T0141357

FLORA OF WEST TROPICAL AFRICA

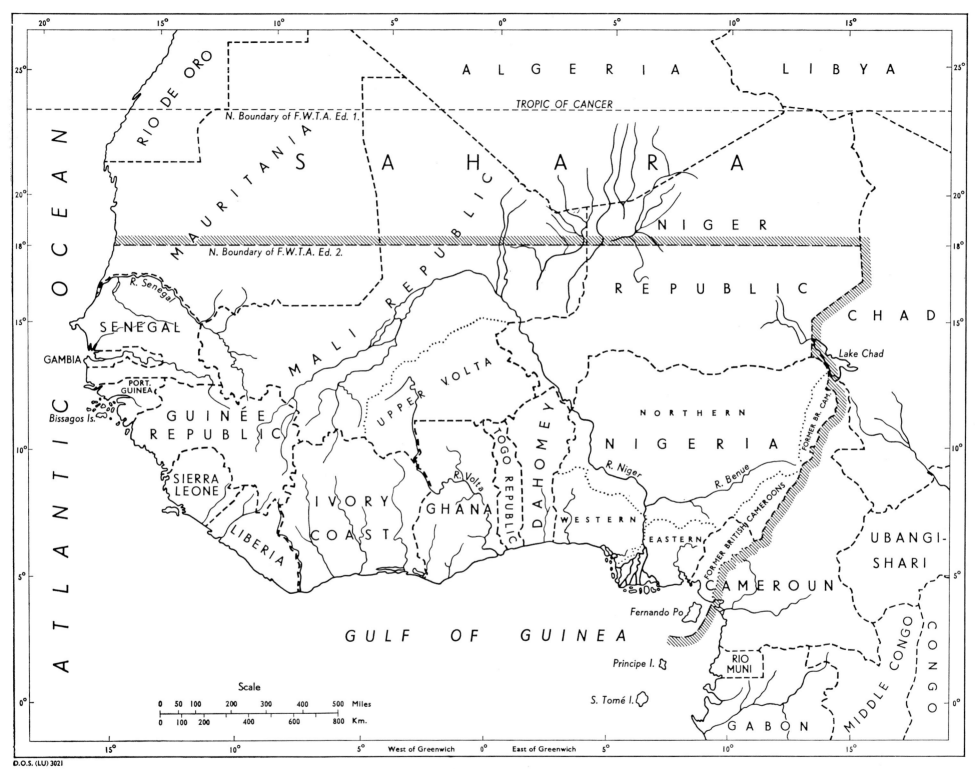

SKETCH-MAP OF WEST AFRICA, SHOWING THE AREA DEALT WITH IN THIS FLORA.

See p. ix for the list of abbreviations used.

D.O.S. (LU) 3021

FLORA

OF

WEST TROPICAL AFRICA

ALL TERRITORIES IN WEST AFRICA SOUTH
OF LATITUDE 18°N. AND TO THE WEST OF
LAKE CHAD, AND FERNANDO PO

BY

J. HUTCHINSON, LL.D., F.R.S., V.M.H., F.L.S.

FORMERLY KEEPER OF THE MUSEUMS OF BOTANY, ROYAL BOTANIC GARDENS, KEW

AND

J. M. DALZIEL, M.D., B.Sc., F.L.S.

FORMERLY OF THE WEST AFRICAN MEDICAL SERVICE, AND
ASSISTANT FOR WEST AFRICA, ROYAL BOTANIC GARDENS, KEW

SECOND EDITION

EDITED BY

F. N. HEPPER, B.Sc., F.L.S.

SENIOR SCIENTIFIC OFFICER, ROYAL BOTANIC GARDENS, KEW

PREPARED AND REVISED AT THE
HERBARIUM, ROYAL BOTANIC GARDENS,
KEW, UNDER THE SUPERVISION OF THE
DIRECTOR

VOL. II

18th October, 1963

PUBLISHED ON BEHALF OF THE GOVERN-
MENTS OF NIGERIA, GHANA, SIERRA
LEONE AND THE GAMBIA

BY THE

CROWN AGENTS FOR OVERSEA GOVERNMENTS   ©
AND ADMINISTRATIONS
MILLBANK, LONDON S.W.1

First printed 1963; © Crown Copyright

Reprinted 1994 by permission of the copyright holders.

This reprint copyright © the Trustees of The Royal Botanic Gardens, Kew

ISBN 0 947643 75 3

For up-to-date prices and availability of other parts, write to:

Publications Sales,
Royal Botanic Gardens,
Kew,
Richmond,
Surrey TW9 3AE,
U.K.

Printed in Great Britain by
Whitstable Litho Printers Ltd, Whitstable, Kent

EDITOR'S PREFACE TO VOLUME II

This second volume, covering the same families as appeared in Part 1 of Volume II in the first edition, concludes the account of the Dicotyledons of West Tropical Africa. The Monocotyledons will be treated separately in a third volume.[1] Owing to the great increase in the content of this edition, it was felt that a division of the Flora into three volumes rather than into two would be more convenient for handling. The Key to the Families of Dicotyledons appeared at the beginning of Volume I of the first edition, but the revised Key is now placed at the end of the present volume, and the index to all the names mentioned in Volume II is also included.

Mr R. W. J. Keay, who revised the bulk of the first volume of the new edition,[2] left Kew to return to Nigeria in 1957, and I succeeded him as Editor. Dr. H. Huber also worked on the Flora at Kew for seven months in 1957–58 and Dr. H. Heine for three years, 1958–61. In all seventeen botanists have contributed to this volume and thus a certain variation of treatment is apparent. A few minor changes have been made, such as the adoption of the spelling " savanna " (in place of " savannah ") in accordance with a recent international recommendation. The full reference to the place of original publication of each accepted species is now given, in order to lessen dependence on the Flora of Tropical Africa. In some of the more complicated families a fuller synonymy and more references to literature are given than was the practice in the first volume.

Political changes have recently taken place with great rapidity within the area of this Flora. To avoid gross anachronism at the time of publication, the names of certain countries have been changed from those used in Volume I. In other cases a compromise between the old and the new has been made in order to avoid confusion, and it is hoped that this will be accepted as the best solution at the time of preparation of this volume. A new map has been prepared as the frontispiece to show these changes, and a full list of abbreviations used is given.

I am most grateful to all who have contributed, to others who have given advice and to the staff of the various herbaria for their continued co-operation.

F. N. HEPPER.

ROYAL BOTANIC GARDENS, KEW,
March 1963.

---

[1] There is also The Ferns and Fern-Allies of West Tropical Africa by A. G. H. Alston, which was published in 1959 as a Supplement to this Revised Edition.
[2] Volume I, part 1 (pp. 1-295) was published in 1954 and part 2 (pp. 296-828, including an index) in 1958.

CONTENTS

LIST OF ABBREVIATIONS

Literature. Only the more commonly cited works are listed here.

Adansonia = continuation of Notulae Systematicae, from 1961.

Aké Assi Contrib. 1 = L. Aké Assi, Contribution à l'Étude Floristique de la Côte d'Ivoire et des Territoires Limitrophes (Thèse) (1961).

Aubrév. Fl. For. C. Iv. = A. Aubréville, La Flore Forestière de la Côte d'Ivoire, ed. 1 (1936), ed. 2 (1959).

Aubrév. Fl. For. Soud.-Guin. = A. Aubréville, Flore Forestière Soudano-Guinéenne (1950).

Appendix = J. M. Dalziel, Useful Plants of West Tropical Africa (Appendix to the Flora of West Tropical Africa) (1937).

Baill. Adansonia = H. Baillon, Adansonia, Paris (1860–79).

Berhaut Fl. Sén. = Berhaut, Flore du Sénégal (1954).

Bull. I.F.A.N. = Bulletin de l'Institut Française d'Afrique Noire, Dakar.

Bull. Jard. Bot. Brux. = Bulletin du Jardin Botanique de l'État, Bruxelles.

Bull. Mus. Hist. Nat. = Bulletin du Muséum National d'Histoire Naturelle, Paris.

Bull. Soc. Bot. Fr. = Bulletin de la Société Botanique de France.

Cat. Talb. = A. B. Rendle (and others), Catalogue of the Plants collected by Mr. & Mrs. P. A. Talbot (1913).

Cat. S. Tomé = A. W. Exell (and others), Catalogue of the Vascular Plants of S. Tomé (with Principe and Annobon) (1944). Supplement (1956).

Cat. Welw. = Catalogue of the African Plants collected by Dr. F. Welwitsch (1896–1901).

Chev. Bot. = A. Chevalier, Exploration Botanique de l'Afrique Occidentale Française, (1920).

Consp. Fl. Ang. = Conspectus Florae Angolensis.

DC. Prod. = De Candolle, Prodromus Systematis Naturalis Regni Vegetabilis.

Engl. Bot. Jahrb. = Engler, Botanische Jahrbücher für Systematik, Pflanzengeschichte und Pflanzengeographie.

Engl. Monogr. Afr. = Engler, Monographieen Afrikanischer Pflanzen-Familien und -Gattungen.

Engl. Pflanzenr. = Engler, Das Pflanzenreich.

Engl. Pflanzenw. Afr. = Engler & Drude, Die Vegetation der Erde, IX:— Engler, Die Pflanzenwelt Afrikas insbesondere seiner tropischen Gebiete.

E. & P. Pflanzenfam. = Engler & Prantl, Die Natürlichen Pflanzenfamilien.

F.T.A. = Oliver (later Dyer, Prain and Hill), Flora of Tropical Africa.

F.T.E.A. = W.B. Turrill (later C. E. Hubbard) & E. Milne-Redhead, Flora of Tropical East Africa.

F.W.T.A., ed. 1 = J. Hutchinson & J. M. Dalziel, Flora of West Tropical Africa, edition 1 (Vol. 1, part 1, 1927 ; part 2, 1928 ; Vol. II, part 1, 1931, part 2, 1936).

Fl. Congo [Belge] = Flore du Congo [Belge] et du Ruanda-Urundi.

Fl. Nigrit. = J. D. Hooker & G. Bentham, Flora Nigritiana, in W. J. Hooker, Niger Flora (1849).

Fl. Oware = A.M.F.J. Palisot de Beauvois, Flore d'Oware et de Benin, en Afrique (1804–20).

Fl. Seneg. = J. A. Guillemin, S. Perrottet and A. Richard, Florae Senegambiae Tentamen (1831–33).

Holl. = J. H. Holland, The Useful Plants of Nigeria, Kew Bulletin, Additional Series IX (1908–22).

J. Agric. Trop. = Journal d'Agriculture Tropicale et de Botanique Appliquée, Paris. (See Rev. Bot. Appliq. below.)

J. Bot. = The Journal of Botany, London.

J. Linn. Soc. = The Journal of the Linnean Society of London—Botany.

Johnston Lib. = H. Johnston, Liberia. Reference is usually to Stapf's List of the Known Plants of Liberia in Appendix IV to Vol. 2 (1906).

Kew Bull. = Kew Bulletin (formerly the Bulletin of Miscellaneous Information).

Mildbr. Wiss. Ergebn. 1910–11 = Wissenschaftliche Ergebnisse der Zweiten Deutschen Zentral-Afrika-Expedition 1910–11 unter Führung Adolf Freidricks, Herzogs zu Mecklenburg, Band II : Botanik (1922).

Notizbl. Bot. Gart. Berl. = Notizblatt des Botanischen Gartens und Museums zu Berlin-Dahlem.

Notulae Syst. = Lecomte (later Humbert), Notulae Systematicae, Paris (1909–61), continued as Adansonia.

Rev. Bot. Appliq. = Revue Internationale de Botanique Appliquée et d'Agriculture
    Tropicale (1921–53), continued as *J. Agric. Trop.*
Schum. & Thonn. Beskr. Guin. Pl. = Schumacher & Thonning, Beskrivelse af Guineiske
    Planter (1827).
Trans. Linn. Soc. = The Transactions of the Linnean Society of London.

Countries

Maur.   —Mauritania.
Sen.   —Senegal.
Gam.   —Gambia.
Mali   —Mali (formerly **Fr. Sud.** : French Sudan), including, in this Flora, Upper Volta.
Port. G.   —Portuguese Guinea.
Guin.   —Guinée (formerly **Fr. G.** : French Guinea).
S.L.   —Sierra Leone.
Lib.   —Liberia.
Iv. C.   —Ivory Coast.
Ghana   —Ghana (formerly **G.C.:** Gold Coast), including former British Togoland.
Togo Rep.—Togo Republic (formerly French Togoland ; British Togoland is now
    incorporated in Ghana).
Dah.   —Dahomey.
Niger   —Niger Republic (formerly **Fr. Nig.:** French Niger Colony).
N. Nig.   —Northern Nigeria.
S. Nig.   —Southern Nigeria (political Western Nigeria and Eastern Nigeria).
[Br.]Cam. —West Cameroon (ex-British Cameroons ; formerly Southern Cameroons,
    now part of Federal Republic of Cameroun ; and Northern Cameroons,
    now incorporated into N. Nigeria).
F. Po   —Fernando Po.

Collectors and Herbaria

Aubrév. = A. Aubréville.
Chev. = A. Chevalier.
Dalz. = J. M. Dalziel.
Esp. Santo = Espírito Santo.
FH = Forest Herbarium, No.
FHI = Forest Herbarium Ibadan, No.
Hb. = Herbarium.
Hutch. = J. Hutchinson.
Jac.-Fél. = H. Jacques-Félix.
Mildbr. = J. Mildbraed.
Sc. Elliot = G. F. Scott Elliot.

Other abbreviations

! = specimen seen by reviser.
Dist. = District.
Div. = Division.
F.R. = Forest Reserve.
Nom. cons. = *nomen conservandum* (conserved
    name listed in International Code of
    Botanical Nomenclature).
Prov. = Province.
R. = River.

PHYLOGENETIC SEQUENCE OF ORDERS AND FAMILIES CONTAINED IN VOLUME II

(The cross-lines show breaks in affinity)

METACHLAMYDEAE

126. ERICACEAE

By R. Ross

Shrubs or undershrubs, less frequently trees ; leaves alternate or whorled, rarely opposite, simple, usually evergreen ; stipules absent. Flowers hermaphrodite, actinomorphic or slightly zygomorphic. Calyx persistent ; sepals free or united. Corolla hypogynous, gamopetalous, inserted below a fleshy disk ; lobes contorted or imbricate. Stamens usually double the number of the corolla lobes or, if the same number, alternate with them, hypogynous ; filaments and anthers usually free ; anthers 2-celled, opening by pores or pore-like slits. Ovary superior, several-celled, with numerous ovules on axile or rarely basal placentas, rarely 1-ovulate ; style simple. Fruit a capsule, berry, or drupe. Seeds with fleshy endosperm and straight embryo, sometimes winged.

Generally distributed, but less frequent and mostly on the mountains in the tropics ; very numerous in the south-western region of South Africa. Amongst African families recognized by the small flowers, hypogynous stamens within a shortly lobed gamopetalous corolla, and the anthers opening by pores or pore-like slits.

Leaves lanceolate, acuminate, fairly large ; corolla not persistent ; flowers in racemes
1. **Agauria**
Leaves very small and heath-like ; corolla persistent ; flowers not in racemes :
Small tree or shrub ; no bracteoles on pedicel ; calyx zygomorphic ; corolla shortly campanulate, less than 2 mm. long ; anthers without appendages .. 2. **Philippia**
Small wiry undershrub ; 1–3 bracteoles on pedicel ; calyx actinomorphic ; corolla tubular, expanded above, more than 4 mm. long ; anthers with appendages
3. **Blaeria**

1. AGAURIA (DC.) Hook.f.—F.T.A. 3 : 483.

Shrub or small tree ; branchlets purplish ; leaves oblong-lanceolate, acute at both ends, more or less acuminate at the apex, 3–9 cm. long, 1–3 cm. broad, glabrous, glaucous beneath ; petiole about 1 cm. long ; flowers in short axillary racemes ; pedicels 3–4 mm. long ; bracts caducous ; sepals triangular, 5, pubescent outside ; corolla deciduous, greenish-yellow tinged with pink, cylindric, contracted at the

FIG. 202.—AGAURIA SALICIFOLIA (*Comm.*) *Hook. f. ex Oliv.* (ERICACEAE).
A, flower. B and C, stamens. D, calyx and pistil.

1

mouth, with 5 short rounded lobes ; stamens 10, filaments pubescent ; fruit a capsule, 5-angled, 4–6 mm. diam. ; seeds curved .. .. .. *salicifolia*

A. salicifolia (*Comm.*) *Hook. f. ex Oliv.* F.T.A. 3 : 483 (1877) ; Sleumer in Engl. Bot. Jahrb. 69 : 381 (1938). *Andromeda salicifolia* Comm. in Lam. Encycl. Méth., Bot. 1 : 159 (1783) ; Bot. Mag. 60 : t. 3286 (1833). A shrub or small tree, attaining about 40 ft. in height, on the margins of the montane forest and in grassland above it.

[Br.]**Cam.** : Cam. Mt., 4,000–10,000 ft. (Dec.–May) *Mann* 1209 ! 2182 ! *Johnston* 12 ! *Deistel* 74 ! *Thresh* 3 ! *Hambler* 210 ! Lakom, 6,000 ft., Bamenda (fr. Apr.) *Maitland* 1631 ! Lake Oku, 7,700 ft., Bamenda (Jan.) *Keay & Lightbody* FHI 28510 ! Bafut-Ngemba F.R., 6,000–7,000 ft., Bamenda (Jan.) *Keay & Lightbody* FHI 28375 ! Nkambe, 6,000 ft. (Feb.) *Hepper* 1903 ! **F.Po** : Clarence Peak, 8,500–9,000 ft. (Dec.–Mar.) *Mann* 621 ! *Guinea* 2934 !

This species is represented in West Africa by only var. *salicifolia* (var. *pyrifolia* (Pers.) Oliv. ; Sleumer, l.c.)

2. PHILIPPIA Klotzsch in Linnaea 9 : 354 (1835). *Ericinella* of F.T.A. 3 : 484, not of Klotzsch.

Small tree or shrub ; branchlets pubescent with simple and barbellate hairs ; leaves ascending to erect, in whorls of three, oblong-lanceolate, acute, 1·5–4 mm. long, with a dorsal groove, minutely scabrid at the margin ; flowers few at the apex of each branchlet ; pedicels glabrous, 1·5–2 mm. long ; calyx-lobes 4, one free, linear, leaf-like, two-thirds of the corolla to almost equalling it in length, three connate to about half their length, triangular, about half the length of the corolla ; corolla campanulate, shallowly cleft, 4-lobed, about 1·5 mm. long ; stamens 6, included ; anthers without appendages, opening by pore-like slits ; stigma broadly peltate ; ovary 3–4-celled ; capsule densely pubescent with very short hairs .. *mannii*

P. mannii (*Hook. f.*) *Alm & Fries* in Kungl. Svenska Vet. Handl., ser. 3, 4, No. 4 : 37, t. 4a, fig. 10m. *Ericinella mannii* Hook. f. in J. Linn. Soc. 6 : 16 (1861) ; Bot. Mag. 92 : t. 5569 (1866) ; F.T.A. 3 : 484 (1877). A heath-like shrub up to 12 ft. with purplish flowers.

[Br.]**Cam.** : Cam. Mt., 4,500–11,000 ft. (fl. May, fr. Nov.–Mar.) *Mann* 1289 ! 2181 ! *Johnston* 104 ! *Deistel* 81 ! *Gray* FHI 7454 ! *Brenan* 9393 ! Lake Bambuluwe, Bamenda (Sept.) *Ujor* FHI 30201 ! Oku, 6,500 ft. Bamenda (fr. Feb.) *Hepper* 2059 ! **F.Po** : Clarence Peak, 9,000 ft. to summit (fl. Dec., fr. Mar.) *Mann* 287 ! *Guinea* 2796 !

3. BLAERIA Linn.—F.T.A. 3 : 484.

Small wiry undershrub, stems covered with short simple and long bristly hairs often branched below, some, usually most or all, glandular at the tip ; leaves in whorls of 3, occasionally spiral or in whorls of 4, 2–5 mm. long, with a dorsal groove, sometimes with short simple hairs, always with long hairs at the margins, some glandular, some branched at the tip ; flowers axillary in whorls of 3, inflorescence usually a spike-like panicle ; pedicels short, bracteolate ; calyx-lobes 4, linear, about one third the length of the corolla, with long marginal hairs some or all glandular at the tip ; corolla 4–6 mm. long, shortly 4-lobed, somewhat expanded above, slightly curved, persistent and tearing at capsule dehiscence ; stamens 4, scarcely exserted ; anthers with appendages, opening by pores ; ovary 4-celled ; capsule pubescent *mannii*

B. mannii (*Engl.*) *Engl.* Bot. Jahrb. 43 : 366 (1909) ; Alm & Fries in Act. Hort. Berg. 8 : 245, t. 5–6 fig. f–g (1924). *B. spicata* of F.T.A. 3 : 484, 'partly, not of Hochst. *B. spicata* var. *mannii* Engl. Hochgebirgsflora trop. Afr. : 325 (1892). *B. tenuipilosa* Alm & Fries, l.c. 247, partly (*Ledermann* 1625) ; Pichi-Sermolli & Heiniger in Webbia 9 : 40 (1953). *B. nimbana* A. Chev. in Rev. Bot. Appliq. 18 : 113 (1938). A heath-like undershrub with purplish flowers.

Iv.C. : Nimba Mts., 5,300–5,800 ft. (July–Oct.) *Jac.-Fél.* 1938 ! *Schnell* 1575 ! 1877 ! [Br.]**Cam.** : Cam. Mt., 6,000 ft. to summit (Nov.–Apr.) *Mann* 1280 ! 2177 ! *Johnston* 7 ! 91 ! *Preuss* 712 ! 778 ! *Dalziel* 8311 ! *Mildbr.* 10837 ! 10877 ! 10904 ! *Dundas* FHI 20634 ! *Brenan* 9394 ! 9394A ! *Keay* FHI 28593 ! FHI 28624 ! *Thresh* 2 ! Lakom, 5,500 ft., Bamenda (Apr.) *Maitland* 1421 ! Lake Oku, 7,900 ft., Bamenda (Jan.) *Keay & Lightbody* FHI 28497 ! Bambili, Bamenda (Aug.) *Ujor* FHI 29981 ! Bafut-Ngemba F.R., 7,000–7,300 ft., Bamenda (Feb., Sept.) *Savory* UCI 486 ! *Hepper* 2126 ! **F.Po** : Clarence Peak, summit (fl. Dec., fr. Mar.) *Mann* 592 ! *Guinea* 2744 !

127. EBENACEAE

By F. White

Trees or shrubs without milky latex, heartwood sometimes black (Ebony of commerce). Leaves nearly always alternate, exstipulate, entire. Flowers actinomorphic, hypogynous, usually unisexual, but frequently with rudiments of other sex, solitary, fasciculate or cymose, rarely in false racemes, sometimes cauliflorous. Calyx gamosepalous, entire to deeply lobed, always persistent in fruit and usually accrescent. Corolla gamopetalous, shortly to deeply lobed ; tube often fleshy and constricted at throat ; lobes contorted in bud. Stamens from 2 to more than 100, epipetalous or borne on receptacle, exserted or included, filaments often very short, anthers usually apiculate, often 2 or more arising from a single filament. Ovary syncarpous, each locule either with 2 ovules or divided by a false septum into 2 uni-ovulate compartments ; styles distinct or basally connate, very rarely completely united ; stigmas usually large and conspicuous ; ovules pendulous from apex of locule. Fruit a berry. Seeds large ; endosperm abundant, hard, sometimes ruminate ; embryo half as long as the endosperm, with foliaceous cotyledons.

Mainly in the tropics and subtropics ; many of local economic importance, a few of considerable commercial importance. The number of floral parts is very variable and inconstant both between and within species. The only other genus now recognized in the family is *Euclea* Murr. which is widespread in Africa but does not occur in our area.

DIOSPYROS Linn.—F.T.A. 3 : 517 ; White in Bull. Jard. Bot. Brux. 26 : 237–246 (1956). *Maba* Forst.—F.T.A. 3 : 514 ; F.W.T.A., ed. 1, 2 : 6.

KEY TO SPECIMENS WITH MALE FLOWERS.

*Corolla lobed to the middle or beyond ; anthers often exserted beyond the throat (to p. 5) :
　Flowers small, corolla less than 8 mm. long ; anthers usually exserted far beyond the throat :
　　Filaments attached to corolla-tube ; anthers exserted beyond throat for whole or greater part of length :
　　　Calyx 3·5-4·5 mm. long, lobes 5–6, suborbicular, imbricate ; stamens 45–120 ; leaves more or less oblong-elliptic, acuminate, up to 17 cm. long and 7 cm. broad, lateral nerves in 8–10 pairs, tertiary nervation laxly reticulate—(see fig. 202A 1)
　　　　　　　　　　　　　　　　　　　　　　　　　　　　　1. *polystemon*
　　　Calyx up to 2·5 mm. long ; lobes with open aestivation, usually deltoid ; stamens 10–25 :
　　　　Peduncles more than 4 mm. long ; cymes lax, 12–20-flowered ; corolla 7·5 mm. long, 4-lobed ; leaves more or less elliptic, apex usually cuspidate, up to 17 cm. long and 11 cm. broad, drying reddish, upper surface glossy, with conspicuously reticulate, raised venation　..　..　..　..　..　　2. *viridicans*
　　　　Peduncles up to 2 mm. long ; cymes mostly congested ; corolla up to 6 mm. long, 3–5-lobed :
　　　　　Calyx densely fulvous-setulose outside ; leaves mostly elliptic, shortly acuminate, up to 18·5 cm. long and 7 cm. broad, usually drying reddish, tertiary nerves subparallel ; corolla (4–)5-lobed　..　..　..　..　..　3. *dendo*
　　　　　Calyx completely glabrous outside except for a few minute marginal cilia ; leaves drying black ; corolla 3(–4)-lobed :
　　　　　　Leaves elliptic-oblong with almost parallel sides, apex caudate-acuminate, up to 13 cm. long and 5 cm. broad, secondary nerves in 12–20 pairs ; flowers distinctly pedicellate in lax cymes ; anthers minutely setulose on the margin
　　　　　　　　　　　　　　　　　　　　　　　　　　　　　4. *piscatoria*
　　　　　　Leaves more or less elliptic, shortly subacuminate, up to 12 cm. long and 4 cm. broad, secondary nerves in 5–12 pairs ; flowers subsessile in contracted cymes ; anthers glabrous　..　..　..　..　..　..　5. *abyssinica*
　　Filaments attached to receptacle ; anthers not, or only slightly exserted beyond the throat :
　　　Flowers in 1–4-flowered fascicles ; pedicels slender, 3–4 mm. long ; corolla-lobes spreading, glabrous ; calyx with open aestivation, regularly lobed ; leaves elliptic, acuminate, up to 11 cm. long and 4·5 cm. broad, drying black, with 6–9 pairs of prominent lateral nerves, tertiary nerves and veins scarcely visible
　　　　　　　　　　　　　　　　　　　　　　　　　　　　　6. *cooperi*
　　　Flowers subsessile in 1–3(–4)-flowered pedunculate cymules ; corolla-lobes ascending, densely strigulose ; calyx closed in the bud, irregularly lobed at anthesis ; leaves very variable, up to 10 cm. long and 4·5 cm. broad, drying brown, venation closely reticulate on both surfaces..　..　..　..　..　..　..　7. *ferrea*
　Flowers large, corolla more than 11 mm. long ; anthers included or only slightly exserted :
　　Inflorescence a lax, 3–5-flowered axillary pseudo-raceme ; pedicels up to 2 cm. long ; calyx-lobes 4, free almost to base, papery, with marginal cilia and conspicuous venation ; stamens entirely included ; branchlets with setose hairs up to 5 mm. long ; some leaves modified to form boat-shaped shelters for ants　　8. *conocarpa*
　　Inflorescence a congested fascicle or cyme, or a lax, many-flowered cyme borne on the older wood ; calyx-lobes variously united ; hairs on branchlets never more than 2 mm. long ; leaves not modified to form shelters for ants :
　　　Corolla hairy :
　　　　Corolla hypocrateriform, fulvous-tomentellous, tube 9 mm. long, slender, contracted at the throat ; stamens attached to receptacle, entirely included, strigose at base ; leaves elliptic, acuminate, up to 18 cm. long and 8 cm. broad, lower surface fulvous-sericeous and with laxly reticulate, prominent venation　9. *kamerunensis*
　　　　Corolla infundibuliform, glabrous except for densely strigose mid-petaline lines ;

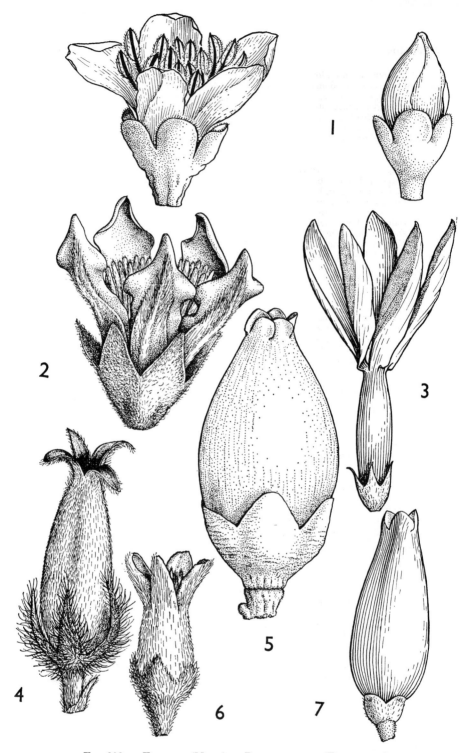

FIG. 202A.—FLOWERS (MALE) OF DIOSPYROS SPP. (EBENACEAE).

1, *D. polystemon* Gürke (*Louis* 3466), ×5. 2, *D. mannii* Hiern (*Tisserant* 1361), ×3. 3, *D. fragrans* Gürke (*Zenker* 1740), ×3. 4, *D. barteri* Hiern (*Zenker* 1798), ×5. 5, *D. crassiflora* Hiern (*Louis* 8910), ×3. 6, *D. chevalieri* De Wild. (*Andoh* FH 5784), ×5. 7. *D. hoyleana* F. White (*Brenan* 9471), ×5.

tube about 5 mm. long, widely open at the throat ; stamens attached to corolla-tube, exserted, densely strigose for entire length ; leaves elliptic or lanceolate-elliptic, acuminate, up to 18 cm. long and 8·5 cm. broad, lower surface pinkish- or greyish-white, strigose-pubescent on nerves, tertiary nervation subparallel— (see fig. 202A 2) .. .. 10. *mannii*
Corolla glabrous :
Leaves pubescent beneath ; flowers always borne on bole and branches, never in axils of leaves ; pedicels more than 3 mm. long ; corolla-lobes 5 :
Flowers in fascicles ; pedicels 3–4 mm. long ; calyx sparsely puberulous, lobes as long as tube, narrowly deltoid ; corolla-tube as long as lobes, constricted at throat ; leaves oblong-elliptic or lanceolate-elliptic, acutely acuminate, up to 11·5 cm. long and 4 cm. broad, drying grey-green, lateral nerves in 4–5 pairs, not impressed above—(see fig. 202A 3) .. .. .. 11. *fragrans*
Flowers in lax, many-flowered cymes ; pedicels 8–13 mm. long ; calyx tomentellous, lobes much shorter than the tube, hemi-orbicular ; corolla-tube much shorter than the lobes, not constricted ; leaves oblong-elliptic, or lanceolate-elliptic, acuminate, up to 14 cm. long and 4·5 cm. broad, lateral nerves in about 7 pairs, leaving midrib at very acute angle, impressed on upper surface
12. *suaveolens*
Leaves glabrous beneath ; some flowers borne in leaf-axils, others on older branch-lets, but never (apparently) on bole and main branches ; pedicels 1 mm. long ; corolla-lobes 4 :
Calyx 8 mm. long, lobes 4 mm. long, narrowly deltoid ; leaves chartaceous, oblong-elliptic or oblanceolate-oblong, caudate-acuminate, up to 26 cm. long and 9 cm. broad, drying olive-brown, venation laxly reticulate.. .. 13. *bipindensis*
Calyx 2·5 mm. long, cup-shaped, margin entire except for 4 minute teeth ; leaves coriaceous, elliptic or oblong-elliptic, caudate-acuminate, up to 19 cm. long and 7 cm. broad, drying grey-green, venation very closely reticulate
14. *physocalycina*
*Corolla lobed only in upper half ; anthers usually completely concealed :
†Corolla tomentose or tomentellous or with large hairs easily visible to naked eye (to p. 6) :
Branchlets densely clothed with long, rusty, spreading hairs about 1·5 mm. long ; leaves cordate at base, usually ovate or lanceolate, up to 16 cm. long and 7 cm. broad, acutely acuminate, glaucous beneath, tertiary nerves subparallel, deeply impressed above ; calyx lobed almost to the base, lobes subulate ; corolla 1 cm. long, tube 7·5 mm. long, gradually narrowed to throat—(see fig. 202A 4)
15. *barteri*
Branchlets without long hairs ; leaves cuneate or rounded at base :
Corolla more than 12 mm. long :
Calyx deeply lobed almost to the base, lobes narrowly deltoid ; corolla glabrous except for a vertical strip of hairs along central axis of each corolla-lobe which is continued down the tube, and a few marginal cilia ; leaves oblanceolate or oblanceolate-elliptic, shortly acuminate, up to 26 cm. long and 11 cm. broad, drying reddish-brown, lower surface sparsely puberulous .. 16. *undabunda*
Calyx only lobed in upper half or not at all ; corolla tomentellous :
Calyx 4 mm. long, cup-shaped, unlobed but sometimes developing irregular splits ; pedicels 3–5 mm. long ; corolla 1·3–1·6 cm. long, tube 1–1·2 cm. long, con-stricted at the throat ; leaves lanceolate or lanceolate-elliptic, shortly acuminate, up to 23 cm. long and 8 cm. broad, glaucous or grey-green beneath, lower surface sparsely and minutely puberulous .. .. 17. *sanza-minika*
Calyx 7–8 mm. long, distinctly lobed ; pedicels less than 2 mm. long :
Calyx sparsely and minutely puberulous, lobes rounded ; corolla 2·5–3 cm. long, tube 2·2–2·8 cm. long, flask-shaped, tapering rapidly from the middle to both ends ; stamens about 90 ; leaves oblong-elliptic or lanceolate-elliptic, acumin-ate, up to 19 cm. long and 7·5 cm. broad, lower surface sparsely and minutely puberulous—(see fig. 202A 5) .. .. 18. *crassiflora*
Calyx densely covered with chocolate-brown or black tomentum, lobes deltoid ; corolla 1·8–2·2 cm. long, hypocrateriform, tube about 1·2 cm. long ; stamens about 20 ; leaves usually oblanceolate-elliptic, very shortly acuminate, up to 38 cm. long and 14 cm. broad, usually drying greenish-brown, lower surface densely setulose-puberulous on nerves and veins, more sparsely so on lamina, venation prominently reticulate .. .. 19. *gabunensis*
Corolla 7–8 mm. long :
Leaves opposite or subopposite, up to 32 cm. long and 10 cm. broad, oblanceolate, apex acute or shortly acuminate, lower surface sparsely and minutely puberulous and glaucous ; first-year branchlets rusty-tomentellous ; stamens exserted— (see fig. 202A 6) .. .. 20. *chevalieri*
Leaves alternate, usually smaller ; branchlets not rusty-tomentellous ; stamens included :

Inflorescence a distinctly pedunculate 3–9-flowered cyme ; leaves more or less elliptic or oblanceolate-elliptic, apex subacuminate, up to 15 cm. long and 5·5 cm. broad, usually drying yellowish- or reddish-brown, lower surface almost glabrous, but with a few appressed hairs   ..   ..   .. 21. *mespiliformis*
Inflorescence a sessile fascicle or congested cyme :
  Corolla narrowly conical, gradually tapering from near base ; leaves rounded or acute at apex, never acuminate, ovate or elliptic, drying reddish above, densely pubescent or tomentose beneath   ..   .. 22. *tricolor*
  Corolla botuliform or urceolate, tapering more or less equally from about the middle ; leaves acuminate :
    Leaves small, up to 14 cm. long and 4·6 cm. broad, minutely and sparsely puberulous beneath, tertiary nerves scarcely visible :
      Calyx-lobes rounded, broader than long ; leaves usually drying greenish-brown above and grey beneath, lower surface with minute, black, strigulose hairs
                                            23. *nigerica*
      Calyx-lobes narrowly deltoid, much longer than broad ; leaves usually drying reddish, lower surface with minute hyaline hairs..   ..   24. *heudelotii*
    Leaves large, up to 20 cm. long and 10 cm. broad, densely puberulous or pubescent beneath except in extreme old age, tertiary nerves very conspicuous, subparallel :
      Calyx 2 mm. long, lobes deltoid with flat margins ; tertiary nerves not or only slightly impressed   ..   .. 25. *liberiensis*
      Calyx 4–5 mm. long, lobes ovate-apiculate with reduplicate margins ; tertiary nerves impressed above   ..   .. 26. *thomasii*
†Corolla glabrous or with minute hairs invisible to naked eye :
  Branches with papery peeling bark and sharp-pointed spines ; flowers subsessile in 3(–5)-flowered cymules at ends of (3–)6–14 mm. long, distally expanded and flattened peduncles ; calyx closed in bud, botuliform, (7–)9–12(–21) mm. long ; corolla rotate, tube 12 mm. long, lobes 7–9 mm. long ; leaves oblanceolate, subacuminate, up to 20 cm. long and 7·5 cm. broad, drying brown   ..   .. 27. *monbuttensis*
  Branches without papery peeling bark and spines ; flowers in fascicles ; calyx with open aestivation, not botuliform :
    Calyx truncate or with only slight indications of lobes :
      Stamens with long strigose hairs (visible to naked eye) clothing filament and proximal half of connective, anthers with apiculus 1 mm. long ; leaves broadly ovate-elliptic to lanceolate-elliptic, acuminate, up to 12 cm. long, and 5·5 cm. broad, drying brownish, secondary nerves not impressed, tertiary nerves and veins invisible   ..   .. 28. *melocarpa*
      Stamens glabrous or with minute setulose hairs, apiculus less than 0·25 mm. long ; secondary nerves deeply impressed above, prominent beneath, tertiary nerves clearly visible :
        Leaves drying reddish, elliptic or lanceolate-elliptic, up to 21 cm. long and 9 cm. broad, lateral nerves in (6–)7–9 pairs, acumen gradually tapering
                                          29. *alboflavescens*
        Leaves drying blackish, elliptic or lanceolate-elliptic, up to 14 cm. long and 6·5 cm. broad, lateral nerves in 4–5 pairs, acumen parallel-sided except for the slightly expanded tip   ..   .. 30. *zenkeri*
    Calyx distinctly lobed :
      Corolla-lobes 4–5 mm. long ; leaves large, venation prominent and very closely reticulate on both surfaces :
        Calyx glabrous, 2·5 mm. long, tightly clasping base of corolla-tube, regularly lobed to middle ; corolla-tube not constricted ; anthers far-exserted ; leaves oblanceolate or oblanceolate-elliptic, shortly acuminate, up to 35 cm. long and 13 cm. broad   ..   .. 31. *preussii*
        Calyx sparsely and minutely strigulose, 4 mm. long, separated from base of corolla-tube by an appreciable space, irregularly lobed in upper third ; corolla-tube constricted at throat ; anthers included ; leaves usually elliptic or oblong-elliptic, acuminate, up to 25 cm. long and 10 cm. broad   .. 32. *canaliculata*
      Corolla-lobes up to 2·5 mm. long ; leaves usually small and without closely reticulate venation :
        Calyx-lobes rounded, or glabrous inside :
          Calyx-lobes acute ; leaves elliptic, oblong-elliptic or oblanceolate-oblong, up to 14 cm. long and 6 cm. broad, apex gradually tapering to a narrowly deltoid acumen, lower surface glabrous, drying dark brown or black   33. *soubreana*
        Calyx-lobes rounded ; leaves not or scarcely acuminate :
          Leaves lanceolate, up to 12 cm. long and 5 cm. broad, apex not emarginate, lateral nerves conspicuous ; calyx 3 mm. long, puberulous, lobed almost to the base ; corolla-lobes 2 mm. long, much longer than broad   34. *elliotii*
          Leaves obliquely rhombic, up to 6 cm. long and 2 cm. broad, apex emarginate, lateral nerves invisible ; calyx 1·5 mm. long, minutely puberulous, lobed only

in upper third ; corolla-lobes 1 mm. long, scarcely longer than broad—(see
fig. 202A 7) .. .. **35. *hoyleana***
Calyx-lobes acute, puberulous inside :
Leaves small, up to 3·5 cm. long, drying greenish or blackish, very asymmetric,
apex with a filiform tip.. .. .. **36. *obliquifolia***
Leaves larger, more than 7 cm. long, drying reddish, more or less symmetric,
without a filiform tip :
Leaves large, up to 19 cm. long and 9 cm. broad, lanceolate or lanceolate-
elliptic, acuminate, tertiary nerves and veins clearly visible on lower surface,
darker than lamina ; corolla glabrous .. .. .. **37. *simulans***
Leaves medium-sized, up to 17 cm. long and 6·5 cm. broad, but usually much
smaller, tertiary nerves and veins scarcely visible ; corolla with a few strigulose hairs :
Flowers subsessile ; corolla 7 mm. long ; leaves tapering gradually to the
rather blunt acumen, drying greyish-green beneath .. **38. *vignei***
Flowers distinctly pedicellate ; corolla 11 mm. long ; leaves tapering rather
suddenly to the acute acumen, drying pinkish beneath .. **39. *cinnabarina***

KEY TO FRUITING SPECIMENS.

Calyx-lobes undulate ; calyx more than 1 cm. long :
Calyx less than half as long as fruit, 1–1·2 cm. long, fulvous-puberulous outside, lobes
(4–)5, ovate-deltoid, 5–6 mm. long, margin undulate with at least one large fold
about one third of the way from the base ; fruit subglobose, 2·8 cm. long, 2·6 cm.
broad, verruculose, almost glabrous at maturity .. .. .. **21. *mespiliformis***
Calyx almost as long as or longer than the fruit :
Calyx glabrous, lobed only in upper half, lobes 4 :
Calyx leathery, without distinct venation, about 3·8 cm. long, lobes (at least in young
fruit) reduplicate-valvate ; tertiary nerves and veins of leaves very closely reticulate and prominent on both surfaces .. .. **31. *preussii***
Calyx papery with distinct longitudinal venation, 3–3·8 cm. long, lobes not redupli-
cate-valvate ; tertiary nerves and veins laxly reticulate and slightly prominent
on both surfaces—(see fig. 202B 3) .. .. **13. *bipindensis***
Calyx puberulous or pubescent, lobed almost to the base, lobes usually 5 :
Young branchlets, leaves and calyces pubescent, hairs visible to naked eye ; calyx
about 4 cm. long, venation not parallel ; fruit depressed-globose, about 3·8 cm.
long and 4 cm. broad .. .. **16. *undabunda***
Young branchlets, etc., puberulous, hairs visible with a ×20 lens ; calyx 2·5–3 cm.
long, venation parallel ; fruit ovoid-conic, about 1·5 cm. long and 1·2 cm. broad
—(see fig. 202B 1) .. .. **3. *dendo***
Calyx-lobes not undulate (or if undulate as in No. 26, then calyx less than 7 mm. long) :
Indumentum of fruit consisting at least in part of long setose hairs clearly visible to
naked eye :
Calyx cup-shaped, the tube surrounding the whole of the lower half of the fruit,
tomentose with chocolate-brown or black hairs, lobes 5, scarcely developed or well-
developed and up to 1·5 cm. long ; fruit up to 2·2 cm. long and 2·6 cm. broad,
chocolate-brown tomentose—(see fig. 202B 6) .. .. .. **19. *gabunensis***
Calyx not cup-shaped, the tube not surrounding the lower half of fruit :
Leaves opposite ; fruit subsessile, about 4·2 cm. long and 2·5 cm. broad ; calyx
4–5 mm. long .. .. **20. *chevalieri***
Leaves alternate ; pedicels up to 3 cm. long ; calyx 1·2–1·5 cm. long (but note that
the fruiting calyx of No. 11 is unknown) :
Fruit sparsely setose :
Fruit ovoid-conic, up to 4 cm. long and 1·7 cm. broad ; calyx papery, sparsely
setose, with conspicuous venation ; some leaves modified to form boat-shaped
shelters for ants .. .. **8. *conocarpa***
Fruit depressed-globose, 3 cm. long and 3·5 cm. broad ; leaves not modified to
form shelters for ants .. .. **11. *fragrans***
Fruit tomentose :
Fruit up to 4 cm. long and 2·8 cm. broad, tomentose with an understorey of short
chocolate-brown hairs and with longer setose hairs—(see fig. 202B 2)
12. *suaveolens*
Fruit about 4·8 cm. long and 4·4 cm. broad, ferrugineous-setose-tomentose
10. *mannii*
Indumentum of fruit consisting of minute hairs visible with a ×20 lens, or fruit
glabrous :
Calyx longer than the fruit, 3–4 cm. long :
Calyx cordiform, broadest just above the base and gradually tapering to apex, lobes
much shorter than the tube, deltoid .. .. **14. *physocalycina***
Calyx broadest near the middle at the junction of lobes and tube, lobes rounded with

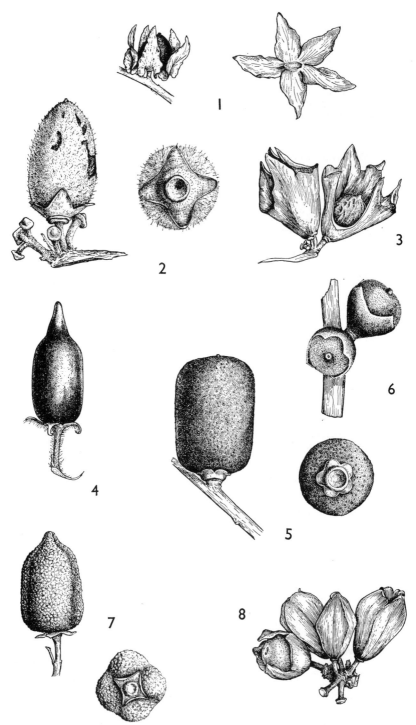

Fig. 202b.—Fruits of Diospyros spp. (Ebenaceae).

1, *D. dendo* Welw. ex Hiern (*Dubois 523, Onochie* FHI 19665). 2, *D. suaveolens* Gürke (*Keay* FHI 37042). 3, *D. bipindensis* Gürke (*Louis 9516*). 4, *D. barteri* Hiern (*Meikle 1393*). 5, *D. sanza-minika* A. Chev. (*Vigne* FH 106). 6, *D. gabunensis* Gürke (*Edwardson 43*). 7, *D. thomasii* Hutch. & Dalz. (*Wallace 33*). 8, *D. canaliculata* De Wild. (*Enti* FH 6851) All ×¾.

8

reduplicate margins forming 4 acute ridges which are continued in the upper part of the tube—(see fig. 202B 8) ..   ..   32. *canaliculata*

Calyx much shorter than the fruit :

  Fruit ovoid-conic or conic, tapering to a pointed apex ; calyx lobed almost to the base, lobes deltoid :

  Branchlets densely clothed with long rusty spreading hairs about 1·5 mm. long ; leaves cordate at base, up to 16 cm. long and 7 cm. broad ; calyx 9 mm. long, rusty pilose ; fruit up to 4·2 cm. long and 1·8 cm. broad—(see fig. 202B 4)
15. barteri

  Branchlets without long hairs ; leaves not cordate at base :

  Leaves very small, up to 3·5 cm. long and 1·6 cm. broad, markedly asymmetric, apex with a filiform tip ; fruit more or less 1·8 cm. long and 0·7 cm. broad
36. obliquifolia

  Leaves larger, more than 6 cm. long, not or scarcely asymmetric, apex not filiform :

  Leaves rounded or acute at apex, densely strigose-pubescent beneath ; fruit 2·2–2·8 cm. long, 1·3–1·6 cm. broad ; shrub of coastal scrub   22. *tricolor*

  Leaves acuminate, sparsely strigulose beneath ; shrubs or small trees of rain forest :

  Fruit conic, tapering from the base, sometimes asymmetric, up to 4 cm. long and 2·5 cm. broad ; seeds 4 or fewer by abortion..   ..   38. *vignei*

  Fruit ovoid-conic, tapering from some distance above the base ; seeds 6 :

  Tertiary nerves and veins clearly visible on lower surface, darker than lamina ; fruit about 4·2 cm. long and 3·2 cm. broad ..   ..   37. *simulans*

  Tertiary nerves and veins scarcely visible ; fruit about 3·5 cm. long and 3 cm. broad ..   ..   39. *cinnabarina*

Fruit of various shapes, never tapering to a pointed apex (but sometimes with an umbo at base of style) :

  Fruit tomentellous (except in extreme old age) :

  Fruit subglobose, up to 2·8 cm. long and 2·4 cm. broad   ..   24. *heudelotii*

  Fruit ellipsoid, cylindric, ovoid- or obovoid-cylindric, 3·5–4·5 cm. long, 2·8–3·8 cm. broad :

  Seeds 6–8 ; tertiary nerves and veins forming an indistinct, lax reticulum
9. kamerunensis

  Seeds 3–4 ; tertiary nerves subparallel :

  Pedicels 9–18 mm. long ; calyx 6–7 mm. long, lobes suborbicular-apiculate, the margins folded back and undulate—(see fig. 202B 7) ..   26. *thomasii*

  Pedicels 2–3 mm. long ; calyx 1 cm. long, lobes narrowly deltoid, the margins flat ..   ..   25. *liberiensis*

Fruit glabrous or almost so :

  Fruit very large, about 10 cm. long and 6·5 cm. broad, obovoid, sparsely strigulose-puberulous, especially near apex, otherwise glabrous ; calyx 1·5 cm. long, with 5–6 wrinkled and minutely puberulous rounded lobes ..   18. *crassiflora*

  Fruit smaller, never more than 5 cm. long and 4·5 cm. broad :

  Fruit cylindric, up to 5 cm. long and 4·5 cm. broad—(see fig. 202B 5)
17. sanza-minika

  Fruit globose, ellipsoid or ovoid, smaller than last :

  Leaves up to 6 cm. long and 2 cm. broad, obliquely rhombic, apex subacuminate and emarginate, nerves and veins invisible ; fruit up to 2·5 cm. long and 2 cm. broad, ovoid or ovoid-cylindric, verruculose ..   ..   35. *hoyleana*

  Leaves nearly always larger, never emarginate, at least the secondary nerves always visible :

  Fruit ellipsoid, small, up to 1·4 cm. long and 0·9 cm. broad ; seeds 1–2 :

  Calyx glabrous outside except for a few minute marginal cilia, deeply lobed almost to base, lobes suborbicular ; fruit drying black, glabrous
5. abyssinica

  Calyx tomentellous, shallowly and irregularly lobed, lobes broadly deltoid ; fruit drying pale brown, with a few hairs near apex ..   ..   7. *ferrea*

  Fruit globose, larger, never less than 1·5 cm. long and 1·5 cm. broad ; seeds 6–8, or sometimes fewer by abortion :

  Calyx cup-shaped, surrounding the lower half of fruit, margin subtruncate, but usually irregularly broken away ; fruit up to 3 cm. long and 3·4 cm. broad
27. monbuttensis

  Calyx not cup-shaped :

  Endosperm ruminate :

  Calyx-lobes 4, about 4 mm. long and 2·5 mm. broad, reflexed ; fruit globose, about 3 cm. long and 3 cm. broad ; tertiary nerves and veins conspicuously raised and reticulate on upper surface   2. *viridicans*

  Calyx-lobes 3(–4), 3 mm. long, or calyx irregularly lobed, lobes not reflexed, tertiary nerves and veins inconspicuous or invisible :

Leaves caudate-acuminate ; fruit about 3·2 cm. long and 3·5 cm. broad ;
    seeds dark brown, surface transversely wrinkled   ..   28. *melocarpa*
Leaves subacute to subacuminate ; fruit up to 2 cm. long and 2·4 cm.
    broad ; seeds black, surface smooth ..   ..   ..   34. *elliotii*
Endosperm uniform :
Calyx-lobes 5–8 mm. long, distinctly longer than broad :
Leaves abruptly tapering at apex to parallel-sided acumen, lateral nerves
    in 12–20 pairs ; pedicels more than 3 mm. long ; fruiting calyx about
    8 mm. long   ..   ..   4. *piscatoria*
Leaves gradually tapering to the deltoid acumen ; lateral nerves in 7–10
    pairs ; fruits subsessile, fruiting calyx 1–1·2 cm. long   33. *soubreana*
Calyx-lobes up to 3 mm. long, not longer than broad, or calyx unlobed :
Leaves minutely strigulose beneath :
Fruiting calyx unlobed, dish-shaped ; fruit about 2·6 cm. long and 3 cm.
    broad ; leaves with tertiary nerves and veins scarcely visible
                                                            23. *nigerica*
Fruiting calyx with 4–5 well-defined hemi-orbicular lobes ; fruit about
    2 cm. long and 2 cm. broad ; leaves with a prominent, lax reticulum
                                                            1. *polystemon*
Leaves glabrous :
Leaves drying reddish ; acumen gradually tapering ; lateral nerves in
    (6–)7–9 pairs, tertiary nerves prominent ; calyx 4–5 mm. long,
    irregularly 3(–4) lobed ; fruit about 2·8 cm. long and 2·8 cm. broad
                                                    29. *alboflavescens*
Leaves drying blackish or greenish-black ; tertiary nervation scarcely
    visible :
Leaves suddenly tapering to the parallel-sided acumen ; lateral nerves
    in 4–5 pairs   ..   ..   30. *zenkeri*
Leaves gradually tapering to the deltoid acumen ; lateral nerves in 6–9
    pairs   ..   ..   6. *cooperi*

1. **D. polystemon** *Gürke* in Engl. Bot. Jahrb. 43 : 210 (1909). Forest shrub or tree up to 90 ft. high ; bole
slightly fluted at base ; flowers white, fragrant.
    [Br.]**Cam.**: Likomba Plantation *Mildbr.* 10754. Also in Cameroun, the Congos and Angola.
    [I have not seen this specimen which is cited in the first edition ; it is probably now destroyed.—F.W.]
2. **D. viridicans** *Hiern* in J. Bot. 59 : 129 (1921). *D. kekemi* Aubrév. & Pellegr. in Bull. Soc. Bot. Fr. 83 :
621, t. 1, 1 (1937) ; Aubrév. Fl. For. C. Iv., ed. 2, 3 : 164, t. 309, 1–5. Small or medium-sized forest
tree 25–60 ft. high ; bole tall, straight ; bark black, smooth at first, becoming scaly ; fruit yellow at
first, turning red, then black.
    S.L.: Njala, Moyamba (fr. Nov.) *Deighton* 6014 ! **Iv.C.**: Abidjan (fr. Jan.) *Aubrév.* 188 ! Rasso (fr. Dec.)
Aubrév. 545 ! Agboville (fr. June) *Aubrév.* 1385 ! Guiglo (fr. Jan.) *Serv. for.* 4042 ! **Ghana**: Pra Anum
F.R., Bekwai (fr. Oct.) *Annan* 122 ! Kwahu Praso (fr. Feb.) *Vigne* FH 1613 ! S. Fomang Su F.R. (♂ fl.
Jan.) *Vigne* FH 2692 ! **S.Nig.**: Omo & Shasha F.R. *Jones & Onochie* FHI 16712 ! 16753 ! Shasha F.R.
(fr. Feb.) *Lamb* 185 ! Njujua R., Obudu (♂ fl. May) *Latilo* FHI 30930 ! [Br.]**Cam.**: Victoria (fr. Nov.)
Maitland 768 ! Bambuko F.R., Kumba *Keay* FHI 37422 ! Also in the Congos, Gabon and Cabinda.
3. **D. dendo** *Welw. ex Hiern* in Trans. Camb. Phil. Soc. 12 : 195, t. 10 (1873) ; F.T.A. 3 : 523. *D. atro-
purpurea* Gürke in Engl. Bot. Jahrb. 26 : 67 (1899) ; op. cit. 43 : 211, t. 4 (1909) ; Mildbr. in Notizbl.
Bot. Gart. Berl. 9 : 1051 (1926) ; F.W.T.A., ed. 1, 2 : 5 ; Kennedy For. Fl. S. Nig. 191. Small or
medium-sized forest tree 25–50 ft. high ; bole slightly fluted with a blackish Guava-like scaly bark ;
slash black outside, bright yellow inside ; sapwood white ; heartwood sometimes black in the centre ;
corolla cream ; fruit and fruiting sepals dark red.
    S.Nig.: Benin (fr. Nov.) *Baldwin* 13730a ! Ojogba-Ugun F.R., Benin (fr. June) *Olorunfemi* FHI 38080 !
Osun F.R., Ijebu (♂ fl. Apr.) *Ejiofor* FHI 26115 ! Akure F.R., Ondo (fr. July) *Symington* FHI 4149 !
Sapoba F.R. (fr. Aug.) *Okeke* FHI 30133 ! [Br.]**Cam.**: Bambuko F.R., Kumba *Keay* FHI 37474 ! Also
in Cameroun, Gabon, Congo and Angola. (See Appendix, p. 346.)
4. **D. piscatoria** *Gürke* in Engl. Bot. Jahrb. 46 : 155 (1911) ; White in Bull. Soc. Bot. Brux. 26 : 280, tt. 76
A–E, 77 A–B ; F.W.T.A., ed. 1, 2 : 6 (excl. syn. *Maba soubreana*) ; Kennedy For. Fl. S. Nig. 191.
M. chrysantha Kennedy l.c. 193 (1936), English descr. only. Shrub or medium-sized forest tree up to
90 ft. high ; bole long, straight, slender, often fluted and sometimes with very small buttresses at base ;
bark black, rough, exfoliating in large scales ; slash black outside, yellow or orange inside ; sapwood
white or pinkish white, heartwood often black, sometimes with greenish-brown streaks ; fruits dull
crimson.
    Guin.: Frecariah (♀ fl. June) *Jac.-Fél.* 1789 ! **S.L.**: Kofiu Mt. (fr. Jan.) *Sc. Elliot* 4600 ! Kurusu (♂ fl.
Apr.) *Sc. Elliot* 5538 ! Matuta (♂ fl. June) *Thomas* 455 ! Bumbuna, Koinadugu (fr. Oct.) *Thomas* 3697 !
Ghana: Dunkwa (♂ fl. Nov.) *Vigne* FH 238 ! Jimira F.R., Kumasi (fr. June) *Vigne* FH 2962 ! **S.Nig.**:
Sapoba *Kennedy* 773 ! 774 *bis* ! Shasha R., Ijebu (♂ fl., fr. Mar.) *Richards* 3212 ! 3331 ! Ekosogo, Ogoja
(♂ fl. Apr.) *Jones* FHI 1539 ! [Br.]**Cam.**: Bambuko F.R., Kumba (fr. Jan.) *Keay* FHI 37472 ! Also in
Cameroun, Rio Muni, the Congos and Cabinda. (See Appendix, p. 349.)
5. **D. abyssinica** (*Hiern*) *F. White* in Bull. Jard. Bot. Brux. 26 : 294, tt. 76 F–K, 77 C–D (1956). *Maba
abyssinica* Hiern in Trans. Camb. Phil. Soc. 12 : 132 (1873) ; F.T.A. 3 : 516. *M. warneckei* Gürke
(1911)—F.W.T.A., ed. 1, 2 : 7. Medium-sized evergreen forest tree, to 90(–120) ft. high, with long
straight slender bole and small dense crown ; bark thin, dark, rough, reticulate and scaling on old trees ;
slash black outside, bright yellow inside ; flowers cream ; fruit black.
    Guin.: Kouroussa *Pobéguin* ! Gangan (♂ fl. Mar.) *Roberty* 17722 ! **Mali**: Kita (fr. Dec.) *Jaeger* 3778 !
(♂ fl. Aug.) *Jaeger* 5776 ! *Jaeger* 5776 ! **Iv.C.**: Anoumaba, Dimbokro *Chev.* 22394 ! **Ghana**: Odumasi, Volta R. (fr.
Nov.) *Irvine* 1700 ! 2109 ! Achimota (♂ fl. June) *Irvine* 2779 ! Aburi Scarp (fr. Oct.) *Scholes* 303 ! **Togo
Rep.**: Kodjeloa (♂ fl. Feb., fr. July) *Kersting* 500 ! 689 ! Lomé *Mildbr.* 7506 ! *Warnecke* 220 ! **S.Nig.**:
Idanre Hills (♂ fl. Apr.) *Symington* FHI 3359 ! Aku Rock (fr. Nov.) *Hambler* 5176 ! Widespread in Africa
to Eritrea, Angola and S. Rhodesia.
6. **D. cooperi** (*Hutch. & Dalz.*) *F. White* in Bull. Jard. Bot. Brux. 26 : 243 (1956). *Maba cooperi* Hutch. &
Dalz. F.W.T.A., ed. 1, 2 : 7 (1931) ; Kew Bull. 1937 : 55. *M. gavi* Aubrév. & Pellegr. in Bull. Soc. Bot.
Fr. 83 : 622, t. 1, 2 (1937) ; Aubrév. Fl. For. C. Iv., ed. 2, 3 : 174, t. 315, 1–3. *Diospyros gavi* (Aubrév.
& Pellegr.) F. White l.c. (1956). Small forest tree 18–40 ft. high ; bark greenish-brown, smooth ; flowers
borne in clusters on stem, mustard-yellow ; fruit yellow.

S.L.: Kambui F.R. (♂ fl. Oct.) *Edwardson* 169 ! Kafago (fr. Apr.) *Sc. Elliot* 5613 ! Mamaha (♀ fl. Nov.) *Thomas* 4585 ! 4653 ! Yonibana, Tonkolili (♂ fl. Nov.) *Thomas* 4666 ! **Lib.:** Ba, Mano R., Boporo (fr. Dec.) *Baldwin* 10697 ! Monrovia (♂ fl. Oct.) *Cooper* 92 ! 316 ! Peahtah *Linder* 1029 ! 1066 ! **Iv.C.:** Agboville to Abenguru *Aubrév.* 186 ! Dakpadu (fr. Jan.) *Aubrév.* 874 ! Man to Danané (fr. Mar.) *Aubrév.* 1095 ! 60 km. N. of Sassandra (fr. Feb.) *Leeuwenberg* 2812 ! 2845 ! (See Appendix, p. 349.)

[*Pobéguin* 910 from Langofome River, Guinée, may be a variant of this species ; its fruit, however, is ovoid-conic.]

7. **D. ferrea** (*Willd.*) *Bakh.* in Gard. Bull. Str. S. 7 : 162 (1933) ; in Bull. Jard. Bot. Buitenz., sér. 3, 15 : 50 (1936), 431 (1941). *Ehretia ferrea* Willd. Phytogr. 1 : 4, t. 2, 2 (1794). *Maba buxifolia* (Rottb.) A. L. Juss. (1804)—Pers. Syn. Pl. 2 : 606 (1807) ; Hiern in Trans. Camb. Phil. Soc. 12 : 116 (1873) ; F.T.A. 3 : 515 ; Thonner Blütenfl. Afr. t. 126 ; F.W.T.A., ed. 1, 2 : 7 ; not *D. buxifolia* (Bl.) Hiern. *Pisonia buxifolia* Rottb. (1783). *Ferreola guineensis* Schum. & Thonn. Beskr. Guin. Pl. 448 (1827). *M. guineensis* (Schum. & Thonn.) A. DC. (1844). *M. smeathmannii* A. DC. (1844). *M. secundiflora* Hutch. in Kew Bull. 1921 : 384, t. 8. *M. lancea* of Chev. Bot. 395 ; F.W.T.A., ed. 1, 2 : 7 (excl. syn. *M. sudanensis*) ; Aubrév. Fl. For. Soud.-Guin. 422, t. 92, 4, not of Hiern. *M. ferrea* (Willd.) Aubrév. Fl. For. C. Iv., ed. 2, 3 : 174, t. 310, 6–7 (1959), nom. illegit. Forest shrub or small under-storey tree 5–45 ft. high with spreading and arching branches ; bark dark grey, slash black outside, red inside, sapwood white or pale straw-coloured ; young shoots densely rusty pilose ; flowers white.

Sen.: Casamance (fl.) *Leprieur* ! **Mali:** Kita (fr. Apr.) *Dubois* 185 ! Timbo (fr. Apr.) *Chev.* 13278 ! Fouta Djalon (fl. Apr.) *Chev.* 13607 ! Kouroussa (fr.) *Pobéguin* 851 ! **Port.G.:** Bissau (fr. Jan.) *Esp. Santo* 1633 ! Farim, Begene (♂ fl. Apr.) *Esp. Santo* 2461 ! **S.L.:** (♀ fl.) *Smeathmann* ! **Iv.C.:** Abidjan (fr.) *Aubrév.* 903 ! Sassandra (fl. May) *Chev.* 17946 ! Cavally (fl.) *Chev.* 19932 ! **Ghana:** Pamu-Berekum F.R. (fr. Sept.) *Vigne* FH 2518 ! **Dah.:** Cotonou (fr. Mar.) *Debeaux* 345 ! **N.Nig.:** Dogon Dawa, Zaria (fr. Feb.) *Daggash* FH 31412 ! Katagum (fr.) *Dalz.* 413 ! Jos Plateau (♂ fl. May) *Lely* P266 ! **S.Nig.:** Akure F.R., Ondo (♂ fl. Oct.) *Keay* FHI 21556 ! Ibadan *Keay* FHI 22548 ! Iyamayong F.R., Ogoja (fr. Apr.) *Binuyo* FHI 41231 ! [**Br.]Cam.:** Johann-Albrechtshöhe (= Kumba) (♂ fl.) *Staudt* 617 ! Gangume, Gashaka Dist. (♂ fl. Dec.) *Latilo & Daramola* FHI 28911 ! Scattered through the palaeotropics as far as Hawaii.

[*D. ferrea* is a very variable species. Some of the variation in our area can be correlated with ecology and distribution, but insufficiently precisely to warrant taxonomic recognition.]

8. **D. conocarpa** *Gürke & K. Schum.* in Engl. Bot. Jahrb. 14 : 311 (1892) ; Mildbr. in Notizbl. Bot. Gart. Berl. 9 : 1048 (1926). *D. staudtii* Gürke (1899)—F.W.T.A., ed. 1, 2 : 6, nom. illegit. Forest shrub or small slender tree up to 30 ft. high ; branches in pseudo-whorls ; bole black with thin papery scales falling to expose powdery purple surface ; slash black outside, pale yellow and fibrous inside ; sapwood pale cream-yellow, hard and tough with blackish markings at centre ; inflorescence drooping ; corolla white or pinkish.

S.Nig.: Afi River F.R., Ogoja (♂ fl. May) *Jones & Onochie* FHI 17398 ! Ojogba-Ugun F.R., Benin (♂ fl. June) *Olorunfemi* FHI 38068 ! Eket *Talbot* 3290 ! 3320 ! [**Br.]Cam.:** Kumba (♂ fl. Mar.) *Binuyo & Daramola* FHI 35630 ! S. Bakundu F.R. *Brenan* 9424 ! Also in Cameroun, Rio Muni, Gabon, Congo and Cabinda.

9. **D. kamerunensis** *Gürke* in Engl. Bot. Jahrb. 26 : 69 (1899) ; op. cit. 43 : 208, t. 3 M-O (1909) ; Aubrév. Fl. For. C. Iv., ed. 2, 3 : 162, t. 308, 1–3. *D. pallescens* A. Chev. Bot. 397, name only. Small or medium-sized forest tree 25–50 ft. high ; bark rough ; wood pink ; flowers white or pale yellow ; fruit yellow or orange.

Lib.: Monrovia (♂ fl. Apr., ♀ fl. Nov.) *Cooper* 90 ! 116 ! 304 ! **Iv.C.:** Abidjan (♂ fl. Oct., Nov.) *Aubrév.* 129 ! 171 ! Yapo *Aubrév.* 588 ! Songan *Chev.* 17686 ! 17728 ! Teké (♂ fl. Sept.) *de Wilde* 572 ! **Ghana:** Wasaw-Aowin, Benso (fr. Apr.) *Andoh* FH 5477 ! Kumasi *Cummins* 201 ! Pra Anum F.R. (♂ fl. July) *Darko* 196 ! Ankasa River F.R., Ahanta-Nzima (fr. Feb.) *Enti* FH 6921 ! Banka, Bekwai (♂ fl. Sept.) *Vigne* FH 1363 ! Also in Cameroun and Gabon. (See Appendix, p. 347.)

10. **D. mannii** *Hiern* in Trans. Camb. Phil. Soc. 12 : 255 (1873) ; F.T.A. 3 : 524 ; Chev. Bot. 396 ; Aubrév. Fl. For. C. Iv., ed. 2, 3 : 160, t. 308, 4–6. *D. talbotii* Wernham in Cat. Talb. 57 (1913) ; F.W.T.A., ed. 1, 2 : 6. *D. aggregata* Gürke (1909). *D. pseudaggregata* Mildbr. (1926). *D. ivorensis* Aubrév. & Pellegr. in Bull. Soc. Bot. Fr. 83 : 621 (1937) ; Aubrév. Fl. For. C. Iv., ed. 2, 3 : 160, t. 308, 7–8. Medium-sized forest tree 40–60 ft. high ; bark black, smooth, hard ; slash black outside, yellow inside ; sapwood citron-yellow, heartwood sometimes with a black centre ; branchlets densely brown-villous ; leaves pale and glaucous beneath ; fruits orange, borne on old branches.

S.L.: Dodo Hills F.R., Kono (fr. Jan.) *Sawyerr* FHI 13569 ! Gola F.R., Kenema *Small* 670 ! **Lib.:** Mano R., Boporo (fr. Dec.) *Baldwin* 10701 ! **Iv.C.:** Taï, Man *Aubrév.* 1214 ! Guiglo, Man *Aubrév.* 2069 ! Byanouan (fr. Mar.) *Chev.* 17695 ! Yapo *Chev.* 22319 ! Divo to Lakota (fr. Aug.) *de Wilde* 311. **Ghana:** Krokosua Hills, Sefwe (♀ fl. Apr.) *Enti* FH 6700 ! **S.Nig.:** Okomu F.R., Benin *Brenan* 8993 ! 9048 ! Oban *Gray* 2/11 ! Also in Ubangi-Shari, Cameroun, Gabon, the Congos and Cabinda.

11. **D. fragrans** *Gürke* in Engl. Bot. Jahrb. 46 : 154 (1912) ; Mildbr. in Notizbl. Bot. Gart. Berl. 9 : 1050 (1926). Shrub or small forest tree up to 50 ft. high, sometimes multiple-stemmed ; bole shallowly and irregularly fluted at base ; bark dark brown, peeling in thin papery flakes up to 2 by 1 in. ; slash blackish outside, deep ochre inside ; sapwood creamy-buff ; leaves glaucous-green beneath ; flowers fragrant, borne in fascicles all up the trunk ; corolla white ; fruit orange.

[**Iv.]Cam.:** S. Bakundu F.R., Kumba (♂ fl. Mar.) *Binuyo & Daramola* FHI 35629 ! *Brenan* 9405 ! Also in Cameroun and Gabon.

[*Onochie* FHI 33426 from Ute Ogboji Forest, Benin may be this species, but as the specimen is sterile and was collected from a stool shoot, there is some doubt.]

12. **D. suaveolens** *Gürke* in Engl. Bot. Jahrb. 26 : 68 (1899). *D. confertiflora* Gürke ex Kennedy, For. Fl. S. Nig. 192 (1936), English descr. only ; not of (Hiern) Bakh. (1933). *D. barteri* of F.W.T.A., ed. 1, 2 : 4, partly (*Mildbr.* 10522), not of Hiern. Small or medium-sized forest tree 30–90 ft. high ; bole fluted at the base ; bark black, rough, scaling ; slash black outside, deep yellow, turning orange inside ; sapwood white, turning pale yellow ; heartwood pinkish ; leaves ashy-grey beneath ; calyx pink ; corolla ivory white ; fruits with irritant hairs, borne on upper part of bole and main branches.

S.Nig.: Owo F.R. (♂ fl. Apr.) *Jones* FHI 3481 ! Idanre F.R. (fr. June) *Keay* FHI 37042 ! Sapoba *Kennedy* 771 ! 778 ! Omo Sawmill, Ijebu (fr. Aug.) *Okeke & Daramola* FHI 36891 ! [**Br.]Cam.:** Bambuko F.R., Kumba *Keay* FHI 37438 ! Likomba Plantation *Mildbr.* 10522 ! Also in Cameroun and Gabon.

13. **D. bipindensis** *Gürke* in Engl. Bot. Jahrb. 26 : 70 (1899) ; Mildbr. in Notizbl. Bot. Gart. Berl. 9 : 1051 (1926). *D. busgenii* Gürke (1911). *D. flavovirens* Gürke (1914). Forest shrub 8–20 ft. high or tree up to 60 ft. high ; bole shallowly fluted with rounded ridges ; bark dark brown, with very close anastomosing fissures ; slash : outer bark jet black, inner bark cream outside, ochre-yellow inside ; sapwood white darkening to pale yellow ; heartwood sometimes black ; flowers strongly scented ; corolla yellowish-cream ; ripe fruit whitish-pruinose, borne on branches.

[**Br.]Cam.:** Banga, Kumba (♂ fl. May) *Binuyo & Daramola* FHI 35595 ! S. Bakundu F.R., Kumba (♂ fl. Mar., fr. Aug.) *Brenan* 9418 ! *Olorunfemi* FHI 30703 ! Also in Cameroun, Gabon, Congo and Uganda.

14. **D. physocalycina** *Gürke* in Engl. Bot. Jahrb. 26 : 68 (1899) ; Mildbr. in Notizbl. Bot. Gart. Berl. 9 : 1052 (1926). *D. xanthochlamys* Gürke op. cit. 43 : 210 (1909), not of F.W.T.A., ed. 1. *Heisteria winkleri* Engl. (1909). Small forest tree up to 30 ft. high ; bark blackish ; slash jet black outside, pinkish-cream inside and immediately turning dark brown, buttercup yellow at inner edge ; calyx pale purplish-green ; corolla white or yellow ; fruit red, surrounded by green calyx.

S.Nig.: Okomu F.R., Benin (♂ fl. Feb., ♀ fl. Mar.) *Brenan* 9158 ! *Ross* 137 ! Idanre F.R. (fr. Aug.) *Okafor & Daramola* FHI 35283 ! Also in Cameroun and Gabon.

15. **D. barteri** *Hiern* in Trans. Camb. Phil. Soc. 12 : 187 (1873) ; F.T.A. 3 : 521 ; Oliv. in Hook. Ic. Pl. 23 : t. 2300 (1894) ; F.W.T.A., ed. 1, 2 : 4 ; (excl. *Mildbr.* 10522). *D. rubicunda* Gürke in Engl. Bot. Jahrb. 43 : 206, t. 3A (1909). Forest shrub or small tree 3–22 ft. high, sometimes scrambling, young parts covered with ferrugineous hairs ; leaves glaucous beneath ; fruit pale yellow or orange.
   Ghana: Akwapim (fr. Apr.) *Irvine* 1153 ! Aburi (♂ fl. July) *Johnson* 1086 ! **S.Nig.:** Lagos (fr.) *Barter* 2194 ! (♀ fl.) *Dalz.* 1099 ! (♂ fl. May) *Dalz.* 1180 ! Ibadan (fr. Apr.) *Meikle* 1393 ! Eket *Talbot* 3331 ! Also in Cameroun. (See Appendix, p. 346.)

16. **D. undabunda** *Hiern ex Greves* in J. Bot. 67, Suppl. Gamopet. : 80 (1929). Forest shrub or tree up to 40 (–120) ft. high ; bole dark grey or black, smooth, with long shallow fissures ; slash deep chocolate-brown outside, pale yellow-brown inside ; leaves deep green and very glossy above ; flowers pale yellow, fragrant ; fruit bright red.
   S.Nig.: Omo and Shasha F.R. *Jones & Onochie* FHI 17589 ! Okomu F.R., Benin (fr. June) *Onochie* FHI 34605 ! Also in Ubangi-Shari, Congo, Cabinda, Angola and N. Rhodesia.
   [The Nigerian plant probably belongs to a distinct sub-species].

17. **D. sanza-minika** *A. Chev.* Vég. Ut. 5 : 155 (1909) ; Bot. 397 ; Aubrév. Fl. For. C. Iv., ed. 2, 3 : 168, t. 314. *D. nsambensis* Gürke (1909). Medium-sized forest tree 40–80 ft. high ; bole long, slender without buttresses ; bark black, with deep longitudinal ridges and fissures, very hard and more resistant to decay than the wood ; heartwood sometimes with a black centre ; leaves greyish beneath ; flowers white ; fruit yellow with a mucilaginous white pulp.
   S.L.: Kambui, Kenema (fr. May) *Lane-Poole* 224 ! Gola F.R., Kenema (♀ fl. Apr., fr. Jan.) *Small* 612 ! 659 ! *Unwin & Smythe* 40 ! Kenema (fr. Jan.) *Thomas* 7654 ! **Lib.:** Monrovia (fr. Apr., ♀ fl. Mar., ♂ fl. May) *Cooper* 71 ! 271 ! 422 ! **Iv.C.:** Fort Binger, Cavally R. (♂ fl., fr. July) *Chev.* 19431 ! 19436 ! 56 km. N. of Sassandra (fr. Jan.) *Leeuwenberg* 2620 ! Abidjan (♂ fl.) *Aubrév.* 39 ! Zaranou, Abengourou (fr. Mar.) *Chev.* 16284 ! **Ghana:** Benso, Wasaw-Aowin (fr. Oct.) *Osei* FH 5245 ! Dunkwa (♀ fl., fr. Apr.) *Vigne* FH 106 ! 1109 ! Ateiku (♂ fl. May) *Vigne* FH 2004 ! Also in Cameroun and Gabon. (See Appendix, p. 349.)

18. **D. crassiflora** *Hiern* in Trans. Camb. Phil. Soc. 12 : 260 (1873) ; F.T.A. 3 : 525 ; Kennedy For. Fl. S. Nig. 191. *D. incarnata* Gürke (1909). Forest tree up to 60 ft. high and 12 ft. girth, but often hollow ; bole slightly fluted at the base ; bark black, flaking to expose a black surface ; slash black and brittle outside, pale salmon pink with cream streaks inside, somewhat granular ; sapwood cream with black streaks ; heartwood black, a source of commercial ebony ; flowers white.
   S.Nig.: Sapoba F.R. (fr. Aug.) *Keay* FHI 37945 ! Okomu F.R., Benin (♂ fl. Oct., Mar.) *Akpabla* 1133 ! *Oyebade* FHI 33655 ! Afi River F.R. *Jones & Onochie* FHI 18727 ! Calabar *Thomson* 47 ! [Br.]**Cam.:** Victoria (♂ fl. Feb.) *Maitland* 407 ! Mamfe (♂ fl. Apr.) *Rosevear* 1/34 ! Also in Ubangi-Shari, Cameroun, Gabon and the Congos. (See Appendix, p. 347.)

19. **D. gabunensis** *Gürke* in Engl. Bot. Jahrb. 26 : 72 (1899) ; Aubrév. Fl. For. C. Iv., ed. 2, 3 : 166, t. 311. *D. castaneifolia* A. Chev. in Journ. de Bot. sér. 2, 2 : 116 (1909) ; Aubrév. tom. cit. : 168, t. 312, 4–5. *D. gilgiana* Gürke op. cit. 43 : 207, t. 3, K-L (1909). *D. mamiacensis* Gürke l.c. 205 (1909). Small or medium-sized forest tree 20–65 ft. high ; bole long, straight, slender, without buttresses ; bark black, smooth, hard and brittle like glass ; slash black outside, pale biscuit-brown inside, yellow towards inner edge, not darkening ; sapwood pale yellow ; heartwood sometimes with black veins ; flowers fragrant ; calyx blackish-brown ; corolla white ; fruits sometimes borne on stem.
   S.L.: Kambui (♂ fl. Sept.) *Aylmer* 600 ! Kambui Hills F.R. (fr. Dec.) *Edwardson* 43 ! **Lib.:** St. John R., Tappita (fr. Aug.) *Baldwin* 9137 ! Monrovia (♂ fl. Mar., Apr., fr. Oct.) *Cooper* 78 ! 87 ! 275 ! Gbanga (♀ fl., fr. Sept.) *Linder* 641 ! **Iv.C.:** Abidjan (♂ fl. Nov.) *Aubrév.* 169 ! Yapo (fr. Dec.) *Aubrév.* 609 ! Grabo (♂ fl. July) *Chev.* 19665 ! 61 km. N. of Sassandra (fr. Jan.) *Leeuwenberg* 2508 ! **Ghana:** Benso (♀ fl., fr. Nov.) *Andoh* FH 5387 ! 5812 ! Tarkwa *Chipp* 247 ! **S.Nig.:** Uquo, Eket (♀ fl. May) *Onochie* FHI 33186 ! [Br.]**Cam.:** S. Bakundu F.R., Kumba (♂ fl. Mar.) *Brenan* 9408 ! Bambuko F.R., Kumba (fr. Sept.) *Olorunfemi* FHI 30752 ! Johann-Albrechtshöhe (=Kumba) (♂ fl. Apr.) *Staudt* 958 ! Also in Cameroun, Gabon and Congo. (See Appendix, p. 347.)

20. **D. chevalieri** *De Wild.* Pl. Bequaert. 3 : 538 (1926) ; Aubrév. Fl. For. C. Iv., ed. 2, 3 : 170, t. 312, 1–3, not of Lecomte (1928). *D. macrophylla* A. Chev. in Journ. de Bot. sér. 2, 2 : 116–117 (1909) ; F.W.T.A., ed. 1, 2 : 4, not of Blume (1826). *D. linderi* Hutch. & Dalz. ex Cooper & Record in Yale Univ. For. Bull. 31 : 99 (1931). Forest shrub up to 10 ft. high ; wood bright yellow ; flowers white, becoming hessian-brown ; fruit chocolate-brown and tomentose at first with bristly hairs in tufts, red and glabrous when ripe.
   S.L.: Bunumbu (fr. Apr.) *Deighton* 3938 ! Kambui F.R. (fr. Mar.) *Jordan* 2012 ! **Lib.:** Tawata, Boporo (fr. Nov.) *Baldwin* 10302 ! Duo, Sinoe (fr. Mar.) *Baldwin* 11362 ! Monrovia (fr. Feb., Nov.) *Cooper* 130 ! 200 ! Gbanga (fl. Sept.) *Linder* 562 ! 567 ! **Iv.C.:** Sassandra R. to Mid. Cavally (♂ fl. July) *Chev.* 19289 ! Grabo, Tabou *Chev.* 19659 ! Keéta, Man (♂ fl. Sept.) *Chev.* 19318 ! **Ghana:** Benso, Tarkwa (fr. Nov.) *Andoh* FH 5411 ! Ankasa F.R., Axim (fr. Feb.) *Enti* FH 6922 ! Wasaw, Prestea *Vigne* FH 3087 ! (See Appendix, p. 347.)

21. **D. mespiliformis** *Hochst. ex A. DC.* Prod. 8 : 672 (1844) ; Hiern in Trans. Camb. Phil. Soc. 12 : 165 (1873) ; F.T.A. 3 : 518 ; Chev. Bot. 397 ; Kennedy For. Fl. S. Nig. 190 ; Aubrév. Fl. For. Soud.-Guin. 422–424, t. 92, 1–3 ; Fl. For. C. Iv., ed. 2, 3 : 164, t. 313, 3–4. *D. senegalensis* Perr. ex A. DC. tom. cit. : 234 (1844) ; Benth. Fl. Nigrit. : 442, partly. Forest and savanna tree up to 90 ft. high ; bark rough, dark brown or black, muricate ; slash black outside, pink inside ; flowers white, sweet-scented fruit yellowish.
   Sen.: Dakar (♂ fl. Mar.) *Adam* 938 ! Baol, Bambey (fr. May) *Dubois* 279 ! Dagana (fr. Apr.) *Leprieur* 15 ! Bellevue, Thiès (fr. Aug.) *Trochain* 4180 ! **Gam.:** on the R. Gambia *Whitfield* ! **Mali:** Fangala, Kita (♂ fl. Apr.) *Dubois* ! Bamako (♀ fl. Apr.) *Hagerup* 18 ! Dogo (♀ fl. Apr.) *Davey* 578 ! Fafa, Gao (fr. Mar.) *de Wailly* 5358 ! Ouagadougou (♂ fl. May) Aubrév. 2426 ! **Guin.:** Kouroussa *Pobéguin* 680 ! 694 ! Sankatan to Kankan (♂ fl. Mar.) *Pobéguin* 842 ! **Iv.C.:** Touba *Aubrév.* 1240 ! **Ghana:** Achimota (♂ fl. Jan.) *Irvine* 1748 ! Bamboi, Sunyani (fr. May) *Vigne* FH 3841 ! Tamale to Yendi (♂ fl. Mar.) *Adams* GC 3883 ! Tamale (♂ fl. July) *Dalz.* 10 ! Lisse to Lambusie (♂ fl. Mar.) *Kitson* 780 ! **Togo Rep.:** Basari to Sokodé *Kersting* A552 ! Lomé *Warnecke* 325 ! 330 ! **Dah.:** Abbo, Zagnanado (fr. Feb.) *Chev.* 22961 ! Dassa-Zoumé (fr. May) *Chev.* 23640 ! Atacora Mts. (fr. June) *Chev.* 24067 ! **Niger:** Maradi to Madoua (♂ fl. Feb.) *Chev.* 43710 ! **N.Nig.:** Zamfara F.R., Sokoto (♂ fl. Apr.) *Keay* FHI 15673 ! Katagum *Dalz.* 219 ! Assob Falls (fr. Oct.) *Hepper* 1043 ! Emiworo, Share F.R., Ilorin (fr. Jan.) *Ujor* FHI 31611 ! **S.Nig.:** Lagos *Foster* 28 ! 35 ! Owo F.R., Ondo *Jones* FHI 3616 ! Olokemeji F.R. (fr. Sept.) *Prov. For. Off.* Oyo FHI 14590 ! Bebi, Ogoja (♂ fl. June) *Catteral* ! Mamu River F.R., Onitsha *Kennedy* 2536 ! Widespread in Africa to the Red Sea, Transvaal and S.W. Africa. Also in Arabia. (See Appendix, p. 347.)

22. **D. tricolor** (*Schum. & Thonn.*) *Hiern* in Trans. Camb. Phil. Soc. 12 : 183, t. 5 (1) (1873) ; F.T.A. 3 : 521 ; Gürke in Engl. Bot. Jahrb. 43 : 203, t. 2 (1909). *Noltia tricolor* Schum. & Thonn. Beskr. Guin. Pl. 189 (1827). Much branched shrub of coastal thicket about 5 ft. high ; fruit yellow or orange.
   Iv.C.: Abouabou, Abidjan (fr. Feb.) *Leeuwenberg* 1471 ! **Ghana:** Axim (fr. Feb., ♀ fl. Nov.) *Irvine* 2157 ! *Vigne* FH 4817 ! **Togo Rep.:** Lomé (fr. Jan., ♂ fl. May) *Warnecke* 26 ! 310 ! **Dah.:** Cotonou (fr. Mar.) *Chev.* 23361 ! *Debeaux* 338. *Poisson* 144 ! **S.Nig.:** Lagos *Barter* 2199 ! 2234 ! *Dalz.* 1181 ! Also in Gabon. (See Appendix, p. 349.)

23. **D. nigerica** *F. White* in Bull. Jard. Bot. Brux. 33 : ined. (1963). *D. heudelotii* of F.W.T.A., ed. 1, 2 : 6, partly (*Mildbr.* 10531), not of Hiern. Small or medium-sized forest tree 20–80 ft. high ; bole slightly fluted ; bark dark brown, rough, scaling in large slabs ; slash black outside, pale pink or dull red inside with yellow inner edge ; sapwood cream ; leaves glaucous beneath ; flowers yellow.

Fig. 203.—DIOSPYROS MESPILIFORMIS *Hochst. ex A. DC.* (EBENACEAE).

A, male flower. B, longitudinal section of male flower. C, stamens. D, longitudinal section of female flower. E, cross-section of ovary. F, shoot with fruit.

W.E.T.

S.Nig.: Shasha F.R. (♂ fl. Mar., fr. Feb.) *Jones & Onochie* FHI 16963! *Ross* 74! Sapoba *Kennedy* 2551! 2611! Owena, Ondo (♂ fl. May) *Darko* 113! Aponmu, Ondo (fr. Sept.) *Symington* FHI 5056! **[Br.]Cam.:** Likomba *Mildbr.* 10595! S. Bakundu, Kumba *Onochie* FHI 32057! Also in Cameroun and Gabon.

24. **D. heudelotii** *Hiern* in Trans. Camb. Phil. Soc. 12 : 215 (1873); F.T.A. 3 : 524; Chev. Bot. 396; F.W.T.A., ed. 1, 2 : 6, partly (excl. *D. gracilescens*, *D. apiculata & Maba cinnabarina*); Aubrév. Fl. For. C. Iv., ed. 2, 3 : 168, t. 313, 1–2. *Maba lancea* Hiern op. cit. 118 (1873); F.T.A. 3 : 516, not of F.W.T.A., ed. 1, 2 : 7 (excl. *Smeathmann*). *D. guineensis* A. Chev. Bot. 396, name only. Shrub or small or medium-sized forest tree up to 60 ft. high, sometimes cauliflorous; slash black outside, wood pink; flowers white or pale yellow; fruit yellow or orange.
   Sen.: Sinédou, Casamance (♂ fl. Jan.) *Chev.* 2818! **Port.G.:** Cantanhez (fr. Feb.) *d'Orey* 289! Bissalanca, Bissau (♀ fl. Jan.) *Esp. Santo* 1660! Bafata (♂ fl. Jan.) *Esp. Santo* 2871! Catio (fr. June) *Esp. Santo* 2998! **Guin.:** Iles de Los, Conakry (fr. Feb.) *Chev.* 12115! Kissidougou *Chev.* 20706! Fouta Djalon *Heudelot* 638! Ditinn *Jac.-Fél.* 607! **S.L.:** *Smeathmann*! Newton, Colony (♂ fl. Nov.) *Deighton* 1475! Kambui F.R., Kenema *Edwardson* 233! Njala (fr. Apr.) *Deighton* 1911! Kambia (fr. Dec.) *Sc. Elliot* 5367! **Iv.C.:** Rasso (fr. Dec.) *Aubrév.* 159! 583! Taï (fr. Apr.) *Aubrév.* 1220! Dimbokro (fr. Nov.) *Chev.* 22400! **Ghana:** Benso, Tarkwa *Andoh* FH 5891! Pamu-Berekum F.R. *Vigne* FH 2509! Akumadan (♂ fl. Oct.) *Vigne* FH 2568! Atuna (fr. Dec.) *Vigne* FH 3512! (See Appendix, p. 347.)

25. **D. liberiensis** *A. Chev. ex Hutch. & Dalz.* F.W.T.A., ed. 1, 2 : 4; Kew Bull. 1937 : 54; Aubrév. Fl. For. C. Iv., ed. 2, 3 : 170, t. 313, 5; Chev. Bot. 396, name only. Forest shrub up to 10 ft. high.
   Lib.: Jabroke, Webo (♂ fl. July) *Baldwin* 6474! **Iv.C.:** Lower Cavally R. (♂ fl. Aug.) *Chev.* 19857! Tabou (fr. Dec.) *Aubrév.* 1669! Yapo (fr. Oct.) *de Wilde* 711!

26. **D. thomasii** *Hutch. & Dalz.* F.W.T.A., ed. 1, 2 : 6; Kew Bull. 1937 : 54. Small forest tree 10–30 ft. high; bark smooth, dark green, finely fissured; leaves bluish-grey beneath; flowers white; fruit yellow.
   Guin.: Badabon *Schnell* 7570! **S.L.:** Njala (♀ fl. Oct., fr. Jan.) *Deighton* 2245! 2830! Kenema (♂ fl. Nov.) *Edwardson* 137! Bumbuna (♂ fl. Oct.) *Thomas* 3293! Pujehun, Bo *Thomas* 8437! **Lib.:** Yila, Gbanga (fr. Aug.) *Baldwin* 9122! Beiden, Boporo (fr. Nov.) *Baldwin* 10268! Monrovia (fr. Nov.) *Cooper* 129! Gbanga (♂ fl. Sept.) *Linder* 481! (See Appendix, p. 349.)

27. **D. monbuttensis** *Gürke* in Engl. Bot. Jahrb. 26 : 66 (1899); op. cit. 43 : 208 (1909); Chev. Bot. 397; Kennedy For. Fl. S. Nig. 190; F. White in Bull. Jard. Bot. Brux. 27 : 518, t. 54 (1957); Aubrév. Fl. For. C. Iv., ed. 2, 3 : 162, t. 310, 1–5. *D. senensis* of F.T.A. 3 : 520, partly (W. Afr. specs. only) not of Klotzsch. Forest shrub or small tree up to 30 ft. high; bole fluted; branches of young trees with large, spreading, sharp-pointed spines; bark papery, peeling, usually purple-brown, sometimes yellow-brown or green; slash pale yellow; sapwood white; flowers fragrant, corolla creamy-white; ripe fruit yellow or orange.
   Iv.C.: Agboville (fr. June, ♂ fl.) *Aubrév.* 1389! 1808 *bis*! Nzi Valley, Baouké (fr. July) *Chev.* 22214! Toumodi, Dimbokro (fr. Oct.) *Pobéguin* 239! **Ghana:** Sefwi Bekwai (fr. Oct.) *Akpabla* 896! New Tafo *Lovi* WACRI 3958! Banka, Obuasi (♂ fl. Mar.) *Vigne* FH 1897! Pamu-Berekum F.R. (fr. Sept.) *Vigne* FH 2481! **Togo Rep.:** Kue stream *Kersting* A637! **Dah.:** Lahama, Allada *Chev.* 23258! **S.Nig.:** Lagos *Rowland*! Ibadan (♂ fl. Apr.) *Meikle* 1394! Aboh *Barter* 290! Abeokuta *Barter* 3390! Mamu R., Onitsha *Kennedy* 2531! **[Br.]Cam.:** Wum (♀ fl. Mar.) *Johnstone* 87/31! Bambui, Bamenda *Maitland* 1634! Also in Cameroun, Ubangi-Shari and Congo. (See Appendix, p. 348.)

28. **D. melocarpa** *F. White* in Bull. Jard. Bot. Brux. 33 : ined. (1963). Forest tree 20–30 (–90) ft. high; slash green then black outside, brown and dry inside; sapwood white, hard; flowers yellowish; fruit yellow.
   S.Nig.: Afi River F.R. *Jones & Onochie* FHI 8302! **[Br.]Cam.:** Kembong F.R., Mamfe (fr. Dec.) *Tiku* FHI 41895! **F.Po:** *T. Vogel*! Also in Cameroun, Gabon and Congo.

29. **D. alboflavescens** (*Gürke*) *F. White* in Bull. Jard. Bot. Brux. 26 : 241 (1956). *Maba alboflavescens* Gürke in Engl. Bot. Jahrb. 43 : 199 (1909). *D. insculpta* Hutch. & Dalz. F.W.T.A., ed. 1, 2 : 4; Kew Bull. 1937 : 54, not of Buch.-Ham. (1827). Small or medium-sized forest tree up to 60 ft. high, sometimes multiple-stemmed; bole fluted; bark black, smooth; slash black outside, reddish-brown inside, crumbling; flowers fragrant, corolla cream; fruit orange, borne on branches and main stem.
   S.Nig.: Okomu F.R., Benin (♂ fl. Jan.) *Brenan & E. W. Jones* 8750! Cross River F.R., Ogoja (fr. June) *Latilo* FHI 31838! Omo, Ijebu (fr. Aug.) *Okeke & Binuyo* FHI 36892! Obom Itiet to Atan Eki, Calabar (fr. May) *Onochie* FHI 33227! Oban *Talbot* 1609! **[Br.]Cam.:** Badshu Abagbe, Mamfe *Smith* 3! Also in Cameroun, Ubangi-Shari, the Congos and Angola. (See Appendix, p. 347.)

30. **D. zenkeri** (*Gürke*) *F. White* in Bull. Jard. Bot. Brux. 26 : 244 (1956). *Maba zenkeri* Gürke in Engl. Bot. Jahrb. 26 : 63 (1899); Mildbr. in Notizbl. Bot. Gart. Berl. 9 : 1047 (1926); F.W.T.A., ed. 1, 2 : 7. *D. longicaudata* Gürke ex Hutch. & Dalz. F.W.T.A., ed. 1, 2 : 4 (1931); Kew Bull. 1937 : 54. Shrub or forest tree up to 60 ft. high; bark black; slash black outside, pale pink or reddish-brown inside with yellow inner edge, granular; sapwood white, heartwood sometimes black; leaves slightly glaucous beneath; corolla cream, fleshy; fruit yellow, turning black, borne on older stems.
   S.Nig.: Afi River F.R. *Jones & Onochie* FHI 16549! 18750! Cross River North F.R., Ogoja (fl. May), *Latilo & Olorunfemi* FHI 43938! **[Br.]Cam.:** Bambuko F.R. *Keay* FHI 37443! *Olorunfemi* FHI 30767! Victoria *Maitland* 619! Johann-Albrechtshöhe (= Kumba) (♂ fl.) *Staudt* 943! Also in Cameroun, Gabon and Congo.

31. **D. preussii** *Gürke* in Engl. Bot. Jahrb. 14 : 313 (1892); op. cit. 26 : 71 (1899). *D. le-testui* Pellegr. in Bull. Mus. Hist. Nat. 4 : 328 (1924). Shrub or small forest tree up to 30 ft. high; bark black; flowers borne on older branchlets and main stem in clusters of up to 30 or more, corolla white; fruiting calyx deep purple.
   S.Nig.: Buden Dunlop Estate, Calabar (fr. July) *Binuyo* FHI 41396! Akampka Rubber Estate, Calabar (♂, ♀ fl. Mar.) *Latilo* FHI 41335! Afi River F.R., Ogoja *Jones & Onochie* FHI 18623! Oban (♂ fl. Mar.) *Richards* 5149! *Talbot* 1329! **[Br.]Cam.:** S. Bakundu F.R. (♂ fl. Feb., fr. Mar.) *Binuyo & Daramola* FHI 35548! *Brenan* 9415! Buenga, Victoria (fr. May) *Motuba* FHI 15068! Badshu Abagbe, Mamfe *Smith* IV! Also in Cameroun and Gabon.

32. **D. canaliculata** *De Wild.* Pl. Bequaert, 3 : 537 (1926). *Maba coriacea* Cummins in Kew Bull. 1898 : 76; Chev. Bot. 395, not *D. coriacea* Hiern (1873). *D. cauliflora* De Wild. in Bull. Jard. Bot. Brux. 5 : 63 (1915), not of Blume (1825), nor of Martius ex Miq. (1856). *D. chlamydocarpa* Mildbr. in Notizbl. Bot. Gart. Berl. 9 : 1052 (1926). *D. xanthochlamys* of F.W.T.A., ed. 1, 2 : 4; of Aubrév. Fl. For. C. Iv., ed. 2, 3 : 159, t. 309, 6–8, not of Gürke. *Cyclostemon gabonensis* of Chev. Bot. 561, not of Pierre. Small or medium-sized forest tree 15–60 ft. high; bark black, thin, smooth, breaking like flint when struck; slash black outside, pale pink turning ochrous-yellow inside; sapwood white, rapidly becoming bright yellow; flowers fragrant; corolla white; fruits borne in fascicles on the branches and trunk; fruiting calyx red.
   Lib.: Gbanga (fr. Jan.) *Harley* 1108! **Iv.C.:** Abidjan *Aubrév.* 216! Agboville (♀ fl., fr. Aug.) *Aubrév.* 1498! Bériby, Cavally basin (fr. Aug.) *Chev.* 19993! Banco (cult., fr. Sept.) *de Wilde* 534! **Ghana:** Anonum (♂ fl. Aug.) *Chipp* 539! Amentia, Ashanti (fr. Mar.) *Vigne* FH 1858! *Irvine* 451! Pra Anum F.R., Bekwai (♂ fl. June) *Mooney* FH 7180! 7199! Banka (fr. Sept.) *Vigne* FH 1362! **N.Nig.:** Dogon Kurmi, Jemaa Div. *Keay & Onochie* FHI 21534! **S.Nig.:** Akilla, Ijebu *Thornewill* 234! Idanre F.R. (♂ fl. Oct.) *Iheuwa* FHI 22372! Omo F.R., Ijebu (♀, ♂ fl. Aug.) *Okeke* FHI 36896! *Okeke & Binuyo* FHI 36897! 36898! **[Br.]Cam.:** Gangume, Gashaka Dist. (fr. Dec.) *Lomax* FHI 32318! Also in Gabon, Ubangi-Shari, Congo, Cabinda and Angola. (See Appendix, p. 349.)

33. **D. soubreana** *F. White* in Bull. Jard. Bot. Brux. 26 : 228, tt. 76 M-R, 77 E-F (1956). *Maba soubreana* A. Chev. Bot. 395, name only; Aubrév. Fl. For. C. Iv., ed. 1, 3 : 146 (1936); ed. 2, 3 : 176, t. 315, 4–8, French descr. only. *D. piscatoria* of F.W.T.A., ed. 1, 2 : 6, partly (syn. *M. soubreana* only), not of

Gürke. Forest shrub or treelet up to 15 ft. high ; bark black ; slash black outside, pale orange-brown inside ; flowers cream ; fruit dark red, turning black, calyx green.

Iv.C.: Davo to Boutobré, Sassandra (♂, ♀ fl. May) *Chev.* 16331 ! 16332 ! Man (♂ fl. July) *Chev.* 19326 ! Bouroukrou, Agboville (fr. Dec.) *Chev.* 16898 ! 16912 ! **Ghana:** Abofaw, Kumasi (♂ fl. May) *Andoh* FH 5491 ! Suhuma F.R., Sefwi (♀ fl. Oct.) *Cansdale* 9 ! Mampong (♀ fl. Apr.) *Vigne* FH 1925 ! Jimira F.R., Kumasi (♂ fl. May) *Vigne* FH 2908 ! 2909 ! **Dah.:** Pedjile to Pobé, Zagnanado (fr. Feb.) *Chev.* 22931 ! **S.Nig.:** Aponmu, Ondo (♀ fl. June) *Ejiofor* FHI 32003 ! Owo F.R. (fr. Apr.) *Jones* FHI 3486 ! Awba Hills F.R., Ibadan (fr. Oct.) *Jones* FHI 5975 ! Ibadan South F.R. (♀ fl. & fr. Apr.) *Keay* FHI 22546 ! 22547 ! Ilaro F.R. (♂ fl. June) *Onochie* FHI 3427 !

34. **D. elliotii** (*Hiern*) *F. White* in Bull. Jard. Bot. Brux. 26 : 243 (1956). *Maba elliotii* Hiern in Sc. Elliot in J. Linn. Soc. 30 : 85 (1894). *M. mannii* Hiern in Trans. Camb. Phil. Soc. 12 : 129 (1873) ; F.W.T.A. ed. 1, 2 : 7, not *Diospyros mannii* Hiern (1873). *D. senegalensis* of Benth. in Fl. Nigrit. 442, partly (Stirling loc.) not of A. DC. Small forest tree 12–35 ft. high ; bark smooth, dark brown ; slash orange-yellow ; flowers white ; fruit yellow on older parts of stem below the leaves, bright orange.

   Sen.: (♂ fl. Apr.) *Perrottet* 453 ! **Port.G.:** Bissau (♂ fl. Mar.) *Esp. Santo* 1894 ! **Guin.:** Kogou *Maclaud* ! **S.L.:** Bendu, Bonthe *Adames* 11 ! Taiama, Moyamba (fr. July) *Deighton* 1948 ! Longo (♀ fl. Apr.) *Deighton* 3722 ! Bagroo R. *Mann* 839 ! Falaba *Sc. Elliot* 5101 ! **Ghana:** Pong Tamale (fr. June) *Vigne* FH 3863 ! **N.Nig.:** Nupe *Barter* 1220 ! Irendu swamp, Kabba (fr. Sept.) *Daramola & Adebusuyi* FHI 38402 ! Gurara R., near Gornapara *Elliot* 172 ! Stirling *T. Vogel* 205 ! **S.Nig.:** Anamba F.R., Onitsha *Jones* FHI 4810 ! (See Appendix, p. 349.)

35. **D. hoyleana** *F. White* in Bull. Jard. Bot. Brux. 26 : 245 (1956). *Maba kamerunensis* Gürke in Engl. Bot. Jahrb. 46 : 150 (1912) ; F.W.T.A., ed. 1, 2 : 6, not *Diospyros kamerunensis* Gürke (1899). Forest shrub or small tree up to 45 ft. high ; bark dark brown or black, smooth ; slash black and brittle outside, reddish inside like that of *D. alboflavescens*, crumbling ; sapwood pinkish-white with black longitudinal lines ; leaves dark green and glossy above, margins coarsely undulate ; flowers pale yellow or white, borne down the leafless twigs but not on the leaves ; fruit red, turning black.

   S.Nig.: Cross R. *Jones* FHI 6882 ! Oban *Richards* 5163 ! [Br.]**Cam.:** S. Bakundu F.R., Kumba (♂, ♀ fl. Mar.) *Brenan* 9288 ! 9309 ! 9471 ! *Onochie* FHI 30853 ! Also in Cameroun, Gabon, Cabinda, Congo, N. Rhodesia and Angola.

36. **D. obliquifolia** (*Hiern ex Gürke*) *F. White* in Bull. Jard. Bot. Brux. 33 : ined. (1963). *Rhaphidanthe obliquifolia* Hiern ex Gürke in E. & P. Pflanzenfam. 4, 1 : 165 (1891) ; Stapf in Hook. Ic. Pl. 31 : t. 3030 (1915). Forest shrub or small tree up to 18 ft. high ; bark blackish, almost smooth ; slash black outside, red as in *D. alboflavescens* inside ; sapwood creamy white ; leaves deep green and glossy above, glaucous beneath ; flower buds yellowish.

   [Br.]**Cam.:** S. Bakundu F.R., Kumba (♂ & ♀ fl. Mar.) *Brenan* 9448 ! 9449 !

37. **D. simulans** *F. White* l.c. (1963). Forest shrub or small tree up to 30 ft. high ; flowers pinkish-white, mostly borne on older branchlets and main stem.

   [Br.]**Cam.:** Banga F.R., Kumba (♂ fl. May) *Binuyo & Daramola* FHI 35602 ! Also in Gabon.

38. **D. vignei** *F. White* l.c. (1963). *D. apiculata* A. Chev. Bot. 395, name only, not of Hiern (1873). *D. heudelotii* of F.W.T.A., ed. 1, 2 : 6, partly (syn. *D. apiculata* and *Chev.* 17560 & 19202 only). *Maba graboensis* Aubrév. Fl. For. C. Iv., ed. 2, 3 : 176 (1959), French descr. only. Forest shrub 6–10 ft. high ; fruit orange or red, borne on the older branchlets.

   Lib.: Webo *Baldwin* 6073 ! **Iv.C.:** Abiati, Lower Comoe (fr. Mar.) *Chev.* 17540 ! Mid. Sassandra to Mid. Cavally (♂ fl. June) *Chev.* 19202 ! 60 km. N. of Sassandra (fr. Feb.) *Leeuwenberg* 2797 ! Amitioro (fr. Sept.) *de Wilde* 259 ! **Ghana:** Opon Mansi F.R., *Akpabla* 925 ! Mamiri F.R., Dunkwa (♂, ♀ fl., fr. May–June) *Enti* FH 6718 ! 6718a ! 6718b ! 6718c ! 7177 ! 7178 ! Wasaw-Aowin, Opon (fr. Jan.) *Irvine* 1096 ! Ankasa F.R. (fr. Dec.) *Vigne* FH 3169 !

39. **D. cinnabarina** (*Gürke*) *F. White* in Bull. Jard. Bot. Brux. 26 : 242 (1956). *Maba cinnabarina* Gürke in Engl. Bot. Jahrb. 43 : 199 (1909). *Diospyros heudelotii* of F.W.T.A., ed. 1, 2 : 6, partly (*Zenker* 3466) not of Hiern. Small forest tree up to 30 ft. high ; bark black, smooth with shallow vertical anastomosing fissures ; slash jet black outside, red and fibrous inside, sapwood white ; leaves dull above, pale glaucous-green beneath.

   [Br.]**Cam.:** S. Bakundu F.R. (♂ fl. Mar.) *Brenan* 9320 ! Likomba Plantation *Mildbr.* 10531 ! Also in Cameroun.

Besides the above, *D. discolor* Willd., a native of the Philippines, widely planted in the tropics for its edible fruits, is grown in our area. *D. heterotricha* (B. L. Burtt) F. White, a native of Central Africa, is also in cultivation.

Imperfectly known species.

1. **D.** sp. Small forest tree 15 ft. high ; differs from *D. obliquifolia* (Hiern ex Gürke) F. White (No. 36 above) in having more acute leaves with more distinctly filiform apices and more densely pubescent lower leaf-surfaces with secondary and tertiary nerves clearly visible.

   S.Nig.: Afi River F.R., Ogoja *Jones & Onochie* FHI 17387 ! *Keay* FHI 28186 !

2. **D.** sp. Medium-sized tree about 50 ft. high ; possibly *D. cinnabarina*, but flowers and fruits are required for confirmation.

   S. Nig.: Ohumbe F.R., Abeokuta *Oladoyinbo* FHI 15006 ! Ogun River F.R., Lagos *Onochie* FHI 18653 !

128. HOPLESTIGMATACEAE

By F. N. Hepper

Trees. Leaves alternate, entire ; stipules absent. Flowers hermaphrodite, in terminal cymes. Calyx closed in bud, bursting irregularly. Corolla shortly tubular, with about 3 series of overlapping corolla lobes. Stamens numerous (about 21), in several rows inserted at the base of the corolla ; anthers 4-celled, opening by longitudinal slits. Ovary superior, 1-celled with 2 intrusive parietal placentas ; style bilobed nearly to the base, branches sharply bent ; stigmas capitate. Ovules pendulous, one on each side of the bilobed placentas. Fruit with a deep groove up each side, the calyx forming a plate at the base, endocarp hard. Seeds with scanty endosperm and long thick straight embryo.

HOPLESTIGMA Pierre in Bull. Soc. Linn. Paris, n. sér., 116 (1899) ; emend. Gilg in
  Engl. Bot. Jahrb. 40, Beibl. 93 : 76 ; Hallier f. in
  Meded. Rijks Herb. Leiden 1 : 35 (1910).

Leaves densely pubescent beneath at least on midrib and nerves with rather long,
  crisped hairs ; lamina broadly obovate-elliptic, rounded and slightly apiculate,
  narrowed to base, 15–30 cm. long, 5–15 cm. broad, venulose-reticulate above ; lateral
  nerves about 10 on each side of midrib ; petiole about 2 cm. long ; cymes terminal,
  softly tomentose all over ; calyx about 1 cm. long ; corolla 1·5 cm. long ; fruits
  broadly ellipsoid, 3 cm. long, 2·6–2·9 cm. broad, glabrous   ..   1. *pierreanum*
Leaves beneath with small, stiff and straight appressed hairs, sometimes sparse, lateral
  nerves about 8 on each side of midrib ; petioles 1–1·5 cm. long ; cymes covered
  with dense stiff hairs ; fruits 2 cm. long, 2·2–2·4 cm. broad   ..   2. *kleineanum*

1. **H. pierreanum** *Gilg* l.c., with fig. (1908). A tree with long slender stem and small crown ; inflorescences like
    Cordia.
      [**Br.]Cam.**: Likomba Plantation, Cam. Mt. (Dec.) *Mildbr.* 10774! Also in Cameroun.
2. **H. kleineanum** *Pierre* l.c. (1899) ; Aubrév. Fl. For. C. Iv., ed. 2, 3 : 155, t. 307. A tree, 65–80 ft. high.
      Iv.C.: Tabou *Aubrév.* 1683! 2809! San Pedro *Aubrév.* 1270! Also in Gabon.

129. SAPOTACEAE

By H. Heine

  Trees, shrubs, or rarely climbers, often with milky juice. Leaves alternate,
entire, leathery ; stipules usually absent. Flowers hermaphrodite, actinomorphic,
usually small. Calyx 4–8-lobed. Corolla 4–8-lobed, lobes 1–2-seriate, imbricate,
sometimes with petaloid external appendages. Stamens inserted on the corolla,
the fertile ones equalling the corolla-lobes and opposite to them, or more numerous
and 2– or more-seriate ; · staminodes sometimes present ; anthers opening
lengthwise. Ovary several-celled, superior ; style simple. Ovules solitary in
each cell, ascending from the inner axil. Fruit 1– many-celled, often a rather
hard berry, rarely a capsule. Seeds with a bony, often shining testa, and a large
broad hilum ; endosperm mostly scanty ; embryo large, with small radicle and
broad foliaceous cotyledons.

  Mainly in the tropics and subtropics ; some of considerable economic value. Distinguished by the woody habit, alternate leaves without or with caducous stipules, and
regular flowers, mostly with petaloid staminodes.

  To assist future workers on this complicated family, fuller references and synonymies
are given than is usual in this Flora. Since this account was completed, two important
additions to the taxonomic literature on the family have been published : Aubréville
in Notulae Syst. 16 : 223–279 (1961), and Fl. Gabon No. 1 : Sapotac. (1961).

Stamens several (3–5) opposite each lobe of the corolla (sterile in the female flowers) ;
  fruits 6–26 cm. diam., globose to depressed-globose :
  Strictly cauliflorous trees, i.e. flowers borne in fascicles on the main trunk, pedicellate ;
    styles thickened or slightly inflated beyond the stigma ; fruits depressed-globose,
    up to 26 cm. diam., with a very thick and hard exocarp and about 20 elliptic,
    flattened seeds up to 4 cm. long and 2·5 cm. broad ..   ..   **1. Omphalocarpum**
  Flowers borne mainly on the younger branches, sometimes (in No. 2) also on the main
    trunk ; fruits globose, up to 8 cm. diam. :
    Flowers in fascicles on slender branches below the leaves on scars of older fallen
      leaves, few in each cluster, pedicels up to 8 mm. long, pedicels and calyx rusty-
      tomentose ; fruits with about 20 flattened seeds, up to 1·5 cm. long
                                                              2. Ituridendron
    Flowers single, but sometimes densely together, sessile, borne on the younger branches
      on scars of older fallen leaves ; calyx nearly glabrous, ciliate ; fruits with 10 seeds,
      up to 3·5 cm. long   ..   ..   **3. Tridesmostemon**
Stamens solitary opposite to each lobe of the corolla ; fruits globose or ellipsoid, up
  to 5 cm. diam. :
  Corolla-lobes with petaloid external appendages (these often appearing to be 3 series
    of petals) ; staminodes also petaloid :
    Staminodes opposite to the stamens ; flowers 5-merous ; fruits dehiscent   **4. Gluema**
    Staminodes alternate to the stamens ; fruits indehiscent :
      Flowers 5-merous ; lobes of the corolla-trilobate ; leaves and flowers in whorl-like
        fascicles at the nodes   ..   ..   **5. Neolemonniera**
      Flowers 6–8-merous, lobes of the corolla entire ; leaves and flowers never in whorl-
        like fascicles at the nodes :

Sepals 3 + 3 ; corolla with 6 lobes .. .. **6. Manilkara**
Sepals 4 + 4 ; corolla with 8 lobes :
  Seeds with a small navel-like scar at the base .. .. **7. Mimusops**
  Seeds with a scar covering more than a third of the base :
    Scars on the seeds basiventral and subrectangular, rather less than half the length
      of the seed ; petals pubescent inside .. .. **8. Autranella**
    Scars on the seeds nearly as long as the length of the seed, broadly elliptic :
      Testa of the seeds woody, very thick ; plants without stipules ; leaf-nerves not
        prominent beneath .. .. **9. Tieghemella**
      Testa of the seeds thin ; plants with stipules ; leaf-nerves very prominent beneath
                10. Baillonella
Corolla-lobes without external appendages :
  Leaves and flowers relatively densely clustered at the top of the shoots ; ovary 8–
    10-celled ; seeds usually 8 ; staminodes petaloid, apiculate .. **11. Butyrospermum**
  Leaves usually not clustered at the top of the shoots ; ovary rarely more than
    5-celled ; petals 4–6 :
    Staminodes well-developed, petaloid :
      Sepals united beyond the middle .. .. **12. Synsepalum**
      Sepals not united beyond the middle ; stamens with filaments about 2 mm. long
                13. Vincentella
    Staminodes very small and subulate, or absent :
      Disk present, glabrous :
        Ovary 5-celled ; leaves oblong-lanceolate, cuneate at base, with 9–15 pairs of
          nerves, glabrous .. .. **14. Kantou**
        Ovary 8-celled :
          Leaves cordate at the base ; corolla long-tubular ; style contorted ; leaves
            golden silky-velutinous beneath, nerves very numerous and close
                  15. Breviea
      Leaves oblong, cuneate at the base ; corolla short ; style not contorted ; leaves
        glabrous beneath, nerves 7–9 pairs .. .. .. **16. Aubregrinia**
    Disk absent ; ovary mostly 5-celled :
      Stamens inserted on the corolla between and at the base of the lobes :
        Fruits ellipsoid, olive- or prune-like, mostly with 1–2 seeds :
          Leaves with pellucid dots, with about 12–20 pairs of very prominent lateral
            nerves, leaf-colour in drying mostly dark brown or greenish-brown, not very
            different on both sides :
            Flowers without staminodes .. .. **17. Malacantha**
            Flowers with subulate staminodes .. .. .. **18. Aningeria**
          Leaves without pellucid dots, with very numerous and dense, not prominent,
            lateral nerves, leaf-colour in drying cream-greyish, light olive-green or light
            rusty-brown, with the lower surface mostly much lighter than the upper
                19. Bequaertiodendron
        Fruits globose, egg-shaped or apple-like, mostly with 3–5 seeds :
          Shrub or small tree ; flowers single or in pairs in the axils of the leaves, all
            parts covered with 3 mm. long rusty-brown hairs, nerves of the leaves 15–20
            pairs, very conspicuous .. .. **20. Delpydora**
          Mostly very big and tall forest trees, sometimes lianes ; flowers fascicled ;
            plants generally not covered with rusty-brown long hairs (if so, leaves
            irregular-rotundate at the base) .. .. **21. Chrysophyllum**
      Stamens inserted in the corolla-tube and alternate with the lobes :
        Stipules persistent ; flowers without staminodes ; leaves with about 16 pairs of
          conspicuous lateral nerves .. .. **22. Pachystela**
        Stipules caducous ; flowers with small staminodes ; leaves with rather numerous
          very obscure lateral nerves .. .. **23. Afrosersalisia**

1. OMPHALOCARPUM P. Beauv.—F.T.A. 1 : 171.

Flowers pedicellate, pedicels not bracteate except at the base :
  Leaves elongate-obovate, long-attenuated to the base, rounded at the apex, about
    40 cm. long and up to 18 cm. broad, with numerous (over 20) pairs of prominent
    lateral nerves, glabrous ; flowers several in each cluster ; pedicels up to 1 cm.
    long, tomentose ; sepals ovate, 1–1·5 cm. long, tomentose ; corolla 3 cm. long,
    the lobes slightly pubescent on the margin ; fruits depressed-globose, about 14 cm.
    diam., the outer shell about 1·5 cm. thick, openly reticulate in section ; seeds nearly
    flat, 4 cm. long, 2·5 cm. broad, with a narrow scar on one side .. .. 1. *ahia*
  Leaves obovate or oblanceolate, attenuated to the base, rounded or obtusely acumin-
    ate at the apex, 12–25 cm. long, 4–8 cm. broad, glabrous, with about 10 pairs of
    lateral nerves ; flowers more or less as above but calyx and pedicels soon glabres-
    cent ; fruits (according to a sketch) depressed-globose .. .. 2. *elatum*
Flowers subsessile, with several imbricate bracts below the calyx ; leaves oblanceolate,

much narrowed to the base, not acuminate at apex, 8–13 cm. long, 3–5 cm. broad, glabrous, with 6–7 pairs of lateral nerves ; bracts very broadly ovate, tomentose outside, about 8 mm. long ; young fruits depressed-globose, glabrous, with a persistent style ; mature fruit depressed-globose, about 15 cm. diam. ; seeds as in *O. ahia*

3. *procerum*

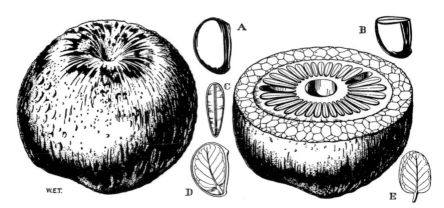

FIG. 204.—FRUITS OF OMPHALOCARPUM AHIA *A. Chev.* (SAPOTACEAE).

A, seed.   B, part of seed.   C, cross-section of seed.   D, seed opened showing one cotyledon and radicle.   E, embryo.

1. **O. ahia** *A. Chev.* in Vég. ut. Afr. trop.f r. 5 : 244 (1909) ; Chev. Bot. 389 ; Aubrév. Fl. For. C. Iv. ed. 2, 3 : 110, t. 288 fig. 1–3. A tall tree with clear bole and large leaves ; conspicuous by the clustered fruits on the main stem.
    S.L.: Jomu, Nongowa (Oct.) *Deighton* 4368 ! Bagbe (fr. Mar.) *Small* 737 ! **Lib.:** *Harley* 920 ! Ganta (Mar.) *Harley* 1120 ! **Iv.C.:** Yapo *Chev.* 22357 ! Songan *Chev.* 16287 ! **Ghana:** Nokeira (Sept.) *Smythe* 242 ! Tarkwa (July) *Chipp* 242 ! Agogo (Apr.) *Vigne* FH 1121 ! *Irvine* 936 ! Bonsa Su (fr. May) *Vigne* FH 1993 ! (See Appendix, p. 359.)

2. **O. elatum** *Miers* in Trans. Linn. Soc., ser. 2, 1 : 16, t. 4 (1875). *O. radlkoferi* Pierre in Bull. Soc. Linn. Paris 1 : 577, 580 (1886) ; Engl. Monogr. Afr. 8 : 16, fig. 4 and t. 4, 5 (1904), incl. var. *pluriloculare* Engl., l.c. *O. anocentrum* Pierre ex Engl. Monogr. Afr. 8 : 15, t. 6 (1904) ; Chev. Bot. 390 ; Aubrév. Fl. For. C. Iv., ed. 2, 3 : 110, t. 289. *O. procerum* of F.T.A. 1 : 171 (1868), partly (*Mann* 712, 815, *Thomson* 128), not of P. Beauv. *O. talbotii* Wernham ex Hutch. & Dalz. F.W.T.A., ed. 1, 2 : 13 (1931), name only in syn. A large tree with white waxy flowers and fruits clustered on the lower part of the stem.
    S.L.: Tunkia (fl. & fr. May) *Jordan* 2067 ! Guma *King-Church* 2 ! Bagroo R. (fl. & fr. Apr.) *Mann* 815 ! **Lib.:** Dukwai *Cooper* 81 ! 383 ! **Iv.C.:** Dabou *Jolly* 162. Accrédou *Chev.* 16194 ! Yapo *Chev.* 22311 ! (Oct.) *de Wilde* 694 ! Abidjan *Aubrév.* 93 ! **Ghana:** Fomang Su F.R. *Vigne* FH 2687 ! Prestea (July) *Vigne* FH 1265 ! Akwantambra (Nov.) *McAinsh* 390 ! **S.Nig.:** Degema *Rosevear* O.C. 116 ! Eket *Talbot* 3142 ! Calabar *Thomson* 128 ! **[Br.]Cam.:** *Mann* 712 ! **F.Po:** Musola (Jan.) *Guinea* 1484 *bis* ! (See Appendix, p. 359.)

3. **O. procerum** *P. Beauv.* Fl. Oware 1 : 7, t. 5, 6 (1805) ; Pierre in Bull. Soc. Linn. Paris 1 : 579 (1886) ; F.T.A. 1 : 171, partly ; Engl. Monogr. Afr. 8 : 14, t. 3B. A tall tree resembling the last ; flowers white.
    Ghana: Agogo (Apr.) *Vigne* FH 1120 ! Pra River Station (Feb.) *Vigne* FH 1035 ! Fomang Su F.R. (Feb.) *Vigne* FH 1822 ! **S.Nig.:** Oware, Benin *P. Beauvois.* Sapoba *Kennedy* 787 ! (See Appendix, p. 360.)

Doubtful species.

O. pierreanum *Engl.* Monogr. Afr. 8 : 14 (1904). *O. laurentii* De Wild. (1907).
    [Br.]Cam.: Johann-Albrechtshöhe *Preuss* 282b. Also in Cameroun and Gabon. Not seen, also no other material from our area. The only difference between this species and *O. procerum* P. Beauv. which the present writer could observe (on Gabon specimens) are the two teeth at the base of the staminodes ; it is therefore highly probable that this is only an occasional form without taxonomic value.

2. ITURIDENDRON De Wild. Pl. Bequaert. 4 : 100 (1926).

A big tree up to 25 m. high, with light brown bark and oblong-oblanceolate leaves, obtusely acuminate, 10–24 cm. long and 4–10 cm. broad, with about 9–12 pairs of lateral nerves, rather stiff and thick coriaceous, glabrous ; petioles up to 4 cm. long ; calyx rusty-tomentose, sepals elliptic, up to 9 mm. long and 7 mm. broad ; corolla up to 5·5 mm. long, staminodes ciliate, with two acumens up to 4 mm. long ; style glabrous, about 2 mm. long

bequaertii

I. bequaertii *De Wild.* l.c. (1926). *Omphalocarpum pachysteloides* Mildbr. ex Hutch. & Dalz. F.W.T.A., ed. 1, 2 : 13 (1931) ; Kew Bull. 1937 : 59 ; Aubrév. Fl. For. C. Iv. ed. 2, 3 : 112, t. 288, fig. 4–10. A tree with white flowers few, in clusters below the leaves.
    S.L.: Kabusa (Apr.) *Sc. Elliot* 5471 ! Samuel Town (fl. Feb., fr. May) *Deighton* 2755 ! 3112 ! Loma Mts. *Jaeger* 4014 ! **Lib.:** Bilipia *Harley* 2082 ! **Ghana:** Akumadan (Oct.) *Vigne* FH 2559 ! **S.Nig.:** Okhuesan F.R., Ishan Dist. (fr. Aug.) *Onochie* FHI 33264 ! Ojogba F.R., Ishan Dist. *Olorunfemi* FHI 38069 ! Also in Cameroun, Gabon, and the Congo.

3. TRIDESMOSTEMON Engl. Bot. Jahrb. 38 : 99, with fig. (1905) ; Pellegr. in Bull. Soc. Bot. Fr. 85 : 179 (1958).

Leaves narrowly oblong-lanceolate, abruptly acuminate, narrowed from well above the upper third down to the base and there long-acuminate, decurrent along the petiole which, therefore, seems to be slightly winged in its upper part, coriaceous, glabrous and shining on both sides, up to 17 cm. long and 5 cm. broad, with slightly revolute and undulate margin ; real petiole (without the very narrow basal part of the leaf) up to 1·5 cm. long ; flowers sessile, mostly solitary in the axils of fallen leaves ; calyx glabrous, the 5 sepals imbricate, ciliate, pink ; corolla, about 8 mm. long, 5-parted, petals oblong-spathulate, imbricate, very finely pilose ; stamens 15, ternate (3 united stamens opposite to each petal), staminodes tridenticulate, the central tooth well-developed, subulate, the 2 lateral minute ; ovary globose, 10-celled, pilose ; fruit subglobose, up to 8 cm. diam., subsessile ; 10-seeded ; seeds flat, obovate-ellipsoid, 5 cm. long, 2·5 cm. broad, 11 mm. thick, with shining testa and a 4 mm. broad scar along the whole median side   ..   ..   *omphalocarpoides*

T. omphalocarpoides *Engl.* l.c. (1905) ; Pellegrin l.c. 180. *T. claessensii* De Wild. in Rev. Zool. Afr. 7, Suppl. Bot. 23 (1919). *T. mortehanii* De Wild. l.c. 24 (1919). *Sideroxylon gossweileri* Greves in J. Bot. 64, Suppl. 2 : 71 (1927). A tree up to 110 ft. high, with rather clustered leaves at the end of the branches, in habit much resembling *Omphalocarpum* trees ; flowers white ; fruits very similar to a small or medium-sized orange.
[Br.]**Cam.:** Victoria to Kumba *Ejiofor* FHI 15254 ! (fr. Apr.) *Olorunfemi* FHI 30541 ! Bonakoko, Kumba Div. (fr. Mar.) *Jeme* FHI 12011 ! Kumba, mile 47 of the road (fl. & fr. Apr.) *Daramola* FHI 29845 ! Banga, S. Bakundu F.R. (Mar.) *Brenan & Onochie* 9434 ! Also in Cameroun, Gabon, Cabinda and Congo.

4. GLUEMA Aubrév. & Pellegr. in Bull. Soc. Bot. Fr. 81 : 797 (1935) ; Aubrév. in Notulae Syst. 16 : 274, fig. 12 (1961).

Leaves glabrous, clustered at the end of the branches, narrowly oblong-lanceolate, long-acuminate, 12–32 cm. long, 3–8 cm. broad, with 12–18 pairs of lateral nerves ; petiole 2–5 cm. long ; flowers 2–3 together, 1 cm. long ; pedicels up to 1·5 cm. long, finely red-tomentose ; ovary covered with rather long hairs ; young fruits apiculate, squamulous-hairy, with persistent calyx   ..   ..   *ivorensis*

G. ivorensis *Aubrév. & Pellegr.* l.c. 798, fig. 2 (1935) ; Aubrév. Fl. For. C. Iv., ed. 2, 3 : 112, t. 290. A large forest tree, in habit resembling *Omphalocarpum*, with small axillary flowers and almond-shaped dehiscent fruits ; along streams in rain forest.
Iv.C.: Mé Yakassé *Aubrév.* 1790 ! 1791 ! Byanouan *Chev.* 16289 ! Bassam *Aké Assi* IA 1760 ! **Ghana:** Dompim (fl. Mar., fr. Nov.) *Vigne* FH 1499 ! 1978 ! Neung F.R. (fr. Jan.) *Bassaw* FH 4842 ! Also in Gabon. (See Appendix, p. 356.)
[This genus is certainly closely allied with *Lecomteodoxa* Pierre (syn. *Walkeria* A. Chev. not of Miller ex Ehret) ; see Aubrév. in Bull. Soc. Bot. Fr. 103 : 8 (1956)].

5. NEOLEMONNIERA Heine in Kew Bull. 14 : 301 (1960). *Le Monniera* H. Lecomte in Notulae Syst. 3 : 337 (1918), not *Lemonniera* De Wild. (1894).

Leaves lanceolate-elliptic, acuminate, acute at the base, 7–15 cm. long, 2–2·5 cm. broad on fertile branches, but sometimes on sterile shoots up to 35 cm. long and 10 cm. broad, glabrous, with about 10 pairs of very indistinct lateral nerves ; petiole 1·5–3 cm. long ; flowers in dense fascicles at the distant nodes, numerous ; pedicels 1·5–3 cm. long ; calyx-segments ovate, obtuse, 4–5 mm. long ; corolla-lobes membranous, trilobate, central lobe ovate-acuminate, lateral lobes lanceolate-acuminate ; fruit 7–8 cm. diam. ; seeds narrowly and obliquely ellipsoid, brown, shining, 3·5 cm. long, 1·5 cm. broad, with the scar reaching nearly the full length ..   ..   *clitandrifolia*

N. clitandrifolia (*A. Chev.*) *Heine* l.c. (1960). *Mimusops clitandrifolia* A. Chev. in Vég. ut. Afr. trop. fr. 5 : 242 (1909) ; Chev. Bot. 392 (as "*clitandrifolium*") ; Chev. in Rev. Bot. Appliq. 23 : 134 (1943). *Le Monniera clitandrifolia* (A. Chev.) H. Lecomte l.c. 342 (1918) ; Aubrév. Fl. For. C. Iv. 3 : 96, t. 278 ; ed. 2, 3 : 116, t. 292. *Sideroxylon aylmeri* Scott in Kew Bull. 1915 : 45 ; F.W.T.A., ed. 1, 2 : 12. *Pouteria aylmeri* (Scott) Baehni in Candollea 9 : 322 (1942). A large forest tree, up to 90 ft. high, with buttressed stem up to 3 ft. diam., glabrous, with white latex ; flowers clustered, white, scarcely ¼ in. long ; in rain forest.
S.L.: Gumati (May) *Lane-Poole* 235 ! Colony F.R. *Vigne* 17 ! Sokei *Edwards* 51 ! Falaba (Apr.) *Aylmer* 57 ! **Lib.:** Dukwia R. (Mar.) *Cooper* 311 ! **Iv.C.:** Nzo *Aubrév.* 1160. Mt. Momi *Aubrév.* 1170 ! Malamalasso (Mar.) *Chev.* 16247 ! Mt. Capé *Chev.* 19699. (See Appendix, p. 361.)
[Note : *N. batesii* (Engl.) Heine l.c. (1960) is cited by Aubréville for Ivory Coast (Fl. For. C. Iv., ed. 2, 3 : 116, t. 293, 1–5 : Tabou *Aubrév.* 2180), but I have not seen any specimen from our area. The species was described from Cameroun.—H.H.]

6. MANILKARA Adans. Fam. pl. 2 : 166 (1763) ; Dubard in Ann. Mus. col. Marseille, sér. 3, 3 : 6 (1915) ; Aubrév. in Notulae Syst. 16 : 223 (1961). *Nom. cons.*

Leaves oblong or narrowly oblong-obovate, rounded or shortly acuminate, usually shortly narrowed at base, 7–14 cm. long, 4–6 cm. broad, greyish, finely silky beneath when young, with numerous pairs of faint lateral nerves ; petiole up to 3 cm. long ; flowers densely clustered amongst the leaves ; pedicels about 1 cm. long, puberulous ; calyx lobed to the base, lobes oblong-elliptic, 3–4 mm. long, shortly tomentellous outside ; corolla-lobes glabrous ; fruits subglobose, about 2·5 cm. diam., several-seeded ; seeds 1 cm. long, shining   ..   ..   1. *multinervis*
Leaves broadly obovate or obovate-elliptic, rounded or slightly acuminate, 6–16 cm. long, 4–8 cm. broad, silky beneath when young, often glaucous-grey when older, with

numerous pairs of obscure lateral nerves ; petiole up to 3 cm. long ; flowers clustered
with the leaves ; pedicels 1–1·5 cm. long, nearly glabrous ; calyx-lobes ovate-triangular, 5 mm. long, finely puberulous ; fruits obovoid, 3 cm. long, finely puberulous
                                                            2. *obovata*

1. **M. multinervis** (*Bak.*) *Dubard* l.c. 24 (1915), incl. var. *poissonii* Dubard ; Aubrév. Fl. For. C. Iv., ed. 2, 3 :
120, t. 294, fig. 8. *Mimusops multinervis* Bak. in F.T.A. 3 : 506 (1877) ; Engl. Monogr. Afr. 8 : 57. *M.
densiflora* Bak. in Kew Bull. 1893 : 148. *M. chevalieri* Pierre in Bull. Mus. Hist. Nat. 7 : 139 (1901).
M. atacorensis A. Chev. Bot. 392 and *M. djalonensis* A. Chev. Bot. 393, names only. *M. poissonii* Pierre
ex Dubard l.c. (1915) ; Aubrév. Fl. For. Soud.-Guin. 425. *Manilkara maclaudii* Pierre ex Lecomte in
Bull. Mus. Hist. Nat. 26 : 647 (1920), incl. var. *membranacea* Pierre ex Lecomte l.c. 648 ; Aubrév. Fl. For.
Soud.-Guin. 425. A fairly large tree, often branched from the base and stem, shrub-like, up to 40 ft. high,
in habit much resembling *Mimusops kummel* Bruce in A.DC. and the following species ; the leaves without
stipules, more or less crowded towards the end of branchlets, with clustered white or brown flowers amongst
them ; in gallery forest.
    Guin.: Donbato N. (July) *Roberty* 17152 ! Sangorola (Feb.) *Chev.* 344 ! Kollangui (Mar.) *Chev.* 12191 !
Iv.C.: *Aubrév.*! Bondoukou *Aubrév.* 749 ! **Ghana:** Pong (Mar.) *Kitson* 662 ! Prang (Jan.) *Vigne* FH 1552 !
Datoka (Jan.) *Vigne* FH 4656 ! Anyima (Jan.) *Vigne* FH 3547 ! R. Tafo to Nchenenchene (fr. Apr.)
Kitson 1048 ! Saku (May) *Vigne* FH 4700 ! **Togo Rep.:** Sokodé (Dec.) *Kersting* A229 ! **Dah.:** Natitingou
to Bocorona *Chev.* 24192 ! **N.Nig.:** Nupe *Barter* 1123 ! Jos Plateau (Jan.) *Lely* P46 ! Jagindi *Keay* FHI
21546 ! **S.Nig.:** Olokemeji F.R. (Dec.) *McGregor* 438 ! *Ross* R42 ! *Keay* FHI 14612 ! Idanre (Jan.) *Brenan
& E. W. Jones* 8638 ! Bebi, Ogoja Prov. (Dec.) *Savory & Keay* FHI 25007 ! (See Appendix, p. 357.)
2. **M. obovata** (*Sabine & G. Don*) *J. H. Hemsley* in Kew Bull. 17 : 171 (1963). *Chrysophyllum obovatum* Sabine
& G. Don in Trans. Hort. Soc. Lond. 5 : 458 (1824) ; F.T.A. 3 : 501 ; Engl. Monogr. Afr. 8 : 43 (1904):
F.W.T.A., ed. 1, 2 : 9. *Mimusops lacera* Bak. in F.T.A. 3 : 507 (1877) ; Engl. l.c. 59, incl. var. *longi-
petiolata* Engl. and var. *newtonii* Engl. (1904). *Manilkara lacera* (Bak.) Dubard in Ann. Mus. col. Marseille,
sér. 3, 3 : 24 (1915) ; Aubrév. Fl. For. C. Iv., ed. 2, 3 : 120. *M. welwitschii* (Engl.) Dubard l.c., 27 (1915).
Mimusops welwitschii Engl. Bot. Jahrb. 12 : 524 (1890). *Manilkara sublacera* A. Chev. Bot. 393, name only.
M. sylvestris Aubrév. & Pellegr. in Bull. Soc. Bot. Fr. 104 : 279 (1957) ; Aubrév. l.c. 122, t. 294, fig. 1–7.
M. angolensis (Engl.) Lecomte ex Pellegr. (1928). A tree up to 80 ft. high, stem up to 4 ft. diam.,
frequently only as a shrub, old trees often resembling poplars ; along lagoons and streams in small groups.
    S.L.: *T. Vogel* 71 ! Kent *Deighton* 5532 ! Kumrabai (Dec.) *Thomas* 6942 ! 6949 ! Dunabaye *King* 66 !
Kambui F.R. (May) *Jordan* 2082 ! Makali *Deighton* 4058 ! Gola *Dawe* 501 ; 569 ! **Lib.:** *Linder* 53 ! 796 !
Gbanga (fr. Sept.) *Linder* 7196 ! Dukwia R. *Cooper* 113 ! **Iv.C.:** Azaguié *Chev.* 22298 ! **Ghana:** Kommendah
(fr. Jan.) *Andoh* 5548 ! 5599 ! Mampong (May) *Vigne* FH 1184 ! Ejura *Chipp* 759 ! E. Nyima (Aug.)
Mensah FH 4901. **Togo Rep.:** Lomé *Warnecke* 106 ! 385 ! **Dah.:** *Poisson* 151 (?) ! Porto-Novo (Mar.)
Chev. 23363 ! **N.Nig.:** Nupe *Barter* 1270 ! Bako R., Bida *Lamb* 16 ! Abinsi *Dalz.*! **S.Nig.:** Lagos *Barter* !
Millen 65 ! Nun R. (Sept.) *Mann* 489 ! Miniata, Onitsha (Feb.) *Jones* FHI 515 ! Onitsha F.R. (June)
Jones FHI 1028 ! Karipi *Ainslie* 404 ! [Br.]**Cam.:** Dieka, Kumba Div. *Akuo* FHI 15198 ! Bamenda (fl.
& fr. May) *Ujor* FHI 30408 ! Also in Cameroun, Gabon, and Angola. (See Appendix, p. 355.)

7. MIMUSOPS Linn.—F.T.A. 3: 305.

Pedicels 5–6 mm. long, shortly tomentose ; leaves tomentose at first, glabrescent ;
petiole about 1·5 cm. long, 3·5 cm. broad, with numerous fine lateral nerves ; flowers
axillary, few ; calyx tomentose, 5–6 mm. long   ..   ..   ..   1. *andongensis*
Pedicels 2–4 cm. long, tomentose ; leaves tomentose beneath, especially on or near the
midrib, elliptic or slightly obovate, obtusely acuminate, 8–15 cm. long, 4–7 cm. broad,
with numerous fine lateral nerves ; petiole thinly tomentose, 2 cm. long ; flowers
few in the leaf-axils ; calyx-lobes narrowly lanceolate, 7–15 cm. long ; ovary 8-
celled ; fruit ellipsoid, mucronate, up to 2·5 cm. long and 1·5 cm. diam. ; seeds with
a small basal scar, 1·5–1·8 cm. long and 1 cm. broad   ..   ..   2. *kummel*

1. **M. andongensis** *Hiern* in Cat. Welw. 1 : 649 (1898). *M. warneckei* Engl. (1904)—F.W.T.A., ed. 1, 2 : 14 ;
Aubrév. Fl. For. C. Iv., ed. 2, 3 : 124. A tree about 60 ft. high with straight bole and light crown, bark
dark brown with shallow longitudinal fissures, slash red with white marks ; fruits organge-red.
    Port.G.: Canchungo (fr. Apr.) *Esp. Santo* 1943 ! Bissau (fr. Jan.) *Esp. Santo* 1704 ! Farim-Corjambarim
(fr. Nov.) *Esp. Santo* 2421 ! Umpacaca (Jan.) *d'Orey* 153 ! **Togo Rep.:** Lomé *Warnecke* 311 ! 326 ! **S.Nig.:**
Miniata, Onitsha (Feb.) *Jones* FHI 5018 ! 5035 ! Mamu River F.R., Akwa Dist. (fl. & fr. Sept.) *Onochie*
FHI 34069 ! Extending to Angola.
2. **M. kummel** *Bruce ex A. DC.* in DC. Prod. 8 : 203 (1844) ; F.T.A. 3 : 508 ; Engl. Monogr. Afr. 8 : 75 ;
Eggeling & Dale, Indig. Trees of Uganda ed. 2, 400 (1952). *M. fragrans* (Bak.) Engl. in E. & P. Pflanzenfam.
4, 1 : 151, fig. 82 N–S (1891) ; Monogr. Afr. 8 : 77 ; F.W.T.A., ed. 1, 2 : 14 ; Aubrév. Fl. For. C. Iv.,
ed. 2, 3 : 124, t. 295 ; Chev. in Rev. Bot. Appliq. 23 : 125 (1943) ; Aubrév. Fl. For. Soud.-Guin. 427.
Imbricaria fragrans Bak. in F.T.A. 3 : 509 (1877). *Mimusops longipes* Bak. in Kew Bull. 1895 : 149 ;
Engl. Monogr. Afr. 8 : 74, t. 28, fig. A. *M. kerstingii* Engl. Monogr. Afr. 8 : 78 (1904). A rather small tree
with rusty-tomentose, long-stalked white flowers 2–4 together in the axils of the upper leaves.
    Ghana: Ejura (May) *Vigne* FH 1165 ! 2023 ! Akuse to Kpong, Volta R. (fr. Oct.) *Morton* ! Gambaga
Scarp (Apr.) *Morton* GC 9000 ! **Togo Rep.:** Sokodé-Basari (Apr.) *Kersting* A46 ! **N.Nig.:** Little Koriga R.
Dalz. 415 ! Jebba *Barter* ! Heipang (Apr.) *Keay, E. W. Jones & Wimbush* FHI 37601 ! Kwarra, Plateau
Prov. (fr. Jan.) *Wimbush* FHI 41822 ! Anara F.R., Zaria *Keay* FHI 21113 ! 22961 ! Alagbede F.R.,
Ilorin (Feb.) *Ejiofor* FHI 19821 ! **S.Nig.:** Lagos *Rowland* ! Yoruba country *Barter* 1217 ! Oyo to Iseyin
(Mar.) *Keay & Stanfield* FHI 37805 ! Bikote, Obudu Div. *Catterall* FHI 9542 ! [Br.]**Cam.:** Mayo Wombo,
Adamawa (Feb.) *Latilo & Daramola* FHI 34489 ! Also in Ethiopia and eastern Africa. (See Appendix
p. 358.)

    M. elengi Linn., a native of India, is cultivated as a shade-tree in gardens and public parks throughout
the area.

8. AUTRANELLA A. Chev. in Vég. ut. Afr. trop. fr. 9 : 271 (1917).

Leaves glabrous beneath, slightly clustered at the end of the branches, obovate-oblong,
shortly acuminate, 10–15 cm. long, 3–4·5 cm. broad, with numerous fine lateral nerves ;
petiole about 4–5 cm. long ; flowers axillary ; pedicels up to 2 cm. long, shortly
tomentose ; calyx tomentose, 2·5 mm. long ; fruit 7·5 cm. long and 5·5 cm. diam. ;
seeds triangular-ovate, relatively flat, 5 cm. long, 3·5 cm. broad, 2·6 cm. thick, with
a basal scar up to 3 cm. long and 1 cm. broad. ..   .. *congolensis*

A. congolensis (*De Wild.*) *A. Chev.* l.c. t. 29, fig. 2 (1917) ; Rev. Bot. Appliq. 23 : 132 (1943). *Mimusops congolensis* De Wild. Miss. E. Laurent 1 : 434, figs. 82, 83 (1907). *M. ebolowensis* Engl. & Krause in Engl. Bot. Jahrb. 49 : 396 (1913). *M. letestui* Lecomte in Bull. Mus. Hist. Nat. 26 : 534, figs. 1, 2 (1920) ; Pellegr. in Mém. Soc. Linn. Normandie, n. sér., Sect. Bot. 1, 3 : 8, fig. 1 (1928). *Autranella letestui* (Lecomte) A. Chev. in Rev. Bot. Appliq. 23 : 133 (1943). *Mimusops boonei* De Wild. in Rev. Zool. Afr. 7, Suppl. Bot. 27 (1919) ; Pl. Bequaert. 4 : 148, incl. var. *acuminata* De Wild. l.c. 150, and var. *abrupte-acuminata* De Wild. l.c. 151 (1926). Tree about 40 ft. high ; flowers white.
S.Nig.: Onitsha *Kennedy* 2423 ! 2424 ! Afi River F.R. (fr. May) *Jones & Onochie* FHI 18636 ! Adazi-Ani, Awka (fl. & fr. July) *Onyeagocha* FHI 16565 ! [Br.]Cam.: Victoria Bot. Gard. (Dec.) *Ogu* FHI 50311 ! Also in Cameroun, Gabon and Congo.

9. TIEGHEMELLA Pierre Not. bot. Sapot. 18 (1890). *Dumoria* A. Chev. in Compt. Rend. Acad. Sci. Paris 145 : 267 (1907).

Leaves 7–15 cm. long, 3–6 cm. broad, with lateral nerves fewer and less prominent than *Autranella*; outer calyx-lobes glabrous, broadly triangular, 5 mm. long, the inner ones softly tomentellous ; leaves obovate or narrowly obovate-elliptic, with numerous faint lateral nerves, glabrous ; petiole slender, up to 4 cm. long ; pedicels 2 cm. long, glabrous ; calyx-lobes ovate, 3 mm. long ; staminodes ovate-lanceolate ; fruits ovoid, 8–10 cm. diam., yellow when ripe, with 2 or 3 seeds ; seeds 5 cm. long and 3 cm. broad, two-thirds (testa) hard and shining, other third (scar) rough   *heckelii*

T. heckelii *Pierre ex A. Chev.* in Vég. ut. Afr. trop. fr. 2 : 172 (1907) ; l.c. 5 : 237 (1909). *Dumoria heckelii* A. Chev. in Compt. Rend. Acad. Sci. Paris 145 : 267 (1907) ; Vég. ut. Afr. trop. fr. 5 : 237 (1909) ; Aubrév. Fl. For. C. Iv., ed. 2, 3 : 126, t. 296 ; Rev. Bot. Appliq. 23 : 129, t. 3 (1943). *Tieghemella heckeliana* [in error] Pierre ex Dubard in Ann. Mus. col. Marseille, sér. 3, 3 : 33 (1915). *Mimusops heckelii* (A. Chev.) Hutch. & Dalz. F.W.T.A., ed. 1, 2 : 14 (1931) ; Aubrév. Fl. For. C. Iv., ed. 1, 3 : 102, t. 280 ; Hutch. & Dalz. in Kew Bull. 1937 : 60. A large forest tree up to 150 ft. high with horizontal branches near the top.
S.L.: *Thomas* 36 ! Kenema (May) *Jordan* 2095 ! Kambui Hills (fr. Mar.) *Small* 528 ! Waanje valley *Deen* SLFD 20033 ! **Lib.:** Wanau *Harley* 2107 ! **Iv.C.:** Bettié *Chev.* 16253 ! Azaguié *Chev.* 22304 ! Sassandra (Feb.) *Leeuwenberg* 2525 ! 2752 ! **Ghana:** Kwahu Prasu (Feb.) *Vigne* FH 1596 ! Aburi *Irvine* 1881 ! Prestea *Green* 897 ! Ashanti *Imp. Inst.* 38765 ! S. Fomang Sur F.R. (Apr.) *Vigne* FH 1213 ! Ateiku *Krukoff* 42K ! **S.Nig.:** Ishan (Mar.) *Kennedy* 2165 ! 2426 ! *Ainslie* 16 ! (See Appendix, p. 358.)

10. BAILLONELLA Pierre Not. bot. Sapot. 13 (1890).

Leaves 20–40 cm. long, 7–12 cm. broad, elongate-obovate, abruptly acuminate, with very numerous prominent lateral nerves, rust-pilose beneath, at length nearly glabrous ; petiole 3–5 cm. long, at first densely villous ; flowers clustered among the leaves ; pedicels up to 4 cm. long ; calyx-lobes very small ; staminodes spoon-shaped ; ovary 8-celled ; fruit subglobose, 6 cm. diam. or more ; seeds obliquely ellipsoid, about 6–7 cm. long, two-thirds (testa) hard and shining, other third (scar) rough ..   ..   *toxisperma*

B. toxisperma *Pierre* l.c. 14 (1890) ; Chev. in Vég. ut. Afr. trop. fr. 9 : 242 (1917) ; Rev. Bot. Appliq. 23 : 126 (1943) ; Bois & For. Trop. 45 : 27, with fig. *Mimusops toxisperma* (Pierre) A. Chev. in Rev. Bot. Appliq. 23 : 126 (1943), in syn. *Mimusops djave* Engl. in E. & P. Pflanzenfam. Nachtr. 1 : 279 (1897) ; Monogr. Afr. 8 : 81, t. 32, 33 (1904) ; Chev. Vég. ut. Afr. trop. fr. 2 : 160, figs. 25, 26, 27 (1907) ; F.W.T.A., ed. 1, 2 : 14. *Baillonella djave* Pierre ex Dubard in Ann. Mus. col. Marseille, sér. 3, 3 : 37 (1915) ; H. Lecomte in Bull. Mus. Hist. Nat. 24 : 143, 145 (1918). *Baillonella obovata* Pierre, mimeograph MS. (1895) ; Pellegrin in Mém. Soc. Linn. Normandie, n. sér. 1 : 10 (1928). *Baillonella obovata* Pierre ex Engl. in E. & P. Pflanzenfam. Nachtr. 1 : 279 (1897), not of Sonder (1850). *M. pierreana* Engl. Monogr. Afr. 8 : 82, t. 33B. *Baillonella pierreana* (Engl.) A. Chev. in Rev. Bot. Appliq. 23 : 127 (1943). A very large forest tree with prominently nerved leaves and rather long-stalked flowers crowded at the ends of stout branchlets.
S.Nig.: Degema *Talbot* ! Boje, Afi River F.R., Ikom (fr. May) *Jones & Onochie* FHI 18627 ! Boshi-Okwangwo F.R., Obudu (fr. May) *Latilo* FHI 30935 ! [Br.]Cam.: Johann-Albrechtshöhe *Staudt* 899. S. Bakundu F.R., Kumba *Ejiofor* FHI 15115 ! Also in Cameroun, Gabon and Congo. (See Appendix, p. 357.)

11. BUTYROSPERMUM Kotschy in Sitzungsber. Acad. Wien, Math. Naturw. Cl. 1, Abth. 1 : 357 (1865) ; F.T.A. 3 : 504 ; Hepper in Taxon 11 : 226 (1962). *Vitellaria* Gaertn. f. Fruct. 3 : 131, t. 205 (1805) ; Pierre in Bull. Soc. Linn. Paris 1 : 578 (1888) ; Baill. Hist. Pl. 11 : 288 (1811).

Small deciduous tree ; leaves clustered at the ends of the shoots, oblong or obovate-oblong, 10–25 cm. long, 5–8 cm. broad, with numerous slender parallel lateral nerves spreading almost at right angles, glabrous, except when young ; petiole up to 10 cm. long ; flowers white, clustered at the end of the shoots ; pedicels about 1·5–2 cm. long, pubescent ; sepals 8 or 10, in two rows, oblong-lanceolate, nearly 1 cm. long, softly tomentose outside ; corolla as long as the calyx, glabrous, lobes 8 or 10, ovate-oblong ; staminodes petaloid, elliptic, apiculate, toothed ; stamens 8 or 10 ; ovary 8–10-celled, tomentose ; fruit ellipsoid, 4–5 cm. long, 1-seeded, style persistent   ..   ..   *paradoxum* subsp. *parkii*

B. paradoxum (*Gaertn. f.*) *Hepper* in Taxon 11 : 227 (1962). *Vitellaria paradoxa* Gaertn. f. l.c. (1805) ; Baill. l.c. 289 (1891). [*Lucuma paradoxa* Gaertn. f. A.D.C. in DC. Prod. 8 : 173 (1844) ; Baehni in Candollea 9 : 427 (1942).
B. paradoxum subsp. **parkii** (*G. Don*) *Hepper* l.c. 227 (1962). *Bassia parkii* G. Don Gard. Dict. 4 : 36 (1838) ; A. DC. in DC. Prod. 8 : 199. *Butyrospermum parkii* (G. Don) Kotschy l.c. 359 (1865) ; Pl. Tinn. 20, t. 8B (1867) ; F.T.A. 3 : 504, partly ; Engl. Monogr. Afr. 8 : 22 (1904) ; Chev. in Vég. ut. Afr. trop. fr. 2 : 27, figs. 1, 2, incl. vars. *mangifolium* A. Chev. and *poissonii* A. Chev. l.c. 32 (1907) ; F.W.T.A., ed. 1, 2 : 8 ; Aubrév. Fl. For. Soud.-Guin. 425 ; Chev. in Rev. Bot. Appliq. 23 : 100 (1943) and 28 : 241 (1948), both

3

with several varieties of no taxonomic importance, excl. var. *nilotica* (Kotschy) Pierre ex Engl. (1904) ; Ruyssen, Agron. Trop. 12 ; 143–172 (1957). The Shea butter tree. A small tree about 30 ft. high with tufted leaves ; bark cracking in squares ; flowers white.
Port.G.: Pirada *d'Orey* 308 ! **Ghana:** Dagonkaddi (Apr.) *Kitson* 899 ! Krepi Plains (Jan.) *Johnson* 545 ! Bukabuka *Thompson* LXX ! Wa (Jan.) *Mehamer* 5–8 ! Yendi (Jan.) *Coull* 1–9 ! *Poisson* 90 ! **Dah.:** Massé to Kétou (Feb.) *Chev.* 23008 ! Koulomba *Vuillet* 570 ! **N.Nig.:** Ilorin to Jebba (Feb.) *Richards* 5028 ! Jebba *Barter* 987 ! Nupe *Barter* 721 ! 1178 ! Jos Plateau (Feb.) *Lely* P143 ! Baro *Yates* ! **S.Nig.:** *Ross* 73 ! Lagos *Rowland* ! *Phillips* 2 ! Abbeokuta *Irving* ! [Br.]**Cam.:** Tapare, Adamawa (Jan.) *Latilo & Daramola* FHI 34413 ! Gurum, Adamawa *teste* Hepper. Abundant in savanna extending to Ubangi-Shari. (See Appendix, p. 350.)
The conservation of the generic name *Butyrospermum* has been proposed (Hepper l.c.), but Dr. H. Heine would have preferred to abandon the name for his revision of this family in favour of *Vitellaria* had not this proposal been made.—ED.

12. SYNSEPALUM (A. DC.) Daniell (1852)—Baill. Hist. Pl. 11 : 286 (1892) ; Chev. in Rev. Bot. Appliq. 23 : 291 (1943). *Sideroxylon* Sect. *Synsepalum* A. DC. (1844).

Leaves closely silky beneath, elliptic-oblanceolate, acuminate, up to 15 cm. long and 5 cm. broad, with 10–12 pairs of lateral nerves ; petiole 1·5 cm. long ; flowers axillary, clustered ; buds finely appressed-pubescent ; fruits subsessile, broadly ellipsoid, beaked, about 2·5 cm. long, puberulous .. .. 1. *glycydorum*
Leaves glabrescent beneath :
Leaves with 8–12 pairs of lateral nerves, rounded or shortly acuminate at apex, markedly attenuated from the upper third to the base, 5–10 cm. long, 2–4 cm. broad ; petiole short, up to 5 mm. long ; flowers axillary, subsessile, pedicels up to 2 mm. long ; calyx ribbed, 5 mm. long, tomentellous outside, lobes short, triangular ; corolla with a narrow tube as long as calyx, lobes exserted, glabrous ; style exserted ; fruit ellipsoid, 1-seeded, 2 cm. long, nearly glabrous ; seeds nearly as large as the fruit .. .. 2. *dulcificum*
Leaves with about 18 pairs of lateral nerves, oblong-lanceolate, long-acuminate, and obtusely pointed at apex, cuneate from the middle to the very narrow petiole-like base, 10–19 cm. long, 4·5–6 cm. broad ; real petiole only up to 3 mm. long, but with the petiole-like base of the leaf apparently 1–1·5 cm. long and winged ; flowers axillary, very numerous ; pedicels up to 3 mm. long ; calyx pilosulous outside, lobes lanceolate-ovate, acute, 2·5 mm. long ; corolla with a narrow tube up to 3 mm. long, lobes exserted, up to 3 mm. long, glabrous ; style exserted ; fruit ellipsoid, 2 cm. long, smooth, 1-seeded .. .. 3. *stipulatum*

1. **S. glycydorum** *Wernham* in J. Bot. 55 : 82 (1917) (as *glycydora*). A shrub or small tree, about 25 ft. high, in leaf characters much resembling a *Chrysophyllum*, especially *C. subnudum*.
S.Nig.: Owasin, Bende Dist. (Jan.) *Mallam* FHI 5801 ! Degema Dist. *Talbot* 3720 ! Akamkpa Rubber Estate, Calabar (Mar.) *Latilo* FHI 41350 ! (See Appendix, p. 361.)
2. **S. dulcificum** (*Schum. & Thonn.*) *Daniell* in Bell Pharm. Journ. Trans. 11 : 445 (1852) ; Baillon l.c. 287 ; Engl. Monogr. Afr. 8 : 32, t. 7 ; Aubrév. Fl. For. C. Iv., ed. 1, 3 : 106 ; Chev. l.c. 23. *Bumelia dulcifica* Schum. & Thonn. in Beskr. Guin. Pl. 130 (1827). *Sideroxylon dulcificum* (Schum. & Thonn.) A. DC. l.c. (1844). *Bakeriella dulcifica* (Schum. & Thonn.) Dubard in Ann. Mus. col. Marseille, sér. 2, 10 : 28 (1912). *Pouteria dulcifica* (Schum. & Thonn.) Baehni in Candollea 9 : 276 (1942) (excl. syn. *S. glycydora* Wernham) A shrub or small tree, young leaves and fruits purplish. Often cultivated ; the "Miraculous Berry".
Ghana: Kibbi (fr. Jan.) *Brent* 404 ! Essuasu (Mar.) *Vigne* FH 168 ! Puru R. *Darko* 374 ! Aburi (fl. & fr. Jan.) *Johnson* 147 ! Komenda (Aug.) *Andoh* 5550 ! Afram Mankrong F.R., Mpraeso Dist. (Nov.) *Enti* FH 6509 ! **Dah.:** Zagnanado to Za (Feb.) *Chev.* 23104 ! **S.Nig.:** Agege *Foster* 221 ! Lagos (Apr.) *Moloney* ! Abeokuta (fl. & fr. Oct.) *Onochie* FHI 13517 ! [Br.]**Cam.:** Victoria Botanic Garden (fl. & fr. Feb.) *Dalz.* 8174 ! Also in Cameroun and Congo. (See Appendix, p. 361.)
3. **S. stipulatum** (*Radlk.*) *Engl.* Monogr. Afr. 8 : 33 (1904). *Stironeurum stipulatum* Radlk. in Ann. Mus. Congo, Bot., sér. 2, 1 : 31 (1899). *Synsepalum longecuneatum* De Wild. in Fedde Rep. 13 : 377 (1914). *S. attenuatum* Hutch. & Dalz. in Aubrév. Fl. For. C. Iv., ed. 1, 3 : 106 (1936), name only ; Dalz. Appendix 361 (1937), English descr. only ; Baehni in Candollea 9 : 429 (1942) ; Chev. in Rev. Bot. Appliq. 23 : 292 (1943). A small tree, up to 50 ft. high.
S.Nig.: Sapoba (Jan.–Mar.) *Kennedy* 198 ! 528 ! 1734 ! *Ross* 222 ! Okomu F.R. Benin (Dec.) *Brenan* 8550 ! *Richards* 9021 ! Omo F.R. (Feb.) *Onochie* FHI 17501 ! 17522 ! *Jones & Onochie* FHI 17212 ! 17377 ! Eme River F.R., Bende Dist. (Aug.) *Olorunfemi* FHI 34206 ! Also in Cameroun and Congo. (See Appendix, p. 361.)

Imperfectly known species.

S. tsounkpe *Aubrév. & Pellegr.* in Notulae Syst. 16 : 265 (1961). *S.* sp. near *congolense* H. Lecomte in Aubrév. Fl. For. C. Iv., ed. 2, 3 : 128, t. 128 (1959).
Iv. C.: Port Bouët *Aubrév.* 1577. (No material seen.)

13. VINCENTELLA Pierre Not. bot. Sapot. 37 (1891) ; Aubrév. & Pellegr. in Bull Soc. Bot. Fr. 81 : 800 (1935).

Young branches and petioles minutely puberulous and becoming very early glabrescent ; leaves oblong-lanceolate, 7–20 cm. long and 3–7 cm. broad, with about 10–13 pairs of lateral nerves ; stipules up to 1·8 cm. long ; flowers numerous, fasciculate on the second season's wood ; pedicels slender, 1 cm. long ; sepals narrowly lanceolate, 1·5 mm. long, puberulous, petals a little longer than the sepals ; ovary densely pilose ; style glabrous or pilose up to $\frac{2}{3}$ of its length, 2–4 mm. long .. .. 1. *revoluta*
Young branches and petioles densely rusty-brown lanate or velvety-puberulous, tomentum becoming purplish-brown or blackish in the second year and then slowly becoming glabrescent :
Leaves spathulate-lanceolate, very variable in size, shape and colour, but nearly

always becoming greenish, olive or olive-brown on drying, cuneate from the upper third to the base and there never rotundate, 5–10(–13) cm. long and up to 5 cm. broad, with about 7–9 pairs of lateral nerves ; leaf-margin sometimes slightly undulate ; stipules up to 2 mm. long ; staminodes as long as the petals, not very conspicuous; ovary densely pilose ; style glabrous, up to 3 mm. long .. .. **2. passargei**

Leaves obovate-lanceolate, 10–25 cm. long, 4·5–10 cm. broad, becoming rusty or dark brown on drying, cuneate from the upper third to the base and there mostly slightly and abruptly rotundate, with about 10–13 pairs of lateral nerves ; stipules 3–5 mm. long ; flowers fasciculate, 3–6 together in the axils of the fallen leaves ; pedicels rusty-velvety-tomentose, 2–4 mm. long ; sepals oblong-lanceolate, 2 mm. long ; petals 3·5 mm. long ; staminodes 5·5 mm. long, very conspicuous ; ovary densely pilose ; style glabrous, about 1 mm. long .. .. **3. brenanii**

1. **V. revoluta** (*Bak.*) *Pierre* Not. bot. Sapot. 37 (1891) ; Aubrév. Fl. For. Soud.-Guin. 427 ; Chev. in Rev. Bot. Appliq. 23 : 284 (1943) ; Heine in Kew Bull. 14 : 302 (1960), for full synonymy. *Sideroxylon revolutum* Bak. in F.T.A. 3 : 503 (1877). *Bakerisideroxylon revolutum* (Bak.) Engl. Monogr. Afr. 8 : 34, t. 11 (1904), incl. var. *brevipetiolulatum* Engl. l.c. ; Chev. Bot. 390 ; F.W.T.A., ed. 1, 2 : 12. *Bakeriella revoluta* Dubard in Ann. Mus. col. Marseille, sér. 2, 10 : 28 (1912). *Pouteria revoluta* (Bak.) Baehni in Candollea 9 : 385 (1942). *Bakerisideroxylon bruneelii* De Wild. in Fedde Rep. 13 : 375 (1914). *Vincentella camerounensis* Pierre ex Aubrév. & Pellegr. in Bull. Soc. Bot. Fr. 81 : 800 (1935) ; Aubrév. & Pellegr. in Fl. For. C. Iv., ed. 2, 3 : 130, t. 306, fig. 5–9 ; Chev. in Rev. Bot. Appliq. 23 : 284 (1943) ; Aubrév. Fl. For. Soud.-Guin. 427. *Pouteria camerounensis* (Pierre ex Aubrév. & Pellegr.) Baehni l.c. 383 (1942). *Vincentella impressa* Sprague in Chipp Gold Coast Trees and Shrubs, 26 (1913), name only. A tree 30–60 ft. high, with fascicles of small whitish flowers.
    Iv.C.: Byanouan (Mar.) *Chev.* 17809. Fort-Binger to Mount Nienokoué (July) *Chev.* 19515! **Ghana:** Boundary Post, Tano R. (Sept.) *Chipp* 340! Benso (Sept.) *Andoh* 5881! Dunkwa (July) *Mensah* FH 6742! **S.Nig.:** Cross R. (fl. & fr. Nov.) *Catterall* 53! **[Br.]Cam.:** Buea (fl. Jan., fr. Apr.) *Maitland* 44! 250! **F.Po:** 2,000 ft. (Dec.) *Mann* 635! Also in Cameroun and Congo.
2. **V. passargei** (*Engl.*) *Aubrév.* in Fl. For. Soud.-Guin. 427 (1950) ; Bull. Soc. Bot. Fr. 104 : 281 (1957), as *passagei*. *Bakerisideroxylon passargei* Engl. Monogr. Afr. 8 : 35, t. 11 (1904). *Pouteria passargei* (Engl.) Baehni in Candollea 9 : 382 (1942). *Sersalisia microphylla* A. Chev. Bot. 388, name only. *Bakerisideroxylon sapini* De Wild. in Rev. Zool. Afr. 7, Suppl. Bot. 16 (1919). *Vincentella sapini* (De Wild.) Brenan in Mem. N.Y. Bot. Gard. 8 : 498 (1954) ; Meeuse in Bothalia 7 : 342 (1960). *Pouteria ligulata* Baehni l.c. 386 (1942). *P. afzelii* (Engl.) Baehni l.c. 320 (1942), partly (*Chev.* 12682). *Sersalisia afzelii* of F.W.T.A., ed. 1, 2 : 11, not of Engl. A tree 15–20 ft. high, with white flowers.
    Guin.: Kollangui (Mar.) *Chev.* 12218! Diaguissa Plateau *Chev.* 12682! Timbicounda, Upper Niger (Oct.) *Jaeger* 104! Bambaya to Kéréssa-dju (Jan.) *Jaeger* 3886! **S.L.:** Loma Mts. (Jan.) *Jaeger* 4015! 4121! **Ghana:** below Shiare, Krachi Dist. (Nov.) *Morton* A4110! **N.Nig.:** Jos Plateau (fr. Jan.) *Lely* P24! Jos F.R. (Nov.) *Keay* FH 21433! FHI 20196! Assob waterfall (Oct.) *Hepper* 1040! **[Br.]Cam.:** Serti, Adamawa (fr. Dec.) *Latilo & Daramola* FHI 28960! Also in Cameroun, Ubangi-Shari, Gabon, Congo and tropical E. Africa. (See Appendix, p. 360.)
    [The plants from Guinea, Sierra Leone, and Ghana are slightly different in some vegetative characters (thickish-woolly young shoots, flowers usually, but not always, subsessile, leaves larger), but it is impossible to separate them as a different taxon as was done by Baehni (l.c. 1942)—H.H.]
3. **V. brenanii** *Heine* in Hook Ic. Pl. t. 3630 (1964). A shrub or small tree, up to 15 ft. high, with small and densely fascicled greenish-cream flowers in the axils of fallen leaves.
    [Br.]Cam.: Banga, S. Bakundu F.R., Kumba Div. (Mar.) *Brenan* 9273! *Binuyo & Daramola* FHI 35589!

14. KANTOU Aubrév. & Pellegr. in Bull. Soc. Bot. Fr. 104 : 276 (1957).

Leaves crowded at the top of the branches, oblong-lanceolate, 15–25 cm. long, 6–7 cm. broad, with about 10–15 lateral nerves, glabrescent ; stipules minute ; petiole 4–5 cm. long ; flowers fasciculate, pedicels about 1 cm. long, pubescent ; bracts ovate, 5–6 mm. long ; sepals about 6 mm. long, pubescent ; corolla glabrous, 7–8 mm. long, 5-lobate, lobes ovate, obtusely tripartite, corolla-tube about 1·5 mm. long ; ovary 5-loculate, conic, hirsute, about 4 mm. long ; fruit globose, reddish, only one well-developed seed ; seed oblong, about 5 cm. long, compressed, with brown and shining testa and a large ovate hilum .. .. *guereensis*

K. guereensis *Aubrév. & Pellegr.* l.c. (1957) ; Aubrév. Fl. For. C. Iv., ed. 2, 3 : 114, t. 291. A tall tree up to 160 ft. high, with small fascicled flowers ; in dense evergreen forest.
    Lib.: Wanau (fr. Oct.) *Harley* 2049! **Iv.C.:** Oumé *Aubrév.* 4147! Guiglo *Aubrév.* 4049! Tai *Aubrév.* 4103!

15. BREVIEA Aubrév. & Pellegr. in Bull. Soc. Bot. Fr. 81 : 792 (1935).

Branches and leaves underneath golden-yellow-brown silky-villous ; leaves elongate-oblong, abruptly caudate-acuminate, cordate at base, 11–22 cm. long and 5–8 cm. broad, with 12–18 pairs of prominent lateral nerves ; petioles very short, up to 5 mm. long ; flowers axillary in small fascicles, 3–5 together ; pedicels 3–4 mm. long ; calyx with 5 ovate imbricate sepals, silky-tomentose outside ; corolla greenish-white, tubular, up to 2 cm. long and 5 mm. diam. ; staminodes subulate, up to 2 mm. long, style up to 7 cm. long ; fruits depressed-globose, 6–7 cm. diam., 6 cm. high, with 8 flat, elliptic seeds 2 cm. long and 1·4 cm. broad .. .. *leptosperma*

B. leptosperma (*Baehni*) *Heine* in Kew Bull. 14 : 302 (1960). *Pouteria leptosperma* Baehni in Candollea 9 : 388 (1942). *Breviea sericea* (A. Chev.) Aubrév. & Pellegr. l.c. 793, fig. 1 (1935) ; Aubrév. Fl. For. C. Iv., ed. 1, 3 : 108, t. 282 ; ed. 2, 3 : 130, t. 297 ; Chev. in Rev. Bot. Appliq. 23 : 284 (1943). *Chrysophyllum sericeum* A. Chev. in Mém. Soc. Bot. Fr. 2, 8 : 296 (1917) ; Chev. Bot. 386 ; F.W.T.A., ed. 1, 2 : 9 ; Hutch. & Dalz. in Kew Bull. 1937 : 55 ; not of Salisb. (1796) and A. DC. (1844). *Chrysophyllum leptospermum* (Baehni) Roberty in Bull. I.F.A.N. 15 : 1417 (1953). A tree up to 100 ft. high ; stem up to 5 ft. girth ; in transition forest between rain and deciduous forest.
    Iv.C.: Zaranou to Bébou *Chev.* 22627! Acuabo *Aubrév.* 642! **Ghana:** Mprisu, Ashanti *Vigne* FH 2059! Ofin Headwaters F.R. (fr. Aug.) *Andoh* FH 4231! 4290!

16. AUBREGRINIA Heine in Kew Bull. 14 : 301 (1960). *Endotricha* Aubrév. & Pellegr. in Bull. Soc. Bot. Fr. 81 : 794 (1935), not *Endotrichia* Suringar (1870).

Leaves oblong, obtuse, 13–25 cm. long, 5–10 cm. broad, membranous, with about 7–9 pairs of lateral nerves ; petioles 2–3 cm. long ; flowers axillary, pedicels glabrous, about 12 mm. long ; flowers 5-merous, unisexual (only female flowers known) ; calyx with 5 imbricate, ciliate lobes, glabrous outside, densely covered with long, silky hairs inside ; corolla 6–7 mm. long, with 5 suborbiculate lobes, lobes 1·5 mm. long and 2 mm. broad ; 5 staminodes opposite the lobes, fixed at base of the corolla, about 3·5 mm. long, 5 staminodes more or less abortive, alternate with the corolla-lobes and affixed in the axils of the lobes ; disk glabrous at base, stiff-strigose at top ; ovary 8-loculate ; style about 5 mm. long, glabrous ; fruit baccate, subglobose, depressed, 8-seeded, 8 cm. diam., 6·5 cm. long ; seeds 5·5 cm. long, 3 cm. broad and 1·8 cm. thick, glabrous, dark-brown, with a long, deep-impressed hilum and woody testa   *taiensis*

A. taiensis (*Aubrév. & Pellegr.*) *Heine* l.c. (1960). *Endotricha taiensis* Aubrév. & Pellegr. l.c. (1935) ; Aubrév. Fl. For. C. Iv., ed. 1, 3 : 110 (as *Entotricha*), t. 283, fig. 1–3 ; ed. 2, 3 : 132, t. 298 ; Aubrév. & Pellegr. in Bull. Soc. Bot. Fr. 104 : 277 (1957). A very large tree, up to 150 ft. high and 8–9 ft. girth with greenish flowers ; fruits greenish-yellow.
Iv.C.: Taï *Aubrév.* 1212. Soubré *Aubrév.* 4776. **Ghana:** Kade Univ. College Farm, Oda Dist. (Jan.) *Enti* FH 6871 !
[The genera 14–16 are closely allied with *Chrysophyllum* Linn. in a broad sense and further taxonomic studies might perhaps make it necessary to unite them with it.]

17. MALACANTHA Pierre Not. bot. Sapot. 60 (1891) ; Engl. Monogr. Afr. 8 : 47 (1904).

Leaves tomentose, becoming more or less glabrescent, slightly cuneate to rounded at base, obovate-elliptic, up to 23 cm. long, 6–12 cm. broad, with 15–20 pairs of very prominent lateral nerves ; flowers sessile, clustered ; calyx light brown-rusty tomentose outside, lobes elliptic ; corolla-tube 4 mm. long, lobes 3 mm. long, glabrous ; ovary tomentose ; fruits ellipsoid, apiculate, up to 2·5 cm. long, pubescent

alnifolia

M. alnifolia (*Bak.*) *Pierre* l.c. 61 (1891) ; Engl. l.c. 49 ; Chev. Bot. 387 ; F.W.T.A., ed. 1, 2 : 12, partly ; Aubrév. Fl. For. C. Iv., ed. 1, 3 : 110 ; ed. 2, 3 : 133 ; Fl. For. Soud.-Guin. 427. *Chrysophyllum? alnifolium* Bak. in F.T.A. 3 : 499 (1877). *Pouteria alnifolia* (Bak.) Roberty in Bull. I.F.A.N. 15 : 1417 (1953). *M. heudelotiana* Pierre l.c. 60 (1891) ; Aubrév. Fl. For. C. Iv., ed. 2, 3 : 134, t. 299. *M. warneckeana* Engl. l.c. 48 (1904). *M. obtusa* C. H. Wright in Kew Bull. 1908 : 58. *M. acutifolia* A. Chev. Bot. 387, name only. A shrub or small tree with scaly bark ; flower-clusters dense in the axils of fallen leaves ; in deciduous and gallery forests.
Sen.: *Perrottet* 452 ! 454 ! Kombo country *Heudelot* 52 ! **Iv.C.:** Anoumaba *Chev.* 22397 ! **Ghana:** Cape Coast *Dalz.* 8277 ! Shai Plains (Jan.) *Johnson* 529 ! Kpedsu (Dec.) *Howes* 1070 ! Aburi (fr. Feb.) *Dalz.* 8269 ! Komenda *Vigne* FH 5598 ! Lawra *Adams* 4028 ! **Togo Rep.:** Lomé *Warnecke* 413 ! *Mildbr.* 7514 ! Törrimaberg *Kersting* A236 ! **Dah.:** Pobé to Adjaouéri (Feb.) *Chev.* 22941 ! **N.Nig.:** *Thornewill* 105 ! Lokoja (Apr.) *Elliott* 51 ! Yola (fl. & fr. Mar.) *Dalz.* 192 ! **S.Nig.:** Lagos (fl. & fr. Mar.) *Foster* 37 ! *Rowland* ! Olokemeji F.R. (fr. Apr.) *Keay* FHI 37598 ! Oyo (Feb.) *Keay* FHI 21076 ! Owo F.R. *Jones* FHI 3622 ! Onitsha *Barter* 1788 ! Widespread in tropical Africa. (See Appendix, p. 356.)

18. ANINGERIA Aubrév. & Pellegr. in Bull. Soc. Bot. Fr. 81 : 795 (1935).

Young shoots and leaves covered with fine greyish tomentum, glabrescent, leaves elliptic, rounded at base, 5–16 cm. long, 3–7 cm. broad, with 12–15 prominent pairs of lateral nerves covered with fine greyish tomentum in young state ; petiole 1–2 cm. long, glabrescent ; flowers fasciculate, 4–5 together in the axils of the terminal leaves ; pedicels tomentellous, up to 5 cm. long ; calyx softly pubescent outside ; corolla white, with 5 ciliate lobes ; staminodes subulate, ⅔ of the length of the corolla-lobes ; anthers not exceeding the corolla ; fruits obovoid-globose olive-shaped, finely pubescent ; seeds ovoid, 1·3 cm. long, with a broad elliptic scar nearly as long   1. *altissima*
Young shoots and leaves underneath densely pubescent, leaves elliptic to oblong-obovate, sometimes emarginate at apex, rounded at base, 8–15 cm. long, 4–6 cm. broad, with 15–20 pairs of lateral orange-coloured nerves, covered with a dense tomentum ; flowers fasciculate, in the axils of fallen leaves ; pedicels tomentellous, up to 3 cm. long ; calyx and corolla like the last but staminodes only ⅓ of the length of the corolla-lobes and anthers exceeding the corolla ; fruit and seeds like the last   2. *robusta*

1. A. altissima (*A. Chev.*) *Aubrév. & Pellegr.* l.c. 796 ; Aubrév. Fl. For. C. Iv., ed. 1, 3 : 111 ; ed. 2, 3 : 136, t. 300, fig. 1–4, under "*Aningueria*" ; Fl. For. Soud.-Guin. 429. *Hormogyne altissima* A. Chev. in Mém. Soc. Bot. Fr. 2, 8 : 265 (1917). *Sideroxylon altissimum* (A. Chev.) Hutch. & Dalz. F.W.T.A., ed. 1, 2 : 12 (1931) ; Kew Bull. 1937 : 59. *Hormogyne gabonensis* A. Chev. l.c. 267 (1917). *Sideroxylon gabonense* (A. Chev.) Lecomte ex Pellegr. in Mém. Soc. Linn. Normandie, n. sér. 1 : 16 (1928). *Pouteria altissima* (A. Chev.) Baehni in Candollea 9 : 292 (1942). Tall forest tree up to 120 ft. high with straight bole ; flowers creamy-white.
Guin.: Kaba *Chev.* 13129 ! 13141 ! **S.L.:** Loma Mts. *Jaeger* 1845 ! **Iv.C.:** Abidjan (Feb.) *Aubrév.* 86 ! 842 ! **Ghana:** Abetifi, Kwahu (Mar.) *Johnson* 624 ! *Brent* 543 ! L. Bosumtwi (Dec.) *Vigne* FH 1498 ! Ahiayom (Dec.) *Andoh* 582 ! Opro River F.R., Ashanti (fr. Apr.) *Keay & Aubrév.* FHI 37594 ! **N.Nig.:** Zaria *Lamb* FHI 3215 ! Esie, Ilorin Dist. (Jan.) *Latilo* FHI 18225 ! [**Br.**]**Cam.:** *Mildbr.* 10 ! Bamenda, 5,000 ft. *Johnstone* 265/32 ! *Kennedy* 2161 ! Extends to Congo, E. Africa, Ethiopia and Sudan. (See Appendix, p. 360.)
2. A. robusta (*A. Chev.*) *Aubrév. & Pellegr.* in Bull. Soc. Bot. Fr. 81 : 797 (1935). *Malacantha robusta* A. Chev. in Vég. ut. Afr. trop. fr. 5 : 241 (1909) ; Chev. Bot. 387 ; Fl. For. C. Iv., ed. 1, 3 : 111, 112 ; ed. 2, 3 : 137, t. 300, fig. 5–9, under "*Aningueria*." *M. heudelotiana* of F.W.T.A., ed. 1, 2 : 12, partly, not of Pierre (1891). *Pouteria aningeri* Baehni l.c. 289 (1942). *A. pierrei* (A. Chev.) Aubrév. & Pellegr. l.c. 796 (1935) ; Aubrév. Fl. For. Soud.-Guin. 429 ; Fl. For. C. Iv., ed. 1, 3 : 111 ; ed. 2, 3 : 134 ; Chev. in Rev. Bot. Appliq.

23 : 136 (1943). *Hormogyne pierrei* A. Chev. Etud. Fl. Afr. centr. 1 : 183 (1913), name only ; Mém. Soc. Bot. Fr. 2, 8 : 265 (1917) ; Vég. ut. Afr. trop. fr. 9 : 262 (1916). *Pouteria pierrei* (A. Chev.) Baehni l.c. 292 (1942). Tall forest tree up to 150 ft. high with clear straight bole, buttressed at base ; flowers whitish, fruits red.

Port.G.: Catio to Cumebú (fr. June) *Esp. Santo* 2095 ! **S.L.:** Heddle's Farm, Freetown *Lane-Poole* 374 ! Peri Ganra *King* 333 ! Kasewe F.R. *King* 113 ! **Iv.C.:** Bouroukrou *Chev.* 16134 ! Taï *Aubrév.* 1206. Sassandra (fl. & fr. Feb., Mar.) *Leeuwenberg* 2809 ! 2922 ! **Ghana:** Adeambra (Mar.) *Vigne* FH 863 ! Juaso (Feb.) *Vigne* FH 1828 ! Bobiri F.R. *Taylor* 5182 ! Kwahu Prasu (Feb.) *Vigne* FH 1614 ! **S.Nig.:** Lagos *Punch* 120 ! Benin (May) *Cousens* FHI 3838 ! Gambari F.R. (Jan.) *Idahosa* FHI 23860 ! Omo F.R. *Jones & Onochie* FHI 16821 ! 17587 ! Owena R. Pilot Mill, Ondo *Jones* FHI 19533 ! **[Br.]Cam.:** Kumba Div. *Ejiofor* FHI 14037 ! Bambui, Bamenda *Bilderbeck* FHI 22207 ! Also in Ubangi-Shari. (See Appendix, p. 357.)

19. BEQUAERTIODENDRON De Wild. in Rev. Zool. Afr. 7, Suppl. Bot. 22 (1919) ; Pl. Bequaert. 4 : 143 (1926) ; emend Heine & Hemsley in Kew Bull. 14 : 306 (1960).

Flowers with pedicels 1·5–2 cm. long ; leaves oblong or obovate-oblong, slightly cuneate from above the broadest upper third, cuneate-rotundate at base, 11–24 cm. long, 8 cm. broad, on drying always becoming brownish or rusty-brown beneath
1. *magalismontanum*

Flowers sessile ; leaves obovate, broadly acuminate, long-cuneate at base, 10–17 cm. long, 4–6 cm. broad, on drying always light greyish olive-green 2. *oblanceolatum*

1. **B. magalismontanum** (*Sond.*) *Heine & J. H. Hemsley* in Kew Bull. 14 : 307 (1960), (q.v. for full synonymy)· *Chrysophyllum magalismontanum* Sond. in Linnaea 23 : 72 (1850). *Zeyherella magalismontana* (Sond.) Aubrév. & Pellegr. in Bull. Soc. Bot. Fr. 105 : 37 (1958). *Pouteria magalismontana* (Sond.) A. Meeuse in Bothalia 7 : 335 (1960). *Tisserantiodoxa oubanguiensis* Aubrév. & Pellegr. in Bull. Soc. Bot. Fr. 104 : 277 (1957) ; Fl. For. C. Iv., ed. 2, 3 : 109, t. 304, fig. 8–15 (1959). *Chrysophyllum laurentii* De Wild. Miss. E. Laurent 1 : 429, t. 133 (1907) ; F.W.T.A., ed. 1, 2 : 10. *C. farannense* A. Chev. Bot. 286 (1920), name only. An extremely variable tree or shrub, 16–20 ft. high, with long subcoriaceous leaves, crowded towards the end of the branches, the leaves shiny above, minutely rusty-greyish, pubescent beneath, with numerous flowers in dense fascicles ; in our region typical of wet places, always near water, along streams, etc. **Guin.:** Faranna *Chev.* 13395. Socourala *Chev.* 20498. **Lib.:** Duport (Nov.) *Linder* 1479 ! **Iv.C.:** Cavally R. *Chandler* ! **Ghana:** Amoma (Jan.) *Vigne* FH 3548 ! **S.Nig.:** Sapoba *Symington* FHI 5693 ! *Kennedy* 1962 ! 2048 ! 2313 ! *Onochie* FHI 23425 ! 34267 ! Stubbs Creek F.R., Calabar *Onochie* FHI 32913 ! Akpan, Calabar Dist. (July) *Ujor* FHI 31621 ! Bonny *Kalbreyer* 59 ! Also in Cameroun, Gabon, Ubangi-Shari, Congo, E. and S. Africa. (See Appendix, p. 355.)

2. **B. oblanceolatum** (*S. Moore*) *Heine & J. H. Hemsley* l.c. 309 (1960). *Sideroxylon oblanceolatum* S. Moore in J. Bot. 14 : 47 (1907). *Chrysophyllum obovatum* of Engl. Monogr. Afr. 8 : 43 (1904), not of Sabine & G. Don (1824). *Pseudoboivinella oblanceolata* (S. Moore) Aubrév. & Pellegr. in Notulae Syst. 16 : 260 (1961). *Manilkara* (as *Manilhora*) *dahomeyensis* of Chev. Bot. 394, not of Pierre ex Dubard (1915). *Chrysophyllum glomeruliferum* Hutch. & Dalz. F.W.T.A., ed. 1,2 : 9 (1931) ; Aubrév. Fl. For. C. Iv., ed. 1,3 : 122, t, 288, fig. 5 ; Hutch & Dalz. in Kew Bull. 1937 : 56. *Boivinella glomerulifera* (Hutch. & Dalz.) Aubrév. & Pellegr. in Bull. Soc. Bot. Fr. 105 : 37 (1958) ; Aubrév. Fl. For. C. Iv., ed. 2, 3 : 148, t. 304, fig. 1–7. *Neoboivinella glomerulifera* (Hutch. & Dalz.) Aubrév. & Pellegr. l.c. 106 : 23 (1959). A tree up to 60 ft high, with pale leaves and small flowers clustered mostly below the leaves ; in deciduous forest. **S.L.:** Mt. Gonkwi (Feb.) *Sc. Elliot* 4867 ! **Iv.C.:** *Chev.* 23789 ! *Aubrév.* 1256 ! Hiré *Aubrév.* 4133 ! **Ghana:** Atewa Range (Feb.) *Vigne* FH 4327 ! Ateiso F.R., E. Reg. (fr. Mar.) *Irvine* 2414 ! Akumadan (Oct.) *Vigne* FH 2566 ! Afram Headwaters F.R. (Apr.) *Andoh* 4736 ! Also in E. Africa.

20. DELPYDORA Pierre in Bull. Soc. Linn. Paris 2 : 1275 (1896) ; Engl. Monogr. Afr. 8 : 49 (1904).

Branchlets hispid-pilose ; leaves narrowly obovate, long and acutely acuminate, slightly unequal-sided at base, up to 30 cm. long and 9 cm. broad, pilose on the midrib beneath, lateral nerves 18–20 pairs ; flowers in axillary fascicles ; pedicels 3 mm. long, pilose ; sepals about 6 mm. long, thinly strigose-pilose ; petals about as long as sepals
gracilis

D. gracilis *A. Chev.* in Mém. Soc. Bot. Fr. 2, 8 : 263 (1917) ; Chev. Bot. 394 ; Hutch. & Dalz. in Kew Bull. 1937 : 58. A shrub or small tree. **Lib.:** Dukwia R. (fr. Feb.) *Cooper* 175 ! 29–46 ! Suen (fr. Nov.) *Baldwin* 10481 ! Yratoke (fr. July) *Baldwin* 6275 ! Boporo *Baldwin* 10383 ! Gbawia (July) *Baldwin* 6715 ! **Iv.C.:** Toula to Nékaougnié *Chev.* 19583 ! Tou *Chev.* 19672 ! **Ghana:** Prestea *Vigne* FH 3083 ! Simpa (fr. May) *Vigne* FH 1973 ! Awiabo (fl. & fr. Nov.) *Vigne* FH 1474 !

21. CHRYSOPHYLLUM Linn.—F.T.A. 3 : 498.

Leaves glabrous beneath, with very numerous, non-prominent, parallel lateral nerves ; leaf-colour on drying becoming on both sides dark blackish-green or dark brownish-green (Sect. *Donella* (Pierre) Engl.) :
Trees :
  Fruits up to 11 cm. long and 8·5 cm. diam. ; leaves oblong, shortly cuneate or obtuse at base, abruptly and obtusely acuminate, 10–16 cm. long, 4–7 cm. broad ; petioles 8–12 mm. long ; flowers axillary, numerous ; pedicels up to 4 mm. long, puberulous at first ; calyx pubescent ; fruits with a fleshy exocarp, well-marked 5-angled ; seeds up to 5·5 cm. long, 3 cm. broad and up to 1·6 cm. diam. 1. *pentagonocarpum*
  Fruits up to 4·5 cm. long and 3 cm. diam. ; leaves very similar to the preceding species, very shortly cuneate at base, abruptly and obtusely acuminate, 5–10 cm. long, 3–5 cm. broad ; petiole up to 5 mm. long ; flowers axillary, numerous ; pedicels up to 5 mm. long, pubescent ; fruit depressed-globose, 5-angled, with a fleshy exocarp ; seeds up to 2·2 cm. long and 1·2 cm. broad .. 2. *pruniforme*
Climbing shrub ; leaves like the preceding species, but broadly rounded at the base, shortly acuminate, 5–8 cm. long, 2·5–3·5 cm. broad ; petiole up to 3 mm. long ;

flowers axillary ; pedicels 2–3 mm. long ; fruit about 4·5 cm. long and 3 cm. in diam., more or less beaked ; seeds 2·5 cm. long, 1 cm. broad and 5–6 mm. diam.   **3. welwitschii**
Leaves always with an indumentum beneath, but sometimes glabrescent ; lateral nerves not very dense and numerous, always clearly prominent ; leaves on drying seldom becoming dark-greenish colour on both sides (Sect. *Gambeya* (Pierre) Engl.) :
Leaves asymmetrical, irregularly rounded at base, rusty-tomentose or pilose beneath with rather long and conspicuous hairs ..   ..   **4. beguei**
Leaves symmetrical regular-cuneate at base :
Fruits clearly beaked, 5–6 cm. long, 3–4 cm. diam. ; leaves very sparsely and shortly rusty-pilose beneath, mainly along the nerves ; oblong-elliptic, 14–18 cm. long, 4–6 cm. broad, with about 7–10 pairs of lateral nerves ; petioles 6–8 mm. long, tomentose ; flowers fascicled, axillary, sessile, 3–4 mm. long ; calyx about 3 mm. long ; sepals hirsute outside, inside glabrous ; petals up to 3 mm. long, oblong-lanceolate ; seeds 25–28 mm. long, 10–13 mm. broad, black ..   **5. azaguieanum**
Fruits not clearly beaked ; leaves silky-tomentose beneath, but sometimes glabrescent :
Leaves usually silvery-silky beneath, indumentum persistent or becoming glabrescent :
Indumentum persistent, i.e. leaves still silvery beneath in adult state, elongate-obovate-elliptic, shortly acuminate, cuneate at base, 12–30 cm. long, 5–9 cm. broad, with 12–15 pairs of distant lateral nerves, finely appressed beneath ; petiole 1–3 cm. long, hoary ; flowers shortly pedicellate ; calyx 3 mm. long, silky ; fruit mostly depressed-globose, about 3 cm. diam...   ..   **6. albidum**
Indumentum not persistent, i.e. leaves at first silvery-silky beneath, becoming glabrescent in adult state :
Leaves oblanceolate, apex rather elongated, obtusely acuminate, 8–16 cm. long, 2·5–5·5 cm. broad, shortly pubescent beneath, with 10–15 pairs of prominent lateral nerves ; petiole 1–2 cm. long, pubescent ; flowers subsessile ; calyx-lobes broadly ovate, pubescent, 3–4 mm. long ; corolla-lobes densely white-ciliate ; fruit about 3 cm. diam., slightly 5-lobed   ..   **7. subnudum**
Leaves elliptic-lanceolate, apex not elongated, sharply acuminate, 10–15 cm. long, 3–5·5 cm. broad, shortly silvery-pubescent beneath, with about 15 pairs of lateral nerves, petiole 1–1·5 cm. long, pubescent ; flowers unknown ; fruit 2–2·5 cm. diam., globose, not lobed ..   ..   **8. taiense**
Leaves fulvous or golden beneath :
Flowers in short racemes   ..   ..   **9. giganteum**
Flowers in fascicles :
Leaves small, usually below 10 cm. long, and 4–5 cm. broad, broadly elliptic-obovate, mucronate-acuminate, with 12–14 pairs of lateral nerves, rusty-silky beneath ; petiole 1 cm. long ; flowers sessile, very small ; calyx appressed-tomentose   ..   ..   **10. prunifolium**
Leaves large, usually well over 10 cm. long, lateral nerves about 15 pairs or more :
Lateral nerves 15–20 pairs ; leaves oblong-elliptic, obtuse or shortly pointed, 15–22 cm. long, 6–8 cm. broad, dark brown-reddish, rusty-tomentose beneath but hairs not appressed ; petiole 2–3 cm. long, tomentellous ; flowers clustered, sessile ; calyx rusty-tomentose ; fruits sessile, globose, 2·5 cm. diam., rusty-tomentose   ..   ..   **11. perpulchrum**
Lateral nerves 20–30 pairs ; leaves elongate-obovate-elliptic, acuminate, 15–30 cm. long ; indumentum always clearly appressed, tawny to rusty-brown :
Mature crown leaves 7–15 cm. long, averaging about 12 cm., up to 3 cm. broad, primary lateral nerves 10–17 with loose reticulum of veins ; fruits up to 3 cm. diam., usually with persistent ferrugineous indumentum   **12. gorungosanum**
Mature crown leaves 13–25 cm. long, averaging about 20 cm., primary lateral nerves 15–30, with rather close reticulum of veins ; fruits large, up to 6 cm. diam.   ..   ..   **13. delevoyi**

1. **C. pentagonocarpum** *Engl. & K. Krause* in Engl. Bot. Jahrb. 49 : 387, fig. 2 (1913). *C. letestuanum* A. Chev. in Mém. Soc. Bot. Fr. 2, 8 : 269 (1917) ; Aubrév. Fl. For. C. Iv., ed. 1, 3 : 122, t. 288, fig. 6–9 ; ed. 2, 3 : 146, t. 303, fig. 6–11 ; Chev. in Rev. Bot. Appliq. 23 : 137 (1943). *Donella letestuana* (A. Chev.) A. Chev. ex Pellegr. in Mém. Soc. Linn. Normandie, n. sér. 1 : 15 (1928). *Chrysophyllum belemba* De Wild. Pl. Bequaert. 4 : 122 (1926). *C. claessensii* De Wild. l.c. 125 (1926). A large tree, with small fascicled flowers and globose lemon- or orange-like, yellow fruits with 5 seeds ; in dense rain forest.
   Iv.C.: Man *Aubrév.* 1165. Oumé *Aubrév.* 4170. **Ghana:** Kwahu Prasu (Feb.) *Vigne* FH 1595 ! **S.Nig.:** Ukpon F.R., Ogoja (fr. May) *Jones & Onochie* FHI 18886 ! Also in Cameroun, Gabon and Congo.
2. **C. pruniforme** *Pierre ex Engl.* Monogr. Afr. 8 : 42 (1904) ; Aubrév. Fl. For. C. Iv., ed. 1, 3 : 122 ; ed. 2, 3 : 146, t. 303, fig. 2–4 ; Chev. in Rev. Bot. Appliq. 23 : 138 (1943). *C. dubardii* Pierre in Agr. prat. pays chauds, Bull. Jard. col. 5e année, n. 28 : 88 (1905). *C. gracile* A. Chev. in Mém. Soc. Bot. Fr. 2, 8 : 268 (1917) ; Chev. Bot. 386. *C. mortehanii* De Wild. Pl. Bequaert. 4 : 137 (1926). A tree up to about 100 ft. high, with small flowers in axillary clusters, like the preceding species, but with much smaller fruits.
   S.L.: *Deighton* 3919 ! *Unwin & Smythe* 30 ! Mt. Aureol, Freetown (June, Dec.) *Deighton* 2718 ! *Dalz.* 958 ! Moselelo (fr. Nov.) *Deighton* 2388 ! Falaba *Aylmer* 34 ! Gola Forest (May, fr. June) *Small* 654 ! 734 ! *Deighton* 5213 ! Kambui F.R. (fr. June) *Jordan* 2121 ! **Iv.C.:** Dabou *Jolly* ! **Ghana:** Axim (Feb.) *Vigne* FH 2830 ! *Irvine* 2152 ! **Dah.:** Bokoutou F.R. *Chev.* 22863 ! **S.Nig.:** Eba (Jan.) *Thornewill* 248 ! Forcados (July) *Unwin* 59 ! Benin *Kennedy* 1703 ! Omo F.R. *Jones & Onochie* FHI 17355 ! Stubbs Creek F.R., Eket *Amachi* FHI 24308 ! Also in Gabon and Congo. (See Appendix, p. 356.)

3. **C. welwitschii** *Engl.* Bot. Jahrb. 12 : 521 (1890) ; Sc. Elliot in J. Linn. Soc. 30 : 84 ; Engl. Monogr. Afr. 8 : 41, t. 13, fig. A. *Micropholis angolensis* Pierre Not. bot. Sapot. 41 (1891). *Donella welwitschii* Pierre ex Engl. Monogr. Afr. 8 : 41 (1904), in syn. *C. klainei* Pierre ex Engl. l.c. 42 (1904). *C. ellipticum* A. Chev. Bot. 386 (1920), name only. *C. ealaense* De Wild. Pl. Bequaert. 4 : 128 (1926). A climbing shrub, with small greenish flowers in axillary clusters.
   S.L.: Njala (fl. June, fr. Nov.) *Deighton* 1956 ! 3102 ! Bwedu to Kangama (Apr.) *Deighton* 3192 ! Bendu (Apr.) *Deighton* 1598 ! Kumrabai *Smythe* 202 ! Bumban to Loko (Apr.) *Sc. Elliot* 5673 ! Rokupr (May) *Jordan* 254 ! **Lib.:** Ganta (May) *Harley* 1158 ! Jabroke (fr. July) *Baldwin* 6627 ! Robertsfield (fr. July) *Baldwin* 6650a ! Mano, Grand Cape Mount Co. (fr. Dec.) *Baldwin* 10812 ! Brewersville (fr. Dec.) *Baldwin* 10970 ! **Iv.C.:** Tingouéla to Assikasso *Chev.* 22576 ! Sassandra (Dec.) *Leeuwenberg* 2252 ! Adiopodoumé (Dec.) *Leeuwenberg* 2267 ! **Ghana:** Akropong, Akwapim (fr. Jan.) *Adams* 4726 ! Elmina (Apr.) *Hall* 1340 ! **S.Nig.:** Ibadan *Keay, Richards & Jones* FHI 22460 ! Abeokuta to Ibadan (Jan.) *Keay* FHI 22465 ! Mamu River F.R., Onitsha (Aug.) *Olorunfemi* FHI 34183 ! Cross River North F.R. (June) *Latilo* FHI 31839 ! Also in Gabon, Congo and in Angola. (See Appendix, p. 356.)

4. **C. beguei** *Aubrév. & Pellegr.* in Bull. Soc. Bot. Fr. 81 : 795 (1935) ; Fl. For. C. Iv., ed. 1, 3 : 118, t. 288, fig. 1 ; ed. 2, 3 : 142, t. 303, fig. 1–2. A shrub or tree, with rusty tomentose branches and glabrous globose fruits, up to 1½ ins. across.
   Iv.C.: Yapo *Aubrév.* 1808. 2270. **Ghana:** Adeambra *Vigne* FH 857 ! Bobiri F.R. *Keay & Aubrév.* FHI 37586 ! Tabeso to Antobea (Apr.) *Enti* FH 6714 ! Also in Congo and N. Rhodesia.

5. **C. azaguieanum** *Miège* in Bull. Soc. Bot. Fr. 103 : 145 (1956) ; Aubrév. Fl. For. C. Iv., ed. 2, 3 : 145. A small tree, 40–50 ft. high, with sessile, fascicled small white flowers ; in rain forest.
   Iv.C.: Azaguié (fl. & fr. July, Aug.) *Hb. I.D.E.R.T.* 1896 ! **Ghana:** Axim to Sekondi (Mar.) *Morton* A391 ! Simpa (Feb.) *Vigne* FH 2805 ! Bomassu (May) *Vigne* FH 2012 ! Ankassa (Aug.) *Chipp* 329 !

FIG. 205.—CHRYSOPHYLLUM ALBIDUM *G. Don* (SAPOTACEAE).

A, flowering shoot. B, open flower. C, stamen. D, longitudinal section of ovary. E, fruit.
F, interior of fruit with seed. G, seeds.

6. **C. albidum** *G. Don* Gen. Syst. 4 : 32 (1837) ; F.T.A. 3 : 500 (1877) ; Engl. Monogr. Afr. 8 : 45 (1904); Chev. Bot. 386 ; Aubrév. Fl. For. C. Iv., ed. 2, 3 : 144. *Sapota mammosa* Gaertn. De Fruct. 2 : 104, t. 104, figs. h, i (1791), nom. ambig. *Gambeya mammosa* (Gaertn.) Pierre Not. bot. Sapot. 63 (1891). *Achras sericea* Schumach. Beskr. Guin. Pl. 179 (1827). *Chrysophyllum millenianum* Engl. Monogr. Afr. 8 : 44 (1904). *C. kayi* S. Moore in J. Bot. 47 : 412 (1909). *C. obovatum* of Chev. in Vég. ut. Afr. trop. fr. 5 : 237 (1909), not of Sabine. A tall tree, with the branchlets and leaves beneath whitish fine-tomentose, and small flowers amongst or below the upper leaves.
   S.L.: Dambaye *King* 71 ! Dodo Hills F.R. *Sawyerr* FHK 13588 ! **Ghana:** *Thonning* (photo) ! Abofaw *Vigne* FH 1180 ! Aburi Gardens *Dalz.* 8264 ! Akwapim *Irvine* 1529 ! Sewerako-Dede F.R. *Osafo* FH 6586 ! **N.Nig.:** Ita, Benue Prov. *Thomas* 1939 ! **S.Nig.:** Lagos *Foster* 107 ! *Moloney* ! Sapoba *Kennedy* 2173 ! Olokemeji F.R. *Ross* 85 ! Nawfia, Awka *Jones* FHI 6632 ! Degema Dist. *Talbot* 3676 ! Etomi, Ikom (June) *Latilo* FHI 31857 ! **[Br.]Cam.:** Gangumi, Adamawa *Latilo & Daramola* FHI 28851 ! Widespread in tropical Africa, described under many names, of which only those described from or used for plants within our area are mentioned above. (See Appendix, p. 354.)

7. **C. subnudum** *Bak.* in F.T.A. 3 : 499 (1877) ; Engl. Monogr. Afr. 8 : 45, t. 15, fig. E (1904) ; Aubrév. Fl. For. C. Iv., ed. 1, 3 : 120, t. 287, fig. 4–8 ; ed. 2, 3 : 144, t. 302, fig. 4–8. *Gambeya subnuda* (Bak.) Pierre Not. bot. Sapot. 63 (1891). *Chrysophyllum renieri* De Wild. in Fedde Rep. 13 : 376 (1914). *C. brieyi* De Wild. in Rev. Zool. Afr. 7, Suppl. Bot. 28 (1919). *C. metallicum* Hutch & Dalz. F.W.T.A., ed. 1, 2 : 9 (1931) ; Kew Bull. 1937 : 56. A tree up to 30 ft. high, the leaves at first silvery-silky beneath, becoming glabrous ; flowers whitish, in small axillary clusters.
   S.L.: Kambui via Hangha (June) *King* 106 ! Solanyé Valley, Kambui (Mar.) *King* 167 ! **Lib.:** Dukwia R. *Cooper* 282 ! Ganta (Aug.) *Harley* 633 ! **Iv.C.:** Abidjan (Oct.) *Aubrév.* 126 ! Brafouédi (fr. Dec.) *de Wit* 504 ! **Ghana:** Banka (Sept.) *Vigne* FH 1364 ! Simpa (May) *Vigne* FH 1956 ! Sefwi Bekwai (fr. Oct.) *Akpabla* 894 ! Axim *Irvine* 2377 ! **S.Nig.:** Calabar (Mar.) *Brenan & Jones* 9255 ! Dukwe *Latilo* FHI 40348 ! Ogoja *Jones & Onochie* FHI 17397 ! 18792 ! **[Br.]Cam.:** S. Bakundu F.R., Kumba Div. *Ejiofor* FHI 29324 ! Also in Cameroun, Gabon, Rio Muni and Congo.

8. **C. taiense** *Aubrév. & Pellegr.* in Bull. Soc. Bot. Fr. 104 : 279 (1957) ; Aubrév. Fl. For. C. Iv., ed. 2, 3 144, t. 303, fig. 10, 11. A medium sized tree, very similar to No. 7 and perhaps only a local race of it.

Iv.C.: Taï *Aubrév.* 4090*bis*. (According to Aubréville widespread in the forests of Taï, Upper Niouniourou, Soubré, etc.)

9. **C. giganteum** *A. Chev.* in Mém. Soc. Bot. Fr. 2, 8 : 267 (1917) ; Bot. 386 ; Aubrév. Fl. For. C. Iv., ed. 1, 3 : 120, t. 287, fig. 103 ; ed. 2, 3 : 142, t. 302, fig. 1–3 ; Hutch. & Dalz. in Kew Bull. 1937 : 57. Tree up to 70 ft. high, sharply buttressed, bark with long narrow fissures, slash pink ; fruit dull orange.
S.L.: *Crichton* 128 ! **Iv.C.**: Akabélé, Krou (fr. Dec.) *Chev.* 22512 ! **Ghana**: Nfaense (Feb.) *Chipp* 115 ! Benchema (Apr.) *Foggie* FH 4457 ! Juaso to Ogbogu Road (Feb.) *Akpabla* 277 ! Antubia, W.P. (Apr.) *Cansdale* 3977 ! **S.Nig.**: Shasha F.R., Ife Dist. (Mar.) *Cooper* FHI 43823 ! (See Appendix, p. 355.)

10. **C. prunifolium** *Bak.* in F.T.A. 3 : 499 (1877). A tree 30–60 ft. high, rusty-silky on the branchlets and under the surface of the leaves.
S.Nig.: *Kennedy* (*fide* Dalz. Appendix, no specimen seen). **F.Po**: (Mar.) *Mann* 2344 ! (See Appendix, p. 356.)

11. **C. perpulchrum** *Mildbr. ex Hutch. & Dalz.* F.W.T.A., ed. 1, 2 : 10 (1931), and in Kew Bull. 1937 : 57 ; Aubrév. Fl. For. C. Iv., ed. 1, 3 : 118, t. 286, fig. 6–8 ; ed. 2, 3 : 140, t. 301, fig. 6–8. *C. africanum* A. DC. var. *orientale* Engl. Bot. Jahrb. 48 : 390 (1913). *C. africanum* A. DC. var. *multinervatum* De Wild. Pl. Bequaert. 4 : 121 (1926). A tree up to 100 ft. high, with subcoriaceous leaves red-brown velvety beneath.
Lib.: Ganta (fr. Feb., Mar.) *Baldwin* 10996 ! *Harley* 326 ! 807 ! 1391 ! **Iv.C.**: Oroumba-Boca *de Wilde* 612 ! Sassandra (fl. & fr. Mar.) *Leeuwenberg* 3115 ! **Ghana**: S. Scarp F.R. *Beveridge* FH 3280 ! Ntakem, Sepuri (Mar.) *Vigne* FH 1185 ! Kumasi (Mar.) *Andoh* FH 5632 ! Awoso, near Dunkwai (fr. Oct.) *Akpabla* 884 ! L. Bosumtwe (fr. Dec.) *Vigne* FH 1503 ! Agogo (fr. May) *Chipp* 444 ! **S.Nig.**: Osanko *Sankey* ! Ekpuro (fr. Mar.) *Thompson* 8 ! Benin *Hitchens* ! Oria, Benin *Umana* FHI 29125 ! Aponmu, Ondo Prov. (fr. July) *Olorunfemi* FHI 34157 ! Also in Congo and E. Africa.

12. **C. gorungosanum** *Engl.* Monogr. Afr. 8 : 44 (1904) ; Brenan in Mem. N.Y. Bot. Gard. 8 : 498 (1954) ; Meeuse in Bothalia 7 : 329 (1960). *C. fulvum* S. Moore in J. Linn. Soc. 40 : 131 (1911). A tree, sometimes reaching 120 ft. high, but usually much smaller.
[Br.]Cam.: Buea, 3,000 ft. (Mar.) *Maitland* 462 ! Mimbia to Lyonga, 4,500 ft. (Apr.) *Brenan* ! Also in Angola, Mozambique and E. Africa ; a typical highland species.

13. **C. delevoyi** *De Wild.* Pl. Bequaert. 4 : 126 (1926). *C. macrophyllum* Sabine & G. Don[1] in Trans. Hort. Soc. Lond. 5 : 458 (1824) ; not of Lam. (1797) nor Gaertn. f. (1805). *C. africanum* A. DC. in DC. Prod. 8 : 163 (1844), invalid name ; F.T.A. 3 : 500 ; Engl. Monogr. Afr. 8 : 43, t. 15A ; Chev. in Vég. ut. Afr. trop. fr. 5 : 236 (1909) ; l.c. 9 : 246 (1917) ; Bot. 385 ; Rev. Bot. Appliq. 23 : 137 ; F.W.T.A., ed. 1, 2: 10 ; Aubrév. Fl. For. C. Iv. 3 : 116, incl. var. *aubrevillei* (Pellegr.) Aubrév. (1936) ; l.c., ed. 2, 3 : 140, t. 301, fig. 1–5. *Gambeya africana* (A. DC. [as "G. Don"]) Pierre Not. bot. Sapot. 63 (1891). *G. africana* var. *aubrevillei* Pellegr. in Bull. Soc. Bot. Fr. 78 : 682 (1932). *C. edule* Hoyle in Kew Bull. 1932 : 269. *C. omumu* Kennedy For. Fl. S. Nig. 194 (1936), name only. A medium-sized or large tree, the rather large leaves covered beneath with a very close light brown or chestnut-coloured tomentum.
S.L.: Kambui F.R. *King* 145 ! 255 ! Kambui Hills *Small* 855 ! Kenema (May) *Small* 77 ! *King* 16 ! Kasewe F.R. *King* 116 ! Dambaye to Gengelu (May) *King* 90 ! Belebu *King* 132 ! **Lib.**: Bili (fr. Mar.) *Harley* 1783 ! **Iv.C.**: Agboville (fr. Nov.) *Chev.* 22344 ! **Ghana**: Boundary Post, Tano R. *Chipp* 339 ! Essuasu, Bonsa R. (fr. Aug.) *Vigne* FH 165 ! 917 ! Bonsasu *Vigne* FH 1991 ! **S.Nig.**: Sapoba *Kennedy* 1613 ! Sunmoga, Ijebu *Jones & Onochie* FHI 17420 ! **[Br.]Cam.**: Victoria *Maitland* 416 ! 1146 ! Mimbia, 4,500 ft. (Apr.) *Brenan* ! **F.Po**: *Mann* 1154 ! *T. Vogel* 116 ! Widespread in tropical Africa. (See Appendix, p. 354.)

C. cainito Linn., Star Apple, a native of C. America, is often cultivated in our area as an ornamental tree.

Doubtful species.

C. akuase *A. Chev.* in Mém. Soc. Bot. Fr. 2, 8 : 267 (1917), and Chev. Bot. 385 as "akuasi," excl. *Chev.* 22669): Aubrév. Fl. For. C. Iv., ed. 1, 3 : 124 ; ed. 2, 3 : 146 (as "C.sp.") ; Chev. in Rev. Bot. Appliq. 23 : 139, fig. 1 (1943) (as "akuasi"). Perhaps conspecific with No. 1.
Iv.C.: Yakassé to Adzopé *Chev.* 22663. *Aubrév.* 647. No material seen.
[Note : F.W.T.A., ed. 1, 2 : 10 states that this is *Uvaria anonoides* Bak. f.; but this is only true in respect of *Chev.* 22669, *vide* Chev. l.c. 141 (1943).]

22. PACHYSTELA Baill. ex Engl. Monogr. Afr. 8 : 35 (1904) ; Baill. in Bull. Soc. Linn. Paris 2 ; 946 (1891), provisional name.

Leaves 15–35 cm. long, 6–16 cm. broad, very thick-coriaceous, on drying upperside mostly light olive-green, rusty-brown beneath, narrowed to a broadish slightly cordate base, abruptly acuminate, elongate-obovate, covered at first with a thin felt of hairs beneath, becoming glabrous, with about 16 pairs of prominent lateral nerves ; stipules subulate-filiform, up to 2 cm. long ; flowers pedicellate   ..   ..   1. *msolo*

Leaves up to 22 cm. long and 7 cm. broad, thick-coriaceous, on drying becoming greyish, glabrous beneath, narrowly cuneate below the middle and acute or slightly cordate at base, with about 10 pairs of lateral nerves ; stipules subulate, up to 1·5 cm. long ; flowers subsessile :

Leaves acute at base, narrowly acuminate at apex, lanceolate ; petiole 5–7 mm. long
      2. *brevipes*

Leaves slightly cordate or rounded at base, shortly acuminate at apex, obovate ; petiole 3–4 mm. long   ..   ..   3. *pobeguiniana*

1. **P. msolo** (*Engl.*) Engl. Monogr. Afr. 8 : 38 (1904) ; Baehni in Candollea 9 : 428 (1942). *Chrysophyllum msolo* Engl. Pflanzenw. Ost-Afr. C : 306, t. 37 (1895). *Pachystela robusta* Engl. Bot. Jahrb. 49 : 386 (1913); Baehni l.c. 428. *P. argentea* A. Chev. in Mém. Soc. Bot. Fr. 2, 8 : 264 (1917) ; Bot. 391 ; Rev. Bot. Appliq. 23 : 289 (1943) ; F.W.T.A., ed. 1, 2 : 10 ; Hutch. & Dalz. in Kew Bull. 1937 : 58 ; Baehni l.c. 427. *P. bequaertii* De Wild. Pl. Bequaert. 4 : 106 (1926) ; Baehni l.c. 427. *Pouteria msolo* (Engl.) Meeuse in Bothalia 7 : 341 (1960). *P. zenkeri* Meeuse l.c. (1960). A tall tree, up to 150 ft. high ; flowers in the axils of fallen leaves, but sometimes also occurring on the stem.
Ghana: Kibi (Feb.) *Vigne* FH 4323 ! Jumpapu (Sept.) *Vigne* FH 4251 ! **Dah.**: Bokoutou F.R. *Chev.* 22865 ! **[Br.]Cam.**: Bamuko F.R., Kumba Dist. *Keay* FHI 37452 ! Mamfe *Lobe Babute* Cam. 7/37 ! Bali-Ngemba F.R. (fr. May) *Ujor* FHI 30348 ! Also in Cameroun, Congo and in tropical E. Africa.

2. **P. brevipes** (*Bak.*) *Baill. ex Engl.* Monogr. Afr. 8 : 37 (1904) ; Baill. l.c. 947 (1891), provis. name ; F.W.T.A., ed. 1, 2 : 10, partly ; Aubrév. Fl. For. C. Iv. 3 : 124 ; Fl. For. Soud.-Guin. 425, 427 ; Chev. in Rev. Bot. Appliq. 23 : 286 (1943). *Sideroxylon brevipes* Bak. in F.T.A. 3 : 502 (1877). *Sersalisia brevipes* (Bak.) Baill. l.c. (1891). *Bakeriella brevipes* (Bak.) Dubard in Ann. Mus. col. Marseille 20 : 27 (1912). *Pouteria brevipes* (Bak.) Baehni in Candollea 9 : 290 (1942) (excl. syn. *Pachystela pobeguiniana* Pierre) ; Meeuse in

[1] This is not "*C. macrophyllum* G. Don" of Hutch. & Dalz., F.W.T.A., ed. 1, 2 : 9 (1931), referring to a very doubtful specimen of a cultivated plant, which is quite different from *C. delevoyi*.

Bothalia 7 : 333 (1960). *Sideroxylon longistylum* Bak. in F.T.A. 3 : 502 (1877). *Vincentella longistyla* Pierre Not. bot. Sapot. 37 (1891). *Pachystela longistyla* Radlk. in Ann. Mus. Congo, Bot., sér. 2, 1 : 33 (1899) ; Engl. Monogr. Afr. 8 : 38. *Bakeriella longistyla* Dubard in Ann. Mus. col. Marseille 20 : 27 (1912). *Chrysophyllum cinereum* Engl. Bot. Jahrb. 12 : 522 (1890). *Pachystela cinerea* (Engl.) Pierre ex Engl. Monogr. Afr. 8 : 36 (1904), incl. var. *undulata* Engl., var. *cuneata* Engl., var. *ogowensis* Engl., var *batangensis* Engl. (1904). *Bakeriella cinerea* Dubard l.c. (1912). *Sideroxylon sacleuxii* Baill. l.c. 911, 946 (1891). *Pachystela sacleuxii* Pierre ex Baill. l.c. (1891). *Pachystela conferta* Radlk., *P. lenticellosa* Radlk. & *P. cuneata* Radlk. l.c. (1899). *P. liberica* Engl. Bot. Jahrb. 49 : 385 (1913). *P. macrocarpa* A. Chev. in Rev. Bot. Appliq. 23 : 288 (1943). A fair-sized tree, up to 45 ft. high, with small whitish flowers in dense clusters along the older branches in the axils of fallen leaves.

Fig. 206.—Pachystela brevipes (*Bak.*) *Baill.* (Sapotaceae).

A, flower.   B, longitudinal section of flower.   C, part of corolla laid open.   D, stamen.   E, stigma.
F, fruits.

Port.G.: Empada (Oct.) *Esp. Santo* 2194 ! **Guin.**: *Heudelot* 731 ! Kouria (Sept.) *Chev.* 14904 ! Tagania to Mangeta (fr. Jan.) *Chev.* 20418 ! Dubreka (Apr.) *Bonery* ! Nzérékoré (Sept.) *Jac.-Fél.* 1122 ! **S.L.**: Kambia (fr. Dec.) *Sc. Elliot* 4365 ! Bumbuna (Oct.) *Thomas* 3840 ! Wallia (Jan.) *Sc. Elliot* 4459b ! Musaia (Mar.) *Sc. Elliot* 5132 ! Gola (fr. Mar.) *Unwin & Smythe* 65 ! Njala (fr. Nov.) *Deighton* 2559 ! 2601 ! **Lib.**: Peahtah (fl. & fr. Oct.) *Linder* 908 ! 935 ! Gbanga (Sept.) *Linder* 644 ! *Baldwin* 10522 ! Nekabozu (Oct.) *Baldwin* 9966 ! **Iv.C.**: Dabou *Chev.* 16200 ! Oumé (fr. Mar.) *Aubrév.* 4171 ! Divo *de Wilde* 462 ! Adiopodoumé *de Wilde* 193 ! Tiassalé (fr. Dec.) *Leeuwenberg* 2149 ! **Ghana**: Agua (Aug.) *Vigne* FH 1302 ! Sa (Aug.) *Vigne* FH 979 ! Mpraeso, Afram R. *Enti* FH 6524 ! Ancobra R. (fl. & fr. Mar.) *Irvine* 2382 ! Nkroful (fl. & fr. Nov.) *Vigne* FH 1406 ! **N.Nig.**: Birnin Gwari (Nov.) *Thornewill* 153 ! **S.Nig.**: Lagos *Millen* 215 ! Ikoyi Plains (July) *Dalz.* 1273 ! Mamu F.R., Ibadan *Jones* FHI 4953 ! Awba Hills F.R., Ibadan *Jones* FHI 7323 ! Olokemeji *Symington* FHI 5084 ! *Foster* 108 ! Ohumbe F.R., Abeokuta *Onochie* FHI 26658 ! **[Br.]Cam.**: Mamfe *Roche* FHI 24739 ! Gangumi, Adamawa *Latilo & Daramola* FHI 28813 ! Mai Idoanu (fr. Feb.) *Latilo & Daramola* FHI 34464 ! Vogel Peak, Adamawa (Nov., Dec.) *Hepper* 1471 ! 2777 ! Widespread in tropical Africa. (See Appendix, p. 360.)

3. **P. pobeguiniana** *Pierre ex Lecomte* in Bull. Mus. Hist. Nat. 25 : 191 (1919)[1] ; Chev. Bot. 301 ; Chev. in Rev. Bot. Appliq. 23 : 288 (1943) ; Aubrév. Fl. For. Soud.-Guin. 427 ; Fl. For. C. Iv., ed. 2, 3 : 150. *Bakeriella pobeguiniana* Dubard in Notulae Syst. 2 : 91 (1911). *Pachystela albida* A. Chev. Bot. 391 (1920), name only.   A tree up to 45 ft. high, leaves with whitish tomentum when young, but never silvery-silky and shining, nerves very prominent and distinct beneath, ascending at an angle of about 45° ; flowers like those of *P. brevipes*.

Sen.: Niokolo-Koba *Berhaut* 1490. **Mali**: Bamako (May) *Hagerup* 53 ! Bougmoumi to Sasso *Aubrév.* 30581. Kolouba *Vuillet* 648 ! Koulikoro *Vuillet* 159 ! Birgo *Dubois* 17. **Guin.**: Kouroussa *Pobéguin* 193 ! 890. Ditinn *Chev.* 13531. Kouroussa to Kankan *Aubrév.* 3059 ! Farannah (fr. May) *Chev.* 13176 ! Kadé (Apr.) *Pobéguin* 2070 ! **Iv.C.**: Dinderesso *Serv. For.* 1952 !

23. AFROSERSALISIA A. Chev. in Rev. Bot. Appliq. 23 : 292 (1943).

Branchlets brownish-tomentose ; leaves oblanceolate to narrowly oblong-oblanceolate, shortly and obtusely pointed, narrowed from well above the middle of the base and there obtuse, 10–16 cm. long, 3–6 cm. broad, villous only on the midrib beneath ; lateral nerves 12–15 pairs, slightly prominent on both sides ; petiole 1 cm. long, stout, tomentose ; flowers clustered, axillary, subsessile ; pedicels up to 2 mm. long ; sepals villous, united only at base ; fruits glabrescent, ellipsoid olive-like, about 1·8 cm. long and 1 cm. thick, tipped by the short persistent slender style

1. *afzelii*

"*P. pobeguiniana* Pierre MSS" ex Dubard (l.c. 91 (1911)) is a name invalidly published in synonymy.

Branchlets glabrescent, with few appressed hairs ; leaves small, oblanceolate, laurel-like in shape, narrowly and obtusely acuminate, not pointed, narrowed from the middle to the base and there cuneate-acuminate, 8–16 cm. long, 2·5–5 cm. broad, glabrous ; lateral nerves 8–12 pairs, canaliculate above, prominent beneath ; petiole 5–8 mm. long, stout, glabrous ; flowers densely clustered, axillary, pedicellate ; pedicels 2–4 mm. long ; sepals pubescent, united beyond the middle ; fruit puberulous, ellipsoid to long cylindric-ellipsoid, 2–3(–4) cm. long, up to 2 cm. thick, tipped by the short persistent slender style   ..   ..   *2. cerasifera*

1. **A. afzelii** (*Engl.*) *A. Chev.* l.c. 293 (1943), partly (excl. *Chev.* 12218 [as "16218"], 12605*bis*, 12829*bis*, 20587, 23932, 24189) ; Aubrév. Fl. For. C. Iv., ed. 2, 3 : 150, t. 304, fig. 1–6 ; Fl. For. Soud.-Guin. 427 ; Bull. Soc. Bot. Fr. 104 : 280 (1957). *Sersalisia afzelii* Engl. Monogr. Afr. 8 : 30, t. 10, fig. B (1904) ; F.W.T.A., ed. 1, 2 : 11, partly. *Mimusops micrantha* A. Chev. in Vég. ut. Afr. trop. fr. 5 : 244 (1909), nom. subnudum; Bot. 393. *Pachystela micrantha* (A. Chev.) Hutch. & Dalz. F.W.T.A., ed. 1, 2 : 11 ; Kew Bull. 1937 : 58. *Sersalisia micrantha* (A. Chev.) Aubrév. & Pellegr. in Bull. Soc. Bot. Fr. 81 : 798 (1935) ; Aubrév. Fl. For. C. Iv., ed. 1, 3 : 126. *Pouteria afzelii* (Engl.) Baehni in Candollea 9 : 320 (1942), partly (excl. syn. *Sersalisia microphylla* A. Chev., *Chev.* 12682). *Pouteria akuedo* Baehni l.c. 384 (1942). *Vincentella micrantha* A. Chev. in Rev. Bot. Appliq. 23 : 284 (1943). *Afrosersalisia micrantha* A. Chev. l.c. 293 (1943). A shrub or tree, up to 20 ft. high, with white flowers and red fruits.
   S.L.: Njala *Deighton* 696 ! 2617 ! *Lane-Poole* 458 ! Taiama *Deighton* 3098 ! Kenema (June) *Jordan* 2130 !
   Lib.: Totokwelli, Medina *Linder* 1294 ! Karmadhun, Kolahun Dist. *Baldwin* 10190 ! Butaw (Mar.) *Baldwin* 11488 ! Zine (fr. Nov.) *Baldwin* 10258 ! **Iv.C.:** Abidjan *Aubrév.* 55 ! Aboisso *Chev.* 16307 ! Abouabou (fr. Apr.) *Leeuwenberg* 2661 ! 3325 ! **Ghana:** Abofaw *Vigne* FH 1179 ! **S.Nig.:** Owerri *Olorun-femi* FHI 34204 ! Sapoba (May) *Kennedy* 334 ! 1671 ! Calabar (Mar.) *Brenan* 9225 ! **[Br.]Cam.:** Kembong F.R., Mamfe Dist. *Onochie* FHI 31180 ! Also in Cameroun and Gabon. (See Appendix, p. 360.)

2. **A. cerasifera** (*Welw.*) *Aubrév.* in Bull. Soc. Bot. Fr. 104 : 281 (1957).[1] *Sapota cerasifera* Welw. Apont. 585, No. 17 (1859). *Chrysophyllum cerasiferum* (Welw.) Hiern Cat. Welw. 3 : 643 (1898). *Sersalisia cerasifera* (Welw.) Engl. Monogr. Afr. 8 : 30 (1904). *Chrysophyllum disaco* Hiern l.c. 3 : 642 (1898). *Sersalisia disaco* (Hiern) Engl. l.c. 30 (1904). *Afrosersalisia disaco* (Hiern) Aubrév. l.c. (1957). *Sersalisia edulis* S. Moore in J. Bot. 44 : 86 (1906). *S. chevalieri* Engl. Bot. Jahrb. 49 : 385 (1913). *Pouteria chevalieri* (Engl.)Baehni in Candollea 9 : 320 (1942). *Rogeonella chevalieri* (Engl.) Chesnais in Rev. Bot. Appliq. 23 : 294 (1943). *Afrosersalisia chevalieri* (Engl.) Aubrév. l.c. (1957) ; Fl. For. C. Iv., ed. 2, 3 : 152, t. 306. *Bakerisideroxylon djalonense* Chev. Bot. 390, name only. *Sersalisia djalonensis* Aubrév. & Pellegr. in Bull. Soc. Bot. Fr. 81 : 798 (1935) ; Aubrév. Fl. For. C. Iv., ed. 1, 3 : 127 ; Fl. For. Soud.-Guin. 427. *Pachystela brevipes* of Hutch. & Dalz. F.W.T.A., ed. 1, 2 : 10, partly, not of Baill. *Pouteria cerasifera* (Welw.) Meeuse in Bothalia 7 : 341 (1960). *P. disaco* (Hiern) Meeuse l.c. (1960). A large tree, up to 90 ft. high, with very hard wood ; white flowers and red fruits, like the last, but larger.
   Guin.: Dalaba to Souguéta *Chev.* 20168.   Téliko Plateau *Maclaud* ! Labé, Fouta Djalon *Chev.* 12309 ! 12321 ; 12356 ! **S.L.:** Loma Mts. (Jan.) *Jaeger* 4128 ! **Iv.C.:** *Chev.* 34194. **[Br.]Cam.:** Binkas, Wum Dist. (fr. June) *Ujor* FHI 29266 ! Also in Ubangi-Shari and in Angola.

130. MYRSINACEAE

By F. N. Hepper

Trees, shrubs or rarely subherbaceous. Leaves alternate, rarely subopposite, simple, punctate or with schizogenous lines. Flowers small, usually herma-phrodite, in clusters, racemes or panicles. Sepals free or connate, often gland-dotted, valvate, imbricate or contorted, persistent. Corolla gamopetalous or rarely petals free ; lobes contorted, imbricate or valvate. Stamens the same number as and opposite the petals, the filaments more or less adnate to the corolla ; anthers introrse, opening by slits or apical pores. Ovary superior to half-inferior, 1-celled ; style simple ; ovules numerous, on a free central placenta. Fruit a berry or drupe, rarely irregularly dehiscent. Seeds with smooth or rarely ruminate endosperm, with the embryo sometimes placed transversely.

Mainly in the tropics. The woody habit and the stamens the same number as and opposite to the petals are characteristic ; leaves usually gland-dotted.

Calyx free from the ovary :
   Flowers in axillary fascicles or below the leaves ; petals united at the base into a short tube :
   Trees ..   ..   1. **Rapanea**
   Herbs or small shrubs   ..   ..   2. **Afrardisia**
   Flowers racemose ; petals free   ..   ..   3. **Embelia**
Calyx-tube adnate to the ovary ; petals united into a short tube   ..   4. **Maesa**

Besides the above indigenous genera, *Ardisia humilis* Vahl (syn. *A. solanacea* Roxb.), a native of India, is reported from Victoria, [Br.] Cameroons.

1. **RAPANEA** Aubl. Hist. Pl. Guian. Fr. 1 : 121, t. 46 (1775) ; Mez in Engl. Pflanzenr. 4, 236, Myrsinac. : 342 (1902). *Myrsine* of F.T.A. 3 : 493, not of Linn.

Leaves obovate-oblanceolate, obtuse at apex, gradually cuneate at the base, 8–12 cm. long, 3–4·5 cm. broad, glabrous, with numerous conspicuous lateral nerves ; flowers in axillary clusters or below the leaves ; calyx 2 mm. long, lobes small and rounded, minutely ciliate ; corolla-tube very short, lobes oblong, 3 mm. long, with black

---

[1] The combination is made there without citation of the basionym.

longitudinal markings ; fruits globose, up to 5 mm. diam., tipped by the short persistent style, longitudinally striate ; veins conspicuous when dry .. *neurophylla*

R. neurophylla (*Gilg*) *Mez* l.c. 374 (1902). *Myrsine neurophylla* Gilg in Engl. Bot. Jahrb. 19, Beibl. 47 : **45** (1894). *M. melanophlaeos* of F.T.A. 3 : 494 (excl. distrib.), not of R. Br. A tree or tall shrub 15–50 ft. high with pinkish petioles and leaf-veins inconspicuous when fresh ; flowers white ; fruits bright bluish-purple ; in moist upland forest.
 [**Br.**]**Cam.**: Cam. Mt., 4,000–7,600 ft. (fl. Dec., fr. Jan.–Mar.) *Mann* 1200 ! 2175 ! *Maitland* 995 ! 1352 ! *Brenan* 9368 ! Bafut-Ngemba F.R., 7,000 ft. (fl. Apr., fr. Feb.) *Ejiofor* FHI 29384 ! *Hepper* 2854 ! Bamenda (young fr. June) *Lightbody* FHI 26346 ! **F.Po** : summit of S. Isabel Peak (fr. Mar.) *Guinea* 2702 ! Also in Uganda.

2. AFRARDISIA Mez in Engl. Pflanzenr. 4, 236, Myrsinac. : 183 (1902) ; de Wit in Blumea, Suppl. 4 : 243 (1958).

Leaf-margin distinctly and irregularly serrate, entire towards base, 10–15(–18) cm. long, 4–5(–8) cm. broad ; petioles up to 1 cm. long ; pedicels 6–14 mm. long   3. *cymosa*
Leaf-margin entire or shallowly denticulate or crenulate:
 Lamina at least 3 times as long as broad :
  Leaves elliptic or oblong-elliptic, 15–20(–26) cm. long, 5–8(–12) cm. broad ; petioles 10–17 mm. long :
   Leaves with 12–16 lateral nerves, reticulations (when dry) lax and delicate ; margins more or less entire ; petioles 10–17 mm. long ; pedicels 6–9 mm. long ; style punctate .. .. 1. *buesgenii*
   Leaves with at least 20 lateral nerves, densely and conspicuously reticulate (when dry), margin usually shallowly wavy ; pedicels 11–12 mm. long ; style not punctate .. .. 2. *conraui*
  Leaves lanceolate, long-cuneate at each end, (8–)10–14(–23) cm. long, 2–3(–5) cm. broad ; petioles slender, 8–13 mm. long ; inflorescences axillary ; pedicels 2–3 mm. long .. .. 7. *zenkeri*
 Lamina 1½–2½ times as long as broad :
  Petals and sepals mucronate ; leaves elliptic, 8–14 cm. long, 3·5–5·5 cm. broad ; petioles 1–1·5 cm. long ; inflorescences nodding, with 10–12 flowers, equalling petioles .. .. 4. *oligantha*
  Petals and sepals not mucronate, acute or obtuse :
   Peduncle absent or nearly so ; pedicels 3–5 mm. long ; leaves oblong, obovate-elliptic or broadly elliptic, 8–12 cm. long, 4·5–5 cm. broad ; petioles 10–13 mm. long ; undershrub .. .. 5. *schlechteri*
   Peduncle up to 1 cm. long ; pedicels 5–8 mm. long ; leaves ovate-elliptic or oblong, 13–18 cm. long, 4·5–7(–8) cm. broad ; petioles 4–6(–15) mm. long ; shrub or tree 6. *staudtii*

1. **A. buesgenii** *Gilg & Schellenb.* in Engl. Bot. Jahrb. 48 : 516 (1912) ; de Wit in Blumea, Suppl. 4 : 247, fig. 3 (1958). A shrub or small tree, 6 ft. or more high, with horizontal more or less whorled branches ; flowers pale pinkish brown.
 S.Nig.: R. Ata, below Mt. Koloishe, 4,000 ft., Obudu Div. (Dec.) *Keay & Savory* FHI 25057 ! Sonkwala, 4,000 ft., Obudu Div. (Jan.) *Carpenter* 731 ! Also in Cameroun.
2. **A. conraui** (*Gilg*) *Mez* in Engl. Pflanzenr. 4, 236, Myrsinac. : 184 (1902) ; de Wit l.c. 249, fig. 5. *Ardisia conraui* Gilg in Engl. Bot. Jahrb. 30 : 98 (1901). *Afrardisia ledermannii* Gilg & Schellenb. (1912). *A. rosacea* Gilg & Schellenb. (1912). A shrub or small tree ; flowers and fruits red.
 [**Br.**]**Cam.**: Mayo Ndaga, Bello, Gashaka Dist. (fr. Jan.) *Latilo & Daramola* FHI 28988 ! Also in Cameroun.
3. **A. cymosa** *Mez* l.c. 186 (1902) ; Exell Cat. S. Tomé 233 ; de Wit l.c. 252, fig. 6–7. *Ardisia cymosa* Bak. in F.T.A. 3 : 495 (1877), not of Blume (1826). *Tinus cymosa* (Bak.) O. Ktze. (1891). *Afrardisia dentata* Gilg & Schellenb. (1912). Erect shrub or small tree, flowers and fruits red ; in upland forest.
 [**Br.**]**Cam.**: Bamenda Dist. : Bamenda (Apr.) *Ujor* FHI 30029 ! Bafut-Ngemba F.R. (Mar.) *Onochie* FHI 34869 ! *Hepper* 2118 ! Bali (fr. May) *Ujor* FHI 30372 ! Kishong (Jan.) *Keay & Russell* FHI 28442 ! Lakom (Apr.) *Maitland* 1408 ! Also in S. Tomé, Congo and Uganda.
4. **A. oligantha** *Gilg & Schellenb.* in Engl. Bot. Jahrb. 48 : 517 (1912) ; de Wit l.c. 256. Probably a shrub.
 [**Br.**]**Cam.**: Cam. Mt. *Weberbauer* 48. Type specimen destroyed.
5. **A. schlechteri** (*Gilg*) *Mez* l.c. 185 (1902) ; de Wit l.c. 259. *Ardisia schlechteri* Gilg in Engl. Bot. Jahrb. 30 : 97 (1901). A subherbaceous shrublet about 1 ft. high, with fibrous rhizomatous stems ; inflorescences very short, 3–6-flowered.
 [**Br.**]**Cam.**: Bibundi (Apr.) *Schlechter* 12417. Type specimen destroyed.
 [*Olorunfemi* FHI 30692 (N. Korup F.R., Kumba) may also belong to this species.]
6. **A staudtii** (*Gilg*) *Mez* l.c. 187 (1902) ; de Wit l.c. fig. 13. *Ardisia staudtii* Gilg l.c. 98 (1901). *A. brunneo-purpurea* Gilg l.c. (1901). *A. haemantha* Gilg l.c. 99 (1901). *Afrardisia brunneo-purpurea* (Gilg) Mez l.c. 186 (1902). *A. haemantha* (Gilg) Mez l.c. (1902). A shrub 1–6 ft. tall, with pink petioles ; flowers in small clusters along the branches, purple or reddish ; fruits bright red ; in forest.
 N.Nig.: Dogon Kurmi, Sanga F.R., Jemaa Div. (Aug.) *Killick* 6 ! **S.Nig.**: Okomu F.R., Benin (fl. & fr. Jan.) *Brenan* 8894 ! Boshi-Okwangwo F.R., Obudu Dist. (May) *Latilo* FHI 30954 ! Also in Cameroun and Congo.
7. **A. zenkeri** (*Gilg*) *Mez* l.c. 186 (1902) ; de Wit l.c. 262, fig. 14. *Ardisia zenkeri* Gilg in Engl. Bot. Jahrb. 30 : 100 (1901). *Afrardisia hylophila* Gilg & Schellenb. (1912). A simple shrub 1–3 ft. high ; fruits reddish.
 S.Nig.: Sapoba F.R., Benin (fr. Feb., Aug.) *Olorunfemi* FHI 34174 ! *Ujor* FHI 32990 ! Also in Cameroun and Gabon.

3. EMBELIA Burm.—F.T.A. 3 : 496 ; Mez in Engl. Pflanzenr. 4, 236, Myrsinac. : 295 (1902). *Nom. cons.*

Inflorescences on leafless part of previous year's branchlets ; leaves glabrous :
 Leaves (when dry) with numerous prominent very slender lateral nerves ; coriaceous when fresh :
  Upland shrub ; veins not markedly reticulate ; leaves elliptic to obovate-elliptic, obtuse at each end, 5–9 cm. long, 2–4 cm. broad, black gland-dots often obscure

except towards apex ; racemes 2–5·5 cm. long, sometimes numerous, flowers about
  5 mm. diam.   ..   ..   1. *schimperi*
Lowland shrub ; veins prominently reticulate ; leaves obovate, rounded at apex,
  cuneate at base, about 5 cm. long and 3 cm. broad, with gland-dots beneath ;
  racemes 2–3 cm. long, few   ..   ..   2. sp. *A.*
Leaves (when dry) with few impressed lateral nerves, veins very obscure, lamina
  obovate, broadly rounded at apex, cuneate at base, 5–8 cm. long, 2·5–4 cm. broad,
  with numerous black gland-dots ; racemes 2–3 cm. long, few, flowers about 4 mm.
  diam.   ..   ..   3. *guineensis*
Inflorescences arising from leaf-axils of current year's growth :
Racemes 1–1·5 cm. long, subumbellate with flowers about 3 mm. diam. clustered towards apex, slender ; leaves elliptic 5–6 cm. long, 2·5–3 cm. broad, margin distinctly
  dentate, glabrous   ..   ..   4. sp. nr. *welwitschii*
Racemes 2–3 cm. long, not subumbellate ; leaves entire, more or less pubescent
  beneath :
Leaves without visible black dots, obovate-oblong to obovate-elliptic, obtuse or very
  shortly apiculate at apex, cuneate at base, 6–13 cm. long, 3·5–7 cm. broad, midrib
  irregularly pubescent or minutely so ; flowers 4 mm. diam. in loose axillary
  racemes ; pedicels 4–6 mm. long in fruit   ..   ..   ..   5. *rowlandii*
Leaves with conspicuous black gland-dots :
Midrib rather densely puberulous beneath, lateral nerves about 15 pairs, glabrous
  and raised on both surfaces, lamina obovate-elliptic, 8–9 cm. long, 4–5 cm. broad
                                                                 6. *djalonensis*
Midrib, the 6–8 pairs of lateral nerves and young branchlets densely ferruginous-
  puberulous ; leaves obovate to obovate-elliptic, 6–9 cm. long, 3·5–6 cm. broad
                                                          7. *nilotica*

1. **E. schimperi** *Vatke* in Linnaea 40 : 206 (1876) ; Mez in Engl. Pflanzenr. 4, 236 : 329 ; Brenan in Mem.
  N.Y. Bot. Gard. 8 : 498 (1954). *E. abyssinica* Bak. in F.T.A. 3 : 497 (1877). Straggling shrub or climber
  with smooth leathery leaves tufted towards ends of the branches ; flowers whitish ; fruits red ; in upland thickets.
  [**Br.]Cam.**: Bafut-Ngemba F.R., 6,200–7,200 ft., Bamenda (fl. Feb., Apr., fr. Jan.) *Ujor* FHI 30049 ! *Keay
  & Lightbody* FHI 28368 ! *Hepper* 2133 ! Bum, 4,000 ft., Bamenda Dist. (fr. June) *Maitland* 1609 ! Widespread in the mountains of tropical Africa.
2. **E. sp. A.** *E. guineensis* of F.W.T.A., ed. 1, 2 : 16, partly (*Thomas* 9111), not of Bak. Climbing shrub ; " white
  berries with black flecks " (*Harley* 1201).
  S.L. : Gbanbama (fr. Mar.) *Thomas* 9111 ! **Lib.** : Ganta (fr. May) *Harley* s.n. ! 1201 ! **S.Nig.** : Onitsha
  (Feb.) *Jones* FHI 4506 ! [*E. tessmannii* of Aké Assi, Contrib. 1 : 115 (Ivory Coast) ; Yapo *Aké Assi* IA
  4911, Mt. Tonkoui *Aké Assi* IA 5474) may also belong here.]
3. **E. guineensis** *Bak.* in F.T.A. 3 : 496 (1877) ; Chev. Bot. 385. *E. gilgii* Mez l.c. 330 (1902). A climbing shrub
  with short rather crowded racemes on the older parts of the branchlets ; flowers white.
  Guin. : R. Mafin, Dalaba (Apr.) *Caille* in Hb. Chev. 18132. **S.L.**: Bagroo R. (Apr.) *Mann* 861 ! Dodo
  (Apr.) *Deighton* 3922 ! near Berria, Falaba (Mar.) *Sc. Elliot* 5229 ! **Iv.C.**: Gankoué to Bampleu, Upper
  Nuon (Apr.) *Chev.* 21165 *bis* ! (leaves not seen). **Togo Rep.**: Misahöhe *Baumann* 107. **Dah.**: Tohoué, near
  Porto-Novo *Chev.* 22798.
  [It is possible that the specimens I have not seen may belong to sp. *A.* above—F. N. H.].
4. **E. sp. nr. welwitschii** (*Hiern*) *K. Schum.* in Just Bot. Jahresb. 26 : 390 (1900) ; Gilg in Engl. Bot. Jahrb.
  30 : 96 (1901) ; Mez l.c. 317. *Pattara welwitschii* Hiern (1898). A climbing shrub, in upland forest undergrowth.
  [**Br.]Cam.**: Cam. Mt., Buea to Musaka Camp (May) *Maitland* 692 !
5. **E. rowlandii** *Gilg* in Engl. Bot. Jahrb. 30 : 96 (1901) ; Mez l.c. 317. *E. dahomensis* A. Chev. Bot. 385, name
  only. Climbing shrub, the older branches with grey bark, the younger ones with fine brown indumentum
  and pale lenticels ; flowers creamy-yellow ; besides streams in savanna.
  Port.G. : Catio (fl. July, fr. Aug.) *Esp. Santo* 2143 ! 2170 ! **Dah.** : Bassila to Péssésoulou (May) *Chev.*
  23825. **N.Nig.**: Anara F.R., Zaria Prov. (May) *Keay* FHI 22868 ! FHI 25954 ! **S.Nig.**: Lagos interior
  Rowland ! Enyinawsa, Epe Dist. (fr. Jan.) *Onochie* FHI 35259 !
6. **E. djalonensis** *A. Chev. ex Hutch. & Dalz.* F.W.T.A., ed. 1, 2 : 16 (1931) ; Kew Bull. 1937 : 60 ; Chev. Bot.
  384, name only. *E. guineensis* of Mez l.c. 331, not of Bak. A climber in woods, the leaves with pubescent
  midrib and numerous lateral nerves.
  Guin. : Kollangui (Mar.) *Chev.* 13537 ! Dantilia (fr. Mar.) *Sc. Elliot* 5266 ! **Iv.C.** : Boundiali *Aké Assi*
  IA 5510.
7. **E. nilotica** *Oliv.* in Trans. Linn. Soc. 29 : 105, t. 71 (1875) ; F.T.A. 3 : 496 (1877) ; Mez l.c. 318. Climbing
  shrub about 20 ft. high ; flowers cream, fragrant ; beside streams in savanna.
  N.Nig.: Anara F.R., Zaria Prov. (May) *Keay* FHI 22894 ! Also in Uganda.

4. MAESA Forsk.—F.T.A. 3 : 491 ; Mez in Engl.
Pflanzenr. 4, 236, Myrsinac. : 15 (1902).

Leaves more or less entire, subcordate to rounded at base, acuminate at apex, broadly
  ovate, 11–17 cm. long, 7–12 cm. broad (much smaller near the inflorescences), glabrous;
  inflorescences lax and spreading panicles up to 25 cm. long   ..   1. *kamerunensis*
Leaves serrate to crenate, cuneate to rounded at base, ovate-elliptic to lanceolate,
  8–14 cm. long, 3–7 cm. broad, glabrous or softly pubescent beneath :
Calyx glabrous, 2 mm. long ; corolla little longer than calyx ; inflorescences about
  11 cm. long, with few branches about 2 cm. long, glabrous ; flowers laxly arranged
  on pedicels 2 mm. long   ..   ..   2. *nuda*
Calyx ciliate, 1 mm. long ; corolla longer than calyx ; inflorescences profusely
  branched, 5–19 cm. long ; flowers usually very numerous and subsessile
                                                                   3. *lanceolata*

1. **M. kamerunensis** *Mez* in Engl. Pflanzenr. 4, 236 : 27 (1902). *M. cordifolia* Bak. in F.T.A. 3 : 492 (1877),

not of Miq. (1859). *M. lanceolata* of Exell Suppl. Cat. S. Tomé, 31 (1956), not of Forsk. A shrub 15–20 ft. high, with large leathery leaves ; flowers white ; in upland forest.
[Br.]**Cam.**: Cam. Mt., 2,500–3,100 ft. (fl. & fr. Dec., Mar., Apr.) *Mann* 1198 ! *Maitland* 1077 ! *Hutch. & Metcalfe* 91 ! *Brenan* 9502 ! above L. Oku, 7,800 ft. (Jan.) *Keay & Lightbody* FHI 28474 ! **F. Po :** Musola (Jan.) *Guinea* 1236 ! 1250 ! Moka, 4,000–5,000 ft. *Exell* 791 ! Also in Cameroun and S. Tomé.

2. **M. nuda** *Hutch. & Dalz.* F.W.T.A., ed. 1, 2 : 16 (1931) ; Kew Bull. 1937 : 60. Apparently a shrub, quite glabrous and with slender inflorescences.
Guin.: *Farmar* 327 !

Fig. 207.—Maesa lanceolata *Forsk.* (Myrsinaceae).
A, flower. B, corolla-lobe and stamen from within. C, stamen. D, fruit.
E, cross-section of fruit.

3. **M. lanceolata** *Forsk.* Fl. Aegypt.-Arab. CVI & 66 (1775) ; F.T.A. 3 : 492 ; Mez l.c. 26 ; Chev. Bot. 384. *M. djalonis* A. Chev. Bot. 384, name only. A shrub or small tree up to 20 ft. high ; flowers greenish white ; beside upland streams and in grassland.
Guin. : Longuery *Caille* in *Hb. Chev.* 14831 !   Diaguissa to Boulivel *Chev.* 18611 ; 18678. Labé *Chev.* 12238. Ditinn *Chev.* 13439. **Iv.C.**: Mt. Do, Upper Cavally (May) *Chev.* 21432 ! Mt. Gbon *Chev.* 21410. Mt. Tonkoui (fr. Aug.) *Schnell* 6370 ! *Boughey* GC 18333 ! **N.Nig.**: Rafin Bauna North F.R., 4,300 ft., Jos (fl. & fr. Oct.) *Hepper* 1178 ! **S.Nig.**: Mt. Koloishe, 5,000 ft., Obudu Div. (Dec.) *Keay & Savory* FHI 25072 ! [Br.]**Cam.**: Cam. Mt., 3,000–9,000 ft. (fl. & fr. Dec., Jan., Mar., May) *Mann* 1208 ! 2176 ! *Dunlap* 149 ! *Brenan* 9392 ! Bamenda (fl. Apr., fr. Sept.) *Ujor* FHI 30058 ! 30219 ! Mambila Plateau, 6,000 ft. (fl. & fr. Jan.) *Latilo & Daramola* FHI 28991 ! 34361 ! **F.Po**: Clarence Peak, 5,000 ft. (Nov.) *Mann* 578 ! Pico Geramo (Jan.) *Guinea* 1954 ! Widespread in tropical Africa, Arabia and Madagascar, and in S. Africa.
[Although *Mann* 578 and *Guinea* 1954 have serrated leaves, cuneate at the base, the inflorescences are more like those of *M. kamerunensis*].

131. STYRACACEAE

By F. N. Hepper

Trees and shrubs, often with stellate or lepidote indumentum. Leaves alternate, stipules absent. Flowers actinomorphic, hermaphrodite, racemose, axillary or terminal. Calyx tubular, more or less adnate to the ovary ; lobes or teeth valvate or open. Corolla gamopetalous or rarely of free petals, lobes 4–7, valvate or imbricate. Stamens equal and alternate with or double the number of the corolla-lobes, adnate to the corolla-tube or rarely free ; anthers 2-celled, opening lengthwise. Ovary superior to inferior, 1–5-celled ; style slender, 3–5-lobed ; ovules 1 to many in each cell, axile, anatropous. Fruit drupaceous or capsular, calyx persistent. Seed with copious endosperm and straight or slightly curved embryo.

A. Chevalier (in Rev. Bot. Appliq. 27 : 26–29, 402 (1947)) transferred *Afrostyrax* to the family *Huaceae*, but the genus differs widely from *Hua* and it should be retained in *Styracaceae* as originally placed by its authors–(Hutchinson, Fam. Fl. Pl., ed. 2, 1 : 263 (1959)).

AFROSTYRAX Perkins & Gilg in Engl. Bot. Jahrb. **43** : 216 (1909).

Small trees ; leaves alternate, usually long-acuminate. Flowers axillary, small ; calyx closed in bud ; petals 5, free ; stamens 10 ; anthers 4-celled ; ovary 1-celled, superior. Fruit dry, indehiscent.

Leaves beneath glabrous or at most sparsely hairy on the midrib, lamina elliptic to ovate, rounded at base, long-caudate-acuminate at apex, 10–16 cm. long, 4·5–7 cm. broad ; petiole 6–10 mm. long, densely brown-tomentellous ; flowers axillary usually paired, pedicels 5–8 mm. long .. .. 1. *kamerunensis*
Leaves beneath densely pale-brown-lepidote, lamina oblong-elliptic to ovate-elliptic, rounded at base, acuminate and often long-acuminate at apex, 9–18 cm. long, 3–6 cm. broad ; petiole 6–10 mm. long, brown-lepidote ; flowers in axillary clusters, pedicels 8–12 mm. long .. .. 2. *lepidophyllus*

1. **A. kamerunensis** *Perkins & Gilg* in Engl. Bot. Jahrb. **43** : 217, with fig. (1909). Medium-sized tree ; flowers pale yellow ; in forest.
    [Br.]**Cam.**: Likomba, Cam. Mt. *Mildbr.* 10888 ! Also in Cameroun.
2. **A. lepidophyllus** *Mildbr.* in Engl. Bot. Jahrb. **49** : 556 (1913). Tree 50–70 ft. high, straight bole, slash cream coloured, smelling of onions.
    Ghana : Benso, Tarkwa Dist. (fl. Sept., fr. Nov.) *Quao* FH 5611 ! 5673 ! *Andoh* FH 5818 ! Also in Cameroun.

132. LOGANIACEAE

By C. F. A. Onochie and A. J. M. Leeuwenberg

Mostly trees and shrubs ; stem usually woody. Leaves opposite, simple ; stipules present or absent. Flowers hermaphrodite, usually actinomorphic, paniculate, corymbose or in globose heads. Calyx-lobes valvate or imbricate. Corolla tubular, lobes 4–16, contorted, imbricate or valvate. Stamens epipetalous, as many as the corolla-lobes and alternate with them, or rarely reduced to 1 ; anthers 2-celled, opening lengthwise. Ovary superior, 2–4-celled ; style single ; ovules several to numerous, rarely solitary, axile or ascending from the base of each cell. Fruit a capsule, berry or drupe. Seeds sometimes winged ; embryo straight, in the middle of fleshy or cartilaginous endosperm.

Tropics and subtropics generally ; resembling *Rubiaceae*, but usually exstipulate and ovary always superior.

Sepals almost completely connate ; calyx-tube about as long as the corolla-tube or longer ; capsule small, about as long as the calyx .. .. 6. **Nuxia**
Sepals free or connate up to two-thirds of their length ; calyx-tube if present shorter than the corolla-tube ; fruit conspicuously larger than the calyx :
  Annual herbs with one-sided (cincinnous) apical spikes .. .. 4. **Spigelia**
  Trees, shrubs or climbers :
    Inflorescence with 2 large approximately orbicular bracts covering the sepals (*M. hirsuta*) .. .. 3. **Mostuea**
    Inflorescence with small bracts never covering the sepals :
      One sepal much larger than the others ; fruit an oblong capsule, about 3–4 times as long as wide .. .. 5. **Usteria**
      Sepals equal or unequal, the largest up to twice as long as the others ; fruit up to twice as long as wide :
        Sepals 4, decussate, orbicular, rounded ; corolla-lobes 8–16, contorted ; fruit a berry .. .. 1. **Anthocleista**
        Sepals 5 or 4, mostly imbricate ; corolla-lobes 5 or 4, valvate or imbricate :
          Fruit a berry, globose or nearly so ; corolla-lobes valvate ; shrubs, trees, or lianes with tendrils ; leaves mostly coriaceous and distinctly 3- or 5-nerved
            2. **Strychnos**
          Fruit an obcordate or bilobed capsule ; corolla-lobes imbricate ; shrubs ; leaves mostly thin and never 3- or 5-nerved .. .. .. 3. **Mostuea**

The weed *Cynoctonum mitreola* (Linn.) Britt. has been recorded from Ghana (Gambaga Scarp (fr. Nov. 1957) *Morton* A2728). It is native in America and has been widely introduced into Asia (see Heine in Kew Bull. 17 : 171 (1963)).

1. **ANTHOCLEISTA** Afzel. ex R. Br.—F.T.A. **4**, 1 : 537 ; E. A. Bruce in Kew Bull. **10** : 45 (1955) ; Leeuwenberg in Acta Bot. Neerl. **10** : 1–53 (1961).

Secondary veins conspicuous ; leaves crowded at the apices of the branchlets ; candelabrum-shaped trees or few-stemmed shrubs :
  Branches armed with short usually paired spines ; flower buds uniformly rounded or subtruncate at the apex :

Branches conspicuously spiny ; spines divergent, confluent at the base, paired or
sometimes 3–4 together ; leaves conspicuously discolorous, dark green above, more
or less glaucous beneath, usually sessile, (6–)15–45(–150) cm. long, (4–)6–20(–45) cm.
broad, margin usually revolute and undulate ; mature corolla in bud usually less
than 5 times as long as the calyx :

Corolla-tube 2–3 times as long as the lobes and 2·5–3·5 times as long as the calyx ;
fruit when dry with 4 more or less irregular dents or irregularly shrivelled ; leaves
often drying black   ..   .. 1. *nobilis*

Corolla-tube 0·9–1·5 times as long as the lobes and 1·25–2 times as long as the calyx ;
fruits drying smooth, neither shrivelled nor dented ; leaves never drying black,
mostly obovate and often comparatively wider than in No. 1 ..   .. 2. *vogelii*

Branches not conspicuously spiny ; spines parallel or slightly divergent, confluent
for at least half their length, paired ; leaves not conspicuously discolorous, petiolate,
margin not revolute nor undulate ; mature corolla in bud 5·5–6 times as long as the
calyx (see also below)   ..   .. 3. *djalonensis*

Branches unarmed or if with incipient or occasional spines, then flower buds not rounded
but tapering at the apex :

Leaves petiolate and buds usually uniformly rounded at the apex :

Leaves obovate, usually rounded at base, 9–35(–115) cm. long, 4·5–17(–50) cm.
broad ; calyx narrowed, not definitely constricted at the mouth ; branches with
incipient or occasional spines ; berry drying smooth, 35–50 mm. long, 25–35 mm.
broad when mature, with a thick wall (about 3 mm.)   .. 3. *djalonensis*

Leaves elongate-oblanceolate, always cuneate at base, 11–75 cm. long, 3–15 cm.
broad ; calyx definitely constricted at the mouth ; branches never spiny ; berry
irregularly shrivelled when dry, 15–27 mm. long, 10–18 mm. broad when mature,
with a thin wall (about 1 mm.)   ..   .. 4. *liebrechtsiana*

Leaves sessile or subsessile, if petiolate then buds not rounded, but tapering to apex,
oblong-obovate to oblanceolate :

Sepals drying smooth, clasping the corolla-tube at anthesis, only spreading under the
mature fruit ; leaves, at least the upper ones, usually petiolate, lower ones often
sessile or subsessile, 7–45(–100) cm. long, 3·5–18(–30) cm. broad, usually subcoriaceous, and with inconspicuous tertiary veins ; corolla-tube about 1–1·5 times
as long as the lobes which are large and reflexed   ..   .. 5. *schweinfurthii*

Sepals drying rugulose, outer pair at least more or less spreading, not closely clasping
the corolla-tube at anthesis, widely spreading under the fruit ; leaves sessile,
40–45(–145) cm. long, 20(–45) cm. broad, usually membranaceous, and with
conspicuous tertiary veins ; corolla-tube 1·5–2·5 times as long as the lobes, which
are smaller and spreading ..   .. 6. *procera*

Secondary veins inconspicuous ; leaves not crowded at the apices of the branchlets ;
climbers, shrubs, or sometimes (?) trees (see also No. 4) :

Outer sepals about as long as broad or broader ; leaves oblong-elliptic to oblong-
obovate, 6–20 cm. long, 2·5–11 cm. broad :

Sepals 4–8 mm. long ..   .. 7. *microphylla*
Sepals 20–30 mm. long   ..   .. 8. *scandens*

Outer sepals 1·5–2 times as long as broad, 12–17 mm. long, 8–11 mm. broad, often
apically torn ; leaves oblong-obovate to oblong-lanceolate, 4·5–19 cm. long, 1–7 cm.
broad ..   .. 9. *obanensis*

1. **A. nobilis** *G. Don* Gen. Syst. 4 : 68 (1838) ; F.T.A. 4, 1 : 538, partly ; Chev. Bot. 441 ; E. A. Bruce in Kew
Bull. 10 : 47 (1955) ; Aubrév. Fl. For. C. Iv. ed. 2, 3 : 186, t. 318 ; Leeuwenberg in Acta Bot. Neerl. 10 :
14, figs. 4–6 (1961). *A. parviflora* Bak. (1895)—F.T.A. 4 : 539. *A. procera* var. *umbellata* and var. *parvi-*
flora of Chev. Bot. 441, not of Bak. A tree about 25–60 ft. high with spiny stem ; flowers white ; in
secondary forest.
Sen.: Bignona to Sindialone, Casamance (Feb.) *Chev.*! **Guin.:** Conakry *Chev.* 12250! *Maclaud* 28! Ymbo
Caille in *Hb. Chev.* 14759! Santa R. to Timbo *Chev.* 12613! Macenta *Roberty* 7150! **S.L.:** Freetown
(Feb.) *Dalz.* 984! Bagroo R. *Mann*! Talla Hills, 3,500 ft. (Mar.) *Sc. Elliot* 5053! Njala (Feb.) *Deighton*
512! Loma Mts. (Sept.) *Jaeger* 1577! **Lib.:** Dukwia R. (Feb.) *Cooper* 190! 380! 428! Buchanan (Mar.)
Baldwin 11197! Gbanga (fl. & fr. Sept.) *Linder* 542! Peahtah *Linder* 1021! **Iv.C.:** Abidjan *Chev.* 17327!
Adiopodoumé *Leeuwenberg* 1921! Banco *Aubrév.* 407! Makougnié *Chev.* 16156 *bis*! **Ghana:** Axim (fl. bud
Feb.) *Irvine* 2198! *Morton* A2221! Jemma (Sept.) *Chipp* 354! near Prestea (July) *Vigne* FH 1270!
(See Appendix, p. 362.)
2. **A. vogelii** *Planch.* in Hook. Ic. Pl. 8 : t. 793–4 (1848) ; E. A. Bruce l.c. 48 ; Bruce & Lewis in F.T.E.A.
Loganiac. 8 (1960) ; Leeuwenberg l.c. 16, fig. 7. *A. talbotii* Wernham in Cat. Talb. 68 (1913). *A. kalbreyeri*
Bak. (1895)—F.T.A. 4 : 540. *A. nobilis* of F.T.A. 4, 1 : 538, partly (*T. Vogel* 51), not of G. Don. A tree
up to 60 ft. high, bark pale grey, inner bark yellowish ; flowers creamy or yellowish-brown in erect
inflorescences ; infructescences drooping ; in wet forest.
S.L.: near Kolia *Deighton* 3204! Kangahun (Aug.) *Jordan* 2131! Manjehun (fr. Nov.) *Deighton* 6148!
Baoma *Deighton* 3203! **Lib.:** Gbanga (fr. Sept.) *Linder* 533! Mecca (fr. Nov.) *Baldwin* 10435! **Iv.C.:**
Cosrou (fr. May) *Leeuwenberg* 4257! 40 km. S. of Taï (fl. & young fr. Mar.) *de Wilde & Leeuwenberg* 3537!
Ghana: Kwaben *Vigne* FH 1295! **N.Nig.:** Anara F.R., Zaria (May) *Keay* FHI 22895! **S.Nig.:** Ibu *T.*
Vogel 51! Nun R. (fr. Sept.) *Mann* 471! Ibadan *Onochie* FHI 40448! Oyo to Ibadan *Meikle* 1311! Oban
Talbot 177! 2027! [**Br.]Cam.:** Ambas Bay (Feb.) *Mann* XVI! Victoria *Maitland* 362! Also in
Cameroun, Ubangi-Shari, Gabon, the Congos, Sudan, Kenya, Tanganyika, N. Rhodesia and Angola. (See
Appendix, p. 362.)
3. **A. djalonensis** *A. Chev.* in Mém. Soc. Bot. Fr. 2, 8 : 47 (1908) ; E. A. Bruce l.c. 49 ; Aubrév. Fl. For. C. Iv.,
ed. 2, 3 : 184 ; Leeuwenberg l.c. 20, figs. 8–10. *A. kerstingii* Gilg ex Volkens in Notizbl. Bot. Gart. Berl.,
App. 22 : 33 (1909) ; F.W.T.A., ed. 1, 2 : 18 ; Aubrév. Fl. For. Soud.-Guin. 436, t. 95, 3. *A. nobilis* of
F.W.T.A., ed. 1, 2 : 18, partly (syn.). *A procera* of Chev. Bot. 441, not of Lepr. ex Bureau. A tree, 30–

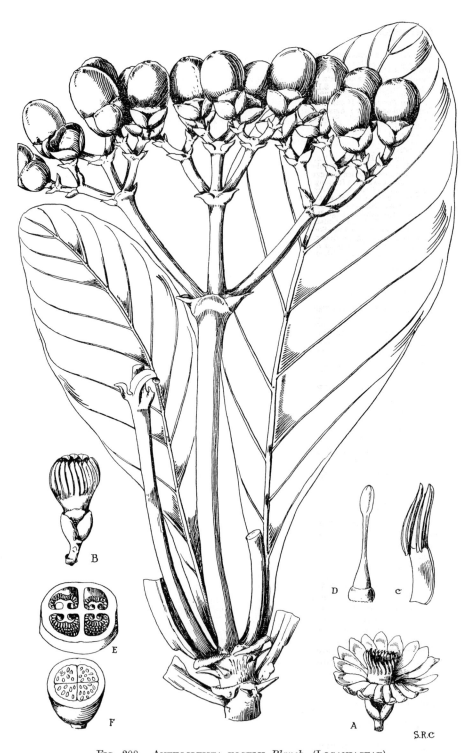

Fig. 208.—Anthocleista vogelii *Planch*. (Loganiaceae).

A, flower.   B, flower-bud.   C, two stamens.   D, pistil.   E, cross-section of ovary.   F, cross-section of fruit.—After Hook. Ic. Pl.

36

45 ft. high with blunt spines on the unbranched pale grey trunk, and widespreading crown ; flowers white or creamy ; in rather dry places, in savanna or thickets.
Mali: Guiri (May) *Chev.* 886 ! **Port. G.:** Teixeira Pinto *d'Orey* 130 ! Gabu *Esp. Santo* 274 ! Pinche-Pansor *Esp. Santo* 3425 ! Fulacunta *d'Orey* 211 ! **Guin.:** Kollangui (fl. & fr. Mar.) *Chev.* 12221 ! 12222 ! 12873 ! Santa Valley *Chev.* 12773 ! **S.L.:** Musaia (fr. July) *Deighton* 4215 ! 4827 ! **Iv.C.:** Séguéla *Leeuwenberg* 3273 ! near Bouaké *Leeuwenberg* 3285 ! 3315 ! Tafiré *Aubrév.* 1393 ! Ferkéssédougou *Aubrév.* 1538 ! **Ghana:** Mpraeso *Beveridge* FH 3288 ! Ejura *Vigne* FH 2029 ! Agogo *Vigne* FH 1125 ! Kwahu *Irvine* 1669 ! **Togo Rep.:** Sokodé (Mar.) *Kersting* A18 ! **Dah.:** Adja, Ouéré *Le Testu* 296 ! Godomey *Poisson* ! **N.Nig.:** Tangale Waja, Bauchi Prov. (July) *Kennedy* FHI 7275 ! Abinsi *Dalz.* 638 ! **S.Nig.:** Rosevear 50/29 ! Olokemeji F.R. *Ross* 39 ! Ibadan (fl. & fr. Mar., Apr.) *Meikle* 1309 ! *Ladipo* FHI 3086 ! *Keay* FHI 26734 ! Hambler 7 ! Onitsha to Uke (fl. & fr. May) *Onochie* FHI 35775 ! Akpaka F.R., Onitsha *Onochie* FHI 40433 ! **[Br.]Cam.:** Jamtari, Adamawa (fr. Dec.) *Latilo & Daramola* FHI 28928 ! Also in Cameroun. (See Appendix, p. 362.)

4. **A. liebrechtsiana** *De Wild. & Th. Dur.* in Compt. Rend. Soc. Bot. Belg. 38 : 96 (1899) ; F.T.A. 4, 1 : 540 ; E. A. Bruce l.c. 51 ; Leeuwenberg l.c. 22, fig. 11. A slender tree without spines, 30–40 ft. high, with open spreading crown and leaves aggregated at branch ends, bole grey, slash cream ; flowers pale creamy white ; in *Raphia—Cynometra* fringing forests.
Dah.: Massi to Goutyssa, Abomey *Chev.* 23269 ! Lower Ouémé R., Dogba to Affamé *Chev.* 23472 ! **S.Nig.:** Omo (formerly part of Shasha) F.R. (Apr.) *Jones & Onochie* FHI 17264 ! Akilla, Oni R. *Kennedy* 2364 ! Siluko (Nov.) *Onochie* FHI 40423 ! Also in Cameroun, Gabon, Congo, N. Rhodesia and Angola.

5. **A. schweinfurthii** *Gilg* in Engl. Bot. Jahrb. 17 : 579 (1893) ; F.T.A. 4, 1 : 541 ; E. A. Bruce l.c. 51 ; Bruce & Lewis in F.T.E.A. Loganiac. 11 ; Leeuwenberg l.c. 24, fig. 12 (for full synonymy). *A. oubanguiensis* Aubrév. & Pellegr. in Bull. Soc. Bot. Fr. 100 : 25 (1953) ; Aubrév. Fl. For. Soud.-Guin. 436, t. 95, 1–2. Tree about 60 ft. high, with dense umbrella crown and grey bole ; flowers white or creamy white.
S.Nig.: Jamieson R., Sapoba *Kennedy* 2136 ! Also in Cameroun, Ubangi-Shari, Gabon, the Congos, Sudan, Uganda, Tanganyika, N. Rhodesia and Angola.

6. **A. procera** *Lepr. ex Bureau* Thèse Loganiac. 74–77, figs. 60–62 (1856) ; F.T.A. 4, 1 : 539 (under *A. parviflora*); Aubrév. Fl. For. Soud.-Guin. 436, t. 95, 4–6 ; E. A. Bruce l.c. 56 ; Aubrév. Fl. For. C. Iv., ed. 2, 3 : 184 ; Leeuwenberg l.c. 31, fig. 13. *A. frezoulsii* A. Chev. in Mém. Soc. Bot. Fr. 2, 8 : 47 (1908) ; F.W.T.A., ed. 1, 2 : 18 ; Chev. Bot. 441. *A. nobilis* of F.T.A. 4, 1 : 538, partly (*Leprieur*), not of G. Don. A tree about 60 ft. high, with spreading branches, without spines ; flowers white ; in orchard bush, usually in swampy places.
Sen.: Koular *Berhaut* 891 ! Tambacounda *Hb.I.F.A.N.* 8608 ! Casamance *Trochain* 1396 ! **Gam.:** Albreda *Leprieur* ! *Perrottet* ! Kombo (June) *Heudelot* 109 ! Lamin *Rosevear* 1 ! **Port.G.:** Bissau *Esp. Santo* 1700 ! Gabú *Esp. Santo* 284 ! Fulacunda to Bedanda *d'Orey* 254 ! Cabuchangue to Quebu *Esp. Santo* 2071 ! **Guin.:** Conakry *Chev.* 12162 ! Kouroussa *Pobéguin* 841 ! Fouta Djalon *Pobéguin* 1918 ! Konkauré to Timbo *Chev.* 12453 ! **S.L.:** Musaia (fr. Apr., fl. July) *Deighton* 5481 ! *Miszewski* 6 ! Njala (July) *Deighton* 5960 ! **Iv.C.:** Agnéby, Dabou *Hb. I.D.E.R.T.* 3140 ! 3672 ! *Leeuwenberg* 3178 ! **S.Nig.:** Lagos *Imp. Inst.* 10 ! (See Appendix, p. 362.)

7. **A. microphylla** *Wernham* in Cat. Talb. 67 (1913) ; Leeuwenberg l.c. 32, fig. 14. *A. micrantha* Gilg & Mildbr. ex Hutch. & Dalz. F.W.T.A., ed. 1, 2 : 18 (1931). A glabrous tree or climber ; flowers white ; in secondary forest.
S.Nig.: Oban *Talbot* 304 ! **F.Po:** S. Isabel Peak, 3,000 ft. (Aug.) *Mildbr.* 6434 ! Also in Principe and S. Tomé.

8. **A. scandens** *Hook. f.* in J. Linn. Soc. 6 : 16 (1861) ; F.T.A. 4, 1 : 542 ; Leeuwenberg l.c. 34, fig. 15, 1–3. *A. exelliana* Monod in Bull. I.F.A.N. sér A, 19 : 347, f. 1–30 (1957). Climbing shrub or tree 20–50 ft. high with square branchlets ; flowers white ; in montane forest.
[Br.]Cam.: L. Bambuluwe, 6,000 ft., Bamenda (fl. & fr. Mar.) *Onochie* FHI 34852 ! *Daramola* FHI 40516 ! **F.Po:** Clarence Peak, 5,000 ft. (fl. & fr. Dec.) *Mann* 623 ! Moka, 4,000–6,000 ft. (Jan.) *Mildbr.* 7111. *Exell* 814 ! Also in S. Tomé.

9. **A. obanensis** *Wernham* in Cat. Talb. 67 (1913) ; Leeuwenberg l.c. 34, fig. 16. Climbing shrub ; flowers yellowish ; in wet forest.
S.Nig.: Jamieson R., Sapoba (fl. bud Sept., Nov.) *Onochie* FHI 34272 ! *Keay* FHI 28079 ! Eket Dist. *Talbot* 3025 ! 3105 ! Oban *Talbot* 305 ! Also in Cameroun and Congo.

2. **STRYCHNOS** Linn.—F.T.A. 4, 1 : 517 ; Duvigneaud in Bull. Soc. Bot. Belg. 85 : 9 (1952); E. A. Bruce in Kew Bull. 10 : 35, 127 (1955), 627 (1956) ; 11 : 153, 267 (1956). *Scyphostrychnos* S. Moore (1913)—F.W.T.A., ed. 1, 2 : 24.

Inflorescence terminal, large, lax, paniculate, and many-flowered, or if not so then sepals lanceolate, acuminate, and not distinctly imbricate, axillary spines and/or linear stipules present, and large and thick-walled fruits :

Savanna trees with axillary spines, without tendrils ; leaves rounded and slightly mucronate or emarginate to acuminate at apex, very variable in shape and size, suborbicular to elliptic, (1·5–)3–8 cm. long, (0·8–)1·3–7 cm. broad, glabrous or pubescent beneath ; branches usually with curved, reflexed spines ; inflorescence seemingly umbellate ; sepals lanceolate ; fruits large, thick-walled .. .. 1. *spinosa*

Climbers or shrubs, never trees, in the forest, with tendrils, sometimes with spines and then usually with linear stipules (No. 3) :

Plant brown-hirsute ; leaves oblong-obovate, oblong-oblanceolate, oblong-elliptic, or sometimes oblong-ovate, 2·5–16·5 cm. long, 1·4–8 cm. broad ; inflorescence lax, paniculate ; corolla-tube very short ; filaments curved and provided with a bilobed pubescent gland ; large colleters (glands) near the base of the petioles ; fruit about 2–3 cm. diam., thin-walled, about 3–7-seeded .. .. .. 2. *phaeotricha*

Plant glabrous or pubescent ; leaves ovate, or elliptic, or nearly so :

Inflorescence seemingly umbellate, rather congested ; sepals lanceolate, acuminate ; plant often with 4 or 8 linear stipules on each node ; fruit large, about 7 cm. diam., thick-walled, many-seeded ; leaves rounded and abruptly acuminate at apex
3. *congolana*

Inflorescence lax, paniculate ; sepals orbicular, ovate, or triangular, rounded to acute ; plant without stipules ; fruits small, about 1–1·5 cm. diam., thin-walled, one-seeded :

Branchlets densely pubescent ; leaves dull beneath ; sepals connate for about half their length, acute, pubescent outside ; pistil glabrous .. 4. *dolichothyrsa*

4

Branchlets glabrous or sparsely pubescent ; leaves shining on both sides ; sepals
   free, obtuse or rounded, minutely pubescent outside ; pistil sparsely pubescent
                                                                    5. *dinklagei*
Inflorescence axillary or both axillary and terminal, mostly on the same branchlet ;
   sepals up to twice as long as wide, rounded to acute, usually distinctly imbricate ;
   plants without spines, occasionally with internodal prickles (No. 15) :
Leaf-apex rounded or obtuse ; lamina elliptic to obovate-elliptic, 2–15 cm. long,
   1–7 cm. broad ; flowers subfasciculate ; corolla in mature bud 6–9 mm. long ;
   sepals ciliate ; fruit globose, 5–12·5 cm. diam., thick-walled, many-seeded ; small
   trees in savanna woodland :
Branchlets glabrous ; leaves glabrous on both sides   6. *innocua* subsp. *innocua*
                                                                var. *innocua*
Branchlets pubescent ; leaves pubescent, at least at the base of the midrib beneath
                                    6a. *innocua* subsp. *innocua* var. *pubescens*
Leaf-apex acuminate or acute ; if obtuse or rounded, a shrub or climber in forest :
Branchlets, petioles, and usually also leaves beneath (at least at the base of the midrib)
   hairy ; fruits small, 1–3 cm. diam., thin-walled, usually 1-seeded :
Pistil pilose or sparsely pubescent (in No. 5) :
Pistil sparsely pubescent ; inflorescence distinctly longer than the leaves, lax, many-
   flowered, 4–7 times branched ; corolla in mature bud 2–3 mm. long ; leaves
   shining and coriaceous (other characters as above) ..   ..   5. *dinklagei*
Pistil distinctly pilose ; inflorescence up to half as long as the leaves, few-flowered,
   if lax up to 3 times branched :
Branchlets and inflorescence-axis puberulous with very short appressed hairs ;
   lamina entirely glabrous, coriaceous ; inflorescence lax ; corolla in mature bud
   3–4 mm. long, lobes 6–7 times as long as the tube ; anthers glabrous 7. *splendens*
Branchlets and inflorescence-axis sparsely pubescent as is the midrib beneath ;
   hairs erect or nearly so :
Leaves coriaceous, large, oblong-elliptic (7·5–)12–19·5 cm. long, 3·5–7 cm. broad,
   sparsely pubescent on the distinctly prominent midrib and main secondary
   veins beneath ; inflorescence rather congested, much shorter than the leaves ;
   corolla in mature bud 4 mm. long, the lobes 1·7 times as long as the tube ;
   anthers bearded at the base ; fruit 2–3 cm. diam., 1–3-seeded 8. *memecyloides*
Leaves papyraceous to thinly coriaceous, smaller, 3–11 cm. long, 1–5 cm. broad :
Inflorescence congested or rather so ; corolla in mature bud 7·5–8·5 mm. long,
   outside more or less distinctly pubescent, lobes about as long as the tube ;
   anthers bearded ; leaves obtuse to acuminate at apex   9. *ngouniensis*
Inflorescence lax ; corolla in mature bud 4·5–5 mm. long, outside glabrous or
   nearly so, the lobes 1·4–2 times as long as the tube ; anthers glabrous ; leaves
   acuminate at apex, acumen often long   ..   .. 10. *boonei*
Pistil glabrous, sometimes some minute hairs on the style (in No. 13) :
Tall tree without tendrils ; leaves variously shaped, oblong-ovate, oblong-lanceolate, or oblong-elliptic, mostly acuminate, never rounded at apex ; petioles
   smooth ; inflorescence not really congested, often rather lax ; corolla in mature
   bud 3·5–4 mm. long, glabrous outside, inside with a broad brush on each lobe ;
   tube 0·6–1·2 times as long as the lobes ; anthers bearded at the base ; fruit
   1–1·5 cm. diam., 1(–2)-seeded   ..   .. 11. *mitis*
Climber or scandent shrub with tendrils ; inflorescence-axis appressed-pubescent
   (see also No. 26 ; if inflorescence-axis glabrous or with ranked hairs see No. 24) :
Sepals sparsely pubescent outside ; leaves usually distinctly discolorous and
   above with conspicuous tertiary venation when dry :
Leaves extremely variable in shape and size, usually broadest above the middle,
   rounded, apiculate, or acuminate at apex ; pubescence not regularly appressed,
   hairs rather long ; corolla in mature bud 1·8–2 mm. long, sparsely pubescent
   outside, inside with a flat brush of white hairs on each lobe ; tube 0·5–1 times
   as long as the lobes ; anthers bearded ; fruit about 1–1·3 cm. diam., 1-seeded
                                                                    12. *afzelii*
Leaves very variable but less than in the preceding species, usually broadest
   below the middle, acuminate, sometimes obtuse or acute at apex, pubescence
   regularly appressed, with very short hairs ; petioles often transversely rugose
   when dry ; corolla in mature bud 2–2·5 mm. long, often very sparingly pubescent outside, inside pilose on the lobes ; lobes 1·5–2 times as long as the tube ;
   anthers glabrous   ..   ..   13. *malacoclados*
Sepals glabrous outside ; leaves usually not discolorous and with inconspicuous
   tertiary venation above, mostly obtuse and apiculate ; petioles mostly transversely rugose when dry ; corolla in mature bud 2–2·5 mm. long, subrotate,
   glabrous outside, inside densely pilose on the base of the lobes, lobes 3–4 times
   as long as the tube ; anthers glabrous ; fruit about 1–2 cm. diam.
                                                                    14. *angolensis*
Branchlets, petioles and leaves glabrous ; some hairs may occur on the stipular line :

Branches and branchlets with internodal prickles ; large climber ; tendrils in 2 or 3 pairs above each other at the apex of short lateral branchlets ; inflorescence much shorter than the leaves, congested ; corolla in mature bud 4–4·7 mm. long, tube 1–1·7 times as long as the lobes, lobes thick ; fruit large, subglobose, about 8–12·5 cm. diam. .. .. 15. *aculeata*

Branches and branchlets unarmed :

Pistil hairy, immature fruit often topped by the hairy base of the style :

Flowers small ; corolla in mature bud 2–4 mm. long ; fruit small, up to 3 cm. diam., thin-walled, usually 1(–6)-seeded :

Inflorescence distinctly longer than the leaves, large, many-flowered, 4–7 times branched (other characters as above, in two places) .. 5. *dinklagei*

Inflorescence up to half as long as the leaves, 1–3 times branched :

Inflorescence lax, few-flowered ; corolla glabrous outside or with some minute hairs ; style 2·5–3·5 mm. long (other characters as above) 7. *splendens*

Inflorescence congested, many-flowered ; corolla pubescent outside ; style 1 mm. long .. .. 16. *talbotiae*

Flowers larger ; corolla in mature bud 5–11 mm. long ; fruit large, 5 cm. diam. or more, thick-walled, many-seeded (if not, see No. 17) :

Corolla in mature bud 5–6 mm. long and rounded at apex, lobes 1·2–2 times as long as the tube ; anthers bearded ; style 1·5 mm. long ; sepals 1·5 mm. long ; fruit 4–5 cm. diam., thin-walled, 3–6-seeded ; leaves about 8–20 cm. long, 4–10 cm. broad, mostly pale greenish-brown when dry .. 17. *chrysophylla*

Corolla in mature bud 7·5–11 mm. long and tapering at apex ; anthers glabrous ; style 4–6 mm. long ; sepals 2–5 mm. long ; fruit 5 cm. diam. or more, thick-walled, many-seeded ; leaves mostly smaller, drying medium or dark brown, sometimes slightly glaucous :

Branchlets usually distinctly lenticellate ; old leaves often thinly coriaceous, usually drying with a flat margin and with rather conspicuously reticulate tertiary venation above ; inflorescence mostly congested ; inflorescence-axis and pedicels often sparsely pubescent ; corolla in mature bud 7·5–8 mm. long, tube 1·5 times as long as the lobes, lobes about twice as long as wide, 3–3·2 mm. long .. .. 18. *densiflora*

Branchlets lenticellate or not ; old leaves often thickly coriaceous, usually drying with a recurved margin and smooth above ; inflorescence lax ; inflorescence-axis and pedicels glabrous ; corolla in mature bud 7·5–11 mm. long, tube 1–1·2 times as long as the lobes, lobes 2·5–3 times as long as wide, 4–5 mm. long .. .. 19. *nigritana*

Pistil glabrous :

Inflorescence, sepals and corolla outside, or some of these organs often partially hairy, if not then inflorescence large, lax, and provided with tendrils, and leaves dull beneath (in No. 21) :

Corolla minutely pubescent outside, inside with 2 rings of white brush-like hairs ; anthers bearded ; inflorescence-axis, pedicels, and sepals outside shortly pubescent all over ; fruits small, up to 4 cm. diam. ; forest tree, 7–20 m. high

20. *staudtii*

Corolla glabrous or sometimes hairy outside, inside with only one ring of hairs, sometimes composed of brushes of which one is on or below each lobe ; fruits small, up to 3 cm. diam. ; climbers or shrubs with tendrils :

Inflorescence lax, about 0·5–1·5 times as long as the leaves, many-flowered, provided with tendrils ; leaves dull beneath ; sepals obtuse, glabrous outside ; corolla in mature bud 6·5–8 mm. long, lobes 1·2–1·3 times as long as the tube ; anthers glabrous .. .. 21. *melastomatoides*

Inflorescence lax or congested, without tendrils, if more than half as long as the leaves corolla in mature bud up to 5 mm. long or leaves shining or matt beneath :

Tall tree without tendrils ; inflorescence-axis pubescent (other characters as above) .. .. 11. *mitis*

Climbers, shrubs, or trees (?), with tendrils or not (?) (habit of No. 22 unknown ; No. 26 may be a tree as far as known from E. African material) :

Leaves dull beneath ; midrib and secondary veins hardly or not prominent, tertiary venation inconspicuous ; inflorescence congested ; peduncle conspicuous, with 2 stripes of pubescence and a pair of bracteoles ; sepals glabrous outside, rounded, distinctly ciliate ; corolla in mature bud 3·5–4·5 mm. long, densely pilose inside ; lobes 3·5 times as long as the tube

22. *gnetifolia*

Leaves shining or matt, never dull beneath ; at least the midrib distinctly prominent beneath, peduncle without bracteoles ; sepals acute or obtuse, if ciliate less distinctly so :

Branches and branchlets green or pale greenish-brown when living, drying usually paler than in the next species, more or less distinctly quadrangular,

not lenticellate ; leaves mostly conspicuously 3-nerved, drying pale
greenish-brown ; inflorescence-axis shortly pubescent all over or with 2
stripes of pubescence ; corolla in mature bud 3·5–5 mm. long, lobes
1–1·5 times as long as the tube ; style 2–2·2 mm. long   23. *johnsonii*
Branches and branchlets often dark brown, terete, very often lenticellate,
if not then corolla in mature bud 4–7·5 mm. long (in No. 25) or stamens
included and leaves distinctly acuminate (in No. 24) ; leaves less conspicuously 3-nerved, drying pale or dark greenish-brown or brown :
Stamens included ; apex of anther reaching the hairy ring in throat ;
corolla in mature bud 4–5 mm. long, with a narrow brush-like ring of
stiff hairs in the throat ; anthers bearded ; leaves papyraceous to subcoriaceous, distinctly acuminate ; branches and branchlets not lenticellate ; branchlets often with some ranked hairs as on the inflorescence-
axis   ..   .. 24. *longicaudata*
Stamens exserted ; anthers glabrous :
Branches lenticellate or not ; branchlets glabrous, not lenticellate ;
corolla in mature bud 4–7·5(–8·5) mm. long ; tube 0·7–1·5(–3) times as
long as the lobes, 2–4(–5) mm. long ; style 2–6 mm. long ; fruit
1–2 cm. diam., sometimes lenticellate ; leaves usually more shining,
thicker, and less distinctly acuminate than in No. 26   25. *floribunda*
Branches usually dark brown, lenticellate ; branchlets glabrous or shortly
pubescent, lenticellate near the base, often covered by a pale skin which
later splits and peels off ; corolla in mature bud 2·5–3·5 mm. long, tube
1–1·4 mm. long, lobes 1·5 times as long as the tube ; style 1–1·5(–2) mm.
long ; fruit 1–1·8 cm. diam., sometimes shortly stipitate
26. usambarensis
Inflorescence, sepals and corolla outside entirely glabrous :
Corolla with an entire or 5-lobed corona ; whole plant drying dark brown or
black ; corolla in mature bud 8–11·5 mm. long ; leaves large, 6–22 cm. long,
3–10 cm. broad ; basal secondary veins mostly not more conspicuous than
the others ; fruits large, thick-walled   ..   ..   .. 27. *camptoneura*
Corolla without corona ; if plant drying dark brown or black then flowers smaller
than in No. 27 ; corolla in mature bud up to 6 mm. long, and fruits small and
thin-walled :
Fruits large, thick-walled ; inflorescence distinctly pedunculate, congested ;
corolla-lobes thick, inside with a flat brush of white hairs ; anthers bearded ;
leaves large, 5–24 cm. long, 3–9 cm. broad, usually drying pale greenish-
brown ; stigma sessile (other characters as above)   .. 15. *aculeata*
Fruits small, thin-walled, 1–3 cm. diam. ; corolla-lobes not very thick ; leaves
if large drying darker (in No. 28) :
Leaves usually large, 5·5–21 cm. long, 2–10 cm. broad, mostly distinctly acuminate, strongly 3-nerved from above the base to the apex, papery to parchment-like, rarely coriaceous ; inflorescence lax ; flowers small, 4-merous ;
corolla in mature bud 2·5 mm. long, lobes 2·6 times as long as the tube ;
style 0·4–0·8 mm. long ; sepals small, 0·4–1 mm. long, nearly halfway
connate ; branches often umbellately branched and shortly below the
branching point provided with a tendril ..   .. 28. *icaja*
Leaves smaller, main secondary veins not very distinct near the apex if
reaching it ; inflorescence mostly congested ; sepals connate at the base ;
tendrils arranged otherwise :
Leaves parchment-like, distinctly acuminate, acumen 7–12 mm. long, mostly
dark brown or black when dry ; branches hardly or not lenticellate ;
inflorescence congested ; flowers small :
Stamens included ; anthers bearded ; branchlets and inflorescence sometimes sparsely pubescent ; tendrils pubescent (other characters as above)
24. longicaudata
Stamens exserted ; anthers glabrous ; branchlets, inflorescence and tendrils
glabrous ; leaves mostly dull and shrivelled when dry ; corolla in mature
bud about 3–4 mm. long, pilose on the base of the lobes inside and at the
tube apex, lobes nearly twice as long as the tube ; style about 1·7 mm.
long   ..   .. 29. *odorata*
Leaves coriaceous or thinly coriaceous, if distinctly acuminate branches mostly
distinctly lenticellate (in Nos. 25 and 26, below) or drying paler, mostly
greenish-brown (in No. 31) :
Leaves rounded, apiculate, or shortly acuminate at apex, often oblong-
obovate, coriaceous, shining on both sides, especially above ; corolla in
mature bud about 6 mm. long, pilose on the base of the lobes inside and at
the tube apex, lobes about as long as the tube, narrow, strap-shaped,
recurved ; style 3–5 mm. long ..   .. 30. *barteri*
Leaves acuminate, never rounded at the apex :

Flowers small ; corolla in mature bud 2·5–3·5 mm. long (other characters
as above) .. .. 26. *usambarensis*
Flowers larger ; corolla in mature bud 4–7·5(–8·5) mm. long :
Flowers 4-merous ; corolla very densely pilose inside, usually also on the
apices of the lobes, in mature bud 6·5 mm. long and tapering at the
apex ; branches and branchlets not lenticellate, pale brownish-green
when dry ; fruit about 2·5–3 cm. diam. ; leaves often distinctly
acuminate .. .. 31. *ndengensis*
Flowers usually 5-, rarely 4-merous ; sepals approximately triangular ;
corolla densely pilose inside, but glabrous on apices of the lobes, in
mature bud mostly rounded at the apex (other characters as above)
25. *floribunda*

1. **S. spinosa** *Lam.* Illustr. 2 : 38 (1794); Aubrév. Fl. For. Soud.-Guin. 438, t. 96, 1–4 ; E. A. Bruce in Kew
Bull. 10 : 42 (1955); Bruce & Lewis in F.T.E.A. Loganiac. 20, fig. 3, 15–16 (1960); Verdoorn in Fl. S.
Afr. 147. *S. lokua* A. Rich. Tent. Fl. Abyss. 2 : 53 (1851). *S. laxa* Solered. in Engl. Bot. Jahrb. 17 :
573 (1893). *S. buettneri* Gilg (1893)—F.T.A. 4, 1 : 535 (1903). *S. spinosa* var. *pubescens* Bak. in F.T.A.
4, 1 : 537. *S. djalonis* A. Chev. Bot. 442, name only ; Rev. Bot. Appliq. 27 : 358 (1947). A small tree
up to 20 ft. high with light spreading crown and branches armed with spines ; branchlets and leaves
sometimes softly pubescent ; flowers greenish-white, in short compound cymes ; corolla not spit ; fruit
with a hard yellow shell ; in savanna woodland.
Sen.: *Perrottet*! **Gam.**: Georgetown to Kuntaur (fr. Jan.) *Dalz.* 8072! **Port. G.**: Canchungo, Empacaca
(Apr.) *Esp. Santo* 1945! Falacunda (fr. May) *Esp. Santo* 2035! **Guin.**: Moria (Feb.) *Sc. Elliot* 4801!
Farana, 3,200 ft. (Mar.) *Sc. Elliot* 5384! Rio Nunez (Mar.–Apr.) *Heudelot* 813! Kaba to Mamou (May)
Chev. 12716! Ditinn (Apr.) *Chev.* 12179 bis! Konkouré to Timbo (Mar.) *Chev.* 12461! **S.L.**: *Thomas*
10261! Newton (Mar.–June) *Garner* in *Hb. Deighton* 2491! *Deighton* 2992! **Iv. C.**: Bounoukou *Aubrév.*
1609. Ferkéssédougou *Aubrév.* 1871. Bouaké (fl. & fr. Apr.) *Leeuwenberg* 3288! **Ghana**: Yendi (Apr.)
Vigne FH 1693! Tamale to Yendi (Mar.) *Adams* 3869! Bonkwe Hill, Ashanti (fr. June) *Chipp* 489!
Kulmasa to Jagalo (Feb.) *Kitson* 665! Pan to Bujan (Apr.) *Kitson* 908! **Togo Rep.**: *Kersting* 102!
N. Nig.: Nupe *Barter* 1140! 1705! Zamfara F.R. (Apr.) *Keay* FHI 15608! Katagum *Dalz.* 373! Anara
F.R., Zaria (fr. Sept.) *Olorunfemi* FHI 24385! Agaie (fl. & young fr. May) *Yates* 54! Abinsi (Mar.) *Dalz.*
924! **S.Nig.**: Ijaiye F.R., Oyo Prov. (Mar.) *Keay & Ladipo* FHI 21190! Ado Ischu *Rowland* 5! Through-
out tropical and S. Africa, Madagascar and Mauritius. (See Appendix, p. 363).
2. **S. phaeotricha** *Gilg* in Engl. Bot. Jahrb. 36 : 105, t. 3 (1905). *S. thyrsiflora* Gilg in Wiss. Ergeb. Deutsch.
Zent.-Afr. Exped. 2 : 532, t. 74 (1914). A liane with paired tendrils ; tertiary veins below forming
characteristic mussel shell markings ; in forest.
Ghana: Foso (Oct.) *West-Skinn* 45! **S.Nig.**: Usonigbe F.R., Benin *Onochie* FHI 35670! Also in Cameroun,
Gabon and Congo.
3. **S. congolana** *Gilg* in Engl. Bot. Jahrb. 28 : 120 (1893); F.T.A. 4, 1 : 521 ; E. A. Bruce in Kew Bull. 10 :
38 (1955) (excl. syn. *S. djalonis* A. Chev.) ; Bruce & Lewis in F.T.E.A. Loganiac. 16 ; Leeuwenberg in
Act. Bot. Neerl. 11 : 48–49, fig. 2 (1962). *S. lecomtei* A. Chev. ex Hutch. & Dalz. F.W.T.A., ed. 1, 2 :
22 (1931) ; and in Kew Bull. 1937 : 333 ; Chev. in Rev. Bot. Appliq. 27 : 368.4 *S. viridiflora*
De Wild. in Pl. Bequaert. 2 : 101 (1923). *S. chloropetala* A. W. Hill in Kew Bull. 1930 : 175. Large
climber, up to 130 ft. high in trees with paired tendrils, when regrowing from stump shrub-like and often
flowering, rarely spiny ; corolla-tube split above the insertion of the stamens ; in forest.
Iv. C.: Adiopodoumé (fl. & fr. Apr.) *Leeuwenberg* 3146! 3701! (June) *Leeuwenberg* 4467! Bingerville
Chev. 15402! **S. Nig.**: Calabar *Thomson* 121! Also in Cameroun, Congo and E. Africa.
4. **S. dolichothyrsa** *Gilg ex Onochie & Hepper* Kew Bull. 16 : 385 (1962). Climber with tomentellous inflorescence-axis and terminal cymes ; in forest.
S.Nig.: Afl River F.R., Ogoja Prov. (fl. & fr. May) *Latilo* FHI 31811! Also in Cameroun.
5. **S. dinklagei** *Gilg* in Engl. Bot. Jahrb. 28 : 121 (1899) ; F.T.A. 4, 1 : 520. A climber with flowers in
terminal and axillary cymes ; in forest.
Lib.: Grand Bassa *Dinklage* 2100! **Iv.C.**: Abouabou (Apr.) *Leeuwenberg* 3332! Brafouédi (fl. Apr.,
fr. Mar.) *Leeuwenberg* 3326! 3706!
6. **S. innocua** *Del.* subsp. **innocua** var. **innocua**—Cent. Pl. Afr. 53 (1826) ; F.T.A. 4, 1 : 532 ; Bullock &
Bruce in Kew Bull. 1938 : 45 ; Bruce & Lewis in Kew Bull. 11 : 270 ; and in F.T.E.A. Loganiac. 25 ;
Aubrév. Fl. For. Soud.-Guin. 440 ; t. 96, 5–7. *S. alnifolia* Bak. in Kew Bull. 1895 : 150 ; F.T.A. 4,
1 : 532. Small tree up to 30 ft. high with axillary clusters of creamy-white or greenish-yellow flowers
and thick-walled spherical fruits ; in savanna woodland.
Mali: Ouassana to Dendéla (Mar.) *Chev.* 623! **Ghana**: various localities (May) *Kitson* 664! 775! 30
miles from Navrongo on Tumu road (fr. Dec.) *Adams & Akpabla* GC 4359! **N.Nig.**: Katagum *Dalz.*
421! Jos (Aug.) *Keay* FHI 12713! **S.Nig.**: Lagos *Rowland*! Ijaiye F.R., Oyo Prov. (Mar.) *Keay* FHI
21182! Also in Congo, Sudan, Ethiopia, E. Africa and S. Rhodesia.
6a. **S. innocua** subsp. **innocua** var. **pubescens** Solered. in Engl. Bot. Jahrb. 17 : 556 (1893) ; Bruce & Lewis in
F.T.E.A. Loganiac. 26. *S. triclisioides* Bak. in Kew Bull. 1895 : 98 ; F.T.A. 4, 1 : 533. A small tree
up to 30 ft. high (other characters as in No. 6) ; in savanna woodland.
Guin.: Kankan (Mar.) *Chev.* 571. **Iv.C.**: Ferkéssédougou *Hb. I.D.E.R.T.* 1284! **Ghana**: Gurumbele to
Bantala (fr. Apr.) *Kitson* 909! **N.Nig.**: Nupe *Barter* 1160! Sokoto (fl. & fr. May) *Lely* 836! Anara F.R.,
Zaria Prov. (May) *Keay* FHI 22965! Jos Plateau (Feb.) *Lely* P115! Yola (Apr.) *Dalz.* 197! Also in
Cameroun, Ubangi-Shari, Ethiopia, Sudan, E. Africa, N. Rhodesia, Congo and Angola. (See Appendix,
p. 363.)
7. **S. splendens** *Gilg* in Engl. Bot. Jahrb. 17 : 571 (1893) ; F.T.A. 4, 1 : 524. *S. chrysocarpa* Bak. in Kew Bull.
1895 : 98 ; F.T.A. 4, 1 : 529. (?) *S. acutissima* Gilg in Engl. Bot. Jahrb. 23 : 200 (1896) ; F.T.A. 4, 1 :
524. *S. odorata* A. Chev. in Rev. Bot. Appliq. 27 : 373 (1947), partly (*Dinklage* 2914 ; *Chev.* 19273).
A climber with paired tendrils and fragrant creamy-white flowers in axillary cymes ; in forest.
Sen.: Sanoukou (fr. Feb.) *Roberty* 16747! **Port.G.**: Ponte de Timbo, Catio (Oct.) *Esp. Santo* 2191!
2192! Prabis, Bissau (fr. Mar.) *Esp. Santo* 1856! Cumura, Bissau (fr. Nov.) *Esp. Santo* 2230! **S.L.**:
Afzelius! Bayabaya, Scarcies R. (fr. Feb.) *Sc. Elliot* 4292! Baiima (Aug.) *Deighton* 6108! Wilberforce
(fr. Mar.) *Johnston* 98! **Lib.**: Monrovia (fr. Oct.) *Dinklage* 2914! Peahtah (fr. Oct.) *Linder* 1084! Jenne,
Loffa R. (fr. Jan.) *Bequaert* 31! Kondessu, Boporo Dist. (fr. Dec.) *Baldwin* 10662b! Mecca (fr. Dec.)
Baldwin 10791! **Iv.C.**: Brafouédi (July) *Aké Assi* 3101! (fr. Apr.) *Leeuwenberg* 3331! **Ghana**: *Burton
& Cameron*! Prang, N.T. (June) *Vigne* FH 3901! Abetifi (Aug.) *Plumptre* GC 205! **S.Nig.**: Ibadan
South F.R. (July–Aug.) *Ahmed & Chizea* FHI 19782! *Keay* FHI 25366! Owo F.R. (fr. Apr.) *Jones*
FHI 3133! Ukpon F.R., Obubra Div. (July) *Latilo* FHI 31874!
8. **S. memecyloides** *S. Moore* in Cat. Talb. 69 (1913) ; F.W.T.A., ed. 1, 2 : 24 (1931), partly (excl. syn.
S. talbotiae S. Moore). *S. memecyloides* var. *effusior* S. Moore l.c. 70 (1913). An erect or climbing shrub
with comparatively large leaves and whitish flowers in short pubescent axillary cymes.
S.Nig.: Afl River F.R., Ogoja (fr. May) *Latilo* FHI 31806! Stubbs Creek F. R., Eket Dist. (fr. May)

W. E. T.

Fig. 209.—Strychnos spinosa *Lam.* (Loganiaceae).

A, leafy, and B, flowering shoot. C, flower. D, longitudinal section of flower. E, stamen (filament should appear bearded all round). F, cross-section of ovary. G, fruit. H, cross-section of fruit.

42

Onochie FHI 32910! Oban *Talbot* 2078! 2079! [**Br.**]**Cam.**: Mombo, S. Bakossi (fr. May) *Olorunfemi* FHI 30563!

9. **S. ngouniensis** *Pellegr.* in Bull. Mus. Hist. Nat. 32 : 394 (1926) ; in Mém. Soc. Linn. Norm., n. sér., 1 : 37, fig. 8 (1928) (as *ngounyensis*). *S. soubrensis* Hutch. & Dalz., F.W.T.A., ed. 1, 2 : 22 (1931) ; Kew Bull. 1937 : 333 ; Chev. in Rev. Bot. Appliq. 27 : 368, t. 17A (1947). *S. jollyana* Pierre ex A. Chev. l.c. 364, t. 14 (1947), partly, excl. fruits. A climber with paired tendrils and white flowers in axillary and terminal cymes ; in forest.
S.L.: *Afzelius*! Kumrabai (fr. Oct.) *Thomas* 7080! **Lib.:** Mecca, Boporo Dist. (fr. Nov.) *Baldwin* 10412! Woeme, Vonjama Dist. (fr. Oct.) *Baldwin* 10097! Soplima, Vonjama Dist. (fr. Nov.) *Baldwin* 10062! **Iv.C.:** Dabou *Jolly* 203! Baléko, 76 km. N. of Sassandra (Apr.) *Leeuwenberg* 3216! Soubré, Sassandra Valley (May) *Chev.* 17994! **S.Nig.:** Sapoba, Benin (fr. Nov.) *Meikle* 636! (fr. Apr.) *Onyeagocha* FHI 7130! Onipanu, Ondo Prov. (fr. Dec.) *Onochie* FHI 5227! Ibadan South F.R. (Apr.) *Keay* FHI 22809! Also in Gabon and Congo.

10. **S. boonei** *De Wild.* in Bull. Jard. Bot. Brux. 5 : 45 (1915) ; Bruce & Lewis in F.T.E.A. Loganiac. 30. A liane with solitary tendrils and fragrant white flowers in rather few-flowered axillary cymes ; leaves brown-pubescent at base beneath ; in forest.
S.Nig.: Ojogba-Ugun F.R., Ishan (June) *Olorunfemi* FHI 38067! Ugo to Jesse, Sapoba F.R. (May) *Keay* FHI 37032! Ugo, Usonigbe F.R. (fl. & fr. July) *Charter* FHI 38704! Also in Cameroun, Congo and Uganda.

11. **S. mitis** *S. Moore* in J. Linn. Soc. 40 : 146 (1911) ; Duvigneaud in Bull. Soc. Bot. Belg. 85 : 24 (1952) ; Bruce & Lewis in F.T.E.A. Loganiac. 21. A tall tree without spines or tendrils ; branchlets pubescent ; flowers in short axillary cymes ; fruits yellow or orange.
S.L.: Falaba, near Berria (fr. Mar.) *Sc. Elliot* 5418! Also in Angola, Uganda, Kenya, Tanganyika, Sudan, S. Rhodesia and Mozambique.
[Note : In E. Africa this species is a tree of upland and lowland rain-forest and fringing forest, 30 to more than 100 ft. high, with branchlets glabrous or more rarely pubescent. The specimen cited here has densely pubescent branchlets with the V-shaped hairs which are also found in the pubescent form of the species in E. Africa.]

12. **S. afzelii** *Gilg* in Engl. Bot. Jahrb. 17 : 572 (1893) ; F.T.A. 4, 1 : 522 ; Chev. in Rev. Bot. Appliq. 27 : 367, t. 17C (1947). *S. zizyphoides* Bak. in F.T.A. 4, 1 : 522 (1903). *S. erythrocarpa* Gilg in Engl. Bot. Jahrb. 23 : 199 (1896). *S. caryophyllus* A. Chev. Bot. 442, name only. *S. odorata* A. Chev. in Rev. Bot. Appliq. 27 : 373 (1947), partly (*Chev.* 17404 ! 17765 ! 17895). A liane with solitary tendrils and very small white flowers in many-flowered axillary cymes ; in forest.
Sen.: *Leprieur*! *Perrottet* 489! *Brene*, Bissau (fl. & fr. Jan.) *Esp. Santo* 1720! Prabis, Bissau (Feb.) *Esp. Santo* 1811! **S.L.:** Sugarloaf Mt. (fr. Dec.) *Sc. Elliot* 4015! Moria, Scarcies (fr. Feb.) *Sc. Elliot* 4480! Mayoso (Aug.) *Thomas* 1442! Jigaya, 1,100 ft. (Sept.) *Thomas* 2583! Bumpe (fl. & fr. July-Nov.) *Deighton* 4839! 4928! **Lib.:** Du R. (July) *Linder* 176! Peahtah (fl. & fr. Oct.) *Linder* 909! Dukwia R. (Feb.) *Cooper* 180! **Iv.C.:** Alépé *Chev.* 17404! 17895! Aboisso *Chev.* 17765! Guidéko to Soubré, Mid. Sassandra R. (June) *Chev.* 19100! Mt. Kouen (Apr.) *Chev.* 21272! **Ghana:** *Burton & Cameron*! Wurubong to Okraji (May) *Kitson* 1138! Alabadi (Apr.) *Scholes* 289! **Togo Rep.:** *Baumann* 558. **S.Nig.:** Ilaro F.R. (fl. & fr. Dec.) *Onochie* FHI 31878! Gambari F.R. (fr. Oct.) *Onochie* FHI 32235! Ogbesse R., Owo Dist. (Apr.) *Jones* FHI 3529! (See Appendix, p. 363.)

13. **S. malacoclados** *C. H. Wright* in F.T.A. 4, 1 : 523 (1903). *S. pansa* S. Moore in Cat. Talb. 68 (1913). A liane with solitary tendrils and bright orange flowers in lax axillary cymes ; fruit small.
Iv.C.: Bianouan, Bia R. *Leeuwenberg* 3938! about 30 km S.W. of Guéyo *Leeuwenberg* 4114! 56 km N. of Sassandra *Leeuwenberg* 3994! Hana R., Taï to Tabou (fr. Mar.) *de Wilde & Leeuwenberg* 3613! **Ghana:** Pamu, Berekum F.R. (Sept.) *Vigne* FH 2515! **S.Nig.:** Oban *Talbot* 1661! Also in Cameroun and Gabon.

14. **S. angolensis** *Gilg* in Engl. Bot. Jahrb. 17 : 571 (1893) ; F.T.A. 4, 1 : 522 ; E. A. Bruce in Kew Bull. 11 : 157. *S. cinnabarina* Gilg ex Hutch. & Dalz. F.W.T.A., ed. 1, 2 : 24 (1931) ; Kew Bull. 1937 : 335. A liane with solitary tendrils and flowers in short puberulous axillary cymes ; in forest.
S.Nig.: Eket Dist. *Talbot*! Also in Cameroun, Gabon, Congo, Angola and E. & S. tropical Africa.

15. **S. aculeata** *Solered.* in Engl. Bot. Jahrb. 17 : 544 (1893) ; F.T.A. 4, 1 : 520 ; Chev. Bot. 442. *S. mortehani* De Wild. in Bull. Jard. Bot. Brux. 5 : 50 (1915). *S. pseudo-jollyana* A. Chev. in Rev. Bot. Appliq. 2 : 366 (1947), partly (*Chev.* 16351 ! 16395). A lofty liane with large fruits and prominent internodal prickles; in forest.
S.L.: Bagroo R. (Apr.) *Mann* 863! Njala (fl. & fr. May) *Deighton* 5052! **Iv.C.:** Guidéko, Sassandra Valley (May) *Chev.* 16351! 16368! 16395! Alépé (Mar.) *Chev.* 17894! **S.Nig.:** Abe, Sapoba F.R., Benin (Nov.) *Onochie* FHI 35969! Oban *Talbot* 1381! 2013! **F.Po:** (Jan.) *Mann* 175! Also in Cameroun, Gabon, the Congos, Angola and tropical E. Africa. (See Appendix, p. 363.)
[Note : The ultimate branchlets sometimes lack the characteristic internodal prickles.]

16. **S. talbotiae** *S. Moore* in Cat. Talb. 69 (1913). An erect shrub with flowers in short dense axillary cymes.
S.Nig.: Oban *Talbot* 2077! Also in Congo.

17. **S. chrysophylla** *Gilg* in Engl. Bot. Jahrb. 28 : 119 (1899) ; F.T.A. 4, 1 : 525. *S. eketensis* S. Moore in J. Bot. 52 : 29 (1914). A liane with ribbed branches, shining, rather leathery leaves with very prominent midribs ; in forest.
S.Nig.: Kwa Ibo R., Eket *Talbot* 3237! Also in Cameroun and Gabon.

18. **S. densiflora** *Baill.* Adansonia 12 : 369 (1879) ; F.T.A. 4, 1 : 528 (1903) ; Chev. in Rev. Bot. Appliq. 27 : 372, t. 16A (1947). *S. suaveolens* Gilg in Engl. Bot. Jahrb. 17 : 566 (1893). *S. chlorocarpa* Gilg l.c. 28 : 120 (1899). *S. martreti* A. Chev. Etudes Fl. Afr. cent. fr. 204 (1913), name only. *S. hirsutostylosa* De Wild. (1915). A liane with paired tendrils and with yellowish flowers in dense axillary cymes ; in forest.
Guin.: Rio Nunez *Heudelot* 861! Erimakuna (= Hérémakon) (Mar.) *Sc. Elliot* 5406! **S.Nig.:** Oban *Talbot*! Also in Cameroun, Gabon, Ubangi-Shari and Congo. (See Appendix, p. 363.)

19. **S. nigritana** *Bak.* in Kew Bull. 1895 : 97 ; F.T.A. 4, 1 : 523. *S. vogelii* Bak. in Kew Bull. 1895 : 96. *S. ciliicalyx* Gilg in Schlechtr. Westafr. Kautschuk.-Exped. 304 (1900), name only ; Gilg & Busse in Engl. Bot. Jahrb. 36 : 95 (1905). *S. imbricata* A. W. Hill, name only. *S. fleuryana* A. Chev. Bot. 443, name only. A liane with few-flowered axillary cymes ; in fringing forests.
Iv.C.: Ahinta, Sanvi (Mar.) *Chev.* 17754! Prolo to Blérion, Lower Cavally (Aug.) *Chev.* 19888! **Ghana:** Kintampo (Mar.) *Dalz.* 19! **Togo Rep.:** Quamikrum (Mar.) *Schlechter* 12957! **Dah.:** Little Popo (July) *Cole*! **N.Nig.:** Idda (Sept.) *T. Vogel* 49! Nupe *Barter*! R. Bahago, Anara F.R., Zaria Prov. (fr. Oct.) *Keay* FHI 20121! Katabua, Anara F.R. (May) *Keay* FHI 22909! **S.Nig.:** Lagos (Apr.) *Barter* 2232! *Dalz.* 1154! Epe *Barter* 3249! Warri (Mar.) *Freeman* 319! Owenna R., Shasha F.R. (Feb.) *Jones & Onochie* FHI 16969! Ijaiye F.R., Oyo Prov. (Apr.) *Onochie* FHI 21957! Also in Cameroun and Ubangi-Shari.

20. **S. staudtii** *Gilg* in Notizbl. Bott. Gart. Berl. 1 : 182 (1896) ; F.T.A. 4, 1 : 528. A tree up to 65 ft. high and 6 ft. girth with whitish flowers in short axillary cymes ; slash light brown ; in forest.
[**Br.**]**Cam.:** Koto Barombi (Feb.) *Lobe Babute* Cam./37/36! Baimayan to Barombi (Feb.) *Binuyo & Daramola* FHI 35511! Ebonji, Kumba Dist. (fr. May) *Olorunfemi* FHI 30598! Johann-Albrechtshöhe (= Kumba) *Staudt* 616! 966!

21. **S. melastomatoides** *Gilg* in Engl. Bot. Jahrb. 23 : 201 (1896) ; F.T.A. 4, 1 : 527. *S. syringiflora* A. Chev. in Mém. Soc. Bot. Fr. 2, 8 : 48 (1908) ; Chev. Bot. 444. Shrub or straggly tree with white fragrant flowers ; tendrils mostly on inflorescences.

Guin.: Kaba Valley, Farana (May) *Chev.* 13181! **S.L.:** *Afzelius*! Kahreni (Apr.) *Sc. Elliot* 5592! "Konnoh Country" *Burbridge* 485! Taiama (Mar.) *Deighton* 3375!

22. **S. gnetifolia** *Gilg ex Onochie & Hepper* in Kew Bull. 16 : 385 (1962). Erect shrub or liane with leathery leaves and flowers in compact few-flowered axillary cymes ; in forest.
S.Nig.: Oban *Talbot* 574! Also in Cameroun.

23. **S. johnsonii** *Hutch. & M. B. Moss* F.W.T.A., ed. 1, 2 : 24 (1931) ; Kew Bull. 1937 : 335, partly (excl. *Sc. Elliot* 4292). *S. goniodes* Duvign. in Bull. Séanc. Inst. Roy. Col. Belg. 19 : 221 (1948), name only. Liane with flowers in loose axillary cymes ; tendrils paired ; in forest.
S.L.: Madina (Sept.) *Adames* 74! Rowalla (fr. July) *Thomas* 1188! **Iv.C.:** Abidjan (Sept.) *de Wilde* 298! **Ghana:** Akropong (Oct.) *Johnson* 802! **Togo Rep.:** Misahöhe (Nov.) *Mildbr.* 7336! **S.Nig.:** Etemi Odo, Omo F.R. (fl. & fr. Nov.) *Tamajong* FHI 20739! Benin *Kennedy* 2195! Also in Cameroun, Congo and Angola.

24. **S. longicaudata** *Gilg* in Engl. Bot. Jahrb. 17 : 570 (1893) ; F.T.A. 4, 1 : 527 (both as *longecaudata*). *S. brevicymosa* De Wild. in Rev. Zool. Afr. 10, Suppl. Bot. 5 : 6 (1922). *S. nigrovillosa* De Wild. Pl. Bequaert. 2 : 95 (1923). Liane with solitary pubescent tendrils ; branchlets and inflorescence-axis often with ranked hairs ; in forest.
Iv.C.: Davo R., 34 km N. of Sassandra (Apr.) *Leeuwenberg* 4018! Oumé (May) *Leeuwenberg* 4144! Sassandra R., 38 km E. of Duékoué (Apr.) *Leeuwenberg* 3901! **S.Nig.:** Omo F.R., Ijebu Prov. (fl. & fr. May) *Tamajong & Latilo* FHI 16782! Also in Cameroun and Congo.

25. **S. floribunda** *Gilg* in Engl. Bot. Jahrb. 17 : 566 (1893) ; F.T.A. 4, 1 : 527. *S. welwitschii* Gilg l.c. 573 ; F.T.A. 4, 1 : 524 ; E. A. Bruce in Kew Bull. 10 : 627, fig. 1A. *S. moloneyi* Bak. in Kew Bull. 1895 : 97 ; F.T.A. 4, 1 : 527 ; F.W.T.A., ed. 1, 2 : 24, partly (excl. syn. *S. chrysocarpa* Bak.). *S. togoensis* Gilg & Busse in Engl. Bot. Jahrb. 36 : 96 (1905). *S. littoralis* A. Chev. ex Hutch. & Dalz. F.W.T.A., ed. 1, 2 : 22 (1931) ; Kew Bull. 1937 : 334 ; Chev. Bot. 443, name only ; Chev. in Rev. Bot. Appliq. 27 : 370, t. 19A (1947). *S. warneckei* Gilg, name only. Climbing shrub or liane ; tendrils solitary ; flowers in axillary cymes ; fruits orange-yellow, globose ; in forest.
S.L.: Yonibana (fr. Nov.) *Thomas* 5032! Berria (fr. Mar.) *Sc. Elliot* 5431! **Lib.:** Duport (fr. July) *Dinklage* 3067! **Iv.C.:** Tabou to Bériby (Aug.) *Chev.* 19953! Abouabou (fr. Feb.) *de Wilde & Leeuwenberg* 3441! **Ghana:** Kwase, Cape Coast Dist. (fr. Nov.) *Chipp* 371! Adeiso, E. Prov. (fr. Mar.) *Irvine* 2412! **Togo Rep.:** Lomé *Warnecke* 369! Tohoué (fl. & fr. Jan.) *Chev.* 22776! Sokodé (fr. Nov.) *Kersting* 691. **S.Nig.:** Lagos *Dalz.* 1415! *Rowland*! Etemi, Omo F.R. (fr. Mar.) *Jones & Onochie* FHI 16632! Idanre (fr. Jan.) *Brenan & Keay* 8652! Onitsha *Barter* 1813! Akpaka F.R., Onitsha (fr. Sept.) *Onochie* FHI 34052! Abakaliki to Obubra (fr. Apr.) *Kitson*! [Br.]**Cam.:** Tiko (fr. Jan.) *Dunlap* 168! Likomba (fl. & fr. Oct.) *Mildbr.* 10521! 10527! Gangumi, Adamawa (fr. Dec.) *Latilo & Daramola* FHI 28849! Also in Cameroun, Gabon, the Congos and Angola.

26. **S. usambarensis** *Gilg* in Engl. Pflanzenw. Ost.-Afr. C : 311 (1895) ; F.T.A. 4, 1 : 526 ; E. A. Bruce in Kew Bull. 10 : 627, fig. 1B (1956) ; Bruce & Lewis in F.T.E.A. Loganiac. 34. *S. micans* S. Moore in J. Linn. Soc. 40 : 146 (1911). *S. cooperi* Hutch. & M. B. Moss F.W.T.A., ed. 1, 2 : 24 (1931) ; Kew Bull. 1937 : 335. Liane with solitary tendrils, shrub or small tree (known in E. Africa as a tree up to 80 ft. high) ; branchlets conspicuously lenticellate, ultimate ones often covered with a pale skin which later splits and peels off ; flowers greenish-yellow, in few-flowered axillary cymes ; in forest thickets.
S.L.: *Thomas* 10493! Bumbuna (fr. Oct.) *Thomas* 3898! **Lib.:** Dukwia R. (Mar.) *Cooper* 300! **Iv.C.:** Sassandra R., 14 km. N. of Sassandra (Apr.) *Leeuwenberg* 3241! **Ghana:** Cape Coast (Feb.-Mar.) *Hall* 1031! 1080! Achimota (Mar.-June ; fl. & fr. July) *Irvine* 343! 1456! *Milne-Redhead* 5123! **N.Nig.:** Sanga River F.R. (fl. & fr. Apr., May) *E. W. Jones* 60! **S.Nig.:** Abeku, Omo F.R. (fr. Feb.) *Jones & Onochie* FHI 17152! Olokemeji F.R. (fr. Sept.-Nov.) *Onochie* FHI 14548! Also in Uganda, Kenya, Tanganyika, Mozambique, S. Rhodesia and S. Africa.

27. **S. camptoneura** *Gilg & Busse* in Engl. Bot. Jahrb. 36 : 93 (1905) ; in Wiss. Erg. Zw. Zentr.-Afr. Exp. 62 (1922) ; Leeuwenberg in Act. Bot. Neerl. 11 : 48, fig. 1 (1962). *Scyphostrychnos talbotii* S. Moore in Cat. Talb. 71, t. 10 (1913) ; F.W.T.A., ed. 1, 2 : 24. *Scyphostrychnos psittaconyx* Duvign. in Bull. Class Sci. sér. 5, 34 : 98, fig. 1 (1948). Large liane, drying dark brown or black ; tendrils in 2 pairs above each other ; fruit very large, up to 6½ in. long, 5 in. broad.
Iv.C.: Mé R. *Leeuwenberg* 4171! Bianouan, Bia R. (fr. Apr.) *Leeuwenberg* 3974! **S.Nig.:** Omo F.R., Ijebu Dist. *Onochie* FHI 15529! Sapoba, Benin (fr. Sept.) *Kennedy* 2220! *Onochie* FHI 34318! Oban *Talbot* 1664! [Br.]**Cam.:** Victoria (immature fr. July) *Buchholz*. Also in Cameroun, Gabon and Congo.

28. **S. icaja** *Baill.* Adansonia 12 : 368 (1879) ; Bull. Soc. Bot. Fr. 58 : 528, t. 18 (1911) ; Rev. Bot. Appliq. 27 : 206, t. 11 (1947). *S. dewevrei* Gilg in Engl. Bot. Jahrb. 28 : 119 (1899). *S. kipapa* Gilg in Notizbl. Bot. Gart. Berl. 2 : 256 (1899), and in Engl. Bot. Jahrb. 28 : 118 (1899). *S. pusilliflora* S. Moore in Cat. Talb. 70 (1913). *S. venulosa* Hutch. & M. B. Moss F.W.T.A., ed. 1, 2 : 24 (1931) ; Kew Bull. 1937 : 334. A stout liane with branchlets bearing solitary tendrils ; flowers minute, borne in slender rather open axillary panicles ; fruit globose ; in forest.
S.L.: Falaba (Apr.) *Aylmer* 58! **Lib.:** Peahtah (fr. Oct.) *Bequaert* 1073! **Iv.C.:** Sassandra *Leeuwenberg* 4026! about 30 km. S.W. of Guéyo *Leeuwenberg* 3732! Hana R., Taï to Tabou (Mar.) *de Wilde & Leeuwenberg* 3521! Nimba Mts., 1,200 ft. *Schnell* 1046! **S.Nig.:** Oban *Talbot* 157! 1256! Also in Cameroun, Ubangi-Shari, Gabon and Congo.

29. **S. odorata** *A. Chev.* in Rev. Bot. Appliq. 27 : 372 (1947), partly (only the type). An entirely glabrous liane with paired tendrils ; leaves like parchment, often drying dark brown or black.
Lib.: Ganta (Aug.) *Harley* 989! Peahtah (Oct.) *Linder* 1065! **Iv.C.:** Abidjan *Chev.* 15435! Aboisso (Apr.) *Chev.* 17824! 20 km S.W. of Guéyo (May) *Leeuwenberg* 3737!

30. **S. barteri** *Solered.* in Engl. Bot. Jahrb. 27 : 556 (1839) ; F.T.A. 4, 1 : 523. A liane with tendrils in 1–3 pairs above each other and fragrant white flowers in dense-flowered axillary cymes ; fruits globose, orange, one-seeded ; in forest.
S.L.: Madina, Limba country (fl. & fr. Apr.) *Sc. Elliot* 5569! 5659! Rowalla (fr. July) *Thomas* 1055! **Iv.C.:** Bianouan, Bia R. *Leeuwenberg* 3952! 31 km. S. of Gagnoa, Davo R. *Leeuwenberg* 4116! 56 km. N. of Sassandra *Leeuwenberg* 3996! Mt. Tonkouï (fr. Apr.) *Leeuwenberg* 3855! **S.Nig.:** Sumoge, Oni R., Ijebu Prov. (Apr.) *Keay* FHI 16060! Shasha F.R. (Feb.) *Richards* 3094! Idanre (fl. & fr. Apr.) *Symington* FHI 3363! Onitsha *Barter* 1247! 1759! Ubulubu (fr. Jan.) *Thomas* 2265!

31. **S. ndengensis** *Pellegr.* in Bull. Mus. Hist. Nat. 33 : 267 (1927) ; Mém. Soc. Linn. Norm., n. sér., 1 : 37 (1928). *S. tricalysioides* Hutch. & M. B. Moss F.W.T.A., ed. 1, 2 : 24 (1931) ; Kew Bull. 1937 : 334. A liane with yellowish-white flowers in condensed axillary cymes ; fruit globose, bright orange ; midrib impressed above ; in forest.
S.Nig.: Afi River F.R., Ikom Dist. (fr. May) *Jones & Onochie* FHI 18929! Omo F.R. (fr. Feb.) *Tamajong* FHI 16972! [Br.]**Cam.:** Likomba (Oct.) *Mildbr.* 10576! Also in Gabon.

3. MOSTUEA

3. MOSTUEA Didr.—F.T.A. 4, 1 : 504 ; Leeuwenberg in Meded. Landb. Wageningen 61, 4 : 1–31 (1961). *Coinochlamys* T. Anders. ex Benth. (1876)—F.W.T.A., ed. 1, 2 : 20.

Inflorescence with two large orbicular bracts covering the sepals ; leaves more or less pilose, ovate, acuminate to subacute at apex, more or less rounded and unequal-sided at base, 1–4·5 cm. long, 6–20 mm. broad ; fruits 2-lobed, pubescent to pilose   3. *hirsuta*

Inflorescence with small bracts never covering the sepals :
Plant pubescent, pilose, or glabrous ; hairs never erect :
Leaves shining when dry, entirely glabrous, equal-sided at base, elliptic to oblong-ovate, 2·5–7·5 cm. long, 1–3·5 cm. broad ; sepals obtuse or acute, usually shorter than the ovary, connate for about half their length ; capsule obcordate, pale brown, conspicuously veined ; fruits retuse and apiculate at apex   ..   1. *adamii*
Leaves dull when dry, usually with some hairs, variable in shape and size ; sepals acute to subulate, usually longer than the ovary ; capsule not conspicuously veined, mostly dark brown :
Capsule dull, usually dark brown, mostly bilobed ; branchlets densely appressed-pubescent to glabrous, if with hairs in two lines then lamina lanceolate ; leaves extremely variable, orbicular to lanceolate   ..   ..   ..   2. *brunonis*
Capsule shining, medium-brown, obcordate ; branchlets apically with two lines of pubescence below the stipules ; leaf often abruptly narrowed into the cuneate base, acuminate at apex, 1·5–5 cm. long, 8–24 mm. broad   4. *hymenocardioides*
Plant hirsute with stiff erect hairs, especially on the branchlets ; leaves variable, elliptic to oblong-obovate, 6–65 mm. long, 4–27 mm. broad ; sepals ovate-lanceolate to linear ; inflorescences very short, 1–3-flowered   ..   ..   ..   5. *batesii*

1. **M. adamii** *Sillans* in Rev. Bot. Appliq. 33 : 546, t. 5 (1953) ; Leeuwenberg in Meded. Landb. Wageningen 61, 4 : 30, fig. 6, 1–5 (1961). Shrub, 1–3 ft. high ; leaves subcoriaceous ; corolla white ; fruit with longitudinal anastomosing veins ; in rain forest.
   Guin.: Bala, 3,600 ft., Macenta Dist. *Adam* 4190! **S.L.:** Namuyei R. *Lane-Poole* 331! **Lib.:** Peahtah *Linder* 978!
2. **M. brunonis** *Didr.* in Vidensk. Meddel. Nat. Foren. Kjoeb. 1853 : 87 (1854) ; F.T.A. 4, 1 : 505 ; Leeuwenberg l.c. 14. *M. thomsonii* (Oliv.) Benth. in Hook. Ic. Pl. 12 : 83 (1876) ; F.T.A. 4, 1 : 505 ; F.W.T.A., ed. 1, 2 : 20. *M. angustifolia* Wernham in Cat. Talb. 66 (1913) ; F.W.T.A., ed. 1, 2 : 20. Extremely variable shrub, 1–23 ft. high ; sepals ovate to ovate-linear ; corolla mostly white with a yellow base. Only var. *brunonis* occurs in our area (Leeuwenberg l.c. 19, figs. 3 & 4).
   Ghana: Bonsasu (fl. & fr. May) *Vigne* FH 1972! **S.Nig.:** Afikpo (fl. & fr. Apr.) *Jones* FHI 1761! Cross R. North F.R., Ikom (fl. & fr. Dec.) *Okafor* FHI 41718! Oban (fr. May) *Talbot* 306! 1035! Calabar (fl. & fr. Dec.) *Thomson* 44! *Robb.! Ujor* FHI 30841! *Brenan* 9247! **[Br.]Cam.:** Mokonyong (fr. Nov.) *Tamajong* FHI 22112! Through much of tropical Africa, and in Madagascar.
3. **M. hirsuta** (*T. Anders. ex Benth.*) *Baill. ex Bak.* in F.T.A. 4, 1 : 509 (1903) ; Chev. in Rev. Bot. Appliq. 27 : 108, t. 6, 4 (1947) ; Leeuwenberg l.c. 7, fig. 1. *Coinochlamys hirsuta* T. Anders. ex Benth. in Benth. & Hook. f. Gen. Pl. 2 : 1091 (1876) ; F.W.T.A., ed. 1, 2 : 20. Shrub or undershrub, 1–6 ft. high, with hairy branchlets ; corolla white, sometimes pale yellow or yellow-striped at the base ; in open places, in forest or savanna, sometimes in moist places ; fire resistant.
   Port.G.: Catio (June) *Esp. Santo* 2067! **Guin.:** Karkandy *Heudelot* 880! Sementa *Maclaud* 280! Séguéa (Oct.) *Adam* 12634! Farmoreah (Apr.) *Roberty* 17336! *Dollaga Roberty* 17602! **S.L.:** Kamakuia (May) *Thomas* 265! Kamalu (May) *Thomas* 353! Binkolo (fr. Aug.) *Thomas* 1750! 1824! Bagroo R. (Apr.) *Mann* 811! Yamandu, near Baoma (fl. & fr. July) *Deighton* 5804! **Iv.C.:** Aboisso *Hb. I.D.E.R.T.* 3816! Aboisso to Krinjabo *Hb. I.D.E.R.T.* 3838! **N.Nig.:** Dogon Kurmi, Sanga River F.R. (Apr.–May) *E. W. Jones* FHI 42232! Samban Kwoi, Zaria *Wimbush* FHI 39620! **S.Nig.:** Engenni, Brass Dist. (fl. & fr. May) *Mallam* FHI 18302! Oron to Eket *Talbot*! **[Br.]Cam.:** Su, Bamenda (May) *Maitland* 1692! S. Bakundu F.R., Kumba (fl. & fr. Mar.) *Brenan* 9453! Also in Cameroun, Rio Muni, Ubangi-Shari, Sudan, Gabon, the Congos, Cabinda and Angola.
4. **M. hymenocardioides** *Hutch. & Dalz.* F.W.T.A., ed. 1, 2 : 20 (1931) ; Kew Bull. 1937 : 61 ; Chev. in Rev. Bot. Appliq. 27 : 106 ; Leeuwenberg l.c. 26, fig. 5, 1–5. *Rytigynia concolor* of Robyns in Bull. Jard. Bot. Brux. 11 : 190, partly (*Chev.* 15705) ; F.W.T.A., ed. 1, 2 : 111, not of (Hiern) Robyns. *Vangueria concolor* of Chev. Bot. 330, not of Hiern. Much-branched slender shrub up to 5 ft. high ; flowers white, red inside.
   Guin.: Kankan (Mar.) *Chev.* 15967! Doubato (Mar.) *Roberty* 17167! Kouroussa *Brossart* in *Hb. Chev.* 15691! 15705! Farana *Sc. Elliot* 5308! *Roberty* 17231! Dantilia R. (fr. Mar.) *Sc. Elliot* 5268! **S.L.:** Tassin, Scarcies R. *Sc. Elliot* 4515! Njala (May) *Deighton* 3524! Ndilajula (Apr.) *Deighton* 3735! Laminaiya (Apr.) *Thomas* 123! (See Appendix, p. 362.)
5. **M. batesii** *Bak.* in F.T.A. 4, 1 : 506 (1903) ; Leeuwenberg l.c. 10, fig. 2, 10–20. Shrub or undershrub 1–5 ft. high ; corolla white, yellow at the base ; in secondary or rain forest.
   [Br.]Cam.: Nyasoso, N.E. of Kumba *Schlechter* 12936! Also in Cameroun, Gabon and Congo.

4. SPIGELIA Linn. Sp. Pl. 149 (1753) ; Leeuwenberg in Acta Bot. Neerl. 10 : 460–465 (1961).

A herb up to 40 cm. high ; stems glabrous ; leaves 4 in a whorl at the end of the stem, broadly lanceolate, gradually acute, subsessile, 5–7 cm. long, 1·5–2·5 cm. broad, slightly scabrid on both surfaces ; stipules connate ; flowers in terminal secund spikes (cincinni) up to 10 cm. long ; calyx 2 mm. long, segments narrow, acute ; corolla narrowly funnel-shaped, 1 cm. long, with 5 small lobes ; fruits 5 mm. diam., 2-lobed, the upper part warted, base of fruit boat-shaped and persisting after the fall of the fruit   ..   ..   *anthelmia*

S. anthelmia *Linn.* Sp. Pl. 149 (1753) ; Leeuwenberg in Acta Bot. Neerl. 10 : 461, fig. 1 (1961). Herb up to 2 ft. high, with pink flowers ; common weed of waste places.
   Guin.: Conakry (May) *Chillou* 1453! **S.L.:** Binkolo (fl. & fr. Aug.) *Deighton* 1284! Makump (fl. & fr. May, Aug.) *Deighton* 1708! *Dawe* 505! Njala (fl. & fr. Sept.) *Payne* 5! Rokupr (fl. & fr. May) *Jordan* 258! **Lib.:** Monrovia (fl. & fr. Nov.) *Linder* 1540! Kakata (fl. & fr. Dec.–Apr.) *Bequaert* 189! Sinkor (fl. & fr. Jan.) *Barker* 1069! **Iv.C.:** Abidjan (Nov.) *Leeuwenberg* 1925! **Ghana:** Benso (fl. & fr. July) *Andoh* FH 5516! Achimota (fl. & fr. June) *Irvine* 1446! Asokwa-Kumasi (fl. & fr. Aug.) *Darko* 708! **N.Nig.:** Jebba (fl. & fr. Dec.) *Hagerup* 718! **S.Nig.:** Lagos (fl. & fr. Oct.) *Dalz.* 1042! Olokemeji (fl. & fr. Apr.) *W. D. MacGregor* 289! Ibadan (fl. & fr. Mar.) *Jones* FHI 753! Port Harcourt (fr. June) *Maitland* A! Native of tropical America and naturalised in W. Africa, Congo and Ethiopia.

5. USTERIA Willd. in Cothenius, Disp. 1 (Jan.–May 1790) ; F.T.A. 4, 1 : 517 ;
F.W.T.A., ed. 1, 2 : 24 ; Leeuwenberg & Stafleu in Taxon 10 : 212 (1961), not of
Medik. (Mar.–May 1790).

A climber with long internodes and smooth glabrous branchlets ; leaves broadly
elliptic, subacute to rounded at the apex, about 10 cm. long and 6 cm. broad, with
3–4 pairs of conspicuous lateral nerves ; inflorescence a lax terminal panicle, minutely
puberulous ; enlarged calyx-lobe narrowly obovate, almost 1 cm. long ; corolla
1–1·5 cm. long ; stamen 1 ; fruit a capsule, oblong, 2-valved, about 3·5 cm. long,
1 cm. broad   ..   ..   *guineensis*

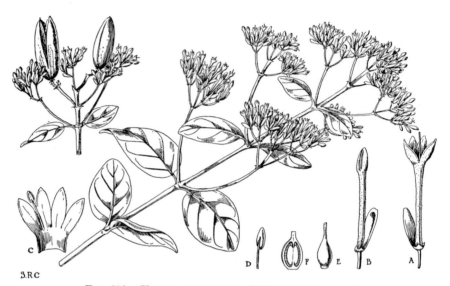

S.R.C

FIG. 210.—USTERIA GUINEENSIS *Willd.* (LOGANIACEAE).

A, flower. B, the same in bud. C, upper part of corolla opened to show the solitary stamen.
D, anther. E, ovary. F, longitudinal section of ovary.

U. guineensis *Willd.* in Schr. Berlin Ges. Naturf. Fr. 10 : 51 (1790) ; F.T.A. 4, 1 : 517 ; F.W.T.A., ed. 1, 2 :
25 ; Chev. Bot. 442. A low climber with whitish or lilac flowers with one large pale calyx-lobe ; in climber
tangles in forests and roadsides.
Sen.: *Heudelot* 597 ! Ziguinchor (Jan.) *Chev.* 2841 ! **Gam.:** *Brown-Lester* 837 ! **Port.G.:** Calequisse,
Teixeira Pinto (Jan.) *d'Orey* 107 ! **Guin.:** *Farmar* 169 ! Conakry (fl. & fr. Jan.) *Dalz.* 8088 ! Kouroussa
(July) *Pobéguin* 280 ! **Mali:** Oualia (Dec.) *Chev.* 98 ! **S.L.:** Banana Is. (Dec.) *Dawe* 423 ! Heddle's Farm
(Jan.) *Lane-Poole* 380 ! Kamalu (fl. & fr. May) *Thomas* 327 ! Widaro (Oct.) *Fisher* in *Hb. Deighton* 54 !
Samu (Dec.) *Sc. Elliot* 4243 ! **Lib.:** Du R. (Aug.) *Linder* 245 ! Ganta (fl. & fr. Dec.) *Barker* 1141 ! Jabroke,
Webo Dist. (July) *Baldwin* 6454 ! Mano (fl. & fr. Dec.) *Baldwin* 10811 ! **Iv.C.:** Abidjan (Oct.) *Leeuwenberg*
1744 ! Danané *Collenette* 44 ! **Ghana:** Lante (Jan.) *Johnson* 828 ! Aburi (Sept., Nov.) *Johnson* 1079 !
Adams 1921 ! Axim (fl. & fr. Feb.) *Irvine* 2123 ! Anum, Trans-Volta (fl. & fr. Nov.) *Morton* GC 7975 !
Dah.: (May) *Le Testu* 153 ! **N.Nig.:** Kaduna (Dec.) *Meikle* 780 ! Zurzufa stream, Nasarawa Div. (Nov.)
Peal FHI 42783 ! **S.Nig.:** Ebute Metta (fl. & fr. Jan.) *Millen* 23 ! Agbede (Sept.) *Farquhar* 34 ! Sapoba
(Mar.) *Thompson* ! Aguku (Oct.) *Thomas* 952 ! 1172 ! Nun R. (Aug.) *Mann* 460 ! Degema *Talbot* ! Oban
Talbot 303 ! Also in Cameroun, Rio Muni, Gabon, the Congos and Angola. (See Appendix, p. 364.)

6. NUXIA Lam.—F.T.A. 4, 1 : 511. *Lachnopylis* Hochst. in Flora 26 : 77 (1843);
F.W.T.A., ed. 1, 2 : 20.

Leaves mostly ternate, elliptic or nearly so, 5–13 cm. long, 2–6 cm. broad, glabrous or
hairy, may be tomentose beneath, entire to serrate, herbaceous to coriaceous, usually
more hairy, more serrate and thinner on young shoots and in the shade, subacute to
rounded at apex, cuneate at base ; petiole 0·5–2 cm. long; peduncles glabrous to
densely pubescent ; calyx about 5 mm. long, nearly cylindric, minutely papillose to
densely pubescent outside, pubescent inside, with erect lobes ; stamens long-exserted ;
capsule bivalved, with apically torn valves; seeds small, fusiform .. *congesta*

1. **N. congesta** *R. Br. ex Fresen.* in Flora 21 : 606 (1838) ; F.T.A. 4, 1 : 512 (1903) ; Bruce & Lewis in F.T.E.A.
Loganiac. 44, fig. 8, 7–8 (1960). *Lachnopylis mannii* (Gilg) Hutch. & M. B. Moss in F.W.T.A., ed. 1, 2 :
20 (1931). *Nuxia mannii* Gilg in Engl. Bot. Jahrb. 32 : 140 (1902) ; F.T.A. 4, 1 : 512 ; Chev. Bot. 440.
Lachnopylis guineensis Hutch. & M. B. Moss in F.W.T.A., ed. 1, 2 : 20 (1931), and in Kew Bull. 1937 :
61. Tree up to 80 ft. high, sometimes with a grey fluted stem ; flowers creamy-white and scented ; in
montane forest.

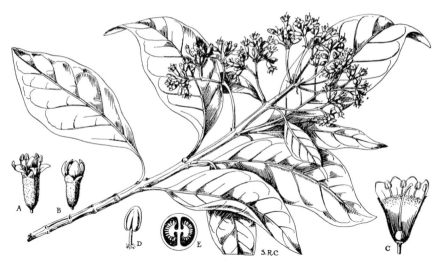

FIG. 211.—NUXIA CONGESTA *R. Br. ex Fresen.* (LOGANIACEAE).
A and B, flowers. C, flower laid open. D, anther. E, cross-section of ovary.

Guin.: Diaguissa, Fouta Djalon (Dec.) *Chev.* 18011! **S.L.:** Mt. Gonkwi, 3,000 ft., Talla (Feb.) *Sc. Elliot* 4824! Bintumane Peak, 5,400 ft. (Jan.) *T. S. Jones* 144! *Jaeger* 4101! **Iv.C.:** Mt. Boho, above 2,500 ft., Upper Sassandra (May) *Chev.* 21494. **Ghana:** Bana Hill, Krobo (Mar.) *Irvine* 893! Amedzofe, 2,000 ft. (Jan.) *de Wit & Morton* A2871! Gbadzeme to Amedzofe (fl.) *Enti* FH 6875! Senchi (Oct.) *Vigne* FH 4052! Odomi River F.R. (Oct.) *St. Cl.-Thompson* FH 3644! **S.Nig.:** Koloishe Mt., 6,000 ft., Ogoja Prov. (Dec.) *Savory & Keay* FHI 25099! **[Br.]Cam.:** Cam. Mt., 5,000–9,000 ft. (Jan., Mar., Apr., Dec.) *Mann* 1206! *Maitland* 458! 1194! *Keay* FHI 28630! *Brenan* 9571! Mondani, Mamfe Dist. (Jan.) *Lobe Babute* Cam. 15/35! Bafut-Ngemba F.R., Bamenda (Jan.) *Lightbody* FHI 26281! **F.Po:** Moka (Jan., May) *Guinea* 2150! 2154! Also in Congo, Sudan, Ethiopia, E. Africa, Nyasaland, Angola, S. Rhodesia, Mozambique and S. Africa.
[The range of variation is extreme in this genus.]

133. OLEACEAE

By P. S. Green

Trees, shrubs or climbers. Leaves opposite or very rarely alternate, simple or pinnate ; stipules absent. Flowers hermaphrodite or rarely unisexual, actinomorphic. Calyx lobed or dentate. Petals present, free or connate, often 4, imbricate or induplicate-valvate. Stamens hypogynous or epipetalous, usually 2 ; anthers apiculate, 2-celled, cells back to back, opening lengthwise. Disk absent. Ovary superior, 2-celled, style simple with a capitate or bifid stigma ; ovules usually 2 in each cell, axile, pendulous or ascending. Fruit baccate or drupaceous ; seeds usually with endosperm ; embryo straight, the radicle sometimes hidden within the base of the cotyledons.

A small family easily recognised by the woody habit, opposite leaves without stipules and the two stamens.

Corolla not exceeding 7 mm. long, with a short tube or petals nearly free ; fruit a drupe ; shrubs or small trees :
Petals very narrow, paired and free almost to the base ; inflorescences axillary ; leaves with or without pits or tufts of hairs in the axils of the main lateral nerves beneath, lamina up to 25 cm. long   ..   ..   **1. Linociera**
Petals more or less ovate, united, with a distinct tube ; inflorescences terminal ; leaves without pits or tufts of hairs in the axils of the main lateral nerves beneath, lamina up to 13 cm. long   ..   ..   **2. Olea**
Corolla not less than 10 mm. long, hypocrateriform with a well developed tube ; fruit a berry or hard and woody; trees, shrubs or climbers :
Fruit hard and woody, 2-valved, seeds winged ; trees ; leaves simple (in our species) ; petioles simple ; calyx-lobes blunt-triangular or very small   ..   **3. Schrebera**

Fruit a berry, seeds not winged ; climbers or shrubs ; leaves simple or trifoliolate (in our species) ; petioles trifoliate or articulate ; calyx-lobes filiform, or often blunt-triangular   ..   ..   **4. Jasminum**

1. LINOCIERA Sw.—F.T.A. 4, 1 : 19 ; Gilg & Schellenberg in Engl. Bot. Jahrb. 51 : 67 (1913). *Nom. cons.*

Rhachis of inflorescence (1·3–)2·5–4(–7) cm. long ; leaves with glabrous pits or tufts of hair beneath in the axils of the majority of the (6–)9–12(–13) main lateral nerves on each side of the midrib ; lamina oblong, oblong-oblanceolate or oblong-lanceolate, (6–)9–15(–25) cm. long, (2·5–)3–6(–9·5) cm. broad ; petioles 6–20 mm. long :

Main lateral nerves with tufts of hair in their axils ; inflorescence rhachis densely appressed-tomentose ; calyx appressed-tomentose often densely so ; petioles thickened, flaky rugulose in dried specimens, even when young   ..   1. *africana*

Main lateral nerves with glabrous pits in their axils ; inflorescence rhachis glabrous or sparsely appressed-tomentose ; calyx glabrous, ciliolate or rarely with a few hairs outside ; petioles slightly thickened, not flaky rugulose   ..   ..   2. *nilotica*

Rhachis of inflorescence 0·2–2 cm. long ; leaves with or without pits in the axils of the (5–)6–7(–8) main lateral nerves on each side of the midrib ; lamina oblanceolate, oblong or oblong-lanceolate (often narrowly so in *L. mannii*) ; petioles 5–10 mm. long, rugulose, sometimes flaky with age in dried specimens :

Inflorescence rhachis 10–20 mm. long, appressed-tomentose ; corolla 3–5 mm. long ; leaves gradually acuminate, tip acute or blunt, lamina (6–)7–17(–19) cm. long, (1·5–)2–5(–7) cm. broad, without or with small pits glabrous (or hairy) beneath in a few of the axils of the 6–7 main lateral nerves on each side of midrib ; main lateral nerves slightly raised, venation more or less obscure beneath ..   ..   3. *mannii*

Inflorescence rhachis 2–7 mm. long, glabrous or appressed-tomentose ; corolla 4·5–7 mm. long ; leaves abruptly caudate-acuminate, tip rounded ; lamina 9·5–24 cm. long, 3·7–8·5 cm. broad, without pits beneath in the axils of the (5–)6–7(–8) main lateral nerves on each side of midrib ; main lateral nerves and venation not prominent beneath   ..   ..   4. *congesta*

1. **L. africana** (*Welw. ex Knobl.*) *Knobl.* in Bot. Centralbl. 61 : 129 (1895). *Mayapea africana* Welw. ex Knobl. in Engl. Bot. Jahrb. 17 : 529 (1893). *Linociera angolensis* Bak. in F.T.A. 4, 1 : 20 (1902). *L. johnsonii* Bak. l.c. (1902) ; Gilg & Schellenb. in Engl. Bot. Jahrb. 51 : 70 ; F.W.T.A., ed. 1, 2 : 26 ; Turrill F.T.E.A. Oleac. 12. *L. fragrans* Gilg & Schellenb. l.c. 71 (1913). *L. oreophila* Gilg & Schellenb. l.c. (1913). Shrub or small tree, to 30 ft. high ; flowers white ; fruits blue-black when ripe ; in rain forest.
   S.L.: Kafoko (Sept.) *Thomas* 2134 ! **Ghana:** (Dec.) *Johnson* 148 ! Aburi Hills (Nov.) *Johnson* 234 ! 453 ! Ejura (Nov.) *Vigne* FH 3456 ! Okrabi, Hohoe Dist. (Nov.) *St. C. Thompson* FH 3698 ! **N.Nig.:** Esie F.R., Ilorin *Latilo* FHI 38015 ! **S.Nig.:** Ibadan North F.R. (fl. Dec., fr. Mar.) *Chizea* FHI 23969 ! 23975 ! Oluwa F.R. *Kennedy* 2475 ! Oban *Talbot* ! [**Br.]Cam.:** Buea (Jan.) *Deistel* 85 ! *Maitland* 306 ! Bali-Ngemba F.R., Bamenda (fr. May) *Ujor* FHI 30335 ! Also in Angola, Uganda and Tanganyika.
2. **L. nilotica** *Oliv.* in Trans. Linn. Soc. 29 : 106, t. 117 (1875) ; F.T.A. 4, 1 : 19 ; Gilg & Schellenb. in Engl. Bot. Jahrb. 51 : 71 ; Chev. Bot. 399 ; Aubrév. Fl. For. C. Iv., ed. 2, 3 : 180, t. 317 ; Aubrév. Fl. For. Soud.-Guin. 442 ; Turrill F.T.E.A. Oleac. 14 ; Berhaut Fl. Sén. 117, 118. *L. sudanica* A. Chev. Bot. 399, name only. Shrub or small tree to 25 ft. high ; flowers white or cream, fragrant ; in fringing forest in the savanna woodland regions.
   Sen.: Niokolo-Koba *Berhaut* 1520. **Mali:** Banfara (fl. & young fr. Mar.) *Chev.* 535 ! **Iv.C.:** Bobo-Dioulasso (May) *Aubrév.* 1848 ! *Chev.* 903 ! Bondoukou *Aubrév.* 737. Tiengara *Aubrév.* 1581. **Ghana:** Fuller Fallst Kintampo Dist. (Nov.) *Andoh* FH 5417 ! Saku, N.T. (May) *Vigne* FH 4704 ! Bahare, Gambaga Dis., (young fr. June) *Akpabla* 735 ! **Togo Rep.:** Sokodé *Kersting* a176 ! **Dah.:** Kouandé to Konkobiri (June) *Chev.* 24265 ! Tankiéta *Aubrév.* 92d. Natitingou *Aubrév.* 80d. **N.Nig.:** 51 miles S.E. of Ilorin *Clarke* 12 ! Jebba (Jan.) *Meikle* 1009 ! Anara F.R., Zaria Prov. (May) *Keay* FHI 25780 ! 25787 ! Jos Plateau (Jan., Nov.) *Lely* 724 ! P53 ! Katagum Dist. *Dalz.* 366 ! [**Br.]Cam.:** foot of Vogel Peak, Adamawa (Nov.) *Hepper* 1463 ! Also in Cameroun, Ubangi-Shari, Angola, Sudan, Uganda, Kenya and Tanganyika.
3. **L. mannii** *Solereder* in Bot. Centralbl. 46 : 17 (1891) ; F.T.A. 4, 1 : 19 ; Chev. in Vég. Util. Afr. Trop. Fr. 5 : 222 (1909) ; Chev. Bot. 399. *L. congesta* Bak. in F.T.A. 4, 1 : 20 (1902), partly (*Mann* 2214). *L. lingelsheimiana* Gilg & Schellenb. in Engl. Bot. Jahrb. 51 : 72 (1913) ; F.W.T.A., ed. 1, 2 : 26 ; Aubrév. Fl. For. C. Iv., ed. 2, 3 : 180, t. 317. *L. macroura* Gilg & Schellenb. l.c. (1913). Tree with spreading crown to 35 ft. high ; flowers white, fragrant ; in rain forest.
   S.L.: Yonibana (Oct.) *Thomas* 4139 ! Kukuna, Scarcies R. (fr. Jan.) *Sc. Elliot* 4717 ! **Lib.:** Monrovia (fl. Oct., fr. Jan.) *Dinklage* 2818 ! 2915 ! **Iv.C.:** coastal thickets *Aubrév.* 901. Zaranou, Indénié (fr. Mar.) *Chev.* 16285 ! **S.Nig.:** Akilla, Ijebu Ode Prov. (fl. Nov., fr. Apr.) *Jones & Onochie* FHI 17305 ! *Kennedy* 2024 ! *Onochie* FHI 8176 ! Kajola to Imope, Ijebu Prov. *Tamajong* FHI 16906 ! [**Br.]Cam.** (?): Cameroons R. (fr. Jan.) *Mann* 2214 ! Also in Gabon. (See Appendix, p. 364.)
   [This species appears to occur in three variants : the specimens from the Gabon having occasional minutely hairy pits in the axils of the main lateral nerves beneath, those from the Cameroons having occasional glabrous pits and those from Sierra Leone and the Ivory Coast without them.]
4. **L. congesta** *Bak.* in F.T.A. 4, 1 : 20 (1902), (excl. *Mann* 2214). Shrub or small tree to 12 ft. high ; flowers white ; in rain forest.
   [**Br.]Cam.:** S. Bakundu F.R., Kumba Div. (Jan.) *Binuyo & Daramola* FHI 35187 ! Also in Gabon and Rio Muni.

2. OLEA Linn.—F.T.A. 4, 1 : 17.

Leaves oblong-elliptic or elliptic, (4·5–)6–10(–13) cm. long, (2–)3–4·5(–5·5) cm. broad, glabrous but, under the hand lens, punctate with peltate appressed scales on both surfaces, coriaceous, base acute-cuneate, apex subacuminate or acute, (5–)6–8(–9) main lateral nerves on each side of the midrib, venation obscure or occasionally slightly raised on both surfaces ; petiole glabrous, 1–1·5(–2) cm. long, more or less rugulose in dried specimens ; inflorescence terminal, 3·5–5(–8) cm. long, 4–5(–8) cm.

broad, with the calyx glabrous except for small peltate appressed scales ; calyx with 4 shallow acute lobes ; corolla 3–3·5 mm. long, glabrous, valvate in bud ; fruit rounded ellipsoid, 1·5 cm. long, 1·1 cm. diam. (? mature)   ..   *hochstetteri*

O. hochstetteri *Bak.* in F.T.A. 4, 1 : 17 (1902); Chev. Bot. 399 ; Aubrév. Fl. For. Soud.-Guin. 442 ; Turrill F.T.E.A. Oleac. 10. *O. guineensis* Hutch. & C. A. Smith in F.W.T.A., ed. 1, 2 : 26 (1931) ; Kew Bull. 1937 : 336 ; Aubrév. Fl. For. C. Iv. ed. 2, 3 : 178, t. 317. Small tree to 40 (–70) ft. in montane forest ; flowers white ; lenticels prominent.
    Guin.: Ziama Massif, 3,800–4,000 ft. (May) *Schnell* 2626 ! *Adam* 50 ; 51. **S.L.:** Picket Hill, 2,300–2,800 ft. (bud & fr. Nov.) *T. S. Jones* 202 ! 213 ! Sugar Loaf Mt. 1,800 ft. *Deighton* 5649 ! **Iv.C.:** Mt. Momy, Upper Cavally, 2,700–2,900 ft. (young fr. Apr.) *Chev.* 21358 ! *Aubrév.* 1177. **[Br.]Cam.:** Cam. Mt., 5,000 ft. (bud & fr. Mar.) *Maitland* 498 ! Bamenda *Johnstone* FHI 356 ! Bali-Ngemba F.R., 5,600 ft., Bamenda (June) *Ujor* FHI 30428 ! Bafut-Ngemba F.R., Bamenda (fr. Jan.) *Lightbody* FHI 26283 ! 26315 ! Also in Congo, Ethiopia, Sudan, Uganda, Kenya, Tanganyika and N. Rhodesia. (See Appendix, p. 365.)

    [*O. welwitschii* (Knobl.) Gilg & Schellenb. known from S. Tomé, Annobon, Angola and tropical E. Africa is very close and, as pointed out by Verdoorn (Bothalia 6 : 581 (1956)) the two species may possibly be regarded as subspecies of the S. African *O. capensis* Linn. *Maitland* 498 and *Lightbody* FHI 26283 approach *O. welwitschii* in their longer slender petioles and slightly acuminate leaf-apices.]

    Besides the above indigenous species, *O. europaea* Linn., native of the Mediterranean region, is cultivated in Senegal (Berhaut Fl. Sén. 113).

3. SCHREBERA Roxb.—F.T.A. 4, 1 : 13. *Nom. cons.*

Leaves broadly oblong, elliptic or ovate-elliptic, (6–)8–14(–18) cm. long, (3·5–)4·5–9 (–11) cm. broad, more or less chartaceous, glabrous except at the base of the midrib, and occasionally of the main lateral nerves, beneath, which range from soft-tomentellous to glabrate or rarely glabrous, base obtuse to rounded slightly decurrent into the petiole, apex abruptly and shortly acuminate to acuminate, with (5–)6–8(–9) main lateral nerves on each side of the midrib ; petiole 1–3 cm. long, glabrate to soft-tomentellous ; inflorescence terminal, glabrate to soft and shortly tomentellous ; calyx campanulate, 4–5 mm. long, glabrate to minutely tomentellous, with 4 shallow ciliolate teeth up to 1 mm. long; corolla hypocrateriform, 12–20 mm. long, tube 2–3 mm. diam., lobes 5–6 mm. long, each with a patch of reddish-brown hairs at the base ; fruit woody, club-shaped, blunt, lenticular, 5–6 cm. long when mature ; seeds 4 cm. long, including the wing   ..   ..   *arborea*

S. arborea *A. Chev.* in Mém. Soc. Bot. Fr. 2, 8 : 180 (1912) ; Chev. Bot. 399 ; Aubrév. Fl. For. C. Iv., ed. 2, 3 : 177, t. 316 ; Turrill F.T.E.A. Oleac. 2 ; Berhaut Fl. Sén. 117, 119. *S. chevalieri* Hutch. & Dalz. F.W.T.A., ed. 1, 2 : 26 (1931) ; Kew Bull. 1937 : 336. *S. golungensis* of Thompson Rep. For. Gold Coast 72, 190 (1910) ; Kennedy For. Fl. S. Nig. 202. Forest tree to 100 ft. high, with clear bole ; flowers white or greenish white with crimson or purplish-brown markings at the base of the petals ; sweetly fragrant ; often planted.
    Sen.: Casamance *Chev.* 3532 ! *Berhaut* 1316. **Port.G.:** Fulacunda *Esp. Santo* 2027. Ilhas das Cobras *Esp. Santo* 1927. **Guin.:** Mamou *Aubrév.* 31 ; 59. **Iv.C.:** Anoumaba (fr. Nov.) *Chev.* B22352 ! *Aubrév.* 72 ; 1340 ; 1812 ! 2008. **Ghana:** Odumase, Krobo (Apr.) *Odonkor* ! Ajena, Volta Gap (fr. Nov.) *Morton* GC 9451 ! **Dah.:** Sakété *fide* Chev. *l.c.* **N.Nig.:** Wana Town (Apr.) *Hepburn* 163 ! Kafanchan (fr. Sept.) *A. F. A. Lamb* FHI 3168 ! **S.Nig.:** Forestry Hill, Ibadan (Mar.) *Jones* FHI 4900 ! *Keay* FHI 25704 ! Okeho, Oyo (Mar.) *Keay* FHI 37742 ! Also from the Congo, Sudan and Uganda. (See Appendix, p. 365.)
    [The Sudan and Uganda specimens are completely glabrous in all parts.]

4. JASMINUM Linn.—F.T.A. 4, 1 : 1 ; Gilg & Schellenberg in Engl. Bot. Jahrb. 51 : 77 (1913).

Leaves trifoliolate, glabrous (often including the nerves), ovate to ovate-lanceolate ; inflorescence many-flowered, more or less densely corymbose-paniculate ; pedicels 1–3(–4) mm. long :
  Indumentum of stem and inflorescence contrasting, the stem and petioles glabrous, the inflorescence densely tomentose ; corolla tube 1 cm. long, with 5 lobes 0·5 cm. long, more or less ovate ; shallow pits with tufts of hairs in most of the axils of the main lateral nerves on the leaf beneath, lamina (5–)8–11 cm. long, (3·5–) 4–7 cm. broad, apex acute to acuminate, base acute to almost rounded ; petioles 20–35 mm. long ; calyx tomentose, lobes acute, 1–2 mm. long, slightly accrescent in fruit   ..   ..   *1. bakeri*
  Indumentum of stem and inflorescence similar, and, with the petioles, crisped pubescent, sometimes minutely so ; corolla tube 2–2·5 cm. long, with 5–7 lobes 1 cm. long, rounded elliptic ; tufts of hairs confined to the axils of the main lateral nerves at the base of the leaf beneath, lamina 3–5 cm. long, 2–3·7 cm. broad, apex acute, base rounded to subcordate ; petioles 5–20 mm. long ; calyx puberulous, lobes blunt, less than 1 mm. long ..   ..   *2. fluminense*
Leaves simple, glabrous or hairy, nerves usually hairy, lamina ovate to narrowly lanceolate :
  Stem and petioles more or less glabrous or if puberulous, minutely so ; leaves without tufts of hairs in the axils of the main lateral nerves beneath :
    Inflorescence with numerous flowers more or less densely corymbose, pedicels 1–3 mm. long, accrescent and thickened in fruit, especially towards the top, 7–15 mm. long; lamina ovate or occasionally broadly lanceolate, (4–)5–11 cm. long, (2·5–) 3–6·5 cm. broad, apex acute, acuminate or rounded, the tip often mucronate occasionally slightly retuse, base acute or rounded ; petioles 7–20 mm. long ; leaves on some shoots often ternate ; calyx glabrous or sparsely puberulous, lobes

0·5–2(–4) mm. long, ciliolate ; corolla tube 1·8–2·5 cm. long, with 5–9 oblanceolate lobes 0·7–1·6 cm. long   ..   ..   3. *dichotomum*

Inflorescence 1–3(–4) flowered, pedicels 2–8(–17) mm. long ; lamina narrow-lanceolate or rarely lanceolate, 3·5–7 cm. long, 1–2·5 cm. broad, apex acute, tip rounded, mucronate, occasionally retuse, base cuneate ; petiole 2–7 mm. long ; calyx puberulous or sparsely puberulous, sometimes minutely so, lobes 1–4 mm. long ; corolla tube 2–2·5 cm. long with 7–8 oblong to narrowly oblong lobes 1·4–2 cm. long

                                                   4. *kerstingii*

Stem pubescent or puberulous with crisped or spreading hairs, often densely so ; leaves with or without tufts of hairs in the axils of the main lateral nerves beneath ; inflorescence 1–12-flowered from the axils of the uppermost leaves :

Tufts of hairs in the axils of the main lateral nerves of the leaf beneath and occasionally at the nerve-junctions, sometimes almost masked by hairs on the midrib ; pedicels 10–20 mm. long, accrescent and thickened in fruit especially towards the top and up to 25 mm. long, petioles 3–18 mm. long ; stem and petioles pubescent, often densely so ; leaves glabrous or scattered pubescent, ovate, ovate-elliptic or rarely oblong, (2–)2·5–6(–11) cm. long, (1–)2–3(–6) cm. broad, base rounded or subcordate, rarely subacute, apex acute or subacuminate, tip mucronate ; calyx pubescent or glabrate, lobes 3–8 mm. long, slightly accrescent in fruit ; corolla tube 1·6–2 cm. long, with 7–9 linear lobes 1–2 cm. long..   ..   5. *pauciflorum*

Tufts of hairs in the axils of the main lateral nerves of the leaf absent beneath ; pedicels 2–20 mm. long, petioles 2–7 cm. long :

Apex of leaf rounded, usually more or less retuse ; stem, pedicels and calyx pubescent, sometimes densely so ; pedicels 2–6(–10) mm. long, slightly accrescent in fruit and thickened especially toward the top ; calyx lobes 1·5–4 mm. long, slightly accrescent in fruit ; leaf surface glabrous or puberulous along the nerves on both surfaces, lamina ovate to lanceolate or rarely elliptic, (1·5–)2–5 cm. long, (0·8–) 1·5–3(–4) cm. broad ; petioles 2–4 mm. long ; corolla tube 2–3 cm. long with 7–9 oblong-lanceolate lobes, 1–1·7 cm. long   ..   ..   ..   6. *obtusifolium*

Apex of leaf subacuminate to cuspidate ; stem and especially the calyx and pedicel with long spreading hairs, often densely so ; pedicels 15–20 mm. long, slightly accrescent in fruit and thickened towards the top ; calyx lobes (4–)6–7 mm. long, linear ; leaf surface with scattered hairs on both surfaces especially on the nerves, densely so beneath, lamina obovate or oblanceolate, rarely ovate, (3–) 4–8 cm. long, (1·7–)2–5 cm. broad ; corolla tube and lobes 2–2·3 cm. long, lobes 6–8, linear   ..   ..   7. *preussii*

1. **J. bakeri** *Sc. Elliot* in J. Linn. Soc. 30 : 86 (1894) ; F.T.A. 4, 1 : 12 ; Chev. Bot. 398. A climber ; flowers white, fragrant.
  Guin.: Bonhouri (Feb.) *Pobéguin* 899 ! Bangadou to Dioromandou (Feb.) *Chev.* 20730 ! Labé (Apr.) *Chev.* 12318. **S.L.:** near Berria (Mar.) *Sc. Elliot* 5409 ! **Iv.C.:** Sogui to Koualé (May) *Chev.* 21697. Also in Congo and N. Rhodesia.

2. **J. fluminense** *Vell.* Fl. Flumin. 10 (1825) ; op. cit. Atlas 1, t. 23 (1827) ; Dandy in Kew Bull. 1950 : 368 ; Turrill F.T.E.A. Oleac. 19 ; Exell Cat. S. Tomé Suppl. 33. *J. mauritianum* Bojer ex DC. Prod. 8 : 310 (1844) ; F.T.A. 4, 1 : 10 ; F.W.T.A., ed. 1, 2 : 27. A scandent shrub ; flowers cream-coloured or white tinged with pink, fragrant ; fruits black.
  S.L.: Njala (cult. fl. Jan.) *Deighton* 3962 ! **N.Nig.:** near Jos (Aug., Dec.) *Batten-Poole* 133 ! *Keay* FHI 20070 ! *Lely* 61 ! *W. D. MacGregor* 422 ! Gindiri, Plateau Prov. (Oct.) *Hepper* 1089 ! Distributed throughout Africa and in Mauritius and the Seychelles ; introduced to S. America and the W. Indies. Several infraspecific groups have been recognised (see Turrill F.T.E.A. Oleac. 19) ; all the tropical West African specimens examined fall within var. *fluminense*.

3. **J. dichotomum** *Vahl* Enum. Pl. 1 : 26 (1804) ; F.T.A. 4, 1 : 9 ; Gilg & Schellenb. in Engl. Bot. Jahrb. 51 : 89 ; Chev. Bot. 398 ; Esta de Sousa in Anais Junta Invest. Col. 4 : 27 ; Turrill F.T.E.A. Oleac. 23 ; Berhaut Fl. Sén. 100, 117. *J. noctiflorum* Afzel. Rem. Guin. 25 (1815). *J. guineense* G. Don Gen. Syst. 4 : 61 (1838). *J. gardeniodorum* Gilg ex Bak. F.T.A. 4, 1 : 8 (1902). *J. nigericum* A. Chev. Bot. 398, name only. *J. dinklagei* of Hoyle Gold Coast Check-list 87. A scrambling shrub or woody climber reaching to 25 ft. or more in trees ; corolla with red tube and pure white lobes, very fragrant.
  Sen.: *Heudelot* 578 ! *Perrottet* 450. **Mali:** R. Bani, San (June) *Chev.* 1082 ! Diafarabé to Sansanding (fl. & fr. Sept.) *Chev.* 2836 ! 2837 ! **Port.G.:** Cacine *Esp. Santo* 2116. Fulacunda to Bubatambom *Esp. Santo* 2224. **Guin.:** *Paroisse* 30 ! Conakry (Feb.) *Chev.* 12094. **S.L.:** *Afzelius* ! *Smeathmann* ! *T. Vogel* 182 ! Heddle's Farm (Oct.) *Lane-Poole* 83 ! Kafogo, Limba (Apr.) *Sc. Elliot* 5485 ! Kumrabai (fr. Dec.) *Thomas* 7067 ! **Lib.:** Monrovia *Dinklage* 3281 ! Congotown (Aug.) *Barker* 1405 ! **Ghana:** Sampa, Ashanti (fl. & fr. Dec.) *Vigne* FH 3486 ! Banda, Wenchi Dist. (fl. & fr. Dec.) *Morton* GC 25127 ! Aburi Scarp *Morton* ! Shai Plains (Jan.) *Johnson* 528 ! Accra (Aug.) *Dalz.* 157 ! **Togo Rep.:** Lome (Nov.) *Mildbr.* 7483 ! *Warnecke* 15 ! 371 ! **Dah.:** Whydah *Don* ! **N.Nig.:** Zaria (Feb.) *Dalz.* 365 ! *Kennedy* 2872 ! Jos Plateau (Jan.) *Lely* P50 ! Ropp (Dec.) *W. D. MacGregor* 393 ! **S.Nig.:** Ilaro *Millen* 160 ! Olokemeji (Feb.) *Foster* 135 ! Igbetti (fr. Feb.) *Keay* FHI 14633 ! **[Br.]Cam.:** Bamenda, 5,700 ft. (Jan.) *Keay* FHI 28382 ! Nguroje, 5,600 ft., Mambila Plateau (Jan.) *Hepper* 1753 ! Vogel Peak, Adamawa (Dec.) *Hepper* 1519 ! Also in Ubangi-Shari, Congo, Angola, Sudan, Uganda, Kenya, Tanganyika, N. Rhodesia and Mozambique. (See Appendix, p. 364.)
  [*Sc. Elliot* 5689 from near Makunde, Sierra Leone cited as this species in ed. 1, is *Didymosalpinx abbeokutae* (Hiern) Keay—see p. 130.]

4. **J. kerstingii** *Gilg & Schellenb.* in Engl. Bot. Jahrb. 51 : 92 (1913). A suffrutex, to 2 ft. high ; flowers white ; in savanna.
  Ghana: Kulpawn R. flats, Bantala (Apr.) *Kitson* 667 ! Bimbila (Mar.) *Hepper & Morton* A3066 ! Tamale to Yendi (Mar.) *Adams* 3868 ! Kumbungu Farm, Tamale (May) *Kitson* ! Yendi (Apr.) *Vigne* FH 1695 ! **Togo Rep.:** Sokobe-Basari (Feb.) *Kersting* 121 ; 557. Endemic to W. Africa but closely allied to the S. African *J. glaucum* (L.f.) Ait.

5. **J. pauciflorum** *Benth.* in Fl. Nigrit. 443 (1849) ; F.T.A. 4, 1 : 6 ; Gilg & Schellenb. in Engl. Bot. Jahrb. 51 : 98 ; Chev. Bot. 398 ; Ester de Sousa l.c. 28 ; Turrill F.T.E.A. Oleac. 28. *J. talbotii* Wernham in Cat. Talb. Nig. Pl. 58 (1913). *J. angustilobum* Gilg & Schellenb. in Engl. Bot. Jahrb. 51 : 95 (1913). *J. calli-*

Fig. 212.—Jasminum pauciflorum *Benth.* (Oleaceae).
A, flower, longitudinal section. B, anther. C, pistil. D, cross-section of ovary. E, fruit.
F, calyx.

anthum Gilg & Schellenb. l.c. 96 (1913). *J. warneckei* Gilg & Schellenb. l.c. 103 (1913). A slender scandent
shrub ; flowers pure white, fragrant ; fruits black.
Port.G.: Catió *Esp. Santo* 2130. **Guin.:** Macenta (May) *Collenette* 16 ! Between R. Kaba and Haut-
Mamou (Apr.) *Chev.* 12760. **S.L.:** *Afzelius* ! Bagroo R. (Apr.) *Mann* 849 ! Kamalu (May) *Thomas* 381 !
Njala (May) *Deighton* 689 ! Nerekoro, Simiria (Apr.) *Glanville* 201 ! Bafodeya, Limba (Apr.) *Sc. Elliot*
5639 ! **Lib.:** Gbanga *Linder* 1155 ! 1172 ! **Iv.C.:** Bampleu to Kouanhoulé (Apr.) *Chev.* 21182. Foot of Mt.
Kouan, Danané (Apr.) *Chev.* 21258. **Ghana:** Cape Coast (fr. July) *Brass* 217 ! *T. Vogel* 10 ! Agogo (Apr.)
Adams 2554 ! Mile 16, Accra to Aburi (Nov.) *Adams* 1920 ! Accra (May) *Dalz.* 5 ! Pamu, Ashanti (May)
Irvine 2524 ! **Togo Rep.:** Lomé *Warnecke* 143 ! 390 ! **N.Nig.:** Nupe *Barter* ! Anara F.R., Zaria Prov. (May)
Keay FHI 22891 ! Jos Plateau (June) *Dent Young* 169 ! *Lely* 251 ! *P364* ! **S.Nig.:** Lagos *Rowland* ! Ibadan
(Apr.) *Brenan* 9610 ! *Meikle* 1469 ! Owo F.R. (May) *Jones* FHI 3587 ! Eket *Talbot* 3262 ! Oban *Talbot*
336 ! [**Br.]Cam.:** Bali, Bamenda (May) *Ujor* FHI 30382 ! Lakom, 6,000 ft. Bamenda (Apr.) *Maitland* 1676 !
Also from Congo, Uganda and Kenya and possibly from other parts of tropical Africa.
    [The S. African *J. streptopus* E. Mey. (1837) is very close, differing in minor characters as pointed out by
Verdoorn (Bothalia 6 : 571 (1956)). However numerous forms described as species from different parts
of Africa are involved and until the group can be revised as a whole the W. African representatives are
best treated as a separate species.]
6. **J. obtusifolium** *Bak.* in Kew Bull. 1895 : 93 ; F.T.A. 4, 1 : 4 ; Chev. Bot. 398. A scandent shrub reaching
to 25 ft. in trees ; flowers white or cream, fragrant.
Mali: Fô (June) *Chev.* 967. **Ghana:** Kologu, Navrongo Dist. (Apr.) *Vigne* FH 3767 ! **Niger:** Fada, Gourma
Dist. (July) *Chev.* 21182 ; 24537. **N.Nig.:** Kawgan, R. Niger *Barter* 3435 ! Sokoto Prov. (Apr., June)
Dalz. 538 ! *Ryan* 38 ! 67 ! Lemme (May) *Lely* 157 ! Mokwa (Apr.) *Keay* FHI 25717 ! Yola (Mar.) *Dalz.*
207 ! **S.Nig.:** Yoruba country *Barter* 1338 !
7. **J. preussii** *Engl. & Knobl.* in Engl. Bot. Jahrb. 17 : 536 (1893) ; F.T.A. 4, 1 : 9. *J. monticola* Gilg &
Schellenb. in Engl. Bot. Jahrb. 51 : 99 (1913). *J. zenkeri* Gilg & Schellenb. l.c. (1913), incl. var. *glabrata.*
A rambling climber or erect shrub, to 3 ft. high ; flowers white.
Ghana: Abra, W. Prov. (Jan.) *Akpabla* 801 ! **S.Nig.:** Omo F.R., Ijebu Ode (June) *Keay* FHI 37054 !
Sapoba (May) *Okafor* FHI 36583 ! Oban *Talbot* 335 ! [**Br.]Cam.:** Cam. Mt., 5,000–6,000 ft. (Mar.-June,
Aug.) *Brenan* 9354 ! *Deistel* 203 ; *Maitland* 39 ! 653 ! Barombi (Apr.) *Preuss* 122 ! Also in Cameroun.
    A small-flowered form, with corolla tube 4 mm., and lobes 6 mm. long, has been described from
Kohoumbo, W. of Toumodi, Ivory Coast, by Roberty (in Bull. I.F.A.N. 15 : 1423 (1953)) as forma
minutiflorum.

Imperfectly known species.

J. dinklagei *Gilg & Schellenb.* in Engl. Bot. Jahrb. 51 : 97 (1913) ; F.W.T.A., ed. 1, 2 : 28. No authentic
material of this species has been seen, but it is suspected that it is closely allied to *J. dichotomum.*
Lib.: White Plains, Monrovia (May) *Dinklage* 2193.

    Besides the above indigenous species, *J. multiflorum* (Burm.f.) Andr. (syn. *J. pubescens* (Retz.) Willd.),
and *J. sambac* (Linn.) Ait., natives of tropical Asia, and *J. volubile* Jacq. (*J. simplicifolium* Auct.), native
of Australia are cultivated in our area.

134. APOCYNACEAE
By H. Huber

    Trees, shrubs or climbers, rarely perennial herbs, with latex. Leaves opposite
or verticillate, rarely alternate, simple, entire ; stipules usually absent. Flowers
hermaphrodite, actinomorphic. Calyx often glandular inside ; lobes 5 or rarely

4, imbricate. Corolla tubular, variously shaped ; lobes contorted-imbricate, very rarely valvate. Stamens 5 or rarely 4, inserted in the tube ; filaments free or rarely united ; anthers often sagittate, free or connivent around the stigma, rarely adherent to the latter, 2-celled, opening lengthwise, connective often produced at the apex ; pollen granular ; disk usually present, annular, cupular or of separate glands. Ovary superior, 1-celled with 2 parietal placentas or 2-celled with the placentas adnate to the septa, or carpels 2 and free or connate only at the base with ventral placentas in each carpel. Style 1, split at the base or entire, thickened and stigmatose below the apex ; ovules 2 or more in each carpel. Fruit entire and indehiscent or of 2 separate carpels, baccate, drupaceous or follicular. Seeds mostly with endosperm and large straight embryo, often winged or appendaged with long silky hairs.

A well-marked family with usually opposite exstipulate entire leaves, latex, contorted gamopetalous corolla without a corona, and superior often apocarpous ovary.

Leaves or leaf-scales spirally arranged ; stem succulent, erect ; fruits dry follicles, the seeds with an apical tuft .. .. **34. Adenium**
Leaves opposite or verticillate (very exceptionally also alternate, but then stems climbing and never fleshy) ; stems not succulent :
*Corolla-lobes overlapping to the left :
  Carpels completely fused ; fruits fleshy :
    Calyx with numerous, or at least with 3, glandular scales on each sepal :
      Leaves with 4–6 pairs of distant lateral nerves ; flowers conspicuous, the corolla-lobes 14–22 mm. long .. .. **8. Vahadenia**
      Leaves with more than 20 pairs of dense lateral nerves ; flowers small, corolla-lobes 2–3 mm. long .. .. **12. Cyclocotyla**
    Calyx without glandular scales (or with 5 solitary scales alternating with the sepals in *Landolphia utilis*) :
      Inflorescences of elongate, branched, terminal panicles :
        Anthers not keeled ; ovary and fruit glabrous ; stipules interpetiolar, early caducous .. .. **7. Dictyophleba**
        Anthers keeled along the back ; ovary and fruit pubescent ; stipules absent
                                                                            9. Ancylobotrys
      Inflorescences contracted, short and often clustered, terminal or axillary :
        Length of the anthers up to ⅓ of the length of the corolla-tube, frequently less :
          Walls of the corolla-tube thickened above the anthers ; anthers not keeled, 0·5 mm. or more long :
            Seeds with albumen .. .. **2. Landolphia**
            Seeds without albumen (use key to *Landolphia* spp., p. 54) **11. Cylindropsis**
          Walls of the corolla-tube not thickened above the anthers :
            Only terminal inflorescences developed .. .. **10. Saba**
            Inflorescences axillary or axillary and terminal :
              Clavuncula (swelling beneath stigma) cylindric, ovate or globose :
                Anthers not keeled, less than 0·5 mm. long inserted above the middle of the corolla-tube .. .. **3. Clitandra**
                Anthers keeled along the back, more than 0·5 mm. long, inserted beneath the middle of the corolla-tube .. .. **4. Orthopichonia**
              Clavuncula (swelling beneath stigma) ring-shaped ; anthers not keeled, 0·8–1 mm. long .. .. **5. Anthoclitandra**
        Length of the anthers almost ½ or more of the length of the corolla-tube ; clavuncula ring-shaped .. .. **6. Aphanostylis**
  Carpels free, or connate only at the very base ; fruits fleshy or dry ;
    Anthers not tailed, fertile to the base :
      Fruits fleshy, indehiscent ; trees or shrubs :
        Disk absent ; leaves opposite, rarely in whorls of 3 ; branches and leaves glabrous ; inflorescences contracted, short and often clustered, terminal or axillary ; corolla-lobes 2–25 mm. long :
          Calyx with glandular scales within ; stigma usually with well-developed apiculus :
            Corolla 2–4 cm. long ; ovary with 70–130 ovules .. .. **13. Picralima**
            Corolla 5–16 mm. long ; ovary with 1–30 ovules .. .. **14. Hunteria**
          Calyx without glandular scales ; apiculus of stigma hardly developed
                                                                            15. Pleiocarpa
        Disk present ; leaves in whorls of 3, 4 or more ; branches and leaves glabrous or puberulous ; flowers in ample, verticillately-branched cymes ; corolla-lobes 1–2 mm. long .. .. **25. Rauvolfia**
      Fruits a pair of dry follicles up to 4 cm. long, opening lengthwise ; seeds glabrous ;

corolla-lobes when fully expanded 1–2 cm. broad ; introduced herbs or under-
shrubs ..   ..   23. **Catharanthus**
Anthers tailed or auriculate, sterile towards the base :
  Corolla with 5-toothed corona-lobes in the throat of the tube ; anthers entirely
    exserted ; fruit a dry follicle ; seeds with a basal tuft ..   32. **Pleioceras**
  Corona absent ; anthers not or only slightly exserted ; seeds glabrous :
    Corolla-tube very short, up to 2 mm. long ; disk absent ; fruit a dry follicle ;
      seeds exarillate   ..   ..   22. **Pycnobotrya**
    Corolla-tube long ; disk frequently present ; fruit fleshy, usually dehiscent ;
      seeds arillate :
      Sepals persistent, free or very shortly connate at base :
        Clavuncula (swelling beneath stigma) circular in cross-section ; flowers rather
          small, corolla-tube 6–9 mm. long ; fruit winged with 3 longitudinal ridges
                16. **Pterotaberna**
        Clavuncula (swelling beneath stigma) ribbed ; fruit not ridged lengthwise :
          Anthers basifixed, topped with a minute mucro   ..   17. **Callichilia**
          Anthers dorsifixed, with a well-developed acumen :
            Disk free ; anthers fixed with a big dorsal callus   ..   18. **Hedranthera**
            Disk indistinct, completely connate with the ovary ; anthers without a dorsal
              callus ..   ..   19. **Tabernaemontana**
      Sepals caducous, more or less connate into a tube, circumscissile at the base
                20. **Voacanga**
*Corolla-lobes overlapping to the right :
  Anthers not tailed, fertile to the base :
    Ovary syncarpous ; fruit fleshy ; branches usually armed with axillary spines
                1. **Carissa**
    Ovary apocarpous ; fruit dry follicles ; branches never armed :
      Leaves whorled, with more than 25 pairs of lateral nerves ; seeds with a tuft of
        hairs at each end   ..   ..   21. **Alstonia**
      Leaves opposite ; seeds with one apical tuft of hairs :
        Corolla-tube up to 9 mm. long ; anthers up to 2 mm. long ; branches and inflor-
          escences mostly puberulous to tomentellous   ..   24. **Holarrhena**
        Corolla-tube more than 10 mm. long ; anthers 5–6 mm. long ; branches and
          inflorescences glabrous   ..   ..   30. **Farquharia**
  Anthers sterile towards the base, caudate or auriculate :
    Corolla with paired corona-appendages between the (frequently long-tailed) corolla-
      lobes ; fruits dry follicles ; seeds produced into a long, plumose apical beak
                27. **Strophanthus**
    Corolla without corona-appendages ; seeds with or without a basal beak :
    Corolla-tube more than 2 cm. long ; fruits fleshy ; seeds arillate   17. **Callichilia**
    Corolla-tube up to 1·5 cm. long, usually less ; fruits dry-follicles ; seeds not arillate :
      Anther-cone exserted from the corolla-tube :
        Flowers in narrow-terminal panicles ; seeds with an apical tuft of hairs
                26. **Isonema**
        Flowers in axillary clusters ; seeds without a tuft of hairs   33. **Malouetia**
      Anther-cone included in the corolla-tube :
        Corolla-lobes folded lengthwise (conduplicated) in bud ; corolla-tube constricted
          markedly in the lower half or ⅓ ; seeds with an apical tuft ; small tree, possibly
          introduced   ..   ..   28. **Mascarenhasia**
        Corolla-lobes not induplicated in bud :
          Trees ; flowers in dense, axillary clusters ; seeds produced into a plumose basal
            beak   ..   ..   31. **Funtumia**
          Climbing shrubs ; inflorescences terminal or axillary, but then elongated or
            branched ; seeds with an apical tuft of hairs :
            Disk absent ; corolla-tube narrowed towards the mouth :
              Interpetiolar stipules present ; no supra-axillary glands on the petiole ;
                anther-tails free   ..   ..   29. **Alafia**
              Interpetiolar stipules absent ; supra-axillary glands on the petiole developed ;
                corolla-tube quite glabrous on the inside ; anther-tails connate with the
                connective   ..   ..   30. **Farquharia**
            Disk present ; corolla-tube wide at the mouth :
              Anther-tails straight, acute ; stigma up to 0·5 mm. long ; corolla-lobes
                2·5–20 mm. long ; follicles narrowly cylindric, 30–50 cm. long, up to 1 cm.
                diam. ..   ..   35. **Baissea**
              Anther-tails curved, obtuse ; stigma 0·5–1 mm. long ; corolla-lobes 2–5 mm.
              long :
                Corolla-tube with 5 corona-scales alternating with the petals in the mouth ;
                  anther-tips glabrous ; follicles 10–25 cm. long, from less than 1 cm. to
                  2 cm. in diam. narrowly cylindrical to spindle-shaped   .. 37. **Oncinotis**
                Corolla-tube without scales in the mouth ; anthers tipped with a minute

plumose mucro ; follicles narrowly spindle-shaped, about 12 cm. long and
1·5–2 cm. in diam. ..   ..   **36. Motandra**

Besides the above, the following have been recorded as cultivated plants in our area : *Allamanda cathartica*
Linn., *A. neriifolia* Hook., *Chonemorpha macrophylla* G. Don, *Ervatamia coronaria* (Jacq.) Stapf, *Nerium oleander*
Linn. (Common oleander), *Plumeria rubra* Linn. (The frangipani), *Thevetia neriifolia* Juss.

1. CARISSA Linn.—F.T.A. 4, 1 : 88. *Nom. cons.*

Spiny shrub up to 5 m. high ; branches glabrous or pubescent with axillary spines
1–3 cm. long ; leaves ovate, usually acute at apex, shortly cuneate or rounded at base,
2–6 cm. long, 2–6 cm. broad, with 3–5 pairs of lateral nerves, glabrous or pubescent ;
petiole short ; inflorescences dense, shortly pedunculate, terminal or axillary cymes ;
calyx 3–5 mm. long, with subulate teeth ; corolla tube 7–20 mm. long, lobes 3–10 mm.
long, always shorter than the tube   ..   ..   *edulis*

C. edulis *Vahl* Symb. Bot. 1 : 22 (1790) ; F.T.A. 4, 1 : 89 ; Aubrév. Fl. For Soud.-Guin. 445, t. 97, 5–6.
C. pubescens A. DC. (1844). *C. edulis* forma *pubescens* (A. DC.) Pichon (1949). *C. edulis* subsp. *continentalis*
Pichon in Mém. Inst. Sci. Madag., sér. B, 2 : 129 (1949). An erect shrub up to 16 ft. high ; flowers fragrant,
white or purplish ; fruits syncarpous berries about ½ in. diam.
Mali: Ouassana *Chev.* 616. Kita *Dubois* 152. **Guin.:** Labé (Mar.) *Chev.* 12405! Fouta Djalon (Mar.)
Schnell 4815! **Iv.C.:** Ouangolo *Aubrév.* 1404! **Ghana:** Accra *T. Vogel* 31! Nsawam (Feb.) *Dalz.* 8289!
Wahaku to Fausi (Apr.) *Kitson* 824! Krobo Plains (Dec.) *Johnson* 510! **Togo Rep.:** Lomé *Warnecke* 266!
Dah.: Cotonou *Chev.* 4497. **N.Nig.:** Kaduna to Zaria (Feb.) *Meikle* 1225! Kogin Kano F.R. (Mar.)
Onwudinjoh FHI 22354! Vom (Feb.) *McClintock* 215! Maiduguri (July) *Ujor* FHI 21947! **S.Nig.:**
A. F. Ross R121! [**Br.**]**Cam.:** Mayo Wombo, Gashaka Dist. (fr. Dec.) *Latilo & Daramola* 28971! Through-
out the drier parts of tropical Africa and Madagascar, to Arabia, India and Indochina. (See Appendix,
p. 367.)

2. LANDOLPHIA P. Beauv. (1804)—F.T.A. 4, 1 : 30 ; Pichon in Mém. I.F.A.N. 35 : 40

(1953). *Nom. cons. Carpodinus* R. Br. ex G. Don (1837)—F.T.A. 4, 1 : 72 ;
F.W.T.A., ed. 1, 2 : 35, mostly.

Ovary glabrous ; cymes axillary and terminal :
   Corolla-tube 2–3 mm. long :
     Young branches glabrous ; leaves acuminate, elliptic to oblong, 5–17 cm. long,
     2–6 cm. broad, with 5–11 pairs of lateral nerves ; corolla-lobes 2–3 mm. long
                                 1. *micrantha*
     Young branches mostly pubescent ; leaves not or hardly acuminate, elliptic to ovate,
     4–13 cm. long, 2–6 cm. broad, with 7–12 pairs of lateral nerves ; corolla-lobes
     1·5–2 mm. long ..   ..   2. *togolana*
   Corolla-tube 13–16 mm. long, lobes 11–13 mm. long ; branches glabrous ; leaves
   elliptic to elliptic-oblong, broadly acuminate, 4–12 cm. long, 2·5–5 cm. broad, with
   9–14 pairs of lateral nerves   ..   ..   3. *macrantha*
Ovary pilose at least towards the apex :
   Inflorescences only terminal ; corolla-lobes pilose outside towards the apex :
     Corolla-tube glabrous in the throat, 3–17 mm. long :
       Corolla-tube 3–5 mm. long, lobes 2–8 mm. long, markedly pubescent on their auricles ;
       young branches glabrous or laxly pubescent ; leaves oblong to elliptic, 6–12 cm.
       long, 2·5–4 cm. broad, with 8–28 pairs of lateral nerves   4a. *parvifolia* var. *johnstonii*
       Corolla-tube 11–17 mm. long, lobes 7–22 mm. long ; young branches pubescent ;
       leaves elliptic or oblong, mostly subcordate at the base, 5–14 cm. long, 2·5–5 cm.
       broad, with 5–11 pairs of lateral nerves   ..   ..   5. *calabarica*
     Corolla-tube pilose in the throat, not exceeding 10 mm. in length :
       Anthers 0·9–1·5 mm. long, inserted distinctly above the middle of the corolla-tube ;
       young branches pubescent or rarely glabrous ; leaves oblong, elliptic or lanceolate,
       5–25 cm. long, 2·5–12 cm. broad, with 6–20 pairs of lateral nerves ; corolla-tube
       4–8 mm. long, lobes 1·7–9 mm. long :
         Pedicels and sepals mostly pubescent; sepals ciliate   6. *owariensis* var. *owariensis*
         Pedicels and sepals glabrous ; sepals not ciliate   ..   6a. *owariensis* var. *leiocalyx*
       Anthers 1·5–1·9 mm. long, inserted at about the middle of the corolla-tube ; young
       branches densely pubescent ; leaves similar to the above, but often smaller ;
       3–11 cm. long, 1·5–5 cm. broad, with 5–12 pairs of lateral nerves ; corolla-tube
       5–9 mm. long, lobes 3·7–7 mm. long   ..   ..   ..   7. *heudelotii*
   Inflorescences axillary or axillary and terminal (but then corolla-lobes glabrous outside):
     Inflorescences terminal and axillary ; corolla-lobes glabrous outside ; leaves not
     gland-dotted beneath :
       Calyx with 5 glandular scales within alternating with the sepals ; young branches
       pubescent ; leaves 3–17 cm. long, 1–9 cm. broad, with 7–11 pairs of lateral nerves ;
       mature flowers unknown   ..   ..   8. *utilis*
       Calyx without glandular scales :
         Corolla-tube 8–13 mm. long, equalling the lobes ; branches glabrous or pubescent ;
         leaves oblong to elliptic, 8–26 cm. long, 3–10 cm. broad, with 10–16 pairs of lateral
         nerves ..   ..   9. *landolphioides*
         Corolla-tube 3–5 mm. long, lobes 1·8–2·5 mm. long ; branches glabrous
                      Cylindropsis parvifolia (see p. 61)

Inflorescences only axillary ; leaves gland-dotted beneath (except in No. 17) :
Leaves gland-dotted beneath ; young branches glabrous or with short hairs (not
more than 1·3 mm. long) :
Corolla-tube glabrous in the throat ; anthers glabrous, less than ¼ of the length of
corolla-tube :
Inflated part of the corolla-tube either glabrous or laxly pubescent outside :
Young branches staying smooth during lignification :
Young branches and leaves quite glabrous ; leaves oblong-elliptic to obovate,
7–25 cm. long, 2·5–9 cm. broad, with 10–20 pairs of lateral nerves ; corolla-
tube 12–22 mm. long, lobes 7–23 mm.   ..   ..   ..   10. *violacea*
Young branches more or less pubescent or pilose ; leaves elliptic, oblong or
ovate, 4–20 cm. long, 1·5–10 cm. broad, with 4–10 pairs of lateral nerves ;
corolla-tube 7–20 mm. long, lobes 5–22 mm. long :
Young branches densely pilose ..   ..   ..   11. *dulcis* var. *dulcis*
Young branches laxly to moderately pilose   ..   11a. *dulcis* var. *barteri*
Young branches becoming rough with small scales during lignification :
Branches and leaves glabrous ; leaves usually cuneate at base, elliptic-oblong,
5–23 cm. long, 1·5–9 cm. broad, with 6–11 pairs of lateral nerves ; net of veins
prominent ; sepals frequently only 4 ; corolla-tube 12–28 mm. long, lobes
8–21 mm. long   ..   ..   12. *congolensis*
Branches glabrous or pubescent ; leaves cordate at base, elliptic-oblong to
broadly lanceolate, 5–24 cm. long, 2–12 cm. broad, with 5–11 pairs of lateral
nerves ; net of veins not prominent ; corolla-tube 14–16 mm. long, lobes
16–20 mm. long ..   ..   13. *membranacea*
Inflated part of the corolla-tube densely pubescent outside :
Branches laxly or irregularly densely pubescent ; leaves opposite, oblong, 9–24 cm.
long, 2·5–7·5 cm. broad, with 8–14 pairs of lateral nerves ; cymes 1–2-flowered ;
sepals 2·6–5 mm. long ; corolla-tube 13–16 mm. long, lobes 13–18 mm. long,
attenuate towards the apex   ..   ..   14. *uniflora*
Branches equally densely pubescent ; leaves opposite or partly alternate or in
whorls of 3, elliptic to obovate, 7·5–32 cm. long, 2·5–18 cm. broad, with
6–19 pairs of lateral nerves ; cymes 3–30-flowered ; sepals 1·2–2·2 mm. long ;
corolla-tube 6–12 mm. long, lobes 2·8–9 mm. long, rounded towards the apex
15. *foretiana*
Corolla-tube pilose in the throat ; anthers pilose on their back about ¼ or ⅓ of the
length of corolla-tube ; branches pubescent ; leaves oblong-elliptic, 4–32 cm.
long, 1·5–12 cm. broad, with 8–17 pairs of lateral nerves ; corolla-tube 5–8 mm.
long, lobes 3–7 mm. long   ..   ..   16. *subrepanda*
Leaves not gland-dotted beneath ; young branches clothed with long hairs
(1–3 mm.) ; leaves elliptic to oblong, 5–20 cm. long, 3·5–10 cm. broad, with
6–10 pairs of lateral nerves, usually hirsute beneath ; corolla-tube 3–4 mm. long,
lobes 2·8–4·5 mm. long   ..   ..   · 17. *hirsuta*

1. **L. micrantha** (*A. Chev.*) *Pichon* in Mém. I.F.A.N. 35 : 84, t. 2, 3–4 (1953). *Clitandra micrantha* A. Chev.
in Mém. Soc. Bot. Fr. 2, 8 : 44 (1908) ; Chev. Bot. 401. *C. elastica* var. *micrantha* (A. Chev.) A. Chev.
ex Dalz., Appendix 369 (1937). *C. leptantha* of F.W.T.A., ed. 1, 2 : 34, partly (*Chev.* 17691), not of
Pierre. Climber up to 60 ft. high ; flowers small, yellow or yellowish-white, in very lax inflorescences.
Lib.: Sinoe Basin (Mar.) *Whyte* 8 ! 15 ! Dukwia R. (Feb.) *Cooper* 235 ! **Iv.C.:** Soubiré to Namouéflo
(Mar.) *Chev.* 17691 ! Noé *Chev.* 19841. Guidéko *Chev.* 16391. Abidjan *Chev.* 15619.
2. **L. togolana** (*Hallier f.*) *Pichon* l.c. 87 (1953). *Cylindropsis togolana* Hallier f. (1900). *Clitandra togolana*
(Hallier f.) Stapf in F.T.A. 4, 1 : (1902) ; F.W.T.A., ed. 1, 2 : 34 (1931). *C. alba* Stapf in F.T.A. 4, 1 :
64 (1902). *C.* " *togoensis* " of Chev. Bot. 402. Climber 15 ft. or more high with tendrils and cream-
coloured or yellow flowers in axillary and terminal inflorescences.
Iv.C.: Sassandra *Chev.* 17938. Abidjan *Chev.* Grand Bassam *Thollon* 167. *Chev.* 17909. Abouabou
Mangenot. **Ghana:** Okroase (May) *Johnson* 753 ! **Togo Rep.:** Lomé *Warnecke* 46 ! 331 ! **Dah.:** Zounou
Poisson 35. Adja-Ouéré *Le Testu* 114. Near Savé *Poisson* 51. **S.Nig.:** Mamu F.R., Lagos (May) *Foster*
2 ! Ogun R. (fr. Nov.) *Jones* FHI 4913 ! Olokemeji F.R. (Mar.) *Keay* FHI 21199 !
3. **L. macrantha** (*K. Schum.*) *Pichon* l.c. 146, t. 4, 3–6 (1953). *Carpodinus macrantha* K. Schum. (1896)—
F.T.A. 4, 1 : 85 ; F.W.T.A., ed. 1, 2 : 36. Glabrous climber with tendrils and white flowers in 1–5
flowered inflorescences.
Guin.: Kouria *Pobéguin* 1530. **S.L.:** *Afzelius* ! Heddle's Farm, Freetown (May) *Lane-Poole* 250 !
Ninia, Talla Hills (fr. Feb.) *Sc. Elliot* 4924 ! Makali (fr. Feb.) *Deighton* 4087 !
4. **L. parvifolia** *K. Schum.* in Engl. Bot. Jahrb. 15 : 409 (1893).
4a. **L. parvifolia** var. **johnstonii** (*A. Chev.*) *Pichon* l.c. 96 (1953). *L. subterranea* var. *johnstonii* A. Chev. in
Mém. Soc. Bot. Fr. 2, 8 : 46 (1908). *L. talbotii* Wernham in Cat. Talb. 60 (1913). Lofty climber with
tendrils ; flowers white or yellow, in many-flowered terminal inflorescences.
S.Nig.: Oban *Talbot* 1038 ! **[Br.]Cam.:** Rio del Rey *Johnston* 1. Also in Cameroun, Gabon, Ubangi-
Shari, the Congos, Cabinda.
5. **L. calabarica** (*Stapf*) *E. A. Bruce* in Kew Bull. 2 : 28 (1947) ; Pichon l.c. 138, t. 4, 1–2. *Carpodinus cala-
baricus* Stapf (1894). *Landolphia bracteata* Dewèvre (1895)—F.T.A. 4, 1 : 41 ; F.W.T.A., ed. 1, 2 : 32 ;
Chev. Bot. 403. Climber with tendrils, up to 22 ft. high ; flowers white in dense or lax terminal cymes.
S.L.: Matotoka (July) *Thomas* 1236 ! Bonjema (fr. Dec.) *Deighton* 6153 ! Wiima (Apr.) *Deighton*
3947 ! **Lib.:** Vahon, Kolahun (fr. Nov.) *Baldwin* 10252. **Iv.C.:** Abidjan *Chev.* 15615. N'Zida forest
Mangenot. **Ghana:** Cape Coast *Brass* ! Ateiku (Mar.) *Vigne* FHI 2001 ! Aburi *Brown* 391 ! Assuantsi
(fr. Jan.) *Baldwin* 14026 ! Sefwi Bekwai (fr. Oct.) *Akpabla* 905 ! **Dah.:** Allada (Mar.) *Chev.* 23406 !
Koussi *Chev.* 23231. Bokoutou *Chev.* 22867. **S.Nig.:** Idanre *Foster* 200 ! Ibadan (fl. & fr. Apr.) *Meikle*
1396 ! Ogbesse F.R., Ondo (July) *Onochie* FHI 33359 ! Calabar (Feb.) *Mann* 2242 !
6. **L. owariensis** *P. Beauv.* var. **owariensis**—Fl. Oware 1, 54 : t. 62 (1804) ; F.T.A. 4, 1 : 49 (1902) ; Chev.
Bot. 406 ; Pichon l.c. 109, t. 2, 12. *L. heudelotii* A. DC. var. *djenge* Stapf (1894), partly (*Sc. Elliot*

4675). *L. stapfiana* Wernham (1913). Climber up to 300 ft. long or small suberect shrub, with only terminal inflorescences ; flowers white to yellow, orange or brownish-yellow.
Guin.: Konkou to Timbo *Chev.* 12478 ! Fouta Djalon (Sept.) *Schnell* 7215 ! Farana *Chev.* 20455. Camaradou *Adam* 3753. **S.L.:** Heddle's Farm, Freetown (Mar.) *Lane-Poole* 442 ! Jigaya (Sept.) *Thomas* 2782 ! Kukuna (fr. Jan.) *Sc. Elliot* 4675 ! Loma Mts. (Sept.) *Jaeger* 1448 ! **Lib.:** Gbanga (Sept.) *Linder* 761 ! Peahtah (Oct.) *Linder* 942 ! Ganta *Baldwin* 9233 ! Greenville *Sim* 3 ! **Iv.C.:** Bingerville (July) *Jolly* 320 ! Prolo (Aug.) *Chev.* 19873 ! Nzi (Aug.) *Chev.* 22270 ! Aboisso *Chev.* 17732. **Ghana:** Bamba, S. Ashanti (Sept.) *Vigne* FH 1366 ! Akropong, Akwapim (Aug.) *Irvine* 765 ! Aburi Hills (Nov.) *Johnson* 282 ! Krepi Hills (Jan.) *Johnson* 524 ! Larte Hills (Nov.) *Johnson* 811 ! **Togo Rep.:** Misahöhe *Baumann* 517. Bismarckburg *Büttner* 325. **Dah.:** Adja-Ouéré *Le Testu* 170 ! Zagnanado (Apr.) *Rudier* ! Sakété to Pedjilé *Chev.* 22909 ! **N.Nig.:** Lokoja (Nov.) *Dalz.* 100 ! Omu, Ilorin *Lamb* 5 ! Bassa (fr. Aug.) *Elliott* 102 ! Agaie (Dec.) *Yates* 46 ! **S.Nig.:** Okomu F.R., Benin (Feb.) *Brenan* 9082 ! Ibadan South F.R. (Apr.) *Keay* FHI 22807 ! Onitsha *Barter* 1810 ! Milliken Hill, Enugu (Mar.) *Hepper* 2226 ! Calabar (Feb.) *Mann* 2311 ! Oban *Talbot* 1617 ! **[Br.]Cam.:** Rio del Rey *Johnston* 3. Johann-Albrechtshöhe *Staudt* 883. Vogel Peak, Adamawa (Dec.) *Hepper* 1517 ! Extends through Cameroun and equatorial Africa to Egypt, Sudan, Uganda and Southern Tanganyika. (See Appendix, p. 374.)

FIG. 213.—LANDOLPHIA OWARIENSIS *P. Beauv.* (APOCYNACEAE).
A, flower. B, same with corolla opened. C, anthers. D, fruit.

6a. **L. owariensis** var. **leiocalyx** (*Pichon*) H. *Huber* in Kew Bull. 15 : 437 (1962). *L. leiocalyx* Pichon l.c. 129, t. 3, 2–4 (1953).
    Guin.: Fouta Djalon *Pobéguin* 1929 !
7. **L. heudelotii** *A. DC.* in DC. Prod. 8 : 320 (1844) ; F.T.A. 4, 1 : 53 ; Chev. Bot. 405 ; Pichon l.c. 129. *L. heudelotii* var. *djenge* Stapf (1894)—F.T.A. 4, 1 : 55 (partly, *Sc. Elliot* 4650, 5450). Climber, sometimes very long, with mostly white flowers in terminal, dense or lax cymes.
    Sen.: M'bidjem to Cayor *Trochain* 644. Sangalkam *Berhaut* 894. Bignona *Trochain* 1457. Itou *Chev.* 2073. **Gam.:** *Dawe* 8 ! Kombo *Heudelot.* **Mali:** Guélia (Feb.) *Chev.* 317 ! **Port. G.:** Prabis, Bissau (Feb.) *Esp. Santo* 1831 ! Naga, Bissora (Jan.) *d'Orey* 89 ! **Guin.:** Rio Nunez *Heudelot* 602 ! Conakry (Feb.) *Chev.* 12726 ! Farana (Jan.) *Chev.* 20459 ! Timbo (Jan.) *Pobéguin* 84 ! **S.L.:** Freetown (May) *Deighton* 3989 ! Wallia, R. Scarcies (fl. & fr. Jan.) *Sc. Elliot* 4650 ! Mano (Sept.) *Deighton* 3789 ! Falaba (fr. Apr.) *Sc. Elliot* 5450 ! **Iv.C.:** Tafiré *Aubrév.* 1455. **Ghana:** Lambussie to Belong (fr. Mar.) *Kitson* 785 ! Also on the Cape Verde Islands. (See Appendix, p. 374.)
    L. utilis (*A. Chev.*) Pichon l.c. 200, t. 9, 5–6 (1953). *Carpodinus utilis* A. Chev. in J. Agric. Trop. 6 : 159 (1906)—Chev. Bot. 411. *Clitandra laurifolia* A. Chev. (1908) and Chev. Bot. 401, name only—F.W.T.A., ed. 1, 2 : 35. Climber 6–10 ft. high with 1–4-flowered cymes.
    Iv.C.: Prolo *Chev.* 19856. Grabo *Chev.* 19739 ; 19767. Davo *Chev.* 17966. Bingerville to Abidjan *Chev.* 15549 ; 15552. Dabou *Chev.* 15617.
    [Note : I have not seen any material of this imperfectly known and aberrant species which is the only species with glandular scales in the calyx—Huber.]
9. **L. landolphioides** (*Hallier f.*) *A. Chev.* in Rev. Bot. Appliq. 28 : 401 (1948) ; Pichon l.c. 53, t. 1, 10. *Clitandra landolphioides* Hallier f. (1900). *Carpodinus landolphioides* (Hallier f.) Stapf in F.T.A. 4, 1 : 80 (1902) ; F.W.T.A., ed. 1, 2 : 36. Climber up to 80 ft. long with tendrils and dense or lax axillary and terminal inflorescences ; corolla white, with the tube sometimes red at the base.
    S.Nig.: Cross R. area *Meyer.* **[Br.]Cam.:** Buea, 1,900–4,000 ft. *Dalz.* 8242 ! *Hutch. & Metcalfe* 108 ! *Deistel* 144 ! 556 ! Barombi *Preuss* 217. Extends to Cameroun, Gabon, Ubangi-Shari and Uganda.
10. **L. violacea** (*K. Schum. ex Hallier f.*) Pichon l.c. 157, t. 5, 7 (1953). *Carpodinus violacea* K. Schum. ex Hallier f. (1900)—F.T.A. 4, 2 : 80 ; F.W.T.A., ed. 1, 2 : 36. Glabrous climber with white, pink or violet flowers in few-flowered, dense axillary cymes.
    S.Nig.: Oban *Talbot* 1445 ! Also in Cameroun, Gabon, and the Congos.
11. **L. dulcis** (*R. Br. ex Sabine*) Pichon var. **dulcis**—l.c. 169, t. 6, 6 (1953). *Carpodinus dulcis* Sabine ex G. Don Gen. Syst. 4 : 101 (1837) ; F.T.A. 4, 1 : 76 ; F.W.T.A., ed. 1, 2 : 36 ; Chev. Bot. 410. *Carpodinus acida* Sabine ex G. Don (1837)—F.T.A. 4, 1 : 86. Climber up to 30 ft. with tendrils and white, yellowish or violet flowers in few-flowered axillary cymes.
    Sen.: *fide* Pichon *l.c.* **Mali:** Kangola *Chev.* 860 ! Sitakoto *Paroisse* 31. **Port.G.:** Prabis, Bissau (fr. Feb.) *Esp. Santo* 1827 ! Mansoa *Esp. Santo* 742. Cacim *Esp. Santo* 631. **Guin.:** N'gebela (May) *Collenette* 32 ! Kouroussa (June) *Pobéguin* 703 ! Diaguissa to Timbo *Chev.* 13433. Youkounkoun *Berhaut* 1472.

S.L.: Roruks (fr. Nov.) *Thomas* 5712! Batkanu (fr. Jan.) *Deighton* 2842! Njala (fr. Apr.) *Deighton* 2899! Musaia (Apr.) *Deighton* 5483! **Iv.C.:** Korhogo to Badiakaha *Mangenot*. Also in Cape Verde Islands.

11a. **L. dulcis** var. **barteri** (*Stapf*) *Pichon* l.c. 166, t. 6, 4–5 (1953). *Carpodinus barteri* Stapf (1894)—F.T.A. 4, 1 : 77 ; F.W.T.A., ed. 1, 2 : 36. *C. parviflora* Stapf (1894)—F.T.A. 4, 1 : 78. *C. oocarpa* Stapf in F.T.A. 4, 1 : 598 (1904) ; F.W.T.A., ed. 1, 2 : 36 ; Chev. Bot. 411. *C. baumannii* Hutch. & Dalz. F.W.T.A., ed. 1, 2 : 36 (1931), and in Kew Bull. 1937 : 337.
Guin.: Fouta Djalon *Heudelot* 624! Kissidougou *Chev.* 20704. Ténémadou *Adam* 3391. Nzérékoré *Adam* 4887. Lola to Nzo *Chev.* 20982. **Lib.:** Ganta *Harley*! Sinoe Basin *Whyte* 3! 9! Kakatown *Whyte*! **Iv.C.:** Sassandra *Chev.* 16330. Guidéko *Chev.* 16386! Dabou *Chev.* 15621. Alépé *Chev.* 17374. **Ghana:** Kumasi (May) *Jackson* FH 1935! Axim *Chev.* 13817. Aburi *Chev.* 13873. **Togo Rep.:** Misahöhe (Mar.) *Mildbr.* 7392! *Baumann* 304! R. Djégo *Kling* 51! **Dah.:** Pahou *Poisson*. Tohoué *Chev.* 22770. **S.Nig.:** Lagos *Barter* 2138! Idanre Hills (Oct.) *Keay* FHI 25505! Ibadan (fl. & fr. Mar.) *Meikle* 1252! Calabar (Feb.) *Mann* 2261! Oban *Talbot* 297! Also in Cameroun, Gabon and Middle Congo.

12. **L. congolensis** (*Stapf*) *Pichon* l.c. 171 (1953). *Carpodinus congolensis* Stapf (1898)—F.T.A. 4, 1 : 76. *C. turbinata* Stapf (1898)—F.T.A. 4, 1 : 83. *C. schlechteri* K. Schum. ex Stapf in F.T.A. 4, 1 : 75 (1902). *C. oxyanthoides* Wernham in Cat. Talb. 61 (1913). Glabrous climber 30 ft. or more high with tendrils ; flowers variable in colour, in mostly 1-flowered axillary inflorescences.
S.Nig.: Oban Dist. *Talbot* 1443! Also in Cameroun, Gabon, Ubangi-Shari, the Congos, and Cabinda.

13. **L. membranacea** (*Stapf*) *Pichon* l.c. 176, t. 7, 2–3 (1953). *Clitandra membranacea* Stapf in F.T.A. 4, 1 : 597 (1904) ; F.W.T.A., ed. 1, 2 : 34. Climber with tendrils and greenish-white flowers mostly solitary in leaf-axils ; sometimes flowering from older, leafless branches.
Guin.: Moussadougou to Lola *Chev.* 20969. **S.L.:** Njala (fr. Dec.) *Deighton* 2932! Yonibana *Thomas* 4087! Mamaka *Thomas* 4576! **Lib.:** Jabroke (fr. July) *Baldwin* 1947! Sinoe Basin *Johnston* 19! Greenville *Sim* 20! **Iv.C.:** Grabo *Chev.* 19664! Kéeta *Chev.* 19329. Nimba to Danané *Mangenot*. Bingerville *Chev.* 16049.

14. **L. uniflora** (*Stapf*) *Pichon* l.c. 178 (1953). *Carpodinus uniflora* Stapf (1894) — F.T.A. 4, 1 : 74 ; F.W.T.A., ed. 1, 2 : 36. Climber with tendrils and lemon-yellow flowers.
S. Nig.: Oban Dist. *Talbot* 298! 1480! Also in Cameroun and Gabon.

15. **L. foretiana** (*Pierre ex Jumelle*) *Pichon* l.c. 179, t. 7, 6 (1953). *Carpodinus foretiana* Pierre ex Jumelle (1898). *C. klainei* Pierre ex Stapf in F.T.A. 4, 1 : 79 (1902) ; Chev. Bot. 410. *C. rufinervis* Pierre ex Stapf in F.T.A. 4, 1 : 79 (1902) ; F.W.T.A., ed. 1, 2 : 36. Climber with tendrils, up to 240 ft. long with white or yellow flowers.
Lib.: Greenville, Sinoe *Sim* 31! **Iv.C.:** Guidéko to Zozro *Chev.* 19019. **S.Nig.:** Sapoba *Thompson* 19! Akwebe *Farquhar* 75! Oban *Talbot* 299! 2030! Extends through Cameroun and Equatorial Africa to Congo and Cabinda.

16. **L. subrepanda** (*K. Schum.*) *Pichon* l.c. 184, t. 7, 7–8 (1953). *Carpodinus subrepanda* K. Schum. (1895)—F.T.A. 4, 1 : 81 ; F.W.T.A., ed. 1, 2 : 36. Climber up to 20 ft. high with tendrils and white flowers in dense, axillary inflorescences.
S.Nig.: Cross R. *McLeod*! [Br.]**Cam.:** Barombi *Preuss* 187! S. Bakosi, Kumba (fr. May) *Olorunfemi* FHI 30570! Also in Cameroun, Gabon and the Congos.

17. **L. hirsuta** (*Hua*) *Pichon* l.c. 193, t. 9, 1–4 (1953). *Carpodinus hirsuta* Hua (1900)—F.W.T.A., ed. 1, 2 : 36, partly (excl. syn. *C. nigerina* A. Chev. and Chev. specs.) ; Chev. Bot. 410. Climber up to 60 ft. high with tendrils and small, yellowish or greenish-yellow flowers, in dense, often many-flowered cymes.
Sen.: Bignona *Chev.* Marsassoum *Portères* 3012. Sinedone *Harcens*. Adéane *Chev.* 2068. **Port.G.:** Fulacunda (fr. Apr.) *Esp. Santo* 2254! Cabuchanque to Guébu *Esp. Santo* 2072! **Guin.:** Conakry *Lecerf*! Kandiafara *Paroisse* 10 (partly). Bambaya *Paroisse* 8. Timbo *Chev.* 12432. **S.L.:** Newton (Feb.) *T. S. Jones* 309! Pelewahun (Jan.) *Deighton* 5736a! Mange (Feb.) *Deighton* 3608! Njala (fl. Feb., fr. May) *Deighton* 3111! 5046! **Lib.:** Kondessu, Boporo (Dec.) *Baldwin* 10666! **Iv.C.:** Alépé to Malamalasso *Chev.* 17515! **Ghana:** Akwapim *Horton*! Fiapere *Chipp* 83! Aburi *Chev.* 13876. **Togo Rep.:** Tonugbe *Baumann* 520. **Dah.:** Djougou *Chev.* 23884 ; 23922. **N.Nig.:** Ankpa, Kabba Prov. *Lamb* 18! **S.Nig.:** Asaba Div. *Unwin* 4! N. Ibadan F.R. *Mackay*! [Br.]**Cam.:** S. Bakundu F.R., Kumba (Mar.) *Brenan* 9431!

3. CLITANDRA Benth.—F.T.A. 4, 1 : 60 ; Pichon in Mém. I.F.A.N. 35 : 205 (1953).

Robust climber with branches and leaves glabrous ; leaves elliptic to oblong, 4·5–20 cm. long, 1·6–9 cm. broad, with 9–21 pairs of lateral nerves ; cymes many-flowered, axillary, dense to lax ; corolla-tube 2·5–7 mm. long, lobes 1·3–6 mm. long *cymulosa*

C. cymulosa Benth. in Fl. Nigrit. 445 (1849) ; F.T.A. 4, 1 : 65 ; Pichon l.c., t. 9, 12. *C. orientalis* K. Schum. (1895)—Chev. Bot. 401. *C. elastica* A. Chev. (1906). Climber up to 300 ft. long, with tendrils and fairly small white, greyish or yellowish flowers, the tube mostly pink.
Guin.: Bouillé to Bangadou (fl. & fr. Feb.) *Chev.* 20728! Korodou (Feb.) *Chev.* 20726! Kaba *Chev.* 13241. Voroa *Adam* 3298. Moussadougou to Lola *Chev.* 20962. **S.L.:** *Don*! Jepihun (fr. Dec.) *Smythe* 230! Kenema *Thomas* 7893! Panguma *Imp. Inst.* 5! **Lib.:** Peahtah (fr. Oct.) *Linder* 900! **Iv.C.:** Fort Binger *Chev.* 19524! **Ghana:** ? locality! **Dah.:** Boukoutou *Chev.* 22866! Sakété to Pédjilé *Chev.* 22892. **N.Nig.:** Bassa (fr. Aug.) *Elliott* 98! **N.Nig.:** Sapoba *Thompson* 15! Ondo (Nov.) *Onochie* FHI 34238! Degema *Talbot* 1657! Also in Gabon, Ubangi-Shari, the Congos, Cabinda, Uganda and Tanganyika.

4. ORTHOPICHONIA H. Huber in Kew Bull. 15 : 437 (1962). *Orthandra* Pichon in Mém. I.F.A.N. 35 : 211 (1953), not of Burret (1940). *Clitandra* of F.T.A., 4, 1 : 60, and of F.W.T.A., ed. 1, 2 : 33, partly.

Corolla-lobes attenuate towards the apex :
Branches and petioles quite glabrous ; corolla-tube 11–16 mm. long, the anthers inserted in the lower ⅓ of the tube ; sepals glabrous outside :
Corolla-lobes 4–5 mm. long, always less than ⅓ length of the tube and frequently ¼ length, pubescent outside towards base, tube 11–15 mm. long ; leaves elliptic to oblong, 8–16 cm. long, 3–8 cm. broad, with 16–35 pairs of lateral nerves
1. *indeniensis*
Corolla-lobes 6–9 mm. long, about half as long as the tube ; leaves oblong to broadly lanceolate, cuneate or almost rounded at base, 5–13 cm. long, 2–6 cm. broad, with 14–35 pairs of lateral nerves ..   .. 2. *longituba*
Corolla-lobes 4–6 mm. long, at least ⅔ length and up to about twice as long as the tube ; leaves broadly obovate, elliptic, oblong or almost orbicular, rounded or indistinctly cuneate at base, 4–16 cm. long, 2–7 cm. broad, with 16–32 pairs of lateral nerves ..   .. 3. *schweinfurthii*

Branches and petioles puberulous when young ; corolla-tube 4–7 mm., lobes 4·5–8 mm. long ; anthers inserted in the middle of the tube ; leaves elliptic to oblong, 8–16 cm. long and 3–8 cm. broad, with 16–35 pairs of lateral nerves ; sepals pubescent at least towards the base ..   ..   4. *staudtii*

Corolla-lobes rounded towards the apex :

Corolla-tube 3–5 mm. long, anthers inserted about 1 mm. above the base of the tube ; corolla-lobes 1·5–2·7 mm. long ; branches glabrous also when young ; leaves elliptic, 6–16 cm. long, 3–7 cm. broad, with 28–44 pairs of lateral nerves   ..   5. *barteri*

Corolla-tube 7–15 mm. long, anthers inserted 2–3 mm. above the base of the tube ; corolla-lobes 3–9 mm. long ; ovary about 1 mm. high, attenuate at the apex ; leaves elliptic, long-acuminate, 4–10 cm. long, 2–5 cm. broad, with 25–45 pairs of lateral nerves, glabrous ..   ..   6. *nigeriana*

1. **O. indeniensis** (*A. Chev.*) *H. Huber* in Kew Bull. 15 : 437 (1962). *Clitandra indeniensis* A. Chev. in Mém. Soc. Bot. Fr. 2, 8 : 43 (1908). *Clitandra ivorensis* A. Chev. ex Hutch. & Dalz. F.W.T.A., ed. 1, 2 : 34 (1931) ; Kew Bull. 1937 : 336. *Orthandra indeniensis* (A Chev.) Pichon in Mém. I.F.A.N. 35 : 218 (1953). Climber with tendrils 15–50 ft. high ; corolla white with tube yellow.
   Lib.: Greenville, Sinoe *Sim* 14 ! Sinoe (fr. Mar.) *Whyte* 4 ! **Iv.C.:** Mid. Cavally to Mid. Sassandra *Chev.* 19276. Abidjan *Chev.* 15183 ! 15616. Yapo *Mangenot.* Aboisso *Chev.* 17751.
2. **O. longituba** (*Wernham*) *H. Huber* l.c. (1962). *Clitandra longituba* Wernham in Cat. Talb. 60 (1913) ; F.W.T.A., ed. 1, 2 : 34. *Orthandra longituba* (Wernham) Pichon l.c. 216, t. 10, 1–4 (1953). Climber with tendrils ; flowers in axillary fascicles.
   S.Nig.: Sapoba *Kennedy* 2156 ! Eket *Talbot* ! Oban *Talbot* 1577 ! **[Br.]Cam.:** Victoria (Feb.) *Maitland* 405 ! S. Bakundu F.R., Kumba (Mar.) *Brenan* 9498 !
3. **O. schweinfurthii** (*Stapf*) *H. Huber* l.c. (1962). *Clitandra schweinfurthii* Stapf in Kew Bull. 1894 : 20 ; F.T.A. 4, 1 : 68. *Clitandra visciflua* K. Schum. ex Hallier f. (1900)—F.T.A. 4, 1 : 66 ; F.W.T.A., ed. 1, 2 : 34. *Orthandra schweinfurthii* (Stapf) Pichon l.c. 213 (1953). Climber with the flowers in axillary clusters ; corolla white or with a yellowish tube.
   S.Nig.: Benin *Unwin* 51 ! Sapoba F.R. (fr. June) *Onochie* FHI 36652 ! Also in Cameroun, Ubangi-Shari, Gabon, Congo and Sudan.
4. **O. staudtii** (*Stapf*) *H. Huber* l.c. (1962). *Clitandra staudtii* Stapf in F.T.A. 4, 1 : 67 (1902) ; F.W.T.A., ed. 1, 2 : 34. *Orthandra staudtii* (Stapf) Pichon l.c. 223, t. 10, 5–8 (1953). Climber with tendrils and white flowers in axillary fascicles.
   [Br.]Cam.: Johann-Albrechtshöhe (=Kumba) (Feb.) *Staudt* 860 ! Also in Cameroun, Gabon and Congo.
5. **O. barteri** (*Stapf*) *H. Huber* l.c. (1962). *Clitandra barteri* Stapf in Kew Bull. 1894 : 20 ; F.T.A. 4, 1 : 66; F.W.T.A., ed. 1, 2 : 34. *Orthandra barteri* (Stapf) Pichon l.c. 223, t. 10, 12, t. 11, 1–3 (1953). Climber, 12 ft. high, similar to the above.
   Ghana: Dome (fr. May) *Thompson* 76 ! **S.Nig.:** Epe *Barter* 3310 ! Sapoba *Kennedy* 2322 ! *Thompson* 21 ! Asaba *Unwin* ! Anwai R. *Unwin* 15 !
6. **O. nigeriana** (*Pichon*) *H. Huber* l.c. (1962). *Orthandra nigeriana* Pichon l.c. 226, t. 11, 7–9 (1953). Climber with tendrils and glabrous branches ; flowers densely clustered, white.
   S.Nig.: Oban *Talbot* 1351 ! 1353 ! 1537 ! **[Br.]Cam.:** Balange, Kumba (Jan.) *Binuyo & Daramola* FHI 35089 !

5. ANTHOCLITANDRA (Pierre) Pichon in Mém. I.F.A.N. 35 : 230 (1953). *Clitandra* of F.T.A. 4, 1 : 60, and of F.W.T.A., ed. 1, 2 : 33, partly.

Lateral nerves 11–15 pairs ; leaves broadly elliptic to obovate, 5–10 cm. long, 2·5–6 cm. broad ; pedicels up to 3 or 4 mm. long ; corolla-tube about 3 mm. long, pubescent in the throat, lobes 4–6 mm. long ; ovary glabrous   ..   ..   1. *nitida*

Lateral nerves 4–10 pairs ; leaves oblong, 7–21 cm. long, 2·5–8 cm. broad ; pedicels very short, not exceeding 1·5 mm. ; corolla-tube 4–6 mm. long, glabrous in the throat, lobes 3–10 mm. long ; ovary pubescent on the apex ..   ..   ..   2. *robusta*

1. **A. nitida** (*Stapf*) *Pichon* in Mém. I.F.A.N. 35 : 230, t. 12, 2 (1953). *Clitandra nitida* Stapf in F.T.A. 4, 1 : 595 (1904) ; F.W.T.A., ed. 1, 2 : 34. Glabrous climber with tendrils and very lax axillary cymes.
   Lib.: Kakatown *Whyte* ! Monrovia (fr. Oct., Nov.) *Cooper* 43 ! Sinoe *Whyte* 1 ! Greenville, Sinoe *Sim* 24 ! Mecca, Boporo Dist. (fr. Nov.) *Baldwin* 10411 ! **Iv.C.:** Kééta *Chev.* 19338.
2. **A. robustior** (*K. Schum.*) *Pichon* l.c. 231, t. 12, 3–5 (1953). *Clitandra robustior* K. Schum. in E. & P. Pflanzenfam. 4, 2 : 130 (1895) ; F.T.A. 4, 1 : 71 ; F.W.T.A., ed. 1, 2 : 34. Climber up to 60 ft. high with very dense, axillary inflorescences ; corolla very variable in colour : white, yellow, ochre, purple or violet.
   S.Nig.: Afi River F.R., Ikom (Dec.) *Keay* FHI 28218 ! 28266 ! **[Br.]Cam.:** Kumba to Ikiliwindi *Preuss* 390. S. Bakundu F.R., Kumba (Jan.) *Binuyo & Daramola* FHI 35095 ! Also in Cameroun, Gabon, the Congos and Cabinda.

6. APHANOSTYLIS Pierre in Bull. Soc. Linn. Paris, n. sér., 89 (1898) ; Pichon in Mém. I.F.A.N. 35 : 235 (1953). *Clitandra* of F.T.A. 4, 1 : 60, and of F.W.T.A., ed. 1, 2 : 33, partly.

Lateral nerves 7–10 pairs ; only axillary inflorescences developed :

Cymes very dense ; pedicels 1–3 mm. long ; corolla-tube 1·3–1·8 mm. long, the anthers inserted in the lower $\frac{1}{3}$; corolla-lobes about 5 mm. long ; branches glabrous or pubescent, leaves oblong, 5–18 cm. long, 2·5–6 cm. broad ..   ..   1. *leptantha*

Cymes lax ; pedicels 4–7 mm. long ; corolla-tube 2·5–2·7 mm. long, the anthers inserted at about the middle ; corolla-lobes about 10 mm. long ; leaves elliptic to oblong, 8–14 cm. long, 3–7 cm. broad   ..   ..   2. *flavidiflora*

Lateral nerves in 11–20 pairs ; axillary and terminal inflorescences developed ; corolla-tube 0·8–2 mm. long, the anthers inserted above the middle ; corolla-lobes 3–10 mm. long ; branches glabrous ; leaves elliptic-oblong, 4–11 cm. long, 2–5 cm. broad

                                                                     3. *mannii*

1. **A. leptantha** (*K. Schum.*) *Pierre* in Bull. Soc. Linn. Paris, n. sér., 89 (1898) ; Pichon in Mém. I.F.A.N. 35 : 237, t. 12, 6–7. *Carpodinus leptantha* K. Schum. (1895). *Clitandra leptantha* (K. Schum.) Hallier f. (1900)

—F.T.A. 4, 1 : 70 ; F.W.T.A., ed. 1, 2 : 34 (excl. syn. *C. micrantha* A. Chev. and *Chev.* 17091). Climbing shrub 12–25 ft. high with tendrils and small white flowers in dense axillary cymes.

   S.Nig.: Oban *Talbot* 1526 ! 2047 ! Eket *Talbot* 3190 ! Also in Cameroun and Gabon.

2. **A. flavidiflora** (*K. Schum.*) Pierre l.c. 90 (1898) ; Pichon l.c. 239, t. 13, 1–2. *Carpodinus flavidiflora* K. Schum. (1896). *Clitandra flavidiflora* (K. Schum.) Hallier f. (1900)—F.T.A. 4, 1 : 70 ; Brenan in Kew Bull. 7 : 450 (1952). A large climber very similar to *A. leptantha* but with laxer cymes.

   [Br.]Cam.: near Banga, Kumba Dist. (Mar.) *Brenan* 9322 ! Also in Cameroun.

3. **A. mannii** (*Stapf*) Pierre l.c. 89 (1898) ; Pichon l.c. 241, t. 13, 3–5. *Clitandra mannii* Stapf (1894)—F.T.A. 4, 1 : 69 ; F.W.T.A. ed. 1, 2 : 34 ; Chev. Bot. 401. Climbing shrub, up to 100 ft. long, with tendrils ; flowers white or yellowish in lax inflorescences.

   Port.G.: Guébu, Catió (fl. July, fr. June) *Esp. Santo* 2080 ! 2136 ! Cacine *Esp. Santo* 692. **Guin.**: Bayan-Bayan *Sc. Elliot* 4547. **S.L.**: *Afzelius. Dudgeon* 1b ! Bagroo R. *Mann* 848 ! Mano (Mar.) *Deighton* 5324 ! Hierakohun *Smythe* 79 ! **Lib.**: Du R. (fr. July) *Linder* 89 ! **Iv.C.**: Bettié *Chev.* 17578 ! Mid. Cavally to Sassandra *Chev.* 19233. Guidéko to Zozro *Chev.* 19037. Nzi *Chev.* 22272. Ahinta to Aboisso *Chev.* 17761. **Togo Rep.**: Misahöhe *Baumann* 514. Katchenki *Kling* 7. **Dah.**: Bokoutou *Chev.* 22873. Sakété to Pédjilé *Chev.* 22910. **S.Nig.**: Ilaro (May) *Onochie* FHI 8233 ! Also in Cameroun, Gabon, Ubangi-Shari, the Congos and Cabinda.

7. DICTYOPHLEBA Pierre in Bull. Soc. Linn. Paris, n. sér. : 92 (1898) ; Pichon in Mém. I.F.A.N. 35 : 250 (1953). *Landolphia* of F.T.A. 4, 1 : 30, and of F.W.T.A., ed 1, 2 : 31, partly.

Anthers inserted about or below the middle of the corolla-tube ; ovary abruptly distinct from the style ; axis of the panicle robust, usually 2–7 mm. diam. near the base :
  Stipules small, markedly caducous, entire, deltoid ; corolla-lobes shorter or longer than the tube :
    Corolla-tube 5–13 mm. long, anthers inserted at about the middle of the tube ; corolla-lobes 4–18 mm. long, longer than ½ of the tube and sometimes longer than the tube ; branches when young more or less puberulous ; leaves broadly elliptic, cordate at base, 8–44 cm. long, 3–21 cm. broad with 7–24 pairs of lateral nerves   1. *ochracea*
    Corolla-tube 11–28 mm. long, anthers inserted in the lower ⅓ of the tube :
      Corolla-lobes 5–10 mm. long, less than ½ length of the tube ; branches pilose when young ; leaves broadly elliptic to obovate, 7–22 cm. long, 5–12 cm. broad, with 5–11 pairs of lateral nerves .. ..   2. *leonensis*
      Corolla-lobes 30–41 mm. long, longer than the tube ; branches glabrous ; leaves ovate, shortly and abruptly acuminate, 10–17 cm. long, 5·5–8·5 cm. broad, with 8–10 pairs of lateral nerves ..   ..   3. *rudens*
  Stipules persistent, lacerate in setiform lobes 2–5 cm. long ; corolla-tube 7–10 mm. long, anthers inserted in the lower ¼ or ⅓ of the tube ; corolla-lobes 14–22 mm. long, always distinctly longer than the tube ; branches laxly pilose when young ; leaves oblong-elliptic, long-acuminate 14–25 cm. long, 6–12 cm. broad, with 8–12 pairs of lateral nerves ..   ..   4. *stipulosa*
Anthers inserted above the middle ; corolla-tube 8–16 mm. long, lobes 5–15 mm. long, usually shorter than the tube ; ovary long-attenuate, gradually passing into the style ; branches glabrous or pilose ; stipules small, deltoid, entire ; leaves elliptic to oblong, 7–18 cm. long, 3–10 cm. broad, with 6–13 pairs of lateral nerves ; axis of the panicle slender, usually less than 2 mm. diam. near the base ..   5. *lucida*

1. **D. ochracea** (*K. Schum. ex Hallier f.*) Pichon in Mém. I.F.A.N. 35 : 251 (1953). *Landolphia ochracea* K. Schum. ex Hallier f. in Jahrb. Hamb. Wiss. Anst. 17, 3, Beih. 86 (1900) ; F.T.A. 4, 1 : 40 ; F.W.T.A., ed. 1, 2 : 33. Climbing shrub up to 130 ft. high, with tendrils ; cymes dense or lax, arranged in terminal panicles ; corolla white, calyx violet.
   S.Nig.: Oban *Talbot* 292 ! Also in Cameroun, Gabon, Ubangi-Shari, Congo and Cabinda.
2. **D. leonensis** (*Stapf*) Pichon l.c. 256, t. 14, 1–4 (1953). *Landolphia leonensis* Stapf in F.T.A. 4, 1 : 36 (1902) ; F.W.T.A., ed. 1, 2 : 33. *Carpodinus macrophyllus* A. Chev. ex Hutch. & Dalz. F.W.T.A., ed. 1, 2 : 35 (1931) ; Kew Bull. 1937 : 337. Very similar to the last.
   Guin.: Tembikounda *Chev.* 20588. **S.L.**: *Smythe* 42 ! *Barter* ! Heddle's Farm, Freetown (May) *Lane-Poole* 268 ! **Lib.**: Dukwia R. (May) *Cooper* 423 ! Sinoe *Whyte* s.n. ! 2 ! Greenville *Sim* 31 ; 35. **Iv.C.**: Jolly. Guidéko *Chev.* 16369 ; 16405. **Ghana**: *Dudgeon* 108a.
3. **D. rudens** Hepper in Kew Bull. 16 : 451 with fig. (1963). A lofty climber on trees in streamside forest in savanna ; flowers white, fragrant.
   [Br.]Cam.: foot of Vogel Peak, Adamawa (Nov.) *Hepper* 1470 !
   [This species has been inserted since Dr. Huber completed his MS.—Ed.]
4. **D. stipulosa** (*S. Moore ex Wernham*) Pichon l.c. 259, t. 15, 1–3 (1953). *Landolphia stipulosa* S. Moore ex Wernham in Cat. Talb. 59 (1913) ; F.W.T.A. ed. 1, 2 : 33. A hairy climber with long setiform stipules.
   S.Nig.: Oban *Talbot* 346 ! 1296 ! Also in Cameroun and Gabon.
5. **D. lucida** (*K. Schum.*) Pierre in Bull. Soc. Linn. Paris, n. sér. : 93 (1898) ; Pichon l.c. 262, t. 15, 4–5. *Landolphia lucida* K. Schum. in Notizbl. Bot. Gart. Berl. 1 : 24 (1895) ; F.T.A. 4, 1 : 59. Climber up to 100 ft. with tendrils ; flowers white, yellow, greenish or orange-yellow, in terminal panicles.
   S.Nig.: Benin *Unwin* 8. Also in Gabon, Ubangi-Shari, Congo, Cabinda, Tanganyika, Mozambique and S. Rhodesia.
   [This species is included here on the record of *Unwin* 8, a sterile specimen at Paris (*vide* Pichon l.c.) ; it was not seen for this revision. *Unwin* 8 at Kew is *Ancylobotrys amoena* Hua].

8. VAHADENIA Stapf in F.T.A. 4, 1 : 29 (1902) ; Pichon in Mém. I.F.A.N. 35 : 266 (1953).

Anthers inserted about the middle of the corolla-tube (i.e. 6–11 mm. above the base of the tube) ; leaves elliptic to lanceolate, 6–20 cm. long, 4–10 cm. broad ; orange-brown beneath when dry, with 5–12 pairs of lateral nerves ; sepals large, 7–10 mm. long ; corolla-tube 1·5–2·5 cm. long, lobes 1–3 cm. long ..   ..   1. *laurentii*
Anthers inserted at the lower ⅓ of the corolla-tube (i.e. about 6 mm. above the base of the

tube) ; leaves obovate, elliptic or oblong, 6–15 cm. long, 3–7·5 cm. broad, pale greyish-green or greyish-brown beneath when dry, with 5–9 pairs of lateral nerves ; sepals 4–7 mm. long ; corolla as above, although frequently slightly smaller   2. *caillei*

1. **V. laurentii** (*De Wild.*) *Stapf* in F.T.A. 4, 1 : 29 (1902) ; Pichon in Mém. I.F.A.N. 35 : 266, t. 16, 1–2. *Landolphia laurentii* De Wild. in Rev. Cult. Col. Paris 8 : 229 (1901) ; F.T.A. 4, 1 : 30. *Vahadenia talbotii* Wernham in Cat. Talb. 58 (1913). *V. caillei* of F.W.T.A., ed. 1, 2 : 31, partly (syn. *V. talbotii* Wernham and *Talbot* 1634). Climber with tendrils, quite glabrous on branches and leaves also when young, up to 150 ft. long ; flowers white or sometimes tinged with purple, in cymes arranged in mostly lax, terminal, elongated panicles.
S.Nig.: Sapoba F.R., Benin (fr. June) *Onochie* FHI 31242 ! Oban *Talbot* 1634 ! **[Br.]Cam.** : Banga F.R., S. Bakundu, Kumba (Mar.) *Binuyo & Daramola* FHI 35618 ! Also in Gabon, the Congos, Ubangi-Shari and Angola.
2. **V. caillei** (*A. Chev.*) *Stapf ex Hutch. & Dalz.* F.W.T.A., ed. 1, 2 : 31 (1931) (excl. syn. *V. talbotii* Wernham and *Talbot* 1634) ; Pichon l.c. 271, t. 16, 3. *Landolphia caillei* A. Chev. in Mém. Soc. Bot. Fr. 2, 8 : 181 (1912). Glabrous climber up to 100 ft. with tendrils ; flowers yellowish-white and pink, fruits greenish-yellow.
Guin.: Kouria (Dec.) *Caille* in *Hb. Chev.* 14895 ! 14938. Bilima *Caille* in *Hb. Chev.* 14901. Sokourala *Chev.* 20499. **S.L.:** Lane-Poole 448 ! Yonibana (fl. & fr. Oct.) *Thomas* 4023 ! 4157 ! Kwaoma (Oct.) *J.D.F.* 84 ! Kasawe F.R. (Dec.) *King* 41b ! **Iv.C.:** Azaguié (fl. & fr. Dec.) *Fleury* in *Hb. Chev.* 22287 ! Guidéko to Zozro *Chev.* 19060. Bingerville *Vasson-Bandiougou* in *Hb. Chev.* 15186. *Chev.* 17340. Cosrou to Lahou *Mangenot.* Yapo *Mangenot.*

9. ANCYLOBOTRYS Pierre in Bull. Soc. Linn. Paris, n. sér. : 91 (1898), as *Ancylobothrys* ; Pichon in Mém. I.F.A.N. 35 : 272 (1953). *Landolphia* of F.T.A. 4, 1 : 30, and of F.W.T.A., ed. 1, 2 : 31, partly.

Corolla-tube 5–17 mm. long :
  Indumentum of young branches persistent and relatively long (usually 0·4–0·8 mm.) ; leaves elliptic to oblong, 4–20 cm. long, 2–9 cm. broad, with 7–17 pairs of lateral nerves ; corolla-lobes 3–19 mm. long .. .. 1. *amoena*
  Indumentum of young branches very short (up to 0·2 mm. long) :
    Petioles 1–3 cm. long; peduncles stout ; leaves ovate to ovate-oblong, 8–28 cm. long, 3–12 cm. broad, with 8–20 pairs of lateral nerves ; corolla-lobes 7–16 mm. long 2. *pyriformis*
    Petioles 4–6 mm. long ; peduncles fairly thin ; leaves elliptic-oblong, 6–10 cm. long, 2·5–4 cm. broad, with 12–22 pairs of lateral nerves ; corolla-lobes 3–6 mm. long 3. *brevituba*
Corolla-tube 18–44 mm. long ; young branches with short, persistent indumentum (up to 0·2 mm. long) ; leaves oblong to broadly lanceolate, 4–15 cm. long, 2–7 cm. broad, with 12–20 pairs of lateral nerves ; corolla-lobes 6–20 mm. long .. 4. *scandens*

1. **A. amoena** *Hua* in Bull. Mus. Hist. Nat. 5 : 186 (1899) ; Pichon in Mém. I.F.A.N. 35 : 274, t. 16, 4, t. 17, 1–2 (1953). *Landolphia amoena* (Hua) Hua & A. Chev. in Journ. de Bot. 15 : 8 (1900) ; F.T.A. 4, 1 : 46 ; F.W.T.A., ed. 1, 2 : 33. *L. scandens* var. *ferruginea* Hallier f. (1900). *L. ferruginea* (Hallier f.) Stapf in F.T.A. 4, 1 : 46 (1902). Climber up to 60 ft. long, with white or yellowish to purplish flowers in very dense cymes arranged in terminal panicles.
Mali: Mamahiza, Bamako *Roberty* 10389. **Guin.:** Pobéguin 170 ! Kouroussa (fr. Feb.) *Pobéguin* ! Farmar 264 ! *Chev.* 384 ! *Paroisse* 24 ! **Iv.C.:** Gafuoa (fr. Aug.) *Schnell* 5921 ! Issia rock (Nov.) *Mangenot.* Koualé to Kouroukoro (fr. May) *Chev.* 21753. Mankono (fr. June) *Chev.* 21899. Kangala to Kindali (fr. May) *Chev.* 861. **Ghana:** *Evans* 7 ! Prang (fl. & fr. Jan.) *Vigne* FH 1577 ! Northern Scarp F.R. (Nov.) *Enti* FH 5606 ! Akwapim Mts. (Nov.) *Johnson* 782 ! **Dah.:** Kouba to Farfa, Atacora Mts. *Chev.* 24039 ! Djougou (fr. June) *Chev.* 23921. Bocorona to Kouandé *Chev.* 24222. Farfa to Tounkountouna (fr. June) *Chev.* 24050 ; 24051. **N.Nig.:** Lokoja (Nov.) *Dalz.* 98 ! Wazeta to Karshe (Feb.) *Dalz.* 370 ! Jebba (fl. & fr. Jan.) *Meikle* 844 ! 1005 ! Anara F.R., Zaria Prov. (fl. & fr. May) *Keay* FHI 22960 ! Naraguta *Lely* 50 ! **S.Nig.:** *Unwin* 2 ! 8 ! 9 ! *Rowland* ! R. Isahin (fr. Mar.) *Rowland* 20 ! **[Br.]Cam.:** Vogel Peak, 4,000 ft. (Dec.) *Hepper* 1555 ! Also in Cameroun, Ubangi-Shari, Congo, Uganda, Tanganyika and Mozambique.
2. **A. pyriformis** *Pierre* in Bull. Soc. Linn. Paris, n. sér.: 127 (1899) ; Pichon l.c. 280, t. 17, 3–4. *A. robusta* Pierre (1899). *Landolphia robusta* (Pierre) Stapf in F.T.A. 4, 1 : 43 (1902) ; F.W.T.A., ed. 1, 2 : 33. *L. pyriformis* (Pierre) Stapf in F.T.A. 4, 1 : 60 (1902). Climber up to 90 ft. long with tendrils.
S.Nig.: Eket *Talbot* s.n. ! 3119 ! Also in Gabon, Ubangi-Shari, and the Congos.
3. **A. brevituba** *Pichon* l.c. 283, t. 17, 5–6 (1953). Climber with tendrils ; flowers white, few, in terminal panicles.
Dah.: Kétou (Jan.) *Le Testu* 110.
4. **A. scandens** (*Schum. & Thonn.*) *Pichon* l.c. 286, t. 18, 4–7 (1953). *Strychnos scandens* Schum. & Thonn. (1827). *Landolphia scandens* (Schum. & Thonn.) F. Didr. (1855)—F.T.A. 4, 1 : 46 ; F.W.T.A., ed. 1, 2 : 33 ; *Chev. Bot.* 408. Climber up to 60 ft. long with tendrils ; flowers white, yellowish, purple or violet in dense cymes arranged in terminal panicles.
Iv.C.: Danané to Mt. Goula (fr. Apr.) *Chev.* 21232. Danané (fr. Apr.) *Chev.* 21286. Guidéko *Chev.* 16478. Boukourou (fr. Dec.) *Chev.* 16682 ; 16806. **Ghana:** *Farmar* 398 ! Krobo Plains (fl. & fr. Dec.) *Johnson* 494 ! Achimota (Apr.) *Irvine* 1424 ! Legon Hill (fr. Jan.) *Adams* 3656 ! Akwapim (Mar.) *Irvine* 1545 ! **Togo Rep.:** Lomé *Warnecke* 252 ! Cabolé (fl. & fr. June) *Annet* 66 ; 68. **Dah.:** (fr. Mar.) *Poisson* ! Adja-Ouéré (Mar.) *Le Testu* 712 ! Adjara *Chev.* 22742. Ouidah (fr. Apr.) *Chev.* 23456. **S.Nig.:** Lagos (Oct.) *Foster* 68 ! *Moloney* ! *Dalz.* 1104 ! Ilaro (Jan.) *Millen* 105 ! Asaba (fr. Apr.) *Unwin* 14 ! Also in Cameroun, Gabon, the Congos, and Angola.

10. SABA Pichon in Mém. I.F.A.N. 35 : 302 (1953). *Landolphia* of F.T.A. 4, 1 : 30, and of F.W.T.A., ed. 1, 2 : 31, partly.

Corolla-tube 15–28 mm. long ; anthers 2·1–2·7 mm. long ; corolla-lobes 1–3 cm. long ; leaves elliptic or ovate, not or only shortly and indistinctly acuminate, 7–24 cm. long, 4–12 cm. broad, with 7–14 pairs of lateral nerves ; ovary densely pubescent to villous .. .. 1. *florida*
Corolla-tube 8–14 mm. long ; anthers 1·5–2 mm. long :
  Ovary glabrous ; leaves not or only shortly and indistinctly acuminate, oblong, 4–17 cm. long, 2–7 cm. broad, with 7–14 pairs of lateral nerves ; corolla-lobes 8–22 mm. long :

Pedicels equally pubescent ; sepals and corolla-tube puberulous outside
$\qquad$ 2. *senegalensis* var. *senegalensis*
Pedicels, sepals and corolla-tube glabrous outside   2a. *senegalensis* var. *glabriflora*
Ovary densely pubescent to villous ; leaves mostly long-acuminate, oblong-elliptic,
6–19 cm. long, 2–8 cm. broad, with 8–13 pairs of lateral nerves ; corolla-lobes
6–12 mm. long .. .. 3. *thompsonii*

1. **S. florida** (*Benth.*) *Bullock* in Kew Bull. 13 : 391 (1959). *Landolphia florida* Benth. (1849)—F.T.A. 4, 1 :
   38 ; F.W.T.A., ed. 1, 2 : 32 ; Chev. Bot. 404. *L. florida* var. *leiantha* Oliv. (1875)—F.T.A. 4, 1 : 39.
   L. comorensis K. Schum. (1893). *L. comorensis* var. *florida* (Benth.) K. Schum. (1893). *Saba comorensis*
   (K. Schum.) Pichon in Mém. I.F.A.N. 35 : 303 (1953), incl. var. *florida* (Benth.) Pichon l.c. 309 (1953).
   A strong woody climber with red-purplish stems and jasmine-like flowers yellow in the throat, and edible
   yellow fruit.
   Mali: Kati (Oct.) *Chev.* 2000 ! Sitakoto (fr. May) *Dubois* 188. Kangaba *Chev.* 269. **Guin.:** Friguiagbé
   to Bambaya (Dec.) *Pobéguin* 39. Loungouri (Dec.) *Caille* in *Hb. Chev.* 14868. Labé (fr. Apr.) *Chev.*
   12314. Tagania (Jan.) *Fleury* in *Hb. Chev.* 20409. Dalaba to Souguéta (Oct.) *Chev.* 20191. **S.L.:** *Smythe*
   7 ! Musaia (fl. & fr. Feb., Mar.) *Sc. Elliot* 5109 ! *Deighton* 4177 ! Ninia, Talla Hills (fr. Feb.) *Sc. Elliot*
   4896 ! Yalamba (Feb.) *Jaeger* 4340 ! **Lib.:** Murphy-Town *Whyte* 4 ! (drawings). Batomba-Town (fr.
   Dec.) *Whyte* 8 ! (drawing). **Iv.C.:** Sindou (fr. May) *Chev.* 863. Sindou & Bobo-Dioulasso, Assakra (Oct.)
   Mangenot. Katiola *Mangenot.* **Ghana:** *Dudgeon. Kitson* ! Okroase (May) *Johnson* 749 ! N. Territories
   (Feb.) *Northcott* 151 ! **Togo Rep.:** Bismarckburg to Misahöhe *Büttner* 756. Bismarckburg *Kling* 172.
   Kirkri *Kersting* 4. **Dah.:** *Poisson* s.n.; 20 ! 54 ! (Oct.) *Le Testu* 124 ! **N.Nig.:** Niger *T. Vogel* 101 ! Jos
   Plateau (Oct.) *Lely* P832 ! Kontagora (fl. & fr. Dec.) *Dalz.* 8 ! Anara F.R., Zaria Prov. (Oct.) *Keay* FHI
   21102 ! Agaie (Oct.) *Yates* 44 ! **S.Nig.:** *Unwin* 4. Orile Ilugu, Egba Dist. (Dec.) *Olorunfemi* FHI 19188 !
   Ibadan *Millen* 118. Benin *Chev.* [**Br.**]**Cam.:** Victoria *Deistel* 130. Barombi *Preuss* 441. Johann-Albrechtshöhe (Mar., Dec.) *Staudt* 487 ; 889. R. Metschum, Bamenda (Jan.) *Keay & Russell* FHI 28529 ! Throughout tropical Africa to the Comores and Madagascar.

2. **S. senegalensis** (*A. DC.*) *Pichon* var. **senegalensis**—in Mém. I.F.A.N. 35 : 316 (1953). *Vahea senegalensis*
   A.DC. (1844). *Landolphia senegalensis* (A. DC.) Kotschy & Peyr. (1831)—F.T.A. 4, 1 : 36 ; F.W.T.A.,
   ed. 1, 2 : 32 ; Chev. Bot. 408 ; Aubrév. Fl. For. C. Iv., ed. 1, 3 : 159. Climber, often more than 120 ft.
   long, with tendrils ; flowers white, yellowish- or greenish-white, cymes in short terminal inflorescences
   Sen.: *Roger* ! Thiès (Mar.) *Poisson.* Sangalkam *Adam* 8361. Karang to Koumbeng (July) *Trochain*
   3995. Kombo *Heudelot* 29 ! **Gam.:** *Skues* ! *Saunders* 11 ! Genieri, Kaiaff (July) *Fox* 166 ! Albreda
   Leprieur. Kan to Kousann *Perrottet* 792. **Mali:** (Mar.) *Davey* 548 ! Sandsanding (Apr.) *Chudeau.* Soumpi
   to Sébi (July) *Chev.* 2069. Niafounké to Diré (Apr.) *Rogeon* 105. El-Oualadji (July) *Chev.* 1199. Bobo-
   Dioulasso (May) *Chev.* 918 ; 919. **Port.G.:** Acóco, Formosa (June) *Esp. Santo* 1995 ! Begene, Farim
   (May) *Esp. Santo* 2384 ! Pussubé, Bissau (Apr.) *Esp. Santo* 1173. Sama, Bafata (fl. & fr. Jan.) *Esp. Santo*
   434. Dandum *Noury.* **Guin.:** Kadé (Feb.) *Maclaud* 321. Sareya (Feb.) *Chev.* 452. Kouroussa (Feb.)
   Chev. 386. Diaragouéla (Feb.) *Chev.* 470. Siguiri (fr. Feb.) *Chev.* 305. **Ghana:** *Evans* 2 ! Kulmasa to
   Iagalo (Feb.) *Kitson* 668 ! Izigi, Lorha (Mar.) *McLeod* 832 ! Burufo (Dec.) *Adams* 4438 !

2a. **S. senegalensis** var. **glabriflora** (*Hua*) *Pichon* l.c. 322 (1953). *Landolphia senegalensis* var. *glabriflora* Hua
    (1899)—F.T.A. 4, 1 : 37 ; F.W.T.A., ed. 1, 2 : 32.
    Sen.: Niafunké (May) *Hagerup* 85a ! Koungheul (July) *Trochain* 173. Nicolo Koba (May) *Trochain*
    3506. **Gam.:** *Dawe* 54 ! near R. Gambia *Whitfield.* **Mali:** Sébékoro (Apr.) *Dubois* 66. Bamako (Jan.)
    Chev. 235. Sébi (July) *Chev.* 1172. San to Douma (Mar.) *Chev.* 43975. Bandiagara to Mopti (fl. & fr.
    Sept.) *Chev.* 24913. Bobo-Dioulasso *Chev.* Ouahigouya (Aug.) *Chev.* 24767. **Guin.:** Kadé (Mar.) *Maclaud*
    377. Diala, near Ouassaya (Mar.) *Chev.* 517. Guiankorokora (Feb.) *Chev.* 352. Siguiri *Pobéguin* 659 !
    Kankan (Mar.) *Chev.* 581. Farana (Mar.) *Sc. Elliot* 5345 ! **Iv.C.:** Niangbo (June) *Mangenot.* Tafiré
    (fl. & fr. May) *Aubrév.* 1444.

3. **S. thompsonii** (*A. Chev.*) *Pichon* l.c. 322, t. 21, 4–6 (1953). *Landolphia thompsonii* A. Chev. in Mém. Soc.
   Bot. Fr. 2, 8 : 182 (1912) ; F.W.T.A., ed. 1, 2 : 32 ; Chev. Bot. 409. *L. senegalensis* of F.W.T.A., ed. 1,
   2 : 32, partly (*Warnecke* 421), not of (A.DC.) Kotschy & Peyr. Climber up to 90 ft. high, with tendrils ;
   flowers white in short terminal inflorescences.
   Iv.C.: near M'Bayiakro, Nzi (Aug.) *Chev.* 22266 ! Bouroukrou *Chev.* 16648 ; 16823. Yabouakrou to
   Tingouéla *Chev.* 22569. **Ghana:** Accra *Traun* 1892. **Togo Rep.:** Lomé *Warnecke* 421 ! **Dah.:** Pira,
   Savalou *Chev.* 23753 ! Ekpé *Poisson.* Adja-Ouéré (Aug.) *Le Testu* 28. Koussi to Aévédji *Chev.* 23250.
   Djougou *Chev.* 23886. **S.Nig.:** Ilushi, Asaba *Unwin* 5 ! Benin *Thompson* 12 ! Ibadan (Dec.) *Punch* 45 !
   Owo (July) *Jones* FHI 4392 ! Ibadan (Sept.) *Onochie* FHI 7664 !

11. CYLINDROPSIS Pierre in Bull. Soc. Linn. Paris, n. sér. : 38 (1898) ; Pichon in Mém. I.F.A.N. 35 : 329 (1953).

Climbing shrub with branches and leaves glabrous ; leaves oblong to elliptic, short- to
sometimes long-acuminate, 4–13 cm. long and 2–5 cm. broad with 6–13 pairs of
lateral nerves ; cymes axillary and terminal ; corolla-tube 3–5 mm. long, the lobes
1·8–2·5 mm. long .. .. *parvifolia*

C. parvifolia *Pierre* l.c. (1898) ; Pichon l.c. *Clitandra parvifolia* (Pierre) Stapf in F.T.A. 4, 1 : 63 (1902).
   C. talbotii Wernham in Cat. Talb. 60 (1913). Climber up to 120 ft. high with small white, yellowish or
   greenish flowers.
   S.Nig.: Oban *Talbot* 1039 ! s.n. ! Also in Gabon, Ubangi-Shari, Congo and Cabinda.

12. CYCLOCOTYLA Stapf in Kew Bull. 1908 : 259.

A climbing shrub ; leaves opposite, elliptic or oblong, shortly acuminate, acute at the
base, about 8–10 cm. long and 4 cm. broad, with numerous spreading lateral nerves ;
petiole 1–1·3 cm. long ; cymes axillary, few-flowered ; peduncle 2 cm. long ; pedicels
5 mm. long ; calyx 3·5 mm. long, lobed to the middle ; corolla-tube 8 mm. long,
lobes 6 mm. long ; anthers inserted about 2 mm. above the base of the corolla-tube ;
style very short .. .. *congolensis*

C. congolensis *Stapf* in Kew Bull. 1908 : 260. *C. oligosperma* Wernham in J. Bot. 52 : 28 (1914) ; F.W.T.A.,
   ed. 1, 2 : 35. Glabrous climber with short-lobed white flowers in paniculate to corymbose inflorescences.
   S.Nig.: Shasha F.R., Ijebu Prov. (Feb.) *Ross* 39 ! Omo F.R., Ijebu Prov. (fl. & fr. Apr.) *Onochie* FHI
   15533 ! Eket *Talbot* 3052 ! Also in Cameroun and Congo.

13. PICRALIMA Pierre—F.T.A. 4, 1 : 96 ; Pichon in Bol. Soc. Brot. 27 : 81 (1953).

Leaves and branches glabrous ; leaves oblong or elliptic, 9–22 cm. long, 3–10 cm. broad, with 14–21 pairs of lateral nerves ; sepals 3–9 mm. long ; corolla-tube 13–20 mm. long, lobes 8–25 mm. long . . . . *nitida*

P. nitida (*Stapf*) *Th. & H. Dur.* in Bull. Jard. Bot. Brux. 2 : 338 (1910) ; Aubrév. Fl. For. C. Iv., ed. 2, 3 : 206, t. 326. *Tabernaemontana nitida* Stapf in Kew Bull. 1895 : 22. *Picralima klaineana* Pierre l.c. 1279 (1897) ; F.T.A. 4, 1 : 96 (1902). *Picralima macrocarpa* A. Chev. Bot. 427, name only. A glabrous tree or shrub, 9–75 ft. high with white to yellow flowers in terminal, mostly densely contracted, inflorescences.
Iv.C.: Bingerville-Abidjan-Dabou *Chev.* 15200 ; 15600. Agboville (fl. & fr. June) *Aubrév.* 1383. Anoumaba (fr. Nov.) *Chev.* 22391 ; 22440. **Ghana:** Adeambra (young fr. May) *Vigne* FH 864 ! Abroase (June) *Johnson* 917 ! Aburi (fr. Sept.) *Imp. Inst.* ! Kumasi (July) *Vigne* FH 3956 ! *Andoh* 4222 ! **S.Nig.:** Agege *Foster* 220 ! Idumuje (Dec.) *Thomas* 2101 ! Okomu F.R., Benin (fr. Jan.) *Brenan & Richards* 8910 ! Oban *Talbot* 219 ! 1690 ! [**Br.**]**Cam.:** Victoria (Nov.–Jan.) *Maitland* 284 ! 764 ! 787 ! Ambas Bay (Jan.) *Mann* 710 ! Wum, Bamenda (fr. June) *Ujor* FHI 29271 ! Also in Cameroun, Gabon, Ubangi-Shari, the Congos and Uganda.

14. HUNTERIA Roxb.—Pichon in Bol. Soc. Brot. 27 : 88 (1953). *Polyadoa* Stapf in F.T.A. 4, 1 : 103 (1902).

Corolla 5–7 mm. long ; tube with 5 minute tufts of hairs inside beneath the anthers, otherwise glabrous ; corolla-tube 3–4 mm. long, lobes 2–3 mm. long ; leaves elliptic to oblong-elliptic, short- or long-acuminate, 8–25 cm. long, 3–9 cm. broad, with 10–18 pairs of lateral nerves ; fruits with a short curved beak   1. *camerunensis*
Corolla 8–28 mm. long, the tube within with a continuous zone of pubescence beneath the anthers :
  The tops of the anthers ending 2–4 mm. beneath the mouth of the corolla-tube ; corolla 15–28 mm. long, corolla-tube up to 17 mm. long, lobes 10–12 mm. ; leaves elliptic to oblong-elliptic, 7–18 cm. long, 3–7 cm. broad, with 7–13 pairs of lateral nerves ; fruits with a straight beak   . .   . .   2. *simii*
  The tops of the anthers ending 0–1 mm. beneath the mouth of the corolla-tube ; corolla 8–16 mm. long ; fruits not beaked :
    Anthers inserted 2–2·5 mm. above the base of the corolla-tube ; style 0·6–1·7 mm. long ; corolla-tube 3–5 mm. long, lobes 5–9 mm. long ; leaves elliptic to oblong, 9–16 cm. long, 3–9 cm. broad, with 12–24 pairs of lateral nerves . .   3. *eburnea*
    Anthers inserted 4–5 mm. above the base of the corolla-tube ; style 3–4 mm. long :
      Axis and branches of inflorescence not thickened towards their apex ; leaves mostly oblong-elliptic, 5–16 cm. long and 2–6 cm. broad, with 11–25 pairs of lateral nerves ; corolla-tube 5–7 mm. long, lobes 4–8 mm. long . .   . .   . .   4. *elliotii*
      Axis and branches of inflorescence markedly thickened at their apex ; leaves broadly elliptic to oblong, 11–23 cm. long, 5–9 cm. broad, with 11–18 pairs of lateral nerves ; corolla-tube 5–7 mm. long, lobes 6–10 mm. long   . .   5. *umbellata*

1. **H. camerunensis** *K. Schum. ex Hallier f.* in Jahrb. Hamb. Wiss. Aust. 17, 3, Beih. 187 (1900). *Pleiocarpa camerunensis* (K. Schum. ex Hallier f.) Stapf in F.T.A. 4, 1 : 102 (1902). *Polyadoa camerunensis* (K. Schum. ex Hallier f.) Brenan in Kew Bull. 7 : 451 (1953). *Comularia camerunensis* (K. Schum. ex Hallier f.) Pichon in Bol. Soc. Brot. 27 : 116 (1953). Glabrous shrub or small tree, 3–9 ft. high, with minute white flowers in dense terminal clusters.
  [**Br.**]**Cam.:** S. Bakundu F.R., Kumba (Jan., Mar.) *Brenan* 9401 ! 9401a ! 9401b ! *Binuyo & Daramola* FHI 35090 ! Also in Cameroun.
2. **H. simii** (*Stapf*) *H. Huber* in Kew Bull. 15 : 437 (1962). *Polyadoa* (?) *simii* Stapf in Johnston Lib. 2 : 624 (1906). *Picralima laurifolia* A. Chev. (1914), name only. *Pleiocarpa simii* (Stapf) Stapf ex Hutch. & Dalz. F.W.T.A., ed. 1, 2 : 37 (1931) ; Aubrév. Fl. For. C. Iv., ed. 1, 3 : 174. *Tetradoa simii* (Stapf) Pichon l.c. 120 (1953). Glabrous shrub with white flowers in mostly terminal, few-flowered clusters.
  S.L.: Gola Forest (fl. & fr. Mar., May) *Small* 539 ! 649 ! Kambui F.R., (fr. Apr.) *Jordan* 2037 ! 2046 ! **Lib.:** Sim 16 ! Dukwia R. (fr. Feb.) *Cooper* 169 ! Mecca, Boporo Dist. (fr. Nov.) *Baldwin* 10443 ! Genna Tanyehun (fr. Dec.) *Baldwin* 10736 ! Ganta (fr. Sept.) *Baldwin* 12561 ! **Iv.C.:** Fort Binger (July) *Chev.* 19535. Kéeta (July) *Chev.* 19311 ; 19367 ; 19368. Guidéko to Zozro (June) *Chev.* 19036.
3. **H. eburnea** *Pichon* l.c. 91 (1953) ; Aubrév. Fl. For. C. Iv., ed. 2, 3 : 208, t. 327. *Picralima gracilis* A. Chev. (1914), name only. *Picralima elliotii* of F.W.T.A., ed. 1, 2 : 40, partly (syn. *P. gracilis* A. Chev. and *Chev.* specimens), not of (Stapf) Stapf. A glabrous tree 20–120 ft. high, with white flowers in dense, short, terminal inflorescences.
  Guin.: Macenta *Adam* 6470. **S.L.:** Gola Forest (fr. Mar.) *Small* 502 ! Kenema *Edwardson* 41. **Iv.C.:** Abidjan (June) *Aubrév* 23 ! Aboisso *Chev.* 16310 ! Bériby (fr. Aug.) *Chev.* 20013. Guidéko (fr. May) *Chev.* 16373. Bouroukrou (Dec., Jan.) *Chev.* 16519. **Ghana:** Simpa (Apr.–May) *Vigne* FH 1957 ! *Enti* FH 6920 ! **S.Nig.:** Afi River F.R., Ikom Dist. (Dec.) *Keay* FH 28250 !
4. **H. elliotii** (*Stapf*) *Pichon* l.c. 97 (1953). *Polyadoa elliotii* Stapf in F.T.A. 4, 1 : 104 (1902). *Picralima elliotii* (Stapf) Stapf in Kew Bull. 1908 : 302 ; F.W.T.A., ed. 1, 2 : 40 (excl. syn. *P. gracilis* A. Chev. and *Chev.* specimens). Glabrous shrub or tree 10–30 ft. high with white to yellowish flowers in terminal, mostly sub-globose inflorescences.
  Sen.: Goloumbo, Gambia Hills (fr. Sept.) *Berhaut* 2082. Ouassadou *Trochain* 3467. *Berhaut* 1669 ; 2081. **Port.G.:** Jamberém, Cantanhez (Apr., June) *d'Orey* 376 ! *Esp. Santo* 3001. Pobusa, Cubisseque (fl. & fr. June) *Esp. Santo* 3016. Empacaca, Canchungo (Apr.) *Esp. Santo* 1940 ! Jangada, Contubo (June) *Esp. Santo* 2257 ! **Guin.:** Boké (Mar.) *Pobéguin* 2015. **S.L.:** Batkanu (Apr.) *Thomas* 19 ! Kasewe (Apr.) *King* 231 ! Njala (Mar.) *Deighton* 5010 ! Kambui F.R. (June) *Jordan* 2120 ! Makunde, Limba (Apr.) *Sc. Elliot* 5690 !
5. **H. umbellata** (*K. Schum.*) *Hallier f.* in Jahrb. Hamb. Wiss. Aust. 17, 3, Beih. 190 (1900). *Carpodinus umbellatus* K. Schum. (1896). *Polyadoa umbellata* (K. Schum.) Stapf in F.T.A. 4, 1 : 103 (1902). *Picralima umbellata* (K. Schum.) Stapf (1908)—F.W.T.A., ed. 1, 2 : 40. Small glabrous tree 25–40 ft. high ; flowers as above, in dense, terminal heads.
  Ghana: Axim (Mar.) *Vigne* FH 1660. Essuboni *Jumah* 1. **S.Nig.:** Lagos *Foster* 5 ! Okomu F.R., Benin (Mar.) *Akpabla* 110 ! Ibadan F.R. (fr. May, Nov.) *Punch* 138 ! *Bakare, Obaseki & Idahosa* FHI 22857 ! Also in Cameroun and Ubangi-Shari.

15. PLEIOCARPA Benth.—F.T.A. 4, 1 : 97 ; Pichon in Bol. Soc.
Brot. 27 : 123 (1953).

Anthers 0·4–1·7 mm. long, their tops 0–1·7 mm. beneath the mouth of the corolla-tube ;
tube 4–20 mm. long, lobes 0·7–9 mm. long ; carpels with 1–2 ovules :
Carpels 2 ; corolla-tube 4–12 mm. long:
  Leaves all opposite, oblong-elliptic to broadly lanceolate, mostly long-acuminate,
  8–15 cm. long, 2·5–5 cm. broad, with 13–30 pairs of lateral nerves ; corolla-tube
  6–12 mm. long, lobes 2–9 mm. long ..   ..   ..   1. *bicarpellata*
  Leaves all or partly in whorls of 3 or 4, oblong to oblong-elliptic, mostly not or only
  short-acuminate, 6–16 cm. long, 2–6 cm. broad, with 11–23 pairs of lateral nerves ;
  corolla-tube 4–9 mm. long, lobes 0·7–3 mm. long   2a. *pycnantha* var. *tubicina*
Carpels 3–5 ; corolla-tube 7–20 mm. long, lobes 3–8 mm. long ; leaves oblong to
  lanceolate or elliptic, 8–20 cm. long, 2·5–8 cm. broad, with 8–23 pairs of lateral
  nerves   ..   ..   3. *mutica*
Anthers 2·3–3·5 mm. long, their tops ending 3 mm. beneath the mouth of the corolla-
tube or deeper ; tube 15–22 mm. long, lobes 7–13 mm. long ; carpels 5, each with
4–8 ovules ; leaves elliptic to oblong, 10–21 cm. long, 4–10 cm. broad, with 6–12
pairs of lateral nerves   ..   ..   4. *talbotii*

1. **P. bicarpellata** *Stapf* in Kew Bull. 1894 : 21 ; F.T.A. 4, 1 : 99. *Hunteria ambiens* K. Schum. in Engl. Bot.
  Jahrb. 23 : 223 (1896). A glabrous shrub or tree 3–46 ft. high, rarely climbing ; flowers pure white, in
  sessile, axillary clusters.
  [**Br.**]**Cam.**: Cam. Mt., 4,000 ft. (Feb.) *Mann* 1213 ! Kumba (Mar.) *Brenan* 9458 ! Johann-Albrechtshöhe
  Staudt 573 ! 683 ! 794. Barombi (Feb.) *Preuss* 44.
2. **P. pycnantha** (*K. Schum.*) *Stapf* in F.T.A. 4, 1 : 99 (1902). *Hunteria pycnantha* K. Schum. l.c. 222 (1896).
2a. **P. pycnantha** var. **tubicina** (*Stapf*) *Pichon* in Bol. Soc. Brot. 27 : 132 (1953). *P. tubicina* Stapf in Kew Bull.
  1898 : 304. *Hunteria breviloba* Hallier f. (1900). *Pleiocarpa micrantha* Stapf in F.T.A. 4, 1 : 100 (1902) ;
  F.W.T.A., ed. 1, 2 : 38. *P. flavescens* Stapf in F.T.A. 4, 1 : 101 (1902) ; F.W.T.A., ed. 1, 2 : 38.
  P. breviloba (Hallier f.) Stapf in F.T.A. 4, 1 : 102 (1902). *P. microcarpa* Stapf in F.T.A. 4, 1 : 102 (1902).
  A glabrous shrub or tree 4–100 ft. high ; flowers white, yellowish or greenish in sessile, mostly axillary
  clusters.

Fig. 214.—Pleiocarpa mutica *Benth*. (Apocynaceae).
A, flower.   B, pistil.   C, upper part of corolla laid open.   D, calyx and style.   E, anther.

Mali: Kerfomouria (Mar.) *Chev.* 597 ! **Port.G.**: Empacaca, Canchungo (Apr.) *Esp. Santo* 1941 ! Poncom,
Fulacunda (fl. & fr. May) *Esp. Santo* 2042 ! Entachá, Bolama (Apr.) *Esp. Santo* 1919 ! **Ghana:** Aburi
Hills (Feb., Mar.) *Johnson* 616 ! 623 ! Akropong, Akwapim (Oct.) *Johnson* 903 ! Legon (fl. Mar., fr. May)
Adams 3789 ! 4331 ! **Togo Rep.**: Lomé *Warnecke* 481 ! *Mahaux* in Hb. *Alleizette* 4582. **Dah.**: Bokoutou
Chev. 22844. **S.Nig.**: *Foster* 182 ! Ikirun to Igbajo, Ibadan (Jan.) *Latilo* FHI 31754 ! Iressi, Ibadan (Jan.)
Latilo FHI 31758 ! Also in Ubangi-Shari, Congo, Angola, Sudan, Uganda and Zanzibar.
3. **P. mutica** *Benth*. in Hook. Ic. Pl. 12 : 71, t. 1181 (1876) ; F.T.A. 4, 1 : 98 ; Pichon l.c. 138 ; Aubrév. Fl.
  For. C. Iv., ed. 2, 3 : 206, t. 325. *P. salicifolia* Stapf in F.T.A. 4, 1 : 99 (1902) ; F.W.T.A., ed. 1, 2 : 38.
  P. bakueana A. Chev. (1914), name only. *P. ternata* A. Chev. (1914), name only. *P. tricarpellata* Stapf
  in Kew Bull. 1915 : 47 ; F.W.T.A., ed. 1, 2 : 38. A glabrous shrub or small tree 2–25 ft. high ; flowers
  white, inflorescences as above.
  S.L.: *Barter* ! Falaba (Apr.) *Aylmer* 35 ! Gbinti, N. Prov. (fr. July) *Deighton* 2501 ! Faiama (Jan.)
  Deighton 3869 ! 3871 ! Gola Forest (fr. May) *Small* 701 ! **Lib.:** Betandu (Mar.) *Bequaert* 148 ! Dukwia R.
  (Feb.) *Cooper* 248 ! Zwedru, Tchien Dist. (fr. Aug.) *Baldwin* 7013 ! Mecca (Dec.) *Baldwin* 10814a !

Greenville, Sinoe R. (fr. Mar.) *Baldwin* 11552! **Iv.C.:** Fort Binger (fr. July) *Chev.* 19432. Bériby (fr. Aug.) *Chev.* 19994. Port Bouet (May) *Bégué* 3110. Bango (Oct.) *Martineau* 350. Aboisso (fr. Apr.) *Chev.* 17862. **Ghana:** Princes *Akpabla* 792! Axim (Feb.) *Irvine* 2173! Asanwinso (Feb.) *Vigne* FH 986! Asenanyo River F.R. (Feb.) *Andoh* 4305! Akwa Range, 1,500 ft., (fl. & fr. Feb.) *Vigne* FH 4345! **S.Nig.:** Cross R. (Jan.-Mar.) *Johnston*! Calabar (fl. & fr. Feb., Mar.) *Mann* 2277! *Holland* 106! Oban *Talbot* 1565! 1654! Agbagbana to Ndiakoro, Itu Dist. (fr. June) *Ujor* FHI 30169! **[Br.]Cam.:** *Staudt* 322! Rio del Rey *Johnston*! Also in Cameroun and Congo.

4. **P. talbotii** *Wernham* in Cat. Talb. 62 (1913). *Carpodinopsis talbotii* (Wernham) Pichon l.c. 144, t. 5, 1–3 (1953). A small glabrous tree or shrub, often with semi-scrambling habit ; flowers pure white in sessile, few-flowered, axillary clusters.

    S.Nig.: Oban *Talbot* 1037! s.n.! Oban F.R. (Feb.) *Onochie* FHI 36293x! British Obokum, Ikom (Dec.) *Keay* FHI 28277! **[Br.]Cam.:** S. Bakundu F.R., Kumba (fr. Jan.) *Keay* FHI 28571!

16. PTEROTABERNA Stapf in F.T.A. 4, 1 : 125 (1902).

Shrub ; leaves elliptic-lanceolate or elliptic-acuminate, 7–18 cm. long, 2·5–7 cm. broad, with 5–10 pairs of lateral nerves ; inflorescence a small, terminal, laxly branched cyme ; corolla-tube 6–9 mm. long, lobes 3–5 mm. long, overlapping to the left ; anthers 3 mm. long ; mericarps paired, with 3 longitudinal wings   *inconspicua*

P. inconspicua (*Stapf*)*Stapf* in F.T.A. 4, 1 : 126 (1902) ; Pichon in Bull. Soc. Bot. Fr. 100 : 172 (1953). *Tabernaemontana inconspicua* Stapf in Kew Bull. 1894 : 120. A glabrous shrub, 3–10 ft. high, with small yellowish-green flowers in lax cymes.

    S.L.: Bumbuna (fr. Oct.) *Thomas* 3868! Also in Cameroun, Gabon, Rio Muni and Congo.

    [Note : our W. African specimen may be distinct, however the material (a single sheet with fruit only) is insufficient to be sure—Huber, *fide* Pichon *l.c.*]

17. CALLICHILIA Stapf in F.T.A. 4, 1 : 130 (1902).

Calyx 2·5–3·5 mm. long ; leaves petiolate, oblong-obovate, 14–30 cm. long and up to 20 cm. broad, with 7–12 pairs of lateral nerves ; corolla-tube 6–9 cm. long, the anthers inserted in about the lower ¼ of the tube, corolla-lobes 4–6 cm. long, overlapping to the left .. .. 1. *macrocalyx*
Calyx 5–12 mm. long ; corolla-tube 2–6 cm. long :
  Corolla-lobes overlapping to the left ; anthers inserted above the middle of the tube :
    Petiole 5–15 mm. long ; leaves elongate-oblong or oblanceolate, blade 15–26 cm. long, 3·5–10 cm. broad, long-cuneate at base, with 6–12 pairs of lateral nerves ; sepals narrowly lanceolate, acute ; corolla-tube 2–6 cm. long, lobes about 2·5 cm. long   2. *monopodialis*
    Leaves subsessile or with a very short petiole up to 5 mm. long, oblong-elliptic to lanceolate, 6–17 cm. long, 3–7 cm. broad, with 5–9 pairs of lateral nerves, rounded or cuneate at base ; sepals oblong, subobtuse ; corolla-tube about 3 cm. long, lobes mostly 2·5–3 cm. long .. .. 3. *mannii*
  Corolla-lobes overlapping to the right :
    Sepals broadly lanceolate or elliptic, obtuse or mucronate at apex ; corolla-tube 2–3 cm. long, the anthers inserted below the middle ; corolla-lobes 2–3 cm. long ; leaves elliptic to lanceolate, 7–20 cm. long, 2–8 cm. broad, with 7–12 pairs of lateral nerves .. .. 4. *subsessilis*
    Sepals narrowly lanceolate, very acute ; corolla-tube 2·5–3·5 cm. long, with the anthers inserted above the middle ; corolla-lobes about 1·5 cm. long ; leaves elliptic to lanceolate, 9–16 cm. long, 3–8 cm. broad with 7–12 pairs of lateral nerves .. .. 5. *stenosepala*

1. **C. macrocalyx** *Schellenb. ex Markg.* in Notizbl. Bot. Gart. Berl. 8 : 310 (1923). *C. magnifica* R. Good in J. Bot. 67, Suppl. 2 : 86 (1929). A glabrous shrub about 10 ft. high with large white flowers in terminal pendulous inflorescences.
    S.Nig.: Oban *Talbot*! Also in Cameroun and Rio Muni.
2. **C. monopodialis** (*K. Schum.*) *Stapf* in F.T.A. 4, 1 : 131 (1902). *Tabernaemontana monopodialis* K. Schum. in Engl. Bot. Jahrb. 23 : 225 (1896). A glabrous, erect or decumbent shrub up to 7 ft. high, with white flowers.
    S.Nig.: Usonigbe F.R., Benin (May) *Keay* FHI 37016! **[Br.]Cam.:** Johann-Albrechtshöhe *Staudt* 832. Also in Cameroun.
3. **C. mannii** *Stapf* in F.T.A. 4, 1 : 131 (1902). Glabrous climbing shrub with white flowers in few-flowered, terminal inflorescences.
    S.Nig.: Oban (May) *Ujor* FHI 30819! **[Br.]Cam.:** Ambas Bay (Dec.) *Mann* 2152! Also in Cameroun.
4. **C. subsessilis** (*Benth.*) *Stapf* in F.T.A. 4, 1 : 132 (1902). *Tabernaemontana subsessilis* Benth. in Fl. Nigrit. 448 (1849). Glabrous shrub 2–8 ft. high with white flowers in terminal inflorescences.
    Guin.: Macenta (fl. May, fr. Oct.) *Collenette* 13! *Baldwin* 9805! Dyeke (fr. Oct.) *Baldwin* 9651! 9673! **S.L.:** Karheni, Limba (fr. Apr.) *Sc. Elliot* 5590! Falaba (May) *Aylmer* 81! Njala (May) *Deighton* 703! Mayogbo (Jan.) *Marmo* 155! Gola Forest (Dec.) *King* 94b! *Small* 586! **Lib.:** T. *Vogel* 4! 5! *Dinklage* 6263! Nikabuzu to Zigida (Mar.) *Bequaert* 126! Ganta (Dec.) *Barker* 1123! Peahtah (Dec.)*Baldwin* 10607! **Ghana:** Popokyere, near Tarquah (Apr.) *Chipp* 195! E. Akim (June) *Johnson* 762! 765! Assin-Yan-Kumasi *Cummins* 235! Pra R. (fr. Feb.) *Vigne* FH 1603! **S.Nig.:** Oban *Talbot*! Also in Cameroun.
5. **C. stenosepala** *Stapf* in F.T.A. 4, 1 : 602 (1904). A glabrous shrub 3–5 ft. high, erect or half-straggling, with white flowers, in terminal inflorescences.
    Lib.: Grant's Farm, Sinoe *Whyte*! **S.Nig.:** Ala (Nov.) *Thomas* 1973! Benin to Owam (fr. Jan.) *Onochie* in *Hb. Brenan* 8944! Agbadi, Benin Div. *Meikle* 545! Degema *Talbot* 3752!

18. HEDRANTHERA Pichon in Mém. Mus. Hist. Nat. Paris, n. sér., 27 : 225 (1948).
Callichilia of F.T.A. 4, 1 : 131, partly.

Leaves distinctly petiolate, elliptic to lanceolate, 6–20 cm. long, 3–9 cm. broad, with 5–12 pairs of lateral nerves ; calyx-lobes broadly oblong, subobtuse to truncate or mucronate ; corolla-tube 3–5 cm. long, lobes 2·5–4 cm. long, overlapping to the left   ..   .. *barteri*

H. barteri (*Hook. f.*) *Pichon* l.c. (1948). *Tabernaemontana barteri* Hook. f. (1870). *Callichilia barteri* (Hook. f.) Stapf in F.T.A. 4, 1 : 133 (1902) ; F.W.T.A., ed. 1, 2 : 39. A glabrous erect shrub 3 ft. high with conspicuous white flowers, mostly in terminal inflorescences.
Ghana: *Buxton*! **N.Nig.:** Lokoja (Mar.) *Shaw* 17! Lafia to Jemaa (Apr.) *Hepburn* 44! Samban Kwoi, Zaria (Apr.) *Wimbush* FHI 39619! **S.Nig.:** Epe (Feb.) *Barter* 3284! *Millen* 170! Ogwashi (Dec.) *Thomas* 2068! Okomu F.R., Benin (Jan.) *Brenan* 8771! Ibadan (Apr.) *Meikle* 1404! Enugu (fl. & fr. Mar.) *Irvine* 3618! Calabar (fl. & fr. Feb.) *Mann* 2271! **[Br.]Cam.:** Mamfe (Mar.) *Richards* 5216! Also in Congo.

19. TABERNAEMONTANA Linn. (1753)—Pichon in Notulae Syst. 13 : 230 (1948).

Conopharyngia G. Don (1837)—F.T.A. 4,1 : 139 ; F.W.T.A., ed. 1, 2 : 40. *Gabunia* K. Schum. ex Stapf in F.T.A. 4, 1 : 136 (1902) ; F.W.T.A., ed. 1, 2 : 39.

Erect trees or shrubs ; leaves with 6–16 pairs of distinct lateral nerves (only in No. 6 smaller leaves frequently with 5 pairs) ; calyx with glandular scales within (Subgen. *Sarcopharyngia*) :
  Corolla-tube 8–12 mm. long in the mature flowers :
    Cymes sessile, axillary, or rarely with a peduncle up to 1·5 cm. long ; corolla-tube pilose within above the anthers ; leaves elliptic to broadly lanceolate, 6–24 cm. long, 4–9 cm. broad, with 6–14 pairs of lateral nerves   ..   .. 1. *penduliflora*
    Cymes all long-pedunculate ; leaves with 9–15 pairs of lateral nerves :
      Corolla-tube densely pilose within above the anthers ; leaves broadly elliptic to elliptic-oblong, 18–50 cm. long, about 10–25 cm. broad   .. 2. *brachyantha*
      Corolla-tube glabrous within above the anthers or with 5 faintly puberulous lines ; leaves oblong, linear-oblong or lanceolate, 8–24 cm. long, 3–10 cm. broad
3. *ventricosa*
  Corolla-tube 15–100 mm. long in the mature flower ; cymes pedunculate :
    Calyx (12–)14–20 mm. long ; corolla-tube 6–9 cm. long, very stout ; anthers 1·6–2·4 cm. long, inserted about 2·5–4 cm. above the base of the tube ; leaves elliptic to oblong, 12–40 cm. long, 5–15 cm. broad, with 9–14 pairs of lateral nerves
4. *chippii*
    Calyx 2–10 mm. long :
      Corolla-tube markedly pilose within above the anthers :
        Corolla-tube 1·5–3 cm. long when fully developed, very stout, the anthers inserted about 1 cm. above the base of the tube ; leaves broadly elliptic to ovate or obovate, 12 to about 50 cm. long, 7–25 cm. broad, with 7–16 pairs of lateral nerves :
          Corolla-tube 2·5–3 cm. long in the mature flower 5. *pachysiphon* var. *pachysiphon*
          Corolla-tube 1·5–2 cm. long in the mature flower 5a. *pachysiphon* var. *cumminsii*
        Corolla-tube 6–12 cm. long, slender in the upper half, the anthers inserted 2–3·5 cm. above the base of the tube ; leaves elliptic, oblong or broadly lanceolate, 7–20(–24) cm. long, 3–9 cm. broad, with 5–10 pairs of lateral nerves   6. *longiflora*
      Corolla-tube glabrous or almost glabrous within above the anthers :
        Anthers 14–16 mm. long, inserted 1–2 cm. above the base of the long corolla-tube ; calyx usually 8–10 mm. long ; leaves elliptic, 15–35 cm. long, 10–20 cm. broad, with 9–16 pairs of lateral nerves   ..   .. 7. *contorta*
        Anthers 6–10 mm. long, inserted 5–12 mm. above the base of the corolla-tube ; calyx 2–8(–10) mm. long ; leaves elliptic, obovate or oblong, 13–40 cm. long, 6–20 cm. broad, with 6–12 pairs of lateral nerves   ..   .. 8. *crassa*
Climbing shrubs ; leaves with 4–9 pairs of lateral nerves or lateral nerves quite indistinct ; calyx without glandular scales (except in No. 10) (Subgen. *Gabunia*) :
  Corolla-tube 8–18 mm. long, corolla-lobes about 1·5 cm. long ; leaves elliptic to lanceolate, 7–12 cm. long, 1·5–4 cm. broad, the lateral nerves quite indistinct ; peduncle usually 1–4 cm. long   ..   .. 9. *psorocarpa*
  Corolla-tube 3·5–10 cm. long, lobes 2–4 cm. long ; leaves with the lateral nerves distinct :
    Calyx 2–4 mm. long, with glandular scales within ; peduncle very short, 0–1·5(–2) cm. long ; leaves oblong, lanceolate or elliptic, 5–16 cm. long, 2–6 cm. broad, with 6–9 pairs of lateral nerves   ..   .. 10. *glandulosa*
    Calyx 3–10 mm. long, without glandular scales ; peduncle about 2–5 cm. long ; leaves elliptic to lanceolate, 8–20 cm. long, 3–8 cm. broad, with 4–6 pairs of lateral nerves
11. *eglandulosa*

1. **T. penduliflora** *K. Schum.* in Engl. Bot. Jahrb. 23 : 225 (1896) ; Pichon in Notulae Syst. 13 : 252 (1948). *Conopharyngia* (?) *penduliflora* (K. Schum.) Stapf in F.T.A. 4, 1 : 149 (1902) ; F.W.T.A., ed. 1, 2 : 41. Glabrous shrub or small tree 3–10 ft. high with white, greenish or pink flowers in sessile, subumbellate inflorescences ; in high forest.
S.Nig.: Okomu F.R., Benin (Feb., Mar.) *Brenan* 8575! 9051! Omo (formerly part Shasha) F.R. (Mar.) *Jones & Onochie* FHI 17207! Degema *Talbot* 3711! Oban *Talbot* 1056! Cross R. Div. (Feb.) *McLeod*! **[Br.]Cam.:** S. Bakundu F.R., Kumba (Jan.) *Binuyo & Daramola* FHI 35079! Also in Cameroun, Congo and Sudan.
2. **T. brachyantha** *Stapf* in Kew Bull. 1894 : 22. *Conopharyngia brachyantha* (Stapf) Stapf in F.T.A. 4, 1 : 148 (1902) ; F.W.T.A., ed. 1, 2 : 41. Glabrous tree up to 50 ft. high with numerous small white flowers in long-peduncled cymes.

S.Nig.: Oban *Talbot!* **[Br.]Cam.:** Victoria (Feb.) *Maitland* 429! Cam. Mt., 4,800 ft. *Keay* FHI 37507! **F.Po:** (Jan.) *Mann* 221! Also in Cameroun.

3. **T. ventricosa** *Hochst. ex A. DC.* in DC. Prod. 8 : 366 (1844). *T. usambarensis* Engl. (1895). *Conopharyngia usambarensis* (Engl.) Stapf in F.T.A. 4, 1: 148 (1902). Small tree with a compact crown ; flowers white with yellow centre, fragrant ; in montane forests.
Ghana: Aburi Gardens (? cult., Jan.) *Irvine* 1902! **S.Nig.:** *Rosevear* 9/30! Ikwette to Balegette, Obudu (Dec.) *Savory* FHI 25214! **[Br.]Cam.:** Cam. Mt. (Jan.-May) *Dundas* FHI 15335! *Dalz.* 8346! *Maitland* 273! 506! 676! Nkambe (Feb.) *Hepper* 1898! Also in Congo, Uganda, Kenya, Tanganyika, Nyasaland, S. Rhodesia, Mozambique to Natal.

4. **T. chippii** *(Stapf) Pichon* l.c. 251 (1948). *Conopharyngia chippii* Stapf in Kew Bull. 1913 : 77 ; F.W.T.A. ed. 1, 2 : 41 ; Aubrév. Fl. For. C. Iv., ed. 2, 3 : 212, t. 328. Glabrous tree 15–20 ft. high with large, white or cream, fragrant, fleshy flowers.
Lib.: *Sim* 18! 26! **Ghana:** Kumasi (Mar.) *Andoh* 3743! Anwhiaso F.R. (Oct.) *Vigne* FH 233! Fwirem, Ashanti (May) *Chipp* 454! Bunso (Aug.) *Darko* 952! Jemma (Sept.) *Chipp* 353!

5. **T. pachysiphon** *Stapf* var. **pachysiphon**—in Kew Bull. 1894 : 22. *Conopharyngia pachysiphon* (Stapf) Stapf in F.T.A. 4, 1: 145 (1902) ; F.W.T.A., ed. 1, 2 : 41. Glabrous shrub or medium-sized tree, 9–32 ft. high with large smooth leaves and fleshy white flowers.
Togo Rep.: Klingfall *Kersting* A634! Okposso *Doering* 296! **Dah.:** *Poisson*! **S.Nig.:** Okomu F.R., Benin (Jan.) *Brenan* 8918! Onitsha *Barter* 1328! Sapoba *Kennedy* 2341! Ogwashi-iku, Asaba (Sept.) *Dept. of Agric.*! Oban *Talbot* 1397! **[Br.]Cam.:** Bamenda (Feb.) *Lightbody* FHI 26302!

5a. **T. pachysiphon** var. **cumminsii** *(Stapf) H. Huber* in Kew Bull. 15 : 438 (1962). *Conopharyngia cumminsii* Stapf in F.T.A. 4, 1: 145 (1902) ; F.W.T.A., ed. 1, 2 : 41.
Ghana: Begoro (Dec.) *Plumptre* 89! Aburi *Akpabla* 951! Assin-Yan-Kumasi (Apr.) *Scholes* 236! *Cummins* 114! Tafo (Sept.) *Darko* 997! **Togo Rep.:** Kutio (Apr.) *Scholes* 36! **S.Nig.:** Onitsha *Chesters* 194!

6. **T. longiflora** *Benth.* in Fl. Nigrit. 447 (1849), and in Bot. Mag. 75 : t. 4484 (1849). *Conopharyngia longiflora* (Benth.) Stapf in F.T.A. 4, 1 : 142 (1902) ; F.W.T.A., ed. 1, 2 : 41 ; Aubrév. Fl. For. C. Iv., ed. 2, 3 : 212, t. 328. Glabrous shrub or small tree 5–25 ft. high with very long, white, fragrant flowers.
Port.G.: Empandja, Bissau (Mar.) *Esp. Santo* 1510! L. Cufada (Jan.) *d'Orey* 224! **Guin.:** *Heudelot* 726! *Farmar* 151! Conakry (Mar., June) *Debeaux* 140! 324! Kouria *Chev.* 15033! **S.L.:** *T. Vogel* 151! Sugar Loaf Mt. (fr. May) *Barter*! Waterloo (fl. & fr. Mar.) *Kirk* 23! Rokupr (May) *Jordan* 257! Loma Mts. (Jan.) *Jaeger* 4054! **Lib.:** Ganta (Sept.) *Harley*! *Baldwin* 9332! Totokwelli, Medina (Oct.) *Linder* 1301! Moala, Suen (Nov.) *Linder* 1383!
[Note : in Fl. *Tabernaemontana grandiflora* Hook. (in Gray & Dochard Trav. West. Afr. App. 389, t. B (1825), not *T. grandiflora* Linn.) which was described from a drawing, was cited as a synonym of *T. chippii* ; I cannot agree with this and prefer to follow Stapf who expected this to belong to *T. longiflora.*—Huber.]

7. **T. contorta** *Stapf* in Kew Bull. 1894 : 23. *Conopharyngia contorta* (Stapf) Stapf in F.T.A. 4, 1 : 142 (1902) F.W.T.A., ed. 1, 2 : 41. Tree 30–40 ft. high with large leaves and fleshy, white, fragrant flowers.
S.Nig.: Oban *Talbot*! **[Br.]Cam.:** Ambas Bay (Jan.) *Mann* 703! Victoria (Jan.) *Maitland* 283! *Kalbreyer.* Mayuko (Feb.) *Dalz.* 8223! Johann-Albrechtshöhe *Staudt* 796. Extends to Cameroun.

8. **T. crassa** *Benth.* in Fl. Nigrit. 447 (1849). *T. durissima* Stapf (1894). *Conopharyngia durissima* (Stapf) Stapf in F.T.A. 4, 1 : 143 (1902) ; F.W.T.A., ed. 1, 2 : 41 ; Aubrév. Fl. For. C. Iv., ed. 2, 212, t. 329. *C. jollyana* Stapf in F.T.A. 4, 1 : 144 (1902) ; Chev. Bot. 415. *C. crassa* (Benth.) Stapf in F.T.A. 4, 1 : 144 (1902) ; F.W.T.A., ed. 1, 2 : 41. *Gabunia dorotheae* Wernham in J. Bot. 52 : 25 (1914) ; F.W.T.A., ed. 1, 2 : 40. A small to medium-sized tree ; flowers white, fragrant, opening in the evening ; in wet forest.
S.L.: Sefalu, Bandajuma (fl. & fr. Nov.) *Deighton* 5271! Pujehun (Apr.) *Aylmer* 62! *Thomas* 8050! Zimi (May) *Deighton* 3727! Gola F.R. (May) *Small* 657! **Lib.:** *T. Vogel* 21! Dukwai R. (Nov.) *Cooper* 117! Brewersville (Jan.) *Barker* 1202! Genna Tanyehun *Baldwin* 10769! Totokwelli, Medina (fl. & fr. Oct.) *Linder* 1302! **Iv.C.:** Assinie (fl. & fr. Apr.) *Chev.* 16314! Little Bassam, Dabou, *Jolly* 168! Bingerville *Chev.* 15193! Abidjan (Feb.) *Aubrév.* 82! **Ghana:** *Cameron*! Tanoso (Dec.) *Chipp* 33! Axim (Dec., Feb.) *Johnson* 893! *Irvine* 2145! Simpa (May) *Vigne* FH 1958! **S.Nig.:** Calabar (Apr., fr. Sept.) *Holland* 4! 150! *Thomson* 73! Oban *Talbot* s.n.! 218! Eket *Talbot* 3387! **[Br.]Cam.:** S. Bakundu F.R., Kumba (Jan.) *Binuyo & Daramola* FHI 35172! FHI 35474! Also in Cameroun, Ubangi-Shari, Gabon and Congo.
[Note : typical *T. crassa* seems to differ from *C. durissima* only in having the corolla-tube shorter than the fully developed specimens of the other species mentioned, which evidently is due to their immaturity.]

9. **T. psorocarpa** *(Pierre ex Stapf) Pichon* in Notulae Syst. 13 : 253 (1948). *Gabunia psorocarpa* Pierre ex Stapf in F.T.A. 4, 1 : 137 (1902) ; F.W.T.A., ed. 1, 2 : 9. Glabrous climbing shrub with small flowers in few-flowered terminal cymes.
Lib.: Yeh R. (Oct.) *Linder* 999! Ganta (Sept.) *Baldwin* 9300! **Ghana:** Awiaho, Axim (Nov.) *Vigne* FH 1467! Axim (fr. Mar.) *Irvine* 2395! **S.Nig.:** Eket *Talbot* 3228a! Also in Gabon.

10. **T. glandulosa** *(Stapf) Pichon* l.c. (1948). *Gabunia glandulosa* Stapf in F.T.A. 4, 1 : 138 (1902) ; F.W.T.A., ed. 1, 2 : 40. Glabrous climbing shrub in closed forest with white, fragrant flowers in sub-umbullate cymes.
S.L.: Regent (Apr.) *Sc. Elliot* 5821! York Pass (July) *Lane-Poole* 310! Roruks (July) *Deighton* 3248! Rokupr (July) *Jordan* 48! Fwendu to Potoru, Pelewahun (Apr.) *Deighton* 1617! **Lib.:** Ganta (May) *Harley* 1150! Sinoë Basin *Whyte*! Nyaake, Webo Dist. (June) *Baldwin* 6191! Kulo (Mar.) *Baldwin* 11404! Fortsville (fr. Mar.) *Baldwin* 11149! **Iv.C.:** several localities. **Ghana:** Assuantsi (Mar.) *Irvine* 1158! Aburi (July) *Johnson* 1085! Kwahu (Apr.) *Johnson* 669! Kumasi (Apr.) *Irvine* 2821! near Kwabeng (Apr.) *Adams* 2696! **Dah.:** Lower Ouémé ? *Coll.* **S.Nig.:** Oban *Talbot* 1481. Oban F.R. (Feb.) *Latilo* FHI 36415z! Also in Congo.

11. **T. eglandulosa** *Stapf* in Kew Bull. 1894 : 24. *Gabunia eglandulosa* (Stapf) Stapf in F.T.A. 4, 1 : 138 (1902) incl. var. **macrocalyx** Stapf l.c. 139 (1902) ; F.W.T.A., ed. 1, 2 : 39. *Gabunia longiflora* Stapf in F.T.A. 4, 1 : 138 (1902). Woody climber in forest tangles ; flowers white, fruits yellowish.
S.Nig.: Ikirum (Feb.) *Millson* 12! Osun F.R., Ijebu-Igbo (fr. May) *Ejiofor* FHI 26130! Omo (formerly part Shasha) F.R. (Apr.) *Onochie* FHI 17280! Calabar (Feb.) *Mann* 2253! Oban *Talbot* 1525! **[Br.] Cam.:** Kumba (Jan.) *Binuyo & Daramola* FHI 35160! **F.Po:** (Jan.) *Mann* 239! Also in Cameroun, Gabon and Congo.

Beside the above, *T. donnell-smithii* Roxb., an American species, is cultivated at Victoria.

20. VOACANGA Thouars—F.T.A. 4, 1 : 151 ; Pichon in Bull. Mus. Hist. Nat. sér. 2, 19 : 409 (1947).

Leaves rounded at apex, with 12–16 pairs of lateral nerves, obovate to oblong, 12–25 cm. long, 4–10 cm. broad ; calyx 14–18 mm. long, early circumscissile ; corolla-tube 1·5–2·5 cm. long, lobes 2–3 cm. long   ..   ..   1. *thouarsii*
Leaves acute and acuminate at apex ; calyx not or tardily circumscissile :

Bracts caducous ; calyx-lobes mostly (but not necessarily) spreading or reflexed at anthesis ; leaves with 8–16 pairs of lateral nerves :
　Corolla-lobes 12–24 mm. long when fully expanded, oblong to obovate, forming in bud an elongate-conical structure 1·5–2·5 times as long as broad ; corolla-tube 6–9 mm. long ; calyx 5–10 mm. long ; leaves broadly elliptic to broadly lanceolate or obovate, triangular at apex, not or shortly acuminate, 6–30 cm. long, 3–16 cm. broad ; petiole glabrous or puberulous　..　..　..　.. 　2. *africana*
　Corolla-lobes 4–7 mm. long, broadly oblong to obovate, in bud forming a shortly conical structure about as long as broad ; corolla-tube about 1 cm. long ; calyx 8–12 mm. long, leaves elliptic-lanceolate, 10–20 cm. long, 3–9 cm. broad ; petiole puberulous　..　..　..　..　..　..　.. 　3. *diplochlamys*
Bracts persistent ; calyx-lobes mostly erect ; leaves usually long-acuminate, with 5–10 pairs of lateral nerves :
　Petiole minutely puberulous　..　..　..　..　.. 　3. *diplochlamys*
　Petiole glabrous :
　　Calyx 6–12 mm. long ; corolla-lobes broadly oblong to obovate, forming in bud a shortly conical head 1–1·5 times as long as broad ; leaves elliptic-acuminate to lanceolate, 9–24 cm. long, 3–8 cm. broad :
　　　Corolla-lobes 3–6 mm. long, about half as long as the tube　4. *bracteata* var. *bracteata*
　　　Corolla-lobes 8–12 mm. long, almost or about as long as the tube
　　　　　　　　　　　　　　　　　　　　　4a. *bracteata* var. *zenkeri*
　　Calyx 12–18 mm. long ; corolla-lobes 10–20 mm. long :
　　　Corolla-lobes oblong to ovate-oblong, 3–5 mm. broad, forming in bud an elongate-conical head 2–3 times longer than broad ; leaves elliptic-lanceolate, 9–24 cm. long, 3–8 cm. broad　..　..　..　..　..　.. 　5. *psilocalyx*
　　　Corolla-lobes narrowly linear, about 2 mm. long, forming in bud a cylindric structure 4–6 times longer than broad, which is not dilated towards the base ; leaves lanceolate or oblanceolate, 12–30 cm. long, 2·5–7 cm. broad　.. 　6. *caudiflora*

1. **V. thouarsii** *Roem. & Schult.* Syst. Veg. 4 : 439 (1819) ; F.T.A. 4, 1 : 154 (1902). *V. obtusa* K. Schum. (1895)—F.T.A. 4, 1 : 153 ; F.W.T.A., ed. 1, 2 : 42. *V. thouarsii* var. *obtusa* (K. Schum.) Pichon in Bull. Mus. Hist. Nat., sér. 2, 19 : 415 (1947) ; Aubrév. Fl. For. C. Iv., ed. 2, 3 : 214, t. 330. Glabrous tree 20–30 ft. high with the leaves crowded at the ends of the branches ; flowers white waxy.
　Gam.: *Hayes* 531! Albreda *Leprieur.* **Mali:** Sirakoro *Dubois* 189. **Port.G.:** Coiada, Gabu (June) *Esp. Santo* 2521! **Guin.:** Kaba valley *Chev.* 13188. Kindia (Mar.) *Chev.* 13213 ; 13214. Sineya *Pobéguin* 804. Laya *Sc. Elliot* 4481! **S.L.:** Batkanu to Port Loko (fl. & fr. July) *Deighton* 1970! Sefadu (July) *Deighton* 3744! Kanya (fr. Oct.) *Thomas* 2929! Rokupr to Mange (fr. July) *Jordan* 59! Gberia Timbako (Oct.) *Small* 459! **Lib.:** Dukwia R. (May) *Cooper* 172! 431! Du R. (July) *Linder* 47! Monrovia (Aug.) *Baldwin* 9206! Paynesville (Apr.) *Bequaert* 182! Ganta (May) *Harley*! **Iv.C.:** Buandougou to Marabadiassa, Mankono (July) *Chev.* 21995. Man *Aubrév.* 955. Bioune *Aubrév.* 1126. Guigla *Serv. For* 2021 ; 2029. **Dah.:** Abomey (Feb.) *Chev.* 23201. **N.Nig.:** Kontagora to Ibeto (Jan.) *Meikle* 1088!. Kontagora (Nov.) *Dalz.* 9! Amo, Jos Plateau (Apr.) *Wimbush* FHI 39624! Ewan R., Niger Prov. (Aug.) *Onochie* FHI 35144! Forest areas of tropical Africa and Madagascar, extending to Natal.
2. **V. africana** *Stapf* in J. Linn. Soc. 30 : 87 (1893) ; F.T.A. 4, 1 : 157 (1902). Aubrév. Fl. For. C. Iv., ed. 2, 3 : 216, t. 331. *V. glabra* K. Schum. (1895). *V. schweinfurthii* var. *parviflora* K. Schum. (1897). *V. magnifolia* Wernham in Cat. Talb. 62 (1913). *V. talbotii* Wernham l.c. 63 (1913). *V. eketensis* Wernham in J. Bot. 52 : 25 (1914). *V. glaberrima* Wernham l.c. (1914). *V. africana* var. *glabra* (K. Schum.) Pichon in Bull. Mus. Hist. Nat., sér. 2, 19 : 412 (1947). A shrub or tree, 5–30 ft. high ; flowers white or greenish.
　Sen.: *Heudelot* 89! Casamance *Heudelot* 568. **Gam.:** *Skues*! **Guin.:** Conakry (Apr.) *Debeaux* 320! La Konkoré (Sept.) *Pobéguin* K16! Timbo (Mar.) *Chev.* 12521! Kaloum *Maclaud* 361. Lola, Guerzés *Chev.* 20974. **S.L.:** Kambui F.R. (Apr.) *Lane-Poole* 251! Ninia, Talla Hills *Sc. Elliot* 4903! Bafodeya (Apr.) *Sc. Elliot* 5484! Mano (Apr.) *Deighton* 3382! Bandakarifaia to Yalumbu (May) *Deighton* 5084! **Lib.:** Karmadhun, Kolahun Dist. (fr. Nov.) *Baldwin* 10193! **Iv.C.:** Banco *Serv. For.* 379. Rasso *Aubrév.* 576. Mt. Tongui *Aubrév.* 1027. Bingerville *Chev.* 16033. Alépé *Chev.* 17422. **Ghana :** Axim (fl. & fr. Jan.) *Chipp* 60! Assin-Yan-Kumasi (Apr.) *Scholes* 235! Dunkwa (fr. Aug.) *Chipp* 716! Essiama (Apr.) *Williams* 436! Kumasi (Mar.) *Vigne* FH 3741! **Togo Rep.:** Misahöhe *Baumann* 464! **Dah.:** Ouidah *Poisson.* Adja Ouéré *Poisson* 9; *Le Testu* 279. Boguila, Abomey *Chev.* 23201. **N.Nig.:** Patti Lokoja (Apr.) *Elliott* 48! 222! Nupe *Barter* 1327! Abinsi (Apr.) *Dalz.* 704! Anara F.R., Zaria (May) *Keay* FHI 22940! Birnin Gwari (Apr.) *Meikle* 1376! **S.Nig.:** Onikan, Lagos (Dec.) *Onochie* FHI 34654! Ibadan (Mar., Apr.) *Meikle* 1478! *Hambler* 24! Idanre, Ondo Prov. (Jan.) *Brenan* 8712! Oban *Talbot* 1053! 2071! Eket *Talbot*! **[Br.]Cam.:** Victoria (Feb., Apr.) *Maitland* 422! 759! Buea, 4,000 ft. (Jan., Apr.) *Hutch. & Metcalfe* 107! *Dunlap* 220! S. Bakundu F.R., Kumba (Mar.) *Binuyo & Daramola* FHI 35645! **F.Po:** (Jan.) *Mann* 243! Also in Cameroun, Congo, Egypt, Sudan and (?) Tanganyika. (See Appendix, p. 383.)
3. **V. diplochlamys** *K. Schum.* in E. & P. Pflanzenfam. 4, 2 : 149 (1895) ; F.T.A. 4, 1 : 160 ; Brenan in Kew Bull. 7 : 452 (1953). A shrub ; flowers yellow.
　S.Nig.: Calabar (Mar.) *Thomson* 39! *Holland* 108! Oban *Talbot* 1330! Eket *Talbot*! **[Br.]Cam.:** *Rudatis* 14! Barombi (Mar.) *Preuss* 14! S. Bakundu F.R., Kumba (Jan., Mar.) *Brenan* 9300! *Binuyo & Daramola* FHI 35051! Also in Cameroun and Rio Muni.
4. **V. bracteata** *Stapf* var. *bracteata*—in Kew Bull. 1894 : 22 ; F.T.A. 4, 1 : 160. A glabrous shrub or small tree, 2–10 ft. or more high ; flowers white or more frequently yellowish to brown, in drooping cymes.
　S.L.: Bagroo R. (Apr.) *Mann* 858! Matotoka (July) *Thomas* 1368! Magburaka (fr. Nov.) *Deighton* 4948! Pujahun *Thomas* 8353! Mesima (fr. Dec.) *Deighton* 5295! **Lib.:** Sinoë Basin *Whyte*! Banga (Oct.) *Linder* 1188! Nekabozu, Vonjama Dist. (Oct.) *Baldwin* 10035! Kitoma, Sanokwele Dist. (fr. Jan.) *Baldwin* 14067! Yratoke, Webo Dist. (July) *Baldwin* 6276! **Ghana:** Axim (Feb.) *Irvine* 2358! Awiabo (Nov.) *Vigne* FH 1465! S. Scarp F.R., Kwahu (Dec.) *Adams* 5167! **S.Nig.:** Oban F.R. (Feb.) *Onochie* FHI 36425x! Also in Cameroun and Congo. (See Appendix, p. 383.)
4a. **V. bracteata** var. **zenkeri** (*Stapf*) *H. Huber* in Kew Bull. 15 : 438 (1962). *V. zenkeri* Stapf in F.T.A. 4, 1 : 159 (1902) ; F.W.T.A., ed. 1, 2 : 42. *V. obanensis* Wernham in Cat. Talb. 62 (1913).
　S.Nig.: Oban (Mar.) *Talbot* 290! 1290! *Richards* 5162! **[Br.]Cam.:** Hele, Mamfe Dist. (Nov.) *Tamajong* FHI 22133! Mamfe (fr. Mar.) *Richards* 5225! Also in Cameroun.
5. **V. psilocalyx** *Pierre ex Stapf* in F.T.A. 4, 1 : 159 (1902). Similar to the last species ; peduncle usually very long.
　S.Nig.: Oban *Talbot* s.n.! 302! Also in Cameroun and Gabon.
6. **V. caudiflora** *Stapf* in F.T.A. 4, 1 : 603 (1906). Shrub or small tree 5–20 ft. high ; flowers cream or greenish.

S.L.: Blama (Dec.) *Deighton* 315 ! **Lib.:** Kakatown *Whyte* ! Mecca, Boporo Dist. (Nov.) *Baldwin* 10403 ! Jabroke, Webo Dist. (July) *Baldwin* 6452 !

21. ALSTONIA R. Br.—F.T.A. 4, 1 : 120 ; Monachino in Pacif. Sci. 3, 2 : 139 (1949). *Nom. cons.*

Petiole 1(–2) cm. long ; leaves in whorls of 4–8, obovate to elongate-oblanceolate, rounded to acuminate at apex, 7–25 cm. long, 3–11 cm. broad, with about 30–55 pairs of lateral nerves ; calyx outside densely villous ; corolla-tube usually 6–12 mm. long, lobes 3–6 mm. long, the lobes about as long as broad ; fruits with 2 linear follicles up to 50 cm. long and 5 mm. broad, villous .. ..   1. *boonei*

Petiole 0–5 mm. long ; leaves as above ; calyx slightly pubescent or glabrous outside; corolla-tube usually 4–6 mm., lobes 5–9 mm. long, the lobes much longer than broad ; fruits glabrous .. ..   2. *congensis*

1. **A. boonei** *De Wild.* in Fedde Rep. 13 : 382 (1914). *A. congensis* of F.W.T.A., ed. 1, 2 : 42 (1931) ; Chev. Bot. 414 ; Aubrév. Fl. For. C. Iv., ed. 2, 194, t. 320, not of Engl. Tree 30–120 ft. high with whorled leaves and branches ; many-flowered terminal inflorescences and paired, narrowly-cylindric fruiting follicles 12–20 in. long.
    Sen.: Sinédone, Casamance (Jan.) *Huay* ! *Chev.* 2690 ! **Gam.:** *Dawe* 37 ! **Port.G.:** Fancati, Teikeira Pinto-Caió (Jan.) *d'Orey* 122 ! Bafata *d'Orey* 198 ! **Guin.:** Labé (Jan.) *Roberty* 16401 ! **S.L.:** Bumbuna (Oct.) *Thomas* 3934 ! Njala (fl. Oct., fr. Jan.) *Deighton* 2993 ! 5610 ! 5722 ! Loma Mts. *Jaeger* 1636 ! **Iv.C.:** Bouroukrou *Chev.* 16114 ! **Ghana:** Ankobra Junction *Kitson* 1019 ! Kumasi (Nov.) *Irvine* 1851 ! Sindura *Armitage* 2 ! Aburi *Gady* ! *Deighton* 3416 ! Offinso (Oct.) *Darko* 736 ! **S.Nig.:** Ibadan (fl. Nov., fr. Dec., Jan.) *Punch* 145 ! *Hambler* 86 ! Ibuzo (Nov.) *Thomas* 1986 ! Sapoba *Kennedy* 2084 ! Aboh *Barter* 490 ! Oban *Talbot* 1488 ! **[Br.]Cam.:** Victoria (Nov.) *Maitland* 765 ! Likomba (Nov.) *Mildbr.* 10708 ! Also in Cameroun, Congo, Egypt, Sudan, Uganda and Cabinda.
2. **A. congensis** *Engl.* Bot. Jahrb. 8 : 64 (1887). *A. congensis* var. *glabrata* Hutch. & Dalz. F.W.T.A., ed. 1, 2 : 42 (1931) ; Kew Bull. 1937 : 337 ; Aubrév. Fl. For. C. Iv., ed. 2, 3 : 196.
    S.L.: Gola Forest *Small* 712 ! **S.Nig.:** Lagos (Oct.) *Dalz.* 1256 ! Ikeja *Onochie* FHI 26678 ! Also in Congo and Cabinda. (See Appendix, p. 366.)

22. PYCNOBOTRYA Benth.—F.T.A. 4, 1 : 202 ; Pichon in Mém. Mus. Hist. Nat. Paris, n. sér., B, 1 : 155 (1951).

Leaves and branches glabrous ; leaves verticillate or opposed, elliptic-oblong to oblong-lanceolate, 6–14 cm. long, 1·5–4·5 cm. broad, with very numerous lateral nerves ; corolla-tube 1·5–2 mm. long, the lobes 3–5 mm. long ; anthers about 1 mm. long   *nitida*

P. nitida *Benth.* in Hook. Ic. Pl. 12 : 72, t. 1183 (1876) ; F.T.A. 4, 1 : 202. *P. multiflora* K. Schum. ex Stapf in F.T.A. 4, 1 : 203 (1902). Woody climber up to 50 ft. high with numerous small flowers in ample, terminal cymes.
    S.Nig.: Sapoba *Kennedy* 1951 ! 2085 ! Extends to Cameroun, Gabon and Congo.

23. CATHARANTHUS G. Don (1838)—Pichon in Mém. Mus. Hist. Nat. Paris 27 : 237 (1948). *Lochnera* Reichenb. f. (1828), name only—F.T.A. 4, 1 : 118.

Stem erect, puberulous, herbaceous or more or less woody, especially towards the base ; leaves opposite, obovate to oblanceolate, 3–8 cm. long, 1–4 cm. broad ; flowers terminal or axillary, the corolla-tube 1·5–3 cm. long, the lobes broadly obovate, 1–3 cm. long ; fruits paired narrow-fusiform follicles 2–4 cm. long .. ..   *roseus*

C. roseus (*Linn.*) G. Don Gen. Syst. 4 : 95 (1838). *Vinca rosea* Linn. (1759). *Lochnera rosea* Reichenb. f. (1828) name only—F.T.A. 4, 1 : 118 ; F.W.T.A., ed. 1, 2 : 37 (1931). A perennial (or occasionally annual) erect herb to small shrub with pink or white flowers. " Madagascar Periwinkle ", a native of tropical America, often planted for ornament and now more or less naturalised in the tropics.

24. HOLARRHENA R. Br.—F.T.A. 4, 1 : 161 ; Pichon in Mém. Mus. Hist. Nat. Paris, n. sér., B, 1 : 157 (1951).

Leaves mostly ovate-acuminate, or ovate-lanceolate, 5–18 cm. long, 2–8 cm. broad, with 6–12 pairs of lateral nerves ; inflorescence a many-flowered, globose or flattened cyme ; corolla-tube 5–9 mm. long, lobes 3·5–8 mm. long :

Leaves glabrous or almost glabrous even when young   1. *floribunda* var. *floribunda*
Leaves densely pubescent beneath at least when young   1a. *floribunda* var. *tomentella*

1. **H. floribunda** (*G. Don*) *Dur. & Schinz* var. **floribunda**—Etud. Fl. Congo 1 : 190 (1896). *Rondeletia floribunda* G. Don (1834). *Holarrhena africana* A. DC. (1844)—F.T.A. 4, 1 : 164 ; Aubrév. Fl. For. C. Iv., ed. 2, 3 : 204, t. 324 ; Chev. Bot. 417. *H. wulfsbergii* Stapf in F.T.A. 4, 1 : 164 (1902) ; Aubrév. Fl. For C. Iv., ed. 2, 3 : 204 ; Chev. Bot. 418. Shrub to medium-sized tree 5–50 ft. high ; flowers white in almost umbel-like inflorescences and paired narrowly cylindrical fruiting follicles 1–2 ft. long ; in deciduous forest.
    Mali: Ouassana to Dendela (fl. & fr. Mar.) *Chev.* 620 ! **Port.G.:** Abu, Formosa (Apr.) *Esp. Santo* 1978 ! **Guin.:** Friguiagbé (fr. Aug.) *Chillou* 606 ! Conakry (fr. Jan.) *Dalz.* 8079 ! Kouroussa (Apr.) *Pobéguin* 140 ! Dalaba (Mar.) *Dalz.* 8402 ! Macenta (fr. Oct.) *Baldwin* 9793 ! **S.L.:** *Don* ! Freetown (Jan.) *Dalz.* 971 ! Karina, N. Prov. (Feb.) *Glanville* 181 ! Kambui F.R. (Apr.) *Jordan* 2033 ! Rokupr (fr. Apr.) *Jordan* 226 ! Musaia (Feb.) *Deighton* 4218 ! **Lib.:** Ganta (Apr.) *Harley* 1136 ! **Iv.C.:** Abidjan (fr. Feb.) *Aubrév.* 64 ! Man *Serv. For.* 406 ! **Ghana:** Aburi (fr. Nov.) *Gould* ! Mampong (Apr.) *Vigne* FH 1920 ! L. Bosumtwe (Apr.) *Adams* 2470 ! Kumasi (Apr.) *Andoh* 4884 ! Ho (Mar.) *Dalz.* 156 ! **Togo Rep.:** Badja *Schlechter* 12909 ! Lomé *Warnecke* 131 ! Waya *Hornberger.* Agatineo, near Todjie R. *Merz.* **Dah.:** Le *Testu* 122 ! Zagnanado (Feb.) *Chev.* 22978. Abomey (Feb.) *Chev.* 23192. **N.Nig.:** Jos Plateau (May, fr. July) *Lely* P311 ! P502 ! Keana (Apr., May) *Hepburn* 49 ! **S.Nig.:** Ibadan (Mar., Dec.) *Meikle* 869 ! *Adekunle* FHI 3007 ! Olokemeji F.R. (fr. Dec.) *Ross* R36 ! Abeokuta *Barter* 3334 ! Idanre, Ondo Prov. (fl. & fr. Jan.) *Brenan* 8655 ! Also in Cameroun, Ubangi-Shari and Congo.

1a. **H. floribunda** var. **tomentella** *H. Huber* in Kew Bull. 15 : 437 (1962). *H. ovata* A. DC. (1844)—F.T.A. 4, 1 : 163.
    Gam.: *Dawe* 40! *Hayes* 509! *Brown-Lester*! R. Gambia *Ozanne* 9! **Mali:** Kangola to Kindolé (May) *Chev.* 857! **Port.G.:** Begene, Farim (June) *Esp. Santo* 2393! **Guin.:** Rio Nunez *Heudelot* 795! **S.L.:** Musaia (fl. & fr. Apr.) *Miszewski* 36! **N.Nig:** Tureta to Damri, Sokoto Prov. (June) *Dalz.* 534! Sokoto to Kebbi (fr. Oct.) *Dalz.* 370! Kanbuja to Yashi, Katsina Prov. (June) *Keay* FHI 7571! Katsina (Mar.) *Meikle* 1355! Anara F.R., Zaria Prov. (May) *Keay* FHI 25788!

25. RAUVOLFIA Linn.—F.T.A. 4, 1 : 108, as *Rauwolfia*

Branches of inflorescences distinctly puberulous ; leaves elliptic-acuminate to broadly lanceolate, 8–24 cm. long, 3–10 cm. broad, with 8–16 pairs of lateral nerves, with a narrowly triangular acute acumen ; corolla 8–10 mm. long ; carpels (if both developed) quite free in fruit ..   ..   1. *vomitoria*
Branches of inflorescences glabrous :
  Leaves with 5–12 pairs of lateral nerves ; pedicels 1–4 mm. long ; flowers usually 6–10 mm. long :
    Petiole short, about 5(–10) mm. long ; leaves elliptic-acuminate to lanceolate, pointed with a triangular acumen ; blade 8–20 cm. long, 2·5–6 cm. broad ; carpels (if both developed) quite free in fruit ..   ..   ..   2. *cumminsii*
    Petiole about 10(–20) mm. long ; leaves elliptic to broadly lanceolate, pointed with a narrowly linear acumen, 10–26 cm. long, 3–9 cm. broad ; fruiting carpels (if both developed) connate to about half of their length ..   ..   ..   3. *mannii*
  Leaves with 16 to about 30 pairs of lateral nerves ; flowers sessile or very shortly pedicellate ; pedicels not exceeding 1 mm. in length ; flowers 4–6 mm. long :
    Petiole about 2–4 cm. long ; leaves oblanceolate-oblong, 7–22 cm. long, 3–6 cm. broad, short-cuneate or almost rounded at base, not or scarcely decurrent on the petiole, pointed with a triangular acumen ; branches 4-angled ; flowers about 6 mm. long 4. *caffra*
    Leaves subsessile or with a petiole up to about 2 cm. long, obovate to obovate-oblong, 10–20 cm. long, 4–9 cm. broad, long-cuneate at base and gradually passing into the petiole, not or scarcely acuminate at apex ; branches almost 4-winged ; flowers about 4 mm. long   ..   ..   5. *macrophylla*

1. **R. vomitoria** *Afzel.* Stirp. Med. Sp. Nov. 1 (1818) ; F.T.A. 4, 1 : 115 ; Chev. Bot. 412 ; Aubrév. Fl. For. C. Iv., ed. 2, 3 : 196, t. 321. *Hylacium owariense* P. Beauv. (1819)—Hepper in Kew Bull. 16 : 338 (1962). *Rauvolfia senegambiae* A. DC. (1844). Small shrub or tree up to 50 ft. high with whorled leaves and small white to greenish flowers in many-flowered, terminal, ample, verticillate cymes.
  Sen.: Rio Nunez *Heudelot* 910. **Mali:** Folo (May) *Chev.* 832! **Port.G.:** Pussubé, Bissau (Oct.) *Esp. Santo* 984! Acóco, Formosa (Apr.) *Esp. Santo* 1960! Catió (Aug.) *Esp. Santo* 2175! **Guin.:** Farmar 162a! Sulimana *Sc. Elliot* 5318! Duyania (Feb.) *Sc. Elliot* 4823! **S.L.:** Bagroo R. *Mann*! Njala (May) *Deighton* 677! Kambia (Dec.) *Sc. Elliot* 4356! Kambui Hills (Mar.) *Small* 513! Loma Mts. (fr. Sept.) *Jaeger* 1541! **Lib.:** Monrovia (Nov.) *Linder* 1419! Karmadhun, Kolahun Dist. (fr. Nov.) *Baldwin* 10194! Ganta (fr. Sept.) *Baldwin* 9281! Gbanga (Mar.) *Blickenstaff* 30! Gletown, Tchien Dist. (fr. July) *Baldwin* 6960! **Ghana:** Asientien (July) *Chipp* 286! Aburi (fr. Oct.) *Ankrah* GC 20106! Kumasi (Jan.) *Irvine* 59! Axim (fl. & fr. Feb.) *Irvine* 2158! Mampong, Akwapim (fr. May) *Adams* 4341! **Dah.:** Cotonou *Debeaux* 173! **N.Nig.:** Nupe *Barter* 1704! **S.Nig.:** Abeokuta *Irving* 64! Okomu F.R., Benin (Jan.) *Brenan* 8831! Ibadan (Feb.) *Meikle* 1165! Sapoba (Nov.) *Meikle* 632! Onitsha (Feb.) *Jones* FHI 575! Calabar (fr. July) *Holland* 64! **[Br.]Cam.:** *Preuss* 903! Mimbia, Buea (Apr.) *Hutch. & Metcalfe* 93! S. Bakundu F.R., Kumba (Mar.) *Binuyo & Daramola* FHI 35594! Bafut-Ngemba F.R. (Feb.) *Tiku* FHI 22159! Also in Cameroun, Gabon, S. Tomé, Congo, Sudan, Uganda and Tanganyika.

2. **R. cumminsii** *Stapf* in F.T.A. 4, 1 : 114 (1902) ; Aubrév. Fl. For. C. Iv., ed. 2, 3 : 196. *R. liberiensis* Stapf in F.T.A. 4, 1 : 601 (1906). *R. ivorensis* A. Chev. Bot. 412, name only. Shrub or small tree, 2–7 ft. high with red fruits.
  Lib.: Mecca, Boporo Dist. (Nov.) *Baldwin* 10428! Monrovia *Whyte*! Totokwelli, Medina (Oct.) *Linder* 1297! Greenville (Mar.) *Baldwin* 11564! Sinoe *Whyte*! **Ghana:** Assin-Yan-Kumasi *Cummins* 216! Awiabo (Nov.) *Vigne* FH 100! Benso, Tarkwa Dist. (Nov.) *Andoh* FH 5408! 5877! Abra, W. Prov. (fr. Jan.) *Akpabla* 812!

3. **R. mannii** *Stapf* in Kew Bull. 1894 : 21 ; F.T.A. 4, 1 : 113 ; Brenan in Kew Bull. 7 : 451. *R. preussii* K. Schum. (1895)—F.T.A. 4, 1 : 114. Small shrub or tree 2–30 ft. high with greenish-white and pink flowers and red berries.
  S.Nig.: Ibadan (fr. Nov.) *Newberry & Etim* 163! Oban F.R. (May) *Ujor* FHI 31782! **[Br.]Cam.:** S. Bakundu F.R., Kumba (fl. & fr. Mar.) *Brenan* 9422! Mamfe (Mar.) *Richards* 5220! Bamenda (fr. June) *Lightbody* FHI 26327! Extends to Cameroun, Gabon and Congo.

4. **R. caffra** *Sond.* in Linnaea 23 : 77 (1850) ; F.T.A. 4, 1 : 110. *R. welwitschii* Stapf in F.T.A. 4, 1 : 110 (1902). Small to large tree up to 70 ft. high with verticillate leaves and small tubular flowers in ample inflorescences.
  N.Nig.: Auchang, Zaria *Dalz.* 358! Zaria (July) *Lamb*! Katsina Allah (June) *Dalz.* 796! **S.Nig.:** Foriku to Agbabu, Ondo (sterile) *Onochie* FHI 33376! Idodo R., Enugu to Abakaliki (fr. May) *Onochie* FHI 35838! Lower Enyong F.R., Calabar (fr. May) *Onochie* FHI 33212! **[Br.]Cam.:** Mai Idoanu, Gashaka Dist. (fr. Feb.) *Latilo & Daramola* FHI 34472! Widespread in tropical and S. Africa.

5. **R. macrophylla** *Stapf* in Kew Bull. 1894 : 20 ; F.T.A. 4, 1 : 110. Tree 40–75 ft. high with a broad crown and the leaf-whorls mostly crowded at ends of the branches ; flowers minute, whitish.
  S.Nig.: Mamu River F.R., Onitsha *Keay* FHI 22300! (sterile, probably this sp.) **[Br.]Cam.:** Likomba (fr. Dec.) *Mildbr.* 10783! Tiko (Jan.) *Dunlap* 248! Ambas Bay (Dec.) *Mann* 1328! Also in Cameroun and S. Tomé.

26. ISONEMA R. Br.—F.T.A. 4, 1 : 187.

Leaves glabrous, broadly elliptic, 5–9 cm. long, 2·5–5 cm. broad, with 4 or 5 pairs of lateral nerves ; corolla-tube 8–10 mm. long, lobes 5–9 mm. long ; panicle lax, with long, spreading branches ..   ..   1. *buchholzii*
Leaves pilose beneath especially on the nerves ; elliptic, obovate or oblong, 4–11 cm. long, 2–5 cm. broad, with 5–7 pairs of lateral nerves ; corolla-tube 8–10 mm. long,

lobes 5–6 mm. long ; panicle narrow, with fairly to very short branches ; fruit with 2 carpels up to 20 cm. long, and 7–9 mm. diam.   ..   ..   *2. smeathmannii*

1. **I. buchholzii** *Engl.* Bot. Jahrb. 7 : 340 (1886) ; F.T.A. 4, 1 : 189. Glabrous woody climber with white to brownish-cream, tubular flowers in ample panicles ; in swamp forest.
 S.Nig.: Orlu, Owerri Prov. (Apr.) *Thompson* 16 ! Eket *Talbot* 3027 ! Cross River North F.R., Ikom Dist. (June) *Latilo* FHI 31844 ! [Br.]**Cam.:** Johann-Albrechtshöhe (= Kumba) *Staudt* 481 ! Also in Cameroun.

2. **I. smeathmannii** *Roem. & Schult.* Syst. Veg. 4 : 401 (1819) ; F.T.A. 4, 1 : 188. Climbing shrub with mostly rusty-tomentose indumentum, red-brown tubular flowers and horizontally diverging cylindrical follicles ; in swamp forest.
 Port.G.: Cumura to Bór, Bissau (Mar.) *Esp. Santo* 1879 ! Safim, Bissau (Mar.) *Esp. Santo* 1905 ! **Guin.:** *Heudelot* 912 ! Kindia (Sept.) *Pobéguin* 1285 ! **S.L.:** Kahreni to Port Loko (fl. & fr. Apr.) *Sc. Elliot* 5814 ! Bagroo R. (fl. & fr. Apr.) *Mann* 824 ! Bonthe, Sherbro (fl. & fr. Feb.) *Dalz.* 943 ! Njala *Deighton* 1747 ! Baiima (May) *Deighton* 5772 ! **Lib.:** Grand Bassa (July) *T. Vogel* 65 ! Dobli Is., St. Paul R. *Bequaert* 24 ! near Mt. Barclay (Jan.) *Dalz.* 8120 ! Sangwin (fr. Mar.) *Baldwin* 11308 ! Dimeh (June) *Barker* 1338 ! **Iv.C.:** Farmar 370 ! **Ghana:** *Burton* !

27. STROPHANTHUS A. DC.—F.T.A. 4, 1 : 167 ; Gilg in Engl. Monogr. Afr. Apocynac. 7 (1903).

Corolla-lobes not produced in tails, not more than 3 times as long as broad at base, not longer than the corolla-tube ; bracts scarious ; leaves glabrous, smooth ; tertiary venation indistinct :
Corolla-tube 3–4·5 cm. long, lobes broadly ovate or orbicular, obtuse at apex, 1·5–3 cm. long and about as broad at their base ; inflorescence with several flowers ; calyx 10–15 mm. long ; leaves elliptic, elliptic-lanceolate or ovate, 7–19 cm. long, 3–9 cm. broad, with 5–9 pairs of lateral nerves   ..   ..   ..   *1. gratus*
Corolla-tube 2·5–3 cm. long, lobes narrowly triangular, acute at apex, 2–3 cm. long, 1·5–3 times longer than broad at their base ; inflorescence 1–few-flowered ; calyx 13–15 mm. long ; leaves elliptic to lanceolate, 8–14 cm. long, 2–5 cm. broad with (4–)6 pairs of lateral nerves   ..   ..   *2. thollonii*
Corolla-lobes produced into long, linear tails, markedly longer than the tube ; bracts herbaceous :
Leaves smooth and glabrous on both surfaces ; branches glabrous :
Corolla-tube 2–3·5 cm. long, lobes 4–16 cm. long, produced into linear tails 2–4 times longer than the oblong-deltoid base ; calyx 10–15 mm. long, lobes lanceolate, very acute ; leaves elliptic or ovate, acuminate, 4–11 cm. long, 2·7 cm. broad, with 4–7 pairs of lateral nerves and frequently with distinct tertiary venation beneath
                                                      3. sarmentosus
Corolla-tube up to 2 cm. long ; lobes produced into subfiliform tails many times longer than the deltoid base ; calyx 8–17 mm. long ; leaves usually with indistinct tertiary venation :
Sepals ovate-lanceolate, broadest below the middle, very acute ; corolla-tube 1·5–2 cm. long, lobes 12–32 cm. long ; leaves elliptic or elliptic-oblong, 3·5–13 cm. long and 2·5–8 cm. broad, with 5–8 pairs of lateral nerves   ..   *4. preussii*
Sepals oblanceolate, broadest towards the apex, subobtuse ; corolla-tube 1·2–1·8 cm. long, lobes 3–8 cm. long ; leaves elliptic to ovate, acuminate, 3–10 cm. long, 2·5 cm. broad, with 4–6 pairs of lateral nerves   ..   ..   ..   *5. barteri*
Leaves scabrid or hairy, at least beneath :
Leaves scabrid or minutely puberulous, especially beneath, elliptic or ovate, 4–12 cm. long and 1·5–4 cm. broad, with 4–6 pairs of lateral nerves ; calyx 10–14 mm. long ; corolla-tube 1–1·5 cm. long, lobes 8–18 cm. long   ..   ..   ..   *6. gracilis*
Leaves hairy beneath, with long patent hairs on midrib and branches :
Corolla-lobes 8–18 cm. long, 5–12 times longer than the tube ; tube 8–15 mm. long ; calyx 14–24 mm. long ; leaves elliptic or ovate, 5–15 cm. long, 2–9 cm. broad, with 5–9 pairs of lateral nerves ; tertiary venation very distinct, deeply impressed above, prominent beneath   ..   ..   *7. hispidus*
Corolla-lobes up to 2 cm. long, about 1·5–2 times longer than the tube ; tube 10–14 mm. long ; calyx 7–12 mm. long ; leaves elliptic, acuminate, 7–12 cm. long, 2·5–5 cm. broad, with 6–12 pairs of lateral nerves ; tertiary venation very indistinct   ..   ..   *8. bullenianus*

1. **S. gratus** *(Hook.) Franch.* in Journ. de Bot. 7 : 321 (1893) ; F.T.A. 4, 1 : 170. *Roupellia grata* Hook. in Bot. Mag. 75 : t. 4466 (1849). Glabrous, climbing or erect shrub or small tree with large, campanulate, white to purple flowers.
 S.L.: *Whitfield* ! *Lane-Poole* 147 ! Pujehun (Dec.) *Deighton* 275 ! Waterloo to York *Smythe* 245 ! Gowana (Dec.) *Burbridge* 425 ! **Lib.:** Monrovia (Nov.) *Linder* 1413 ! Kakatown *Whyte* ! Belleyella, Boporo Dist. (Dec.) *Baldwin* 12554 ! Ganta (Dec.) *Barker* 1147 ! Suacoco, Gbanga (Apr.) *Konneh* 151 ! **Iv.C.:** Abouabou *Leeuwenberg* 2353 ! **Ghana:** Aburi Gardens (cult. Nov.) *Irvine* 1863 ! *Baldwin* 13403 ! Akwaseho, Kwahu (Dec., Jan.) *Irvine* 1681 ! Assin-Yan-Kumasi *Cummins* 44–55 ! Akwapim (Dec.) *Dalz.* 158 ! **N.Nig.:** Benue R. *Flegel.* **S.Nig.:** Olokemeji *Foster* 139 ! Ibadan (Feb.) *Meikle* 1159 ! Ubuluku (Dec.) *Thomas* 2079 ! Awka to Onitsha (Nov.) *Baldwin* 13749 ! Bendiga Ayuk, Ikom (Dec.) *Keay* FHI 28166 ! Oban *Talbot* 2054 ! [Br.]**Cam.:** Victoria (cult. Mar.) *Preuss* 1301. Mamfe (Dec.) *Baldwin* 13814 ! 13815 ! Also in Cameroun, Gabon and Congo. (See Appendix, p. 378.)

2. **S. thollonii** *Franch.* l.c. 299 (1893) ; F.T.A. 4, 1 : 171. Glabrous, woody climber with white and purple flowers similar to the above.
 S.Nig.: Oban *Talbot* ! [Br.]**Cam.:** Rio del Rey *Johnston* 44 ! Extends to Cameroun and Gabon.

3. **S. sarmentosus** *DC.* in Bull. Soc. Philom. Paris 3 : 123 (1802) ; F.T.A. 4, 1 : 180. A stout, glabrous climber up to about 20 ft. high with long-tailed flowers, pale yellow to almost white ; fruiting follicles horizontally spreading, thick and woody, about 8 in. long and 1–1½ in. broad.

Fig. 215.—Strophanthus sarmentosus *DC.* (Apocynaceae).
A, leafy shoot. B, flowers. C, stamen. D, one valve of fruit from within. E, seed.

71

Sen.: *Heudelot* 337 ! *Leprieur* 457 ; 467. *Boivin* 1408 ! Richard-Toll *Leprieur.* Cayor *Leprieur.* **Gam.:** *Hayes* 510 ! 526 ! *Brown-Lester* 36 ! *Skues* ! *Ingram* 66 ! Genieri (Feb.) *Fox* 96 ! **Mali:** Bamako (Apr.) *Hagerup* 30 ! Siguire (Feb.) *Chev.* 288 ! **Guin.:** *Farmar* 166 ! Boké (Apr.) *Maclaud* 386 ! Pela, Nzérékoré (fr. Sept.) *Baldwin* 13320 ! Dalaba, Fouta Djalon (Mar.) *Schnell* 4640 ! Nimba Mts. (Feb.) *Schnell* 4263! **S.L.:** Bunce Is. (Mar.) *Kirk* 39 ! Njala (Feb.) *Deighton* 1928 ! Port Loko (Jan.) *Deighton* 1048 ! Kasawe F.R. (Jan.) *King* 207*B* ! Kukuna (Jan.) *Sc. Elliot* 4512 ! **Lib.:** Monrovia (Nov.) *Linder* 1517 ! Gbanga (fr. Sept.) *Linder* 596 ! Dobli Is., St. Paul R. *Bequaert* 29 ! Suacoco (Dec.) *Barker* 1185 ! St. John R., Sanokwele Dist. *Baldwin* 13209 ! **Iv.C.:** Dapola (young fr. July) *Katz & Schmutz* H72 ! Odienné to Touba *Katz & Schmutz* H73 ! Korogo to Odienné *Katz & Schmutz* H71 ! Bobo-Dioulasso *Katz & Schmutz* H74 ! Man *Katz & Schmutz* H68 ! **Ghana:** Dodowah Plains (Feb.) *Irvine* 1500 ! Saltpond (Jan.) *Baldwin* 14017 ! Dunkwa (Apr.) *Vigne* FH 147 ! Kumasi (fl. & young fr. May) *Katz & Schmutz* H10 ! Saboba (Mar.) *Hepper & Morton* A3119 ! Gambaga (June) *Saunders* 5 ! **Togo Rep.:** *Baumann* 93 ! Lomé *Warnecke* 476 ! Agome Mts. *Schlechter* 12962. Pewa *Katz & Schmutz* H70 ! **N.Nig.:** Nupe *Barter* 1325 ! Mokwa (Mar.) *Meikle* 1238 ! Jos (Mar.) *Hill* 15 ! Sokoto (May) *Lely* 825 ! Daddara, Katsina (Feb.) *Meikle* 1479 ! Kaduna to Zaria (Dec.) *Meikle* 838 ! **S.Nig.:** Lagos (Feb.) *Millson* ! Abeokuta *Irving* ! Idanre (Jan.) *Brenan & Keay* 8692 ! Ogun River F.R. (Dec.) *Ejiofor & Latilo* FHI 32026 ! Okomu F.R., Benin (Feb.) *Brenan & E. W. Jones* 9133 ! Oban *Talbot* 1349 ! **[Br.]Cam.:** Gashaka to Mayo Wombo (fr. Feb.) *Latilo & Daramola* FHI 34457 ! foot of Vogel Peak, Adamawa (fr. Nov.) *Hepper* 1466 ! Also Cameroun, Ubangi-Shari, Gabon, Congo and Cabinda.

4. **S. preussii** *Engl. & Pax* in Engl. Bot. Jahrb. 15 : 369 (1892) ; F.T.A. 4, 1 : 176. Glabrous climber or small shrub with white and purple flowers turning yellow.
   Guin.: *Collenette* 19 ! Nimba Mts. (Apr.) *Schnell* 4902 ! 5228 ! **S.L.:** Freetown (Aug.) *Burbridge* 520 ! Gorahun (Apr.) *Deighton* 3684 ! Jahama *Smythe* 77 ! Yumbuma (Apr.) *Deighton* 3943 ! Mayiera (July) *Marmo* 170 ! **Lib.:** Bobei (Feb.) *Baldwin* 14189 ! Ganta (Apr.) *Harley* ! **Iv.C.:** Adiopodoumé (Nov.) *Roberty* 15511 ! **Ghana:** Oda, W. Akim (fl. & fr. Jan.) *Darko* 497 ! Essiama (Apr.) *Williams* 437 ! Aburi (May) *Johnson* 150 ! Kumasi to Agogo (Apr.) *Bally* 181 ! Assuantsi (fr. Jan.) *Baldwin* 14027 ! **Togo Rep.:** *Baumann* 591*a* ! Mamu F.R., Ibadan (Mar.) *Keay* FHI 22538 ! Okomu F.R., Benin (Dec.) *Brenan* 8447 ! Atta *Barter* 3322 ! Calabar (Apr.) *Holland* 100 ! Isobendige, Ikom Dist. (May) *Latilo* FHI 31820 ! Degema *Talbot* 3834 ! **[Br.]Cam.:** *Preuss* 116 ! Victoria (Mar.) *Preuss* 1114. Johann-Albrechtshöhe (Mar.) *Staudt* 664. Banyu, Kumba Dist. *Olorunfemi* FHI 30699 ! **F.Po:** (Jan.) *Mann* 177 ! Also in Cameroun, Congo, Uganda, Angola and Tanganyika.

5. **S. barteri** *Franch.* in Journ. de Bot. 7 : 323 (1893) ; F.T.A. 4, 1 : 177 ; Chev. Bot. 419. Glabrous climbing shrub with white to yellow flowers turning yellow.
   Guin.: Nimba Mts. (fl. & fr. Apr.) *Schnell* 5249 ! **Iv.C.:** Mt. Kouan to Danané *Chev.* 21252. Man to Zagoué *Chev.* 21556. **Ghana:** Kumasi (Apr.) *Bally* 115 ! Otrokpe (Apr.) *Vigne* FH 4380 ! Agogo, Ashanti *Vigne* FH 1083 ! Mensah Dawa (May) *Darko* 579 ! Kokofu to Kuntanase (Apr.) *Adams* 2488 ! **S.Nig.:** Lagos *Dalz.* 1106 ! Abeokuta *Barter* 3346 ! Ikire (fr. Jan.) *Katz & Schmutz* H83 ! Ibadan (fr. Feb.) *Meikle* 1129 ! Gambari F.R., Ibadan Dist. (Mar.) *Onochie* FHI 21700 ! **[Br.]Cam.:** Fang, Bamenda (May) *Maitland* 1691 !

6. **S. gracilis** *K. Schum. & Pax* in Engl. Bot. Jahrb. 15 : 370 (1892) ; F.T.A. 4, 1 : 175. Slender climbing shrub with minutely scabrid branches and fairly small pale lemon-yellow flowers spotted with crimson or brown.
   S.Nig.: Nun R. (Sept.) *Mann* 499 ! Bonny *Kalbreyer* 70 ! Sapoba, Benin (Nov.) *Meikle* 540 ! Eket *Talbot* 3083 ! Calabar (Nov.) *Baldwin* 13756 ! Also in Cameroun and Gabon.

7. **S. hispidus** *DC.* in Bull. Soc. Philom. Paris 3 : 123 (1802) ; F.T.A. 4, 1 : 174. A coarsely hairy climbing shrub with long-caudate white flowers turning to yellow, spotted with red or brown inside ; fruit woody, 12–18 in. long and about ½ in. thick. Sometimes cultivated.
   Sen.: Louga *Joret.* Casamance R. *Perrottet.* **Mali:** Banan (Mar.) *Chev.* 522 ! Kangala (May) *Chev.* 859 ! **Port.G.:** Pussubé, Bissau (Mar.) *Esp. Santo* 1481 ! Bissoram to Mansoa (Sept.) *Esp. Santo* 853 ! **Guin.:** Labé (Mar.) *Schnell* 4757 ! Boké (Apr.) *Maclaud* 381 ! Nimba Mts. (Feb.) *Schnell* 4487 ! Nzérékoré (fr. Sept.) *Baldwin* 13321 ! Dantilia (Mar.) *Sc. Elliot* 5300 ! Rio Nunez *Heudelot* 829 ! **S.L.:** Sherboro R. (Mar.) *Mann* 793 ! Messima (Apr.) *Adames* 28 ! Njala (Mar.) *Deighton* 2865 ! Loma Mts. (Aug.) *Jaeger* 935 ! **Lib.:** Totata, Salala Dist. (Feb.) *Baldwin* 14175 ! **Ghana:** Afram Plains (fl. & fr. Mar.) *Johnson* 594*b* ! Tuna (Feb.) *Kitson* 894 ! Pokoase, Accra Plains (Mar.) *Vigne* FH 2 ! Tamale (Apr.) *Bally* 143 ! Bossuso (Feb.) *Andoh* 5106 ! **N.Nig.:** Nupe *Barter* 749 ! Kontagora (Jan.) *Dalz.* 11 ! (cult.) *Meikle* 1054 ! Kufana, Kaduna Div. (cult. Dec.) *Meikle* 789 ! Lokoja (May) *Elliott* 233 ! Atakati (June) *Elliott* 236 ! **S.Nig.:** Abeokuta *Irving* 188 ! Olokemeji F.R. (Feb.) *Ejiofor* FHI 26898 ! Omo F.R. (Mar.) *Jones & Onochie* FHI 16070 ! Isa *W. D. MacGregor* 180 ! Oban *Talbot* 1482 ! **[Br.]Cam.:** S. Bakundu F. R., Kumba (Mar.) *Binuyo & Daramolo* FHI 35617 ! Mandesi, Kumba (Apr.) *Olorunfemi* FHI 30529 ! Mamfe to Bamenda *Baldwin* 13730 ! Also in Cameroun, Gabon, Cabinda, Congo & Uganda. (See Appendix, p. 379.)

8. **S. bullenianus** *Mast.* in Gard. Chron. 1870 : 1471, excl. fruit ; F.T.A. 4, 1 : 175. *S. schlechteri* K. Schum. & Gilg in Engl. Bot. Jahrb. 32 : 158 (1902) ; Climber with pilose branches and fairly small yellow and brown or pink flowers.
   S.Nig.: Calabar (fl. & fr. Feb.) *Mann* 2247 ! *Thomson* 22 ! Oban *Talbot* 211 ! Oban to Calabar (May) *Ujor* FHI 31798 ! **[Br.]Cam.:** Victoria (June) *Maitland* 56 ! Kumba (fr. Nov.) *Krukoff* 064 (photo !) **F.Po:** *Mann* 1444 ! Also in Gabon.

28. MASCARENHASIA A. DC. (1844)—F.T.A. 4, 1 : 193
(1902). Lanugia N. E. Br. (1927).

Leaves elliptic, obovate or lanceolate, 6–15 cm. long and 3–6 cm. broad, with up to 12 pairs of lateral nerves ; corolla-tube 8–12 mm. long, constricted at about the middle, wider above ; lobes ovate-acuminate, 6–8 mm. long, in bud lengthwise folded together ; disk with 5 distinct segments   ..   ..   ..   *arborescens*

M. arborescens *A. DC.* in DC. Prod. 8 : 488 (1844) ; Pichon in Mém. Inst. Sci. Madag., sér. B, 2 : 81 (1949). *M. variegata* Britten & Rendle (1894)—F.T.A. 4, 1 : 193 ; Brenan in N.Y. Bot. Jard. 8 : 503 (1954). *M. elastica* K. Schum. in Notizbl. Bot. Gart. Berl. 2 : 270, with fig. (1899) ; F.T.A. 4, 1 : 194. *Lanugia variegata* (Britten & Rendle) N. E. Br. (1927). A large shrub or small tree with creamy-white flowers in axillary and terminal, several to many-flowered fascicles.
   Togo Rep.: Lomé *Warnecke* 481 *bis.* (*vide* Pichon *l.c.*). Possibly cultivated or subspontaneous. Also in Congo, Kenya, Tanganyika, Pemba, Zanzibar, Mozambique and Madagascar.

Imperfectly known species.

Savory & Keay FHI 25205 from E. Nigeria belongs to a genus unrepresented in this Flora, but the key to the genera may lead to the above genus. It differs from *Mascarenhasia* by having oblong corolla-lobes, flat in bud, the absence of the disk and the inflorescence being 1-few-flowered. Fruits and further flowering material are required.—Huber.

29. ALAFIA Thouars. (1806)—F.T.A. 4, 1 : 195 ; Pichon in Mém. Inst. Sci. Madag., sér. B, 2 : 45 (1949). Holalafia Stapf in Kew Bull. 1894 : 123 ; F.T.A. 4, 1 : 201.

Carpels and meric*r* rps connate ; corolla-tube 12–15 mm. long, lobes broadly ovate,

10–12 mm. long ; leaves glabrous, broadly elliptic to elliptic-oblong, 7–18 cm. long, 4–10 cm. broad, with 6–10 pairs of lateral nerves, rounded or truncate at base
1. multiflora

Carpels and mericarps free ; corolla-tube up to 12 mm. long :
Leaves glabrous above, densely tomentellous beneath, broadly elliptic to elliptic-oblong, large, 18–26 cm. long and 9–14 cm. broad, with about 10–12 pairs of lateral nerves ; corolla-tube 6–7 mm. long, pilose within immediately above the insertion of the anthers ; corolla-lobes broadly ovate, 5–7 mm. long   ..   2. grandis
Leaves glabrous on both sides, usually small or medium-sized ; corolla-tube glabrous within immediately above the insertion of the anthers :
Corolla-lobes forming in bud a broadly ovoid-conical head not more than 1½ times longer than broad ; the lobes rarely exceeding 5 mm. in length and always distinctly shorter than the tube :
Leaves obovate-elliptic, 12–20 cm. long and 6–8 cm. broad, with 4–6 pairs of lateral nerves, cuneate at base ; corolla-tube 9–11 mm. long   ..   ..   3. whytei
Leaves usually 4–13 cm. long and 2–6·5 cm. broad, lateral nerves usually 6–11 pairs ; corolla-tube 3·5–7·5 mm. long :
Corolla-tube 3·5–5·5 mm. long ; leaves elliptic, elliptic-oblong or obovate, often rounded at base ; corolla-tubes almost orbicular, ciliate   ..   4. barteri
Corolla-tube 6–7·5 mm. long ; leaves oblong or obovate, acuminate, shortly cuneate at base ; corolla-lobes 4·5–7 mm. long   ..   ..   4a. klaineana
Corolla-lobes forming in bud a conical or mostly oblong head 2–5 times as long as broad (if rather short the corolla-lobes as long or longer than the tube) ; lobes usually more than 5 mm. long when fully developed :
Inflorescences many-flowered and mostly dense ; corolla-lobes ovate or elliptic, about 3 mm. broad, or broader :
Corolla-tube 3·5–5·5 mm. long, the lobes usually longer than the tube :
Corolla-lobes forming in bud a conical head ; flowers pale yellow with red centre ; leaves elliptic to obovate, 4–12 cm. long, 2·5–6 cm. broad, with 4–6(–7) pairs of lateral nerves   ..   ..   5. lucida
Corolla-lobes forming in bud an oblong or ovate-oblong structure ; flowers white, with throat and tube purple ; leaves elliptic or elliptic-oblong, 7–13 cm. long, 2·5–5·5 cm. broad, with 8–12 pairs of lateral nerves   ..   6. scandens
Corolla-tube 6–10 mm. long, the lobes shorter than the tube, forming an oblong structure in bud ; flowers white or white with a greenish tube ; leaves as above but with 4–8 pairs of lateral nerves only   ..   7. schumannii
Inflorescences few-flowered (1–10, rarely up to 15 flowers) ; corolla-lobes oblong, less than 3 mm. broad :
Leaves elliptic to obovate or obovate-lanceolate, 3·5–9 cm. long, 1–4 cm. broad ; lateral nerves 6–10 pairs, very oblique, set at an angle of 30°–40° ; pedicels 4–10 mm. long ; corolla-tube 3–4 mm. long   ..   ..   8. benthamii
Leaves elliptic-oblong, 5–10 cm. long, 1·7–4·5 cm. broad ; lateral nerves 5–7 pairs, spreading, set at an angle of 60°–70° ; pedicels 0·5–3 mm. long ; corolla-tube 8–10 mm. long   ..   ..   ..   9. parciflora

1. **A. multiflora** (Stapf) Stapf in Kew Bull. 1908 : 303 ; Pichon in Mém. Inst. Sci. Madag., sér. B, 2 : 60, t. 3, fig. 12, 29 ; t. 4, fig. 7. Holalafia multiflora Stapf (1894)—F.T.A. 4, 1 : 201. A stout woody glabrous climber with many-flowered, terminal, corymbose inflorescences ; corolla white ; follicles in fruit up to 2½ ft. long.
   Ghana: Kunumoasi, Kumasi Dist. Andoh FH 5583! Mampong to Koforidua (Oct.) Katz & Schmutz 48! **S.Nig.:** Osayolu (Nov.) Farquhar 9! Ibadan F.R. (Nov.) Punch 49! 51! [Br.]**Cam.:** Rio del Rey Johnston 44! Victoria (Nov.) Maitland 754! 791! Kumba Sale! **F.Po:** Mann 1164! Also in Gabon, Ubangi-Shari, Congo, Egypt and Sudan.
2. **A. grandis** Stapf in F.T.A. 4, 1 : 196 (1902) ; Pichon l.c. 60, t. 3, fig. 11, 28 ; t. 4, fig. 6. Stout scrambling shrub over 20 ft. high with dense, many-flowered corymbose inflorescences ; fruiting follicles over 18 in. long.
   S.Nig.: Ilaro F.R., Egbado Dist. Onochie FHI 31890! Also in Cameroun, Congo and Uganda.
3. **A. whytei** Stapf in F.T.A. 4, 1 : 610 (1904) ; Pichon l.c. 60. Climbing glabrous shrub with rather large leaves and subsessile corymbose inflorescences.
   Lib.: Sinoe Basin Whyte!
4. **A. barteri** Oliv. in Hook Ic. Pl. 20 : t. 1992 (1890) ; F.T.A. 4, 1 : 197 ; Pichon l.c. 55, t. 3, fig. 7 ; t. 4, fig. 5. A. giraudii Dubard (1912). High-climbing glabrous shrub with small, pure white or pink flowers in mostly fairly lax corymbose inflorescences ; fruiting follicles over 1 ft. long, slender, paired.
   S.L.: Mayoso (Aug.) Thomas 1423! Mayombo, near Roruks (July) Deighton 4675! **Lib.:** Jaurazon, Sinoe (Mar.) Baldwin 11460! Sanokwele (fr. Feb.) Baldwin 14184! Gletown, Tchien Dist. (July) Baldwin 6743! **Iv.C.:** Bouroukrou Chev. 16742. Dinébo to Yabouakrou Chev. 22568. **Ghana:** Adiembra to Domiabra, Ashanti (June) Kitson 1227! Mampong (Apr.) Vigne FH 1919! Bunso (fr. Sept.) Darko 981! Kumasi (fr. Oct.) Baldwin 13455! Agogo (Apr.) Adams 2600! **Dah.:** Porto-Novo Chev. 22841. **S.Nig.:** Oshogbo (Apr.) Millson! Ibadan S.F.R. (fl. & fr. Feb.) Keay FHI 22543! Osun F.R., Ijebu Igbo (Apr.) Ejiofor FHI 26116! Onitsha Barter 1321! Ekoson, Ikom (Apr.) Jones FHI 6480! [Br.]**Cam.:** S. Bakundu F.R., Kumba (fl. & fr. Mar.) Brenan 9428! Binuyo & Daramola FHI 35091! (See Appendix, p. 366.)
5. **A. lucida** Stapf in Kew Bull. 1894 : 122 ; F.T.A. 4, 1 : 198 ; Pichon l.c. 54, t. 3, fig. 6, 24, 25, t. 4, fig. 4, 9. A. cuneata Stapf in Kew Bull. 1894 : 122. A. reticulata K. Schum. (1895). A climbing glabrous shrub 15–20 ft. high with fairly small, cream to yellow flowers in corymbose inflorescences.
   Lib.: Genna Tanyehun (Dec.) Baldwin 10764! **Ghana:** Mumeromemo Armitage 1! 2! Assuantsi Baldwin 14023! **Togo Rep.:** (July) Debeaux 389! **S.Nig.:** Nun R. Mann 491! Calabar (fl. & fr. Feb.) Mann 2245! Degema Talbot 3643! Also in Cameroun, the Congos, Gabon, Angola and Uganda. (See Appendix, p. 366.
6. **A. scandens** (Thonning) De Wild. Not. Apoc. laticif. Congo 15 (1903) ; Pichon l.c. 50, t. 3, fig. 31, 32. Nerium scandens Thonning (1827). Holarrhena landolphioides A. DC. (1844). Alafia landolphioides (A. DC.) K.

Schum. (1895)—F.T.A. 4, 1 : 197. Climbing shrub of 50–65 ft. with many-flowered corymbose inflorescences ; follicles in fruit 14–20 in. long ; in moist forest.

Sen.: Casamance (Jan.) *Chev.* 2671 ! **Gam.:** *Whitfield.* **Port.G.:** Cantankez (Feb.) *d'Orey* 284 ! Bedanda, Fulacunda (Jan.) *d'Orey* 246 ! Safim, Bissau (fr. Mar.) *Esp. Santo* 1909 ! Gadamael, Cacine (July) *Esp. Santo* 2114 ! Catio to Saucunda, Catio (Dec.) *Esp. Santo* 2235 ! **Guin.:** Rio Nunez *Heudelot* 626 ! Bénary Is. *Heudelot* 776. Bayas country *Heudelot* 777. Lamiah *Paroisse* 29. Koukoré (Sept.) *Pobéguin* 1198 ! **S.L.:** Mano (May) *Deighton* 4624 ! Bunce Isl. (fr. Mar.) *Kirk* 5 ! Port Loko (fl. & fr. Jan.) *Deighton* 1047 ! Kambia, Magbema (Feb.) *Jordan* 391 ! Ninia, Talla Hills (Feb.) *Sc. Elliot* 4809 ! **Ghana:** Sekondi (Dec.) *Johnson* 895 ! Achimota (Apr.) *Irvine* 1422 ! Krepi, Volta R. (Jan.) *Johnson* 561 ! Puliano to Asare (Apr.) *Thompson* 59 ! Abokobi to Biabase (fl. & fr. Dec.) *Adams* 3637 ! **Togo Rep.:** Lomé *Warnecke* 110 ! 333 ! **Dah.:** (Jan.) *Poisson* 01·3–1 ! L. Abémé, Séborhoué *Poisson* 03·2–74. Adja Ouéré (Sept.) *Pobéguin* 93 ; 264. Abomey *Chev.* 23127. **N.Nig.:** Patti Lokoja (fl. & fr. Mar.) *Elliott* 33 ! **S.Nig.:** Lagos *Barter* 2218 ! Abeokuta *Irving* 145 ! Asaba (fl. & fr. Apr.) *Unwin* 7 ! Owo (fr. Nov.) *Meikle* 503 ! Idanre, Ondo Prov. (fl. & fr. Jan.) *Brenan* 8648 ! Also in the Congos.

7. **A. schumannii** *Stapf* in F.T.A. 4, 1 : 197 (1902) ; Pichon l.c. 52, t. 3, fig. 3, 17, 18, 36, 37. A climbing glabrous white-flowered shrub with abruptly acuminate leaves and dense corymbose inflorescences.

S.L.: Toma *Thomas* 8774 ! **Iv.C.:** Bouroukrou *Chev.* 17001. **S.Nig.:** Ukunzu (Jan.) *Thomas* 2213 ! [**Br.**] **Cam.:** S. Bakundu F.R., Kumba (Feb.) *Binuyo & Daramola* FHI 35519 ! Also in Cameroun, Gabon, Congo, Cabinda, Uganda and Tanganyika.

8. **A. benthamii** (*Baill. ex Stapf*) Stapf in F.T.A. 4, 1 : 199 (1902) ; Pichon l.c. 53, fig. 19, 38. *Ectinocladus benthamii* Baill. ex Stapf in J. Linn. Soc. 30 : 88 (1894), and in Hook. Ic. Pl. 24 : t. 2341 (1894). A woody, glabrous climber with stiff leaves, greenish-yellow flowers with a red centre and narrow-cylindrical follicles 5–9 in. long.

Guin.: *Farmar* 324 ! *Chev.* 20316. Badi R. *Pobéguin* 897. **S.L.:** Mofari, Scarcies R. (Jan.) *Sc. Elliot* 4440 ! Batkanu (Feb.) *Deighton* 4186 ! Njala (Feb.) *Deighton* 517 ! Baiima (May) *Deighton* 5177 ! Pujahun *Thomas* 8333 ! **Lib.:** Greenville, Sinoe *Sim.* **S.Nig.:** Ikeja, Lagos (Dec.) *Onochie* FHI 26672 ! Osse R., Benin Prov. (Mar.) *Onochie* FHI 38322 ! Calabar *Thomson* 117 ! Also in Gabon and the Congos.

9. **A. parciflora** *Stapf* in F.T.A. 4, 1 : 610 (1904) ; Pichon l.c. 60. Glabrous climbing shrub with yellow flowers with a red centre, in about 2–3-flowered inflorescences.

Lib.: Greenville, Sinoe *Sim* 25 !

Imperfectly known genus.

Vilbouchevitchia atro-purpurea *A. Chev.* in Boissiera 7 : 254, t. 20 (1943). A monotypic genus recorded from Aboisso, Ivory Coast, but no type specimen cited. Pichon was unable to find the specimen and considered (in Mém. Inst. Sci. Madag., sér. B, 2 : 46 (1949)) that *Vilbouchevitchia* should be treated as part of *Alafia*.

30. FARQUHARIA Stapf in Kew Bull. 1912 : 279.
Aladenia Pichon in Mém. Inst. Sci. Madag., sér. B, 2 : 61 (1949).

Leaves and branches glabrous ; leaves elliptic, elliptic-oblong or ovate-acuminate, 6–12 cm. long, 2·5–6 cm. broad, with 5–9 pairs of lateral nerves ; corolla-tube 10–18 mm. long, lobes ovate, 8–13 mm. long ; follicles about 30 cm. long when ripe, paired, covered with a rusty felt .. .. *elliptica*

F. elliptica *Stapf* in Kew Bull. 1912 : 279 ; F.W.T.A., ed. 1, 2 : 607 ; Brenan in Kew Bull. 7 : 453. *Alafia jasminiflora* A. Chev. Bot. 423, name only. *A. mirabilis* A. Chev. Bot. 423, name only. *Holalafia jasminiflora* Hutch. & Dalz. F.W.T.A. 2 : 43 (1931), and in Kew Bull. 1937 : 388. *Aladenia jasminiflora* (Hutch. & Dalz.) Pichon in Mém. Inst. Sci. Madag., sér. B, 2 : 61 (1949). A stout woody climber with dense terminal corymbose inflorescences of white fleshy flowers.

Iv.C.: Mbasso (Mar.) *Chev.* 17606 ! Morénou, near Akabilékrou (Dec.) *Chev.* 22507 ! Bingerville *Chev.* 16583. Bouroukrou (Dec., Jan) *Chev.* 16644. Assikasso (Dec.) *Chev.* 22582. **Ghana:** Kumasi (May) *Vigne* FH 2011 ! **S.Nig.:** Onitsha *Chesters* OBS156 ! Awka to Onitsha (fr. Nov.) *Baldwin* 13748 ! Central Prov. (June) *Rosevear* BR10 ! Okuni, Ikom Dist. (fl. & fr. July) *Latilo* FHI 31860 ! Mogumu, Benin (Mar.) *Farquhar* 8 ! Also in Cameroun, Gabon and Congo.

31. FUNTUMIA Stapf—F.T.A. 4, 1: 189.

Leaves without pits in the axils of the lateral nerves, elliptic or elliptic-oblong, acuminate, 12–28 cm. long, 3·5–12 cm. broad, with 6–12 pairs of lateral nerves ; flower-buds more or less cylindrical ; corolla-lobes longer than or only slightly shorter than the tube ; tube 6–9 mm. long, lobes 8–13 mm. long .. .. 1. *africana*
Leaves with pits in the axils of the lateral nerves beneath, otherwise as above ; flower-buds conical, acute ; corolla-tube 6–8 mm. long, lobes 3–4 mm. long ; the lobes about ½ as long as the tube .. .. 2. *elastica*

1. **F. africana** (*Benth.*) *Stapf* in Hook. Ic. Pl. 27 : tt. 2696–7 (1901) ; F.T.A. 4, 1 : 190 ; Aubrév. Fl. For. C. Iv. ed. 2, 3 : 200. *Kickxia africana* Benth. in Hook. Ic. Pl. 13 : 59, t. 1276 (1879). *K. latifolia* Stapf in Kew Bull. 1898 : 307. *Funtumia latifolia* (Stapf) Schlechter (1900)—F.T.A. 4, 1 : 192 ; Aubrév. Fl. For. C. Iv., ed. 2, 3 : 200, t. 323, 1. *Kickxia zenkeri* K. Schum. in Notizbl. Bot. Gart. Berlin 3 : 81 (1900). A glabrous tree, frequently about 30 ft. high but sometimes up to 100 ft., with the flowers in axillary clusters ; corolla fleshy, yellowish- to greenish-white or orange.

Port.G.: Empada, Cubisseco (fl. & fr. Aug.) *Esp. Santo* 2166 ! **Guin.:** Boubouni, Macenta (May) *Collenette* 17 ! **S.L.:** Bagroo R. (fl. & fr. Apr.) *Mann* 817 ! Kambui F.R. (May) *Lane-Poole* 239 ! Njala (fr. Sept.) *Small* 391 ! Kukuna (fr. Jan.) *Sc. Elliot* 4506 ! Kabala (fr. July) *Glanville* 249 ! **Lib.:** Dukwia R. (May) *Cooper* 373 ! 468 ! Peahtah (fr. Oct.) *Linder* 888a ! Gbanga (fr. Sept.) *Linder* 743 ! Ganta (fr. Feb.) *Baldwin* 11034 ! Diebla, Webo Dist. (July) *Baldwin* 6279 ! **Iv.C.:** Banco *Ser. For.* 369 ! Zaranou, Indénié *Chev.* 17617 ! Dabou (fr. May) *Jolly* 169 ! 174 ! **Ghana:** E. Akim (Apr.) *Johnson* 692 ! Owabi, Ashanti (Apr.) *Andoh* FH 4180 ! Prah R. (fl. & fr. Dec.) *Johnson* 925 ! Koforidua (fr. Dec.) *Johnson* 434 ! Assuantsi (Mar.) *Irvine* 1551 ! **Togo Rep.:** Agome Mts. *Schlechter.* **Dah.:** Le *Testu* 103 ! Porto Novo (cult.) *Poisson* ! **S.Nig.:** Okomu F.R., Benin (Dec.) *Brenan* 8532 ! Sapoba *Kennedy* 345 ! Ekure, Cross R. (fr. Jan.) *Holland* 160 ! Obazogbon (Apr.) *Farquhar* 24 ! Ogoja *Rosevear* 79/29 ! Oban *Talbot* 1466 ! [**Br.**]**Cam.:** Victoria (Dec.) *Preuss* 1382 ! Mungo R. *Schlechter.* **F.Po:** *Mann* ! Also in Cameroun, S. Tomé, Gabon, Congo, Uganda, Kenya, Tanganyika, Mozambique and Angola. (See Appendix, p. 371.)

.**elastica** (*Preuss*) *Stapf* l.c. tt. 2694–5 (1901) ; F.T.A. 4, 1 : 191 ; Aubrév. Fl. For. C. Iv., ed. 2, 3 : 200, t. 323, 2. *Kickxia elastica* Preuss (1899). Up to 60ft. high and very similar to the above ; flowers mostly white.

Guin.: Nzérékoré (fr. Sept.) *Baldwin* 13308 ! **S.L.:** (cult.) *Smythe* 24 ! Nyago *Smythe* 78 ! Njala (Apr.) *Deighton* 1159 ! Konero, Panguma Dist. *Imp. Inst.* 1 ! **Lib.:** Sinoe R. *Sim* ! **Iv.C.:** Aubrév. 646 ! Zaranou, Indénié (Mar.) *Chev.* 16282 ! **Ghana:** *Vigne* FH 1513 ! Mampong Hills *Johnson* 255 ! Assin-Yan-Kumasi

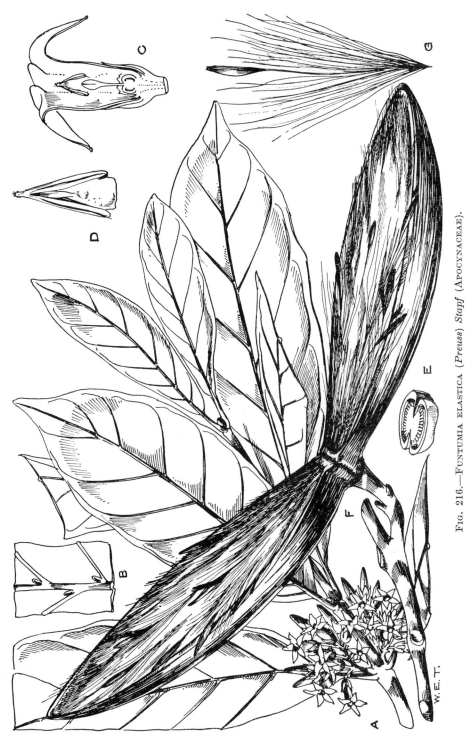

Fig. 216.—Funtumia elastica (*Preuss*) *Stapf* (Apocynaceae).

A, flowering shoot. B, part of under surface of leaf showing pits. C, longitudinal section of flower. D, anther. E, cross-section of ovary. F, fruit. G, seed.

Cummins 217! Agogo *Adams* 2615! **S.Nig.:** Lagos *Millen* 178! Ijebu-Ode to Ibadan (May) *Chukwuogo* FHI 4652! Ife to Ibadan (Dec.) *Meikle* 867! *Keay* FHI 28291! Calabar (Mar.) *Holland* 162! Abarogba, Cross R. (fr. Jan.) *Holland* 158! 159! 161! **[Br.]Cam.:** Victoria (Jan.) *Maitland* 319! Malende to Nyoke, Victoria (Dec.) *Preuss* 1381! Also in the Congos, Rio Muni, Egypt, Sudan & Uganda. (See Appendix, p. 371.)

32. PLEIOCERAS Baill.—F.T.A. 4, 1 : 165.

Inflorescences puberulous, usually pyramidal panicles ; pedicels 1–5 mm. long ; corolla-tube 3–5 mm. long, lobes 2–4 mm. long ; leaves elliptic-oblong, elliptic-lanceolate or ovate, 6–16 cm. long, 2·5–5 cm. broad, with 5–10 pairs of lateral nerves ; follicles up to 50 cm. long :

Leaves minutely puberulous on and between the secondary nerves on the under-surface   ..   ..   **1.** *barteri* var. *barteri*

Leaves glabrous beneath or puberulous only on the midrib and sparsely so on the secondary nerves, glabrous between them   ..   ..   **1a.** *barteri* var. *zenkeri*

Inflorescences glabrous, corymbose ; pedicels 1–2 cm. long ; corolla-tube 5–7 mm. long, lobes about as long as the tube ; leaves quite glabrous, ovate- or elliptic-acuminate, 5–10 cm. long, 3·5 cm. broad, with 5–10 pairs of lateral nerves   ..   **2.** *afzelii*

1. **P. barteri** *Baill.* var. **barteri**—in Bull. Soc. Linn. Paris 1 : 759 (1888) ; F.T.A. 4, 1 : 166 (1902) ; Aubrév. Fl. For. C. Iv., ed. 2, 3 : 198, t. 322. *Wrightia parviflora* Stapf in Kew Bull. 1894 : 121. Usually an erect or climbing shrub ; inflorescence a many-flowered, terminal panicle with small creamy-yellow flowers with a red tube.
   Lib.: Mnalulu, Webo Dist. (fr. June) *Baldwin* 7027! **Iv.C.:** Bingerville, Abidjan to Dabou *Chev.* 15190! Adiopodoumé (Oct.) *Roberty* 15309! **Ghana:** Atwabo (fr. Feb.) *Irvine* 2347! Abakrampa (Mar.) *Irvine* 1604! Assuantsi (Mar.) *Irvine* 1543! Sekondi (Nov.) *Vigne* FH 1402! Tikwabo (fl. & fr. July) *Chipp* 267! **S.Nig.:** Lagos *Barter* 2170! Epe *Barter* 3278! Ikorodu, Ikeja Dist. (fl. & fr. Dec.) *Onochie* FHI 26700! Orosun Mt., Idanre (Jan.) *Brenan* 8696! Ibadan (fl. & fr. Mar.) *Newberry* 45! *Meikle* 899! Adiabo, Calabar (Mar.) *Holland* 86! (See Appendix, p. 377.)

1a. **P. barteri** var. **zenkeri** (*Stapf*) *H. Huber* in Kew Bull. 15 : 438 (1962). *P. zenkeri* Stapf in F.T.A., 4, 1 : 166 (1902). *P. glaberrima* Wernham in J. Bot. 52 : 26 (1914). *P. oblonga* Wernham l.c. 27 (1914). *P. stapfiana* Wernham l.c. 27 (1914). *P. talbotii* Wernham l.c. 26 (1914). A small tree.
   [Br.]Cam.: S. Bakundu F.R., Kumba (Aug.) *Olorunfemi* FHI 30735! Also in Cameroun.

2. **P. afzelii** (*K. Schum.*) *Stapf* in F.T.A. 4, 1 : 166 (1902). *Wrightia afzelii* K. Schum. (1896). *Pleioceras whytei* Stapf in F.T.A. 4, 1 : 604 (1904). Erect, glabrous shrub or small tree 9–30 ft. high ; inflorescence many-flowered, corymbose, with flowers red outside, yellow within, and a fringe of filiform appendages in the throat.
   S.L.: *Afzelius.* Regent (Mar.) *Lane-Poole* 217! Newton (Feb., Apr.) *Deighton* 4745! 5000! Taiama (Apr.) *Deighton* 3127! **Lib.:** Sinoe R. *Whyte*!

33. MALOUETIA A. DC.—F.T.A. 4, 1 : 194.

Leaves and branches glabrous ; leaves elliptic-acuminate to oblong, 7–14 cm. long, 3–7 cm. broad, with 6–10 pairs of lateral nerves ; flowers very few in subsessile axillary cymes ; corolla-tube narrowly cylindrical, 12–15 mm. long, lobes obovate, 7–10 mm. long ; anthers exserted   ..   ..   *heudelotii*

M. heudelotii *A. DC.* in DC. Prod. 8 : 380 (1844) ; F.T.A. 4, 1 : 195. Erect, glabrous shrub 3–5 ft. high with the axillary tubular flowers white to pale mauve, and paired, narrowly cylindric follicles ; seeds linear-lanceolate, not tufted ; in swamps or along streams.
   Guin.: *Heudelot* 714! 890! Kindia (June) *Jac.-Fél.* 1670! Bayabaya *Sc. Elliot* 4782! Sasseni (fl. & fr. Jan.) *Sc. Elliot* 4524! **S.L.:** Roruks (July) *Deighton* 2512! Roken (July) *Jordan* 58! Tendekom, Yoni Mabanta (May) *Jordan* 898! Magbile (Dec.) *Thomas* 6295! Jama to Magibisi (Sept.) *Deighton* 3063! Njala (June) *Deighton* 1762! Also indicated from Congo, but identification doubtful.

34. ADENIUM Roem. & Schult.—F.T.A. 4, 1 : 226.

Branches succulent, leaves spirally arranged, glabrous (in our area), obovate-oblong to oblanceolate (in our area), cuneate at base, sessile, up to 10 cm. long (but often smaller) and 2–3 cm. broad with 6–12 pairs of lateral nerves ; pedicels and calyx puberulous ; corolla-tube 3·5–4 cm. long, constricted in the lower $\frac{1}{4}-\frac{1}{3}$, campanulate above, puberulous outside ; lobes ovate or orbiculate, 15–25 mm. long ; anthers provided with long appendages, which are shortly exserted from the tube ; follicles spindle-shaped, minutely puberulous ; seed with an apical tuft   ..   *obesum*

A. obesum (*Forsk.*) *Roem. & Schult.* Syst. 4 : 411 (1819) ; Aubrév. Fl. For. Soud-Guin. 445, t. 97, 7–10 ; M. Pichon in Mém. Mus. Hist. Nat. Paris, n. sér. B, 1 : 78 (1950). *Nerium obesum* Forsk. Fl. Aegypt. Arab. 205 (1775). *Adenium honghel* A. DC (1844)—F.T.A. 4, 1 : 229 (1902) ; F.W.T.A., ed. 1, 2 : 29. An erect shrub 4–6 ft. high, with a stout almost conical trunk and smooth, fleshy branches ; flowers rather large, pink to purple, campanulate, 1½–2½ in. long in short, subsessile, terminal cymes, appearing before or with the leaves ; in drier parts of our area and often grown for ornament.
   Sen.: Dagana *Leprieur*! *Roger*! Oualo *Roger* 133! Baké *Vuillet* 264. **Mali:** Birgo, Kaarta *Dubois* 133. **Port.G.:** Madina de Boé, Gabu (Feb.) *d'Orey* 310! Boé, Jangada to Madina (Jan.) *Esp.* Santo 2364! **Guin.:** Kouroussa *Pobéguin* 603 ; 809. **Iv.C.:** Banfora (Mar.) *Chev.* 530! **N.Nig.:** Onisaja, Ilorin (fl. & fr. May) *Ejiofor* FHI 19828! Borgu *Barter* 711! Zaria (cult. May) *Lamb* 46! Jos Plateau (Mar.) *Lely* P188! Kano (Dec.) *Hagerup* 669! Katagum Dist. *Dalz.* 430! Yola (cult.) *Dalz.* 97! **[Br.]Cam.:** R. Senchu, Alantika Mts., Adamawa (Dec.) *Hepper* 1609! Gwoza, Dikwa Div. (Jan.) *McClintock* 141! Also in Egypt, Sudan, Somaliland, Arabia, Ethiopia, Kenya, Uganda, and (?) Tanganyika.

   Adenium is doubtfully native in our area ; the W. African plant usually known as *A. honghel* is possibly merely an introduced form of the widespread and variable E. African and Arabian species which can all be called *A. obesum*.

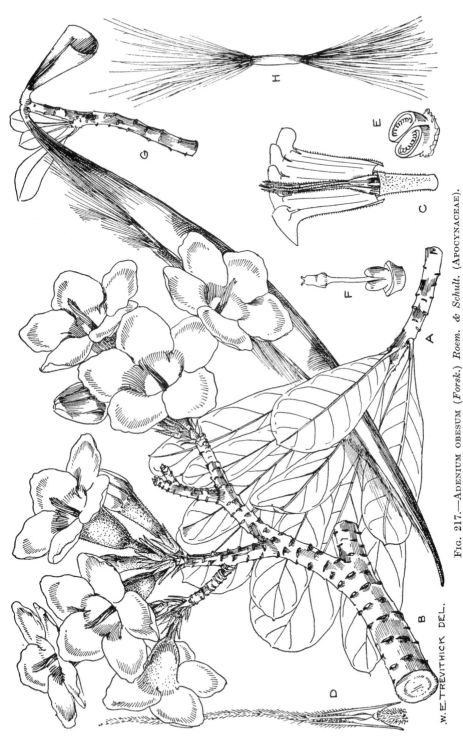

.W.E.TREVITHICK DEL.

Fig. 217.—Adenium obesum (*Forsk.*) *Roem. & Schult.* (Apocynaceae).

A, leafy shoot. B, flowering shoot. C, corolla laid open showing staminal cone. D, stamen. E, cross-section of ovary. F, pistil. G, one carpel in fruit. H, seed.

35. BAISSEA A. DC. (1844)—F.T.A. 4, 1 : 203.

Zygodia Benth. (1876)—F.T.A. 4, 1 : 217. *Codonura* K. Schum. (1896).

Petioles 5–15 mm. long (when only about 5 mm. long, then with 7 or more pairs of lateral nerves) ; leaves cuneate at base, glabrous beneath or with a tuft of hairs in the nerve-axils ; branches glabrous or minutely puberulous ; lateral nerves 4–15 pairs :

Corolla-lobes 7–16 mm. long, mostly linear :

Calyx 2·5–4 mm. long, glabrous as well as the pedicels ; corolla-lobes 12–16 mm. long, tube about 6 mm. long ; leaves obovate-oblong to ovate-acuminate, 8–20 cm. long, 4–7·5 cm. broad, with 9–15 pairs of lateral nerves   ..   ..   1. *calophylla*

Calyx 1–1·5 cm. long, minutely puberulous as well as the pedicels ; leaves 4–12 cm. long, 2–5·5 cm. broad :

Leaves elliptic- or ovate-acuminate, with 5–8 pairs of lateral nerves ; pedicels 1–4 mm. long ; corolla dark violet ; corolla-lobes 7–14 mm. long, narrowed in their apical half into an almost threadlike beak ; tube much shorter than the lobes   ..   ..   2. *tenuiloba*

Leaves with 7–15 pairs of lateral nerves ; corolla white, corolla-lobes linear-oblong or linear-lanceolate, not threadlike :

Pedicels 5–10 mm. long ; corolla-lobes 8–16 mm. long, much longer than the tube ; leaves ovate- or elliptic-acuminate   ..   ..   3. *odorata*

Pedicels 1–5 mm. long ; corolla-lobes (exceptionally) up to 8 mm. long
                                                  4. *leonensis*

Corolla-lobes 3–6 mm. long, mostly ovate or deltoid-oblong ; pedicels not exceeding 5 mm. in length :

Leaves elliptic, elliptic-acuminate or sometimes obovate, 4–10 cm. long, 2–5·5 cm. broad, with 8–10 pairs of lateral nerves ; pedicels thin, about 0·2–0·3 mm. diam.
                                                  4. *leonensis*

Leaves elliptic-obovate, with 4–6 pairs of lateral nerves ; pedicels mostly stout, 0·3–0·7 mm. diam.   ..   ..   5. *lane-poolei*

Petioles 1–5 mm. long ; leaves rounded, subcordate or shortly cuneate at base, frequently pilose on the midrib beneath ; branches frequently puberulous to tomentellous ; lateral nerves in 2–8 pairs :

Corolla-lobes 6–20 mm. long, deltoid-linear to broadly linear :

Sepals about 1 mm. long ; pedicels 5–20 mm. long, slender, 0·2–0·3 mm. in diam. ; corolla-lobes 6–10 mm. long ; leaves elliptic, elliptic-lanceolate or oblanceolate, 4–8 cm. long, 1·8–3 cm. broad, with 2–5 pairs of main lateral nerves, cuneate or rarely rounded at base, glabrous beneath except for tuft of hairs in nerve-axils
                                                  6. *laxiflora*

Sepals 2–5 mm. long ; pedicels 5–10 mm. long, 0·5–0·7 mm. diam. ; leaves with 4–8 pairs of lateral nerves :

Leaves rounded or truncate, rarely slightly cuneate at base ; corolla-lobes 6–20 mm. long :

Leaves acuminate at apex ; elliptic, ovate or broadly lanceolate, 5–13 cm. long, 2–5·5 cm. broad, midrib puberulous beneath when young, later frequently glabrous except in the nerve-axils ; corolla-lobes 12–20 mm. long   7. *multiflora*

Leaves not acuminate, rounded or shortly deltoid at apex ; usually oblong, sometimes ovate- or obovate-oblong, 2·5–6 cm. long, 1·2–4 cm. broad, midrib mostly pubescent beneath ; corolla-lobes 6–10 mm. long   ..   ..   8. *zygodioides*

Leaves slightly cordate at base, rounded or deltoid at apex, oblong or obovate-oblong, 3–9 cm. long, 1·5–3·5 cm. broad ; corolla-lobes 6–9 mm. long   9. *subsessilis*

Corolla-lobes sharply triangular, 2–4 mm. long ; leaves subcordate or truncate at base, up to 7 cm. long :

Leaves gradually acuminate at apex, oblong, ovate or obovate, 2·5–7 cm. long, 1·2–3 cm. broad ; inflorescence 1–5 cm. long ; peduncle 8–20 mm. long
                                                 10. *breviloba*

Leaves rounded or mostly deltoid at apex, not acuminate-oblong, 2·5–5 cm. long, 1·2–2·7 cm. broad ; inflorescence very short, 0·5–1·5 cm. long ; peduncle about 2–5 mm. long   ..   ..   11. *axillaris*

1. **B. calophylla** (*K. Schum.*) *Stapf* in F.T.A. 4, 1 : 205 (1902) ; Brenan in Kew Bull. 7 : 454. *Codonura calophylla* K. Schum. in Engl. Bot. Jahrb. 23 : 229 (1896). Robust liane up to 32 ft. high with leaves very glossy above ; flowers white to yellow, with red inside the tube.
   S.Nig.: Sapoba (Feb.) *Kennedy* 1959 ! *Meikle* 542 ! Ekeji-Ipetu F.R., Ijesha Dist. (Dec.) *Onochie* FHI 5232 ! Usonigbe F.R., Benin (Nov.) *Ejiofor* FHI 24632 ! *Meikle* 559 ! *Chizea* FHI 24475 ! Okomu F.R., Benin (Feb.) *Brenan* 9162 ! Also in Cameroun.

2. **B. tenuiloba** *Stapf* in Kew Bull. 1894 : 124 ; F.T.A. 4, 1 : 214 (1902). *Guerkea uropetala* K. Schum. l.c. 228 (1896). A climbing shrub with long-tailed violet flowers in terminal and axillary cymes.
   S.Nig.: Bonny R. (Feb., Oct.) *Mann* 504 ! *Kalbreyer* 73 ! Also in Cameroun and (?) Uganda.

3. **B. odorata** *K. Schum. ex Stapf* in F.T.A. 4, 1 : 212 (1902). A glabrous woody climber with white, purple-streaked flowers ; inflorescences as above.
   Lib.: Greenville, Sinoe *Sim* 32 ! **S.Nig.:** Lagos (Mar., Dec.) *Dalz.* 1258 ! *Hagerup* 806 ! Also in Cameroun.

4. **B. leonensis** *Benth.* in Fl. Nigrit. 452 (1849) ; F.T.A. 4, 1 : 213 (1902) ; Chev. Bot. 424. *B. brachyantha* Stapf in Kew Bull. 1894 : 125 ; F.T.A. 4, 1 : 213. A climbing shrub with small white flowers in axillary and terminal cymes ; follicles in fruit up to 2 ft. long.
   Port.G.: Catio (July) *Esp. Santo* 2134 ! **Guin.:** Diaguissa *Chev.* 12421 ; 12693 ; 12930 ; 13613. Dalaba

Caille in *Hb. Chev.* 18117. Forécariah (June) *Jac.-Fél.* 1732! **S.L.:** (June) *T. Vogel* 119! Freetown (fr. Mar.) *Lane-Poole* 208! Bagroo R. (Apr.) *Mann* 854! Kenema (Apr.) *Jordan* 2051! Mano *Thomas* 9976! Bafodeya (Apr.) *Sc. Elliot* 5500! **Lib.:** Roberts Field (May) *Baldwin* 5800! Ba, Little Kola R. (Mar.) *Baldwin* 11188! Grand Bassa *Dinklage*! Suacoco, Gbanga (Apr., June) *Konneh* 163! Ganta (Apr.) *Harley* 1163! **Ghana:** *Farmar* 419! Aburi Hills (Nov.) *Johnson* 283! Axim (Feb.) *Irvine* 2165! Assuantsi (fr. Jan.) *Baldwin* 14021! Akwapim (Mar.) *Murphy* 677! **S.Nig.:** Oban *Talbot*! Eket (May) *Talbot* 3324! 3326! *Onochie* FHI 32935! Also in Cabinda.

5. **B. lane-poolei** *Stapf* in Kew Bull. 1915 : 46. *B. concinna* Stapf ex Hutch. & Dalz. in Kew Bull. 1937 : 338. Very similar to the last.
    S.L.: York Pass (July) *Lane-Poole* 322! Sugar Loaf Mt. (Apr.) *Roberty* 17297! Sambaia (June) *Thomas* T68! **Lib.:** Grand Bassa *Dinklage* 1630!

6. **B. laxiflora** *Stapf* in Kew Bull. 1894 : 124 and in Hook. Ic. Pl. 24 : t. 2342 (1894) ; F.T.A. 4, 1 : 208. Stout woody climber with pale cream flowers.
    S.Nig.: Ubulubu (Jan.) *Thomas* 2266! Calabar (Feb.) *Mann* 2258! Eket *Talbot* 3164! Oban *Talbot* 2053! **[Br.]Cam.:** S. Bakundu F.R., Kumba (Mar.) *Latilo & Ujor* FHI 31190! Also in Cameroun, Gabon, Rio Muni and the Congos.

FIG. 218.—Baissea multiflora *A. DC.* (Apocynaceae).

A, flowering shoot. B, anther. C, pistil. D, cross-section of ovary. E, fruit. F, seed.

7. **B. multiflora** *A. DC.* in DC. Prod. 8 : 424 (1844) ; F.T.A. 4, 1 : 207 ; Chev. Bot. 424. *B. caudiloba* Stapf in J. Linn. Soc. 30 : 90 (1894). *B. heudelotii* Hua (1898). *B. multiflora* var. *caudiloba* (Stapf) Stapf in F.T.A. 4, 1 : 208 (1902) ; Chev. Bot. 425. Woody climber 10-25 ft. high, usually rusty pubescent ; flowers white or yellow, in axillary and terminal cymes ; follicles narrowly cylindrical, up to 18 in. long when ripe.
    Sen.: Ferlo *Heudelot* 186! Cape Nase (Feb.) *Döllinger* 45! **Gam.:** *Dawe* 53! Albreda *Perrottet* 480. Kuntaur (fl. & fr. Jan.) *Dalz.* 8043! Boroba *Imp. Inst.* 8! S. bank of R. Gambia *Brown-Lester* S14! **Mali:** Badinko *Chev.* 118! Kati *Chev.* 175. **Port.G.:** Pelaque, Bissau (Feb.) *Esp. Santo* 1723! Bissau to Prabis (Dec.) *d'Orey* 15! Bissalanca, Bissau (Jan.) *Esp. Santo* 1620! **Guin.:** *Farmar* 319a! Fouta Djalon *Heudelot* 597! 601! Kouroussa (May, Dec.) *Pobeguin* 242! Kindia *Chev.* 13371. **S.L.:** Jama (Mar.) *Deighton* 4726! Ninia, Talla Hills (Feb.) *Sc. Elliot* 4807! Kenema (Apr.) *Jordan* 2050! Gola Forest (Apr.) *Small* 584! Njala (Feb.) *Deighton* 4147! **Iv.C.:** Zaranou *Chev.* 17603 ; 17616. **Niger:** Diapaga *Chev.* 24436. Extends to Cabinda. (See Appendix, p. 367.)

8. **B. zygodioides** (*K. Schum.*) Stapf in F.T.A. 4, 1 : 210 (1902) ; Chev. Bot. 425. *Oncinotis zygodioides* K. Schum. l.c. 227 (1896). *Baissea aframensis* Hutch. & Dalz. in Kew Bull. 1937 : 338. Woody climber with fairly small leaves and axillary cymes of white to yellow flowers ; follicles more than 1 ft. long when ripe.
    Guin.: Macenta (Apr.) *Jac.-Fél.* 829! Nimba Mts. (Apr.) *Schnell* 5250! **S.L.:** *Afzelius*! Bwedu to Kangama (Apr.) *Deighton* 3155! Njala (fl. May, fr. Jan.) *Deighton* 3435! 4577! Kenema (Apr.) *Jordan* 2042! **Lib.:** Sanokwele (fr. Sept.) *Baldwin* 9528! Gbanga (Mar.) *Blickenstaff* 32! Dobli Is., St. Paul R. (Apr.) *Bequaert* 167! Ganta (May) *Harley* 1169! Mnanulu (fr. June) *Baldwin* 6042! **Iv.C.:** *vide* Chev. l.c. **Ghana:** Kumasi (Apr.) *Vigne* FH 1107! Volta River F.R. (Mar.) *Vigne* FH 4373! Mayara, Akwapim (Mar.) *Irvine* 1534! Wuruboug to Okraji, Afram Plains (May) *Kitson* 1140! Aburi Hills (June) *Patterson* 287! Dodowa Plains (fr. Jan.) *Johnson* 558!

9. **B. subsessilis** (*Benth.*) Stapf ex Hutch. & Dalz. F.W.T.A., ed. 1, 2 : 46 (1931). *Zygodia subsessilis* Benth. in Hook. Ic. Pl. 12 : 73 sub. t. 1184. *Oncinotis subsessilis* (Benth.) K. Schum. in Schlechter Westafr. Kautsch.-Exp. 307 (1900) ; F.T.A. 4, 1 : 224. Climbing, reddish-brown puberulous shrub up to 20 ft. high with pale yellow flowers in axillary and terminal cymes.
    S.Nig.: Ishagama (Mar.) *Schlechter* 12310! Osun F.R., Ijebu Igbo (Apr.) *Ejiofor* FHI 26114! Gambari to Ibadan (Mar.) *Onochie* FHI 21801!

10. **B. breviloba** *Stapf* in Kew Bull. 1912 : 278. *B. ivorensis* A. Chev. Bot. 424, name only. Climbing shrub up to 10 ft. high very similar to the above.
    Guin.: Nzérékoré (June) *Collenette* 34! **Iv.C.:** Makougnié, Agniéby valley *Chev.* 17191! **Ghana:** Ateiku (May) *Vigne* FH 1949! Abura Dunkwa (Apr.) *Scholes* 275! Manso (Apr.) *Scholes* 265! **S.Nig.:** Ikpe (Apr.) *Farquhar* 1! Igoubazowa, Benin Prov. (May) *Onochie* FHI 34894! Okomu F.R., Benin (June) *Onochie* FHI 31231!

11. **B. axillaris** (*Benth.*) *Hua* in Compt. Rend. Acad. Paris 134 : 857 (1902) ; F.T.A. 4, 1 : 210. *Zygodia axillaris* Benth. in Hook. Ic. Pl. 12 : t. 1184 (1876). A climbing shrub similar to the last.
S.Nig.: Lagos *Punch* 77 ! Sapoba, Benin Prov. (June) *Ejiofor* FHI 32010 ! Abeokuta (fr. Oct.) *Baldwin* 13648 ! Ijaiye F.R., Ibadan (fr. Feb.) *Onochie* FHI 20691 ! Calabar (Feb.) *Mann* 2301 ! **[Br.]Cam.:** Johann-Albrechtshöhe *Staudt* 518 ! Also in Cameroun, Congo and Angola.

36. MOTANDRA A. DC.—F.T.A. 4, 1 : 224.

Leaves elliptic- or obovate-oblong, acuminate, 6–13 cm. long and 2–5 cm. broad, with 5–8 pairs of lateral nerves, thickly pubescent in the nerve-axils ; flowers in oblong, terminal panicles ; corolla-tube 2·5–4 mm. long, the lobes about 5 mm. long
guineensis

M. guineensis (*Thonning*) *A. DC.* in DC. Prod. 8 : 423 (1844) ; F.T.A. 4, 1 : 224. *Echites guineensis* Thonning (1827). *Motandra rostrata* K. Schum. (1903). A climbing or sometimes suberect shrub, more or less yellowish- or fulvo-tomentose, with small white flowers in tomentose inflorescences ; fruiting carpels spindle-shaped, about 5 in. long and ¾ in. across, densely brown-scaly.
Guin.: Dalaba (Jan.) *Jac.-Fél.* 754 ! **S.L.:** Smythe 266 ! Konnoh Country *Burbridge* 481 ! Bendu to Mamaima (Apr.) *Deighton* 3942 ! **Lib.:** Cess R., near Bahtown (fr. Aug.) *Baldwin* 9013 ! **Ghana:** Aburi Hills (Mar.) *Brown* 684 ! Kumasi (fl. Mar., fr. Oct.) *Baldwin* 13462 ! *Vigne* FH 3729 ! Zome (Mar.) *Thomas* D174 ! Atonso (fr. Oct.) *Baldwin* 13503 ! Tain II F.R. (fr. Dec.) *Adams* 2785 ! **Togo Rep.:** Misahöhe (fr. Nov.) *Mildbr.* 7379 ! **Dah.:** Adja Ouéré (Mar.) *Le Testu* 115 ! Bassila *Poisson* ! **S.Nig.:** Ilaro F.R., Abeokuta Prov. (fr. Oct.) *Onochie* FHI 13613 ! Ishagama (Mar.) *Schlechter* 12312 ! Mamu F.R., Ibadan Dist. (Mar.) *Keay & Foggie* FHI 22529 ! Onitsha *Barter* 1636 ! Oban *Talbot* ! **[Br.]Cam.:** Nkom-Wum F.R. (fr. July) *Ujor* FHI 29296 ! R. Sulli, Gashaka Dist. (fr. Dec.) *Latilo & Daramola* FHI 28838 ! Also in Cameroun, Congo, Uganda and Angola.

37. ONCINOTIS Benth.—F.T.A. 4, 1 : 220.

Branches and undersurface of leaves rusty-pubescent ; leaves obovate-elliptic, broadest mostly above the middle, 5–12 cm. long, 2·5–5·5 cm. broad ; lateral nerves 4–7 pairs ; corolla-tube about 3 mm. long, lobes 4–5 mm. long ; ripe follicles 12–20 cm. long, narrowly spindle-shaped to almost cylindrical, up to about 1 cm. diam.   1. *gracilis*
Branches and leaves glabrous ; leaves with 6–16 pairs of lateral nerves ; ripe follicles 1–2 cm. diam. :
    Leaves obovate, broadest above the middle, usually rather long-cuneate at base, 4–9 cm. long, 2·5–6 cm. broad, main lateral nerves mostly 6 mm. apart from each other or less ; corolla-tube about 4 mm. long, lobes about the same ; ripe follicles cylindrical, 20–25 cm. long, 1–1·5 cm. broad .. ..   2. *nitida*
    Leaves elliptic or ovate-elliptic, broadest at or below the middle, short-cuneate or rounded at base, 8–12 cm. long, 3–5 cm. broad ; main lateral nerves more than 5 mm. apart from each other ; fruiting follicles spindle-shaped, 10–20 cm. long, 1·5–2 cm. broad .. ..   3. *glabrata*

1. **O. gracilis** *Stapf* in Kew Bull. 1894 : 124 ; F.T.A. 4, 1 : 223. *O. chlorogena* K. Schum. in Schlechter Westafr. Kautschuk-Exp. 307 (1900), name only. A rusty-pubescent climbing shrub up to 40 ft. high ; flowers greenish or yellowish in terminal and axillary cymes.
Guin.: Pela, Nzérékoré (fr. Sept.) *Baldwin* 13312 ! **S.L.:** Yonibana (fr. Oct.) *Thomas* 4128 ! **Lib.:** Javajai, Boporo Dist. (fr. Nov.) *Baldwin* 10269a ! Nikabuzu (Mar.) *Bequaert* 119 ! Zigida (Mar.) *Bequaert* 135 ! Sanokwele (fr. Sept.) *Baldwin* 9537 ! Tappita (fr. Aug.) *Baldwin* 9074 ! **Ghana:** Akropong (Mar.) *Johnson* 902 ! Kumasi (fr. Oct.) *Baldwin* 13454 ! Agogo (Apr.) *Adams* 2584 ! Kwahu (Apr.) *Johnson* 658 ! Atonso, Ashanti Dist. (fr. Oct.) *Baldwin* 13502 ! **N.Nig.:** Lokoja (Mar.) *Shaw* 18 ! **S.Nig.:** Igbessa, Lagos *Millen* 189 ! Ijoko (fr. Nov.) *Baldwin* 13705 ! Sapoba (fr. Nov.) *Keay* FHI 25572 ! Mamu F.R., Ibadan (Mar.) *Keay* FHI 22531 ! Oban *Talbot* 1675 ! Also in Cameroun.
2. **O. nitida** *Benth.* in Fl. Nigrit. 451 (1849) ; F.T.A. 4, 1 : 221. Climbing glabrous shrub up to 25 ft. high ; in gallery forest.
Port.G.: Empada (fr. Dec.) *Esp. Santo* 2337 ! Catió (Aug.) *Esp. Santo* 2174 ! **S.L.:** Sc. *Elliot* ! T. *Vogel* 63 ! Waterloo (July) *Lane-Poole* 300 ! **Lib.:** Sinoe, Greenville *Sim* ! **Togo Rep.:** Lomé *Warnecke* 450 ! **S.Nig.:** Lagos (June) *Dalz.* 1347 ! *Batten-Poole* 48 ! Ibadan (fr. Dec.) *Baldwin* 13677 !
3. **O. glabrata** (*Baill.*) *Stapf ex Hiern* in Cat. Welw. 1 : 674 (1898) ; F.T.A. 4, 1 : 222. *Motandra glabrata* Baill. (1888). *Oncinotis campanulata* K. Schum. (1896)—F.T.A. 4, 1 : 222. *O. batesii* Stapf in F.T.A. 4, 1 : 221 (1902). *O. glandulosa* Stapf in F.T.A. 4, 1 : 221 (1902). A glabrous woody climber.
Guin.: Nzérékoré (fr. Sept.) *Baldwin* 13324 ! **Lib.:** Peahtah (Oct.) *Linder* 903 ! Dobli Is. *Bequaert* 22 ! Kondessu, Boporo Dist. (fr. Dec.) *Baldwin* 10684a ! Sanokwele (fr. Dec.) *Baldwin* 13271 ! Tappita (fr. Aug.) *Baldwin* 9066a ! **Ghana:** Aburi (Apr.) *Johnson* 883 ! 1064 ! Ateiku (May) *Vigne* FH 1951 ! Tafo (fr. July) *Darko* WACRI 914 ! Assuantsi (fr. Jan.) *Baldwin* 14031 ! **S.Nig.:** Sapoba to Benin *Thompson* 20 ! Oban *Talbot* 1692 ! **[Br.]Cam.:** above Buea (Dec.) *Migeod* 256 ! Also in Cameroun, Gabon ,Congo, Angola and (?) Tanganyika.

135. PERIPLOCACEAE [1]
By A. A. Bullock

Climbing, twining or erect, softly woody or occasionally with wiry stems, or herbaceous perennials with a tuberous rootstock ; leaves opposite, entire, linear to very broad, pinnately nerved ; stipules absent, but sometimes with a stipular frill around the stem at the nodes which may become woody ; inflorescence a terminal cyme often lateral by sympodial growth of the axis, sometimes racemiform or umbelliform ; bracts very small ; flowers bisexual, sometimes functionally unisexual and then dioecious ; calyx-tube very short or obsolescent,

[1] In Ed. 1 regarded as a tribe of *Asclepiadaceae*.

segments 5, valvate or imbricate but opening very early ; corolla gamopetalous, 5-lobed, tube short or sometimes as long as or longer than the lobes, lobes contorted or very rarely valvate ; corona of 5 lobes arising from the base of the stamen filaments, of varied form, or rarely absent or reduced to tubercles ; stamens 5, inserted in the corolla-tube or at its throat, free from each other, but with the anthers pressed together and to the expanded apex of the style ; anthers basifixed, introrse, usually with a scarious apical appendage inflexed over the style ; pollen granular, in tetrads discharged on to 5 spathulate pollen carriers derived from the expanded part of the style and with a glandular base ; pollen carriers concealing the 5 stigmatic surfaces ; carpels two, joined only by their styles, which unite to form a flat or variously elongated head ; ovules numerous, multiseriate on a single adaxial placenta ; fruit of two (or one by abortion or pollination failure) divergent or reflexed sessile follicles ; follicles more or less fusiform and often greatly elongated, or sometimes broadly ovoid, smooth, glabrous or hairy, sometimes warted or winged, finally dehiscing lengthwise adaxially ; seeds flat, often winged, and crowned with a coma of soft silky hairs ; endosperm present ; embryo straight, almost as long as the seed, cotyledons flat.

Distinguished from *Asclepiadaceae*, in which the genera were formerly placed, by the granular pollen carried on spathulate glandular carriers. Confined to the tropics and warm temperate regions of the Old World and reaching its greatest diversity in genera and species in tropical and S. Africa.

Corolla-tube a mere ring, or saucer-shaped :
  Anthers more or less villous   ..   ..   **1. Parquetina**
  Anthers glabrous :
    Corona very small, of 5 indistinct emarginate lobules adnate to the bases of the filaments ; pollen carrier rhombic, deeply cleft, attached to its gland by a short slender stalk ..   ..   **2. Batesanthus**
    Corona-lobes relatively large ; pollen-carriers spathulate :
      Corona-lobes broadly obcordate, with a linear dorsal appendage ; stems with conspicuous stipular frills at the nodes ..   ..   **3. Mondia**
      Corona-lobes linear, simple or divided :
        Leaves with numerous straight horizontal parallel lateral veins looped near the margin ; flowers in congested lateral cymes ; pedicels long and filiform
                                  4. Zacateza
        Leaves with fewer, curved, ascending lateral nerves ; flowers in lax paniculate cymes ; pedicels relatively short..   ..   ..   **5. Tacazzea**
Corolla-tube well developed ; anthers glabrous :
  Climbing shrubs :
    Corolla-tube cylindrico-campanulate, less than half as long as the lobes ; leaves not glaucous beneath   ..   ..   **6. Cryptolepis**
    Corolla-tube narrowly urceolate, much longer than the lobes ; leaves glaucous beneath   ..   ..   **7. Mangenotia**
  Erect herbs or subshrubs with fleshy or woody tubers :
    Corolla-tube cylindric or cylindrico-campanulate, not saccate at the base, corolla-lobes lanceolate or ovate-lanceolate, white, cream or purplish-pink :
      Subshrubs with wiry, erect stems and linear to narrowly oblong leaves ; rootstock woody ; flowers greenish-cream ..   ..   **8. Ectadiopsis**
      Herbs with erect simple stems from a tuberous rootstock ; flowers pale violet-purple to deep crimson ; corona of erect filiform lobes ; follicles elongate-linear, erect
                                  9. Raphionacme
Corolla-tube broadly campanulate, as broad as long, 5-angled and 5-saccate at the base ; flowers blue, in 2–5-flowered terminal cymes ; stamens with a horny callus at the base of the anthers ..   ..   **10. Pentagonanthus**

Cryptostegia grandiflora (Roxb.) R. Br. ex Lindley is frequently cultivated and may have escaped in various places ; it is a climbing or erect glabrous shrub with oblong-elliptic to almost rotund leathery leaves 5–10 cm. long, and terminal or lateral cymes of large funnel-shaped pink or lilac flowers 6–12 cm. across. It is a native of Madagascar.

1. PARQUETINA Baill. in Bull. Soc. Linn. Paris 2 : 806 (1889) ; Bullock in Kew Bull. 15 : 204 (1960). *Periploca*[1] of F.W.T.A., ed. 1, 2 : 52, not of Linn. *Omphalogonus* Baill. l.c. (1889).

Scrambling woody climber, glabrous ; leaves coriaceous, oblong, elliptic, ovate, obovate or subrotund, abruptly acuminate, rounded to cordate at base, up to 15 cm. long and

---

[1] It is surprising that *Periploca linearifolia* Dillon & A. Rich. has never been recorded from the area of this Flora.

8 cm. broad, dark shining green above, glaucous beneath ; petioles 1·5–8 cm. long ;
flowers in lateral cymes ; pedicels rarely exceeding 10 mm. long ; sepals broadly
ovate, 3 mm. long ; corolla rotate, the lobes broadly ovate, about 10 mm. long ;
follicles widely spreading, woody, up to 45 cm. long but often much shorter   *nigrescens*

P. nigrescens (*Afzel.*) *Bullock* in Kew Bull. 15 : 205 (1961). *Periploca nigrescens* Afzel. (1817)—F.T.A. 4,
1 : 256 ; F.W.T.A., ed. 1, 2 : 52. *Parquetina gabonica* Baill. l.c. (1889). *Omphalogonus nigritanus* N. E.
Br. in Kew Bull. 1912 : 279 ; F.W.T.A., ed. 1, 2 : 53. *O. calophyllus* Baill. in Bull. Soc. Linn. Paris 2 :
812 (1889) ; F.T.A. 4, 1 : 256 ; Chev. Bot. 430. *Periploca calophylla* (Baill.) Roberty in Bull.
I.F.A.N. 15 : 1429 (1953). A twiner often herbaceous but becoming woody, with leathery glossy leaves ;
flowers greenish outside, deep red inside ; in regrowth forest.
Sen. : Diouloulou (fl. and fr. Sept., Oct.) *Adam* 17248. **Port.G. :** Cubisseco, Pobresa (fr. Feb.) *Esp.
Santo* 2379 ! Cacheu (fl. & fr. Oct.) *Esp. Santo* 1266 ! **Guin. :** Kindia (July) *Jac.-Fél.* 1803 ! Massadou,
Macenta *Collenette* 26 ! **S.L. :** *Barter* ! Bagroo R. (Apr.) *Mann* 845 ! Njala (Apr.) *Deighton* 5724 !
Kambia Dist. (Apr.) *Hepper* 2649 ! Limba (Apr.) *Sc. Elliot* 5631 ! **Lib. :** Nyaake (June) *Baldwin* 6141 !
Zwedru (Aug.) *Baldwin* 7037 ! Sinoe Basin *Johnston* ! Dukwia R. (May) *Cooper* 453 ! Gbanga (Sept.)
Linder 529 ! Ganta (Mar.) *Harley* 195 ! **Iv.C. :** Abouabou Forest, Abidjan to Grand Bassam (Jan.)
Leeuwenberg 2384 ! **Ghana :** Assuantsi (Mar.) *Irvine* 1562 ! Ancobra Junction (fr. Jan.) *Irvine* 1079 !
Eijan, Ashanti (Apr.) *Vigne* FH 1123 ! Bibiani (Dec.) *Adams* 1951 ! **Togo Rep. :** Lomé *Warnecke* 216 !
Dah. : *Poisson* ! **N.Nig. :** Kontagora *Dalz.* 50 ! **S.Nig. :** *Thomas* 1011 ! Agbadi, Sapoba (Nov.) *Meikle* 547 !
Ile Ife, Lagos (fr. Dec.) *Millen* 70 ! Ibadan (July, Aug.) *Keay & Latilo* FHI 37131 ! *Newberry* 49 !
Degema (Sept.) *Holland* 127 ! (See Appendix, p. 389.)

2. BATESANTHUS N. E. Br.—F.T.A. 4, 1 : 253 ; Bullock in Kew Bull. 15 : 203 (1961).

Woody climber ; leaves glabrous, elliptic to oblong-obovate or subrotund, apex abruptly
acuminate, base cordate, usually 7–16 cm. long, 4–10 cm. broad, or sometimes larger ;
petioles more or less 2 cm. long ; stipular fringes interpetiolar, reflexed, leathery,
subpersistent ; flowers in lax lateral panicles ; calyx small ; corolla rotate, lobes
elliptic-ovate up to 18 mm. long and 11 mm. broad, rounded at apex ; corona annular,
shortly 5-lobed, lobes emarginate, adnate to the base of the stamen-filaments ;
follicles more or less woody, about 12 cm. long and 2 cm. diam., suberect, shortly
beaked   ..   ..   *purpureus*

B. purpureus *N. E. Br.* in Hook. Ic. Pl. 25 : t. 2500 (1896) ; F.T.A. 4, 1 : 254 ; Bullock l.c. 204. *B. talbotii*
S. Moore in Cat. Talb. 63 (1913) ; F.W.T.A., ed. 1, 2 : 52. A twiner in forest with rather thick leaves ;
flowers greenish outside and (?) red inside.
S.Nig. : Oban *Talbot* 63 ! 2021 ! Eket (Apr.) *Talbot* ! [**Br.**]**Cam. :** Buea (May) *Maitland* 666 ! Bamenda
(June) *Maitland* 1565 ! Bafut-Ngemba F.R., Bamenda (Feb.) *Hepper* 2173 !

3. MONDIA Skeels in Bull. U.S. Dep. Agric. Bur. Pl. Ind. No. 223 : 45 (1911). *Chlorocodon*
Hook. f. (1871), not of Fourreau (1869)—F.T.A. 4, 1 : 254.

Softly woody climber with a large tuberous rootstock ; stems puberulous, glabrescent,
with persistent lignescent stipular frills at the nodes and prominent lenticels ; leaves
petiolate, broadly oblong, ovate, obovate or almost rotund, shortly acuminate,
rounded to deeply cordate at base, up to about 17 cm. long and 15 cm. broad, but
often smaller, glabrescent ; petiole up to 7 cm. long but usually shorter ; flowers in
lax lateral pedunculate panicles up to 15 cm. long and 10 cm. across ; pedicels
slender, 0·5–2 cm. long ; corolla subrotate, 5-lobed almost to the base, lobes 10–12
mm. long and about 6 mm. broad, ovate-oblong, glabrous on both sides but minutely
ciliate along one (the outside in bud) margin ; follicles somewhat woody, lanciform,
horizontally divaricate, about 9 cm. long and 1·5–2 cm. diam.   ..   ..   *whitei*

M. whitei (*Hook. f.*) *Skeels* l.c. (1911) ; Bullock in Kew Bull. 15 : 203 (1961). *Chlorocodon whitei* Hook. f. in
Bot. Mag. 97 : t. 5898 (1871) ; F.T.A. 4, 1 : 255. *Tacazzea viridis* A. Chev. ex Hutch. & Dalz. in Kew Bull.
1937 : 339 ; F.W.T.A., ed. 1, 2 : 52. A woody climber of drier forest country, with pale greenish-white
or cream flowers.
Port.G. : Bissau *Esp. Santo* 1760 ! Fulacunda (Aug.) *Esp. Santo* 2161 ! Bolola to Buba (Aug.) *Esp. Santo*
2155 ! Bafatá (Nov.) *Esp. Santo* 2977 ! **Guin. :** Bafing R., Man to Touba road (July) *Collenette* 55 ! Kindia
(May, June) *Jac.-Fél.* 1632 ! 1750 ! **S.L. :** Musaia (July, Aug.) *Deighton* 4405 ! 4805 ! *Jordan* 494 !
Kabala (July) *Glanville* 267 ! **Lib. :** Tappita (fr. Aug.) *Baldwin* 9051 ! **Iv.C. :** Mankono, Dialakoro to
Kénégoué (July) *Chev.* 21975 ! **N.Nig. :** Ankpa to Acharane, Kabba Prov. (June) *Daramola* FHI 38043 !
S.Nig. : Lagos *Batten-Poole* ! Ibadan North F.R. (Aug.) *Keay* FHI 25361 ! Oke Eleyele, Ibadan (July)
Sutton ! *Jones & Keay* FHI 16223 ! Gambari F.R. (fr. Nov.) *Onochie* FHI 34943 ! [**Br.**]**Cam. :** Aba-ajia,
Wum Dist. (July) *Ujor* FHI 30462 ! Widely distributed in tropical Africa, and replaced in eastern coastal
districts by the very closely allied *M. ecornuta* (N. E. Br.) Bullock.

4. ZACATEZA Bullock in Kew Bull. 9 : 361 (1954).

Slender twining shrub, internodes 3–15 cm. long, completely glabrous ; leaves shortly
petiolate, oblong, abruptly and shortly acuminate, broadly rounded to truncate at
base, up to 20 cm. long and 5·5 cm. broad, but usually about 10 cm. long and 3 cm.
broad, with numerous slender, parallel, horizontal lateral veins looped near the margin
and with fine reticulation between them ; cymes lateral, with very short peduncle
and main branches, but with slender pedicels 3–4 cm. long ; flowers buds oblong,
rounded at apex, 6 mm. long just before anthesis, tube very short, lobes oblong,
spreading, rounded at apex ; follicles falcate, reflexed, about 5·5 cm. long, 6–7 cm.
diam. near the base, with two ventro-lateral longitudinal ridges ..   *pedicellata*

Z. pedicellata (*K. Schum.*) *Bullock* l.c. (1954). *Tacazzea pedicellata* K. Schum. in Engl. Bot. Jahrb. 17 : 115
(1893) ; F.T.A. 4, 1 : 262, incl. var. *occidentalis* N. E. Br. l.c. 263 (1902). Climber up to 50 ft. ; flowers
cream.
S.Nig. : Lagos (Jan.) *Dalz.* 1348 ! Eglessa (Jan.) *Millen* 130 ! Eket Dist. *Talbot* 3265 ! Also recorded from
N.E. Congo and Angola.

5. **TACAZZEA** Decne.—F.T.A. 4, 1 : 253 ; Bullock in Kew Bull. 9 : 350 (1954).

Woody climber ; leaves oblong, ovate or subrotund, obtuse to acuminate at apex, rounded to cordate at base, 4–10 cm. long or more, 1·5–6 cm. broad, more or less glabrous, or pubescent, or tomentose beneath, glabrous or pubescent above ; flowers in terminal or lateral diffuse panicles ; calyx-lobes ovate, 1·5 mm. long ; corolla-lobes 6–7 mm. long ; follicles divaricate, narrowly or broadly ovoid, acuminately tapered to the apex, glabrous or densely hirsute ..   .. *apiculata*

T. apiculata *Oliv.* in Trans. Linn. Soc. 29 : 108 (1875) ; F.T.A. 4, 1 : 267 ; Bullock l.c. 354 ; Berhaut Fl. Sén. 60. *T. barteri* Baill. (1889)—F.T.A. 4, 1 : 266 ; F.W.T.A., ed. 1, 2 : 52. *T. nigritana* N. E. Br. (1895)—F.T.A. 4, 1 : 265 ; F.W.T.A., ed. 1, 2 : 52. *T. apiculata* var. *benedicta* Sc. Elliot (1894)—F.T.A. 4, 1 : 267 ; F.W.T.A., ed. 1, 2 : 52. A luxuriant dry forest climber with masses of creamy-white flowers. **Sen.:** *Berhaut* 1342. **Mali:** *Sole* 589 ! **Port.G.:** Gabu (July) *Esp. Santo* 2708 ! **Guin.:** Macenta (Apr.) *Jac.-Fél.* 831 ! Kindia (May) *Jac.-Fél.* 1607 ! Dantilia R. (Mar.) *Sc. Elliot* 5269 ! **S.L.:** Rowalla (fr. July) *Thomas* 1061 ! Mattru (June) *Deighton* 4773 ! Musaia (fr. July) *Deighton* 4797 ! **Ghana:** Elmina (June) *Hall* 1469 ! Kumasi (May) *Vigne* FH 1709 ! Wenchi (Aug.) *Darko* 937 ! Kpong (June) *Irvine* 1789 ! Pong Tamale (June) *Vigne* FH 3865 ! Gambaga (fl. May) *Morton* GC 7413 ! **N.Nig.:** Kuzata R., Niger Prov. (June) *Keay* FHI 25877 ! Jemaa (Feb.) *McClintock* 183 ! Mairabo plantation, Zaria (Jan.) *Lowe* FHI 48408 ! Kachia to Kaduna (Feb.) *Meikle* 1204 ! Gaya, Kano (July) *Daggash* FHI 22393 ! Sokoto (fl. & fr. May) *Dalz.* 525 ! Widely distributed in tropical and S. Africa. (See Appendix, p. 390.)

6. **CRYPTOLEPIS** R. Br.—F.T.A. 4, 1 : 242, partly.

Twining and scrambling thin-stemmed shrub with blood-red juice ; leaves thinly herbaceous, elliptic, oblong-elliptic or ovate, acutely and shortly acuminate, rounded or more rarely acutely cuneate at base, 2·5–7 cm. or more long, 1–3 cm. or more broad, glabrous ; cymes lateral on lateral shoots, laxly few-flowered ; pedicels 6–15 mm. long ; corolla-tube 5 mm. long, lobes contorted to the left in bud, about 12 mm. long, lanceolate-acuminate ; follicles linear, 18 cm. or more long, spreading or somewhat reflexed ; seeds 12 mm. long, with a coma of long silky hairs .. *sanguinolenta*

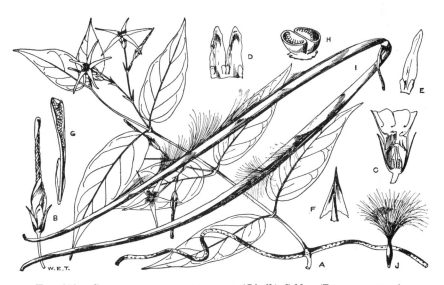

FIG. 219.—CRYPTOLEPIS SANGUINOLENTA (*Lindl.*) *Schltr.* (PERIPLOCACEAE).

A, flowering shoot. B, flower-bud.   C, longitudinal section of lower part of flower.   D, corona-lobes.   E, corona-lobe.   F, appendage.   G, carpel, inverted.   H, cross-section of ovary. I, fruit. J, seed.

C. sanguinolenta (*Lindl.*) *Schltr.* Westafr. Kautschuk-Exped. 308 (1900) ; Berhaut Fl. Sén. 82 ; Bullock in Kew Bull. 10 : 280 (1955). *Pergularia sanguinolenta* Lindl. in Bot. Mag. t. 2532 (1825). *Cryptolepis triangularis* N. E. Br. (1894)—F.T.A. 4, 1 : 245 ; Berhaut l.c. 59. Slender climber up to 25 ft. high with greenish-yellow flowers. **Sen.:** Sangalkam, Niayes *Berhaut* 1811. Tambacounda *Berhaut* 1190. **Port.G.:** Bolama (fl. & fr. Apr.) *Esp. Santo* 1925 ! Bissau (fr. Feb.) *Esp. Santo* 1739 ! Bafatá (Aug.) *Esp. Santo* 3277 ! **Guin.:** Macenta (Mar.) *Jac.-Fél.* 1581 ! 1582 ! **S.L.:** *Don* ! Lumbaraya (Feb.) *Sc. Elliot* 5013 ! Jigaya (Sept.) *Thomas* 2771 ! **Ghana:** Aburi Hill (June) *Williams* 289 ! L. Bosumtwe (Apr.) *Adams* 2473 ! Ejura (June) *Vigne* FH 2050 ! **N.Nig.:** *Barter* 3359 ! Nupe *Barter* 1333 ! Kabba (June) *Daramola* FHI 38040 ! Sokoto *Moiser* ! **S.Nig.:** Lagos *Rowland* ! Abeokuta (May) *Keay & Onochie* FHI 37013 ! Ibadan (Mar.) *Meikle* 1296 ! Ogwashi (Nov.) *Thomas* 2041 ! (See Appendix, p. 387.)

7. MANGENOTIA Pichon in Bull. Soc. Bot. Fr. 101 : 246 (1954).

Scrambling and twining thin-stemmed shrub, internodes uniformly 2–3 cm. long ; leaves oblong, acutely and abruptly acuminate, rounded or subacute at base, very uniform in size, about 3–5 cm. long, 1–2 cm. broad, glabrous, strongly discolorous, dark green above, glaucous beneath ; cymes laxly few-flowered, lateral ; peduncles 1–1·5 cm. long ; pedicels 3–6 mm. long ; corolla-tube urceolate, 6–8 mm. long, 1·5 mm. diam. in the lower part, narrower above, lobes spreading, lanceolate, acute, 5 mm. long ; follicles straight, horizontally spreading, about 13 cm. long, 8 mm. diam., tapering to an acute tip   ..   ..   *eburnea*

M. eburnea *Pichon* l.c. (1954) ; Bullock in Kew Bull. 9 : 587 (1955). A shrub, erect at first, later climbing and twining ; flowers creamy white.
　　S.L.: Port Loko (fr. Dec.) *Thomas* 6585 ! Mattru to Gbangbama (Nov.) *Deighton* 2349 ! Mamaka (fl. & fr. Nov.) *Thomas* 4377 ! 4544 ! Yonibana (Oct.) *Thomas* 4119 ! Sukudu *Dawe* 543 ! **Ghana:** near Mampong (July) *Darko* 689 ! Mampong Scarp (May) *Vigne* FH 1739 ! **S.Nig.:** Lagos, Ikoyi Plains (Aug.) *Dalz.* 1405 !

8. ECTADIOPSIS Benth. in Benth. & Hook. f. Gen. Pl. 2 : 741 (1876) ; Bullock in Kew Bull. 10 : 267 (1955).

Erect subshrub with annual or perennial simple or sparingly branched stems 30–60 cm. high, glabrous ; leaves linear, linear-lanceolate or oblanceolate, or more rarely oblong or ovate, apex subacutely narrowed and more or less mucronulate, cuneate at base, very variable in size, up to 10 cm. long ; cymes lateral, few to many-flowered, lax or somewhat congested, branched from the base ; pedicels 1–4 mm. long ; corolla-tube 2–4 mm. long, lobes 3–4 mm. long ; follicles erecto-patent, straight, tapering to a blunt apex, up to 14 cm. long, 1 cm. diam. but frequently smaller ; seeds oblong-elliptic, about 1 cm. long and 2·5–3 mm. broad, crowned with a dense coma of long silky hairs   ..   ..   *oblongifolia*

E. oblongifolia (*Meisn.*) *Schltr.* in Engl. Bot. Jahrb. 18, Beibl. 45 : 14 (1894) ; Bullock l.c. 268. *Ectadium* ? *oblongifolium* Meisn. (1843). *Ectadiopsis nigritana* Benth. in Hook. Ic. Pl. 12 : 75, t. 1187 (1876). *Cryptolepis nigritana* (Benth.) N. E. Br. in F.T.A. 4, 1 : 251 (1902) ; F.W.T.A., ed. 1, 2 : 53. An erect shrub to about 2 ft. high with cream flowers ; in savanna woodland.
　　Port.G.: Pitche, Pansor (fr. Oct.) *Esp. Santo* 3429 ! Gabu, Calicunda to Paunca (fr. Sept.) *Esp. Santo* 2791 ! **Guin.:** Dabola (fl. & fr. June) *Pobéguin* 352 ! **Ghana:** Bole (May) *Vigne* FH 3828 ! Kintampo (July) *Dalz.* 46 ! Amedzofe (fl. Mar., fr. Nov.) *Morton* 9417 ! *Hepper* 2330 ! Shiare (Apr.) *Hall* 1424 ! Gambaga (July) *Vigne* FH 4536 ! **Togo Rep.:** (Feb.) *Büttner* 404 ! **N.Nig.:** Nupe *Barter* 1324 ! Zungeru (fl. & fr. July) *Dalz.* 7 ! Anara F.R., Zaria (May) *Keay* FHI 22939 ! Mande F.R., Zaria Prov. (June) *Keay* FHI 25841 ! Vom *Dent Young* 174 ! Nabardo (May) *Lely* 216 ! **S.Nig.:** Upper Ogun F.R., Oyo Prov. (Feb.) *Keay* FHI 22509 ! Yala Hill (May) *Hambler* 450 ! [**Br.**]**Cam.:** Abakpa, Bamenda (May) *Ujor* FHI 30096 ! Bamenda (June) *Maitland* 1694 ! Vogel Peak, Adamawa (fr. Dec.) *Hepper* 1556 ! Widely distributed in tropical and S. Africa. (See Appendix, p. 387.)

9. RAPHIONACME Harvey—F.T.A. 4, 1 : 268.

Leaves linear, up to 20 times as long as broad ; inflorescence a lax terminal or lateral cyme with pedicels 8–14 mm. long ; flowers frequently precocious, corolla-tube 3 mm. long, lobes 5–7 mm. long, oblong, spreading or reflexed, pink ; corona-lobes erect, filiform 5–6 mm. long ; follicles erect, often arcuate, linear-terete, 10–30 cm. long ; dwarf 10–30 cm. high, with a fleshy tuberous rootstock   ..   ..   1. *brownii*
Leaves lanceolate, elliptic, or more or less rotund-obovate :
Inflorescence lateral from upper leafy nodes :
　Inflorescence 2–3-flowered, peduncles very short, pedicels 1–2 cm. long ; flowers sometimes precocious ; corolla-tube more or less 3 mm. long, lobes oblong, about 13 mm. long, pink inside, green outside, spreading or somewhat reflexed ; leaves shortly petiolate, lanceolate to broadly elliptic or obovate, apex rounded, apiculate, base cuneate to rounded, up to 12 cm. long and 4 cm. broad, the lower relatively broader and smaller ; follicles "linear, erect 8–10 cm. long" (Berhaut) ; herb 10–30 cm. high, with a tuberous rootstock..   ..   ..   2. *daronii*
　Inflorescence cymose, multiflorous ; pedicels 4–6 mm. long ; corolla-tube 2 mm. long, lobes triangular, 5–6 mm. long, spreading or at length reflexed, green with a deep red zone at the base, minutely papillate inside ; corona-lobes rectangular at the base, with a flexuose whip-like apical appendage intertwined and connivant over the style and stamens ; stems simple, numerous, erect from a woody tuber, up to 1 m. high ; leaves shortly or obsoletely petiolate, oblong-lanceolate, up to 12 cm. long and 2 cm. broad, apex and base acute, glaucous below and crispate-pubescent on both surfaces   ..   ..   3. *keayii*
Inflorescence a branched, congested, terminal cyme, more or less cylindrical in outline, with a bibracteate peduncle 2–3 cm. long ; the congested lateral cymules racemosely arranged along the erect rhachis ; bracts and bracteoles minute, pedicels 1–2 mm. long ; corolla-lobes ovate-lanceolate, about 3·5 mm. long, the tube cylindrico-campanulate, 2 mm. long ; corona-lobes filiform, erect, 5 mm. long ; stems simple, erect, about 6–16 cm. high, puberulous ; leaves obovate-oblanceolate, the upper-most largest, 4–10 cm. long, 1–3 cm. broad, glabrescent above, scabrido-puberulous beneath, apex rounded, cuneate at base and shortly petiolate   ..   4. *vignei*

1. **R. brownii** *Sc. Elliot* in J. Linn. Soc. 30 : 91 (1894) ; F.T.A. 4, 1 : 273 ; Bullock in Kew Bull. 8 : 60 (1953).
R. excisa Schltr. in J. Bot. 33 : 301 (1895) ; F.T.A. 4, 1 : 272. *R. brownii* var. *longifolia* A. Chev. Bot. 430, name only. *R. juvensis* N. E. Br. in F.T.A. 4, 1 : 272 ; Bullock in Kew Bull. 8 : 60 (1953). A precocious herb 6–12 in. high with diffuse cymes of pink flowers about 1–2 in. diam.
Port.G.: Boé, Madina (June) *Esp. Santo* 2921 ! Bissoran, Lalaconquili (June) *Esp. Santo* 2391 !
Guin.: Labi (fr. Apr.) *Chev.* 12269 ! Beyla (fl. & fr. Apr.) *Collenette* 2 ! Telimélé (June) *Jac.-Fél.* 1777 !
S.L.: Kasanko (May) *Adames* 228 ! Falaba (fl. & fr. Mar.) *Sc. Elliot* 5179 ! **Ghana:** Achimota (fr. Apr.)
Irvine 1834 ! Berekum to Sampa (fl. & fr. Apr.) *Morton* A3241 ! Sunyani (Jan.) *Chipp* 76 ! Okroso to
Otiso Ferry (Mar.) *Hepper & Morton* A3038 ! Tamale to Bolgatanga (fr. Apr.) *Adams* 4163 ! **N.Nig.:**
Ago-Are to Aha, Oyo Prov. (Mar.) *Meikle* 22525 ! Lokoja (Mar.) *Dalz.* 6 ! Jebba (Jan.) *Meikle* 1104 !
Birnin Gwari (Feb.) *Meikle* 1230 ! Abinsi *Dalz.* 688 ! Widely distributed in the Sudan, northern Congo
and E. Africa. (See Appendix, p. 389.)
2. **R. daronii** *Berhaut* in Bull. Soc. Bot. Fr. 101 : 374 (1955) ; Fl. Sén. 82. A pink-flowered herb of stony
savanna, said to be less precocious than the preceding.
Sen.: Vélor, Thiès to N'Gazobil *Berhaut* 1638 ! 4777. **Ghana:** W. of Abene, Kwahu (Jan.) *Chipp* 633 !
Kwahu Tafo to Mankrong (Dec.) *Adams* 5091 ! Laboni F.R., Tamale to Damongo (Mar.) *Adams* 3935 !
Bamboi to Kintampo (Mar.) *Hepper & Morton* A3191 !
3. **R. keayii** *Bullock* in Kew Bull. 8 : 63 (1953). An erect herb to 3 ft. high, of open *Isoberlinia-Uapaca* woodland.
N.Nig.: Mando F.R., Zaria Prov. (July) *Keay* FHI 25983 ! 25990 ! Also in Cameroun (Beretum,
Mildbr. 9664).
4. **R. vignei** *Bruce* in Kew Bull. 1936 : 477. A dwarf erect geophytic herb from a fleshy edible disciform tuber,
with pale green flowers.
Ghana: *Dalz.* 55 ! Sao Forest, Wa (May) *Vigne* FH 3823 ! Bole (Apr.) *Morton* GC 3301 !

10. PENTAGONANTHUS Bullock in Hook. Ic. Pl. 36 : sub. t. 3583 (1961).

Erect herb 1–2 ft. high with simple stems from a large woody tuber, more or less hirsute
but soon glabrescent ; leaves subsessile, reduced to scales in the lower part of the
stem, increasing in size upwards, linear, up to about 12 cm. long and about 5 mm.
broad, with recurved edges and a prominent midrib, apex acute, narrowed to the base ;
flowers blue, in sessile fascicles 2–5 cm. apart of 2–4 flowers racemosely arranged
along the upper leafless part of the stem ; pedicels 6–10 mm. long, elongating to about
2 cm. long as the flower ages ; corolla-tube 5-angled, shallowly 5-saccate at the base,
4 mm. long and 4 mm. diam., lobes oblong-lanceolate, acute or subacute, nearly 2 cm.
long and about 7 mm. broad ; follicles not known ..   ..   *caeruleus*

P. caeruleus (*E. A. Bruce*) *Bullock* l.c. (1961). *Raphionacme caerulea* E. A. Bruce in Kew Bull. 1935 : 279.
(?) *R. sudanica* A. Chev. ex Hutch. & Dalz. F.W.T.A., ed. 1, 2 : 54 (1931) ; Chev. Bot. 430, name only.
A blue-flowered herb, unique in the family.
Guin.: Kindia (July) *Jac.-Fél.* 1793 ! *Jaeger* ! **S.L.:** Bumban (Aug.) *Deighton* 1246 !

136. ASCLEPIADACEAE

By A. A. Bullock

Twining or erect shrubs or perennial herbs ; leaves opposite, linear to
orbicular-obovate, entire, or rarely lobed or toothed. Inflorescence cymose,
most frequently umbelliform but sometimes with flowers more or less racemosely
fasciculate along a simple or branched rhachis. Flowers regular, pentamerous ;
calyx-tube very short or obsolete ; corolla contorted, imbricate or valvate,
gamopetalous, the lobes sometimes connivent at the apex ; filaments united
into a tube, anthers connivent throughout their length and united with the
expanded style apex, introrse, bilocular and provided with lateral horny wings.
Corona arising from the staminal column, very variable in form or rarely absent
or reduced to five small fleshy tubercles. Pollen in waxy masses (pollinia)
attached in pairs by caudicles of various form to 5 corpuscular usually horny
sutured pollen carriers arising from the style apex and concealing the stigmatic
surfaces. Ovary of two separate carpels ; styles free to the apex where they are
united in a peltate disk which is convex, conical or beaked ; ovules multiseriate,
on a single adaxial placenta in each carpel. Fruit of two (by abortion often one)
erect or divergent follicles which may be linear to ovoid or ellipsoidal, membranous
to woody, smooth or winged or variously armed with soft or indurated prickles,
always dehiscent lengthwise adaxially. Seeds compressed, often with a membranous or thickened margin and nearly always crowned with a coma of silky
hairs.

A mainly tropical family of some 130 genera, most of its members occupying savanna,
monsoon or semi-arid country, more rarely found in rain forest. The tribe *Ceropegieae*
is confined to the Old World. The gamopetalous corolla, staminal column and corona
and peltate style-apex, together with follicular fruits, flattened seeds with silky coma
and opposite exstipulate leaves, are characteristic. Except in the *Secamonoideae*, woody
plants are rare, and the wood itself is very soft ; milky latex is universally present, but
sometimes small in quantity.

KEY TO THE SUBFAMILIES

Pollen-contents of each anther-loculus consisting of two minute waxy masses ; pollen carrier pale-coloured, not of horny consistency, caudicles subobsolete ; style often produced beyond the anthers .. .. SECAMONOIDEAE

Pollen-contents of each anther-loculus consisting of a single waxy mass, often hyaline along one edge ; pollen carrier dark in colour, horny, caudicles well-developed, filiform or variously flattened ; style usually not produced beyond the anthers
ASCLEPIADOIDEAE

KEY TO THE TRIBES AND GENERA

Subfamily—SECAMONOIDEAE

A single tribe .. .. I. SECAMONEAE

Tribe I—SECAMONEAE

Style apex truncate or bilobed, scarcely exceeding the anthers ; corona-lobes laterally flattened or subulate ; inflorescence a diffuse dichotomous cyme.. 1. **Secamone**

Style apex subulate-conical to clavate, much exceeding the anthers ; corona-lobes dorsally flattened:

Inflorescence racemiform, much longer than the leaves ; corona much exceeding the anthers ; corolla-lobes about as long as the tube or a little longer
2. **Rhynchostigma**

Inflorescence cymose, shorter than the leaves ; corona scarcely exceeding the anthers ; corolla-lobes several times as long as the tube .. .. .. 3. **Toxocarpus**

Subfamily—ASCLEPIADOIDEAE

Pollen masses pendulous in the anther-loculi, suspended by caudicles from horny reddish-brown to black corpuscular carriers, the latter with a dorsal suture
II. ASCLEPIADEAE

Pollen masses horizontal, ascending or erect in the anther loculi, attached by caudicles to horny dark or light-coloured sutured carriers:

Corolla-lobes imbricate or contorted, never valvate .. .. III. MARSDENIEAE

Corolla-lobes valvate, sometimes adherent to each other at their tips
IV. CEROPEGIEAE

Tribe II—ASCLEPIADEAE[1]

Leafy twiners, scramblers or erect herbs or shrubs ; leaves sometimes caducous :

Corona of 5 free lobes arising from the junction between the corolla and staminal column ; herbs of semi-arid areas .. .. 4. **Glossonema**

Corona simple or double, arising from the staminal column, sometimes at its extreme base :

Corona tubular, membranous, variously toothed or laciniate ; inflorescence sub-umbelliform or flowers in fascicles racemosely arranged along an elongating axis ; leaves usually cordate at the base .. .. 5. **Cynanchum**

Corona never tubular, but the lobes sometimes united near the base to form an annulus or shallow cup :

Twining climbers :

Flowers about 3 cm. diam., saucer-shaped ; corona simple, of 5 erect lanceolate lobes with a wrinkled gibbosity at the base .. .. .. 6. **Oxystelma**

Flowers not exceeding 1·5 cm. diam., usually much smaller :

Corona double, the outer membranous, annular with very short subquadrate, often denticulate lobes ; inner corona of 5 fleshy erect lobes adnate to the staminal column, free above with subulate horns incurved over the staminal column and produced below into spreading or deflexed spurs ; corolla-tube cylindric, the lobes spreading .. .. 7. **Pergularia**

Corona simple ; corolla lobed almost to the base :

Corona-lobes fleshy, apex rounded or truncate, slipper-shaped, margins infolded, with an apical reflexed horn on the inner face ; follicles usually softly echinate, ovoid .. .. 8. **Pentarrhinum**

Corona-lobes laterally compressed, adnate to the staminal column, with a free triangular apex ; follicles narrowly lanceolate in outline, smooth
9. **Pentatropis**

Erect herbs, or shrubs, occasionally tree-like :

Corona-lobes laterally flattened, with a conspicuous curved basal spur ; large, softly woody tree-like shrub .. .. 10. **Calotropis**

---

[1] The tribe *Asclepiadeae* has been until recently referred to as *Cynancheae* ; this change is necessary under the International Code of Botanical Nomenclature. The *Ceropegieae* include the *Stapelieae* of the previous edition of this work and of the Flora of Tropical Africa ; I find no justification for the separation of the usually leafless stapeliads. *Frerea*, an Indian genus now sometimes united with *Caralluma*, has the fleshy habit of other stapeliads, but large long-persistent leaves such as are found in *Ceropegia*. On the other hand, all other stapeliads have minute caducous leaves, whilst the flowers of many of them recall *Brachystelma*.

Corona-lobes without a spur at the base :
Corona-lobes petaloid, complicate and claw-like at the base, expanded above into
a broad limb, variously toothed or fimbriate, or almost entire
11. Margaretta
Corona-lobes not petaloid and without a claw-like basal part :
Flowers densely spirally arranged along the thickened apical part of the peduncle
which elongates as the flowers develop ; leaves narrowly lanceolate, with
minute bristles in their axils ..   ..   **12. Kanahia**
Flowers in a false umbel ; leaves without axillary bristles :
Corona-lobes dorsally flattened, without an expanded free part ; erect herbs
with linear or linear-lanceolate leaves and flowers not exceeding 5 mm. diam.
13. Aspidoglossum
Corona-lobes laterally flattened or not at all compressed, sometimes with a
strap-shaped free part :
Corona-lobes cucullate (i.e. hooded), the inner face densely beset with erect
papillae   ..   ..   **14. Trachycalymma**
Corona-lobes not cucullate and not internally papillate :
The corona-lobes complicate :
Corona-lobes with a conspicuous horn arising from the base of the inner face
15. Asclepias
Corona-lobes quadrate in outline, but without an inner basally affixed horn
16. Gomphocarpus
The corona-lobes not complicate, fleshy and solid :
Corona-lobes arising from the staminal column on short horizontal stipes,
with a fleshy apical ovoid-conical apical part tapering to a subulate point ;
follicles erect, long-fusiform   ..   ..   ..   **17. Xysmalobium**[1]
Corona-lobes dorsally flattened, sometimes with slightly infolded margins,
with or without a strap-shaped apical appendage, and with or without
wing-like keels on the inner face..   ..   ..   ..   **18. Pachycarpus**
Leafless scrambling shrubs with flowers in lateral umbelliform inflorescences at the
nodes ; corolla rotate or rotate-campanulate ; corona-double, the outer annular, the
inner of 5 erect fleshy laterally compressed lobes   ..   ..   ..   **19. Sarcostemma**

Tribe III—Marsdenieae

Corona with three series of lobes ; inflorescence racemiform, with a long naked peduncle
and long filiform pedicels   ..   ..   **20. Neoschumannia**
Corona simple or double ; inflorescence not as above :
*Corona simple, often reduced to minute tubercles :
Corolla-lobes with reflexed and more or less laciniate flaps arising from their bases,
projecting into the tube and closing its throat ; corona-lobes arising high on the
staminal column, ovate, acuminate, coherent at their tips over the truncate style-
apex and much swollen dorsally ; flowers densely spirally arranged along the
swollen apex of the peduncle   ..   ..   **21. Dalzielia**[2]
Corolla not as above, at most hairy in the throat :
Corona arising from the sinus between the corolla and staminal column and partly
adnate to the corolla, consisting of 5 fleshy lobes channelled down the face and with
free incurved tips ; a climbing or sub-erect shrub with rusty indumentum
22. Gymnema
Corona arising from the staminal column, not at all adnate to the corolla :
Corolla subrotate ; corona of 5 minute tubercles ; climbing herbaceous or softly
woody lianes, or erect herbs from a woody rootstock ; inflorescence usually a
zigzag cyme   ..   ..   **23. Tylophora**
Corolla globose, campanulate or tubular :
The corolla globose or globose campanulate, the lobes shorter than the tube ;
erect or semi-twining herbs from a woody rootstock..   .. **24. Sphaerocodon**
The corolla tubular-campanulate or with a short tube :
Corolla tubular-campanulate, inflated at the base, densely hairy in the throat ;
corona-lobes with the free part expanded and with a linear appendage arising
from its inner face ; a twining shrub..   ..   ..   ..   **25. Telosma**
Corolla with a short cylindric tube ; corona-lobes without an appendage on the
inner face ; corolla-lobes spreading, reflexed, or suberect :
Flowers 1 cm. or more diam., the lobes spreading or reflexed ; follicles woody,
ovoid, often winged ..   ..   **26. Dregea**
Flowers not more than 5 mm. diam., the lobes suberect ; follicles membranous,
lanceolate in outline, never winged ..   ..   **27. Gongronema**

[1] For the purpose of this Flora I am retaining the circumscription of F.T.A. and Ed.1, but this genus is in
fact more properly restricted to *X. undulatum* E. Mey. The species enumerated here under *Xysmalobium*
will be transferred to another genus at present undescribed.—A.A.B.
[2] The corona of *Dalzielia* was originally described as double, but the so-called outer corona consists of
outgrowths of the corolla and has nothing to do with the staminiferous corona.—A.A.B.

*Corona double, the inner arising from the staminal column, the outer from the sinus between the corolla and staminal column ; high-climbing glabrous shrub with umbelliform inflorescences and very long widely divergent follicles   28. **Anisopus**

Tribe IV—Ceropegieae

Leafy erect or twining shrubs or herbs, or if leafless, not at all succulent :
 Corolla-tube short, scarcely longer than broad, or almost none :
  Twining leafy climbers or leafless shrubs ; flowers small, campanulate
                                                              29. **Leptadenia**
  Erect herbs with a tuberous rootstock ; flowers more or less rotate, the limb with a
   broad disk, the lobes often long-caudate ..   .. 30. **Brachystelma**
 Corolla-tube much longer than broad, the limb spreading or with the lobes variously
  connate at their tips ..   .. 31. **Ceropegia**
Succulent plants, the leaves reduced to minute triangular caducous scales :
 Corolla with the sinus between the lobes produced into small triangular teeth
                                                              32. **Huernia**
 Corolla without teeth at the sinuses between the lobes   ..   .. 33. **Caralluma**

1. SECAMONE R. Br.—F.T.A. 4, 1 : 276.

Style with a truncate apex, not protruding beyond the anthers ; flowers very small, rotate, about 3 mm. diam. ; follicles horizontally spreading ; leaves abruptly and fairly long-acuminate, ovate to oblong-elliptic, up to 5 cm. long and 2 cm. broad
                                                              1. *afzelii*
Style apex ovoid-clavate, most often somewhat bilobed, protruding beyond the anthers ; flowers larger than above, corolla-lobes erecto-patent, 3 mm. long, the tube about 1·5 mm. long ; follicles reflexed ; leaves gradually acuminate or merely acute, some- what narrowly ovate or elliptic, up to 7·5 cm. long but often less   2. *africana*

1. **S. afzelii** (*Schultes*) *K. Schum.* in Engl. Bot. Jahrb. 23 : 234 (1896) ; Bullock in Kew Bull. 9 : 586 (1955). *Ichnocarpus afzelii* Schultes in Roem. & Schult. Syst. Veg. 4 : 399 (1819). *Secamone myrtifolia* Benth. (1849)—F.W.T.A., ed. 1, 2 : 54 ; Berhaut Fl. Sén. 57. *Toxocarpus leonensis* Sc. Elliot (1894). *Secamone leonensis* (Sc. Elliot) N. E. Br. (1902)—F.W.T.A., ed. 1, 2 : 54.[1] Climbing or scrambling shrub with yellowish or orange flowers ; in secondary forest.
 Sen.: (May) *Thierry* 58 ! Niayes *Berhaut* 413. **Port.G.:** Cumura, Bissau (fr. Nov.) *Esp. Santo* 2233 ! Bijinita, Bissau (Sept.) *Esp. Santo* 2183 ! **Guin.:** Péla (Sept.) *Baldwin* 13317 ! **S.L.:** Sugar Loaf Mt. (Apr.) *Sc. Elliot* 5773 ! Kissy Brook, Freetown (Aug.) *Melville & Hooker* 204 ! Njala (fr. Jan.) *Deighton* 2437 ! Jigaya (Sept.) *Thomas* 2538 ! Scarcies (fr. Feb.) *Sc. Elliot* 4781 ! Miligi, Tonko-Limba (Oct.) *Jordan* 928 ! **Lib.:** Bonuta (Oct.) *Linder* 883 ! Tappita (Aug.) *Baldwin* 9045 ! Suacoco (Sept.) *Traub* 273 ! Gbanga (Sept.) *Linder* 699 ! Belleyella (fr. Dec.) *Baldwin* 10644 ! **Ghana:** Ancobra Junction (fr. Jan.) *Irvine* 1077 ! Kumasi (Aug.) *Darko* 711 ! Aburi Scarp (fr. Sept.) *Adams* 1780 ! Somanya (Aug.) *Johnson* 1077 ! Oda (Aug.) *Howes* 963 ! **Togo Rep.:** Lomé *Warnecke* 203 ! **Dah.:** *Poisson* ! **N.Nig.:** Dogon Kurmi, Jemaa Div. (Sept.) *Killick* 70 ! **S.Nig.:** Ijebu Ode (fr. Oct.) *Tamajong* FHI 20970 ! Ibadan (fr. Feb.) *Meikle* 1163 ! Ishan (Aug.) *Onochie* FHI 33276 ! Onitsha (Sept.) *Onochie* FHI 33444 ! Calabar (fr. Jan.) *Onochie* FHI 36172 ! Also in Cameroun and Gabon.

2. **S. africana** (*Oliv.*) *Bullock* in Kew Bull. 8 : 362 (1953), and 15 : 195 (1961). *Toxocarpus africanus* Oliv. in Trans. Linn. Soc. 29 : 109, t. 108A (1875). *Secamone platystigma* K. Schum. (1893)—F.T.A. 4, 1 : 280. Climbing shrub about 15 ft. high ; flowers white with yellow centre.
 [Br.]Cam.: L. Nyos, 3,500–4,000 ft., Bamenda (Aug.) *Savory* UCI 344 ! Widely distributed in tropical Africa.

2. RHYNCHOSTIGMA Benth. in Benth. & Hook. f. Gen. Pl. 2 : 771 (1876) ; Bullock in Kew Bull. 15 : 194 (1961).

Twining climber ; leaves coriaceous, oblong, cuspidate, 6–10 cm. long, 2·5–4 cm. broad, with numerous parallel obscure lateral veins ; inflorescence lateral, distantly racemi- form, 5–15 cm. long and with 4–10 flowers in pairs ; pedicels 1–2 cm. long and the pairs of flowers 1–2·5 cm. apart ; corolla-tube 3–4 mm. long, lobes 4–4·5 mm. long, recurved-spreading, oblong, obtuse, apex of style exserted, slender ; corona-lobes incurved, somewhat clavate ; follicles linear, about 15 cm. long, 5 mm. diam., widely divaricate   ..   ..   *racemosum*

R. racemosum *Benth.* in Hook. Ic. Pl. 12 : 77, t. 1189 (1876) ; Bullock l.c. *Toxocarpus racemosus* (Benth.) N. E. Br. in F.T.A. 4, 1 : 287 ; F.W.T.A., ed. 1, 2 : 54. Climber over bushes.
 [Br.]Cam.: Cam. Mt., 4,500 ft. (Feb.) *Mann* 1273 ! **F.Po:** Moka, 6,000 ft. (fr. Sept.) *Wrigley & Melville* 592 !

3. TOXOCARPUS Wight & Arn.—F.T.A. 4, 1 : 268, partly ; Bullock in Kew Bull. 15 : 193 (1961).

Twining climber ; young parts and inflorescences covered with conspicuous red pubes- cence ; leaves discolorous, narrowly elliptic, oblong, lanceolate or oblanceolate through ovate or obovate to almost rotund, abruptly and shortly acuminate, cuneate at base, 3–10 cm. long, rarely more than 4 cm. broad, lateral veins conspicuous, slightly ascending, parallel and numerous ; cymes lateral, 6–10-flowered, shorter than the leaves ; pedicels 4–12 mm. long ; corolla 5-lobed almost to the base, lobes narrow, oblong, 5–7 mm. long, widely spreading, conspicuously contorted in bud ; style

------

[1] See Kew Bull. 15 : 195 (1961) ; the doubt expressed there as to the status of *S. leonensis* is confirmed. I now regard the type as a poor specimen of *S. afzelii*—A.A.B.

protruding about 2 mm. beyond the anthers, the apical part acuminately conical, obtuse, spirally grooved ; anther appendages fimbriate .. .. .. *brevipes*

T. brevipes (*Benth.*) *N. E. Br.* in F.T.A. 4, 1 : 287 ; Bullock l.c. *Rhynchostigma brevipes* Benth. in Hook. Ic. Pl. 12 : 78, sub. t. 1189 (1876). Slender climbing shrub ; flowers white turning yellow.
S.Nig.: Ikoyi Plains, Lagos (Aug.) *Dalz.* 931 ! Agulu, Onitsha Prov. (Nov.) *Keay* FHI 25590 ! Milliken Hill, Enugu (Mar.) *Hepper* 2235 ! Nun R. *Barter* 2120 ! *Mann* 484 ! Bendiga Afi, Ogoja Prov. (Dec.) *Keay* FHI 28271 ! Also in Cameroun, Congo and N. Rhodesia.

4. GLOSSONEMA Decne.—F.T.A. 4, 1 : 290, partly ; Bullock in Kew Bull. 10 : 613 (1956).

Spreading herb up to 30 cm. high, stems shortly pubescent ; leaves linear or linear-lanceolate, up to 12 cm. long, 3–10 mm. broad, pubescent, at least beneath ; flowers few, in subsessile extra-axillary cymes ; corolla 4 mm. diam. ; fruits fleshy when young, narrowly elongate-ovoid, acuminate, echinate, pubescent, up to 6 cm. long

boveanum subsp. *nubicum*

G. boveanum (*Decne.*) *Decne.* subsp. **nubicum** (*Decne.*) Bullock in Kew Bull. 10 : 617 (1956). *G. nubicum* Decne. (1844)—F.T.A. 4, 1 : 291 ; Chev. Bot. 431 ; F.W.T.A., ed. 1, 2 : 55. A herb a few inches to 1 ft. high, woody at the base ; flowers greenish-white ; a weed of cultivation in dry country.
Mali: Kolé (Feb.) *Vaillant* 900 ! Macina (Nov.) *Davey* 1411 Dioura (fl. & fr. Jan.) *Davey* 211! Niami-Niama, near Bore *O.I.C.M.A.* 44/1957 ! Timbuktu (fl. & fr. July) *Hagerup* 214 ! **Ghana:** Navrongo (fl. & fr. Mar.) *Vigne* FH 4487 ! 4682 ! **N.Nig.:** *Thornewill* 35 ! Fodama *Moiser* 187 ! Ibeto (fl. & fr. Jan.) *Meikle* 1048 ! Katagum (Dec.) *Dalz.* 99 ! Sokoto *Dalz.* 368 ! **S.Nig.:** Lagos *Barter* ! [Br.]**Cam.:** Bama, Dikwa Div. (Jan.) *McClintock* 155 ! Also in the Sudan: *G. boveanum* subsp. *boveanum* occurs throughout N. Africa, extending into the Sudan, Ethiopia, Somalia, Aden and Arabia.

5. CYNANCHUM Linn.—F.T.A. 4, 1 : 390.

Slender climbers ; cymes lateral at leafy nodes ; leaves 5-nerved, long-petiolate, the lamina deeply reniform-cordate at the base :
Corolla-lobes 2–2·5 mm. long, erecto-patent ; pedicels 4–6 mm. long ; corona 2 mm. long :
Peduncles 0–6 mm. long .. .. 1a. *adalinae* subsp. *adalinae*
Peduncles 2–3 cm. long .. .. 1b. *adalinae* subsp. *mannii*
Corolla-lobes 5–6 mm. long, widely spreading ; pedicels 1·5–2·5 cm. long ; corona 4 mm. long .. .. 2. *longipes*
Erect, precocious herb, the flowers arising in cushioned cymes at ground level, 3–5 cm. high ; leafy shoots[1] arising later, the leaves more or less lanceolate, shortly petiolate, 2–4 cm. long, cuneate at the base .. .. 3. *praecox*

1. **C. adalinae** (*K. Schum.*) *K. Schum.* in E. & P. Pflanzenfam. 4, 2 : 253 (1895) ; F.T.A. 4, 1 : 394 ; F.W.T.A., ed. 1, 2 : 54. *Vincetoxicum adalinae* K. Schum. (1893).
1a. **C. adalinae** (*K. Schum.*) *K. Schum.* subsp. **adalinae**—Bullock in Kew Bull. 17 : 185 (1963). *C. macinense* A. Chev. Bot. 435, name only. Slender twiner with yellowish flowers ; in secondary forest.
Mali: Macina (Sept.) *Chev.* 24861. **Ghana:** Aburi Scarp (June, Sept.) *Morton* ! *Adams* 1776 ! **S.Nig.:** E. Lagos (fl. & fr. May) *Lamborn* 303 ! Aweba, Idanre (fl. & fr. Jan.) *Brenan* 8691 ! Ijebu Dist. (Apr.) *Tamajong* FHI 23259 ! Sapoba F.R., Benin Prov. (fl. & fr. June) *Onochie* FHI 36654 ! S.W. Boje, Ogoja Prov. (May) *Jones & Onochie* FHI 18918 ! [Br.]**Cam.:** Ambas Bay *Mann* 765 ! Victoria *Kalbreyer* 3 ! **F.Po:** *Mann* !
1b. **C. adalinae** subsp. **mannii** (*Sc. Elliot*) Bullock in Kew Bull. 17 : 186 (1963). *Vincetoxicum mannii* Sc. Elliot (1894). *Cynanchum mannii* (Sc. Elliot) N. E. Br. in F.T.A. 4, 1 : 394 (1903) ; F.W.T.A., ed. 1, 2 : 54. *C. acuminatum* (Benth.) K. Schum. (1895), not Humb. & Bonpl. ex Schultes.
S.L.: Bagroo R. *Mann* ! Bumban to Port Loko (Apr.) *Sc. Elliot* 5672 ! Binkolo (Aug.) *Thomas* 1816 ! Njala (fr. May) *Deighton* 2522 ! Kabusa *Sc. Elliot* 5479 ! Vevehun, Fwendu to Potoru (Apr.) *Deighton* 1649 ! **Lib.:** Duport (July) *Dinklage* 3064 ! Du R. (July) *Linder* 258 ! Beiden, Boporo Dist. (fr. Nov.) *Baldwin* 12602 ! Peahtah (Oct.) *Linder* 914 ! 1026 !
2. **C. longipes** *N. E. Br.* in Kew Bull. 1897 : 273 ; F.T.A. 4, 1 : 393 ; Bullock in Kew Bull. 10 : 286 (1955). Slender twiner with greenish-yellow flowers.
Port.G.: Bafatá, Geba to Mato de Cao (fr. Sept.) *Esp. Santo* 3370 ! Bafatá to Capé (Aug.) *Esp. Santo* 3311 ! **Guin.:** Kindia (July) *Jac.-Fél.* 1810 ! **S.L.:** Ronieta (Nov.) *Thomas* 5377 ! Mamaha (Nov.) *Thomas* 4553 ! Njala (Sept., Oct.) *Deighton* 3229 ! 5188 ! **Lib.:** Gbanga *Linder* 653 ! **S.Nig.:** Papalayito, E. Lagos *Millen* 48 ! Also in Congo and Uganda.
3. **C. praecox** *Schltr. ex S. Moore* in J. Bot. 40 : 256 (1902) ; Bullock in Kew Bull. 8 : 354 (1953) and 9 : 362 (1954). *C. pygmaeum* Schltr. (1913). Precocious herb with inflorescence about 2 in. high bearing brownish flowers, greenish inside with a white corona ; appearing after grass fires in upland.
S.L.: Loma Mts. (Jan.) *Jaeger* 4168 ! **N.Nig.:** N. of Zaria (Feb.) *Milne-Redhead* 5030 ! [Br.]**Cam.:** Kumbo to Oku, 6,000 ft. (Feb.) *Hepper* 2011 ! Gembu, 4,500 ft., Mambila Plateau (Jan.) *Hepper* 1831 ! Also in Cameroun, Congo, Tanganyika and N. Rhodesia.

Additional species.

C. hastifolium *N. E. Br.* in Kew Bull. 1895 : 257 ; F.T.A. 4, 1 : 397 ; Berhaut Fl. Sén. 57. A twining plant. Previously recorded only from Ethiopia and Eritrea.
Sen.: *Adam* 9996 (not seen).

6. OXYSTELMA R. Br.—F.T.A. 4, 1 : 382.

Twiner ; stems wiry, glabrous ; leaves oblong-lanceolate, mucronate at the apex, rounded, truncate or subcordate at base, 3–8 cm. long, 1·5–3·5 cm. broad, glabrous ; cymes 2–4-flowered ; pedicels slender, about 1 cm. long ; calyx-lobes lanceolate, 4 mm. long, glabrous ; corolla 3 cm. diam., lobes broadly triangular, ciliate at the

---

[1] Only very young leafy shoots have been collected hitherto ; at a later stage they may climb by twining, and the leaves then may be much larger and of different shape.

sinus, veined dark purple inside ; follicles about 5 cm. long, obliquely ellipsoid, glabrous .. .. *bornouense*

O. bornouense *R. Br.* in Denham & Clapp. Travels in N. & C. Afr., App. 239 (1826) ; F.T.A. 4, 1 : 383 ; Berhaut Fl. Sén. 57. Twiner with white flowers suffused crimson.
Sen.: (Nov.) *Roger* ! *Heudelot* 423 ! *Berhaut* 1406 ! **Gam.**: *Saunders* 59 ! **Port.G.**: Farim (Nov.) *Esp. Santo* 3634 ! **Guin.**: Faranah (Mar.) *Roberty* 17227 ! **S.L.**: *Dawe* 510 ! Laminaia (fl. & fr. Apr.) *Thomas* 145 ! **Ghana**: Aburi *Anderson* 15 ! Kete Krachi (Dec.) *Adams* 4594 ! Bawku (Apr.) *Vigne* FH 3744 ! **N.Nig.**: Nupe *Barter* 1131 ! Minna (Dec.) *Meikle* 731 ! Lemme (May) *Lely* 144 ! Katagum Dist. *Dalz.* 98 ! L. Chad *Talbot* 1014 ! **S.Nig.**: *Thomas* ! Lagos *Rowland* ! Abeokuta *Irving* ! Olokemeji (Aug.) *Foster* 298 ! [Br.]**Cam.**: Bama, Dikwa Div. (Jan.) *McClintock* 136 !

7. PERGULARIA Linn. Mant. 8 : 53 (1767) ; N. E. Br. in Kew Bull. 1907 : 323 (1907), and in This.-Dyer, Fl. Cap. 4, 1 : 757 (1908). *Daemia* R. Br. (1810)—F.T.A. 4, 1 : 385.

Stems scrambling and twining, densely and harshly tomentose when young ; leaves petiolate, ovate-orbicular, cordate at base, apiculate but not acuminate at apex, up to about 5 cm. diam. but often smaller, very densely strigose-tomentose on both surfaces when young, more or less glabrescent in age ; peduncles lateral, short ; pedicels slender, about 1 cm. long ; corolla white and purple, tube narrowly cylindric, lobes elliptic, bearded within the margin ; follicles paired, divaricate, more or less echinate, very shortly tomentose, at least when young, ovoid in shape, beaked, up to about 7 cm. long.. .. .. 1. *tomentosa*
Stems twining or trailing over the ground, more or less hirsute with stiff spreading hairs ; leaves broadly ovate or more rarely almost orbicular, widely cordate at base, cuspidate-acuminate, up to 15 cm. or more long and 12 cm. or more broad, but usually smaller, thinly mesophytic, shortly pubescent on both surfaces, often early glabrescent, long-petiolate ; peduncles slender, elongate, lateral ; flowers in a terminal racemiform corymb ; pedicels up to 3 cm. long, filiform ; corolla white or greenish-white, about 1 cm. long, lobes bearded on the margin ; follicles paired, widely divaricate, lanceolate in outline, smooth or variously echinate, sometimes densely so, up to 8 or 9 cm. long, densely pubescent to almost glabrous .. .. 2. *daemia*

1. **P. tomentosa** *Linn.* Mant. 53 (1767). *Asclepias cordata* Forsk. (1775). *Daemia cordata* (Forsk.) R. Br. (1810)— F.T.A. 4, 1 : 386. *D. incana* Decne. (1838). *D. tomentosa* (Linn.) Pomel (1874)—K. Schum. in E. & P. Pflanzenfam. 4, 2 : 258 (1895). A low shrubby climber of very dry country, with purplish-white sweet-scented flowers.
Mali: Timbuktu *Hagerup* 232 ! 238a ! Gao (Aug.) *de Wailly* 5162 ! Taga-aras (Sept.) *Hagerup* 319 ! **N.Nig.**: Sokoto (Nov.) *Moiser* 46 ! *Lely* 147 ! 154 ! Ngzuru (fl. & fr. June) *Onochie* FHI 23325 ! Katagum Dist. (fl. & fr. Jan.) *Dalz.* 322 ! Molomawa (Feb.) *Foster* 14 ! Extending to N. Africa, Egypt, Arabia, Pakistan and India.

2. **P. daemia** (*Forsk.*) *Chiov.* Res. Sci. Miss. Stefan.-Paoli Somal. Ital. 1 : 115 (1916). *Asclepias daemia* Forsk. (1775). *Pergularia extensa* (R. Br.) N. E. Br. in This.-Dyer Fl. Cap. 4, 1 : 758 (1908), partly ; F.W.T.A., ed. 1, 2 : 55 ; Berhaut Fl. Sén. 57. *Daemia extensa* R. Br. (1810)—F.T.A. 4, 1 : 387, partly.[1] A high climbing herbaceous or semi-woody climber of forest edges and damp savanna, with white or greenish sweet-scented flowers.
Sen.: *Heudelot* 416 ! *Farmar* 125 ! *Berhaut* 518 ! Fasena R. (Feb.) *Döllinger* ! **Gam.**: *Don* 8 ! (July) *Brooks* 20 ! **Port.G.**: Praia Varela, Suzana (fl. Mar., fl. & fr. Oct.) *Esp. Santo* 2248 ! 2281 ! Suzana to Casselol (fl. & fr. Nov.) *Esp. Santo* 3138 ! **Lib.**: Ganta (Oct.) *Harley* ! Flumpa, Sanokwele Dist. (Sept.) *Baldwin* 9355 ! Nakabozu, Vonjama Dist. (Oct.) *Baldwin* 9978 ! **Ghana**: Cape Coast (July) *T. Vogel* 36 ! Accra *T. Vogel* 15 ! Nungua (fl. & fr. Mar.) *Ankrah* GC 20165 ! Aburi (fl. & fr. Nov.) *Adams* 1923 ! Jiasi (fl. & fr. Oct.) *Vigne* FH 1384 ! Akayao, Kwamikrom, Togo (Dec.) *Adams* 4503 ! **N.Nig.**: Lokoja *Barter* ! *Ansell* ! Nupe *Barter* 1320 ! Pandiaki *Ansell* ! Gwari, Gawu Hills, Niger Prov. (Aug.) *Onochie* FHI 35916 ! Bossa, Minna (Dec.) *Meikle* 1232 ! Yola & Vango, Malabu Hills (fl. & fr. May) *Dalz.* 95 ! **S.Nig.**: Lagos (fl. Nov., fl. & fr. Apr.) *Dalz.* 1257 ! *Moloney* ! Makori to Araromi, Ijebu Ode (fl. & fr. Jan.) *Tamajong* FHI 21053 ! Abeokuta to Ibadan (fl. & fr. Dec.) *Burtt* B3 ! Busogboro, Ibadan Dist. (fl. & fr. Nov.) *Okafor* FHI 34951 ! Oban *Talbot* 997 ! [Br.]**Cam.**: Buea (Nov.) *Migeod* 93 ! Widely distributed in tropical Africa, extending to Arabia and further east.

8. PENTARRHINUM E. Mey.—F.T.A. 4, 1 : 378.

Twining climber ; leaves glabrous or nearly so, ovate-oblong, deeply cordate with a wide rounded sinus at the base, caudate-acuminate, up to about 10 cm. long and 6 cm. broad ; petioles slender, 2–5 cm. long ; inflorescence umbelliform, lateral at the nodes; peduncle about as long as the petioles, thinly pubescent ; pedicels very slender, 1–5 cm. long, pubescent ; corolla saucer-shaped, about 1 cm. diam., the lobes ovate, rounded at the apex ; corona-lobes with an abruptly inflexed apical horn ; follicles narrowly ovoid, narrowed to a beak at apex, more or less softly tuberculate-echinate or nearly smooth and glabrous .. .. *insipidum*

P. insipidum *E. Mey.* Comm. Pl. Afr. Austr. 200 (1837) ; F.T.A. 4, 1 : 378. Twining herbaceous climber 4–6 ft. or more high ; corolla-lobes grey-green with faint purple lines, corona-lobes white, gynostegium brownish ; in open parts of montane forest.
[Br.]**Cam.**: Cam. Mt., near Mann's Spring, 7,000 ft. (Mar.) *Brenan* 9510 ! Tropical and S. Africa generally.

9. PENTATROPIS Wight & Arn.—F.T.A. 4, 1 : 380 ; Bullock in Kew Bull. 10 : 283 (1955).

Stems twining, wiry, glabrescent ; leaves elliptic to ovate, slightly cordate at base, acute, 2–5 cm. long, 1–2 cm. broad, glabrescent ; flowers 1–4 on a very short extra-

---

[1] An extensive synonymy is given by N. E. Brown, but it seems likely that the species as understood by him is heterogeneous. In W. Africa, I include most of Brown's synonymy, but the species is evidently polymorphic and may merit division into infraspecific units at least.—A.A.B.

axillary peduncle ; pedicels about 1 cm. long, very slender ; calyx-lobes ovate, very small ; corolla deeply lobed, lobes lanceolate, acuminate, about 1 cm. long, puberulous inside .. .. *spiralis*

P. spiralis (*Forsk.*) *Decne.* in Ann. Sci. Nat., sér. 2, 9 : 327, t. 11E (1838) ; Bullock l.c. 284. *Asclepias spiralis* Forsk. Fl. Aegypt.-Arab. 49 (1775). *Pentatropis senegalensis* Decne. l.c. 328 (1838). *P. cynanchoides* var. *senegalensis* (Decne.) N. E. Br. in F.T.A. 4, 1 : 381 ; F.W.T.A., ed. 1, 2 : 54. A woody twiner with slightly fleshy leaves ; flowers dull purplish-green ; in thickets, sandy places and by seasonal streams. **Sen.:** *Roger* ! *Heudelot* 530 ! *Berhaut* 86. **Mali:** Koubita (Feb.) *Davey* 452 ! Rharous (Sept.) *Hagerup* 321 ! **N.Nig.:** Katagum Dist. (Aug.) *Dalz.* 94 ! Extending across the northern dry savanna to Kenya, Ethiopia and Somalia, the Orient, Pakistan and India ; also in Madagascar and the Comoro Islands.

10. CALOTROPIS R. Br.—F.T.A. 4, 1 : 294.

Shrub or small tree ; leaves oblong-obovate to broadly obovate, cordate at base, abruptly and shortly acuminate, up to 30 cm. long and up to 15 cm. or more broad, very glaucous ; cymes umbelliform, pedunculate, extra-axillary, several-flowered ; pedicels 1–3 cm. long ; calyx segments 6 mm. long ; corolla campanulate, about 2 cm. diam. ; follicles inflated, subglobose to obliquely ovoid, up to 10 cm. or more long
procera

C. procera (*Ait.*) *Ait. f.* Hort. Kew. ed. 2, 2 : 78 (1811) ; F.T.A. 4, 1 : 294 ; Berhaut Fl. Sén. 82. A shrub or small tree with large glaucous leaves ; flowers green with blue or violet markings ; follicles bladder-like ; seeds with a dense coma used as kapok ; found abundantly in semi-arid conditions. **Sen.:** *Berhaut* 37. **Gam.:** *Ozanne* 3 ! *Pirie* 52 ! *Dawe* 71 ! Kuntaur *Ruxton* ! **Port.G.:** Bissau (fl. & fr. Mar.) *Esp. Santo* 1158 ! **S.L.:** Yoni, Bonthe Isl. (cult., Mar.) *Deighton* 2493 ! **Ghana:** Achimota (Apr.) *Adams* 2717 ! Kpedsu (fl. & fr. Dec.) *Howes* 143 ! Samdu (May) *Kitson* 719 ! Navrongo (May) *Andoh* 5187 ! **N.Nig.:** Jebba (fl. & fr. Feb.) *Richards* 5037 ! Ilorin Div. (fl. & fr. Dec.) *Meikle* 871 ! Katagum Dist. *Dalz.* 304 ! Sokoto (Mar.) *Moiser* ! Jos Plateau (Mar.) *Lely* P189 ! [**Br.**]**Cam.:** near Vogel Peak, Adamawa (Dec.) *Hepper* 1558 ! **F.Po:** *Mann* 238 ! Widely distributed (? introduced) in tropical Africa, extending to N. Africa, Orient, Pakistan and India.

11. MARGARETTA Oliv.—F.T.A. 4, 1 : 372 ; Bullock in Kew Bull. 7 : 411 (1952).

Herb ; stems simple or subsimple, from a woody tuberous rhizome, pubescent, rarely exceeding 30 cm. high ; leaves broadly linear or lanceolate, acute, rounded or sub-cordate at base, sessile, 3–7 cm. long, up to 1 cm. broad, scabrid-pubescent on both surfaces ; cymes umbelliform, terminal or becoming lateral ; flowers numerous, corolla-lobes reflexed, corona-lobes conspicuous and petaloid .. .. *rosea*

M. rosea *Oliv.* in Trans. Linn. Soc. 29 : 111, t. 76 (1875) ; F.T.A. 4, 1 : 373 ; Bullock l.c. *M. inopinata* Hutch. in Kew Bull. 1921 : 387 ; F.W.T.A., ed. 1, 2 : 58. Perennial with annual shoots up to 1 ft. high from a tuberous rootstock ; flowers with white, yellow, pink or red petaloid corona, corolla-lobes white, small, ligulate, reflexed ; in upland savanna. **N.Nig.:** between Hepham and Ropp (July) *Lely* 366 ! Jos Plateau (Apr.) *Lely* P226 ! Heipang (Apr.) *Keay*, *Wimbush & E. W. Jones* FHI 37610 ! [**Br.**]**Cam.:** Bamenda (fl. & fr. May) *Maitland* 171 ! Ndu, 6,000 ft. (Feb.) *Hepper* 2844 ! Mambila Plateau, 5,600 ft. (Jan.) *Hepper* 1749 ! Vogel Peak Massif, Adamawa (Dec.) *Hepper* 1563 ! Widespread in tropical Africa.

12. KANAHIA R. Br.—F.T.A. 4, 1 : 295 ; Bullock in Kew Bull. 7 : 421 (1952).

Stems erect, woody ; leaves linear-lanceolate, ascending, acute, up to 15 cm. long and 1–1·5 cm. broad, glabrous ; flowers in extra-axillary multiflorous pedunculate umbelli-form racemes ; pedicels up to 2·5 cm. long ; corolla deeply 5-lobed, about 1 cm. long, lobes ovate-triangular, villous inside ; follicles ovoid, up to 5 cm. long .. *laniflora*

K. laniflora (*Forsk.*) *R. Br.* in Mem. Wern. Soc. 1 : 40 (1810) ; F.T.A. 4, 1 : 296 ; Bullock in Kew Bull. 7 : 421 (1952). *K. consimilis* N. E. Br. (1902)—F.T.A. 4, 1 : 298 ; F.W.T.A., ed. 1, 2 : 55. *Asclepias fluviatilis* A. Chev. Bot. 434, name only. A woody herb, stiffly erect, to 6 ft. high with creamy white sweet-scented flowers ; along seasonal rivers. **Iv.C.:** Alépé to Malamalasso *Chev.* 17514. Marabadiassa *Chev.* 22021. **Dah.:** Savé to Agouagon *Chev.* 23594. **N.Nig.:** Lokoja (Mar.) *Shaw* 33 ! Jira (May) *Lely* 125 ! Plateau Prov. (Feb.) *McClintock* 249 ! Jos Plateau (fl. & fr. July) *Lely* P345 ! Kaura Namoda, Sokoto Prov. (May) *Keay* FHI 18082 ! **S.Nig.:** Idagun *Rowland* ! Ijebu (Apr.) *Tamajong* FHI 23255 ! Etemi (Mar.) *Jones & Onochie* FHI 16692 ! Abeokuta (fr. Dec.) *Meikle* 923 ! Oban *Talbot* 1551 ! [**Br.**]**Cam.:** *Preuss* 1364 ! Johann-Albrechtshöhe (=Kumba) *Staudt* 603 ! Bamenda (Apr.) *Maitland* 1508 ! Kimbi Bridge, Wum Div. (Feb.) *Hepper* 1933 ! near Vogel Peak, Adamawa (Nov.) *Hepper* 1403 ! Widely distributed throughout tropical Africa, extending southwards to Mozambique, the Transvaal and S.W. Africa, and northwards into Egypt and Arabia.

13. ASPIDOGLOSSUM E. Mey. Comm. Pl. Afr. Austr. 200 (1837).

Corolla-lobes conspicuously tomentose with long white hairs inside, pubescent outside ; herb up to 60 cm. or more high, stems strictly erect, slender, simple or sparsely branched in the upper third, softly pubescent, arising from a carrot-like woody rootstock ; internodes 5–10 cm. long ; leaves linear, stiffly erecto-patent, 5–7 cm. long, margins incurved, rather thinly and shortly pubescent on both surfaces ; flowers in dense fascicles about 2 cm. diam., lateral at the upper nodes ; pedicels uniform in length in each fascicle, 5–7 mm. long .. .. 1. *angustissimum*
Corolla-lobes with a few scattered hairs inside and outside ; herb similar in habit to above, but rarely exceeding 40 cm. in height ; stems slender, very thinly pubescent or almost glabrous ; flowers few, in fascicles, lateral at the upper nodes ; pedicels of variable length in each fascicle, 2–8 mm. long .. .. .. 2. *interruptum*

1. **A. angustissimum** (*K. Schum.*) *Bullock* in Kew Bull. 7 : 418 (1952). *Schizoglossum angustissimum* K. Schum. in Engl. Bot. Jahrb. 17 : 123 (1893) ; F.T.A. 4, 1 : 357 ; F.W.T.A., ed. 1, 2 : 58, partly (only the name). Erect herb from a woody base.

[Br.]Cam.: Nchan, 5,000 ft., Bamenda (June) *Maitland* 1384 ! Widely distributed in savanna grasslands of tropical Africa.

2. **A. interruptum** (*E. Mey.*) *Bullock* l.c. 419 (1952). *Lagarinthus interruptus* E. Mey. l.c. 208 (1837). *Schizoglossum interruptum* (E. Mey.) Schltr. in Engl. Bot. Jahrb. 18, Beibl. 45 : 4 (1894) ; N. E. Br. in This.-Dyer Fl. Cap. 4, 1 : 660 (1907). *S. angustissimum* of F.W.T.A., ed. 1, 2 : 58, partly (specimens cited), not of K. Schum. Similar to the last ; flowers dull purplish.

Port.G.: Gabu, Pitche (Aug.) *Esp. Santo* 2723 ! **Ghana**: Winneba Plains (Apr.) *Keay & Adams* FHI 37764 ! Legon Hill (Nov.) *Adams* 3528 ! Akuse (fl. & fr. June) *Dalz.* 154 ! Ayafie to Atafie (Dec.) *Adams* 4671 ! **Niger**: Diapaga to Fada, Gourma (fl. & fr. July) *Chev.* 24492 ! **N.Nig.**: Zungeru (June) *Dalz.* 574 ! Abinsi (May, June) *Dalz.* 686 ! Maska, Katsina (June) *Keay* FHI 25890 ! [Br.]Cam.: Basenako, 5,000 ft., Bamenda (fl. & fr. June) *Maitland* 1453 ! Ngong, Bamenda (June) *Maitland* 1775 ! Widely distributed through the savanna of tropical and S. Africa.

14. TRACHYCALYMMA Bullock in Kew Bull. 8 : 348 (1953).

Corona-lobes truncate, the internal papillae exserted ; leaves linear-lanceolate, up to 10 cm. long and scarcely 5 mm. broad, with prominent midrib, narrowed to the base and sessile or nearly so ; flowers 4–6, in umbelliform lateral cymes, saucer-shaped ; follicles up to 20 cm. long, erect, narrowly fusiform, the stipe more than half the total length .. .. 1. *cristatum*

Corona-lobes produced upwards to a rounded distal horn, the internal papillae not exserted ; leaves, etc., very much as above .. .. 2. *pulchellum*

1. **T. cristatum** (*Decne.*) *Bullock* l.c. 349 (1953). *Gomphocarpus cristatus* Decne. in Ann. Sci. Nat. sér. 2, 9 : 325 (1838). *Asclepias palustris* (K. Schum.) Schltr. (1895)—F.T.A. 4, 1 : 349. A weak erect perennial herb of open savanna, from a small turnip-like tuber, a few inches to 1 ft. high, with white flowers suffused with pink and very long erect fusiform follicles.

N.Nig.: Mongu, 4,300 ft. (Aug.) *Lely* 442 ! Rabo'utu R., Birnin Gwari Dist. (June) *Keay* FHI 25885 ! [Br.]Cam.: Lakom, Bamenda (Apr.) *Maitland* 1494 ! Widely distributed in savanna country of tropical Africa.

2. **T. pulchellum** (*Decne.*) *Bullock* l.c. 350 (1953). *Gomphocarpus pulchellus* Decne. l.c. (1838). *Asclepias foliosa* (K. Schum.) Hiern (1898)—F.T.A. 4, 1 : 349 ; F.W.T.A., ed. 1, 2 : 56. A weak erect herb very like the preceding but differing markedly in the form of the corona.

N.Nig.: Naraguta (fl. & fr. July) *Lely* 317 ! Vom *Dent Young* 172 ! Anara F.R., Zaria Prov. (May) *Keay* FHI 22915 ! Mando F.R., Birnin Gwari Dist. (June) *Keay* FHI 25816 ! Kaciya to Zonkwa, Zaria Prov. (June) *G. V. Summerhayes* 96 ! Widely distributed in savanna country of tropical Africa.

[Note : It is possible that the two species enumerated above represent no more than ecotypic genotypes of the same species. Although both are well-represented in herbaria, collectors' observations are not sufficient for a decision to be made.]

15. ASCLEPIAS Linn.—F.T.A. 4, 1 : 313, partly (introduced New World species).

Erect more or less glabrous herb up to 1·5 m. high ; leaves lanceolate, acuminate, 1·5–2·5 cm. broad ; corolla red-orange, the lobes soon reflexed ; follicles erect, fusiform, smooth, about 8 cm. long .. .. *curassavica*

A. curassavica *Linn.* Sp. Pl. 215 (1753) ; F.T.A. 4, 1 : 328 ; Berhaut Fl. Sén. 88, 95. An erect, shrubby herb up to 5 ft. high with umbelliform inflorescences of flowers with scarlet reflexed corolla-lobes and yellow corona.

Widely cultivated in our area as an ornamental plant and recorded from most countries from Senegal to Fernando Po. A native of tropical America and an escape from cultivation in our area, but not known to have established itself among the native flora.

16. GOMPHOCARPUS R. Br. in Mem. Wern. Soc. 1 : 37 (1810). *Asclepias* of F.T.A. 4, 1 : 294 ; of F.W.T.A., ed. 1, 2 : 56, partly, not of Linn.

Follicles densely invested with filiform bristle-like outgrowths ; peduncles and pedicels densely pubescent ; flowers more or less cream-coloured ; leaves linear to lanceolate up to 15 cm. long :

Follicles inflated, bladder-like when ripe, apex rounded ; calyx-lobes up to 1 cm. long ; corona-lobes shorter than the staminal column .. .. .. 1. *physocarpus*

Follicles not inflated, more or less ovoid, apex rostrate ; calyx-lobes not exceeding 5 mm. long ; corona-lobes at least as long as the staminal column 2. *fruticosus*

Follicles without outgrowths, narrowly fusiform, membranous, glabrous ; peduncles and pedicels glabrous or with lines of sparse short hairs ; leaves linear, up to 20 cm. long or more ; flowers greenish-blue :

Leaves linear, very sharply ascending, 15–20 cm. long, scarcely 5 mm. wide ; the whole plant glabrous ; flowers more or less nodding, 2 cm. or more in diam.

3. *solstitialis*

Leaves linear-lanceolate, about 10 cm. long, 5–6 mm. wide, glabrous or with a few very short hispid hairs beneath ; flowers not more than 1·5 cm. diam., pedicels and peduncles with opposite lines of sparse short hairs .. .. 4. *kamerunensis*

1. **G. physocarpus** *E. Mey.* Comm. Pl. Afr. Austr. 202 (1837) ; Bullock in Kew Bull. 7 : 408 (1952). *G. semilunatus* A. Rich. (1851). *Asclepias semilunata* (A. Rich.) N. E. Br. in F.T.A. 4, 1 : 328 (1902) ; F.W.T.A., ed. 1, 2 : 56, partly ; Berhaut Fl. Sén. 265. A perennial herb up to 6 ft. high with cream flowers and bladder-like fruits ; in upland seasonal swamp grassland.

Sen.: Kayar & Niayes *Berhaut* 1386. N.Nig.: Jos Plateau (fl. & fr. Apr.) *Lely* P248 ! [Br.]Cam.: Bamenda, 5,000 ft. (fl. & fr. Dec.) *Baldwin* 13863 ! Nchan, 5,000 ft., Bamenda (May) *Maitland* 1708 ! Widespread in tropical and S. Africa.

2. **G. fruticosus** (*Linn.*) *Ait. f.* Hort. Kew ed. 2, 2 : 80 (1811) ; Bullock l.c. 406. *Asclepias fruticosa* Linn. (1753)—F.T.A. 4, 1 : 330. *A. foliosa* of F.W.T.A., ed. 1, 2 : 56, partly, not of N. E. Br. *A. semilunata* of

F.W.T.A., ed. 1, 2 : 56, partly (synonymy). *A. euphorbioides* A. Chev. (1909). Perennial herb up to 6 ft. high, with cream flowers ; the fruits are not inflated ; in savanna grassland.
   Guin.: Mali (Sept., Oct.) *Schnell* 7267 ! *Jac.-Fél.* 1970 ! Dalaba-Diaguissa Plateau *Chev.* 18372 ! Ditinn to Diaguissa *Chev.* 12849 ! Widespread in tropical and S. Africa.
3. **G. solstitialis** (*A. Chev.*) *Bullock* l.c. 340 (1953). *Asclepias solstitialis* A. Chev. in Mém. Soc. Bot. Fr. 2, 8 : 270 (1917) ; Chev. Bot. 434 ; F.W.T.A., ed. 1, 2 : 56. A perennial herb about 3 ft. high with small bluish flowers ; in stream beds.
   Guin.: Soarella (June) *Pobéguin* 1007 ! **Iv.C.**: Mt. Kamouéniboka, Marabadiassa (July) *Chev.* 22017 ! **Ghana**: Yeji (Aug.) *Pomeroy* 1343 ! **N.Nig.**: Aguji, Ilorin *Thornton* ! Zungeru (fl. & fr. July) *Dalz.* 3 ! **S.Nig.**: Ogbuga, Enugu (June) *Stevenson* FHI 23643 !
4. **G. kamerunensis** (*Schltr.*) *Bullock* l.c. 8 : 340 (1953). *Asclepias kamerunensis* Schltr. in Engl. Bot. Jahrb. 51 : 137 (1913). A perennial herb about 3 ft. high, with bluish-green flowers ; in stream beds.
   Ghana: Shiare, Buen-Krachi Dist. (Apr.) *Hall* 1448 ! **N.Nig.**: Mando, Zaria Prov. (June) *Keay* FHI 25857 ! **S.Nig.**: Nsuka, Onitsha Prov. (June) *Thompson* ! [**Br.**]**Cam.**: Ngong, Bamenda (fl. & fr. June) *Maitland* 1747 ! Also in Cameroun.

17. XYSMALOBIUM R. Br.—F.T.A. 4, 1 : 298, partly.

Leaves elliptic-oblong-oblanceolate, long-cuneate at base, with a petiole up to 1 cm. long, the lamina broadly acute, up to 8 cm. long and 1·5 cm. broad ; inflorescences sessile, 6–8-flowered, extra-axillary at the upper nodes, umbelliform ; pedicels 8–15 mm. long ; corolla-lobes spreading-campanulate, oblong, 7 mm. long, 3 mm. broad
   1. *membraniferum*

Leaves narrowly oblong-linear, narrowly oblong, lanceolate, oblanceolate or rarely ovate or obovate, never long-cuneate at the base, usually rounded, very shortly petiolate, up to 11 cm. long, but usually 4–7 cm. long ; inflorescences 10–12-flowered, extra-axillary at the upper nodes, umbelliform on very short or obsolete peduncles ; pedicels 8–12 cm. long ; corolla-lobes abruptly reflexed, oblong-lanceolate, 4–5 mm. long   ..   .. 2. *heudelotianum*

1. **X. membraniferum** *N. E. Br.* in Kew Bull. 1895 : 251 ; F.T.A. 4, 1 : 304. An erect herb 12–18 in. high.
   S.L.: Port Loko *Sc. Elliot* 5183 ! Falaba (Mar.) *Sc. Elliot* 5184 !
2. **X. heudelotianum** *Decne.* in DC. Prod. 8 : 520 (1844) ; F.T.A. 4, 1 : 304. An erect herb from a fusiform tuber a few inches to 4 ft. high ; flowers greenish-yellow, follicles erect, long-stipulate, narrowly fusiform, up to nearly 1 ft. long but often much less ; in grassy savanna.
   Sen.: Komba *Heudelot* ! **Gam.**: Genieri (Feb.) *Fox* 84 ! **S.L.**: Samaia (fr. May) *N. W. Thomas* 233 ! **Ghana**: Achimota (May) *Irvine* 1617 ! Aburi *Anderson* ! Nangodi to Bongo (Nov.) *Morton* A3802 ! **N.Nig.**: Nupe *Barter* 1323 ! Anara F.R., Zaria (May) *Keay* FHI 19199 ! Assob (Feb.) *McClintock* 237 ! Mongu (July) *Lely* 439 ! Abinsi (fl. & fr. June–July) *Dalz.* 692 ! 693 ! 694 ! Kyana, Nassarawa Prov. (Feb.) *Hepburn* 3 ! Widely distributed in savanna country of tropical and S. Africa.
   [It is very doubtful whether *X. membraniferum* can be maintained as a species distinct from *X. heudelotianum*. The distinctions made above are somewhat poor and can be explained as artefactal or developmental. I have seen only the two original specimens of *X. membraniferum*—A.A.B.]

18. PACHYCARPUS E. Mey. Comm. Pl. Afr. Austr. 200 (1837).

Follicles inflated, broadly rounded at base and apex ; corona-lobes arising from the base of, and shorter than the gynostegium ; leaves oblong-elliptic, rounded to subacute at base and apex, 6–15 cm. long, 2–6 cm. broad, with numerous parallel patent lateral veins ; petiole 4–10 mm. long ; stems simple, up to 1 m. high ; inflorescence pedunculate, lateral from the upper nodes, umbelliform ; the whole plant more or less hispid hairy..   ..   .. 1. *lineolatus*

Follicles broadly fusiform, narrowed to a beak above and a stipe below ; corona-lobes from above the base of, and slightly overtopping the gynostegium ; habit and leaves as in the preceding   ..   .. 2. *schweinfurthii*

1. **P. lineolatus** (*Decne.*) *Bullock* in Kew Bull. 8 : 333 (1953). *Gomphocarpus lineolatus* Decne. in Ann. Sci. Nat. Paris, sér. 2, 9 : 326 (1838). *Asclepias lineolata* (Decne.) Schltr. in J. Bot. 33 : 336 (1895), partly ; F.T.A. 4, 1 : 322 ; F.W.T.A., ed. 1, 2 : 56 ; Berhaut Fl. Sén. 82. An erect simple-stemmed perennial, hispid-hairy all over, with inflated bladder-like follicles.
   Sen.: Niokolo-Koba *Berhaut* 1555. **Guin.**: Beyla (June) *Jac.-Fél.* 985 ! Fabala, Kérouané (July) *Collenette* 75 ! **Ghana**: Aburi *Anderson* 3 ! Yeji (Aug.) *Vigne* FHI 3310 ! Tolon (June) *Williams* 822 ! Logba Lota (June) *Howes* 925 ! **N.Nig.**: Aguji, Ilorin *Thornton* ! Nupe *Barter* 1323a ! Zungeru (July) *Dalz.* 2 ! Jos Plateau (May, July) *Lely* P329 ! *Dent Young* 171 ! *Keay & King* FHI 37100 ! Maska, Katsina Prov. (June) *Keay* FHI 25897 ! [**Br.**]**Cam.**: Bamenda, 5,000 ft. (fl. & fr. May) *Maitland* 1428 ! Nchan, Bamenda (fl. & fr. June) *Maitland* 1748 ! 1789 ! Widely distributed in the savanna of tropical Africa.
2. **P. schweinfurthii** (*N. E. Br.*) *Bullock* l.c. 330 (1953). *Asclepias schweinfurthii* N. E. Br. in Kew Bull. 1895 : 253 ; F.T.A. 4, 1 : 323. A hispid hairy perennial very similar to the preceding, but the follicles fusiform, not inflated.
   N.Nig.: Mando F.R., Birnin Gwari Dist. (June) *Keay* FHI 25837 ! [**Br.**]**Cam.**: Nchan, 5,000 ft., Bamenda (fr. May) *Maitland* 1521 ! Ngong, Bamenda (June) *Maitland* 1746 ! Nguroje, Mambila Plateau (Jan.) *Hepper* 2819 ! Widely distributed in the savanna of tropical Africa.

   [A specimen collected by T. D. Maitland (No. 1437) at Bamenda in May 1931 may represent a third species for our area.]

19. SARCOSTEMMA R. Br.—F.T.A. 4, 1 : 383 ; Bullock in Kew Bull. 11 : 504 (1957).

Scrambling and twining shrub ; branchlets leafless, terete, pubescent at the nodes ; flowers numerous in sessile umbelliform fascicles, lateral at the nodes ; pedicels 6–10 mm. long, more or less woolly ; corolla cream or greenish, 5–6 mm. long ; follicles up to 10 cm. long, narrowly linear-lanceolate, thinly puberulous   ..   .. *viminale*

S. viminale (*Linn.*) *R. Br.* in Mem. Wern. Soc. 1 : 51 (1810) ; F.T.A. 4, 1 : 384. Forming dense scrambling bushes with intricately interlaced branches ; flowers cream or greenish, sweetly scented ; a plant of dry savanna country, or frequently found in semi-arid conditions.

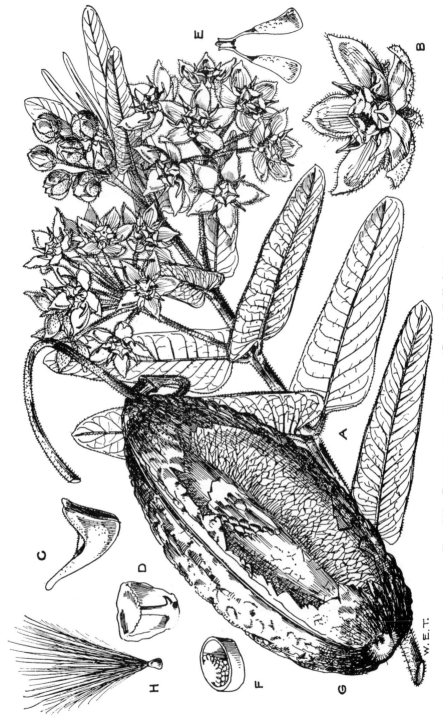

Fig. 220.—Pachycarpus lineolatus (*Decne.*) *Bullock* (Asclepiadaceae).

A, flowering shoot. B, open flower. C, outer corona-lobe. D, inner corona-lobe. E, pollinia. F, cross-section of carpel. G, fruit. H, seed.

Ghana: Komenda (Sept.) *Chipp* 577! Accra (Jan.) *Dalz.* 71! Accra Plains *Vigne* FH 2911! Achimota (fr. Jan.) *Irvine* 1969! Krobo Plains (Jan.) *Johnson* 604! **N.Nig.**: Takwara *Lely* 109! **S.Nig.**: Iseyin (Apr.) Onochie FHI 34687! Idanre (Jan.) *Brenan, Jones, Richards & Keay* 8649! Widely distributed in tropical Africa.

20. NEOSCHUMANNIA Schltr. in Engl. Bot. Jahrb. 38 : 38 (1905).

Twining glabrous climber with very slender stems and long internodes ; leaves elliptic or oblong elliptic, shortly acuminate, rounded at the base, 9–12 cm. long, 5–6·5 cm. broad ; petiole flexuous, 2–2·5 cm. long ; inflorescence extra-axillary, racemiform, with a slender naked peduncle as long as or longer than the leaves, the floriferous upper part bracteate, stouter, with numerous flowers borne on filiform pedicels about 4 cm. long ; corolla lobed almost to the base, lobes about 11 mm. long, lanceolate, ciliate with clavate hairs ; corona in three series, the outer lobes reflexed, glabrous, oblong, 2 mm. long, the intermediate lobes sickle-shaped, erect distally and bidentate, about 1 mm. long, maculate, the inner lobes lanceolate, erect, from a narrow base, elliptic, obtusely acuminate, 3 mm. long, ciliate on the margins and villous inside

kamerunensis

N. kamerunensis *Schltr.* l.c., fig. 4 (1905). A slender twining climber among shrubs at the forest edge ; flowers dark purple outside, greenish inside.
[**Br.]Cam.**: near Man O' War Bay (Apr.) *Schlechter* 12384.
[I have not seen a specimen of this remarkable plant which is unique in the family in its inflorescence and corona.]

21. DALZIELIA Turrill in Hook. Ic. Pl. 31 : t. 3061 (1916).

A shrub or woody herb 60–100 cm. high with simple shortly pubescent stems ; leaves narrowly linear-oblanceolate, 10–12 cm. long, 1–1·5 cm. broad, glabrous; flowers racemosely umbelliform, numerous, at the apex of a short extra-axillary; peduncle ; pedicels 2 cm. long ; calyx-lobes subulate-lanceolate, 4–5 mm. long, minutely ciliate ; corolla lobed to the middle, the lobes ovate, 8 mm. long   ..   ..   *oblanceolata*

D. oblanceolata *Turrill* l.c. (1916). *Xysmalobium graniticola* A. Chev. Bot. 432, name only. A shrubby perennial herb with simple stems up to 3 ft. high, crowded willow-like leaves and cream flowers, bearded on the inner face of the lobes ; along seasonal rivers.
Guin.: Macenta (fl. & fr. May) *Jac.-Fél.* 894! Mt. Nimba (Apr.) *Schnell* 5010! Upper Niger (Apr.) *Chev.* 13596! **S.L.**: Bandajuma (May) *Aylmer* 74! York (May) *Adames* 204! York Pass (June) *Lane-Poole* 50! Makump (July) *Thomas* 933! Yifin (Apr.) *Deighton* 5085! **Iv.C.**: Guinea frontier (Aug.) *Boughey* 18135!

22. GYMNEMA R. Br.—F.T.A. 4, 1 : 413.

Shrubby scrambling twiner, young parts more or less tomentose ; leaves more or less ovate, acuminate, cuneate to rounded or cordate at base, densely pubescent to glabrescent, up to about 7 cm. long and 4 cm. broad, but often smaller ; flowers numerous, in sessile or shortly pedunculate lateral umbelliform cymes ; pedicels about 6 mm. long ; corolla about 4 mm. long, lobes spreading-recurved ; follicles lanceolate, up to 10 cm. long ..   ..   *sylvestre*

G. sylvestre *(Retz.) Schultes* Syst. Veg. 6 : 57 (1820) ; F.T.A. 4, 1 : 413 ; Chev. Bot. 436 ; Bullock in Kew Bull. 8 : 66 (1953) ; Berhaut Fl. Sén. 719. *Periploca sylvestris* Retz. Obs. Bot. 2 : 5 (1781). A scrambling shrub, leaves and young shoots densely hairy ; flowers greenish, inconspicuous, sweetly scented ; a plant of savanna or deciduous forest.
Maur.: *Webb* ! **Sen.**: (fl. & fr. Mar.) *Roger* 49! *Sieber* 22! Dagana (fr. Feb.) *Roberty* 16878! *Berhaut* 719. **Mali**: Bone (Aug.) *Demange* 39/1957! **Port.G.**: Carache (fr. May) *Esp. Santo* 1999! Farim (Aug.) *Esp. Santo* 3061! Fulacunda (July) *Esp. Santo* 2145! Bafata (Aug.) *Esp. Santo* 3279! **Guin.**: Nzérékoré (Sept.) *Jac.-Fél.* 1142! Kouroussa (July) *Pobéguin* 277! **Ghana**: Komenda (Aug.) *Andoh* FH 5549! Winneba Plains (fl. & fr. Oct.) *Darko* 1022! Accra (Aug.) *Dalz.* 155! Krobo Plains (Aug.) *Johnson* 1083! **Togo Rep.**: Lomé *Warnecke* 248! **N.Nig.**: Jebba (fl. & fr. Dec.) *Hagerup* 701! Kaduna (fr. Dec.) *Meikle* 837! Katagum (Sept.) *Keay* FHI 25401! Ogun R., Oyo (Sept.) *Latilo* FHI 23536! Oyo (Sept.) *Keay* FHI 28040! Ukpon F.R., Obubra Dist. (Aug.) *Adebusuyi* FHI 43961! Widely distributed throughout tropical Africa, and in S. Africa and Madagascar, extending eastwards into India and beyond.

23. TYLOPHORA R. Br.—F.T.A. 4, 1 : 404.

*Twining climbers ; leaves petiolate, spreading :
   Leaves large, about 15 cm. diam. or more, suborbicular, rounded subcordate at base, very shortly cuspidate at apex ; flowers purple :
     Stems and leaves glabrous ; leaves green   ..   ..   ..   1. *cameroonica*
     Stems and inflorescences golden-tomentose ; leaves glaucous, tomentose on the veins beneath   ..   ..   2. *glauca*
   Leaves much smaller, oblong-obovate to linear-lanceolate, usually more than twice as long as broad, often acuminate, but not, or only rarely, cuspidate at apex :
     Flowers dark reddish-brown, 1·5 cm. diam., in crowded racemiform spikes at the ends of the inflorescence branches ; pedicels 1 cm. long ; leaves oblong-obovate, deeply cordate at base, broadly acuminate, 10–16 cm. long, up to 8 cm. broad
       3. *conspicua*
     Flowers not exceeding 1 cm. diam., usually 3–5 mm. diam. :
       Flowers in very short racemiform clusters scattered diffusely along the inflorescence branches ; pedicels about 5 mm. long ; leaves oblong-ovate to obovate, rarely*

almost rotund, acutely acuminate, deeply cordate at base, with broad sinus, up
to 16 cm. long and 8 cm. broad, thinly herbaceous, petiole 2–3 cm. long ; follicles
usually solitary, somewhat falcate, linear, about 7–9 cm. long   4. *sylvatica*
Flowers in umbelliform or racemiform groups at the ends of the inflorescence
branches, or in simple cymes :
Leaves 3–5-nerved from the cordate base, oblong, ovate or somewhat obovate,
acutely acuminate, up to 8 cm. long and 4 cm. broad ; petiole 1–2 cm. long
                                                                    5. *gilletii*
Leaves pinninerved :
Corolla-lobes broadly rounded, the flower about 1 cm. diam. :
Leaves with a definite narrow sinus at the cordate base, oblong, ovate or
elliptical, apex acute, often apiculate but scarcely acuminate, up to 10 cm.
long and 5 cm. broad, but usually smaller, glabrous ; corona of 5 small,
separate, fleshy tubercles   ..   ..   6. *oculata*
Leaves obtusely cuneate, truncate or cordate with a broad shallow sinus at base,
oblong, ovate, elliptical or obovate, somewhat larger than above ; corona of
5 laterally elongated, contiguous fleshy tubercles   ..   ..   7. *oblonga*
Corolla-lobes narrow, acute, the flower 2–3 mm. diam. ; leaves rounded or obtuse
at base, not at all cordate, ovate to oblong-ovate, scarcely acuminate but with
fine mucronate tip, lamina usually less than 5 cm. long and about half as broad ;
petiole 0·5–1 cm. long ; inflorescence a lateral cyme about as long as the leaves
                                                                    8. *dahomensis*
*Erect perennial herb or sub-shrub with a woody rhizome and simple stems ; leaves
ascending, linear-lanceolate, sessile, apex acute, 3–7 cm. long, 5 mm. broad, glabrous ;
inflorescence lateral, longer than the leaves, with scattered fascicles of a few flowers
in the upper half ; flowers brown, about 3 mm. diam., with narrowly triangular acute
lobes ; follicles very slightly divergent, linear-acute, 3–4 cm. long   9. *congolana*

1. **T. cameroonica** *N. E. Br.* in Kew Bull. 1895 : 258 ; F.T.A. 4, 1 : 407 ; Bullock in Kew Bull. 9 : 585 (1955).
A glabrous climber with very broad leaves and small purple flowers in clusters along the branches of large
diffuse paniculate cymes.
[Br.]Cam.: Rio del Rey *Johnston* ! Buea (fl. & fr. Feb.) *Dalz.* 8236 *bis* ! Also in Uganda.

2. **T. glauca** *Bullock* l.c. 585 (1955). A climber very like the above, but with less expanded panicles, the purple
flowers larger and fewer, stems, inflorescences and veins golden-tomentose and the leaves very glaucous.
S.Nig.: Obomkpa, Benin Prov. (fl. & fr. Aug.) *Onochie* FHI 33430 ! Also in Congo.

3. **T. conspicua** *N. E. Br.* in Kew Bull. 1895 : 298 ; F.T.A. 4, 1 : 405 ; Bullock l.c. 583. Twiner with dark
reddish-brown flowers.
Lib.: Ganta (May) *Harley* 1203 ! **Ghana:** Aburi (May) *Johnson* ! Tafo (July) *Owen* 572 ! Amentia *Irvine* 518 !
W. Ashanti (May) *Irvine* 2501 ! Kumasi (fr. Oct.) *Baldwin* 13473 ! Extending to Cameroun, Congo and
E. Africa.

4. **T. sylvatica** *Decne.* in Ann. Sci. Nat. sér. 2, 9 : 273 (1838) ; F.T.A. 4, 1 : 407 ; Berhaut Fl. Sén. 57 ; Bullock
l.c. 582. *T. adalinae* K. Schum. (1895)—F.T.A. 4, 1 : 410 ; Bullock in Kew Bull. 15 : 202 (1961). *Mars-
denia profusa* N. E. Br. (1895)—F.T.A. 4, 1 : 425 ; F.W.T.A., ed. 1, 2 : 60. Herbaceous twiner ; flowers
dull purple-brown ; in thickets.
Sen.: *Berhaut* 1554. **Guin.:** *Heudelot* 816 ! Nzérékoré (fr. Oct.) *Baldwin* 9731 ! **S.L.:** Leicester (June)
Lane-Poole 288 ! Kuntaia (fl. & fr. June) *Thomas* 418 ! Rokupr (fl. Apr., fr. May) *Jordan* 229 ! 256 ! **Lib.:**
Gbanga (fl. & fr. Sept.) *Linder* 704 *bis* ! Cess R. (fr. Aug.) *Baldwin* 9010 ! Cape Palmas *T. Vogel* 3 ! Zolopla
(Sept.) *Baldwin* 9380 ! **Ghana:** Essuanaw *Fishlock* 9 ! Agogo (fl. & fr. Apr.) *Adams* 2612 ! Aburi (Aug., Oct.)
Johnson 770 ! 946 ! Nsawam (fl. & fr. May) *Irvine* 2837 ! **Togo Rep.:** Misahöhe *Baumann* 511 ! **N.Nig.:**
Nupe *Barter* 1321 ! **S.Nig.:** Ikoyi Plains, Lagos (fr. Aug.) *Dalz.* 1411 ! Omo F.R., Ijebu Prov. (fl. & fr. Apr.)
Onochie FHI 15528 ! Oke Are, Oyo Prov. *Latilo* FHI 21011 *bis* ! Brass *Barter* 16 ! Akassa *Dutton* 29 !
Nun R. (Sept.) *Mann* 479 ! Onitsha (Apr.) *Killick* 113 ! Oban *Talbot* 374 *bis* ! 375 ! **[Br.]Cam.:** Victoria
Rosevear ! Kumba road (fl. & fr. Feb.) *Maitland* 432 ! **F.Po:** (Nov.) *T. Vogel* 237 ! Extending to Congo,
E. Africa, N. Rhodesia, Angola and Madagascar.

5. **T. gilletii** *De Wild.* in Ann. Mus. Congo, sér. 5, 1 : 193 (1904) ; Bullock in Kew Bull. 9 : 580. *T. smilacina*
S. Moore (1914)—F.W.T.A., ed. 1, 2 : 59. Climber, unique on account of its 3–5-nerved leaves.
S.Nig.: near Sacred Lake, Ikotobo, Eket Dist. *Talbot* 3252 ! Also in Congo.

6. **T. oculata** *N. E. Br.* in Kew Bull. 1895 : 112 ; F.T.A. 4, 1 : 406 ; Bullock l.c. 580. *T. deightonii* Hutch.
& Dalz. F.W.T.A., ed. 1, 2 : 59 (1931), and in Kew Bull. 1937 : 340. Twiner ; flowers with crimson
centre ; in secondary forest.
Guin.: Nzérékoré to Macenta (Oct.) *Baldwin* 9744 ! **S.L.:** *Sc. Elliot* ! Hill Station, Freetown (Oct.) *Deighton*
219 ! Kanya (Oct.) *Thomas* 3007 ! Matotaka (fr. July) *Thomas* 1277 ! Toma (fr. Feb.) *Thomas* 8781 ! **Lib.:**
Peahtah (Oct.) *Linder* 1070 ! **Ghana:** Jimira F.R., Ashanti (Oct.) *Vigne* FH 3112 ! **S.Nig.:** Ibadan (Sept.)
Keay FHI 37676 !

7. **T. oblonga** *N. E. Br.* in Kew Bull. 1895 : 257 ; F.T.A. 4, 1 : 408 ; Bullock l.c. 581. *T. anfracta* N. E. Br.
in F.T.A. 4, 1 : 408 (1903). *T. liberica* N. E. Br. in J. Linn. Soc. 37 : 109 (1905) ; F.W.T.A., ed. 1, 2 : 59.
Secamone conostyla S. Moore in Cat. Talb. 64 (1913).
Lib.: Sinoe Basin *White* ! Ganta (Oct.) *Harley* 696 ! **N.Nig.:** Benue R. *Talbot* 1013 ; **S.Nig.:** *Talbot* 3360 !
Ibadan (Sept.) *Newberry* 70 ! Sapoba, Jamieson R. (fl. & fr. Nov.) *Keay* FHI 28064 ! Also in Cameroun.

8. **T. dahomensis** *K. Schum.* in Engl. Bot. Jahrb. 33 : 329 (1903) ; F.T.A. 4, 1 : 619. *T. minutiflora* A. Chev.
Bot. 436, name only. Twiner in thickets ; flowers reddish inside and greenish outside.
Ghana: Legon Bot. Gard. (May) *Adams* 4783 ! Legon Hill (Nov., fl. & fr. Dec.) *Adams* 3543 ! 3565 !
Achimota (fl. & fr. June, Oct.) *Morton* ! *Milne-Redhead* 5109 ! **Togo Rep.:** Lomé *Warnecke* 190 *bis* !
S.Nig.: Oyo road, Ibadan (Apr.) *Meikle* 1444 !

9. **T. congolana** (*Baill.*) *Bullock* in Kew Bull. 9 : 585 (1955). *Nanostelma congolanum* Baill. Hist. Pl. 10 : 248
(1890) ; F.T.A. 4, 1 : 411. *Tylophora orthocaulis* K. Schum. (1896)—F.T.A. 4, 1 : 410 ; Chev. Bot. 436 ;
F.W.T.A., ed. 1, 2 : 59. *Schizoglossum glanvillei* Hutch. & Dalz. F.W.T.A., ed. 1, 2 : 58 (1931), and in
Kew Bull. 1937 : 340. Erect herb from a woody base, 1–2 ft. high ; flowers green and brown ; amongst
grass.
Guin.: Kouria *Caille* in *Hb. Chev.* 15064 ! **S.L.:** Port Loko to Lungi (June) *Jordan* 1047 ! Rowala (fr. July)
Thomas 1075 ! 1096 ! Kulufaga (Apr.) *Glanville* 192 ! Also in Cameroun and Congo.

24. SPHAEROCODON Benth.—F.T.A. 4, 1 : 411.

Stems subsimple from a woody rootstock, 30–100 cm. tall, softly pubescent ; leaves oblong, lanceolate or oblanceolate to elliptic, 5–10 cm. long, 1·5–6 cm. broad, softly pubescent on the nerves ; peduncles lateral at the upper nodes, about 2 cm. long ; pedicels about 1 cm. long ; flowers very dark purple, very variable in size, rotate, 1–2 cm. diam., glabrous or minutely pubescent outside ..   ..   ..   *caffrum*

S. caffrum (*Meisn.*) *Schltr.* in J. Bot. 37 : 339 (1895). *Tylophora caffra* Meisn. (1843). *Sphaerocodon obtusifolium* Benth. in Hook. Ic. Pl. 12 : 78, t. 1190 (1876) ; F.T.A. 4, 1 : 412 ; F.W.T.A., ed. 1, 2 : 61. Suberect perennial 1–3 ft. high, branched from the base only, with lateral umbelliform cymes of dark purple (almost black) flowers.
S.L.: Musaia (July) *Deighton* 4796 ! **N.Nig.:** Kontonkoro (June) *Dalz.* 552 ! Abinsi (July) *Dalz.* 687 ! **[Br.]Cam.:** Bamenda (May) *Maitland* 1558 ! Mt. Lakom (May) *Maitland* ! Widespread in tropical Africa, extending into Natal.

25. TELOSMA Coville in U.S. Dept. Agric. Contrib. Nat. Herb. 9 : 384 (1905). *Pergularia* R. Br.—F.T.A. 4, 1 : 426, not of Linn.

Stems glabrous ; leaves ovate-oblong to broadly ovate, shortly cuspidate, cuneate to cordate at base, 5–10 cm. long, up to 7 cm. broad ; cymes umbelliform, many-flowered, extra-axillary, sessile or shortly pedunculate ; pedicels 4–6 mm. long ; calyx-lobes ovate to lanceolate, 4 mm. long ; corolla-lobes linear, 7–12 mm. long, spreading, sinuous, strongly contorted in bud ; villous on the inner face, tube inflated in the lower half, 6–8 mm. long, conspicuously villous in the throat ..   *africanum*

T. africanum (*N. E. Br.*) *Colville* l.c. (1905) ; Berhaut Fl. Sén. 57. *Pergularia africana* N. E. Br. (1895)—F.T.A. 4, 1 : 426 ; Chev. Bot. 437. *P. sanguinolenta* Britten (1894), not of Lindley. A slender climber with yellow flowers, found in riverine or damper deciduous forest.
Guin.: *Berhaut* 392 ! Moussaya (May) *Roberty* 17851 ! **S.L.:** Nongowa (Apr.) *Deighton* 3930 ! Karga (Mar.) *Thomas* 9833 ! Bafodeya (Apr.) *Sc. Elliot* 5498 ! Kafogo (Apr.) *Sc. Elliot* 5489 ! Kamalu (May) *Thomas* 321 ! **Ghana:** Accra (June) *Adams* 4342 ! Assuantsi (Mar.) *Irvine* 1559 ! Kumasi (Mar.) *Vigne* FH 3739 ! Dawa Mato Kole (Mar.) *Thomas* D177 ! L. Bosumtwe (Apr.) *Adams* 2474 ! **N.Nig.:** Nupe *Barter* ! Zaria (June) *Keay* FHI 25892 ! **S.Nig.:** Ijaye *Barter* 3332 ! W. Lagos *Rowland* ! Ibadan (Apr.) *Meikle* 1427 ! Oyo (Apr.) *Keay* FHI 19165 ! Calabar *Thomson* 113 ! **[Br.]Cam.:** Kumba (Feb.) *Binuyo & Daramola* FHI 35556 ! Widely distributed in tropical Africa, extending southwards into Natal.

26. DREGEA E. Mey. Comm. Pl. Afr. Austr. 199 (after Apr. 1837), nom. cons. ; Bullock in Kew Bull. 11 : 512 (1957) ; not of Ecklon & Zeyher (Apr. 1837).

Corolla about 12 mm. diam., rounded in bud, lobes villous only near their margins, ovato-rotund, spreading or reflexed ; follicles about 7 cm. long, with many intricately convolute, often cristate wings forming a spongy mass more than 1 cm. thick ; leaves broadly acuminate, broadly cuneate at base, 6–11 cm. long, 3·5–7 cm. broad with 4–5 pairs of lateral nerves ; inflorescence pedunculate, peduncle stout, woody, up to 2 cm. long ; pedicels crowded—(Subgen. *Dregea*) ..   ..   1. *abyssinica*
Corolla markedly acute in bud, lobes narrowly oblong, parallel-sided and twisted ; follicles without wings—(Subgen. *Traunia*) :
Leafy branches densely appressed-pubescent ; follicles thinly to densely pubescent, the surface shining and wrinkled, 7–14 cm. long ; leaves ovate or ovate-elliptic, shortly acuminate, rounded-truncate at the base, 5–12 cm. long and half as broad ; petioles 1–2·5 cm. long   ..   ..   2. *schimperi*
Leafy branches clothed with long tawny spreading hairs ; leaves acutely acuminate, cordate or rounded at base, 9–12 cm. long, 3·5–8 cm. broad, loosely pilose above and densely pilose on the nerves beneath ; cymes pedunculate, subumbellate ; lower bracts linear, leafy ; sepals lanceolate, pilose   ..   ..   3. *crinita*

1. D. abyssinica (*Hochst.*) *K. Schum.* in Engl. Pflanzenw. Ost-Afr. C : 326 (1895) ; Bullock l.c. 516. *Pterygo-carpus abyssinicus* Hochst. in Flora 26 : 78 (1843). *Marsdenia abyssinica* (Hochst.) Schltr. (1913)— F.W.T.A., ed. 1, 2 : 60 ; Berhaut Fl. Sén. 59. *M. spissa* S. Moore (1901)—F.T.A. 4, 1 : 420. Climber ; flowers white, fragrant.
Sen.: *Berhaut* 759. **Ghana:** Shai Hills, Accra Plains (June) *Harris* ! Tamale *Anderson* 14 ! Tumu (Apr.) *Vigne* FH 3794 ! **N.Nig.:** Kontagora (Apr.) *Dalz.* 369 ! **S.Nig.:** Abeokuta *Barter* 3369 ! Ibadan (Apr.) *Meikle & Keay* 1456 ! Latilo FHI 21006 ! Ado rock, Oyo (fr. Feb.) *Onochie* FHI 35279 ! Olla Hills F.R., Ogbomosho (Mar.) *Binuyo* FHI 36903 ! Widely distributed in tropical Africa.
2. D. schimperi (*Decne.*) Bullock l.c. 518 (1957). *Mardenia schimperi* Decne. (1844)—F.T.A. 4, 1 : 419. *Traunia albiflora* K. Schum. (1895). Climber with whitish flowers.
[Br.]Cam.: Buea (May) *Maitland* 654 ! Bamenda (May) *Daramola* FHI 41071 ! Also in Sudan, eastwards to Somalia and Aden, and in Kenya and Tanganyika.
3. D. crinita (*Oliv.*) Bullock l.c. 519 (1957). *Marsdenia crinita* Oliv. in Hook. Ic. Pl. 20 : t. 1993 (1891) ; F.T.A. 4, 1 : 418 ; F.W.T.A., ed. 1, 2 : 60. Climber with brown-hairy branches ; flowers white.
Port.G.: Catió (June) *Esp. Santo* 2094 ! **S.L.:** Baoma (Apr.) *Deighton* 3653 ! Tiama (fr. Jan.) *Dalz.* 8095 ! Yumbuma (Apr.) *Deighton* 3944 ! Bendembu (Apr.) *Sc. Elliot* 5651 ! **Ghana:** Apla, E. Prov. (Oct.) *Vigne* FH 4044 ! **S.Nig.:** Ibadan (Apr.) *Meikle* 1430 ! Oyo (May, Sept.) *Millson* ! *Barter* 3426 ! *Patel* FHI 51345 ! Also recorded from Gabon and Angola.

27. GONGRONEMA Decne. in DC. Prod. 8 : 624 (1844) ; Bullock in Kew Bull. 15 : 197 (1961).

Climbing shrubs ; leaves broadly ovate or ovate-rotundate, apex acutely acuminate : Flowers in racemosely arranged fascicles along the branches of the inflorescence ; pedicels 2–4 mm. long ; follicles narrowly lanceolate, about 8 cm. long, glabrous, 1·5 cm. diam. near the base ..   ..   1. *latifolium*

Flowers umbellately disposed at the ends of the inflorescence branches ; pedicels 8–10 mm. or more long ; follicles linear, 10–11 cm. long, more or less densely pilose, 4 mm. diam. ..   ..   2. *angolense*
Erect shrub ; leaves triangular-ovate or more or less oblong, apex apiculate

3. *obscurum*

1. **G. latifolium** *Benth.* in Fl. Nigrit. 456 (1849) ; Bullock l.c. 198. *Marsdenia leonensis* Benth. l.c. 198. F.T.A. 4, 1 : 424 ; F.W.T.A., ed. 1, 2 : 60. *M. glabriflora* Benth. l.c. (1849) ; F.T.A. 4, 1 : 424 ; F.W.T.A., ed. 1, 2 : 60. *M. latifolia* (Benth.) K. Schum. (1900)—F.W.T.A., ed. 1, 2 : 60. *M. racemosa* K. Schum. (1893)—F.T.A. 4, 1 : 425. Climber, woody below, with hollow glabrous stems ; flowers greenish-yellow. **Port.G.:** Catió (June, July) *Esp. Santo* 2093 ! 2137 ! **Guin.:** Macenta (Apr.) *Jac.-Fél.* 830 ! **S.L.:** Waterloo (July) *Lane-Poole* 316 ! Yonibana (fr. Oct.) *Thomas* 4131 ! Makali (June) *Marmo* 49 ! Njala (May) *Deighton* 681 ! Mabonto (fr. Oct.) *Thomas* 3507 ! **Lib.:** Sanokwele Dist. (Sept.) *Baldwin* 9378 ! Ganta *Harley* 1171 ! **Iv.C.:** Nimba Mt. (Aug.) *Schnell* 6208 ! **Ghana:** Agogo (Apr.) *Adams* 2595 ! *Irvine* 965 ! Amentia *Irvine* 506 ! New Tafo (Feb.) *Lovi* WACRI 3894 ! Kwahu (Apr.) *Johnson* 649 ! **Togo Rep.:** *Baumann* 465 ! **N.Nig.:** Ankpa (June) *Daramola* FHI 38041 ! **S.Nig.:** W. Lagos *Rowland* ! Sapoba (fr. Nov.) *Ejiofor* FHI 24664 ! Iyekohiomwon, Benin (Apr.) *Emwiogbon* FHI 45501 ! Udi Dist. (May) *Onochie* FHI 35853 ! Ibadan (Apr.) *Latilo* FHI 34985 ! Oban *Talbot* 2072 ! [Br.]**Cam.:** Buea (Mar., May) *Maitland* 494 ! 705 ! Kumba *Daramola* FHI 29841 ! Bulijambo (Mar.) *Maitland* 581 ! Widely distributed in tropical Africa.

2. **G. angolense** (*N. E. Br.*) *Bullock* l.c. 199 (1961). *Marsdenia angolensis* N. E. Br. in Kew Bull. 1895 : 258 ; F.T.A. 4, 1 : 423 ; F.W.T.A., ed. 1, 2 : 60. A softly hairy climber with greenish-white or cream flowers on short dense cymes. **S.L.:** Yakala (Sept.) *Thomas* 2365 ! **N.Nig.:** (July) *Lely* P515 ! Widely distributed in tropical Africa.

3. **G. obscurum** *Bullock* l.c. 200 (1961). A small erect shrub, softly hairy, with small dense extra-axillary cymes of cream flowers. **Ghana:** Salaga (June) *Vigne* FH 3891 ! Yeadi *Akpabla* 559 ! Wa (June) *Adams* 768 ! Navrongo (June) *Andoh* FH 5905 !

28. ANISOPUS N. E. Br.[1]—F.T.A. 4, 1 : 415 ; Bullock in Kew Bull. 11 : 510 (1957).

Glabrous twining shrub ; leaves petiolate, oblong, elliptic, ovate or more or less rotund, acute to cordate at base, shortly cuspidate at apex, up to 15 cm. or more long and 12 cm. broad, but usually 6–8 cm. long and 3–4 cm. broad, petiole 1–2 cm. long ; cymes about 10–20-flowered, umbelliform, often 2 at a node and then one sessile and one shortly pedunculate ; pedicels about 1 cm. long ; corolla-lobes tomentose inside, rotund-ovate, 3–5 mm. long ; follicles 15–20 cm. long, linear, widely divaricate

mannii

A. mannii *N. E. Br.* in Kew Bull. 1895 : 259 ; F.T.A. 4, 1 : 416 ; Bullock l.c. 511. *Marsdenia efulensis* N. E. Br. in Hook. Ic. Pl. 25 : t. 2497 (1896) ; F.T.A. 4, 1 : 416 (1903) ; F.W.T.A., ed. 1, 2 : 60. *M. bicoronata* K. Schum. (1896). *Anisopus bicoronatus* (K. Schum.) N. E. Br. in F.T.A. 4, 1 : 416 (1903) ; F.W.T.A., ed. 1, 2 : 61. *Marsdenia rostrifera* N. E. Br. (1906)—F.W.T.A., ed. 1, 2 : 60. *Anisopus rostriferus* (N. E. Br.) Bullock l.c. 512 (1957), and l.c. 15 : 202 (1961). *Marsdenia rhynchogyna* K. Schum. (1896). A strong climber with greenish flowers in globose lateral umbelliform cymes and horizontally opposed follicles 6–8 in. long and about ½ in. thick, tapering to a slightly hooked point at the apex. **Ghana:** Akim (Mar.) *Irvine* 1813 ! Banka (Mar.) *Vigne* FH 1865 ! Aburi (Sept.) *Johnson* 1078 ! **S.Nig.:** Akassa *Dutton* 32 ! Okomu, Benin (Mar.) *Akpabla* 1114 ! Oban *Talbot* 1684 ! [Br.]**Cam.:** Kumba (Feb.) *Binuyo & Daramola* FHI 35579 !

29. LEPTADENIA R. Br.—F.T.A. 4, 1 : 430 ; Bullock in Kew Bull. 10 : 287 (1955).

Erect, leafless shrubs, or with a few small, linear, fugacious leaves ; peduncles very short, lateral ; pedicels 1–2 mm. long, tomentellous ; calyx and corolla outside shortly pubescent ; follicles very narrow, long-beaked, about 12 cm. long, closely grooved ..   ..   1. *pyrotechnica*
Twining shrubs with broad, well-developed and persistent petiolate leaves :
Calyx longer than the corolla-tube ; corolla-lobes 4–5 mm. long ..   2. *hastata*
Calyx shorter than the corolla-tube ; corolla-lobes 2–2·5 mm. long   3. *arborea*

1. **L. pyrotechnica** (*Forsk.*) *Decne.* in Ann. Sci. Nat. sér. 2, 9 : 270 (1838) ; Berhaut Fl. Sén. 8 ; Bullock l.c. 289. *Cynanchum pyrotechnicum* Forsk. Fl. Aeg.-Arab. 53 (1775). *Leptadenia spartium* Wight & Arn. (1834) —F.T.A. 4, 1 : 432 ; F.W.T.A., ed. 1, 2 : 63 ; Chev. Bot. 438. Erect, leafless shrub with green rather succulent stems, 3–6 ft. high ; flowers yellowish ; in dry sandy places. **Sen.:** *Roger* ! *Leprieur* ! *Heudelot* 482 ! Oualo *Berhaut* 1397 ! **Mali:** Timbuktu (fl. & fr. June) *Hagerup* 103 ! **Niger:** Konna (July) *Lean* 49 ! **N.Nig.:** Chad (fl. & fr. Jan.) *Foster* 93 ! Bure (Dec.) *Elliott* 116 ! Kukawa *Rosevear* FHI 26608 ! Throughout the northern part of this Flora area, extending in semi-desert areas to Arabia and Palestine, to Baluchistan and Sind.

2. **L. hastata** (*Pers.*) *Decne.* in DC. Prod. 8 : 551 (1844) ; Berhaut Fl. Sén. 58 ; Bullock l.c. 289. *Cynanchum hastatum* Pers. Syn. Pl. 1 : 273 (1805). *C. lancifolium* Schum. & Thonn. (1827). *Leptadenia lancifolia* (Schum. & Thonn.) Decne. (1838)—F.T.A. 4, 1 : 430 ; F.W.T.A., ed. 1, 2 : 63 ; Chev. Bot. 437. Twiner with corky bark on the older stems ; flowers dull yellowish ; in dry savanna. **Sen.:** *Brunner* 158 ! *Roger* ! *Döllinger* 60 ! Dakar (May) *Baldwin* 5704 ! *Debeaux* 181 ! **Gam.:** *Saunders* 20 ! **Mali:** Goumal, Dioura (Jan.) *Davey* 232 ! Timbuktu (July) *Hagerup* 227 ! **Port.G.:** Bissau (Feb.) *Esp. Santo* 1819 ! Suzana to Casselol (Nov.) *Esp. Santo* 3137 ! **Guin.:** Kindia (May) *Jac.-Fél.* 1623 ! Dinguiraye (Dec.) *Jac.-Fél.* 1422 ! **Ghana:** Accra (Oct.) *Brown* 309 ! *Dalz.* 7 ! Accra Plains *Irvine* 417 ! Navrongo (May) *Vigne* FH 4505 ! Dedora Tankara (June) *Andoh* 5899 ! **Niger:** *Gaillard* 5 ! **N.Nig.:** Bida (Jan.) *Meikle* 1017 ! Zamfara (Apr.) *Keay* FHI 15636 ! Jibiya, Katsina (May) *Onwudinjoh* FHI 22373 ! near L. Chad. (Dec.) *Elliott* 110 ! Yola (Dec.) *Hepper* 1615 ! **S.Nig.:** W. Lagos *Rowland* ! Yoruba country *Barter* 3336 ! Iseyin (Feb.) *Onochie* FHI 35277 ! Ibadan (May) *Onochie* FHI 19175 ! [Br.]**Cam.:** Bama, Dikwa Div. (Nov.) *McClintock* 7 ! Extending to Uganda, Kenya and southern Ethiopia.

3. **L. arborea** (*Forsk.*) *Schweinf.* Arab. Pflanzenn. 167 (1912) ; Bullock l.c. 290. *Cynanchum arboreum* Forsk. Fl. Aeg.-Arab. 53 (1775). *C. heterophyllum* Del. (1826). *Leptadenia heterophylla* (Del.) Decne. (1838)— F.T.A. 4, 1 : 432 ; F.W.T.A., ed. 1, 2 : 63. Twiner over shrubs ; flowers greenish-yellow. **Mali:** Selingourou *Davey* 27 ! **Niger:** Zinder (Nov.) *Hagerup* 602 ! **N.Nig.:** Katagum (Nov.) *Dalz.* 96 !

---

[1] I am now reluctantly obliged to regard this genus as monotypic ; there is considerable variation in the specimens examined but I can find no characters sufficient to warrant specific segregation.—A.A.B.

Gajibo, Bornu (Nov.) *Johnston* N87! Bure, near L. Chad (Dec.) *Elliott* 112! **[Br.]Cam.**: Bama, Dikwa Div. (Dec., fl. & fr. Jan.) *McClintock* 126! 161! Gulumba, Dikwa Div. (Dec.) *McClintock* 106! Extending to Egypt and Arabia.

30. BRACHYSTELMA R. Br. in Bot. Mag. t. 2343 (1822).

Flowers several, in umbelliform fascicles at the apex of leafy branches ; corolla with a broad flat or funnel-shaped disk and triangular lobes ; stems stout, branched sparingly, leafy :

Pedicels about 4 cm. long ; corolla purple, densely tomentose inside, with a very short tube and broad flat disk, nearly 2 cm. diam., the lobes very short, deltoid ; leaves oblong-oblanceolate, up to 6 cm. long and 1·5 cm. broad, more or less hispid on both sides .. .. 1. *omissum*

Pedicels from less than 10 to 14 mm. long ; corolla glabrous or pubescent inside, yellow, green or purple, the tube widely funnel-shaped, 5 mm. long, the lobes triangular, 5–10 mm. long and about 4 mm. broad at the base ; leaves oblong-oblanceolate, up to 11 cm. long and 4 cm. broad, more or less pubescent on both sides

2. *togoense*

Flowers solitary at the upper nodes of leafless stems ; pedicels filiform, up to 2 cm. long ; corolla divided almost to the base into linear segments more than 4 cm. long and about 1 mm. wide at the base, beset with long purple clavate (? vibratile) hairs particularly towards the base ; stems arising from a small flattened tuber, slender, with 2–3 long internodes below the flowering upper third, leaves reduced to scale-like cataphylls ; stems extending after flowering to become leafy and flexuous ; leaves lanceolate, sessile, subacute, subcordate at base, up to about 6·5 cm. long and 1 cm. wide, thinly scaberulous-puberulous on both sides .. .. 3. *exile*

1. **B. omissum** *Bullock* in Kew Bull. 17 : 193, fig. 2 (1963). A perennial herb up to 1 ft. high, from a fleshy disciform tuber, with purple flowers.
   [Br.]Cam.: Jua, Bamenda (Apr.) *Maitland* 1712!
2. **B. togoense** *Schltr.* in Engl. Bot. Jahrb. 38 : 40, fig. 5H–O (1905) ; Bullock l.c. 191 *B. atacorense* A. Chev. (1917)—F.W.T.A., ed. 1, 2 : 65. A perennial erect herb up to 1 ft. high, from a fleshy disciform tuber, with green flowers, or greenish-yellow and often flecked with purple.
   Ghana: Accra Plains (Feb., Mar.) *Cox* 130! **Togo Rep.**: Foot of Agomé Mts. (Mar.) *Schlechter* 12961. **Dah.**: Atacora Mts., Kovandé, 1,200–1,500 ft. (June) *Chev.* 24013. **N.Nig.**: Mongu, 4,300 ft. (July) *Lely* 443! Cece F.R., Niger Prov. (June) *E. W. Jones* 188! Mando, Birnin Gwari Dist. (June) *Keay* FHI 25856!
3. **B. exile** *Bullock* l.c. fig. 1 (1963). A precocious erect geophytic herb of savanna grassland, about 18 in. high, the purple flowers with long spidery corolla-lobes beset with vibratile purple hairs.
   N.Nig.: Ancho, Plateau Prov. (May) *Hepburn* 141! Mando F.R., Birnin Gwari Dist. (June) *Keay* FHI 25848! Also in Cameroun.

Doubtful species.

B. bingeri *A. Chev.* in Rev. Cult. Colon. 8 : 67, with fig. (1901) ; F.T.A. 4, 1 : 469. I have been unable to identify this plant from the description. Chevalier's description makes reference to the genus *Brachystelma* very doubtful, whilst his figure shows a turnip-like tuber unlike that of any other species of the genus. The small cyme is also strange, whilst the rather poor sketch of the flowers recalls the genus *Pentagonanthus* (*Periplocaceae*) rather than *Brachystelma*. The plant is recorded from the dry savanna of Mali and Dahomey (*Chev.* 992 ; 1015 ; 2703 ; 24251).

31. CEROPEGIA Linn.—F.T.A. 4, 1 : 435 ; H. Huber in Mem. Soc. Brot. 12 : 5–203 (1958).

Slender erect herbs from a few to 40 cm. high ; leaves linear to very narrowly lanceolate, sometimes absent at flowering time ; corolla-lobes more or less coherent :

Corolla with subulate spreading appendage 3–6 mm. long at the sinus between corolla-lobes ; tube 4–7 cm. long, mottled in upper half, lobes linear, 4–6 cm. long ; leaves 5–8 cm. long, rarely as much as 8 mm. broad .. .. .. 1. *ledermannii*

Corolla without appendages at the sinus between corolla-lobes, or these reduced to very short broadly deltoid teeth :

Corolla-tube about 2 cm. long ; leaves linear, about 4 cm. long ; corolla-lobes linear, minutely pubescent but with a few long purple hairs

2. *deightonii*

Corolla-tube 4 cm. long or usually more :

Corolla-tube broadly funnel-shaped and 5-angled at the throat and somewhat flask-shaped at the base, glabrous outside, lobes very variable in length, up to 5 cm. long

3. *campanulata*

Corolla-tube almost cylindrical and very slender, pubescent outside, plants usually leafless at flowering time ; corolla-lobes up to 9 cm. long .. 4. *porphyrotricha*

Twining climbers ; leaves lanceolate to broadly ovate or elliptic :

Corolla-lobes quite free from each other, short and spreading in the open flower, triangular to oblong, somewhat longer than broad ; tube 8–10 mm. long, thinly to densely pubescent, more or less globose at the base ; leaves ovate or ovate-lanceolate, more or less cordate at base, thinly setulose above, and thinly pubescent beneath

5. *nigra*

Corolla-lobes connate into a slender column about 8 mm. long terminating in an ellipsoid head, tube 1·5–2 cm. long, swollen at the base ; leaves ovate-lanceolate, ovate or elliptic ; flowers in umbelliform lateral cymes ; stems and leaves somewhat fleshy .. .. 6. *rhynchantha*

Corolla-lobes coherent at the tips only :

Calyx-lobes linear-subulate, about 2 cm. long ; corolla-tube cylindric for 2–5 cm., expanded above, lobes 1·5 cm. long ; leaves ovate, caudate-acuminate, almost glabrous   ..   ..   7. *talbotii*
Calyx-lobes more or less triangular, not more than about 5 mm. long :
   Leaves ovate or elliptic, not linear or linear-lanceolate :
      Leaves more or less hairy, at least on the veins beneath :
         Corolla-tube cylindric, 2·5 cm. long, the lobes 1·4 cm. long ; leaves densely pilose on the veins beneath, subulate-acuminate, mostly cordate at base 8. *johnsonii*
         Corolla-tube curved, 2 cm. long, the lobes 4 cm. long ; leaves thinly pubescent on the veins beneath, subulate-acuminate   ..   ..   ..   9. *peulhorum*
      Leaves glabrous :
         Corolla-lobes with a fringe of long hairs at the apex, about 8 mm. long, the tube 1·5 cm. long ; sepals 2·5 mm. long ; leaves obovate-oblanceolate, acutely acuminate, 4–7 cm. long   ..   ..   10. *sankuruensis*
         Corolla-lobes without a fringe of long hairs at the apex :
            Corolla-lobes glabrous inside, broadly lanceolate :
               Corolla-tube 3 cm. long ; leaves elliptic, acutely long-mucronate-acuminate, about 6 cm. long and 3·5 cm. broad   ..   ..   ..   11. *fusiformis*
               Corolla-tube 1·5 cm. long, gibbous at the base ; leaves elliptic, narrowly long-triangular-acuminate ..   ..   12. *aristolochioides*
            Corolla-lobes pilose inside, linear, 1 cm. long, the tube 1·5–2·5 cm. long, mottled towards the apex ; calyx-lobes narrowly lanceolate, glabrous :
               Leaves small, ovate, apex acute, base rounded or acute, not cordate ; stems with fleshy deciduous axillary jointed gemmae   ..   ..   13. *gemmifera*
               Leaves ovate-rounded, very acute, cordate at base ; stems without gemmae
                                   14. *yorubana*
  Leaves linear or linear lanceolate :
    Corolla-lobes densely covered with purple setae, about 6 mm. long ; leaves linear
                                   15. *tourana*
    Corolla-lobes without purple setae :
      Leaves linear, somewhat fleshy ; corolla-tube 8–14 mm. long, the lobes narrowly linear, 5–7 mm. long   ..   ..   16. *linophyllum*
      Leaves usually linear-lanceolate, rarely broader ; corolla-tube 1–2 cm. long, the lobes oblong or ovate, 5–12 mm. long, thinly white-pilose inside
                                   17. *racemosa*

1. **C. ledermannii** *Schltr.* in Engl. Bot. Jahrb. 51 : 154 (1913) ; Hutch. in Kew Bull. 1921 : 387 ; H. Huber in Mem. Soc. Brot. 12 : 131 (1958). Erect herb 6–12 in. high with an underground discoid tuber about 2 in. across, and purple-mottled tubular flowers with long green thread-like lobes.
   N.Nig.: Takwara (May) *Lely* 110 ! Vom, Jos Plateau *Dent Young* ! **S.Nig.**: Oyo (June) *Burton* H1616/38 ! Also in Cameroun.

2. **C. deightonii** *Hutch. & Dalz.* F.W.T.A., ed. 1, 2 : 61, 62 (1931) ; Kew Bull. 1937 : 340 ; H. Huber l.c. 133. From a few to 18 in. high, from a tuber up to ¾ in. diam. ; flowers green to deep red or purple with very hairy lobes ; often frequent in grassland on wet lateritic soils.
   S.L.: Waterloo (fl. & fr. June, fl. Aug.) *Deighton* 2057 ! 2743a ! *Melville & Hooker* 279 ! Kambia (fl. Aug., fr. Sept.) *Jordan* 292 ! *Adames* 183 ! Binkolo (Aug.) *Deighton* 1289 ! Gbap (June) *Adames* 48 ! Kakama, Gbo (Aug.) *Deighton* 6100 ! **Ghana**: Bolgatanga (fl. & fr. June) *Vigne* FH 4511 ! **N.Nig.**: Bichikki (May) *Lely* 179 ! Zaria (Mar.) *Milne-Redhead* 5044 ! Randa (May) *Hepburn* 58 !

3. **C. campanulata** *G. Don* Gen. Syst. 4 : 112 (1837) ; F.T.A. 4, 1 : 462 ; Bullock in Kew Bull. 9 : 592 (1955), partly ; H. Huber l.c. 129, partly, incl. var. *abinsica* (N. E. Br.) H. Huber l.c. 130 (1958). *C. kerstingii* K. Schum. (1903)—F.W.T.A., ed. 1, 2 : 63. *C. abinsica* N. E. Br. (1914)—F.W.T.A., ed. 1, 2 : 62 ; Bullock l.c. *C. dalzielii* N. E. Br. (1913)—F.W.T.A., ed. 1, 2 : 62. *C. hepburnii* Hutch. & Dalz. F.W.T.A., ed. 1, 2 : 62 (1931) ; Kew Bull. 1937 : 341. *C. tamalensis* W. W. Sm. (1922). Erect herb, 6 in. high, from a tuber, with dark purple-brown flowers 2 in. long including the corolla lobes.
   Mali: *Mowan. Don* ! **Ghana**: *Don* ! Tamale *Dalz.* 60 (partly) ! Achimota (cult. July,) *Adams* 967 ! **Togo Rep.**: *Kersting* 307 ! **N.Nig.**: Naraguta (Apr.) *King in Hb. Hepper* 2869 ! Abinsi (fl. June, fl. & fr. July) *Dalz.* 689 ! 690 ! Randa (fl. & fr. May) *Hepburn* 57 ! 62 ! **S.Nig.**: Old Oyo F.R. (Feb.) *Keay* FHI 23434 !

4. **C. porphyrotricha** *W. W. Sm.* in Notes Roy. Bot. Gard. Edinb. 13 : 307 (1922). *C. campanulata* of Bullock l.c. 592, partly, not of G. Don. *C. campanulata* var. *porphyrotricha* (W. W. Sm.) H. Huber l.c. 130 (1958). Erect herb about 1 ft. high, simple, from a discoid tuber ; flowers 3–4 in. long, green and deep purple including the erect linear lobes, the lower part of the tube only slightly dilated, the base of the lobes fringed with dark purple hairs ; in the leafy state one pair of leaves is much larger than the others.
   Ghana: Accra (Apr.) *Adams* 230 ! Tamale (Apr.) *Dalz.* 60 (partly) ! Kales Hill, Wa (Apr.) *Adams* 4004 ! 4007 ! **N.Nig.**: Katagum Dist. *Dalz.* 418 !

5. **C. nigra** *N. E. Br.* in Kew Bull. 1895 : 261 ; F.T.A. 4, 1 : 440 ; Bullock l.c. 591 ; H. Huber l.c. 159. *C. kroboensis* N. E. Br. in F.T.A. 4, 1 : 440 (1903) ; F.W.T.A., ed. 1, 2 : 63 ; Berhaut Fl. Sén. 59. A pubescent twiner ; flowers ½–⅔ in. long, the corolla-tube curved above the obliquely inflated base, the lobes dark blue-green, turning black.
   Sen.: Tambacounda *Berhaut* 871. **Port.G.**: Bafatá (Aug.) *Esp. Santo* 2747 ! **S.L.**: Rokupr (Oct.) *Adames* 255a ! **Ghana**: Krobo Plains (fl. & fr. Oct.) *Johnson* 790 ! Apla, E.P. (Oct.) *Vigne* FH 4047 ! **N.Nig.**: Niger *Baikie* ! **S.Nig.**: Gambari, Oshun Dist. (Oct.) *Onochie* FHI 34936 !

6. **C. rynchantha** *Schltr.* in Engl. Bot. Jahrb. 51 : 155 (1913) ; Berhaut Fl. Sén. 58 ; H. Huber l.c. 100. A slender perennial twiner with glabrous stems and peduncled cymes of green and red-mottled flowers 1–1¼ in. long including the club-topped connate corolla-lobes.
   Sen.: Tambacounda *Berhaut* 859. **Mali**: Bamako *Waterlot* 1281. **Guin.**: Kouroussa *Pobéguin* 438. **N.Nig.**: *Lely* 599 ! Naraguta (Aug.) *Lely* 523 ! 550 ! Anara F.R., Zaria Prov. (July) *Keay* FHI 25951 ! Also in Cameroun.

7. **C. talbotii** *S. Moore* in Cat. Talb. 65 (1913) ; H. Huber l.c. 86. A nearly glabrous twiner usually leafy, with flowers 2 in. long, dirty white with purple blotching.
   S.L.: Masingbe (Oct.) *Deighton* 4401 ! Bobuabu (Sept.) *Deighton* 2249 ! Kanya (Oct.) *Thomas* 2987 ! **Lib.**:

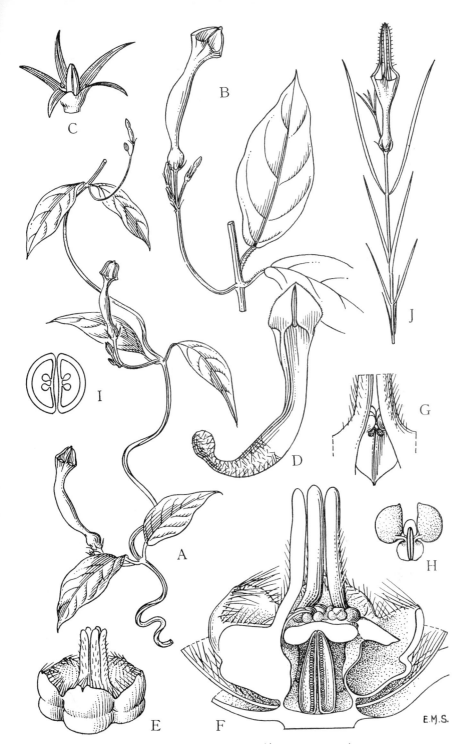

Fig. 221.—Ceropegia spp. (Asclepiadaceae).

C. *fusiformis* N. E. Br.—A, habit, × ⅔. B, inflorescence, × 1. C, calyx with carpels, × 4.
D, longitudinal section of corolla, × 1½. E, corona, × 8. F, longitudinal section of flower,
× 16. G, dehisced anther, stigmatic surfaces and part of corona, × 16. H, pollen-carrier
and pollinia, × 40. I, transverse section of ovary (diagrammatic). Drawn from FHI
25382.
C. *campanulata* G. Don—J, habit, × ½. Drawn from *Adams* 967.

8

Gbanga (Sept.) *Linder* 817! Peahtah (Oct.) *Linder* 1032! **S.Nig.:** Ibadan (Sept.) *Newberry* 69! Eket Dist. *Talbot* 3357! Oban *Talbot* 116!

8. **C. johnsonii** *N. E. Br.* in F.T.A. 4, 1 : 451 (1903) ; H. Huber l.c. 85. A twiner with fairly stout, half-succulent glabrous stems ; flowers 1½–2 in. long, whitish or green with purple blotches, the tube slightly curved.
   Lib.: Ganta (Sept.) *Harley* 656! **Ghana:** Aburi (Nov.) *Johnson* 768! 1055! Ho to Adaklu (fl. & fr. Nov.) *Adams* 5376! Also recorded from Congo and Uganda.

9. **C. peulhorum** *A. Chev.* in Journ. de Bot. 22 : 118 (1909) ; H. Huber l.c. 87, incl. var. *breviloba* Huber l.c. (1958) ; Adam in Bull. I.F.A.N. 24 : 944 (1962). Habit of the last, rather slender stemmed, glabrous, with dull mottled purple or almost black-purple flowers 2 in. or more long including the linear lobes ; fruits narrow, paired, 3–5 in. long.
   Sen.: Sélety, Diouloulou (Oct.–Nov.) *Adam* 18189. **Mali:** *Chev.* 18395! *Macleod* 73! **Port.G.:** Cumura, Bissau (Nov.) *Esp. Santo* 2231! Balanasinho, Fulacunda (July) *Esp. Santo* 2147! **S.L.:** Freetown Peninsula (Aug., Sept.) *Melville & Hooker* 83! 394! 470! *Litchford*! Rokupr (Aug.) *Adames* 115! Masasa (Sept.) *Jordan* 322! Mabum (Aug.) *Thomas* 1608! Also in Ubangi-Shari.

10. **C. sankuruensis** *Schltr.* in Engl. Bot. Jahrb. 51 : 155 (1913), as " *sankurnensis* " ; H. Huber l.c. 79. *C. anceps* S. Moore in Cat. Talb. 66 (1913) ; F.W.T.A., ed. 1, 2 : 63. *C. degenensis* S. Moore (1919). *C. batesii* S. Moore (1926). *C. aristolochioides* of F.W.T.A., ed. 1, 2 : 63 (1931), partly, not of Decne. A glabrous leafy climber, with flowers in short-pedunculate cymes, creamy-white and blotched on the tube, the lobes dark purple.
    S.L.: Ronietta (Nov.) *Thomas* 5561! Yonibana (Oct.) *Thomas* 3982! Kanya (Oct.) *Thomas* 2981! Mamaka (Nov.) *Thomas* 4434! Mabonto (Oct.) *Thomas* 3561! **Lib.:** Bumbuma (Oct.) *Linder* 1323! Ganta (Oct.) *Harley* 718! **Ghana:** Nsemse F.R., Borku to Sunyani (fr. Dec.) *Adams* 5237! Wenchi (Sept.) *Irvine* 4769! **S.Nig.:** Degema *Talbot* 3652! Oban *Talbot* 174! Calabar (Nov.) *Lloyd* (drawing)! Also in Cameroun, Congo, E. Africa and extending to S. Rhodesia.

11. **C. fusiformis** *N. E. Br.* in Kew Bull. 1897 : 273 ; F.T.A. 4, 1 : 450 ; H. Huber l.c. 91. A glabrous twiner with curved flowers 1¾ in. long, subumbellate (at first) on a peduncle 1–1½ in. long, greenish-white, purple at the base ; fruits narrow, 8 in. long.
    Iv.C.: Baoulé *Chev.* 22216! *Pobéguin* 181! **Ghana:** Krobo Plains (fl. & fr. Dec.) *Johnson* 499! Odumasi (Aug.) *Johnson* 1075! Kumawu (Aug.) *Cansdale* 123! **S.Nig.:** Lagos (Sept.) *Millen* 89! Idanre (Oct.) *Keay* FHI 25501! Olokemeji F.R. (Sept.) *Keay* FHI 25391! Ibadan (Mar., Sept.) *Latilo* FHI 31775! *Keay* FHI 25382!

12. **C. aristolochioides** *Decne.* in Ann. Sci. Nat., sér. 2, 9 : 263 (1838) ; F.T.A. 4, 1 : 464 ; F.W.T.A., ed. 1, 2 : 63, partly ; Berhaut Fl. Sén. 57 ; H. Huber l.c. 91. *C. perrottetii* N. E. Br. (1898)—F.T.A. 4, 1 : 448. A glabrous half-succulent twiner, with flowers clustered at the end of a peduncle scarcely 1 in. long, dark purple and mottled, about 1 in. long.
    Sen.: *Heudelot* 477! *Perrottet* 791! *Berhaut* 386. **Mali:** *Chev.* 24499! Extending eastwards to Ethiopia, Somalia, E. Africa and Congo.

13. **C. gemmifera** *K. Schum.* in Engl. Bot. Jahrb. 33 : 328 (1903) ; F.T.A. 4, 1 : 620. *C. nilotica* Kotschy var. *simplex* H. Huber l.c. 104 (1958). A half-succulent twiner, usually with few leaves, the axillary jointed gemmae acting as propagules and the species is unique in this respect ; flowers 1½ in. long, lurid-purple mottled ; fruits 4–5 in. long.
    Ghana: Accra (July) *Dalz.* 153! Achimota (May) *Milne-Redhead* 5086b! Pawmpawm R. (May) *Morton* 7261! **Togo Rep.:** Lomé *Warnecke* 242!
    [The writer cannot agree with Huber in uniting this with *C. denticulata* K. Schum. as a variety of *C. nilotica* Kotschy.—A.A.B.]

14. **C. yorubana** *Schltr.* in Engl. Bot. Jahrb. 38 : 48 (1905) ; Milne-Redhead in Kew Bull. 3 : 465 (1948) ; H. Huber l.c. 93. A glabrous climber, usually leafy with flowers about 1 in. long.
    Ghana: Amosima, Cape Coast (Mar.) *Hall* 1737! Abokobi (Feb.) *Irvine* 1989! Anum (Dec.) *Plumptre* 31! Kumasi (Aug., Nov.) *Darko* 617! 705! **S.Nig.:** Ibadan (fl. & fr. Mar.) *Meikle* 1256!

15. **C. tourana** *A. Chev.* in Mém. Soc. Bot. Fr. 2, 8 : 61 : 274 (1917) ; H. Huber l.c. 132. A climber with deep purple flowers ¾ in. long, and slender fruit follicles 10 in. long.
    Iv.C.: Toura, Sassandra *Chev.* 21653. **S. Nig.:** near Ado-Awaive, Oyo Dist. *Hambler* 650!

16. **C. linophyllum** *H. Huber* l.c. 115 (1958).[1] *C. senegalensis* H. Huber l.c. 116 (1958). A slender twiner with small tubers, corolla dull purple outside, the lobes and inside very dark crimson.
    Sen.: Near Dakar *Pitot*! **Ghana:** Navrongo (Sept.) *Vigne* FH 4602! Boro, near Wa *Adams* 1751! **S.Nig.:** Idanre Hills *Keay* FHI 22588! 37908!

17. **C. racemosa** *N. E. Br.* in Kew Bull. 1895 : 262 ; F.T.A. 4, 1 : 456 ; Werderman in Engl. Bot. Jahrb. 70: 207 (1939) ; Bullock in Kew Bull. 9 : 591 (1955) ; H. Huber l.c. 93. *C. pedunculata* Turrill in Kew Bull. 1921 : 389 ; F.W.T.A., ed. 1, 2 : 62. *A. atacorensis* A. Chev. & *A. gourmaca* A. Chev. Bot. 439, names only. A slender twiner with glabrous stems and flowers 1½ in. long, orange or brown-purple with paler corolla-lobes ; fruit follicles very narrow.
    Mali: Ouagadougou to Ouahigouya, Mossi, Upper Volta (Aug.) *Chev.* 24689! 24732! **Port.G.:** Bafata, Tamtam Cossé (fr. Sept.) *Esp. Santo* 3340! Farim (Aug.) *Esp. Santo* 3056! Fulacunda (Aug.) *Esp. Santo* 2179! Gabu, Gitche (fr. Oct.) *Esp. Santo* 3488! **Ghana:** Boro, Wa (June) *Adams* 681! **N.Nig.:** Anara F.R., Zaria Prov. (July) *Keay & Clayton* FHI 37084! Bonu Hill, Gwari, Niger Prov. (fl. & fr. Aug.) *Onochie* FHI 18262! Jos (Aug.) *Keay* FHI 20183! Naraguta (Aug.) *Lely* 495! Kilba country, near Yola (Aug.) *Dalz.* 96! A highly polymorphic species ; widely distributed through tropical and S. Africa, and in Madagascar. Four subspecies are recognized by Huber (l.c. 1958), but I am unable to agree with his circumscriptions.

Imperfectly known species.

C. achtenii *De Wild.* subsp. **togoensis** *H.* Huber l.c. 157 (1958). *C. achtenii* De Wild. is widely distributed in central equatorial Africa, and this subspecies is based on a single specimen (Ghana: Waribo Mt., N. of Wurupong (*Adams* 1834) preserved in the East African Herbarium, Nairobi.

   In the first edition of this work no fewer than 22 species were recognized ; in this treatment several of them are reduced to synonymy following the work by Huber and by the present writer. There is no doubt, however, that several further species will be found in our area.

32. HUERNIA R. Br.—F.T.A. 4, 1 : 495 ; White & Sloane, Stapelieae, ed. 2, 3 : 819 (1937).

Stems succulent, glabrous, erecto-patent, about (3–)5(–8) cm. long and up to 1·5 cm. thick, 4–5 angled, glabrous, bluntly toothed ; leaves opposite, triangular, 2 mm. long, at the tips of the stem teeth ; flower solitary, extra-axillary, near the base of young stems ; pedicel about 1 cm. long ; corolla campanulate, tube 7–8 mm. long, 8–9 mm. diam., lobes broadly triangular, about 10 mm. long, sinus teeth deltoid, about 1·5 mm. long ; flower pink outside, crimson-blotched on yellow papillate ground inside, corona very dark purple .. .. *nigeriana*

[1] The epithet was published as " *linophyllum* " ; I have treated this as a noun in apposition.—A.A.B.

H. nigeriana *Lavranos* in Journ. S. Afr. Bot. 27 : 233 (1961). Succulent herb with campanulate flowers, pink outside and with crimson blotches inside ; corona-lobes blackish-purple.
N.Nig.: Mongu F.R., Plateau Prov. *Keay* FHI 37098 ! 48 km. W.S.W. of Bauchi (July) *Lavranos* 1058.

33. CARALLUMA R. Br.—F.T.A. 4, 1 : 477 ; White & Sloane, Stapelieae, ed. 2, 1 : 151 (1937).

Stems cylindric, not angled or winged, 20–30 cm. long, slender, much branched ; leaves minute, caducous ; flowers 1–2 together, extra-axillary near the stem apex, on slender curved pedicels 4–6 mm. long ; calyx-lobes ovate-lanceolate, 3–4 mm. long, glabrous ; corolla quite glabrous, campanulate urceolate, 7–8 mm. long and as broad, yellowish-green, longitudinally ridged inside ; corolla-limb dark purple with a narrow disk and 5 erect oblong-triangular lobes 5 mm. long and 2·5 mm. broad at the base ; outer corona lobes bipartite, forming 10 filiform horns 5 mm. long   ..   **1. mouretii**
Stems winged or angled :
Stems with 4 wing-like ribs up to 2 cm. wide, indurated at the margins and with recurved or spreading indurated teeth from which the minute triangular leaves are early caducous ; the stems branched, erect, up to about 60 cm. high ; flowers in large apparently sub-terminal umbels forming a compound globose head 7–10 cm. or more diam. and with a hundred or more flowers ; pedicels 2–4 cm. long, glabrous ; calyx-lobes narrowly lanceolate, 2–5 mm. long ; corolla 12–16 mm. diam., rotate, lobed to about half-way, the lobes ovate-triangular, 5–6 mm. long and as broad at the base, sparsely covered with purple hairs or glabrous, ciliate with vibratile club-shaped hairs ; outer corona cup-shaped with 5 pairs of arcuate-subulate teeth 1–1·5 mm. long ; follicles erecto-divergent, terete, 12–18 cm. long   ..   **2. retrospiciens**
Stems 4 or 5 angled but not winged :
Corolla-lobes 4–5 mm. long, lanceolate, with a constriction near the base, pilose inside with purple multicellular hairs, distinctly maculate below the constriction ; outer corona lobes deeply bipartite into subulate horns ; corolla almost rotate, the tube shortly campanulate ; flowers in extra-axillary fascicles of 1–3 scattered along the upper part of the stem ; calyx-lobes subulate-lanceolate, about 4 mm. long ; stems 30 cm. or more high, with minute, caducous triangular leaves   **3. dalzielii**
Corolla-lobes 10–12 mm. long, ovate-oblong, acute ; outer corona annular, truncate, with a reddish rim ; corolla 2·5 cm. diam., rotate, with a very short tube ; flowers in fascicles of 2–4 near the apex of the stem ; calyx-lobes oblong-subulate, 4–5 mm. long, glabrous ; stems 10–40 cm. high, up to 1·5 cm. thick at base, the minute triangular leaves borne at the apex of stout conical teeth, shield-shaped at the base
                                                4. decaisneana

1. **C. mouretii** *A. Chev.* in Rev. Bot. Appliq. 14 : 270 (1934) ; White & Sloane l.c. 181. *C. edulis* A. Chev. ex Hutch. & Dalz. F.W.T.A., ed. 1, 2 : 65 (1931) ; Chev. Bot. 440, name only, not of Benth. (1876). A succulent herb with yellowish-green flowers.
Maur.: Tidjikja (Sept.) *Charles* in *Hb. Chev.* 28733. Also in Morocco.
[There is no specimen of this plant at Kew, and I rely on the identification given by White and Sloane.—A.A.B.]
2. **C. retrospiciens** (*Ehrenb.*) *N. E. Br.* in Gard. Chron. 12 : 370 (1892) ; F.T.A. 4, 1 : 480 ; White & Sloane l.c. 236. *Desmidorchis retrospiciens* Ehrenb. in Abh. Acad. Berlin 33 (1831). *Boucerosia tombuctuensis* A. Chev. (1900). *Caralluma tombuctuensis* (A. Chev.) N. E. Br. in F.T.A. 4, 1 : 622 ; F.W.T.A., ed. 1, 2 : 65 ; Berhaut Fl. Sén. 3. *C. retrospiciens* var. *tombuctuensis* (A. Chev.) White & Sloane l.c. 240 (1937). An erect sparsely branched greyish-green shrubby succulent, perhaps the largest of the stapeliads ; the winged stems indurated on the margins and with spreading or reflexed hard teeth ; flowers almost black in large globose heads at the tips of the branches.
Sen.: Guiers *Berhaut*. **Mali**: near Timbuktu and Arnassay (July) *Chev.* 1317 ; 1318.
[White and Sloane admit five varieties, all except one having been regarded as distinct species at one time or another. The species as here understood is widespread in semi-arid areas from the west of our area to the Red Sea coast and islands, Ethiopia, Kenya and Somalia. Within our area, White and Sloane record it from " many points in Mauretania, Senegal, French Sudan (i.e. Mali), Nigeria, and the Algerian Sahara, up to 6,000 ft. elevation ", but I have not seen specimens from most of these areas. It is likely that some other plants which have been treated as distinct species, may belong here.—A.A.B.]
3. **C. dalzielii** *N. E. Br.* in Kew Bull. 1912 : 280 ; White & Sloane l.c. 185. An erect sparsely branched succulent herb with green stems and scattered dark reddish-purple star-like flowers.
Ghana: Gambaga (Dec.) *Morton* A1408 ! **N.Nig.**: Sokoto (July) *Dalz.* 317 ! Katagum (cult.) *Dalz.* 367 ! Found in cultivation in many localities.
4. **C. decaisneana** (*Lem.*) *N. E. Br.* in Gard. Chron. 12 : 369 (1892) ; F.T.A. 4, 1 : 488 ; F.W.T.A., ed. 1, 2 : 65 ; Bullock in Kew Bull. 17 : 195 (1963). *Boucerosia decaisneana* Lem. in Herb. Gén. Amat. sér. 2, 4 : t. 21 (1844). *Stapelia ango* A. Rich. (1851). *Caralluma ango* (A. Rich.) N. E. Br. (1892). *C. sprengeri* N. E. Br. (1895). *C. commutata* Berger (1910). *C. hesperidum* Maire (1922). *C. venenosa* Maire (1934). *Stapelia decaisneana* (Lem.) A. Chev. (1934)—Berhaut Fl. Sén. 2, 3. A slender much-branched succulent, the stems decurved and rooting freely in contact with the soil ; flowers blackish-purple, frosted with minute white papillae.
Maur.: Tichoten *Charles* in *Hb. Chev.* 28690. **Mali**: Dienné (July) *Chev.* 1145 ! **Sen.**: *fide* Berhaut l.c. St. Louis *Chev.* 3489. A widespread and variable species extending through the Sahara and Sudan to Ethiopia, Somalia and southern Arabia.

137. RUBIACEAE

By F. N. Hepper[1] and R. W. J. Keay[2]

Trees, shrubs or rarely herbs. Leaves opposite or rarely verticillate, entire ; stipules inter- or intra-petiolar, often connate, rarely leafy and not distinguishable from the leaves. Flowers usually hermaphrodite, actinomorphic or very rarely slightly zygomorphic, solitary to capitate. Calyx adnate to the ovary. Corolla epigynous, more or less tubular, rarely campanulate ; lobes 4–12, contorted, imbricate or valvate. Stamens epipetalous, as many as and alternate with the corolla-lobes ; anthers mostly separate, 2-celled, opening lengthwise or rarely by terminal pores, rarely transversely septate. Ovary inferior or rarely superior, 2- or more-celled, with axile, apical or basal placentas, or rarely 1-celled with parietal placentas ; style often slender ; ovules 1 to many. Fruit a capsule, berry or drupe. Seeds rarely winged, mostly with endosperm, the latter rarely ruminate ; embryo straight or curved.

A very large and mostly tropical family, the herbaceous representatives largely in temperate regions. Recognised at once in the *Metachlamydeae* (*Gamopetalae*) by the usually inferior ovary, regular corolla, separate anthers, and stipulate simple *entire* leaves.

KEY TO THE TRIBES[3]

Corolla-lobes contorted or imbricate :
Flowers not in globose heads ; corolla contorted (rarely imbricate) ; trees, shrubs, climbers, or rarely subherbaceous (use combined Key for genera Nos. 4–42)
    II. GARDENIEAE, III. ALBERTEAE, IV. IXOREAE, V. HAMELIEAE
Flowers in globose heads ; corolla imbricate ; trees or shrubs   VI. NAUCLEEAE
Corolla-lobes valvate ; trees, shrubs or herbs :
Ovules 2 or more in each ovary-cell :
Fruit a capsule or dicoccous ; seeds winged or appendaged :
Trees or shrubs :
Flowers in lax panicles or spike-like cymes ..   ..   I. CINCHONEAE
Flowers in globose heads   ..   ..   ..   VI. NAUCLEEAE
Herbs or very small shrublets ..   ..   XII. HEDYOTIDEAE
Fruit a berry or dry and indehiscent   ..   ..   ..   VII. MUSSAENDEAE
Ovule solitary in each ovary-cell :
Ovule pendulous from near the top of the cell :
Woody plants :
Stigma capitate ; cystoliths absent on leaves etc. ..   VIII. VANGUERIEAE
Stigma bilobed ; cystoliths present on leaves and calyx ; leaves often yellowish
    IX. CRATERISPERMEAE
Herbs   ..   ..   XII. HEDYOTIDEAE
Ovule erect from near the base or peltately attached towards the middle of the cell :
Stipules not leaf-like :
Calyx-tubes confluent ; flowers in heads ; fruits more or less united into a mass ; trees or shrubs   ..   ..   X. MORINDEAE
Calyx-tubes not confluent :
Fruits fleshy, indehiscent ; ovule attached to base of the cell ; mostly trees and shrubs ; flowers cymose or capitate, rarely axillary (in *Lasianthus*)
    XI. PSYCHOTRIEAE
Fruits dry, usually dehiscent ; herbs or undershrubs often with small flowers in axillary or terminal clusters :
Ovule attached to the septum of the ovary ; flowers hermaphrodite
    XIII. SPERMACOCEAE
Ovule attached to base of the ovary ; flowers sometimes unisexual (*Antho-spermum*) ..   ..   XIV. ANTHOSPERMEAE
Stipules leaf-like, the " leaves " verticillate ; herbs or herbaceous climbers
    XV. RUBIEAE

Tribe I—CINCHONEAE

Flowers in simple or slightly branched spike-like racemes, sometimes subtended by large leaf-like white or coloured bracts ; corolla-lobes without appendages ; anthers included ; style long-exserted ; capsules loculicidally dehiscent   **1. Hymenodictyon**

[1] Genera 32, 41–91.
[2] Genera 1–31, 33–40.
[3] See Verdcourt in Bull. Jard. Bot. Brux. 28 : 209–290 (1958) for a discussion on the classification of this family.

Flowers in panicles ; corolla-lobes with appendages :
 Anthers and style exserted ; capsule loculicidally dehiscent ; flowers (in our species) 4-merous ; corolla-appendages (in our species) short, subspherical, close together in bud .. .. **2. Corynanthe**
 Anthers and style included ; capsule septicidally (and sometimes partially loculicidally) dehiscent ; flowers 5-merous ; corolla-appendages long, filiform or subulate, erect and often divergent in bud .. .. **3. Pausinystalia**

Tribes II—GARDENIEAE, III—ALBERTEAE, IV—IXOREAE & V—HAMELIEAE

(Corolla-lobes contorted or rarely imbricate, overlapping to the left (except in some species of *Rothmannia*) ; fruit indehiscent (except *Crossopteryx*) ; trees, shrubs and lianes).

Fruit capsular ; seeds winged ; ovary 2-celled ; ovules numerous ; style long-exserted, style-head clavate, slightly 2-lobed ; stamens exserted ; flowers (4-)5(-6)-merous ; flowers numerous in dense terminal corymbs .. .. .. **4. Crossopteryx**[1]
Fruit indehiscent ; seeds not winged :
 Ovary (3-)6-celled ; ovules 2 to numerous per cell, placentation axile ; inflorescences several- to many-flowered, appearing laterally :
 Corolla-lobes 5, longer than the short tube ; ovules 2-4 per cell ; stamens 5, exserted ; style-head fusiform, undivided ; flowers pedicellate in pedunculate cymes borne laterally at alternate nodes ; leaves petiolate, up to 14 cm. long and 8 cm. broad
  5. Morelia
 Corolla-lobes (6-)12, shorter than the long tube ; ovules numerous in each cell ; stamens (6-)13, included ; style-head 4-6-lobed ; flowers sessile in a dense sessile cluster subtended by broad bracts and borne at ends of shoots opposite a single leaf and just above a pair of leaves ; leaves up to 120 cm. long and 45 cm. broad; ovary (3-)6-celled .. .. **10. Schumanniophyton**
Ovary 1-2-celled :
 *Solitary flowers or inflorescences terminal (sometimes on lateral branchlets), or appearing laterally by sympodial growth of stem, never truly axillary, if apparently in axils then never on both sides of stem at same node (to p. 107) :
 †Ovules several to numerous ; placentas on walls or septum of ovary (to p. 107):
  ‡Style-head usually capitate, clavate or fusiform, undivided or sometimes shortly cleft but never with spreading arms ; calyx-lobes mostly open (but see *Pseudogardenia* and *Sherbournia*) (to p. 107):
  Inflorescences several- to many-flowered appearing laterally at alternate nodes ; pollen grains not in tetrads ; ovary 2-celled with placentas on the septum, rarely 1-2-celled with parietal placentation (*Brenania* only) :
  Stipules not forming a truncate sheath ; laminas rounded to acuminate at apex usually with a fringe of hairs around the node inside them ; ovary 2-celled, placentas borne on the septum ; small or medium sized trees :
  Style well exserted from the cylindrical corolla-tube ; corolla-lobes often reflexed :
   Corolla villous in the throat, glabrous or pubescent outside ; anthers well exserted ; pollen grains single ; style-head entire ; fruits subglobose, up to 1·5 cm. diam. .. .. **6. Aidia**
   Corolla glabrous inside and outside ; anthers included ; pollen grains aggregated into spindle-shaped masses about 250 μ long ; style-head divided ; fruits ovoid, beaked, 5-9 cm. long .. .. .. **7. Massularia**
  Style and anthers included in the funnel-shaped corolla-tube ; corolla-lobes mostly erect ; corolla silky pilose outside except at base, glabrous inside except for a band of hairs near the base ; pollen grains single ; style-head entire ; fruits ellipsoid, up to 3 cm. long .. .. .. **8. Porterandia**
  Stipules forming a truncate sheath, ciliate at the margin ; corolla-tube broadly cylindrical, glabrous except for a ring of hairs inside, lobes broad, rounded, erect ; anthers included ; style-head divided ; placentas parietal, ovary 1-celled or partially 2-celled ; pollen grains single ; a large tree **9. Brenania**
 Inflorescences terminal, or appearing laterally at successive nodes, or flowers solitary or paired :
 The inflorescences several- to many-flowered always terminal (sometimes on lateral branchlets) :
 Placentas attached to the septum of the distinctly 2-celled ovary :
  Inflorescence either an elongated panicle of small cymes, or flowers in axils few or congested ; fruits globose, not exceeding 1 cm. diam. ; ovules numerous .. .. **41. Bertiera**
 Inflorescences not as above ; trees, shrubs or lianes :
  Fruits oblong or fusiform, 3-11 cm. long ; ovules numerous ; flowers in small cymes ; lianes .. .. **40. Atractogyne**

---

[1] *Crossopteryx* is usually placed in the tribe *Cinchoneae*, but it appears here for convenience.

Fruits globose :
  Corolla with 1 lobe imbricated ; calyx-lobes foliaceous ; fruit 1·5 cm.
    diam. ; ovules numerous ; inflorescences few-flowered ..   42. **Heinsia**
  Corolla contorted ; calyx small ; fruit not more than 1 cm. diam. ; ovules
    1–6 ; inflorescences mostly corymbose ; trees or shrubs, rarely climbers ;
    see also below ..   ..   28. **Tarenna**
Placentas parietal ; ovules numerous ; ovary 1–2-celled :
  Anthers not included in the tubular lower portion of the corolla but completely
    exserted at the top between the lobes ; style-head subglobose, very long-
    exserted ; cymes 1- to many-flowered, terminating long and short shoots ;
    corolla hairy outside ; ovary 1-celled   ..   ..   11. **Macrosphyra**
  Anthers included wholly or mostly in the corolla-tube ; style-head usually
    longer than broad, not long-exserted :
    Calyx-lobes distinct ; corolla-tube cylindrical or infundibuliform, without a
      large campanulate upper portion :
      Cymose panicles subsessile at the ends of long shoots :
        Leaves and stem hirsute, the latter hollow ; calyx-lobes subulate or
          lanceolate ; corolla infundibuliform, pubescent outside, glabrous inside
          except for a ring of hairs near base ; style-head clavate or obovate ;
          ovary 1-celled   ..   ..   12. **Calochone**
        Leaves and stems glabrous, the latter solid ; calyx-lobes leafy, persistent
          in fruit ; corolla-tube cylindrical, glabrous outside, slightly pubescent
          inside ;   style-head linear-ellipsoid ;   ovary 2-celled, sometimes only
          partially so   ..   ..   13. **Preussiodora**
      Cymose panicles borne terminally on young relatively short lateral shoots
        with 1–3 pairs of new leaves and conspicuous stipules ; stem and leaves
        glabrous ; corolla very narrowly infundibuliform, glabrous inside and
        outside ; style-head obovate-spathulate ; ovary 1-celled   14. **Polycoryne**
    Calyx truncate ; lower portion of corolla shortly and narrowly tubular,
      abruptly expanded into the relatively large upper campanulate portion ;
      corolla glabrous except for a dense band of hairs inside at the top of the
      narrow tubular portion ; ovary 1-celled ; style-head elongate fusiform ;
      cymes short, few-flowered ; stem and leaves glabrous   15. **Oligocodon**
The inflorescences appearing laterally at successive nodes, or if terminal then
  flowers 1–2 (very rarely 3) :
  Branches usually armed with spines formed from modified lateral branchlets ;
    flowers 1–2 (–3) together, terminating opposite pairs of very much abbreviated
    leafy lateral shoots ; ovary 2-celled, placentas attached to the septum which
    divides the seed-masses in fruit ; pollen grains single   ..   16. **Xeromphis**
  Branches not armed with spines (but if, in some species of *Gardenia*, lateral
    branchlets tend to be spine-like, then pollen grains in tetrads and ovary
    1-celled with 3–9 placentas) ; placentas 2–9, parietal ; ovary 1–2-celled ;
    seeds forming a single mass in fruit :
    Flowers solitary or paired, terminal, or appearing lateral by sympodial growth ;
      corolla usually more or less funnel-shaped and often large :
      Stipules chaffy, persistent ; leaves deciduous, crowded below the solitary
        flowers, with long almost leafless internodes between the clusters ; calyx-
        tube very short, lobes spreading ; ovary 2-celled ; pollen grains in tetrads
                                                                  17. **Euclinia**
      Stipules not chaffy, leaves not as above :
        Stipules sheathing, persistent, often truncate ; young parts usually
          glutinous ; ovary 1-celled (or partially 2-celled in *G. sokotensis*), placentas
          2–9 ; calyx-tube well developed ; pollen grains in tetrads   18. **Gardenia**
        Stipules not as above, mostly soon caducous ; young parts not glutinous
          (except in *Rothmannia lujae*) ; ovary 1–2-celled ; placentas 2 ; pollen
          grains single :
          Calyx-lobes in bud open or overlapping to the left ; flowers terminal,
            solitary (except *Rothmannia lujae*), or appearing laterally on abbreviated
            shoots on account of the sympodial growth :
            Calyx-tube glabrous and minutely papillose inside, truncate, with broad
              leafy overlapping lobes attached outside ; corolla-lobes overlapping to
              left ; ovary 1-celled ; scandent shrubs   ..   19. **Pseudogardenia**
            Calyx-tube velutinous inside, with dentate, linear or filiform lobes not
              attached outside a truncate tube and not overlapping in bud ; calyx-
              tube sometimes split at anthesis ; corolla-lobes overlapping to left or
              right ; ovary 2-celled or partially 2-celled ; erect shrubs or trees
                                                                  20. **Rothmannia**
          Calyx-lobes in bud contorted and over-lapping to the right (in species with
            short lobes, see very young bud or the thin margin on the left of each
            lobe) ; cymes 1 to several-flowered normally axillary on one side of

stem at successive nodes, rarely terminal ; calyx-tube finely velutinous inside and outside ; corolla velutinous outside except for a glabrous portion at the base, lobes short and broad, overlapping to the left ; lianes ; see also below .. .. 21. **Sherbournia**

Flowers 3 or more together in cymes appearing laterally or axillary (on only one side of the stem) :

Stipules rounded at apex, caducous and exposing a fringe of hairs at the nodes ; ovary and fruit often distinctly 10-ribbed ; ovary 2-celled ; corolla funnel-shaped or campanulate, with short broad lobes ; anthers and style included ; cymes 1- to several-flowered ; climbing shrubs 21. **Sherbournia**

Stipules acute at apex, persistent:

Anthers not included in the long narrowly cylindrical corolla-tube ; style well-exserted .. .. 22. **Oxyanthus**

Anthers included (or nearly so) in the funnel-shaped corolla-tube ; style included or just exserted .. .. 23. **Mitriostigma**

‡Style with 2 recurved spreading arms :

Calyx-lobes open in bud ; flowers in terminal and axillary cymes ; ovules several in each cell, pendulous from the upper part of the septum .. 25. **Pouchetia**

Calyx-lobes contorted in bud ; flowers in terminal corymbs ; placentas adnate to the septum of the 2-celled ovary :

Anther cells divided into small compartments by longitudinal and numerous transverse septa .. .. 26. **Dictyandra**

Anther cells not divided into compartments .. .. .. 27. **Leptactina**

†Ovules solitary or paired ; ovary 2-celled, placentation not parietal :

Flowers normally 4-merous ; ovules attached to the septum (if ovules basally attached, see *Rutidea*) ; flowers mostly in terminal corymbs :

Style-head clavate, subentire ; bracts stipular in origin (hence at right angles to the lateral branches of the inflorescence) ; floral bracts and bracteoles absent or very small ; leaves and stems often with bacterial nodules ; ovule solitary (rarely 2) .. .. 29. **Pavetta**

Style with 2 recurved spreading arms ; bracts foliar in origin (hence subtending the lateral branches of the inflorescence) ; floral bracts and bracteoles well-developed (in our species) ; bacterial nodules absent ; ovule solitary

30. **Ixora**

Flowers normally 5-merous (but 4-merous in 2 spp. of *Rutidea*, ovules then basally attached) ; bacterial nodules absent :

Flowers 1-several subtended by a single leaf, clustered together at the ends of very short terminal branchlets often appearing laterally at the nodes owing to sympodial growth of the stem ; ovules paired in each ovary-cell, immersed in a pendulous placenta ; style long and very narrowly clavate, sometimes with 2 short arms .. .. 31. **Aulacocalyx**

Flowers not arranged as above, mostly in terminal panicles or corymbs ; style clavate, subentire :

Ovules 1–6, attached to the septum ; endosperm not ruminate ; see also above

28. **Tarenna**

Ovule 1, attached to base of ovary ; endosperm ruminate .. 32. **Rutidea**

*Solitary flowers or subsessile clusters of flowers, axillary on both sides of the node ; growth of stem monopodial :

Style-head fusiform or clavate, entire ; ovules numerous on 2 parietal placentas which often meet to make the ovary partially 2-celled ; flowers solitary, supra-axillary ; corolla trumpet-shaped, longitudinally ribbed ; stems often armed with spinescent abbreviated lateral shoots .. .. .. 24. **Didymosalpinx**

Style more or less filiform, entire or with 2 arms ; ovary 2-celled, ovules 1–16 in each cell ; placentas not parietal :

‖Style exserted ; anthers exserted or included ; corolla-tube glabrous or hairy inside :

Style undivided or shortly bifid ; ovules solitary and pendulous, or 2–4 and attached to the septum :

Flowers opposite and solitary, appearing on lateral branchlets usually before these elongate and while they are still more or less covered by the bud-scales ; bracteoles lanceolate ; ovules 2–3, attached to the septum ; corolla puberulous or glabrous inside .. .. 33. **Feretia**

Flowering not as above ; ovules solitary and pendulous ; corolla densely hairy in the throat :

Flowers 5(–6)-merous ; bracteoles not cupular ; anthers exserted ; calyx-teeth distinct ; endosperm not ruminate .. .. 34. **Cremaspora**

Flowers 4-merous ; bracteoles cupular ; anthers included ; calyx truncate or subtruncate ; seeds with ruminate endosperm .. 35. **Polysphaeria**

Style distinctly divided into two spreading arms ; ovules solitary and attached to the septum, or 1–16 and inserted in the placentas ; bracteoles often cupular :

Corolla pubescent or pilose inside ; ovules 1–16, inserted on the outer face of
each placenta ; placentas peltately attached to the septum ; bracteoles often
cupular, but only rarely with foliaceous appendages ..   **36. Tricalysia**
Corolla glabrous inside ; ovules solitary in each cell, attached to the lower half
of the septum, very rarely 2 immersed in each placenta ; bracteoles usually
cupular and often with foliaceous appendages ; see also below   **37. Coffea**
||Style and anthers included in the corolla-tube, style not reaching to the level of the
anthers ; corolla-tube glabrous inside ; flowers sessile or subsessile, solitary or a
few together ; bracteoles often cupular :
  Ovules solitary in each cell, attached to the septum ; calyx-lobes open in bud :
    Calyx-lobes not accrescent ; seeds without a longitudinal groove ; (*Coffea*
    rupestris and *C. ebracteolata*), for other species see above   **37. Coffea**
    Calyx-lobes accrescent in fruit ; seeds with a longitudinal groove
                                           38. Psilanthus
  Ovules 2 in each cell, inserted on the inner face of a pendulous placenta ; fruits
  marked at the apex by the wide remains of the disk and the calyx-lobes ; calyx-
  lobes imbricate in bud, not enlarged in fruit ; seeds 2–4 in each fruit, without
  longitudinal grooves   ..   ..   **39. Belonophora**

Tribe VI—Naucleeae

Flowers entirely free from one another ; fruit a capsule :
  Corolla-lobes valvate ; seeds shortly winged ; bracteoles present ; stigma caplike,
  longitudinally grooved ..   ..   **43. Mitragyna**
  Corolla-lobes imbricate ; seeds with tail-like wings ; stigma more or less clavate,
  smooth :
    Scandent shrubs ; branchlets more or less 4-angled, armed with axillary pairs of
    recurved spines ; leaves not whorled ; bracteoles absent ..   ..   **44. Uncaria**
    Trees ; branchlets terete, without spines ; leaves whorled ; bracteoles present
                                             45. Adina
Flowers fused together by their ovaries and calyx-tubes, syncarpous in fruit ; seeds
not winged ; corolla-lobes imbricate ; stigma fusiform, with two papillose areas in the
lower half ..   ..   **46. Nauclea**

Tribe VII—Mussaendeae

Inflorescences terminal, often one of the calyx-lobes of the outer flowers enlarged and
petaloid ; straggling shrubs   ..   ..   **47. Mussaenda**
Inflorescences axillary (terminal and axillary in *Pauridiantha viridiflora*) ; calyx-
lobes never petaloid :
  Calyx-teeth minute or calyx truncate (up to 3 mm. long in *P. stipulosa*) ; bracts not
  enclosing inflorescence ; erect shrubs or sometimes trees   ..   **48. Pauridiantha**
  Calyx-teeth well-developed ; bracts often present and sometimes enclosing the
  inflorescence ; mostly climbing shrubs, others erect shrubs ..   ..   **49. Sabicea**

Tribe VIII—Vanguerieae

Corolla-tube curved, broad and about 2·5 cm. long, glabrous ; stems annual from a
rhizome ; calyx-limb truncate ..   ..   **50. Temnocalyx**
Corolla-tube always straight, mostly short (except *Hutchinsonia* & *Robynsia*) ; trees or
shrubs (herbs in *Fadogia*) :
  Corolla about 2 cm. long, densely pubescent, narrowly tubular ; shrubs :
    Flowers solitary ; corolla-lobes with long filiform apex ; stigma vertically lobed,
    shortly exserted   ..   ..   **51. Hutchinsonia**
    Flowers cymose, numerous ; corolla-lobes without appendages ; stigma laterally
    lobed, long exserted   ..   ..   **52. Robynsia**
  Corolla small (up to about 1 cm. long), usually glabrous ; flowers rarely solitary :
    Bracts and bracteoles foliaceous and conspicuous ; calyx-lobes usually foliaceous or
    very rarely truncate ; ovary 2–5-locular ; corolla very acute ; straggling shrubs
                                             53. Cuviera
    Bracts and bracteoles not foliaceous :
      Leaves in whorls of 3–5 ; stems mostly herbaceous, angular, rarely trees or shrubs ;
      calyx-limb truncate, irregularly dentate or lobed ; stigma 3–5-lobed   **54. Fadogia**
      Leaves opposite ; shrubs or trees :
        Ovary 2-locular ; fruits didymous or by abortion 1-celled :
          Inflorescences with the ultimate branches always elongated, cymose-racemose ;
          corolla-throat glabrous ; calyx deeply lobed   ..   ..   **55. Vangueriopsis**
          Inflorescences with the ultimate branches not elongated, mostly cymose-corymbose
          or umbellate ; corolla-throat bearded or glabrous ; calyx truncate or minutely
          denticulate :
            Flowers numerous, inflorescences mostly cymose-corymbose or rarely clustered
                                             56. Canthium

Flowers few, in clusters (*R. affine*) .. .. ..   **57. Rytigynia**
Ovary (3–)4–5-locular ; fruits mostly of 5 pyrenes :
    Flowers in clusters or very short cymes :
        Calyx truncate or minutely denticulate .. ..   **57. Rytigynia**
        Calyx deeply lobed .. ..   **58. Globulostylis**
    Flowers in cymose racemose inflorescences with ultimate branches always
        elongated ; calyx lobed ; corolla throat villous .. ..   **59. Vangueria**

Tribe IX—Craterispermeae

Shrubs or small trees ; leaves often yellowish ; flowers in axillary clusters ; fruits free
    from one another .. ..   **60. Craterispermum**

Tribe X—Morindeae

Trees or climbing shrubs ; flowers in heads ; fruits fused .. ..   **61. Morinda**

Tribe XI—Psychotrieae

Inflorescences strictly axillary, flowers fasciculate and sessile :
    Ovules solitary .. ..   **62. Schizocolea**[1]
    Ovules 6–10 .. ..   **63. Lasianthus**
Inflorescences terminal (rarely also axillary), paniculate, capitate or spicate :
    The inflorescences paniculate or if capitate not surrounded by an involucre ; mainly
        shrubs :
        Ovary superior ; stipules markedly ochreate .. .. ..   **64. Gaertnera**[2]
        Ovary inferior :
            Inner surface of seeds deeply and broadly concave, endosperm not ruminate ;
                corolla-tube often curved, rather long .. .. .. **65. Chassalia**
            Inner surface of seeds more or less flat :
                Erect shrubs with panicles or if undershrubs inflorescences not capitate ; endo-
                    sperm either ruminate or not .. .. **66. Psychotria**
                Undershrubs with capitate inflorescences ; endosperm not ruminate
                                                                    69. Trichostachys
    The inflorescences capitate and involucrate, or spicate and without an involucre :
        Inflorescences capitate and involucrate or flowers 1–few :
            Shrubs or creeping undershrubs ; leaves cuneate ; flowers numerous **67. Cephaëlis**
            Herbs with soft, creeping stems ; leaves usually cordate ; flowers numerous or
                1–few .. ..   **68. Geophila**
        Inflorescences spicate, sometimes shortly so .. .. ..   **69. Trichostachys**

Tribe XII—Hedyotideae[3]

Ovules solitary ; ovary 2-celled ; uncommon plants :
    Pyrophyte, 5–25 cm. high ; fruit indehiscent ; calyx-lobes unequal ; in upland
        savanna .. ..   **70. Pentanisia**
    Half-succulent herb, 50–200 cm. high ; fruit dehiscent ; one calyx-lobe enlarged and
        leafy ; in crevices of rock outcrops .. ..   **71. Neobaumannia**
Ovules numerous :
    Stigma capitate :
        Anthers opening by terminal pores, connivent ; fruit dehiscing irregularly ; small
            annual herbs in wet places with a basal rosette of leaves   **72. Argostemma**
        Anthers opening by longitudinal slits ; fruit with one valve remaining attached by a
            pedicel, the other deciduous ; large herbs .. .. **73. Virectaria**
    Stigma lobed :
        Flowers mainly 4-merous :
            Stigma over-topped by anthers, all inserted ; common weeds .. **74. Kohautia**
            Anthers over-topped by stigma which is often exserted :
                Capsule with a horny wall, with or without a beak, but never with a solid beak,
                    early dehiscent ; common weeds .. .. **75. Oldenlandia**
                Capsule with a thick woody wall and a solid conical beak, dehiscence delayed ;
                    rare .. ..   **76. Lelya**
        Flowers mainly 5-merous :
            Plant glabrous ; leaves half-succulent ; inflorescences lax axillary cymes ; in wet
                places .. ..   **77. Pentodon**
            Plants not glabrous ; leaves not succulent :
                Climbing plants :

---

[1] Bremekamp placed this genus in the tribe *Coussareeae*, which is otherwise entirely American. As there is
some doubt as to its true position, the genus is included here in the *Psychotrieae* for convenience.
[2] *Gaertnera* was placed in *Loganiaceae* in F.W.T.A., ed. 1, 2 : 21.
[3] Adapted from Verdcourt in Bull. Jard. Bot. Brux. 23 : 250–252 (1953), (q.v. for a very useful key to the
herbaceous genera of *Rubiaceae* occurring in tropical Africa). *Hedyotideae* here includes genera of the tribes
Knoxieae and *Argostemmeae* which are recognised by Bremekamp and Verdcourt.

Corolla-tube 5–6 mm. long ; inflorescences terminal lax or congested panicles
    slender shrubs   ..   ..   78. **Sacosperma**
Corolla-tube 25–40 mm. long ; flowers in a terminal cluster ; herb ; (*O. volubilis*)
                                                  83. **Otomeria**
Erect, ascending or procumbent plants :
  Inflorescences of long branched or un-branched spikes from middle axils ; flowers
    minute ; slender shrub   ..   ..   79. **Hekistocarpa**
  Inflorescences terminal, axillary and confined to upper axils, or flowers solitary or
    paired :
    Flowers solitary or paired, sessile in axils :
      Annual herb ; style inserted   ..   ..   ..   80. **Thecorchus**
      Perennial herb ; style exserted ..   ..   ..   81. **Batopedina**
    Flowers in terminal inflorescences (with or without additional solitary flowers in
      lower axils) or in axillary clusters :
      Inflorescences axillary, clustered, not elongating in fruit   82. **Parapentas**
      Inflorescences terminal :
        Flowering inflorescences capitate, later elongating into a long, single spike,
          rarely with axillary spikes from upper axils and frequently with solitary
          flowers at lower nodes ; fruits oblong to turbinate ..   83. **Otomeria**
        Flowering inflorescences capitate or lax, much-branched, complicate-cymose,
          individual branches becoming spicate in fruit ; fruits globose, obtriangular
          or oblong (in *P. herbacea*)   ..   ..   84. **Pentas**

Tribe XIII—Spermacoceae

Ovary 3-celled ; stigmas 3 ; introduced prostrate hairy herb ..   ..   85. **Richardia**
Ovary 2-celled ; stigmas 2 or style bilobed :
  Fruit separating into 2 parts, each part remaining indehiscent ; trailing or scandent
    herbs ; stipules multi-setose ; flowers 1–few at nodes   ..   86. **Diodia**
  Fruit with both loculi dehiscent ; usually erect herbs ; flowers in dense terminal or
    axillary clusters ; common weeds :
    Fruit with loculi dehiscent at apex   ..   ..   87. **Borreria**
    Fruit circumscissile   ..   ..   88. **Mitracarpus**

Tribe XIV—Anthospermeae

Fruits with 1(–2) persistent foliaceous calyx-lobe ; flowers hermaphrodite, homostylous ;
  prostrate herb   ..   ..   89. **Otiophora**
Fruits without foliaceous persistent calyx-lobe ; flowers often unisexual, heterostylous ;
  heath-like undershrubs   ..   ..   90. **Anthospermum**

Tribe XV—Rubieae[1]

Climbers with hooked hairs, and whorls of linear leaves and leaf-like stipules 91. **Galium**

    Some additional genera have been introduced into our area :—several species of *Cinchona* Linn., quinine
plants, from tropical S. America, have been introduced especially to the Cameroons, including *C. calisaya* Wedd.,
C. ledgeriana Moens. ex Trimen and *C. succioubra* Pav. ex Klotzsch ; *Warscewiczia coccinea* (Vahl) Klotzsch
(from tropical America) and *Pseudomussaenda flava* Verdc. (from E. Africa) are cultivated as ornamentals.
    Benzona corymbosa Schumach. Beskr. Guin. Pl. 114 (1827) must remain an imperfectly known genus and
species since the whereabouts of the type specimen is not known and it has not been possible to determine it
from the description alone.

1. HYMENODICTYON Wall.—F.T.A. 3 : 42 ; K. Schum. in E. & P. Pflanzenfam. 4,
                      4 : 47 (1891).

Corolla 3 mm. long, consisting of a tube 1·5 mm. long and 0·6 mm. diam. and an abruptly
  expanded upper part 1·5 mm. long and 3 mm. diam., densely pubescent outside ;
  calyx-lobes 4·5–7 mm. long, subulate ; anthers 0·75–1 mm. long ; free part of filament 0·25 mm. long ; inflorescence simple or branched, up to 32 cm. long, pubescent,
  without large leaf-like bracts ; branchlets stout, glabrous, with the leaves crowded
  at the apices ; leaves obovate or obovate-oblong, 8–18 cm. long, 5–10 cm. broad,
  glabrous or with hairs by the midrib beneath, venation obscure beneath ; fruits
  ellipsoid, 5–5·5 cm. long, 2–2·2 cm. broad ; seeds with broad wings   1. *pachyantha*
Corolla 4–7·5 mm. long, obovate to clavate or clavate-ellipsoid gradually widened from
  the base, glabrous or papillose outside ; calyx-lobes 1–3 mm. long ; anthers 1–2 mm.
  long ; free part of filament 1·25–2·25 mm. long ; inflorescences subtended by large
  leaf-like bracts ; fruits 0·8–2 cm. long :
  Leaves quite glabrous, venation very obscure beneath ; inflorescences often branched ;
    bracteoles caducous or persistent, up to 15 mm. long and 1·5 mm. broad ; calyx-
    lobes 2–3 mm. long ; corolla 4–7 mm. long ; fruits 1–2 cm. long ; seeds with linear
    tails, one tail split to the base ; leaves obovate to elongate-elliptic, 9–17 cm. long,
    3·5–7·3 cm. broad   ..   ..   2. *biafranum*

    This tribe was previously known as *Galieae*.

Leaves pubescent beneath, sometimes densely so, venation distinct beneath ; inflorescences not branched ; bracteoles subulate, up to 5 mm. long, more or less caducous ; calyx-lobes about 1·5 mm. long ; corolla 4–5·25 mm. long ; fruits 0·8–1 cm. long ; seeds with a triangular wing deeply bilobed at one end ; leaves obovate, 6–16 cm. long, 3–7·5 cm. broad .. .. 3. *floribundum*

1. **H. pachyantha** *K. Krause* in Engl. Bot. Jahrb. 57 : 26 (1920). *H. bracteatum* of F.W.T.A., ed. 1, 2 : 70, not of *K.* Schum. *H. gobiense* Aubrév. & Pellegr. in Bull. Soc. Bot. Fr. 105 : 34 (1958) ; Aubrév. Fl. For. C. Iv., ed. 2, 3 : 294, t. 361, 1–3. A tree, to 70 ft. high with clear, rough bole and rather small, pyramidal crown ; leaves all green ; flowers green ; in forest.
   Iv.C.: Gobia, N.W. of Oumé (Mar.) *Aubrév.* 4153 ! **S.Nig.:** Oke Igbo, Ondo Div. (Feb.) *Tamajong* FHI 14676 ! Benin City (Apr.) *Farquhar* 18 ! 33 ! Auchi *Kennedy* 2356 ! Onitsha Dist. *Rosevear* 26/28 ! Mamu River F.R. *Kennedy* 2529 ! Iyamoyong F.R., Ogoja Prov. (fr. Apr.) *Binuyo* FHI 41271 ! [Br.]**Cam.:** Aiyomojok, Mamfe Div. (fr. Feb.) *Keay & Tiko* FHI 41899 ! Also in Cameroun. (See Appendix, p. 400.)
2. **H. biafranum** *Hiern* in F.T.A. 3 : 42 (1877). *H. bracteatum* K. Schum. in Engl. Bot. Jahrb. 23 : 424 (1896), not of F.W.T.A., ed. 1, 2 : 70. *H. epidendron* Mildbr. ex Hutch. & Dalz. F.W.T.A., ed. 1, 2 : 70 (1931). *H. oreophyton* Hoyle in J. Bot. 75 : 168 (1937). *H. reflexum* Hoyle l.c. 169 (1937). A shrub or tree to 30 ft. or more high, deciduous ; flowers pale purple, in erect racemes, subtended by white leaf-like bracts which turn red with age ; in open places, very abundant on the 1922 lava flow of Cameroon Mt., sometimes epiphytic.
   S.Nig.: Oban *Talbot* 213 ! 256 ! [Br.]**Cam.:** Cam. Mt., 4,500 ft. (fr. Feb.) *Mann* 1194 ! Litoka, Cam. Mt., 4,500 ft., bordering 1922 crater (Apr.) *Maitland* 1074 ! Ebang, Cam. Mt., 5,000 ft. (May) *Johnstone* 320/32 ! Buea, 3,000 ft. (fr. Jan.) *Maitland* 201 ! Mopanya (Apr.) *Kalbreyer* 183 ! Isobi to Bibundi, 1922 lava flow, near sea level (fl. June, fr. Jan.) *Dundas* FHI 8466 ! *Keay* FHI 28652 ! *Rosevear* Cam. 68/37 ! **F.Po:** N. side of St. Isabel's Peak *Mildbr.* 6431. Moka (fr. Sept.) *Wrigley* 581 ! Also in Cameroun and Principe.
3. **H. floribundum** (*Steud. & Hochst.*) *B. L. Robinson* in Proc. Amer. Acad. 45 : 404 (1910) ; Aubrév. Fl. For. C. Iv., ed. 2, 3 : 294, t. 349, 3–5 ; Fl. For. Soud.-Guin. 455. *Kurria floribunda* Steud. & Hochst. (1842). *Hymenodictyon kurria* Hochst. (1843)—F.T.A. 3 : 42 ; Chev. Bot. 308. A shrub or tree, to 25 ft. high, in open places, commonly on rocky hills ; racemes erect, subtended by a pair of red leaf-like bracts ; calyx yellow ; corolla yellow, turning crimson-brown.
   Guin.: Labé *Pobéguin* 2081. Diaguissa *Chev.* 12919. Ditinn to Dalaba *Chev.* 18543. Dalaba *Chev.* 20279. **S.L.:** summit of Sugar Loaf Mt. (Apr.) *Sc. Elliot* 5779 ! Lumley, by river (May) *Deighton* 2730 ! **Iv.C.:** Mt. Momy (Apr.) *Chev.* 21352 ! S. of Taï (young fr. Aug.) *Boughey* GC 14973 ! Mt. Nimba, Mt. Dou, Niangbo & Guiglo *fide* Aubrév. l.c. **Ghana:** Obuasi (young fr. Sept.) *Vigne* FH 927 ! **Togo Rep.:** Sokode-Basari *Kersting* A741 ! **N.Nig.:** Kargi Hill, Birnin Gwari, Zaria Prov. (June) *Keay* FHI 25884 ! Neill's Valley, Jos (June) *Lely* 266 ! Jos Plateau (May) *Lely* P321 ! **S.Nig.:** Lagos *MacGregor* 163 ! *Rowland* ! Idanre Hills (fr. Apr.) *Symington* FHI 3366 ! [Br.]**Cam.:** Fonfuka, Bamenda (Apr.) *Maitland* 1418 ! Bapinyi, Bamenda (May) *Ujor* FHI 30364 ! Also in Cameroun, Ubangi-Shari, Congo, Sudan, Ethiopia, E. Africa, Nyasaland, N. & S. Rhodesia and Angola.

2. CORYNANTHE[1] Welw. in Trans. Linn. Soc. 27 : 37 (1869) ; W. Brandt in Archiv der Pharmazie 260 : 55 (1933). *Pseudocinchona* A. Chev. ex E. Perrot in Compt. Rend. Acad. Sci. Paris 148 : 1466 (1909).

Capsules 0·7–1 cm. long, glabrous ; seeds flat, thin, with a long membranous tail at each end ; branchlets glabrous, with very short internodes ; stipules early deciduous, leaving a very short ciliolate ring ; leaves obovate-elliptic, cuneate at base, bluntly acuminate, 14–20 cm. long, 4–8 cm. broad, glabrous, with 6–10 main lateral nerves on each side of midrib and numerous parallel tertiary nerves ; petiole up to 4 cm. long ; flowers in sessile clusters, numerous, in a crowded panicle ; ovary glabrous ; flowers 4-merous ; calyx-lobes very short and thick ; corolla infundibuliform, with a tube 2·5 mm. long and 4 lobes 2 mm. long each with a subspherical appendage about 1 mm. long ; anthers and style exserted, stigma ellipsoid, entire .. 1. *pachyceras*
Capsules 2·5–3 cm. long ; leaves oblong, cuneate at base, acuminate, 15–25 cm. long, 6–8 cm. broad, with 12–16 main lateral nerves on each side of midrib ; flowers not known ; otherwise similar to above .. .. .. 2. *dolichocarpa*

1. **C. pachyceras** *K. Schum.* in Notizbl. Bot. Gart. Berl. 3 : 96 (1901) ; W. Brandt in Archiv der Pharmazie 260 : 62 ; Aubrév. Fl. For. C. Iv., ed. 2, 3 : 296, t. 362 ; Kennedy For. Fl. S. Nig. 210. *Pausinystalia pachyceras* (K. Schum.) De Wild. (1922). *Pseudocinchona africana* A. Chev. ex E. Perrot (1909)—Chev. Vég. Util. 5 : 229 ; Bot. 308. *P. pachyceras* (K. Schum.) A. Chev. (1926). A lower storey forest tree to 70 ft. high and 6 ft. girth, with deeply fluted, twisted bole, branching low ; flowers white, in copious panicles, sweetly scented.
   S.L.: Njala (Oct.) *Deighton* 1333 ! Dodo Hills F.R. (fr. Jan.) *Sawyerr* FHK 13598 ! **Lib.:** Moylakwelli (Oct.) *Linder* 1271 ! Suacoco, Gbanga (Oct.) *Okeke* 22 ! Kolahun (Nov.) *Baldwin* 10146 ! **Iv.C.:** Nzi, Baoulé-Nord (fl. buds Aug.) *Chev.* 22271 ! Bouroukrou (fr. Dec.–Jan.) *Chev.* 16141 ! Abidjan *Aubrév.* 188 ! Banco *Martineau* 344 ! Rasso *Aubrév.* 144 ! **Ghana:** Amentia (Sept.) *Vigne* FH 1368 ! Sefwi Bekwai (Oct.) *Akpabla* 886 ! Jabo, Upper Wassaw F.R. (young fr. June) *Vigne* FH 966 ! Pammer, Kilelu Dist. (fr. Dec.) *Vigne* FH 1569 ! **S.Nig.:** Akilla, Ijebu Ode Prov. (fl. buds & fr. Dec.) *Mitchell* ! Okomu F.R., Benin Prov. (fr. Dec.–Mar.) *Brenan* 8609 ! *A. F. Ross* 195 ! Benin City *Dennett* 25 ! Bendiga Ayuk, Ikom Div. (fl. & fr. Dec.) *Keay* FHI 28164 ! Oban FHI 32444 ! [Br.]**Cam.:** Likomba (Nov.) *Mildbr.* 10606 ! Tiko (fr. Jan.) *Dunlap* 165 ! Kumba (fr. Dec.) *Bamenda* FHI 8437 ! Also in Cameroun, Rio Muni and Congo. (See Appendix, p. 395.)
2. **C. dolichocarpa** *W. Brandt* l.c. 62, t. 1, 1 & 8, t. 2, 3 (1922). A forest tree, to 100 ft. high.
   [Br.]**Cam.:** Likomba (young fr. Dec.) *Mildbr.* 10773 ! Also in Cameroun.

---

[1] The type species of *Corynanthe* is *C. paniculata* Welw. (from Angola) which has 5-merous flowers and filiform corolla-appendages, erect and divergent in bud. *C. pachyceras* with 4-merous flowers and short subspherical corolla-appendages has been separated as a distinct genus, *Pseudocinchona*. These two species are however, linked by a third species, originally described as *Pausinystalia mayumbensis* Good (from Cabinda), which has 4-merous flowers and long corolla-appendages. Three other species of *Corynanthe* were described by W. Brandt (1922), but are not yet known in flower. R. Hamet in Compt. Rend. Acad. Sci. Paris 212 : 305 (1941) maintained *Pseudocinchona* for *Corynanthe pachyceras* (syn. *Ps. africana*) and *Ps. mayumbensis* (Good) R. Hamet, mainly because these two species contain alkaloids not found in *Corynanthe paniculata*. I consider, however, that the differences are best represented by regarding *Pseudocinchona* A. Chev. as a subgenus of *Corynanthe* Welw.—R.W.J.K.

3. PAUSINYSTALIA Pierre ex Beille in Act. Soc. Linn. Bordeaux 61 : 130 (1906) ;
W. Brandt in Archiv der Pharmazie 260 : 64 (1922).

Leaves elongate-obovate or oblanceolate, narrowed to the usually auriculate base, very
shortly and obtuse acuminate, 13–35 cm. long, 5–11·5 cm. broad, with 10–16 main
lateral nerves on each side of midrib, prominently venose-reticulate above, margins
often undulate ; petioles up to 0·7 cm. long ; panicles up to 18 cm. long, glabrous,
flowers in clusters ; corolla-appendages filiform, about 1·5 cm. long ; capsules
1–1·6 cm. long   ..   ..   1. *johimbe*
Leaves oblong or elliptic, not more than 20 cm. long ; petioles up to 2·5 cm. long :
Ovary and inflorescence-axes glabrous or nearly so ; corolla-appendages filiform,
about 1 cm. long, drying blackish ; leaves thinly coriaceous or papery, elliptic,
cuneate at base, acuminate, 6·5–15 cm. long, 2·5–6 cm. broad, with 5–10 main lateral
nerves on each side of midrib ; petioles 0·8–2·5 cm. long ; panicles up to 15 cm. long ;
capsules 1–1·7 cm. long   ..   ..   2. *macroceras*

FIG. 222.—PAUSINYSTALIA MACROCERAS (*K. Schum.*) *Pierre ex Beille* (RUBIACEAE).
A, flower.   B, part of corolla laid open.   C, seed.

Ovary and inflorescence-axes more or less densely pubescent ; leaves coriaceous,
cuneate to rounded or cordate at base, with (7–)10–15 main lateral nerves on each
side of midrib :
Corolla-appendages angular, dark brown, blackish or crimson and not usually wrinkled
when dry, up to 8 mm. long ; capsules 1·7–2·3 cm. long ; leaves oblong to obovate-
oblong, 9–20 cm. long, 4·2–11·5 cm. broad ; panicles up to 15 cm. long   3. *talbotii*
Corolla-appendages terete, pale brown and wrinkled when dry, up to 15 mm. long ;
capsules 0·8–1·4 cm. long ; leaves oblong, obovate-oblong or elliptic, 7–20 cm. long,
3·2–9·5 cm. broad ; panicles up to 25 cm. long   ..   ..   ..   4. *lane-poolei*

1. **P. johimbe** (*K. Schum.*) *Pierre ex Beille* in Act. Soc. Linn. Bordeaux 61 : 130 (1906), as *P.* "*yohimba*" ;
   W. Brandt in Archiv der Pharmazie 260 : 70 ; Kennedy For. Fl. S. Nig. 209. *Corynanthe johimbe* K.
   Schum. in Notizbl. Bot. Gart. Berl. 3 : 94 (1901). *Pausinystalia macroceras* of Kennedy l.c. 210. A forest
   tree, to 90 ft. high ; bole straight, without buttresses.
   S.Nig.: Sapoba , Benin Prov. (Nov.) *Chukwuogo* FHI 4666 ! *Kennedy* 291 ! 331 ! 2071 ! 2117 ! 2143 ! Also
   in Cameroun and Rio Muni. (See Appendix, p. 407.)
2. **P. macroceras** (*K. Schum.*) *Pierre ex Beille* l.c. (1906), as *P.* "*microceras*" by error ; W. Brandt l.c. 65.
   Corynanthe macroceras K. Schum. in Engl. Bot. Jahrb. 23 : 424 (1896).   *Pausinystalia brachythyrsa* of
   F.W.T.A., ed. 1, 2 : 71 ; of Kennedy For. Fl. S. Nig. 209 ; (?) not of (K. Schum.) De Wild. A forest tree,
   to 135 ft. high, with clean straight bole up to 6 ft. in girth ; flowers white, scented ; slash white, quickly
   turning yellow or brown.
   S.Nig.: Ijebu Ode Prov. : Oni *Sankey* 2 ! Omo (formerly part of Shasha) F.R. (fr. Feb., Mar.) *Richards*
   3047 ! *Ross* 67 ! Akilla *Kennedy* 367 ! Benin Prov. : Sapoba (June) *Kennedy* 409 ! 1423 ! 1516 ! 1740 !
   2633 ! Okomu F.R. (Sept.) *Iriah* FHI 23074 ! Ogoja Prov. : Cross River North F.R. *Latilo* FHI 31847 !
   Boshi-Okwango F.R. (May) *Latilo* FHI 30958 ! [**Br.**]**Cam.:** Mbalange, Kumba Div. *Olorunfemi* FHI
   30701 ! Also in Cameroun, Rio Muni, Gabon and Congo. (See Appendix, p. 406.)
3. **P. talbotii** *Wernham* in Cat. Talb. 49 (1913). *P. sankeyi* Hutch. & Dalz. F.W.T.A., ed. 1, 2 : 71 (1931). A
   forest tree, to 90 ft. high ; corolla white to crimson.

S.Nig.: Oni R., near Alafara (fr. Feb.) *Jones & Onochie* FHI 17529! Ijebu Ode Prov. : Oni *Sankey*!
Akilla (young fr. Dec.) *Mitchell*! Omo (formerly part of Shasha) F.R. (fl. & fr. Apr.) *Richards* 3356! *Ross*
240! Calabar Prov. : Oban *Talbot* 1493! *Gray* 2/1! Odot (Mar.) *Unwin* 10! (See Appendix, p. 406.)
4. **P. lane-poolei** (*Hutch.*) Hutch. ex Lane-Poole Trees, Shrubs etc. of Sierra Leone 74 (1916). *Corynanthe lane-*
poolei Hutch. in Kew Bull. 1912 : 98. New genus mentioned on page 381 of Hook. Fl. Nigrit. *Pausinystalia*
reticulata Hutch., name only. A forest tree, up to 50 ft. high, with pitted bark ; flowers white, turning to
purplish pink, very numerous with a sickly scent.
S.L.: *Don*! York Pass (June, Dec.) *Aylmer* 1! *Lane-Poole* 46! Kenema *Lane-Poole* 468! Hangha (fr.
Mar.) *Lane-Poole* 334! Kambui Hills South (Apr.) *Small* 897! Gola Forest (fl. & fr. Mar.) *Small* 508!
Lib.: Dukwia R. (Nov.) *Cooper* 143! 278! **Ghana:** Ankasa F.R. (Dec.) *Vigne* FH 3171! (See Appendix,
p. 406.)

4. CROSSOPTERYX Fenzl—F.T.A. 3 : 43 ; K. Schum. in E. & P. Pflanzenfam. 4, 4 : 51 (1891).

Shoots and leaves varying from glabrous to shortly and softly pubescent ; stipules
small ; leaves elliptic to suborbicular, rounded to shortly acuminate, averaging about
7 cm. long, mostly pubescent beneath ; flowers very numerous in rather dense terminal
corymbs ; bracts small, linear ; calyx very small ; corolla tomentose, about 8 mm.
long ; stamens shortly exserted ; style long-exserted, glabrous ; fruit subglobose,
slightly didymous, reticulate, about 1 cm. diam. ; seeds thin, flat, with a jagged wing
all round .. .. *febrifuga*

C. febrifuga (*Afzel. ex G. Don*) Benth. in Fl. Nigrit. 381 (1849) ; Chev. Bot. 309 ; Aubrév. Fl. For. Soud.-Guin.
455, t. 98, 1–2. *Rondeletia febrifuga* Afzel. ex G. Don (1834). *Crossopteryx kotschyana* Fenzl (1839)—
F.T.A. 3 : 44 ; Chev. Bot. 309. *Rondeletia africana* T. Winterb. (1803), name only. A savanna tree,
or shrub, up to 30 ft. high, with scaly bark ; variable in leaf-shape and indumentum ; flowers creamy-
white, sickly scented ; fruits blackish.
Sen.: *Heudelot* 357! Kolda, Casamance *Serv. For.* 9. **Mali:** Koulikoro *Vuillet* 658. Sareya (Feb.) *Chev.*
445*bis*! Kita *Dubois* 15. **Port.G.:** *Esp. Santo* 519! Tabanca de Forol, Bissoram (May) *Esp. Santo* 2388!
Guin.: *Heudelot* 766! Sineia (Mar.) *Pobéguin* 199! Yorogama *Pobéguin* 904! Diaguissa *Chev.* 18067.
S.L.: *Afzelius*! *Don*! Musaia (Apr.) *Deighton* 5478! Mt. Gonkwi, Talla Hills *Sc. Elliot* 4983! Batkanu
(Apr.) *Thomas* 93! Magbile, Maforki (May) *Jordan* 891! 894! **Iv.C.:** Banco *Martineau* 250! Dimbokro
Aubrév. 431! And other localities *fide* Aubrév. *l.c.* **Ghana:** Achimota (Apr.) *Irvine* 1423! Kintampo (fl.
& fr. May) *Stevenson* FH 4225! Salaga *Anderson* 56! Afram Plains (May) *Kitson* 1118! Tamale (fl. & fr.
Apr.) *Williams* 119! Kpandu (May) *Robertson* 136! **Togo Rep.:** *Baumann* 384! Lomé *Warnecke* 445!
Sokode (Apr.) *Kersting* 45! **Dah.:** Atacora Mts. *Chev.* 24137. **N.Nig.:** Nupe *Barter* 1197! 1233! Sokoto
Ryna 12! Zungeru (fl. & fr. May *Dalz.* 220! Jos Plateau (Feb.) *Lely* P150! **S.Nig.:** Onitsha Dist. *Rosevear*
10/29! Iva Valley, Enugu (Feb.) *Ainslie*! **[Br.]Cam.:** Kentu, Bamenda (fl. & fr. June) *Maitland* 1550!
1580! Bamenda Prov. (May) `ohnstone 123/31! Gashaka Dist. (fr. Dec., fl. Jan.) *Latilo & Daramola*
FHI 28934! 34407! Widespread in tropical Africa. (See Appendix, p. 396.)

5. MORELIA A. Rich. ex DC.—F.T.A. 3 : 112 ; K. Schum. in E. & P. Pflanzenfam. 4, 4 : 79 (1891).

Branchlets glabrous ; leaves oblong-elliptic, rather abruptly and obtusely acuminate,
about 12 cm. long and 5–6 cm. broad, glabrous, with 5–6 pairs of lateral nerves rather
sharply raised beneath ; venation very obscure ; petiole about 1 cm. long ; cymes
somewhat extra-axillary, short and many-flowered ; bracts small, ciliate ; calyx
truncate and undulately lobed, shortly ciliate, 4·5 mm. long, glabrous outside or
rarely pubescent ; corolla 1·5 cm. long, the lobes reaching down to the top of the
calyx and overlapping to the left in bud ; fruit globose, 1·5 cm. diam., crowned by the
short calyx .. .. *senegalensis*

M. senegalensis *A. Rich. ex DC.* Prod. 4 : 617 (1830) ; F.T.A. 3 : 113 ; Chev. Bot. 322 ; Aubrév. Fl. For.
C. Iv., ed. 2, 3 : 282, t. 355, 1–4. *Lamprothamnus fosteri* Hutch. (1907). An evergreen shrub or tree, up
to 40 ft. high, branching low, with white fragrant flowers in abundant cymes ; common by streams,
especially in the savanna regions.
Sen.: *Leprieur*! Niokolo-Koba *Berhaut* 1469. **Mali:** Kita (Dec.) *Chev.* 105. Siguri (Jan.) *Chev.* 287!
Port.G.: R. Jumbembem, Farim (Feb.) *d'Orey* 329! Cacine to Buba, Fulacunda (fr. July) *Esp. Santo*
2109! **Guin.:** *Heudelot* 737! Fouta Djalon *Maclaud* 353! Dentilia (Mar.) *Sc. Elliot* 5298! Boké *Chillou*!
Kouroussa (Feb.) *Pobéguin* 649! **S.L.:** Pujehun (Apr.) *Dawe* 472! *Thomas* 8404! Falaba (Mar.) *Aylmer*
26! Mawele (Feb.) *Deighton* 4192! Freetown (Dec.) *Deighton* 495! **Iv.C.:** Viala (fr. May) *Bodard* 1785!
And various localities *fide* Aubrév. *l.c.* **Ghana:** Sungbo, Pong (Mar.) *Kitson* 872! Pong Tamale (fr. June)
Vigne FH 3866! Atebubu (Jan.) *Vigne* FH 1526! Lissa (Mar.) *Kitson* 817! **Togo Rep.:** Kpedsu (Feb.)
Howes 1124! **Dah.:** *Poisson*! Djougou *Chev.* 23870. Konkobiri *Chev.* 24349. **N.Nig.:** Zungeru (Jan.)
Elliott 25! Kontagora (Dec.) *Dalz.* 225! Nupe *Barter* 1180! R. Taraba, Adamawa Prov. (Jan.)
Latilo & Daramola FHI 34427! **S.Nig.:** Lagos *Foster* 4! Oyo to Iseyin, Ogun R. (Feb.) *Brenan & Keay*
8960! Ibo Country *T. Vogel* 29! Okpo (Jan.) *Unwin* 39! Calabar R. (Feb.) *Mann* 2274! Also in Cameroun,
Rio Muni, Gabon, Congo and Sudan. (See Appendix, p. 403.)

6. AIDIA Lour. Fl. Cochinch. 143 (1790) ; G. Taylor in Cat. S. Tomé 197 (1944).

Calyx-tube about 2 mm. long, glabrous outside, with thick, oblong-ovate, recurved
lobes about 2 mm. long ; corolla-tube 5–6 mm., lobes 7–9 mm. long, broadly subacute ; anthers 4·5–5 mm. long, with filaments 1·5–2·5 mm. long ; style about 11 mm.
long ; leaves coriaceous, oblong-elliptic, cuneate or obtuse at base, shortly and
broadly acuminate, 9–16 cm. long, 3–6 cm. broad, with 7–8 main lateral nerves on
each side of midrib, venation obscure beneath, glabrous, drying reddish brown ;
flowers in short paniculate cymes, appearing laterally opposite a very small leaf just
above a pair of normal leaves ; fruits globose, not crowned by the calyx, glabrous,
smooth .. .. *1. rubens*

Calyx-tube 6–7 mm. long, more or less pubescent outside, with narrow teeth up to 1·5 mm. long ; corolla-tube 16–17·5 mm., lobes 11–16 mm. long, gradually very acute ; anthers about 7 mm. long, with filaments 1–2·5 mm. long ; style about 26 mm. long ; leaves thin, oblong-elliptic or elliptic, cuneate or obtuse at base, rather long-acuminate, 5–17 cm. long, 2–7 cm. broad, with 4–5(–7) main lateral nerves on each side of midrib, glabrous or very nearly so beneath, drying a dull or blackish olivaceous colour ; flowers in short cymes, appearing laterally at alternate, equally spaced nodes ; fruits globose, crowned by the persistent calyx-tube, more or less glabrous, 1–1·2 cm. diam.
2. *genipiflora*

1. **A. rubens** (*Hiern*) *G. Tayl.* in Cat. S. Tomé 200 (1944).   *Randia rubens* Hiern in F.T.A. 3 : 95 (1877) ; F.W.T.A., ed. 1, 2 : 79.   *R. refractiloba* K. Krause (1907). An understorey tree, to 18 ft. high ; flowers white ; in forest.
S.Nig.: Eket Dist. *Talbot* 3304 ! Calabar (Mar.) *Thomson* 87 ! Anyankene stream, Okwango F.R., Obudu Div. (May) *Latilo* FHI 30959 ! Also in Cameroun.

2. **A. genipiflora** (*DC.*) *Dandy* in F.W. Andr. Fl. Pl. A.-E. Sud. 2 : 424 (1952) ; Aubrév. Fl. For. C. Iv., ed. 2, 3 : 282, t. 354, t. 354.   *Randia genipaeflora* DC. (1830)—F.T.A. 3 : 95 ; F.W.T.A., ed. 1, 2 : 27.   *Gardenia genipaeflora* (DC.) Roberty (1954).   *Randia sassandrae* A. Chev. Bot. 322, name only.   *R. rhacodosepala* of Chev. Bot. 321, not of K. Schum. An understorey tree, to 40 ft., rarely to 70 ft. high ; or sometimes only a shrub ; flowers white, turning yellow, fragrant ; in forest.
Port.G.: Fulacunda to Bedanda (Jan.) *d'Orey* 253 ! **Guin.:** Heudelot 780 ! Macenta (Mar.) *Jac.-Fél.* 1548 ! Nzo (Mar.) *Schnell* 811 ! **S.L.:** *Afzelius* ! Bagroo R. (Apr.) *Mann* 804 ! Kambia (Dec.) *Sc. Elliot* 4358 ! Kambui F.R. (Apr.) *Lane-Poole* 220 ! Goderich (Apr.) *Deighton* 4753 ! Moyamba (Mar.) *Deighton* 1915 ! **Lib.:** Dukwia R. (fr. Aug.) *Cooper* 403 ! *Linder* 260 ! Gbanga (fr. Sept.) *Linder* 690 ! Mecca, Boporo (Nov.) *Baldwin* 10426 ! Ganta (fl. buds Feb.) *Baldwin* 11010 ! **Iv.C.:** Yapo (Oct.) *Chev.* 22371 ! Bingerville *Chev.* 15170 ! Abidjan *Aubrév.* 90 ! Danané *Aubrév.* 1037 ! **Ghana:** Cape Coast *Brass* ! Banka, Ashanti (Mar.) *Vigne* FH 1879 ! Ampunyasi, Obuasi (July) *Andoh* FH 4413 ! Axim (Mar.) *Irvine* 2394 ! Pamuso Bibiani (Mar.) *Darko* 856 ! Togo Plateau F.R. *St. C. Thompson* FH 3588 ! **S.Nig.:** Western Lagos *Rowland* ! Olokemeji (Apr.) *A. F. Ross* 155 ! Degema Dist. *Talbot* 3728 ! Oban *Talbot* 1553 ! [**Br.]Cam.:** Johann-Albrechtshöhe (= Kumba) *Staudt* 615 ! **F.Po:** *Barter* ! *Mann* 228 ! Also in Sudan. (See Appendix, p. 409.)

7. **MASSULARIA** (K. Schum.) Hoyle, Check List of the Gold Coast 110, 115 (1937). *Randia* Linn. sect. *Massularia* K. Schum. in E. & P. Pflanzenfam. 4, 4 : 76 (1891).

Branchlets puberulous ; leaves obliquely elongate elliptic-obovate, asymmetrically subcordate and auriculate at base, acuminate, 15–32 cm. long, 5–12·5 cm. broad, pubescent on midrib and nerves beneath, with 12–16 main lateral nerves on each side of midrib ; petioles very short ; flowers in pseudolateral inflorescences ; peduncles up to 3·5 cm. long, with 2–5 cymose branches up to 4·5 cm. long, glabrous, except for the ciliate bracts ; calyx 7–10 mm. long, glabrous except towards the top and on the short teeth ; corolla glabrous, very acute in bud, tube 11–15 mm. long, lobes acutely and narrowly lanceolate, reflexed, 8–12 mm. long ; anthers included, pollen adhering in tiny spindle-shaped masses ; style grooved, stigma 2-cleft, well exserted ; fruits glabrous, ovoid, beaked, 5–9 cm. long, up to 6·5 cm. diam. ..   ..   *acuminata*

M. acuminata (*G. Don*) *Bullock ex Hoyle* l.c. (1937) ; Aubrév. Fl. For. C. Iv., ed. 2, 3 : 282, t. 353.   *Gardenia? acuminata* G. Don (1834).   *Randia acuminata* (G. Don) Benth. (1849)—F.T.A. 3 : 95 ; K. Schum. l.c. fig. 28, E–F ; Chev. Bot. 320 ; F.W.T.A., ed. 1, 2 : 79. A shrub or small tree, 2–30 ft. high ; corolla-tube greenish-white at base, shading into pink above, lobes pink or reddish-purple, often with white margins ; in forest.
Guin.: Nimba Mts. *Schnell* 3481 ! **S.L.:** *Afzelius* ! *Don* ! Njala (Jan., June) *Dalz.* 8091 ! *Deighton* 1748 ! Fwendu, Perri (Apr.) *Deighton* 1625 ! Pujehun *Thomas* 8620 ! 8647 ! Gola Forest (fr. Apr.) *Small* 579 ! **Lib.:** Gbanga (Sept.) *Linder* 449 ! Kakatown *Whyte* ! Dukwia R. (Jan., July) *Cooper* 201 ! 252 ! *Linder* 188 ! Brewersville (June) *Barker* 1331 ! **Iv.C.:** Massa Mé *Aubrév.* 132 ! Banco *Aubrév.* 232. And other localities *fide* Chev. *l.c.* **Ghana:** Achimkrom, N.W. of Prestea (July) *Vigne* FH 1266 ! Kwahu Prasu (Feb.) *Vigne* FH 1945 ! Aburi (May) *Johnson* 954 ! **S.Nig.:** Epe (Feb.) *Millen* 164 ! Okomu F. R., Benin (Dec.– Jan.) *Brenan* 8473 ! 8830 ! R. Nun *T. Vogel* 40 ! Aguku *Thomas* 1148 ! 1234 ! Oban *Talbot* 210 ! 1328 ! [**Br.]Cam.:** Victoria (Nov.) *Maitland* 767 ! Johann-Albrechtshöhe *Staudt* 506 ! Abonando (Mar.) *Rudatis* 23 ! Oyomojok (July) *Johnstone* 160/31 ! **F.Po:** *Mann* 211 ! Boloko (Jan.) *Guinea* 907 ! 909 ! Also in Cameroun, Rio Muni, Gabon, Cabinda and Congo. (See Appendix, p. 408.)

8. **PORTERANDIA** Ridley in Kew Bull. 1939 : 593 (1940). *Randia* Linn. sect. *Anisophyllea* Hook. f. (1880)—K. Schum. in E. & P. Pflanzenfam. 4, 4 : 75 (1891).

Branchlets densely pubescent, later glabrescent, with a ring of villous hairs at each node exposed on the falling of the elliptic or obovate up to 3·5 cm. long stipules ; leaves obovate, cuneate at base, rounded at apex, 11–38 cm. long, 8–18 cm. broad, densely appressed-pubescent beneath and on midrib and nerves above, with 10–19 main lateral nerves on each side of midrib, venation obscure beneath ; flowers in short cymes borne pseudolaterally at alternate nodes, mostly on leafless parts of the stem ; ovary and calyx velutinous outside (rusty-brown when dry) ; calyx-tube about 5 mm. long, truncate or 5-denticulate, velutinous inside ; narrow part of corolla-tube about 5 mm. long, glabrous outside, expanded part (18–20 mm. long) and lobes (about 15 mm. long and up to 8 mm. broad) silky-pilose outside, corolla glabrous within except at base ; anthers included, 7–7·5 mm. long ; style included ; ovary 2-celled ; fruits ellipsoid, about 1·5 cm. long and 1 cm. broad, with 10 faint longitudinal lines crowned by the persistent calyx ..   ..   *cladantha*

P. cladantha (*K. Schum.*) *Keay* in Bull. Jard. Bot. Brux. 28 : 24 (1958).   *Randia cladantha* K. Schum. in Engl. Bot. Jahrb. 28 : 62 (1899) ; F.W.T.A., ed. 1, 2 : 79 ; Kennedy For. Fl. S. Nig. 212. A tree to 50 ft. high ; flowers creamy-white or pink, yellowish inside ; ripe fruits yellow or orange ; in the forest regions, especially in regrowth vegetation.

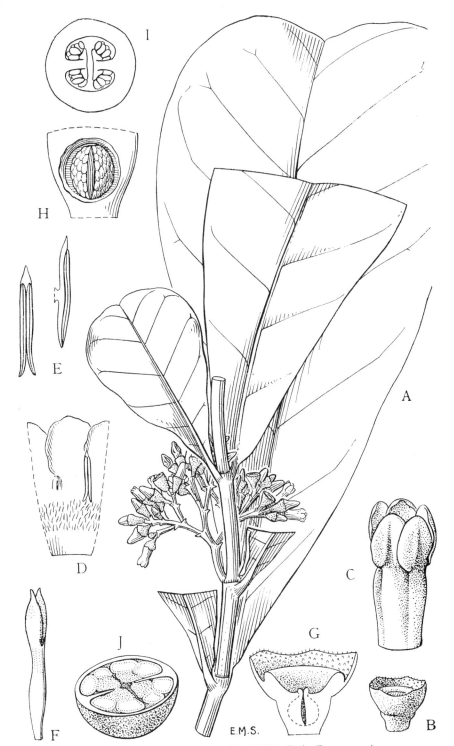

FIG. 223.—BRENANIA BRIEYI (*De Wild.*) *Petit* (RUBIACEAE).

A, flowering shoot, × ⅔.   B, calyx, × 3.   C, corolla, × 3.   D, section of corolla, × 3.
    E, stamen, × 6.   F, style, × 3.   G, section through calyx, receptacle and ovary, × 6.
    H, dissected ovary, × 8.   I, transverse section of ovary, × 8.   J, half a fruit, × ⅔.
    A drawn from *Mildbr.* 8148.   B–I from *Mildbr.* 7720.   J from *Brenan* 9053.

115

S.Nig.: Agege *Foster* 223 ! Okomu F.R., Benin Div. (fr. Dec.) *Brenan* 8482 ! Sapoba, Benin Div. (fl. May-Aug., fr. Apr.) *Kennedy* 353 ! 1391 ! Onitsha (fr. Nov.) *Keay* FHI 22283 ! Ikom (Aug.) *Catterall* 35 ! Oban *Talbot* 145 ! 214 ! [Br.]**Cam.**: Mombo, Kumba Div. (fl. & young fr. May) *Olorunfemi* FHI 30555 ! Also in Cameroun, Rio Muni, Cabinda and Congo.

9. BRENANIA Keay in Bull. Jard. Bot. Brux. 28 : 26 (1958).

A glabrous tree ; young parts glutinous ; branchlets stout ; stipules united, truncate, ciliate on the margin, persistent ; petioles up to 5 mm. long ; leaves obovate-oblanceolate, subcordate to cuneate at base, rounded at apex and very shortly and bluntly acuminate, 30–62 cm. long, 12–25 cm. broad, with 13–18 main lateral nerves on each side of midrib ; inflorescences about 5 cm. long, cymose, borne laterally opposite a single leaf at alternate nodes, axis becoming woody in fruit ; flowers glabrous except for inside of corolla-tube ; ovary about 5 mm. long ; calyx-tube less than 3 mm. long, truncate or obscurely 5-toothed ; corolla-tube 6–8 mm. long, 4 mm. diam. at throat, lobes about 5 mm. long and 4–5 mm. broad ; anthers about 4·5 mm. long, inserted near the top of the tube ; style about 6 mm. long, stigma 2-lobed, acuminate, longitudinally grooved, about 7 mm. long ; fruits ellipsoid-subglobose, 3–6 cm. diam., glabrous, with a hard woody shell ..   ..   *brieyi*

B. brieyi (*De Wild.*) *Petit* in Bull. Jard. Bot. Brux. 31 : 5 (1961). *Anthocleista brieyi* De Wild. Miss. de Briey 210 (1920). *Randia spathulifolia* Good in J. Bot. 64, Suppl. Gamopet. 13 (1926). *R. walkeri* Pellegr. in Rev. Bot. Appliq. 18 : 499 (1938). *Brenania spathulifolia* (Good) Keay l.c. t. 1 (1958). *Gardenia voacangoides* Mildbr. (1922), name only. A forest tree to 90 ft. high, with cylindrical stem up to 8 ft. girth, and dense evergreen crown ; corolla yellow with purple-red margins inside.
S.Nig.: Okomu F.R., Benin Div. (fr. Feb., Mar.) *A. F. Ross* 190 ! *Brenan* 9053 ! Akpa Ntong, N. of Odoro Nkot, Eket Div. (May) *Onochie* FHI 33163 ! [Br.]**Cam.**: Mbo F.R. (fr. May) *Tiku* FHI 24744 ! Also in Cameroun, Gabon, Cabinda and Congo.

10. SCHUMANNIOPHYTON Harms in E. & P. Pflanzenfam. Nachtr. 1 : 313 (1897). *Tetrastigma* K. Schum. (1896), not of Planch. (1887). *Assidora* A. Chev. (1948).

Leaves sessile with lamina cuneate to base, elongate-obovate, 60–120 cm. long, 30–45 cm. broad, with about 30 pairs of lateral nerves, minutely pubescent on the nerves beneath ; flowers sessile in a cluster ; corolla-tube 6–7 cm. long, densely tomentellous, lobes 7–10 narrowly lanceolate, 2 cm. long ; ovary 3–4-celled   .. 1. *magnificum*
Leaves with a thick petiole about 2 cm. long, lamina rounded to cuneate at base, 30–40 cm. long, 18–36 cm. broad, with 12–18 pairs of lateral nerves, nerves more or less pubescent beneath ; flowers sessile clustered in a spathe ; corolla-tube 5–6 cm. long, densely puberulous, lobes 8–10, 1 cm. long ; ovary (3–)6-celled ; fruits ovoid, about 5·5 cm. long   ..   .. 2. *problematicum*

1. S. magnificum (*K. Schum.*) *Harms* l.c. (1897). *Tetrastigma magnificum* K. Schum. in Engl. Bot. Jahrb. 23 : 445 (1896). *Randia immanifolia* Wernham (1913). *Schumanniophyton klaineanum* Pierre ex A. Chev. (1917). A shrub or small tree, 12–16 ft. high, with soft-wooded stems and very large leaves ; flowers white or yellow, in a dense cluster subtended by broad bracts and borne at ends of shoots opposite a single leaf and just above a pair of leaves ; in forest.
S.Nig.: Idumuje (Dec.) *Thomas* 2104 ! Degema *Talbot* 3681 ! Oban *Talbot* 189 ! Boje, Ikom Div. (May) *Latilo & Okeke* FHI 31807 ! Calabar to Mamfe (Jan.) *Onochie & Okafor* FHI 35999 ! Also in Cameroun, Gabon and Cabinda. (See Appendix, p. 413.)
2. S. problematicum (*A. Chev.*) *Aubrév.* Fl. For. C. Iv., ed. 2, 3 : 255, t. 361, 4–7 (1959). *Assidora problematica* A. Chev. in Compt. Rend. Acad. Sci. Paris 228 : 1115 (1948). A small tree 20–40 ft. high with large deciduous leaves grouped in threes at the ends of the branches ; flower yellowish-white, fragrant ; in forest.
S.L.: Bagbe, Gola Forest (fr. Mar.) *Small* 679 ! Kpuabu, Gaura (fr. Apr.) *Pyne* 169 ! **Iv.C.**: Kassa (fl. July, fr. June) *Aké Assi* 1178 ! 5408 ! **Ghana**: Ateiku (fr. Nov.) *Vigne* FH 1506 !

11. MACROSPHYRA Hook. f. in Benth. & Hook. f. Gen. Pl. 2 : 86 (1873) ; F.T.A. 3 : 105 ; K. Schum. in E. & P. Pflanzenfam. 4, 4 : 77 (1891).

Flowers white, cream or yellowish ; corolla-tube 20–50 mm. long, lobes 10–15 mm. long ; flowers several to many in a terminal sessile cluster ; ovary and calyx densely pubescent, tube 1–2 mm. long, lobes 3–8 mm. long ; branchlets and leaves densely pubescent, almost tomentose at first ; leaves obovate to ovate-lanceolate, cuneate to subcordate at base, acuminate, 9–19 cm. long, 5·5–11 cm. broad, main lateral nerves 5–8 on each side of midrib ; petioles 1·5–4·5 cm. long ; fruits subglobose, about 4·5 cm. diam.   ..   .. 1. *longistyla*
Flowers red or pink (with yellow markings) ; corolla-tube less than 20 mm. long, lobes 45–66 mm. long ; flowers 1–3 in a terminal cluster ; leaves up to 8 cm. long and 4·5 cm. broad ; petioles up to 2 cm. long ; fruits not known ; otherwise similar to the above
2. *brachysiphon*

1. M. longistyla (*DC.*) *Hiern* in F.T.A. 3 : 106 (1877) ; Chev. Bot. 318 ; Aubrév. Fl. For. Soud.-Guin. 458, t. 98, 3–6. *Randia longistyla* DC. (1830). A scandent shrub ; corolla-tube and exposed parts of lobes in bud cream or yellow, inner part of lobes white or cream ; flowers with a strong, sweet scent.
Sen.: *Boivin* ! Cayor *Leprieur* ! R. Salum *Brunner* 110 ! 133 ! Thiès de *Wailly* 4622. **Gam.**: Albreda *Leprieur* ; *Perrottet*. Koto (May) *Fox* 108 ! Kerewan (May) *Rosevear* ! **Mali**: Birgo *Dubois* 184. **Port.G.**: Bissau Dist. (Mar.) *Esp. Santo* 1880 ! Bubaque (May) *Esp. Santo* 2017 ! **Guin.**: Kouroussa *Pobéguin* 221 ! **S.L.**: *Don* ! R. Bagroo (Apr.) *Mann* 821 ! Waterloo (Mar.) *Kirk* ! Mando (Apr.) *Jordan* 213 ! Helewa, Dase (Apr.) *Deighton* 6046 ! **Lib.**: Ganta (Mar.) *Harley* 412 ! **Iv.C.**: Tafiré *Aubrév.* 1420. Ferkéssédougou *Aubrév.* 1532. **Ghana**: Kumasi (Mar.) *Vigne* FH1641 (partly) ! Birrifwa (Mar.) *Kitson* 807 ! Burufo, Lawra (Apr.) *Adams* 4016 ! Kpong (Apr.) *A. S. Thomas* M11 ! Kpandu (fr. June) *Andoh* FH 5284 ! **Togo**

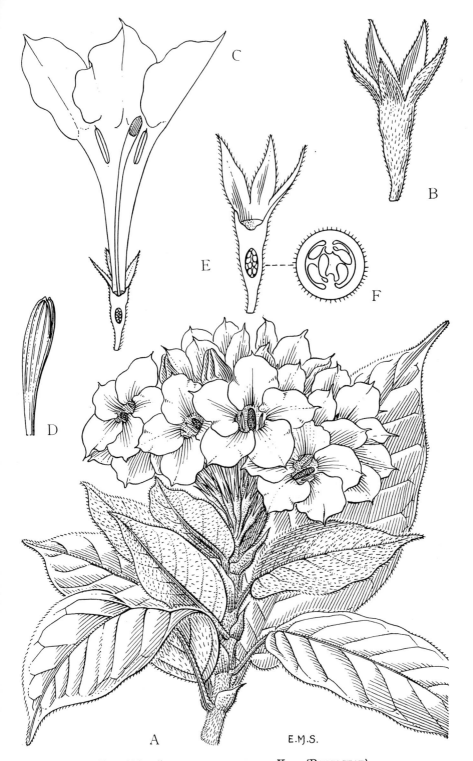

FIG. 224.—CALOCHONE ACUMINATA *Keay* (RUBIACEAE).

A, flowering shoot, × ⅔. B, calyx, × 2. C, longitudinal section of flower, × 1. D, stigma, × 2. E, longitudinal section of calyx and ovary. F, transverse section of ovary (diagrammatic). Drawn from *Keay* FHI 28554.

E.M.S

Fig. 225.—POLYCORYNE FERNANDENSIS (Hiern) Keay (RUBIACEAE).

A, flowering shoot, × ⅓. B, flower bud, × 2. C, ovary and section of calyx, × 6. D, longitudinal section of flower, × 2. E, stigma, × 6. F, transverse section of ovary (diagrammatic) × 6 ....

118

Rep.: *Baumann* 451! **Dah.**: *Burton*! Zagnanado *Chev.* 23068. Abomey *Chev.* 23147. **N.Nig.**: Zurmi, Sokoto Prov. (Apr.) *Keay* FHI 16128! Lokoja (Apr.) *Elliott* 49! Vom (Feb.) *McClintock* 229! Naraguta *Lely* 83! **S.Nig.**: Ebute Metta (Jan.) *Millen* 91! Ado Rock *Rowland*! Abeokuta *Irving* 166! Agulu *Thomas* 169! [Br.]**Cam.**: Fonfaka, Bamenda (May) *Maitland* 1700! Extends to Sudan and Uganda. (See Appendix, p. 401.)

2. **M. brachysiphon** *Wernham* in J. Bot. 55 : 79 (1917). A scandent shrub ; corolla " azalea-pink ", or " brick-red, and yellow near the tube " ; in forest.

S.Nig.: *Maitland* 16! Ogba R., Benin (Dec.) *Farquhar* 45! Sapoba (Dec.) *Collier* FHI 4667! Abua-Owere-were Road, Degema *Talbot* 3764!

12. CALOCHONE Keay in Bull. Jard. Bot. Brux. 28 : 28 (1958).

Branchlets, petioles and leaves densely hirsute ; inflorescence densely many-flowered ; ovary and calyx densely pubescent, tube 1 mm. long, lobes lanceolate, about 8 mm. long ; corolla-tube pubescent outside, gradually expanded above, about 40 mm. long, lobes about 30 mm. long and 25 mm. broad, acuminate ; anthers 10·5–12 mm. long ; leaves oblong-ovate to oblong-obovate, cordate at base, gradually long-acuminate, 12–22 cm. long, 5·5–8 cm. broad, with 8–13 main lateral nerves on each side of midrib ; petioles up to 2·8 cm. long   ..   ..   *acuminata*

C. acuminata *Keay* l.c. 30, t. 2 (1958). A shrub, to 10 ft. high ; midrib and nerves red beneath ; calyx green ; corolla bright orange-pink ; in secondary forest.

[Br.]**Cam.**: 45 miles from Kumba, on Mamfe Road (Jan.) *Keay* FHI 28554! Wone *Keay* FHI 37356! Also in Cabinda.

13. PREUSSIODORA Keay in Bull. Jard. Bot. Brux. 28 : 31 (1958).

A glabrous shrub ; stipules ovate to lanceolate, long-acuminate, united at base, persistent, 7–9 mm. long ; petioles 1·5–5·5 cm. long ; leaves membranous, oblong-oblanceolate or elongate-obovate, cuneate at base, acuminate, 18–30 cm. long, 6·5–13 cm. broad, with 11–13 main lateral nerves on each side of midrib ; inflorescences terminal, short, about 6-flowered ; ovary glabrous, about 3 mm. long ; calyx mostly glabrous, tube about 3 mm. long, lobes oblanceolate, ciliate at margins, up to 15 mm. long and 7 mm. broad ; corolla-tube glabrous outside, slightly pubescent inside, 12–20 mm. long, lobes ovate, 12–17 mm. long and broad, spreading ciliate at margins ; anthers included in corolla-tube or slightly exserted ; style-head linear-ellipsoid, longitudinally grooved, entire ; fruits ellipsoid about 2·5 cm. long and 2 cm. broad, crowned by the persistent calyx   ..   ..   *sulphurea*

P. sulphurea (*K. Schum.*) *Keay* l.c. 32 (1958). *Randia sulphurea* K. Schum. in Engl. Bot. Jahrb. 23 : 441 (1896) ; F.W.T.A., ed. 1, 2 : 79. *R. exserta* K. Schum. in Engl. Bot. Jahrb. 33 : 343 (1903). A forest shrub, to 6 ft. high with sulphur-yellow or yellowish-white flowers.

S.Nig.: Degema *Talbot*! Ndiumo to Orora Road (Nov.) *Rosevear* C32! Oban *Talbot* 1544! [Br.]**Cam.**: Rio del Rey *Johnston*! Victoria (fl. Feb.-May, July, fr. Dec.) *Preuss* 1296 ; 1308 ; 1327. *Maitland* 790! Tiko (Feb.) *Dunlap* 254! Mbalange to Bombe Ndifo, Kumba Div. (Mar.) *Onochie* FHI 30881! Kumba *Preuss* 34. Mokundange to Ngeme (Mar.) *Preuss* 1386! **F.Po**: *Mann* 288!

14. POLYCORYNE Keay in Bull. Jard. Bot. Brux. 28 : 32 (1958).

Branchlets and leaves glabrous ; leaves broadly oblong-elliptic, cordate, rounded, obtuse or cuneate at base, shortly acuminate, 4·5–25 cm. long, 3–14 cm. broad, with 5–6 main lateral nerves on each side of midrib ; petioles 1–4·8 cm. long ; cymes many-flowered, borne terminally on lateral branches, with 1–5 pairs of new leaves and large stipules ; calyx-tube 0·5–1 mm. long, with ovate-lanceolate, ciliate lobes about 2 mm. long ; corolla glabrous outside and inside, tube about 30 mm. long, with spreading oblong lobes about 10 mm. long and 6 mm. broad ; anthers 4–5 mm. long, partially exserted ; style-head obovate-spathulate ; fruits globose, about 3 cm. diam.   *fernandensis*

P. fernandensis (*Hiern*) *Keay* l.c. 34, t. 3 (1956). *Gardenia fernandensis* Hiern in F.T.A. 3 : 105 (1877) ; F.W.T.A., ed. 1, 2 : 75. *Randia psychotrioides* K. Schum. (1896). *R. chloroleuca* K. Schum. (1899). *Gardenia ramentacea* K. Schum., MS. name only. *R. exserta* K. Schum. in ... — A scrambling shrub or liane ; flowers fragrant, white or with green corolla-tube and white lobes ; in forest.

S.L.: *Litchford* in Hb. *Lane-Poole* 343! Freetown (Feb.) *Deighton* 3991 Bumban, N. Prov. (Feb.) *Glanville* 163! Ninia, Talla Hills (Feb.) *Sc. Elliot* 4914! Falaba (Mar.) *Sc. Elliot* 5184! **Lib.**: Kakatown *Whyte*! **Iv.C.**: Abouabou (fr. Feb.) *Leeuwenberg* 2689! Adiopodoumé (Nov.) *Roberty* 15490! **Ghana**: Bunsu to Koforidua (Jan.) *Morton* GC 8336! Kumasi (Feb., Apr.) *Darko* 523! *Vigne* FH 1052! 1116! Worawora, Kabo River F.R. (Dec.) *St. C. Thompson* FH 3705! **N.Nig.**: Loko, R. Benue (Nov.) *Dalz.* 44! **S.Nig.**: Degema *Talbot*! Eket Dist. *Talbot* 3030! Oban *Talbot* 2041! Ikom (Dec.) *Keay* FHI 28159! [Br.]**Cam.**: Johann-Albrechtshöhe (=Kumba) (Nov.) *Staudt* 445! **F.Po**: (Dec.) *Mann* 22! Also in Cameroun, Gabon, Cabinda and Congo.

15. OLIGOCODON Keay in Bull. Jard. Bot. Brux. 28 : 36 (1958).

Branchlets and leaves glabrous ; leaves obovate-elliptic, cuneate at base, rounded at apex, 6–8·5 cm. long, 3–4 cm. broad, with 4–5 main lateral nerves on each side of midrib ; petioles 1–3 cm. long ; cymes about 3–6-flowered, short, terminating short lateral branchlets ; calyx-tube subtruncate, scarcely 1 mm. long ; corolla glabrous except for a band of hairs inside near the base, bell-shaped, with a narrow tube 5 mm. long, an abruptly expanded part about 35 mm. long and 30 mm. diam., and rounded lobes about 18 mm. long and 12–15 mm. broad ; anthers 18 mm. long, included or

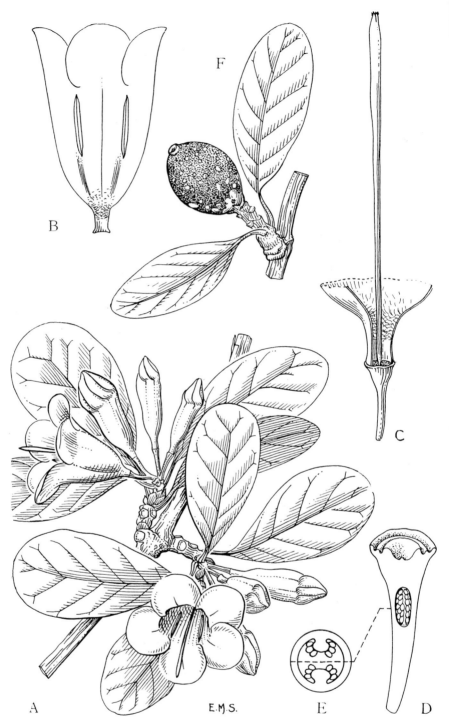

FIG. 226.—OLIGOCODON CUNLIFFEAE (*Wernham*) *Keay* (RUBIACEAE).

A, flowering shoot, × ⅔.   B, section of corolla,   × 1.   C, gynoecium and section of corolla,   × 2.
D, longitudinal section of ovary,   × 4.   E, transverse section of ovary (diagrammatic).
F, fruit, × ⅔.   A–E drawn from *Talbot* 3149.   F from *Onochie* FHI 26675.

only slightly exserted ; style-head linear-fusiform ; fruits ellipsoid, up to 3·5 cm.
long and 2·5 cm. diam. .. .. *cunliffeae*

O. **cunliffeae** (*Wernham*) *Keay* l.c. 36, t. 4 (1958). *Gardenia cunliffeae* Wernham in J. Bot. 52 : 5 (1914) ;
F.W.T.A., ed. 1, 2 : 75. A glabrous climbing shrub with rough splitting bark ; corolla-tube very pale-
green externally, dark green inside towards the base, lobes white above, with vivid purple splashes ;
flowers fragrant, turning black in drying as do the leaves ; fruits vivid orange ; in forest vegetation.
S.Nig.: Lagos (fl. Feb., Dec., fr. Dec.) *Hagerup* 754 ! *Millen* 144 ! *Onochie* FHI 26675 ! Ubium R., Eket
Dist. *Talbot* 3149 ! Akampa, Calabar (fr. Mar.) *Latilo* FHI 41312 ! [**Br.**]**Cam.**: R. Meme, Kumba *Keay*
FHI 37379 !

16. XEROMPHIS Raf. Sylva Tellur. 21 (1838). *Lachnosiphonium* Hochst. (1842).

Branches whitish, pubescent, armed with sharp spines 1–3 cm. long ; spines usually
solitary and alternate, but sometimes in unequal pairs ; leaves and flowers on much
abbreviated lateral branchlets borne in pairs just below the spines ; leaves obovate,
cuneate at base, rounded at apex, 2–7 cm. long, 1–3 cm. broad, shortly pubescent
especially beneath, with 3–4 main lateral nerves on each side of midrib ; flowers
solitary or paired, terminating the lateral branchlets ; pedicels 5–10 mm. long ; calyx
about 7 mm. long, shortly toothed, glabrous outside ; corolla about twice as long as
the calyx, villous outside ; fruits ellipsoid, 1·8–2 cm. long, 1–1·5 cm. broad, glabrous,
2-celled .. .. *nilotica*

X. **nilotica** (*Stapf*) *Keay* in Bull. Jard. Bot. Brux. 28 : 39 (1958). *Randia nilotica* Stapf (1906)—F.W.T.A.,
ed. 1, 2 : 78 ; Aubrév. Fl. For. Soud.-Guin. 462, t. 100, 5–6. *Lachnosiphonium niloticum* (Stapf) Dandy
(1952). *Randia dumetorum* of F.T.A. 3 : 94, not of (Retz.) Lam. A stiff, spiny shrub, often many-stemmed,
with whitish bark ; corolla white ; fruits yellowish ; in arid types of savanna woodland.
Guin.: Gaya *Aubrév.* 19n. **N.Nig.**: Sokoto (fl. & fr. May, July) *Lely* 824 ! *Ryan* 39 ! 68 ! *Keay* FHI 37787 !
Katagum *Dalz.* 298 ! Sherifuri (fr. May) *Thornewill* 182 ! Lantewa, Bornu Prov. (June) *Onochie* FHI 23369 !
Extends to Sudan, Ethiopia and E. Africa.

17. EUCLINIA Salisb. Parad. Lond. index sexualis & errata sub. t. 93 (1808).

Stipules chaffy, persistent, 5–16 mm. long ; leaves crowded at the ends of almost leafless branchlets, deciduous, membranous, oblong or oblong-obovate, cuneate at base,
acuminate, 7–28 cm. long, 3·5–10 cm. broad, with 8–10 main lateral nerves on each
side of midrib, pubescent beneath on midrib, nerves and in axils of nerves ; petioles
up to 4 cm. long ; flowers 5-merous ; ovary and calyx glabrous, tube up to 1·5 mm.
long only, lobes triangular-lanceolate, spreading, 9–16 mm. long ; corolla-tube
16–24 cm. long, glabrous outside, pubescent inside, lobes 1·5–5 cm. long and 1·5–3·5 cm.
broad ; fruits ellipsoid-globose, 2·6–4 cm. diam., with 15–16 faint longitudinal lines,
crowned with the persistent, spreading calyx-lobes .. .. .. *longiflora*

E. **longiflora** *Salisb.* l.c. (1808) ; Aubrév. Fl. For. C. IV., ed. 2, 3 : 280. *Randia longiflora* Salisb. l.c. t. 93
(1808), not of Lam. (1789). *Gardenia longiflora* (Salisb.) Dryander (1810), not of Ruiz & Pav. (1799).
G. macrantha Schultes (1819). *Randia macrantha* (Schultes) DC. (1830)—F.T.A. 3 : 97 ; Chev. Bot. 320 ;
F.W.T.A., ed. 1, 2 : 78. *Rothmannia macrantha* (Schultes) Robyns Fl. Sperm. Parc Nat. Alb. 2 : 341
(1947) ; F.W. Andr. Fl. Pl. A.-E. Sud. 2 : 462. *Gardenia ? longifolia* G. Don (1834). *Randia bowieana*
Hook. (1835). *Rothmannia bowieana* (Hook.) Benth. (1849). *Gardenia devoniana* Lindl. (1846). A shrub
or small tree, to 20 ft. high, deciduous, with membranous leaves and conspicuous, persistent stipules ;
flowers fragrant, corolla white, turning to cream, with red markings between the lobes ; in forest and
forest regrowth vegetation.
Port.G.: Catió (fl. Aug., fr. May) *Esp. Santo* 2061 ! 2159 ! **Guin.**: Kissidougou (Apr.) *Martine* 295 ! Nimba
Mts. (fr. Oct.) *Schnell* 3775 ! **S.L.**: *Don* ! *Smeathmann* ! Regent (fr. Dec.) *Sc. Elliot* 4137 ! Njala (fl. May,
Dec., fr. Nov.) *Deighton* 671 ! 2587 ! 3099 ! Blama (Dec.) *Deighton* 314 ! Kurusu, Limba (Apr.) *Sc. Elliot*
5511 ! Gola Forest (Apr.) *Small* 608 ! **Lib.**: Gbanga (Sept.) *Linder* 643 ! **Iv.C.**: various localities *fide*
Chev. l.c. **Ghana**: Kibi (Aug.) *Miles* 541 ! Banka, Ashanti (Mar.) *Vigne* FH 1869 ! Tukobo, Axim (Nov.)
Vigne FH 1446 ! Arikasa F.R. (Dec.) *Vigne* FH 3178 ! New Janbin (Mar.) *Johnson* 694 ! **S.Nig.**: Lagos
Millson ! *Rowland* ! Epe *Barter* 3291 ! Sapoba *Kennedy* 2213 ! Oban *Talbot* 1672 ! [**Br.**]**Cam.**: Likomba
(Oct.) *Mildbr.* 10568 ! Bamenda (young fr. May) *Ujor* FHI 30357 ! Also in Cameroun, Sudan, Uganda,
Congo and Angola. (See Appendix, p. 409.)

18. GARDENIA Ellis—F.T.A. 3 : 99. *Nom. cons.*

Corolla-tube 1–1·2 cm. long, campanulate, with 5 lobes about 5 mm. long and broad ;
calyx-tube 2–3 mm. long, with 5(–6) linear lobes, 7–8 mm. long ; placentas 2, often
joining to make the ovary spuriously 2-celled ; flowers puberulous outside, borne in
terminal pairs (rarely solitary) usually appearing at the nodes on account of sympodial
growth ; pedicels slender, 3–12 mm. long in flower, up to 18 mm. long in fruit ; fruits
ellipsoid-subglobose, about 10 mm. long and 7 mm. diam., with a thin shell ; leaves
oblong-elliptic, obtuse or sometimes cuneate at base, subacute to rounded at apex,
2–10 cm. long, 1–5 cm. broad, with (6–)9–13 main lateral nerves on each side of
midrib, puberulous beneath .. .. 1. *sokotensis*
Corolla-tube 1·2–22·5 cm. long, relatively slender and gradually expanded, with 5–9
(–11) lobes ; placentas (2–)3–9 ; flowers sessile or on short stout pedicels ; fruits at
least 2 cm. long, with thick walls :
Corolla-tube scabrid-puberulous outside, 11–22·5 cm. long, much widened towards the
3–4 cm.-wide top, with 5 lobes 2·5–5 cm. long ; flowers in pairs, terminal but often
appearing at the nodes on account of sympodial growth ; ovary and calyx-tube each
about 10 mm. long, calyx-lobes 12–15 mm. long ; placentas 2–3 ; fruits ellipsoid,

4·5–5 cm. long, 3·5–5 cm. broad ; leaves more or less sessile, obovate, with small pouches at the base, shortly acuminate, 20–50(–80) cm. long, 10–22·5(–40) cm. broad, with 13–29 main lateral nerves on each side of midrib, venation parallel, scabrid puberulous beneath ; tree to 20 m. high   ..   ..   *2. imperialis*

Corolla-tube 1·2–15 cm. long, less than 1 cm. wide at top, with 5–9 lobes ; flowers usually solitary (but see Nos. 3 & 4), terminal (but see No. 3) ; placentas 3–6(–9) ; leaves up to 30 cm. long and 11 cm. broad ; shrubs or small trees to 5 m. high :

Calyx-lobes broadly spathulate or obovate, 12–25 mm. long, 3–13 cm. broad, tube 8–11 mm. long ; flowers solitary or rarely paired, terminal but often appearing laterally at the nodes by sympodial growth ; corolla-tube 7·5–10·5 cm., lobes 2·5–4·5 cm. long ; placentas 4 ; fruits clavate, woody, 5–8 cm. long, 1·5–2·5 cm. broad ; leaves elliptic, cuneate at base, acuminate, 3·3–14·5 cm. long, 1·3–7 cm. broad, with 6–12 main lateral nerves on each side of midrib, glabrous   *3. nitida*

Calyx-lobes filiform to linear-oblong, 1–15 mm. long ; flowers terminal :

Corolla-tube 12–15 cm. long, with 5(–6) lobes 3–8·5 cm. long ; calyx-tube 15–18 mm. long, very sparsely setose outside, lobes 1–15 mm. long ; flowers solitary or some-times 3 together, the 2 lateral ones borne on very short lateral shoots and sub-tended by short broad bract-like leaves ; placentas 3 ; fruits elongate, up to 11 cm. long and 1·3 cm. diam. ; leaves oblong-oblanceolate, cuneate at base, acuminate, margin sometimes lobulate, 8·5–30 cm. long, 3–10·5 cm. broad, with 10–12 main lateral nerves on each side of midrib, glabrous ; a forest shrub or small tree, with relatively slender branchlets   ..   ..   *4. vogelii*

Corolla-tube 1·2–9 cm. long with 6–9 lobes ; flowers solitary, terminal ; placentas 6(–9) ; savanna shrubs or small trees with stout branchlets ; leaves and branching often ternate :

Plants puberulous, pubescent or tomentose at least on the branchlets and outside of calyx :

Leaves pubescent on both surfaces ; fruits fibrous or woody :

Corolla-tube 1·2–3(–4) cm. long, pubescent outside, lobes 0·9–2 cm. long ; calyx-lobes subulate, 1·5–3 mm. long ; leaves oblanceolate or obovate, rather densely pubescent especially beneath, nerves and veins very prominent and more or less impressed above at maturity, 2–6 cm. long, 1–2 cm. broad on flowering shoots ; fruits ellipsoid, ribbed, 2–3(–4) cm. long, 1·2–1·6(–2) cm. broad   ..   ..   *5. aqualla*

Corolla-tube 4–6(–9) cm. long, glabrescent outside, lobes 3–5 cm. long ; calyx lobes linear-oblong, 3–15 mm. long ; leaves obovate, scabrid-puberulous, nerves and veins prominent on both surfaces, 3·5–12 cm. long, 1·2–5·5 cm. broad ; fruits ellipsoid, ribbed when young, 4–8 cm. long, 1·8–3·5 cm. broad   *6. triacantha*

Leaves glabrous, except occasionally for a few hairs near the very short petiole and rarely in the axils of the nerves beneath :

Calyx-tube and branchlets densely tomentose ; leaves drying with the upper surface reddish and the lower surface glaucous green with reddish nerves, broadly obovate, up to 21 cm. long and 11 cm. broad ; calyx-lobes subulate. ; corolla-tube 2–8 cm. long ; fruits ellipsoid or subglobose, up to 8 cm. long, rather fleshy, edible   ..   ..   *7. erubescens*

Calyx-tube puberulous, branchlets pubescent ; leaves drying uniformly greenish, obovate to oblanceolate, up to 20 cm. long and 8 cm. broad ; calyx-lobes linear-oblong, often very short ; corolla-tube 2·5–6(–9) cm. long ; fruits ellipsoid, up to 6 cm. long, fibrous or woody   ..   ..   ..   *8. lutea*

Plants quite glabrous (except for the style) ; calyx subtruncate or with 1–10 mm. long linear-oblong lobes ; corolla-tube 4–9 cm., lobes 2–4 cm. long ; fruits more or less ellipsoid, up to 10 cm. long, fibrous or woody ; leaves obovate, to 19 cm. long and 7 cm. broad, prominently reticulate with more or less parallel venation on both surfaces   ..   ..   *9. ternifolia*

1. **G. sokotensis** *Hutch.* in Kew Bull. 1912 : 99 ; in Hook. Ic. Pl. t. 2991 (1913) ; Aubrév. Fl. For. Soud.-Guin. 460–462, t. 99, 10. *G. mossica* A. Chev. Bot. 323, name only. *Randia lucida* A. Chev. Bot. 320, name only. *R. mossica* A. Chev. Bot. 321, name only. A shrub, or small tree to 8 ft. high, with tough branchlets, glutinous on young parts ; flowers white ; on dry rocky hills in the drier savanna regions.
Sen.: Ouassadou *Berhaut* 1457. **Mali**: Kangala (May) *Chev.* 829. Fô (June) *Chev.* 965. Koulikoro *Chev.* 2118. Bamako *Waterlot* 1138. **Port.G.**: Madina, Boé (fr. Jan.) *Esp. Santo* 2368 ! **Guin.**: Kankan to Kouroussa *Chev.* 15640. **Iv.C.**: Yako to Gourpé, Mossi (fl. & fr. Aug.) *Chev.* 24733 ! Banfara *Serv. For.* 286. Kaya *Serv. For.* 2211. Black Volta *Serv. For.* 2601. **Ghana**: Gradaw, Banda Hills (Oct.) *Rose Innes* GC 31645 ! Gambaga Dist. (fl. & young fr. June) *Akpabla* 703 ! Burufo, N.T. (fl. Apr., fr. Dec.) *Adams* GC 4018 ! 4405 ! **N.Nig.**: Sokoto (fl. & young fr. July) *Dalz.* 402 ! Zurmi (fr. Apr.) *Keay* FHI 16181 ! Rimi Dist., Katsina (Aug.) *Onwudinjoh* FHI 24006 ! Bichi Dist., Kano *Keay* FHI 211421 ! (See Appendix, p. 399.)

2. **G. imperialis** *K. Schum.* in Engl. Bot. Jahrb. 23 : 442 (1896) ; Aubrév. Fl. For. C. Iv., ed. 2, 3 : 274, t. 351 ; Kennedy For. Fl. S. Nig. 210. *Randia physophylla* K. Schum. (1899). *Gardenia physophylla* (K. Schum.) De Wild. (1904). *G. viscidissima* S. Moore (1905)—Chev. Bot. 324. A tree, to 60 ft. high, with stout branchlets and large shining leaves, viscid on the younger parts ; flowers white ; in swamp forest and by streams.
Sen.: Badi *Berhaut* 1489. Bignona to Massasoun *Chev.* 2120. **Mali**: Nono *Chev.* 431. Siguiri (May) *Chev.* 887 ! **Port.G.**: Pecixe Isl., Canchungo *Esp. Santo* 2046 ! **Guin.**: Mamou (fr. Mar.) *Dalz.* 8405 ! Kollangui

Chev. 13501. **S.L.:** Falaba (fr. Mar.) *Sc. Elliot* 5141! Jama to Yamadu (July) *Aylmer* 91! Kyema (Apr.) *Aylmer* 92! Samaia *Thomas* 241! **Iv.C.:** Dabou (fr. Feb.) *Chev.* 16210! Bondoukou, Tiengara, Port Bouët & Ferkéssédougou *fide* Aubrév. *l.c.* **Ghana:** *Burton & Cameron*! Essiama (fr. Feb.) *Irvine* 2323! Neung F.R., Tarkwa Dist. (fr. June) *Enti* FH 6243! **N.Nig.:** Jos Plateau (Jan.) *Lely* P117! Vom (fl. & fr. Feb.) *McClintock* 205! **S.Nig.:** Eba Isl. (Aug.) *Sankey*! Sapoba (July) *Kennedy* 1409! Mamu River F.R., Awka (Sept.) *Onochie* FHI 34070! Degema *Talbot* 3616! Oban *Talbot* 175! 278! Also in Cameroun, Congo, Uganda, Tanganyika, Nyasaland, N. Rhodesia and Angola. (See Appendix, p. 398.)

3. **G. nitida** Hook. Bot. Mag. t. 4343 (1847); F.T.A. 3 : 102. *G. assimilis* Afzel. ex Hiern in F.T.A. 3 : 102 (1877). *G. fragrantissima* Hutch. in Kew Bull. 1916 : 39, & fig. *G. lane-poolei* Hutch. ex Lane-Poole (1916), name only. A glabrous shrub or small tree, to 12 ft. high; flowers white, turning yellow, fragrant, lasting only a day; branchlets and fruits ashen-grey; in drier types of forest; sometimes planted in villages.

Port.G.: Catió *Esp. Santo* 3004. Cantanhez (fr. Apr.) *d'Orey* 380! **S.L.:** *Afzelius*! *Whitfield*. *Wilford*! Sugar Loaf Mt. (fl. & fr. May) *Tindall* 66! Kessewe (Apr.) *Lane-Poole* 127! Kambia, Scarcies R. (fl. Nov., fr. Jan.) *Jordan* 374! *Sc. Elliot* 4410! Makump (fr. July) *Thomas* 934! **Ghana:** Cape Coast *Brass*! Nsawam (Jan.) *Brown* 132! Atrokpe, Krobo (fl. & fr. Nov.) *Irvine* 1702! Ofuman, Wenchi Dist. (fr. Jan.) *Vigne* FH 3558! Afram Plains (May) *Kitson* 1134! Dawa Matokole (fl. & fr. Jan.) *A. S. Thomas* D67! **S.Nig.:** Foster 190! Olokemeji (fr. Feb.) *Ejiofor* FHI 26899! Ibadan South F.R. (fl. Apr., fr. Apr., Aug.) *Keay* FHI 22549! 25370! Gambari F.R. (fl. & fr. Mar.) *Latilo* FHI 34982! (See Appendix, p. 399.)

4. **G. vogelii** Hook. f. ex Planch. in Hook. Ic. Pl. t. 782–3 (1848); F.T.A. 3 : 103; Chev. Bot. 324. A shrub or small tree, to 15 ft. high; flowers white, fragrant; fruits often curved; in forest.

Lib.: Dukwia R. (fr. Aug.) *Bequaert* in *Hb. Linder* 297! **Iv.C.:** Bouroukrou *Chev.* 16515! 16517. Taté *Chev.* 19796. **Ghana:** Akyease (fl. & fr. June) *Fishlock* 79! Benso (fr. Sept.) *Andoh* FH 5888! **N.Nig.:** Kurmin Damisa, Jemaa Div. (fr. Nov.) *Keay & Onochie* 21728! **S.Nig.:** Okomu F.R., Benin (fl. & fr. June) *Onochie* FHI 31234! Aboh *Barter* 156! Ibo country (fl. & fr. Aug.) *T. Vogel* 58! Oban *Talbot* 1243! **[Br.]Cam.:** Bamenda (young fr. May) *Maitland* 1536! Also in Sudan, Uganda, Congo, Angola and S. Rhodesia. (See Appendix, p. 399.)

5. **G. aqualla** Stapf & Hutch. in J. Linn. Soc. 38 : 427 (1909); Chev. Bot. 322; Aubrév. Fl. For. Soud.-Guin. 460–461, t. 99, 1–2. *G. thunbergia* of F.T.A. 3 : 100, partly (*Barter* 1234). A savanna shrub, to 9 ft. high; flowers yellow, fragrant.

Sen.: *Berhaut* 130. **Mali:** Zorgongo to Ouagnan, Mossi *Chev.* 24636. **Iv.C.:** Lalérabah *Aubrév.* 1445. Bâtié *Aubrév.* 2496. **Ghana:** Sambisi to Hamboi (May) *Kitson* 670! Lawra to Lissa (fl. & fr. Mar.) *Kitson* 811! Bawku (Apr.) *Vigne* FH 3752! Kugri, N.T. (Apr.) *Vigne* FH 4706! **N.Nig.:** *Thornewill* 102! Morai, Ilorin Prov. (Mar.) *J. A. D. Jackson* FHI 27746! Sokoto *Ryan* 16! Zaria (Feb.) *Dalz.* 352! Gari Gwenchi, E. of Zaria (Apr.) *Ryan* 46! Anara F.R., Zaria Prov. (fr. Oct.) *Keay* FHI 20106! Faran, N. of Yola (fr. Dec.) *Hepper* 2896! **S.Nig.:** hills in Yoruba country *Barter* 1234! Ago-Are, Oyo *Keay* FHI 37558! Also in Ubangi-Shari and Sudan. (See Appendix, p. 398.)

6. **G. triacantha** DC. Prod. 4 : 382 (1830); Stapf & Hutch. l.c. 426; Aubrév. l.c. 460–461, t. 99, 11–12. *G. nigerica* A. Chev. Bot. 323, name only. *G. thunbergia* of F.T.A. 3 : 100, partly. *G. ternifolia* of F.W.T.A., ed. 1, 2 : 73, partly (*Talbot* 271 (partly), *Dalz.* 161). A savanna shrub, to 12 ft. high; flowers white, fragrant.

Sen.: *Roger*! Tivaouane to Mbara *de Wailly* 4625. Bakel *Carrey* 57! *Collin* 63. **Gam.:** *Leprieur*; *Mungo Park*! *Perrottet*; *Whitfield*! N. bank (July) *Ozanne* 7! **Mali:** Moriquéniéba (Mar.) *Chev.* 447 *bis*. Tiédiana (June) *Chev.* 993. Sansanding (fr. Sept.) *Chev.* 2025. Mamba, Diafarabé (Apr.) *Davey* 606! **Port.G.:** *Esp. Santo* 1883! **Guin.:** Boké *Paroisse* 27. Kadé *Pobéguin*. Ditinn (fl. & fr. Mar.) *Langdale-Brown* 2619! **S.L.:** *Glanville* 184 (partly)! *Smythe*! *Thomas*! Yana, Tambaka *Miszewski* 29a! Musaia (fl. & fr. Feb.) *Deighton* 4214! **Iv.C.:** *Aubrév.* 2592. **Ghana:** Kuluissa to Tagalo (Feb.) *Kitson* 868! **N.Nig.:** Katagum Dist. (fr. Feb.) *Dalz.* 161 (partly)! Near L. Chad (Jan., Feb.) *Talbot* 271 (partly)! Also in Chad and Sudan. (See Appendix, p. 399.)

7. **G. erubescens** Stapf & Hutch. l.c. 428 (1909); Chev. Bot. 322; Aubrév. l.c. 460–462, t. 99, 3–4. *G. triacantha* var. *parvilimbis* F. N. Williams (1907). *G. ternifolia* of F.W.T.A., ed. 1, 2 : 73, partly (*Anderson* 50, *Chipp* 493, 630). A savanna shrub, or small tree, to 10 ft. high, rarely more; flowers white, turning yellow, fragrant; fruits yellow when ripe, edible.

Sen.: Bondou *Heudelot* 181! Fouladou, Casamance *Etesse* 59! **Gam.:** *Whitfield*! Georgetown (Jan.) *Dalz.* 8044! Bulak (Apr.) *Rosevear* 52! Genieri (fl. & fr. Feb.) *Fox* 78! 79! **Mali:** Nigalia (fl. & fr. Jan.) *Chev.* 160! Bamako *Waterlot* 1139. Arbala *Dubois* 17. Koulikoro *Vuillet* 655. **Guin.:** Mamou (Mar.) *Dalz.* 8408! Timbo *Pobéguin* 121. Tombia, Samu Country (Dec.) *Sc. Elliot* 4370! Wulia, Scarcies R. (Feb.) *Sc. Elliot* 4574! **S.L.:** *Glanville* 184 (partly)! Batkanu *Thomas* 91! Yana, Tambaka (Jan.) *Miszewski* 29! **Iv.C.:** Ghighi *Aubrév.* 2422! **Ghana:** Sissu, N. Ashanti (young fr. June) *Chipp* 493! W. Abene, Kwahu (Jan.) *Chipp* 630! Yendi (Apr.) *Vigne* FH 1696! Yipala to Kulmassa (Feb.) *Kitson* 857! Tumu to Lawra (Dec.) *Adams* GC 4377! Salaga *Anderson* 50! **Dah.:** Natitingou *Aubrév.* 100d. **N.Nig.:** Kontagora (fl. & young fr. Jan.) *Dalz.* 224! Jos Plateau (Jan.) *Batten-Poole* 238! *Lely* P93! Katagum Dist. (Feb.) *Dalz.* 161 (partly)! Also in Ubangi-Shari, Uganda and Sudan. (See Appendix, p. 398.)

8. **G. lutea** Fres. in Mus. Senckenb. 2 : 167 (1837); Stapf & Hutch. l.c. 425. *G. thunbergia* of F.T.A. 3 : 100, partly. A savanna shrub, like the preceding.

[Br.]Cam.: Bamumkumbit, Bamenda (Jan.) *Johnstone* 278/32! Also in Sudan, Ethiopia and (?) E. Africa.

9. **G. ternifolia** Schum. & Thonn. Beskr. Guin. Pl. 147 (1827); Stapf & Hutch. l.c. 425; Chev. Bot. 323; Aubrév. l.c. 460–462, t. 99, 5–9. *G. medicinalis* Vahl ex Schumach. (1827). *G. thunbergia* of F.T.A. 3 : 100, partly. *G. jovis-tonantis* of F.W.T.A., ed. 1, 2 : 73, not of (Welw.) Hiern, see note below. A savanna shrub, 5–15 ft. high; flowers fragrant, white, opening at night, turning yellow next day; fruits fibrous, grey-green, long-persistent; extending into rather moister regions than the preceding four species.

Sen.: *Berhaut* 802. *Irvine* 3239! Niokolo-koba *Adam* 14178! **Mali:** Ouala *Chev.* 35. Ouagadougou *Aubrév.* 2392. **S.L.:** Kabala *Thomas* 2334! **Guin.:** Timbo *Miquel* 13. **Iv.C.:** Dotou *Chev.* 21778. Fétékro *Aubrév.* 800. Bâtié *Aubrév.* 2835. **Ghana:** Gah & Adampi *Thonning*. Gira (fr. Aug.) *Chipp* 737! Afram Plains *Kitson* 1131! Abofaw (Feb.) *Vigne* FH 1837! Jema, Ashanti *Brown* FH 2153! Navrongo (fl. & fr. Mar., Apr.) *Vigne* FH 4684! *Hughes* FH 5276! Kpedsu (fl. & fr. Jan.) *Howes* 1078 [1] **Togo Rep.:** Kenre *Kersting* A408! **Dah.:** Tanguéta, Atacora Mts. (fr. June) *Chev.* 24105. Allada *Chev.* 23230. **N.Nig.:** Nupe *Barter* 1205! *Baikie*! Zungeru (Jan.) *Elliott* 30! Jemaa (fl. & fr. Feb.) *McClintock* 194! Katagum *Dalz.* 162! Near L. Chad (Jan., Feb.) *Talbot* 271 (partly)! Kukawa (Apr.) *E. Vogel* 92! **S.Nig.:** Olokemeji (fl. & fr. Jan.) *Dodd* 401! Old Oyo F.R. (Feb.) *Keay* FHI 16274! S. Ugboha (Jan., Feb.), Ishan (Feb.) *Umana* FHI 29122! **[Br.]Cam.:** Gashaka Dist. (Jan.) *Latilo & Daramola* FHI 34437! Also in Ubangi-Shari and Sudan. (See Appendix, p. 399.)

[*G. jovis-tonantis* (Welw.) Hiern is widespread in tropical Africa south of the Equator. It differs from *ternifolia* in the calyx which is puberulous and in the normally 8–9-merous ovary and corolla. Aubréville (l.c.) treats it as *G. ternifolia* var. *jovis-tonantis* (Welw.) Aubrév.—R.W.J.K.]

---

[1] The Kew specimen *Howes* 1078 consists of two sheets, one of which was named *G. jovis-tonantis* and the other *G. ternifolia*; the number *Howes* 1078 was accordingly cited under both species in F.W.T.A., ed. 1. There is, however, no reason for believing this specimen to be a mixture of two species and I regard both sheets as *G. ternifolia*. Both sheets were in fact probably used in preparing Fig. 199 in the first edition, but as this drawing has two errors it has been omitted from the second edition. The pubescent shoot in F.W.T.A., ed. 1, 2 : 74, fig. 199 is not of *G. ternifolia*, a species with glabrous branchlets as the description on the previous page correctly states; the fruits in Fig. 199 are wrongly drawn with longitudinal ridges.—R.W.J.K.

Imperfectly known species.

G. alba *J. D. Kennedy* For. Fl. S. Nig. 211 (1936), name only. Specimen not traced.
S.Nig.: Olokemeji *Kennedy* 73.

19. PSEUDOGARDENIA Keay in Bull. Jard. Bot. Brux. 28 : 45 (1958).

Young stems, petioles and underside of leaves rusty- setose -pilose ; leaves obovate, cuneate or obtuse at base, shortly acuminate, 9–25 cm. long, 4–11 cm. broad, with 8–14 main lateral nerves on each side of midrib ; flowers solitary, terminal ; calyx sparingly or more or less densely setose outside, glabrous and minutely papillose inside, lobes 2·5–3 cm. long, 1–2 cm. broad, attached outside the truncate 2·5–3·2 cm. long tube ; corolla-tube 10·5–15 cm. long, gradually expanded, densely silky outside, lobes 6–8 cm. long ; anthers 2·5 cm. long, partially exserted ; style-head fusiform, about 4 cm. long ; fruits obovoid-cylindrical, 10-ribbed, about 6 cm. long and 4·5 cm. diam., crowned with the persistent calyx-tube .. .. .. *kalbreyeri*

P. kalbreyeri (*Hiern*) *Keay* l.c. 46 (1958). *Gardenia kalbreyeri* Hiern in J. Bot. 16 : 97, t. 195 (1878) ; F.W.T.A., ed. 1, 2 : 73. *G. gossleriana* J. Braun & K. Schum. (1889). *Randia purpureo-maculata* C. H. Wright (1901). A forest shrub, usually scrambling, 5–40 ft. high ; corolla cream or yellow, streaked dark red inside. **S.Nig.:** Akilla, Ijebu Ode Prov. (Jan.) *Emumwen* FHI 20684 ! Omo (formerly part of Shasha) F.R. (Apr.) *Ross* 239 ! Owena, Ondo Prov. (Mar.) *Darko* 189 ! Uquo, Eket Div. (May) *Onochie* FHI 33156 ! Calabar (Jan., Feb.) *Holland* 8 ! *Kalbreyer* 212 ! Oban *Talbot* 200 ! 1311 ! 1317 ! **[Br.]Cam.:** Abonando, Mamfe Div. (May) *Rudatis* 56 ! Kembong, Mamfe Div. (May) *Ndep Enoh* Cam. 49/38 ! Also in Cameroun, Rio Muni, Gabon, Congo and Cabinda. (See Appendix, p. 399.)

20. ROTHMANNIA Thunb. in Vet. Acad. Handl. Stockh. 65 (1776) ; Fagerlind in Arkiv för Bot. Stockh. 30A, No. 7 : 39 (1943).

Corolla-lobes overlapping to the right :
  Flowers 7–8-merous ; calyx-tube 3–5 mm. long, not split, lobes 2–4·5 cm. long, setose-pilose ; corolla-tube 14–18 cm. long, gradually expanded towards apex, spreading-pubescent outside, lobes 3·5–8 cm. long, 1–1·5 cm. broad, acute ; branchlets and leaves hirsute ; leaves oblanceolate or elliptic-oblanceolate, cuneate at base, caudate or shortly acuminate, 9–22 cm. long, 2·5–8 cm. broad, with 5–12 main lateral nerves on each side of midrib ; fruits cylindrical-club-shaped, about 13·5 cm. long and 3·5 cm. broad, nearly glabrous .. .. *1. octomera*
  Flowers 5-merous ; calyx usually split down one side :
    Calyx-tube 2·5–4 cm. long, markedly spathaceous, lobes linear or lanceolate, 3–4 cm. long ; corolla-tube 13–17 cm. long, densely shaggy villous outside, lobes 2·5–3 cm. long ; fruits ellipsoid and markedly 5-ridged, each ridge longitudinally grooved, 6–11 cm. long, 3·5–6 cm. broad ; branchlets and leaves spreading-pilose ; leaves elliptic, more or less obtuse at base, long-acuminate, 9–20 cm. long, 3·5–8 cm. broad, with 7–8 main lateral nerves on each side of midrib .. .. .. *2. hispida*
    Calyx-tube 1–1·5 cm. long, lobes filiform, 0·3–1 cm. long ; corolla-tube 4·5–6·5 cm. long, pilose outside, lobes 1·5–3 cm. long ; fruits ellipsoid or subglobose, not ridged but with 10–12 longitudinal lines, 3·5–7·5 cm. long, 3·3–5·5 cm. broad ; branchlets and leaves pubescent and glabrescent or quite densely pilose ; leaves elliptic, cuneate or obtuse at base, acuminate, 8–13 cm. long, 2·8–5·5 cm. broad, with 5–8 main lateral nerves on each side of midrib .. .. .. *3. urcelliformis*
Corolla-lobes overlapping to the left ; flowers 5-merous :
  Corolla glabrous outside, consisting of a 11–18 cm. long narrow lower portion and an abruptly expanded 1·5–3 cm. long campanulate upper portion, lobes 2–3 cm. long ; flowers normally in pairs ; calyx-tube 5-angled, glabrous, 7–9 mm. long, teeth 1–2 mm. long ; fruits 15–20 cm. diam., woody, taking several years to grow ; leaves obovate-oblong, cuneate at base, acuminate, 11–28 cm. long, 4·5–14 cm. broad, with 6–15 main lateral nerves on each side of midrib ; a tree 10–30 m. high .. *4. lujae*
  Corolla appressed-puberulous (at least when young) or densely tomentellous or velutinous outside, tube gradually expanded and funnel-shaped ; flowers solitary ; fruits up to 12 cm. long :
    Calyx glabrous outside, teeth usually much shorter than tube ; leaves glabrous except for tufts in nerve-axils :
      Leaves more or less elliptic, cuneate at base, acuminate, 5–13 cm. long, 2–5·2 (–6·5) cm. broad with 4–6 main lateral nerves on each side of midrib ; corolla appressed-puberulous outside, sometimes glabrescent, tube 12–25 cm. long, lobes 1·5–4 cm. long ; fruits subglobose, 4–5 cm. diam. .. .. *5. longiflora*
      Leaves obovate-oblanceolate, broadly cuneate to subcordate at base, very shortly and broadly acuminate, 19–35 cm. long, 9·5–21·5 cm. broad, with 7–10 main lateral nerves on each side of midrib ; fruits ellipsoid, 10–12 cm. long, 6·5–9 cm. broad
6. megalostigma
    Calyx densely brown-tomentellous outside, lobes linear, (3–)5–66 mm. long ; leaves rusty pubescent on midrib and nerves beneath when young ; fruits up to 7 cm. long :
      Ovary and fruit 5-ridged, the latter ellipsoid, about 6 cm. long and 3·5 cm. diam. ; calyx-tube inside finely silvery velutinous above, sparsely so below, lobes up to 3–15 mm. long ; corolla-tube densely and shortly velutinous outside, 20–25 cm.

long, very gradually widened and only about 2·5 cm. wide at mouth, lobes about 4·5 cm. long ; leaves obovate-oblanceolate, cuneate at base, acuminate, 12–24 cm. long, 4–9 cm. broad, with 7–9 main lateral nerves on each side of midrib, venation very obscure beneath .. .. 7. *talbotii*
Ovary 10-ridged ; fruit 10-ridged or nearly smooth, subglobose, up to 7 cm. diam. ; calyx-tube densely and evenly brown-velutinous inside, lobes 15–66 cm. long ; corolla-tube densely long-velutinous outside, 3–17 cm. long, with expanded part (including lobes) 3·7–12 cm. long, lobes 1·7–6 cm. broad ; leaves oblong-obovate, oblanceolate or narrowly elliptic, cuneate at base, acuminate, 6–30 cm. long, 2–11 cm. broad, with 7–12 main lateral nerves on each side of midrib, venation obvious beneath .. .. 8. *whitfieldii*

1. **R. octomera** (*Hook.*) *Fagerlind* in Arkiv för Bot. Stockh. 30A, 7 : 39 (1943). *Gardenia octomera* Hook. Bot. Mag. t. 5410 (1863). *Randia octomera* (Hook.) Hook. f. (1873)—F.T.A. 3 : 98 ; F.W.T.A., ed. 1, 2 : 78. *R. galtonii* Wernham (1914). *R. cunliffeae* Wernham (1914). A forest shrub or small tree, about 7 ft. high ; " calyx very dark green, corolla-tube bright green, lobes white within and divided on the back longitudinally into a bright green half and a creamish-white half, stamens light drab, style and stigma cream " ; " fruits green with pale green speckles ".
S.Nig.: Degema *Talbot* 3631 ! Oron to Eket (Feb.) *Talbot* 3219 ! 3220 ! 3385 ! Calabar *Goldie* ! *Robb* ! *Thomson* 140 ! Lower Enyong F.R., Itu (fr. May) *Onochie* FHI 33205 ! [Br.]**Cam.**: Banga, Kumba Div. (fl. buds & fr. Mar.) *Brenan* 9318 ! **F.Po**: *Mann* ! Also in Cameroun, Gabon, Cabinda and Congo.
2. **R. hispida** (*K. Schum.*) *Fagerlind* l.c. (1943) ; Aubrév. Fl. For. C. Iv., ed. 2, 3 : 278, t. 352A. *Randia hispida* K. Schum. in Engl. Bot. Jahrb. 23 : 437 (1896) ; F.W.T.A., ed. 1, 2 : 78. *R. pynaertii* De Wild. (1907). *Gardenia spathicalyx* K. Schum. ex Wernham (1919). An erect shrub or small tree, to 35 ft. high ; corolla white, with purple markings inside, densely silky outside ; in forest.
Guin.: Nimba Mts. (Feb.) *Schnell* 4267 ! **Lib.**: Gbanga (fr. Sept.) *Linder* 785 ! Belefanai, Gbanga Dist. (Dec.) *Baldwin* 10542 ! Karmadhun, Kolahun (fr. Nov.) *Baldwin* 10188 ! **Iv.C.**: Danané to Mt. Kouan *Chev.* 21261 ! Banco *Aubrév.* 366 ! **Ghana**: Tukobo, Axim *Vigne* FH 1438 ! Nyaso, Ashanti (Dec.) *Vigne* FH 1481 ! Benso, Tarkwa (fl. Sept., fr. Apr.) *Andoh* FH 5479 ! *Vigne* FH 4746 ! **S.Nig.**: Sapoba (Feb.) *Kennedy* 724 ! 2026 ! 2070 ! 2575 ! Awka Dist. (Nov.) *Onwudinjoh* FHI 21914 ! Degema *Talbot* 3638 ! Eket *Talbot* 3137 ! Aboabam, Ikom Div. (Dec.) *Keay* FHI 28204 ! Oban *Talbot* 201 ! 2004 ! 2026 ! [Br.]**Cam.**: Ndebanja, Mamfe (Nov.) *Ndep Enoh* 8 ! Okoyong, Mamfe (Feb.) *Enjong* 4 ! Also in Cameroun, Cabinda and Congo. (See Appendix, p. 409.)
3. **R. urcelliformis** (*Hiern*) *Bullock ex Robyns* Fl. Sperm. Parc Nat. Alb. 2 : 340 (1947) ; Aubrév. Fl. For. C. Iv., ed. 2, 3 : 278, t. 355, 5–6. *Gardenia urcelliformis* Hiern in F.T.A. 3 : 104 (1877). *Randia urcelliformis* (Hiern) Eggeling (1940). *Gardenia tigrina* Welw. ex Hiern (1898)—Aubrév. Fl. For. C. Iv., ed. 1, 3 : 248. *Randia stenophylla* K. Krause (1909)—F.W.T.A., ed. 1, 2 : 78. *R. chevalieri* Aubrév. l.c. 244 (1936), French descr. only. A shrub or small tree, to 25 ft. or more high ; corolla white, mottled purple within ; in forest.
Guin.: Dalaba (Jan.) *Roberty* 16472 ! Mt. Benna, Kindia (Dec.) *Jac.-Fél.* 2156 ! **S.L.**: Mano, Luawa (fr. Apr.) *Deighton* 3217 ! Kukuna (Jan.) *Sc. Elliot* 4673 ! **Lib.**: Vahon, Kolahun (fr. Nov.) *Baldwin* 10246 ! **Iv.C.**: Man *Aubrév.* 987 ! 1086. Orumbo-Bocca *Aubrév.* 839. Bouroukrou *Chev.* 17003 ! **Ghana**: Akwapim Hills (Jan.) *Johnson* 555 ! Suhum *West-Skinn* ! Ejan, N. of Agogo (Jan.) *Vigne* FH 1785 ! Ejura *Vigne* FH 3457 ! Koforidua (fl. & fr. Jan.) *Vigne* FH 4322 ! Kudje, Hohoe Dist. (Nov.) *St. C. Thompson* FH 3689 ! **Togo Rep.**: Agaua (fr. Jan.) *Kersting* A285. **Dah.**: *Chev.* 22836 ; 22932 ; 23696. **N.Nig.**: Sanga River F.R., Jemaa Div. (Nov.) *Keay* FHI 37238 ! Gidan-Anju, S. Muri Dist., Adamawa (Nov.) *Latilo & Daramola* FHI 28726 ! **S.Nig.**: Olokemeji *Foster* 131 ! *Hepper* 2300 ! Awba Hills F.R., Ibadan Prov. (fr. Apr.) *Onochie* FHI 32078 ! Idumuje (fl. buds Jan.) *Thomas* 2181 ! [Br.]**Cam.**: Cam. Mt., 5,700 ft. (fr. Apr.) *Hutch. & Metcalfe* 44 ! Bamenda (Jan.) *Johnstone* 1831 ! Also in Sudan, Congo, Ethiopia, Kenya, Uganda, Tanganyika, Angola, S. Rhodesia and Mozambique. (See Appendix, p. 410.)
4. **R. lujae** (*De Wild.*) *Keay* in Bull. Jard. Bot. Brux. 28 : 53 (1958). *Randia lujae* De Wild. (1904). *R. coriacea* K. Schum. ex Hutch. & Dalz. F.W.T.A., ed. 1, 2 : 78 (1931) ; Kennedy For. Fl. S. Nig. 211 ; not of Benth. (1849). *Rothmannia coriacea* Fagerlind l.c. (1943). *R. giganthrophaera* Fagerlind l.c., name only. A forest tree, 30–105 ft. high, with clean cylindrical bole up to 5 ft. girth and relatively small dense crown of deep green glossy leaves ; splash reddish with fine white lines, exuding dark orange sap ; corolla white ; fruits brown, heavy, pendulous, apparently taking several years to mature as the twigs on which they are borne become woody and up to 2·5 cm. diam.
S.Nig.: Freeman 350 ! Okomu F.R. (fr. Feb.) *Brenan* 9111 ! Sapoba *Kennedy* 189 ! 235 ! 934 ! 1625 ! Usonigbe F.R., Benin (fl. buds & fr. Nov.) *Keay & Onochie* FHI 21605 ! Degema to Isoba (Oct.) *King-Church* 48 ! Oban *Talbot* 1267 ! [Br.]**Cam.**: mile 48, Kumba to Victoria (fr. Apr.) *Daramola* FHI 29817 ! Also in Cameroun, Rio Muni, Cabinda and Congo. (See Appendix, p. 409.)
5. **R. longiflora** *Salisb.* Parad. Lond. t. 65 (1807) ; Aubrév. Fl. For. C. Iv., ed. 2, 3 : 278, t. 353, 6. *Randia maculata* DC. (1830)—F.T.A. 3 : 96 ; Chev. Bot. 321 ; F.W.T.A., ed. 1, 2 : 78, excl. distrib. Gambia. *Rothmannia maculata* (DC.) Fagerlind (1943). *Gardenia speciosa* A. Rich. (1834). *G. stanleyana* Hook. ex Lindl. (1845). *Rothmannia stanleyana* (Hook. ex Lindl.) Hook. ex Benth. (1849). *Randia stanleyana* (Hook. ex Lindl.) Walp. (1851–52). *R. thomasii* Hutch. & Dalz. F.W.T.A., ed. 1, 2 : 78 (1931). A shrub or small tree, up to 15 ft. high, with shiny foliage, often much-branched ; flowers more or less erect, corolla greenish or reddish outside, the lobes white with purplish or reddish markings within ; mostly in forest regrowth vegetation.
Port.G.: Dandua, Bafata (Oct.) *Esp. Santo* 190 ! **Guin.**: Pobéguin 2197 ! Santa to Timbo *Chev.* 12627. **S.L.**: *Whitfield* ! Kambia (Dec., Jan.) *Deighton* 839 ! *Sc. Elliot* 4392 ! Serabu (Jan.) *Smythe* 238 ! Zimmi (Nov.) *Deighton* 384 ! Rokupr (Nov.) *Jordan* 378 ! Yetaya (fl. buds Sept.) *Thomas* 2311 ! **Lib.**: Kakatown *Whyte* ! Zorzor, Vonjama (Oct.) *Baldwin* 10071 ! Ganta (May) *Harley* ! Javajai, Boporo (Nov.) *Baldwin* 10263a ! **Iv.C.**: Danané *Collenette* 447 ! Mt. Tonkoui (Dec.) *Roberty* 15740 ! Agboville *Aubrév.* 1921 ! **Ghana**: Cape Coast *Brass* ! L. Bosumtwi (Jan.) *Dalz.* 65 ! Asuansi Road (May) *Fishlock* 21 ! Nkwantanan, Kwahu (Nov.) *Irvine* 1717 ! Nkroful (Nov.) *Vigne* FH 1417 ! Togo Plateau F.R. (Nov.) *Adogla* FH 4228 ! **Togo Rep.**: Lomé *Warnecke* 472 ! **Dah.**: *Poisson* ! *Le Testu* 101 ! **N.Nig.**: Nupe *Barter* ! **S.Nig.**: Lagos *Punch* 79 ! *Rowland* 16 ! Ibadan (Dec.) *Keay* FHI 37195 ! Ejiofor FHI 23632 ! Ogwashi (Nov.) *Thomas* 2028 ! Degema Dist. *Talbot* 3678 ! Oban *Talbot* 181 ! [Br.]**Cam.**: Victoria (Jan.) *Maitland* 183 ! Abonando, Mamfe (Apr.) *Rudatis* 51 ! **F.Po**: *Mann* 178 ! Also in Cameroun, Congo, Sudan, Uganda and Angola. (See Appendix, p. 409.)
6. **R. megalostigma** (*Wernham*) *Keay* l.c. 54 (1958) ; Aubrév. Fl. For. C. Iv., ed. 2, 3 : 278, t. 354 B. *Randia megalostigma* Wernham in J. Bot. 57 : 279 (1919). *R. lane-poolei* Hutch. & Dalz. F.W.T.A., ed. 1, 2 : 78 (1931). (?) *R. munsae* Schweinf. ex Hiern in F.T.A. 3 : 99 (1877), partly. A shrub or tree, 8–50 ft. high, with straight bole ; calyx deep purplish-green ; corolla brownish, lobes pale green outside and white inside, with pink spots in throat, stigma yellow ; in swamp forest.
S.L.: Taninahun, Tunkia (fr. Dec.) *Deighton* 3831 ! Wandei, Kayamba (fr. July) *Jordan* 2132 ! Quia (fl. & fr. July) *Lane-Poole* 309 ! Bumbuna *Thomas* 3793 ! **Lib.**: *Johnston* ! Peahtah (fr. Oct.) *Linder* 967 ! Ganta (May) *Harley* 514 ! **Iv.C.**: Danipleu *Aubrév.* 1155. Abidjan *Aubrév.* 1648. **Ghana**: Neung F.R., Tarkwa (Jan.) *Vigne* FH 4837 ! Subiri F.R. (fr. Sept.) *Andoh* FH 5235 ! **S.Nig.**: Abeku, Ijebu Ode Prov.

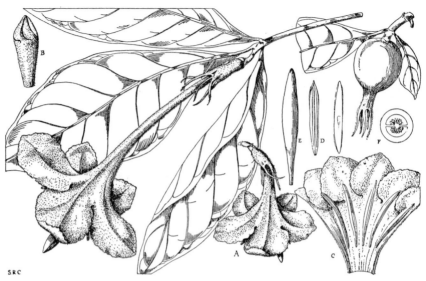

FIG. 227.—ROTHMANNIA WHITFIELDII (*Lindl.*) *Dandy* (RUBIACEAE).
A, a short-tubed flower.   B, corolla-bud.   C, upper part of corolla laid open.   D, anthers.
E, stigma.   F, cross-section of ovary.

Jones & Onochie FHI 17172! Okomu F.R., Benin (Feb.) *Brenan & Onochie* 9011! [Br.]**Cam.**: Serti,
Gashaka Dist. (Feb.) *Latilo & Daramola* FHI 34453! Also in Cameroun and Congo.

7. **R. talbotii** (*Wernham*) *Keay* l.c. (1957). *Randia talbotii* Wernham in Cat. Talb. 45 (1913); F.W.T.A., ed. 1,
2 : 78. *R. tubaeformis* Pellegr. (1936). A small tree; leaves whitish beneath; in forest.
S.Nig.: Oban *Talbot* 217! Boje, Afi River F.R., Ikom Div. (fl. buds May) *Jones & Onochie* FHI 5814!
[Br.]**Cam.**: Ekoneman, Mamfe (Aug.) *Johnstone* 147/31! Also in Cameroun, Gabon, Cabinda and Congo.

8. **R. whitfieldii** (*Lindl.*) *Dandy* in F. W. Andr. Fl. Pl. A.-E. Sud. 2 : 461, fig. 165 (1952); Aubrév. Fl. For. C.
Iv., ed. 2, 3 : 276, t. 352B. *Gardenia whitfieldii* Lindl. Bot. Reg. 31 : sub. t. 47 (1845). *G. malleifera* Hook.
Bot. Mag. t. 4307 (1847). *Rothmannia malleifera* (Hook.) Hook. ex Benth. (1849). *Randia malleifera* (Hook.)
Hook. f. (1873)—F.T.A. 3 : 98; Chev. Bot. 32; F.W.T.A., ed. 1, 2 : 78; Aubrév. Fl. For. Soud.-Guin.
462–464, t. 100, 1–4. A shrub or tree, 6–40 ft. high; flowers fragrant with the corolla white inside brownish-
velvety outside, pendulous; fruit 10-ridged or nearly smooth, yielding an inky black dye; in forest
vegetation, including forest outliers in the savanna regions.
Sen.: Mangacounda, Casamance (fl. bud Jan.) *Chev.* 2099! **Mali**: Banancoro (fl. bud Feb.) *Chev.* 505!
Port.G.: Prabis, Bissau (fl. bud Mar.) *Esp. Santo* 1854! **Guin.**: Rio Nunez *Heudelot* 809! Mamou (May)
Schnell 5409! **S.L.**: *Whitfield*! *Turner*! Bagroo R. (Apr.) *Mann* 913! Lomaburu, Scarcies R. (Feb.)
Sc. Elliot 5025! Giehun (Mar.) *Aylmer* 565! Njala (Apr.) *Deighton* 2887! Mafokoya, Roruks (May)
Deighton 3003! **Lib.**: Kakatown *Whyte*! Dukwia R. (fr. May) *Cooper* 450! Gbanga (fr. Sept.) *Linder*
482! Taninewa (Mar.) *Bequaert* 150! *Irvine* 1688! *Irvine* 1688! **N.Nig.**: Kurmin Damisa, Jemaa (fr. Nov.)
Ghana: Kumasi (Mar.) *Dalz.* 36! Prestea (July) *Vigne* FH 1264! Banka, Ashanti (Dec.) *Vigne* FH 1487!
Pra R. (Feb.) *Vigne* FH 1597! Kwahu (Jan.) *Irvine* 1688! **N.Nig.**: Kurmin Damisa, Jemaa (fr. Nov.)
Keay FHI 21661! **S.Nig.**: Lagos *Foster* 76! Agbemia *Barter* 3367! Okomu F.R., Benin (Feb.) *Onochie*
in *Hb. Brenan* 9179! Awka *Thomas* 75! Degema *Talbot* 3697! Insofan (Jan.) *Holland* 252! [Br.]**Cam.**:
Mamfe (Mar.) *Collier* in *Hb. Maitland* 1060! Fonfuka, Bamenda (May) *Maitland* 1585! Babessi (Oct.)
Ledermann 5797! Lus, Nkambe Div. (Feb.) *Hepper* 1874! R. Sulli, Gashaka Dist. (fr. Dec.) *Latilo &
Daramola* FHI 28837! Also in Cameroun, Rio Muni, Gabon, Congo, Ubangi-Shari, Sudan, E. Africa,
N. Rhodesia and Angola. (See Appendix, p. 409.)

Doubtfully recorded species.

R. capensis *Thunb.* in Vet. Acad. Handl. Stockh. 67, t. 2 (1776). *Gardenia rothmannia* Linn. f. (1781). *Randia
maculata* of F.T.A. 3 : 96 and F.W.T.A., ed. 1, 2 : 78, partly (*Boteler* ex Gambia). This is a South African
species and it is very unlikely that Boteler's specimen was indigenous in the Gambia. I suggest that either
the specimen was not in fact collected in the Gambia, or that it was from an introduced cultivated plant.

21. **SHERBOURNIA** G. Don in Loudon Encycl. Pl., Suppl. 2, 1303 (1855). *Amaralia*
Welw. ex Hook. f.—F.T.A. 3 : 112; Wernham in J. Bot. 55 : 1 (1917); Keay in
Bull. Jard. Bot. Brux. 28 : 57 (1958).

Calyx-lobes 6–28 mm. long, much longer than the calyx-tube :
Main lateral nerves (4–)5–7 on each side of midrib; lamina appressed-puberulous to
nearly glabrous beneath, midrib and main nerves usually more densely puberulous
or pubescent; flowers solitary, rarely paired :
Leaves mostly cordate and often unequal at base, ovate-elliptic or obovate-elliptic,
acuminate, 9–21 cm. long, 5–11·5 cm. broad; calyx-tube 3–10 mm. long, lobes
17–28 mm. long; corolla-tube 36–55 mm. long, lobes 10–18 mm. long and broad;
fruits ellipsoid-subglobose, faintly ribbed about 7·5 cm. long and 5·5 cm. diam.

1. *bignoniiflora*

Leaves cuneate to obtuse or rarely rounded at base, narrowly elliptic or obovate-elliptic, gradually acuminate, 7–17 cm. long, 2·5–7·5 cm. broad ; calyx-tube up to 4 mm. long ; corolla-tube up to 30 mm. long :

Calyx-tube very short, up to 1 mm. long, lobes 15–28 mm. long, 6–10 mm. broad ; corolla-tube 15–30 mm. long, lobes 10–11 mm. long, 12–14 mm. broad ; fruits campanulate or broadly ellipsoid, only very slightly ribbed, 3·2–4·5 cm. long, 2·5–3 cm. broad ..   ..   2. *calycina*

Calyx-tube 3–4 mm. long, lobes 6–10 mm. long, 3·5–4·5 mm. broad ; corolla-tube 24–28 mm. long, lobes 4–7 mm. long, 5–8 mm. broad ; fruits cylindrical, often slightly curved, somewhat ribbed, 2·5–3·2 cm. long, 1–1·2 cm. broad   3. *millenii*

Main lateral nerves 10–12 on each side of midrib ; leaves rather densely appressed-pubescent beneath, elliptic, ovate-elliptic or obovate-elliptic, acute or obtuse at base, acuminate, 9–14 cm. long, 3·3–7 cm. broad ; cymes 1–3(–5)-flowered ; calyx-tube about 3 mm. long, lobes 12–15 mm. long, 5–7 mm. broad ; corolla-tube 26–32 mm. long, lobes about 10 mm. long and broad ; fruits cylindrical, strongly ribbed, about 3·5 cm. long and 1·5 cm. broad   ..   ..   4. *zenkeri*

Calyx-lobes broadly triangular, up to 2·5 mm. long, much shorter than the 9–10 mm. long tube ; cymes up to 7-flowered ; fruit oblong-ellipsoid, strongly ribbed :

Leaves densely setose-pubescent beneath, with 9–15 main lateral nerves on each side of midrib, broadly elliptic or obovate-elliptic, unequal and rounded at base, abruptly acuminate, 14–31 cm. long, 7–15 cm. broad ; corolla-tube about 26 mm. long, lobes about 8 mm. long and broad ; fruits about 2·3 cm. long and 2 cm. broad
5. *hapalophylla*

Leaves pubescent only on midrib beneath, with 5–6 main lateral nerves on each side of midrib, broadly elliptic, acute or somewhat obtuse at base, shortly acuminate, 10–17 cm. long, 5–8 cm. broad ; corolla-tube about 35 mm. long, lobes about 6 mm. long and 5 mm. broad ; fruits about 3·5 cm. long and 1·7 cm. broad
6. *amaraliocarpa*

1. **S. bignoniiflora** (*Welw.*) *Hua* in Bull. Soc. Hist. Nat. Autun 14 : 396 (1901). *Gardenia bignoniaeflora* Welw. (1859). *Amaralia bignoniiflora* (Welw.) Hiern in F.T.A. 3 : 112 (1877), partly ; Wernham in J. Bot. 55 : 5. *Amaralia heinsioides* Wernham l.c. 5 (1917) ; F.W.T.A., ed. 1, 1 : 84. *A. buntingii* Wernham l.c. 4 (1917). A scandent shrub or liane ; calyx green with crimson markings, turning purplish within ; in forest. **S.L.:** Kafogo (Apr.) *Sc. Elliot* 5602! Mano *Thomas* 10553! Yalumbu to Kurobonia (young fr. May) *Deighton* 5083! **Lib.:** S.W. of Jiu, Gola Forest (fl. & fr. Apr.) *Bunting*! Kulo, Sinoe Co. (Mar.) *Baldwin* 11432! 11438! **Iv.C.:** Adiopodoumé (July) *de Wilde* 82! Sassandra (Dec.) *Leeuwenberg* 2240! **Ghana:** Finsenase, Obuasi (Mar.) *Chipp* 149! Bawdia (Apr.) *Vigne* 164! Mrenya, Ashanti *Irvine* 484! **S.Nig.:** Lagos *Rowland*! Ijaiye F.R., Oyo (fl. & fr. Apr.) *Onochie* FHI 21959! Etemi, Omo F.R., Ijebu Ode Prov. (May) *Tamajong & Latilo* FHI 16776! Owo (May) *Jones* FHI 3623! Ukpon F.R. (Apr.) *Jones* 1533! Also in Cameroun, Sudan, Congo, N. Rhodesia and Angola. (See Appendix, p. 391.)

2. **S. calycina** (*G. Don*) *Hua* l.c. 398 (1901), partly (excl. *Barter ex Eppah*) ; Chev. Bot. 324. *Gardenia calycina* G. Don (1834). *Amaralia calycina* (G. Don) K. Schum. in E. & P. Pflanzenfam. 4, 4 : 78, fig. 28, K (1891) ; Wernham l.c. 4. *Gardenia sherbourniae* Hook. Bot. Mag. t. 4044 (1843). *Randia sherbourniae* (Hook.) Hook. (1849). *Amaralia sherbourniae* (Hook.) Wernham l.c. (1917), partly (excl. distrib. Cameroons) ; F.W.T.A., ed. 1, 2 : 84, partly (excl. distrib. Nigeria & Cameroons). *Randia doniana* Benth. (1849). *Sherbournia foliosa* G. Don (1855)—Chev. Bot. 324. *Randia amaralioides* K. Schum. ex Hutch. & Dalz. F.W.T.A., ed. 1, 2 : 79 (1931), partly (*Soward* 650). *Amaralia bignoniaeflora* of F.T.A. 3 : 112, partly ; *A. huana* of Wernham l.c. 5, partly. A scandent shrub or liane ; corolla white, fading to cream, sometimes pink, with deep pink markings inside ; ripe fruits red or orange ; mostly in forest. **Guin.:** Karkandy (Apr.) *Heudelot* 839! Lola to Nzo (Mar.) *Chev.* 20985! Kaba valley, *Chev.* 13127. Kouria *Chev.* 14712. **S.L.:** *Barter*! *Don*! Bagroo R. (Apr.) *Mann* 819! Regent (Dec.) *Sc. Elliot* 4110! Kurusa, Kafogo (Apr.) *Sc. Elliot* 4602! Njala (fl. May, fr. Feb.) *Deighton* 676! 2459! 2634! Moyamba (June) *Lane-Poole* 51! **Lib.:** Monrovia *Whyte*! Sinoe Basin *Whyte*! Dukwia R. (Mar.) *Cooper* 328! Brewersville (Mar.) *Barker* 1231! Ba, Little Kolu R. (Mar.) *Baldwin* 11186! **Iv.C.:** Bingerville Region *Chev.* 15290. Guidéko *Chev.* 16445. Aboisso (Apr.) *Chev.* 17841. Davo confluence, Sassandra R. (May) *Chev.* 17976. **Ghana:** Bonsa Su (May) *Vigne* FH 1987! Obuasi, Ashanti (young fr. Jan.) *Soward* 650! Kumasi *Cummins* 143! (See Appendix, p. 391.)

3. **S. millenii** (*Wernham*) *Hepper* in Kew Bull. 16 : 459 (1963). *Amaralia millenii* Wernham in J. Bot. 55 : 6 (1917). *A. micrantha* Wernham l.c. (1917) ; F.W.T.A., ed. 1, 2 : 84, partly (excl. *Talbot* 202). *A. huana* Wernham l.c. 5 (1917), partly (excl. *Mann* 819). *A. sherbourniae* of F.W.T.A., ed. 1, 2 : 84, partly (distrib. Nigeria & Cameroons only). *Randia amaralioides* K. Schum. ex Hutch. & Dalz. F.W.T.A., ed. 1, 2 : 79 (1931), partly (*Rowland* s.n.) ; not *R. amaralioides* K. Schum. ex Hua (1901). A liane, to about 20 ft. ; calyx purplish, corolla and style red ; in forest. **S.Nig.:** Lagos (Nov.) *Millen* 143! *Rowland*! Ebute Metta (Dec.) *Millen* 31! Epe *Barter*! Olokemeji *Foster* 295! Oyo (May) *Millson*! Omo F.R., Ijebu Ode Prov. (Apr.) *Jones & Onochie* FHI 16841! *Onochie* FHI 15541! Okomu F.R., Benin Div. (Feb.) *Brenan* 9172! Also in Cameroun.

4. **S. zenkeri** *Hua* (1901) ; Kennedy For. Fl. S. Nig. 213. *Amaralia zenkeri* (Hua) Wernham l.c. 6 (1917). *A. millenii* of F.W.T.A., ed. 1, 2 : 84, partly (*Talbot* 202). Scandent shrub or liane with dull purplish branchlets and petioles ; calyx and corolla dull purple with whitish-grey indumentum (*fide* Onochie), or corolla white with reddish-purple throat (*fide* Kennedy) ; in forest. **S.Nig.:** Ipopon, Benin Div. (Apr.) *Farquhar* 198! Sapoba, Benin Div. (fl. Apr., May, fr. June) *Ejiofor* FHI 32008! *Kennedy* 1134! 2631! 2632! Oban Dist. (Mar.) *Onochie* FHI 34810! *Talbot* 2021 s.n.! [Br.]**Cam.:** Ambas Bay (Feb.) *Mann* 1329! Victoria (Mar.) *Maitland* 1063! Bulifambo, Buea (Mar.) *Maitland* 539! Also in Cameroun and Cabinda.

5. **S. hapalophylla** (*Wernham*) *Hepper* l.c. 459 (1963). *Randia hapalophylla* Wernham in J. Bot. 55 : 9 (1917); F.W.T.A., ed. 1, 2 : 79. *Amaralia hapalophylla* (Wernham) Keay in Bull. Jard. Bot. Brux. 28 : 58 (1958). *R. streptocaulon* of Cat. Talb. 131, not of K. Schum. A climbing shrub ; in forest. **S.Nig.:** Oban *Talbot* 211a! [Br.]**Cam.:** Abonando, Mamfe (May) *Rudatis* 66! Banga, Kumba Div. (fr. Jan.) *Binuyo & Daramola* FHI 35185! Also in Cameroun.

6. **S. amaraliocarpa** (*Wernham*) *Hepper* l.c. 459 (1963). *Randia amaraliocarpa* Wernham l.c. 8 (1917); F.W.T.A., ed. 1, 2 : 79. *Amaralia amaraliocarpa* (Wernham) Keay l.c. 59 (1958). A scandent shrub ; calyx " bronzy-green " with greyish indumentum ; corolla-tube creamy yellow, lobes with dark purplish to black spots ; in farm clearings in the forest regions. **S.Nig.:** Oron to Eket *Talbot* 3021!

22. OXYANTHUS DC.—F.T.A. 3 : 106 ; K. Schum. in E. & P. Pflanzenfam. 4, 4 : 78 (1891).

Branchlets and leaves more or less densely puberulous, pubescent or pilose ; corolla-tube[1] 10–17 cm. long :

  Leaves 18–40 cm. long, 10–25 cm. broad, very asymmetrical at base with one side broadly auriculate, lamina ovate-elliptic, shortly acuminate, spreading pilose beneath and on midrib above, with 8–14 main lateral nerves on each side of midrib ; inflorescences widely paniculate, axes (including pedicels) up to 8 cm. long, with very numerous flowers ; calyx-lobes filiform, about 7 mm. long ; corolla-tube 13–15·5 cm. long, lobes about 2 cm. long ; fruits about 2·5 cm. diam. ..   1. *unilocularis*

  Leaves not exceeding 24 cm. long and 10 cm. broad ; inflorescence-axes (including pedicels) up to 2 cm. long, with not more than about 12 flowers ; calyx-lobes triangular-lanceolate, 1·5–2·5 mm. long ; main lateral nerves 7–10 on each side of midrib :

    Branchlets, petioles and undersurface of leaves setose-pilose ; corolla-tube 16–17 cm. long, glabrous outside ; leaves oblong, more or less asymmetrical and rounded to cuneate at base, acuminate, 14–22 cm. long, 5·3–10 cm. broad ; inflorescence-axes (including pedicels) up to 0·8 cm. long .. .. 2. *setosus*

    Branchlets, petioles and undersurface of leaves softly puberulous or pubescent ; corolla-tube 11–14 cm. long, pubescent or glabrescent ; leaves oblong, very asymmetrical, cuneate and rounded to subcordate at base, broadly acuminate, (5–)9–20 cm. long, (1·5–)2–8 cm. broad ; inflorescence-axes (including pedicels) up to 2 cm. long .. .. 3. *racemosus*

Branchlets and leaves quite glabrous, or leaves sparingly pubescent or puberulous beneath mainly by midrib (Nos. 4, 5 & 8) :

  Corolla-tube[1] 10–17·5 cm. long, 2–3 mm. wide (when flattened), lobes (1·5–)2–3·2 cm. long :

    Calyx-lobes well-developed, 1·5–2·5 mm. long ; leaves symmetrical and usually cuneate at base, oblong, acuminate ; flowers up to about 15, but usually much fewer :

      Leaves pubescent on midrib and especially in axils of main nerves beneath, rarely minutely puberulous on lamina, 5–18 cm. long, 2–7 cm. broad ; inflorescence-axes (including pedicels) less than 1 cm. long, with conspicuous persistent bracts :

        Main lateral nerves 5–8 on each side of midrib ; corolla-tube 11–15 cm., lobes 2–2·5 cm. long ; fruits ellipsoid, about 3·2 cm. long and 1·8 cm. broad

                4. *subpunctatus*

        Main lateral nerves (8–)10–12 on each side of midrib ; corolla-tube 10–10·5 cm., lobes 1·5–2 cm. long ; immature fruits ovoid, about 5 cm. long ..   5. sp. *A*

    Leaves quite glabrous, 11–25 cm. long, 4–10 cm. broad, with 6–8 main lateral nerves on each side of midrib ; inflorescence-axes (including pedicels) up to 2·5 cm. long, with minute bracts ; fruits ellipsoid, about 3·5 cm. long and 1·8 cm. broad

                6. *pallidus*

  Calyx subtruncate or with very short teeth ; leaves usually asymmetrical, cuneate to obtuse and obtuse to subcordate at base, with 9–11 main lateral nerves on each side of midrib, oblong, shortly and broadly acuminate, 15–30 cm. long, 4–12 cm. broad ; inflorescence-axes (including pedicels) up to 10 cm. long, flowers up to about 30, but usually much fewer, bracts minute ..   .. .. 7. *formosus*

Corolla-tube 2·5–7 cm. long, up to 1·5 mm. wide (when flattened), lobes 0·5–1·2 cm. long ; leaves more or less symmetrical and often cuneate at base :

  Calyx 1 mm. wide or less, teeth less than 1 mm. long ; corolla-tube 0·75–1·5 mm. wide (when flattened) ; inflorescence a lax corymbose cyme, with axes (including the slender pedicels) up to 2 cm. long, bracts inconspicuous, usually 1 mm. or less long :

    Corolla-tube 5·2–7 cm., lobes 1–1·2 cm. long ; branchlets slender ; leaves membranous, slightly pubescent by midrib beneath, especially in the axils of the 5–7 main lateral nerves ; lamina oblong-elliptic, acuminate, 8·5–16 cm. long, 3–6·5 cm. broad .. .. 8. *gracilis*

    Corolla-tube 2·7–3·2 cm., lobes 0·5–0·9 cm. long ; branchlets rather stout ; leaves subcoriaceous, quite glabrous, with 4–6 main lateral nerves on each side of midrib ; lamina oblong or oblong-elliptic, acuminate, 9–20 cm. long, 3–8·5 cm. broad

                9. *laxiflorus*

  Calyx about 1·5 mm. wide, teeth 1–2 mm. long ; corolla-tube about 1·5 mm. wide (when flattened) ; inflorescence a dense, many-flowered, narrow panicle, with the main axis up to 8 cm. long and conspicuous bracts up to 7 mm. long ; corolla-tube (2·5–)4–5·5(–6·5) cm., lobes 0·6–1·2 cm. long ; leaves oblong, rounded, obtuse or cuneate at base, shortly and broadly acuminate, 7–25 cm. long, 2–13 cm. broad, with 7–10 main lateral nerves on each side of midrib ; fruits ellipsoid, stipitate, about 2·5 cm. long and 1·5 cm. broad, black and shiny when dry .. 10. *speciosus*

---

[1] Measurements of the corolla given in this key are taken only from fully expanded flowers. Measurements of unexpanded corollas can be most misleading ; e.g. the type specimen of *O. breviflorus* Benth. is simply *O. formosus* Hook. f. in young bud stage.

1. **O. unilocularis** *Hiern* in F.T.A. 3 : 110 (1877) ; Aubrév. Fl. For. C. Iv., ed. 2, 3 : 286, t. 357 ; F. W. Andr. Fl. Pl. A.-E. Sud. 2 : 455. *O. macrophyllus* Schweinf. ex Hiern l.c. (1877). *O. litoreus* S. Moore (1905). A shrub or tree, to 25 ft. high, with stout, more or less angled branchlets sometimes becoming hollow ; leaves dark glossy green above, coarsely hairy beneath ; calyx green, corolla white ; in forest, including fringing forest in the savanna regions.
S.L.: Kahreni to Port Loko *Sc. Elliot* 5827 ! Makeni (young fr. Jan.) *Deighton* 2837 ! Njala (fl. buds Apr.) *Deighton* 2888 ! Bandajuma (May) *Aylmer* 80 ! **Lib.:** Ganta (June) *Harley* 1211 ! Flumpa (fl. buds Sept.) *Baldwin* 9361 ! Dobli Isl., St. Paul R. *Bequaert* 17 ! Moylakwelli *Linder* 1317 ! **Iv.C.:** Abidjan *Aubrév.* 195 ; 469. **Ghana:** Tukobo, Axim Dist. (Nov.) *Vigne* FH 1437 ! Adeiso (Mar.) *Irvine* 1612 ! Bansu (Mar.) *Vigne* FH 1900 ! Aburi Hills (Oct.) *Johnson* 480 ! Kumasi (Mar., Apr.) *Darko* 383 ! *Vigne* FH 1646 ! **N.Nig.:** Jebba *Barter* 1075 ! Acharane, Kabba Prov. (June) *Daramola* FHI 38030 ! Anara F.R., Zaria Prov. (fr. Oct.) *Keay* FHI 5463 ! Katsina Ala *Dalz.* 644 ! **S.Nig.:** Ibadan (fl. buds Jan.) *Meikle* 1119 ! Sapoba *Kennedy* 2269 ! 2317 ! Bonny R. (Oct.) *Mann* 506 ! Oban *Talbot* 289 ! 1416 ! [**Br.**]**Cam.:** Serti, Gashaka Dist. (fr. Feb.) *Latilo & Daramola* FHI 34452 ! Also in Cameroun, Ubangi-Shari, Sudan, Gabon, Cabinda, Congo and Uganda. (See Appendix, p. 406.)

2. **O. setosus** *Keay* in Bull. Jard. Bot. Brux. 28 : 44 (1958). A forest shrub, to 8 ft. high ; calyx pale green ; corolla-tube greenish-white below, white at apex, lobes white, spreading or somewhat reflexed.
[**Br.**]**Cam.:** near Banga, S. Bakundu F.R., Kumba Div. (Jan., Mar.) *Brenan* 9283 ! 9438 ! *Latilo* FHI 43828 !

3. **O. racemosus** (*Schum. & Thonn.*) *Keay* l.c. 42 (1958). *Ucriana racemosa* Schum. & Thonn. (1827). *Oxyanthus hirsutus* DC. (1830). *O. thonningii* Benth. (1849). *O. rubriflorus* Hiern in F.T.A. 3 : 108 (1877) ; **F.W.T.A.**, ed. 1, 2 : 80. *O. ? sulcatus* Hiern l.c. (1877) ; Chev. Bot. 325. *O. tubiflorus* of F.T.A. 3 : 107 ; of Chev. Bot. 325 ; of F.W.T.A., ed. 1, 2 : 80 ; not of (Andr.) DC. A shrub or small tree, 2½–15 ft. high ; flowers white, fragrant ; ripe fruits orange or red ; in coastal scrub and in forest vegetation well inland.
Port.G.: Catió (fl. buds June) *Esp. Santo* 2097 ! Cacine Guileje (young fr. Aug.) *Esp. Santo* 2152 ! **Guin.:** Kindia (Mar.) *Chev.* 13070. Doubaya, Forécariah (June) *Jac.-Fél.* 1728 ! **S.L.:** *Afzelius* ! *Stormont* ! Bagroo R. *Mann* ! Yoni, Sherbro Isl. (Mar.) *Deighton* 2474 ! Goderich (Apr.) *Deighton* 4751 ! Mesima (Apr.) *Deighton* 3700 ! Mando (May) *Jordan* 249 ! **Lib.:** Brewersville (fl. & fr. Dec.) *Baldwin* 11956 ! **Iv.C.:** Bouroukrou *Chev.* 16997. **Ghana:** Cape Coast to Saltpond (Apr.) *Scholes* 207 ! Aburi *Brown* 431 ! Nchenenchene (Apr.) *Kitson* 1072 ! Accra *T. Vogel* ! Akwapim (May) *Irvine* 1765 ! *Thonning* ! **Togo Rep.:** Lomé *Warnecke* 460 ! **N.Nig.:** near Gerki, Abuja Dist. (Apr.) *Keay* FHI 37648 ! **S.Nig.:** Lagos *Rowland* ! Elugu *Barter* 3417 ! Abeokuta *Barter* 3325 ! Ibadan (Apr.) *Meikle* 1395 ! Onitsha *Barter* 1794 ! Also in Angola. (See Appendix, p. 406.)

4. **O. subpunctatus** (*Hiern*) *Keay* l.c. 44 (1958). *Mitriostigma ? subpunctatum* Hiern in F.T.A. 3 : 111 (1877 ; **F.W.T.A.**, ed. 1, 2 : 72. *O. tenuis* Stapf in J. Linn. Soc. 37 : 107 (1905) ; **F.W.T.A.**, ed. 1, 2 : 80. A forest shrub 2–10 ft. high, with dark green leaves ; corolla white or with pale green tube, fragrant ; fruits orange.
S.L.: Madina, Limba Country (Apr.) *Sc. Elliot* 5571 ! Nyandehun (Apr.) *Deighton* 3669 ! Dodo (Apr.) *Deighton* 3910 ! Gola Forest (Mar.) *Small* 549 ! **Lib.:** Sinoe Basin *Whyte* ! **Ghana:** Agogo, Ashanti (Apr.) *Vigne* FH 1079 ! Kumasi *Cummins* 1 ! **Togo Rep.:** Badja (Mar.) *Schlechter* 12975 ! **S.Nig.:** Lagos (May) *Dalz.* 1052 ! *Foster* 83 ! Olokemeji (fl. Apr., fr. Nov.) *Onochie* FHI 8138 ! *A. F. Ross* 154 ! Mamu F.R., Ibadan Prov. (Mar.) *Keay & Foggie* FHI 22527 ! Osho, Omo F.R., Ijebu Ode Prov. (Apr.) *Jones & Onochie* FHI 17259 ! [**Br.**]**Cam.:** Mamfe (Apr.) *Ndep Enoh* Cam. 39/38 ! Also in Congo. (See Appendix, p. 406.)

5. **O. sp. A.** A species of montane forest, close to the preceding ; more material required.
[**Br.**]**Cam.:** Nyanga Camp, 4,000 ft. (Dec.) *Maitland* 1208 ! **F.Po:** Marcelino Puente's estate (Jan.) *Guinea* 1695 ! Moka, 4,600 ft. (fl. & fr. Sept.) *Wrigley & Melville* 559 !

6. **O. pallidus** *Hiern* in Cat. Welw. 1 : 465 (1898). *O. tubiflorus* of F.W.T.A., ed. 1, 2 : 80, partly (*Johnson* 603). A shrub or small tree to 25 ft. high, with dark green foliage usually drying paler ; flowers white, fragrant ; in forest.
Iv.C.: Anguededou Forest (Mar.) *Aké Assi* 1509 ! Abidjan to Dabou (fr. July) *de Wilde* 104 ! **Ghana:** Simpa, W. Prov. (Feb.) *Kinloch* FH 3228 ! *Vigne* FH 2792 ! Kukerentumi, E. Akim (Mar.) *Johnson* 603 ! S. Scarp F.R., E. Prov. (Mar.) *Vigne* FH 4353 ! Banka (Mar.) *Irvine* 477 ! *Vigne* FH 1883 ! Begoro Dist. (Mar.) *Irvine* 1815 ! Also in Congo and Angola.

7. **O. formosus** *Hook. f. ex Planch.* in Hook. Ic. Pl. tt. 785–6 (1848) ; F.T.A. 3 : 109 ; Chev. Bot. 325. *O. breviflorus* Benth. (1849)—F.T.A. 3 : 109 ; F.W.T.A., ed. 1, 2 : 80. A forest shrub, to 15 ft. high, with dark green foliage, and stout but soft-wooded branchlets ; flowers white, fragrant.
Guin.: Nimba Mts. (June) *Schnell* 2893 ! **S.L.:** Njala (July) *Deighton* 1168 ! **Lib.:** Cape Palmas (July) *T. Vogel* 27 ! Jabroke, Webo Dist. (fr. July) *Baldwin* 6677 ! **Iv.C.:** various localities *fide* Chev. *l.c.* Banco (July) *de Wilde* 132 ! **Ghana:** Kumasi (Apr., May) *Akwa* FH 1706 ! *Vigne* FH 2924 ! Beyera stream, New Tafo (fr. Nov.) *Lovi* WACRI 3954 ! Accra *Moloney* ! **S.Nig.:** Omo F.R., Ijebu Ode Prov. (fr. Mar.) *Jones & Onochie* FHI 16836 ! 17210 ! Akure F.R. (fr. Aug.) *Jones* FHI 19531 ! Eket Dist. *Talbot* 3310 ! Ikwette Plateau, 5,200 ft., Obudu Div. (fl. buds Dec.) *Keay & Savory* FHI 25259 ! Aboabam (May) *Jones & Onochie* FHI 14150 ! [**Br.**]**Cam.:** Buea to Musake, 5,500 ft., Cam. Mt. (fl. & fr. Apr.) *Hutch. & Metcalfe* 68 ! **F.Po:** (Apr.) *Mann* 390 ! Gutwok Bay (Nov.) *T. Vogel* 209 ! Also in Cameroun, Gabon, Congo and Sudan.

8. **O. gracilis** *Hiern* in F.T.A. 3 : 109 (1877). A forest shrub, 2–10 ft. high ; flowers white ; ripe fruits yellow.
S.Nig.: Omo (formerly part of Shasha) F.R., Ijebu Ode Prov. (Feb.-Apr.) *Richards* 3217 ! 3354 ! *Tamajong* FHI 23257 ! Nikrowa, Benin Div. (Mar.) *Ross* 117 ! Obaritin F.R., Benin Div. (May) *Chizea* FHI 8299 ! Sapoba (fr. Aug.) *Kennedy* 118 ! 2241 ! *Olorunfemi* FHI 34163 ! **F.Po:** (Apr.) *Mann* 389 !

9. **O. laxiflorus** *K. Schum. ex Hutch. & Dalz.* F.W.T.A., ed. 1, 2 : 80 (1931). A glabrous, forest shrub.
S.Nig.: Eket Dist. *Talbot* ! Oban *Talbot* 1314 ! 1359 ! 2034 ! Orem, Oban F.R. (fl. & fr. Jan., Feb.) *Onochie & Okafor* FHI 36107x ! 36147x ! 36272x ! [**Br.**]**Cam.:** Banga, Kumba Div. (June) *Akuo* FHI 15105 ! *Ejiofor* FHI 15257 !¼ Also in Cameroun.

10. **O. speciosus** *DC.* in Ann. Mus. Paris 9 : 218 (1807) ; F.T.A. 3 : 108 ; Chev. Bot. 325 ; Aubrév. Fl. For. C. Iv., ed. 1, 3 : 248, t. 340. *O. racemosus* of Aubrév. l.c. ed. 2, 3 : 286, t. 356. A shrub or tree, to 50 ft. high, with dark shining foliage and fragrant white flowers ; in forest.
Port.G.: Fulacunda (young fr. Aug.) *Esp. Santo* 587 ! Bafata (fr. Aug.) *Esp. Santo* 3298 ! **Guin.:** *Heudelot* 849 ! Dantilia (Mar.) *Sc. Elliot* 5301 ! Pela, Nzérékoré (young fr. Sept.) *Baldwin* 13316 ! **S.L.:** *Smeathmann* ! Heddle's Farm, Freetown (May) *Deighton* 5927 ! Berria, Falaba (fl. buds Mar.) *Sc. Elliot* 5231 ! Rolal, Mambolo (May) *Jordan* 248 ! Sefadu (May) *Small* 69 ! **Lib.:** Moylakwelli (fl. & fr. Oct.) *Linder* 1265 ! Yeh R., 2 miles above junction with St. Paul R. (fr. Oct.) *Linder* 1000 ! Browntown, Tchien (fr. Aug.) *Baldwin* 7087 ! Ganta (Apr.) *Harley* 1904 ! **Iv.C.:** Abidjan *Aubrév.* 342 ; 1353. Djibi *Aubrév.* 535. Rasso *Aubrév.* 568. Malamalasso (Mar.) *Chev.* 17543. **Ghana:** Bia F.R., Sunyani Dist. (Apr.) *Enti* FH 6694 ! Kumasi (May) *Cummins* 213 ! *Vigne* FH 1720 ! Kwahu (Apr.) *Johnson* 466 ! Aburi (June) *Patterson* 298 ! Akropong, Akwapim (fr. Aug.) *Irvine* 795 ! **S.Nig.:** Lagos *Millen* 33 ! 202 ! *Rowland* ! Ibadan (Nov.) *Jones & Keay* FHI 13897 ! Urhehue, Benin Div. (Nov.) *Keay* FHI 25585 ! Bonny R. (Oct.) *Mann* 509 ! Brass *Barter* 1852 ! Oban *Talbot* 288 ! [**Br.**]**Cam.:** Rio del Rey *Johnston* ! Banga, Kumba Div. (fr. Apr.) *Olorunfemi* FHI 30527 ! Also in S. Tomé and widespread in tropical Africa. (See Appendix, p. 406.)

Imperfectly known species.

O. tubiflorus (*Andr.*) *DC.* Prod. 4 : 376 (1830) ; not of F.T.A. 3 : 107, nor of F.W.T.A., ed. 1, 2 : 80. *Gardenia* *tubiflora* Andr. Bot. Repos. 3 : t. 183 (1801). Andrews' plate is of a plant grown in a hot-house at Hammersmith (London) from seed sent from Sierra Leone ; I have not been successful in tracing a specimen of this plant. The plate differs from *O. racemosus* (the plant referred to as *O. tubiflorus* in the F.T.A. and F.W.T.A.) as follows :—(i) stem, leaves and flowers glabrous, (ii) leaves cuneate at base, (iii) inflorescence shorter and ebracteate, (iv) calyx-lobes broader. I cannot, moreover, satisfactorily match Andrews' plate with any other specimen. It is perhaps nearest to *O. subpunctatus* but differs in the broadly triangular calyx-lobes. Since the name *O. tubiflorus* cannot certainly be associated with any known species of *Oxyanthus* it seems best to cease using it.—R.W.J.K.

23. MITRIOSTIGMA Hochst.—F.T.A. 3 : 111.

Branchlets glabrous ; leaves broadly oblanceolate, very acute at base, acutely acuminate, 10–15 cm. long, 3–4·5 cm. broad, glabrous, with about 6–8 pairs of looped lateral nerves and lax venation between ; stipules subulate from a broad base, glabrous ; cymes axillary, very short and few-flowered ; calyx-lobes subulate, 1–1·5 mm. long ; corolla 1 cm. long, gradually widened from a cylindric base ; fruits oblong-lanceolate in outline, beaked, 3 cm. long ; seeds with a fibrous testa ..   ..   *barteri*

M. barteri *Hook.f. ex Hiern* in F.T.A. 3 : 111 (1877) ; Gen. Pl. 2 : 90 (1873), name only. *Randia barteri* (Hook. f. ex Hiern) K. Schum. in E. & P. Pflanzenfam. 4, 4 : 75 (1891). Half-woody undershrub 3 ft. high ; " calyx green, corolla brown ".
F.Po: *Barter* ! *Mann* 234 !

24. DIDYMOSALPINX Keay in Bull. Jard. Bot. Brux. 28 : 61 (1958).

Calyx glabrous, tube scarcely 1 mm. long, lobes triangular 2–4 mm. long ; corolla-tube 3·5–8 cm. long, lobes ovate, acuminate, 1·5–3 cm. long ; anthers 1–1·4 cm. long ; style-head dilated, fusiform ; pedicels 1·2–2·4 cm. long, without bracteoles ; branchlets, stipules and leaves glabrous ; arrested lateral branchlets sometimes becoming spines ; petioles up to 8 mm. long ; leaves drying greenish, elliptic to elliptic-lanceolate, cuneate at base, acuminate at apex, 3·5–18 cm. long, 1·7–8 cm. broad, with 5–7 main lateral nerves on each side of midrib, usually pitted in the axils ; fruits subglobose, about 2·5 cm. diam. ..   ..   1. *abbeokutae*
Calyx puberulous, tube about 2 mm. long, lobes triangular, 1·5–2 mm. long ; corolla-tube 1·8–2·5 cm. long, lobes obovate-elliptic, rounded at apex, 1·5–2 cm. long ; anthers about 1 cm. long ; style-head attenuate ; pedicels about 1·2 cm. long, with puberulous bracteoles ; stipules and young leaves pubescent ; branchlets not becoming spines ; petioles 8–16 mm. long ; leaves drying reddish beneath, oblong or obovate-oblong, usually obtuse at base, acuminate at apex, 8·5–22 cm. long, 4–8 cm. broad, with 5–10 main lateral nerves on each side of midrib, not pitted in axils ; young fruits fusiform ..   ..   2. *parviflora*

1. **D. abbeokutae** *(Hiern) Keay* in Bull. Jard. Bot. Brux. 28 : 62, t. 5 (1958) ; Aubrév. Fl. For. C. Iv., ed. 2, 3 : 274. *Gardenia abbeokutae* Hiern in F.T.A. 3 : 104 (1877) ; F.W.T.A., ed. 1, 2 : 75. *G. hiernii* Sc. Elliot (1894). A scrambling glabrous shrub with shining leaves and cream-white or greenish-yellow flowers ; in forest vegetation.
Guin.: Kissidougou (Apr.) *Martine* 297 ! **S.L.:** Bagroo R. (Apr.) *Mann* 840 ! Kafogo, Limba (Apr.) *Sc. Elliot* 5516 ! Freetown (May) *Deighton* 2984 ! Njala (fl. Apr., May, fr. Mar.) *Deighton* 3434 ! 4744 ! 5749 ! Magbile (fr. Dec.) *Thomas* 5946 ! 6475 ! **Lib.:** Jaurazon, Sinoe County (Mar.) *Baldwin* 11451 ! 11459 ! Dukwia R. (Feb.) *Cooper* 290 ! 327 ! 388 ! Taninewa (Mar.) *Bequaert* 151 ! Mt. Barclay (May) *Bunting* ! **Iv.C.:** Mt. Tonkoui (fr. Jan.) *Schnell* 4065 ! Man *Chev.* 21240. Bériby *Pobéguin* 45. **Ghana:** Amentia (Mar., Apr.) *Vigne* FHI 1873 ! 2889 ! Bosusu, Begoro (July) *Moor* 91 ! Cape Coast (May) *Dalz.* 51 ! Asakraka to Nteso (Apr.) *Kitson* 1044 ! **S.Nig.:** Abeokuta *Irving* 32 ! Ibadan South F.R. (Apr.) *Keay* FHI 22545 ! Oban *Talbot* 1519 ! **[Br.]Cam.:** Victoria to Bimbia (May) *Preuss* 1259 ! Also in Cameroun and Congo. (See Appendix, p. 398.)
2. **D. parviflora** *Keay* l.c. 64 (1958). An erect shrub, 6–12 ft. high, or small tree ; flowers slightly fragrant, calyx pale green, corolla white ; in forest.
S.Nig.: Boje, Afi River F.R., Ikom Div. (May) *Jones & Onochie* FHI 18905 ! 18912 ! **[Br.]Cam.:** Mabeta, Victoria Div. (young fr. May) *Motuba* FHI 15067 ! Bopo, S. Bakundu F.R. (Mar.) *Brenan* 9425 !

25. POUCHETIA A. Rich. ex DC. Prod. 4 : 393 (1830) ; A. Rich. in Mém. Soc. Hist. Nat. Paris 5 : 251 (1834) ; F.T.A. 3 : 116 ; K. Schum. in E. & P. Pflanzenfam. 4, 4 : 79 (1891).

Upper bracteate leaves imbricate, markedly decreasing in size towards apex of shoot and broadly ovate with cordate base, 1–4 cm. long, 1–2·5 cm. broad, normal leaves ovate-elliptic, rounded to cuneate at base, long-acuminate, 6–9(–15) cm. long, 2–4(–6) cm. broad, glabrous ; corolla about 1 cm. long, tubular ; flowers in axillary clusters or shortly pedunculate with peduncle up to 2·5 cm. long ..   1. *gilletii*
Leaves all similar, cuneate to obtuse at base, not as above :
Corolla funnel-shaped, the 3·5 mm. long tube gradually widened into the 3·5 mm. long, more or less erect lobes ; anthers 4 mm. long ; inflorescence and outside of calyx glabrous, except for the ciliate calyx-teeth ; flowers in lax axillary and terminal racemes or panicles, up to 11 cm. long ; fruits ellipsoid, acute, nearly 1 cm. long ; leaves oblong-elliptic, obtuse or cuneate at base, broadly acuminate, 2–11 cm. long, 1–5·5 cm. broad, pubescent only in the axils of the 4–7 main lateral nerves beneath   2. *africana*
Corolla with narrow tube 2 mm. long, abruptly widened into the reflexed, 3 mm. long lobes ; anthers 2·75 mm. long ; inflorescence and outside of calyx more or less densely puberulous ; flowers in axillary and terminal usually narrow panicles, up to 13 cm.

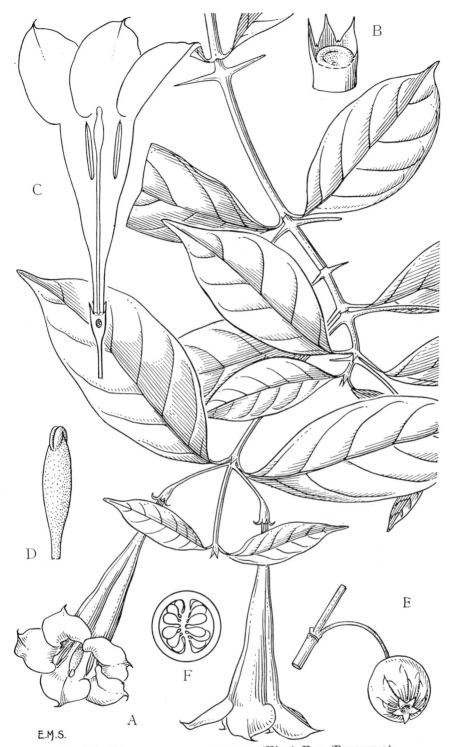

E.M.S.

FIG. 228.—DIDYMOSALPINX ABBEOKUTAE (*Hiern*) KEAY (RUBIACEAE).

A, flowering shoot, × ⅔. B, ovary and section of calyx, × 4. C, longitudinal section of flower, × 1. D, apex of style and stigma, × 2. E, fruit, × ⅔. F, transverse section of ovary (diagrammatic). A drawn from *Deighton* 2660. B–D & F from *Keay* FHI 22545. E from *Thomas* 6473.

long ; leaves broadly elliptic, cuneate at base, gradually acuminate, 6–14 cm. long, 2·5–6 cm. broad, puberulous on the midrib and 6–9 main lateral nerves beneath .... 3. *parviflora*

1. **P. gilletii** *De Wild.* in Ann. Mus. Congo, sér. 5, 3 : 289 (1910). A shrub about 10 ft. high with pendulous inflorescences ; flowers pale cream and berries dark red ; in swamp forest.
    S.Nig.: Imo River Village, Port Harcourt to Aba (fl. & fr. Dec.) *Onochie* FHI 40442 ! Also in Congo and Angola.

2. **P. africana** *A. Rich. ex DC.* Prod. 3 : 393 (1830) ; F.T.A. 3 : 117, excl. var. ; Chev. Bot. 325 ; Aubrév. Fl. For. Soud.-Guin. 464, t. 101, 6–7. A shrub or small tree, up to 15 ft. or more high, glabrous, with greenish or white, scented flowers ; berries blackish-purple when ripe ; in coastal vegetation and in forest, including fringing forest in savanna regions.
    Sen.: Casamance & Cayor *fide* DC. *l.c.* **Gam.**: *Heudelot* 28 ! *fide* DC. *l.c.* **Mali**: Kouroukouto (fr. Mar.) *Roberty* 17082 ! Katou (Mar.) *Chev.* 541. Nguer to Bir, Tialam (Dec.) *Chev.* 2097 ! **Port.G.**: Prabis, Bissau (young fr. Feb.) *Esp. Santo* 1813 ! Bissalanca, Bissau (Jan.) *Esp. Santo* 1661 ! **Guin.**: *Heudelot* 682 ! Conakry *Chev.* 12018. Kadé, R. Badi, Timbo & Bramaya *fide* Aubrév. *l.c.* **S.L.**: *Don* ! *Smeathmann*. *T. Vogel* 85 ! Freetown (Oct.) *Deighton* 248 ! Bonthe (Apr.) *Deighton* 4303 ! Surinuia, Talla Hills (Apr.) *Sc. Elliot* 5527 ! Rosino (Mar.) *Deighton* 4974 ! Mabonto (Oct.) *Thomas* 3527 ! **Lib.**: Monrovia (Nov.) *Linder* 1420 ! 1559 ! Sangwin (fl. & fr. Dec.) *Baldwin* 11440 ! Congotown (Aug.) *Barker* 1398 ! **Iv.C.**: Diénougou *Aubrév.* 2737 ! **Ghana**: Bou, W. Prov. (Nov.) *Vigne* FH 1419 ! Prang (Jan.) *Vigne* FH 1562 ! Sambisi to Hamboi (May) *Kitson* 793 ! Wa, N.T. (Dec.) *Adams & Akpabla* 4469 ! **Togo Rep.**: Dgaba *Kersting* A270 ! **N.Nig.**: Borgu *Barter* 714 ! Also in Ubangi-Shari (*fide* Aubrév. *l.c.*).

3. **P. parviflora** *Benth.* in Hook. Fl. Nigrit. 395 (1849) ; F.T.A. 3 : 117. *Feretia ? virgata* K. Schum. (1905). A shrub or small tree, with white flowers ; in forest.
    S.L.: Lumbaraya, Talla Hills, up to about 3,000 ft. (Feb.) *Sc. Elliot* 4989 ! **S.Nig.**: Gambari Group, Ibadan (young fr. Feb.) *Chizea* FHI 23972 ! *W. D. MacGregor* 579 ! Ibadan (fl. & young fr. Jan.) *Meikle* 954 ! **F.Po**: (Jan., Nov.) *Barter* 2065 ! *T. Vogel* 183 ! 252 ! Also in Principe and S. Tomé.

26. DICTYANDRA Welw. ex Hook. f.—F.T.A. 3 : 85 ; K. Schum. in E. & P. Pflanzenfam. 4, 4 : 73 (1891).

Corolla about 2½ times as long as the calyx, silky-tomentose outside, 3 cm. long ; calyx-lobes broadly elliptic, about 1 cm. long, with one thickened recurved margin ; leaves obovate-elliptic, shortly-acuminate, 12–20 cm. long, 5–9 cm. broad, slightly pubescent in the axils of the nerves beneath ; stipules broadly triangular, 5 mm. long ; flowers in lax cymes ; style hairy ..   .. 1. *arborescens*
Corolla about 8 times as long as the calyx, slender, appressed-pilose, about 8 cm. long ; calyx-lobes lanceolate, hardly 1 cm. long, with slightly curled edges ; leaves obovate, sharply cuneate at the base, 12–23 cm. long, 6–12 cm. broad, glabrous ; stipules semi-orbicular, leafy, 1–1·5 cm. long ; flowers rather crowded ..   .. 2. *involucrata*

1. **D. arborescens** *Welw. ex Hook.f.* in Benth. & Hook. f. Gen. Pl. 2 : 85 (1873) ; F.T.A. 3 : 86 ; Aubrév. Fl. For. C. Iv., ed. 2, 3 : 272, t. 350. A shrub or small tree, up to 25 ft. high, with shining dark green foliage, white silky fragrant flowers 1–1½ in. long in cymes, and small guava-like fruits crowned by the leafy calyx ; bole with short, stout, blunt, conical spines ; in forest.
    Guin.: Nzo (Mar.) *Schnell* 869 ! **S.L.**: Limba, Bafodeya (Apr.) *Sc. Elliot* 5584 ! Koflu Mt. (Jan.) *Sc. Elliot* 4760 ! Kenema *Thomas* 7837 ! 7923 ! Jama (fr. Sept.) *Deighton* 3081 ! **Lib.**: Zigida (Mar.) *Bequaert* 131 ! **Iv.C.**: Bouroukrou *Aubrév.* 712 ! Bouaké *Aubrév.* 119 ! **Ghana**: Aburi Hills (Oct.) *Johnson* 465 ! Banka (Dec.) *Vigne* FH 1484 ! Bunso, Akim (Oct.) *Scholes* 312 ! Awoso, Dunkwa (Oct.) *Akpabla* 875 ! **Dah.**: Le Testu 67 ! **S.Nig.**: Abakurudu, Shasha F.R. (Mar.) *Jones & Onochie* FHI 16635 ! Idanre Hills (Jan.) *Brenan* 8676 ! Idumuje (Dec.) *Thomas* 2112 ! 2137 ! Oban *Talbot* 1515 ! Calabar *Thomson* 77 ! **[Br.]Cam.**: Kiliwindi to Mamfe (fr. July) *Akuo* FHI 15155 ! **F.Po**: Guinea 1512 ! Extends to Uganda, Congo and Angola. (See Appendix, p. 396.)

2. **D. involucrata** (*Hook. f.*) *Hiern* in F.T.A. 3 : 86 (1877). *Leptactina involucrata* Hook. f. (1871). A shrub 15–20 ft. high or more or less climbing, with rather broad leaves and long tubular white flowers 3–4 in. long, crowded in terminal cymes ; in forest.
    S.L.: Lumbaraya (young fr. Feb.) *Sc. Elliot* 4935 ! **Ghana**: Kudje, Hohoe Dist. (Nov.) *St. C.-Thompson* FH 3688 ! **N.Nig.**: Patti Lokoja (Nov.) *Dalz.* 99 ! **S.Nig.**: Awba Hills, Ibadan (Oct.) *Jones* FHI 5972 ! Sapoba F.R., Benin (Nov.) *A. F. Ross* 231 ! Okwashi (young fr. Nov.) *Thomas* 2035 ! Oban *Talbot* 283 ! Degema Dist. *Talbot* ! **[Br.]Cam.**: Ambas Bay (Dec.) *Mann* 2156 ! S. Bakundu, Kumba Dist. (fr. Jan.) *Binuyo & Daramola* FHI 35469 ! Ossidinge (Sept.) *Rudatis* 73 ! R. Sulli, Gashaka Dist. (fr. Dec.) *Latilo & Daramola* FHI 28840 ! Also in Cameroun and Rio Muni.

27. LEPTACTINA Hook. f.—F.T.A. 3 : 87 ; K. Schum. in E. & P. Pflanzenfam. 4, 4 : 73 (1891).

Style included in the corolla-tube ; leaves more than 7·5 cm. long and 3 cm. broad, with 7–10 main lateral nerves on each side of midrib ; petioles more than 5 mm. long ; calyx-lobes 1·3–3 cm. long ; corolla-tube (2–)4–8(–10·5) cm., lobes (1·2–)1·8–3 (–3·5) cm. long ; branchlets densely or sparingly pilose, or nearly glabrous ; leaves broadly elliptic to obovate, obtuse or cuneate at base, more or less acuminate at apex, 7·5–20 cm. long, 3–11 cm. broad :
Ovary densely pubescent ; leaves usually with long weak hairs and tufts of shorter hairs by the midrib and nerves beneath ..   .. 1. *densiflora* var. *densiflora*
Ovary glabrous or sparingly pubescent ; leaves with only short hairs beneath .. 1a. *densiflora* var. *glabra*
Style exserted ; leaves narrowly elliptic, acute at both ends, 3–8 cm. long, 1–3·5 cm. broad, softly pubescent beneath, with 5–8 main lateral nerves on each side of midrib ; petioles 3–4 mm. long ; calyx-lobes about 1·3 cm. long ; corolla-tube 2–5 cm., lobes 1·5–2·5 cm. long, ovary densely pubescent ..   ..   .. 2. *senegambica*

1. **L. densiflora** *Hook. f.* var. *densiflora*—Hook. Ic. Pl. sub t. 1092 (1871) ; F.T.A. 3 : 87. A large bushy shrub with rather large leaves, and white fragrant tubular flowers 1–3 in. long, in dense heads ; fruits black, ribbed and wrinkled, ½ in. diam., crowned by the long calyx.
    S.L.: Mabonto *Thomas* 3542 ! Nganyama, Tikonko (Sept.) *Deighton* 4028 ! Yonibana (Nov.) *Thomas*

4861 ! 4908 ! 5031 ! Mamaha *Thomas* 4396 ! Batkanu (fr. Jan.) *Deighton* 2845 ! **Lib.**: Gbanga (Sept.) *Linder* 617 ! Gonatown, Grand Cape Mount *Baldwin* 10783 ! **Iv.C.**: Adiopodoumé *Giovannetti* 51 ! **Ghana:** Krobo Plains (Dec.) *Johnson* 512 ! Bana Hill, Krobo (fr. Jan.) *Irvine* 1916 ! Volta River F.R. *Morton* GC 6072 ! Alavanyo, Hohoe Dist. (Nov.) *St. C.-Thompson* FH 3654 ! **S.Nig.**: Lagos (Dec.) *Millen* 23 ! 181 ! Abeokuta (Jan.) *Irving* 94 ! *Rowland* ! Eba Isl. *Kennedy* 1705 ! Sapoba (July) *Onochie* FHI 23421 ! Uwet Odot F.R., Itu (Jan.) *Jones* FHI 6861 ! Oban *Talbot* 220 ! 1613 !

1a. **L. densiflora** var. **glabra** *Hutch. & Dalz.* F.W.T.A., ed. 1, 2 : 84 (1931), excl. *Johnson* 512. Similar to above but less hairy.

Ghana: Atauaso, N.E. of Tarkwa (May) *Chipp* 230 ! Kumasi *Cummins* 150 ! Jiasi (fl. & fr. Oct.) *Vigne* FH 1391 ! Essiama (Nov.) *Vigne* FH 1407 ! Central Prov. *Fishlock* 76 ! **S.Nig.**: Sapoba, Benin Prov. (fr. Nov.) *Ejiofor* FHI 24614 ! *Kennedy* 1921 !

2. **L. senegambica** *Hook. f.* l.c. (1871) ; F.T.A. 3 : 88 ; Chev. Bot. 318 ; Aubrév. Fl. For. Soud.-Guin. 469, t. 101, 1–3. A shrub with rather stiff subquadrangular branchlets, dark shining foliage, and greenish flowers 1–2½ in. long, softly hairy outside in small dense cymes.

Mali: Moussaia (Feb.) *Chev.* 402. **Guin.**: 126 km. S. of Kouroussa (Dec.) *Roberty* 16194 ! Kouroussa (Feb., Apr.) *Chev.* 375 ! *Pobéguin* 220 ! 678 ! Rio Pongo *Heudelot* 893 ! And other localities *fide* Aubrév. l.c. **S.L.**: *Afzelius* ! Koflu Mt., Scarcies (Jan.) *Sc. Elliot* 4759 ! Kukuna (Apr.) *Hepper* 2643 ! Makumre (June) *Thomas* 504 ! (See Appendix, p. 401.)

28. TARENNA Gaertn.—F.T.A. 3 : 88.

Style pilose in the lower half, always well-exserted ; ovules several in each ovary cell :
Calyx glabrous outside, lobes very short, often broader than long ; inflorescences glabrous or puberulous ; corolla-tube glabrous outside, pilose within :
Corolla-tube 9–30 mm. long ; pedicels 4–35 mm. long, glabrous or puberulous :
Pedicels 25–35 mm. long ; cymes 1–3-flowered ; corolla-tube 10–13 mm. long, lobes 7–8 mm. long ; leaves elliptic, rounded or obtuse at base, abruptly spathulate-acuminate, 3·5–8·5 cm. long, 1·2–3·8 cm. broad, glabrous, with 2–4 main lateral nerves looped on each side of the midrib .. .. .. 1. *flavo-fusca*
Pedicels 4–10 mm. long ; cymes several- to many-flowered ; leaves more or less asymmetrically long-cuneate at base, acumen not spathulate ; main lateral nerves 4–9 on each side of midrib :
Corolla-tube 25–30 mm., lobes 8–9 mm. long ; leaves oblong-oblanceolate or oblong-elliptic, acuminate, 12–24 cm. long, 4–9·5 cm. broad, glabrous ; fruits sub-globose, 10–12 mm. diam., 10-ribbed, glabrous .. .. 2. *grandiflora*
Corolla-tube 9–10 mm., lobes 6–8 mm. long ; leaves oblong-oblanceolate, long-acuminate, 5·5–17 cm. long, 1·7–5 cm. broad ; fruits as above, but scarcely ribbed :
Leaves rather densely pubescent beneath .. .. 3. *vignei* var. *vignei*
Leaves glabrous or nearly so beneath .. .. 3a. *vignei* var. *subglabra*
Corolla-tube 2·5–4 mm., lobes 4–6(–7) mm. long ; pedicels 1–2 mm. long, puberulous ; leaves oblong-elliptic, rather asymmetrically cuneate at base, acuminate, 8–17 cm. long, 2·5–6·5 cm. broad, glabrous, with 4–6 main lateral nerves on each side of midrib, prominent beneath .. .. 4. *pallidula*
Calyx pubescent or setose outside, lobes short or long, mostly longer than broad ; corolla-tube glabrous, puberulous or pubescent outside, glabrous or pilose inside ; inflorescences mostly densely puberulous, pubescent or setose ; pedicels 0·5–10 mm. long :
Corolla-tube glabrous inside :
Outside of corolla-tube puberulous ; leaves glabrous ; calyx-lobes ovate-triangular, up to 0·5 mm. long, densely puberulous ; corolla-tube 6–8 mm., lobes 4–6 mm. long ; pedicels 3–5 mm. long ; leaves ovate, elliptic or obovate-elliptic, cuneate at base, acuminate, 7–18 cm. long, 2–7·5 cm. broad, with 7–10 main lateral nerves on each side of midrib ; fruits slightly bilobed, 8–9 mm. diam., glabrous, wrinkled and shining when dry .. .. 5. *gracilis*
Outside of corolla-tube glabrous ; leaves pubescent on midrib and nerves and appressed-puberulous on lamina beneath ; calyx-lobes triangular-lanceolate, 1–2 mm. long, densely setose-pubescent ; corolla-tube 3–5 mm., lobes 5–7 mm. long ; pedicels 1–3 mm. long ; leaves oblanceolate, oblong-elliptic or lanceolate, cuneate at base, 2·8–12 cm. long, 1·5–5 cm. broad, with 6–9 main lateral nerves on each side of midrib ; fruits subglobose, 5 mm. diam., glabrous, wrinkled and shining when dry .. .. 6. *thomasii*
Corolla-tube pilose inside :
Outside of corolla-tube pubescent, at least when young ; inflorescence-axes, the 4–10 mm. long pedicels, ovary and the 1 mm. long ovate-acute calyx-lobes densely golden-brown (when dry) silky pubescent ; corolla-tube 6 mm., lobes 4–4·5 mm. long ; leaves coriaceous, drying brown, elliptic to obovate, cuneate at base, acuminate, 8·5–15 cm. long, 2·7–6 cm. broad, pubescent beneath especially on the midrib, with 5–8 main lateral nerves on each side of midrib .. 7. *eketensis*
Outside of corolla-tube glabrous ; pedicels 1–4 mm. long :
Calyx-lobes oblong, rounded at apex, 2 mm. long, 1·5 mm. broad, margins ciliate, often overlapping at anthesis ; leaves rounded or obtuse at base, oblong, 10–17 cm. long, 4–5 cm. broad, rather abruptly acuminate, glabrous, with 4–6 main lateral nerves on each side of midrib ; corolla-tube 7–8 mm., lobes 8 mm. long ; pedicels 1–1·5 mm. long .. .. 8. *baconioides*

Calyx-lobes triangular, ovate or lanceolate, narrower and/or shorter than above, not overlapping at anthesis ; leaves cuneate at base, mostly puberulous beneath :
Calyx-lobes lanceolate, 1·5–3·5 mm. long, up to 1 mm. broad ; inflorescence-axes, pedicels and ovary densely setose-pilose ; corolla-tube 7–10 mm., lobes 5–8 mm. long ; leaves elliptic or obovate, obtuse or cuneate at base, acuminate, 8–20 cm. long, 3–10 cm. broad, appressed-puberulous on midrib above and on midrib, nerves and lamina beneath ; main lateral nerves 6–13 on each side of midrib ; fruits subglobose, glabrous, about 7 mm. diam., wrinkled, black and shining when dry   ..   ..   9. *lasiorhachis*
Calyx-lobes broadly triangular or ovate, up to 1·5 mm. long ; inflorescence-axes, pedicels and ovary more or less densely appressed-pubescent :
Inflorescence up to 3·5 cm. diam., densely flowered, with pedicels up to 2 mm. long, closely subtended by 1–2 pairs of small leaves near the apex of long lateral shoots ; calyx-lobes 1–1·5 mm. long ; corolla-tube and lobes each about 5 mm. long ; leaves obovate or elliptic, distinctly and often abruptly acuminate, 8–17·5 cm. long, 3–7·8 cm. broad, rather densely puberulous beneath, with 6–9 main lateral nerves on each side of midrib ; fruits subglobose, about 8 mm. diam., glabrous, shining   ..   ..   10. *conferta*
Inflorescence 4–10 cm. diam., pedicels 1–4 mm. long ; leaves sparingly puberulous, especially in axils of nerves, or glabrous beneath ; calyx-lobes 0·5–1 mm. long :
Inflorescence dense ; leaves obovate, acuminate, 7–16 cm. long, 3–6·5 cm. broad, with 6–9 main lateral nerves on each side of midrib ; corolla-tube 4–5 mm., lobes 4–5·5 mm. long   ..   ..   11. *pavettoides*
Inflorescence lax ; leaves narrowly elliptic or elliptic-oblanceolate, very gradually long-acuminate, 6–14 cm. long, 2–5·5 cm. broad, with 5–8 main lateral nerves on each side of midrib ; corolla-tube and lobes each 5–8 mm. long
                                                                           12. *nitidula*
Style glabrous in the lower half, often only slightly exserted or included ; leaves usually drying blackish ; pedicels (2–)4–7 mm. long :
Calyx and ovary glabrous ; corolla-tube glabrous inside :
Calyx-lobes broad and imbricate, overlapping to the left, 1 mm. long ; corolla-tube short (specimens available only in bud) ; pedicels setose ; leaves oblanceolate-elliptic, cuneate at base, acuminate, 7–12 cm. long, 2–5 cm. broad, with 6–8 main lateral nerves on each side of midrib ; ovules 4 per cell   ..   13. *hutchinsonii*
Calyx-lobes triangular, 0·5 mm. long ; corolla-tube 12–19 mm. long, glabrous inside and outside, lobes 6–10 mm. long ; pedicels glabrous ; leaves ovate, lanceolate, elliptic or obovate, obtuse, cuneate or rounded at base, acuminate, 6–16 cm. long, 2–9 cm. broad, with 3–6 main lateral nerves on each side of midrib ; ovules 1 per cell ; fruit usually with a single seed or with a second imperfectly developed seed
                                                                           14. *soyauxii*
Calyx and ovary puberulous outside ; corolla-tube pilose inside :
Corolla-tube 4–7(–10) mm. long, glabrous or puberulous outside, lobes 7–8(–9) mm. long ; calyx puberulous, lobes lanceolate to subulate, 1–2 mm. long ; pedicels puberulous ; leaves narrowly elliptic, elliptic-lanceolate or oblanceolate, usually unequal at base, obtuse and/or cuneate, gradually obtusely acuminate, 6·5–18 cm. long, 3–8 cm. broad, with 7–11 main lateral nerves on each side of midrib ; ovules usually 3 per cell ; fruit with 2–6 seeds   ..   ..   ..   15. *bipindensis*
Corolla-tube 4 mm. long, glabrous outside, lobes 4–4·5 mm. long ; calyx minutely puberulous ; lobes triangular, 0·75 mm. long ; pedicels glabrous or puberulous ; leaves similar to above   ..   ..   16. *brachysiphon*

1. **T. flavo-fusca** (*K. Schum.*) *S. Moore* in J. Linn. Soc. 37 : 302 (1906). *Chomelia flavo-fusca* K. Schum. in Engl. Bot. Jahrb. 33 : 339 (1903). *C. laxissima* K. Schum. l.c. 340 (1903). *Ixora laxissima* (K. Schum.) Hutch. & Dalz. F.W.T.A., ed. 1, 2 : 87 (1931). A slender climber, or small tree to 12 ft. high ; flowers white ; in forest.
Lib.: Yeh R., 2 miles above junction with St. Paul R. (Oct.) *Linder* 993! **Ghana:** Konongo (Apr.) *Vigne* FH 1929! Abofaw (June) *Vigne* FH 1175! **S.Nig.:** Oban *Talbot* 232! [Br.]**Cam.:** Kumba (June) *Preuss* 312! Mbalange F.R., Kumba Div. (fr. Jan.) *Binuyo & Daramola* FHI 35479! Also in Cameroun, Angola and Uganda.
2. **T. grandiflora** (*Benth.*) *Hiern* in F.T.A. 3 : 91 (1877). *Stylocoryne grandiflora* Benth. (1849). *Chomelia grandiflora* (Benth.) O. Ktze. (1891). *C. neurocarpa* K. Schum. (1903). A shrub or small tree to 50 ft. high ; corolla green outside and white inside ; fruits pale orange ; in forest.
S.Nig.: Eket Dist. *Talbot* 3366! Oban *Talbot* 279! Ikom (fr. May) *Jones & Onochie* FHI 17444! [Br.] **Cam.:** Mile 49, Victoria to Kumba road (Aug.) *Olorunfemi* FHI 30745! Korup F.R., Kumba (fr. July) *Olorunfemi* FHI 30670! **F.Po:** (June, Dec.) *Barter*! *Mann* 25! *T. Vogel* 262! Also in Cameroun.
3. **T. vignei** *Hutch. & Dalz.* var. vignei—F.W.T.A., ed. 1, 2 : 76 (1931). A forest shrub.
Ghana: Amentia (fl. buds Nov.) *Vigne* FH 1461!
3a. **T. vignei** var. **subglabra** *Keay* in Bull. Jard. Bot. Brux. 27 : 98 (1957). *Tarenna* sp. of Aubrév. Fl. For. C. Iv., ed. 1, 3 : 237. A forest shrub or small tree, to 12 ft. high ; flowers white, fragrant ; fruits yellow or orange when ripe.
S.L.: Gola Forest *Small* 554! Gorahun to Zimi, Gola Forest (fl. & fr. Apr.) *Deighton* 3640! 3685! **Lib.:** Vahon, Kolahun Dist. (fr. Nov.) *Baldwin* 10248! Ba, Mano R. (fr. Sept.) *Baldwin* 10724! Peahtah, Salala Dist. (fr. Dec.) *Baldwin* 10610! Zeahtown, Tchien Dist. (fr. Aug.) *Baldwin* 6930! Sarbo, Webo Dist. (fr. July) *Baldwin* 6409! **Iv.C.:** Man to Danané *Aubrév.* 1033. Kissidougou *Chev.* 20758. **Ghana:** Tano-Nimri F.R. (fr. July) *Foggie* FH 4459! Ankasa Su (fr. July) *Chipp* 304! Ankassa F.R. (fl. buds & fr. Dec.) *Vigne* FH 3179! Enchi (fr. Oct.) *Tolmie* FH 2380!

4. **T. pallidula** *Hiern* in F.T.A. 3 : 91 (1877). *Chomelia oligoneura* K. Schum. (1896). *Tarenna patula* Hutch. & Dalz. F.W.T.A., ed. 1, 2 : 76 (1931). A glabrous shrub ; in forest.
S.Nig.: Eket Dist. *Talbot* 3257! Stubb's Creek F.R., Eket Dist. (young fr. May) *Onochie* FHI 33161!
Also in Cameroun, Rio Muni and Gabon.

5. **T. gracilis** (*Stapf*) *Keay* l.c. 99 (1957) ; Chev. Bot. 319, name only. *Webera gracilis* Stapf in J. Linn. Soc. 37 : 106 (1905). *Tarenna thomasii* of F.W.T.A., ed. 1, 2 : 76, partly (*Sc. Elliot* 5554, only). *T. nitidula* of F.W.T.A., ed. 1, 2 : 76, partly. A shrub, apparently deciduous, 3–9 ft. high ; flowers white, fragrant ; fruits white, borne on leafless lower lateral branchlets.
S.L.: Bafodeya, Limba (Apr.) *Sc. Elliot* 5554! Kenema (Apr.) *Lane-Poole* 284! Njala (May) *Deighton* 3437! 3438! 3968! Nyeminga (Apr.) *Deighton* 3680! Panguma (Apr.) *Deighton* 3907! Bumpe, Peri (fr. Aug.) *Deighton* 5134! **Lib.:** Sinoe Basin *Whyte*! Kulo, Sinoe County (fl. buds Mar.) *Baldwin* 11426! Ganta (Feb.) *Harley* 867! Wanau (Mar.) *Harley* 1389! **Iv.C.:** Zaranou, Indénié (Mar.) *Chev.* 17623! **Ghana:** Simpa, W. Prov. (Feb.) *Andoh* FH 2789! Mpraeso Scarp (fr. Apr.) *Morton* A 694!

6. **T. thomasii** *Hutch. & Dalz.* F.W.T.A., ed. 1, 2 : 76 (1931), excl. *Sc. Elliot* 5554. *T. flexilis* A. Chev. Bot. 319 (1920), name only. *T.* sp. ? nov. of F.T.A. 3 : 90. *T. conferta* of Chev. Bot. 319, partly (*Chev.* 578) ; of F.W.T.A., ed. 1, 2 : 76, partly (Guin. & Iv.C. records). A shrub, 5–12 ft. high ; flowers white, fragrant; fruits red ; in forest vegetation, especially by streams.
Guin.: R. Milo, Kankan (Mar., Apr.) *Adam* 12134! *Chev.* 578! R. Niger, Farana (Mar.) *Sc. Elliot* 5344! **S.L.:** *Thomas* 10463! 10496! 10539! 10614! Laminaiya (Apr.) *Thomas* 126! Njala (Mar., May) *Deighton* 1582! 5042! Mesima (Apr.) *Deighton* 3695! Diamain (Apr.) *Adames* 36! **Iv.C.:** Ouaguié to Capiécrou, Agniéby Valley (Jan.) *Chev.* 17177! **Ghana:** Prang (fr. June) *Vigne* FH 3897! Suguri, N.T. (Apr.) *Pomeroy* FH 1232! **N.Nig.:** Abinsi (fr. May) *Dalz.* 636! **S.Nig.:** Onitsha *Barter* 1245! Obubra (Jan.) *Jones* FHI 5878!

7. **T. eketensis** *Wernham* in J. Bot. 52 : 4 (1914). *Rutidea degemensis* Wernham in op. cit. 55 : 80 (1917). *R. talbotiorum* Wernham l.c. 81 (1917). A scandent shrub to 30 ft. high, or a tree ; inflorescence light golden-silky pubescent, flowers white ; in forest, especially in regrowth.
Lib.: Karmadhun, Kolahun (fr. Nov.) *Baldwin* 10164! **Ghana:** Bou, W. Prov. (Nov.) *Vigne* FH 1411! Bogosu (Dec.) *Vigne* FH 3147! **S.Nig.:** Sapoba (Mar., Nov.) *Jones* FHI 547! *Keay* FHI 28076! Oron-Eket Road *Talbot* 3024! Degema Dist. *Talbot* 3827! 3828! s.n.! Oban Dist. *Talbot*! [**Br.]Cam.:** Abonando (May) *Rudatis* 61! Also in Cameroun, Rio Muni and Gabon.
[*Chomelia gilletii* De Wild. & Th. Dur. (1900) from Congo is very similar but has membranous instead of coriaceous leaves.]

8. **T. baconioides** *Wernham* in Cat. Talb. 44 (1913). A shrub, glabrous except on the inflorescence ; flowers white, crowded in terminal trichotomous cymes ; in forest.
S. Nig.: Oban *Talbot* 1595!

9. **T. lasiorhachis** (*K. Schum. & K. Krause*) *Bremek.* in Fedde Rep. 37 : 196 (1934). *Pavetta lasiorhachis* K. Schum. & K. Krause in Engl. Bot. Jahrb. 39 : 550 (1907). A forest shrub or small tree, to 10 ft. high ; flowers white.
[**Br.]Cam.:** Victoria (fl. Nov., fr. Apr.) *Maitland* 87! 620! Mamfe (Mar.) *Richards* 5221! Also in Cameroun.

10. **T. conferta** (*Benth.*) *Hiern* in F.T.A. 3 : 90 (1877) ; F.W.T.A., ed. 1, 2 : 76, partly (S. Nig. & F. Po spec. only). *Stylocoryne conferta* Benth. (1849). *Tarenna talbotii* Wernham (1913). A shrub or liane ; flowers white, fragrant ; in swampy and riverside forest.
S.Nig.: Lagos (Feb., Apr., July, Oct., fr. July) *Dalz.* 1019! 1054! 1202! *Jones* FHI 18846! 19428! Sapoba *Kennedy* 1795! Brass *Barter* 28! R. Nun (fl. & fr. Aug.) *Barter* 2131! *T. Vogel* 23! Oban *Talbot* 1548! **F.Po:** (fl. & fr. Jan.) *Mann* 207! Also in Cameroun and Gabon. (See Appendix, p. 413.)

11. **T. pavettoides** (*Harv.*) *Sim* For. & For. Fl. Col. Cape of Good Hope 239 (1907). *Kraussia pavettoides* Harv. Fl. Cap. 3 : 22 (1865). A shrub, in fringing forest.
[**Br.]Cam.:** Bamenda Prov.: Bum, 4,000 ft. (May) *Maitland* 1530! Lakom, 6,000 ft. (May) *Maitland* 1776! Also in Uganda, Sudan, Tanganyika, N. & S. Rhodesia and S. Africa.
[A specimen (*Deighton* 3211) from Jaiama, Sierra Leone, is, perhaps also this species.]

12. **T. nitidula** (*Benth.*) *Hiern* in F.T.A. 3 : 90 (1877), excl. var. *afzelii* Hiern ; Chev. Bot. 319 ; F.W.T.A., ed. 1, 2 : 76, partly. *Stylocoryne nitidula* Benth. (1849). *Pavetta striatula* Hutch. & Dalz. F.W.T.A., ed. 1, 2 : 91 (1931) ; Bremek. in Fedde Rep. 37 : 205. *Tarenna nimbana* Schnell in Bull. I.F.A.N. 16 : 87, fig. 5 (1954). A shrub 6–7 ft. high, with shining leaves ; flowers white, fragrant ; fruits white ; in forest.
Port.G.: Cumura, Bissau (fr. Nov.) *Esp. Santo* 2229! **Guin.:** Koundian to Ouria, Kissi (Feb.) *Chev.* 20758! Nimba Mts. (Apr.) *Schnell* 4361! 4955! 4988! 5331! **S.L.:** *Barter*! *Smythe* 48! 131! 256! *Whitfield*! *T. Vogel* 147! Hills near Freetown (fl. Feb., Apr., Dec., fr. Nov., Dec.) *Deighton* 523! 5639! *Johnston* 4! *Sc. Elliot* 3869! 5769! Heddle's Farm (fl. Apr., fr. Dec.) *Lane-Poole* 279! Limba Country (Apr.) *Sc. Elliot* 5473! 5618! Kukuna, Scarcies (fl. & fr. Jan.) *Sc. Elliot* 4700! **Lib.:** Monrovia (Feb.) *Barker* 1215! Taninewa (Mar.) *Bequaert* 153! Dukwia R. (Feb.) *Cooper* 206! **Iv.C.:** Tonkoui Mt. *Schnell* 4172! Oroumbo Boka (fr. Aug.) *Schnell* 6461!

13. **T. hutchinsonii** *Bremek.* in Fedde Rep. 37 : 200 (1934). *Pavetta nigrescens* Hutch. & Dalz. F.W.T.A., ed. 1, 2 : 91 (1931), not *Tarenna nigrescens* Hiern. A shrub with shining foliage turning blackish in drying and rather small white flowers numerous in branched terminal cymes.
S.L.: Kenema (Jan.) *Thomas* 7500! 7527! 7554! 7834!

14. **T. soyauxii** (*Hiern*) *Bremek.* in Fedde Rep. 37 : 7 (1934). *Ixora soyauxii* Hiern in F.T.A. 3 : 166 (1877). *I. thomsonii* Hiern l.c. 164 (1877) ; F.W.T.A., ed. 1, 2 : 89. *I. asteriscus* K. Schum. (1899). *Tarenna asteriscus* (K. Schum.) Bremek. l.c. (1934). *Pavetta melanophylla* K. Schum. (1899)—F.W.T.A., ed. 1, 2 : 91. *Ixora atrata* Stapf (1905)—F.W.T.A., ed. 1, 2 : 87. *Tarenna lagosensis* Hutch. & Dalz., F.W.T.A., ed. 1, 2 : 76 (1931). *T. nigrescens* Good (1926), not of Hiern (1877). *T. nigroviridis* Good (1926). A much-branched shrub, sometimes scandent, or tree to 16 ft. high ; leaves deep green, drying black ; flowers white or partly pale green, very fragrant.
S.L.: Mt. Aureol *Lane-Poole* 362! Giewahun (fl. buds & fr. Dec.) *Deighton* 3836! Kenema *Lane-Poole* 456! **Lib.:** Monrovia *Whyte*! Zuie, Boporo Dist. (fr. Nov.) *Baldwin* 10236a! **Ghana:** Bobiri F.R. (fr. Aug.) *Tsawe* FH 6258! **S.Nig.:** Lagos (Feb., Dec.) *Dalz.* 1043! 1203! 1375! Akilla, Ijebu-Ode Prov. (Jan.) *A. F. A. Lamb* FHI 4526! Okomu F.R., Benin Div. (Dec.) *Brenan* 8556! Calabar *Thomson* 45! Mbot to Akpan, Calabar (fr. July) *Ujor* FHI 31464! [**Br.]Cam.:** Likomba (Dec.) *Mildbr.* 10793! N. Korup F.R., Kumba Div. (fr. July) *Olorunfemi* FHI 30675! Also in Cameroun, Gabon, Cabinda and Congo.

15. **T. bipindensis** (*K. Schum.*) *Bremek.* in Fedde Rep. 37 : 7, 208 (1934). *Chomelia bipindensis* K. Schum. (1903). *Pavetta warburgiana* De Wild. & Th. Dur. (1901) ; not *Tarenna warburgiana* Valeton (1924). *Ixora bipindensis* K. Schum. ex Hutch. & Dalz. F.W.T.A., ed. 1, 2 : 87 (1931). *Tarenna adamii* Schnell in Bull. I.F.A.N. 16 : 79, fig. 2 (1954), incl. var. *nigeriana* Schnell. *T. mangenotii* Schnell l.c. 77, fig. 1 (1954). A shrub, often scandent ; leaves usually drying black ; flowers green ; in forest.
Guin.: Macenta (May) *Collenette* 12! *Adam* 5282 ; 5951 ; s.n.! **S.L.:** Gomdama, Maje (May) *Deighton* 5782! **Iv.C.:** *Mangenot*! Sampleu (Apr.) *Chev.* 21093! *Adam* 13494! **Ghana:** Abra, W. Prov. (Jan.) *Akpabla* 807! **S.Nig.:** Lagos (Aug.) *Rowland*! Oban *Talbot* 1048! Oban F.R. (fr. May) *Ujor* FHI 30847! **F.Po:** (Jan.) *Mann* 212! Also in Cameroun and Gabon.

16. **T. brachysiphon** (*Hiern*) *Keay* in Bull. Jard. Bot. Brux. 27 : 99 (1957). *Ixora brachysiphon* Hiern in F.T.A. 3 : 165 (1877) ; F.W.T.A., ed. 1, 2 : 89. *Tarenna ombrophila* Schnell in Bull. I.F.A.N. 16 : 85, fig. 4

(1954). *T. tomensis* Schnell l.c. 81, fig. 3 (1954). A glabrous shrub with shining leaves and white flowers in lax cymes.
Guin.: Nimba Mts. (Apr.) *Schnell* 5114! Macenta (Feb.) *Adam* 3885! **S.L.:** *Thomas* 10045! Bagroo R. (Mar.) *Mann* 792! Mabonto (fr. Oct.) *Thomas* 3546! 3657!

29. PAVETTA Linn.—F.T.A. 3 : 167 ; Bremekamp in Fedde Rep. 37 : 1–208 (1934).

*Corolla-tube 2·5–7 mm. long, usually shorter than the lobes but sometimes equalling them ; *throat of corolla bearded* (i.e. with a dense ring of long hairs) ; calyx often imbricate in bud—(Subgen. *Baconia*) :
  Each ovary cell with 2 collateral ovules ; midrib, nerves and veins of leaves impressed above ; leaves more or less elliptic, 10–22 cm. long, 4–10 cm. broad, softly pubescent beneath, with 11–15 main lateral nerves on each side of midrib, prominent beneath ; branchlets, inflorescence-axes and calyces densely pubescent ; flowers numerous in corymbose cymes ; corolla-tube 5 mm., lobes 8 mm. long .. .. 1. *lasioclada*
  Each ovary cell with a solitary ovule ; nerves and veins not impressed above ; leaves mostly glabrous or nearly so beneath (but see Nos. 2a and 3) :
    Stipules and bracts pilose inside :
    Calyx-lobes persistently imbricate at anthesis, 1–1·5 mm. long, 1·5–2 mm. broad, rounded or truncate at apex, puberulous outside ; corolla-tube about 5 mm., lobes about 10 mm. long ; cymes many-flowered, widely corymbose ; leaves oblanceolate to elliptic, 7·5–20 cm. long, 2–8 cm. broad, with 7–9 main lateral nerves on each side of midrib, bacterial nodules usually conspicuous :
      Leaves puberulous or nearly glabrous ; branchlets and inflorescence puberulous
                                                  2. *corymbosa* var. *corymbosa*
      Leaves more or less densely pubescent beneath ; branchlets and inflorescence pubescent .. .. 2a. *corymbosa* var. *neglecta*
    Calyx-lobes open or imbricate in bud only, or calyx truncate ; corolla (tube and lobes together) 5·5–17 mm. long :
      Leaves softly and densely pubescent all over the lower surface, especially on the midrib and nerves ; branchlets, inflorescence and calyx shortly tomentose ; leaves elliptic to obovate, 8–20 cm. long, 4–10 cm. broad, with 9–14 main lateral nerves on each side of midrib ; corolla about 10 mm. long ; flowers numerous, in rather dense, widely corymbose cymes .. .. 3. *mollissima*
      Leaves glabrous or puberulous beneath mainly on midrib and nerves or in axils of nerves ; branchlets glabrous or puberulous :
        Corolla 8–17 mm. long :
          Calyx-lobes less than 0·5 mm. long, or calyx merely 4-lobulate ; corolla about 8 mm. long ; inflorescences puberulous, up to 5 cm. diam. ; branchlets and leaves glabrous or nearly so ; leaves elliptic, 9–22 cm. long, 2·5–9 cm. broad, with 4–9 main lateral nerves on each side of midrib, venation rather obscure
                                              4. *brachycalyx*
        Calyx-lobes distinct, 1–3 mm. long :
          Leaves sessile or subsessile, oblong or oblanceolate, shortly cuneate or rounded at base, 8–15 cm. long, 1·8–4·8 cm. broad, puberulous beneath mainly on midrib and nerves, with 7–8 main lateral nerves on each side of midrib ; branchlets and inflorescence puberulous, the latter less than 4 cm. diam. ; corolla about 11 mm. long .. .. 5. *oblongifolia*
          Leaves distinctly petiolate, mostly long-cuneate at base :
            Calyx-lobes longer than broad, 1·5–3 mm. long, slightly keeled ; corolla 15–17 mm. long ; inflorescence puberulous, up to 10 cm. diam., rather lax ; branchlets and leaves glabrous or nearly so ; leaves oblanceolate to elliptic, 9–24 cm. long, 2·5–7 cm. broad, with 5–8 main lateral nerves on each side of midrib .. .. 6. *glaucescens*
           Calyx-lobes as long as broad or shorter, 1–2 mm. long, not keeled ; corolla 8–12 mm. long :
              Corolla 8–9 mm. long ; plant usually drying blackish ; calyx-lobes broad and rounded ; branchlets glabrous ; leaves oblanceolate to obovate, 7–12 cm. long, 2·5–5 cm. broad, with 7–11 main lateral nerves on each side of midrib
                                            7. *hookeriana*
              Corolla 10–12 mm. long ; plant drying greenish-brown ; calyx-lobes obtuse to subacute ; branchlets minutely puberulous ; leaves elliptic to oblanceolate or obovate, 7–22 cm. long, 3–9 cm. broad, with 6–12 main lateral nerves on each side of midrib .. .. .. 8. *owariensis*
      Corolla about 5·5 mm. long ; calyx-lobes distinct ; inflorescence puberulous, up to 7 cm. diam. ; branchlets and leaves glabrous ; leaves oblong-elliptic, 11–17 cm. long, 2·5–6 cm. broad, with 6–8 main lateral nerves on each side of midrib, venation slightly prominent, lower surface shining, bacterial nodules conspicuous .. .. 9. *obanica*

Stipules and bracts entirely glabrous ; branchlets and leaves glabrous or nearly so ; bacterial nodules usually present :
Corolla 6–8 mm. long ; inflorescences 1–3 cm. diam., subtended by a single pair of leaves at the ends of long axillary shoots :
Venation of leaves obscure, main lateral nerves 8–11 on each side of midrib ; calyx and inflorescence glabrous ; calyx-lobes distinct, imbricate at anthesis ; leaves coriaceous, elongate-oblong-elliptic, 14–25 cm. long, 6–10 cm. broad
10. *mannioides*
Venation of leaves prominently reticulate on both surfaces, main lateral nerves 10–14 on each side of midrib ; calyx and inflorescence puberulous ; calyx sub-truncate or with short broad lobes not imbricate at anthesis ; leaves stiffly chartaceous (when dry), elliptic to lanceolate, 9–25 cm. long, 2·5–10 cm. broad
11. *ixorifolia*
Corolla 10–16 mm. long ; inflorescences 3–13 cm. diam., subtended by 1–3 pairs of leaves :
Calyx-lobes persistently imbricate, subquadrate, very short ; corolla about 13 mm. long ; leaves oblong or lanceolate, 10–18 cm. long, 3–5·5 cm. broad, with 7–9 main lateral nerves on each side of midrib .. .. .. 12. *staudtii*
Calyx-lobes imbricate in bud only, triangular to ovate :
Corolla about 11 mm. long ; leaves oblong-elliptic, 15–22 cm. long, 6–8·5 cm. broad, with 9–13 main lateral nerves on each side of midrib ; stipules not seen
13. *neurocarpa*
Corolla 12–16 mm. long ; leaves lanceolate, elongate-elliptic or broadly oblance-olate, 12–28 cm. long, 3·5–9 cm. broad, with 12–14 main lateral nerves on each side of midrib ; stipules with points up to 10 mm. long .. .. 14. *bidentata*
*Corolla-tube 8–40 mm. long (but only 3–4 mm. long in No. 20), usually longer than the lobes but sometimes as long as the lobes ; *throat of corolla not bearded* ; calyx very rarely imbricate—(Subgen. *Pavetta*) :
Calyx-lobes 12–20 mm. long, linear-filiform, densely plumose-pilose ; corolla 32–38 mm. long ; branchlets and leaves densely spreading pilose :
Leaves elongate-obovate or oblanceolate, 20–35 cm. long, 6–11 cm. broad, with (12–)15–16 main lateral nerves on each side of midrib ; corolla glabrous outside, tube 17–18 mm., lobes 15–20 mm. long .. .. 15. *hispida*
Leaves oblong-elliptic, 15–22 cm. long, 5–8 cm. broad, with 9–11 main lateral nerves on each side of midrib ; corolla-tube about 22 mm. long, lobes pilose outside, about 13 mm. long .. .. 16. *plumosa*
Calyx-lobes up to 6 mm. long, or if up to 12 mm. then oblong, obovate or spathulate and up to 5 mm. broad ; corolla 7–50 mm. long ; branchlets and leaves pilose, pubescent or glabrous :
Calyx-lobes oblong, obovate or spathulate, accrescent and reaching 9–12 mm. long and 3–5 mm. broad in fruit ; corolla 24–32 mm. long ; leaves narrowly elliptic, with 7–12 main lateral nerves on each side of midrib :
Flowers in a dense subglobose cluster ; calyx-lobes oblong to obovate-oblong, up to 11 mm. long and 3 mm. broad ; upper surface of leaves glabrous except near the petiole ; branchlets pubescent to pilose ; leaves 11–28 cm. long, 3·5–8·5 cm. broad, more or less densely pubescent beneath ; corolla-tube 16–22 mm., lobes about 10 mm. long .. .. .. ; .. .. 17. *genipifolia*
Flowers in corymbose cymes ; calyx-lobes obovate or spathulate ; corolla-tube 14–14·5 mm., lobes 10–11 mm. long :
Leaves pilose on both surfaces ; calyx-lobes spathulate, up to 12 mm. long and 5 mm. broad ; leaves 8–23 cm. long, 2·5–7·5 cm. broad, rather densely pilose beneath .. .. 18. *leonensis*
Leaves glabrous or sparsely pubescent ; calyx-lobes obovate, about 9 mm. long and 5 mm. broad ; leaves 10–25 cm. long, 3·5–10 cm. broad .. 19. *platycalyx*
Calyx-lobes 0–6 mm. long, broadly triangular to linear-filiform :
Each ovary cell with 2 collateral ovules ; glabrous shrub, not exceeding 1 m. high ; leaves oblanceolate, attenuate at base into the 0·5–3 cm. long petiole, acuminate, 9·5–20 cm. long, 2·5–7·5 cm. broad, with 8–11 main lateral nerves on each side of midrib ; inflorescences up to 2 cm. long ; calyx-lobes broad and rounded, scarcely 1 mm. long ; corolla-tube 5–8 cm. long, pubescent inside, lobes about 5 mm. long and 3–4 mm. broad ; style about 30 mm. long, far-exserted .. 20. *tetramera*
Each ovary cell with a solitary ovule ; shrubs exceeding 1 m. high :
Corolla-tube 3–4 mm. long, approximately equal in length to the lobes ; calyx-lobes broadly triangular, up to 2 mm. long ; branchlets and leaves glabrous ; leaves elliptic-oblong, up to 32 cm. long and 9 cm. broad, petiole up to 7 cm. long, with about 12 main lateral nerves on each side of midrib ; cymes up to 2 cm. long, appearing axillary .. .. 21. *talbotii*
Corolla-tube (8–)10–40 mm. long, longer than the lobes :
Calyx-lobes (3–)5–6 mm. long, linear-lanceolate to filiform ; corolla-tube 24–40 mm. long ; main lateral nerves 6–8 on each side of midrib :

Inflorescence subumbellate, congested ; corolla-tube about 24 mm. long, lobes about 9 mm. long ; leaves broadly elliptic, 12–20 cm. long, 6–12·5 cm. broad, pubescent beneath..   ..   ..   22. *dolichosepala*
Inflorescence paniculate, lax ; corolla-tube 35–40 mm. long, lobes about 10 mm. long, acute-acuminate ; leaves elliptic, 5–11 cm. long, 2·5–4 cm. broad, densely and softly pubescent beneath   ..   ..   23. *mollis*
Calyx-lobes 0–2 mm. long, or if up to 3 mm. long (species No. 28) then corolla not exceeding 16 mm. long and leaves, branchlets, inflorescence and outside of flowers glabrous or sparingly pubescent :
Flowering shoots greenish ; corolla-lobes 10–16 mm. long, tube 13–19 mm. long ; branchlets and leaves glabrous or nearly so ; inflorescence laxly corymbose, minutely puberulous ; calyx-lobes broadly triangular, 0·2–1 mm. long ; leaves elliptic, 15–30 cm. long, 7·5–14 cm. broad, with 8–11 main lateral nerves on each side of midrib   ..   ..   24. *rigida*
Flowering shoots covered with whitish corky bark ; corolla-lobes 5–8 mm. long ; leaves with 5–8 main lateral nerves on each side of midrib :
Stipules and bracts glabrous inside ; branchlets stout ; leaves often in threes, subsessile, glabrous, narrowly elongate-oblong-oblanceolate, rounded at apex, 15–25 cm. long, 3–5 cm. broad, corymbs many-flowered, glabrous ; calyx subtruncate, glabrous outside ; corolla-tube 12–18 mm. long, lobes about 5 mm. long..   ..   ..   25. *crassipes*
Stipules and bracts pilose inside ; branchlets more slender ; leaves elliptic, oblong or obovate, 3–11(–16) cm. long, 1·5–4·5(–7) cm. broad ; calyx-lobes mostly distinct :
Flowers subsessile, in clusters on very short lateral shoots ; corolla-tube about 20 mm. long, 3–4 times as long as the lobes ; calyx densely puberulous outside, lobes triangular ; branchlets and leaves densely pubescent ; leaves elliptic, rounded or obtuse at apex, 4–7 cm. long, 2–4 cm. broad 26. *subcana*
Flowers rather long-pedicellate, in lax corymbs on relatively long shoots ; corolla-tube not more than twice as long as the lobes :
Leaves rather densely puberulous, especially beneath when young, obtuse at apex, cuneate at base, 10–16 cm. long, 3–7 cm. broad ; calyx-lobes triangular ; branchlets, inflorescence and calyx puberulous ; corolla glabrous, tube 7 mm., lobes 5 mm. long ; fruits 6 mm. diam.   ..   ..   27. *cinereifolia*
Leaves, branchlets, inflorescence and outside of flowers glabrous or sparingly pubescent :
Leaves elliptic or oblong, 5–8 cm. long, 2·5–4 cm. broad, with a distinct pubescent petiole 4–15 mm. long ; calyx-lobes to 3 mm. long ; corolla-tube 14–15·5 mm., lobes 7–8 mm. long ; style 40–45 mm. long
28. *subglabra*
Leaves obovate or oblanceolate, 3–6·5(–8) cm. long, 1·2–2(–3) cm. broad, long-attenuate and almost sessile at base, quite glabrous ; calyx-lobes less than 0·5 mm. long ; corolla-tube 8–12 mm., lobes 5–7 mm. long ; style 20–22 mm. long   ..   ..   29. *saxicola*

1. **P. lasioclada** (*K. Krause*) *Mildbr. ex Bremek.* in Fedde Rep. 37 : 62 (1934) ; Aubrév. Fl. For. Soud.-Guin. 471–474, t. 104, 1. *Chomelia lasioclada* K. Krause (1909). *Pavetta viburnoides* A. Chev. ex Hutch. & Dalz. F.W.T.A., ed. 1, 2 : 91 (1931) ; Chev. Bot. 334, name only. A shrub 8–15 ft. high, pale brownish-pubescent, with white flowers.
Mali: Moussaia (Feb.) *Chev.* 403 ! **Guin.:** Timbo (Feb.) *Pobéguin* 158 ! Kollangui (Mar.) *Chev.* 12193 ! Mali (Jan.) *Roberty* 16554 ! Labé *Chev.* 12351. Timbo *Chev.* 12523. Kankan to Kouroussa *Chev.* 15649. **S.L.:** Falaba (Mar.) *Sc. Elliot* 5142 ! Bintumane Peak, 3,000 ft. (bud Jan.) *Glanville* 453 ! **Iv.C.:** Tafiré *Aubrév.* 1829 ! Dyolo country *Chev.* 21376. Mt. Dou *Aubrév.* 1075. Ferkéssédougou *Aubrév.* 2313. **Togo Rep.:** Sokode (fr. Dec.) *Kersting* A56. Bismarckburg *Büttner* 392. Also in Cameroon.

2. **P. corymbosa** (*DC.*) *F. N. Williams* var. **corymbosa**—in Bull. Herb. Boiss. sér. 2, 7 : 378 (1907) ; Bremek. l.c. 69, as var. *glabra* ; Aubrév. Fl. For. Soud.-Guin. 471–474, t. 104, 2. *Baconia corymbosa* DC. (1807). *Ixora nitida* Schum. & Thonn. (1827). *Pavetta nitida* (Schum. & Thonn.) Hutch. & Dalz., F.W.T.A., ed. 1, 2 : 91 (1931), partly ; Aubrév. Fl. For. C. Iv., ed. 1, 3 : 219. *P. baconia* Hiern in F.T.A. 3 : 176 (1877) ; Kennedy For. Fl. S. Nig. 213 (as " *P. baconi* "). *P. rhombifolia* Bremek. l.c. 64 (1934). *P. hygrophytica* Bremek. l.c. 68 (1934), partly (*Chipp* 420). A shrub or small tree, usually less than 20 ft. high ; corolla greenish-white outside, white inside ; fruits whitish when young, black when ripe ; in forest regrowth and drier types of forest including fringing forest and forest margins.
Sen.: Cape Verde *Leprieur.* Casamance *Perrottet.* **Gam.:** *Heudelot* 62 ! **Port.G.:** Fonte, Bubaque (May) *Esp. Santo* 2020 ! Brene, Bissau (fr. Jan.) *Esp. Santo* 1681 ! **Guin.:** Rio Nunez *Heudelot* 772 ! Koloiea *Paroisse* 4 ! Toumbo *Paroisse* 80. Souguékourou *Paroisse* 144. **S.L.:** *Hormont* ! *Smeathmann. Whitfield* ! Heddle's Farm (May) *Lane-Poole* 270 ! Goderich (Apr.) *Deighton* 4763 ! Musaia (Mar.) *Deighton* 5454 ! Sellakuri (Mar.) *Sc. Elliot* 5093 ! Bafodeya (Apr.) *Sc. Elliot* 5506 ! Falaba (Mar.) *Sc. Elliot* 5129 ! **Iv.C.:** *fide* Aubrév. l.c. **Ghana:** Axim (young fr. Apr.) *Chipp* 420 ! Cape Coast *Brass* ! Achimota (Feb.) *Irvine* 1997 ! Aboma F.R., Mampong Dist. (fl. & fr. Jan.) *Andoh* FH 5452 ! 5454 ! Kpeve (fl. buds Mar.) *Williams* 68 ! Amedzofe (Mar.) *Williams* 96 ! 292 ! **Togo Rep.:** Lome *Warnecke* 49 ! **Dah.:** *Le Testu* 250 ! **N.Nig.:** Anara F.R., Zaria Prov. (fl. May, fr. Oct.) *Keay* FHI 21133 ! 25783 ! **S.Nig.:** Lagos *Barter* 2200 ! *Dalz.* 1053 ! *Millen* 209 ! Abeokuta *Irving* 140 ! Oyo *Barter* 3402 ! Onitsha *Barter* 1257 ! [Br.]**Cam.:** Gangumi, Gashaka Dist. (fr. Dec.) *Latilo & Daramola* FHI 28846 ! (See Appendix, p. 407.)

2a. **P. corymbosa** var. **neglecta** *Bremek.* l.c. 70 (1934). As above, but leaves densely pubescent.
Iv.C.: Dimbokro *Aubrév.* 1811 ! **Ghana:** Volta River F.R. (Feb.) *Vigne* FH 4351 ! Akropong (Mar.) *Johnson* 899 ! Achimota (Feb.) *Irvine* 285 ! 2000 ! Inchaban (Jan.) *Vigne* FH 4129 ! Kpandu *Robertson* 102 ! **Dah.:** *Burton* ! Boguila, Abomey (Feb.) *Chev.* 23194 ! **N.Nig.:** Dogon Kurmi, Jemaa Div. (May) *E. W. Jones* 47 ! **S.Nig.:** Lagos *Millen* 75 ! Ibadan (fl. Jan.-Apr., fr. Aug.-Jan.) *Meikle* 949 !

1307 ! *Newberry* 46 ! 190 ! Ijaiye F.R. (Mar.) *Onochie* FHI 20694 ! Idanre Hills (Jan.) *Brenan* 8680 ! Also in Ubangi-Shari. (See Appendix, p. 407.)

3. **P. mollissima** *Hutch. & Dalz.* F.W.T.A., ed. 1, 2 : 91 (1931) ; Bremek. l.c. 70. A forest shrub, 6 ft. high, with profuse white flowers.
Iv.C.: Gagnoa (fl. buds & young fr. Feb.) *Kerharo & Bouquet* 981 ! **Ghana:** Kwahu Praso (Feb.) *Vigne* FH 1601 ! Banso, Ashanti (Mar.) *Vigne* FH 1866 ! S. Fomang Su F.R. (Jan.) *Andoh* FH 2680 !

4. **P. brachycalyx** *Hiern* in F.T.A. 3 : 169 (1877) ; Bremek. l.c. 76. *Exechostylus flaviflorus* K. Schum. (1899). *Pavetta flaviflora* (K. Schum.) Hutch. & Dalz. F.W.T.A., ed. 1, 2 : 92 (1931), partly. *P. hierniana* Bremek. l.c. 76, partly (*Dunlap* 20). A shrub up to 10 ft. high ; in forest.
[Br.]Cam.: Cam. Mt., 2,000–4,800 ft. (Dec., Jan.) *Dunlap* 20 ! *Maitland* 213 ! 908 ! *Mann* 2159 ! Johann-Albrechtshöhe (=Kumba) (Jan.) *Staudt* 565. Also in Cameroun.

5. **P. oblongifolia** (*Hiern*) *Bremek.* l.c. 65 (1934). *P. baconia* Hiern var. *oblongifolia* Hiern in F.T.A. 3 : 176 (1877). *P. schweinfurthii* Bremek. var. *oblongifolia* (Hiern) Aubrév. Fl. For. Soud.-Guin. 471–474, t. 103, 4 ; Berhaut Fl. Sén. 106. *P. nitida* of F.W.T.A., ed. 1, 2 : 91, partly (*Heudelot* 673). A shrub.
Sen.: *Berhaut* 1229. Niokolo-koba *Adam* 14236 ! **Mali:** Kerfomouria (Mar.) *Chev.* 600 ! Niayes to Koulikoro *Chev.* 2020 ! **Port.G.:** Contabani to Guilege, Boé (June) *Esp. Santo* 3196 ! **Guin.:** Landouma country *Heudelot* 673 ! Kadé *Pobéguin*. **S.L.:** Laminaiya (Apr.) *Thomas* 147 ! Also in Ubangi-Shari (*fide* Aubrév.). (See Appendix, p. 407.)

6. **P. glaucescens** *Hiern* in F.T.A. 3 : 171 (1877) ; Bremek. l.c. 71. *P. chionantha* K. Schum. & K. Krause (1907)—Bremek. l.c., excl. syn. *P. owariensis. P. hygrophytica* Bremek. l.c. 68 (1934), excl. *Chipp* 420. A small tree ; corolla white.
S.Nig.: Eket *Talbot* 3120 ! Calabar (Mar.) *Thomson* 68 ! Oban *Talbot* 1451 ! 2042 ! Orem, Calabar to Mamfe (Jan.) *Onochie* FHI 36075x ! [Br.]Cam.: Rio del Rey *Johnston* ! S. Bakundu F.R. (Mar., Apr.) *Ejiofor* FHI 29306 ! *Onochie* FHI 31199 ! **F.Po:** *Mann* ! Also in Cameroun, Rio Muni, Gabon, Cabinda and Congo.

7. **P. hookeriana** *Hiern* in F.T.A. 3 : 176 (1877) ; Bremek. l.c. 63 (excl. spec. ex Annobon). *Baconia montana* Hook. f. (1864) ; not *Pavetta montana* Reinw. ex Blume (1826). *Pavetta exellii* Bremek. l.c. 73 (1934). *P. molundensis* of Bremek. l.c. 64, partly (*Johnstone* 102/31). A well-branched shrub, to 10 ft. high, with white flowers ; in montane forest.
[Br.]Cam.: Cam. Mt., 4,000–7,600 ft. (fl. Dec.-Apr., fr. Dec.-Feb.) *Mann* 2166 ! *Maitland* 206 ! 992 ! 1139 ! 1312 ! Bafut-Ngemba F.R., Bamenda (Feb.-Apr.) *Onochie* FHI 34850 ! *Ujor* FHI 30050 ! *Hepper* 2160 ! Kishaw, Bamenda (Apr.) *Johnstone* 102/31 ! L. Oku, Bamenda (Jan.) *Keay* FHI 28514 ! **F.Po:** Moka, 4,000–5,000 ft. (fl. & young fr. Sept., Jan.) *Exell* 787 ! 790 ! *Wrigley & Melville* 529 !
[*P. monticola* Hiern, from S. Tomé and Annobon is closely related, but appears to differ in having considerably longer corolla-lobes.]

8. **P. owariensis** *P. Beauv.* Fl. Oware 1 : 87, t. 52 (1806) ; F.T.A. 3 : 170, excl. *Vogel* spec. *P. smythei* Hutch. & Dalz. F.W.T.A., ed. 1, 2 : 92 (1931) ; Bremek. l.c. 76 (1934). *P. eketensis* Bremek. l.c. 75 (1934). *P. hierniana* Bremek. l.c. 76 (1934), (excl. *Dunlap* 20). *P. intermedia* Bremek. l.c. 71 (1934), partly (*Kalbreyer* 92 & ? *Dinklage* 2635). *P. nigritana* Bremek. in Fedde Rep. 47 : 20 (1939). *P. staudtii* of Bremek. in Fedde Rep. 37 : 77, partly (S.L. spec.). *P. flaviflora* of F.W.T.A. ed. 1, 2 : 92, partly. A forest shrub or small tree, 6–20 ft. high ; flowers white.
Guin.: Nimba Mts. *Schnell* 426 ! **S.L.:** Fundu (Jan.) *Smythe* 214 ! Sinja (Feb.) *Lane-Poole* 153 ! Gorahun (Apr.) *Deighton* 3657 ! Bumbuna (fr. Oct.) *Thomas* 3439 ! Ronietta (fr. Nov.) *Thomas* 5514 ! **Lib.:** *Dinklage* 2635. Batwa, Gola Forest (Dec.) *Bunting* 87 ! **Iv.C.:** R. Davo, Béyo (Jan.) *Leeuwenberg* 2599 ! **Ghana:** Sui F.R. (Mar.) *Foggie* FH 4449 ! Donkoto, Bibiani (Mar.) *Darko* 857 ! **S.Nig.:** Iddu, Benin Div. (Mar.) *Farquhar* 32 ! Warri to Buonopozo *Beauvois* ! Okomu F.R. (Dec., Jan.) *Brenan* 8569 ! 8876 ! Sapoba *Kennedy* 1776 ! 1990 ! 2072 ! 2346 ! 2620 ! Eket Dist. *Talbot* ! Degema Dist. *Talbot* 3806 ! 3807 ! [Br.]Cam.: Victoria (Mar.) *Kalbreyer* 92 ! **F.Po:** *Mann* 195 ! Clarence (fl. buds Nov.) *T. Vogel* 114 ! Also in Cameroun.

9. **P. obanica** *Bremek.* l.c. 77 (1934). *P. flaviflora* of F.W.T.A., ed. 1, 2 : 92, partly. A forest shrub.
S.Nig.: Oban *Talbot* s.n. ! 349 ! 359 ! Obom Itiat, Itu Div. (Jan.) *Jones* FHI 6871 ! (See Appendix, p. 407.)

10. **P. mannioides** *Hutch. & Dalz.* F.W.T.A., ed. 1, 2 : 91 (1931) ; Bremek. l.c. 80. A glabrous forest shrub.
S.Nig.: Oban *Talbot* !

11. **P. ixorifolia** *Bremek.* in Fedde Rep. 37 : 79 (1934). A forest shrub, to 15 ft. high ; flowers green or cream.
Guin.: Dalaba (Feb.) *Chev.* 18137 ! **Iv.C.:** Sassandra to Gagnoa (fr. Aug.) *de Wilde* 338 ! Béyo (fl. & fr. Dec.) *Leeuwenberg* 2216 ! Abouabou Forest (fl. & fr. Jan.) *Leeuwenberg* 2367 ! Grand Bassam (Dec.) *Schnell* 3957 ! **Ghana:** Princes, W. Prov. (Jan.) *Akpabla* 781 ! Tano Anwia, Enchi Dist. (fr. May) *Enti* FH 6722 ! S. Scarp F.R., Mpraeso Dist. (Dec.) *Moor* FH 2117 ! *Adu* FH 6583 ! **S.Nig.:** Okomu F.R., Benin (Feb.) *Richards* 3938 !

12. **P. staudtii** *Hutch. & Dalz.* F.W.T.A., ed. 1, 2 : 91 (1931) ; Bremek. l.c. 77, excl. S.L. spec. A glabrous forest shrub.
[Br.]Cam.: Johann-Albrechtshöhe (=Kumba) *Staudt* 599 ! Also in Cameroun.

13. **P. neurocarpa** *Benth.* in Hook. Fl. Nigrit. 414 (1849) ; F.T.A. 3 : 172 ; F.W.T.A., ed. 1, 2 : 91, excl. syn. *P. macrostemon* K. Schum. ; Bremek. l.c. 79. *P. mannii* Hiern in F.T.A. 3 : 169 (1877) ; F.W.T.A., ed. 1, 2 : 91, excl. Oban spec. A glabrous forest shrub about 3 ft. high ; corolla white.
[Br.]Cam.: Cameroon R. (Dec.) *Mann* ! NE. of Bafia, Victoria Dist. (Feb.) *Keay* FHI 37533 ! **F.Po:** (fr. Nov.) *T. Vogel* 151 !

14. **P. bidentata** *Hiern* in F.T.A. 3 : 176 (1877) ; Bremek. l.c. 78. *P. deistelii* K. Schum. (1903). *P. longistipulata* Bremek. l.c. 78 (1934). A forest shrub.
S.Nig.: Eket Dist. *Talbot* 3316 ! [Br.]Cam.: Victoria (Apr., Dec.) *Deistel* 127. *Maitland* 1180 ! *Smith* Cam. 34/37 ! Njoke to Malende *Schlechter* 12872 ! **F.Po:** (Apr.) *Mann* 395 ! Also in Congo.

15. **P. hispida** *Hiern* in F.T.A. 3 : 175 (1877) ; Bremek. l.c. 135. A forest shrub, to 8 ft. high ; flowers white.
S.Nig.: Oban *Talbot* 26 ! 1490 ! [Br.]Cam.: Kembong F.R., Mamfe Div. (fl. buds Mar.) *Onochie* FHI 32053 ! **F.Po:** (Jan.) *Mann* 227 ! Also in Cameroun and Cabinda.

16. **P. plumosa** *Hutch. & Dalz.* F.W.T.A., ed. 1, 2 : 91 (1931). A forest shrub, about 3 ft. high.
[Br.]Cam.: Likomba (Dec.) *Mildbr.* 10743 !

17. **P. genipifolia** *Schumach.* Beskr. Guin. Pl. 78 (1827) ; F.W.T.A., ed. 1, 2 : 91, excl. spec. ex S.L. ; Keay in Bull. Jard. Bot. Brux. 27 : 100 (1957). *P. megistocalyx* K. Krause (1909)—Bremek. l.c. 136. A shrub, to 10 ft. high ; flowers white ; in forest and thickets.
Ghana: *Farmar* 431 ! Mampong Scarp (May) *Vigne* FH 1129 ! Aburi (May) *Johnson* 131 ! Kwahu Hills (fl. buds Apr.) *Johnson* 905 ! Jimira F.R. (June) *Vigne* FH 2966 ! Akropong, Akwapim (fl. May, fr. Jan.) *Irvine* 1760 ! **Togo Rep.:** *Baumann* 278 ! Misahöhe (fr. Dec.) *Busse* 3424. **S.Nig.:** Oni River F.R., Ife-Ilesha Dist. (Apr.) *Onochie* FHI 15508 ! Oluwa F.R., Ondo Dist. (young fr. July) *Kennedy* 2481 ! *Onochie* FHI 33413 !

18. **P. leonensis** *Keay* in Bull. Jard. Bot. Brux. 27 : 101 (1957). *P. genipifolia* of Hiern in F.T.A. 3 : 175 ; F.W.T.A., ed. 1, 2 : 91, partly (spec. ex S.L.) ; Bremek. l.c. 136 ; not of Schum. A shrub, to 10 ft. high. **Guin.:** Nimba Mts. (June) *Schnell* 3014 ! **S.L.:** Barter ! Near top of Sugar Loaf Mt. (fr. Dec.) *Sc. Elliot* 4024 ! Kent (fr. Nov.) *Tindall* 13 ! Nyandehun (young fr. July) *Deighton* 5792 ! Benikoro (fr. Oct.) *Thomas* 2886 ! Masankoi (June) *Marmo* 24 !

19. **P. platycalyx** *Bremek.* l.c. 137 (1934). A forest shrub ; flowers white.
Guin.: Forocontan (fr. Oct.) *Jaeger* 86 ! **S.L.:** Sankan Biriwa Massif (fr. Jan.) *Cole* 178 ! **Lib.:** Bilimu

(fr. Dec.) *Harley* 1629 ! **Iv.C.**: Sassandra *Chev.* 17982. Foot of Mt. Momy, Haut-Cavally (Apr.) *Chev.* 21353 !

20. **P. tetramera** (*Hiern*) *Bremek.* l.c. 134 (1934). A low shrub, 1–3 ft. high, with white flowers ; in forest.
[Br.]**Cam.**: S. Bakundu F.R., Banga, Kumba Div. (Jan.) *Binuyo & Daramola* FHI 35156 ! *Keay* FHI 28559 ! Also in Cameroun, Rio Muni, Gabon, Congo and Angola.

21. **P. talbotii** *Wernham* in Cat. Talb. 52 (1913) ; Bremek. l.c. 133.
S.Nig.: Oban *Talbot* 1638 !

22. **P. dolichosepala** *Hiern* in F.T.A. 3 : 174 (1877) ; Bremek. l.c. 141. A forest shrub, to 6 ft. high.
[Br.]**Cam.**: Likomba (young fr. Nov.) *Mildbr.* 10602 ! Also in the Congos.
[Although cited, without comment, by Bremekamp, the specimen from our area has the sepals considerably more pubescent than in the type from Middle Congo.]

23. **P. mollis** *Afzel. ex Hiern* in F.T.A. 3 : 174 (1877) ; Bremek. in Fedde Rep. 37 : 178. *P. apiculata* Hutch. & Dalz. F.W.T.A., ed. 1, 2 : 91 (1931) ; Bremek. l.c. 139. A shrub, to 15 ft. high, with pale branchlets ; flowers whitish, scented.
Guin.: Kouroussa (July) *Pobéguin* 274 ! **Iv.C.**: Lomo (fr. Sept.) *Aké Assi* 5947 ! **Ghana**: Cape Coast *Brass* ! Okroase (fl. buds May) *Johnson* 754 ! Achimota (fl. buds Apr.) *Irvine* 330 ! 756 ! Afram to Mankrong (Apr.) *Morton* A731 !
[Hiern cites a specimen from Hb. Afzelius, and Bremekamp cites *Afzelius* s.n. in Hb. Berlin, both from Sierra Leone ; I have not been able to trace these specimens.]

24. **P. rigida** *Hiern* in F.T.A. 3 : 178 (1877) ; Bremek. l.c. 188. *P. macrostemon* K. Schum. (1899)—Bremek. l.c. 132. *P. grandiflora* K. Schum. & K. Krause (1907). *P. neurocarpa* of F.W.T.A., ed. 1, 2 : 91, partly (*Kalbreyer* 171). A glabrous shrub, to 12 ft. high ; corolla white except for a green stripe down back of lobes ; inflorescence, ovary and calyx green ; in forest.
S.Nig.: Afi River F.R., Ikom Div. (May) *Jones & Onochie* FHI 18761 ! [Br.]**Cam.**: Cam. Mt., 4,000 ft. (Mar.) *Deistel* 134 ! *Kalbreyer* 171 ! *Maitland* 509 ! *Rudatis* 164. Barombi *Preuss* 293. Tinto Banyang *Conrau* 111. Nkom-Wum F.R., Bamenda Prov. (Apr.) *Ujor* FHI 30065 ! **F.Po**: (Jan., Nov.) *Mann* 183 ! *Guinea* 2612 ! Also in Cameroun.

25. **P. crassipes** *K. Schum.* in Engl. Pflanzenw. Ost-Afr. C : 389 (1895) ; Bremek. l.c. 169 ; Aubrév. Fl. For. Soud.-Guin. nr. 104, 4–5. *P. barteri* Dawe (1906). *P. utilis* Hua (1907)—Chev. Bot. 334. A glabrous shrub with stout squarish branchlets covered with pale corky bark which splits and falls off ; leaves often in threes ; flowers greenish-white ; fruits black ; in savanna.
Sen.: Koulaye, Casamance *Chev.* 2021. **Mali**: Bongouni (Apr.) *Chev.* 684 ! Kéméne *Chev.* 653. Linaré to Gampéla, Mossi Dist. *Chev.* 24608 ! Birgo *Dubois* 199. **Port.G.**: Gabu (fr. Oct.) *Esp. Santo* 3471 ! 3558 ! **Guin.**: Kouroussa *Pobéguin* 272. **Iv.C.**: Touba *Aubrév.* 1253. Ouangolo *Aubrév.* 1472. Ferkéssé-dougou *Aubrév.* 1524. **Ghana**: Bussie, Wa (Mar.) ? *coll.* 823 ! Parria to Nabingo (May) *Kitson* 739 ! 963 ! Han (Apr.) *Vigne* FH 3799 ! Asi, Nante (June) *Lamptey* FH 2423 ! Bosomoa F. R. (fr. June) *Vigne* FH 2999 ! **Dah.**: Atacora Mts. *Chev.* 24130. **N.Nig.**: Sokoto (May) *Lely* 829 ! Kontagora (Feb.) *Dalz.* 217 ! Lokoja *Barter* ! *Elliott* 194 ! Meringa, Biu (May) *Rosevear* FHI 26606 ! Yola (May) *Dalz.* 42 ! **S.Nig.**: Aboh *Barter* 324 ! Also in Cameroun, Congo, Sudan, E. Africa, N. Rhodesia, Nyasaland and Mozambique. (See Appendix, p. 407.)

26. **P. subcana** *Hiern* in F.T.A. 3 : 172 (1877) ; Bremek. l.c. 175. A more or less scrambling shrub pubescent on the young parts, with whitish branches and slender white flowers crowded in clusters on short lateral branchlets ; on hills in savanna nowhere.
N.Nig.: Malabu Hills, near Yola (May) *Dalz.* 40 ! Also in Ubangi-Shari, Sudan, Kenya and Uganda.

27. **P. cinereifolia** *Berhaut* in Mém. Soc. Bot. Fr. 1953–54 : 8 (1954) ; Fl. Sén. 105 (1954). A shrub 3–8 ft. high with rather thick white bark and hairy leaves drying dark ; in savanna woodland.
Sen.: Tambacounda, Goudiry (fr. Sept.-Nov.) *Berhaut* 1456 ! 2921 ; 3148 ! 3162 ! Niokolo-Koba to Kédougou *Adam* 14405 ! **Mali**: Kita (July-Aug.) *Duong* 403 !

28. **P. subglabra** *Schumach.* Beskr. Guin. Pl. 78 (1827) ; F.T.A. 3 : 174 ; Bremek. in Fedde Rep. 37 : 207 ; op. cit. 47 : 92. *P. warneckei* K. Schum. & K. Krause (1907)—F.W.T.A., ed. 1, 2 : 91 ; Bremek. in op. cit. 37 : 179. A nearly glabrous shrub, to 12 ft. high ; flowers white ; fruits black.
Ghana: *Thonning* (photo !) Abetifi (June) *Box* 3452 ! Nungua and Aflenya, Accra Plains (May) *Rose Innes* GC 30048 ! *Adams* 4771 ! Achimota *Irvine* 751 ! **Togo Rep.**: Lomé (fl. & fr. May) *Warnecke* 301 !

29. **P. saxicola** *K. Krause* in Engl. Bot. Jahrb. 48 : 421 (1912) ; Bremek. l.c. 180. *P. thorbeckei* K. Krause in Engl. Bot. Jahrb. 57 : 40 (1920) ; Bremek. l.c. 179. An erect shrub, to 6 ft. high, with glossy leaves ; flowers yellow ; fruits black ; on rocky hills.
Mali: Guimel, Douentza (fr. Oct.) *Jaeger* 5515 ! **Iv.C.**: Geriereni, Haut Sassandra (May) *Fleury* in Hb. *Chev.* 21644 ! **Togo Rep.**: Ssodu-Dako Plateau, Kumonde (May) *Kersting* A740 ! Kodjelua (May) *Kersting* A613 ! **N.Nig.**: Nupe *Yates* ! Naraguta (June) *D. E. S. King* FHI 38861 ! **S.Nig.**: Ado Rock, Oyo Prov. (fl. Apr., fr. Oct.) *Keay & Savory* FHI 25444 ! Onochie FHI 34686 ! Idanre Hills (fr. Aug.) *Jones* FHI 20727 ! [Br.]**Cam.**: Vogel Peak, 4,700 ft., Adamawa (Nov.) *Hepper* 2739 ! Also in Cameroun.

Imperfectly known species.

1. **P. oresitropha** *Bremek.* in Fedde Rep. 37 : 73 (1934). Type specimens presumably destroyed in Berlin. Near to *P. owariensis* and *P. hookeriana*.
F.Po: *Mildbr.* 7100 ; 7135.

2. **P. bangweensis** *Bremek.* l.c. 79 (1934). Type specimen presumably destroyed in Berlin.
[Br.]**Cam.**: Bangwe *Conrau* 66.

In addition to the above, I have seen a number of sterile and fruiting specimens which probably belong to Pavetta but which cannot at present be satisfactorily determined.—R.W.J.K.

30. IXORA Linn.—F.T.A. 3 : 162.

Leaves oblanceolate or oblanceolate-obovate, subsessile, with conspicuous pouches (1–1·5 cm. long) on either side at the base, acuminate, 20–33 cm. long, 7·5–10 cm. broad, with 10–12 main lateral nerves ; peduncle 1·5–16 cm. long ; flowers sessile, in clusters ; calyx with short triangular lobes ; corolla-tube about 20 mm., lobes about 6 mm. long   ..   ..   1. *hippoperifera*
Leaves without large pouches at their base :
Corolla-tube not exserted from the calyx, about 0·5 mm. long, lobes 3·5–4 mm. long ; calyx-lobes short and broad ; inflorescence and flowers glabrous ; peduncle 4·5–8·5 cm. long, pedicels 5–15 mm. long ; leaves glabrous, obovate or obovate-oblong, cuneate, distinctly acuminate, 4·5–16·5 cm. long, 1·8–6 cm. broad, with 10–16 main lateral nerves   ..   ..   2. *divaricata*
orolla-tube well exserted from the calyx, 5–60 mm. long :
*Corolla-tube 5–35(–37) mm. long (to p. 142) :

†Leaves mostly glabrous or sometimes puberulous beneath ; peduncle 0–13(–15) cm. long (to p. 142) :

Corolla-tube 5–11 mm. long :
  Peduncle 5–15 cm. long ; lateral pedicels 8–17 mm. long ; corolla-tube 6–11 mm., lobes 7–10 mm. long ; leaves oblong or obovate-oblong, obtuse to cuneate at base, abruptly acuminate, 11–20 cm. long, 3–8 cm. broad, with about 12 main lateral nerves .. ..   3. *nematopoda*
  Peduncle 0–2 cm. long ; lateral pedicels 0–3·5 mm. long :
    Corolla-lobes 9–11 mm., tube 10–11 mm. long ; calyx with short broad lobes ; inflorescence up to 3 cm. long, with the primary branches up to 1·2 cm. long, minutely and sparingly puberulous ; leaves elliptic, oblong-elliptic or lance-olate, cuneate or obtuse at base, acuminate, 6–16 cm. long, 2–5·8 cm. broad, thinly membranous, drying pale green, with 10–16 main lateral nerves spreading at a wide angle .. ..   4. *delicatula*
    Corolla-lobes 4–5·5 mm. long :
      Inflorescences 4–11 cm. long, with the primary branches 1·5–7 cm. long, puberulous :
        Corolla-tube 10–15 mm. long ; inflorescence up to 7 cm. long, with the primary branches 1·5–3 cm. long ; leaves elliptic or oblong-elliptic, cuneate, acuminate, 6–17(–19) cm. long, 3–8 cm. broad, with 6–8 main lateral nerves ; at low altitudes .. ..   5. *guineensis*
        Corolla-tube 4·5–8 mm. long ; inflorescence up to 11 cm. long, with the primary branches 2–7 cm. long ; leaves 9–19 cm. long, 3·5–6·5 cm. broad, with 7–10 main lateral nerves ; at higher altitudes .. ..   6. *breviflora*
      Inflorescences up to 3 cm. long, with the primary branches not exceeding 1·5 cm., minutely puberulous ; corolla-tube 5·5–11 cm., lobes 4–5 mm. long ; leaves oblanceolate to elongate-elliptic, cuneate, usually acutely acuminate, 6·5–15 cm. long, 2·5–4·5 cm. long, with 8–10 main lateral nerves ; whole plant not more than 1 m. high .. ..   7. *talbotii*

Corolla-tube 13–37 mm. long (but occasional flowers in No. 14 have corolla-tubes 7–13 mm. long) :
  Peduncle 0–3·5 cm. long ; bracteoles minute, caducous ; corolla rounded or sub-acute in bud :
    Inflorescence glabrous ; corolla-tube 13–17 mm. long :
      Primary branches of inflorescence not exceeding 1 cm., lateral pedicels 0–1 mm. long ; leaves thin, elliptic, cuneate or obtuse at base, broadly acuminate, 4·5–9·5 cm. long, 2–4 cm. broad, with 8–10 main lateral nerves ; corolla-tube 14·5 mm., lobes 5·5 mm. long .. ..   8. *sp.* A
      Primary branches of inflorescence 1·5–4 cm., lateral pedicels 1–6 mm. long :
        Leaves membranous, very narrowly elliptic, attenuate at base, gradually acutely acuminate, 5–13 cm. long, 1·2–2·4 cm. broad, with 6–13 main lateral nerves ; corolla-tube 13–14 mm., lobes 6–7 mm. long .. ..   9. *sp.* B
        Leaves subcoriaceous, elliptic to oblanceolate, cuneate at base, subacute or obtuse at apex, 6–17 cm. long, 2·2–5 cm. broad, with 8–15 main lateral nerves ; corolla-tube 13–17 mm., lobes 6 mm. long   10. *bauchiensis*
    Inflorescence densely puberulous, with the primary branches 1·7–4 cm. long ; corolla-tube 10–30 mm. long :
      Corolla-tube 10–15 mm., lobes 4·5–5·5 mm. long ; inflorescence very shortly puberulous ; see also above .. ..   5. *guineensis*
      Corolla-tube 16–30 mm., lobes 6–8 mm. long ; inflorescence puberulous ; leaves subcoriaceous, elongate-elliptic, acute at each end, 12–20 cm. long, 4·2–6·8 cm. broad, with 10–15 main lateral nerves .. ..   11. *hiernii*
  Peduncle 2·2–13 cm. long ; bracteoles ovate or triangular ¼ to ¾ length of calyx, often persistent ; corolla rounded, subacute or acuminate in bud :
    Main lateral nerves 2–7, looped well within the margin ; peduncle 3·5–4·5 cm., lateral pedicels 3–6 mm. long ; corolla-tube 30 mm., lobes 11 mm. long, acuminate in bud ; leaves elliptic or oblong-elliptic, rounded or obtuse at base, long-acuminate, 3–12 cm. long, 1·5–5·5 cm. broad .. 12. *baldwinii*
    Main lateral nerves 8–21 :
      Corolla white or pale pink, tube (7–)15–24(–34) mm. long, lobes at least ¼ length of tube :
        Peduncle 9–13 cm. long ; flowers numerous and congested in a flat corymbose cyme, pedicels 0–1 mm. long ; corolla-tube 16–20 mm., lobes 8 mm. long, conspicuously acuminate in bud ; leaves oblanceolate, cuneate into a distinct petiole, long-acuminate, 18–35 cm. long, 4–12 cm. broad, with 15–21 main lateral nerves ; leaves subtending the inflorescences subcordate and sub-sessile .. ..   13. *aggregata*
        Peduncle 2·2–8·5 cm. long ; flowers in usually lax inflorescences, pedicels 0–12 mm. long ; corolla subacute or rounded in bud ; leaves up to 25 cm. long and 9 cm. broad, with up to 18 main lateral nerves :

Calyx-lobes ovate-triangular, about 1 mm. long ; bracteoles ¾ length of calyx ;
corolla-tube (7–)13–18 mm., lobes 6 mm. long ; lateral pedicels 0–2·5 mm.
long ; leaves obovate or elliptic, cuneate, shortly and obtusely acuminate,
7–14 cm. long, 3–6 cm. broad, with about 12 main lateral nerves ; leaves
subtending inflorescences often smaller and relatively broad ; stipules with
long points   ..   ..   14. *foliosa*
Calyx subtruncate or with short broad lobes ; bracteoles about ¼ length of
calyx ; corolla-tube (10–)15–25(–34) mm., lobes 8–12 mm. long ; lateral
pedicels (1–)4–12 mm. long ; leaves elliptic-oblanceolate, oblong-obovate,
oblong-elliptic, ovate-oblong or lanceolate, cuneate to rounded at base,
shortly acuminate, 4·5–20 cm. long, 2–7·5 cm. broad, with 8–18 main lateral
nerves ..   ..   15. *laxiflora*
Corolla-tube and outside of lobes red, tube 24–37 mm. long, lobes 5–7 mm. long
(i.e. about ¼ length of tube), rounded or subacute in bud ; peduncle 6–13 cm.
long, lateral pedicels 0–4 mm. long ; calyx-lobes ovate-triangular to lanceolate ; leaves oblong-lanceolate or oblong-oblanceolate rounded, obtuse or
cuneate at base, acuminate, 10–25 cm. long, 3–9 cm. broad, with 7–12 main
lateral nerves   ..   ..   16. *nimbana*
†Leaves densely covered with short hairs beneath ; peduncle 13–31 cm., pedicels
0–4·5 mm. long, both densely pubescent ; corolla densely pubescent outside, tube
red, 19–20 mm. long, lobes 7–8 mm. long, acuminate in bud ; calyx-lobes triangular ; bracteoles nearly as long as the calyx ; leaves oblong-elliptic, cuneate,
abruptly acuminate, 12–30 cm. long, 3·5–8·5 cm. broad, with 12–15 main lateral
nerves rather conspicuous beneath ..   ..   17. *nigerica*
*Corolla-tube (25–)40–60 mm. long ; pedicels 0–1(–2) mm. long ; flowers in rather
dense terminal corymbs ; leaves glabrous :
Peduncle 0–1 cm. long ; upper and lower leaves with petioles about 1 cm. long,
elliptic or obovate-elliptic, cuneate at base, gradually and acutely acuminate,
12–15 cm. long, 4–6 cm. broad, with about 8 main lateral nerves ; corolla-tube
40–45 mm., lobes 6–7 mm. long   ..   ..   18. *degemensis*
Peduncle 3–7 cm. long ; upper leaves subsessile, rounded or subcordate at base,
lower leaves petiolate ; leaves oblong, elliptic, ovate or oblanceolate, shortly
acuminate, 11–27 cm. long, 4·4–11(–13·5) cm. broad, with 9–13 main lateral
nerves ; corolla (37–)40–60 mm. long, lobes 8–9 mm. long   19. *brachypoda*

1. **I. hippoperifera** *Bremek.* in Hook. Ic. Pl. t. 3241 (1934). A glabrous shrub, with white flowers ; in forest.
   S.Nig.: Kwa Falls, Calabar (fr. Mar.) *Brenan* 9241 ! [**Br.**]**Cam.**: S. Bakundu F.R., Kumba Div. (Jan.,
   Mar.) *Binuyo & Daramola* FHI 35163 ! *Onochie* FHI 30868 ! Also in Cameroun.
2. **I. divaricata** *Hutch. & Dalz.* F.W.T.A., ed. 1, 2 : 89 (1931). A forest shrub or small tree, 10–35 ft. high,
   with small flowers in a lax panicle ; fruits red.
   S.L.: Gola Forest (Apr.) *Small* 619 ! **Lib.:** Dukwia R. (May) *Cooper* 58 ! 307 ! 308 ! Zwedru, Tchien
   (fr. Aug.) *Baldwin* 7014 ! (See Appendix, p. 400.)
3. **I. nematopoda** *K. Schum.* in Engl. Bot. Jahrb. 33 : 357 (1903). *I. rosea* K. Schum. l.c. 358 (1903) ; F.W.T.A.,
   ed. 1, 2 : 89 ; not of Wall. (1820). *I. tetramera* K. Schum. ex Wernham in Cat. Talb. 132 (1913), name
   only. A forest shrub.
   S.Nig.: Oban *Talbot* 231 ! 2037 ! Eyeyin, Ikom Div. (July) *Mason* FHI 4209 (partly) ! [**Br.**]**Cam.**:
   Preuss 1338 ! Barombi to Ninga (June) *Preuss* 345 ! Mungo (May) *Buchholz*. Kumba to Barombi (Aug.)
   Preuss 338. **F.Po:** *Mann* !
   [The specimen *Preuss* 1338 which I have seen at the British Museum is quite possibly a duplicate of
   the specimen cited by K. Schumann as *Preuss* 338.]
4. **I. delicatula** *Keay* in Bull. Jard. Bot. Brux. 27 : 96 (1957). Erect shrub, to 6 ft. high ; corolla white ; in
   forest.
   S.Nig.: Degema Dist. *Talbot* 3822 ! [**Br.**]**Cam.**: S. Bakundu F.R., Banga (Jan.) *Keay* FHI 28568 !
5. **I. guineensis** *Benth.* in Fl. Nigrit. 414 (1849) ; F.T.A. 3 : 165. *I. riparia* Hiern in F.T.A. 3 : 164 (1877) ;
   F.W.T.A., ed. 1, 2 : 89. (?) *I. buchholzii* Engl. (1886). *I. obanensis* Wernham (1913). A small shrub ;
   corolla pale pinkish-white ; near the sea coast and by streams.
   S.Nig.: Oban *Talbot* 230 ! [**Br.**]**Cam.**: Victoria (Apr., Nov.) *Brenan* 9592 ! *Maitland* 771 ! Tiko (Jan.)
   Dunlap 169 ! Mungo R. *Buchholz*. Bai to Bokosso (Jan.) *Keay* FHI 37367 ! S. Bakundu F.R., Kumba
   (fl. & fr. Dec.) *Latilo* FHI 43810 ! **F.Po:** (Dec.) *Mann* 19 ! Also in Cameroun and Rio Muni.
   [The type of *I. guineensis* Benth. is a specimen at Kew from Hb. Hooker ; it has the locality " Guinea
   Coast " but no collector's name.]
6. **I. breviflora** *Hiern* in F.T.A. 3 : 165 (1877). (?) *I. carniflora* K. Krause (1917). Erect shrub or small tree,
   to 25 ft. high ; corolla creamy-white, fragrant, becoming pinkish (or red *fide* Mann).
   S.Nig.: Ikwette Plateau, 5,200 ft., Obudu Div. (Dec.) *Keay & Savory* FHI 25213 ! [**Br.**]**Cam.**: Bafut-
   Ngemba F.R., 7,200 ft., Bamenda (Feb.) *Hepper* 2132 ! **F.Po:** 1,000 ft. *Mann* 315 ! N. side of St. Isabel
   Peak, 3,000–3,200 ft. (Aug.) *Mildbr.* 6391. Also in Cameroun.
7. **I. talbotii** *Wernham* in Cat. Talb. 51 (1913). Low shrub, with red calyx and white corolla ; in forest.
   S.Nig.: Oban *Talbot* 1413 ! 2038 ! Ikom (Dec.) *Rosevear* 63/30a ! Cross River North F.R. (Dec.) *Keay*
   FHI 28152 ! Afi River F.R. (Jan.) *Motuba* FHI 28685 ! [**Br.**]**Cam.**: Mamfe (Dec.) *Baldwin* 13818 !
8. **I. sp. A.**—Flowers cream ; in forest ; further material required.
   Lib.: Mt. Bili (Dec.) *Barker* 1153 ! Ganta (Dec.) *Harley* 1465 !
9. **I. sp. B.**—A forest shrub ; flowers white ; further material required.
   Ghana: Esem Epan F.R., W. Prov. (Oct.) *Darko* 1013 ! Southern Scarp F.R. (fr. Feb.) *Moor* FH 2229 !
10. **I. bauchiensis** *Hutch. & Dalz.* F.W.T.A., ed. 1, 2 : 87 (1931). A shrub, to 20 ft. high ; flowers white,
    sweetly scented ; in fringing forest by streams.
    N.Nig.: Naraguta (Nov.) *Lely* 723 ! Jos Plateau (fl. Jan., Oct., fr. Jan.) *Lely* P102 ! 845 ! **S.Nig.:** Kundeve,
    Obudu Div. (Dec.) *Keay & Savory* FHI 25138 ! 25148 ! [**Br.**]**Cam.**: Mai-Idoanu, Gashaka Dist. (Dec.)
    Latilo & Daramola FHI 28972 !
11. **I. hiernii** *Sc. Elliot* in J. Linn. Soc. 30 : 82 (1894) ; Aubrév. Fl. For. C. Iv., ed. 2, 3 : 292, t. 360 B. *I.
    breviflora* of F.W.T.A., ed. 1, 89, partly (S.L. records). A shrub or tree, to 15 ft. high, with white flowers
    and red berries.
    Guin.: *Chev.* 12980. *Pobéguin* 833. Eremakuna *Sc. Elliot* 5405 ! **S.L.:** Freetown *Sc. Elliot* 5909 ! Koflu,

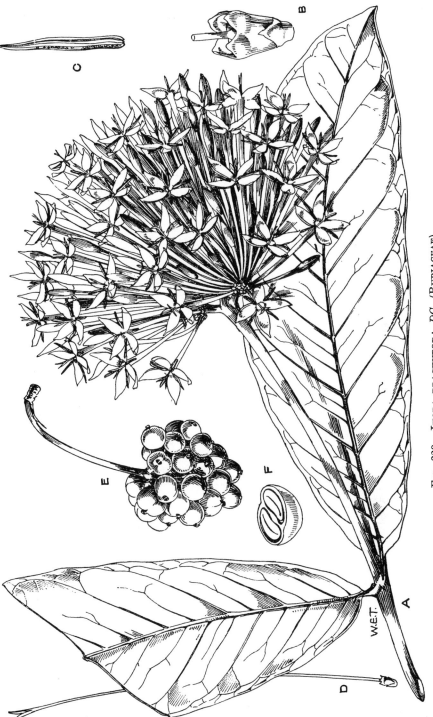

Fig. 229.—Ixora brachypoda *DC.* (Rubiaceae).

A, flowering shoot. B, ovary. C, anther. D, pistil. E, fruits. F, cross-section of fruit.

W.E.T.

143

R. Scarcies (Jan.) *Sc. Elliot* 4596 ! Ninia, Talla Hills (fr. Feb.) *Sc. Elliot* 4890 ! Musaia (Dec.) *Deighton* 4538 ! **Iv.C.:** Man *Aubrév.* 1059 ! Mt. Dou *Aubrév.* 2113. **Ghana:** Adamsu (Dec.) *Vigne* FH 3502 ! Kadjakpe, Hohoe Dist. (Nov.) *St. C. Thompson* FH 3669 ! **N.Nig.:** Kurmin Damisa, Jemaa (Nov.) *Keay & Onochie* FHI 21727 ! Kwakuti, near Minna (Dec.) *Keay* FHI 37309 ! [**Br.]Cam.:** W. of Nkambe (fl. & fr. Feb.) *Hepper* 1915 ! Vogel Peak, 3,850 ft., Adamawa (Nov.) *Hepper* 1337 !

12. **I. baldwinii** *Keay* in Bull. Jard. Bot. Brux. 27 : 97 (1957). A forest shrub with red fruits.
Lib.: Jaurazon (Mar.) *Baldwin* 1146 ! Mnanulu, Webo (fr. June) *Baldwin* 6047 ! Nyaake, Webo (fr. June) *Baldwin* 6095 ! Diebla, Webo (fr. July) *Baldwin* 6312 !

13. **I. aggregata** Hutch. F.W.T.A., ed. 1, 2 : 607 (1936). *I. congesta* Stapf (1905)—F.W.T.A., ed. 1, 2 : 87; not of Roxb. (1820). A shrub to 4 ft. high, or small tree ; flowers white.
Lib.: Sinoe Basin *Whyte* ! Dukwia R. (fl. buds Feb.) *Cooper* 201 ! About 15 miles inland from River Cess (Mar.) *Baldwin* 11248 ! Nyaake, Webo (June) *Baldwin* 6181 ! Jabroke, Webo (fr. July) *Baldwin* 6500 ! 6608 ! 6609 !

14. **I. foliosa** *Hiern* in F.T.A. 3 : 166 (1877). A tree to 30 ft. high ; corolla white, tinged pink ; fruits black and glossy ; in montane forest and woodland.
[**Br.]Cam.:** Cam. Mt., 4,000–7,200 ft. (fl. Dec.-Apr., fr. Apr.) *Brenan* 9562 ! *Maitland* 198 ! 208 ! 514 ! *Mann* 2154 ! *Preuss* 665 ! *Thorold* CM42 ! Bafut-Ngemba F.R., 5,000–6,700 ft., Bamenda (Dec.-Apr.) *Keay* FHI 28357 ! *Onochie* FHI 34868 ! Lakom, 6,000 ft., Bamenda (Apr.) *Maitland* 1380 ! Also in Kenya.

15. **I. laxiflora** *Sm.* in Rees Cycl. 19, No. 8 (1819) ; Bot. Mag. t. 4482 ; F.T.A. 3 : 164, partly (excl. syn. *I. brachypoda* DC.) ; Aubrév. Fl. For. C. Iv., ed. 2, 3 : 292, t. 360 A. *I. odoratissima* Klotzsch in Abh. Acad. Berl. 227, t. 1 (1856). *I. brachypoda* of F.W.T.A., ed. 1, 2 : 87, not of DC. A shrub or small tree, to 12 ft. high ; flowers pale pink in lax drooping cymes ; by streams and in sandy coastal situations. Specimens from coastal areas tend to have broader, thicker leaves.
Port.G.: R. Colufe, Gabu (June) *Esp. Santo* 2509 ! Cacine *Esp. Santo* 643 ! **Guin.:** Kouroussa (May) *Pobéguin* 893 ! Macenta (May) *Jac.-Fél.* 936 ! **S.L.:** *Afzelius* ! *Whitfield* ! Sugar Loaf Mt. *Barter* ! Bagroo R. *Mann* ! Tombo (May) *Deighton* 2670 ! York (Apr.) *Jordan* 863 ! Gbap (June) *Adames* 52 ! **Lib.:** Grand Bassa (July) *T. Vogel* 73 ! *Middleton* ! Cape Palmas (Nov.) *Schoenlein* ; *T. Vogel* 6 ! 61 ! Sinoe Basin *Whyte* ! Monrovia (May) *Baldwin* 5829 ! *Linder* 1535 ! *Whyte* ! Ba, Little Kola R. (Mar.) *Baldwin* 11190 ! **Iv.C.:** Tabou *Aubrév.* 1284. **Ghana:** Kikum (Jan.) *Brent* 1dd ! Apremdo, Takoradi (fl. buds May) *Hughes* 509 ! 533 ! Essiama (June) *Fishlock* 16 ! (See Appendix, p. 401.)
[*I. linderi* Hutch. & Dalz. F.W.T.A., ed. 1, 2 : 87 (1931) appears to be a coastal form of this species with very short pedicels.—**Lib.:** Monrovia (Nov.) *Linder* 1536 !]

16. **I. nimbana** *Schnell* in Rev. Gén. Bot. 57 : 282 (1950). *I. laxiflora* of F.W.T.A., ed. 1, 2 : 87, partly. Shrub to 6 ft. high, or climber ; corolla-tube and outside of lobes red, inside of lobes pinkish-white ; fruit red ; in forest.
Guin.: lower slopes of N.E. Nimba Mts. (Aug.) *Schnell* 3334 ! **S.L.:** Mayoso (Aug.) *Thomas* 1462 ! Njagbahuna, Kowa (Aug.) *Deighton* 6103 ! Roruks (July) *Deighton* 2508 ! 3251 ! Sewa R., Jaiama (July-Aug.) *Dawe* 529 ! Levuma (Aug.) *Deighton* 2210 ! **Lib.:** Gletown, Tchien (July) *Baldwin* 6793 ! Peahtah (fr. Dec.) *Baldwin* 10612 ! Ganta (fl. & fr. July) *Harley* 622 ! **Iv.C.:** Taï (Aug.) *Boughey* GC 14968 ! **Ghana:** E. Akim (May) *Johnson* 767 ! Kwahu Praso (June) *Vigne* FH 1751 ! Kumasi (June) *Vigne* FH 1204 ! 1764 ! Jimira F.R. (June) *Vigne* FH 2963 !

17. **I. nigerica** *Keay* in Bull. Jard. Bot. Brux. 27 : 96 (1957). A forest shrub ; corolla-tube and outside of lobes red, inside of lobes white.
S.Nig.: Umon-Ndealichi F.R., Itu Div. (June) *Ujor* FHI 30176 ! Ukpon F.R., Obubra Div. (July) *Latilo* FHI 31875 !

18. **I. degemensis** *Hutch. & Dalz.* F.W.T.A., ed. 1, 2 : 87 (1931). A glabrous shrub with slender flowers in a terminal corymb ; in forest.
S.Nig.: Degema *Talbot* 3767 ! 3773 !

19. **I. brachypoda** *DC.* Prod. 4 : 488 (1830) ; not of F.W.T.A., ed. 1, 2 : 87 ; Keay l.c. 95. *I. radiata* Hiern in F.T.A. 3 : 163 (1877) ; Chev. Bot. 333 ; Aubrév. Fl. For. C. Iv., ed. 1, 3 : 254 ; Fl. For. Soud.-Guin. 469, t. 103, 3. *I. capitata* A. Chev. Bot. 332. A shrub, 10–20 ft. high, with terminal corymbs of flowers ; corolla white, with the lobes pink at the tips ; fruits red ; in swamps and other moist situations.
Sen.: Casamance (fl. buds & fr. Apr.) *Perrottet & Leprieur* ! Sangalkam *Berhaut* 1624. **Gam.:** *Hayes* 524 ! **Mali:** Bougouri to Sikasso (Dec.) *Roberty* 13319 ! Bobo-Dioulasso (May) *Chev.* 904. **Port.G.:** Nova Cintra, Bolama (Apr.) *Esp. Santo* 1923 ! Buruntuma, Gabu (fl. & fr. July) *Esp. Santo* 2711 ! **Guin.:** Santa to Timbo *Chev.* 12802. **S.L.:** Sherbro (Feb.) *Dalz.* 952 ! Gegbwema (Nov.) *Deighton* 438 ! Mabonto (Oct.) *Thomas* 3610 ! Kokoru, Gaura (Oct.) *Deighton* 5212 ! Daragbe, Samu (Dec.) *Sc. Elliot* 4306 ! **Lib.:** *Millen* 196 ! Monrovia (Aug.) *Baldwin* 13063 ! Road to Brewersville (Sept.) *Barker* 1057 ! Buchanan (Mar.) *Baldwin* 11209 ! **Iv.C.:** Ferkéssédougou *Aubrév.* 1627 ; 1836. Gaoua to Banfora *Aubrév.* 2209. Bouaké *Chev.* 22081 ; 2219. **Ghana:** Ejura (fl. Oct., fr. Nov.) *Vigne* FH 3466 ! 4871 ! **Togo Rep.:** Wiawia, Hohoe Dist. (Oct., Nov.) *St. C. Thompson* FH 3619 ! **N.Nig.:** Loin Nupe *Barter* 1236 ! Kontagora (Dec.) *Dalz.* 214 ! Kaduna (Dec.) *Meikle* 750 ! 775 ! **S.Nig.:** Lagos *Barter* 2210 ! *Moloney* ! Aligere, Lagos (Feb.) *Millen* 149 ! Sapoba *Kennedy* 1974 ! 2757 ! Mamu River F.R., Awka (Sept.) *Onochie* FHI 34079 ! [**Br.]Cam.:** Gangumi, Gashaka Dist. (fr. Dec.) *Latilo & Daramola* FHI 28808 ! Extends to Congo and the Sudan. (See Appendix, p. 401.)

Besides the above, the following Asiatic species are cultivated in our area : *I. coccinea* Linn. and *I. parviflora* Vahl.

Imperfectly known species.

I. sp.—Near to *I. inundata* Hiern and *I. euosmia* K. Schum. ; further material of this group needed. Shrub of river banks with elliptic-lanceolate leaves, attenuate at base and acutely acuminate at apex.
[**Br.]Cam.:** R. Marie, Anyu-Lipundu path, Kumba Div. *Olorunfemi* FHI 30617 !

31. AULACOCALYX Hook. f.—F.T.A. 3 : 129 ; K. Schum. in E. & P. Pflanzenfam. 4, 4 : 87 (1891). *Dorothea* Wernham (1913)—F.W.T.A., ed. 1, 2 : 76.

Corolla-tube not split at anthesis :
Anthers 13–15 mm. long, exserted for at least half their length ; main lateral nerves 2–4 on each side of midrib, not conspicuously looped, with numerous closely parallel fine transverse tertiary nerves spreading at right angles to the midrib ; leaves elliptic, long-cuneate to obtuse or rarely rounded at base, gradually long-acuminate, 6·5–16 cm. long, 2·2–6(–9) cm. broad, thinly pilose on midrib and main nerves beneath ; flowers silky outside, 1–8 together ; corolla-tube 1·5–2·8 cm., lobes 1–1·5 cm. long
　　　　　　　　　　　　　　　　　　　　　　1. *jasminiflora*
Anthers 10–12 mm. long, only very slightly exserted ; main lateral nerves 6–10 on each side of midrib, looped well within the margin, with rather irregular tertiary nerves between ; leaves elliptic or ovate-elliptic or obovate-elliptic, subcordate to

auriculate and often asymmetric at base, long-acuminate, 7·5–19 cm. long, 3·5–9 cm. broad, thinly pilose on midrib and main nerves beneath ; flowers pilose outside, 1–3 together ; corolla-tube 4–5 cm., lobes 1·5–1·7(–2·4) cm. long ..   2. *caudata*

Corolla-tube split down one side at anthesis ; style recurved through the slit in the corolla-tube ; anthers exserted well beyond the corolla-tube :

Calyx divided into subulate teeth, 4–5 mm. long ; tube about 3·5 mm. long ; corolla-tube 0·7–1·5 cm., lobes 1–2 cm. long ; anthers 1–1·3 cm. long ; flowers 3–4 together, pilose outside ; branchlets pubescent ; leaves elliptic to narrowly obovate, cuneate or obtuse at base, gradually long-acuminate, 6–10 cm. long, 2–4 cm. broad, with 5–7 main lateral nerves on each side of midrib, pubescent beneath   3. *divergens*

Calyx subtruncate, sulcate, about 5 mm. long ; corolla-tube 3·5–4·5 cm., lobes 1·5–2 cm. long ; anthers 1·8–2 cm. long ; flowers 2–3 together, pilose outside ; branchlets glabrous ; leaves elliptic to obovate-elliptic, cuneate at base, gradually acuminate, 8–18 cm. long, 3·5–8·5 cm. broad, with 3–5 main lateral ascending nerves on each side of midrib, conspicuous and pubescent beneath ..   ..   ..   4. *talbotii*

1. **A. jasminiflora** *Hook.f.* in Hook. Ic. Pl. t. 1126 (1873) ; F.T.A. 3 : 129 ; Aubrév. Fl. For. C. Iv., ed. 2, 3 : 290. An understorey forest tree, to 40 ft. high, rarely more, with fluted bole ; calyx pale green, corolla greenish below, white above.
S.L.: Kafogo, Limba (young fr. Apr.) *Sc. Elliot* 5611 ! Dansogoia, Tonkolili Dist. (fl. buds Jan.) *King* 153b ! **Lib.**: Kakatown *Whyte* ! Dukwia R. (Nov.) *Cooper* 90 ! 142 ! **Ghana**: Pusupusu (Nov.) *McAinsh* FH 1572 ! Begoro, Akim (Jan.) *Irvine* 1161 ! S. Scarp F.R. (fl. Dec., fr. Feb.) *Moor* FH 1149 ! 2224 ! Pinpinsu, Begoro (Dec.) *Tinsley* WACRI 4936 ! Buaku, Wiaso (Nov.) *Talmie* FH 3037 ! **S.Nig.**: Omo F.R., Ijebu Ode Prov. (young fr. Mar.) *Jones & Onochie* FHI 17201 ! Okomu F.R., Benin Prov. (Feb., Dec.) *Brenan* 8562 ! 8602 ! *Hide* 39/37 ! [**Br.**]**Cam.**: Buea (Mar.) *Maitland* 456 ! 493 ! 687 ! Mopanya (fr. Apr.) *Maitland* 1126 ! Banga, Kumba Div. (Jan.) *Keay & Russell* FHI 28681 ! Also in Cameroun, Rio Muni, Gabon, Congo and Angola. (See Appendix, p. 391.)

2. **A. caudata** (*Hiern*) *Keay* in Bull. Jard. Bot. Brux. 28 : 60 (1958). *Randia ? caudata* Hiern in F.T.A. 3 : 96 (1877). *Aulacocalyx leptactinoides* K. Schum. in Engl. Bot. Jahrb. 23 : 453 (1897). *Randia leptactinoides* K. Schum. ex Hutch. & Dalz. F.W.T.A., ed. 1, 2 : 78 (1931). *Rothmannia leptactinoides* (K. Schum. ex Hutch. & Dalz.) Fagerlind (1943). A forest shrub ; flowers white.
S.Nig.: Oban *Talbot* 285 ! Orem (Jan.) *Onochie* FHI 36060x ! Akamkpa Rubber Estate, Calabar R. Div. (Mar.) *Latilo* FHI 40930 ! Ukpon River F.R., Obubra Dist. (Aug.) *Amachi* FHI 38282 ! [**Br.**]**Cam.**: Kumba Div. (fr. June) *Akuo* FHI 15107 ! Nta Ali F.R., near Bakebe, Mamfe Div. (Jan.) *Umana* FHI 28551 ! Also in Cameroun, Rio Muni and Gabon.

3. **A. divergens** (*Hutch. & Dalz.*) *Keay* l.c. 61 (1958). *Dorothea divergens* Hutch. & Dalz. F.W.T.A., ed. 1, 2 : 76 (1931). *D. minor* Wernham ex A. Chev. Bot. 324, name only. A shrub or tree, 3–25 ft. high, with white flowers, sickly scented ; in forest.
Guin.: *Farmar* 317 ! Kindia (Feb.) *Jac.-Fél.* 1544 ! Kaba Valley (May) *Chev.* 13180 ! Farana (Apr.) *Chev.* 13198. S.L.: Mt. Horton (Jan.) *T. S. Jones* 305 ! Hangha (Apr.) *Lane-Poole* 219 ! Musaia (young fr. Apr.) *Deighton* 5407 !

4. **A. talbotii** (*Wernham*) *Keay* l.c. 61 (1958). *Dorothea talbotii* Wernham in Cat. Talb. 46, t. 7 (1931) ; F.W.T.A., ed. 1, 2 : 76. A small forest tree, to 15 ft. high, with cream-coloured flowers.
S.Nig.: Oban *Talbot* 1546 ! Aking, Calabar Prov. (Apr.) *Onochie* FHI 21887 ! Dukwe, Calabar R. Div. (fr. Mar.) *Latilo* FHI 40342 ! Also in Cameroun.

32. RUTIDEA DC.—F.T.A. 3 : 187.

Stipules divided (at least near the inflorescence) into 3–7 filiform segments ; corolla-tube glabrous, 6–8 mm. long ; branchlets bristly-pubescent :

Bracteoles linear, as long as calyx, 2–3 mm. long ; stipules with 3–5 segments ; bracts linear-subulate, about 5 mm. long ; leaves elliptic, 6–10 cm. long, 3–5·5 cm. broad, rounded to subcordate at base, acutely acuminate, subglabrous above, pilose on nerves beneath often with tuft of pale coloured hairs in nerve axils   1. *parviflora*

Bracteoles subulate, twice as long as calyx, about 6 mm. long ; stipules with 5–7 segments ; bracts subulate, about 1 cm. long, spreading-pilose ; leaves elliptic, 8–14 cm. long, 4–6 cm. broad, cuneate to rounded at base, acute or long mucronate at apex, setose above and below, pilose on nerves beneath with dense tuft of brown hairs in nerve axils   ..   ..   2. *syringoides*

Stipules undivided :

Leaves setose-pilose above and beneath, ciliate on margins, elliptic to ovate-elliptic, 7–14 cm. long, 4–7·5 cm. broad, subcordate at base, acuminate at apex, nerves impressed above and prominent beneath without hair clusters in the axils ; branchlets pilose ; bracts and bracteoles linear-lanceolate, much longer than calyx, about 1 cm. long, pilose ; corolla-tube 13 mm. long, pubescent, corolla-lobes 5   3. *pavettoides*

Leaves glabrous, subglabrous or softly tomentose beneath ; bracts and bracteoles inconspicuous :

Leaves softly tomentose beneath, densely so in the axils of the nerves which loop distinctly round the margin, 7–15 cm. long, 4–8 cm. broad, rounded at base, acute or subacute at apex, all nerves impressed above ; branchlets softly tomentose ; corolla-tube 4 mm. long, pubescent, corolla-lobes 5   ..   4. *olentricha*

Leaves beneath and branchlets glabrous or subglabrous :

Corolla-tube densely pubescent, 3 mm. long, corolla-lobes 5 ; leaves ovate-elliptic, 8–15 cm. long, 5–7 cm. broad, cuneate to rounded at base, acutely acuminate, with 8–10 pairs of minutely pubescent nerves ; panicles spreading   5. *smithii*

Corolla-tube glabrous or sparsely pubescent :

Reflexed spinescent arrested branchlets, 1·5–4 cm. long, present in some axils ; inflorescence about 11 cm. diam. of spreading corymbs ; corolla-tube glabrous,

3 mm. long, corolla-lobes 4 ; leaves elliptic or broadly so, 7–13 cm. long, 4–7 cm.
broad, 8–10 pairs of nerves   ..   ..   6. *membranacea*
Modified branchlets absent :
Inflorescences dense terminal corymbs about 3 cm. long and 4 cm. diam. ; corolla-
tube 5 mm. long, glabrous, corolla-lobes 4, bud-swelling ovoid, obtuse, 2 mm.
long ; leaves elliptic, 8–15 cm. long, 3–7 cm. broad, with 8–10 pairs of nerves,
long acuminate   ..   ..   7. *glabra*
Inflorescences lax axillary panicles 4–9 cm. long ; corolla-tube 3 mm. long ;
glabrous or sparsely pubescent, corolla-lobes 5, bud-swelling ellipsoid, acute,
3 mm. long ; leaves broadly elliptic, 7–13 cm. long, 3–7·5 cm. broad with about
6 pairs of nerves, long acuminate   ..   ..   8. *decorticata*

1. **R. parviflora** *DC.* in Ann. Mus. Paris 9 : 219 (1807) ; F.T.A. 3 : 188 ; Chev. Bot. 377. *R. kerstingii* K.
Krause (1909). *Tarenna pobeguinii* Pobéguin (1906). Scandent shrub with shiny leaves ; flowers white,
fragrant, fruits yellow-red ; in regrowth forest.
Sen.: *Perrottet* ! **Mali:** Bambanatoumba (fr. Mar.) *Chev.* 558 ! **Port.G.:** Biombo, Bissau (fr. Feb.) *Esp.
Santo* 1795 ! Prabis, Bissau (fr. Feb.) *Esp. Santo* 1825 ! **Guin.:** Kissidougou (fr. Mar.) *Martine* 278 !
Timbo (buds & fr. Feb.) *Pobéguin* 157 ! Nzérékoré to Macenta (fr. Oct.) *Baldwin* 9747 ! Konkouré to
Timbo *Chev.* 12465 ; 12510. Kindia *Chev.* 12793. **S.L.:** Lumbaraya, Talla Hills (Feb.) *Sc. Elliot* 4945 !
Falaba to Pujehun (Mar.) *Aylmer* 19 ! Loma Mts. (bud Jan.) *Jaeger* 4041 ! Njala (fl. Mar., fr. Jan.) *Deighton*
2881 ! 5020 ! Newton (fl. & fr. Feb.) *Deighton* 5360 ! **Lib.:** Robertsport (fr. Dec.) *Baldwin* 10927 ! Jabrocca
(Dec.) *Baldwin* 10854 ! Gbanga (Dec.) *Baldwin* 10534 ! Cape Palmas (fr. July) *T. Vogel* 26 ! 39 ! Monrovia
White ! **Iv.C.:** Potou Lagoon to Alépé *Chev.* 17380. Grabo, Cavally basin *Chev.* 19728. Mt. Tonkoui
(Mar.) *Leeuwenberg* 2980 ! **Ghana:** Kumasi (fl. & fr. Jan.) *Vigne* 1641 ! *Darko* 502 ! Bibiani (Mar.) *Darko*
858 ! Essiama (fr. Oct.) *Fishlock* 4 ! **N.Nig.:** Kaduna (fl. & fr. Feb.) *Meikle* 1206 ! Nupe *Barter* ! Assob,
Plateau (Feb.) *McClintock* 240 ! **S.Nig.:** Lagos *Rowland* !
2. **R. syringoides** (*Webb*) *Bremek.* in Fedde Rep. 37 : 206 (1934). *Pavetta syringoides* Webb (1850). *Rutidea
rufipilis* Hiern in F.T.A. 3 : 188 (1877) ; F.W.T.A., ed. 1, 2 : 93. Scandent shrub covered with brown
hairs ; flowers white ; fruits yellow-red.
S.L.: Heddle's Farm, Freetown (fr. Dec.) *Sc. Elliot* 3927 ! Ganta (fr. Sept.) *Baldwin* 9250 ! *Harley*
1522 ! **Iv.C.:** Adiopodoumé *Mangenot & Aké Assi* IA 1229. N'Zida *Aké Assi* IA 1603. **Ghana:** Kubeasi
(Mar.) *Andoh* 4953 ! Bosumkese F.R. (fr. Dec.) *Adams* 5263 ! **S.Nig.:** Lagos *Moloney* 7 ! Jamieson R.,
Sapoba (fr. Feb.) *Onochie* FHI 31942 ! Nikrowa Creek, Okomu F.R. (fr. Feb.) *Brenan* 9185 ! Calabar (fl.
& fr. Feb., Mar.) *Mann* 2295 ! *Brenan* 9220 ! Oban *Talbot* 223 ! **F.Po:** Musola (fr. Jan.) *Guinea* 1143 !
Also in Cameroun, Congo, Uganda, Kenya, Tanganyika and Mozambique.
3. **R. pavettoides** *Wernham* in J. Bot. 57 : 345 (1919). Scandent shrub or liane, densely hairy with stiff brown
hairs ; flowers white, fragrant ; fruits yellow or yellow-red ; in forest.
S.Nig.: Lagos (Dec.) *Dalz.* 1201 ! *Hagerup* 809 ! Ilaro F.R., Egbado Dist. (Dec.) *Onochie* FHI 31889 !
Etemi to Atikiriji, Ijebu-Ode Dist. (Dec.) *Tamajong* FHI 20278 ! Oban *Talbot* !
4. **R. olentricha** *Hiern* in F.T.A. 3 : 189 (1877). Scandent shrub with dense velvety-brown indumentum and
dark green shiny leaves ; flowers white ; fruits yellowish.
S.L.: *Don* ! Heddle's Farm, Freetown (May, young fr. Dec.) *Sc. Elliot* 3929 ! *Deighton* 1192 ! Kenema
(fr. Jan.) *Thomas* 7663 ! Leicester (May) *Lane-Poole* 281 ! **Lib.:** Monrovia to Ganta (fr. Jan.) *Harley*
1105 ! Bobei (Mar.) *Harley* 1791 ! **Iv.C.:** Soubiré to Yaou *Chev.* 17803 ! **Ghana:** Aburi (May) *Brown* 325 !
Abetifi, Kwahu (young fr. June, Dec.) *Irvine* 313 ! 1655 ! Begoro, Akim (Apr.) *Irvine* 1188 ! Simpa (Feb.)
Vigne FH 2788 ! **S.Nig.:** Ibadan South F.R. (Apr.) *Keay* FHI 22832 ! Aguku Dist. *Thomas* 980 ! Ekeji-
Ipetu F.R., Ilesha Dist. (fr. Dec.) *Onochie* FHI 5242 ! **[Br.]Cam.:** S. Bakundu, Kumba (fr. Mar.) *Binuyo
& Daramola* FHI 35641 ! R. Labati, Serti, Gashaka Dist. (fr. Feb.) *Latilo & Daramola* FHI 34449 ! Also
in Cameroun, Rio Muni, Sudan, Congo, Angola and N. Rhodesia.
5. **R. smithii** *Hiern* in F.T.A. 3 : 189 (1877). *R. albiflora* K. Schum. (1899). Scandent shrub becoming a liane
20 ft. or more high ; flowers whitish, strongly fragrant ; fruits yellow-red ; in secondary forest or forest
margins.
S.L.: Ronietta (bud & fr. Nov.) *Thomas* 5509 ! 5661 ! Berria, Tambakka (Feb.) *Sc. Elliot* 5003 ! Joru
(Jan.) *Deighton* 3853 ! Tikonka (Jan.) *Deighton* 3886 ! **Lib.:** Kakatown *Whyte* ! Du R. (fr. Aug.) *Linder*
280 ! Karmahun (fr. Nov.) *Baldwin* 10157 ! **Iv.C.:** Guédéyo to Soubré (fr. Apr.) *Leeuwenberg* 3220 !
Ghana: Kumasi (fr. Oct.) *Andoh* FH 5577 ! New Jauben (Mar.) *Johnson* 629 ! Mpraeso (fl. & fr. Feb.)
Foggie FH 4857 ! Assuantsi Road (Jan.) *Fishlock* 25 ! **S.Nig.:** Lagos *Moloney* ! *Schlechter* 12298 ! Ijaiye
F.R., Oyo (Feb.) *Keay* FHI 21156 ! Ibadan (Jan.) *Keay* FHI 19804 ! *Meikle* 1123 ! Okomu F.R., Benin
(bud & fr. Dec., fl. Jan.) *Brenan* 8496 ! 8907 ! Oban *Talbot* 2039 ! **[Br.]Cam.:** Buea, 3,200 ft. Cam. Mt.
(Feb., Apr.) *Maitland* 1016 ! *Hutch. & Metcalfe* 94 ! Lakom, Bamenda, 6,000 ft. (Apr.) *Maitland* 1656 !
Also in Annobon, Cameroun, Gabon, Congo, Sudan, Uganda, Kenya, Tanganyika, N. Rhodesia and
Angola.
6. **R. membranacea** *Hiern* in F.T.A. 3 : 190 (1877). *R. glabra* of F.W.T.A., ed. 1, 2 : 93, partly (*Thomson*
126 ; *Johnson* 650 ; *Irvine* 382), not of Hiern. *R. obtusata* K. Krause (1920). Rambling shrub usually on
edge of forest ; flowers white in spreading panicles, fragrant, fruits yellow-red.
S.L.: Gorahun (Apr.) *Deighton* 3643 ! Daru (Apr.) *Deighton* 3886 ! **Lib.:** Peahtah (fr. Oct.) *Linder* 1503 !
Zwedru, Tchien Dist. (fr. Aug.) *Baldwin* 7028 ! Vahon, Kolahun Dist. (fr. Nov.) *Baldwin* 10208 ! **Iv.C.:**
Adiopodoumé *Aké Assi* IA 1009. Anguédédou (fr. Aug.) *de Wilde* 199 ! **Ghana:** *Burton* ! *Irvine* 382 !
Osino (June) *Arnold* 373 ! Kwahu, 2,000 ft. (Apr.) *Johnson* 650 ! Ebin-Mamiri F.R., Tarkwa Dist.
(fr. June) *Enti* FH 6247 ! **S.Nig.:** Ogwashi-Uku F.R., Asaba Dist. (fr. Aug.) *Onochie* FHI 33296 ! Okomu
F.R., Benin (fr. June) *Onochie* FHI 34614 ! Calabar (Apr.) *Thomson* 126 ! **[Br.]Cam.:** Buea, 3,000 ft. (fr.
Jan.) *Maitland* 652 ! Nkom-Wum F.R. (fr. July) *Ujor* FHI 29286 !
[The two Nigerian FHI specimens cited have larger fruits than is usual for this species and lack the
axillary hooks.]
7. **R. glabra** *Hiern* in F.T.A. 3 : 190 (1877) ; F.W.T.A., ed. 1, 2 : 93, partly (excl. syn. *R. membranacea* Hiern).
Pavetta brachycoryne K. Schum. (1899). Slender rambling shrub with papery outer bark sometimes
peeling when dry ; flowers white in compact terminal panicles.
S.Nig.: Calabar (Apr.) *Thomson* 107 ! Oban *Talbot* 1573 ! **[Br.]Cam.:** Barombi *Preuss* 37 ! Abonando
(Mar.) *Rudatis* 11 ! Buea, 2,500 ft. (Mar.) *Maitland* 556 ! Also in Cameroun. (See Appendix, p. 410.)
8. **R. decorticata** *Hiern* in F.T.A. 3 : 190 (1877). *R. landolphioides* Wernham (1917). A climbing shrub up to
30 ft. high, with the papery outer bark peeling off ; flowers in slender, lax, divaricately branched panicles
2–3 in. long.
S.Nig.: Oban *Talbot* 248 ! Degema *Talbot* 3829 ! **[Br.]Cam.:** Cam. Mt., 2,000–3,000 ft. *Mann* 2157 !
Bamenda (fr. Feb.) *Daramola* FHI 40499 ! Also in Cameroun and Gabon.

Imperfectly known species.

R. sp. A (?) straggling shrub densely pilose with brown or purplish hairs, very long subulate bracts and bracteoles,
and large entire stipules : further material required.
Lib.: Sinoe Basin *Whyte* ! Sanokwele (bud Sept.) *Baldwin* 9509 !

33. FERETIA Del.—F.T.A. 3 : 115 ; K. Schum. in E. & P. Pflanzenfam. 4, 4 : 79 (1891).

Leaves appearing with or usually after the flowers, ovate-elliptic to obovate, mucronate, 4–6 cm. long, 2–3 cm. broad, pubescent on the nerves beneath when young ; flowers clustered on the usually leafless shoots ; bud-scales (perulae) broadly ovate, dry ; pedicels up to 2·5 cm. long, with one or two bracteoles towards the apex ; calyx glabrous, lobes linear-lanceolate, 2·5 mm. long ; corolla widely expanded from a short cylindrical base, 1·5–2 cm. long ; anthers wholly exserted, nearly 1 cm. long ; fruit globose, stalked, about 8 mm. diam. ..   ..   *apodanthera*

FIG. 230.—FERETIA APODANTHERA *Del.* (RUBIACEAE).

A, flowers. B, pistil. C, fruiting shoot. D, fruit. E, stipules.

F. apodanthera *Del.* in Ann. Sci. Nat., sér. 2, 20 : 92, t. 1, 4 (1843) ; F.T.A. 3 : 115 ; Chev. Bot. 326. *F. ? canthioides* Hiern in F.T.A. 3 : 116 (1877) ; Chev. Bot. 326 ; F.W.T.A., ed. 1, 2 : 81 ; Aubrév. Fl. For. Soud.-Guin. 464, t. 101, 4–5. A bushy shrub, up to 15 ft. high, with sweetly scented flowers abundant on short lateral shoots usually appearing before the leaves ; calyx pink ; corolla-tube pink and white, lobes white within and half pink and half white outside ; in savanna.
Sen.: Dagana *Heudelot* 436 ! **Gam.:** *Hayes* 551 ! **Mali:** Dialacoro (Apr.) *Chev.* 705 ! Kouroukoto (Mar.) *Roberty* 17087 ! Famsala (Mar.) *Davey* 262 ! Bobo-Dioulasso (May) *Chev.* 917. **Iv.C.:** Ouangolo *Aubrév.* 1415. Kaya *Aubrév.* 2375. **Ghana:** Bawku (Apr.) *Vigne* FH 3749 ! Pan to Bujan (Apr.) *Kitson* 682 ! Samboro to Basisau (fl. & fr. May) *Kitson* 766 ! Kologa, Navrongo Dist. (Apr.) *Vigne* FH 4499 ! **Dah.:** Boukombé *Aubrév.* 86d. **N.Nig.:** Nupe *Barter* 1242 ! Sokoto *Ryan* 24 ! Zungeru (Apr.) *Dalz.* 91 ! *Keay* FHI 25720 ! Katagum *Dalz.* 164 ! Yola (May) *Dalz.* 38 ! Also in Ubangi-Shari, Sudan, Ethiopia, Kenya and Uganda. (See Appendix, p. 397.)

[Experimental field studies are needed to determine how far the variation in length of pedicel is significant taxonomically. Specimens with long pedicels have been named *F. canthioides* and are commoner in western than in eastern Africa, but there are many intermediates. It would appear that when the flowers first appear on leafless shoots the pedicels are very short, but as flowering continues in the rainy season they are then much longer].

34. CREMASPORA Benth.—F.T.A. 3 : 126 ; K. Schum. in E. & P. Pflanzenfam. 4, 4 : 87 (1891).

Underside of leaves pubescent with relatively short hairs more or less appressed and directed away from the base, scattered on the lamina as well as on the midrib and nerves ; branchlets pubescent ; leaves broadly or narrowly ovate, or elliptic rounded or obtuse at base, broadly acuminate, 3–10 cm. long, 1·8–5 cm. broad, with 4–6 main lateral nerves on each side of midrib ; flowers in dense axillary clusters ; calyx setose-pilose, teeth about 1 mm. long ; corolla-tube 4–6 mm. long, pubescent or glabrous outside, lobes 3–5 mm. long, pubescent outside ; fruits ellipsoid, ultimately glabrous, about 1 cm. long, tipped with the persistent calyx ..   ..   1. *triflora*
Underside of leaves pilose with 1–1·5 mm.-long spreading hairs mainly on midrib and nerves ; branchlets long-spreading pilose ; leaves oblong-ovate, rounded at base, abruptly acuminate, 2·5–11 cm. long, 1·3–5 cm. broad, with 3–6 main lateral nerves on each side of midrib: fruits spreading-pilose ; otherwise similar to above   2. *thomsoni*

1. **C. triflora** (*Thonn.*) *K. Schum.* in E. & P. Pflanzenfam. 4, 4 : 88, fig. 31, F–G (1891) ; Hutch. & Dalz. F.W.T.A., ed. 1, 2 : 85 (1931), excl. *Warnecke* 415. *Psychotria triflora* Thonn. (1827). *Coffea ? microcarpa* DC. (1830). *Cremaspora microcarpa* (DC.) Baill. (1879). *Coffea hirsuta* G. Don (1834). *Cremaspora africana* Benth. (1849)—F.T.A. 3 : 126 ; Chev. Bot. 327. *C. heterophylla* F. Didr. (1854). *C. thomsonii* of Chev. Bot. 327. A fair-sized shrub with rusty-pubescent, usually spreading and more or less straggling branches, fragrant white subsessile flowers ½ in. long and usually several together in the axils, and scarlet berries.
Sen.: Cape Rouge, Casamance *Leprieur* ; *Perrottet*. **Mali:** Kouroukoto (Mar.) *Roberty* 17084 ! **Port.G.:** Catió (fr. July) *Esp. Santo* 2103 ! Brene, Bissau (fr. Jan.) *Esp. Santo* 1719 ! **Guin.:** Rio Nunez *Heudelot* 654. Konkouré to Timbo (Mar.) *Chev.* 12505 ! **S.L.:** *Don* ! Hill Station (fl. & fr. May) *Deighton* 2981 ! Musaia (Mar.) *Sc. Elliot* 5120 ! Yonibana (fr. Nov.) *Thomas* 5079 ! Batkanu (fr. May) *Deighton* 3006 ! **Lib.:** Sanokwele (fr. Sept.) *Baldwin* 13264 ! Gletown, Tchien (fr. July) *Baldwin* 6742 ! **Iv.C.:** Nandala to Mankono (fr. June) *Chev.* 21866 ! Dixcove (fr. Mar.) *Morton* A465 ! Kwahu (Mar.) *Johnson* 699 ! Nabingo Waterside (May) *Kitson* 964 ! Asiama *Thonning.* **N.Nig.:** Yates 2 ! **S.Nig.:** Lagos *Foster* 74 ! *Millen* 56 ! Ijaiye (Apr.) *Barter* 3333 ! Onochie FHI 21954 ! Ibadan (fl. Apr., fr. Jan.) *Meikle* 962 ! 1431 ! Oban *Talbot* 1507 ! **[Br.]Cam.:** Su, Bamenda (Apr.) *Maitland* 1622 ! Bum, Bamenda (May) *Maitland* 1670 ! Gangumi, Gashaka Dist. (fr. Dec.) *Latilo & Daramola* FHI 28898. Widespread in tropical Africa. (See Appendix, p. 395.)

2. **C. thomsoni** *Hiern* in F.T.A. 3 : 126 (1877). A climbing shrub with brown-hirsute branchlets with white flowers ½ in. long, few, sessile in axillary clusters ; fruits deep red.
S.Nig.: Eket *Talbot* 3189 ! Oban *Talbot* 1049 ! Calabar (fl. & fr. Mar.) *Thomson* 94 ! Also in Cameroun.

35. POLYSPHAERIA Hook. f.—F.T.A. 3 : 127 ; K. Schum. in E. & P. Pflanzenfam. 4, 4 : 201 (1891) ; Brenan in Kew Bull. 4 : 854 (1949).

Axillary flower-clusters with peduncle up to 9 mm. long ; flowering branchlets subterete ; leaves lanceolate or narrowly ovate-lanceolate, shortly cuneate or obtuse at base, gradually tapered at apex, 7·5–19 cm. long, 1·6–6·2 cm. broad, with 8–10 looped lateral nerves on each side of midrib, glabrous, shining above when dry ; calyx obconic, undulately lobed, 1·5 mm. long, puberulous or nearly glabrous outside, corolla-tube 4 mm. long, densely villous inside, lobes 4 mm. long ; anthers included ; style long-exserted, densely pilose ; fruits ellipsoid-globose, about 1 cm. diam., capped by the cupular persistent calyx ..   ..   *1. arbuscula*
Axillary flower-clusters sessile ; flowering branchlets angled ; leaves oblong-elliptic, obtuse or acute at base, gradually acuminate, 13–25 cm. long, 5–8 cm. broad, with 8–12 looped lateral nerves on each side of midrib, glabrous, often drying rather dull and blackish ; calyx glabrous outside ; corolla-tube and lobes each about 3 mm. long ; fruits nearly 1·5 cm. long ; otherwise similar to above ..   ..   *2. macrophylla*

1. **P. arbuscula** *K. Schum.* in Engl. Bot. Jahrb. 33 : 349 (1903) ; F.W.T.A., ed. 1, 1 : 86, excl. syn. *P. pedunculata* K. Schum. ex De Wild. ; Aubrév. Fl. For. Soud.-Guin. 469, t. 103, 1–2. *P. lagosensis* A. Chev. Bot. 328 (1920), name only. A shrub or small tree, up to 12 ft. high ; flowers cream, sweet-scented ; ripe fruits red ; especially common in fringing forest in savanna regions.
Ghana: *Anderson* 44 ! Afram Plains (May) *Kitson* 1083 ! 1112 ! 1122 ! Ejura (June) *Vigne* FH 2058 ! Fuller Falls, Kintampo (fr. Nov.) *Andoh* FH 5416 ! **Togo Rep.:** Sokode-Basari (Aug.) *Kersting* 385. **Dah.:** Ouessé *Poisson.* Djougou *Chev.* 23917. Konkobiri, Atakora *Chev.* 24318. **N.Nig.:** Giwa Dist., Zaria Prov. (July) *Keay* FHI 25941 ! Anara F.R., Zaria Prov. (fr. Oct.) *Keay* FHI 21127 ! Zungeru (June) *Dalz.* 213 ! Abinsi (July) *Dalz.* 637 ! **S.Nig.:** Lagos *Foster* 85 ! Olokemeji (fl. & fr. Aug.) *Symington* FHI 5076 ! **[Br.]Cam.:** Kwagiri to Gangumi, Gashaka Dist. (fr. Dec.) *Latilo & Daramola* FHI 28782 ! foot of Vogel Peak, Adamawa (fr. Nov.) *Hepper* 1472 ! Also in Cameroun.

2. **P. macrophylla** *K. Schum.* l.c. (1903). A forest shrub or small tree ; flowers cream tinged pink, with red calyx ; fruits blue-black.
Ghana: Bobiri F.R. (fr. Nov.) *Andoh* FH 5841 ! Pra-Anum (fr. Oct.) *Amediwole* FH 5846 ! **S.Nig.:** Ibadan North F.R. (fr. Dec.) *Jones* FHI 7179 ! Eluju, Omo (formerly part of Shasha) F.R. (Apr.) *Ross* 226 ! Mamu River F.R., Onitsha Prov. (fr. Sept.) *Onochie* FHI 34068 ! **[Br.]Cam.:** Isongo to Bakingele, Victoria Div. (Mar.) *Preuss* 1382. Bulifambo, Buea (Mar.) *Maitland* 599 ! Ekumbe Ndene waterfalls, Meme R., Kumba Dist. (fr. Jan.) *Keay* FHI 37375 !

36. TRICALYSIA A. Rich. ex DC.—F.T.A. 3 : 117.

*Calyx not covering the corolla in bud and not splitting into 2 paleaceous pieces at anthesis ; corolla-tube glabrous, or nearly so outside (but densely appressed-pilose in No. 21) (to p. 150) :
Calyx consisting of a rim less than 0·25 mm. high and teeth up to 1 mm. long, glabrous ; ovules 1 on each placenta ; style glabrous ; bracteoles not cupular ; corolla-tube and lobes each 3–3·5 mm. long, glabrous outside, lobes acute in bud, reflexed at anthesis, throat very densely villous ; leaves oblong or narrowly elliptic, margins revolute at the cuneate or obtuse base, apex acuminate, 7·5–16 cm. long, 2–6·3 cm. broad, midrib and nerves (5–6 on each side) and reticulate venation prominent on both surfaces, glabrous ..   ..   *1. reflexa*
Calyx with a distinct campanulate tube or with well-developed lobes ; placentas attached to the upper part of the septum back to back, with 1–16 ovules immersed or partially immersed on the face :
†Style glabrous (to p. 150) :
‡Branchlets densely pubescent (to p. 149) :
Leaves cordate or auriculate at base, petiole very short ; calyx-tube about 4 mm. long, with short teeth ; flowers 8-merous, solitary ; corolla densely appressed-pilose outside, lobes subglabrous inside, tube 9 mm., lobes 8 mm. long ; leaves oblong-elliptic or oblong-obovate, sometimes subfalcate, (5–)12–19·5 cm. long, 2·8–4·8 cm. broad, long-acuminate, with conspicuous parallel venation ; fruits ellipsoid, about 3 cm. long and 2 cm. diam. ; ovules 4 on each placenta
                                                                                            21. auriculata

Leaves cuneate at base ; calyx-tube 1 mm. long or less ; corolla-tube glabrous or nearly so outside ; fruits up to 1·5 cm. long :
Flowers solitary, on peduncles 4–5 mm. long ; calyx-tube about 1 mm. long, with 4–5 filiform lobes about 2 mm. long ; calyx densely puberulous, closely subtended by the connate calyx-like bracteoles ; leaves oblong-lanceolate, cuneate at base, caudate-acuminate, 7–14 cm. long, 3–4·5 cm. broad, glabrous beneath, with about 5 main lateral nerves and very obscure venation ; fruits fusiform, about 15 mm. long and 6 mm. diam. ; ovules 1 on each placenta   .. 2. *wernhamiana*
Flowers several ; calyx not as above :
Calyx-tube about 1 mm. long, with short teeth ; calyx puberulous outside, closely subtended by the connate calyx-like bracteoles ; corolla-lobes densely pubescent outside ; inflorescences rather lax, with branches up to 10 mm. long ; leaves ovate- or obovate-elliptic or oblong-elliptic, cuneate at base, acuminate, 5–18 cm. long, 2·3–6·5 cm. broad, with 3–5 main lateral nerves on each side of midrib, nearly glabrous except for the midrib beneath ; ovules 2 on each placenta ..   .. 3. *oligoneura*
Calyx deeply divided into ovate or triangular lobes, densely pubescent outside ; bracteoles connate but scarcely calyx-like ; flowers in dense sessile clusters ; corolla-lobes glabrous or nearly so outside ; ovules 2(–3) on each placenta :
Leaves shortly appressed-pubescent mainly on midrib and nerves beneath ; corolla-tube about 4 mm. long ; leaves oblong-elliptic or slightly obovate-elliptic, cuneate or obtuse at base, shortly and obtusely acuminate, 2–9 cm. long, 1·2–4 cm. broad, with usually 4 main lateral nerves on each side of midrib   ..   .. 4. *deightonii*
Leaves spreading-pilose mainly on midrib and nerves beneath ; corolla-tube 3–3·5 mm. long ; leaves elliptic or obovate-elliptic, cuneate or obtuse at base, acutely acuminate, 4–12 cm. long, 2–4·7 cm. broad, with 4–8 main lateral nerves on each side of midrib, impressed above   ..   .. 5. *talbotii*
‡Branchlets glabrous, or with very minute hairs when young :
Calyx closely subtended by the connate calyx-like bracteoles ; calyx subtruncate, glabrous outside :
Flowers with short but distinct pedicels and peduncles ; bracteolar calyculus with two processes ; calyx up to 1 mm. long ; corolla-tube 2–4 mm., lobes about 4 mm. long ; leaves elongate-elliptic, ovate-elliptic or obovate-elliptic, cuneate at base, gradually long-acuminate, 8–18 cm. long, 3–7 cm. broad, glabrous, with 5–8 main lateral nerves on each side of midrib ; ovules (1–)2 on each placenta
6. *biafrana*
Flowers sessile in the leaf-axils ; bracteolar calyculus with undulate margin ; calyx 1–2 mm. long ; corolla-tube about 5 mm., lobes 6–7 mm. long ; leaves oblong or broadly elliptic, cuneate, obtuse or rounded at base, broadly acuminate, 13–25 cm. long, 4·5–11 cm. broad, glabrous, with 6–8 main lateral nerves on each side of midrib ; fruits globose, 7–8 mm. diam. ; ovules 3 on each placenta
7. *vignei*
Calyx separated from the connate bracts by pedicels which bear usually free, ovate or lanceolate bracteoles :
Corolla-tube 9–12 mm. long ; calyx-tube 3 mm. long, subtruncate, with short teeth, appressed-pubescent outside ; ovules 5–8 on each placenta ; leaves elongate-elliptic, cuneate (or obtuse or rounded) at base, gradually long-acuminate, 8–25 cm. long, 3–7 cm. broad, glabrous, except for tufts in the axils of the 5–7 pairs of main lateral nerves ..   .. 8. *elliotii*
Corolla-tube 2·5–7 mm. long ; calyx-tube distinctly and rather deeply lobed :
Leaves glabrous beneath ; inflorescence and outside of calyx pubescent :
Calyx-lobes rounded-oblong, overlapping at anthesis, about 1 mm. long ; corolla-tube and lobes each 3·5–4 mm. long ; ovules 12–16 on each placenta ; leaves broadly lanceolate, rather abruptly narrowed to the cuneate base, gradually acuminate, 11–22 cm. long, 4–8 cm. broad, with 6–10 main lateral nerves prominent on each side of midrib beneath, strongly reticulate 9. *macrophylla*
Calyx-lobes ovate, not overlapping at anthesis, about 1·5 mm. long ; corolla-tube 3·5–4 mm., lobes 4·5–5 mm. long ; ovules 2(–5) on each placenta ; leaves elliptic, oblong-elliptic, ovate-elliptic or obovate-elliptic, obtuse or cuneate at base, distinctly acuminate, 6·5–16 cm. long, 1·5–7 cm. broad, with 4–6 main lateral nerves on each side of midrib, venation reticulate ; fruits 5–7 mm. diam.   ..   .. 10. *bracteata*
Leaves minutely puberulous on midrib and nerves beneath ; inflorescence and outside of calyx glabrous or pubescent ; ovules 2 on each placenta :
Corolla-tube 5–7 mm. long, densely pilose in the throat, lobes 4·5 mm. long ; leaves oblong-lanceolate or oblong-elliptic, cuneate at base, acuminate, 13–26 cm. long, 4–9 cm. broad, with 8–11 main lateral nerves on each side of midrib ; seeds with ruminate endosperm   ..   ..   .. 11. *discolor*
Corolla-tube 3·5–4 mm. long, sparingly pubescent in the throat, lobes 3–4 mm.

long ; leaves narrowly elliptic or oblong-lanceolate, 12–20 cm. long, 4·5–
6 cm. broad, with about 7 main lateral nerves on each side of midrib
                                                                     12. *mildbraedii*
†Style pilose or pubescent ; calyx truncate or subtruncate or with short teeth, closely
subtended by the connate calyx-like bracteoles :
Flowers 11–14-merous ; corolla-tube about 10 mm., lobes about 6 mm. long ;
anthers densely pubescent, long-exserted ; calyx puberulous outside, 3–3·5 mm.
diam. ; leaves oblong or oblong-lanceolate, obtuse or rounded at base, broadly
acuminate, about 22 cm. long and 7·5 cm. diam., glabrous with 8–9 main lateral
nerves on each side of midrib ; ovules several on each placenta ..   13. *pleiomera*
Flowers 5–8-merous ; corolla-tube 2–5·5 mm. long ; anthers glabrous, or pubescent
mainly at the tips :
Calyx glabrous outside, 2–2·5 mm. long, margin undulate ; corolla-tube and lobes,
each 3–4 mm. long ; flower-cluster crowded and sessile ; leaves coriaceous,
broadly elliptic to oblong-obovate, cuneate or obtuse at base, broadly and shortly
acuminate, 9–16 cm. long, 3–9 cm. broad, glabrous, venation rather obscure between the 4–6 main lateral nerves ; ovules about 4 on each placenta ; fruits
about 10 mm. diam.   ..   ..   14. *coriacea*
Calyx pubescent outside :
Branchlets glabrous ; corolla-tube 4–5 mm. long ; calyx-tube 1·5–2·5 mm. long ;
leaves glabrous except often in nerve-axils :
Calyx with subulate teeth 1–1·5 mm. long ; corolla-lobes 3·5–4 mm. long ; midrib
of leaves only very slightly prominent beneath, venation and the 3–4 pairs of
main lateral nerves obscure ; leaves elliptic, cuneate at base, long-acuminate,
6·5–9 cm. long, 2·5–3·5 cm. broad ; ovules 2 on each placenta   15. *parva*
Calyx subtruncate ; corolla-lobes 5–7 mm. long ; midrib venation and the 4–6
pairs of main lateral nerves prominent beneath ; ovules 3–4 on each placenta :
Outside of calyx and inflorescence rather densely setose-pubescent ; outside of
corolla-lobes pubescent ; leaves more or less elliptic, cuneate at base, broadly
and slightly pointed at apex, 6–14·5 cm. long, 2–7 cm. broad
                                                       16. *okelensis* var. *okelensis*
Outside of calyx and inflorescence puberulous ; outside of corolla-lobes
appressed-puberulous or subglabrous ; leaves oblanceolate, cuneate at base,
obtuse at apex, 4·5–12 cm. long, 1·8–4·5(–5·5) cm. broad
                                                       16b. *okelensis* var. *oblanceolata*
Branchlets pubescent or puberulous, sometimes very minutely so :
Leaves densely and softly pubescent beneath ; branchlets densely and softly
puberulous ; otherwise similar to var. *okelensis* (see above)
                                                       16a. *okelensis* var. *pubescens*
Leaves glabrous or sparsely puberulous beneath :
Venation of leaves strongly reticulate ; corolla-tube 4–5·5 mm., lobes 5–6 mm.
long ; calyx-tube 2 mm. long ; leaves oblong-elliptic, lanceolate-elliptic or
obovate-elliptic, cuneate at base, rather abruptly acuminate, 5·5–17 cm. long,
1·8–6·2 cm. broad, with 4–6 main lateral nerves on each side of midrib :
Young branchlets very minutely puberulous ; pedicels very short ; ovules
3–6 on each placenta ..   ..   17. *reticulata*
Young branchlets appressed-puberulous or shortly pubescent ; pedicels 2–4 mm.
long ; ovules 2 on each placenta   ..   ..   ..   18. *faranahensis*
Venation of leaves obscure ; corolla-tube 2–3 mm. long ; branchlets puberulous :
Lateral nerves very obscure, 5–6 on each side of midrib ; leaves papery,
oblanceolate to elliptic-lanceolate or obovate-elliptic, cuneate or obtuse at
base, distinctly acuminate, 5·5–16 cm. long, 2–5·5 cm. broad ; fruits about
5 mm. diam. ; ovules 2 on each placenta ..   ..   ..   19. *pallens*
Lateral nerves distinct, 3–4(–5) on each side of midrib ; leaves subcoriaceous,
elliptic or obovate-elliptic, cuneate at base, acuminate, 4–12·5 cm. long,
2–4·5 cm. broad ; ovules 3–4 on each placenta   ..   ..   20. *obanensis*

*Calyx covering the corolla in bud and splitting at anthesis into usually 2 paleaceous
pieces 5–7 mm. long, densely pilose outside ; corolla-tube densely appressed-pilose
outside ; flowers 6–8-merous ; ovules 1–3 on each placenta ; fruits about 1–2 cm.
diam. :
Leaves glabrous or sparingly pubescent on midrib and nerves beneath, but sometimes
with tufts of hair in the nerve-axils, oblanceolate, cuneate at base, often undulate
towards the apex :
Leaves with 7–8 main lateral nerves spreading on each side of midrib, lamina abruptly
acuminate, 12–17 cm. long, 4·5–8 cm. broad ; ovules solitary on each placenta ;
fruits about 2 cm. diam. ; trees, up to 20 m. high ..   ..   22. *toupetou*
Leaves with 4–5 main lateral nerves ascending on each side of midrib, lamina acute
or obtuse at apex, 4–12 cm. long, 1·2–4·5 cm. broad ; ovules 2–3 on each placenta ;
fruits about 1 cm. diam. ; shrubs, up to 4 m. high :

Branchlets densely pubescent ; leaves sparingly pubescent on midrib and nerves
beneath .. .. 23. *trilocularis*
Branchlets glabrous ; leaves glabrous on midrib and nerves beneath 24. *paroissei*
Leaves densely pilose beneath, obovate-oblong, rounded or subcordate at base,
acuminate, 6–17 cm. long, 2·5–6·5 cm. broad, with 7–10 main lateral nerves on each
side of midrib .. .. 25. *chevalieri*

1. **T. reflexa** *Hutch.* in Kew Bull. 1915 : 44. *Coffea arabica* var. *leucocarpa* Hiern (1876)—F.T.A. 3 : 181.
T. reticulata of F.W.T.A., ed. 1, 2 : 83, partly (*Chipp* 368). A forest shrub 4–6 ft. high ; flowers white.
S.L.: *T. Vogel* 174 ! Kessewe (Apr.) *Lane-Poole* 131 ! Gola Forest (Jan.) *King* 1 ! Waima, Kori (young
fr. June) *Deighton* 5333 ! Giewahun (young fr. Apr.) *Deighton* 3661 ! **Lib.:** Dukwia R. (July) *Linder*
113 ! Nikabuzu to Zigida (Mar.) *Bequaert* 121 ! **Iv.C.:** Mt. Tonkoui *Aubrév.* 994 ! Bingerville (cult.,
fl. Mar.) *Bodard* 1360 ! **Ghana:** Nkwanta, W. Prov. (fr. Sept.) *Chipp* 368 ! Dompim, Tarkwa Dist.
(Aug.) *Darko* 978 ! Benso, Tarkwa Dist. (Sept.) *Andoh* FH 5774 ! **S.Nig.:** Lagos (young fr. Apr.) *Dalz.*
1379 ! Ilaro F.R., Abeokuta Prov. *Onochie* FHI 8198 ! 13542 ! [Br.]**Cam.:** Bova, Kumba Div. (young
fr. Sept.) *Olorunfemi* FHI 30778 ! (See Appendix, p. 413.)
2. **T. wernhamiana** (*Hutch. & Dalz.*) *Keay* in Bull. Jard. Bot. Brux. 28 : 291 (1958). *Cremaspora wernhamiana* Hutch. & Dalz. F.W.T.A., ed. 1, 2 : 85 (1931). *Coffea talbotii* Wernham in Cat. Talb. 52 (1913).
A forest shrub ; fruits red.
S.Nig.: Oban *Talbot* 1620 !
3. **T. oligoneura** *K. Schum.* in Engl. Bot. Jahrb. 23 : 448 (1896). A shrub or small tree, 5–20 ft. high, with
smooth grey bark ; leaves papery, deep green and glossy above, pale beneath ; flowers white, corolla-
lobes reflexed ; fruits orange when ripe.
S.Nig.: Abeku, Omo F.R. (Mar.) *Jones & Onochie* FHI 17163 ! Oshun F.R., Ijebu-Ode (young fr. Apr.)
Chizea FHI 23954 ! Omo (formerly part of Shasha) F.R. (Feb.) *Richards* 3073 ! Idanre Hills (fl. buds
Jan.) *Brenan* 8673 ! Okomu F.R., Benin (fl. Jan., fr. Feb.) *Brenan* 8877 ! 9024 ! Also in Cameroun.
4. **T. deightonii** *Brenan* in Kew Bull. 8 : 112 (1953). A forest shrub, 4–7 ft. high ; flowers white.
S.L.: *Thomas* 7964 ! Jama, Moyamba Dist. (Mar., Dec.) *Deighton* 4692 ! 4723 ! Faiama (Jan.) *Deighton*
3872 ! 3873 ! Nyandehun, Koya (Dec.) *Deighton* 3659 ! 3804 ! Koribandu, Jaiama (Feb.) *Deighton*
5907 ! Gola Forest (Dec.) *King* 80b !
5. **T. talbotii** (*Wernham*) *Keay* l.c. 291 (1958). *Cremaspora talbotii* Wernham in Cat. Talb. 49 (1913) ; F.W.T.A.,
ed. 1, 2 : 85. A forest shrub, erect or half-straggling, to 6 ft. high ; flowers white ; fruits " white and
purple ".
S.Nig.: Oban *Talbot* 287 ! Boli to Bateriko, Obudu Div. (fr. June) *Jones & Onochie* FHI 17350 !
6. **T. biafrana** *Hiern* in F.T.A. 3 : 122 (1877). *T. hookeri* Hutch. & Dalz. F.W.T.A., ed. 1, 2 : 83 (1931). A
shrub or small tree, 5–30 ft. high ; flowers white, fragrant ; in forest.
Lib.: Kakatown *Whyte* ! **Iv.C.:** Davo R., Béyo (Jan.) *Leeuwenberg* 2598 ! **Ghana:** 7 miles N. of Abofuo
(Feb.) *Vigne* FH 1805 ! **S.Nig.:** Ibadan (Feb., Mar.) *Meikle* 1132 ! 1240 ! Onda, Omo F.R. (Feb.)
Jones & Onochie FHI 16951 ! Okomu F.R., Benin (Feb.) *Brenan* 9070 ! Calabar (Feb.) *Mann* 2305 !
[Br.]**Cam.:** *Preuss* 1310 ! 1311 ! Ambas Bay (Feb.) *Mann* 7 ! Victoria (Feb.) *Maitland* 981 ! Cam. Mt.,
500 ft. (Jan.) *Mann* 1212 ! Mai Idoanu, Gashaka Dist. (Feb.) *Latilo & Daramola* FHI 34471 !
7. **T. vignei** *Aubrév. & Pellegr.* in Bull. Soc. Bot. Fr. 83 : 41, fig. 2A (1936) ; Aubrév. Fl. For. C. Iv., ed. 2,
3 : 288, t. 358, A. A shrub or small tree, with large coriaceous leaves ; flowers white ; fruits red ; in
forest.
Iv.C.: Danané *Aubrév.* 1103 ! **Ghana:** Beyin, W. Prov. (Nov.) *Vigne* FH 1469 !
8. **T. elliotii** (*K. Schum.*) *Hutch. & Dalz.* F.W.T.A., ed. 1, 2 : 83 (1931). *Probletostemon elliotii* K. Schum. in
Engl. Bot. Jahrb. 23 : 450 (1897). *Tricalysia batesii* A. Chev. Caf. du Globe 3 : 243 (1947) ; op. cit.
2 : t. 146, name and plate only (1942). A tree, to 20 ft. high ; flowers white, fragrant ; in forest.
S.L.: Lumbaraya (Feb.) *Sc. Elliot* 4937 ! **Ghana:** Akwapim Hills (Nov.) *Johnson* 775 ! Aburi Hills
(Feb.) *Johnson* 941 ! Kumasi (Jan.) *Vigne* FH 1000 ! 1576 ! **S.Nig.:** Ugo, Benin (Jan.) *Latilo & Olorunfemi* FHI 33692 ! Also in Cameroun.
9. **T. macrophylla** *K. Schum.* in Engl. Bot. Jahrb. 28 : 66 (1899) ; Brenan in Kew Bull. 2: 73 (1947). *T. pluri-
ovulata* K. Schum. ex Hoyle in Gold Coast Check-list 119 (1937), English descr. only. A shrub or tree,
to 50 ft. high, with long drooping branches ; fruits black, red and white.
Iv.C.: Abengourou (young fr. Mar.) *Renaud* in Hb. *Bodard* 1380 ! **Togo:** Asuokoko R. (fr. Mar.)
Beveridge FH 2943 ! **S.Nig.:** Jamieson R., Sapoba (young fr. Nov.) *Kennedy* 1822 ! 1854 ! 1922 !
Meikle 549 ! Also in Cameroun.
10. **T. bracteata** *Hiern* in F.T.A. 3 : 120 (1877) ; Chev. Bot. 327 ; Aubrév. Fl. For. Soud.-Guin. 466–469,
t. 102, 4–7. *T. syrmanthera* Hiern l.c. (1877). A shrub or tree, 5–20 ft. high ; flowers white ; fruits white
at first, turning red and finally brownish-purple ; in forest.
Port.G.: Balanasinho, Fulacunda (young fr. June) *Esp. Santo* 2083 ! **Guin.:** Labé (Mar.) *Chev.* 12414 !
Between R. Konkouré and Timbo (Mar.) *Chev.* 12482 ! Karkandy *Heudelot* 855 ! Bafing (fr. Apr.) *Adam*
11863 ! **S.L.:** *Afzelius* ! Talla Hills (Feb.) *Sc. Elliot* 4908 ! 5054 ! Falaba (young fr. Mar.) *Sc. Elliot*
5425 ! 5428 ! Bintumane (Jan.) *Glanville* 458 ! Kambui F.R., Kenema (young fr. Mar.) *Jordan* 2006 !
Gola Forest (fr. Apr.) *Small* 589 ! **Ghana:** Suku Suku *Chipp* !
11. **T. discolor** *Brenan* in Kew Bull. 2 : 72 (1947). A tree, 20–30 ft. high ; flowers white ; in forest.
Lib.: Beiden, Boporo Dist. (fr. Nov.) *Baldwin* 10265 ! **Iv.C.:** Kodopleu *Bodard* 1247 ! **Ghana:** Krokosu
Hills F.R., Sefwi Wiawso Dist. (fr. Apr.) *Enti* FH 6701 ! Mampong Scarp (Feb.) *Vigne* FH 2748 ! Bobiri
F.R., Juaso (young fr. July) *Andoh* FH 5538 ! Kwajo Nkwanta to Sikamang (young fr. June) *Kitson*
1248 !
12. **T. mildbraedii** *Keay* in Bull. Jard. Bot. Brux. 28 : 293 (1958). A tree to 25 ft. high.
[Br.]**Cam.:** Dikome, Mamfe Rd., Kumba Div. (young fr. July) *Akuo* FHI 15158 ! Also in Cameroun.
13. **T. pleiomera** *Hutch.* in Cat. Talb. 48 (1913). A shrub or small tree ; in forest.
S.Nig.: Oban *Talbot* 1277 !
14. **T. coriacea** (*Benth.*) *Hiern* in F.T.A. 3 : 120 (1877) ; Chev. Bot. 327 ; Aubrév. Fl. For. Soud.-Guin. 468–
469. *Randia coriacea* Benth. in Fl. Nigrit. 387 (1849). A shrub, 6–12 ft. high ; flowers white, fragrant
in undergrowth of forest in swampy places, including brackish areas.
Guin.: Télimélé *Pobéguin* 2190 ! **S.L.:** Mofari, Scarcies R. (Jan.) *Sc. Elliot* 4441 (Kew spec.) ! Njala
(Jan.) *Deighton* 1519 ! 2440 ! Gbanbama *Thomas* 8973 ! Ronietta (fl. bud Nov.) *Thomas* 5419 ! Moselelo
to Sembehun (Nov.) *Deighton* 2389 ! **Lib.:** Grand Bassa (July, Aug.) *T. Vogel* 172 ! *Dinklage* 1696 !
Duport (Nov.) *Linder* 1501 ! 1504 ! **Iv.C.:** Aboisso *Chev.* 17862. Kéeta, Mid. Cavally (July) *Chev.* 19309.
Ghana: Beyin, W. Prov. (Nov.) *Vigne* FH 1454 ! **S.Nig.:** Okomu F.R., Benin (fr. Feb.) *Brenan* 9109 !
Jamieson R., Sapoba *Kennedy* 2800 ! 2818 ! Also in Cameroun, Gabon and Congo.
15. **T. parva** *Keay* l.c. 292 (1958). Small shrub ; ripe fruit red, with persistent waxy red calyx ; in forest.
S.L.: S. Prov. (Jan.) *Smythe* 239 ! Gola Forest (fr. May) *Small* 675 ! 691 !
16. **T. okelensis** *Hiern* var. **okelensis**—F.T.A. 3 : 122 (1877) ; Aubrév. Fl. For. Soud.-Guin. 466–468, t. 102,
1–3 ; Fl. For. C. Iv., ed. 2, 3 : 288, t. 358B ; Berhaut Fl. Sén. 103. *T. pobeguinii* Hutch. & Dalz.
F.W.T.A., ed. 1, 2 : 83 (1931). A shrub or small tree, to 15 ft. high ; flowers white ; in savannah wood-
land and fringing forest.
Sen.: Niokolo-Koba *Berhaut* 1458. **Mali:** Bobo Dioulasso, Upper Volta *Aubrév.* 1820. **Guin.:** Kouroussa
(Jan.) *Pobéguin* 638 ! 650 ! 822 ! Kissidougou (Jan.) *Martine* 197 ! Timbo *Pobéguin* 112. Dinguiraye
(Dec.) *Jac.-Fél.* 1457 ! **S.L.:** Musaia (fl. Jan.-Feb., fr. Apr.) *Deighton* 4158 ! 4180 ! 5418 ! *Miszewski*

27! **Iv.C.**: Ferkéssédougou *Aubrév.* 1470 ; 2309! Bondoukou *Aubrév.* 762! **Ghana**: Grube (Jan.) *Vigne* FH 1580! Ejura (Nov.) *Vigne* FH 3463! Bosomoa F.R. (Jan.) *Lyon* FH 2740! Kpandu *Robertson* 52! Ho (Feb.) *Howes* 1105! Also in Ubangi-Shari and Sudan.

16a. **T. okelensis** var. **pubescens** *Aubrév.* & *Pellegr. ex Keay* l.c. 292 (1958) ; Aubrév. Fl. For. Soud.-Guin. 466, 468 (1950), French descr. only.
  Guin.: Farana *Chev.* 20454. **Iv.C.**: Oumé (young fr. Apr.) *Aubrév.* 1261! Tafiré *Aubrév.* 1582! Ferkéssédougou *Aubrév.* 1629!

16b. **T. okelensis** var. **oblanceolata** (*Hutch.* & *Dalz.*) *Keay* l.c. 292 (1958). *T. oblanceolata* Hutch. & Dalz. F.W.T.A., ed. 1, 2 : 83 (1931). A shrub or small tree to 30 ft. high ; flowers white ; fruits red ; in fringing forest in the savanna regions.
  N.Nig.: 12 miles N. of Zaria (young fr. Feb.) *Thornewill* 37! Anara F.R., Zaria Prov. (fl. Nov., fr. May) *Keay* FHI 21685! 25784! 28123! Naraguta (Nov.) *Lely* 725! Jos Reservoir Catchment Area (Nov.) *Keay* FHI 21435! [**Br.**]**Cam.**: Bambui, Bamenda (young fr. Apr.) *Maitland* 1664! Bayango, Bamenda (young fr. Apr.) *Ujor* FHI 30043! Vogel Peak, Adamawa (Nov., Dec.) *Hepper* 1464! 1543!

17. **T. reticulata** (*Benth.*) *Hiern* in F.T.A. 3 : 121 (1877) ; Chev. Bot. 327 ; Aubrév. Fl. For. Soud.-Guin. 466. *Randia reticulata* Benth. in Fl. Nigrit. 386 (1849). *Tricalysia pseudoreticulata* Aubrév. & Pellegr. in Bull. Soc. Bot. Fr. 100 : 25 (1953) ; Aubrév. Fl. For. Soud.-Guin. 466. A shrub 3–7 ft. high ; flowers white ; in forest, especially near water.
  Port.G.: Bubaque (May) *Esp. Santo* 2016! Formosa (Apr.) *Esp. Santo* 1973! Balanasinho, Fulacunda (young fr. July) *Esp. Santo* 2106! **Guin.**: Karkandy *Heudelot* 884! Bouma, Fouta Djalon (Mar.) *Pobéguin* 2283! Between R. Konkouré and Timbo (Mar.) *Chev.* 12484! Boké *Chillou*! **S.L.**: *Afzelius*! *Don*! Freetown and environs (fl. Dec.-Mar., fr. June) *Dalz.* 963! *Deighton* 468! 1052! 4748! Bagroo R. (Apr.) *Mann* 887! Ninia, Talla Hills (Feb.) *Sc. Elliot* 4810! Mange, Bure Makonte (Mar.) *Adames* 223! Gola Forest (May) *King* 127b! **Lib.**: Peahtah, Salala Dist. (Dec.) *Baldwin* 10571! **Iv.C.**: Mt. Tonkoui (buds Nov.) *de Wilde* 916! (See Appendix, p. 413.)
  [In the first edition this species was recorded from Ghana, but the specimen (*Chipp* 368) is *T. reflexa* Hutch., q.v.]

18. **T. faranahensis** *Aubrév.* & *Pellegr.* in Bull. Soc. Bot. Fr. 100 : 25 (1953). *T. chevalieri* Aubrév. Fl. For. Soud.-Guin. 468 (1950), French descr. only ; not of K. Krause (1909). A shrub.
  Guin.: Timbikounda to Farakoro, Farana (Jan.) *Chev.* 20632! **Iv.C.**: Abengourou (fr. Feb., Apr.) *Bodard* 1399! 1592!

19. **T. pallens** *Hiern* in F.T.A. 3 : 121 (1877). A shrub or small tree, 7–30 ft. high, with profuse white flowers ; fruits red ; in forest.
  Iv.C.: Akabossué to Ebrinahoué, Mid. Comoé (Dec.) *Chev.* 22607! Abengourou (young fr. Mar.) *Renaud* in *Hb. Bodard* 1391! Dieouzon to Sebazon (fl. buds Jan.) *Bodard* 1244! **Ghana**: L. Bosumtwi (Dec.) *Vigne* FH 1497! Kumasi *Cummins* 136! Aburi Hills (young fr. Feb.) *Johnson* 942! Asenano F.R., Akota (Nov.) *Darko* 1059! Nkwatia to Mpraeso, Kwahu Plateau (fr. May) *Kitson* 1183! Kabo River F.R., Okrabi, Hohoe Dist. (Nov.) *St. C. Thompson* FH 3695! **S.Nig.**: Okomu F.R., Benin (Dec.) *Brenan* 8577! Sapoba, Benin (Mar.) *Ejiofor* FHI 24615! *Keay* FHI 25571! *Kennedy* 1864! 1902! Calabar (fr. Apr.) *Holland* 103! [**Br.**]**Cam.**: Buea (young fr. Mar.) *Maitland* 492! Nyanga camp, 4,500 ft., Cam. Mt. (fl. buds Dec.) *Maitland* 1206! **F.Po**: *Mann*! (See Appendix, p. 413.)

20. **T. obanensis** *Keay* in Bull. Jard. Bot. Brux. 28 : 294 (1958).
  S.Nig.: Oban *Talbot* 1324! s.n.!

21. **T. auriculata** *Keay* l.c. 295 (1958).
  S.Nig.: near waterside, between Budeng and Ewen Road, Calabar Prov. (fl. & fr. Nov.) *Rosevear* C33! C34! Also in Cameroun.

22. **T. toupetou** *Aubrév.* & *Pellegr.* in Bull. Soc. Bot. Fr. 100 : 26 (1953) ; Aubrév. Fl. For. Soud.-Guin. 466, 468 (1950), French descr. only. Tree to about 65 ft. high and 4 ft. girth ; in forest.
  Iv.C.: Tabou *Serv. For.* 2797! 2800!

23. **T. trilocularis** (*Sc. Elliot*) *Hutch.* & *Dalz.* F.W.T.A., ed. 1, 2 : 83 (1931). *Aulacocalyx triloculare* Sc. Elliot in J. Linn. Soc. 30 : 81 (1894). Shrub to 12 ft. high, with fragrant white flowers ; leaves usually with bacterial nodules ; on river banks.
  S.L.: Moria (fl. buds & fr. Jan.-Feb.) *Sc. Elliot* 4435! 4563! Mange (Apr.) *Hepper* 2609! Njala (Feb.) *Deighton* 1815! 4154! 4158! Rowalla (young fr. July) *Thomas* 1063! Laminaiya (Apr.) *Thomas* 138!

24. **T. paroissei** *Aubrév.* & *Pellegr.* in Bull. Soc. Bot. Fr. 100 : 26 (1953) ; Aubrév. Fl. For. Soud.-Guin. 466, 468 (1950), French descr. only. A shrub, about 3 ft. high.
  Guin.: Konkouré Falls, Lahaya *Paroisse* 112! R. Badi, Toudou (Apr.) *Roberty* 17482!

25. **T. chevalieri** *K. Krause* in Engl. Bot. Jahrb. 43 : 141 (1909) ; not of Aubrév. (1950). *Feretia coffeoides* A. Chev. in Mém. Soc. Bot. Fr. 8 : 179 (1912) ; Chev. Bot. 326. *Tricalysia coffeoides* (A. Chev.) Hutch. & Dalz. F.W.T.A., ed. 1, 2 : 83 (1931) ; Aubrév. Fl. For. Soud.-Guin. 466, 469, t. 102, 8–10 ; Berhaut Fl. Sén. 102, 105 ; not of Good (1926). *T. sudanica* A. Chev. Caf. du Globe 3 : 238 (1947) ; Brenan in Kew Bull. 5 : 221 (1950). *T. jasminiflora* of Chev. Bot. 327. A shrub, erect or scandent, 3–12 ft. high, flowers white, appearing when the reddish twigs are almost leafless ; ripe fruit red and glossy.
  Sen.: Tambacounda *Berhaut* 1244 ; 4068. **Mali**: Birgo *Dubois* 196. Chute de la Conisé *Adam* 15282! **Port.G.**: Madina, Boé (fl. buds Jan.) *Esp. Santo* 2357! **Guin.**: Dalaba (fl. & fr. Dec.-Jan.) *Chev.* 20285! *Roberty* 16482! 16842! Kouroussa (Feb.) *Pobéguin* 641! **S.L.**: Loma Mts. (fr. Aug.) *Jaeger* 1281! **Iv.C.**: Pâ *Aubrév.* 2216. Boromo *Aubrév.* 2215. Mt. Dourou, Koualé *Chev.* 21735! **Ghana**: Shiare, Buem-Krachi Dist. *Hall* 1411! Chai River F.R., Kete-Krachi Dist. (fr. Nov.) *Enti* FH 7061! Nalerugu, Gambaga (young fr. June) *Akpabla* 686! **Dah.**: Atacora Mts. *Chev.* 24062. **N.Nig.**: Wimbush FHI 42216! **S.Nig.**: *Foster* 165! Olokemeji (fr. Aug.) *Symington* FHI 5079! R. Ogun, Lanlate F.R. (Mar.) *Keay* FHI 21169! Itesi, Oyo Prov. (Mar.) *Onochie* FHI 20695! Idanre Hills (fl. bud & fr. Jan.) *Brenan* 8729! Asata, Enugu (Feb.) *Jones* FHI 709! Also in Ubangi-Shari. (See Appendix, p. 413.)

Imperfectly known species.

1. **T.** sp. Small tree or shrub ; fruits bright red ; in forest. Probably near *T. biafrana*, but with narrowly elliptic-lanceolate leaves, 6–9 cm. long, 1–2 cm. broad. More material required.
  [**Br.**]**Cam.**: banks of Mainya R., Mundani, Mamfe Div. (fl. & fr. Sept.) *Eyeku* FHI 22306!
2. **T.** sp. Shrub or small tree, 5–15 ft. high ; flowers white fragrant. More material required.
  Ghana: Tumfa Hills, 1,000 ft., Akim (Dec.) *Johnson* 274! Opon-Mansi F.R., Atobiasi (fr. Mar.) *Foggie* 189!

37. COFFEA Linn.—F.T.A. 3 : 179 ; A. Chev. Les Caféiers du Globe 1 (1929), 2 (1942), 3 (1947) ; Lebrun[1] Inst. Roy. Colon. Belge, sect. Sci. Nat. et Méd., Mém. 11 (1941).

*Anthers wholly exserted on distinct filaments ; leaves evergreen ; corolla-tube less than twice as long as the lobes ; style exserted (to p. 154) :

[1] Lebrun has separated several species of *Coffea* into new genera. For the purposes of this Flora, however, I have felt it wise to retain the traditional broad view of the genus *Coffea*, thus including several anomalous

Calyx consisting of a mere rim or very small teeth, shorter than the disk ; seeds with a longitudinal groove on the flat inner face ; flowers solitary to numerous in each axil :

Leaves usually cuneate at base or sometimes obtuse, rarely rounded, never long-attenuate to a subcordate or rounded base ; petioles 3 mm. or more long ; trees or shrubs, often exceeding 2 m. high :

Stipules obtuse or subacute at apex ; leaves rounded, obtuse, or shortly and broadly acuminate at apex, obovate, oblanceolate or broadly elliptic, (9–)14–37(–47) cm. long, (4–)6–13·5(–20) cm. broad, usually coriaceous, with 6–16 main lateral nerves on each side of midrib, domatia (pits) usually on the base of the nerves ; flowers 4–50 in each axil, 5–6(–9)-merous, exserted from the connate bracteoles which lack foliar appendages ; corolla-tube (6–)10–15(–18) mm., lobes (6–)10–21 (–25) mm. long ; disk distinctly lobed at apex ; fruits 12–25 mm. long, 9–21 mm. broad, 8–16 mm. thick, disk generally prominent at the apex .. 1. *liberica*

Stipules cuspidate or long-acuminate at apex ; leaves distinctly and often long-acuminate at apex ; connate bracteoles often with foliar appendages, disk not lobed at apex :

Flowers (4–)10–24(–48) in each axil, 5–6(–7)-merous ; calyx subtruncate ; main lateral nerves 9–14 on each side of the midrib ; fruits red, orange, or yellow at maturity :

Foliar appendages of the bracteoles conspicuous, 2–22 mm. long, 1–9 mm. broad ; calyces more or less hidden by the bracteoles at anthesis ; leaves broadly elliptic or oblong or obovate, 12–35(–40) cm. long, 5–12 cm. broad ; puberulous domatia (pits) usually present in the axils of the main lateral nerves ; corolla-tube (5–)9–14(–16) mm., lobes (8–)9–14(–21) mm. long ; fruits 9–17 mm. long, 7–13 mm. broad, 6–12 mm. thick .. .. 2. *canephora*

Foliar appendages of the bracteoles small, 0·5–5 mm. long, 0·3–1·5 mm. broad ; calyces well exserted from the bracteoles at anthesis ; leaves elliptic or oblong or ovate, (6–)8–16(–18) cm. long, 3–7(–8·5) cm. broad ; domatia (pits) in the axils of the main lateral nerves usually glabrous ; corolla-tube 5–11 mm., lobes 8–13 mm. long ; fruits 13–17 mm. long, 9–13 mm. broad, 7–10 mm. thick
3. *arabica*

Flowers 1–2(–4) in each axil, (4–)5–8(–9)-merous ; calyx with more or less distinct teeth ; main lateral nerves 6–10 on each side of midrib :

Calyx well exserted from the bracteoles at anthesis ; corolla-tube 6–8 mm., lobes 10–15 mm. long ; leaves with glabrous domatia (pits) usually present in the axils of the 6–10 main lateral nerves on each side of midrib :

Branchlets minutely puberulous ; leaves oblong-elliptic, 6–11 cm. long, 2–4 cm. broad ; fruits 10–14 mm. long, 7–10 mm. broad, colour not recorded
4. *togoensis*

Branchlets glabrous ; leaves narrowly oblong-oblanceolate, 5–14 cm. long, 0·7–4·5 cm. broad ; ripe fruits violet-black, 12–13 mm. long, 8–10 mm. broad and 6–8 mm. thick .. .. 5. *stenophylla*

Calyx hidden by the bracteoles at anthesis :

Stipules with the point not exceeding 1 mm. long ; leaves obovate, obovate-elliptic or oblanceolate, 7–14 cm. long, 2·8–6·5 cm. broad, with glabrous domatia (pits) usually present in the axils of the 6–8 main lateral nerves on each side of midrib ; corolla-tube 3·5–4(–6) mm., lobes 10(–19) mm. long
6. sp. nr. *carrissoi*

Stipules with points about 2 mm. long ; leaves oblanceolate or obovate-elliptic, 10–21 cm. long, 3·8–8 cm. broad, with 7–8 main lateral nerves on each side of midrib, domatia absent ; corolla-tube about 7 mm., lobes about 10 mm. long ; ripe fruits red .. .. 7. *brevipes*

Leaves oblanceolate, long-attenuate to a rounded or subcordate base, acuminate, 15–37 cm. long, 5–11 cm. broad, with 10–15 main lateral nerves on each side of midrib, domatia (pits) present or absent ; petiole 1–3 mm. long ; stipules with subulate points ; whole plant seldom exceeding 2 m. high ; flowers 1–3 in each leaf axil ; foliar appendages of bracteoles up to 15 mm. long ; fruits 15–25 mm. long, 8–12 mm. broad, red when ripe .. .. 8. *humilis*

Calyx well-developed ; seeds (where known) without a longitudinal groove ; flowers 1–few in each axil :

Flowers shortly pedicellate ; calyx-tube 2–7 mm. long ; erect shrubs or trees, to 8 m. high ; leaves 7–24 cm. long, 2–9 cm. broad :

Calyx-tube about 2 mm. long, not split at anthesis ; corolla-tube 7–10 mm. long, lobes 10–12 mm. long, 5–8 mm. broad ; bracteoles with foliar appendages up to 10 mm. long ; leaves elliptic, obtuse at base, long-acuminate, 17–24 cm. long, 6·5–9 cm. broad with 9–11 main lateral nerves on each side of midrib ; stipules

groups. A thorough revision of *Coffea*, *Tricalysia* and related genera is much needed, but this would be a major undertaking involving many Asiatic as well as African genera and species. From what I have seen of these genera in preparing this account for the F.W.T.A., I would suggest that detailed studies in the placentation of the ovules and the structure of the seeds would be essential.—R.W.J.K.

with a point about 1 cm. long ; each ovary cell with 2 collateral ovules completely immersed in the placenta   ..   ..   9. *macrochlamys*

Calyx-tube 3–7 mm. long, split down one side at anthesis ; corolla-tube 9–14 mm. long, lobes 10–12 mm. long, 3–5·5 mm. broad ; bracteoles with small narrow foliar appendages up to 2·5 mm. long ; leaves elliptic narrowly oblong-elliptic or lanceolate, obtuse or cuneate at base, abruptly long-acuminate, 7–19 cm. long, 2–6·5 cm. broad, with 4–9 main lateral nerves on each side of midrib ; stipules with a point 2·5–4 mm. long ; each ovary cell with 1 ovule completely immersed in the placenta
                                                                              10. *spathicalyx*

Flowers sessile ; calyx deeply divided into distinct lobes ; lianes or low shrubs ; leaves 3–10 cm. long, 1·5–5·5 cm. broad, oblong-elliptic or elliptic or more or less obovate or ovate, usually rather abruptly long-acuminate :

Branchlets and leaves glabrous ; leaves cuneate, obtuse or sometimes rounded at base, main lateral nerves 3–5 on each side of midrib with glabrous domatia (pits) at their bases or in their axils :

Corolla-tube 8–16 mm., lobes 10–15 mm. long ; foliar appendages of bracteoles more or less ovate, up to 10 mm. long and 6 mm. broad   ..   ..   11. *afzelii*

Corolla-tube 2–3 mm., lobes 4–5 mm. long ; foliar appendages of bracteoles ovate-orbicular, about 2 mm. long and broad   ..   ..   12. *pulchella*

Branchlets puberulous or densely hispid ; leaves puberulous or pubescent mainly on midrib and nerves beneath ; main lateral nerves 4–6 on each side of midrib with pubescent domatia (pits) :

Branchlets closely puberulous ; leaves obtuse or rounded at base ; calyx-lobes rounded ; corolla-tube 5–7 mm., lobes 5–6·5 mm. long ..   ..   13. *lemblini*

Branchlets densely hispid ; leaves subcordate or rounded at base ; calyx-lobes triangular ; corolla-tube 7–17 mm., lobes 5–10 mm. long   ..   14. *subcordata*

*Anthers included in the corolla-tube for most of their length, sessile or on very short filaments ; leaves deciduous, with 4–6 main lateral nerves on each side of midrib ; calyx very short ; flowers solitary to few :

Flowers axillary, subtended by several glumaceous or foliaceous bracts and bracteoles, sessile, corolla-tube at least twice as long as the lobes ; fruits subglobose 6–10 mm. diam. ; seeds without a longitudinal groove ; young branchlets densely pubescent or minutely and rather sparingly puberulous :

Flowers subtended by rigid dry bracts and bracteoles with inrolled margins and sharp setose points ; style exserted ; corolla-tube (10–)18–30 mm., lobes (5–)7–11 mm. long ; leaves elliptic or ovate or obovate, rounded or subcordate or obtuse at base, long-acuminate, 3–11 cm. long, 1·5–6 cm. broad, pubescent beneath especially on midrib and main lateral nerves   ..   ..   15. *eketensis*

Flowers subtended by soft, often rather leafy, bracts and bracteoles ; style included, much shorter than corolla-tube ; corolla-tube 28–45 mm., lobes 7–18 mm. long ; leaves broadly or narrowly elliptic or obovate-elliptic, cuneate or obtuse at base, gradually acuminate, 3–12 cm. long, 1·5–4·5 cm. broad, glabrous or pubescent beneath ..   ..   16. *rupestris*

Flowers subtended only by a pair of leaves, solitary or paired, terminal ; corolla-tube (7–)13–20 mm., lobes 7–15 mm. long ; style included ; fruits 2-lobed, 7–8 mm. long and broad, 4 mm. thick ; seeds with a longitudinal groove ; young branchlets minutely puberulous or glabrous ; leaves elliptic or ovate, cuneate at base, long-acuminate, 2·5–10·5 cm. long, 1–5·3 cm. broad, glabrous   ..   17. *ebracteolata*

1. C. liberica *Bull ex Hiern* in Trans. Linn. Soc., ser. 2, 1 : 171, t. 24 (1876) ; F.T.A. 3 : 181 ; Aubrév. Fl. For. C. Iv., ed. 2, 3 : 291, t. 359, 5–9 ; Lebrun Inst. Roy Colon. Belge. sect. sci. Nat. et Méd., Mém. 11, 3 : 153, tt. 15–19 (q.v. for syn.) ; Chev. Caf. du Globe 2 : t. 1 ; 3 : 170. *C. dewevrei* De Wild. & Th. Dur. (1899)—Chev. op. cit. 2 : t. 2 ; 3 : 179. *C. excelsa* A. Chev. (1903). *C. abeokutae* Cramer (1913)— F.W.T.A., ed. 1, 2 : 96 ; Chev. op. cit. 2 : t. 42 ; 3 : 175. *C. excelsoidea* Portères (1937). An evergreen tree, to 60 ft. high, spontaneous in the understorey of rain forest ; also commonly cultivated. For the varieties and forms reference should be made to the special literature. **Port.G.:** Bissalanca, Bissau (cult., fr. Jan.) *Esp. Santo* 1627 ! **Guin.:** spont. & cult. *fide* Chev. *l.c.* **S.L.:** Njala (cult., fl. Mar.) *Deighton* 3376 ! Kambema, Joru (cult., fl. Jan.) *Deighton* 3854 ! Kassewe F.R. (Apr.) *Lane-Poole* 128 ! **Lib.:** Cape Palmas *T. Vogel* 63 ! Bangee, Boporo (cult., fr. Nov.) *Baldwin* 10369 ! Grand Bassa (Feb.) *Dinklage* 1778 ! **Iv.C.:** Assikasso, Mid. Comoe (fr. Dec.) *Chev.* 22588 ! Mt. Momy *Aubrév.* 1175. **S.Nig.:** Lagos *Millen* 192 ! Omo (formerly part of Shasha) F.R., Ijebu Prov. (Feb.) *Jones & Onochie* FHI 17214 ! 17229 ! *Richards* 3070 ! *Ross* 251 ! Sapoba, Benin Div. (Jan.) *Onyeagocha* FHI 7605 ! **[Br.]Cam.:** Lakom, 6,000 ft., Bamenda (fr. May) *Maitland* 1587 ! Also in Cameroun, Gabon, Cabinda, the Congos, Ubangi-Shari, Sudan and Uganda ; and widely cultivated in the moister tropics. (See Appendix, pp. 393–394.)

[*C. abeokutae* Cramer was described from living material cultivated at Buitenzorg, Indonesia, specimens of which are preserved in the Kew herbarium. This material was presumably obtained originally from one of the several forms which were known in W. Nigeria as " Abeokuta " coffee early in this century.]

2. C. canephora *Pierre ex Froehner* in Notizbl. Bot. Gart. Berl. 1 : 237 (1897) ; Aubrév. Fl. For. C. Iv., ed. 1, 2 : 250, 252 ; Lebrun l.c. 122, tt. 11–14 (q.v. for syn.) ; Chev. Caf. du Globe 2 : tt. 28–39 ; 3 : 186 (q.v. for vars.). *C. robusta* Linden Cat. Pl. Econ. Hort. Colon. 64 (1900). *C. canephora* var. *robusta* (Linden) A. Chev. op. cit. 3 : 191 (1947). *C. maclaudi* A. Chev. (1905)—F.W.T.A., ed. 1, 2 : 96. *C. canephora* var. *maclaudi* (A. Chev.) A. Chev. op. cit. 2 : 30, t. 34 (1942) ; 3 : 194. An evergreen shrub or small tree, to 20 ft. high ; spontaneous in the understorey of rain forest ; also cultivated ; flowers white. For the varieties and forms reference should be made to the special literature. **Guin.:** Labé (Apr.) *Chev.* 12330 ! Mt. Bilima, Upper Konkouré (May) *Chev.* 12332 ! **S.L.:** Njala (cult., fl. Mar.) *Deighton* 3377 ! **Iv.C.:** Assikasso, Mid. Comoé (Dec.) *Chev.* 22590 ! **Dah.:** Allada (cult., fr. Mar.) *Chev.* 23389 ! **S.Nig.:** Shasha F.R., Oyo Prov. (Feb.) *Jones & Onochie* FHI 16968 ! Akure F.R., Ondo

Fig. 231.—Coffea liberica *Bull ex Hiern* (Rubiaceae).

A, longitudinal section of flower. B, part of cymule. C, stamen. D, fruits.

S.R.C.

Prov. (fr. Nov.) *Keay* FHI 25542! Okomu F.R., Benin Prov. (Jan., Feb.) *Brenan* 8871! 9028! *Onochie* FHI 19727! Degema *Talbot*! **[Br.]Cam.:** Likomba (fl. buds Dec.) *Mildbr.* 10766! Also in Cameroun, Gabon, the Congos, Sudan, Uganda and Angola ; and widely cultivated in the moister tropics. (See Appendix, p. 394.)

3. **C. arabica** *Linn.* Sp. Pl. 1 : 172 (1753) ; F.T.A. 3 : 180 (excl. var. *leucocarpa* Hiern) ; Lebrun l.c. 114 ; Chev. Caf. du Globe 2 : tt. 17–24 ; 3 : 196 (q.v. for vars.). Not indigenous in W. Africa, where it thrives in cultivation only at altitudes above 3,000–4,000 ft. Numerous varieties and forms have been described. **Sen.:** Sedhiou, Casamance (cult.) *Chev.* 2019. **Guin.:** Conakry (cult.) *Chev.* 12025. Macenta (cult.) *Barthe* 24. **S.L.:** Njala (cult.) *Deighton* 3378! **S.Nig.:** Lagos (cult.) ? *Millen*! Calabar (cult.) *Chev.* 13724. **[Br.]Cam.:** Gembu, Mambila Plateau, Adamawa (cult., fr. Jan.) *Latilo & Daramola* FHI 34371! **F.Po:** (cult.) *Barter* 2077! Indigenous in Arabia and N.E. tropical Africa, now cultivated widely in the tropics. (See Appendix, p. 393.)

4. **C. togoensis** *A. Chev.* in Rev. Bot. Appliq. 19 : 402 (1939) ; Chev. Caf. du Globe 2 : t. 49 ; 3 : 169. A shrub or small tree, 6–10 ft. high ; flowers white or pale pink, sweetly scented.
Ghana: near R. Asuokoko (Mar.) *Beveridge* FH 2945! **Togo Rep.:** lateritic plateaux near Lomé (fr. Sept.) *Warnecke* 415!

5. **C. stenophylla** *G. Don* Gen. Syst. 3 : 581 (1834) ; F.T.A. 3 : 182 : Bot. Mag. t. 7475 ; Chev. Caf. du Globe 2 : t. 46 ; 3 : 210. ? *C. affinis* De Wild. (1904). A glabrous shrub or small tree, to 10 ft. high. Indigenous in Guinée, Sierra Leone and Ivory Coast, but also cultivated in these and other countries.
Guin.: spont. & cult. *fide* Chev. l.c. **S.L.:** *Barter*! *Don*! *Morson*! *Sc. Elliot* 3909! Kassewe Hills (fl. Apr., fr. Nov.) *Lane-Foole* 125! *Gbanja*! **Iv.C.:** Aniasvé, R. Comoé *Court.* Abengourou *Chev.* And other localities *fide* Chev. l.c. **Ghana:** Aburi (cult., fr. Feb.) *Johnston* 611! (See Appendix, p. 395.)

6. **C. sp. nr. carrissoi** *A. Chev.* in Rev. Bot. Appliq. 19 : 401 (1939). A shrub or small tree, to 12 ft. high, with white flowers. The type specimen of *C. carrissoi* from Angola is in fruit only.
N.Nig.: Sanga River F.R., Jemaa Div. (Nov.) *Keay* FHI 22263! **S.Nig.:** Ohosu F.R., Benin (Dec.) *Hide* 5/38! Okomu F.R., Benin (fl. Jan., young fr. June) *Brenan* 8854! 8878! *Onochie* FHI 34608! Sapoba *Kennedy* 2167! 2352! **[Br.]Cam.:** Barombi (Sept.) *Preuss* 517! 520!

7. **C. brevipes** *Hiern* in Trans. Linn. Soc., ser. 2, 1 : 172 (1876) ; F.T.A. 3 : 182 ; Lebrun l.c. 147 ; Chev. Caf. du Globe 2 : t. 53 ; 3 : 166. *C. staudtii* Froehner (1897). ? *C. montana* K. Schum. ex A. Chev. op. cit. 2 : t. 45 ; 3 : 169 (1947). A forest shrub to 6 ft. high ; flowers white.
[Br.]Cam.: 2,000–3,000 ft., Cam. Mt. (fl. & fr. Dec.) *Mann* 2158! *Keay* FHI 37420! Also in Cameroun and Congo. (See Appendix, p. 394.) *Preuss* 1383. Johann-Albrechtshöhe (= Kumba) *Staudt* 548! Bambuko F.R., Kumba Dist. (fl. & fr. Jan.) *Keay* FHI 37420! Also in Cameroun and Congo. (See Appendix, p. 395.)

8. **C. humilis** *A. Chev.* in C.R. Acad. Sci. Paris 145 : 349 (1907) ; Chev. Caf. du Globe 2 : tt. 55 & 56 ; 3 : 165. A forest undershrub, to 6 ft. high, rarely more ; ripe fruits red ; flowers seldom collected.
Lib.: Dukwia R. (fr. Oct., Nov.) *Cooper* 28! 137! Ganta (fr. Sept.) *Linder* 628! Tawata, Boporo (fr. Nov.) *Baldwin* 10295! Ganta (fr. Sept.) *Baldwin* 9327! **Iv.C.:** Guidéko, Mid. Sassandra (fr. May) *Chev.* 16406! Sampleu to Ganhoué, Danané *Chev.* 21153. And other localities *fide* Chev. l.c.

9. **C. macrochlamys** *K. Schum.* in Engl. Bot. Jahrb. 23 : 463 (1897). *Tricalysia macrochlamys* (K. Schum.) A. Chev. in C.R. Acad. Sci. Paris 210 : 358 (1940) ; Caf. du Globe 2 : 36, t. 143 (1942) ; 3 : 240 (1947). A shrub or small tree, to 25 ft. high ; seeds not known.
Ghana: *Bunting* 9 (partly)! **F.Po:** *Guinea*! Also in Cameroun.

10. **C. spathicalyx** *K. Schum.* l.c. 587 (1897). *Calycosiphonia spathicalyx* (K. Schum.) Lebrun l.c. 69, t. 7 (1941). *Tricalysia spathicalyx* (K. Schum.) A. Chev. Caf. du Globe 2 : 36, t. 144 (1942) ; 3 : 240 (1947). A straggling shrub or small tree, to 25 ft. high ; flowers white ; in forest.
Iv.C.: Abengourou (young fr. Feb.) *Bodard* 1400! **Ghana:** Ofin Headwaters F.R. (Jan.) *Andoh* FH 4289! Oda (Jan.) *Enti* FH 6874! **S.Nig.:** Omo (formerly part of Shasha) F.R., Ijebu Ode Prov. (Feb.) *Jones & Onochie* FHI 16960! 17525! *Richards* 3103! 3164! Ohosu F.R., Benin Div. (Dec.) *Hide* 6/38! **[Br.]Cam.:** Kuke Bova, Kumba Div. (Jan.) *Keay* FHI 37463! Fonfuka, 6,000 ft., Bamenda (fr. May) *Maitland* 1376! Also in Cameroun, Gabon, Congo, Sudan and Angola.

11. **C. afzelii** *Hiern* in Trans. Linn. Soc., ser. 2, 1 : 174 (1876) ; F.T.A. 3 : 184 ; Chev. Caf. du Globe 2 : t. 102 ; 3 : 134. *C. scandens* K. Schum. (1897). *Argocoffeopsis scandens* (K. Schum.) Lebrun l.c. 56 (1941). *Coffea ligustrifolia* Stapf (1909)—F.W.T.A., ed. 1, 2 : 96 ; Aubrév. Fl. For. C. IV. 3 : 252. A forest liane ; flowers white, very fragrant ; leaves shining.
S.L.: *Afzelius*. Victoria *Thomas* 9059! 9062! Zimmi (Nov.) *Deighton* 365! Kenema (Oct.) *Deighton* 5229! Mamansu, Kumrabai (Nov.) *Marmo* 143! **Lib.:** Gola Forest (Nov.) *Bunting* 63! Kakatown *Whyte*! Sinoe Basin *Whyte*! Bonuta (Oct.) *Linder* 885! Beiden, Boporo (Nov.) *Baldwin* 10277! **Iv.C.:** *fide* Aubrév. l.c. Angédédou Forest (Nov.) *Leeuwenberg* **Ghana:** Bou, Axim Dist. (Nov.) *Vigne* FH 1587! **[Br.]Cam.:** Victoria (young fr. Mar.) *Maitland* 527! Bai, Kumba (Feb.) *Lobe Babute* Cam. 48/36! Kumba, near Boviongo (Jan.) *Keay* FHI 37389! Also in Cameroun and Congo.

12. **C. pulchella** *K. Schum.* in Engl. Bot. Jahrb. 23 : 462 (1897) ; Chev. Caf. du Globe 2 : t. 98 ; 3 : 132. *Cremaspora glabra* Wernham (1913)—F.W.T.A., ed. 1, 2 : 85. A forest liane, attaining more than 70 ft. in height ; flowers white.
S.Nig.: Eket *Talbot* 3200! Oban *Talbot* 1536! Also in Gabon.

13. **C. lemblini** (*A. Chev.*) *Keay* in Bull. Jard. Bot. Brux. 28 : 296 (1958) ; Chev. Bot. 336 (1920), name only. *Randia lemblini* A. Chev. in Rev. Bot. Appliq. 18 : 415, t. 10 (1938). A much branched shrub, to 20 in. high ; flowers white ; in forest.
Iv.C.: Voguié, Agnéby valley (Jan.) *Chev.* 17180!

14. **C. subcordata** *Hiern* in Trans. Linn. Soc., ser. 2, 1 : 174 (1876) ; F.T.A. 3 : 184 ; Chev. Caf. du Globe 2 : t. 95 ; 3 : 130. *Argocoffeopsis subcordata* (Hiern) Lebrun l.c. 61, t. 6 (1941). A small liane or scrambling shrub ; flowers white, fragrant ; fruits red ; in forest.
S.Nig.: Eket *Talbot*! Calabar (Mar.) *Thomson* 35! Oban *Talbot* 243! 1523! *Onochie* FHI 36348x! **[Br.]Cam.:** Mbalange to Ndifo, S. Bakundu (Jan., Mar.) *Onochie* FHI 30854! 35477! Also in Cameroun, Gabon, Cabinda and Congo.

15. **C. eketensis** *Wernham* in J. Bot. 52 : 8 (1914). *C. jasminoides* Welw. ex Hiern (1876)—F.T.A. 3 : 185 ; Aubrév. Fl. For. C. IV. 3 : 252 ; Chev. Caf. du Globe 2 : t. 99 ; 3 : 131 ; not of Cham. (1834). *Argocoffea jasminoides* (Welw. ex Hiern) Lebrun l.c. 40, t. 4 (1941). A scandent shrub ; flowering when leafless ; flowers white, fragrant ; fruits in forest regrowth and margins.
Guin.: Farmar 252! Dalaba Chev. 18115. **S.L.:** *Smythe* 35! Mofari *Sc. Elliot* 4443! Sendugu (young fr. June) *Thomas* 561! 600! Hallawa, Dasse (young fr. May) *Jordan* 2102! **Iv.C.:** *fide* Aubrév. l.c. **S.Nig.:** Onitsha *Barter* 1249! Orlu (Feb.) *Thompson*! Awka *Thomas* 16! Eket *Talbot* 3064! 3171! Calabar *Thomson* 37! Also in Cameroun, Gabon, the Congos, Angola and N. Rhodesia. (See Appendix, p. 394.)

16. **C. rupestris** *Hiern* l.c. 174 (1876) ; F.T.A. 3 : 184 ; Chev. Caf. du Globe 2 : t. 16 ; 3 : 135. *Argocoffea rupestris* (Hiern) Lebrun l.c. 48 (1941). *Coffea divaricata* K. Schum. (1897) ; not of DC. (1830). *C. nudiflora* Stapf (1905)—Chev. op. cit. 2 : t. 104 ; 3 : 137. *C. nudicaulis* A. Chev. Bot. 336 (1920), name only. *C. brenanii* J. F. Leroy in J. Agric. Trop. 7 : 714 (1961), French descr. only. *Psilanthus jasminoides* Hutch. & Dalz. F.W.T.A., ed. 1, 2 : 92 (1931) ; Chev. Caf. du Globe 2 : t. 136 ; 3 : 225. An erect shrub to 6 ft. high, with stiff branches ; flowers white, very sweet scented ; fruits yellow, turning red ; in forest and also in open situations, especially rocky hills. There is a tendency for the specimens from forest to have longer, narrower leaves and minutely puberulous branchlets, and for those from open situations to have smaller, broader leaves and more densely spreading pubescent branchlets ; these distinctions break down however, especially in W. Nigeria, and do not seem to be of sufficient importance for specific recognition.

Guin.: Fossakoidou to Kesséridou, Koniankè *Chev.* 20816. **S.L.**: Faya (Feb.) *Marmo* 164! **Lib.**: Monrovia *Whyte*! Sinoe Basin *Whyte*! Ganta (fr. Apr.) *Harley*! **Iv.C.**: Mid. Sassandra to Mid. Cavally *Chev.* 19262. Mt. Kouan, Danané *Chev.* 21255. Bontrou to Bampleu (fr. Apr.) *Chev.* 21170. **Ghana:** near Larte (Jan.) *Johnson*! Bou, Axim Dist. (Nov.) *Vigne* FH 1473! Kintampo (Mar.) *Dalz.* 27! Kpong (fl. Feb., Apr., fr. Apr.) *Johnson* 964! *A. S. Thomas* M2! Accra Plains (fl. Jan.-Mar., young fr. Mar., Apr.) *Adams* 3733! *Irvine* 1425! *Rose Innes* GC 30102! **Togo Rep.**: Lomé *Warnecke* 89! Agome (Mar.) *Schlechter* 12963! Misahöhe (Mar.) *Baumann* 377. Kpedsu (Jan.) *Howes* 1112! **Dah.**: L. Azri to Zagnanado (Feb.) *Chev.* 23053! **N.Nig.**: Esie, Ilorin (Jan.) *Latilo* FHI 18224! **S.Nig.**: Abeokuta *Barter* 3343! *Rowland*! Ogbomosho *Rowland*! Owo *G. Burton*! Sapoba (Jan.) *Brenan* 8932! Obu *Thomas* 356! (See Appendix, p. 394.)

[The Congo plant (e.g. *Louis* 1234) referred to by Lebrun (l.c.) as *Argocoffea rupestris* is a distinct species, note in particular the exserted style.]

17. **C. ebracteolata** (*Hiern*) *Brenan* in Kew Bull. 8 : 115 (1953). *Psilanthus ? ebracteolatus* Hiern in F.T.A. 3 : 186 (1877); *Chev.* Caf. du Globe 2 : t. 135 ; 3 : 226. *P. ? tetramerus* Hiern l.c. 187 (1877). *Coffea nigerina* A. Chev. in Rev. Bot. Appliq. 19 : 402 (1939). A forest shrub, 1–8 ft. high ; flowers white.

Guin.: Kaba valley (May) *Chev.* 13245! Farana (Mar.) *Sc. Elliot* 5382! Farmoreah (Apr.) *Roberty* 17342! Kissidougou *Chev.* 20718! **S.L.**: Bagroo R. (Apr.) *Mann* 846! Lomaburn, Scarcies (fl. & young fr. Feb.) *Sc. Elliot* 5028! Kuntaia (June) *Thomas* 450! Njala (fl. Mar., Apr., fr. Nov.) *Deighton* 1136! 1908! 5659! York (May) *Deighton* 5526! **Lib.**: Peahtah (fr. Oct.) *Linder* 1018! Gletown (fr. July) *Baldwin* 6945! Ganta (Mar.) *Harley* 451! **Iv.C.**: Bouroukrou *Chev.* 16914! Sea coast between Tabou and Bériby *Chev.* 19991 ; 20010. **Ghana:** Bia F.R. (Apr.) *Foggie* FH 4455! S. Scarp F.R. (young fr. Dec.) *Moor* FH 2111! Sekasua (Mar.) *A.S. Thomas* D172! Togo Plateau F.R. (Apr.) *Beveridge* FH 2944! **N.Nig.**: Irenedu, Kabba Prov. (fr. Sept.) *Daramola & Adebusuyi* FHI 38411! **S.Nig.**: Olokemeji *Foster* 275! 282! Ibadan South F.R. (Apr.) *Keay* FHI 22829! Akure F.R. (fr. Oct.) *Adekunle* FHI 24230! Idanre *Foster* 199! Oji R., Udi Dist. (Apr.) *Kitson*! [**Br.**]**Cam.**: Ambas Bay (Jan.) *Mann* 740! Johann-Albrechtshöhe (= Kumba) *Staudt* 562! S. Bakundu F.R., Kumba (Jan.) *Latilo* FHI 43830!

38. PSILANTHUS Hook. f.—F.T.A. 3 : 185. *Nom. cons.*

A shrub or small tree ; branchlets glabrous or minutely puberulous ; stipules very short ; leaves elliptic, oblong-elliptic, obovate-elliptic or elliptic-oblanceolate, cuneate at base, acuminate, 7–18 cm. long, 2·5–7 cm. broad, with 4–6 main lateral nerves on each side of midrib, glabrous ; flowers axillary, solitary or paired, subsessile ; calyx-lobes lanceolate, about 2 mm. long at anthesis, glabrous ; corolla glabrous, tube 45–67 mm., lobes 26–34 mm. long ; anthers mostly included ; style much shorter than corolla-tube ; fruits obovoid-ellipsoid, 12–20 mm. long, 10–14 mm. broad, ribbed and crowned by the persistent, enlarged, 20–30 mm. long calyx-lobes .. .. *mannii*

P. mannii *Hook. f.* in Hook. Ic. Pl. t. 1129 (1873) ; F.T.A. 3 : 186 ; Chev. Caf. du Globe 2 : tt. 131–132 ; 3 : 223 ; Rev. Bot. Appliq. 27 : 499, t. 23. *Heinsia jasminiflora* of Chev. Bot. 317 (*fide* Hutch. & Dalz.). A forest shrub or small tree, to 12 ft. high ; corolla white ; fruits black when ripe, with conspicuous green persistent calyx lobes.

Guin.: Kaba *Chev.* 13245. Farana *Chev.* 13401. **S.L.**: Pujehun (Apr.) *Aylmer* 67! Yonibana (fr. Oct., Nov.) *Thomas* 4108! 4254! 4971! **Iv.C.**: Grand Bassam *Vest.* Guidéko (fr. May) *Chev.* 16420. Bouroukrou (Dec.-Jan.) *Chev.* 16914. Bériby *Chev.* 20010. **Ghana:** Akwapim (fr. Apr.) *Irvine* 1152! S. Scarp F.R. (Feb.) *Beveridge* FH 4160! Ofin Headwaters F.R. (fr. Nov.) *Vigne* FH 3413! **S.Nig.**: Oshogbo *Millson*! Akure F.R. (fr. Oct.) *Adekunle* FHI 24228! Okomu F.R., Benin (fr. Jan., Feb.) *Brenan* 8783! 9191! Oban (Mar.) *Richards* 5203! *Talbot* 1414! [**Br.**]**Cam.**: Bimbi road, Victoria *Maitland* 612! Bambuko F.R., Kumba Div. (fr. Sept.) *Olorunfemi* FHI 30769! **F.Po:** on the beach *Mann* 266! Also in Cameroun, Ubangi-Shari, Gabon, Congo and Angola.

39. BELONOPHORA Hook. f.—F.T.A. 3 : 129 ; Keay in Bull. Jard. Bot. Brux. 28 : 297 (1958). *Kerstingia* K. Schum. (1903). *Diplosporopsis* Wernham (1913).

Leaves cordate-auriculate and sometimes pouched at base, up to 35 cm. long, and 20 cm. broad, with 9–12 main lateral nerves on each side of midrib :
Branchlets scurfy puberulous ; stipules lanceolate and long-acuminate, up to 3 cm. long and 0·5 cm. broad, with a strong dorsal nerve ; petiole up to 7 mm. long ; leaves obovate or obovate-elliptic, narrowed to the markedly-pouched auriculate base, gradually long-acuminate, 16–30 cm. long, 6–13 cm. broad ; calyx-lobes triangular-ovate, about 4 mm. long, conspicuous ; corolla-tube 15–20 mm. long, lobes 10–13 (–18) mm. long, 3·5–5 mm. broad ; anthers about 6 mm. long, their tips 1·5 mm. below the apex of the corolla-tube .. .. 1. *wernhamii*
Branchlets glabrous ; stipules broadly ovate, acuminate, up to 4·5 cm. long and 2·5 cm. broad, without a strong dorsal nerve ; petiole up to 15 mm. long ; leaves elliptic or obovate-elliptic, broadly cordate-auriculate at base, rather abruptly acuminate, 22–35 cm. long, 9–19·5 cm. broad ; calyx-lobes small ; corolla-lobes 18 mm. long, anthers 8–9 mm. long ; otherwise similar to above .. .. 2. *talbotii*
Leaves cuneate or obtuse at base, elliptic, oblong-elliptic or obovate-elliptic, up to 27 cm. long and 12·5 cm. broad ; petiole distinct ; branchlets glabrous ; calyx-lobes 1–2 mm. long :
Stipules linear-oblong, oblong-spathulate, or oblong, rounded at apex, 8–14(–20) mm. long, 1·5–6(–10) mm. broad ; leaves shortly acuminate, (5–)10–27 cm. long, (2·5–) 4–11 cm. broad, with 6–7 main lateral nerves ; corolla-tube about 13 mm. long, lobes about 8 mm. long and 3 mm. broad ; anthers about 6 mm. long, their tips about 1 mm. from the apex of the corolla-tube ; fruits about 15 mm. diam.
3. *coriacea*
Stipules lanceolate or triangular, with long points, up to 18 mm. long ; leaves gradually and often long acuminate, 12–23 cm. long, 3–9(–12·5) cm. broad, with 4–6(–7) main lateral nerves on each side of midrib ; corolla-tube 9–17 mm. long, lobes 4–9 mm.

long and 2–4 mm. broad ; anthers 4·5–5·5 mm. long, with their tips 1–4 mm. from the apex of the corolla-tube ; fruits ellipsoid-subglobose, about 12 mm. long

4. *hypoglauca*

1. **B. wernhamii** *Hutch. & Dalz.* F.W.T.A., ed. 1, 2 : 85 (1931). *Diplosporopsis coffeoides* Wernham in Cat. Talb. 47, t. 8 (1913), not *B. coffeoides* Hook. f. (1873). Treelet, 3–8 ft. high ; corolla white ; in forest.
    S.Nig.: Oban *Talbot* 1649 ! [Br.]**Cam.**: S. Bakundu F.R., Banga, Kumba Div. (Mar.) *Brenan* 9292 ! 9437 !
2. **B. talbotii** (*Wernham*) *Keay* in Bull. Jard. Bot. Brux. 28 : 298 (1958). *Diplosporopsis talbotii* Wernham in Cat. Talb. 47 (1913). A small tree ; in forest.
    S.Nig.: Oban *Talbot* 1056 !
3. **B. coriacea** *Hoyle* in Kew Bull. 1935 : 263. A tree, 4–40 ft. high ; flowers white ; in forest.
    S.Nig.: Lagos *Rowland* 45 ! Okomu F.R., Benin (fr. Jan.) *Brenan* 8844 ! Sapoba, Benin (fl. Aug., fr. Apr.) *Kennedy* 1730 ! 1852 ! 2279 ! Mamu River, Onitsha Prov. (fr. Feb.) *Kennedy* 2542 ! [Br.]**Cam.**: Kumba Div. (young fr. Apr.) *Olorunfemi* FHI 30530 ! Nkom-Wum F.R., Bamenda (July) *Ujor* FHI 29283 ! Also in Cameroun. (See Appendix, p. 391.)
4. **B. hypoglauca** (*Welw. ex Hiern*) *A. Chev.* in Rev. Bot. Appliq. 19 : 398 (1939) ; Chev. Caf. du Globe 3 : 247. *Coffea hypoglauca* Welw. ex Hiern (1876)—F.T.A. 3 : 183. *Kerstingia lepidopoda* K. Schum. (1903). *Belonophora lepidopoda* (K. Schum.) Hutch. & Dalz. F.W.T.A., ed. 1, 2 : 85 (1931) ; Aubrév. Fl. For. C. Iv., ed. 2, 3 : 290, t. 359, 1–4 ; Chev. Caf. du Globe 3 : 248. *B. morganae* Hutch. (1921)—F.W.T.A., ed. 1, 2 : 85 (1931). *B. glomerata* M. B. Moss (1929). A shrub or small tree, to 30 ft. high ; flowers white, strongly scented ; in evergreen forest, especially by streams in the savanna woodland regions and on hills.
    Guin.: Dalaba *Chev.* 18124. Pita, 4,000 ft. *Chev.* 34064. **S.L.:** Smythe 132 ! Kambui F.R., Kenema (young fr. June) *Jordan* 2122 ! Bintumane, 3,000 ft. (Jan.) *Glanville* 451 ! **Lib.**: Mt. Wolagwisa, 4,500 ft., Pandami (Mar.) *Bequaert* 102 ! Mt. Bobei, Sanokwele (fr. Sept.) *Baldwin* 9603 ! 12596 ! **Iv.C.:** Mt. Tonkoui (fr. Dec.) *Aubrév.* 1006 ! *Roberty* 15878 ! Abengourou *Bodard* 1394 ! 1674 ! **Ghana:** Atuna, N.W. Ashanti (fl. buds & fr. Dec.) *Vigne* FH 3504 ! **Togo Rep.:** Loso, Sokode-Basari (Apr.) *Kersting* 362 ! 398 ! **N.Nig.:** Gwaha Kurmi, Kuta Dist., Niger Prov. (fl. buds Mar.) *Jibrin Jia* FHI 5572 ! Zaria Prov. (fl. buds Dec.) *Taylor* FHI 9870 ! Jos *Morgan* ! *Kennedy* FHI 11885 ! Also in Ubangi-Shari, Sudan, Uganda and Angola.

40. ATRACTOGYNE Pierre in Bull. Soc. Linn. Paris, 2 : 1261 (1896) ; F. Hallé in Adansonia 2 : 309–311 (1962). *Afrohamelia* Wernham (1913).

Fruit oblong, about 3 cm. long and 1–2 cm. diam. ; cymes about 1·5 cm. long, pedicel and peduncle elongating up to 6 cm. long in fruit ; corolla about 1 cm. long ; leaves ovate, rounded or subcordate at base, acuminate, 7–20 cm. long, 3–11 cm. broad, with 6–7 main lateral nerves on each side of midrib, puberulous on midrib and nerves beneath ..   ..   1. *bracteata*
Fruit fusiform, 9–11 (–15) cm. long and about 1 cm. wide ; leaves ovate, obtuse, rounded or subcordate at base, acuminate, 6–16 cm. long, 2·5–6·2 cm. broad, puberulous only on the pits in the axils of the main lateral nerves beneath ; otherwise similar to above ..   ..   2. *gabonii*

1. **A. bracteata** (*Wernham*) *Hutch. & Dalz.* F.W.T.A., ed. 1, 2 : 98 (1931) ; F. Hallé l.c. 313, t. 3–5. *Afrohamelia bracteata* Wernham in Cat. Talb. 43, t. 6 (1913). *Atractogyne melongenifolia* A. Chev. Bot. 350 (1920), name only. A shrub, climbing by recurved branches ; in forest.
    Iv.C.: Guidéko (May) *Chev.* 16388 ! Bingerville Dist. *Chev.* 15377 ! Taté, Cavally Basin (Aug.) *Chev.* 19797 ! **S.Nig.:** Oban *Talbot* 262 ! 1662 ! Also in Cameroun and Gabon.
2. **A. gabonii** *Pierre* in Bull. Soc. Linn. Paris 2 : 1262 (1896) F. Hallé l.c. 311, t. 2. *A. batesii* Wernham (1919). A liane ; fruits red when ripe ; in forest.
    [Br.]**Cam.**: S. Bakundu F.R., Kumba Dist. (fr. Mar., Apr.) *Olorunfemi* FHI 30547 ! *Binuyo & Daramola* FHI 35598 ! Also in Cameroun and Gabon.

41. BERTIERA Aubl.—F.T.A. 3 : 82 ; Wernham in J. Bot. 50 : 110 (1912) ; N. Hallé in Notulae Syst. 16 : 280 (1960).

Flowers on elongated inflorescences :
  Inflorescence with stalked lateral branchlets, lax flowered :
    Inflorescence with a few scorpioid branches at length recurved, like that of *Helio-tropium*, pubescent ; leaves elliptic to broadly oblanceolate, acuminate, 7–12 cm. long, 3–6 cm. broad, sparingly pubescent beneath, lateral nerves about 5 pairs

1. *breviflora*

  Inflorescence more or less elongated and raceme-like :
    Stipules 5–8 mm. long, connate and shortly sheathing, acuminate ; leaves rather broadly and shortly elliptic, pubescent on the 5–6 pairs of nerves beneath, glabrous between them, abruptly acuminate, 8–12 cm. long, 3–6 cm. broad ; inflorescence slender, many flowered ; bracteoles linear, about 5 mm. long ; corolla 7 mm. long ; fruits globose, 4 mm. diam., strongly ribbed ..   ..   2. *bracteolata*
    Stipules 20–35 mm. long, free ; corolla about 2 cm. long ; fruits globose, 5–7 mm. diam., strongly ribbed ; leaves with 7–9 pairs of nerves, sparsely appressed-pubescent beneath :
      Leaves, branchlets and stipules glabrous, inflorescence-rhachis pubescent ; stipules lanceolate, very acute, 2–2·5 cm. long, 7–8 mm. broad ; leaves elliptic, abruptly acuminate, 10–17 cm. long, 4–7 cm. broad, with 7–8 pairs of nerves ; inflorescence lax ..   ..   3a. *laxa* var. *bamendae*
      Leaves (at least the nerves beneath), branchlets and stipules pubescent ; stipules broadly lanceolate to ovate, shortly and very acutely acuminate, 2–2·5(–3·5) cm. long, 7–11 mm. broad ; leaves narrowly elongated-ovate, acuminate, 10–20 cm. long, 4–8 cm. broad, with 7–9 pairs of nerves ; inflorescence rather few flowered, lax ..   ..   3. *laxa* var. *laxa*
Inflorescence with sessile or subsessile clusters of crowded flowers, or subsimply racemose :

Branchlets densely clothed with long spreading hairs ; leaves ovate-oblong, acutely acuminate, rounded or subcordate at base, 15–25 cm. long, 4–7 cm. broad, pilose beneath, especially on the 10–12 pairs of lateral nerves ; inflorescence up to 20 cm. long ; calyx pubescent with short triangular lobes ; corolla nearly 1 cm. long, sparsely pubescent   ..   ..   **4.** *spicata*
Branchlets glabrous or with closely appressed hairs :
Flowers on simple pedicels on the axis ; corolla 4 mm. long, minutely pubescent ; calyx 1 mm. long ; leaves oblong-elliptic, abruptly acuminate, 6–9 cm. long, 3–4 cm. broad, pubescent on the 4–5 pairs of nerves beneath ; stipules connate and shortly sheathing, acuminate, about 5 mm. long   ..   ..   **5.** *chevalieri*
Flowers in crowded cymules on the main axis ; stipules 1–3 cm. long :
Stipules very acutely long-acuminate, sheathing, thin, 1·5–1·8 mm. long, 4 mm. broad, median nerve prominent with pubescence limited to it ; corolla-limb ovoid in bud, nearly as broad as long, corolla-tube 6–7 mm. long, all densely pubescent ; flowers densely clustered in whorls on the rhachis ; calyx 1 mm. long, densely pubescent ; leaves oblong-lanceolate, gradually long-acuminate, cuneate at base, 10–15 cm. long, 4–5 cm. broad, pubescent on the 8–10 pairs of lateral nerves beneath   ..   ..   **6.** *subsessilis*
Stipules subacute at apex, thick, 1–2·5 cm., about 7 mm. broad at base, more or less evenly appressed-pubescent, 1-nerved ; flowers in subsessile clusters, not whorled ; corolla limb narrowly ellipsoid in bud, pubescent or glabrous :
Calyx glabrous or slightly pubescent ; corolla-limb glabrous at apex or entirely so ; inflorescence-branches 1–2 cm. long ; leaves oblong-elliptic, acuminate, rounded at base, distinctly petiolate, 15–20 cm. long, about 10 cm. broad, with about 8 pairs of lateral nerves :
Corolla pubescent   ..   ..   **7.** *racemosa* var. *racemosa*
Corolla glabrous   ..   ..   **7a.** *racemosa* var. *glabrata*
Calyx densely pubescent ; corolla-limb entirely pubescent, acutely acuminate at apex ; inflorescence branches up to 1 cm. long ; leaves slightly oblanceolate, shortly acuminate, slightly cuneate at base, subsessile, 10–35 cm. long, 4–11 cm. broad, with about 9 pairs of lateral nerves   ..   ..   ..   **8.** *retrofracta*
Flowers axillary, not in elongated inflorescences :
Stoloniferous subherbaceous plant of the forest floor ; stems about 3 cm. high above ground ; flowers solitary or almost so, sessile ; leaves elliptic, 5–7 cm. long, 2·5–4 cm. broad, rather densely appressed-pilose, spreading on the margins ; petiole slender   ..   ..   **9.** *adamsii*
Shrub up to 1·2 m. high, upper nodes congested, internodes about 1·5 cm. long obscured by large sheathing stipules with fimbriate margins ; flowers concealed within the stipules ; leaves elongate-oblanceolate and slightly pandurate, 30–42 cm. long, 6–7 cm. broad, young leaves sericeous beneath ; petiole short and stout
10. *fimbriata*

1. **B. breviflora** *Hiern* in F.T.A. 3 : 85 (1877) ; Wernham in J. Bot. 50 : 156 (1912). Small single-stemmed shrub 3–12 ft. high ; flowers greenish-white on the spreading branches of terminal and axillary cymes, fruits red ; in forest.
   S.L.: Kukuna, R. Scarcies (fr. Jan.) *Sc. Elliot* 4646 ! Mabonto (fr. July) *Deighton* 3266 ! Kenema *Thomas* 7892 ! Gola F.R. (fr. May) *Small* 686 ! **Lib.:** Peahtah (fr. Oct.) *Linder* 965 ! Genna Tanyehun (fr. Dec.) *Baldwin* 10751 ! Gletown (July) *Baldwin* 6909 ! **Iv.C.:** Béyo (fr. Dec.) *Leeuwenberg* 2206 ! **Ghana:** Assin Nyankumasi *Cummins* 124 ! Otumi, Kibbi Dist. (fr. Apr.) *Andoh* 5149 ! Bunso Plantation (Oct.) *Darko* 1046 ! **S.Nig.:** Lagos *Rowland* ! Epe *Barter* 3292 ! Omo F.R., Ijebu (Feb.) *Jones & Onochie* FHI 17151 ! Umon-Ndealichi F.R., Itu Dist. (fl. & fr. June) *Ujor* FHI 30181 ! Aboabam, Ikom Dist. (fl. & fr. Dec.) *Keay* FHI 28205 ! **[Br.]Cam.:** Victoria (fl. & fr. Feb.) *Maitland* 393 ! Johann-Albrecht-shöhe (= Kumba) *Staudt* 550 ! Barombi *Preuss* 324 ! S. Bakundu F.R., Kumba (fl. Apr., fr. Jan.) *Ejiofor* FHI 29344 ! *Binuyo & Daramola* FHI 35071 ! Also in Cameroun, Gabon and Congo.

2. **B. bracteolata** *Hiern* in F.T.A. 3 : 84 (1877) ; Wernham l.c. 160. *B. africana* of Chev. Bot. 317, partly, not of A. Rich. Scandent shrub ; flowers white in slender racemes 6–12 ins. long ; in forest.
   S.L.: Bonganema (Oct.) *Deighton* 6132 ! Gbangbama (Nov.) *Deighton* 2337 ! Ronietta (fl. & fr. Nov.) *Thomas* 5455 ! Kafogo (fr. Apr.) *Sc. Elliot* 5619 ! **Lib.:** Vahon (Nov.) *Baldwin* 10232 ! Gbanga (Sept.) *Linder* 675 ! Peahtah (Oct.) *Linder* 985 ! Gletown (July) *Baldwin* 6919 ! Tappita (Aug.) *Baldwin* 9111 ! **Iv.C.:** Agniéby Valley *Chev.* 17086 ; 17208. Alépé *Chev.* 17456. Taté to Tabou *Chev.* 19812. **Ghana:** Atroni (Aug.) *Vigne* FH 2445 ! Asankrangwa (Aug.) *Vigne* FH 1288 ! Mt. Ejuanema, Kwahu (fl. & fr. Dec.) *Adams* 5128 ! Benso (fl. & fr. Nov.) *Andoh* FH 5412 ! **S.Nig.:** Ibadan to Ife *Meikle* 1458 ! Oban *Talbot* 1288 ! Onochie FHI 36090x ! **[Br.]Cam.:** S. Bakundu F.R., Kumba (Jan.) *Binuyo & Daramola* FHI 35491 ! Also in Cameroun, Rio Muni, Gabon. (See Appendix, p. 391.)

3. **B. laxa** *Benth.* var. **laxa**—in Fl. Nigrit. 394 (1849) ; F.T.A. 3 : 85 ; Wernham l.c. 157. *B. maitlandii* Hutch. & Dalz. F.W.T.A., ed. 1, 2 : 97 (1931). A shrub 12–15 ft. high ; flowers green ; fruits bluish.
   S.Nig.: Nikrowa, Benin (fr. Dec.) *Brenan* 8401 ! Oban *Talbot* 51 ! **[Br.]Cam.:** Johann-Albrechtshöhe (= Kumba) *Staudt* 579 ! Bulifumbo, 2,500 ft., Buea (Mar.) *Maitland* 544 ! Bafla (fr. Feb.) *Keay* FHI 37540 ! **F.Po:** *T. Vogel* 148 ! *Mann* 200 ! Ureka (fl. & fr. Feb.) *Guinea* 2445 ! 2503 ! Also in Cameroun.

3a. **B. laxa** var. **bamendae** *Hepper* in Kew Bull. 13 : 405 (1959). Shrub 4 ft. high, leaves very dark green above, paler beneath ; fruits bluish ; in high forest.
   [Br.]Cam.: Wum Fuel Plantation, Bamenda (fr. Apr.) *Ujor* FHI 30087 !

4. **B. spicata** (*Gaertn. f.*) *Wernham* in J. Bot. 50 : 160 (1912). *Pomatium spicatum* Gaertn. f. (1805). *Bertiera africana* A. Rich. (1834)—F.T.A. 3 : 84 ; Chev. Bot. 317, partly. A hairy shrub 6–12 ft. high ; flowers white, yellowish-green or pinkish, fruits brown ; in forest, often beside water.
   Port.G.: Bangacia (fr. Dec.) *Esp. Santo* 3178 ! Pussubé (fr. Jan.) *Esp. Santo* 1114 ! Bissalanca (fr. Jan.) *Esp. Santo* 1665 ! **Guin.:** *Heudelot* 719 ! Kindia *Chev.* 13003 ; 13083. Kouria to Ymbo *Chev.* 14998. **S.L.:** Gbinti (Jan.) *Deighton* 980 ! Njala (Jan., July) *Deighton* 1740 ! *Dalz.* 8074 ! Kenema (fr. Jan.) *Smythe* 235 ! Kasawe F.R. (fr. Dec.) *King* 49b ! Rokupr (Nov.) *Jordan* 693 ! *Hepper* 2620 ! **Lib.:**

Belleyella (fr. Dec.) *Baldwin* 10650! Genna Tanyehun (fr. Dec.) *Baldwin* 10737! Ganta (Sept.) *Baldwin*
9241! Rippue's Town (Aug.) *Linder* 361! Bahtown (Aug.) *Baldwin* 9019!

5. **B. chevalieri** *Hutch. & Dalz.* F.W.T.A., ed. 1, 2 : 97 (1931). Shrub with the appearance of *B. bracteolata*
except for the simpler inflorescences.
    Guin.: Bérezia to Diorodougou (Feb.) *Chev.* 20786! **Lib.:** Bilimu (fr. Aug.) *Harley* 1448!

6. **B. subsessilis** *Hiern* in F.T.A. 3 : 83 (1877). A shrub about 12 ft. high, with wide-angled brittle branches;
flowers white or pale mauve in pendant inflorescences; in *Rhizophora* swamp forest.
    S.Nig.: Brass *Barter* 1831! Port Harcourt (fl. & fr. Jan.) *Jones* FHI 6198! Degema *Talbot*! Eket
Talbot 3067! Stubbs Creek F.R., Eket (Jan.) *Keay* FHI 37713! Also in Gabon.

7. **B. racemosa** (*G. Don*) *K. Schum.* var. **racemosa**—in Bol. Soc. Brot. 10 : 127 (1892); Wernham l.c. 160;
Aubrév. Fl. For. C. Iv., ed. 2, 3 : 294. *Wendlandia racemosa* G. Don (1834). *Bertiera macrocarpa*
Benth. (1849)—F.T.A. 3 : 84. *B. racemosa* var. *glabrata* Hutch. & Dalz. F.W.T.A., ed. 1, 2 : 97, partly.
B. montana Hiern in F.T.A. 3 : 83 (1877); F.W.T.A., ed. 1, 2 : 97; Aubrév. l.c. *B. laxa* of Chev. Bot.
318, not of Benth. Erect (or scandent, *Chizea* FHI 12386) shrub or small tree 10–25 ft. high, with thin
papery scales on the bole; flowers white in pendulous racemes, fruits greenish-brown; in high forest.
    S.L.: Kenema (Oct.) *Deighton* 5235! Kambui Hills (fr. Mar.) *Small* 518! Sasseni, R. Scarcies (fr.
Jan.) *Sc. Elliot* 4523! Vevehun (fr. Apr.) *Deighton* 1618! **Lib.:** Monrovia (Oct.) *Cooper* 55! Kakatown
Whyte! Gletown, Tchien Dist. (fr. July) *Baldwin* 6959! **Iv.C.:** Abidjan *Aubrév.* 477. Dabou *Chev.*
15288. Mbago *Chev.* 17076. Grabo *Chev.* 19730. Alépé *Chev.* 17509. **Ghana:** Tano R. (fl. & fr. Sept.)
Chipp 349! Tarkwa (Feb.) *Thompson* 42! Kumasi (fl. & fr. Sept.) *Vigne* FHI 1353! Jiasi (fl. & fr. Oct.)
Vigne FHI 1392! **Dah.:** *Debeaux* 374! **S.Nig.:** Lagos *Millen* 13! Ehor F.R., Benin (fr. Feb.) *Chizea*
FHI 12386! Brass *Barter* 10! Calabar (fr. Dec.) *Thomson* 101! Olorunfemi FHI 34208! Eket (fr. May)
Onochie FHI 33184! **[Br.]Cam.:** Rio del Rey *Johnston*! S. Bakosi F.R., Kumba (fr. May) *Olorunfemi*
FHI 30595! **F.Po:** 7,000 ft. *Mann* 292! Also in Cameroun, Gabon, Congo, Cabinda, Principe, Uganda
and Tanganyika. (See Appendix, p. 392.)

7a. **B. racemosa** var. **glabrata** (*K. Schum.*) *Hutch. &. Dalz.* F.W.T.A., ed. 1, 2 : 97 (1931). *B. glabrata* K.
Schum. (1897)—Wernham l.c. 161.
    S.L.: Kambui F.R. (Mar.) *Lane-Poole* 191! **Lib.:** Sinoe Basin *Whyte*! Gbanga (Sept.) *Linder* 582!
Flumpa, Sanokwele Dist. (Sept.) *Baldwin* 9366! Zigida, Vonjama (Oct.) *Baldwin* 10021a!
    [Apart from the corolla being glabrous in this variety it appears to be indistinguishable from var.
racemosa, hence some of the fruiting specimens cited under the type variety might ultimately prove to
be var. *glabrata*.]

8. **B. retrofracta** *K. Schum.* in Engl. Bot. Jahrb. 23 : 452 (1897). *B. obversa* K. Krause (1917). A shrub or
small tree 8–10 ft. high, with obtusely quadrangular branchlets; flowers white, fruits clustered; in
high forest.
    S.Nig.: Oban *Talbot* 236! Boshi-Okwangwo F.R., Obudu Dist. (May) *Latilo* FHI 30942! **[Br.]Cam.:**
Victoria to Bimbia (May) *Preuss* 1279. Barombi (June) *Preuss* 337! Mbalange, S. Bakundu F.R. (fr.
Jan.) *Binuyo & Daramola* FHI 35452! Kumba F.R. (Aug.) *Okereke* FHI 8372! **F.Po:** Bokoko (Oct.)
Mildbr. 6833.

9. **B. adamsii** (*Hepper*) *N. Hallé* in Adansonia 3 : 177 (1963). *Sabicea adamsii* Hepper in Kew Bull. 13 : 291, fig.
2 (1958); Petit in Bull. Jard. Bot. Brux. 32 : 193 (1962). A hairy plant with slender stolons above and
below ground level; flowers white, fruits red; amongst leaves on the forest floor.
    Ghana: Mpameso F.R., W. Ashanti (fl. & fr. Dec.) *Adams* 2967! Also in Rio Muni and Congo.

10. **B. fimbriata** (*A. Chev. ex Hutch. & Dalz.*) *Hepper* in Kew Bull. 16 : 329 (1962). *Psychotria fimbriata* A.
Chev. ex Hutch. & Dalz. F.W.T.A., ed. 1, 2 : 123 (1931); Chev. Bot. 341. A small shrub 2–4 ft. high
with greenish-white flowers.
    Lib.: Boporo (Nov.) *Baldwin* 10372! **Iv.C.:** Toula to Nékaougnié, Cavally (July) *Chev.* 19582! Mt.
Tou, Cavally (July) *Chev.* 19683! Tabou (Oct.) *Aké Assi* 6079!

42. HEINSIA DC.—F.T.A. 3 : 80.

Branchlets very slightly pubescent; leaves elliptic-lanceolate, acutely acuminate,
acute at base, 5–10 cm. long, 2–4 cm. broad, slightly pubescent on the nerves beneath;

Fig. 232.—Heinsia crinita (*Afzel.*) G. Tayl. (Rubiaceae).
A, anther. B, style. C, longitudinal section of ovary. D, seed.

stipules short and subulate ; flowers usually solitary at the end of short branchlets ; ovary setulose ; calyx-lobes foliaceous, oblanceolate, 1–1·5 cm. long ; corolla-tube 1·5–2·5 cm. long, appressed-tomentose, very setose-tomentose in the mouth, with style slightly exserted, limb 5–6 cm. diam., the lobes ovate to oblanceolate with crispate margins ; fruit globose, 1–1·5 cm. diam., crowned with the calyx-lobes ..   *crinita*

H. crinita *(Afzel.) G. Tayl.* in Exell Cat. S. Tomé 209 (1944). *Gardenia crinita* Afzel. (1829). *G. pulchella* G. Don (1824), name only. *Heinsia jasminiflora* DC. (1830)—F.T.A. 3 : 81 ; Chev. Bot. 317. *H. pulchella* K. Schum. (1891)—F.W.T.A., ed. 1, 2 : 98. A branching shrub in secondary forest or a small understorey tree in high forest, 8–25 ft. sometimes up to 40 ft. high ; flowers white with yellow throat-hairs, fruits yellow.
Guin.: Kaba Valley *Chev.* 13245. Farana *Chev.* 13401 ; 13431. **S.L.:** Yonibana (fr. Oct.) *Thomas* 4192 ! Mt. Omi *Lane-Poole* 116 ! Njala (May, Sept.) *Deighton* 654 ! 675 ! SLH 2130 ! Gola Forest (Mar.) *Small* 560 ! **Lib.:** Monrovia (Nov.) *Whyte* ! *Linder* 1564 ! Congotown (fl. & fr. Mar.) *Barker* 1222 ! Javajai (fr. Nov.) *Baldwin* 10275 ! Bobei Mt. (fr. Sept.) *Baldwin* 9604 ! Fortsville (Mar.) *Baldwin* 11146 ! **Iv.C.:** Guidéko *Chev.* 16420. Bouroukrou *Chev.* 16914. Kéeta *Chev.* 19317. Bériby *Chev.* 19991 ; 20010 ; 20043. **Ghana:** Bonsasu, Tarkwa Dist. (fr. June) *Enti* FH 6246 ! **S.Nig.:** Lagos *Punch* 46 ! Oban *Talbot* 276 ! 281 ! Boshi-Okwangoro F.R., Ogoja (fr. May) *Okeke* FHI 30943 ! **[Br.]Cam.:** Victoria (Jan.) *Kalbreyer* 12 ! Abonando (May) *Rudatis* 35 ! S. Bakundu F.R., Kumba (Jan., Apr.) *Olorunfemi* FHI 30535 ! *Binuyo & Daramola* FHI 35077 ! Korup F.R., Kumba (fr. July) *Olorunfemi* FHI 30663 ! Mamfe (fr. Nov.) *Tamajong* FHI 22109 ! Mbikas, Wum (fr. July) *Ujor* FHI 30477 ! **F.Po:** *T. Vogel* 177 ! *Barter* 2064 ! *Mann* 402 ! Also in Cameroun, Gabon, Congo, Principe, Kenya, Tanganyika, S. Rhodesia, Nyasaland, Mozambique, Cabinda and Angola. (See Appendix, p. 400.)

43. MITRAGYNA Korth.—F.T.A. 3 : 40 (as *Mitragyne*). *Nom. cons.*

Corolla-lobes glabrous outside ; anthers exserted and pendulous between corolla-lobes ; calyx truncate, glabrous ; globose inflorescences 2–2·5 cm. diam., about 1·5 cm. diam. in fruit ; fruits with a callous ring at apex ; leaves obovate-elliptic, acute to shortly acuminate, rounded to subcordate at base, 6–9 cm. long, 3·5–5 cm. broad, ciliate on nerves beneath ; petioles about 1 cm. long ; stipules oblong-lanceolate, thin, 1·5–2 cm. long, acute, early caducous ; savanna tree .. .. 1. *inermis*
Corolla-lobes densely pubescent outside ; anthers included ; leaves large ; forest trees :
Calyx glabrous, truncate forming an erect cupule in fruit ; globose inflorescences (1·5–) 2–2·5 cm. diam., about 1·5 cm. diam. in fruit ; leaves broadly elliptic, 15–45 cm. long, 8–15 cm. broad, more or less pubescent on nerves beneath ; stipules obovate up to 10 cm. long and 7 cm. broad, tomentose especially towards the base   2. *stipulosa*
Calyx ciliate on the 5 distinct lobes which persist and usually curve inwards in fruit ; globose inflorescences 1·5–2 cm. diam., about 1 cm. diam. in fruit ; leaves broadly elliptic, 15–62 cm. long, 8–44 cm. broad, subglabrous beneath with tufts of hairs in nerve axils ; stipules ovate-elliptic to obovate, 4–10 cm. long, 3–7 cm. broad, tomentose .. .. 3. *ciliata*

1. **M. inermis** *(Willd.) O. Ktze.* Rev. Gen. Pl. 1 : 288 (1891) ; K. Schum. in E. & P. Pflanzenfam. 4, 4 : 56 ; Aubrév. Fl. For. Soud.-Guin. 474 ; Berhaut Fl. Sén. 104. *Uncaria inermis* Willd. (1793). *Mitragyna africana* (Willd.) Korth. (1839)—F.T.A. 3 : 40 ; Chev. Bot. 307. *Nauclea africana* Willd. (1798). Shrub or low-branching tree 20–40 ft. high with scaly bark ; flower-heads white ; usually on wet heavy clay in savanna.
Sen.: Bondu *Heudelot* 456 ! Walo country (Sept.) *Roger* ! Samandiniéry, Casamance *Chev.* 2100 ! Tambou-kané *Chev.* 2103 ! **Gam.:** *Saunders* 39 ! Kuntaur *Ruxton* 54 ! Genieri (July) *Fox* 174 ! **Mali:** San to Bani *Chev.* 1102 ! Sebi *Chev.* 1180 ; 2112 ! Labézanga (Sept.) *Hagerup* 445 ! **Port.G.:** Prabis, Bissau (fr. Feb.) *Esp. Santo* 1807 ! **Guin.:** Kouroussa *Pobéguin* 431 ! Farana (fr. Mar.) *Sc. Elliot* 5378 ! **S.L.:** *Don* ! **Lib.:** Grand Bassa *Ansell* ! Ouangolo *Aubrév.* 1398. Groumania *Aubrév.* 781. Red Volta *Aubrév.* 2429. **Ghana:** Accra Plains (Feb.) *Brown* 929 ! Wenchi Dist. (fl. & fr. July) *Enti* FH 6245 ! Yeji (Aug.) *Pomeroy* 1341 ! Kpedsu (fr. Jan.) *Howes* 1085 ! **Togo Rep.:** Lomé (Nov.) *Warnecke* 247 ! *Mildbr.* 7507 ! **Dah.:** Djougou *Chev.* 23878 ! Konkobiri to Diapaga *Chev.* 24389 ! **Niger:** Zinder *Arnaud* in *Hb. Chev.* 25180 ! **N.Nig.:** Nupe *Barter* 1189 ! Share F.R., Ilorin (fr. Jan.) *Ujor* FHI 31604 ! Zungeru (Aug.) *Elliott* 5 ! Bornu (Nov.) *Elliott* 120 ! *E. Vogel* 70 ! Yola (Oct.) *Shaw* 62 ! *Hepper* 1614 ! **S.Nig.:** Osomari F.R. *Kennedy* 2494 ! Orle F.R., Kukuruku, Afenmai Div. (Aug.) *Onochie* FHI 33290 ! Nkalago, Enugu to Abakaliki (Jan.) *Cons. of For.* 216 ! Bansara (Oct.) *Smith* 27 ! **[Br.]Cam.:** Gulumba, Dikwa Div. (fr. Dec.) *McClintock* 45 ! Also in Mauritania, Cameroun, Ubangi-Shari, Sudan and Congo. (See Appendix, p. 401.)

2. **M. stipulosa** *(DC.) O. Ktze.* Rev. Gen. Pl. 1 : 289 (1891) ; F.W.T.A., ed. 1, 2 : 98, is mainly *M. ciliata* and *Nauclea diderrichii* except *Sc. Elliot* 5014 ; Aubrév. Fl. For. C. Iv., ed. 2, 3 : 260 ; Berhaut Fl. Sén. 104. *Nauclea stipulosa* DC. (1830). *N. stipulacea* G. Don (1834). *Mitragyna macrophylla* Hiern in F.T.A. 3 : 41 (1877), partly ; Chev. Bot. 307. Tree 20–100 (–140) ft. high with grey bark ; slash dull white ; flower-heads white ; in swamp forest.
Sen.: Leprieur ! *Serv. For.* 56. Koulage-Harage *Chev.* 2016. Badi *Berhaut* 1847. **Guin.:** Timbo *Pobéguin* 1505 ! *Chev.* 12430 ! 12822 ! Kouria *Dumas* in *Hb. Chev.* 18201 ! **S.L.:** Kenema (Apr.) *Lane-Poole* 214 ! Likuru (fr. Feb.) *Sc. Elliot* 5014 ! Rokupr (fr. Jan.) *Deighton* 2964 ! Commendi (Nov.) *Aylmer* 615 ! **Iv.C.:** Bondoukou *Aubrév.* 748 ! Man *Aubrév.* 989 ; 2103. Konilou *Sargos* 119. *fide* Chev. *l.c.* **N.Nig.:** Agaie (Dec.) *Yates* 2 ! Mker F.R., Benue Prov. (fr. Feb.) *Jones* 929 ! **S.Nig.:** Udi Plateau *Cons. of For.* 170 ! Enugu F.R. (Dec.) *Cons. of For.* 184 ! 192 ! 193 ! **[Br.]Cam.:** Bakaw Dist. *Maitland* FHI 10017 ! Mai Idoam, Gashaka Dist. (Jan., Feb.) *Latilo & Daramola* FHI 28985 ! 34480 ! Also in Ubangi-Shari, Gabon, Congo, Angola, Sudan, Uganda and N. Rhodesia. (See Appendix, p. 402.)

3. **M. ciliata** *Aubrév. & Pellegr.* in Bull. Soc. Bot. Fr. 83 : 36 (1936) ; Aubrév. Fl. For. C. Iv., ed. 2, 3 : 262, t. 345. *M. stipulosa* of F.W.T.A., ed. 1, 2 : 98, mainly. *M. macrophylla* Hiern in F.T.A. 3 : 41 (1877), partly. A tree 40–100 ft. or more high with low buttresses and clear bole ; flower-heads white ; in swamp forest and beside streams.
Lib.: Dukwia R., Monrovia (bud Oct.) *Cooper* 104 ! 312 ! Ganta (Feb.) *Harley* 880 ! **Iv.C.:** Alépé (fr. Mar.) *Chev.* 16234 ! Banco *Aubrév.* 877 ! Man *Aubrév.* 1633 ! Azaguié *Chev.* 22301 ! San Pedro *Thoiré* 40 ! **Ghana:** Esherasu (Jan.) *Thompson* 5 ! Imbraim (Mar.) *Thompson* 21 ! Benso (Nov.) *Quao* FH 4920 ! Axim (fr. Mar.) *Chipp* 417 ! **S.Nig.:** Sapoba (May) *Kennedy* 1272 ! Nun (fr. Sept.) *Mann* 494 ! Oni *Sankey* 7 ! Akilla (fr. June) *Owoseje* FHI 25352 ! Oban *Talbot* 2023 ! Uwet Odot F.R., Calabar (July) *L. G. Cooper* FHI 39688 ! **[Br.]Cam.:** Victoria (fr. Feb.) *Maitland* 363 ! **F.Po:** *T. Vogel* 138 ! Also in Cameroun, Gabon, Rio Muni, the Congos and Cabinda.

44. UNCARIA Schreb.—F.T.A. 3 : 41. *Nom. cons.*

Calyx shortly 5-toothed with the obtuse tips usually dark brown and glabrous ; flower-heads about 4 cm. diam. ; fruit-heads 5–8 cm. diam. ; flowers usually sessile, becoming pedicellate in fruit with pedicel about 1 cm. long ; fruit linear-elliptic, ribbed, about 2 cm. long ; seed with long hyaline tails, one of them deeply divided ; corolla-tube about 1 cm. long, densely yellow-pubescent outside ; leaves elliptic or ovate-elliptic, 7–15 cm. long, 3–6·5 cm. broad, sub-glabrous to sparsely pilose on nerves beneath ; larger branchlets 4-angled, glabrous　..　　..　　..　　1a. *africana* var. *africana*

Calyx deeply toothed or if rather shortly so then flower-head 9–11 cm. diam. ; flowers and fruits pedicellate in the head :

Corolla-tube up to 1 cm. long with style exserted about 5 mm., appressed-pubescent outside ; flower-head about 4 cm. diam. ; calyx deeply 5-toothed with the tips usually glabrous ; leaves elliptic to ovate-elliptic, 7–13 cm. long, 3–6 cm. broad, subglabrous or sparsely pilose on nerves beneath ; larger branchlets 4-angled, subglabrous　..　　..　　..　　..　　..　　1b. *africana* var. *angolensis*

Corolla-tube 2–3 cm. long with style exserted about 2 cm., densely appressed·pubescent outside ; flower-heads 9–11 cm. diam. ; calyx rather deeply 5-toothed, wholly pubescent ; leaves ovate-elliptic 11–14 cm. long, 5–8 cm. broad, densely pilose on nerves beneath ; all branchlets markedly 4-angled and pubescent on the angles

2. *talbotii*

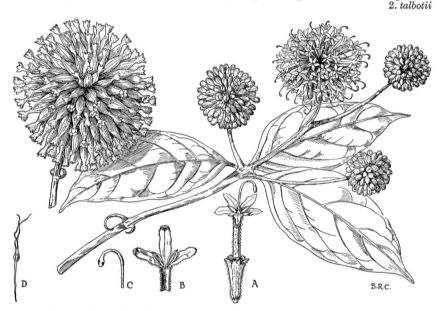

FIG. 233.—UNCARIA AFRICANA *G. Don* (RUBIACEAE).

A, flower.　B, part of corolla laid open.　C, stigma.　D, seed.

1. **U. africana** *G. Don* Gen. Syst. 3 : 471 (1834) ; F.T.A. 3 : 41 ; Haviland in J. Linn. Soc. 33 : 76 ; Chev. Bot. 308 ; Petit in Bull. Jard. Bot. Brux. 27 : 445 (1957).
1a. **U. africana** *G. Don* var. **africana.** *U. talbotii* of F.W.T.A., ed. 1, 2 : 99, partly (*Talbot* 39). A climbing shrub 7–60 ft. high with paired axillary hooks ; flowers in dense heads, yellow-silky outside ; in secondary forest.
  Port.G. : Ponte de Daba, Fulacunda (Oct.) *Esp. Santo* 2201 ! Sare N'gana Geba (Nov.) *Esp. Santo* 3118 ! **Guin.** : Nzo (fr. Dec.) *Roberty* 16084 ! Kindia *Chev.* 13001. Bilima *Caille* in *Hb. Chev.* 14879. Kouria *Caille* in *Hb. Chev.* 14954. Bembaya, Farana *Chev.* 20676. **S.L.** : *Don* ! Njala (July) *Deighton* 5984 ! Kasewe F.R. (Jan.) *King* 206b ! Kambia (fl. & fr. Jan.) *Sc. Elliot* 4709 ! Rokupr (Feb.) *Jordan* 192 ! **Lib.** : Kakatown *Whyte* ! Banga (Oct.) *Linder* 1258 ! Tappita (Aug.) *Baldwin* 9083 ! **Iv.C.** : Dabou *Chev.* 15347 ; 15463. **Ghana** : Obuasi, Ashanti (fr. Jan.) *Soward* 649 ! Shiare (Nov.) *Morton* A4113 ! **N.Nig.** : Patti Lokoja (Oct.) *Dalz.* 39 ! Abinsi (fr. Feb.) *Dalz.* 778 ! **S.Nig.** : Oban *Talbot* ! Also in Congo, Sudan, Uganda, Tanganyika, Angola and Annobon. (See Appendix, p. 413.)
1b. **U. africana** var. **angolensis** *Havil.* in J. Linn. Soc. 33 : 76 (1897) ; Petit l.c. 447. *U. angolensis* (Havil.) Welw. ex Hutch. & Dalz. F.W.T.A., ed. 1, 2 : 99 (1931). A climbing shrub like the preceding.
  Iv.C. : Gagnoa to Sassandra (July) *de Wilde* 169 ! **S.Nig.** : Onitsha (Oct.) *Onochie* FHI 34065 ! Oban *Talbot* 274 (partly) ! Also in Congo and Angola.
2. **U. talbotii** *Wernham* in Cat. Talb. 40 (1913). A climbing shrub with hooks and acutely 4-angled stem hairy on the angles ; flowers white with yellow pubescence outside ; in fringing forest.
  S.L. : Kamabai to Kabala (Sept.) *Deighton* 3980 ! **Lib.** : Zigida, Vonjama Dist. (fr. Oct.) *Baldwin* 9997 ! Ganta (July) *Harley* 1225 ! Gbau, Sanokwele Dist. (Sept.) *Baldwin* 9407 ! Tappita (Aug.) *Baldwin* 9110 ! Yamoussokro to Bouaflé *Mangenot & Aké Assi* IA 721. Man to Danané *Aké Assi* IA 3246. **Ghana** : Konongo (Oct.) *Adjie* FH 1394 ! Bobiri (Sept.) *Tsawe* FH 6266 ! Nkawkaw (Sept.) *Darko* 1002 ! Togo Plateau F.R. (Sept.) *St. C. Thompson* FH 3576 ! **S.Nig.** : Oban *Talbot* 168 ! 274 (partly) !

45. ADINA Salisb.—F.T.A. 3 : 39.

Branchlets glabrous, dull ; stipules caducous ; leaves lanceolate, cuneate at base, acute or slightly acuminate at apex, 9–15 cm. long, 2·5–5 cm. broad, with numerous lateral nerves slightly prominent beneath, glabrous ; flowers sessile in axillary, solitary, globose heads, 1·5–2 cm. diam. ; peduncle 3 cm. long with a pair of ovate, membranous bracts above the middle ; calyx-lobes oblong, 1·5 mm. long, pubescent ; corolla 5 mm. long, silky-pubescent, style long-exserted with an ellipsoid stigma
microcephala

A. **microcephala** (*Del.*) *Hiern* in F.T.A. 3 : 40 (1877) ; Chev. Bot. 307 ; Aubrév. Fl. For. Soud.-Guin. 474. *Nauclea microcephala* Del. (1826). Tree 15–50 ft. high with grey fissured and scaly bark and shiny leaves ; flowers yellowish ; beside streams in savanna.
Mali: Douiana to Sikasso (Sept.) *Jaeger* 5136 ! U. Volta : Banfora *Aubrév.* 1850 ! Kampti *Aubrév.* 2500. **Iv.C.:** Tourmi *Vuillet* 704. **Ghana:** ⟨Gurumbele to Bantala (Apr.) *Kitson* 833 ! Gambaga (Oct.) *Vigne* FH 4569 ! **Togo Rep.:** Syori (Apr.) *Kersting* 57 ! **Dah.:** Atacora Mts. (June) *Chev.* 24144 ! Konkobiri *Chev.* 24339 ! Tankiéta *Aubrév.* 9*d.* **N.Nig.:** Ilorin (Sept.) *Adejumo* FHI 5737 ! Kaduna (Feb.) *Meikle* 1203 ! Zamfara F.R., Sokoto (fl. & fr. Apr.) *Keay* 15619 ! 18011 ! Assob (Feb.) *McClintock* 247 ! Beli, Muri Dist. (bud Jan.) Latilo & Daramola FHI 34426 ! [Br.]**Cam.:** Karamti, Gashaka Dist. *Latilo & Daramola* FHI 28944 ! Vogel Peak Massif *fide* Hepper. Also in Cameroun, Ubangi-Shari, Sudan, Kenya, Tanganyika, N. & S. Rhodesia, Nyasaland, Mozambique and Angola. (See Appendix, p. 391.)

Adina cordifolia Hook. f. from tropical Asia has been introduced into Olokemeji F.R., S. Nigeria. It has rather large ovate and cordate leaves.

46. NAUCLEA Linn. Sp. Pl., ed. 2, 1 : 243 (1762) ; Merrill in J. Wash. Acad. Sci. 5 : 530 (1915). *Sarcocephalus* Afzel. ex Sabine (1824)—F.T.A. 3 : 38; F.W.T.A., ed. 1, 2 : 99.

Stipules broadest at the base, rounded or subacute at apex, 3–7 mm. long ; corolla 8–10 mm. long :
  Calyx-lobes pyramidal, 0·5–1 mm. long, pubescent ; corolla glabrous inside ; syncarps 2–5 cm. diam., the surface pitted ; leaves broadly elliptic to rounded-ovate, abruptly and shortly acuminate at apex, shortly cuneate to rounded or subcordate at base, 10–21 cm. long, 7–12 cm. broad ; petioles 1–2 cm. long ; pubescent bracteoles present ; small tree or shrub with arching branchlets, in savanna .. 1. *latifolia*
  Calyx-lobes oblong-spathulate, 2–4 mm. long, pubescent all over ; corolla pubescent inside ; syncarps 2–3 cm. diam., the surface densely papillose with the hairy, long calyx-lobes ; leaves broadly elliptic, rounded to subacute at apex, rounded to truncate at base, 12–18 cm. long, 8–11 cm. broad ; petioles slender, 2–4·5 cm. long ; bracteoles absent ; tall tree in forest .. .. .. 2. *pobeguinii*
Stipules broadest above the base, elliptic or obovate, often keeled, rounded or obtuse at apex, up to 50 mm. long ; calyx-lobes not exceeding 1 mm. long :
  Leaves elliptic (to oblong-elliptic)[1], 8–12 (–40) cm. long, 4–10 (–18) cm. broad, with 5–8 (–11) nerves on each side, cuneate (to rounded) at base ; corolla about 6 mm. long, densely pilose inside ; syncarp about 3 cm. diam., pits on surface of mature fruit, 7–9 mm. diam. ; peduncle about 2 cm. long ; tall tree, in forest 3. *diderrichii*
  Leaves oblong, 20–44 cm. long, 10–20 cm. broad, with 8–11 nerves on each side, truncate at base ; corolla about 12 mm. long, glabrous inside ; syncarp sessile, about 7 cm. diam., pits on surface of mature fruit 3–4 mm. diam. ; small tree, in forest
4. *vandeguchtii*

1. **N. latifolia** *Sm.* in Rees Cyclop. 24, No. 5 (1813) ; Milne-Redhead in Kew Bull. 3 : 459 (1948) ; Petit in Bull. Jard. Bot. Brux. 28 : 8 (1957). *Sarcocephalus esculentus* Afzel. ex Sabine (1824)—F.T.A. 3 : 38 ; F.W.T.A. ed. 1, 2 : 100 ; Chev. in Rev. Bot. Appliq. 18 : 179 (incl. vars.); Aubrév. Fl. For. Soud.-Guin. 477, t. 105, 3–6 ; Berhaut Fl. Sén. 103. *S. russeggeri* Kotschy ex Schweinf. (1868)—F.T.A. 3 : 39 ; Chev. Bot. 306. *S. sambucinus* K. Schum. (1891). *Nauclea esculenta* (Afzel. ex Sabine) Merrill (1915). *Sarcocephalus sassandrae* A. Chev. Bot. 307, name only. Straggling shrub or small tree about 10 ft. high ; flowers white, fragrant ; fruits reddish ; in savanna woodland.
Sen.: *Heudelot* ! Ugazobil (July) *Berhaut* 167 ! **Gam.:** *Saunders* 22 ! *Ozanne* 23 ! **Mali:** Ouacoro *Chev.* 92 ! Couroula *Chev.* 734. Soubaraniédougou to Kountseni *Chev.* 894 ! **Port.G.:** Uno (May) *Esp. Santo* 2012 ! **Guin.:** Kouroussa (Apr.) *Pobéguin* 238 ! Dalaba *Chev.* 18140 ! **S.L.:** Bagroo R. (Apr.) *Mann* 801 ! Njala (May) *Deighton* 655 ! Batkanu (Apr.) *Thomas* 5 ! Bumbuna (fr. Oct.) *Thomas* 3719 ! Falaba (Apr.) *Aylmer* 51 ! **Lib.:** Monrovia (fl. July, fr. June) *Linder* 12 ! *Baldwin* 5874 ! Suacoco (Mar., May) *Blickenstaff* 27 ! 29 ! *Daniel* 174 ! Vonjama Dist. (fr. Oct.) *Baldwin* 9993 ! **Iv.C.:** Dabou *Chev.* 15286 ! Bériby *Chev.* 2000 ! Brafouédi (Dec.) *Leeuwenberg* 2309 ! Diamancrou, Nzi *Chev.* 20126 ! Kangoroma Mt. *Chev.* 22170 ! **Ghana:** Cape Coast (July) *T. Vogel* 93 ! Ejura (fl. May, fr. Aug.) *Chipp* 722 ! *Vigne* FH 1199 ! Aburi (May) *Johnson* 952 ! Kpedsu (fr. Jan.) *Howes* 1142 ! Tamale (Apr.) *Williams* 117 ! **Togo Rep.:** Sokode (Apr.) *Kersting* 83 ! **Dah.:** Atacora Mts. *Chev.* 23950 ! **N.Nig.:** Nupe *Barter* 1244 ! Kontagora (Nov.) *Dalz.* 219 ! Abuja to Badeggi (Feb.) *Meikle* 1210 ! Assob (Feb.) *McClintock* 245 ! **S.Nig.:** Lagos (Aug.) *Phillips* 28 ! Shaki Road (fr. May) *Denton* 14 ! Enugu (Mar.) *Ainslie* 104 ! Mamu River F.R. *Kennedy* 3085 ! [Br.]**Cam.:** (cult.) *Maitland* 97 ! Also in Cameroun, the Congos, Cabinda and Uganda. (See Appendix, p. 411.)
2. **N. pobeguinii** (*Pobéguin ex Pellegr.*) *Petit* in Bull. Jard. Bot. Brux. 32 : 191 (1962), q.v. for discussion on valid combinations ; Aubrév. Fl. For. C. Iv., ed. 2, 3 ; 265, t. 346. *Sarcocephalus pobeguini* Pobéguin ex Pellegr. in Bull. Soc. Bot. Fr. 79 : 222 (1932) ; Berhaut Fl. Sén. 103. *Mitragyna stipulosa* of F.W.T.A., ed. 1, 2 : 98, partly. *Sarcocephalus diderrichii* of F.W.T.A., ed. 1, 2 : 607, partly. A tree 25–100 ft. high ; flowers white ; in marshes and flooded forest.

[1] Sapling trees of *N. diderrichii*, e.g. *Edwardson* 224, have much larger and more oblong-shaped leaves than the mature trees. These can be confused with adult leaves of *N. vandeguchtii*. The characters given in brackets in the key refer to such saplings.

Sen.: Bantancountou *Trochain* 3543! **Port.G.:** Gabú (June) *Esp. Santo* 2513! **Guin.:** Kouroussa *Pobéguin* 433! **S.L.:** Njala (June) *Deighton* 3522! **Iv.C.:** Abidjan *Aubrév.* 170! Dabou *Aubrév.* 892! **Ghana:** Kotokrom, Ofin R. (June) *Vigne* FH 1200! Pamu-Berekum F.R. (fr. Sept.) *Vigne* FH 2520! **N.Nig.:** Kurmin Damisa, Zaria Prov. (fr. Nov.) *Keay* FHI 21545! **S.Nig.:** Shasha F.R., Oyo (Apr., June) *Jones & Onochie* FHI 17311! 17594! Akilla (June) *Lamb* 222/37! *Kennedy* 2017! Awna R., Ibadan (fr. Nov.) *Keay & Jones* FHI 14208! 14226! Boje, Afi River F.R. (May) *Jones & Onochie* FHI 18933! Also in Cameroun and N. Rhodesia.

3. **N. diderrichii** (*De Wild. & Th. Dur.*) *Merrill* l.c. 535 (1915); Petit in Bull. Jard. Bot. Brux. 28 : 10. *Sarcocephalus diderrichii* De Wild. & Th. Dur. in Rev. Cult. Colon. 9 : 7 (1901); F.W.T.A. ed. 1, 2 : 608, partly (excl. syn. *S. pobeguini*); Pellegr. l.c. 223 ; Aubrév. l.c. 232, t. 330. *S. trillesii* Pierre—A. Chev. in Vég. Util. 9 : 229, 230, t. 24 (1917) ; Rev. Bot. Appliq. 18 : 185 (1938). Aubrév. Fl. For. Soud.-Guin. 475. *S. badi* Aubrév. l.c. 475, name only. Tree 30–130 ft. high with clear bole and low buttresses.
S.L.: Njala (Aug.) *Deighton* 3025! Kambui F.R. (fr. Mar.) *Jordan* 2003! *Edwardson* 224! **Lib.:** Gletown, Tchien Dist. (fr. July) *Baldwin* 6937! Dukwia R. (Feb.) *Cooper* 279! **Iv.C.:** Azaguié (fr. Sept.) *Chev.* 22302! Alépé *Chev.* 17235! Dakpadou F.R. *Aubrév.* 871! 872! Abidjan *Aubrév.* 1369! Guiglo *Aubrév.* 2047! **Ghana:** Kankan (June) *Thompson* 89! Dunkwa *Foggie* 200! Abofaw (June) *Vigne* FH 1211! **S.Nig.:** Ondo *Thornewill* 226! Omo River F.R., Ijebu-Ode (fr. Nov.) *Tamajong* FHI 20994! Onysanya to Owesemele, Shasha F.R. (fr. Mar.) *Jones & Onochie* FHI 16677! Udi Plateau *Kennedy* 3094! Degema to Abua (fr. Sept.) *King Church* 41! [Br.]**Cam.:** Likomba, Victoria (fr. Nov.) *Mildbr.* 10605! Mamfe *Eyong* 3! Badsu Akagbe, Mamfe *Johnstone* 152/31! Also in Cameroun, Ubangi-Shari, Gabon, Congo, Uganda and Mozambique.

4. **N. vanderguchtii** (*De Wild.*) *Petit* in Bull. Jard. Bot. Brux. 28 : 12 (1958). *Sarcocephalus vanderguchti* De Wild. (1923). *S. nervosus* Hutch. & Dalz. F.W.T.A., ed. 1, 2 : 100 (1931) ; Chev. in Rev. Bot. Appliq. 18 : 189, 190 (1938), incl. var. *cordifolia* A. Chev. A tree 50 ft. high with ant-holes in the branchlets ; flowers white ; beside rivers and streams in forest.
S.Nig.: Akilla (Jan.) *Kennedy* 2363! 2579! Oluwa F.R., Ondo (Apr.) *Symington* 3389! Shasha F.R. (Apr.) *Jones & Onochie* FHI 17282! Oban *Talbot* 1604! Also in Cameroun, Congo and Cabinda.

47. MUSSAENDA Linn.—F.T.A. 3 : 65 ; Wernham in J. Bot. 51 : 233 (1913) ; F. Hallé in Adansonia 1 : 266–298 (1961).

Upper leaf-surface glabrous or at most ciliate on midrib (see also No. 14) ; one enlarged white calyx-lobe present on some flowers ; corolla-tube about 2 cm. long :
 Corolla and calyx pilose outside ; calyx-lobes linear, about 1 cm. long ; leaves broadly elliptic, shortly acuminate at apex, rounded at base, setose on nerves beneath, 13–16 cm. long, 8–9 cm. broad, with about 8 nerves on each side ; petiole 6 mm. long, setose .. .. *1. conopharyngiifolia*
Corolla and calyx minutely pubescent or glabrous :
 Corolla-tube $1\frac{1}{2}$–$2\frac{1}{2}$ times as long as corolla-lobes ; calyx glabrous (rarely pubescent), lobes subulate, 2–10 mm. long, erect or slightly recurved, enlarged lobe glabrous ; inflorescence usually glabrous ; leaves elliptic to ovate, acuminate, cuneate to rounded at base, 7–16 cm. long, 4–8 cm. broad, glabrous or ciliate on the nerves beneath .. .. *2. arcuata*
 Corolla-tube 5–7 times as long as corolla-lobes ; calyx appressed-pubescent, lobes triangular-subulate, 1–2 (–3) mm. long, slightly incurved, enlarged lobe shortly pubescent on the nerves ; inflorescence puberulous ; leaves elliptic, caudate-acuminate, rounded to cuneate at base, 7–12 cm. long, 3·5–6 cm. broad, slightly ciliate on the nerves beneath .. .. *3. polita*
Upper leaf-surface pubescent (at least on nerves) or scabrid ; enlarged red, white or yellow calyx-lobe present or absent :
 Leaves covered with white appressed hairs on both surfaces, specially above when young, ovate to elliptic, slightly acuminate, cuneate at base, 7–9 cm. long, 4–5 cm. broad, with about 10 pairs of lateral nerves ; petiole about 1·5 cm. long ; corolla densely pubescent, 1·5 cm. diam., tube 2 cm. long ; calyx-lobes 1 cm. long, oblanceolate, densely pubescent outside, glabrous inside, enlarged lobe absent *4. nivea*
Leaves variously more or less pubescent beneath, but not white-appressed :
 Calyx-lobes (both the normal and enlarged) red, normal lobes lanceolate, about 1 cm. long, pubescent and ciliate ; corolla 1·5 cm. diam., tube 2 cm. long, pilose ; leaves broadly ovate to elliptic, truncate to cuneate at base, shortly acuminate, 7–15 cm. long, 5–11 cm. broad, densely to slightly tomentose beneath .. *5. erythrophylla*
 Calyx-lobes not red :
  Lobes of calyx obtuse, mucronate, obovate, 3–4 mm. long, all softly pubescent ; corolla-tube 2 cm. long, densely reflexed appressed-pubescent ; leaves ovate, ovate-elliptic or oblanceolate, rounded or subcordate at base, 10–14 cm. long, 4·5–7 cm. broad, with about 13 closely parallel nerves, shortly scabrid-hispid on both sides ; fruit about 2·5 cm. long and 1·5 cm. broad, densely pubescent .. .. *6. afzelii*
  Lobes of calyx acute or acuminate, not mucronate :
   Corolla-tube 4–5 (–6) mm. diam. in middle, about 2·5 cm. long, lobes 6 (–7) ; stamens 6 (–7), style-lobes (2–) 3 (–4) ; calyx-lobes 7 mm. broad at base :
    Hairs on outside of corolla-tube reflexed, lobes ovate, apiculate, 1·5 cm. long ; flowers terminal, 1-several ; calyx-lobes triangular-lanceolate, about 1 cm. long, about 5 mm. broad at base ; leaves ovate to elliptic, 10–15 cm. long, 4–7 cm. broad, pilose on nerves beneath .. .. *7. grandiflora*
    Hairs on outside of corolla-tube ascending, lobes suborbicular, apiculate, 5 mm. long ; leaves obovate, 8–11 cm. long, 3–5·5 cm. broad, pilose on nerves beneath
                     8. tristigmatica

Corolla-tube 1·5–3 mm. diam. in middle ; calyx-lobes up to 1·5 mm. broad :
  Corolla-lobes 1–4 cm. long, 1–2·5 cm. broad, broadly obovate to elliptic, densely
  pubescent outside, tube 2–3 cm. long ; calyx-lobes linear to linear-oblanceolate,
  5–10 mm. long, not enlarged ; leaves elliptic to obovate-elliptic, subacute to
  rounded at base, 7–11 cm. long, 3–6 cm. broad, shortly pubescent beneath ; fruits
  oblong-ellipsoid, glabrescent, with persistent calyx-lobes .. .. 9. *elegans*
Corolla-lobes 2–4 mm. long :
  Leaves scabrid above with many minute acute papillae and short curved hairs :
    Lower surface of leaves densely to moderately covered with long, soft, brown
    hairs, oblong-elliptic, shortly and abruptly acuminate ; calyx-lobes triangular
    to triangular-subulate, densely appressed-pubescent, enlarged lobe 1·5–
    3 cm. diam. on stalk about 1 cm. long ; corolla densely appressed-pubescent
    outside, lobes not tailed .. .. 10. *linderi*
  Lower surface of leaves glabrous and papillose (for other characters see below)
            12. *chippii*
  Leaves hispid or subglabrous above :
    Calyx-lobes subulate (to linear-triangular) 3–6 mm. long ; corolla-lobes distinctly
    tailed, tube pubescent with rather long more or less appressed brown hairs ;
    leaves elliptic, acutely acuminate, 9–12 cm. long, 5–8 cm. broad, setulose on
    the 8–10 lateral nerves beneath .. .. 11. *tenuiflora*
    Calyx-lobes shortly triangular, 1–2 mm. long :
      Leaves beneath (at least when dry) papillose ; corolla-tube minutely and
      sparsely pubescent, hairs often in longitudinal lines, corolla-lobes acute but
      not tailed ; branchlets more or less appressed-pubescent · 12. *chippii*
      Leaves beneath not papillose ; corolla-tube pubescent :
        Corolla-tube densely pubescent with erect spreading hairs, 2–3 cm. long,
        lobes slightly tailed ; branchlets hispid with recurved hairs ; leaves with
        about 10 nerves on each side .. .. .. 13. *landolphioides*
        Corolla-tube appressed-pubescent :
          Leaves with about 6 nerves widely spaced on each side ; corolla-tube
          3–4 cm. long, lobes slightly tailed ; branchlets shortly pubescent,
          glabrescent .. .. 14. *isertiana*
          Leaves with 11–14 nerves, closely parallel, on each side ; corolla-tube
          2–2·5 cm. long ; branchlets pubescent .. .. .. 15. *afzelioides*

1. **M. conopharyngiifolia** *Stapf* in J. Linn. Soc. 37 : 104 (1905) ; Wernham in J. Bot. 51 : 275 (1913) ; Chev.
Bot. 313, partly. A straggling shrub with rather large leaves and hirsute corymbs.
Lib.: Sinoe Basin *Whyte*! **Iv.C.:** *fide* Chev. *l.c.*

2. **M. arcuata** *Lam. ex Poir.* in Lam. Encycl. Méth. Bot. 4 : 392 (1797) ; F.T.A. 3 : 68 ; Wernham l.c. 274.
Aubrév. Fl. For. Soud.-Guin. 478, t. 108, 5–7 ; F. W. Andr. Fl. Pl. Sud. 3 : 446 (incl. var. *pubescens*
Wernham). *M. laurifolia* A. Chev. Bot. 314. Erect or scrambling shrub ; flowers yellow with yellow-
hairy centre which turns brown later, fruits orange, enlarged calyx white ; in forest and beside streams.
Guin.: Dalaba to Diaguissa *Chev.* 18582 ; 18847. Macenta (Oct.) *Baldwin* 9794! **S.L.:** Jigaya (Sept.)
Thomas 2743! Warantamba (Oct.) *Small* 351! Firana (fr. Mar.) *Sc. Elliot* 5353! Loma Mts. (Sept.)
Jaeger 1943! **Lib.:** Kolahun (fl. & fr. Nov.) *Baldwin* 10132! **Iv.C.:** Bouaké *Chev.* 22134. **Ghana:** Gambaga
to Nakpanduri (fr. Mar.) *Hepper & Morton* A3142! **N.Nig.:** Kontagora Prov. (Apr.) *Dalz.* 403! Nupe
Barter 1239! Vom (Feb.) *McClintock* 219! Katsina Alla *Dalz.* 645! **S.Nig.:** Agbadi, Sapoba (fl. & fr.
Nov.) *Meikle* 548! Okomu F.R., Benin (fl. & fr. Mar.) *Akpabla* 1117! Oban (Apr.) *Ejiofor* FHI 21873!
Talbot 203! Boshi, Ogoja Dist. *Rosevear* 62/29! [**Br.**]**Cam.:** Victoria to Kumba (fr. Apr.) *Ejiofor* FHI
29367! Johann-Albrechtshöhe (= Kumba) *Staudt* 458! Mambila Plateau (Jan.) *Latilo & Daramola*
FHI 34386! *Hepper* 1647! **F.Po:** *Mann*! Widespread in tropical Africa and Madagascar.

3. **M. polita** *Hiern* in F.T.A. 3 : 67 (1877) ; Wernham l.c. 239. Scrambling shrub ; flowers yellow, about
¼ in. across, enlarged calyx-lobe white ; in forest.
S.Nig.: Oban *Talbot*! Calabar *Williams* 26! 36! Cross R. (Dec.) *Holland* 228! [**Br.**]**Cam.:** Kumba
(Oct.) *Dundas* FHI 15326! Also in Cameroun and Gabon.

4. **M. nivea** *A. Chev. ex Hutch. & Dalz.* F.W.T.A., ed. 1, 2 : 101 (1931) ; Chev. Bot. 315, name only. Straggling
shrub with silvery hairy leaves ; flowers red.
Iv.C.: Guidéko, Mid. Sassandra (May) *Chev.* 16467! **Ghana:** Dunkwa (fl. & fr. July) *King-Church* 903!
Osino *Arnold* 371! Manso (Apr.) *Scholes* 266! Assin *West-Skinn* 16!

5. **M. erythrophylla** *Schum. & Thonn.* Beskr. Guin. Pl. 116 (1827) ; F.T.A. 3 : 69 ; Wernham l.c. 275 ; Chev.
Bot. 314. Climbing shrub to 40ft. high ; flowers cream, yellow or orange, enlarged sepal pink or red ;
in forest and beside streams. The " Ashanti Blood ".
Guin.: Labé *Chev.* 12279. Kollangui *Chev.* 12545. Kouria to Irébéléya *Chev.* 18241. Dalaba-Diaguissa
Plateau *Chev.* 18367. **S.L.:** Kofiu Mt. (Jan.) *Sc. Elliot* 4608! Kaballa (Sept.) *Thomas* 2227! Jaiama (Apr.)
Deighton 3160! Gbinti (Jan.) *Deighton* 5898! **Lib.:** Ganta (Apr.) *Harley* 2047! Bili (Aug.) *Harley*
1265! Sakimpa (Mar.) *Harley* 1797! Sodu (Jan.) *Bequaert* 44! Dubo (Sept.) *Baldwin* 9479! **Iv.C.:**
Nzi *Croux* in *Hb. Chev.* 20144. Danané to Kouan Mt. *Chev.* 21263. **Ghana:** Winnebah (fl. & fr. Sept.)
Fishlock 53! Akwapim *Reade*! Prasu *Cummins* 3 *p*! Kumasi (Apr.) *Vigne* FH 1681! Kpeve (June)
Howes 1008! **N.Nig.:** Jemaa to Kafanchan (Apr.) *Mutch* FHI 21861! Naraguta (June) *Lely* 288! Vom
Dent Young 110! Jos (fl. & fr. Oct.) *Thornewill* 121! **S.Nig.:** Boshi, Ogoja, 3,000 ft. *Rosevear* 60/29!
[**Br.**]**Cam.:** Cam. Mt., 1,600–4,500 ft. (Feb.-Apr.) *Kalbreyer* 93! *Deistel* 642! *Mann* 1278! Bali (May)
Ujor FHI 30353! Widdekum, 3,000 ft. (Jan.) *Keay* FHI 28544! Widespread in tropical Africa. (See
Appendix, p. 405.)

6. **M. afzelii** *G. Don* Gen. Syst. 3 : 490 (1834) ; F.T.A. 3 : 66 ; Wernham l.c. 237 ; Chev. Bot. 313. Small
tree or shrub sometimes climbing, up to 20 ft. high ; flowers yellow, foliaceous sepal white ; in moist
forest.
Guin.: Konkouré to Timbo *Chev.* 12512. Kindia *Chev.* 13030. SW. of Dalaba (fr. Mar.) *Langdale-Brown*
2629! **S.L.:** Mahela (Dec.) *Sc. Elliot* 4166! Kumrabai (fr. Dec.) *Thomas* 7043! Port Loko (Dec.) *Thomas*
6627! Ronietta (Nov.) *Thomas* 5422! **Lib.:** Monrovia (Dec.) *Okeke* 75! *Baldwin* 10987! Suen to Brewer's
Landing (Nov.) *Linder* 1405! Zorzor (Jan.) *Harley* 1524! **Iv.C.:** San Pedro (Feb.) *Aké Assi* 2861! Danané
(Nov.) *Aké Assi* 5499! Also in (?) Angola. (See Appendix, p. 404.)

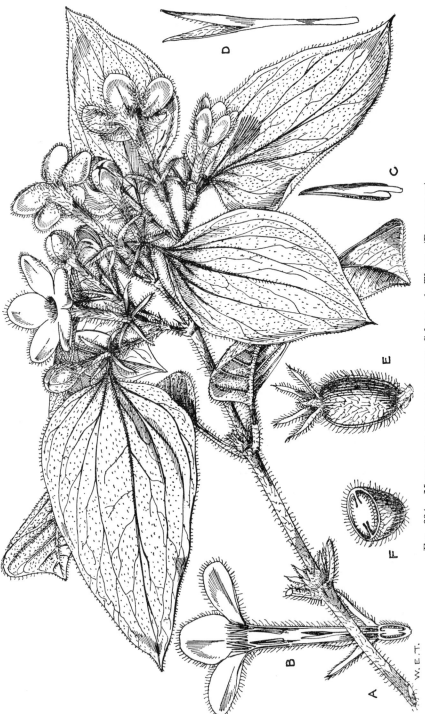

Fig. 234.—Mussaenda erythrophylla *Schum. & Thonn.* (Rubiaceae).

A, flowering shoot. B, longitudinal section of flower. C, stamen. D, style. E, young fruit. F, cross-section of ovary.

W. E. T.

7. **M. grandiflora** *Benth.* in Fl. Nigrit. 392 (1849) ; F.T.A. 3 : 70 ; Wernham l.c. 277. *M. macrosepala* Stapf (1905)—Chev. Bot. 315. *M. uniflora* Hutch. & Dalz. F.W.T.A., ed. 1, 2 : 101 (1931), not of Wall. ex G. Don (1834). *M. collenettei* Hutch. F.W.T.A., ed. 1, 2 : 608 (1936). Scrambling shrub with coarsely hairy stems and leaves ; flowers yellow, enlarged calyx-lobe pure white, sometimes absent ; in forest.
    Guin.: Zoubouroumai (May) *Collenette* 28 ! Nzo (Apr.) *Schnell* 1102 ! **S.L.:** Njala (July, Aug.) *Deighton* 722 ! 1833 ! Kambui Hills (fr. Mar.) *Small* 512 ! Bumpe to Mano (Aug.) *Deighton* 6112 ! Gola F.R. (May) *Small* 682 ! **Lib.:** Monrovia (July) *Linder* 14 ! Kakatown *Whyte* ! Roberts Field (May) *Baldwin* 5801 ! Ganta (May) *Harley* 1181 ! White Plains (Apr.) *Barker* 1280 ! **Iv.C.:** Dabou *Chev.* 16231. Agnieby Valley *Chev.* 17049. Bingerville *Chev.* 17313. Abidjan *Chev.* 17320. Cavally Valley *Chev.* 19625.
8. **M. tristigmatica** *Cummins* in Kew Bull. 1898 : 74 ; Wernham l.c. 276 ; Chev. Bot. 315. Straggling hairy shrub ; flowers yellow, enlarged calyx-lobe white.
    Lib.: Kasono (fl. & fr. Sept.) *Harley* 1914 ! **Iv.C.:** Dabou *Chev.* 15173 ! Adiopodoumé *Leeuwenberg* 1777 ! Brafouedi (fr. Dec.) *de Wit* 7220 ! **Ghana:** Asin-Nyankumasi *Cummins* 41 ! 113 ! Axim (fl. Dec., fr. Mar.) *Johnson* 877 ! *Chipp* 169 ! Simpa (May) *Vigne* FH 1960 ! Tarkwa (fl. & fr. Jan.) *Vigne* FH 4123 ! Also in (?) Gabon. (See Appendix, p. 405.)
9. **M. elegans** *Schum. & Thonn.* Beskr. Guin. Pl. 117 (1827) ; F.T.A. 3 : 69 ; Wernham l.c. 276 ; Chev. Bot. 313. *Gardenia coccinia* G. Don (1824), not *Mussaenda coccinia* Poir. (1797). Shrub or scrambler in thickets 10 to 30ft. high, with slender branches ; flowers scarlet or orange with yellow throat-hairs.
    Mali: Kangala *Chev.* 830 ! 858. **Port.G.:** Catio (June) *Esp. Santo* 2091 ! **Guin.:** *Heudelot* 806 ! Kouroussa *Chev.* 15707 ; 15726. Conakry *Debeaux* 423 ! Macenta, 2,000 ft. (May) *Collenette* 14 ! **S.L.:** Sherbro *Sc. Elliot* 5795 ! No. 2 River F.R. (Apr.) *Hepper* 2514 ! Moyamba (June) *Lane-Poole* 47 ! Njala (May) *Deighton* 688 ! **Lib.:** Ganta (May) *Harley* 1148 ! **Iv.C.:** Bingerville *Chev.* 16055. Abidjan *Chev.* 17316. Sassandra Port *Chev.* 17996. Nzi *Chev.* 20146. **Ghana:** Accra to Aburi (Apr.) *Baldwin* 11988 ! Tarkwa (May) *Chipp* 231 ! Kumasi (Mar., Apr.) *Vigne* FH 1680 ! 3738 ! Dawa (May) *Howes* 909 ! **Dah.:** Dassa-Zoumé *Chev.* 23631. Savalou *Chev.* 23799. **N.Nig.:** Aguji, Ilorin *Thornton* ! Nupe *Barter* ! Mada Hills (June) *Hepburn* 72 ! Ruka, Zaria (Feb.) *Dalz.* 397 ! Jos Plateau (Aug.) *Lely* P640 ! **S.Nig.:** Lagos *Phillips* 51 ! Okomu F.R. (Mar.) *Chev.* ! Ejiofor FHI 21875 ! [Br.]Cam.: Victoria (fl. & fr. Nov.) *Mildbr.* 10598 ! Kumba (Apr.) *Hutch. & Metcalfe* 152 ! Ekwe, Bamenda, 3,000–4,000 ft. (June) *Maitland* 1573 ! Also in Cameroun, Congo, Sudan and Uganda. (See Appendix, p. 404.)
    [This is a species variable in several characters: several varieties of doubtful taxonomic value have been described.—F.N.H.]
10. **M. linderi** *Hutch. & Dalz.* F.W.T.A., ed. 1, 2 : 103 (1931). *M. punctulata* Hutch. & Dalz. F.W.T.A., ed. 1, 2 : 103 (1931). Brown-pubescent shrub 3–5 ft. high ; flowers yellow, enlarged calyx-lobe white.
    Lib.: Miamu (Aug.) *Linder* 368 ! Gbanga (Sept.) *Linder* 536 ! Tubman Bridge (Sept.) *Barker* 1424 ! R. Cess (fr. Mar.) *Baldwin* 11262 ! **Iv.C.:** Guiglo to Taï *Mangenot, Miège & Aké Assi* IA 5350 ! **Ghana:** Prestea (Sept.) *Vigne* FH 3103 !
11. **M. tenuiflora** *Benth.* in Fl. Nigrit. 392 (1849) ; F.T.A. 3 : 69 ; Wernham l.c. 274. *M. entomophila* Wernham (1916). Shrub ; flowers creamy yellow, enlarged calyx-lobe white.
    Guin.: Macenta, 2,000–2,500 ft. (Oct.) *Baldwin* 9838 ! **S.L.:** Jau, Tunkia Chiefdom (Oct.) *Deighton* 5196 ! **S.Nig.:** Degema *Talbot* ! Calabar (Aug.) *Holland* 73 ! Oban *Talbot* 1050 ! Nkfuru to Abaragba, Ikom (Apr.) *Jones* 1547 ! [Br.]Cam.: Victoria (Jan.) *Kalbreyer* 71 ! Victoria to Kumba (Apr., June) *Hutch. & Metcalfe* 141 ! *Dundas* FHI 8351a ! S. Bakundu, Kumba (Jan.) *Binuyo & Daramola* FHI 35174 ! **F.Po:** (fl. & fr. Nov.) *T. Vogel* 67 ! *Barter* ! *Mann* 60 ! Plaza de Ureka (Nov.) *Guinea* 2505 ! Moka, 5,000 ft. (Sept.) *Boughey* 89 ! Also in Cameroun, Congo, Angola and S. Tomé and Principe.
    [*Baldwin* 9838 and *Deighton* 5196 cited above have broader calyx-lobes than usual but appear to be referable to this sp.—F.N.H.]
12. **M. chippii** *Wernham* l.c. 237 (1913). *M. buntingii* Wernham l.c. (1913). *M. conopharyngiifolia* of Chev. Bot. 313, not of Stapf. A slender straggling shrub ; flowers yellow or orange, enlarged calyx-lobe white ; on forest margins and roadsides in forest.
    Guin.: Macenta, 2,000–2,500 ft. (Oct.) *Baldwin* 9838a ! **S.L.:** Malema (Nov.) *Deighton* 328 ! Kenema (fl. Oct. & fr. Nov.) *Deighton* 5278 ! Joru (Oct.) *Deighton* 5195 ! **Lib.:** Monrovia (June) *Baldwin* 5877 ! Begwai, Gola *Bunting* 144 ! Gbanga (Sept.) *Linder* 535 ! Ganta (Sept.) *Harley* ! Kakatown *Whyte* ! **Iv.C.:** Yapo (fl. & fr. Oct.) *Chev.* B22366 ! Prolo (Aug.) *Chev.* 19859 ! Abidjan (Dec.) *Chev.* 16503 ! **Ghana:** Amokokrom (Aug.) *Chipp* 323 ! Dixcove to Busua (fr. Mar.) *Morton* A264 ! Ankobra Junction (Jan.) *Kitson* 1008 ! Nsuaem (fl. & fr. Sept.) *Fishlock* 51 ! Tarkwa (fl. & fr. Jan.) *Vigne* FH 4122 !
13. **M. landolphioides** *Wernham* l.c. 238 (1913). *M. scabrida* Wernham l.c. 236 (1913). *M. brachyantha* Wernham l.c. 238 (1913). A woody climber with harsh reflexed hairs on the younger stems ; flowers yellow, enlarged calyx-lobe white or pale yellow ; in swamp forest.
    Lib.: Begwai, Gola *Bunting* 147 ! **Iv.C.:** N'Zida *Aké Assi* IA 5494. **S.Nig.:** Lagos *Millen* 95 ! 100 ! Sapoba *Kennedy* 2647 ! *Meikle* 611 ! Nikrowa, Okomu F.R. (Sept.) *Iriah* FHI 23090 ! Ibadan *Foster* 157 !
14. **M. isertiana** *DC.* Prod. 4 : 371 (1830) ; F.T.A. 3 : 67 ; Wernham l.c. 238 ; Chev. Bot. 314. *M. macrophylla* Schum. & Vahl (1827), not of Wall. (1824). Climbing shrub up to 30 ft. high ; flowers yellow, strongly scented, enlarged calyx-lobe white ; in forest.
    S.L.: *Don* ! **Iv.C.:** Banco (June) *Aké Assi* 5409 ! **Ghana:** Essiama (Oct.) *Fishlock* 5 ! Axim *Fishlock* 40 ! **Dah.:** Ouida (= Whydah) *Isert*. Porto Novo *Chev.* 22728 ! **N.Nig.:** Bassa (Aug.) *Elliott* 94 ! Anara F.R., Zaria (May) *Keay* FHI 22889 ! **S.Nig.:** Lagos *Barter* 3236 ! Ogun River F.R. (Mar.) *Hepper* 2257 ! Jamieson R., Sapoba *Kennedy* 2038 ! Shasha F.R. (Mar.) *Richards* 3202 ! Abeokuta *Irving* 95 ! Eket *Talbot* 3000 ! 3124 ! **F.Po:** *T. Vogel* 226 ! Also in Cameroun.
15. **M. afzelioides** *Wernham* in Cat. Talb. 40 (1913) ; J. Linn. Soc. 51 : 238. A climbing shrub.
    S.Nig.: Oban *Talbot* 212 ! 275 !

Imperfectly known species.

M. lancifolia *K. Krause* in Engl. Bot. Jahrb. 57 : 58 (1920). The type specimen was destroyed at Berlin and it has not been possible to key out the species satisfactorily from the description alone.
    F.Po: Bokoko (Oct.) *Mildbr.* 6813.

48. PAURIDIANTHA Hook. f.—F.T.A. 3 : 71 ; Bremkamp in Engl. Bot. Jahrb. 71 : 200–227 (1940). *Urophyllum* of F.T.A. 3 : 72 ; of F.W.T.A., ed. 1, 2 : 103, not of Wall.

Inflorescences (6–) 10–20 cm. long, sometimes as long as or longer than the leaves, openly cymose ; stipules large and foliaceous :
  Calyx glabrous, shortly toothed ; inflorescences always axillary, the branches puberulous with very short crispate hairs ; leaves broad at base, oblong-elliptic, acuminate, up to 20 cm. long and 7 cm. broad, glabrous beneath, except in the axils of the 15–20 pairs of lateral nerves, reticulate between them ; stipules broadly ovate, abruptly mucronate .. .. 1. *floribunda*

Calyx shortly pubescent outside, undulate ; inflorescences terminal with some **axillary,** the branches pubescent with loose hairs ; leaves similar to above but acute at base ; stipules broadly ovate, gradually acute ; fruits globose, wrinkled when dry about 5 mm. diam., capped by cupular calyx .. .. 2. *viridiflora*
Inflorescences axillary clusters at most 4 cm. long (in No. 11), often dense :
Branchlets more or less glabrous :
　Calyx undulate or with triangular teeth :
　　Stipules spathulate, at least 1·5 mm., and up to 6 mm., broad, early caducous ; leaves reddish beneath, rather coriaceous, 10–28 cm. long, 4–12 cm. broad, with distinct intra-marginal nerve ; inflorescences 2–3 cm. long ; branchlets glabrous :
　　　Intra-marginal nerve of leaves beneath clearly looped, lateral nerves about 7, apex acute ; calyx about 2 mm. diam., truncate ; fruits 5–6 mm. diam. ; stipules ovate, 4–6 mm. broad .. .. 3. *rubens*
　　　Intra-marginal nerve of leaves beneath almost parallel with margin, lateral nerves about 12, apex acuminate ; calyx about 4 mm. diam., slightly lobed ; stipules linear-lanceolate, 1·5–3 mm. broad .. .. 4. *ziamaeana*
　　Stipules filiform ; branchlets sparsely and minutely pubescent ; inflorescences few-flowered, 1 cm. long ; leaves light green, long-acuminate, mucronate, shortly cuneate at base, oblong-elliptic, 4–14 cm. long, 1·5–4·5 cm. broad, with 5–6 pairs of lateral nerves .. .. 9. *sylvicola*
　Calyx-teeth subulate, 1 mm. long ; leaves long-acuminate, cuneate at base, lanceolate to oblong-elliptic, 7–11 cm. long, 2–3(–4) cm. broad, with about 8 pairs of lateral nerves ; stipules linear-triangular .. .. 5. *paucinervis*
Branchlets conspicuously pubescent :
　Stipules filiform or subulate :
　　Branchlets hirsute with spreading hairs ; leaves oblong-elliptic, broadly acuminate, about 12 cm. long and 5 cm. broad, pilose on nerves beneath, tertiary nerves inconspicuous .. .. 6. *insculpta*
　　Branchlets shortly and densely pubescent (see also No. 9 above) ; calyx-teeth shortly subulate or triangular :
　　　Tertiary nerves numerous (obvious at least when dry) and at right angles to the midrib, leaves elliptic to oblong-elliptic, broadly acuminate, 3–6 cm. long, 1·5–3 cm. broad ; calyx-lobes very shortly subulate .. 7. *canthiiflora*
　　　Tertiary nerves inconspicuous and anastomosing, leaves oblong-elliptic, acuminate, 4–8 cm. long, (1·5–) 2–2·5 cm. broad ; calyx-lobes triangular 8. *afzelii*
　Stipules ovate to lanceolate ; branchlets hirsute :
　　Inflorescence congested axillary cluster ; calyx-lobes lanceolate about 3 mm. long, densely hairy ; stipules lanceolate 1·5–2 cm. long, 5–7 mm. broad ; leaves narrowly oblong-elliptic, acuminate, 15–18 cm. long, 5–6 cm. broad, pilose beneath, about 18 pairs of lateral nerves .. .. 10. *stipulosa*
　　Inflorescence shortly pedunculate ; calyx truncate, glabrous ; stipules ovate-lanceolate, 1·3–2·5 cm. long, 7–11 mm. broad ; leaves oblanceolate, acute or abruptly acuminate, 5–8 cm. broad, pubescent beneath, about 20 pairs of lateral nerves .. .. 11. *hirtella*

1. **P. floribunda** (*K. Schum. & K. Krause*) *Bremek.* in Engl. Bot. Jahrb. 71 : 216 (1940). *Urophyllum floribundum* K. Schum. & K. Krause (1907)—F.W.T.A., ed. 1, 2 : 104. Erect shrub 12–25 ft. high ; flowers small and whitish ; in high forest.
　　S.Nig.: Sapoba *Kennedy* 354 ! 2308 ! Eket Dist. *Talbot* ! Ikom (July) *Catterall* 12 ! Boshi-Okwangwo F.R., Ogoja (May) *Latilo* FHI 30950 ! **[Br.]Cam.:** S. Bakundu F.R., Kumba (fr. Aug.) *Akuo* FHI 15170 ! Mungo F.R., Kumba (May) *Olorunfemi* FHI 30603 ! Also in Cameroun, Gabon and S. Tomé. (See Appendix, p. 413.)

2. **P. viridiflora** (*Schweinf. ex Hiern*) *Hepper* in Kew Bull. 13 : 405 (1959). *Urophyllum viridiflorum* Schweinf. ex Hiern in F.T.A. 3 : 74 (1877) ; F.W.T.A., ed. 1, 2 : 104. *U. eketense* Wernham (1914). *Pamplethantha viridiflora* (Schweinf. ex Hiern) Bremek. l.c. 217 (1940). Shrub or small tree 10–20 ft. high, with yellow-green leaves ; flowers greenish or grey ; at edges of forest.
　　S.Nig.: Eket Dist. *Talbot* 3321 ! Degema Dist. *Talbot* ! **[Br.]Cam.:** Buea and Mopanya, Cam. Mt., 2,400–3,200 ft. (Mar.-Apr.) *Maitland* 502 ! 1075 ! Bali (May) *Ujor* FHI 30378 ! Wum (June) *Daramola* FHI 41093 ! Also in Cameroun, Ubangi-Shari, Congo, Uganda and Tanganyika.

3. **P. rubens** (*Benth.*) *Bremek.* l.c. 215 (1940). *Urophyllum rubens* Benth. (1849)—F.T.A. 3 : 73 ; F.W.T.A., ed. 1, 2 : 104. A glabrous shrub with the branches and leaves beneath reddish ; at forest edge.
　　[Br.]Cam.: Bulifambo, Buea, 2,500 ft. (fr. Mar.) *Maitland* 576 ! Johann-Albrechtshöhe (= Kumba) *Staudt* 860a. **F.Po:** (Oct.) *T. Vogel* 48 ! *Arabin* ! Musola, 1,600 ft., *Mildbr.* 6933 ; 7006. Also in Cameroun, Congo and S. Tomé. (See Appendix, p. 414.)

4. **P. ziamaeana** (*Jac.-Fél.*) *Hepper* in Kew Bull. 13 : 405 (1959). *Urophyllum ziamaeanum* Jac.-Fél. in Bull. I.F.A.N. 16 : 990 (1954). *U. rubens* Benth. (1849), partly (*Afzelius*). Shrub up to 15 ft. high.
　　Guin.: Ziama, near Macenta (Mar.) *Jac.-Fél.* 1559 ! **S.L.:** *Afzelius* ! Kambui Hills (Feb.) *Edwardson* 9 ! **Lib.:** Truo, Sinoe Co. (fr. Mar.) *Baldwin* 11390 ! Kulo, Sinoe Co. (fr. Mar.) *Baldwin* 11414 !

5. **P. paucinervis** (*Hiern*) *Bremek.* l.c. 212 (1940). *Urophyllum paucinerve* Hiern in F.T.A. 3 : 74 (1877) ; F.W.T.A., ed. 1, 2 : 10. A shrub 10–20 ft. high with arching branches and slender green branchlets ; flowers white ; fruits red ; in understorey of upland forest.
　　[Br.]Cam.: Cam. Mt., 4,000–5,000 ft. (Feb.) *Maitland* 1308 ! 1320 ! Bali-Ngemba F.R. (May) *Ujor* FHI 30342 ! Nkambe (fl. & fr. Feb.) *Hepper* 1896 ! **F.Po:** *Mann* 577 ! Mioko, 4,500 ft. (Feb.) *Exell* 844 ! Las Cascadas (Jan.) *Guinea* 2172 ! Finca Puente (Jan.) *Guinea* 1791 !

6. **P. insculpta** (*Hutch. & Dalz.*) *Bremek.* l.c. 215 (1940). *Urophyllum insculptum* Hutch. & Dalz. F.W.T.A., ed. 1, 2 : 104 (1931). A shrub with shortly hirsute branches.
　　S.Nig.: Oban *Talbot* ! Orem, Oban F.R. (fr. Feb.) *Onochie* FHI 36257x !

7. **P. canthiiflora** *Hook. f.* in Benth. & Hook. f. Gen. Pl. 2 : 70 (1871) ; F.T.A. 3 : 71 (as *P. canthiifolia*) ;

Bremek. l.c. 210. *Urophyllum canthiiflorum* (Hook. f.) Hutch. & Dalz. F.W.T.A., ed. 1, 2 : 104 (1931).
U. talbotii Wernham in Cat. Talb. 41 (1913). A shrub 6–15 ft. high ; flowers few, yellowish.
S.Nig.: Oban *Talbot* 225 ! **[Br.]Cam.:** Victoria (Oct.) *Winkler* 548 ! Johann-Albrechtshöhe (= Kumba) *Staudt* 566 ! 804 ! Banga F.R., Kumba (fr. Mar.) *Binuyo & Daramola* FHI 35591 ! Mamfe (fr. Mar.) *Richards* 5222 ! **F.Po:** *Mann* 167 ! Also in Congo.

8. **P. afzelii** (*Hiern*) *Bremek.* l.c. 212 (1940). *Urophyllum afzelii* Hiern in F.T.A. 3 : 73 (1877) ; F.W.T.A., ed. 1, 2 : 104 ; Chev. Bot. 315 ; Aubrév. Fl. For. C. Iv., ed. 2, 3 : 298. *U. micranthum* of Chev. Bot. 315, not of Hiern. A shrub up to 15 ft. high (or tree, in Ghana, to 40 ft. high), with slightly pubescent green branchlets ; flowers greenish-white ; fruits red ; in high forest.
Port.G.: Bissau (fl. & fr. Feb., Mar.) *Esp. Santo* 1833 ! 1888 ! **Guin.:** Labé (Apr.) *Chev.* 12301 ! Fiffa (fr. Dec.) *Roberty* 16221 ! Erimakuna (Mar.) *Sc. Elliot* 5263 ! **S.L.:** Berria, Falaba (fr. Mar.) *Sc. Elliot* 5413 ! Njala (Mar.) *Deighton* 1581 ! Baoma (fr. Apr.) *Deighton* 3168 ! Pujehun (Apr.) *Lane-Poole* 59 ! **Lib.:** Ganta (fr. Jan.) *Baldwin* 14056 ! *Harley* 292 ! **Iv.C.:** Yapo *Serv. For.* 1717 ! Ouaguié *Chev.* 17175. Béyo (fl. & fr. Jan.) *Leeuwenberg* 2453 ! **Ghana:** Afram Plains (fl. & fr. May) *Kitson* 1135 ! Tano Nimiri F.R. (fr. July) *Agbley* 6257 ! **S.Nig.:** near Lagos *Rowland* 51 ! (See Appendix, p. 413.)

9. **P. sylvicola** (*Hutch. & Dalz.*) *Bremek.* l.c. 212 (1940). *Urophyllum sylvicola* Hutch. & Dalz. F.W.T.A., ed. 1, 2 : 104 (1931). *U. afzelii* of F.W.T.A., ed. 1, 2 : 104, partly (*Cooper* 48 ; 250), not of Hiern. Shrub or small tree ; flowers whitish ; in understorey of high forest.
Lib.: Dukwai R. (fr. Nov., Feb.) *Cooper* 48 ! 250 ! Medina, Bumbuna (Oct.) *Linder* 1307 ! Karmadhun, Kolahun Dist. (fl. & fr. Nov.) *Baldwin* 10204 ! Ba, Boporo Dist. (fr. Dec.) *Baldwin* 10716 ! **Ghana:** Esem Epan F.R. (Oct.) *Darko* 1015 ! Sefwi Bekwai (fr. Oct.) *Akpabla* 899 ! Benso, Tarkwa Dist. (fl. & fr. Aug.) *Andoh* 5745 ! 5753 ! **[Br.]Cam.:** Banga, S. Bakundu F.R. (Mar.) *Brenan* 9452 !
[Close to *P. afzelii* and perhaps a less pubescent form of it ; the leaves are usually broader and longer acuminate—F.N.H.]

10. **P. stipulosa** (*Hutch. & Dalz.*) *Hepper* in Kew Bull. 13 : 405 (1959). *Urophyllum stipulosum* Hutch. & Dalz. F.W.T.A., ed. 1, 2 : 104 (1931). *Pentaloncha* ? *stipulosa* (Hutch. & Dalz.) Bremek. l.c. 226 (1940). A shrub with white flowers.
Lib.: Dukwai R. (Feb., Nov.) *Cooper* 37 ! 205 ! Du R. (July) *Linder* 86 ! Duo (Mar.) *Baldwin* 11347 !

11. **P. hirtella** (*Benth.*) *Bremek.* l.c. 216 (1940). *Urophyllum hirtellum* Benth. (1849)—F.T.A. 3 : 73 ; F.W.T.A., ed. 1, 2 : 104 ; Aubrév. Fl. For. C., ed. 2, 3 : 298. Shrub up to 15 ft. high, usually hairy ; flowers greenish-white ; in secondary swamp forest and moist ground.
Port G.: Bissau (fl. Feb., fr. Mar.) *Esp. Santo* 1776 ! 1832 ! 1861 ! **S.L.:** Bagroo R. *Mann* 868 ! Gbanbama *Thomas* 8951 ! Zimi (fl. & fr. Apr.) *Deighton* 3636 ! Dodo (fr. Apr.) *Deighton* 3923 ! Kassa, 4,000 ft. (Mar.) *Sc. Elliot* 5058 ! **Lib.:** Dobli Isl. (fr. Apr.) *Bequaert* 175 ! Barclayville (fr. Mar.) *Baldwin* 11108 ! Mecca (Dec.) *Baldwin* 10822 ! Gbanga (Sept.) *Baldwin* 528 ! **Iv.C.:** Port Bouët *Aubrév.* 1637 ! Danipleu *Serv. For.* 1154. **Ghana:** Benso (fr. Nov.) *Andoh* FHI 5413 ! Ankobra R., Axim *Morton* A373 ! Nkroful (fl. & fr. Nov.) *Vigne* FH 1409 ! Essiama (fr. Oct.) *Unknown Collector* No. 7 ! **S.Nig.:** Lagos (July) *Dalz.* 1200 ! Omo (formerly part of Shasha) F.R. (Mar.) *Jones & Onochie* FHI 17001 ! Iva Valley, Enugu (Feb.) *Jones* FHI 567 ! Calabar (Dec.) *Mann* 2237 ! **[Br.]Cam.:** Mamfe (Mar.) *Onochie* FHI 34833 ! Man o' War Bay (Apr.) *Schlechter* 12380 ! Also in Cameroun. (See Appendix, p. 414.)

49. SABICEA Aubl. (1775)—F.T.A. 3 : 74 ; Wernham Monogr. Sabicea (1914) ; Hepper in Kew Bull. 13 : 289 (1958) ; F. Hallé in Adansonia 1 : 266–298 (1961).
Stipularia P. Beauv. (1807)—F.T.A. 3 : 79 ; F.W.T.A. 2 : 107.

*Leaves more or less thinly pilose or pubescent beneath, not completely covered with a felt of hairs ; (to p. 171):
Inflorescences cauliflorous, i.e. on older stems and separated from the leaves (see also No. 8) :
  Lateral nerves of leaves 8–10 (–15) pairs ; leaves up to 15 cm. long ; stipules ovate, about 1 cm. long ; corolla hairy outside :
    Inflorescences on aerial shoots, sessile, few-flowered ; calyx-lobes linear-filiform, 1·5 cm. long, whole calyx densely pilose ; corolla densely pilose ; bracts wanting or very small ; leaves broadly elliptic, rounded or subcordate at base, slightly acuminate, 10–14 cm. long, 4–7·5 cm. broad, thinly pilose above and densely spreading-pilose on midrib beneath   ..   ..   ..   1. *speciosa*
    Inflorescences on rooting prostrate shoots, shortly pedunculate, many-flowered ; calyx-lobes broadly oblong to lanceolate, 1·5 cm. long, whole calyx long-ciliate ; corolla villose ; bracts forming an open involucre ; leaves obovate, cuneate at base, long-caudate-acuminate, 12–15 cm. long, 4–7 cm. broad, thinly pilose above and on nerves beneath   ..   ..   2. *pilosa*
  Lateral nerves of leaves 20–25 pairs ; leaves 30 or more cm. long ; stipules 3–6 cm. long ; corolla glabrous ; calyx-lobes filiform, about 1 cm. long :
    Leaves shortly pubescent on nerves beneath, weakly pilose above, ciliate ; stipules ovate about 3 cm. long ; young branchlets densely hispid with long spreading hairs ..   ..   3. *xanthotricha*
    Leaves long-pilose on nerves beneath, pilose above ; stipules broadly lanceolate about 6 cm. long ; young branchlets densely villose ..   ..   4. *urbaniana*
Inflorescences not cauliflorous, in axils of leaves, flowers solitary or clustered :
  Prostrate undershrub ; flowers in bracteate heads 2–3 cm. diam. ; calyx and corolla villose ; peduncle about 1·5 cm. long ; stems far-creeping, rooting at nodes ; leaves elliptic, 3–7 cm. long, 2–4 cm. broad, laxly pilose ..   ..   ..   19. *geophiloides*
  Climbing or erect shrubs :
    Flowers solitary in leaf axils :
      Calyx-lobes 2·5–5 mm. long, triangular ; leaves cuneate at base, elliptic, about 8 cm. long and 4 cm. broad, with about 11 pairs of lateral nerves, pubescent on nerves beneath ; corolla densely pilose ; bracts small and inconspicuous at base of calyx :
        Corolla 14 mm. long ; calyx-lobes long-triangular, about 5 mm. long   5. *bracteolata*

Corolla 20 mm. long ; calyx-lobes short-triangular, about 2·5 mm. long   6. *rosea*
Calyx-lobes 10–15 mm. long, subulate or linear ; bracts united into a cup 5 mm.
    long closely surrounding ovary :
    Calyx-lobes subulate, 0·5 mm. broad ; corolla about 3 cm. long, densely pilose
        with spreading pink hairs ; leaves cordate to cuneate at base, ovate-elliptic,
        10–13 cm. long, 4–6·5 cm. broad, pubescent beneath on the 9–12 pairs of lateral
        nerves ; branchlets spreading-pilose .. ..   7. *cordata*
    Calyx-lobes linear, 2 mm. broad ; corolla 2 cm. long, densely appressed-pilose ;
        leaves cuneate at base, elliptic, 8–10 cm. long, 3–4·5 cm. broad, pilose all over
        lower surface, lateral nerves about 12 pairs ; branchlets appressed-pilose,
        becoming subglabrous .. ..   10. *liberica*
Flowers clustered :
    Stout, erect, shrub ; stipules free, foliaceous (3–) 7 cm. long, (1–) 4 cm. broad ;
        leaves oblanceolate, long-cuneate to base, 20–50 cm. long, 8–20 cm. broad
                                                                    24. *gigantistipula*
    Climbing shrubs :
    Inflorescences lax, not distinctly surrounded by bracts :
        Flowers in small cymes, 2–3 cm. long, with about 2 small often reflexed bracts
            at base ; corolla 9–10 mm. long ; calyx-lobes about 3 mm. long ; leaves oblong-
            elliptic to ovate, acutely acuminate, rounded to subcordate at base, 8–10 cm.
            long, 4–5 cm. broad, distinctly setulose above .. ..   12. *venosa*
        Flowers in large cymes, up to 20 cm. long, with numerous bracteoles ; corolla
            about 4 mm. long ; calyx-lobes about 2 mm. long ; leaves oblong-elliptic to
            elliptic, acute to acuminate, cuneate to subcordate at base, 10–15 cm. long,
            5–7 cm. broad, almost glabrous above except on nerves ..   13. *floribunda*
    Inflorescences close, usually globose, surrounded by distinct bracts :
    Flower-heads long-pedunculate, (2–) 4–8 cm. long :
        Peduncles glabrous ; calyx-lobes 5–8 mm. broad, foliaceous, ovate ; ovary
            glabrous ; corolla-lobes pubescent, tube glabrous ; leaves oblong-elliptic,
            acutely acuminate, rounded to cordate at base, 5–8 cm. long, 3–4 cm. broad,
            thinly pilose beneath, with 8–12 pairs of lateral nerves ..   14. *calycina*
        Peduncles pubescent ; calyx-lobes linear (narrowly lanceolate in No. 15) ;
            ovary pubescent :
            Calyx-lobes spreading-pilose, linear, 9–12 mm. long, 1 mm. broad ; (see other
                characters below) .. ..   20. *vogelii*
            Calyx-lobes appressed- or sparsely-pubescent :
                Peduncle 4–8 cm. long :
                    Leaves glabrous above, ovate, 5–8 cm. long, 3–6 cm. broad, acuminate,
                        evenly appressed-pubescent and densely so on the midrib and lateral
                        nerves beneath ; calyx-lobes linear, 1·5–2 mm. broad, 10–12 mm. long,
                        shortly pubescent inside, ciliate on margins, pubescent along midrib
                        outside ; corolla densely pubescent .. .. ..   16. *harleyae*
                    Leaves pubescent above, narrowly elliptic, cuneate at both ends, 8–12 cm.
                        long, 2·5–5 cm. broad, with 14–18 pairs of appressed-pubescent lateral
                        nerves, otherwise nearly glabrous ; corolla appressed-pilose 15. *schaeferi*
                Peduncle 2–2·5 cm. long ; leaves oblong-elliptic, shortly cuneate at base,
                    6–10 cm. long, 3–4 cm. broad, pubescent only on the 10 pairs of lateral
                    nerves ; calyx campanulate, lobes linear ; bracts leafy at apex 23. *robbii*
    Flower-heads sessile or shortly pedunculate (with peduncle up to 1 cm. long) :
    Flowers 2–4 together, sessile :
        Bracts an obscure cup about 1 mm. long at base of each distinct flower ;
            calyx shortly pubescent ; (see also above) .. .. ..   6. *rosea*
        Bracts fused into an open or urn-shaped cup :
            Bracts fused into an urn-shaped cup about 1 cm. long completely enclosing
                the calyx ; corolla glabrous, about 2·5 cm. long ; leaves obovate, 7–11 cm.
                long, 3–5 cm. broad, hispid on both sides, with about 15 pairs of arching
                lateral nerves .. ..   11. *urceolata*
            Bracts forming an open involucral cup ; corolla densely pilose:
                Branchlets and leaves on both surfaces spreading-pilose :
                    Leaves shortly and abruptly acuminate, cuneate at base, ovate to obovate,
                        7–11 cm. long, 4–6 cm. broad, shortly pilose above and on midrib and
                        nerves beneath, with 9 pairs of lateral nerves ; calyx-lobes long-triangular,
                        4–5 mm. long, tube densely pilose at base ; involucre about 7 mm. diam. ;
                        stipules ovate about 7 mm. long and broad .. ..   8. sp. nr. *cordata*
                    Leaves long-acuminate at apex, cuneate to rounded at base, elliptic to
                        oblanceolate, 10–17 cm. long, 3–7 cm. broad spreading-pilose above and
                        on the midrib and nerves beneath, with about 11 pairs of lateral nerves ;
                        calyx-lobes linear, 10 mm. long, densely pilose ; involucre about 1·5 cm.
                        diam. ; stipules broadly ovate, up to 12 mm. long and broad 9. *neglecta*

Branchlets shortly and densely pubescent ; leaves glabrous above except
the nerves, pubescent on the nerves beneath .. .. 5. *bracteolata*
Flowers numerous (at least 4) in the head :
Bracts usually obscure except when young, reflexed beneath the globose
inflorescence, ovate-lanceolate about 1 cm. long ; corolla-lobes pubescent,
tube glabrous ; calyx-lobes linear, 9–12 mm. long, 1 mm. broad ; leaves
ovate, about 8 cm. long and 4 cm. broad, thinly pilose ; petioles up to 1 cm.
long .. .. 20. *vogelii*
Bracts conspicuous at all stages of inflorescence ; most of corolla pubescent :
Inflorescences aggregated near ends of shoots, compact ; 1 cm. diam. ;
bracts ovate, about 1 cm. long ; leaves ovate-elliptic, acuminate, 8–9 cm.
long, 3·5–4 cm. broad, scabrid above, sparingly pubescent beneath
22. *talbotii*
Inflorescences axillary, not specially towards ends of shoots :
Bracts ovate to ovate-lanceolate, up to 1·5 cm. long and at most 1 cm. broad
at base ; leaves ovate 6–9 (–12) cm. long, 3–5 cm. broad, long-cuneate to
acute at apex, rounded at base, rather densely pubescent on both sides ;
petiole about 1 cm. long .. .. 21. *brevipes*
Bracts very broadly ovate, broadest above the base, up to 2 cm. broad,
enclosing the inflorescences :
Petioles 3–6 cm. long ; leaves obovate to ovate-elliptic, 9–17 cm. long ;
5–9 cm. broad, with 12–15 pairs of lateral nerves, acute or shortly
acuminate at apex, often long-cuneate at base, glabrous or shortly
pubescent above, pubescent beneath ; in lowland forest   17. *gabonica*
Petioles about 1 cm. long ; leaves elliptic to ovate-elliptic, 4–11 cm. long,
2–7 cm. broad, with about 13 pairs of lateral nerves, acute or shortly
acuminate at apex, rounded to cuneate at base, pubescent above and often
densely so beneath ; upland species .. .. .. 18. *efulenensis*
*Leaves completely covered with a dense felt of usually white hairs :
Flowers in a loose shortly pedunculate cyme ; leaves oblong or ovate-elliptic, acutely
acuminate, 6–9 cm. long, 3–4·5 cm. broad, shortly pilose above, with 12–18 pairs of
lateral nerves ; stipules ovate, about 7 mm. long ; corolla appressed-pubescent
outside, 1 cm. long .. .. 26. *discolor*
Flowers in a compact head or solitary (then also with creeping habit) ; involucre of
bracts usually conspicuous :
The flowers solitary or a few together ; creeping and with erect stems up to 15 cm.
high ; lower leaves markedly smaller than upper, normal leaves about 9 cm. long
and 5 cm. broad, appressed-pilose above ; stipules 8–14 cm. long, margins densely
brown-pilose .. .. 27. *medusula*
The flowers numerous ; climbing or erect shrubs :
Stipules very large, about 30 mm. long and broad, enclosing sessile inflorescence in
urn-shaped fused bracts 2–3 cm. long ; leaves lanceolate-elliptic or broadly
oblanceolate, subacute at apex, 10–15 cm. long, 4–7 cm. broad, glabrous above,
margins and petioles ciliate ; erect shrub .. .. .. 25. *africana*
Stipules 5–20 mm. long (when over 10 mm. long then inflorescence pedunculate) :
Bracts fused into a sessile urn-shaped involucre enclosing the flowers ; stipules
about 5 mm. long ; leaves broadly lanceolate to obovate or elliptic, acute, 10–
15 cm. long, 4–7 cm. broad, with about 11 pairs of distinct lateral nerves ; erect
shrub .. .. 28. *elliptica*
Bracts not fused into urn-shaped involucre ; climbers:
Flowers pedicellate in the heads ; leaves ovate-elliptic, acutely acuminate,
8–13 cm. long, 4–6 cm. broad :
Calyx-lobes 5–6 mm. long, narrowly lanceolate, spreading, ciliate ; pedicels about
6 mm. long ; leaves with up to 20 pairs of lateral nerves ; corolla a little longer
than the calyx, pubescent on lobes outside .. .. .. 29. *johnstonii*
Calyx-lobes 1·5–2·5 mm. long, narrowly triangular, reflexed against ovary,
woolly outside ; pedicels about 6 mm. long, up to 13 mm. long in fruit ; leaves
with about 10 pairs of lateral nerves ; corolla much longer than the calyx,
pubescent on lobes outside .. .. 30. *pedicellata*
Flowers sessile or subsessile in the head :
Heads sessile or subsessile :
Leaves yellow rusty-brown beneath ; (see other characters below) 33. *ferruginea*
Leaves white or greyish-brown beneath :
Calyx-lobes lanceolate, acute, more or less brown appressed-pubescent, about
3 mm. long ; inflorescences sessile or with peduncle up to 1·5 cm. long ;
leaves ovate-elliptic to oblanceolate, 5–15 cm. long, 2–8 cm. broad, sub-
glabrous above, greyish-brown beneath .. .. 31. *capitellata*
Calyx-lobes markedly acuminate from ovate base, densely clothed with long
white hairs, about 7 mm. long ; inflorescences sessile ; leaves ovate to

obovate, rounded to subcordate at base, 5–12 cm. long, 3–5 cm. broad, rather
densely pubescent above, white beneath (at least when young)
                                                             32. *arachnoidea*
Head pedunculate, peduncle 4–10 cm. long :
    Leaves yellow rusty-brown beneath, broadly ovate subcordate to rounded or
       subacute at base, 10–15 cm. long, 5–11 cm. broad, sparsely hairy above ;
       stipules broadly ovate, 1–1·5 (–2) cm. long, striate ; inflorescences 3–5 cm.
       diam. ; calyx-lobes filiform, about 1 cm. long .. .. 33. *ferruginea*
    Leaves grey or light brownish beneath, elliptical, subacute at both ends, 8–
       12 cm. long, 4–6·5 cm. broad, glabrous or loosely woolly above when young ;
       calyx-lobes oblong, 4 mm. long .. .. 34. *lanuginosa*

1. **S. speciosa** *K. Schum.* in Engl. Bot. Jahrb. 23 : 429 (1896) ; Wernham Monogr. Sabicea 44 (1914). Slender
    climbing shrub ; flowers densely pink-hairy, in dense clusters.
    Togo Rep.: *Büttner* 263. **S.Nig.:** Oban *Talbot* 1040. Afi River F. R., Ogoja Prov. (May) *Jones & Onochie*
    FHI 18906! 18936! [**Br.]Cam.:** Victoria *Winkler* 27. Barombi (Nov.) *Preuss* 150! Abonando *Rudatis*
    57! Lakka, Bamenda, 4,000 ft. (June) *Maitland* 1569! Also in Cameroun.
2. **S. pilosa** *Hiern* in F.T.A. 3 : 76 (1877) ; Wernham l.c. 61. A creeping and climbing shrub up to 10 ft. or
    more high, flowering on the fairly stout older stems which root at the nodes ; flowers pink.
    [**Br.]Cam.:** Barombi *Preuss* 15 ; 278. Also in Cameroun and Gabon.
       [*Vigne* FH 2807 (Simpa, Ghana) and *Aké Assi* 2830 (Gabo to Gaouri, Iv.C.) are closely allied but have
    narrow calyx-lobes and sparsely pilose corolla.—F.N.H.]
3. **S. xanthotricha** *Wernham* in Cat. Talb. 42, t. 11, 1–4 (1913), and Monogr. Sabicea 28. Hairy erect shrub with
    fascicles of flowers on older branches.
    S.Nig.: Oban *Talbot* 249! 251!
4. **S. urbaniana** *Wernham* Monogr. Sabicea 28 (1914). Undershrub up to 8 ft. high ; flowers white, densely
    clustered on older wood.
    F.Po: Musola, 1,950–3,900 ft., *Mildbr.* 7041. [This specimen was apparently destroyed in Berlin and the
    species needs to be re-collected. It appears to be a shrub 8 ft. high and not up to 70 ft. as stated in
    F.W.T.A., ed. 1.]
5. **S. bracteolata** *Wernham* l.c. 73 (1914). *S.* aff. *segregata* of Chev. Bot. 316, partly (*Chev.* 12390). A shrub,
    the branchlets densely rusty-tomentose when young ; flowers densely silky-pubescent outside, solitary.
    Guin.: Labé (Apr.) *Chev.* 12367! 12390! Pita *Adam* 11651!
       [*Vigne* FH 4841 from Tarkwa, Ghana, has filiform calyx-lobes about 1 cm. long.—F.N.H.]
6. **S. rosea** *A. C. Hoyle* in Kew Bull. 1935 : 264. Slender scrambling or climbing shrub ; flowers red, very
    hairy ; in forest.
    Iv.C.: Yapo (Dec.) *Aké Assi* 3889! Gouro to Gouékangouiné *Chev.* 21397! Taï (Mar.) *Leeuwenberg* 3013!
    Ghana: Boinsu (Dec.) *Vigne* FH 3190! Axim (fr. Feb.) *Irvine* 2376! **S.Nig.:** Omo (formerly part
    Shasha) F.R. (fl. & fr. Apr.) *Jones & Onochie* FHI 17279!
7. **S. cordata** *Hutch. & Dalz.* F.W.T.A., ed. 1, 2 : 106 (1931). Slender climbing shrub up to 10 ft. high ;
    flowers scarlet, densely hairy, fruits red ; in secondary deciduous forest.
    Iv.C.: Daoukrou to Morénou (Dec.) *Chev.* 22499! **Ghana:** Mampong Scarp (Apr.) *Dalz.* 54! Nsuta (fl.
    & fr. May) *Vigne* FH 1738! Abetifi (Jan.) *Irvine* 1691a! Odomi River F.R., Hohoe Dist. (Oct.) *St. C.
    Thompson* 1569!
8. **S. sp. nr. cordata** *Hutch. & Dalz.* l.c. (1931). Climbing shrub.
    S.Nig.: Lagos *Batten Poole* 1! [This is probably a form of *S. cordata* with short calyx-lobes like
    those of *S. bracteolata*.]
9. **S. neglecta** *Hepper* in Kew Bull. 14 : 255 (1960). Climbing shrub ; flowers red ; in forest.
    S.Nig.: Oban F.R. (May) *Ujor* FHI 30836!
10. **S. liberica** *Hepper* in Kew Bull. 13 : 293, fig. 1 (1958). A climbing shrub ; flowers pink, fruits red.
    Lib.: Moala (Nov.) *Linder* 1380! Bahtown, Tchien Dist. (fr. Aug.) *Baldwin* 8030!
11. **S. urceolata** *Hepper* in Kew Bull. l.c., fig. 4 (1958). Slender climbing shrub about 4 ft. high ; flowers white,
    bracts green sometimes tipped pink ; in secondary forest.
    S.Nig.: Okomu F.R., Benin (Feb.) *Brenan* 9032! Oban F.R. (Feb.) *Onochie* FHI 36325! **F.Po:** Musola
    (Jan.) *Guinea* 1053!
12. **S. venosa** *Benth.* in Fl. Nigrit. 399 (1849) ; Wernham l.c. 32 ; Chev. Bot. 317. *S. discolor* of Chev. Bot.
    316 (*Chev.* 17493), not of Stapf. *S. schumanniana* of Wernham l.c. 42, partly ; of Aké Assi Contrib. 1 :
    151, not of Büttner. A climbing shrub with branchlets pubescent or subglabrous ; flowers whitish, fruits
    red.
    Port.G.: Bissau (Mar.) *Esp. Santo* 1876! Fulacunda (fr. July) *Esp. Santo* 2105! **Guin.:** Karkandy
    Heudelot 821! Timbo to Conakry (fr. July) *Pobéguin* 770! Timbo *Chev.* 12423! **S.L.:** Hill Station *Smythe*
    58! Sulimania (Mar.) *Sc. Elliot* 5276! Mabum (fr. Aug.) *Thomas* 1616! Njala (fr. May) *Deighton* 2518!
    Lib.: Yratoke, Webo Dist. (fr. July) *Baldwin* 6257! **Iv.C.:** Guidéko *Chev.* 16453! Dabou *Chev.* 17160!
    17222! Lamé (Oct.) *Leeuwenberg* 1730! **Ghana:** Axim (Feb.) *Irvine* 2136! Essiama *Fishlock* 11!
    Tarkwa (fl. fr. Dec.) *Thompson* 46! *Miles* 7! [**Br.]Cam.:** Buea (fl. Apr., fr. July) *Maitland* 32!
    Hutch. & Metcalfe 95! Also in Cameroun and the Congos.
13. **S. floribunda** *K. Schum.* in Engl. Bot. Jahrb. 23 : 428 (1897) ; Wernham l.c. 72. Slender climbing shrub ;
    flowers white or greenish-white, fruits whitish.
    S.Nig.: Oban *Talbot* 228! New Ndebiji, Calabar (Apr.) *Ejiofor* FHI 21883! Also in Cameroun, Gabon
    and the Congos.
14. **S. calycina** *Benth.* in Fl. Nigrit. 399 (1849) ; F.T.A. 3 : 76 ; Wernham l.c. 63 ; Chev. Bot. 316. *S. barteri*
    Wernham l.c. 64, t. 11, 5–7 (1914). Slender climbing shrub, sometimes creeping ; flowers white, calyx and
    stipules pinkish, fruits red ; in thickets chiefly in transitional savanna to forest regions.
    S.L.: Njala (May) *Deighton* 4765! Kasewe Hills F.R. (June) *Deighton* 5829! Bumbuna (fr. Oct.) *Thomas*
    3338! **Lib.:** Cess River, Tchien Dist. (fr. Aug.) *Baldwin* 9021! **Iv.C.:** Alépé *Chev.* 17421! Indénié *Chev.*
    17667! Sanvi *Chev.* 17722! Danané (fl. & fr. June) *Collenette* 46! **Ghana:** Dajou, near Axim (fl. & fr.
    Apr.) *Chipp* 167! Amentia *Irvine* 514! Nsuta (fr. May) *Vigne* FH 1742! Mampong (Apr.) *Vigne* FH 1922!
    N.Nig.: Awton (Apr.) *Elliott* 77! **S.Nig.:** Epe (= Eppah) (fl. & fr. May) *Barter* 3282! *Hambler* N48!
    Ikene (Jan.) *Burtt* B30! Eket Dist. *Talbot*! Ikom (May) *Jones & Onochie* FHI 17442! [**Br.]Cam.:**
    Victoria (fr. June) *Maitland* 54! Victoria to Kumba ,2,000 ft. (fr. Apr.) *Hutch. & Metcalfe* 147! Kumba
    to Mamfe (Jan.) *Keay* FHI 28555! Wum Div. (fr. July) *Ujor* FHI 30476! **F.Po:** Clarence Peak (Oct.)
    T. Vogel 35! Las Corteras (Jan.) *Guinea* 2072! Musola (Jan.) *Guinea* 973! Also in Cameroun, Gabon and
    Congo. (See Appendix, p. 410.)
15. **S. schaeferi** *Wernham* l.c. 59 (1914). Slender climbing shrub up to 12 ft. high ; flowers white, buds pink.
    F.Po: Moka (Jan.) *Guinea* 2241! L. Biao (Sept.) *Wrigley* 656! Also in Cameroun.
16. **S. harleyae** *Hepper* in Kew Bull. 13 : 292, fig. 3 (1958). A slender climbing shrub ; flowers white, bracts and
    calyx-lobes pinkish.
    S.L.: Kambui Hills (May) *Bakshi* 186! **Lib.:** Ganta (May) *Harley* 1167! s.n. !
17. **S. gabonica** (*Hiern*) *Hepper* l.c. (1958). *Stipularia gabonica* Hiern in F.T.A. 3 : 80 (1877). A slender climbing
    shrub in forest.

S.Nig.: Okomu F.R., Benin (fr. Dec., Feb.) *Brenan* 8410! 8997! **[Br.]Cam.:** Mamfe (fr. Jan.) *Maitland* 1161! S. Bakundu F.R., Kumba (fr. Jan.) *Binuyo & Daramola* FHI 35067! Also in Gabon.

18. **S. efulenensis** (*Hutch.*) *Hepper* l.c. (1958). *Stipularia efulenensis* Hutch. in Kew Bull. 1908 : 291. *Sabicea tchapensis* K. Krause (1912), and var. *glabrescens* Wernham l.c. 59 (1914). Climbing shrub up to 15 ft. high ; fruits red ; upland forest margins and stream sides.
S.Nig.: Sonkwala, Obudu Div. (Dec.) *Savory & Keay* FHI 25217! **[Br.]Cam.:** Bamenda (Jan., Apr.) *Migeod* 433! *Maitland* 1706! *Daramola* FHI 40615! Mambila Plateau (Jan.) *Hepper* 1800! 2807! Vogel Peak, Adamawa (Nov.) *Hepper* 1439! Also in Cameroun.

19. **S. geophiloides** *Wernham* in Cat. Talb. 41 (1913), and Monogr. Sabicea 60. Prostrate herb with long reddish stems giving rise to erect leafy shoots ; flowers white, calyx lobes pink, fruits red ; on stream banks in forest.
Ghana: Kumasi (Apr.) *Piening* 2345! Mpraeso Scarp (Apr.) *Morton* A707! **S.Nig.:** Sapoba *Kennedy* 1098! Ibadan North F.R. (fl. & fr. June) *Jones* FHI 3643! Olokemeji F.R. (June) *Jones & Keay* FHI 18813! Oban *Talbot* 255! **[Br.]Cam.:** Abonando *Rudatis* 44!

20. **S. vogelii** *Benth.* in Fl. Nigrit. 398 (1849) ; F.T.A. 3 : 76 ; Wernham Monogr. Sabicea 43 ; Chev. Bot. 317. *S. globifera* Hutch. & Dalz., F.W.T.A., ed. 1, 2 : 106 (1931). Slender climbing shrub ; flowers white or pale pink.
Guin.: Labé *Chev.* 12348! Diaguissa *Chev.* 12687! Nimba Mts. (Nov.) *de Wilde* 873! **S.L.:** Kumrabai (Jan.) *Thomas* 6855! Kenema (Apr.) *Lane-Poole* 227! Yonibana to Moyamba (Apr.) *Deighton* 6049! Bintumane (fr. Aug.) *Jaeger* 1058! (See Appendix, p. 411.)

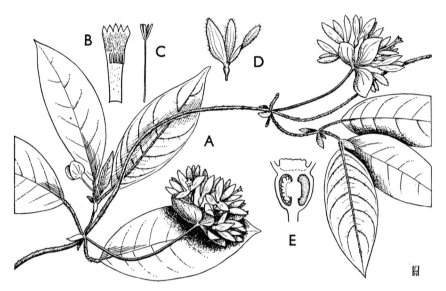

FIG. 235.—SABICEA CALYCINA *Benth.* (RUBIACEAE).

A, flowering branch, × ⅓. B, corolla laid open, ×1. C, style, × 1. D, calyx and ovary, × 1. E, longitudinal section of ovary, × 6. Drawn from *Burtt.* 30.

21. **S. brevipes** *Wernham* l.c. 58 (1914). Climbing or erect shrub 3 or 4 ft. high ; flowers white or mauvish, fruits and calyx red.
Iv.C.: Yapo (Aug.) *Aké Assi* 3122! **Ghana:** Volta River F.R. (Oct.) *Morton* GC 6069! Kpeve (May) *Darko* 567! Trawa (Jan.) *Vigne* FH 4396! **N.Nig.:** Gawu Hills, Niger Prov. (Aug.) *Onochie* FHI 35931! Anara F.R., Zaria (May) *Keay* FHI 22912! Mande F.R., Zaria (June) *Keay* FHI 25821! Naraguta (June) *Lely* 284! **S.Nig.:** Milliken Hill, Enugu (Feb.) *Jones* FHI 560!

22. **S. talbotii** *Wernham* in Cat. Talb. 43 (1913), and Monogr. Sabicea 66.
S.Nig.: Oban *Talbot* 259!

23. **S. robbii** *Wernham* Monogr. Sabicea 69, t. 10, 1–2 (1914). A (? climbing) shrub 10–15 ft. high ; flowers whitish.
S.Nig.: Calabar *Robb*! Also in Gabon.

24. **S. gigantistipula** *K. Schum.* in Engl. Bot. Jahrb. 33 : 337 (1903), as *S. gigantostipula* ; Wernham l.c. 27. Erect shrub 10 ft. high, soft-wooded and with large leaves ; in forest.
S.Nig.: Oban *Talbot* 259! s.n.! **F.Po:** Mt. Balen *Guinea* 427! El Pico, 3,000 ft. (Dec.) *Boughey* 192! Also in Cameroun.

25. **S. africana** (*P. Beauv.*) *Hepper* in Kew Bull. 13 : 292 (1958). *Stipularia africana* P. Beauv. (1807)—F.T.A. 2 : 80 ; F.W.T.A., ed. 1, 2 : 108 ; Chev. Bot. 317. Soft-stemmed shrub 3–8 ft. high ; flowers white, involucre reddish, fruits red ; in swamps and wet places.
S.L.: Luseniya *Sc. Elliot* 4077! Samuel Town (Dec.) *Deighton* 3306! Bagroo R. (Apr.) *Mann* 877! Mano Bonjema *Deighton* 3708! **Lib.:** Paynesville (Apr.) *Bequaert* 165! Duport (Nov.) *Linder* 1482! Jabrocca (Dec.) *Baldwin* 10850! Bushrod Isl. (Feb.) *Baldwin* 11080! **Iv.C.:** Dabou (Feb.) *Chev.* 17230. **Ghana:** Awiabo (Nov.) *Vigne* FH 1435! **S.Nig.:** Lagos Isl. *Dalz.* 1055! Sapoba *Kennedy* 1801! Eket Dist. *Talbot* 3292! Oban Dist. (Mar.) *Onochie* FHI 7710! Also in Cameroun, Gabon and Congo.

26. **S. discolor** *Stapf* in J. Linn. Soc. 37 : 105 (1905) ; Wernham l.c. 35. *S. loxothyrsus* K. Schum. & Dinkl. ex Stapf (1906). *S. discolor* var. β *laxothyrsa* Wernham l.c. (1914). Climbing shrub with leaves white beneath ; flowers pink or white.
S.L.: Njala (May, Sept.) *Deighton* 2125! 4766! Mayaso (Aug.) *Thomas* 1381! Falaba (Apr.) *Aylmer* 31! **Lib.:** Monrovia *Whyte* 1! Grand Bassa (May) *Dinklage* 1903! Ganta *Harley* 425! Brewersville (fr. June) *Barker* 1329! **Iv.C.:** Abidjan *Chev.* 15343! Borobo *Chev.* 17664! **Ghana:** Akwapim (Mar.) *Murphy* 679! Essiama (fr. Apr.) *Williams* 433! Kumasi (Mar.) *Vigne* FH 1642! Tarkwa (Jan.) *Vigne* FH 4118!

27. **S. medusula** *K. Schum. ex Wernham* l.c. 44, t. 6, 1–3 (1914). Creeping herb about 6 ins. high, leaves white-felted beneath ; fruits white ; in high forest.
[Br.]**Cam.**: Mamfe (fr. Mar.) *Coombe* 198 ! 237 ! Also in Cameroun.

28. **S. elliptica** (*Schweinf. ex Hiern*) *Hepper* in Kew Bull. 13 : 292 (1958). *Stipularia elliptica* Schweinf. ex Hiern in F.T.A. 3 : 80 (1877) ; F.W.T.A., ed. 1, 2 : 108. Soft-stemmed shrub about 4 ft. high, with quadrangular branches, hoary on the young parts and underleaf ; flowers white, enclosed in an urn-shaped involucre of red-tinged bracts ; in wet places.
S.Nig.: Onitsha *Barter* 166 ! Also in Sudan.

29. **S. johnstonii** *K. Schum. ex Wernham* l.c. 66 (1914). A climbing shrub up to 10 ft. high, with leaves pale-tomentose beneath ; flower whitish.
S.Nig.: Ologbo *Kennedy* 2103 ! Ulasi, Onitsha Prov. (Mar.) *Brenan* 9204 ! Calabar (fr. Mar.) *Holland* 95 ! *Brenan* 9217 ! Eket *Talbot* ! Cross R. *Johnston* ! Also in Congo.

30. **S. pedicellata** *Wernham* in Cat. Talb. 42 (1913), and Monogr. Sabicea 67. Climbing shrub ; fruits red.
S.Nig.: Oban (Mar.) *Talbot* 1367 ! 2033. *Onochie* FHI 36477x ! Afi River F.R., Ogoja Prov. (fr. May) *Jones & Onochie* FHI 18940 ! Boshi-Okwangwo F.R., Obudu Dist. (fr. May) *Latilo* FHI 30945 ! [Br.]**Cam.**: Kumba to Mamfe (Jan.) *Keay* FHI 28556 !

31. **S. capitellata** *Benth.* in Fl. Nigrit. 398 (1849) ; Wernham Monogr. Sabicea 65. *S. brunnea* Wernham l.c. (1914). A climbing shrub with rufose-pilose branchlets ; flowers pinkish-brown ; fruits red, yielding a red juice, with stiff grey hairs.
S.Nig.: Oban F.R. (fr. Jan.) *Onochie* FHI 36101x ! Okuni and Ikom *Holland* 262 ! [Br.]**Cam.**: S. Bakundu F.R. (fr. Mar.) *Brenan* 9299 ! Kembong F.R. (fl. & fr. Mar.) *Richards* 5210 ! Ikom to Mamfe (Mar.) *Onochie* FHI 30894 ! **F.Po**: *Mann* 41 ! *T. Vogel* 88 ! Also in Cameroun, Gabon, Congo and S. Tomé.

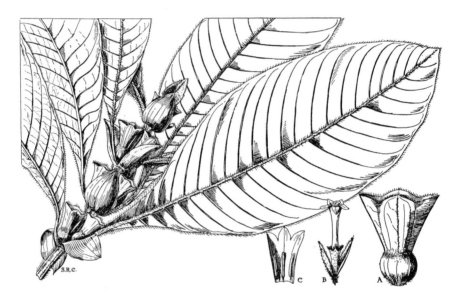

FIG. 236.—SABICEA AFRICANA (*P. Beauv.*) *Hepper* (RUBIACEAE).
A, involucre. B, flower. C, part of corolla laid open.

32. **S. arachnoidea** *Hutch. & Dalz.* F.W.T.A., ed. 1, 2 : 106 (1931). A rather slender-branched pubescent climber with pale underleaf ; flowers in woolly heads.
S.L.: Roruks (Nov.) *Thomas* 5703 ! Yonibana (Nov.) *Thomas* 5062 ! 5275 !

33. **S. ferruginea** (*G. Don*) *Benth.* in Fl. Nigrit. 397 (1849) ; F.T.A. 3 : 75 ; Wernham Monogr. Sabicea 60. *Cephaelis ferruginea* G. Don (1834). *Sabicea lasiocalyx* Stapf (1905)—F.W.T.A., ed. 1, 2 : 107. *S. ferruginea* var. *lasiocalyx* (Stapf) Wernham l.c. (1914). *S. salmonea* A. Chev. Bot. 316, name only. A climbing shrub with fairly large rusty-tomentose leaves.
S.L.: *Don* ! Jigaya (fr. Sept.) *Thomas* 2842 ! Bagroo R. *Mann* 862 ! Jaiama (Apr.) *Deighton* 3161 ! **Lib.**: Tappita (fr. Aug.) *Baldwin* 9072 ! Monrovia *Whyte* ! Banga *Linder* 1193 ! Dukwia R. (May) *Cooper* 459 ! **Iv.C.**: Guébo *Chev.* 17031 ! Alépé *Chev.* 17418 ! Mamba (fr. Oct.) *Aké Assi* 287 ! Gouékangouiné (May) *Chev.* 21452 ! Lamé (Nov.) *Leeuwenberg* 1898 ! **Ghana**: Assin Juaso *West-Skinn* 24 ! (See Appendix, p. 410.)
[Specimens from Sierra Leone and Liberia have pedunculate inflorescences while those from Ivory Coast and Ghana have them sessile.]

34. **S. lanuginosa** *Wernham* l.c. 67, t. 9, 1–3 (1914). A climbing shrub up to 12 ft. high ; flowers white ; in mangrove swamps and swamp forest.
S.Nig.: Lagos *Millen* 68 ! *Hagerup* 808 ! Ikoyi Plains *Dalz.* 1377 ! Apapa (Nov.) *Keay & Savory* FHI 22451 !

50. TEMNOCALYX Robyns in Bull. Jard. Bot. Brux. 11 : 23, 317 (1928).

Branches glabrous, with rather long internodes ; leaves in whorls of three, obovate, shortly and bluntly pointed, 3–5 cm. long, 2–3 cm. broad, glabrous, with about 4 pairs of lateral nerves ; flowers supra-axillary, paired ; pedicels 1 cm. long ; calyx truncate, rim-like ; corolla-tube curved, 1·5–2 cm. long ; anthers partly exserted ; stigma long-exserted, short and cylindric-capitate ; ovary 5-celled ; fruit globose

obovatus

T. obovatus (*N.E. Br.*) *Robyns* l.c. 320, figs. 31, 32 (1928). *Fadogia obovata* N.E. Br. in Kew Bull. 1906 : 105.
    An undershrub sometimes in clumps, 2–3 ft. high, with 3-angled slightly pubescent stems ; flowers greenish with darker apex, solitary or a few together in the axils.
    N.Nig.: Kontagora Prov. (June) *Dalz.* 554 ! Naraguta, Jos Plateau (June) *Lely* 326 ! Extends in savanna regions to E. Africa and south to S. Rhodesia.

51. HUTCHINSONIA Robyns in Bull. Jard. Bot. Brux. 11 : 24 (1928).

Branchlets covered with long slender spirally twisted hairs ; leaves ovate-oblong or oblong-elliptic, slightly cordate at base, acuminate, 3·5–6 cm. long, 1·5–2 cm. broad, laxly pilose on both surfaces with setose hairs, with 5–6 pairs of lateral nerves ; stipules linear from a broad base, 3–5 mm. long ; flowers solitary, axillary ; pedicels 3–4 mm. long, setose ; calyx-lobes filiform, 2·5–3 mm. long ; corolla cylindric, tube 2·3 cm. long, lobes ovate, with long ciliate tails ; ovary 3–4-celled .. 1. *barbata*

Fig. 237.—HUTCHINSONIA BARBATA *Robyns* (RUBIACEAE).

A, flower.   B, same with part of corolla removed.   C, upper part of corolla laid open.   D, anther. E, longitudinal section of ovary.   F, young fruit.

Branchlets nearly glabrous ; leaves ovate-oblong or oblong-elliptic, slightly cordate at base, gradually acuminate, 3·5–5·5 cm. long, 1·3–2 cm. broad, glabrous except the midrib and the axils of the nerves ; lateral nerves about 6 pairs ; stipules linear-subulate from a broad base, 5 mm. long, sparingly ciliate ; flowers axillary or terminal ; pedicels 4 mm. long ; calyx-lobes linear, 2·5–3 mm. long ; corolla-tube 2·2 cm. long, lobes narrowly triangular, appendiculate ; ovary 3-celled   ..   2. *glabrescens*

1. **H. barbata** *Robyns* l.c. 25, figs. 5 & 6 (1928). A straggling shrub up to 10 ft. high with occasional spines on the lower stems ; flowers yellow or orange ; undershrub in forest.
    Guin.: Kolamba, Gueckedou *Adam* 5773 ! **S.L.**: Kambui F.R. (Feb.) *Lane-Poole* 341 ! *Small* 864 ! Nongowa Forest, Kenema (fl. & fr. Mar.) *Jordan* 1007 ! Baiima (May) *Deighton* 5773 ! Makump (May) *Deighton* 1715 ! Gola F.R. (Apr.) *Small* 604 ! **Lib.**: Peahtah (Dec.) *Baldwin* 10605 ! Kitoma (Feb.) *Adam* 16648 ! **Iv.C.**: confluence of R. Davo and R. Sassandra (May) *Chev.* 17954 !
2. **H. glabrescens** *Robyns* l.c. 28 (1928). A straggling shrub up to 6 ft. high ; flowers yellow or orange.
    Lib.: Monrovia (July) *Linder* 40 ! at the mouth of Sinoe R. (Dec.) *Dinklage* 2144 ! Tappita (Aug.) *Baldwin* 9109 !

52. ROBYNSIA Hutch. F.W.T.A., ed. 1, 2 : 108 (1931). *Nom. cons.*

Branchlets shortly pubescent ; stipules with a long tail about 2 cm. long, the sheath at the base villous within ; leaves elliptic, subcordate at the base, acuminate, 10–14 cm. long, 5–6 cm. broad, nearly glabrous beneath, lateral nerves about 8 pairs ; petiole up to 1 cm. long, pubescent ; cymes axillary, shortly pedunculate, many-flowered ; bracts linear, 1–1·5 cm. long ; calyx-lobes linear, 1 cm. long, pubescent ;

FIG. 238.—ROBYNSIA GLABRATA *Hutch.* (RUBIACEAE).

A, flower.   B, lower part of corolla, ovary and calyx.   C, upper part of corolla laid open.   D anthers.   E, cross-section of ovary.   F, longitudinal section of ovary.

corolla long-tubular, 2 cm. long, pubescent, lobes broadly linear ; style exserted for 1 cm. ; fruit globose   ..   ..   *glabrata*

R. glabrata *Hutch.* l.c. (1931) ; Hepper in Kew Bull. 16 : 157 (1962). *Cuviera bolo* Aubrév. & Pellegr. in Bull. Soc. Bot. Fr. 83 : 38 (1936) ; Aubrév. Fl. For. C. Iv., ed. 2, 3 : 300 ; N. Hallé in Bull. Soc. Bot. Fr. 106 : 345. A shrub or tree 40–60 ft. high with a drooping crown ; flowers greenish-white. **Iv.C.:** Mudjika (fr. Aug.) *Aubrév.* 1798 ! **Ghana:** Amumuniso to Enchi (fl. & fr. May) *Enti* FH 6707 ! Mim, Ashanti (fl. & fr. June) *Andoh* FH 5512 ! **S.Nig.:** Sapoba *Kennedy* 177 ! 2552 !

53. CUVIERA DC.—F.T.A. 3 : 156. *Nom. cons.*

Style pubescent ; ovary 5-locular :
  Bracteoles ovate-lanceolate to linear-lanceolate, 1·5–4 cm. long, 4–8 mm. broad; calyx-lobes 5, linear to oblanceolate, 1–1·5 cm. long, 2·5–4 mm. broad ; leaves oblong to oblong-elliptic, 15–27 cm. long, 5–10 cm. broad, with about 10 pairs of lateral nerves, subcordate or upper ones cuneate at base ..   ..   1. *longiflora*
  Bracteoles and calyx-lobes subulate to linear, 1–2 (–4) mm. broad :
    Calyx-lobes and bracteoles subulate, about 1 cm. long and 1 mm. broad ; inflorescences congested ; leaves oblong-elliptic to narrowly oblong, 18–20 cm. long, 3–6 cm. broad, with 6–8 pairs of lateral nerves, cuneate to rounded or rarely subcordate at base
                                                   2. *macroura*
    Calyx-lobes and bracteoles linear, 2–4 cm. long, 2 (–4) mm. broad ; inflorescences branched ; leaves oblong, 20–35 cm. long, 6–8 cm. broad, subcordate at base
                                                   3. *subuliflora*
Style glabrous ; ovary (1–) 2–5-locular :
  Calyx-lobes spathulate, 1 cm. long ; ovary 3-celled ; leaves oblong to oblong-elliptic, broadly acuminate at apex, rounded at base, about 15 cm. long and 5 cm. broad, with about 8 pairs of lateral nerves ; petiole 1 cm. long ; flowers few in a small cyme ; corolla glabrous ; fruits obliquely ellipsoid, ribbed, 3 cm. long
                                                   4. *trilocularis*
  Calyx-lobes never spathulate :
    Calyx truncate or calyx-lobes 1–5, unequal ; bracteoles oblanceolate, 3·5–4 cm. long ; leaves oblong, about 25 cm. long and 10 cm. broad, coriaceous   5. *truncata*
    Calyx-lobes 5, equal, calyx never truncate :
      Corolla glabrous ; inflorescences much-branched paniculate cymes :
        Ovary and fruit 5-locular ; calyx-lobes linear-lanceolate, about 10 mm. long, 1–2 mm. broad ; bracteoles similar, 1·5–2·5 cm. long, 2–3 mm. broad ; leaves coriaceous, oblong, 12–20 cm. long, 4–9·5 cm. broad, shortly acuminate at apex, rounded to subcordate at base   ..   ..   6. *acutiflora*
        Ovary and fruit (1–) 2-locular ; calyx-lobes lanceolate, acute, 5 mm. long, 1·5 mm. broad at base ; bracteoles similar, about 1 cm. long ; leaves hardly coriaceous,

oblong to lanceolate, 11–14 cm. long, 3·5–7 cm. broad, acuminate at apex,
cuneate to rounded at base .. .. 7. sp. *A*
Corolla sparsely pilose ; leaves not coriaceous, elliptic, 6–12 cm. long, 2·5–5·5 cm.
broad, long-apiculate at apex :
Calyx-lobes lanceolate, long-attenuate at apex, longer than corolla, about 2 cm. long
and 6 mm. broad at base .. .. 8. *calycosa*
Calyx-lobes narrowly oblong-elliptic, acute at apex, shorter than corolla, about
8 mm. long and 3 mm. broad .. .. 9. *nigrescens*

1. **C. longiflora** *Hiern* in F.T.A. 3 : 157 (1877) ; Hepper in Kew Bull. 16 : 453 (1963). *C. angolensis* Welw.
ex Hiern (1898). A small glabrous tree up to 30 ft. high, with arching branches usually infested with ants ;
flowers green, with pale yellowish bracts ; in montane forest.
[Br.]**Cam.**: Cam. Mt., 2,000–3,200 ft. (Dec.-Jan.) *Mann* 1211 ! *Maitland* 1212 ! Bafut-Ngemba F.R.
(Mar.) *Richards* 5307 ! *Tiku* FHI 22247 ! Bali-Ngemba F.R., 6,500 ft. (June) *Ujor* FHI 30415 ! Lakom,
6,000 ft., Bamenda (May) *Maitland* 1361 ! Nkambe, 3,600 ft. (Feb.) *Hepper* 1916 ! Also in Cameroun,
Congo and Angola.

2. **C. macroura** *K. Schum.* in Engl. Bot. Jahrb. 33 : 352 (1903). *C. subuliflora* of Chev. Bot. 331, not of Benth.
C. djalonensis A. Chev. Bot. 331, name only. A glabrous, straggling tree 20–40 ft. high ; flowers white,
becoming yellow, fragrant, crowded in short axillary cymes.
Guin.: Diaguissa *Chev.* 12417 ! Konkouré to Timbo *Chev.* 12496. Bilima *Caille* in *Hb. Chev.* 14821 ! **S.L.**:
Binkolo (bud Aug.) *Thomas* 1857 ! Bumbuna (Oct.) *Thomas* 3184 ! Komorobai (Dec.) *Thomas* 6914 !
Loma Mts. (Sept.) *Jaeger* 1959 ! **Lib.**: Peatah (fl. & fr. Oct.) *Linder* 1035 ! Zeahtown, Tchien Dist.
(Aug.) *Baldwin* 6967 ! **Iv.C.**: Prolo, Lower Cavally *Chev.* 19869 ! Bliéron to Tabou *Chev.* 19921 ! **Ghana**:
Kumasi *Vigne* FH 3008 ! Wassaw F.R. (June) *Vigne* FH 181 ! Anyinasin, Akim (fl. & fr. July) *Darko*
898 ! Boin River F.R., Enchi (July) *Agbley* FH 6263 ! **Dah.**: *Spire* 87 !

3. **C. subuliflora** *Benth.* in Fl. Nigrit. 407 (1849) ; F.T.A. 3 : 157 (1877). *C. plagiophylla* K. Schum. (1903). A
small tree 15–25 ft. high, with horizontal branches ; flowers reddish and green with yellow anthers ; at
the edge of forest.
Ghana: Kumasi (July) *Vigne* FH 1242 ! **S.Nig.**: Degema *Talbot* 3383 ! Oban *Talbot* 222 ! 2036 ! [Br.]**Cam.**:
Victoria (Feb., Oct.) *Maitland* 417 ! 740 ! Barombi (Feb.) *Binuyo & Daramola* FHI 35530 ! **F.Po**: *T. Vogel*
83 ! *Mann* 87 !

4. **C. trilocularis** *Hiern* in F.T.A. 3 : 157 (1877). A small glabrous shrub ; flowers few in axillary cymes, green,
bracts greenish-white.
S.Nig.: Calabar *Thomson* 122 ! Oban *Talbot* 221 ! 272 !

5. **C. truncata** *Hutch. & Dalz.* F.W.T.A., ed. 1, 2 : 118 (1931). *C. subuliflora* of Wernham in Cat. Talb. 131, not
of Benth. A scrambling shrub with occasional sharp supra-axillary spines ; corolla greenish outside and
orange-yellow inside, with conspicuous pale yellow-green bracts.
N.Nig.: Sanga River F.R., Jemaa Dist. (Nov.) *Keay & Onochie* FHI 21535 ! *Keay* FHI 37246 ! **S.Nig.**:
Onitsha *Unwin* 64 ! Degema *Talbot* 3686 ! Oban *Talbot* 286 ! [Br.]**Cam.**: Gangumi, Adamawa (Dec.)
Latilo & Daramola FHI 28861 ! Vogel Peak, Adamawa (Nov.) *Hepper* 1448 !
[Note : most of these specimens have irregular calyx-lobes and are intermediate between true *C. truncata*
(*Talbot* 286, 3686) and *C. acutiflora*—F.N.H.]

6. **C. acutiflora** *DC.* Ann. Mus. Paris 9 : 222, t. 15 (1807) ; F.T.A. 3 : 156 ; Chev. Bot. 331 ; Aubrév. Fl.
For. C. Iv., ed. 2, 3 : 300, t. 363. *Vangueriopsis coriacea* Robyns in Bull. Jard. Bot. Brux. 22 : 319 (1952).
Shrub or small tree up to 30 ft. high with leathery shining leaves ; flowers whitish-green, sometimes with
a pink centre, in spreading cymes with conspicuous pale bracteoles and calyx-lobes.
Guin.: R. Lanfofomé (Mar.) *Pobéguin* 907 ! **S.L.**: Leicester Peak (Dec.) *Sc. Elliot* 3898b ! Kambui Hills
(Mar.) *Lane-Poole* 463 ! Njala (Nov.) *Small* 808 ! Rokupr (Apr.) *Jordan* 240 ! **Lib.**: Monrovia (Nov.)
Linder 1539 ! Grand Bassa (July) *T. Vogel* 72 ! Brewersville (Dec.) *Baldwin* 10955 ! Fisherman's L.
(Dec.) *Baldwin* 10893 ! Gbanga (Oct.) *Traub* 299 ! **Iv.C.**: Yapo (Oct.) *Roberty* 15338 ! Dabou *Chev.* 15380 !
15456 ! Assinie *Chev.* 17865 ! 17877 ! Bliéron *Chev.* 19905 ! **Ghana**: Alavango, Hohoe Dist. (Nov.) *St.
C. Thompson* FH 3658 ! **N.Nig.**: Lokoja (Oct.) *Dalz.* 45 ! **S.Nig.**: Sapoba *Kennedy* 2079 ! Owena F.R.,
Ondo (Oct.) *Ujor* FHI 23920 ! Ibadan South F.R. (Dec.) *Onochie* FHI 7495 ! Olokemeji (Dec.) *Foster* 25 !
[Br.]**Cam.**: Victoria (Nov.) *Maitland* 784 ! *Kalbreyer* 29 ! Ambas Bay (Feb.) *Mann* 776 ! Also in Cameroun
and Gabon.

7. **C. sp. A.** A small tree about 15 ft. high, with straight spines 1½ in. long on some branches ; flowers greenish-
white.
Lib.: Geo, Gbanga (Dec.) *Baldwin* 10573 ! Ganta (fl. Apr., fr. May) *Harley* 513 !

8. **C. calycosa** *Wernham* in J. Bot. 52 : 7 (1914). A small tree about 20 ft. high ; flowers white turning yellow
and with a red centre ; in forest.
S.Nig.: Esuk Ekkpo Abassi, Eket *Talbot* 3300 ! [Br.]**Cam.**: S. Bakundu F.R., Kumba (Mar.) *Binuyo &
Daramola* FHI 35613 ! Also in Cameroun and Gabon.

9. **C. nigrescens** (*Sc. Elliot ex Oliv.*) *Wernham* in J. Bot. 49 : 321 (1911) ; A.F.I. For. C. Iv., ed. 2, 3 :
302, t. 363. *Vangueria nigrescens* Sc. Elliot ex Oliv. in Hook. Ic. Pl. t. 2283 (Jan. 1894) ; Sc. Elliot in J.
Linn. Soc. 30 : 81 (1 Feb. 1894). *Cuviera minor* C. H. Wright in Kew Bull. 1906 : 105. *C. trichostephana*
K. Schum. (1897). Shrub or small tree up to 25 ft. high, with pale grey bark and sharp spines on the bole,
slash green and brown, branches arching, leaves dry black ; flowers white, fragrant ; in forest.
Port.G.: Guiledje, Boé (June) *Esp. Santo* 2923 ! Gadamael, Cacine (July) *Esp. Santo* 2113 ! **Guin.**: Kaba
R. (May) *Chev.* 13134 ! 13232 ! **S.L.**: York Pass (June) *Lane-Poole* 52 ! Mesima (Apr.) *Deighton* 3721 !
Falaba (Mar.) *Sc. Elliot* 5736 ! Vevehun, Fwendu to Potoru (Apr.) *Deighton* 1615 ! **Lib.**: Jui, Gola (Apr.)
Bunting ! **Iv.C.**: *Aubrév.* 91 ! 277 ! Mt. Kouan (Apr.) *Chev.* 21267 ! **Ghana**: Kwahu (Apr.) *Johnson* 646 !
Kumasi (Apr.) *Vigne* FH 1113 ! 1712 ! Boankra, Kumasi Dist. (Mar.) *Andoh* 5484 ! S. Fomang Su F.R.,
Ashanti (Feb.) *Vigne* FH 1834 ! **S.Nig.**: Lagos *Foster* 73 ! Sapoba (Apr.) *Kennedy* 2319 ! Onyeagocha
FHI 7124 ! Omo F.R., Ijebu Ode (May) *Tamajong* FHI 16942 ! Owo F.R. (May) *Jones* FHI 3610 ! Ijaiye
F.R., Ibadan (Apr.) *Onwudinjo* FHI 23400 !

54. FADOGIA Schweinf.—F.T.A. 3 : 152, partly ; Robyns in Bull. Jard. Bot. Brux. 11 : 41 (1928).

Leaves glabrous beneath except sometimes (in No. 4) on the midrib and lateral nerves :
Small tree or shrub ; leaves in whorls of 3, elliptic, 4–7 cm. long, 2–4 cm. broad,
glabrous, with paler conspicuous lateral nerves ; flowers in dense clusters on the
previous season's growth ; calyx-lobes oblong-lanceolate, 2·5 mm. long ; corolla
4·5 mm. long, 5–6-lobed ; fruit oblique, 1 cm. long .. 1. *erythrophloea*
Perennial herbs or undershrubs :
Calyx truncate ; flowers in few-flowered cymes without bracteoles or solitary ; pedicels
about 1 cm. long ; corolla 5–6 mm. long, glabrous ; leaves usually in threes, elliptic-
lanceolate, 3·5–6 (–10) cm. long, 1–2 (–2·5) cm. broad, glabrous 2. *ledermannii*

Calyx toothed :
  Branchlets glabrous ; leaves paired or in threes, oblanceolate, 4–5 cm. long, about
    1·5 cm. broad, closely venose beneath ; flowers few, in very shortly pedunculate
    cymes ; calyx-lobes ovate ; corolla 1 cm. long, with acuminate lobes **3. andersonii**
  Branchlets pubescent ; leaves in whorls of 3 or 4, obovate-elliptic, abruptly
    apiculate, subsessile, 6–8 cm. long, 3–4·5 cm. broad, setulose above, at length
    glabrous, setulose on the midrib beneath, glaucous-green ; flowers about 3 in
    shortly pedunculate axillary cymes ; pedicels about 3 mm. long ; calyx-lobes
    linear-lanceolate ; corolla about 8 mm. long, lobes acuminate ; fruit globose
                                                                              4. pobeguinii
Leaves pubescent or tomentose more or less all over the lower surface :
  Calyx subulate-dentate :
    Leaves lanceolate, acute at both ends, 5–6 cm. long, 1–1·5 cm. broad, sparingly
      setulose above, softly tomentellous beneath, with 4–5 pairs of lateral nerves ;
      flowers in small 2–3-flowered cymes   ..   ..   **5. rostrata**
    Leaves obovate or elliptic-obovate, subacute, 3·5–8 cm. long and about 2·5 cm. broad,
      softly tomentellous beneath ; flowers about 3 in each axillary cyme ; corolla-
      lobes sparingly pubescent outside ; fruits globose, hardly 1 cm. diam.
                                                                              6. cienkowskii
  Calyx with broad ovate-triangular lobes ; leaves obovate, rounded at apex, sub-
    cuneate at base, 4–8 cm. long, 2–4 cm. broad, densely softly tomentose beneath ;
    flowers several in short axillary cymes, tomentose all over ; corolla about 8 mm.
    long   ..   ..   **7. agrestis**

1. **F. erythrophloea** (*K. Schum. & K. Krause*) Hutch. & Dalz. F.W.T.A., ed. 1, 2 : 109 (1931) ; Hepper in Kew
   Bull. 16 : 453. *Vangueria erythrophloea* K. Schum. & K. Krause in Engl. Bot. Jahrb. 39 : 533 (1907).
   V. dalzielii Hutch. (1913). *Fadogia djalonensis* A. Chev. ex Robyns in Bull. Jard. Bot. Brux. 11 : 88
   (1928) ; Chev. Bot. 331, name only. *F. cinerascens* Robyns l.c. (1928). *F. leucophloea* Hiern var. *djalonensis*
   Aubrév. Fl. For. Soud.-Guin. 480, name only. Shrub or small tree 4–19 ft. high, with the younger branches
   grey, the bark later flaking and becoming red ; flowers yellow-green or white, usually on leafless branches ;
   in savanna woodland.
   Mali: Kobale *Chev.* 334 ! **Guin.:** Bissikrima (June) *Pobéguin* 801 ! Kankan *Chev.* 572 ! Labé (fr. Apr.)
   Chev. 12307 ! Mamou to Timbo (Dec.) *Chev.* 14639 ! **S.L.:** *Sc. Elliot* 5159b ! **Iv.C.:** Boundiali *Aké Assi*
   IA 4693. **Ghana:** Dutukpene (fl. & fr. Mar.) *Hepper & Morton* A3058 ! Shiare (Apr.) *Hall* 1420 ! **Togo Rep.:**
   Basari *Kersting* 539. **Dah.:** Atacora Mts. *Chev.* 24065 ! **N.Nig.:** Naraguta *Lely* 64 ! Katagum *Dalz.* 379 !
   Katsina to Zaria (Dec.) *Meikle* 823 ! Katabu, Zaria (fl. & fr. May) *Keay* FHI 25790 ! Zuru, Niger Prov.
   (fl. & fr. Mar.) *Daley* FHI 32254 ! **[Br.]Cam.:** Gangumi, Adamawa (fl. & fr. Dec.) *Latilo & Daramola*
   FHI 28893 ! Also in Cameroun. (See Appendix, p. 397.)
2. **F. ledermannii** *K. Krause* in Engl. Bot. Jahrb. 48 : 416 (1912) ; Robyns in Bull. Jard. Bot. Brux. 11 : 59 ;
   Hepper in Kew Bull. 16 : 454. An erect herb 1½–3 ft. high with several stems arising from a woody rootstock ; flowers greenish-white with white hairs in the throat ; in savanna.
   [Br.]Cam.: Vogel Peak, Adamawa (Nov.) *Hepper* 1368 ! Also in Cameroun and Uganda.
3. **F. andersonii** *Robyns* l.c. 86 (1928). Erect undershrub 1–2 ft. high, stems more or less 3-angled from a stout
   woody rootstock ; flowers greenish-yellow, fruits yellow ; in savanna woodland.
   Ghana: Volta R. *Anderson* 43 ! Damongo (Apr., July) *Adams* 3942 ! *Andoh* 5204 ! **N.Nig.:** Samaru, Zaria
   (fl. & fr. June) *Keay* FHI 25862 !
   [Note — *Keay* FHI 25862 has more spreading inflorescences and the corolla is more pubescent than the
   type—F.N.H.]
4. **F. pobeguinii** *Pobéguin* in Ess. Fl. Guin. Franç. 316 (1906) ; Robyns l.c. 74. *F. dalzielii* Robyns l.c. 66 (1928).
   F. cienkowskii of Robyns l.c. 79, partly (*Chev.* 12832). Erect undershrub 1–4 ft. high, with several pubescent
   stems arising from a woody rootstock ; flowers yellow ; in savanna.
   Guin.: *Farmar* 270 ! Timbo *Pobéguin* 89 ! Fouta Djalon (fr. Mar.) *Chev.* 12832 ! **Ghana:** Banda F.R.,
   Wenchi Dist. (July) *Enti* FH 6264 ! Damongo Scarp (May) *Enti* FH 6981 ! Jema (fr. June) *Brown* 2279 !
   N.Nig.: Kontagora (fl. & fr. Feb.) *Dalz.* 218b ! Anara F.R., Zaria (May, Oct.) *Keay* FHI 21135 ! 22929 !
   Gidan Mudi, Zaria (July) *Keay* FHI 37123 ! **S.Nig.:** Ngwo, Udi Dist. (May) *Jones* FHI 1694 ! Also in
   Cameroun.
5. **F. rostrata** *Robyns* l.c. 78 (1928). Erect herb or undershrub about 1½ ft. high from a woody base.
   Dah.: Pelebina to Djougou *Chev.* 23855 ! Tomba, Atacora *Chev.* 24143 !
6. **F. cienkowskii** *Schweinf.* in Reliq. Kotsch. 47, t. 32 (1868) ; F.T.A. 3 : 154 ; Robyns l.c. 79 ; Aubrév. Fl.
   For. Soud.-Guin. 480, t. 106, 4–5 ; Chev. Bot. 330. *Cuviera cienkowskii* (Schweinf.) Roberty (1954). Erect
   undershrub from stout base 1½–3 ft. high with pale under surfaces to the leaves ; flowers greenish-yellow ;
   in savanna.
   Mali: Bambonatounba (Mar.) *Chev.* 561 ! **Guin.:** Faranna *Chev.* 13407. Timbo *Caille* in Hb. *Chev.* 14672 !
   Iv.C.: Ouango-Fitini *Aké Assi* IA 4692. **Ghana:** Karega (Apr.) *Williams* 147 ! **Dah.:** Kouandé to
   Konkobiri, Atacora Mts. *Chev.* 24274 ! **N.Nig.:** Nupe *Barter* 1116 ! Naraguta *Lely* 48 ! Shere Hills, Plateau
   (fr. Dec.) *Coombe* 111 ! Birnin Gwari (Mar.) *Meikle* 1340 ! Cece F.R., Lapai Dist. (Apr.) *Abiai* FHI 7044 !
   S.Nig.: Ngwo, Udi Dist. (Dec.) *Jones* FHI 6844 ! Milliken Hill, Enugu (Feb.) *Abrahall* FHI 27558 ! Widespread in the drier parts of tropical Africa. (See Appendix, p. 397.)
7. **F. agrestis** *Schweinf. ex Hiern* in F.T.A. 3 : 154 (1877) ; Robyns l.c. 92 ; Aubrév. Fl. For. Soud.-Guin. 480,
   t. 106, 1–3 ; Chev. Bot. 330. *F. cienkowskii* of Chev. Bot. 330, partly (*Chev.* 24274). An erect undershrub
   with yellowish tomentellous stems and leaves, 1–3 ft. high ; flowers yellow-green ; in savanna.
   Mali: Koumantou *Chev.* 714 ! Bobo-Dioulasso, U. Volta (fr. June) *Serv. For.* 1963 ! **Guin.:** Dafila *Pobéguin*
   798 ! Banko *Pobéguin* 802 ! Dantilia (Mar.) *Sc. Elliot* 5301 ! **Ghana:** Damongo *Andoh* FH 1963 ! 5207 !
   Yeji (fr. Aug.) *Vigne* FH 3357 ! Tamale to Yendi (fl. & fr. Mar.) *Adams* 3853 ! Nakpanduri (fl. & fr. Mar.)
   Hepper & Morton A3134 ! Salaga *Krause* ! **Dah.:** Abomey *Chev.* 23132 ! Gouka to Bante *Chev.* 23724 !
   Djougou *Chev.* 23871 ! **N.Nig.:** Jebba (Jan.) *Meikle* 1004 ! Samaru (fl. & fr. May) *Keay* FHI 25723 !
   Sanga River F.R., Zaria (July) *Keay* FHI 37125 ! Vom, 3,000–4,500 ft. *Dent Young* 121 ! **S.Nig.:** Lagos
   (fr. Mar.) *Dennett* 485 ! Also in Cameroun, Ubangi-Shari and Sudan. (See Appendix, p. 397.)

55. VANGUERIOPSIS Robyns in Bull. Jard. Bot. Brux. 11 : 23, 248 (1928).

Corolla glabrous outside or at most shortly puberulous :
  Young branches glabrous :
    Petioles 4–5 mm. long ; leaves obovate-elliptic, shortly abruptly acuminate, 10–

14 cm. long, 4·5–7 cm. broad, leathery, with about 5 pairs of lateral nerves ; pedicels and ovary puberulous ; calyx-lobes suborbicular, conspicuous ; branchlets without spines .. .. 1. *calycophila*
Petioles about 1 cm. long ; pedicels and ovary glabrous :
  Branchlets greenish, without spines ; leaves broadly oblong-elliptic, abruptly long-acuminate, rounded at base, 10–15 cm. long, 5–7 cm. broad, with 6–7 pairs of lateral nerves ; calyx-lobes 1–2 mm. long, linear-lanceolate ..   2. *chlorantha*
  Branchlets grey, with sharp slightly curved spines ; leaves broadly elliptic to obovate, acutely acuminate, rounded and subcordate to acute at base, 6–12 cm. long, 2·5–7 cm. broad, very thin, glabrous ; calyx-lobes inconspicuous
                                            3. *spinosa*
Young branches pubescent or tomentose :
  Branchlets often spiny ; leaves much paler beneath, broadly elliptic, abruptly and obtusely acuminate, 6–10 cm. long, 3–4 cm. broad, puberulous on the nerves beneath ; cymes many-flowered ; calyx-lobes broadly ovate, small ; corolla 5 mm. long ; fruit very oblique, 1·5 cm. long .. ..   4. *discolor*
Corolla tomentose or (at least the lobes) pilose outside ; branchlets pubescent or setose :
Corolla tomentose outside ; calyx-lobes ovate, 3 mm. long :
  Corolla-buds with long subulate apices ; leaves ovate-elliptic, acutely acuminate, rounded at base, 6–12 cm. long, 3·5–7 cm. broad, pubescent only on the nerves beneath ; veinlets very close and parallel beneath ; fruit mostly kidney-shaped, 2·5–3 cm. long .. .. 5. *vanguerioides*
  Corolla-buds not apiculate or very shortly so ; leaves ovate-elliptic to lanceolate, shortly acuminate, more or less rounded at base, 8–15 cm. long, 4·5–7 cm. broad, shortly pubescent on the nerves and veins beneath, the latter close and fine ; cymes laxly few-flowered, with rather large persistent bracts and broad calyx-lobes ; fruit about 2·5 cm. long, 1–2-seeded .. .. 6. *nigerica*
Corolla-lobes pilose ; calyx-lobes subulate, 1–5 mm. long :
  Leaves ovate-oblong or elliptic, never over 5·5 cm. broad, caudate-acuminate, very thin, rather densely pilose on both surfaces ; branchlets setose and with curved sharp spines ; cymes subracemose, long-pilose ; pedicels slender, about 5 mm. long ; upper part of corolla long-pilose ; calyx-lobes 3–5 mm. long, subulate 7. *subulata*
  Leaves broadly ovate-rounded up to 8·5 cm. long, 6·5–7 cm. broad, abruptly and shortly acuminate, shortly pilose on the nerves beneath ; branchlets with rather long spines ; cymes subracemose, shortly pilose ; pedicels 3 mm. long ; corolla-lobes very thinly pilose ; calyx-lobes 1 mm. long, subulate .. ..   3. *spinosa*

1. **V. calycophila** (*K. Schum.*) *Robyns* in Bull. Jard. Bot. Brux. 11 : 256 (1928). *Plectronia calycophila* K. Schum. (1899). Erect or climbing shrub ; flowers greenish, few in short loose cymes ; in forest.
  S.Nig.: Oban *Talbot* 246 ! 1657 ! Eket *Talbot* ! Also in Cameroun and Gabon.
2. **V. chlorantha** (*K. Schum.*) *Robyns* l.c. 255 (1928). *Plectronia chlorantha* K. Schum. (1897). A glabrous shrub 6–10 ft. high with smooth greenish branchlets and thorns on the bole ; flowers greenish-white ; in forest. **S.Nig.:** Oban *Talbot* 1549 ! 1653 ! Aningeje to Mamfe, Calabar Dist. (fr. Apr.) *Ujor* FHI 31778 ! [Br.]**Cam.:** Banga, S. Bakundu F.R. (Mar.) *Binuyo & Daramola* FHI 35607 ! Also in Cameroun.
3. **V. spinosa** (*Schum. & Thonn.*) *Hepper* in Kew Bull. 17 : 170 (1963). *Vangueriopsis leucodermis* (K. Krause) Hutch. & Dalz. F.W.T.A., ed. 1, 2 : 117 (1931). *Phallaria spinosa* Schum. & Thonn. Beskr. Guin. Pl. 133 (1827). *Canthium thonningii* Benth. (1849)—F.T.A. 3 : 134. *Chomelia leucodermis* K. Krause (1909). *Vangueriopsis membranacea* Robyns l.c. 271 (1928). *V. violacea* Robyns l.c. 270 (1928). A shrub or small tree up to 12 ft. high ; flowers greenish, the corolla appearing inflated ; in shady savanna woodland.
  Iv.C.: Ouango-Fitini (Apr.) *Mangenot & Aké Assi* 4266 ! **Ghana:** Accra Plains (Jan., Apr., fr. Mar.) *Irvine* 207 ! *Adams* 3741 ! 3790 ! Damongo Scarp (Mar.) *Hepper & Morton* A3179 ! Akim, Ashanti (fr. Apr.) *Irvine* 960 ! Kpandu (Feb.) *Robertson* 108 ! **Togo Rep.:** Lomé *Warnecke* 349 ! Basari (fr. June) *Kersting* A69. **S.Nig.:** Lagos *Rowland* ! Olokemeji *Foster* ! *Hepper* 2305 ! Ijebu (Apr.) *Tamajong* FHI 23251 !
4. **V. discolor** (*Benth.*) *Robyns* l.c. 264 (1928) ; Aubrév. Fl. For. Soud.-Guin. 484 ; Berhaut Fl. Sén. 99. *Canthium discolor* Benth. (1849)—F.T.A. 3 : 138 ; Chev. Bot. 329. *Canthium venosissimum* Hutch. & Dalz. F.W.T.A., ed. 1, 2 : 115, partly (*Linder* 267). Small tree or scrambling shrub, often spiny, the young branchlets rusty-pubescent, with leaves pale beneath ; flowers greenish in short branched cymes often at nodes below the leaves.
  Sen.: Diouloulou, Casamance (fr. Sept.) *Adam* 18139 ! Badi *Berhaut* 1807 ; 4736. **Port.G.:** Fonte, Bubaque (May) *Esp. Santo* 2021 ! 2052 ! Pessubé, Bissau (Mar.) *Esp. Santo* 1153 ! **Guin.:** Kouroussa (Mar.) *Pobéguin* 664 ! Ditinn (Apr.) *Chev.* 12993 ! Fouta Djalon *Heudelot* 802 ! Bafing (Apr.) *Adam* 11864 ! **S.L.:** Freetown (Apr.) *Deighton* 4747 ! *Hepper* 2481 ! Bagroo R. (Apr.) *Mann* 878 ! Njala (Apr.) *Deighton* 1120 ! Rokupr (Apr.) *Jordan* 233 ! **Lib.:** Dukwia R. (May) *Cooper* 425 ! Du R. (fr. Aug.) *Linder* 267 ! White Plains (Apr.) *Barker* 1279 ! Kulo, Sinoe (Mar.) *Baldwin* 11406 ! **Iv.C.:** Danané *Chev.* 21185 ! (See Appendix, p. 414.)
5. **V. vanguerioides** (*Hiern*) *Robyns* l.c. 260 (1928) ; Aubrév. Fl. For. C. Iv., ed. 2, 3 : 302. *Canthium vanguerioides* Hiern in F.T.A. 3 : 146 (1877), mostly ; Chev. Bot. 329. *Plectronia vanguerioides* (Hiern) K. Schum. (1897). Shrub or small tree up to about 25 ft. high ; flowers greenish in lax branched cymes mostly at nodes below the leaves ; in secondary forest.
  Guin.: Kouria *Dumas* 18194. Kaba to Mamou *Chev.* 12754 ! **S.L.:** Sugar Loaf Mt. (Mar., Apr.) *Sc. Elliot* 5776 ! *Hepper* 2544 ! Kenema (May) *Lane-Poole* 280 ! Baoma (Apr.) *Deighton* 3218 ! Roruks (fr. July) *Deighton* 3256 ! **Lib.:** Ganta *Harley* 1905 ! **Iv.C.:** Zoanhé, Upper Sassandra (May) *Chev.* 21462 ! Nimba Mt. *Aubrév.* 1136 ! Mt. Tonkoui (Mar.) *Leeuwenberg* 2968 ! *Aubrév.* 998 ! (See Appendix, p. 414.)
6. **V. nigerica** *Robyns* l.c. 261, figs. 25, 26 (1928). *Canthium vanguerioides* Hiern in F.T.A. 3 : 138 (1877), partly ; Robyns l.c. 260, partly (*Chev.* 23710, *Baumann* 308) ; F.W.T.A., ed. 1, 2 : 116, partly. *Vangueriopsis lanceolata* Robyns l.c. 271 (1928). A shrub or tree with arching branches ; flowers greenish in lax branched cymes mostly at older nodes ; in streamside forest.
  Ghana: Afram Plains (Mar.) *Johnson* 710 ! Neung F.R., Tarkwa *Enti* FH 6229 ! **Togo Rep.:** Misahöhe *Baumann* 243 ! 308 ! **Dah.:** Savalou (May) *Chev.* 23710 ! **N.Nig.:** Lokoja *Elliott* 229 ! Katsina Allah

(June) *Dalz.* 756 ! Mando F.R., Zaria (June) *Keay* FHI 25820 ! **S.Nig.:** Abeokuta *Irving* 155 ! Idu *Barter* 1251 ! Ibadan (Apr.) *Meikle* 1442 ! Ijaiye F.R., Oyo (Apr.) *Onochie* FHI 21970 ! Also in Cameroun.
7. **V. subulata** *Robyns* l.c. 268 (1928). *V. setosa* Robyns l.c. 268 (1928). A scrambling or climbing shrub with spines ; flowers greenish ; in forest.
    Guin.: Mt. Boola (Mar.) *Chev.* 20928 ! **S.L.:** Berria, Falaba (Mar.) *Sc. Elliot* 5400 ! **S.Nig.:** Oban *Talbot* 1688 ! Degema *Talbot.* [Br.]**Cam.:** Mbalange F.R., Kumba (Mar.) *Binuyo & Daramola* FHI 35649 ! Fonfuka, Bamenda (Apr.) *Maitland* 1768 ! Also in Cameroun and Ubangi-Shari.

56. CANTHIUM Lam.—F.T.A. 3 : 132.

Inflorescences sessile or subsessile umbellate fascicles ; calyx truncate :
  Flower-buds very acute :
    Corolla 1–1·3 cm. long ; pedicels slender, 8 mm. long, about 6 or more on a short peduncle ; leaves elliptic, shortly and broadly acuminate, 9–13 cm. long, 4–6 cm. broad, with about 6 pairs of lateral nerves raised on both surfaces, glabrous, tertiary nerves obscure   ..   ..   1. *henriquesianum*
  Corolla up to 7 mm. long, or if 10 mm. long (in No. 3) then the tertiary veins prominent on the leaves above :
    Veins between the 4–5 pairs of lateral nerves obscure or invisible when dry ; leaves broadly elliptic, acuminate, rounded to the acute base, 6–10 cm. long, 3–5 cm. broad, with very prominent sharp nerves beneath, glabrous ; corolla about 7 mm. long ..   ..   2. *acutiflorum*
    Veins between the lateral nerves prominent and raised on the upper surface when dry ; pedicels elongated and slender in fruit ; leaves oblong-elliptic, acuminate, shortly cuneate at the base, 7–10 cm. long, 4–6 cm. broad, glabrous ; corolla about 8 mm. long ; fruits didymous or obliquely globose, the lobes about 7 mm. diam., black   ..   ..   3. *horizontale*
  Flower-buds rounded at apex ; leaves acutely cuneate at base, ovate, gradually acuminate, 4–10 cm. long, 2–4·5 cm. broad, glabrous beneath, glossy above ; pedicels of flower about 7 mm. long, densely fasciculate   ..   ..   4. *schimperianum*
Inflorescences shortly pedunculate cymes ; corolla-buds obtuse :
  Branchlets conspicuously hispid-pilose, but not pubescent or glabrous :
    Stipules large and persistent, broadly lanceolate about 1 cm. broad and 2 cm. long ; stems, leaves beneath and inflorescences densely pilose with reddish hairs ; leaves oblong-elliptic, rounded at base, shortly acuminate at apex, 10–14 cm. long, 4–5 cm. broad, glabrous above including midrib ; calyx-teeth acute, about 2 mm. long
         5. sp. *A*
    Stipules inconspicuous and often caducous ; leaves pilose above, at least the midrib :
      Leaves with 8–9 parallel nerves on each side of midrib, oblong-elliptic, shortly cordate at base, acuminate at apex, 5–12 cm. long, 3–5·5 cm. broad, pilose or scabrid above, usually densely pubescent beneath ; calyx shortly lobed, densely pilose ; stems densely pilose   ..   ..   6. *rufivillosum*
      Leaves with 5–6 widely spaced nerves on each side of midrib ; calyx pilose or nearly glabrous :
        Hairs on stems few and near the nodes ; leaves 8–15 cm. long, 4–8 cm. broad ; (see other characters below)   ..   ..   10. *setosum*
        Hairs evenly spaced on stems ; leaves 5–11(–13) cm. long, 2–4(–6) cm. broad, rounded to slightly cordate at base, hispid ; calyx rather densely pubescent
           7. *hispidum*
  Branchlets glabrous or pubescent, but not hispid-pilose :
    Branchlets glabrous or almost so :
      Pedicels densely and rather long-pubescent :
        Stipules linear-lanceolate, with a broad base and linear above, about 6 cm. long, caducous and only seen at the younger nodes ; leaves elliptic or obovate-elliptic, cuneate at base, shortly acuminate at apex, 8–15 cm. long, 2·5–6 cm. broad, entirely glabrous on both sides (rarely a few hairs in the axils of the lateral nerves beneath)   ..   ..   8. *mannii*
        Stipules ovate-lanceolate, 9–16 mm. long, 5–13 mm. broad, persistent ; (other characters as below)   ..   ..   21. *rubens*
      Pedicels glabrous or more or less shortly pubescent :
        *Pedicels of flowers 5–8 mm. long ; inflorescences not spreading-branched :
          Inflorescences subsessile ; leaves cordate or truncate at base :
            Armed with 1 cm.-long supraxillary spines ; leaves oblong-ovate, cuneate at base, caudate at apex, 5–10 cm. long, 2–4 cm. broad, glabrous or sparsely pubescent; pedicels more or less pubescent ; flowers few ; calyx very shortly toothed
               9. *orthacanthum*
            Unarmed ; flowers numerous :
              Leaves setose on both surfaces, especially beneath, glabrescent above later, subcordate at base, shortly and abruptly acuminate, broadly ovate-elliptic, 8–15 cm. long, 4–8 cm. broad ; pedicels setose ; calyx shortly lobed
                 10. *setosum*
              Leaves glabrous except for the nerves and clusters of brown hairs in some

axils, rather long acutely and abruptly acuminate, broadly obovate, 9–13 cm.
long, 5–8 cm. broad ; pedicels glabrous ; calyx truncate   11. *rubrinerve*

Inflorescences pedunculate ; flowers congested on the peduncle :

Leaves broadly ovate-rounded, unequally rounded at base, broadly acuminate,
4–5 cm. long, 2·5–3·5 cm. broad, glabrous ; calyx-lobes glabrous
    12. *inaequilaterum*

Leaves elliptic to ovate, shortly acuminate, cuneate or rarely rounded at base,
(5–)7–10 cm. long, 3–4·5 cm. broad, glabrous ; calyx-lobes usually with a tuft
of hairs ..   ..   13. *multiflorum*

*Pedicels of flowers up to 3 mm. long (4 mm. in No. 15) ; inflorescences spreading-
branched :

Stipules persistent, densely pilose inside ; leaves ovate, 4–7 cm. long, 2·5–3·5 cm.
broad, glabrous ; peduncle and pedicels slightly puberulous ; calyx glabrous
    14. *pobeguinii*

Stipules caducous, glabrous inside :

Calyx, ovary, pedicels and peduncle minutely pubescent :

Inflorescences 1–2 cm. long, congested ; leaves oblong-elliptic, 6–9 cm. long,
3–4 cm. broad, midrib and lateral nerves raised above with lamina smooth
between them   ..   ..   15. *arnoldianum*

Inflorescences about 5 cm. long, rather lax and branched ; leaves ovate or ovate-
elliptic, 11–16 cm. long, 5–7 cm. broad, with raised reticulation above (at
least when dry)   ..   ..   16. *vulgare*

Calyx glabrous :

Leaves broadly ovate, rounded, truncate or slightly cordate at base, 9–20 cm.
long, 5–17 cm. broad, sparsely setose or glabrous on both sides ; lowland
tree   ..   ..   17. *subcordatum*

Leaves oblong or oblong-elliptic, rounded to cuneate at base, 15–20 cm. long,
about 6 cm. broad, glabrous ; highland tree   ..   18. *dunlapii*

Branchlets conspicuously pubescent :

Leaves harshly scabrid above and beneath with short stiff hairs, lanceolate, cuneate
to rounded at base, acuminate at apex, 9–15 cm. long, 3·5–7 cm. broad ; calyx,
ovary, pedicels and peduncle densely pubescent   ..   19. *scabrosum*

Leaves pubescent or glabrous above, not scabrid :

Leaves long-oblong, truncate at base, long-acuminate at apex, 9–15 cm. long, 2·5–
4 cm. broad, shortly pubescent beneath ; petiole about 3 mm. long ; pedicels
etc. pubescent   ..   ..   20. *palma*

Leaves more or less ovate or elliptic :

Stipules ovate, sometimes broadly so, more or less persistent ; bracteoles ovate at
apex of the peduncles ; leaves ovate, rounded at base, shortly acuminate at
apex, 6–13 cm. long, 3·5–7·5 cm. broad, glabrous above, pubescent beneath ; (see
other characters above) ..   ..   21. *rubens*

Stipules lanceolate usually with a linear apex :

Corolla-lobes 4 :

Leaves densely and softly pubescent beneath, ovate-elliptic, 6–8 cm. long,
3·4 cm. broad, tertiary nerves parallel   ..   ..   22. sp. *B*

Leaves pubescent on the nerves beneath :

Tertiary nerves markedly parallel, numerous and prominent especially beneath
23. *venosum*

Tertiary nerves reticulate, few and not prominent   ..   24. *zanzibaricum*

Corolla-lobes 5 ; tertiary nerves reticulate :

Pedicels glabrous, or sparsely setose ; (other characters as above) 10. *setosum*

Pedicels densely pubescent :

Leaves with about 10 lateral nerves on each side of the midrib, impressed above
and prominent beneath, rounded at base, shortly acuminate at apex, ovate,
7–14 cm. long, 3–7 cm. broad, pubescent on the nerves beneath, more or less
glabrous above ; inflorescences slightly branched   ..   25. *gueinzii*

Leaves with about 7 lateral nerves on each side of the midrib :

Flowers clustered at the end of the peduncle ; leaves subcordate to rounded at
base, acuminate at apex, ovate, 6–8·5 cm. long, 2–5·4 cm. broad, lateral
nerves impressed above ; calyx shortly toothed   ..   26. *manense*

Flowers in a closely branched inflorescence ; leaves usually cuneate, sometimes
rounded or subcordate at base, acute or slightly acuminate at apex, 4–
10 cm. long, 2–5 cm. broad, pubescent beneath or on the nerves only ;
calyx deeply lobed ..   ..   27. *cornelia*

1. **C. henriquesianum** (*K. Schum.*) *G. Tayl.* in Cat. S. Tomé 210 (1944). *Plectronia henriquesiana* K. Schum.
(1893). *C. kraussioides* Hiern in Cat. Welw. 1 : 473 (1898) ; F.W.T.A., ed. 1, 2 : 113; Aubrév. Fl. For.
Soud.-Guin. 484. A glabrous climbing or scrambling shrub with shiny coriaceous leaves ; flowers yellow-
green ; beside upland streams.
Guin.: Pita *Pobéguin* 2185! Boulivel *Pobéguin* 18042. Mali *Schnell* 2301! **S.L.:** Bintumane Peak,
5,400 ft. (Jan.) *T. S. Jones* 141! *Jaeger* 4170! Sankan Biriwa summit, 5,500 ft. (fr. Jan.) *Cole* 150!
N.Nig.: Jos Plateau (Nov., Jan.) *Lely* 719! P51! *Batten-Poole* 255! **S.Nig.:** Ikwette Plateau, 5,000 ft.,
Obudu Div. (Dec.) *Savory & Keay* FHI 25132! 25241! [**Br.**]**Cam.:** Sabga, 6,200 ft., Bamenda (fl. &

fr. Jan.) *Keay* FHI 28462! Mambila Plateau, 6,000 ft., Adamawa (fr. Jan.) *Latilo & Daramola* FHI 34398! Vogel Peak, 4,000 ft., Adamawa (Nov.) *Hepper* 1487! Also in S. Tomé, Angola and Nyasaland.

2. **C. acutiflorum** *Hiern* in F.T.A. 3 : 136 (1877). *Plectronia acutiflora* (Hiern) K. Schum. (1907). A glabrous shrub with coriaceous shiny leaves.
  Ghana: Gradaw, Banda Hills (fl. & fr. Oct.) *Rose Innes* GC 31620! Damongo Scarp *Harris*! **N.Nig.:** Anara F.R., Zaria Prov. (fr. May) *Keay* FHI 22989! **S.Nig.:** Stubbs Creek F.R., Eket Dist. (fr. May) *Onochie* FHI 33174! Calabar *Thomson* 97! Oban *Talbot* 1543! **[Br.]Cam.:** foot of Cam. Mt. (Feb.) *Mann* 1179! Mopanya, Cam. Mt. 4,000–5,000 ft. (July) *Kalbreyer* 182!
  [Species 2, 3 and 4 are easily confused when not in flower.]

3. **C. horizontale** (*Schum. & Thonn.*) *Hiern* in F.T.A. 3 : 137 (1877). *Phallaria horizontalis* Schum. & Thonn. (1827). *Canthium anomocarpum* DC. (1830)—F.T.A. 3 : 136 ; F.W.T.A., ed. 1, 2 : 113 ; Aubrév. Fl. For. Soud.-Guin. 484 ; Berhaut Fl. Sén. 101. *C. caudatiflorum* Hiern in F.T.A. 3 : 137 (1877) ; Chev. Bot. 328. *Plectronia anomocarpa* (DC.) K. Schum. (1900). Glabrous shrub, more or less scrambling ; flowers greenish-white, fruits black ; beside streams usually in savanna.
  Sen.: *Heudelot* 567! *Berhaut* 1862! Floup-Fedyan *Chev.* 2093! **Mali:** Tabacco *Chev.* 3025. **Port.G.:** Biombo, Bissau (fr. Feb.) *Esp. Santo* 1797! Prabis, Bissau (fr. Mar.) *Esp. Santo* 1855! **Guin.:** Santa to Timbo *Chev.* 12583! Kouria *Caille* in Hb. *Chev.* 14733! **S.L.:** York (Sept.) *Melville & Hooker* 402! Kangahun, Gandima (Mar.) *Deighton* 5993! Njala (fr. Feb.) *Deighton* 3996! Musaia (fr. Feb.) *Deighton* 4243! **Lib.:** Monrovia (Nov.) *Linder* 1570! Mecca (fr. Nov., Dec.) *Baldwin* 10397! 10826! **Iv.C.:** Mt. Niénoué, 1,600 ft. *Chev.* 19445! Ferkéssédougou *Aubrév.* 2743! Brafouédi (fl. & fr. Dec.) *Leeuwenberg* 2299! **Ghana:** Sekondi (Aug., Oct.) *Vigne* FH 4787! *Howes* 975! Achimota (fr. Mar.) *Irvine* 422! Sekodamasi (Oct.) *Vigne* FH 3405! Prang (fr. Jan.) *Vigne* FH 1561! **Togo Rep.:** Lomé *Warnecke* 120! Badja (Mar.) *Schlechter* 12976! **Dah.:** Allada (Mar.) *Chev.* 23400! **N.Nig.:** Nupe *Barter* 1034! Osi to Oke-Opin, Ilorin (fr. Dec.) *Ajayi* FHI 19277! Anara F.R., Zaria (Oct.) *Keay* FHI 5468! 20123! **S.Nig.:** Lagos Isl. *Barter* 2208! Iddo Isl. (fr. Dec.) *Millen*! Ezi (fr. Feb.) *Thomas* 2337! Idanre, Akure Div. (Jan.) *Brenan & Keay* 8654! Agulu, Onitsha (Oct.) *Onochie* FHI 34130! (See Appendix, p. 329.)

4. **C. schimperianum** *A. Rich.* Tent. Fl. Abyss. 1 : 350 (1847) ; F.T.A. 3 : 385 ; Bullock in Kew Bull. 1932 : 385 ; Aubrév. Fl. For. Soud.-Guin. 488, t. 107, 3. *C. euryoides* Bullock ex Hutch. & Dalz. F.W.T.A., ed. 1, 2 : 113 (1931) ; Bullock l.c. 384 ; F.W. Andr. Fl. Pl. A.-E. Sud. 2 : 429. *C. nitens* Hiern in F.T.A. 3 : 135 (1877), not of DC. Evergreen shrub, climber or small tree ; flowers whitish.
  Dah.: Atacora Mts. *Chev.* 24049. **S.Nig.:** R. Nun *Barter* 2114! Also in Cameroun, Sudan, Ethiopia, Uganda, Kenya, Tanganyika and Congo.

5. **C. sp. A.** Climbing shrub with long brown hairs ; in high forest and edge of savanna.
  S.Nig.: Sapoba (Sept.) *Onochie* FHI 34311! Mamu F.R., Akwa Dist. *Onochie* FHI 34067!
  [Apparently closely related to *C. rubens* Hiern].

6. **C. rufivillosum** *Robyns ex Hutch. & Dalz.* F.W.T.A., ed. 1, 2 : 113 (1931). A scrambling shrub densely covered with rust-coloured hairs ; in forest.
  S.L.: Yonibana (Nov.) *Thomas* 4740! 4946! **Lib.:** Kakatown *Whyte*! **S.Nig.:** Calabar (Nov.) *Baldwin* 13790! **[Br.]Cam.:** mile 42, Victoria to Kumba (Apr.) *Olorunfemi* FHI 30513!

7. **C. hispidum** *Benth.* in Fl. Nigrit. 409 (1849) ; F.T.A. 3 : 140 ; Chev. Bot. 329, partly. Climbing or straggling shrub ; flowers white ; in forest.
  S.L.: *Don*! Sugar Loaf summit, 2,490 ft. (fr. Nov.) *Deighton* 5619! Gbangbama *Thomas* 9467! **Lib.:** Dukwia R. (fr. Mar.) *Cooper* 309! Harbel (fr. July) *Baldwin* 6619! Nekabozu (Oct.) *Baldwin* 9959! Kolahun (Nov.) *Baldwin* 10142! **Iv.C.:** Sassandra Port *Chev.* 17916! *Leeuwenberg* 2233! Assinie *Chev.* 17873! 17884! **Ghana:** Essiama (Nov.) *Vigne* FH 1475! Aburi *Brown* 683! Kumasi (fr. Jan.) *Irvine* 40! Abotia Hills F.R., Ho Dist. *Enti* FH 6876! **Dah.:** Ouidah to Adjounja (Apr.) *Chev.* 23468! Allada to Niaouli *Chev.* 23394! **S.Nig.:** Lagos *Barter* 2182! *Dalz.* 1051! Oshun River F.R. (fr. May) *Chizea* FHI 23950! Olokemeji (fr. Nov.) *Latilo* FHI 8102! Ibadan (fl. May, fr. Jan.) *Keay* FHI 16209! *Meikle* 966! Also in Cameroun and (?) Gabon. (See Appendix, p. 393.)

8. **C. mannii** *Hiern* in F.T.A. 3 : 143 (1877). *C. nervosum* Hiern in F.T.A. 3 : 143 (1877) ; Aubrév. Fl. For. Soud.-Guin. 484. *Plectronia flaviflora* K. Schum. & K. Krause (1907). Climbing shrub up to 20 ft. high ; flowers white ; beside streams in forest.
  Guin.: R. Koumi *Pobéguin* 1630! Mamou *Pobéguin* 1622! Kissi *Chev.* 20759! **S.L.:** Bagroo R. (Apr.) *Mann* 814! Makunde (Apr.) *Sc. Elliot* 5695! Njala (Apr.) *Deighton* 4743! Tabor (Apr.) *Deighton* 3723! **S.Nig.:** Shasha F.R., Ijebu-Ode (fl. & fr. Apr.) *Keay* FHI 16098! Orle F.R., Benin (fr. Aug.) *Onochie* FHI 33287! Carter's Peak, Idanre (Jan.) *Brenan & Keay* 8690! Ikpan, Calabar (fr. July) *Ujor* FHI 31635! **[Br.]Cam.:** Tiko, Victoria (Feb.) *Maitland* 974! Johann-Albrechtshöhe (= Kumba) *Staudt* 511! Also in Cameroun, and Ubangi-Shari.

9. **C. orthacanthum** (*Mildbr.*) *Robyns* in Bull. Jard. Bot. Brux. 11 : 334 (1928) ; Aké Assi Contrib. 1 : 154 (excl. syn. *C. thonningii* and *Plectronia spinosa*). A spiny shrub ; in moist forest.
  Iv.C.: Banco Forest *Aké Assi* IA 5318 ; 5325. Also in Cameroun and Congo.

10. **C. setosum** *Hiern* in F.T.A. 3 : 141 (1877). *Plectronia macrophylla* K. Schum. (1899). A climbing shrub, sometimes up to 50 ft. high ; flowers white, fragrant ; at edge of forest.
  S.L.: Yonibana (Nov.) *Thomas* 4733! Mabonto (Oct.) *Thomas* 3576! 3606! **Lib.:** Dukwia R. (fr. Apr.) *Cooper* 389! Bushrod Isl. (fr. Aug.) *Baldwin* 13092! Beiden, Boporo Dist. (fr. Nov.) *Baldwin* 10271! Bobei Mt., Sanokwele Dist. (fr. Sept.) *Baldwin* 9632! **Ghana:** Kibbi to Potroase (fr. Oct.) *Glover* 9! Owabi (fl. & fr. Dec.) *Lyon* 2630! **S.Nig.:** Benin (Nov., Dec.) *Brenan* 8890! *Ajayi* FHI 26925! Afi River F.R., Ikom (Dec.) *Keay* FHI 28249! Oban *Talbot*! Degema Dist. *Talbot* 3680! **[Br.]Cam.:** Bimbia Forest, Victoria (fr. Feb.) *Maitland* 392! Cam. Mt., 2,500 ft. (Dec.) *Mann* 1190! Buea *Deistel* 628! **F.Po:** Musola (Jan.) *Guinea* 242! 978! 979! Also in Ubangi-Shari and Gabon.

11. **C. rubrinerve** (*K. Krause*) *Hepper* in Kew Bull. 16 : 157 (1962). *Plectronia rubrinervis* K. Krause (1917). *Canthium favosum* Hutch. & Dalz. F.W.T.A., ed. 1, 2 : 113 (1931). *C. setosum* of F.W.T.A., ed. 1, 2 : 113, partly (syn. *Plectronia rubrinervis* K. Krause), not of Hiern. A climber ; in forest.
  S.Nig.: Oban *Talbot* 1284! **[Br.]Cam.:** Victoria (Oct.) *Mildbr.* 10564! 10585! Barombi (fr. Feb.) *Binuyo & Daramola* FHI 35533! Also in Cameroun.

12. **C. inaequilaterum** *Hutch. & Dalz.* F.W.T.A., ed. 1, 2 : 115 (1931). A glabrous climbing shrub.
  S. Nig.: Oban *Talbot* 277!

13. **C. multiflorum** (*Schum. & Thonn.*) *Hiern* in F.T.A. 3 : 144 (1877). *Psychotria multiflora* Schum. & Thonn. (1827). *C. afzelianum* Hiern in F.T.A. 3 : 142 (1877) ; F.W.T.A., ed. 1, 2 : 113 ; Aubrév. Fl. For. Soud.-Guin. 483 ; Berhaut Fl. Sén. 105. *C. kitsoni* S. Moore in J. Bot. 48 : 220 (1910). *C. heudelotii* Hiern in F.T.A. 3 : 139 (1877) ; F.W.T.A., ed. 1, 2 : 114 ; Berhaut Fl. Sén. 101. A more or less straggling shrub with whitish flowers.
  Mali: Kita (July) *Duong* 353! 521! **Guin.:** *Heudelot* 653! Pita *Pobéguin* 2050! **S.L.:** Kumrabai (Dec.) *Thomas* 6944! 7068! Kukuna, Scarcles R. (fl. & fr. Jan.) *Sc. Elliot* 4503! 4685! Musaia to Faranna (bud Mar.) *Sc. Elliot* 5178! Njala (fr. Sept.) *Deighton* 2916! Mano *Thomas* 14069! **Ghana:** *Thonning*! Cape Coast (fl. Feb., fr. July) *T. Vogel* 63! *Hall* 1045! Accra (fl. & fr. Apr.) *Irvine* 277! *Rose Innes* GC 30344! **Dah.:** Atacora Mts. (June) *Chev.* 2399! 24128! Also in Cameroun, Gabon, Ubangi-Shari and Sudan.

14. **C. pobeguinii** *Hutch. & Dalz.* F.W.T.A., ed. 1, 2 : 115 (1931). A shrub or tree to 20 ft. high with dark reddish branchlets and leaves paler beneath.
  Guin.: Kouroussa (Apr., May) *Pobéguin* 229! 240! **Ghana:** N. Scarp F.R., Mpraeso Dist. (fr. May) *Enti* FH 7162!
  [Note : this may be a small-leaved form of *C. crassum* Hiern (in F.T.A. 3 : 145 (1877) ; Aubrév. Fl. For. Soud.-Guin. 484, t. 107, 1, 2) known from Ubangi-Shari, Sudan and other parts of Africa—F.N.H.].

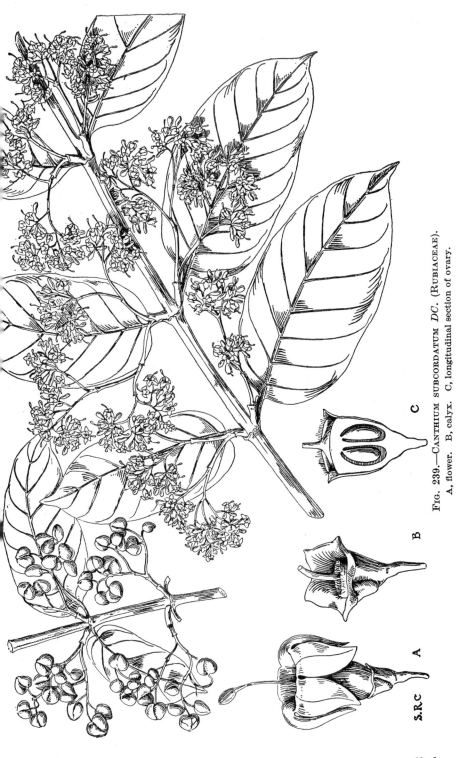

S.RC

FIG. 239.—CANTHIUM SUBCORDATUM *DC.* (RUBIACEAE).
A, flower. B, calyx. C, longitudinal section of ovary.

15. **C. arnoldianum** (*De Wild. & Th. Dur.*) *Hepper* in Kew Bull. 16 : 156 (1962). *Plectronia arnoldiana* De Wild. & Th. Dur. (1900). *Canthium tekbe* Aubrév. & Pellegr. in Bull. Soc. Bot. Fr. 83 : 37, fig. 1 (1936); Aubrév. Fl. For. C. Iv., ed. 2, 3 : 306, t. 365. *C. anomocarpum* of F.W.T.A., ed. 1, 2 : 113, partly (*Chev.* B22333), not of DC. A tree up to 80 ft. high ; in forest.
   Iv.C.: Yapo (fr. Oct.) *Chev.* B22333 ! Djibi F.R. *Aubrév.* 1362 ! Abidjan *Aubrév.* 1363 ! Banco F.R. *Aubrév.* 1371 ! **S.Nig.:** Ekiadolor F.R., Benin *Wardrop* FHI 293 ! Also in Cameroun and Congo.
16. **C. vulgare** (*K. Schum.*) *Bullock* in Kew Bull. 1932 : 374. *Plectronia vulgaris* K. Schum. in Pflanzenw. Ost-Afr. C : 386 (1895). *C. glabriflorum* of F.W.T.A., ed. 1, 2 : 115, partly (*Chev.* 23843), not of Hiern. *C. horizontale* of Chev. Bot. 329, (Chev. 23743), not of Schum. & Thonn. Tree 20–100 ft. high or sometimes a shrub, bark rough, slash brown with grey spots turning greenish ; flowers creamy-yellow or greenish-white, with an unpleasant smell ; beside rivers.
   Sen.: Basse Casamance (fr. Sept.) *Adam* 18038 ! **Port.G.:** Granja, Catio (June) *Esp. Santo* 2087 ! **Guin.:** Kouroussa (June) *Pobéguin* 704 ! Dalaba (fr. Sept.) *Chev.* ! Forécariah (June) *Jac.-Fél.* 1672 ! **S.L.:** Heddle's Farm, Freetown (May) *Deighton* 1199 ! Loma Mts. (fr. Sept.) *Jaeger* 1396 ! **Iv.C.:** Daloa to Oumé *Aubrév.* 1257 ! **Ghana:** Awura F.R., Ejura (fr. June) *Enti* FH 6708 ! Kade, Oda Dist. (fr. Dec.) *Enti* FH 6852 ! Bana Hill, Krobo (Apr.) *Irvine* 2895 ! Mampong (Apr.) *Vigne* FH 1928 ! **Dah.:** Savalou (May) *Chev.* 23743 ! **N.Nig.:** Gindiri, Jos Plateau (Oct.) *Hepper* 1109 ! **S.Nig.:** Olokemeji F.R. (May) Okeke & *Adebusuyi* FHI 28299 ! Owo F.R. (May) *Jones* FHI 3611 ! [**Br.**]**Cam.:** Bamenda to Kentu, 1,500 ft. (June) *Maitland* 1721 ! Widespread in tropical Africa.
   [Note : *Esp. Santo* 2041 & 2052 from Portuguese Guinea have the calyx slightly and minutely pubescent like *C. vulgare* but leaves setose above like some forms of *C. subcordatum* DC.]
17. **C. subcordatum** DC. Prod. 4 : 473 (1830) ; F.T.A. 3 : 141 ; Aubrév. Fl. For. Soud.-Guin. 480 ; Fl. For. C. Iv., ed. 2, 3 : 306, t. 366B. *C. glabriflorum* Hiern in F.T.A. 3 : 140 (1877) ; Chev. Bot. 329. *Plectronia glabriflora* (Hiern) K. Schum. (1892). A small or medium-size tree about 30 ft. or up to 80 ft. high with horizontal branches near the crown, slash whitish, buttresses at the base ; flowers white, heavily scented ; in the more open parts of high forests.
   Gam.: Albreda *Leprieur* ! **Guin.:** Mt. Benna, Kindia (Dec.) *Jac.-Fél.* 2120 ! Labé *Chev.* 12347 ! 12353 ! Dalaba (fl. Sept., fr. Mar.) *Chev.* A ! *Langdale-Brown* 2609 ! **S.L.:** Waterloo (July) *Lane-Poole* 293 ! Bonthe (fr. June) *Deighton* 4840 ! Njala (Apr.) *Deighton* 1137 ! Sefadu (May) *Small* 96 ! Loma Mts. (Nov.) *Jaeger* 415 ! **Lib.:** Kakatown *Bequaert* ! **Iv.C.:** Abidjan *Chev.* 15281 ! 15384. Bettié, Bas-Comoé *Chev.* 17581 ! **Ghana:** E. Akim (June) *Johnson* 736 ! Kumasi (fr. July) *Andoh* FH 5520 ! Burufo Plateau (Apr.) *Adams* 4029 ! Juaso *Dalz.* 159 ! Obuasi (fr. June) *Andoh* 4214 ! **S.Nig.:** Usonigbe F.R., Benin (Oct.) *Keay & Onochie* FHI 19668 ! Jekri Isl., Sapoba R. (Oct.) *A. F. Ross* 217 ! Awka to Imofia, Onitsha (May) *Jones* FHI 496 ! Calabar *Thomson* 112 ! Eket *Talbot* 3041 ! Also in Cameroun, Angola and S. Tomé.
18. **C. dunlapii** Hutch. & Dalz. F.W.T.A., ed. 1, 2 : 115 (1931). A medium-sized tree ; flowers white ; in upland forest.
   [**Br.**]**Cam.:** Cam. Mt. (Dec., Jan., Mar., Apr.) *Dunlap* 93 ! *Maitland* 795 ! *Brenan* 9553 ! *Hutch. & Metcalfe* 10 ! Bamenda (fl. Mar., fr. Jan.) *Johnstone* 264/32 ! *Daramola* FHI 40530 ! **F.Po:** Moka, 4,000–6,000 ft. (fl. Jan., Feb., fr. Sept.) *Exell* 810 ! 851 ! *Wrigley* 572 !
19. **C. scabrosum** *Bullock* in Kew Bull. 1932 : 367. Climbing shrub ; flowers whitish.
   [**Br.**]**Cam.:** S. Bakundu F.R., Kumba Dist. (Mar.) *Onochie* FHI 30852 ! *Binuyo & Daramola* FHI 35612 !
20. **C. palma** (*K. Schum.*) *Good* in J. Bot. 64, Suppl. 2 : 22 (1926). *Plectronia palma* K. Schum. (1899). *P. oddoni* De Wild. (1907). Small tree with spreading branches ; beside rivers.
   S.Nig.: Sapoba *Kennedy* 2240 ! 2329 ! *A. F. Ross* 237 ! Ikom (July) *Catterall* 19 ! *Mackay* 15 ! Afl River F.R., Ikom *Jones & Onochie* FHI 18717 ! Eket Dist. *Talbot* 3069 ! Also in Cameroun, Gabon and Congo.
21. **C. rubens** *Hiern* in F.T.A. 3 : 142 (1877). A climbing shrub up to 20 ft. high, rusty-pubescent on the young branchlets, becoming glabrous.
   Guin.: Diaguissa (Oct.) *Jac.-Fél.* 2061 ! Mali (Jan.) *Roberty* 16566 ! **S.L.:** Bagroo R. (Apr.) *Mann* 818 ! Berria (? = Gberia Fotombu) (fr. Mar.) *Sc. Elliot* 5416 ! Gberia Fotombu (fr. Sept.) *Small* 386 ! **Ghana:** Krokosua F.R., Kwakrom (Feb.) *Cansdale* FH 386 !
   [The Guinée specimens may be specifically distinct from true *C. rubens*.]
22. **C. sp. B.** A climbing shrub about 10 ft. high, branchlets and nerves of leaves rusty-pubescent ; it may prove to be a form of the next.
   Ghana: Princes (fl. bud Jan.) *Darko* 790 ! **N.Nig.:** Omo (formerly part of Shasha) F.R. (Feb.) *Jones & Onochie* FHI 17584 !
23. **C. venosum** (*Oliv.*) *Hiern* in F.T.A. 3 : 144 (1877), incl. var. *pubescens* Hiern ; Aubrév. Fl. For. Soud.-Guin. 483, t. 107, 4 ; Berhaut Fl. Sén. 102. *Plectronia venosum* Oliv. in Trans. Linn. Soc. 29 : 85, t. 49 (1873). *Canthium barteri* Hiern in F.T.A. 3 : 143 (1871). *C. heudelotii* of F.W.T.A., ed. 1, 2 : 115, partly (*Sc. Elliot* 5434), not of Hiern. Straggling shrub or small tree up to 20 ft. high with rough square scaly bark, slash reddish brown ; flowers whitish ; in savanna woodland and beside streams.
   Sen.: *Berhaut* 1858. **Mali:** Oualia, Kati (fr. Dec.) *Chev.* 103 ! **Port.G.:** Prabis, Bissau (fr. Feb.) *Esp. Santo* 1804 ! **Guin.:** Kouroussa (July) *Pobéguin* 282 ! Fouéboudougou, Beyla (July) *Collenette* 70 ! Timbo to Ditinn (Sept.) *Chev.* 18380 ! Dalaba (fr. Feb.) *Langdale-Brown* 2583 ! **S.L.:** Berria (? = Gberia Fotombu) (fr. Mar.) *Sc. Elliot* 5434 ! Gberia Fotombu (Oct.) *Small* 336 ! Kambui F.R. (fr. Apr.) *Jordan* 2055 ! Sini-Koro to Zanda Kaifaia (Sept.) *Jaeger* 1896 ! Musaia (fl. Oct., fr. Feb.) *Thomas* 2628 ! *Deighton* 4179 ! **Lib.:** Dukwia R. (fr. May) *Cooper* 448 ! Mecca (fr. Nov., Dec.) *Baldwin* 10448 ! 10817 ! Bobei Mt., Sanokwele Dist. (fr. Sept.) *Baldwin* 9627 ! **Iv.C.:** Tiengara *Aubrév.* 1563. Bondoukou *Aubrév.* 764. Marabadissa *Chev.* 21998. **Ghana:** New Jantin (Mar.) *Johnson* 633 ! Kintampo (fr. Nov.) *Andoh* FH 5427 ! Mampong to Ejura (July) *Darko* 927 ! Awura F.R. (Aug.) *Andoh* FH 5319 ! **Dah.:** Bimbéréké *Aubrév.* 13d ! Koundé *Chev.* 24293 ! Tchaourou *Poisson* 41 ! **N.Nig.:** Atakati, Bassa (June) *Elliot* 235 ! Cece-Boku F.R., Lapai Dist. (Aug.) *Onochie* FHI 35110 ! Jos (Aug.) *Keay* FHI 20198 ! Abinsi (Aug.) *Dalz.* 768 ! **S.Nig.:** Lagos *MacGregor* 220 ! Aguku Dist. *Thomas* 1157 ! Asaba *Barter* 285 ! Onitsha *Barter* 1800 ! Mamu F.R., Awka Dist. (Aug.) *Olorunfemi* FHI 34185 ! Oban *Talbot* 2075 ! [**Br.**]**Cam.:** Gangumi, Gashaka Dist. (fr. Dec.) *Latilo & Daramola* FHI 28799 ! Also in Cameroun, Ubangi-Shari, Congo, Angola, Sudan and E. Africa to N. Rhodesia.
24. **C. zanzibaricum** *Klotzsch* in Peters Reise Mossamb. Bot. 291 (1861) ; F.T.A. 3 : 138 ; Bullock l.c. 373. *C. cornelia* of F.W.T.A., ed. 1, 2 : 115, partly (*Kitson* 679), not of Cham. & Schlecht. Straggling shrub 10–30 ft. high ; flowers white.
   S.L.: Kambaia (Oct.) *Small* 471 ! Mabonto (Oct.) *Thomas* 3539 ! Mafombo (Sept.) *Deighton* 3058 ! **Ghana:** Black Volta R. (fl. & fr. Mar.) *Kitson* 679 ! [**Br.**]**Cam.:** Bafachu, Bamenda *Johnstone* 261 ! Bamenda (fr. May) *Ujor* FHI 30363 ! Also in Sudan, Uganda, Kenya, Tanganyika, Zanzibar and Angola.
25. **C. gueinzii** *Sond.* in Linnaea 23 : 54 (1850) ; Bullock in Kew Bull. 1932 : 368 ; Hepper in Kew Bull. 16 : 157 (1962). *C. venosissimum* Hutch. & Dalz. F.W.T.A., ed. 1, 2 : 115 (1931), partly (*Vigne* FH 1336). A shrub with white flowers.
   Ghana: Kumasi (Sept.) *Vigne* FH 1336 ! [**Br.**]**Cam.:** Buea, 3,000 ft. (fr. Jan.) *Maitland* 194 ! Bamenda Nkwe (Aug.) *Ujor* FHI 29987 ! Lakom, 6,000 ft., Bamenda (Apr.) *Maitland* 1658 ! Nkambe, 6,000 ft. (Feb.) *Hepper* 1897 ! Also in E. Africa.
26. **C. manense** *Aubrév. & Pellegr.* in Bull. Soc. Bot. Fr. 83 : 38 (1936) ; Aubrév. Fl. For. C. Iv., ed. 2, 3 : 306, t. 366A. Small tree, with horizontal branches.
   Iv.C.: Man to Danané (Mar.) *Aubrév.* 1099 ! Ferkessédougou (Apr.) *Aubrév.* 2310.
27. **C. cornelia** *Cham. & Schlecht.* in Linnaea 4 : 14 (1829) ; F.T.A. 4 : 140 ; F.W.T.A., ed. 1, 2 : 115, partly ; Aubrév. Fl. For. Soud.-Guin. 483 ; Berhaut Fl. Sén. 102. *C. ruminatum* Baill. (1878). *C. heudelotii* of

F.W.T.A., ed. 1, 2 : 115 partly (*Heudelot* 191), not of Hiern. Straggling shrub ; flowers greenish-yellow ; in moist thickets in northern savanna.
Sen.: *Sieber* 21 ! *Roger* 37 ! *Leprieur* ! *Heudelot* 101 (= 191) ! Ouassadou *Berhaut* 1537 ! Niokolo-Koba (fr. Nov.) *Adam* 17192 ! **Gam.:** Genieri (July) *Fox* 138 ! *Hayes* 555 ! **Mali:** Kita *Duong* 543 ! *Dubois* 31 ! Samandini *Chev.* 948 ! L. Debo *Vuillet* 217. **Port.G.:** Sonaco, Gabu (fr. Nov.) *Esp. Santo* 3582 ! **Guin.:** Kouroussa *Pobéguin* 229 ; 240. Farana (Apr.) *Pitot* ! Dantilia (Mar.) *Sc. Elliot* 5296 ! **Ghana:** Suguri (Apr.) *Pomeroy* 1234 ! Saboba (Mar.) *Hepper & Morton* A3118 ! Tumu (Mar.) *Morton* GC 8837 ! Kumbungu (May) *Kitson* 677 ! 717 ! **Niger:** Diapaga *Chev.* 24392. Also in Cameroun and Ubangi-Shari (*fide* Aubrév. *l.c.*).

This genus presents considerable taxonomic difficulties owing to great variability and often wide distribution of the species.

57. RYTIGYNIA Blume Mus. Bot. Lugd. Bat. 1 : 178 (1850) ; Robyns in Bull. Jard. Bot. Brux. 11 : 132 (1928).

Leaves covered all over the lower surface by a fine whitish tomentellous indumentum, ovate-elliptic, narrowly rounded at base, 7–10 cm. long, 3–4·5 cm. broad, with about 5 pairs of lateral nerves ; petioles 3 mm. long ; flowers axillary, solitary or paired ; pedicels 4–5 mm. long ; corolla-tube 3–3·5 mm. long, lobes appendiculate ; fruit 6–7 mm. diam. ..   ..   1. *argentea*
Leaves from densely tomentose to glabrous, not whitish tomentellous, beneath :
  Flowers in short branched cymes, rather numerous ; branchlets glabrous or pubescent ; leaves subacute at base, obtusely acuminate at apex, broadly elliptic, 8–12 cm. long, 3–5 cm. broad, thin and membranous, bright green when dry, with 6–7 pairs of lateral nerves ; pedicels nearly 1 cm. long ; branchlets villous at the nodes within the stipules ; corolla-tube 3–4 mm. long ..   ..   ..   2. *membranacea*
  Flowers fasciculate on a short simple peduncle, few or subsolitary :
    Peduncle longer than the pedicels ; flowers in pairs :
      Branchlets glabrous, lenticellate ; leaves broadly acuminate at apex, ovate-elliptic, 5–6 cm. long, 2–3 cm. broad, with about 6 pairs of lateral nerves often with tufts of hairs in their axils, otherwise glabrous ; pedicels about 5 mm. long, glabrous ; bracts subulate ; corolla-tube 3·5 mm. long, lobes slightly apiculate ; fruits obliquely ellipsoid, 1 cm. long   ..   ..   3. *senegalensis*
      Branchlets thinly tomentose ; leaves broadly acuminate at apex, ovate-lanceolate, 3·5–6 cm. long, 1–2·5 cm. broad, discolorous, thinly pubescent beneath ; pedicels 2 mm. long, tomentose ; corolla-tube 2 mm. long, lobes acutely apiculate
                                                         4. *rhamnoides*
    Peduncle much shorter than the pedicels :
      Pedicels of flowers at least 7 mm. long, in fruit up to 2·5 cm. long ; flowers paired ; branchlets glabrous ; leaves oblong-elliptic, acuminate to caudate, rounded at base, 5–8 cm. long, 2–3·5 cm. broad, paler beneath, with repeatedly looped lateral nerves, glabrous ; bracts ovate-apiculate ; calyx denticulate ; corolla-tube 5 mm. long ..   ..   5. *rubra*
      Pedicels of flowers up to 5 mm. long (up to 1 cm. long in fruit in some species) ; inflorescences 1-several flowered :
        Flower-buds long-apiculate at apex ; corolla-lobes 5 ; 1–3 flowers at nodes :
          Leaves densely pale tomentose beneath, shortly acuminate at apex, rounded at base, oblong-elliptic, 3–6 cm. long, 1·5–2·5 cm. broad ; corolla densely pubescent outside   ..   ..   6. *leonensis*
          Leaves setose-pilose beneath, broadly acuminate at apex, rounded at base, ovate to elliptic, 4–8 cm. long, 2–4 cm. broad ; corolla-lobes sparsely setose outside
                                                       7. *nigerica*
        Flower-buds at most shortly mucronate at apex, glabrous outside (except rarely in Nos. 8 & 13) ; corolla-lobes 4–5 :
          Young branchlets glabrous or only slightly pubescent ; leaves glabrous beneath or slightly pubescent on the midrib :
            Flowers at nodes on the new-growth branchlets ; corolla-lobes 4–5 :
              Corolla-lobes 3 mm. long and half as long as the tube ; flowers 2–several in short stout peduncle ; calyx-teeth acute, distinct ; ovary 2-celled   8. *affinis*
              Corolla-lobes 4 as long as the tube ; flowers several in subsessile clusters ; calyx very shortly toothed ; ovary 4-celled   ..   ..   ..   9. *liberica*
            Flowers on one-year-old (or previous season's) branchlets ; corolla-lobes 5 :
              Corolla 8–10 mm. long, tube about 2·5 mm. diam ; leaves long acuminate at apex ; upland shrub   ..   ..   10. *neglecta*
              Corolla about 5 mm. long, tube about 1·5 mm. diam. ; leaves acuminate, rounded to subacute at base, elliptic to ovate-elliptic, 5–10 cm. long, 3–5 cm. broad, with tufts of short hairs in the axils of the 5 pairs of lateral nerves
                                                       11. *umbellulata*
        Young branchlets densely pubescent :
          Leaves rather densely puberulous beneath, less so above, elliptic to ovate, rather long-acuminate at apex, rounded at base, 4–6 cm. long, 1·5–3 cm. broad ; flowers mostly in the axils of leaves on one-year-old branchlets   12. *laurentii*
          Leaves slightly pubescent on nerves beneath and on upper surface, oblong-

elliptic ;   more or less rounded at base, rather long-acuminate at apex,
5–9(–11) cm, long, 2–3·5(–5) cm. broad ; flowers in the axils of leaves on the
young branchlets . .   . .   **13.** *canthioides*

1. **R. argentea** (*Wernham*) *Robyns* in Bull. Jard. Bot. Brux. 11 : 205 (1928). *Vangueria argentea* Wernham in
Cat. Talb. 49 (1913). A shrub, with rusty-pubescent branchlets, leaves silvery-grey beneath and small
single or paired axillary flowers.
    S.Nig.: Oban *Talbot* 215 !
2. **R. membranacea** (*Hiern*) *Robyns* l.c. 217 (1928). *Vangueria membranacea* Hiern in F.T.A. 3 : 151 (1877).
Rytigynia viridissima (Wernham) Robyns l.c. 213 (1928). *Canthium viridissimum* Wernham (1914). A
shrub or small tree, mostly glabrous, with small yellow flowers in subumbellate cymes.
    S.Nig.: Sapoba *Kennedy* 2553 ! Eket Dist. *Talbot* 3121 ! Calabar *Thomson* 114 ! [**Br.**]**Cam.:** R. Kindong,
S. Bakundu, Kumba Dist. (Jan.) *Binuyo & Daramola* FHI 35494 !
3. **R. senegalensis** *Blume* in Mus. Bot. Lugd. Bat. 1 : 179 (1850) ; Robyns l.c. 163, figs. 19 & 20 ; F.W.
Andr. Fl. Pl. A.-E. Sud. 2 : 462 ; Berhaut Fl. Sén. 102. *Vangueria senegalensis* (Blume) Hiern in F.T.A.
3 : 149 (1877) ; Chev. Bot. 330, *V. euonymoides* Schweinf. ex Hiern in F.T.A. 3 : 150 (1877) ; Chev. Bot.
330. A glabrous shrub with slender branchlets, more or less scrambling, with elegant foliage ; flowers
white or greenish, paired on slender axillary peduncles, with reflexed petals ; moist shady places in
savanna.
    Sen.: *Leprieur* ! *Park* 255 ! *Berhaut* 645 ! **Mali:** Bafing *Lécard* 82 ! Kita (July) *Duong* 544 ! **Guin.:**
Kaba to Haut Mamou *Chev.* 12711. Kouroussa *Pobéguin* 266 ; 667. **Lib.:** Grand Bassa *Dinklage* 2085.
Iv.C.: Yaou to Ahiamé *Chev.* 17698. Bya-nouan to Soubiré *Chev.* 17744. **Ghana:** Kumbungu (fl, & fr.
May) *Kitson* 678 ! Zantana to Lunbungu (May) *Kitson* 962 ! Yeji *Anderson* 23 ! Wa (May) *Vigne* FH
3826 ! **N.Nig.:** Zungeru (July) *Dalz.* 221 ! Abinsi (June) *Dalz.* 757 ! Extends to the Sudan.
4. **R. rhamnoides** *Robyns* l.c. 207 (1928). *Vangueria canthioides* of F.T.A. 3 : 149, partly (*Barter* 1243). A
slender shrub, about 12 ft. high ; flowers greenish-white.
    N.Nig.: Nupe *Barter* 1243 !
5. **R. rubra** *Robyns* l.c. 162 (1928). A climbing or straggling shrub about 8 ft. high, with wiry branches and
thorns ; flowers green ; in forest.
    S.Nig.: Ikoyi Plains, Lagos (Mar.) *Dalz.* 1050 ! Ishan (fr. Apr.) *Daramola* FHI 31265 ! Degema Dist.
Talbot 3823 ! [**Br.**]**Cam.:** Barombi L., Kumba (Mar.) *Brenan* 9457 ! Also in Cameroun and Gabon.
6. **R. leonensis** (*K. Schum.*) *Robyns* l.c. 205 (1928). *Vangueria leonensis* K. Schum. in Engl. Bot. Jahrb. 23 :
457 (1897). *V. oxyantha* K. Schum. (1899), excl. *Rowland* s.n. Shrub or small tree in forest.
    Guin.: Kouroussa *Pobéguin* 666 ! Farana (Mar.) *Sc. Elliot* 5358 ! **S.L.:** Da-Ouley, Loma Mts. (fr. Aug.
Jaeger 1283 !
7. **R. nigerica** (*S. Moore*) *Robyns* l.c. 208 (1928). *Vangueria nigerica* S. Moore (1910), *V. oxyantha* K. Schum.
in Engl. Bot. Jahrb. 28 : 72 (1899), partly (*Rowland* s.n.). Shrub or small tree with horizontal branches ;
flowers greenish-white. Similar to *R. canthioides*.
    Ghana: Afram Plains (fl. & fr. May) *Kitson* 1141 ! Achimota *Milne-Redhead* 5057 ! Legon Hill (Apr.)
Adams 4194 ! **N.Nig.:** Bonu (June) *E. W. Jones* 151 ! Gornapara, Gurara R. *Elliott* 181 ! **S.Nig.:** Lagos
Rowland ! Olokemeji F.R. (Apr.) *Keay* FHI 37003 ! Ibadan (Apr.) *Meikle* 1467 ! *Brenan* 9605 ! Eket
Dist. *Talbot* 3289 !
8. **R. affinis** (*Robyns*) *Hepper* in Kew Bull. 17 : 171 (1963). *Dinocanthium affine* Robyns in Bull. Jard. Bot.
Brux. 17 : 94 (1943). *D. bequaertii* Robyns l.c. 95 (1943). *Canthium affine* (Robyns) Hepper in Kew Bull.
16 : 338 (1962). A climbing shrub with greenish-yellow, sweet-scented flowers.
    Ghana: Donkolo, near Bibiani (Feb.) *Darko* 854 ! Atewa Range F.R. (Feb.) *Vigne* FH 4330 ! Also in
Congo.
9. **R. liberica** *Robyns* l.c. 182 (1928). *R. claviflora* of F.W.T.A., ed. 1, 2 : 111, partly (syn. *R. liberica* Robyns ;
Whyte) not of Robyns. A shrub.
    Lib.: Kakatown *Whyte* ! Suen, Montserrado (Nov.) *Baldwin* 10484 !
10. **R. neglecta** (*Hiern*) *Robyns* l.c. 183 (1928). *Canthium neglectum* Hiern in F.T.A. 3 : 183 (1877). *R. rubra*
of F.W.T.A., ed. 1, 2 : 111, partly (*Maitland* 648), not of Robyns. Scrambling shrub or small understorey
tree about 15 ft. high ; flowers greenish-white ; in montane forest.
    [**Br.**]**Cam.:** below Musaka, about 5,000 ft., Cam. Mt. (Apr., May) *Maitland* 648 ! *Hutch. & Metcalfe* 30 !
Bamenda : Mbakakeka Mt., 7,800 ft. (fr. July) *Lightbody* FHI 30112 ! Lakom, 6,000 ft. (May) *Maitland*
1359 ! L, Oku, 7,200 ft. (Jan.) *Keay & Lightbody* FHI 28490 ! Also in Sudan, Ethiopia, Uganda, Kenya,
Tanganyika and Congo.
11. **R. umbellulata** (*Hiern*) *Robyns* l.c. 184 (1928). *Vangueria umbellulata* Hiern in F.T.A. 3 : 150 (1877). *V.
canthioides* of K. Schum. in Schlechter W.-Afr. Kautschuk Exped. 320 (1901), partly (*Schlechter* 12322).
V. concolor Hiern in F.T.A. 3 : 150, partly (*Barter* 3418). Scrambling or erect shrub or small tree up to
15 ft. high ; flowers pale greenish or yellowish ; fruits reddish ; in secondary forest.
    Port.G.: Catio (fr. July) *Esp. Santo* 2135 ! **S.L.:** Sugar Loaf Mt. (Apr.) *Sc. Elliot* 5780 ! Mt. Aureol (fr.
June) *Deighton* 2717 ! Messima (Apr.) *Deighton* 3699 ! **Ghana:** Axim (Feb.) *Irvine* 2273 ! Begoro (Mar.)
Vigne FH 4349 ! Kumasi (Apr.) *Vigne* FH 1114 ! Aboma F.R., Mampong Dist. (Jan.) *Andoh* FH 5453 !
N.Nig.: Anara F.R., Zaria (May) *Keay* FHI 22974 ! Sanga River F.R., Jemaa Div. (May) *E. W. Jones*
43 ! Cece F.R., Bida (fr. June) *E. W. Jones* 191 ! **S.Nig.:** Ikoyi Plains, Lagos (fl. & fr. Mar.) *Dalz.* 1050a !
Ibadan (Mar., Apr.) *Schlechter* 12322 ! *Latilo* FHI 22744 ! Ijaye *Barter* 3418 ! Onitsha Akpaka F.R.
(fr. Mar.) *Onochie* FHI 35752 ! Akame-Oghe, Udi Dist. (fr. May) *Onochie* FHI 35849 ! Also in Cameroun,
Angola and N. Rhodesia.
12. **R. laurentii** (*De Wild.*) *Robyns* l.c. 192 (1928). *Vangueria laurentii* De Wild. in Mission Laurent 292 (1906).
Shrub or small tree up to 15 ft. high ; flowers greenish-yellow.
    N.Nig.: Sanga River F.R., Jemaa Dist. (Apr.) *Keay* FHI 37627 ! [**Br.**]**Cam.:** Bafut-Ngemba F.R.,
6,000 ft., Bamenda (Apr., May, fr. June) *Ujor* FHI 30309 ! *Richards* 5316 ! *Lightbody* FHI 26338 !
Johnstone 306/32 ! Also in Congo.
13. **R. canthioides** (*Benth.*) *Robyns* l.c. 188 (1928). *Vangueria canthioides* Benth. (1849)—F.T.A. 3 : 149, partly.
Canthium benthamianum Baill. (1878). *Rytigynia claviflora* Robyns l.c. 187 (1928) ; F.W.T.A., ed. 1, 2 :
111, partly (excl. syn.). A straggling shrub or small tree with rusty-pubescent branchlets and horizontal
branches ; flowers green or yellowish, fragrant ; in secondary forest.
    Guin.: Macenta (fl. & fr. Apr., May) *Jac.-Fél.* 907 ! *Adam* 12077 ! **S.L.:** *Don* ! Sugar Loaf Mt.,
2,500 ft, (Apr.) *Hepper* 2541 ! Bagroo R. (Apr.) *Mann* 866 ! Kenema (Apr., May) *Lane-Poole* 240 ! *Small*
43 ! Njala (May) *Deighton* 6061 ! **Lib.:** Mecca (fl. & fr. Dec.) *Baldwin* 10815 ! Kakatown *Whyte* ! Moala-
Suen (Nov.) *Linder* 1385 ! Ganta (May) *Harley* 1156 ! Gletown (July) *Baldwin* 6907 ! **Iv.C.:** Tonkouy
(Dec.) *Roberty* 15844 ! Sassandra (Dec.) *Leeuwenberg* 2231 ! **Ghana:** Abra, W. Prov. (Jan.) *Akpabla*
810 ! Kumasi (Mar.) *Vigne* FH 1643 ! Ashanti Akim (Apr.) *Irvine* 959 ! Aboma F.R., Mampong Dist.
(fl. & fr. Jan.) *Andoh* FH 5450 ! **Togo Rep.:** Misahöhe *Baumann* 433. **S.Nig.:** Eba Isl., Ijebu Prov.
(Mar.) *Richards* 3250 !
    [Note : *Jac.-Fél.* 907, *Adam* 12077 and *Harley* 1156 are intermediate between this species and *R.
nigerica* : they also possess pubescent corolla and calyx.]

Additional species,

R. gracilipetiolata (*De Wild.*) *Robyns* in Bull. Jard. Bot. Brux. 11 : 176 (1928) ; Berhaut Fl. Sén. 102 ; Aké
Assi Contrib. 1 : 153.
    Sen.: Sangalkam *Berhaut* 1625. **Iv.C.:** Adiopodoumé *Aké Assi* IA 5298 ; 5299.

58. GLOBULOSTYLIS Wernham in Cat. Talb. 49, t. 9 (1913).

Branchlets glabrous ; leaves oblong-elliptic, subabruptly acuminate, subacute at base, with 5–7 pairs of lateral nerves, glabrous ; tertiary nerves few and distant ; calyx-lobes ovate, leafy, about 8 mm. long, nearly enclosing the corolla :

Corolla sharply keeled in bud, pilose on keels, tailed at apex ; calyx-lobes ciliate ; flowers axillary, few ; leaves up to 15 cm. long and 6 cm. broad ; petioles 5–10 mm. long   ..   ..   1. *talbotii*

Corolla neither pilose on keels nor tailed at apex ; calyx-lobes not ciliate ; flowers subumbellate on a short peduncle ; leaves up to 12 cm. long ; petioles 2–4 mm. long
2. *minor*

1· **G. talbotii** *Wernham* in Cat. Talb. 49, t. 9 (1913). An almost glabrous shrub.
   S.Nig.: Oban *Talbot* 2051 !
2. **G. minor** *Wernham* l.c. 50 (1913).
   S.Nig.: Oban *Talbot* 247 !

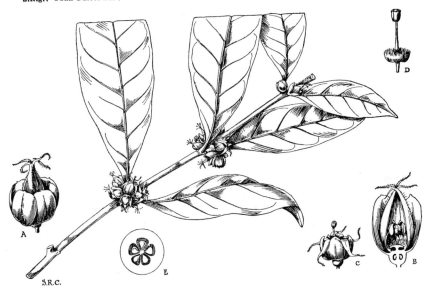

FIG. 240.—GLOBULOSTYLIS TALBOTII *Wernham* (RUBIACEAE).

A, flower. B, same in longitudinal section. C, same with corolla removed. D, style. E, transverse section of ovary.

59. VANGUERIA Juss.—F.T.A. 3 : 146, partly ; Robyns in Bull. Jard. Bot. Brux. 11 : 273 (1928).

Inflorescence glabrous or nearly so ; flowers sessile or subsessile ; corolla villous at the throat, tube 3–4 mm. long, lobes ovate, acute ; leaves deciduous, ovate-oblong, narrowed at the base, gradually acuminate, 7–15 cm. long, 3·5–7 cm. broad, with 8–12 pairs of lateral nerves, glabrous or finely pubescent ; fruit stalked, globose, slightly ribbed, 1·5–2·5 cm. diam., mostly 5-seeded ..   ..   ..   1. *venosa*

Inflorescence pilose ; flowers pedicellate ; corolla long-villous at the throat, tube 3 mm. long, lobes ovate, obtuse, pilose outside ; leaves ovate, rounded at the base, rather obtuse at the apex, 3–5 cm. long, 1·5–3 cm. broad, with 5–7 pairs of lateral nerves ; fruit globose, subsessile, about 1 cm. diam., slightly pubescent, 5-seeded
2. *kerstingii*

1. **V. venosa** *Hochst. ex Del.* in Ferret & Galinier Voy. Abyss. 3 : 140 (1847) ; Robyns in Bull. Jard. Bot. Brux. 11 : 290. *V. edulis* of F.T.A. 3 : 148, excl. var., not of Vahl. A shrub or small tree glabrous in most parts ; flowers greenish in branched cymes usually on the young leafy shoots.
   Mali: Kita *Duong* 489 ! **Ghana:** Aburi Gardens (cult., fl. & fr. May) *Irvine* 3036 ! **N.Nig.:** Sanga River F.R., Jemaa Dist. (May) *E. W. Jones* 50 ! **S.Nig.:** Aboh *Barter* 362 ! Ida *Barter* 1250 ! Also in Sudan, Ethiopia and Uganda. (See Appendix, p. 414.)
2. **V. kerstingii** *Robyns* l.c. 305 (1928). A shrub with red-brown striate branchlets flaking with age ; flowers in short pilose cymes.
   Togo Rep.: Basari *Kersting* A620 !
   [Probably only a densely pubescent form of *V. venosa*.]

Imperfectly known species.

V. sp. A. A glabrous shrub ; fruit globose, about 1½ in. across ; flowers not known.
   S.L.: *Thomas* 1378 ! Mabum (fr. Aug.) *Thomas* 1645 ! **Lib.:** Wanau (fr. July) *Harley* 2041 !

60. CRATERISPERMUM Benth.—F.T.A. 3 : 160.

Inflorescences sessile or nearly so in axils :
  Bracts long-aristate up to 8 mm. long ; leaves broadly oblong-elliptic, broadly acuminate at apex, cuneate at base, 15–20 cm. long, 6–9 cm. broad, laxly reticulate beneath, with about 7 pairs of lateral nerves .. .. ..   1. *aristatum*
  Bracts small, about 2 mm. long ; leaves elliptic to oblanceolate, shortly acuminate at apex, cuneate at base, 5–8 cm. long, 2–3·5 cm. broad, finely reticulate beneath
                                                   2. sp. *A*
Inflorescences pedunculate, markedly supra-axillary :
  Stipules early caducous ; fruits with a pedicel up to 7 mm. long on the peduncle ; leaves oblong or oblong-elliptic, shortly cuneate at base, caudate-acuminate at apex, 8–14 cm. long, 2·5–4·5 cm. broad, finely reticulate on both sides ; calyx denticulate
                                                  3. *caudatum*
  Stipules more or less persistent ; leaves yellow-green :
    Leaves closely and strongly reticulate beneath, coriaceous, acute or obtuse and very shortly and abruptly acuminate at apex, 10–20 cm. long, 4–9 cm. broad ; calyx cupular, truncate ; fruits shortly pedicellate .. .. ..   4. *laurinum*
    Leaves laxly reticulate beneath, hardly coriaceous, distinctly acuminate at apex, cuneate at base, elongate oblong-obovate, 7–15 cm. long, 3–6 cm. broad ; calyx denticulate ; fruits subsessile .. ..   5. *cerinanthum*

1. C. aristatum *Wernham* in Cat. Talb. 51 (1913). A glabrous shrub with large leaves ; flowers congested in the axils or slightly supra-axillary.
  S.Nig.: Oban *Talbot* 251 !
2. C. sp. A. Probably a shrub, with yellowish leaves ; further material required.
  Lib.: Greenville, Sinoe (Mar.) *Baldwin* 11572 !
  [This may be a depauperate specimen of *C. cerinanthum* Hiern.]
3. C. caudatum *Hutch.* in Kew Bull. 1920 : 23. *C. montanum* of K. Schum. in Engl. Bot. Jahrb. 23 : 460, partly (*Sc. Elliot* 4987), not of Hiern. *C. gracile* A. Chev. ex Hutch. & Dalz. F.W.T.A., ed. 1, 2 : 116 (1931) ; Chev. Bot. 332, name only ; Aubrév. Fl. For. C. Iv., ed. 2, 3 : 310, t. 367. A small tree about 15 ft. high, with green, glabrous branchlets ; flowers white, fruits white, yellow or purple ; understorey tree in forest.
  S.L.: Likuru, Talla Hills (Feb.) *Sc. Elliot* 4987 ! Njala (fr. May) *Deighton* 5763 ! Faiama (Jan.) *Deighton* 3870 ! Dia (fr. Apr.) *Deighton* 3206 ! Bandajuma (May) *Aylmer* 78 ! **Lib.:** Gbanga (fl. & fr. Sept.) *Linder* 574 ! 713 ! Ganta (fr. May) *Harley* 1192 ! **Iv.C.:** Bouroukrou (Jan.) *Chev.* 17017 ! Banco F.R. *Aubrév.* 178 ! Danané *Aubrév.* 1105 ! **Ghana:** Asenanyo F.R. (Feb.) *Andoh* 4306 ! Pamu-Berekum F.R. (Sept.) *Vigne* FH 2500 ! **S.Nig.:** Akamkpa, Calabar (fr. Mar.) *Latilo* FH 40903 !
4. C. laurinum (*Poir.*) Benth. in Fl. Nigrit. 411 (1849) ; F.T.A. 3 : 160 ; Aubrév. Fl. For. Soud.-Guin. 484 ; Fl. For. C. Iv., ed. 2, 3 : 308, t. 367. A glabrous shrub or small tree up to 25 ft. high with yellow-green coriaceous leaves ; flowers white ; beside streams.
  Sen.: " gallery forests " *Berhaut* 1550. **Port.G.:** Balanasinho, Fulacunda (fr. May) *Esp. Santo* 2146 ! **Guin.:** Rio Pongo *Heudelot* 908 ! Conakry *Maclaud* ! Iles de Los *Pobéguin* 1210 ! Mamou *Pobéguin* 1557 ! Timbo *Pobéguin* 166 ! **S.L.:** Bagroo R. (Apr.) *Mann* 808 ! Mange, Port Loko Dist. (Apr.) *Hepper* 2605 ! Njala (Mar.) *Deighton* 2880 ! Loma Mts. (Feb.) *Jaeger* 4207 ! Gola F.R. (fr. Apr.) *Small* 576 ! **Lib.:** Monrovia *Whyte* ! Beiden, Boporo Dist. (Nov.) *Baldwin* 10272 ! Zigida, Vonjama Dist. (Oct.) *Baldwin* 9989 ! Gbanga (Dec.) *Baldwin* 10513 ! Ganta (fl. & fr. Feb.) *Harley* 470 ! **Iv.C.:** Mt. Dou *Aubrév.* 1080 ! **Ghana:** Amedzofe (Feb.) *Irvine* 179 ! **N.Nig.:** Jos Plateau (Jan.) *Lely* P 47 ! *Batten-Poole* 258 ! Kaduna to Zaria (Dec.) *Meikle* 761 ! Also in Congo, Uganda, Kenya, Tanganyika, Nyasaland, N. & S. Rhodesia, Angola, Mozambique and Madagascar.
5. C. cerinanthum *Hiern* in F.T.A. 3 : 161 (1877). *C. brachynematum* Hiern in F.T.A. 3 : 161 (1877). *C.* sp. of Aubrév. Fl. For. C. Iv., ed. 2, 3 : 310. *C. laurinum* of Chev. Bot. 332, not of (Poir.) Benth. Small tree up to 50 ft. high, usually much less and sometimes a shrub ; flowers white ; understorey tree in forest, sometimes in open woodland.
  Iv.C.: Bingerville *Jolly* 261 ! 309 ! *Aubrév.* 919 ! Dabou (fl. & fr. Feb.) *Chev.* 17226 ! **Ghana:** Aburi (Feb., Nov.) *Brown* 886 ! *Johnson* 292 ! South Fomang Su F.R. (fl. & fr. Feb.) *Vigne* FH 1823 ! Pamuso (May) *Darko* 868 ! Sansome, Nsawam (fr. Dec.) *Harris* ! **N.Nig.:** Ankpa to Acharane, Kabba Prov. (Mar.) *Okafor* FHI 36885 ! **S.Nig.:** Lagos *Barter* 2175 ! Idanre (fr. Jan.) *Brenan* 8671 ! Omo (formerly part of Shasha) F.R. (Apr.) *Jones & Onochie* FHI 17233 ! 17352 ! Bikote, Obudu (fr. Dec.) *Catterall* ! [**Br.]Cam.:** Kembong F.R., Mamfe (fr. Feb., Mar.) *Onochie* FHI 31179 ! *Keay* FHI 37554 ! S. Bakundu F.R., Kumba (Mar.) *Binuyo & Daramola* FHI 35620 ! Johann-Albrechtshöhe (= Kumba) *Staudt* 569 ! Esu, 5,000 ft. (Feb.) *Johnstone* 59/31 ! Also in Congo.
  [Some specimens from S. Nigeria (FHI 17233 ; 17352 ; *Catterall* s.n.) and from [Br.] Cameroons (FHI 35620) show papillose stems when dry and faint venation : they may be distinct.—F.N.H.]

61. MORINDA Linn.—F.T.A. 3 : 191.

Climbing shrubs ; peduncles terminal or terminating short lateral branchlets, never leaf-opposed ; flowers 6–7-merous :
  Peduncles in pairs ; corolla-tube elongated, slender, (3–)4–8 cm. long, densely villous or pubescent in the throat ; calyx undulate ; leaves oblong-elliptic or obovate-elliptic, shortly cuneate at the base, acuminate, 6–12 cm. long, 2–7 cm. broad, glabrous, with 5–6 pairs of lateral nerves ; ovary glabrous ; fruit turgid (not lobed), 2–3 cm. diam., horned by the persistent calyces .. ..   1. *longiflora*
  Peduncles solitary ; corolla-tube short and stout in bud, up to 3 cm. long, glabrous in the throat ; calyx-tube truncate, 4 mm. long ; leaves elliptic or oblong-elliptic, shortly cuneate at the base, shortly and obtusely acuminate, 6–15 cm. long, 3–8 cm. broad, glabrous, with about 6 pairs of lateral nerves ; ovary mostly puberulous ; fruit lobulate, 4 cm. diam. .. ..   2. *morindoides*
Trees or shrubs, much-branched ; peduncles leaf-opposed and often also terminal ; flowers mostly 5-merous :
  Branchlets slender, more or less terete ; stipules large and foliaceous, soon falling off ; peduncles usually 3 together, elongated and slender ; flower-buds slender ; leaves

broadly elliptic to broadly ovate, more or less acuminate, rounded to broadly cuneate at the base, 8–16 cm. long, 4–8 cm. broad, often dark-purplish when dry, glabrous beneath, with 6–9 pairs of lateral nerves ; heads globose ..   **3.** *lucida*
Branchlets stout, quadrangular ; stipules small and persistent ; peduncles stout ; flower-buds stout ; leaves broadly elliptic, triangular at the apex, up to 20 cm. long and 12 cm. broad, glabrous, dark when dry ; fruit mostly subturgid, transversely ellipsoid or globose   ..   ..   **4.** *geminata*

1. **M. longiflora** *G. Don* Gen. Syst. 3 : 545 (1834) ; F.T.A. 3 : 192 ; Hutch. in Kew Bull. 1916 : 9 ; Aubrév. Fl. For. Soud.-Guin. 485, t. 108, 1–3. A glabrous climbing shrub ; flowers white, fragrant ; fruits yellow composed of connate berries ; in secondary forest.
Guin.: Kisosso (July) *Pobéguin* 793 ! Macenta (Apr.) *Jac.-Fél.* 847 ! **S.L.:** Kessewe (Apr.) *Lane-Poole* 132 ! Mano *Thomas* 9969 ! Rosino (fl. & fr. Feb.) *Deighton* 5009 ! Bweda to Kangama (Apr.) *Deighton* 3154 ! Gola F.R. (May) *Small* 671 ! **Lib.:** Timbo, Grand Bassa (Mar.) *Baldwin* 11215 ! Grand Cess (Mar.) *Baldwin* 11622 ! Sanokwele (fr. Sept.) *Baldwin* 9636 ! Gbanga (fr. Mar.) *Blickenstaff* 33 ! Vonjama (Feb.) *Bequaert* 85 ! **Iv.C.:** Abouabou Forest (fl. & fr. Jan.) *Leeuwenberg* 2343 ! **Ghana:** Kwahu (Apr.) *Johnson* 667 ! Benso (July) *Andoh* FH 5560 ! **S.Nig.:** Idumuye (Dec.) *Thomas* 2111 ! Eket *Talbot* 3255 ! Oban *Talbot* 204 ! 205 ! 1667 ! **F.Po:** *Barter* 2066 ! (June) *Mann* 411 ! 2341 ! Also in Gabon and Congo. (See Appendix, p. 403.)

FIG. 241.—MORINDA LUCIDA *Benth.* (RUBIACEAE).
A, flowering shoot. B, corolla laid open. C and D, anthers.

2. **M. morindoides** (*Bak.*) *Milne-Redh.* in Kew Bull. 1944 : 31 (1947). *Gaertnera morindoides* Bak. in Kew Bull. 1892 : 83. *Morinda confusa* Hutch. l.c. 11, with fig. ; F.W.T.A., ed. 1, 2 : 119 ; Aubrév. Fl. For. Soud.-Guin. 485, t. 108, 4. *M. longiflora* of F.T.A. 3 : 192, partly (*Mann* 810 ; *Barter* 3272 ; *T. Vogel* 188) ; of Chev. Bot. 338, not of G. Don. A glabrous climber ; flowers white, fragrant ; fruits yellow ; in forest.
Port.G.: Guebu to Guileje, Catio (June) *Esp. Santo* 2082 ! **Guin.:** N'zo (Dec.) *Roberty* 16024 ! Bilima *Chev.* 14690 ! **S.L.:** Gegbwema (Nov.) *Deighton* 437 ! Bagroo R. (fl. & fr. Apr.) *Mann* 810 ! Ninia, Talla Hills (Feb.) *Sc. Elliot* 4901 ! Njala (fr. Mar.) *Pyne* 120 ! Gola F.R. (fr. Apr.) *Small* 575 ! **Lib.:** Monrovia *Whyte* ! Mecca, Boporo Dist. (Nov.) *Baldwin* 10425 ! Ganta (fr. Feb.) *Baldwin* 11017 ! Tappita (Aug.) *Baldwin* 9039 ! **Iv.C.:** Sassandra (Dec.) *Leeuwenberg* 2249 ! Gonokrom to Techikrom, Dormaa (Dec.) *Adams* 2974 ! **Ghana:** Dixcove (fr. Apr.) *Chipp* 178 ! Otumi, Kibbi Dist. (Apr.) *Andoh* FH 5154 ! Ancobra Junction (fr. Dec.) *Irvine* 1050 ! Obuom (Dec.) *Vigne* FH 1486 ! Kumasi *Cummins* 43 ! **S.Nig.:** Lagos (Jan.) *Moloney* ! *Dalz.* 1383 ! Eppah *Barter* 3272 ! Ilaro F.R., Egbado Dist. (Dec.) *Onochie* FHI 32441 ! Ibadan to Ife (fr. Mar.) *Meikle* 1268 ! Akpaka F.R., Onitsha (Feb.) *Jones* FHI 7435 ! Oban *Talbot* 1335 ! **[Br.]Cam.:** Banga to Victoria (fr. Mar.) *Binuyo & Daramola* FHI 35600 ! **F.Po:** *T. Vogel* 188 ! Also in Cameroun, Congo, Sudan and Angola. (See Appendix, p. 403.)
3. **M. lucida** *Benth.* in Fl. Nigrit. 406 (1849) ; Hutch. l.c. 12, with fig. ; Aubrév. Fl. For. C. Iv., ed. 2, 3 : 270, t. 348, and Fl. For. Soud.-Guin. 485 ; Berhaut Fl. Sén. 104. *M. citrifolia* of F.T.A. 3 : 191, partly ; Chev. Bot. 338, partly, not of Linn. A medium-sized tree with scaly grey bark, short crooked branches and shining foliage ; flowers whitish in small heads, fruits green with a white interior ; in forest.
Sen.: *fide* Berhaut *l.c.* **S.L.:** Musaia (Mar.) *Deighton* 5441 ! **Lib.:** Tappita (fr. Aug.) *Baldwin* 9056 ! **Iv.C.:** Bingerville (May) *Chev.* 20149 ! Bouaké *Aubrév.* 118 ; 834. Agboville *Aubrév.* 1382 ! **Ghana:** Axim (Nov.) *Chipp* 21 ! Dixcove (fr. Mar.) *Morton* A478 ! Dunkwa (Feb.) *Vigne* FH 214 ! Achimota (fr. June) *Akpabla* GC 511 ! Obuasi (fl. & fr. Apr.) *King-Church* 871 ! **Togo Rep.:** Lomé *Warnecke* 177 ! Sokode (fr. Apr.) *Kersting* 64 ! **Dah.:** Gouka, Savalou *Chev.* 23721 ! **N.Nig.:** Nupe *Barter* ! Abinsi (May-July) *Dalz.* 641 ! Oturkpo to Akwana, Benue Prov. (Feb.) *Jones* FHI 550 ! **S.Nig.:** Lagos *Moloney* ! Ibadan (fl. & fr. Mar., fr. Dec.) *Meikle* 908 ! *Ayewoh* FHI 3010 ! Awka *Thomas* 8 ! Degema *Talbot* 3709 ! Eket *Talbot* 3148 ! 3233 ! **F.Po:** *Barter* 2039 ! *T. Vogel* 77 ! Also in Cameroun, the Congos, Sudan, Uganda, Tanganyika, Cabinda and Angola. (See Appendix, p. 403.)
4. **M. geminata** *DC.* Prod. 4 : 447 (1830) ; Hutch. l.c. 14, with fig. ; Aubrév. Fl. For. C. Iv., ed. 2, 3 : 272, t. 249, and Fl. For. Soud.-Guin. 485 ; Berhaut Fl. Sén. 106. *M. citrifolia* of F.T.A. 3 : 191, partly ; Chev.

Bot. 330, partly, not of Linn. A small tree 20–30 ft. high, with 4-angled branchlets ; flowers white, fragrant; fruits fleshy ; in secondary forest and sometimes planted.
Sen.: Caniag, Cayor *Döllinger* 52 ! Komboi *Heudelot* 376 ! **Gam.:** *Hayes* 530 ! Albreda *Perrottet* 420 ! **Mali:** Sareya (Feb.) *Chev.* 464 ! **Port. G. :** Manooa (fl. & fr. Sept.) *Esp. Santo* 854 ! **Guin.:** Kaba Valley (May) *Chev.* 13186 ! Kouroussa (Apr.) *Pobéguin* 693. Kolenté (May) *Chillou* 1380 ! **S.L.:** (June) *T. Vogel* 145 ! Freetown *Dalz.* 957 ! Batkanu (fr. Apr.) *Thomas* 1 ! Mano *Thomas* 10373 ! Njala (Feb.) *Deighton* 1062 ! Makene (fr. Apr.) *Deighton* 5495 ! **Lib.:** Dukwia R. (fr. May) *Cooper* 445 ! Marshall, Montserrado (fr. Feb.) *Baldwin* 11049 ! Ganta (Mar.) *Harley* 427 ! **Iv.C.:** Danipleu *Aubrév.* 1156. (See Appendix, p. 403.)

62. SCHIZOCOLEA Bremek. in Hook. Ic. Pl. t. 3482 (1950).

Shrub ; branches slender, hirsute ; leaves oblanceolate, 8–12 cm. long, 2·5–4·5 cm. broad, more or less cuneate at base, long-acuminate, midrib hirsute above and beneath ; petiole 3–5 mm. long, hirsute ; stipules fimbriate above, segments subulate, hirsute, up to 2 cm. long ; flowers few in sessile axillary clusters, or solitary ; calyx about 8 mm. long, lobes subulate, hirsute ; corolla-tube about 2 cm. long, expanded above, sparsely pubescent outside, lobes about 1 cm. long, valvate ; filaments 1 mm. long, anthers 4–5 mm. long ; style glabrous, 12 mm. long ; fruit ovoid, 10–12 mm. long, 6–7 mm. diam., surmounted by persistent calyx, 1-seeded .. .. *linderi*

S. linderi (*Hutch. & Dalz.*) Bremek. l.c. (1950). *Urophyllum linderi* Hutch. & Dalz. F.W.T.A., ed. 1, 2 : 104 (1931). *Sabicea ? linderi* (Hutch. & Dalz.) Bremek. (1940). A shrub 4–10 ft. high ; flowers white, fragrant, fruits red ; understorey shrub in high forest.
S.L.: No. 2. River F.R., Freetown (Dec.) *Vigne* 21 ! York Pass (Dec.) *Deighton* 3311 ! Kambui Hills Lane-Poole 469 ! Gola Forest (fl. Jan., fr. Apr.) *King* 10 ! *Deighton* 3686 ! *Small* 577 ! **Lib.:** Kakatown *Whyte* ! Dukwia R. (fr. Apr.) *Cooper* 151 ! Harbel (Dec.) *Bequaert* 14 ! Boporo Dist. (Nov.) *Baldwin* 10362 ! Peahtah (Oct.) *Linder* 1072 ! **Iv.C.:** Nimba Mts. *Schnell* 5156 ! Nouba Mts. (fr. Apr.) *Chev.* 21138 ! 21167 ! Man to Danané (Nov.) *de Wilde* 854 ! Tiapleu Forest (Nov.) *de Wilde* 898 !

63. LASIANTHUS Jack—F.T.A. 3 : 228. *Nom. cons.*

Creeping herb or undershrub rooting at nodes ; stems pilose ; leaves elliptic to obovate-elliptic, 6–10 cm. long, 2·5–5 cm. broad, cuneate to subcordate at base, shortly acuminate or acute at apex, pilose on both surfaces ; petioles 2–3 cm. long, pilose ; flowers congested in leaf axils ; fruits depressed-globose, 6–10 seeded, about 6 mm. diam. .. .. 1. *repens*
Small erect shrub ; stems pilose ; leaves long-elliptic to long-obovate-elliptic, 15–20 cm. long, 5–9 cm. broad, cuneate at base, shortly and acutely acuminate, midrib more or less pilose on both sides ; petioles 2–6 cm. long, pilose ; flowers congested in leaf axils ; fruits depressed-globose, 8–12 seeded, about 8 mm. diam. .. 2. *batangensis*

1. **L. repens** *Hepper* in Kew Bull. 16 : 332 (1962) (excl. F. Po spec. *Melville* 629). Creeping herb with runners and pilose stems ; flowers white, fruits at first orange, becoming blue ; in forest.
Guin.: Sérédou (fl. & fr. May) *Aké Assi* ! **S.L.:** Monomo, Peninsula (fr. Sept.) *Melville & Hooker* 627 ! Nyandehun (Apr.) *Deighton* 3654 ! **Lib.:** Ganta (fl. & fr. May) *Harley* 537 ! **Ghana:** Kibi to Apapam (Dec.) *Morton* GC 8166 ! Also in Gabon, Ubangi-Shari, Congo and Cabinda.
[Since describing this species I have seen several Fernando Po specimens (*Guinea* 2156, 2157, *Boughey* 72) which match *Melville* 629. I do not now regard this as *L. repens* and these specimens, distinct in themselves, are close to *L. mannii* Wernham (see note below)—F.N.H.]

2. **L. batangensis** *K. Schum.* in Engl. Bot. Jahrb. 28 : 107 (1899), incl. var. *longepetiolata* K. Schum. ? *L. mannii* Wernham in Cat. Talb. 56 (1913). *L. guineensis* A. Chev. Bot. 347, name only. An erect undershrub 1–5(–10) ft. high, little branched from the base, sometimes ascending, with pilose stems ; flowers blue or white, fruits pale blue ; in closed forest.
S.L.: Zimi to Gorahun, Gola (fr. Apr.) *Deighton* 3638 ! Gola (fr. Dec.) *King* 96b ! Jepihun *Smythe* 229 ! **Lib.:** Bumbuna to Moala (fl. & fr. Nov.) *Linder* 1331 ! Ba, Mano R. (fr. Dec.) *Baldwin* 10698 ! Zuie, Boporo Dist. (fl. & fr. Nov.) *Baldwin* 10249 ! Kulo, Sinoe (fr. Mar.) *Baldwin* 11428 ! Tchien (Aug.) *Baldwin* 7015 ! **Iv.C.:** Abidjan *Chev.* 15280 ! 15283 ! Guidéko *Chev.* 16384 ! 16451 ! Alépé (Apr.) *Chev.* 17466 ! Yapo (Sept.) *Roberty* 12117 ! **Ghana:** Asamankese F.R. (Sept.) *Vigne* FH 4255 ! Opon Mansi F.R. (fr. Oct.) *Akpabla* 920 ! Benso (Aug.) *Andoh* FH 5403 ! **S.Nig.:** Oban *Talbot* 266 ! [**Br.]Cam.:** Lipundu to Banyu, Kumba (fr. July) *Olorunfemi* FHI 30698 ! Extends southwards to Cabinda.
[Talbot 266 (type) and FHI 30698 are referable to *L. mannii*, if this is to be regarded as a distinct species. Detailed study is required of this difficult genus.]

64. GAERTNERA Lam.—F.T.A. 4, 1 : 542. *Nom. cons.*

Inflorescence a large loosely paniculate cyme ; peduncle glabrous ; nerves of leaves shortly pilose beneath or almost glabrous ; leaves chartaceous, elliptic to oblong, acutely acuminate, more or less cuneate at the base, up to 18 cm. long and 7 cm. broad ; calyx about 1 mm. long, bracteole shorter .. .. 1. *paniculata*
Inflorescence compact or small ; peduncle puberulous ; nerves of leaves puberulous beneath :
Bracteoles ovate or broadly triangular, not longer than the calyx :
  Calyx, with the lobes, 3 mm. long ; bracteoles about as long ; corolla-tube about 3 mm. wide ; leaves chartaceous, oblong, truncate to acute at the base, shortly acuminate, 15–25 cm. long, 7–10 cm. broad, tertiary nerves prominent ; inflorescence many flowered, compact .. .. 2. *cooperi*
  Calyx 1·5–2 mm. long ; bracteoles not half as long ; corolla about 2 mm. wide ; leaves thin, oblong to lanceolate, cuneate, long-acuminate, 8–11 cm. long, 2–3 cm. broad ; tertiary nerves faint ; inflorescence few-flowered, loose 3. *liberiensis*
Bracteoles narrowly triangular or lanceolate, longer than the calyx ; calyx about

3 mm. long ; leaves thin, elliptic to oblong or lanceolate, more or less cuneate, acuminate, up to 12 cm. long and 4·5 cm. broad ; inflorescence very compact
                                              4. *longevaginalis*

1. **G. paniculata Benth.** in Fl. Nigrit. 459 (1849) ; F.T.A. 4, 1 : 543 ; Chev. Bot. 444, partly ; Petit in Bull. Jard. Bot. Brux. 29 : 42 (1959). *G. occidentalis* H. Baill. (1880), not of F.W.T.A., ed. 1, 2 : 21. A shrub or tree 10–20 ft. high with low branches ; flowers whitish in rather lax pyramidal panicles ; in forest. **Guin.**: *Heudelot* 888. Macenta (Oct.) *Baldwin* 9833 ! Faranna (fr. Mar.) *Sc. Elliot* 5326 ! Kouria (Oct.) *Caille* in *Hb. Chev.* 14794 ! 14992 ! **S.L.**: Leicester Peak (Dec.) *Sc. Elliot* 3860 ! York (Sept.) *Hooker & Melville* 613 ! Mamansu (Feb.) *Deighton* 4060 ! Rokupr (Feb.) *Deighton* 3604 ! Loma Mts. (Nov., Apr.) *Frith* 23 ! *Jaeger* 597 ! **Lib.**: Grand Bassa (July) *T. Vogel* 20 ! Monrovia (Nov.) *Linder* 1537 ! Brewersville (June) *Barker* 1335 ! Greenville (fr. Mar.) *Baldwin* 11554 ! **Iv.C.**: *fide* Chev. *l.c.* **Ghana**: Essiama (Feb.) *Irvine* 2339a ! *Cudjoe* 138 ! Assin-Yan-Kumasi *Cummins* 194 ! Amedzofe (Nov.) *Enti* FH 6882 ! Toga Mt. (Nov.) *Morton* A3580 ! **S.Nig.**: Lagos *Dalz.* 1045 ! Brass *Barter* 1877 ! Jamieson R., Sapoba *Kennedy* 1569 ! *Brenan* 8934 ! Akpaka F.R., Onitsha (Mar.) *Duruh* FHI 15466 ! Eket (May) *Talbot* 3391 ! Onochie FHI 33168 ! **[Br.]Cam.**: Bali-Ngemba F.R., Bamenda (fr. Feb.) *Daramola* FHI 40471 ! Bum, Bamenda (fr. Apr.) *Maitland* 1410 ! Vogel Peak, 4,900 ft., Adamawa (Dec.) *Hepper* 1514 ! Also in Cameroun, Gabon, Congo and N. Rhodesia. (See Appendix, p. 362.)

2. **G. cooperi Hutch. & M. B. Moss** in F.W.T.A., ed. 1, 2: 21 (1931), Kew Bull. 1937 : 62, partly (excl. *Chev.* 12420 ; 12664 ; 12936). A shrub or small tree 5–25 ft. high ; flowers white, fruits blue ; in moister parts of forest. **Lib.**: Dukwai R. (Aug.) *Cooper* 202 ! 287 ! Duport (fl. May, Nov., fr. Nov.) *Dinklage* 3056 ! *Linder* 1487a ! Paynesville, Monrovia (Apr.) *Bequaert* 176 ! Cape Palmas (May) *Cooper* 465 ! Sinoe *Whyte* ! **Iv.C.**: Banco Serv. For. 381 ! Abouabou, Abidjan to Grand Bassam (fr. Jan.) *Leeuwenberg* 2366 ! **Ghana**: Ateiku (May) *Vigne* FH 1948 ! Ankobra R., Axim (Mar.) *Morton* GC 8473 ! (See Appendix, p. 362.)

3. **G. liberiensis** *Petit* in Bull. Jard. Bot. Brux. 29 : 40 (1959). *G. salicifolia* Hutch. & J. B. Gillett in F.W.T.A., ed. 1, 2 : 21 (1931) ; Kew Bull. 1937 : 62, not of C. H. Wright ex Bak. (1903). A tree 20 ft. high with reddish bark. **Lib.**: Dukwai R. (June) *Cooper* 277 ! (See Appendix, p. 362.)

4. **G. longevaginalis** *(Schweinf. ex Hiern)* Petit in Bull. Jard. Bot. Brux. 29 : 45 (1959). *Psychotria longe-vaginalis* Schweinf. ex Hiern in F.T.A. 3 : 201 (1877). *Gaertnera plagiocalyx* K. Schum. in Schlechter W. Afr. Kautsch.-Exped. 322, name only. *G. cooperi* of F.W.T.A., ed. 1, 2 : 21, partly, and *G. paniculata* of Chev. Bot. 144, partly (*Chev.* 12420 ; 12664 ; 12936). *G. occidentalis* of F.W.T.A., ed. 1, 2 : 21, not of H. Baill. A weak shrub, or rarely a small tree, a few feet high ; flowers white, fruits blue-purple ; in forest beside streams. **Guin.**: Ditinn (Apr.) *Chev.* 12984 ! Diaguissa *Chev.* 12420 ; 12664 ; 12936. **S.L.**: Bagroo R. (fl. & fr. Apr.) *Mann* 802 ! Makali (Feb.) *Deighton* 4086 ! Kambui F.R. (May) *Lane-Poole* 246 ! Falaba (Apr.) *Jaeger* 45 ! Loma Mts. (Sept.) *Jaeger* 1465 ! **Lib.**: Gbanga (fl. & fr. Sept.) *Linder* 758 ! Maa, Makona R. (fl. & fr. Feb.) *Bequaert* 55 ! Mt. Wolagwisi (Mar.) *Bequaert* 106 ! Ganta (fr. Oct.) *Harley* 1035 ! Baila (Feb.) *Harley* 1910 ! **Iv.C.**: Tonkoui (Aug.) *Boughey* GC 18334 ! Also in Cameroun, Congo and Sudan.

65. CHASSALIA Comm. ex Poir. in Lam. Encycl. Méth. Bot., Suppl. 2 : 450 (1812). *Psychotria* of F.T.A. 3 : 193, partly (sect. *Chasalia*).

Inflorescences subsessile, congested, unbranched except in fruit ; leaves glabrous :
  Corolla 2 cm. long, the tube slightly curved ; calyx denticulate ; flowers in several congested whorls ; leaves elongate-oblanceolate, about 25 cm. long, 4–6 cm. broad, with about 15 pairs of lateral nerves   ..   ..   1. *elongata*
Corolla up to 1 cm. long :
  Climbing shrub ; corolla curved   ..   ..   11. *zenkeri*
  Undershrubs ; corolla straight ; fruits round :
    Fruiting pedicels thick and fleshy, about 3 mm. diam. ; leaves elliptic, rounded to cuneate at base, acuminate, 10–19 cm. long, 4–8 cm. broad, with about 11 pairs of lateral nerves, venation prominent ; petioles up to 3 cm. long   2. *corallifera*
    Fruiting pedicels not thickened ; leaves obovate-elliptic, rather long-cuneate at base, acuminate, 8–14 cm. long, 3·5–6 cm. broad, with about 7 pairs of lateral nerves, venation obscure ; petioles 1–2 cm. long :
      Stems with a very narrow longitudinal ridge from the stipules on each side
                                     3. *subherbacea*
      Stems terete ..   ..   4. *subnuda*
Inflorescences pedunculate, branched and sometimes very lax :
  Inflorescences minutely tomentose ; flowers subsessile :
    Midrib minutely tomentose beneath, leaves broadly oblanceolate, shortly subcordate at base, abruptly acuminate, 8–31 cm. long, 4–9 cm. broad, with 11–13 pairs of lateral nerves ; internodes usually short ; inflorescence terminal, with peduncle about 5 cm. long, shortly branched above   ..   ..   5. *ischnophylla*
    Midrib glabrous beneath :
      Corolla-lobes keeled, tube about 5 mm. long, straight ; inflorescences branched spreading panicles, 3–7 cm. diam., with numerous flowers ; leaves elliptic to oblong-elliptic, cuneate at base, long-acuminate, 8–14 cm. long, 3–5(–8) cm. broad, with about 11 pairs of closely arched lateral nerves   ..   6. *cristata*
      Corolla-lobes not keeled :
        Fruits spherical, smooth ; inflorescences shortly branched, about 2 cm. long, few-flowered   ..   ..   7. *kolly*
        Fruits ovoid, angular (at least when dry) ; inflorescences lax and spreading, more or less 6 cm. long and diam.   ..   ..   8. *afzelii*
  Inflorescences glabrous :
    Flowers subsessile ; shrubs ; corolla keeled :
      Corolla straight, the small lobes keeled ; (other characters as above)   6. *cristata*

Corolla curved, at least 1 cm. long, expanded above and keeled ; inflorescences about
    2 cm. long ; [young] leaves elliptic, cuneate at base, rather long-acuminate,
    6–8 cm. long, 2–3 cm. broad   ..   ..   9. sp. nr. *umbraticola*
Flowers pedicellate ; climbers ; corolla not keeled :
Pedicels 2–3 mm. long :
Inflorescences lax and spreading panicles about 6 cm. long and diam. ; corolla
    about 9 mm. long, nearly straight ; fruits ovoid, angular (when dry) ; leaves
    oblong, cuneate at base, long-acuminate, 8–14 cm. long, 3–4·5 cm. broad
                                                                        10. *laxiflora*
Inflorescences 2–3 cm. long, very shortly branched ; corolla about 11 mm. long,
    curved ; leaves elliptic, caudate at apex, cuneate at base, 9–17 cm. long, 3–8 cm.
    broad   ..   ..   11. *zenkeri*
Pedicels 7–13 mm. long ; inflorescences lax, about 6 cm. long, few-flowered ; corolla
    sharply curved at base, 1·5 cm. long ; leaves elliptic to oblong-elliptic, cuneate to
    rounded at base, caudate at apex, 8–15 cm. long, 2·5–8 cm. broad   12. *cupularis*

1. **C. elongata** *Hutch. & Dalz.* F.W.T.A., ed. 1, 2 : 127 (1931). A shrub with rather long leaves ; flowers
    yellow with the inflorescence rhachis red-purple tinged.
    Lib.: Dukwai R., Monrovia (Oct., Nov.) *Cooper* 33 !
    [*Rosevear* 63/20A from S. Nigeria, has smaller flowers and leaves. It appears to be closely related to
    this sp. and to *C. simplex* K. Krause (1920) in Cameroun.]
2. **C. corallifera** (*A. Chev. ex De Wild.*) *Hepper* in Kew Bull. 16 : 329 (1962). *Psychotria corallifera* A. Chev.
    ex De Wild. in Pl. Bequaert. 2 : 354 (1924). An undershrub 1–3 ft. high ; flowers white ; fruits deep
    violet on fleshy coral-pink pedicels ; in forest.
    S.L.: *Thomas* 3857 ! **Lib.:** Gbanga (fr. Sept.) *Linder* 566 ! Zolopla, Sanokwele Dist. (fr. Sept.) *Baldwin*
    9463 ! Ganta (fr. Sept.) *Harley* 659 ! *Baldwin* 12520 ! Gletown, Tchien Dist. (fr. July) *Baldwin* 6916 !
    Monroviatown, Tchien Dist. (fr. Aug.) *Baldwin* 7095 ! **Iv.C.:** Montézo to Alépé (fl. & fr. Feb.) *Chev.*
    17368 ! 17417 ! 17431 ! Koléahinou forest (fl. buds Jan.) *Aké Assi* 3735 !
3. **C. subherbacea** (*Hiern*) *Hepper* l.c. 331 (1962). *Psychotria subherbacea* Hiern in F.T.A. 3 : 208 (1877).
    P. coralloides A. Chev. ex De Wild. l.c. 356 (1924). An undershrub 1–3 ft. high ; flowers white, open in
    the morning ; on forest floor.
    S.L.: Waterloo to York (fr. Sept.) *Melville & Hooker* 626 ! **Iv.C.:** Gouléako (fl. & fr. Jan.) *Aké Assi*
    3737 ! Yapo (Dec.) *Aké Assi* 3738 ! 3739 ! Taï (Mar.) *Leeuwenberg* 3020 ! **Ghana:** Princes Town (fr.
    Mar.) *Morton* A361 ! **S.Nig.:** Omo (formerly part Shasha) F.R. (Apr.) *Jones & Onochie* FHI 17406 !
    Ross 208 ! **[Br.]Cam.:** Kembong F.R., Mamfe (Mar.) *Onochie* FHI 31177 ! **F.Po:** *Mann* 1420 !
4. **C. subnuda** (*Hiern*) *Hepper* l.c. 331 (1962). *Psychotria subnuda* Hiern in F.T.A. 3 : 209 (1877). A shrub
    4–6 ft. high.
    S.Nig.: Cross R., Calabar (fl. & fr. Feb.) *Mann* 2270 ! .
5. **C. ischnophylla** (*K. Schum.*) *Hepper* l.c. 330 (1962). *Psychotria ischnophylla* K. Schum. in Engl. Bot. Jahrb.
    28 : 93 (1899). A small simple-stemmed shrub, erect or ascending, 1–1½ ft. high ; flowers white ; fruits
    black ; in wet forest.
    S.Nig.: Usonigbe F.R., Benin (fr. Oct.) *Keay & Onochie* FHI 19675 ! Okomu F.R., Benin (Feb.) *Brenan*
    9089 ! Eket *Talbot* ! Oban (Mar.) *Richards* 5181 ! 5199 ! Okarara, Calabar (fr. May) *Ujor* FHI 30837 !
    [Br.]Cam.: Korup F.R., Kumba (fr. June) *Olorunfemi* FHI 30645 !
6. **C. cristata** (*Hiern*) *Bremek.* in Bull. Jard. Bot. Brux. 22 : 104 (1952). *Psychotria cristata* Hiern in F.T.A.
    3 : 205 (1877). A shrub up to 10 ft. high ; flowers white ; in forest.
    S.Nig.: Degema Dist. *Talbot* 3672 ! 3743 ! **[Br.]Cam.:** Cam. Mt. : Buca to Musaka, 4,500 ft. (fr. May)
    Maitland 706 ! Buea *Deistel* 645 ! Lakom, Bamenda (Apr.) *Maitland* 1665 ! Bafut-Ngemba F.R.,
    Bamenda (fr. June) *Lightbody* FHI 26326 ! **F.Po:** above Moka, 4,800 ft. (fl. & fr. Sept.) *Wrigley* 578 !
    Also in Congo, Sudan, Uganda, Kenya, Tanganyika and Angola.
7. **C. kolly** (*Schumach.*) *Hepper* in Kew Bull. 16 : 330 (1962). *Psychotria kolly* Schumach. (1827). *P. parviflora*
    Benth. (1849). *P. benthamiana* Hiern in F.T.A. 3 : 204 (1877) ; F.W.T.A., ed. 1, 2 : 125. *P. warneckei*
    K. Schum. & K. Krause (1907)—F.W.T.A., ed. 1, 2 : 125. *P. infundibularis* of Chev. Bot. 341, not of
    Hiern. A soft-stemmed shrub 3–10 ft. high ; flowers white or pink ; in forest.
    Guin.: Macenta (Apr., May) *Collenette* 6 ! 18 ! **S.L.:** Kukuna (Jan.) *Sc. Elliott* 4713 ! Makene (Apr.)
    Deighton 5502 ! Musaia (Mar.) *Deighton* 5437 ! Batkanu (fl. & fr. Apr.) *Thomas* 54 ! Binkolo (fr. Aug.)
    Thomas 1676 ! **Lib.:** Gbanga (Sept.) *Linder* 607 ! Nekaboyu, Vonjama Dist. (fl. & fr. Oct.) *Baldwin*
    9983 ! Ganta (fr. Sept.) *Baldwin* 9325 ! Yratoke (July) *Baldwin* 6274 ! St. John R., Baila (May) *Harley*
    1424 ! **Iv.C.:** Bingerville, Abidjan, Dabou *Chev.* 15270 ; 15335 ; 15361. Accrédiou (fr. Feb.) *Chev.*
    17073. Malamalasso to Daboïssué *Chev.* 17558. Tabou to Bériby *Chev.* 19992. Béyo (fl. & fr. Jan.)
    Leeuwenberg 2610 ! **Ghana:** Nsawam (Oct.) *Morton* GC 7806 ! Achimota (fl. & fr. July) *Irvine* 746 !
    Kakumdu (July) *Hall* 443 ! Aburi Scarp (fl. & fr. Jan.) *Morton* A2844 ! Kpedsu (fl. & fr. Jan.) *Howes*
    1084 ! **Togo Rep.:** Lomé (fl. Nov., fl. & fr. Aug.) *Mildbr.* 7485 ! *Warnecke* 44 ! 381 ! **N.Nig.:** Katsina
    Allah (Dec.) *Dalz.* 648 ! **S.Nig.:** Olokemeji F.R. (Mar.) *Hepper* 2294 ! Abeokuta (Jan.) *Rowland* !
    Ibadan (fl. & fr. Apr.) *Meikle* 1403 ! Ibadan to Abeokuta (Mar.) *Schlechter* 13029 ! Okomu F.R., Benin
    (Jan.) *Brenan* 8818 ! **[Br.]Cam.:** Cam. Mt. (Mar., Apr.) *Hambler* 144 ! Vogel Peak, Adamawa (Nov.)
    Hepper 1455 ! **F.Po:** *T. Vogel* !
8. **C. afzelii** (*Hiern*) *K. Schum.* in Engl. Bot. Jahrb. 23 : 469 (1897). *Psychotria afzelii* Hiern in F.T.A. 3 : 205
    (1877) ; Chev. Bot. 339. *Chassalia laxiflora* Benth. (1849), partly (*Don*). A straggling shrub or tree up
    to 30 ft. high ; flowers white with yellow centre ; in forest.
    Port. G.: Catio (Aug.) *Esp. Santo* 2177, partly ! **Guin.:** Kindia *Chev.* 13004 ; 13125. Diaguissa to Timbo
    Chev. 13447. **S.L.:** *Afzelius* ! *Don* ! Hastings (Sept.) *Melville & Hooker* 429 ! No. 2. River F.R. (July)
    Small 163 ! Bagroo R. (Apr.) *Mann* 843 ! Yonibana (fl. & fr. Nov.) *Thomas* 4734 ! Njala (Sept.) *Deighton*
    104 ! **Lib.:** Tubman Bridge (Sept.) *Barker* 1411 ! Ganta (fl. & fr. Sept.) *Harley* 628 ! Gbau, Sanokwele
    Dist. (Sept.) *Baldwin* 9402 ! **Iv.C.:** Adiopodoumé (fr. Nov.) *Leeuwenberg* 2095 ! **Ghana:** Mpameso F.R.
    (fl. & fr. Dec.) *Adams* 2941 ! Assin (Sept.) *West-Skinn* 53 ! 139 !
9. **C. sp. nr. umbraticola** *Vatke* in Oestr. Bot. Zeitschr. 25 : 230 (1875). A shrub about 3 ft. high ; flowers
    white ; in secondary upland forest.
    [Br.]Cam.: Bamenda-Nkwe (Apr.) *Ujor* FHI 30027 ! Bafut-Ngemba F.R. (May) *Daramola* FHI 41068 !
    [Very similar to plants widely spread in E. Africa which are named *C. umbraticola* Vatke, *C. discolor*
    K. Schum. and *C. albiflora* K. Krause. Further work is required to elucidate this complex.]
10. **C. laxiflora** *Benth.* in Fl. Nigrit. 416 (1849), partly (*Ansell*). *Psychotria ansellii* Hiern in F.T.A. 3 : 214
    (1877), partly (*Ansell*). Scrambling shrub ; flowers white or pinkish with yellow throat, and the inflores-
    cence-rhachis often purple.
    Lib.: Monrovia *Farmar* 348 ! *Whyte* ! Congotown (Mar.) *Barker* 1224 ! Ganta (Sept.) *Baldwin* 9301 !
    Grand Bassa *Ansell* ! **Iv.C.:** Azaguié (Sept.) *Chev.* 22280 ! 22281 !
11. **C. zenkeri** *K. Schum. & K. Krause* in Engl. Bot. Jahrb. 39 : 565 (1907). *Psychotria ansellii* Hiern in
    F.T.A. 3 : 214 (1877), partly (*Thomson* 105) ; *Chasalia laxiflora* of F.W.T.A., ed. 1, 2 : 127, partly
    (*Thomson* 105), not of Benth. A climber ; flowers waxy, crimson and yellow ; in forest.

S.Nig.: Shakwa, Obudu (May) *Latilo* FHI 30929 ! Degema Dist. *Talbot* 3821 ! Eket Dist. *Talbot* 3056 ! Calabar *Thomson* 105 ! Also in Cameroun.

12. **C. cupularis** *Hutch. & Dalz.* F.W.T.A., ed. 1, 2 : 127 (1931). A glabrous climbing shrub ; flowers scarlet, yellow inside ; in forest.

S.Nig.: Oban (fr. Feb.) *Talbot* 233 ! Onochie FHI 36256x ! Afi River F.R., Ikom (fl. & fr. Dec.) *Keay* FHI 28253 !

Imperfectly known species.

C. garretii *K. Schum. & K. Krause* in Engl. Bot. Jahrb. 39 : 564 (1907). Apparently close to *C. laxiflora* Benth. **W.Africa.**: without locality (Oct. 1893) *Garret.*

Besides the above, several species at present under *Psychotria* may belong to *Chassalia*, e.g. *P. pteropetala* K. Schum.

66. PSYCHOTRIA Linn.—F.T.A. 3 : 193. *Nom. cons. Grumilea* Gaertn.—F.T.A. 3 : 215 ; F.W.T.A., ed. 1, 2 : 216.

*Bracts and bracteoles conspicuous and persistent ; inflorescences either small pilose and congested, or spreading (to p. 194) :

Inflorescences more or less globose or congested, subsessile :

Ascending undershrub with prostrate rooting stems ; flowers very few, apical, calyx-lobes linear ; sheathing stipules drawn out into 2 filiform lobes, about 10 mm. long, on each side, pilose ; leaves ovate, acute at apex, rounded or subcordate at base, about 6 cm. long and 3 cm. broad, pilose at least beneath   **1. brenanii**

Erect undershrubs :

Leaves glabrous or pubescent on the nerves and midrib, lateral nerves about 15, close, lamina oblanceolate to ovate, acuminate, cuneate to rounded at base, 10–17 cm. long, 3–6 cm. broad ; stems more or less glabrous but nodes pilose ; inflorescence usually 1 globose head (rarely 3 heads on a short common peduncle), 1·5–2 cm. diam. ; calyx-lobes linear, pilose on the margins ; fruits ellipsoid, 1 cm. long ..   ..   **2. globosa**

Leaves pilose all over the lower surface ; lateral nerves 8–10 on each side of midrib :

Petioles (5–)10–17(–40) mm. long, slender, pilose ; leaves thinly pilose above, more densely beneath, oblanceolate to ovate, long-cuneate base, 8–17 cm. long, 3·5–6·5 cm. broad ; inflorescence globose, solitary, 1–2 cm. diam. ; calyx-lobes linear, 3 mm. long, pilose on the margins ; fruits ellipsoid, 1 cm. long   **3. nigerica**

Petioles 5 mm. long, stout, densely pilose ; leaves glabrous above, densely pilose beneath, oblanceolate, caudate at apex, cuneate at base, 7–9 cm. long, 2·5–3 cm. broad ; inflorescence congested, 8 mm. diam. ; calyx-lobes acute, 1·5 mm. long, densely pilose ; fruits spherical, 6 mm. diam. ..   ..   **4. sp.** *A*

Inflorescences more or less pyramidal branched cymes or rarely fairly long-pedunculate clusters :

Branchlets tomentose or pubescent all round :

Ascending or creeping undershrubs :

Leaves gradually acutely acuminate, oblong-elliptic about 10 cm. long and 3–4 cm. broad, reddish-brown when dry, rather thin ; internodes not very short ; inflorescence small with rather small bracts ; fruits about 1 cm. diam. when fleshy, drying ribbed and about 5 mm. long ..   ..   ..   **5. reptans**

Leaves rounded to subacute at apex, oblong-elliptic, 5–7 cm. long, 2–3 cm. broad, shortly pubescent on midrib and nerves beneath ; internodes short with stipules often overlapping ; inflorescence very small on peduncle about 3 cm. long   **6. kitsonii**

Erect small shrubs :

Midrib pubescent above ; stems and peduncles rusty-pilose :

Leaves not bullate, with 12–15 nerves on each side of midrib, acuminate, cuneate to rounded at base, 10–12 cm. long, 5–6 cm. broad, rusty-pilose on both sides ; inflorescences small, up to 6 cm. long, flowers few, not in distinct heads ; calyx-lobes 2–3 mm. long, pilose   ..   ..   **7. coeruleo-violacea**

Leaves bullate with about 20 nerves on each side of midrib, acuminate, more or less cuneate at base, 9–13 cm. long, 3·5–7 cm. broad, rusty-pilose beneath ; inflorescences up to 13 cm. long, flowers in distinct heads ; bracteoles broad, 3-toothed ; calyx-lobes about 1 mm. long, pilose   **8. rufipilis** var. *rufipilis*

Midrib glabrous above :

Leaves bullate, midrib beneath and peduncles rusty-pilose ; (see also below)   **8a. rufipilis** var. *konkourensis*

Leaves not bullate ; midrib beneath and peduncles pubescent ; lamina obovate to obovate-elliptic ; broadly acuminate, cuneate at base, 10–20 cm. long, 5–10 cm. broad, with about 20 lateral nerves on each side of midrib conspicuously looped near margin ; inflorescences pedunculate and branched, with small heads ; fruits ribbed when dry, about 8 mm. long   ..   **9. vogeliana**

Branchlets glabrous or only with longitudinal stripes of hairs :

Midrib above pilose with spreading hairs ; leaves bullate, midrib impressed above   **8a. rufipilis** var. *konkourensis*

Midrib above glabrous or with a few hairs :
  Leaves not coriaceous ; inflorescences lax, reflexed or erect :
    Stipules shortly acuminate ; inflorescences reflexed later :
      Bracts entire or very slightly toothed, 1–1·5 cm. long ; inflorescences lax, flowers
        few .. .. **10.** *latistipula*
      Bracts 3-lobed, 6 mm. long ; inflorescences lax with flowers in small tight heads
                                                        11. *subglabra*
    Stipules long-acuminate, 1·2–1·5 cm. long :
      Inflorescences reflexed ; leaves oblanceolate to obovate, acuminate, cuneate at
        base, 12–22 cm. long, 5–7 cm. broad, glabrous .. .. **12.** *strictistipula*
      Inflorescences not reflexed ; leaves elliptic to ovate-elliptic, acuminate, cuneate
        at base, 7–16 cm. long, 2·5–4 cm. broad, shortly pubescent to glabrescent beneath
                                                        13. *obscura*
  Leaves coriaceous ; lateral nerves 15–20 on each side of midrib, looped near margin :
    Midrib beneath entirely glabrous, leaves obovate, acuminate, cuneate at base,
      12–21 cm. long, 5–8 cm. broad ; inflorescences long-pedunculate, peduncle
      8–15 cm. long with usually 3 distinct heads at apex, more or less glabrous
                                                        14. *bidentata*
    Midrib beneath minutely or rusty-pubescent :
      Leaves with midrib minutely pubescent beneath, lamina oblanceolate to obovate,
        acuminate, cuneate at base, 15–21 cm. long, 6–7 cm. broad ; inflorescence
        pyramidal on a peduncle about 4 cm. long, rather close and not divided into
        distinct heads .. .. **15.** *maliensis*
      Leaves with midrib rusty-pubescent beneath ; peduncle up to 10 cm. long ;
        inflorescences formed of a number of distinct small heads .. **9.** *vogeliana*
*Bracts and bracteoles inconspicuous :
  Leaves with numerous bacterial nodules showing as black or pellucid dots (round or
    sometimes irregular) on the leaves beneath [1] (to p. 195) :
  Flowers 5-merous :
    Peduncles partly flattened and 2-winged ; branchlets glabrous ; inflorescences very
      lax, about 12 cm. long ; pedicels about 4 mm. long ; calyx truncate ; stipules
      deeply lobed into 2 broad twisted and reflexed lobes, nearly glabrous ; leaves
      elliptic, acuminate, cuneate and unequal sided at base, 11–15 cm. long, 4–6 cm.
      broad, glabrous .. .. **16.** *ceratalabastron*
    Peduncles not flattened or winged, filiform ; branchlets pubescent ; stipules
      acuminate into 2 subulate lobes ; inflorescences 2–3 cm. long, very few-flowered ;
      peduncle and pedicels filiform :
      Branchlets with opposite longitudinal stripes of pubescence ; leaves obovate-
        oblanceolate, rather long-acuminate, 5–8 cm. long, 1·5–3 cm. broad, with about
        8 pairs of lateral nerves ; petiole 5 mm. long, strigose above .. **17.** *bifaria*
      Branchlets pubescent all round :
        Corolla-lobes ovate-triangular :
          Leaves with about 8 pairs of lateral nerves, obovate to obovate-oblong, rather
            long-acuminate, cuneate at base, 5–10 cm. long, 2–4·5 cm. broad, glabrous ;
            inflorescences very slender 2–6 cm. long, branched, rather few flowered
                                                        18. *mannii*
          Leaves with about 15 pairs of lateral nerves, oblanceolate, acute to shortly
            acuminate, long-cuneate to a slightly rounded base, 12–23 cm. long, 3·5–8 cm.
            broad, slightly pubescent ; flowers rather clustered at the end of the 3–7 cm.
            long peduncle .. .. **19.** *konguensis*
        Corolla-lobes narrow, horn-like ; (for other characters see below) **21.** *humilis*
  Flowers 4-merous :
    Midrib pubescent beneath :
      Leaves subcordate to rounded at base, acuminate, oblong-lanceolate, 9–20 cm. long,
        3·5–7 cm. broad ; erect shrub ; corolla with ovate-triangular lobes
                                                        20. *brachyanthoides*
      Leaves cuneate to rounded at base ; corolla-lobes narrow and horn-like ; (for other
        characters see below) .. .. **21.** *humilis*
    Midrib glabrous beneath :
      Corolla-lobes narrow and horn-like, very small :
        Leaves acutely acuminate, cuneate at base into slender petiole 1–2 cm. long,
          lamina elliptic to obovate, 5–7 cm. long, 2·3–5 cm. broad, glabrous ; inflorescence
          1·5 cm. long with few, rather clustered flowers ; undershrub **22.** *cornuta*
        Leaves acute to slightly acuminate, rounded to cuneate at base, ovate-elliptic,
          3–6(–10) cm. long, 1·5–3(–6) cm. broad, pubescent or glabrous ; inflorescences
          few-flowered 1(–6) cm. long ; more or less herbaceous .. **21.** *humilis*

[1] Beware of superficial fungal spots and other blemishes. Bacterial nodules seem to be very consistently present in certain species and, although they are often visible on the upper surface, they are best seen on the under surface of the leaves.

Corolla-lobes ovate-triangular, shrubs or undershrubs :
  Branchlets pubescent all round ; leaves with about 8 nerves on each side of
    midrib ; (for other characters see above) .. .. .. 18. *mannii*
  Branchlets, inflorescences and leaves glabrous ; stipules more or less pubescent ;
    drying brownish ; bacterial nodules small and in irregular clusters :
    Branchlets terete, without longitudinal ridges :
      Leaves with about 9 or fewer nerves each side of midrib, oblong-elliptic, 5–12 cm.
        long, 2·5–4·5 cm. broad, long-acuminate, cuneate at base ; inflorescences very
        lax and few-flowered ; peduncle filiform, 2–4 cm. long ; corolla 4 mm. long
                                                    23. *linderi*
      Leaves with about 13 nerves on each side of midrib, obovate to oblanceolate,
        10–17 cm. long, 3–7 cm. broad, more or less acuminate, cuneate at base ;
        inflorescences 5–9 cm. long, well-branched ; peduncle about 3 cm. long,
        longer than flowering part of inflorescence ; corolla 2 mm. long
                                              24. *brachyantha*
    Branchlets with 2 longitudinal ridges ; leaves with about 15 lateral nerves on
      each side of midrib, oblanceolate to oblong-elliptic, 13–20 cm. long, 4·5–9 cm.
      broad, abruptly and obtusely acuminate, cuneate at base ; drying green with
      bacterial nodules appearing as distinct black dots ; inflorescences 3–8 cm. long,
      well-branched ; peduncle shorter than or equal to the flowering part of the
      inflorescence ; corolla 2 mm. long .. .. 25. *leptophylla*
Leaves without black dots beneath :
  Midrib beneath with linear black bacterial nodules (several mm. long) at irregular
    intervals on each side :
    Leaf margins strongly undulate, acute at each end, narrowly elliptic-lanceolate,
      7–11 cm. long, 1·5–3 cm. broad, with about 7 nerves on each side of midrib ;
      branchlets with 2 conspicuous ridges ; stipules deeply divided into 2 reflexed
      setae 2 mm. long ; inflorescences terminal with peduncle 1–2 cm. long, few
      flowered and rather congested at apex .. .. 26. *recurva*
    Leaf margins not undulate ; stems terete ; stipules bilobed at apex :
    Leaves with 4–5(–7) nerves on each side of midrib, linear-elliptic to lanceolate-
      elliptic, acute, narrowly cuneate at base, 5–14 cm. long, 1–5 cm. broad ; inflores-
      cences pedunculate, branched above, reflexed in fruit .. 27. *talbotii*
    Leaves with 7–11 nerves on each side of midrib ; inflorescences branched-
      paniculate :
      Apex of leaves acute or obtuse, not acuminate, lanceolate-elliptic, 4–13 cm. long,
        1–4·5 cm. broad ; petioles up to 2 cm. long ; in savanna .. .. 28. *huae*
      Apex of leaves acuminate, very variable in shape, lanceolate-elliptic to broadly
        ovate-elliptic, 8–15(–20) cm. long, 2–6(–10) cm. broad ; petioles 1–5 cm. long ;
        usually in forest .. .. 29. *calva*
Midrib without linear nodules :
†Endosperm not ruminate (to p. 197) :
  Inflorescences sessile, unbranched, congested :
    Ascending undershrub with prostrate rooting stems ; (for other characters see
      above)   1. *brenanii*
    Shrubs with erect stems :
      Calyx-lobes broadly ovate 1–2 mm. long ; branchlets without thick soft bark,
        terete ; (for other characters see below) .. .. .. 30. *subobliqua*
      Calyx-lobes shortly triangular or calyx truncate :
        Petioles 2–6 cm. long :
          Stipules about 15 mm. long, broadly ovate, ferrugineous, caducous ; calyx
            truncate ; (for other characters see below) .. .. 31. *dorotheae*
          Stipules about 5 mm. long, thick, straw-coloured :
            Leaves elliptic, 17–20 cm. long, 8–10 cm. broad, minutely pubescent beneath ;
            petioles up to 6 cm. long ; calyx distinctly lobed .. 32. *obanensis*
            Leaves oblong-oblanceolate, narrowly cuneate at base, 22–31 cm. long,
            4–7·5 cm. broad, glabrous ; petioles up to 8 cm. long ; calyx obscurely
            lobed .. .. 33. *longituba*
        Petioles up to 1 cm. long ; leaves elliptic, acuminate, cuneate at base, 7–15 cm.
        long, 2·5–7 cm. broad ; branchlets with thick cracking bark :
          Branchlets with 2 longitudinal ridges ; leaves glabrous beneath ; corolla 4-lobed,
            about 3 mm. long .. .. 34. *lophoclada*
          Branchlets terete ; leaves minutely pubescent beneath ; corolla 5-lobed,
            about 6 mm. long .. .. 35. *insidens*
  Inflorescences pedunculate (capitate or branched) :
    Peduncles pilose, pubescent or puberulous :
      Indumentum of peduncle conspicuously pilose to puberulous ; inflorescences
        capitate :
        Leaves glabrous beneath, obovate, 10–15 cm. long, 4·5–7 cm. broad ; stems
          glabrous ; stipules and peduncles densely pilose with purple-brown hairs ;

peduncles 2–3 cm. long with a capitate inflorescence at apex ; calyx-lobes
ovate, 2 mm. long ; fruits ellipsoid, 9 mm. long ..   ..   36. *globiceps*
Leaves beneath, stipules and stems puberulous :
   Leaves oblanceolate, cuneate at base, 12–19 cm. long, 4·5–7 cm. broad,
   glabrous above   ..   ..   37. *liberica*
   Leaves ovate to ovate-elliptic, rounded to broadly cuneate at base, 4–7 cm.
   long, 2·5–5 cm. broad, puberulous above ..   .. 38. sp. nr. *tarambassica*
Indumentum of peduncle very short and dense (lens needed) :
   Inflorescences capitate ; (for other characters see above) ..   31. *dorotheae*
   Inflorescences branched ; leaves elliptic, 8–14 cm. long, 3–5 cm. broad, nearly
   glabrous ; flowers 4(–5)-merous ; indumentum of inflorescences black
                                                 39. *nigrescens*
Peduncles glabrous :
   Calyx-teeth acute, distinct, ovate, 1–2 mm. long, often reflexed :
      Inflorescences branched and paniculate, (1–)3 cm. diam. ; branchlets with 2
      longitudinal ridges, bark becoming thick, pale yellow and cracking ; stipules
      thick ; leaves oblanceolate to elliptic, acutely acuminate, cuneate at base,
      7–17 cm. long, 3–5·5 cm. broad   ..   ..   ..   40. sp. nr. *abrupta*
      Inflorescences capitate, 1 cm. diam. on peduncle up to 3 cm. long ; branchlets
      terete, bark thin ; stipules membranaceous ; leaves elliptic, acuminate,
      cuneate at base, 9–13 cm. long, 3–5 cm. broad   ..   ..   30. *subobliqua*
   Calyx-teeth obtuse, short, or calyx truncate :
      Branchlets with thick outer bark, cracking and often breaking off, usually with
      2 longitudinal ridges ; corolla 5-merous :
         Style pubescent ; hairs deep in throat of corolla ; leaves broadly elliptic,
         rather long-acuminate, cuneate at base, 5–11 cm. long, 3–6 cm. broad, with
         5–6 nerves on each side of midrib ..   ..   41. *abrupta*
         Style glabrous ; hairs exserted from throat of corolla ; leaves oblong-oblan-
         ceolate, acuminate, cuneate at base, 8–15 cm. long, 3–4·5 cm. broad, with
         7–9 nerves on each side of midrib ..   ..   42. *sciadephora*
      Branchlets with thin bark, mainly terete :
         Flowers 4-merous :
            Leaves obovate to oblanceolate, 10–17 cm. long, 3–7 cm. broad, with about
            13 nerves on each side of midrib ; (for other characters see above)
                                             24. *brachyantha*
           Leaves linear-lanceolate, 5–8·5 cm. long, 1–2·5 cm. broad, with 8–9 nerves on
           each side of midrib ..   ..   43. *adafoana*
         Flowers 5-merous :
           Branchlets and stipules puberulous ; leaves broadly lanceolate, acute, 7–10
           cm. long, 2–3 cm. broad, with 6–7 nerves on each side of midrib ; inflores-
           cences few-flowered, shortly pedunculate, flowers subsessile 44. *albicaulis*
           Branchlets and stipules glabrous :
              Stipules divided to the base into 2 subulate lobes about 4 mm. long ;
              branchlets longitudinally 2 ridges ; leaves oblong to oblong-lanceolate,
              3–8 cm. long, 1–3·5 cm. broad, with 4–5 nerves on each side of midrib
                                        45. *pleuroneura*
             Stipule-lobes not subulate from the base :
               Leaves narrowly obovate to elongate-oblanceolate or oblong, cuneate at
               base :
                  Inflorescence-branches small and rather few-flowered, 2–2·5 cm. diam. :
                     Leaves narrowly obovate to elongate-oblanceolate, acutely acuminate,
                     10–20 cm. long, 3–7 cm. broad, with 12–14 pairs of lateral nerves ;
                     petiole 2 cm. long ; flowers small and narrow   ..   46. *potanthera*
                     Leaves oblong, shortly cuneate and unequal-sided at the base, 16 cm. long,
                     4–6 cm. broad, abruptly acuminate, with about 10 pairs of lateral
                     nerves spreading at a right angle ; petiole 6–7 cm. long   47. *viticoides*
                Inflorescences rather large and spreading, 4–8 cm. diam. ; corolla-tube
                2 mm. diam. ; leaves oblong-elliptic, acuminate, cuneate at base, with
                10–12 pairs of lateral nerves ; petiole 4–6 cm. long   48. *malchairei*
             Leaves broadly ovate, shortly cuneate at base :
                Inflorescences long-pedunculate, peduncle 3–11 cm. long with or without
                 a pair of small leaves at the first branches ; leaves broadly elliptic,
                 obtusely pointed at apex, subdecurrent on the petiole, 12–16 cm. long,
                 6–10 cm. broad, glabrous ; lateral nerves about 10 pairs ; veins
                 conspicuous ; corolla 7 mm. long   ..   ..   ..   49. *dalzielii*
                Inflorescences shortly pedunculate, peduncle about 2 cm. long :
                  Veins very conspicuous between the 7–11 nerves on each side of midrib :
                     Nerves about 7 on each side of midrib, lamina elliptic, very shortly
                     acuminate, 14–18 cm. long, 6–9 cm. broad ; inflorescences about
                     3 cm. long   ..   ..   50. *piolampra*

Nerves 10–11 on each side of midrib :
Leaves elliptic, very shortly acuminate, about 25 cm. long and 12 cm.
broad, with a very stout midrib ribbed beneath ; petiole 4 cm.
long ; inflorescence nearly sessile ; corolla 7 mm. long, densely
hairy within the throat   ..   ..   ..   51. *arborea*
Leaves obovate-elliptic, broadly acuminate, about 30 cm. long and
14 cm. broad, with a stout smooth midrib beneath ; petiole 5 cm.
long ; inflorescence subsessile, few-flowered ; corolla over 1 cm.
long, thick, densely hairy within the throat   ..   52. *crassicalyx*
Veins invisible or nearly so :
Cymes 6–10 cm. diam. with many spreading branches and numerous
flowers ; peduncle about 2 cm. long ; leaves ovate-elliptic, obtuse at
apex, about 30 cm. long and 12 cm. broad, with 12–14 pairs of
lateral nerves ; petiole 3–4 cm. long ; fruits 8 mm. long   53. *limba*
Cymes 2–3 cm. diam. with few branches and few flowers :
Lobes of corolla not winged, corolla-tube thick, 5 mm. long ; leaves
broadly oblong-elliptic, very shortly pointed, very abruptly cuneate
at base, 16–25 cm. long, 6–10 cm. broad, with 12–14 pairs of
lateral nerves ; petiole 1·5–2 cm. long   ..   54. *rowlandii*
Lobes of corolla winged, corolla-tube slender, 5 mm. long ; leaves
elliptic, acuminate, cuneate at base, 6–14 cm. long, 3–6 cm. broad,
with about 8 pairs of lateral nerves ; petiole 1–2 cm. long
   55. *pteropetala*
†Endosperm ruminate :
Inflorescences branched-paniculate :
Calyx deeply lobed, lobes 2–3 mm. long, acutely deltoid ; corolla about 9 mm.
diam., conspicuously hairy in the throat ; fruits about 1 cm. long ; inflorescences
bracteolate ; leaves obovate-elliptic, acuminate, subcordate to cuneate at base,
8–14 cm. long, 3–7 cm. broad, with about 11 pairs of lateral nerves ; petioles up
to 2 cm. long   ..   ..   56. *elongato-sepala*
Calyx-teeth up to 1 mm. long or obscure :
Leaves closely reticulate beneath, coriaceous, oblong-elliptic, abruptly and
shortly acuminate, more or less rounded at base, about 12 cm. long and 5 cm.
broad with about 12 lateral nerves on each side, often drying yellow-green ;
calyx shortly toothed ; stipules membranous, apex irregular and obscurely
bilobed, about 1 cm. long and nearly as broad, early caducous
   57. *succulenta*
Leaves laxly reticulate beneath, hardly coriaceous, not drying yellow-green ;
calyx subtruncate :
Lateral nerves 8–14 on each side of midrib, strongly looped near margin :
Inflorescences laxly branched, up to 14 cm. long (including petiole) and 11 cm.
broad ; corolla-tube about 1·5 mm. diam. ; fruits about 3 mm. diam. :
Leaves with 8–11 lateral nerves on each side and ant-holes (dermatia)
localised towards the apex, lamina oblong-elliptic, up to 17 cm. long and
7 cm. broad, shiny above and with raised tertiary venation   58. *venosa*
Leaves with about 15 lateral nerves on each side and ant-holes (dermatia) in
many of their axils, lamina obovate to oblong-elliptic, 8–17 cm. long, 4–8
cm. broad, dull above :
Inflorescences usually glabrous ; young branchlets more or less pubescent ;
leaves drying purplish-brown, usually pubescent   ..   59. *articulata*
Inflorescences usually puberulous ; young branchlets usually glabrous ;
leaves drying dull brown, usually glabrous   ..   ..   60. *guineensis*
Inflorescences up to 6 cm. long and broad ; corolla-tube 2–3 mm. diam. ;
leaves broadly elliptic, abruptly acuminate, cuneate at base, 9–13 cm. long,
4·5–7 cm. broad   ..   ..   61. *chalconeura*
Lateral nerves 6–7 on each side of midrib, slightly looped near margin, lamina
ovate-elliptic, acuminate, cuneate at base, 3–12 cm. long, 3–8 cm. broad ;
petioles 1–3 cm. long ; corolla appearing inflated   ..   62. *djumaensis*
Inflorescences clustered on a peduncle or sessile or shortly branched :
Inflorescences sessile, densely clustered, terminal ; leaves variable in shape from
narrowly obovate to suborbicular, obtuse at apex, 10–20 cm. long, 4–14 cm.
broad, glabrous ; fruits ellipsoid, strongly ribbed, about 1 cm. long and 8 mm.
broad, with calyx persistent at apex ..   ..   ..   63. *psychotrioides*
Inflorescences pedunculate or shortly branched :
Midrib pilose or puberulous beneath :
Midrib pilose beneath ; (other characters as above) ..   56. *elongato-sepala*
Midrib puberulous beneath ; leaves ovate-elliptic, acuminate, cuneate at base,
10–15 cm. long, with about 15 pairs of lateral nerves ; branchlets puberulous ;
inflorescences with several short branches ; fruits spherical, 6 mm. diam.,
smooth ..   ..   64. *fernandopoensis*

14

Midrib glabrous beneath (or sometimes minutely puberulous in No. 66) :
Calyx subtruncate ; leaves ovate-elliptic to obovate, shortly acuminate, cuneate
at base, 10–25 cm. long, 4–9 cm. broad ; stipules about 3 mm. long, glabrous
or slightly pubescent, ant-holes usually absent from nerve axils beneath ;
petioles 1–2(–3) cm. long ; inflorescences shortly branched, sometimes shortly
pedunculate, 3–4 cm. diam. ; fruits about 1 cm. long, not ribbed
                                                    65. *gabonica*
Calyx shortly 5-lobed ; leaves usually with ant-holes in the axils of the nerves
beneath   ··   ··   ··   ··   ··   ··   66. *brassii*

1. **P. brenanii** *Hepper* in Kew Bull. 16 : 333 (1962). *Geophila pilosa* A. Chev. Bot. 346, partly (*Chev.* 19222 *bis*), name only. A creeping herb with wiry hispid stems ; flowers white.
    Iv.C.: *Aké Assi* 4863 ! Mid. Sassandra to Mid. Cavally (fr. July) *Chev.* 19222 *bis* ! **S.Nig.:** Shasha F.R., Ijebu (Mar.) *Richards* 3219 ! Okomu F.R., Benin (Dec., Jan.) *Brenan* 8416 ! 8621 ! 8761 ! Also in Cameroun.

2. **P. globosa** *Hiern* in F.T.A. 3 : 208 (1877). *P. ionantha* K. Schum. (1903). *Uragoga hexamera* K. Schum. (1899)—F.W.T.A., ed. 1, 2 : 128. *Cephaëlis hexamera* (K. Schum.) Wernham (1919). Undershrub 1–3 ft. high, decumbent and rooting at the base ; flowers white ; in forest.
    S. Nig.: Okomu F.R. (Nov., Jan., Feb.) *Onochie* FHI 40298 ! *Brenan* 8429 ! 9063 ! Oban *Talbot* ! Kwa Falls (Mar.) *Brenan* 9244 ! **[Br.]Cam.:** *Preuss* 1159 ! Ambas Bay (Feb.) *Mann* 1330 ! Also in Cameroun.
    [*Mann* 1330, the type of this species, is described as a shrub 3 ft. high ; all the other specimens cited are decumbent and rooting at the base.]

3. **P. nigerica** *Hepper* in Kew Bull. 16 : 337 (1962). *Trichostachys talbotii* Wernham in Cat. Talb. 55 (1913) ; F.W.T.A., ed. 1, 2 : 126, not *Psychotria talbotii* Wernham l.c. 54 (1913). A more or less simple-stemmed erect undershrub about 2 ft. high ; flowers whitish ; in forest.
    S.Nig.: Degema *Talbot* 3361 ! Oban (fr. Jan., Apr.) *Talbot* 1041 ! 1368 ! *Ejiofor* FHI 21885 ! *Onochie & Latilo* FHI 36123x ! Afi River F.R., Ogoja (fr. May) *Jones & Onochie* FHI 14146 ! **[Br.]Cam.:** Etende, 4,000 ft., Cam. Mt. (Apr.) *Maitland* 1110 ! Mopanya (Mar.) *Kalbreyer* 153 ! Barombi Lake F.R., Kumba (Apr.) *Daramola* FHI 41008 ! Also in Cameroun.

4. **P. sp. A.** A shrub about 4 ft. high with densely yellowish-hairy young stems and leaves ; "flowers pinkish ", fruits deep red ; in forest.
    S.Nig.: Okarara to New Ndebiji, Calabar (May) *Ujor* FHI 30850 ! Oban F.R. (fr. Jan.) *Onochie* FHI 36083x ! A very distinct species of which more material with flowers is required.

5. **P. reptans** *Benth.* in Fl. Nigrit. 418 (1849) ; F.T.A. 3 : 211 ; Schnell in Mém. I.F.A.N. 50 : 74, fig. 4. *Myrstiphyllum reptans* (Benth.) Hiern (1898). *P. rufipilis* of F.W.T.A., ed. 1, 2 : 124, partly (*Sc. Elliot* 5824), not of A. Chev. ex De Wild. A creeping undershrub up to 3 ft. high ; flowers greenish-white, fruit white and fleshy, drying ribbed ; in forest.
    Guin.: Dalaba (fr. Sept.) *Schnell* 6847 ! Badabou R. *Schnell* 7662 ! Bafing (fr. Oct.) *Adam* 12650 ! Kindia to Conakry (May) *Schnell* 5635 ! **S.L.:** Sugar Loaf Mt. (Jan.) *Sc. Elliot* 5824 ! Leicester (Mar. Apr.) *Hepper* 2486 ! *Brenan* 9617 ! Rokupr (May) *Jordan* 263 ! Musaia (fr. Dec.) *Deighton* 4476 ! Njala (May) *Deighton* 664 ! Bagroo R. *Mann* ! Also in Angola.

6. **P. kitsonii** *Hutch. & Dalz.* F.W.T.A., ed. 1, 2 : 124 (1931). *P. juglasiana* Aké Assi in Bull. Jard. Bot. Brux. 29 : 359, t. 5 (1959). A herb with half-woody stems creeping and rooting ; flowers white, fruit red ; in forest.
    Iv.C.: Abengourou (fl. & fr. Aug.) *Miège & Aké Assi* IA 1927 ! Amitioro Forest *Hallé* IA 4843 ! Kassa Forest *Mangenot & Aké Assi* IA 4844 ! Abaky to Dedesa, Afram Plains (May) *Kitson* 1129 ! Mpameso F.R. (fr. Dec.) *Adams* 2932 ! **S.Nig.:** Iguobazowa F.R., Benin (fr. Feb.) *Onochie* FHI 27140 ! Owam River F.R., Benin (fr. Nov.) *Meikle* 631 !

7. **P. coeruleo-violacea** *K. Schum.* in Engl. Bot. Jahrb. 33 : 363 (1903). *P. ciliata* De Wild. Pl. Bequaert. 2 : 351 (1924). An undershrub with slender reddish-hairy stems ; flowers blue ; in forest.
    S.Nig.: Okomu F.R., Benin (Feb.) *Brenan* 9008 ! 9008a ! **[Br.]Cam.:** *Preuss* 1123 ! Victoria to Bimbia (Apr.) *Maitland* 631 ! Also in Gabon.

8. **P. rufipilis** *A. Chev. ex De Wild.* var. **rufipilis**—Pl. Bequaert. 2 : 415 (1924) ; Chev. Bot. 342, name only ; Schnell in Mém. I.F.A.N. 50 : 73, photo. 13. *P. nimbana* Schnell l.c. 68, photo. 9, figs. 2, 3 (1957), incl. var. *djalonensis* Schnell l.c. 70, photo. 10, f. *vallicola* Schnell l.c. photo. 11, var. *gaidensis* Schnell l.c. 72, photo. 12. A shrub 4–10 ft. high, tinged red-brown ; flowers white ; in forest.
    Port.G.: Brene, Bissau (fr. Jan.) *Esp. Santo* 1708 ! **Guin.:** Dalaba (fl. Sept., fr. Dec.) *Chev.* 20332 ! *Schnell* 6834 ! 6951 ! Fon Massif (fr. Aug.) *Schnell* 6596 ! Kounounkan R. (fr. Oct.) *Schnell* 7618 ! Nimba Mts. (fr. Aug.) *Schnell* 6223 ! **S.L.:** Kumrabai (fr. Dec.) *Thomas* 6982 ! Yonibana (fl. & fr. Nov.) *Thomas* 5063 ! Njala (July) *Deighton* 1758 (partly) ! 5820 ! Kenema (fr. Jan.) *Thomas* 7559 ! Loma Mts. (Aug., Sept.) *Jaeger* 1226 ! 1524 ! **Lib.:** Mt. Wolagwisi, Pandamai (fr. Mar.) *Bequaert* 112 ! Bobei Mt., Sanokwele (fr. Sept.) *Baldwin* 9614 ! Ganta (fr. Oct.) *Harley* 743 ! *Baldwin* 9257 ! **Iv.C.:** Nimba Mts. (fr. Aug.) *Boughey* GC 18159 !

8a. **P. rufipilis** var. **konkourensis** (*Schnell*) *Hepper* in Kew Bull. 16 : 337 (1962). *Cephaëlis konkourensis* Schnell in Bull. Jard. Bot. Brux. 30 : 365 (1960), not *C. psychotrioides* Valeton (1913). *Uragoga psychotrioides* Schnell in Bull. I.F.A.N. 15 : 118, fig. 12 (1953). *Psychotria psychotrioides* (Schnell) Schnell in Mém. I.F.A.N. 50 : 73 (1957).
    Guin.: Konkouré (June) *Pobéguin* 1628 ! **S.L.:** Njala (fl. July, fr. Sept.) *Deighton* 1758 (partly) ! 2124 ! Kamaranka *Thomas* 486 ! Samaia (May) *Thomas* 237 ! **Lib.:** Kakatown *Whyte* ! Du R. (July) *Linder* 102 ! Banga (fr. Oct.) *Linder* 1214 ! Dukwia R. (May) *Cooper* 447 ! Brewersville (June) *Barker* 1330 !

9. **P. vogeliana** *Benth.* in Fl. Nigrit. 420 (1849) ; F.T.A. 3 : 210 ; Chev. Bot. 342, partly ; Schnell in Mém. I.F.A.N. 50 : 76, photos. 1–5, incl. var. *horhogoensis* Schnell l.c. 62, photo. 2 (1957). *P. multinervis* De Wild. Pl. Bequaert. 2 : 391 (1924) ; Schnell l.c. 68. *P. reptans* of Chev. Bot. 342, not of Benth. *Cephaëlis cornuta* Hiern in F.T.A. 3 : 224 (1877). *C. talbotii* Wernham (1914). *Psychotria pilifera* Hutch. & Dalz. F.W.T.A., ed. 1, 2 : 123, partly (excl. descr. and *Talbot* 234). A small shrub up to about 6 ft. high ; flowers cream or greenish ; in savanna woodland.
    Mali : Soukouroba (fr. May) *Chev.* 883 ! **Iv.C.:** Touleleu R. (fr. Aug.) *Schnell* 6156 ! Korhogo to Badikaha (June) *Mangenot* ! **Ghana:** Asafo to Cape Coast (June) *Morton* GC 7826 ! Krepi (Jan.) *Johnson* 545 ! Ejura (fr. Aug.) *Chipp* 765 ! Sampa (fr. Jan.) *Enti* FH 6259 ! Yendi (Apr.) *Vigne* FH 1703 ! **N.Nig.:** Mande F.R., Zaria Prov. (June) *Keay* FHI 25844 ! Abuja (May) *E. W. Jones* 141 ! Jos Plateau (May) *Lely* W280 ! Abinsi (May) *Dalz.* 642 ! **S.Nig.:** Nun R. (fl. & fr. Sept.) *Mann* 502 ! Ubiaja F.R., Benin (Apr.) *Umana* FHI 29101 ! Olokemeji F.R. (Apr.) *Keay* FHI 37001 ! Onitsha *Barter* 1253 ! Eket *Talbot* 3386 ! Calabar *Robb* ! **[Br.]Cam.:** Rio del Rey Johnston ! Malende to Muyuka, Victoria (Mar.) *Brenan* 9325 ! Abonando (Mar.) *Rudatis* 32 ! Also in Cameroun and Congo.

10. **P. latistipula** *Benth.* in Fl. Nigrit. 419 (1849) ; F.T.A. 3 : 210 ; F.W.T.A., ed. 1, 2 : 123 ; Chev. Bot. 342 ; Schnell in Mém. I.F.A.N. 50 : 66. *P. pilifera* Hutch. & Dalz. F.W.T.A., ed. 1, 2 : 123 (1931). partly (*Talbot* 234), not *Cephaëlis talbotii* Wernham. A slender shrub or small tree ; flowers white ; in forest.
    S.Nig.: Olokemeji (Aug.) *Jones* FHI 5609 ! Oban *Talbot* 234 ! Afi River F.R., Ikom Dist. (Dec.) *Keay* FHI 28239 ! Obubra Dist. (Apr.) *Jones* FHI 7397 ! **[Br.]Cam.:** Victoria (May) *Maitland* 721 ! S

Bakundu F.R., Victoria (Apr.) *Daramola* FHI 29839! **F.Po:** *T. Vogel*! *Barter*! *Mann* 40! 1425!
Bokoko (fr. Jan.) *Guinea* 840! Also in Cameroun and Gabon.

11. **P. subglabra** *De Wild.* Pl. Bequaert. 2 : 425 (1924) ; Schnell l.c. *P. latistipula* of F.W.T.A., ed. 1, 2 :
123, partly (*Chev.* 19216 and syn.), not of Benth. A shrub with white flowers ; in forest.
    Iv.C.: Mid. Cavally to Mid. Sassandra (July) *Chev.* 19216! Sérédou (May) *Aké Assi* 2389!

12. **P. strictistipula** *Schnell* in Bull. I.F.A.N. 15 : 131, fig. 15 (1953) ; Mém. I.F.A.N. 50 : 67. A small shrub.
    Guin.: Nimba Mts. (Apr.) *Schnell* 5165! **Lib.:** Ba, Boporo Dist. (fr. Dec.) *Baldwin* 10719!

13. **P. obscura** *Benth.* in Fl. Nigrit. 419 (1849) ; F.T.A. 3 : 212 ; Hepper in Kew Bull. 16 : 457. *P. sodifera*
De Wild. Pl. Bequaert. 2 : 421 (1924) ; Schnell in Mém. I.F.A.N. 50 : 74. *P. farmari* Hutch. & Dalz.
F.W.T.A., ed. 1, 2 : 123 (1931). *Grumilea sodifera* A. Chev. (1913), name only. Small shrub up to 4 or
5 ft. high ; flowers white, fragrant, fruits red ; in forest and in savanna woodland.
    Iv.C.: Toumodi (Aug.) *Chev.* 22414! Bouaké (Apr.) *Leeuwenberg* 3292! **Ghana:** *Farmar* 385! Accra
T. Vogel! Odumasi (Dec.) *Johnson* 513! Cape Coast (Mar.) *Hall* 339! Asikuma to Anum (Nov.)
Morton GC 7934! **N.Nig.:** *Shaw* 54! Bonu, Niger Prov. (May) *E. W. Jones* 142! Afaka F.R., Zaria
(June) *Keay* FHI 37079! Sanga River F.R., Jemaa Div. (fl. May, fr. Nov.) *E. W. Jones* 71! *Keay* FHI
37221! **S.Nig.:** Owo (Feb.) *Brenan* 8990! Olokemeji F.R. (fl. Apr., fr. Aug.) *Jones* FHI 5609! *Keay*
FHI 37002! Idanre Hills (fl. & fr. Oct.) *Keay* FHI 25517! Also in Ubangi-Shari.

FIG. 242.—PSYCHOTRIA VOGELIANA *Benth.* (RUBIACEAE).
A, lower surface of leaf. B, flower. C, calyx and bracts. D, longitudinal section of ovary.

14. **P. bidentata** (*Thunb. ex Roem. & Schult.*) *Hiern* in F.T.A. 3 : 209 (1877) ; Schnell in Mém. I.F.A.N. 50 :
67. *Cephaëlis bidentata* Thunb. ex Roem. & Schult. (1819). *P. garrettii* K. Schum. (1903). *Uragoga
sangalkamensis* Schnell in Bull. I.F.A.N. 15 : 116, fig. 11 (1953). *Cephaëlis sangalkamensis* (Schnell)
Schnell in Mém. I.F.A.N. 50 : 68 (1957). *Psychotria sangalkamensis* (Schnell) Schnell l.c. (1957). A
small shrub 2–5 ft. high with leathery leaves ; flowers white ; in moist sandy soil, sometimes epiphytic
on palms.
    Sen.: Sangalkam, Dakar *Adam* 420 ; 1619! **Port.G.:** Mato de Amedi, Formosa (Apr.) *Esp. Santo*
1971! **S.L.:** *Smeathmann*! Murray Town (May) *Deighton* 5933! Kent (fr. Aug.) *Melville & Hooker*
363! Yungeru (fr. Jan.) *Thomas* 7386! Bendu, Kente Creek (Mar.) *Adames* 10! **Lib.:** Grand Bassa
(July) *T. Vogel* 40! *Dinklage* 214! Buchanan (Mar.) *Baldwin* 11196! Monrovia *Whyte*! Sinoe Basin
Whyte!

15. **P. maliensis** *Schnell* in Mém. I.F.A.N. 50 : 76, photo. 14, figs. 5, 6 (1957). A rather robust shrub 6–10 ft.
high ; flowers white ; in forest beside streams.
    Guin.: Mali (Sept.) *Schnell* 7041! 7159! s.n.! Pita (Mar.) *Adam* 11654!

16. **P. ceratalabastron** *K. Schum.* in Engl. Bot. Jahrb. 33 : 362 (1903). *P. alatipes* Wernham in Cat. Talb. 53
(1913). A shrub about 5 ft. high ; flowers white ; in forest.
    S.Nig.: Oban *Talbot* 241! Eket *Talbot*! Okarara (May) *Ujor* FHI 30152! **[Br.]Cam.:** Victoria to
Bimbia *Preuss* 1271. Kebo *Conrau* 211.

17. **P. bifaria** *Hiern* in F.T.A. 3 : 198 (1877). A shrub 2–3 ft. high ; flowers white ; in forest.
    [Br.]Cam.: Buea, 2,500 ft. (Mar.) *Maitland* 541! Rio del Rey *Johnston*! **F.Po:** (June) *Mann* 419!
Also in Cameroun.

18. **P. mannii** *Hiern* in F.T.A. 3 : 197 (1877). A shrub 2–5 ft. high, branching near the base with a spreading
habit ; flowers white ; in deep shade in forest.
    S.Nig.: Okomu F.R., Benin (fl. & fr. Jan.) *Brenan* 8759! Afi River F.R., Ogoja (fr. May) *Jones &
Onochie* FHI 18743! Oban *Talbot* 2043! **[Br.]Cam.:** S. Bakundu F.R., Kumba (Jan., Mar.) *Daramola*
FHI 35504! *Brenan* 9269! Also in Gabon.

19. **P. konguensis** *Hiern* in F.T.A. 3 : 200 (1877). A shrub 2–3 ft. high.
    [Br.]Cam.: S. Bakundu F.R., Kumba (fr. Sept.) *Akua* FHI 15178! Bombe, Kumba (Apr.) *Daramola*
FHI 29811! Also in Cameroun, Gabon and Cabinda.

20. **P. brachyanthoides** *De Wild.* Pl. Bequaert. 2 : 337 (1924). *P. brachyantha* of Chev. Bot. 340, not of
Hiern. Shrub or small tree with white flowers.
    Iv.C.: Makouguié *Chev.* 17108! Apouassou to Soubiré (Mar.) *Chev.* 17801!

21. **P. humilis** *Hiern* in F.T.A. 3 : 198 (1877). A creeping herb with rooting stems just below the soil and
erect leafy shoots up to 1 ft. high, leaves very dark green with a white margin ; flowers white, fruits
scarlet ; in forest.

S.Nig.: Okomu F.R., Benin (Dec.) *Brenan* 8392! 8417! 8518! Usonigbe F.R., Sapoba (Nov.) *Keay* FHI 25624! *Meikle* 629! Oban *Talbot* 362! Also in Gabon.

[Note: *Brenan* 8392 & 8518 are ascending and have leaves and inflorescences larger than the others cited ; the plants may be pilose or almost glabrous ; very near *P. cornuta*.]

22. **P. cornuta** *Hiern* in F.T.A. 3 : 198 (1877). An ascending undershrub about 1 ft. high ; flowers few and rather congested.
    F.Po: *Mann* 1429!

23. **P. linderi** *Hepper* in Kew Bull. 16 : 335, fig. 1, 2 (1962). A shrub about 3 ft. high ; flowers white in pendent delicate panicles ; in forest.
    Guin.: Ziama, Macenta (Aug.) *Adam* 5969! 6054! **S.L.:** Mabonto (fr. July) *Deighton* 3259! **Lib.:** Gbanga (Sept.) *Linder* 649! Bobei Mt., Sanokwele (fr. Sept.) *Baldwin* 9580! Ganta (fr. Dec.) *Harley* 791! Bilimu (fr. Sept.) *Harley* 1532!

24. **P. brachyantha** *Hiern* in F.T.A. 3 : 196 (1877). *P. aledjoensis* De Wild. l.c. 328 (1924), incl. var. *glabra* De Wild. l.c. 330 (1924) ; F.W.T.A., ed. 1, 2 : 125. A shrub about 6 ft. high ; flowers greenish-yellow.
    Guin.: Yagedou, Boola (Mar.) *Chev.* 20929! Aledjo (Oct.) *Chev.* 27047! **Iv.C.:** Amitioro (fr. Nov.) *Aké Assi* 5452! Banco (July) *de Wilde* 131! **Ghana:** Banka (Mar.) *Vigne* FH 1884! L. Bosumtwe (Apr.) *Adams* 2463! Amentia, Ashanti *Irvine* 509! **Dah.:** R. Ouémé *Le Testu* 300! [**Br.**]**Cam.:** Barombi L. (Mar.) *Brenan* 9466! Johann-Albrechtshöhe (=Kumba) *Staudt* 920! **F.Po:** *Mann* 1423!

25. **P. leptophylla** *Hiern* in F.T.A. 3 : 200 (1877) ; Hepper in Kew Bull. 16 : 334. *Pavetta ? tenuifolia* Benth. (1849), not *Psychotria tenuifolia* Sw. (1788). *Psychotria setacea* Hiern in F.T.A. 3 : 197 (1877) ; F.W.T.A., ed. 1, 2 : 124. A shrub 2–8 ft. high ; fruits shining red ; in forest.
    S.Nig.: Oban *Talbot* 2002! 2024! [**Br.**]**Cam.:** Victoria (Apr.) *Maitland* 1166! Buea (fl. & fr. Jan., Feb., fr. May) *Mann* 1191! *Dalz.* 8208! *Dunlap* 21! *Maitland* 704! S. Bakundu F.R., Kumba (Apr.) *Daramola* FHI 29830! **F.Po:** (fl. & fr. Nov.) *T. Vogel* 167! *Guinea* 2618! Also in Cameroun.

26. **P. recurva** *Hiern* in F.T.A. 3 : 206 (1877). *P. potamogetonoides* Wernham (1917). *P. bicarinata* Mildbr., name only. *P. talbotii* of F.W.T.A., ed. 1, 2 : 124, partly (syn. and *Talbot* 3817), not of Wernham. A glabrous shrub 4–5 ft. high, the narrow leaves with crinkled margins.
    S.Nig.: Degema *Talbot* 3817! **F.Po:** *Mann* 1430! Bogogo *Mildbr.* 6825.

27. **P. talbotii** *Wernham* in Cat. Talb. 54 (1913) ; F.W.T.A., ed. 1, 2 : 124, partly (excl. syn. and *Talbot* 3817) ; Brenan in Kew Bull. 5 : 222. Small shrub 2–5 ft. high, often well-branched ; flowers white, fruits and pedicels deep red ; in forest.
    S.Nig.: Idanre, Akure Div. (fl. & fr. Jan., fl. Apr.) *Brenan* 8641! 8651! 8679! *Symington* FHI 3360! Oban *Talbot* 1054! Kwa Falls (Mar., Apr.) *Brenan* 9244! *Ejiofor* FHI 21882! Akampa Estate, Calabar (fl. & fr. Mar.) *Latilo* FHI 41346! [**Br.**]**Cam.:** Rio de Rey *Johnston*!

28. **P. huae** *De Wild.* Pl. Bequaert. 2 : 372 (1924). *P. williamsii* Hutch & Dalz. F.W.T.A., ed. 1, 2 : 124 (1931). *P. kolly* of Chev. Bot. 341, partly, not of Schum. & Thonn. Shrub ; flowers white or yellowish, fragrant ; in savanna woodland.
    Guin.: Ditinn to Diaguissa *Chev.* 12660! 12675! **Ghana:** *Johnson* 143! Ejura (fl. & fr. Apr.) *Williams* 228! Ampoti to Adumasia, Afram Plains (May) *Kitson*! Opro River F.R. (Apr.) *Keay* FHI 37591! **N.Nig.:** Vom, Jos Plateau *Dent Young* 123!

29. **P. calva** *Hiern* in F.T.A. 3 : 199 (1877) ; Berhant Fl. Sén. 106. *P. umbellata* Schum. & Thonn. (1827), not of Vell. (1825). *Pavetta ? laevis* Benth. (1849), not *Psychotria laevis* DC. (1830). *Grumilea micrantha* Hiern in F.T.A. 3 : 217 (1877) ; Chev. Bot. 343. *Psychotria micrantha* var. *floribunda* A. Chev. Bot. 343, name only. *P. floribunda* De Wild. l.c. 366 (1924). *P. porto-novensis* De Wild. l.c. 405 (1924). *P. warneckei* of F.W.T.A., ed. 1, 2 : 125, partly (*Chev.* 20937), not of K. Schum. & K. Krause. An erect shrub 2–6 ft. high, or higher when straggling, with leaves very variable in width ; flowers whitish, fruits red ; beside streams in forest.
    Sen.: "gallery forests" *Berhaut* 1670. **Port.G.:** Brene, Bissau (fr. Jan.) *Esp. Santo* 1694! **Guin.:** Boola to Moribadou (Mar.) *Chev.* 20937! Fouta Djalon (Mar.) *Chev.* 12833! Farana (Mar.) *Sc. Elliot* 5261! **S.L.:** Kundala (fr. Mar.) *E. Macdonald* 21! Musaia (Apr.) *Deighton* 5423! Pejehun *Thomas* 8253! Makump (May) *Deighton* 1722! Mange (Apr.) *Hepper* 2617! **Lib.:** Monrovia *Whyte*! Sinoe Basin *Whyte*! **Iv.C.:** Tabou to Bériby (Apr.) *Chev.* 19951! Agniéby *Chev.* 17200! Sanvi *Chev.* 17817! Assinie *Chev.* 17888! **Ghana:** Cape Coast (Mar.) *Hall* 1064! Aburi (May) *Hughes* 823! Afram Plains (May) *Kitson* 1123! Banda, Wenchi Dist. (fr. Dec.) *Morton* GC 25080! **S.Nig.:** Lagos *Barter* 2204! 3245! Obu *Thomas* 493! Sapoba (Feb.) *Onochie* FHI 31946! Okomu F.R., Benin (Mar.) *Brenan* 9181! Shasha F.R. (May) *Richards* 3468! [**Br.**]**Cam.:** Bafut-Ngemba F.R., Bamenda (June) *Lightbody* FHI 26327! **F.Po:** *T. Vogel* 190! Bahia Venus (fr. Dec.) *Guinea* 238!
    [A complex aggregate is included here. *P. hallei* Aké Assi & Bouton in Bull. Jard. Bot. Brux. 30 : 393, t. 12 (1960), from Ivory Coast (*Aké Assi* 4404!) may also belong to the aggregate—F.N.H.]

30. **P. subobliqua** *Hiern* in F.T.A. 3 : 206 (1877). *P. refractiloba* K. Schum. (1899). *P. caduciflora* De Wild. Pl. Bequaert. 2 : 346 (1924). *P. copeensis* De Wild. l.c. 353 (1924). *P. yabaensis* De Wild. l.c. 440 (1924). *P. anetoclada* of Chev. Bot. 339, (incl. var. *angustifolia*), names only, not of Hiern. A nearly glabrous shrub a few feet high ; flowers pale green or whitish, soon falling off ; in forest.
    Guin.: Boula, Macenta (Apr.) *Adam* 12040! **Iv.C.:** Tébo, Mid. Cavally (July) *Chev.* 19412! Anyama (Feb.) *Chev.* 20098! Mt. Copé, Grabo Dist. (July) *Chev.* 19677! **Ghana:** Fomena (Apr.) *Adams* 2490! S. Fomangsu F.R. (Apr.) *Adams* 2657! Atewa Range F.R. (Jan.) *de Wit & Morton* A2935! **S.Nig.:** Ikorodu (Mar.) *Schlechter* 12305! Oban (Mar.) *Richards* 5185! [**Br.**]**Cam.:** Victoria (Mar., Apr.) *Maitland* 701! 1142! **F.Po:** *Mann* 1421! Also in Cameroun, S. Tomé and Principe.

31. **P. dorotheae** *Wernham* in Cat. Talb. 53 (1913). A shrub about 6–12 ft. high ; flowers white ; fruits red ; in forest.
    Lib.: Du R. (fr. July) *Linder* 93! Bobei Mt., Sanokwele Dist. (fr. Sept.) *Baldwin* 9629! **Iv.C.:** Tiapleu (fr. May) *Aké Assi*! **Ghana:** Princes Town (Mar.) *Morton* GC 6623! **S.Nig.:** Oban F.R. (Feb.) *Onochie & Latilo* FHI 36312! Akamkpa, Calabar (fl. & fr. Mar.) *Latilo* FHI 38096! Oban *Talbot* 1561!

32. **P. obanensis** *Wernham* in Cat. Talb. 53 (1913). A shrub with rather large, long-petiolate leaves ; flowers densely clustered in 1 to 3 subsessile heads.
    S.Nig.: Oban *Talbot* 244!

33. **P. longituba** *A. Chev. ex De Wild.* Pl. Bequaert. 2 : 382 (1924) ; Chev. Bot. 342 ; Hepper in Kew Bull. 16 : 337 (1962). *P. capitellata* A. Chev. ex De Wild. l.c. 348 (1924) ; Chev. Bot. 340, name only, not of DC. (1830), nor Benth. & Hook. f. (1873). A shrub up to 5 ft. high, with long, rather narrow leaves ; fruits red, angular when dry.
    Lib.: Kulo (fr. Mar.) *Baldwin* 11409! Jaurazon (fr. Mar.) *Baldwin* 11477! **Iv.C.:** Grabo, Cavally (fl. & fr. July) *Chev.* 19618! 19779! s.n.! Malamalasso (fr. Mar.) *Chev.* 17541!

34. **P. lophoclada** *Hiern* in F.T.A. 3 : 197 (1877). A stiffly branched glabrous undershrub 1–2 ft. high with pale yellowish branchlets ; flowers yellowish, small and crowded in sessile heads, fruits red ; in forest.
    S.L.: *Don*! Sugar Loaf Mt. (May, Dec., fr. Nov.) *Sc. Elliot* 3959! *Tindall* 63! *Deighton* 5621! **Lib.:** Bobei Mt., Sanokwele (fr. Sept.) *Baldwin* 9568! Ganta (May) *Harley* 1174! Bilima (fl. & fr. Feb.) *Harley* 1775! Gbanga *Linder* 565! **Iv.C.:** *Béyo* (Dec.) *Leeuwenberg* 2221! Nimba Mts. *Schnell* 3553!

35. **P. insidens** *Hiern* in F.T.A. 3 : 208 (1877). Undershrub 1–3 ft. high ; flowers white or yellow ; in forest.
    S.Nig.: Omo F.R., Ijebu (Apr.) *Onochie* FHI 15522! Shasha F.R., Ijebu (Apr.) *Ross* 204! Oban (Mar.) *Richards* 5184! **F.Po:** *Mann* 310!

36. **P. globiceps** *K. Schum.* in Engl. Bot. Jahrb. 28 : 93 (1899). *P. avakubiensis* De Wild. Pl. Bequaert. 6: 74 (1932). A slender shrub 2 ft. high ; flowers white ; in forest.

[Br.]Cam.: Victoria *Schlechter* 12374! S. Bakundu F.R., Kumba (fr. Jan., Mar.) *Binuyo & Daramola* FHI 35453! *Brenan* 9481! Also in Cameroun and Congo.

37. **P. liberica** *Hepper* in Kew Bull. 16 : 335, fig. 1, 1 (1962). An undershrub ; in forest.
 Lib.: Ganta (fl. Sept., fr. Oct.) *Harley* 710! 1912! *Baldwin* 12515! Duo (fr. Mar.) *Baldwin* 11344!
  [Note : *Aké Assi* 3185 (Ivory Coast) and *Brenan* 9294 ([Br.] Cameroons) are related to this species and to *P. globiceps*, but appear to be distinct from them and from each other—F.N.H.]

38. **P. sp. nr. tarambassica** *Bremek.* in J. Bot. 71 : 280 (1933). A compact shrub about 4 ft. high ; flowers whitish ; in upland savanna.
 [Br.]Cam.: Vogel Peak, Adamawa (Dec.) *Hepper* 1539!
  [This specimen lacks bacterial dots on the leaves which are typical of *P. tarambassica* in E. Africa.]

39. **P. nigrescens** *De Wild.* Pl. Bequaert. 2 : 393 (1924). A shrub about 2 ft. high, with small white flowers.
 Guin.: Boola to Moribadou (Mar.) *Chev.* 20938! 20940! Bérézia (Feb.) *Chev.* 20777! **S.L.**: Bunumbu to Sembehun, Male R. (Apr.) *Deighton* 3937! **Lib.**: Gbau (fr. Sept.) *Baldwin* 9455! Kitoma (Mar.) *Adam* 16753! 16764! 16769! **Iv.C.**: Taï (Mar.) *Leeuwenberg* 3034!

40. **P. sp. nr. abrupta** *Hiern* in F.T.A. 3 : 205 (1877). An undershrub about 1 ft. high, or a shrub up to 15 ft. high ; in moist forest.
 Lib.: Dukwai R. (fl. & fr. July) *Cooper* 163! 360! Du R. (July) *Linder* 108! Duo (Mar.) *Baldwin* 11335!
  [These plants belong to a complex of species that appears to be widespread in tropical Africa : further elucidation is necessary.]

41. **P. abrupta** *Hiern* in F.T.A. 3 : 205 (1877). A small tree or shrub ; flowers white ; beside streams in forest.
 Ghana: Adiembra, Ashanti (Feb.) *Andoh* FH 4297! **[Br.]Cam.**: S. Bakundu F.R., Kumba (Mar.) *Brenan* 9277! Also in Cameroun, Kenya, Tanganyika, N. Rhodesia and Mozambique.

42. **P. sciadephora** *Hiern* in F.T.A. 3 : 202 (1877). *P. brevistipulata* De Wild. l.c. 341 (1924). *P. albifaux* K. Schum. ex Hutch. & Dalz. F.W.T.A., ed. 1, 2 : 125 (1931). A shrub up to 10 ft. high, the leaves and inflorescence turning blackish on drying ; flowers whitish or brown in bud ; in forest.
 Guin.: Nzo to Sokonante (Mar.) *Chev.* 21081! near Mt. Benna *Schnell* 7671! **S.L.**: Roruks (fr. July) *Deighton* 4676! Kasewe Hills F.R. (June) *Deighton* 5827! **Lib.**: Jaurazon (fr. Mar.) *Baldwin* 11448! Vahon (fr. Nov.) *Baldwin* 10234! Ganta (fl. Apr., fr. July, Sept.) *Harley* 913! 1226! **Iv.C.**: Tonkoui Mt. (fr. Aug.) *Schnell* 6335! Sakonanta to Samphea *Chev.* **Ghana**: Bibiani (Apr.) *Darko* 865! Mpraeso Scarp (Apr.) *Morton* A695! **S.Nig.**: Okomu F.R. (Jan.) *Brenan* 8784! 8841! Sapoba *Kennedy* 1773! Usonigbe F.R., Benin (fl. Feb., fr. Sept.) *Onochie* FHI 27697! 34282! Akure F.R. (fr. Oct., Nov.) *Keay* FHI 25549! *Ejiofor* FHI 24608! Ojo to Nyenekosum, Calabar (fr. June) *Ujor* FHI 27998! **[Br.]Cam.**: Ambas Bay (Jan.) *Mann* 729! Victoria (fr. Oct.) *Maitland* 741! Banga (Mar.) *Brenan* 9281! Cam. Mt., 4,000 ft. (Feb.) *Mann* 1192! Also in Cameroun.
  [Note : the specimens from Guinée to Ghana have slightly smaller flowers than true *P. sciadephora* Hiern, and they have been distinguished as *P. brevistipulata* De Wild.—F.N.H.]

43. **P. adafoana** *K. Schum.* in Engl. Bot. Jahrb. 28 : 90 (1899). *P. ivorensis* De Wild. Pl. Bequaert. 2 : 375 (1924). *P. recurva* of Chev. Bot. 342, not of Hiern. A low-branched shrub, nearly glabrous ; flowers small, white in small shortly pedunculate cymes.
 Iv.C.: Abradines, Lower Comoé (Mar.) *Chev.* 17569! Assinie (Apr.) *Chev.* 17879! **Ghana**: Adafo (Nov.) *Krause* 55.

44. **P. albicaulis** *Sc. Elliot* in J. Linn. Soc. 30 : 83 (1894). An undershrub, slightly hairy on the young branchlets ; in forest beside rivers.
 Guin.: Erimakuna to Farana *Sc. Elliot* 5312! **S.L.** Musaia to Falaba (Mar.) *Sc. Elliot* 5124!

45. **P. pleuroneura** *K. Schum.* in Engl. Bot. Jahrb. 33 : 368 (1903). An undershrub ; in forest.
 [Br.]Cam.: Johann-Albrechtshöhe (=Kumba) (Apr.) *Staudt* 218!

46. **P. potanthera** *Wernham* in Cat. Talb. 54 (1913). A glabrous shrub with pithy branchlets ; flowers narrow and whitish in short lax cymes.
 S.Nig.: Oban *Talbot* 240!

47. **P. viticoides** *Wernham* in Cat. Talb. 54 (1913). A shrub with rather larger flowers than the last.
 S.Nig.: Oban *Talbot* 2080!

48. **P. malchairei** *De Wild.* Pl. Bequaert. 2 : 386 (1924). A shrub 6–8 ft. high ; flowers rather large, white ; in forest.
 Ghana: *Burton*! **S.Nig.**: Shasha F.R., Ijebu Prov. (Apr., May) *Ross* 205! 265! *Richards* 3381! **F.Po**: Moka (Sept.) *Melville* 658! Also in Congo.
  [The W. African specimens have smaller flowers and fewer lateral nerves. The fruits appear to be galled frequently.]

49. **P. dalzielii** *Hutch.* in F.W.T.A., ed. 1, 2 : 125 (1931). A perennial herb or small shrub ; flowers yellow ; in savanna woodland.
 N.Nig.: Mando F.R., Zaria Prov. (Aug.) *Keay* FHI 28004! Abinsi (Aug.) *Dalz.* 643!

50. **P. piolampra** *K. Schum.* in Engl. Bot. Jahrb. 28 : 97 (1899). A glabrous undershrub about 2 ft. high ; flowers white with orange calyx ; in forest.
 [Br.]Cam.: Johann-Albrechtshöhe (=Kumba) *Staudt* 738! Mamfe (Mar.) *Richards* 5223!

51. **P. arborea** *Hiern* in F.T.A. 3 : 202 (1877). A small tree, nearly glabrous with large leaves ; flowers white in a short branching terminal cyme.
 S.Nig.: Calabar *Thomson* 106!

52. **P. crassicalyx** *K. Krause* in Engl. Bot. Jahrb. 57 : 46 (1921). Erect shrub with stout branchlets shortly pilose when young ; flowers white ; in forest.
 F.Po: above Basilé, 1,800–2,450 ft. (Nov.) *Mildbr.* 7124. (This specimen was seen for ed. 1.)

53. **P. limba** *Sc. Elliot* in J. Linn. Soc. 30 : 83 (1894). A shrub with large leaves ; beside streams and in forest.
 Guin.: Nimba Mts. (fr. Sept.) *Schnell* 3666! **S.L.**: Kahreni to Port Loko (Apr.) *Sc. Elliot* 5815! Mabonto (fr. Oct.) *Thomas* 3566! Yonibana (fr. Nov.) *Thomas* 4919! Lumbaraya (fl. & fr. Feb.) *Sc. Elliot* 4947! Lomaburu, Talla (Feb.) *Sc. Elliot* 5023! **Lib.**: Bobei Mt., Sanokwele (fr. Sept.) *Baldwin* 9575!

54. **P. rowlandii** *Hutch. & Dalz.* F.W.T.A., ed. 1, 2 : 124 (1931). A nearly glabrous shrub with fairly large leaves and flowers in stoutish terminal cymes. Fruits are required to ascertain whether this species has ruminate endosperm.
 S.Nig.: Lagos *Rowland*! Emege, Brass Dist. (Jan.) *Akpota* FHI 3943! Omo (formerly part Shasha) F.R. (Mar., Apr.) *Jones & Onochie* FHI 16083! 17307!

55. **P. pteropetala** *K. Schum.* in Engl. Bot. Jahrb. 33 : 369 (1903). A small shrub.
 [Br.]Cam.: Barombi *Preuss* 1201! **F.Po**: *Mann* 310!
  [See note on p. 193 at end of *Chassalia* spp.]

56. **P. elongato-sepala** (*Hiern*) *Petit* in Bull. Jard. Bot. Brux. 33 : ined. (1963). *Grumilea elongato-sepala* De Wild. Pl. Bequaert. 2 : 457 (1924). *G. rufo-pilosa* De Wild. l.c. 478 (1924). *G. glabrifolia* De Wild. l.c. 460 (1924). *G. puberulosa* De Wild. l.c. 468 (1924). *Psychotria vogeliana* of Chev. Bot. 345, partly, not of Benth. A climber with pilose branches ; flowers greenish-white ; fruits pink.
 Guin.: Farana *Sc. Elliot* 5346! Pela, Nzérékoré (fr. Sept.) *Baldwin* 13322! **Lib.**: Ganta (Feb., Apr.) *Harley* 832! 1401! Sanokwele (fr. Sept.) *Baldwin* 9639! Zeahtown, Tchien Dist. (Aug.) *Baldwin* 6954! **Iv.C.**: Mt. Nienkué, Cavally (July) *Chev.* 19442! 19444! Soubré to Taï (Mar.) *Chev.* 17697! Mt. Kouan to Danané (Apr.) *Chev.* 21274! Taï *Hallé* 518! Alépé (Mar.) *Chev.*! Yapo (fr. Jan.) *Aké Assi* 3423! **S. Nig.**: Omo (formerly part of Shasha) F.R. (fl. & fr. Apr.) *Jones & Onochie* FHI 17363! (not quite typical).

[Note : The plants described as *Grumilea elongato-sepala*, *G. glabrifolia*, *G. puberulosa* and *G. rufo-pilosa* by De Wildeman are very variable in indumentum and in the shape of the calyx-lobes, but there does not appear to be more than one species involved.—F.N.H.]

57. **P. succulenta** (*Hiern*) *Petit* l.c. (1963). *Grumilea succulenta* Hiern in F.T.A. 3 : 216 (1877) ; Hepper in Kew Bull. 16 : 455 (1963). A glabrous shrub or small tree up to 30 ft. high, with coriaceous reticulate-veined leaves ; flowers white, fruits yellow turning red ; beside upland streams.
  [Br.]Cam.: Bamenda-Nkwe (Feb.) *Daramola* FHI 40468 ! Banso (fl. & fr. Jan.) *Keay* FHI 28454 ! *Lightbody* FHI 26296 ! Ngel Nyaki, Mambila Plateau (fr. Jan.) *Latilo & Daramola* FHI 34399 ! Bellel, Mambila Plateau (fl. & fr. Jan.) *Hepper* 1961 ! Also in Cameroun, Sudan, Uganda, Tanganyika and N. Rhodesia.

58. **P. venosa** (*Hiern*) *Petit* l.c. (1963). *Grumilea venosa* Hiern in F.T.A. 3 : 217 (1877) ; Chev. Bot. 344 ; Aubrév. Fl. For. C. Iv., ed. 2, 3 : 311, t. 368. A shrub or small tree with shining foliage ; flowers white, in spreading paniculate cymes.
  S.Nig.: Oban *Talbot* 235 ! 2040 ! **F.Po:** *Mann* 31 ! 254 ! 1440 ! Also in Cameroun, Gabon and Congo.

59. **P. articulata** (*Hiern*) *Petit* l.c. (1963). *Grumilea articulata* Hiern in F.T.A. 3 : 218 (1877). A shrub or small tree about 20 ft. high with rounded crown and young branchlets sometimes pubescent ; flowers white ; in swamp forest.
  S.Nig.: Lagos *Moloney* ! Epe *Barter* 3259 ! Okomu F.R., Benin (Jan., Feb.) *Brenan* 9184 ! Also in Cameroun.

60. **P. guineensis** *Petit* l.c. (1963). *Grumilea ivorensis* De Wild. in Bull. Jard. Bot. Brux. 9 : 38 (1923), not *Psychotria ivorensis*. *G. venosa* of F.W.T.A., ed. 1, 2 : 126, partly (*Chev.* 16315) not of Hiern. Small tree or shrub up to 25 ft. high ; flowers white.
  Iv. C.: Assinie (Apr.) *Chev.* 16315 ! 17861 ! Banco *Aubrév.* 1370 ! Abidjan (Feb.) *Aubrév.* 888 ! **Ghana:** Sekondi to Axim (Feb., Mar.) *Irvine* 2339b ! 2404 ! **S.Nig.:** Sapoba (Feb.) *Onochie* FHI 27682 ! Umudike, Umuahia (June) *Oseni* FHI 18307 ! Cross R. *Johnston* ! N. of Calabar (Feb.) *Mcleod* ! **[Br.] Cam.:** Batibo (May) *Johnstone* ! Also in Cameroun.

61. **P. chalconeura** (*K. Schum.*) *Petit* l.c. (1963). *Grumilea chalconeura* K. Schum. (1903). A shrub ; in forest.
  [Br.]Cam.: Lakom, Bamenda (fl. & fr. May) *Maitland* 1744 ! Also in Cameroun.

62. **P. djumaensis** *De Wild.* Miss. E. Laurent 1 : 349 (1906), not of De Wild. (1907). *Grumilea albiflora* De Wild. in Bull. Jard. Bot. Brux. 9 : 25 (1923) ; Pl. Bequaert. 2 : 450. Climber on tree ; flowers white or yellow ; fruits yellow.
  S.L.: Bumbuna (fr. Oct.) *Thomas* 3406 ! 3673 ! **Lib.:** Bahtown, Tchien Dist. (Aug.) *Baldwin* 9014 ! ? Ganta (fl. & fr. May, fr. Sept.) *Harley* 1178 ! *Baldwin* 9317 ! **N.Nig.:** *Thornewill* 119 ! Also in Congo.
  [The identity of these specimens with the Congo species is at present tentative.]

63. **P. psychotrioides** (*DC.*) *Roberty* in Bull. I.F.A.N. 16 : 62 (1954). *Grumilea psychotrioides* DC. (1830)—F.T.A. 3 : 216 ; Chev. Bot. 344 ; F.W.T.A., ed. 1, 2 : 126 ; Berhaut Fl. Sén. 106. *Uragoga tumbaensis* De Wild. (1936), French desc. A glabrous shrub or small tree, 4–20 ft. high with shining leathery leaves ; flowers white, fruits red ; beside streams in savanna.
  Sen.: Itou *Perrottet & Leprieur* ! Niayes *Chev.* 2012. **Mali :** Sikasso (May) *Chev.* 798 ! **Port.G.:** Prabis, Bissau (fr. Feb.) *Esp. Santo* 1834 ! Gabu (June) *Esp. Santo* 2498 ! **Guin.:** Mali (fr. Sept.) *Chev.* D ! Dalaba to Mamou (fr. Jan.) *Chev.* 20346 ! **S.L.:** Sugar Loaf Mt. (fl. May, fr. Nov.) *Barter* ! *Deighton* 5650 ! Bagroo R. *Mann* 871 ! Sasseni, Scarcies R. (fr. Jan.) *Sc. Elliot* 4540 ! Ninia, Talla Hills (Feb.) *Sc. Elliot* 4873 ! 4928 ! Sulimania (Mar.) *Sc. Elliot* 5349 ! Loma Mts. (fr. Aug.) *Jaeger* 1152 ! **Lib.:** Monrovia (fl. Apr., fr. Aug.) *Baldwin* 13064 ! Bequaert 180 ! Robertsport (fr. Dec.) *Baldwin* 10954 ! Garaway (fr. Mar.) *Baldwin* 11629 ! Baila (May) *Harley* 1426 ! **Iv.C.:** Tabou to Bériby *Chev.* 19990 ! Moossou (fr. Sept.) *Aké Assi* 4650 ! Zoanlé (May) *Chev.* 21461 (partly) ! Bouaké (Apr.) *Leeuwenberg* 3294 ! **Ghana:** Berekum to Techiman (Mar.) *Morton* GC 8566 ! Banda, Wenchi Dist. (fr. Dec.) *Morton* GC 25109 ! *Adams* 3131 ! Fuller Falls, Kintampo Dist. (fr. Nov.) *Andoh* FH 5414 ! Yapei (Sept.) *Goodall* GC 15773 ! Gambaga (Apr.) *Morton* GC 8987 ! **Dah.:** Djougou to Pobégou (June) *Chev.* 23931 ! Natitingou *Chev.* 24197 ! **N.Nig.:** Birnin Gwari (fr. Feb.) *Daggash* FHI 31423 ! Anara F.R., Zaria (Oct.) *Keay* FHI 5462 ! 20147b ! Jos Plateau (May) *Lely* P 281 ! **S.Nig.:** Ibadan (fr. Jan.) *Meikle* 950 ! Ibu *T. Vogel* 24 ! Onitsha *Barter* 1235 ! Eket (May) *Onochie* FHI 33183 ! Calabar (Mar.) *Latilo* FHI 41345 ! Boshi-Okwangwo F.R., Obudu Dist. (May) *Latilo* FHI 30946 ! **[Br.]Cam.:** Gangumi, Gashaka Dist. (fr. Dec.) *Latilo & Daramola* FHI 28807 ! Also in Cameroun, Ubangi-Shari, Chad and Sudan.

64. **P. fernandopoensis** *Petit* in Bull. Jard. Bot. Brux. 33 : ined. (1963). *Grumilea sphaerocarpa* Hiern in F.T.A. 3 : 218 (1877) ; F.W.T.A., ed. 1, 2 : 125. Shrub or small tree 8–15 ft. high ; flowers white.
  Lib.: Dukwia R. (fr. May) *Cooper* 410 ! Ba, Boporo Dist. (Dec.) *Baldwin* 10718 ! **Iv.C.:** Niapidou, Sassandra (Jan.) *Leeuwenberg* 2481 ! near Béyo, Sassandra (Jan., Feb.) *Leeuwenberg* 2537 ! 2823 ! **F.Po:** (fl. & fr. Jan.) *Mann* 223 !
  [If *Mann* 223, the type of *P. fernandopoensis*, does not prove to be conspecific with the other specimens, a name will be required for them.]

65. **P. gabonica** *Hiern* in F.T.A. 3 : 201 (1877) ; Chev. Bot. 341. *Grumilea lehmbachii* K. Schum. in Engl. Bot. Jahrb. 28 : 101 (1899) ; F.W.T.A., ed. 1, 2 : 126. A shrub or small tree 6–11 ft. high ; flowers whitish ; fruits yellow.
  S.L.: Yonibana (fr. Oct., Nov.) *Thomas* 4005 ! 4782 ! Njala (Apr., May) *Deighton* 3132 ! 3530 ! **Lib.:** Gbanga (fr. Sept.) *Linder* 744 ! Ganta (fl. Mar., fr. Feb., Sept.) *Baldwin* 12518 ! *Harley* 418 ! **Iv.C.:** Agniéby R. *Chev.* 17085 ! Sanvi *Chev.* 17734 ! **Ghana:** Bompata (Feb.) *Vigne* FH 1838 ! Fomang Su F.R. (Apr.) *Adams* 2650 ! **S.Nig.:** Epe *Barter* 3260 ! Onitsha *Barter* 1246 ! Omo (formerly part of Shasha) F.R. (Apr., May) *Jones & Onochie* FHI 16083 ! 17288 ! **[Br.]Cam.:** Cam. Mt., 2,000–4,000 ft. (fl. & fr. Dec.) *Mann* 2160 ! *Maitland* 1217 ! *Keay* FHI 28575 ! S. Bakundu F.R., Kumba (Mar., Apr.) *Olorunfemi* FHI 30542 ! *Brenan* 9454 ! Also in Gabon.

66. **P. brassii** *Hiern* in F.T.A. 3 : 204 (1877). Erect shrub about 5 ft. high ; flowers whitish ; in forest.
  Ghana: Cape Coast *Brass* ! **S.Nig.:** W. Lagos *Rowland* ! Gambari F.R. (Mar.) *Hepper* 2275 ! Ibadan South F.R. (Apr.) *Keay* FHI 22833 ! Omo (formerly part of Shasha) F.R. (Mar.) *Jones & Onochie* FHI 17004 ! *Ross* 66 ! *Richards* 3394 !

Imperfectly known species.

P. epiphytica *Mildbr.* in Wiss. Ergebn. 1910–11, 2 : 194, (1922) name only. An epiphytic shrub (in fruit only)
  F.Po: 2,400–3,000 ft. (fr. Aug.) *Mildbr.* 6433.

P. erythropus *K. Schum.* in Engl. Bot. Jahrb. 28 : 92 (1899). Shrub up to 12 ft. high.
  [Br.]Cam.: Cam. Mt., above Buea, 6,000 ft. (fr. Oct.) *Preuss* 1044.

P. minimicalyx *K. Schum.* in Engl. Bot. Jahrb. 28 : 94 (1899). A small shrub 1–2 ft. high, dichotomously branching, brown-red pubescent on the young parts, with small white flowers in small subsessile panicles ; in forest.
  [Br.]Cam.: Johann-Albrechtshöhe (= Kumba) (Feb.) *Staudt* 206.

67. CEPHAËLIS Sw. (1788)—F.T.A. 3 : 222 ; Hepper in Kew Bull. 16 : 153 (1962).
    Nom. cons. Uragoga Baill. (1879)—F.W.T.A., ed. 1, 2 : 127.

Peduncles very long, 40–400 cm. long ; ovary 3–4 locular :
  Leaves with 9–12 nerves on each side of midrib, broadly obovate, 15–26 cm. long, 8–14

cm. broad, cuneate at base, shortly acuminate or mucronate at apex, glabrous ;
   petioles 2–5 cm. long ; stipules ovate, 1–1·5 cm. long, acute, glabrous ; inflorescences
   axillary ; peduncle up to 45 cm. long ; flowers in 2–3 involucrate close heads ;
   involucral bracts glabrous ..   .. **1.** *mannii*
Leaves with 15–19 (–22) nerves on each side of the midrib, oblong-obovate about
   30 cm. long and 11 cm. broad, long-cuneate at base, shortly acuminate, glabrous ;
   petioles 2–3 cm. long ; stipules ovate, 2–4 cm. long, bifid, glabrous ; peduncle up to
   400 cm. long ; flowers in 3–5 involucrate heads with 1 head sometimes up to 20 cm.
   beyond the remainder ; involucral bracts pubescent ..   ..   **2.** *densinervia*
Peduncles up to 15 cm. long or absent ; ovary 2-locular :
Inflorescence heads with peduncles (2–) 5 cm. or more long (3. *peduncularis* aggregate) :
  Peduncles pubescent :
    Peduncles pubescent all round :
      Leaves shortly pubescent beneath, at least on the nerves
                              3b. *peduncularis* var. *suaveolens*
      Leaves glabrous beneath :
        Calyx-teeth 3–4 mm. long, distinct ..   ..   **3c.** *peduncularis* var. *A*
        Calyx-teeth 1 mm. long, acute or very short and rather obscure :
          Calyx-teeth pubescent ..   ..   ..   **3d.** *peduncularis* var. *palmetorum*
          Calyx-teeth glabrous ..   ..   **3e.** *peduncularis* var. *B*
    Peduncles with hairs in 2 lines on opposite sides :
      Involucral bracts free or deeply lobed ; peduncles solitary (and occasionally 3-
        headed at *apex* in No. 3a) :
        Calyx-teeth pubescent ..   ..   ..   **3a.** *peduncularis* var. *peduncularis*
        Calyx-teeth glabrous ..   ..   ..   **3f.** *peduncularis* var. *guineensis*
      Involucral bracts entire ; peduncles grouped in 3's
                              3h. *peduncularis* var. *ivorensis*
  Peduncles glabrous :
    Involucral bracts free or deeply lobed ..   **3g.** *peduncularis* var. *hypsophila*
    Involucral bracts entire :
      Heads in groups of 3 (rarely 2), terminal ..   **3h.** *peduncularis* var. *ivorensis*
      Heads solitary ..   ..   **3i.** *peduncularis* var. *tabouensis*
Inflorescence heads sessile or nearly so (peduncle up to 1 cm. long) :
  Involucral bracts more or less free :
    Leaves on both sides and stems pilose with reddish hairs ; stipules fused, tubular,
      about 1 cm. long ; leaves ovate, 3–5 cm. long, 2–3 cm. broad, with about 11
      nerves on each side of midrib ; inflorescences terminal or axillary, enclosed by
      imbricated bracts about 8 mm. long ; calyx-lobes linear, 3 mm. long, pilose ;
      peduncle 1 cm. long ..   ..   **4.** *mangenotii*
    Leaves glabrous above, glabrous or shortly tomentose beneath ; stipules free :
      Leaves shortly tomentose on the nerves beneath, ovate-elliptic, 8–11 cm. long,
        4–6 cm. broad, with about 13 nerves on each side of midrib looped near margin ;
        petioles up to 1 cm. long ; inflorescences subsessile, imbricated bracts broad,
        about 1 cm. long ; calyx-lobes triangular, 1·5 mm. long ..   **5.** *schnellii*
      Leaves glabrous on both sides :
        Petioles 3–4 mm. long ; leaves oblanceolate, cuneate at base, shortly acuminate
          at apex, 28–35 cm. long, 7–11·5 cm. broad, with 25–30 nerves on each side of
          midrib looped near margin ; inflorescences stout with numerous imbricated
          bracts 1·5–3 cm. long (head may contain 3 inflorescences) ..   **6.** *yapoensis*
        Petioles about 1 cm. long ; leaves obovate to elliptic, cuneate or slightly rounded
          at base, more or less shortly acuminate at apex, 8–15 cm. long, 3–8 cm. broad,
          with 11–16 nerves on each side of midrib looped near margin ; inflorescences
          similar to the last but smaller ..   ..   **7.** *ombrophila*
  Involucral bracts united ; leaves glabrous (rarely pilose in No. 10) on both sides :
    Involucre 6–15 mm. long, 6–40 mm. diam., not bilobed :
      Petioles less than 1 cm. long ; leaves obovate-elliptic, 6–8 cm. long, 2·5–3·5 cm.
        broad, with about 12 nerves on each side of midrib looped near margin ; inflores-
        scence sessile, involucre about 6 mm. long ; calyx subtruncate **8.** *abouabouensis*
      Petioles 4–5 cm. long ; leaves broadly obovate to oblanceolate, 18–26 cm. long,
        7–11 cm. broad, with about 20 nerves on each side of midrib looped near margin ;
        inflorescences subsessile, involucre about 1·5 cm. long, 2–4 cm. diam. ; calyx
        distinctly toothed ..   ..   **9.** *spathacea*
    Involucre (3–) 4–5 cm. long, 5–8 cm. diam., bilobed ; leaves obovate-elliptic rather
      long-cuneate at base, broadly and shortly acuminate at apex, about 15 cm. long
      and 7 cm. broad, with about 18 nerves on each side of midrib   **10.** *biaurita*

1. **C. mannii** (*Hook f.*) *Hiern* in F.T.A. 3 : 225 (1877). *Camptopus mannii* Hook. f. (1869)—K. Krause in
    Notizbl. Bot. Gart. Berl. No. 68 : 40 (=384) (1920). *Uragoga mannii* (Hook. f.) Hutch. & Dalz.
    F.W.T.A., ed. 1, 2 : 127 (1931) ; Schnell in Bull. I.F.A.N. 25 : 123 (1953). *U. peduncularis* of
    F.W.T.A., ed. 1, 2 : 127, partly (syn. *Camptopus mannii* Hook. f.), not of (Salisb.) K. Schum. A shrub
    or small tree 6–15 ft. high ; flowers white in pendulous heads with red involucre.
    S.Nig.: *Talbot* 154 ! Calabar *Hewan* ! Obudu, 5,500 ft. (Aug.) *Stone* 76 ! [**Br.**]**Cam.:** Cam. Mt. *Lehmbach*

21. *Deistel* 147! 453. *Winkler* 1262. Dikume Barue to Madie (Jan.) *Smith* FHI 10230! Nyanga, Kumba (Feb.) *Jeme* FHI 12003! Bamenda to Nkewe (Apr.) *Ujor* FHI 30028! **F.Po:** 2,000 ft. *Mann* 312! Also in Cameroun and Congo.

2. **C. densinervia** (*K. Krause*) *Hepper* in Kew Bull. 16 : 153 (1962). *Camptopus densinervius* K. Krause in Notizbl. Bot. Gart. Berl. No. 68 : 42 (=386) (1920). Small tree about 20 ft. high ; flowers yellow, clustered at the end of a peduncle 10 ft. or more long which reaches to, or near to, the ground ; in forest. [Br.]Cam.: Mungo Ndaw, near R. Loh, Kumba Div. (Oct.) *Dundas* FHI 15324! Also in Cameroun.

3. **C. peduncularis** *Salisb.* Parad. Lond. t. 99 (1808) ; F.T.A. 3 : 223 ; Hepper in Kew Bull. 16 : 154 (1962). *Uragoga peduncularis* (Salisb.) K. Schum. (1891)—F.W.T.A., ed. 1, 2 : 127. *? U. thorbeckei* K. Krause (1912), type destroyed. A shrub about 3 ft. high ; flowers white ; in forest. Very variable.

3a. **C. peduncularis** *Salisb.* var. **peduncularis**—*C. debauxii* De Wild. (1936), name only ; Schnell in Bull. I.F.A.N. 25 : 114. *Uragoga guerzeensis* Schnell l.c. 112 (1953), incl. f. *puberula* Schnell and f. *saouroana* Schnell. *Cephaëlis guerzeensis* (Schnell) Schnell in Mém. I.F.A.N. 50 : 57 (1957). A shrub 2–6 ft. high ; flowers white ; in forest.
Sen.: Kombo country *Heudelot* 33! Gam.: *Ingram*! Port.G.: Fancati (fr. Jan.) *d'Orey* 115! Guin.: Conakry *Debeaux*! Nimba Mts. (fr. Aug.) *Schnell* 5174! 6595! S.L.: Fundombaia (fr. Oct.) *Thomas* 2880! Rowala (fr. July) *Thomas* 1097! Lib.: Cape Palmas *T. Vogel* 62! Gbanga (fr. Sept.) *Linder* 711! Ghana: Bunso (Apr.) *Morton* A505! Mpraeso (fr. Apr.) *Morton* A709! S.Nig.: Ojogba-Ugun F.R., Ishan Dist. (June) *Olorunfemi* FHI 38062! Ibadan (Apr.) *Meikle* 1407! Keay FHI 22839! Widespread in tropical Africa.

3b. **C. peduncularis** var. **suaveolens** (*Schweinf. ex Hiern*) *Hepper* in Kew Bull. 16 : 156 (1962). *C. suaveolens* Schweinf. ex Hiern in F.T.A. 3 : 224 (1877). *C. nimbana* (Schnell) Schnell in Mém. I.F.A.N. 50 : 57 (1957). *Uragoga lecomtei* De Wild. var. *nimbana* Schnell in Rev. Gén. Bot. 57 : 284 (1950), illegitimate name. *U. nimbana* Schnell in Bull. I.F.A.N. 15 : 106 (1953). A shrub 3–6 ft. high ; flowers white, fruits fleshy blue-black or waxy-white ; in forest beside upland streams.
Guin.: Nimba Mts., 1,600 ft. (Apr., June, Aug.) *Schnell* 2941! 5349! 6224! Ghana: Togo Plateau F.R. (Dec.) *St. C.-Thompson* 3574! [Br.]Cam.: Bamenda, 4,000 ft., *Lightbody* FHI 26317! Vogel Peak, 4,900 ft., Adamawa (fr. Dec.) *Hepper* 1575! 2730! Also in Sudan, Congo, Uganda, Tanganyika and N. Rhodesia.

3c. **C. peduncularis** var. **A.**
S.Nig.: Oban *Talbot* 237!

3d. **C. peduncularis** var. **palmetorum** (*DC.*) *Hepper* l.c. (1962). *Morinda palmetorum* DC. Prod. 4 : 448 (1830). *M. geminata* of Aubrév. Fl. For. C. Iv., ed. 2, 3 : 272 (as to syn. *M. palmetorum*). *Uragoga nimbana* Schnell in Bull. I.F.A.N. 15 : 106 (1953), partly (*Schnell* 5349).
Sen.: St. Louis to Cape Verde (Mar.) *Döllinger* 53! Port.G.: Formosa (Apr.) *Esp. Santo* 1971! Guin.: Nimba (Apr.) *Schnell* 5349! S.L.: Freetown (May) *Deighton* 1182! Ghana: Hohoe (Apr.) *Morton* GC 9183! Axim *Irvine* 2362! S.Nig.: Carpenter 453! [Br.]Cam.: Cam. Mt. (Jan.) *Dunlap* 244!

3e. **C. peduncularis** var. **B.**
[Br.]Cam.: Johann-Albrechtshöhe (=Kumba) (June) *Staudt* 334! Bamenda (fr. Jan.) *Daramola* FHI 40631! Lakom (May) *Maitland* 1360! Oku (Jan., Feb.) *Keay* FHI 28517! *Hepper* 2008! Nkambe to Binka (Sept.) *Savory* UCI 382!

3f. **C. peduncularis** var. **guineensis** (*Schnell*) *Hepper* l.c. 154 (1962). *Uragoga peduncularis* var. *guineensis* Schnell in Rev. Gén. Bot. 57 : 285 (1950). *U. guineensis* (Schnell) Schnell in Bull. I.F.A.N. 15 : 103 (1953), incl. var. *bindelyensis* Schnell. *Cephaëlis guineensis* (Schnell) Schnell in Mém. I.F.A.N. 50 : 57 (1957). *C. coriacea* G. Don (1834). *C. bidentata* of Benth. in Fl. Nigrit. 421, partly (*Don*), not of Thunb. ex Roem. & Schult. *Psychotria bidentata* of F.T.A. 3 : 209, partly (*Afzelius*).
Port.G.: Buba (fr. Oct.) *Esp. Santo* 1269! Guin.: Nimba Mts. (Apr.) *Schnell* 5324! 5360! S.L.: *Afzelius*! Don! Bagroo R. *Mann*! Fwendu to Pujehun (Apr.) *Deighton* 1608! Bumban (June) *Glanville* 434! Njala (Apr.) *Deighton* 4770! Musaia (July) *Deighton* 4818! Lib.: Banga (Oct.) *Linder* 1221! Tchien (fr. Aug.) *Baldwin* 7040! Ganta (fl. Mar., fr. Apr.) *Harley* 419! Taninewa (Mar.) *Bequaert* 154! Gbanga (Apr.) *Konneh* 154! Ghana: Amentia (Dec.) *Vigne* FH 1489! Kwahu Prahou (fr. June) *Vigne* FH 1750! Kumasi *Cummins*! Also in Congo.

3g. **C. peduncularis** var. **hypsophila** (*K. Schum. & K. Krause*) *Hepper* l.c. 155 (1962). *Psychotria hypsophila* K. Schum. & K. Krause (1907). (?) *Uragoga calathea* K. Schum. & K. Krause (1907). *Psychotria bidentata* of Chev. Bot. 340, partly (*Chev.* 19179), not of (Thunb. ex Roem. & Schult.) Hiern.
Guin.: Macenta (May) *Collenette* 22! Lib.: Kakatown *Whyte*! Bushrod Isl. (Apr.) *Barker* 1289! Gbanga (Sept.) *Linder* 695! Iv.C.: Mid. Sassandra *Chev.* 19179! Ghana: Farmar 435! N.Nig.: Abinsi (June) *Dalz.* 635! S.Nig.: Koloishe Mt., 6,000 ft., Obudu (fr. Dec.) *Savory & Keay* FHI 25097! [Br.] Cam.: Victoria to Bimbia (Apr.) *Maitland* 617! Cam. Mt., above Buea (fl. & fr. Apr., fr. Nov.) *Migeod* 136! *Dundas* FHI 15336! *Brenan* 9548! F.Po: *Mann* 222! *Barter*! El Pico, 5,000 ft. (fr. Dec.) *Boughey* 179! Also in Cameroun.

3h. **C. peduncularis** var. **ivorensis** (*Schnell*) *Hepper* l.c. 156 (1962). *Uragoga ivorensis* Schnell in Bull. I.F.A.N. 15 : 109, fig. 7 (1953), incl. var. *puberula* Schnell. *Cephaëlis ivorensis* (Schnell) Schnell in Mém. I.F.A.N. 50 : 57, 88 (1957).
Iv.C.: Abidjan (Dec.) *Schnell* 3942! Agnéby *Chev.* 17184 ; 17085. Adiopodoumé (Nov.) *Leeuwenberg* 1935! Béyo (Feb.) *Leeuwenberg* 2734! Ghana: Simpa *Vigne* FH 2815! Subiri F.R., Benso (fr. Apr.) *Andoh* FH 5434! S.Nig.: Shasha F.R., Ijebu (Mar.) *Richards* 3223!

3i. **C. peduncularis** var. **tabouensis** (*Schnell*) *Hepper* l.c. 156 (1962). *C. tabouensis* Schnell in Mém. I.F.A.N. 50 : 87 (1957). *C. baillehachei* Aké Assi in Bull. Jard. Bot. Brux. 30 : 16, t. 2 (1960).
Lib.: Cape Palmes (July) *T. Vogel* 9! Iv.C.: Tabou (Feb.) *Mangenot & Miège* 2236! Ahouabo (Adzopé) (Apr.) *Aké Assi* 1871!
[*Uragoga thorbeckei* K. Krause (in Engl. Bot. Jahrb. 48 : 429 (1912)) appears to belong to the *C. peduncularis* aggregate : type presumably destroyed. [Br.]Cam.: Bamenda *Thorbecke* 266.]

4. **C. mangenotii** *Aké Assi* in Bull. Jard. Bot. Brux. 29 : 364, fig. 7 (1959). *Geophila rotundifolia* A. Chev. Bot. 346, partly (*Chev.* 19205), name only. A herb with wiry creeping stems rooting on the forest floor ; flowers white.
Iv.C.: Tiapleu (fr. Aug.) *Mangenot & Aké Assi* 4845! Sérédou (fr. May) *Aké Assi* 2426! Mid. Sassandra to Mid. Cavally (June) *Chev.* 19205! Ghana: Kade (May) *Morton* A4023!

5. **C. schnellii** *Aké Assi* l.c. 362, fig. 6 (1959). An undershrub with ascending stems rooting near the lower nodes.
Iv.C.: Guého to Soubré (Dec.) *Mangenot, Miège & Aké Assi* 4846! Koleahinou (fr. Jan.) *Aké Assi* 5378! Adiopodoumé (Mar.) *Aké Assi* 4848!

6. **C. yapoensis** (*Schnell*) *Schnell* in Mém. I.F.A.N. 50 : 57, 91 (1957). *Uragoga yapoensis* Schnell in Bull. I.F.A.N. 15 : 101 (1953). A shrub up to 5 ft. high.
Lib.: Jabroke, Webo Dist. (July) *Baldwin* 6435! Iv.C.: Yapo (Aug., Oct., Dec.) *Schnell* 3944! 5868! *Roberty* 15337!

7. **C. ombrophila** (*Schnell*) *Schnell* in Mém. I.F.A.N. 50 : 57 (1957). *Uragoga ombrophila* Schnell in Rev. Gén. Bot. 57 : 283, fig. 3 (1950). An undershrub, about 1 ft. high, with stems creeping and rooting ; flowers white ; in forest.
Guin.: Nimba Mts. *Schnell* 3475! 5102! S.L.: Gola F.R. (fr. May) *Small* 84! 694! Lib.: Vahun, Kolahun Dist. (fr. Nov.) *Baldwin* 10210! Ganta (fr. Oct.) *Harley* 1248! Bobei Mt., Sanokwele Dist. (Sept.) *Baldwin* 9609! Tchien (Aug.) *Baldwin* 7016! Ghana: Ancobra (bud Dec.) *Johnson* 976!

8. **C. abouabouensis** *Schnell* in Mém. I.F.A.N. 50 : 85, fig. 8 (1957). An undershrub about 1 ft. high, from a creeping base ; flowers white ; in deciduous forest.
   Iv.C.: Abouabou, Abidjan *Miège* 2248 !
9. **C. spathacea** *Hiern* in F.T.A. 3 : 225 (1877) ; Chev. Bot. 347. *Uragoga spathacea* (Hiern) Hutch. & Dalz. F.W.T.A., ed. 1, 2 : 128 (1931). A shrub with rather large leaves, 3–12 ft. high ; flowers white in close heads ; in forest.
   Lib.: River Cess (Mar.) *Baldwin* 11299 ! Sangwin (Mar.) *Baldwin* 11307 ! **Iv.C.:** Abidjan *Chev.* 15338 ! Fort Binger to Nienokué Mt. *Chev.* 19506 ! Grabo, Cavally R. (July) *Chev.* 19638 ! Also in S. Tomé.
10. **C. biaurita** (*Hutch. & Dalz.*) Hepper in Kew Bull. 16 : 153 (1962). *Uragoga biaurita* Hutch. & Dalz. F.W.T.A., ed. 1, 2 : 128 (1931). *Psychotria biaurita* A. Chev. Bot. 340, name only. *Cephaëlis condensata* A. Chev. Bot. 346, name only. A shrub 1–6 ft. high with rather large leaves ; flowers white or bluish, enclosed in a large cup, fruits black on white pedicels ; in forest.
    Guin.: Lolo to Nzo (Mar.) *Chev.* 20996 ! **S.L.:** Masingbe (fr. Jan.) *Deighton* 4579 ! Masimo *Marmo* 27 ! Njagbela *Deighton* 5815 ! Loma Mts. (Sept.) *Jaeger* 2055 ! Gola F.R. (Apr.) *Deighton* 3629 ! **Lib.:** Bobei Mt., Sanokwele Dist. (Sept.) *Baldwin* 9566 ! Wohmen (fr. Oct.) *Baldwin* 10095 ! Bilimu (Jan.) *Harley* 1330 ! **Iv.C.:** Abouassou to Soubiré *Chev.* 17743 ! Danané *Chev.* 21285 ! **Ghana:** Axim (Mar.) *Irvine* 2407 ! Kwahu Plateau (Apr.) *Kitson* 1040 !
    [*C. castaneo-pilosa* Aké Assi (in Bull. Jard. Bot. Brux. 31 : 314, t. 2 (1961)) appears to be only a form of this with pilose leaves.]

68. GEOPHILA D. Don[1]—F.T.A. 3 : 220.

Flowers 1 (–3), subsessile or with a peduncle up to 2 cm. long ; calyx-lobes subulate, 2–3 mm. long, often reflexed ; corolla 1·5 cm. long ; leaves broadly ovate, reniform to suborbicular and deeply cordate at base, 1–5 cm. long, glabrous ; petioles 1–5 cm. long, pubescent ..   .. 1. *repens*
Flowers several together, subsessile or pedunculate and conspicuously bracteate :
Inflorescences without conspicuous leafy bracts, subsessile :
Leaves hirsute (at least on the midrib above), ovate to ovate-oblong, cordate, 2–5 cm. long, 1–4 cm. broad ; petiole 0·5–3·5 cm. long, hirsute ; stems prostrate, rooting at nodes and internodes ; stipules deeply divided   ..   .. 2. *hirsuta*
Leaves glabrous above :
Leaves shortly and sparingly tomentose on the nerves beneath, ovate, deeply cordate at base, 7–8 cm. long, 3–5 cm. broad ; petiole 1–3 cm. long, appressed-pubescent ; stems prostrate and ascending ; stipules bilobed to base   .. 3. *liberica*
Leaves usually densely pubescent on the nerves beneath often accentuating the reticulate venation, oblong or obovate to oblanceolate, usually rounded at each end or subcordate at base, 4–10 cm. long, 1·5–4 cm. broad   .. 4. *neurodictyon*
Inflorescences subtended by leafy bracts or with a cupular involucre :
Inflorescences subtended by leafy ovate bracts :
Stipules entire ; involucral bracts glabrous ; flowers without bracteoles ; inflorescences long-pedunculate, peduncles 2–7 cm. long, glabrous ; leaves ovate-deltoid, deeply cordate at base, subacute at apex, 2–5 cm. long, 1·5–4·5 cm. broad ; petioles slender, densely pubescent on the sides towards apex 5. *obvallata*
Stipules bifid ; involucral bracts pubescent ; flowers with linear bracteoles ; inflorescences usually shortly pedunculate, peduncles up to 3 cm. long, pubescent ; leaves ovate, cordate at base, subacute at apex, 3–6 cm. long, 2–5·5 cm. broad ; petioles pubescent all round   ..   .. 6. *afzelii*
Inflorescences surrounded by a cupular involucre about 1 cm. diam., toothed and pubescent on margin ; leaves orbicular to ovate, obtuse at apex, cordate to cuneate at base, 4·5–6 (–10) cm. long, 3–6 cm. broad, more or less pubescent on both surfaces; petiole 1–2 (–4) cm. long, pubescent   ..   .. 7. *rotundifolia*

1. **G. repens** (*Linn.*) *I. M. Johnston* in Sargentia 8 : 281 (1949) ; Brenan in Mem. N.Y. Bot. Gard. 8 : 453 (1954). *Rondeletia repens* Linn. (1759). *G. reniformis* D. Don (1825)—F.T.A. 3 : 220 ; Chev. Bot. 346. *G. uniflora* Hiern in F.T.A. 3 : 221 (1887) ; F.W.T.A., ed. 1, 2 : 128. *G. lancistipula* Hiern in F.T.A. 3 : 221 (1877) ; F.W.T.A., ed. 1, 2 : 128. *Carinta uniflora* (Hiern) G. Tayl. in Exell Suppl. Cat. S. Tomé 25 (1956). Creeping prostrate herb ; flowers white, fruits scarlet ; in forest.
   Port.G.: Cacine to Buba, Fulacunda (July) *Esp. Santo* 2149 ! **Guin.:** La Kolenté *Chillou* 941 ! Friguiagbé (fr. July) *Chillou* 2106 ! **S.L.:** Gahun, Nando (fr. Oct.) *Deighton* 3768 ! Lengekoro (July) *Glanville* 269 ! **Iv.C.:** Bouroukrou *Chev.* 16558 ! 16674 ! Mid. Sassandra to Mid. Cavally *Chev.* 19221 ! 19385 ! **Ghana:** Swedru (fr. Feb.) *Dalz.* 8287 ! Mankrong (Apr.) *Morton* A556 ! Amentia *Irvine* 441 ! Bobiri F.R., Juaso (fl. & fr. Dec.) *Foggie* FH 5120 ! Okomfokrom, Togo Plateau (Apr.) *Morton* GC 9157 ! **N.Nig.:** Nupe *Barter* 1240 ! Sanga River F.R., Zaria (May) *E. W. Jones* 136 ! **S.Nig.:** Lagos *Punch* ! Nun R. *T. Vogel* 57 ! Brass *Barter* 60 ! Okomu F.R. (fl. & fr. Nov.) *Brenan* 8622 ! Akure F.R. (fr. Nov.) *Keay* FHI 25600 ! Pan-tropical. (See Appendix, p. 400.)
2. **G. hirsuta** *Benth.* in Fl. Nigrit. 422 (1849) ; F.T.A. 3 : 221 ; Chev. Bot. 345, partly. *G. pilosa* A. Chev. Bot. 346, partly, name only. Creeping, prostrate hairy plant ; flowers white, fruits orange coloured ; in forest.
   Guin.: Kouria *Caille* in *Hb. Chev.* 14883 ! **S.L.:** Kenema *Lane-Poole* 446 ! Boama (fr. Apr.) *Deighton* 3666 ! Kambui Hills South (fr. Nov.) *Small* 829 ! Loma Mts., 2,600 ft. (Aug.) *Jaeger* 960 ! **Lib.:** Monrovia (fr. Oct., Nov.) *Cooper* 41 ! Du R. (July) *Linder* 106 ! Tai (fr. July) *Harley* 1227 ! Duo (Mar.) *Baldwin* 11354 ! Zuie, Boporo Dist. (fr. Nov.) *Baldwin* 10240 ! **Iv.C.:** Taï (fr. Aug.) *Boughey* GC 14971 ! Bingerville *Chev.* 15383 ! Tébo, Cavally *Chev.* 19379 ! **Ghana:** Apapam (Aug.) *Chipp* 557 ! **Dah.:** Roubé to Sakété (Aug.) *Le Testu* 18–23 ! **S.Nig.:** Nun R. *T. Vogel* 44 ! Brass *Barter* ! Usonigbe F.R., Benin (fl. & fr. Sept., Nov.) *Meikle* 568 ! *Keay* FHI 25552 ! 34288 ! Shasha F.R. (fr. Mar.) *Richards* 3167 ! [**Br.]Cam.:** Victoria (fr. Oct.) *Mildbr.* 10581 ! 10586 ! Also Cameroun, Ubangi-Shari, Congo, Uganda and Tanganyika.
3. **G. liberica** *A. Chev. ex Hutch. & Dalz.* F.W.T.A., ed. 1, 2 : 128 (1931); Chev. Bot. 345, name only. A semi-prostrate herb.

---

[1] *Geophila* D. Don (1825) has been proposed as a conserved name against *Geophila* Berger (1803), *Carinta* W. F. Wight (1905) and *Geocardia* Standley (1914) ; see Hepper in Taxon 9 : 88 (1960).

Lib.: Sanokwele (Sept.) *Baldwin* 9554! **Iv.C.:** Tébo, Cavally *Chev.* 19382! **Ghana:** Fomena (Apr.) *Adams* 2522! Siresotinpom (Apr.) *Darko* 864! Also in Cameroun.

4. **G. neurodictyon** (*K. Schum.*) *Hepper* in Kew Bull. 16 : 331 (1962). *Psychotria neurodictyon* K. Schum. in Engl. Bot. Jahrb. 33 : 368 (1903). *Geophila fissistipula* K. Krause (1907). *G. aurantiaca* A. Chev. Bot. 334, name only. More or less decumbent herb, or creeping and sending up erect shoots, with softly pubescent stems 6–18 in. high ; lower surface of leaves with a prominent network of hairy nerves or almost glabrous ; flowers white, fruits orange ; on forest floor.

Guin.: Kounounkau valley (fr. Oct.) *Schnell* 7647! **S.L.:** Kasewe Hills F.R. (June) *Deighton* 5826! **Lib.:** Bobei Mt. (fr. Sept.) *Baldwin* 12500! Ganta (Apr.) *Harley* 1128! Kitomu *Harley* 1129! **Iv.C.:** Taï (fr. Aug.) *Boughey* GC 14853! Malamalasso *Chev.* 17538! Tébo, Mid. Cavally *Chev.* 19380! **Ghana:** Atewa Range F.R. (fl. Apr., fr. Jan.) *Adams* 2675! *de Wit & Morton* A2928! Begoro, Akim (fr. Jan.) *Irvine* 1174! **S.Nig.:** Omo F.R. (Apr.) *Onochie* FHI 15520! Okomu F.R., Benin (fr. Dec.) *Brenan* 8461! *Richards* 3300! **[Br.]Cam.:** Victoria (Apr.) *Schlechter* 12364! *Mildbr.* 10589! Barombi *Preuss* 1196! S. Bakundu F.R. (Mar.) *Brenan* 9286! *Binuyo & Daramola* FHI 35652! Nkom to Wum, Bamenda (June) *Ujor* FHI 29264! Also in Principe, Cameroun and Uganda.

5. **G. obvallata** (*Schumach.*) *F. Didr.* in Vidensk. Medd. Naturhist. Foren. Kjöbenh. 1854 : 186 (1855) ; F.T.A. 3 : 222 ; F.W.T.A., ed. 1, 2 : 128, partly (excl. syn. *G. afzelii* Hiern) ; Chev. Bot. 345. *Psychotria obvallata* Schumach. (1827). *Carinta obvallata* (Schum.) G. Tayl. in Exell Suppl. Cat. S. Tomé 25 (1956). *Geophila afzelii* of Chev. Bot. 344, mostly. Creeping prostrate herb ; flowers white, fruits blue ; on the forest floor, often around the base of trees.

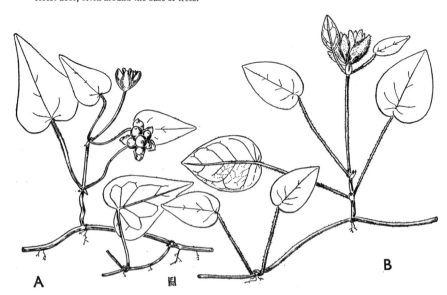

FIG. 243.—GEOPHILA SPP. (RUBIACEAE).

G. obvallata (Schumach.) F. Didr. A, habit, × ¼. Drawn from *Brenan* 8628.
G. afzelii Hiern B, habit, × ¼. Drawn from *Melville & Hooker* 489.

Port.G.: Bafata (fr. Sept.) *Esp. Santo* 3372! **Guin.:** *Heudelot* 602! Friguiagbé *Chillou* 2149! **S.L.:** Kinsita *Sc. Elliot* 4240! Gorahun (Nov.) *Deighton* 362! Njala (fl. Aug., fr. Jan.) *Dalz.* 8045! Loma Mts. (fr. Oct., Jan.) *Jaeger* 235! 4976! **Lib.:** Ganta (fl. & fr. Aug.) *Harley* 204! Nyaake, Webo (June) *Baldwin* 6084! Zule, Boporo Dist. (fr. Nov.) *Baldwin* 10259! **Iv.C.:** Adiopodoumé (Oct.) *Leeuwenberg* 1766! **Ghana:** Aquapim *Isert*! Asamang (fl. & fr. Apr.) *Chipp* 155! Agogo (Apr.) *Irvine* 939! Kwahu-Tafo (Apr.) *Morton* A527! Kumasi *Cummins* 67! **S.Nig.:** Brass *Barter* 4! Lagos *Millen* 6! Okomu F.R. (Dec.) *Brenan* 8628! Sapoba (fl. & fr. Nov.) *Meikle* 635! *Ujor* FHI 23929! Afi River F.R., Ogoja (May) *Jones & Onochie* FHI 17320! **[Br.]Cam.:** Binkas, Nkom-Wum F.R. (June) *Ujor* FHI 29263! **F.Po:** *Barter*! *Mann* 173! Also in Cameroun, Gabon and Congo.

6. **G. afzelii** *Hiern* in F.T.A. 3 : 221 (1877) ; Chev. Bot. 344, partly (*Chev.* 16727) ; Hepper in Kew Bull. 16 : 454 (1963). *G. cordiformis* A. Chev. ex Hutch. & Dalz. F.W.T.A., ed. 1, 2 : 128 (1931) ; Exell Cat. S. Tomé 217 ; Chev. Bot. 344, name only. *G. speciosa* K. Schum. (1897). *Carinta cordiformis* (A. Chev. ex Hutch. & Dalz.) G. Tayl. in Exell Suppl. Cat. S. Tomé 25 (1956). *Geophila lutea* A. Chev. Bot. 345, name only. *G. obvallata* of F.W.T.A., ed. 1, 2 : 128, partly (syn., *MacGregor* 305) ; of Exell Cat. S. Tomé 217 (*Mann*), not of (Schum.) F. Didr. *G. flaviflora* Aké Assi in Bull. Jard. Bot. Brux. 31 : 316, t. 3 (1961). Prostrate hairy herb ; flowers white or yellow, soon falling, fruits red ; in forest.

Guin.: Nzérékoré (fr. Jan.) *Adam* 3202! **S.L.:** *Afzelius*! No. 2 River F.R., Freetown *Hepper* 2517! Waterloo (fl. & fr. Sept.) *Melville & Hooker* 489! Lowoma (Oct.) *Fisher* 76! Rokupr (Feb.) *Deighton* 3606! Kruto to Sini-Koro (fr. Sept.) *Jaeger* 1812! **Lib.:** Dukwai R., Monrovia (fr. Oct., Nov.) *Cooper* 50! Brewersville (June) *Barker* 1334! Du R. (July) *Linder* 183! Ganta (fl. & fr. July) *Harley* 204a! Firestone Cavalla, Maryland (fl. & fr. June) *Baldwin* 6003! **Iv.C.:** Bouroukrou *Chev.* 16727! Potou L. to R. Mé *Chev.* 17437! Abidjan (fr. Dec.) *Boughey* GC 13505! Copé Mt. *Chev.* 19692! Yapo (fr. Oct.) *Roberty* 15354! Taï (fl. & fr. Aug.) *Boughey* GC 14855d! **Ghana:** Subri F.R., Benso (fr. Mar.) *Foggie* FH 4950! Awaso (fr. Dec.) *Morton* A3619! **S.Nig.:** Lagos *MacGregor* 305! *Kennedy* 2697! Akilla, Ijebu (fl. & fr. Nov.) *Onochie* FHI 8168! Usonigbe F.R., Benin (fl. & fr. Nov.) *Meikle* 563! Okomu F.R., Benin (fl. & fr. Dec.) *Brenan* 8493! **[Br.]Cam.:** N. Kembong F.R. (fr. Sept.) *Tiku* FHI 22188! Also in Cameroun, Gabon, Congo, Cabinda and Principe.

7. **G. rotundifolia** A. *Chev. ex Hepper* in Kew Bull. 16 : 331 (1962). Chev. Bot. 346, partly (*Chev.* 17789 *bis*), name only. *G. hirsuta* of F.W.T.A., ed. 1, 2 : 128, partly (*Chev.* 17789 *bis*) not of Benth. A prostrate trailing herb with ascending shoots ; flowers white, fruits yellow or orange ; in forest.

S.L.: Waterloo to York (fr. Sept.) *Melville & Hooker* 621! **Lib.:** Bilimu (fr. Aug.) *Harley* 1452! **Iv.C.:** Soubré to Guidéko, Sassandra (May) *Chev.* 17789 *bis*! **Ghana:** Kwahu (Mar.) *Johnson* 659! **S.Nig.:** Shasha F.R., Ijebu-Ode (Apr., May) *Tamajong* FHI 16764! *Richards* 3352! 3384! **[Br.]Cam.:** S. Bakundu F.R., Kumba (fr. Aug.) *Olorunfemi* FHI 30717! N. Korup F.R., Kumba (July) *Olorunfemi* FHI 30690! Nkom-Wum F.R., Bamenda (fr. June) *Ujor* FHI 29265! Also in Cameroun and Uganda.

69. TRICHOSTACHYS Hook f.—F.T.A. 3 : 215. *Nom. cons.*

Inflorescence spicate, interrupted, about 4 cm. long ; leaves narrowly obovate, about 15 cm. long and 6 cm. broad, acutely cuneate at base, midrib glabrous above and beneath, lateral nerves looped, about 7 on each side of midrib ..   1. *interrupta*
Inflorescences capitate, about 2 cm. long ; peduncle 1–2 cm. long :
  Midrib pubescent above, more or less appressed-pilose beneath, upper surface of leaves more or less pilose, obovate to shortly oblanceolate, 6–20 cm. long, 4–9 cm. broad, cuneate at base, shortly acuminate, acute or rounded at apex ; calyx-teeth acutely deltoid, about 1 mm. long, pilose ; fruits ellipsoid, about 4 mm. long, sparsely pilose
    2. *aurea*
  Midrib and lamina glabrous above, densely spreading-pilose beneath, oblanceolate, about 35 cm. long and 12 cm. broad, long-cuneate at base, acuminate ; calyx subtruncate, ciliate on margin .. .. 3. sp. *A*

1. **T. interrupta** *K. Schum.* in Engl. Bot. Jahrb. 33 : 360 (1903). *T. krausiana* Wernham in Cat. Talb. 55 (1913). Undershrub or herb with thick decumbent stem ; flowers white ; in forest.
  S.Nig.: Oban *Talbot* 1045! Kwa Falls (Mar.) *Brenan* 9238! **[Br.]Cam.:** Barombi *Preuss* 466.
2. **T. aurea** *Hiern* in F.T.A. 3 : 227 (1877). Erect undershrub 1–2 ft. high ; flowers purple, fruits blue, on reflexed peduncle ; in forest.
  S.L.: Jala (July) *Bunting* 65! **Lib.:** Genna Loffa, Kolahun Dist. (fr. Nov.) *Baldwin* 10077! Soplima, Vonjama Dist. (fr. Nov.) *Baldwin* 10026a! Boporo (fr. Nov.) *Baldwin* 10386! Ganta (May) *Harley* 534! Sanokwele (fr. Sept.) *Baldwin* 9560! **Iv.C.:** Adiopodoumé (June) *Aké Assi* 4344! Yapo (fr. Oct.) *Aké Assi* 4403! **Ghana:** Puso Puso Ravine (June) *Adams* 4824! Atewa Range F.R., Pra R. (Mar.) *Morton* A3872! **S.Nig.:** Sapoba F.R. (June) *Onochie* FHI 36653! Omo F.R., Ijebu (May) *Tamajong & Latilo* FHI 16779! Okomu F.R., Benin (fr. Dec.) *Brenan* 8557! Oluwa F.R., Ondo (July) *Onochie* FHI 33411! Oban F.R. (fr. Nov.) *Onyeagocha* FHI 7767! Afi River F.R., Ogoja Prov. (May) *Jones & Onochie* FHI 17318! **[Br.]Cam.:** Bakosi F.R., Kumba (May) *Olorunfemi* FHI 30590! Also in Cameroun and Gabon.
  [Note : large-leaved specimens (FHI 33411, FHI 16779) are similar to the type (*Zenker* 4626, from Cameroun) of *T. zenkeri* De Wild. (1932), which may be only a form of this species—F.N.H.]
3. **T. sp. A.** Unbranched undershrub 2 ft. high with greenish stem ; flower buds red-purple ; in forest.
  [Br.]Cam.: S. Bakundu F.R., Banga (Mar.) *Brenan* 9284! 9284a!

Imperfectly known species.

T. lehmbachii *K. Schum.* in Engl. Bot. Jahrb. 23 : 467 (1897). Erect herb about 25 cm. high ; inflorescence sessile, 2·5 cm. diam. ; stipules bilobed 2·5 cm. long.
  [Br.]Cam.: Victoria *Lehmbach.* (Probably destroyed.)

70. PENTANISIA Harvey—F.T.A. 3 : 131.

Leaves narrowly oblanceolate, acute, 1–2·5 cm. long, glabrous ; cymes shortly pedunculate, rather few-flowered ; calyx-lobes very unequal ; corolla-tube narrow and slender, bright blue, thinly and minutely setulose outside ; ovary densely white-setulose, tomentose .. .. *schweinfurthii*

P. schweinfurthii *Hiern* in F.T.A. 3 : 131 (1877) ; Verdcourt in Bull. Jard. Bot. Brux. 22 : 254 (1952). Perennial herb with a tuft of stems 2–12 in. high from a woody rootstock ; flowers bright blue ; flowering after fires.
  N.Nig.: Jos Plateau : Naraguta *Hill* 29! *Kennedy* 2914! Jos (Dec.) *MacGregor* 414! Bokkos (Dec.) *Coombe* 80! Bukuru *Savory* UCI 148! Extends to the Sudan.

71. NEOBAUMANNIA Hutch. & Dalz. F.W.T.A., ed. 1, 2 : 132 (1931). *Baumannia* K. Schum. (1896), not of DC. (1833).

Indumentum short and sparse on leaves and stems, more or less appressed on stems above, stems from a woody base, glabrous and subsucculent below ; leaves subsessile, ovate to broadly lanceolate, 4–10 cm. long, 2–4 cm. broad, with several oblique nerves, puberulous on the nerves beneath ; stipules divided into numerous filiform segments ; inflorescences terminal panicles, the flowers congested at the apex of the erect branches ; calyx-lobes 4, one narrowly enlarged ; corolla 3 mm. long, 4(–5) lobed ; fruits dehiscent, the valves caducous .. .. .. 1. *hedyotoidea* var. *hedyotoidea*
Indumentum abundant and spreading on leaves and stems
    1a. *hedyotoidea* var. *longipila*

1. **N. hedyotoidea** (*K. Schum.*) *Hutch. & Dalz.* var. **hedyotoidea**—F.W.T.A., ed. 1, 2 : 132 (1931). *Baumannia hedyotoidea* K. Schum. in Engl. Bot. Jahrb. 23 : 455 (1896). Stems 2–8 ft. high, tinged pink or purple, often numerous, fleshy ; flowers blue or purple, enlarged calyx whitish ; on rock outcrops.
  Ghana: Banda, Wenchi *Morton* GC 25269! Aboma F.R. (Nov.) *Vigne* FH 3442! Krobo Hill (Oct., Nov.) *Rose Innes* GC 30870! 31165! Kapjakpe, Hohoe Dist. (Oct.) *St. C.-Thompson* 1503! Maliato (Nov.) *Morton* A3538! **Togo Rep.:** Misahöhe *Baumann* 323. **N.Nig.:** Jos Plateau : Naraguta, 4,000 ft. (Aug.) *Lely* 528! Amo, 4,100 ft. (Oct.) *Hepper* 1031!
1a. **N. hedyotoidea** var. **longipila** *Brenan* in Kew Bull. 5 : 228 (1950). A hairy variety.
  S.Nig.: Carter's Peak, Idanre Hills (fl. & fr. Oct.) *Keay & Onochie* FHI 21561! *Keay* FHI 22586! Akure F.R. (Oct.) *Keay* FHI 22664! Aponmu F.R., Akure (fr. Nov.) *Ajayi* FHI 26917!

72. ARGOSTEMMA Wall.—F.T.A. 3 : 44.

Leaves obliquely elliptic-lanceolate, 2–3·5 cm. long, up to 1·3 cm. broad, pilose with long weak hairs above and on the midrib beneath ; flowers from solitary to about 5 in a

loose cyme ; pedicels pubescent, up to 1·3 cm. long ; calyx slightly toothed ; corolla divided nearly to the base, lobes oblong-elliptic, 5–6 mm. long ; anthers long-exserted, opening by pores　　..　　..　　..　　..　　..　　..　　1. *pumilum*
Leaves oblanceolate, subacute, 2–4·5 cm. long, 0·5 cm. broad, slightly pilose beneath ; flowers up to 5 in each cyme ; pedicels glabrous ; calyx undulate ; corolla divided almost to the base, lobes broadly lanceolate, 6 mm. long ; anthers long-exserted
　　　　　　　　　　　　　　　　　　　　　　　　　　2. *africanum*

1. **A. pumilum** *Benn.* Pl. Jav. Rar. 95 (1838) ; F.T.A. 3 : 44, partly (*Afzelius* spec.). A small herb, 2–3 in. high ; flowers white ; on wet rocks.
　　Guin.: Conakry to Kindia *Jac.-Fél.* 7018 ! **S.L.:** *Afzelius* ! Sugar Loaf Mt., 2,000 ft. (Nov.) *T. S. Jones* 285 ! Bumban, R. Seli (Aug.) *Deighton* 1308 ! Kulafaga to Bumbuna (Aug.) *Deighton* 5155 ! Kenema *Lane-Poole* 465 ! Loma Mts. (Oct.) *Jaeger* 233 ! **Lib.:** Kitomu (Sept.) *Harley* 1601 ! Sanokwele (Sept.) *Baldwin* 9503 ! **Iv.C.:** Nimba Mts. (Aug.) *Schnell* 6161 !
2. **A. africanum** *K. Schum.* in Engl. Bot. Jahrb. 23 : 423 (1896). A small herb similar to the last.
　　S.Nig.: Oban *Talbot* 345 ! [**Br.]Cam.:** Tiko (Oct.) *Schlechter* 15793 ! Also in Rio Muni.

73. VIRECTARIA Bremek. in Verh. K. Nederl. Akad. Wetensch., Afd. Natuurk., sect. 2, 48 : 21 (1952). *Virecta* of Sm., partly—F.T.A. 3 : 47, not of Linn. f.

Disk composed of two cones ; style longer than the exserted stamens ; corolla-tube about 10 mm. long, lobes 6 mm. long ; calyx-lobes linear, 3–4 mm. long ; stems usually densely long-pilose ; leaves elliptic to elliptic-lanceolate, acute at apex, narrowly cuneate at base, 1·3–13 cm. long, 7–27 mm. broad, with about 7 pairs of lateral nerves ; stipules with 2–3 filiform segments up to 5 mm. long, erect
　　　　　　　　　　　　　　　　　　　　　　　　　　1. *multiflora*
Disk cylindrical, not divided into 2 cones :
Leaves broadly lanceolate to ovate-lanceolate :
　Erect herbs ; stamens conspicuously exserted, style longer than stamens :
　　Calyx-lobes subulate, about 6 mm. long, variable in length in the same flower, pilose ; inflorescences many-flowered, dense ; corolla 15–20 mm. long ; stems densely pubescent ; leaves lanceolate to ovate-lanceolate, 2·3–9·5 cm. long, 7–45 mm. broad, with up to 11 pairs of nerves markedly arcuate beneath, pubescent ; stipules deltoid 1–2 mm. long or with 2 lanceolate lobes 4–8 mm. long, often reflexed　　..　　..　　..　　..　　..　　..　　..　　2. *major* var. *major*
　　Calyx-lobes often spathulate at apex, very unequal, 3 or 4 mm. long, slightly pilose ; inflorescences few-flowered, rather lax ; corolla about 10 mm. long ; stems pubescent ; leaves with about 7 pairs of nerves, sparingly pubescent
　　　　　　　　　　　　　　　　　　　　　　　　　2a. *major* var. *spathulata*
　Straggling herb ; calyx-lobes subequal, spathulate at apex, 1–6 mm. long ; inflorescences very few-flowered ; corolla up to 1 cm. long ; style shorter than stamens which are shortly exserted ; stems pubescent, hairs often in 2 opposite longitudinal lines ; leaves ovate-oblong, elliptic or subspathulate, 1–6 cm. long, 6–34 mm. broad ; stipules deltoid or bifid, 3 mm. long　　..　　..　　..　　3. *procumbens*
Leaves narrowly elliptic or oblanceolate, subacute at apex, 3·7–5 cm. long, 6·5–12 cm. broad ; stipules narrowly lanceolate, 1–3 mm. long ; inflorescences very few-flowered ; calyx-lobes lanceolate-oblong, about 1 mm. long ; corolla-tube 4–4·5 mm. long, lobes 3·5–4 mm. long ; stamens slightly exserted　　..　　..　　4. *angustifolia*

1. **V. multiflora** (*Sm.*) *Bremek.* in Verh. K. Nederl. Akad. Wetensch., Afd. Natuurk., sect. 2, 48 : 21 (1952) ; Verdc. in Bull. Jard. Bot. Brux. 23 : 40, fig. 5A, etc. *Virecta multiflora* Sm. (1817)—F.T.A. 3 : 48 ; F.W.T.A., ed. 1, 2 : 130 ; Chev. Bot. 309 ! Berhaut Fl. Sén. 77. Erect herb, a few inches to 2–4 ft. high flowers white or pinkish ; in wet places and rock outcrops.
　　Mali: Fantiéla *Chev.* 745 ! Badi *Berhaut* 884 ; 4700. **Port.G.:** Bijimita (Feb.) *Esp. Santo* 1770 ! Gabu (Sept.) *Esp. Santo* 2785 ! Abreno (May) *Esp. Santo* 2006 ! S. Domingos (Aug.) *Esp. Santo* 3077 ! **Guin.:** Conakry (fr. Oct.) *Adam* 12610 ! Friguiagbé *Chillou* 656 ! Maneah *Pobéguin* 778 ! Macenta (Oct.) *Baldwin* 9831 ! **S.L.:** Sugar Loaf Mt., 2,400 ft. (Oct.) *T. S. Jones* 252 ! Rokupr (Oct.) *Jordan* 655 ! Njala (Sept.) *Deighton* 2126 ! *Pyne* 2 ! Musaia (Sept.) *Small* 226 ! **Lib.:** Monrovia (Nov.) *Dinklage* 3240 ! Gbanga (Oct.) *Daniel* ! Brewersville (Nov.) *Barker* 1086 ! Flumpa, Sanokwele Dist. (Sept.) *Baldwin* 9338 ! **Iv.C.:** Mt. Dou, Upper Sassandra *Chev.* 21488 ! Maniko to Tiégouakro (Aug.) *Chev.* 22319 ! Nimba Mts. (Aug.) *Boughey* GC 18081 ! Tonkoui (Dec.) *Roberty* 15857 ! **Ghana:** Kumasi (Dec.) *Adams* 4444 ! Atwabo *Fishlock* 63 ! Bou (fl. & fr. Nov.) *Vigne* FH 1462 ! Pepease (Nov.) *Irvine* 1710 ! **N.Nig.:** Patti Lokoja (Nov.) *Dalz.* 86 ! **S.Nig.:** Idanre Hills (Oct.) *Keay & Onochie* FHI 20242 ! Ondo (fl. & fr. Nov.) *Onochie* FHI 34232 ! Onitsha *Barter* 1792 ! Oban *Talbot* 282 ! Eket *Talbot* 3015 ! Also in Congo and Cabinda.
2. **V. major** (*K. Schum.*) *Verdc.* var. **major**—l.c. 42, fig. 5E, etc. (1953). *Virecta major* K. Schum. (1895). *V. kaessneri* S. Moore (1910). Tall herb with half-woody lower stems, up to 6 ft. high ; flowers with the corolla pink outside, paler or white inside ; in moist upland grassy places.
　　N.Nig.: Jos Plateau (June) *Lely* P385 ! Vom, 4,000 ft. *Dent Young* 117 ! Monguna, 4,700 ft. (June) *King* FHI 50765 ! **S.Nig.:** Obudu Plateau, 5,500 ft., Ogoja Prov. (Aug.) *Stone* 64 ! [**Br.]Cam.:** above Buea, 6,500 ft., Cam. Mt. (Nov.) *Migeod* 239 ! Bafut-Ngemba F.R. (Aug.) *Ujor* FHI 29964 ! Jakiri, 6,000 ft. (Feb.) *Hepper* 2067 ! Ndu, 6,500 ft. *Savory* UCI 394 ! Vogel Pk., 4,900 ft., Adamawa (Dec.) *Hepper* 1573 ! **F.Po:** Tams & Exell 758 ! Moka, 4,000 ft. (Dec., Jan.) *Guinea* 1923 ! *Boughey* 33 ! Also in Congo, Uganda, Tanganyika and N. and S. Rhodesia.
2a. **V. major** var. **spathulata** *Verdc.* l.c. 46, fig. 5F, etc. (1953).
　　S.L.: Jigaya (fl. & fr. Sept.) *Thomas* 2534 ! 2823 ! **Lib.:** Ganta (fl. & fr. May) *Harley* 225 ! Vonjama (Oct.) *Baldwin* 9929 ! **Iv.C.:** Tonkoui (Aug.) *Boughey* GC 13328 ! Also in Congo.
3. **V. procumbens** (*Sm.*) *Bremek.* l.c. 21 (1952) ; Verdc. l.c. 46, fig. 5B, etc. *Virecta procumbens* Sm. (1817)—F.T.A. 3 : 48 ; F.W.T.A., ed. 1, 2 : 130 ; Chev. Bot. 310. Straggling herb often rooting at nodes, up to 1 ft. high ; flowers white ; a weed and in forest.
　　Port.G.: Bafata (fl. & fr. June) *Esp. Santo* 3010 ! **Guin.:** Friguiagbé (Nov.) *Chillou* 544 ! 898 ! Mamou *Schnell* 6763 ! Tossekré, Fouta Djalon *Adam* 12749 ! **S.L.:** Freetown (Aug.) *Deighton* 2070 ! Bagroo R.

Mann 812! Rokupr (Sept.) *Jordan* 96! Njala (Oct.) *Deighton* 1405! Loma Mts. (Oct.) *Jaeger* 222!
Lib.: Dukwai R., Monrovia (Oct., Nov.) *Cooper* 6! 18! Kakatown *Whyte*! Sinkor (fl. & fr. June) *Barker*
1355! Gletown, Tchien Dist. (July) *Baldwin* 6747! **Iv.C.:** Tai to Tabou (Aug.) *Boughey* GC 14950!
Tonkoui, 3,500 ft. (Aug.) *Boughey* GC 18328! Malamalasso to Aboissué *Chev.* 17560! **Ghana:** Begoro,
Akim (fr. Apr.) *Irvine* 1193! Dawa Mate Kole (Dec.) *A. S. Thomas* D33! Kibbi (fl. & fr. Jan.) *Morton*
GC 6402! Dutukpene (Mar.) *Hepper & Morton* A3055! **S.Nig.:** Okomu F.R., Benin (fl. & fr. Dec.)
Brenan 8523! 8615! Oban (fl. & fr. Feb.) *Talbot* 253! *Onochie* FHI 36306! Stubbs Creek F.R., Eket
(fl. & fr. May) *Onochie* FHI 32091! **F.Po:** *Guinea* 1095! Also in Cameroon, Cabinda and Congo.

4. V. angustifolia (*Hiern*) *Bremek*. l.c. 21 (1952) ; Verdc. l.c. 47. *Virecta angustifolia* Hiern in F.T.A. 3 : 48
(1879). *V. heteromera* K. Schum. (1896)—F.W.T.A., ed. 1, 2 : 130. *Virectaria heteromera* (K. Schum.)
Bremek. l.c. (1952). Erect undershrub or half-woody herb nearly 1 ft. high ; flowers white ; on rocks in
rivers.
S.Nig.: Kwa R., Oban (fl. Aug., fr. Nov.) *Onyeagocha* FHI 7766! *Maggs* 155! **[Br.]Cam.:** Bibundi (fr.
Nov.) *Mildbr.* 10659! Korup F.R., Kumba (fl. & fr. June) *Olorunfemi* FHI 30642! Also in Cameroun
and Gabon.

[*V. angustifolia* var. *schlechteri* Verdc. l.c. 48 (1953) with oblong-lanceolate leaves, rounded at apex,
up to 2·9 cm. long and 5 mm. broad, was collected in the Cameroons " between Mafura and Mundame "
(*Schlechter* 12926), which may be in our area.]

74. KOHAUTIA Cham. & Schlecht.—Bremek. in Verh. K. Nederl. Akad. Wetensch., Afd., Natuurk., sect. 2, 48 : 56 (1952). *Oldenlandia* of F.T.A. 3 : 51 ; of F.W.T.A., ed. 1, 2 : 130, partly

Fruiting pedicels (2–)10–32 mm. long, slightly shorter in flower :
Corolla-lobes acutely acuminate, linear-subulate, up to 2 mm. broad, corolla-tube
about 10 mm. long, glabrous outside ; fruits globose, 2·5–3 mm. diam., smooth ;
leaves linear, 3–5 cm. long, 1–3 mm. broad, glabrous ; stems smooth
                                                                        1. *senegalensis*
Corolla-lobes obtuse, corolla-tube about 5 mm. long ; fruits more or less globose :
Corolla-lobes 2–2·5 mm. broad, corolla-tube finely tomentellous outside ; fruits 4–5
mm. diam., scabrid ; stems and leaves finely scabrid ; leaves linear, 1·5–3·5 cm.
long   2. *ubangensis*
Corolla-lobes up to 1 mm. broad, corolla-tube glabrous outside ; fruits 2–3 mm. diam. :
Stipules with only two lateral filiform segments adnate to the base of the leaf ;
leaves linear-subulate, 1–2 mm. broad ; stems glabrous ; flowers few in laxly
branched inflorescences ; fruiting pedicels 20–32 mm. long   3. *confusa*
Stipules with central filiform segments free from the leaf ; leaves linear to linear-
lanceolate, 2–5 mm. broad, ciliate ; stems shortly pubescent below, glabrous
above ; flowers numerous ; fruiting pedicels 2–17 mm. long ..   4. *virgata*
Fruiting pedicels all 1–3 mm. long (if also some flowers with longer pedicels in the
inflorescences see Nos. 3 and 4 above) :
Fruits ellipsoid, 5–6 mm. long, 3–4 mm. broad, scabrid, surmounted by 4 persistent
erect calyx-lobes 4–6 mm. long ; corolla-lobes 1–2 mm. broad, inflorescences attenuated ; stems shortly scabrid-pubescent ; leaves linear-lanceolate, 1–6·5 cm. long,
1–8 mm. broad..   ..   ..   5. *coccinea*
Fruits globose, 3 mm. diam., glabrous ; calyx-lobes 1 mm. long ; corolla-lobes 3–6 mm.
broad ; inflorescences subcorymbose, congested ; stems very finely scabrid ; leaves
linear, about 5 cm. long and 2 mm. broad ..   ..   ..   6. *grandiflora*

1. K. senegalensis *Cham. & Schlecht.* in Linnaea 4 : 156 (1829) ; Bremek. l.c. 92. *Oldenlandia senegalensis*
(Cham. & Schlecht.) Hiern in F.T.A. 3 : 56 (1877) ; F.W.T.A., ed. 1, 2: 131 ; Chev. Bot. 312 ; Berhaut
Fl. Sén. 77. *O. confusa* Hutch. & Dalz. F.W.T.A., ed. 1, 2 : 131 (1931), partly (*Schlechter* 12978). Erect,
slender branched annual, 1–3 ft. high ; flowers dull brownish, white or pink ; beside roads and in cultivated
ground.
Sen.: *Roger*! Sédhiou (fl. & fr. Feb.) *Chev.* 2113a! Niayes *Chev.* 2011! Walo *Perrottet* 382! **Gam.:**
Brown-Lester N3! *Brooks* 7! 105! 106! **Mali:** Timbuktu (Aug.) *Hagerup* 256! Dioura
(Sept.) *Davey* 112! Koulikoro *Chev.* 2113b! **Port.G.:** Pirada, Gabú (fl. & fr. Nov.) *Esp. Santo* 3150!
Canguelifá, Gabú (fl. & fr. Oct.) *Esp. Santo* 2958! **Guin.:** Siguiri *Pobéguin* 1146. Faranah *Chev.* 20552!
S.L.: Talla Hills, 4,000 ft. (fl. & fr. Mar.) *Sc. Elliot* 5087b! Falaba, 4,000 ft. *Sc. Elliot* 5168! **Iv.C.:** Baoulé
Pobéguin 129! **Ghana:** Achimota (fl. & fr. May) *Irvine* 1618! Legon Hill (Oct.) *Adams* 3401! Krobo
Plains (Dec.) *Johnson* 491! Kintampo (fl. & fr. Mar.) *Dalz.* 31! Navrongo (fl. & fr. Oct.) *Vigne* FH 4593!
Togo Rep.: Badja (fr. Mar.) *Schlechter* 12978! **Dah.:** Zagnanado *Chev.* 23062! **Niger :** Niamey *Hagerup*
523! **N.Nig.:** Nupe *Barter* 956! Minna (fl. & fr. Dec.) *Meikle* 755! Pankshin *MacGregor* 427! Sokoto
(fl. & fr. Nov.) *Moiser* 10! **S.Nig.:** Oyo to Ogbomosho (Oct.) *Onochie* FHI 35297! Oyo to Ileshin (Feb.)
Brenan 8965! **[Br.]Cam.:** Su, 3,500 ft., Bamenda (fl. & fr. May) *Maitland* 1698! Gurum, Adamawa
(Nov.) *Hepper* 1316! Bama, Dikwa Div. (fl. & fr. Dec.) *McClintock* 92! Also in Cameroun, Ubangi-Shari,
Sudan, Ethiopia, Eritrea, Arabia and Cape Verde Islands. (See Appendix, p. 406.)

2. K. ubangensis *Bremek*. l.c. 90 (1952). Slender erect herb 1–2 ft. high, with grey scabrid stems ; flowers
bright scarlet or light mauve ; in cultivated ground.
N.Nig.: Kogon Gandu, 3,200 ft., Bauchi Prov. (fl. & fr. Dec.) *G. V. Summerhayes* 48! **[Br.]Cam.:** Vogel
Peak, 1,600 ft., Adamawa (Nov.) *Hepper* 1410! Gangumi, Gashaka Dist. (fl. & fr. Dec.) *Latilo & Daramola*
FHI 28824! Also in Ubangi-Shari.

3. K. confusa (*Hutch. & Dalz.*) *Bremek*. l.c. 89 (1952). *Oldenlandia confusa* Hutch. & Dalz. F.W.T.A., ed. 1, 2 :
131 (1931), partly ; Berhaut Fl. Sén. 77. *O. effusa* of F.T.A. 3 : 59, partly (*Heudelot* 220), not of Oliv.
Erect slender-branched herb 1–2 ft. high ; flowers dull red or whitish.
Sen.: Galam *Heudelot* 220! Walo *Leprieur*! Malème to Hoddar (fr. Nov.) *Berhaut* 1817! **Mali:** Nyammo
to Koulikoro *Chev.* 2024! **Guin.:** Kouroussa (fl. & fr. Sept.) *Pobéguin* 436! Port. Guinea frontier *Maclaud*!
Also in Tanganyika (?) and Nyasaland.

4. K. virgata (*Willd.*) *Bremek*. l.c. 77 (1952). *Hedyotis virgata* Willd. (1798). *Oldenlandia virgata* (Willd.) DC.
(1830)—F.T.A. 3 : 59 ; F.W.T.A., ed. 1, 2 : 132 ; Chev. Bot. 313, partly (*Chev.* 4434). Herb a few inches
to 1 ft. high, usually with numerous erect branches ; flowers pink, sometimes white, red or mauve.
Ghana: Achimota *Irvine* 660! Accra (fl. & fr. Nov.) *Morton* GC 6114! Shai Plains (fl. & fr. Mar.) *Irvine*
947! Labadi (Jan.) *Adams* 372! **Togo Rep.:** Lomé *Warnecke* 174! Talerni *Mahout* 529. **Dah.:** Cotonou
Chev. 4434. **S.Nig.:** Lagos Isl. *Dalz.* 1048! Attah *T. Vogel* Widespread in tropical and S. Africa.

5. **K. coccinea** *Royle* Ill. Himal. 241, t. 53, fig. 1 (1839) ; Bremek. l.c. 82. *Oldenlandia abyssinica* (Hochst. ex A. Rich.) Hiern in F.T.A. 3 : 57 (1877) ; F.W.T.A., ed. 1, 2 : 131. Slender erect herb a few inches to 1 ft. high, with few branches ; flowers few in lax inflorescences, bright scarlet, sometimes pink, mauve or white ; in cultivated ground.
Sen.: M'Bidgem *Hb. Franqueville* ! **N.Nig.:** Jos Plateau (fl. & fr. Sept.) *Lely* P730 ! Vom *Dent Young* 116 ! Pankshin (Oct.) *Hepper* 1119 ! [Br.]**Cam.:** Banso (fl. & fr. Oct.) *Tamajong* FHI 23456 ! Widespread in tropical Africa ; also in India.

6. **K. grandiflora** *DC.* Prod. 4 : 430 (1830) ; Bremek. l.c. 88. *Oldenlandia grandiflora* (DC.) Hiern in F.T.A. 3 : 57 (1877) ; F.W.T.A., ed. 1, 2 : 131 ; Chev. Bot. 311 ; Berhaut Fl. Sén. 77. Slender erect herb 1 ft. or more high ; flowers scarlet ; in savanna and as a weed of cultivation.
Sen.: *Roger* ! *Farmar* 133 ! Dagana *Leprieur* ! Dakar (May) *Baldwin* 5710 ! **Gam.:** *Brown-Lester* 64 ! *Dawe* 5 ! **Port.G.:** Buba (Oct.) *Esp. Santo* 1273 ! Apilho (fl. & fr. Dec.) *Esp. Santo* 3726 ! **Mali:** Bamako (fl. & fr. Jan.) *Chev.* 202 ! between Senegal and Niger *Bellamy* 445 ; 446 ! Ouacoro *Chev.* 81 ! Tivaouane *Chev.* 2095 ! **S.L.:** *Purdie* ! **Ghana:** Zuarungu (Nov.) *Darko* 446 ! Tamale *Williams* 494 ! Navrongo (Oct.) *Vigne* FH 4594 ! **Togo Rep.:** Sokode *Kersting* A685 ! **Dah.:** Savé *Annet* 23. **Niger:** Niamey (Oct.) *Hagerup* 501 ! **N.Nig.:** Aguji, Ilorin *Thornton* ! Nupe *Barter* 856 ! Anchau, Zaria (Nov.) *Keay* FHI 21666 ! Sokoto *Ryan* 27 ! Biu, Bornu (Sept.) *Noble* 32 ! **S.Nig.:** Lagos *Rowland* ! Sapoba *Kennedy* 2873 ! Shaki, Oyo (fl. & fr. Jan.) *Hambler* 113 ! Oban *Talbot* 257 ! [Br.]**Cam.:** Buea *Mildbr.* 9618. Garbabi to Balli, Gashaka Dist. (fl. & fr. Jan.) *Latilo & Daramola* FHI 34421 ! Also in Ubangi-Shari, Sudan and Uganda. (See Appendix, p. 405.)

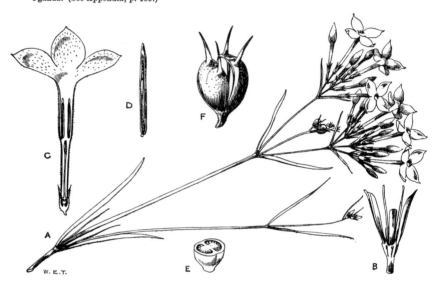

FIG. 244.—KOHAUTIA GRANDIFLORA *DC.* (RUBIACEAE).
A, flowering shoot.   B, stipules.   C, longitudinal section of flower.   D, anther.   E, cross-section of ovary.   F, fruit.

75. OLDENLANDIA Linn.—F.T.A. 3 : 51, partly ; Bremek. in Verh. K. Nederl. Akad. Wetensch., Afd. Natuurk., sect. 2, 48 : 20 (1952).

Flowers congested in terminal and axillary heads :
Leaves elliptic or ovate-elliptic, 3–5-nerved at base, 1·5–2·5 cm. long, 5–13 mm. broad, minutely setulose along midrib above ; ovary and calyx-lobes pubescent ; prostrate trailing herb   ..   ..   1. *goreensis*
Leaves linear ; usually erect, not trailing :
Leaves thin, glabrous or slightly scabrid ; stipular sheath shortly fimbriate ; flowers shortly pedicellate :
Calyx-lobes 4 ..   ..   2. *capensis* var. *capensis*
Calyx-lobes more than 4 and up to 8   ..   ..   2a. *capensis* var. *pleiosepala*
Leaves thickened, midrib conspicuous beneath, scabrid with short aculeate hairs ; stipular sheath long-fimbriate   ..   ..   3. *sclerophylla*
Flowers not congested, often long-pedicellate, solitary in axils or in terminal and axillary panicles :
Leaves ovate ; stems square :
Calyx glabrous ; leaves 1·5–3 cm. long, 7–15 mm. broad, with 4–5 pairs of nerves, glabrous ; petiole 2–7 mm. long ; inflorescences terminal, lax ..   4. *chevalieri*
Calyx hispid ; leaves 2–3 cm. long, 7–17 mm. broad, with much smaller bracteate leaves, setulose above, subglabrous beneath ; petiole 2–5 mm. long ; flowers axillary and terminal, a few together, shortly pedicellate   ..   5. *echinulosa*
Leaves linear to lanceolate :
Pedicels and/or peduncle about 2 mm. long :

Leaves thin (for other characters see above)   ..   2. *capensis* var. *capensis*
Leaves thickened (for other characters see above)..   ..   ..   3. *sclerophylla*
Pedicels much longer than 2 mm. :
   Flowers with distinct peduncle, pedicels forming 2(–4)-flowered umbels :
     Leaves narrowly linear or filiform, less than 1 mm. broad, 1–4 cm. long ; seed
      testa with straight-walled cells ..   ..   6. *linearis*
     Leaves linear or linear-lanceolate, 2–7 mm. broad ; seed testa with wavy-walled
      cells :
       Style glabrous ; common herb ; corolla about as long as calyx   7. *corymbosa*
       Style hairy ; uncommon herb   ..   ..   12a. *caespitosa* var. *lanceolata*
   Flowers solitary or paniculate :
     Capsule 3–3·5 mm. diam., externally with 2 obvious loculi apparent :
      Leaves narrowly lanceolate, 2–6 cm. long, 2–7 mm. broad ; stems square, nearly
       glabrous, more or less prostrate with long branches ; flowers axillary ; pedicels
       1–1·5 cm. long   ..   ..   8. *lancifolia*
      Leaves narrowly linear, about 5 cm. long, 1–2 mm. broad ; stems subterete
       scabrid, erect, profusely branched ; flowers axillary or more or less paniculate ;
       pedicels 1·5–3 cm. long   ..   ..   10. *rhabdina*
     Capsule 1–2(–2·5) mm. diam., spherical :
      Flowers in mainly terminal panicles ; corolla-tube about 3 mm. long, lobes 2 mm.
       long ; leaves lanceolate, 2–4·5 cm. long, 6–16 mm. broad, setulose   9. *affinis*
      Flowers axillary, solitary or paired without a peduncle ; corolla 1–2 mm. long :
       Leaves narrowly linear, 1·5–4 cm. long, 1–2 mm. broad ; diffusely branched erect
       herb, common ; seed testa cell-walls straight   ..   ..   11. *herbacea*
       Leaves oblong, 5–10 mm. long, 1–2 mm. broad ; small decumbent or erect herb,
       rare ; seed testa cell-walls wavy   ..   ..   12. *caespitosa* var. *caespitosa*

Oldenlandia callitrichoides Griseb., a West Indian species with slender prostrate stems and orbicular leaves about 2 mm. diam. has been recorded from Njala, Sierra Leone. As it does not belong to *Oldenlandia* in the strict sense (see Bremek. l.c. 269) it has recently been transferred to *Hedyotis* (*H. callitrichoides* (Griseb.) W. H. Lewis in Rhodora 63 : 222 (1961)).

1. **O. goreensis** (*DC.*) *Summerh.* in Kew Bull. 1928 : 392 ; Bremek. l.c. 48 : 192 ; Berhaut Fl. Sén. 78. *Hedyotis goreensis* DC. (1830). *Oldenlandia trinervia* of F.T.A. 3 : 63 ; of Chev. Bot. 313 ; not of Retz. Prostrate herb with slender stems up to 18 in. long, with bright green leaves drying yellowish ; flowers white with mauve corolla-lobes, or pink ; in moist, muddy places.
   Sen.: *Berhaut* 1049 ! 4737. Kounoum, Cape Verde *Perrottet* 484 ! Dakar *Boivin* 405 ! **Gam.:** Albreda *Leprieur* ! Kombo country *Heudelot* 72. **Mali:** Dendéla *Chev.* 633 ! Sikasso *Chev.* 789. **Port.G.:** Antula, Bissau (Jan.) *Esp. Santo* 1451 ! **Guin.:** Friguiagbé (Mar.) *Chillou* 79 ! Manèa (Feb.) *Chillou* 2527 ! **S.L.:** Pendembu (July) *Thomas* 833 ! Rokupr (Jan.) *Deighton* 2952 ! Hepper 2630 ! Konta (Aug.) *Deighton* 1254 ! Kangahun (May) *Deighton* 4137 ! Kabala (Feb.) *Deighton* 4195 ! **Iv.C.:** Manikro to Tiégouakro *Chev.* 22317 ! **Ghana:** Damongo (Mar.) *Morton* GC 8718 ! **Togo Rep.:** Lomé *Warnecke* 239 ! **N.Nig.:** Nupe *Barter* 1241 ! Jebba (Dec.) *Meikle* 846 ! Kontagora *Dalz.* 215 ! Vom, 4,000 ft., Jos Plateau *Dent Young* 114 ! **[Br.]Cam.:** Wum, 3,500 ft. (Apr.) *Maitland* 1591 ! Jakiri, 6,000 ft. (Feb.) *Hepper* 2063 ! Ngel Nyaki, 5,500 ft., Mambila Plateau (Jan.) *Hepper* 1714 ! Widespread in tropical Africa also in Madagascar, Comores and Mascarene Islands. (See Appendix, p. 405.)
   [*Warnecke* 239, cited above, is the var. *trichocaula* Bremek. l.c. 198 (1952) which is only distinguishable from var. *goreensis* by its hairy stems, leaves and corolla-lobes.]

2. **O. capensis** *Linn. f.* var. **capensis**—Suppl. Pl. 127 (1781) ; F.T.A. 3 : 62 ; Bremek. l.c. 265 ; Berhaut Fl. Sén. 77. *O. sabulosa* DC. (1830)—Chev. Bot. 312. Annual herb 3–4 to 1 ft. high ; flowers white ; on river banks and in wet places.
   Sen.: *Roger* ! Senegal R. *Perrottet* 388 ! Walo *Perrottet* 389. Foula Toro *Leprieur* ! Niokolo-Koba *Berhaut* 1421 ! **Mali:** San *Chev.* 1067 ! Djenné *Chev.* 1115 ! Gorinuta, near Dogo (Apr.) *Davey* 611 ! **Dah.:** Hétinn to Aguégué *Chev.* 23309. Savé to Agouagon *Chev.* 23596. **N.Nig.:** Nupe *Barter* ! Mongu, 4,300 ft. (July) *Lely* 440 ! Taura (May) *Lely* 121 ! Daura (Apr.) *Meikle* 1360 ! Sokoto (Mar.) *Moiser* ! **[Br.]Cam.:** Bamenda *Maitland* ! Widespread in drier regions of Africa, also in S. Europe and Persia.

2a. **O. capensis** var. **pleiosepala** Bremek. l.c. 267 (1952).
   Port.G.: Bissau (fr. Apr.) *Esp. Santo* 1538 ! **Mali:** San *Chev.* 1079 ! L. Tengrela, Upper Volta *Aké Assi* ! **S.L.:** Port Loko (Apr.) *Sc. Elliot* 5876 ! Falaba (Mar.) *Sc. Elliot* 5169 ! Bumpe (Apr.) *Deighton* 5513 ! **Ghana:** White Volta R. (Mar.) *Hepper & Morton* A3156 ! Yendi (Mar.) *Hepper* 2372 ! **Dah.:** Atacora Mts. *Chev.* 24282 ! **N.Nig.:** Jebba Isl. (Mar.) *Meikle* 1290 ! Mande, Zaria Prov. (June) *Keay* FHI 25869 ! Also in Algeria, Egypt, Sudan, Ubangi-Shari, Tanganyika, Nyasaland, N. and S. Rhodesia, Angola and Congo.

3. **O. sclerophylla** *Bremek.* l.c. 268 (1952). Small, branched herb.
   Mali: *Collin* 167 ! Timbuktu *Hagerup* 163a ! Gao to Berra (Mar.) *de Wailly* 5008 !

4. **O. chevalieri** Bremek. l.c. 216 (1952). Straggling herb with stems 1–1½ ft. high ; flowers white ; in forest.
   Lib.: Jaurazon, Sinoe (Mar.) *Baldwin* 11453 ! **Iv.C.:** Alépé *Chev.* 17470 ! Alépé to Potou Lagoon *Chev.* 17383 ! Bianco, Abidjan (Dec.) *Boughey* GC 13516 ! **Ghana:** Fomena (Apr.) *Adams* 2502 ! Adansi Scarp (fl. & fr. Oct.) *Hall* 1606 ! Mt. Ejuanema, Kwahu (Dec.) *Adams* 5133 !

5. **O. echinulosa** *K. Schum.* in Engl. Pflanzenw. Ost-Afr. C : 375 (1895) ; Bremek. l.c. 213. Erect herb 3–18 in. high ; flowers white ; amongst rocks.
   S.L.: Loma Mts. (fr. Nov.) *Jaeger* 681 ! **N.Nig.:** Jos Plateau (Sept.) *Lely* P 737 ! Jos (Aug.) *Keay* FHI 12712 ! **[Br.] Cam.:** Vogel Peak *Hepper* 1342 ! Also in Congo, Angola, Rhodesia, Nyasaland and Tanganyika.

6. **O. linearis** *DC.* Prod. 4 : 425 (1830) ; Bremek. l.c. 258. *O. heynei* of F.T.A. 3 : 59, partly (syn. ? *O. linearis* DC.). A slender erect herb usually a few inches high and often simple ; flowers white or pale mauve.
   Sen.: St. Louis *Perrottet* ! Also in Sudan, Eritrea, Uganda, Kenya, Tanganyika, N. and S. Rhodesia and Congo.

7. **O. corymbosa** *Linn.* Sp. Pl. 119 (1753) ; F.T.A. 3 : 62 ; Chev. Bot. 311 ; Bremek. l.c. 254, incl. var. *microcarpa* Bremek. (1952) ; Berhaut Fl. Sén. 77. An erect and diffusely branched herb, nearly glabrous, up to 1 ft. or more high ; flowers white or mauve ; a weed of cultivation and beside paths.
   Sen.: *Berhaut* 451. Bondou *Heudelot* 136 ! Dakar *Thiebaut* 188 ! Goree *Talmy* 46 ! **Gam.:** *Saunders* 114 ! **Mali:** Oualia *Chev.* 32 ! Bafaga *Chev.* 702 ! **Guin.:** Kouria *Caille* in *Chev.* 15030 ! Conakry *Maclaud* 14. **S.L.:** Freetown (fl. & fr. Sept.) *Hepper* 925 ! Tisana, Bonthe Isl. (fl. & fr. Nov.) *Deighton* 2325 ! Njala (fl. & fr. Oct.) *Deighton* 1354 ! Rokupr (fl. & fr. Nov.) *Jordan* 684 ! **Lib.:** Zolopla, Sanokwele Dist. (fl. & fr. Sept.) *Baldwin* 9395 ! Fishtown *Dinklage* 2079. **Iv.C.:** Adiopodoumé *Leeuwenberg*

2337! **Ghana:** Achimota (fl. & fr. Oct.) *Morton* GC 6007! Krobo Plains (fl. & fr. Dec.) *Johnson* 492! Legon Hill (fl. & fr. May) *Adams* 4784! Atwabo *Fishlock* 7! **Dah.:** Cotonou *Debeaux* 154! **Niger:** Niamey (fl. & fr. Oct.) *Hagerup* 505! **N.Nig.:** Ilorin (fl. & fr. Oct.) *Adejumo* FHI 5746! Anara F.R., Zaria (fl. & fr. Sept.) *Olorunfemi* FHI 24377! Vom, 3,000–4,500 ft. *Dent Young* 111! Maiduguri (fl. & fr. Oct.) *Noble* A5! **S.Nig.:** Ibadan (fl. & fr. Jan.) *Meikle* 1001! Stirling (fl. & fr. Sept.) *T. Vogel* 133! Enugu (fl. & fr. Oct.) *Jones* FHI 6761! Onyadama, Ogoja Prov. (fl. & fr. Apr.) *Jones* FHI 7398! **[Br.]Cam.:** Victoria *Winkler* 42a. Gurum, Adamawa (fl. & fr. Nov.) *Hepper* 1234! **F.Po:** *T. Vogel* 201! *Milne*! Widespread in tropical Africa from Sudan to Transvaal and in tropical and subtropical countries throughout the world. (See Appendix, p. 405.)

[*O. praetermissa* Bremek. l.c. 253 (1952) has the appearance of *O. corymbosa* Linn. and is said to differ from it in the hairy style and exserted anthers and stigmata. Heterostyly is frequent in Rubiaceae and I doubt whether it is more than a form of *O. corymbosa*. Several specimens are cited under both names by Bremekamp, e.g. *Fishlock* 7.—F.N.H.]

8. **O. lancifolia** (*Schumach.*) *DC.* Prod. 4 : 425 (1830) ; F.T.A. 3 : 61 ; Chev. Bot. 312, partly (excl. *Chev.* 2116) ; Bremek. l.c. 230 (1952), incl. var. *grandiflora* Bremek., var. *microcarpa* Bremek., var. *scabridula* Bremek., var. *longipes* Bremek ; Berhaut Fl. Sén. 78. *Hedyotis lancifolia* Schumach. (1827). Prostrate or straggling, branched near the base and branches almost simple, 1–2 ft. long ; flowers white, sometimes pale pink or mauve ; in moist places.
    Sen.: Cape Verde *Maille.* Casamance *Perrottet*! Tamboukané *Chev.* 2104. Badi *Berhaut* 777. **Port.G.:** Bafata (fl. & fr. Oct.) *Esp. Santo* 3167! **Mali:** Koulikoro *Chev.* 2022! **S.L.:** Kundita, Talla *Sc. Elliot* 5045! Kailahun (fl. & fr. Feb.) *Deighton* 4008! Njala (fl. & fr. Oct.) *Deighton* 752! Yonibana (fl. & fr. Nov.) *Thomas* 4700! **Lib.:** Du R. (fl. & fr. Oct., Nov.) *Cooper* 4! *Linder* 216! Gbau (fl. & fr. Sept.) *Baldwin* 9419! Gbanga (fl. & fr. Aug.) *Traub* 244! Dubo, Sanokwele (fl. & fr. Sept.) *Baldwin* 9467! **Iv.C.:** Alangouassou to Mbayakro *Chev.* 22232! Dabou (fl. & fr. Oct.) *Howes* 986! Assuantsi (fl. & fr. Mar.) *Irvine* 1573! New Tafo (fl. & fr. Sept.) *Lovi* WACRI 3912! Bimbila (fl. & fr. Mar.) *Hepper & Morton* A3099! Togo **Rep.:** *Mahoux* 516. **N.Nig.:** Nupe *Barter* 1659! 1714! Vom, 4,000 ft. *Dent Young* 112! **S. Nig.:** Okomu F.R., Benin (fl. & fr. Dec.) *Brenan* 8627! Idanre, Akure (fl. & fr. Jan.) *Brenan & Keay* 8660! Ibadan (fr. Mar.) *Schlechter* 12344! Port Harcourt (fl. & fr. Sept.) *Taylor* 10! *Talbot* 226! **[Br.]Cam.:** Buea, 2,500 ft. (fl. & fr. Mar.) *Maitland* 580! Mamfe (fl. & fr. Dec.) *Migeod* 268! Wum, 3,500 ft. (fl. & fr. Apr., July) *Maitland* 1592! *Ujor* FHI 29293! Jakiri, 6,000 ft. (fl. & fr. Feb.) *Hepper* 2065! Nguroje, 5,000 ft., Mambila Plateau (fl. & fr. Jan.) *Hepper* 1739! **F.Po:** Moka, 4,000–4,500 ft. (fl. & fr. Jan.) *Exell* 822a! Widespread in tropical and S. Africa, introduced into S. America. (See Appendix, p. 405.)

9. **O. affinis** (*Roem. & Schult.*) *DC.* Prod. 4 : 428 (1830) ; Bremek. l.c. 226. *Hedyotis affinis* Roem. & Schult. (1818). *Oldenlandia decumbens* (Hochst.) Hiern in F.T.A. 3 : 54 (1877) ; F.W.T.A., ed. 1, 2 : 132. A glabrous decumbent or straggling herb, stems about 2 ft. long, branched ; flowers deep blue ; amongst grass.
    Lib.: Nana Kru (fl. & fr. Mar.) *Baldwin* 11583! **Iv.C.:** Dabou *Chev.* 17157! Abidjan *Scaëtta* 3008! Bingerville *Jolly.* Adiopodoumé (fl. & fr. Oct.) *Roberty* 12290! **Ghana:** Princes Town (Feb.) *Morton* GC 8454! Cape Coast (fr. Oct.) *Hall* 1138! Anaji, W. Prov. (Apr.) *Lloyd* 30! Atwabo *Fishlock* 66! 77! **Togo Rep.:** Lomé *Warnecke* 261! Noépe *Mahout* 513. **Dah.:** Agoué *Ménager.* **S.Nig.:** Lagos *Barter* 2181! *Dalz.* 1049! Sobo Plains, Benin (fl. & fr. May) *Keay* FHI 37030! Onitsha (fl. & fr. Apr.) *Killick* 109! Calabar (fl. & fr. Mar., May) *Brenan* 9224! Onochie FHI 32938! Also in Cameroun, Gabon, Congo, Angola, Uganda, Kenya, Tanganyika, Mozambique, S. Rhodesia and Transvaal and eastwards to Malay Peninsula.

10. **O. rhabdina** *Bremek.* in Kew Bull. 14 : 317 (1960). An erect herb about 1 ft. high ; in cultivated ground.
    N.Nig.: Bangwele, Lame Dist. (fl. & fr. Sept.) *G. V. Summerhayes* 44!

11. **O. herbacea** (*Linn.*) *Roxb.* Hort. Bengal 11 (1814) ; Bremek. l.c. 244, incl. var. *flaccida* Bremek. l.c. 248 ; Berhaut Fl. Sén. 78. *Hedyotis herbacea* Linn. (1753). *Oldenlandia heynei* G. Don (1834)—F.T.A. 3 : 59 ; Chev. Bot. 311 (incl. var. *djalonis* A. Chev., name only). *O. virgata* of Chev. Bot. 313, partly (*Chev.* 18561). Erect, diffusely branched weed a few inches to 1 ft. high.
    Sen.: *Berhaut* 606. Kounoum *Perrottet*! Bakel *Collins* 163! Casamance *Trochain* 1430! **Gam.:** Pirie 38! Kombo *Dawe* 4! **Mali:** Bamako *Waterlot* 1134! Koulikoro *Chev.* 2116! **Port.G.:** Bissau (fl. & fr. Oct.) *Esp. Santo* 2590! **Guin.:** Dalaba-Diaguissa Plateau (fl. & fr. Oct.) *Chev.* 18708! Ditinn to Dalaba *Chev.* 18561! Kouria *Chev.* 14990! Dyeke (fl. & fr. Oct.) *Baldwin* 9663! Nzérékoré (fl. & fr. Oct.) *Baldwin* 9707! **S.L.:** Freetown (Oct.) *Hepper* 924! Rokupr (fl. & fr. Mar.) *Jordan* 656! Njala (fl. & fr. Nov.) *Deighton* 1497! Kenema *Thomas* 7918! Musaia (Dec.) *Deighton* 4518! **Lib.:** Monrovia (fl. & fr. Nov.) *Dinklage* 3276! *Linder* 1557! Tappita (fl. & fr. Aug.) *Baldwin* 9077! Robertsport (fl. & fr. Dec.) *Baldwin* 10936! Sinkor (fl. & fr. June, Sept.) *Barker* 1065! 1358! **Iv.C.:** Manikro to Tiégouakro *Chev.* 22322! Nimba, 1,000 ft. (fl. & fr. Aug.) *Boughey*! **Ghana:** Accra *Don*! Kete Krachi to Dutukpene (fl. & fr. Dec.) *Adams* 4655! Nsawkaw (fl. & fr. Dec.) *Adams* 3145 4655! Navrongo (fl. & fr. Nov.) *Darko* 451! **N.Nig.:** Jebba (fl. & fr. Feb.) *Meikle* 1190! Bida (fl. & fr. Mar.) *Meikle* 1322! Sokoto *Lely* 123! Naraguta (fl. & fr. Aug.) *Lely* 501! Yola (fl. & fr. Jan.) *Dalz.* 46! **S.Nig.:** Lagos (fl. & fr. May) *Dalz.* 1374! Stirling (Sept.) *T. Vogel* 135! Nanka, Awka Dist. (Nov.) *Keay* FHI 22706! Achalla F.R., Onitsha (Sept.) *Onochie* FHI 34057! Eket *Talbot*! **[Br.]Cam.:** Bali, 4,000 ft. (May) *Ujor* FHI 30368! *Migeod* 287! Belo, 4,500 ft. (fr. Apr.) *Maitland* 1707! Gembu, 6,000 ft., Mambila Plateau (fl. & fr. Jan.) *Hepper* 1807! near Vogel Peak, Adamawa (fl. & fr. Nov.) *Hepper* 1377! Widespread in tropical and S. Africa, also in Cape Verde Islands, Socotra and Egypt and in tropical Asia. (See Appendix, p. 405.)

12. **O. caespitosa** (*Benth.*) *Hiern* var. **caespitosa**—in F.T.A. 3 : 61 (1877) ; Bremek. l.c. 262. *O. herbacea* (?) var. *caespitosa* Benth. (1849). *O. parva* Trochain in Bull. Mus. Hist. Nat., sér. 2, 4 : 604, fig. 1 (1932). A small herb.
    Sen.: Cape Verde *Perrottet* 387. Ziguinchor *Trochain* 1536! **Lib.:** Cape Palmas *T. Vogel* 51! **S.Nig.:** Ikom (fl. & fr. Jan.) *Rosevear* 10/31! Eket Dist. *Talbot*! Also in Cameroun, Congo, Angola, N. and S. Rhodesia, Mozambique, Kenya, Uganda and Tanganyika.

12a. **O. caespitosa** var. **lanceolata** *Bremek.* l.c. 264 (1952). A straggling herb having the appearance of *O. corymbosa*. **S.Nig.:** Aguku Dist. *Thomas* 1003! Also in Cameroun and N. Rhodesia.

76. LELYA Bremek. in Verh. K. Nederl. Akad. Wetensch., Afd. Natuurk., sect. 2, 48 : 181 (1952).

Perennial herb with decumbent branches about 10 cm. long, shortly pubescent ; leaves elliptic-lanceolate or lanceolate, with 2–3 lateral nerves on each side, acute at each end, 5–14 mm. long, 3·5–5 mm. broad, glabrous above, pubescent beneath ; flowers in terminal clusters of 2 or 3, or solitary in axils, subsessile ; calyx-tube 0·4 mm. long, lobes 4, oblong, about 2 mm. long ; corolla-tube about 2 mm. long, lobes 4, 3–5 mm. long ; fruit with a thick woody wall and solid conical beak ..   *osteocarpa*

L. osteocarpa *Bremek.* l.c. 181 (1952). A small herb decumbent and forming a small mat ; flowers white ; on bare soil, appearing after fires.
N.Nig.: Jos Plateau (Jan.) *Lely* P96! P663! Naraguta, 4,000 ft. (May) *Lely* 249! Heipang (fl. & fr. Apr.) *Keay & Wimbush* FHI 37612! Also in Congo, N. Rhodesia and Tanganyika.

77. PENTODON Hochst.—Bremek. in Verh. K. Nederl. Akad. Wetensch., Afd. Natuurk., sect. 2, 48 : 175 (1952).

Decumbent herb, somewhat fleshy ; leaves sessile, linear-lanceolate to lanceolate, rounded to cuneate at base, subacute at apex, 3–6 cm. long, 1–2 cm. broad, glabrous ; panicles axillary, 6–17 cm. long ; pedicels about 1 cm. long ; calyx with very short fine teeth ; corolla 2–3 mm. long .. .. *pentandrus*

P. pentandrus (*Schum. & Thonn.*) *Vatke* in Oest. Bot. Zeit. 25 : 231 (1875) ; Bremek. l.c. 176 (as *P. pentander*). *Hedyotis pentandra* Schum. & Thonn. (1827). *Oldenlandia macrophylla* DC. (1830)—F.T.A. 3 : 63 ; Chev. Bot. 312 ; F.W.T.A., ed. 1, 2 : 132 ; Berhaut Fl. Sén. 77. A straggling half-succulent herb, stems often tinged purplish ; flowers white, pinkish or blue ; in wet, muddy places.
Sen.: Leprieur. *Berhaut* 837. **Gam.:** Kuntaur *Ruxton* 106 ! 124 ! **Port.G.:** Pussubé (June) *Esp. Santo* 1181 ! Bafatá (Aug.) *Esp. Santo* 3280 ! **Guin.:** Nzérékoré (Oct.) *Baldwin* 9727 ! R. Kolenté *Chillou* 1390 ! **S.L.:** Rokupr (Mar.) *Jordan* 18 ! Bumbuna (Oct.) *Thomas* 3228 ! Gbinti (Sept.) *Deighton* 4850 ! Pelewahun (May) *Deighton* 5943 ! **Lib.:** Gola *Bunting* ! Bushrod Isl. (Apr.) *Barker* 1288 ! Nyaake (June) *Baldwin* 6198 ! Du R. (Aug.) *Linder* 262 ! **Iv.C.:** Yapo (Oct.) *Chev.* B22365 ! Grabo to Taté *Chev.* 19753. Toumodi to Dimbroko *Chev.* 22251. **Ghana:** Ancobra Junction (Jan.) *Irvine* 1052 ! Pra R. *Irvine* 435 ! Owrabi (Sept.) *Andoh* FH 4552 ! Kpandu *Robertson* 36 ! **Togo Rep.:** Lomé *Warnecke* 240 ! **Dah.:** Hétinn to Aguégué *Chev.* 23309. Savé to Agouagon *Chev.* 23596. **N.Nig.:** Sokoto (Mar., June) *Moiser* 86 ! *Dalz.* 401 ! **S.Nig.:** Angiama *Barter* 91 ! Ogun F.R., Lagos (Mar.) *Hepper* 2248 ! Okeluse F.R. (July) *Onochie* FHI 33368 ! Ibadan (Nov.) *Meikle* 670 ! **[Br.]Cam.:** Yoke R. (Mar.) *Brenan* 9326 ! L. Barombi (Apr.) *Daramola* FHI 41020 ! Widespread in tropical Africa, also in Cape Verde Islands, S. Tomé and Arabia.

78. SACOSPERMA G. Tayl. in Exell, Cat. S. Tomé 218 (1944) ; Bremek. in Verh. K. Nederl. Akad. Wetensch., Afd. Natuurk., sect. 2, 48 : 43 (1952).

Stipules broadly triangular or ovate, ending in a simple or occasionally bifid awn ; inflorescence lax, branched, flowers sessile ; calyx-lobes equal, ovate, about 1 mm. long ; corolla-tube 5–9 mm. long, lobes 1–2 mm. long, glabrous ; leaves ovate to elliptic, acuminate at apex, cuneate at base, 6–10 cm. long, 2–4 cm. broad, subglabrous to pubescent .. .. 1. *paniculatum*
Stipules deeply divided into 2 or 3 filiform awns ; inflorescences contracted, with few short branches, flowers sessile ; calyx-lobes linear-oblong, 3 mm. long ; corolla-tube 5–6 mm. long, lobes 1–2 mm. long, glabrous ; leaves ovate-lanceolate, acuminate at apex, cuneate at base, 5–7 cm. long, 2–2·5 cm. broad, nerves pubescent beneath 2. *parviflorum*

1. **S. paniculatum** (*Benth.*) *G. Tayl.* in Exell, Cat. S. Tomé 218 (1944) ; Bremek. l.c. 45, incl. var. *pubescens* Bremek. (1952). *Peltospermum paniculatum* Benth. (1849). *Oldenlandia peltospermum* Hiern in F.T.A. 3 : 53 (1877) ; F.W.T.A., ed. 1, 2 : 132 ; Chev. Bot. 312 ; Berhaut Fl. Sén. 104. A slender climbing shrub up to 15 ft. high ; flowers red, white or blue ; in secondary forest, sometimes beside water.
Sen.: Niokolo-Koba *Berhaut* 1509. **Port.G.:** Brene, Bissau (fr. Feb.) *Esp. Santo* 1732 ! Catió (Aug.) *Esp. Santo* 2177a ! **Guin.:** Rio Nunez *Heudelot* 628 ! Fouta Djalon *Pobéguin* 1930 ! Kouria *Caille* in *Hb. Chev.* 14698 ! **S.L.:** Waterloo (Oct.) *Lane-Poole* 378 ! Fundembaia (Oct.) *Thomas* 2860 ! Rokupr (Oct.) *Jordan* 139 ! Njala (fl. Sept., fr. Oct.) *Deighton* 2128 ! 2409 ! Bintumane (fr. Aug.) *Jaeger* 1189 ! **Lib.:** Kakatown *Whyte* ! Gbanga (Sept.) *Linder* 648 ! Ganta (Sept.) *Harley* 629 ! Gbau *Baldwin* 9432 ! **Iv.C.:** Dabou *Chev.* 15329 ! Guidéko *Chev.* 16479 ! Makougnié *Chev.* 16994 ! Bériby to Tabou *Chev.* 20033 ! **Ghana :** Axim (fr. Feb.) *Irvine* 2148 ! Essiama (fr. Oct.) *Fishlock* 2 ! Nkroful (fr. Nov.) *Vigne* FH 1418 ! **S.Nig.:** Sapoba, Jamieson R. (fl. Nov., fr. Dec.) *Kennedy* 1798 ! Ajayi & Odukwe FHI 26942 ! Onochie FHI 31238 ! Oban *Talbot* 242 ! **[Br.]Cam.:** Buea, 3,000 ft. (fl. & fr. Jan.) *Deistel* 624 ! *Maitland* 240 ! **F.Po:** (fl. & fr. Nov.) *T. Vogel* 74 ! *Barter* 2059 ! *Mann* 58 ! Also in Ubangi-Shari, Cameroun, Gabon, Cabinda, Congo and S. Tomé.

2. **S. parviflorum** (*Benth.*) *G. Tayl.* l.c. 218 (1944) ; Bremek. l.c. 46. *Pentas parviflora* Benth. in Bot. Mag. sub. t. 4086 (1844) ; F.T.A. 3 : 47 ; F.W.T.A., ed. 1, 2 : 129. Twining shrub in thickets ; flowers bluish-purple.
Iv.C.: near Béyo, Sassandra (fl. Feb., fr. Dec.) *Leeuwenberg* 2210 ! 2736 ! **Ghana:** Accra *T. Vogel* ! Aburi Scarp (June, Nov.) *Johnson* 971 ! 1916 ! *Morton* GC 6126 ! Ajena to Akwamu (June) *Adams* 4794 ! **N.Nig.:** Mando F.R., Zaria (July) *Keay* FHI 25994 ! Abinsi (fl. & fr. June) *Dalz.* 632 ! **S.Nig.:** Lower Enyong F.R., Calabar (May) *Onochie* FHI 33211 ! **[Br.]Cam.:** Lakom (fl. & fr. May) *Maitland* 1653 !

79. HEKISTOCARPA Hook f.—F.T.A. 3 : 65.

Stems pubescent ; leaves obovate to broadly oblanceolate, acutely acuminate, acute at the base, 8–14 cm. long, 3–6 cm. broad, appressed-setulose-pubescent on both surfaces ; petiole about 1·5 cm. long, setulose ; stipules leafy, ovate-lanceolate, up to 1 cm. long ; flowers very small ; sessile in axillary secund sparsely branched cymes ; calyx-teeth shortly subulate ; ovary setulose ; corolla glabrous, about twice as long as the calyx ; anthers included .. .. *minutiflora*

H. minutiflora *Hook f.* Ic. Pl. t. 1151 (1873) ; F.T.A. 3 : 65. Herb or undershrub 2–5 ft. high, trailing and forming thickets, midrib beneath often purplish ; flowers white, calyx purplish ; in forest, often beside water.
S.Nig.: Omo F.R. (fl. & fr. Feb., Mar.) *Jones & Onochie* FHI 16996 ! 17199 ! Calabar (Feb.) *Mann* 2314 ! Oban (May) *Ujor* FHI 31786 ! Aboabam, Ikom (Dec.) *Keay* FHI 28177 ! **[Br.]Cam.:** *Preuss* 1295 ! S. Bakundu F.R., Kumba (Jan.) *Binuyo & Daramola* FHI 35188 ! *Olorunfemi* FHI 30613 ! Mamfe (Dec.) *Baldwin* 13824 ! Also in Cameroun.

80. THECORCHUS Bremek. in Verh. K. Nederl. Akad. Wetensch., Afd. Natuurk., sect. 2, 48 : 54 (1952).

Annual herb with erect branches or simple, 15–25 cm. high, glabrous and slightly scabrid; stems with 4 slight ribs ; leaves linear, 1–3 cm. long, 1·2–3 mm. broad ; flowers 1–3 in axils, sessile, becoming shortly pedicellate in fruit ; calyx-lobes narrow-triangular, about 1 mm. long, accrescent and becoming foliaceous, about 5 mm. long in fruit ;

15

corolla about 2 mm. long, anthers and style included ; fruits oblong, about 5 mm. long and 2 mm. broad ; ovary with 2 loculi and numerous minute seeds   *wauensis*

T. wauensis (*Hiern*) *Bremek.* l.c. 55 (1952), incl. var. *scabrida* Bremek. l.c. *Oldenlandia wauensis* Hiern in F.T.A. 3 : 64 (1877). *O. abyssinica* of F.W.T.A., ed. 1, 2 : 131, partly (*Chev.* 23598), not of Hiern. An erect herb 6–10 in. high ; flowers white ; on sandy river banks.
Sen.: Niokolo Koba (fr. May) *Trochain* 3477 ! **Mali:** L. Debo (May) *Hagerup* 155 ! Bani R. *Chev.* 1068 ! **Guin.:** Dabola (fr. Apr.) *Pitot* ! **Iv.C.:** Ouango-Fitini *Aké Assi* 5332 ! **Ghana:** Yapei (fl. & fr. Mar.) *Adams* 3918 ! White Volta R., near Lumbunga (Apr.) *Morton* A3895 ! Tumu (fl. & fr. Mar.) *Morton* GC 8889 ! **Dah.:** Savé to Agouagon (fl. & fr. May) *Chev.* 23598 ! Also in Ubangi-Shari and Sudan.

81. BATOPEDINA Verdc. in Bull. Jard. Bot. Brux. 23 : 29 (1953).

Perennial herb ; stems simple or branched ; leaves ovate to ovate-elliptic, narrow-cuneate at base, acute at apex, 2·2–2·5 cm. long, 6–9 mm. broad, with about 4 pairs of lateral nerves, pubescent on both sides, sessile ; flowers axillary, solitary or paired ; calyx-lobes unequal, about 1 mm. long with one lobe ovate-lanceolate, 11 mm. long, 2·7 mm. broad, 3-nerved ; corolla-tube 1·5–2·2 cm. long, pilose outside, lobes 6·5 mm. long, 1·5 mm. broad ; style slightly exserted, stigma bifid, anthers included ; capsule ovoid, 5 mm. long   ..   ..   *tenuis*

B. tenuis (*A. Chev. ex Hutch. & Dalz.*) *Verdc.* in Bull. Jard. Bot. Brux. 23 : 30 (1953). *Otomeria tenuis* A. Chev ex Hutch. & Dalz., F.W.T.A., ed. 1, 2 : 129 (1931) ; Chev. Bot. 310, name only. Stems about 6 in. high from a woody perennial rootstock ; flowers white ; in rock crevices.
Iv.C.: Sindou (May, Sept.) *Chev.* 866 ! *Jaeger* 5202 ! Bamfora (Aug.) *Bouquet.* **Ghana:** Jema, Ashanti (Oct.) *Collector unknown* 2538 !

82. PARAPENTAS Bremek. in Verh. K. Nederl. Akad. Wetensch., Afd. Natuurk., sect. 2, 48 : 50 (1952).

Straggling herb rooting at nodes, pubescence crisped and dense on young parts ; leaves ovate to elliptic, 2·7–3·7 cm. long, 1–2 cm. broad, almost glabrous ; petiole 2–11 mm. long ; stipules with broad bases and 3–6 filiform segments up to 2·5 mm. long ; inflorescences axillary, sessile and congested ; corolla-tube 5 mm. long, lobes deltoid, about 2 mm. long ; stigma bifid ; fruits about 3 mm. long, splitting into 4 valves along the septa ; seed numerous   ..   ..   *setigera*

P. setigera (*Hiern*) *Verdc.* in Bull. Jard. Bot. Brux. 23 : 57 (1953). *Virecta setigera* Hiern in F.T.A. 3 : 48 (1877) ; F.W.T.A., ed. 1, 2 : 130. *Parapentas gabonica* Bremek. l.c. 57 (1952). A creeping or straggling herb rooting at the nodes and sending up lateral shoots ; flowers white ; on the forest floor.
Guin.: *Heudelot* 866 ! **S.L.:** Bagroo R. (Apr.) *Mann* 874 ! Makondi, Njala (Feb.) *Deighton* 1571 ! Gbashama (Mar.) *Marmo* 204 ! Gola Forest (Nov.) *Deighton* 372 ! **Lib.:** Sangwin (Mar.) *Baldwin* 11313 ! Tappi (Dec.) *Baldwin* 70 ! **Iv.C.:** Kéeta *Chev.* 19372 ! Sanvi *Chev.* 17778 ! Tiapleu *Aké Assi* 5971 ! **S.Nig:** Okomu F.R. (Jan.) *Brenan* 8529 ! Shasha F.R. (Apr.) *Ross* 216 ! 299 ! Oban *Talbot* 253 ! Cross R. North F.R. (Dec.) *Keay* FHI 28154 ! Afi River F.R. (May) *Jones & Onochie* FHI 18612 ! **F.Po:** *Barter* 2048 ! *Mann* 215 ! Also in Cameroun, Gabon, Cabinda, Congo and Nyasaland.

83. OTOMERIA Benth.—F.T.A. 3 : 49 ; Verdcourt in Bull. Jard. Bot. Brux. 23 : 6 (1953). Hepper in Kew Bull 14 : 253 (1960). *Tapinopentas* Bremek. in Verh. K. Nederl. Akad. Wetensch., Afd. Natuurk., sect. 2, 48 : 49 (1960).

Corolla-tube 15–28 mm. long ; corolla red ; erect or climbing plants :
  Erect herb ; leaves rounded at base and very shortly petiolate, lanceolate to ovate-lanceolate, 4–9 cm. long, 1–4 cm. broad, densely pubescent to subglabrous ; stems more or less pilose ; inflorescence at first capitate, elongating in fruit up to 30 cm. or more long ; calyx-lobes unequal, about 2 mm. long, triangular, with one linear-lanceolate lobe about 1 cm. long in fruit ; corolla-tube about 1·5 cm. long, dilated above, pubescent outside ; fruits turbinate, about 9 mm. long and 6 mm. broad at the top   ..   ..   1. *elatior*
  Climbing slender shrub ; leaves cuneate at base with slender petioles about 1·5 cm. long, elliptic to slightly ovate-elliptic, 3·5–10 cm. long, 1·5–3·5 cm. broad, sub-glabrous ; stems glabrous below, shortly pubescent above ; inflorescence remaining congested ; calyx-lobes foliaceous, unequal, 6–10 mm. long, 2·5–3 mm. broad ; corolla-tube 1·8–2·8 cm. long, slender and slightly dilated above, minutely pubescent ; fruits oblong, 8–10 mm. long, 3·5–5 mm. broad   ..   ..   ..   2. *volubilis*
Corolla-tube 3–5 mm. long ; corolla white ; inflorescence congested, elongating in fruit ; leaves ovate-elliptic or ovate-lanceolate, (0·9–)2–8 cm. long, 4–40 mm. broad ; erect or ascending herbs :
  Inflorescence in fruit 15–34 cm. long, rather stout, simple :
    Fruits oblong, about 6 mm. long ; calyx-lobes lanceolate   ..   3. *guineensis*
    Fruits obconical about 2 mm. long ; calyx-lobes broadly ovate ..   ..   4. *sp. A*
  Inflorescence in fruit 3–5 cm. long, slender, with 1–3 small branches ; fruits ovoid   5. *cameronica*

1. O. elatior (*A. Rich. ex DC.*) *Verdc.* in Bull. Jard. Bot. Brux. 23 : 18, fig. 3 A–D (1953). *Sipanea elatior* A. Rich. ex DC. (1830). *Otomeria dilatata* Hiern in F.T.A. 3 : 50 (1877) ; F.W.T.A., ed 1, 2 : 129 ; Chev. Bot. 310. Erect herb, 2–3 ft. high, sparingly branched ; flowers scarlet to pink ; in open marshes amongst grass.
Mali: Sikasso *Chev.* 789. **Guin.** Bouhong (June) *Pobéguin* 736 ! Férédougouoa R. (July) *Collenette* 68 ! Kouria *Caille* in *Hb. Chev.* 15026. Irébéléya to Timbo *Chev.* 18339. Erimakuna (fl. & fr. Mar.) *Sc. Elliott* 5239 ! 5392 ! **S.L.:** Jigayo (fl. & fr. Sept.) *Thomas* 2503 ! **Iv.C.:** Marabadiassa to Gottoro (July) *Chev.*

22027 ! **Ghana**: Busunu (fl. & fr. May) *Vigne* FH 3849 ! Damongo (Mar.) *Morton* GC 8714 ! Gambaga (Mar.) *Hepper & Morton* A3140 ! **N.Nig.**: Abinsi (June) *Dalz.* 751 ! Andaha (fl. & fr. Feb.) *Hepburn* 137 ! Nupe *Barter* 1237 ! [**Br.**]**Cam.**: Bamenda (fl. & fr. May) *Maitland* 1426 ! Mambila Plateau *Chapman* 57 ! Also in Cameroun, Congo, Angola, Mozambique, Nyasaland, N. and S. Rhodesia, Tanganyika, Kenya, Uganda and Sudan.

2. **O. volubilis** (*K. Schum.*) *Verdc.* in Kew Bull. 7 : 361 (1952) ; Bull. Jard. Bot. Brux. 23 : 26, fig. 3E. *Pentas volubilis* K. Schum. (1897)—F.W.T.A., ed. 1, 2 : 129. *Otomeria batesii* Wernham (1916)—F.W.T.A., ed. 1 2 : 129. A wiry-stemmed twining plant, nearly glabrous ; flowers red ; in forest.
S.Nig.: Degema *Talbot* 3608 ! [**Br.**]**Cam.**: Barombi (Sept.) *Preuss* 471 ! Kumba to Ikiliwinda *Preuss* 393 Also in Cameroun, Congo and Uganda.

3. **O. guineensis** *Benth.* in Fl. Nigrit. 405 (1849) ; F.T.A. 3 : 49 ; F.W.T.A., ed. 1, 2 : 129, partly (*T. Vogel* 39 ; *Ansell* ; *Linder* 1558). Verdc. in Bull. Jard. Bot. Brux. 23 : 14, fig. 3F, G (excl. *A. S. Thomas* D140) Erect, more or less branched herb 1-2 ft. high ; flowers white ; in grassy places.
Lib.: Grand Bassa (fl. & fr. July) *T. Vogel* 39 ! Monrovia *Farmar* 355 ! White Plains (fl. & fr. June) *Barker* 1344 ! R. Sanou *McWilliam* ! Sinoe Basin *Whyte* ! **S.Nig.**: Isuofla, Awka Dist. (fl. & fr. May) *Jones* FHI 6526 ! Port Harcourt (fl. Sept., fl. & fr. Mar., June) *Brenan* 9223 ! *Maitland* D ! Ideno, Eket Dist. (fl. & fr. May) *Onochie* FHI 32093 ! Also in Gabon, Congo and S. Tomé.

4. **O. sp. A.** *O. micrantha* (?) of Verdc. l.c. 12, partly (*Jones & Onochie* FHI 18993), not of K. Schum. Herb, more or less erect.
Guin.: Bambaya (fl. & fr. Aug.) *Jaeger* 859 ! **S.Nig.**: Afi River F.R., Ogoja (fl. & fr. June) *Jones & Onochie* FHI 18993 !

5. **O. cameronica** (*Bremek.*) *Hepper* in Kew Bull. 14 : 253 (1960). *Tapinopentas cameronica* Bremek. in Verh. K. Nederl. Akad. Wetensch., Afd. Natuurk., Sect. 2, 48 : 49 (1952). *T. latifolia* Verdc. in Bull. Jard. Bot. Brux. 23 : 60 (1953). *Otomeria guineensis* of F.W.T.A., ed. 1, 2 : 129, partly *Williams* 247 ; *Dunlap* 140) ; of Verdc. l.c. 14, partly (*A. S. Thomas* D140). Creeping herb from a strong rootstock, much-branched and with erect shoots 1-3(-4) ft. high ; flowers whitish ; in or near upland forest.
S.L.: Loma Mts. summit, 6,390 ft. (fl. Aug., fr. Nov.) *Jaeger* 402 ! 991 ! *Bakshi* 257 ! **Iv.C.**: Nimba Mts. (Aug.) *Boughey* 18085 ! *Schnell* 1503 ! Mt. Dou *Chev.* 21488 ! **Ghana**: Bompata (May) *Chipp* FH 442 ! Sesiamang (Feb.) *A. S. Thomas* D140 ! Ijang F.R. (fr. Apr.) *Williams* 247 ! Amedzofe (fl. & fr. Apr.) *Hall* 1758 ! **Dah.**: Atacora Mts. (June) *Villiers* ! [**Br.**]**Cam.**: Buea, 3,000 ft. (Jan.) *Maitland* 277 ! *Dunlap* 140 ! Bamenda, 6,000 ft. (Jan., Apr.) *Maitland* 1754 ! *Migeod* 354 ! Bafut-Ngemba F.R., 7,000 ft. (Feb., Apr.) *Hepper* 2083 ! Ujor FHI 30020 ! Kakara, 4,700 ft., Mambila Plateau (Jan.) *Hepper* 1795 ! **F.Po**: Moka (Sept.) *Boughey* 2 ! *Melville* 679 ! Also in Cameroun.
[Note : The Sierra Leone specimens are not quite uniform with the other specimens cited.—F.N.H.]

84. PENTAS Benth.—F.T.A. 3 : 45 ; Verdcourt in Bull. Jard. Bot. Brux. 23 : 237 (1953).

Corolla-tube 35–80 mm. long, pubescent, lobes 6–11 mm. long ; inflorescences densely capitate ; leaves lanceolate, subsessile, acute, 6–9 cm. long, 2–3 cm. broad, minutely pubescent above and on the 7–8 nerves beneath ..   1a. *decora* var. *triangularis*
Corolla-tube 4–13 mm. long :
  Inflorescences densely capitate, 1·3–2·5 cm. diam., sessile and subtended by 2 leafy bracts ; corolla-tube about 5 mm. long, finely pubescent ; leaves lanceolate, ovate-lanceolate or oblong-lanceolate, subsessile, acute, 4·5–11·5 cm. long, 1–3·2 cm. broad, reticulate beneath   ..   ..   2. *purpurea*
  Inflorescences more or less lax, often spreading, not sessile :
    Corolla-tube 9–13 mm. long ; calyx-lobes about 8 mm. long, subequal, subulate ; leaves ovate-elliptic, acute, 8–15 cm. long, 3–6·3 cm. broad, nerves ferrugineous beneath ; distinctly petiolate   ..   ..   5a. *schimperiana* subsp. *occidentalis*
    Corolla-tube 4–5 mm. long ; calyx-lobes not subulate, unequal, usually with one lobe extended :
      Tufted herb appearing after fires ; leaves obtuse or subacute at apex, oblong-lanceolate, 2·5–8 cm. long, 4–20 mm. broad, subsessile, with about 6 nerves on each side ; calyx-lobes oblong, up to 2·5 mm. long, subequal. .   4. *arvensis*
      Tall half-woody herbs ; leaves acuminate or acute, lanceolate, with about 10 nerves on each side :
        Calyx and corolla outside subglabrous ; leaves petiolate, 3–15 cm. long, 1·2–5 cm. broad, with rather thin texture ..   ..   6a. *pubiflora* subsp. *bamendensis*
        Calyx and corolla outside densely and minutely puberulous ; leaves sessile, 4–10 cm. long, 1–2·5 cm. broad, with rather thick texture ..   ..   3. *nervosa*

1. **P. decora** *S. Moore* in J. Bot. 48 : 219 (1910) ; Verdc. in Bull. Jard. Bot. Brux. 23 : 287 (1953).
1a. **P. decora** var. **triangularis** (*De Wild.*) *Verdc.* l.c. 291 (1953). *P. triangularis* De Wild. (1914). *P. globiflora* Hutch. in Kew Bull. 1921 : 374, fig. 5 ; F.W.T.A., ed. 1, 2 : 129. Erect herb 1–2 ft. high ; flowers white, fragrant ; in upland grassland.
Guin.: Beyla (June) *Jac.-Fél.* 966 ! Bondou (July) *Martine* 367 ! Macenta Milo (Aug.) *Adam* 5901 ! **N.Nig.**: Jos Plateau (June) *Lely* P396 ! Mongu, 4,300 ft. (July) *Lely* 386 ! Hoss road (Aug.) *Keay* FHI 12715 ! [**Br.**]**Cam.**: Ninong, Kumba (May) *Collier* FHI 10111 ! Bamenda, 5,500 ft. (Apr., May) *Ujor* FHI 30317 ! *Maitland* 1433 ! Maisamari, Mambila Plateau (June) *Chapman* 50 ! Also in Cameroun, Congo, Uganda, Kenya and Tanganyika.

2. **P. purpurea** *Oliv.* in Trans. Linn. Soc. 29 : 83 (1873) ; F.T.A. 3 : 46 ; Verdc. l.c. 330. Erect simple herb 1–2 ft. high, from a woody rootstock ; flowers mauve ; in upland grassland.
[**Br.**]**Cam.**: Bamenda Prov. : Bum, 4,000 ft. (June) *Maitland* 1602 ! Lakom 6,000 ft. (May) *Maitland* 1371 ! Bafut-Ngemba F.R., 6,000–7,000 ft. (Feb., Mar.) *Hepper* 2102 ! *Richards* 5303 ! Vogel Peak, 5,400 ft., Adamawa (fr. Nov.) *Hepper* 1501 ! Also in Congo, Tanganyika, Mozambique and S. Rhodesia.
[A specimen (*Migeod* 449) from Bamenda with leaves in whorls of 3 is considered by Verdc. l.c. 334 to be a possible variety of the above, or even a new species. *Pitot* 1154 *bis* (in Hb. I.F.A.N.) also has whorled leaves and a laxer, more ciliate inflorescence—F.N.H.]

3. **P. nervosa** *Hepper* in Kew Bull. 14 : 254 (1960), and in Kew Bull. 16 : 457, fig. 2 (1963). Erect slightly-branched herb about 2½ ft. high, from a woody rootstock ; flowers pale mauve ; in upland grassland.
[**Br.**]**Cam.**: Vogel Peak, 4,000 ft., Adamawa (Dec.) *Hepper* 1535 !

4. **P. arvensis** *Hiern* in F.T.A. 3 : 47 (1877) ; Verdc. l.c. 335, fig. 35E, F. Tufted herb up to 1 ft. high, from perennial rootstock ; flowers white or pink ; appearing in upland grassland after fires.
N.Nig.: Vom, Jos Plateau (Jan.) *Lely* P26 ! *Dent Young* 120 ! Also in Ubangi-Shari, Sudan, Uganda and Kenya.

5. **P. schimperiana** (*A. Rich.*) *Vatke* subsp. **schimperiana**—in Linnaea 40 : 192 (1876) : Verdc. l.c. 263.
5a. **P. schimperiana** subsp. **occidentalis** (*Hook f.*) Verdc. l.c. 266 (1953). *Vignaldia occidentalis* Hook. f. (1864).
   Pentas occidentalis (Hook. f.) Benth. & Hook. ex Hiern in F.T.A. 3 : 46 (1877). Half-woody herb 4–7 ft.
   high ; flowers pale pink, fragrant ; in moist places in montane grassland.
   [**Br.**]**Cam.**: Cam. Mt., 4,400–8,500 ft. (Nov.–May) *Mann* 1227 ! *Maitland* 988 ! *Keay* FHI 28596 !
   Brenan 9385 ! Mbakakeka Mt., 6,500 ft. (Jan.) *Keay* FHI 28389 ! Kumbo to Oku, 6,500 ft. (fl. & fr. Feb.)
   Hepper 2007 ! Ndu (fl. & fr. Jan.) *Keay & Russell* FHI 28435 ! **F.Po:** 8,500 ft. *Mann* ! Pico Guinea
   2724 ! L. Biao (Sept.) *Melville* 674 ! Also in S. Tomé, Cameroun and Congo.
6. **P. pubiflora** *S. Moore* in J. Linn. Soc. 38 : 254 (1908) ; Verdc. l.c. 327.
6a. **P. pubiflora** subsp. **bamendensis** *Verdc.* l.c. 329 (1953). Erect herb with reddish hairs ; flowers bluish with
   white lobes ; at edge of montane forest.
   S.Nig.: Ikwette, Obudu Div. (fl. & fr. Dec.) *Savory & Keay* FHI 25231 ! [**Br.**]**Cam.**: Lakom, Bamenda
   (Apr.–May) *Maitland* 1422 ! 1653 ! 1743 ! Oku (fl. & fr. Feb.) *Hepper* 2051 !

Besides the above, a form of *P. lanceolata* (Forsk.) Deflers is cultivated in Sierra Leone and Ghana.
It is an undershrub with pink flowers ; probably native on the Comoros but long cultivated.

85. RICHARDIA Linn.—F.T.A. 3 : 242.

Herb ; stems pubescent to pilose ; ovate to obovate, 1–3 cm. long, 7–20 mm. broad,
coarsely pubescent on both sides, shortly petiolate or subsessile ; flowers clustered in
terminal heads subtended by leafy sheathing bracts ; corolla 4–5-lobed, valvate in
bud ; stamens 4–5, shortly exserted, inserted in throat of corolla ; ovary 3-celled ;
fruit splitting longitudinally into 3 indehiscent cocci, muricate and pubescent outside,
broadly obovoid, 2 mm. long .. .. *brasiliensis*

R. brasiliensis *Gomez* Mem. Ipecac. 31, t. 2 (1801). *R. scabra* of F.T.A. 3 : 242, not of Linn. A hairy, branched
   prostrate herb with stems up to a foot long; flowers white ; introduced weed.
   Ghana: near Achimota (fl. & fr. Apr.) *Akpabla* 1868 ! **S.Nig.:** Apapa, Lagos (fl. & fr. June) *Onochie* FHI
   23303 ! Introduced from S. America into many places in Africa and other parts of the tropics and subtropics.

86. DIODIA Linn.—F.T.A. 3 : 230.

Calyx-lobes 2, 3 mm. long in fruit ; corolla about 7 mm. long ; fruit 2·5 mm. long ;
leaves ovate to obovate 1·5–3 cm. long, 7–18 mm. broad, with about 4 pairs of lateral
nerves, rounded or subacute at apex, cuneate into a short petiole at base, not fleshy,
glabrous with scabrid margins above.. .. .. 1. *rubricosa*
Calyx-lobes 4 :
   Leaves without visible lateral nerves (even in dry state), fleshy, linear-lanceolate to
   ovate-oblong, 2–4 cm. long, 6–9 mm. broad, glabrous, margin ciliate towards base,
   internodes usually congested or with congested short-shoots arising at more widely
   spaced nodes of long-shoots ; flowers solitary ; corolla 1 cm. long ; fruits 5–6 mm.
   long ; coastal species .. .. 2. *vaginalis*
   Leaves with distinct lateral nerves, not or hardly (in No. 3) fleshy :
   Flowers solitary or subsolitary ; fruits 5–6 mm. long, glabrous ; leaves lanceolate to
   elliptic, 2–4 cm. long, 6–15 mm. broad, stiffly acute, margins scabrid, glabrous or
   scabrid-pubescent on both sides, subsessile ; internodes congested or distinct ;
   stems stout ; coastal species .. .. 3. *serrulata*
   Flowers several together in axillary clusters ; fruits 3–4 mm. long, glabrous
   or pubescent ; leaves ovate to ovate-lanceolate, 2·5–5 cm. long, 1–2·5 cm. broad,
   more or less pubescent beneath and scabrid above, shortly petiolate ; internodes
   long ; stems slender ; inland species .. .. 4. *scandens*

1. **D. rubricosa** *Hiern* in F.T.A. 3 : 231 (1877) ; Chev. Bot. 348, partly. *D. scandens* of F.W.T.A., ed. 1, 2 : 133,
   partly (*Whyte*), not of Sw. A prostrate creeping herb, stems rooting at the nodes, a few inches to 2 ft.
   long ; flowers whitish or pale pink.
   S.L.: *Afzelius* ! Yele, Turtle Isl. (fl. & fr. Nov.) *Deighton* 2301 ! **Lib.:** Sinkor (fr. June) *Barker* 1354 ! Sinoe
   Basin *Whyte* ! Gletown, Tchien Dist. (fl. & fr. July) *Baldwin* 6750 ! **Iv.C.:** Bingerville Chev. 16080 !
   Lamé (Oct.) *Leeuwenberg* 1731 ! Adiopodoumé (Oct.) *Leeuwenberg* 1703 ! **Ghana:** Prah R., Ashanti *Irvine*
   664 ! Aburi (fl. & fr. Mar., Aug.) *Deighton* 605 ! *Johnson* ! Princes Town (Feb.) *Morton* A1641 ! **S.Nig.:**
   Badagry (fl. & fr. Aug.) *Onochie* FHI 33468 ! Oron to Eket *Talbot* !
2. **D. vaginalis** *Benth.* in Fl. Nigrit. 424 (1849) ; F.T.A. 3 : 232. Prostrate herb with fleshy stems and leaves ;
   flowers white or pink ; on the seashore near high-water mark.
   S.L.: York *Smythe* ! Lumley (Aug.) *Jordan* 781 ! Tlsana, Bonthe Isl. (Nov.) *Deighton* 2352 ! **Lib.:** Monrovia *Baldwin* 5828 ! 10991 ! *Harley* 1093 ! Grand Bassa *T. Vogel* 68 ! **Ghana:** Half Assinie (July) *Chipp*
   273 ! **Togo Rep.:** *Schinz* ! **S.Nig.:** Brass *Barter* 50 ! Nun R. *T. Vogel* 21 ! Badagry (Aug.) *Onochie* FHI
   33483 ! Also in Cameroun.
3. **D. serrulata** (*P. Beauv.*) *G. Tayl.* in Exell Cat. S. Tomé 220 (1944). *Spermacoce serrulata* P. Beauv. Fl. Oware
   1 : 39, t. 23 (1805). *Diodia maritima* Thonning (1827)—F.T.A. 3 : 231 ; F.W.T.A., ed. 1, 2 : 133 ;
   Chev. Bot. 347 ; Berhaut Fl. Sén. 78. Prostrate herb with stems 1 to several ft. long rooting at the nodes,
   rarely scandent, or somewhat tufted ; flowers white ; on the seashore near high-water mark.
   Sen.: *Farmar* 102 ! Casamance *Heudelot* 572 ! Niayès *Chev.* 2012. Dakar (fr. Jan.) *Adam* 275 ! **Port.G.:**
   Ilha de Pecixe, Canchungo (fr. May) *Esp. Santo* 2050 ! Conakry *Paroisse* 220 ! **S.L.:** Mahera (fl. & fr. Oct.)
   Glanville 446 ! Turtle Isl. (fl. & fr. Nov.) *Deighton* 2312 ! Mano Salija (fl. & fr. Nov.) *Deighton* 277 ! **Lib.:**
   Monrovia (fl. & fr. Oct.–Dec.) *Linder* 1427 ! *Baldwin* 10915 ! *Harley* 1676 ! **Iv.C.:** Sassandra port Chev.
   17924 ! Petit Bassam *Hedin* 2532 ! **Ghana:** Atwabo *Fishlock* 91 ! Esiama (fl. & fr. May) *C. J. Taylor*
   FH 5295 ! Axim (Feb.) *Irvine* 2311 ! **Dah.:** Cotonou (fl. & fr. Apr.) *Debeaux* 164 ! **S.Nig.:** Oware *Beauvois* !
   Lagos *MacGregor* 359 ! *Dalz.* 1380 ! Nun R. (fl. & fr. Aug.) *T. Vogel* 2 ! *Mann* 465 ! Stubbs Creek F.R.,
   Eket Dist. (fl. & fr. May) *Onochie* FHI 32941 ! [**Br.**]**Cam.:** Victoria (fl. & fr. Feb.) *Maitland* 1347 ! Also
   in Cameroun, Angola and C. America.
4. **D. scandens** *Sw.* Prod. Veg. Ind. Occ. 30 (1788) ; Berhaut Fl. Sén. 78. *D. breviseta* Benth. (1849)—F.T.A.
   3 : 231 ; Chev. Bot. 347. *D. rubricosa* of Chev. Bot. 348, partly, not of Hiern. *Spermacoce pilosa* (Schum.
   & Thonn.) DC. (1830)—F.T.A. 3 : 235 ; F.W.T.A., ed. 1, 2 : 135. (?) *S. palmetorum* DC. (1830)—Hepper
   in Kew Bull. 14 : 257. *Diodia pilosa* Schum. & Thonn. (1827). Straggling herb with slender stems
   several feet long and scabrid leaves ; flowers whitish ; in thickets.

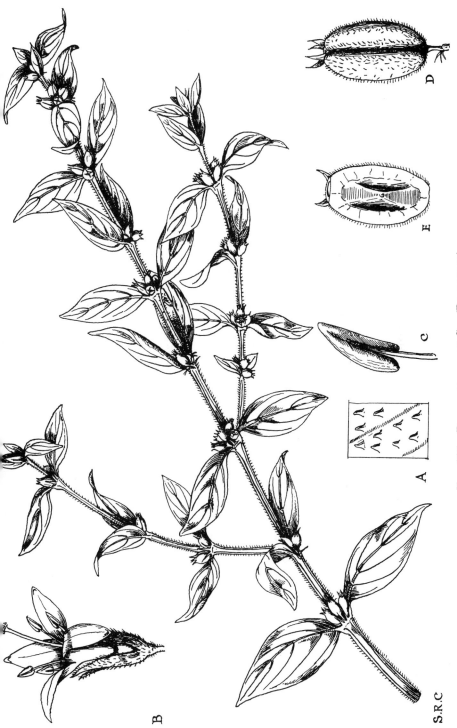

FIG. 245.—Diodia scandens *Sw.* (Rubiaceae).

A, surface of leaf. B, flower. C, anther. D, fruit. E, closed half of fruit.

S.R.C

217

Sen.: Niayes *Berhaut* 1323. Cayor *Leprieur*! **Fr.Sud.:** Médinani (fr. Feb.) *Chev.* 507! **Port.G.:** Antula, Bissau (fr. Feb.) *Esp. Santo* 1474! **Guin.:** Conakry *Chev.* 12160! **S.L.:** Freetown (June) *T. Vogel* 107! Magbile (fr. Dec.) *Thomas* 6300! Yonibana (fr. Nov.) *Thomas* 5047! Njala (fl. & fr. July) *Deighton* 723! **Lib.:** Monrovia *Whyte*! Grand Bassa (fl. & fr. July) *T. Vogel* 84! Ganta (fr. Aug.) *Harley* 811! **Iv.C.:** Sanvi *Chev.* 17819! **Ghana:** Sekondi (fl. & fr. Oct.) *Howes* 985! Akropong, Akwapim (fl. & fr. Aug.) *Irvine* 768! Dormaa (fl. & fr. Dec.) *Adams* 3001! Kpandu *Asamany* 118! **Dah.:** Cotonou (July) *Debeaux* 155! **N.Nig.:** Nupe *Barter* 1722! Tilde Filani, 3,300 ft. *Lely* 225! Vom, 4,000 ft. *Dent Young* 124! Abinsi (fr. Mar.) *Dalz.* 633! **S.Nig.:** Lagos (fl. & fr. May) *Dalz.* 1381! Okomu F.R., Benin (fl. & fr. Dec.) *Brenan* 8440! Sapoba (fl. & fr. Nov.) *Meikle* 591! Milliken Hill, Enugu (fl. & fr. Mar.) *Hepper* 2214! Oban *Talbot* 206! **[Br.]Cam.:** Victoria, on 1922 lava (fl. & fr. June) *Rosevear* 48/37! Bellel, Mambila Plateau (fl. & fr. Jan.) *Hepper* 1696! **F.Po:** *T. Vogel* 63! *Mann* 403! Widespread in tropical Africa, also in Mascarene Islands, tropical Asia and tropical America.

Doubtfully recorded species.

D. arenosa *DC.* Prod. 4 : 564 (1830). Tentatively referred to that species in Fl. Nigrit. 423 and F.T.A. 3 : 231. ? **S.L.:** *Don* (not traced).

87. BORRERIA G.F.W. Mey. Prim. Fl. Esseq. 79 (1818), *nom. cons. Spermacoce* of F.T.A. 3 : 233, not of Linn. *Octodon* Thonn.—F.T.A. 3 : 241; F.W.T.A., ed. 1, 2 : 135; Hepper in Kew Bull. 14 : 260 (1960).

Stems pilose or densely pubescent (hairs not merely in vertical lines) :
  Leaves with a very thick nerve along the margin, lamina linear, 3–5 cm. long, 4–6 mm. broad, sharply acute at apex, glabrous, lacking visible lateral nerves ; stipules pubescent outside, with several filiform segments 6–10 mm. long ; flowers in congested terminal or subterminal heads, surrounded by numerous long leaves
        1. *radiata*
  Leaves without a thick marginal nerve :
    Corolla very large, 3–5 cm. long, lobes and tube above sparsely pilose ; stems densely pilose ; leaves lanceolate, 3–9 cm. long, 8–18 mm. broad, with about 6 lateral nerves on each side of and almost parallel to the midrib, deeply impressed above and prominent beneath ; scabrid and pubescent, sessile ; calyx-lobes about 5 mm. long .. .. 14. *macrantha* var. *macrantha*
    Corolla up to 1 cm. long :
      Stems narrowly 4-winged, shortly pubescent ; leaves ovate to obovate, 3–4·5 cm. long, 1·5–2·5 cm. broad, shortly pubescent on both sides, subsessile ; flowers in axillary clusters ; calyx-lobes 1 mm. long ; corolla 3 mm. long   2. *latifolia*
      Stems terete or 4-angled, not winged :
        Perennial herbs, straggling, or tufted and ascending ; corolla 1 cm. long :
          Leaves broadly ovate, 2–5 cm. long, 1·5–2·5 cm. broad, with 5–6 pairs of lateral nerves, more or less rounded at base and acuminate at apex, shortly petiolate ; stipule segments filiform, 9–11 mm. long ; flowers few in axils
            17a. *princeae* var. *pubescens*
        Leaves lanceolate or linear-lanceolate :
          Leaves pubescent above not harshly scabrid, lanceolate, 2–6 cm. long, 8–25 mm. broad, with 3–4 pairs of lateral nerves, cuneate at base, acute at apex, shortly petiolate ; stipule segments filiform, 5 mm. long ; flowers in terminal or subterminal clusters .. .. 3. *saxicola*
          Leaves harshly scabrid above, linear to linear-lanceolate ; (for other characters see below) .. .. 6. *stachydea* var. *stachydea*
        Annual herbs more or less erect :
          Calyx-lobes about 1 mm. long ; flowers numerous in spherical, terminal and axillary heads ; corolla 3 mm. long ; fruits 1·5–2 mm. long, densely pubescent above ; leaves linear 2–5 cm. long, 3–8 mm. long, scabrid above ; stems sparsely pilose .. .. 4. *filiformis*
          Calyx-lobes and fruits 2–5 mm. long :
            Leaves pubescent above, not harshly scabrid ; (for other characters see above)
              3. *saxicola*
           Leaves harshly scabrid above, linear to linear-lanceolate, 3–9 cm. long, 5–16 mm. broad :
            Bracteate leaves narrowed at base ; fruits 3 mm. long ; calyx-lobes 3 mm. long
              5. *scabra*
            Bracteate leaves broadened at base, subtending most flowers of each inflorescence or at least reduced to a broad bract with a short leafy part ; inflorescence in fruit very distended ; fruits 4–5 mm. long ; calyx-lobes 4 mm. long
              6. *stachydea* var. *stachydea*
Stems glabrous or hairy on the angles (in No. 7 sometimes pilose below the stipular sheath) :
  Flowers small and numerous, often in globose heads, with corolla up to 3 mm. long ; calyx-lobes 0·5–1 mm. long ; fruits 1–3 mm. long :
    Stems square, hairy or ciliate on the angles ; stipular sheaths pubescent :
      Leaves linear-lanceolate, 3–7 cm. long, 2–15 mm. broad, slightly pubescent on both sides and scabrid above ; inflorescences in fruit up to 2 cm. diam. ; fruits 2–3 mm. long ; filiform bracteoles between flowers few :
        Main stem stout ; leaves 4–15 mm. broad   7. *chaetocephala* var. *chaetocephala*

Main stem slender ; leaves 2–4 mm. broad   7a. *chaetocephala* var. *minor*
Leaves ovate to ovate-lanceolate, 1·5–3 cm. long, 5–12 mm. broad, glabrous and smooth on both sides, midrib more or less pubescent beneath ; inflorescences in fruit up to 5 mm. diam. ; fruits 1 mm. long   ..   ..   ..   8. *ocymoides*
Stems obscurely angled, glabrous ; stipular sheaths glabrous or scabrid :
Terminal inflorescence globose, 1–1·5 cm. diam., usually with 2 leafy bracts about 1 cm. long reflexed beneath ; leaves oblanceolate, 3–5 cm. long, 5–10 mm. broad, glabrous, with faint lateral nerves   ..   ..   ..   9. *verticillata*
Terminal inflorescence, if any, overtopped by leafy bracts :
Perennial herbs :
Inflorescences mainly terminal about 1 cm. diam., hard, with the flowers almost obscured by numerous filiform bracteoles ; (for other characters see below)
     13. *octodon*
Inflorescences numerous, at most nodes, about 7 mm. diam. in fruit, with few bracteoles ; fruits glabrous ; stems slender, branching near the base, 30–60 cm. long ; leaves linear to linear-lanceolate, 1–4 cm. long, 3–10 mm. broad ; upland plant   ..   ..   10. *natalensis*
Annual herbs with main stems and lateral branches ; usually lowland plants :
Inflorescences with numerous filiform bracteoles between, and the same length as, the flowers :
Calyx-lobes acute, lanceolate, about 1 mm. long ; fruits more or less pubescent at apex, 1 mm. long ; leaves linear, 2–4 cm. long, 3–6 mm. broad, glabrous minutely scabrid above   ..   ..   11. *pusilla*
Calyx-lobes obtuse, minute :
Stipular sheath glabrous ; leaves filiform, 4–7 cm. long   ..   12. *filifolia*
Stipular sheath pubescent ; leaves linear or linear-lanceolate, 6–10 cm. long
     13. *octodon*
Inflorescences with few bracteoles between, and much shorter than, the flowers ; calyx-lobes minute ; fruits glabrous, 2 mm. long ; leaves linear (3–) 5–9 cm. long, 4–7 mm. broad, glabrous, scabrid above   ..   ..   15. *paludosa*
Flowers rather few and corolla 4–50 mm. long ; calyx-lobes 3–5 mm. long ; fruits 2–5 mm. long :
Corolla very large, 4–5 cm. long ; whole plant glabrous ; leaves lanceolate, 6–8 cm. long, 1·5–2 cm. broad, with scabrid margins ; calyx-lobes about 5 mm. long
     14a. *macrantha* var. *glabra*
Corolla 4–15 mm. long :
Leaves linear-lanceolate ; plants glabrous or nearly so :
Calyx-lobes denticulate or hispid on margins, 3–5 mm. long ; fruits 2–3 (–5) mm. long ; leaves 4–9 cm. long, 3–10 (–21) mm. broad, scabrid above ; bracteate leaves more or less reflexed ; corolla 7–15 mm. long ..   ..   16. *compressa*
Calyx-lobes entire and glabrous, 3 mm. long ; fruits 5 mm. long ; leaves 3–5·5 cm. long, 3–10 mm. broad, smooth above ; bracteate leaves erect, broad at base
     6a. *stachydea* var. *phyllocephala*
Leaves ovate to ovate-lanceolate, 2–5 cm. long, 1–2 cm. broad, more or less hairy above and on the nerves beneath :
Stipular sheaths longer than broad, (5–) 7–9 mm. long, with filiform segments rather longer ; calyx-lobes 4 mm. long ; leaves ovate to ovate-lanceolate, shortly acuminate, slightly scabrid near margins above, with about 6 pairs of lateral nerves deeply impressed above ; short petiole adnate to stipular sheath
     17. *princeae* var. *princeae*
Stipular sheaths broader than long, 2–3 mm. long, obscure when in fruit, with filiform segments about 2 mm. long ; calyx-lobes 1·5 mm. long ; leaves ovate, long-acuminate, smooth above, with about 5 pairs of lateral nerves hardly impressed above ; petiole rather long ..   ..   ..   18. *intricans*

1. **B. radiata** *DC.* Prod. 4 : 542 (1830) ; F.W. Andr. Fl. Pl. A.–E. Sud. 2 : 427 ; Berhaut Fl. Sén. 79. *Spermacoce radiata* (DC.) Sieber ex Hiern in F.T.A. 3 : 237 (1877) ; Chev. Bot. 349. An erect hispid herb half-woody below, stems often reddish-purple, about 1 ft. high ; flowers very small, white or mauve, with leaves radiating from the congested terminal heads ; in savanna.
Sen.: *Heudelot* 416 ! *Sieber* 8 ! *Berhaut* 435. **Gam.:** *Hayes* 572 ! *Saunders* 89 ! **Mali:** Timbuktu *Chev.* 1307 ! Koulikoro *Chev.* 2117 ! Ségou *Chev.* 2811 ! **Ghana:** Dutukpene (Dec.) *Adams* 4651 ! Sandema (Sept.) *Hughes* FH 5342 ! **Togo Rep.:** Basari *Kersting* A668 ! **N.Nig.:** Nupe *Barter* 989 ! Bida to Zungeru (Oct.) *Hepper & Keay* 954 ! Katagum *Dalz.* 166 ! Sokoto *Moiser* 83 ! Naraguta (Sept.) *Lely* 567 ! **S.Nig.:** Olokemeji to Iseyin (Nov.) *Hambler* 78 ! Awba Hills F.R., Ibadan (Oct.) *Jones* FHI 7066 ! Eruwa to Lanlate (Oct.) *Keay & Savory* FHI 25294 ! **[Br.]Cam.:** Gurum, Adamawa *Hepper* 1290 ! Extends east to the Sudan and Uganda.
2. **B. latifolia** (*Aubl.*) *K. Schum.* in Mart. Fl. Bras. 6, 6 : 61 (1888). *Spermacoce latifolia* Aubl. (1775). Straggling herb with square slightly winged stems 1–2 ft. long, leaves and stems yellow-green ; flowers blue or pink ; introduced weed.
S.L.: Newton (fl. & fr. Sept.) *Deighton* 4879 ! 6001 ! **Lib.:** Ganta *Harley* 2171 ! Kitoma *Adam* 16552 ! **Iv.C.:** Adiopodoumé (fl. & fr. Oct.) *Leeuwenberg* 1702 ! **Ghana:** Kumasi (fl. & fr. Sept., Nov.) *Darko* 346 ! 473 ! A native of tropical S. America.
3. **B. saxicola** *K. Schum.* in Engl. Bot. Jahrb. 28 : 112 (1899). Herb with ascending or hanging, tufted stems 1–2 ft. long ; flowers white ; in rock crevices.
N.Nig.: Wana, Jos Plateau (Sept.) *Hepburn* 127 ! Naraguta (Aug.) *Lely* 491 ! Dogon Kurmi, Jemaa (Aug.) *Killick* 36 ! Gawu Hills, Niger Prov. (Aug.) *Onochie* FHI 35929 ! **[Br.]Cam.:** Bafut-Ngemba

F.R., 6,000 ft., Bamenda (Sept., Oct.) *Lightbody* FHI 26254! 26260! Binka, 6,500 ft., Nkambe Div. (Sept.) *Savory* UCI 378! Vogel Peak, 4,700 ft., Adamawa (Nov.) *Hepper* 1349! Also in Cameroun.

4. **B. filiformis** (*Hiern*) *Hutch. & Dalz.* F.W.T.A., ed. 1, 2 : 135 (1931). *Spermacoce filiformis* Hiern in F.T.A. 3 : 234 (1877). Erect annual herb with pilose stems 6–18 in. high, often tinged pink or purple ; flowers white ; in crevices of rock out-crops.

N.Nig.: Nupe *Barter* 1720! Gwari Hills, Niger Prov. (Aug.) *Onochie* FHI 35911! Kufena, Zaria (Aug.) *Keay* FHI 28020! Naraguta (Aug.) *Lely* 503! **S.Nig.:** Fashola to Oyo (fr. Oct.) *Onochie* FHI 34923! Ado Rock, Oyo (Oct.) *Keay & Savory* FHI 25445!

5. **B. scabra** (*Schum. & Thonn.*) *K. Schum.* in Pflanzenw. Ost.-Afr. C : 385 (1895). *Diodia scabra* Schum. & Thonn. (1827). *Spermacoce ruelliae* DC. (1830)—F.T.A. 3 : 238. *Borreria ruelliae* (DC.) K. Schum. ex H. Thoms (1909)—F.W.T.A., ed. 1, 2 : 135 ; Chev. Bot. 349 ; F.W. Andr. Fl. Pl. A.–E. Sud. 2 : 428 ; Berhaut Fl. Sén. 80. *B. stachydea* of F.W.T.A., ed. 1, 2 : 135, partly (*Chev.* 22285) ; Chev. Bot. 349 not of (DC.) Hutch. & Dalz. Erect herb with or without branches, scabrid and coarsely pubescent about 1 ft. high ; flowers white, sometimes pink ; a weed.

Sen.: *Berhaut* 1137. Galam *Heudelot* 278! **Mali:** *fide* Chev. *l.c.* Dioura (Aug.) *Davey* 28! **Port.G.:** Gabu (Oct.) *Esp. Santo* 3494! **Guin.:** Pobéguin 342! Nzérékoré *Baldwin* 9735! **S.L.:** Freetown (June, Sept.) *T. Vogel* 72! *Hepper* 920! Kitchom, R. Scarcies (fr. Dec.) *Sc. Elliot* 4338! Gberia Fotombu (Sept.) *Small* 291! Musaia (Sept.) *Small* 239! **Iv.C.:** Nzi (Aug.) *Chev.* 22285! Toumodi *Boughey* GC 18555! Lopou (Dec.) *Leeuwenberg* 2323! **Ghana:** Accra *T. Vogel*! Nungua, Accra Plains (fr. June) *Ankrah* GC 20185! Sunyani (Sept.) *Vigne* FH 2461! Zuarungu (Sept.) *Hughes* FH 5336! **Togo Rep.:** Lomé *Warnecke* 47! **Dah.:** *Le Testu* 138! *Burton*! **N.Nig.:** Stirling, Lokoja (fl. & fr. Sept.) *T. Vogel* 157! Nupe *Barter* 442! Lemu, Niger Prov. (Oct.) *Hepper & Keay* 951! Kontagora *Dalz.* 216! Naraguta (Aug.) *Lely* 474! **S.Nig.:** Attah (fr. Sept.) *T. Vogel* 53! Lagos *Rowland*! Ibadan (Aug.) *Newberry* 57! Olokemeji F.R. (Aug.) *Keay* FH 37149! Obudu (Aug.) *Stone* 71! **[Br.]Cam.:** Bum, Bamenda Div. (May) *Maitland* 1358! **F.Po:** Moka (Aug.) *Melville* 404! Also in Cameroun, the Congos, Sudan, Uganda, Kenya, N. and S. Rhodesia, Nyasaland and Angola.

6. **B. stachydea** (*DC.*) *Hutch. & Dalz.* var. **stachydea**—F.W.T.A., ed. 1, 2 : 135 (1931) ; Berhaut Fl. Sén. 80. *Spermacoce stachydea* DC. (1830)—F.T.A. 3 : 237. *S. leucadea* Hochst. ex Hiern in F.T.A. 3 : 237 (1877). *Borreria leucadea* (Hochst. ex Hiern) K. Schum. (1891)—F.W. Andr. Fl. Pl. A.–E. Sud. 2 : 427. Erect or spreading, hairy herb, about 1 ft. high ; flowers mauve ; in savanna.

Sen.: Bakel (fr. Aug.) *Leprieur*! Vélor (fr. Apr.) *Berhaut* 3847! **Gam.:** Yundum (fr. Dec.) *Austin* 38! **Port.G.:** Antula, Bissau (fr. Oct.) *Esp. Santo* 2586! **Ghana:** Bole (Nov.) *Harris*! Gushiago to Pigu (fr. Oct.) *Rose Innes* GC 30726! Banda, Trans-Volta (fr. Dec.) *Adams & Akpabla* 4020! Kete-Krachi to Dutukpene (fr. Dec.) *Adams* 4627! **N.Nig.:** Nupe *Barter* 948! Share F.R., Ilorin (fr. Jan.) *Ujor* FHI 30250! Fodama (fr. Dec.) *Moiser* 261! Jos (fr. Dec.) *Coombe* 24! **[Br.]Cam.:** Gurum, Adamawa (Nov.) *Hepper* 2897! Also in Sudan.

6a. **B. stachydea** var. **phyllocephala** (*DC.*) *Hepper* in Kew Bull. 14 : 256 (1960). *Spermacoce phyllocephala* DC. Prod. 4 : 553 (1830). *Borreria galeopsidis* of Berhaut Fl. Sén. 79, not *Spermacoce galeopsidis* DC. A glabrous variety.

Sen.: Walo country (fr. Sept.) *Perrottet*! Kouma (Oct.–Dec.) *Perrottet*(?)! *Perrottet & Leprieur* (photo)! Senegal *R. Bose*! Tivaouane (fr. Oct.) *Trochain* 854!

7. **B. chaetocephala** (*DC.*) *Hepper* in Kew Bull. 14 : 256 (1960). *Spermacoce chaetocephala* DC. Prod. 4 : 554 (1830). *S. hebecarpa* (Hochst. ex A. Rich.) Oliv. [not of DC.] var. *major* Schweinf. Hiern in F.T.A. 3 : 237 (1877). *B. stachydea* of F.W.T.A., ed. 1, 2 : 135, partly (*Chev.* 24959), not of (DC.) K. Schum. *B. compacta* of Berhaut Fl. Sén. 79, not of (Hochst.) K. Schum. An erect annual herb with a rather stout stem, slightly branched, about 1 ft. high ; in savanna.

Sen.: Tambacounda (fr. Dec.) *Adam* 17336! Diohine (fr. Apr.) *Berhaut* 1461! **Mali:** Sansanding (Sept.) *Chev.* 24959! Ansongo (fr. Sept.) *Hagerup* 383! **N.Nig.:** Katagum *Dalz.* 165! Sokoto (Sept.) *Dalz.* 399! Also in Cameroun and Sudan.

7a. **B. chaetocephala** var. **minor** *Hepper* l.c. (1960). *B. hebecarpa* Hochst. ex A. Rich. Tent. Fl. Abyss. 1 : 347 (1848) ; F.T.A. 3 : 326 ; F.W. Andr. Fl. Pl. A.–E. Sud. 2 : 427. A slender erect annual up to 18 in. high ; flowers pink.

N.Nig.: Nupe *Barter* 1259! Also in Uganda and Sudan.

8. **B. ocymoides** (*Burm. f.*) *DC.* Prod. 4 : 544 (1830). *Spermacoce ocymoides* Burm. f. (1768). *Borreria ramisparsa* DC. (1830)—F.T.A. 3 : 238 ; F.W.T.A., ed. 1, 2 : 135 ; Chev. Bot. 349 ; Berhaut Fl. Sén. 79. *Tardavel ocymoides* (Burm. f.) Hiern (1898). A weak herb, erect or decumbent, 6–9 in. high ; flowers very small, white ; wayside weed.

Sen.: *Heudelot* 558! *Berhaut* 1125 ; 4559. **Guin.:** Conakry (Oct.) *Adam* 12580! Fouta Djalon (Oct.) *Adam* 12748! Bilima *Chev.* 15041! **S.L.:** Freetown (Sept.) *Hepper* 910! Rokupr (Sept.) *Jordan* 559! Yonibana (Oct.) *Thomas* 4271! Njala (Sept.) *Pyne* 8! **Lib.:** Du R. (Aug.) *Linder* 222! Brewersville (Sept.) *Barker* 1432! Ganta (July) *Harley* 1443! Jabroke (fr. July) *Baldwin* 6635! **Iv.C.:** Bouroukrou *Chev.* 16751! Guidéko *Chev.* 19075! **Ghana:** Kpandu *Robertson* 2! **N.Nig.:** Patti Lokoja *T. Vogel* 181! Jos Plateau *Lely* P738! **S.Nig.:** Okomu F.R., Benin (Feb.) *Brenan* 9155! Idanre (Jan.) *Brenan* 8738! Sapoba *Kennedy* 2675! Onitsha *Barter* 1765! **[Br.]Cam.:** Victoria, 1922 lava flow (fr. Sept.) *Ngongi* FHI 15089! Mamfe (fr. Dec.) *Migeod* 275! **F.Po:** El Pico (Dec.) *Boughey* 186! Also in Cameroun, Gabon, Ubangi-Shari, Congo, Uganda and S. Tomé.

9. **B. verticillata** (*Linn.*) *G. F. W. Mey.* Prim. Fl. Esseq. 83 (1818) ; F.W. Andr. Fl. Pl. A.–E. Sud. 2 : 428. *Spermacoce verticillata* Linn. (1753). *S. globosa* Schum. & Thonn. (1827)—F.T.A. 3 : 240 ; Chev. Bot. 248. Erect, glabrous branched undershrub or half-woody herb 1–2 ft. high ; flowers white in compact spherical heads ; a weed of cultivation and waste places.

Sen.: Richard-Tol (fr. Jan.) *Döllinger*! Rosso (Apr.) *Popov* 2! Dakar (fr. May) *Baldwin* 5705! St. Yago Isl. *Brunner*! **Gam.:** Brown-Lester 5! 6! Yundum (Sept.) *Wallace* 8! **Mali:** San *Chev.* 1026. Korienza (July) *Lean* 51! **Guin.:** Conakry *Chev.* 12066! Kouria *Caille* in Hb. Chev. 15094! Kankan to Kouroussa *Brossart* in Hb. Chev. 15671! **S.L.:** Freetown (Sept.) *Hepper* 909! Kambia (Dec.) *Deighton* 880! Mano *Thomas* 10335! **Lib.:** Monrovia *Whyte*! Kakatown *Whyte*! Nyaake, Webo (fr. June) *Baldwin* 6161! Ganta (July) *Harley* 1441! **Iv.C.:** Prolo *Chev.* 19864! **Ghana:** Ada (May) *Adams* 2733! **Togo Rep.:** Lomé *Warnecke* 217. **N.Nig.:** Jos (Apr.) *Mutch* FHI 21867! Kano (fr. May) *Ejiofor* FHI 26150! Tilde Filani (May) *Lely* 228! **S.Nig.:** Badagry (Aug.) *Onochie* FHI 33478! Ibadan (Jan.) *Onochie* FHI 8184! Port Harcourt (June) *Maitland* B! Also in Gabon, S. Tomé, Congo, Angola, Sudan and Madagascar.

10. **B. natalensis** (*Hochst.*) *K. Schum. ex S. Moore* in J. Linn. Soc. 40 : 103 (1911). *Spermacoce natalensis* Hochst. (1844). *S. stricta* of F.T.A. 3 : 236, partly (in syn.), not of Linn. f. Straggling or more or less erect perennial herb with yellow-green branched stems 1–2 ft. long ; flowers white, tinged mauve ; wayside weed.

[Br.]Cam.: Bamenda : Bafut-Ngemba F.R., 6,000 ft. *Hepper* 2143! *Onochie* FHI 34861! Bambui *Pedder* 29! Basenako *Maitland* 1672! Nguroje, Mambila Plateau (Jan.) *Hepper* 1760! Widespread in tropical Africa.

11. **B. pusilla** (*Wall.*) *DC.* Prod. 4 : 543 (1830) ; F.W. Andr. Fl. Pl. A.–E. Sud. 2 : 427. *Spermacoce pusilla* Wall. (1820). *S. stricta* Linn. f. (1781)—F.T.A. 3 : 236 ; Chev. Bot. 349. *Borreria stricta* of F.W.T.A., ed. 1, 2 : 135 ; Berhaut Fl. Sén. 79, not of G. F. W. Mey. (1818). *B. hebecarpa* of F.W.T.A., ed. 1, 2 : 135, partly (*Linder* 1498), not of Hochst. ex A. Rich. Small erect herb more or less branched, 6–16 in. high ; flowers whitish ; in sandy places.

Sen.: *Roger*! Niokolo-Koba *Berhaut* 1558. **Guin.:** Mt. Loura, Fouta Djalon (Sept.) *Schnell* 7135!
S.L.: Lumley (Oct.) *Deighton* 1788! Bonthe Isl. (Nov.) *Deighton* 2295! Gberia Fotombu (Oct.) *Small*
292! 365! Kolahun (Oct.) *Glanville* 14! **Lib.:** Duport (Nov.) *Linder* 1498! **N.Nig.:** Vom *Dent Young*
127! **S.Nig.:** Sobo Plain, Warri Prov. (Sept.) *Butler-Cole* 23! Throughout tropical Africa and in
tropical Asia.
  [A closely related species seems to have been introduced into our area (Freetown *Deighton* 231 ;
Ibadan *Onochie* FHI 3316 ; Victoria *Maitland* 51). They match unnamed specimens from Zanzibar and
E. Tanganyika, and others from India where they are regarded as "*Spermacoce stricta* Linn. f.". The
correct name is at present in some doubt—F.N.H.]
12.   **B. filifolia** (*Schum. & Thonn.*) *K. Schum.* in E. & P. Pflanzenfam. 4, 4 : 144 (1891). *Octodon filifolium*
Schum. & Thonn. in Beskr. Guin. Pl. 74 (1827) ; F.T.A. 3 : 241 ; F.W.T.A., ed. 1, 2 : 136 ; Chev.
Bot. 349 ; F.W. Andr. Fl. Pl. A.–E. Sud. 2 : 448. Erect slender glabrous herb, simple or branched,
1–2 ft. high or sometimes a few inches high ; flowers white, the corolla-lobes sometimes purple ;
amongst grass in moist places.
Sen.: *Heudelot* 249! **Mali:** San (fl. & fr. Sept.) *Chev.* 2114! Ansongo (fl. & fr. Sept.) *Hagerup* 389!
Dioura (Sept.) *Davey* 85! Bamako (Sept.) *Waterlot* 1390! **Port.G.:** Pitche to Burumtuma (July) *Esp.
Santo* 2710! Dandum, Bafata (Sept.) *Esp. Santo* 3379! Lenquerem, Gabu (Aug.) *Esp. Santo* 2736!
Guin.: Ditinn (fr. Sept.) *Schnell* 7402! Mamou (Sept.) *Schnell* 6782! Sambailo (fr. Oct.) *Adam* 12715!
Ghana: Prampram (fl. & fr. Jan.) *Robertson* 74! Pepease (Nov.) *Irvine* 1709! Bole to Wa (fl. & fr. Sept.)
Rose Innes GC 30180! Kpedsu (fl. & fr. Dec.) *Howes* 1034! **N.Nig.:** Lokoja (fl. & fr. Sept.) *T. Vogel*
118! Bida to Badeggi (fl. & fr. Aug.) *Onochie* FHI 35413! Anara F.R., Zaria (fl. & fr. Oct.) *Hepper* 993!
Jos Plateau (July) *Lely* P537! Abinsi (Aug.) *Dalz.* 640! Also in Congo ?, Sudan and N. and S. Rhodesia.

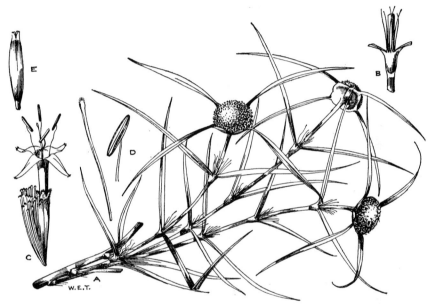

FIG. 246.—BORRERIA OCTODON *Hepper* (RUBIACEAE).
A, flowering shoot.   B, stipules.   C, flower and bracts.   D, anther.   E, fruit.

13.   **B. octodon** *Hepper* in Kew Bull. 17 : 171 (1963). *Octodon setosum* Hiern in F.T.A. 3 : 242 (1877) ;
F.W.T.A., ed. 1, 2 : 136 ; Chev. Bot. 349. *Borreria setosa* (Hiern) K. Schum. (1891), not of Mart. & Gal.
(1844). Erect, usually branched, scabrid herb about 2 ft. high ; flowers white, sometimes tinged pale
mauve ; in open sandy savanna.
Sen.: Faracounda to Tambanaba *Chev.* 2066! **Mali:** Koumantou (Apr.) *Chev.* 723! Toukoto *Chev.* 70!
Port.G.: Piche (Oct.) *Esp. Santo* 3418! Gabu (Oct.) *Esp. Santo* 3503! **Guin.:** Bounalol (Oct.) *Adam*
12569! **Iv.C.:** Mt. Orumbo-Boka, Toumodi (Dec.) *Boughey* GC 14449! Angouakoukro (Aug.) *Chev.*
22401! Béoumi (Dec.) *Lowe*! N. of Ferkéssédougou *Leeuwenberg* 1986! **Ghana:** Shai Plains (Jan.)
Johnson 527! Attabubu (Nov.) *Morton*! Menji to Banda (Dec.) *Adams* 3116! Kpedsu (Dec.) *Howes*
1036! Bimbila (Mar.) *Hepper & Morton* A3103! **Dah.:** *Burton*! Ouidah to Adjounga (Apr.) *Chev.*
23470! **N.Nig.:** Nupe *Barter* 1258! Bida to Zungeru (Oct.) *Hepper & Keay* 952! Anara F.R., Zaria
(Sept.) *Olorunfemi* FHI 24367! Jos Plateau *Batten-Poole* 306! Maifula, Muri Dist. (Nov.) *Latilo &
Daramola* FHI 28712! **S.Nig.:** Lagos *W. MacGregor* 184! 269! Eruwa to Lanlate (Oct.) *Savory &
Keay* FHI 25295! Oyo (fl. & fr. Jan.) *Hambler* 130! Enugu (Sept.) *Onochie* FHI 34113! [**Br.**]**Cam.:**
Gurum, Adamawa (Nov.) *Hepper* 1309!
14.   **B. macrantha** *Hepper* var. **macrantha**—in Kew Bull. 14 : 258 (1960). Straggling, densely pubescent,
perennial herb with subsimple stems 2–4 ft. long ; flowers white ; in upland grassland.
Guin.: Macenta, 2,000–2,500 ft. (Oct.) *Baldwin* 9780! **S.L.:** Loma Mts. 3,500 ft. (Sept.) *Jaeger* 1433!
Sankan Biriwa massif (fr. Jan.) *Cole* 128! **Iv.C.:** Nimba Mts. (Aug., Sept.) *Boughey* GC 18077! 18157!
Schnell 1852! Mt. Tonkoui (Oct.) *Aké Assi* 5972!
14a. **B. macrantha** var. **glabra** *Hepper* l.c. (1960). Straggling herb with simple stem 2½ ft. long ; flowers white.
Guin.: Nzérékoré, Mt. Koire (Sept.) *Baldwin* 13299! Guékedou (July) *Adam* 5594! **Iv.C.:** Mt. Tonkoui
(Oct.) *Aké Assi* 5973! *Villiers*!
15.   **B. paludosa** *Hepper* l.c. 259 (1960). *B. compressa* of F.W.T.A., ed. 1, 2 : 135, partly (*Barter* 1231 ; *Sc.
Elliot* 5264), not of Hutch. & Dalz. *Spermacoce compressa* of Chev. Bot. 348, not of Afzel. ex Hiern.
Erect herb 1–2 ft. high ; flowers pink or mauve ; in moist places.
Guin.: Timbo to Ditinn (fr. Sept.) *Chev.* 18470! 18471! Kouria to Ymbo (fr. July) *Caille* in *Hb. Chev.*

14617! R. Dantilia (fl. & fr. Mar.) *Sc. Elliot* 5264! **S.L.**: Kambia (fr. Mar., fl. & fr. May) *Jordan* 433! 873! **N.Nig.**: Nupe *Barter* 1231! **S.Nig.**: Afikpo, Ogoja Prov. (fl. & fr. June) *Stone* 9!

16. **B. compressa** *Hutch. & Dalz.* F.W.T.A., ed. 1, 2 : 135 (1931); Hepper in Kew Bull. 14 : 258 (1960). *Spermacoce compressa* Afzel. ex Hiern in F.T.A. 3 : 235 (1877), not of Wall. ex G. Don (1834). *Borreria velorensis* Berhaut Fl. Sén. 78 (1954), French desc. only. More or less erect herb, glabrous, 1–2 ft. high; flowers white, yellowish or sometimes tinged mauve; in moist places on rock out-crops.
Sen.: Vélor (Oct., fr. Nov.) *Berhaut* 692! 4057! **Mali:** Diafarabé (Oct.) *Davey* 372! **Guin.**: Conakry (fr. Oct.) *Adam* 12599! Kouroussa *Pobéguin* 440! Filicoundji (Oct.) *Adam* 12589! Macenta (Oct.) *Baldwin* 9764! **S.L.**: *Afzelius*! Hill Station, Freetown (Oct.) *Deighton* 2151! Wonkifu (Sept.) *Jordan* 346! Bafodia (Sept.) *Small* 265! Musaia (Sept.) *Small* 211! **Lib.**: Vonjama (fl. & fr. Oct.) *Baldwin* 9860! **Iv.C.**: Mbayakro (fr. Aug.) *Chev.* 22287! **Ghana:** Ayafie (fr. Dec.) *Adams* 4689! Attabubu (Nov.) *Morton* A2734! Damongo (Mar.) *Morton* GC 8675! Yendi to Kete Krachi (Apr.) *Morton* GC 9098!

17. **B. princeae** *K. Schum.* var. **princeae**—in Engl. Bot. Jahrb. 34 : 341 (1904). '*Diodia scandens* of F.W.T.A., ed. 1, 2 : 133, partly (*Johnston* 97), not of Sw. Straggling herb with several subsimple branches about 2 ft. long; flowers whitish; at the edge of montane forest.
N.Nig.: Cam. (Apr.) *Lely* P224! **S.Nig.**: Obudu Plateau, 5,500 ft. *Stone* 87! [**Br.**]**Cam.**: Cam. Mt., 5,000–7,000 ft. (fl. & fr. Feb.–Apr.) *Johnston* 97! *Maitland* 1327! *Brenan* 9552! Bafut-Ngemba F.R., 7,000 ft., Bamenda (Sept., Feb.) *Hepper* 2107! *Ujor* FHI 30212! Binka, 6,500 ft., Nkambe Div. (Sept.) *Savory* UCI 366! Also in Congo, Uganda, Kenya and Tanganyika.

17a. **B. princeae** var. **pubescens** *Hepper* in Kew Bull. 14 : 259 (1960). Straggling herb with few branches about 1 ft. long; flowers white; in montane forest.
[**Br.**]**Cam.**: Buea, 4,000 ft. (Sept.) *Dundas* FHI 15316! **F.Po**: Moka, 4,000 ft. (Dec.) *Boughey* 32! *Melville* 404! Also in Uganda, Kenya and Tanganyika.

18. **B. intricans** *Hepper* l.c. 257 (1960). *Spermacoce phyllocephala* of F.T.A. 3 : 240, partly (*Don*), not of DC. A more or less erect, branched herb about 1 ft. high: flowers white; in forest.
Sen.: Bafing (Oct.) *Adam* 12655! **Guin.**: Soumbalako to Boulivel (fl. & fr. Sept.) *Chev.* 18644! Nzérékoré (Sept.) *Adam* 6321! **S.L.**: *Don*! Sugar Loaf Mt., 1,000 ft. (fl. & fr. Nov.) *Deighton* 5624! Baiima (fl. & fr. Sept.) *Deighton* 3054! Yonibana (fr. Oct.) *Thomas* 4133! Bintumane Peak (fl. & fr. Nov.) *Jaeger* 539! **Ghana:** Begoro (fr. May) *Morton* A3664!

Imperfectly known species.

Spermacoce galeopsidis *DC.* Prod. 4 : 554 (1830); Hepper in Kew Bull. 14 : 256 (1960).
Sen.: Dagana *Leprieur & Perrottet.*

88. MITRACARPUS Zucc.—F.T.A. 3 : 243.

A herb; stems puberulous; leaves lanceolate, subacute, 3–6 cm. long, 0·7–1·5 cm. broad, glabrous below, scabrous above or nearly smooth; flowers densely crowded at the nodes within the pectinately divided stipular sheath; calyx-lobes unequal, two oblong and two subulate smaller ones; corolla-tube a little longer than the calyx; bracteoles filiform ..   .. *scaber*

M. scaber *Zucc.* in Schultes Mant. 210, 399 (1827)[1]; F.T.A. 3 : 243; Chev. Bot. 350; Berhaut Fl. Sén. 79 *M. verticillatus* (Schum. & Thonn.) Vatke (1876)—F.W.T.A., ed. 1, 2 : 136. *Staurospermum verticillatum* Schum. & Thonn. Beskr. Guin. Pl. 73 (1827).[1] Annual herb, half-woody at the base, erect, with terete branches, from a few inches to 2 ft. high; flowers white; a weed.
Maur.: Dikel *Charles* in *Hb. Chev.* 25486. **Sen.**: St. Louis (fl. & fr. May) *Roger*! *Chev.* 3493. Bondou *Heudelot* 136! M'bidgem *Chev.* 2096; 3534. Kaolak *Kaichinger* in *Hb. Chev.* **Gam.**: Brown-*Lester* 42! Genieri (fr. Feb.) *Fox* 43! Kuntaur *Ruxton* 156! **Mali:** Dioura, Macina (Nov.) *Davey* 162! Timbuktu (July) *Hagerup* 211! Kabarah *Chev.* 1339. **Guin.**: Kouroussa (June) *Pobéguin* 300! **S.L.**: Freetown (Dec. *Deighton* 1536! Kabala (Feb.) *Deighton* 826! Njala (Feb.) *Deighton* 1063! Falaba (Apr.) *Deighton* 1650! **Lib.**: Grand Cess (Mar.) *Baldwin* 11627! **Iv.C.**: Bingerville *Chev.* 16005. **Ghana:** Prampram (May) *Irvine* 1433! Kumasi (Sept.) *Darko* 343! Amedzofe (Oct.) *Morton* GC 6040! Tamale (Nov.) *Williams* 410! **Dah.**: Cotonou (Mar.) *Debeaux* 156! **N.Nig.**: Nupe *Barter* 1132! Jos Plateau (Jan.) *Lely* P124! Sokoto *Lely* 127! Maiduguri (Oct.) *Noble* A1! **S.Nig.**: Lagos *Dawodu* 10! Sapoba *Kennedy* 1779! Ibadan (Nov.) *Newberry & Etim* 164! [**Br.**]**Cam.**: Adamawa Prov. : Gangumi (Nov.) *Latilo & Daramola* FHI 28780! Gurum (Nov.) *Hepper* 1266! Also in Ubangi-Shari, the Congos, Sudan, Uganda, Kenya, Tanganyika, N. Rhodesia and Cape Verde Islands.

89. OTIOPHORA Zucc.—Verdcourt in J. Linn. Soc. 53 : 383 (1950).

Prostrate perennial herb; stems branched, densely pubescent; stipules small, fimbriate; internodes 5–25 mm. long; leaves ovate, truncate or subcordate at base, obtuse or subacute at apex, 8–15 mm. long, 5–13 mm. broad, nearly glabrous above, long hairs on midrib beneath, sessile; flowers in dense terminal clusters; calyx with 1 (–2) foliaceous elliptic lobe about 4 mm. long and 2 mm. broad, ciliate on margin and beneath, glabrous above; corolla-tube 4 mm. long, lobes 1·5 mm. long, sparsely ciliate; stamens 4–5, exserted; style exserted, stigma bifid; fruits ellipsoid, 2 mm. long, 2-celled; 1 seed in each cell ..   .. *latifolia* var. *bamendensis*

O. latifolia *Verdc.* in J. Linn. Soc. 53 : 406 (1950). Var. *latifolia* occurs in E. Africa.
O. latifolia var. **bamendensis** *Verdc.* l.c. 407 (1950). Prostrate perennial herb with occasionally-rooting branches about 6 in. long; flowers pink or light mauve; in rocky montane grassland.
[**Br.**]**Cam.**: Bamenda, 5,000 ft. (May) *Maitland* 1515! Nchan, 5,000 ft., Bamenda (May) *Maitland* 1445! Mayo Daga, Mambila Plateau (July) *Chapman* 58! Vogel Peak, 4,900 ft., Adamawa (fr. Nov.) *Hepper* 1511!

90. ANTHOSPERMUM Linn.—F.T.A. 3 : 229.

Fruits densely and softly tomentose; stems softly tomentellous, stout; leaves narrowly oblanceolate, acute, about 8 mm. long, glabrous; stipules sheathing, divided into several setiform segments, pubescent outside the sheath ..   .. 1. *cameroonense*
Fruits glabrous; stems scabrid-puberulous; leaves oblanceolate, markedly apiculate,

---

[1] As there is still some dispute as to whether the true publication date of Schumacher & Thonning's Beskr. Guin. Pl. was 1827 or 1829, priority for the epithet of this species is given to Zuccarini who certainly published his in 1827—F.N.H.

about 6 mm. long, slightly scabrid on the margin ; stipules divided into 2 segments
from a broad sheathing base ..   ..   2. *asperuloides*

1. **A. cameroonense** *Hutch. & Dalz.* F.W.T.A., ed. 1, 2 : 136 (1931). *A. asperuloides* of F.T.A. 3 : 230, partly
   (Cam. Mt. spec.). A heathy undershrub 1–4 ft. high ; flowers purplish-green ; in montane vegetation.
   [Br.]**Cam.**: Cam. Mt., 8,500–12,000 ft. (fl. & fr. Jan., Mar.) *Mann* 1290 ! *Brenan* 9531 !
2. **A. asperuloides** *Hook. f.* in J. Linn. Soc. 6 : 11 (1861) ; F.T.A. 3 : 230, partly (excl. Cam. Mt. spec.).
   Straggling heathy undershrub about 1 ft. high ; flowers creamy ; in montane vegetation.
   [Br.]**Cam.**: hut II, 8,100 ft., Cam. Mt. (fl. & fr. Dec., Jan.) *Keay* FHI 28592 ! *Morton* K796 ! **F.Po**: summit
   of Clarence Peak, 9,348 ft. (fr. Dec.) *Mann* 593 !

91. GALIUM Linn.—F.T.A. 3 : 244.

Leaves (and stipules) about 8 in a whorl, acute, linear-oblanceolate, 10–40 mm. long
  2–4 mm. broad, margin with recurved hooks ; flowers solitary, axillary ; pedicels
  elongating to 1·5 cm. long, recurved ; fruits glabrous or with hooked bristles, black,
  about 4 mm. diam.   ..   ..   1. *simense*
Leaves (and stipules) 4 in a whorl, apiculate, ovate-elliptic, 5–8 mm. long, 3–5 mm.
  broad, margin and surfaces setulose-pilose ; flowers cymose, cymes forming an
  oblong panicle ; pedicels 4–5 mm. long ; fruits densely clothed with hooked bristles,
  greenish, about 1·5 mm. diam.   ..   ..   2. *thunbergianum*

1. **G. simense** *Fresen.* in Mus. Senckenb. 2 : 165 (1837). *G. aparine* Linn. var. *spurium* Hiern in F.T.A. 3 : 245
  (1877), partly. *G. spurium* of F.W.T.A., ed. 1, 2 : 137, not of Linn. (?) *G. deistelii* K. Krause (1909),
  type not seen. Straggling herb with slender stems about 6 ft. long, readily clinging to clothes, skin, etc. ;
  flowers yellowish-green or white, fruits black ; in montane thickets.
  [Br.]**Cam.**: Cam. Mt., 5,000–9,000 ft. (fr. Nov.–Jan.) *Mann* 1324 ! *Dunlap* 29 ! *Keay* FHI 28634 ! Mt.
  Mbakakeka summit, 8,000 ft. (fl. & fr. Feb.) *Hepper* 2168 ! L. Oku, 7,200 ft. (fr. Jan.) *Keay & Lightbody*
  FHI 28489 ! Kumbo to Oku, 6,000 ft. (fl. & fr. Feb.) *Hepper* 2002 ! **F.Po**: 6,000–8,000 ft., *Guinea* 2887 !
  Also in Ethiopia.
2. **G. thunbergianum** *Eckl. & Zeyh.* Enum. 369 (1837) ; Hedberg in Sym. Bot. Upsal. 15, 1 : 178, 329 (1957).
  G. biafrae Hiern in F.T.A. 3 : 245 (1877). Scrambling, erect or ascending herb up to 3 ft. high ; flowers
  yellowish ; in rocky montane grassland.
  [Br.]**Cam.**: Cam. Mt., 7,000–12,000 ft. (fl. & fr. Nov., Jan., Apr.) *Mann* 1284 ! 2001 ! *Maitland* 1265 !
  Brenan 9560 ! **F.Po**: summit of Clarence Peak (Pico de S. Isabel), 9,348 ft. (fl. & fr. Dec., Mar.) *Mann* 605 !
  Guinea 2813 ! Also on other tropical African mountains, and in S. Africa.

138. DIPSACACEAE

By F. N. Hepper

Perennial or annual herbs. Leaves opposite or verticillate, entire or pinnately
divided ; stipules absent. Flowers hermaphrodite, zygomorphic, often crowded
into heads with an involucre of leafy bracts. Calyx epigynous, cupular or divided
into pappus-like segments. Corolla gamopetalous, lobes imbricate. Stamens
usually 4, rarely 2–3, alternate with the corolla-lobes and inserted usually
towards the bottom of the tube ; anthers 2-celled, opening lengthwise. Ovary
inferior, 1-celled ; style slender ; ovule solitary, pendulous from the top. Seeds
with large straight embryo in scanty endosperm.

Prickly herb ; calyx limb cup-shaped   ..   .. 1. **Dipsacus**
Herb without prickles ; calyx with 5 setose teeth ..   .. 2. **Succisa**

1. DIPSACUS Linn.—F.T.A. 3 : 249.

A prickly perennial herb ; stems stout, hollow, more or less ridged and with recurved
prickles ; leaves lanceolate, up to 35 cm. long, more or less pubescent above, prickly
on the midrib beneath, dentate, lower leaves petiolate, upper leaves sessile and auriculate ; inflorescences globose about 4 cm. diam. with acute deflexed involucral
bracts about 1·5 cm. long ; bracts of the receptacle equal to or exceeding the flowers
                                                     pinnatifidus

D. pinnatifidus *Steud. ex A. Rich.* Tent. Fl. Abyss. 1 : 367 (1848) ; Hedberg in Sym. Bot. Upsal. 15 : 183
  (1957). A stout erect perennial herb up to 5 ft. high, with prickly stems and midribs ; flowers white, in
  prickly, long-pedunculate heads ; in montane grassland.
  [Br.]**Cam.**: ridge above L. Oku, 7,700 ft., Bamenda Dist. (Jan.) *Keay & Lightbody* FHI 28515 ! Also in the
  mountains of eastern Africa.

2. SUCCISA Haller Hist. Stirp. Helv. 1 : 87 (1768). *Scabiosa* Linn.—F.T.A. 3 : 251, partly ; F.W.T.A., ed. 1, 2 : 137.

A perennial herb ; stems sparingly leafy, pubescent ; leaves lanceolate, subacute,
gradually narrowed into the winged petiole, 10–12 cm. long, 2–4 cm. broad, entire,
pilose above, glabrous or pilose only on the midrib beneath ; flowers in pedunculate
heads ; bracts narrowly lanceolate, about 1 cm. long, densely setose-pilose ; bracts
of the receptacle shorter than the flowers ; corolla pubescent outside   *trichotocephala*

S. trichotocephala *Baksay* in Ann. Mus. Nat. Hungar., n. ser. 2 : 249 (1952). *Scabiosa succisa* Linn. (1753)—
  F.T.A. 3 : 252 ; F.W.T.A., ed. 1, 2 : 137. Erect sparingly branched herb up to 4 ft. high from a perennial
  stock ; flowers white or pale mauve ; in montane grassland.
  [Br.]**Cam.**: Cam. Mt., 6,000–10,500 ft. (Jan.–Mar.) *Mann* 1309 ! *Dalz.* 8326 ! *Dundas* FHI 20356 ! Lakom,
  6,000 ft., Bamenda (Apr.) *Maitland* 1667 ! above L. Oku, 7,700 ft. (Jan.) *Keay* FHI 28479 ! (See Appendix,
  p. 414).

Fig. 247.—Succisa trichotocephala *Baksay* (Dipsacaceae).

A, habit, × ⅙. B, compound inflorescence, × ⅓. C, inflorescences, × 1½. D, flower with bract pulled away, showing calyx, × 4. E, two bracts with apices of calyx segments, × 4. F, receptacular bracts, × 6. G, involucel, × 6. Drawn from FHI 28479.

224

139. COMPOSITAE

By C. D. Adams

Herbs, shrubs or rarely small trees or climbers. Leaves alternate or opposite, simple or variously divided ; stipules absent. Flowers (florets) crowded into heads (capitula) surrounded by an involucre of one or more series of free or connate bracts ; sometimes the heads compound with the capitula few- or single-flowered ; receptacle paleate, setose, pitted or naked, usually convex, sometimes elongated or concave. Florets of one or two kinds in each capitulum, herma-phrodite, unisexual or neuter, rarely dioecious, the outer ones often ligulate (ray-florets), the inner ones tubular (disk-florets), or all tubular, or all ligulate. Calyx epigynous, reduced to a pappus of persistent or caducous hairs, bristles or scales, or absent. Corolla sympetalous, 4–5-fid (actinomorphic disk-florets), filiform, ligulate or rarely bilabiate (zygomorphic ray-florets). Stamens 5, rarely 4, epipetalous ; filaments free ; anthers connate into a tube, rarely free, 2-locular, opening lengthwise, often appendaged at the apex and tailed at the base. Ovary inferior, 1-locular, 1-ovuled ; style of the hermaphrodite or female florets mostly 2-fid, the style-arms smooth, papillose or hairy, tapered, rounded, deltoid or truncate, with or without a terminal appendage. Ovule erect from the base. Fruit (achene) sessile, sometimes beaked. Seed without endosperm ; embryo straight with plano-convex cotyledons.

A highly advanced, easily recognised family with world-wide distribution.

KEY TO THE TRIBES.

*Florets not all ligulate ; sap usually clear :
†Florets of two kinds in each capitulum, the outer usually ligulate or filiform and female or neuter, the inner tubular and hermaphrodite, or the capitula unisexual :
Capitula bisexual :
Receptacle with paleae subtending the florets :
Leaves opposite .. .. I. HELIANTHEAE
Leaves alternate :
Anther-base obtuse or shortly mucronate :
Outer florets ligulate .. .. I. HELIANTHEAE
Outer florets eligulate .. .. IV. ASTEREAE
Anther-base sagittate, tailed ; outer florets ligulate .. .. .. V. INULEAE
Receptacle without paleae ; leaves mostly alternate :
Involucral bracts in one main series, sometimes with an outer series of smaller ones at the base (calyculus) .. .. II. SENECIONEAE
Involucral bracts in two or more series :
Anther-base not tailed :
Involucral bracts in two series ; leaves pinnately divided, or small and entire or toothed ; style-arms of the hermaphrodite florets truncate or obtuse
III. ANTHEMIDEAE
Involucral bracts usually in several series :
Style-arms of the hermaphrodite florets with a lanceolate or deltoid tip (see Fig. 253) ; leaves various .. .. IV. ASTEREAE
Style-arms of the hermaphrodite florets narrowly linear, papillose ; leaves subentire .. .. V. INULEAE
Anther-base tailed or sagittate ; style-arms of the hermaphrodite florets club-shaped or obtuse, smooth (see Fig. 255) :
Style of both types of floret divided .. .. V. INULEAE
Style of the sterile florets entire ; sap milky ; leaf-margin spiny
XI. ARCTOTIDEAE
Capitula unisexual ; leaves deeply pinnately divided ; male capitula in a terminal raceme ; female capitula axillary, clustered .. .. .. VI. AMBROSIEAE
†Florets all of one kind, all tubular and hermaphrodite :
Leaves opposite :
Receptacle paleaceous .. .. I. HELIANTHEAE
Receptacle not paleaceous :
Florets yellow .. .. X. HELENIEAE
Florets not yellow :
Style-arms tapering, minutely hairy all over .. .. .. VII. VERNONIEAE
Style-arms obtuse or clavate, usually glabrous .. VIII. EUPATORIEAE
Leaves alternate :
Involucral bracts in a single main series .. .. .. II. SENECIONEAE

Involucral bracts in 2-several series :
  Style-arms tapering, minutely hairy all over ..   ..   ..   VII. Vernonieae
  Style-arms truncate, obtuse or clavate :
    Anther-base rounded, not tailed   ..   ..   ..   VIII. Eupatorieae
    Anther-base sagittate, usually tailed at the base :
      Receptacle naked in simple capitula, or if heads compound capitula with 3 or
      more florets :
        Style-arms clavate without an apical appendage ..   ..   V. Inuleae
        Style-arms obtuse or broadened at the tip with a short glandular appendage ;
        anther-base tails very long   ..   ..   ..   IX. Mutisieae
      Receptacle setose, or if naked capitula 1-flowered in globose compound heads ;
      involucral bracts spinous ..   ..   XII. Cynareae
 * Florets all ligulate and hermaphrodite ; sap usually milky   ..   XIII. Cichorieae

Tribe I.—Heliantheae

Paleae of the receptacle flat or nearly so, not folded around the achenes :
  Involucral bracts free, the outer herbaceous, the inner merging into the paleae of the
  receptacle ; leaves simple :
    Achenes of two kinds, the outer flat, pectinate-winged, the inner angular, crowned
    with 2–3 bristles ..   ..   **1. Synedrella**
    Achenes all more or less alike, angled :
      Pappus present ; low weedy herbs ; ligules about 5, white or cream :
      Pappus paleaceous ..   ..   **2. Galinsoga**
      Pappus of spreading alternately long and short pennately plumose bristles
                                                3. Tridax
      Pappus absent ; tall herbs ; corolla-tube densely pubescent especially at the base ;
      ligules yellow ..   ..   **4. Guizotia**
  Inner involucral bracts connate at the base, the outer fewer and usually smaller ;
  leaves mostly divided :
    Leaves opposite ; erect herbs mostly with conspicuous flower-heads :
      Achenes flat or concavo-convex with or without lateral wings ; aristae 2, antrorsely
      barbellate or smooth, or absent ..   ..   ..   **5. Coreopsis**
      Achenes angled, several-ribbed, wingless ; aristae 2–4, retrorsely barbellate
                                                6. Bidens
    Leaves alternate ; pappus absent ; low herb with small flower-heads
                                             7. Chrysanthellum
Paleae of the receptacle concave or folded around the achenes :
  Inner involucral bracts not embracing the outermost achenes :
    Leaves alternate, ovate ; flower-heads terminal ; involucral bracts leafy ; achenes
    smooth without a pappus ..   ..   **8. Sclerocarpus**
    Leaves opposite :
      Ligulate florets absent :
        Achenes compressed ; capitula long-pedunculate, erect ; florets white
                                              9. Spilanthes
        Achenes 4-angled ; capitula very short-pedunculate, at length nodding ; florets
        yellow ..   ..   **10. Eleutheranthera**
      Ligulate florets present :
        Pappus forming a small cup or annulus with or without bristles :
          Ray-florets fertile (with style and stigmas and producing mature achenes) ; outer
          achenes more or less triquetrous :
            Ligules conspicuous, yellow ; leaves lobulate ; petioles short ; capitula long-
            pedunculate ; achenes tuberculate with caducous aristae   ..   **11. Wedelia**
            Ligules small, white ; leaves simple ; petioles slender ; capitula subsessile or
            shortly pedunculate ; achenes minutely transversely rugulose with persistent
            barbellate aristae ..   ..   **12. Blainvillea**
          Ray-florets sterile (although style and stigmas rarely present) ; all achenes more
          or less alike, bilaterally convex or obscurely 4-angled   ..   ..   **13. Aspilia**
        Pappus composed of separate bristles or reduced to teeth, not united into a cup or
        annulus :
          Achenes thick or angular, not compressed :
            Scales of the receptacle numerous and conspicuous, broad ; flower-heads fairly
            large ; stem often grooved and appressed-strigose   ..   **14. Melanthera**
            Scales of the receptacle few and narrow ; ray-florets numerous, white, with
            narrow ligules ..   ..   **15. Eclipta**
          Achenes compressed :
            Achenes winged ..   ..   **16. Verbesina**
            Achenes often marginate but not winged ..   ..   ..   **9. Spilanthes**
  Inner involucral bracts embracing the fertile ray-achenes :
    Flower-heads axillary, sessile or subsessile :
      Achenes armed with hooked spinules ; ray-florets in one series 17. **Acanthospermum**

Achenes not spinose, with a pappus ; ray-florets in several series   ..   18. **Enydra**
Flower-heads paniculate ; achenes curved, glabrous ; ray-florets uniseriate ; involucral bracts glandular   ..   ..   19. **Sigesbeckia**

Tribe II.—Senecioneae

Involucral bracts connate, the free tips shortly triangular ; capitula discoid   20. **Bafutia**
Involucral bracts free to the base at least in fruit ; capitula discoid or radiate :
  Shrubby climbers ; leaves rather fleshy, broadly rounded, repand-dentate, with a
    crooked petiole-base ; capitula discoid, heterogamous ; florets long-tubular below,
    campanulate distally   ..   ..   21. **Mikaniopsis**
  Herbs, herbaceous climbers, shrubs or small trees :
    Plants succulent ; flowering shoots leafless or with reduced leaves ; capitula discoid ;
      calyculus absent but peduncles bracteolate often close to the involucre   22. **Kleinia**
    Plants not or hardly succulent ; flowering shoots leafy :
      Style-arms long-tapered, flexuous ; capitula discoid   ..   ..   ..   23. **Gynura**
      Style-arms truncate or obtuse, with or without a terminal appendage :
        Ray-florets absent :
          Calyculus absent ..   ..   24. **Emilia**
          Calyculus present :
            Style-arms with a short or long awl-shaped appendage   .. 25. **Crassocephalum**
            Style-arms not appendaged   ..   ..   26. **Senecio**
        Ray-florets present   ..   ..   26. **Senecio**

Tribe III.—Anthemideae

Capitula heterogamous, discoid, the outer florets female in several series; corolla of the
  hermaphrodite florets 4-toothed ; pappus absent :
  Leaves pinnatifid or pinnatisect ; capitula shortly stalked   ..   ..   27. **Cotula**
  Leaves entire, subentire or toothed ; capitula subsessile   ..   28. **Centipeda**

Tribe IV.—Astereae

Pappus well-developed, setaceous :
  Ligulate florets present though sometimes small :
    Shrubs or climbers   ..   ..   29. **Microglossa**
    Herbs :
      Ligule narrow, dull yellow or greenish ; biennial or perennial herbs   30. **Erigeron**
      Ligule oblong, bright yellow ; annual herb ; lower leaves opposite ..   31. **Felicia**
  Ligulate florets absent ; herbs :
    Style of the hermaphrodite florets 2-lobed ; achenes flattened ; florets dull or pale
      yellow   ..   ..   32. **Conyza**
    Style of the hermaphrodite florets entire or emarginate ; achenes terete ; florets
      purplish   ..   ..   33. **Adelostigma**
Pappus poorly developed or absent :
  Receptacle paleaceous, flat ; pappus minutely setiform ..   ..   ..   34. **Ceruana**
  Receptacle naked, convex to columnar :
    Pappus absent ; florets greenish or mauve   ..   ..   ..   35. **Dichrocephala**
    Pappus present ; florets yellow :
      Pappus of a few minutely hooked caducous setae   ..   ..   36. **Microtrichia**
      Pappus of minute curved subspathaceous teeth   ..   ..   37. **Grangea**

Tribe V.—Inuleae

Capitula separate, not crowded into compound heads :
  Receptacle with paleae subtending the florets ; ray-florets ligulate :
    Involucre at length subglobose, hard ; branches dichotomous, broadly winged ;
      leaves subentire   ..   ..   38. **Geigeria**
    Involucre hemispherical ; branches not winged ; leaves crenately toothed or pinnately
      divided :
      Pappus uniform of scales as long as the ovary ; leaves sessile ; low undershrub
                                                                     39. **Bubonium**
      Pappus usually unequal of jagged short scales with or without intermediate bristles ;
        leaves petiolate ; erect or bushy herbs   ..   ..   ..   40. **Anisopappus**
  Receptacle without paleae :
    Female (outer) florets ligulate :
      Pappus an inconspicuous ring ; achenes barrel-shaped, strongly ribbed   41. **Mollera**
      Pappus of at least the inner series setaceous :
        Outer pappus present :
          Outer pappus very short, paleaceous, connate at the base into a cup
                                                                     42. **Pulicaria**
          Outer pappus setaceous ..   ..   43. **Inula**

Outer pappus absent .. .. **44. Vicoa**
Female florets filiform or tubular, or all florets hermaphrodite and tubular :
  Pappus absent ; style terete, undivided or minutely emarginate ; stem narrowly
    winged .. .. **45. Epaltes**
  Pappus present :
    Style-arms of the hermaphrodite florets filiform or broadened and rounded at the
      apex, not truncate :
      Shrubs or undershrubs ; involucral bracts ovate .. .. .. **46. Pluchea**
      Herbs :
        Anther-base tailed :
          Outer florets tubular :
            Pappus uniform, setaceous :
              Perennial herbs .. .. **43. Inula**
              Annual herb .. .. **44. Vicoa**
            Pappus 2-seriate, the outer short, multifid, the inner longer, plumose ; annual
              herb .. .. **47. Pegolettia**
          Outer florets filiform :
            Pappus bristles 5–15 .. .. **48. Porphyrostemma**
            Pappus bristles numerous .. .. **49. Blumea**
        Anther-base not tailed :
          Pappus bristles 5–10 .. .. **50. Nicolasia**
          Pappus bristles numerous .. .. **51. Laggera**
    Style-arms of the hermaphrodite florets truncate ; herbs usually with woolly stems
      and leaves, especially when young ; involucral bracts often thin, scarious,
      hyaline or coloured :[1]
      Plants perennial .. .. **52. Helichrysum**
      Plants annual .. .. **53. Gnaphalium**
Capitula crowded into globose or hemispherical compound heads :
Herbs ; achenes subterete or slightly compressed ; pappus absent **54. Sphaeranthus**
Scrambling shrub with a thorn at the base of the petiole of each leaf ; achenes of the
  female florets winged ; pappus of few stiff bristles .. **55. Blepharispermum**

Tribe VI.—Ambrosieae

Capitula monoecious, rather small in racemes ; female capitula 1-flowered in axillary
  clusters ; bushy aromatic herb .. .. **56. Ambrosia**

Tribe VII.—Vernonieae

Capitula few-flowered in glomerules subtended by leafy bracts ; pappus bristly
                                           57. Elephantopus
Capitula 1- to many-flowered, not tightly clustered :
  Pappus present :
    Pappus more or less persistent :
    Pappus cupular or coroniform :
      Flower-heads sessile in the leaf-axils ; tubular pappus obscurely toothed
                                          58. Struchium
      Flower-heads small in lax terminal corymbs ; pappus coroniform, pectinately
        toothed .. .. **59. Triplotaxis**
    Pappus segments free :
      Flower-heads large, solitary, terminal, much-overtopped by the elongated upper
        leaves ; pappus-setae numerous in many series .. .. .. **60. Aedesia**
      Flower-heads not much-overtopped by elongated upper leaves ; pappus-setae in
        2 to few series :
        Inner involucral bracts connate ; outer pappus of scales, inner pappus of 3–5
          longer barbellate bristles ; low spreading herb .. .. .. **61. Herderia**
        Involucral bracts free ; pappus uniform or if outer paleaceous, inner of numerous
          bristles ; erect herbs, shrubs or trees .. .. .. **62. Vernonia**
    Pappus early caducous, uniform :
      Achenes obovate, 4–7-ribbed ; leaves mostly opposite or verticillate **63. Erlangea**
      Achenes linear-oblong, closely 10-ribbed ; leaves alternate .. **64. Centratherum**
  Pappus absent :
    Achenes broadly truncate at the apex .. .. **65. Ethulia**
    Achenes rounded at the apex .. .. **66. Gutenbergia**

Tribe VIII.—Eupatorieae

Pappus-setae numerous ; anther-connective produced at the apex :
  Herbs or shrubs ; involucral bracts several to numerous ; leaves opposite or alternate
                                      67. Eupatorium

---

[1] *Gnaphalium* and *Helichrysum* together form a natural group within which no clear generic distinctions
can be drawn.

Climbers ; involucral bracts 4–5　..　..　..　..　..　.. 68. **Mikania**
Pappus-setae 5 or fewer ; leaves opposite :
Pappus gland-tipped, adhesive ; apex of the anther truncate or minutely apiculate
　　　　　　69. **Adenostemma**
Pappus of 5 attenuated scales, not gland-tipped ; apex of the anther with an ovate
　appendage　..　..　..　..　..　..　..　..　.. 70. **Ageratum**

Tribe IX.—Mutisieae

Caulescent branched herbs :
Involucral bracts about 8-seriate, not pungent-pointed ; florets bright red
　　　　　　71. **Pleiotaxis**
Involucral bracts in 5–6 series, rigid, pungent-pointed　..　..　.. 72. **Dicoma**
Acaulescent scapigerous herb with a solitary capitulum　..　..　.. 73. **Gerbera**

Tribe X.—Helenieae

Capitula homogamous ; involucral bracts in few series, rather large ; anthers appen-
　daged at the apex, not tailed at the base ; style arms obtuse　　74. **Hypericophyllum**

Tribe XI.—Arctotideae

Capitula heterogamous, radiate ; involucral bracts connate at the base, spinous ;
　achenes villous ; pappus of denticulate obtuse paleae in several rows　75. **Berkheya**

Tribe XII.—Cynareae

Capitula 1-flowered, crowded into globose heads　..　..　..　.. 76. **Echinops**
Capitula many-flowered, not crowded into regular globose heads :
Pappus-setae connate into a ring at the base and deciduous together ; setae barbellate
　　　　　　77. **Carduus**
Pappus-setae more or less connate, persistent, paleaceous at the base, plumose above
　　　　　　78. **Atractylis**
Pappus-setae free or absent　..　..　..　..　..　.. 79. **Centaurea**

Tribe XIII.—Cichorieae

Pappus-setae plumose　..　..　..　..　..　..　..　.. 80. **Picris**
Pappus-setae smooth or barbellate :
Achenes beaked :
　Achenes compressed, ribbed or smooth, usually reddish or blackish-purple, glabrous,
　　or if terete greyish and minutely tuberculate ; pappus white ; involucral bracts
　　not setulose..　..　..　..　..　..　..　..　.. 81. **Lactuca**
　Achenes fusiform, many-ribbed, yellowish-brown, tapered into the beak, finely
　　spiculate ; pappus buff ; involucral bracts setulose　..　..　.. 82. **Crepis**
Achenes not beaked ; pappus mostly white :
Involucral bracts without or with only narrow scarious margins ; achenes oblong-
　terete or more or less flattened, few or many-ribbed, rough or smooth　83. **Sonchus**
Involucral bracts with broad scarious margins ; achenes fusiform, few-ribbed, rough
　　　　　　84. **Launaea**

1. SYNEDRELLA Gaertn.—Benth. & Hook. f. Gen. Pl. 2 : 383.

Stems erect, appressed-pubescent ; leaves opposite, ovate, cuneate at base, up to 8 cm.
long and 5 cm. broad, finely crenate-serrate, shortly appressed-setulose on both
surfaces ; petiole winged, bristly-ciliate ; flower-heads crowded and subsessile in the
leaf axils ; involucral bracts herbaceous, broadly oblong-lanceolate, finely nervose,
8–10 mm. long, inner at length glumaceous ; ligulate florets 3–5, ligules short ; outer
achenes flat with pectinate wings, inner 2–3-sided, scabrid, crowned with 2 or 3
aristae ..　..　..　..　..　..　..　..　..　.. *nodiflora*

S. nodiflora *Gaertn.* Fruct. et Sem. Pl. 2 : 456, t. 171 (1791) ; F.T.A. 3 : 386, in obs. ; Chev. Bot. 372. *Blainvillea gayana* of Chev. Bot. 369, partly (*Chev.* 18639), not of Cass. A branched, usually annual herb, half-
woody below, a few inches to 4 ft. or more high, with narrow oblong clustered flower-heads and small
golden-yellow ray-florets.
Sen.: *Berhaut* 1340. Niokolo (Oct.) *Adam* 15674 ! **Mali:** Kita (Nov.) *Jaeger* ! **Port.G.:** Bissau (Nov.) *Esp.
Santo* 1018 ! **Guin.:** *Farmar* 323 ! Kouria (Nov.) *Caille* in *Hb. Chev.* 14848 ! Soumbalako to Boulivel,
Fouta Djalon (Sept.) *Chev.* 18639 ! **S.L.:** Leicester Peak (Dec.) *Sc. Elliot* 3888 ! Bumbuna (Oct.) *Thomas*
3224 ! Njala (Nov.) *Deighton* 1870 ! Rokupr (Nov.) *Jordan* 674 ! **Lib.:** Suacoco (Jan.) *Daniel* 86 ! Gbanga
(Sept.) *Linder* 553 ! Gletown, Tchien (July) *Baldwin* 6949 ! Cape Palmas (June) *Hale* 103 ! **Iv.C.:**
Adiopodoumé (Dec.) *Leeuwenberg* 2117 ! **Ghana** : Aburi (Nov.) *Howes* 1011 ! Asuansi (Feb.) *Williams* 10 !
Koforidua (Mar.) *Johnson* 627 ! Kumasi (Mar.) *Irvine* 109 ! Wenchi, Ashanti (Dec.) *Adams* 3228 ! Amedzofe
(Oct.) *Morton* GC 6028 ! **N.Nig.** : Jebba (Dec.) *Hagerup* 686 ! **S.Nig.** : *Rowland* ! Olokemeji (Nov.) *Obasheki*
FHI 23833 ! Ibadan (Oct.) *Keay* FHI 28103 ! Oban *Talbot* 392 ! [Br.]**Cam.:** Victoria (Jan., Sept.)
Maitland ! *Boughey* X1179 ! Buea, 3,000 ft. (Jan.) *Maitland* 278 ! Likomba (Dec.) *Mildbr.* 10784 !
Kumba (Apr.) *Hutch. & Metcalfe* 162 ! **F.Po:** S. Isabel (Dec.) *Guinea* 540 ! *Boughey* 189 ! A common
tropical weed. (See Appendix, p. 421.)

2. GALINSOGA Ruiz & Pav.—Benth. & Hook. f. Gen. Pl. 2 : 390 ;
Adams in J.W. Afr. Sci. Assoc. 3 : 113 (1957).

Stems glabrous or nearly so ; peduncle-hairs short, appressed-ascending, mixed with
shortly stalked spreading glands ; leaves ovate, subentire ; receptacle-scales trifid ;
achene-paleae about as long as the body of the achene, longer than the corolla of
the disk florets　　..　　　..　　　..　　　..　　　..　　1. *parviflora*
Stems and peduncles hispid with long spreading hairs and long-stalked glands ; leaves
broadly ovate, serrate-crenate ; receptacle-scales simple or shortly laciniate ; achene-
paleae about half the length of the body of the achene, shorter than the corolla of the
disk florets　　..　　..　　..　　..　　..　　..　　..　　..　　2. *ciliata*

1. **G. parviflora** *Cav.* Icon. 3 : 41, t. 281 (1794). An annual weed up to 18 in. high with white ray and yellow
disk florets in rather small heads.
[Br.]**Cam.**: Cam. Mt., 3,200–6,000 ft. (Jan., Mar., Apr.) *Dunlap* 119 ! *Brenan* 9587 ! *Boughey* GC 7031 !
Bamenda, 5,000 ft. (Jan., Mar.) *Migeod* 291 ! *Morton* K66 ! Native of S. America and introduced into
several localities in tropical Africa, mainly at medium elevations, in, for example, Sudan, Congo and
S. Tomé.

2. **G. ciliata** (*Raf.*) *Blake* in Rhodora 24 : 35 (1922) ; Lousley in Watsonia 1 : 240 (1950). *Adventina ciliata*
Raf. (1836). An annual weed like the last up to about 1 ft. high and distinctly more hairy.
[Br.]**Cam.**: Cam. Mt., 3,000–7,500 ft. (Nov.-Apr.) *Dunlap* 118 ! *Migeod* 9 ! *Dalz.* 8228 ! *Maitland* 309 !
Brenan 9263 ! *Akpabla* GC 10495 ! *Morton* GC 7094 ! Bambui, Bamenda, 6,300 ft. (Dec.) *Adams* GC 11208 !
Mambila Plateau, 4,600 ft. (Jan.) *Hepper* 1653 ! Native of S. America.

3. TRIDAX Linn.—Benth. & Hook. f. Gen. Pl. 2 : 392.

Annual (or perennial when mown) hispid herb with trailing stems branching low from
a tap root, sometimes rooting from the lower nodes ; leaves opposite, up to 7 cm.
long and 3 cm. broad, ovate-rhomboid, coarsely serrate-dentate, soft, cuneate at the
base to a slender petiole 1–2 cm. long ; flowering branches ascending ; capitula
solitary on slender peduncles up to 20 cm. long ; involucral bracts in 2 series, 3–5 mm.
long, the outer broadly ovate, the inner oblong, apiculate, green, shortly setulose ;
receptacle-scales glumaceous, linear, striate, acutely apiculate, glabrous except at
the tip, persistent ; ligulate florets female (4–)5(–7) in each head, limb about 5 mm.
long and 4 mm. broad, usually 3-dentate, spreading horizontally, with the style-arms
rolled in the mouth of the slender pubescent corolla tube ; disk florets tubular, hermaphrodite, the tube linear, 5 mm. long with 5 slightly spreading pubescent lobes ;
achenes grey-silky, 2 mm. long ; pappus-setae slightly longer of about 20 alternately·
long and short pennately plumose bristles, at length spreading ..　　..　*procumbens*

T. procumbens *Linn.* Sp. Pl. 900 (1753) ; Hutch. in Kew Bull. 1921 : 381 ; Trochain Mém. I.F.A.N. 2 : 385
(1940) ; Berhaut Fl. Sén. 70. A low herb with weak trailing branches and long-stalked flower-heads about
⅓ in. long ; ray-florets cream-coloured, drying yellow ; disk-florets yellow, often tinged reddish ; common
in waste places throughout the area, but apparently of quite recent introduction.
Sen. : Hann (Oct.) *Roberty* 15039 ! **S.L.** : Freetown (Sept., Oct.) *Deighton* 2144 ! *Hepper* 917 ! Njala (June)
Deighton 6071 ! Roruks (July) *Deighton* 2505 ! **Iv.C.** : Adiopodoumé (Oct.) *Leeuwenberg* 1753 ! **Ghana** :
Achimota (Oct.) *Irvine* 818 ! Accra (May) *Irvine* 1444 ! Kumasi (June) *Vigne* FH 3051 ! Wenchi, Ashanti
(Dec.) *Adams* 3277 ! Lawra (Apr.) *Adams* 4141 ! Takoradi (Apr.) *Brenan* 9614 ! **N.Nig.**: Zungeru (Sept).
Dalz. 204 ! Fodama (Dec.) *Moiser* 204 ! Kano (Dec.) *Hagerup* 631 ! Samaru, Zaria (May) *Baldwin*
12008 ! Jos Plateau (Feb.) *Lely* P137 ! **S.Nig.**: Olokemeji (Apr.) *W. D. MacGregor* 288 ! *Obaseki*
FHI 23825 ! Lagos (Dec.) *Dalz.* 1164 ! Yaba (Nov.) *Meikle* 501 ! Port Harcourt *Maitland* ! [Br.]**Cam.**:
Gurum, Adamawa (Nov.) *Hepper* 1228 ! Gwoza, 3,500 ft., Dikwa (Jan.) *McClintock* 145 ! Native to central
America, now spread to many warm countries.

4. GUIZOTIA Cass.—F.T.A. 3 : 384. *Nom. cons.*

Outer involucral bracts linear to lanceolate, usually longer than the inner and leafy ;
stems, upper surfaces of leaves, peduncles and involucral bracts mostly scabrid-
setulose ; leaves opposite, lanceolate, 6–18 cm. long, 2–5 cm. broad, serrate-crenate,
sessile, auriculate, sometimes perfoliate ; ray-florets with a broad several-nerved limb ;
corolla-tube densely pubescent especially at base ; achenes glabrous, quadrangular
1. *scabra*
Outer involucral bracts ovate or broadly lanceolate, hardly longer than the inner, not
leafy ; stems glabrescent ; leaves minutely scabrid above with prominent veins
beneath, serrulate to subentire ; peduncles and involucral bracts pubescent ; (other
characters as above) ..　　..　　..　　..　　..　　..　　..　　..　　2. *abyssinica*

1. **G. scabra** (*Vis.*) *Chiov.* in Ann. Ist. Bot. Roma 8 : 184 (1904). *Veslingia scabra* Vis. (1840). *G. schultzii*
Hochst. ex Sch. Bip. (1848)—F.T.A. 3 : 385 ; F.W.T.A., ed. 1, 2 : 141. An erect harshly pubescent
soft-wooded herb 1–6 ft. high with bright yellow florets in numerous heads 1 in. or more across ; ligules
of ray-florets 3–4-fid at the tip.
N.Nig.: Jos Plateau (Jan., July, Sept., Oct.) *Lely* 413 ! P113 ! P704 ! *Summerhayes* 49 ! Jos (Sept., Dec.)
Lely 589 ! *Hepper* 1160 ! *Monod* 9665 ! Birnin Gwari (Mar.) *Meikle* 1382 ! Kontagora (Nov.) *Dalz.* 194 !
Kurmin Damisa (Nov.) *Keay* FHI 21539 ! Kaduna (Nov.) *Keay* FHI 28124 ! **S.Nig.**: Ikwette, 5,200 ft.,
Ogoja (Dec.) *Savory & Keay* FHI 25234 ! [Br.]**Cam.**: Bamenda, 3,600–7,500 ft. (Dec., Jan.) *Migeod* 379 !
Boughey GC 10412 ! 10774 ! 11110 ! *Adams* GC 11097 ! 11285 ! Wum (Dec., Apr.) *Boughey* GC 11571 !
Morton K294 ! Kumbo (Nov.) *Egbuta* FHI 3772 ! Ndu, 6,400 ft. (Jan.) *Keay & Russell* FHI 28440 ! Vogel
Peak, 3,800 ft., Adamawa (Nov.) *Hepper* 1339 ! Extends to Sudan, Ethiopia, Eritrea, Congo, Uganda,
Kenya, Tanganyika, N. Rhodesia and Nyasaland.

2. **G. abyssinica** (*Linn. f.*) *Cass.* in Dict. Sci. Nat. 59 : 248 (1829) ; F.T.A. 3 : 384 ; Adams in J. W. Afr. Sci.
Assoc. 3 : 114 (1957). *Polymnia abyssinica* Linn. f. (1781). A nearly glabrous shrubby herb up to about
4 ft. high with numerous heads of bright yellow flowers 1–1½ in. across.
[Br.]**Cam.**: Gashaka, Adamawa (Dec.) *Latilo & Daramola* FHI 28966 ! From Ethiopia to Nyasaland
sporadically, and also in India and as a casual in Europe.

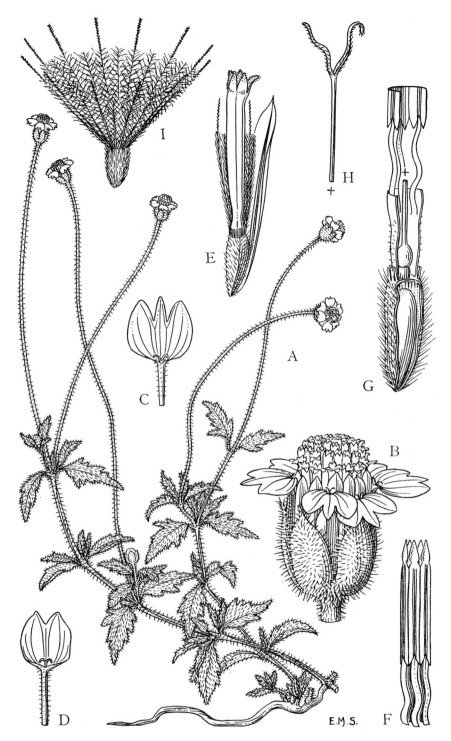

FIG. 248.—TRIDAX PROCUMBENS *Linn.* (COMPOSITAE).

A, habit, × ⅔. B, flower-head, × 4. C, ray floret—ovary removed. D, ray floret—
ovary removed, × 4. E, disk floret, × 8. F, stamens from disk floret, × 16. G, section
of ray floret, × 16. H, continuation of style, with stigmas, × 16. 1, fruit, × 6.
A, D, I, drawn from *Robinson* 811; B, C, E–H, from *Adams* s.n.

5. COREOPSIS Linn.—F.T.A. 3 : 387 ; Sherff in Publ. Field
Mus. 11 : 6 (1936).

Achenes with at least some outer ones in each capitulum broadly and distinctly winged ;
annual herbs :
 Leaves typically 3-partite, segments linear, serrate with mucronate teeth ; outer
  involucral bracts linear, inner broadly lanceolate, acuminate, with scarious margins ;
  winged achenes suborbicular about 12 mm. long and 8 mm. broad ; capitula up to
  7 cm. diam. ..   ..   1. *borianiana*
 Leaves typically simple, elliptic-lanceolate, coarsely toothed or rarely 3-lobed, irregularly serrate-crenate ; capitula as above but smaller ; winged achenes oblong or
  scutelliform, 6–7 mm. long, 3–5 mm. broad   ..   ..   ..   2. *barteri*
Achenes not or very narrowly or indistinctly winged :
 Aristae present :
  Involucral bracts broadly lanceolate in 2–3 series, about 1·5 cm. long and up to 8 mm.
   broad ; ligules 1 cm. or more broad ; achenes flat, oblong, antrorsely scabrid on
   both surfaces, stiffly ciliate on the margin, with 2 short divergent aristae smooth
   or sparingly antrorsely barbellate   ..   ..   3. *asperata*
  Involucral bracts not as above, smaller :
   Segments of leaves linear, very long, remotely serrate, rarely as much as 5 mm.
    broad ; outer involucral bracts linear, as long as or longer than the inner ; inner
    involucral bracts ovate-lanceolate, 5 mm. long, glabrous ; achenes 3 mm. long,
    scabrid ; perennial herb..   ..   ..   4. *camporum*
   Segments of leaves ovate, serrately cut, 1 cm. or more broad ; achenes 3–6 mm. long :
    Ligules about 2 cm. long, conspicuous ; capitulum in flower 3·5–4 cm. diam. ;
     achenes glabrous except for ciliate wing-margin :
     Stems and leaves glabrous   ..   ..   ..   5. *monticola* var. *monticola*
     Stems, leaves and outer involucral bracts pilose   5a. *monticola* var. *pilosa*
    Ligules 7–8 mm. long ; capitulum in flower not exceeding 2 cm. diam. ; achenes
     pubescent   ..   ..   6. *setigera*
 Aristae absent ; achenes convex with a marked midrib on the inner surface, glabrous,
  black, 2·5–3·5 mm. long, 1·5 mm. broad, not ciliate ; leaves up to about 5 cm. long,
  with short oblong apiculate segments 2–3 mm. broad ; annual herb 7. *occidentalis*

1. **C. borianiana** *Sch. Bip.* in Verh. Zool.-Bot. Ges. Wien 18 : 684 (1868) ; Sherff in Publ. Field Mus. 11 : 394.
C. guineensis Oliv. & Hiern in F.T.A. 3 : 390 (1877) ; Chev. Bot. 372, partly (excl. specimens from Guinée and Sierra Leone) ; F.W.T.A., ed. 1, 2 : 143, partly (excl. syn. *C. camporum* Hutch. & *Lely* 383); Berhaut Fl. Sén. 11, 70, 85. *C. togensis* Sherff in Bot. Gaz. 76 : 87 (1923). A nearly glabrous erect branched annual herb up to about 6 ft. high with golden-yellow flower-heads 2–3 in. across.
 Sen. : Niokolo (Oct.) *Adam* 15875 ! **Gam. :** *Saunders* 109 ! **Mali:** Tiguiberri (Jan.) *Chev.* 293 ! Koulikora (Apr.) 2083 ! San (Sept.) *Chev.* 2084 ! Bandiagara to Mopti, Macina (Sept.) *Chev.* 24949 ! Kita Massif (Sept.) *Jaeger* ! **Port.G.:** Pitche to Paiama (Oct.) *Esp. Santo* 3487 ! Sonaco (Nov.) *Esp. Santo* 3587 ! **Iv.C.:** Bouakrou to Alangouassou, Baoulé-Nord (July) *Chev.* 22228 ! **Ghana:** Tamale (Nov.) *Williams* 387 ! Ejura (Nov.–Jan.) *Darko* 748 ! *Morton* GC 9536 ! A1528 ! A2559 ! Ho (Nov., Dec.) *A. S. Thomas* K8 ! *Plumptre* 29 ! *Morton* A2296 ! Kete Krachi to Dutukpene (Dec.) *Adams* 4629 ! Paga *Vigne* 4592 ! **Togo Rep.:** *fide* Sherff *l.c.* **Dah.:** Agouagon (May) *Chev.* 23493 ! **N.Nig. :** Nupe *Barter* 933 ! Zungeru (Oct.) *Dalz.* 192 ! Abinsi (Nov.) *Dalz.* 652 ! Nabardo (Sept.) *Lely* 616 ! Osi, Ilorin (Dec.) *Ajayi* FHI 19299 ! Tula, Bauchi, 3,300 ft. (Nov.) *Summerhayes* 77 ! **S.Nig.:** *Rowland* ! (Oct.) *Foster* 136 ! Lagos *MacGregor* 283 ! Abeokuta (Nov.) *Irving* ! [Br.]**Cam.:** Bama, Dikwa Div. (Jan., Nov.) *McClintock* 3 ! 138 ! Also in Ubangi-Shari and Sudan.
2. **C. barteri** *Oliv. & Hiern* in F.T.A. 3 : 390 (1877) ; Sherff in Publ. Field Mus. 11 : 372. *C. badia* Sherff in Bot. Gaz. 76 : 90 (1923). An erect glabrous branched annual herb up to 3 ft. high with simple serrate upper leaves and golden-yellow flower-heads 1–2 in. across ; a weed of fields and grasslands in savanna.
 Ghana: Zuarungu (Dec.) *Adams & Akpabla* GC 4270 ! Navrongo (Dec.) *Adams & Akpabla* GC 4347 ! Bawku (Dec.) *Morton* A1352 ! **Togo Rep.:** *fide* Sherff *l.c.* **N.Nig.:** Borgu *Barter* 870 ! Biu, Bornu (Sept.) *Noble* 33 ! Naraguta (Aug., Sept.) *Lely* 43 ! *Keay* FHI 12732 ! Jos (Mar.) *Hill* 12 ! *Batten-Poole* 357 ! Buruku, Zaria (Aug.) *Barry* 116 ! Kufena Rock, Zaria (Aug.) *Keay* FHI 28019 ! **S.Nig.:** *Baikie* ! Lagos *Rowland* ! [Br.]**Cam.:** Kishong, Bamenda, 6,700 ft. (Jan.) *Keay & Russell* FHI 28447 ! Ndop to Kumbo (Dec.) *Boughey* GC 11162 ! 11170 !
3. **C. asperata** *Hutch. & Dalz.* F.W.T.A., ed. 1, 2 : 143 (1931) ; Sherff in Bot. Gaz. 93 : 219. *Bidens asperata* (Hutch. & Dalz.) Sherff *l.c.* 220 (1932) ; Sherff in Publ. Field Mus. 16 : 548, t. 139 (1937). *C. guineensis* of Chev. Bot. 372, partly (*Chev.* 20554 ; 20568), not of Oliv. & Hiern. A perennial herb up to about 10 ft. high with thick ribbed stems and deep yellow flower-heads 3 in. across ; in hill grassland and open savanna woodland.
 Guin. : Sambadougou to Boria, Faranna (Jan.) *Chev.* 20554 ! Beyla (Oct.) *Jac.-Fél.* 2031 ! **S.L.:** *Glanville* 384 ! Source of R. Niger, Sorémoudou (Jan.) *Chev.* 20568 ! **Iv.C.:** Gouékouma, Toura, Upper Sassandra (May) *Chev.* 21675 ! **Ghana :** Dutukpene 600–1,600 ft. (Dec.) *Adams* 4543 ! *Morton* ! Shiare (Apr.) *Hall* !
4. **C. camporum** *Hutch.* in Kew Bull. 1921 : 381 ; Sherff in Publ. Field Mus. 11 : 367. *C. guineensis* of F.W.T.A., ed. 1, 2 : 143, partly (*Lely* 383), not of Oliv. & Hiern. A perennial herb with several stems 2–4 ft. high from a woody stock ; ray-florets bright golden-yellow in heads about 1 in. across ; in hill grassland.
 S.L.: Bintumane Peak, 6,000 ft. (May, Aug.) *Deighton* 5095 ! *Jaeger* 516 ! 1243 ! **N.Nig. :** Guduma, near Minna (Oct.) *Hepper* 959 ! Vom *Dent Young* 153 ! Hepham to Ropp (July) *Lely* 383 ! 449 ! Jos Plateau (July, Aug.) *Lely* P516 ! P632 ! [Br.]**Cam.:** Bambulue, Bamenda, 6,000–7,000 ft. (Sept.) *Savory* 470 ! Bambili (Aug.) *Ujor* FHI 29966 ! Kumbo (Dec.) *Boughey* GC 17397 ! Bamenda to Ndop, 5,000–6,300 ft. (Dec.) *Adams* GC 11238 ! 11275 ! *Boughey* GC 10843 ! 11182b ! Also in Cameroun.
5. **C. monticola** (*Hook. f.*) *Oliv. & Hiern* var. **monticola**—in F.T.A. 3 : 390 (1877) ; Sherff in Publ. Field Mus. 11 : 372. *Verbesina monticola* Hook. f. (1864). A shrubby almost glabrous herb 3 ft. or more high ; young stems and leaves reddish ; ray-florets pale yellow ; heads 2–3 in. across in flower ; in mountain grassland.
 [Br.]**Cam.:** Cam. Mt., 5,000–9,000 ft. (Nov.-Jan., Apr.) *Mann* 1219 ! 1922 ! *Johnston* 26 ! *Maitland* 1191 ! *Brenan* 9558 ! *Boughey* GC 12511 !
6. **C. monticola** var. **pilosa** *Hutch. & Dalz.* F.W.T.A., ed. 1, 2 : 143 (1931) ; Sherff *l.c.* 373.
 [Br.]**Cam.:** Cam. Mt., 7,000–10,000 ft. (Nov., Jan., Apr.) *Migeod* 211 ! *Dunlap* 204 ! *Maitland* 450 !

W.E.T.

Fig. 249.—Coreopsis borianiana *Sch. Bip.* (Compositae).

A, ray-flower.   B, achene and subtending palea.   C, disk floret and palea.   D, achene.   E, anthers.   F, style-arms.

233

Irvine 1452 ! *Morton* GC 7074 ! Kumbo (Oct.) *Tamajong* FHI 23493 ! Mba Kokeka Mt., 6,500 ft. (Jan.)
Keay FHI 28390 ! Bafut-Ngemba F.R. (Aug.) *Ujor* FHI 29975 ! Santa Mt. (Dec.) *Boughey* GC 11027 !
[Note : Intermediates between vars. *monticola* and *pilosa* are represented by the following specimens :—
Bamenda, 7,000 ft. (Feb., Mar.) *Migeod* 489 ! *Morton* K139 ! These have slightly smaller capitula and the
young parts only are pubescent—C. D. A.]

6. **C. setigera** *Sch. Bip.* in Walp. Rep. 6 : 163 (1846). *Bidens setigera* (Sch. Bip.) Sherff in Bot. Gaz. 90 : 390
(1930) ; Publ. Field Mus. 16 : 627, t. 188. An annual herb 2–3 ft. high with yellow flowers about ¼ in.
across ; growing on rocks.
N.Nig.: Jos Plateau (Sept.) *Lely* P736 ! Also in Ethiopia.
[Note : The West African plant is generally puberulous ; Ethiopian plants are almost glabrous.]

7. **C. occidentalis** (*Hutch. & Dalz.*) *C. D. Adams* in J.W. Afr. Sci. Assoc. 6 : 149 (1961). *Microlecane occidentalis* Hutch. & Dalz. F.W.T.A., ed. 1, 2 : 143 (1931), partly (*Chev.* 18296). An erect annual herb up to
about 1 ft. high with yellow flower-heads 1 in. across ; in hill grassland.
Guin.: Irébéléya to Timbo, 2,000 ft. (Sept.) *Chev.* 18296 ! **Mali** (Sept.) *Schnell* 7060 ! [**Br.**]**Cam.**: Bafut-
Ngemba F.R., 5,000–6,000 ft. (Oct.–Nov.) *Lightbody* FHI 26275 !

6. BIDENS Linn.—F.T.A. 3 : 392 ; Sherff in Publ. Field Mus. 16 (1937).

Ray-florets present :
  Outer involucral bracts ovate to lanceolate or linear, shorter than the inner :
    Ligules yellow, often rather short and few ; leaves bipinnatisect, usually pubescent
    at least towards the base of the petiole ; leaflet-margin crenate-serrate ; peduncle
    ebracteate ..   ..   1. *bipinnata*
    Ligules white, usually conspicuous ; leaves pinnate (rarely simple), glabrescent ;
    leaflet-margin regularly serrate ..   ..   2. *pilosa*
  Outer involucral bracts linear-spathulate, up to 2 cm. long, much longer than the
  inner ; leaves bipinnatifid, thinly pilose ; leaflet-margin deeply serrate ; peduncle
  with a single bracteole below the involucre, similar to an outer involucral bract
                                          3. *biternata*
Ray-florets absent ; leaves pinnatifid, mostly alternate ; inner involucral bracts 3,
8–9 mm. long ; outer 2 or 3, 1·5–2·5 mm. long ; achenes narrow, slightly compressed,
8–9 mm. long, pilose, biaristate ..   ..   4. *minuta*

1. **B. bipinnata** *Linn.* Sp. Pl. 832 (1753) ; F.T.A. 3 : 393 ; *Chev.* Bot. 373, partly (Guinée specimens) ; Sherff
in Publ. Field Mus. 16 : 366, t. 89. An erect, sometimes bushy annual herb 1–4 ft. high, nearly glabrous or
pubescent on the stems and leaves when young ; ray-florets yellow, few or rarely absent.
Guin.: *fide* Chev. *l.c.* **S.L.**: Musaia (Dec.) *Deighton* 4508 ! **Ghana**: Nwereme, Ashanti (Dec.) *Adams* 5268 !
Sunyani (Dec.) *Adams* 5351 ! **N.Nig.**: Neill's Valley, Naraguta (June) *Lely* 267 ! Kadaura *Lely* 600 !
Kachia to Kaduna (Dec.) *Meikle* 316 ! Widespread in warm countries. (See Appendix, p. 416.)

2. **B. pilosa** *Linn.* Sp. Pl. 832 (1753) ; F.T.A. 3 : 392 ; *Chev.* Bot. 373 ; Sherff l.c. 412, t. 99 & 102 ; Berhaut
Fl. Sén. 11, 27, 48 ; Schnell in Ic. Pl. Afr. I.F.A.N. 5 : t. 99. *B. abortiva* Schum. & Thonn. (1827). An
erect annual herb with yellow disk-florets and white ray-florets ; fruits strongly adherent ; a common
weed of disturbed ground.
Sen.: *fide* Berhaut *l.c.* **Mali**: Kita Massif (Oct.) *Jaeger* ! **Port.G.**: Brandao to Fulacunda (Oct.) *Esp. Santo*
2208 ! Gabu (Oct.) *Esp. Santo* 3553 ! **Guin.**: (July) *Collenette* 72 ! Diaguissa (Apr.) *Chev.* 12681 *bis* ! Kouria
(Oct.) *Caille* in Hb. *Chev.* 14716 ! Mali (Dec.) *Schnell* 2410 ! **S.L.**: (June) *T. Vogel* 71 ! 113 ! Freetown (Aug.)
Burbridge 518 ! *Deighton* 72 ! Leicester Peak (Dec.) *Sc. Elliot* 3972 ! Pendembu (July) *Thomas* 850 !
Loma Mts. (Nov.) *Jaeger* 2035 ! **Lib.**: Monrovia (May) *Barker* 1317 ! Nyaake, Webo (June) *Baldwin*
6162 ! Gletown, Tchien (July) *Baldwin* 6781 ! Ganta (May, Sept.) *Harley* ! **Iv.C.**: Bouroukrou (Jan.)
Chev. 17021 ! Tonkoui (Aug.) *Boughey* GC 18329 ! Nimba Mts. *Schnell* 2929 ! **Ghana**: Aburi (Oct.) *Brown*
405 ! *Morton* GC 7872 ! Kpedsu (Dec.) *Howes* 1038 ! Mampong, Ashanti (Sept.) *Rattray* 27 ! *Rose Innes*
GC 30137 ! Wenchi (Dec.) *Adams* 3250 ! **Niger**: Aïr *De Miré & Gillet* MN2/10–11. **N.Nig.**: Naraguta *Lely*
42 ! 46 ! Mongu (July) *Lely* 396 ! Jos Plateau (Feb.) *Lely* P167 ! Vom *Dent Young* 154 ! **S. Nig.**: *T. Vogel* !
Lagos *Millen* 142 ! Ibadan (July, Oct., Dec.) *Ahmed & Chizea* FHI 20003 ! *Newberry* 109 ! 127 ! Sobo
Plains, Benin (Mar.) *Jones* FHI 6294 ! Calabar (May) *Holland* 47 ! [**Br.**]**Cam.**: Cam. Mt., 3,200–7,500 ft.
(Jan., Mar., Apr.) *Dunlap* 76 ! *Boughey* GC 7038 ! *Morton* GC 7078 ! Bamenda (Dec., Jan., Apr.) *Migeod*
314 ! *Ujor* FHI 30072 ! *Boughey* GC 10395 ! **F.Po**: (Dec.) *Mann* 68 ! S. Isabel (Dec.) *Boughey* GC 17372 !
Moka, 4,000 ft. (Dec.) *Boughey* 197 ! Musola *Guinea* 1077. Widespread in the tropics. (See Appendix,
p. 416.)

3. **B. biternata** (*Lour.*) *Merrill & Sherff* in Bot. Gaz. 88 : 293 (1929) ; Sherff in Publ. Field Mus. 16 : 388, t.
99. *Coreopsis biternata* Lour. (1790). *B. bipinnata* of Chev. Bot. 373, partly (*Chev.* 21959), not of Linn. An
annual weedy herb with yellow flowers.
Iv.C.: R. Béré to Dialokoro, Mankono (June) *Chev.* 21959 ! **Ghana**: Cape Coast (Nov.) *Morgan* GC 17363 !
N. Nig. : Naraguta (June) *Lely* 276 ! A tropical weed, also reported from Ethiopia and from Sudan to
Angola.

4. **B. minuta** *De Miré & Gillet* in J. Agric. Trop. 3 : 703, fig. 3 (1956). A therophyte only a few inches high
with rayless flower-heads nearly ½ in. long.
Niger: Mt. Baguesane, Aïr, 4,000 ft. (Nov.) *De Miré & Gillet* MN3–30.

7. CHRYSANTHELLUM Rich.—F.T.A. 3 : 394.

A low herb with branches radiating from the stem-base ; leaves alternate, deeply
pinnatisect, up to about 5 cm. long including the subamplexicaul petiole, segments
oblong-lanceolate, subacute, apiculate ; flower-heads small, numerous, pedunculate ;
involucral bracts 1–2-seriate, about 4 mm. long, obtuse, with scarious margins ;
ligulate florets usually 8–12 with very short ligules ; receptacle-scales linear, persistent;
achenes compressed, oblong, mostly winged with a thick straw-coloured slightly hairy
margin, emarginate or truncate at the apex ..   ..   *americanum*

C. americanum (*Linn.*) *Vatke* in Abh. Nat. Brem. 9 : 122 (1885) ; F.W.T.A., ed. 1, 2 : 608 ; Berhaut Fl. Sén.
173. *Anthemis americana* Linn. (1753). *Chrysanthellum procumbens* Pers. (1807)—F.T.A. 3 : 395 ; Chev.
Bot. 373 ; F.W.T.A., ed. 1, 2 : 143. A faintly aromatic branching annual herb, usually less than 1 ft. high.
with yellow disk-florets and paler very small rays ; a common weed of roadsides and waste places.
Sen.: *fide* Berhaut *l.c.* Niokolo-Koba (Oct.) *Adam* 15583 ! **Gam.**: *Hayes* 575 ! **Mali**: Bamako (Jan., Nov.)
Chev. 199 ! *Jaeger* ! **S.L.**: Kabala (July) *Glanville* 265 ! Musaia (Oct., Dec.) *Deighton* 4460 ! *Thomas* 2641 !

Iv.C. : N. of Bouaké (Nov.) *Leeuwenberg* 2069! **Ghana**: Senchi (June) *Morton* GC 7244! *Boughey* GC 13021!
Agogo, Ashanti (May) *Irvine* 592! *Williams* 272! Dedoro-Tankara F.R., Navrongo (June) *Andoh* 5906!
Zuarungu (Sept.) *Hughes* FH 5330! Ho to Amedzofe (Nov.) *Morton* A2291! **Dah.**: *Burton*! **Niger**: Aïr,
Gal (Aug.) *De Miré & Gillet in Hb. Bouillon.* **N.Nig.**: Quorrah (R. Niger) (Sept.) *T. Vogel* 192! Nupe
Barter 1298! Kontagora (Nov.) *Dalz.* 209! Lokoja *Parsons* 8! Sokoto *Lely* 125! **S.Nig.**: Lagos (Apr.)
MacGregor 92! *Dalz.* 1351! Ibadan to Abeokuta (Mar.) *Schlechter* 12349! Ibadan (Oct.) *Newberry &
Etim* 143! Batun, Obudu (Aug.) *Stone* 62! A common tropical weed, but apparently not often found near
the sea in W. Africa (See Appendix, p. 417.)

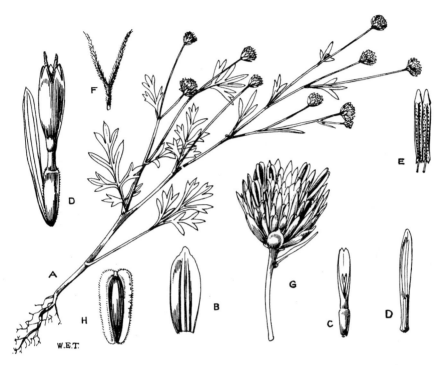

FIG. 250.—CHRYSANTHELLUM AMERICANUM (*Linn.*) *Vatke* (COMPOSITAE).

A, whole plant.   B, involucral bract.   C, ray floret.   D, disk floret with palea and separate palea.
E, anthers.   F, style-arms.   G, fruiting head.   H, achene.

8. SCLEROCARPUS Jacq. f.—F.T.A. 3 : 373.

An erect branched annual herb with ribbed stems ; leaves alternate or rarely opposite,
ovate, acute, abruptly narrowed at base, 7–12 cm. long, 3·5–6 cm. broad, crenate-
serrate except at base, scabrid-setulose above, pilose beneath, petiole slender ;
flower-heads terminal, sessile, subsessile or pedunculate with leafy bracts below ;
paleae of the receptacle large, boat-shaped, ribbed, enveloping the fertile florets ;
ray-florets usually 3, small, sterile ; achenes smooth without a pappus   *africanus*

S. africanus *Jacq. ex Murr.* Syst. Veg., ed. 14, 783 (1784) ; F.T.A. 3 : 374 ; Chev. Bot. 369 ; Berhaut Fl. Sén.
68. A weedy annual herb 1–4 ft. high, more or less roughly pubescent, with mostly alternate leaves and
yellow florets in heads ½–⅔ in. across.
Sen.: Niokolo-Koba (Oct.) *Adam* 15753! **Mali**: Boré (Aug.) *Demange* 32/1957! **Lib.**: Cape Palmas (July)
Ansell! **Ghana**: *Thonning*! Kpeve *Westwood*! Akuse to Kpong (Oct.) *Morton* A26! Yendi (Dec.) *Adams &
Akpabla* GC 4069! *Morton* A1411! **Dah.**: *Le Testu* 41! Zagnanado (Feb.) *Chev.* 23079! **N.Nig.**: Nupe
Barter 1294! Sokoto *Lely* 129! Panyan, 4,500 ft. (July) *Lely* 438! Abinsi (Feb.) *Dalz.* 657! R. Niger
(Dec.) *Hagerup* 736! **S.Nig.**: Lagos *Rowland*! Abeokuta *Millen* 92! Ogwashi (Nov.) *Thomas* 2044!
Throughout tropical Africa and also in India. (See Appendix, p. 420.)

9. SPILANTHES Jacq. Enum. Pl. Carib. 8 (1760) ; F.T.A. 3 : 383
(as of Linn.) ; Adams in Webbia 12 : 325 (1956).

Ligulate florets absent ; disk-florets white ; achenes ribbed, biaristate, 2·8–3 mm. long ;
capitula up to 1 cm. long and 9 mm. broad on long stout peduncles ; leaves oblong-
ovate, subentire, up to 5 cm. long and 2 cm. broad.   ..   ..   ..   1. *costata*
Ligulate florets present ; disk-florets yellow or orange ; achenes unribbed, 1·8–2 mm.
long :

Stem suberect, rooting only from the lower nodes ; leaves lanceolate ; peduncles elongated, slender, exceeding the upper leaves ; capitula ovoid, about 6 mm. long ; achenes ciliate on the margins, shortly biaristate   ..   ..   2. *uliginosa*

Stem creeping, rooting from the nodes ; leaves ovate ; achenes not ciliate, scarcely aristate :

Peduncles shorter than the leaves ; ligules not or hardly exceeding the involucre ; leaves subentire or remotely and shallowly serrate-dentate, up to 4 cm. long including the petiole ; glabrescent ; achenes sparsely pubescent along distinct margins
           3. *filicaulis*

Peduncles about twice as long as the leaves ; ligules usually exceeding the involucre ; leaves distinctly serrate-dentate, up to 8 cm. long including the petiole, sparsely pubescent ; achenes almost glabrous, without thickened margins. ..   4. *africana*

1. **S. costata** *Benth.* in Fl. Nigrit. 436 (1849) ; Adams in Webbia 12 : 329. *S. acmella* of Chev. Bot. 372, partly (*Chev.* 20051), not of Murr. *S. acmella* of F.W.T.A., ed. 1, 2 : 147, partly (*T. Vogel* 41), not of Linn. A glabrescent fleshy decumbent herb with greenish-white disk-florets ; mostly near the sea.
Lib.: Cape Palmas (July) *T. Vogel* 41 ! **Iv.C.**: Tabou, Cavally basin (Aug.) *Chev.* 20051 ! Adiopodoumé (July) *Boughey* GC 14575 ! **Ghana**: Cape Coast (July) *T. Vogel* 34 ! Half Assini (Feb.) *Morton* A1633 !

2. **S. uliginosa** *Sw.* Prod. 110 (1788) ; Adams l.c. 326. *S. acmella* of Chev. Bot. 372, partly (*Chev.* 2082 ; 12140), not of Murr. *S. acmella* of F.W.T.A., ed. 1, 2 : 146, partly (*Sc. Elliot* 3891 ; *Barter* ; *Whyte*), not of Linn. An annual herb a few inches to nearly 1 ft. high with erect branches ; disk-florets yellow and ray-florets pale yellow or white ; in swampy or low-lying ground.
Sen.: Mangacounda, Casamance (Jan.) *Chev.* 2082 ! Santamba (Oct.) *Monod* 8530 ! **Port.G.**: Bafata (Nov.) *Esp. Santo* 2833 ! **Guin.**: Conakry (Feb., Apr.) *Maclaud* ! *Chev.* 12140 ! *Martine* 57 ! **S.L.**: (May) *Barter* ! Leicester Peak (Dec.) *Sc. Elliot* 3891 ! Mayoso (Aug.) *Thomas* 1393 ! Makene (Apr.) *Deighton* 5497 ! Rokupr (Nov.) *Jordan* 683 ! Kruto (Sept.) *Jaeger* 1780 ! **Lib.**: Kakatown *Whyte* ! Ganta *Harley* ! Jabrocca, Grand Cape Mount Co. (Dec.) *Baldwin* 10867 ! Queentown (Dec.) *Adam* 16263 ! **Iv.C.** : Azaguié (Dec.) *Giovannetti* 390 ! **Ghana**: Aiyinasi, W. Region (Dec.) *Rose Innes* GC 30320 ! In the West Indies and also reported from Tanganyika.

3. **S. filicaulis** (*Schum. & Thonn.*) *C. D. Adams* in Webbia 12 : 326 (1956). *Eclipta filicaulis* Schum. & Thonn. Beskr. Guin. Pl. 390 (1827). *S. acmella* of Chev. Bot. 372, partly (*Chev.* 19330), not of Murr. *S. acmella* of F.W.T.A., ed. 1, 2 : 146, partly (*Millen* 196 ; *Foster* 345 ; *Lely* 233), not of Linn. A creeping herb with flower-heads on short ascending peduncles ; disk-florets orange, ray-florets yellow.
Guin.: Kissidougou (June) *Martine* 314 ! **Lib.**: Gletown, Webo (July) *Baldwin* 6767 ! **Iv.C.**: Kéeta, Middle Cavally (July) *Chev.* 19330 ! Nimba Mts. (Apr.) *Schnell* 5334 ! Man (July) *Kerharo-Bouquet* 630 ! **Ghana** : Akwapim *Thonning* 227 ! Aburi (Mar., May) *Johnson* 145 ! *Deighton* 616 ! Winneba Plains (Apr.) *Morton* A2022 ! Amentia, Ashanti (Apr.) *Irvine* 511 ! Vane, Trans-Volta Togo (Nov.) *Morton* GC 9379 ! **N.Nig.**: Tilde Filani, 3,300ft. (May) *Lely* 233 ! Jos Plateau *Lely* 859 ! **S.Nig.**: (Sept.) *Foster* 345 ! Lagos *Millen* 196 ! Okomu F.R., Benin (Feb.) *Brenan* 9154 ! Sapoba (Sept.) *Onochie* FHI 34279 ! Oban *Talbot* 974 ! **[Br.]Cam.**: Cam. Mt., Buea, 3,200–3,800 ft. (Mar., Dec.) *Boughey* GC 7036 ! *Adams* GC 11809 ! Mamfe (Mar.) *Onochie* FHI 30888 ! Bamenda, 5,000 ft. (Feb.) *Migeod* 498 ! Gembu, 5,300 ft., Mambila Plateau (Jan.) *Hepper* 1812 ! Serti, Adamawa (Dec.) *Latilo & Daramola* FHI 28954 ! Vogel Peak, 1,200 ft., Adamawa (Nov.) *Hepper* 1434 ! **F.Po**: Basilé (Dec.) *Boughey* GC 10971 ! Moka, 3,500–4,000 ft. (Dec.) *Boughey* 77 ! Also in S. Tomé and Principe. (See Appendix, p. 420.)

4. **S. africana** *DC.* Prod. 5 : 623 (1836). Like the last but more hairy and more robust with the flower-heads on longer peduncles.
[Br.]Cam.: Cam. Mt., 4,400 ft. (Apr.) *Maitland* 1130 ! Bafut-Ngemba F.R., Bamenda (Jan., Feb.) *Keay & Lightbody* FHI 28369 ! *Hepper* 2098 ! Lakom, 6,000 ft. (Apr.) *Maitland* 1655 ! Ndu, 5,900 ft. (Feb.) *Hepper & Charter* 1947 ! Extending to Natal.

10. ELEUTHERANTHERA Poit. ex Bosc—Lawalrée in Bull. Jard. Bot. Brux. 17 : 55 (1943).

An erect branched annual herb ; stems ribbed, pubescent ; leaves opposite, ovate, 3–5(–10) cm. long, 1–2(–4) cm. broad, narrowed to a slender petiole about 5 mm. long, serrate-crenate, obtusely pointed, thinly pilose on both surfaces, minutely glandular ; capitula on slender axillary peduncles hardly exceeding 1 cm. in length ; outer involucral bracts 6–8 mm. long, longer than the inner, oblong and rounded at apex ; ray-florets almost always absent ; disk-florets 4–5-merous ; receptacle-scales persistent, ciliate ; achenes 4-angled, tuberculate, glabrescent, without pappus
           ruderalis

E. ruderalis (*Sw.*) *Sch. Bip.* in Bot. Zeit. 24 : 165 (1866) ; Meikle in Kew Bull. 8 : 118 (1953). *Melampodium ? ruderale* Sw. (1806). *Eleutheranthera ovata* Poit. ex Steud. (1840)—Berhaut Fl. Sén. 68. A weedy, thinly hispid annual herb up to about 2 ft. high with small golden-yellow shortly stalked axillary flower-heads nodding in fruit ; in wet areas mostly near the sea ; flowering more frequently during the rains.
Sen. : Ouassadou *Berhaut* 1267. **Guin.**: Conakry (Apr.) *Martine* 52 ! Moussaya (Nov.) *Schnell* 2091 ! Gangan (Apr.) *Adam* 11942 ! **S.L.**: *Thomas* 8293 ! 9885 ! 10438 ! Freetown (July, Aug., Sept.) *Deighton* 2020 ! *Hepper* 911 ! *Melville & Hooker* 104 ! Roruks (July) *Deighton* 3257 ! **Lib.**: *Delafosse* (see note below). Monrovia (May, June) *Barker* 1318 ! *Baldwin* 5882 ! Cape Palmas (Apr.) *Todd* 37 ! Siatown (Dec.) *Adam* 16365 ! **Ghana**: Axim (Mar., Apr., June, Dec.) *Morton* GC 6574 ! A419 ! 2223 ! 2458 ! Cape Coast (Aug.) *Hall* 572 ! Dunkwa (Feb.) *Whiting* 12 ! **S.Nig.**: Lagos (July) *Dalz.* 1352 ! **[Br.]Cam.**: Victoria (Sept.) *Boughey* X1199 ! X1294 ! **F.Po**: S. Isabel (Jan.) *Boughey* GC 10972 ! In the American, Asian and Australasian tropics, and apparently now spreading in Africa.

[Note: According to Berhaut (*in litt.*) the collection by *Delafosse*, which I have not seen, was made in 1897 and is probably the earliest record for this species in our area—C. D. A.].

11. WEDELIA Jacq.—Adams in Webbia 12 : 229 (1956). *Nom. cons.*

Stems procumbent, rooting at the nodes ; leaves opposite, about 4 cm. long and 3 cm. broad, trilobed, broadly cuneate in the lower half, lobes broadly triangular, dentate, sparingly setulose on the nerves ; petiole up to 5 mm. long, sparingly setulose ; flower-heads axillary, solitary on long peduncles ; involucral bracts leafy, sparingly setulose ; ray-florets spreading, fertile; achenes irregularly ovoid, warted, minutely puberulous in the upper part, with short early-caducous aristae   ..   *trilobata*

W. trilobata (*Linn.*) *Hitchc.* in Rep. Missouri Bot. Gard. 4 : 99 (1893). *Silphium trilobatum* Linn. (1759). *Wedelia carnosa* Pers. (1807). A prostrate weedy herb with yellow flower-heads up to 1 in. across ; introduced from tropical America.
Guin.: Conakry (? Apr.) *Martine* 55 ! **S.L.:** Freetown (May, Sept.) *Deighton* 1186 ! *Hepper* 912 ! Makump (Aug.) *Deighton* 1380 ! Njala, introduced from Freetown (June) *Deighton* 6070 !

[Note : No other species of *Wedelia* is known to occur in W. Africa (see *Aspilia africana*) ; those listed in F.T.A. 3 : 376–377 for other parts of Africa may, when revised, prove to belong also to *Aspilia* or other genera—C. D. A.]

12. BLAINVILLEA Cass.—F.T.A. 3 : 374 ; Adams in Webbia 12 : 229 (1956).

An erect hirsute annual herb with striate angular stems and branches ; leaves opposite or alternate above, broadly ovate, acuminate, very shortly cuneate and 3-nerved at base, 5–12 cm. long, 3–6 cm. broad, thin, shortly pilose on the nerves and veins, shallowly crenate-serrate ; petiole slender, pilose ; flower-heads pedunculate at the ends of the shoots, or axillary ; peduncles variable, up to 4 cm. long ; capitula oblong in flower, hemispherical in fruit, about 1 cm. long ; outer involucral bracts lanceolate, acute, thin, striate-hairy outside, inner glabrous ; ligules of outer female florets about 1 mm. long, white ; achenes blackish, sharply angled, the outer 3-angled with 3 persistent but fragile barbellate aristae, the inner 2- or 3-angled, longer than the outer, with 2 aristae, all transversely rugulose, truncate at apex   ..   ..   *gayana*

B. gayana *Cass.* in Dict. Sci. Nat. 47 : 90 (1827) ; F.T.A. 3 : 375 ; Chev. Bot. 369 partly (*Chev.* 24864) ; *Berhaut* Fl. Sén. 69, 70. An erect hairy annual 1–3 ft. high with white flower-heads about ½ in. long ; a weed of open ground in semi-arid regions.
Sen.: *Heudelot* 248 ! M'Bidjem *Perrottet* 50 ! Dakar *Vermoesen* 1013 ! Kaolak (Sept.) *Berhaut* 411 *bis* ! Ngazobil *Berthollet* 23 ! Niokolo-Koba (Oct.) *Adam* 15798 ! **Gam.:** Albreda *Perrottet.* **Mali:** Macina (Sept.) *Chev.* 24864 ! Labézanga *Hagerup* 449 ! Bore (Sept.) *Demange* 38/1959 ! **Niger** : Niamey (Oct.) *Hagerup* 511 ! **N.Nig.:** Katagum *Dalz.* 168 ! Also in Cape Verde Islands, Ethiopia and Arabia.

13. ASPILIA Thouars—F.T.A. 3 : 378 ; Adams in Webbia 12 : 230 (1956).

Ligules of the ray-florets yellow (cream to orange) :
Ray-florets pale yellow or cream :
Ligules pale sulphur-yellow ; densely hispid erect herbs of mostly perennial habit ; receptacle-scales appendaged and hairy at apex ; achene-aristae absent :
Capitula mostly involucrate by the upper leaves ; stem-leaves lanceolate ; achenes not more than 5 mm. long ; appendage of the receptacle-scale not exceeding 1·5 mm. long   ..   ..   *1. rudis* subsp. *rudis*
Capitula distinctly pedunculate ; stem-leaves ovate ; achenes 6 mm. or more long ; appendage of the receptacle-scale up to 4 mm. long 1a. *rudis* subsp. *fontinaloides*
Ligules cream ; sparsely setulose straggling-branched annual herb ; receptacle-scales obtuse, glabrous ; achene-aristae very short and inconspicuous   11. *helianthoides*
Ray-florets yellow to orange :
Leaves ovate to lanceolate with a pair of distinct lateral nerves arising near the base :
Leaves sessile or subsessile :
Capitula distinctly pedunculate ; leaves ovate to ovate-lanceolate, acuminate ; involucral bracts recurved at apex ; indumentum coarse ..   2. *spenceriana*
Capitula subsessile or shortly pedunculate ; leaves lanceolate, acute ; involucral bracts erect, acute ; indumentum very fine with copious short appressed hairs   3. *chevalieri*
Leaves petiolate ; peduncles 4–10 cm. long ; involucral bracts erect or spreading, as long as or longer than the disk-florets ; indumentum of short whitish appressed or spreading hairs :
Leaves ovate to ovate-lanceolate :
Hairs on peduncles spreading ; rudimentary styles in ray-florets few or absent :
Capitula 2 cm. or more in diam. (in flower) ; ray-florets 10 or more ; leaves up to about 10 cm. long and 5 cm. broad   ..   4a. *africana* var. *africana*
Capitula about 1·5 cm. diam. ; ray-florets 5–6 ; leaves up to 6 cm. long and 2 cm. broad, mostly much smaller   ..   ..   ..   4b. *africana* var. *minor*
Hairs on peduncles appressed-ascending ; rudimentary styles usually present in most of the ray-florets   ..   ..   ..   4c. *africana* var. *ambigua*
Leaves lanceolate ; hairs on peduncles spreading   4d. *africana* var. *guineensis*
Leaves linear to linear- or elliptic-lanceolate, without a pair of distinct lateral veins arising near the base ; annual herbs :
Receptacle-scales shortly spiculate or triangular-tipped ; central disk-florets fertile ; outer involucral bracts as long as the ray-florets or longer ; leaves subsessile :
Leaves narrowly oblong-lanceolate ; low much-branched herb ; internodes 3–4 cm. long ; peduncles 1·5–6 cm. long   ..   ..   *5. bracteosa*

Leaves linear-lanceolate to linear ; tall straggling herb ; internodes up to about 15 cm. long ; peduncles up to 10 cm. long .. . . 6. *paludosa*
Receptacle-scales (at least of the disk-florets) with a terminal appendage 1·5–3 mm. long ; central disk-florets mostly sterile ; outer involucral bracts rarely exceeding the rays :
Outer involucral bracts mostly longer than the inner, acute or triangular-tipped :
Leaves tapered to the base ; pappus-aristae 2(–3), up to 1·5 mm. long, unequal ; stems and leaves scabrid with short appressed hairs .. .. 7. *mortonii*
Leaves abruptly narrowed at the base ; pappus-aristae very short or absent ; stems and leaves hispid, the hairs on the midrib below spreading 8. *linearifolia*
Outer involucral bracts shorter than the inner, obtuse ; leaves tapered to a rather broad base, subsessile ; pappus-aristae very short, curved ; stems hispid
9. *angustifolia*
Ligules of the ray-florets white, violet or purple :
Capitula more or less clustered and sessile ; leaves linear-lanceolate, often pandurate towards the base ; stems rough with spreading curved hairs :
Ray-florets purple .. .. 10. *kotschyi* var. *kotschyi*
Ray-florets white .. .. 10a. *kotschyi* var. *alba*
Capitula pedunculate :
Involucral bracts in 2–4 rows, the outer mostly as long as or longer than the inner, with pointed tips ; leaves ovate to ovate-lanceolate or rarely lanceolate :
Flower-heads large, 2 cm. or more in diam. ; leaves ovate, 5–12 cm. long, 1·5– 4·5 cm. broad, shortly petiolate, more or less rounded at base, acute at apex, coarsely serrate with a thick texture ; stems often purplish, with stiff rather sparse spreading or ascending hairs ; peduncles up to 9 cm. long, with spreading hairs ; pappus-aristae very short or absent ; ligules about 8, white or cream
11a. *helianthoides* subsp. *helianthoides*
Flower-heads less than 1·5 cm. diam. ; leaves ovate to lanceolate, acute, remotely serrate or subentire ; peduncles short or long, with numerous spreading or ascending hairs ; pappus-aristae short or long or rarely absent ; ligules 5–8, white, violet or purplish :
Leaves ovate to ovate-lanceolate, remotely serrate or subentire, of thin texture ; peduncles short ; pappus-aristae 2, unequal, rarely absent ; outer involucral bracts acute :
Hairs on peduncles spreading ; pappus-aristae very short and caducous, or absent ; leaves distinctly petiolate, cuneate at base.. 11b. *helianthoides* subsp. *ciliata*
Hairs on peduncles more or less ascending ; pappus-aristae up to 2·5 mm. long, unequal, more or less persistent ; leaves shortly petiolate
11c. *helianthoides* subsp. *prieuriana*
Leaves lanceolate to linear-lanceolate, remotely serrate, of firm texture ; peduncles variable in length, with ascending hairs ; pappus-aristae 1·5–2 mm. long, more or less persistent ; outer involucral bracts slender, linear, obtuse
11d. *helianthoides* subsp. *papposa*
Involucral bracts in 4–5 rows, the outer much shorter than the inner, with spathulate obtuse or rounded tips ; leaves lanceolate to linear-lanceolate ; peduncles slender with appressed hairs .. .. 12. *bussei*

1. **A. rudis** *Oliv. & Hiern* subsp. **rudis**—in F.T.A. 3 : 380 (1877) ; Chev. Bot. 371, partly ; F.W.T.A., ed. 1, ·2 : 145 (excl. syn. ; *Chev.* 27174 ; *Irvine* 887) ; Adams in Webbia 12 : 233. A coarse erect, often perennial hispid herb about 3 ft. high, with pale yellow or almost white flowers ; in savanna grassland or woodland.
   Mali: Baguineda, Bamako (Aug.) *Roberty* 2580 ! **Guin.:** Yagadou (Mar.) *Chev.* 20932 ! **Ghana:** Banda Hills F.R., Ashanti (Dec.) *Morton* GC 25186 ! **N.Nig.:** Nupe *Barter* 1292 ! Biu, 2,500 ft., Bornu (Aug.) *Noble* 3 ! Yola (July) *Dalz.* 30 ! Abinsi (Oct.) *Dalz.* 653 ! Benue R. *Talbot* ! Also in Principe.

1a. **A. rudis** subsp. **fontinaloides** *C. D. Adams* in Webbia 12 : 233 (1956). *A. rudis* of Chev. Bot. 371, partly (*Chev.* 23992). A robust erect coarsely hispid herb up to 4 ft. high with pale yellow flowers ; leaves rather broad and sometimes in whorls of 3 or 4 ; in savanna.
   Guin.: Kinkou (June) *Adam* 14671 ! **Ghana:** Afram Plains (May) *Morton* GC 7250 ! Nsawkaw, Ashanti (fr. Dec.) *Adams* 3153 ! Aframso, Ashanti (July) *Darko* 919 ! Ejura Scarp (Dec.) *Morton* GC 9599 ! **Dah.:** Kouandé, Atacora Mts. (June) *Chev.* 23992 !

2. **A. spenceriana** *Muschl.* in Engl. Bot. Jahrb. 50, Suppl. : 334 (1914) ; Adams l.c. 234. *A. baoulensis* A. Chev. Bot. 370, name only. *A rudis* of F.W.T.A., ed. 1, 2 : 145, partly (*Chev.* 22174), not of Oliv. & Hiern. *A. helianthoides* of Chev. Bot. 370, partly (*Chev.* 24496), not of Oliv. & Hiern. An erect herb with stalked heads nodding in fruit, with the outer involucral bracts recurved ; florets pale yellow ; in savanna.
   Iv.C.: Kangoroma Mt. (July) *Chev.* 22174 ! **Ghana:** Kete Krachi *Graf Zech* 92. Kpeve *Westwood* GC 5578 ! Ejura Scarp (Dec.) *Morton* A2564 ! **Niger:** Diapaga to Fada, Gourma (July) *Chev.* 24496 !

3. **A. chevalieri** *O. Hoffm. & Muschl.* in Mém. Soc. Bot. Fr. 2, 8 : 115 (1910) ; Chev. Bot. 370 ; Adams l.c. 234 (excl. *Roberty* 2968). An erect yellow-flowered herb up to 2 ft. high ; stems and leaves covered with a fine whitish indumentum.
   Mali: Bamako (Jan.) *Chev.* 202 *bis* ! Nyamina to Koulikoro (Oct.) *Chev.* 2027 ! **Guin.:** Farana (Mar.) *Sc. Elliot* 5371 ! **S.L.:** Buyabuya, Scarcies (Feb.) *Sc. Elliot* 4268 ! Loma Mts., 4,000 ft. (Nov.) *Jaeger* 480 !

4. **A. africana** (*Pers.*) *C. D. Adams* l.c. 236 (1956). *Wedelia africana* Pers. (1807)—F.T.A. 3 : 376 ; F.W.T.A., ed. 1, 2 : 145. *Aspilia latifolia* Oliv. & Hiern in F.T.A. 3 : 379 (1877) ; Chev. Bot. 371 ; F.W.T.A., ed. 1, 2 : 145 ; Berhaut Fl. Sén. 70. A weed widespread in Africa usually with deep yellow flowers ; very variable and separable into at least the following varieties :—

4a. **A. africana** (*Pers.*) *C. D. Adams* var. **africana**. *A. helianthoides* of Chev. Bot. 370, partly (*Chev.* 16076). A spreading herb or scrambling shrub up to 6 ft. high.

Sen.: *Adam* 6022. **Gam.:** Genieri-Kaiaaf (July) *Fox* 162! **Guin.:** Nzo Mt. (Mar.) *Chev.* 21031! Dyeke (July) *Baldwin* 9658! Beyla (Oct.) *Jac.-Fél.* 2047! **S.L.:** *Winwood Reade!* Leicester Peak (Dec.) *Sc. Elliot* 3893! Newton (June) *Deighton* 5546! Musaia (Sept.) *Small* 236! Bintumane Peak, 5,400 ft. (Jan.) *T. S. Jones* 70! Loma Mts. (Sept., Nov.) *Jaeger* 534! 1587! **Lib.:** *Bunting!* Suacoco *Okeke* 9! Ganta (May) *Harley!* **Iv.C.:** Bingerville (Dec.) *Chev.* 16076! Man (Aug.) *Boughey* GC 18416! **Ghana:** Accra (Aug.) *T. Vogel* 26! Asuansi (Feb.) *Williams* 7! Kumasi (July) *Chipp* 537! Banka *Irvine* 478! Amedzofe (Nov.) *Morton* GC 9426! **N.Nig.:** *Baikie!* Lokoja (Sept.) *Parsons* 4! Abinsi (Oct.) *Dalz.* 658! **S.Nig.:** Lagos *Moloney!* *Millen!* Ibadan (Jan., Oct.) *Meikle* 994! *Keay & Jones* FHI 13738! Akure F.R. (Oct.) *Keay* FHI 25464! Calabar (Feb.) *Mann* 2325! *Kalbreyer* 210! **[Br.]Cam.:** Bamenda, 5,000 ft. (Jan.) *Migeod* 317! 430! Nkambe, Bamenda (Sept.) *Ujor* FHI 30228! (See Appendix, p. 415.)

4b. **A. africana** var. **minor** *C. D. Adams* l.c. 237 (1956). *A. helianthoides* of Chev. Bot. 370, partly (*Chev.* 18783). A small bushy herb ; at higher altitudes.
 Guin. : Dalaba Plateau, Fouta Djalon, 3,000–4,000 ft. (Oct.) *Chev.* 18783! Timbi-Madina (Mar.) *Adam* 11705! **[Br.]Cam.:** Wum, Bamenda (Dec.) *Boughey* GC 11565! 11567! Ndop, 3,600 ft.(Dec.) *Boughey* GC 11099!

4c. **A. africana** var. **ambigua** *C. D. Adams* l.c. 238 (1956). *Wedelia africana* of F.W.T.A., ed. 1, 2 : 145, partly (*Johnston* 5 ; *Chipp* 622 ; *Lely* 60 ; *Carpenter*). A weed of cleared ground in drier forest country ; ligules golden-yellow.
 Guin.: Bafing (Oct.) *Adam* 12651! Timbo (Oct.) *Jac.-Fél.* 1915! **S.L.:** Kortright Hill (Mar.) *Johnston* 5! Wallia, Scarcies *Sc. Elliot* 4268! Makumri (June) *Thomas* 540! Gberia Fotumbu (Sept.) *Small* 287! **Ghana:** Senya Beraku (May) *Morton* GC 9215! Abetifi (Dec.) *Adams* 1926! Dukwesein, Agogo (Dec.) *Chipp* 622! Dzodze (Oct.) *A. S. Thomas* K3! **N.Nig.:** Liruwen-Kano Hills *Carpenter!* Birnin Gwari, Zaria (June) *Keay* FHI 25842! Naraguta *Lely* 60! Jos *Batten-Poole* 358! Bula, Bauchi (June) *Summerhayes* 5! **S.Nig.:** Olokemeji *Foster* 301! Ibadan (Feb.) *Keay & Meikle* 1136!

4d. **A. africana** var. **guineensis** (*O. Hoffm. & Muschl.*) *C. D. Adams* l.c. 238 (1956). *A. guineensis* O. Hoffm. & Muschl. (1910)—Chev. Bot. 370. *A. latifolia* of F.W.T.A., ed. 1, 2 : 145, partly (*Chev.* 14622 ; 14650 ; 18244). A perennial herb with narrow leaves ; ligules yellow.
 Guin.: Kouria to Longuery (Aug.) *Caille* in Hb. *Chev.* 14622! 14650! Kouria to Irébéléya (Sept.) *Chev.* 18244!

5. **A. bracteosa** *C. D. Adams* l.c. (1956). A bushy annual herb up to 2 ft. high with yellow flowers about ¾ in. across.
 Ghana: Yendi (Dec.) *Morton* A1475!

6. **A. paludosa** *Berhaut* in Bull. Soc. Bot. Fr. 101 : 375 (1954) ; Fl. Sén. 69, 71 ; Adams l.c. 239. A branched annual herb up to 4 ft. high with pale yellow florets ; in low-lying ground in savanna. **Mali:** Kokry, Macina (Nov.) *Roberty* 2968! **Ghana:** Bimbila (Dec.) *Morton* A1422! Tamale (Oct.) *Baldwin* 13557b! Pong-Tamale (Dec.) *Morton* GC 8930! **N.Nig.:** (May) *Thornewill* 180!
 Sen.: Tambacounda (Sept.-Oct.) *Berhaut* 1271! Bambousaie (Sept.) *Berhaut* 3274! 4047.

7. **A. mortonii** *C. D. Adams* l.c. 240 (1956). A straggling annual herb up to 6 ft. or more high, with slender branches and orange-yellow flowers ; in low-lying savanna.
 Ghana: Damongo, Gonja (Mar., July, Sept., Dec.) *Morton* GC 8747! 9963! 25028! *Adams* 3969! *Andoh* 5214! *Rose Innes* GC 30188! Kwahu Tafo (Apr., Dec.) *Morton* A662! *Adams* 4919!

8. **A. linearifolia** *Oliv. & Hiern* in F.T.A. 3 : 380 (1877) ; Adams l.c. 240. An erect bushy herb up to 6 ft. high with harshly hispid stems and leaves ; flower-heads golden-yellow, up to 1½ in. across.
 Ghana: Burbulakofe, Wurupong (Sept.) *Adams* 1848! Adidome to Ho (May) *Morton* A2059! A2061! Yeji (Nov.) *Morton* A2705! **N.Nig.:** ? Lokoja *Baikie!*

9. **A. angustifolia** *Oliv. & Hiern* in F.T.A. 3 : 380 (1877) ; Adams l.c. 241. An annual herb up to 6 ft. high with orange-yellow flowers about 1 in. across ; in wet places in savanna.
 Ghana : Kpandai (Dec.) *Adams & Akpabla* GC 4029! Bimbila (Dec.) *Irvine* 2718! Kete Krachi (Jan., May, Dec.) *Morton* GC 6295! 7165! A1516! Yabrasu to Kintampo (Dec.) *Morton* A1189! Ejura to Atebubu (Nov.) *Morton* A2752! **N.Nig.:** Nupe *Barter* 1007! 1185! Kontagora (Nov.) *Dalz.* 195! **S.Nig.:** Ado Rock, Oyo (Oct.) *Savory & Keay* FHI 25349! Ibadan (Oct.) *Keay & Jones* FHI 13777!

10. **A. kotschyi** (*Sch. Bip.*) *Oliv.* var. **kotschyi**—in Trans. Linn. Soc. 29 : 98 (1873) ; F.T.A. 3 : 381 ; Adams l.c. 241 ; Berhaut Fl. Sén. 70. *Dipterotheca kotschyi* Sch. Bip. (1842). *A. kotschyana* of Chev. Bot. 371. An erect hispid herb up to 4 ft. high with clustered purple or brownish-red flowers.
 Sen.: Diohine *Berhaut* 2269! **Mali:** San (Sept.) *Chev.* 2081! Ségou (Jan., Nov.) *Roberty* 2737! 3174! **N.Nig.:** Attah (Sept.) *T. Vogel!* Aboh *Barter* 338! Zungeru (Aug.) *Dalz.* 198! Fodama (Dec.) *Moiser* 164! Jos Plateau (Aug.) *Lely* 557! Bornu *Parsons!* **[Br.]Cam.:** Gurum, Adamawa (Nov.) *Hepper* 1288! Extends from Sudan through E. Africa to Angola. (See Appendix, p. 415.)

10a. **A. kotschyi** var. **alba** *Berhaut* in Bull. Soc. Bot. Fr. 101 : 375 (1954) ; Fl. Sén. 70. *A.* sp. aff. *kotschyana* of Chev. Bot. 371. A herb similar to the last, with white flowers.
 Sen.: Bargny *Berhaut* 1726 ; 4778. **Mali:** Yatenga-Nord (Aug.) *Chev.* 24805! **N.Nig.:** Katagum (Sept.) *Dalz.* 167!

11. **A. helianthoides** (*Schum. & Thonn.*) *Oliv. & Hiern* in F.T.A. 3 : 381 (1877) ; Chev. Bot. 370, partly ; Berhaut Fl. Sén. 69, 70. *Coronocarpus helianthoides* Schum. & Thonn. (1827). Annual herbs with white or cream-coloured ligulate florets, sometimes turning purplish on drying ; widespread and variable. The following subspecies occur in our area :—

11a. **A. helianthoides** (*Schum. & Thonn.*) *Oliv. & Hiern* subsp. **helianthoides**—Adams l.c. 244. A robust herb with white or cream ligules ; in savanna and hill grassland.
 Guin.: Bambaya, Upper Niger (Oct.) *Jaeger* 4! Kouroussa *Brossart* in Hb. *Chev.* 15701! **S.L.:** Yalamba (Jan.) *Jaeger* 3949! **Ghana:** Thonning! Accra (Aug.) *T. Vogel* 27! Menji to Sampa, Ashanti (Oct.) *Morton* A2612! Togo Plateau F.R. (Oct.) *Boughey* GC 14391! Hohoe (Apr.) *Morton* GC 9160! **S.Nig.:** Lagos *Phillips* 9! Newi *Kitson!* (See Appendix, p. 415.)

11b. **A. helianthoides** subsp. **ciliata** (*Schumach.*) *C. D. Adams* l.c. 245 (1956). *Verbesina ciliata* Schumach. (1827) *Aspilia smithiana* Oliv. & Hiern in F.T.A. 3 : 380 (1877) ; F.W.T.A., ed. 1, 2 : 145. *Blainvillea prieuriana* of F.W.T.A., ed. 1, 2 : 144, partly (*Foster* 317), not of DC. A bushy weed herb, with flower-heads about ¾ in. across and white or purple-tinged ligules.
 Sen.: Niokolo (Jan.) *Adam* 15658! **Iv.C.:** Techikrom (Dec.) *Adams* 2980! **Ghana:** *Thonning!* Accra *Don* 23! Achimota (June) *Irvine* 704! Accra Plains (Oct.) *Baldwin* 13417! Kpandu (Sept.) *Adams* 1786! Yeji *Anderson* 36! **Togo Rep.:** *Büttner* 124! Atikpui (Apr.) *Schlechter* 12983! Lomé *Warnecke* 256! **Dah.:** *Burton!* **N.Nig.:** Quorrah (= Niger) (Sept.) *T. Vogel* 20! Aboh *Barter* 370! Nupe *Barter* 1293! Abinsi (Aug.) *Dalz.* 660! R. Benue *Talbot!* **S. Nig.:** Olokemeji *Foster* 317! Abeokuta (Sept.) *Odukwe* FHI 33914! Oyo (Oct.) *Onochie* FHI 34914! Also in Congo and Angola.

11c. **A. helianthoides** subsp. **prieuriana** (*DC.*) *C. D. Adams* l.c. 246 (1956). *Blainvillea prieuriana* DC. (1836)—F.W.T.A., ed. 1, 2 : 144, partly (*Thomas* 7279 ; *Hayes* 580). *Aspilia helianthoides* of Chev. Bot. 370, partly (*Brossart* in Hb. *Chev.* 15674, and var. *minor*, *Chev.* 207). An annual weed like the last ; in more inland localities.
 Sen.: Kaolak (Oct.) *Berhaut* 2276! **Gam.:** *Hayes* 580! Albreda *Perrottet!* **Mali:** Bamako (Jan.) *Chev.* 207! Yatenga-Nord, Ouahigouya to Koro (Aug.) *Chev.* 24805 *bis.* Kita (Oct.) *Jaeger* 23! **Port. G.:** Bissau (Oct.–Nov.) *Esp. Santo* 1031! **Guin.:** Kouroussa *Pobéguin* 468! Kankan *Brossart* in Hb. *Chev.* 15674! **S.L.:** Yungeru (Jan.) *Thomas* 7279! Rosino (Feb.) *Jordan* 8! Batkanu (Jan.) *Deighton* 2838! Rokupr (Jan., Apr., Oct.) *Deighton* 2951! 4575! *Jordan* 132! **Ghana:** Bagabaga to Tamale (Nov.) *Darko* 536! Bame Pass (Dec.) *Adams* 4486! Dutukpene (Dec.) *Adams* 4538! (See Appendix, p. 416.)

11d. **A. helianthoides** subsp. **papposa** (*O. Hoffm. & Muschl.*) *C. D. Adams* l.c. 247 (1956). *A. helianthoides* var.

papposa O. Hoffm. & Muschl. (1914). *Blainvillea prieuriana* of F.W.T.A., ed. 1, 2 : 144, partly (*Sc. Elliot* 4593 ; *Hayes* 573), not of DC. An uncommon herb of grassland.

    Gam.: *Hayes* 573 ! **Mali**: Kita Massif (Sept.) *Jaeger* ! **S.L.**: Kora, Scarcies R. (Feb.) *Sc. Elliot* 4593 ! **Togo Rep.**: Bassari *Kersting* 127. **N.Nig.**: Naraguta (Aug.) *Lely* 507 ! *Keay* FHI 20099 ! Jos (July, Oct.) *Lely* P531 ! 812 ! *Batten-Poole* 138 ! 160 !

12. **A. bussei** *O. Hoffm. & Muschl.* in Engl. Bot. Jahrb. 50 : 341 (1914). *A. angustifolia* A. Chev. Bot. 370, name only, not of Oliv. & Hiern. *A. kitsonii* S. Moore in J. Bot. 65 : 14 (1927). *Blainvillea prieuriana* of F.W.T.A., ed. 1, 2 : 144, partly (*Chev.* 23557), not of DC. A slender-branched straggling herb of savanna grassland with flower-heads about ⅓–¾ in. across ; ligules white or pale purple, rarely lemon-yellow.

    Iv.C.: *Lowe* ! Toumodi (Dec.) *Boughey* GC 14448 ! Séguéla (Aug.) *Boughey* GC 18453 ! Asakra (Aug.) *Boughey* GC 18620 ! **Ghana**: Ejura (Mar., Apr., Aug.) *Williams* 230 ! *Andoh* 5043 ! *Darko* 553 ! Tamale (Dec.) *Morton* GC 6223 ! Kumawu, Ashanti *Chipp* 465 ! **Togo Rep.**: Madse *Busse* 3502. **Dah.**: Agouagon to Savé (May) *Chev.* 23557 !

14. MELANTHERA Rohr—F.T.A. 3 : 381.

Scales of the receptacle elongated into long stiff barbellate setae, the outer ones pectinate-ciliate ; leaves petiolate, ovate-triangular, 4·5–7 cm. long, about 3 cm. broad, pilose on both surfaces ; peduncles slender ; involucral bracts lanceolate, hispid-setulose outside..   ..   ..   1. *gambica*
Scales of the receptacle rather abruptly acuminate or triangular at apex :
  Leaves petiolate :
    Stems, petioles and veins of the leaves beneath densely fulvous-pilose with matted hairs 3–4 mm. long ; leaves broadly ovate, truncate-subcordate, crenate-dentate, acuminate, up to 15 cm. long and 10 cm. broad ; capitula 1·5–2 cm. broad ; involucral bracts lanceolate, 1 cm. long, shortly setose outside with bulbous-based hairs ; achenes obovate, glabrous with 2 unequal caducous bristles   ..   2. *felicis*
    Stems, petioles and leaf-veins shortly hispid to minutely scabrid-setulose :
      Receptacle-scales acutely long-acuminate ; leaves acute or acuminate at apex :
        Leaf-bases abruptly and broadly cuneate ; lamina 8–15 cm. long, 3–10 cm. broad, rhomboid, coarsely dentate in the upper half ; petiole about 1·5 cm. long ; achenes obovate, glabrous, minutely rugulose ; pappus of 1 or 2 very short caducous bristles   ..   ..   3. *rhombifolia*
        Leaf-bases various, usually truncate to subcordate but rarely rounded and shortly cuneate or subhastate, margin crenate-serrate to subentire ; petiole slender 2–4 cm. long ; capitula rather few on long slender peduncles ; achenes broad at apex, pitted, coronate, quadrangular or slightly flattened ; pappus of several short caducous bristles   ..   ..   4. *scandens*
      Receptacle-scales triangular at apex ; leaves oblong to ovate-elliptic, shortly cuneate at base, not acuminate, 4–8 cm. long, 2–4 cm. broad, closely scabrid-setulose on both surfaces, prominently 3-nerved above the base ; petiole 0·5–1·5 cm. long ; capitula rather numerous ; involucral bracts densely setulose with short appressed whitish hairs ; achenes quadrangular ; pappus of few caducous bristles 5. *elliptica*
  Leaves sessile or subsessile :
    Leaves oblong-lanceolate, rounded at base, 6–8 cm. long, 1·5–2·5 cm. broad, scabrid-setulose on both surfaces, shortly crenate-serrate ; involucral bracts in 2–3 series, ovate, obtuse, not exceeding 5 mm. long, in heads about 1 cm. diam. in fruit ; achenes oblong-obovate, bilateral, rounded above, glabrous, with few minute caducous bristles   ..   ..   6. *abyssinica*
    Leaves linear, tapered to base, 4–10 cm. long, 0·5–1 cm. broad, minutely scabridulous on both surfaces, subentire ; involucral bracts in 2 series, ovate-lanceolate, acuminate, 1–1·5 cm. long, 3–5 mm. broad, in heads 2–2·5 cm. broad in fruit on rather long peduncles ; achenes as above but sparsely pubescent   ..   7. *elegans*

1. **M. gambica** *Hutch. & Dalz.* F.W.T.A., ed. 1, 2 : 146 (1931) ; Berhaut Fl. Sén. 70. A hispid slender-stemmed perennial herb 1–2 ft. high, often branching near the ground ; flower-heads ¾–1 in. across ; ligules yellow ; pappus-bristles early caducous.
    Sen.: Kaolak (Aug.) *Berhaut* 2143 ! Niokolo (Oct.) *Adam* 15546 ! **Gam.** : *Hayes* 586 ! Kuntaur *Ruxton* 50 ! Genieri (July) *Fox* 142 ! **Port. G.** : Pirada, Gabu (June) *Esp. Santo* 3032 ! Also in Mauritania.
2. **M. felicis** *C. D. Adams* in J.W. Afr. Sci. Assoc. ined. (1963).
    A robust herb with a thick deeply grooved very hairy stem up to 5 ft. high ; florets yellow in heads about ¾ in. across.
    Guin.: Faranah (Oct.) *Jac.-Fél.* 1868 !
3. **M. rhombifolia** *O. Hoffm. & Muschl.* in Mém. Soc. Bot. Fr. 2, 8 : 117 (1910) ; Chev. Bot. 372. *M. elliptica* of F.W.T.A., ed. 1, 2 : 146, partly (*Chev.* 999), not of O. Hoffm. An erect perennial herb 1–2 ft. high with much-branched stems ; leaves rather large, finely scabrid, sharply serrate ; flower-heads up to 1 in. across with bright orange-yellow rays.
    Mali: Tiédiana (June) *Chev.* 999 ! Kita Massif (July) *Jaeger* 1 ! 3 ! **Iv.C.**: Tonkoui (Dec.) *Roberty* 15827 ! **Ghana**: Han to Lawra (May) *Morton* GC 7601 ! Gambaga Scarp (Apr., Sept.) *Morton* GC 9003 ! *Hall* 740 ! **N.Nig.**: Mando F.R., Birnin Gwari, Zaria (June) *Keay* FHI 25825 ! 25840 !
4. **M. scandens** (*Schum. & Thonn.*) *Roberty* in Bull. I.F.A.N. 16 : 68 (Jan. 1954) ; Brenan in Mem. N.Y. Bot. Gard. 8 : 480 (Feb. 1954). *Buphthalmum scandens* Schum. & Thonn. (1827). *M. brownei* (DC.) Sch. Bip. (1844)—F.T.A. 3 : 382 ; Chev. Bot. 371 ; F.W.T.A., ed. 1, 2 : 146, partly (excl. syn. *M. djalonensis*) ; Trochain Mém. I.F.A.N. 2 : 378 ; Guinea in Ann. Jard. Bot. Madrid 10 : 304, t. 331. *Lipotriche brownei* DC. (1836). A branched scabrid herb usually scrambling or scandent with rather broad leaves and orange radiate flower-heads about 1 in. across.
    Sen.: *fide* Trochain *l.c.* **Guin.**: Kouroussa (July) *Pobéguin* 329 ! Kouria (Oct.) *Caille* in Hb. Chev. 15024 ! Source of R. Niger (Oct.) *Jaeger* 99 ! Bafing (Oct.) *Adam* 12654 ! **S.L.**: Njala (July) *Deighton* 762 ! Jigaya (Sept.) *Thomas* 2505 ! Musaia (Dec.) *Deighton* 4556 ! **Lib.**: Nyaake, Webo (June) *Baldwin* 6188 ! Flumpa, Sanokwele (Sept.) *Baldwin* 9350 ! Ganta (Sept.) *Harley* ! **Iv.C.**: Bingerville *Chev.* 15405 ! Middle Cavally

Basin (July) *Chev.* 19396! Nzo (July) *Schnell* 1515. Oroumba-Boka Mt. (Aug.) *Boughey* GC 18578!
Ghana: *Thonning*! Nsuaem, Oda (Oct.) *Fishlock* 54! Asin-Yan-Kumasi *Cummins* 244! Cape Coast
(July) *T. Vogel* 42! Aburi (Nov.) *Howes* 1014! Wenchi, Ashanti (Dec.) *Adams* 3240! Kete Krachi to
R. Oti (Dec.) *Morton* GC 6431! **N.Nig.:** Nupe *Barter* 1297! **S.Nig.:** Lagos (Oct.) *Dalz.* 1169! Abeokuta
Irving! Ibo country (Aug.) *T. Vogel* 19! Akure F.R., Ondo (Oct.) *Keay* FHI 25486! Oban *Talbot* 389!
986! [**Br.**]**Cam.:** Rio del Rey *Johnston*! Cam. Mt., 3,000–6,500 ft. (Sept., Nov.-Jan., Mar.) *Dunlap* 86!
Mildbr. 10752! *Maitland* 217! *Migeod* 229! *Brenan* 9580! *Boughey* GC 6755! Lus, Nkambe Div. (Feb.)
Hepper 1867! Kwagiri-Gangumi, Adamawa (Dec.) *Latilo & Daramola* FHI 28783! **F.Po:** (Oct., Dec.)
T. Vogel 42! *Mann* 59! *Monod* 10485! Moka, 4,000 ft. (Dec.) *Boughey* 52! *fide* Guinea *l.c.* Widespread
in tropical Africa. (See Appendix, p. 418.)

5. **M. elliptica** *O. Hoffm.* in Engl. Bot. Jahrb. 24 : 474 (1898) ; F.W.T.A., ed. 1, 2 : 146, partly (excl. syn. *M. rhombifolia* and *Chev.* 999). *M. chevalieri* O. Hoffm. & Muschl. (1910). An erect perennial bushy herb 2–4 ft. high with numerous yellow flower-heads $\frac{3}{4}$–1 in. across ; in savanna.
Mali: Kita Massif (Oct.) *Jaeger* 30! **Ghana:** Akuse (May, Oct., Nov.) *Morton* A28! 2204! 2373! Kete
Krachi to R. Oti (May) *Morton* GC 7283! Jasikan (Apr.) *Darko* 252! Banda (Dec.) *Adams & Akpabla*
GC 4011! Gambaga (Sept.) *Hall* 780! **Togo Rep.:** *Büttner* 34. **N.Nig.:** Zaria (Feb.) *Dalz.* 361! Anara
F.R., Zaria (May) *Keay* FHI 25785! Birnin Gwari, Zaria (June) *Keay* FHI 25822! Hepham to Ropp,
4,600 ft., Jos Plateau (July) *Lely* 379! Biu, 2,500 ft., Bornu (Aug.) *Noble* 5! [**Br.**]**Cam.:** Vogel Peak.
3,600 ft., Adamawa (Nov.) *Hepper* 1401! Also in Ubangi-Shari.

6. **M. abyssinica** (*Sch. Bip.*) *Oliv. & Hiern* in F.T.A. 3 : 382 (1877). *Wuerschmittia abyssinica* Sch. Bip. (1846),
M. djalonensis A. Chev. Bot. 372, name only. *M. sokodensis* Muschl. ex Hutch. & Dalz. F.W.T.A., ed.,
1, 2 : 146. A perennial scabrid herb with erect conspicuously grooved stems (at least when dry) ; flower-heads about 1 in. across with orange-yellow rays ; in mountain grassland.
Guin.: Dalaba-Diaguissa Plateau, 3,000–4,000 ft. (Sept., Oct.) *Chev.* 18699! 18829! Mt. Loura (Sept.)
Schnell 7100! **S.L.:** Bintumane Peak (Loma Mts.), 5,000 ft. (Jan., Sept.) *T.S. Jones* 94! *Jaeger* 1987!
Sereleu to Kouka, Loma Mts. (Sept.) *Jaeger* 1588! 1703! **Togo Rep.:** Sokode (Oct.) *Schroeder* 73! Also in
Sudan and Ethiopia.

7. **M. elegans** *C. D. Adams* l.c. ined. (1963.) A many-grooved slender-stemmed perennial herb 2–3 ft. high
from a rhizome or decumbent base ; leaves pale green, narrow ; ray-florets rich yellow in heads about
1½ in. across ; on outcropping rocks.
Guin.: Tossékré (Oct.) *Adam* 12728! Labé (June) *Adam* 14765!

15. **ECLIPTA** Linn.—F.T.A. 3 : 373. *Nom. cons.*

A herb covered all over with short scabrid appressed hairs ; leaves opposite, lanceolate
or narrowly ovate-lanceolate, subacute, shortly narrowed at base, 4–10 cm. long,
1–3 cm. broad, distantly serrate, 3-nerved from well above the base ; flower-heads
axillary and terminal, about 1 cm. diam. ; peduncle slender, 1–7 cm. long ; involucral
bracts few, imbricate, ovate-orbicular ; ray-florets numerous, very small, white ;
achenes dentate at apex, finely tuberculate ..   ..   *prostrata*

E. prostrata (*Linn.*) *Linn.* Mant. Pl. Alt. 286 (1771). *Verbesina prostrata* Linn. (1753). *Eclipta alba* (Linn.)
Hassk. (1848)—F.T.A. 3 : 373 ; Chev. Bot. 369 ; F.W.T.A., ed. 1, 2 : 146 ; Schnell in Ic. Pl. Afr. I.F.A.N.
1 : t. 18 ; Berhaut Fl. Sén. 69. A decumbent or erect annual herb up to 2 ft. or more high, with rough leaves
and stems and white flower-heads about ¼ in. across ; whole plant becoming dark on drying ; a common
tropical weed especially in damp places.
Sen.: Richard Tol (Jan.) *Döllinger*! Dakar (Jan., May) *Chev.* 2085! *Baldwin* 5717! **Gam.:** *Saunders* 4!
Kuntaur *Ruxton* 17! **Mali:** Timbuktu (July) *Chev.* 1262! Kabarah (Aug.) *Chev.* 1342! Kobala, Bamako
(July) *Roberty* 2453! Ouasagouna (Sept.) *Hagerup* 433! **Guin.:** Heudelot 786! Kouroussa *Brossart* in
Hb. *Chev.* 15636! **S.L.:** Sumbuya (Apr.) *Deighton* 1694! Njala (May) *Deighton* 2519! Musaia (Dec.)
Deighton 4468! Makuma (Sept.)*Jordan* 921! **Lib.:** Kakatown *Whyte*! Ganta (May) *Harley*! Harbel (July)
Baldwin 6649! Gletown, Tchien (July) *Baldwin* 6780! **Iv.C.:** Bingerville (Dec.) *Chev.* 16073! **Ghana:**
Shai Plains (Jan.) *Johnson* 570! Keta (Mar.) *Morton* GC 6535! Asamankese (Aug.) *Howes* 941! Amentia,
Ashanti *Irvine* 467! Tumu (May) *Morton* GC 7575! **Dah.:** Aguégué, Porto Novo (Mar.) *Chev.* 23321!
Togo Rep.: Lomé *Warnecke* 256a! **Niger:** Kolo (Nov.) *Vaillant* 838! **N.Nig.:** Nupe *Barter* 865! Jebba
(Mar.) *Meikle* 1285! Sokoto (Nov.) *Moiser* 36! Jira (May) *Lely* 128! Jos Plateau (Feb.) *Lely* P147!
Abinsi (July) *Dalz.* 662! **S.Nig.:** Attah (Sept.) *T. Vogel* 68! Nun R. (Aug., Sept.) *T. Vogel* 29! *Mann*
470! Lagos *W. MacGregor* 133! Ibadan (Feb., Mar.) *Brenan* 8953! *Meikle* 1262! Onitsha (May) *Onochie*
FHI 7200! Oban *Talbot* 393! [**Br.**]**Cam.:** Gurum, Adamawa (Nov.) *Hepper* 1262! Common throughout
the tropics. (See Appendix, p. 417.)

16. **VERBESINA** Linn.—Benth & Hook. f. Gen. Pl. 2 : 379.

Stems pubescent ; leaves ovate-triangular, petiolate, about 7 cm. long, dentate, whitish
beneath with short appressed hairs ; flower-heads about 4 cm. diam. on long peduncles ; involucral bracts linear, 1 cm. long ; ligules deeply lobed at apex ; achenes
obovate, pilose, with a broad white wing ; pappus-bristles 2, smooth   *encelioides*

V. encelioides (*Cav.*) *A. Gray* in Bot. Calif. 1 : 350 (1876) ; Berhaut Fl. Sén. 167. *Ximenesia encelioides* Cav.
Ic. 2 : 60, t. 178 (1793) ; F.T.A. 3 : 383. An annual herb with yellow flower-heads ; probably escaped from
cultivation.
Sen.: *Perrottet. Etesse*! Introduced from tropical America ; also in the Sudan.

17. **ACANTHOSPERMUM** Schrank—Benth. & Hook. f. Gen. Pl. 2 : 349.

Leaves petiolate ; achenes obliquely elliptic, about 8-ribbed, sulcate, with lateral hooked
spinules only ..   ..   1. *brasilum*
Leaves sessile or the lower shortly petiolate ; achenes triangular with numerous lateral
hooked spinules and two large straight or hooked erect apical spines   2. *hispidum*

1. **A. brazilum** *Schrank* Pl. Rar. Hort. Monac. 2 : 53 (1822). An erect annual weed, about 1 ft. high, with
yellow flowers.
Ghana: Cape Coast (July) *Morgan* GC 17335! Also in tropical America.

2. **A. hispidum** *DC.* Prod. 5 : 522 (1836) ; Berhaut Fl. Sén. 68. *A. humile* of Chev. Bot. 368, not of DC. A bushy
annual weed about 12–18 in. high with small sessile flower-heads ; florets pale greenish-yellow ; achenes
5–10, spreading widely in fruit.
Sen.: Kaolak *Chev.* 47! Dakar (May) *Baldwin* 5730! **Gam.:** Kudang (Jan.) *Dalz.* 8121! **Mali:** Saninkoura,
Ségou (Aug., Oct., Nov.) *Roberty* 56! 448! 2867! Kabala, Bamako (July) *Roberty* 2446! **Port.G.:** Bissau
Esp. Santo 846! **Guin.:** Conakry *Debeaux* 310! **S.L.:** Freetown (Dec.) *Deighton* 476! Kumrabai (Dec.)

Thomas 7029 ! **Lib.**: Monrovia (Nov.) *Linder* 1525 ! Ganta (Sept.) *Baldwin* 9295 ! **Iv.C.**: Garango *Prost* ! Bingerville (Dec.) *Chev.* 16077 ! Nzo (Oct.) *Schnell* 3860. **Ghana:** Achimota (June) *Irvine* 692 ! Wenchi, Ashanti (Dec.) *Adams* 3253 ! Tumu (May) *Morton* GC 7587 ! **Niger:** Niamey (Dec.) *Hagerup* 543 ! **N.Nig.**: Jos (Dec.) *Coombe* 2 ! Kano (Aug.) *Barry* 104 ! **S.Nig.**: Lagos (Aug.) *Dalz.* 1413 ! Native of tropical America and now widely spread. (See Appendix, p. 414.)

18. ENYDRA[1] Lour.—F.T.A. 3 : 372.

Stems elongated, rooting in the lower part, glabrous ; leaves opposite, sessile, linear or linear-lanceolate, 5–12 cm. long, up to 2 cm. broad, subauriculate at base, entire or distantly serrate, finely puberulous and minutely glandular beneath ; flower-heads few, axillary, sessile, about 1 cm. diam. ; involucral bracts 3–4, broad, leafy ; receptacle with ribbed paleae closely sheathing the florets ; achenes oblong, without a pappus.. .. .. *fluctuans*

E. fluctuans *Lour.* Fl. Cochinch. 511 (1790) ; F.T.A. 3 : 372 ; Berhaut Fl. Sén. 73, 89, 92. A perennial spreading herb of swampy ground, especially in coastal areas ; florets white or yellowish-white in sessile heads ¼–⅓ in. across.
Sen.: *fide* Berhaut *l.c.* **Ghana:** Adabraka, Accra (July) *Irvine* 739 ! Prampram (Jan.) *Robertson* 72 ! Accra Plains (Jan., Dec.) *Morton* GC 6323 ! *Adams* 3615 ! Sekondi to Axim (Mar.) *Irvine* 2403 ! **S.Nig.**: Calabar (Feb.) *Mann* 2320 ! **F.Po** : (June, Nov.) *T. Vogel* 218 ! *Barter* ! *Mann* 428 ! Widespread in the tropics.

19. SIGESBECKIA Linn.—F.T.A. 3 : 371.

Outer involucral bracts usually 5, very narrowly elongate-spathulate, 0·5–1 cm. long, with numerous stalked glands ; leaves opposite, the lower with a long, winged petiole, lamina triangular-rhomboid, up to 12 cm. long and 5 cm. broad, crenate-dentate, thinly hispid on both surfaces, the undersurfaces with shining sessile glands ; ray-florets few with very short deeply divided ligules ; disk-florets with corolla 1·5 mm. long, broadly campanulate in the upper half ; achenes almost enclosed by the adherent-glandular receptacle-scales, curved-pyriform, striate and minutely transversely rugulose, glabrous ; pappus absent .. .. 1. *orientalis*
Outer involucral bracts ovate to oblong, 3–4 mm. long with occasionally a few sessile glands but without stalked glands ; leaves as above but remotely serrate and with a few sessile glands below or glands lacking ; florets and achenes as above ; receptacle-scales glabrous or with a few sessile glands .. .. 2. *abyssinica*

1. S. orientalis *Linn.* Sp. Pl. 900 (1753) ; F.T.A. 3 : 372. An erect branched annual glandular herb about 3 ft. high with minute yellow ray- and disk-florets in heads ⅛ in. across, including the spreading outer involucre ; a weed.
[Br.]Cam.: Bamenda, 4,000–5,800 ft. (Apr.–June) *Maitland* 1390 ! *Morton* K280 ! Throughout tropical Africa on higher ground and generally in tropical and subtropical regions.
2. S. abyssinica (*Sch. Bip.*) *Oliv. & Hiern* in F.T.A. 3 : 372 (1877) ; Robyns Fl. Sperm. Parc. Nat. Alb. 2 : 519, t. 47. *Limnogenneton abyssinicum* Sch. Bip. (1846). A diffusely branched herb like the last but much less glandular in all its parts ; florets yellow in heads ¼ in. broad ; in damp mountain grassland.
[Br.]Cam.: Mba Kokeka Mt., 7,400 ft., Bamenda (Jan.) *Keay & Lightbody* FHI 28395 ! Also in Congo, Ethiopia, Kenya, Uganda, Tanganyika and Nyasaland.

20. BAFUTIA C. D. Adams in Kew Bull. 15 : 439 (1962).

Slender-branched erect glabrous herb ; leaves mostly basal or on the lower part of the stem, linear-oblanceolate, narrowed to the half-amplexicaul base, up to 5 cm. long and 1 cm. broad ; flower-heads numerous, solitary at the ends of slender peduncles, 1–2 cm. long ; involucre campanulate about 4 mm. long with 8 short triangular teeth minutely puberulous at the tips ; florets all hermaphrodite, about 20, with slender exserted corollas ; achenes about 2 mm. long shallowly ribbed or smooth ; pappus of five or six short minutely plumose caducous setae .. .. .. *tenuicaulis*

B. tenuicaulis *C. D. Adams* l.c., fig. 1 (1962). An annual herb up to about 1 ft. high with pinkish-purple florets in numerous flower-heads less than ⅛ in. long, forming a diffuse corymb ; in mountain grassland.
[Br.]Cam.: Bafut-Ngemba F.R., 5,000–6,000 ft., Bamenda (Sept.–Nov.) *Lightbody* FHI 26264 ! *Ujor* FHI 30216 !

21. MIKANIOPSIS Milne-Redhead in Exell Suppl. Cat. S. Tomé 27 (1956).

Involucral bracts of the inner series about 8, 5–6 mm. long, obtusely narrowed at apex, rounded dorsally, much exceeded by the florets and pappus in flower ; capitula in rounded clusters in axillary branched inflorescences 15–25 cm. long ; stems greyish-woolly-tomentose ; leaves rounded to openly cordate at base, acuminate, repand-dentate, up to about 10 cm. long and 8 cm. broad .. .. 1. *paniculata*
Involucral bracts of the inner series about 12, about 1 cm. long, long-acute, flat, slightly exceeded by the florets and hardly by the pappus in flower ; stems glabrescent, striate :
Inflorescences slender, 10–15 cm. long, with long-pedunculate capitula irregularly scattered along the glabrescent axis ; leaves rather fleshy, shortly petiolate, rounded at the base, shallowly repand-dentate, acuminate, up to 6 cm. long and 5 cm. broad
2. *tedliei*
Inflorescences congested, up to about 8 cm. long with capitula more or less contiguous on short lateral branches ; axis and branches grey-pubescent ; leaves thick and

---

[1] The correct spelling is *Enydra* not *Enhydra* as in F.T.A., F.W.T.A. ed. 1, and other works.

fleshy, broadly ovate, shallowly cordate at base, long caudate-acuminate, up to 8 cm. long, remotely repand-dentate, 5–7-nerved above the base, glabrous   3. *maitlandii*

1. **M. paniculata** *Milne-Redhead* in Exell Suppl. Cat. S. Tomé 28, fig. 1 (1956). A shrubby climber to about 20 ft. with numerous pale yellow flowers.
    S.Nig.: Ikwette Plateau, 5,200 ft., Ogoja (Dec.) *Savory & Keay* FHI 25238! Also in S. Tomé.
2. **M. tedliei** (*Oliv. & Hiern*) *C. D. Adams* in J.W. Afr. Sci. Assoc. 6 : 153 (1961). *Senecio tedliei* Oliv. & Hiern in F.T.A. 3 : 420 (1877). *Gynura tedliei* (Oliv. & Hiern) S. Moore ex Hutch. & Dalz. F.W.T.A., ed. 1, 2 : 608 (1936). A shrubby climber like the last.
    Ghana: *Tedlie*! Hemang, Pra R. (Sept.) *Chipp* 572! Pra Suhein F.R. *Morton* A3681! **[Br.]Cam.:** Cam. Mt.: Litoka, 3,000–4,000 ft. (Apr.) *Maitland* 1122! 1311! Mann's Springs, 7,400 ft. (Dec.) *Boughey* GC 10590! *Morton* K870!
3. **M. maitlandii** *C. D. Adams* in J.W. Afr. Sci. Assoc. 6 : 152 (1961). A shrubby climber with purplish stems, glossy leaves and dull yellow florets in clustered flower-heads about ½ in. long.
    [Br.]Cam.: Cam. Mt.: Ukele, 7,300–7,600 ft. (Feb.) *Maitland* 1001! Mann's Springs, 7,400 ft. (Mar.) *Brenan* 9513!

22. KLEINIA Mill. Gard. Dict. Abridg., ed. 4 (1754).

Capitula 1–3, rarely more numerous, up to 4 cm. broad ; inflorescence subscapose, terminating an erect leafy or bracteate stem from a tuberous rootstock ; leaves towards the base of the flowering shoot or on a separate shoot, obovate-lanceolate, long-tapered to a sessile base, dentate or denticulate in the upper half or subentire, 8–15 cm. long, 3–5(–8) cm. broad, glabrous, apiculate ; involucral bracts 12–16, about 2 cm. long and up to 0·5 cm. broad with scarious margins ; whole plant glabrous except for a minute tuft of hairs at the tip of each involucral bract ; bracteoles lanceolate ; achenes glabrous or sparsely pubescent ; pappus-setae up to about 2 cm. long, smooth or minutely plumose, white or pale buff ..   ..   ..   1. *abyssinica*
Capitula 20 or more, oblong, less than 1 cm. broad, shortly pedunculate in rounded umbelliform clusters at the ends of short leafless branches ; peduncles about 1 cm. long, with small subulate bracteoles not forming a distinct calyculus ; leaves (when present) rhomboid-elliptic, obtuse, up to about 10 cm. long and 5 cm. broad, long-cuneate at base into a petiole ; involucral bracts about 10, up to 1·5 cm. long, oblong-lanceolate with narrow scarious margins and a minute tuft of hairs at the tip ; achenes ribbed, glabrous ; pappus-setae minutely barbellate, white ; a succulent shrub with deciduous leaves   ..   ..   2. *cliffordiana*

1. **K. abyssinica** (*A. Rich.*) *A. Berger* in Monatsschr. Kakt. 15 : 11 (1905). *Notonia abyssinica* A. Rich. (1848)—F.T.A. 3 : 407. *N. dalzielii* Hutch. in F.W.T.A., ed. 1, 2 : 149 (1931). A glaucous succulent herb with erect stems up to 18 in. high from a tuberous rootstock ; florets deep purplish-red in a few large heads with pale involucres.
    Guin.: Mali (June) *Adam* 14449! **N.Nig.:** Gimi, Katagum (Feb.) *Dalz.* 351! **S.Nig.:** Igboho, Oyo (Feb.) *Keay* FHI 23438! 23448! Also in Ubangi-Shari, Sudan, Ethiopia, Uganda, Kenya and Tanganyika.
2. **K. cliffordiana** (*Hutch.*) *C. D. Adams* in J.W. Afr. Sci. Assoc. 6 : 150 (1961). *Senecio cliffordianus* Hutch. in Kew Bull. 1921 : 250 ; F.W.T.A., ed. 1, 2 : 151. A glaucous succulent shrub up to 8 ft. high, leafless at flowering time, with numerous clustered flower-heads about ¾ in. long in terminal umbelliform corymbs ; involucres pale green ; florets white ; amongst rocks.
    Mali: Summit of Ouari Tondo, near Hombori *Jaeger* 5482! Guimel, Douentza (fl. & fr. Oct.) *Jaeger* 5518! **N.Nig.:** Zaria to Kaduna (Dec.) *Meikle* 836! Jos Plateau (Jan., Nov.) *Lely* P103! *Hill* 14! *Batten-Poole* 104! Vom *Dent Young* 161! Naraguta (Dec.) *Coombe* 30! Ropp to Mongu (Apr.) *Morton* K405!

23. GYNURA Cass.—F.T.A. 3 : 401, partly. *Nom. cons.*

Climbing robust herb with striate glabrescent stems ; leaves elliptic, acute, shortly cuneate at base, shortly petiolate, 4–10 cm. long, 2–5 cm. broad, subentire or remotely toothed, with about 5 pairs of lateral nerves, thinly pubescent ; inflorescences of few capitula on short lateral branches ; capitula with several narrow calycular bracts ; involucral bracts about 15, glabrous or nearly so, 1·5 cm. long   1. *sarmentosa*
Erect or straggling herb with pubescent grooved stems ; leaves lanceolate to obovate in plan, the lower broadly petiolate, the upper sessile, coarsely dentate to deeply pinnatifid, up to about 15 cm. long and 7 cm. broad, the segments remotely toothed, crisped-pubescent on both surfaces ; capitula several in a terminal corymbose inflorescence, each with several calycular bracts ; involucral bracts 12–14, pubescent, 1–1·3 cm. long, with broad scarious margins ..   ..   2. *miniata*

1. **G. sarmentosa** (*Blume*) *DC.* Prod. 6 : 298 (1838). *Cacalia sarmentosa* Blume (1826). *Gynura buntingii* S. Moore (1916). *Senecio baoulensis* A. Chev. in Mém. Soc. Bot. Fr. 2, 8 : 260 (1917) ; Chev. Bot. 375, not *Crassocephalum baoulense* (A. Chev.) Milne-Redhead (1951). A wide-spreading robust glabrous herbaceous climber with rather fleshy leaves ; leaves often purple beneath on young plants ; florets greenish-yellow in heads ¾ in. long ; pappus white ; in forest margins and thickets.
    S.L.: Freetown (Oct.) *Deighton* 240! **Lib.:** Begwai (Sept.) *Bunting* 122! Zinkor (Apr.) *Dinklage* 3035! Duport (Oct.) *Barker* 1438! Diebla, Webo (July) *Baldwin* 6368! Grand Bassa (Mar.) *Baldwin* 11236! **Iv.C.:** Toumodi, Baoulé-Sud (Aug.) *Chev.* 22395! **Ghana:** Akwapim (Oct.) *Johnson* 823! Aburi Scarp, 1,000 ft. (Nov.) *Adams* 1914! Volta River F.R. (Nov.) *Morton* GC 6071! Ajena (Nov.) *Morton* GC 9456! **S.Nig.:** Lagos (July) *Dalz.* 1410! Olokemeji F.R. (July) *Keay* FHI 22414! Also in tropical Asia, from where it has possibly been introduced into our area. (See Appendix, p. 418.)
2. **G. miniata** *Welw.* Apont. 586 (1858) ; F.T.A. 3 : 403 ; Adams in J.W. Afr. Sci. Assoc. 3 : 114 (1957). *Crassocephalum miniatum* (Welw.) Hiern (1898). *Gynura amplexicaulis* of F.W.T.A., ed. 1, 2 : 148, partly, (*Dalz.* 656, *Migeod* 481), not of Oliv. & Hiern Note 1. (?) *Gynura* sp. Bally in J.E. Afr. & Uganda Nat. Hist. Soc. 22 : 124, t. 20, fig. 4 (1945) Note 2. A perennial rather succulent herb with straggling leafy stems 2–4 ft. high from a mass of hard tubers ; stems and leaves pale green, crisped-pubescent ; leaves very variable, the lowest usually being deeply pinnatifid, but the upper often only toothed ; florets orange-brown or rarely yellow in rather broad heads up to ¾ in. long ; in rocky hill grassland or planted near villages for medicinal purposes.

Guin.: Mali, 5,000 ft. (June) *Adam* 14500 ! Note ² **S.L.:** Bintumane Peak, 6,400 ft. (Jan., Aug.) *T. S. Jones* 103 ! *Jaeger* 982 ! **N.Nig.:** Abinsi (June) *Dalz.* 656 ! Kakangi, Zaria (July) *Keay* FHI 25947 ! **[Br.]Cam.:** Bamenda Dist., 5,000–6,000 ft. (Feb.–May) *Maitland* 1416 ! *Migeod* 481 ! *Ejiofor* FHI 30079 ! *Ujor* FHI 30095 ! *Morton* K193 ! *Hepper* 2066 ! Mambila Plateau, 4,800 ft., Adamawa (Jan.) *Hepper* 1801 ! Also recorded from Kenya, Tanganyika and Angola. (See Appendix, p. 417.)

[Note 1 : The type of *G. amplexicaulis* Oliv. & Hiern (*Schweinfurth* 3770 from Sudan) superficially resembles *Crassocephalum picridifolium* (DC.) S. Moore, but has the style-arms of *Gynura*—C. D. A.]

[Note 2 : *Adam* 14500 and the plant described by Bally (*Coryndon Museum* 6799) have yellow florets and more divided leaves than is typical of this species—C. D. A.].

24. EMILIA Cass.—F.T.A. 3 : 405.

Upper leaves ovate, minutely pubescent beneath ; involucral bracts distinctly shorter than the florets :

Leaves repand-dentate ; involucral bracts about 15 ; florets orange-yellow 1. *coccinea*
Leaves dentate ; involucral bracts about 10 ; florets pale yellow ; styles pink
        2. *praetermissa*

Upper leaves lanceolate, sparsely setulose beneath ; involucral bracts nearly as long as the florets, about 8 ; florets mauve, rarely white .. .. 3. *sonchifolia*

1. **E. coccinea** (*Sims*) *G. Don* in Sweet Hort. Brit., ed. 3, 382 (1839). *Cacalia coccinea* Sims (1802). *E. sagittata* DC. (1838)—F.T.A. 3 : 405 ; Chev. Bot. 374 ; F.W.T.A., ed. 1, 2 : 149. A straggling glaucous sparingly pubescent herb with weak stems up to about 3 ft. long ; leaves often purple beneath ; florets deep yellow in long-stalked heads about ½ in. long ; a weed along roadsides and in clearings in forest country.
Guin.: Labé (Mar.) *Chev.* 12366 ! Nzérékoré to Macenta (Oct.) *Baldwin* 9745 ! Macenta, 2,000–2,500 ft. (Oct.) *Baldwin* 9841 ! Gangan (Apr.) *Adam* 11927 ! Nimba Mts. (Sept.) *Schnell* 3532 ! **S.L.:** Leicester Peak (Dec.) *Sc. Elliot* 3852 ! Picket Hill, 2,800 ft. (Nov.) *T. S. Jones* 212 ! Kenema (Oct.) *Deighton* 5227 ! Da-Oulen, 5,000 ft., Loma Mts. (Aug.) *Jaeger* 1349 ! Makoni (Apr.) *Marmo* 217 ! **Lib.:** Monrovia *Whyte* ! Dukwai R., Monrovia (Aug., Oct.–Nov.) *Cooper* 8 ! *Linder* 224 ! Brewersville (Sept.) *Barker* 1431 ! Ganta (May, Oct.) *Harley* ! Nimba Mts. (Sept.) *Adam* 16476 ! **Iv.C.:** Grabo, Cavally Basin (July) *Chev.* 19612 ! Tonkoui Mt. (Aug., Sept.) *Schnell* 1718 ! *Boughey* GC 18338 ! Nimba Mts., 2,500–5,000 ft. (Aug.) *Boughey* GC 18050 ! 18080 ! Téké (Dec.) *Giovannetti* 310 ! **Ghana:** Aburi (Feb., June) *Johnson* 1101 ! *Williams* 299 ! Axim (Feb.) *Irvine* 2199 ! Tarkwa (Dec.) *Andoh* FH 5445 ! Mampong, Ashanti (Dec.) *Morton* GC 9651 ! Amedzofe (July, Oct., Nov.) *Kwami* 80 ! *Morton* GC 6027 ! 9427 ! *Boughey* GC 13038 ! **Dah.:** Sakété (Jan.) *Chev.* 22808 ! **N.Nig.:** Lokoja (Oct.) *Parsons* 99 ! Abinsi (Feb.) *Dalz.* 664 ! **S.Nig.:** Lagos (Mar.) *Dalz.* 1163 ! *Hagerup* 835 ! Idanre Hills (Oct.) *Keay* FHI 25506 ! 25507 ! Okomu F.R., Benin (Feb., Mar.) *Akpabla* 1119 ! *Brenan* 9150 ! Sapoba (Nov.) *Meikle* 584 ! Abeokuta *Irving* 79 ! **[Br.]Cam.:** Cam. Mt., Buea, 3,000–5,000 ft. (Nov., Jan., Mar.) *Preuss* 1156 ! *Dunlap* 96 ! *Migeod* 57 ! *Boughey* GC 7015 ! Ambas Bay (Jan.) *Mann* 731 ! Bamenda Dist., 3,600–6,000 ft. (Jan., Mar., Apr., July, Dec.) *Migeod* 320 ! *Maitland* 1756 ! *Ujor* FHI 30470 ! *Boughey* GC 11139 ! *Morton* K31 ! Vogel Peak, 4,000 ft., Adamawa (Dec.) *Hepper* 1542 ! **F.Po :** Moka, 4,000 ft. (Dec.) *Boughey* 22 ! *Monod* 10490 ! Also in Sudan. (See Appendix, p. 417.)

[As at present construed this species has scarlet-flowered forms in central Africa and Asia, and crimson-flowered forms in the New World tropics as well as the yellow-flowered West African plants. Some of these forms are known to be diploid, others tetraploid (cf. Baldwin in Bull. Torr. Bot. Club 76 : 346 (1949)). Similarly with *Emilia sonchifolia* the smaller mauve-flowered plants of West Africa and America are diploid, while larger red-flowered plants of eastern Asia are tetraploid.—C. D. A.]

2. **E. praetermissa** *Milne-Redhead* in Kew Bull. 5 : 375 (1951). An erect herb 2–3 ft. high with pale flesh-coloured florets in heads about ½ in. long ; in clearings in forest.
S.L.: Freetown (Apr., May, Oct.) *King* 182 ! *Deighton* 1188 ! *Tindall* 15 ! Wellington (Feb.) *Deighton* 4984 ! Mt. Horton, 1,800 ft. (Jan.) *T. S. Jones* 299 ! Hastings (Sept.) *Melville & Hooker* 431 ! Rokupr (June) *Jordan* 448 ! **Iv.C.:** Lamé (Nov.) *Leeuwenberg* 1904 ! **Ghana:** Kanawirebo Hill, 1,700 ft., Awaso, Sefwi (Dec.) *Adams* 2078 ! Dunkwa (Dec.) *Morton* A3607 ! **S.Nig.:** Lagos (Apr., Nov.) *Baldwin* 11981 ! *Meikle* 502 ! Ibadan (Dec.) *Keay* FHI 25595 ! Sapoba (Nov.) *Meikle* 621 ! *Keay* FHI 28065 ! Enugu (Mar.) *Hepper* 2222 ! Port Harcourt (Jan.) *Gregory* 74 ! Calabar (Nov.) *Baldwin* 13754 !

3. **E. sonchifolia** (*Linn.*) *DC.* Prod. 6 : 302 (1838) ; F.T.A. 3 : 405 ; Chev. Bot. 374 ; Berhaut Fl. Sén. 171. *Cacalia sonchifolia* Linn. (1753). A bushy annual herb usually about 1 ft. high with rather glaucous sparingly pubescent stems and leaves ; florets mauve or rarely creamy-white in narrow erect heads nearly ½ in. long.
Sen.: *fide* Berhaut *l.c.* **Guin.:** Conakry (Feb.) *Chev.* 12167 ! **S.L.:** Freetown (May) *Deighton* 1189 ! Njala (Sept.) *Deighton* 2080 ! Musaia (Apr.) *Deighton* 5380 ! **Iv.C.:** Lamé (Nov.) *Leeuwenberg* 1905 ! **Ghana:** Achimota (Mar., May, Oct.) *Irvine* 400 ! *Akpabla* 27 ! *Morton* A1571 ! Bantroasi (Nov.) *Morton* GC 7898 ! Axim to Sekondi (May) *Cudjoe* 132 ! Koforidua (June, fl. white) *Baker* in Hb. *Morton* A2237 ! Asuansi (Feb.) *Williams* 3 ! **S.Nig.:** Lagos (Oct.) *Dalz.* 1165 ! Enugu (Oct.) *Jones* FHI 6171 ! **F.Po:** S. Isabel (Dec.) *Guinea* 196. A common tropical weed. (See Appendix, p. 417.)

Imperfectly known species.

Senecio tenuicaulis *Muschl.* in Engl. Bot. Jahrb. 43 : 60 (1909). A slender glabrous herb about 1 ft. high with a single terminal flower-head ; pappus white.
? **Togo Rep.:** *Büttner* 456. Type specimen destroyed.

25. CRASSOCEPHALUM Moench—S. Moore in J. Bot. 50 : 209 (1912). *Gynura* of F.T.A.3 : 401, and of F.W.T.A., ed. 1, 2 : 147, partly, not of Cass. *Senecio* of F.T.A.3 : 408, and of F.W.T.A., ed 1, 2 : 148, partly, not of Linn.

Corolla yellow, orange or brownish-red :

Flower-heads small, up to about 1 cm. long, numerous in panicles ; involucral bracts glabrous ; florets pale yellow :

Small tree ; leaves closely serrate, up to 40 cm. long and 10 cm. broad, pubescent on the nerves beneath ; achenes pubescent .. .. 1. *mannii*

Climber ; leaves more or less triangular, subhastate and sometimes with a pair of lateral segments, long-petiolate, 5–8 cm. long, 3–5 cm. broad, repand-dentate, glabrous ; achenes glabrous .. .. 2. *biafrae*

Flower-heads usually exceeding 1 cm. in length ; erect or straggling herbs :

Capitula several to numerous :

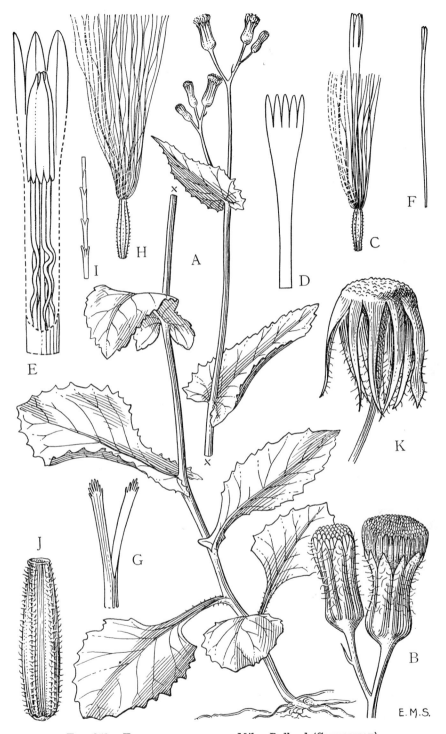

Fig. 251.—Emilia praetermissa *Milne-Redhead* (Compositae).

A, habit, × ⅔. B, capitula, × 3. C, floret, × 6. D, corolla-tube opened out, × 6. E, section of corolla-tube showing staminal tube, × 16. F, style, × 6. G, stigma, × 24. H, fruit with pappus, × 6. I, section of pappus hair, × 60. J, fruit without pappus, × 16. K, empty receptacle, × 3. A drawn from *Baldwin* 11981. B–K from *King* 182.

Leaves ovate in plan, pinnatifid and sharply dentate ; capitula rather numerous ; erect herbs :
Florets brownish-red ; involucral bracts 9–10 mm. long ; style-arm appendages distinct .. .. **3. *crepidioides***
Florets yellow ; involucral bracts 7–8 mm. long ; style-arm appendages very short
                                                                                       4. *montuosum*
Leaves lanceolate, subentire, shallowly repand-dentate, the upper sessile and mostly broadly auriculate ; capitula few on peduncles up to 15 cm. long ; straggling or erect herbs ; florets golden-yellow ; involucral bracts 11–12 mm. long
                                                                                       5. *picridifolium*
Capitula solitary on long peduncles :
Heads up to 2 cm. broad ; involucral bracts 7·5–9 mm. long, pubescent ; leaves coarsely dentate with rounded stipuliform auricles at the base :
Leaves ovate-elliptic, cuneate down to point of attachment ; involucre shortly pubescent .. .. **6. *gracile***
Leaves broadly ovate, rounded to truncate or subcordate at base ; involucre densely setulose-pubescent .. .. **7. *vitellinum***
Heads about 2·5 cm. broad ; involucral bracts 10–12 mm. long, glabrescent ; leaves finely serrate without basal stipuliform auricles .. .. **8. *bougheyanum***
Corolla purple, pink, mauve, blue or rarely white ; more or less erect annual herbs :
Flower-heads up to about 1 cm. long :
Capitula several on short peduncles :
Leaves linear, shallowly repand-dentate ; involucral bracts 4–6 mm. long ; stems slender, erect, sparingly branched ; style-arm appendages distinct **9. *guineense***
Leaves pinnately lobed ; involucral bracts 7–9 mm. long ; stems much-branched ; style-arm appendages very short .. .. **10. *bauchiense***
Capitula solitary ; leaves oblanceolate, cuneate, dentate, glabrous **11. *libericum***
Flower-heads 1·5 cm. or more long :
Capitula solitary on long peduncles ; leaves mostly subentire, dentate **12. *rubens***
Capitula several on long or short peduncles ; at least the upper leaves distinctly pinnatifid or deeply toothed :
Peduncles up to 8 cm. long ; flower-heads rarely clustered ; upper leaves doubly dentate, pinnatifid with 2–4 pairs of segments ; lower leaves subentire
                                                                                       13. *sarcobasis*
Peduncles up to 2 cm. long ; flower-heads mostly clustered ; leaves usually pinnatifid with 4 or more pairs of segments :
Leaves linear-lanceolate in plan with numerous linear segments not cut down to the midrib and smaller towards the apex ; involucral bracts densely setulose
                                                                                       14. *baoulense*
Leaves ovate to obovate in plan with 4–5 pairs of large oblanceolate or elliptic segments cut down nearly to the midrib and larger towards the apex, and with smaller paired linear lobes towards the base ; involucral bracts sparsely setulose
                                                                                       15. *togoense*

1. **C. mannii** (*Hook. f.*) *Milne-Redhead* in Kew Bull. 5 : 377 (1951). *Senecio mannii* Hook. f. (1862)—F.T.A. 3 : 418 ; F.W.T.A., ed. 1, 2 : 151. A shrub or soft-wooded tree up to about 25 ft. high, nearly glabrous, with yellow florets in dense panicles of heads ⅓–½ in. long ; usually in clearings in mountain forest ; sometimes planted as fences.
S.Nig.: Enugu (cult.) *Tamajong* FHI 26840 ! **[Br.]Cam.:** Cam. Mt., 3,500–8,000 ft. (Dec.–Apr.) *Mann* 1935 ! *Maitland* 236 ! 985 ! *Dalz.* 8226 ! *Hutch. & Metcalfe* 73 ! *Brenan* 9511 ! Bamenda, 6,500 ft. (Apr.) *Morton* K227 ! Bafut-Ngemba F.R., 5,200 ft. (Jan.) *Lightbody* FHI 26279 ! *Migeod* 374 ! L. Oku, 7,000 ft., Bamenda (Jan.) *Keay* FHI 28486 ! Nkambe, 6,000 ft. (Feb.) *Hepper* 1900 ! Mandaga, 6,000 ft. Adamawa (cult. Jan.) *Latilo & Daramola* FHI 34353 ! **F.Po:** 6,000 ft. (Dec., Feb.) *Mann* 282 ! *Boughey* 115 ! Pico Serrano, Moka, 6,000 ft. (Jan.) *Guinea* 2045 !
[*C. multicorymbosum* (Klatt) S. Moore, which is very closely allied if indeed distinct (see Milne-Redhead l.c.), occurs in Sudan, E. Africa, Congo and Angola.]

2. **C. biafrae** (*Oliv. & Hiern*) *S. Moore* in J. Bot. 50 : 211 (1912). *Senecio biafrae* Oliv. & Hiern in F.T.A. 3 : 420 ; F.W.T.A., ed. 1, 2 : 151. *S. gabonicus* of Chev. Bot. 375, not of Oliv. & Hiern. A glabrous climbing herb with numerous pale yellow flower-heads about ½ in. long in rounded pedunculate clusters ; along roadsides and in secondary forest in hilly forest country.
Guin.: Bilima (Dec.) *Caille* in Hb. Chev. 14880 ! **S.L.:** Mt. Gonkwi, Talla Hills (Feb.) *Sc. Elliot* 4837 ! Njala (cult. Apr.) *Deighton* 2995 ! Koindu (cult. Feb.) *Deighton* 3508 ! **Ghana:** Aburi (Jan.–Feb.) *Johnson* 961 ! *Akpabla* 93 ! *Morton* GC 6337 ! Tafo (Nov.) *Robertson* 48 ! Begoro (Feb.) *Morton* A1848 ! Abetifi (Jan.) *Irvine* 1677 ! **Dah.:** Sakété to Pedjilé, Porto Novo (Feb.) *Chev.* 22913 ! **S.Nig.:** Lagos *Moloney* ! Ebute Metta (Jan.) *Millen* 96 ! Ibadan (Apr.) *Meikle* 1406 ! Shasha F.R. (Mar.) *Jones & Onochie* FHI 17033 ! Oban *Talbot* 1549 ! **[Br.]Cam.:** Victoria (Mar.) *Maitland* 504 ! Cam. Mt., 7,000 ft. (Jan.) *Mann* 1325 ! L. Bambulue, Bamenda, 5,500 ft. (Jan.) *Keay* FHI 28349 ! **F.Po:** *Boughey* GC 17406 ! S. Isabel (Jan.) *Guinea* 782. Musola (Jan.) *Guinea* 1275. Also in S. Tomé. (See Appendix, p. 420.)

3. **C. crepidioides** (*Benth.*) *S. Moore* in J. Bot. 50 : 211 (1912). *Gynura crepidioides* Benth. (1849)—F.T.A. 3 : 403 ; Chev. Bot. 374 ; F.W.T.A., ed. 1, 2 : 148, partly (excl. *Mann* 51 ; *Dunlap* 87 ; *Maitland* 268 ; 477 ; *Millen* 51). A stout erect herb up to about 3 ft. high with bright brownish-red florets in heads ¼–⅓ in. long ; a weed of farms and waste places.
Guin.: *Heudelot* 785 ! Sérédou (Apr.) *Adam* 11987 ! **S.L.:** Njala (Feb.) *Deighton* 1871 ! Makump (Sept.) *Glanville* 410 ! Kaballa (Sept.) *Thomas* 2168 ! Musaia (Mar.) *Deighton* 5374 ! Loma Mts. (Sept., Nov.) *Jaeger* 569 ! 1994 ! **Lib.:** Dukwai R. (Oct.–Nov.) *Cooper* 1 ! Brewersville (Sept.) *Barker* 1430 ! Toroke, Webo (July) *Baldwin* 6733 ! Robertsfield (July) *Baldwin* 6657 ! Ganta (Sept.) *Harley* 12 ! **Iv.C.:** Makouguié (Jan.) *Chev.* 16993 ! **Ghana:** Assuansi (Mar.) *Irvine* 1563 ! Anyinam (Dec.) *Dalz.* 103 ! Mampong, Ashanti. (Dec.) *Thorold* 47 ! Wenchi (Dec.) *Adams* 3239 ! Zolo-Kpuita, Trans-Volta Togo (Nov.) *Morton* GC 9366 ! Legon Hill (Oct.) *Adams* 3387 ! **N.Nig.:** Jos Plateau (July) *Lely* P390 ! Kwakwi, Zaria (Aug.)

Fig. 252.—Crassocephalum mannii (*Hook. f.*) *Milne-Redhead* (Compositae).
A, involucre. B, flower. C, anthers. D, style-arms.

247

Keay FHI 20157! **S.Nig.**: Lagos *Moloney*! Idanre F.R., Ondo (July) *Onochie* FHI 33377! Usonigbe F.R., Benin (Nov.) *Meikle* 580! Onitsha *Barter* 1809! Calabar (Jan.) *Holland* 19! Ikwette Plateau, 5,500 ft., Obudu (Aug.) *Stone* 56! [**Br.**]**Cam.**: Ambas Bay (Apr.) *Hutch. & Metcalfe* 133! Cam. Mt. 3,000 ft. (Jan.) *Dunlap* 150! Bamenda, 6,000 ft. (Jan.) *Migeod* 439! Wum (Dec.) *Boughey* GC 10947! 11374! Mambila Plateau, 4,600 ft. (Jan.) *Hepper* 1683! **F.Po:** (Nov.) *T. Vogel* 139! Mioko, 5,000 ft. (Dec.) *Boughey* 152! In tropical Africa generally and in S. Tomé and the Mascarene Islands. (See Appendix, p. 418.)

4. **C. montuosum** (*S. Moore*) *Milne-Redhead* in Kew Bull. 5 : 376 (1951) ; Adams in J.W. Afr. Sci. Assoc. 1 : 27 (1954). *Senecio montuosus* S. Moore in J. Linn. Soc. 35 : 354 (1902). *Gynura crepidioides* of F.W.T.A. ed. 1, 2 : 148, partly (*Mann* 51 ; *Dunlap* 87 ; *Maitland* 268 ; 477). An erect much-branched herb up to about 4 ft. high with yellow florets in numerous heads usually less than ½ in. long ; a weed of clearings in hilly districts.

[**Br.**]**Cam.**: Rio del Rey *Johnston*! Cam. Mt., 3,000–6,000 ft. (Nov., Jan., Mar.) *Maitland* 268! 300! 477! *Migeod* 188! *Dunlap* 87! Mann's Springs (Dec.) *Boughey* GC 10733! *Adams* GC 11723! Mamfe to Bamenda (Jan.) *Keay* FHI 28545! Bamenda (Mar., Aug.) *Morton* K80! *Maitland*! *Savory* UCI 319! Bali-Ngemba F. R. (May) *Ujor* FHI 30332! **F.Po:** *Mann* 51! *Monod* 10473! 10474! Monte Balea (Dec.) *Guinea* 396! Moka, 4,000–6,000 ft. (Dec.–Jan.) *Boughey* 12! *Exell* 805! Pico de S. Isabel (Dec.) *Boughey* 10804a! Also in Sudan, Congo and S. Tomé.

5. **C. picridifolium** (*DC.*) *S. Moore* in J. Bot. 50 : 212 (1912). *Senecio picridifolius* DC. (1838)—F.T.A. 3 : 413 ; Chev. Bot. 375, partly (*Chev.* 222, 635). *Gynura amplexicaulis* of F.W.T.A., ed. 1, 2 : 148, partly (excl. *Dalz.* 656 ; *Migeod* 481), not of Oliv. & Hiern (see note 1 under *Gynura miniata*). *G. gracilis* of F.W.T.A., ed. 1, 2 : 148, partly (*Chev.* ; 635), not of Hutch. & Dalz. A scrambling or erect herb up to 2 ft. high in damp open ground, with usually several flower-heads over ½ in. long ; florets golden-yellow.

Gam.: *Brown-Lester* 40! **Mali:** Sicoro (Jan.) *Chev.* 222! Dendéla (Mar.) *Chev.* 635! **Port.G.:** Brene, Bissau (Jan.) *Esp. Santo* 1697! **S.L.:** (July) *Glanville* 264! Falaba (Mar.) *Sc. Elliot* 5143! Kaballa (Sept.) *Thomas* 2198! **N.Nig.:** Sokoto (Sept.) *Dalz.* 406! Matyoro (Oct.) *Thornewill* 130! Vom (Jan.) *Dent Young* 157! Bokkos, 4,000 ft. (Dec.) *Coombe* 76! Maiduguri *Kennedy* 2949! **F.Po** : summit of the Peak, 9,200 ft. (Dec., Feb.) *Boughey* GC 10839! *Guinea* 2738. Also in Mauritania and extending to Sudan, Congo and Angola.

[This species has a strong superficial resemblance to *Gynura amplexicaulis* Oliv. & Hiern, in which the style-arms are uniformly tapered.]

6. **C. gracile** (*Hook. f.*) *Milne-Redhead ex Guinea* in Anal. Jard. Bot. Madrid 10, 1 : 307 (1951). *Gynura vitellina* var. *gracilis* Hook. f. in J. Linn. Soc. 7 : 202 (1864). *G. gracilis* Hook. f. ex Hutch. & Dalz. F.W.T.A., ed 1, 2 : 147, 148, (1931) partly (excl. *Chev.* 222 ; 635). *Senecio picridifolius* of Chev. Bot. 375, partly, not of DC. An erect or straggling herb up to 3 ft. high with orange-yellow florets in heads ½–¾ in. across ; leaves sessile, narrowed to the base ; in mountain grassland.

Guin.: *fide* Chev. *l.c.* Timbi-Madina (Mar.) *Adam* 11702! Timbi (June) *Adam* 14652! [**Br.**]**Cam.**: Cam. Mt., 6,000–9,000 ft. (Dec., Jan., Apr., May) *Mann* 1317! *Johnston* 35! *Maitland* 650! 847! 955! *Morton* GC 6945! Litoka, 4,500 ft. (Apr.) *Maitland* 1073! Mann's Springs, 7,200–8,700 ft. (Dec., Mar.) *Boughey* GC 10608! 12583! A528b! *Brenan* 9374! **F.Po:** Musola (Jan.) *Guinea* 1369!

7. **C. vitellinum** (*Benth.*) *S. Moore* in J. Bot. 50 : 212 (1912). *Gynura vitellina* Benth. (1849)—F.T.A. 3 : 402 ; F.W.T.A., ed. 1, 2 : 148, partly (*T. Vogel* 188 ; *Mann* 38 ; *Barter* ; *Maitland* 307 ; *Migeod* 455). A weak straggling herb 2–3 ft. high with deep yellow to orange florets in heads ½–¾ in. across ; in grassy clearings in mountain areas.

S.Nig.: Koloishe Mt., 5,200 ft., Obudu (Dec.) *Savory & Keay* FHI 25068! Ikwette Plateau, 5,200 ft., Obudu (Dec.) *Savory & Keay* FHI 25237! [**Br.**]**Cam.**: Cam. Mt., 3,000–4,000 ft. (Nov.–Mar.) *Akpabla* GC 10494! *Boughey* GC 7017! *Maitland* 307! *Mildbr.* 10213! *Migeod* 58! Bamenda, 5,000 ft. (Feb.) *Migeod* 455! Bamenda to Santa (Mar.) *Morton* K36! Bamenda to Bambui (Jan.) *Keay* FHI 28332! Nkambe, 6,000 ft. (Feb.) *Hepper* 1891! Mambila Plateau, 4,600 ft. (Jan.) *Hepper* 1663! **F.Po:** *T. Vogel* 188!198! *Barter*! *Mann* 38! Moka, 4,000 ft. (Dec.) *Boughey* 23! Generally distributed in the mountains of tropical Africa.

8. **C. bougheyanum** *C. D. Adams* in J.W. Afr. Sci. Assoc. 3 : 111 (1957). *Gynura vitellina* of F.T.A. 3 : 402 ; F.W.T.A., ed. 1, 2 : 148, partly (*Mann* 610 ; 1931 ; *Dunlap* 33 ; 85 ; *Maitland* 478 ; *Migeod* 234 ; *Mildbraed* 10807), not of Benth. (1849). *C. vitellinum* of Guinea in Anal. Jard. Bot. Madrid. 10 : 305 (1951), not of (Benth.) S. Moore. A straggling herb up to 8 ft. or more high with glabrescent stems and involucral bracts, and yellow florets in heads 1 in. across ; in grassy clearings in upper margins of mountain forest. [**Br.**]**Cam.**: Cam. Mt., 5,000–7,000 ft. (Nov.–Jan.) *Mann* 1931! *Dunlap* 33! 85! *Irvine* 1461a! *Maitland* 5,000–6,000 ft. (Nov., Dec., Mar., Apr.) *Migeod* 234! *Mildbr.* 10807! *Maitland* 478! *Hutch. & Metcalfe* 17! *Thorold* 36! Mann's Springs, 6,000–7,500 ft. (Dec., Mar., Apr.) *Adams* GC 11729! *Boughey* GC 11545! *Brenan* 9514! *Morton* GC 7079! **F.Po:** Pico de S. Isabel, 8,500 ft. (Dec., Mar.) *Mann* 610! *Guinea* 2884 ; 2885!

9. **C. guineense** *C. D. Adams* in J. W. Afr. Sci. Assoc. ined. (1963). *Lactuca tenerrima* of Chev. Bot. 379, partly (*Chev.* 18291 ; 18449), not of Pourret (1788). *Emilia guineensis* of F.W.T.A., ed. 1, 2 : 149, partly (*Chev.* 18291 ; 18449), not of Hutch. & Dalz. (1931). *Senecio perrottetii* of F.W.T.A., ed. 1, 2 : 151, partly (*Thomas* 3090 ; *Deighton* 1252 ; 1253), not of DC. (1838). A slender erect annual herb 1–2 ft. high with reddish-purple, blue, mauve, pink or white florets in heads about ¼ in. long ; in grassy places on rock-outcrops.

Guin. : Fouta Djalon, 3,000–4,000 ft. (Sept., Oct.) *Chev.* 18291! 18449! 18559! 18820! Boumalol (Oct.) *Adam* 12570! Mt. Bilima, 3,000 ft., Mamou (Sept.) *Jac.-Fél.* 1830! **S.L.** : Bumban, 1,000 ft. (Aug.) *Deighton* 1252! 1253! Kanya, 1,350 ft. (Oct.) *Thomas* 3090! Bumbuna to Farangbaia (Sept.) *Deighton* 5148! Sasa, Tonko Limba (Sept.) *Jordon* 334a! 334b! Loma Mts., 5,000 ft. (Sept.) *Jaeger* 1704!

10. **C. bauchiense** (*Hutch.*) *Milne-Redhead* in Kew Bull. 5 : 376 (1951). *Gynura bauchiensis* Hutch. F.W.T.A. ed. 1, 2 : 608 (1936). *G. caerulea* Hutch. & Dalz. F.W.T.A., ed. 1, 2 : 147, 148 (1931), not *G. coerulea of* O. Hoffm. (1893). An erect bushy annual herb about 1 ft. high with bright blue florets in numerous heads ½ in. long ; in rough open ground.

N.Nig.: Vom, 3,000–4,500 ft. Dent *Young* 158! Jos Plateau *Batten-Poole* 165! [**Br.**]**Cam.**: Bambui, Bamenda, 6,300 ft. (Dec.) *Adams* GC 11239!

11. **C. libericum** *S. Moore* in J. Bot. 54 : 282 (1916). *Gynura liberica* (S. Moore) Hutch. & Dalz. F.W.T.A., ed. 1, 2 : 148 (1931). A prostrate nearly glabrous herb with ascending slender branches, shortly petiolate leaves and pink florets in heads ½ in. long.

Lib.: Begwai (Oct.) *Bunting* 142!

[This obscure plant may be merely a depauperate form of *C. rubens*—C. D. A.]

12. **C. rubens** (*Juss. ex Jacq.*) *S. Moore* in J. Bot. 50 : 212 (1912). *Senecio rubens* Juss. ex Jacq. (1777). *S. cernuus* Linn. f. (1781). *Cacalia uniflora* Schum. & Thonn. (1827). *Gynura cernua* Benth. (1849)—F.T.A. 3 : 402 ; Chev. Bot. 374, partly (excl. *Chev.* 22399) ; F.W.T.A., ed. 1, 2 : 148. An erect branched herb with heads ½–⅔ in. long on rather long peduncles drooping in flower, erect in fruit ; florets usually bright purple, but occasionally white, pink or blue ; a common weed of disturbed ground.

Mali: Guélia *fide* Chev. *l.c.* **Guin.:** Diomandou (Jan.) *Adam* 12070! Macenta, 2,000–2,500 ft. (Oct.) *Baldwin* 9762! **S.L.:** Leicester Peak (Dec.) *Sc. Elliot* 3879! Kasokora (Aug.) *Deighton* 1257! Bonthe (Nov.) *Adames* 107! Kabala (Dec.) *Deighton* 4513! Loma Mts. (Sept.) *Jaeger* 1554! 1764! **Lib.:** Buchanan (Nov.) *Adam* 16052! 16440! **Iv.C.:** Toumodi to Nzakro, Baoulé-Sud (Aug.) *Chev.* 22421 (partly)! Nimba Mts., 2,500–5,000 ft. (Aug.) *Boughey* GC 18064! 18066! **Ghana:** Aburi (Nov.) *Brown* 786! *Johnson* 784! Legon Hill (Oct.) *Adams* 3449! Techiman, Ashanti (Dec.) *Adams & Akpabla* GC 4479! Banda, Ashanti

(Dec.) *Morton* GC 25213 ! 25216 (white fl.) ! Damongo (Mar.) *Morton* GC 8715 ! **Togo Rep.:** Lomé *Warnecke* 296 ! **N.Nig.**: Abinsi *Dalz.* 655 ! Kontagora (Nov.) *Dalz.* 207 ! Jos Plateau (Jan.) *Lely* P32 ! Anara F.R., Zaria (Nov.) *Keay* FHI 28121 ! **S.Nig.:** *T. Vogel* ! Lagos (Dec.) *Dalz.* 1160 ! Ibadan *Newberry* 84 ! Ikwette Plateau, 5,500 ft., Obudu (Aug.) *Stone* 41 ! **[Br.]Cam.**: Bamenda, 4,500–6,000 ft. (Jan, Mar., Apr.) *Migeod* 447 ! *Maitland* 1466 ! *Morton* K77 ! Bafut-Ngemba F.R., 5,000–6,000 ft. (Aug., Oct.–Nov.) *Lightbody* FHI 26269 ! *Tamajong* FHI 26892 ! Bambui, 6,300 ft. (Dec.) *Boughey* GC 10421 ! Mambila Plateau, 4,600 ft. (Jan.) *Hepper* 1682 ! 1684 ! Widespread in warm countries. (See Appendix, p. 418.)

13. **C. sarcobasis** (*Boj. ex DC.*) *S. Moore* in J. Bot. 50 : 211 (1912). *Senecio sarcobasis* Boj. ex DC. (1838). An annual herb about 1 ft. high with mauve to deep pink or magenta florets ; in hilly districts.
N.Nig.: Vom *Dent Young* 156 ! **[Br.]Cam.**: Lakom, Bamenda, 6,000 ft. (Apr.) *Maitland* 1482 ! Belo, Bamenda, 4,500 ft. (Apr.) *Maitland* 1769 ! Kishong, 6,700 ft. (Jan.) *Keay & Russell* FHI 28446 ! Bambui, 6,300 ft. (Dec.) *Boughey* GC 10420 ! Santa, 5,700 ft. (Mar.) *Morton* K103 ! Throughout tropical Africa and in Madagascar.
[This species comes near to *C. rubens* in the W. African area, but the heads are always smaller, more numerous and, according to Milne-Redhead, do not nod in flower or fruit. This last character may also serve to distinguish this species from *C. crepidioides*, with which it has been united (e.g. Robyns, 1947) —C. D. A.].

14. **C. baoulense** (*Hutch. & Dalz.*) *Milne-Redhead* in Kew Bull. 5 : 377 (1951) ; Adams in J.W. Afr. Sci. Assoc. 2 : 62 (1956). *Gynura baoulensis* Hutch. & Dalz. F.W.T.A., ed. 1, 2 : 148 (1931). *G. cernua* of Chev. Bot. 374, partly (*Chev.* 22399), not of Benth. An annual herb 2–3 ft. high with pilose stems ; florets bright purple ; in low-lying savanna grassland.
Iv.C.: Toumodi to Dimbokro, Baoulé-Sud (Aug.) *Chev.* B22259 ! Touminiané to Angouakoukrou (Aug.) *Chev.* 22399 ! **Ghana:** Afram Plains (Jan.) *Irvine* 612 ! Opro F.R., Ashanti (Jan.) *Andoh* FHI 2403 ! Nwereme, Ashanti (May, Dec.) *Irvine* 2476 ! *Gillett* 24 ! Ejura (Nov.–Jan.) *Vigne* FH 3470 ! *Morton* GC 9492 ! A1522 ! Sampa, Ashanti (Dec.) *Morton* A2624 !

15. **C. togoense** *C. D. Adams* in J.W. Afr. Sci. Assoc. 1 : 27 (1954). *Gynura crepidioides* of F.W.T.A., ed. 1, 2 : 148, partly (*Millen* 51). An erect annual herb 2–3 ft. high ; florets mauve ; in savanna.
Ghana: Amedzofe (Nov.) *Morton* GC 9428 ! 9429 ! A2299 ! **N.Nig.:** Kogigiri, Jos, 4,100 ft. (Oct.) *Hepper* 1055 ! **S.Nig.:** Lagos *Millen* 51 ! Olokemeji F.R. (Aug., Oct.) *Keay* FHI 25387 ! *Hepper* 934 ! Ibadan (Oct.) *Keay* FHI 25435 ! Ilora junction, Oyo (Sept.) *Keay* FHI 28035 !

26. SENECIO Linn.—F.T.A. 3 : 408.

Annual herbs ; capitula cymose-corymbose, rather small ; florets yellow :
Ray-florets absent ; leaves linear, subentire, 2–6 cm. long ; involucres campanulate, calyculate, 3–4 mm. long, glabrous, on peduncles up to 1 cm. long ; stems and leaves glabrous ; achenes minutely setulose on the ribs, glabrescent ; achene-setae conspicuously zig-zag and barbellate, caducous ..   ..   1. *schimperi*
Ray-florets present ; achenes shortly pubescent on the ribs ; involucre 5–6 mm. long :
Calyculus absent ; capitula oblong on slender peduncles ; leaves spathulate, repand-dentate, slightly setulose, up to 4 cm. long ..   ..   ..   2. *abyssinicus*
Calyculus present ; capitula campanulate, more or less clustered on peduncles up to 1 cm. long ; leaves lanceolate, pinnatifid or irregularly lobed, up to 10 cm. long and 3 cm. broad, glabrous ..   ..   3. *perrottetii*
Perennial herbs ; ray-florets mostly present :
Leaves entire, sessile, glabrous or glabrescent :
Capitula solitary on naked peduncles 10–15 cm. long ; leaves linear-lanceolate, narrowed to base, 2–4 cm. long, up to 8 mm. broad, at first thinly villous, glabrescent, with 3–5 ascending nerves ; involucral bracts 1·5 cm. long ; achenes pubescent ..   ..   4. *baberka*
Capitula several, corymbose, calyculate ; leaves ovate to obovate, rounded at base, up to 8 cm. long and 1–3 cm. broad, glabrous, with pinnate nerves ; involucral bracts 8–10 mm. long ; achenes glabrous ..   ..   ..   5. *ruwenzoriensis*
Leaves toothed or lobed :
Stem-leaves broad and auriculate at base ; involucral bracts 1 cm. long, woolly or pubescent ; capitula numerous, corymbose ; achenes glabrous :
Leaves oblong, up to 15 cm. long and 4 cm. broad, coarsely and doubly dentate or lobulate, obtuse, apiculate, pubescent beneath ..   ..   6. *clarenceanus*
Leaves broadly oblanceolate, up to 20 cm. long and 6 cm. broad, closely denticulate, with numerous faint lateral nerves, acuminate, woolly-tomentose with white hairs beneath or rarely glabrescent ..   ..   7. *burtonii*
Stem-leaves at most narrowly rounded at base, sessile ; lower leaves oblanceolate, stalked ; achenes pubescent :
Leaf-margins regularly minutely denticulate ; stems and involucral bracts glabrous or nearly so ; capitula few, with involucres 10–12 mm. long ..   8. *lelyi*
Leaf-margins irregularly and coarsely crenate-dentate, remotely denticulate ; stems and involucral bracts pubescent ; capitula numerous with involucres 7–9 mm. long ..   ..   9. *hochstetteri*

1. **S. schimperi** *Sch. Bip. ex A. Rich.* Tent. Fl. Abyss. 1 : 435 (1848) ; F.T.A. 3 : 412. *Lactuca tenerrima* of Chev. Bot. 379, partly (*Chev.* 14702 ; 18458), not of Pourret. *Emilia guineensis* Hutch. & Dalz. F.W.T.A., ed. 1, 2 : 149, partly (*Chev.* 14702 ; 18458). A slender annual herb with yellow tubular florets in heads less than ¼ in. broad.
Guin.: Bouria (Nov.) *Caille* in *Hb. Chev.* 14702 ! Timbo to Ditinn (Sept.) *Chev.* 18458 ! Mt. Loursa, 3,000 ft., Dabola (Sept.) *Jaeger* 4913 ! Dalaba (Sept.) *Schnell* 6838 ! Madina Tossékré, 3,500 ft. (Oct.) *Adam* 12531 ! Also in Ethiopia, Eritrea and Congo.

2. **S. abyssinicus** *Sch. Bip. ex A. Rich.* l.c. 438 (1848) ; F.T.A. 3 : 410. An annual herb a few inches to 18 in. high with striate glabrous or sparsely pilose stems, and yellow ray- and disk-florets in narrow heads ¼ in. long.
N.Nig.: Ilorin (Dec.) *Clarke* 26a ! *Keay* FHI 37293 ! Jos Plateau (Jan., Oct.) *Lely* P31 ! *Batten-Poole* 109 ! Jos (Oct.) *Hepper* 1003 ! Naraguta *Lely* 32 ! Vom *Dent Young* 159 ! **S.Nig.:** Lagos *Dodd* 433 ! Ogun

R. *Millen* 121! **[Br.]Cam.:** Bamenda, 5,000 ft. (Jan.) *Migeod* 293! Kumbo (June, Dec.) *Boughey* GC 17485! *Gregory* 131! Lus, Nkambe Div. (Feb.) *Hepper* 1862! Widespread in central and east tropical Africa. (See Appendix, p. 420.)

3. **S. perrottetii** *DC.* Prod. 6 : 343 (1838) ; F.T.A. 3 : 412 ; F.W.T.A., ed. 1, 2 : 151, partly (excl. records from Sierra Leone) ; Berhaut Fl. Sén. 171. An annual herb with glabrous leafy stems 1–2 ft. high ; ray- and disk-florets yellow in campanulate flower-heads about ¼ in. long ; along sandy riverbanks and as a weed of cultivation in northerly districts.
Sen.: (Jan.) *Perrottet*! *Roger*! *Adam*! Richard Tol (Feb., May) *Döllinger* 65! *Roberty* 16842! **Mali:** Fafa rapids, Gao (Mar.) *De Wailly* 5357! Fetodie, near Dogo (Jan.) *Davey* 078!

4. **S. baberka** *Hutch.* in Kew Bull. 1913 : 180. *S. fallax* Mattf. (1924). Stems leafy, from a few inches to about 1 ft. high, appearing after grass fires from a perennial woody stock ; florets bright yellow in solitary heads ½–⅝ in. long.
N.Nig.: Jos Plateau (Dec., Jan.) *Lely* P37! Ropp (Dec.) *Coombe* 53! Miango to Jos (Dec.) *Coombe* 92! Naraguta *Lely* 59! Jos (Mar., Apr.) *Hill* 13! *Morton* K356! Vom, 4,000 ft. (Feb., Apr.) *Dent Young* 160! *McClintock* 208! *Morton* K396! Zaria (Feb., Mar.) *Dalz.* 390! *Milne-Redhead* 5041! Anara F.R., Zaria (Nov.) *Keay* FHI 21683! Birnin Gwari (Feb.) *Daggash* FHI 31417! Also in Cameroun and Sudan.

5. **S. ruwenzoriensis** *S. Moore* in J. Linn. Soc. 35 : 355 (1902). *S. thorbeckei* Mattf. (1924). A perennial herb with rather fleshy glaucous stems and leaves, 1–2 ft. high, drying black ; ray- and disk-florets yellow in heads ⅓ in. long ; in montane grassland.
[Br.]Cam.: Bamenda Dist., 6,000–7,500 ft. (Dec., Feb., Mar.–May) *Maitland* 1367! *Ujor* FHI 30012! *Adams* GC 11206! *Morton* K23! Bafut-Ngemba F.R. *Ujor* FHI 30098! *Onochie* FHI 34867! *Richards* 5329! *Hepper* 2157! Also in Congo, Uganda, Kenya and Tanganyika.

6. **S. clarenceanus** *Hook. f.* in J. Linn. Soc. 6 : 14 (1862) ; F.T.A. 3 : 418. An erect or low bushy perennial herb from a few inches to about 5 ft. high with ribbed stems and reddish-purple, mauve or pink florets in numerous flower-heads ⅓ in. long ; in montane grassland.
[Br.]Cam.: Cam. Mt., 7,600 ft. to the summit (Dec.-Apr.) *Mann* 1237! 1923! *Deistel* 561! *Dundas* FHI 13913! *Maitland* 815! 972! 1284! *Keay* FHI 28590! *Irvine* 1454! *Boughey* GC 12501! *Morton* GC 6900! **F.Po:** Summit of the peak (Feb., Mar.) *Mann* 609! *Guinea* 2758! Also in Congo.

7. **S. burtonii** *Hook. f.* in J. Linn. Soc. 7 : 202 (1864) ; F.T.A. 3 : 416. A robust herb up to about 4 ft. high with grooved stems and leathery leaves ; stems usually greyish-white and leaves variably so beneath ; flower-heads numerous about ½ in. long with conspicuous yellow rays ; in montane grassland.
[Br.]Cam.: Cam. Mt., 7,000–12,000 ft. (Nov.–Apr.) *Mann* 1246! 1929! *Deistel* 613! *Maitland* 844! 920! *Brenan* 9386! *Keay* FHI 28629! *Irvine* 1471! *Boughey* GC 10656! *Morton* GC 6862! Bamenda Dist. 6,000–8,500 ft. (Dec., Jan., Mar., Apr., June) *Maitland* 1395! 1722! *Keay* FHI 28406! *Boughey* GC 11039! *Morton* K140!

8. **S. lelyi** *Hutch.* in Kew Bull. 1921 : 382 ; F.W.T.A., ed. 1, 2 : 151, partly (*Lely* 356). *S. graciliserra* Mattf. (1924). An erect almost glabrous perennial herb with slender grooved stems up to about 2 ft. high from a woody stock ; ray-florets yellow in rather few heads ½–⅝ in. long.
Ghana: Kyilinga, 2,000 ft., Trans-Volta Togo (Apr.) *Hall* 1451! **N.Nig.:** Jos Plateau (May) *Lely* P278! Hepham to Ropp, 4,600 ft. (July) *Lely* 356! Fusa Jarawa, 3,200 ft. (June) *Summerhayes* 61! **[Br.]Cam.:** Bamenda Dist., 5,000 ft. (Mar.–May) *Maitland* 1431! 1544! *Morton* K146! Also in Cameroun.
[Note : *S. lygodes* Hiern is very near to *S. lelyi*, but has glabrous achenes and a tendency for the stem-leaf bases to be decurrent ; it occurs in Sudan, Congo, Rhodesia, Nyasaland and Angola—C. D. A.]

9. **S. hochstetteri** *Sch. Bip. ex A. Rich.* Tent. Fl. Abyss. 1 : 435 (1848) ; F.T.A. 3 : 414. *S. lelyi* of F.W.T.A. ed. 1, 2 : 151, partly (*Lely* 364 ; P89 ; *Migeod* 364 ; 479), not of Hutch. An erect, pubescent perennial herb like the last with more numerous flower-heads ¼–⅓ in. long and pale yellow or white ray-florets.
Guin.: Kinkon (June) *Adam* 14673! **N.Nig.:** Jos Plateau (Aug.) *Lely* P565! Naraguta (Jan.) *Lely* P89! Hepham to Ropp (July) *Lely* 364! **[Br.]Cam.:** Bamenda Dist., 5,000–6,000 ft. (Feb.–May) *Migeod* 364! 479! *Maitland* 1481! *Ujor* FHI 30315! *Coombe* 218! *Morton* K53! 265! *Hepper* 2113! Mambila Plateau, 4,900 ft., Adamawa (Jan.) *Hepper* 1796! Also in Ethiopia, Congo, Kenya, Uganda and Tanganyika.

De Miré and Gillet report *S. coronopifolius* Desf. from Aïr, Niger (J. Agric. Trop. 3 : 710 (1956)).

27. COTULA Linn.—F.T.A. 3 : 397

A small herb with woolly or sparsely hairy stems and leaves, the latter deeply pinnately divided, acute ; flowers-heads very small and numerous, shortly stalked, 4–5 mm. diam. ; involucral bracts 2-seriate, oblong, with thin scarious margins ; achenes narrowly winged ; receptacle nearly flat and prominently tuberculate after the fall of the flowers.. .. .. *anthemoides*

C. anthemoides *Linn.* Sp. Pl. 891 (1753) ; F.T.A. 3 : 397 ; Berhaut Fl. Sén. 171. A spreading or prostrate herb with branches a few inches long, with small spherical yellow flower-heads scarcely ¼ in. diam.
Sen.: *Perrottet. fide* Berhaut *l.c.* Common in the warmer parts of the Old World.

28. CENTIPEDA Lour. Fl. Cochinch. 492 (1790) ; Benth. & Hook. f. Gen. Pl. 2 : 430.

An annual herb with numerous spreading prostrate slender leafy branches ; leaves alternate, obovate-oblong, 5–12 mm. long, subentire or with few teeth ; capitula globose, solitary, axillary, subsessile, 3–5 mm. broad ; involucral bracts in two series, spreading in fruit ; receptacle naked ; achenes 4-angled, the angles with simple hairs *minima*

C. minima (*Linn.*) *A. Br. & Aschers.* Ind. Sem. Hort. Berol. 1867, App. 6 (1868) ; Merrill in Trans. Amer. Phil. Soc., n. ser. 241 : 392 (1935). *Artemisia minima* Linn. (1753). A low weedy herb with small yellow flower-heads.
Lib.: Robertsfield (Aug. 1960) *Aké Assi* 5542! This is the first record for West Africa of a common Asian plant ; also reported from Australia, Pacific islands and Madagascar.

29. MICROGLOSSA DC.—F.T.A. 3 : 308.

Petioles distinct, 1 cm. or more long ; leaves ovate to broadly lanceolate :
Leaves ovate, rounded or broadly cuneate at base, acutely acuminate, 6–10 cm. long, 3–5 cm. broad, more or less pubescent on both surfaces, glabrescent above ; achenes ribbed, pubescent ; pappus pale or reddish :
Flower-heads densely crowded in terminal corymbs ; leaves coarsely serrate-dentate 1. *densiflora*

Flower-heads in laxly branched terminal corymbs ; leaves entire or nearly so
                                                                    2. *pyrifolia*
Leaves lanceolate, cuneate at base, long-tailed-acuminate, entire, 5–8 cm. long, 1·5–
2·5 cm. broad, thinly puberulous on the nerves beneath ; inflorescence laxly branched
                                                                    3. *caudata*
Petioles not exceeding 5 mm. long, or leaves sessile ; leaves narrowly elliptic to linear-
oblanceolate, acute or shortly acuminate ; achenes pubescent :
Outer involucral bracts ovate ; leaves glabrous or nearly so, with long-subulate tips,
  5–8 cm. long, 1·5–2·5 cm. broad, very thin ; flower-heads rather few, terminal or
  terminating short lateral branches :
  Leaves rather abruptly narrowed at the base, entire or subentire ; branchlets glabre-
    scent   ..   ..   4a. *afzelii* var. *afzelii*
  Leaves tapered to a slender petiole, serrate in at least the upper half ; branchlets
    puberulous ..   ..   4b. *afzelii* var. *serratifolia*
Outer involucral bracts linear ; leaves pubescent beneath, acute, lanceolate, narrowed
  to the base, sessile, shallowly serrate-dentate, 5–8 cm. long, 1–1·5 cm. broad, scabrid
  above ; flower-heads pedunculate, numerous in a terminal corymbose inflorescence
                                                                    5. *angolensis*

1. **M. densiflora** *Hook. f.* in J. Linn. Soc. 7 : 200 (1864) ; F.T.A. 3 : 308 ; Guinea in Anal. Jard. Bot. Madrid
   10 : 313 (1951). A climbing shrub up to 25 ft. high with white ligulate and yellow disk florets in heads
   ⅛ in. long densely clustered in rounded terminal corymbs.
   [Br.]Cam.: Cam. Mt., 3,000–9,000 ft. (Dec.–Mar.) *Mann* 1320 ! 1327 ! 1915 ! *Kalbreyer* 123 ! *Dunlap* 201 !
   Maitland 244 ! 503 ! 513 ! *Keay* FHI 28641 ! *Boughey* GC 6809 ! *Morton* GC 7100 ! Bamenda, 6,500 ft.
   (Jan., Apr.) *Keay* FHI 28337 ! *Morton* K230 ! Bafut-Ngemba F.R. (Feb.) *Lightbody* FHI 26299 ! **F.Po:**
   Musola to Moka (Jan.) Guinea 1922. Also in Congo.
   [Note : This species is only really distinct from *M. pyrifolia* at the stations of higher altitude listed
   above. In the following localities, plants with densely clustered heads but otherwise intermediate between
   M. densiflora and *M. pyrifolia*, occur :—**Port.G.**: Pessubé, Bissau (Mar.) *Esp. Santo* 1496 ! **S.Nig.**: Ibadan
   North F.R. (Feb.) *Ejiofor* FHI 19805 ! Ilora, Oyo (Jan.) *Keay* FHI 25645 ! [Br.]Cam.: Cam. Mt., 4,400 ft.
   (Apr.) *Maitland* 1080 !—C. D. A.]
2. **M. pyrifolia** (*Lam.*) *O. Ktze.* Rev. Gen. Pl. 1 : 353 (1891). *Conyza pyrifolia* Lam. (1786). *Microglossa volu-*
   bilis DC. (1836)—F.T.A. 3 : 309 ; Chev. Bot. 362 ; F.W.T.A., ed. 1, 2 : 152. *Pluchea subumbellata* Klatt
   (1873). *Conyza heudelotii* Oliv. & Hiern in F.T.A. 3 : 317 (1877) ; F.W.T.A., ed. 1, 2 : 154. An erect or
   straggling shrub up to 10 ft. or more high with more or less pubescent stems and leaves ; florets pale
   yellow or white in numerous small heads.
   Sen.: Youni-Fogny, Casamance (Feb.) *Chev.* 2057 ! **Port.G.:** Bissalanca, Bissau (Jan.) *Esp. Santo* 1667 !
   Mato de Amedi, Formosa (Apr.) *Esp. Santo* 1968 ! **Guin.:** *Heudelot* 609 ! Bambaya, Upper Niger (Jan.)
   Jaeger 3860 ! Pofodara (Mar.) *Adam* 11596 ! **S.L.:** Sugar Loaf Mt., 2,500 ft. (Oct.) *Tindall* 18 ! Free-
   town (Jan.) *Dalz.* 991 ! Magbema (May) *Jordan* 443 ! Kukuna (Jan.) *Sc. Elliot* 4715 ! Loma Mts. (Jan.)
   Jaeger 4160 ! **Lib.:** Kakatown *Whyte* ! Dukwia R. (May) *Cooper* 449 ! Vonjama (Oct.) *Baldwin* 9921 !
   Gondolahun, Kolahun (Nov.) *Baldwin* 10109 ! Duo, Sinoe (Mar.) *Baldwin* 11361 ! **Iv.C.:** *Farmar* 376 !
   Bouroukrou (Jan.) *Chev.* 17003 *bis* ! Sassandra Port (May) *Chev.* 17918 ! Orimbo-Boka Mt., Toumodi
   (Dec.) *Boughey* GC 14432 ! Bingerville, Banco (July) *Chev.* ! *Chev.* ! **Ghana:** Aburi Scarp (Sept., Oct.) *Johnson* 468 ! *Adams* 1783 ! *Morton* GC
   9286 ! Dixcove to Busua (Mar.) *Morton* A280 ! Accra (Aug.) *T. Vogel* ! Wenchi, Ashanti (Dec.) *Adams*
   3246 ! Amedzofe (Feb., Oct.) *Morton* GC 6038 ! 8401 ! **Togo Rep.:** *Baumann* 425 ! Lomé *Warnecke* 418 !
   Dah.: Sakété to Pedjilé, Porto-Novo (Feb.) *Chev.* 22890 ! **N.Nig.:** Ilorin (Dec.) *Ajayi* FHI 19268 ! Mongu,
   4,300 ft. (July) *Lely* 397 ! Naraguta (Nov.) *Lely* 710 ! Jemaa, 2,500 ft. (Feb.) *McClintock* 181 ! Shere
   Hills, 4,000 ft. (Dec.) *Coombe* 107 ! **S.Nig.:** Lagos (Jan.) *Dalz.* 1161 ! Ibadan (Dec.) *Meikle* 856 ! 868 !
   Cam. Mt., Buea, 3,000–3,480 ft. (Mar.) *Boughey* GC 7026 ! *Maitland* 503a ! Lakom, Bamenda, 6,000 ft.
   (Apr.) *Maitland* 1583 ! Bafut-Ngemba F.R. (Feb.) *Tiku* FHI 22160 ! *Hepper* 2192 ! Mambila Plateau,
   5,200 ft., Adamawa, *Hepper* 1833 ! **F.Po:** Moka (Feb.) Guinea 2271. Widespread
   in tropical Asia and Africa and also in S. Tomé. (See Appendix, p. 419.)
3. **M. caudata** *O. Hoffm. & Muschl.* in Mém. Soc. Bot. Fr. 2, 8 : 113 (1910) ; Chev. Bot. 362. A shrub with
   pubescent stems, peduncles and undersurfaces of leaves ; inflorescence paniculate ; pappus pinkish.
   Guin.: Bowali to Kouria (July) *Caille* in Hb. Chev. 15069 !
4. **M. afzelii** *O. Hoffm.* in Engl. Bot. Jahrb. 24 : 469 (1898) ; Chev. Bot. 362. *M. sessilifolia* of F.T.A. 3 : 309,
   not of DC.
4a. **M. afzelii** *O. Hoffm.* var. **afzelii**. A scrambling shrub 12–15 ft. high with the young branchlets and peduncles
   thinly pubescent ; flower-heads about ⅛ in. across with conspicuous white ligules to the outer florets ;
   in thickets and fringing forest.
   Guin.: Kaba R. to Upper Mamou (Aug.) *Chev.* 12762 ! Bafing R., Mau to Touba (July) *Collenette* 54 !
   Gangan (Apr.) *Adam* 11931 ! **S.L.:** Duunia, Talla (Feb.) *Sc. Elliot* 4818 ! Kafogo (Apr.) *Sc. Elliot* 5491 !
   Kenema (Apr.) *Lane-Poole* 223 ! Moyamba (Feb., Mar.) *Deighton* 1921 ! *Smythe* 54 ! Kangahun (Apr.)
   Deighton 6043 ! **Lib.:** Gbanga (Jan., Aug., Sept., Dec.) *Daniel* 82 ! *Linder* 376 ! *Okeke* 11 ! *Baldwin*
   10520 ! Browntown, Tchien (Aug.) *Baldwin* 7077 ! Ganta (May) *Harley* ! Nimba Mts. (May) *Schnell* 1362 ! **Ghana:** Aburi
   (Mar., May, Sept., Oct.) *Adams* 398 ! *Johnson* 467 ! *Howes* 1183 ! *Morton* GC 9278 ! Elmina Plains
   (Mar.) *Morton* A1880 ! Agogo, 1,400 ft., Ashanti (Apr., Dec.) *Deakin* 5 ! *Adams* 2550 ! Ahinsa, Ashanti
   (Jan.) *Darko* 507 ! Amedzofe, 2,500 ft. (Feb., Mar., Nov.) *Irvine* 157 ! *Morton* GC 9410 ! *Williams* 104 !
   Togo Rep.: Misahöhe (Mar.) *Baumann* 445 ! **S.Nig.:** Ado Ekiti, Ondo *Hoskyns-Abrahall* FHI 20434 ! Oban
   Talbot 1671 ! [Br.]Cam.: Bamenda, 6,000 ft. (Jan.) *Migeod* 411 ! Also in Sudan. (See Appendix, p. 418.)
4b. **M. afzelii** var. **serratifolia** *C. D. Adams* in J.W. Afr. Sci. Assoc. 3 : 115 (1957). A slender-branched scrambling shrub with greenish-yellow flowers in heads about ⅛ in. across ; in secondary growth.
   Lib.: Nyandamolahun (Feb.) *Bequaert* 77 ! Tappita (Aug.) *Baldwin* 9075 ! **Ghana:** Aburi Scarp (Jan.)
   Morton GC 6338 !
5. **M. angolensis** *Oliv. & Hiern* in F.T.A. 3 : 309 (1877). An erect undershrub 2–4 ft high with pale yellow or
   greenish-yellow florets in numerous stalked heads about ⅛ in. broad, clustered at the ends of leafy branches;
   along streams and forest margins in hilly districts.
   N.Nig.: Jos Plateau (May) *Lely* 347 ! Bukuru to Hepham, 4,300 ft. (July) *Lely* 347 ! Sho (Nov.) *Keay*
   FHI 21448 ! Jos (Apr.) *Savory* UCI 109 ! *Batten-Poole* 210 ! 360 ! *Morton* K414 ! [Br.]Cam.: Bamenda
   Dist., 4,500–6,500 ft. (Dec., Jan., Mar., June) *Maitland* 1685 ! *Migeod* 300 ! *Boughey* GC 11406 ! *Adams*
   GC 11341 ! *Morton* K38 ! *Hepper* 2189 ! Vogel Peak, 4,800 ft., Adamawa (Nov.) *Hepper* 1358 !
   [Note: This species is almost indistinguishable from *Conyza pyrrhopappa* Sch. Bip. ex A. Rich. (1848),
   which seems only to differ in having shorter ligules to the outer florets. If taken together, the distribution
   extends from Ethiopia through Uganda and Congo to Angola—C. D. A.]

FIG. 253.—MICROGLOSSA PYRIFOLIA (*Lam.*) *O. Ktze.* (COMPOSITAE).

A, female flower. B, achene. C, hermaphrodite flower. D, seta of pappus. E, stamens.
F, style-arms. G, part of involucre.

252

30. ERIGERON Linn.—F.T.A. **3** : 307 ; B.L. Burtt in Kew Bull. 3 : 371 (1949).

Leaves linear, up to 5 mm. broad, the upper filiform ; pappus pinkish-red ; achenes pubescent ..   .. 1. *bonariensis*
Leaves linear-lanceolate to oblanceolate, usually 1 cm. or more broad ; pappus buff ; achenes minutely pubescent, glabrescent ..   .. 2. *floribundus*

1. **E. bonariensis** *Linn.* Sp. Pl. 863 (1753) ; Burtt in Kew Bull. 3 : 371. *E. sumatrensis* Retz. (1789). *E. linifolius* Willd. (1804)—F.T.A. 3 : 308. *Conyza bonariensis* (Linn.) Cronquist in Bull. Torr. Bot. Club 70 : 632 (1943). An erect herb up to 18 in. high ; in waste ground.
    Iv.C. : Adiopodoumé (Oct.) *Leeuwenberg* 1762 ! **Ghana**: Abetifi, 2,000 ft. (Jan., July) *Irvine* 1697 ! *Akpabla* 157 ! Also in S. Tomé, Principe and E. Africa.
2. **E. floribundus** (*H.B. & K.*) *Sch. Bip.* in Bull Soc. Bot. Fr. 12 : 81 (1865). *Conyza floribunda* H.B. & K. (1820). A herb up to about 6 ft. high ; florets dull yellow in heads about ¼ in. across ; a weed of roadsides and cultivation in mountainous regions and lowland areas transitional between forest and savanna.
    Guin.: Timbi Madina (Mar.) *Adam* 11698 ! Nzérékoré (Oct.) *Baldwin* 9724 ! **S.L.**: Njala (June) *Deighton* 730 ! Dia (Apr.) *Deighton* 3158 ! Musaia (Apr.) *Deighton* 5385 ! **Lib.**: Gbanga (Sept.) *Linder* 816 ! Ganta (May) *Harley* ! Jabroke, Webo (July) *Baldwin* 6428 ! **Iv.C.**: Yapo Forest, near Abidjan (July) *Boughey* GC 14577 ! Adiopodoumé (Oct.) *Roberty* 15322 ! Bianco Forest (Aug.) *Boughey* GC 13502 ! Gonokrom to Techikrom (Dec.) *Adams* 2979 ! **Ghana**: Techiman, Ashanti (Dec.) *Adams & Akpabla* GC 4499 ! Kumasi (Oct.) *Lovi* WACRI 3933 ! Agogo, Ashanti, 1,400 ft. (Apr.) *Adams* 2607 ! Amedzofe (Nov.) *Morton* GC 9423 ! Elmina (Mar.) *Morton* A1878 ! **S.Nig.**: Owo (Nov.) *Meikle & Keay* 504 ! Akure F.R., Ondo (Oct.) *Keay* FHI 25459 ! Afi River F.R., Ogoja (June) *Jones & Onochie* FHI 18980 ! **[Br.]Cam.**: Cam. Mt., 3,400–9,000 ft. (Mar.–Apr.) *Boughey* GC 7016 ! *Morton* GC 6879 ! 7078 ! Buea (Jan., June) *Maitland* 40 ! FHI 10335 ! Bambui, Bamenda, 6,300 ft. (Dec.) *Boughey* GC 10396 ! Mambila Plateau, 4,500 ft., Adamawa (Jan.) *Hepper* 1830 ! **F.Po**: Musola (Jan.) *Guinea* 1384. Widespread in warm countries.

31. FELICIA Cass. in Bull. Sci. Soc. Philom. Paris 1818 : 165 ; Dict. Sci. Nat. 16 : 314. *Nom. cons.*

Stems erect, branched, leafy ; leaves opposite below, alternate above, linear-lanceolate, 1–5 cm. long, 2–5 mm. broad ; stems and leaves thinly pubescent ; peduncles slender 2–7 cm. long ; involucral bracts in 3 series the outer shorter, the inner up to 7 mm. long, linear, with a narrow dark median nerve and broad scarious margins, abruptly acuminate, mucronate, glabrescent ; receptacle papillose, 4–5 mm. diam. ; outer florets shortly ligulate, inner filiform, minutely 5-lobed ; style-arms tapered ; achenes 1 mm. long, elliptic, flattened, minutely puberulous, greyish-buff ; pappus-setae slender, white, caducous   ..   .. *homochroma*

F. homochroma *S. Moore* in J. Bot. 59 : 229 (1921). A slender erect annual herb ½–2 ft. high ; short ray- and disk-florets bright yellow in slender-stalked heads about ½ in. across.
    N.Nig.: Naraguda F.R., Jos (Nov.) *Keay* FHI 37207 ! **[Br.]Cam.**: Vogel Peak 3,800 ft., Adamawa (Nov.) *Hepper* 1429 ! Also in Congo, Tanganyika, Nyasaland and N. Rhodesia.

    Felicia richardii Vatke is reported by De Miré and Gillet from Aïr, Niger (J. Agric. Trop. 3 : 707 (1956)).

32. CONYZA Less.—F.T.A. **3** : 311. *Nom. cons.*

Herbs or undershrubs, erect or scandent, usually 1 m. or more high ; capitula numerous (more than 20) on slender peduncles in terminal branched inflorescences ; stems leafy ; leaves distinctly serrate or serrulate but not divided :
Stems and leaves glabrous ; lower stem-leaves petiolate, elliptic, serrate, upper sessile subauriculate linear-lanceolate, serrulate ; involucral bracts oblong, acute, 3–4 mm. long, glabrous ; peduncles slender ; achenes glabrous..   .. 1. *persicifolia*
Stems and leaves pubescent ; upper stem-leaves gradually narrowed to a clasping base ; involucral bracts lanceolate, thinly pubescent in the mid-line ; achenes minutely puberulous :
Capitula about 5 mm. diam. in flower ; scrambling or weakly erect herb ; leaf-base auricled ; stem and leaves thinly pubescent ; leaves glabrescent above, shortly setulose on the veins beneath ..   .. 2. *steudelii*
Capitula 6–8 mm. diam. in flower ; erect robust herb ; leaf-bases not auricled ; stem densely setulose-pubescent ..   .. 3. *gigantea*
Herbs rarely reaching 1 m. high, often scapose or tufted from a short stock :
Leaves pinnately divided ; leaves and stems villous-pilose ; involucral bracts linear, slender ; annual herbs :
Uppermost stem-leaves subentire, linear ; inflorescence spreading 4. *aegyptiaca* var. *aegyptiaca*
Uppermost stem-leaves pinnatifid, divided into narrow lobes ; inflorescence contracted ..   .. 4a. *aegyptiaca* var. *lineariloba*
Leaves entire or subentire, often serrate but not divided ; stems and leaves more or less pubescent, rarely glabrescent ; mostly perennial herbs :
Stems usually 5–30 cm. high ; leaves mostly in a basal rosette, broadly oblanceolate, serrate ; capitula about 1 cm. diam., clustered at the apex of an almost leafless scape ..   .. 5. *subscaposa*
Stems usually 30–60 cm. high ; leaves scattered along the flowering stems, linear to linear-oblanceolate :
Flower-heads 6–12 mm. diam. ; stem-leaves more than 3 cm. long ; stems more leafy below than above :

Capitula in a dense terminal cluster ; outer involucral bracts obtuse, purplish
                                                           6. *clarenceana*
Capitula borne on branches in the axils of the upper leaves :
  Involucral bracts purplish, acute ; flower-heads 9–12 mm. diam. ; stems sparsely
    setulose, glabrescent ; leaf-margins at first ciliate   ..   ..   7. *theodori*
  Involucral bracts green, obtuse ; flower-heads 6–8 mm. diam. ; stems appressed-
    pilose   ..   ..   8. *gouanii*
Flower-heads 6 mm. or less diam. ; stem-leaves less than 3 cm. long ; stems leafy
  only in the upper part at flowering time :
  Capitula numerous, 3–4 mm. diam. in rounded clusters ; leaves oblanceolate,
    toothed near apex, about 2 cm. long and 4–5 mm. broad ; stem indistinctly
    striate when dry ; pubescent ..   ..   9. *stricta*
  Capitula few, 5–6 mm. diam. ; leaves linear, about 2 mm. broad ; stem deeply
    grooved when dry, glabrescent   ..   ..   ..   10. *spartioides*

1. **C. persicifolia** (*Benth.*) *Oliv. & Hiern* in F.T.A. 3 : 312 (1877) ; Chev. Bot. 363. *Erigeron persicaefolius* Benth. (1849). *Eschenbachia persicifolia* (Benth.) Exell Cat. S. Tomé 224 (1944). An erect glabrous annual herb 2–6 ft. high ; lower leaves stalked ; upper leaves clasping the stem but hardly auricled ; flower-heads small and numerous with pale yellow florets ; a weedy plant of open ground.
    Guin.: *fide* Chev. l.c. **S.L.:** Kofiu (Jan.) *Sc. Elliot* 4610 ! Kurusu (Apr.) *Sc. Elliot* 5649 ! **Iv.C.:** Droupleu to Zouanlé, Upper Sassandra (May) *Chev.* 21453 ! Gouréni to Gouékouma, Upper Sassandra (May) *Chev.* 21650 ! **Ghana:** Tarkwa (Sept.) *Dalz.* 107 ! **N.Nig.:** Naraguta *Lely* 45 ! *Kennedy* 2953 ! Vom, 4,000 ft. (Feb., Apr.) *Dent Young* 147 ! *McClintock* 201 ! *Morton* K328 ! 340 ! Katagum *Dalz.* 368 ! Tilde Filani, 3,300 ft. (May) *Lely* 234 ! **S.Nig.:** Abeokuta *Irving* 80 ! Lagos *Millen* 128 ! Afi River F.R., 2,500 ft., Ogoja (June) *Jones & Onochie* FHI 18801 ! **[Br.]Cam.:** Buea, 3,500 ft. (Feb., June, Nov.) *Dalz.* 8188 ! *Maitland* 43 ! *Migeod* 43 ! Widekum, 2,000 ft. (Jan.) *Keay* FHI 28327 ! Bamenda, 5,000 ft. (June) *Maitland* 1612 ! Mambila Plateau, 5,200 ft., Adamawa (Jan.) *Hepper* 1832 ! **F.Po:** Musola (Nov.) *Guinea* 1080 ! Moka (Feb.) *Guinea* 2269. Widespread in tropical Africa and in S. Tomé.

2. **C. steudelii** *Sch. Bip. ex A. Rich.* Tent. Fl. Abyss. 1 : 388 (1848) ; F.T.A. 3 : 313. *C. volkensii* of Adams in J.W. Afr. Sci. Assoc. 1 : 26 (1954), not of O. Hoffm. An erect or scandent much-branched shrubby herb 3–6 ft. high with thin pale green leaves ; flower-heads with green involucral bracts and pale yellow florets ; at the margins of montane woodland.
    [Br.]Cam.: Cam. Mt., 8,000–9,000 ft. (Jan., Mar., Apr.) *Keay* FHI 28632 ! *Brenan & Onochie* 9545 ! *Morton* GC 6875 ! Also in the upland regions of Ethiopia, Congo, Kenya, Uganda and Tanganyika.

3. **C. gigantea** *O. Hoffm.* in Engl. Pflanzenw. Ost-Afr. C : 408 (1895). An erect robust bushy herb 5 ft. or more high with hispid stems and yellow florets in heads clustered at the ends of leafy branches ; in montane grassland and woodland margins.
    S.L.: Loma Mts., 4,500 ft. (Feb., Nov.) *Jaeger* 371 ! 462 ! 4194 ! 4269 ! Bintumane Peak, 6,390 ft. (Jan.) *T.S. Jones* 104 ! Also in the mountains of Uganda and Tanganyika.
    [The W. African plants are slightly smaller in all their parts than specimens so named from E. African localities, but the type (*Volkens* 1141) has the smaller leaves of the W. African specimens. Plants with glabrous stems have been determined as the closely related *C. montigena* S. Moore (1908), but the type of this species (*Wollaston* s.n.) has a pubescent stem—C. D. A.].

4. **C. aegyptiaca** (*Linn.*) *Ait.* var. **aegyptiaca**—Hort. Kew. 3 : 183 (1789) ; F.T.A. 3 : 314 ; Chev. Bot. 363 ; Berhaut Fl. Sén. 172. *Erigeron aegyptiacus* Linn. (1767). An annual or biennial herb from a few inches to 3 ft. high with numerous pinnately lobed leaves ; flower-heads about ½ in. diam. in terminal clusters, opening widely and exposing masses of silky pappi in fruit ; florets dull yellow ; a weed of cultivated ground and waste places outside the forest areas.
    Sen.: *Richard* ! (Jan.) *Roger* ! Bandia (June) *Berhaut* 5297 ! **Mali:** Koba (Mar.) *Chev.* 609 ! Koldry (Apr.) *Roberty* 3597 ! **Port.G.:** Cumura, Bissau (Apr.) *Esp. Santo* 2383 ! **Guin.:** Yagadou (Mar.) *Chev.* 20931 ! Mamou (Apr.) *Adam* 11870 ! Kindia (May) *Jac.-Fél.* 1614 ! **S.L.:** Sulimania (May) *Sc. Elliot* 5286 ! Kamalo (May) *Thomas* 386 ! Musaia (Feb., Apr., Dec.) *Deighton* 4263 ! 4555 ! 5384 ! **Ghana:** *Williams* 495 ! Salaga *Krause* ! Nwereme, Ashanti (May, Dec.) *Irvine* 2492 ! *Morton* A2635 ! Techiman, Ashanti (Dec.) *Adams & Akpabla* GC 4518 ! Damongo (Mar.) *Morton* GC 8717 ! Kete Krachi to Oti R. (May) *Morton* GC 7162 ! **Dah.:** Konbongou, Atacora Mts. (June) *Chev.* 24156 ! **N.Nig.:** Vom *Dent Young* 148 ! Naraguta (Nov.) *Lely* 717 ! Bauchi, 2,200 ft. (Jan., May) *Lely* 170 ! Baradau, 3,000 ft. (May) *Lely* 92 ! Kaduna to Zaria (May) *Keay* FHI 25756 ! Abinsi (June) *Dalz.* 654 ! **[Br.]Cam.:** Wum Crater L., Bamenda (Apr.) *Morton* K297 ! Mapeo, 3,000 ft., Alantika Mts. (Dec.) *Hepper* 2791 ! Throughout tropical Africa and Asia.

4a. **C. aegyptiaca** var. **lineariloba** (*DC.*) *O. Hoffm.* in Bol. Soc. Brot. 13 : 23 (1896) ; Chev. Bot. 363. *C. lineariloba* DC. (1836). An annual herb like the last but more slender with narrower inflorescences and leaves.
    Guin.: Kouria to Ymbo (July) *Chev.* 15017 ! Timbi (June) *Adam* 14678 ! **S.L.:** Rowala (July) *Thomas* 1130 ! **N.Nig.:** Bokkos, 4,000 ft., Plateau Prov. (Dec.) *Coombe* 65 ! **[Br.]Cam.:** Bamenda, 5,000–7,000 ft. (Feb., Apr., May) *Maitland* 1447 ! *Morton* K216 ! Bafut-Ngemba F.R. (Feb., Apr.) *Hepper* 2697 ! *Morton* K252 ! Also in Congo and Angola.

5. **C. subscaposa** *O. Hoffm.* in Engl. Bot. Jahrb. 20 : 225 (1894) ; Robyns Fl. Sperm. Parc. Nat. Alb. 2 : 469, t. 44. *C. hochstetteri* of Adams in J.W. Afr. Sci. Assoc. 1 : 26 (1954), not of Sch. Bip. A perennial herb with a basal rosette of stiff leaves and a weak flowering shoot from a few inches to 1 ft. high ; florets dull yellow; involucral bracts often purplish ; in montane grassland.
    [Br.]Cam.: Cam. Mt., in grassland up to the summit (Dec.–Mar.) *Maitland* 675 ! 1100 ! *Bumpus* FHI 12312 ! *Dundas* FHI 20361 ! *Keay* FHI 28623 ! *Brenan* 9503 ! 9528 ! 9573 ! *Boughey* GC 6805 ! 12683 ! Bamenda, 6,000–7,500 ft. (Dec., Mar.) *Richards* 5306 ! *Adams* GC 11310 ! *Boughey* GC 10880 ! 11414 ! Maisamari, 5,300 ft., Mambila Plateau (Jan.) *Hepper* 1734 ! Also in Ethiopia, Sudan, Congo, Kenya, Uganda, Tanganyika and Rhodesia.

6. **C. clarenceana** (*Hook. f.*) *Oliv. & Hiern* in F.T.A. 3 : 316 (1877). *Vernonia clarenceana* Hook. f. (1862). An erect slightly branched sparsely pubescent perennial herb with stems up to 2 ft. high, narrow clasping leaves and a terminal cluster of flower-heads ; florets yellow ; involucral bracts rich purple.
    F.Po: Clarence Peak, 8,500 ft. (Dec.) *Mann* 599 ! *Guinea* 2898. *Boughey* GC 17383 ! Pico Serrano (Dec., Jan.) *Guinea* 1969 ! *Monod* 10480 !

7. **C. theodori** *R. E. Fries* in Acta Hort. Berg. 9 : 122, t. 1, fig. 1 (1928). A perennial herb with reddish stems 12–18 in. high ; flower-heads in small clusters in the branched terminal inflorescence ; florets yellow ; involucral bracts long-pointed, purplish ; in upland grasslands.
    [Br.]Cam.: Cam. Mt., 4,400 ft. (Apr.) *Maitland* 1098a ! Lakom, 6,000 ft. (Apr.–May) *Maitland* 1378 ! 1480 ! 1493 ! Kumbo, 5,000 ft. (Feb.) *Hepper* 2698 ! Also in Ethiopia, Congo and Kenya.

8. **C. gouanii** (*Linn.*) *Willd.* Sp. Pl., ed. 4, 3 : 1928 (1803) ; F.T.A. 3 : 316 (incl. syn. *C. gnaphalioides* Sch. Bip. (1848), but not of H.B.K. (1820)). *Erigeron gouani* Linn. (1771). A perennial herb like the last with shorter greenish involucral bracts forming smaller heads ; stems often purple ; in upland grass-lands.
    S.Nig.: Ata R., Obudu (Dec.) *Keay & Savory* FHI 25011 ! **[Br.]Cam.:** Cam. Mt., 4,000 ft. (Apr.) *Maitland*

1098! Bamenda, 6,000 ft. (Mar.) *Morton* K16! Bamenda to Ndop, 5,000–5,400 ft. (Dec.) *Adams* GC 11279! *Boughey* GC 10454! 11372! Bambui, Bamenda, 6,500 ft. (Dec.) *Boughey* GC 10846! Also in Canary Islands, Ethiopia, Tanganyika and Rhodesia.

9. **C. stricta** *Willd.* Sp. Pl., ed. 4, 3 : 1922 (1803) ; F.T.A. 3 : 318. An annual herb 1–3 ft. high often with several ascending branches from the main stem, white-pubescent with hooked hairs ; flower-heads about ¼ in. across ; florets yellow ; pappus deciduous in one piece.
Guin.: Mali Peak, 5,000 ft. (Jan.) *Chev.*! Mt. Loura, 5,000 ft. (June) *Adam* 14515! **N.Nig.:** Jos Plateau *Lely* 860! Gindiri, 3,500 ft., Jos Plateau (Oct.) *Hepper* 1102! Throughout tropical Africa.

10. **C. spartioides** *O. Hoffm.* in Engl. Bot. Jahrb. 20 : 224 (1894). *Nidorella spartioides* (O. Hoffm.) Cronquist in Bull. Jard. Bot. Brux. 22 : 310 (1952). *N. stricta* O. Hoffm. (1901). A perennial herb 1–2 ft. high with slender flexuous, sometimes almost leafless, branches from a woody stock ; flower-heads rather few in terminal clusters ; florets yellow ; involucral bracts green.
[Br.]Cam.: Fang, Bamenda, 3,000–5,000 ft. (Apr.) *Maitland* 1472! Bafut-Ngemba F.R., 7,200 ft. (Feb., Mar.) *Hepper* 2119! *Morton* K41! Santa Mt., 8,500 ft., Bamenda (Apr.) *Morton* K313! Also in Kenya, Uganda, Tanganyika, Angola and Rhodesia.

33. ADELOSTIGMA Steetz—F.T.A. 3 : 320.

Erect annual, sparingly branched in the upper part ; stems thinly pilose, glabrescent ; leaves linear in outline, deeply pinnately partite, up to about 5 cm. long, with linear segments about 5 mm. long, glabrescent ; flower-heads terminal, surrounded closely by the upper leaves, 1·5–2 cm. diam. ; involucral bracts about 3-seriate, linear, about 1 cm. long, silky-pubescent ; outer florets with a filiform corolla, inner narrowly campanulate ; achenes pubescent ; pappus-bristles smooth, pale buff   *senegalense*

A. senegalense *Benth.* in Hook. Ic. Pl. 12 : t. 1144 (1873) ; F.T.A. 3 : 320 : Chev. Bot. 364. An erect annual herb a few inches to 18 in. or more high with soft, thick whitish roots and slightly angled stems ; florets mauve or bluish-purple in thistle-like heads ; in damp grassy flats.
Guin.: Fouta Djalon *Heudelot* 677! Timbo to Ditinn (Sept.) *Chev.* 18462! 18513! Bouria (Nov.) *Chev.* 14963! Bilima (Sept.) *Chev.* 15039! Mali (Sept.) *Schnell* 7057! Timbi *Adam* 12846! **S.L.:** Waterloo (Jan., Aug.) *Lane-Poole* 408! Melville & *Hooker* 323! Freetown (Oct.) *Johnston* 19! Kanya (Oct.) *Thomas* 2959! Wellington (Nov.) *Deighton* 866! Wonkifu (Oct.) *Jordan* 931! No. 2 River, Colony (Nov.) *T. S. Jones* 374!

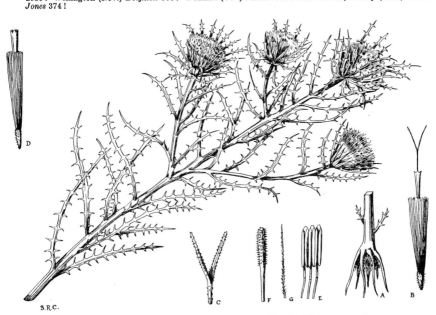

S.R.C.

FIG. 254.—ADELOSTIGMA SENEGALENSE *Benth.* (COMPOSITAE).

A, roots.   B, female flower.   C, style-arms.   D, hermaphrodite flower.   E, anthers.   F, style before opening.   G, seta of pappus.

34. CERUANA Forsk.—F.T.A. 3 : 304.

Branchlets pilose ; leaves obovate-oblanceolate, narrowly auriculate at base, coarsely crenate in the broader part of the lamina, 4–6 cm. long, about 1·5 cm. broad, thinly pilose on both sides ; flower-heads very shortly pedunculate, campanulate, 6–8 mm. long ; outer involucral bracts leafy, inner dry and chaffy ; outer florets female, narrowly tubular, with a rudimentary ligule ; receptacle paleaceous ; achenes slightly angular, glabrous, crowned with a minute setiform pappus   ..   *pratensis*

C. pratensis *Forsk.* Fl. Aegypt.-Arab. LXXIV (1775) ; F.T.A. 3 : 305 ; Chev. Bot. 362 ; Berhaut Fl. Sén. 172. An erect branched half-woody annual herb 1–2 ft. high, with dull yellow or greenish-yellow florets in leafy-bracteate heads ; in drying marshy ground and on river banks.

Sen. : Richard Tol (May) *Döllinger* ! **Gam.**: *Hayes* 548 ! **Mali** : Djenné (June) *Chev.* 1116 ! Kolby, Macina (July) *Roberty* 2554 ! **Guin.**: R. Niger, Kouroussa (Apr.) *Adam* 12169 ! **Ghana**: Red Volta R., Bolgatanga to Bawku (Apr.) *Morton* GC 8932 ! 8952 ! **N.Nig.**: Nupe *Barter* 1128 ! R. Benue, Yola (May) *Dalz.* 35 ! Extends to Sudan and Egypt.

35. DICHROCEPHALA L'Hérit. ex DC.—F.T.A. 3 : 302.

Flower-heads subsolitary, 6–8 mm. diam., rather few on elongated bracteate peduncles ; stems scabrid or glabrescent ; leaves broadly auriculate at base, coarsely dentate or pinnately lobulate, the lower leaves spathulate, up to 10 cm. long and 3 cm. broad, more or less scabrid-setulose above, pilose on the midrib beneath   1. *chrysanthemifolia*
Flower-heads numerous in panicles, 3–4 mm. diam. on peduncles up to about 5 mm. long ; stems more or less scabrid ; leaves petiolate, not auriculate, pinnately partite, the terminal lobe usually broadly ovate, serrate, setulose-pilose on both surfaces
                                                                                   2. *integrifolia*

1. **D. chrysanthemifolia** (*Blume*) *DC.* in Guill. Arch. Bot. 2 : 518 (1833) ; F.T.A. 3 : 303. *Cotula chrysanthemi-folia* Blume (1826). An aromatic erect or straggling branched herb 1–3 ft. or more high with more or less pubescent stems and leaves, and spherical red-purple to mauve flower-heads ; in montane grasslands. [**Br.**]**Cam.**: Cam. Mt., 7,000–10,500 ft. (Nov.–Apr.) *Mann* 1267 ! 1927 ! *Mildbr.* 10850 ! *Maitland* 907 ! 973 ! 1278 ! *Dalz.* 8321 ! *Keay* FHI 28601 ! *Morton* K538 ! *Boughey* GC 12647 ! Mbakokeka Mt., 7,400 ft., Bafut-Ngemba F.R. (Dec., Feb.) *Hepper* 2130 ! Bambui Mt. (Dec.) *Boughey* GC 10759 ! Bamenda (Dec.) *Boughey* GC 11413 ! **F.Po** : summit of the Peak, 9,000 ft. (Dec., Mar.) *Mann* 606 ! *Guinea* 2847 ! Through-out tropical Africa, mostly on high ground, and in Asia and Madagascar.
   [The larger more glabrous forms are probably shade plants.]
2. **D. integrifolia** (*Linn. f.*) *O. Ktze.* Rev. Gen. Pl. 1 : 333 (1891). *Hippia integrifolia* Linn. f. (1781). *Cotula bicolor* Roth (1800). *Dichrocephala latifolia* DC. (1833)—F.T.A. 3 : 303 ; *Chev.* Bot. 361. *D. bicolor* (Roth) Schlechtend. (1852)—F.W.T.A., ed. 1, 2 : 155. An erect branched annual herb 1–3 ft. or more high, with scabrid angular stems and roughly pubescent leaves ; the flower-heads have an involucre of green bracts with whitish tips, and greenish-yellow central florets, the outer florets may be mauve ; a weed of high ground.
   Guin.: Ditinn Fall, Fouta Djalon (Sept.) *Chev.* 18535 ! Labé (Mar.) *Adam* 11534 ! **Iv.C.**: Soucourala to Sanrou, Upper Sassandra (May) *Chev.* 21586 ! **Ghana**: Amedzofe, 2,000 ft. (Apr.) *Morton* A1997 ! **N.Nig.**: Hepham to Ropp, 4,600 ft. (July) *Lely* 368 ! Werram R., Jos (Aug.) *Keay* FHI 12722 ! **S.Nig.**: Obudu Plateau, 5,500 ft. (Aug.) *Stone* 50 ! [**Br.**]**Cam.**: Cam. Mt., 3,000–8,000 ft. (Nov.–Jan., Mar.) *Mann* 1268 ! 1926 ! *Migeod* 111 ! *Boughey* GC 10703 ! *Morton* GC 6705 ! Bamenda Dist. (Mar., June, Dec.) *Maitland* 1388 ! *Adams* GC 11309 ! *Boughey* GC 10949 ! *Morton* K35 ! Ntim, Nkambe Div. (Feb.) *Hepper* 1860 ! Mambila Plateau, 5,600 ft., Adamawa (Jan.) *Hepper* 1736 ! **F.Po**: refuge on the Peak, 6,000 ft. (Dec.) *Boughey* GC 17419 ! Widespread in tropical and subtropical Africa and also in Asia and southern Europe.

36. MICROTRICHIA DC.—F.T.A. 3 : 302.

A small annual herb ; leaves obovate, rounded at apex, narrowly cuneate at base, the lowermost long-petiolate, up to 6 cm. long and 2 cm. broad, coarsely crenate, thinly and minutely pubescent ; flower-heads in terminal subsessile clusters, 3–4 mm. diam. ; involucral bracts in 2–3 rows, oblanceolate or linear, slightly pubescent ; outer florets female, tubular ; inner florets hermaphrodite ; achenes pubescent, subterete ; pappus of a few denticulate caducous bristles   ..   ..   *perrottetii*

M. perrottetii *DC.* Prod. 5 : 366 (1836) ; F.T.A. 3 : 302 ; Chev. Bot. 361 ; Berhaut Fl. Sén. 174, 176, 294. A diffuse or tufted herb from a few inches to over 1 ft. high with the odour of *Artemisia* ; flower-heads ⅛–¾ in. broad ; florets yellow ; in damp sandy places in savanna areas.
Sen.: Ouassadou *Berhaut* 1691. **Mali**: *fide* Chev. *l.c.* **Port.G.**: Antula, Bissau (Feb.) *Esp. Santo* 1473 ! **Guin.**: *Heudelot* 859 ! Timbi-Madina Apr.) *Adam* 11817 ! Kankan *Adam* 12135 ! **S.L.**: Port Loko *Sc. Elliot* 5858 ! Njala (May) *Deighton* 640 ! Waima (June) *Deighton* 5334 ! Yonibana (June) *Deighton* 5848 ! Kasanko (May) *Adames* 227 ! **Iv.C.**: between the banks of Bandama R. and Marabadiassa R., Mankono (July) *Chev.* 22002 ! **Ghana**: White Volta R., Yapei (Mar.) *Morton* GC 8767 ! Tumu (Apr.) *Morton* GC 8887 ! Volta R., Kete Krachi (May) *Morton* GC 7293 ! **Dah.**: Gouka to Banté, Savalou (May) *Chev.* 23734 ! Kouandé to Konkobiri, Atacora Mts. (June) *Chev.* 24281 ! **N.Nig.**: Aboh *Barter* 181 ! Abinsi (July) *Dalz.* 651 ! Anara F.R., Zaria (May) *Keay* FHI 19193 ! Birnin Gwari, Zaria (June) *Keay* FHI 25867 ! Extends eastwards through Cameroun to Sudan and Uganda.

37. GRANGEA Adans.—F.T.A. 3 : 304.

Prostrate herb or shrub, clothed everywhere with long white hairs ; leaves variably pinnately divided, often nearly to the midrib, sessile, up to 5–7 cm. long, segments oblong, subacute ; flower-heads depressed-globose, up to 1 cm. diam. on short peduncles ; involucral bracts about 3-seriate, obovate-elliptic, 3–4 mm. long, gland-ular ; outer florets female, in 1 or more series ; florets persistent ; achenes sub-terete, scabrid, with very short curved pappus-bristles ..   ..   *maderaspatana*

G. maderaspatana (*Linn.*) *Poir.* in Lam. Encycl. Méth. Bot., Suppl. 2 : 825 (1812) ; F.T.A. 3 : 304 ; Chev. Bot. 361 ; Hagerup in Biol. Medd. 9, 4 : 55 ; Berhaut Fl. Sén. 172. *Artemisia maderaspatana* Linn. (1753). A hairy annual or perennial herb or shrub with prostrate branches from a woody root, odorous ; florets yellow ; in damp places.
Sen.: *Döllinger* ! *fide* Chev. & Berhaut *l.c.* Dakar (May) *Baldwin* 5701 ! **Mali**: Badumbé (Dec.) *Chev.* 51 ! Bandiagara to Mopti (Sept.) *Chev.* 24928 ! Timbuktu (July) *Hagerup* 148 ! 159 ! R. Niger, Macina (May) *Lean* 16 ! Diafarabé to Koumbé (May) *Lean* 25 ! **Port.G.**: Antula, Bissau (June) *Esp.Santo* 1512 ! Peluba, Bissau (Jan.) *Esp. Santo* 1707 ! **Ghana**: Kakundu, Cape Coast (Mar.) *Hall* 1301 ! **N.Nig.**: Nupe *Barter* 1200 ! Katagum *Dalz.* 174 ! Gashua, Bornu (June) *Onochie* FHI 23366 ! Maiduguri (June) *Ujor* FHI 21931 ! Dapchi, Bornu (Nov.) *Noble* A16 ! [**Br.**]**Cam.**: Bama, Dikwa Div. (Jan.) *McClintock* 174 ! Widely spread through tropical Africa and Asia.

38. GEIGERIA Griessel.—F.T.A. 3 : 367 ; Merxmüller in Mitt. Bot. Staatss. München 1, 7 : 239 (1953).

An erect divaricately branched glabrous annual herb with broadly winged stems ; leaves opposite, sessile and decurrent, oval-oblong, subacute, mucronate, entire or minutely denticulate, 2–8 cm. long, 0·5–2·5 cm. broad, glandular ; flower-heads campanulate, about 0·5 cm. long, sessile in the forks of branches, 10–12-flowered ; involucral bracts rigid in a few series, oval-oblong, spine-tipped, thinly puberulous, glandular ; receptacle with numerous slender setaceous paleae ; ray-florets in a single series, ligulate ; disk-florets tubular ; scales of pappus aristate, hyaline   ..   *alata*

G. **alata** (*DC.*) *Oliv. & Hiern* in F.T.A. 3 : 368 (1877) ; F.W. Andr. Fl. Pl. Sudan 3 : 31, fig. 5. *Diplostemma alatum* DC. (1838). A herb up to about 1 ft. high with a 3-winged stem ; leaves of two sorts occur at each node, two with decurrent bases and one without ; flower-heads about ¼ in. long with the tips of the outer involucral bracts spreading and leafy ; florets yellow ; a weed of semi-desert.
Mali: Niami-Niama, near Boré (Nov.) *Demange* 33/1957! **Niger**: Aïr *De Miré & Gillet*. Zinder (Dec.) *Vaillant* 792! Kolo *Vaillant* 864! Also in Mauritania, Cameroun, Sudan, Egypt, Somaliland, Arabia, Ethiopia, Eritrea, Kenya, Angola and S.W. Africa.

39. BUBONIUM Hill Veg. Syst. 2 : 74 (1761).

Shrubby up to 50 cm. high ; branches rigid, whitish, thinly velvety ; leaves pinnately lobed or toothed, up to 5 cm. long, sessile, amplexicaul, viscid-pubescent ; involucres rounded, terminal and subsessile in the branch-axils, the outer bracts foliaceous, mucronate or apiculate, the inner ovate, puberulous ; receptacle with palea ; ray-florets short, 2–3-toothed, yellow ; disk-florets tubular, acutely 5-lobed ; achenes costate, hairy on the ribs, the outer somewhat compressed, the inner terete ; pappus-scales scarious, dissected at the apex, equalling the ovary   ..   ..   *graveolens*

B. **graveolens** (*Forsk.*) *Maire* in Bull. Soc. Hist. Nat. Afr. Nord. 27 : 233 (1936) ; De Miré & Gillet in J. Agric. Trop. 3 : 706. *Buphthalmum graveolens* Forsk. (1775). *Odontospermum graveolens* (Forsk.) Sch. Bip. (1844)—F.T.A. 3 : 370. A tough bushy herb up to about 18 in. high with yellow florets in heads ½–¾ in. across.
Niger: Aïr *fide* De Miré & Gillet *l.c.* Extends from Mauritania through Sudan to Egypt and Arabia.

40. ANISOPAPPUS Hook. & Arn.—F.T.A. 3 : 369.

Pappus a single row of jagged scales ; leaves ovate to triangular-lanceolate, truncate or subcordate at base, crenate, 4–7 cm. long, 2–3 cm. broad, slightly scabrid above, puberulous beneath ; petiole slender, up to 2 cm. long ; flower-heads several, usually 4–8, rarely 1 ; involucral bracts about 3-seriate, up to 8 mm. long, scabridulous ; exserted ligules longer than the involucre   ..   ..   1. *africanus*
Pappus a row of jagged scales with intermediate bristles :
Intermediate pappus-bristles about twice as long as the jagged scales ; leaves very

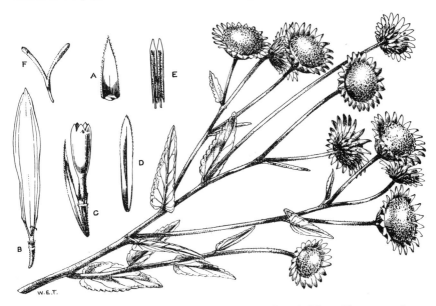

Fig. 255.—ANISOPAPPUS AFRICANUS (*Hook. f.*) *Oliv. & Hiern* (COMPOSITAE).
A, outer involucral bract. B, ray floret. C, disk floret and palea. D, involucral bract
E, anthers. F, style-arms.

variable from shortly triangular to lanceolate, crenate and often distinctly lobed towards the base, cuneate to cordate, up to 8 cm. long and 2·5 cm. broad, slightly scabrid ; petiole slender ; flower-heads 1–6, usually 4 or 5 ; involucral bracts grey-puberulous outside ; exserted ligules about equal to the involucre   .. 2. *dalzielii*
Intermediate pappus-bristles about equal in length to the jagged scales ; leaves ovate to suborbicular, up to 3 cm. long and broad, cordate with a blunt tip and rounded basal lobes, densely scabridulous above, puberulous beneath ; petiole 0·5–1 cm. long ; flower-heads 1–4 ; involucral bracts densely puberulous, green, the lower recurved ; exserted ligules exceeding the involucre   ..   .. 3. *suborbicularis*

1. **A. africanus** (*Hook. f.*) *Oliv. & Hiern* in F.T.A. 3 : 369 (1877). *Telekia africana* Hook. f. (1864). A robust herb up to 4 ft. high with striate pubescent pithy stems and yellow flower-heads about 1 in. across ; in savanna woodland or montane grassland.
   Ghana: Ejura, Ashanti (Dec.–Jan.) *Morton* GC 9529! A1529! Techiman to Nkoranza (Dec.) *Morton* A1154! **S.Nig.:** Mt. Koloishe, 5,000–5,400 ft., Ogoja (Aug.) *Stone* 49! [Br.]**Cam.:** Cam. Mt., 4,800–9,000 ft. (Sept., Nov.–Feb., Apr.) *Mann* 1313! 1920! *Irvine* 1468! *Dalz.* 8317! *Brenan* 9384! *Boughey* A445! GC 6845! *Keay* FHI 28639! Bamenda to Ndop, 4,800 ft. (Dec.) *Adams* GC 11257! Throughout tropical Africa.
2. **A. dalzielii** *Hutch.* F.W.T.A., ed. 1, 2 : 156 (1931). *A. africanus* of Chev. Bot. 368, not of Oliv. & Hiern. *A. chinensis* of Chev. Bot. 368, not of Hook. & Arn. An erect perennial herb 1–3 ft. high with yellow ray- and disk-florets in few heads ; in rough open grassland or savanna woodland.
   Mali: Tabacco (Oct.) *Chev.* 2075! **Guin:** Kouria (Dec.) *Chev.* 14821! Timbo to Ditinn (Sept.) *Chev.* 18453! 18512! Kaoulendougou, Nzo (Mar.) *Chev.* 21054! Mali, 5,000 ft. (Sept.) *Schnell* 7059! 7226! Madina Tossékré (Oct.) *Adam* 12558! **S.L.:** *Glanville* 377! Bintumane Peak, 5,000 ft. (Jan., Oct.) *T. S. Jones* 115! *Jaeger* 308! Yalamba (Jan.) *Jaeger* 3962! **Ghana:** Nwereme, Ashanti (Dec.) *Morton* A2634! Dutukpene (Dec.) *Adams* 4549! Amedzofe (Nov.) *Morton* GC 9432! A2305! **N.Nig.:** Kontagora (Oct.) *Dalz.* 196! Anara F.R., Zaria (Oct.) *Keay* FHI 5479! Kwakwi, Zaria (Nov.) *Onochie & Keay* FHI 21739! Jos Plateau (Oct.) *Keay* P823! Amo, Jos Plateau (Oct.) *Hepper* 2894! **S.Nig.:** Unya, 2,500 ft., Ogoja, (Dec.) *Savory & Keay* FHI 25019! [Br.]**Cam.:** Madaki, Adamawa (Dec.) *Latilo & Daramola* FHI 28834! Vogel Peak, 4,300 ft., Adamawa (Nov.) *Hepper* 1345! Also in Uganda and N. Rhodesia.
3. **A. suborbicularis** *Hutch. & B. L. Burtt* in Rev. Zool. & Bot. Afr. 23 : 41 (1932). *A. africanus* of F.W.T.A., ed. 1, 2 : 156, partly (*Migeod* 453). A straggling or erect bushy herb up to 3 ft. high with bright yellow flowers ; in montane grassland.
   [Br.]**Cam.:** Bamenda, 6,000 ft. (Jan.) *Migeod* 453! Nchan, Bamenda, 5,000 ft. (June) *Maitland* 1741! Kumbo to Oku, 6,000 ft. (Feb.) *Hepper* 2006! Mambila Plateau, 4,900 ft., Adamawa (Jan.) *Hepper* 1709! Also in Congo.

41. MOLLERA O. Hoffm. in E. & P. Pflanzenfam. 4, 5 : 205 (1890) ; Bol. Soc. Brot. 10 : 174 (1892).

Stem narrowly winged from the decurrent leaf-bases ; lower leaves opposite, upper alternate, oblanceolate, 1–3·5 cm. long, 3–6 mm. broad, remotely serrate-crenate towards the apex, minutely scabrid ; peduncles setulose 1–2 cm. long ; involucral bracts in 2 series, narrowly ovate, acuminate, scabridulous, 3–4 mm. long, the tips reflexed at maturity ; outer ligulate florets prominent with the tube exserted from the involucre, 2–3-toothed ; achenes barrel-shaped, strongly 8–10-ribbed, minutely glandular between the ribs ; pappus a minute inconspicuous ring ..   .. *angolensis*

M. angolensis *O. Hoffm.* l.c. (1890). *M. punctulata* Hiern Cat. Welw. 1 : 569 (1898). An annual herb up to about 1 ft. high with yellow ray- and disk-florets in heads ½–⅓ in. diam. ; a weed of fields and open ground.
   N.Nig.: Jos Plateau (Aug.) *Lely* P588! Bukuru to Hepham, 4,300 ft. (July) *Lely* 346! Ropp to Mongu (Apr.) *Morton* K363! Anara F.R., Zaria (Oct.) *Keay* FHI 20142! Naraguta racecourse (Nov.) *Keay* FHI 37205! Also in Angola.

42. PULICARIA Gaertn.—F.T.A. 3 : 363.

Involucral bracts in 4–5 rows ; capitula numerous ; achenes dark brown, oblong, sulcate, glabrous ; pappus 2-seriate, outer of short scales, inner of 10–12 slender setae, plumose towards the tips, the two series caducous together ; leaves rather small, oblong, with usually strongly crisped involute margins ..   .. 1. *crispa*
Involucral bracts in 2–3 rows ; capitula usually rather few ; achenes pale buff, smooth, shallowly grooved, glabrescent ; pappus 2-seriate, outer of short scales, inner of 12–15 slender setae, minutely barbellate along their entire length and slightly broader towards the tips, the inner series caducous, the outer persistent ; leaves oblanceolate, sinuate-margined, usually flat   ..   .. 2. *undulata*

1. **P. crispa** (*Forsk.*) *Oliv.* in Trans. Linn. Soc. 29 : 96 (1873) ; F.T.A. 3 : 366 ; Chev. Bot. 366 ; Hagerup in Biol. Medd. 9, 4 : 57 ; Berhaut Fl. Sén. 172 ; F. W. Andr. Fl. Pl. Sudan 3 : 46, fig. 8. *Aster crispus* Forsk. (1775). An annual herb with decumbent branches spreading and ascending to 1 or 2 ft. high ; stems and leaves silvery, but sometimes glabrescent ; flower-heads ¼–⅓ in. across with light yellow ray and disk florets ; an aromatic weed of fields and arid country.
   Sen.: Richard Tol (Feb.) *Roberty* 16858! **Mali:** Bamako (Jan.) *Chev.* 197! Timbuktu (July) *Hagerup* 168! Bendofo, Kita (July) *Jaeger* 6! Uburu, Lawra (Dec.) *Adams & Akpabla* GC 4067! Kaleo, Wa (Apr.) *Adams* 4009! **Dah.:** Djougou (June) *Chev.* 23903! **Niger:** Zinder (Nov.) *Hagerup* 598! **N.Nig.:** Borgu *Barter* 798! Kontagora (Jan., Nov.) *Dalz.* 201! *Meikle* 1036! Kano (Oct.) *Keay* FHI 21136! Bichikki (Sept.) *Lely* 630! Jengre, 3,000 ft., Zaria (Dec.) *Coombe* 84! [Br.]**Cam.:** Bama, Dikwa Dist. (Dec.) *McClintock* 51! Also in Mauritania, Egypt, Sudan, Ethiopia and Arabia. (See Appendix, p. 419.)
2. **P. undulata** (*Linn.*) *C. A. Mey.* Verzeichn. Pfl. 79 (1831) ; F.T.A. 3 : 365. *Inula undulata* Linn. (1767). *Pulicaria undulata* var. *alveolosa* (Batt. & Trab.) Maire —Berhaut Fl. Sén. 176 ; De Miré & Gillet in J. Agric. Trop. 3 : 709. An aromatic weedy herb like the last, but generally smaller and more silky-hairy ; flower-heads ¼–⅓ in. across ; florets yellow.
   Sen.: *fide* Berhaut *l.c.* **Mali:** Toguère de Banguita (Jan.) *Davey* 12! Niondjé (Jan.) *Davey* 212! Fenadjé (Jan.) *Davey* 254! **Ghana:** Tongo (Dec.) *Adams & Akpabla* GC 4313! Zuarungu (Apr.) *Morton* GC 8915! **Niger:** Aïr *fide* De Miré & Gillet *l.c.* **N.Nig.:** Fodama (Nov., Dec.) *Moiser* 186! 214! 230! 238!

Katagum *Dalz.* 314! Yola (Jan.) *Dalz.* 34! Areje, Bornu (Dec.) *Elliott* 152! Kalkala, L. Chad *Golding* 52! [Br.]Cam.: Belel, Adamawa (Jan.) *Hepper* 1632! Also in Mauritania, Egypt, Sudan and Arabia.

[De Miré and Gillet in J. Agric. Trop. 3 : 709, 710 (1956) report *P. arabica* Coss. subsp. *inuloides* Maire and *P. volkonskyana* Maire from Aïr, Niger.]

43. INULA Linn.—F.T.A. 3 : 357.

Flower-heads numerous, up to 1 cm. long ; stems closely ribbed, becoming nearly glabrous ; radical leaves elongate-lanceolate :

  Capitula crowded into rather broad corymbs ; peduncles slender, 1 cm. long ; involucral bracts about 4-seriate, lanceolate, pubescent, nearly 1 cm. long ; stem-leaves sessile, ovate-lanceolate, mucronate, auriculate at base, up to 25 cm. long and 8 cm. broad, pilose-pustulate above, woolly beneath, denticulate .. .. .. 1. *mannii*

  Capitula in tight terminal clusters ; peduncles with gland-tipped hairs ; involucral bracts about 3-seriate, linear, up to 0·5 cm. long ; stem-leaves rather small, scabrid above, scabridulous on the veins beneath, otherwise glabrous, shallowly dentate

                                        2. *subscaposa*

Flower-heads 2–5 or rarely solitary, up to 5 cm. diam. ; stems densely clothed with silky hairs ; radical leaves broadly oblong, dentate, up to 30 cm. long and 15 cm. broad, scabrid above, densely silky-villous beneath, with petioles up to 20 cm. long ; upper stem-leaves sessile, amplexicaul ; involucral bracts 6–7-seriate, the outer silky-pubescent .. .. 3. *klingii*

1. **I. mannii** (*Hook. f.*) *Oliv. & Hiern* in F.T.A. 3 : 358 (1877). *Vernonia ? mannii* Hook. f. (1864). A stout soft-stemmed herb 2–10 ft. high, with rather large leaves pale-tomentose beneath and campanulate flower-heads nearly ⅓ in. long in dense corymbs ; florets yellow.
  [Br.]Cam.: Cam. Mt., 6,700–7,500 ft. (Dec.–Feb.) *Mann* 1314! 1933! *Maitland* 975! 1004! Bamenda, 7,400–8,200 ft. (Dec.) *Boughey* GC 10771! 11008! Also in Ethiopia and Congo.

2. **I. subscaposa** *S. Moore* in J. Linn. Soc. 35 : 341 (1902). An erect perennial herb with slender stems 3–4 ft. high arising from a silky-hairy stock ; ray- and disk-florets yellow in flower-heads about ½ in. long.
  [Br.]Cam.: Bamenda Dist., 4,000–6,000 ft. (May–June) *Maitland* 1451! 1457! 1683! 1684! Gembu, 5,000 ft., Mambila Plateau, Adamawa (Jan.) *Hepper* 1819! Also in Rhodesia and Nyasaland.

3. **I. klingii** *O. Hoffm.* in Engl. Bot. Jahrb. 24 : 472 (1898). A perennial herb with an erect stout stem 1–6 ft. high ; inner involucral bracts dull yellow ; disk-florets yellow.
  Guin.: Mali, 4,500 ft. (Sept.) *Schnell* 7096! Pita (Oct.) *Jac.-Fél.* 1965! *Adam* 12546! Togo Rep.: Bismarckburg *Kling* 153! N.Nig.: Jagindi, Jemaa Div. (Nov.) *Keay & Onochie* FHI 21715! Naraguta, 4,000 ft. (Oct.) *Hepper* 1010! [Br.]Cam.: Wum crater lake, Bamenda (Apr.) *Morton* K295! Also in Congo and Angola.
  [This species is very near to *I. shirensis* Oliv. (1882). *I. klingii* is described as rayless ; *I. shirensis* as rayed. Further field observations are needed to establish the variability of this character—C. D. A.]

44. VICOA Cass.—F.T.A. 3 : 361.

Stem leafy, sparsely pubescent, glabrescent, brownish when dry ; leaves linear-lanceolate, sessile, auriculate, acute, 5–8 cm. long, 0·8–1·5 cm. broad, denticulate, scabridulous on both surfaces ; flower-heads at the ends of long branches ; peduncles slender ; involucral bracts 3–4-seriate, linear, acute, up to 7 mm. long, glandular outside ; florets all tubular (outer ligulate in Asian species) ; achenes very small, pubescent at least in the upper part ; pappus-bristles few, almost smooth, white

                                            leptoclada

V. leptoclada (*Webb*) *Dandy* in F.W. Andr. Fl. Pl. Sudan 3 : 62 (1956). *Inula leptoclada* Webb (1849). *Vicoa indica* of F.W.T.A., ed. 1, 2 : 608 and of Berhaut Fl. Sén. 172, not of DC. *Inula indica* of Chev. Bot. 366, not of Linn. *V. auriculata* of F.T.A. 3 : 362 and of F.W.T.A., ed. 1, 2 : 157, not of Cass. An erect annual herb 1–4 ft. high ; leaves dark green above, paler beneath ; florets yellow ; a weed of cultivated ground and open places mostly outside the high-forest areas.
  Sen.: Thiès (Dec.) *Chev.* 2033! Kaolak (Nov.) *Berhaut* 1169! Gam.: *Boteler!* *Brown-Lester* 21! Kudang (Jan.) *Dalz.* 8032! Genieri (Feb.) *Fox* 45! Mali: Koulouba (Nov.) *Jaeger* 19! Diona, Boré (Oct.) *Demange* 36/1957! Port.G.: Bafatá (Dec.) *Esp. Santo* 2857! Guin.: Faranna to Tindo (Jan.) *Chev.* 20476! Erimakuma (Mar.) *Sc. Elliot* 5253! S.L.: Musaia (Feb.) *Deighton* 4157! Ghana: Achimota (June, Oct., Nov.) *Irvine* 743! 1511! *Milne-Redhead* 5173! Accra Plains (Oct.) *Baldwin* 13434! Lumbunga, Tamale (Dec.) *Morton* GC 6236! Bolgatanga to Bawku (Apr.) *Morton* GC 8946! Dah.: Massé to Kétou (Feb.) *Chev.* 22994! Niger: Maradi (Dec.) *Vaillant* 856! N.Nig.: Nupe *Barter* 855! Ilorin (Jan.) *Ejiofor* FHI 19823! 30239! Kontagora (Nov.) *Dalz.* 200! Jos Plateau (Feb.) *Lely* P145! Kano (Dec.) *Hagerup* 638! Biu, Bornu, 2,500 ft. (Nov.) *Noble* 37! S.Nig.: Lagos *Phillips* 59! Ogurude (Jan.) *Holland* 265! Oluwatedo, Oyo (Jan.) *Keay* FHI 25636! [Br.]Cam.: Bama, Dikwa Div. (Dec.) *McClintock* 95! Vogel Peak, 2,100 ft., Adamawa (Nov.) *Hepper* 1304! Widespread in tropical Africa.

45. EPALTES Cass.—F.T.A. 3 : 331.

A small much-branched herb with a narrowly winged stem and alternate lanceolate leaves 2–4 cm. long and about 6(–10) mm. broad, remotely and shortly serrate, scabridulous ; capitula 3–4 mm. long in flower, campanulate, on peduncles not exceeding 5 mm. long in terminal umbelliform clusters ; involucral bracts in several series, ovate, acuminate, pubescent ; outer florets female, inner hermaphrodite, all tubular ; achenes oblong ; pappus absent .. .. *alata*

E. alata (*Sond.*) *Steetz* in Peters Reise Mossamb. Bot. 452 (1864). *Ethulia alata* Sond. (1850). A bushy herb usually about 1 ft. high with mauve florets in small clustered flower-heads.
  Mali: Boré (Aug., Nov.) *Demange* 35! Niger: Zinder (Oct.) *Hagerup* 582! Dakoro (Dec.) *Vaillant* 854! Also in Sudan, Ubangi-Shari, Tanganyika, Rhodesia, Nyasaland and S.W. Africa.

46. PLUCHEA Cass.—F.T.A. 3 : 327.

Leaves oblong to oblong-ovate, sessile and auriculate at base and sometimes decurrent on
  the stem with toothed wings, rounded at apex, 4–7 cm. long, 1·5–3 cm. broad, repand-
  dentate, rather densely pubescent and minutely glandular ; flower-heads several in
  a spreading terminal corymb ; involucral bracts about 5-seriate, ovate to lanceolate,
  thinly pubescent, straw-coloured ..   .. 1. *ovalis*
Leaves lanceolate to elliptic-oblanceolate, sessile but not auriculate at base, not decurrent, obtuse, apiculate, entire, 4–5 cm. long, up to 1 cm. broad, prominently nerved,
  softly appressed-tomentellous ; flower-heads crowded on short terminal branches,
  shortly pedunculate ; involucral bracts about 5-seriate, ovate-rounded, pubescent,
  purplish ..   .. 2. *lanceolata*

1. **P. ovalis** (*Pers.*) *DC.* Prod. 5 : 450 (1836) ; F.T.A. 3 : 328 ; Fl. Sperm. Parc Nat. Alb. 2 : 483, t. 45 ; Berhaut Fl. Sén. 146, 170. *Baccharis ovalis* Pers. (1807). *P. dioscoridis* of Chev. Bot. 365, not of DC. An
  aromatic glandular shrub 4–5 ft. high with pale grey-green stems and leaves ; florets yellowish-white or
  pale mauve in heads ¼ in. broad.
  Sen.: *Sieber* 61 ! M'Bidjem (Dec.) *Chev.* 2037 ! Dakar (Jan.) *Hagerup* 771 ! **Mali:** Cotaga, Niger R. (July)
  Chev. 1159 ! Timbuktu (July) *Chev.* **Ghana:** Achimota (May, Nov., Dec.) *Milne-Redhead* 5088 ! *Akpabla*
  1804 ! *Morton* A2834 ! Ada (June) *Morton* GC 9262 ! Weija (Nov.) *Morton* A63 ! **Togo Rep.:** Lomé
  Warnecke 223 ! 312 ! Also in Mauritania, Sudan, Congo, Kenya, Uganda, Tanganyika, Angola and the
  Cape Verde Islands.
2. **P. lanceolata** (*DC.*) *Oliv. & Hiern* in F.T.A. 3 : 329 (1877) ; Berhaut Fl. Sén. 204. *Berthelotia lanceolata* DC.
  (1836). A small shrub covered with a fine grey indumentum with clusters of flower-heads about ¼ in. broad
  at the tips of striate leafy twigs ; florets pink.
  Sen.: *Perrottet* ! *Heudelot* 400 ! Fann rocks, Dakar (Feb.) *Adam* 11087 ! Toumba (July) *Chev.* 4082 !

47. PEGOLETTIA Cass.—F.T.A. 3 : 360.

Stems pubescent, ribbed ; leaves linear, 1–3 cm. long, up to 5 mm. broad, entire or
  slightly toothed, densely crisped-setulose, glandular-punctate on both surfaces ;
  flower-heads numerous, terminal on bracteate branches 4–7 cm. long, 4–6 mm. broad
  in flower and up to 1 cm. broad in fruit, campanulate with 2-seriate pubescent involucral bracts 1 cm. long, the outer linear, the inner lanceolate ; florets all tubular ;
  achenes closely ribbed, thinly pubescent, 5 mm. long ; pappus 2-seriate, the outer
  short, multifid, the inner longer plumose..   ..   .. *senegalensis*

P. senegalensis *Cass.* Dict. Sci. Nat. 38 : 232 (1826) ; F.T.A. 3 : 361 ; Berhaut Fl. Sén. 204. A branched
  aromatic annual herb, woody below, up to about 1 ft. high with yellow florets turning purplish, in heads
  ¼ in. long.
  Sen.: *Roger* 86 ! Sor Isl. *Brunner* 187 ! **Niger:** Aïr *De Miré & Gillet.* **N. Nig.:** Kuka (Feb.) *E. Vogel* 60 !
  Kalkala, Bornu, *Golding* 68 ! Extends in northerly latitudes from Mauritania eastwards to central Sudan,
  Arabia and western India ; also in Angola and the drier parts of south tropical Africa.

48. PORPHYROSTEMMA Grant ex Benth.—F.T.A. 3 : 367.

An annual herb ; stems appressed-pilose ; leaves sessile, linear, up to 12 cm. long and
  1 cm. broad, thinly hispid, glabrescent ; capitula corymbose, few, shortly pedunculate,
  about 2 cm. diam. ; involucres rounded at base, with numerous linear, very acute
  shortly pilose bracts up to 1·5 cm. long in about 4 series ; peduncles villous ; corolla-
  limb pubescent ; achenes thinly pubescent ; pappus-bristles 10–15, uniform, barbellate
  chevalieri

P. chevalieri (*O. Hoffm.*) *Hutch. & Dalz.* F.W.T.A., ed. 1, 2 : 158 (1931) ; Berhaut Fl. Sén. 204. *P. grantii*
  Benth. ex Oliv. var. *chevalieri* O. Hoffm. in Mém. Soc. Bot. Fr. 2, 8 : 41 (1908) ; Chev. Bot. 367. A herb
  from a few inches to 2 ft. high with pink florets in solitary flower-heads ¾–1 in. across, terminal on the
  branches ; a weed of sandy soil in arid districts.
  Sen.: Sinédone (Jan.) *Chev.* 2091 ! Kaolak (Oct.) *Berhaut* 3374 ! **Port.G.:** Bissoram (Oct.) *Esp. Santo* 2409 !
  Suzana (Nov.) *Esp. Santo* 3680 ! **Ghana:** Navrongo (Nov.) *Akpabla* 387 ! Burufo, near Lawra (Dec.)
  Adams & Akpabla GC 4445 ! **N.Nig.:** Sokoto *Lely* 144 ! Jos Plateau (Oct.-Dec.) *Lely* P840 ! *Keay* FHI
  21024 ! *Coombe* 70 ! Yola (Nov.) *Hepper* 1193 ! Also in N. Rhodesia.
  [Except by the greater number of pappus-bristles this species is hardly to be distinguished from *P. grantii*
  Benth. ex Oliv. and *P. cuanzense* O. Hoffm. ,which together extend the range to Tanganyika and Angola—
  C. D. A.].

49. BLUMEA DC.—F.T.A. 3 : 322. *Nom. cons.*

Leaf-bases decurrent forming wings on the stem :
  Stem-wings entire, long ; whole plant densely whitish tomentose ; leaves entire,
    linear ..   .. 1. *gariepina*
  Stem-wings short, interrupted, stipuliform ; leaves dentate or lobed :
    Peduncles 2–3 cm. long ; inflorescence spreading ; glandular hairs long-stalked, few
      or absent ; glands colourless or yellow ; stems often violet-tinged 2. *aurita* var. *aurita*
    Peduncles 1–1·5 cm. long ; inflorescence compact ; glandular hairs short-stalked,
      numerous ; glands brown ; stems always green   2a. *aurita* var. *foliolosa*
Leaf-bases not decurrent ; stems without wings :
  Capitula about 1 cm. diam. on peduncles 1–1·5 cm. long ; branches of the inflorescence
    with numerous bracts :
    Leaves dentate ; inflorescence compact ; florets pink or mauve
                                                          2a. *aurita* var. *foliolosa*
    Leaves deeply divided, doubly dentate with acutely pointed mucronate lobes, thin ;
      inflorescence diffuse ; florets yellow or orange-yellow   ..   .. 3. *laciniata*

Capitula about 0·5 cm. diam. ; peduncles mostly short ; inflorescences dense with capitula clustered ; leaves not divided, denticulate ; florets pink to pale purple :

Lower leaves long-petiolate ; capitula on slender peduncles ; glands few, pale ; leaves glabrescent .. .. 4. *mollis*

Lower leaves narrowed to the base but not distinctly petiolate ; capitula subsessile in terminal and lateral bracteate clusters ; glands numerous, long-stalked, brown ; leaves and stems densely villous .. .. 5. *perrottetiana*

1. **B. gariepina** *DC.* Prod. 5 : 448 (1836) ; De Miré & Gillet in J. Agric. Trop. 3 : 706. A densely silky bushy herb or undershrub about 2 ft. high, with stalked flower-heads about ¼ in. long.
Niger: Aïr *vide* De Miré & Gillet *l.c.* Also in Sudan and south tropical Africa.

2. **B. aurita** (*Linn. f.*) *DC.* var. **aurita**—in Wight Contrib. Bot. Ind. 16 (1834); DC. Prod. 5 : 449 ; F.T.A. 3 : 322 ; Chev. Bot. 364. *Conyza aurita* Linn. f. (1781). *C. senegalensis* Willd. (1803)—F.T.A. 3 : 314, partly (*Don*), excluding var. β. An erect glandular aromatic densely silky-hairy herb 2–3 ft. high forming a basal rosette of leaves at first ; inflorescence spreading ; florets vary from greenish-yellow or white to pale mauve or pink ; a common weed in drier areas.
Sen.: Dakar (Apr., May) *Jaeger* ! *Baldwin* 5703 ! Bandia (Feb.) *Berhaut* 5349 ! **Gam.**: *Don* ! **Mali**: Bamako (Jan.) *Chev.* 194 ! Dogo (Apr.) *Davey* 571 ! 572 ! **Guin.**: Kouroussa (Dec.) *Pobéguin* 609 ! **S.L.**: Freetown (June) *Deighton* 2704 ! Sasseni (Jan.) *Sc. Elliot* 4395 ! Kambia (Dec.) *Deighton* 815 ! Musaia (Feb.) *Deighton* 4226 ! **Ghana**: *Thonning* 351 ! Achimota (June) *Irvine* 711 ! Sekodumase *Kitson* ! Wenchi (Dec.) *Adams* 3242 ! Pong Tamale (Dec.) *Morton* GC 9869 ! Kete Krachi to Dutukpene (Dec.) *Adams* 4638 ! **Togo Rep.**: Lomé *Warnecke* 274 ! **Dah.**: Ouidah *Isert* ! **N.Nig.**: Nupe *Baikie* ! Oke Opin, Ilorin (Dec.) *Ajayi* FHI 19266 ! Sokoto *Dalz.* 413 ! Fodama (Dec.) *Moiser* 165 ! Jos Plateau (Jan.) *Lely* P138 ! **S.Nig.**: Abeokuta (Jan.) *Rowland* ! Awka *Thomas* 22 ! Widespread in warm countries. (See Appendix, p. 416.)

2a. **B. aurita** var. **foliolosa** (*DC.*) *C. D. Adams* in J.W. Afr. Sci. Assoc. 6 : 149 (1961). *B. guineensis* var. *foliolosa* DC. Prod. 5 : 449 (1836). *Conyza senegalensis* of F.T.A. 3 : 314 ; of F.W.T.A., ed. 1, 2 : 154, partly (*Sieber* 42). *Blumea lacera* of Chev. Bot. 364, partly (*Chev.* 13225), not of DC. *B. guineensis* of Berhaut Fl. Sén. 171, not of DC. A green-stemmed densely glandular aromatic herb.
Sen.: *Perrottet* ! *Sieber* 42 ! (May) *Berhaut* 820 ! Dakar (May) *Baldwin* 5744 ! Mbao (June) *Berhaut* 5265 ! Bandia (Feb.) *Berhaut* 5348 ! **Port.G.**: Pessubé, Bissau (Mar.) *Esp. Santo* 1505 ! Tor, Bissau (Feb.) *Esp. Santo* 1781 ! **Guin.**: Kindia (Mar.) *Chev.* 13225 !

3. **B. laciniata** (*Roxb.*) *DC.* Prod. 5 : 436 (1836). *Conyza laciniata* Roxb. Fl. Ind. 3 : 427 (1832). *C. senegalensis* of F.W.T.A., ed. 1, 2 : 154, partly (*Sc. Elliot* 4956), not of Willd. A slender annual weed 2–3 ft. high with yellow or orange-yellow florets in long-stalked heads.
Sen.: Niokolo-Koba (Jan.) *Berhaut* 1552 ! Kayar *Adam* 12207 ! **S.L.**: Likuru, Talla (Feb.) *Sc. Elliot* 4956 ! Wara Wara, Yagala (Jan.) *King* 185b ! **Ghana**: Red Volta R., Bolgatanga to Bawku (Apr.) *Morton* GC 8934 ! **N.Nig.**: Birnin Gwari (Apr.) *Meikle* 1377 ! Also in India and eastern Asia.

4. **B. mollis** (*D. Don*) *Merrill* in Philipp. J. Sci. 5 : 395 (1910). *B. lacera* of F.T.A. 3 : 322, partly. A slender weed up to 3 ft. high with many small capitula in terminal inflorescences ; basal and lower stem-leaves long-stalked ; florets mauve or pale purple ; in low-lying ground.
Sen.: *Heudelot* 10 ! Sangalkam (June) *Berhaut* 4770 ! **Gam.**: Kombo country *Heudelot* ! **Guin.**: Rio Nunez *Heudelot* 670 ! Sala Falls (Apr.) *Adam* 11776 ! **S.L.**: Kamaranka (Feb.) *Deighton* 4043 ! Musaia (Feb.) *Deighton* 4204 ! **Ghana**: Legon (Jan., Oct., Dec.) *Boughey* GC 4646 ! *Baker* ! *Morton* A3640 ! **S.Nig.**: Ogurude (Jan.) *Holland* 269 ! Scattered through the Old World tropics from India to Ethiopia, Madagascar and Angola.

5. **B. perrottetiana** *DC.* Prod. 5 : 443 (1836). *B. lacera* of F.T.A. 3 : 322, partly ; of F.W.T.A., ed. 1, 2 : 158 partly ; of Chev. Bot. 364, partly (*Chev.* 128), not of DC. *Pluchea odorata* of Berhaut Fl. Sén. 266, not of Cass. An erect, densely velvety aromatic herb up to 4 ft. high ; florets pale pink to reddish-purple in numerous clustered flower-heads.
Sen.: *Leprieur* ! *Perrottet*. Kayar (Apr.) *Berhaut* 1381 ! Niokolo-Koba (Jan.) *Berhaut* 4684 ! Sangalkam *Adam* 9409 ! **Gam.**: *Pirie* 44 ! **Mali**: Tabacco (Jan.) *Chev.* 158 ! **Port.G.**: Brene, Bissau (Jan.) *Esp. Santo* 1721 ! **Guin.**: *Heudelot* 677 *bis* ! **S.L.**: Laya (Jan.) *Sc. Elliot* 4484 ! Wara Wara, Yagala (Jan.) *King* 184b ! Rosino (Feb.) *Deighton* 4976 ! Rokupr (Mar.) *Jordan* 420 ! Musaia (Feb., Mar.) *Deighton* 4156 ! 5394 ! **Lib.**: Kakatown *Whyte* ! **Ghana**: Winneba Plains (Mar.) *Morton* A1856 ! Menji, Ashanti (Dec.) *Adams* 3112 ! *Morton* A2611 ! Wenchi, Ashanti (Dec.) *Adams* 3236 ! near Sorri Lake, Damongo (Dec.) *Morton* GC 25023 ! **Togo Rep.**: *Baumann* 420 ! **N.Nig.**: Nupe *Barter* 1164 ! Sokoto (Nov., July) *Moiser* 44 ! *Dalz.* 404 ! Jos Plateau (Jan.) *Lely* P76 ! Vom (Apr.) *Morton* K400 ! Alagbade F.R., Ilorin (Feb.) *Ejiofor* FHI 19816 ! [Br.]**Cam.**: Lakom, Bamenda, 5,500 ft. (May) *Maitland* 1439 ! Metschum R., Bamenda, 2,100 ft. (Jan.) *Keay & Russell* FHI 28526 ! Nkambe, 4,000 ft. (Feb.) *Hepper* 1912 ! Gwoza, 3,000 ft., Dikwa Div. (Jan.) *McClintock* 143 ! Throughout tropical Africa.
[Small, often perennial, forms of this species are found in upland pastures, for example *King* 184b and *Hepper* 1912, cited above. These may be separable as *B. dregeanoides* Sch. Bip. or *B. solidaginoides* (Poir.) DC.]

50. NICOLASIA S. Moore in J. Bot. 38 : 458 (1900) ; Merxmüller in Mitt. Bot. Staatss. München, 2, 11, 1 (1954).

An erect herb with spreading branches, 20–50 cm. high ; stem and leaves with scattered glandular hairs ; leaves linear-oblanceolate, entire, decurrent at the base into a narrow stem-wing, the midrib prominent on both surfaces ; capitula with the outer florets female, the inner hermaphrodite, all tubular, 10–15 mm. long ; involucral bracts imbricate in several series, narrow, the inner more or less scarious, the outer shorter ; peduncles slender ; receptacle naked ; style arms 2, linear, mostly distinct but sometimes appressed ; anther-base shortly lobed, not tailed ; achenes slightly flattened ; pappus-setae 5–10, stiff .. .. *stenoptera*

N. stenoptera (*O. Hoffm.*) *Merxm.* l.c. 9/10 : 402 (1954). *Laggera stenoptera* O. Hoffm. in Bull. Herb. Boiss. 1 : 76 (1893). An annual herb with pinkish-white florets in red-tinged involucres ; by streams in rocky hill country.
[Br.]**Cam.**: Vogel Peak, 4,000 ft., Adamawa (Dec.) *Hepper* 1554 ! Mambila Plateau, 4,600 ft. (Jan.) *Hepper* 1650 ! Also in S.W. Africa and S. Rhodesia.

51. LAGGERA Sch. Bip. ex Benth.—F.T.A. 3 : 323.

Stem-wings toothed, at least in the upper part ; inflorescence paniculate, spreading or contracted ; capitula numerous on short or long slender peduncles .. 1. *pterodonta*

Stem-wings entire :

18

Capitula few, usually less than 12 ; perennial herbs with shoots arising from a woody
   stock :
   Flower-heads sessile or subsessile, terminal or axillary, 2–3 cm. diam...   2. *braunii*
   Flower-heads borne at the ends of short winged branches, about 1·5 cm. diam. ;
     involucral bracts uniformly puberulous, the outer reflexed ..   ..   3. *oloptera*
Capitula numerous, usually more than 20 in a spreading paniculate inflorescence ;
   annual or short-lived perennial herbs, rarely woody at base ; flower-heads up to
   1·5 cm. diam., but usually smaller :
   Stem-leaves linear or narrowly oblong-elliptic, up to 20 cm. long ; inflorescence-leaves
     linear ; leaves sparsely pubescent or glabrous beneath, scabrid above ; veins
     prominent ; stem-wings 4–6 mm. broad, scabrid; peduncles up to 3 cm. long ;
     capitula 1–1·5 cm. diam. ; middle involucral bracts pubescent in the mid-line and
     towards the apex outside ; stems often reddish ..   ..   4. *heudelotii*
   Stem-leaves broadly oblong to elliptic, rarely exceeding 15 cm. in length ; inflores-
     cence-leaves oblong to elliptic ; leaves pubescent to tomentose beneath :
     Capitula 1 cm. or more diam. ; robust herbs :
       Leaves in the inflorescence elliptic-lanceolate ; stem-leaves broadly oblong-elliptic,
       obtuse ; stem and undersurface of leaves puberulous ; venation clearly visible ;
       stem-wings about 3 mm. broad ; involucral bracts puberulous ; whole plant
       aromatic ..   ..   5. *alata* var. *alata*
       Leaves in the inflorescence elliptic ; stem-leaves broadly elliptic ; stem and under-
       surface of leaves densely tomentose with crisped yellowish-brown hairs, with the
       venation scarcely visible ; stem-wings 5–6 mm. broad ; involucral bracts woolly-
       tomentose ..   ..   5a. *alata* var. *montana*
     Capitula 5–7 mm. diam. ; nodding slender herb ; leaves oblong, rounded at apex,
     whitish-puberulous with conspicuously articulated crisped hairs and yellow glands
     beneath, hairs and glands fewer above ; stem-wings about 3 mm. broad ;
     peduncles slender, not exceeding 1·5 cm. long ; involucral bracts in about
     3 series, the outer reflexed ..   ..   6. *gracilis*

1. **L. pterodonta** (*DC.*) *Sch. Bip. ex Oliv.* in Trans. Linn. Soc. 29 : 94 (1873) ; F.T.A. 3 : 324 ; Chev. Bot.
   365. *Blumea pterodonta* DC. (1834). *Laggera appendiculata* of Adams in J. W. Afr. Sci. Assoc. 3 : 115,
   not of Robyns (1943). A strongly aromatic glandular weedy herb up to 5 ft. high with purple or greenish
   florets in numerous flower-heads about ½ in. long.
   Sen.: Sinédone, Casamance (Jan.) *Chev.* 2039. **S.L.**: Kabala (Dec.) *Deighton* 4514! Rokupr (Feb.) *Jordan*
   390! Magbema (Jan.) *Jordan* 763! Scarcies R., Wallia (Jan.) *Sc. Elliot* 4458! **Ghana**: Ejura Scarp (Dec.)
   Morton A2561! Wenchi, Ashanti (Dec.) *Adams* 3231! Kete Krachi to Atebubu (Dec.) *Adams* 4616!
   Kpedsu (Jan.) *Howes* 1123! **N.Nig.**: Pankshin (Dec.) *MacGregor* 416! Naraguta (Nov.) *Lely* 708! Vom
   Dent Young 149! **S.Nig.**: Sonkwala, Ogoja, 5,000–5,400 ft. (Dec.) *Savory & Keay* FHI 25091! **[Br.]Cam.**:
   Cam. Mt., Mann's Springs, 7,500 ft. (Dec.) *Boughey* GC 10597! Bamenda, 4,500 ft. (Mar., May) *Maitland*
   1695! *Morton* K124! Ndop to Bamenda (Dec.) *Boughey* GC 11371! Vogel Peak, Adamawa (Nov.)
   Hepper 2879! Throughout tropical Africa and Asia.
2. **L. braunii** *Vatke* in Linnaea 39 : 486 (1875) ; F.T.A. 3 : 326. An erect perennial herb 2–3 ft. high with large
   reddish-purple sessile flower-heads ; in hill grassland.
   N.Nig.: Naraguta, 4,000 ft. (Dec.) *Coombe* 32! **[Br.]Cam.**: Bamenda to Ndop, 5,000 ft. (Dec.) *Adams*
   GC 11270! Also in Ethiopia.
3. **L. oloptera** (*DC.*) *C. D. Adams* in J.W. Afr. Sci. Assoc. 6 : 152 (1961). *Blumea oloptera* DC. Prod. 5 : 448
   (1836). *Laggera oblonga* Oliv. & Hiern in F.T.A. 3 : 327 (1877) ; F.W.T.A., ed. 1, 2 : 158 ; Berhaut
   Fl. Sén. 170. *L. macrorrhiza* O. Hoffm. & Muschl. (1910)—Chev. Bot. 365. A perennial herb of grassland
   with shoots 1–2 ft. high branching near the base from a woody stock ; flower-heads about ¾ in. long on
   short peduncles, rather few and not clustered, with white, pink or purple florets.
   Sen.: *fide* Berhaut *l.c.* **Gam.**: Goaal, near Gambia R. (Mar.) *Perrottet*! **Mali**: Médinani (Dec.) *Chev.* 504!
   S.L.: Scarcies R., Buyabuya (Feb.) *Sc. Elliot* 4545! **Ghana**: Achimota (Apr., May) *Irvine* 1766! *Morton*
   A1982! Legon Hill (Nov.) *Adams* 3549! Wenchi, Ashanti (Dec.) *Adams* 3264! Zuarungu (Dec.) *Akpabla*
   409! Burufu (Dec.) *Adams & Akpabla* GC 4418! 4430! **N.Nig.**: Kafanchan (May) *Jones* FHI 14105!
   Kontagora (Nov.) *Dalz.* 205! Naraguta *Lely* 54! Jos Plateau (Feb.) *Lely* P162! Abinsi (Oct.) *Dalz.* 666!
   Also in Sudan, Congo and Angola.
4. **L. heudelotii** *C. D. Adams* in J. W. Afr. Sci. Assoc. 6 : 151 (1961). *L. alata* of F.W.T.A., ed. 1, 2 : 158, partly
   (*Heudelot* 639 ; *Barter* 1287 ; *Sc. Elliot* 4044). An erect herb 3–6 ft. high with linear, sparsely pubescent
   strongly veined leaves ; stem often reddish ; florets pale purple in heads about ⅜ in. long.
   Sen.: *Heudelot* 37! **Guin.**: *Heudelot* 639! *Jaeger* 2129! **S.L.**: Wilberforce (Dec.) *Sc.*
   Elliot 4044! Musaia (Dec.) *Deighton* 4418! Mange, Bure (Dec.) *Jordan* 729! **Iv.C.**: Tchumkrou to
   Akakoumoëkrou, Ano (Dec.) *Chev.* 22929! **Ghana**: Kintampo, Ashanti (Dec.) *Morton* A1164! 1180!
   2588! Wenchi, Ashanti (Dec.) *Adams* 3245! Zolo-Kpuita (Nov.) *Morton* GC 9368! **N.Nig.**: Nupe *Barter*
   1287! **S.Nig.**: Lagos *W. D. MacGregor* 178! Ibadan (Aug.) *Tamajong* FHI 19511! Abeokuta (Jan.)
   Keay FHI 21154! **[Br.]Cam.**: Bamenda, 5,000 ft. (Jan.) *Migeod* 324! Also in Angola.
5. **L. alata** (*D. Don*) *Sch. Bip. ex Oliv.* var. *alata*—in Trans. Linn. Soc. 29 : 94 (1873) ; F.T.A. 3 : 326 (excl.
   syn. *Blumea oloptera* DC.) ; Chev. Bot. 364. *Erigeron alatus* D. Don (1825). A strongly aromatic erect
   weedy herb up to 10 ft. high with broad obtuse round-sided leaves ; florets usually mauve in numerous
   heads about ½ in. long.
   Guin.: Dalaba (Apr.) *Caille* in Hb. *Chev.* 18138! Mali (Dec.) *Jac.–Fél.* 633! **S.L.**: Sugar Loaf Mt., 2,000 ft.
   (Dec.) *Sc. Elliot* 3971! **Lib.**: Mpaka Fossa Mt., Kolahun (Jan.) *Bequaert* 33! **Iv.C.**: Bouroukrou (Jan.)
   Chev. 17020! Gonokrom to Techikrom (Dec.) *Adams* 2976! **Ghana**: Aburi Hills (Feb.) *Brown* 933!
   Breppaw (Jan.) *A. S. Thomas* D97! New Tafo (Dec.) *Lovi* WACRI 3974! Pamu, Ashanti (May)
   Irvine 2506! Dormaa-Ahenkro (Dec.) *Adams* 3036! Sakogu, Gambaga Scarp (Dec.) *Adams & Akpabla*
   GC 4241! **N.Nig.**: (Nov.) *Lely* 693! **S.Nig.**: Lagos *Rowland*! Oban *Talbot*! **[Br.]Cam.**: Buea, 3,000 ft.
   (Dec.–Feb.) *Dunlap* 160! *Maitland* 126! 204! Bamenda (Mar.) *Morton* K124! Bamenda to Ndop,
   5,400 ft. (Dec.) *Boughey* GC 10455! 10456! Vogel Peak, 4,000 ft., Adamawa (Nov.) *Hepper* 1532! (See
   Appendix, p. 418.)
5a. **L. alata** var. **montana** *C. D. Adams* in J. W. Afr. Sci. Assoc. 6 : 150 (1961). *L. alata* of F.W.T.A., ed. 1,
   2 : 158, partly (*Mann* 1269, 1914 ; *Dunlap* 196). An erect tomentose herb up to 6 ft. high with a spreading inflorescence of heads about ½ in. long ; florets variable in colour from greenish-white to reddish-
   brown.
   [Br.]Cam.: Cam. Mt., 5,000–8,000 ft. (Dec.–Feb., Apr.) *Dunlap* 196! *Hutch. & Metcalfe* 26! *Keay* FHI

28642 ! *Irvine* 1466 ! *Mann* 1269 ! 1914 ! *Maitland* 304 ! Mann's Springs, 7,500 ft. (Mar., Apr.) *Brenan* 9382 ! *Morton* GC 6787 ! 6798 ! Bamenda, 5,000-7,000 ft. (Dec.–Feb.) *Migeod* 303 ! 486 ! *Boughey* GC 10746 ! Bafut-Ngemba F.R., 7,000 ft. (Feb.) *Hepper* 2185 ! *Tiku* FHI 22165 ! **F.Po:** Pico Serrano (Jan.) *Guinea* 1975. Moka (Dec.) *Monod* 10498 !

6. **L. gracilis** (*O. Hoffm. & Muschl.*) *C. D. Adams* in J. W. Afr. Sci. Assoc. 6 : 151 (1961). *Blumea alata* var. *gracilis* O. Hoffm. & Muschl. in Mém. Soc. Bot. Fr. 2, 8c : 113 (1910). A graceful erect herb 4-5 ft. high with slender branches and peduncles ; florets pink in numerous small heads ; although glandular, apparently not aromatic.

Guin.: Bambaya (Oct.) *Jaeger* 829 ! **S.L.:** Musaia (Dec.) *Deighton* 4479 ! **Ghana:** Kintampo, Ashanti (Dec.) *Morton* A1165 ! 1194 ! Banda, Ashanti (Dec.) *Morton* A1103 ! **N.Nig.:** Jos, 4,000 ft.(Dec.) *Coombs* 4 ! **[Br.]Cam.:** Vogel Peak, Adamawa, (Nov.) *Hepper* 1399 ! Also in Ubangi-Shari.

52. HELICHRYSUM Mill. corr. Pers. Syn. 2 : 414 (1807). *Nom. cons.*

Plant leafy at flowering time :
   Leaves small, 1-2 cm. long, at least in the upper part of the plant, linear-oblanceolate to lanceolate, densely tomentose on both surfaces ; flower-heads small in rounded clusters :
      Involucral bracts golden-brown fading to buff, blunt, membranous with frayed margins ; capitula about 2 mm. long and 1 mm. broad in flower ; leaves woolly-tomentose, often with recurved margins, ericoid .. .. .. 1 *cymosum*
      Involucral bracts silvery-white, subacute, imbricate, glumaceous, entire ; capitula 4-5 mm. long and 2 mm. broad ; leaves smooth, the margins not recurved but whole leaf folded lengthwise at least when dry .. .. .. 2. *glumaceum*
   Leaves larger, various, not ericoid :
      Upper part of shoot leafy with usually several distinct inflorescence branches ; basal leaves mostly withered or absent at time of flowering :
         Involucral bracts silvery-white or pinkish, broadly ovate, subacute :
            Flower-heads 1·5-2 cm. broad ; leaves numerous, clustered on the stem, lanceolate to ovate-lanceolate, acutely acuminate, up to 12 cm. long and 3 cm. broad, scabrid above, woolly-tomentose beneath .. .. 3. *mannii*
            Flower-heads about 1 cm. broad ; leaves few, spread along the stem, obovate to obovate-lanceolate, obtuse, up to 8 cm. long and 2·5 cm. broad, yellowish-woolly-tomentose on both surfaces .. .. .. 4a. *antunesii* var. *latifolium*
      Involucral bracts bright lemon-yellow or buff coloured :
         Involucral bracts bright lemon-yellow :
            Leaves ovate-lanceolate, broadly subauriculate at base, 6-12 cm. long, 2·5-4 cm. broad, acute, densely woolly-tomentose on both surfaces ; leaves on inflorescence branches broad, numerous ; flower-heads 2-2·5 cm. across   5. *cameroonense*
            Leaves linear to linear-lanceolate, less than 2 cm. broad, scabridulous or glabrescent above :
               Stems not winged ; flower-heads 1-1·5 cm. broad ; leaves narrowly lanceolate, acutely acuminate, sessile-auriculate, up to 12 cm. long, glabrescent above ; leaves on inflorescence branches several, conspicuous .. .. 6. *foetidum*
               Stems narrowly winged ; leaves scabridulous above ; leaves on inflorescence branches narrow, few or absent :
                  Capitula 6-7 mm. broad .. .. 7. *biafranum*
                  Capitula 1-2·5 mm. broad .. .. 8. *odoratissimum*
         Involucral bracts buff, acute or acuminate, membranous ; flower-heads more or less campanulate, about 3·5 mm. broad ; leaves flat, linear-lanceolate ; stem not winged .. .. [see *Gnaphalium undulatum*]
      Upper part of shoot usually simple with few leaves or leafless ; basal leaves petiolate ; flower-heads up to 4 mm. broad, numerous and tightly clustered in a subglobose very shortly branched terminal inflorescence :
         Flower-heads oblong, 4-5 mm. long and 2·5-3 mm. broad ; involucral bracts shiny-brown ; leaves persistently white-tomentose beneath .. .. 9. *globosum*
         Flower-heads more or less rounded in bud, up to 4 mm. long :
            Involucral bracts pale dull yellow, white or white and tinged with purple ; leaves persistently greyish-tomentose beneath ; stem-leaves acuminate :
               Stem-leaves woolly above, sessile ; lower leaves more or less cordate, acuminate, long-petiolate ; involucral bracts acute, dull silvery-white or yellowish, at length spreading .. .. 10. *albiflorum*
               Stem-leaves glabrescent above, very long-pointed, ½-amplexicaul ; lower leaves with 3-5 conspicuous veins beneath ; outer involucral bracts tinged reddish   11. *rhodolepis*
            Involucral bracts bright rich yellow ; at least the lowest leaves glabrescent on both surfaces ; stem-leaves acute .. .. .. 12a. *nudifolium* var. *leiopodium*
Plant leafless at flowering time, except for two or three reduced leaves on the densely woolly stems, the latter from a rhizome ; leaves when present basal, spathulate-elliptic or almost orbicular up to 20 cm. long and 7 cm. broad, mucronate, hairy on both surfaces ; petioles slender, up to 10 cm. long ; flower-heads 7-10 mm. long and 5·6 mm. broad, several in a terminal cluster ; involucral bracts oblong-elliptic, tinged reddish .. .. 13. *mechowianum*

1. **H. cymosum** (*Linn.*) *Less.* Syn. Gen. Comp. 302 (1832); F.T.A. 3 : 353. *Gnaphalium cymosum* Linn. (1753) *H. cymosum* subsp. *fruticosum* Hedberg in Symb. Bot. Upsal. 15, 1 : 203, 339 (1957). *H. fruticosum* Vatke (1875)—Moeser in Engl. Bot. Jahrb. 44 : 257; F.W.T.A., ed. 1, 2 : 159; F. W. Andr. Fl. Pl.Sudan 3 : 36. An erect perennial aromatic herb or undershrub, usually woody at the base, from a few inches to 4 ft. high, thinly woolly; with very numerous small golden-brown flower-heads; very variable in leaf-size and spacing and in the shape of capitulum; in montane grassland.
[Br.]Cam.: Cam. Mt., 7,000 ft. to summit (Nov.–Apr.) *Mann* 1287! 1930! *Mildbr.* 10835! 10847! *Dalz.* 8322! *Maitland* 945! 1000! 1241! *Migeod* 156! *Boughey* GC 6969! *Morton* GC 6870! 7062! *Keay* FHI 28595! Bamenda Dist., 5,500–8,000 ft. (Feb.) *Migeod* 341! *Savory* 398! *Morton* K99! 206! *Adams* GC 11094! 11209! Bafut-Ngemba F.R. (Sept., Jan., Feb.) *Hepper* 2096! *Keay & Lightbody* FHI 28388! *Ejiofor* FHI 29381! **F.Po:** summit of the Peak (Dec., Mar.) *Mann* 284 (partly)! *Boughey* GC 10981! *Guinea* 2895. Also in Ethiopia, Sudan, Uganda, Kenya, Tanganyika, Congo, Angola and Arabia.

[An extreme form with widely spaced larger leaves, longer more straggling stems and rather larger flower-heads, which may be a plant of shady habitats, is represented by *Brenan* 9543 and *Boughey* GC 10644 from Cameroon Mt., *Migeod* 442 from Bamenda and *Hepper* 2815 from Mambila Plateau.]

2. **H. glumaceum** DC. Prod. 6 : 197 (1838). *Achyrocline glumacea* (DC.) Oliv. & Hiern in F.T.A. 3: 340 (1877), partly; F.W.T.A., ed. 1, 2 : 159; Berhaut Fl. Sén. 205. *A. luzuloides* of Chev. Bot. 366, not of Vatke. An annual or perennial bushy herb from a few inches to about 1 ft. high with small whitish-tomentose leaves; flower-heads narrow, pointed, silvery-white; on littoral dunes.
Sen.: Dakar (Jan., Oct.) *Chev.* 2045! *Adam* 9176! Also in Mauritania, Ethiopia, Eritrea, Socotra, Kenya and Tanganyika.

3. **H. mannii** *Hook. f.* in J. Linn. Soc. 6 : 12 (1862). A stout herb 1–2 ft. high with a pithy stem, crowded sessile half-amplexicaul leaves, and several clustered silvery-white flower-heads ¾ in. across; florets cream-coloured to pale yellow.
[Br.]Cam.: Cam. Mt., in grasslands from 8,000 ft. to the summit and in the crater lip (Nov.–Jan., Mar., Apr.) *Mann* 1937! *Johnston* 51! *Kingsley*! *Dundas* FHI 20362! *Migeod* 194! *Maitland* 1285! *Boughey* GC 6989! 12641! **F.Po:** Summit of Clarence Peak (Dec.) *Mann* 615!

4. **H. antunesii** *Volkens & O. Hoffm.* in Engl. Bot. Jahrb. 32 : 149 (1902).

4a. **H. antunesii** var. **latifolium** *C. D. Adams* in J. W. Afr. Sci. Assoc. 2 : 63 (1956). A perennial herb with densely yellowish-woolly leaves and stems 2–3 ft. high; flower-heads about ⅜ in. across in several terminal corymbose clusters, pinkish when dry; in montane grassland.
[Br.]Cam.: Nchan, Bamenda, 5,000 ft. (May–June) *Maitland* 1446! 1516! 1709 *bis*! Bamungo, Bamenda (May) *Ujor* FHI 30303! Also in Angola.

5. **H. cameroonense** *Hutch. & Dalz.* F.W.T.A., ed. 1, 2 : 159 (1931). An aromatic, glandular-viscid stout-stemmed herb up to 4 ft. high with lemon-yellow flower-heads ¾–1 in. across, florets darker than the involucral bracts.
[Br.]Cam.: Cam. Mt., in grasslands up to about 13,000 ft. (Dec.–Feb., Apr.) *Mann* 1306! 1932! *Mildbr.* 10852! *Dalz.* 8307! *Dundas* FHI 13917! *Hutch. & Metcalfe* 59! *Maitland* 986! *Keay* FHI 28627! *Plumptre* 244! *Boughey* GC 6923! Bafut-Ngemba F.R., 5,000–6,000 ft. (Dec.) *Tamajong* FHI 22209! Bambui, 7,400 ft. (Dec.) *Boughey* GC 10770! L. Oku, 7,800 ft. (Jan.) *Keay & Lightbody* FHI 28511!

6. **H. foetidum** (*Linn.*) *Moench* Meth. Pl. 575 (1794); F.T.A. 3 : 352, partly (excl. syn. *H. mannii*). *Gnaphalium foetidum* Linn. (1753). An erect aromatic branched herb with a pithy stem up to about 5 ft. high; leaves narrow, whitish beneath; flower-heads bright yellow, about ½ in. across.
S.Nig.: Calabar *Hewan*! Koloishe Mt., 6,000 ft., Ogoja (Dec.) *Savory & Keay* FHI 25095! Ikwette, Ogoja (Dec.) *Savory & Keay* FHI 25235! **[Br.]Cam.:** Cam. Mt., 5,000–10,000 ft. (Dec.–Apr.) *Mann* 1916! *Dunlap* 194! *Dalz.* 8308! *Maitland* 251! 969! *Keay* FHI 28635, 28643! *Brenan* 9575! *Boughey* GC 6810! Bamenda Dist., 5,400–8,200 ft. (Dec., Jan., Mar., Apr.) *Migeod* 336! *Adams* GC 11329! *Boughey* GC 10563! *Morton* K79! Bambui (Dec.) *Maitland* 1465! Kumbo to Oku, 6,000 ft. (Jan.) *Hepper* 1993! **F.Po:** Clarence Peak *Mann* 281! Pico Serrano (Jan.) *Guinea* 1995. Also in Ethiopia, Sudan, Congo and S. Tomé.

7. **H. biafranum** *Hook. f.* in J. Linn. Soc. 7 : 202 (1862); F.T.A. 3 : 351. An erect slender branched herb up to 10 ft. high with narrowly winged stems and numerous bright yellow flower-heads about ½ in. broad.
[Br.]Cam.: Cam. Mt., in ravine above Buea, and at edge of forest near Mann's Springs, 7,000 ft. (Dec.) *Mann* 1934 (partly)! *Boughey* GC 12656!

8. **H. odoratissimum** (*Linn.*) *Less.* Syn. Gen. Comp. 301 (1832); Moeser in Engl. Bot. Jahrb. 44 : 242 (1910); Hedberg in Symb. Bot. Upsal. 15, 1 : 201, 338 (1957). *Gnaphalium odoratissimum* Linn. (1753). *Achyrocline hochstetteri* Sch. Bip. ex A. Rich. (1848)—F.T.A. 3 : 339; F.W.T.A., ed. 1, 2 : 159. A perennial herb with slender white-woolly erect or straggling winged branches 1–5 ft. high; flower-heads pale golden-yellow, very small and numerous in terminal corymbs.
S.Nig.: Ikwette, 5,200 ft., Ogoja (Dec.) *Savory & Keay* FHI 25228! **[Br.]Cam.:** Cam. Mt., 5,000–9,000 ft. (Dec.–May) *Mann* 1228! 1934 (partly)! *Dalz.* 8323! *Maitland* 660! 922! 970! *Brenan* 9387! *Thorold* CM 24! *Keay* FHI 29631! *Boughey* GC 11553! *Morton* GC 6869! Bamenda, 5,000–8,200 ft. (Dec., Jan., Mar.) *Migeod* 331! *Boughey* GC 11557! *Morton* K120! *Adams* 1516! Mba Kokeka Mt., 7,400 ft. (Jan.) *Keay & Lightbody* FHI 28394! Kumbo (Feb.) *Hepper* 1974! Maisamari, 4,600 ft., Mambila Plateau (Jan.) *Hepper* 1676! **F.Po:** Clarence Peak, 8,500 ft. (Dec.) *Mann* 607! Extends from Ethiopia through Sudan and Uganda to Congo, Angola and Transvaal.

9. **H. globosum** *Sch. Bip. ex A. Rich.* Tent. Fl. Abyss. 1 : 425 (1848); F.T.A. 3 : 354. A whitish-tomentose perennial herb a few inches to 1 ft. or more high with stiff leaves mostly in a basal rosette; inflorescence scapose, probably elongating during flowering, with numerous pale brown clustered flower-heads and pale yellow florets; in montane grassland.
[Br.]Cam.: Cam. Mt., 5,000–10,000 ft. (Nov.–Jan., Mar.) *Mann* 1912! *Johnston* 82! *Maitland* 482! *Brenan* 9504! 9574! *Boughey* GC 6806! 12526! Bamenda, 5,000 ft. (Jan., June) *Migeod* 294! *Maitland* 1613! Ngel Nyaki, 5,400 ft., Mambila Plateau (Jan.) *Hepper* 1721! **F.Po:** summit of the Peak *Mann* 284 (partly)! Pico Serrano (Jan.) *Guinea* 1974! Moka (Dec., Jan.) *Guinea* 2188. *Monod* 10499! Also in Congo, Rhodesia and Angola.

10. **H. albiflorum** *Moeser* in Engl. Bot. Jahrb. 44 : 269 (1910). An erect perennial herb 2–3 ft. high with the stems and the leaves on both surfaces pale greyish-tomentose; flower-heads numerous, about ⅛ in. broad, tightly clustered in a dense terminal corymb, pale silvery-yellow or white.
S.Nig.: Obudu Plateau, 5,500 ft., Ogoja (Aug.) *Stone* 48! **[Br.]Cam.:** Bamenda Dist., 5,400–7,500 ft. (Dec., Jan., Mar., Apr.) *Boughey* GC 10467! 11156! 11431! *Morton* K125! 207! Jakiri (Jan.) *Keay* FHI 28430! Mayo Binka, 4,000 ft., Nkambe (Feb.) *Hepper* 1884! Also in Congo, Uganda and Tanganyika.

11. **H. rhodolepis** *Bak.* in Kew Bull. 1898 : 150. An erect perennial herb with simple whitish-tomentose stems 2–3 ft. high; flower-heads numerous in a dense terminal cluster, about ⅓ in. broad, with the inner involucral bracts silvery and the outer reddish; florets cream-coloured.
[Br.]Cam.: Kuk, 4,000 ft., Bamenda (Apr.) *Maitland* 1701! Ndop to Kumbo (Dec.) *Boughey* GC 11167! Crater L., Wum (Apr.) *Morton* K291! Also in Cameroun, Sudan, Congo, Uganda, Kenya, Tanganyika and Nyasaland.

12. **H. nudifolium** (*Linn.*) *Less.* Syn. Gen. Comp. 299 (1832). *Gnaphalium nudifolium* Linn. (1753).

12a. **H. nudifolium** var. **leiopodium** (*DC.*) *Moeser* l.c. 266 (1910). *H. leiopodium* DC. (1838). A perennial herb with a rosette of long-stalked strongly veined leaves, variably tomentose below, glabrescent above; flower-heads bright yellow in a dense terminal corymb; in montane grassland.

Fig. 256.—Helichrysum cameroonense *Hutch. & Dalz.* (Compositae).
A, involucral bract.   B, flower.   C, pappus-bristle.

265

S.L.: Bintumane Peak, 5,000–6,000 ft. (Jan., May) *Glanville* 470! *Deighton* 5094! Loma Mts. (Feb.) *Jaeger* 4191! [**Br.**]**Cam.**: Bamenda Dist., 5,000 ft. *Migeod* 478! *Morton* K55! Lakom (Apr.) *Maitland* 1483! Banso (Jan.) *Keay* FHI 28456! Bafut-Ngemba F.R. (Dec., Feb., Apr.) *Ujor* FHI 30008! *Hepper* 2156! *Boughey* GC 11450! Maisamari, 5,000 ft., Mambila Plateau (Jan.) *Hepper* 2800! Also in Ethiopia, Congo and extending to Nyasaland and S. Africa.

13. **H. mechowianum** *Klatt* in Ann. Nat. Hofmus. Wien 7 : 101 (1892). *H. hoepfnerianum* Vatke (1896). *H. congolanum* Schltr. & O. Hoffm. (1903). A perennial herb with flowering scapes often branched at or below ground level, 3–8 in. high, appearing before the leaves ; leaves in a rosette ; flower-heads about ⅓ in. long, numerous in a dense terminal cyme, rich yellow, often tinged red ; in grassland, flowering after fires.

Togo Rep.: Agome (Mar.) *Schlechter* 12960! **N.Nig.**: Jos Plateau (Jan.) *Lely* P21! Jos (Mar.) *Hill* 11! Vom *Dent Young* 151! [**Br.**]**Cam.**: Ndop to Kumbo (Dec.) *Boughey* 11160! Kakara, 4,700 ft., Mambila Plateau (Jan.) *Hepper* 1782! Also in Cabinda, Congo, Tanganyika, Rhodesia and Angola.

53. GNAPHALIUM Linn.—F.T.A. 3 : 341.

Involucre spreading widely in fruit exposing the receptacle ; annual herbs :
  Capitula in rounded clusters, mostly at the ends of upper branches ; receptacle 1·5 mm. diam. ; stem usually erect, but occasionally branched low down and spreading from the base ; lower leaves spathulate ; upper leaves linear-oblanceolate, up to 8 cm. long ..   ..   1. *luteo-album*
  Capitula in elongated clusters at the ends of terminal and numerous lateral ascending branches ; receptacle 1 mm. diam. ; whole plant usually less than 10 cm. high with decumbent branches spreading radially ; rosette and stem-leaves spathulate, 2–3 cm. long   ..   ..   2. *indicum*
Involucre campanulate in fruit ; perennial herb ; capitula about 3·5 mm. broad with acute or acuminate buff-coloured involucral bracts ; leaves linear-lanceolate, about 6 cm. long, glabrescent above, absent from the lower parts of the shoot at flowering time   ..   ..   3. *undulatum*

1. **G. luteo-album** *Linn.* Sp. Pl. 851 (1753) ; F.T.A. 3 : 343 ; Chev. Bot. 366 ; Berhaut Fl. Sén. 205. A herb of variable habit, usually annual, with erect or straggling grey-green cottony stems up to 2 ft. high ; florets pale dull yellow, involucral bracts silvery, in heads about ⅛ in. across, diffuse or tightly clustered ; in sandy waste places, especially by streams and on riverbanks, or as a weed of farms in hilly districts, or in periodically burnt grassland (where the plants are often larger and tend to a perennial habit).

Sen.: Dakar (June) *Adam* 12229! **Mali**: Toguéré de Saredina (May) *Davey* 95! Dogo (Apr.) *Davey* 580! Gao (Feb.) *de Wailly* 5337! **Guin.**: Kaba to Upper Mamou (May) *Chev.* 13266! Macenta (Mar.) *Jac.-Fél.* 1593! Kankan (Apr.) *Adam* 12150! **Ghana**: White Volta R., near Bawku (Apr.) *Morton* GC 8951! **N.Nig.**: Sokoto (Jan.–Feb.) *Dalz.* 408! 409! 410! Vom (Apr.) *Dent Young* 150! *Morton* K327! Kano (Mar.) *Onyeagocha* FHI 7639! *Onwudinjoh* FHI 22361! Aboh *Barter* 194! Birnin Gwari (Mar.) *Meikle* 1334! [**Br.**]**Cam.**: Cam. Mt., Buea, 3,000–3,500 ft. (Nov., Feb.–Apr.) *Dalz.* 8229! *Migeod* 110! *Maitland* 513! *Morton* K471! *Brenan & Jones* 9584! Mamfe (Jan.) *Lobe Babute* FHI 12979! Bamenda Dist., 5,800–6,600 ft. (Dec., Mar., Apr.) *Boughey* GC 10875! *Adams* GC 11224! *Morton* K284! Bama, Dikwa Dist. (Jan.) *McClintock* 176! **F.Po**: (Jan.) *Guinea* 2189. Moka L. (Dec.) *Boughey*! Widespread in warm countries.

2. **G. indicum** *Linn.* Sp. Pl. 852 (1753) ; Chev. Bot. 366. *G. niliacum* Spreng. (1826)—F.T.A. 3 : 344 ; F.W.T.A., ed. 1, 2 : 161 ; Berhaut Fl. Sén. 205. *G. spathulatum* Lam. (1797)—Berhaut Fl. Sén. 205. A small spreading herb with numerous silvery-woolly flower-heads crowded in spike-like clusters ; florets pale yellow ; on river-sand and in waste places in northern districts.

Sen.: *Perrottet*! *Roger* 34! Niokolo-Koba (Jan.) *Roberty* 16659! **Mali**: Petal (Feb.) *Davey* 431! Sangara (Jan.) *Chev.* 264! [**Br.**]**Cam.**: Bama, Dikwa Dist. (Jan.) *McClintock* 165! Widespread in the Old World tropics.

3. **G. undulatum** *Linn.* Sp. Pl. 852 (1753). *Helichrysum steudelii* Sch. Bip. ex A. Rich. (1848). *G. steudelii* Sch. Bip. ex Hochst. (1841). A perennial herb with woolly-tomentose glabrescent shoots up to 15 in. high, rather woody at the base ; florets dull yellow ; in hill grassland.

[**Br.**]**Cam.**: Santa Mt., 8,500 ft., Bamenda (Apr.) *Morton* K312! Bamenda to Ndop, 5,000 ft. (Dec.) *Adams* GC 11283! Nguroje to Kakara, 5,400 ft., Mambila Plateau (Jan.) *Hepper* 1761! Also in Ethiopia, Uganda, Tanganyika, Nyasaland, Rhodesia, Natal and Madagascar.
[In the southern localities this species is apparently normally annual.]

54. SPHAERANTHUS Linn.—F.T.A. 3 : 332 ; Robyns in Kew Bull. 1924 : 177–199 ; Ross-Craig in Hook. Ic. Pl., ser. 5, 6 : 1 (1954).

Stems not winged ; common involucre not visible ; leaves oblong, mostly about 2 cm. long and 1 cm. broad, auriculate, subentire, densely pilose ; heads about 1 cm. diam.
                                                                                    1. *flexuosus*
Stems winged ; leaves decurrent :
  Common involucre conspicuous ; stem-wings entire or nearly so ; leaves subentire :
    Receptacle more or less flat ; common bracts enclosing more than half of the inflorescence, broadly obovate, sharply acuminate, pubescent, with a strong midrib ; heads 6–8 mm. broad ; leaves about 4 cm. long, minutely scabrid and sparsely glandular ..   ..   2. *angustifolius*
    Receptacle subspherical ; common bracts enclosing less than half of the inflorescence, broadly obovate, mucronate, puberulous, without a strong midrib ; heads about 10 mm. broad ; leaves up to 8 cm. long, glabrescent, punctate-glandular on both surfaces ..   ..   3. *talbotii*
[Common involucre not visible ; stem-wings sharply toothed ; receptacle orbicular ;
  ⌊ heads 9–13 mm. diam. ; leaves linear to oblong-oblanceolate, subulate-mucronate, up to 10 cm. long and 2 cm. broad, with sharply subulate marginal teeth, woolly
                                                                                    4. *senegalensis*

1. **S. flexuosus** *O. Hoffm.* in Ann. Mus. Congo, sér. 4, p. x (1903) ; Ross-Craig in Hook. Ic. Pl., ser. 5, 6 : 34, t. 3505B. *S. brounae* Robyns (1924)—F.W.T.A., ed. 1, 2 : 162. A much-branched softly tomentose annual herb ; flowers purple, in heads ¼–⅓ in. across.

N.Nig.: Jos Plateau (Jan.) *Lely* P136! Gajibo, Bornu (Nov.) *H. B. Johnston* N99! [Br.]**Cam.:** Gwoza, 1,100 ft., Dikwa Div. (Dec.) *McClintock* 56! Also in Sudan and Congo.

2. **S. angustifolius** *DC.* Prod. 5 : 370 (1836) ; Ross-Craig l.c. 56, t. 3513A. *S. nubicus* Sch. Bip. ex Oliv. & Hiern in F.T.A. 3 : 335 (1877) ; F.W.T.A., ed. 1, 2 : 161. *S. lelyi* Robyns (1924). A glandular-pubescent, erect or diffuse annual herb 1–2 ft. high ; flowers pale red in heads about ⅓ in. across.
   Sen.: Casamance *Perrottet* 26. **Mali:** west of Agades *Brodie*! **Ghana:** White Volta R., Bawku to Bolga-tanga (Apr., Dec.) *Morton* GC 8961! A1354! **N.Nig.:** Kuka *E. Vogel* 39! Sokoto *Lely* 149! **Niger:** Maradi (Dec.) *Vaillant* 865! Also in Sudan. (See Appendix, p. 420.)

3. **S. talbotii** *S. Moore* in Macleod Chiefs & Cities Centr. Afr. 303 (1912), and in Cat. Talb. 56 (1913) ; Ross-Craig l.c. 65, t. 3516B. An annual herb 1–3 ft. high with glabrous, whitish stems ; flowers purple in de-pressed-globose heads about ⅓ in. across.
   N.Nig.: Dapchi to Damaturu (Nov.) *Noble* A19! Bornu (fr. June) *Onochie* FHI 23380! Birnin Kudu, Kano (Nov.) *Onwudinjoh* FHI 24022! Also in Cameroun, Sudan, Tanganyika and Nyasaland.

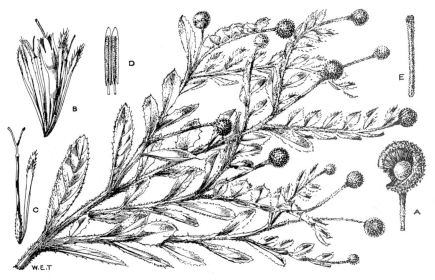

FIG. 257.—SPHAERANTHUS SENEGALENSIS *DC.* (COMPOSITAE).
A, portion of compound head. B, partial head. C, female flower. D, anthers. E, style.

4. **S. senegalensis** *DC.* Prod. 5 : 370 (1836) ; Ross-Craig l.c. 25, t. 3502B ; Berhaut Fl. Sén. 170. *S. hirtus* of F.T.A. 3 : 334 ; Holl. 3 : 384 ; Chev. Bot. 365, not of Willd. *S. lecomteanus* O. Hoffm. & Muschl. (1910)—Chev. Bot. 365. A branched annual herb 1–3 ft. high or half-decumbent ; softly whitish-hairy ; flower-heads reddish-purple about ⅓–½ in. across.
   Sen.: *Perrottet*! *Roger*! *Brunner* 113! *Farmar* 130! Labgehar (Feb.) *Roberty* 16791! Dakar (May) *Baldwin* 5702! **Gam.:** *Brown-Lester* N2! S8! *Ozanne*! *Pirie* 31! Basse (Mar.) *Todd* 11! Genieri (Feb.) *Fox* 55! **Mali:** Sienso, near San (June) *Chev.* 1042! Bamako (Apr.) *Hagerup* 26! Koussilé (Mar.) *Roberty* 17017! **Guin.:** *Heudelot* 771! **Ghana:** Dedoro-Tankara, near Navrongo (Dec.) *Adams & Akpabla* GC 4349! Burufo, near Lawra (Dec.) *Adams & Akpabla* GC 4420! **Dah.:** *Poisson*! **N.Nig.:** Kano (Nov.) *Hagerup* 623! Kontagora (Nov.) *Dalz.* 189! Kaduna (Dec.) *Meikle* 745! Jengre, 3,000 ft., Zaria (Dec.) *Coombe* 85! Also in Mozambique, N. Rhodesia and tropical Asia. (See Appendix, p. 420.)

   [Note : *Morton* GC 8796 from Lumbunga, Tamale, Ghana, bears a close resemblance to *S. gazaensis* Bremek., but it may be merely a regrowth on recently burnt-over ground from persistent stocks of *S. senegalensis*—C. D. A.]

55. BLEPHARISPERMUM Wight ex DC. in Wight Contrib. Bot. Ind. 11 (1834); F.T.A. 3 : 335.

A scrambling shrub with finely striate rusty-pubescent branchlets and a persistent deflexed thorn below the insertion of each leaf ; leaves alternate, up to about 8 cm. long and 4 cm. broad, ovate to elliptic-lanceolate, rounded or broadly cuneate at base and slightly unequal-sided, acuminate, minutely apiculate, subentire, with about 5 pairs of lateral pubescent nerves ; petiole slender, about 1 cm. long ; flower-heads compound in a spherical cluster, terminating short, stiff lateral branches ; each floret enclosed in a scale, the outer female with 4 corolla-lobes and a large style with thick tongue-shaped style arms, the inner hermaphrodite or male with 5 corolla-lobes and a style with a minutely bifid apex ; achene (formed by outer florets) flat, winged, obcordate-emarginate, ciliate, with about 6 unequal ascending-barbellate pappus-bristles in the sinus .. .. *spinulosum*

B. spinulosum *Oliv. & Hiern* in F.T.A. 3 : 335 (1877). A scrambling deciduous shrub with persistent climbing-accessory thorns at the nodes ; flowers greenish-white in spherical heads ⅓–¾ in. across, sweet-scented in thickets of secondary forest.

Iv.C.: Morénou, Eterokrou to Tchumkrou (Dec.) *Chev.* 22517 ! **Ghana:** Aburi scarp (fl. Nov.–Dec., fr. Feb.) *Morton* GC 6132 ! A1610 ! *Adams* 1917 ! Afram R., Mankrong (Apr.) *Morton* A723 ! Also in Congo and Angola.

56. AMBROSIA Linn.—F.T.A. 3 : 370.

An annual or short-lived herb ; stems pilose with whitish hairs ; leaves petiolate, deeply bipinnately divided with lobulate segments, up to about 12 cm. long, softly pubescent; flower-heads unisexual ; male involucres cup-shaped with the bracts united and crenately lobed, about 5 mm. diam., in a terminal raceme ; male florets with a corolla ; female capitula 1-flowered in axillary clusters, the involucre angular in fruit with 4–6 horns ; female floret without a corolla ; achene obovoid, smooth, without a pappus..   ..   ..   *maritima*

A. **maritima** *Linn.* Sp. Pl. 988 (1753) ; F.T.A. 3 : 370 ; Berhaut Fl. Sén. 27, 48. *A. senegalensis* DC. (1836)— F.T.A. 3 : 371 ; Chev. Bot. 368 ; Trochain Mém. I.F.A.N. 2 : 364 (1940). An erect, branched aromatic herb 1–4 ft. high, often woody below and sometimes almost shrubby, with striate stems, more or less hoary ; flower-heads small, greenish-yellow forming rather dense spikes ; flowering during the rains. **Sen.:** *Perrottet* ! **Mali :** San (June) *Chev.* 1023 ! Djenné (June) *Chev.* 1114 ! Kabarah (Aug.) *Chev.* 1540 ! Timbuktu *Hagerup* 198a ! **Ghana:** White Volta R., near Bawku (June) *Hepper & Morton* A3159 ! *T. M. Harris* ! **N.Nig.:** Borgu *Barter* 837 ! Bornu *E. Vogel* 87 ! Fodama (Nov.) *Moiser* 205 ! 244 ! Jebba (June) *Meikle* 1280 ! *Onochie* FHI 18674 ! Maiduguri (June) *Ujor* FHI 21939 ! Distributed from the Mediterranean region through Sudan, eastern and S. Africa to the Mascarene Islands ; also Principe and S. Tomé. (See Appendix, p. 415).

57. ELEPHANTOPUS Linn.—F.T.A. 3 : 298.

Flower-heads narrow in slender spikes ; involucral bracts narrowly lanceolate, keeled, acute ; achenes slightly pubescent ; pappus-bristles few, minutely barbellate, one on each side of the achene longer and vertically folded ; leaves linear to oblanceolate, those on the upper part of the stem about 2 cm. long and 2–3 mm. broad with reflexed margins, closely and minutely gland-dotted beneath, pilose on the nerves   1. *spicatus*
Flower-heads broad, surrounded by divergent leaf-like bracts, paniculate or capitate ; leaf-veins impressed above ; leaves shallowly crenate ; achenes pubescent :
Inflorescence a leafy panicle of pedunculate heads ; common involucral bracts leafy, broadly ovate, bluntly acuminate ; leaves more or less obovate-oblanceolate, acutely acuminate, sheathing at base, up to 15 cm. long and 5 cm. broad, scabrid above, thinly and shortly pilose beneath and minutely glandular ; involucral bracts with sharp setose points ; pappus of 5–6 straight bristles   ..   ..   2. *mollis*
Inflorescence a single terminal cluster of heads surrounded and exceeded by leaves ; leaves linear-lanceolate, 10–15 cm. long, 1·5–3 cm. broad, pilose on both surfaces or almost glabrous beneath, minutely glandular beneath ; involucral bracts ribbed, gradually pointed ; pappus of numerous straight bristles   ..   3. *senegalensis*

1. **E. spicatus** *B. Juss. ex Aubl.* Pl. Guian. 2 : 808 (1775). *Pseudelephantopus spicatus* (B. Juss. ex Aubl.) Vahl (1792). An erect branched stiff herb 1–2 ft. or more high, more or less pilose, with small white or mauve florets in narrow oblong heads forming spikes ; a weed.

FIG. 258.—ELEPHANTOPUS MOLLIS *Kunth* (COMPOSITAE).
A, portion of plant.   B, partial flower-head.   C, flower.   D, achene.   E, anthers.   F, style-arms.

Guin.: *Farmar*! Gangan (Apr.) *Adam* 11920! Bambaya, Upper Niger (Jan.) *Jaeger* 3861! **S.L.:** Freetown (Feb.) *Johnston* 61! Leicester Peak (Dec.) *Sc. Elliot* 3863! Mt. Aureol (Dec.) *Dalz.* 945! Bo (Jan.) *Thomas* 7410! Njala (Feb.) *Deighton* 1925! **Lib.:** Dimei (Feb.) *Barker* 1216! A native of tropical America, introduced into our area and Asia.

2. **E. mollis** *Kunth* Nov. Gen. & Sp. Pl. 4 : 26 (1820) ; Philipson in J. Bot. 76 : 303–304 (1938). *E. scaber* of F.T.A. 3 : 299, partly ; of Chev. Bot. 359 ; of F.W.T.A., ed. 1, 2 : 170 ; of Berhaut Fl. Sén. 204, not of Linn. An erect branched coarsely hairy perennial herb with wrinkled leaves ; usually 2–4 ft. high, but exceptionally up to 8 ft. ; florets small, white, in bracteate heads up to ¾ in. across ; in open grassy places in woodlands, savanna, fringing forest, and sometimes cultivated land.
Sen.: *fide* Berhaut *l.c.* **Mali:** Tabacco (Oct.) *Chev.* 2817! **Guin.:** *Heudelot* 674! Kouroussa (Oct.) *Pobéguin* 572! Macenta, 2,000–2,500 ft. (Oct.) *Baldwin* 9752! **S.L.:** *Winwood Reade*! Leicester Peak (Dec.) *Sc. Elliot* 3880! Kanya (Oct.) *Thomas* 3056! Kent *Deighton* 3302! Sefadu (Nov.) *Deighton* 4663! **Lib.:** Gbanga (Oct.) *Linder* 1196! Tawata, Boporo (Nov.) *Baldwin* 10308! Cess R. (Mar.) *Baldwin* 11232! **Ghana:** Chirano to Afao, Sefwi (Dec.) *Adams* 1994! Aduamoa, Kwahu (Dec.) *Morton* A2434! Techiman, Ashanti (Dec.) *Adams & Akpabla* GC 4491! Amedzofe (Nov.) *Morton* GC 9411! Dutukpene (Dec.) *Adams* 4547! **N.Nig.:** Damisa, Madaki (Nov.) *Keay & Onochie* FHI 21716! Jos Plateau (Oct.) *Lely* P837! **S.Nig.:** Okuni, Cross R. (Jan.) *Holland* 169! Calabar (May) *Ujor* FHI 30158! Oban *Talbot*! **[Br.]Cam.:** Cam. Mt., Buea, 3,000–3,400 ft. (Jan., Mar., June, Dec.) *Dunlap* 106! *Maitland* 61! 128! 249! *Boughey* GC 7014! Victoria (Jan.) *Winkler* 676! Kumba (Jan., Mar.) *Ejiofor* FHI 29301! *Binuyo & Daramola* FHI 35463! Vogel Peak, Adamawa (Nov.) *Hepper* 1443! **F.Po:** (Jan., June) *Mann* 225! *Barter*! Extends to Sudan and Angola, and also in S. Tomé and Principe ; a tropical American plant also introduced into Asia.

3. **E. senegalensis** (*Klatt*) *Oliv. & Hiern* in F.T.A. 3 : 299 (1877) ; Chev. Bot. 359 ; Berhaut Fl. Sén. 204. *Synchodendron senegalense* Klatt (1873). An erect perennial herb 1–2 ft. high with hairy striate stems ; leaves long, narrow and rather wrinkled ; flower-heads in a terminal cluster 1–1½ in. across ; in rocky savanna grassland or woodland, uncommon.
Sen.: Niokolo-Koba *Berhaut* 1589! **Guin.:** *Heudelot* 646! Sambadougou, Faranna (Jan.) *Chev.* 20545! Fouta Djalon, 5,000 ft. (Sept.) *Schnell* 7168! **S.L.:** Musaia (Oct.) *Thomas* 2659! **Ghana:** Banda, Ashanti (Dec.) *Morton* GC 25148! Nwereme, Ashanti (Dec.) *Morton* A2637! Adibu *Goodall* GC 16590! Amedzofe (Nov.) *Morton* A2298! **N.Nig.:** Jos Plateau *Batten-Poole* 176! **S.Nig.:** Lagos *W. D. MacGregor* 158! Also in Angola.

58. STRUCHIUM P. Br. Hist. Jamaic. 312, t. 34 (1756). *Sparganophorus* Crantz (1766) —F.T.A. 3 : 262.

Stems thinly pubescent ; leaves obovate-oblanceolate, acute, narrowed into the winged petiole, 7–15 cm. long, 2·5–6 cm. broad, crenate-serrate, minutely gland-dotted and slightly pubescent beneath ; flower-heads sessile, clustered in the leaf axils, 5–7 mm. broad ; involucral bracts in several rows, ovate, very acute, pilose, about 5 mm. long ; achenes 3–5-angled, glabrous, glandular, crowned by a short tubular obscurely toothed cartilaginous pappus   ..   ..   *sparganophora*

S. sparganophora (*Linn.*) *O. Ktze.* Rev. Gen. Pl. 1 : 366 (1891). *Ethulia sparganophora* Linn. (1763). *Sparganophorus vaillantii* Crantz (1766)—F.T.A. 3 : 262 ; Chev. Bot. 350 ; F.W.T.A., ed. 1, 2 : 171 ; Berhaut Fl. Sén. 175. *Struchium africanum* P. Beauv. Fl. Oware 1 : 81, t. 48 (1806). An erect or decumbent half-succulent herb 1–4 ft. high, with white, mauve or pink florets ; in damp places.
Sen.: Casamance (Jan.) *Chev.* 2055. **Mali:** Négala (Jan.) *Chev.* 164. **Port.G.:** Pessubé, Bissau (Mar.) *Esp. Santo* 1498! **Guin.:** *Heudelot* 656! **S.L.:** Mofari, Digisinn (Jan.) *Sc. Elliot* 4405! Tombo (Jan.) *Deighton* 958! Kambia, Magbema (Dec.) *Jordan* 760! Port Loko (Oct.) *Tindall* 45! Ronietta (Nov.) *Thomas* 5341! **Lib.:** Gletown, Tchien (July) *Baldwin* 6802! Genna Tanyehun, Grand Cape Mount Co. (Dec.) *Baldwin* 10767! Tappita (Aug.) *Baldwin* 9085! **Iv.C.:** Orumbo-Boka Mt., Toumodi (Dec.) *Boughey* GC 14443! Pokoase (Apr.) *Irvine* 1593! Nsawkaw, Ashanti (Dec.) *Adams* 3175! Kpandu *Robertson* 39! **Togo Rep.:** Lomé *Warnecke* 412! **Dah.:** Porto-Novo (Mar.) *Chev.* 23310! **S.Nig.:** Ogun R., Abeokuta *Barter* 3341! Okomu F.R., Benin (Dec.) *Brenan & Jones* 8584! Ibadan (Jan.) *Coombe* 120! Ijebu (Oct.) *Tamajong* FHI 20269! Oban *Talbot* 399! **[Br.]Cam.:** *Preuss* 1351! Victoria (Feb.) *Maitland* 358! Tiko (Feb.) *Maitland* 977! **F.Po:** (June) *Barter*! *Mann* 408! Ureka (Feb.) *Guinea* 2369. Throughout tropical Africa and also in S. Tomé, Principe, the Mascarene Islands, S.E. Asia and tropical America. (See Appendix, p. 420.)

59. TRIPLOTAXIS Hutch. in Kew Bull. 1914 : 355.

Stems herbaceous, slender, puberulous ; leaves shortly petiolate, ovate or ovate-lanceolate, acute or subacute, 2–6 cm. long, 1–3 cm. broad, serrate-crenulate to sub-entire, crispate-puberulous especially on the nerves ; flower-heads very small in lax divaricately branched corymbs ; peduncles slender 5–15 mm. long, shortly pubescent ; involucral bracts 3-seriate, free, pubescent ; achenes turgid, sparingly pilose ; pappus cupular, pectinately toothed   ..   ..   *stellulifera*

T. stellulifera (*Benth.*) *Hutch.* in Kew Bull. 1914 : 356 (1914). *Herderia stellulifera* Benth. (1849)—F.T.A. 3 : 298 ; Chev. Bot. 358. An erect or decumbent annual herb 1–2 ft. high with mauve or rarely white florets in flower-heads less than ¼ in. long ; a weed of pathsides and clearings in forest areas.
Guin.: Boula (Apr.) *Adam* 12065! **S.L.:** Heddle's Farm, Freetown (Dec.) *Sc. Elliot* 3933! Kenema (Nov.) *Deighton* 396! Njala (June) *Deighton* 3235! Rowala (July) *Thomas* 1144! Rokupr (Sept.) *Jordan* 315! **Lib.:** Du R. (July) *Linder* 127! Ganta (May) *Harley*! Gbawia, Webo (July) *Baldwin* 6716! Gletown, Tchien (July) *Baldwin* 6738! Tawata, Boporo (Nov.) *Baldwin* 10309! **Iv.C.:** *fide* Chev. *l.c.* Adiopodoumé (July) *Boughey* GC 14523! Abidjan (Dec.) *Boughey* GC 13516! Guitry (Dec.) *Boughey* GC 13550! Guiglo (Aug.) *Boughey* GC 14795! **Ghana:** Princes Town (Mar.) *Morton* A350! Amentia, Ashanti *Irvine* 533! Abetifi, 2,000 ft. (Jan.) *Irvine* 1660! Enchi (Dec.) *Adams* 2218! Amedzofe, Trans-Volta Togo (Nov.) *Morton* GC 9422! **S.Nig.:** Ishagama (Mar.) *Schlechter* 12309! Sapoba (Apr., June) *Onyeagocha* FHI 7141! *Ejiofor* FHI 32013! Ubiaja F.R., (Aug.) *Onochie* FHI 33257! Okomu F.R., Benin (Jan., Feb.) *Brenan* 8776! 9151! **[Br.]Cam.:** Rio del Rey *Johnston*! Tiko (Feb.) *Maitland*! Buea, 3,000–3,600 ft. (Mar., Apr., June, Nov., Dec.) *Maitland* 34! *Migeod* 99! *Hutch. & Metcalfe* 125! *Akpabla* GC 10493! Path to Mann's Springs (Dec.) *Adams* GC 11798! Ndop, 3,600 ft., Bamenda (Dec.) *Boughey* GC 11125! **F.Po:** (June, Dec.) *Barter*! *T. Vogel* 265! *Boughey* GC 10976b! S. Isabel (Dec.) *Guinea* 549! Musola (Jan.) *Guinea* 1316. Also in Uganda and through Gabon and Congo to Angola ; and in Principe and S. Tomé. See Appendix, p. 421.

Fig. 259.—Aedesia baumannii *O. Hoffm.* (Compositae).
A, flower. B, anthers. C, style-arms. D, achene. E, pappus-bristle.

60. AEDESIA O. Hoffm. in E. & P. Pflanzenfam. Nachtr. 1 : 321 (1897), and in Engl. Bot. Jahrb. 24 : 467 (1898).

Leaves lanceolate, rather abruptly acute, about 15 cm. long and 2·5 cm. broad at base, serrulate, glabrous, with about 15 pairs of lateral nerves ; flower-head about 5 cm. diam., solitary, closely surrounded and over-topped by the upper leaves ; involucral bracts 1– 2-seriate, linear-spathulate, 3–4 cm. long, glabrous ; achenes ribbed, pubescent ; pappus buff, 2·5–3 cm. long   ..   ..   1. *glabra*
Leaves very similar to above but much narrower and elongated to a very narrow apex ; flower-heads about 3 cm. diam.   ..   ..   2. *baumannii*

1. **A. glabra** (*Klatt*) O. *Hoffm.* in Engl. Bot. Jahrb. 24 : 468 (1898) ; Chev. Bot. 381 ; Berhaut Fl. Sén. 170. *Bojeria glabra* Klatt (1873). *A. latifolia* A. Chev. Bot. 381, name only. A perennial herb with erect simple stems 1–2 ft. high from a stout knotted rootstock ; florets purple ; in damp grassland.
    Sen.: Tombana, Casamance (Feb.) *Chev.* 3538 ! **Guin.:** Kouria to Ymbo (July) *Chev.* 14613 *bis* ! Tonflli to Tagania, Faranna (Jan.) *Chev.* 20399 ! **S.L.:** Gonkwi Mt., Talla Plateau (Feb.) *Sc. Elliot* 4985 ! Kumrabai (Sept.) *Glanville* 442 ! Musaia (Sept.) *Deighton* 4893 ! Kakama (Aug.) *Adames* 197 ! Loma Mts. (Sept.) *Jaeger* 2015 ! **Iv.C.:** Siana to Nandala, Mankono (June) *Chev.* 21858 ! Keoulenta (July) *Schnell* 1542. **N.Nig.:** Vom *Dent Young* 143 ! Also in Cameroun and Congo.
2. **A. baumannii** O. *Hoffm.* l.c. (1898) ; Chev. Bot. 380. An erect perennial herb about 1 ft. high ; florets deep reddish-purple or (?) yellow in a solitary terminal head much overtopped by the involucre ; in savanna grassland.
    Mali: *fide* Chev. *l.c.* **Iv.C.:** Dotou (May) *Chev.* 21765 ! Languira, Baoulé-Nord (July) *Chev.* 22202 ! **Ghana:** Salaga *Easmon* ! Yeji *Anderson* 49 ! Bosomoa F.R. (June) *Vigne* FH 3056 ! Banda, Ashanti (Dec.) *Morton* GC 25081 ! Burufu, Lawra (Dec.) *Adams & Akpabla* GC 4429 ! Yendi (July) *Akpabla* GC 17347 ! **N.Nig.:** Borgu *Barter* 713 ! Yola (July) *Dalz.* 28 ! Naraguta (July) *Lely* 253 ! Kogigiri, 4,100 ft., Jos (Oct.) *Hepper* 1062 ! **S.Nig.:** Lagos *W. MacGregor* 174 ! Also in Cameroun.

61. HERDERIA Cass.—F.T.A. 3 : 297, partly (excl. *H. stellulifera*).

Stems numerous, procumbent or ascending, variably woolly-pubescent ; leaves spathulate-obovate, narrowed to base, rounded or subtruncate at apex and there sparingly dentate, 1–3(–4) cm. long, 5–15 mm. broad, thinly pubescent or villous when young, glandular beneath ; flower-heads broadly campanulate, about 6 mm. long and broad ; outer involucral bracts leafy, oblanceolate, inner connate to above the middle ; florets all tubular ; achenes 4-angled, glabrous ; pappus-setae in two rows, the outer scale-like, the inner 3–5, longer and barbellate ..   ..   *truncata*

H. truncata *Cass.* Dict. Sci. Nat. 60 : 599 (1830) ; F.T.A. 3 : 298 ; Hutch. in Kew Bull. 1914 : 354, with fig. ; Chev. Bot. 358 ; Berhaut Fl. Sén. 173, 198. *H. truncata* var. *chevalieri* O. Hoffm. (1908)—Chev. Bot. 359. *Vernonia yatesii* S. Moore (1914). A spreading annual herb, woody at the base ; florets reddish-purple to mauve ; in damp sandy places.
    Sen.: *fide* Berhaut *l.c.* **Gam.:** *Hayes* 541 ! **Mali:** Sébi (July) *Chev.* 1165 ! Guguber *Hagerup* 318 ! Korienza (July) *Lean* 54 ! **Port.G.:** Boé, Vehetche (June) *Esp. Santo* 2933 ! **Guin.:** Kankan *Adam* 12136 ! **S.L.:** Batkanu (Jan.) *Glanville* 151 ! Pujehun (Apr.) *Deighton* 1654 ! Makeni (May) *Deighton* 4141 ! Mange, Bure (May) *Jordan* 868 ! **Iv.C.:** *fide* Chev. *l.c.* **Ghana:** Yeji *Anderson* 28 ! Babile, near Lawra (Oct.) *Hinds* 5001 ! Yendi (Dec.) *Adams & Akpabla* GC 4085 ! Akuse to Kpong (June) *Irvine* 1778 ! Tefle (Jan., May) *Morton* GC 8313 ! A2069 ! **N.Nig.:** Aboh *Barter* 308 ! Abinsi (May) *Dalz.* 661 ! Yola (May) *Dalz.* 37 !
    A variety (var. *villosa* DC. Prod. 5 : 14 (1836) ; Berhaut Fl. Sén. 174, 204) with the leaves densely villous with appressed hairs, has been recorded from Senegal (*Perrottet*).

62. VERNONIA Schreb.—F.T.A. 3 : 266. *Nom. cons.*

*Trees, shrubs or woody climbers (to p. 272) :
Involucral bracts with coloured or whitish appendages ; capitula 1·5–3 cm. or more diam. ; achenes often black and ribbed ; pappus-setae more or less uniform, usually deciduous and buff in colour ; leaves often toothed ; erect shrubs or small trees :
  Leaves sessile :
    Leaves narrowed to base without auricles ; pappus persistent with the outer setae shorter ; achenes minutely pubescent   ..   ..   ..   1. *kotschyana*
    Leaves auricled ; pappus caducous, more or less uniform ; achenes glabrous :
      Capitula 3–4 cm. diam. ; leaves up to 30 cm. long and 10 cm. broad, thinly pubescent beneath ; branches shortly velvety   ..   ..   2. *calvoana* var. *calvoana*
      Capitula 1–2 cm. diam. ; leaves up to 15 cm. long and 7 cm. broad, densely pubescent beneath ; branches thinly tomentose ..   2a. *calvoana* var. *microcephala*
  Leaves petiolate :
    Capitula large, 3 cm. or more diam. in flower ; pappus caducous :
      Leaves thinly pubescent beneath ; achenes glabrous ; pappus uniform ; lamina cordate-auriculate at junction with petiole   ..   ..   3. *insignis*
      Leaves woolly-tomentose beneath ; achenes pubescent ; outer pappus very short :
        Involucral bracts softly tomentellous outside, linear ; leaves rather small, irregularly dentate   ..   ..   4. *stenostegia*
        Involucral bracts puberulous, lanceolate ; leaves up to 20 cm. long and 6 cm. broad, serrate ..   ..   5. *tenoreana*
    Capitula smaller, hardly exceeding 2·5 cm. diam. in flower :
      Lamina tapered to base ; leaf-margin serrate :
        Middle involucral bracts obtuse or rounded :
          Leaves tomentose beneath ; lamina-base decurrent on the petiole ; achenes glabrous or nearly so   ..   ..   6. *mokaensis*

Leaves pubescent and glandular beneath ; lamina-base very narrowly cordate at the junction with the petiole ; achenes sparsely pubescent

7. *leucocalyx* var. *leucocalyx*

Middle involucral bracts acute ; leaves thinly tomentose beneath ; lamina-base broadly cordate at the junction with the petiole ; achenes minutely puberulous

7a. *leucocalyx* var. *acuta*

Lamina broadly rounded at base ; leaf-margin coarsely dentate ; leaves tomentose beneath, puberulous only above ; middle involucral bracts acute ; achenes pubescent or minutely puberulous ; pappus-setae of variable length, the outer shorter .. .. 8. *iodocalyx*

Involucral bracts not appendaged :

Capitula about 2 cm. long, sessile or nearly so, each subtended by a tomentose foliaceous bract ; involucral bracts lanceolate, acute ; leaves very large, sessile, fimbriately lobed near base, undulate-margined, densely tomentose beneath ; achenes glabrous ; pappus uniform ; a tree .. .. 9. *frondosa*

Capitula with involucres hardly exceeding 1 cm. in length, without foliaceous bracts ; involucral bracts more or less ovate or oblong :

Climbers or scramblers :

Leaves ovate, acutely acuminate ; involucral bracts coloured ; florets purple :

Inner involucral bracts not exceeding 5 mm. in length, more or less narrowed to the tip ; outer pappus-setae very short, distinctly fimbriate, scale-like and spreading at maturity.. .. .. 10. *biafrae*

Inner involucral bracts 6–7 mm. long, broadly rounded at the tip ; outer pappus-setae very little broader than the inner and hardly scale-like ; leaves often sinuate-margined .. .. 11. *tufnelliae*

Leaves broadly ovate, rounded at apex, entire ; petiole slender, about 1 cm. long, leaving a peg on the stem after leaf-fall ; achenes glandular, glabrous or nearly so; involucral bracts green ; florets white, 10–11 in each capitulum .. 12. *andohii*

Erect shrubs or small trees :

Leaves petiolate :

Pappus 2-seriate, the outer setae shorter ; florets mauve or white :

Achenes pubescent ; leaf-margin serrate :

Petiole auricled at base ; capitula with a single floret .. 13. *auriculifera*

Petiole not auricled at base ; florets several to each capitulum ; leaves up to 30 cm. long and 10 cm. broad, softly pubescent when young, glabrescent :

Leaves long-cuneate to base ; florets 3–5 in each capitulum ..14. *myriantha*

Leaves abruptly narrowed to the petiole ; florets about 7 in each capitulum

15. *ampla*

Achenes glabrous ; involucres up to 6 mm. long :

Branches of inflorescence short, flexuous, ascending, bearing capitula in dense rounded clusters ; leaves abruptly narrowed to base, acute at apex, puberulous only on the veins beneath .. .. 16. *thomsoniana*

Branches of inflorescence spreading laterally with capitula distinctly separated, or not in large clusters ; leaves cuneate, densely pubescent beneath :

Lateral branches of inflorescence hardly exceeding 8 cm. in length ; capitula on peduncles over 5 mm. long or also a few clustered at the tips of branches ; leaves oblanceolate, up to 15 cm. long and 5 cm. broad, entire, obtuse

17. *doniana*

Lateral branches of inflorescence wide-spreading, 15 cm. or more long ; capitula subsessile or on peduncles not exceeding 5 mm. in length ; leaves very large, sinuate-margined .. .. 18. *conferta*

Pappus more or less uniform ; florets white :

Lamina rather abruptly narrowed to base, entire, undulate ; achenes glabrous, shallowly 10-ribbed, densely glandular between the ribs .. 19. *colorata*

Lamina long-cuneate to base, remotely toothed, flat ; achenes pubescent

20. *amygdalina*

Leaves sessile, the lamina continuous with stem-clasping auricles ; capitula mostly sessile along the inflorescence branches or clustered at their tips ; achenes pubescent :

Inflorescence subumbellate with numerous small capitula clustered at branch-tips ; florets 3 ; involucral bracts 3-seriate, glumaceous, glabrescent ; outer pappus-setae shorter and more or less united at base .. .. 21. *subuligera*

Inflorescence paniculate ; involucral bracts 4–5-seriate, puberulous ; pappus-setae uniform :

Florets 5–8 ; involucres about 5 mm. long ; inner involucral bracts mostly obtusely rounded at apex without broad hyaline margins 22. *theophrastifolia*

Florets about 10 ; involucres about 7 mm. long ; inner involucral bracts mostly acute at apex, with broad hyaline margins .. .. 23. *richardiana*

*Herbs or undershrubs :

Flower-heads appearing before or separately from the leaves ; stem arising from a perennial stock :
> Apex of stock at most whitish-tomentose and scaly with reduced leaves ; roots fusiform, fleshy ; involucral bracts greyish-green, 4–5-seriate, thinly woolly outside, up to 1·3 cm. long, rounded at apex ; leaves lanceolate, variable in size up to 15 cm. long and 6 cm. broad, serrulate, pustulate-scabrid above, densely tomentose beneath ; pappus-bristles flat, the outer shorter ; achenes strongly ribbed, pubescent and glandular   ..   ..   51. *pumila*
> Apex of stock densely brownish-woolly ; roots fibrous ; involucral bracts purplish, slightly pubescent outside ; leaves (when present) elongate-oblanceolate, 20 cm. or more long, 5–6 cm. broad, petiolate, minutely denticulate or repand-dentate, with numerous lateral nerves ; pappus with the outer bristles successively shorter but not much broader or flatter than the inner ; achenes pubescent :
>> Capitula solitary or tightly clustered on very short peduncles arising directly from the stock ; involucral bracts linear-lanceolate, up to 1·5 cm. long, the inner with long-acute crisped apex ; leaves softly pubescent beneath ..   24. *chthonocephala*
>> Capitula several, long-pedunculate in a branched almost leafless inflorescence ; involucral bracts orbicular to oblong and rounded at apex, less than 1 cm. long ; leaves puberulous beneath   ..   ..   25. *subaphylla*

Flower-heads accompanying the leaves :
> Capitula solitary or subsolitary, terminal on simple leafy or leafless scapes ; achenes pubescent :
>> Head subsessile, closely surrounded and overtopped by 5 or 6 of the uppermost leaves   ..   ..   26. *chapmanii*
>> Head distinctly pedunculate :
>>> Leaves mostly in a basal rosette, oblanceolate, up to 12 cm. long and 4 cm. broad, tomentose when young ; apex of the stock brown-woolly :
>>>> Involucral bracts 2–3-seriate, ovate, acuminate, up to 1·5 cm. long and 7–8 mm. broad, purplish ; peduncle terete ; achenes 3–4 mm. long, ribbed ; outer pappus of short fimbriate scales, the inner shortly plumose   27. *acrocephala*
>>>> Involucral bracts 3–4-seriate, lanceolate, acute, up to 2 cm. long and 4 mm. broad, not purplish ; peduncle ribbed ; pappus uniform, the outer setae shorter
>>>>   28. *gerberiformis*

Leaves mostly on the subaerial shoot :
> Capitulum 3·5–5 cm. diam. ; involucral bracts 4–5-seriate, lanceolate or narrowly oblong :
>> Leaves densely tomentose beneath   ..   ..   ..   52c. *guineensis* var. *procera*
>> Leaves glabrous or puberulous beneath :
>>> Leaves oblong-obovate, rounded at apex, up to 20 cm. long and 7 cm. broad, obscurely denticulate, glabrous ; involucral bracts glabrous, long-apiculate ; outer pappus shorter and slightly broader than the inner   29. *macrocyanus*
>>> Leaves elliptic, acute, up to 6 cm. long and 2 cm. broad, serrate, puberulous ; outer pappus of slender setae, shorter than inner ; inner pappus-setae distinctly broadened and fimbriate towards apex   ..   ..   30. *nimbaensis*
> Capitulum up to about 2 cm. diam. ; involucral bracts 2–3-seriate, broadly ovate, purplish ; leaves elliptic, up to 4 cm. long and 1·5 cm. broad, finely serrate ; pappus as in No. 30   ..   ..   31. *jaegeri*

Capitula several to numerous in branched inflorescences :
> Innermost involucral bracts very long and tinged crimson on the inner surface, about twice as long as the pappus and achene ; outer involucral bracts spreading down the peduncle ; stems woolly at base ; leaves obovate or obovate-lanceolate, cuneate at base, up to 15 cm. long and 7 cm. broad, denticulate, thick, harshly scabrid on both surfaces ; achenes densely tomentose   ..   ..   32. *nigritiana*
> Innermost involucral bracts not much longer than the others or the pappus, not coloured :
>> Flower-heads sessile subtended by broad leafy bracts ; leaves scabrid on both surfaces, glandular-punctate beneath, linear-lanceolate to elliptic, acute, entire or nearly so, up to 18 cm. long and 4 cm. broad ; achenes villous :
>>> Capitula exceeding 1 cm. diam., widely campanulate, few ; involucral bracts sharply apiculate, arachnoid-pubescent   ..   33. *purpurea* var. *purpurea*
>>> Capitula smaller, rounded at the base ; leafy bracts rather small ; involucral bracts tomentose   ..   ..   33a. *purpurea* var. *schnellii*
>> Flower-heads not closely subtended by leafy bracts :
>>> Capitula sessile on elongated spike-like inflorescence branches ; involucre tomentose but occasionally glabrescent, about 1 cm. long ; leaves oblong or elliptic, apiculate, 5–15 cm. long, 2·5–6 cm. broad, scabrid above, woolly-tomentose beneath, nervose ; outer pappus of very short setae ; achenes pubescent
>>>   34. *camporum*

> Capitula corymbose or paniculate :
>> †Outermost involucral bracts linear-filiform ; pappus uniform ; achenes pubescent :

Outer bracts nearly as long as the intermediate ones, slender ; leaves narrowly
  lanceolate, up to 2 cm. broad, distantly serrate..   ..   .. 35. *pauciflora*
Outer bracts rather short ; leaves elliptic, up to 18 cm. long and 6 cm. broad,
  serrate or dentate ; involucres up to 2 cm. long :
Stems glabrescent or setulose ; peduncles glabrescent except at the extreme
  apex ; leaves elliptic-lanceolate, up to 10 cm. long and 2·5 cm. broad,
  abruptly narrowed into a very short petiole, sharply but not deeply serrate,
  harshly crisped-setulose on the veins beneath, rigid
                                                36a *glabra* var. *occidentalis*
Stems pubescent ; peduncles dark-brownish pubescent ; leaves elliptic, up to
  18 cm. long and 6 cm. broad, long-cuneate to a short petiole, remotely and
  shallowly dentate, setulose on the midrib and veins beneath, minutely
  glandular and sparsely pubescent on the lamina, flaccid ; capitula usually
  few ..   ..   36b. *glabra* var. *hillii*
†Outermost involucral bracts not linear-filiform, rather short :
Leaves linear to linear-lanceolate, up to 5 mm. broad; achenes pubescent ;
  pappus whitish or buff :
Leaves linear-acicular with recurved margins ; flower-heads campanulate,
  1–1·5 cm. long ; involucral bracts acute, villous :
Pappus 2-seriate, the outer of short scales, the inner shortly plumose ; leaves
  1–3 cm. long, scabrid-setulose ; achenes 5–6-ribbed, pubescent on the ribs,
  glandular between ; annual herb   ..   ..   ..   37. *perrottetii*
Pappus more or less uniform, the outer setae shorter but not broader than the
  inner ; leaves 1–1·5 cm. long, pubescent ; perennial herb   38. *bauchiensis*
Leaves linear-lanceolate with numerous gland-pits beneath, 1–3 cm. long,
  more or less flat ; achenes uniformly pubescent ; pappus 2-seriate, inner
  setae shortly plumose ; flower-heads up to 1 cm. long :
Outer pappus of very short scales ; capitula rather few on slender peduncles
  1–2 cm. long ; involucral bracts lanceolate, acute, keeled with a distinct
  median nerve ; leaves glabrescent beneath ; annual herb
                                                39a. *poskeana* var. *elegantissima*
Outer pappus of short setae ; capitula numerous on peduncles up to 4 cm.
  long ; involucral bracts oblanceolate, obtuse or triangular-tipped, 3-nerved ;
  leaves crisped-pubescent beneath ; perennial herb   40. *plumbaginifolia*
Leaves more than 5 mm. broad :
Pappus purple-tipped ; leaves broadly oblanceolate, acute, 6–7 cm. long,
  2–2·5 cm. broad, serrate at least in the upper half, glabrous, coarsely reticulate-veined ; involucral bracts broadly linear-lanceolate, villous ; capitula
  densely corymbose ; achenes silky ; outer pappus inconspicuous
                                                41. *djalonensis*

Pappus white, yellow, buff or brown :
Capitula elongate-ovoid or turbinate, distinctly longer than broad, with
  involucral bracts in many series ; shrubby plants with numerous upper
  stem-leaves and a leafy inflorescence ; leaves narrowly elliptic to lanceolate,
  4–8 cm. long ; pappus 2-seriate ; achenes pubescent :
Involucral bracts rounded and villous at apex, the outer ovate to orbicular,
  glumaceous below ; leaves lanceolate with inrolled margins, densely
  gland-pitted and pubescent beneath ; inner pappus-setae plumose, pale
  buff ..   .. 42. *oocephala*
Involucral bracts acute, tomentose especially near the margins, glabrescent,
  purplish-brown ; leaves elliptic to oblanceolate, 4–6 cm. long, 1–2 cm.
  broad, rigidly leathery, closely reticulate-veined, shiny, glabrous, minutely
  gland-dotted beneath, flat ; inner pappus-setae almost smooth, outer
  with a scale-like base, brown ..   .. 43. *saussureoides*
Capitula not much longer than broad :
‡Outer pappus of short scales distinctly broader than the inner bristles ;
  achenes 4(–5)-angled, often glandular between the ribs (to p. 274) :
Achenes pubescent at least on the angles :
Achenes uniformly pilose ; outer pappus-scales rather narrow, setaceous,
  stiff ; capitula 2–3 cm. diam. ; involucral bracts 4–6-seriate, acutely
  setose-acuminate, villous ; leaves sessile, lanceolate, 5–6 cm. long and
  about 1 cm. broad, thinly scabrid-pubescent ; perennial herb
                                                44. *philipsoniana*
Achenes pubescent only on the angles, glandular between :
Capitula 10–15 mm. broad, more or less clustered at the ends of spreading
  leafy branches ; stem leafy at flowering time ; upper leaves stem-
  clasping, oblanceolate, serrate ; annual herb   ..   .. 45. *ambigua*
Capitula 5–8 mm. broad, densely clustered at the ends of long erect
  almost leafless branches ; leaves at flowering time in a sub-basal
  rosette ; seasonal ' biennial ' herb   ..   .. 46. *bambilorensis*

Achenes glabrous or glabrescent, glandular between the angles ; perennial
  herbs :
Capitula mostly shortly stalked :
  Flower-heads 5–8 mm. broad, numerous in a compact terminal corymb ;
    upper stem leafy ; leaves shortly petiolate, elliptic-lanceolate, acute,
    or obtuse, 3–7 cm. long, 1–2 cm. broad, pubescent and minutely
    glandular beneath, scabrid above   ..   ..   47. *blumeoides*
  Flower-heads 10–15 mm. broad, 4 to 12 in a terminal corymb, or some-
    times in a single tight cluster ; upper stem subscapose with a few
    sessile leaves ; leaves rhomboid-elliptic, up to about 10 cm. long and
    4 cm. broad, crisped-pubescent, minutely glandular and venose
    beneath, scabridulous above, margin crenate-denticulate   48. *klingii*
Capitula mostly sessile, spread along the upper sides of unequally divari-
  cate inflorescence-branches :
  Flower-heads 1·5–2 cm. broad ; upper leaves sessile, ½-amplexicaul,
    densely tomentose beneath, glabrescent above ..   .. 49. *bamendae*
  Flower-heads 1–1·5 cm. broad ; upper leaves rounded at base but not
    amplexicaul, densely pubescent beneath, scabrid above   50. *rugosifolia*
‡Outer pappus of bristles shorter but not broader than the inner, or if broader
  at base produced into a simple bristle, or pappus uniform :
Leaves woolly- or silky-tomentose beneath ; achenes pubescent :
  Leaves coarsely toothed, lanceolate to narrowly obovate, up to 15 cm.
    long and 5 cm. broad, white- or buff-woolly beneath ; achenes ribbed ;
    pappus-bristles flat, uniform, scarcely barbellate :
    Capitula numerous in a branched corymb (1–)1·5–2·5 cm. broad in flower :
      Appendage to the involucral bracts blunt, smoothly obtuse or ragged
                    52a. *guineensis* var. *guineensis*
      Appendage to the involucral bracts acute, coloured, papery
                      52b. *guineensis* var. *cameroonica*
    Capitula very few or solitary, terminal on long simple branches, 3–6 cm.
    broad in flower   ..   ..   52c. *guineensis* var. *procera*
  Leaves entire or minutely toothed ; involucral bracts with dark tips ;
    outer pappus mostly of shorter bristles ; perennial herbs :
  Leaves sessile and rounded at base, ovate-lanceolate, 3–6 cm. long,
    1–1·5 cm. broad ; capitula 7–8 mm. long on leafy branches   53. *nestor*
  Leaves narrow at base :
    Inflorescence a broad panicle ; capitula sessile along spreading branches,
    about 6–8 mm. long ; involucral bracts pubescent outside ; leaves
    broadly lanceolate ; achenes 4-ribbed ..   ..   54. *cistifolia*
    Inflorescence a small corymb ; capitula 5–8 mm. diam. ; involucral
    bracts appressed-silky ; leaves lanceolate, mucronate, 2–10 cm. long,
    1–2 cm. broad, distinctly veined and uniformly densely silky beneath,
    thinly silky and glabrescent above ; pappus more or less uniform
                      55. *smithiana*
Leaves at most rather densely pubescent beneath ; herbs or undershrubs :
  Leaves mostly basal, oblanceolate, up to 25 cm. long and 4 cm. broad ;
    involucral bracts obtuse in heads 1–1·5 cm. broad   25. *subaphylla*
  Leaves not mostly basal :
    Achenes glabrous, glabrescent or glandular only :
    Leaves sessile, rounded and clasping at base, lanceolate, up to
      8(–12) cm. long and 2(–2·5) cm. broad, very shallowly dentate,
      apiculate, minutely glandular-punctate beneath ; involucral bracts
      thinly pubescent, glabrescent ; outer pappus-setae about ⅓ the length
      of the inner, broadened into a scale-like base ; achenes 1·5 mm. long,
      rounded on one side, smooth, hardly angled ..   .. 56. *migeodii*
    Leaves petiolate, narrowed to base, obovate to oblanceolate, rigidly
      leathery, up to 10 cm. long and 3 cm. broad, remotely serrulate,
      closely reticulate, sparsely glandular-punctate beneath, rusty-
      pubescent when young, glabrescent ; capitula very small with about
      5 florets, less than 1 cm. long and 5 mm. broad ; involucral bracts
      obtuse ; pappus uniform ; achenes 3 mm. long, strongly 8–10-ribbed,
      glandular ; woody undershrub   ..   ..   .. 57. *glaberrima*
    Achenes pubescent or tomentose ; pappus 2-seriate, the outer bristles
      shorter but not broader than the inner :
      Involucral bracts long-tapered, acute ; capitula about 5 mm. broad in
        flower ; leaves broadly ovate, obtuse, crenate-dentate, abruptly
        narrowed to the petiole ..   .. 58. *cinerea*
      Involucral bracts abruptly triangular-tipped or obtuse, apiculate :
        Capitula about 1 cm. broad in flower, pedunculate in broad corymbs ;
        stem simple or sparingly branched, striate, thinly pubescent ; leaves

ovate-rhomboid, up to 8 cm. long and 2 cm. broad, very acute, mucronate, remotely serrate-dentate, tapered at the base to a short petiole, thinly pubescent and gland-dotted beneath, scabridulous above   ..   ..   59. *undulata*

Capitula 5–8 mm. broad in flower, subsessile or pedunculate in a divaricately branched inflorescence ; stem woody, profusely and shortly branched, terete, minutely but densely white-tomentose ; leaves obovate to oblanceolate, 1–2 cm. long, 5–10 mm. broad, rounded at apex, tapered at base, subentire, minutely puberulous, glabrescent, gland-dotted beneath   ..   ..   60. *cinerascens*

1. **V. kotschyana** *Sch. Bip.* in Walp. Rep. 2 : 947 (1843) ; F.T.A. 3 : 380 ; Chev. Bot. 355 ; Berhaut Fl. Sén. 176. A shrub with erect, pithy, striate, sparsely pubescent stems 2–5 ft. high ; florets pale mauve in heads 1–1½ in. across.
Sen.: *fide* Chev. *l.c.* **Mali**: *fide* Chev. *l.c.* **N.Nig.** : Kano (Dec.) *Hagerup* 641! Sokoto *Lely* 151! Katagum (Sept.) *Dalz.* 173! Mongu, 4,300 ft. (July) *Lely* 385! [Br.]**Cam.**: Gurum, Vogel Peak, Adamawa (Nov.) *Hepper* 1318! Extends through Cameroun and Ubangi-Shari to Ethiopia. (See Appendix, p. 423.)

2. **V. calvoana** (*Hook. f.*) *Hook. f.* var. **calvoana**—in Bot. Mag. 94 : t. 5698 (1868) ; F.T.A. 3 : 293 ; Adams in J.W. Afr. Sci. Assoc. 3 : 117. *Stengelia calvoana* Hook. f. (1864). A shrub up to about 15 ft. high with ribbed velvety pithy branches and rather large sessile leaves ; florets purple to pale mauve in heads nearly 2 in. across which have white appendages to the middle and inner involucral bracts ; in clearings in montane forest.
[Br.]**Cam.**: Cam. Mt., 2,500–7,000 ft. (Nov.–Feb.) *Mann* 1238! *Dalz.* 8325! *Dunlap* 112! *Maitland* 124! 242! *Migeod* 38! 49! *Steele* 5! Moyuko, 2,000 ft. (Apr.) *Hutch. & Metcalfe* 84! (See Appendix, p. 422.)

2a. **V. calvoana** var. **microcephala** *C. D. Adams* l.c. 118 (1957). A shrub 2–6 ft. high ; flower-heads up to 1 in. across ; florets mauve ; in sheltered places in montane grassland.
[Br.]**Cam.**: Sagbo, 5,400 ft., Bamenda to Ndop (Dec.) *Boughey* GC 10444! Nkambe to Misage, 6,000–7,000 ft. (Sept.) *Lightbody* FHI 26259! Mambila Plateau, 4,600–6,000 ft., Adamawa (Jan.) *Hepper* 1671! *Latilo & Daramola* FHI 28998!
[A variety intermediate between *V. calvoana* var. *calvoana* and var. *microcephala*, occurring in Kenya and Uganda, has been described (Adams l.c. 118). Low regrowth from the stem-bases, and presumably also young plants of *V. calvoana* in W. Africa, may have small leaves lacking basal auricles, which, therefore, appear to be petiolate rather than sessile—C. D. A.]

3. **V. insignis** (*Hook. f.*) *Oliv. & Hiern* in F.T.A. 3 : 292 (1877). *Stengelia insignis* Hook. f. (1864). A shrub or small tree 15–20 ft. high with flower-heads about 2 in. across ; florets pale purple to mauve ; in clearings in montane forest.
[Br.]**Cam.**: Cam. Mt., 3,000–8,200 ft. (Sept., Dec.–Feb., Apr.) *Mann* 1925! *Maitland* 1027! *Keay* FHI 28647! *Adams* GC 11725! *Boughey* A5025! GC 17461! *Irvine* 1465! *Morton* GC 7102!

4. **V. stenostegia** (*Stapf*) *Hutch. & Dalz.* F.W.T.A., ed. 1, 2 : 166 (1931). *Candidea stenostegia* Stapf in Bot. Mag. 149 : t. 8981 (1923). A shrub 2–6 ft. high, sometimes leafless below ; flower-heads 1–1½ in. across with linear, velvety involucral bracts ; florets pale bluish-purple.
N.Nig.: Nassarawa *Robins*! Vom (Apr., Dec.) *Dent Young* 136! *Morton* K334! Jos, 4,000 ft. (Feb., Mar., Dec.) *Hill* 10! *Batten-Poole* 211! *Coombe* 15! *Lely* P160! Pankshin (Dec.) *W. D. MacGregor* 418!

5. **V. tenoreana** *Oliv.* in Trans. Linn. Soc. 29 : 92 (1873) ; F.T.A. 3 : 290. *V. lasiolepis* O. Hoffm. (1898)— Chev. Bot. 355. An erect shrub up to 6 ft. high with flower-heads 1–1½ in. across ; florets cream-coloured.
Sen.: *fide* F.T.A. *l.c.* **Port.G.**: R. Buba (Dec., Jan.) *Esp. Santo* 2335! *d'Orey* 214! Chitole (Dec., Feb.) *Esp. Santo* 2882! 3170! Bambadinca, Bafata (Dec.) *Esp. Santo* 3833! **Guin.**: Timbo (Oct.) *Jac.-Fél.* 1911! **Dah.**: Ouidah *Isert*! Porto-Novo (Jan.) *Chev.* 22786! **N.Nig.**: Tama, 2,800 ft., Bauchi Prov. (Aug.) *Summerhayes* 34! Makurdi (Mar.) *Morton*! **S.Nig.**: Lagos (Mar.) *Millen* 129! Abeokuta *Millen* 94! Ibadan (Jan., July–Sept.) *Chev.* 14062! *Newberry* 65! *Jones* FHI 14831! *Tamajong* FHI 19512! Abo *Barter* 368! Idogun *Rowland*! Also in Cameroun. (See Appendix, p. 423.)

6. **V. mokaensis** *Mildbr. & Mattf.* in Engl. Bot. Jahrb. 59, Beibl. 133 : 5 (1924) ; Guinea in Anal. Inst. Bot. Cav. 10, 1 : 319 (fig.), 320. A shrub or small tree up to 10 ft. high with serrate leaves narrowed to both ends ; flower-heads up to about 1 in. across with purple florets.
F.Po: Moka, scrub and grassland, 4,000–4,500 ft. (Sept.–Nov., Dec. without fl.) *Mildbr.* 7077. *Tessman* 2807. *Boughey* 26! *Wrigley & Melville* 678! Finca Puente, Musola (Jan.) *Guinea* 1638!

7. **V. leucocalyx** *O. Hoffm.* var. **leucocalyx**—in Engl. Bot. Jahrb. 30 : 422 (1901). A shrub 6–9 ft. high with elliptic, acute, long-cuneate leaves ; flower-heads about 1 in. across with broad white involucral bract appendages and blue florets ; in upland grassland.
S.Nig.: Mt. Koloishe, 5,000 ft., Obudu (Dec.) *Savory & Keay* FHI 25069! [Br.]**Cam.**: Batibo to Bali, Bamenda (Jan., May) *Keay* FHI 28331! *Ujor* FHI 30380! Also in Tanganyika.

7a. **V. leucocalyx** var. **acuta** *C. D. Adams* in J.W. Afr. Sci. Assoc. 3 : 120 (1957). An erect shrub 9–12 ft. high ; flower-heads ½–1 in. broad with acutely appendaged white involucral bracts and purple florets ; in the open margins of montane forest.
[Br.]**Cam.**: Bamenda, 6,000–8,200 ft. (Dec.–Jan., Apr.–June) : Lakom *Maitland* 1485! Bambui *Boughey* GC 10776! Santa *Boughey* GC 11036! Bafut-Ngemba F.R. *Keay & Lightbody* FHI 28366!

8. **V. iodocalyx** *O. Hoffm.* in Engl. Pflanzenw. Ost-Afr. C : 403 (1895). An erect shrub with coarsely dentate, petiolate leaves ; flower-heads under 1 in. broad, distinctly stalked but clustered at the ends of velvety branches, with white-appendaged acute involucral bracts ; florets exserted, purple fading to mauve.
Guin.: Mali, 5,000 ft., Fouta Djalon (Sept., Oct.) *Schnell* 7169! *Jac.-Fél.* 1997! Also in E. Africa.

9. **V. frondosa** *Oliv. & Hiern* in F.T.A. 3 : 294 (1877). *V. aff. frondosa* of Schnell in Mém. I.F.A.N. 22 : 512 (1952). A little-branched shrub or small tree 15–20 ft. high and 1 ft. girth ; leaves sessile, up to 3 ft. long and 1½ ft. broad, crowded near the top of the stem ; flower-heads sessile or subsessile, about 1 in. across, in clusters along the branches of a panicle 2–3 ft. broad ; florets with purple lobes and white tubes ; in forest clearings.
Iv.C.: Nimba Mts., 3,000 ft. (Dec., Apr.) *Adam* 16464! *Schnell* 5189. **S.Nig.**: Oban *Sankey* 9! *Kennedy* 1979! 2078! Epe *Barter* 3294! Calabar to Oban (Jan.) *Onochie & Okafor* FHI 36000a! Omo F.R., Ijebu (Dec.) *Onochie* FHI 36593! [Br.]**Cam.**: R. Kindong F.R., Kumba (Jan.) *Binuyo & Daramola* FHI 35072! Also in Cameroun. (See Appendix, p. 422.)

10. **V. biafrae** *Oliv. & Hiern* in F.T.A. 3 : 270 (1877) ; F.W.T.A., ed. 1, 2 : 168, partly (excl. *Migeod* 381 ; *Mildbr.* 10800). *V. senecioides* A. Chev. in Mém. Soc. Bot. Fr. 28 : 259 (1917) ; Chev. Bot. 357. A climbing or scrambling shrub with ribbed puberulous branches, and copious panicles of purple campanulate flower-heads ½ in. wide.
Guin.: Tonfili, Faranna (Jan.) *Chev.* 20436! **S.L.**: Njala (Feb.) *Deighton* 2441! Makali (Feb.) *Deighton* 4057! **Ghana**: Aburi Hills (Mar.) *Johnson* 622! Mt. Toga, 2,000 ft., Togo Plateau *Morton* A3832! **S.Nig.**: Sonkwala, 2,500 ft., Ogoja (Dec.) *Savory & Keay* FHI 25021! [Br.]**Cam.**: Cam. Mt., 2,000–4,500 ft. (Nov.–Apr.) *Dalz.* 8187! *Dundas* FHI 20390! *Maitland* 294! *Mann* 1296! *Migeod* 63! *Lehmbach* 135! *Schlechter* 12849! *Morton* K465! **F.Po**: Musola to Moka (Jan.) *Guinea* 1284! Moka, 4,000 ft. (Dec.) *Boughey* 5! Also in Congo. (See Appendix, p. 422.)

11. **V. tufnelliae** *S. Moore* in J. Bot. 46 : 292 (1908). *V. biafrae* of F.W.T.A., ed. 1, 2 : 168, partly (*Migeod*

381 ; *Mildbr.* 10800), not of Oliv. & Hiern. A scandent shrub like the last with slightly larger more brightly coloured flower-heads and coarsely toothed or sinuate-margined leaves; at the margins of montane forest.

[Br.]Cam.: Cam. Mt., 3,000 ft. (Dec.) *Mildbr.* 10800! Bangwa, Mamfe (Jan.) *Lobe Babute* Cam 21/37! Bamenda, 6,000–7,500 ft. (Sept., Dec., Jan.) *Adams* GC 11325! *Keay & Lightbody* FHI 28355! *Migeod* 381! Kumbo (Feb.) *Hepper* 1968! Binka, Nkambe Div. *Savory* UCI 365! Also in Rio Muni, Congo, Uganda and Tanganyika.

12. **V. andohii** *C. D. Adams* in J.W. Afr. Sci. Assoc. 3 : 115 (1957). *Microglossa volubilis* var. *laxiflora* A. Chev. Bot. 363, name only. A scrambling shrub up to 30 ft. high in open forest ; florets white in numerous shortly stalked heads less than ½ in. long.

Iv.C.: Bouroukrou (Dec.–Jan.) *Chev.* 16804! **Ghana:** Asuansi (Jan.) *Fishlock* 20! Abetifi, 2,000 ft., Kwahu (Jan.) *Irvine* 1672! Pepease, 2,000 ft. (Jan.) *Irvine* 1827! Mt. Ejuanema, Kwahu (Dec.) *Adams* 5138! Akropong, Akwapim (Jan.) *Irvine* 2619! Obuasi, Ashanti (Feb.) *Andoh* FH 4137!

13. **V. auriculifera** *Hiern* in Cat. Welw. 1 : 539 (1898). *V. uniflora* Hutch. & Dalz. F.W.T.A., ed. 1, 2 : 169 (1931). A shrub or small tree 6–15 ft. high with broad corymbs of minute mauve flowers ; in open grassland or at the margin of forest in mountainous country.

[Br.]Cam.: Bamenda, 6,000–7,500 ft. (Dec., Jan., Mar., Apr.) *Boughey* GC 11421! *Migeod* 356! *Maitland* 1629! *Morton* K47! *Johnston* FHI 10438! Bafut-Ngemba F.R. (Jan.) *Tiku* FHI 22234! Bamenda to Ndop, 5,000 ft. (Dec.) *Adams* GC 11093! Ndu (Feb.) *Hepper* 1945! Also in Congo, Uganda, Kenya and Angola.

14. **V. myriantha** *Hook. f.* in J. Linn. Soc. 7 : 198 (1864) ; F.T.A. 3 : 297 ; F.W.T.A., ed. 1, 2 : 169 (excl, *Chev.* 20330, 20689 and *Migeod* 484). A shrub or small soft-wooded tree 10–20 ft. high with numerous small heads of white florets in dense terminal corymbose panicles.

[Br.]Cam.: Cam. Mt., 5,000–6,000 ft. (Dec.–Mar.) *Mann* 1913! 1936! *Dalz.* 8195! *Dunlap* 80! *Keay* FHI 28646! *Dundas* FHI 13907! *Maitland* 256! 516! 954! 1227! *Adams* GC 11763! **F.Po:** 4,000–5,000 ft. (Dec.) *Mann* 622!

15. **V. ampla** *O. Hoffm.* in Engl. Bot. Jahrb. 30 : 423 (1901) ; Adams in J.W. Afr. Sci. Assoc. 2 : 64 (1956). *V. conferta* of Chev. Bot. 353, partly (Chev. 20330 ; 20689), not of Benth. *V. myriantha* of F.W.T.A., ed. 1, 2 : 169, partly (*Chev.* 20330, 20689) ; of Aubrév. Fl. For. C. Iv., ed. 2, 3 : 313, not of Hook. f. A shrub or small tree 15–20 ft. high with mauve florets in numerous flower-heads each about ½ in. long; in clearings of upland forest.

Guin.: Dalaba Plateau, 3,600 ft. (Dec.) *Chev.* 20330! Soundia, Kissi (Feb.) *Chev.* 20689! Timbi-Madina (fr. Mar.) *Adam* 11637 *bis* ! **S.L.:** Duunia, Talla (Feb.) *Sc. Elliot* 4833! Moria, Scarcies (Feb.) *Sc. Elliot* 4839! Loma Mts. (Jan.) *Jaeger* 4122! Also in Ethiopia and south to Transvaal and Zululand.

16. **V. thomsoniana** *Oliv. & Hiern* in Trans. Linn. Soc. 29 : 91 (1873) ; F.T.A. 3 : 295. *V. amygdalina* of Chev. Bot. 352, partly (Chev. 20398). *V. pobeguinii* Aubrév. Fl. For. Soud.-Guin. 487, t. 109, fig. 3 (1950), French descr. only. A shrub up to about 10 ft. high with numerous flower-heads ½ in. long densely packed in rounded terminal clusters ; florets mauve ; in upland savanna.

Guin.: Timbo (Jan.) *Pobéguin* 52! 1431! Dalaba to Pitou (Dec.) *Pobéguin* 2043![Tonfili, Faranna (Feb.) *Chev.* 20398! [Br.]Cam.: R. Metschum, Bafut, 3,000 ft., Bamenda (Jan.) *Keay* FHI 28521! Extends through Cameroun and Congo to Sudan, east tropical Africa and Angola.

17. **V. doniana** *DC.* Prod. 5 : 23 (1836) ; F.T.A. 3 : 295. A shrub 6 ft. high, rusty-pubescent on the branchlets and inflorescence, with small campanulate-turbinate flower-heads ¼ in. across in short racemes or pedunculate clusters ; florets ? purple.

S.L.: *Don*! Leicester Peak *Miss Turner*! Freetown (Aug.) *Burbridge* 522! **Lib.:** Bilimu (Jan.) *Harley* 1350!

18. **V. conferta** *Benth.* in Fl. Nigrit. 427 (1849) ; F.T.A. 3 : 294 ; Chev. Bot. 353, partly (excl. *Chev.* 20330 ; 20689) ; Aubrév. Fl. For. C. Iv., ed. 2, 3 : 314, t. 369. A tree with a slender stem up to 25 or 30 ft. high, pale rusty-tomentose on branchlets and inflorescence, with large distinctly petiolate sinuate-margined leaves crowded near the ends of the branches ; florets white or pale mauve in numerous small flower-heads forming a very large terminal erect panicle ; in secondary forest.

Guin.: Korodou (Feb.) *Chev.* 20725! **S.L.:** *Don*! Kukuna, Scarcies (Jan.) *Sc. Elliot* 4501! Njala (Feb., Mar.) *Deighton* 1070! *Aylmer* 571! **Lib.:** Cape Palmas (May) *Huntting* 71! Palilah, Gbanga (Feb.) *Baldwin* 11037! **Iv.C.:** Abidjan *Aubrév.* 866! Morénou (Dec.) *Chev.* 22483! **Ghana:** Aburi (Dec.) *Johnson* 882! Bunso (Sept.) *Scholes* 336! Ankobra Junction (Feb.) *Kitson* 1027! Sunyani, Ashanti (Jan.) *Chipp* 75! Kumasi (Feb.) *Irvine* 99! Togo Plateau F.R. (Feb.) *Beveridge* FH 2926! **Togo Rep.:** *Baumann* 432! **S.Nig.:** Epe (Feb.) *Millen* 168! Idumuje (Jan.) *Thomas* 2152! Oban *Talbot* 398! 1498! [Br.]Cam.: Johann-Albrechtshöhe *Zenker & Staudt* 559! Buea, 2,500–3,000 ft. (Dec., Jan.) *Maitland* 190! 368! Pete-Mutoko, Kumba (Feb.) *Binuyo & Daramola* FHI 35559! **F.Po:** *Mann* 255! Musola to Moka (Jan.) *Guinea* 1023. Extends through central Africa to Uganda and Angola. (See Appendix, p. 422.)

19. **V. colorata** *(Willd.) Drake* in Bull. Soc. Bot. Fr. 46 : 230 (1899) ; Hist. Pl. Madag. Atlas t. 466 ; Aubrév. Fl. For. C. Iv., ed. 2, 3 : 313, and Fl. For. Soud.-Guin. 488, t. 109, 1 ; Berhaut Fl. Sén. 141. *Eupatorium coloratum* Willd. (1803). *V. senegalensis* Less. (1829)—F.T.A. 3 : 283 ; Chev. Bot. 357, partly (excl. *Chev.* 15105). A shrub or small tree up to about 15 ft. high, with soft pale rusty-tomentose branchlets and white sweet-scented florets in heads about ½ in. broad ; inflorescence a broadly rounded corymbose panicle ; pappus cream, later becoming tawny ; common in derived savanna.

Sen.: *Michelin*! *Richard* 168! " N'doute T. Vigne " (Dec.) *Chev.* 2034! **Gam.:** *Ingram*! *Brown-Lester* 73! Georgetown (Jan.) *Dalz.* 8034! **Mali:** Tabacco (Jan.) *Chev.* 133. Diafarabe (Jan.) *Davey* 402! **Port.G.:** Pessubé, Bissau (Dec.) *Esp. Santo* 1084! **Guin.:** Timbo (Jan.) *Pobéguin* 97! **S.L.:** Kambia (Dec.) *Sc. Elliot* 4201! Musaia (Dec.) *Deighton* 4548! Saiama (Dec.) *Deighton* 5284! Bantoro (Nov.) *Koroma* 5! Wara Wara (Jan.) *King* 191b! **Lib.:** (Jan.) *Dinklage* 3359! **Iv.C.:** Agboville *Aubrév.* 637! **Ghana:** Keta *Thonning* 131! Cape Coast (Jan.) *Vigne* FH 947! Achimota (Feb., Apr.) *Irvine* 147! *Milne-Redhead* 5089! Dutukpene, near Kete Krachi (Dec.) *Adams* 4542! Han to Wa (Mar.) *Kitson* A15! **Togo Rep.:** Lomé *Warnecke* 29! **Dah.:** Kétou, Zaganando (Feb.) *Chev.* 22022. **N.Nig.:** Ilorin (Jan.) *Meikle* 1007! Nupe *Barter* 810! Jebba (Dec.) *Hagerup* 755! Zungeru to Kagura (Oct.) *Hepper* 982! Jos Plateau *Batten-Poole* 208! **S.Nig.:** Olokemeji (Dec.) *Jones* FHI 4873! [Br.]Cam.: Gwoza, 3,500 ft., Dikwa Div. (Jan.) *McClintock* 148! Mapeo, Alantika Mts., 3,200 ft. (Dec.) *Hepper* 1604! Throughout central and south tropical Africa. (See Appendix, p. 421.)

[This is a variable species ; the leaves of many plants, particularly from the drier and more northerly localities, have more acute tips and the undersurfaces much less woolly with the secondary and tertiary veins clearly visible ; in these plants the middle and outer involucral bracts may be mucronate or acuminate, rather than rounded, with the tips slightly reflexed ; there is also variation in the number and distribution of glands on the involucral bracts.—C. D. A.].

20. **V. amygdalina** *Del.* Cent. Pl. Afr. 41 (1826) ; F.T.A. 3 : 284 ; Chev. Bot. 352, partly (excl. *Chev.* 20398) ; Aubrév. Fl. For. C. Iv., ed. 2, 3 : 313, and Fl. For. Soud.-Guin. 488, t. 109, fig. 2. *V. senegalensis* of Chev. Bot. 357, partly (*Chev.* 15105), not of Less. A shrub or small tree 6–15 ft. high with striate, pubescent branchlets becoming glabrous ; florets sweet-scented, white, in heads ½ in. broad ; pappus buff or reddish ; common and often planted.

Mali: *fide* Aubrév. (1950) *l.c.* **Guin.:** Kouria (Jan.) *Caille* in Hb. *Chev.* 15105! Timbi-Madina (Mar.) *Adam* 11643 *bis* ! **S.L.:** *Sc. Elliot* 5753! Njala (Jan.) *Deighton* 544! Freetown (Mar.) *Kirk* 40! Yiraia (Jan.) *Glanville* 483! **Lib.:** Owensgrove, Grand Bassa (Mar.) *Baldwin* 11101! **Iv.C.:** *fide* Chev. *l.c.* **Ghana:** Dunkwa (Aug.) *Sampeney* 712! Achimota *Irvine* 46a! Ayafie, near Kete Krachi (Dec.) *Adams* 4708! Amedzofe (Feb.) *Dokosi* in Hb. *Adams* 4747! *Morton* A2687! Bole to Bamboi (Dec.) *Morton* GC 25116!

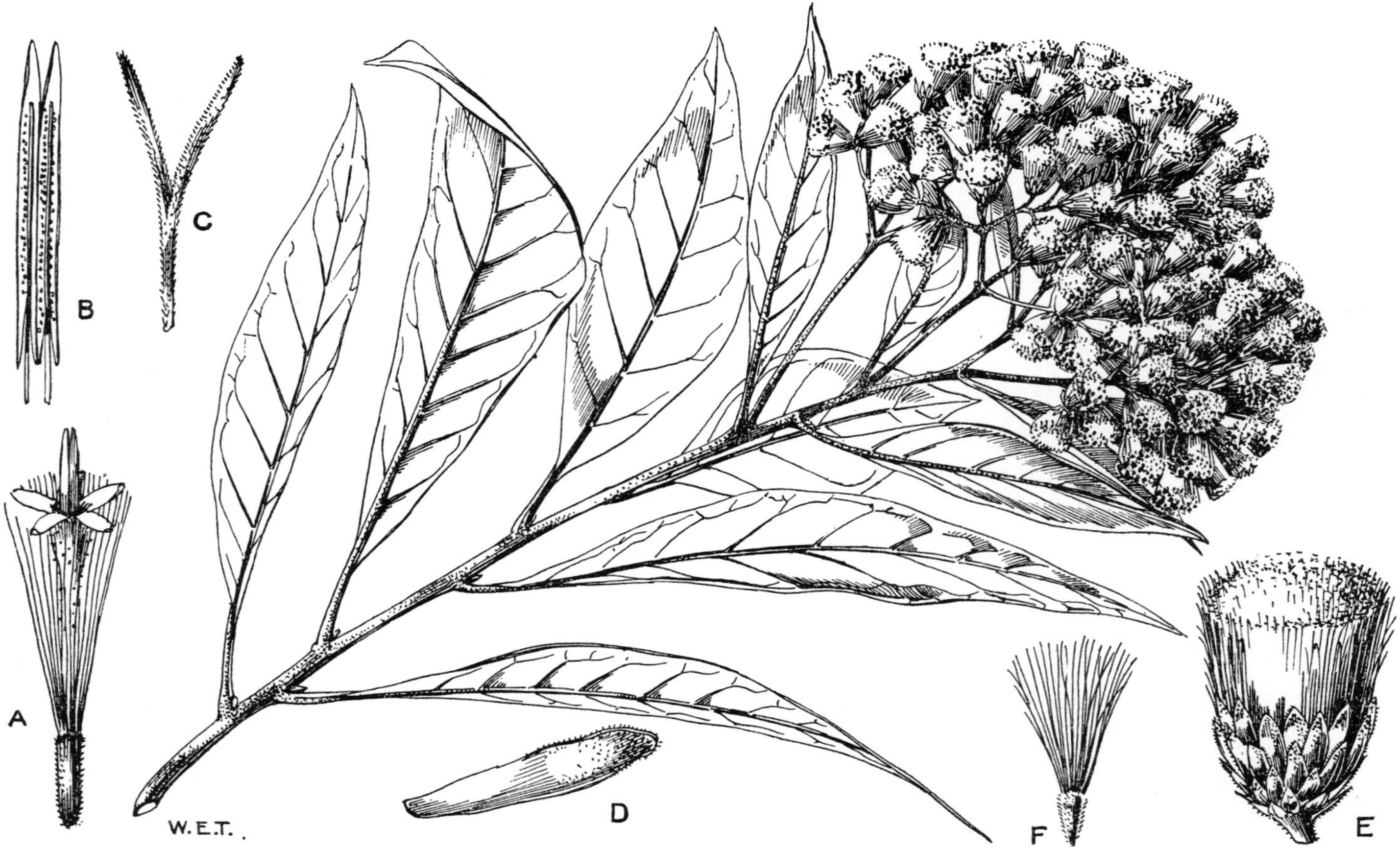

FIG. 260.—VERNONIA AMYGDALINA *Del.* (COMPOSITAE).

A, flower. B, anthers. C, style-arms. D, involucral bract. E, flower-head. F, achene and pappus.

W.E.T.

278

Dah.: *fide* Chev. *l.c.* **N.Nig.:** Nupe *Barter* 1172! Wawa, Borgu *Barter* 722! Sokoto *Lely* 138! Jos Plateau (Dec., Jan.) *Lely* P3! *W. D. MacGregor* 397! *Coombe* 16! Abuja (Jan.) *Savory* 140! **S.Nig.:** Ibadan (Jan.) *Newberry* 29! Abeokuta (Feb.) *Irving* 68! Olokemeji F.R. (Apr.) *Ross* R149! Enugu *Smith* 39! Oji R. (Feb.) *Todd* 230! Oban *Talbot*! [Br.]**Cam.:** Victoria (June) *Maitland* 53! Cam. Mt., 3,000 ft. (Jan.) *Dunlap* 125! *Maitland* 316! Mamfe (Jan.) *Maitland* 1165! Ntim, 4,400 ft., Nkambe Div. (Feb.) *Hepper* 1858! Gashaka, Adamawa (Dec.) *Latilo & Daramola* FHI 28786! **F.Po:** *T. Vogel* 257! (Dec.) *Mann* 28! S. Isabel (Dec., Feb.) *Dalz.* 8175! *Boughey* GC 10974! Musola to Moka (Jan.) *Guinea* 1302. Throughout tropical Africa from Ethiopia and Sudan to Rhodesia, Angola, S. Tomé and Annobon. (See Appendix, p. 421.)

21. **V. subuligera** *O. Hoffm.* in Engl. Pflanzenw. Ost-Afr. C: 403 (1895). *V. myriantha* of F.W.T.A., ed. 1, 2: 169, partly (*Migeod* 484), not of Hook. f. A shrub about 12 ft. high with flower-heads in dense rounded clusters; florets bluish-mauve.
    [Br.]**Cam.:** Bamenda, 5,500–7,000 ft. (Jan., Feb.) *Keay & Lightbody* FHI 28353! *Migeod* 484! Bafut-Ngemba F.R., 6,500 ft. (Jan.) *Tiku* FHI 22235! Also in Uganda and Tanganyika.

22. **V. theophrastifolia** *Schweinf. ex Oliv. & Hiern* in F.T.A. 3: 294 (1877). A shrub or small tree up to 15 ft. high with sessile auriculate oblanceolate leaves and small sessile or subsessile flower-heads in a divaricately branched inflorescence; florets reddish-purple.
    [Br.]**Cam.:** Lekitaba, Mambila Plateau, Adamawa (Jan.) *Latilo & Daramola* FHI 34376! Extends to Sudan.

23. **V. richardiana** (*O. Ktze.*) *P.-Sermolli* in Webbia 7: 340 (1950). *Cacalia richardiana* O. Ktze. (1891). *V. seretii* De Wild. in Ann. Mus. Congo, sér. 5, 2: 207 (1907); Adams in J.W. Afr. Sci. Assoc. 2: 66 (1956). *V. myriocephala* A. Rich. (1848)—F.T.A. 3: 296, not of DC. (1836). A shrub 6–10 ft. high, unbranched below, with sessile undulate-margined leaves; flower-heads subsessile in divaricately branched panicles; florets bluish-purple to mauve.

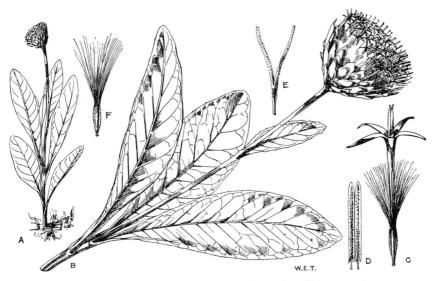

FIG. 261.—VERNONIA MACROCYANUS *O. Hoffm.* (COMPOSITAE).

A, whole plant. B, portion of same. C, flower. D, anthers. E, style-arms. F, achene and pappus.

Ghana: Amedzofe Hill, 2,300 ft. (Feb., Apr.) *Morton* GC 8410! A1999! A2690! **N.Nig.:** Jos Plateau *Batten-Poole* 212! Kurra Falls, Jos (Dec.) *Savory* 119! Naraguta, 4,000 ft. (Dec.) *Coombe* 41! Ropp (Dec.) *W. D. MacGregor* FHI 10503! [Br.]**Cam.:** Mayo Selbe, Gashaka Dist., Adamawa (Jan.) *Hepper* 1634! Also in Ethiopia, Sudan and across central Africa to Congo.

24. **V. chthonocephala** *O. Hoffm.* in Bol. Soc. Brot. 13: 17 (1896). A dwarf perennial herb with flower-heads about 1 in. across, having purplish involucral bracts, usually appearing near the ground after grass burning and separately from the leaves; leaves oblanceolate, up to about 8 in. long; florets purple.
    Togo Rep.: Bismarckburg *Büttner* 379. **N.Nig.:** Vom, 3,000–4,500 ft. (Jan.) *Dent Young* 133! Jos (Dec.) *Monod*! [Br.]**Cam.:** Kakara, 4,900 ft., Mambila Plateau (Jan.) *Hepper* 1775! Also in Sudan, Uganda, Kenya, Congo, Nyasaland and Angola.

25. **V. subaphylla** *Bak.* in Kew Bull. 1895: 290. A perennial herb with few round-based flower-heads, about ½ in. across, terminal on the branches of an almost leafless stem about 2 ft. high, arising from a densely woolly stock; florets reddish-purple; in upland grassland, flowering usually after burning.
    [Br.]**Cam.:** Nchan, Bamenda, 5,000 ft. (Apr.) *Maitland* 1471! Bambui, 3,500–4,000 ft. (Apr.) *Maitland* 1715! Bamenda (leaf & fl. Mar.) *Morton* K123! Ndop to Kumbo (Dec.) *Boughey* GC 11151! Ngong, Wum Div. (Feb.) *Hepper* 2837! Also in Congo.

26. **V. chapmanii** *C. D. Adams* in J.W. Afr. Sci. Assoc. ined. (1963). A perennial herb with leafy stems about 1 ft. high arising from a woody stock; florets purple in terminal solitary heads partly hidden by the uppermost leaves.
    [Br.]**Cam.:** between Chabbel Zanche and Kan Iyaka, 5,500 ft., Mambila Plateau (July) *Chapman* 42!

27. **V. acrocephala** *Klatt* in Ann. Nat. Hofmus. Wien 7: 100 (1892). A perennial herb 2–8 in. high with single or rarely double-headed scapes arising from a basal rosette of leaves; involucre of rather few broad acuminate purplish bracts about ⅓ in. long; florets pale purple; in montane grassland.
    [Br.]**Cam.:** Bamenda, 5,000 ft. (Mar., Apr., June) *Maitland* 1417! 1549! *Morton* K118! *Onochie* FHI 34876! Bafut-Ngemba F.R., 7,200 ft. (Feb.) *Hepper* 2181! Also in Cameroun and Angola.

28. **V. gerberiformis** *Oliv. & Hiern* in F.T.A. 3: 285 (1877). An acaulescent herb with a basal rosette of

oblanceolate leaves and a solitary-headed scape arising from a densely woolly perennial stock ; scape 1–5 in. long in flower, longer in fruit ; florets purplish-blue.
N.Nig.: Jos Plateau : Bokkos, 4,000 ft. (Dec.) *Coombe* 73 ! N. of Ropp (Apr.) *Morton* K377 ! Extends through Sudan and Congo to Uganda, Kenya and Tanganyika and south to Angola.

29. **V. macrocyanus** *O. Hoffm.* in Bol. Soc. Brot. 13 : 20 (1897). *V. primulina* of Chev. Bot. 356, and of F.W.T.A., ed. 1, 2 : 166, not of O. Hoffm. A perennial herb with smooth ribbed simple or sparingly branched stems 1–2 ft. high in clumps from a fleshy rhizome and large solitary flower-heads 1½–2 in. across with rich blue or purple florets ; in burnt savanna.
Mali: Bougouni (Apr.) *Chev.* 681. **N.Nig.:** Sokoto (Sept.) *Ryan* 2 ! Naraguta (Nov.) *Lely* 695 ! Ropp, 4,000 ft. (Dec.) *Coombe* 47 ! Zaria (Feb.–Mar.) *Bunny & Ryan* 31 ! *Dalz.* 363 ! *Onyeagocha* FHI 7642 ! [Br.]Cam.: Belel, N. Adamawa (Jan.) *Hepper* 1629 ! Also in Uganda, Tanganyika and Angola. (See Appendix, p. 423.)

30. **V. nimbaensis** *C. D. Adams* in J.W. Afr. Sci. Assoc. ined. (1963). A perennial herb with erect usually branched, striate puberulous stems 1 ft. or more high ; leaves sessile, serrate, prominently reticulate-veined ; involucre solitary and terminal to the main stem on a stout peduncle, up to 1½ in. across in fruit ; in montane grassland.
Guin.: Nzo Mt., 4,000–4,500 ft. (fr. Mar.) *Chev.* 21009 ! 21023 ! ridges of Nimba Mts. (fr. Apr.) *Schnell* 925 ! **S.L.:** Loma Mts. (sterile Aug.) *Jaeger* 1086 !

31. **V. jaegeri** *C. D. Adams* in l.c. ined. (1963). A perennial herb up to about 9 in. high with a leafy stem arising from a woody stock ; heads solitary or few, about ¾ in. across.
S.L.: savanna at about 5,600 ft., Loma Mts. (fr. Feb.) *Jaeger* 4242 !

32. **V. nigritiana** *Oliv. & Hiern* in F.T.A. 3 : 288 (1877) ; Chev. Bot. 355 ; Berhaut Fl. Sén. 177 (as *V. nigritana*). *V. baoulensis* A. Chev. Bot. 352, name only. An erect, branched herb or undershrub with annual regrowth of pithy stems 1–2 ft. or more high arising from a perennial stock ; leaves harshly scabrid ; inner involucral bracts coloured bright red on the inner surface, spreading in heads up to 2 in. across ; in savanna.
Sen.: Kaolak *Kaichinger* 78 ! *Berhaut* 610 ! **Gam.:** *Saunders* 69 ! *Brown-Lester* 60 ! (July) *Brooks* 15 ! Niambai Forest (Oct.) *Frith* 131 ! **Mali:** Ouacoro (Dec.) *Chev.* 2063 ! **Port.G.:** Canguelifé to Pitche, Gabu (Sept.) *Esp. Santo* 2778 ! Geba to Mato de Cao, Bafatá (Sept.) *Esp. Santo* 3353 ! Pitche to Rio Bidigor, Gabu (Oct.) *Esp. Santo* 3397 ! Pitche (Oct.) *Esp. Santo* 3430 ! **Guin.:** Fouta Djalon *Maclaud* 272 ! Douné to Timbo (Dec.) *Caille* in *Hb. Chev.* 14636 ! **S.L.:** Lomaburu, N. Tambakka (Feb.) *Sc. Elliot* 5034 ! Warantamba (Oct.) *Small* 354 ! Knokcba (Oct.) *Small* 445 ! Musaia (Dec.) *Deighton* 4470 ! Loma Mts. (Sept.) *Jaeger* 2016 ! **Iv.C.:** Bouaké to Langouassou (July) *Chev.* 22147 ! Askara (Aug.) *Boughey* GC 18640 ! Orumbo Boka Mt., near Toumodi (Dec.) *Boughey* GC 14447 ! Abaka (Dec., Jan.) *Irvine* 623 ! 628 ! Ejura, Ashanti (Nov., Dec.) *Vigne* FH 3453 ! Deakin 99 ! *Morton* GC 9579 ! Paga (Sept.) *Vigne* FH 4582 ! Kete Krachi to Dutukpene (Dec.) *Adams* 4627 ! Amedzofe Hill (Feb.) *Morton* GC 8416 ! Ho (Dec.) *Plumptre* 24 ! **N.Nig.:** Nupe *Barter* 753 ! Niger *Baikie* 19 ! Kontagora (Nov.) *Dalz.* 208 ! Zungeru (Oct.) *Elliott* 13 ! Share F.R., Ilorin (Jan.) *Ujor* FHI 31613 ! **S.Nig.:** Abeokuta *Irving* ! Ishoka (Oct.) *Unwin* 157 ! [Br.]Cam.: Gashaka, Adamawa (Dec.) *Latilo & Daramola* FHI 28930 ! Gurum, near Vogel Peak, Adamawa, (Nov.) *Hepper* 1299 ! (See Appendix, p. 423.)

33. **V. purpurea** *Sch. Bip.* var. **purpurea**—in Walp. Rep. 2 : 946 (1843); F.T.A. 3 : 281 ; Chev. Bot. 357 ; Berhaut Fl. Sén. 177, 204. A stiff erect scabrid perennial herb with a ribbed stem ; stems simple or branched above, a few inches to 4 ft. high ; florets purple or reddish-purple in heads up to 1 in. across.
Sen.: Niokolo (Oct.) *Adam* 15556 ! 15766 ! **Mali:** *fide* Chev. *l.c.* **Port. G:** Gabu (Sept.) *Esp. Santo* 2769 ! **Guin.:** Kindia (Oct.) *Jac.-Fél.* 1858 ! Pita (Oct.) *Jac.-Fél.* 2026 ! Fon Massif (Aug.) *Jaeger* 4857 ! **Iv.C.:** *fide* Chev. *l.c.* **Ghana:** Pong Tamale (Nov.) *Akpabla* 37 ! Yeji *Anderson* 27 ! Damongo (July) *Vigne* 371 ! *Andoh* 5199 ! *Morton* GC 25003 ! 25040 ! Gambaga Scarp (Dec.) *Adams & Akpabla* GC 4247 ! Kete Krachi to Atebubu (Dec.) *Adams* 4608 ! **N.Nig.:** Nupe *Barter* 1288 ! Hepham to Ropp (July) *Lely* 375 ! 376 ! Ropp, 4,600 ft. *Lely* 452 ! Vom, 4,000 ft. (Oct.) *Dent Young* 142 ! *Hepper* 1158 ! Jos (Aug., Dec.) *Batten-Poole* 380 ! *Coombe* 104 ! *Keay* FHI 104 ! Naraguta F.R. (Aug.) *Keay* FHI 20062 ! **S.Nig.:** Obudu, Ogoja (Aug.) *Stone* 81 ! [Br.]Cam.: Bamenda Dist., 5,000–8,200 ft. (Feb., Apr.–June, Sept., Dec.) *Maitland* 1454 ! 1456 ! 1768 ! *Ujor* FHI 30210 ! *Boughey* GC 11042 ! *Adams* GC 11317 ! *Morton* K264 ! *Hepper* 2103 ! Extends to Sudan, Congo, Uganda and Mozambique.

33a. **V. purpurea** var. **schnellii** *C. D. Adams* in J.W. Afr. Sci. Assoc. 3 : 121 (1957). ? *V. purpurea* forma *poly-cephala* O. Hoffm. of Chev. Bot. 357. A perennial herb of montane grassland, like the last but with smaller more numerous heads and less conspicuous leafy bracts ; florets mauve.
Guin.: Mali, Fouta Djalon (Sept., Oct.) *Schnell* 7200 ! *Jac.-Fél.* 1996 ! Kouria (Sept.) *Chev.* 14769 ! **S.L.:** Musaia (Sept.) *Deighton* 4874 ! **Iv.C.:** *fide* Chev. *l.c.*

34. **V. camporum** *A. Chev.* in Mém. Soc. Bot. Fr. 2, 8e : 259 (1917) ; Chev. Bot. 353. An erect woody herb up to 7 ft. high with a terete softly pale-pubescent little-branched stem ; undersurface of leaves and inflorescence tomentose ; inflorescence much-branched with numerous sessile narrowly turbinate flower-heads crowded in long spikes ; florets purple ; in savanna.
Sen.: Niokolo-Koba (Oct.) *Adam* 15702 ! **Guin.:** Kouroussa (Jan.) *Pobéguin* 630 ! Farako to Tagania (Jan.) *Chev.* 20408 ! **S.L.:** Falaba (Mar.) *Sc. Elliot* 5222 ! Musaia (Feb., Dec.) *Deighton* 4224 ! 4455 ! **Ghana:** Dukwesein, near Agogo, Ashanti (Dec.) *Chipp* 620 ! Ejura, Ashanti (Nov.–Dec.) *Vigne* FH 3464 ! *Morton* GC 9474 ! A2563 ! Banda Hills (Dec.) *Adams* 3135 ! *Morton* GC 25136 ! Jeketi to Ajena (Feb.) *Morton* GC 8531 ! Sakogu, Gambaga Scarp (Dec.) *Adams & Akpabla* GC 4248 ! **N.Nig.:** Nupe *Barter* 809 ! Ilorin (Dec., Jan.) *Ajayi* FHI 19263 ! *Ejiofor & Ujor* FHI 30242 ! Bonu, Niger Prov. (June) *E. W. Jones* 220 ! [Br.]Cam.: Serti, Adamawa (Dec.) *Latilo & Daramola* FHI 28958 !

35. **V. pauciflora** (*Willd.*) *Less.* in Linnaea 4 : 292 (1829) ; F.T.A. 3 : 283 ; Chev. Bot. 356 ; Berhaut Fl. Sén. 177. *Conyza pauciflora* Willd. (1803). An erect annual herb up to about 4 ft. high with striate, pubescent stems ; florets bluish-mauve or rarely white in heads ¾–1 in. across.
Sen.: *Roger* 123 ! Niokolo-Koba (Oct.) *Adam* 15873 ! **Gam.:** Kombo *Dawe* 3 ! Kudang (Jan.) *Dalz.* 8036 ! Genieri (Feb.) *Fox* 46 ! **Mali:** Goundam (Oct.) *Chev.* 2080. Lébézénga (Nov.) *Hagerup* 436 ! Kita Massif (Sept.) *Jaeger* ! Boré (Sept.) *Demange* 34/1957. **Port.G.:** Farim (Nov.) *Esp. Santo* 2411 ! Bafata (Nov.) *Esp. Santo* 2832 ! Bissura to Mansaba (Dec.) *Esp. Santo* 3743 ! **Guin.:** Mangata (Jan.) *Chev.* 20420 ! Kouroussa (Nov.) *Pobéguin* 599 ! Dantilia (Mar.) *Sc. Elliot* 5291 ! **Ghana:** Yendi (Apr., Dec.) *Adams & Akpabla* GC 4082 ! *Morton* GC 9082 ! Sakogu, Gambaga Scarp (Dec.) *Adams & Akpabla* GC 4238 ! White Volta R. (Dec.) *Adams & Akpabla* GC 4291 ! **Dah.:** *Burton* ! **Niger:** Zinder (Nov.) *Hagerup* 612 ! Aïr *De Miré & Gillet.* **N.Nig.:** Borgu *Barter* 870 ! Nupe *Baikie* ! Sokoto *Lely* 124 ! Katagum (Dec.) *Elliott* 166 ! Anchau, Zaria (Nov.) *Keay* FHI 21667 ! **S.Nig.:** Oyo (Jan.) *Keay* FHI 25643 ! Lagos (Aug.) *Phillips* 57 ! [Br.]Cam.: Bama, Dikwa Div. (Jan.) *McClintock* 10 ! Extends through Ubangi-Shari to Sudan, Ethiopia, Eritrea and east tropical Africa. (See Appendix, p. 423.)

36. **V. glabra** (*Steetz*) *Vatke* in Oestr. Bot. Zeitschr. 27 : 194 (1877) ; F.T.A. 3 : 286 ; Adams in J.W. Afr. Sci. Assoc. 3 : 118 (1957). *Linzia glabra* Steetz (1863). A widespread species in tropical Africa of which two varieties occur in our area.

36a. **V. glabra** var. **occidentalis** *C. D. Adams* l.c. 119 (1957). An erect herb 2–4 ft. high with blue florets ; a weed of disturbed ground in hilly districts.
N.Nig.: Jos (Nov.–Jan., Apr.) *Lely* P87 ! *Monod* ! *Savory* UCI 137 ! Naraguta *Lely* 696 ! Vom Dent Young 135 ! *Morton* K341 ! [Br.]Cam.: Nchan, 5,000 ft., Bamenda (Apr.) *Maitland* 1476 ! Bafut-Ngemba F.R., 6,500 ft. (Jan., Mar.) *Tiku* FHI 22237 ! *Onochie* FHI 34855 ! Ndop, 3,600 ft. (Dec.) *Boughey* GC 11100 ! Maisamari, 4,600 ft., Mambila Plateau (Jan.) *Hepper* 1672 !

36b. **V. glabra** var. **hillii** (*Hutch. & Dalz.*) *C. D. Adams* l.c. (1957). *V. hillii* Hutch. & Dalz. F.W.T.A., ed. 1, 2 : 168 (1931). A coarse erect herb up to 5 ft. high with blue florets.
[Br.]Cam.: Cam. Mt., 2,000–4,000 ft. (Nov.–Jan.) *Hill*! *Mildbr.* 10799! *Maitland* 234! *Migeod* 61! *Dalz.* 8225! *Akpabla* GC 10519! *Keay* FHI 28651!

37. **V. perrottetii** *Sch. Bip.* in Walp. Rep. 2 : 947 (1843) ; F.T.A. 3 : 272 ; Chev. Bot. 356 ; Berhaut Fl. Sén. 203. An erect much-branched annual herb 1–2 ft. or more high, with bright magenta florets in heads ½–¾ in. across, solitary at the ends of the branches ; in fields.
Sen.: *fide* Chev. *and* Berhaut *l.c.* **Gam.:** Bathurst (Jan.) *Irvine* 2796! Kudang (Jan.) *Dalz.* 8031! Genieri (Feb.) *Fox* 26! **Mali:** Oualia (Dec.) *Chev.* 38! Kita (Nov.) *Jaeger*! Dioura (Nov.) *Davey* 161! **Port.G.:** Pussube, Bissau (fl. & fr. Dec.) *Esp. Santo* 1085. **Guin.:** Dalaba to Mamou (Jan.) *Chev.* 20353! **S.L.:** Wallia (Feb.) *Sc. Elliot* 4263! Musaia (Apr.) *Deighton* 5466! Bumban (Feb.) *Glanville* 164! **Ghana:** Zuarungu (Nov.) *Akpabla* 411! Nalerigu (Dec.) *Adams & Akpabla* GC 4212! Tumu (Mar.) *Morton* GC 8846! Gambaga (Apr., May) *Morton* GC 7352! 8982! **Dah.:** *Burton*! **Niger:** Zinder (Dec.) *Vaillant* 872! **N.Nig.:** Kontagora (Nov.) *Dalz.* 191! Kano (Dec.) *Hagerup* 637! Jos, 4,000 ft. (Dec., Jan., Apr.) *Lely* P35! *Morton* K355! *Coombe* 5! Ropp (July) *Lely* 448! Bornu (Nov.) *Elliott* 125! Maiduguri (Oct.) *Noble* A12! **S.Nig.:** Lagos *Rowland*! Agulu, Awka (Nov.) *Keay* FHI 21523! **[Br.]Cam.:** Bama, Dikwa Div. (Nov.) *McClintock* 1! Extends to Sudan, Uganda, Nyasaland, Rhodesia and Angola. (See Appendix, p. 423.)

38. **V. bauchiensis** *Hutch. & Dalz.* F.W.T.A., ed. 1, 2 : 168 (1931). A perennial, slightly pubescent herb with stems about 6 in. high from a woody base ; florets purple in heads ½ in. long, solitary at the ends of the branches.
N.Nig.: Jos Plateau (Jan.) *Lely* P86!

39. **V. poskeana** *Vatke & Hildebrandt* in Oestr. Bot. Zeitschr. 25 : 324 (1875) ; F.T.A. 3 : 274 ; Berhaut Fl. Sén. 176.

39a. **V. poskeana** var. **elegantissima** (*Hutch. & Dalz.*) *C. D. Adams* in J.W. Afr. Sci. Assoc. 3 : 121 (1957). *Vernonia elegantissima* Hutch. & Dalz. F.W.T.A., ed. 1, 2 : 167 (1931). A slender erect slightly branched annual herb, usually about 1 ft. high ; florets purple, mauve or white in heads about ⅓ in. long, fairly numerous on slender stalks.
Sen.: Niokolo-Koba (Jan.) *Berhaut* 1553! **Ghana:** Dutukpene Scarp, 1,600 ft. (Dec.) *Adams* 4527! **N.Nig.:** Patti Lokoja (Nov.) *Dalz.* 36! Anara F.R., Zaria (Oct.) *Keay* FHI 5498! Igabi, Zaria (Nov.) *Keay* FHI 28112! Jos (Oct.) *Lely* P824! *Hepper* 1006! Other varieties and forms occur in Congo, Angola and E. Africa.

40. **V. plumbaginifolia** *Fenzl ex Oliv. & Hiern* in F.T.A. 3 : 279 (1877) ; Chev. Bot. 356, partly (*Chev.* 525 ; 560). Stems herbaceous slender striate thinly pubescent, ½–2 ft. high from a perennial woody stock ; florets bright purple in subcampanulate heads ¼–½ in. across.
Mali: Banfara (Mar.) *Chev.* 525! Bambanatoumba (Mar.) *Chev.* 560! **N.Nig.:** Birnin Gwari (Mar.) *Meikle* 1343! Also in Sudan.

41. **V. djalonensis** *A. Chev. ex Hutch. & Dalz.* F.W.T.A., ed. 1, 2 : 167 (1931) ; Chev. Bot. 354, name only. An erect robust perennial soft-stemmed undershrub, branched above, with crowded leaves and broadly turbinate flower-heads ¾ in. across when open ; florets blue- or reddish-purple ; middle and inner involucral bracts densely whitish- or yellow-woolly except at the tips, which are deep purple.
Guin.: Fouta Djalon (Mar., Aug.–Oct.) ; Diaguissa to Boulivel *Chev.* 18043! Timbo to Ditinn *Chev.* 18394! Dalaba Plateau, 3,000–4,000 ft. *Chev.* 18568! *Jac.-Fél.* 2063! 7065! *Adam* 12666! Mali *Schnell* 7228!

42. **V. oocephala** *Bak.* in Kew Bull. 1895 : 68. An erect undershrub 1–4 ft. high, almost unbranched or branched from near the base, with whitish florets in numerous crowded straw-coloured heads about ½ in. long.
N.Nig.: Jos Plateau *Lely* 250! Naraguta (May) *Lely* 67! Shere Hills (Dec.) *Savory* 128! Jos (bud Nov., fl. Dec.) *Onochie* FHI 21743! *Coombe* 3! *Morton* K350! Kaduna to Zaria (Dec.) *Meikle* 764! Zaria to Wazeta (Feb.) *Dalz.* 367! **[Br.]Cam.:** Vogel Peak, 4,000 ft., Adamawa (Dec.) *Hepper* 1534! Also in Congo, Tanganyika, Nyasaland, Rhodesia and Angola.

43. **V. saussureoides** *Hutch.* in Kew Bull. 1921 : 378, fig. 6. A perennial herb with erect glabrous pithy stems 2–3 ft. high arising several together from a woody stock ; florets rich blue in top-shaped heads ½–¾ in. long ; in hill grassland.
N.Nig.: Jos (Apr., June, July) *Lely* P236! *Batten-Poole* 117! Naraguta *Lely* 286! Ropp to Mongu *Morton* K364! Mongu F.R., Pankshin *Keay & King* FHI 37087! **[Br.]Cam.:** Bamenda to Bambui, 4,500 ft. (June) *Maitland* 1611! Nkom-Wum F.R. (July) *Ujor* FHI 30466! Kumbo (Dec.) *Boughey* GC 17477! Vogel Peak, 4,000 ft., Adamawa (Dec.) *Hepper* 2782!

44. **V. philipsoniana** *Lawalrée* in Inst. Sci. Nat. Belg. Expl. Hydrobiol. L. Tang. 4 : 59 (1955). *V. lappoides* O. Hoffm. (1897), not of Bak. (1873). *V. hoffmanniana* Hutch. & Dalz. F.W.T.A., ed. 1, 2 : 167, not of S. Moore (1900). A perennial herb woody at the base ; stems ribbed and loosely hairy, mostly simple or sparingly branched below the inflorescence ; florets mauve or purple in rather few flower-heads about 1 in. across.
N.Nig.: Jos Plateau : Ropp, 4,600 ft. (July) *Lely* 453! Vom *Dent Young* 132! Extends through Cameroun, Congo, Nyasaland and Rhodesia to Angola.

45. **V. ambigua** *Kotschy & Peyr.* Pl. Tinn. 35, t. 17B (1867) ; F.T.A. 3 : 272 ; Chev. Bot. 352 ; F.W.T.A., ed. 1, 2 : 167, partly (excl. syns. *V. klingii* and *V. courtetii* and *Chev.* specimens) ; Berhaut Fl. Sén. 172. *V. benthamiana* of Chev. Bot. 353, not of Oliv. & Hiern. An erect coarse annual ½–2 ft. high with ribbed pubescent stems ; florets magenta or mauve in heads ½ in. across.
Sen.: *fide* Berhaut *l.c.* Badi (Oct.) *Adam* 15826! **Mali:** Dendela (Mar.) *Chev.* 634! **Port.G.:** Pessube, Bissau (Nov.) *Esp. Santo* 1056! Antula, Bissau (Nov.) *Esp. Santo* 2621! Ticoe to Sare Bacar (Nov.) *Esp. Santo* 3616! **Guin.:** Labé (Mar.) *Adam* 11535! **S.L.:** Berria (Mar.) *Sc. Elliot* 5410! Masasa (Oct.) *Glanville* 1! Mabasike (Dec.) *Deighton* 5715! **Iv.C.:** Mankono (June) *Chev.* 21936! **Ghana:** Yendi (Dec.) *Adams & Akpabla* GC 4075! Tamale (Nov., Dec.) *Adams & Akpabla* GC 4176! *Darko* 433! *Morton* GC 9925! Damongo (Dec.) *Morton* GC 9945! 9996! 25058! Gambaga (Apr.) *Morton* GC 9053! Daboya ferry (Dec.) *Morton* GC 6194! **N.Nig.:** Nupe *Barter* 824! 942! Kontagora (Oct.) *Dalz.* 199! Kabama, Zaria (Nov.) *Sule* FHI 7016! Ilorin (Jan., Mar.) *Ejiofor* FHI 19843! 30238! Jos (Oct., Dec.) *Hepper* 1024! *Coombe* 103! **S.Nig.:** Lagos (Nov., Dec.) *Onochie* FHI 34657! *Dalz.* 1162! **[Br.]Cam.:** Gurum, near Vogel Peak, Adamawa (Nov.) *Hepper* 1267! Bama, Dikwa Div. (Dec.) *McClintock* 105! Extending through the drier parts of tropical Africa to Sudan and Angola. (See Appendix, p. 421.)
[The first-formed flower-heads have more numerous series of involucral bracts than those formed later on the same plants.]

46. **V. bambilorensis** *Berhaut* in Mém. Soc. Bot. Fr. 1953–54 : 7 (1954) ; Fl. Sén. 172. An erect biennial herb with an ascending-branched almost leafless stem 15–30 in. high ; leaves mostly in a basal rosette ; flower-heads clustered, about ½ in. broad with purple florets.
Sen.: Bambilor (Mar.–May, Sept.–Oct.) *Berhaut* 434 ; 2942 ; 3200 ; 3220! Sangalkam (May) *Berhaut* 5229!

47. **V. blumeoides** *Hook. f.* in J. Linn. Soc. 7 : 198 (1864) ; F.T.A. 3 : 281 ; Guinea in Anal. Inst. Bot. Cav. 10, 1 : 322. An erect little-branched pithy-stemmed undershrub 2–4 ft. high with shortly pubescent striate stems and mauve to reddish-purple florets in heads ¼–⅓ in. broad in dense terminal corymbs.
N.Nig.: Vom, 4,000 ft. (Oct.) *Hepper* 1155! **S.Nig.:** Sonkwala, 5,200 ft., Ogoja (Dec.) *Savory & Keay* FHI 25230! Koloishe, 4,000 ft., Ogoja (Dec.) *Savory & Keay* FHI 25053! **[Br.]Cam.:** Cam. Mt., 4,000–9,000 ft. (Nov.–Jan., Mar.–May) *Mann* 1241! 1921! *Maitland* 305! 828! 971! *Mildbr.* 10830! *Brenan*

9399 ! *Keay* FHI 28638 ! *Boughey* GC 6812 ! *Morton* GC 6786 ! K600 ! Bamenda to Ndop, 4,000–5,400 ft. (Dec.) *Adams* 1524 ! GC 11268 ! *Boughey* GC 10462*a* ! Maisamari, 4,600 ft., Mambila Plateau (Jan.) *Hepper* 1642 ! Vogel Peak, 3,500 ft., Adamawa (Nov.) *Hepper* 1485 !

48. **V. klingii** *O. Hoffm. & Muschl.* in Mém. Soc. Bot. Fr. 2, 8c : 112 (1910) ; Chev. Bot. 355 ; Adams in J.W. Afr. Sci. Assoc. 2 : 65 (1956). *V. ambigua* of F.W.T.A., ed. 1, 2 : 167, partly (*Chev.* 20895), not of Kotschy & Peyr. *V. conyzoides* Hutch. & Dalz. F.W.T.A., ed. 1, 2 : 168 (1931), not of DC. (1834). *V. thomasii* Hutch. (1936). Stems herbaceous, striate, pubescent with whitish curled hairs, 1–3 ft. high from a perennial woody stock ; florets purple in heads ⅓–½ in. across ; in grassy hill savanna.
Guin.: Kaba to Upper Mamou (Apr.) *Chev.* 12758 ! Diédédou to Nionssomoridou (Feb.) *Chev.* 20851 ! Boula (Apr.) *Adam* 12042 ! Dalaba (May) *Roberty* 1932 ! Kindia (June) *Jac.-Fél.* 1775 ! **S.L.:** Falaba (Mar.) *Sc. Elliot* 5223 ! Bafodeya (Apr.) *Sc. Elliot* 5684 ! Makumri (June) *Thomas* 546 ! Kabala (Feb.) *Deighton* 4247 ! Bintumane Peak, 6,000 ft. (May, Sept.) *Jaeger* 1990 ! *Deighton* 5096 ! **Iv.C.:** Gouékan-gouiné to Droupleu, Upper Sassandra (May) *Chev.* 21451 ! **Ghana:** Banda ravine, Ashanti *Morton* A3279 ! Shiare, 2,500 ft., Trans Volta Togo (Apr.) *Hall* 1410 ! **Dah.:** Birni to Nioro, Atacora Mts. (June) *Chev.* 23966 !

49. **V. bamendae** *C. D. Adams* in J.W. Afr. Sci. Assoc. 3 : 116 (1957). A robust herb up to 5 ft. high with a brownish-tomentose ribbed stem and broad sessile leaves ; inflorescence spreading with flower-heads ⅜ in. broad sessile or subsessile along the upper sides of the branches ; florets purple.
[Br.]Cam.: Bamenda, 5,000–8,200 ft. (Dec.–Jan., May–June) : Basenako *Maitland* 1455 ! Lakom *Maitland* 1514 ! Mba Kokeka Mt. *Keay* FHI 28410 ! Santa Mt. *Boughey* GC 11013 !

50. **V. rugosifolia** *De Wild.* in Bull. Jard. Bot. Brux. 5 : 89 (1915). An erect herb of grassland 2–3 ft. high with numerous shortly stalked flower-heads about ⅓ in. long spread along the branches of a compact inflorescence : florets purple.
Guin.: Mali (Jan.) *Roberty* 16570 ! **N.Nig.:** Zaria *Kennedy* 2899 ! **[Br.]Cam.:** Bambui, 6,300 ft., Bamenda (Dec.) *Boughey* GC 10411 ! Nchan, 5,000 ft., Bamenda (June) *Maitland* 1457 !

51. **V. pumila** *Kotschy & Peyr.* Pl. Tinn. 37, t. 17A (1867) ; F.T.A. 3 : 292 ; Chev. Bot. 357 ; Berhaut Fl. Sén. 4, 174, 185. A perennial herb with pale-pubescent flowering stems from a few inches to 2 ft. high, appearing before the leaves from a woody stock with thickened spindle-shaped roots ; florets reddish or bluish-purple fading to white in heads ¾–1in. across ; stamens white ; on bare sandy or stony ground in savanna.
Sen.: Niokolo-Koba *Berhaut* 1503. **Mali:** Sarankoro (Feb.) *Chev.* 363. **Guin.:** *Chillou* 16 ! Ditinn to Labé (Apr.) *Chev.* 12896 ! **Dah.:** Atacora Mts., 1,700 ft. (June) *Chev.* 24064 ! **N.Nig.:** Ilorin *Rowland* ! Zungeru (Jan.) *Elliott* 26 ! Kontagora (Jan.) *Dalz.* 190 ! Minna (Dec., Jan.) *Meikle* 690 ! *Keay* FHI 37325 ! Zaria (Feb.) *Dalz.* 386 ! Vom (Jan.) *Dent Young* 139 ! **S.Nig.:** Oyo (Jan., Feb.) *Keay* FHI 14770 ! *Brenan* 8967 ! Extends to Sudan and Uganda. (See Appendix, p. 423.)

52. **V. guineensis** *Benth.* in Fl. Nigrit. 427 (1849) ; F.T.A. 3 : 285 : Chev. Bot. 354 ; Adams in J.W. Afr. Sci. Assoc. 6 : 153 (1961). A widespread species with the following varieties in our area.

52a. **V. guineensis** *Benth.* var. **guineensis.** *V. firma* Oliv. & Hiern in F.T.A. 3 : 290 (1877). A herb with strong erect stems arising from a perennial stock having numerous thick fusiform roots ; florets bluish-purple fading to pink or mauve in heads ½–1 in. across, generally smaller towards the western end of the range ; in savanna.
Mali: *fide* Chev. *l.c.* **Guin.:** Timbo, Fouta Djalon (Mar.) *Chev.* 12431 ! Labé (Apr.) *Adam* 11763 ! Beyla (Feb.) *Jac.-Fél.* 1546 ! **S.L.:** Batkanu (Feb.) *Glanville* 168 ! Rowala (July) *Thomas* 1126 ! Mambolo (Jan.) *Deighton* 979 ! Musaia (Apr.) *Deighton* 5388 ! Kamaro to Baleo (Aug.) *Jaeger* 902 ! **Iv.C.:** Bouaké, Baoulé-Nord (July) *Chev.* 22106 ! Asakra (Aug.) *Boughey* GC 18621 ! **Ghana:** Salaga *Krause* ! Afram Plains (Mar.) *Johnson* 715 ! Tamale (Mar., Dec.) *Adams & Akpabla* GC 4168 ! *Morton* GC 8778 ! Ejura Scarp (Mar.) *Adams* 2453 ! Amedzofe Hill (Feb., Apr.) *Morton* GC 6513 ! A2008 ! **Togo Rep.:** *Baumann* 92 ! Agome (Mar.) *Schlechter* 12964 ! **N.Nig.:** Jos Plateau (Jan.) *Lely* P88 ! Naraguta, 4,000 ft. (May) *Lely* 242 ! Shere Hills, 4,000 ft. (Dec.) *Coombe* 106 ! Jenre, Zaria (Dec.) *Coombe* 83 ! Extends to Sudan. (See Appendix, p. 422.)

52b. **V. guineensis** var. **cameroonica** *C. D. Adams* l.c. 154 (1961). An erect herb with densely grey- or brownish-pubescent stems 1–5 ft. high ; florets purple fading to pale mauve in heads about ¾ in. across with the tips of the involucral bracts papery and acute, coloured mauve or pink fading to brown.
[Br.]Cam.: Bamenda, 5,000–6,000 ft. (Jan., Mar., Apr., June) *Migeod* 295 ! *Maitland* 1389 ! *Ujor* FHI 30038 ! *Richards* 5294 ! Ndop, 3,600 ft. (Dec.) *Adams* GC 11064 ! Ndop to Kumbo (Dec.) *Boughey* GC 11166 ! Gashaka, Adamawa (Feb.) *Latilo & Daramola* FHI 34467 !

52c. **V. guineensis** var. **procera** (*O. Hoffm.*) *C. D. Adams* l.c. 154 (1961). *V. procera* O. Hoffm. in Mém. Soc. Bot. Fr. 2, 8 : 39 (1908) ; Chev. Bot. 357. Flower-heads solitary, about 2 in. across with mauve florets turning white ; leaves often yellowish-tomentose beneath at least when dry.
Dah.: Ady, near Savé (May) *Chev.* 23569. **N.Nig.:** Jos Plateau (Mar., Apr., Sept., Nov.) *Lely* P179 ! *Kennedy* FHI 11882 ! *Morton* K345 ! Naraguta *Lely* 70 ! 721 ! Abinsi (Feb.) *Dalz.* 659 ! Bonu, Niger Prov. (June) *E. W. Jones* 157 ! *Onochie* FHI 18700 ! **S.Nig.:** Olokemeji *Foster* 280 ! Upper Ogun F.R., Oyo (Feb.) *Keay* FHI 22508 ! Also in Cameroun and Ubangi-Shari.
[*Dalz.* 659 and *Keay* FHI 22508 are somewhat intermediate between this and var. *guineensis.*]

53. **V. nestor** *S. Moore* in J. Linn. Soc. 35 : 317 (1902). *V. chariensis* O. Hoffm. (1908). A perennial herb with stiff stems 1–2 ft. high branched only at the top ; leaves densely crowded and rather small, these and the stems densely clothed with long silky hairs ; florets mauve in numerous heads ⅓ in. long in rounded clusters in a broad corymb.
Guin.: Boumalol, Fouta Djalon (Oct.) *Adam* 12568 ! **N.Nig.:** Kontagora (Nov.) *Dalz.* 202 ! Jos, 4,000 ft. (Dec.) *Lely* P796 ! *Coombe* 26 ! *Monod* 9688 ! Vom *Dent Young* 141 ! Maiduguri *Kennedy* 2989 ! **[Br.] Cam.:** Bamenda, 4,500–5,000 ft. (Jan., Apr.) *Migeod* 301 ! *Maitland* 1582 ! Nchan (June) *Maitland* ! Bamenda to Ndop, 5,000 ft. (Dec.) *Adams* GC 11271 ! Extends on high ground to Nyasaland.

54. **V. cistifolia** *O. Hoffm.* in Engl. Pflanzenw. Ost-Afr. C : 404 (1895). A shortly pubescent undershrub up to 6 ft. high with red-purple florets in heads ¼–⅓ in. long, in cymes forming a panicle about 1 ft. long.
N.Nig.: Jos (Oct.) *Lely* P801 ! Naraguta (Nov.) *Keay* FHI 21026 ! Also in Congo and E.Africa.

55. **V. smithiana** *Less.* in Linnaea 6 : 638 (1831) ; F.T.A. 3 : 276 ; Chev. Bot. 358. A perennial herb with erect slender pale-silky stems 1–3 ft. high from a woody stock ; florets pinkish- or reddish-purple in hemispherical heads about ½ in. across, crowded in small dense terminal corymbs ; in hill grassland.
Mali: Morigueya (Feb.) *Chev.* 434 ! **Guin.:** Timbikounda to Farakoro (Jan.) *Chev.* 20626 ! Diédédou to Nionssomoridou (Feb.) *Chev.* 20850 ! **[Br.]Cam.:** Bamenda Dist., 5,000–7,400 ft. (Feb.–May, Dec.) *Maitland* 1370 ! 1755 ! *Ujor* FHI 30004 ! *Onochie* FHI 34857 ! *Hepper* 2153 ! *Adams* GC 11216 ! *Boughey* GC 10786 ! *Morton* K9 ! K56 ! Gembu, 4,600 ft., Mambila Plateau (Jan.) *Hepper* 1837 ! Also in Congo, Sudan and through E. Africa to Angola.

56. **V. migeodii** *S. Moore* in J. Linn. Soc. 35 : 319 (1902). *V. courtetii* O. Hoffm. & Muschl. (1910)—Chev. Bot. 354. *V. plumbaginifolia* of Chev. Bot. 356, partly (*Chev.* 24083), not of Fenzl. *V. ambigua* of F.W.T.A., ed. 1, 2 : 167, partly (*Chev.* 23676 ; 24083), not of Kotschy & Peyr. A perennial herb with several erect striate sparsely pubescent stems from a woody stock, 2–4 ft. high ; leaves sessile, rounded at the base, minutely glandular-punctate beneath ; florets purple in heads about ½ in. long, with the outer involucral bracts much shorter than the inner.
Ghana: Yendi (Apr.) *Akpabla* 514 ! Dutukpene (Jan.) *Akpabla* 1806 ! **Dah.:** Agouagon to Savalou (May) *Chev.* 23676 ! Toukountouna to Tanguéta (June) *Chev.* 24083 ! **N.Nig.:** Lokoja (Oct.) *Dalz.* 32 ! Anara F.R., Zaria (May) *Keay* FHI 22952 ! Igabi, Zaria (May) *Keay* FHI 25768 ! R. Benue *Talbot* ! **S.Nig.:** Lagos *Phillips* 50 ! Awka *Thomas* 33 ! Obu *Thomas* 367 ! 428 ! Ogbomosho *Rowland* ! Onitsha to

Enugu (Mar.) *Morton* K5 ! Ikom (May) *Jones* FHI 10484 ! [Br.]**Cam.**: Kentu, Bamenda (June) *Maitland* 1574 ! Also in Ubangi-Shari. (See Appendix, p. 423.)

57. **V. glaberrima** *Welw. ex O. Hoffm.* in Bol. Soc. Brot. 13: 15 (1896) ; Chev. Bot. 354. An erect slender-branched undershrub 1–6 ft. high, scabrid-pubescent only on the younger parts, soon glabrous ; leaves finely reticulate-veined ; florets white or pale pink in numerous heads ¼ in. long ; in open hillside grass-land.

Guin.: Kaba Valley (May) *Chev.* 13237 ! **S.L.**: Falaba (Mar.) *Sc. Elliot* 5152 ! *Burbridge* 543 ! 554 ! Musaia (Apr., Aug.) *Deighton* 4352 ! 5417 ! **Iv.C.**: Zoanlé, Upper Sassandra (May) *Chev.* 21493 ! **Ghana:** Togo Plateau F.R. (Oct.) *St. C. Thompson* 1523 ! Amedzofe (Feb., Apr.) *Morton* GC 6517 ! A2011 ! 2684 ! **Dah.**: Pobégou to Birni, Atacora Mts. (June) *Chev.* 23953 ! **N.Nig.**: Samaru, Zaria (May) *Keay* FHI 25731 ! Birnin Gwari (Mar.) *Meikle* 1341 ! Auchang (Feb.) *Dalz.* 346 ! Naraguta *Lely* 65 ! 75 ! Jos (Apr.) *Morton* K346 ! [Br.]**Cam.**: Bamenda, 3,000–6,000 ft. (Mar.–June, Dec.) *Maitland* 1402 ! 1540 ! 1696 ! *Boughey* GC 11098 ! *Morton* K197 ! Also in Angola. (See Appendix, p. 422.)

58. **V. cinerea** (*Linn.*) *Less.* in Linnaea 4 : 291 (1829) ; F.T.A. 3 : 275 ; Chev. Bot. 353 ; F.W.T.A., ed. 1, 2 : 168 (excl. syn. *V. undulata*) ; Berhaut Fl. Sén. 176, 204. *Conyza cinerea* Linn. (1753). An erect branched herb 1–3 ft. or rarely more high with shortly pubescent striate stems and mauve or reddish-purple florets in numerous heads about ¼ in. broad ; a common weed of farms, roadsides and waste places.

Sen.: *Roger* ! M'Bidgem *Hb.* Franqueville 56 ! Niombato (Nov.) *Berhaut* 670 ! **Mali:** Kara (Dec.) *Davey* 180 ! Fenadjé (Jan.) *Davey* 256 ! **Port.G.**: Pessubé, Bissau (Nov.) *Esp. Santo* 1058 ! **Guin.**: Labé (Apr.) *Adam* 11790 ! **S.L.**: (June) *T. Vogel* 108 ! Kumbagwe (Sept.) *Thomas* 2095 ! Kambia (Dec.) *Deighton* 854 ! Njala (Sept.) *Deighton* 2118 ! Musaia (Apr.) *Deighton* 5379 ! Newton (Apr.) *Deighton* 5923 ! **Lib.:** Monrovia (Sept.) *Baldwin* 9213a ! Buchanan (Nov.) *Adam* 16054 ! **Iv.C.**: Mankono (Aug.) *Boughey* GC 18497 ! **Ghana:** Accra (Oct.) *Brown* 306 ! Elmina (Oct.) *Andoh* 5074 ! Jasikan (Dec.) *Box* 251 ! Ejura, Ashanti (Aug.) *Andoh* 5054 ! Lawra (Apr.) *Adams* 4130 ! **Togo Rep.**: Lomé *Warnecke* 233 ! **Dah.**: *fide* Chev. *l.c.* **N.Nig.**: *Kennedy* 1768 ! Nupe *Barter* 1081 ! Lokoja (Sept.) *Parsons* 3 ! **S.Nig.**: Aguku *Thomas* 582 ! Lagos (Oct.) *Dalz.* 1166 ! Ibadan (Aug.) *Newberry* 58 ! [Br.]**Cam.**: Victoria (Apr.) *Morton* GC 7130 ! Widespread in tropical countries. (See Appendix, p. 422.)

[A diffusely branched form of this species with abnormally small capitula is represented by *Kennedy* 1768 listed above. *Deighton* 5923 and *Baldwin* 9213a are similar but the divergence from normal is less exaggerated.]

59. **V. undulata** *Oliv. & Hiern* in F.T.A. 3 : 276 (1877) ; Chev. Bot. 358. A perennial herb with erect leafy stems up to 4 ft. high from a woody stock ; leaves smaller towards the stem-base ; florets magenta in distinctly stalked heads ¼ in. long ; in grassland.

Guin.: Nzo (Mar.) *Chev.* 21050 ! 21055 ! **S.L.**: *Burbridge* 538 ! Musaia (Apr.) *Deighton* 5443 ! **Ghana:** Kpandu *Robertson* 109 ! Amedzofe (Feb., Apr., Nov.) *Morton* GC 6511 ! 9425 ! A2010 ! **N.Nig.**: Kaduna (Dec.) *Meikle* 791 ! Jos Plateau (Jan.) *Lely* 10 ! 90 ! *Morton* K336 ! Samaru, Zaria (July) *Keay* FHI 25924 ! Kwakuti, Niger (Dec.) *Keay* FHI 37310 ! [Br.]**Cam.**: Nchan, Bamenda, 5,000 ft. *Maitland* 1682 ! Cheddar Gorge, 6,000 ft., Bamenda (Mar.) *Morton* K196 ! Wum crater lake (Apr.) *Morton* K292 ! Gembu, 4,600 ft. Mambila Plateau (Jan.) *Hepper* 1838 ! Also in Sudan, Congo, Uganda and Angola.

60. **V. cinerascens** *Sch. Bip.* in Schweinf. Beitr. Fl. Aethiop. 162 (1867) ; F.T.A. 3 : 275 ; De Miré & Gillet in J. Agric. Trop. 3 : 711. A shrub 1–3 ft. high with whitish-pubescent glandular stems and leaves ; florets deep reddish-purple in stalked heads scarcely ¼ in. broad.

Sen.: Niokolo (June) *Adam* 14230 ! **Niger:** central and southern Afr, among rocks and talus in mountains above 5,000 ft. *vide* De Miré & Gillet *l.c.* Extends through Sudan to the Red Sea, Ethiopia and north-western India, and also in Angola and S.W. Africa.

63. ERLANGEA Sch. Bip.—F.T.A. 3 : 265, not of F.W.T.A., ed. 1, 2 : 170.

Involucres 1–1·3 cm. broad on velvety-tomentose peduncles 1·5 cm. or more long, few and hardly contiguous ; involucral bracts ovate, acuminate with spreading tips ; leaves opposite, verticillate or alternate, shortly petiolate, ovate to lanceolate, up to 12 cm. long and 3 cm. broad, serrate, with 8–12 pairs of ascending lateral nerves .. .. 1. *fruticosa*

Involucres 5–8 mm. broad on peduncles 1–5 mm. long, numerous and tightly clustered in broad cymes at the ends of slender pubescent branches ; involucral bracts ovate with erect acute tips and broad scarious margins ; leaves opposite or alternate, petiolate, blade elliptic, up to 15 cm. long and 6 cm. broad, serrate-dentate, with 10–14 pairs of curved-ascending lateral nerves .. .. .. 2. *schimperi*

1. **E. fruticosa** *C. D. Adams* in J.W. Afr. Sci. Assoc. ined. (1963). A shrub up to about 8 ft. high with velvety striate branchlets scarred with the bases of old leaves ; flower-heads about ¼ in. across, rather few in flat terminal corymbs ; florets purple or mauve.

Guin.: Mt. Benna, Kindia (fl. & fr. Nov.) *Jac.–Fél.* 2123 ! Madina Tossékré (fr. Oct., fr. Mar.) *Adam* 11680 ! 12532 ! **S.L.**: Loma Mts. (fr. Nov.) *Jaeger* 489 !

2. **E. schimperi** (*Oliv. & Hiern ex Benth.*) *S. Moore* in J. Linn. Soc. 35 : 313 (1902). *Bothriocline schimperi* Oliv. & Hiern ex Benth. in Hook. Ic. Pl. t. 1133 (1873) ; F.T.A. 3 : 266. Erect and shrubby up to about 4 ft. high with white, pale blue or mauve florets in clustered heads ¼ in. across.

S.L.: Bintumane Peak (Nov.) *Jaeger* 363 ! **S.Nig.**: Ikwette Plateau, 5,200 ft., Ogoja (Dec.) *Savory & Keay* FHI 25220 ! [Br.]**Cam.**: Bamenda Dist. (Dec., Jan., Mar.) *Keay* FHI 28330 ! *Boughey* GC 11123 ! *Morton* K150 ! Also in E. Africa.

64. CENTRATHERUM Cass. in Bull. Sci. Soc. Philom. Paris 1817 : 31 ; Dict. Sci. Nat. 7 : 383 (1817).

Stems densely villous, glabrescent ; leaves alternate narrowly oblanceolate, acute, up to 8 cm. long and 1·5 cm. broad, thinly setulose above, woolly-tomentose beneath, sharply serrulate ; flower-heads few in a loose terminal corymb, shortly pedunculate, 1–1·5 cm. diam. ; involucral bracts very numerous, subulate-filiform, very acute, up to 1 cm. long, pubescent, the outermost sometimes leafy or the flower-head closely subtended by a single linear leafy bract ; florets all tubular ; achenes glabrous, closely 10-ribbed ; pappus caducous, subplumose towards base .. .. *angustifolium*

C. angustifolium (*Benth.*) *C. D. Adams* in J.W. Afr. Sci. Assoc. 6 : 149 (1961) *Gymnanthemum angustifolium* Benth. (1849). *Vernonia benthamiana* Oliv. & Hiern in F.T.A. 3 : 282 (1877). *Erlangea angustifolia* (Benth.) Hutch. & Dalz. F.W.T.A., ed. 1, 2 : 170 (1931). An erect simple or slightly branched annual herb with

fairly stout striate stems 1–2 ft. high ; florets pale mauve or pink to bluish-purple in subglobose flower-heads about ⅓ in. across.
Guin.: Dalaba-Diaguissa Plateau (Oct.) *Chev.* 18765 ! 18793 ! *Jac.-Fél.* 2019 ! Kindia (Nov.) *Jac.-Fél.* 2073 ! Sondomoli Mt. (Sept.) *Schnell* 7314 ! Madina Tossékré (Oct.) *Adam* 12517 ! 12562 ! **S.L.:** *Barter* ! Makene (Oct.) *Glanville* 6 ! Bumban (Aug., Sept.) *Thomas* 2009 ! *Deighton* 1219 ! Njala (Sept.) *Deighton* 3443 ! Loma Mts. (Nov.) *Jaeger* 623 ! 640 ! 664 ! Mayumba (Dec.) *King* 13b ! **Iv.C.:** Danané to Mt. Goula, Dyolas country (Apr.) *Chev.* 21234 ! **N.Nig.:** *Hepburn* ! (See Appendix, p. 417.)

65. ETHULIA Linn. f.—F.T.A. 3 : 262.

Stems ribbed, shortly pubescent ; leaves oblanceolate to obovate-oblanceolate, acute, 6–10 cm. long, 1–3·5 cm. broad, serrate, shortly pubescent and glandular beneath, with conspicuous parallel lateral veins ; petiole narrowly winged ; flower-heads discoid, very small, in lax corymbose terminal cymes ; involucral bracts in about 3 rows, up to 2·5 mm. long, obovate, thinly pilose ; achenes 4–5-ribbed ; pappus a cartilaginous undulate ring .. .. *conyzoides*

E. conyzoides *Linn. f.* Decas Prima Pl. Rar. 1, t. 1 (1762) ; F.T.A. 3 : 262 ; *Chev. Bot.* 351 ; Berhaut Fl. Sén. 176. An erect branched aromatic herb 2–4 ft. high, with numerous small flower-heads with pink, mauve or rarely white florets ; usually in wet grassland or by rivers.
Sen.: *Perrottet* ! *Berhaut* 1067 ; 4259. **Gam.:** *Pirie* 33 ! Georgetown (Jan.) *Dalz.* 8035 ! **Mali:** Sikasso (May) *Chev.* 786 ! Dio (Nov.) *Jaeger* 7 ! **Port.G.:** Bafata (Feb., Oct., Dec.) *Esp. Santo* 2673 ! 2799 ! 3799 ! Gabu (Nov.) *Esp. Santo* 3579 ! **Guin.:** Kouroussa *Pobéguin* 457 ! Nzérékoré (Oct.) *Baldwin* 9733 ! Dantilia (Mar.) *Sc. Elliot* 5273 ! **S.L.:** Pendembu (July) *Thomas* 763 ! Yonibana (June) *Deighton* 5844 Rokupr (Sept.) *Jordan* 351 ! Musaia (Feb.) *Deighton* 4208 ! **Lib.:** Du R. (Aug.) *Linder* 243 ! Ganta (May, Dec.) *Harley* ! *Barker* 1127 ! **Iv.C.:** Prolo (Aug.) *Chev.* 19865. Tiassalé, Bandama R. (Dec.) *Leeuwenberg* 2135 ! **Ghana:** Nwereme, Ashanti (May) *Irvine* 2472 ! Akuse to Kpong (Oct.) *Morton* A27 ! White Volta R., Bolgatanga to Bawku (Apr.) *Morton* GC 8954 ! Oti R., near Kete Krachi (May) *Morton* GC 7285 ! **Dah.:** Djougou (June) *Chev.* 23897. **N.Nig.:** Nupe *Barter* 1291 ! Aboh *Barter* 76 ! Kontagora (Nov.) *Dalz.* 193 ! Katagum *Dalz.* 170 ! Sokoto (Nov.), Dec.) *Moiser* 177 ! 222 ! Ropp, 4,000 ft. (Dec.) *Coombe* 59 ! **S.Nig.:** Lagos (Mar.) *Akpabla* 1085 ! Ibadan (Apr.) *Keay & Jones* FHI 19439 ! *Meikle* 1411 ! Eket *Talbot* ! Calabar (Apr.) *Akpata* 3929 ! **[Br.]Cam.:** N. Korup F.R., Kumba (July) *Olorunfemi* FHI 30696 ! Bali, Bamenda (May) *Ujor* FHI 30385 ! Bama, 1,000 ft., Dikwa Dist. (Nov.) *McClintock* 28 ! In tropical and extra-tropical Africa including S. Tomé, and Asia and the Mascarene Islands. (See Appendix, p. 417.)

66. GUTENBERGIA Sch. Bip.—F.T.A. 3 : 263.

Flower-heads about 5 mm. diam., numerous ; involucral bracts in 3–4 series, acute, loosely pubescent outside ; leaves sessile, auricled at base, linear-lanceolate, 4–6 cm. long, up to 1 cm. broad, discolorous, thinly pubescent above, white-tomentose beneath; achenes 1·5 mm. long, about 8-ribbed, glabrous ; corolla 3·5 mm. long 1. *rueppellii*
Flower-heads 1 cm. or more diam., few or solitary :
    Capitula several, pedunculate ; involucral bracts in 6 or more series, acutely acuminate or mucronate ; leaves subsessile, oblong-lanceolate ; corolla 6–8 mm. long ; achenes 2–2·5 mm. long, 10-ribbed, glabrous :
        Middle involucral bracts acuminate, pubescent ; capitula about 1 cm. diam. ; leaves mostly narrowed to base or shortly rounded, thinly greyish-pilose beneath
                          2. *nigritana*
        Middle involucral bracts subtruncate-mucronate, glabrescent ; capitula 2–2·5 cm. diam. ; leaves mostly broadly rounded at base, usually white woolly pilose beneath
                          3. *macrocephala*
    Capitula solitary, sessile, 2·5–3 cm. diam., closely surrounded by clustered leaves ; stem-leaves small, linear-lanceolate .. .. 4. *foliosa*

1. G. rueppellii *Sch. Bip.* in Gedenkb. 4 Jubelf. Buchdr. 120, t. 4 (1840) ; F.T.A. 3 : 263. *Ethulia rueppellii* (Sch. Bip.) Hochst. ex A. Rich. (1848). An erect much-branched annual herb 1–2 ft. high with the leaves whitish beneath ; florets reddish-purple or white in heads about ⅓ in. across ; a local weed of waysides and open land.
N.Nig.: Jos Plateau (Sept.) *Lely* P696a ! Naraguta, common (July, Nov.) *Lely* 464 ! 711 ! Also in Sudan, Ethiopia, Eritrea, Uganda and Kenya.
2. G. nigritana (*Benth.*) *Oliv. & Hiern* in F.T.A. 3 : 264 (1877) (incl. var. *scabra* Oliv. & Hiern). *Oiospermum nigritanum* Benth. (1849). *Gutenbergia macrocephala* of *Chev. Bot.* 351, not of Oliv. & Hiern. An erect herb, 1–4 ft. high with puberulous striate branches, wrinkled leaves and mauve to reddish-purple florets in heads ⅓ in. across ; in open grassland.
Guin.: Manankoro to Boola, Beyla (Mar., Apr.) *Chev.* 20904 ! *Adam* 12064 ! Nzérékoré (Oct.) *Jac.-Fél.* 2045 ! Timbo (Oct.) *Jac.-Fél.* 1872 ! 2084 ! Bafing (Oct.) *Adam* 12656 ! Nimba Mts. (Apr., Sept.) *Schnell* 1873 ! 3582 ! **S.L.:** *Glanville* 371 ! Betagya (Oct.) *Small* 493 ! Falaba (Nov.) *Miszewski* 18 ! Musaia (Dec.) *Deighton* 4437 ! **Ghana:** Wanki, Ashanti (June) *Chipp* 485 ! Pepease, 2,000 ft., Kwahu (July, Nov.) *Irvine* 1715 ! *Akpabla* 179 ! Adamsu, Ashanti (Dec.) *Adams* 3085 ! Amedzofe Hill (May) *Morton* GC 6512 ! Kpandu Fessi (July) *Asamany* 84 ! **N.Nig.:** Kontagora (June, Dec.) *Dalz.* 206 ! 667 ! Kaduna (Dec.) *McClintock* 196 ! *Morton* K380 ! *Hepper* 2876 ! **S. Nig.:** Niger (Sept.) *T. Vogel* 172 ! Lagos W. MacGregor 286 ! 291 ! Abeokuta (July) *Irving* 30 ! Chizea FHI 7871 ! Aboh *Barter* 435 ! Yoruba country *Barter* 1296 ! Oke-Iho, Oyo (July) *Savory* UCI 258 !
3. G. macrocephala *Oliv. & Hiern* in F.T.A. 3 : 264 (1877), not of Chev. Bot. 351. A robust herb up to about 4 ft. high with leaves usually wrinkled ; flower-heads at least 1 in. across, florets bright purple ; in grassy places.
Iv.C.: Tchumkrou to Akakoumoëkrou, Ano (Dec.) *Chev.* 22541 ! **Ghana:** Bame Pass (Oct.) *Morton* GC 9314 ! Mampong, Ashanti (July) *Darko* 686 ! Kukuramoa, Nsawkaw to Wenchi (Dec.) *Adams* 3182 ! Nyankpala (Dec.) *Adams & Akpabla* GC 4189 ! Vane (Nov.) *Morton* GC 9349 ! **S.Nig.:** Lagos *Millen* 62 ! Lagos to Abeokuta (July) *Savory* UCI 185 ! Abeokuta *Irving* !
[Both this species and *G. nigritana* are very variable in respect of leaf-margin, which may be serrated or entire, and the shape of the leaf-base, which may be broad or narrow. The density of the tomentum on the undersurface of the leaf in *G. macrocephala* is also variable, and there is a great range of shape between the inner and outer involucral bracts—C. D. A.]

4. **G. foliosa** *O. Hoffm.* in Engl. Bot. Jahrb. 24 : 462 (1897). A simple-stemmed herb up to 4 ft. high with scabrid crowded narrow leaves, woolly-tomentose beneath.

Togo Rep: Misahöhe, Angomé Tongbe (July) *Baumann* 34.

[*Baumann* 34 was seen for the first edition, but the specimen presumably was destroyed in Berlin ; apparently no further gatherings have been made—C. D. A.]

67. EUPATORIUM Linn.—F.T.A. 3 : 300.

Shrubby scrambler with pithy, finely striate branches ; leaves opposite, ovate, grossly dentate to subentire, cuneate at base, acute, 3–10 cm. long, 2–6 cm. broad, with a pair of curved lateral veins arising above the base, densely gland-dotted and puberulous beneath ; petiole slender, about 1 cm. long ; flower-heads pedunculate in spreading inflorescences at the ends of short stiff lateral branches ; involucre oblong-turbinate, 1 cm. long ; involucral bracts numerous, strongly nerved, imbricate, the outer shorter ; achenes ribbed with cartilaginous margins, glabrous or nearly so ; pappus-setae buff, in a single series, barbellate, about 6 mm. long   ..   1. *odoratum*

Erect herbs :

Stem-leaves mostly alternate, sessile or shortly petiolate, ovate to narrowly lanceolate, subacute, 5–7 cm. long, up to 3 cm. broad (frequently much smaller and narrower), gland-dotted beneath ; shoots several usually arising from a woody perennial stock ; stems, leaves and inflorescence curly-pubescent ; flower-heads crowded in a terminal inflorescence, few-flowered ; involucral bracts few, oblong, about 6 mm. long, membranous on the margins, not nerved ; achenes sulcate, pubescent ; pappus as above   ..   ..   2. *africanum*

Stem-leaves mostly opposite ; annual herbs with a single main glabrous or glabrescent stem :

Leaves elliptic-lanceolate, tapered to and distinctly 3-veined at base ; flower-heads up to 1 cm. broad ; involucral bracts linear, acute   ..   ..   3. *triplinerve*

Leaves ovate-rhomboid, abruptly narrowed to a slender petiole 1–3 cm. long, 3-veined, the lateral pair of veins submarginal at base ; flower-heads up to 5 mm. broad ; involucral bracts 2-seriate, linear-oblong, 3–4 mm. long, equalling or exceeding the florets, glabrous, glumaceous ; achenes blackish, sulcate ; pappus-setae very fine, about 20, spreading, white   ..   ..   4. *microstemon*

1. **E. odoratum** *Linn.* Syst. Nat., ed. 10, 2 : 1205 (1759). A diffuse shrub with angular pubescent branched stems up to about 8 ft. high ; florets white, mauve or pale blue in heads arranged in sub-umbellate inflorescences at the ends of short branches ; a rank-scented weed.

S.Nig.: Ubiaja, Benin (Apr.) *Daramola* FHI 31258 ! Udi, Enugu pitwood plantations (June, Nov.) *Jones* FHI 18804 ! *Keay & Mallam* FHI 22721 ! mile 34 Enugu to Abakaliki (Feb.) *Hepper* 2237 ! Native of Central and tropical S. America ; also in S. E. Asia ; spreading rapidly in our area.

2. **E. africanum** *Oliv. & Hiern* in F.T.A. 3 : 301 (1877) ; Chev. Bot. 360 ; Robyns Fl. Sperm. Parc. Nat. Alb. 2 : 457, t. 43. *Vernonia humilis* C. H. Wright (1897). *V. malosana* Bak. (1898). A herb or undershrub with usually several erect curled-pubescent stems 6 in. to 3 ft. high arising from a perennial woody stock ; florets cream to greyish-mauve, rather few in numerous small heads clustered at the branch-tips ; in grassland, mostly in hilly districts.

Guin.: Fassakouidou to Késséridou (Feb.) *Chev.* 20809 ! Kindia (Dec.) *Jac.-Fél.* 2199 ! Boula to Baizia (? Apr.) *Adam* 12057 ! Mali (June, Sept.) *Schnell* 7109 ! *Adam* 14451 ! **S.L.:** near Falaba (Mar.) *Sc. Elliot* 5148 ! Musaia to Sinkunia (Apr.) *Deighton* 5420 ! **Iv.C.:** Nimba Mts., 2,500–5,000 ft. (Apr., Aug.) *Schnell* 926 ! *Boughey* GC 18048 ! **Ghana:** Kintampo (Mar.) *Dalz.* 41 ! Trumeso, near Wenchi (Mar., Dec.) *Adams* 2873 ! *Morton* GC 8557 ! Kpandu (Apr.) *Asamany* 164 ! Amedzofe Hill (Mar., Apr.) *Morton* GC 6502 ! A2001 ! **N.Nig.:** Zaria (Feb.) *Dalz.* 403 ! Mando F.R., Zaria (June) *Keay* FHI 25836 ! Vom *Dent Young* 145 ! Sho (Nov.) *Keay* FHI 22523 ! Zaranda Mt., 5,800 ft. (May) *Lely* 191 ! **S.Nig.:** Upper Ogun F.R., Oyo (Mar.) *Keay* FHI 21021 ! [**Br.**]**Cam.:** Bamenda Dist., 4,500–8,000 ft. (Jan.–June) *Migeod* 328 ! 450 ! 493 ! *Maitland* 1412 ! 1492 ! *Ujor* FHI 30040 ! *Morton* K126 ! Nkambe, 6,100 ft. (Feb.) *Hepper* 1888 ! 2129 ! Kakara, 4,900 ft., Mambila Plateau (Jan.) *Hepper* 2818 ! Gashaka, Adamawa (Feb.) *Latilo & Daramola* FHI 34462 ! Throughout tropical Africa ; the only indigenous species.

3. **E. triplinerve** *Vahl* Symb. Bot. 3 : 97 (1794). A herb 2–3 ft. high with almost glabrous stems and leaves ; florets mauve in rather few heads about ⅓ in. broad ; a weed.

Dah.: Abomey (Feb.) *Chev.* 23174 ! Widely spread in the tropics of Asia and America and also in the Mascarene Islands.

4. **E. microstemon** *Cass.* in Dict. Sci. Nat. 25 : 432 (1822). *E. guadalupense* Spreng. (1826). An erect annual herb, branched above, 1–3 ft. high with terete striate glabrescent stems ; florets pale mauve to white in small numerous more or less clustered heads on long branches ; a common weed of farms and roadsides in forest areas, at least in Ghana.

Iv.C.: *teste* Mangenot. **Ghana:** Bosuso (May) *Adams* 1760 ! Asiakwa (Nov.) *Boughey* GC 13125 ! Techiman to Ejura (Dec.) *Morton* GC 9561 ! Wenchi to Chiraa (Dec.) *Adams* 3198 ! Aburi to Nsawam (Oct.) *Morton* A3440 ! Pramkese (Jan.) *De Wit & Morton* A2957 ! New Tafo (Oct.) *Lovi* WACRI 3926 ! Dunkwa (Feb.) *Whiting* 14 ! Also in Central America, Mexico and the West Indies and almost certainly introduced in our area.

[The first gathering of this weed was apparently made as recently as May 1953 ; it should be sought in other parts of the area of this Flora—C. D. A.]

68. MIKANIA Willd.—F.T.A. 3 : 301. *Nom. cons.*

Leaves coarsely crenate-dentate to subentire, ovate-triangular, acuminate, cordate to rounded-truncate at base, 5–10 cm. long, up to 8 cm. broad ; involucral bracts variable in length up to 8 mm. ; achenes black, 3 mm. long, mostly cartilaginous-ribbed on the angles, glandular ; pappus-setae minutely barbellate, often pinkish, up to 5 mm. long :

Capitula mostly sessile or shortly stalked in tightly rounded clusters ; involucral bracts about 6 mm. long   ..   ..   1. *cordata* var. *cordata*

Capitula on stalks 5–15 mm. long in an open panicle ; involucral bracts 7–8 mm. long
                             1a. *cordata* var. *chevalieri*
Leaves 3–5(–7)-lobed, up to about 5 cm. long and 4 cm. broad ; lobes oblong-lanceolate,
mucronate ; leaves openly cordate ; involucral bracts about 4 mm. long ; achenes
black, 2–2·5 mm. long, angled but margins not cartilaginous, eglandular, minutely
transversely rugulose ; pappus-setae barbellate, buff, less than 2 mm. long   2. *carteri*

1. **M. cordata** (*Burm. f.*) *B. L. Robinson* var. **cordata**—in Contrib. Gray Herb. 104 : 65 (1934). *Eupatorium
cordatum* Burm. f. (1768). *M. chenopodifolia* Willd. (1803). *M. scandens* of F.T.A. 3 : 301 ; Chev. Bot. 361 ;
F.W.T.A. ed. 1, 2 : 172, not of Willd. Scrambling up to 25 ft. or more in old farms and clearings ; florets
white or rarely bluish in numerous small few-flowered clustered heads.
  Sen.: Sedhiou, Casamance (Feb.) *Chev.* 2038 *bis* ! **Gam.:** *Hayes* 557 ! **Port.G.:** Pessube, Bissau (Oct.)
Esp. Santo 982 ! **Guin.:** *Heudelot* 610 ! *Farmar* 267 ! Kouria (Oct.) *Caille* in *Hb. Chev.* 14914 ! **S.L.:**
Don ! Sugar Loaf Mt. (Dec.) *Sc. Elliot* 3865 ! Mt. Horton, 1,500 ft. (Jan.) *T. S. Jones* 297 ! Pujehun (Dec.)
Deighton 255 ! Kitchom (Jan.) *Deighton* 961 ! Bintumane Peak, 5,400 ft. (Jan., Nov.) *T. S. Jones* 139 !
156 ! *Jaeger* 532 ! **Lib.:** Kakatown *Whyte* ! Gbanga (Sept.) *Linder* 751 ! Monrovia (June, Nov.) *Baldwin*
5876 ! *Dinklage* 3266 ! Genna Tanyehun (Dec.) *Baldwin* 10772 ! Dukwai R. (Oct.–Nov.) *Cooper* 40 !
Nimba Mts. (Dec.) *Adam* 16463 ! **Iv.C.:** Bingerville *Chev.* 15458 ! Nimba Mts., 5,000 ft. (Feb.) *Schnell*
446. **Ghana:** Ewisa, near Saltpond (Aug.) *Chipp* 567 ! Aburi (Apr.) *Brown* 322 ! Asin-Nyankumasi
Cummins 220 ! Wenchi, Ashanti (Dec.) *Adams* 3249 ! Togo Plateau F.R. (Sept.) *St. Clair Thompson* 1479 !
  Togo Rep.: *Baumann* 452 ! Lomé *Warnecke* 407 ! **N.Nig.:** Nupe *Barter* 186 ! Sokoto (June) *Dalz.* 550 !
Naraguta (June) *Lely* 321 ! Jebba *Hagerup* 754 ! Vom *Dent Young* 146 ! **S.Nig.:** Abeokuta *Irving* !
Nun R. (Aug.) *T. Vogel* 22 ! Lagos (Dec.) *Dalz.* 1168 ! Ibadan (Nov., Dec.) *Umana* FHI 26757 ! *Fakunle*
FHI 27671 ! *Ujor* FHI 29397 ! Oban *Talbot* 395 ! **[Br.]Cam.:** Cam. Mt., up to 7,000 ft. (Nov.–Feb.)
Mann 1326 ! 1924 ! *Preuss* 1097 ! *Mildbr.* 10809 ! Johann-Albrechtshöhe *Staudt* 554 ! Bamenda, 5,000 ft.
(Jan.) *Migeod* 322 ! Oku L., Bamenda (Jan.) *Keay & Lightbody* FHI 28670 ! Vogel Peak, 4,900 ft.,
Adamawa (Dec.) *Hepper* 1568 ! Ngel Nyaki, 5,400 ft., Mambila Plateau (Jan.) *Hepper* 1715 ! **F.Po:**
(June, Dec.) *T. Vogel* 165 ! *Barter* ! *Mann* 33 ! 7,000 ft. *Mann* 283 ! Monte Balea, 1,800 ft. (Dec.)
Guinea 468 ! Moka, 4,000 ft. (Dec.) *Boughey* 6 ! Widespread in the Old World tropics. (See Appendix,
p. 419.)

1a. **M. cordata** var. **chevalieri** *C. D. Adams* in J.W. Afr. Sci. Assoc. ined. (1963). *M. laxa* A. Chev. Bot.
360, name only, not of DC. This plant differs from typical *M. cordata* in having a more open inflorescence
with slightly larger capitula ; the stems and undersurface of the leaves are often purplish and the whole
plant is usually glabrous.
  S.L.: Jigaya (Nov.) *Thomas* 2839 ! Bumbuna (Oct.) *Thomas* 3937 ! **Lib.:** Gbanga (Sept.) *Linder* 448 !
Flumpa, Sanokwele (Sept.) *Baldwin* 9364 ! **Iv.C.:** Mbasso to Abongoua (Dec.) *Chev.* 22651 ! **S.Nig.:**
Lagos (Oct.) *Moloney* ! Sapoba, Benin (Sept., Nov.) *Jones* FHI 4981 ! *Meikle & Keay* 524 ! *Onochie*
FHI 34275 ! Nun R. (Sept.) *Mann* 496 ! Extending probably into Congo and Angola.

2. **M. carteri** *Bak.* in Kew Bull. 1895 : 106 ; Robinson l.c. 67. *M. scandens* Willd. var. *laciniata* Hutch. &
Dalz. F.W.T.A., ed. 1, 2 : 172. *M. tropaeolifolia* O. Hoffm. (1898). A scrambling herb with rather small
divided leaves ; involucres green ; florets white with purple styles.
  S.Nig.: Lagos *Rowland* ! Idanre Hills, Ondo (Oct.) *Keay & Onochie* FHI 20241 ! Okelife Hill, Ondo (Nov.)
Onochie FHI 34344 ! Ado Rock, Oyo (Sept., Oct.) *Savory & Keay* FHI 25450 ! *Keay* FHI 37182 ! Onitsha
(Oct.) *Onochie* FHI 34126 ! *Okafor* FHI 35871 ! Also in Cameroun ; this species closely resembles
M. corydalifolia Griseb. (1863) from Cuba.

69. ADENOSTEMMA J.R. & G. Forst.—F.T.A. 3 : 299.

Achenes smooth ; leaves long-petiolate, ovate to ovate-rhomboid, cuneate and subentire towards the base, 8–15 cm. long, 3–7 cm. broad, glabrous or nearly so beneath,
prominently trinerved from above the base, rather sharply serrate-dentate ; flower-
heads numerous, 6–8 mm. long ; achenes with a club-like pappus   1. *mauritianum*
Achenes closely warted ; pappus as above :
Leaves ovate, abruptly narrowed at base into the long petiole, crenate or coarsely
  serrate, 5–12 cm. long, 3–6 cm. broad, subscabrid on the nerves beneath ; flower-
  heads small and numerous   ..   ..   2. *perrottetii*
Leaves oblong-lanceolate to linear, subsessile or shortly petiolate, shallowly serrate, up
  to 12 cm. long and 1–3 cm. broad, glabrous ; flower-heads large and few   3. *caffrum*

1. **A. mauritianum** *DC.* Prod. 5 : 110 (1836). *A. viscosum* of F.T.A. 3 : 300, partly, not of J.R. & G. Forst.
An erect usually branched herb 2–4 ft. high, with rather rough stems and white florets in small heads in
loose dichotomous terminal panicles.
  S.Nig.: Eket *Talbot* 3371 ! [Br.]Cam.: Cam. Mt., 2,500–7,500 ft. (Nov.–Apr.) *Mann* 1218 ! 1958 ! *Dalz.*
8338 ! *Dunlap* 84 ! *Maitland* 481 ! *Migeod* 42 ! 129 ! *Hutch. & Metcalfe* 8 ! *Adams* 1643 ! Oku, 6,500 ft.,
Bamenda Dist. (Feb.) *Hepper* 2047 ! **F.Po:** 4,000–8,000 ft. (Dec.) *Mann* 575 ! *Boughey* 183 ! Extends to
Sudan, Congo, Rhodesia and the Mascarene Islands.

2. **A. perrottetii** *DC.* l.c. 110 (1836) ; F.T.A. 3 : 300, in syn. ; Berhaut Fl. Sén. 69. *A. viscosum* of Chev. Bot.
359, partly, not of J.R. & G. Forst. *A. mauritianum* of F.W.T.A., ed. 1, 2 : 172, partly (Liberian record),
not of DC. *Siegesbeckia abyssinica* of Stapf in Johnston Lib. 615, not of Oliv. & Hiern. A weak-stemmed
annual weed of damp shady places, 1–3 ft. high ; florets white or pale mauve.
  Sen.: Niokolo-Koba (Oct.) *Adam* 15796 ! **Mali:** Gaïbala Spring, Kita Massif (Oct.) *Jaeger* ! **Port.G.:** Safim,
Bissau (Jan.) *Esp. Santo* 1657 ! Brene, Bissau (Jan.) *Esp. Santo* 1692 ! **Guin.:** Kouria to Ymbo (Nov.)
Chev. 14725 ! **S.L.:** Freetown (Oct.) *Deighton* 2133 ! Mabonto (Oct.) *Thomas* 3637 ! Kenema (Nov.)
Deighton 448 ! Njala (Nov.) *Deighton* 1785 ! Gbinti (Jan.) *Deighton* 5895 ! **Lib.:** Bumbumi-Moala (Nov.)
Linder 1332 ! Kakatown *Whyte* ! Dukwia R. (Oct.–Nov.) *Cooper* 2 ! Du R. (Aug.) *Linder* 291 ! Suacoco,
Gbanga (Apr.) *Konneh* 156 ! Jabrocca, Webo (July) *Baldwin* 6681 ! **Iv.C.:** Tébo, Middle Cavally (July)
Chev. 19395 ! Orumbo Mt., Toumodi (Dec.) *Boughey* GC 14442 ! Brenase, Akim (Apr.) *Irvine* 554 !
S. Fomangsu F.R. (Apr.) *Adams* 2658 ! Ofinso (Nov.) *Darko* 642 ! Vane, Amedzofe (Nov.) *Morton* GC
9381 ! Dutukpene (Dec.) *Adams* 4532 ! **N.Nig.:** Jos Plateau (Sept.) *Lely* P750 ! **S.Nig.:** Aboh *Barter* 1289 !
Lagos *W. MacGregor* 303 ! Akure F.R., Ondo (Aug., Oct.) *Jones* FHI 20701 ! *Keay* FHI 25530 ! [Br.]Cam.:
Cam. Mt., Buea, 5,000 ft. (Jan.) *Maitland* 225 ! Ndop, Bamenda, 3,600 ft. (Dec.) *Boughey* GC 11113 !
Vogel Peak, 4,900 ft., Adamawa (Dec.) *Hepper* 1577 ! **F.Po:** Venus Bay (Dec.) *Guinea* 328 ! Moka, 3,500–
4,000 ft. (Dec.) *Boughey* 79 ! Also Sudan, Eritrea, Uganda, Tanganyika and S. Tomé. (See Appendix,
p. 414.)

3. **A. caffrum** *DC.* l.c. 112 (1836). *A. viscosum* of F.T.A. 3 : 299, partly ; of Chev. Bot. 359, partly ; of F.W.T.A,
ed. 1, 2 : 172 ; not of J.R. & G. Forst. A very variable erect or decumbent herb up to about 2 ft. high
with white florets in rather few large heads.
  Mali: Tabacco (Jan.) *Chev.* 127 ! **Guin.:** Konkouré to Timbo (Mar.) *Chev.* 12503 ! Ditinn to Dalaba (Mar.
Chev. 18085 ! Pita (Dec.) *Jac.-Fél.* 673 ! **S.L.:** Banda Karfaia to Sivi-Kozo (Feb.) *Jaeger* 4307 ! Farangbaya

Tonkolili *Crisp* C! G! **Ghana:** Techiman (Dec.) *Adams & Akpabla* GC 4486! Wenchi (Dec.) *Adams* 3193! **Dah.:** Guiliméro to Quétécou (June) *Chev.* 24256! **N.Nig.:** Jos Plateau (Aug.) *Lely* P676! Naraguta (Nov.) *Lely* 707! **S.Nig.:** Koloishe Mt., 5,000 ft., Obudu (Dec.) *Keay & Savory* FHI 25133! 25248! [**Br.]Cam.:** Vogel Peak, 1,200–4,000 ft., Adamawa (Dec.) *Hepper* 1533! 2770! Also in Sudan, Congo and extending to southern Africa.

[Plants of wet places in the western part of the range tend to be smaller, decumbent and have linear leaves, possibly divided when submerged.]

70. AGERATUM Linn.—F.T.A. 3 : 300.

A softly hispid herb with opposite leaves ; petioles slender 1–3 cm. long, blades ovate up to 6 cm. long and 4 cm. broad ; flower-heads in compact terminal cymose inflorescences ; involucral bracts in about 3 rows, oblong, shortly pointed, strongly 2-nerved, glabrous or with a few setose hairs ; achenes sharply 4-angled, black when mature, sparsely pubescent, glabrescent ; pappus of 5(–6) spreading fimbriate scales each terminating in a barbellate hair .. .. *conyzoides*

A. conyzoides *Linn.* Sp. Pl. 839 (1753) ; F.T.A. 3 : 300 ; Chev. Bot. 360. A faintly aromatic erect or procumbent annual herb 1–3 ft. high with mauve florets in heads about ¼ in. long ; a common weed.
Sen.: (Jan.) *Roger*! Dakar (Jan.) *Chev.* 2043! Cape Verde (Dec.) *Berhaut* 725! **Gam.:** Genieri (Feb.) *Fox* 59! Basse (Mar.) *Todd* 12! **Mali:** Badumbé (Dec.) *Chev.* 52! **Guin.:** (Apr.) *Bouery* 21! Kouroussa (July) *Pobéguin* 325! **S.L.:** *Winwood Reade*! *Don*! (June) *T. Vogel* 298! Picket Hill, 2,300 ft. (Nov.) *T. S. Jones* 198! Musaia (Apr.) *Deighton* 5460! Loma Mts. (Sept.) *Jaeger* 1955! **Lib.:** *Whyte*! Dukwia R. (Oct.–Nov.) *Cooper* 24! Monrovia (July) *Linder* 11! Ganta (Sept.) *Harley*! **Iv.C.:** Adiopodoumé (July) *Boughey* GC 14576! Nimba Mts., 2,500–5,000 ft. (Aug.) *Boughey* GC 18082! **Ghana:** Cape Coast (July) *T. Vogel* 43! Amuni, Ashanti (Dec.) *Chipp* 45! Kumawu (Sept.) *Whiting* 4! Tamale (Dec.) *Adams & Akpabla* GC 4171! 4172! Amedzofe (Oct.) *Morton* GC 6041! **N.Nig.:** Borgu *Barter* 1030! Sokoto (Nov.) *Lely* 133! *Moiser* 69! Fodama (Dec.) *Moiser* 155! Jos Plateau *Batten-Poole* 368! **S.Nig.:** Lagos (Aug.) *Dalz.* 1409! Ibadan (Jan.) *Newberry* 17! Calabar (Jan.) *Holland* 20! Obudu Plateau, 5,000 ft. (Aug.) *Stone* 49! [**Br.]Cam.:** Buea, 3,000–6,000 ft. (Jan., Apr., June) *Dunlap* 74! *Hutch. & Metcalfe* 119! *Maitland* 43! 293! Bamenda, 5,000 ft. (Jan.) *Migeod* 427! **F.Po:** (Oct.) *T. Vogel* 17! Monte Balea (Dec.) *Guinea* 463! Widespread through the tropics and subtropics. (See Appendix, p. 414.)

A. houstonianum Mill. (1768) with long-pointed linear-lanceolate shortly hispid involucral bracts and white or bluish-purple florets in larger more round-based heads is reported from Mali (*Roberty* 1035! 1036! cult.) and from Buea (*teste* Morton, ? escape from cultivation). A native of Mexico.

71. PLEIOTAXIS Steetz—F.T.A. 3 : 440.

Stems and lower surfaces of the leaves woolly-tomentose ; leaves broadly lanceolate, acutely acuminate, acute at base, 10–15 cm. long, 2–4 cm. broad, denticulate with about 10–15 pairs of parallel lateral nerves ; petiole about 1·5 cm. long, encircling the stem ; flower-heads narrowly campanulate, about 4 mm. long, the innermost broadly lanceolate and 2 cm. long, with a broad darkened grooved midrib ; achenes pubescent ; pappus straw-coloured, 1·5 cm. long .. .. *newtonii*

P. newtonii *O. Hoffm.* in Engl. Bot. Jahrb. 15 : 537 (1893) ; Chev. Bot. 377 ; Berhaut Fl. Sén. 175, 203. An erect perennial herb with striate stems up to 4 or 5 ft. high and leaves softly whitish-tomentose beneath ; florets bright red ; in savanna.
Sen.: Niokolo-Koba *Berhaut* 1536 ; 4631. **Port.G.:** Gabu (Oct., Nov.) *Esp. Santo* 2773! 2951! 2963! Souaco to Bafatá (Nov.) *Esp. Santo* 3595! **Guin.:** Kouroussa (July) *Pobéguin* 328! Irébéléya to Timbo, 1,800–2,150 ft. (Sept.) *Chev.* 18341! Mamou to Irébéléya (Sept.) *Chev.* 18616! Siguiri (Nov.) *Jac.-Fél.* 513! Dabola (Sept.) *Jaeger* 4907! Also in Angola. (See Appendix, p. 419.)

72. DICOMA Cass.—F.T.A. 3 : 442 ; C. F. Wilson in Kew Bull. 1923 : 377.

Outer pappus of barbellate bristles, inner of lanceolate 1-nerved membranous scales ; achenes long-villous ; leaves oblanceolate to linear, 1·5–6 cm. long, less than 1 cm. broad, white-woolly tomentose on both surfaces but paler beneath, mucronate, minutely and sharply serrate ; flower-heads sessile or subsessile, numerous, mostly axillary ; involucral bracts about 5-seriate, up to 1·5 cm. long, terminating in a long sharp bristle from a lanceolate base, the outer woolly .. .. 1. *tomentosa*
Outer and inner pappus of plumose bristles ; achenes densely silky-villous ; leaves oblanceolate to elliptic, 6–12 cm. long and 1–2 cm. broad ; stems and leaves beneath at first silky, glabrescent ; flower-heads subsessile, terminal or axillary, few ; involucral bracts about 6-seriate, up to about 3 cm. long, narrowly lanceolate with a bristle tip, glabrous .. .. 2. *sessiliflora*

1. **D. tomentosa** *Cass.* in Bull. Sci. Soc. Philom. Paris 1818 : 47 ; F.T.A. 3 : 443 ; Wilson in Kew Bull. 1923 : 377 ; Chev. Bot. 377 ; Berhaut Fl. Sén. 200 ; De Miré & Gillet in J. Agric. Trop. 3 : 706. An annual weedy herb, usually branched from a woody base, up to 2 ft. high, with cottony stems and leaves and greenish-white to pale pink florets in spinose heads ¾ in. or more long ; in dry fields.
Sen.: (June) *Leprieur*! **Mt.** Rolland (Dec.) *Berhaut* 738! **Mali:** Sicoro (Jan.) *Chev.* 221a! Kita (Oct.) *Jaeger* 5! **Iv.C.:** *Poisson*! **Ghana:** Zuarungu (Apr., Nov., Dec.) *Akpabla* 396! *Adams & Akpabla* GC 4262! *Morton* GC 8925! Nandom to Lambusie (Apr.) *Adams* 4108! **Dah.:** Adjara, near Porto-Novo (Jan.) *Chev.* 22755! **Niger:** Ansogo (Sept.) *Hagerup* 396! Aïr *vide* De Miré & Gillet *l.c.* **N.Nig.:** Borgu *Barter* 1299! Kuka, Bornu (Jan.) *E. Vogel* 30! Sokoto *Lely* 111! Auchang (Feb.) *Dalz.* 171! Gaya, Kano (July) *Daggash* FHI 22378! Extends through Sudan and Ethiopia to tropical Asia and south to Angola and Mozambique.

2. **D. sessiliflora** *Harv.* in Fl. Capensis 3 : 518 (1865) ; F.T.A. 3 : 444 ; Wilson l.c. 386 ; Berhaut Fl. Sén. 198, 204. A perennial herb with pithy stems from a few inches to 4 ft. high from a woody stock ; florets yellowish in heads 1–1½ in. long ; the leaves often tinged wine-red ; in stony or rocky places in savanna.
Sen.: *fide* Berhaut *l.c.* **Ghana:** Burufo, near Lawra (Dec.) *Adams & Akpabla* GC 4462! Damongo Scarp, Gonja (Dec.) *Morton* GC 10000! Wa (July) *Adams* GC 17421! **N.Nig.:** Borgu *Barter* 717! Abinsi (Oct.) *Dalz.* 663! Naraguta (Nov.) *Lely* 701! Anara F.R., Zaria (Nov.) *Keay* FHI 22703! Share to Ilorin (Jan.) *Ejiofor & Ujor* FHI 30245! *Ejiofor* FHI 30804! Extends to Sudan and Angola.

FIG. 262.—DICOMA SESSILIFLORA *Harv.* (COMPOSITAE).
A, portion of plant. B, outer flower. C, pappus-bristle. D, anther. E, style. F, inner flower.

73. GERBERA Linn. ex Cass. in Bull. Sci. Soc. Philom. Paris 1817 : 34 ; F.T.A. 3 : 445 ;
Brenan in Mem. N.Y. Bot. Gard. 8 : 488 (1954) ; Taxon 9 : 160 (1960). *Nom. cons.*

An acaulescent perennial herb with a silky-tomentose stock ; leaves elliptic, rounded at
the apex, narrowed into a slender petiole, 5–20 cm. long, 2–6 cm. broad, entire or
denticulate, sparingly pilose above, softly pilose beneath ; flower-heads solitary on a
tomentose peduncle up to 30 cm. long ; peduncle thickened towards the apex and
lengthening further in fruit ; involucral bracts linear, 1·5–2 cm. long, shaggy outside,
glabrous and shiny-black within ; ray-florets tubular, about 2 cm. long, variable in
colour ; disk-florets shorter, usually white ; achenes spreading at maturity, long-
beaked, setulose ; pappus of numerous smooth pinkish-buff spreading setae
piloselloides

G. **piloselloides** (*Linn.*) *Cass.* in Dict. Sci. Nat. 18 : 461 (1820) ; F.T.A. 3 : 445 ; Chev. Bot. 377. *Arnica
piloselloides* Linn. (1760). A perennial herb with solitary flower-heads on leafless peduncles 6–18 in. high,
arising from a rosette of leaves ; outer florets pale crimson to greenish-yellow ; in hill grassland.
Guin.: Diaguissa (Apr.) *Chev.* 12944 ! **Ghana:** Amedzofe *Dokosi* ! **[Br.]Cam.:** Bamenda Dist., 4,500–
6,500 ft. (Dec., Jan., Mar., Apr., June) *Maitland* 1470 ! 1710 ! *Migeod* 365 ! *Boughey* GC 10886 ! *Morton*
K60 ! 96 ! 242 ! *Hepper* 2110 ! 2838 ! Gembu to Likitaba, Mambila Plateau (June) *Chapman* 48 ! From
Eritrea and Ethiopia in high grassland through east and central Africa to Angola, S.W. Africa and
Transvaal, and also in Madagascar.

74. HYPERICOPHYLLUM Steetz in Peters Reise Mossamb. Bot. 49, t. 50 (1863).
Jaumea of F.T.A. 3 : 395, not of Pers.

Stems several from a woody rhizome, slightly setulose but soon glabrous, ribbed ; leaves
lanceolate to ovate-elliptic, obtuse at apex, more or less rounded at base, 5–8 cm. long,
1·5–4 cm. broad, faintly gland-dotted, reticulate on both sides, opposite, sessile, more
or less scale-like below ; flower-heads solitary or subsolitary, about 2·5 cm. diam. ;
peduncle up to 15 cm. long ; involucral bracts 2–3-seriate, broadly ovate-lanceolate,
acutely acuminate, about 2 cm. long, glabrous ; florets all tubular ; achenes pubescent ; pappus in a single series with golden-buff, densely and shortly plumose setae
1–1·3 cm. long   ..   ..   *multicaule*

H. **multicaule** *Hutch.* in Kew Bull. 1907 : 50.   An erect perennial herb with several stiff pale green stems,
pinkish at the base, up to 2 ft. high, arising from a woody rhizome ; leaves bright green, rather fleshy and
glossy ; involucre pale green ; florets greenish-white in heads 1 in. long ; flowering after grass burning.
S.L.: Loma Mts. (sterile, Aug.) *Jaeger* 1085 ! **Iv.C.:** Koualé to Kouroukoro (May) *Chev.* 21756 ! **N.Nig.:**
Kontagora (Jan.) *Dalz.* 52 ! *Meikle* 1058 ! Birnin Gwari, Zaria (May) *Keay* FHI 25807 ! Jos Plateau
(Feb.) *Lely* P79 !

75. BERKHEYA Ehrh.—F.T.A. 3 : 428. *Nom. cons.*

Stems erect, ribbed, with crisped hairs ; leaves linear-oblong, up to about 15 cm. long,
sparsely and minutely scabrid-pubescent above, white-woolly tomentose beneath,
with an acute spinescent apex and numerous spine-tipped teeth on the margin, sessile,
auriculate and amplexicaul ; flower-heads several, pedunculate, 5–7 cm. diam. ;

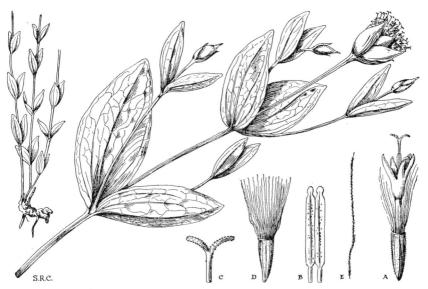

Fig. 263.—Hypericophyllum multicaule *Hutch.* (Compositae).
A, flower. B, anthers. C, style-arms. D, achene and pappus. E, pappus-bristle.

involucral bracts margined with long spines, longer than the disk-florets ; ligules of ray-florets linear ; pappus-scales oblanceolate, serrulate.   ..   *spekeana*

B. spekeana *Oliv.* in Trans. Linn. Soc. 29 : 100, t. 66 (1873) ; F.T.A. 3 : 429. A leafy thistle-like herb up to 4 or even 6 ft. high, with yellow florets in heads nearly 3 in. across and milky latex ; in montane grassland. [Br.]Cam.: Bamenda, 5,000–7,500 ft. (Jan., Mar., June, Aug.) *Migeod* 380 ! *Morton* K29 ! *Savory* UCI 286 ! Njinikom (June) *Maitland* 1718 ! Jakiri (Feb.) *Hepper* 2081 ! Banja, Bamenda to Nkwa (Aug.) *Ujor* FHI 29955 ! Shwai, Banso (Dec.) *Egbuta* FHI 3776 ! Also in Congo, Sudan, Kenya, Uganda and Tanganyika. [The West African plants have the upper leaf surface more sparsely appressed-setulose than the central and eastern African forms.]

76. ECHINOPS Linn.—F.T.A. 3 : 430.

Leaves very small, not exceeding 3 cm. long, usually 1·5–2 cm. long, shortly pinnately toothed ; compound heads 3–5 cm. diam. :
  Stem-wings continuous between the nodes from the decurrent leaf-bases ; stems and leaves beneath thinly cottony when young, glabrescent ; involucral bracts subentire, rhomboid-elliptic   ..   ..   1. *gracilis*
  Stems winged only for a short distance in two lines below each leaf ; whole plant glabrous ; involucral bracts lanceolate to rhomboid-elliptic, serrulate in the upper half   ..   ..   2. *guineensis*
Leaves much larger than above, usually white-woolly-tomentose beneath :
  Stem-leaves not divided, entire or at most dentate :
    Leaves smooth and glabrous above ; stem smooth ; leaves linear, shortly auriculate, more or less distinctly dentate, 10–25 cm. long, 0·5–2(–3) cm. broad (incl. the spine-tipped teeth) ; heads terminal on rather short peduncles, 5–6 cm. broad
                                          3. *longifolius*
    Leaves densely setose above ; stem setulose, deeply grooved ; leaves lanceolate to oblong-elliptic, more or less rounded at base, sessile, entire ; heads terminal, 4–5 cm. broad, those on lateral branches smaller :
      Lamina up to about 12 cm. long and 2 cm. broad, acutely acuminate
                                         4. *mildbraedii*
      Lamina 8–10 cm. long and 3–3·5 cm. broad, shortly acute   ..   5. *lanceolatus*
  Stem-leaves divided, lanceolate to ovate or elliptic in plan, with segments spine-tipped :
    Outer involucral bracts of some of the partial capitula developed into hard spines up to 3 cm. long, projecting beyond the spherical head which is about 6 cm. diam. in flower ; stems grooved, woolly-tomentose, very thinly setulose ; leaves woolly-tomentose on both surfaces, also scabrid above, dentate-pinnatifid, with marginal and terminal spines 1 cm. or more long ; partial heads about 2·5 cm. long with numerous buff setae 1·5 cm. long arising near the base ; involucral bracts minutely ciliate on the shoulders of the broader basal portion below a spinose apex
                                         6. *spinosissimus*

Outer involucral bracts uniform, not developed into long spines ; stems grooved, setose ; leaves woolly-tomentose beneath ; heads 6–12 cm. diam. :

Leaves ovate in plan, doubly dentate in deeply pinnatifid segments, sessile, amplexicaul and auriculate, thinly tomentose and scabridulous above ; partial heads (capitula) up to 4 cm. long with the basal setae buff about 1 cm. long ; involucral bracts pectinate-ciliate, smoothly tapered at apex ..   ..   7. *amplexicaulis*

Leaves obovate in plan, the lower nearly entire, the upper deeply pinnatisect, shortly stalked, the bases of the petioles half-amplexicaul but not auriculate, shortly and densely setose above and on the veins beneath ; partial heads (capitula) 1·5–3 cm. long with the basal setae buff 3–7 mm. long ; involucral bracts expanded at apex into a spinulose tuft .. .. 8a. *giganteus* var. *lelyi*

1 . **E. gracilis** *O. Hoffm.* in Mém. Soc. Bot. Fr. 2, 8 : 42 (1908). A bushy perennial herb or undershrub usually about 2–3 ft., but sometimes up to 6 ft. high ; stem woody below ; florets pale mauve or fading white in sessile or shortly stalked spherical heads about 2 in. across ; in hill grassland.

N.Nig.: Jos Plateau (Apr.) *Lely* P219 ! Vom, 4,000 ft. (Feb.) *Dent Young* ! *McClintock* 213 ! Bokkos (Dec.) *Coombe* 62 ! [Br.]Cam.: Bamenda Dist., 4,800–6,300 ft. (Dec., Jan., Mar., Apr.) *Maitland* 1539 ! *Migeod* 298 ! *Onochie* FHI 34841 ! *Boughey* GC 10403 ! *Adams* GC 11352 ! *Morton* K121 ! Gembu, Mambila Plateau, 5,300 ft. (Jan.) *Hepper* 1844 ! *Latilo & Daramola* FHI 34363 ! Vogel Peak, 4,000 ft., Adamawa (Dec.) *Hepper* 1531 ! Also in Cameroun, Ubangi-Shari, Sudan, Congo and Uganda.

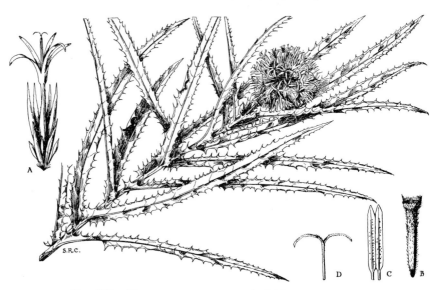

FIG. 264.—ECHINOPS LONGIFOLIUS *A. Rich.* (COMPOSITAE).

A, partial flower-head. B, achene. C, anthers. D, style-arms.

2. **E. guineensis** *C. D. Adams* in J.W. Afr. Sci. Assoc. 3 : 111 (1957). *E. elegans* Hutch. & Dalz. F.W.T.A., ed. 1, 2 : 177 (1931), not of Bertol. (1853). *E. gracilis* of Chev. Bot. 375, not of O. Hoffm. A glabrous shrubby perennial herb with short spiny leaves and spherical compound heads 1–1½ in. across.

Guin.: Beyla to Sadougou (Mar.) *Chev.* 20871 ! Beyla (Mar., Aug.) *Jaeger* 4890 ! *Chev.* 20864 ! Iv.C.: Toura country, Upper Sassandra (May) *Chev.* 21703 !

3. **E. longifolius** *A. Rich.* Tent. Fl. Abyss. 1 : 452, t. 61 (1848) ; F.T.A. 3 : 431 ; Chev. Bot. 375. *E. otarus* Mattf. (1924). *E. bathrophyllus* Mattf. (1924). A perennial herb 1–3 ft. high with long narrow regularly spinose-margined leaves closely whitish-tomentose beneath ; florets pale blue fading to mauve or white in spherical compound heads about 2 in. across ; in savanna.

Mali: Sikasso (fr. Sept.) *Adam* 15094 ! Guin.: Kouroussa (July) *Pobéguin* 318 ! Mt. Loursa, near Dabola (Sept.) *Jaeger* 4897 ! Ghana: Amansare, Ashanti (July) *Chipp* 523 ! Yeji (Aug.) *Anderson* 48 ! *Vigne* FH 3368 ! Damongo, Gonja (July, Dec.) *Andoh* 5198 ! *Morton* GC 9999 ! 25038 ! Yendi (May) *Akpabla* 580 ! Kete Krachi (Dec.) *Adams* 4513 ! 4552 ! Sakogu, Gambaga (Dec.) *Adams & Akpabla* GC 4251 ! Dah.: Cabolé to Bassila, Savalou (May) *Chev.* 23802 ! N.Nig.: Jos Plateau (June, Nov.) *Lely* P325 ! *Keay* FHI 21021 ! Ropp Hills, 4,000 ft. (Dec.) *Coombe* 51 ! Gindiri (Oct.) *Hepper* 1150 ! Kilba country (July) *Dalz.* 29 ! Bornu (July, Aug.) *Parsons* ! *Daggash* FHI 24862 ! S.Nig.: Lagos *W. MacGregor* 152 ! [Br.]Cam.: Vogel Peak area, Adamawa (Nov.) *Hepper* 1286 ! Also in Cameroun, Ubangi-Shari, Sudan, Ethiopia, Uganda, Kenya and Tanganyika. (See Appendix, p. 417.)

4. **E. mildbraedii** *Mattf.* in Engl. Bot. Jahrb. 59, Beibl. 133 : 49 (1924). A much-branched herb with stout stems 2–3 ft. high from a woody stock ; florets blue or white in spherical heads up to 2 in. across ; in open grassland.

[Br.]Cam.: Ndop to Kumbo, Bamenda Dist. (Dec.) *Boughey* GC 4701 ! Also in Cameroun.

5. **E. lanceolatus** *Mattf.* l.c. 51 (1924). A perennial herb very similar to and perhaps not distinct from the last.

[Br.]Cam.: Vogel Peak, 4,000 ft., Adamawa (Dec.) *Hepper* 1549 ! Also in Cameroun.

6. **E. spinosissimus** *Turra* Farset. Nov. Gen. 13 (1765). *E. spinosus* Linn. (1767)—F.T.A. 3 : 341. An erect branched herb up to 5 ft. high with pinnatifid spiny leaves ; stems tomentose, glabrescent ; florets white and involucral bracts green in spherical spiny heads up to 4 in. across ; a weed of waste ground in dry open country.

Mali: Karseni (Apr.) *Davey* 264! Niafounké (Feb.) *Davey* 467! [Br.]**Cam.**: Bama, 1,000 ft., Dikwa Div. (Jan.) *McClintock* 177! Also in Sudan and Ethiopia.

7. **E. amplexicaulis** *Oliv.* in Trans. Linn. Soc. 29 : 101, t. 67 (1873) ; F.T.A. 3 : 431. A stout herb 6–8 ft. high with the stems and midribs of the leaves smooth or very sparingly setulose ; corollas and anthers deep crimson in spherical heads 3–5 in. across.
[Br.]**Cam.**: Mambila Plateau, 5,600 ft., Adamawa (Jan.) *Hepper* 1750! Also in Ethiopia, Sudan, Congo, Kenya, Uganda and Tanganyika.

8. **E. giganteus** *A. Rich.* Tent. Fl. Abyss. 1 : 449 (1848) ; F.T.A. 3 : 342. *E. velutinus* O. Hoffm. (1901). A stout herb with large compound heads and densely setulose stems and leaves recorded from Sudan and Tanganyika.

8a. **E. giganteus** var. **lelyi** (*C. D. Adams*) *C. D. Adams* in J.W. Afr. Sci. Assoc. ined. (1963) *E. velutinus* var. *lelyi* C. D. Adams 3 : 113 (1957). An erect branched herb 2–5 ft. high with white flowers in spherical heads up to about 3 in. across ; stems and midribs of leaves beneath setulose ; in rough open country and cultivated land.
N.Nig.: Jos Plateau (Aug.) *Lely* P671! *Batten-Poole* 374! Vom, 4,000 ft. (Dec.) *Coombe* 105! [Br.]**Cam.**: Bamenda (May) *Ujor* FHI 30365! Vogel Peak, 3,500 ft., Adamawa (Dec.) *Hepper* 1525!

77. CARDUUS Linn.—F.T.A. 3 : 433.

Stem erect, simple or branched, striate, with spinous wings ; leaves elliptic or lanceolate, pinnately divided into ovate or lanceolate segments ; segments dentate, with the teeth and apex spine-tipped ; flower-heads clustered in dense cymes 3–5 cm. diam. ; involucral bracts up to 5 mm. long terminating in a long spreading spine ; receptacle densely setose ; achenes glabrous ; pappus setae scabrid, united at base, finally caducous   ..   ..   *nyassanus*

C. nyassanus (*S. Moore*) *R. E. Fries* in Acta Hort. Berg. 8 : 25 (1925). *C. leptacanthus* Fresen. var. *nyassana* S. Moore in J. Linn. Soc. 37 : 326 (1906). A perennial herb of damp montane grassland, up to 4 ft. high ; florets white.
[Br.]**Cam.**: Lakom, 6,000 ft., Bamenda (Apr.) *Maitland* 1649! Kumbo to Oku, 7,000 ft., Bamenda (leaf Feb.) *Hepper* 2719! Nguroje, 5,600 ft., Mambila Plateau (leaf Jan.) *Hepper* 2820! Also in Uganda, Ethiopia, Congo, Tanganyika and Nyasaland.

78. ATRACTYLIS Linn.—Benth. & Hook. f. Gen. Pl. 2 : 465.

A very prickly herb ; stems slightly pubescent ; leaves linear-lanceolate, about 2·5 cm. long, divided into pungent points, woolly when young ; flower-heads 1·5 cm. long, long, sessile, terminal, surrounded by the upper leaves ; inner involucral bracts with membranous margins, thinly pubescent outside ; achenes silky-villous ; pappus plumose   ..   ..   *aristata*

A. aristata *Battandier* in Bull. Soc. Bot. Fr. 49 : 291 (1903) ; De Miré & Gillet in J. Agric. Trop. 3 : 703. A herb with very prickly leaves and yellow flower-heads.
Niger: Aïr *vide* De Miré & Gillet *l.c.* Also in Mauritania, Central Sahara and Algeria.

79. CENTAUREA Linn.—F.T.A. 3 : 436 ; Philipson in J. Bot. 77 : 227 (1939).

Stems winged by the decurrent leaves ; leaves linear-lanceolate, acute, 7–9 cm. long, about 1 cm. broad, thinly woolly-pubescent, entire or nearly so with leaf and wing-margins finely setulose ; involucral bracts pectinate with several fine bristles, at first erect, later reflexed, the innermost with spathulate toothed tips ; achenes glabrous, with a short pappus ..   ..   1. *nigerica*
Stems not winged :
  Bristles of the involucral bracts short ; leaves linear to oblong, up to 12 cm. long, 0·5–2 cm. broad, thinly pubescent, glabrescent, scabrid ; flower-heads usually near the ground ; achenes glabrous with a short pappus   ..   ..   ..   2. *praecox*
  Bristles of the involucral bracts long, spinescent, straw-coloured ; stem-leaves ovate, lanceolate or oblanceolate, semi-amplexicaul, toothed or pinnately lobed, scabrid ; flower-heads borne at the ends of the branches :
    Achenes with a well-developed pappus ; spinules of the involucral bracts 3–5 on each side ; stem-leaves up to 5 cm. long   ..   ..   3. *senegalensis*
    Achenes without a pappus (the receptacle in *Centaurea* is setaceous, and when dissecting heads these hairs may be mistaken for deciduous pappus) ; spinules of the involucral bracts 1–3 on each side ; stem-leaves up to 10 cm. long   4. *perrottetii*

1. **C. nigerica** *Hutch.* in Kew Bull. 1921 : 383, fig. 7. An erect perennial herb 2–3 ft. high, sparingly branched above, slightly woolly, with globose-campanulate flower-heads about 1 in. long ; florets pale purple to rose.
Guin.: Bara, Fouta Djalon (Sept.) *Schnell* 7321! Labé (June) *Adam* 14763! **N.Nig.:** Jos *Keay* FHI 20177! Naraguta *Lely* 252! Pankshin *W. D. MacGregor* 426! Ropp to Pankshin, 4,000 ft. *Coombe* 48! Amo, 4,100 ft. (Oct.) *Hepper* 1028! Federe-Fobur, Jarawa *Summerhayes* 12!

2. **C. praecox** *Oliv. & Hiern* in F.T.A. 3 : 438 (1877) ; Chev. Bot. 376. *C. rhizocephala* Oliv. & Hiern in F.T.A. 3 : 438 (1877) ; F.W.T.A., ed. 1, 2 : 176, not of Trautv. (1873). *C. atakorensis* A. Chev. Bot. 376, name only. An erect branched herb 1–2 ft. high from a perennial woody stock ; florets reddish-purple or rose-pink the outer fading first to pale mauve, in prickly heads ¾–1 in. long, usually appearing close to the ground, the leafy shoots elongating later ; leaves very variable in breadth.
Mali: Kéméné (Mar.) *Chev.* 655! Kita massif (July) *Jaeger*! **Ghana:** Yendi (Apr., Dec.) *Williams* 164! Akpabla 416! *Adams & Akpabla* GC 4058! Salaga *Krause*! Bole to Sikiri (Mar.) *Todd* 310! Yeji *Anderson* 46! Tumu (Apr., May, Dec.) *Vigne* FH 3781! *Morton* GC 7525! *Adams & Akpabla* GC 4362! Kintampo, Ashanti (Mar.) *Dalz.* 11! **Dah.:** Djougou (June) *Chev.* 23892! Djougou to Pobégou (June) *Chev.* 23930! **Niger:** Diapaga to Fada (July) *Chev.* 24497! **N.Nig.:** Jebba (Jan.) *Meikle* 1105! Sokoto (Dec.) *Lely* 116! *Dalz.* 405! Anara F.R., Zaria (Nov.) *Keay* FHI 21682! Nassarawa (Mar.) *Hepburn* 15! Naraguta, 4,000 ft. (Dec.) *Lely* 58! *Coombe* 12! **S.Nig.:** Yoruba country *Barter* 1223! Extends to Sudan. (See Appendix, p. 416.)

3. **C. senegalensis** *DC.* Prod. 6 : 598 (1838) ; F.T.A. 3 : 437 ; Chev. Bot. 376 ; Hagerup in Biol. Medd. 9, 4 : 55 ; Schnell in Ic. Pl. Afr. I.F.A.N. t. 37 ; Berhaut Fl. Sén. 170. An annual or chamaephytic branched

scabrid herb, erect 1–2 ft. high or prostrate ; stem-leaves semi-amplexicaul, up to 2 in. long, the basal leaves of early growth larger ; florets purple or mauve in spiny heads about ¾ in.broad terminating leafy branches.

Sen.: *Leprieur!* *Roger!* *Berhaut* 1295. **Mali:** Timbuktu (June, July) *Chev.* 1258! *Hagerup* 118! **N.Nig.:** Kuka, Bornu (Jan.) *E. Vogel* 17! Nguru, Bornu (June) *Onochie & Ibrahim* FHI 23346! **[Br.] Cam.:** Bama, Dikwa Div. (Nov.) *McClintock* 5! Extends from Mauritania through Tibesti to Sudan.

4. **C. perrottetii** *DC.* l.c. (1838) ; Arènes in Bull. I.F.A.N. 14 : 32, t. 1–3 ; Schnell in Ic. Pl. Afr. I.F.A.N. t. 38 ; Berhaut Fl. Sén. 169. *C. calcitrapa* of F.T.A. 3 : 437 and of Chev. Bot. 376, not of Linn. *C. alexandrina* of F.W.T.A., ed. 1, 2 : 176, not of Del. A herb like the last up to 18 in. high, with rather smaller heads and larger leaves ; florets purple.

Sen.: *Perrottet!* Dakar (May, Sept.–Nov.) *Chev.* 33834! 33997! *Farmar* 146a! *Baldwin* 5711! **Gam.:** Sukuko (Jan.) *Dalz.* 8033! Bathurst (Jan., July) *Irvine* 2793! *Deighton* 5583! Cape St. Mary (Mar.) *Todd* 24! Fajara (fl. & fr. Jan.) *Wallace* AXA! **Mali:** Bamako (Jan.) *Chev.* 193! Kita (Oct.) *Jaeger* 22! Dioura (Mar.) *Davey* 260! **Port.G.:** Suzana (Aug.) *Esp. Santo* 3067! **Niger:** Niamey (Oct.) *Hagerup* 493! Tanout (Aug.) *Barry* 88! **N.Nig.:** Kuka, Bornu (Feb.) *E. Vogel* 51! Katagum (Sept.) *Dalz.* 176! Fodama (Nov.) *Moiser* 228! Maiduguri (June) *Ujor* FHI 21923! Kano (Mar., Dec.) *Bally* 7! *Meikle* 801! **[Br.] Cam.:** Bama, 1,000 ft., Dikwa Div. (Jan.) *McClintock* 156! Extends from Mauritania through central Sahara. (See Appendix, p. 416.)

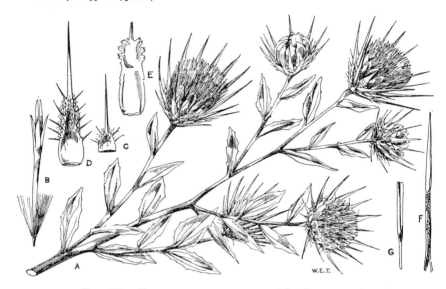

FIG. 265.—CENTAUREA SENEGALENSIS *DC.* (COMPOSITAE).

A, portion of plant. B, flower. C, outer, D, intermediate, E, inner involucral bract. F, stamen. G, upper part of style.

80. PICRIS Linn.—F.T.A. 3 : 447.

Perennial acaulescent herb with a strong tap-root ; leaves mostly radical, linear-oblanceolate, repand-dentate with mucronate teeth, up to 20 cm. long and 2 cm. broad, setose with forked-tipped hairs on both surfaces and on the margin ; flower-heads 1–3 on long peduncles, the scapes usually with a few oblanceolate leaves ; involucral bracts unequal up to 1·5 cm. long, pointed, sparingly setose outside, spreading widely in fruit ; achenes 5 mm. long, shortly beaked, scabrid on the ribs ; pappus-setae buff, rigid, plumose or the outer shorter and simple   ..   ..   ..   *humilis*

P. humilis *DC.* Prod. 7 : 130 (1838) ; F.T.A. 3 : 447, in obs. ; Berhaut Fl. Sén. 173. Flowering stems erect from a firm stock, 6–30 in. high, with yellow or white latex ; florets yellow in oblong heads ⅔ in. long ; often appearing after grass fires.

Sen.: Senegal R. *Berhaut* 1395. **Mali:** San (Feb.) *Roberty* 3386! **N.Nig.:** N. of Zaria (Feb.) *Dalz.* 392! Naraguta, 4,000 ft. (May, June, Dec.) *Lely* 24! 341! *Coombe* 44! *Kennedy* FHI 1198! Jos (Apr., Aug.) *Keay* FHI 12714! *Morton* K351! *Monod!* Vom *Dent Young* 163! Gindiri, 3,500 ft. (Oct.) *Hepper* 1133!

81. LACTUCA Linn.—F.T.A. 3 : 450.

Scandent herbs ; inflorescences axillary and terminal, broadly paniculate :

Stems, peduncles and bracts pilose with gland-tipped hairs ; leaves lyrate-pinnatifid with a broadly winged rachis, with usually one pair of lanceolate lateral lobes and a terminal triangular or ovate-deltoid lamina, up to 7 cm. long, repand-denticulate ; flower-heads up to 1·5 cm. long ; achenes with a long slender beak, flattened with several wing-like ribs   ..   ..   ..   1. *glandulifera* var. *glandulifera*

Stems, peduncles and bracts glabrous or nearly so, eglandular, otherwise as above

1a. *glandulifera* var. *calva*

Erect herbs ; inflorescence terminal, spreading or contracted :
Achenes with a single median rib on each flat side ; flower-heads 1 cm. or more long :
Stem-leaves linear or linear-lanceolate with acutely downward-pointed auricles, simple or runcinate-pinnatifid, very acute, margin entire or irregularly denticulate, up to 20 cm. long and 1·5 cm. broad ; stems glabrous, smooth ; achenes narrowed into a long beak, without a conspicuously thickened margin ; pappus usually bright yellow at base .. .. 2. *capensis*
Stem-leaves broadly oblanceolate with more or less rounded auricles, shallowly dentate, remotely denticulate ; stems setose, glabrescent ; achenes with a very short beak and conspicuously thickened margins .. .. 3. *schweinfurthii*
Achenes several-ribbed, narrowed to a slender beak ; stems or peduncles glabrous or glabrescent :
Beak distinct ; body of achene rather flat, reddish ; stems sparingly setose :
Involucres 6–8 mm. long ; stem-leaves numerous, lanceolate, shallowly dentate, denticulate, with rounded auricles .. .. 4. *schulzeana*
Involucres 1–1·2 cm. long ; stem-leaves very few and small, exauriculate ; basal leaves absent at flowering time ; involucral bracts curled at apex, purplish
5. *tuberosa*

Beak short, indistinct, body of achene fusiform, grey ; stems glabrous :
Inner involucral bracts 5–6, two to three times as long as the outer ones ; leaves mostly deltoid-pinnatifid ; panicle much-branched ; florets deep yellow
6. *taraxacifolia*

Inner involucral bracts 7–8 ; leaves dentate, sinuate or subentire, denticulate ; panicle elongated with few ascending branches ; florets pale yellow 7. *intybacea*

1. **L. glandulifera** *Hook. f.* var. glandulifera—in J. Linn. Soc. 7 : 203 (1864) ; F.T.A. 3 : 454 ; Stebbins in Bull. Jard. Bot. Brux. 14 : 350. A herb with straggling green or purplish stems up to about 4 ft. high with gland-tipped hairs ; florets yellow.
S.L.: Loma Mts. (Feb.) *Jaeger* 4276 ! **[Br.]Cam.:** Cam. Mt., 5,000–8,000 ft. (Dec., Feb., Mar.) *Mann* 1240 ! 1917 ! *Maitland* 1316 ! *Brenan* 9534 ! L. Bambalue, Bamenda (Mar.) *Morton* K73 ! Mambila Plateau, 5,400 ft., Adamawa (Jan.) *Hepper* 1730 ! **F.Po:** Musola to Moka (Jan.) *Guinea* 1903 ; 1905. Also in Ethiopia, Congo, Uganda, Kenya and Tanganyika.
1a. **L. glandulifera** var. **calva** (*R. E. Fries*) *Robyns* Fl. Sperm. Parc Nat. Alb. 2 : 606 (1947). *L. glandulifera* forma *calva* R. E. Fries (1928). *L. dunlapii* Hutch. & Dalz. F.W.T.A., ed. 1, 2 : 177 (1931). Like the last, but with the peduncles and involucral bracts at most minutely setulose without gland-tipped hairs.
[Br.]Cam.: Cam. Mt., 5,000–9,000 ft. (Dec., Jan.) *Dunlap* 47 ! *Mildbr.* 10814 ! *Maitland* 226 ! *Boughey* GC 17256 ! Also in Congo, Uganda, Kenya, Tanganyika and Nyasaland.
[The distinction of these two varieties is not wholly satisfactory. The following may be regarded as intermediates :—Bamenda Dist.: Mbakokeke Mt., 6,300 ft. (Jan.) *Keay & Lightbody* FHI 28387 ! Bafut-Ngemba F.R., 7,000 ft. (Feb.) *Hepper* 2195 ! Bambui, 6,000 ft. (Dec.) *Boughey* GC 10877 ! Kumbo to Oku, 6,000 ft. (Feb.) *Hepper* 2014 !].
2. **L. capensis** *Thunb.* Prod. Fl. Cap. 2 : 139 (1800) ; F.T.A. 3 : 452 ; Chev. Bot. 378. *L. holophylla* Bak. (1895). *L. sassandrensis* A. Chev. Bot. 378, name only. Very variable perennial herbs from a few inches to over 5 ft. high from a robust woody stock.
Mali: *fide* Chev. *l.c.* **Guin.:** *fide* Chev. *l.c.* Mali, 4,800 ft. (June) *Adam* 14464 ! Dantilia (Mar.) *Sc. Elliot* 5303 ! **S.L.:** Musaia (Feb.) *Deighton* 4234 ! Kangahun (Apr., May) *Deighton* 3525 ! 6045 ! Yalamba (Jan.) *Jaeger* 3954 ! **Iv.C.:** Gouékouma, Upper Sassandra (May) *Chev.* 21651 ! 21667 ! **Ghana:** Kpong (June) *Irvine* 1775 ! Kintampo (Mar.) *Dalz.* 39 ! Wenchi, Ashanti (Dec.) *Adams* 3316 ! Gambaga (May) *Morton* GC 7368 ! Yendi (Dec.) *Adams & Akpabla* GC 4068 ! Amedzofe Hills (Nov.) *Morton* GC 9430 ! **Dah.:** Zagnanado (Feb.) *Chev.* 23071 ! Tchitopa to Koussi, Allada (Feb.) *Chev.* 23229 ! **N.Nig.:** Kontagora (Jan., Apr.) *Meikle* 1039 ! Naraguta, 4,000 ft. (June) *Lely* 338 ! Jos (Nov.) *Keay & Onochie* FHI 21746 ! Vom (Jan., Apr.) *Dent Young* 164 ! *Morton* K412 ! **S.Nig.:** Lagos *Rowland* ! Olokemeji F.R. (Mar.) *Keay* FHI 25699 ! Sonkwala, 4,000 ft., Ogoja (Dec.) *Savory & Keay* FHI 25030 ! **[Br.]Cam.:** Cam. Mt., 5,000 ft. to summit (Nov.–Apr.) *Mann* 1300 ! 1928 ! *Mildbr.* 10839 ! *Dalz.* 8313 ! *Maitland* 932 ! 1256 ! *Keay* FHI 28584 ! *Boughey* GC 6984 ! *Brenan* 9379 ! Bamenda Dist., 4,000–6,500 ft. (Dec., Jan., Mar., Apr.) *Migeod* 296 ! 407 ! *Maitland* 1543 ! 1673 ! Nkambe, 5,800 ft. (Feb.) *Hepper* 1934 ! Mambila Plateau, 4,700 ft., Adamawa (Jan.) *Hepper* 1686 ! **F.Po:** summit of the peak (Mar.) *Guinea* 2786. Widespread in tropical and subtropical Africa and also in the Mascarene Islands. (See Appendix, p. 418.)
[The small high-mountain forms are branched rosette plants with blue flowers ; the plants of medium altitudes have numerous basal leaves, few scape leaves, and the flowers are usually blue or mauve ; the tallest plants of lowland localities usually lack basal leaves at flowering time, the leaves are glabrous or rarely setulose, the leaf-margins are entire or rarely minutely denticulate and the flowers are deep rich purple or rarely yellow. In view of the wide distribution of *L. capensis* in tropical and south Africa, it is not considered appropriate formally to establish any subdivisions of the species, for the purpose of this Flora, on the basis of the West African material only.—C. D. A.]
3. **L. schweinfurthii** *Oliv. & Hiern* in F.T.A. 3 : 452 (1877). A perennial herb erect hispid-setulose stems up to about 6 ft. high ; inflorescence spreading with numerous flower-heads about ½ in. long ; florets mauve or yellow.
[Br.]Cam.: Fonfuka, 3,000 ft., Bamenda (June) *Maitland* 1396 ! Also in Cameroun, Sudan, Congo, Tanganyika and Angola.
4. **L. schulzeana** *Büttner* in Verh. Bot. Verein Brandenburg 31 : 72 (1889). An erect herb up to about 6 ft. high with yellow or white florets in rather small heads in a wide diffuse panicle.
[Br.]Cam.: Lakom, 6,000 ft., Bamenda (Apr.) *Maitland* 1479 ! Bum to Nchan, 5,000 ft., Bamenda (May) *Maitland* 1749 ! Also in Congo, Uganda and Angola.
5. **L. tuberosa** *A. Chev.* in Mém. Soc. Bot. Fr. 2, 8 : 261 (1917) ; Chev. Bot. 379. A herb with erect shoots about 1 ft. high from a massive perennial stock ; leaves mostly basal but absent at flowering time ; inflorescences rather narrow with erect flower-heads ; florets pale blue or white.
Guin.: Laya-Santo, Faranna (Jan.) *Chev.* 20485 ! **[Br.]Cam.:** Nguroje to Kakara, Mambila Plateau, 5,400 ft., Adamawa (Jan.) *Hepper* 1764 !
6. **L. taraxacifolia** (*Willd.*) *Schum. ex Hornemann* De Ind. Pl. Guin 17 (1819) ; Schum. & Thonn. Beskr. Guin. Pl. 380 (1827) ; F.T.A. 3 : 451 ; Chev. Bot. 378 ; F.W.T.A., ed. 1, 2 : 197 ; Berhaut Fl. Sén. 171. *Sonchus taraxacifolius* Willd. (1803). A weedy herb with erect stems 2–4 ft. high arising after the formation of a basal rosette of leaves from an underground branching and proliferating rhizome ; sap milky ; florets yellow.
Sen.: Richard Tol (Jan.) *Perrottet* 2 ! *Roger* ! Dakar (Apr.) *Chev.* 15803 ! **S.L.:** Freetown (Dec.) *Hagerup* 766 ! *Deighton* 1533 ! **Iv.C.:** Ouodé to Gouréni, Upper Sassandra (May) *Chev.* 21628 ! **Ghana:** Aburi

20

T. W. Brown 402 ! Legon Hill (Oct.) *Adams* 3374 ! Achimota (June) *Irvine* 734 ! *Morton* A1564 ! Koforidua (Nov.) *Dalz.* 104 ! Dawa̱(May) *Howes* 911 ! **Dah.**: Sakété, Porto-Novo (Jan.) *Chev.* 22809 ! Zagnanado (Feb.) *Chev.* 23086 ! **Niger:** Aïr *De Miré & Gillet.* **N.Nig.:** Katagum *Dalz.* 169 ! Sokoto *Lely* 120 ! Kouka, Bornu (Mar.) *E. Vogel* 47 ! Kalkala, near L. Chad *Golding* 59 ! **S.Nig.:** *Dodd* 379 ! Lagos (Aug.) *Rowland* ! *Phillips* 3 ! *Millen* 133 ! Ijaye *Barter* 3414 ! Ibadan (Nov., Dec., Feb.) *Gwynn* 86 ! *Newberry* 128 ! *Meikle* 917 ! *Latilo* FHI 13647 ! Also in Sudan and Ethiopia. (See Appendix, p. 418.)
[This species behaves in habitat and location as a casual or introduced weed. There is, however, at present no further evidence to suggest or confirm its presence or origin outside tropical Africa.]

7. **L. intybacea** *Jacq.* Ic. Pl. Rar. 1 : 16, c. 162 (1781). *Lactuca goraeensis* (Lam.) Sch. Bip. (1842)—F.T.A. 3 : 452 ; F.W.T.A., ed. 1, 2 : 177, partly (excl. syn. *L. tuberosa* A. Chev. and *Rowland, Millen* 133, Dodd 379) ; Trochain Mém. I.F.A.N. 2 : 376 ; Berhaut Fl. Sén. 171. *Sonchus goraeensis* Lam. (1792). *Launaea goraeensis* (Lam.) O. Hoffm. (1893)—Chev. Bot. 380. An annual herb with erect stems up to 6 ft. high from a taproot ; florets pale lemon-yellow ; leaves yellowish-green, glossy.
Sen.: Dakar (Jan.) *Chev.* 2045. *fide* Trochain *l.c.* **Ghana:** Ada road *Morton* A2183 ! **Niger:** Aïr *De Miré & Gillet.* **N.Nig.:** (Nov.) *Noble* A27 ! Thinly scattered through the Old and New World tropics in semi-arid regions.

82. CREPIS Linn.—F.T.A. 3 : 448 ; Babcock in Univ. Calif. Publ. 22 (1947).

Scape leafy ; inner involucral bracts lanceolate, acute, pubescent within, 1-nerved in the outer midline, not changing at maturity, ultimately reflexed ; achenes 8–10 mm. long, fusiform, 10-ribbed, finely setulose ; pappus buff, 5 mm. long   1. *cameroonica*
Scape bracteate ; inner involucral bracts lanceolate, obtuse, glabrous within, becoming carinate dorsally and spongy-thickened at the base in mature heads ; achenes about 7 mm. long ; pappus pale buff or white, 6 mm. long   .. 2a. *newii* subsp. *kundensis*

1. **C. cameroonica** *Babc. ex Hutch. & Dalz.* F.W.T.A., ed. 1, 2 : 178 (1931) ; Babcock l.c. 346, fig. 63. *C. hookeriana* Oliv. & Hiern in F.T.A. 3 : 450 (1877), not of Ball (?) (1873) or of C. B. Clarke (?) (1876). *Anisorhamphus hypochaeroides* Hook. f. (1864), not of DC. (1838). A perennial herb with loosely hairy stems 6 to 20 in. high arising in a basal rosette of pale green leaves from a thick stock ; florets bright yellow ; involucral bracts green with blackish hairs, forming heads about ¾ in. broad.
[Br.]**Cam.**: Cam. Mt., common from 7,000 ft. to near summit (Nov.–Apr.) *Mann* 1318 ! 1918 ! *Maitland* 848 ! 888 ! 1002 ! 1243 ! *Migeod* 199 ! *Dundas* FHI 20360 ! *Keay* FHI 28608 ! *Brenan* 9559 ! *Morton* GC 6903 ! *Boughey* GC 12665 !

2. **C. newii** *Oliv. & Hiern* in F.T.A. 3 : 449 (1877) ; Babcock l.c. 369.
2a. **C. newii** subsp. **kundensis** *(Babc.)* Babc. l.c. 374, fig. 77 (1947). *Crepis bumbensis* Hiern subsp. *kundensis* Babc. in Bull. Jard. Bot. Brux. 15 : 300 (1937). A perennial herb with a branched scape 1–2 ft. high from a rosette of leaves ; florets yellow in heads ⅓ in. broad ; latex white ; in open hill pasture.
[Br.]**Cam.**: Bamenda Dist., 4,500–7,500 ft. (Dec., Jan., Mar., Apr.) *Maitland* 1469 ! 1481a ! 1678 ! *Migeod* 443 ! *Keay* FHI 28384 ! *Morton* K98 ! *Adams* GC 11212 ! *Boughey* GC 11433 ! *Hepper* 2115 ! 2713 ! Binka, Nkambe (Oct.) *Hepper* 1887 ! Maisamari, 4,700 ft., Mambila Plateau (Jan.) *Hepper* 1688 ! The range of *C. newii*, including the several subspecies of Babcock, extends to Tanganyika and Angola.

Additional Species.

Davey 493 (**Mali**): Diafarabé) has a rosette with a few small lobulate, sparsely setulose leaves on a short scape 6 in. high and yellow florets in an involucre of sparsely setulose scarious-margined bracts. This plant may be *C. juvenalis* F. Schulz (in Flora 23 : 719 (1840)), a native of western N. Africa also recorded from the Atlantic Islands, and from Europe as an alien.

83. SONCHUS Linn.—F.T.A. 3 : 457.

Flowering stems leafy :
  Leaves distinctly auriculate ; bracteoles on peduncles absent, those subtending inflorescence-branches linear-subulate like miniature leaves ; base of involucre more or less woolly-pubescent, at least when young :
    Leaves obovate to oblanceolate in plan, 10–20 cm. long, coarsely and irregularly lobed, sharply triangular-denticulate ; involucre glabrescent ; annual weedy herbs :
      Achenes smooth, about 10-ribbed, winged, flat, elliptic, rounded at both ends ; leaf-base auricles more or less rounded   ..   ..   1. *asper*
      Achenes rugulose, grooved, not winged, obovate, narrowed to the base ; leaf-base auricles more or less pointed   ..   ..   2. *oleraceus*
    Leaves linear to linear-lanceolate, up to 40 cm. long, entire or sparsely runcinate-pinnatifid, minutely denticulate, long-acute, tailed-auriculate ; peduncles and bases of the involucres persistently buff or whitish woolly-pubescent ; achenes oblong, 4–8-ribbed, minutely transversely rugulose ; tall annual or perennial hollow-stemmed herbs of hill grassland :
      Outer involucral bracts densely setose, especially in the midline or at base and apex with ascending bronze-coloured hairs ..   ..   3. *angustissimus*
      Outer involucral bracts glabrous or with a very few minute setae at apex
                                                                4. *schweinfurthii*
  Leaves shortly or not auriculate ; bracteoles on peduncles and at inflorescence-branches ovate to ovate-lanceolate, small, more or less clasping ; base of involucre glabrous, glandular or thinly pubescent, not woolly :
    Leaves linear-lanceolate, mostly basal, runcinate-pinnatifid and coarsely dentate, shortly auriculate at base, the lower 10–12 cm. long ; achenes pubescent or glabrous, 4- or many-ribbed in the same capitulum   ..   ..   ..   5. *chevalieri*
    Leaves linear to linear-oblanceolate, mostly cauline, simple or nearly so, entire or remotely denticulate, not distinctly auriculate at base :
      Stems spreading horizontally ; leaves 5–10 cm. long, 4–7 mm. broad, obtuse, apiculate ; achenes as above ; pappus pale buff ..   ..   6. *brunneri*
      Stems erect ; leaves up to 15 cm. long and 1 cm. broad, acute ; achenes 10–12-ribbed,

Fig. 266.—Sonchus angustissimus *Hook. f.* (Compositae).

A, involucral bract. B, flower. C, style-arms. D, achene and pappus. E, pappus-bristle.

smooth or tuberculate ; pappus white ; involucral bracts and peduncles puberulous   ..   ..   7. *exauriculatus*

Flowering stems leafless ; leaves when present in a basal rosette, obovate to oblanceolate, rounded at apex, 6–20 cm. long, 2–8 cm. broad, denticulate or rarely smooth-margined; achenes oblong, ribbed, smooth, glabrous ; pappus white :

Stems dwarf, hardly exceeding 10 cm. in length, from a massive tuberous rootstock ; involucres 1–1·5 cm. long, on bracteate peduncles up to 1 cm. long, contiguous in a bunched inflorescence :

Inner involucral bracts ovate to lanceolate, successively longer from outside to inside, twisted at apex ; stem-apex and bracteoles within densely clothed with long matted buff hairs ; bracteoles linear   ..   ..   8. *ledermannii*

Inner involucral bracts linear, distinct from and about four times longer than the outer without intermediates, more or less flat ; stem-apex and bracteoles within glabrous or thinly pubescent ; bracteoles ovate   ..   ..   9. *elliotianus*

Stems 25–50(–75) cm. long from a stout taproot ; inflorescence divaricately branched into a spreading panicle ; involucres narrowly oblong, 1·5–2 cm. long on peduncles up to 3 cm. long, each with a single minute bracteole or ebracteolate   10. *rarifolius*

1. **S. asper** (*Linn.*) *Hill* Herb. Brit. 1 : 47 (1769). *Sonchus oleraceus* var. *asper* Linn. (1753)—F.T.A. 3 : 457 ; Hiern in Cat. Welw. 1 : 622. A coarse erect or decumbent herb 1–3 ft. high with sharply toothed undulate leaves, pale green beneath and glossy above ; florets yellow.
[Br.]**Cam.**: Cam. Mt., Upper Farm, Buea, 3,300 ft. (Mar., Apr.) *Boughey GC* 7032 ! *Brenan & Jones* 9582 ! A casual weed ; cosmopolitan.

2. **S. oleraceus** *Linn.* Sp. Pl. 794 (1753) ; F.T.A. 3 : 457 ; *Chev.* Bot. 379 ; Berhaut Fl. Sén. 171. A coarse erect annual herb 1–3 ft. high with a milky sap and occasionally with glandular hairs on stems and peduncles ; florets yellow.
Sen.: *fide* Berhaut *l.c.*   **Mali**: Kankan (Mar.) *Chev.* 564 ! Timbuktu (Aug.) *Hagerup* 248 ! **Guin.**: Mali (Jan.) *Roberty* 16551 ! **N.Nig.**: Jos, 4,100 ft. (fr. Oct.) *Hepper* 1027 ! [Br.]**Cam.**: Bamenda, 5,000 ft. (Feb.) *Migeod* 457 ! **F.Po**: estate of Marcelino Puente (Jan.) *Guinea* 1736 ; 1737. A cosmopolitan weed.

3. **S. angustissimus** *Hook. f.* in J. Linn. Soc. 7 : 203 (1864) ; F.T.A. 3 : 458. An erect herb 2–10 ft. high with glabrous striate hollow stems and campanulate flower-heads ¾ in. long with pale yellow florets.
S.Nig.: Ikwette, 5,200 ft., Ogoja (Dec.) *Savory & Keay* FHI 25261 ! [Br.]**Cam.**: Cam. Mt., 5,000–9,500 ft. (Nov.–May) *Mann* 1311 ! 1919 ! *Deistel* 200 ! *Deistel* 616 ! *Mildbr.* 10843 ! *Brenan & Onochie* 9390 ! Bamenda Dist., 6,000–8,200 ft. (Jan., Feb., Apr., May) *Migeod* 342 ! 491 ! *Maitland* 1750 ! *Morton* K255 ! *Boughey* GC 11044 ! Also in Sudan, Uganda, Kenya, Tanganyika and S. Rhodesia.

4. **S. schweinfurthii** *Oliv. & Hiern* in F.T.A. 3 : 458 (1877). An erect herb like the last, 3–6 ft. high, with white latex ; ligules of florets yellow within, pinkish outside.
[Br.]**Cam.**: Bamenda, 3,600–5,000 ft. (Dec., Jan., Mar., Apr.) *Migeod* 366 ! *Keay* FHI 28416 ! *Ujor* FHI 30053 ! *Boughey* GC 11108 ! *Morton* K39 ! Kumbo, 5,500 ft. (Feb.) *Hepper* 1988 ! Maisamari, 4,600 ft., Mambila Plateau (Jan.) *Hepper* 1656 ! Also in Ubangi-Shari, Congo, Sudan, Kenya, Uganda, Tanganyika, Nyasaland, N. Rhodesia and Angola.

5. **S. chevalieri** (*O. Hoffm. & Muschl.*) *Dandy* in F. W. Andr. Fl. Pl. Sudan 3 : 51 (1956). *Launaea chevalieri* O. Hoffm. & Muschl. (1910)—Chev. Bot. 380. *Sonchus prenanthoides* Oliv. & Hiern in F.T.A. 3 : 459 (1877) ; F.W.T.A., ed. 1, 2 : 178, not of Bieb. (1808). *Launaea integrifolia* Hagerup (1930). A palestemmed diffusely branched annual or perennial herb up to about 2 ft. high, woody at the base, with a rosette or dentate leaves ; florets white with pale yellow ligules often reddish outside.
Sen.: Rufisque (Dec.) *Chev.* 2051 ! Nianing (Feb.) *Berhaut* 5350 ! **Mali**: Timbuktu (July) *Chev.* 1259 ! *Hagerup* 134 ! Selingourou (Feb.) *Davey* 26 ! **Ghana**: Gambaga (Dec.) *Adams & Akpabla* GC 4224 ! Tongo Hills, Zuarungu (Apr.) *Morton* GC 8926 ! **Niger**: Fada to Koupéla, Gourma (July) *Chev.* 24534 ! **N.Nig.**: Kouka, Bornu (Feb.) *E. Vogel* 45 ! Sokoto (Feb.) *Dalz.* 411 ! Kano (Dec.) *Hagerup* 661 ! Also in Mauritania, Sudan, Ethiopia, Kenya and Tanganyika. (See Appendix, p. 420.)

6. **S. brunneri** (*Webb*) *Oliv. & Hiern* in F.T.A. 3 : 459 (1877) ; F.W.T.A., ed. 1, 2 : 178 ; Berhaut Fl. Sén. 4, 198. *Rhabdotheca brunneri* Webb in Fl. Nigrit. 147 (1849). *S. bipontini* of Chev. Bot. 379, not of Aschers. A bushy spreading stiff-branched herb of littoral sands about 8–12 in. high at most ; ligules white, pale yellow at the base and tinged pink outside.
Sen.: Sôr Isl. *Brunner* 26 ! Dakar (Jan.) *Chev.* 2046 ! Kayar (Mar.) *Berhaut* 5661 ! Also in Mauritania.

7. **S. exauriculatus** (*Oliv. & Hiern*) *O. Hoffm.* in Engl. Pflanzenw. Ost-Afr. C : 421 (1895). *S. bipontini* Aschers. var. *exauriculatus* Oliv. & Hiern in F.T.A. 3 : 459 (1877). An erect perennial herb with glaucous stems up to about 4 ft. high ; ligules of florets pale yellow, reddish outside.
[Br.]**Cam.**: Gurum, Vogel Peak area, Adamawa (Dec.) *Hepper* 1598 ! From Sudan and Somaliland to Congo, Nyasaland and Zanzibar.

8. **S. ledermannii** *R. E. Fries* in Act. Hort. Berg. 8 : 118 (1925). A perennial herb with a thick taproot yielding white latex ; the leaves appear separately from the tufted inflorescences which are about 2 in. high with yellow florets in purplish involucres ; ligules reddish outside.
[Br.]**Cam.**: Ngel Nyaki, 5,400 ft., Mambila Plateau (Jan.) *Hepper* 2878 ! Vogel Peak, 4,900 ft., Adamawa (Dec.) *Hepper* 2877 ! Also in Cameroun, Tanganyika and N. Rhodesia.

9. **S. elliotianus** *Hiern* in Cat. Welw. 1 : 623 (1898) ; Chev. Bot. 379. *S. nanus* O. Hoffm. (1895), not of Sond. (1865). *Lactuca nana* Bak. (1895). A perennial herb with leafless flowering shoots a few inches high from a thick laticiferous rhizome ; florets yellow, often tinged reddish.
Guin.: Diaguissa (Apr.) *Chev.* 12910 ! Timbikounda, Faranna (Jan.) *Chev.* 20604 ! **S.L.**: *Sc. Elliot* 5010 ! York (May) *Deighton* 5530 ! Makene (Jan.) *Deighton* 3590 ! Kabala (Feb.) *Deighton* 4245 ! Bintumane Peak, Loma Mts. (Sept.) *Jaeger* 1998 ! **Iv.C.**: *fide* Chev. 339 ! **Ghana**: Kintampo (fl. & leaves Mar.) *Dalz.* 40 ! Abafuo (Feb.) *Vigne* FH 1808 ! Tumu (May) *Morton* GC 7563 ! Legon Hill (Feb., May, Nov.) *Morton* A2168 ! *Adams* 3554 ! 3750 ! **Togo Rep.**: Misahöhe (Mar.) *Baumann* 131. **Dah.**: Zagnanado (Feb.) *Chev.* 23058 ! **N.Nig.**: Abinsi (fl. & leaves Oct.) *Dalz.* 665 ! Vom, 3,000–4,500 ft. (Apr.) *Dent Young* 165 ! *Morton* K383 ! **S.Nig.**: Ibadan (Feb., Mar.) *Meikle* 1151 ! *Keay* FHI 25698 ! Ogbomosho to Igbetti (Mar.) *Hambler* 133 ! [Br.]**Cam.**: Jua, 3,500 ft., Bamenda (Apr.) *Maitland* 1589 ! Bambui, 6,300 ft. (Dec.) *Adams* GC 11245 ! Kumbo to Oku, 6,000 ft. (Feb.) *Hepper* 2013 ! Ngel Nyaki, 5,400 ft., Mambila Plateau (Jan.) *Hepper* 1732 ! Vogel Peak, 4,900 ft., Adamawa (Dec.) *Hepper* 1591 ! Also in Congo, Sudan, Uganda, Kenya, Tanganyika, Nyasaland, Rhodesia, Mozambique, Transvaal and Angola.
[*Keay* FHI 28413 from Mbakokeka Mt., Bamenda is intermediate between *S. ledermannii* and *S. elliotianus*.]

10. **S. rarifolius** *Oliv. & Hiern* in F.T.A. 3 : 460 (1877) ; Chev. Bot. 380 ; R. E. Fries l.c. 113, t. 3, fig. 1–2. *Lactuca stenocephala* Bak. (1895). An erect stiff-branched herb, the flowering stems leafless, 1–2½ ft. high from a laticiferous perennial stock ; involucres narrow ½–¾ in. long with yellow florets, reddish outside.
Mali: Dougoura (Feb.) *Chev.* 339 ! **Guin.**: *fide* Chev. l.c.   **S.L.**: Musaia (Mar.) *Sc. Elliot* 5115 ! Loma Mts. (Feb.) *Jaeger* 4263 ! **Iv.C.**: Zoanlé, Upper Sassandra (May) *Chev.* 21491 ! Alangouassou to Mbayakro, Baoulé-Nord (Aug.) *Chev.* 22244 ! **Ghana**: Wenchi, Ashanti (Dec.) *Adams* 3247 ! Amedzofe Hill (Apr.,

May) *Morton* A2009! GC 7246! **N.Nig.**: Nupe *Barter* 1224! Giwa, Zaria (May) *Keay* FHI 25805! **S.Nig.**: Ilorin *Rowland*! Iseyin, Oyo (Feb.) *Brenan* 8966! Igbobo, Oyo (Feb.) *Keay* FHI 23447! Bansara, Abakaliki to Ikom (Mar.) *Morton* K1! [**Br.**]**Cam.**: Bamenda Dist., 5,000 ft. (fl. Mar., leaves Dec.) *Maitland* 1615! *Boughey* GC 11368! *Morton* K49! Lus, 2,500 ft., Nkambe (Feb.) *Hepper* 1863! Maisamari, 4,700 ft., Mambila Plateau (Jan.) *Hepper* 1687! Also in Congo, Uganda, Rhodesia, Angola, Transvaal and Madagascar.

84. LAUNAEA Cass.—F.T.A. 3 : 460.

Involucral bracts with distinct broad hyaline margins ; involucres cylindrical ; leaves mostly basal, coarsely dentate and denticulate with sharp whitish cartilaginous teeth ; outer achenes wrinkled, muricate, few or several-ribbed ; bracteoles on peduncles very few or absent ..   ..   *nudicaulis*

L. nudicaulis (*Linn.*) *Hook. f.* in Fl. Brit. Ind. 3 : 416 (1881) ; Muschl. Man. Fl. Egypt 2 : 1059. *Chondrilla nudicaulis* Linn. (1771)—De Miré & Gillet in J. Agric. Trop. 3 : 708 (1956). A rosette herb with a branched scapose inflorescence up to about 1 ft. high ; ligules of florets yellow within, crimson outside. **Niger**: Aïr *vide* De Miré & Gillet *l.c.* Also in Cape Verde Islands, Mauritania, Tibesti, through N. Africa and Egypt to Arabia and India, and in S. W. Africa.

140. GENTIANACEAE

By P. Taylor

Annual or perennial herbs. Leaves opposite, often connate at the base ; stipules absent. Flowers hermaphrodite, actinomorphic or rarely zygomorphic. Calyx tubular or of separate sepals, imbricate. Corolla contorted. Stamens the same number as the corolla-lobes and alternating with them, inserted on the corolla ; anthers 2-celled, opening lengthwise. Disk present or absent. Ovary superior, mostly 1-celled with 2 parietal placentas, sometimes 2-celled with the placentas adnate to the septa ; style simple. Ovules often numerous. Fruit usually a capsule. Seeds with copious endosperm and small embryo.

Mainly in temperate and subtropical regions, abundant in the mountains of the northern hemisphere. Recognised by the herbaceous habit, opposite exstipulate leaves, gamopetalous corolla, the stamens the same number as and alternate with the corolla-lobes, and the superior ovary with usually parietal placentas.

Leaves well developed, green :
  Flowers solitary in both leaf axils all along the stems ; calyx-tube ribbed
                                         1. Neurotheca

  Flowers in axillary or terminal cymes or clusters :
    Flowers in terminal dichotomous cymes :
      Ovary 2-celled :
        Corolla blue   ..   ..   **2. Exacum**
        Corolla white or yellow   ..   ..   **3. Sebaea**
      Ovary 1-celled with parietal or sometimes intrusive placentas :
        Anthers all fertile :
          Corolla-lobes with 1–2 fringed nectaries near their base   ..   **4. Swertia**
          Corolla-lobes without nectaries near their base :
            Calyx shortly 4-lobed ; anthers straight ; corolla yellow   ..   **5. Schultesia**
            Calyx deeply 5-lobes ; anthers twisted ; corolla pink..   ..   **6. Centaurium**
        Anthers not all (usually only one) fertile ; corolla-lobes without nectaries
                                      7. Canscora

    Flowers in dense axillary clusters all along the stem or in dense terminal clusters or heads :
      Anthers exserted from the corolla-tube :
        Flowers 4-merous :
          Anthers not all (usually only one) fertile ; corolla white   ..   **8. Schinziella**
          Anthers all fertile ; corolla blue or mauve :
            Stigma 2-lobed, throat of corolla with a fringe of hairs   ..   **9. Djaloniella**
            Stigma simple, throat of corolla without a fringe of hairs   ..   **10. Faroa**
        Flowers 3-merous   ..   ..   **11. Pycnosphaera**
      Anthers not exserted ; flowers 5-merous   ..   ..   ..   **12. Enicostema**
Leaves rudimentary, membranous ; flowers solitary, terminal ; stamens included in the corolla-tube ; calyx-lobes subulate   ..   ..   **13. Voyria**

1. NEUROTHECA Salisb.—F.T.A. 4, 1 : 559.

Calyx-tube as long as or shorter than the teeth, unequally 10–12-ribbed, lobes subulate-lanceolate ; leaves oblanceolate, 1–2·5 cm. long, narrowed to the base ; flowers small and inconspicuous   ..   ..   **1.** *longidens*

Calyx-tube longer than the teeth, regularly 8-ribbed ; lobes about ⅓ as long as the tube, triangular-lanceolate, very acute ; leaves up to 2·5 cm. long ; flowers numerous and conspicuous .. .. 2. *loeselioides*

1. **N. longidens** *N. E. Br.* in F.T.A. 4, 1 : 560 (1904). A prostrate or suberect herb with slender angular branches up to 12 in. long, glabrous, with small white or pale lilac flowers solitary in the leaf axils.
   S.Nig.: Lagos (Nov.) *Dalz.* 1240 ! (Apr., July) *A. P. D. Jones* FHI 19409 ! *Moloney* ! Also in the Congo and Uganda.

2. **N. loeselioides** (*Spruce ex Prog.*) *Baill.* Hist. 10 : 138 (1888) ; F.T.A. 4, 1 : 560 ; Chev. Bot. 446 (incl. var. *grandiflora* Knobl.). *Octopleura loeselioides* Spruce ex Prog. in Mart. Fl. Bras. 4, 1 : 212 (1865). *N. robusta* Hua (1906)—Chev. Bot. 446. *N. rupicola* Hua (1906). An usually branched herb with slender angular stems up to 18 in. high, with blue or violet flowers ⅓–½ in. long solitary in the leaf axils.
   Sen.: Niayes (Oct.) *Chev.* 2830 ! Casamance (Feb.) *Chev.* 2827. **Mali:** Koulikoro (Oct.) *Chev.* 2828. **Port.G.:** Suxana to Bulol (Nov.) *Esp. Santo* 3719 ! **Guin.:** Conakry (Oct.) *Adam* 12613 ! Kouria (Oct.) *Chev.* 14921. Boulivel to Dalaba (Sept.) *Chev.* Mt. Benna (Oct.) *Jac.-Fél.* 7156 ! Koba (Nov.) *Jac.-Fél.* 7202 ! Macenta (Oct.) *Baldwin* 9792 ! **S.L.:** Newton (Nov.) *Deighton* 1487 ! Waterloo (Aug.) *Melville & Hooker* 281 ! Rokupr (Nov.) *Jordan* 155 ! Sefadu (Nov.) *Deighton* 4662 ! Gberia Fotumbu (Oct.) *Small* 417 ! Topan (Oct.) *Deighton* 5036 ! **Lib.:** Paynesville (Oct.) *Harley* 1689 ! Monrovia (Nov.) *Linder* ! Kolahun (Nov.) *Baldwin* 10090 ! Grand Cape Mount (Dec.) *Baldwin* 10908 ! Duport (Oct.) *Barker* 1449 ! **Iv.C.:** Akabilékrou to Etérokrou (Dec.) *Chev.* 22520. **Ghana:** Kwahu Tafo (Dec.) *T. M. Harris* ! Atwabo (Feb.) *Irvine* 2297 ! Kintampo (Nov.) *Morton* A3818 ! Senchi (May) *Morton* 7236 ! Sampa (Apr.) *Morton* 3258 ! Mampong (Oct.) *Rose Innes* GC 31167 ! **N.Nig.:** Lokoja (Oct.) *Dalz.* 122 ! Guduma, Minna (Oct.) *Hepper* 970 ! Zaria (Sept.) *Olorunfemi* FHI 24391 ! Dogon Kurmi, Jemaa Dist. (Aug.) *Killick* 26 ! Jos *Batten-Poole* ! Naraguta (Sept.) *Lely* 530 ! **S.Nig.:** Lagos (Oct.) *Dalz.* 1239 ! Akure F.R., Ondo (Nov.) *Latilo & Olorunfemi* FHI 24414 ! Onitsha *Barter* 1761 ! Agulu Enu, Awka Dist. (Nov.) *Keay* FHI 21531 ! Ukpor-Nzagba, Onitsha Prov. (Oct.) *Okafor* FHI 35873 ! [Br.]**Cam.:** Mamfe (Dec.) *Morton* K677 ! **F.Po:** (Feb.) *Moore* ! Also in Uganda.

2. EXACUM Linn.—F.T.A. 4, 1 : 545.

A slender annual up to 75 cm. high, simple or branched above ; leaves linear to lanceolate-ovate, acute, up to 6 cm. long, 3–5-nerved ; flowers in terminal cymes ; pedicels up to 1·4 cm. long ; sepals ovate, 4 mm. long, broadly winged on the keel ; corolla campanulate, tube 3 mm. long, the lobes shorter than the tube ; anthers exserted ; capsule globose, 4 mm. diam. .. .. *quinquenervium*

E. quinquenervium *Griseb.* Gen. et Sp. Gent. 112 (1838) ; F.T.A. 4, 1 : 546 ; Chev. Bot. 445. Erect herb of wet places with pale blue flowers.
   Sen.: Casamance (Jan.) *Chev.* 2843. **Port.G.:** Bafata to Saltinho *Esp. Santo* 3823 ! **Guin.:** Kankan *Chev.* 15681. Tossékré (Oct.) *Adam* 12752 ! **S.L.:** Mateboi (Nov.) *Jordan* 837 ! Musaia (Dec.) *Deighton* 5700 ! **Ghana:** Achimota (Sept.) *Milne-Redhead* 5133 ! Accra to Winneba (Oct.) *Cudjoe* ! **N.Nig.:** Kano (Dec.) *Hagerup* 664 ! Abo *Barter* 140 ! **S.Nig.:** Lagos *Rowland* ! Apapa (Oct.) *Dalz.* 1254 ! Aguku *Thomas* 1057 ! Enugu (Aug.) *Olorunfemi* FHI 34198 ! [Br.]**Cam.:** Vogel Peak, Adamawa (Nov.) *Hepper* 1436 ! Widespread in tropical Africa and Madagascar.

3. SEBAEA R. Br.—F.T.A. 4, 1 : 546.

Stamens inserted in the corolla sinuses, exserted ; leaves orbicular, up to 1·5 cm. diam.; flowers shortly pedicellate, numerous, in dense terminal cymes :
   Corolla about 6 mm. long .. .. 1. *brachyphylla*
   Corolla about 12 mm. long .. .. 2. *longicaulis*
Stamens inserted in the corolla-tube, included ; leaves lanceolate to elliptic ; flowers very long-pedicellate in lax terminal cymes or more or less solitary :
   Corolla-tube 2–3 times as long as the calyx, lobes 10–15 mm. long, white ; leaves lanceolate .. .. 3. *teuszii*
   Corolla-tube about equal to the calyx, lobes cream or pale yellow :
      Leaves lanceolate ; plant up to 20 cm. high :
         Corolla more than 2 cm. long, lobes 5–10 mm. long ; leaves up to 3 cm. long
                                                                                  4. *grandis*
         Corolla 1–1·5 cm. long, lobes 2·5–5 mm. long ; leaves up to 1·5 cm. long
                                                                                  5. *luteo-alba*
      Leaves elliptic ; plant 5–10 cm. high ; corolla about 8 mm. long, lobes about 1·5 mm. long .. .. 6. *pumila*

1. **S. brachyphylla** *Griseb.* Gen. et Sp. Gent. 170 (1838). *S. multinodis* N. E. Br. in F.T.A. 4, 1 : 548 (1904) ; F.W.T.A., ed. 1, 2 : 182. Erect herb of wet places up to 60 cm. high ; flowers bright yellow.
   [Br.]**Cam.:** Cam. Mt., 8,000–9,800 ft. (Nov., Jan., Mar.) *Brenan* 9522a ! *Migeod* 158 ! *Steele* 39 ! Kumbo, 5,500 ft. (Feb.) *Hepper* 1969 ! Mambila Plateau, 4,900 ft. (Jan.) *Hepper* 1787 ! **F.Po:** Moka, 4,900 ft. (Dec.) *Boughey* 153 ! Clarence Peak (Pico de S. Isabel), 8,300–9,000 ft. (Mar., Dec.) *Mann* 598 ! *Guinea* 2921 ! Widespread in the mountains of tropical Africa.

2. **S. longicaulis** *Schinz* in Bull. Herb. Boiss. 2 : 219 (1894). *S. oreophila* Gilg (1901)—F.T.A. 4, 1 : 547. Very similar to the above but with much larger flowers.
   [Br.]**Cam.:** Kumbo to Oku, 6,000 ft., Bamenda Div. (Feb.) *Hepper* 2809 ! Widespread in the mountains of tropical Africa and extending into S. Africa.

3. **S. teuszii** (*Schinz*) *P. Tayl.* in Taxon 12 : ined. (1963). *Belmontia teuszii* Schinz in Viertelj. Nat. Gesells. Zürich 36 : 335 (1891) ; F.T.A. 4, 1 : 554. *Exochaenium teuszii* (Schinz) Schinz (1906). *Belmontia chevalieri* des Abbayes & Schnell in Bull. Soc. Bot. Fr. 96 : 204 (1950). Herb of wet places with white flowers.
   Guin.: Conakry (Oct.) *Schnell* 7720 ! Mt. Benna (Oct.) *Jac.-Fél.* 7159 ! *Schnell* 2135 ! Dalaba, 4,200 ft. (Nov.) *des Abbayes* ! **S.L.:** Bumban (Oct.) *Deighton* 3273 ! Also in E. and S. tropical Africa.

4. **S. grandis** (*E. Mey.*) *Steud.* Nom., ed. 2, 2 : 550 (1841). *Belmontia grandis* E. Mey. (1836)—F.T.A. 4, 1 : 553. *Exochaenium grande* (E. Mey.) Griseb. (1845). Herb with pale yellow flowers.
   N.Nig.: Kagoro Hills, Zaria Prov. (Aug.) *Killick* 20 ! Naraguta (July) *Lely* 463 ! Abinsi (Sept.) *Dalz.* 724 ! [Br.]**Cam.:** Bamenda (June) *Maitland* 1790 ! Also in E. and S. tropical Africa and S. Africa.

5. **S. luteo-alba** (*A. Chev.*) *P. Tayl.* in Taxon 12 : ined.] (1963). *Belmontia luteo-album* A. Chev. in Journ. de Bot. 1909 : 119 (1909) ; Chev. Bot. 445 ; F.W.T.A., ed. 1, 2 : 183. Similar to the above but with smaller flowers.

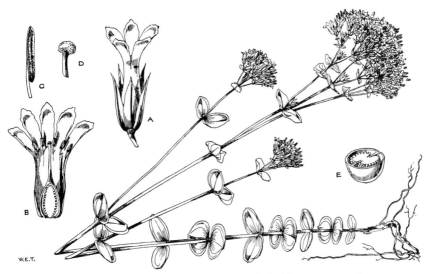

FIG. 267.—SEBAEA BRACHYPHYLLA *Griseb.* (GENTIANACEAE).

A, flower.   B, longitudinal section of flower.   C, stamen.   D, stigma.   E, cross-section of ovary.

Guin.: Kouria to Irébéléya (Sept.) *Chev.* 18239. Timbo to Ditinn (Sept.) *Chev.* 18425. Ditinn to Dalaba (Sept.) *Chev.* 18556 ! Dalaba (Sept.) *Chev.* 20206. *Schnell* 6811 ! **S.L.:** Loma Mts. (Nov.) *Jaeger* 574 !

6. **S. pumila** (*Bak.*) *Schinz* in Bull. Herb. Boiss., sér. 2, 6 : 731 (1906). *Belmontia pumila* Bak. in Kew Bull. 1894 : 25 ; F.T.A. 4, 1 : 552 ; F.W.T.A., ed. 1, 2 : 183. *Exochaenium pumilum* (Bak.) A. W. Hill (1908). Minute herb of wet places.

N.Nig.: Nupe *Barter* 1680 ! Naraguta (Aug.) *Lely* 452 ! Abinsi (Sept.) *Dalz.* 788 !

4. SWERTIA Linn.—F.T.A. 4, 1 : 570.

Leaves ovate to orbicular, up to 2·5 cm. long and 1·5 cm. wide ; sepals oblanceolate to obovate-spathulate, 3–6 mm. long ; corolla 0·8–1·3 cm. long ; capsule about as long as the corolla   ..   ..   1. *abyssinica*

Leaves linear-lanceolate to elliptic ; sepals lanceolate to narrowly lanceolate :

Annual ; leaves linear-lanceolate to narrowly elliptic, 1–3 cm. long ; corolla-lobes narrowly lanceolate, 0·5–1 cm. long, up to 3 mm. broad, acute ; cymes rather lax, elongated   ..   ..   2. *mannii*

Perennial ; leaves narrowly lanceolate to elliptic, acute, 1·5–5 cm. long ; corolla-lobes oblong-obovate, 1–1·3 cm. long, up to 5 mm. broad ; cymes congested, corymbiform 3. *quartiniana*

1. **S. abyssinica** *Hochst.* in Flora 1844 : 28 (1844) ; F.T.A. 4, 1 : 578. *S. clarenceana* Hook. f. (1861)—F.T.A. 4, 1 : 578 ; F.W.T.A., ed. 1, 2 : 183. *S. subalpina* N. E. Br. in F.T.A. 4, 1 : (1904) ; F.W.T.A., ed. 1, 2: 183. *S. dissimilis* N. E. Br. in F.T.A. 4, 1 : 579 (1904) ; F.W.T.A., ed. 1, 2 : 183. Erect, often much-branched herb up to 18 ins. high, with numerous white flowers. **[Br.]Cam.:** Cam. Mt., 6,000–10,000 ft. (Nov.–Jan.) *Mann* 1216 ! 1994 ! *Mildbr.* 10880 ! L. Oku, 7,800 ft., Bamenda Div. (Jan.) *Keay & Lightbody* FHI 28495 ! **F.Po:** Clarence Peak, 8,500 ft. (Dec.) *Mann* 596 ! Also in Ethiopia and E. Africa.

2. **S. mannii** *Hook. f.* in J. Linn. Soc. 7 : 206 (1864) ; F.T.A. 4, 1 : 581. Erect, slender annual with numerous small white flowers. **Guin.:** Madina (Oct.) *Adam* 12529 ! Dalaba, 4,000 ft. (Oct.) *des Abbayes* 683 ! Yambéring (Sept.) *Schnell* 7014 ! **S.L.:** Bintumane, 5,400 ft. (Jan.) *T. S. Jones* 85 ! **N.Nig.:** Vom, 3,000–4,500 ft. *Dent Young* ! Jos, 4,000–5,000 ft. *Saunders* ! **[Br.]Cam.:** Cam. Mt., 7,000–8,000 ft. (May, Nov.) *Mann* 2000 ! *Maitland* 642 ! Bafut-Ngemba F.R., 5,000–6,000 ft., Bamenda (Sept.) *Lightbody* FHI 26253 ! Nkambe, 6,500 ft. (Sept.) *Savory* UCI 374 !

3. **S. quartiniana** *Hochst. ex A. Rich.* Tent. Fl. Abyss. 2 : 56 (1851) ; F.T.A. 4, 1 : 574. Simple or sparsely branched perennial from 6 in. to 2 ft. high, with mauve flowers. **[Br.]Cam.:** Bafut-Ngemba F.R., 6,500 ft., Bamenda (Feb., May) *Keay* FHI 37921 ! *Hepper* 2084 ! Likitaba to Gembu, 5,500 ft., Mambila Plateau (June) *Chapman* 76 ! Also in Ethiopia, E. and S. tropical Africa.

5. SCHULTESIA Mart.—F.T.A. 4, 1 : 559. *Nom. cons.*

Erect annual up to 50 cm. high ; stems 4-angled ; leaves sessile, ovate-lanceolate to narrowly oblong, 1·5–4 cm. long, thin, glabrous ; flowers large, 1·5–2 cm. diam., few, cymose ; pedicels ascending ; calyx 1·5 cm. long ; segments subulate-acuminate, broadly winged and reticulate, dry ; corolla-tube longer than the calyx, lobes broadly obovate ; capsule enclosed in the persistent calyx   ..   *stenophylla* var. *latifolia*

S. stenophylla *Mart.* Nov. Gen. et Sp. 2 : 106, t. 182 (1826). An American species with the following variety in tropical Africa.

S. stenophylla var. **latifolia** *Mart. ex Prog.* Fl. Bras. 6 : 203 (1865) ; F.T.A. 4, 1 : 559 ; Chev. Bot. 446. Erect, simple or branched herb of wet places, with yellow flowers ¾–1 in. long.
Sen.: *Heudelot* 551 ! Niomoum (Jan.) *Chev.* 2821 ! 2823 ! Samandiny (Feb.) *Chev.* 2822. **Gam.**: Brown-*Lester* 39 ! **Port.G.**: Bissau (Nov.) *Esp. Santo* 1750 ! Sedengal to S. Domingos (Nov.) *Esp. Santo* 3666 ! **Guin.**: Koba (Nov.) *Jac.-Fél.* 7215 ! **S.L.**: Aberdeen (Nov.) *Small* 796 ! Lumley (Aug.) *Melville & Hooker* 252 ! Batbai (Jan.) *Deighton* 3551 ! Totenbana (Mar.) *Adames* 157 ! Konta (Mar.) *Jordan* 409 !

6. CENTAURIUM Hill Brit. Herb. 62 (1756). *Erythraea* Borck. (1796)—F.T.A. 4, 1 : 556.

Erect branched annual up to 12 cm. high ; leaves narrowly lanceolate-oblong, acute, up to 3 cm. long, 3–5-nerved ; flowers in terminal cymes ; pedicels 2–5 mm. long ; sepals linear-lanceolate, about 6 mm. long ; corolla-tube slender, about 8 mm. long, lobes about half as long as the tube, narrowly ovate ; stamens inserted at the top of the tube, anthers twisted ; capsule cylindrical, about 7 mm. long   .. *pulchellum*

C. pulchellum (*Sw.*) *E. H. L. Krause* in Sturm, Fl. Deutsch., ed. 2, 10 : 14 (1903). *Gentiana pulchella* Sw. (1783). *Erythraea ramosissima* (Vill.) Pers. (1805)—F.T.A. 4, 1 : 556. Herb of damp depressions among sand dunes ; flowers bright pink.
Sen.: Kayan (July) *Raynal* 7162 ! **Niger**: Bilma *Le Coeur* 19 ! Widespread in the Mediterranean region.

7. CANSCORA Lam.—F.T.A. 4, 1 : 557.

Calyx about 1 cm. long, tube swollen towards the base, teeth 4, short, ovate-lanceolate ; corolla half as long again as the calyx ; leaves lanceolate, acute, 1·5–3 cm. long, 3–5-nerved from the base, glabrous ; flowers few in leafy cymes ; pedicels stout, winged   1. *decussata*
Calyx about 5 mm. long, tube narrow scarcely swollen towards the base, teeth 4, very short, subulate ; corolla a little longer than the calyx ; leaves elliptic or ovate-elliptic, acute, 1–4 cm. long, 3–5-nerved from the base, glabrous ; flowers very numerous in leafy cymes ; pedicels filiform, not winged ..   ..   ..   2. *diffusa*

1. **C. decussata** (*Roxb.*) *Roem. & Schult.* Syst. Mant. 3 : 229 (1827) ; F.T.A. 4, 1 : 557 ; Chev. Bot. 445. *Pladera decussata* Roxb. Fl. Ind. 1 : 418 (1820). Erect, much branched herb up to 1 ft. high, with small white flowers.
Sen.: Sedhiou to Tambanaba (Feb.) *Chev.* 2851. **Port.G.**: Bafata (Dec.) *Esp. Santo* 2842 ! **S.L.**: Kamaranka (Feb.) *Deighton* 4042 ! Musaia (Dec.) *Deighton* 4462 ! **Ghana**: Krachi (Oct.) *Morton* A 3727 ! Sunyani (Sept.) *Vigne* FH 2464 ! **Togo Rep.**: Lomé *Warnecke* 298 ! **N.Nig.**: Anara F.R., Zaria (Sept.) *Olorunfemi* FHI 24393 ! Dogon Kurmi, Jemaa Dist. (Sept.) *Killick* 64 ! Jos Plateau (Oct.) *Lely* 803 ! **S.Nig.**: Oyo (Nov.) *Savory & Keay* FHI 25300 ! Owo *Burton* 1 ! Ogurude, Ogoja Prov. (Jan.) *Holland* 266 ! **[Br.]Cam.**: Vogel Peak, Adamawa (Nov.) *Hepper* 1387 ! Throughout tropical Africa, Madagascar, also in tropical Asia and Australia.

2. **C. diffusa** (*Vahl*) *R. Br. ex Roem. & Schult.* Syst. 3 : 301 (1820) ; F.T.A. 4, 1 : 558 ; Chev. Bot. 445. *Gentiana diffusa* Vahl Symb. 3 : 47 (1794). Erect herb up to 2 ft. high with a much-branched many-flowered inflorescence ; flowers white or lilac.
Sen.: *Heudelot* 170 ! **Port.G.**: Boé, Danduru (Jan.) *Esp. Santo* 2365 ! **Mali**: Badumbe (Dec.) *Chev* 62 ! **Guin.**: Sambadougou, Faranna (Jan.) *Chev.* 20533 ! **S.L.**: York (Dec.) *Deighton* 3307 ! Makump (Jan.) *Glanville* 149 ! Kambia (Dec.) *Deighton* 5690 ! Musaia (Feb.) *Deighton* 4277 ! **Lib.**: Ganta (Jan.) *Harley* 1333 ! **Ghana**: Krachi (Nov.) *Morton* 4076 ! Bongo (Nov.) *Morton* 3798 ! Kamba, Lawra (Nov.) *T. M. Harris* ! **N.Nig.**: Minna (Dec.) *Meikle* 699 ! Dogon Kurmi, Jemaa Dist. (Nov.) *Keay* FHI 37252 ! Kaduna (Nov.) *Keay* FHI 28114 ! Kontagora (Nov.) *Dalz.* 57 ! Jos Plateau (Oct.) *Lely* 836 ! **S.Nig.**: Bebi, Obudu Div. (Dec.) *Savory & Keay* FHI 25010 ! **[Br.]Cam.**: Vogel Peak, Adamawa (Nov.) *Hepper* 1334 ! Throughout tropical Africa, also in tropical Asia and Australia.

8. SCHINZIELLA Gilg—F.T.A. 4, 1 : 557.

Erect perennial herb 30–60 cm. high, stems simple or very sparsely branched from the base, 4-angled, broadly winged ; leaves sessile, the lowermost elliptic, up to 1·5 cm. long, those above narrowly lanceolate, much reduced ; flowers sessile or subsessile in compact terminal heads ; calyx 4 mm. long, lobes acuminate about equal to the tube ; corolla about 6 mm. long, lobes acute, about equal to the tube   *tetragona*

S. tetragona (*Schinz*) *Gilg* in E. & P. Pflanzenfam. 4, 2 : 74 (1892) ; F.T.A. 4, 1 : 557. *Canscora tetragona* Schinz. in Viertelj. Nat. Gessells. Zürich 37 : 388 (1891). Herb of wet places ; flowers white rapidly turning brown.
Guin.: Ditinn, 2,600 ft. (Nov.) *des Abbayes* 882 ! Also in Congo and E. and S. tropical Africa.

9. DJALONIELLA P. Tayl. in Taxon 12 : ined. (1963).

Dwarf annual up to 6 cm. high, stem simple or sparsely branched, 4-angled, very narrowly winged ; leaves lanceolate to linear-lanceolate acute or acuminate, up to 2 cm. long ; flowers in congested terminal leafy cymes ; calyx about 3 mm. long, sepals linear-lanceolate ; corolla 6 mm. long, lobes narrowly lanceolate, acuminate, shorter than the tube ..   ..   *ypsilostyla*

D. ypsilostyla *P. Tayl.* in l.c. (1963). *Swertia caerulea* A. Chev. in Mém. Soc. Bot. Fr. 2, 8 : 183 (1912) ; Chev. Bot. 446 ; F.W.T.A., ed. 1, 2 : 183, not of Royle (1835). Small herb of wet places, flowers blue.
Guin.: Dalaba, Diaguissa Plateau, 3,200–4,300 ft. (Oct.) *Chev.* 18759 ! *Schnell* 7375 !

10. FAROA Welw.—F.T.A. 4, 1 : 565.

A small annual 2–10 cm. high ; stems simple or sparsely branched, 4-angled, distinctly winged ; leaves elliptic to lanceolate, 1·5–3·5 mm. long, membranous, shortly petiolate ; flowers pedicellate in axillary and terminal clusters ; calyx 2·5 mm. long, lobes

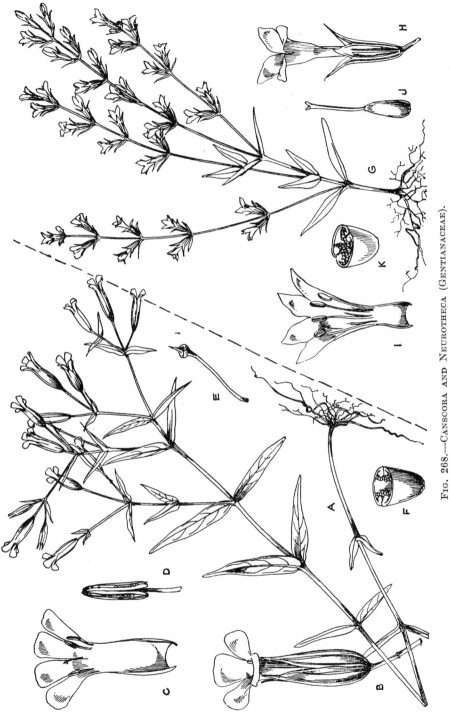

Fig. 268.—Canscora and Neurotheca (Gentianaceae).

Canscora decussata (Roxb.) Roem. & Schult.—A, habit. B, flower. C, corolla laid open. D, stamens. E, style. F, cross-section of ovary.
Neurotheca loeselioides (Spruce ex Prog.) Baill.—G, habit. H, flower. I, corolla laid open. J, pistil. K, cross-section of ovary.

ovate, acute, longer than the tube ; corolla slightly longer than the calyx, lobes
lanceolate, acute, shorter than the tube   ..   ..   *pusilla*

F. **pusilla** *Bak.* in Kew Bull. 1894: 26 (1894); F.T.A. 4, 1 : 568. Herb of damp places, flowers white or pale
mauve.
    Ghana : Gambaga (Sept.) *Hall* 7481 **N. Nig.:** Nupe *Barter* 10081 17101 Lokoja (Nov.) *Dalz.* 411 Guduma,
Minna (Oct.) *Hepper* 9571 [**Br.**]**Cam.**: Maifula, Adamawa (Nov.) *Latilo & Daramola* FHI 287061 Also in
Sudan.

11. PYCNOSPHAERA Gilg—F.T.A. 4, 1 : 564.

Erect perennial herb up to 1 m. high, usually unbranched ; stems terete, leaves sessile,
narrowly oblanceolate to linear-lanceolate, up to 5 cm. long ; flowers in dense
bracteate, corymbosely arranged heads ; bracts obovate, shortly acuminate ; sepals
3, lateral pair lanceolate, carinate, 2·5 mm. long, dorsal sepal minute ; corolla 5 mm.
long, lobes ovate-oblong, acute, shorter than the tube ; anthers apiculate, stigma
bifid   ..   ..   *buchananii*

P. **buchananii** (*Bak.*) *N. E. Br.* in F.T.A. 4, 1 : 565 (1904). *Faroa buchananii* Bak. in Kew Bull. 1894 : 26.
Herb of wet places ; flowers mauve.
    Guin.: Ditinn (Nov.) *des Abbayes* 8811 Timbo (Oct.) *Jac.-Fél.* 18951 Diélela *Pobéguin.* Also in the Congo
(Katanga), E. and S. tropical Africa.

12. ENICOSTEMA Blume—F.T.A. 4, 1 : 563. *Nom. cons.*

A glabrous herb ; leaves sessile, linear to narrowly oblong-lanceolate, acute to obtuse,
up to 8 cm. long ; flowers in axillary clusters all along the stem, sessile ; calyx-tube
2 mm. long, lobes ovate-lanceolate to linear-lanceolate, acuminate, 2 mm. long with
spreading tips, margins submembranous ; corolla-tube 5 mm. long, lobes about
3 mm. long ; capsule oblong   ..   ..   *hyssopifolium*

E. **hyssopifolium** (*Willd.*) *Verdoorn* in Bothalia 7 : 462 (1961). *Exacum hyssopifolium* Willd. Sp. 1 : 640 (1798).
E. littorale Blume (1826)—F.T.A. 4, 1 : 563 ; F.W.T.A., ed. 1, 2 : 184. Herb with angled or narrowly
winged stems and small green and white flowers.
    Gam.: *Brown-Lester.* Throughout tropical Africa, S. Africa and tropical Asia.

13. VOYRIA Aubl.—F.T.A. 4, 1 : 569.

Stems erect, slightly flexuose, up to 6 cm. long, bearing about 3 pairs of very small
rudimentary membranous leaves ; flower terminal, solitary ; calyx about 7 mm. long,
lobes linear-lanceolate, setaceous at the apex ; corolla-tube 2 cm. long, lobes 5,
shortly spathulate-orbicular ; anthers included in the corolla-tube ; capsule elliptic,
membranous   ..   ..   *platypetala*

V. **platypetala** *Bak.* in Kew Bull. 1894 : 26 (1894) ; F.T.A. 4, 1 : 570. A tiny pallid saprophytic herb of forest
floors.
    Lib.: Bobei Mt. (Sept.) *Baldwin* 95871 **Ghana:** (Dec.) *Vigne* FH 32171 **S.Nig.:** Nun R. *Mann* 5141

141. MENYANTHACEAE

By P. Taylor

Aquatic or marsh herbs. Leaves alternate, entire or 3-foliolate, sometimes
orbicular and peltate. Flowers hermaphrodite, actinomorphic. Corolla-lobes
valvate or induplicate-valvate, margins on inside sometimes fimbriate. Stamens
inserted towards the base of the tube or between the lobes ; anthers 2-celled,
sagittate or versatile. Ovary superior 1-celled, with 2 parietal placentas. Style
1, bifid at the apex. Ovules few to numerous. Fruit a 4 or 2-valved capsule, or
fleshy and indehiscent. Seeds with copious endosperm and small embryo.

Mainly in the temperate regions of both hemispheres. Formerly included in *Gentianaceae* but differing in the aquatic habit, alternate leaves and valvate aestivation.

NYMPHOIDES Hill Brit. Herb. 77 (1756). *Limnanthemum* Gmel. (1770)—F.T.A. 4,
1 : 583.

Leaves simple, ovate-orbicular, deeply cordate at the base, 1–20 cm. long, glabrous,
entire ; petiole 1–2 cm. long ; flowers fasciculate, usually numerous, variable in size ;
pedicels 3–6 cm. long ; sepals ovate-lanceolate ; corolla campanulate, 5–6-lobed,
fimbriate ; stamens 5–6 ; fruits indehiscent ; seeds lenticular, smooth or more or less
tuberculate, the margin sometimes carinate..   ..   ..   *indica*

N. **indica** (*Linn.*) *O. Ktze.* Rev. Gen. 2 : 429 (1891). *Menyanthes indica* Linn. (1753). *Villarsia senegalensis*
G. Don (1837). *Limnanthemum indicum* (Linn.) Griseb. (1839). *L. niloticum* Kotschy & Peyr. (1867)
—F.T.A. 4, 1 : 585. *L. senegalense* (G. Don) N. E. Br. in F.T.A. 4, 1 : 586 (1904) ; Chev. Bot. 447 ;
F.W.T.A., ed. 1, 2 : 184. Aquatic plant with rounded leaves floating on the surface and emergent clusters
of white or yellow flowers.
    Sen.: Richard Toll (Dec., Jan.) *Roger* 1 *Leprieur* 1 Séléki, Casamance *Chev.* 2824 ; 2825. **Gam.:** *Pirie* 1
Kuntaur (July) *Ruxton* 28 1 **Port.G.:** Mansoa (Feb.) *Esp. Santo* 23811 Bafata (Dec.) *Esp. Santo* 37561

Mali: Gao (Nov.) *de Wailly* 4898! Fafa (Mar.) *de Wailly* 5353! Ansongo (Sept.) *Hagerup* 422! Guin.: Koba (Nov.) *Jac.-Fél.* 7209! Dalaba (Mar.) *Langdale-Brown* 2636! S.L.: Kambia (Nov.) *Adames* 199! Rhombe (Mar.) *Adames* 143! Mange (Dec.) *Jordan* 720! Sherbro Isl. (July) *T. S. Jones* 165! Iv.C.: Ouellé (Sept.) *des Abbayes* 1130! Ghana: Sesiamang (Feb.) *A. S. Thomas* D147! Bimbila (Jan.) *Akpabla* 1864! N.Nig.: Nupe *Barter* 1329! Katagum *Dalz.* 197! S.Nig.: Okomu F.R., Benin (Dec.) *Brenan* 8595! Widespread in Old World tropics.

142. PRIMULACEAE

By P. Taylor

Annual or perennial herbs or rarely shrubs ; stems erect or procumbent and rooting at the nodes. Leaves mostly basal, rarely cauline, alternate, opposite or verticillate, simple or lobate, often dentate ; stipules absent. Flowers solitary to paniculate, actinomorphic, hermaphrodite. Calyx persistent, often leafy. Corolla hypocrateriform, campanulate or tubular, lobes 5, imbricate. Stamens inserted on the corolla, the same number as and opposite to the lobes. Ovary superior, rarely semi-inferior, 1-celled with a free basal placenta and numerous, or very rarely few ovules. Fruit a capsule, many-seeded or very rarely 1-seeded. Seeds angular with a small straight embryo in copious endosperm.

Mostly in the mountainous parts of the north temperate zone, rare in the Southern Hemisphere. Recognised at once amongst the *Gamopetalae* by the united petals with the stamens inserted opposite to them and the peculiar placentation. *Ardisiandra* is endemic to the mountains of west and east tropical Africa.

Ovary superior :
  Leaves entire, glabrous :
    Capsule circumscissile .. .. 1. **Anagallis**
    Capsule valvate .. .. 2. **Lysimachia**
  Leaves lobate-dentate pubescent .. .. 3. **Ardisiandra**
Ovary half inferior ; capsule valvate ; leaves glabrous, entire .. .. 4. **Samolus**

1. ANAGALLIS Linn.—F.T.A. 3 : 490 ; Pax & Knuth in Engl. Pflanzenr. 4, 237, Primulac. : 321 (1905) ; P. Taylor in Kew Bull. 10 : 321 (1955) and 13 : 133 (1958).

Corolla bright blue or red, fringed with glands ; leaves always opposite ; fruiting pedicel strongly recurved ; corolla not persistent..   .. .. .. 1. *arvensis*
Corolla white or pink, not fringed with glands ; leaves alternate (sometimes the lower-most subopposite) ; fruiting pedicel erect ; corolla persistent :
  Flowers pedicellate, usually 5-merous, corolla as long as or longer than the calyx :
    Plant branched above or simple ; flowers in terminal racemes occupying the upper half of the stem or branches ; leaves sessile, narrowly elliptic to broadly ovate, base cuneate to subtruncate, apex acute ; corolla equal in length to or exceeding the calyx ; filaments filiform ; seeds 0·3–0·45 mm. long :
      Filaments glabrous .. .. 2a. *pumila* var. *pumila*
      Filaments bearded .. .. 2b. *pumila* var. *barbata*
    Plant branched from the base or simple ; flowers axillary almost to the base of the stem ; leaves petiolate, obovate-spathulate, apex rounded or acute ; corolla shorter than the calyx ; filaments dilated and flattened ; seeds 0·45–0·6 mm. long   3. *djalonis*
  Flowers sessile, usually 4-merous, corolla shorter than the calyx..   .. 4. *minima*

1. **A. arvensis** *Linn.* Sp. Pl. 148 (1753) ; F.T.A. 3 : 490 ; Pax & Knuth l.c. 323. Stems erect or ascending winged ; a weed of cultivation.
  N.Nig.: Bornu *Oudney*! Native of W. Europe and the Mediterranean region now widespread as a weed in both hemispheres.
2. **A. pumila** *Sw.* Prod. Veg. Ind. Occ. 1 : 40 (1788).
2a. **A. pumila** *Sw.* var. **pumila**—P. Taylor in Kew Bull. 10 : 343 (1955). Erect annual herb to 1 ft. high ; flowers white, corolla persisting and becoming reddish-brown in fruit ; in damp places.
  Guin.: Kouloundala (Jan.) *Chev.* 20384! Ghana: Elmina (Oct.) *Hall* 852! N.Nig.: Lokoja (Nov.) *Dalz.* 52! Jos Plateau (Sept.) *Lely* P728! Anara F.R., Zaria Prov. (Sept.) *Olorunfemi* FHI 24394! S.Nig.: Olokemeji F.R., Ibadan Prov. (Aug.) *Keay* FHI 37167! [Br.]Cam.: Gurum, Adamawa (Nov.) *Hepper* 1313! Widespread throughout the tropics.
2b. **A. pumila** var. **barbata** *P. Tayl.* l.c. 345 (1955). Erect annual herb to 18 in. high ; flowers white ; in damp places.
  Mali: Kita (Sept.) *Jaeger*! N.Nig.: Borgu *Barter* 784! [Br.]Cam.: Gurum, Adamawa (Nov.) *Hepper* 1250! Throughout tropical Africa.
3. **A. djalonis** *A. Chev.* in Journ. de Bot., sér. 2, 22 : 115 (1909) ; Chev. Bot. 384. *A. pumila* var. *djalonis* (A. Chev.) P. Tayl. l.c. 346 (1955). Erect annual herb to 6 in. ; flowers white, usually at high altitudes.
  Guin.: Fouta Djalon (Sept.–Oct.) *Chev.* 18668! 18876! 20193! *Adam* 12584! [Br.]Cam.: Bamenda (Dec.) *Daramola* FHI 40564! Gembu, Mambila Plateau (Jan.) *Hepper* 1810! Throughout tropical Africa.
4. **A. minima** (*L.*) *E. H. L. Krause* in Sturm. Fl. Deutsch. ed. 2, 9 : 251 (1901) ; P. Tayl. l.c. 347 (1955). *Centunculus minimus* Linn. Sp. Pl. 116 (1753). Minute annual herb to about 2 in. ; flowers white or pink ; in damp places at high altitudes.
  [Br.]Cam.: Cam. Mt., Mann's Spring, 9,500 ft. (fl. & fr. Dec.) *Morton* K867! Also in Ethiopia, India, N., C. & S. America and Europe.

2. LYSIMACHIA Linn.—F.T.A. 3 : 489 ; Pax & Knuth in Engl. Pflanzenr. 4, 237, Primulac. : 256 (1905) ; P. Taylor in Kew Bull. 13 : 142 (1958).

A perennial herb with numerous fibrous roots and erect or decumbent stems ; leaves elliptic, entire, petiolate, with numerous black punctate glands ; flowers numerous in terminal racemes, shortly pedicellate ; calyx segments oblong, obtuse, about 3 mm. long ; corolla about as long as calyx ; fruit globose, about 3 mm. in diam. ; style persistent, about 1 mm. long   ..   .. *ruhmeriana*

L. ruhmeriana *Vatke* in Linnea 40 : 204 (1876) ; F.T.A. 3 : 489 ; Pax & Knuth l.c. 292. A branched herb to 4 ft. high ; flowers white or pink ; in grassland.
[Br.] **Cam.** : Bamenda (fl. & fr. Apr., May) *Ujor* FHI 30383! *Maitland* 1406! Throughout tropical and southern Africa and Madagascar.

3. ARDISIANDRA Hook. f.—F.T.A. 3 : 488 ; Pax & Knuth in Engl. Pflanzenr. 4, 237, Primulac. : 222 (1905) ; P. Taylor in Kew Bull. 13 : 146 (1958).

A creeping pubescent herb ; stems rooting at the nodes ; leaves suborbicular, cordate at base, deeply lobed, the lobes coarsely toothed ; petiole as long as or longer than lamina ; flowers in few-flowered extra-axillary corymbs or racemes ; pedicels up to 1 cm. long ; calyx-segments broadly ovate-lanceolate, somewhat accrescent ; corolla campanulate, slightly longer than the calyx ; capsule globose, dehiscing by 5 valves
sibthorpioides

A. sibthorpioides *Hook. f.* in J. Linn. Soc. 7 : 205 (1864) ; F.T.A. 3 : 488 ; Pax & Knuth l.c. 222. Stems slender, up to 2 ft. long ; flowers white ; in montane forest.
[Br.]**Cam.**: Cam. Mt., 3,500–7,600 ft. (fl. & fr. Dec., Feb., fr. Aug.) *Mann* 2022! *Rosevear* Cam. 101/37! *Maitland* 1323! *Schlechter* 12846! Bafut-Ngemba F.R., 7,000 ft., Bamenda (fr. Feb.) *Hepper* 2863! **F.Po:** 6,500–7,500 ft. (fl. & fr. Mar., Sept.) *Mann* 1458! *Wrigley & Melville* 584! Also in mountains of eastern Africa.

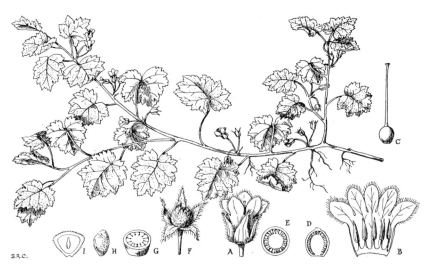

Fig. 269.—Ardisiandra sibthorpioides *Hook. f.* (Primulaceae).
A, flower. B, corolla laid open. C, pistil. D and E, longitudinal and cross-sections of ovary. F, fruit. G, cross-section of fruit. H, seed. I, cross-section of seed.

4. SAMOLUS Linn.—F.T.A. 3 : 490 ; Pax & Knuth in Engl. Pflanzenr. 4, 237, Primulac. : 336 (1905).

Herb with erect stems ; leaves obovate-spathulate, rounded at apex, entire, up to 8 cm long and 3 cm. broad, the basal ones often forming a rosette ; flowers in slender terminal racemes ; pedicels up to 1·5 cm. long, with a bract in the middle ; calyx campanulate, 3–4 mm. long, 5-toothed ; corolla campanulate, slightly longer than the calyx, 5-lobed   ..   ..   *valerandi*

S. valerandi *Linn.* Sp. Pl. 1 : 171 (1753) ; F.T.A. 3 : 490 ; Pax & Knuth l.c. 337. A glabrous herb 1–3 ft. high, with small white flowers ; in damp places.
N.Nig.: Bornu *Denham & Clapperton.* Also in many other parts of tropical Africa and widespread throughout the world.

Fig. 270.—Plumbago zeylanica *Linn.* (Plumbaginaceae).

A, longitudinal section of flower. B, stamen. C, style-arms. D, calyx in fruit. E, sepal. F, fruit. H, stalked gland.

W.E.T.

143. PLUMBAGINACEAE

By F. N. Hepper

Herbs, undershrubs or climbers. Leaves in a basal rosette or alternate; stipules absent. Flowers hermaphrodite, actinomorphic, often in unilateral inflorescences or subumbellate ; bracts often sheathing, dry and membranous. Calyx often ribbed, mostly membranous between the ribs. Corolla gamopetalous, lobes imbricate, mostly persistent. Stamens 5, opposite the corolla-lobes ; anthers 2-celled, opening lengthwise. Disk absent. Ovary superior 1-celled ; styles 5, free or connate. Ovule 1, pendulous from a basal funicle. Fruit various. Seed with or without endosperm, and with a straight embryo.

Mainly halophytes ; numerous around the Mediterranean ; some highly decorative species.

PLUMBAGO Linn. —F.T.A. 3 : 486.

Perennial herbs or shrubby and occasionally climbing ; leaves alternate ; calyx tubular with stalked glands ; flowers racemose.

A climbing undershrub ; branches ribbed, glabrous ; leaves ovate or ovate-lanceolate, acute, shortly cuneate at the base, 6–10 cm. long, 3–6 cm. broad, glabrous ; petiole amplexicaul at the base ; racemes mostly branched ; calyx about 1 cm. long, clothed with stiff stalked glands ; corolla longer than the calyx ; capsule membranous, included in the persistent calyx   ..   .. *zeylanica*

P. zeylanica *Linn.* Sp. Pl. 151 (1753) ; F.T.A. 3 : 486 ; Chev. Bot. 383 ; van Steenis in Fl. Males., ser. 1, 4 : 109 (1949). Climbing shrub about 3 ft. high, viscid above ; flowers white, ¾ in. long ; in thickets and near villages.
Sen. : *Heudelot* 98 ! Tivaouane (Dec.) *Chev.* 2835 ! Séléki, Casamance *Chev.* 2833. **Guin. :** Beyla to Manankoro *Chev.* 20891. **S.L. :** Regent (Aug.) *Deighton* 2048 ! Hastings (Aug.) *Melville & Hooker* 449 ! Yungeru (Jan.) *Thomas* 7153 ! Njala (Oct.) *Deighton* 1772 ! **Lib. :** Ganta (May) *Harley* ! Gouréni, Upper Sassandra *Chev.* 21646. **Ghana :** Aburi Scarp (Jan.) *Morton* GC 6341 ! Akropong, Akwapim (Aug.) *Irvine* 779 ! Abokobi (Oct.) *Vigne* FH 4262 ! **Togo Rep. :** Lomé *Warnecks* 210 ! **N.Nig. :** Kabba road (Oct.) *Parsons* L91 ! Nupe *Barter* 289 ! Katagum *Dalz.* 315 ! **S.Nig. :** Lagos (Dec.) *Dalz.* 1276 ! Idumuye (Dec.) *Thomas* 2128 ! Calabar (Feb.) *Mann* 2318 ! Oban *Talbot* ! **[Br.]Cam. :** Cam. Mt., 2,400 ft. (Apr.) *Maitland* 1113 ! **F.Po :** *T. Vogel* 108 ! Widely spread in the tropics. (See Appendix, p. 424.)

Besides the above *P. capensis* Thunb. is cultivated in gardens ; it has blue flowers and stalked glands only on the upper part of the calyx.

144. PLANTAGINACEAE

By F. N. Hepper

Herbs. Leaves all radical or alternate or opposite, simple, sometimes much reduced. Flowers usually hermaphrodite, spicate, actinomorphic. Calyx herbaceous. Corolla gamopetalous, scarious, 3–4-lobed, lobes imbricate. Stamens usually 4, inserted on the corolla-tube and alternate with the lobes or hypogynous ; anthers 2-celled, opening lengthwise. Ovary superior, 1–4-celled ; style simple. Ovules 1 or more in each cell, axile or basal. Fruit a circumscissile capsule or nut. Seeds peltately attached, with fleshy endosperm and straight or curved embryo.

A small family represented in our area by one very distinct species.

PLANTAGO Linn.—F.T.A. 5 : 504.

Herbs with usually radical leaves and long-pedunculate spikes of small flowers ; stamens usually 4 ; anthers versatile ; fruit a circumscissile capsule.

A perennial herb with numerous roots ; leaves all radical, very long-petiolate, ovate-orbicular in outline, widely cordate at the base, coarsely crenately lobed, about 9 cm. diam., with 3 principal nerves and several spreading side nerves, slightly pubescent on the nerves beneath, gradually narrowed into the petiole ; spikes long-pedunculate, about 2 cm. long in flower, soon lengthening and slender   ..   *palmata*

P. palmata *Hook. f.* in J. Linn Soc. 6 : 19 (1861) ; F.T.A. 5 : 504 ; Pilger in Engl. Pflanzenr. 4, 269, Plantaginac. : 77 (1937). Sessile herb with a tuft of long-petiolate leaves ; flowers green ; in montane grassland. **[Br.]Cam. :** Cam. Mt., 4,400–7,000 ft. (fl. & fr. Dec., Apr.) *Mann* 1962 ! *Maitland* 1129 ! **F.Po :** Clarence Peak, 8,000 ft. (Dec.) *Mann* 611 ! Also on the mountains of E. Africa, from Ethiopia to Lake Nyasa.

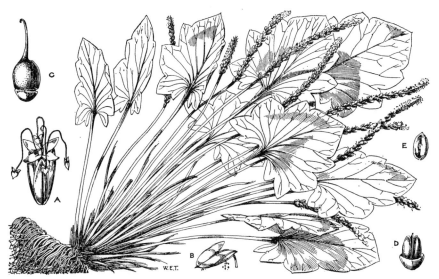

FIG. 271.—PLANTAGO PALMATA *Hook. f.* (PLANTAGINACEAE).
A, flower. B, anther. C, pistil. D, open fruit. E, seed.

145. SPHENOCLEACEAE[1]

By F. N. Hepper

Annual herbs. Stems erect or decumbent, with aerenchymous tissue at the base. Leaves spirally arranged, simple, entire, exstipulate. Inflorescences terminal, densely spicate, acropetal. Flowers subtended by a bract and two bracteoles, bisexual, regular. Calyx-tube adnate to the ovary, lobes 5. Corolla campanulate-urceolate, perigynous, lobes 5, imbricate. Stamens 5, epipetalous, alternating with corolla-lobes; filaments short; anthers rounded, 2-locular, dehiscing longitudinally. Ovary semi-inferior, 2-locular; style short, stigma capitate. Fruit a capsule, circumscissile, seeds numerous, minute.

A monogeneric family with one pantropical species and one species endemic to our area. Separable from *Campanulaceae* by the circumscissile capsule and imbricate corolla lobes on the regular flowers.

SPHENOCLEA Gaertn.—F.T.A. 3 : 480. *Nom. cons.*

Leaves lanceolate, more or less acute, up to 10 cm. long and 3 cm. broad, entire, glabrous; petiole up to 1·3 cm. long; spikes terminal, cylindric, up to 7 cm. long, about 1 cm. diam.; calyx-lobes ovate-rounded, imbricate; corolla divided nearly to the base, as long as the calyx, white; anthers 5, subsessile, suborbicular; style very short; stigma shortly bifid; capsule depressed-globose, dehiscing transversely .. 1. *zeylanica*
Leaves obovate-elliptic, sessile, rounded at the apex, 2–4 cm. long, 1–2 cm. broad; spikes very short, oblong; corolla twice as long as the calyx-lobes, pink; filaments as long as the anthers ..   .. 2. *dalzielii*

1. **S. zeylanica** *Gaertn.* Fruct. 1 : 113, t. 24, fig. 5 (1788); F.T.A. 3 : 481; Chev. Bot. 382; Berhaut Fl. Sén. 197; Airy Shaw in Fl. Males., ser. 1, 4 : 27. *Phytolacca octandra* of F.T.A. 6, 1 : 98 (*Rowland*), not of Linn. An erect glabrous herb with spongy stems 1–4 ft. high; flowers greenish-white in congested spikes; in rice-fields, beside rivers and open swamps, often gregarious in tidal creeks.
Sen. : *Farmar* 118! *Berhaut* 577; 3975. **Gam.** : Kuntaur *Ruxton* 115! **Mali** : Gao (fl. & fr. Sept.) *Hagerup* 353! **Port.G.** : Brene, Bissau (Jan.) *Esp. Santo* 1681! **S.L.** : Freetown *Sc. Elliot* 5924! Tombo (fl. & fr. Jan.) *Deighton* 965! Gbabai (fl. & fr. Feb.) *Glanville* 253! Rokupr (Feb.) *Jordan* 770! **Iv.C.** : Mankono *Chev.* 21937. Kondroko to Touminiané, Toumodi *Chev.* 22380. **Ghana** : Accra (fl. & fr. July) *Irvine* 681! Achimota (Nov.) *Irvine* 1519! Shai Plains (fl. & fr. May) *Irvine* 2677! Kepetchu (Mar.) *Hepper & Morton*

[1] Airy Shaw in Fl. Males., ser. 1, 4 : 27 (1948). Dr. J. Hutchinson does not recognise this family (Fam. Fl. Pl., ed. 2, 1 : 476 (1959)), but there seems to be a good case for separating it from *Campanulaceae*, though not for allying it with *Phytolaccaceae* as Airy Shaw proposed.

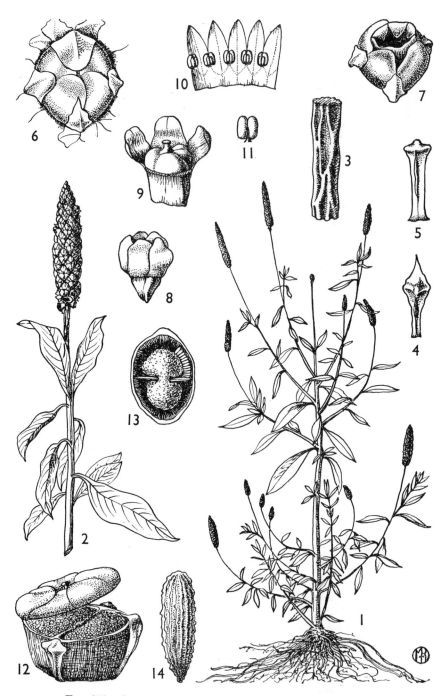

FIG. 272.—SPHENOCLEA ZEYLANICA *Gaertn.* (SPHENOCLEACEAE).

1, habit, × ¼. 2, inflorescence, × ½. 3, inflorescence axis, × 4. 4 & 5, bracteoles, × 4.
6, flower bud, × 4. 7, open flower, × 4. 8, calyx, × 4. 9, pistil (2 calyx-lobes removed),
× 4. 10, corolla laid open, × 4. 11, stamen, × 8. 12, fruit, × 4. 13, transverse
section of ovary, × 4. 14, seed, × 40. 1, 2, 14, drawn from *Milne-Redhead & Taylor*
7463. 3–6, 12, 13, from *Jones* FHI 18808. 7–11, from *Deighton* 132a.

A3034! **Niger** : Bentia to Ansongo (fl. & fr. Sept.) *Hagerup* 416! **N.Nig.** : Nupe *Barter* 1715! Sokoto (fl. & fr. July) *Dalz.* 383! Kano (fl. & fr. Dec.) *Hagerup* 672! Gombe *Lely* 663! **S.Nig.** : Lagos *Rowland* ! *Ejiofor* FHI 32050! Ibadan (fl. & fr. Dec., Jan.) *Ujor* FHI 23944! *Meikle* 995! Onitsha (May) *Onochie* FHI 7518! [**Br.**]**Cam.** : Gurum, Adamawa Div. (fl. & fr. Nov.) *Hepper* 1261! Bama, Dikwa Dist. (fl. & fr. Dec.) *McClintock* 84! In the tropics generally.

2. **S. dalzielii** *N.E. Br.* in Kew Bull. 1912 : 277 ; Berhaut Fl. Sén. 194, fig. 20, 5. Straggling herb with spongy stems ; flowers pink ; on mud beside pools.
   Sen. : Tambacounda *Berhaut* 1649 ; 3262. **Gam.** : Bansang, Macarthy Div. *Duke* 7! **N.Nig.** : Katagum Dist. *Dalz.* 201! Also in Ubangi-Shari.

146. CAMPANULACEAE

By F. N. Hepper[1]

Herbs to small trees, nearly always with milky juice. Leaves alternate, rarely opposite, simple ; stipules absent. Flowers hermaphrodite, actinomorphic. Calyx-tube adnate to the ovary, 3–10-lobed. Corolla gamopetalous, tubular or campanulate, lobes valvate. Stamens as many as the corolla-lobes and alternate with them, inserted towards the base of the corolla or the disk ; anthers free, 2-celled, opening lengthwise. Ovary inferior or rarely superior, 2–10-celled, with axile placentas. Ovules mostly numerous. Fruit capsular or baccate.

A widely distributed family, in Africa mainly confined to the south. The absence of stipules distinguishes it from the herbaceous *Rubiaceae*.

Corolla-lobes short and broad ; fruits obconic or long and narrow   .. 1. **Wahlenbergia**
Corolla deeply divided into narrow lobes ; fruits always shortly obconic :
  Stigma capitate or slightly bilobed ; style not swollen towards apex 2. **Cephalostigma**
  Stigma shortly but distinctly 2–3 lobed ; style swollen towards apex   3. **Lightfootia**

1. WAHLENBERGIA Schrad. ex Roth—F.T.A. 3 : 477 ;
von Brehmer in Engl. Bot. Jahrb. 53 : 9 (1915). *Nom. cons.*

Mature capsules 2–4 mm. long, obconic ; ovary (at time of flowering) 1·5 mm. long:
Corolla about 10 [mm. long ; calyx-lobes triangular-subulate, about 3 mm. long ;
  flowers few on long pedicels ; montane perennial ..   .. 1. *arguta*
Corolla 1–2 mm. long ; calyx-lobes linear, 1–2 mm. long ; flowers very few on short
  pedicels ; lowland annual   ..   .. 2. *campanuloides*
Mature capsules 8–15 mm. long:
Leaves linear, about 5–6 cm. long, glabrous or pubescent, with a thick cartilaginous
  crenulate margin ; flowers several in a long-pedunculate cyme ; capsule campanulate-
  obconic, glabrous, 8–10 mm. long ; lowland annual   ..   ..   .. 3. *riparia*
Leaves lanceolate, about 1 cm. long, sparingly setose-pilose, with a cartilaginous
  minutely crenulate margin ; flowers very few on long pedicels ; capsule subcylindric, about 1·5 cm. long, with several ribs ; montane perennial   .. 4. *mannii*

1. **W. arguta** *Hook. f.* in J. Linn. Soc. 6 : 15 (1861) ; von Brehmer in Engl. Bot. Jahrb. 53 : 99 (1915), incl · var. *parvilocula* v. Brehm. A herb with slender straggling stems about 1 ft. high from a perennial root-stock ; flowers whitish or pale mauve with blue veins ; in montane grassland.
   [**Br.**]**Cam.** : Cam. Mt., 7,000–12,000 ft. (fl. & fr. Nov.-Feb.) *Mann* 1943! 1944! *Dalz.* 8314! *Keay* FHI 28588! *Maitland* 1250! Bafut-Ngemba F.R., Bamenda (fl. & fr. Jan., Feb.) *Hepper* 2086! *Daramola* FHI 40633! near L. Oku (fl. & fr. Jan.) *Keay & Lightbody* FHI 28469! **F.Po** : Clarence Peak (Pico de S. Isabel), 8,500 ft. (fl. & fr. Dec., Mar.) *Mann* 601! *Guinea* 2712! Also in Cameroun.

2. **W. campanuloides** (*Del.*) *Vatke* in Linnaea 38 : 700 (1874). *Cervicina campanuloides* Del. Fl. Egypte 7, t. 5, 2 (1812). *Wahlenbergia cervicina* A. DC. Monogr. Campanulac. 156 (1830) ; Prod. 7 : 440 (1839) ; von Brehmer l.c. 130 ; F.W.T.A., ed. 1, 2 : 190. Small herb, branched from near the base, about 2 in. high ; flowers white.
   Sen. : *Leprieur* ! Also in Egypt and S.W. Africa.

3. **W. riparia** *A. DC.* Monogr. Campanulac. 146 (1830) ; von Brehmer l.c. 110 ; Chev. Bot. 382 ; Berhaut Fl. Sén. 174. *W. humilis* A. DC. (1830). Erect simple or branched annual herb, pilose in the lower parts, 1–2 ft. high ; flowers blue or violet ; in cultivated, moist ground and beside rivers in savanna.
   Sen. : *Roger* 46! *Berhaut* 917. *Leprieur & Perrottet. Lelièvre.* **Port.G.** : Pessubé, Bissau (fl. & fr. Mar.) *Esp. Santo* 2243! **Mali** : Faramana (fl. & fr. June) *Chev.* 971! **N.Nig.** : Nupe *Barter* 1202! Birnin Gwari (fl. & fr. Mar.) *Meikle* 1333! Takwara, 3,200 ft. (fl. & fr. May) *Lely* 108! Naraguta F.R. (fl. & fr. Aug.) *Keay* FHI 20057! Vom (fl. & fr. Nov.) *Dent Young* 168! Abinsi (fl. & fr. Oct.) *Dalz.* 702! Also in (?) Socotra.

4. **W. mannii** *Vatke* in Linnaea 38 : 700 (1874) ; von Brehmer l.c. 128 (1915), incl. var. *intermedia* v. Brehm. A herb with wiry stems from a horizontal perennial base ; flowers pale bluish or white with blue veins ; in montane grassland.
   [**Br.**]**Cam.** : Cam. Mt., 7,000–9,000 ft. (fl. & fr. Dec.–Feb.) *Mann* 1226! 1945! *Dalz.* 8315! *Brenan* 9378! Bafut-Ngemba F.R., Bamenda (fl. & fr. Feb., May) *Hepper* 2144! *Ujor* FHI 30314! Kumbo to Oku (Feb.) *Hepper* 2010! near L. Oku (Jan.) *Keay* FHI 28482! **F.Po** : Clarence Peak (Pico de S. Isabel), 9,000 ft. (fl. & fr. Dec., Mar.) *Mann* 600! *Guinea* 2715!

[1] I am indebted to the late Prof. F. E. Wimmer for his assistance in drawing up the generic and specific de-limitations. However, since this account was prepared P. Tuyn (in Fl. Males., ser. 1, 6 : 111 (1960)) has united *Cephalostigma* and *Lightfootia* under *Wahlenbergia*. Although this may be the best solution, I think it is advisable to wait until the family is revised for southern Africa where the maximum number of species concerned is to be found. According to a note in Baileya 5 : 197 (1957), *Lightfootia* L'Hérit. is antedated by a few months by *Lightfootia* Sw. *Wahlenbergia* is proposed for further conservation by van Steenis in Taxon 9 : 125 (1960).—F. N. H.

S.R.C.

FIG. 273.—CEPHALOSTIGMA AND WAHLENBERGIA (CAMPANULACEAE).

Cephalostigma perottetii A. DC.—A, habit.   B, flower-bud.   C, open flower.   D, the same with corolla removed.   E, stigma.   F, cross-section of ovary.   G, fruit.
Wahlenbergia arguta Hook. f.—H, habit.   I, longitudinal section of flower.   J, stamen.   K, stigmas.   L, fruits.

2. CEPHALOSTIGMA A. DC. Monogr. Campanulac. 117 (1830); F.T.A. 3 : 471.

Herb ; stems more or less terete, usually pilose, sometimes glabrous ; leaves oblong-lanceolate, 1–5 cm. long, 5–15 mm. broad, acute at each end, margin undulate and thickened with a nerve ; flowers very numerous, the cymules forming an oblong panicle ; pedicels slender 1 cm. or more long ; ovary pilose or glabrous ; petals very short and slender ; calyx-lobes about 1 mm. long ; fruits obconical with 5 prominent nerves ; seeds trigonous .. .. *perrottetii*

C. perrottetii *A . DC.* Monogr. Campanulac. 118 (1830) ; Prod. 7 : 420 ; Chev. Bot. 382 ; Berhaut Fl. Sén. 175.
C. prieuri A. DC. (1830). *C. diaguissae* A. Chev. Bot. 382, name only. Annual herb, very variable in size and pubescence, a few inches to 2 ft. high ; flowers very small and pale mauve ; a weed of cultivation and in sandy places.
Sen. : *Heudelot* 554 ! *Berhaut* 1215 ; 4463 ; 4668. Bignona to Sindalone *Chev.* 2833. Cape Verde *Perrottet & Leprieur* ! **Mali :** Ouacoro *Chev.* 80 ! **Guin. :** Kouria *Caille* in Hb. *Chev.* 14682 ! Dalaba-Diaguissa Plateau *Chev.* 18868 ; 18881. Bafing *Adam* 12652 ! **S.L. :** Kumrabai (fl. & fr. Dec.) *Thomas* 6796 ! Rokupr (fl. & fr. Nov.) *Jordan* 692 ! Kenema (Nov.) *Deighton* 394 ! Makump (fl. & fr. Dec.) *Glanville* 117 ! **Lib. :** Monrovia *Whyte* ! **Iv.C. :** Marabadiassa to Gottoro *Chev.* 22023. **Ghana :** Sekondi (fl. & fr. Oct.) *Howes* 984 ! Nkwanta (fl. & fr. Jan.) *Chipp* 68 ! Techiman (fr. Dec.) *Adams* 4480 ! Tamale (fl. & fr. Dec.) *Adams & Akpabla* 4151 ! Yendi (fr. Dec.) *Adams & Akpabla* 4061 ! **Togo Rep. :** Sokode to Basari *Schroeder* 117 ! **N.Nig. :** Borgu *Barter* 826 ! Kontagora (Oct.) *Dalz.* 58 ! Kano (fr. Dec.) *Hagerup* 654 ! Naraguta (Oct.) *Hepper* 1020 ! Abinsi (fl. & fr. Nov.) *Dalz.* 707 ! **S.Nig. :** Lagos *MacGregor* 363 ! Aguku Dist. *Thomas* 1377 ! Ibadan (Nov.) *Newberry & Etim* 171 ! Owo (fl. & fr. Nov.) *Meikle* 507 ! Oron to Eket *Talbot* 3044 ! **[Br.]Cam. :** Bamenda (fl. & fr. Dec.) *Daramola* FHI 40563 ! Gembu, Mambila Plateau (fl. & fr. Jan.) *Hepper* 1809 ! Gurum, Adamawa Prov. (Nov.) *Hepper* 1280 ! Also in Cameroun, Ubangi-Shari, the Congos, Sudan, Uganda, Comoro Isl. and tropical S. America.

3. LIGHTFOOTIA L'Hérit. Sert. Angl. 4 (1788) ; F.T.A. 3 : 473.

Stems glabrous, angular, slender ; leaves subulate, about 1 cm. long and 1 mm. broad, apex acute, decurrent at base, margins thick and with a few minute teeth towards apex ; flowers numerous in a branched inflorescence ; corolla about 2 mm. long ; calyx-lobes 0·5 mm. long ; ovary glabrous ; seeds trigonous **1a. abyssinica var. tenuis**
Stem pilose at least below ; leaves crenulate and undulate ; corolla about as long as calyx lobes :
Leaves lanceolate about 1 cm. long and 3 mm. broad ; flowers rather numerous in lax inflorescences ; pedicels very slender 1–3·5 cm. long ; calyx-teeth of fruit shortly triangular, about 1 mm. long, much shorter than the ovary ; fruit about 3 mm. long ; seeds trigonous .. .. **2. ramosissima**
Leaves ovate or obovate, 1–3 cm. long, 8–20 mm. broad ; flowers numerous in branched inflorescences ; pedicels slender 1–1·5 cm. long ; calyx-teeth of fruit long-triangular, about 2·5 mm. long, as long as the ovary ; fruits about 2·5 mm. long ; seeds trigonous .. .. **3. hirsuta**

1. **L. abyssinica** *Hochst ex A . Rich.* Tent. Fl. Abyss. 2 : 1 (1851).
1a. **L. abyssinica** var. **tenuis** *Oliv.* in J. Linn. Soc. 21 : 401 (1885) ; Hepper in Kew Bull. 15 : 61. Slender, well-branched herb 1–2 ft. high.
　[Br.]Cam. : Mambila Plateau *Chapman* 59 ! Also in Uganda, Kenya and Tanganyika.
2. **L. ramosissima** *(Hemsley) E. Wimm. ex Hepper* in Kew Bull. 15 : 61 (1961). *Cephalostigma ramosissimum* Hemsley in F.T.A. 3 : 472 (1877) ; F.W.T.A., ed. 1, 2 : 191. Erect branched annual with slender ribbed hispid stems 3–10 ins. high ; flowers blue.
　[Br.]Cam. : Cam. Mt., 5,000–7,000 ft. (fl. & fr. Nov., Dec.) *Mann* 1992 ! *Maitland* 1185 !
3. **L. hirsuta** *(Edgew.) E. Wimm. ex Hepper* l.c. (1961). *Cephalostigma hirsutum* Edgew. in Trans. Linn. Soc. 20 : 81 (1851) ; F.T.A. 3 : 472. *C. perotifolium* Hutch. & Dalz. F.W.T.A., ed. 1, 2 : 191 (1931), not *Wahlenbergia perotifolia* Wight & Arn. (1834). Erect much-branched pilose annual a few inches high ; flowers blue, rarely yellow ; in moister places between rocks.
　Ghana : Sunyani (fl. & fr. Dec.) *Adams* 2870 ! Gambaga Scarp (Sept.) *Hall* 811 ! **N.Nig. :** Nabardo *Lely* 623 ! Jos Plateau (fl. & fr. Oct.) *Lely* P802 ! P804 ! Vom *Dent Young* 167 ! Gindiri (fl. & fr. Oct.) *Hepper* 1100 ! 1100a ! **[Br.]Cam. :** Vogel Peak, Adamawa (fl. & fr. Nov.) *Hepper* 1341 ! Extends to India.

147. LOBELIACEAE
By F. E. Wimmer

Herbs, shrubs or small trees, always with milky juice ; leaves alternate, rarely opposite or verticillate, simple ; stipules absent. Flowers hermaphrodite, zygomorphic. Calyx-tube adnate to the ovary, 5-lobed. Corolla gamopetalous, tubular, usually bilabiate, lobes 5, valvate. Stamens 5, epigynous, alternate with the corolla-lobes, inserted on or free from the corolla ; filaments partly free, anthers completely connate in a tube around the filiform style, 2-celled, opening lengthwise. Ovary inferior, 2-celled with axile placentation, rarely 1-celled with parietal placentation ; ovules numerous, very rarely few. Fruit capsular or baccate ; seeds minute, embryo straight in fleshy endosperm.

A widely distributed family in the tropics and subtropics, few species in the temperate zone, in Africa mainly confined to the South, some remarkable species of *Lobelia* in the African mountains, especially in E. Africa, where they are a prominent feature of the vegetation.

Distinguished from *Campanulaceae* by the zygomorphic corolla and the anthers cohering into a tube around the style.

Capsule more or less globose or obovoid (nearly as broad as long), dehiscent with 2 valves:
　Flowers bluish or white ; stigma 2-lobed, lobes more or less rounded to oblong,
　　spreading　.. 　.. 　.. 　.. 　.. 　.. 　.. 　.. 　.. **1. Lobelia**
　Flowers yellowish ; stigma 2-lobed, lobes long-filiform, rolling back　　**2. Monopsis**
Capsule elongate-conic to fusiform, tortuose, at length splitting into 5 narrow-linear
　strips; flowers bluish　.. 　.. 　.. 　.. 　.. 　.. 　.. **3. Dielsantha**

1. LOBELIA Linn.—F.T.A. 3 : 464; E. Wimm. in Engl. Pflanzenr. 4, 276b Campanulac.-
　　　　　　Lobelioid.: 408 (1953).

Tall stiff erect herb, mostly unbranched, 1·5–2·5 m. high ; stem softly tomentellous ;
　leaves sessile, lanceolate, acute, up to 20 cm. long and 3 cm. broad, densely denticulate
　to irregularly dentate, shortly pubescent on both surfaces ; flowers in a dense terminal
　raceme, 20–25 cm. long ; bracts linear-lanceolate, 25 mm. long, 3–5 mm. broad ;
　pedicels 10–15 mm. long, 2-bracteolate in the middle ; calyx-lobes sublinear, entire,
　tomentellous, obtuse and mucronulate, 10–15 mm. long ; corolla 25–35 mm. long,
　puberulous ; filaments glabrescent, up to 28 mm. long ; anther-tube glabrous at the
　apex, 6 mm. long　　.. 　.. 　.. 　.. 　.. 　.. 　.. **1. *columnaris***
Small herbs, erect, decumbent or prostrate, 5–80 cm. high :
*Anthers with only the 2 lower bearded at the apex:
　Leaves basal, few, ovate to elliptic, shortly petiolate, rounded at the base or cuneate,
　　subacute at the apex, densely denticulate on the margin, scarcely pilose above,
　　glabrous beneath ; stem without leaves, 5–20 cm. high, hirtellous below, glabrous
　　above ; inflorescence a one-sided raceme with few flowers ; pedicel 4 mm. long,
　　calyx-tube obconic, lobes subulate with a tooth near the base ; corolla 6 mm. long ;
　　stamens 3 mm. long ; anther-tube glabrous　　.. 　.. 　.. 　.. **2. *lelyana***
　Leaves cauline, alternate:
　　Prostrate, creeping herb, densely leafy, stems glabrous :
　　　Leaves ovate-rounded, 4–5 mm. long and about as broad, glabrous, truncate or
　　　　rounded at base, margin bluntly toothed and mucronulate ; petiole 1–2 mm.
　　　　long ; flowers solitary in the uppermost axils of the leaves ; pedicels about
　　　　twice as long as the leaf, hirtellous ; calyx-tube turbinate, hirtellous, the lobes
　　　　linear, acute, 1·5 mm. long ; corolla 8 mm. long, deep sky blue or blue and white,
　　　　glabrous, the lower 3 lobes oblong ; anther-tube glabrous upwards
　　　　　　　　　　　　　　　　　　　3a. *minutula* var. *kiwuensis*
　　　Leaves ovate-rounded, 5–10 mm. long, 5–12 mm. broad, rounded-truncate at base,
　　　　margin angular-dentate, glabrous ; petiole 2 mm. long ; flowers singly in the
　　　　axils of the uppermost leaves ; pedicels 10–20 mm. long, hirtellous ; calyx-tube
　　　　obconic, hirtellous, the lobes subulate, ciliate, 2 mm. long ; corolla dull blue,
　　　　11 mm. long, subglabrous ; stamens 5 mm. long ; anther-tube upwards hirtellous
　　　　　　　　　　　　　　　　　　　　　　　　　4. *acutidens*
　　Erect or ascending (occasionally decumbent) herbs, stems with scattered leaves :
　　　Leaves (except the lowest) lanceolate to linear :
　　　　Calyx-lobes ciliate ; corolla 5 mm. long, glabrous within ; lower leaves ovate to
　　　　　oblong, 2–4 cm. long, 1·2–2 cm. broad, with petiole 5–10 mm. long, the upper
　　　　　leaves subsessile ; seeds ovoid-globose, 0·4–0·5 mm. long　.. 　**5. *molleri***
　　　　Calyx-lobes glabrous ; corolla about 10 mm. long, upper lobes hirtellous ; lower
　　　　　leaves spathulate, about 3 cm. long, 8 mm. broad, narrowed into a short petiole ;
　　　　　the others subsessile, remotely denticulate, all glabrous ; seeds oblong-ellipsoid,
　　　　　0·5 mm. long　.. 　.. 　.. 　.. 　.. 　.. 　.. 　**6. *senegalensis***
　　　Leaves suborbicular, ovate to elliptic or subrhomboidal, the uppermost often more
　　　　lanceolate ; petioles 3–10 mm. long :
　　　　Corolla 3–6 mm. long ; erect herbs :
　　　　　Stem and branches conspicuously 3-winged ; leaves ovate to elliptic, 1–2 cm. long,
　　　　　　3–15 mm. broad, subcordate truncate to gradually narrowed to base, blunt to
　　　　　　acute at the apex, more or less dentate, scarcely hirtellous above ; bracts
　　　　　　lanceolate, acute ; pedicels 1 cm. long ; calyx-lobes lanceolate to linear, entire,
　　　　　　2 mm. long ; corolla about 5 mm. long, stamens 3 mm. long
　　　　　　　　　　　　　　　　　　　7a. *heyneana* var. *inconspicua*
　　　　　Stem and branches not winged, the leaf at most very narrowly decurrent ; leaf
　　　　　　ovate to ovate-elliptic, 1·5–5 cm. long, 8–32 mm. broad, glabrous or scarcely
　　　　　　hirtellous ; petiole 3–10 mm. long ; flowers in a lax, terminal raceme ; pedi-
　　　　　　cels 3–8 mm. long with short bracts ; calyx-tube turbinate, soon oblong-
　　　　　　cylindrical, 3–4 mm. long, lobes subulate, entire, 2 mm. long ; corolla 3–4 mm.
　　　　　　long ; anther-tube hirtellous above　..　.. 　.. 　.. 　.. **8. *sapinii***
　　　　Corolla 9–10 mm. long ; erect, or often decumbent below and rooting at nodes ;
　　　　　leaves ovate :
　　　　　Leaves bluntly acuminate, 5–40 mm. long, 3–17 mm. broad, rounded to cuneate

at base, margin denticulate ; petiole 3–10 mm. long ; flowers solitary in axils of the upper leaves or in a lax raceme ; bracts more or less lanceolate ; calyx-tube obconic, lobes linear, 3–6 mm. long, sometimes ciliate, spreading ; seeds oblong-ellipsoid .. .. **9. rubescens**
Leaves acute at apex, 15–25 mm. long, 5–10 mm. broad, rounded to cuneate at base, margin serrulate ; petiole 5–7 mm. long, winged ; flowers laxly clustered ; calyx-tube obconic, lobes subulate, 2–4 mm. long, minutely ciliate, spreading ; seeds compressed-elliptic .. .. **10. kamerunensis**

*Anthers all bearded at apex :
Leaves sessile, linear-lanceolate (some of lowermost leaves ovate), acute, denticulate, 1–3·5 cm. long, 4–6 mm. broad, glabrous ; flowers solitary, axillary or in lax racemes; pedicels 2–4 cm. long with 2 divaricate bractlets at the base ; bracts narrow-lanceolate, shorter than the pedicels ; calyx-tube obovoid, lobes subulate, 1 mm. long **11. djurensis**
Leaves petiolate, ovate, about 5 cm. long, 3 cm. broad, acute to acuminate at apex, subcordate or rounded to cuneate at base, sparsely pilose on both sides or only beneath :
Corolla 7–9 mm. long, white or pale blue, with lilac margins ; pedicels with 2 bractlets near the middle ; seeds compressed-ovoid, 0·8 mm. long .. **12. hartlaubii**
Corolla 14–25 mm. long, blue, rarely white ; pedicels with 2 bractlets near the base ; seeds convex-oblong, 1 mm. long .. .. **13. baumannii**

1. **L. columnaris** *Hook. f.* in J. Linn. Soc. 6 : 14 (1862) ; F.T.A. 3 : 466 ; E. Wimm. in Engl. Pflanzenr. 4, 276b, Campanulac.-Lobelioid. : 660. *Tupa columnaris* (Hook. f.) Vatke (1874). Erect simple herb, 6–9 ft. high ; flowers blue or lilac ; in montane grassland.
 [Br.]Cam.: Cam. Mt. : above Ukile 8,500 ft. (Mar.) *Brenan* 9536 ! above Bamenda, 6,000 ft., *Migeod* 360 ! Ndu, 6,000 ft. (fl. & fr. Feb.) *Hepper* 1937 ! F.Po: Clarence Peak (Dec.) *Mann* 1316 ! Moka, 3,800–4,000 ft. (Nov., Jan.) *Mildbr.* 7067 ! *Guinea* 1609 ! Also in Cameroun.
2. **L. lelyana** *E. Wimm.* l.c. 465 (1953). Erect annual herb, up to 8 in. high ; flowers dull blue ; on rocks.
 N.Nig.: Jos Plateau (Sept.) *Lely* P727 !
3. **L. minutula** *Engl.* Bot Jahrb. 19, Beibl. 47 : 50 (1894).
3a. **L. minutula** var. **kiwuensis** (*Engl.*) *E. Wimm.* in Ann. Naturhist. Mus. Wien 56 : 347 (1948) ; Engl. Pflanzenr. l.c. 486. *L. kiwuensis* Engl. Ergebn. Deutschen Zentralafr.-Exped. 1907–08, 2 : 345 (1911). A small caespitose herb with stems creeping ; flowers deep blue or white-spotted.
 [Br.]Cam.: Lakom, 6,000 ft., Bamenda (fl. & fr. Apr.) *Maitland* 1773 !
4. **L. acutidens** *Hook. f.* in J. Linn. Soc. 7 : 204 (1864) ; F.T.A. 3 : 467 ; E. Wimm. l.c. 488. A small prostrate, creeping herb ; flowers dull blue.
 F.Po: 9,000 ft. (Mar.) *Mann* 1452 ! above Basilé, 8,000 ft., St. Isabel's Peak *Mildbr.* 7171 !
5. **L. molleri** *Henriq.* in Bol. Soc. Brot. 10 : 137 (1892) ; E. Wimm. l.c. 472. *L. thomensis* Engl. & Diels (1899). *L. fervens* of F.W.T.A., ed. 1, 2 : 193, not Thunb. A weak erect or partly decumbent herb, glabrous, 1 ft. or more high ; flowers white.
 [Br.]Cam.: above Buea, frequent *Deistel* 39 (652) ! 91 ! F.Po: Moka, about 3,600 ft. *Tessmann* 2873 ! Also in Cameroun, S. Tomé, Uganda and Tanganyika.
6. **L. senegalensis** *A. DC.* Prod. 7 : 372 (1839) ; F.T.A. 3 : 469 ; E. Wimm. l.c. 551 ; Berhaut Fl. Sén. 174. *L. kohautiana* Vatke (1874). *L. chilawana* Schinz (1900). *L. trierarchi* R. Good (1924). A glabrous erect, branched herb, about 1 ft. high ; flowers pale blue with white spots.
 Sen.: *Perrottet* 439 ! *Sieber* 15 ! *Boivin* 403 ! Cayor *Leprieur* ! R. Senegal *Heudelot* 512 ! Sangalkam (Dec.) *Berhaut* 204 ! Also in Sudan, Somaliland and Mozambique.
7. **L. heyneana** *Roem. & Schult.* Syst. Veg. 5 : 50 (1819).
7a. **L. heyneana** var. **inconspicua** (*A. Rich.*) *E. Wimm.* l.c. 475, fig. 79 c. (1953). *L. inconspicua* A. Rich. (1851) —F.T.A. 3 : 468. *L. ilysanthoides* Schlechter (1922), partly. A small erect herb, 2–11 cm. high, stem narrowly winged and sparsely pilose ; flowers 3–5 mm. long, reddish or lilac.
 S.L.: *Jaeger* 451 ! Also in Cameroun, Ethiopia and Tanganyika.
8. **L. sapinii** *De Wild.* in Bull Jard. Bot. Brux. 3 : 261 (1911) ; E. Wimm. l.c. 511. A glabrous erect branched herb, up to 8 in. high ; flowers blue or white with deep blue stripes or bright blue with 2 rows of dark dots across lower lip.
 N.Nig.: Anara F.R., Zaria (Oct.) *Hepper* 1072 ! Naraguta F.R., 3,800 ft., (Nov.) *Hepper* 1072 ! [Br.]Cam.: Vogel Peak, Adamawa (Nov.) *Hepper* 1484 ! Also in Cameroun, Congo, Tanganyika and Nyasaland.
9. **L. rubescens** *De Wild.* in Rev. Zool. Bot. Afr. 8, Suppl. 27 (1920) ; E. Wimm. l.c. 507. Erect, subglabrous herb to 1 ft. high, often decumbent or somewhat trailing and rooting at the base ; flowers blue, about ½ in. long ; in marshes.
 S.L.: Loma Mts. (Feb.) *Jaeger* 4257 ! Also in Congo, Uganda and Tanganyika.
 [Note. : This specimen seems to be an intermediate form between *L. kamerunensis* Engl. and *L. rubescens* De Wild.—E. W.]
10. **L. kamerunensis** *Engl. ex Hutch. & Dalz.* F.W.T.A., ed. 1, 2 : 193 (1931) ; E. Wimm. l.c. 507. Erect or decumbent-ascending herb to 1 ft. high, stems and leaves above slightly pilose or glabrescent ; flowers white, bluish with dull spots on the lower lip, ½ in. long.
 S.L.: Loma Mts. 1,400 ft. (Sept.) *Jaeger* 1979 ! S.Nig.: Sonkwala, 5,500 ft., Obudu Div. (Aug., Dec.) *Stone* 80 ! *Savory & Keay* FHI 25247 ! [Br.]Cam.: Cam. Mt. : Mann's Spring, 7,700 ft., *Preuss* 748 ! Musake *Maitland* 1239 ! Bafut-Ngemba F.R., 7,000 ft., Bamenda (Feb.) *Hepper* 2106 ! Maisamari, 4,600 ft., Mambila Plateau (Jan.) *Hepper* 1667 ! F.Po: Moka, 4,000 ft., and Mioka area *Boughey* 38 ! 90 ! Also in Cameroun.
11. **L. djurensis** *Engl. & Diels* in Engl. Bot. Jahrb. 26 : 116 (1898) ; E. Wimm. l.c. 570, fig. 93 d. *L. baoulensis* A. Chev. in Mém. Soc. Bot. Fr. 2, 8 : 180 (1912) ; F.W.T.A., ed. 1, 2 : 193 ; Berhaut Fl. Sén. 174. A glabrous erect herb, stem 4-angled, succulent and spongy ; flowers pale lilac or purple, ¼ in. long ; in muddy places.
 Sen.: Badi (Dec.) *Berhaut* 1827 ! 4424. Mali: Diafarabé (Dec.) *Davey* 600 ! Guin.: Grandes Chutes (Dec.) *Chev.* 20313 ! Iv.C.: Baoulé (Aug.) *Chev.* 22252 ! 22389 bis ! Ghana: near Burufo (Dec.) *Adams & Akpabla* 4383 ! N.Nig.: Borgu *Barter* 797 ! Anara F.R., Zaria Prov. (Oct.) *Hepper* 994 ! Lokoja and Kontagora (Oct., Nov.) *Dalz.* 59 ! Minna (Dec.) *Meikle* 728 ! Also in Middle Congo and Sudan.
12. **L. hartlaubii** *Buchenau* in Abh. Naturwiss. Ver. Bremen. 7 : 201 (1881) ; E. Wimm. l.c. 586. *L. schaeferi* Schlechter in Engl. Bot. Jahrb. 57 : 624 (1922). Decumbent or ascending herb, 1–2 ft. high, often rooting below, branched, sparsely pilose ; flowers white, blue or mauve with white margins ; in montane forest.
 [Br.]Cam. (eastern border) : Mbo, 5,000 ft. (Nov.) *Ledermann* 6054 ! Also in Cameroun, Uganda and Madagascar.

Fig. 274.—Lobelia columnaris *Hook. f.* (Lobeliaceae).

A, bract. B, flower. C, longitudinal section of flower. D, anther. E, stigma. F cross-section of ovary.

13. **L. baumannii** *Engl.* in Abhandl. Preuss. Akad. Wiss. 46 (1894) ; E. Wimm. l.c. 586. Erect herb, stem decumbent at the base, rooting at the nodes, 1 ft. or more high ; flowers white with purple markings on the throat, anthers blue.

    S.Nig.: Sonkwala, Obudu Div. (Dec.) *Savory & Keay* FHI 25208 ! Widespread in E. Africa.

2. **MONOPSIS** Salisb. in Trans. Hort. Soc. Lond. 2 : 37 (1817) ; E. Wimm. in Engl. Pflanzenr. 4, 276b, Campanulac.-Lobelioid. : 698 (1953).

Small herb ; stems prostrate, scabrid ; leaves opposite, lanceolate, up to 2·5 cm. long and 6 mm. broad, subacute at both ends, midrib scabrid, margins callous and crenate ; flowers solitary in the axils of the leaves ; peduncles 8–25 mm. long, divaricate ; calyx-tube obconic, retrorse-pilose, lobes sublinear, 4 mm. long, ciliate, divaricate or deflexed ; corolla about 8 mm. long, bilabiate ; capsule oblong-obovoid, hirtellous, 6 mm. long ; seeds compressed-globose, scrobiculate, 0·6–0·8 mm. long

                                                      stellarioides var. *schimperiana*

M. stellarioides (*Presl*) *Urb.* in Jahrb. Bot. Gart. Berl. 1 : 275 (1881). *Dobrowskya stellarioides* Presl (1836), partly.

M. stellarioides var. schimperiana (*Urb.*) *E. Wimm.* l.c. 704 (1953). *M. schimperiana* Urban in Jahrb. Bot. Gart. Berl. 1 : 275 (1881). *Lobelia stellarioides* (Presl) Benth. & Hook. f. ex Hemsley in F.T.A. 3 : 470 (1877). *L. mukuluensis* De Wild. (1920). Straggling herb, with weak stems 1–2 ft. long ; flowers yellow with dull red throat and often reddish outside ; in montane marshes.

    [Br.]Cam. : Litoka, 4,400 ft., Cam. Mt. (Apr.) *Maitland* 1087 ! Bamenda (fl. & fr. May) *Daramola* FHI 41185 ! Jakiri, 5,000 ft., Bamenda (Feb.) *Hepper* 1957 ! **F.Po**: Biao, 6,500 ft. (Sept.) *Wrigley & Melville* 469 ! Also in Cameroun, tropical E. Africa, E. Congo.

3. **DIELSANTHA** E. Wimm. in Ann. Naturhist. Mus. Wien 56 : 372 (1948) ; Engl. Pflanzenr. 4, 276b, Campanulac.-Lobelioid.: 743, fig. 111 (1953).

Erect or decumbent, branched, herb ; stems succulent, more or less square, often rooting below ; leaves ovate to oblong-ovate, 3–7 cm. long, 2–3·5 cm. broad, subacuminate to acute at the apex, rounded to cuneate, margin serrate, slightly hirsute on the surfaces or glabrescent ; petioles 5–15 mm. long ; flowers few, sessile in sessile axillary fascicles ; corolla as in *Lobelia*, 1 cm. long, all segments ciliate at the apex ; all anthers bearded ; calyx-tube at first peduncle-like, soon becoming conical or spindle-shaped ; fruit 1·5–2 cm. long, 2 mm. broad, with 5 sepals at apex, splitting into 5 narrow strips

                                                           galeopsoides

D. galeopsoides (*Engl. & Diels*) *E. Wimm.* l.c. (1953). *Lobelia galeopsoides* Engl. & Diels in Engl. Bot. Jahrb. 26: 118 (1899) ; F.W.T.A. ed. 1, 2 : 193. A weak herb, about 8 in. high ; flowers lilac or blue-violet.

    S.Nig. : Oban *Talbot* 1259 ! **[Br.]Cam.** : Barombi gorge (May) *Preuss* 184 ! **F.Po** : St. Isabel's Peak (Aug.) *Mildbr.* 6365 !

148. GOODENIACEAE

By F. N. Hepper

Herbs or undershrubs. Leaves alternate or rarely opposite ; stipules absent. Flowers zygomorphic. Calyx tubular, adnate to the ovary, rarely free. Corolla gamopetalous, bilabiate or rarely 1-lipped, lobes valvate, often induplicate. Stamens 5, alternate with the corolla-lobes, free or rarely shortly adnate to the corolla ; anthers 2-celled, free or connivent around the style. Ovary mostly inferior, 1–4-celled ; stigma indusiate at the top. Ovules 1 or more in each cell, erect or ascending. Fruit drupaceous or nut-like, or capsular. Seeds small, with straight embryo in the middle of copious endosperm.

Mainly Australian ; represented in our area by one widely distributed species.

SCAEVOLA Linn.—F.T.A. 3 : 462. *Nom. cons.*

A succulent low shrub ; leaf-scars oblique, prominent ; leaves narrowly obovate, fleshy, rounded at the apex, broadly petiolate, 5–8 cm. long, 2·5–4 cm. broad, glabrous, faintly nerved when dry ; flowers in short axillary cymes, sessile ; bracts linear ; calyx very short and annular ; corolla about 2 cm. long, glabrous outside, densely woolly inside ; style pubescent, indusiate stigma finely ciliate ; fruit fleshy, about 1·3 cm. long, ellipsoid-globose..   ..   ..   *plumieri*

S. plumieri (*Linn.*) *Vahl* Symb. Bot. 2 : 36 (1791). *Lobelia plumieri* Linn. (1753). *S. lobelia* Murr. (1774) —F.T.A. 3 : 462 ; Chev. Bot. 381. A low shrub 1–3 ft. high, succulent and nearly glabrous, with thick branchlets marked by leaf-scars ; flowers white or greenish-yellow, fruits blue-black ; on sandy shores near high-water mark.

    Sen. : *Sieber* 23 ! *Leprieur* ! *Berhaut* 1012. Dakar (fl. & fr. May) *Baldwin* 5755 ! *Chev.* 2834. **S.L.** : Lumley (Aug.) *Deighton* 2044 ! Hamilton (Apr.) *Hepper* 2531 ! Messima (Apr.) *Adames* 31 ! **Lib.** : Grand Bassa (July) *T. Vogel* 116 ! Robertsport (fl. & fr. Dec.) *Baldwin* 10909 ! **Iv.C.** : Bériby *Chev.* 20017. Abouabou (fl. & fr. Jan.) *Leeuwenberg* 2376 ! Near Assinie (July) *Chipp* 283 ! **Ghana** : Accra *Moloney* ! Tema *Irvine* 948 ! Prampram (fr. June) *Boughey* GC 512 ! **S.Nig.** : Badagry (Aug.) *Onochie* FHI 33488 ! Sea coast of Africa, tropical America and S. India. (See Appendix, p. 424.)

    Scaevola sericea Vahl, a native of Madagascar and tropical Asia, is cultivated at Victoria.

149. HYDROPHYLLACEAE

By F. N. Hepper

Annual or perennial herbs. Leaves radical or alternate, rarely opposite, entire to pinnately or palmately lobed. Flowers usually cymose, hermaphrodite, actinomorphic. Calyx-segments 5, imbricate. Corolla gamopetalous, lobes imbricate or contorted. Stamens the same number as the corolla-lobes and alternate with them, mostly inserted towards the base of the tube ; anthers 2-celled. Ovary superior, 1-celled with 2 parietal placentas, or 2–3-celled ; styles 1 or 2. Ovules numerous. Fruit a loculicidal or rarely septicidal capsule. Seeds with fleshy endosperm and small straight embryo.

Mainly North American.

HYDROLEA Linn.—F.T.A. 4, 2 : 2 ; Brand in Engl. Pflanzenr. 4, 251, Hydrophyllac.: 174 (1913). *Nom. cons.*

Herbs ; leaves alternate, entire, narrow. Flowers blue, in axillary clusters, or in terminal cymes. Sepals 5, free to near the base, imbricate. Corolla 5-lobed to near the base, open. Stamens 5 ; anthers sagittate. Ovary 2- (rarely 3-) celled ; ovules numerous; styles 2, divergent. Capsule septicidally 2-valved.

Flowers in axillary clusters along the stems or in very short axillary racemes ; leaves lanceolate, very acutely acuminate, acute at the base, 4–8 cm. long, 1–1·5 cm. broad, glabrous, with numerous lateral nerves ; sepals oblong-lanceolate, acute, glabrous, about 6 mm. long ; corolla lobed nearly to the base, very short ; capsule broadly ovoid, glabrous ..   ..   1. *glabra*
Flowers in cymes ; leaves linear or linear-lanceolate :
Sepals cordate-ovate, about 7 mm. long ; cymes lax, few-flowered ; pedicels slender ; leaves sessile, linear-lanceolate, acute, long, 0·5–1 cm. broad, with very few lateral nerves ; capsule subglobose, hidden by the slightly accrescent veiny sepals ..   ..   2. *macrosepala*
Sepals lanceolate, 4 mm. long ; cymes rather compact, many-flowered ; leaves linear, up to 9 cm. long, acute, glabrous ; capsule globose, about 4 mm. diam., glabrous
                                                                3. *floribunda*

1. **H. glabra** *Schum. & Thonn.* Beskr. Guin. Pl. 161 (1827) ; Brand l.c. 175. *H. guineensis* Choisy in Ann. Sci. Nat., sér. 2, 1 : 180 (1834) ; F.T.A. 4, 2 : 3 ; Chev. Bot. 447 ; F.W.T.A., ed. 1, 2 : 194. Glabrous herb with spongy stems ascending or weakly erect ; flowers blue ; in pools and wet places.

Fig. 275.—HYDROLEA FLORIBUNDA *Kotschy & Peyr.* (HYDROPHYLLACEAE).
A, flower. B, part of corolla from inside. C, stamen. D, calyx and pistil. E, cross-section of ovary. F, fruit.

Sen. : *Heudelot* 79 ! Niokolo-Koba (fl. & fr. Apr.) *Berhaut* 1500 ! Badi (fl. & fr. Dec.) *Berhaut* 2163 ! **S.L.** : Pendembu (July) *Thomas* 735 ! Kahreni to Port Loko (Apr.) *Sc. Elliot* 5805 ! Mapaki (Aug.) *Deighton* 1207 ! Mano (Jan.) *Deighton* 2447 ! **Lib.** : Monrovia *Whyte* ! Tauiné (Sept.) *Linder* 825 ! **Iv.C.** : Bliéron *Chev.* 19919. Grabo to Taté *Chev.* 19787. **Ghana** : Sesiamang (Feb.) *A. S. Thomas* D149 ! Dormaa-Ahenkro (Dec.) *Adams* GC 3009 ! **N.Nig.** : Lokoja (Oct.) *Dalz.* 217 ! Agiare, Bida (Dec.) *Savory* UCI 102 ! Kontagora (Dec.) *Dalz.* 56 ! **S.Nig.** : Abo *Barter* 185 ! Abeokuta *Irvine* 46 ! Badagry (Aug.) *Onochie* FHI 33464 ! Adani (Mar.) *Tuley* 71 ! Also in Cameroun, Congo and Madagascar.

2. **H. macrosepala** *A. W. Bennett* in J. Linn. Soc. 11 : 277, t. 1, fig. 13 (1870) ; F.T.A. 4, 2 : 4. *H. djalonensis* A. Chev. Bot. 447, name only. Erect branched glabrous herb, stems up to 2 ft. high, spongy in the lower part ; flowers blue in very lax panicles ; in swamps and moist ground in savanna.
Sen. : Niokolo-Koba *Berhaut* 1144 ! **Guin.** : Kouroussa (Dec.) *Pobéguin* 612 ! Kouria *Caille* in Hb. *Chev.* 14962 ! **S.L.** : Kambia (Nov.) *Jordan* 376 ! **Ghana** : Lawra (Oct.) *Hinds* 5011 ! **N.Nig.** : Nupe *Barter* 902 ! Kontagora (Nov.) *Dalz.* 54 ! Mada (Nov.) *Hepburn* 515 ! **[Br.]Cam.** : Gurum, Adamawa (Nov.) *Hepper* 1242 ! Also in Sudan.

3. **H. floribunda** *Kotschy & Peyr.* Pl. Tinn. 22, t. 9B (1867) ; F.T.A. 4, 2 : 5 ; Brand l.c. 177, fig. 37. *H. graminifolia* A. W. Bennett l.c. t. 1, fig. 12 (1870) ; F.T.A. 4, 2 : 5 ; Chev. Bot. 447 ; F.W.T.A., ed. 1, 2 : 194. Erect herb branched above at the corymbose inflorescence, glabrous, stems spongy below, up to 3 ft. high ; flowers rich blue ; in swamps in savanna.
Sen. : Niokolo-Koba (Jan.) *Roberty* 16636 ! **Mali** : Tiguiberri (fr. Jan.) *Chev.* 296 ! Dogo (Mar., May) *Davey* 99 ! 007 ! **Guin.** : Kollangui *Chev.* 12190 ; 13530. Kebeya *Pobéguin* 717 ! Férédougouba R. (July) *Collenette* 69 ! Erimakuna (= Hérémakon) (Mar.) *Sc. Elliot* 5235 ! **Iv.C.** : Alangouassou to Mbayrakro *Chev.* 22240 ! **Ghana** : Yeji (fl. & fr. Aug.) *Vigne* FH 3318 ! Yeji to Ejura (Apr.) *Dalz.* 48 ! Damongo Scarp (Sept.) *Rose Innes* GC 30201 ! Yendi to Zabzugu (Oct.) *Rose Innes* GC 30696 ! **N.Nig.** : Jebba *Barter* 888 ! Bida (Dec.) *Savory* UCI 103 ! *Keay* FHI 37296 ! Kontagora (Nov.) *Dalz.* 55 ! Abinsi (June) *Dalz.* 750 ! **[Br.]Cam.** (?) : Tuburi marshes *Talbot* 332 ! Also in Cameroun, Uganda and Sudan.

150. BORAGINACEAE

By H. Heine

Trees, shrubs or herbs, often roughly scabrid or hispid. Leaves alternate or rarely opposite, simple ; stipules absent. Flowers often in scorpoid cymes, actinomorphic or rarely oblique, mostly hermaphrodite. Calyx-lobes imbricate or rarely valvate. Corolla with contorted or imbricate lobes. Stamens the same number as the corolla-lobes and alternate with them, inserted on the corolla. Anthers 2-celled, opening lengthwise. Disk present or absent. Ovary superior, 2-celled or 4-celled by spurious septa, entire or deeply 4-lobed ; style terminal or gynobasic. Ovules paired, erect or spreading from the central axis. Fruit a drupe or of 4 nutlets. Seeds with or without endosperm.

Generally distributed ; very numerous in the drier parts of the Mediterranean regions. Recognized by the usually rough hairs and often lobed ovary with gynobasic style.

Style terminal on top of the ovary (ovary not vertically lobed) :
Trees or shrubs :
Style once divided :
Stamens inserted at the throat of the corolla-tube ; flowers small, cymose
            1. Ehretia
Stamens inserted at the bottom of the corolla-tube ; flowers few, in leafy cymes
            2. Rotula
Style twice divided .. .. **3. Cordia**
Herbs :
Styles two, separate ; annual procumbent herb densely setose all over ; flowers tetra-merous, solitary, axillary .. .. **4. Coldenia**
Style one ; herbs with scorpoid cymes (in one species flowers appearing solitary) ; flowers pentamerous (except the gynaecium) .. .. .. **5. Heliotropium**
Style inserted between the lobes of the ovary (ovary vertically lobed) :
Calyx large and conspicuous, accrescent in fruit .. .. .. **6. Trichodesma**
Calyx rather small, not or only slightly accrescent :
Nutlets depressed-globose, covered with glochidiate bristles .. **7. Cynoglossum**
Nutlets ovoid, without glochidiate bristles :
Corolla-tube rather long and narrow ; plants densely hispid with long stiff white hairs ; flowers yellow .. .. **8. Arnebia**
Corolla-tube short, hypocrateriform, plants clothed with short spreading bristly hairs ; flowers white or bright blue .. .. **9. Myosotis**

1. EHRETIA Linn.—F.T.A. 4, 2 : 19.

Leaves glabrous beneath except sometimes on the nerves, broadly elliptic to obovate or almost suborbicular, abruptly acuminate, rounded to subacute at the base, 10–15 cm. long and up to 10 cm. broad, entire ; petiole 1–2 cm. long, slightly pilose ; cymes large, terminal, many-flowered ; flowers small :
Flowers clearly pedicellate, articulated at or near base of the calyx ; inflorescences usually minutely pubescent ; leaves nearly glabrous, sometimes very sparsely pilose

on the midrib and principal nerves of the lower surface ..   1a. *cymosa* var. *cymosa*
Flowers subsessile, articulated at or near the base of calyx ; inflorescences pubescent,
    tomentellous or shortly pilose   ..   ..   ..   lb. *cymosa* var. *zenkeri*
Leaves shortly setulose with broad-based hairs beneath, obovate-elliptic, gradually
    acuminate, rounded at base, 12–16 cm. long, 5–8 cm. broad, with about 5–6 pairs of
    lateral nerves ; petiole extremely short ; cymes few-flowered ; calyx-segments linear-
    lanceolate, glabrous ; fruit narrowly obovoid, ribbed, 1 cm. long   2. *trachyphylla*

1. **E. cymosa** *Thonning* Beskr. Guin. Pl. 129 (1827) ; F.T.A. 2 : 25 ; F.W.T.A., ed. 1, 2 : 195 ; Chev. Bot.
    449 ; Brenan in Mem. N.Y. Bot. Gard. 9, 1 : 4 (1954) ; Aubrév. Fl. For. C. Iv., ed. 2, 3 : 217, t. 332,
    6–9. *E.thonningiana* Exell Suppl. Cat. S. Tomé 34 (1956)[1]. Shrub or small tree up to 20 ft. high, sometimes
    with weak drooping branches ; flowers small, white, fragrant, in copious panicled cymes ; fruits red.
1a. **E. cymosa** *Thonning* var. **cymosa**—Brenan l.c. 5 (1954).
    S.L.: Bafodeya highlands (fr. Apr.) *Sc. Elliot* 5645 ! Duunia (Feb.) *Sc. Elliot* 4856 ! Bumban (Apr.) *Sc.
    Elliot* 5723 ! Kanah (May) *Thomas* 406 ! Njala (Apr.) *Deighton* 2636 ! **Iv.C.:** Sassandra *Chev.* 16338 ;
    17915. Dimbokro *Aubrév.* 436 ; 437. Bondoukou *Aubrév.* 759. Tonkoui Mt. *Aubrév.* 1020. **Ghana:** Accra
    (Feb., May), *T. Vogel* ! *Johnson* 607 ! *Brown* 372 ! Aburi *Patterson* 184 ! Secondi (Jan.) *Dalz.* 67 !
    Kumasi (fl. & fr. Feb.) *Irvine* 9 ! **Togo Rep.:** Lomé *Warnecke* 82 ! 370 ! **S.Nig.:** Abeokuta *Chev.* 13915 ;
    14085 ; 14116. *Irving* 52 ! *Punch* 2 ! Shaki (fl. May) *Denton* 24 ! Aboh *Barter* 287 ! Degema *Talbot* 3839 !
    Eket Dist. *Talbot* ! **[Br.]Cam.:** Rio del Rey *Johnston* ! Buea (May) *Maitland* 694 ! Also in Cameroun,
    Gabon, Congo and Uganda. (See Appendix, p. 426.)
1b. **E. cymosa** var. **zenkeri** (*Gürke ex Bak. & Wright*) Brenan l.c. (1954). *E. zenkeri* Gürke ex Bak. & Wright in
    F.T.A. 4, 2 : 25 (1905). Like the last but with longer and denser indumentum on the inflorescence.
    N.Nig.: *Thornewill* 70 ! 83 ! **S.Nig.:** Lagos *Phillips* 34 ! Also in Cameroun and S. Tomé.
    [Note: Intermediates between var. *cymosa* and var. *zenkeri* are frequent in S. Nigeria, e.g. *Rowland* !
    Lagos (Feb.) *Dalz.* 1018 ! (May) *Foster* 86 ! Abeokuta *Barter* 3388 ! Ibadan *Newberry* 59 ! Aku Rock (fr.
    Nov.) *Hambler* 582 !—H. H.]
2. **E. trachyphylla** *C. H. Wright* in Kew Bull. 1907 : 53 ; Aubrév. Fl. For. C. Iv., ed. 2, 3 : 218, t. 332, 1–5.
    A small tree with lenticellate branches, and small white flowers in lax pubescent cymes.
    Iv.C.: *Aubrév.* 153 ; 797 ; 1378 ; 1787 *bis* ; 1792. **Ghana:** Dunkwa (Aug.) *Vigne* FH 4781 ! Ateiku
    Vigne FH 4775 ! Aburi (June) *Johnson* 974 ! Cape Coast (fr. July) *Hall* 1502 !

2. ROTULA Lour. Fl. Cochinch. 121 (1790) ; Johnston in J. Arn. Arb.
32: 14 (1951). *Rhabdia* Mart.—F.T.A. 4, 2 : 27.

Shrub with numerous setose to glabrescent branchlets ; leaves oblanceolate, up to
3·5 cm. long and 1 cm. broad, rather long-setose-pilose to almost glabrous beneath ;
corymbs lax, lateral ; bracts foliaceous ; calyx-segments linear-oblong, 5 mm. long ;
corolla widely campanulate, 1 cm. long, throat glabrous ; stamens included ; fruit
slightly ribbed, ovoid, surrounded by the persistent calyx   ..   *aquatica*

R. aquatica *Lour.* l.c. (1790) ; Bunting in J. Bot. 47 : 269 (1909) ; Merrill in Trans. Am. Phil. Soc., n. ser., 24
    2 : 330 (1935) ; Johnston l.c. 15 ; Berhaut Fl. Sén. 7 ,134. *Rhabdia lycioides* Mart. (1827)—F.T.A. 4, 2 :
    28 ; Chev. Bot. 449. *Zombiana africana* Baill. (1888)—F.T.A. 5 : 263. A stiff-branched shrub 3–6 ft. high,
    with red flowers ; by streams or half-submerged.
    Sen.: *Heudelot* 116 ! *Berhaut* 1276 ; 4088 ; 4493. M'bidjem (May) *Thierry* 217 ! **Mali :** Badumbé *Chev.* 59.
    Quignaba *Chev.* 255. Kéniégué *Chev.* 273. **Port. G. :** Chitole Saltinko (fl. & fr. Feb.) *Esp. Santo* 2669 !
    Guin. : Tongué (Dec.) *Roberty* 16316 ! Farana *Chev.* 13397. *Sc. Elliot* 5339 ! **S.L.:** Njala (Feb.) *Deighton*
    3116 ! Musaia (Feb.) *Deighton* 4237 ! Makump (Oct.) *Glanville* 137 ! Kukuna (Jan.) *Sc. Elliot* 4718 !
    Batkanu (Jan.) *King* 178b ! **Ghana:** Kete Krachi (Dec.) *Adams* 4554 ! Black Volta, N.T. (Dec.) *Adams &
    Akpabla* 4453 ! Black Volta R., Lawra (Jan.) *Crisp* ! **S.Nig.:** R. Awon, Yoruba country *Barter* 1143 ! Pan-
    tropical.

3. CORDIA Linn.—F.T.A. 4, 2 : 6.

Leaves opposite or subopposite, oblong-oblanceolate, rounded at the apex, 4–8 cm. long,
    1·5–3 cm. broad, slightly scabridulous above, finely reticulate beneath ; petiole 0·5–
    1 cm. long ; flowers few ; pedicels 5 mm. long ; calyx 5 mm. long, slightly setulose ;
    fruit acute, 1 cm. long   ..   ..   1. *rothii*
Leaves alternate :
    Leaves glabrous on the lower surface :
        Calyx-tube not ribbed, contracted at the base, about 7 mm. long, glabrous ; leaves
            long-petiolate, oblong to narrowly ovate, acutely acuminate, very shortly cuneate
            at the base, 8–15 cm. long, 3–6 cm. broad, with about 5 pairs of lateral nerves ; fruit
            1·5 cm. long, acute   ..   ..   2. *senegalensis*
        Calyx-tube strongly ribbed, 1 cm. long, velvety-tomentellous ; corolla shortly exserted,
            lobes reflexed ; leaves elliptic or ovate-elliptic, acutely acuminate, rounded at the
            base, 10–25 cm. long, 5–10 cm. broad, with about 5 pairs of lateral nerves ; petiole
            2 cm. long   ..   ..   3. *aurantiaca*
Leaves pubescent to tomentose beneath :
    Calyx very strongly ribbed :
        Calyx 1·3 cm. long, appressed-setulose ; corolla shortly exserted ; branchlets densely
            pilose ; leaves oblong-lanceolate, broadly acuminate, about 20 cm. long and 5 cm.
            broad, long-pilose on the midrib on both surfaces ; petiole 1 cm. long   4. *vignei*
        Calyx less than 1 cm. long, tomentellous ; corolla long-exserted, funnel-shaped,
            about 2·5 cm. long ; leaves ovate-orbicular, rounded to widely cordate at the base,
            rounded at the apex, about 8 cm. diam., with very prominent parallel tertiary

---

[1] *E. cymosa* Willd. ex Roem. & Schult. Syst. Veg. 4 : 805 (1819) was invalidly published in syn. for *E. laevis*
Roxb., an Indian plant.

W.E.T.

Fig. 276.—Cordia africana *Lam.* (Boraginaceae).

A, calyx. B, open corolla. C, pistil. D, cross-section of fruit. E, longitudinal section of ovary. E, fruit.

319

nerves, shortly tomentellous beneath; cymes many-flowered; fruit 1·3–1·5 cm. long, fleshy .. .. 5. *africana*

Calyx not ribbed or only very obscurely so:

Calyx glabrous outside or minutely puberulous only towards the tip, contracted at the base and widely obconic; cymes about 8 cm. long; pedicels very short and stout; leaves ovate-orbicular, slightly and broadly mucronate, slightly pubescent on the closely reticulate lower surface, flabellately trinerved at the base, 6–12 cm. long, 4–9 cm. broad .. .. 6. *myxa*

Calyx tomentose or tomentellous outside:

Leaves rounded at the apex; calyx apiculate in bud:

Leaves small, at most 7–8 cm. diam. .. .. 7. *guineensis*

Leaves large, 15–25 cm. long, obovate-orbicular .. .. 8. *millenii*

Leaves acuminate, broadly ovate-elliptic, rounded to slightly cuneate at base, up to 18 cm. long and 12 cm. broad, shortly pubescent beneath, with prominent and numerous parallel tertiary nerves; calyx not apiculate in bud, about 7 mm. long; cymes divaricate .. .. 9. *platythyrsa*

1. **C. rothii** *Roem. & Schult.* Syst. Veg. 4: 798 (1819); F.T.A. 4, 2: 18; Chev. Bot. 448; Berhaut in Mém. Soc. Bot. Fr. 1953/54: 4 (1954), and Fl. Sén. 109. *Cornus gharaf* Forsk. (1775), name only. *Cordia gharaf* (Forsk.) Ehrenb. ex Aschers. in Sitzungsber. Ges. Naturf. Fr. Berlin 1879: 46 (1879), and Verh. Bot. Ver. Prov. Brandenburg 21: 69 (1879); Aubrév. Fl. For. C. Iv., ed. 2, 3: 218, and Fl. For. Soud.-Guin. 493, t. 110, 2, 3; Pellegr. in Bull. Soc. Bot. Fr. 81: 272. *Cordia senegalensis* of A. DC. Prod. 9: 480, not of Juss., incl. var. *pelida* A. DC. (1845). A shrub or small tree up to 16 ft., with white flowers ¼ in. long in rather small pubescent cymes and ovoid, pointed fruits on a saucer-shaped calyx.
Sen.: Dagana *Leprieur*. *Perrottet* 131; 544. Fadiout *Ezanno* 41. *Berhaut* 357. **Mali:** Tacadji *Chev.* 3179. Timbukto *Chev.* 3547. Niafounké (fr. Mar.) *Davey* 603! **Ghana:** Winneba Plain (fr. Feb.) *Dalz.* 8275! Accra (Jan.) *Dalz.*! *Adams* 4763! Newningo (Mar.) *Morton* A169! Shai Plains (Feb.) *Irvine* 1981! **Togo Rep.:** Lomé *Warnecke* 344! **N.Nig.:** Yo, Bornu (fl. & fr. Dec.) *Elliott* 149! In the drier parts of tropical Africa generally and throughout Arabia to India.
[*Davey* 603 is a specimen which seems to be an intermediate between this and *C. ovalis* R. Br. ex DC.]

2. **C. senegalensis** *Juss.* in Poir. Encycl. 7: 47 (1806); F.T.A. 4, 2: 15, partly; Pellegr. in Bull. Soc. Bot. Fr. 81: 270; Berhaut Fl. Sén. 14: 1; Aubrév. Fl. For. C. Iv., ed. 2, 3: 220, t. 333, 1–5. *C. heudelotii* Bak. (1894)—F.T.A. 4, 2: 14. *Cordia mannii* C. H. Wright in F.T.A. 4, 2: 15. *Vitex syringaefolia* Bak. (1895)—F.T.A. 5: 350. *Ehretia acutifolia* Bak. (1894). A nearly glabrous shrub or tree up to 30 ft. high, with white flowers ⅓ in. long in lax terminal and axillary cymes.
Sen.: *Perrottet* 541. *Heudelot* 253! 342! *Irvine* 3237! *Berhaut* 209. **Iv.C.:** *Aubrév.* 1128; 1779. **Ghana:** *Rowland*! Mampong *Vigne* FH 1061! Juaso (Feb.) *Vigne* FH 1820! **S.Nig.:** Lagos *Rowland*! Yoruba country *Barter* 3425! Ibadan North F.R. (Feb.) *Chizea* FHI 23970! Eruwa, Oyo (Apr.) *Keay* FHI 37808! **[Br.]Cam.:** Ambas Bay (fl. Feb.) *Mann* 968! Johann-Albrechtshöhe *Staudt* 614! Also in Cameroun.

3. **C. aurantiaca** *Bak.* in Kew Bull. 1894: 26; F.T.A. 4, 2: 10 (1905). *C. dusenii* Gürke (1895). A shrub or small tree 15–25 ft. high, brownish-velvety on the young parts, with orange-yellow flowers, ¾ in. long in terminal cymes.
S. Nig.: Idumuje (Dec.) *Thomas* 2103! 2140! Sapoba *Kennedy* 1971! Oban *Talbot* 318! **[Br.]Cam.:** Buea (fl. & fr. Mar.) *Deistel* 620! *Hutch. & Metcalfe* 102! *Maitland* 570! Victoria (Mar.) *Maitland* 511! Boirongo (fl. Jan.) *Keay* FHI 37390! **F.Po:** (June) *Barter*! *Mann* 4! (See Appendix, p. 425.)

4. **C. vignei** *Hutch. & Dalz.* F.W.T.A., ed. 1, 2: 196 (1931); Aubrév. Fl. For. C. Iv., ed. 2, 3: 220, t. 333, 6–9. A shrub with rather long acutely-pointed leaves, densely pale brown pubescent on the young parts, with flowers ½–⅝ in. long.
S.L.: Levuma (fr. Aug.) *Deighton* 2212! Kaboura-Mamayema (fl. Apr.) *Deighton* 3927! Mano (fl. Apr.) *Deighton* 4619! **Iv.C.:** Agboville *Aubrév.* 1387; 1388. **Ghana:** Kumasi (fl. & fr. Apr., June) *Vigne* FH 1098! 1744! Ankaful *Hall* 1845!

5 **C. africana** *Lam.* Tabl. Encycl. Ill. 1: 420 (1792). *C. abyssinica* R.Br. (1814)—F.T.A. 4, 2: 9; Aubrév. Fl. For. Soud-Guin. 493, t. 110, 4, and Fl. For. C. Iv., ed. 2, 3: 218. *C. platythyrsa* of Chev. Bot. 448, not of Bak. *C. ubanghensis* A. Chev. (1913). A shrub or small tree up to 25 ft. high, with broad leaves strongly-nerved beneath, conspicuous funnel-shaped white flowers in paniculate cymes, and yellow succulent fruits ⅓–½ in. in diam.; often planted.
Guin.: Koumi *Pobéguin* 1486! *Scaëtta* 3348. Labé *Cochet* 64. Kouria (fl. Dec.) *Caille* in *Hb. Chev.* 14666! **N.Nig.:** *Moiser* 61! *Lely* 140! Katagum Dist. *Dalz.* 88! Wase (Nov.) *Lamb* 79! Nafada (fr. Dec.) *Foster* 84! Widespread in tropical Africa. (See Appendix, p. 424.)

6. **C. myxa** *Linn.* Sp. Pl. 190 (1753); F.T.A. 4, 2: 14; Hutch. in Kew Bull. 1918: 217, t. on p. 220, 1; Chev. Bot. 448; Berhaut Fl. Sén. 130; Aubrév. Fl. For. C. Iv., ed. 2, 3: 218, t. 334, 6; Fl. For. Soud.-Guin. 490, 493, t. 112, 5. *Vitex gomphophylla* Bak. in F.T.A. 5: 319 (1900). A shrub or tree up to 40 ft. high with stout stem, nearly glabrous branchlets, and cream-white flowers ⅓ in. long in lax paniculate cymes.
Sen.: Niakoulrab *Berhaut* 192. **Mali:** Birgo *Dubois* 57. Monguiéneba *Chev.* 446. Moussaia (Feb.) *Chev.* 463! Somilia *Chev.* 446. **Port.G.:** *Esp. Santo* 403! 737! **Guin.:** Timbo *Chev.* 13604. Kouroussa *Pobéguin* 837. Kouroussa to Kankan *Aubrév.* 3072. Dantilia (fr. Mar.) *Sc. Elliot* 5389! **Iv.C.:** Langouassou *Chev.* 22152. Ouangolo *Aubrév.* 1476; 1845. **Ghana:** Wenchi (Mar.) *Dalz.* 28! Banda, Wenchi Dist. (fr. Dec.) *Morton* GC 15288! Kunche, Cherepong to Yaga (fr. Mar.) *Kitson* 848! 849! Introduced and cultivated but apparently naturalised; native of Asia Minor and Palestine. (See Appendix, p. 425.)

7. **C. guineensis** *Schum. & Thonn.* Beskr. Guin. Pl. 128 (1827); F.T.A. 4, 2: 17; F.W.T.A., ed. 1, 2: 198 (as imperfectly known species); Aké Assi, Contrib. 1: 176 (1961). *C. johnsonii* Bak. in F.T.A. 4, 2: 13 (1905). *C. warneckei* Gürke ex Bak. & Wright in F.T.A. 4, 2: 13 (1905); Chev. Bot. 449. *C. kabarensis* De Wild. (1923). *C. bequaerti* De Wild. Rev. Zool. Afr. 9, 3, Suppl. Bot. 88 (1921), not *C. bequaerti* De Wild., l.c. 8: 43 (1920). A low shrub with densely pubescent branchlets; flowers ⅓–½ in. long, cream-yellowish, in short dense cymes; fruit pointed, ⅓ in. long.
Iv.C.: Forêt de Singrobo *Aké Assi* IA 4457. Bouna to Bondoukou *Aké Assi* IA 5395. **Ghana:** *Isert*! Afram Plains (fl. Mar.) *Johnson* 705! Ningo (fl. & fr. May) *Irvine* 1435! Ayikuma (fr. May) *Irvine* 2679! Elmina (May) *Morton* A893! Accra Plains (Mar.) *Morton* A229! Banda (fr. Dec.) *Morton* GC 25282! **Togo Rep.:** Lomé *Warnecke* 100! 308! **Dah.:** Djougou *Chev.* 23911. Also in Congo.

8. **C. millenii** *Bak.* in Kew Bull. 1894: 27; F.T.A. 4, 2: 11; Aubrév. Fl. For. C. Iv., ed. 2, 3: 224; Aké Assi, Contrib. 1: 176 (1961). *C. chrysocarpa* Bak. (1894)—F.T.A. 4, 2: 11; F.W.T.A., ed. 1, 2: 198. *C. irvingii* Bak. (1895), partly; F.T.A. 4, 2: 11; Bruce in Kew Bull. 1940: 62 (for full synonymy). A large forest tree pale brownish-tomentose on the young parts; flowers yellowish ½ in. long, crowded in cymes; often planted in towns.
Iv.C.: Kokondékro, near Bouaké *Aubrév.* (?); many localities cited by *Aké Assi*. **Ghana:** East Akim *Johnson* 744! *Thompson* 77! Asafo (Mar.) *Moor* FH 854! Abeambra *Vigne* FH 861! Nfoum (fr. Oct.) *Vigne* FH 109! Asakrakra-Kwahu (Mar.) *Johnson* 625! **S.Nig.:** Lagos (Nov.) *Millen* 12! 210! Okeado *Onyeagocha* FHI 7149! Shagamu (Mar.) *Schlechter* 13010! Abeokuta *Barter* 3400! *Irving* 147! *Chev.*

13898 ; 13950. [Br.]Cam.: Buea (buds Feb.) *Maitland* 369! Widespread in tropical Africa. (See Appendix, p. 425.)

9. **C. platythyrsa** *Bak.* in Kew Bull. 1894 : 27 ; F.T.A. 4, 2 : 12 ; Aubrév. Fl. For. C. Iv., ed. 2, 3 : 222, t. 334, 1–5. *C. irvingii* Bak. (1895), partly. *C. candidissima* A. Chev. Bot. 447, name only. " *C. platyphylla* Bak.*" of Chev. Bot. 448, name only in error. A tree 30–80 ft. high, pale brownish-pubescent on the young parts, with white funnel-shaped flowers in cymes forming a spreading panicle.
S.L.: Freetown *Burbridge* 535 ! Bagroo R. *Mann* 875 ! Moyamba (Mar.) *Lane-Poole* 101 ! **Lib.:** Firestone Plantation (Apr.) *Harley* 1491 ! **Iv.C.:** Anoumaba *Chev.* 22349. Guidéko *Chev.* 16366 ; 19098. Eryma-kougnié *Chev.* 16966. Indénié *Chev.* 17666 ; 17789. **Ghana** : Kumasi (Mar.) *Vigne* FH 3740 ! Swedru (Apr.) *Morton* A500 ! **S.Nig.:** Abeokuta *Irvine* 63 ! Ilugun to Olokemeji (Apr.) *Keay* FHI 37807 ! Ikorodu (Mar.) *Schlechter* 12999 ! Lagos *MacGregor* 179 ! (See Appendix, p. 425.)

Imperfectly known species.

Cordia tisserantii *Aubrév.* Fl. For. Soud.-Guin. 490, t. 110, fig. 1 (1950), French descr. only ; Berhaut Fl. Sén. 108, 130. " A small tree or tall shrub or coppice regrowth in forest clearing " (*Patel* FHI 51343).
Sen.: Niakoulrab (sterile material) *Berhaut* 192. **S.Nig.:** Foster 325 ! Nigerian College, Ibadan *Patel* FHI 51343 ! Onda, Omo F.R., Ijebu *Jones & Onochie* FHI 16978 ! Also in Ubangi-Shari. The identity of this species seems doubtful.

4. COLDENIA Linn.—F.T.A. 4, 2 : 28; Johnston in Journ. Arn. Arb. 32 : 12 (1951).

Annual procumbent herb, densely setose all over ; leaves shortly petiolate, oblong, up to 2·5 cm. long and 1 cm. broad, rounded at each end, dentately lobulate, the hairs converging between and revealing the nerves above ; flowers solitary, axillary, shortly pedicellate ; fruit glandular-pubescent, laterally 4-lobed, apiculate, about 4 mm. diam. .. .. *procumbens*

C. procumbens *Linn.* l.c. (1753) ; F.T.A. 4, 2 : 28 ; Chev. Bot. 449 ; Johnston, l.c. 13. Whitish hairy, spreading from a woody base, with wrinkled leaves and small scarlet or white flowers ; in wet places, by muddy pools (in habit resembling *Chrozophora*, *Forskohlea* and *Neurada*).
Sen.: (fl. & fr. Nov.) *Perrottet* ! Richard Tol (Jan.) *Roger* 144 ! **Gam.:** *Hayes* 542 ! Genieri (Feb.) *Fox* 88 ! **Mali** : Kabarah (fl. & fr. Sept.) *Hagerup* 229 ! Dogo (Apr.) *Davey* 509 ! Day *Chev.* 1351. **Port.G.:** Bissau (fl. & fr. Feb.) *Esp. Santo* 1780 ! **Guin.:** Kankan *Chev.* 587. **Ghana** : (fl. & fr. Mar.) *Kitson* ! Sapoba (fr. Mar.) *Hepper & Morton* A3120 ! Tapei (fr. Apr.) *Adams* 3921 ! Tamale to Damongo (fr. Mar.) *Morton* GC 8764 ! **N.Nig.:** Lokoja *Barter* 1672 ! Jos, 2,200 ft. (fl. & fr. May) *Lely* 173 ! Yola (fl. & fr. Apr.) *Dalz.* 155 ! Extends to the Red Sea, E. Africa and south to Angola. (See Appendix, p. 424.)

5. HELIOTROPIUM Linn.—F.T.A. 4, 2 : 28.

Leaves broadly ovate, more or less cuneate into the winged petiole, more or less tri-angular-pointed at the apex, up to 12 cm. long and 8 cm. broad, more or less undulate-dentate, shortly pubescent ; sometimes bullate ; stems densely pilose or villous ; spikes slender, curled towards the tip, up to 20 cm. long ; corolla about 8 mm. long ; fruit ovoid, ribbed, glabrous .. .. 1. *indicum*
Leaves linear to obovate (never ovate) :
  Corolla-lobes with long linear-filiform points ; leaves lanceolate, acute, 4–6 cm. long, up to 1·4 cm. broad, setulose-pilose ; stems long-setose-pilose and with short gland-tipped hairs ; fruits ovoid, slightly rugose .. .. 2. *subulatum*
  Corolla-lobes without filiform points :
    Leaves rather broadly obovate-oblanceolate to obovate :
      Flowers very small and numerous in a close slender spike-like cyme, appressed-pilose ; calyx 2–3 mm. long ; leaves oblong-elliptic, densely to thinly villous on both sides ; fruits depressed, pubescent .. .. 3. *ovalifolium*
      Flowers few, rather laxly spicate, softly villous ; calyx about 5 mm. long ; leaves elliptic, appressed-pilose between the nerves above, villous beneath ; fruit ovoid, 4-ribbed, rugulose .. .. 4. *supinum*
    Leaves more or less lanceolate to linear :
      Leaves linear or linear-oblanceolate, several times longer than broad :
        Leaves linear, not undulate, 2–2·5 cm. long, appressed-setose-pilose on both sides ; flowers distinctly pedicellate ; fruit depressed, sparingly setulose .. 5. *strigosum*
        Leaves oblanceolate, with strongly undulate margins, up to 8 cm. long and 1·5 cm. broad, setulose-scabrid on both surfaces ; flowers in dense short scorpioid cymes ; stems setose-pilose ; fruit glabrous :
          Nutlets free, equal, pilose at first, finally glabrous, rugose .. 6. *bacciferum*
          Nutlets cohering in pairs ; glabrous and smooth, each pair margined with a broad horny wing, with a round sinus at the apex .. .. .. 7. *pterocarpum*
      Leaves shortly lanceolate, about 3 times as long as broad ; inflorescence leafy, the flowers appearing solitary ; fruit ovoid :
        Fruit beaked up to 2 mm. long ; flowers white with yellow throat
                                    8a. *baclei* var. *baclei*
        Fruit beaked up to 5 mm. long ; flowers mostly entirely bright yellow
                                    8b. *baclei* var. *rostratum*

1. **H. indicum** *Linn.* Sp. Pl. 130 (1753) ; F.T.A. 4, 2 : 32 ; Chev. Bot. 450 ; Berhaut Fl. Sén. 200. (?) *H. africanum* Schum. & Thonn. Beskr. Guin. Pl. 87 (1827) ; A. DC. Prod. 9 : 548 ; F.T.A. 4, 2 : 43 ; F.W.T.A., ed. 1, 2 : 199. Robust annual 1–3 ft. high with small lilac, sometimes white, flowers in long dense curled spikes ; a common weed of waste places near dwellings, throughout the area.
Sen.: *Berhaut* 1013. *Perrottet* 527 ! 537 ! Dakar *Chev.* 15815. Richard Tol (fl. & fr. May) *Döllinger* 63 ! *Baldwin* 5734 ! *Lelièvre* ! **Gam.:** *Hayes* 533 ! Kuntaur *Ruxton* 33 ! Albreda (fl. & fr. June) *Leprieur* ! **Mali:** Bamako *Chev.* 201. Goundam *Chev.* 2883. Kéniéba (fl. & fr. Mar.) *Roberty* 17032 ! **Port.G.:** Bissau (fl. & fr. Mar.) *Esp. Santo* 1131 ! **Guin.:** Timbo *Caille* in Hb. Chev. 14675. Kankan *Brossart* 15680 ! **S.L.:**

Kukuna (fl. & fr. Jan.) *Sc. Elliot* 4719! Batkanu (fl. & fr. Apr.) *Thomas* 16! Yetaya (fl. & fr. Sept.) *Thomas* 2269! Ronietta (fl. & fr. Nov.) *Thomas* 5546! Freetown (fl. Oct.) *Deighton* 2165! Gembahun (fl. & fr. Jan.) *King* 164b! **Lib.:** Monrovia *Johnston*! Ganta (fr. Oct.) *Harley*! Gbanga (fl. & fr. Sept.) *Linder* 659! **Iv.C.:** Alépe to Malamalasso *Chev.* 17506. **Ghana:** Achimota (fr. Mar.) *Morton* GC 25401! Adabraka, Accra (fr. July) *Morton*! Kumasi *Cummins* 52! Tamale *Williams* 157! **Togo Rep.:** Lomé *Warnecke* 277! Misahöhe (fl. & fr. July) *Baumann* 210! **Dah.:** Pédjilé *Chev.* 22916. **N.Nig.:** Aguju (fl. & fr. Aug.) *Thornton*! Lokoja *Parsons* 66! Zaria (fl. & fr. Apr.) *Ryan* 50! Lemme (fl. & fr. July) *Lely* 151! Maifoni (fl. & fr. July) *Parsons*! **S.Nig.:** Agogo-Igun (fl. & fr. May) *MacGregor* 3! Idanre *Brenan & Jones* 8732! Aguku (fl. & fr. Jan.) *Thomas* 1393! Ibazo (fl. & fr. Nov.) *Thomas* 1989! Kuni (fl. & fr. Jan.) *Holland* 209! [Br.]**Cam.:** *Brown* 56! Mamfe (fl. & fr. Jan.) *Maitland* 1179! Bama, Dikwa Div. (fl. Jan.) *McClintock* 163! Throughout the tropics in the Old World. (See Appendix, p. 426.)

2. **H. subulatum** (*Hochst. ex A. DC.*) *Vatke* in Linnaea 43 : 316 (1882) ; Duthie in Fl. Gangetic Plain 2 : 91 (1917). *Tournefortia subulata* Hochst. ex A. DC. in DC. Prod. 9 : 528 (1845). *Heliotropium zeylanicum* of F.T.A. 4, 2 : 31 ; F.W.T.A., ed. 1, 2 : 199 ; Chev. Bot. 451 ; Berhaut Fl. Sén. 201, not of Lam. (1789). A coarse, erect branched perennial with tap-root, more or less glandular hairy, with small whitish tubular flowers in lax scorpioid spikes up to 9 in. long.
Sen.: *Perrottet* 526! Richard Tol *Roger* 109! Maka-Diama *Berhaut* 475. **Mali:** Féreibo (fl. & fr. Aug.) *Chev.* 2884! **Niger:** Zinder (fl. & fr. Dec.) *Hagerup* 583! **N.Nig.:** Katagum (fl. & fr. June) *Dalz.* 76! Kuka, L. Chad (fl. & fr. Feb.) *E. Vogel* 36! Kalkala, L. Chad *Golding* 43! Throughout tropical Africa, and in India. (See Appendix, p. 427.)

3. **H. ovalifolium** *Forsk.* Fl. Aegypt.-Arab., Descr. Pl. 38 (1775) ; F.T.A. 4, 2 : 34 ; Berhaut Fl. Sén. 201. *H. niveum* A. Chev. Bot. 450, name only. Diffuse densely white-villous stems from a woody base and tap-root, with small white flowers in cymes 2–3 in. long.
Sen.: *Berhaut* 1010. M'Bogosse *Chev.* 2885. *Sieber* 12! Dakar (fl. & fr. May) *Baldwin* 5712! Richard Tol (fl. & fr. June) *Döllinger* 56! **Mali:** Sébi *Chev.* 1167. Sompi *Chev.* 2882. Timbuktu (fl. & fr. May) *Hagerup* 209! (fl. & fr. Aug.) 249! Bossobougou (fl. & fr. July) *De Wailly* 5085! Sarédina *Davey* 108! 140! **Ghana:** Accra Plains (fl. & fr. June) *Morton* GC 9231! White Volta (May) *Kitson*! **N.Nig.:** Kuka (July) *E. Vogel*! Nupe *Barter*! Sokoto (fl. & fr. July) *Dalz.* 388! Mamu (fl. May) *Lely* 164! 165! Bani (fl. & fr. June) *Lean* 43! In tropical Africa generally, extending to Transvaal. (See Appendix, p. 426.)

4. **H. supinum** *Linn.* Sp. Pl. 130 (1753) ; F.T.A. 4, 2 : 37 ; Berhaut Fl. Sén. 201. Hairy much-branched herb, decumbent from a woody base.
Sen.: *Berhaut* 1053 ; 2357. **Mali:** Sarédina (fl. & fr. Mar.) *Davey* 94! 95! **Niger:** Zinder *Hagerup* 603! **N.Nig.:** Kuka, L. Chad. *E. Vogel* 12, partly! [Br.]**Cam.:** Bama, Dikwa Div. (fl. Dec.) *McClintock* 85! In N. & S. Africa, Canary Islands, S. Europe and eastern tropics.

5. **H. strigosum** *Willd.* Sp. Pl. 1, 2 : 743 (1798) ; F.T.A. 4, 2 : 41 ; Chev. Bot. 451 ; Berhaut Fl. Sén. 201. Perennial, branched, erect or spreading from a woody base up to 1 ft. high, with appressed white bristly hairs, and small white flowers ; in sandy soil, on farms, waste places etc.
Sen.: *Berhaut* 755 ; 2356. Richard Tol (fl. Sept., Nov.) *Roger* 106! 108! *Lelièvre*! **Gam.:** Kuntaur *Ruxton* 150! **Mali:** Tenétou *Chev.* 656. Sindou *Chev.* 871. Guiébili *Chev.* 1008. Beragungu *Hagerup* 311! **Port.G.:** Bafata (fl. & fr. Aug.) *Esp. Santo* 2721! **Iv.C.:** Mankono *Chev.* 21991! 21992. **Ghana:** Cape Coast *Don*! Accra Plains *Brown* 255! *T. Vogel* 6! 14! Achimota (fl. & fr. May, June) *Irvine* 685! 1430! Aburi (fl. & fr. Dec.) *Anderson* 30! Ashanti *Chipp* 501! Yendi (Mar.) *Hepper & Morton* A3078! **Togo Rep.:** Lomé *Warnecke* 19! **Dah.:** *Burton* s.n.! **N.Nig.:** Zaria (fl. & fr. July) *Keay* FHI 2597! Birnin Gwari (fl. & fr. June) *Keay* FHI 25879! Lokoja *Shaw* 37! Liruwen-Kano Hills *Carpenter*! Lemme (fl. & fr. May) *Lely* 124! Throughout tropical Africa and in Egypt and Arabia ; also in Australia. (See Appendix, p. 426.)

6. **H. bacciferum** *Forsk.* Fl. Aegypt.-Arab., Descr. Pl. 38 (1775). *H. undulatum* Vahl (1790)—F.T.A. 4, 2 : 37, partly ; Chev. Bot. 451 (partly ?) ; F.W.T.A., ed. 1, 2 : 199, partly ; Berhaut Fl. Sén. 201 (?)[2]; Schnell in Ic. Pl. Afr. I.F.A.N. 2, t. 40. Sub-erect or prostrate from a perennial stock, covered with white bulbous-based bristles, with small white flowers, nutlets 4, equal, free, finally glabrous.
Sen.: Dakar *Debeaux* 85! 189! (fl. & fr. May) *Roger* 40! Coté des Maures (fl. & fr. Jan.) *Döllinger*! M'Bidjem *Thierry* 51! **Mali:** Timbuktu *Hagerup* 149! Doura *Davey* 079! **N.Nig.:** Dabehi *Noble* A20! Kalkala *Golding* 73! Common in dry regions in N. Africa, Arabia and tropical Asia.

7. **H. pterocarpum** (*DC. & A. DC.*) *Hochst. & Steud. ex Bunge* in Bull. Soc. Imp. Natural. Moscou 42 (1) : 331 (1869). *Heliophytum pterocarpum DC. & A. DC.* Prod. 9 : 552 (1845). "*Heliotropium pterocarpum* Hochst. ex DC.''—F.T.A. 4, 2 : 35.[1] *H. undulatum* of F.T.A. 4, 2 : 37 ; of F.W.T.A., ed. 1, 2 : 199, partly ; Chev. Bot. 451 (partly ?) ; Berhaut Fl. Sén. 175, 201 (?)[2], not of Vahl. In habit like the last, but nutlets cohering in pairs and each pair margined with a broad horny wing, with a rounded sinus at the apex.
Sen.: *Perrottet*! Dakar (fl. & fr. Mar.) *Dalz.* 8409! **Mali:** Timbuktu (fl. & fr. July) *Hagerup* 155! **N.Nig.:** Kuka (fl. & fr. Jan.) *E. Vogel* 12, partly! Fodama (fl. & fr. Nov.) *Moiser* 229! 237! Bornu (fl. ; fr. Nov.) *Elliott* 122! Katagum *Dalz.* 74! Kalkala (fl. & fr. Jan.) *Parsons*! Jos Plateau (fl. & fr. Dec.) *Kennedy* 2848! Also in N.E. Africa and Arabia. (See Appendix, p. 426.)

8. **H. baclei** *DC. & A. DC.* Prod. 9 : 546 (1845), partly ; F.T.A. 4, 2 : 34 ; Johnston in Contrib. Gray Herb. 92 : 91 (1930) ; Berhaut Fl. Sén. 195, 200. *H. marifolium* of F.T.A. 4, 2 : 40, partly (W. Afr. specimens) ; Chev. Bot. 450, partly, not of Retz. Much-branched, decumbent herbs, slightly woody beneath. The following two varieties are only distinguishable by the key-characters given above.

8a. **H. baclei** *DC. & A. DC.* var. baclei.
Sen.: Ouassadou *Berhaut* 1690. *Heudelot* 874! **Gam.:** (fl. & fr. July) *Hayes* 543! Quoia *Bacle.* **Mali:** San *Chev.* 1063! 1069. Kabara (fl. & fr. Sept.) *Hagerup* 317! Day *Chev.* 1365. **Port.G.:** Pitche to Buruntuma (fl. June) *Esp. Santo* 2713! Catió (fl. & fr. May) *Esp. Santo* 2055! 2474! **Guin.:** Mamou to Kindia *Chev.* 13589. **S.L.:** Mesima (fl. & fr. Apr.) *Deighton* 3704! Makunde (fl. & fr. Apr.) *Sc. Elliot* 5715! Batkanu *Glanville* 152! 190! Gbundapi *McDonald* 4! Madina, Makari Ganta (fr. Apr.) *Bakshi* 152!

8b. **H. baclei** var. rostratum *Johnston* l.c. (1930). *H. katangense* Gürke in De Wild. Ann. Mus. Congo, Bot. Sér. 4, 223 (1903) ; F.T.A. 4, 2 : 34. *H. nigerinum* A. Chev. in Sudania 1 : 24 (1911) and Chev. Bot. 450, name only. *H. marifolium* of F.T.A. 4, 2 : 34, partly ; Chev. Bot. 450, partly, not of Retz.
Mali: " Dunes along streams " *Lécard* 42. Sébi (fl. & fr. July) *Chev.* 1168! Gumguber (fl. & fr. Sept.) *Hagerup* 309! Korienza (fl. & fr. June) *Lean* 52! 53! *Pitot* 2666! Mopti (fr. July) *Pitot* 1135! (fl. & fr. June) *Lean* 36! Scudégéré (fr. Sept.) *Pitot* 5033! Also in Congo, N. Rhodesia and Tanganyika.

6. TRICHODESMA R. Br.—F.T.A. 4, 2 : 44. *Nom. cons.*

Leaves (except the upper ones) fairly long-petiolate, ovate to broadly lanceolate, truncate to acute at the base, up to 12 cm. long and 7 cm. broad, scabrid on both surfaces with short bristles from a discoid base; cymes lax and few-flowered; calyx-segments narrowly lanceolate, gradually acute, about 1 cm. long, with a conspicuously

---

[1] " *Heliotropium pterocarpum* Hochst. & Steud." in A. DC. Prod. 9 : 552 (1845) is a ms. (label)-name, invalidly published in synonymy.
[2] The above species, Nos. 6 and 7, are obviously confused in Berhaut's Fl. Sén.—None of the specimens quoted by Berhaut has been seen for this revision.

bristly midrib and margins ; anthers densely villous, with spirally connivent tips
    1. *africanum*
Leaves all sessile, lanceolate or ovate-lanceolate, narrowed at the base, about 5 cm.
    long and up to 2·5 cm. broad, tuberculate-scabrid ; cymes lax and few-flowered ;
    calyx-segments ovate-lanceolate, acutely acuminate, 1·5 cm. long, broadening in fruit ;
    anthers loosely villous   ..   ..   2. *ledermannii*

1. **T. africanum** (*Linn.*) *Lehm.* Pl. Asperif. 195 (1818)[1] ; F.T.A. 4, 2 : 48 ; Berhaut Fl. Sén. 71. *Borago africana* Linn. Sp. Pl. 138 (1753). *T. africanum* var. *homotrichum* Bornm. & Kneuck. in Allg. Bot. Zeitschr. 22 : 2 (1916) ; Brand in Engl. Pflanzenr. 4, 252 : 30 (1921). *T. fruticosum* Maire in Bull. Soc. Hist. Nat. Afr. Nord 39 : 136 (1949). Erect branched annual 2–3 ft. high, harshly scabrid, with flowers ¼ in. long white, the throat yellow and with brown spots.

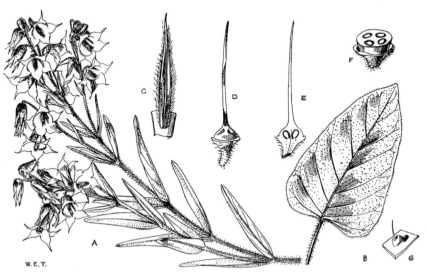

FIG. 277.—TRICHODESMA AFRICANUM (*Linn.*) *Lehm.* (BORAGINACEAE).
A, upper part of plant. B, lower leaf. C, anther. D, pistil. E, longitudinal, and F, cross-section of same. G, portion of leaf showing bulbous-based hair.

Sen.: *Perrottet* 547 ; 2006. Dakar *Maire*. Volcan de Mamelles, Cape Verde peninsula (fl. & fr. Dec.) *Jaeger* 5685 ! **Mali** : Bargoussi, Hombori (fl. & fr. Oct.) *Jaeger* 5485 ! Mopti (fl. & fr. Dec.) *De Wailly* 5305 ! Gao (fl. & fr. Sept.) *Hagerup* 344 ! **Ghana** : Nangodi (Mar.) *Hepper & Morton* A3764 ! **Dah.**: Garimana *Gironcourt* 101. **N.Nig.** : Borgu *Barter* 770 ! Kontagora (fl. & fr. Oct.) *Dalz.* 115 ! Bauchi (fl. & fr. May) *Lely* 169 ! Gujba (fl. Dec.) *Foster* 83 ! L. Chad (fl. Feb.) *E. Vogel* 43 ! Kalkala (fl. Mar.) *Golding* 46 ! Yola (Dec.) *Hepper* 1620 ! [**Br.**]**Cam.**: Bama, Dikwa Div. (fl. Nov.) *McClintock* 11 ! Also in Cape Verde Islands and Mauritania, extending to the coast of the Red Sea and to S. Africa. (See Appendix, p. 427.)
2. **T. ledermannii** *Vaupel* in Engl. Bot. Jahrb. 48 : 529 (1912) ; Brand in Engl. Pflanzenr. 4, 252 : 24 (1921). *T. physaloides* of F.W.T.A., ed. 1, 2 : 200, not of (Fenzl) A. DC. (1846). Stems herbaceous 1–2 ft. high from a woody rootstock, roughly hairy at first, with conspicuous bluish or white flowers 1–1½ in. across, spotted in the throat.
N.Nig.: Kaduna (fl. Nov.) *Lamb* 95 ! Maigana *Hill* 47 ! Jos Plateau (fl. Jan.) *Lely* P74 ! *Young* 176 ! Vom (fl. Dec.) *W. D. MacGregor* 378 ! Also in Cameroun.

Imperfectly known species.

T. uniflorum *Brand* in Fedde Rep. 12 : 504 (1913), and in Engl Pflanzenr. l.c. 38 ([**Br.**] **Cam.**: Johann-Albrechts-Hütte *Winkler* 1255. Victoria *Zahn* 530) is treated in F.W.T.A., ed. 1 as a synonym of *T. indicum* (L.) Lehm. (1818)[1], an introduced species from the Eastern Tropics. No material seen for the Revised Edition.

7. CYNOGLOSSUM Linn.—F.T.A. 4, 2 : 51 ; Brand in Engl. Pflanzenr. 4, 252 (1921).

Nutlets cohering at the apex with the style, not separated from the style at maturity :
    Spines on the nutlets spread all over the surface ; leaves broadly lanceolate, 4–6 cm.
    long, 1·5–2 cm. broad, setulose-scabrid above, sparingly so beneath ; cymes long and
    slender when in fruit ..   ..   ..   1a. *lanceolatum* subsp. *lanceolatum*
    Spines on the nutlet marginal and on the midrib ; leaves lanceolate, acutely acuminate,
    up to 14 cm. long and 3 cm. broad, rather closely pilose with bulbose-based hairs
    above, setose-pilose beneath ; cymes slender ; calyx-lobes oblong ; corolla slightly
    longer than calyx   ..   ..   1b. *lanceolatum* subsp. *geometricum*

[1] *T. africanum* R.Br., Prod. 496 (1810) and *T. indicum* R.Br. l.c. (1810) are binomials wrongly attributed to R. Brown, who (l.c.) only quoted *Borago africana* and *indica* as belonging to his newly established genus *Trichodesma* without making the necessary new combinations.—H. H.

Nutlets free before maturity, 7 mm. diam., smooth between the bristles on margin and
midrib ; leaves ovate, acuminate, narrowed at base with a short winged petiole,
6–12 cm. long, 4–7 cm. broad, pilose with discoid-based hairs on the upper surface,
softly and shortly pubescent beneath ; cymes few-flowered ; calyx-lobes broadly
ovate, setulose ; corolla nearly twice as long as calyx

2a. amplifolium forma macrocarpum

1. **C. lanceolatum** *Forsk.* Fl. Aegypt.-Arab., Descr. Pl. 41 (1775) ; F.T.A. 4, 2 : 54 ; Brand in Engl. Pflanzenr.
    4, 252 : 139 (1921) ; Chev. Bot. 452. A pilose herb 3–6 ft. high, with white or bluish flowers.
1a. **C. lanceolatum** *Forsk.* subsp. **lanceolatum.** *C. lanceolatum* Forsk. subsp. *eu-lanceolatum* Brand, l.c. (1921).
    Lib.: Cape Palmas (fl. & fr. Mar.) *T. Vogel* 18 ! **Iv.C.:** Guidéko *Chev.* 19099. **Ghana:** Lakte (fl. & fr. Oct.)
    Johnson 820 ! Amedzofe (fl. & fr. May) *Morton* A3662 ! **N.Nig.:** Jos Plateau (fl. & fr. May) *Lely* 275 ! **S.
    Nig.:** *Thomas* 1745 ! 1827 ! **[Br.]Cam.:** Bamenda (fl. & fr. Apr.) *Maitland* 1630 ! Jakiri (fl. & fr. Feb.)
    Hepper 1954 ! Widely distributed in the Old World tropics and subtropics.
1b. **C. lanceolatum** subsp. **geometricum** (*Bak. & Wright*) Brand l.c. 140 (1921). *C. geometricum* Bak. & Wright
    in F.T.A. 4, 2 : 52 (1905). *C. mannii* Bak. & Wright in F.T.A. 4, 2 : 52 (1905) ; F.W.T.A., ed. 1, 2 : 200.
    C. lanceolatum subsp. *geometricum* var. *mannii* (Bak. & Wright) Brand l.c. 140 (1921). *C. mannii* Bak. &
    Wright l.c. (1905). Erect herb, scabrid-pilose, up to 4–6 ft. high, with white flowers.
    [Br.]Cam.: Cam. Mt. (fl. & fr. Jan., Feb., June, July) *Deistel* 657 ! *Dalz.* 8274 ! *Migeod* 108 ! *Maitland*
    47 ! 1015 ! *Mann* 2005 ! *Hutch. & Metcalfe* 12 ! Oku (fl. & fr. Feb.) *Hepper* 2026 ! Nguroje, 5,500 ft.,
    Mambila Plateau (fl. & fr. Jan.) *Hepper* 1756 ! *Chapman* 67 ! Also in tropical E. Africa.
2. **C. amplifolium** *Hochst. ex A. DC.* in Prod. 10 : 149 (1846). Widespread, but in our area represented by :
2a. **C. amplifolium** forma **macrocarpum** Brand l.c. 141 (1921). *C. lancifolium* Hook. f. in J. Linn. Soc. 7 :
    207 (1864) ; F.T.A. 4, 2 : 53 ; F.W.T.A., ed. 1, 2 : 200. Erect herb up to 3 or 4 ft. high, hispid, with
    conspicuous bluish or white flowers 1–1½ in. across, spotted in the throat.
    [Br.]Cam.: Cam. Mt., 7,000–8,000 ft. (fl. & fr. Dec., fr. Jan.) *Mann* 1266 ! 2004 ! *Johnson* 10 ! 57 ! *Mait-
    land* 966 ! Bafut-Ngemba F.R., Bamenda (fl. & fr. Jan., Feb.) *Hepper* 2859 ! *Daramola* FHI 40628 !

---

8. **ARNEBIA** Forsk.—F.T.A. 4, 2 : 55 ; Johnston in J. Arn. Arb. 35 : 51 (1954).

Whole plant densely hispid with long stiff white hairs ; leaves linear-lanceolate, subacute, up to 8 cm. long and 1 cm. broad ; cymes scorpoid ; calyx-segments linear-
subulate, about 8 mm. long, densely setose ; corolla-tube narrow, 1 cm. long, lobes
slightly pubescent ; nutlets angular, rugose..   ..   ..   ..   *hispidissima*

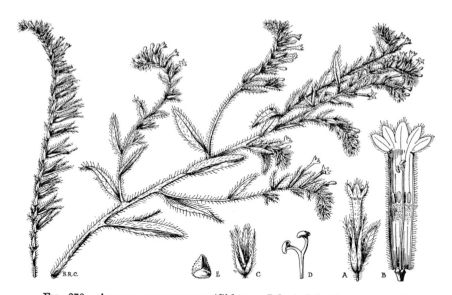

Fig. 278.—ARNEBIA HISPIDISSIMA (*Sieber ex Lehm.*) DC. (BORAGINACEAE).
A, flower. B, flower with corolla laid open. C, calyx and young fruit. D, style-arms. E,
nutlet.

A. hispidissima (*Sieber ex Lehm.*) *DC.* Prod. 10 : 94 (1846) ; F.T.A. 4, 2 : 56 ; Johnston in J. Arn. Arb. 35 :
    55 (1954). *Lithospermum hispidissimum* Sieber ex Lehm. (1821)—Johnston in J. Arn. Arb. 33 : 325
    (1952). *Arnebia asperrima* (Del.) Hutch. & Dalz. F.W.T.A., ed. 1, 2 : 201 (1931). *Anchusa asperrima*
    Del. (1813), name only. Much-branched from a hard woody base, sometimes short and bushy, the hairs
    almost pungent, with yellow tubular flowers ⅓ in. long and a blood-red root ; a plant of dry sandy soil.
    N.Nig.: Deli (fl. & fr. Apr.) *E. Vogel* 97 ! Katagum Dist. (fl. & fr. Feb.) *Dalz.* 75 ! **[Br.]Cam.:** Bama, Dikwa
    Div. (fl. & fr. Feb.) *McClintock* 2 ! Extends through N. tropical Africa to N. India. (See Appendix, p. 424.)

---

9. **MYOSOTIS** Linn.—F.T.A. 4, 2 : 57 ; Stroh in Beih. Bot. Cbl. 61B : 317 (1941).

Racemes with a leafy bract subtending each flower, short; stems dwarf; leaves
narrowly oblong-lanceolate, the larger about 2 cm. long and 6 mm. broad, pilose-

setose on both surfaces ; calyx-lobes narrowly linear-lanceolate, 3–4 mm. long
                                                                        1. *abyssinica*
Racemes without leaf bracts, slender ; stems up to 60 cm. high ; leaves linear-lanceolate,
acute, the larger about 5 cm. long and 1 cm. broad, thinly pilose on both surfaces ;
pedicels up to 5 mm. long in fruit, shortly pubescent ; calyx-lobes lanceolate, pubescent   ..   ..   **2. sp. nr. *vestergrenii***

1. **M. abyssinica** *Boiss. & Reut.* in Boiss. Diagn. pl. or., n. sér. 1, 11 : 122 (1849) ; F.T.A. 4, 2 : 58 ; Stroh
   l.c. 334 ; Hedberg in Symb. Bot. Upsal. 15, 1 : 157 (1957). Herb a few inches high, harshly pubescent,
   with small white flowers.
   [Br.]Cam.: Cam. Mt., 8,000–10,000 ft. (fl. & fr. Nov.) *Mann.* 2033 ! **F.Po** : 8,500 ft. *Mann* 1459 ! Also
   in E. Africa from Ethiopia to Tanganyika.
2. **M. sp. nr. vestergrenii** Stroh l.c. 328 (1941) ; Hedberg l.c. 158, 316. *M. sylvatica* of F.T.A. 4, 2 : 57, partly
   (*Johnston* 72) ; of F.W.T.A., ed. 1, 2 : 201, not of Hoffm. Erect 1–2 ft. high, thinly hairy, with bright
   blue flowers.
   [Br.] **Cam.**: Cam. Mt., 7,000–8,000 ft., upper forest (Dec., Feb.) *Johnston* 72 ! *Maitland* 1342 ! *Morton*
   K852 !

151. SOLANACEAE

By H. Heine

Herbaceous or woody. Leaves alternate, simple ; stipules absent. Flowers
hermaphrodite, mostly actinomorphic. Calyx 4–6-lobed, persistent. Corolla
gamopetalous, usually 5-lobed, lobes folded, contorted or valvate. Stamens
inserted on the corolla-tube and alternate with its lobes ; anthers 2-celled, cells
parallel, opening lengthwise or by apical pores. Ovary 2-celled, the cells sometimes again divided by a false septum ; style terminal. Ovules very numerous,
axile. Fruit a capsule or berry. Seeds with copious endosperm and curved or
annular embryo.

Generally distributed, but most numerous in tropical regions.

Fruit a capsule :
  Capsule 4-valved ; stamens 5 ; leaves and flowers large   ..   ..   .. **1. Datura**
  Capsule 2-valved :
    Stamens 5 ; flowers medium-sized   ..   ..   **2. Nicotiana**
    Stamens 2 ; flowers very small   ..   ..   **3. Schwenckia**
Fruit a berry :
  Anthers opening by slits lengthwise :
    Calyx not enlarged in fruit or not strikingly so :
      Corolla rotate, valvate   ..   ..   **4. Capsicum**
      Corolla urceolate, induplicate-valvate   ..   ..   **5. Discopodium**
    Calyx enlarged and inflated in fruit :
      Flowers solitary   ..   ..   **6. Physalis**
      Flowers fasciculate   ..   ..   **7. Withania**
  Anthers opening by terminal pores   ..   ..   **8. Solanum**

Cyphomandra betacea (Cav.) Sendtner (in Flora 28 : 172 (1845) ; Sandwith in Chron. Bot. 4 : 225 (1938),
for full syn. ; *Solanum betaceum* Cav. (1799)). The tree-tomato. This species, a native of Peru, is cultivated
and often becomes naturalized in many parts of the tropics. In Africa it occurs particularly in the E. and
SE. Attention is drawn to the first record from W. Africa (Cam. Mt., Mann's Spring, fl. & fr. Dec. 1959, *Morton*
K852 : " small tree 15 ft. high in moist forest, fl. very pale pinkish, fr. plum-shaped, orange ") as it may soon
become widely naturalized in our area.

1. DATURA Linn.—F.T.A. 4, 2 : 258 ; Fosberg in Taxon 8 : 52 (1959) ;
  Satina, Avery & Sachet in Blakeslee, The Genus Datura 16–47 (1959).

Flowers 25–30 cm. long, pendulous ; calyx not circumscissile, either persisting like a
  husk appressed to the fruit or spathiform ; fruits unarmed and indehiscent, spheroid,
  ellipsoid or fusiform, terete ; ovate-acuminate ; leaves large, up to 24 cm. long and
  12 cm. broad, ovate-acuminate, leaf-margin mostly entire :
  Calyx spathiform, terminating in a point, finely pubescent outside, appressed to the
    funnel-shaped lower part of the corolla-tube ; principal nerves of the corolla termi-
    nating in 1–2 cm. long tail-like appendages between which the margin is entire or
    rounded ; anthers free ; fruit lemon-shaped   ..   .. **1. candida**
  Calyx 5-toothed at the apex, nearly glabrous, very inflated and not appressed to the
    very thin pipe-like lower part of the corolla-tube ; principal nerves of the corolla
    terminating in very short (0·5 cm. long) appendages ; anthers coherent ; fruit
    fusiform..   ..   ..   **2. suaveolens**
Flowers up to 20 cm. long, erect ; calyx circumscissile near the base, the base persistent
  and expanding like a frill, shield or cup, calyx-tube falling off together with the
  corolla ; fruit ovoid or globular, armed and dehiscent ; leaves up to 16 cm. long and

8·5 cm. broad, ovate or triangular-ovate, leaf-margin frequently sinuate-repand, more or less dentate :

Petioles and branchlets densely and softly grey-pubescent or pilose ; leaves ovate, unequal, rounded or truncate at the base, 8–20 cm. long, 4–14 cm. broad, entire, thinly pubescent beneath ; flowers solitary, usually axillary ; pedicels 1·5 cm. long, densely pilose ; calyx tubular, about 9 cm. long ; corolla narrowly trumpet-shaped, up to 18 cm. long, 10-lobed, lobes broadly triangular ; capsule pendulous on the elongated pedicel, prickly, globose, 3–4 cm. diam., dehiscence irregular

3. *innoxia*

Petioles and branchlets glabrous ; corolla 5–6 lobed, lobes long-cuspidate or long-acuminate :

Capsule pendulous, dehiscing irregularly ; leaves elliptic to broadly ovate or triangular-ovate, up to 20 cm. long, 14 cm. broad, subentire or with a few dentate lobes ; pedicels 1 cm. long ; calyx tubular, 6 cm. long, 1·5 cm. diam. ; corolla often double, up to 20 cm. long, the 5–6 lobes long-cuspidate ; capsule covered with blunt warts .. .. 4. *metel*

Capsule erect, dehiscing by 4 valves ; leaves rather narrowly ovate, up to 24 cm. long, 10 cm. broad, more or less coarsely dentate ; pedicels very short ; calyx tubular, up to 3·5 cm. long, 4 mm. diam. ; corolla about 8 cm. long, the 5–6 lobes long-acuminate ; capsule densely prickly .. .. 5. *stramonium*

1. **D. candida** (*Pers.*) *Safford* in J. Wash. Acad. Sci. 11 : 182 (1921) ; Pasquale in Cat. R. Orto Bot. Napoli 36 (1867), not validly published. *Brugmansia candida* Pers. (1805). *Datura arborea* of Ruiz & Pavon Fl. Peruv. 2, 15 : 127 (1799), not of Linn. (1753) ; Fosberg l.c. 52 ; Sachet l.c. 41. A handsome shrub or small tree ; flowers white, very fragrant.
Ghana : Mampong Scarp (Aug.) *Darko* 712 ! [Br.]**Cam.**: Buea (Feb., Nov.) *Migeod* 148 ! *Keay* FHI 37545 ! Bamenda (Jan., Apr.) *Ujor* FHI 30035 ! *Daramola* FHI 40607 ! Ntim (Feb.) *Hepper* 1858 ! Known only in cultivation ; believed to be native of the Andean region of South America, now spreading widely in tropical and subtropical countries.
[Note : P. C. Standley states in Publ. Field Mus. of Nat. Hist. Bot. 18, 3 : 1055 (1938) : "In recent years this species has been listed frequently as *D. candida* (Pers.) Pasqu., the specific name under which it was treated by Safford, but it now appears probable that the name *D. arborea* L. is the correct one." I cannot accept this brief discussion as sufficient for a taxonomic conclusion.—H. H.]

2. **D. suaveolens** *Humb. & Bonpl. ex Willd.* Enum. Pl. Berol. 227 (1809) ; Safford, l.c. 185 ; Sachet l.c. 46. *D. gardneri* Hook. Bot. Mag. t. 4252 (1846). A shrub or small tree like the last ; flowers white, fragrant.
S.L.: Regent *Sc. Elliott* 4107 ! **Ghana** : Agogo (Apr.) *Vigne* FH 1909 ! [Br.]**Cam.**: Buea (Nov.) *Migeod* 113 ! A native of Brazil, now much cultivated in the tropics.

3. **D. innoxia** *Mill.* Gard. Dict., ed. 8, no. 5 (1768), as *inoxia* ; Safford l.c. 179 (1921) and Ann. Rep. Smithsonian Inst. 1920 : 545, t. 3, fig. 4 (1922) ; Fosberg l.c. 55 ; Satina & Avery l.c. 25 ; Sachet l.c. 44. *D. metel* of F.T.A. 4, 2 : 256 ; F.W.T.A., ed. 1, 2 : 202 ; Chev. Bot. 468 ; Berhaut Fl. Sén. 199, not of Linn. Erect-branched herbaceous undershrub 2–4 ft. high, with long white flowers and long spiny spherical fruit ; in waste places.
Sen.: *Boivin* ! *Chev. Berhaut* 25. **Port.G.**: Cachen Mator (fl. & fr. Feb.) *Esp. Santo* 1312 ! **Guin.**: Farana (fl. & fr. Mar.) *Sc. Elliot* 5387 ! **N.Nig.**: Lokoja *Shaw* 41 ! Katagum *Dalz.* 93 ! Widespread weed in the tropics ; native of tropical S. America.
[Note : The spelling "inoxia", used by Miller (l.c.) and Fosberg and subsequent authors, is orthographically incorrect ; it is not used here as Miller based this epithet upon the pre-Linnean polynomial of Boerhaave (Ind. alt. hort. acad. Lugd.-Bat. 1 : 262, 1720) in which the correct spelling "innoxia" is used. It seems hardly justifiable to perpetuate the incorrect spelling.—H. H.]

4. **D. metel** *Linn.* Sp. Pl. 1 : 179 (1753) ; Safford in J. Wash. Acad. Sci. 11 : 178, t. 1, B, C (1921) and Ann. Rep. Smithsonian Inst. 1920 : 546, t. 1–2 (1922) ; Satina & Avery, l.c. 32 ; Sachet l.c. 44. *D. fastuosa* Linn. (1759)—F.T.A. 4, 2 : 256 ; F.W.T.A., ed. 1, 2 : 202 ; Chev. Bot. 467 ; Berhaut Fl. Sén. 192. *D. alba* Nees in Trans. Linn. Soc. 17 : 73 (1837) ; Dunal in DC. Prod. 13, 1 : 541 (1852) ; Fosberg l.c. 53 ; *D. fastuosa* var. *alba* (Nees) C. B. Cl. in Hook. f., Fl. Brit. Ind. 4 : 243 (1884) ; F.T.A. 4, 2 : 257 ; Chev. Bot. 468. Habit of the last, 2–3 ft. high, with white or wine-purple, sometimes (in cultivation) double flowers.
Sen.: Dakar (Jan.) *Chev.* 2874 ! *Berhaut* 905. **Guin.**: *Poisson* ! **S.L.**: Kanike (Sept.) *Thomas* 2058 ! 9333 ! Roruks (fl. & fr. Nov.) *Thomas* 5754 ! Gorongo (fl. Dec.) *Deighton* 267 ! Serabu (fl. & fr. Apr.) *Deighton* 1724 ! Njala (fl. & fr. Nov.) *Deighton* 2807 ! **Iv.C.**: Abinta *Chev.* 17728. Tabou to Beriby *Chev.* 19509 ; 19510. Grabo to Taté *Chev.* 19781. **Ghana** : Cape Coast (fl. & fr. May) *Morton* A3667 ! Accra *Moloney* ! 113 ! Accra to Taté *Chev.* 19781. **Dah.**: Djougou *Chev.* 23862. **S.Nig.**: Aponmu, Akure Dist. (Oct.) *Ujor* FHI 26195 ! Extends through the Old World tropics. (See Appendix, p. 429.)

5. **D. stramonium** *Linn.* Sp. Pl. 179 (1753) ; F.T.A. 4, 2 : 257 ; Chev. Bot. 468 ; Satina & Avery l.c. 18 ; Sachet l.c. 45 ; Berhaut Fl. Sén. 199 ; Aké Assi Contrib. 1 : 177 . *D. tatula* Linn. Sp. Pl., ed. 2, 1 : 256 (1762). Erect branched annual with smooth stems 1–2 ft. or more high, and rather slender white (*D. stramonium* Linn. proper) or purple-violet (*D. tatula* Linn. proper) flowers. The flower-colour is of no taxonomic importance.
Mali : Kabarah *Chev.* 1363. **Ghana** : Accra (fl. June) *Dalz.* 118 ! (fr. May) *Morton* A884 ! Assin Yan Kumasi *Cummins* 8as ! **Togo Rep.**: Lomé *Warnecke* 364 ! **N.Nig.**: Kuka (fl. & fr. Feb.) *E. Vogel* 56 ! [Br.]**Cam.**: Bamenda (fr. July) *Daramola* FHI 41588 ! Cosmopolitan. (See Appendix, p. 429.)

2. NICOTIANA Linn.—F.T.A. 4, 2 : 259 ; Goodspeed, The Genus Nicotiana (1954).

Corolla about 4 cm. long, 3–4 mm. diam. in lower half of the tube, lobes very acute ; upper leaves oblong-lanceolate to elliptic, apex acute, decurrent at the base, 8–15 cm. long, 1·5–6 cm. broad, sessile or subsessile ; inflorescence a terminal cyme ; pedicels glandular-pubescent, 1·5 cm. long ; calyx about 1·5 cm. long, lobes long-acuminate ; capsule ovoid or ellipsoid, 2 cm. long, surrounded by the persistent calyx, and with a short apical beak .. .. 1. *tabacum*

Corolla not more than 2 cm. long, 6–7 mm. diam., lobes rounded or subacute ; upper leaves broadly ovate or oblong-elliptic rounded or subacute at the apex and base, 4–10·5 cm. long, 2–7 cm. broad, petioles up to 4 cm. long ; inflorescence and pedicels as above ; calyx about 1 cm. long, lobes broadly triangular, acute ; capsule globose, about 1·2 cm. diam., without an apical beak   ..   ..   ..   2. *rustica*

1. **N. tabacum** *Linn.* Sp. Pl. 180 (1753) ; F.T.A. 4, 2 : 259 ; Chev. Bot. 468 ; Berhaut Fl. Sén. 203 ; Goodspeed l.c. 372. Robust annual up to 6 ft. high, with long tubular white, pinkish, or sometimes cream flowers, viscid-glandular outside.
   Sen.: Dakar *Chev.* 2875 ! **Mali:** Faraba *Chev.* 1020. **Guin.:** Kouria *Caille* in *Hb. Chev.* 14757. Koulikoro *Chev.* 20644 ; 25025. **S.L.:** Freetown (fl. & fr. Feb.) *Johnston* 72 ! Kuntara (fr. June) *Thomas* 434 ! Kanya (fl. Oct.) *Thomas* 3005 ! Njala *Deighton* 2454 ! **Iv.C.:** Mankono *Chev.* 21841 ; 21924. Bouaké *Chev.* 22110. **Ghana:** (fl. & fr. Feb.) *Brown* FHI 2164 ! **Dah:** Agouagon *Chev.* 23511 ! **Niger:** Diapaga to Fada *Chev.* 24473. **N.Nig.:** *Robins* ! *Barter* 1345 ! Manga *Dudgeon* 60 ! **S.Nig.:** Ulepelu (fr. June) *Umana* FHI 29131 ! Ukungu (Jan.) *Thomas* 2200 ! **[Br.]Cam.:** Bamenda (fr. May) *Ujor* FHI 30338 ! Kumba (Feb.) *Binuyo & Daramola* FHI 35567 ! Widely cultivated tobacco in warmer parts of the world ; native of S. America. (See Appendix, p. 430.)

2. **N. rustica** *Linn.* l.c. 180 (1753) ; F.T.A. 4, 2 : 260 ; Berhaut Fl. Sén. 203 ; Goodspeed l.c. 351. Herb, branched up to 4 ft. high, with short greenish-yellow flowers villous outside.
   Mali: Gao (fl. & fr. Feb.) *De Wailly* 5343 ! Kangaba *Chev.* 268. Simona *Chev.* 976 ! **Guin.:** Soucourala *Chev.* 25014. **S.L.:** Smythe 128 ! Kabula (fr. Feb.) *Tindall* 56 ! Yetaya *Thomas* 2288 ! 2426 ! **Ghana:** *Johnston* ! **[Br.]Cam.:** Bafawchu (fr. May) *Ujor* 30339 ! Kakara, 4,900 ft., Mambila Plateau (fl. & fr. Jan.) *Hepper* 2823 ! Cultivated tobacco ; native of S. America. (see Appendix, p. 430.)

3. SCHWENCKIA D. van Royen ex Linn.—F.T.A. 4, 2 : 20, as *Schwenkia* Linn.; Heine in Kew Bull. 16 : 465 (1963).

Herb, glabrous or sparingly pubescent, branching freely ; lower leaves petiolate, elliptic to ovate, up to 4 cm. long and 2·5 cm. broad, rounded to subacute, entire ; upper leaves becoming narrowly oblong, sessile ; inflorescence a lax panicle, pedicels 0·5 cm. long ; calyx-teeth triangular, tube cylindric, 3 mm. long ; corolla narrowly tubular, 7–9 mm. long, lobes broadly and obtusely triangular, two rather larger than the rest, alternating with club-shaped processes ; fruit a subglobose capsule 4 mm. diam., surrounded by the persistent calyx and dehiscing by two valves   ..   *americana*

S. americana *Linn.* Gen. Pl., ed. 6, 577 (error " 567 ") (1764) ; F.T.A. 4, 2 : 260 ; Chev. Bot. 469 ; Berhaut Fl. Sén. 203. *S. hirta* of C. H. Wright in F.T.A. 4, 2 : 260, not of Klotzsch ; Chev. Bot. 469. (?) *S. guineensis* Schum. & Thonn. (1827). Erect slender-branched herb 1–3 ft. high ; flowers greenish-yellow ; weed of cultivation.
   Sen.: *Berhaut* 1126. **Mali:** (fl. & fr. Apr.) *Leprieur* ! Kouroussa (Dec.) *Chev.* 390 ! Dakar (fr. May) *Baldwin* 5729 ! Tabacco *Chev.* 135. **Guin.:** Dalaba-Diaguissa *Chev.* 18870 ; 18897. Boola to Moribadou *Chev.* 20942. Conakry (fl. & fr. June) *Debeaux* 421 ! **S.L.:** Kitchom *Sc. Elliot* 4322 ! Leicester *Sc. Elliot* 3887 ! Kumrabai (Dec.) *Thomas* 6783 ! Lumley beach (fl. Dec.) *Deighton* 790 ! Kambia (fl. Dec.) *Deighton* 893 ! Musaia (fr. Apr.) *Deighton* 5381 ! Talla Hills *Sc. Elliot* 5087a ! **Lib.:** Sinkor (fl. & fr. Sept.) *Barker* 1063 ! Bushrod Is. (fl. & fr. Apr.) *Barker* 1295 ! **Iv.C.:** Adiopodoumé (fr. Sept.) *Roberty* 12208 ! Brafouédi (fr. Dec.) *de Wit* 7252 ! Mt. Nimba (fl. & fr. June) *P. Gruys* 42 ! **Ghana :** Cape Coast Castle (fr. July) *T. Vogel* 12 ! Accra *Brown* 308 ! Achimota (fl. & fr. May) *Irvine* 696 ! Pong Tamale (fl. July) *Williams* 827 ! **Dah.:** Allado *Chev.* 23377. **N.Nig.:** R. Niger (Quorra) (fl. Sept.) *T. Vogel* 193 ! Nupe *Barter* ! Sokoto Prov. (June) *Dalz.* 546 ! Zaria (fl. & fr. Mar.) *Hill* 41 ! Abinsi (fl. & fr. July) *Dalz.* 722 ! Yola (fl. & fr. July) *Dalz.* 118 ! **S.Nig.:** Lagos (fr. Apr.) *Millen* 100 ! *Dawodu* ! Ibadan (Mar.) *Schlechter* 12332 ! Aguku Dist. *Thomas* 1392 ! **[Br.] Cam.:** *Mildbr.* 9449 ! Bamenda, 5,000 ft. (fl. & fr. Apr.) *Maitland* 1647 ! Gurum, Adamawa (fl. & fr. Nov.) *Hepper* 1315 ! Throughout Africa, and in tropical America. (See Appendix, p. 432.)

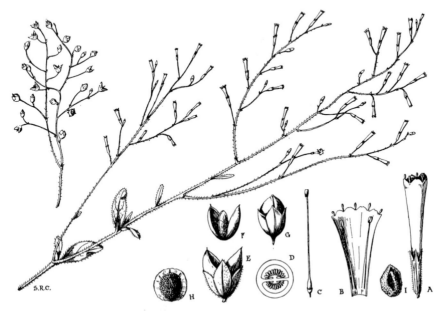

Fig. 279.—Schwenckia americana *Linn.* (Solanaceae).
A, flower.   B, corolla laid open.   C, pistil.   D, cross-section of ovary.   E, F, G, fruit.   H, I, seed.

4. CAPSICUM Linn.—F.T.A. 4, 2 : 250 ; Smith & Heiser in Am. J. Bot. 38 : 362 (1951).

Annual or biennial herbs ; leaves broadly lanceolate to ovate, apex acutely acuminate, cuneate or abruptly acute at the base, 5–8 cm. long, 2–5 cm. broad, entire, sub-glabrous ; flowers in 3–8-flowered axillary clusters, producing usually only one fruit ; pedicels up to 1·5 cm. long, swollen in the upper part ; calyx obscurely 5-toothed, 10-ribbed ; corolla rotate-campanulate, deeply 5-lobed, 5 mm. diam. ; fruit a more or less elongated berry, subglobose to acutely conical, 1–several cm. long   1. *annuum*
Suffruticose perennial ; very similar to the above, but readily distinguished by its more or less shrubby habit ; calyx truncate ; fruits usually 2 or more from each flower-cluster, much smaller and rarely exceeding 2 cm. in length   ..   2. *frutescens*

1. **C. annuum** *Linn.* Sp. Pl. 188 (1753) ; F.T.A. 4, 2 : 251 ; Chev. Bot. 466. Stout herb 2–5 ft. high, much branched, angular, glabrous stems ; flowers white or greenish ; fruits red ; cultivated.
    Mali : San *Chev.* 1035. **S.L.**: Freetown (fr. Jan.) *Johnston* 51 ! Matotoka *Thomas* 1339 ! Njala (Feb.) *Deighton* 1883 ! **Iv.C.**: *fide* Chev. *l.c.* **N.Nig.**: *Barter* 522 ! *Yates* 65 ! Lokoja (fl. & fr. Sept.) *Parsons* 85 ! **S.Nig.**: Sapoba *Kennedy* 2617 ! Obompa (fr. Jan.) *Thomas* 2236 ! Port Harcourt (fr. Sept.) *Stubbins* 46 ! Widely dispersed throughout the tropics, cultivated and sometimes naturalized. (See Appendix, p. 427.)
2. **C. frutescens** *Linn.* l.c. 189 (1753) ; F.T.A. 4, 2 : 251 ; Chev. Bot. 467. Undershrub 2–3 ft. high, much-branched, angular, nearly glabrous stems ; flowers white or pale yellow.
    Sen.: *Roger* ! Cape Verde *T. Vogel* ! **Mali** : Kati *Chev.* 177. Siguiri *Chev.* 300. **S.L.**: *Glanville* 403 ! Kanitole (fr. Sept.) *Thomas* 2067 ! Yonibana (fr. Nov.) *Thomas* 4718 ! Hangha (Jan.) *Thomas* 7780 ! Njala (fl. & fr. Feb.) *Deighton* 1884 ! **Iv.C.**: Sassandra (fr. Feb.) *Leeuwenberg* 2888 ! **Ghana** : *Irvine* 457 ! Kumasi *Cummins* 152 ! **Togo Rep.**: Lomé *Warnecke* 356 ! **N.Nig.**: Sapoba *Kennedy* 2615 ! 2616 ! 2618 ! **S.Nig.**: Obompa (fr. Jan.) *Thomas* 2231 ! Lagos (fl. & fr. Sept.) *Dodd* 439 ! [**Br.**]**Cam.**: Gwoza, Dikwa Div. (fr. Dec.) *McClintock* 67 ! Widely dispersed throughout the tropics, cultivated and sometimes naturalized. (See Appendix, p. 427.)

[It is impossible to give here an account of the very difficult taxonomy of this critical genus. The present writer therefore, feels unable to deal here with the various hybrids and infraspecific taxa (mainly cultivars) within the genus, or to recognize *C. sinense* Jacq. as a distinct species occurring in our area (as stated by J. Y. Wilson in Nature 183 (4) : 1142 (1959)—H. H.]

5. DISCOPODIUM Hochst.—F.T.A. 4, 2 : 253.

Small tree or shrub, 15–20 ft. high ; leaves elliptic to oblong-elliptic, broadly acuminate, cuneate at the base, 10–25 cm. long, 3–10 cm. broad, entire, glabrous or slightly pubescent on the midrib and 10–12 pairs of lateral nerves beneath ; petioles 1–2 cm. long ; flowers fasciculate, axillary ; pedicels 1–1·5 cm. long ; calyx 4 mm. long, lobes broadly triangular, spreading ; corolla cylindrical, about 8 mm. long, tomentellous outside above the calyx, lobes more or less reflexed or spreading, half as long as the tube ; stamens and style included ; berry globose, 6–8 mm. diam.   ..   *penninervium*

D. penninervium *Hochst.* in Flora 1844 : 22 ; F.T.A. 4, 2 : 253 ; Robyns Fl. Parc Nat. Alb. 2 : 203 (1947) *D. penninervium* var. *sparsearaneosum* Bitter in Engl. Bot. Jahrb. 57 : 17 (1920). Flowers white or yellowish fading to brown.

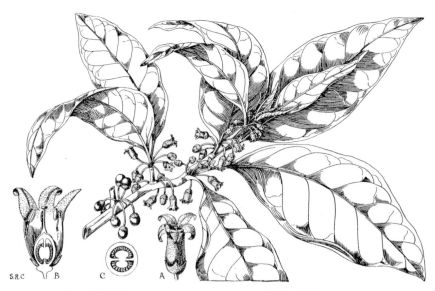

Fig. 280.—DISCOPODIUM PENNINERVIUM *Hocket.* (SOLANACEAE).
A, flower. B, longitudinal section of flower. C, cross section of ovary.

[Br.]Cam.: Cam. Mt., 7,000–7,600 ft. (Feb., Mar., Dec.) *Mann* 1236! 2178! *Brenan* 9375! *Maitland* 1332! Mt. Mbakakeka, 7,900 ft., Bafut-Ngemba F.R., Bamenda (fl. Jan., fl. & fr. Feb.) *Hepper* 2166! *Keay & Lightbody* FHI 26288! 28365! **F.Po:** 4,000 ft. *Mann* 300! Also in Cameroun, Congo and E. Africa from Ethiopia to Nyasaland, a typical highland species. Some of the E. African plants are considered as varietally distinct by Bitter (l.c. 16 (1920)).

6. **PHYSALIS** Linn.—F.T.A. 4, 2 : 246 ; O. E. Schulz in Urban, Symb. Antill. 6 : 140 (1909) ; Waterfall in Rhodora 60: 107–114, 128–142, 152–173 (1958).

Plants glabrous or nearly so (sometimes sparingly villose in No. 1) :
    Leaves subentire, ovate 1·5–3(–5) cm. long, acuminate at apex, cuneate at base ; fruiting calyx sub-globular, up to 15 mm. long and 12 mm. diam., finely puberulous ; pedicels puberulous, up to 3 cm. long; corolla very small, up to 4 mm. long
                                                                                                      1. *micrantha*
    Leaves coarsely sinuate-dentate, ovate, 5–10 cm. long, acuminate at apex, broadly cuneate to rounded at base ; fruiting calyx ovoid, up to 3 cm. long and 2 cm. broad, glabrous or very finely puberulous ; pedicels glabrous, up to 1 cm. long, corolla up to 8·5 mm. long ..   ..   2. *angulata*
Plants villous or appressed-strigose :
    Perennial, the whole plant densely villous ; leaves entire or with a few large teeth, rhomboid to deltoid, acuminate at apex, broadly rounded, truncate or cordate at base, 8–10 cm. long, 6–7·5 cm. broad ; pedicels up to 5–8 mm. long, not becoming longer in fruit ; corolla 15 mm. long ; fruiting calyx ovoid, about 4 cm. long and 3 cm. broad, villous ..   ..   3. *peruviana*
    Annual, the lower parts glabrescent, upper parts and pedicels appressed-strigose ; leaves coarsely and irregularly sinuate-dentate, ovate to suborbicular, acuminate at apex, broadly rounded to cordate at base, 2–5 cm. long ; flower-pedicels 5–8 mm. long, later 2–3 cm. long ; corolla 8–10 mm. long ; fruiting calyx turbinate, i.e. ovoid, abruptly acute, about 3 cm. long and 2·2 cm. diam., nearly glabrous
                                                                                                      4. *pubescens*

1. **P. micrantha** *Link* Enum. Pl. Hort. Berol. 1 : 181 (1821) ; O. E. Schulz l.c. 147 (1909), for full synonymy. (?) *P. divaricata* D. Don Prod. Fl. Nep. 97 (1825) ; Baldwin & Speese in Bull. Torrey Bot. Club 78, 3 : 255 (1951). *P. minima* of F.T.A. 4, 2 : 246, of Chev. Bot. 465, and of Berhaut Fl. Sén. 175, not of Linn. *P. angulata* of F.W.T.A., ed. 1, 2 : 205, partly. A very variable annual, stems more or less prostrate or sometimes erect up to 3 ft. high ; flowers cream ; calyx in fruiting state bladder-like.
Sen.: *Perrottet* 553! 567! *Roger* 132! *Berhaut* 494. **Gam.:** *Saunders* 68! **Mali:** Sikasso *Chev.* 748. Djenné *Chev.* 2881. **S.L.:** Njala (Apr.) *Thomas* 100! (fr. Mar.) *Deighton* 542! Mange Bureh (fl. & fr. May) *Deighton* 1044! Roboli, Rokupr (fl. & fr. July) *Jordan* 286! **Lib.:** Sinoe R. *Johnston*! Dukwai R. (fl. & fr. Oct.) *Cooper* 5! Jabroke (fr. July) *Baldwin* 6433! Begwai (fl. & fr. Oct.) *Bunting* 201! Sanokwele (fr. Sept.) *Baldwin* 13213! **Iv.C.:** Soubré *Chev.* 19137. **Ghana:** " Guinea " *Isert* (*fide* O. E. Schulz *l.c.*) Achimota (fr. Aug.) *Irvine* 854! Akroso (fr. Aug.) *Stowes* 949! Bomase (Apr.) *A. S. Thomas* D182! Nangodi (fr. Dec.) *Adams & Akpabla* 4294! **Togo Rep.:** Lomé *Warnecke* 192! **Niger:** Diapaga to Fada *Chev.* 24493 bis. Niamey (fl. & fr. Oct.) *Hagerup* 536! **N.Nig.:** Barter 1493! Katagum (fr. Sept.) *Dalz.* 90! Sokoto *Lely* 110! Lokoja (fr. Oct.) *Parsons* L47! Idah *T. Vogel* 66! **S.Nig.:** Ibadan (fl. & fr. Aug.) *Onochie* FHI 7586! Owo *Burton*! Onitsha (fr. Oct.) *Thomas* 1867! Agege (fr. Oct.) *Baldwin* 14531! [Br.]Cam.: Dawo, Vogel Peak, Adamawa (fr. Nov.) *Hepper* 1409! Pantropical (?) ; but owing to the complexity of the genus and the lack of a monographic treatment, neither distribution nor taxonomic position properly known.
2. **P. angulata** *Linn.* Sp. Pl. 183 (1753) ; F.T.A. 4, 2 : 249 ; O. E. Schulz l.c. 140, for full synonymy ; Berhaut Fl. Sén. 174, 195 ; Baldwin & Speese l.c. 255. *P. minima* of F.T.A. 4, 2 : 247, partly, not of Linn. Erect glabrous annual up to 3 ft. high ; flowers cream.
Sen.: *Perrottet*! *Berhaut* 450. **Gam.:** *Hayes* 505! **Mali:** Kouroussa (fr. Aug.) *Pobéguin* 369! **S.L.:** Mgamba (fr. Jan.) *King* 136B! Waterloo *Deighton* 2068! Kortright (fr. Feb.) *Gledhill* 10! Jigaya (fr. Sept.) *Thomas* 2763! Yonibana (Nov.) *Thomas* 4866! **Lib.:** Monrovia (fl. & fr. July) *Linder* 17! Ganta *Harley*! Harbel (fr. June) *Baldwin* 5934! Gbanga (Sept.) *Linder* 588! Ghana: *T. Vogel* 51! Accra (fr. Apr.) *Bally* 22! Owabi (fr. Apr.) *Andoh* 4188! Achiasi (fl. & fr. Mar.) *Kitson* 1033! Nkwanta (fl. & fr. Apr.) *Kitson* 1077! White Volta (fr. Dec.) *Adams & Akpabla* 4300! **Togo Rep.:** Lomé *Warnecke* 253! 255! **N.Nig.:** Lokoja *Parsons* 1! Katagum (Sept.) *Dalz.* 89! Ngurno (fl. & fr. Feb.) *T. Vogel* 54! Kano (Dec.) *Hagerup* 648! Kalkala *Golding* 53! **S.Nig.:** Ibadan (Oct.) *Newberry* 119! Owo *Burton*! Attah (fl. & fr. Sept.) *T. Vogel* 37! Calabar (fr. Jan.) *Holland* 14! [Br.]Cam.: Dawo, Vogel Peak, Adamawa (Nov.) *Hepper* 1405! **F.Po:** Botonos (fl. & fr. Feb.) *Exell* 870! Native of tropical America, widely distributed in the tropics. (See Appendix, p. 432.)
3. **P. peruviana** *Linn.* Sp. Pl. ed. 2, 1670 (1763) ; F.T.A. 4, 2 : 248 ; Baldwin & Speese l.c. 255, 257. Erect perennial up to 3 ft. high, densely hairy, from a creeping rootstock. The " Cape Gooseberry."
S.L.: Njala (fr. Oct.) *Deighton* 2419! **Ghana:** Kumasi (fl. & fr. Nov.) *Irvine* 1855! **N.Nig.:** Jos Plateau (fl. & fr. Oct.) *Hepper* 1179! Kura Falls *Kennedy* 2965! [Br.]Cam.: Bamenda (fl. & fr. Dec., Feb.) *Baldwin* 13853! *Migeod* 454! Jakiri (fl. & fr. Feb.) *Hepper* 1966! Native of tropical S. America, occasionally cultivated and naturalized.
4. **P. pubescens** *Irvine* Sp. Pl. 183 (1753) ; O. E. Schulz l.c. 143 ; Waterfall l.c. 164, for discussion. *P. turbinata* Medik. (1780)—O. E. Schulz. An annual herb about 2 ft. high ; flowers yellow with brown centre.
Ghana: Achimota (fl. & fr. Nov.) *Akpabla* 1801! Chilinga, Krachi (fl. & fr. May) *Morton* A3971! Introduced from America.

7. **WITHANIA** Pauq.—F.T.A. 4, 2 : 248. *Nom. cons.*

Shrubby herb, stems stellate-tomentose ; leaves elliptic to broadly ovate-lanceolate, apex acute to rounded, abruptly acute to long-decurrent at the base, 5–10 cm. long, 2–5 cm. broad, entire or sinuate ; petiole 1–2 cm. long ; flowers 2–6 in axillary fascicles ; pedicels about 5–6 mm. long, elongating to 1 cm. or more in fruit ; calyx campanulate, divided to about the middle into 5 acute triangular lobes, inflated in fruit, 1–2 cm. long, up to 1 cm. broad ; corolla campanulate, twice as long as the calyx, about 7–8 mm. long, stamens slightly exserted, style included ; fruit a red, globose berry, enclosed by the inflated calyx   ..   ..   *somnifera*

W. somnifera (*Linn.*) *Dunal* in DC. Prod. 13, 1 : 453 (1852) ; F.T.A. 4, 2 : 249. *Physalis somnifera* Linn. (1753). Much-branched undershrub 3–7 ft. high ; flowers small, pale.
Mali: Onatagouna *Hagerup* 430 ! **N.Nig.:** Ngurno (fr. Feb.) *E. Vogel* 58 ! Katagum (fr. Sept.) *Dalz.* 91 ! Jos Plateau (fl. & fr. Apr.) *Lely* P209 ! Vom (fl. & fr. Feb.) *McClintock* 252 ! Throughout the drier parts of Africa; in the Atlantic Is. and E. to India. (See Appendix, p. 453.)
[A specimen collected by T. Vogel and cited as No. 35 in F.T.A. (l.c.) and F.W.T.A., ed. 1, 2 : 205 from Cape Palmas, Liberia (fl. & fr. July 1841) may not have been collected there due to confusion of labels.]

8. SOLANUM Linn.—F.T.A. 4, 2 : 207.

Inflorescence terminal :
  Leaves entire :
    Inflorescences paniculate or broadly cymose (not spiciform) :
      Leaves glabrous or nearly so :
        Leaves coriaceous, shining above, leaf-nerves 7–9 pairs anastomosing clearly at the margin ; not climbing (?)   ..   ..   1. *clerodendroides*
        Leaves membranous, never becoming coriaceous, not shining, leaf-nerves 11–13 pairs never clearly anastomosing ; climbing   2b. *terminale* subsp. *sanaganum*
      Leaves tomentose beneath with stellate hairs:
        Leaves upperside entirely glabrous   ..   ..   3. *giganteum*
        Leaves with indumentum on both sides :
          Stems with axillary subactive buds bearing small leaves up to 2 cm. diam. ; leaves elliptic to ovate-lanceolate, up to 16 cm. long and 8 cm. broad, densely tomentose on both surfaces ; panicles lax, globose ; calyx about 6 mm. long ; corolla 1 cm. long   ..   ..   4. *mauritianum*
          Stems with subdormant axillary buds ; leaves elliptic to ovate, up to 21 cm. long and 10 cm. broad, pubescent above, tomentose beneath ; inflorescences and flowers as above ..   ..   5. *verbascifolium*
    Inflorescences spiciform, cymules subsessile on the axis ; leaves oblong to obovate, gradually acuminate, 7–15 cm. long, 3–6 cm. broad ; pedicels very short ; calyx cupular, with minute teeth ; corolla 7–8 mm. long, lobes linear-oblong ; fruits about 5 mm. diam. ..   ..   2c. *terminale* subsp. *welwitschii*
  Leaves pinnate or pinnately lobed up to 15 cm. long and 8 cm. broad with 3–4 pairs of leaflets, the leaflets of the upper pairs mostly confluent with the rhachis ; cymes very large, many-flowered ; calyx 1 mm. long ; corolla 9 mm. long ; climber, entirely glabrous and without spines or prickles   ..   6a. *seaforthianum* var. *disjunctum*
Inflorescences lateral, axillary, extra-axillary or leaf-opposed :
  *Indumentum stellate (always fairly copious) :
    Flowers subsolitary to about 4 together :
      Flowers white, corolla 12–16 cm. diam. :
        Plants green ; fruits globose or pear-like, up to 2·5 cm. diam., bright red, shining   8a. *gilo* var. *gilo*
        Plants tinged with violet ; fruits globose, light red-brown and tinged with violet   8b. *gilo* var. *pierreanum*
      Flowers almost purple-violet (rarely white), corolla up to 3 cm. diam. :
        Tomentum on stem floccose ; plants always armed ; leaves mostly up to 10 cm. long and 7 cm. broad ; fruit globose, yellow   ..   ..   9. *incanum*
        Tomentum on stem not floccose ; plants (occurring in Africa) never armed ; leaves up to 16 cm. long and 10 cm. broad ; fruit egg-shaped, sometimes cucumber-like or pyriform, intense violet, white or dirty-yellow   10b. *melongena* var. *inerme*
    Flowers rather numerous, cymose (occasionally also a few solitary flowers) ; corolla up to 2 cm. diam. ; fruits about 1 cm. diam. or less :
      Leaves glabrous above, green, spines sometimes very large, up to 15 mm. long, compressed, base up to 5 mm. long, sharply recurved   ..   11. *aculeastrum*
      Leaves tomentose throughout, spines never longer than 1 cm., straight or slightly recurved :
        Leaves rather small, ovate, 2–7 cm. long, subentire ; cymes about 7 cm. long, lax ; calyx about 3 mm. long ; corolla 1·3 cm. long ; fruit up to 1 cm. diam.   12. *albicaule*
        Leaves large, more or less elliptic, up to 10–16 cm. long, but sometimes smaller ; lobate-sinuate to subentire :
          Corolla about 1 cm. long and up to 3·3 cm. diam. :
            Cymes corymbose, nearly always unarmed ; pedicels about 6 mm. long, with glandular hairs ; calyx 4 mm. long, lobes triangular ; corolla up to 2·5 cm. diam., white ; fruit 1 cm. diam., dirty-brown   ..   ..   13. *torvum*
            Cymes racemose, mostly armed with straight yellow spines ; pedicels 10–15 mm., hairs not glandular ; calyx 6 mm. long ; corolla up to 3·3 cm. diam. (lower flowers of the inflorescence), violet ; fruit up to 3 cm. diam., yellow or olive-yellow, shining :

Leaves pubescent when young, becoming glabrescent and rather thinly scabrid-
pubescent   ..   ..   ..   **14a. *cerasiferum* subsp. *cerasiferum***
Leaves densely yellowish tomentose, not becoming glabrescent
    14b. *cerasiferum* subsp. *crepinii*
Corolla about 5 mm. long ; cymes raceme-like ; leaves very shortly pubescent
above, subtomentose beneath :
Plants unarmed   ..   ..   **15a. *indicum* subsp. *distichum* var. *distichum***
Plants armed :
Plants armed with spines up to 10 mm. long and at the base 6 mm. broad
    15b. *indicum* subsp. *distichum* var. *grandemunitum*
Plants armed with spines up to 5 mm. long and at the base 2·5 mm. broad
    15c. *indicum* subsp. *distichum* var. *modicearmatum*
*Indumentum simple or absent :
Leaves sessile or subsessile, the latter long-decurrent on the midrib at the base :
Calyx armed with long sharp bristles, very accrescent in fruit ; the fruit about 3 cm.
diam. ; leaves pinnately lobed nearly to the middle, elliptic-obovate in outline,
with prickles on the nerves, pilose on both surfaces   ..   ..   **16. *dasyphyllum***
Calyx not armed and not accrescent in fruit, the latter 4 cm. diam. ; leaves less
lobate than above and glabrous   ..   ..   **17. *macrocarpon***
Leaves distinctly petiolate :
Plants armed with sharp prickles :
Fruits 3 cm. diam. or more ; leaves pinnately lobed, shortly pubescent; flowers
subsolitary ; calyx-lobes pubescent, becoming prickly in fruit :
Corolla white ; ripe fruit dirty-yellow or brownish ; seeds about 2 mm. diam.
without conspicuous hyalin margin   ..   ..   ..   **18. *aculeatissimum***
Corolla white or pale violet ; ripe fruit orange or bright red ; seeds about 4 mm.
diam., with a conspicuous hyalin margin about 1 mm. broad   **19. *ciliatum***
Fruits less than 1 cm. diam. ; flowers not subsolitary ; inflorescence subfasciculate,
leaves glabrous or nearly so, with undulate margins   ..   ..   **20. *anomalum***
Plants unarmed :
Woody plants :
Climber ; leaves elliptic, acuminate, entire, up to 12 cm. long and 8 cm. broad,
glabrous or nearly so ; inflorescence few-flowered, subracemose
    2a. *terminale* subsp. *inconstans*
Erect shrub ; branches mealy-pubescent ; leaves obovate-elliptic, hardly acumi-
nate, up to 12 cm. long and 6 cm. broad, glabrous, undulate on the margin ;
flowers subfasciculate on a short peduncle   ..   ..   ..   **20. *anomalum***
Herbs :
Plants pubescent ; stems slightly winged ; leaves ovate, subacute, up to 12 cm.
long and 5 cm. broad, subentire ; inflorescence 4–6-flowered
    21. *pseudospinosum*

Plants glabrous or nearly so :
Flowers umbellate on a common peduncle ; leaves ovate, shortly ciliate, entire
or nearly so   ..   ..   **22. *nigrum***
Flowers fasciculate, without a peduncle ; pedicels about 5 mm. long ; leaves
lobulate, glabrous   ..   ..   **7. *aethiopicum***

1. **S. clerodendroides** *Hutch. & Dalz.* F.W.T.A., ed. 1, 2 : 206 (1931). An undershrub with crowded shining
    leaves, and flowers in a thyrsoid panicle.
    S.Nig.: Eket *Talbot* 3211! Apparently endemic, only known from the type collection.
2. **S. terminale** *Forsk.* subsp. **terminale**—Fl. Aegypt.-Arab. 45 (1775) ; Bitter in Fedde Rep. 18 : 301 (1922).
    S. bifurcatum Hochst. ex A. Rich. Tent. Fl. Abyss. 2 : 98 (1851) ; Exell Cat. S. Tomé 252. *S. bifurcum*
    Hochst. ex Dunal in DC. Prod. 13, 1 : 105 (1852) ; F.T.A. 4, 2 : 213 ; Bitter in Engl. Bot. Jahrb. 54 :
    454 (1917). In Ethiopia, Arabia and tropical N. E. Africa.
2a. **S. terminale** subsp. **inconstans** (*C. H. Wright*) *Heine* in Kew Bull. 14 : 247 (1960). Aké Assi, Contrib. 1 : 177.
    S. inconstans C. H. Wright in Kew Bull. 1894 : 127 ; F.T.A. 4, 2 : 211 (partly, excl. *S. symphostemon* De
    Wild. & Th. Dur.) ; Bitter in Engl. Bot. Jahrb. 54 : 482 (1917) ; F.W.T.A., ed. 1, 2 : 207. *S. togoense*
    Dammer in Engl. Bot. Jahrb. 38 : 59 (1905) ; F.T.A. 4, 2 : 246, 571 ; Bitter l.c. 478 ; F.W.T.A., ed. 1,
    2 : 207. *S. suberosum* Dammer l.c. 182 (1906) ; Bitter in Fedde Rep. 18 : 307 (1922). Inflorescences very
    small, long-peduncled, mostly lateral ; fruit ellipsoid.
    Togo Rep.: Badja (Mar.) *Schlechter* 12974! **S.Nig.:** Ondo Prov. (Aug.) *Jones* FHI 19515! Shasha F.R.
    (fl. & fr. Apr.) *Richards* 3373! Eket *Talbot*! Oban *Talbot*! [**Br.]Cam.:** Cam. Mt. *Kalbreyer* 172! **F.Po :**
    Mann 62! *Guinea* 1144! 1146! El Pico, 3,000 ft. (fl. & fr. Dec.) *Boughey* 190! Also in Cameroun.
2b. **S. terminale** subsp. **sanaganum** (*Bitter*) *Heine* in Kew Bull. 14 : 248 (1960). *S. bansoense* Dammer in Engl.
    Bot. Jahrb. 48 : 237 (1912) ; Bitter in Engl. Bot. Jahrb. 54 : 472 (1917), and in Fedde Rep. 18 : 303
    (1922), incl. var. *episporadotrichum* Bitter l.c. (1922) and subsp. *sanaganum* Bitter l.c. 304 (1922). *S.
    plousianthemum* Dammer (1906)—Bitter in Engl. Bot. Jahrb. 54 : 456 (1917). Inflorescence large,
    terminal, fruits ellipsoid.
    [**Br.]Cam.:** Bamenda, 3,500 ft. (May) *Maitland* 1960! Bafut-Ngemba F.R. (Apr.) *Ujor* FHI 30026!
    Dengdeng (Mar.) *Mildbr.* 8616! Nkom-Wum F.R. (June) *Daramola* FHI 41085! Also in Cameroun,
    Congo and tropical E. Africa. Distribution without revision of the whole group not properly known.
2c. **S. terminale** subsp. **welwitschii** (*C. H. Wright*) *Heine* in Kew Bull. 14 : 248 (1960). Aké Assi, Contrib. 1 : 178 ;
    S. welwitschii C. H. Wright in Kew Bull. 1894 : 126, incl. var. *strictum* and *oblongum* ; F.T.A. 4, 2 : 213 ;
    Bitter in Engl. Bot. Jahrb. 54 : 478 (1911) ; F.W.T.A., ed. 1, 2 : 206. Inflorescence spike-like, terminal,
    fruits globose.
    Guin.: Sérédou (May) *Collenette* 29! Lola to Uzo *Chev.* 20992! Boola *Chev.* 20935! Mt. Tongoui *Schnell*
    4079! **Iv.C.:** Sakota to Sassandra (Aug.) *de Wilde* 306! Yapo (fl. & fr. Oct.) *de Wilde* 715! **Ghana:**
    Bosuso to Begoro (May) *Robertson* K9! Akim (fl. & fr. Jan., Mar.) *Irvine* 1172! 1812! Aburi (June)

Johnson 158! Sutah *Cummins* 45! **S.Nig.:** Sonkwala, Obudu Div. (fl. & fr. Dec.) *Savory & Keay* FHI 25255! **[Br.]Cam.:** Bakosi F.R. (May) *Olorunfemi* FHI 30589! Ambas Bay (fl. & fr. Feb.) *Mann X*! **F.Po:** *Mann* 274! Also in Cameroun, Rio Muni, the Congos and Angola.

3. **S. giganteum** *Jacq.* Collect. 4 : 125 (1790) ; Bot. Mag. 44 : t. 1921 ; F.T.A. 4, 2 : 229 ; Bitter in Engl. Bot. Jahrb. 57 : 256 (1921). A shrub or tree up to 25 ft. high, with white tomentum on all parts, except the surfaces of the leaves ; flowers violet-purple.
[Br.]Cam.: Cam. Mt., Buea 4,000 ft. (fl. & fr. May) *Mildbr.* 9507! 9525! Bali-Ngemba F.R. (fr. May) *Ujor* FHI 30356! Widespread in tropical Africa also in S. Africa, S. India and Ceylon.

4. **S. mauritianum** *Scop.* Delic. Fl. et Faun. Insubr. 3 : 16 (1788) ; Bitter in Engl. Jahrb. 54 : 491 (1917). *S. auriculatum* Ait. (1789)—F.W.T.A., ed. 1, 2 : 206. *S. verbascifolium* of F.T.A. 4, 2 : 221, partly (Angolan specimens), not of Linn. A densely stellate-tomentose tree or shrub up to 20 ft. high, with blue-purple flowers.
S.L.: Njala, cultivated (July) *Deighton* 3961! **[Br.]Cam.:** Cam. Mt., Buea, 3,500–4,000 ft. (fl. Feb., Apr. June, Nov.) *Dundas* FHI 20389! *Dalz.* 8207! *Dunlap* 103! *Hutch. & Metcalfe* 124! *Maitland* 58! *Migeod* 242! Probably a native of S. America (Brazil, Paraguay, Uruguay) ; long-introduced and natura-lized in many tropical and subtropical regions of the Old World. Not a native of Mascarene Islands, as stated in F.W.T.A., ed. 1, 2 : 206. (See Appendix, p. 433.)

5. **S. verbascifolium** *Linn.* Sp. Pl. 184 (1753) ; F.T.A. 4, 2 : 221, partly ; O. E. Schulz in Urban, Symb. Antill. 6 : 182 Bitter l.c. 490. An unarmed shrub, mealy-stellate-pubescent, with white flowers turning blue-purple.
Guin.: Lola (May) *Roberty* 18011! **S.L.:** Kenema (fl. & fr. Feb.) *Deighton* 4142! Tunkia (fl. Dec., fr. Nov., Jan.) *Deighton* 3827! 3877! *Pyne* 18! Gola Forest (Apr.) *Small* 609! **Lib.:** Vahon (fr. Nov. *Baldwin* 10253! Gonatown (fr. Dec.) *Baldwin* 10785! Mecca (fr. Dec.) *Baldwin* 10400! Klay (fl. & fr. Aug.) *Barker* 1895! **Ghana:** Krobo *Williams* 304! Afram, Mankrang F.R. *Foggie* FH 4856! Begoro (fr. Nov.) *Morton*! **N.Nig.:** Jebba (Nov.) *Hagerup* 729! **S.Nig.:** Lagos *Millen* 50! Olokemeji F.R. (Apr.) *Ross* 135/10/3! Akilla *Kennedy* 2376! Nikrowa, Okomu F.R. (fl. & fr. Sept.) *Iriah* FHI 23086! A very common weed in the tropics of the whole world. (See Appendix, p. 435.)

6. **S. seaforthianum** *Andr.* Bot. Reposit. 8 : t. 504 (1808).

6a. **S. seaforthianum** var. **disjunctum** O. E. Schulz in Urban Symb. Antill. 6 : 169 (1909) ; Bitter in Fedde Rep. Beih. 16 : 309 (1923) ; Berhaut Fl. Sén. 47. A very handsome climber, with large panicles of violet-purple flowers ; fruits globose, red.
Sen.: *Berhaut* 1869. **S.L.:** Njala (fl. & fr. May) *Deighton* 5155! **S.Nig.:** Olokemeji (fl. & fr. Nov.) *MacGregor* FHI 364! A native of W. Indies and Central America, introduced for ornamental purposes and naturalized in many parts of tropical Africa.

7. **S. aethiopicum** *Linn.* Amoen. Acad. 4 : 307 (1759) ; F.T.A. 4, 2 : 217, partly ; F.W.T.A., ed. 1, 2 : 208, partly (*Thomas* 146) and excl. syn. ; Bitter in Fedde Rep. 16 : 44 (1923). A glabrous herb up to 2 ft. high, with white or pale violet flowers up to ⅔ in. diam. and globose, red shining fruits up to ¾ in. diam. **Dah.:** Cabolé (fr. May) *Chev.* 23769. Zagnanado *Chev.* 23070! **S.Nig.:** *Thomas* 146! (fr. Mar.) *Irvine* 3604! Various forms cultivated in many parts of tropical Africa, frequently misidentified and confused with other species. (See Appendix, p. 432.)

8. **S. gilo** *Raddi* in Atti Soc. Ital. Sci. Modena 18 : 31 (1820) ; Bitter in Fedde Rep. 16 : 48 (1923). A wide-spread long cultivated plant in many parts of tropical Africa ; taxonomically it was almost entirely mis-understood until Bitter's revision. Introduced (and naturalized ?) in S. America from where it was originally described.

8a. **S. gilo** var. **gilo**—*S. geminifolium* Thonning in Schum. & Thonn. Beskr. Guin. Pl. 121 (1827) ; F.T.A. 4, 2 : 223. *S. naumannii* Engl. (1886)—F.T.A. 4, 2 : 216. An unarmed shrub up to 4 ft. high, with white flowers and intense shining pink, sometimes yellowish, globose fruits.
Guin.: Tonfili (fl. & fr. July) *Chev.* 20446! **S.L.:** Likuru (Feb.) *Sc. Elliot* 4955! Njala (fl. & fr. July) *Deighton* 2225! **Lib.:** Filoke (fr. July) *Baldwin* 6694! **Iv.C.:** Bébon (Dec.) *Chev.* 22643! Baoulé (July) *Chev.* 22179! 22194! 22297! Bouaké (fr. July) *Chev.* 22124! **N.Nig.:** *Yates* 13a! 20! 43! Katagum Dist. *Dalz.* 316! **S.Nig.:** *Thomas* 1237!
[Note : The note on " *S. incanum* Linn." in the Appendix p. 433 includes this species, and the plants called " Yalo " and similar belong to *S. gilo* Raddi var. *gilo*—H. H.]

8b. **S. gilo** var. **pierreanum** (*Paillieux & Bois*) Bitter l.c. (1923). *S. pierreanum* Paillieux & Bois in Rev. Sci. Nat. Appl. 37 : 483, fig. p. 484 (1890) ; André in Rev. Hort. 62 : 343 (1890), 71 : 495 (1899) and 72 : 238, with fig. (1900) ; Chev. Bot. 464. Like the last var., but fruits larger and lengthwise sulcate and like the whole plant tinged with violet.
Mali: Kankan (Mar.) *Chev.* 589! Only known as cultivated plant ; also in Gabon.
[Note : This is clearly a horticultural variety, i.e. " cultivar " (cv.). The present writer, however, feels that it is better not to change the rank of this taxon given by Bitter without modern revision of the whole group.—H. H.]

9. **S. incanum** *Linn.* Sp. Pl. 1 : 188 (1753) ; F.T.A. 4, 2 : 238 ; Bitter l.c. 270, 297 ; F.W.T.A., ed. 1, 2 : 206, partly (fig. 249) ; Berhaut Fl. Sén. 170. A spiny undershrub up to 6 ft. high, pale stellate-tomentose all over, with violet purple, rarely white, 5-parted flowers and globose yellow fruits, up to 1 in. diam.
Sen.: *Berhaut* 147. Tamba-Kounda (Sept.) *Berhaut* 3205! **Mali:** Compongan (July) *Chev.* 24359! Mopti (fl. & fr. Dec.) *de Wailly* 5311! Timbuktu *Hagerup* 222! Iles de Bourem (July) *Chev.* 1291! **Togo Rep.:** Lomé *Warnecke* 351! **Dah.:** Cotonou (fl. & fr. June) *Debeaux* 191! Allada (Feb.) *Chev.* 23238! **N.Nig.:** Jebba *Barter*! Katagum (May) *Dalz.* 92! Daddara (fr. Dec.) *Meikle* 816! Sokoto (fl. & fr. July) *Dalz.* 387! Yola *Dalz.* 184! Kuka (fl. & fr. Jan.) *E. Vogel* 28! Kalkala *Golding* 52! **S.Nig.:** Abeokuta *Millen* 96! **[Br.]Cam.:** Rosevear Cam. 50/37! The typical variety is widespread in drier north tropical Africa and north and eastwards to India. Many other infraspecific taxa accepted by Bitter (l.c. 275–282 (1923)) from other parts of Africa could not be studied in connection with this revision. (See Appendix, p. 433.)
[*Berhaut* 3205 is probably a hybrid between *S. incanum* and *S. dasyphyllum*.]

10. **S. melongena** *Linn.* Sp. Pl. 186 (1753) ; F.T.A. 4, 2 : 242 ; Bitter l.c. 292. The Egg-plant or Aubergine, a native of tropical Asia, widely cultivated in many horticultural varieties and often naturalized in the warmer parts of the world. (See Appendix, p. 433.)

10a. **S. melongena** *Linn.* var. **melongena.** Plants to which this name could be applied do not occur wild in Africa, according to Bitter. Specimens of cultivated plants concerned are for example : **Mali:** Mau *Portères*! **Iv.C.:** Guidéko (May) *Chev.* 16377! **S.Nig.:** Lagos *Rowland*!

10b. **S. melongena** var. **inerme** (*DC.*) *Hiern* in Cat. Welw. 1, 3 : 748 (1898) ; F.T.A. 4, 2 : 242 ; Bitter l.c. 294. *S. esculentum* Dunal var. *inerme* Dunal in DC. Prod. 13, 1 : 355 (1852). *S. incanum* of F.W.T.A., ed. 1, 2 : 206, partly, not of Linn. Unarmed undershrub, flowers mostly 7–8-parted.
Sen.: Tamboukané (Dec.) *Chev.* 2878! **Guin.:** Faranna, Tonfili (fl. & fr. Jan.) *Chev.* 20451! **S.L.:** *T. Vogel* 52! *Garrett* 22! Motenta (June) *Thomas* 456! Mamba (Nov.) *Thomas* 4357! Kangama (Sept.) *Fisher* 91! **Togo Rep.:** Lomé *Warnecke* 351!
[Note: *Garrett* 22, determined by Hiern, has a few very small spines on the calyx—H. H.]

11. **S. aculeastrum** *Dunal* in DC. Prod. 13, 1 : 366 (1852) ; F.T.A. 4, 2 : 243 ; Bitter in Fedde Rep. 16 : 166 (1923). A tree or shrub up to 20 ft. high, with white tomentum on all parts, except the surface of the leaves, and mostly very big and sharp hook-like spines, but sometimes also unarmed.

11a. **S. aculeastrum** *Dunal* var. **aculeastrum:** the armed variety.
[Br.]Cam.: Bafut-Ngemba F.R., 7,100 ft. (fl. & fr. Feb., June) *Ujor* FHI 30449! *Hepper* 2183! Madange (fl. & fr. Jan.) *Latilo & Daramola* FHI 34356! Widely distributed in the tropics and frequently culti-vated as a hedge-plant.

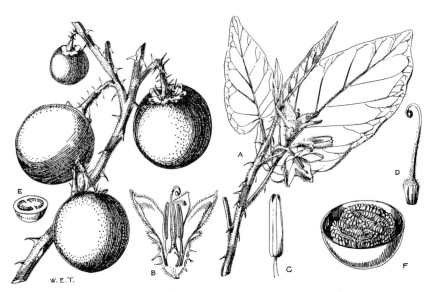

FIG. 281.—SOLANUM INCANUM *Linn.* (SOLANACEAE).

A, flowering shoot. B, longitudinal section of flower. C, anther. D, pistil. E, cross-section of ovary. F, section of fruit.

11b. **S. aculeastrum** var. **albifolium** (*C. H. Wright*) *Bitter* in Fedde Rep., Beih 16 : 170 (1923), including var. *exarmatum* Bitter and var. *conraui* (Dammer) Bitter (1923). *S. albifolium* C. H. Wright in Kew Bull. 1894 : 127 ; F.T.A. 4, 2 : 224. *S. conraui* Dammer in Engl. Bot. Jahrb. 48 : 242 (1912). Unarmed variety.
　　[Br.]Cam.: Bamenda, Lakom, 6,000 ft. (Apr.) *Maitland* 1657 ! Bangwe (June) *Conrau* 255 ! Also in tropical East Africa and Angola.

12. **S. albicaule** *Kotschy ex Dunal* in DC. Prod. 13, 1 : 204 (1852) ; F.T.A. 4, 2 : 255 ; Bitter in Fedde Rep. Beih. 16 : 101 (1923) ; Carvalho & Gillet in J. Agric. Trop.7 : 237 (1960) ; Heine in Kew Bull. 16: 204. *S. heudelotii* Dunal (1852) ; F.T.A. 4, 2 : 224 ; Bitter l.c. 5 (1923) ; F.W.T.A., ed. 1, 2 : 206 ; Monod in Contrib. Et. Sahara Occ. 2 : 108 ; Berhaut Fl. Sén. 122. *S. scindicum* Prain J. As. Soc. Beng. 65 : 542 (1896). An undershrub with slender branches and reflexed spines, shortly hoary-tomentose.
　　Sen.: *Heudelot* 34 ; 417 ! *Perrottet* 340 ! *Leprieur* ! *Huard* ! Oualo *Berhaut* 1376 ! Also in Mauritania, Chad, N.E. Africa (Ethiopia, Sudan, Egypt, Eritrea, Somalia), Arabia and Pakistan.

13. **S. torvum** *Sw.* Prod. 47 (1788) ; F.T.A. 4, 2 : 231 (excl. syn. *S. ferrugineum* Jacq., incl. *S. torvum* var. *compactum* Wright) ; O. E. Schulz, in Urban, Symb. Antill. 6 : 233 (1909), for full synonymy. Bitter in Engl. Bot. Jahrb. 57 : 252(1921). *S. mannii* C. H. Wright, in Kew Bull 1894 : 129, incl. var. *compactum* C. H. Wright, l.c. (1894) ; F.W.T.A., ed. 1, 2 : 207. A shrubby weed attaining 10 ft. or more in height, pale stellate-tomentose, with flattened spines, white or lilac flowers ; fruits ⅓ in. diam.
　　Sen.: *Richard* ! Lib.: Gbanga (fl. & fr. Nov.) *Daniel* 31 ! Lakrata (Apr.) *Bequaert* 188 ! Monrovia (fl. & fr. Dec.) *Dinklage* 3307 ! S.L.: Bafodea (fl. & fr. Apr.) *Deighton* 5470 ! Hill Station (fl. & fr. Nov.) *Deighton* 229 ! Regent (Jan.) *Gledhill* 40 ! Tombo (Jan.) *Deighton* 957 ! Guma (fl. & fr. Mar.) *Hepper* 2511 ! Iv.C.: Bingerville (fl. & fr. Dec.) *Chev.* 16030 ! 16074 ! (16014?). Ghana: Asawasi, Kumasi (Apr.) *Andoh* 5267 ! Bunsu, Akim (Mar., Apr.) *Irvine* 1356 ! 1754 ! Kotonfo (fl. & fr. Aug.) *Andoh* 4419 ! Dunkwa (fl. & fr. Feb.) *Whiting* 16 ! S.Nig.: Gambari F.R. (fl. & fr. Mar.) *Meikle* 1272 ! Lagos (fl. & fr. May) *Dalz.* 1190 ! Sapoba *Kennedy* 1766 ! 2717 ! Yoruba country (Dec.) *Newberry* 13 ! Nikrowa, Okomu F.R. (Sept.) *Iriah* FHI 23085 ! Baba Eko, Shasha F.R. (fl. & fr. Mar.) *Ross* 80 ! 273 ! [Br.]Cam.: Cam. Mt., 3,000–3,500 ft. (Jan., Nov.) *Maitland* 281 ! *Migeod* 241 ! *Dunlap* 104 ! Mamfe to Ikom (fl. & fr. Feb.) *Hepper* 2209 ! S. Bakundu F.R. (fl. & fr. Feb.) *Binuyo & Daramola* FHI 35550 ! F.Po: *Mann* 55 ! Musola *Guinea* 1269 ! *Boughey* 107 ! A very common weed throughout the tropics. (See Appendix, p. 435.)

14. **S. cerasiferum** *Dunal* in DC. Prod. 13, 1 : 365 (1852) ; F.T.A. 4, 2 : 241, partly.

14a. **S. cerasiferum** *Dunal* subsp. **cerasiferum**—Bitter in Fedde Rep., Beih. 16 : 283 (1923) (incl. var. *garuense* Bitter l.c. 285.) *S. xanthocarpum* var. *schraderi* of Schweinf. Pl. nilot. 25, t. 9 (1862) ; F.T.A. 4, 2 : 234, not of DC. *S. yolense* Hutch & Dalz. in F.W.T.A., ed. 1, 2 : 206 (1931). Herbaceous undershrub with blue-purple flowers and yellow fruits ⅓–¾ in. across.
　　N.Nig.: Yola *Dalz.* 183 ! Also in Cameroun, Ubangi-Shari, Sudan and Ethiopia.

14b. **S. cerasiferum** subsp. **crepinii** (*van Heurck*) *Bitter* l.c. 287 (1923) (incl. subsp. *duchartrei* (Heckel) Bitter l.c. 286 (1923)). *S. crepini* van Heurck in Obs. Bot. 1 : 89 (1870). *S. duchartrei* Heckel in Rev. Gén. Bot. 2 : 49, t. 2 (1890).
　　Sen.: Plateau de Thiès, Rufisque *Duclot & Sambuc* ; *Sieber*. Walo (fl. & fr. Oct.) *Perrottet* 557 !

15. **S. indicum** *Linn.* Sp. Pl. 187 (1753) ; F.T.A. 4, 2 : 232, partly. A very critical group of a very wide distribution in the Old World tropics. As far as Bitter's revision is the most up-to-date work on these plants which are extremely difficult to classify, his infraspecific concept is generally accepted here, although it does not correspond well with modern taxonomic methods.
　　S. indicum subsp. **distichum** (*Thonning*) *Bitter* in Fedde Rep. Beih 16 : 13 (1923). *S. distichum* Thonning (1827)—F.T.A. 4, 2 : 223.

15a. **S. indicum** subsp. **distichum** (*Thonning*) *Bitter* var. **distichum.** *S. indicum* Linn. subsp. *distichum* (Thonning) Bitter var. *immunitum* Bitter l.c. 14 (1923). *S. senegambicum* Dunal in DC. Prod. 13, 1 : 194 (1852); F.T.A. 4, 2 : 222. *S. anomalum* of F.T.A. 4, 2 : 232 ; of F.W.T.A., ed. 1, 2 : 207, not of Thonning. *S schroederi* Dammer (1912), partly. *S. scalare* C. H. Wright (1894)—F.T.A. 4, 2 : 224. A coarse tomentose, unarmed undershrub up to 6 ft. high, with white flowers and erect red globose fruits ⅛–¼ in. diam.

Sen.: near R. Nunez *Heudelot* 47 ; 713 ! **S.L.**: *Garrett* 18 ! Yungeru (fr. Jan.) *Thomas* 1321 ! Njala (fl. & fr. Oct., Nov.) *Deighton* 1407 ! 1784 ! Sakaro (fl. & fr. Jan., Feb.) *Sc. Elliot* 4394 ! 4898 ! Bumbuna (fl. & fr. Oct.) *Thomas* 3677 ! Bafodea (fr. Apr.) *Deighton* 5471 ! **Lib.**: *Whyte* ! Ganta (fr. May) *Harley* ! Gbanga (fl. & fr. Sept.) *Linder* 550 ! 655 ! Bobei Mt. (fr. Sept.) *Baldwin* 9638 ! **Iv.C.**: Bouaké, Baoulé Nord (fl. & fr. July) *Chev.* 22127 ! 22194 ! 22274 ! Tonkoui (fr. Aug.) *Boughey* 18321 ! **Ghana**: *Thonning* (photo) ! Kpeve *Williams* 95 ! Akim *Irvine* 556 ! Kumasi (fr. June) *Andoh* FH 2271 ! Aburi (fl. & fr. June) *Brown* 320 ! Achimota (fl. & fr. June) *Irvine* 1770 ! Pramkese (fl. & fr. Jan.) *de Wit & Morton* A2967 ! Asuansi (fr. Oct.) *Darko* 1033 ! **N.Nig.**: Abinsi *Dalz.* 723 ! **S.Nig.**: Yorubaland (fr. Mar.) *Schlechter* 13004 ! Lagos *Rowland* ! Abeokuta *Harrison* 3 ! Sapoba *Kennedy* 2325 ! Jamieson R. *Kennedy* FHI 1958 ! Baba Eko, Shasha F.R. (fl. & fr. May) *Ross* 85 ! 276 ! Cross R. (fl. & fr. Dec.) *Holland* 188 ! **[Br.]Cam.**: Mambila Plateau : Bellei (fl. & fr. Jan.) *Hepper* 1707 ! Gembu (fl. & fr. Jan.) *Hepper* 1828 ! A common weed (sometimes cultivated for its fruits) in tropical Africa. (See Appendix, p. 433, also for both the following varieties [*grandemunitum* and *modicearmatum*], under " *S. anomalum* Thonn.")

15b. **S. indicum** subsp. **distichum** var. **grandemunitum** *Bitter* l.c. 17 (1923). *S. rederi* Dammer in Engl. Bot. Jahrb. 48 : 251 (1912). Stems densely armed with long, sharp spines.

Guin.: Diaguissa (fr. Sept., Oct.) *Chev.* 18368 ! 18595 ! **S.L.**: Musaia (fl. & fr. Apr.) *Deighton* 5469 ! **[Br.]Cam.**: Cam. Mt., 5,800–8,000 ft. (fl. & fr. Nov.—Jan.) *Mann* 1322 ! 1975 ! *Maitland* 895 ! 1330 ! *Mildbr.* 10811 ! *Dunlap* 31 ! Lakom to Basenako (fr. Apr.) *Maitland* ! **F.Po:** (fl. & fr. Jan.) *Guinea* 1746 !

15c. **S. indicum** subsp. **distichum** var. **modicearmatum** *Bitter* l.c. 16 (1923). *S. buettneri* Dammer in Engl. Bot. Jahrb. 38 : 59 (1905) ; F.T.A. 4, 2 : 273. *S. schroederi* Dammer in Engl. Bot. Jahrb. 48 : 250 (1912), partly. *S. indicum* Linn. var. *micranthum* Hook. f. in J. Linn. Soc. 6 : 18 (1861), name only—F.T.A. 4, 2 : 232, in syn. Loosely armed with much shorter and narrower spines.

Sen.: *Roger* ! **Gam.**: Genieri *Fox* 92 ! **Port.G.**: Boé (fl. & fr. Jan.) *Esp. Santo* 2369 ! **Guin.**: Sulimania (fl. & fr. Mar.) *Sc. Elliot* 5335b ! **S.L.**: *Thomas* ! *G. Don* ! *T. Vogel* 87 ! Musaia (Mar., Apr.) *Deighton* 5451 ! 5472 ! Pendambu (fl. & fr. July) *Thomas* 761 ! Yonibana (fl. & fr. Oct.) *Thomas* 4194 ! Njala (fl. & fr. Nov.) *Deighton* 1783 ! **Lib.**: Begwai (fl. & fr. Oct.) *Bunting* 124 ! Grant's Farm, Sinoe R. *Whyte* ! **Togo Rep.**: Sokodé *Schroeder* 92 ! **N.Nig.**: Jos (fl. & fr. June) *Lely* 316 ! 365 ! *Batten-Poole* 339 ! Udonoria (fl. & fr. May) *Umana* FHI 29110 ! **[Br.]Cam.**: *Migeod* 423 ! Cam. Mt., 6,000 ft. (fl. & fr. Apr.) *Hutch. & Metcalfe* 16 ! Bamenda (fl. & fr. Jan.) *Migeod* 306 ! 424 ! Banso, Bamenda (fl. & fr. Oct.) *Tamajong* FHI 23484 ! Vogel Pk., Adamawa (Dec.) *Hepper* 1586 ! **F.Po:** 6,000 ft. (fl. & fr. Dec.) *Mann* 625 ! 1165 ! *T. Vogel* 52 ! 83 ! *Guinea* 2256 ! Moka (fl. & fr. Dec.) *Boughey* 4 ! A common weed in nearly all parts of tropical Africa.

There are many intermediate forms between both var. *modicearmatum* and var. *grandemunitum*, for example : **Guin.**: *Chev.* 20447 ! 20448 ! Pobéguin 1043 ! **Iv.C.**: *Chev.* 19133 ! **Dah.**: Le *Testu* 132 ! *Chev.* 23910 ! 24022 !

16. **S. dasyphyllum** *Schum. & Thonn.* in Beskr. Guin. Pl. 126 (1827) ; Bitter in Fedde Rep., Beih. 16 : 188 (1923), incl. var. *semiglabrum* (C. H. Wright) Bitter l.c. 191, and var. *inerme* Bitter l.c. 192. *S. duplosinuatum* Klotzsch in Peters, Reise nach Mossamb., Bot. 1 : 233 (1862) ; F.T.A. 4, 2 : 243 ; F.W.T.A., ed. 1, 2 : 207. *S. afzelii* Dunal in DC. Prod. 13, 1 : 363 (1852). *S. duplosinuatum* Klotzsch var. *semiglabrum* C. H. Wright in F.T.A. 4, 2 : 244 (1905). *S. macinae* A. Chev. Bot. 463 (1920), name only. Usually a coarsely hairy and spiny undershrub up to 3 ft. high, with bluish-purple flowers 1–1½ in. diam. Unarmed plants occur (var. *inerme* Bitter) but the occurrence of spines cannot be regarded as of taxonomic importance.

Sen.: Marancounda (fr. Feb.) *Chev.* 2880 ! **Gam.**: Kuntaur *Ruxton* 39 ! **Port.G.**: Bissau, Prabis (fr. Feb.) *Esp. Santo* 1818 ! **S.L.**: *Afzelius* ! Njala (fr. Nov.) *Deighton* 2867 ! **Lib.**: Gbanga (fl. & fr. Sept.) *Linder* 571 ! **Iv.C.**: Sassandra (June) *Chev.* 21817 ! Mossi *Chev.* 24667 ! 24680 ! Abidjan (July) *Chev.* 13809 ! **Dah.**: Atacora (fr. June) *Chev.* 24139 ! Porto Novo (Mar.) *Chev.* 23317 ! **N.Nig.**: *Yates* ! Baradan (May) *Lely* 91 ! Gawu, Niger Prov. (fr. Aug.) *Onochie* FHI 35941 ! Lokoja (fl. & fr. Nov.) *Dalz.* 186 ! Nupe *Barter* 1344 ! **S.Nig.**: Lagos (Sept.) *Dawodu* 37 ! *Rowland* ! Gambari (fl. & fr. Sept.) *Onochie* FHI 7681 ! Oban *Talbot* ! Ukunzu (Jan.) *Thomas* 2199 ! Omo F.R. (Feb.) *Jones & Onochie* FHI 16714 ! Obompa (Jan.) *Thomas* 220 ! **[Br.]Cam.**: Bamenda (Feb.) *Migeod* 456 ! Vogel Peak, Adamawa (Dec.) *Hepper* 2788 ! Throughout tropical Africa and in S. Africa. (See Appendix, p. 433.)

17. **S. macrocarpon** *Linn.* Mant. Alt. 205 (1771) ; F.T.A. 4, 2 : 214 ; Bitter in Fedde Rep., Beih. 16 : 195 (1923) incl. var. *calvum* Bitter, l.c. 198 (1923), and other varieties ; Burkill in Kew Bull. 1925 : 333. Half-woody usually unarmed undershrub with rather stout branches up to 5 ft. high, hairy or sometimes glabrescent ; flowers ¾–1 in. diam., white or bluish-purple ; often cultivated.

Mali: between Senegal and Niger *Bellamy* 136 ! 208 ! **Guin.**: Tonfili (fr. Jan.) *Chev.* 20450 ! Timbo *Pobéguin* 1878 ! **S.L.**: Njala, cult. (fl. & fr. Dec.) *Deighton* 1803 ! 1803a ! Kangama (Sept.) *Fisher* 94 ! Kanika (Sept.) *Thomas* 2063 ! Rowala (July) *Thomas* 1162 ! Mabum (Aug.) *Thomas* 1635 ! **Iv.C.**: Attéou *Chev.* 17045 ! Bouaké (July) *Chev.* 22136 ! 22138 ! Arbasso (fl. Dec.) *Chev.* 22642 ! **N.Nig.**: *Thornewill* 95 ! **S. Nig.** : Lagos *Dalz.* 1189 ! Enugu (Mar.) *Irvine* 3605 ! Also in E. Africa. (See Appendix, p. 434.)

[Note : Intermediate forms between both *S. dasyphyllum* and *S. macrocarpon* are not infrequent (for example : **S.L.**: *Deighton* 1803A ! 1804), and it seems to be evident that the two species are very close. No. 17 may be a glabrous form of No. 16, arisen in cultivation, and perhaps there are hybrid swarms between both " species " : see Bitter, l.c. 197 (1923).]

18. **S. aculeatissimum** *Jacq.* Icon. Pl. Rar. 1 : 5, t. 41 (1781) ; F.T.A. 4, 2 : 228, partly (excl. syn. *S. ciliatum* Lam.) ; Bitter l.c. 148. An undershrub 1–2 ft. high, armed with nearly straight spines, with white flowers ½ in. diam. and smooth brown-yellow nodding fruits.

Guin.: Nzérékoré (fr. Oct.) *Baldwin* 12101 ! Diaguissa (Apr.) *Chev.* 12686 ! **S.L.**: Mano Bonjena (fr. May) *Jordan* 2114 ! Kanya (fr. Oct.) *Thomas* 3946 ! Njala (fr. Apr.) *Deighton* 2635 ! (Nov.) *Deighton* 1782 ! Yonibana (Oct.) *Thomas* 4213 ! **Lib.**: *Whyte* ! Ganta (Oct.) *Harley* ! **Iv.C.**: Guidéko *Chev.* 16428 ! Japo *Chev.* 16603 ! Soubré (fl. & fr. Dec.) *Leeuwenberg* 2170 ! **[Br.]Cam.**: Bamenda, 5,500 ft. *Migeod* ! Bamenda : Lakom, 6,000 ft. (May) *Maitland* ! Bali-Ngemba F.R. (fr. Apr.) *Ujor* FHI 30333 ! Jakiri, 5,600 ft. (fl. & fr. Feb.) *Hepper* 1949 ! Throughout tropical Africa and in S. Africa.

19. **S. ciliatum** *Lam.* Illustr. 2 : 21, No. 2360 (1793) ; Encycl. Méth. Bot. 4 : 298 (1796) ; Bitter in Fedde Rep Beih. 16 : 151, partly (excl. syn. *S. campechiense* Dunal (1852) and *S. polycanthum* L'Hér. MS.) ; Dinklage in Fedde Rep. 41 : 265 (1937). *S. aculeatissimum* of F.T.A. 4, 2 : 228 ; F.W.T.A., incl. ed. 1, 2 : 207, partly, and of Exell Cat. S. Tomé 252, not of Jacq. *S. aculeatissimum* var. *purpureum* A. Chev. Bot. 460 (1920), name only. Like the last, but indumentum less dense, particularly in the leaves above ; leaves clearly ciliate (epithet !) ; fruits always red.

Lib.: Fishtown, Grand Bassa (fl. & fr. July) *Dinklage* 1661 ! **Iv.C.**: Taté to Tabou (fr. Aug.) *Chev.* 19830 ! Also in Principe (*Quintas* 45, *Mann* 1110 ! *Exell* 541 !). Pantropical, but mainly in tropical America and East Asia. Most likely (in Africa) an introduced plant. Often confused with No. 18 and perhaps overlooked and more widespread.

20. **S. anomalum** *Thonning* in Beskr. Guin. Pl. 126 (1827) ; Bitter in Engl. Bot. Jahrb. 57 : 272 (1921) ; not of F.T.A. 4, 2 : 232 and F.W.T.A., ed. 1, 2 : 207 (which is No. 15a). *S. mannii* Wright var. *compactum* C. H. Wright in Kew Bull. 1894 : 129. *S. warneckeanum* Dammer in Engl. Bot Jahrb. 38 : 168 (1906) ; F.W.T.A., ed. 1, 2 : 207, not of C. H. Wright (1894). *S. pauperum* of F.T.A. 4, 2 : 217, partly (*Ansell*), of F.W.T.A. ed. 1, 2 : 207, not of C. H. Wright (1894). Unarmed or armed shrub, up to 5 ft. high, branchlets and young leaves densely stellate-tomentose at first, branches sometimes with rather large flattened spines ; fruits globose, ½–¾ in. diam., yellow-brown.

Lib.: Zeahtown (fr. July) *Baldwin* 6999 ! Moylakwelli (fl. & fr. Oct.) *Bequaert* 1286 ! **Iv.C.**: Toumodi to Uzaakro *Chev* .22429 ! Guidéko (fl. & fr. May, June) *Chev* .19003 ! Soubré *Chev.* 19147 ! Sassandra (fl

& fr. May, June) *Chev.* 17937! **Ghana:** *Thonning* 135 (photo)! Accra *Ansell*! Pokoase (May) *Hughes* 832! Asuantsi (Mar., fl. & fr. Oct.) *Irvine* 1579! *Darko* 1031! Newtown (July) *Chipp* 278! **Togo Rep.:** Lomé *Warnecke* 145! **Dah.:** *Poisson* 132! Cotonou *Chev.* 4498! **S.Nig.:** *Thomas* 1679! Ukunzu (fr. Jan.) *Thomas* 2201! Ibazo (fr. Nov.) *Thomas* 2020! Badagry (fr. Aug.) *Onochie* FHI 33481! (See Appendix, p. 433.)

21. **S. pseudospinosum** *C. H. Wright* in F.T.A. 4, 2 : 220 (1906) ; Bitter in Fedde Rep. 10 : 546 (1912). *S. molliusculum* Bitter l.c. (1912). Herb, prostrate or ascending, sometimes slightly woody below, with small white flowers and violet-black berries.
[**Br.**]**Cam.:** Cam. Mt., 6,000–10,000 ft. (fl. & fr. Dec.-Apr.) *Mann* 1321! 1938! *Dalz.* 8335! *Mildbr.* 10888! *Hutch. & Metcalfe* 54! *Maitland* 1301! 1333! A quite easily recognised segregate of No. 22, endemic in our area.

22. **S. nigrum** *Linn.* Sp. Pl. 186 (1753) (incl. vars.) ; F.T.A. 4, 2 : 218 ; Exell in Cat. S. Tomé 253. *S. nodiflorum* Jacq. (1788)—F.T.A. 4, 2 : 218 ; F.W.T.A., ed. 1, 2 : 208 ; Berhaut Fl. Sén. 174, 194. *S. guineense* (Linn.) Lam. (1797), not of Linn. (1753). A weed 1–2 ft. or more high, sometimes cultivated, with small white flowers and berries ⅛–¾ in. diam.
Sen.: *Berhaut* 378. *Roger* 17! **Mali:** Timbuktu *Hagerup* 144! **S.L.:** *Thomas* 4396! Ronietta (Nov.) *Thomas* 5435! Katuna (fl. & fr. Jan.) *Deighton* 1033! Njala (cult. var. *guineense* Linn.; fr. Dec.) *Deighton* 1811! Freetown (fl. & fr. Dec.) *Deighton* 474! Musaia (fr. Dec.) *Deighton* 4571! Bumbuna (fr. Oct.) *Thomas* 3703! Kambia (fr. Sept.) *Glanville* 412! **Ghana:** Amedzofe (fr. Nov.) *Morton*! Asuansi (Oct.) *Darko* 1032! **N.Nig.:** Nupe *Barter* 1054! Jos Plateau (fl. & fr. July) *Lely* P392! Naraguta *Lely* 31! **S.Nig.:** Lagos *Dalz.* 1188! 1188a! 1188b! Ogwashi (fl. & fr. Nov.) *Thomas* 2042! [**Br.**]**Cam.:** Buea, 3,000 ft. (fl. & fr. Nov.) *Migeod* 88! Bamenda, 6,000 ft. (fr. Jan.) *Migeod* 358! Bafut-Ngemba F.R., 7,000 ft. (fl. & fr. Feb.) *Hepper* 2146! *Lightbody* FHI 28492! Kumbo (fl. & fr. Feb.) *Hepper* 1981! **F.Po:** Musola, 3,000 ft. (fl. & fr. Jan.) *Guinea* 1348! Moka, 4,000 ft. (fl. & fr. Dec.) *Boughey* 109! Widespread throughout the world. (See Appendix, p. 435.)
[Note : A broad view of this species has been taken for this revision owing to its complexity throughout the world. For a discussion on *Solanum nigrum* var. *guineense* Linn. Sp. Pl. 186 (1753), see Heine in Kew Bull. 14 : 246 (1960).]

Numerous hybrids have been recorded in this genus e.g. *S. indicum* subsp. *distichum* var. *modicearmatum* × *gilo* var. *gilo*, *S. indicum* subsp. *distichum* var. *distichum* × *gilo* var. *gilo* (**Guin.:** *Chev.* 16449! **Iv.C.:** *Chev.* 16436!) ; *S. dasyphyllum* × *macrocarpon* (see note below *S. macrocarpon*) ; *S. incanum* × *dasyphyllum* (?), etc.

Cultivated species.

S. wrightii *Benth.* in Fl. Hong Kong 243 (1861) ; Heine in Kew Bull. 14 : 248 (1960). *S. macranthum* hort. in Rev. Hort. 1867 : 132, not of Dunal. *S. grandiflorum* auct. not of Ruiz & Pavon Fl. Peruv. 2 : 35, t. 168, fig. b. (1799) ; Bitter in Fedde Rep., Beih. 16 : 180 (1923). A large shrub or tree up to 20 ft. high, with very large, violet " potato-like " flowers up to 3 in. diam., cultivated at Njala, Sierra Leone. A native of Bolivia, frequently cultivated in the tropics and sometimes naturalized. The " Potato-Tree ".

S. hispidum *Pers.* Syn. 1 : 228 (1805) ; Dunal in DC. Prod. 13, 1 : 275 ; C. T. White in Kew Bull. 1939 : 666. *S. stellatum* Ruiz & Pavon Fl. Peruv. 2 : 40, t. 176 (1799), not of Jacq. (1789) or of Moench (1794). *S. warscewiczi* Weick ex Lambertye in Rev. Hort. 1865 : 429 ; Bitter in Engl. Bot. Jahrb. 57 : 253 (1921). *S. antiguense* Coult. in Donnell Smith, Enum. pl. Guatem. 4 : 187 (1895). *S. pynaertii* De Wild. in Miss. E. Laurent 2 : 437, t. 119 (1907). A small shrub or tree up to 30 ft. high, in habit resembling *Solanum torvum* Linn., but with rusty-coloured tomentum of stellate hairs on all parts, mainly the younger ones, and more (and particularly more regularly) pinnatisect leaves.
Guin.: Dalaba, Fouta Djalon (Sept.) *Schnell* 7467! An ornamental plant of Mexican and Guatemalan origin ; apparently naturalized in Congo. Naturalized in Queensland, Ceylon and obviously in Peru, from where it was first described (" habitat in *ruderatis*, ad Pillao et Panao vicos " Ruiz & Pavon, l.c.).

The " European " potato (*Solanum tuberosum* Linn. [1753]) is cultivated in the upland parts of our area, and the tomato (*S. lycopersicum* Linn. (1753) incl. var. ; syn. *Lycopersicum esculentum* Mill. (1768), *L. cerasiforme* Dunal (1813), etc.) is more widely cultivated. Both, originally American, are mainly represented in our area by cultivars introduced from Europe ; *S. lycopersicum* particularly frequently by the var. *cerasiforme* (Dunal) A. Voss (1896) and related cultivars.

Excluded species.

S. guineense *Linn.* Sp. Pl. 184 (1753) ; Heine in Kew Bull. 14 : 245 (1960). *S. aggregatum* Jacq. (1790)—F.T.A. 4, 2 : 214. A S. African plant which, since Linnaeus, has been erroneously indicated from our area.

152. CONVOLVULACEAE

By H. Heine

Herbaceous or woody plants, often climbing, juice usually milky. Leaves alternate, simple ; stipules absent. Flowers hermaphrodite, actinomorphic ; bracts often forming an involucre. Sepals usually free, imbricate, persistent. Corolla gamopetalous, mostly funnel-shaped, lobes 5, contorted. Stamens 5, inserted towards the base of the corolla-tube and alternate with the lobes ; anthers 2-celled, opening lengthwise. Ovary often surrounded by a disk, 1–4-celled ; ovules solitary or paired, erect ; style terminal. Fruit a capsule or fleshy. Seeds sometimes hairy, with rather scanty endosperm and more or less curved embryo. Cotyledons folded or crumpled.

A widely distributed family, mainly in warmer regions ; flowers very delicate and often fugitive. For an account of the classification of the family, with extensive bibliography, see K. A Wilson in J. Arn. Arb. 41 : 298 (1960).

Leafless parasitic plants with filiform stems and clusters of small flowers .. 1. **Cuscuta**
Leafy plants, not parasitic:
 Pollen grains smooth :

The 2 outer sepals much enlarged in fruit ; bracts not enlarged ; woody climbers
                   2. **Calycobolus**
Sepals not enlarged in fruit :
 Bracts enlarged in fruit ; plants shrubby .. .. 3. **Neuropeltis**
 Bracts not enlarged in fruit ; plants generally herbaceous (shrubby in *Bonamia* and
  in *Ipomoea* sp. No. 14, as well as in one cultivated *Ipomoea* sp.) :
  Ovary 2-cleft with 2 ovules in each chamber ; fruit 2-lobed, styles 2, inserted be-
   tween the lobes of the ovary; a small prostrate and creeping herb, with cordate
   to reniform leaves .. .. 4. **Dichondra**
  Ovary not deeply lobed ; fruit not 2-cleft ; style simple or, if styles 2, terminal ;
   plants of various habits :
   Styles 2, or single and rather deeply bilobed :
    Small herbs with very small flowers :
     Styles 2, forked .. .. 5. **Evolvulus**
     Styles 2, not forked .. .. 6. **Cressa**
    Shrubby climbers ; style with 2 unequal branches with globose stigmas 7. **Bonamia**
   Style 1, entire or slightly lobed :
    Outer sepals not much larger than the inner :
     Stigmas filiform or elliptic :
      Stigmas linear-filiform .. .. 8. **Convolvulus**
      Stigmas elliptic .. .. 9. **Jacquemontia**
     Stigmas biglobular :
      Capsule circumscissile ; upper part of the epicarp separating from the lower
       part and from the endocarp ; peduncles and pedicels winged 10. **Operculina**
      Capsule opening by 4 valves or more or less irregularly dehiscent ; peduncles
       and pedicels not winged .. .. 11. **Merremia**
    Outer sepals conspicuously larger than the inner :
     Stigmas 2, ovate-oblong ; capsule globose .. .. .. 12. **Hewittia**
     Stigmas biglobular ; capsule ovoid .. .. 13. **Aniseia**
 Pollen grains spinose :
  Fruit a capsule :
   Stamens inserted at the base of scales .. .. 14. **Lepistemon**
   Stamens not attached to scales :
    Indumentum stellate ; erect shrubs .. .. 15. **Astripomoea**
    Indumentum absent or if present then not stellate ; climbing, trailing or creeping
     plants, very rarely (*I. verbascoidea*) erect shrubs .. .. 16. **Ipomoea**
  Fruit fleshy and indehiscent ; cymes few-flowered and subsessile ; leaves long-petiolate
                  17. **Stictocardia**

1. CUSCUTA Linn.—F.T.A. 4, 2 : 202 ; Yuncker in Mem. Torrey Bot. Club 18 : 2
(1932) ; Verdcourt in E. Afr. Agric. Journ. 18 : 85 (1952).

Capsule not circumscissile :
 Scales in the corolla-tube not reaching the base of filaments, oblong, variously fim-
  briated to almost entire ; corolla-lobes broad, obtuse or rounded, ovate to suborbicular, erect .. .. 1. *australis*
 Scales in the corolla-tube usually reaching the base of the filaments, ovate, abundantly
  fringed with fairly long processes ; corolla-lobes broadly triangular, acute often with
  inflexed tips, rarely obtuse, usually spreading .. .. .. 2. *campestris*
Capsule circumscissile ; scales in the corolla-tube large, broadly oblong, fimbriate, embracing the styles and thus closing the throat of the corolla ; corolla exceeding the
 calyx, tube shortly campanulate, about 2 mm. long, lobes spreading, ovate, obtuse,
 2·5 mm. long, margin slightly and irregularly crenate .. .. 3. *blepharolepis*

1. **C. australis** *R. Br.* Prod. Fl. Nov. Holl., ed. 1, 491 (1810) ; Yuncker in Mem. Torrey Bot. Club 18 : 124,
fig. 1 ; van Oostatr. in Blumea 3 : 66 (1938) ; Fl. Males., ser. 1, 4 : 392 ; Verdcourt in E. Afr. Agric.
Journ. 18 : 85, 86 ; Meeuse in Bothalia 6 : 647 (1958). *C. cordofana* (Engelm.) Yuncker l.c. 127, fig. 2
(1932). *C. chinensis* of F.T.A., 4, 2 : 204, partly (W. Afr. specimens) ; of F.W.T.A., ed. 1, 2 : 219, not
of Lam. (1786). Leafless parasitic twiner with very slender tangled stems bright yellow or reddish-orange
in colour ; flowers small, white ; mostly on marsh plants.
 S.L. : Makomba (fl. & fr. Feb.) *Deighton* 4081 ! Panguma (fl. & fr. Aug.) *Smythe* 88 ! **Lib.** : Gbanga (fl.
& fr. Sept.) *Linder* 544 ! Vahon (fl. & fr. Nov.) *Baldwin* 10238 ! Bobli Mt. (fl. & fr. Sept.) *Baldwin* 9567 !
Geo (fl. & fr. Dec.) *Baldwin* 10562 ! **Iv.C.** : Danané (fl. & fr. Nov.) *de Wilde* 896 ! **Ghana** : Aburi (fl. & fr.
June) *Johnson* 157 ! Koforidua (fl. & fr. July) *Vigne* FH 4249 ! **S.Nig.** : Lagos (fl. & fr. July ; Dec.)
Dalz. 1152 ! 1353 ! [Br.]**Cam.** : Banso (Oct.) *Tamajong* FHI 23459 ! Widespread in the Old World tropics.
2. **C. campestris** *Yuncker* l.c. 138, fig. 14 (1932) ; van Oostatr. in Blumea 3 : 68 ; Verdcourt l.c. 85–86 ;
Meeuse, l.c. 648. A parasite like the last ; very common in some parts of tropical Africa and a serious pest
on cultivated plants.
 [Br.]**Cam.** : Kumbo (fl. & fr. Feb.) *Hepper* 1980 ! A native of America, now a very widespread weed in
warmer parts of the world.
3. **C. blepharolepis** *Welw. ex Hiern* Cat. Welw. 1 : 743 (1898) ; F.T.A. 4, 2 : 205 ; Yuncker l.c. 198, fig. 68.
A saffron coloured parasite.
 Guin.: Nzérékoré (fl. & fr. Sept.) *Baldwin* 13280 ! Also in Angola.

2. CALYCOBOLUS Willd. ex Roem. & Schult. Syst. Veg. 5 : II (1819) ; House in Bull. Torrey Bot. Club 34 143 ; Heine in Kew Bull. 16 : 387 (1963). *Prevostea* Choisy, Ann. Sci. Nat. 4 : 496 (1825) : F.T.A. 4, 2 : 81.

Leaves very shortly acuminate or not acuminate, ovate or ovate-elliptic, rounded at the base, 8–14 cm. long, 5–8 cm. broad, glabrous, pitted in the axils of the 4–7 pairs of lateral nerves ; petiole 2–2·5 cm. long, glabrous ; cymes several-flowered ; sepals softly tomentose ; corolla 1·5–2·5 cm. long, the 5 lobes slightly pubescent outside; style-arms unequal ; two sepals in fruit very unequal, one twice as large as the other, orbicular, cordate at the base, the larger about 5 cm. diam., venose   1. *heudelotii*
Leaves more or less long-acuminate :
  Corolla 1·5 cm. long or more :
    Petiole tomentose ; leaves shortly and broadly narrowed at base, obovate-elliptic, about 15 cm. long, 7 cm. broad, with 7–8 pairs of lateral nerves not pitted in their axils ; pedicels pubescent ; two sepals in fruit orbicular, one ⅔ as large as the other, the larger 5 cm. diam.   ..   ..   2. *insignis*
    Petiole slightly pubescent, 1 cm. long ; leaves oblong-oblanceolate, long-tailed-acuminate, gradually narrowed from above the middle to the obtuse base, 10–15 cm. long, 3·6 cm. broad, minutely glandular-punctate below, with 8–10 pairs of prominently looped lateral nerves ; corolla 2 cm. long, very shortly lobed ; two sepals in fruit very unequal, broadly ovate-orbicular, slightly pointed, one less than ⅓ as large as the other, the larger 7 cm. long   ..   ..   3. *africanus*
  Corolla up to 12 mm. long ; leaves ovate-lanceolate to oblong-elliptic, broadly acuminate, rounded to acuminate at base, 7–15 cm. long, 3–8 cm. broad, with about 8 pairs of lateral nerves prominently looped :
    Leaves minutely pubescent on the nerves beneath, becoming nearly glabrous, petiole pubescent, up to 1 cm. long ; pedicel and calyx minutely pubescent and appearing nearly glabrous ; corolla urceolate, slightly inflated, slightly pubescent only near the margin   ..   ..   ..   4. *micranthus*
    Leaves and the 5–20 mm.-long petioles entirely glabrous ; pedicels and calyx persistently whitish to greyish pubescent ; corolla tubular, straight, pubescent
                                                       5. *parviflorus*

1. **C. heudelotii** (*Bak. ex Oliv.*) *Heine* in Kew Bull. 15 : 390 (1963). *Prevostea heudelotii* (Bak. ex Oliv.) Hallier f. in Bull. Herb. Boiss. 5 : 1009 (1897) ; F.T.A. 4, 2 : 83 (1905), incl. var. *minor* Rendle. *Breweria heudelotii* Bak. ex Oliv. in Hook. Ic. Pl. t. 2276 (1894). *Baillaudea mirabilis* (Bak.) G. Roberty in Candollea 14 : 25 (1952), partly. *P. gilgiana* Pilger (1908). *Stachyanthus nigeriensis* S. Moore (1920) and *Chlamydocarya rostata* Bullock in Kew Bull. 1933 : 469 (1933), as to leaves ; (see F.W.T.A., ed. 2, 1 : 643). Climbing shrub, glabrous, with shining leaves ; flowers in terminal inflorescences, white, corolla variable in length ; in forest.
Port.G. : Catio (fr. May) *Esp. Santo* 2056 ! **Guin.** : Fouta Djalon *Heudelot* 864 ! Labé *Chev.* 12232 ; 12293. Bomboli *Chev.* 13566. **S.L.** : Hill Station, Freetown (Apr.) *Deighton* 5024 ! Rokupr (Apr.) *Jordan* 201 ! **Iv.C.** : Sassandra (fl. & fr. Dec.) *Leeuwenberg* 2208 ! **Ghana** : Owabi (Mar.) *Andoh* 4311 ! Kumasi (Dec.) *Dalz.* 121 ! Kamokra to Boinso (Dec.) *Adams* 2220 ! Aburi Scarp and Hills (fl. & fr. Feb., Apr.) *Adams* 3773 ! *Johnson* 614 ! *Morton* A3644 ! Dodowa (Jan.) *Johnson* 585 ! **S.Nig.:** Lagos (fl. & fr. Feb.) *Dalz.*

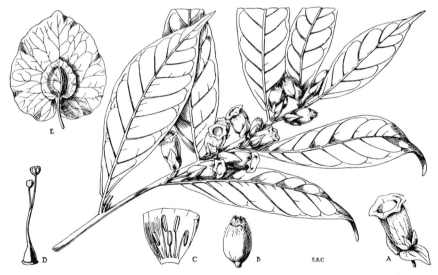

FIG. 282.—CALYCOBOLUS AFRICANUS (*G. Don*) *Heine* (CONVOLVULACEAE).
A, flower. B, corolla in bud. C, open lower part of corolla. D, pistil. E, fruit.

1069! Olokemeji (Feb.) *Richards* 5011! Ibadan to Lagos (fl. & fr. Mar.) *Meikle, Keay & Davey* FHI 25689! *Onochie* FHI 201693! Acharane (Mar.) *Okafor* FHI 36881! Gambari F.R. (Mar.) *Hepper* 2289! [Br.]**Cam.**: Mamfe *Maitland* 1167 (partly, leaves)! Also in Cameroun.

2. **C. insignis** (*Rendle*) *Heine* l.c. 390 (1963). *Prevostea insignis* Rendle in F.T.A. 4, 2 : 571 (1906). A shrub with slightly brown-pubescent branchlets ; flowers unknown, fruits with large membranous veined sepals, sometimes purple-tinged.
    Lib. : Kakatown *Whyte*! Mecca (fr. Nov.) *Baldwin* 10429!

3. **C. africanus** (*G. Don*) *Heine* l.c. 388 (1963). *Prevostea africana* (G. Don) Benth. in Fl. Nigrit. 469, t. 46 (1849) ; F.T.A. 4, 2 : 82 ; Chev. Bot. 459. *Codonanthus africanus* G. Don Gen. Syst. 4 : 166 (1838). *C.? alternifolia* Planch. in Hook. Ic. Pl. t. 796 (1848). *Breweria codonanthus* Bak. ex Oliv. in Hook. Ic. Pl. t. 2276 (1894). *B. alternifolia* (Planch.) Radlk. (1884). *Prevostea nigerica* Rendle Cat. Talb. 72 (1913) ; F.W.T.A., ed. 1, 2 : 209. A woody climber with pubescent branches becoming glabrous ; flowers white. **S.L.:** *G. Don*! Kafogo (fr. Apr.) *Sc. Elliot* 5518! **Iv.C.:** Malamalasso *Chev.* 17491. Niapidou (Feb.) *Leeuwenberg* 2765! **Ghana:** Kimso (Feb.) *Chipp* 110! Pra River Station (Feb.) *Vigne* FH 1038! 1826! *Andoh* 3253! **S.Nig.:** Lagos *Moloney*! Sapoba *Kennedy* 2139! Ibadan (fr. Mar.) *Idahosa* FHI 23862! Okomu F.R. (Jan.) *Brenan* 8882! Oban *Talbot* 1484! Also in Cameroun, Rio Muni, Cabinda and Congo.

4. **C. micranthus** (*U. Dammer*) *Heine* l.c. 390 (1963). *Prevostea micrantha* U. Dammer in Engl. Bot. Jahrb.2 3, Beibl. 57 : 57 (1897) ; F.T.A. 4, 2 : 82. A shrub 7–9 ft. high, with densely puberulous branchlets when young ; flowers white, rather small, few together on short axillary peduncles sometimes below the leaves.
    [Br.]**Cam.** : Johann-Albrecht-Höhe *Staudt* 637!
    [Note : Only known from the type collection and perhaps only an occasional small-flowered form of *C. africanus* (G. Don) Heine—H. H.]

5. **C. parviflorus** (*Mangenot*) *Heine* l.c. (1963). *Prevostea parviflora* Mangenot in Bull. I.F.A.N. 19 : 359 (1957). A glabrous climber up to 30 ft. high ; flowers tubular, yellowish-green up to ½ in. long, in very short axillary cymes ; in swamp forest.
    Lib. : Moala (Nov.) *Linder* 1374! **Iv. C.** : Adiopodoumé (fl. & fr. Dec.) *Mangenot*!

3. NEUROPELTIS Wall.—F.T.A. 4, 2 : 80.

Inflorescences long-spicate, 12–30 cm. long, up to 50-flowered ; flowers hypocrateriform, up to 7 mm. long ; leaves elliptic tending to obovate, up to 10 cm. long and 5 cm. broad, subcoriaceous, older leaves always flat :
    Branches and leaves beneath very finely whitish-pubescent, becoming glabrous
                                              1. *acuminata*
    Branches and leaves beneath densely ferrugineously hairy   ..   2. *velutina*
Inflorescences short-racemose, 2–5 cm. long, 4–10 flowered ; flowers campanulate, up to 16 mm. long ; leaves elliptic, 8–14 cm. long and 4–6 cm. broad, chartaceous, older leaves very often folded towards the upper central part ; fruiting bracts up to 4 cm. diam.   ..   ..   3. *prevosteoides*

1. **N. acuminata** (*P. Beauv.*) *Benth.* in Fl. Nigrit. 469 (1849) ; F.T.A. 4, 2 : 80 ; De Wild. Pl. Bequaert. 1 : 537 (1922) ; Mangenot in Rev. Bot. Appliq. 31 : 521 (1951). *Porana acuminata* P. Beauv. Fl. Oware 1 : 66, t. 39 (1805). *N. vermoesenii* De Wild l.c. 542 (1922). A climbing shrub with shining leaves and small white flowers in narrow racemes 3–6 in. long forming a panicle, with a large papery veined bract in fruit. **S.L.** : *Don*! Mano *Smythe* 119! **Lib.** : Peahtah (Oct.) *Linder* 1048! Gbanga (Sept.) *Linder* 658! **Ghana** : (Dec.) *Vigne* FH 194! Bia F.R. (Feb.) *Foggie* 4447! Bobiri F.R. (Jan.) *Andoh* FH 5460! **N.Nig.** : Dogon Kurmi, Jemaa Div. (fr. Sept.) *Killick* 66! **S.Nig.** : *P. Beauvois*! Sapoba (Nov.) *Ejiofor* FHI 24651! *Kennedy* 1886! 1937! Benin (fl. & fr. Jan.) *Keay* FHI 37350! Okomu F.R. (Jan.) *Onochie* in Hb. *Brenan* 8885! Ibadan South F.R. (Sept.) *Keay* FHI 28042! Calabar *Williams*! [Br.]**Cam.**: Victoria *Kalbreyer* 11! Also in Cameroun, Gabon, Congo and Angola.

2. **N. velutina** *Hallier f.* in Bull. Herb. Boiss. 5 : 374 (1897) ; F.T.A. 4, 2 : 81 ; De Wild., Pl. Bequaert. 1 : 540 ; Mangenot l.c. 521. *N. anomala* Pierre ex De Wild. Pl. Bequaert. 1 : 539 (1922). Like the last, but the whole plant ferrugineously hairy. **S.L.** : Mabomba (Nov.) *Deighton* 3772! Mabonto (Oct.) *Thomas* 3575! **Lib.** : *Harley* 492! Ganta (fr. Jan.) *Baldwin* 14049! Kondessu (fr. Dec.) *Baldwin*! **S.Nig.** : Gambari F.R. (fr. July) *Symington* FHI 4117! Ibadan (fr. July) *Ahmed & Chizea* FHI 20008! Eket *Talbot*! Also in Cameroun, Gabon and Congo.

3. **N. prevosteoides** *Mangenot* l.c. 521 (1951). A lofty climber with relatively small golden-brownish pilose inflorescences and yellowish-white flowers with bracts becoming membranaceous-transparent in the fruiting state.
    Lib.: Gbanga (Sept.) *Linder* 818! **Iv.C.:** Niapidou (fr. Jan.) *Leeuwenberg* 2513! Adiopodoumé *Mangenot*. [According to Mangenot (l.c. 521) common in all rain forests of Ivory Coast.]

4. DICHONDRA J. G. & R. Forster—F.T.A. 4, 2 : 65.

Leaves reniform, 1–3 cm. broad, more or less silky, especially on the lower face, long-petioled ; flowers axillary, solitary ; peduncles shorter than the petiole ; calyx up to 2 mm. long, sepals oblong to oblong-spathulate ; corolla yellow, slightly shorter than the calyx ; utricles of the fruit about 2 mm. broad   ..   ..   ..   *repens*

D. repens *J. G. & R. Forst.* Char. Gen. 40, t. 20 (1776) ; F.T.A., l.c. 65 ; van Ooststr. in Blumea 3 : 72 (1938) ; Meeuse in Bothalia 6 : 657 (1958). A perennial herb with slender trailing pubescent stems rooting from the nodes, in moist places.
    Maur. : Kaédi to Mbagne (Dec.) *Mosnies* in Hb. *Raynal* 7538! **Mali** : " Fleuve Blanc " *Peney*! [Br.]**Cam.**: Cam. Mt., 3,000 ft. *Maitland* 1310! Bamenda, 5,000 ft. *Maitland* 145! Oku (fl. & fr. Feb.) *Hepper* 2024! Lus, Nkambe Div. (fl. & fr. Feb.) *Hepper* 1861! Cosmopolitan and widely spread in the warmer regions of both hemispheres.

5. EVOLVULUS Linn.—F.T.A. 4, 2 : 66 ; van Ooststr. in Meded.
Bot. Mus. Herb. Utrecht 14 : 1–267 (1934). *Volvulopsis* Roberty (1952).

Leaves elliptic, oblanceolate-oblong to lanceolate ; corolla shallowly lobed ; capsule 2-celled   ..   ..   1. *alsinoides*
Leaves orbicular or orbicular-obovate or elliptic ; corolla deeply lobed, capsule 1-celled
                                                  2. *nummularius*

1. **E. alsinoides** (*Linn.*) *Linn.* Sp. Pl., ed. 2, 1 : 392 (1762) ; F.T.A. 4, 2 : 67 ; Chev. Bot. 458 ; van Ooststr. l.c. 26. *Convolvulus alsinoides* Linn. Sp. Pl. 157 (1753). A tufted or spreading hairy herb ; woody at the base, with sky-blue funnel-shaped flowers ¼ in. across.
Sen. : *Leprieur* ! *Roger* 12 ! *Heudelot* 229 ! Dakar *Chev.* 2891. Mt. Roland *Chev.* 2903. **Gam.** : *Hayes* 554 ! **Mali** : Siguiri *Chev.* 289. Sikasso *Chev.* 759. Gao (fl. & fr. Sept.) *Hagerup* 350 ! Ansongo (fl. & fr. Sept.) *Hagerup* 387 ! Karora Hills *Robbie* 18 ! Dioura (fl. & fr. Oct.) *Davey* 087 ! **Port.G.** : Praia de Varela (Aug.) *Esp. Santo* 3071 ! **Guin.** : Kouroussa *Brossart* in *Hb. Chev.* 15750. **Iv.C.** : Orodoungou *Chev.* 21810. Bouaké to Mt. Lémé-libou *Chev.* 22100. Bouaké (fl. & fr. Nov.) *Leeuwenberg* 2087 ! **Ghana** : Accra (fl. & fr. Oct.) *T. Vogel* ! *Brown* 360 ! Achimota (fl. & fr. Dec.) *Andoh* 4473 ! Amansare (fl. & fr. Aug.) *Chipp* 520 ! Tamale (fl. & fr. Feb.) *Williams* 483 ! Navrongo (fl. & fr. June) *Vigne* FH 4531 ! **Togo Rep.** : Lomé *Warnecke* 212 ! **N.Nig.** : Fobur, Bauchi (July) *Summerhayes* 14 ! Kogigiri, Jos (fl. & fr. Oct.) *Hepper* 1053 ! Jos Plateau (fl. & fr. Feb.) *Lely* 164 ! R. Benue *Talbot* 799 ! Baradan (fl. & fr. May) *Lely* 89 ! Katagum Dist. (fl. & fr. July) *Dalz.* 189 ! [**Br.**]**Cam.** : Bama, Dikwa Div. (fl. & fr. Nov.) *McClintock* 26 ! Gurum, Adamawa (fl. & fr. Nov.) *Hepper* 1274 ! Throughout the tropics.

2. **E. nummularius** (*Linn.*) *Linn.* Sp. Pl., ed. 2, 1 : 391 (1762) ; F.T.A. 4, 2 : 68 ; Chev. Bot. 458 ; van Ooststr. l.c. 114. *Convolvulus nummularius* Linn. Sp. Pl. 157 (1753). *Volvulopsis nummularium* (Linn.) G. Roberty (1952). A perennial prostrate herb ; flowers small, white, rarely pale blue.
Iv.C. : Marabadiassa *Chev.* 22012. **Ghana** : Achimota (fl. & fr. Aug.) *Milne-Redhead* 5127 ! Akropong, Ashanti (fl. & fr. June) *Darko* 685 ! Lawra (Apr.) *Adams* 4081 ! Throughout tropical Africa, also in tropical America and in India.

6. CRESSA Linn.—F.T.A. 4, 2 : 72.

A much-branched herb ; branchlets dense, pubescent ; leaves ovate, sessile, subacute, 5–6 mm. long, thinly pilose ; flowers subsessile, crowded at the tops of the branchlets, the latter forming a panicle ; calyx campanulate, pubescent ; corolla 4–5 mm. long, lobes sharply reflexed ; stamens exserted ; stigma capitate ; fruit ovoid, shortly beaked, slightly nerved, 4 mm. long, glabrous ; seed ovoid, dark brown   *cretica*

C. cretica *Linn.* Sp. Pl. 223 (1753) ; F.T.A. 4, 2 : 72 ; Chev. Bot. 459 ; Berhaut Fl. Sén. 5, 6, 203. Stems woody below, slender, spreading or ascending from a few inches to 1 ft. long ; in damp sandy places, especially near the sea.
Sen.: *Leprieur* ! *Roger* 73 ! *Heudelot* 409 ! *Berhaut* 201 ! Dakar *Chev.* 15812. **Mali:** Niayes (fl. & fr. Dec.) *Chev.* 2906 ! **Port.G.:** Bambadinca, Bafáta (fl. & fr. June) *Esp. Santo* 2518 ! **Guin.:** Conakry *Maclaud* ! Also in the Mediterranean region, Socotra, E. Africa, Madagascar, and in many other places of both hemispheres.

7. BONAMIA Thouars—F.T.A. 4, 2 : 78 ; van Ooststr. in Blumea 3 : 75 (1938), and in Fl. Males., ser. 1, 4 : 398 (1953) ; Meeuse in Bothalia 6 : 664 (1958). *Nom. cons.*

Leaves oblong-ovate, mucronate, rounded at the base, 4–6 cm. long, 2–3·5 cm. broad, glabrous or nearly so above, appressed-silky-tomentose beneath to nearly glabrous, with about 8 pairs of lateral nerves impressed above, prominent beneath ; petiole 1 cm. long ; flowers in shortly pedunculate axillary cymes ; pedicels tomentose, about 1 cm. long ; calyx-segments ovate, acuminate, about 8 mm. long, tomentose ; corolla 2 cm. long, pilose outside, slightly pilose to nearly glabrous in the throat, lobes only up to 4 mm. long ; fruit ovoid, apiculate, about 1 cm. long ; seeds black, glabrous   *1. thunbergiana*

Leaves oblong to lanceolate-elliptic or obovate, 5–12 cm. long, 2–5 cm. broad, at first slightly pilose on both sides and with ciliate margin, becoming glabrous or nearly so above, slightly furfuraceous and pilose on the midrib and lateral nerves beneath, with 5–7 pairs of lateral nerves ; petiole 3–7 mm. long, tomentose ; flowers in dense, subsessile terminal cymes ; pedicels tomentose, 4–6 mm. long ; calyx-segments elliptic to broad-elliptic, 4–7 mm. long, and 3–5 mm. broad, tomentose outside, glabrous inside, accrescent in fruit, becoming slightly orbicular and about 7 mm. diam. ; corolla up to 15 mm. long, appressed-tomentose outside, very villose in the throat, lobes half as long as the corolla-tube ; fruit obliquely ovoid, apiculate, 8–10 mm. in diam., seeds dark red and with a persistent blackish aril   ..   *2.vignei*

1. **B. thunbergiana** (*Roem. & Schult.*) *F. N. Williams* in Bull. Herb. Boiss., 2e sér., 7 : 371 (1907); Heine in Kew Bull. 14 : 276 (1960). *Convolvulus thunbergianus* Roem. & Schult. Syst. Veg. 4 : 884 (1819). *C. cymosus* Roem. & Schult. l.c. 303 (1819), not of Desrousseaux in Lam. (1791), or of Ruiz & Pavon (1799). *Bonamia cymosa* (Roem. & Schult.) Hallier f. in Engl. Bot. Jahrb. 18 : 91 (1893) ; F.T.A. 4, 2 : 79 ; Chev. Bot. 459 ; F.W.T.A., ed. 1, 2 : 210. A woody climber and twiner, with wrinkled leaves brownochry beneath, and rather small white flowers ⅜ in. long in dense cymes forming one-sided panicles.
Gam. : *Ingram* ! **Port.G.** : Catio (Dec.) *Esp. Santo* 2236 ! **Guin.** : Kouria to Ymbo *Caille* in *Hb. Chev.* 14754. **Iv.C.** : Kouria *Caille* 14795. **S.L.** : *Afzelius* ! *Don* ! *Hart* ! *Barter* ! Makumpo (Dec.) *Glanville* 108 ! Mamaneu (Nov.) *Marmo* 145 ! Kambui Hills (Dec.) *Small* 857 ! **Lib.** : Dukwia R. (fr. Feb.) *Cooper* 220 ! Monrovia *Whyte* ! Kakatown *Whyte* ! Banga (fl. & fr. Feb.) *Linder* 1190 ! Dobli Island *Bequaert* 19 ! Ganta (Apr.) *Harley* ! **Iv.C.:** Beyo (Jan.) *Leeuwenberg* 2455 ! **Ghana:** Assin *Cummins* ! Sampa (fr. Mar.) *Irvine* 1583 ! Axim (?) (fr. Feb.) *Irvine* 2366 ! Bibiani to Kumasi (Dec.) *Adams* 1941 ! Mpameso F.R. (Dec.) *Adams* 2943 ! **S.Nig.:** Lagos *Barter* 2167 ! 2227 ! *Millen* 34 ! Ikoyi Plains (Dec.) *Dalz.* 1157 ! Oban *Talbot* 88 ! 88a ! 1535 ! Also in Cameroun. (See Appendix, p. 435.)

2. **B. vignéi** *Hoyle* in Kew Bull. 1934 : 188. A woody climber with dark brown pilose stems, inflorescences and petioles, leaves nearly glabrous ; flowers small, white, in corymbose inflorescences (never forming one-sided panicles) ; in forest.
Ghana : Tiasi (Oct.) *Vigne* FH 1387 ! Owabi (fr. Dec.) *Lyon* 2631 !

8. CONVOLVULUS Linn.—F.T.A. 4, 2 : 88 ; Verdcourt in Kew Bull. 12 : 344 (1957) ; Meeuse in Bothalia 6 : 666 (1958).

Leaves linear-oblanceolate, acute at the base, rounded and mucronate at the apex, 2–3 cm. long, 3–6 mm. broad, pilose ; flowers solitary or paired in the leaf-axils ;

calyx-segments lanceolate, acuminate, 8 mm. long, densely pilose ; corolla 1 cm. long
.. .. 1. *microphyllus*
Leaves lanceolate, hastate-lobulate at the base, about 4 cm. long, shortly pubescent ;
flowers paired on a common peduncle ; calyx-segments about 8 mm. long, setulose ;
corolla 1·3-1·5 cm. long   ..   ..   2. *aschersonii*

1. **C. microphyllus** *Sieb. ex Spreng.* Syst. Veg. 1 : 611 (1824) ; Hallier f. in Engl. Bot. Jahrb. 18 : 98 (1893) ;
F.T.A. 4, 2 : 91 ; F.W. Andr. Fl. Pl. Sudan 3 : 106 ; Berhaut Fl. Sén. 200, 287 ; Carvalho & Gillet in
J. Agric. Trop. 7 : 240 ; Heine in Kew Bull. 16 : 205. *Ipomoea microphylla* Roth (1821). *Convolvulus
pluricaulis* Choisy (1834)—Hallier f. l.c. ; F.T.A. 4, 2 : 91 ; F.W.T.A., ed. 1, 2 : 210. *C. scindicus* Boiss.
(1856), not of Stocks (1852). Stems slender, rusty pilose, from a woody stock, erect or spreading, with
whitish shortly funnel-shaped flowers ⅓ in. long.
Sen.: *Heudelot* 403 ! *Leprieur* (*fide* Hallier f. *l.c.*). *Berhaut* 1009. Also in Sudan, Egypt, Arabia, W. Pakistan
and N. India.
2. **C. aschersonii** *Engl.* Hochgebirgsfl. trop. Afr. 349 (1892 ; = Abh. Preuss. Akad. Wiss. 1891, 2) ; Meeuse
in Bothalia 6 : 677 (1958). *C. sagittatus* Thunb. Prod. 35 (1794), var. *abyssinicus* (Hallier f.) Bak. &
Rendle in F.T.A. 4, 2 : 96 (1905) ; F.W.T.A., ed. 1, 2 : 210. *C. sagittatus* var. *parviflorus* subvar. *abyssinicus* Hallier f. in Bull Herb. Boiss. 6 : 533 (1898). *C. sagittatus* var. *aschersonii* (Engl.) Verdc. in Kew
Bull. 12 : 345 (1957). Stems prostrate, grey-greenish, pubescent, with pink-purplish or white flowers
⅓ in. long.
N.Nig. : Hepham to Ropp, 4,600 ft. (July) *Lely* 362 ! Widespread in the drier parts of tropical Africa
from Eritrea and Ethiopia to Angola, Bechuanaland, S. Rhodesia and Transvaal.

9. JACQUEMONTIA Choisy—F.T.A. 4, 2 : 85 ; van Ooststr. in Blumea 3 : 267 (1939),
Fl. Males., ser. 1, 4 : 431 (1953) ; Meeuse in Bothalia 6 : 699 (1958).

Flowers subumbellate, few ; pedicels about 1 cm. long, slender ; bracts subulate ;
leaves oblong-lanceolate, rounded at the apex or emarginate, 3-4 cm. long, 1-2·5 cm.
broad, entire, glabrous ; calyx-segments ovate, 5 mm. long ; corolla 7-8 mm. long ;
capsule about 5 mm. diam.   ..   ..   1. *ovalifolia*
Flowers capitate, surrounded by leafy bracts ; leaves broadly ovate, acuminate, cordate
to rounded at the base, up to 8 cm. long and 5·5 cm. broad, glabrous or nearly so
beneath ; calyx-segments 1 cm. long, linear-lanceolate, long-pilose with brown hairs ;
corolla about 9 mm. long ; capsule 4-5 mm. diam.   ..   ..   2. *tamnifolia*

1. **J. ovalifolia** (*Vahl*) *Hallier f.* in Engl. Bot. Jahrb. 18 : 96 (1893) ; F.T.A. 4, 2 : 87. *Convolvulus ovalifolius*
Vahl Eclog. Am. 2 : 16 (1798). *C. coeruleus* Schum. & Thonn. (1827). Prostrate and ascending from a half-
woody base, glabrous, with pale blue flowers ⅓ in. long.
Ghana : Accra *Thonning.* (June-July) *T. Vogel* 11 ! *Brown* 374 ! *Dalz.* 8 ! *Irvine* 668 ! 689 ! Prampram
(Nov.) *Morton* 6075 ! Nungua, Accra Plains (fl. & fr. May) *Rose Innes* GC 30064 ! **Togo Rep.** : Lomé *Warnecke* 254 ! Also in other coastal regions in tropical Africa (Somaliland, tropical E. Africa and Angola).
[The type is from the West Indies, but the plant has never been recorded from there since the type
collection was made.]
2. **J. tamnifolia** (*Linn.*) *Griseb.* Fl. Brit. W. Ind. 474 (1861) ; Bruce in Kew Bull. 1940 : 63 ; Meeuse in
Bothalia 6 : 700. *Ipomoea tamnifolia* Linn. (1753). *Convolvulus capitatus* Desr. (1791). *Jacquemontia capitata* (Desr.) G. Don (1837)—F.T.A. 4, 2 : 85 ; Chev. Bot. 457 ; F.W.T.A., ed. 1, 2 :
211. *Convolvulus guineensis* Schum. (1827). A slender twiner, pilose, with blue flowers.
Sen. : Richard Tol (fl. & fr. Sept., Oct.) *Roger* 9 ! *Farmar* 38 ! **Mali** : Télé *Chev.* 2901. Sompi *Chev.* 2902.
San *Chev.* 2904. **Port.G.** : Bissau (Oct.) *Esp. Santo* 2533 ! **Lib.** : (July) *T. Vogel* 38 ! **Ghana** : Accra Plains
(Jan., May) *Irvine* 691 ! *Dalz.* 120 ! *Ansell* ! Nungua (May) *Rose Innes* GC 30047 ! Damongo (Nov.) *Harris* !
Togo Rep. : Lomé *Warnecke* 243 ! **Dah.** : *Burton* ! **N.Nig.** : *Baikie* 3 ! Jebba *Barter* ! Sokoto (fl. & fr.
Nov.) *Lely* 158 ! *Moiser* 72 ! Katagum Dist. (fl. & fr. Sept.) *Dalz.* 179 ! Bornu (fl. & fr. Nov.) *Elliott* 130 !
[Br.]Cam. : Bama, Dikwa Div. (fl. & fr. Jan.) *McClintock* 157 ! A native of tropical America, found there
in all warmer parts, naturalized in tropical and S. Africa, and the Mascarene Islands.

J. pentantha (Jacq.) G. Don (syn. *Convolvulus pentanthus* Jacq.), a native of the warmer parts of America, is
sometimes cultivated as an ornamental plant.

10. OPERCULINA Silva Manso, Enum. Subst. Brazil. 16 (1836) ; van Ooststr. in
Blumea 3 : 361 (1939), Fl. Males., ser. 1, 4 : 454 (1953).

Stems stout, hollow, conspicuously 4-winged on the older portions ; leaves up to 13 cm.
long and 18 cm. broad, palmatisect with 5 or more more or less elliptic lobes, with
bluntish mucronulate apex and tapering base, veins beneath winged and conspicuous,
sometimes obsoletely puberulous ; petiole stout, narrowly winged, nearly as long as
the blade ; peduncle about 4 cm. long ; few-flowered ; bracts deciduous, membranous, bluntly ovate, about 7 mm. long ; pedicels stoutly clavate and broadly winged,
about 2·5 cm. long ; calyx broadly cup-shaped, up to 2·5 cm. long and broad, glabrous,
lobes broadly obovate to orbicular ; corolla about 6·5 cm. long and nearly as broad ;
fruits depressed-globose, up to 3·5 cm. diam. (the persistent calyx included), bilocular
(2-)3-4-seeded, epicarp fleshy, in the superior part circumscissile and forming a little
cap (" operculum ") ; seeds about 1 cm. diam., glabrous   ..   ..   *macrocarpa*

O. macrocarpa (*Linn.*) *Urban* Symb. Antill. 3 : 343 (1902) (for full synonymy) ; Heine in Kew Bull. 14 : 397.
Convolvulus macrocarpus Linn. Syst. Nat., ed. 10, 2 : 923 (1759) ; Sp. Pl., ed. 2, 222 (1762). *Operculina
convolvulus* Silva Manso l.c. 49 (1836). *Merremia alata* Rendle in F.T.A. 4, 2 : 102 (1905) ; F.W.T.A., ed.
1, 2 : 212. A stout glabrous climber with winged hollow stems ; flowers widely funnel-shaped, creamy-
white.
Ghana : Takoradi (fl. & fr. Oct.) *Howes* 979 ! Odumassu (Oct.) *Johnson* 155 ! Krobo plains (Dec.) *Johnson*
517 ! Cape Coast (fr. Sept.) *Hall* 1600 ! **Togo Rep.** : Lomé *Warnecke* 273 ! (fl. & fr. Nov.) *Mildbr.* 7502 !
A native of Brazil, also found in Martinique and Guadeloupe, apparently long-introduced and naturalized
in our area. (See Appendix p. 440).

11. MERREMIA Dennst. ex Hallier f.—F.T.A. 4, 2 : 101 ; van Oostst. in Blumea 3 : 292 (1939), Fl. Males., ser. 1, 4 : 439 (1953) ; Meeuse in Bothalia 6 : 700 (1958).

Leaves deeply pinnatipartite, up to 4 cm. long, segments linear, thinly pilose ; peduncles axillary, 2–3 cm. long, pilose, 1–3-flowered ; calyx-segments with long linear points ; corolla narrowly funnel-shaped, a little longer than the calyx ; fruit globose, pilose .. .. 1. *pinnata*
Leaves not pinnately divided :
  Leaves linear-lanceolate or linear, usually hastately-lobulate or toothed at the base, subsessile, 3–5 cm. long, glabrous ; peduncles slender, axillary, 1–3-flowered, bracteate towards the top ; calyx-segments 5 mm. long, mucronate ; corolla 1·5 cm. long ; fruit ovoid-globose, about 8 mm. long   2a. *tridentata* subsp. *angustifolia*
  Leaves neither linear nor lanceolate :
    Leaves more or less ovate, entire or 3-lobed to the middle :
      Corolla less than 1 cm. long ; calyx-segments dry and chaffy, 6 mm. long, obovate-elliptic ; pedicels very short ; leaves small, mostly shortly trilobed ; fruit quadrangular-globose, about 8 mm. diam., glabrous .. .. 3. *hederacea*
      Corolla 3 cm. long ; calyx segments 1 cm. long, leathery ; pedicels 1·5 cm. long ; leaves broadly ovate, deeply cordate, about 12 cm. long and 7–10 cm. broad, glabrous or pubescent on the midrib beneath ; fruit 1·5 cm. long ; seeds covered with short hairs .. .. 4a. *umbellata* subsp. *umbellata*
    Leaves deeply divided :
      Calyx-segments and branchlets densely covered with long slender hairs ; leaves divided completely to the base into 5 obovate acutely acuminate segments up to 8 cm. long and 4 cm. broad ; calyx-segments 1·5–2 cm. long in flower, increasing to 3 cm. in fruit ; corolla 2–2·5 cm. long ; seeds glabrous .. 5. *aegyptia*
      Calyx-segments glabrous ; branchlets usually glabrous :
        Calyx 2–3 cm. long :
          Segments of the leaves sinuate-dentate, acute, 7–10 cm. long, lanceolate ; stems and often the petioles usually with long slender hairs ; corolla 3·5 cm. long ; fruit ovoid, 1·5 cm. long ; seeds glabrous .. .. .. 6. *dissecta*
          Segments of the leaves nearly entire, acute 7–9 cm. long, oblanceolate ; stems and petioles glabrous ; corolla about 5 cm. long .. .. .. 7. *kentrocaulos*
        Calyx up to about 1 cm. long :
          Leaves digitately lobed ; stems winged in the lower part ; leaf-segments variable, from linear- to broadly-lanceolate, long mucronate, up to 10 cm. long and 3 cm. broad, pubescent towards the base of the principal nerves beneath ; corolla-lobes densely setose in bud .. .. 8. *pterygocaulos*
          Leaves subpinnately divided with 2 lobes on each side, the terminal lobe linear-lanceolate, 5–6 cm. long, thinly pilose on the nerves beneath ; peduncle densely pubescent, short ; pedicels 5 mm. long ; calyx 8 mm. long
                          9a. *pes-draconis* var. *nigerica*

1. **M. pinnata** (*Hochst. ex Choisy*) *Hallier f.* in Engl. Bot. Jahrb. 16 : 552 (1893) ; F.T.A. 4, 2 : 113 ; Chev. Bot. 458 ; Berhaut Fl. Sén. 46. *Ipomoea pinnata* Hochst. ex Choisy in DC. Prod. 9 : 353 (1845). Annual, trailing and twining, with slender loosely hairy stems, and cream-white flowers ½–⅓ in. long mostly hidden by the hairy calyx.
  Sen.: *Farmar* 154 ! *Berhaut* 252 ! M'bidgem *Thierry* 18 ! Tivaouane *Chev.* 2887. Kaolak 41 *Chev.* **Gam.:** *Ingram* ! *Brown-Lester* 79 ! **Mali:** Dioura (Oct.) *Davey* 097 ! San *Chev.* 2888. Koulikoro *Chev.* 2889. Sompi *Chev.* 2895. **Port.G.:** Bissau (Oct.) *Esp. Santo* 2540 ! **S.L.:** Kitchom (fr. Dec.) *Sc. Elliot* 1891 ! *Deighton* 981 ! Juring (fl. & fr. Dec.) *Deighton* 287 ! Rokupr *Deighton* 1326 ! **Lib.:** Monrovia (Nov.) *Linder* 1437 ! Sinkor (fr. Dec.) *Harley* 1957 ! **Ghana:** Wa (fr. Nov.) *Harris* ! **N.Nig.:** Jebba *Barter* ! Sokoto (fr. Sept.) *Dalz.* 377 ! Fodama (fr. Jan.) *Moiser* 170 ! Jos Plateau (fr. Nov.) *Lely* P635 ! Bakura Tureta (fr. Jan.) *Yunus Bornu & Jibiringia* FHI 29052 ! Yola (fl. & fr. Nov.) *Hepper* 1222 ! Extends throughout tropical Africa.
2. **M. tridentata** (*Linn.*) *Hallier f.* in Engl. Bot. Jahrb. 16 : 552 (1893). *Convolvulus tridentatus* Linn. Sp. Pl. 157 (1753).
  M. tridentata subsp. **tridentata**—van Oostst. in Fl. Males., ser. 1, 4 : 445 (1953). *M. tridentata* subsp. *genuina* (Hallier f. ex Oostst.) Oostst. in Blumea 3 : 315 (1939). E. tropical Africa, Mascarene Is., tropical Asia and Malaysia.
2a. **M. tridentata** subsp. **angustifolia** (*Jacq.*) *Oostst.* l.c. 323 (1939). *Ipomoea angustifolia* Jacq. (1788). *Merremia angustifolia* (Jacq.) Hallier f.—F.T.A. 4, 2 : 111 ; Chev. Bot. 457 ; F.W.T.A., ed. 1, 2 : 211 ; Berhaut Fl. Sén. 153, 171. A glabrous annual, prostrate and twining amongst grass, etc., with pale yellow funnel-shaped flowers ; in savanna.
  Sen.: *Berhaut* 224 ; 2277. Kaolak 101 *Kaichinger* in *Hb. Chev.* **Gam.:** Saunders 33 ! *Skues* ! **Mali:** Timbuctu (fl. & fr. July) *Hagerup* 223 ! Folo *Chev.* 834. San *Chev.* 2910. Ségou *Chev.* 2911. **Guin.:** Cotonou (fl. & fr. Apr.) *Debeaux* 347 ! **S.L.:** Tisana (fr. Nov.) *Deighton* 2276 ! Yungeru (fl. & fr. Jan.) *Thomas* 7127 ! 7222 ! Mahela (fl. & fr. Dec.) *Sc. Elliot* 4891 ! Aberdeen (fl. & fr. Aug.) *Melville & Hooker* 133 ! **Lib.:** Cape Palmas *T. Vogel* 55 ! **Iv.C.:** *Farmar* 363 ! Bingerville (fl. & fr. Aug.) *de Wilde* 240 ! *Chev.* 17331. **Ghana:** Wa (fr. May) *Vigne* FH 3819 ! Accra Plains (fr. Apr.) *Bally* B25 ! Achimota *Irvine* 340 ! **Togo Rep.:** Lomé *Warnecke* 35 ! **N.Nig.:** Niger *T. Vogel* 112 ! Nupe *Barter* 1266 ! Sokoto (fl. & fr. Jan., July) *Dalz.* 379 ! *Moiser* 41 ! Katagum Dist. *Dalz.* 181 ! Makurdi (Feb.) *Jones* FHI 677 ! **S.Nig.:** Lagos (fl. & fr. Oct.) *Stubbings* 83 ! *Dalz.* 1073 ! *Dawodu* 375 ! Olokemeji F.R. (Mar.) *Keay* FHI 25688 ! [Br.]**Cam.:** Gurum, Adamawa (Nov.) *Hepper* 1270 ! In tropical Africa generally, and in S. Africa.
3. **M. hederacea** (*Burm. f.*) *Hallier f.* in Engl. Bot. Jahrb. 18 : 118 (1894) ; van Oostst. in Blumea 3 : 302, fig. 1, e–f, m–n ; Fl. Males., ser. 1, 4 : 441 ; Meeuse in Bothalia 6 : 700. *Evolvulus hederaceus* Burm. f. (1768). *Merremia convolvulacea* Dennst. ex Hallier f. (1893)—F.T.A. 4, 2 : 114 ; Chev. Bot. 457. A nearly glabrous twiner and climber, the stems often reddish-purple-tinged, with rather small yellow flowers.
  Sen.: *Berhaut* 153. Richard Tol (fr. Feb.) *Roger* 117 ! *Roberty* 16862 ! **Port.G.:** Contubo-el (fl. & fr. Nov.) *Esp. Santo* 2814 ! **S.L.:** Mofari (fr. Jan.) *Sc. Elliot* 4433 ! Mange (fr. Dec.) *Jordan* 726 ! **Iv.C.:** Makougnié

Chev. 16964! Tiassalé (fl. & fr. Dec.) *Leeuwenberg* 2134! **Ghana:** Nungua, Accra Plains (fr. Apr.) *Ankrah* GC 20170! Mankessim (fl. & fr. Dec.) *Hall* 1673! **N.Nig.:** Nupe *Barter* 885! Jebba (fl. Dec.) *Hagerup* 680! Kontagora (fr. Nov.) *Dalz.* 119! Kumu (fr. Oct.) *Lely* 668! Yola (Dec.) *Hepper* 1679! **S.Nig.:** Sasha, Ijebu Ode *Tamajong* FHI 20300! Olokemeji F.R. (Feb.) *Keay* FHI 22491! **[Br.]Cam.:** Jalo, Gashaka Dist. *Latilo & Daramola* FHI 28878! Bama, Dikwa Div. (fr. Jan.) *McClintock* 166! In the tropics of the Old World.

4. **M. umbellata** (*Linn.*) *Hallier f.* in Engl. Bot. Jahrb. 16 : 552 (1893) ; F.T.A. 4, 2 : 106 ; Chev. Bot. 458 ; van Ooststr. in Blumea 3 : 333 (1939) ; Fl. Males., ser. 1, 4 : 449 (1953) ; Verdcourt in Kew Bull. 13 : 186. *Convolvulus umbellatus* Linn. Sp. Pl. 155 (1753).

4a. **M. umbellata** (*Linn.*) *Hallier f.* subsp. **umbellata**—van Ooststr. l.c. (1953) ; Exell Cat. S. Tomé, Suppl. 35 (1956). *M. umbellata* var. *occidentalis* Hallier f. in Verslag 's Lands Pl.-tuin Buitenz. 1895 ; 127 (1896), and in Bull. Herb. Boiss. 5 : 375 (1897) ; van Ooststr. in Blumea 3 : 341, 296 (1939). *Ipomoea fragilis* of F.W.T.A., ed. 1, 2 : 216, partly (*Thomas* 305 ; 7305), not of Choisy in DC. (1845). *I. oenotheriflora* A. Chev. Bot. 455, name only. A perennial twiner, glabrous or sparingly hairy, with sulphur-yellow funnel shaped flowers in umbellate cymes on a peduncle 2–6 in. long ; in swampy places by streams.
Gam.: *Skues*! Kudang (fl. Jan.) *Dalz.* 8047! **Mali:** Banfara *Chev.* 533! **Guin.:** Faranna, Kaba *Chev.* 20391! Kindia *Chev.* 13207! **S.L.:** Njala (Jan.) *Dalz.* 8136! Makump (Dec.) *Glanville* 107! Regent (Dec.) *Sc. Elliot* 4324! Mange Bureh (Jan.) *Deighton* 1039! Kamah (fr. May) *Thomas* 305! Yungeru (Jan.) *Thomas* 7305! **Lib.:** *Harley* 854! Kakatown *Whyte*! Genna Tanyehun (Dec.) *Baldwin* 10776! **Iv.C.:** Sassandra (fl. & fr. Feb.) *Leeuwenberg* 2902! **Ghana:** Axim (Jan.) *Chipp* 59! Cape Coast *Deakin* 61! **N.Nig.:** Ibi (Nov.) *Dalz.* 77! Abutshi *Woodruff*! **S.Nig.:** Ikwette, Obudu (Dec.) *Savory & Keay* FHI 25002! **[Br.]Cam.:** Victoria to Buea (Mar.) *Hambler* 653! Mamfe (Nov.) *Tamajong* FHI 22125! **F.Po:** *Mann* 85! *Barter* 1838! *T. Vogel* 54! Also in tropical America.
[The subsp. *orientalis* (Hallier f.) Ooststr. in Fl. Males., ser. 1, 4 : 449 (1953) is the subspecies occurring in E. Africa and in tropical Asia.]

5. **M. aegyptia** (*Linn.*) *Urban* Symb. Antill. 4 : 505 (1910) ; van Ooststr. in Blumea 3 : 327 ; Fl. Males. ser. 1, 4 : 448 ; *Ipomoea aegyptia* Linn. Sp. Pl. 162 (1753). *Convolvulus pentaphyllus* Linn. (1762) *Merremia pentaphylla* (Linn.) Hallier f. (1893)—F.T.A. 4, 2 : 108 ; F.W.T.A. ed. 1, 2 : 212 ; Schnell in Ic. Pl. Afr. I.F.A.N., 1, t. 19 (1953) ; Berhaut Fl. Sén. 23. A robust annual twiner, hirsute with campanulate funnel-shaped white flowers 1 in. long, the buds covered with spreading yellowish hairs.
Sen.: *Roger* 34! Kanu to Wothan *Baumann*! Richard Tol (Jan.) *Döllinger* 8! Thiès *Chev.* 2865. Tivaouane *Chev.* 2867. Farmar 36! **Gam.:** *Brooks* 26! **Port.G.:** Bissau (fr. Dec.) *Esp. Santo* 1589! **Guin.:** Conakry (Apr.) *Debeaux* 314! **S.L.:** Freetown (fr. Dec.) *Deighton* 1535! Regent (fr. Dec.) *Sc. Elliot* 4060! Wellington (Dec.) *Deighton* 3353! Ninia (fr. Feb.) *Sc. Elliot* 4869! Port Loko (Dec.) *Thomas* 5622! Kumrabai (Jan.) *Thomas* 6866! **Ghana:** Cape Coast (Sept.) *Dalz.* 132! Elmina Castle (Jan.) *Dalz.* 36! Kpedsu (Dec.) *Howes* 1049! Shai Hills (fr. Oct.) *Adams*! **Togo Rep.:** Lomé *Warnecke* 275! **Niger:** Labézanga (Sept.) *Hagerup* 447! **N.Nig.:** Nupe *Barter* 863! Sokoto (Nov.) *Moiser* 8! Katagum Dist. (Sept.) *Dalz.* 184! Bornu *Parsons*! **S.Nig.:** Lagos (Mar.) *Millen* 118! Ibadan (Feb.) *Foster* 155! (Dec.) *Keay* FHI 37289! Widespread throughout the tropics.

6. **M. dissecta** (*Jacq.*) *Hallier f.* in Engl. Bot. Jahrb. 16 : 552 (1893) ; F.T.A. 4, 2 : 104 ; van Ooststr. in Blumea 3 : 328 ; Fl. Males., ser. 1, 4 : 448. *Convolvulus dissectus* Jacq. Obs. 2 : 4, t. 27 (1767). A perennial twiner, the stems, etc., at first coarsely hirsute, becoming glabrous, with widely funnel-shaped flowers 1½ in. long, white with purple throat, opening in the evening.
S.L.: Njala (Oct.) *Deighton* 2818! **Iv.C.:** Abidjan (Mar.) *Leeuwenberg* 3123! **Ghana:** Achimota (fr. Jan.) *Irvine* 1107! **S.Nig.:** Lagos *Cons. of For.* 472! Benin *Dundas* FHI 21480! Ibadan (fr. Oct.) *Newberry & Etim* 145! **F.Po:** *T. Vogel* 238! Cultivated and sometimes naturalized in many parts of the tropics ; a native of tropical America.

7. **M. kentrocaulos** (*C. B. Cl.*) *Rendle* in F.T.A. 4, 2 : 103 (1905) ; Berhaut Fl. Sén. 152 ; Meeuse in Fl. Pl. Afr. 30 : t. 1194, and in Bothalia 6 : 704. *Ipomoea kentrocaulos* C. B. Clarke in Hook. f. Fl. Brit. Ind. 4 : 213 (1883). A glabrous perennial twiner, with widely funnel-shaped flowers 2–2½ in. long, cream-white with dark purple centre.
Sen.: Berhaut 438 ; 2137. **N.Nig.:** Nupe *Barter* 930! Jos Plateau (Nov.) *Lamb* 78! Mokwa (Nov.) *Mutch* FHI 27964! Matyoro (Oct.) *Thornewill* 123! **S.Nig.:** Okeifo, Oyo (Oct.) *Onochie* FHI 40871! **[Br.]Cam.:** Kwagiri, Gashaka Dist. (Nov.) *Latilo & Daramola* FHI 28769! Throughout Africa south of the Sahara, and in tropical Asia.

8. **M. pterygocaulos** (*Steud. ex Choisy*) *Hallier f.* in Engl. Bot. Jahrb. 16 : 552 (1893) ; F.T.A. 4, 2 : 105 ; Meeuse l.c. 702. *Ipomoea pterygocaulos* Steud. ex Choisy in DC. Prod. 9 : 381 (1845). *I. leucantha* of F.W.T.A. ed. 1, 2 : 216, partly (*Hayes* 521), not of Webb (1849). A perennial twiner with glabrous acutely-angled and often winged stems, and campanulate flowers about 1½ in. long, white, occasionally yellow, with a purple or red ring in the throat ; by swamps and streams.
Gam.: *Hayes* 521! Kudang (Jan.) *Dalz.* 8049! **S.L.:** Musaia *Deighton* 4558! (fr. Feb.) 4238! Kaballa *Thomas* 2175! Yalamba (Jan.) *Jaeger* 3956! Mabahdu (Dec.) *Marmo* 102! Rowala (Jan.) *King* 151B! **Lib.:** Ganta (fr. Nov.) *Harley* 304! **N.Nig.:** Jos Plateau *Batten-Poole* 228! Vom *Dent Young* 182! Naraguta (Nov.) *Lely* 709! Kontagora (fr. Dec.) *Dalz.* 121! **S.Nig.:** Abokom to Ikura (fr. Jan.) *Holland* 256! **[Br.]Cam.:** Bamenda, 3,000 ft. (Jan.) *Keay* FHI 28520! Maisamari, 4,600 ft., Mambila Plateau (Jan.) *Hepper* 1677! Vogel Peak, 4,000 ft., Adamawa (Dec.) *Hepper* 1552! Widespread throughout tropical Africa, also in the Mascarene Islands.

9. **M. pes-draconis** *Hallier f.* in Bull. Herb. Boiss. 6 : 537 (1898) ; F.T.A. 4, 2 : 107.

9a. **M. pes-draconis** var. **nigerica** *Rendle* Cat. Talb. 73 (1913). A rather slender twiner, covered with dull yellowish pubescence, with 7-lobed leaves and flowers ¾ in. long.
N.Nig.: R. Benue *Talbot* 832!

M. cissoides (Lam.) Hallier f., and *M. tuberosa* (Linn.) Rendle, both natives of tropical America, are occasionally cultivated in our area.

Imperfectly known species.

M. sp. ; " between *M. palmata* Hallier f. and *M. pterygocaulos* (Choisy) Hallier f." (Verdcourt MS.).
N. Nig. : Katagum Dist. *Dalz.* 313! This specimen is conspecific with *A.W. Cruse* 396 (= 17) from Dambo, N. Rhodesia.

12. HEWITTIA Wight & Arn.—F.T.A. 4, 2 : 100 ; Meeuse in Bothalia 6 : 698 (1958).

Stems twining ; leaves ovate to triangular-ovate, mucronate, cordate at the base, up to 12 cm. long and 9 cm. broad, thinly pubescent on the nerves beneath ; flowers few in a bracteate head on an axillary peduncle ; bracts ovate or ovate-lanceolate, green, about 1 cm. long, pubescent ; calyx-segments ovate-lanceolate, 1 cm. long ; corolla densely pilose in bud, 2·5 cm. long ; fruit subglobose   ..   ..   *sublobata*

H. sublobata (*Linn. f.*) *O. Ktze.* Rev. Gen. Pl. 441 (1892) ; van Ooststr. in Blumea 3 : 286 ; Fl. Males. ser. 1, 4 : 438 ; Meeuse l.c. *Convolvulus sublobatus* Linn. f. Suppl. 135 (1781). *C. bicolor* Vahl (1794). *Hewittia*

bicolor (Vahl) Wight & Arn. (1837)—F.T.A. 4, 2 : 100 ; Chev. Bot. 457. Herbaceous twiner, pubescent, with flowers 1 in. long, cream-white or pale yellow with deep purple centre and leafy bracts and calyx.
Gam. : *Brown-Lester* 26 ! **Mali :** Bambaya to Forboia-Bafé (fl. & fr. Aug.) *Jaeger* 861 ! **S.L. :** Regent (fl. & fr. Dec.) *Sc. Elliot* 4111 ! Kumrabai (fl. & fr. Dec., Jan.) *Thomas* 6766 ! Njala (fl. & fr. Feb.) *Deighton* 1927 ! Makump (fl. & fr. Dec.) *Glanville* 105 ! Kaballa (Sept.) *Thomas* 2341 ! **Lib. :** Cape Palmas *T. Vogel* 14 ! Dukwai R. (July) *Linder* 117 ! **Iv.C. :** Bingerville *Chev.* 17333 ! Tiassalé (Aug.) *de Wilde* 440 ! **Ghana :** *Farmar* 531 ! *Cummins* ! Agogo *Deakin* 3 ! Elmina (Dec.) *Deakin* 56 ! Dodowali (Nov.) *Harris* K20 ! **S. Nig. :** Lagos (fl. & fr. Oct.) *Dalz.* 1158 ! *Barter* 2160 ! Abeokuta *Barter* 3351 ! *Irving* 120 ! Ibadan (Jan., Feb.) *Ejiofor* FHI 31927 ! *Daggash* FHI 22747 ! Oban *Talbot* 798 ! [Br.] **Cam. :** Rio del Rey *Johnston* ! Throughout tropical Africa, Asia and Polynesia.

13. ANISEIA Choisy—F.T.A. 4, 2 : 88.

Stems twining, soon nearly glabrous ; leaves oblong-lanceolate, rounded at the apex and long-mucronate, up to 9 cm. long and 2·5 cm. broad, glabrous or slightly pubescent ; petiole up to 2·5 cm. long ; flowers usually solitary, on rather long axillary peduncles ; bracts subulate, a little below the ovate-lanceolate acute sepals, the latter about 1·5 cm. long ; corolla about 3 cm. long, pilose outside ; fruit 2 cm. long
martinicensis

A. martinicensis (*Jacq.*) *Choisy* in Mém. Soc. Phys. Genève 8 : 66 (1838) ; van Ooststr. in Blumea 3 : 279 ; Fl. Males., ser. 1, 4 : 435. *Convolvulus martinicensis* Jacq. Select. Stirp. Americ. 26, t. 17 (1763). *C. uniflorus* Burm. (1768). *Aniseia uniflora* (Burm.) Choisy (1833)—F.T.A. 4, 2 : 88 ; Chev. Bot. 456. *Ipomoea ryssenii* A. Chev. Rev. Bot. Appliq. 30 : 270, t. 13 (1950). *Ipomaëlla ryssenii* (A. Chev.) A. Chev. l.c. 272 (1950), illegitimate name. A slender twiner thinly pubescent at first, with white flowers 1 in. long with leafy sepals.
Sen. : *Roger* 38 ! **Guin. :** Kouria *Caille* in *Hb. Chev.* 14665. Dantilia *Sc. Elliot* 5271 ! **S.L. :** Rokupr (fl. & fr. Feb.) *Deighton* 4188 ! 5003 ! Mahela (Dec.) *Sc. Elliot* 4051 ! Makunde (Apr.) *Sc. Elliot* 5696 ! Rokel R. bridge (Apr.) *Hepper* 2562 ! Njala (Jan.) *Dalz.* 8046 ! **Lib. :** Experimental Sta., Central Prov. (Mar.) *Blickenstaff* 35 ! Firestone Plantations, Dukwai R. (fl. & fr. Oct.-Nov.) *Cooper* 9 ! *Linder* 120 ! Kakatown *Whyte* ! Ganta (Sept.) *Harley* ! **Iv.C. :** Abradine *Chev.* 17587. Bliéron *Chev.* 19910. Gabon (Apr.) *Irvine* 1592 ! Kpotame (fl. & fr. Dec.) *Adams* 3620 ! Black Volta (fr. Dec.) *Adams & Akpabla* 4452 ! **Niger :** Gaya *Chev.* **N.Nig. :** Benue (Nov.) *Dalz.* 74 ! Ndoni (?) *Barter* 1771 ! **S.Nig. :** Lagos (July) *Dalz.* 1354 ! Abeokuta *Barter* 3340 ! (Dec.) *Keay & Jones* FHI 4871 ! Ibadan (Nov.) *Newberry & Etim* 173 ! Benin (fl. & fr. Mar.) *Onochie* FHI 38515 ! Widespread throughout the tropics.

14. LEPISTEMON Blume—F.T.A. 4, 2 : 115 ; van Ooststr. in Blumea 5 : 340 (1943) ; Fl. Males., ser. 1, 4 : 489 (1953).

Flowers rather few, subfasciculate ; stems long-pilose with reflexed hairs ; leaves very broadly ovate, widely cordate at the base, mucronate at the apex, up to about 15 cm. long, entire or sinuately lobed or dentate ; pedicels 1–1·5 cm. long, setulose-pubescent ; calyx-segments ovate, 7 mm. long ; corolla 2·5–3 cm. long, broadly funnel-shaped, wider towards the base ; fruit densely bristly ..   ..   ..   1. *owariense*
Flowers numerous, in short cymes ; stems glabrous ; leaves as above but smaller, appressed-pilose on the upper surface ; pedicels 1 cm. long, glabrous ; calyx-segments 5 mm. long ; corolla 1 cm. long   ..   ..   2. *parviflorum*

1. **L. owariense** (*P. Beauv.*) *Hallier f.* in De Wild. Etud. Fl. Katanga 1 : 112 (1902). *Ipomoea owariensis* P. Beauv. Fl. Oware 2 : 41, t. 82 (1816). *Lepistemon africanum* Oliv. in Hook. Ic. Pl., ser. 3, 3 : 54, t. 1270 (1878) ; F.T.A. 4, 2 : 115. *L. chevalieri* Trochain (1933). A robust twiner clothed with almost pungent hairs ; yellowish-white flowers, sometimes with purple throat, about 1 in. long in short axillary cymes. **Port.G. :** Farim, Cajambarim (Oct.) *Esp. Santo* 2315 ! **S.L. :** Musaia (fr. Dec.) *Deighton* 4414 ! **Iv.C. :** Férkéssédougou (Nov.) *Leeuwenberg* 2041 ! **Ghana :** Se-Mahem (fr. Jan.) *Irvine* 644 ! Anum (Nov.) *Johnson* 811 ! Aburi (Jan.) *de Wit* 8018 ! Prahu *Cummins* 11 ! Dedeman (fr. Dec.) *Irvine* 2101 ! **N.Nig. :** Lokoja *Dalz.* 195 ! Kabba (Oct.) *Parsons* 83 ! Gindiri, Jos Plateau (fl. & fr. Oct.) *Hepper* 1079 ! **S.Nig. :** Lagos (fl. & fr. Mar.) *Millen* 33 ! 68 ! 180 ! Ibadan (Nov.) *Newberry* 111 ! *Newberry & Etim* 176 ! [Br.]**Cam. :** Victoria (fr. Jan.) *Maitland* 310 ! Jalo (fr. Dec.) *Latilo & Daramola* FHI 28880 ! Also in Congo, Ubangi-Shari, Sudan, E. Africa and south to Angola.

2. **L. parviflorum** *Pilger ex Büsgen* in Mitt. Deutsche Schutzgeb. 23 : 87 (before July 1910) ; Pilger in Engl. Bot. Jahrb. 45 : 219 (Dec. 1910). A slender twiner with smaller and more numerous flowers than the last. **S.L.:** Joru to Daru (Nov.) *Pyne* 31 ! **Lib. :** Genna Tanyehun, Grand Cape Mount Co. (fr. Dec.) *Baldwin* 10771 ! **Ghana :** Kwahu, 2,350 ft. (Dec.) *Adams* 5727 ! **S.Nig. :** Onipanu, Ondo (Dec.) *Onochie* FHI 5237 ! Oban *Talbot* 1289 ! Ikom (Dec.) *Keay* FHI 28161 ! [Br.]**Cam.:** Johann-Albrechtshöhe *Büsgen* 64. *Winkler* 1041. Victoria (fr. Feb.) *Maitland* 396 ! Southern Bakundu F.R., Kumba (fr. Jan.) *Binuyo & Daramola* FHI 35497 !

15. ASTRIPOMOEA Meeuse in Bothalia 6 : 709 (1958) ; Verdcourt in Kew Bull. 13 : 189. *Astrochlaena* Hallier f. (1893)—F.T.A. 4, 2 : 68 ; F.W.T.A. ed. 1, 2 : 213 ; not *Asterochlaena* Corda (1845) nor *Asterochlaena* Garcke (1850).

Calyx-segments not keeled, ovate, about 1 cm. long in fruit, stellate-tomentellous ; flowers few on a very short common peduncle ; corolla about 1·5 cm. long, with a narrow purple throat ; leaves broadly ovate, entire or spreading-undulate, truncate or rounded at the base, up to 12 cm. long and 9 cm. broad, loosely stellate-pubescent on both surfaces ; fruit 1 cm. long, glabrous   ..   ..   ..   1. *lachnosperma*
Calyx-segments keeled, broadly lanceolate, 1 cm. long, stellate-tomentellous ; flowers few on a long common peduncle ; corolla up to 5 cm. long, gradually widened from the middle upwards ; leaves ovate, acute, rounded at the base, mucronate, 7–9 cm. long, 5–6 cm. broad, thinly stellate-pubescent above, more densely so beneath   2. *malvacea*

1. **A. lachnosperma** (*Choisy*) *Meeuse* l.c. 710 (1958) ; Verdcourt l.c. 195. *Ipomoea lachnosperma* Choisy in DC. Prod. 9 : 356 (1845). *Astrochlaena lachnosperma* (Choisy) Hallier f. (1893)—F.T.A. 4, 2 : 119 ; N.E. Br. in Kew Bull. 1909 : 124 ; F.W.T.A. ed. 1, 2 : 213. Erect perennial herb, woody below 1–2 ft. high, covered with yellowish stellate tomentum, with funnel-shaped flowers ¾ in. long, white with red-purple centre.

Mali : Ségouba (fr. Oct.) *Davey* 060 ! Boré *Demange* 42/1957 ! Bargoussi, near Hombori (fr. Oct.) *Jaeger* 5487 ! **Niger** : Zinder (fr. Oct.) *Hagerup* 573 ! **N.Nig.** : Sokoto (fl. & fr. Aug.) *Dalz.* 371 ! Hadeija (fl. & fr. Sept.) *Dalz.* 182 ! Maiduguri (fl. & fr. Nov.) *Noble* A25 ! [Br.]**Cam.** : Bama, Dikwa Div. (fr. Dec.) *McClintock* 93 ! Extends to Eritrea and to S. tropical Africa.

2. **A. malvacea** (*Klotzsch*) *Meeuse* l.c. 710 (1958); Verdcourt in Kew Bull. 13 : 192 (1958), incl. var. *malvacea*, var. *volkensii* (Damm.) Verdc. and var. *floccosa* (Vatke) Verdc. *Breweria malvacea* Klotzsch in Peters Reise Mossamb., Bot. 245, t. 37 (1861). *Astrochlaena malvacea* (Klotzsch) Hallier f. (1893)—F.T.A. 4, 2 : 121 ; F.W.T.A., ed. 1, 2 : 213 ; Brenan in Mem. N.Y. Bot. Gard. 9 : 8 (1954). *A. chariensis* A. Chev. (1913), name only. Habit of the last, 2–3 ft. high, hoary-stellate-tomentose, with red-purple flowers 1½–2 in. long.

N. Nig.: Yola (June, Oct.) *Dalz.* 72 ! *Shaw* 76 ! 104 ! From Ubangi-Shari and Ethiopia to S. Africa.

16. IPOMOEA

Linn. Sp. Pl. 159 (1753) ; F.T.A. 4, 2 : 128 ; van Ooststr. in Blumea 3 : 481 (1940) ; Fl. Males., ser. 1, 4 : 458 (1953), incl. *Calonyction* Choisy (1833)—F.T.A. 4, 2 : 117 ; and *Quamoclit* Moench (1794)—F.T.A. 4, 2 : 127).

Sepals distinctly awned at or below the apex ; awn straight or curved ; corolla salver-shaped with a long and narrow tube ; stamens and style mostly exserted :

    Corolla white with greenish bands or pale bluish-purple, 5–15 cm. long ; outer sepals 5–12 mm. long (without awn), inner ones 7–15 mm. (without awn) :

        Corolla white ; tube not or slightly widened above, 7–12 cm. long ; limb rotate ; stamens and style exserted .. .. 1. *alba*

        Corolla purplish ; the tube distinctly widened above, 3–6 cm. long, the limb funnel-shaped to rotate ; stamens and style not or scarcely exserted   2. *muricata*

    Corolla scarlet, rarely pure white, rather small, 3–4·5 cm. long ; outer sepals 2–4·5 mm. long (without awn) ; inner ones 3–6 mm. (without awn) :

        Leaves pinnately parted into numerous linear or filiform segments, rarely less deeply pinnately cut .. .. 3. *quamoclit*

        Leaves not pinnately cut ; ovate to orbicular, cordate at the base ; margin entire or lobed .. .. 4. *hederifolia*

Sepals obtuse, acute or acuminate, whether mucronulate or not, but not distinctly awned at or below the apex ; corolla mostly funnel-shaped, or campanulate, sometimes salver-shaped ; stamens and style mostly included, sometimes exserted (when leaves pinnately cut, see 3, *I. quamoclit*):

Flowers in involucrate heads:

    Leaves digitately and deeply 7–9-lobed, long-pilose beneath, lobes acutely acuminate ; bracts ovate-acuminate, up to 3 cm. long, pilose ; calyx about 1 cm. long, pubescent; corolla about 3 cm. long .. .. 5. *pes-tigridis*

    Leaves not lobed :

        Bracts not united, lanceolate :

            Leaves ovate-subcordate, up to 8 cm. long and 6 cm. broad, setose-pilose above, silky below ; petiole and branches long-pilose ; petiole up to 5 cm. long ; corolla 3 cm. long ; twining plant .. .. 6. *heterotricha*

            Leaves oblong-lanceolate, mucronate, truncate or subcordate at the base, up to 12 cm. long and 4 cm. broad, pilose above, silky below ; petiole 0.5–1 cm. long ; corolla about 5 cm. long ; plants usually not twining .. 7. *argentaurata*

        Bracts united into an oval doubly acuminate involucre, up to 5 cm. diam. ; leaves ovate, cordate, mucronate-acuminate, up to 7 cm. long and 5 cm. broad, pilose on both surfaces ; corolla up to 4·5 cm. long:

            Corolla 4·5 cm. long ; sepals lanceolate, acute .. .. .. 8. *involucrata*

            Corolla 2·5–3 cm. long ; sepals oblong-spathulate, obtuse .. .. 9. *pileata*

Flowers not in involucrate heads :

    Leaves bilobed at the apex, suborbicular, up to 12 cm. diam.,with numerous nerves radiating from near the base ; petiole up to 8 cm. long ; flowers few in cymes ; calyx 1·5 cm. long ; corolla 5–6 cm. long   10a. *pes-caprae* subsp. *brasiliensis*

    Leaves not bilobed at the apex :

        Leaves more or less ovate to suborbicular, pandurate or triangular, usually cordate at the base, never lobed :

        *Flowers fairly long-pedunculate, not in a subsessile cluster :

            Leaves suborbicular, not acuminate, widely cordate at the base, up to about 12 cm. diam., with few radiating nerves from near the base ; petiole up to 15 cm. long ; flowers cymose ; calyx segments unequal, 1–1·5 cm. long ; corolla 7–8 cm. long ; fruit subglobose, 1·5 cm. diam. ; seeds puberulous   11. *asarifolia*

        Leaves not suborbicular, usually acuminate :

        †Flowers purplish-red, pink or lilac (not yellow or white) :

        ‡Leaves softly tomentose or pubescent, surface sometimes becoming glabrescent :

            Flowers subumbellate or in rather dense cymes with lanceolate bracts :

                Leaves cordate, acuminate, about 9 cm. long and 6 cm. broad, not markedly reticulate beneath ; flowers umbellate ; calyx-lobes about 8 mm. long, rounded at the apex, pilose ; corolla 6 cm. long .. .. 12. *rubens*

                Leaves broadly cordate-ovate, acuminate, about 10 cm. long and 8 cm. broad, entire, thinly pilose beneath ; petiole pubescent ; corolla 3 cm. long ; fruit loosely pilose .. .. 13. *velutipes*

Flowers solitary or geminate in the axils of the leaves :
  Leaves ovate-oblong to suborbicular, 8–12 cm. long and 6–9 cm. broad,
    woolly on the marked venation beneath, rounded or slightly cordate at
    base ; flowers solitary, surrounded by 2 large bracts ; calyx 1·5 cm. long ;
    corolla 8–10 cm. long   ..   ..   14. *verbascoidea*
  Leaves oblong-ovate to triangular-mucronate, base retuse to shallowly cordate,
    ovate, obtuse 2·5–6 cm. long and up to 4 cm. broad ; flowers solitary or
    paired, inflorescences with two linear bracts up to 7 mm. long ; calyx up to
    12 mm. long ; corolla up to 2·5 cm. long   ..   ..   15. *pyrophila*
‡Leaves glabrous or nearly so :
  Stems setose-pilose ; leaves ovate-triangular, broadly and shortly acuminate,
    mucronate, widely cordate at base, up to 14 cm. long and 10 cm. broad ;
    calyx up to 2·5 cm. long, segments oblong-elliptic ; corolla up to 10 cm. long
                                                                    16. *setifera*
  Stems glabrous :
    Corolla 18–20 mm. long ; calyx-segments pilose-ciliate to glabrous, apiculate ;
      (forms with entire leaves, for other characters see below)   ..   24. *triloba*
    Corolla 3·5–9 cm. long :
      Leaves broadly cordate, subacute to obtusely acuminate, with a little mucro,
        4–11 cm. long, and broad ; corolla 3·5–5 cm. long :
        Sepals elliptic, with rounded apex ; corolla-tube very narrow at the base
          and remaining 2–2·5 mm. in diam. for about 5–10 mm. above the base ;
          fruit capsule about 1·5 cm. in diam., seeds brown, 7–8 mm. long, 5–6 mm.
          broad, slightly pilose ; plants never cultivated   ..   17. *stenobasis*
        Sepals oblong, shortly and abruptly acute, mucronate, corolla-tube cam-
          panulate funnel-shaped, about 5–6 mm. diam. at the base ; fruit capsules
          very rarely developed in African specimens, seeds glabrous ; plants
          cultivated, but sometimes apparently wild in abandoned farm-land ;
          (forms with entire leaves ; for other characters see below)   32. *batatas*
      Leaves variable, lanceolate or long-hastate, hastate, cordate or entire at the
        base, always much longer than broad, up to 14 cm. long :
        Leaves broadly triangular to lanceolate, more or less hastate at base, up
          to 14 cm. long and 8 cm. broad, with numerous lateral nerves ; calyx
          1 cm. long ; corolla up to 9 cm. long   ..   ..   18. *aquatica*
        Leaves triangular-ovate and entire to deeply trilobed with linear segments,
          about 7 cm. long with few lateral nerves ; calyx about 6 mm. long ;
          corolla about 5 cm. long   ..   ..   19. *hellebarda*
†Flowers yellow or white (or very rarely purplish in No. 22c) :
  Stems pubescent :
    Leaves bluntly elliptic-ovate to more or less pandurate, widely cordate at base,
      4–6 cm. long, 3–4 cm. broad, pilose on both surfaces ; flowers few, in axil-
      lary fascicles or solitary ; calyx 1 cm. long, pilose ; corolla a little larger
      than the calyx, not exceeding 14 mm.   ..   ..   20. *vagans*
    Leaves acuminate-cordate :
      Leaves, petioles and inflorescences more or less densely clothed with yellowish
        spreading hairs ; leaves never becoming glabrescent, cordate-ovate, acute,
        mucronate, 4–9 cm. long ; petioles 2·5–5 cm. long ; flowers generally in
        dense, many-flowered cymes ; corolla up to 2·5 cm. long   21. *tenuirostris*
      Leaves, petioles, stems and inflorescences finely pubescent, never clothed with
        yellowish spreading hairs ; leaves becoming almost glabrescent except the
        margin :
        Leaves ovate with rounded sides, widely cordate at base, broadly acuminate
          and mucronate, 6–9 cm. long, 3·5–7 cm. broad, shortly ciliate, pubescent
          on the nerves beneath ; petioles 1·5–4 cm. long ; flowers cymose ; calyx-
          segments 7–8 mm. long, mucronate, the outer ones not cordate or auricled
          at base, corolla 4 cm. long ; fruit broadly ovoid, mucronate, 1·5 cm. long
                                                              22. *obscura* aggregate
        Leaves cordate-ovate or cordate-oblong, obtuse or acute, mucronate, basal
          sinus variable, usually deep and rather narrow, 3–8 cm. long, 2–5 cm. broad,
          margin ciliate ; petioles slender, 1–9 cm. long ; flowers in 1–3-flowered in-
          florescences, calyx-segments very unequal, outer ones cordate or auricled
          at the base, long-acuminate, very acute ; corolla not exceeding 2 cm. ;
          fruit globose, apiculate, up to 9 mm. in diam.   ..   23. *sinensis*
    Stems not pubescent ; flowers umbellate, pedicels up to 1 cm. long ; calyx-
      segments long apiculate ; (forms with entire leaves ; for other characters see
      above)   ..   ..   24. *triloba*
*Flowers in a subsessile cluster or subsolitary and subsessile :
  ‖Leaves very variable, from ovate-triangular to narrowly lanceolate, widely cordate,
    mucronate, up to 8 cm. long and 5 cm. broad, pubescent on the nerves beneath ;

flowers generally clustered, but sometimes solitary or paired ; calyx-segments
lanceolate, 7 mm. long, pubescent ; corolla 1 cm. long ; fruit pilose
                                                          25. *eriocarpa*
Leaves ovate, widely cordate at the base, rounded and mucronate at the apex,
  3–5 cm. long, 1·5–3·5 cm. broad, long-petiolate ; flowers solitary or paired,
  shortly pedicellate ; calyx-segments oblong, 3–4 mm. long, thinly pilose ;
  corolla 6–8 mm. long ; fruit glabrous, nearly 1 cm. long   26. *verticillata*
‖Leaves linear or lanceolate and not cordate at the base, or more or less deeply lobed
  and then cordate (in some forms of *I. kotschyana* not properly cordate) :
Leaves linear to lanceolate ; flowers more or less pedicellate :
  Calyx pubescent outside :
    Calyx about 8 mm. long ; corolla slightly longer than the calyx ; leaves oblong-
      lanceolate, pilose beneath ; stems pubescent   .. .. 27. *coscinosperma*
    Calyx 2 cm. long ; corolla thrice as long as the calyx ; leaves linear or oblong-
      linear, nearly glabrous ; stems puberulous ; fruit 1 cm. long
                                                          28. *blepharophylla*
  Calyx glabrous or warted outside :
    Leaves acute at apex ; calyx warted outside ; stems usually pilose   29. *barteri*
    Leaves rounded or emarginate at apex ; calyx not warted ; stems glabrous
                                                          30. *stolonifera*
Leaves more or less deeply lobed :
  Calyx glabrous outside :
    Leaf-lobes deeply divided into small segments ; flowers subsolitary ; calyx
      warted, 5 mm. long ; corolla 1·5 cm. long ; fruit 8 mm. long   31. *coptica*
    Leaf-lobes entire or nearly so :
      Sepals apiculate, 7–10 mm. long :
        Plants cultivated for their edible subterraneous tubers, sometimes escaped
          from cultivation ; stems mostly prostrate, thick ; leaves broad-ovate to
          orbicular in outline, cordate or truncate at base, entire, or angular to pal-
          mately 3–5(–7)-lobed ; petiole up to 15 cm. long ; corolla pale violet, 3–4·5 cm.
          long ..   .. 32. *batatas*
        Plants without tubers and generally not cultivated ; stems mostly twining,
          thinner ; leaves mostly 3-lobed (but sometimes entire) ; corolla pink or red-
          purple, 18–20 mm. long   ..   .. 24. *triloba*
      Sepals not apiculate :
        Leaves broadly hastate, 3-lobed, sometimes nearly to the base, the outer lobes
          linear-lanceolate, about 5 cm. long, glabrous ; flowers about 3 in a cyme ;
          calyx 8 mm. long ; corolla 6 cm. long, contracted at the base ; seeds shortly
          tomentellous ..   .. 19. *hellebarda*
        Leaves palmate or palmately lobed :
          Leaves usually 5–7-lobed to about the middle, lobes acuminate ; flowers
            cymose ; corolla about 7 cm. long, contracted at base ; fruit 1·5 cm. long ;
            seeds long-villous ..   .. 33. *mauritiana*
          Leaves 5-lobed nearly to base, the outer lobes often lobulate, acutely mucro-
            nate ; flowers cymose ; calyx 6 mm. long ; corolla 4·5 cm. long ; fruit
            1 cm. long ; seeds long-pilose   ..   .. 34. *cairica*
  Calyx pilose or pubescent outside :
    Leaves glabrous ; sepals 7–8 mm. long, apiculate, always distinctly fimbriate at
      the margin, with some rather long hairs on the back ; corolla 18–20 mm. long
                                                          24. *triloba*
    Leaves pubescent, pilose or tomentose; sepals lanceolate to ovate, acuminate,
      pubescent or pilose-hairy :
      Leaf-shape orbicular in outline, 1–2-pinnatifid, with linear lobes ; flowers soli-
        tary, axillary, subsessile, about 1 cm. long ; sepals pubescent 35. *kotschyana*
      Leaves trilobed, with broad-angular lobes ; petioles about as long as the lamina ;
        flowers in rather long-peduncled inflorescences (up to 12 cm. long) ; corolla
        2·5–7 cm. long :
        Leaves thinly pilose beneath, cordate, trilobed to about the middle, lobes
          acuminate ; petiole with reflexed hairs ; calyx-segments 2 cm. long, long
          acuminate, densely pilose in the lower half ; corolla up to 7 cm. long
                                                          36. *nil*
        Leaves white-woolly-tomentose beneath, trilobed to the middle, cordate ;
          calyx as above but shorter ; corolla about 2 cm. long ..   .. 37. *aitonii*

1. **I. alba** *Linn.* Sp. Pl. 161 (1753) ; van Ooststr. in Blumea 3 : 547 ; Meeuse in Bothalia 6 : 765. *Convolvulus aculeatus* Linn. (1753). *Ipomoea bona-nox* Linn. (1762). *Calonyction aculeatum* (Linn.) House in
Bull. Torrey Bot. Club 31 : 590 (1904) ; F.W.T.A., ed. 1, 2 : 213 ; Exell Cat. S. Tomé 249. *C. speciosum*
Choisy in Mém. Soc. Phys. Genève 6 : 441 (1833) ; F.T.A. 4, 2 : 117 ; Chev. Bot. 457. A strong twiner
with stems smooth or sparingly tubercled, and large white flowers 4–6 in. across the limb, expanding and
fragrant at night ; wild in secondary vegetation, but often cultivated for ornament.
Guin.: Kouria *Caille* in *Hb. Chev.* 14871. **S.L.**: Hanga (fr. Jan.) *Thomas* 7783 ! 7799 ! Yungeru (Jan.)
Thomas 7251 ! 7287 ! Kumrabai (Dec.) *Thomas* 6741 ! 6867 ! 6959 ! Ronietta (Nov.) *Thomas* 5567 !
Njala (Dec.) *Deighton* 1851 ! **Lib.**: Bakratown (Oct.) *Linder* 877 ! **Ghana**: Akroso (Jan.) *Deakin* 115 !

N.Nig.: near Ibi, R. Benue (Nov.) *Dalz.* 73! **S.Nig.**: Lagos (fr. Mar.) *Millen* 88! Ibu *T. Vogel* 38! Ibadan (fr. Dec.) *Newberry* 124! Ogwashi (Dec.) *Thomas* 7053! Calabar (Feb.) *Mann* 2324! Oban *Talbot* 825! A native of tropical America, often cultivated and naturalized, now widely distributed at low and medium altitudes throughout the tropics.

2. **I. muricata** (*Linn.*) *Jacq.* Hort. Schoenbr. 3 : 40, t. 323 (1798) ; van Oostr. l.c. 551. *Convolvulus muricatus* Linn. Mant. 44 (1767). *Calonyction muricatum* (Linn). G. Don Gen. Syst. 4 : 264 (1837) ; F.T.A. 4, 2 : 118 ; F.W.T.A., ed. 1, 2 : 213 ; Chev. Bot. 456. A strong climber like the last with tubercled stems and funnel-shaped flowers, white or reddish or lilac.
Sen.: Pont *Chev.* 2852. **Gam.**: *Ingram* ! **S.L.**: Mahela (fr. Dec.) *Sc. Elliot* 4074! Musaia (fr. Feb.) *Glanville* 404! **N.Nig.**: Sokoto (Aug.) *Dalz.* 372! **S.Nig.**: Lagos *Dawodu* 257! A native of tropical America, cultivated, and now widespread throughout the tropics, but apparently not as frequent as *I. alba* Linn.

3. **I. quamoclit** *Linn.* Sp. Pl. 159 (1753) ; van Oostr. l.c. 555 ; O'Donell in Lilloa 29 : 74, for full synonymy. *Convolvulus pennatus* Desr. (1791). *Quamoclit pennata* (Desr.) Boj. Hort. Maurit. 224 (1837) ; F.W.T.A., ed. 1, 2 : 213. *Quamoclit vulgaris* Choisy in Mém. Soc. Phys. Genève 6 : 434 (1833) ; F.T.A. 4, 2 : 128 ; Exell Cat. S. Tomé 249. A glabrous twiner, cultivated like the last.
S.L.: Waterloo (fl. & fr. Aug.) *Melville & Hooker* 171! Njala (July) *Deighton* 1162! Bo (fl. & fr. Jan.) *Thomas* 7423! **Lib.**: *Whyte* **N.Nig.**: Sokoto (fl. & fr. Nov.) *Moiser* 25! **S.Nig.**: Sapoba *Kennedy* 152! 1585! Lagos *MacGregor* 85! Port Harcourt (fl. & fr. July) *Stubbings* 191! **[Br.]Cam.**: Man-of-War Bay (Apr.) *Schlechter* 12398! **F.Po**: (fr. July) *Mann* 400! *Barter* 1265! Native of tropical America, often cultivated in warner countries as an ornamental and frequently naturalized.

4. **I. hederifolia** *Linn.* Syst. Nat., ed. 10, 925 (1759) ; O'Donell l.c. 45, for full synonymy. *I. angulata* Lam. Tabl. Encycl. 1 : 464 (1797) ; van Oostr. l.c. 553. *Quamoclit angulata* (Lam.) Boj. Hort. Maurit. 224 (1837). *Ipomoea phoenicea* (Roxb.) Choisy l.c. 433 (1833). *Quamoclit coccinea* of F.T.A. 4, 2 : 128, and of F.W.T.A., ed. 1, 2 : 213, not of Moench (1794) ; (see note by van Oostr. l.c. 555, referring to a note of Hallier f. in Bull. Herb. Boiss. 7 : 415). Annual twiner with brilliant scarlet tubular salver-shaped flowers.
S.L.: (fl. & fr. Dec.) *Sc. Elliot* 3987! 3894! Freetown (fl. & fr. Feb.) *Johnston* 63! Njala (fl. & fr. Dec.) *Deighton* 2568! **Ghana**: Aburi *Brown* 387! Asuansi (fl. & fr. Feb.) *Williams* 1! **N.Nig.**: Kabba *Parsons* 76! Jemaa (Nov.) *Lamb* 87! **S.Nig.**: *Rowland*! Agbor *Kennedy* 1985! 2076! Olokemeji (Nov.) *MacGregor* 362! Port Harcourt (Jan.) *Stubbings* 158! Native of tropical America, widespread in warm countries as an ornamental plant.

5. **I. pes-tigridis** *Linn.* Sp. Pl. 162 (1753) ; F.T.A. 4, 2 : 158 ; Hallier f. in Bull. Herb. Boiss. 6 : 539 (1898), incl. var. *africana* Hallier f. (1898) and var. *indica* Hallier f. (1898) ; van Oostr. l.c. 504 ; Brenan in Mem. N.Y. Bot. Gard. 9 : 7 (1954) ; Roberty in Bull. I.F.A.N. 16 : 1010, incl. forma *subintegrifolia* Roberty (1954) ; Meeuse l.c. 744 (1958). Coarsely hairy twiner with deeply divided leaves, and pink or red-purple flowers 1 in. or more long in pedunculate bracteate heads.
Mali: Dioura (fr. Oct.) *Davey* 171! **Port.G.**: Bissau (Oct.) *Esp. Santo* 2529! **Niger**: Niamey (fl. & fr. Oct.) *Hagerup* 474! 507! **N.Nig.**: *Moiser* 74! Ako (fl. & fr. Oct.) *Lely* 652! Katagum Dist. (fr. Oct.) *Dalz.* 186! Extends to Sudan, Somaliland and S.E. tropical Africa ; also in Mauritius and tropical Asia.

6. **I. heterotricha** *F. Didr.* in Vidensk. Meddel. naturhist. Foren. Kjöbenhavn 1854 : 220 (1855). *I. amoena* Choisy in DC. Prod. 9 : 365 (1845) ; F.T.A. 4, 2 : 154 ; Chev. Bot. 453 ; F.W.T.A., ed. 1, 2 : 218, not of Blume (1825). *I. amoenula* Dandy in F.W. Andr. Fl. Pl. Sudan 3 : 112 (1956). A twiner, densely yellow-hairy, with leaves silvery-silky beneath, and flowers either red-purple or white with a purple centre ¾–1 in. long in pedunculed bracteate heads.
Sen.: *Heudelot* 146! **Gam.**: *Ingram* ! *Brooks* 29! Kuntaur *Ruxton* M16! **Port.G.**: Bissau (fr. Oct.) *Esp. Santo* 2593! **S.L.**: Musaia (Dec.) *Deighton* 5709! Yalamba (fr. Jan.) *Jaeger* 3953! Kabala (fr. Jan.) *King* 188B! (fr. Dec.) *Deighton* 4511! **Iv.C.**: Assakra (Oct.) *de Wilde* 595! Séguéla to Mankono (Nov.) *de Wilde* 928! 959! Bouaké *Leeuwenberg* 2075! **Ghana**: Achimota (fl. & fr. Nov.) *Irvine* 819! Utonso (Oct.) *Darko* 388! Krepi Plains (Jan.) *Johnson* 544! Agogo *Chipp* 608! Yendi (Nov.) *Williams* 415! Sampa (fl. & fr. Dec.) *Vigne* FH 3490! Gambaga (fr. Dec.) *Adams* 4221! **N.Nig.**: R. Benue, *Barter* ! Panshanu (Nov.) *Lely* 608! Fuka, Minna (Oct.) *Hepper* 964! Toro (Oct.) *Lely* 686! Jos Plateau (Oct.) *Lely* P843! **S.Nig.**: Lagos *Rowland*! Owo *Burton* 7! Akure F.R., Ondo (Oct.) *Keay* FHI 25461! **[Br.]Cam.**: foot of Vogel Peak, Adamawa (Nov.) *Hepper* 1446! Extends to Sudan (?) and south to Angola.

7. **I. argentaurata** *Hallier f.* in Engl. Bot. Jahrb. 18 : 132 (1893) ; F.T.A. 4, 2 : 153. Stems prostrate or ascending from a woody base, whitish-tomentose and with spreading yellow hairs ; flowers whitish turning light purplish with darker centre, 2 in. long and as much across the limb.
S.L.: Sumbaraya *Sc. Elliot* 4949! **Iv.C.**: Ourossantiakara (Nov.) *Leeuwenberg* 2005! **Ghana**: Wa (Nov.) *Darko* 1052! Tamale *Williams* 407! Bole *Harris*! **Togo Rep.**: Misahöhe to Bismarckburg *Büttner* 746! **N.Nig.**: R. Niger (Kworra) (Sept.) *T. Vogel* 123! Nupe *Barter* 1031! Zungeru (Feb., Nov) *Elliott* 16! *Dalz.* 118! Naraguta F.R. (fl. & fr. Oct.) *Hepper* 1017! Abinsi (fl. & fr. Nov.) *Dalz.* 684! Yola (Oct., Nov.) *Shaw* 80! *Hepper* 1220! **S.Nig.**: Lagos *Rowland*! Ibadan *Foster* 141! Awba Hills F. R. (Dec.) *Olafusi* FHI 14198! Gadika, Bornu (Oct.) *Daggash* FHI 24786! **[Br.]Cam.**: near Vogel Peak, Adamawa (Nov.) *Hepper* 1378! Also in Cameroun.

8. **I. involucrata** *P. Beauv.* Fl. Oware 52, t. 89 (1817) ; F.T.A. 4, 2 : 150 ; Chev. Bot. 455 ; Berhaut in Bull. Soc. Bot. Fr. 101 : 373 (1955) ; van Oostr. in Fl. Males., ser. 1, 4 : 468 ; Meeuse l.c. 744. Twiner, often widely climbing, pubescent, with red-purple flowers 1½ in. long or more, opening one at a time in head enclosed in a boat-shaped involucre.
Sen.: *Heudelot* ! **Gam.**: *Brown-Lester* 24! **Guin.**: Dyoko (Oct.) *Baldwin* 9681! Bambaya (Jan.) *Jaeger* 3881! **S.L.**: *G. Don*! *Barter*! Leicester (Dec.) *Sc. Elliot* 3845! Panguma *Deighton* 3932! Mange, Bureh (Jan.) *Deighton* 1026! Njala (July) *Sc. Elliot* 754! Loma Mts. (Nov.) *Jaeger* 288! **Lib.**: Cape Palmas (July) *T Vogel* 43! Monrovia (Oct. & Nov.) *Cooper* 11! *Whyte*! Kakatown *Whyte*! Ganta (Apr.) *Harley*! Dukwai R. (July) *Linder* 119! **Iv.C.**: Bingerville *Chev.* 17335! Lamé (fr. Oct.) *Leeuwenberg* 1748! **Ghana**: Axim (fl. & fr. Dec.) *Chipp* 49! Dunkwa (fl. & fr. Feb.) *Whiting* 8! Assene *Irvine* 566! Achimota (Nov.) *Morton* 6136! Aburi (Nov.) *Johnson* 239! **Dah.**: *Burton*! **N.Nig.**: Kontagora (fl. & fr. Feb.) *Dalz.* 120! Naraguta (fl. & fr. Aug.) *Lely* 514! Vom, 3,000–4,500 ft. *Dent.-Young* 179! Jebba (Dec.) *Hagerup* 747! **S.Nig.**: Lagos *Barter* 2185! Carter's Peak, Idanre (Nov.) *Brenan* 8647! Abeokuta *Irving*! Ibadan (Dec.) *Newberry* 12! Port Harcourt *Stubbings* 199! Onitsha *Barter* 1748! Oban *Talbot* 814! **[Br.]Cam.**: Buea *Maitland* 220! (Nov.) *Migeod* 80! Cam. Mt. (Jan.) *Dunlap* 6! **F.Po**: *T. Vogel* 234! Very common throughout tropical Africa.

9. **I. pileata** *Roxb.* Fl. Ind., ed. Carey & Wall. 2 : 94 (1824) ; F.T.A. 4, 2 : 151 ; van Oostr. l.c. 507 ; Fl. Males., ser. 1, 4 : 468 ; Berhaut l.c. *I. involucrata* var. *saxicola* A. Chev. ex Meeuse l.c. 745 (1958). A slender twiner with flowers not more than 1 in. long ; in all parts more slender than No. 8.
Sen.: Vélor, near Saloum *Berhaut* 525 ; 3485; 4779. **Gam.**: (fr. Dec.) *Ingram*! Also in E. and S. Africa, the Mascarenes and in tropical Asia.

10. **I. pes-caprae** (*Linn.*) *Sweet* Hort. Suburb. Londin. 35 (1818) ; van Oostr. l.c. 532 ; Exell Cat. S. Tomé 251 (1944) ; Meeuse l.c. 754 ; Naegelé in Bull. I.F.A.N. 21 : 1186, f. 6 (1959). *Convolvulus pes-caprae* Linn. Sp. Pl. 159 (1753).

I. pes-caprae (*Linn.*) *Sweet* subsp. **pes-caprae**—In Somaliland, tropical Asia and Malaysia.

10a. **I. pes-caprae** (*Linn.*) *Sweet* subsp. **brasiliensis** (*Linn.*) Oostr. l.c. 3 : 533 ; Meeuse l.c. *Convolvulus brasiliensis* Linn. Sp. Pl. 159 (1753). *Ipomoea biloba* of F.T.A. 4, 2 : 172, not of Forsk. (1775) (W. Afr. specimens) ; Chev. Bot. 453. *I. pes-caprae* "Roth" of F.W.T.A., ed. 1, 2 : 215, only partly of (Linn.) Sweet (1818). Glabrous perennial trailing on sand of the seashore, with half-succulent leaves and large red-purple flowers.

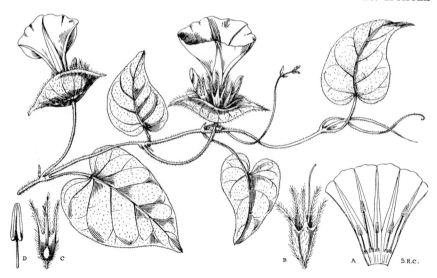

FIG. 283.—IPOMOEA INVOLUCRATA *P. Beauv.* (CONVOLVULACEAE).
A, corolla laid open. B, calyx and style. C, the same with the calyx cut open. D, anther.

Sen.: *Boivin*! St. Louis *Roger*! **Guin.:** Conakry (fl. & fr. July) *de Wit* 7123! *Chev.* 12254! **S.L.:** *Deighton* 2039! Lumley (fl. & fr. Aug.) *Melville & Hooker* 245! River Beach (July) *Small* 157! Yungeru (fl. & fr. Jan.) *Thomas* 7304! **Lib.:** Cape Palmas (July) *T. Vogel* 44! 23! **Iv.C.:** Sassandra (fl. & fr. Feb.) *Leeuwenberg* 2790! **Ghana:** Axim (Mar.) *Chipp* 414! Accra (Jan.) *Irvine* 814! 1962! Ada (Apr.) *Bally* 23! **S.Nig.:** Lagos (Sept.) *Dalz.* 1070! *Stubbings* 9! **[Br.]Cam.:** Victoria (Nov.) *Nditapah* FHI 50289! **F.Po:** (Jan.) *Mann* 241! Clarence beach (fl. & fr. Nov.) *T. Vogel*! Circumtropical.

11. **I. asarifolia** (*Desr.*) *Roem. & Schult.* Syst. Pl. 4 : 251 (1819) ; van Ooststr. l.c. 539. *Convolvulus asarifolia* Desr. in Lam. Encycl. Méth. Bot. 3 : 562 (1789). *Ipomoea repens* Lam. Tabl. Encycl. 1 : 467 (1791) ; F.T.A. 4, 2 : 172 ; Chev. Bot. 456 ; F.W.T.A., ed. 1, 2 : 215 ; Berhaut Fl. Sén. 158 ; F.W. Andr. Fl. Pl. Sudan 3 : 120 (1956), not of Roth (1821) ; Roberty in Bull. I.F.A.N. 16 : 1011 (1954), incl. forma *deltoidea* Roberty and forma *suffrutescens* Roberty. A glabrous perennial, trailing, with hollow stems and large red-purple flowers.
Sen.: *Roger*! *Heudelot* 218! *Brunner*! *Farmar* 59! Dakar *Chev.* 15814. **Gam.:** Genieri to Massembe (Feb.) *Fox* 31! **Mali:** Couroula *Chev.* 753. Djenné *Chev.* 1133. Korioumé (July) *Chev.* 1294! Kabarah (Sept.) *Hagerup* 300! **Guin.:** Ditinn *Chev.* 12987 ; 13526. **Iv.C.:** Bliéron *Chev.* 19898. **Ghana:** Abetifi Kwahu (Jan.) *Irvine* 1668! Navrongo (Apr.) *Vigne* 3759! **N.Nig.:** Nupe *Barter* 841! Sokoto *Moiser* 76! Katagum Dist. *Dalz.* 187! Mopti (June) *Lean* 39! Kano (Dec.) *Meikle* 802! Kuka, L. Chad. (Feb) *E. Vogel* 49! **S.Nig.:** *Foster* 360! Lagos *Rowland*! *Dalz.* 1072! Ibadan (Dec.) *Foster* 132! **[Br.]Cam.:** Bama, Dikwa Div. (Dec.) *McClintock* 50! Also in Cape Verde Islands, tropical Asia and America.

12. **I. rubens** *Choisy* in Mém. Soc. Phys. Genève 6 : 463 (1834) ; Verdcourt in Webbia 13 : 324 (1958). *I. lilacina* Blume Bijdr. 716 (1826) ; F.T.A. 4, 2 : 187 ; F.W.T.A., ed. 1, 2 : 215 ; Berhaut Fl. Sén. 153, not of Schrank (1822). *I. riparia* G. Don (1838)—Exell Cat. S. Tomé 251. *I. senegambica* A. Chev., in Bull. Mus. Hist. Nat. Paris, sér. 2, 5 : 234 (1933). A fairly stout twiner with grey-pubescent striate stems and bright red-purple flowers 1¼–2 in. long.
Sen.: *Roger*! Niayes (Dec.) *Chev.* 2855! *Berhaut* 1308! **Guin.:** Cercle de Tillabéri (Jan.) *de Wailly* 5820! **N.Nig.:** *T. Vogel* 241! Nupe *Barter* 969! Jebba (Dec.) *Hagerup* 679! Yola (May) *Dalz.* 75! Also in Sudan, Uganda, E. Africa, Congo and Angola.

13. **I. velutipes** *Welw. ex Rendle* in J. Bot. 32 : 175 (1894) ; F.T.A. 4, 2 : 155. *I. camporum* A. Chev. ex Hutch. & Dalz. F.W.T.A., ed. 1, 2 : 215 (1931) ; A. Chev. in Bull. Mus. Hist. Nat. Paris, sér. 2, 5 : 231 (1933) ; Verdcourt in Kew Bull. 13 : 208. A twiner with striate pubescent stems and long-pedunculate cymes of pale red-purple flowers.
Guin.: Kaba, Faranna (fl. & fr. Jan.) *Chev.* 20390! Ditinn (fl. & fr. Jan.) *Roberty* 16457! Dalaba to Mamou (fl. & fr. Jan.) *Chev.* 20392! **S.L.:** Musaia (Dec.) *Deighton* 4481! **S.Nig.:** Bendiga Ayuk, Ogoja (Dec.) *Keay* FHI 28136! Also in Congo, Angola and N. Rhodesia.

14. **I. verbascoidea** *Choisy* in DC. Prod. 9 : 356 (1845) ; F.T.A. 4, 2 : 183 ; Meeuse l.c. 769. *I. lukafuensis* De Wild. Ann Mus. Congo, sér. 4, 1 : 112, t. 2 (1903) ; F.T.A. 4, 2 : 184. *I. tessmannii* Pilger in Notizbl. Bot. Gart. Berl. 7 : 542 (1921) ; F.W.T.A., ed. 1, 2 : 215. A stout erect plant, with wrinkled leaves whitish-woolly beneath, and large reddish-purple flowers 3 in. long.
N.Nig.: Jos Plateau (June) *Lely* P366! Naraguta (June) *Lely* 287! Vom, 3,000–4,500 ft. *Dent Young* 180! Also in Cameroun, Congo, Angola, E. Africa, N. and S. Rhodesia, Bechuanaland and S.W. Africa.

15. **I. pyrophila** *A. Chev.* in Bull. Mus. Hist. Nat. Paris, sér. 2, 5 : 235 (1933). A slender trailing plant with a thickened rootstock, densely covered with short spreading hairs ; flowers in pedunculate cymes ; in savanna after fires.
Guin.: Kouredrari[1] (Feb.) *Chev.* 421! **N.Nig.:** Kontagora (fl. & fr. May) *Dalz.* 378! Ntachu to Madengyen (Jan.) *Meikle* 1073! Owudugu, Birnin Gwari (Feb.) *Daggash* FHI 31422! Sokoto *Lely* 132! Jos Plateau (Jan.) *Lely* P43! Naraguta (May) *Lely* 40! 41! 468! Vom *Dent Young* 181!
[This species belongs to the *I. asperifolia-fulvicaulis* complex, which has been studied recently by B. Verdcourt (Kew Bull. 13 : 202, and 14 : 337). It may be treated later as an infraspecific taxon, but as these studies are not yet complete and have not dealt with the plants of our area, it seems unadvisable to establish a new taxon for it.
I. atacorensis A. Chev. (l.c. 232 (1933)) belongs to the same group ; it is only known from the type-

---

[1] This is the original spelling of the locality on the label of Chevalier's specimen ; in the place of publication the locality occurs as " Koundian ".

specimen (Dahomey, Atacora Mts. (June) *Chev.* 24138 !) and may be an extreme form of a taxon already described. Also, the specific delimitation between *I. tenuirostris* Choisy, incl. vars. (No. 21 below) and representatives of the *I. asperifolia-fulvicaulis* complex is neither clear nor satisfactory. Intermediates are represented at Kew from Sierra Leone (Musaia *Deighton* 4206 ! 5399 !) and N. Nigeria(Katagum *Dalz.* 369 ! Birnin Gwari *Meikle* 1350 !).]

16. **I. setifera** *Poir.* in Lam. Encycl. Méth. Bot. 6 : 17 (1804) ; F.T.A. 4, 2 : 199 ; F.W.T.A., ed. 1, 2 : 216, partly (excl. *Plumtre* 65) ; Hill & Sandwith Fl. Trinidad & Tobago 2 : 228 (1953) ; Berhaut Fl. Sén. 156, 158. *Convolvulus ruber* Vahl Eclog. Am. 2 : 12 (1798). *Ipomoea rubra* (Vahl) Millesp. in Field Col. Mus. Bot. 2 : 86 (1900) ; Urban Symb. Antill. 3 : 345 (1902), not of Murray (1774). A hirsute rather robust twiner and trailer with red-purple flowers 2 to nearly 4 in. long, the sepals often plicate, tubercled and apiculate.

Sen.: *Berhaut* 1780. **Gam.:** Kudang (Jan.) *Dalz.* 8050 ! S. bank of Gambia R. *Brown-Lester* 9 ! **Guin.:** *Heudelot* 750 ! **S.L.:** York (Dec.) *Deighton* 3338 ! Yungeru (Jan.) *Thomas* 7207 ! 7278 ! Port Loko (fr. Dec.) *Thomas* 6552 ! 6638 ! Kambia (cult.) (Dec.) *Sc. Elliot* 5902 ! Rokupr (fl. & fr. Jan.) *Deighton* 2945 ! **Iv.C.:** Marahoui (fr. Jan.) *de Wit* 7647 ! A native of tropical America and now naturalized in our area.

17. **I. stenobasis** *Brenan* in Kew Bull. 5 : 228 (1950). General appearance as in *I. asarifolia*, but more slender and graceful, and leaves always clearly pointed or acuminate, with a little mucro up to 1 mm. long (which is always lacking in *I. asarifolia*), and with a much more conspicuous glandular dotting on the lower surface.

S.Nig.: Idanre Hills (fl. & fr. Oct.) *Keay* FHI 22525 ! 22674 ! Ado Rock (Oct.) *Hambler* 342 ! *Savory & Keay* FHI 25442 ! Akure F.R. (Oct.) *Keay* FHI 25468 ! Also in Congo and in Uganda.

18. **I. aquatica** *Forsk.* Fl. Aegypt.-Arab. Descr. Pl. 44 (1775) ; F.T.A. 4, 2 : 170 ; van Ooststr. l.c. 528 ; Meeuse l.c. 753. *I. reptans* Poir. in Lam. Encycl. Méth. Bot. 3 : 460 (1814) ; F.W.T.A., ed. 1, 2 : 215 ; Berhaut Fl. Sén. 158, not *Convolvulus reptans* Linn. (1753). *I. barteri* of Chev. Bot. 453, not of Bak. Nearly glabrous, stems hollow, from a stout woody stock, trailing on mud and rooting at the nodes, with pale or dark red-purple flowers 2–3 in. long.

Sen.: Richard Tol (fl. & fr. Sept.) *Roger* 96 ! Tamboukané *Chev.* 2863 ! *Berhaut* 559, 2271 ! **Gam.:** Kuntaur *Ruxton* 31 ! Pendaba *Gardiner* 17 ! 22 ! **Mali:** Dogo (fl. & fr. Apr., May) *Davey* 131 ! 577 ! Douentza to Hombori (Oct.) *Jaeger* 5497 ! **S.L.:** Ronietta (Nov.) *Thomas* 5527 ! Njala (Feb.) *Deighton* 2877 ! Moselolo (Nov.) *Deighton* 2391 ! L. Yomboro, near Konta (Apr., Dec.) *Jordan* 715 ! *Hepper* 2654 ! Aberdeen (Aug.) *Melville & Hooker* 132 ! **Ghana:** Yendi *Thorold* 312 ! Densu R., near Sekumu Lagoon (Jan.) *Irvine* 1100 ! Tamale (Nov.) *Williams* 399 ! **N.Nig.:** Addacudu *T. Vogel* 126 ! Lokoja (Oct.) *Parsons* 56 ! Fodama (Nov.) *Moiser* 217 ! Yola to Jimeta (Nov.) *Hepper* 1198 ! Katagum Dist. (Aug.) *Dalz.* 188 ! Lemme (May) *Lely* 143 ! **S.Nig.:** Lagos *Rowland* ! Eyinawsa *Keay* FHI 16047 ! Asaba (Dec.) *Keay* FHI 28304 ! Sapoba *Kennedy* 2906 ! [**Br.**]**Cam.:** Victoria *Maitland* 1346 ! Vogel Peak, Adamawa (Nov.) *Hepper* 1438 ! Dikwa (Nov.) *McClintock* 32 ! Throughout the tropics.

19. **I. hellebarda** *Schweinf. ex Hiern* in Cat. Welw. 1, 3 : 737 (1898), incl. var. *sarcopoda* Welw. ex Hiern ; F.T.A. 4, 2 : 170 (excl. syn.) ; Hallier f. in Engl. Bot. Jahrb. 18 : 142, in syn. (1893) ; Engl. Bot. Jahrb. 28 : 43 (1899). *Merremia hastifolia* A. Chev. in Bull. Mus. Hist. Nat. Paris, sér 2, 5 : 236 (1933). A glabrous twiner or trailer with very variable leaves and flowers 2 in. long, lilac or darker red-purple.

Gam.: *Ingram* ! Kambo, Central Prov. (Oct.) *Frith* 232 ! **Mali:** Goumal (fl. & fr. Jan.) *Davey* 230 ! **Guin.:** Mamou *Chev.* 20388 ! **Ghana:** *Thonning* (photo.) ! Krobo *Irvine* 1022 ! Kpeloe (fr. Oct.) *A. S. Thomas* K5 ! Kumasi (Sept.) *Dalz.* 133 ! **Dah.:** Massé to Kétou *Chev.* 22992. **N.Nig.:** Naraguta *Lely* 549 ! Throughout tropical Africa.

20. **I. vagans** *Bak.* in Kew Bull. 1894 : 70 ; F.T.A. 4, 2 : 137. *I. sulphurea* Hochst. ex Choisy in DC. Prod. 9 : 356 (1845) ; F.T.A. 4, 2 : 137 ; F.W.T.A., ed. 1, 2 : 216 ; Roberty in Bull. I.F.A.N. 16 : 1060, incl. forma *deltoidea* Roberty ; Berhaut, Fl. Sén. 157, 155 not of G. Don (1838). Trailing, woody at the base, pale-pilose, with rather small white (or yellow) flowers with a leafy calyx.

Sen.: Perrottet. Leprieur. *Berhaut* 464, 2260, 2834. **Ghana:** *Isert.* **N.Nig.:** Sokoto (fl. & fr. Sept.–Nov.) *Moiser* 17 ! 81 ! Katagum Dist. *Dalz.* 180 ! Also in Sudan.

21. **I. tenuirostris** *Choisy* in DC. Prod. 9 : 379 (1845) ; F.T.A. 4, 2 : 143 ; Verdcourt in Kew Bull. 13 : 202 (on *I. asperifolia* Hallier f.). *I. gracilior* Rendle in J. Bot. 46 : 180 (1908). Climbing in grass ; flowers pale mauve with mauve centre, in cymose inflorescences.

S.L.: Makump (Dec.) *Glanville* 106 ! Jaiama (fr. Dec.) *Deighton* 3468 ! Musaia (fl. & fr. Dec.) *Deighton* 4539 ! Bintumane Peak, 5,400 ft. (fr. Jan.) *T. S. Jones* 61 ! **N.Nig.:** Naraguta (Oct.) *Hepper* 1164 ! **S.Nig.:** Koloishe Mt., 4,900 ft., Obudu (fl. & fr. Dec.) *Savory & Keay* FHI 25102 ! [**Br.**]**Cam.:** Bamenda, 5,000 ft. (fl. & fr. Jan.) *Migeod* 310 ! Also in Cameroun.

[Note: A critical species ; material at Herb. Kew not homogenous and in some specimens (*Adams* 5145 ; *Deighton* 3468 ; 4539 ; etc.) perhaps hybridized with *I. eriocarpa* R. Br.—H. H.]

22. **I. obscura** (*Linn.*) *Ker-Gawl.* in Bot. Reg. 3, t. 239 (1817) ; Verdcourt in Kew Bull. 13 : 210, incl. var. *sagittifolia* Verdc. ; F.T.A. 4, 2 : 164 ; van Ooststr. l.c. 519 ; Meeuse l.c. 746 (1958), incl. var. *fragilis* (Choisy) Meeuse in Fl. Pl. Afr. 31, t. 1222 (1959). *Convolvulus obscurus* Linn. Sp. Pl. ed. 2, 1 : 220 (1762). *I. fragilis* Choisy in DC. Prod. 9 : 372 (1845) ; F.T.A. 4, 2 : 165 ; F.W.T.A., ed. 1, 2 : 216. (?) *C. micrantha* Hallier f. in Bull. Herb. Boiss. 6 : 541 (1898) ; F.T.A. 4, 2 : 166. *I. afra* Choisy l.c. 380 (1845), partly. In typical specimens the corolla is 1·4–2·5 cm. long, whitish-yellow with or without a purplish centre ; extremely common throughout tropical Africa. See note under 22c.

22a. **I. ochracea** (*Lindl.*) *G. Don* Gen. Syst. 4 : 270 (1837) ; F.T.A. 4, 2 : 166 ; Verdcourt in Kew Bull. 13 : 210. *Convolvulus ochraceus* Lindl. in Bot. Reg., 13 : t. 1060 (1827). *Ipomoea ochroleuca* Spanoghe in Linnaea 15 : 340 (1841) (as " ochroleucea "), van Oostr. l.c. 523, Fl. Males., ser. 1, 5 : 561 (1958). *I. kentrocarpa* Hochst. ex A. Rich. Tent. Fl. Abyss. 2 : 70 (1851) ; F.T.A. 4, 2 : 163 ; F.W.T.A., ed. 1, 2 : 216, partly (excl. *Sc. Elliot* 4679). In typical specimens the corolla is 2·7–4 cm. long, bright yellow with a purple centre ; fairly common through our area in savanna. See note under 22c.

22b. **I. trichocalyx** *Schum. & Thonn.* Beskr. Guin. Pl. 91 (1827) ; Verdcourt in Kew Bull. 13 : 210. **Ghana:** *Thonning* (Photo !) (identified by F. Didrichsen as " *I. afra* Choisy," by H. Hallier f. as *I. ochracea*).

22c. **I. acanthocarpa** (*Choisy*) *Ascherson & Schweinf.* in Schweinf. Beitr. Fl. Aethiop. 277 (1867) ; F.T.A. 4, 2 : 166 (1905) ; *Calonyction? acanthocarpum* Choisy in DC. Prod. 9 : 346 (1845). Corolla 1 in. long, whitish or pale purple, peduncles short, 1–3 flowered ; fruits tipped with a sharp persistent style-base.

Sen.: *Perrottet* 521 ; 524. *Lelièvre.* **Port.G.:** Antula, Bissau (fl. & fr. Oct.) *Esp. Santo* 2583 ! **Niger:** Zinder (fl. & fr. Oct.) *Hagerup* 585 ! **N.Nig.:** Katagum Dist. *Dalz.* 183 ! Sokoto (fr. Aug.) *Dalz.* 375 ! [**Br.**]**Cam.:** Dikwa (fl. & fr. Nov.) *McClintock* 16 ! Also in Sudan.

[The taxa *I. obscura*, *I. ochracea* and *I. trichocalyx* are kept nominally by Verdcourt (in Kew Bull. 13 : 209 (1958)) specifically distinct, but *I. ochracea* is considered as " actually little more than a large *I. obscura* " and Verdcourt is " far from satisfied " with his conclusions and says that " cultural work is needed on the members of this complex." *I. acanthocarpa*, *Merremia geophiloides* A. Chev. (in Bull. Mus. Hist. Nat. Paris, sér. 2, 5 : 236 (1933), Sarafinian *Chev.* 20637 ! Koulounda to Kaba *Chev.* 20388 !) and *Ipomoea sudanica* A. Chev. (l.c. 235 (1933), Moriguénieba *Chev.* 416. Koundian *Chev.* 446 !) from Guinée also belong to the *Ipomoea obscura*-complex, and I am quite unable to key satisfactorily these above-mentioned species and to refer properly the W. African material to them ; for this reason no specimens are cited for Nos. 22 and 22a. It is, meanwhile, best to accept the name *I. obscura* (Linn.) Ker.-Gawl. for the whole and to treat it as an aggregate species—H. H.]

23. **I. sinensis** (*Desr.*) *Choisy* in Mém. Soc. Phys. Genève 6 : 459 (1834) ; Meeuse in Bothalia 6 : 728 (1958). *Convolvulus sinensis* Desr. in Lam. Encycl. Méth. Bot. 3 : 557 (1789). *Ipomoea tropica*

Santapau & Patel in J. Bombay Nat. Hist. Soc. 54 : 799 (1957). *I. sinensis* subsp. *blepharosepala* (Hochst. ex A. Rich.) Verdc. ex Meeuse in Bothalia 6 : 729 (1958). *I. cardiosepala* Hochst. ex Bak. & Wright in F.T.A. 4, 2 : 147 (1905), not of (H.B.K.) Meissner (1869). *I. blepharosepala* Hochst. ex A. Rich. Tent. Fl. Abyss. 2 : 72 (1857). A slender perennial, trailing or twining ; flowers white or pinkish.

Niger: Aïr (fl. & fr. Aug.) *Rodd* ! In the tropics of the Old World.

24. **I. triloba** *Linn.* Sp. Pl. 161 (1753) ; Meissner in Martius Fl. Brasil. 7 : 277 (1869), incl. vars. ; Eggers in Bull. U.S. Nat. Mus. 13 : 71 (1879), incl. vars. ; Hallier f. in Engl. Bot. Jahrb. 18 : 138 (1893) ; Britton & Wilson in Sci. Surv. P. Rico & Virgin Isl. 6, 1 : 116 (1925) ; van Ooststr. in Blumea 3 : 509 (1940). *I. leucantha* Webb ex Hook. Fl. Nigrit. 152 (1849) ; F.T.A. 4, 2 : 176 ; F.W.T.A., ed. 1, 2 : 216 ; Berhaut Fl. Sén. 157 ; Eggers l.c. (1879) ; Britton & Wilson l.c. 117 (1925), not of Jacq. (1788). A glabrous twiner and trailer with rather angular stems, sepals long-ciliate, density of indumentum very variable ; flowers white, mauve, pink or pale red-purple ¾–1 in. long.

Sen.: Ouassadou *Berhaut* 1231 ; 2129. Dakar (fl. & fr. Sept.) *Berhaut* 834 ! **Guin.:** *Heudelot* 734 ! **S.L.:** *Glanville* 366 ! Magbile (Dec.) *Thomas* 5930 ! 5932 ! 5981 ! Njala (fl. & fr. Nov.) *Small* 805 (red-fl. form) ! Magbile (Dec.) *Thomas* 6065 ! 6491 ! Port Loko (Dec.) *Thomas* 5822 ! 6535 ! Russell (fl. & fr. Dec.) *Deighton* 3316 (red-fl. form) ! Kambia (Dec.) *Deighton* 1082 ! Musaia (fl. & fr. Dec.) *Deighton* 4466 ! Segbwema (fl. & fr. Dec.) *Deighton* 3476 ! Roboli, near Rokupr (Nov.) *Jordan* 178 ! **Iv.C.:** Béyo (fl. & fr. Dec.) *Leeuwenberg* 2001 ! A circumtropical weed, native of tropical America.

[Note : The W. African specimens are almost white-flowered, but this character is insufficient for a taxonomic distinction between American and African specimens ; such white-flowered forms do occur also in the W. Indies and in S. America (see Meissner and Eggers, etc. l.c.), and they have sometimes been treated unnecessarily in the older literature as varietally or (as with *I. leucantha* Webb ex Hook., not of Jacq.) specifically distinct.—H. H.]

25. **I. eriocarpa** *R. Br.* Prod. Fl. Nov. Holl., ed. 1, 484 (1810) ; F.T.A. 4, 2 : 136 ; Exell Cat. S. Tomé 250 ; van Ooststr. in Fl. Males., ser. 1, 4 : 462 (1953). *I. hispida* (Vahl) Roem. & Schult. Syst. Pl. 4 : 238 (1819) ; F.W.T.A., ed. 1, 2 : 216 ; van Ooststr. in Blumea 3 : 490 (1940), not of Zuccagni (1806). *Convolvulus hispidus* Vahl (1794). *Ipomoea morsonii* Bak. in Kew Bull. 1894 : 91 ; F.T.A. 4, 2 : 140 ; F.W.T.A. ed. 1, 2 : 216. *I. leptocaulos* Hallier f. in Engl. Bot. Jahrb. 18 : 126 (1893) ; F.T.A. 4, 2 : 140. (?) *I. kourankoensis* A. Chev. in Bull. Mus. Hist. Nat. Paris, sér. 2, 5 : 234 (1933). A slender twiner, sparingly pubescent, with white flowers about ½ in. long.

Sen.: *Roger* 116 ! *Ingram* ! **Gam.:** Genieri (fr. Feb.) *Fox* 23 ! **Mali:** Kara (fl. & fr. Dec.) *Davey* 177 ! Dogo (fr. Apr.) *Davey* 570 ! **Guin.:** Kouroussa (fr. Oct.) *Pobéguin* 584 ! **Port.G.:** Bissau (fl. & fr. Oct.) *Esp. Santo* 2588 ! **S.L.:** Kambia (Jan.) *Deighton* 916 ! Falaba (fr. Mar.) *Sc. Elliot* 5186 ! Gbangbama to Senge (Nov.) *Deighton* 2355 ! Kenema (fl. & fr. Dec.) *Deighton* 3454 ! Kasawe F.R. (fl. & fr. Dec.) *King* 52*B* ! **Iv.C.:** Mankono (fl. & fr. Nov.) *de Wilde* 957 ! **Ghana:** Bole (fl. & fr. Nov.) *Harris* ! Yendi (fr. Dec.) *Adams* 4053 ! Kpeve (fr. Jan.) *Robertson* 115 ! **Togo Rep.:** Lomé *Warnecke* 253 ! Bismarckburg *Büttner* 365. **Dah.:** *Burton* ! **N.Nig.:** Nupe *Barter* 1027 ! Kontagora (fl. & fr. Oct.) *Dalz.* ! Sokoto (fl. & fr. Sept.) *Dalz.* 373 ! Katagum Dist. *Dalz.* 177 ! Jos Plateau (Oct.) *Lely* P793 ! Abinsi (fl. & fr. Oct.) *Dalz.* 679 ! **S.Nig.:** Lagos (fr. Mar.) *Millen* 139 ! 167 ! Ibadan (Jan.) *Newberry & Etim* 183 ! **[Br.]Cam.:** Bamenda, 3,000 ft. (fl. & fr. Apr.) *Maitland* 1711 ! Gurum, Adamawa (fl. & fr. Nov.) *Hepper* 1294 ! 1322 ! Also throughout tropical Africa and in Madagascar, tropical Asia and N. Australia. (see Appendix p. 438).

26. **I. verticillata** *Forsk.* Fl. Aegypt.-Arab. Descr. Pl. 44 (1775) ; F.T.A. 4, 2 : 136 ; Berhaut Fl. Sén. 157. A prostrate annual with small white axillary flowers.

Sen.: *Perrottet. Leprieur.* Richard Tol *Berhaut* 1411. **Mali:** Kakarah (fr. Aug.) *Hagerup* 291 ! Also in Sudan, Ethiopia and Angola.

27. **I. coscinosperma** *Hochst. ex Choisy* in DC. Prod. 9 : 354 (1845) ; F.T.A. 4, 2 : 138, incl. var. *glabra* Schimper ex Bak. & Rendle (1905) (= var. *coscinosperma*) and var. *hirsuta* A. Rich. (1851) ; Meeuse l.c. 721 (1958) ; Berhaut Fl. Sén. 155. Trailing herb with loosely hairy slender stems ; flowers red or purplish, only about ⅓ in. long.

Sen.: *Berhaut* 1228 ! 3474. **Mali:** Gao (Sept.) *Hagerup* 342a ! Sompi (fl. & fr. Aug.) *Chev.* 2899 ! **Ghana:** Bongo to Nangodi (fr. Nov.) *Morton* ! **Niger:** Niamey (fl. & fr. Oct.) *Hagerup* 522 ! **N.Nig.:** Sokoto (fl. & fr. Sept.) *Dalz.* 374 ! Extends to E. Sudan and Ethiopia and to S. Africa.

[Note : the W. African plants belong to the var. *hirsuta* A. Rich. Tent. Fl. Abyss. 2 : 66 (1851), but it seems unreasonable to distinguish taxa in varietal rank of this extremely variable and widespread species ; see Meeuse l.c.—H. H.]

28. **I. blepharophylla** *Hallier f.* in Engl. Bot. Jahrb. 18 : 125 (1893) ; F.T.A. 4, 2 : 141 ; Chev. Bot. 454 ; Berhaut Fl. Sén. 156. Stems slender, pubescent, trailing from a woody rootstock, with narrow funnel-shaped flowers 2–2½ in. long, pale- or red-purple with darker lines and throat.

Sen.: Ouassadou *Berhaut* 1226. **Mali:** Sarankoro (Feb.) *Chev.* 366 ! **Ghana:** Afram Plains (Mar.) *Johnson* 728 ! **N.Nig.:** Mongu (fl. & fr. Aug.) *Lely* 392 ! Jos Plateau (May) *Lely* P272 ! Naraguta (June) *Dalz.* 311 ! Abinsi (fl. & fr. July) *Dalz.* 683 ! Yola (July) *Dalz.* 69 !

29. **I. barteri** *Bak.* in Kew Bull. 1894 : 70 (1894) ; F.T.A. 4, 2 : 169. *I. fleuryana* A. Chev. in Bull. Mus. Hist. Nat. Paris, sér. 2, 5 : 233 (1933). Stems twining, very slender, pilose at first, becoming glabrous, with flowers 2–2½ in. long and as much across the limb, pale, almost white, or reddish-purple with darker throat and tube.

Guin.: Fouta Djalon *Chev.* 18294 ! Soumbalato to Boulivel *Chev.* 18640 ! **S.L.:** Tabe (Sept.) *Deighton* 3045 ! Bunkolo (Aug.) *Thomas* 1770 ! Loma Mts. (Sept.) *Jaeger* 2014 ! Warantamba, Gberia Fotombu (Oct.) *Small* 362 ! **N.Nig.:** Jebba *Barter* ! Jos Plateau (fl. & fr. July, Aug.) *Lely* P510 ! P603 ! Vom, 3,000–4,500 ft., *Dent Young* 178 ! Mongu, 4,300 ft. (July) *Lely* 391 ! Yola (fl. & fr. July) *Dalz.* 70 !

30. **I. stolonifera** (*Cyrill.*) *J. F. Gmel.* in Linn. Syst. Nat., ed. 13, 2 : 345 (1791) ; F.T.A. 4, 2 : 171 ; van Ooststr. in Blumea 3 : 540 (1940) ; Naegelé in Bull. I.F.A.N. 21 : 1186, fig. 5 (1959). *Convolvulus stoloniferus* Cyrill. Pl. Rar. Neap. 1 : 14 (1788). Stems slender, glabrous, from a perennial tuberous root, creeping and sending up short leafy branchlets ; flowers pure white with a purple centre, 2 in. across the open limb.

Gam.: Bathurst (Nov.) *Gardiner* 113 ! **S.L.:** *Morson* ! Yele, Turtle Isl. (Nov.) *Deighton* 2315 ! Yoni to Ugepe, Bonthe Isl. (fl. & fr. Nov.) *Deighton* 2275 ! Ruka (Dec.) *Sc. Elliot* 4109 ! Shengbi (Dec.) *King* 74*B* ! **Lib.:** Monrovia (Nov.) *Linder* 1438 ! (fr. Jan.) *Harley* 1098 ! Robertsport (Dec.) *Baldwin* 10911 ! **Iv.C.:** Azuretti (Sept.) *de Wit* 7132 ! Abidjan to Grand Bassam (Jan.) *Leeuwenberg* 2378 ! **Ghana:** Axim (Feb.) *Irvine* 2302 ! Teshi, Accra (fr. Nov.) *Irvine* 810 ! **Togo Rep.:** Lomé *Warnecke* 56 ! Widespread on the shores of warm countries.

31. **I. coptica** (*Linn.*) *Roth ex Roem. & Schult.* Syst. Veg. 4 : 208 (1819) ; Roth Nov. Pl. Spec. 110 (1821) ; van Ooststr. l.c. 544 (1940) ; Meeuse in Fl. Pl. Afr. 31 : t. 1217a (1956) ; Bothalia 6 : 760 (1958). *Convolvulus copticus* Linn. Mant. Alt., App. 559 (1771). *Ipomoea dissecta* Willd. Phytogr. 5, t. 2 (1794) ; F.T.A. 4, 2 : 176 ; F.W.T.A., ed. 1, 2 : 218. A slender glabrous trailing herb with much-divided leaves and small white flowers ⅓–⅝ in. long.

Sen.: (fl. & fr. Sept.) *Roger* ! *Heudelot* 270 ! **Mali:** Kabarah (fl. & fr. Aug.) *Hagerup* 290 ! Sarédina (fl. & fr. May) *Davey* 107 ! 138 ! Gao (fl. & fr. Aug.) *de Wailly* 5120 ! Boré (Sept.) *Demange* 41/1957 ! **Ghana:** *Brown* 305 ! Accra (fl. & fr. July) *Irvine* 710 ! Nungua, Accra Plains (fr. July) *Ankrah* GC 20188 ! **Togo Rep.:** Lomé *Warnecke* 284 ! **N.Nig.:** R. Benue *Talbot* 806 ! Katagum Dist. (fr. Sept.) *Dalz.* 178 ! Maifoni, Bornu *Parsons* ! Bindawa (Aug.) *Grove* 49 ! **S.Nig.:** Lagos *Dawodu* 85 ! Throughout tropical Africa, also in S. Africa, Asia and Australia.

32. **I. batatas** (*Linn.*) *Lam.* Tabl. Encycl. 465 (1791) ; F.T.A. 4, 2 : 175 ; Chev. Bot. 453 ; van Ooststr. l.c. 512 (1490) ; Meeuse l.c. 746 (1958). *Convolvulus batatas* Linn. Sp. Pl. 154 (1753). *Ipomoea setifera* of F.W.T.A., ed. 1, 2 : 216 partly (*Plumptre* 65), not of Poir. in Lam. Trailing and climbing, nearly glabrous

from tuberous root ; flowers campanulate-funnel-shaped with whitish or pink-tinged limb and red-purple centre 1½–2 in. long. The sweet potato.

Sen.: Tiaraye *Chev.* 2857. " Kaolak 167 " *Kaichinger* in *Hb. Chev.* **Port.G.:** Bissau (Nov.) *Esp. Santo* 1370! **Guin.:** Conakry *Chev.* 12017 ; 12131. **S.L.:** Rowala (July) *Thomas* 1185! Bumba (Sept.) *Thomas* 1946! Jigaya (Sept.) *Thomas* 2474! Roruks (Nov.) *Thomas* 5743! 5776! **Iv.C.:** Bingerville *Chev.* 17334. **Ghana:** Achimota (Aug.) *Irvine* 780! Accra Plains (Nov.) *Morton* 6121! Asamankase (Dec.) *Plumptre* 65! Kumasi *Chipp* 709! **N.Nig.:** *Dalz.* 680! *Lely* 713! **S.Nig.:** Lagos (Aug.) *Dodd* 424! Obu *Thomas* 489! Port Harcourt (Dec.) *Stubbings* 166! Commonly cultivated and naturalised near habitations or railways, etc. (See Appendix, p. 436).

33. **I. mauritiana** *Jacq.* Coll. Bot. 4 : 216 (1791) ; Hort. Caes. Schoenbrunn. 2 : 39, t. 200 (1797) ; Exell in Bull. I.F.A.N., sér. 1, 21 : 462 (1959). *I. digitata* of F.T.A. 4, 2 : 189 (1905) incl. vars. ; Chev. Bot. 454, incl. var. *djalonensis* A. Chev., name only and var. *eriosperma* (P. Beauv.) A. Chev. (1920), illegitimate name ; F.W.T.A., ed. 1, 2 : 216, fig. 251 and various other Floras of the Old World Tropics, see van Ooststr. in Fl. Males. ser. 1, 4 : 483 (1953), and in Blumea 3 : 558 (1940), not of Linn. (1759). Perennial with a large tuberous root, with widely climbing glabrous hollow stems, and conspicuous rose-red flowers 2–2½ in. long in pedunculate cymes.

FIG. 284.—IPOMOEA MAURITIANA *Jacq.* (CONVOLVULACEAE).
A, flowering shoot. B and C, stamen. D, fruits. E, seed.

Sen.: *Berhaut* 359 ; 3238. **Gam.:** Kuntaur *Ruxton* 63! **Guin.:** Kouria *Caille* in *Hb. Chev.* 14944. Longuéry *Caille* in *Hb. Chev.* 14945. Golea, near Conakry (fl. & fr. July) *de Wit* 7013! **Port.G.:** Bedanda (fr. Jan.) *d'Orey* 249! **S.L.:** *T. Vogel* 96! Kortright (fl. & fr. Aug.) *Melville & Hooker* 29! Waterloo (July) *Lane-Poole* 294! Kukuna, Scarcies (fl. & fr. Jan.) *Sc. Elliot* 4507! Kaballa (fr. Sept.) *Thomas* 2335! **Lib.:** Kakatown *Whyte*! Robertsport (fr. Dec.) *Baldwin* 10920! Dukwai R. (fl. & fr. Oct.–Nov.) *Cooper* 15! Gbanga (fr. Aug.) *Okeke* 14! Ganta (Oct.) *Harley*! **Iv.C.:** Nékakougnié to Grabo *Chev.* 19601. Bassam (Sept.) *de Wit* 5771! Niapidou (Apr.) *Leeuwenberg* 3198! Adiopodoumé (fl. & fr. June) *de Wilde* 34! **Ghana:** Axim (Feb.) *Irvine* 2301! Achimota (Dec.) *Irvine* 815! Techeri (June) *Chipp* 473! Kumasi (Apr.) *Akwa* 1682! Kpeve *Williams* 44! **Togo Rep.:** Lomé *Warnecke* 244! **N.Nig.:** Nupe *Barter* 1267! Lokoja *Parsons* 120! Abinsi (fl. & fr. Mar.) *Dalz.* 682! Malabu Hills, near Yola (fl. & fr. May) *Dalz.* 71! **S.Nig.:** Lagos *Dalz.* 1074! Sapoba *Kennedy* 2626! Ibadan (Jan.) *Newberry* 24! Port Harcourt (Sept.) *Stubbings* 59! Eket (May) *Onochie* FHI 32948! Calabar *Clarke*! **[Br.]Cam.:** Victoria (Mar.) *Maitland* 470! **F.Po:** (fr. June) *Mann* 422! Cosmopolitan in the tropics.

34. **I. cairica** (*Linn.*) *Sweet* Hort. Brit., ed. 1, 287 (1827) ; van Ooststr. in Blumea 3 : 542 (1940) ; Brenan in Mem. N.Y. Bot. Gard. 9 : 8 (1954) ; Meeuse l.c. 761 (1958). *Convolvulus cairicus* Linn. Syst. Nat., ed. 10, 922 (1759). *I. palmata* Forsk. Fl. Aegypt.-Arab. 43 (1775) ; F.T.A. 4, 2 : 178 ; Chev. Bot. 455 ; incl. var. *ciliolata* A. Chev., name only ; Berhaut Fl. Sén. 152. *I. vesiculosa* P. Beauv. Fl. Oware 2 : 73, t. 106 (1819). A slender twiner with glabrous, sometimes warted stems, and funnel-shaped flowers pale or reddish-purple 1½–2 in. long.

Sen.: *Roger*! *Leprieur*! *Brunner*! *Farmar* 158! *Berhaut* 839. **Gam.:** *Ingram*! *Hayes* 520! **Port.G.:** Bissau (Dec.) *Esp. Santo* 2637! **S.L.:** *T. Vogel* 48! *Sc. Elliot* 4082! Boltum, Turtle Isl. (Nov.) *Deighton* 2366! Shenghi (fr. Dec.) *King* 73B! Luti (Oct.) *Jordan* 941! **Lib.:** Monrovia *Whyte*! **Iv.C.:** San Pedro *Thoiré*! Bingerville (Feb.) *Chev.* 17330! Bériby (Aug.) *Chev.* 20001 *bis*! **Ghana:** Emisano, near Elmina (Dec.) *Deakin* 53! **N.Nig.:** Bornu *Parsons*! **S.Nig.:** Lagos (Apr.) *Dalz.* 1071! **F.Po:** Clarence Pk. (fl. & fr. Nov., Dec.) *T. Vogel* 40! 236! *Mann* 54! Throughout tropical Africa ; also in Egypt, S. Africa and widespread in the tropics.

35. **I. kotschyana** *Hochst. ex Choisy* in DC. Prod. 9 : 354 (1845) ; Hallier f. in Engl. Bot. Jahrb. 18 : 125 (1893) ; F.T.A. 4, 2 : 139 ; Berhaut Fl. Sén. 48, 152 ; Verdcourt in Kew Bull. 13 : 200. A slender annual with small flowers.

Sen.: *Leprieur*. *Berhaut* 948. Dagana *Perrottet*! **Guin.?:** *Heudelot* 734! **Mali:** Timbuktu (Aug.) *Hagerup* 241! Also in Sudan and Eritrea.

[*Heudelot* 734 at the British Muesum is dated 1847 ; but *Heudelot* 734 at Kew is *I. triloba* and dated 1838.]

36. **I. nil** (*Linn.*) *Roth* Cat. Bot. 1 : 36 (1797) ; van Ooststr., in Blumea 3 : 497 (1940) ; Verdcourt in Taxon 6 : 231 (1957) ; Meeuse l.c. 733 (1958). *Convolvulus nil* Linn. Sp. Pl., ed. 2, 219 (1762). *Ipomoea hederacea* of F.T.A. 4, 2 : 159 ; of F.W.T.A., ed. 1, 2 : 218 (and of many other Floras), not of Jacq. (1787). A

twiner, the stems, etc., covered when young with spreading yellow hairs ; the flowers pale blue in the morning turning bright pink or red-purple, with long narrow calyx-lobes ; on fences in villages and in fields.

S.L.: Njala, cult. (July) *Deighton* 3740 ! **Ghana:** Cape Coast *Brass*. **N.Nig.:** Jebba *Barter* ! Lokoja (Oct.) *Dalz.* 76 ! R. Benue *Talbot* 822 ! Sokoto (Dec.) *Dalz.* 376 ! Kadana (fl. & fr. Nov.) *Lely* 592 ! **S.Nig.:** Lagos (Oct.) *Punch* 39 ! Ado Rock, Oyo (Oct.) *Savoury & Keay* FHI 25237 ! Shabe Rock, Oyo (Oct.) *Hambler* 532 ! Widespread in the tropics.

37. **I. aitonii** *Lindl.* Bot. Reg. 21 ; t. 1794 (1836) ; F.W. Andr. Fl. Pl. Sudan 3 : 114 (1956). *I. arachnosperma* Welw. (1858)—Meeuse in Fl. Pl. Afr. 31 : t. 1203 (1956) ; Bothalia 6 : 736 (1958). *I. pilosa* (Roxb.) Sweet Hort. Brit., ed. 1, 289 (1827) ; F.T.A. 4, 2 : 161 ; F.W.T.A., ed. 1, 2 : 218, not of Hottouyn (1777), or of Cavanilles (1797). A perennial twiner, coarsely hairy, with leaves white below and rather small pink or red-purple flower ¾ in. long in pedunculate axillary cymes.

Sen.: *Roger* ! **Gam.:** *Ingram* ! **Ghana:** Abetifi (Dec.) *Irvine* 1654 ! Sunyani (fr. Dec.) *Adams* 2754 ! **N.Nig.:** *Moiser* ! *Gosling* ! Katagum Dist. *Dalz.* 185 ! Nabardo (fl. & fr. Sept.) *Lely* 611 ! R. Benue *Talbot* 808 ! Throughout tropical Africa.

The following herbaceous species are cultivated in our area : *I. acuminata* (Vahl) Roem. & Schult. (syn. *I. congesta* R. Br.) ; *I. leari* Paxton ; *I. purpurea* (Linn.) Roth ; *I. setosa* Ker-Gawl. ; *I. tuba* (Schlechtend.) G. Don ; *I. tricolor* Cav. (syn. *I. rubrocaerulea* Hook.). *I. intrapilosa* Rose, a tree about 12 ft. high with large white flowers, is cultivated at Ibadan, S. Nigeria. *I. arborescens* (Humb. & Bonpl. ex Willd.) G. Don, closely allied, is recorded from gardens in Senegal by Berhaut (Fl. Sén. 126, 130).

Doubtfully recorded species.

I. tiliacea *(Willd.)* Choisy in DC. Prod. 9 : 375 (1845) ; van Ooststr. in Blumea 5 : 233 (1942) ; Fl. Males., ser. 1, 4 : 469 (1953) ; Exell Cat. S. Tomé 252. *Convolvulus tiliaceus* Willd. (1809). *C. fastigiatus* Roxb. (1824). *Ipomoea fastigiata* (Roxb.) Sweet Hort. Brit., ed. 1, 288 (1826) ; Choisy in DC. Prod. 9 : 380 (1845) ; F. Didr. in Vidensk. Medd. Kjöbenhavn 1854 : 226 (1855) ; F.T.A. 4, 2 : 169. Native of tropical America ; found on Principe (*vide* Exell *l.c.*) and in Cameroun (*vide* F.T.A. *l.c.*), also recorded by Didrichsen (l.c.) " e Guinea, *Mortensen* " without further comment. No material seen, but it seems very probable that this species occurs in our area.

Imperfectly known species.

I. ardissima *A. Chev.* in Rev. Bot. Appliq. 30 : 272 (1950). **Niger :** Gaya to Dosso *A. Chev.* s.n. According to Chevalier this species belongs to " the group " of *I. cairica* and *I. coptica*.

17. **STICTOCARDIA** Hallier f. in Engl. Bot. Jahrb. 18 : 158 (1894) ; van Ooststr. in Blumea 5 : 346 (1943) ; Fl. Males., ser. 1, 4 : 491 (1953) ; Meeuse in Bothalia 6 : 772 (1958).

Branches puberulous ; leaves long-petiolate, very widely ovate-orbicular, widely cordate at the base, broadly acuminate, about 15 cm. diam., with numerous nerves from near the base, puberulous beneath, with large glandular areas here and there which perish and leave large holes ; cymes subsessile or shortly pedunculate, few-flowered ; calyx-segments 1·3 cm. long, stellate-pubescent towards the base ; corolla about 4·5 cm. widely tubular   ..   ..   *beraviensis*

S. beraviensis *(Vatke)* Hallier f. l.c. (1894) ; Verdcourt in Kew Bull. 13 : 189. *Ipomoea beraviensis* Vatke in Linnaea 43 : 514 (1882). *Argyreia* (?) *beraviensis* (Vatke) Bak. in F.T.A. 4, 2 : 201 (1906) ; Chev. Bot. 453. *Ipomoea hierniana* Rendle in J. Bot. 39 : 58 (1901) ; Chev. Bot. 453. ; F.T.A. 4, 2 : 188. A stout climber with finely pubescent pithy stems and crimson or red-purple flowers 1½–2 in. long.

Mali : Kita (Nov.) *Jaeger* 5566 ! **Port. G. :** Bafata (fr. Dec.) *Esp. Santo* 2873 ! **Guin. :** Kouria *Caille* in Hb. Chev. 14874. **S.L. :** Njala (Nov.) *Deighton* 2570 ! Maboma (Nov.) *Deighton* 3778 ! **Lib. :** Ganta (Nov.) *Harley* ! Banga (Oct.) *Linder* 1232 ! Kakatown *Whyte* ! Zigida (Oct.) *Baldwin* 10037 ! **Ghana :** Akwapim (Dec.) *Johnson* 759 ! Akim Swedru (Dec.) *Irvine* 2064 ! **N. Nig. :** Lokoja (Oct.) *Parsons* 77 ! Mada Hills (fl. & fr. Oct.) *Hepburn* 87 ! **S. Nig. :** *Foster* 366 ! Lagos (Mar.) *Millen* 55 ! Idumuye (Dec.) *Thomas* 2122 ! Udi F.R. (Oct.) *Smith* S 20 ! Onitsha Olassa *Thomas* 1887 ! Ondo, Okelife (Nov.) *Onochie* FHI 34350 ! (Dec.) *Okereks* FHI 26881 ! Oban *Talbot* 801 ! Also in Cameroun, Congo, E. Africa and Madagascar.

153. SCROPHULARIACEAE

By F. N. Hepper

Herbs or shrubs, rarely small trees. Leaves alternate, opposite or verticillate; stipules absent. Flowers hermaphrodite, mostly zygomorphic. Calyx imbricate or valvate. Corolla-lobes imbricate. Stamens often 4 or 2, rarely 5, inserted on the corolla and alternate with the lobes, sometimes the fifth stamen represented by a staminode; filaments free from each other; anthers 1–2-celled, opening lengthwise. Ovary superior, entire, usually 2-celled ; style terminal. Ovules numerous on axile placentas. Fruit a capsule or berry. Seeds numerous, with fleshy endosperm.

Widely distributed. Recognized by the usually opposite exstipulate leaves, mostly zygomorphic flowers, entire ovary with terminal style, and axile placentas.

Flowers spurred ; lower leaves opposite, upper leaves alternate :
Capsule dehiscing by apical pores ; corolla 1–2 cm. long   ..   ..   .. 1. **Linaria**
Capsule splitting loculicidally and spreading ; corolla 5 mm. long ..   .. 2. **Diclis**
Flowers not spurred :
Leaves alternate, never all radical (but densely congested in No. 3) :
Stamens 2, staminodes absent..   ..   .. 3. **Anticharis**

Stamens 4 (rarely 5 in Nos. 7 and 9) :
　Leaves orbicular ; creeping herb with small rotate axillary flowers　　7, **Sibthorpia**
　Leaves not orbicular ; erect herbs :
　　Corolla rotate ; flowers in long racemes .. 　.. 　.. 　.. 　.. 　5. **Celsia**
　　Corolla with distinct tube :
　　　Leaves densely congested on short stems ; flowers sessile amongst the leaves
　　　　　　　　　　　　　　　　　　　　　　　　　　　　4. **Aptosimum**
　　　Leaves not congested ; stems erect ; flowers axillary on long slender pedicels
　　　　　　　　　　　　　　　　　　　　　　　　　　　　9. **Capraria**
Leaves opposite, or at least subopposite (sometimes the upper leaves alternate) or all
　radical :
　Stamens 2 (2 staminodes also present in No. 20 and very reduced in No. 15) :
　　Leaves much divided ; aquatic (*L. sessiliflora*) ; see also below ..　　13. **Limnophila**
　　Leaves not divided ; creeping or erect :
　　　Creeping herb 1 cm. high ; flowers solitary, slender-pedicellate ; calyx 3-lobed ;
　　　　anther-cells confluent　.. 　.. 　.. 　.. 　.. 　　22. **Glossostigma**
　　　Plant usually much larger, or if only about 2 cm. high and erect (as in No. 23)
　　　　then calyx 5-toothed :
　　　　Corolla-tube short, almost rotate ; stamens exserted .. 　.. 　　6. **Veronica**
　　　　Corolla with distinct tube, usually un-equally 2-lipped ; stamens included :
　　　　　Flowers subsessile in dense axillary clusters, minute ; stems very slender 2–3 cm.
　　　　　　high ; anther cells distinct ; staminodes absent ; stigma spathulate
　　　　　　　　　　　　　　　　　　　　　　　　　　　23. **Psammetes**
　　　　　Flowers axillary or in terminal racemes, not in dense clusters ; stigma bilobed :
　　　　　　Staminodes (inside lower lip) with a boss-like appendage and long slender sterile
　　　　　　　filament ; capsule septicidal　.. 　.. 　.. 　.. 　20. **Ilysanthes**
　　　　　　Staminodes (inside lower lip) reduced to 2 small protuberances or quite obsolete;
　　　　　　　capsule loculicidal　.. 　.. 　.. 　.. 　.. 　15. **Dopatrium**
　Stamens 4 (rarely 3 in No. 4) :
　　Leaves all radical ; small tufted herbs with or without stolons　..　21. **Limosella**
　　Leaves not all radical ; more or less erect herbs :
　　　Calyx-lobes very unequal　.. 　.. 　.. 　.. 　.. 　.. 　14. **Bacopa**
　　　Calyx-lobes not markedly unequal :
　　　　Stamens all (or some of them) inserted at the throat of the corolla :
　　　　　Flowers 2–3 cm. long[1] ; calyx 5-partite, segments broad and imbricate
　　　　　　　　　　　　　　　　　　　　　　　　　　　16. **Artanema**
　　　　　Flowers up to 1 cm. long :
　　　　　　Lower filaments arched with a small boss-like appendage at the base ; calyx
　　　　　　　slightly 2-lipped and narrowly winged along the nerves ..　18. **Torenia**
　　　　　　Lower filaments sharply angled and with a large boss-like appendage at base :
　　　　　　　Calyx deeply 5-partite ; flowers solitary or in lax spikes ..　19. **Lindernia**
　　　　　　　Calyx 5-toothed ; flowers in rather or very dense spikes　..　17. **Craterostigma**
　　　　Stamens all inserted in corolla-tube :
　　　　　Anthers with 2 equal or subequal cells :
　　　　　　Leaves pinnately or variously deeply divided, in at least the submerged ones :
　　　　　　　Corolla-tube short and campanulate, at most 1·5 cm. long, yellow, rarely white
　　　　　　　　　　　　　　　　　　　　　　　　　　　13. **Limnophila**
　　　　　　　Corolla-tube long and narrow, about 2·5 cm. long, white (*R. fistulosa*); see also
　　　　　　　　below .. 　.. 　.. 　.. 　.. 　.. 　.. 　28. **Rhamphicarpa**
　　　　　　Leaves all entire or toothed, not deeply divided :
　　　　　　　Anther-cells distinct and separate　.. 　.. 　.. 　.. 　11. **Stemodia**
　　　　　　　Anther-cells distinct but contiguous :
　　　　　　　　Corolla-lobes 4 ; calyx deeply 4- or 5-lobed .. 　.. 　.. 　8. **Scoparia**
　　　　　　　　Corolla-lobes 5, funnel-shaped or campanulate tube ; calyx 5-toothed or lobed:
　　　　　　　　　Flowers distinctly pedicellate, 5–20 mm. long :
　　　　　　　　　　Fruits markedly reflexing on maturity　.. 　.. 　.. 　12. **Stemodiopsis**
　　　　　　　　　　Fruits not reflexing :
　　　　　　　　　　　Bracteoles absent ; corolla markedly bilabiate ; pedicels 1–2 cm. long ;
　　　　　　　　　　　　plants drying green　.. 　.. 　.. 　.. 　.. 　10. **Mimulus**
　　　　　　　　　　　Bracteoles present beneath calyx ; corolla not bilabiate ; pedicels
　　　　　　　　　　　　5–10 mm. long ; plants drying blackish　..　24. **Micrargeria**
　　　　　　　　　Flowers sessile or subsessile ; plants drying blackish :
　　　　　　　　　　Bracteoles absent ; corolla neither thin nor conspicuously veined ; seeds
　　　　　　　　　　　winged　.. 　.. 　.. 　.. 　.. 　.. 　25. **Bartsia**
　　　　　　　　　　Bracteoles present ; corolla thin, conspicuously longitudinally veined ;
　　　　　　　　　　　each seed enclosed in a transparent or opaque envelope　.. 26. **Alectra**
　　　　　Anthers 1-celled or with 1 fertile cell and 1 variously modified sterile cell or
　　　　　　appendage ; dry plants usually dark ; mainly semi-parasitic :

[1] The introduced *Torenia fournieri* Linden ex Fourn. also has flowers of this size, but its calyx is conspicuously
winged.

Calyx-teeth woolly inside (conspicuous as a white fringe on calyx-margin) ; anthers (at least two of them) 2-celled, with 1 fertile cell and 1 sterile cell or appendage ; corolla subrotate   ..   ..   ..   **27. Sopubia**
Calyx-teeth not woolly inside nor on margins ; anthers all 1-celled ; corolla-tube long and narrow :
  Corolla-tube sharply curved at or above the middle, lobes unequal   **30. Striga**
  Corolla-tube straight or nearly so :
    Fruits obliquely ovoid, long-beaked ; calyx campanulate ; corolla-tube about 2 cm. long ..   ..   **29. Rhamphicarpa**
    Fruits straight, oblong or ellipsoid ; calyx tubular :
      Corolla small, about 5 mm. long, with subequal lobes ; fruits dry and dehiscent   ..   ..   **28. Buchnera**
      Corolla large, about 4 cm. long, upper lobes united and smaller than the lower; fruits fleshy and indehiscent   ..   ..   ..   **31. Cycnium**

Besides the above, *Russelia equisetiformis* Schlecht. & Cham. (syn. *R. juncea* Zucc.) a native of Mexico, is cultivated as an ornamental plant in our area.

1. LINARIA Juss.—F.T.A. 4, 2 : 288.

Stems and leaves glandular-pubescent ; leaves more or less ovate, with a few large teeth or hastately lobed, 5–15 mm. long, 4–10 mm. broad ; petiole up to 9 mm. long, pubescent, or leaves subsessile ; flowers solitary, axillary, corolla about 1 cm. long with spur 3 mm. long, pubescent   ..   ..   1. *aegyptiaca*
Stems and leaves glabrous ; leaves linear to lanceolate-hastate, 1·5–4·5 cm. long, 2–10 mm. broad ; petiole 2–10 mm. long, glabrous ; flowers solitary, axillary, corolla about 2 cm. long with spur 10 mm. long, slightly glandular-pubescent ..   2. *sagittata*

1. **L. aegyptiaca** (*Linn.*) *Dum.-Cours.* Bot. Cult., ed. 1, 2 : 92 (1802) ; De Miré and Gillet in J. Agric. Trop. 3 : 718 (1956), incl. infra-specific forms. *Antirrhinum aegyptiacum* Linn. (1753). *Kickxia aegyptiaca* (Linn.) Nábělek in Publ. Fac. Sci. Univ. Masaryk, Brno 70 : 31 (1926). Straggling under-shrub with many slender branches a few inches to 2 ft. high ; flowers yellow ; in semi-desert.
  Maur.: Kankossa (fl. & fr. Mar.) *Roberty* 16944 ! **Niger:** Aïr, fide *De Miré & Gillet* l.c. Also in N. Africa and Syria.

2. **L. sagittata** (*Poir.*) *Hook. f.* Bot. Mag. t. 6060 (1873) ; F.T.A. 4, 2 : 291 ; Berhaut Fl. Sén. 190. *Antirrhinum sagittatum* Poir. (1816). *L. heterophylla* (Schousb.) Spreng. (1825), not of Desf. (1798). *Kickxia heterophylla* (Schousb.) Dandy in F.W. Andr. Fl. Sudan 3 : 137 (1956). Slender twining herb with stems about 2 ft. long ; flowers yellow.
  Sen.: *Perrottet* 571 ! Wakam (Sept., Oct.) *Berhaut* 1153 ! 1980 ! Also in Sudan, Ethiopia, Somaliland, N. Africa, Arabia and Canary Islands.

2. DICLIS Benth.—F.T.A. 4, 2 : 286.

Prostrate or ascending annual ; leaves mostly alternate, ovate, acute at apex, truncate to cuneate at base, with several large teeth on each side, (5–)10–18 mm. long and nearly as broad, minutely pubescent ; petiole (3–)8–15 mm. long ; flowers solitary, axillary ; pedicels 11–34 mm. long ; calyx deeply 5-lobed ; corolla 2-lipped, upper lip small, lower lip 3 mm. long, 3-lobed, spur 1·5 mm. long ; capsule splitting and spreading, 2 mm. long ; seeds ovoid, longitudinally wrinkled ..   ..   *ovata*

D. ovata *Benth.* in Hook. Comp. Bot. Mag. 2 : 23 (1836) ; F.T.A. 4, 2 : 287 ; Hepper in Kew Bull. 15 : 65. Slender herb at first more or less erect becoming prostrate, with sub-simple branches up to a foot long ; flowers white, streaked reddish, spur yellowish ; in moist, open places.
  [**Br.**] **Cam.:** Buea (fl. & fr. May) *Morton* K943 ! Mambila Plateau : Gembu, 5,300 ft. (fl. & fr. Jan.) *Hepper* 1813 ! Maisamari to Bellel, 5,000 ft. (fl. & fr. Jan.) *Hepper & Daramola* 1710 ! Also in Cameroun (?), Ethiopia, Uganda, Kenya, Tanganyika, N. & S. Rhodesia, Nyasaland, Angola, Madagascar and Mauritius.

3. ANTICHARIS Endl.—F.T.A. 4, 2 : 275.

Leaves narrowly linear, 2–5·5 cm. long, about 3 mm. broad, with glandular hairs, acute ; flowers axillary, solitary ; pedicels very slender, 1–2 cm. long, with a pair of bracteoles above the middle ; fruits ovoid, about 8 mm. long, beaked   ..   1. *linearis*
Leaves ovate-elliptic, about 1 cm. long and 5 mm. broad, densely glandular-pubescent, obtuse ; pedicels 2–3 mm. long, with 2 inconspicuous bracteoles about the middle ; fruits ovoid, about 7 mm. long, beaked   ..   ..   2. *glandulosa*

1. **A. linearis** (*Benth.*) *Hochst. ex Aschers.* in Monatsber. Akad. Wiss. Berl. 1866 : 882 (1867) ; F.T.A. 4, 2 : 276 ; Chev. Bot. 469 ; Berhaut Fl. Sén. 200 ; De Miré & Gillet in J. Agric. Trop. 3 : 717 (1956). *Doratanthera linearis* Benth. (1846). Erect much-branched herb, woody below, up to 1 ft. or more high ; flowers pink or purplish ; in semi-desert.
  Sen.: *Leprieur* ! Walo *Perrottet* 567 ! Richard Tol *Berhaut* 1753. **Mali :** Ansongo (fl. & fr. Sept.) *Hagerup* 380 ! Goundam (Aug.) *Chev.* 3300 ! **Iv. C.:** Ouagadougou (Oct.) *Aubrév.* 2662 ! **Niger :** Fada to Koupéla *Chev.* 24519 ! Aïr Mts. *fide* De Miré & Gillet *l.c.* **N. Nig.:** Fodama (fl. & fr. Nov., Dec.) *Moiser* 213 ! 231 ! Kuka, Bornu Prov. (fl. & fr. Jan.) *E. Vogel* 11 ! From Cape Verde Islands to India.
2. **A. glandulosa** *Aschers.* l.c. 880 (1867) ; De Miré & Gillet l.c. 717 (1956). Erect herb about 6 ins. high, densely glandular ; flowers bluish ; in semi-desert.
  Niger: Aïr Mts. *fide* De Miré & Gillet *l.c.* Also in N. Africa, Somaliland and extends to India.

4. APTOSIMUM Burch.—F.T.A. 4, 2 : 267.

Densely caespitose herb, half woody at base, 3–10 cm. high ; leaves linear to linear-lanceolate, 2–6 cm. long, 2–6 mm. broad, midrib prominent beneath, more or less pubescent on both sides, long-ciliate beneath ; flowers sessile, axillary ; calyx deeply 5-cleft, lobes subulate, about 1 cm. long, long-ciliate ; corolla about 1 cm. long, tube narrow at base, broadening above, lobes subequal ; stamens 4 .. .. *pumilum*

A. pumilum (*Hochst.*) *Benth.* in DC. Prod. 10 : 345 (1846) ; F.T.A. 4, 2 : 268. *Chilostigma pumilum* Hochst. (1841). *Anticharis linearis* of Monod in Contrib. Et. Sahara Occ. 2 (Phanéro.) : 110 (1939), ? partly, not of (Benth.) Hochst. Densely tufted, with 2 or 3 horizontal branches at ground level 2 or 3 in. long ; flowers white outside and blue inside ; in semi-desert.
Maur. : Tagant Region *Monod* 1789 ! Néma to Timbrédra *Rossetti* 61/278 ! **Niger**: Tabello, Aïr *Chopard & Villiers* ! Agadès (fl. & fr. Nov.) *de Wailly* 5263 ! Also in Sudan and Somaliland.

5. CELSIA Linn.—F.T.A. 4, 2 : 280 ; Murbeck in Lunds Univ. Årsskrift. n.f. 22 : 1 (1925).

Leaves tomentose beneath, often very densely so, lanceolate, closely serrulate, 2–7 cm. long, 7–20 mm. broad ; stems densely tomentose ; bracts ovate-lanceolate, nearly as long as pedicels in flower ; pedicels 1 cm. long in flower, 1·5 cm. long in fruit, glandular ; flowers about 2 cm. diam. ; fruits 6–8 mm. long .. .. 1. *densifolia*
Leaves sparsely tomentose on the nerves beneath, oblong-lanceolate, crenate, up to 16 cm. long and 6 cm. broad ; stems sparsely tomentose ; bracts ovate-lanceolate, lower ones cordate, shorter than the pedicels ; pedicels 1–2 cm. long in flower, 1·5–3 cm. long in fruit, glandular ; flowers about 2·5 cm. diam. ; fruits 8–10 mm. long 2. *ledermannii*

1. **C. densifolia** *Hook. f.* in J. Linn. Soc. 7 : 208 (1862) ; F.T.A. 4, 2 : 284 ; Murbeck in Lunds Univ. Årsskrift. n.f. 22 : 63 (1925). Robust erect herb with pithy stems 3 ft. high, woody at the base ; flowers yellow in a long terminal raceme ; in montane grassland and thickets.
[**Br.**]**Cam.**: Cam. Mt., 9,000–10,000 ft. (fl. & fr. Dec.) *Johnston* 9 ! *Maitland* 892 ! *Thorold* CM 20 ! Lakom, Bamenda (fl. & fr. Apr.) *Maitland* 1409 ! Kumbo to Oku, 6,000 ft. (fl. & fr. Feb.) *Hepper* 1996 ! **F.Po** : 8,500 ft. *Mann* 1453 !
2. **C. ledermannii** *Schlechter* in Murbeck l.c. 64 (1925). Stout erect herb woody at the base, 4–5 ft. high ; flowers yellow ; in montane grassland.
[**Br.**]**Cam.**: Muti, Bamenda *Ledermann* 5851. Lakom, 6,000 ft., Bamenda (fl. & fr. June) *Maitland* 1547 ! Mbakakeka Mt., 7,400 ft., Bamenda (fl. & fr. Jan.) *Keay* FHI 28409 ! *Hepper* 2131 ! Ndu, 6,000–7,000 ft. Bamenda (Sept.) *Savory* UCI 397 !

6. VERONICA Linn.—F.T.A. 4, 2 : 356.

Stems and calyx pubescent :
Leaves petiolate, ovate, rounded or truncate at base, (1–)2–4 cm. long, 1–2 cm. broad, shortly pubescent on both surfaces, serrate except towards base ; petiole about 1 cm. long ; flowers paired or a few together on a slender axillary peduncle ; calyx 3–4 mm. long, pubescent ; corolla 8–10 mm. diam. ; fruit bilobed, as long as calyx, pubescent .. .. 1. *abyssinica*
Leaves sessile, lanceolate, narrowed at base, 1–4 cm. long, 2–12 mm. broad, pilose at first, soon becoming glabrous ; flowers in a dense terminal spike-like raceme ; calyx 4–6 mm. long, pubescent ; corolla 10–14 mm. diam. ; fruits retuse, shorter than calyx, pubescent .. .. 2. *mannii*
Plant entirely glabrous ; leaves shortly petiolate, oblong-elliptic, 1·5–3 cm. long, 1–1·5 cm. broad, slightly serrate ; flowers numerous in opposite axillary racemes ; pedicels slender about 5 mm. long ; calyx 2–3 mm. long ; corolla about 4 mm. diam. 3. *beccabunga*

1. **V. abyssinica** *Fres.* in Bot. Zeit. 2 : 356 (1844) ; F.T.A. 4, 2 : 358 ; Römpp in Fedde Rep., Beih. 50 : 141 (1928) ; Hedberg in Sym. Bot. Uppsal. 15 : 169 (1957). *V. africana* Hook. f. in J. Linn. Soc. 7: 208 (1864). Prostrate herb, creeping, branched from the base, more or less pilose ; flowers blue or pinkish ; in montane grassland and scrub beside small streams.
[**Br.**]**Cam.**: Cam. Mt., 5,000–9,000 ft. (fl. & fr. Dec., Jan.) *Mann* 1263 ! *Dunlap* 75 ! Bamenda, 6,000–8,000 ft. (fl. & fr. Jan., Apr., June, Sept.) *Ujor* FHI 30445 ! *Savory* UCI 279 ! *Boughey* GC 10557 ! Ndu (Feb.) *Hepper* 1939 ! Ngel Nyaki, Mambila Plateau (Jan.) *Hepper* 1718 ! Also in the highlands of eastern Africa from Ethiopia to Nyasaland.
2. **V. mannii** *Hook. f.* in J. Linn. Soc. 6 : 19 (1861) ; F.T.A. 4, 2 : 359 ; Hedberg l.c. 169, 323. Erect or ascending perennial herb, stems with 2 lines of hairs, 6–18 in. high, internodes short, inflorescence densely glandular-pubescent ; flowers blue ; in montane grassland.
[**Br.**]**Cam.**: Cam. Mt., 9,000–13,000 ft. (fl. & fr. Nov.-Mar.) *Mann* 1312 ! 2030 ! *Maitland* 798 ! *Keay* FHI 28606 ! **F.Po**: Clarence Peak, 9,350 ft. (fl. & fr. Dec.) *Mann* 604 !
3. **V. beccabunga** *Linn.* Sp. Pl. 12 (1753) ; F.T.A. 4, 2 : 357. *V. anagallis* of F.W.T.A. ed. 1, 2 : 231 ; Chev. Bot. 474 ; not *V. anagallis-aquatica* Linn. A soft-stemmed herb, ascending up to 1 ft. high ; flowers blue ; in wet places.
Niger: Bilma Oasis (fl. & fr. May) *Ducellier* in *Hb. Chev.* 28627 ! *Le Coeur* 29 ! Widely distributed in warm and temperate regions (See Appendix, p. 442).
[*V. anagallis-aquatica* Linn. (as *V. anagallis* Linn.) was included by error in the first edition. However, it is widespread in Africa and may well occur in our area ; easily distinguished from *V. beccabunga* by its sessile leaves.—F. N. H.]

7. SIBTHORPIA Linn.—F.T.A. 4, 2 : 353.

A weak, straggly herb rooting at the nodes ; stems thinly pubescent ; leaves reniform-orbicular, crenate with numerous small teeth, 1·5–2 cm. diam., pilose on both surfaces ; petiole slender, pubescent ; pedicels 3–4 mm. long ; calyx about 2 mm. long, pubescent, shortly lobed ; corolla purple .. .. *europaea*

S. europaea *Linn.* Sp. Pl. 631 (1753) ; Hedberg in Bot. Not. 108 : 168 (1955), and in Sym. Bot. Uppsal. 15 : 168 (1957). *S. europaea* var. *africana* Hook. in J. Linn. Soc. 7 : 208 (1862) ; F.T.A. 4, 2 : 354. *S. australis* Hutch. F.W.T.A., ed. 1, 2 : 221 (1931). A prostrate herb in moist montane habitats.
[Br.]Cam.: Cam. Mt., 7,000–8,500 ft. (Dec., Feb., Apr.) *Mann* 1963 ! *Maitland* 1197 ! 1322 ! *Mildbr.* 7096. *Brenan* 9566 ! Mbakakeka Mt. *Coombe* 236 ! Bafut-Ngemba F.R. (Feb.) *Hepper* 2094 ! Oku, 6,500 ft. (Feb.) *Hepper* 2009 ! **F.Po:** 7,500 ft. *Mann* 1455 ! Widespread in oceanic parts of W. & S. Europe, in the Azores and on high mountains in tropical Africa.

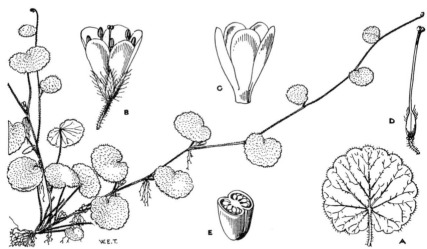

FIG. 285.—SIBTHORPIA EUROPAEA *Linn.* (SCROPHULARIACEAE).

A, leaf, from beneath. B, flower. C, corolla. D, pistil. E, cross-section of fruit.

8. SCOPARIA Linn.—F.T.A. 4, 2 : 354.

Stems ribbed, glabrous ; leaves opposite or whorled, narrowly oblanceolate, narrowed and entire in the lower half, crenulate in the upper part, 2·5–5 cm. long, up to 1·5 cm. broad, glabrous, areolate-punctate below ; flowers 1–2 in the upper reduced leaf-axils forming slender racemes ; pedicels about 6 mm. long, glabrous ; calyx and corolla 2–3 mm. long ; capsule globose, 4 mm. diam. .. .. .. *dulcis*

S. dulcis *Linn.* Sp. Pl. 116 (1753) ; F.T.A. 4, 2 : 354 ; Chev. Bot. 474 ; Berhaut Fl. Sén. 61, 74. Erect shrubby weed 1–3 ft. ; flowers white or bluish, usually 4 petals, densely bearded inside.
Maur.: El Berbera (fl. & fr. Apr.) *Monod* 9531 ! **Sen.:** Richard Tol (fl. & fr. Jan.) *Döllinger* 4 ! **Gam.:** Kuntaur *Ruxton* 143 ! Genieri (fl. & fr. July) *Fox* 133 ! **Mali :** Timbuktu (fl. & fr. June) *Hagerup* 113 ! Macina to Diafarabé (fl. & fr. May) *Lean* 22 ! **Port.G.:** Pussubé, Bissau (fl. & fr. Oct.) *Esp. Santo* 966 ! **Guin.:** Koukoré (fl. & fr. Sept.) *Pobéguin* K 6 ! Kouroussa (fl. & fr. July) *Pobéguin* 301 ! **S.L.:** *Sc. Elliot* 5134 ! Freetown (fl. & fr. Aug.) *Deighton* 2059 ! Njala (fl. & fr. Mar.) *Deighton* 1819 ! Musaia (fl. & fr. Apr.) *Deighton* 5377 ! **Lib.:** Monrovia *Whyte* ! Grand Bassa (fl. & fr. July) *T. Vogel* 47 ! Gbanga (fl. & fr. Aug.) *Freeman* 6 ! **Ghana :** Accra Plains (fl. & fr. Nov.) *Morton* GC 6120 ! Bomase (fl. & fr. Apr.) *A. S. Thomas* D.187 ! Juaso, Ashanti (fl. & fr. June) *Irvine* 272 ! Tamale (fl. & fr. Feb., June) *Williams* 482 ! 821 ! **Togo Rep.:** Lomé *Warnecke* 24 ! 377 ! **N.Nig.:** Lokoja (fl. & fr. Sept.) *Parsons* L.17 ! Nupe *Barter* 1174 ! Sokoto (fl. & fr. Nov.) *Moiser* 40 ! 58 ! Bindawa, Katsina (fl. & fr. Aug.) *Grove* 31 ! Naraguta *Lely* 33 ! **S.Nig.:** Lagos (fl. & fr. Dec.) *Onochie* FHI 32034 ! Ibadan to Abeokuta (fl. & fr. Mar.) *Schlechter* 12337 ! Nun R. *Mann* 475 ! Ogbomosho *Barter* 3405 ! Olokemeji *Foster* 290 ! Calabar (fl. & fr. Jan., Apr.) *Holland* 17 ! 98 ! [Br.]Cam.: Johann-Albrechtshöhe *Staudt* 598 ! Mamfe (fl. & fr. Dec.) *Migeod* 269 ! Gembu, Mambila Plateau (fl. & fr. Jan.) *Latilo & Daramola* FHI 34373 ! Gurum, Adamawa (Nov.) *Hepper* 2742 ! Widespread in the tropics. (See Appendix, p. 441.)

9. CAPRARIA Linn.—F.T.A. 4, 2 : 355.

A small shrub ; branches thinly pilose ; leaves broadly oblanceolate, long-cuneate and entire towards the base, sharply serrate in the upper half, 3–3·5 cm. long, 1–1·5 cm. broad, slightly pubescent towards the margin ; flowers 1–2, axillary ; pedicels nearly 1 cm. long ; calyx-segments linear-lanceolate, about 4 mm. long, ciliate ; corolla widely funnel-shaped, 1 cm. long ; fruit broadly ellipsoid, 5 mm. long, gland-dotted
biflora

C. biflora *Linn.* Sp. Pl. 628 (1753) ; F.T.A. 4, 2 : 355 ; Sprague in Kew Bull. 1921 : 209. Erect branched under-shrub, 2–3 ft. high ; flowers white ; in waste places.
Ghana : Cape Coast (fl. & fr. July) *T. Vogel* 79 ! Accra (fl. & fr. Oct.) *Brown* 377 ! *Dalz.* 115 ! *Irvine* 1004 ! Adaiso (fl. & fr. Nov.) *Morton* GC 6145 ! Asebu, Asuansi (fl. & fr. Oct.) *Lovi* WACRI 3930 ! A tropical American weed introduced into our area, Cape Verde Islands and Mauritius.

10. MIMULUS Linn.—F.T.A. 4, 2 : 310.

Glabrous perennial ; stems quadrangular, slightly winged ; leaves lanceolate or linear-lanceolate, 2–6 cm. long, 5–12 mm. broad, shortly toothed or undulate on margin, acute at apex, rounded or amplexicaul at base ; flowers solitary, axillary ; pedicels about 2 cm. long ; calyx 7 mm. long, 5-toothed ; corolla about 1 cm. long, very unequally 2-lipped ; stamens 4 .. .. *gracilis*

M. gracilis *R. Br.* Prod. 439 (1810) ; F.T.A. 4, 2 : 310. Slender stiffly erect herb, 1–2 ft. high, simple or sparingly branched ; flowers white or mauve with yellowish spots in the throat ; in upland savanna.
  N.Nig. : Jos Plateau (fl. & fr. June) *Lely* P 355 ! Panyam, 4,500 ft. (fl. & fr. July) *Lely* 410 ! Also in Sudan, Ethiopia, Tanganyika, N. & S. Rhodesia, Nyasaland, Mozambique and Angola.

11. STEMODIA Linn.—F.T.A. 4, 2 : 313. *Nom. cons.*

Leaves lanceolate, sessile, broad at the base, up to 5 cm. long, dentate towards the apex, pubescent on the nerves ; flowers in leafy spike-like racemes ; calyx-segments linear, about 6 mm. long, pubescent ; corolla 2-lipped ; stamens 4, included ; fruit narrow, glabrous, as long as the calyx .. .. 1. *serrata*
Leaves more or less ovate, serrate except the cuneate base, petiolate, about 1·5 cm. long ; flowers axillary, subsolitary ; fruit globose .. .. .. 2. *verticillata*

1. S. serrata *Benth.* in DC. Prod. 10 : 381 (1846) ; F.T.A. 4, 2 : 314 ; Berhaut Fl. Sén. 74. An erect, much branched viscid glandular-pubescent aromatic herb, 6–12 in. high ; flowers yellow or white.
  Sen. : *Leprieur* ! *Perrottet* 441 ! *Roger.* *Berhaut* 1307 ! **Ghana** : Nangodi (Dec.) *Morton* A1254 ! Also in Sudan, Ethiopia, Nyasaland and India.
2. S. verticillata *(Mill.) Boldingh* Zakfl. Landb. Java 165 (1916). *Erinus verticillatus* Mill. Gard. Dict. ed. 8, No. 5 (1768). *Stemodia parviflora* Ait. (1789)—F.W.T.A., ed. 1, 2 : 221. A small, much-branched herb, slightly woody below, erect or decumbent, a few inches high ; flowers blue-purple ; weed of waste places.
  Guin.: Lalie (fr. Oct.) *Adam* 12667 ! **S.L.** : Rokupr (fl. & fr. July) *Jordan* 278 ! Bonthe (fl. & fr. Nov.) *Deighton* 2264 ! Njala (fl. & fr. Sept.) *Deighton* 2087 ! *Hepper* 2555 ! Makump (fl. & fr. Aug.) *Deighton* 1387 ! Nyiyama (fl. & fr. Apr.) *Deighton* 1659 ! **Lib.** : ? Ganta (May) *Harley* 923 ! **[Br.]Cam.**: Ebonji, Kumba (fl. & fr. May) *Olorunfemi* FHI 30586 ! **F.Po** : El Pico, 1,700 ft. (fl. & fr. Dec.) *Boughey* 186a ! Introduced from tropical America ; also in Mauritius and Java.

Imperfectly known species.

S. senegalensis *Desf.* Cat. Hort. Paris. ed. 3, 107 (1829), name only. An annual from Senegal without description. It seems more probable that this is *S. serrata*, which is known from Senegal, than *S. verticillata* (syn. *S. parviflora*) under which it was provisionally placed in F.W.T.A., ed. 1, 2 : 221.

12. STEMODIOPSIS Engl.—F.T.A. 4, 2 : 314.

Perennial herb from a small tufted base, pubescent with short stiff whitish hairs ; leaves opposite, ovate, acute at apex, cuneate at base, coarsely dentate, 1–2 cm. long, 6–15 mm. broad ; petiole 5–10 cm. long ; inflorescences few-flowered, cymes in most axils ; calyx deeply 5-fid, segments linear, 3 mm. long ; corolla 5 mm. long, 2-lipped, lower lip shortly 3-lobed ; fruit conical, 5–6 mm. long, slightly hairy .. *humilis*

S. humilis *Skan* in F.T.A. 4, 2 : 316 (1906) ; Hepper in Kew Bull. 15 : 65. A perennial with tufted stems about 5 in. long from cracks in large rocks ; flowers greenish-yellow.
  [Br.]Cam.: Mapeo, Alantika Mts., 2,500–3,500 ft., Adamawa (fl. & fr. Dec.) *Hepper* 1601 ! Also in Sudan, Uganda, Kenya, Tanganyika, Nyasaland, S. Rhodesia and Mozambique.

13. LIMNOPHILA R. Br.—F.T.A. 4, 2 : 316. *Nom. cons.*

Flowers with slender pedicels 6–18 mm. long, axillary or subterminal ; leaves deeply divided, the lower in whorls, the upper whorled or opposite and sometimes only serrate, 1–3 cm. long ; calyx-segments linear, 5 mm. long ; corolla 1 cm. long ; fruit nearly equalling calyx .. .. 1. *indica*
Flowers sessile or subsessile :
  Corolla 5–10 mm. long, about 2 mm. diam. ; flowers in lax spikes or axillary, solitary :
    Bracts and leaves above water ovate, often broadly so and subcordate, 1·5–2·5 cm. long, 1–2 cm. broad, 5–7-nerved, serrulate, softly tomentose on both sides ; submerged leaves much divided and whorled ; stems above densely pubescent ; flowers in interrupted spikes, numerous ; corolla 5 mm. long ; calyx pubescent, 4 mm. long .. .. 2. *barteri*
    Bracts and leaves all pinnatisect and whorled, up to 1 cm. long, glabrous, submerged leaves with finer segments ; stems glabrous or slightly pubescent ; flowers few, solitary, axillary ; corolla 8–10 mm. long ; calyx glabrous, gland-dotted, 5 mm. long .. .. 3. *sessiliflora*
  Corolla 13–15 mm. long, about 6 mm. diam., in a terminal congested spike and often also some axillary flowers ; aerial leaves glabrous, punctate beneath, ovate, rarely all divided, rounded at base, 7–20 mm. long, 4–8 mm. broad, 3-nerved ; calyx 5 mm. long, glabrous .. .. 4. *dasyantha*

1. L. indica *(Linn.) Druce* in Rep. Bot. Exch. Cl. Brit. Isles 3 : 420 (1914). *Hottonia indica* Linn. (1759). *L. gratioloides* R. Br. (1810)—F.T.A. 4, 2 : 319 ; F.W.T.A., ed. 1, 2 : 223 ; Berhaut Fl. Sén. 61. *Ambulia gratioloides* (R. Br.) Baill. ex Engl.—Chev. Bot. 470. Slender glabrous or slightly glandular-pubescent herb, branched or simple, a few inches to 2 ft. long ; flowers small, white ; in mud or submerged sometimes in masses. A dwarf form about 2 in. high with congested leaves and few flowers (*Barter* 1709) has been named *L. gratioloides* var. *nana* Skan in F.T.A. 4, 2 : 319 (1906).
  Sen. : *Roger* 59 ! *Berhaut* 1511. **Mali** : Nafadié (fl. & fr. Jan.) *Chev.* 155 ! Gao (Mar.) *de Wailly* 5355 ! **Togo Rep.**: Lomé *Warnecke* 230 ! **[Br.]Cam.**: Gurum, Adamawa (fl. & fr. Dec.) *Hepper* 1599 ! Widespread in tropical Africa, Asia and Australia.
2. L. barteri *Skan* in F.T.A. 4, 2 : 317 (1906) ; Chev. Bot. 470 ; Berhaut Fl. Sén. 68. *L. fluviatile* A. Chev. in Bull. Mus. Hist. Nat. sér. 2, 4 : 587 (1932), incl. forma *fluviatile* and forma *terrestris*. Erect or semi-aquatic herb 1–2 ft. high, with viscid hairs ; flower white.

Sen. : *Berhaut* 1518. **Mali** : Dogo (fr. Apr.) *Davey* 562 ! **Port.G.** : Bafatá (fl. & fr. Dec.) *Esp. Santo* 2856 ! **Guin.**: Kindia to Télimélé (fl. & fr. Feb.) *Roberty* 10744 ! **S.L.** : Lomaburu (fr. Feb.) *Sc. Elliot* 5032 ! Port Loko (fl. & fr. Dec.) *Thomas* 6549 ! **Iv.C.** : Alangouassou to Mbayakro *Chev.* 22249 ! **Ghana** : Burufo (fl. & fr. Dec.) *Adams & Akpabla* GC 4381 ! Bimbila (fr. Mar.) *Hepper & Morton* A3092 ! **N.Nig.**: *Barter* 751 ! Kontagora (fl. & fr. Nov.) *Dalz.* 161 !

3. **L. sessiliflora** (*Vahl*) *Blume* Bijdr. Fl. Ned. Ind. 749 (1825) ; F.T.A. 4, 2 : 318. *Hottonia sessiliflora* Vahl (1791). Small aquatic herb about 6 ins. long ; flowers yellow.
   S.L. : Makeni to Kabala (fl. & fr. Jan.) *King* 172b ! Also in Tanganyika, N. & S. Rhodesia, Angola, S. Africa, tropical Asia and China and Japan.

4. **L. dasyantha** (*Engl. & Gilg*) *Skan* in F.T.A. 4, 2 : 318 (1906) ; *Chev. Bot.* 470. *Ambulia dasyantha* Engl. & Gilg (1903). *Ceratophyllum demersum* of F.W.T.A., ed. 2, 1 : 65 (partly *Deighton* 2339), not of Linn. Weak-stemmed aquatic herb 2–4 ft. long, leafless in the lower part ; flowers conspicuous yellow ; in stagnant pools.
   Mali : Oualia, Kita *Chev.* 101 ! **Port.G.** : Antula (Dec.) *Esp. Santo* 2638 ! Empandja *Esp. Santo* 1609 ! S. Domingos to Suzana (Nov.) *Esp. Santo* 3677 ! **Guin.** : Koba (Nov.) *Jac.-Fél.* 7223 ! **S.L.** : Bentembu (Apr.) *Sc. Elliot* 5862 ! Songo (Oct.) *Lane-Poole* 379 ! Rokupr (Apr.) *Adames* 169 ! Njala (Oct.) *Deighton* 2810 ! Mattru to Gbangbama *Deighton* 2339 ! Also in Angola. (See Appendix, p. 441.)

14. BACOPA Aubl. Pl. Guian. 1 : 128, t. 49 (1775), *nom. cons. Moniera* Browne (1756).—F.T.A. 4, 2 : 319.

Leaves palmately nerved with 5–7 almost equal nerves, spathulate, undulate on margin towards apex, broadly cuneate into petiole, 1–2·5 cm. long, 5–15 mm. broad, glabrous ; prostrate, rooting at most nodes ; flowers solitary ; pedicels 1–1·5 cm. long ; brac-teoles absent ; sepals and corolla-lobes 3–5, stamens 3–4 .. .. 1. *egensis*
Leaves 1-nerved, midrib prominent :
   Calyx 2–5(–6) mm. long in fruit ; fruiting pedicels 1(–3) mm. long :
      Leaves broad and subcordate at base, ovate-lanceolate, 1–2 cm. long, 4–9 mm. broad, densely glandular-punctate beneath ; younger stems appressed-pubescent
                                                                              2. *occultans*
      Leaves not subcordate at base narrowly lanceolate or linear :
         Fruits and smaller calyx segments obscured by 2 large ovate calyx segments about 4 mm. broad ; leaves narrowly lanceolate, narrowed to base, 2–3 cm. long, 3–7 mm. broad ; stem usually branched below   3. *hamiltoniana* var. *hamiltoniana*
         Fruits not obscured by calyx, segments all visible, at most 2·5 mm. broad ; leaves linear, up to 2·5 cm. long and 2 mm. broad ; stem simple
                                                                        3a. *hamiltoniana* var. *angustisepala*
   Calyx 8–11 mm. long in fruit ; fruiting pedicels (2–)3–13 mm. long :
      Pedicels pubescent, 5–11 mm. long ; largest calyx-segment broadly ovate, about 8 mm. long and 7 mm. broad in fruit, pubescent ; leaves lanceolate, narrowed to base or sessile, 1–3·5 cm. long, 3–6(–15) mm. broad, pubescent   4. *floribunda*
      Pedicels and other parts glabrous :
         Erect, annual herb ; largest calyx-segment oblong-lanceolate, about twice as long as broad, in fruit 8–11 mm. long and 4–6 mm. broad, toothed or entire ; flowers solitary or often in short congested axillary racemes .. .. 5. *decumbens*
         Creeping herbs ; flowers usually solitary in axils :
            Calyx-segments conspicuously venose at least in fruit, broadly ovate, 7–8 mm. long, 5–6 mm. broad ; pedicels shorter than leaves ; leaves lanceolate, crenate-serrate above, subacute, 2–3·5 cm. long, 5–11 mm. broad .. .. .. 6. *crenata*
            Calyx-segments not venose ; pedicels as long as or longer than the leaves :
               Leaves oblanceolate, subentire, obtuse at apex, 5–15 mm. long, 3–8 mm. broad ; calyx-segments narrowly ovate ; stems fleshy .. .. 7. *monniera*
               Leaves ovate-elliptic, coarsely serrate, subacute at apex, about 2 cm. long and 1 cm. broad ; calyx-segments broadly ovate ; stems not fleshy 8. *procumbens*

1. **B. egensis** (*Poepp. & Endl.*) *Pennell* in Proc. Acad. Nat. Sci. Phil. 98 : 96 (1946). *Hydranthelium egense* Poepp. & Endl. (1845)—F.T.A. 4, 2 : 351 ; F.W.T.A., 1, 2 : 229. A creeping herb ; flowers white or bluish, fragrant ; on muddy river and stream sides.
   S.Nig.: Onitsha *Barter* 1252 ! Also in the Congos and Central and S. America.
2. **B. occultans** (*Hiern*) *Hutch. & Dalz.* F.W.T.A., ed. 1, 2 : 222 (1931) ; *Berhaut* Fl. Sén. 74. *Moniera occultans* Hiern (1898). Erect or ascending herb, up to 1 ft. high, sparingly branched near the base : flowers white ; in rice swamps and margins of lakes.
   Sen.: Banbankountia (Dec.) *Berhaut* 1822 ! **S.L.**: Njala (fl. & fr. Oct.) *Deighton* 2811 ! Miligi (Oct.) *Jordan* 926 ! **Ghana**: Kamba, near Lawra (Nov.) *Harris* ! **N.Nig.**: Nupe *Barter* 1689 ! **S.Nig.**: Eket *Talbot* ! Also in Angola.
3. **B. hamiltoniana** (*Benth.*) *Wettst.* var. **hamiltoniana**—in E. & P. Pflanzenfam. 4, 3B : 77 (1891) ; F.W.T.A., ed. 1, 2 : 222, excl. syn. *B. alternifolia* Engl. ; Berhaut Fl. Sén. 74. *Herpestis hamiltoniana* Benth. (1835). *Moniera hamiltoniana* (Benth.) T. Cooke (1905)—F.T.A. 4, 2 : 323. *M. scabrida* A. Chev. in Mém. Soc. Bot. Fr. 2, 8 : 184 (1912) ; *Chev. Bot.* 471. Erect usually well-branched herb about 6 in. high ; in damp places.
   Sen.: *Leprieur* 237 ! *Berhaut* 1462. **Guin.**: Mankoutan (Nov.) *Jac-Fél.* 7349 ! **Iv.C.**: Toumodi *Chev.* 22391 ! **Ghana**: Gambaga Scarp (Sept.) *Hall* 752b ! **N.Nig.**: Kontagora (fl. & fr. Nov.) *Dalz.* 158 ! Anara F.R., Zaria (fr. Oct.) *Hepper* 987 ! Yola (fl. & fr. Nov.) *Hepper* 1192 ! [**Br.**]**Cam.**: Jada, Adamawa (fl. & fr. Nov.) *Hepper* 1226 ! Also in India.
3a. **B. hamiltoniana** var. **angustisepala** *Hepper* in Kew Bull. 15 : 64 (1961). Small half-succulent, simple herb 2–7 in. high ; flowers pale blue ; in moist ground.
   N.Nig.: Guma, Minna (fl. & fr. Oct.) *Hepper* 976 !
4. **B. floribunda** (*R. Br.*) *Wettst.* in E. & P. Pflanzenfam. 4, 3B : (1891) ; Berhaut Fl. Sén. 74. *Herpestis flori-bunda* R. Br. (1810). *Moniera floribunda* (R. Br.) T. Cooke (1905)—F.T.A. 4, 2 : 322. *M. pubescens* Skan in F.T.A. 4, 2 : 322 (1906). *M. calycina* of Chev. Bot. 471, not of (Benth.) Hiern. *Bacopa pubescens* (Skan) Hutch. & Dalz. F.W.T.A., ed. 1, 2 : 222 (1931). *Moniera bicolor* A. Chev. in Mém. Soc. Bot. Fr. 2, 8 : 183 (1912). Erect or ascending variable herb, 3–18 in. high flowers mauve ; in damp places.

Sen.: *Perrottet* 583! 598! 602! Badi (fr. Dec.) *Berhaut* 1839! **Gam.:** *Dawe* 28! **Port. G.:** Pirada (fl. & fr. Nov.) *Esp. Santo* 3161! Bafatá (fl. & fr. Dec.) *Esp. Santo* 2841! **S.L.:** Falaba R. (fl. & fr. Apr.) *Sc. Elliot* 5451! **Iv.C.:** Toumodi *Chev.* 22400! 22410! **Ghana:** Burufo (fl. & fr. Dec.) *Adams & Akpabla* 4382! Lumbaga (fl. & fr. Dec.) *Morton* GC 6263! Tumu (fr. Dec.) *Adams & Akpabla* 4373! **N.Nig.:** Nupe *Barter* 1011! Bida (fl. & fr. Mar.) *Meikle* 1325! Lokoja (Oct.) *Parsons* L60! Anara F.R., Zaria (Oct.) *Hepper* 986! Jos Plateau (Oct.) *Lely* P826! **[Br.]Cam.:** Kwagiri, Gashaka Dist. (Nov.) *Latilo & Daramola* FHI 28766! Jada, Adamawa (fl. & fr. Nov.) *Hepper* 1225! Also in Ubangi-Shari, Sudan, Congo, Kenya, Tanganyika, Pemba, Zanzibar, Mozambique and Madagascar, extending into tropical Asia and Australia.

5. **B. decumbens** (*Fernald*) *F. N. Williams* in Bull. Herb Boiss., sér. 2, 7 : 369 (1907). *Herpestis decumbens* Fernald (1897). *Moniera decumbens* (Fernald) Skan in F.T.A. 4, 2 : 321 (1906). *Bacopa erecta* Hutch. & Dalz. F.W.T.A., ed. 1, 2 : 222 (1931) ; Berhaut Fl. Sén. 74. Erect herb 1–2 ft. high with soft stem ; flowers white ; in brackish swamps and rice fields.
Sen.: *Heudelot* 595! *Perrottet* 589! 607! *Berhaut* 1772. **Gam.:** Dawe 66! Sukuko (fl. & fr. Jan.) *Dalz.* 8060! **Port.G.:** Bissau (fl. & fr. Jan., Feb., Nov.) *Esp. Santo* 1013! 1445! 1751! Sedeugafe to Ingoré (fl. & fr. Dec.) *Esp. Santo* 3738! **Guin.:** Koba (Nov.) *Jac.-Fél.* 7203! **S.L.:** Smeathmann! Mahela (Dec.) *Sc. Elliot* 3847! Rokupr (Feb., Apr.) *Jordan* 769! *Hepper* 2598! Maswari (fl. & fr. Feb.) *Deighton* 4985! Urika (Mar.) *Glanville*! Also in Mexico and C. America.
[In spite of this being described as a new species in the first edition, I cannot see that it is specifically distinct from Fernald's species and I therefore revert to the F.T.A. determination with its apparently inappropriate epithet—F. N. H.]

FIG. 286.—BACOPA CRENATA (*P. Beauv.*) *Hepper* (SCROPHULARIACEAE).
A, flower.   B, corolla laid open.   C, one sepal.   D, another sepal.   E, a third sepal.   F, pistil.
G, calyx enveloping the fruit.

6. **B. crenata** (*P. Beauv.*) *Hepper* in [Kew Bull. 14 : 407 (1960). *Herpestis crenata* P. Beauv. Fl. Oware 2 : 83 t. 112 (1819). *H. calycina* Benth. (1836). *Bacopa calycina* (Benth.) Engl. ex De Wild. in Bull. Herb. Boiss 1 : 832 (1901) ; F.W.T.A., ed. 1, 2 : 222 ; Berhaut Fl. Sén. 74. *Moniera calycina* (Benth.) Hiern (1898)—F.T.A. 4, 2 : 320. *M. cuneifolia* of F.T.A. 4, 2 : 320, partly (*Beauvois* s.n.), not of Michx. *Bacopa monnieria* of F.W.T.A., ed. 1, 2 : 222, not of (Linn.) Wettst. Creeping and ascending herb with soft spongy stems a few inches to 3 ft. long, rooting at the lower nodes ; flowers white ; in swamps and ditches.
Sen.: *Leprieur* 214! *Perrottet* 12! 604! *Berhaut* 603. **Lib.:** Nyaake, Webo Dist. (June) *Baldwin* 6159! Vonjama (Oct.) *Baldwin* 9874! **Iv.C.:** Tadio Lagoon *Tournier*! Bouakro to Alangoussou *Chev.* 22225! **Ghana:** Beyin (fl. & fr. Feb.) *Irvine* 2350! Accra Plains (fl. & fr. July) *Irvine* 725! Atwabo *Fishlock* 54! 81! Bimbila (fl. & fr. Mar.) *Hepper & Morton* A3104! **Togo Rep.:** Lomé *Warnecke* 226! **N.Nig.:** Borgu *Barter* 1690! Lokoja (fl. & fr. Nov.) *Dalz.* 1238! Badagry (fl. & fr. Aug.) *Onochie* FHI 33492! 33497! Ibadan (fl. & fr. Aug.) *Tamajong* FHI 19507! Also in Sudan, the Congos, Angola and Madagascar.

7. **B. monniera** (*Linn.*) *Wettst.* in E. & P. Pflanzenfam. 4, 3B : 77 (1895). *Lysimachia monnieri* Linn. (1756). *Moniera cuneifolia* Michx. (1803)—F.T.A. 4, 2 : 320 and F.W.T.A., ed. 1, 2 : 222, partly (excl. *Beauvois* and syn. *Herpestis crenata* P. Beauv.) A creeping, or sometimes aquatic, herb ; flowers blue or white.
S.Nig.: Ogun F. R., Lagos (Mar.) *Hepper* 2249! Majidun Ilaje, Lagos (Feb.) *Hambler* 410! Throughout the tropics.

8. **B. procumbens** (*Mill.*) *Greenman* in Publ. Field Columb. Mus., Bot. ser., 2 : 261 (1907). *Erinus procumbens* Mill. Gard. Dict., ed. 8, No. 6 (1768). *Herpestis chamaedryoides* H. B. & K. (1817). *Bacopa chamaedryoides* (H.B. & K.) Wettst. in E. & P. Pflanzenf. 4, 3B : 76 (1895). A prostrate creeping herb with large yellow flowers ; plant usually drying black.
[Br.]Cam.: Buea, 4,000 ft. (fl. & fr. Jan.) *Morton* K947! Widespread in tropical America and presumably introduced into our area and recorded here for the first time from Africa ; also in India and Queensland.

15. DOPATRIUM Buch.-Ham. ex Benth.—F.T.A. 4, 2 : 324.
By P. Taylor

Lowermost flowers in inflorescence sessile and cleistogamous ; perfect flowers pedicellate, pedicels 5–10 mm. long ; lower leaves oblanceolate, 1–3 cm. long ; stem leaves ovate, 1–5 mm. long ; calyx 1–2 mm. long ; corolla 5–6 mm. long ; fruits globose or ovoid, 1·5–2 mm. long ..   .. 1. *junceum*
All flowers perfect and pedicellate :
Corolla less than 5 mm. long ; calyx 3–3·5 mm. long, segments about 2 mm. long,

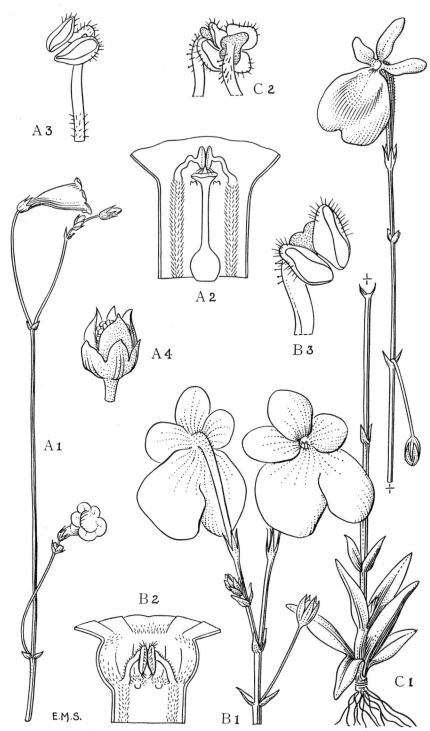

FIG. 287.—DOPATRIUM SPP. (SCROPHULARIACEAE).

D. senegalense Benth.—A1, part of inflorescence, × 2 (*Adames* 121). A2, opened corolla-tube
with stamens and pistil, × 8. A3, stamen, × 20. A4, fruit, × 2. (*Farmar* 199a).
D. macranthum Oliv.—B1, part of inflorescence, × 2. B2, opened corolla-tube with stamens,
× 8. B3, stamen, × 20. (*Hinds* 5004a).
D. longidens Skan—C1, habit, × 2. C2, stamens, × 20. (*Hepburn* 65).

360

narrowly lanceolate, acuminate ; basal leaves elliptic, 10–15 mm. long ; fruits
narrowly ovoid   ..   ..   2. *baoulense*
Corolla more than 10 mm. long :
Calyx 4–6 mm. long ; corolla salver-shaped, tube slender, 1–1·5 cm. long, limb
  2–3 cm. across ; racemes few-flowered :
Pedicels in fruit reflexed, up to 2 cm. long ; basal leaves lanceolate acute, stem
  leaves deltoid   ..   ..   3. *longidens*
Pedicels in fruit erect, up to 2 cm. long ; basal leaves narrowly lanceolate, stem
  leaves ovate ..   ..   4. *macranthum*
Calyx 2–3 mm. long ; corolla subcampanulate, tube widening from base, less than
  1 cm. long, limb less than 1 cm. across ; lower leaves lanceolate ; stem leaves
  minute, deltoid ; racemes usually many flowered ; pedicels up to 2 cm. long
                                                              5. *senegalense*

1. **D. junceum** *Buch.-Ham. ex Benth.* Scroph. Ind. 31 (1835). Erect, usually simple herb, to 1 ft. high ; flowers
white or pale mauve ; in wet places.
   Sen. : Richard Tol (Sept.) *Martine* 111 ! **N.Nig.** : Kumu (Oct.) *Lely* 666 ! Widespread in tropical Africa
and Asia.
2. **D. baoulense** *A. Chev.* in Mém. Soc. Bot. Fr. 2, 8 : 184 (1912) ; Chev. Bot. 471. *Ilysanthes parviflora* of
F.W.T.A., ed. 1, 2 : 230, partly (syn. & *Chev.* 22250), not of (Roxb.) Benth. Erect simple herb to 8 in. high ;
in wet places.
   Iv.C. : Baoulé (Aug.) *Chev.* 22250 !
3. **D. longidens** *Skan* in F.T.A. 4, 2 : 325 (1906). Erect herb often branched from base to 15 in. high ; flowers
large and showy, purple or mauve ; in wet places.
   Mali : Ségou (Aug.) *Roberty* 3756 ! **N.Nig.** : Nupe *Barter* 1683 ! Anara F.R., Zaria Prov. (Sept.) *Olorunfemi*
FHI 24353 ! Jos (Aug.) *Keay* 20072 ! Naraguta (Aug.) *Lely* 526 ! Jos Plateau *Saunders* 9 ! *Dent
Young* ! Gidan-Anju, Muri Div., Adamawa (Nov.) *Latilo & Daramola* FHI 28728 !
4. **D. macranthum** *Oliv.* in Trans. Linn. Soc. 29 : 120 (1875). *D. schweinfurthii* Wettst. in E. & P. Pflanzenfam.
4, 36 : 75 (1895). *D. luteum* Engl. (1897)—F.T.A. 4, 2 : 324. Erect, usually simple herb to 18 ins. high ;
flowers large and showy, purple or yellow ; in wet places.
   Mali : Bamako to Ségou (fl. & fr. Sept.) *Jaeger* 5116 ! Boroma to Ouagadougou (Sept.) *Jaeger* 5271 ! **Ghana** :
Yendi to Gambaga (Oct.) *Rose Innes* GC 30486 ! Lawra (July) *Hinds* s.n. ! 5004 ! 5004a ! Also in Congo,
Uganda and Sudan.
5. **D. senegalense** *Benth.* in DC. Prod. 10 : 407 (1846) ; F.T.A. 4, 2 : 326 ; Berhaut Fl. Sén. 4. *D. nanum*
Sc. Elliot (1895)—F.T.A. 4, 2 : 326. *D. peulhorum* A. Chev. in Mém. Soc. Bot. Fr. 2, 8 : 185 (1912) ;
Chev. Bot. 471. *D. dawei* Hutch. & Dalz. F.W.T.A., ed. 1, 2 : 230 (1931). Erect, usually branched herb,
to 18 in. high ; flowers violet ; in wet places.
   Sen. : *Heudelot* 138 ! Tambacounda *Berhaut* 1261. **Gam.** : Bansang *Duke* 3 ! **Guin.** : Fouta Djalon (Sept.)
Chev. 18554 ! 18623 ! 18657. *Schnell* 7383 ! Kindia *Farmar* 199a ! Dalaba *Abbayes* 844 ! **S.L.** : Scarcies
(Feb.) *Sc. Elliot* 4275b ! Moyamba (Aug.) *Dawe* 568 ! Bunkababe *Dawe* 493 ! Binkolo *Deighton* 1280 !
Kambia *Adames* 185 !

16. **ARTANEMA** D. Don—F.T.A. 4, 2 : 327. *Nom. cons.*

Stems glabrous ; leaves opposite, broadly lanceolate, acutely acuminate, shortly
petiolate, (6–)10–25 cm. long, (1–)2–9 cm. broad, serrulate except the cuneate base,
scabrid above especially towards margin ; racemes 10–15 cm. long ; pedicels 1 cm.
long ; calyx glabrous, lobes ovate to lanceolate, 3–8 mm. long ; corolla 3 cm. long,
slightly irregular ; fruits globose, about 1 cm. diam.   ..   ..   *longifolium*

A. longifolium (*Linn.*) *Vatke* in Linnaea 43 : 307 (1882). *Columnea longifolia* Linn. (1767). *Achimenes
sesamoides* Vahl (1791). *Artanema sesamoides* (Vahl) Benth. (1835)—F.T.A. 4, 2 : 327 ; Chev. Bot. 471.
Herb about 3 ft. high with hollow, succulent stems ; flowers deep red-purple, paler inside ; in water and
marshy places ; drying green.
   Lib. : Browntown, Tchien Dist. (Aug.) *Baldwin* 7081 ! Gipu (Aug.) *Harley* 2194 ! **Iv.C.** : Sassandra Basin
Chev. 17977 ! Bériby *Chev.* 20006 ! 20042 ! Alangouassou *Chev.* 22221 ! 22243 ! **Ghana** : Accra *Brown* 381 !
Cape Coast (June) *Fishlock* 37 ! Swedru (fl. & fr. Sept.) *Dalz.* 146 ! Subiri F.R., Benso (fl. & fr. Sept.)
Andoh FH 5563 ! Kpandu (Sept.) *Robertson* 110 ! *Baumann* 245 ! **Dah.** : Savé *Chev.* 23583 ! **S.Nig.** :
Ndoni *Barter* 1769 ! Abeokuta *Irving* ! Ibadan (Apr.) *Meikle* 1474 ! Sapoba *Kennedy* 2377 ! British
Ogbokum, Ikom (fl. & fr. May) *Jones & Onochie* FHI 18877 ! **[Br.]Cam.** : Mombo to Bakosi, Kumba (fr.
May) *Olorunfemi* FHI 30568 ! Also in Cameroun, Rio Muni, the Congos and Uganda and extends into
tropical Asia. (See Appendix, p. 440.)

17. **CRATEROSTIGMA** Hochst.—F.T.A. 4, 2 : 328.

Lower lip of corolla 5–7 mm. long, tube 6–7 mm. long, densely puberulous ; flowers
congested in terminal heads ; stems quadrangular, at least towards base, glabrous ;
lower leaves ovate to elliptic, obtuse, upper leaves lanceolate, acute, sessile, 1–2·5 cm.
long, 3–11 mm. broad, subentire, glabrous   ..   ..   1. *schweinfurthii*
Lower lip of corolla 2–3 mm. long, tube 5 mm. long, slightly puberulous and gland-dotted ;
flowers several together or solitary in interrupted spikes ; leaves lanceolate, acute,
with 2 or 3 small serrations on each side, 7–20 mm. long, 3–7 mm. broad, glabrous
                                                              2. *guineense*

1. **C. schweinfurthii** (*Oliv.*) *Engl.* in Engl. Bot. Jahrb. 23 : 501 (1897) ; F.T.A. 4, 2 : 332 ; Chev. Bot. 472,
partly. *Torenia schweinfurthii* Oliv. (1878). Stem herbaceous 1–2 ft. high, erect or slightly decumbent at
base with fibrous roots along the internodes, leaves tinged brown or purple beneath ; flowers rich violet-
blue ; in marshes.
   Ghana : Yeji (fl. & fr. July) *Vigne* FH 1256 ! Bosuso (May) *Vigne* FH 3848 ! **Dah.**: Atacora Mts. *Chev.*
23989 ! **N.Nig.** : Kontagora (fl. & fr. Jan., Nov.) *Dalz.* 165 ! *Meikle* 1044 ! Nupe *Barter* ! Bokkos, Plateau
Prov. (fl. & fr. Dec.) *Coombe* 74 ! Also in Cameroun, Sudan, Uganda, N. Rhodesia and Angola.
2. **C. guineense** *Hepper* in Kew Bull. 14 : 407, fig. 2 (1960). *C. schweinfurthii* of Sousa in Est. Bot. Anais 5 :
39 (1950) ; of Chev. Bot. 472, partly ; (?) Berhaut Fl. Sén. 81. Slender erect herb 1½–16 ins. high, sparingly
branched, similar in habit to the last ; small blue flowers with yellow hairs on the lower lip ; in marshes.
   Sen. : Sedhiou to Tambaba (fl. & fr. Feb.) *Chev.* 2849 ! Badi *Berhaut* 1831 (not seen). **Port.G.** : Antula,

FIG. 288.—CRATEROSTIGMA SCHWEINFURTHII (*Oliv.*) *Engl.* (SCROPHULARIACEAE).
A, flower. B, calyx. C, stamen and staminode. D, stigma. E, young fruit.

Bissau (fl. & fr. Dec.) *Esp. Santo* 1424! Prabis, Bissau (Nov.) *Esp. Santo* 1821! **Guin.** : Grandes Chutes (Dec.) *Chev.* 20239! 20318! Timbo to Ditinn (fl. & fr. Sept.) *Chev.* 18422! Dalaba-Diaguissa Plateau, 3,200–4,200 ft. (Oct.) *Chev.* 18764! 18864!

18. TORENIA Linn.—F.T.A. 4, 2 : 334.

Flowers subsessile in leafy spikes ; plant erect ; leaves oblanceolate, dentate, narrowed to base and subsessile, 1–3 cm. long, 3–10 mm. broad, glabrous ; calyx shortly 5-toothed, 3·5 mm. long, in fruit up to 6 mm. long and longitudinally winged ; lower stamens inserted in a small boss .. .. 1. *spicata*
Flowers distinctly pedicellate, pedicel at least 1 cm. long ; plants prostrate :
Lamina lanceolate to ovate-lanceolate, 10–35 mm. long, 7–16 mm. broad, crenate-serrate ; petioles 4–8 mm. long ; bracteoles filiform at base of pedicel 2–5 mm. long ; flowers axillary, solitary or a few together on a short leafless peduncle ; pedicels 1·5–3 cm. long, usually reflexing in fruit ; calyx slightly winged, at least in fruit, 10 mm. long ; seeds pitted, numerous, very small ; filaments with small appendage at base .. .. 2. *thouarsii*
Lamina ovate to suborbicular, (8–)10–12(–17) mm. long, 6–13 mm. broad, crenate ; petioles 1–2(–4) mm. long ; bracteoles absent ; flowers axillary, solitary ; filaments with a large appendage towards base ; pedicels about 1 cm. long, not reflexing ; calyx prominently winged, in fruit 9 mm. long ; seeds pitted, few, twice as large as in No. 2 .. .. 3. *dinklagei*

1. **T. spicata** *Engl.* Bot. Jahrb. 23 : 502, t. 7, figs. G–M. (1897) ; F.T.A. 4, 2 : 334. *Pseudolobelia humilis* A. Chev. Bot. 480, name only. Erect glabrous annual, simple or branched, the stems acutely 4-angular, 2–9 in. high ; flowers blue, sometimes white or pink ; weed of cultivation and waste places in savanna.
Mali : Ségou (Oct.) *Roberty* 2819! **Guin.** : Timbo to Ditinn (fl. & fr. Sept.) *Chev.* 18497! **Ghana** : Tamale (fl. & fr. Dec.) *Adams & Akpabla* 4154! Dedoro to Tankara (fr. Dec.) *Adams* 4331! **N.Nig.** : Keana, Nassarawa Prov. (fl. & fr. Oct.) *Hepburn* 39! Nupe *Yates*! Jos Plateau (Aug.) *Lely* P579! Kadaura (Sept.) *Dalz.* 603! Yola (fl. & fr. July, Nov.) *Dalz.* 127! *Hepper* 1219! Also in Sudan, Tanganyika, N. & S. Rhodesia, Nyasaland, Angola and Transvaal.
2. **T. thouarsii** (*Cham. & Schlechtend.*) *O. Ktze.* Rev. Gen. Pl. 2 : 468 (1891). *Nortenia thouarsii* Cham. & Schlechtend. (1828). *Torenia parviflora* Buch.-Ham. ex Benth. (1835)—F.T.A. 4, 2 : 335 ; F.W.T.A., ed. 1, 2 : 229, mostly ; Chev. Bot. 472 ; Berhaut Fl. Sén. 74. Slender branched procumbent or weakly ascending herb, rooting at nodes and sometimes along internodes ; flowers mauve or white ; in moist places.
Sen. : *Perrottet* 587! 606! *Farmar* 50! *Berhaut* 1014. Adéane, Casamance *Chev.* 3289. **Mali** : Sikasso *Chev.* 787. **S.L.** : Rokupr (fl. & fr. Nov., Apr.) *Jordan* 697! *Hepper* 2624! Kenema (fl. & fr. Nov.) *Deighton* 422! Futa (Sept.) *Deighton* 2238! Gola Forest (fl. & fr. Jan.) *King* 133b! **Lib.** : Firestone Cavalla Plantation (fl. & fr. June) *Baldwin* 6005! **Iv.C.** : Angouakoukro to Toumodi *Chev.* 22408. Mankono *Chev.* 21934. Manikro to Tiégouakro *Chev.* 22321. **Ghana** : Owabi (fl. & fr. Sept.) *Andoh* FH 4553! Sesiamang (fl. & fr. Feb.) *A. S. Thomas* D 151! Bole *Harris*! Otiso Ferry (Mar.) *Hepper & Morton* A3032! Amedzofe (fl. & fr. Mar.) *Morton* GC 6499! **N.Nig.** : Nupe *Barter* 1199! Jebba Island (fl. & fr. Mar.) *Meikle* 1286! Kontagora (fl. & fr. Nov.) *Dalz.* 156! Abinsi (fl. & fr. Mar.) *Dalz.* 701! **S.Nig.** : Lagos (fl. & fr. Feb.) *Dalz.* 1237! Ibadan to Abeokuta (fl. & fr. Mar.) *Schlechter* 12339! Akinjare, Owo (fl. & fr. Mar.) *Jones* FHI 3072! Milliken Hill, Enugu (fl. & fr. Feb.) *Jones* FHI 657! **[Br.]Cam.** : Mamfe (fl. & fr. Dec.) *Migeod* 270! Kakara, Mambila Plateau (Jan.) *Hepper* 1786! Throughout tropical Africa, also in Madagascar, tropical Asia and America. (See Appendix, p. 442.)
[*T. thouarsii* and *T. dinklagei* sometimes grow together when hybrids may occur and appear intermediate between these otherwise quite distinct species.]
3. **T. dinklagei** *Engl.* Bot. Jahrb. 57 : 610 (1922). *T. parviflora* of F.T.A. 4, 2 : 335, partly (*Mann* 1442) ; of F.W.T.A., ed. 1, 2 : 229, partly (*Mann* 1442) ; not of Buch.-Ham. ex Benth. *Lindernia diffusa* of Chev. Bot. 473, partly (*Chev.* 16978 ; 19740), not of (Linn.) Wettst. A slender creeping herb rooting at the nodes ; flowers blue or white ; in waste places and moist sandy ground.
S.L. : Njala (fl. & fr. Sept.) *Deighton* 2104! Puabu, Gaura Chiefdom (fl. & fr. Oct.) *Deighton* 5201! Pujehun, S. Prov. (fl. & fr. Apr.) *Deighton* 1621! Kasawe F.R. (fl. & fr. Dec.) *King* 326! **Lib.** : Diebla, Webo Dist. (fl. & fr. July) *Baldwin* 6277! Begwai (fl. & fr. Oct.) *Bunting* 13! **Iv.C.** : Guitri (fl. & fr. Dec.) *Boughey* GC 13555! Makougnié *Chev.* 16978. Grebo, Cavally *Chev.* 19740. **Ghana** : Oda (fl. & fr. Aug.) *Howes* 966! **Togo Rep.** : *Baumann* 158! **S.Nig.** : Port Harcourt (fl. & fr. Sept.) *T. W. J. Taylor* 9 (partly)! **[Br.] Cam.** : Mamfe (fl. & fr. Dec.) *Migeod* 266! **F.Po** : *Mann* 1442! Moka Plateau, 3,500–4,000 ft. (fl. & fr. Dec.) *Boughey* 80! *Exell* 823! Also in Cameroun, Ubangi-Shari and Congo.

Besides the above, *T. fournieri* Linden ex Fourn. with large flowers is cultivated in our area.

19. LINDERNIA Allioni—F.T.A. 4, 2 : 337.

Flowers terminal or subterminal ; erect herbs 5–13 cm. high ; pedicels 1–1·5 cm. long, erect :
Fruits obtuse, as long as calyx ; calyx shortly 5-toothed, 2–4 mm. long, thin ; leaves ovate to ovate-elliptic, serrate, 5–10 mm. long, upper leaves and bracts decreasing in size and becoming subulate .. .. 1. *crustacea*
Fruits acute, slightly longer than calyx ; leaves and bracts similar and not decreasing in size .. .. 2. *numulariifolia*
Flowers axillary, solitary ; erect or procumbent :
Calyx 2–3 mm. long :
Leaves broadly ovate about 1 cm. long and broad, serrate, subsessile ; flowers slenderly pedicellate, rarely sessile ; calyx shortly triangular-toothed 2. *numulariifolia*
Leaves narrowly lanceolate about 1 cm. long and 3 mm. broad, entire, subsessile ; flowers sessile in lax spikes ; calyx segments linear-filiform .. .. 3. *debilis*
Calyx 5–9 mm. long :
Fruits and flowers sessile or pedicel up to 2 mm. long in fruit ; stems prostrate, pubescent ; leaves ovate to suborbicular, 1–3 cm. long, about 1 cm. broad, serrate,

obtuse at apex, subsessile ; calyx about 7 mm. long, lobes subulate ; fruit narrowly
lanceolate, 11 mm. long, beak beyond seeds 1–2 mm. long 4. *diffusa* var. *diffusa*
Fruits (and usually flowers) distinctly pedicellate ; pedicels about 1–2 cm. long :
  Leaves ovate to suborbicular, or elliptic-oblong, 3–10 mm. broad :
    Leaves ovate to suborbicular, about 1 cm. broad, not fleshy or congested :
      Beak of fruit beyond seeds 1–2 mm. long :
        Leaves broadly ovate to suborbicular, 1–3·5 cm. long, 1–2 cm. broad, obtuse to
        subacute ; fruits up to 11 mm. long .. .. 4a. *diffusa* var. *pedunculata*
        Leaves ovate to ovate-lanceolate, 1–3 cm. long, 1–1·5 cm. broad, acute ; fruits
        up to 19 mm. long .. .. 5. *senegalensis*
      Beak of fruits beyond seeds 2·5–4 mm. long, attenuate ; fruits very narrow, up
      to 18 mm. long ; fruiting calyx 8–9 mm. long ; leaves ovate to suborbicular,
      obtuse, about 1 cm. long and broad, crenate, shortly petiolate .. 6. *vogelii*
  Leaves elliptic-oblong, 3–5 mm. broad, 7–13 mm. long, subentire, ciliate, rather
  fleshy and congested ; calyx-lobes lanceolate ; fruits about 1 cm. long, shortly
  beaked .. .. 7. *abyssinica*
Leaves linear, 1–2·5 cm. long, 1–2 mm. broad, acute, sessile ; corolla about 1 cm.
long, with broad rounded lobes on lower lip .. .. .. 8. *oliverana*

1. **L. crustacea** (*Linn.*) *F. Muell.* Census 97 (1882). *Capraria crustacea* Linn. (1767). Annual, erect or ascending, 2–3 in. high ; flowers blue ; weed of gardens and waste places.
    S.L.: Makump (fl. & fr. Aug.) *Deighton* 1385 ! Rokupr (fl. & fr. Sept.) *Jordan* 558 ! Njala (fl. & fr. June) *Deighton* 2079 ! Mattru (fl. & fr. July) *Deighton* 4841 ! **Iv. C.**: Angedédou (Dec.) *Leeuwenberg* 2272 ! **Ghana** : Kumasi (fl. & fr. June) *Vigne* FH 3058 ! **S.Nig.**: Ibadan (Sept.) *Keay* FHI 37175 ! Ozziza, Ogoja (fl. & fr. July) *Stone* 31 ! Enugu (fl. & fr. Oct.) *Jones* FHI 2217 ! **[Br.]Cam.**: Mamfe (fl. & fr. Dec.) *Migeod* 267 ! Widespread in tropical and sub-tropical countries.
2. **L. numulariifolia** (*D. Don*) *Wettst.* in E. & P. Pflanzenfam. 4, 3B : 79 (1891) ; F.T.A. 4, 2 : 341. *Vandellia numularifolia* D. Don (1825). Erect herb 2–5 in. high, simple or sparingly branched, leaves often tinged purple beneath ; flowers mauve and white or yellow ; weed of paths and damp places, usually in hilly districts.
    S.L.: Hill Station (Oct.) *Deighton* 2152 ! Loma Mts. (fl. & fr. Sept.) *Jaeger* 237 ! 1947 ! **Iv.C.**: Tonkoui, 3,500 ft. (fl. & fr. Aug.) *Boughey* GC 18326 ! **N.Nig.**: Naraguta & Jos (fl. & fr. Aug., Sept.) *Lely* 518 ! 558 ! **S.Nig.**: Akure Div. (fl. & fr. Jan.) *Brenan & E. W. Jones* 8739 ! **[Br.]Cam.**: Buea, 3,500 ft. (Apr.) *Dundas* FHI 15338 ! *Preuss* 1004 ! Gembu, Mambila Plateau (fl. & fr. Jan.) *Hepper* 1814 ! Also in Sudan, Ethiopia, E. Africa, N. Rhodesia and Angola ; extends into tropical Asia.
3. **L. debilis** *Skan* in F.T.A. 4, 2 : 344 (1906). A slender herb, 1–6 in. high ; flowers purplish with yellow on the lower lip ; in damp places.
    N.Nig.: Lokoja (fr. Nov.) *Dalz.* 128 ! Anara F.R. (fl. & fr. Oct.) *Hepper* 989 ! **[Br.]Cam.**: Gurum, Ada-mawa (fl. & fr. Nov.) *Hepper* 1249 ! Also in Sudan.
4. **L. diffusa** (*Linn.*) *Wettst.* var. **diffusa**—in E. & P. Pflanzenfam. 4, 3B : 80 (1891) ; F.T.A. 4, 2 : 338 ; Chev. Bot. 473, partly. *Vandellia diffusa* Linn. (1767). A pubescent much-branched creeping herb with stems about 6 in. long, rooting at nodes ; flowers white ; in damp places and roadsides.
    Mali: Guiri *Chev.* 890 ! **Guin.**: Dalaba-Diaguissa Plateau *Chev.* 18858 ! **S.L.**: Freetown (fl. & fr. Sept.) *Hepper* 927 ! Regent *Sc. Elliot* 4013 ! Rokupr (fl. & fr. Sept.) *Jordan* 112 ! Bumbuna (fl. & fr. June) *Thomas* 3250 ! **Lib.**: Kakatown *Whyte* ! Monrovia (fl. & fr. Nov.) *Barter* 1451 ! Suacoco, Gbanga (fl. & fr. May) *Daniel* 173 ! **Iv. C.**: Dabou (Nov.) *Leeuwenberg* 1955 ! **Ghana**: Begoro (fl. & fr. Mar.) *Johnson* 696 ! New Koforidua (fl. & fr. June) *Darko* 319 ! Psah R. *Irvine* 434 ! Amedzofe (May) *Morton* GC 7223 ! **N.Nig.**: Lokoja *Parsons* ! Kontagora (fl. & fr. Nov.) *Dalz.* 152 ! Nupe *Yates* ! **S.Nig.**: Lagos (fl. & fr. Oct.) *Dalz.* 1241 ! Onochie FHI 32029 ! Sapoba (fl. & fr. Nov.) *Meikle* 594 ! Ibadan to Abeokuta (fl. & fr. Mar.) *Schlechter* 12341 ! Onitsha *Barter* 1766 ! Oban *Talbot* 315 ! **[Br.]Cam.**: *Preuss* 1184 ! Mamfe (fl. & fr. Dec.) *Migeod* 265 ! Abonando (fl. & fr. Mar.) *Rudatis* 6 ! Mayo Selbe, Gashaka Dist. (fl. & fr. Jan.) *Hepper* 1635 ! **F.Po** : *Mann* 1441 ! Also in Cameroun, Gabon, Middle Congo, Uganda and Tanganyika.
4a. **L. diffusa** var. **pedunculata** (*Benth.*) *Skan* in F.T.A. 4, 2 : 338 (1906). *Vandellia diffusa* var. *pedunculata* Benth. (1846). *Ilysanthes stictantha* Hiern (1898). *Lindernia stictantha* (Hiern) Skan in F.T.A. 4, 2 : 339 (1906). Ascending or prostrate herb with white flowers.
    S.L.: *Barter* ! *T. Vogel* 111 ! Njala (fl. & fr. Sept.) *Deighton* 2078 ! Makump (fl. & fr. Aug.) *Deighton* 1386 ! Kambaia (fl. & fr. Oct.) *Small* 487 ! **Lib.**: Monrovia (fl. & fr. July) *Linder* 24 ! Also in S. Tomé, Congo and Angola.
5. **L. senegalensis** (*Benth.*) *Skan* in F.T.A. 4, 2 : 339 (1906) ; Chev. Bot. 473 ; Berhaut Fl. Sén. 67. *Vandellia senegalensis* Benth. (1846). Ascending from a woody rootstock, stems slightly pubescent becoming glabrous, about 1 ft. long ; in wet places.
    Sen.: *Leprieur* ! *Perrottet* 586 ! Niokolo-Koba *Berhaut* 1562. **Guin.**: *Heudelot* 804 ! Timbo to Ditinn *Chev.* 18509. **S.L.**: *Barter* ! *Don* ! Lomaburn, Talla (Feb.) *Sc. Elliot* 5031 ! **Iv.C.**: Nimba Mts., 5,200 ft. *Schnell* 3710 !
6. **L. vogelii** *Skan* in F.T.A. 4, 2 : 339 (1906) ; Chev. Bot. 473, partly. A slender glabrous herb, stems ascending ; flowers white and pink.
    Iv.C.: Agniéby Valley *Chev.* 17061. Cavally Basin *Chev.* 19587. **S.Nig.**: Nun R. (fl. & fr. Aug.) *T. Vogel* 1 ! Brass *Barter* 27 ! **F.Po**: Ureka (fl. & fr. Feb.) *Guinea* !
7. **L. abyssinica** *Engl.* Bot. Jahrb. 23 : 503, t. 9, figs. A–E (1897) ; F.T.A. 4, 2 : 343. Small perennial, erect or ascending, 2–4 in. high ; flowers blue ; a montane plant.
    [Br.]Cam.: Bamenda : Lakom, 6,000 ft. (fl. & fr. May) *Maitland* 1790 ! Bafut-Ngemba F.R., 6,000 ft. (Nov.) *Lightbody* FHI 26268 ! Bali-Ngemba F.R. (May) *Ujor* FHI 30325 ! Also in Sudan, Ethiopia and Uganda.
8. **L. oliveriana** *Dandy* in F.W. Andr. Fl. Pl. Sudan 3 : 139 (1956). *Vandellia lobelioides* Oliv. in Trans. Linn. Soc. 29 : 120, t. 121B (1875), not of F. Muell. (1858). *Lindernia lobelioides* (Oliv.) Wettst. in E. & P. Pflanzenfam. 4, 3B : 80, fig. 36 A, B (1891) ; F.T.A. 4, 2 : 340, not of F. Muell. (1882). An ascending herb about 1 ft. high, with slender, nearly simple stems and few narrow leaves ; flowers blue with white and yellow markings ; in swampy places.
    N.Nig.: Heipang, Jos Plateau (Aug.) *Summerhayes* 155 ! Also in Sudan, Ethiopia and through E. Africa to S. Rhodesia.

20. ILYSANTHES Raf.—F.T.A. 4, 2 : 345.

Calyx deeply divided almost to base ; prostrate or erect ; fruits globose-ellipsoid ;
flowers axillary, solitary ; pedicels slender about 6 mm. long and up to 2 cm. in fruit :
  Prostrate ; leaves suborbicular, sessile, subcordate, 6–12 mm. long and broad, about

7-nerved from base ; calyx 5 mm. long, lobes linear, about 1 mm. broad ; fruits about as long as calyx .. .. **1.** *rotundifolia*
Erect ; leaves ovate, oblanceolate to linear, (3–)6–12 mm. long, (1–)3–8 mm. broad, 3-nerved from base ; calyx 2 mm. long, lobes slender ; fruits twice as long as calyx
2. *parviflora*

Calyx 5-toothed ; plants erect ; fruits linear, acute :
Stems and leaves minutely but distinctly scabrid ; leaves linear to lanceolate, 5–15 mm. long, with distinct lamina, 2–3(–4) mm. broad, midrib distinct ; pedicels 5–12 mm. long, remaining erect in fruit ; fruit 9 mm. long .. .. .. **3.** *gracilis*
Stems and leaves smooth and glabrous (sometimes slightly scabrid towards the base of the stem) ; leaves subulate to linear about 5 mm. long, 0·5–1 mm. broad, distinctly 2-nerved, midrib obscure ; pedicels 1–8 mm. long, reflexing in fruit ; fruit about 5 mm. long .. .. **4.** *schweinfurthii*

1. **I.** **rotundifolia** (*Linn.*) *Benth.* in DC. Prod. 10 : 420 (1846) ; F.T.A. 4, 2 : 346 ; Hepper in Kew Bull. 15 : 65. *Gratiola rotundifolia* Linn. (1771). *Ilysanthes rotundata* Pilger in Engl. Bot. Jahrb. 45 : 214 (1910). Creeping herb rooting at the nodes ; flowers few, white or pale mauve ; in boggy ground. **[Br.]Cam. :** Bafut-Ngemba F.R., 6,000 ft., Bamenda (Jan.) *Keay* FHI 28380 ! Jakiri (Feb.) *Hepper* 1956 ! Nguroje, Mambila Plateau (Jan., June) *Hepper* 1713 ! *Chapman* 25 ! Also in Cameroun, Congo, E. Africa, Madagascar and through India to China.
2. **I.** **parviflora** (*Roxb.*) *Benth.* in DC. Prod. 10 : 419 (1846) ; F.T.A. 4, 2 : 346 ; F.W.T.A., ed. 1, 2 : 230, partly (excl. syn. *Dopatrium baoulensis* A. Chev. & *Chev.* 22250) ; Berhaut Fl. Sén. 54. *Gratiola parviflora* Roxb. (1819). *Lindernia parviflora* (Roxb.) Haines (1922)—F.W. Andr. Fl. Pl. Sudan 3 : 139. Erect or ascending herb, 2–9 in. high with slender usually diffuse branches ; flowers blue and white ; in swamps. **Sen. :** *Perrottet* 609 ! 610 ! Casamance (fr. Apr.) *Perrottet* 590 ! Ross *Berhaut* 1425 ! **Mali :** Gao to Koko-romme (fl. & fr. Mar.) *De Wailly* 4987 ! **Ghana :** Dormaa to Ahenkro (fl. & fr. Dec.) *Adams* 3010 ! **Niger :** Aïr *De Miré & Gillet*. **N.Nig. :** near Gombe (fl . & fr. Oct.) *Lely* 665 ! Also in Sudan, Congo, E. Africa Mozambique, N. Rhodesia and India.

FIG. 289.—ILYSANTHES GRACILIS *Skan* (SCROPHULARIACEAE).
A, whole plant. B, front view of flower and C, side view. D, stamen. E, capsule. F, seed.

3. **I. gracilis** *Skan* in F.T.A. 4, 2 : 349 (1906) ; Chev. Bot. 474, partly. Erect slender herb 3–6 in. high, simple or freely branched, leaves purple beneath ; flower pink and white or lilac ; in moist pockets in rock outcrops. **Mali :** Bamako *Garnier* ! **Lib. :** Atonso (fl. & fr. Oct.) *Baldwin* 13495 ! **Iv. C. :** Séguéla (fl. & fr. Aug.) *Boughey* GC 18485 ! Bouaké *Chev.* 22090. Langouassou to Fetékro *Chev.* 22157. Kodiokoffi *Chev.* 22365. **Ghana :** Mankrong, Kwahu (Dec.) *Adams* 4993 ! **N.Nig. :** Jebba *Barter* ! Jos, 4,000–5,000 ft. (fl. & fr. Aug.) *Saunders* 3 ! *Keay* FHI 20051 ! Naraguta (fl. & fr. Aug.) *Lely* 519 ! **S.Nig. :** Idanre (fl. & fr.Aug.) *Jones* FHI 20725 ! Okelifi, Ondo (fl. & fr. Nov.) *Onochie* FHI 34224 ! Ado Rock, Ibadan Dist. (fr. Oct.) *Hambler* 504 !
4. **I. schweinfurthii** *Engl.* Bot. Jahrb. 23 : 504 (1897) ; F.T.A. 4, 2 : 350, incl. var. *linearifolia* Engl. (1897) ; Berhaut Fl. Sén. 5, 54, 93. *I. barteri* Skan in F.T.A. 4, 2 : 350 (1906) ; F.W.T.A., ed. 1, 2 : 230. Chev. Bot. 473 ; Sousa in An. Junta Invest. Colon. 7 : 58 (1952). *Lindernia schweinfurthii* (Engl.) Dandy in F.W. Andr. Fl. Pl. Sudan 3 : 139 (1956). Slender erect herb like the last and in similar habitats. **Sen. :** Tambacounda (Sept.) *Berhaut* 1648 ! **Port.G :** Bafatá (Aug.) *Esp. Santo* 2743 ! **Mali :** Koulikoro *Chev.* 2787. Banfora (Sept.) *Jaeger* 5256 ! Mt. Zongapignié, Mossi *Chev.* 24642. **Guin. :** Timbo to Ditinn *Chev.* 18421. Dalaba-Diaguissa Plateau *Chev.* 18762. **S.L. :** *Barter* ! Brookfields (Aug.) *Jordan* 65 ! Kambia (Sept.) *Adames* 186 ! Lengekoro (July) *Glanville* 262 ! **Ghana :** Damongo Scarp (Nov.) *Rose Innes* GC 30837 ! **N.Nig. :** Jebba *Barter* ! Lame Dist. (fl. & fr. Aug.) *Summerhayes* 20 ! Fuka, Minna (fl. & fr. Oct.) *Hepper* 963 ! Jos (Aug.) *Saunders* 4 ! *Keay* FHI 20073 ! Yola (Aug.) *Dalz.* 129 ! **S.Nig. :** Ado Rock, Oyo (fl. & fr. Oct.) *Keay & Savory* FHI 25350 ! Fashola, Oyo (fl. & fr. Oct.) *Onochie* FHI 34925 ! Also in Ubangi-Shari, Sudan, Congo and Uganda.

21. LIMOSELLA Linn.—F.T.A. 4, 2 : 352 ; Glück in Engl. Bot. Jahrb. 66 : 488 (1934).

Leaves spathulate with a distinct lamina and long petiole, 1·8–9 cm. long, 2·5–10 mm. broad, all radical ; flowers axillary, solitary on pedicels about 1 cm. long ; calyx 5-toothed, about 2 mm. long, glabrous ; corolla a little longer than calyx ; fruits subglobose, about 3 mm. long ..   .. 1. *africana*

Leaves subulate or acicular, terete or subterete, without a distinct lamina, 2–4·3 cm. long, 1 mm. broad ; otherwise similar to above ..   ..   .. 2. *subulata*

1. **L. africana** *Glück* in Notizbl. Bot. Gart. Berl. 12 : 76 (1934), and in Engl. Bot. Jahrb. 66 : 540, 558 (1934) ; Hedberg in Sym. Bot. Uppsal. 15 : 166, 318 (1957). *L. aquatica* of F.T.A. 4, 3 : 352 ; of F.W.T.A., ed. 1, 2 : 221 ; not of Linn. A small tufted aquatic or marsh herb, often with stolons bearing tufts of leaves and rooting ; flowers small, lilac or pinkish.
   Mali : Bentia, Ansongo *Hagerup* 424 ! [Br.]**Cam.** : Cam. Mt., 9,800–10,800 ft. *Mann* ! Through much of tropical Africa and in S. Africa.

2. **L. subulata** *Ives* in Trans. Phys.-Med. Soc. New York 1 : 440 (1817) ; Glück in Engl. Bot. Jahrb. 66 : 491. *L. aquatica* var. *tenuifolia* (Nutt.) Hook. f. in F.T.A. 4, 2 : 352 (1906). *L. tenuifolia* Nutt. (1818)—F.W.T.A., ed. 1, 2 : 221, not of Wolf ex Hoffm. (1804). Stolons usually absent.
   F.Po : Clarence Peak, 9,350 ft. (Dec.) *Mann* 597 ! Cosmopolitan.

22. GLOSSOSTIGMA Arn. in Nov. Act. Nat. Cur. 18 : 355 (1836). *Nom. cons.*

Small creeping herb ; stems filiform prostrate rooting at nodes ; leaves spathulate, about 1 cm. long, 1–1·5 mm. broad ; flowers axillary, solitary ; pedicels 3–9(–18) mm. long ; calyx 3-lobed ; corolla 5-lobed ; stamens 2 ; stigma compressed ; fruits 2-celled, many-seeded ..   .. *diandra*

G. diandra (*Linn.*) *O. Ktze.* Rev. Gen. Pl. 2 : 461 (1891) ; A. Chev. in Bull. Mus. Hist. Nat. Paris, sér. 2, 4 : 587 (1932). *Limosella diandra* Linn. (1767). *Microcarpaea spathulatum* Hook. (1831). *Glossostigma spathulatum* (Hook.) Arn. (1836)—Berhaut Fl. Sén. 93. In small mats on mud.
   Sen. : *Leprieur* ! *Perrottet* ! Podor (Dec.) *Martine* 33 ! *Berhaut* 1426 ! **Mali** : Timbuktu to Bamba (Mar.) *Chev.* ! Timbuktu to Gao *Chev.* Also in Kenya, India and Ceylon.

23. PSAMMETES Hepper in Hook. Ic. Pl. t. 3582 (1962).

A small erect unbranched herb, 2·5–4 cm. high ; stem narrowly 4-winged, lower internodes about 5 mm. long, upper internodes up to 16 mm. long ; leaves opposite, linear-lanceolate, 3–7 mm. long, 1–1·5 mm. broad, glabrous ; flowers cleistogamous, subsessile in axils of upper leaves and terminal ; calyx (4–)5-lobed, lobes 1 mm. long, linear ; corolla bilabiate, 1·2 mm. long, upper lip emarginate, lower lip 3-lobed ; stamens 2 ; ovary 2-celled, style glabrous, stigma flattened ; fruit ovoid, dry, 1 mm. long with numerous seeds ..   .. *nigerica*

P. nigerica *Hepper* l.c. (1962). A slender annual about 2 in. high ; flowers pale yellow-brown ; " in coconut palm plantation on degraded sands in furrows between ridges."
   S.Nig.: Badagry (fl. & fr. Aug.) *Onochie* FHI 33493 ! Ogoyo-Igbosere, Lagos (Aug.) *Savory* UCI 271a !

24. MICRARGERIA Benth.—F.T.A. 4, 2 : 457.

Plants minutely scabrid all over ; leaves linear-acicular, acute, 1–3·5 cm. long, more or less spreading ; stems much-branched ; flowers numerous in lax racemes ; calyx campanulate, 3–4 mm. long, teeth triangular, acute or slightly acuminate, 1(–2) mm. long ; corolla about 1 cm. long, widely funnel-shaped, puberulous outside ; fruits subglobose, 3–4 mm. diam., smooth ..   .. 1. *filiformis*

Plants glabrous, only slightly scabrid on stem below ; leaves erect ; stems sparingly branched ; calyx-teeth acuminate, 2 mm. long ; otherwise similar to above 2. *barteri*

1. **M. filiformis** (*Schum. & Thonn.*) *Hutch. & Dalz.* F.W.T.A., ed. 1, 2 : 223 (1931) ; Berhaut Fl. Sén. 54, 94. *Gerardia filiformis* Schum. & Thonn. (1827). *Gerardiella scopiformis* Klotzsch (1861). *Micrargeria scopiformis* (Klotzsch) Engl. (1895)—F.T.A. 4, 2 : 457 ; Chev. Bot. 480. *Sopubia parviflora* of F.W.T.A., ed. 1, 224, partly (*Lely* 604, *Glanville* 22). Slender-branched erect annual herb, 1–2 ft. high ; flowers showy, pink or purple ; in moist grassland in savanna and rice-fields ; drying grey-brown.
   Sen. : Casamance *Chev.* 3299. Badi *Berhaut* 1837. **Gam.** : Kombo Prov. *Dawe* 33 ! **Mali** : Kita massif (Nov.) *Jaeger* 5563 ! **Port.G.** : Bafatá (fl. & fr. Nov., Dec.) *Esp. Santo* 2855 ! 2966 ! Suzana (Nov.) *Esp. Santo* 3147 ! Gabu (fl. & fr. Oct.) *Esp. Santo* 3505 ! **Guin.** : Kouria to Irébéléya *Chev.* 18273. Timbo to Ditinn *Chev.* 18493. Dalaba-Diaguissa Plateau *Chev.* 18894. **S.L.** : Mateboi (fl. & fr. Oct.) *Glanville* 22 ! **Iv.C.** : Mankono *Chev.* 21908. **Ghana** : Shai Plains (fl. & fr. Jan.) *Johnson* 576 ! Navrongo (fl. & fr. Nov.) *Vigne* FH 4667 ! **Togo Rep.** : Lomé *Warnecke* 271 (partly) ! **|N.Nig.** : Anara F.R., Zaria (fl. & fr. Sept., Oct.) *Olorunfemi* FHI 24395 ! *Hepper* 988 ! Abinsi (fl. & fr. Dec.) *Dalz.* 710 (partly) ! Panshanu (fl. & fr. Sept.) *Lely* 604 ! Jos Plateau (fl. & fr. Oct.) *Lely* P821 ! [Br.]**Cam.** : Gurum, Vogel Peak area (Nov.) *Hepper* 1244 ! Also in Sudan, E. Africa, Angola, N. Rhodesia and Nyasaland.

2. **M. barteri** *Skan* in F.T.A. 4, 2 : (1906) ; Chev. Bot. 480. Annual herb, 2–3 ft. high, sparingly branched ; flowers ornamental, pink ; in moist grassland in savanna ; drying grey-brown.
   Port.G. : Gabu (fl. & fr. Oct.) *Esp. Santo* 3455 ! **Iv.C.** : Alangouassou to Mbayakro *Chev.* 22241. Kodiokoffi *Chev.* 22327 ! 22370 ! **Ghana** : Kayoro (fl. & fr. Oct.) *Vigne* FH 4610 ! Deloro-Tankara (fr. Dec.) *Adams & Akpabla* 4336 ! Elmina (fl. & fr. Oct.) *Hall* 855 ! **Togo Rep.** : Lomé *Warnecke* 271 (partly)! **N.Nig.** : Abinsi (Dec.) *Dalz.* 710 (partly) ! Zungeru (fr. Oct.) *Dalz.* 174 ! Nupe *Barter* 755 ! 1706 ! Jos Plateau (Oct.) *Lely* P829 ! Makurdi (fl. & fr. Nov.) *Keay* FHI 22275 ! **S.Nig.** : Abeokuta (fl. & fr. Aug.) *Symington* FHI 5095 ! Awba Hills F.R. (fl. & fr. Oct.) *Jones* FHI 6311 ! Enugu Extension F.R. (fl. & fr. Sept.) *Onochie* 34107 ! [Br.]**Cam.** : Vogel Peak, Adamawa (Nov.) *Hepper* 1391 !

25. BARTSIA Linn.—F.T.A. 4, 2 : 458 ; R. E. Fries in Acta Horti Berg. 8 : 63 (1924). *Nom. cons.*

Stems branched, up to 170 cm. long ; flowers numerous in lax terminal inflorescences ; corolla 11 mm. long ; fruit about 12 cm. long ; leaves 1–3·5 cm. long, narrowed to base with very short but distinct petiole .. .. 1. *mannii*
Stems simple or branches arising from near the rootstock, up to 25 cm. long ; flowers few, 2–9, in terminal clusters ; corolla 9 mm. long ; fruit 8(–12) mm. long ; leaves 6–12 mm. long, quite sessile .. .. 2. *petitiana*

1. **B. mannii** *Hemsl.* in F.T.A. 4, 2 : 459 (1906). Straggling perennial herb 1–3 ft. high ; flowers pale yellow turning pink ; in montane grassland ; drying dark brown.
  S.Nig. : Ikwette Plateau, 5,200 ft., Obudu Div. (fl. & fr. Dec.) *Keay & Savory* FHI 25258 ! [Br.]**Cam.** : Cam. Mt., 6,000–9,000 ft. (fl. & fr. Nov.–Feb., fr. May) *Mann* 1264 ! 1986 ! *Dunlap* 198 ! *Maitland* 668 ! Bafut-Ngemba F.R., Bamenda (fl. & fr. Dec.) *Tamajong* FHI 22213 ! *Hepper* 2093 ! Kumbo (Feb.) *Hepper* 2843 ! Maisamari, Mambila Plateau (fl. & fr. Jan.) *Hepper* 1661 !
2. **B. petitiana** (*A. Rich.*) *Hemsl.* in F.T.A. 4, 2 : 460 (1906) ; Hedberg in Sym. Bot. Uppsal. 15 : 171, 325 (1957). *Alectra petitiana* A. Rich. (1851). *Bartsia mannii* of F.W.T.A., ed. 1, 2 : 224, partly (*Johnston* 80). Erect perennial herb 6 in. to 1 ft. high ; flowers carmine or bluish-mauve ; in montane grassland ; drying dark brown.
  [Br.]**Cam.** : Cam. Mt., 8,000–12,000 ft. (fl. & fr. Nov.–Jan.) *Johnston* 80 ! *Migeod* 183 ! *Maitland* 805 ! 1298 ! *Keay* FHI 28607 ! Also in Ethiopia and E. Africa.

26. ALECTRA Thunb.—F.T.A. 4, 2 : 362 ; Melchior in Notizbl. Bot. Gart. Berl. 15 : 423 (1941).

Calyx 13–16 mm. long, pilose on nerves outside, calyx-teeth about 5 mm. long ; anthers apiculate, filaments all glabrous ; flowers few, scattered in axils ; stems slender, with 2 lines of hairs ; leaves spreading, linear to linear-oblong, remotely serrate, sessile, only the midrib prominent, 2–5 cm. long, 3–8 mm. broad ; seeds opaque 1. *linearis*

Calyx 5–10 mm. long :
Calyx glabrous, teeth 2 mm. long ; anthers apiculate, filaments all glabrous ; flowers towards tips of stems ; stems slender, with 2 obscure lines of hairs ; leaves linear, subentire with ciliate margins, sessile, appressed to stem, only the midrib prominent, 1–5 cm. long, 2–4 mm. broad ; seeds opaque .. 2a. *rigida* subsp. *paludosa*
Calyx pubescent, at least on margins and nerves :
  Pubescence of calyx on margins and nerves only, hairs usually long :
    Whole plant densely hispid ; leaves clasping the simple stem, sessile, lanceolate, serrate, 1–2·5 cm. long, 5–10 mm. broad ; inflorescence congested and with leafy bracts rather larger than the stem leaves ; seeds in a transparent envelope 3. sp. nr. *capensis*
    Plants slightly hispid or glabrous :
      Leaves 2–4 mm. broad, 7–11 mm. long, lanceolate, rounded at base, slightly toothed, 1-nerved, ciliate ; stems hispid-pilose ; anthers apiculate, filaments all bearded ; seeds opaque .. .. 4. *virgata*
      Leaves (5–)10–20 mm. broad, 10–50 mm. long, 3-nerved from base ; seed in a transparent envelope 1–1·5 mm. long ; filaments 2 bearded and 2 glabrous ; anthers apiculate :
        Lower leaves ovate, broadly rounded at base or cordate, almost sessile, slightly to deeply serrate .. .. 5a. *sessiliflora* var. *senegalensis*
        Lower leaves lanceolate to elliptic, cuneate at base, distinctly petiolate, often deeply serrate .. .. 5b. *sessiliflora* var. *monticola*
  Pubescence of calyx short and evenly spread ; leaves and bracts subentire ; filaments glabrous ; anthers not apiculate .. .. 6. *vogelii*

1. **A. linearis** *Hepper* in Kew Bull. 14 : 402, fig. 1 (1960). Slender, branched annual herb 6–17 ins. high with spreading leaves ; flowers pale yellow ; in drier parts of swamps ; drying black.
  Iv.C. : Nzi, Baoulé-Nord (fl. & fr. Aug.) *Chev.* 22286 ! **Ghana** : Babile, Lawra (fr. Sept.) *Irvine* 4733 ! **N.Nig.** : Lokoja (fr. Oct.) *Dalz.* 218 ! **S.Nig.** : Mamu R., Awka Dist. (May) *Jones* FHI 6142 !
2. **A. rigida** (*Hiern*) *Hemsl.* in F.T.A. 4, 2 : 369 (1906). *Melasma rigida* Hiern (1898).
2a. **A. rigida** subsp. **paludosa** (*A. Chev.*) *Hepper* in Kew Bull. 14 : 404 (1960). *A. paludosa* A. Chev. Bot. 475 (1920). *A. virgata* of F.W.T.A., ed. 1, 2 : 223 and of Melchior l.c. 437, partly (*Chev.* 18896). Erect simple or branched annual herb, 1–2 ft. high with appressed leaves ; flowers yellow ; amongst grass in moist places ; drying black.
  Port.G. : Bôr, Bissau (fl. & fr. Jan.) *Esp. Santo* 1448 ! Pelubé, Bissau (Dec.) *Esp. Santo* 1613 ! Gabu (Oct., Dec.) *Esp. Santo* 3510 ! 3766 ! **Guin.** : Dalaba-Diaguissa Plateau (Oct.) *Chev.* 18896 ! **S.L.** : Waterloo (fl. & fr. Dec.) *Deighton* 3282 ! Kent *Deighton* 3301 ! Mankara, Maforki Chiefdom (fl. & fr. Oct.) *Jordan* 812 ! **Ghana** : Kwahu Tafo (Dec.) *Harris* !
3. **A. sp. nr. capensis** *Thunb.* Nov. Gen. Pl. 82 (1784) ; Melchior l.c. 432 ; Hepper in Kew Bull. 15 : 64. Stiffly erect, harshly hispid herb about 10 in. high ; in wet grassland.
  Ghana : Elmina (Feb.) *Hall* 1050 ! **N.Nig.** : Naraguta, 3,700 ft., Jos Plateau (fl. & fr. Oct.) *Hepper & Wimbush* 1169 !
  [Note : The group to which these collections belong is in need of further revision—F. N. H.]
4. **A. virgata** *Hemsl.* in F.T.A. 4, 2 : 369 (1906) ; Melchior l.c. 437. A small slender herb, simple or with a few branches, about 6 ins. high ; drying brown.
  S.Nig. : Lagos area *Millson* !
5. **A. sessiliflora** (*Vahl*) O. Ktze. Rev. Gen. Pl. 2 : 458 (1891) ; Melchior l.c. 437 ; Hepper in Kew Bull. 14 : 404 (1960). *Gerardia sessiliflora* Vahl (1794). *Alectra melampyroides* Benth. (1846)—F.T.A. 4, 2 : 371.

5a. **A. sessiliflora** var. **senegalensis** (*Benth.*) *Hepper* l.c. 405 (1960). *A. senegalensis* Benth. (1846)—F.T.A. 4, 2 : 371 ; F.W.T.A., ed. 1, 2 : 223 ; Melchior l.c. 439 ; Berhaut Fl. Sén. 68. *A. communis* of F.W.T.A., ed. 1, 2 : 223, and of Chev. Bot. 474, partly (*Chev.* 3296, 18428, 18435, 18880), not of Hemsl. *A. cordata* Benth. (1846) partly, Afr. spec.,—F.T.A. 4, 2 : 371 ; Melchior l.c. 439. A slender erect usually roughly pilose herb 1 ft. or more high but may be reduced to a few inches high ; flowers yellow ; drying greenish-brown ; parasitic on grasses.
Sen.: *Leprieur* 9 ! 10 ! *Perrottet* 573 ! Casamance *Chev.* 3295 ! 3296 ! **Gam.:** *Brown-Lester* 27 ! **Mali:** Kita (fl. & fr. Oct.) *Chev.* 3297 ! **Guin.:** *Heudelot* 787 ! Timbo to Ditinn *Chev.* 18435 ! Dalaba-Diaguissa Plateau *Chev.* 18880. **S.L.:** Picket Hill, 2,914 ft. (fl. & fr. Nov.) *T. S. Jones* 216 ! 228 ! **Ghana:** Kumasi *Darko* 352 ! **N.Nig.:** Kontagora (fl. & fr. Jan.) *Dalz.* 153 ! Naraguta (Aug.) *Lely* 502 ! Vodni, 4,600 ft. (July) *Lely* 421 ! Sara Hills, Bauchi Prov. (fl. & fr. Oct.) *Hepper* 1122 ! **S.Nig.:** Akure F.R., Ondo (fl. & fr. Nov.) *Latilo & Olorunfemi* FHI 24448 ! [**Br.**]**Cam.:** Cam. Mt., 3,000–10,000 ft. (fl. & fr. Nov., Dec.) *Mann* 1262 ! 1985 ! Bafut-Ngemba F.R., 6,000 ft., Bamenda (fl. & fr. Feb., Oct., Nov.) *Lightbody* FHI 26273 ! *Hepper* 2187 ! Kumbo to Oku (fl. & fr. Feb.) *Hepper* 2005 ! Gembu, 5,300 ft., Mambila Plateau *Hepper* 1806 ! Vogel Peak area, 1,900–5,000 ft. Adamawa (fl. & fr. Nov.) *Hepper* 1276 ! 2766 ! Also in eastern Africa from the Sudan to Mozambique.

5b. **A. sessiliflora** var. **monticola** (*Engl.*) *Melch.* in Notizbl. Bot. Gart. Berl. 15 : 126 (1940) and 438 (1941) ; Hepper l.c. 406. *Glossostylis avensis* Benth. (1835). *A. avensis* (Benth.) Merrill (1917), " *arvensis* " by error. *A. communis* Hemsl. in F.T.A. 4, 2 : 372 (1906) ; F.W.T.A., 4, 2 : 372 (1906) ; F.W.T.A., ed. 1, 2: 223 ; Berhaut Fl. Sén. 68. Erect scabrid herb 1–3 ft. high ; flowers pale yellow ; amongst grass.
Sen.: Simenti *Berhaut* 1609 ! **S.L.:** Zimmi (Nov.) *Deighton* 366 ! Pujehun, near Njala (Dec.) *Deighton* 1515 ! **Lib.:** Kakatown *Whyte* ! **Ghana:** Abetifi (fl. & fr. Jan.) *Irvine* 1666 ! **S.Nig.:** Forest Hill, Ibadan (Oct.) *Jones* FHI 13735 ! *Latilo* FHI 24439 ! Oban *Talbot* 317 ! [**Br.**]**Cam.:** Buea, 3,000–3,500 ft. (Nov.) *Migeod* 56 ! 115 ! Cam. Mt., lava flow (Sept.) *Ngongi* FHI 15087 ! L. Bambuluwe, Bamenda (Sept.) *Ujor* FHI 30205 ! Vogel Peak area, (Nov.) *Hepper* 2743 ! **F.Po:** Moka 3,500–4,000 ft. (Dec.) *Boughey* 82 ! 86 ! Musola (fr. Jan.) *Guinea* 1402 ! Widespread in tropical Africa, in the Mascarenes and tropical Asia to the Philippines and Formosa.
[The division between these two varieties is by no means distinct.—F. N. H.]

6. **A. vogelii** Benth. in DC. Prod. 10 : 339 (1846) ; F.T.A. 4, 2 : 368 ; Melchior l.c. 432. *A. senegalensis* var. *arachidis* A. Chev. Bot. 475, and *A. arachidis* A. Chev. (1936), names only. *A. communis* of F.W.T.A., ed. 1, 2 : 223, partly (syn. and *Chev.* 18285). Stoutly stemmed herb branching from near the base, or simple, 7–18 ins. high, with few leaves that are often deciduous on drying ; flowers yellow ; semi-parasitic weed of cultivation, particularly of Papilionaceae, in savanna ; drying greenish-brown.
Guin.: Irébéléya to Timbo (Sept.) *Chev.* 18285 ! **Ghana:** Ketin (fl. & fr. Oct.) *Vigne* FH 4609 ! **N.Nig.:** Borgu *Barter* 789 ! Sokoto (fl. & fr. Nov.) *Dalz.* 356 ! Oke-Ogin, Ilorin (fl. & fr. Dec.) *Ajayi* FHI 19265 ! Patti Lokoja *T. Vogel* 186 ! Tangale-Waja, Bauchi Prov. (Nov.) *Summerhayes* 75 ! **S.Nig.:** Ebute Metta *Millen* 172 ! Oluwatedo, Oyo (fl. & fr. Jan.) *Keay* FHI 25638 ! Also in E. Africa, Nyasaland, N. & S. Rhodesia, Angola and Cabinda.

27. SOPUBIA Buch.-Ham.—F.T.A. 4, 2 : 444.

Calyx 2–3 mm. broad and long, glabrous and smooth outside, only margin woolly inside, teeth less than 1 mm. long ; bracteoles filiform, nearly as long as calyx ; flowers about 6 mm. diam. ; stems slender, slightly puberulous, almost terete ; leaves and bracts linear-acicular or almost filiform, 1–2(–4) cm. long ; annual   .. 1. *parviflora*
Calyx 4–5 mm. broad, 4–7 mm. long, smooth, scabrid or verrucose ; perennials :
  Stems angular and longitudinally furrowed, almost glabrous ; calyx 4–5 mm. long and broad, scabrid-pubescent outside, teeth 2 mm. long, woolly inside ; bracteoles subulate much shorter than calyx-tube ; flowers about 1·5 cm. diam. ; leaves linear, more or less appressed to stem, 5–30 mm. long, slightly scabrid   2. *simplex*
  Stem not angular, internodes puberulous in alternate strips ; leaves spreading, 1–3 cm. long, scabrid, axillary tufts or branchlets frequent ; calyx-teeth densely woolly inside :
    Calyx verrucose and minutely scabrid-pubescent outside, teeth 3 mm. long, acuminate ; bracteoles linear often leaf-like with a midrib and about as long as calyx ; flowers about 1 mm. diam. ; midrib prominent   ..   ..   .. 3. *ramosa*
    Calyx smooth and glabrous or rarely slightly woolly and verrucose outside but not scabrid, 4–5 mm. long and broad, teeth 2 mm. long, triangular ; bracteoles subulate, nearly as long as calyx-tube ; flowers about 1·5 cm. diam. ; midrib more or less obscure :
      Pedicels 1 cm. long or less ; flowers numerous, often densely arranged ; leaves densely congested on stem   ..   .. 4. *mannii* var. *mannii*
      Pedicels 1·5–2 cm. long ; flowers few and laxly arranged ; leaves laxly arranged or rather congested   ..   .. 4a. *mannii* var. *tenuifolia*

1. **S. parviflora** *Engl.* Bot. Jahrb. 18 : 65 (1893) ; F.T.A. 4, 2 : 452. Erect slender branched annual herb with obtusely 4-angled slightly scabrid stems 1 2 ft. high ; flowers yellow with a purple centre ; in damp grassland ; drying brown.
Mali: Bamako to Ségou (Sept.) *Jaeger* 5108 ! **Guin.:** Kouroussa (Aug.) *Pobéguin* 371 (partly) ! **S.L.:** *Afzelius* ! Kakama (fl. & fr. Aug.) *Deighton* 6097 ! Tabe (Sept.) *Deighton* 3043 ! Roboli (fl. & fr. Nov.) *Jordan* 176 ! Musaia (fl. & fr. Sept., fr. Dec.) *Deighton* 4520 ! 4875 ! **Iv.C.:** Séguéla (fr. Aug.) *Boughey* GC 18481 ! Nimba Mt. (fl. & fr. Aug., Sept.) *Boughey* GC 18141 ! *Schnell* 3571 ! **Ghana:** Kwahu Tafo (Dec.) *Harris* ! Damongo (fr. Sept.) *Goodall* GC 15908 ! Gambaga Scarp (Sept.) *Hall* 812 ! **N.Nig.:** Zaria (fl. & fr. Sept.) *Olorunfemi* FHI 24390 ! Zungeru (fl. & fr. Sept.) *Dalz.* 175 ! Naraguta (Aug.) *Lely* 524 ! Jos (fr. Oct.) *Hepper* 1064 ! Also in Congo and Sudan.

2. **S. simplex** (*Hochst.*) *Hochst.* in Flora 27 : 27 (1844) ; F.T.A. 4, 2 : 45 ; Chev. Bot. 479. *Rhaphidophyllum simplex* Hochst. (1841). Erect perennial herb with slender sparingly branched ribbed stem, 1–2 ft. high, from a stout woody rootstock ; flowers pink or mauve with a purple centre ; in moist grassland in savanna ; drying brown.
Mali: Niana to Namesa (Feb.) *Chev.* 330 ! Guienguénia to Moriguéya *Chev.* 433. **Guin.:** Kankan *Pobéguin* 1014a ! Kebeya (June) *Pobéguin* 1014b ! Erimakuna (Mar.) *Sc. Elliot* 224 ! **S.L.:** Ninia Talla *Sc. Elliot* 4872 ! **Ghana:** Afram Plains (fl. & fr. Mar.) *Johnson* 727 ! Amedzofe (Mar., May) *Scholes* 58 ! *Hepper* & *Morton* A3044 ! **Togo Rep.:** Agome (Mar.) *Schlechter* 12959 ! **N.Nig.:** Nupe *Barter* ! Abinsi (Mar.) *Dalz.* 708 ! Zaria (fl. & fr. Feb.) *Dalz.* 383 ! Tof, 4,000 ft., Jos Plateau (Feb.) *Hepburn* 136 ! Biu, 2,500 ft.,

Bornu Prov. (May) *Noble* 53! **S.Nig.**: Upper Ogun F.R., Oyo (Feb., Mar.) *Keay* FHI 22520! 23443!
[**Br.**]**Cam.**: Bamenda, 3,000–5,000 ft. (Jan., Apr., fl. & fr. May) *Migeod* 367! *Maitland* 1414! 1762! *Ujor*
FHI 30094! Gembu, Mambila Plateau (Jan.) *Hepper* 1817! Mayo Wombo to Mai Idoanu, Gashaka
Dist. (fl. & fr. Feb.) *Latilo & Daramola* FHI 34460! Also in Cameroun, Ubangi-Shari, Congo and
eastern Africa from Sudan to S. Africa. (See Appendix, p. 442.)

3. **S. ramosa** (*Hochst.*) *Hochst.* l.c. (1844) ; F.T.A. 4, 2 : 449 ; Chev. Bot. 479. *Rhaphidophyllum ramosum*
Hochst. (1841). Robust erect shrub or woody herb 1–5 ft. high ; flowers pink or mauve with darker centre ;
in savanna woodland or grassy places ; drying brown.
Guin.: Kouroussa (July, Oct.) *Pobéguin* 565! 1021! **S.L.**: Bumbuna (fr. Jan.) *Deighton* 3597! Rofankma
(fr. Dec.) *Koroma* 1! Lomaburn (fr. Jan.) *Sc. Elliot* 5036! **Ghana**: Kayoro (Oct.) *Vigne* FH 4613!
Kintampo Scarp (Nov.) *Morton* A3814! Wa (fr. Dec.) *Adams* 4474! Techiman to Nkoransa (fl. & fr. Dec.)
Adams 4520! **Togo Rep.**: Alafanyo *Krause*! **Dah.**: *Poisson*! **N.Nig.**: Borgu *Barter* 774! Liruwen-Kano
Hills *Carpenter*! Dogon Kurmi, Jemaa Div. (Nov.) *Keay & Onochie* FHI 21538! Jos (Oct.) *Hepper* 1065!
Maifula, Muri Dist. (fl. & fr. Nov.) *Latilo & Daramola* FHI 28711! **S.Nig.**: Abeokuta *Irving*! Oyo to
Iseyin *Hambler* 114! Oban *Talbot* 316! [**Br.**]**Cam.**: Banso, Bamenda (fr. Dec.) *Egbuta* FHI 3775! Banso
to Bamenda (Oct.) *Tamajong* FHI 23487! Vogel Peak area (Nov.) *Hepper* 1425! Also in Congo, Sudan,
E. Africa, Nyasaland, N. and S. Rhodesia.

4. **S. mannii** *Skan* var. **mannii**—in F.T.A. 4, 2 : 450 (1906) ; Hepper in Kew Bull. 14 : 409 (1960). *S. dregeana*
Benth. (1846), illegitimate name, partly—F.T.A. 4, 2 : 451. Erect undershrub about 1 ft. high, rootstock
woody, stems branched and with short internodes ; flowers purplish or magenta ; in montane grassland
above timber line ; drying brown.
Iv.C.: Nimba Mt., 4,700 ft. *Monod* 10264! [**Br.**]**Cam.**: Cam. Mt., 5,000–11,000 ft. (Jan., Dec., fl. & fr.
Mar.) *Mann* 1287! 2003! *Johnson* 2! *Maitland* 447! *Keay* FHI 28589! Bafut-Ngemba F.R., Bamenda
6,000 ft. *Lightbody* FHI 26255! Basenako, Bamenda, 6,000 ft. (June) *Maitland* 1723! L. Oku, 7,800 ft.,
Bamenda (Jan.) *Lightbody* FHI 28496! Also in Cameroun, Uganda, Tanganyika, Angola, N. & S. Rhodesia
and S. Africa.

4a. **S. mannii** var. **tenuifolia** (*Engl. & Gilg*) *Hepper* l.c. 410 (1960). *S. dregeana* Benth. var. *tenuifolia* Engl. &
Gilg (1903). A few-flowered undershrub of montane grassland.
S.L.: Bintumane, 5,000–6,000 ft. (fl. Feb., fl. & fr. May, fr. Aug.) *Deighton* 5089! *Glanville* 472! *Jaeger*
1135! 4247! Also in Uganda, N. & S. Rhodesia, Angola and S. Africa.

28. BUCHNERA Linn.—F.T.A. 4, 2 : 373.

Spikes head-like, 1–2·5 cm. long ; leaves lanceolate to linear-lanceolate, entire or slightly
dentate, lower ones obtuse, upper ones subacute :
Stems and nerves of leaves beneath hispid with spreading hairs about 1 mm. long ;
lower leaves narrowed to base ; calyx-teeth triangular, acuminate, about 1·5 mm.
long ; bracts ovate, villous   ..   ..   1. *bowalensis*
Stems above and upper leaves pubescent in the younger parts with hairs 0·25 mm. long,
glabrous below ; lower leaves slightly narrowed at base ; calyx-teeth subulate,
2–3 mm. long ; bracts broadly lanceolate, acuminate   ..   ..   2. *capitata*
Spikes elongate, usually lax and discontinuous, not head-like ; leaves lanceolate to
linear-lanceolate, the lowest often oblanceolate, petiolate :
Plant hispid with hairs about 1 mm. long ; calyx densely hispid ; corolla-tube about
5 mm. long, sparsely villous outside ; leaves often coarsely dentate   ..   3. *hispida*
Plant almost glabrous ; calyx glabrous or nearly so, with ciliate teeth ; corolla-tube
about 5 mm. long, glabrous outside ; leaves subentire   ..   4. *leptostachya*

1. **B. bowalensis** *A. Chev.* in Mém. Soc. Bot. Fr. 2, 8 : 185 (1912). Chev. Bot. 475. Slender erect herb, 4–20 in.
high ; flowers blue ; in pools on ironstone plateau ; drying black.
Mali: Banfora, Upper Volta (Sept.) *Jaeger* 5220! **Guin.** : Dalaba-Diaguissa Plateau, 3,200–4,000 ft. (Oct.)
Chev. 18867! Bouria (fl. & fr. Sept.) *Chev.* 18461! Ditinn (Sept.) *Schnell* 7404!

2. **B. capitata** *Benth.* in DC. Prod. 10 : 495 (1846) ; F.T.A. 4, 2 : 381. Erect herb 1–3 ft. high, simple or slightly
branched with bright green leaves ; flowers usually white, or pale mauve ; in wet places amongst grass ;
drying black.
Iv.C. : Alangouassou to Mbayakro, Nzi Valley (fl. & fr. Aug.) *Chev.* 22248! **Ghana** : Banda, Wenchi Dist.
(Dec.) *Morton* GC 25206! **N.Nig.** : Bida (fl. & fr. Mar.) *Meikle* 1320! Tangale (Oct.) *Lely* 659! Yola and
Kilba (fl. & fr. Jan., Aug.) *Dalz.* 116! **S.Nig.**: Akure F.R., Ondo (fl. & fr. Oct., Nov.) *Keay* FH 22663!
25470! [**Br.**]**Cam.** : Vogel Peak, 4,800 ft., Adamawa (fl. & fr. Nov.) *Hepper* 1355! Kakara, 4,900 ft.,
Mambila Plateau (fl. & fr. Jan.) *Hepper* 1769! Widespread in tropical Africa and in Madagascar.

3. **B. hispida** *Buch.-Ham. ex D. Don* Prod. Fl. Nepal. 91 (1825) ; F.T.A. 4, 2 : 397 ; Chev. Bot. 475 ; Berhaut
Fl. Sén. 69. *B. longifolia* Klotzsch (1861)—F.T.A. 4, 2 : 398 ; F.W.T.A., ed. 1, 2 : 225 ; Berhaut l.c.
Variable erect herb 1–2 ft. high ; flowers mauve ; in waste sandy places ; drying black.
Sen. : *Heudelot* 132! *Leprieur* 8! Oualia *Chev.* 42. Thiès *Chev.* 3287! **Gam.** : *Brown-Lester* 62! *Boteler*!
MacCarthy Isl. *Dalz.* 8063! Jappeni Jarra Central (Nov.) *Frith* 263! **Mali** : Sonni Koura (?) *Roberty* 487!
Port. G. : Antula (Jan.) *Esp. Santo* 1454! **Guin.** : Kouroussa *Pobéguin* 566! **Ghana** : Kayoro (Oct.) *Vigne*
FH 4611! Nangodi (fr. Dec.) *Adams & Akpabla* 4275! Deloro-Tankara (fl. & fr. Dec.) *Adams & Akpabla*
4328! Vendi (Dec.) *Adams & Akpabla* 4056! **Niger** : Daddora (fl. & fr. Dec.) *Meikle* 819! **N.Nig.** :
Kano (fl. & fr. Oct., Dec.) *Hagerup* 659! *Keay* FHI 21418! Anchu, Zaria (fl. & fr. Dec.) *Keay* FHI 21668!
Kontagora (Nov.) *Dalz.* 172! Shere Hills, Jos (fl. & fr. Dec.) *Savory* UCI 120! Jos (Oct.) *Hepper* 1044!
Maiduguri (fl. & fr. Oct.) *Noble* A15! [**Br.**]**Cam.** : Gurum, near Vogel Peak (Nov.) *Hepper* 1284! Bama,
Dikwa Div. (fl. & fr. Dec.) *McClintock* 102! Widespread in tropical Africa also in S.W. & S. Africa, Mada-
gascar and India.

4. **B. leptostachya** *Benth.* in DC. Prod. 10 : 497 (1846) ; F.T.A. 4, 2 : 394 ; Chev. Bot. 476 ; Berhaut Fl. Sén.
69. Erect slender-branched herb, stem 1–2 ft. high, from a rosette of leaves, leaf-veins sometimes tinged
purple beneath ; flowers mauve or occasionally white ; roadsides and moist places ; drying black.
Sen. : *Leprieur* 7! Badi *Berhaut* 1557. **Gam.** : *Brown-Lester* S42! **Mali** : Tabacco (fl. & fr. Jan.) *Chev.*
134! **Port.G.** : Cantubo (Nov.) *Esp. Santo* 3602! **Guin.** : *Heudelot* 613! Dalaba-Diaguissa Plateau *Chev.*
18758! 18865! Kissidougou (fr. Dec.) *Martine* 226! **S.L.** : Tower Hill, Freetown (fl. & fr. Dec.) *Deighton*
793! Bonthe (fl. & fr. Nov.) *Adames* 108! Wallia (fl. & fr. Feb.) *Sc. Elliot* 4266! Yiraia (fl. & fr. Jan.)
Glanville 481! **Iv.C.** : Mankono *Chev.* 21994! **Ghana** : Burufo (fr. Dec.) *Adams & Akpabla* 4389! Bimbila
Hepper & Morton A3102! **N.Nig.** : Nupe *Barter* 907! Kontagora (fl. & fr. Nov., Dec.) *Dalz.* 160! Hoss, Jos
(fl. & fr. Nov.) *Keay* FHI 21736! [**Br.**]**Cam.** : Cam. Mt., 5,500 ft. (Feb.) *Keay* FHI 37505! Kumbo to Oku
(Feb.) *Hepper* 2015! Nguroje, Mambila Plateau (fl. & fr. Jan.) *Hepper* 1744! Karamti, Gashaka Dist. (fl.
& fr. Dec.) *Latilo & Daramola* FHI 28938! Gurum, Vogel Peak area (Dec.) *Hepper* 1597! Also in Cameroun, Uganda, Kenya, Tanganyika and Nyasaland. (See Appendix, p. 441.)

29. RHAMPHICARPA Benth.—F.T.A. 4, 2 : 418.

Leaves pinnatisect into slender filiform segments, 1–8 cm. long ; corolla 1·5 cm. diam.
across the lobes, tube very slender but distended above, 2·5–3 cm. long, very slightly
pubescent ; calyx 6 mm. long, lobes ovate, long-acuminate-subulate ; fruits obliquely
ovoid with the erect beak to one side, compressed, dehiscing longitudinally between
flanges, 1 cm. long .. .. 1. *fistulosa*
Leaves linear to narrow-lanceolate, remotely serrate, 3–11 cm. long, 2–9 mm. broad ;
corolla 3–5 cm. diam. across the lobes, tube uniform, 2·5 cm. long, glandular-pubescent ;
calyx 1(–2) cm. long, lobes lanceolate ; fruits very obliquely ovoid with the beak
horizontal and projecting between calyx-lobes, 6 mm. long .. .. 2. *tubulosa*

1. **R. fistulosa** (*Hochst.*) *Benth.* in DC. Prod. 10 : 504 (1846) ; F.T.A. 4, 2 : 419. *Macrosiphon fistulosa* Hochst.
(1841). *Rhamphicarpa longiflora* Benth. in Hook. Comp. Bot. Mag. 1 : 368 (1836), partly (African spec.)—
Berhaut Fl. Sén. 11, not of Benth. in DC. Prod. 10 : 504 (1846). Erect much-branched glabrous herb a
few inches to 2 ft. high, half-woody below, with fibrous roots ; flowers white, rarely yellow ; in swamps and
damp places in savanna ; drying black.
   Sen. : *Perrottet* 568 ! 570 ! St. Louis Isl. *Leprieur* ! **Gam.** : Kuntaur *Ruxton* 84 ! **Mali** : Kabarah (fl. & fr.
Aug.) *Chev.* 1336 ! Bamako *Garnier* ! Kita *Jaeger* 9 ! **Port.G.** : Farim (fl. & fr. Apr.) *Esp. Santo* 2678 !
Gabu (July) *Esp. Santo* 2704 ! **Guin.** : Kouroussa (fl. & fr. Aug.) *Pobéguin* 373 ! Fouta Djalon (fl. & fr.
Sept.) *Schnell* 7064 ! **Ghana** : Mirigu (fl. & fr. Oct.) *Vigne* FH 4595 ! Kete Krachi (Dec.) *Adams* 4568 !
Burufu, near Lawra (Nov.) *Harris* ! **Niger** : Aïr *De Miré & Gillet.* **N.Nig.** : Nupe *Barter* 1713 ! Sokoto
(fl. & fr. Sept., Oct.) *Moiser* 115 ! Gindiri, Plateau (Oct.) *Hepper* 1096 ! Zaria (fl. & fr. July, Sept.)
Olorunfemi FHI 24388 ! *Keay* FHI 25914 ! Yola (fl. & fr. Sept.) *Dalz.* 120 ! L. Chad shore (fl. & fr. Apr.)
Potter ! **S.Nig.** : Ado Rock, near Olokemeji (June) *Hambler* 289 ! *Onochie* FHI 31208 ! **[Br.]Cam.** : Jada,
Adamawa (Nov.) *Hepper* 1224 ! Widespread in tropical Africa, also in S. Africa and Madagascar. (See
Appendix, p. 441.)
2. **R. tubulosa** (*Linn. f.*) *Benth.* in Hook. Comp. Bot. Mag. 1 : 368 (1836) ; F.T.A. 4, 2 : 428. *Gerardia tubulosa*
Linn. f. (1781). An erect herb 1–2 ft. high, sparingly branched ; flowers blue, pink or white, fragrant ; in
wet grassland ; drying black.
   N.Nig. : *Thornewill* 74 ! Nupe *Yates* ! Odeke, Kabba (Feb.) *Jones* FHI 5896 ! **S.Nig.** : Onitsha *Talbot* !
Also in Ethiopia, E. Africa, Congo, N. & S. Rhodesia, Nyasaland, S. Africa and Madagascar.

30. STRIGA Lour.—F.T.A. 4, 4 : 399.

Calyx 10- or more ribbed, 1 rib or more between each calyx-tooth :
   Plant glabrous ; leaves reduced to scales clasping the stem, 2–7 mm. long ; corolla-
tube about 1 cm. long, slightly curved, corolla-lobes 1–2 cm. long ; capsule-lobes
2, linear, reflexing, about 6 mm. long ; calyx 4-toothed, about 15- ribbed
                                                                    1. *baumannii*
   Plants pubescent ; capsule-lobes not reflexing at maturity :
      Flowers imbricated in a very dense spike :
         Bracts ovate, about 7 mm. long and nearly as broad, evenly and shortly glandular-
pubescent ; corolla-tube 1–1·5 cm. long, densely pubescent above, corolla more
or less 1·5 cm. diam. ; leaves linear-lanceolate, coarsely serrate, 2–12 cm. long ;
stems hispid .. .. 2. *macrantha*
         Bracts lanceolate, hispid with hairs about 1 mm. long on the margins and midrib ;
stems hispid :
            Leaves oblong-lanceolate, 3–4 cm. long, 3 mm. broad, very scabrid-setose ; corolla
about 4 mm. diam., lower lobes 2 mm. long, tube about 1 cm. long, shortly
glandular-pubescent ; bracts 8–12 mm. long, about 3 mm. broad, the hispid hairs
arising from bulbous bases .. .. 3. *klingii*
            Leaves linear, 2–3 cm. long, 3 mm. broad, hispid-setose ; corolla nearly 2 cm. diam.,
lower lobes about 1 cm. long, tube 2 cm. long, puberulous ; bracts 15 mm. long,
3 mm. broad, hair-bases inconspicuous .. .. 4. *dalzielii*
      Flowers not imbricated, forming a loose spike :
         Calyx-lobes about 5 mm. long in flower, becoming about 10 mm. long in fruit ; bracts
leafy each subtending 1 flower, opposite ; flowers 1–2 cm. diam., corolla-tube
2–2·5 cm. long, sparsely pubescent .. .. 5. *forbesii*
         Calyx-lobes at most 2 mm. long ; bracts neither leafy nor opposite :
            Calyx-lobes triangular ; calyx rather obscurely about 15-ribbed, pubescent ;
corolla-tube about 2 cm. long, glandular-pubescent, corolla about 1·5 cm. diam. ;
bracts ovate-lanceolate, 2–5 mm. long .. .. 6. *primuloides*
            Calyx-lobes subulate ; calyx distinctly 10-ribbed, usually hispid ; corolla-tube up
to 1 cm. long, minutely pubescent, corolla 5–10 mm. diam. ; bracts linear-lanceolate, about 5 mm. long .. .. 7. *asiatica*
Calyx 4–5-ribbed, each rib ending in a tooth :
   Indumentum closely appressed to the stems, hairs directed upwards ; leaves subulate,
appressed, 5(–10) mm. long ; bracts similar to the leaves ; corolla tomentose-
pubescent, the tube about 1 cm. long, the lobes 2–3 mm. long .. 8. *linearifolia*
   Indumentum of stems more or less erect or spreading, scabrid-hispid or plant glabrous:
      *Stems scabrid-hispid or conspicuously pubescent :
         Corolla-tube conspicuously pubescent or tomentose, often glandular ; flowers usually
alternate :
            Bracts acuminate, obovate to oblanceolate, 3–4 mm. broad, 7–9 mm. long, ciliate,

the lower ones linear, about 8 mm. long, spikes compact, 2·5–3·5 cm. long ; leaves linear, spreading, 4–6 cm. long, basal ones about 1 cm. long and erect, scabrid

9a. *bilabiata* subsp. *jaegeri*

Bracts acute, lanceolate to linear-lanceolate, ciliate :
Stems square, sharply angular with decurrent leaf-bases, branching above ; inflorescences usually only about ½ as long as the rest of the stem ; bracts linear-lanceolate, about 1 cm. long, not imbricate ; leaves linear to linear-lanceolate, 1–5·5 cm. long, usually spreading. .   9b. *bilabiata* subsp. *rowlandii*
Stems terete or obtusely square, branching below ; inflorescences usually about twice as long as the rest of the stem, lax ; bracts linear, 1–1·5(–3) cm. long ; leaves linear, 1–3 cm. long, few   . .   . .   9c. *bilabiata* subsp. *barteri*
Corolla-tube nearly glabrous or very shortly pubescent ; flowers alternate ; stems not sharply angular :
Bracts linear to subulate :
Calyx-tube 1·5–2 mm. long, teeth 1 mm. long, up to 2 mm. in fruit ; mature fruits 2–3 mm. long ; leaves 1–2 cm. long, scabrid . .   . .   . .   10. *brachycalyx*
Calyx-tube 3–8 mm. long ; leaves linear, 1–5 cm. long, scabrid :
Bracts mostly shorter than calyx, about 3 mm. long (lower ones sometimes longer) ; calyx-tube 3–4 mm. long, teeth 1–2 mm. long ; mature fruits about 4 mm. long   . .   . .   11. *aspera*
Bracts leafy, much longer than calyx, up to 3 cm. long, only uppermost about same length ; calyx-tube 6–8 mm. long, teeth 2 mm. long, spreading and lengthening to up to 5 mm. in fruit ; mature fruits 5–6 mm. long

12. *passargei*

Bracts lanceolate to lanceolate-elliptic, markedly ciliate on margin and sometimes on midrib outside, about 1 cm. long and 2(–3) mm. broad ; leaves linear or slightly elliptic-linear, 3–9 cm. long, 3–7 mm. broad, scabrid ; corolla often very large. .   . .   .   13. *hermontheca*
*Stems glabrous or very slightly pubescent, not scabrid ; flowers opposite or sub-opposite:
Leaves subulate to linear, 1–2(–3) cm. long ; corolla-tube densely pubescent

14. *aequinoctialis*

Leaves ovate, scale-like from a broad base, few, about 5 mm. long ; corolla-tube glabrous   . .   . .   15. *gesnerioides*

1. **S. baumannii** *Engl.* Bot. Jahrb. 23 : 515, t. 12 figs. O–T (1897) ; F.T.A. 4, 2 : 414. Tall glaucous herb 2 ft. or more high ; flowers dull-yellow, in long slender spikes ; roots subsucculent from a fairly stout stock ; parasitic on grass roots in moist grassland ; drying grey.
Mali: Oualia, Kita (Dec.) *Chev.* 102 ! **S.L.:** *Sc. Elliot* 5085 ! Kumrabai (fl. & fr. Jan.) *Glanville* 148 ! Brookfields (fl. & fr. Aug.) *Jordan* 784 ! Waterloo (fr. Aug.) *Melville & Hooker* 329 ! **Ghana:** Amedzofe, 2,600 ft. (fl. & fr. Mar.) *Hepper & Morton* 2313 ! Togo Rep.: Quamikrum (Mar.) *Schlechter* (fl. & fr. Mar.) 12954 ! Misahöhe *Baumann* 29. **N.Nig.:** Abinsi (fl. & fr. Mar.) *Dalz.* 709 ! **S.Nig.:** Ago-Are F.R., Oyo (Mar.) *Keay* FHI 37556 ! [Br.]Cam.: Bum-Nchan, 5,000 ft., Bamenda (June) *Maitland* 1636 ! Vogel Peak, 5,200 ft. (fr. Nov.) *Hepper* 1497 ! Also in Cameroun, Kenya and (?) Congo.
2. **S. macrantha** (*Benth.*) *Benth.* in DC. Prod. 10 : 503 (1846) ; F.T.A. 4, 2 : 413 ; Chev. Bot. 477, partly ; Berhaut Fl. Sén. 68, 83. *Buchnera macrantha* Benth. (1836). *B. buettneri* Engl. (1894)—Chev. Bot. 475. Robust herb 2–5 ft. sparingly branched above ; flowers white, tube green ; amongst tall grass ; drying dark brown.
Sen.: *Berhaut* 650 ! **Gam.:** *Brooks* 4 ! *Brown-Lester* 834 ! **Mali:** *Roberty* ! Nafadié to Ouacoro *Chev.* 3275. **Port.G.:** Pussubé, Bissau (fl. & fr. Jan.) *Esp. Santo* 1464 ! Safim, Bissau (fr. Jan.) *Esp. Santo* 1699 ! **Guin.:** *Heudelot* 607 ! *Farmar* 246a ! Farana *Chev.* 20402 ! 20440 ! **S.L.:** Regent (fl. & fr. Oct.) *Deighton* 220 ! Bumban to Port Loko (fl. & fr. Apr.) *Sc. Elliot* 5677 ! Musaia (Dec.) *Deighton* 4494 ! Bintumane Peak, 4,500 ft. (fl. & fr. Jan.) *Nichols* 11 ! **Iv.C.:** Ano, Tchumkrou to Akakoumoékrou *Chev.* 22543 ! **Ghana:** Winneba Plains (Oct.) *Darko* 1026 ! Attabubu (Dec.) *Morton* GC 6183 ! Salaga *Krause* ! Dutukpene (Mar.) *Hepper & Morton* 2355 ! Techiman (fl. & fr. Dec.) *Adams & Akpabla* 4509 ! **Togo Rep.:** Bismarck-burg *Kling* 179. **N.Nig.:** Oke-Opim, Ilorin (Dec.) *Ajayi* FHI 19271 ! Nupe *Barter* 985 ! Kurmin Damisa, Jemaa Div. (Nov.) *Keay* FHI 21541 ! Jos Plateau (Dec.) *Lely* P785 ! **S.Nig.:** Lagos (fl. & fr. Mar.) *Millen* ! Abeokuta (Oct.) *Baldwin* 13639 ! Ibadan (Dec.) *Newberry* 6 ! Jesse, Wari Prov. (fl. & fr. Sept.) *Butler-Cole* 4 ! [Br.]Cam.: Vogel Peak, 3,400 ft., Adamawa (Nov.) *Hepper* 1392 ! Also in Cameroun and Angola.
3. **S. klingii** (*Engl.*) *Skan* in F.T.A. 4, 2 : 413 (1906). *Buchnera klingii* Engl. (1893). *S. macrantha* of Chev. Bot. 477, partly (Chev. 18403, 18612, 20176), not of (Benth.) Benth. Erect, little-branched herb 1 ft. high, with harshly hairy stem ; flowers pink or red with white throat ; in wet places or amongst grass ; drying brown.
Port.G.: Gabu (Oct.) *Esp. Santo* 3447 ! Farim (Nov.) *Esp. Santo* 3622 ! S. Domingo to Suzana (Nov.) *Esp. Santo* 3668 ! Bafatá (Dec.) *Esp. Santo* 3754 ! 3808 ! **Guin.:** Mamou (Sept.) *Schnell* 6785 ! Fon Massif, Beyla (Aug.) *Schnell* 6626 ! Fouta Djalon *Chev.* 18403 ! 18612 ! 20176 ! **Ghana:** Kete Krachi (Dec.) *Morton* GC 6292 ! **Togo Rep.:** Assumafaru *Büttner* 293. Ketschen Timoy *Büttner* 221. **N.Nig.:** Borgu *Barter* 758 ! Rinji (Nov., Dec.) *Lely* 680 ! [Br.]Cam.: Vogel Peak, 3,600 ft., Adamawa (Nov.) *Hepper* 1397 ! Also in Ubangi-Shari.
4. **S. dalzielii** *Hutch.* in F.W.T.A., ed. 1, 2 : 226 (1931). Similar to the last ; in wet places ; drying brown.
N.Nig.: Zungeru (Sept.) *Dalz.* 168 !
5. **S. forbesii** *Benth.* in Hook. Comp. Bot. Mag. 1 : 364 (1836) ; F.T.A. 4, 2 : 410 ; Berhaut Fl. Sén. 83. Erect, simple or little-branched scabrid herb a few inches to 2 ft. high ; flowers pink, or sometimes scarlet or yellow, subtended by leafy bracts ; in wet places ; drying greenish-brown.
Sen.: *Leprieur* 5 ! *Perrottet* 566 ! **Gam.:** Kuntaur *Ruxton* 67 ! 149 ! **Mali:** Mopti (June) *Lean* 37 ! **Port.G.:** Farim to Bigene (fl. & fr. Aug.) *Esp. Santo* 3063 ! **Guin.:** Kouroussa (Dec.) *Pobéguin* 1018 ! **Ghana:** Navrongo (Aug.) *Vigne* FH 4580 ! **N.Nig.:** Zungeru (Aug.) *Dalz.* 167 ! Neill's Valley, Naraguta (June) *Lely* 255 ! Vom, 3,000–4,500 ft. *Dent Young* 184 ! Yola (June) *Dalz.* 109 ! Ruma, Katsina (fl. & fr. Aug.) *Onwudinjoh* FHI 24005 ! **S.Nig.:** Oban *Talbot* 308 ! [Br.]Cam.: Mambila Plateau (July) *Chapman* 64 ! Widespread in tropical Africa and in Madagascar and S. Africa.
6. **S. primuloides** *A. Chev.* in Mém. Soc. Bot. Fr. 2, 8 : 185 (1912) ; Chev. Bot. 478. Erect leafless parasite about 1 ft. high ; flowers primrose-yellow ; drying brown.

Iv.C.: Anoumaba (fl. & fr. Nov.) *Chev.* 22404! Mt. Dourou, 1,800–2,500 ft., Upper Sassandra *Chev.* 21714! Thiassalé (Aug.) *Pobéguin* 120! **S.Nig.:** Awkaka (Jan.) *Birkett-Smith in Hb. Hambler* 49!

7. **S. asiatica** (*Linn.*) *O. Ktze.* Rev. Gen. Pl. 2 : 466 (1891) ; Berhaut Fl. Sén. 83 (incl. vars.). *Buchnera asiatica* Linn. (1753). *Striga lutea* Lour. (1790)—F.T.A. 4, 2 : 409 ; Chev. Bot. 477. *S. hirsuta* Benth. (1846)—Chev. Bot. 477. Extremely variable slender erect herb, scabrid-pubescent or almost glabrous, a few inches to 1 ft. high ; flowers usually brilliant scarlet inside, yellow outside, sometimes all red, yellow or white ; in moist shallow sandy soil overlying rock amongst short grass ; drying greenish.
Sen.: *Berhaut* 555. Ouassadou *Berhaut* 1673. **Port.G.:** Bafatá (Aug.) *Esp. Santo* 2740! 3329! **Guin.:** Macenta (fl. & fr. Oct.) *Baldwin* 9787! **S.L.:** *Morson*! Waterloo (fl. & fr. Aug.) *Deighton* SLH 2042! Rokupr (Nov.) *Jordan* 151! Kasewe Hills (fl. & fr. June) *Deighton* 5841! Konjo-ngiehum, Bagbe (July) *Deighton* 5812! **Lib.:** Harper (fl. & fr. June) *Baldwin* 5988! **Iv.C.:** Sabodougou, Touba (fl. & fr. July) *Collenette* 59! **Ghana:** Aburi *Anderson* 53! Birifu, Lawra (fl. & fr. Aug.) *Thorold* 213! Damongo (fl. & fr. July) *Andoh* FH 5202! 5206! Banda (fl. & fr. Dec.) *Adams & Akpabla* 4021! **N.Nig.:** Nupe *Barter* 538! Anara F.R., Zaria (fl. & fr. Sept.) *Olorunfemi* FHI 24389! Vom, 3,000–4,500 ft. *Dent Young* 184a! Naraguta (fl. & fr. June, Aug.) *Lely* 254! 525! Yola (fl. & fr. Aug., Sept.) *Dalz.* 111! 113! **S.Nig.:** Apapa, Lagos (fl. & fr. Feb.) *Dalz.* 1390! Calabar *Robb*! Ogoja Dist. *Rosevear* 44/29! 45/29! [**Br.**]**Cam.:** Bum, 4,000 ft., Bamenda (May) *Maitland* 1599! Vogel Peak, Adamawa (Nov.) *Hepper* 2745! Widespread in tropical Africa and Asia, S. Africa and the Mascarenes.

8. **S. linearifolia** (*Schum. & Thonn.*) *Hepper* in Kew Bull. 14 : 416 (1960). *Buchnera linearifolia* Schum. & Thonn. (1827)—F.T.A. 4, 2 : 415. *Striga canescens* Engl. (1895)—F.T.A. 4, 2 : 406 ; Chev. Bot. 477. *S. strictissima* Skan in F.T.A. 4, 2 : 407 (1906) ; F.W.T.A., ed. 1, 2 : 227. Erect herb, simple or little-branched from near the base, stems ribbed and square, 1–2 ft. high ; flowers pale pink, pale lilac or white in rather dense spikes 3–12 ins. long ; moist ground in savanna grassland ; drying grey.
Mali: Macina *Roberty* 2676! Bamako *Waterlot* 1240! **Guin.:** *Pobéguin* 364! Kouroussa *Pobéguin* 389! **Iv.C.:** Dotou *Chev.* 21775! **Ghana:** Accra (Jan.) *Dalz.* 145! Nangodi (fr. July) *Williams* 534! Vigne FH 3355! Busa (fl. & fr. Aug.) *Thorold* 209! Amedzofe (May) *Scholes* 59! *Hepper* 2314! Dutukpene (Mar.) *Hepper & Morton* A3050! **N.Nig.:** Nupe *Barter* 1263! Lokoja (Apr.) *Parsons* L117! Maska, Katsina Prov. (fl. & fr. June) *Keay* FHI 25896! Lemme, 2,600 ft. (May) *Lely* 146! **S.Nig.:** Ibadan F.R. (May) *Keay* FHI 22864! Also in Ethiopia, East Africa and Angola.

9. **S. bilabiata** (*Thunb.*) *O. Ktze.* Rev. Gen. Pl. 3, 2 : 240 (1898) ; Hepper in Kew Bull. 14 : 411 (1960). *Buchnera bilabiata* Thunb. (1800). *Striga thunbergii* Benth. (1836)—F.T.A. 4, 2 : 404.

9a. **S. bilabiata** subsp. **jaegeri** *Hepper* l.c. 415 (1960). *S. welwitschii* Engl. var. *longifolia* Berhaut Fl. Sén. 84 (1954), French descr. only. Erect simple perennial herb about 7 in. high ; in scrub on ironstone plateau and laterite hills ; drying greenish.
Sen.: Ouassadou (fl. & fr. Sept.) *Berhaut* 1681! **Mali:** Kita Massif (July, Sept.) *Jaeger* K2! 2541!

9b. **S. bilabiata** subsp. **rowlandii** (*Engl.*) *Hepper* l.c. 415 (1960). *S. rowlandii* Engl. (1897)—F.T.A. 4, 2 : 405 ; F.W.T.A., ed. 1, 2 : 227 ; Berhaut Fl. Sén. 84. *S.* sp. of Berhaut l.c. Herb about 1 ft. or more high with mauve, pink or white flowers ; in burnt savanna grassland ; drying greenish.
Sen.: Niokolo-Koba (fl. & fr. Jan.) *Berhaut* 1630! Jouat (Mar.) *Perrottet* 649! **Gam.:** *Hayes* 539! **Mali:** Dougoura (fl. & fr. Feb.) *Chev.* 336! Pessubé, Bissau (Mar.) *Esp. Santo* 1504! **Guin.:** *Pobéguin* 364! **S.L.:** Mange *Glanville* 241! Falaba (fl. & fr. Mar.) *Sc. Elliot* 5207! Rhombe (Mar.) *Adames* 146! Robis (Jan.) *Jordan* 190! **Ghana:** Sesiamang (fl. & fr. Feb.) *A. S. Thomas* D138! Afram Plains (fl. & fr. Mar.) *Johnson* 709! Lumbuga (May) *Morton* GC 8795! Yendi (Mar.) *Hepper & Morton* A3110! **N.Nig.:** Nupe *Barter* 1169! Liruwen-Kano Hills *Carpenter*! Zaria (Mar.) *Dalz.* 407! *Milne-Redhead* 5043! Mando F.R., Birnin Gwari (June) *Keay* FHI 25833! Vom, 3,000–4,500 ft. *Dent Young* 185! **S.Nig.:** *Rowland*! Upper Ogun F.R., Oyo (Feb.) *Keay* FHI 23442!

9c. **S. bilabiata** subsp. **barteri** (*Engl.*) *Hepper* l.c. 414 (1960). *S. barteri* Engl. (1897)—F.T.A. 4, 2 : 406 ; F.W.T.A., ed. 1, 2 : 227 ; Berhaut Fl. Sén. 84 ; F.W. Andr. Fl. Pl. Sud. 3 : 143. *S. glandulifera* Engl. (1897)—F.T.A. 4, 2 : 406. Perennial herb 6–8 ins. high ; flowers pinkish or violet ; in savanna ; drying dark brown or greenish.
Sen.: *fide* Berhaut *l.c.* **Port.G.:** Contabane to Guilege (June) *Esp. Santo* 3199! **Guin.:** Mali (Jan.) *Roberty* 16608! Kouroussa (Mar.) *Pobéguin* 208! **N.Nig.:** Nupe *Barter* 1170! Jebba (Jan.) *Meikle* 1011! Alagbe F.R., Ilorin (Feb.) *Ejiofor* FHI 19822! Also in Ubangi-Shari, Sudan and Uganda.

10. **S. brachycalyx** *Skan* in F.T.A. 4, 2 : 403 (1906). *S. varneckei* Engl. ex Skan in F.T.A. 4, 2 : 414 (1906). Erect usually diffusely branched perennial herb, 1–1½ ft. high, scabrid ; flowers purple ; weed of cultivation, drying brown.
Ghana: Bole (Apr.) *Morton* A3295! Wa (fl. & fr. Nov.) *Harris*! Tili, Kusai Dist. (July) *Vigne* FH 4534! **Togo Rep.:** Lomé *Warnecke* 201! Basari to Heppe *Kersting* A665! **N.Nig.:** Jebba *Barter*! Jos Plateau (Aug.) *Lely* P567! Samaru, Zaria (Oct.) *Harris* 4! Abinsi (fl. & fr. Nov.) *Dalz.* 790! **S.Nig.:** Enugu Extension F.R. (fl. & fr. Sept.) *Onochie* FHI 34106! Oban *Talbot* 1033! Also in Cameroun.

11. **S. aspera** (*Willd.*) *Benth.* in Hook. Comp. Bot. Mag. 1 : 362 (1836) ; F.T.A. 4, 2 : 403, incl. var. *filiformis* Benth. (1846) ; Chev. Bot. 476 ; Berhaut Fl. Sén. 84. *Euphrasia aspera* Willd. (1801). Slender erect slightly hispid herb, 6–18 in. high ; flowers pink or purple, sometimes white ; weed of waste places and cultivation in savanna ; drying blackish.
Sen.: *Heudelot* 165! *Roger*! **Gam.:** *Mungo Park*! Kuntaur *Ruxton* 65! 152! **Mali:** Dogo *Davey* 039! Ségou *Chev.* 3282! Koulikoro *Chev.* 3284! **Guin.:** Kouroussa (fl. & fr. Aug.) *Pobéguin* 371 (partly)! Kankan (May) *Pobéguin* 1020! **Iv.C.:** Gampéla *Chev.* 24616. Ouadougou to Ouahigonya *Chev.* 24718. **Ghana:** *Williams* 548! Cape Coast *Brass*! Zuarungu (Sept., Dec.) *Adams & Akpabla* 4272! Jiripa (Sept.) *Thorold* 252! **Niger:** Diapaga *Chev.* 24415. **N.Nig.:** Jebba *Barter*! Zaria (June) *Keay* FHI 25911! Kontagora (Nov.) *Dalz.* 170! Vom *Dent Young* 186! Biu, Bornu (Aug.) *Noble* 1! **S.Nig.:** Onitsha (fl. & fr. Feb.) *Jones* FHI 4557! [**Br.**]**Cam.:** Gulumba, Dikwa Div. *McClintock* 42! Also in Sudan.

12. **S. passargei** *Engl.* Bot. Jahrb. 23 : 515, t. 12, figs. M, N (1897) ; Chev. Bot. 478, partly ; Berhaut Fl. Sén. 83. *S. aspera* of F.T.A. 4, 2 : 403, in note ; F.W.T.A., ed. 1, 2 : 226, partly (*Dalz.* 110, 355 ; *Lely* 532). Erect slightly scabrid herb, 4–18 in. high ; flowers creamy-white to pale yellow with pink centre, rarely wholly pale pink (*Lely* 532) ; in rocky places ; drying black.
Sen.: Ouassadou (fl. & fr. Sept.) *Berhaut* 1666! **Ghana:** Jiripa (fl. & fr. Sept.) *Thorold* 251! **N.Nig.:** Sokoto (fl. & fr. July) *Dalz.* 355! Kakangi F.R., Zaria (fl. & fr. July) *Keay* FHI 25936! Kufena Rock, Zaria (fl. & fr. July) *Keay* FHI 25977! Naraguta (fl. & fr. Aug.) *Lely* 532! Yola (fl. & fr. July) *Dalz.* 110! *Passarge* 48. Also in Sudan and Tanganyika.

13. **S. hermontheca** (*Del.*) *Benth.* l.c. 365 (1836) ; F.T.A. 4, 2 : 407. *Buchnera hermontheca* Del. (1813). *S. senegalensis* Benth. (1846)—F.T.A. 4, 2 : 408 ; F.W.T.A., ed. 1, 2 : 226. An erect semi-parasitic herb about 1 ft. high ; flowers bright pink ; usually parasiting *Sorghum* roots.
Sen.: *Heudelot*! **Gam.:** *Ingram*! *Saunders* 60! 62! **Mali:** Ségou *Chev.* 3281! Sompi *Chev.* 3286! Bamako (Nov.) *Waterlot* 1451! **Port.G.:** Bafatá (Oct.) *Esp. Santo* 2796! Gabu (Oct.) *Esp. Santo* 3520! **Ghana:** *Saunders* 7! Burufo (Dec.) *Adams* 4419! Paliba (fl. & fr. Dec.) *Adams & Akpabla* 4107! Birifu, near Lawra (Aug.) *Thorold* 254! **Niger:** Niamey (Oct.) *Hagerup* 478! Dosso (Oct.) *Monod* 7138! **N.Nig.:** Nupe *Barter* 866! Kontagora (fl. & fr. Oct.) *Dalz.* 171! Onisaja, Ilorin (fl. & fr. Jan.) *Ejiofor* FHI 30811! 30240! Naraguta (Oct.) *Hepper* 1162! Biu, Bornu (fl. & fr. Sept.) *Noble* 30! Lassa, Chad (Oct.) *Royer* 136! **S.Nig.:** Stirling *T. Vogel* 130! Aboh *Barter* 136! Ogoja (fl. & fr. Dec.) *Talbot*! [**Br.**]**Cam.:** R. Donga, Mambila Dist. (Feb.) *Hepper* 1847! Gangumi, Adamawa Prov. (fl. & fr. Dec.) *Latilo & Daramola* FHI 28792! Gurum, Vogel Peak area (Nov.) *Hepper* 1278! Gwoza, Dikwa Div. (fl. & fr. Dec.) *McClintock* 74! Widespread in the drier parts of tropical Africa and in Madagascar.

Fig. 290.—Striga hermontheca (*Del.*) *Benth.* (Scrophulariaceae).
A, habit. B, corolla opened out. C, pistil. D, upper bract.

[Extreme variability in corolla size and in inflorescence compactness occurs in this species. A specimen from Biu, N. Nigeria (*Noble* 31) has a congested, imbricated inflorescence and flowers with corolla-lobes about 2 mm. long, whilst *Davey* FHI 27126 from Bornu, N. Nigeria, has a laxer inflorescence with corolla-lobes about 13 mm. long : there is no satisfactory point of separation in the range.—F. N. H.]

14. **S. aequinoctialis** *A. Chev. ex Hutch. & Dalz.* F.W.T.A., ed. 1, 2 : 227 (1931) ; Chev. Bot. 476. Erect, slender, simple or little-branched herb, 6–18 ins. high ; flowers purple to almost white ; in moist montane grass-land ; drying black.
Guin.: Mt. Nzo, 5,300 ft. *Chev.* 21035. Siredougou, 2,500 ft. (Apr.) *Collenette* 3 ! Maneah (fl. & fr. July) *Pobéguin* 782 ! Nimba Mts. (Apr.) *Schnell* 930 ! **S.L.:** Freetown (Aug.) *Burbridge* 510 ! Bumbuna to Farangbaia (Sept.) *Deighton* 5146 ! **Iv.C.:** Nimba Mts., 2,500–5,000 ft. (Aug.) *Boughey* GC 18052 ! 18067b ! 18076 ! 18095b !

15. **S. gesnerioides** (*Willd.*) *Vatke* in Oest. Bot. Zeitschr. 25 : 11 (1875). *Buchnera gesnerioides* Willd. (1801). *S. orobanchoides* Benth. (1836)—F.T.A. 4, 2 : 402 ; Chev. Bot. 477. Stiff erect perennial herb with usually rather stout stems a few inches to 1 ft. or more high ; flowers pink or purple or creamy white ; in waste places and cultivated land ; drying black.
Maur.: Dendari to Bou Zériba (Oct.) *Monod* 4007 ! **Sen.:** *Sieber* 52 ! *Heudelot* 540 ! *Leprieur* 6 ! *Farmar* 131 ! **Gam.:** Kuntaur *Ruxton* 53 ! **Mali:** Ségou (Sept.) *Chev.* 3280 ! **S.L.:** Buyabuya, Scarcies (fl. & fr. Feb.) *Sc. Elliot* 4279 ! **Ghana:** Zuarungu (fl. & fr. Sept.) *Thorold* 210 ! Manga (fl. & fr. Sept.) *Thorold* 248 ! **Togo Rep.:** Lomé *Warnecke* 189 ! **Dah.:** Cotonou (Nov.) *Le Testu* 229 ! **Niger:** Niamey (fl. & fr. Oct.) *Hagerup* 479 ! **N.Nig.:** Sokoto (fl. & fr. Aug., Nov.) *Dalz.* 357 ! *Moiser* 63 ! Naraguta (fl. & fr. Aug.) *Lely* 489 ! Kilba Hills and Yola (fl. & fr. Aug.) *Dalz.* 300 ! **S.Nig.:** Addarundu *T. Vogel* ! Ado Rock, Oyo (fl. & fr. Oct.) *Keay & Savory* FHI 25446 ! Also in Cape Verde Islands, throughout tropical and S. Africa and through Arabia to India. (See Appendix, p. 442.)

31. CYCNIUM E. Mey. ex Benth.—F.T.A. 4, 2 : 430.

Leaves narrowed to the base, often slightly petiolate, oblong-oblanceolate, serrulate, 4–8 cm. long, 1·5–3 cm. broad, scabrid ; flowers supra-axillary ; pedicels about 1·5 cm. long, glandular-pubescent ; calyx 1·5 cm. long, lobes unequal, ovate ; corolla-tube 4·5 cm. long, pubescent outside ; limb about 3 cm. diam. .. 1. *camporum*

Leaves sessile, ovate, rounded at the base, 3–4·5 cm. long, 1–2·5 cm. broad, densely glandular-scabrid, irregularly serrulate ; flowers slightly supra-axillary ; pedicels 2 cm. long, with two leafy bracts at the top ; calyx 2 cm. long, glandular-tomentellous, lobes ovate ; corolla-tube 3·5–4 cm. long, unilaterally constricted above the middle, glandular-pilose, limb 3–4 cm. diam. .. .. 2. *petunioides*

1. **C. camporum** *Engl.* Bot. Jahrb. 18 : 73 (1893) ; F.T.A. 4, 2 : 432 ; Chev. Bot. 479. Much-branched perennial herb from a woody base, 1–3 ft. high, pilose and scabrid ; flowers pure white, occasionally with pink spots or outside mauve fading to indigo or black ; amongst grass ; drying bluish-brown.
Mali : Penia, Sonfarasso (May) *Chev.* 817 ! **Guin. :** Siredougou, 2,000 ft. (Apr.) *Collenette* 4 ! 5 ! Erimakuna

(Mar.) *Sc. Elliot* 5729 ! **S.L.** : *Deighton* 2002 ! Lengekoro to Kabala (Apr.) *Glanville* 200 ! **Ghana** : Kpong (Feb.) *Johnson* 962 ! Kintampo (Mar.) *Dalz.* 29 ! Shai Plain *Irvine* 946 ! Sesiamang (Feb.) *A. S. Thomas* D136 ! Gambaga (May) *Vigne* FH 3855 ! **Togo Rep.** : Misahöhe *Baumann* ! Gbiu (fr. Mar.) *Schlechter* 12949 ! **N.Nig.** : Sokoto (May) *Dalz.* 361 ! Anara F.R., Zaria (May) *Keay* FHI 22923 ! Vom, Jos Plateau *Dent Young* 183 ! Fusa, Jarawa Dist., 3,200 ft. (June) *Summerhayes* 60 ! **S.Nig.** : Ogbomosho *Barter* 3398 ! Obba *Foster* 187 ! Ado to Iseyin *Rowland* ! Owo to Benin (Mar.) *Richards* 5351 ! **[Br.]Cam.** : Bamenda Prov. : Bafut-Ngemba F.R., 5,000–6,000 ft. (Jan., Mar.) *Tiku* FHI 22248 ! Onochie 34847 ! *Hepper* 2858 ! Bali-Ngemba F.R. (fr. May) *Ujor* FHI 30330 !, Bamenda, 5,000 ft. (Jan., Apr.) *Maitland* 1541 ! *Migeod* 309 ! Kumbo (Apr.) *Ejiofor* FHI 30080 ! Also in Cameroun, Congo, Cabinda, Sudan, Uganda and Kenya. (See Appendix, p. 441.)

2. **C. petunioides** *Hutch.* in Kew Bull. 1921 : 251. Erect perennial herb 9–18 ins. high ; flowers white, fading black ; in grassland ; drying green to brown.
N.Nig. : Jos Plateau (Jan.) *Lely* P135 ! Naraguta (Dec.) *Kennedy* 2850 ! *Lely* 17 ! Vom, 4,000 ft. (Feb.) *McClintock* 214 ! Jos (Mar.) *Hill* 17 ! **[Br.]Cam.** : Ndop, Bamenda (Jan.) *Keay* FHI 28427 ! Gembu, Mambila Plateau (Jan.) *Hepper* 1820 !

154. OROBANCHACEAE

By F. N. Hepper

Herbs, parasitic on roots, often covered with scales at the base, never green ; stems with alternate, often crowded, scales. Flowers solitary in the axils of bracts, often crowded, hermaphrodite, zygomorphic. Calyx 4–5-toothed or lobed or variously split, lobes open or valvate. Corolla gamopetalous, often curved ; limb oblique or 2-lipped, lobes 5, imbricate, the adaxial 2 interior. Stamens 4, didynamous, inserted below the middle of the corolla-tube, alternate with the lobes, the fifth (adaxial) one reduced to a staminode or absent; anthers often connivent in pairs, opening lengthwise. Ovary superior, 1-celled, with 4 parietal placentas; style terminal; ovules numerous. Capsule often enveloped by the calyx, opening by 2 valves. Seeds very numerous, small, with fleshy endosperm and minute embryo.

A small family recognised by the parasitic habit, absence of leaves, and the zygomorphic flowers usually crowded into spikes.

Calyx-lobes rounded ; corolla 2·5–5 cm. long ; stems fleshy   ..   1. **Cistanche**
Calyx-lobes acute ; corolla about 1 cm. long ; stems not fleshy ..   2. **Orobanche**

1. CISTANCHE Hoffmgg. & Link—F.T.A. 4, 2 : 463.

A parasite, destitute of chlorophyll ; stem stout, fleshy, glabrous ; leaves reduced to fleshy scales, lurid-purple or yellowish, lower crowded, triangular, caudate-acuminate to acute, upper lanceolate, 1–2·5 cm. long, with thinner margins ; spikes up to 30 cm. long, dense ; bracts like the leaf-scales ; calyx 5-lobed to about the middle ; corolla-tube 2·5–5 cm. long, at length curved, wider in the upper half, villous in the lower part ; lobes broader than long ; anthers woolly ..   .. *phelypaea*

C. phelypaea (*Linn.*) *Cout.* Fl. Port. 571 (1913) ; Graham in F.T.E.A. Orobanchac. 2 (q.v. for discussion) ; F.W. Andr. Fl. Pl. Sudan 3 : 150, fig. 40 ; Berhaut Fl. Sén. 5. *Lathraea phelypaea* Linn. (1753). *Cistanche lutea* (Desf.) Hoffmgg. & Link Fl. Port. 1 : 319, t. 63 (1809) ; F.T.A. 4, 2 : 463 ; Chev. Bot. 480. *Phelipaea lutea* Desf. (1798). Stem 6–18 in. high, swollen at base ; flowers bright yellow ; parasite on plants in maritime and sandy inland habitats.
Sen. : *Berhaut* 999. *Leprieur* ! Dakar (Dec., Jan.) *Chev.* 3317 ! *Debeaux* ! St. Louis (June) *Roger* ! Kayar (July) *Kesby* 17 ! **Mali** : Goundam, parasitic on *Salvadora persica*, Chev. 3316 ! Gao de Wailly 5396 ! **Dah.** : Haut Dahomey *Gironcourt* 76 ! Also in Cape Verde Isl., Spain, along southern shore of Mediterranean to Egypt, Sudan, Arabia and possibly beyond.
[G. Beck in Engl. Pflanzenr. 4, 261, Orobanchac. 33, 35 (1930), records the species *C. senegalensis* (Reuter) G. Beck (*Phelipaea senegalensis* Reuter (1847)) ; *C. hesperugo* (B. Webb) G. Beck (*Phelipaea hesperugo* B. Webb (1849)) and *C. brunneri* (B. Webb) G. Beck (*Phelipaea brunneri* B. Webb (1849)) from Senegal. These may all be *C. phelypaea* (Linn.) Cout., but available material is insufficient to be sure.—F. N. H.]

2. OROBANCHE Linn.—F.T.A. 4, 2 : 464.

A parasitic herb, without chlorophyll ; stems simple or slightly branched, not fleshy ; leaves reduced to a few small ovate scales, up to 1 cm. long ; spikes elongated or short and congested ; bracts subtending each sessile flower, lanceolate to ovate-lanceolate as long as the calyx ; bracteoles narrower and nearly as long as the bracts ; calyx 4-lobed, each lobe acutely deltoid ; corolla hairy (or glabrous in W. Afr. specimen)
                                                                                                                    ramosa

O. ramosa *Linn.* Sp. Pl. 633 (1753) ; F.T.A. 4, 2 : 465 ; G. Beck in Engl. Pflanzenr. 4, 261 Orobanchac. 66 (1930) ; Graham in F.T.E.A. Orobanchac. 3, fig. 1, 1–2. A herb with pale yellow stems about 1 ft. high ; flowers mauve-blue or pale yellow with a blue lip ; parasitic on a wide range of hosts.
Mali : Timbuktu *Hagerup* 200a ! Widespread in Europe ; in Africa and America probably introduced.

155. LENTIBULARIACEAE

By P. Taylor

Annual or perennial herbs, terrestrial, epiphytic or aquatic, with much modified parts for the capture of small organisms. Leaves entire or much divided, often bearing traps. Inflorescence racemose, bracteate, those of some aquatic species supported on the water by a whorl of modified spongy leaves (floats). Flowers zygomorphic. Calyx-lobes 2, 4 or 5. Corolla 2-lipped, spurred or saccate. Stamens 2, attached at the base of the corolla, filaments often winged ; anthers 2-celled, dorsifixed, the cells sometimes confluent. Ovary superior ; style usually very short, persistent ; stigma unequally 2-lipped. Ovules 2—numerous on a free central placenta. Fruit a capsule, usually globose opening by valves or pores, circumscissile or indehiscent. Seeds without endosperm, sometimes winged.

A small and very remarkable family of great morphological interest. Habitat usually aquatic or damp places; flowers mostly small or very small but usually brightly coloured.

Calyx-lobes 5 ; traps tubular, with 2 divergent helically twisted arms   **1. Genlisea**
Calyx-lobes 2, free to or near the base ; traps bladder-like, ovoid or globose
                                              2. Utricularia

1. GENLISEA St. Hil. Voy. Dist. Diam. 2 : 428 (1833) ; F.T.A. 4, 2 : 497.

Inflorescence and ovary more or less densely covered with gland-tipped hairs ; leaves spathulate, numerous in a basal rosette, 0·5–3 cm. long ; scape up to 35 cm. tall, 2–12-flowered ; corolla violet, rarely white or yellow, lower lip deeply 3-lobed, about as long as spur .. .. *1. africana*
Inflorescence and ovary without gland-tipped hairs ; ovary, calyx and pedicels hispid ; leaves spathulate, up to 5 cm. long ; scape up to 30 cm. tall, 2–6-flowered ; corolla violet, lower lip shortly 3-lobed, shorter than the spur .. .. *2. hispidula*

1. **G. africana** *Oliv.* in J. Linn. Soc. 9 : 145 (1865) ; F.T.A. 4, 2 : 497 ; Chev. Bot. 484. *G. stapfii* A. Chev. in Mém. Soc. Bot. Fr. 2, 8 : 188 (1912) ; Chev. Bot. 484. Slender, erect herb of wet grassland with remarkable subterranean inverted Y-shaped traps for the capture of small organisms ; flowers pedicellate, $\frac{1}{4}$–$\frac{1}{3}$ in. long.
Mali: Bamako *Garnier*! **Port.G.:** Suzana to Cassalal (Nov.) *Esp. Santo* 3125! Gabu (Oct.) *Esp. Santo* 3525! **Guin.:** Timbo to Ditinn (Sept.) *Chev.* 18415! Friguiagbé (July) *Chillou* 581! Macenta (Oct.) *Baldwin* 9798! Nimba Mts. (Aug.) *Schnell* 6199! **S.L.:** Waterloo (Aug.) *Melville & Hooker* 275! Bonthe (Nov.) *Deighton* 2297! Bumban (Aug.) *Deighton* 1303! Mateboi (Oct.) *Glanville* 19! Loma Mts. (Nov.) *Jaeger* 1704! **Lib.:** Monrovia (Aug.) *Baldwin* 9187! Paynesville (Oct.) *Harley* 1685! Duport (Nov.) *Linder* 1448! **Iv.C.:** Mt. Dou (May) *Fleury* in *Hb. Chev.* 21499! Nimba Mts. (Aug.) *Boughey* GC 18161! [**Br.]Cam.:** Mamfe (Mar.) *Onochie* FHI 34870! Also in E. and S.E. tropical Africa.
2. **G. hispidula** *Stapf* in Fl. Cap. 4, 2 : 437 (1904) ; F.T.A. 4, 2 : 498. Similar to the above but lacking the gland-tipped hairs and with fewer, larger leaves.
N.Nig.: Jos Plateau (Apr.) *Morton* K404! [**Br.]Cam.:** Buea (May) *Mildbr.* 9468! Jakiri, 5,800 ft., Bamenda (Feb.) *Hepper* 1964! Also in S.E. tropical and S. Africa.

2. UTRICULARIA Linn.—F.T.A. 4, 2 : 469 ; Pellegr. in Bull. Soc. Bot. Fr. 61 : 13–21 (1914) ; P. Taylor in Kew Bull. 18 (1) (1963).

*Terrestrial or epiphytic herbs with entire leaves or apparently leafless or nearly so :
Inflorescence and 1–2 obovate-spathulate or broadly linear leaves arising from a small globose tuber ; epiphyte ; corolla yellow, 12–25 mm. long .. .. *1. mannii*
Inflorescence and leaves not arising from a tuber :
Bracts not produced below the point of attachment :
Fruiting pedicel as long as or longer than the calyx, flattened or winged :
Seeds verrucose, ovoid ; leaves usually present at flowering time, linear, 3-nerved, thalloid, rosulate at the base of the scape ; corolla yellow, less than 10 mm. long
                                                2. andongensis
Seeds smooth or reticulate :
Corolla yellow ; seeds ovoid, smooth with a prominent hilum ; leaves 1-nerved often not present at flowering time :
Corolla 12 mm. or more long, spur thick, straight, conical, obtuse or subacute :
Upper lip of corolla shorter than the calyx, lower lip about $\frac{1}{4}$ as long as the spur
                                   3a. *micropetala* var. *micropetala*
Upper lip of corolla longer than the calyx, lower lip $\frac{1}{2}$ as long as the spur
                                 3b. *micropetala* var. *macrocheilos*
Corolla less than 10 mm. long, spur slender, curved, acute
                                 4. *scandens* var. *schweinfurthii*
Corolla violet or mauve :
Seeds globose, reticulate, the cells of the testa isodiametric ; fruiting pedicels erect :

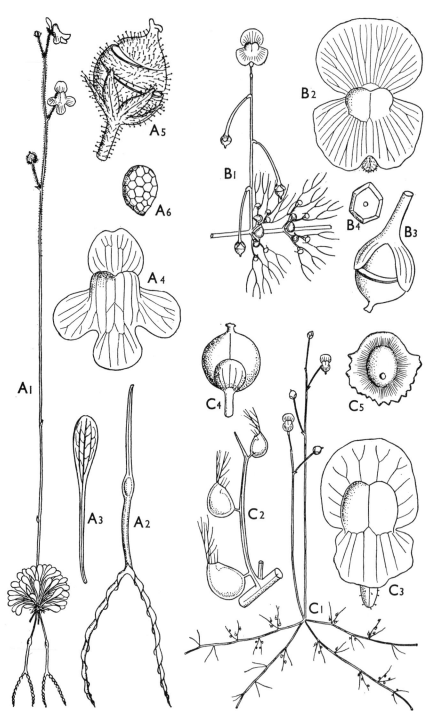

Fig. 291.—Genlisea and Utricularia spp. (Lentibulariaceae).

Genlisea africana Oliv.—A1, habit, × 1. A2, trap, × 4. A3, leaf, × 4. A4, corolla, × 4. A5, fruit, × 8. A6, seed, × 30.
Utricularia reflexa Oliv.—B1, habit, × 1. B2, corolla, × 4. B3, fruit, × 4. B4, seed, × 12.
U. gibba Linn. subsp. *exoleta* (R. Br.) P. Tayl.—C1, habit, × 1. C2, leaf and traps, × 8. C3, corolla, × 8. C4, fruit, × 8. C5, seed, × 15.
A1–6 drawn from *Baldwin* 13003. B1–4 from *Hepper* 1626. C1–5 from *T. S. Jones* 310.

Spur curved, acute ; inflorescence usually twining :
  Corolla 12–30 mm. long, upper lip broad, longer and wider than the calyx
                                                    5a. *spiralis* var. *spiralis*
  Corolla up to 10 mm. long, upper lip shorter and narrower than the calyx
                                                    5b. *spiralis* var. *tortilis*
Spur straight, obtuse ; inflorescence not twining ; corolla 10–12 mm. long,
  upper lip broad, longer and wider than the calyx 5c. *spiralis* var. *pobeguinii*
Seeds ovoid, reticulate, the cells of the testa distinctly longer than broad ; fruit-
  ing pedicels recurved ..   ..   6. *baoulensis*
Fruiting pedicel much shorter than the calyx, terete :
Calyx, bracts and bracteoles pectinate ; leaves narrowly linear, rosulate at the base
  of the scape ; scape 3–12 cm. high ; corolla deep yellow ..   7. *fimbriata*
Calyx, bracts and bracteoles entire ; leaves narrowly spathulate :
Corolla yellow, sub-persistent ; margin of calyx-lobes strongly recurved ; flowers
  numerous   ..   ..   8. *firmula*
Corolla mauve or white, not persistent ; margin of calyx-lobes not recurved ;
  flowers few (2–4)   ..   ..   9. *arenaria*
Bracts produced below the point of attachment :
Bracteoles present, similar to the bracts but smaller ; flowers white or mauve :
Leaves inconspicuous, peltate ; calyx-lobes more or less equal, upper acute ;
  corolla lower lip entire ; seeds smooth   ..   ..   ..   10. *pubescens*
Leaves numerous and conspicuous at flowering time, orbicular or reniform, long-
  petiolate ; calyx-lobes very unequal, upper emarginate ; corolla lower lip 5-
  crenate ; seeds glochidiate   ..   ..   11. *striatula*
Bracteoles absent ; flowers yellow :
Inflorescence zigzag ; all bracts on the inflorescence-axis fertile ; calyx-lobes
  obtuse, not strongly nerved at flowering time   ..   ..   12. *subulata*
Inflorescence straight ; some bracts on the inflorescence-axis sterile ; calyx-
  lobes acute, strongly nerved at flowering time   ..   ..   13. *stanfieldii*
*Aquatic herbs with leaves divided into filiform segments :
Plants attached by thickened claw-like rhizoids ; spur of corolla absent :
Leaves up to 100 cm. long, bearing few or no traps ; corolla 4–6 mm. long, lower lip
  emarginate ; fruits about 1·6 mm. long ..   ..   14. *rigida*
Leaves up to 6 cm. long, traps numerous on the rachis ; corolla 2–3 mm. long, lower
  lip 4-lobed ; fruits about 1·2 mm. long ..   ..   15. *tetraloba*
Plant free floating ; corolla with well developed spur :
Inflorescence with a whorl of spongy floats :
Corolla upper lip entire, lower lip about equal to the spur ; calyx accrescent, longer
  than the fruit :
Inflorescence floats narrowly cylindrical up to 15 mm. long, 5–10 times as long as
  wide ; corolla usually mauve   ..   ..   ..   16a. *inflexa* var. *inflexa*
Inflorescence floats ellipsoid up to 40 mm. long, 2–3 times as long as wide ; corolla
  usually yellow   ..   ..   16b. *inflexa* var. *stellaris*
Corolla upper lip deeply bilobed, lower lip much shorter than the inflated cylindrical
  spur ; calyx not accrescent, smaller than the fruit   ..   17. *benjaminiana*
Inflorescence without a whorl of floats :
Traps arising in the angle of bifurcation of the leaf segments ; inflorescence 1–4-
  flowered ; corolla yellow, hairy ; fruiting pedicels reflexed   ..   18. *reflexa*
Traps arising laterally on the leaf segments above the point of bifurcation :
Scape many-flowered ; fruiting pedicels recurved ; leaves very large, repeatedly
  pinnately divided ; seeds with a narrow regular wing ..   ..   ..   19. *foliosa*
Scape 1–4-flowered ; fruiting pedicels erect ; leaves very small 1–2 times forked
  only ; seeds with a broad irregular wing :
Corolla 10–20 mm. long, upper lip broader than the lower
                                                    20a. *gibba* subsp. *gibba*
Corolla up to 6 mm. long, upper lip about as wide as the lower
                                                    20b. *gibba* subsp. *exoleta*

1. **U. mannii** *Oliv.* in J. Linn. Soc. 9 : 149 (1865) ; F.T.A. 4, 2 : 484. *U. bryophila* Ridl. in Ann. Bot. 2 :
    306 (1889) ; F.T.A. 4, 2 : 484. Epiphyte on mossy tree trunks with slender scape 1–4 in. high arising
    from small tuber and bearing 1–3 yellow flowers ⅜–1 in. long.
    [Br.]Cam.: Cam. Mt., 4,000–6,000 ft. (Nov.) *Mann* 2112 ! Buea (Aug.) *Gregory* 321b ! Bamenda, 7,000 ft.
    (May) *Keay* FHI 37937 ! **F.Po**: Ureka, 2,000–3,000 ft. (Aug.) *Thorold* 80 ! Moka, 4,600 ft. (Sept.)
    Wrigley & Melville 434a ! 434b ! Also in S. Tomé and Principe.
2. **U. andongensis** *Welw. ex Hiern* Cat. Welw. 787 (1900) ; F.T.A. 4, 2 : 481. Terrestrial herb of wet rocks
    with erect or twining scape arising from a rosette of thalloid linear leaves and bearing a few small yellow
    flowers.
    Guin.: Kindia (Apr.) *Chillou* 1306 ! Nzérékoré (Sept.) *Baldwin* 13296 ! **S.L.**: Tonkolili Dist. (Nov.)
    Marmo 132 ! Bintumane *Jaeger* 1100 ! **Lib.**: Sanokwele (Sept.) *Baldwin* 9556 ! **N.Nig.**: Jos (Aug.)
    Keay FHI 20092 ! **S.Nig.**: Ife (Sept.) *Stanfield* 120 ! **[Br.]Cam.**: Mamfe (Sept.) *Gregory* 321a ! Onochie
    FHI 34870 ! Bamenda (May) *Maitland* ! Bum, Wum Div. (Feb.) *Hepper* 1920 ! Mambila Plateau,
    Adamawa Div. (Jan.) *Hepper* 1770 ! 1793 ! Throughout tropical Africa.
3. **U. micropetala** *Sm.* in Rees Cyclop. 37 : No. 58 (1819) ; F.T.A. 4, 2 : 483.
3a. **U. micropetala** *Sm.* var. **micropetala**. Erect or twining herb to about 6 in. high ; pedicels longer than the
    calyx, winged ; corolla yellow with a thick conical spur and very short upper and lower lips.

Guin.: Friguiagbé (Nov.) *Boismare* 482! Fouta Djalon *Heudelot* 676! Ditinn (Jan.) *Roberty* 16461! Mt. Gangan (Nov.) *Schnell* 2207! **S.L.**: *Afzelius*! Waterloo (Oct.) *Lane-Poole* 405! **N.Nig.**: Vom Dent *Young* 187! Naraguta (Aug.) *Lely* 553! [**Br.**]**Cam.**: Vogel Peak, Adamawa Div. (Nov.) *Hepper* 1486!

3b. **U. micropetala** var. **macrocheilos** *P. Tayl.* Taxon 12: ined. (1963). Similar to above but upper and lower lips of the corolla well developed.

Guin.: Kindia (Oct.) *Pobéguin*! *Farmar* 199! Macenta (Oct.) *Baldwin* 9796! **S.L.**: Bintumane (Sept.) *Jaeger* 1994!

4. **U. scandens** *Benj.* in Linnaea 20 : 309 (1847). *U. wallichiana* Wight (1850). *U. gibbsiae* Stapf in F.T.A. 4, 2 : 574 (1906). A widespread species, subsp. *scandens* occurring throughout E. & S. tropical Africa, Madagascar and tropical Asia, and the following subspecies in W. Africa.

4a. **U. scandens** *Benj.* subsp. **schweinfurthii** (*Bak. ex Stapf*) *P. Tayl.* l.c. (1963). *U. schweinfurthii* Bak. ex Stapf in F.T.A. 4, 2 : 482 (1906). *U. prehensilis* of Pellegr. in Bull. Soc. Bot. Fr. 61 : 17 (1914), partly ; of F.W.T.A., ed. 1, 2 : 234, not of E. Mey. Erect or twining ; corolla yellow ; seeds smooth with a prominent hilum.

Guin.: Benna (Nov.) *Schnell* 2144! **Iv.C.**: Mankono (Jan.-Nov.) *Chev.* 21888! *de Wilde* 943! **N.Nig.**: Kontagora (Nov.) *Dalz.* 145! [**Br.**]**Cam.**: Bamenda (Jan.-Feb.) *Hepper* 1963! *Migeod* 418! *Keay* FHI 28533! Also in Ubangi-Shari, Congo, Sudan and Ethiopia.

5. **U. spiralis** *Sm.* in Rees Cyclop. 37 : No. 5 (1819).

5a. **U. spiralis** *Sm.* var. **spiralis**. *U. spiralis* Sm.—F.T.A. 4, 2 : 482, partly ; F.W.T.A., ed. 1, 2 : 234, partly (excl. syn. *U. baoulensis*). *U. baumii* Kam. (1902)—F.T.A. 4, 2 : 480 ; Pellegr. in Bull. Soc. Bot. Fr. 61 : 16, incl. var. *leptocheilos* Pellegr. l.c. (1914). Slender twining herb to about 1 ft. high ; corolla violet, ⅓–1 in. long, spur curved, slender, acute.

Guin.: Conakry *Maclaud*! Kindia *Farmar* 201! Telémélé (Jan.) *Roberty* 16468! Benna (Nov.) *Schnell* 2152! **S.L.**: Mano, Bonjema (Jan.) *Capstick* in Deighton 5308! Sherbro Is. (Feb.) *Dalz.* 1027! Wellington (Nov.) *Deighton* 1865! Mateboi (Oct.) *Glanville* 26! **Lib.**: Monrovia (Aug.) *Baldwin* 10535! Paynesville (Jan.) *Barker* 1187! Duport (Nov.) *Linder* 1442! **Iv.C.**: Mankono (June) *Chev.* 21886! Also in tropical Africa.

5b. **U. spiralis** var. **tortilis** (*Welw. ex Oliv.*) *P. Tayl.* l.c. (1963). *U. spiralis* of F.T.A. 4, 2 : 482, partly, not of Sm. ; Pellegr. in Bull. Soc. Bot. Fr. 61 : 17 (1914) ; F.W.T.A., ed. 1, 2 : 234, partly. *U. tortilis* Welw. ex Oliv. in J. Linn. Soc. 9 : 150 (1865) ; F.T.A. 4, 2 : 483. Slender twining herb similar to above but with smaller corolla (up to ⅔ in. long) with a relatively smaller and much narrower upper lip.

Sen.: Badi (Dec.) *Berhaut* 1828! **Port.G.**: Gabu (Oct.) *Esp. Santo* 3531! Bissau (Jan.) *Esp. Santo* 1444! **Mali**: Bamako (Mar.) *Raynal* 5641! Sansanding (Feb.) *Chev.* 3301! **Guin.**: Friguiagbé (Dec.) *Chillou* 1065! Conakry *Maclaud*! Mali (Dec.) *Schnell* 2355! **S.L.**: Waterloo (Aug.) *Melville & Hooker* 282! Bonthe (Oct.) *Adames* 85! Kambia (Sept.) *Adames* 182! Roboli (Nov.) *Jordan* 172! **Lib.**: Monrovia (Aug.) *Baldwin* 9188! Paynesville *Harley* 1681! **Iv.C.**: Mossou (Aug.) *de Wilde* 223! Grand Bassam (Sept.) *de Wit* 1198! **Ghana**: Sanpa (Apr.) *Morton* A3261 Yendi *Morton* A1465! **N.Nig.**: Kontagora (Nov.) *Dalz.* 144! **S.Nig.**: Badagry (Aug.) *Thorold* 2007! Lagos *Barter* 2214! Sapoba (Nov.) *Kennedy* 2726! Throughout tropical Africa.

5c. **U. spiralis** var. **pobeguinii** (*Pellegr.*) *P. Tayl.* l.c. (1963). *U. pobeguinii* Pellegr. in Bull. Soc. Bot. Fr. 61 : 16 (1914). Usually erect, not twining ; upper lip of corolla broad, spur obtuse.

Guin.: Kindia (Aug.) *Pobéguin* 1361! Friguiagbé (Sept.) *Chillou* 649! 2119!

6. **U. baoulensis** *A. Chev.* in Mém. Soc. Bot. Fr. 2, 8 : 186 (c. 21 Sept. 1912) ; Chev. Bot. 481. *U. scandens* Oliv. in J. Linn. Soc. 3 : 181 (1859), not of Benj. (1847). *U. tenerrima* Merrill in Philipp. Journ. Sci., Bot. 7 : 247 (30 Sept. 1912). Very slender twining herb with minute mauve flowers and recurved fruiting pedicels.

Mali: Koulouba, Balasoko (Nov.) *Duong*! **Iv.C.**: Baoulé (Aug.) *Chev.* 22247! **Ghana**: Navrongo (Oct.) *Vigne* FH 4603 (partly)! **S.Nig.**: Owo (Sept.) *Stanfield* 122! Also in E. tropical Africa, Madagascar and tropical Asia (Madras and Philippines).

7. **U. fimbriata** *H.B.K.* Nov. Gen. et Sp. 2 : 225 (1818). *Polypompholyx laciniata* Benj. in Mart. Fl. Bras. 10 : 251 (1847), partly ; Pellegr. in Bull. Soc. Bot. Fr. 61 : 20. Inflorescence stiffly erect bearing a few small orange-yellow flowers with deeply fimbriate calyx-lobes.

Port.G.: Gabu (Oct.) *Esp. Santo* 3522! **Lib.**: Duport (Nov.) *Linder* 1452! Paynesville (Oct.) *Harley* 2159! Monrovia (Aug.) *Baldwin* 13037! Also in Gabon, Congo, N. Rhodesia and Angola, and from peninsula Florida southwards to Brazil.

8. **U. firmula** *Welw. ex Oliv.* F.T.A. 4, 2 : 479 (1906) ; F.W.T.A., ed. 1, 2 : 234, partly (excl. *Lane-Poole* 404 ; *Chev.* 21865) ; Pellegr. in Bull. Soc. Bot. Fr. 61 : 16 ; Chev. Bot. 481. Stiffly erect terrestrial herb with numerous very small subsessile yellow flowers.

Sen.: *Trochain* 1466! Koulaye (Feb.) *Chev.* 3302! **Gam.**: Kombo *Dawe* 24! **Guin.**: Kadé (Jan.) *Pobéguin* 1981! *Chev.* 20319! **Lib.**: Johnsonville (Feb.) *Dinklage* 2960! **N.Nig.**: Kontagora (Dec.) *Dalz.* 147! Zaria (Dec.) *Milne-Redhead* 5004! Jos (Oct.) *Hepper* 1013! **S.Nig.**: Owo (Dec.) *Stanfield* 24! Throughout tropical Africa and in Madagascar.

9. **U. arenaria** *A. DC.* in DC. Prod. 8 : 20 (1844). *U. tribracteata* Hochst. ex A. Rich. Tent. Fl. Abyss. 2 : 18 (1851) ; F.T.A. 4, 2 : 475, partly (excl. syn. and *Schimper* 1149 ; *Edith Cole*) ; F.W.T.A., ed. 1, 2 : 234. *U. kirkii* Stapf in F.T.A. 4, 2 : 476 (1906). *U. exilis* Oliv. in J. Linn. Soc. 9 : 154 (1865) ; F.T.A. 4, 2 : 477 ; Pellegr. in Bull. Soc. Bot. Fr. 61 : 16. Erect, delicate herb 2–6 ins. tall bearing 2–4 white or mauve flowers ; leaves oblanceolate-spathulate.

Sen.: St. Louis *Perrottet*! *Leprieur*! **Mali**: Bamako *Garnier*! **S.L.**: Lumley (Aug.) *Melville & Hooker* 157! **Iv.C.**: Mankono (Aug.) *Schnell* 6440! **Ghana**: Atwabo *Fishlock* 1! 3! Navrongo (Oct.) *Vigne* FH 4603a! **N.Nig.**: Anara F.R., Zaria (Oct.) *Hepper* 1001! Naraguta (Aug.) *Lely* 521! Vom Dent *Young* 188! [**Br.**]**Cam.**: Gurum, Adamawa (Nov.) *Hepper* 1251! Throughout tropical Africa and in Madagascar.

10. **U. pubescens** *Sm.* in Rees Cyclop. 37 : No. 53 (1819). *U. papillosa* Stapf in Kew Bull. 1916 : 41 ; F.W.T.A., ed. 1, 2 : 234. *U. graniticola* A. Chev. & Pellegr. in Mém. Soc. Bot. Fr. 2, 8 : 276 (1917) ; Chev. Bot. 482 ; F.W.T.A., ed. 1, 2 : 234. *U. peltatifolia* A. Chev. & Pellegr. l.c. 276 (1917) ; Chev. Bot. 482. *U. hydrocotyloides* Lloyd & G. Tayl. in Contrib. Gray Herb. 165 : 85 (1947). *U. fernaldiana* Lloyd & G. Tayl. l.c. 87 (1947). *U. thomasii* Lloyd & G. Tayl. l.c. 88 (1947). *U. deightonii* Lloyd & G. Tayl. l.c. 89 (1947). Erect terrestrial herb with small peltate leaves and slender, few-flowered scape ; flowers white or mauve, very variable in size, calyx and scape often pubescent.

Guin.: Fouta Djalon (Aug.-Oct.) *Chev.* 18414! 18416! 20207! Nimba Mt. (Aug.) *Schnell* 6200! Nzérékoré (Sept.) *Baldwin* 13297! **S.L.**: Bonthe (Oct.) *Adames* 96! Sugar Loaf Mt. (Nov.) *T. S. Jones* 238! Waterloo (Aug.) *Melville & Hooker* 284! Rokupr (Nov.) *Jordan* 174! Bintumane (Sept.) *Jaeger* 1771! **Lib.**: Paynesville (Oct.) *Harley* 1686! Monrovia (Aug.) *Dinklage* 2830! Sanokwele (Sept.) *Baldwin* 9557! Gbanga (Aug.) *Baldwin* 9153! **Iv.C.**: Mt. Dou (May) *Chev.* 24497! Mossou (Sept.) *Aké Assi* 3155! Mt. Tonkoui (Oct.) *Aké Assi* 3821! Mankono (June) *Chev.* 21918! **N.Nig.**: Abinsi (Nov.) *Dalz.* 732! Bida (Mar.) *Meikle* 1328! **S.Nig.**: Idanre Hills (Oct.) *Keay* FHI 22659! Owo (Sept.) *Stanfield* 123! [**Br.**]**Cam.**: Mamfe (Mar.) *Onochie* FHI 38798! Also in Uganda, Tanganyika, N. Rhodesia, Angola, and tropical S. America.

11. **U. striatula** *Sm.* in Rees Cyclop. 37 : No. 17 (1819). F.T.A. 4, 2 : 486 ; Pellegr. in Bull. Soc. Bot. Fr. 61 : 18 ; F.W.T.A., ed. 1, 2 : 234, partly (excl. syn. *U. peltatifolia* A. Chev. & Pellegr.). A delicate epiphyte of mossy trees or rocks with numerous orbicular-spathulate leaves and small pink or white flowers, very unequal calyx-lobes and glochidiate seeds.

Guin.: Dalaba (Sept.) *Schnell* 6841! **S.L.**: Regent (Aug.) *Deighton* 2071! Sugar Loaf Mt. (Nov.)

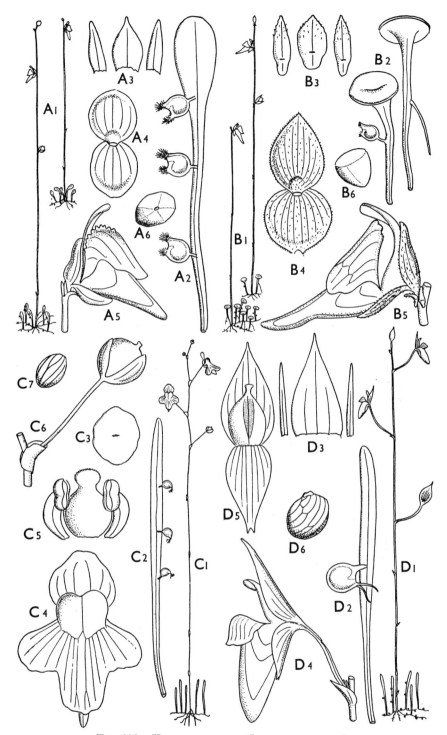

FIG. 292.—UTRICULARIA SPP. (LENTIBULARIACEAE).

Utricularia arenaria A. DC.—A1, habit, × 1. A2, leaf and traps, × 8. A3, bract and bracteoles, × 15. A4, calyx, × 8. A5, flower, × 8. A6, seed, × 40.
U. pubescens Sm.—B1, habit, × 1. B2, leaves and trap, × 8. B3, bract and bracteoles,
× 15. B4, calyx, × 8. B5, flower, × 8. B6, seed, × 40.
U. subulata Linn.—C1, habit, × 1. C2, leaf and traps, × 8. C3, bract, × 15. C4, corolla,
× 4. C5, pistil and stamens, × 15. C6, fruit, × 8. C7, seed, × 40.
U. micropetala Sm. var. *micropetala*—D1, habit, × 1. D2, leaf and trap, × 8. D3, bract and
bracteoles, × 15. D4, flower, × 6. D5, calyx with fruit, × 6. D6, seed, × 60.
A1 drawn from *Killick* 61. A2 from *Hepper* 1000. A3, 4, 6 from *Melville & Hooker* 157.
A5 from *Thorold* 2013. B1–6 from *Baldwin* 13004. C1–7 from *Baldwin* 13014. D1, 2, 6
from *Hepper* 1486. D3–5 from *Roberty* 16461.

T. S. Jones 288! Bumban (Aug.) *Deighton* 1307! Bintumane (Sept.) *Jaeger* 1948! **Lib.:** Ganta (Aug.) *Harley* 1535! Sanokwele (Sept.) *Baldwin* 9621! **Iv.C.:** Mt. Tonkoui (Sept.) *Aké Assi* 3335! **Ghana:** Boti Falls (Oct.) *Hall* 1644! **S.Nig.:** Ogoja (Aug.) *Stone* 40! Calabar (Mar.) *Richards* 9253! [**Br.**]**Cam.:** Cam. Mt. (Nov.) *Mann* 1964! Buea (Sept.) *Dundas* FHI 15307! Vogel Peak, Adamawa (Nov.) *Hepper* 1477! **F.Po:** Ureka (Aug.) *Thorold* 84a! Moka (Sept.) *Wrigley & Melville* 641! Also in Gabon, Uganda, Tanganyika, N. Rhodesia, Angola and throughout tropical Asia.

12. **U. subulata** *Linn.* Sp. Pl. 18 (1753) ; F.T.A. 4, 2 : 485 ; Pellegr. in Bull. Soc. Bot. Fr. 61 : 18 ; Chev. Bot. 484. Inflorescence erect 1–10 in. high, scape capillary, flexuous or zigzag, bearing small yellow flowers on pedicels about ¼ in. long with obtuse, obscurely nerved calyx-lobes.
Sen.: Djongouts (Feb.) *Trochain* 1465! **Mali:** Sikasso (May) *Chev.* 794! 795! 3304! Bamako (Sept.) *Waterlot* 1340! **Port.G.:** Bissau (Sept.) *Esp. Santo* 1343! **Guin.:** Conakry *Maclaud*! Fon (Aug.) *Schnell* 6604 *bis*! Fouta Djalon (Sept., Oct.) *Chev.* 18426 ; 18489 ; 20205! Macenta (Oct.) *Baldwin* 9797! Mt. Nimba (Sept.) *Baldwin* 13296! **S.L.:** Waterloo (Aug.) *Deighton* 2073! Bonthe (Oct.) *Adames* 93! Mandu (Oct.) *T. S. Jones* 415! Kambia (Sept.) *Adames* 180! Lunsar (Mar.) *Jordan* 29! Bintumane, 5,000 ft. (Nov.) *Jaeger* 575! **Lib.:** Monrovia (Aug.) *Baldwin* 9178! Gbanga (Dec.) *Baldwin* 10544! Boporo (Nov.) *Baldwin* 10347! Miamu (Aug.) *Linder* 415! Paynesville (Apr.) *Barker* 1246! **Iv.C.:** Mt. Tonkoui (Nov.) *Aké Assi* 3476! Nimba (Aug.) *Boughey* GC 18162! Mossou (Aug.) *de Wilde* 222! **Ghana:** Ntonso (Oct.) *Baldwin* 13493! Atwabo *Fishlock* 2! Banda (Dec.) *Morton* GC 25211! **N.Nig.:** Kontagora (Jan.) *Meikle* 1095! Bida (Mar.) *Meikle* 1331! Dogon Kurmi, Jemaa Dist. (Aug.) *Killick* 46! Naraguta (Aug.) *Lely* 522! **S.Nig.:** Lagos *Barter* 2213! Carter's Peak, Idanre Hills (Oct.) *Keay* FHI 25508! Enugu to Agbami (May) *Onochie & Awa* FHI 35808! Owo (Aug.) *Stanfield* 114! Throughout tropical Africa and Madagascar, eastern N. America southwards to S. Brazil ; also in tropical Asia, Siam and Borneo.

13. **U. stanfieldii** *P. Tayl.* l.c. (1963). Inflorescence erect, scape capillary, straight bearing 2–6 small yellow flowers with 3-lobed lower corolla lip and acute, strongly nerved calyx-lobes.
S.L.: Roboli (Nov., Oct.) *Jordan* 173! 1053! **Lib.:** Sanokwele (Sept.) *Baldwin* 9483! Paynesville (Jan.) *Barker* 1192! **S.Nig.:** Oba, Ondo Prov. (Sept.) *Stanfield* 121! Idogun Hill, Benin Prov. (Oct.) *Stanfield*!

14. **U. rigida** *Benj.* in Linnaea 20 : 303 (1847) ; F.T.A. 4, 2 : 486 ; Pellegr. in Bull. Soc. Bot. Fr. 61 : 19 ; Chev. Bot. 483. Aquatic attached to rocks in flowing water with emerging stiffy erect inflorescences bearing numerous small white flowers and long submerged leaves divided into capillary segments.
Port.G.: Nova Lamego to Canjadude (Oct.) *Esp. Santo* 3451! Gabu (Dec.) *Esp. Santo* 3032! Nhampassare (Dec.) *Esp. Santo* 3785! **Mali:** Sikasso (May) *Chev.* 793! Koulikoro (Oct.) *Chev.* 3305! Kita Massif (Nov.) *Jaeger* 5558! Bamako (Nov.) *Raynal* 5192! **Guin.:** Ymbo to Kouria (Oct.) *Chev.* 14996! Boffa *Jac.-Fél.* 7360! Dalaba (Sept.) *Schnell* 6840! Mali (Dec.) *Schnell* 2345! **S.L.:** Sugar Loaf Mt. (Aug., Nov.) *T. S. Jones* 275a! 239b! *Melville & Hooker* 70! Kambia (Sept.) *Adames* 187! Bumban (Aug.) *Deighton* 1302! **N.Nig.:** Jos (Oct.) *Hepper* 1029! Naraguta (Aug.) *Lely* 555!

15. **U. tetraloba** *P. Tayl.* l.c. (1963). Similar to the above but much shorter and more slender with shorter leaves bearing numerous traps.
Guin.: Kanea (Oct.) *Adam* 12624! Friguiagbé (Oct.) *Boismare* 420! Mt. Benna *Jac.-Fél.* 7172! **S.L.:** Sugar Loaf Mt. (Sept., Oct., Nov.) *T. S. Jones* 246! 239a! *Milne-Redhead* 5157! Loma Mts. (July) *Jaeger* 431!

16. **U. inflexa** *Forsk.* Fl. Aegypt. Arab. Descr. Pl. 9 (1775).
16a. **U. inflexa** *Forsk.* var. **inflexa**—P. Tayl. in Mitt. Bot. Staatss. München 4 : 95 (1961). *U. thonningii* Schumach. in Schumach. & Thonn. Beskr. Guin. Pl. 12 (1827) ; F.T.A. 4, 2 : 487 ; Pellegr. in Bull. Soc. Bot. Fr. 61 : 19 ; F.W.T.A., ed. 1, 2 : 234. Diffuse aquatic half submerged with emergent inflorescences supported by a whorl of long narrowly cylindrical spongy floats ; flowers usually mauve.
Sen.: Sangalkam (Jan.) *Berhaut* 222! Telel (Apr.) *Trochain* 3032! Kankossa (Mar.) *Roberty* 16988! **Mali:** Soye (Nov.) *Monod* 4! Gao (Nov.) *de Wailly* 4889! **Ghana:** Mankessim (Oct.) *Box* FH 3314! Atiawi (Sept.) *Akpabla* 345! Agbosome *Thorold* 259! **Dah.:** Dogba (Oct.) *Le Testu* 209! **N.Nig.:** Amabo, Kabba Prov. (Feb.) *A. P. D. Jones* 2800 (= FHI 632)! Keana *Hepburn* 50! Yola (Jan.) *Dalz.* 130! **S.Nig.:** Ijebu-Ode (Aug.) *Horne* FHI 39717! Ologun to Owoseni (Dec.) *Onochie* FHI 26689! Throughout tropical Africa southwards to Natal, also in Madagascar and India.

16b. **U. inflexa** *Forsk.* var. **stellaris** (*Linn.f.*) P. Tayl. in Mitt. Bot. Staatss. München 4 : 96 (1961). *U. stellaris* Linn. f. Suppl. 86 (1781) ; F.T.A. 4, 2 : 488 ; Pellegr. in Bull. Soc. Bot. Fr. 61 : 19 ; F.W.T.A., ed. 1, 2 : 234. *U. trichoschiza* Stapf in F.T.A. 4, 2 : 488 ; F.W.T.A., ed. 1, 2 : 234. Similar to the above, but with shorter ellipsoid floats and small yellow flowers.
Maur.: Kankossa (Mar.) *Roberty* 17005! **Sen.:** N'Tal (Dec.) *Trochain* 2231! N'Diaël (Dec.) *Trochain* 2025! Matam (Nov.) *Trochain* 1002! Kaedi (Dec.) *Chev.* 3306! **Gam.:** Bansang *Duke* 5! Kuntaur *Ruxton* 103! 111! Fulladu (Nov.) *Frith* 259! **Mali:** Macina *Roberty* 2948! San (June) *Chev.* 1032! Gao (Dec.) *de Wailly* 5269! 5271! **Port.G.:** Bafata to Capé (Aug.) *Esp. Santo* 3320a! Bissau (Nov.) *Esp. Santo* 1477! **Guin.:** Koba (Nov.) *Jac.-Fél.* 7205! Faranah (Jan.) *Chev.* 20461! **S.L.:** Njala (Sept.) *Deighton* 2093! 5855! Gbap (Sept.) *Adames* 78! Gbundapi (Sept.) *Adames* 70! Kambia (Aug.) *Jordan* 961! Ronietta (Nov.) *Thomas* 5322! **Iv.C.:** (Jan.) *Hedin* 2563! Baoule Nord (Aug.) *Chev.* 22374! **Ghana:** Achimota (Nov.) *Irvine* 1518! Tamale (Dec.) *Adams & Akpabla* 4136! 4137! Jirapa (Nov.) *Harris*! Navrongo (Oct.) *Vigne* FH 4604! **Togo Rep.:** (Sept.) *Unknown Collector* 425! **Dah.:** Logozohé (July) *Annet* 168! Niger: Niamey to Gao *Ryff*! **N.Nig.:** Isoburi (Dec.) *Elliott* 139! Kontagora (Dec.) *Dalz.* 142! Nupe *Barter* 1553! Maiduguri (Nov.) *Noble* 30! **S.Nig.:** Benin (Dec.) *Stanfield* 160! [**Br.**] **Cam.:** Gulumba, Dikwa Div. (Dec.) *McClintock* 47! Widespread in tropical Africa, also in S. Africa, Madagascar and tropical Asia to N. Australia.

17. **U. benjaminiana** *Oliv.* in J. Linn. Soc. 4 : 176 (1860) ; Kam in Engl. Bot. Jahrb. 33 : 109. *U. gilletii* De Wild. & Th. Dur. in Compt. Rend. Soc. Bot. Belg. 38 : 40 (1900). *U. villosula* Stapf in F.T.A. 4, 2 : 490 (1906) ; F.W.T.A., ed. 1, 2 : 234 ; Pellegr. in Bull. Soc. Bot. Fr. 61 : 19. Aquatic half submerged with emergent slender racemes of mauve and white flowers supported by a whorl of slender fusiform floats.
Sen.: Casamance *Trochain* 1551! **Mali:** Macina (Nov.) *Monod* 2! Gao (Mar.) *de Wailly* 4974! **Port.G.:** Bafata to Capé (Aug.) *Esp. Santo* 3320! **Guin.:** Mankoutan *Jac.-Fél.* 7348! Benna (Nov.) *Schnell* 2164! **S.L.:** Newton (Oct.) *T. S. Jones* 247! Njala (Dec.) *Deighton* 5973! Sherbro Is. (Feb.) *Dalz.* 1025! **Lib.:** Monrovia (Nov.) *Dinklage* 3277! Duport (Nov.) *Linder* 1475! **Iv.C.:** Mossou (Aug.) *Schnell* 6557! **Ghana:** Atwabo (Jan.) *Thorold* 293! **S.Nig.:** Sapoba (Nov., Feb.) *Meikle* 543! *Onochie* FHI 27679! Throughout tropical Africa and Madagascar, also in Brazil and Guiana.

18. **U. reflexa** *Oliv.* in J. Linn. Soc. 9 : 146 (1867) ; F.T.A. 4, 2 : 492 ; Pellegr. in Bull. Soc. Bot. Fr. 61 : 19. *U. platyptera* Stapf in F.T.A. 4, 2 : 492 (1906) ; F.W.T.A., ed. 1, 2 : 234. *U. charoidea* Stapf in F.T.A. 4, 2 : 492 (1906) ; F.W.T.A., ed. 1, 2 : 234. *U. pilifera* A. Chev. in Mém. Soc. Bot. Fr. 2, 8 : 187 (1912) ; Pellegr. in Bull. Soc. Bot. Fr. 61 : 19. *U. kalmaloensis* A. Chev. in Bull. Mus. Hist. Nat. sér. 2, 4 : 588 (1932). Aquatic with long stolons bearing short lateral scapes, with 1 to 4 yellow flowers, the pedicels elongating and reflexing in fruit.
Sen.: Richard Toll (Jan.) *Trochain* 2310! **Gam.:** *Pirie* 41! Kuntaur *Ruxton* 32! **Mali:** Macina (Nov.) *Monod* 3! Gao (Mar., Dec.) *de Wailly* 4973! 5270! **S.L.:** Madina, Port Loko Dist. (Aug.) *Jordan* 906! Kambia Dist. (Sept.) *Adames* 191! **Iv.C.:** Baoule-Nord (Aug.) *Chev.* 22297! Mankono (June) *Chev.* 21881! **Ghana:** Viume (Dec.) *Foote*! Yendi to Tamali (Oct.) *Akpabla* 4138! **Dah.:** Porto-Novo (Jan.) *Chev.* 22764! **Niger:** Birni N'Koni (Feb.) *Chev.* N.Nig.: Kontagora (Dec.) *Dalz.* 143! Zaria (Dec.) *Milne-Redhead* 5001! Nupe *Barter* 890! Lokoja *Barter*! **S.Nig.:** Badagry (Aug.) *Thorold* 2009! Sapoba

(Nov.) *Meikle* 520! *Keay & Onochie* FHI 21518! [Br.]Cam.: Gulumba, Dikwa Div. (Dec.) *McClintock*
107! 108! Throughout tropical Africa and Madagascar.
19. **U. foliosa** Linn. Sp. Pl. 18 (1753) ; F.T.A. 4,2 : 491 ; Pellegr. in Bull. Soc. Bot. Fr. 61 : 19. Submerged
aquatic forming large masses in still or slowly flowing deep water, with erect many-flowered scapes up
to 1 ft. long ; flowers yellow.
Sen.: (Dec.) *Trochain* 1203! **Mali:** Gao (Jan.-Mar.) *de Wailly* 4952! 4973 *bis*! Macina (Nov.)
Monod 5! **Guin.:** Koba *Jac.-Fél.* 7214! **S.L.:** Sembehun (Dec.) *Deighton* 301! Sherbro Is. (July)
T. S. Jones 166! Subu (Oct.) *Jordan* 592! L. Mape (Jan.) *Adames* 5! L. Mabesi (Jan.) *Capstick* in *Hb.
Deighton* 5303! **Iv.C.:** *Jolly*! **Niger:** Niamey to Gao *Ryff*! **N.Nig.:** Bukwami (Mar.) *Clayton* 1538
(= FHI 41999)! Birnin Kebbi (Dec.) *Philcox* 212 (= FHI 43152)! Throughout tropical Africa and
Madagascar, also from southern U.S.A. to Brazil.
20. **U. gibba** Linn. Sp. Pl. 18 (1753).
20a. **U. gibba** subsp. **gibba**—P. Tayl. in Mitt. Bot. Staatss. München 4 : 98 (1961). *U. obtusa* Sw. (1788)—
F.T.A. 4, 2 : 495. Creeping on wet mud with yellow flowers about ½ in. long, 1–3 on a slender scape 1–6 ins.
S.Nig.: Okomu F.R., Benin (Jan.) *Brenan & Jones* 8598a! Throughout central, E. and S. tropical
Africa, and widespread in N. and S. America.
20b. **U. gibba** subsp. **exoleta** (*R. Br.*) P. Tayl. in Mitt. Bot. Staatss. München 4 : 101 (1961). *U. exoleta* R. Br.
(1810)—F.T.A. 4, 2 : 495 ; Pellegr. in Bull. Soc. Bot. Fr. 61 : 20 ; F.W.T.A., ed. 1, 2 : 234 ; Chev.
Bot. 481. *U. riccioides* A. Chev. in Mém. Soc. Bot. Fr. 2, 8 : 187 (1912). Similar to above but flowers
much smaller, less than ¼ in. long.
Sen.: *Perrottet*! Casamance (Feb.) *Chev.* 3303! Dagana (Mar.) *Leprieur*! Galam *Heudelot* 117! Sangalkam (Jan.) *Berhaut* 570! **Mali:** Bamako (Apr.) *Waterlot* 1085! Gao (Mar.) *de Wailly* 4972! **Port.G.:**
Bissau (Oct.) *Esp. Santo* 1317! Gabu (July) *Esp. Santo* 2703! **Guin.:** Koba (Nov.) *Jac.-Fél.* 7208!
Kindia (Mar.) *Chev.* 13372! **S.L.:** Kumrabai (Dec.) *Thomas* 6761! Newton (Mar.) *Adames* 201! Waanje
R. (Dec.) *T. S. Jones* 386! Lunsar (Mar.) *Jordan* 28! Rokupr (Oct.) *Jordan* 145! Njala (Feb.) *Deighton*
2957! **Lib.:** Clay (May) *Harley* 2113! Ganta (Sept.) *Baldwin* 13225! Monrovia (Feb.) *Baldwin* 11044!
Vonjama (Oct.) *Baldwin* 9856! **Iv.C.:** Mossou (Aug.) *Schnell* 6559! Songon Agban to Dabou (Feb.)
Leeuwenberg 2637! Mankono (June) *Chev.* 21865! **Ghana:** Gambaga (Mar.) *Hepper & Morton* 2413!
Nungua (Feb.) *Rose Innes* GC 30096! Amisana (Dec.) *Hall* 1249! **N.Nig.:** Bida (Jan.) *Meikle* 1021!
Bukwium, Sokoto (Mar.) *Clayton* FHI 42000! Katsina (Feb.) *Meikle* 1224! Vom *Dent Young* 189!
Yola (Jan.) *Dalz.* 131! **S.Nig.:** Lagos *Barter* 2212! Badagry (Aug.) *Thorold* 2008! Calabar *Mann*
2333! Ogoja (June) *Stone* 13! [Br.]Cam.: Vogel Peak, Adamawa (Dec.) *Hepper* 1545! Gulumba,
Dikwa Div. (Dec.) *McClintock* 109! Throughout tropical and S. Africa, Madagascar, N. Africa, Portugal
and tropical Asia to Australia.

156. GESNERIACEAE[1]

By B. L. Burtt

Acaulescent or caulescent herbs, or rarely shrubs. Leaves opposite (rarely
alternate), those of a pair equal or unequal ; plants sometimes 1-foliate and the
leaf cotyledonary in origin. Inflorescence generally of open axillary cymes, the
flowers at each dichotomy being paired and opening serially ; occasionally
much congested and sub-capitate, or pseudoracemose. Flowers hermaphrodite
(very rarely unisexual) often protandrous, usually zygomorphic, often large and
showy, sometimes cleistogamous with reduced corolla. Calyx tubular and 5-lobed
or divided to the base or 3 upper lobes only united. Corolla gamopetalous with
distinct tube, often 2-lipped, proportion of lobes to tube variable ; lobes imbricate,
and adaxial pair often interior. Stamens rarely 5, usually 4 or 2, inserted on
corolla-tube ; anthers free or variously connate, 2-celled, opening lengthwise.
Disk annular or cup-like, often lobed or undulate, rarely oblique. Ovary superior,
1-celled with 2 parietal bilamellate placentas, occasionally 2-celled by their
union centrally. Ovules numerous. Fruit a capsule (often linear, sometimes
spirally twisted) or a more or less fleshy berry. Seeds numerous, small, more or
less ellipsoid, sometimes tailed with hair-like appendages at either end ; endosperm absent or very slight.

Mainly in the tropics and subtropics, often favouring shady places (at least for their
roots) and sometimes epiphytic. Comparatively few in Africa. The African genera
either occur in Asia or have affinity with genera there. They are not closely related to
the tropical American genera.

Flowers in open pedunculate cymes arising from leaf-axil or leaf-base, or subsolitary ;
fertile stamens rarely 4, usually only the 2 anticous ones ; fruit dehiscing longitudi-
nally, straight or spirally twisted:
Capsule many times longer than broad ; filaments without appendages ; caulescent
(in W. African species) :
Capsule valves flat ; flowers subsolitary ; stems creeping .. 1. **Didymocarpus**
Capsule valves twisting spirally ; flowers cymose ; stems erect   2. **Streptocarpus**
Capsule ovoid or oblong ; filaments with distinct tooth-like appendage ; inflorescence
arising from the base of the midrib of the single leaf ; acaulescent 3. **Acanthonema**
Flowers in densely congested, bracteate, cymose heads ; 2 posticous stamens only
fertile ; capsule circumscissile .. .. 4. **Epithema**

[1] The family description does not include characters found only in the New World genera.

In addition to the above indigenous genera the E. African *Saintpaulia ionantha* and members of the American genus *Achimenes* are sometimes cultivated.

1. DIDYMOCARPUS Wall.—F.T.A. 4, 2 : 503. *Nom. cons.*

Stems prostrate and rooting at and between nodes, glabrous ; leaves ovate, rounded at base, subobtuse, 4–6 cm. long, 3–4 cm. broad, glabrous, with about 6 pairs of lateral nerves ; petiole 2·5–3·5 cm. long, nearly glabrous ; flowers subsolitary ; pedicels 1 cm. long ; calyx-segments linear, 8 mm. long, slightly pilose ; corolla urceolate, 8 mm. long, lobes very short, acute ; fruit linear .. .. .. *kamerunensis*

D. kamerunensis *Engl.* Bot. Jahrb. 18 : 79, t. 4–5 (1894) ; F.T.A. 4, 2 : 503. A weak-branched herb 1 ft. or more long, with white axillary flowers.
[Br.]Cam.: Barombi *Preuss* 951 ! **F.Po**: Moka, 4,550 ft. (Jan., Sept.) *Guinea* 2094 ! *Wrigley* 619 !

2. STREPTOCARPUS Lindl.—F.T.A. 4, 2 : 504.

Ovary glabrous or nearly so ; ripe capsule glabrous, 4–5 cm. long ; branchlets hispid with gland-tipped hairs ; lower leaves very long-petiolate, upper sessile, ovate-elliptic, acuminate, rounded or subcordate at base, 8–12 cm. long, 4–6 cm. broad, setulose above, nearly glabrous beneath ; corolla 1 cm. long .. **1. elongatus**
Ovary and ripe capsule pubescent ; corolla 2·5–3·5 cm. long, the tube glandular-pubescent outside ; leaves ovate-elliptic, rounded to subcordate at base, shortly and obtusely pointed, up to 12 cm. long and 7 cm. broad, pubescent to nearly glabrous on both surfaces ; lower flowers small and cleistogamous, the tiny corolla narrowly hood-like and persistent on the young fruit :

Capsule at most 5 cm. long, pubescent .. .. **2. nobilis**
Capsule 12 cm. long, glandular-pubescent .. .. **3. insularis**

1. S. elongatus *Engl.* Bot. Jahrb. 18 : 76, t. 4–5, fig. A (1894) ; F.T.A. 4, 2 : 509. Erect herb 1–1½ ft. high, with white flowers in lax pedunculate cymes.
[Br.]Cam.: Buea, 4,000 ft. (Jan.) *Maitland* ! *Preuss* 1010 ! Buea to Mann's Spring, 4,500 ft. (Dec.) *Morton* K 889 !
[Note : *Jaeger* 236 & 455 from Loma Mts., Sierra Leone, are very closely allied to *S. elongatus* but the material is not yet sufficient to say if they are really conspecific—B. L. B.]

2. S. nobilis *C.B. Cl.* in DC. Monogr. Phan. 5, 1 : 155 (1883) ; F.T.A. 4, 2 : 511. *S. balsaminoides* Engl. (1894)—F.W.T.A., ed. 1, 2 : 236. *S. lagosensis* C.B. Cl. (1906). *S. atroviolaceus* Engl. (1921). *S. princeps* Mildbr. & Engl. (1921). (?) *S. violascens* Engl. (1921—fruiting spec. only). Stems up to 3 ft. high, half-succulent ; chasmogamous flowers deep violet-purple, 1–1¼ in. long. Sometimes only a few inches high with all the flowers cleistogamous.
Guin.: Landouman country (fr. Jan.) *Heudelot* 728 ! **S.L.**: Bumban, (Oct.) *Deighton* 3277 ! **Iv.C.**: Donané, 1,100 ft. (Oct.) *Chev.* 4179 ! **Ghana**: Bobiri F.R. (Oct.) *Andoh* 4915 ! Mpraeso scarp (Nov.) *Morton* A2766 ! Adiambra to Domiabra (June) *Kitson* 1231 ! **Togo Rep.**: Misahöhe, (Nov.) *Mildbr.* 7329 ! **N.Nig.**: Patti Lokoja (Oct.) *Dalz.* 121 ! Mada Hills *Hepburn* 80 ! Gindiri, 3,700 ft., Jos Plateau (fr. Oct.) *Hepper* 1099 ! **S.Nig.**: Akure F.R., Ondo (Oct.) *Keay* FHI 21573 ! Olokemeji F.R. (Oct.) *Ogbeni* FHI 5199 ! [Br.]Cam.: Victoria (Oct.) *Maitland* 743 ! Likomba *Mildbr.* 10509 ! 10619 ! Vogel Peak, 2,500 ft., Adamawa (fr. Nov.) *Hepper* 2759 ! Also on S. Tomé.

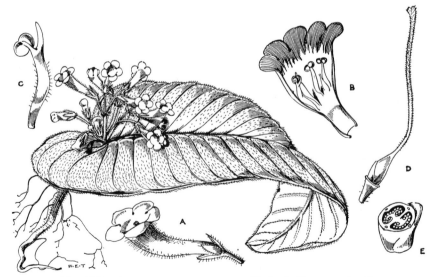

Fig. 293.—Acanthonema strigosum *Hook. f.* (Gesneriaceae).
A, flower. B, corolla laid open. C, stamen. D, pistil. E, cross-section of ovary.

3. **S. insularis** *Hutch. & Dalz.* F.W.T.A., ed. 1, 2 : 237 (1931). *S. denticulatus* Engl. Bot. Jahrb. 57 : 213 (1921), not of Turrill (1915). A handsome plant 3 ft. high with dark violet-blue flowers ; in mountain grassland. **F.Po:** Moka, 4,000–5,500 ft. (Nov.) *Mildbr.* 7113.

Imperfectly known species.

S. kerstingii *Engl.* Bot. Jahrb. 57 : 209 (1921). Specimen in fruit and leaf only. **Togo Rep.:** Sokode to Basari (fr. Nov.) *Kersting* 501 (22).

3. ACANTHONEMA Hook. f.—F.T.A. 4, 2 : 502. *Nom. cons.*

Stems short, tomentose, bearing one leaf and a short cyme ; leaf ovate-oblong, cordate at base, up to 20 cm. long and 10 cm. broad, with numerous lateral nerves, setulose-pilose above and on the nerves beneath ; cymes several-flowered ; pedicels 1 cm. long ; calyx-segments lanceolate, 5 mm. long, pilose ; corolla slightly curved, 2 cm. long ; fruit ovoid, sharply beaked, 6 mm. long, glabrous   ..   ..   *strigosum*

A. strigosum *Hook. f.* in Bot. Mag. t. 5339 (1862) ; F.T.A. 4, 2 : 502. *Carolofritschia diandra* Engl. (1899). Flowers bluish-purple, in short cymes borne 1–4 together at the base of the leaf ; epiphytic or on rocks, etc. **S.Nig.:** Calabar *Hewan* ! [Br.]**Cam.:** Cam. Mt., 2,000 ft. (Dec.) *Mann* 1948 ! **F.Po:** 1,000–2,000 ft. (Nov.) *Mann* 569 ! Also in Cameroun.

[This plant normally has two stamens : 4 or 5 stamens may occur exceptionally, associated with a more or less regular corolla, in many diandrous genera of Gesneriaceae].

4. EPITHEMA Blume—F.T.A. 4, 2 : 501.

Nearly stemless herb ; leaves ovate-elliptic, cordate at the base, up to 12 cm. long and 3–8 cm. broad, thinly pilose on both surfaces, serrulate ; petiole 2–2·5 cm. long, pilose ; mature flowers not seen   ..   ..   *tenue*

E. tenue *C. B. Cl.* in DC. Monogr. Phan. 5, 1 : 181 (1883) ; F.T.A. 4, 2 : 502. *E. graniticolum* A. Chev. in Mém. Soc. Bot. Fr. 2, 8 ; 189 (1912) ; Chev. Bot. 484. Somewhat succulent, with a single leaf at flowering ; calyx hirsute-velvety ; on rocks in forest. **Guin.:** Boola Mt., 3,200 ft., Guerzé country (Mar.) *Chev.* 20924. **S.L.:** Denkali, Loma Mt. (Oct.) *Jaeger* 227 ! **Lib.:** Vahon, Kolahun Dist. (Nov.) *Baldwin* 10215 ! **Iv.C.:** Do Mt., 2,900 ft., Upper Cavally (May) *Chev.* 21141. Droupleu to Zoanlé, Upper Sassandra (May) *Chev.* 21418 ! Digoualé to Gouané (May) *Chev.* 21457. Sampleu to Ganhouie, Upper Nuon *Chev.* 21506. [Br.]**Cam.:** Metschum Falls, Bamenda (Aug.) *Savory* UCI 299 ! Nbika's by-pass, Wum (Aug.) *Ujor* FHI 29260 ! **F.Po:** *Mann* 2345 ! Also in Uganda.

157. BIGNONIACEAE

By H. Heine

Trees or shrubs, sometimes scandent, very rarely herbs. Leaves opposite, rarely alternate, mostly compound, digitate or pinnate, sometimes the terminal leaflet tendril-like (but not in the native genera of W. Africa) ; stipules absent, pseudostipules present in some genera. Flowers often showy, hermaphrodite, more or less zygomorphic. Calyx campanulate, closed or open in bud, truncate or 5-toothed, sometimes spathaceous. Corolla with 5 imbricate lobes sometimes forming 2 lips, the upper of 2, the lower of 3 lobes. Stamens alternate with the corolla-lobes, only 4 or 2 perfect. Anthers connivent in pairs or rarely free, 2-celled, opening lengthwise ; staminode representing the fifth stamen often short, sometimes absent, often 3 present when only two stamens. Disk usually present. Ovary superior, 2-celled with 2 placentas in each cell or 1-celled with parietal bifid placentas. Style terminal, 2-lipped ; ovules numerous. Fruit capsular or fleshy and indehiscent. Seeds often winged, without endosperm ; embryo straight.

Climbing shrub (if, exceptionally, not climbing, then with very long slender shoots), ovary 2-celled, ovules in 8 series (each placenta bearing four rows of ovules) ; fruit dehiscent, cylindrical, apparently very similar to those of *Newbouldia* (not yet known in ripe state) ; calyx irregularly 5-lobed, sometimes slightly bilabiate or subspathaceous   ..   ..   1. **Dinklageodoxa**
Erect trees or shrubs :
  Calyx not spathaceous :
    Ovary 1-celled ; fruit indehiscent, sausage-shaped ; seeds not winged .. 2. **Kigelia**
    Ovary 2-celled ; fruit dehiscent ; seeds winged   ..   ..   3. **Stereospermum**
  Calyx spathaceous :
    Calyx split on the adaxial side ; ovules in 4 or more series :
      Ovules in many series ; valves of the fruit boat-shaped ; calyx recurved, acuminate, ribbed   ..   ..   4. **Spathodea**
      Ovules in 4–6 series ; valves of the fruit flat ; calyx beaked in bud   5. **Markhamia**
    Calyx split on the abaxial side, bilobed at the apex ; ovules in 2 series   6. **Newbouldia**

W. E. T.

Fig. 294.—Kigelia africana (*Lam.*) *Benth.* (Bignoniaceae).

A, flowers. B, ovary. C, cross-section of ovary. D, leaf. E, immature fruit (the mature fruit is turgid and sausage-shaped).

The following ornamentals have been introduced into our area : *Bignonia capreolata* Linn. (S.E. North America) ; *Crescentia cujete* Linn. (Tropical America) ; *Haplophragma adenophyllum* (Wall. ex G. Don) P. Dop (India) ; *Jacaranda mimosifolia* D. Don (N.W. Argentine) ; *Pandorea pandorana* (Andr.) van Steenis (Indonesia, New Guinea) ; *Parmentiera cereifera* Seem. (C. America) ; *Pyrostegia venusta* (Ker-Gawl.) Miers (Brazil) ; *Saritaea magnifica* (Sprague ex van Steenis) Dugand (syn. *Arrabidaea magnifica* Sprague ex van Steenis) (Colombia) ; *Tabebuia rosea* (Bertol.) DC. (C. America) ; *Tecoma stans* (Linn.) H.B.K. (C. America) ; *Tecomaria capensis* (Thunb.) Spach (S. Africa).

1. DINKLAGEODOXA Heine & Sandwith in Kew Bull. 16: 223 (1962).

Branchlets glabrous ; leaflets 1–2 pairs, long-elliptic, entire, acuminate on both sides (sometimes in big leaflets leaf-base rounded), 7–18 cm. long and 2·5–6 cm. broad ; minutely punctate and sparsely glandulose beneath ; inflorescence terminal, 10–20 flowered, racemose-paniculate ; peduncles 5–7 mm. long, 1–3-flowered, pedicels 7–9 mm. long ; calyx about 15 mm. long, and at the top 15 mm. in diam., glabrous, covered with small flattened glands ; corolla campanulate funnel-shaped, 5–6 cm. long, glabrous outside, slightly puberulous within, with small immersed glands in the inferior part ..   ..   *scandens*

D. **scandens** *Heine & Sandwith*, l.c. 223, figs. 1 & 2 (1962). *Kigelia dinklagei* Aubrév. & Pellegr. in Fl. For. C. Iv. 3 : 210 (1936), invalidly published (short French descr. only). *Dinklageanthus volubilis* Melchior ex Dinklage & Mildbr. in Fedde Rep. 41 : 266 (1937), name only. A climber, up to 11 ft. high, in trees of coastal savanna, with white flowers, corolla outside lilac at the base, inside with purplish lines and dots ; ripe fruits unknown.
Lib.: Monrovia (Nov.) *Dinklage* 2926 ! 3282 ! Duport (Nov.) *Bequaert* in *Hb. Linder* 1511 ! Mt. Barclay (Nov.) *Bunting* 22 ! Arboretum of the College of Forestry, Paynesville (Nov.) *Voorhoeve* 76 ! 82 !

2. KIGELIA DC. Bibl. Univ. Genève, n.s., 17 : 135 (1838) ; F.T.A. 4, 2 : 533.

A tree, extremely variable in all its parts, with imparipinnate leaves up to 50 cm. long, with 3–6 pairs of leaflets ; leaflets elliptic, elliptic-oblong, to oblong-lanceolate, up to 20 cm. long and 6 cm. broad, papery to subcoriaceous, rounded to apiculate, entire to slightly dentate-serrate in the upper part, sometimes scabrid above, indumentum on the rhachis and on the leaves beneath present or absent, with 7–12 pairs of lateral nerves, sessile or stalked up to 1 cm. long ; inflorescences in long-peduncled hanging panicles, up to 50 cm. long ; flowers 6–13 cm. long ; calyx 1–3 cm. long and 1–2 cm. broad ; corolla 5–10 cm. long ; fruits sausage-like, up to 45 cm. long and 15 cm. diam. ..   ..   *africana*

K. **africana** (*Lam.*) *Benth.* in Fl. Nigrit. 463 (1849) ; F.T.A. 4, 2 : 536 ; Chev. Bot. 487 ; Aubrév. Fl. For. C. Iv. ed. 2, 3 : 242, t. 231 ; Fl. For. Soud.-Guin. 494, t. 111 ; Sillans in Not. Syst. 14 : 323 (1953), incl. var. *aethiopica* (Decne.) Aubrév. ex Sillans[1] and var. *elliptica* (Sprague) R. Sillans l.c. p. 324. *Bignonia africana* Lam. Encycl. 1 : 424 (1785). *Crescentia pinnata* Jacq. Collect. 3 : 203, t. 18 (1789). *Sotor aethiopium* Fenzl (1844). *Kigelia aethopica* Decne. in Deless. Ic. Select. Pl. 5 : 39, t. 93 (1845) ; F.T.A. 4, 2 : 538, incl. vars. *abyssinica* (A. Rich.) Sprague, *bornuensis* Sprague and *usambarica* Sprague (1906). *K. abyssinica* A. Rich. Tent. Fl. Abyss. 2 : 60 (1847). *K. pinnata* (Jacq.) DC. Prod. 9 : 247 (1845). *K. acutifolia* Engl. ex Sprague in F.T.A. 4, 2 : 535 (1906) ; F.W.T.A., ed. 1, 2 : 240. *K. elliottii* Sprague in F.T.A. 4, 2 : 536 (1906) ; Chev. Bot. 487 ; F.W.T.A. ed. 1, 2 : 240. *K. elliptica* Sprague in F.T.A. 4, 2 : 534 (1906) ; F.W.T.A. ed. 1, 2 : 240. *K. impressa* Sprague in F.T.A. 4, 2 : 535 (1906) ; F.W.T.A., ed. 1. 2 : 240. *K. spragueana* Wernham in J. Bot. 52 : 31 (1914). *K. tristis* A. Chev. Bot. 487 (1920), name only. *K. talbotii* Hutch. & Dalz. F.W.T.A., ed. 1, 2 : 238 (1931). *K. aethiopium* (Fenzl) Dandy in F. W. Andr. Fl. Sudan 3 : 156 (1956). A medium-sized tree 20–50 ft. high ; flowers variable in colour, purplish-red, purple-yellow, orange, greenish-yellow, and sometimes spotted with darker reddish colours ; fruits like long sausages. The Sausage Tree.
Sen.: *Adanson* 199a. *Michelin* ! **Gam.:** *Brown-Lester* 815 ! **Mali:** Timbuktu *Chev.* 1208 ! El Massara *Chev.* 3314 ! 3315 ! **Guin.:** Kouria *Caille* in *Hb. Chev.* 14916 ; 14949. **S.L.:** Freetown (Sept.) *Lane-Poole* 357 ! Buyabuya (Feb.) *Sc. Elliot* 4757 ! *Harley* 1406 ! Talla (Feb.) *Sc. Elliot* 5037 ! Musaia (Sept.) *Small* 267 ! Rotifunk (Oct.) *Deighton* 4134 ! **Lib.:** Beiden, Boporo Dist. *Baldwin* 10273 ! **Iv.C.:** Dabou *Chev.* 15559. Agboville *Aubrév.* 427 ; Rasso F.R. *Aubrév.* 563 ; Djibi *Aubrév.* 1361 ; Banco *Aubrév.* 1368. **Ghana:** Cape Coast *T. Vogel* 88 ! Accra & Dodawah Plains (Feb.) *Irvine* 1525 ! Akotaa (Oct.) *Andoh* FH 5579 ! Nangodi *Vigne* FH 4489 ! Kpedsu (fl. & fr. Jan.) *Howes* 1088 ! **Togo Rep.:** *Kersting* 79 ! **N.Nig.:** Jos Plateau (Mar.) *Lely* P198 ! Wawa, Borgu *Barter* 724 ! Katagum Dist. (Aug.) *Dalz.* 105 ! Bornu (fl. & fr. Apr.) *E. Vogel* 83 ! **S.Nig.:** Bonny *Monteiro* ! R. Ossin *Barter* 1145 ! Oban *Talbot* 115 ! 1016 ! Cross R. Div. (fr. Jan.) *Holland* 237 ! Ogoja Dist. *Rosevear* 78/29. **[Br.]Cam.:** Barombi (Mar.) *Preuss* 202. Bali-Ngemba F.R., 5,500 ft. (June) *Tiku, Dioh & Ujor* FHI 30417 ! Serti, Adamawa Prov. *Latilo & Daramola* FHI 28952 ! Mayuko (Feb.) *Dalz.* 8224 ! Lus, Nkambe Div. (Feb.) *Hepper* 1878 ! **F.Po:** (Dec.) *Mann* 2 ! Widespread in tropical Africa. (See Appendix, p. 443.)
[Note : Sprague in F.T.A. accepted for *Kigelia* 14 taxa : 10 in specific, and 4 in infraspecific (varietal) rank. Aubréville & Sillans more recently accepted only 1 sp. with 3 varieties in our area and Ubangi-Shari : i.e. *K. africana* var. *africana*, var. *aethiopica* (Decne.) Aubrév. ex Sillans, var. *elliptica* (Sprague) Sillans. But Sillans states that the characters used in his key are due to ecological adaptations. I find from a taxonomic study of all the characters accepted in Sprague and Sillans that they break down into variability and the only practical conclusion possible is that there is one extremely variable species in Africa. Much of its variability may be accounted for by the fact that it grows in both dense forest and savanna. —H. H.]

3. STEREOSPERMUM Cham. in Linnaea 7 : 720 (1832) ; F.T.A. 4, 2 : 517.

Calyx campanulate, 5–6 mm. long, softly tomentellous to nearly glabrous outside ; corolla about 5 cm. long, tube softly pubescent outside ; leaflets 3–4 pairs, stalked, oblong or oblong-elliptic, shortly acuminate, 5–10 cm. long, entire or crennate-serrate, glabrous or softly pubescent ; panicle large, usually softly pubescent ; capsule cylindrical, elongate, up to 60 cm. long ; seeds winged at each end, 2·5–3 cm. long
1. *kunthianum*

[1] *K. africana* var. *aethiopica* Aubrév., Fl. For. Soud.-Guin. 496 (1950), is not validly published.

Calyx tubular, 1·5–2 cm. long, glabrous outside ; corolla 5–6 cm. long, tube glabrous or very sparingly pubescent outside ; leaflets 5–6 pairs, oblong, long-acuminate, 8–14 cm. long, closely reticulate below, glabrous ; panicle glabrous except for the small bracts ; fruit as above   ..   ..   *2. acuminatissimum*

Fig. 295.—Stereospermum kunthianum *Cham.* (Bignoniaceae).

A, leaf. B, flowers. C, stamen. D, calyx and style. E, ovary. F, cross-section of ovary.
G, seeds. H, fruits.

1. **S. kunthianum** *Cham.* in Linnaea 7 : 721 (1832) ; F.T.A. 4, 2 : 518 ; Chev. Bot. 487 ; Aubrév. Fl. For. C. Iv., ed. 2, 3 : 240 ; Fl. For. Soud.-Guin. 497, t. 112, fig. 1–4. A tree of savanna woodlands, up to 30–40 ft. high, with drooping ample panicles of pink or purplish and darker streaked flowers 1–2 in. long ; flowering in the dry season usually while leafless.
    Sen.: *Heudelot* 202 ! *Perrottet* 499 ! **Gam.:** Genieri (Feb.) *Fox* 75 ! Niani (Feb.) *Frith* 165 ! **Mali:** Macina to Sarro (Mar.) *Davey* 502 ! Kayes (?) (Jan.) *Roberty* 17025 ! **Port. G.:** Farim *Juan d'Orey* 282 ! Geba, Bafata (Nov.) *Esp. Santo* 2675 ! 2667 ! Catauliez *Juan d'Orey* 282 ! **Guin.:** Fouta Djalon *Pobéguin* 180 ! **Ghana:** Agomeda, Shai Plains (Feb.) *Irvine* 1505 ! Adakulu Plains (Mar.) *Lloyd Williams* 58 ! Prang to Attabubu *Kitson* 686 ! Kengerigongo to Bisuma, *Kitson* 822 ! **Togo Rep.:** *Baumann* 372 ! **N.Nig.:** Bida (May) *Lamb* 36 ! Nupe *Barter* 1135 ! Kontagora (fr. Jan., fl. Mar.) *Dalz.* 181 ! *Daley* FHI 32256 ! Bornu E. *Vogel* 89 ! Zungeru (fl. Oct.) *Elliott* 12 ! **S.Nig.:** Lagos *Rowland* ! *Foster* 26 ! Olokemeji *Ross* 15 ! [**Br.**] **Cam.:** Madaki, Adamawa (Dec.) *Latilo & Daramola* FHI 28832 ! Extends to the Red Sea, and south to Nyasaland and Congo Basin.
2. **S. acuminatissimum** *K. Schum.* in E. & P. Pflanzenfam. 4, 3B : 243 (1895) ; F.T.A. 4, 2 : 519 ; Aubrév. Fl. For. C. Iv., ed. 2, 3 : 242, t. 340 ; Fl. For. Soud.-Guin. 494 ; Chev. Bot. 486. A tall tree up to 100 ft. or more high ; with pink or pale purple flowers 1½–2 in. long in ample corymbose-pyramidal panicles ; in forests and often planted for ornamental purposes.
    Guin.: *Collenette* 39 ! Kindia (July) *Jac.-Fél.* 1800 ! **S.L.:** Wilberforce (June) *Lane-Poole* 56 ! Kessewe (Aug.) *Lane-Poole* 133 ! Panguma (June) *Lane-Poole* 90 ! Bumbuna (Aug.) *Dawe* 518 ! Tonkoli Forest (Aug.) *Thomas* 69 ! Panguma (July) *Deighton* 374 ! Loma Mts. *Jaeger* 1568 ! Musaia (July) *Miszewski* 2 ! 13 ! **Lib.:** Zeahtown (Aug.) *Baldwin* 6975 ! Gbau, Sanokwele Dist. (fr. Sept.) *Baldwin* 9425 ! **Iv.C.:** *Aubrév.* 659 ! 1597 ! **N.Nig.:** Mada Hills, (July) *Hepburn* 91 ! **S.Nig.:** Idanre (Aug.) *Olorunfemi* FHI 41527 ! Imo R. (Dec.) *Unwin* 18 ! 31 ! Lagos *Phillips* 41 ! Ukpon River F.R. (Aug.) *Amachi* FHI 38279 ! Ukpon F.R., Ogoja Prov. (July) *Latilo* FHI 38164 ! [**Br.**]**Cam.:** Barombi *Preuss* 332 ! Johann-Albrechts-höhe (=Kumba) *Staudt* 950 ! Bamenda (June) *Ujor* FHI 30441 ! Also in Cameroun. (See Appendix, p. 445.)

4. SPATHODEA P. Beauv. Fl. Oware. 1 : 46 (1805) ; F.T.A. 4, 2 : 528.

Branchlets tomentellous ; leaves opposite, sometimes unequal in each pair ; leaflets 4–8 pairs, elliptic or oblong, rather gradually acuminate, shortly cuneate and with a large gland at the base, up to 14 cm. long and 6 cm. broad, very minutely reticulate beneath, with about 6 pairs of lateral nerves ; racemes terminal, few-flowered ; pedicels stout, up to 6 cm. long, finely tomentellous ; calyx recurved, long-acuminate, ribbed, softly tomentellous, 6–8 cm. long, split down one side ; corolla widely campanulate from a contracted base, 10–12 cm. long ; fruit 15–20 cm. long ; seeds winged all round, 2 cm. broad (including the hyaline wing)   ..   ..   *campanulata*

S. campanulata *P. Beauv.* l.c. 47, t. 27 (1805) ; F.T.A. l.c. 529 ; Chev. Bot. 485 ; Aubrév. Fl. For. C. Iv., ed. 2, 3 : 248, t. 344 ; Fl. For. Soud.-Guin. 494. A tree 20–70 ft. high with conspicuous flaming inflorescences, flowers scarlet or orange-red with yellow margin ; habitat mainly fringing forests.
    Guin.: Kouria *Caille* in *Hb. Chev.* 14916 ! Timbo *Caille* in *Hb. Chev.* 14765. Mamou *Caille* in *Hb. Chev.* 14816. Fouta Djalon *Chev.* 18720. **S.L.:** Yetaya *Thomas* 2268 ! Kabala (Sept.) *Thomas* 2208 ! Panguma (Aug.) *Smythe* 59 ! Kambia (Dec.) *Sc. Elliot* 4502 ! Njala (Oct.) *Deighton* 1415 ! **Lib.:** Gbau, Sanokwele Dist. (Sept.) *Baldwin* 9403 ! **Iv.C.:** Dabou (fr. Feb.) *Chev.* 12616 ! Bériby *Chev.* 20024. Bouaké *Chev.* 22131. Agboville *Chev.* 22335. **Ghana:** Ejura (Aug.) *Chipp* 730 ! Kanyankor (Nov.–Dec.) *Miles* 16 ! Chibbi

(Sept.) *Gent* 934 ! Aburi (Feb.) *Howes* 1012 ! Mampong *Brown* 2064 ! **Togo Rep.**: Lomé *Warnecke* 447 ! Kuië stream *Kersting.* **Dah.**: (Oct.) *Le Testu* 19 ! **N.Nig.**: Stirling Hill *Ansell* ! Tonti, Kurmi Tonduru (Nov.) *Latilo & Daramola* FHI 28741 ! Kabba *Parsons* 157 ! Agaie (Oct.) *Yates* 22 ! **S.Nig.**: *Barter* 560 ! Lagos *Irving* ! Abeokuta *Irving* ! Enugu F.R. *Smith* 13 ! Ibadan (Nov.) *Newberry & Etim* 177 ! *MacGregor* 557 ! Ugwogo, Enugu (Sept.) *Rosevear* 1/28 ! Benin (Aug.) *Unwin* 126 ! Calabar (Sept.) *Holland* 151 ! **F.Po**: (May) *Mann* 387 ! Extending to Congo and Angola. (See Appendix p. 445.)

5. MARKHAMIA Seemann ex K. Schum. in E. & P., Pflanzenfam. 4, 3B : 242 (1895)[1]; Sprague in F.T.A. 4, 2 : 522, and in Kew Bull. 1919 : 309.

Pseudostipules orbicular, foliaceous, about 2 cm. diam. ; calyx densely lepidote out-side, 2·5 cm. long, with a long slender beak in bud, split down one side ; leaflets 3–4 pairs, oblong or oblong-elliptic, gradually acuminate, unequally wedge-shaped at the base, up to 17 cm. long and 7 cm. broad, pubescent only in the axils of the nerves ; panicles small, few-flowered ; corolla about 6 cm. long, contracted for nearly 1 cm. at the base, broadly expanded to the top ; fruit elongate-linear, 1·5 cm. broad, with one nerve down the middle   ..   ..   1. *lutea*

Pseudostipules conical, tomentose ; calyx split down one side, tomentose outside, 2·5 cm. long, shortly beaked in bud ; leaflets 4–5 pairs, oblong-elliptic, rather gradu-ally acuminate, shortly wedge-shaped at the base, 8–15 cm. long, 4–6 cm. broad, more or less pubescent or tomentose beneath ; corolla 5–6 cm. long, gradually expanded from the base ; fruits elongate-linear, 1·5 cm. broad, softly tomentose

                                     2. *tomentosa*

1. **M. lutea** *(Benth.) K. Schum.* l.c. (1895) ; Sprague in F.T.A. 4, 2 : 525 ; Kew Bull. 1919 : 525 ; *lutea* Benth. in Fl. Nigrit. 461 (1844) ; A. Chev. Bot. 486 ; Aubrév. Fl. For. C. Iv., ed. 2, 3 : 246, t. 1343, fig. 8 ; Fl. For. Soud.-Guin. 499. A shrub or tree 15–30 ft. high with finely scaly branchlets becoming glabrous, and yellow flowers marked with purple, over 2 in. long in terminal and axillary corymbose panicles, and long narrow falcate fruits 1½–3 ft. long with seeds 1¼ in. long, including the hyaline wing at each end.

    Ghana: *Brown* 393 ! Anum *Plumptre* 38 ! Odumase (fr. Jan.) *Irvine* 1932 ! Senchi Ferry (Oct.) *Morton* ! **S.Nig.**: Abeokuta (fr. Dec.) *Olorunfemi* FHI 40321 ! **[Br.]Cam.**: *Preuss* 436 ! Victoria *Kalbreyer* 33 ! Buea (Mar.) *Maitland* 475 ! 572 ! (fr. Feb.) *Dalz.* 3237 ! Johann-Albrechtshöhe *Staudt* 486 ! **F.Po**: (fl. Mar., Nov., Dec., fr. Nov.) *T. Vogel* 60 ! *Mann* 5 ! 399 ! *Ansell* !

2. **M. tomentosa** *(Benth.) K. Schum. ex Engl.* in Glied. Veg. Usambara 34 (1894) ; E. & P. Pflanzenfam, 4, 3B : 242 (1895) ; Sprague in F.T.A. 4, 2 : 528 ; Kew Bull. 1919 : 311, incl. var. *gracilis* Sprague in F.T.A. 4, 2 : 528 (1906) ; Chev. Bot. 486 ; Aubrév. Fl. For. C. Iv., ed. 2, 3 : 246, t. 343, fig. 1–7 ; Fl. For. Soud.-Guin. 499. *Spathodea tomentosa* Benth. in Fl. Nigrit. 462 (1849). A tree up to 30 ft. high with yellowish-tomentose brachlets, and yellow flowers with purple lines, 2–2½ in. long in long terminal oblong tomentose racemes, and fruit up to 2 ft. or more long ; habitat commonly fringing forests.

    Guin.: Kouroussa to Conakry *Pobéguin* 746 ! **S.L.**: Sefadu (Nov.) *Deighton* 4665 ! Sulimania (Mar.) *Sc. Elliot* 5281 ! Kumala (June) *Thomas* T63 ! Musaia (fr. Feb. ; fl. & fr. Oct.) *Deighton* 4210 ! *Thomas* 2622 ! Talla Hills (fr. Feb.) *Sc. Elliot* 4992 ! **Ghana**: Larte (Apr.) *Morton* A834 ! Pepeose to Nkwantaneng *Morton* A796 ! Ejura (fl. & fr. Aug.) *Chipp* 770 ! Nwereme (May) *Irvine* 2513 ! **Togo Rep.**: Kuië stream *Kersting* 633 ! **N.Nig.**: Patti Lokoja *Barter* 555 ! Nupe *Barter* 1310 ! Dekina (June ; Aug.) *Elliott* 89 ! 242 ! Wana (July) *Hepburn* 90 ! Acharane F.R. (June) *Daramola* FHI 38042 ! Odoba Forest (Aug.) *Horwood* FHI 38351 ! **S.Nig.**: Lagos (Aug.) *Phillips* 58 ! *MacGregor* 703 ! *Dawodu* 203 ! *Foster* 101 ! Shabe Rock (Oct.) *Hambler* 518 ! Asaba *Leslie* 37 ! Oban (May) *Talbot* 2055 *bis* ! Enugu (Oct.) *Onyeagocha* FHI 16597 ! **[Br.] Cam.**: Bamenda (June) *Maitland* 1576 ! 1610 ! Also in Cameroun. (See Appendix, p. 444.)

Fig. 296.—NEWBOULDIA LAEVIS (*P. Beauv.*) *Seemann ex Bureau* (BIGNONIACEAE).

A, calyx and style. B, stamens and part of corolla-tube. C, pistil.

[1] *Markhamia* Seemann in J. Bot. 1 : 226 (1863), name only typification by citing " *Spathodea stipulata* Wall.'' as "type''of the new genus.

6. NEWBOULDIA Seemann ex Bureau, Mon. Bignon., Atlas t. 15 (1864) ; Seemann in J. Bot. 1 : 225 (1863), name only ; l.c. 8 : 337 (1870) ; F.T.A. 4, 2 : 520.

Branchlets glabrous ; leaflets 3–6 pairs, oblanceolate to broadly elliptic, serrate to entire or nearly so, long-acuminate, glandular at the base beneath, up to 20 cm. long and 10 cm. broad, minutely punctate beneath ; flowers in a terminal racemose panicle ; pedicels 6–8 mm. long, glabrous ; calyx 2·5 cm. long, split down one side and shortly bilobed at the apex, glabrous outside ; corolla nearly regular, narrowly campanulate from a contracted cylindrical base, about 6 cm. long, glabrous outside ; fruit about 30 cm. long, the valves 3-nerved ; seeds winged at each end, 3·5 cm. long ..   ..   *laevis*

N. laevis (*P. Beauv.*) *Seemann ex Bureau* l.c. (1864) ; Seemann in J. Bot. 8 : 337 (1870)[1] ; F.T.A. 4, 2 : 251 ; Chev. Bot. 485 ; Aubrév. Fl. For. C. Iv., ed. 2, 3 : 244, t. 342 ; Fl. For. Soud.-Guin. 494. *Spathodea laevis* P. Beauv. Fl. Oware. 1 : 48, t. 29 (1805). A tree of medium size in forest, commonly slender, 10–40 ft. high in towns, nearly glabrous, with red-purple or pink and white foxglove-like flowers 2 in. or more long, in erect dense raceme-like panicle.

Sen.: (Dec.) *Döllinger* 36 ! *Perrottet* 500 ! *Heudelot* 818 ! **Gam.:** *Brown-Lester* S31 ! S46 ! Tamba Jang Jarra (fl. & fr. Apr.) *Frith* 152 ! **Guin.:** (fr. Sept.) *Farmar* 185 ! Timbo *Pobéguin* 129 ! Kindia *Chev.* 13027 ; 13223 ; 13308. Sangorola (Feb.) *Chev.* 347 ! Kollangui *Chev.* 12559. **S.L.:** *Barter* ! *Garrett* 21 ! *Thomas* 8624 ! 10028 ! Kambia (fr. Dec.) *Sc. Elliot* 4350 ! Mabasike (Feb.) *Lane-Poole* 79 ! Batkanu (Apr.) *Thomas* 37 ! Njala (Feb.) *Deighton* 509 ! Jala (Feb.) *Bunting* 13 ! **Lib.:** *Whyte* ! **Iv.C.:** Bingerville *Chev.* 16061 ; 17287. Bouroukrou *Chev.* 16133. Brafouédi (Dec.) *de Wit* 7878 ! **Ghana:** *Thonning* (photo) ! Ejura (Dec.) *Morton* 9755 ! Sahara, W. Prov. (Feb.) *Vigne* FH 218 ! Aburi (Jan.) *Morton* 6345 ! *Johnson* 297 ! Kpeve (fr. Mar.) *Lloyd Williams* 42 ! Wurubong to Okraji (fr. May) *Kitson* 1145 ! **Togo Rep.:** Lomé *Warnecke* 65 ! **Dah.:** *Burton* ! Sakété *Chev.* 22854. **N.Nig.:** *Rosevear* 1/30 ! 36/28 ! 40/28 ! Kontagora (Jan.) *Dalz.* 184 ! **S.Nig.:** Lagos *Moloney* ! Olokemeji F.R. (Jan.) *Ross* 54 ! Akwa *Thomas* 62 ! Ibadan (Jan.) *Richards* 5003 ! Ala (Nov.) *Thomas* 1949 ! Oban *Talbot* 1342 ! Degema *Talbot* 3689 ! *Millen* 25 ! 59 ! *Lamborn* 29 ! [Br.]**Cam.:** Victoria (Jan.) *Maitland* 282 ! 772 ! Likomba *Mildbr.* 10599 ! Johann-Albrechtshöhe *Staudt* 898. S. Bakundu F.R., Kumba (Jan.) *Daramola* FHI 35500 ! Gashaka, Adamawa (Dec.) *Latilo & Daramola* FHI 28967 ! **F.Po:** (fl. & fr. Dec.) *Mann* 84 ! Also in Cameroun. (See Appendix, p. 444.)

158. PEDALIACEAE

By H. Heine

Annual or perennial herbs. Leaves opposite or the upper alternate, simple ; stipules absent. Flowers hermaphrodite, zygomorphic. Calyx of 5 or 4 segments or 4-fid or spathaceous. Corolla gamopetalous, often oblique ; lobes 5, imbricate. Stamens 4, or rarely only 2 perfect, alternate with the corolla-lobes ; anthers connivent in pairs, 2-celled, cells distinct, parallel or divaricate, opening lengthwise ; fifth stamen often represented by a staminode. Disk hypogynous, fleshy. Ovary sessile, 1-celled with 2 intrusive parietal placentas or 2–4-celled, the cells again often divided by spurious septa ; style terminal. Ovules solitary or numerous on each placenta. Fruit a capsule, nut, or subdrupaceous ; endocarp hardened and often horned or prickly. Seeds without endosperm ; embryo with flat cotyledons.

Flowers in erect terminal racemes ; pedicels not inserted between two small sessile nectarial glands, short   ..   ..   **1. Martynia**

Flowers solitary, axillary ; pedicels inserted between 2 small nectarial glands :
Ovary and fruit 2-celled ; cells undivided ; fruit indehiscent, more or less turbinate, regular and symmetric, 4-angled, up to 2 cm. long, with one lateral spreading spine at each basal angle ..   ..   **2. Pedalium**

Ovary and fruit 2-(rarely 4-) celled ; cells incompletely divided by a false septum ; fruit dehiscent or indehiscent, never turbinate or properly 4-angled, obliquely ovoid (and therefore more or less asymmetric) and rostrate when spined at the base :
Cells of the ovary and capsule very unequal, the posticous one small and indehiscent   **3. Rogeria**

Cells of the ovary and capsule equal-sized :
Capsule acute or beaked, without lateral appendages ..   ..   ..   **4. Sesamum**
Capsule with 2 divergent horns or spines at the apex ..   ..   **5. Ceratotheca**

1. MARTYNIA Linn. Sp. Pl. 618 (1753).

Annual ; clothed with patent, long, glandular-viscid hairs ; main stem almost hollow ; leaves opposite, long-petioled, broad, palmatinerved ; flower with 2 membranous bracteoles at the base ; calyx 5-partite to the base ; corolla obliquely campanulate thimbleshaped, lobes very unequal, broad ; perfect stamens 2 (anterior ones) ; anthers cohering ; ovary 1-celled, divided into 4 compartments by 2 deeply intruding, parietal,

---

[1] Seemann, l.c. 1 : 225 (1863) discusses only the taxonomic position of *Spathodea laevis* P. Beauv. (which he cites as the "type" of *Newbouldia*) but he then omitted to make the necessary new combination under *Newbouldia.*—H. H.

opposite, T-shaped placentas touching in the middle ; fruit a beaked horn-like capsule
about 3 cm. long ..   ..   *annua*

M. annua *Linn.* l.c. (1753) ; Backer in Fl. Males., ser. 1, 4 : 221, fig. 3 (p. 220) (1951). A foetid, erect, rank
herb, frequently widely branched, up to 6 ft. high, with yellowish or greenish-white flowers.
S.Nig.: Ago-Are, Oyo (Apr.–May) *Onochie* FHI 34001 ! *Sofoluwe* FHI 38160 ! Sabongida Ora, Benin Prov.
(July) *Umana* FHI 34507 ! Ibadan (June) *Squire* K1 ! Olokemeji (May) *Foster* 303 ! Okeke &
Adebusuyi FHI 18248 ! Native of Mexico, locally naturalized in tropical Africa and in other parts of the
tropics.

2. PEDALIUM Linn.—F.T.A. 4, 2 : 540.

Annual herb with simple or branched subsucculent stem ; leaves petiolate, elliptic,
obovate or oblong, rounded or truncate at the apex, entire or coarsely dentate, 1·5–
5 cm. long ; flowers solitary in the leaf axils ; calyx small ; corolla-tube subcylindric
or narrowly funnel-shaped, about 2·5 cm. long, pale primrose-yellow ; fruit an
indehiscent 4-angled capsule, subpyramidal, rounded to acute at the apex, with a
spreading spine at each basal angle, and then abruptly contracted into a narrow collar
below, 1–2 cm. long and 0·6–1 cm. broad ..   ..   *murex*

P. murex *Linn.* Syst. Nat., ed. 10, 1123 (1759) ; F.T.A. 4, 2 : 540 (1906) ; F.T.E.A., Pedaliac. 6, fig. 2.
spreading or subprostrate, sparsely glandular herb, 1–2½ ft. high ; a saline soil indicator in sand or lime-
stone in short grass near the coast.
Ghana: Keta (fr. July) *Akpabla* 32 ! **Togo Rep.:** *Mahoux* 511 ! **S.Nig.:** Badagry (fr. Aug.) *Onochie* FHI
33470 ! Also in East tropical Africa, Socotra, Madagascar, and in India and Ceylon.

3. ROGERIA J. Gay ex Delile, Cent. Pl. Afr. 78 (1826)[1] ; F.T.A. 4, 2 : 548.

Stems mealy-glandular in the upper part ; leaves broadly obovate, 3-lobed, truncate or
shortly cuneate at the base, 5–10 cm. long, glandular below and rather glaucous ;
lobes obtuse, dentate, teeth subulate, with glands at the base ; flowers 1–3 in the
leaf-axils, subsessile ; corolla 4–5 cm. long ; capsule 2·5–3 cm. long with 4 spreading
spines below the middle, often with 4 smaller between ..   ..   ..   *adenophylla*

R. adenophylla *J. Gay ex Delile* l.c., t. 2, fig. 3 (1826) ; J. Gay in Ann. Sci. Nat. sér. 1, 1 : 457 (1824)[2] ; F.T.A.
4, 2 : 548 ; Chev. Bot. 488 ; Berhaut Fl. Sén. 64. Stout herb up to 9 ft. high ; flowers white to purplish ;
in dry savanna.
Sen.: *Roger* ! *Perrottet* 548 ! Richard Tol (Jan.) *Döllinger* 68 ! Bondy (fl. & fr. Nov.) *Trochain* 891 ! *Berhaut*
1202. **N.Nig.:** Jos *Kennedy* 2863 ! Keffin Hausa *Dalz.* 100 *bis* ! Kouka, L. Chad (fl. & fr. Feb.) *E. Vogel*
48 ! Mongonu, Baga (fr. Jan.) *Daggash* FHI 24962 ! Sokoto (July) *Dalz.* 365 ! *Moiser* 85 ! Also in
Mauritania, Sudan and extending to Angola. (See Appendix, p. 446.)

4. SESAMUM Linn.—F.T.A. 4, 2 : 550.

Seeds muricate, 2–3 mm. long, with a suborbicular 2–3 mm.-long wing at the base and
apex ; capsule sub-obconical, narrowed to the base, about 5 cm. long, long-beaked,
strigillose, calyx deciduous in fruit ; plants nearly glabrous, lower leaves palmately
3–5-foliolate or -partite with narrow lobes or segments ; corolla up to 3 cm. long,
pubescent ..   ..   1. *alatum*
Seeds not winged ; capsule parallel-sided, rounded at the base ; calyx persistent in
fruit ; plants usually pubescent, leaves when lobed or partite with broad and ovate
segments :
Mature seeds on their broad surfaces with finely reticulate ribs or almost smooth,
up to 3 mm. long and 1·5 mm. broad ; capsule up to 3 cm. long and 12 mm. broad,
crowned by a very conspicuous subulate beak ; lowest leaves often deeply divided or
palmately compound ..   ..   2. *indicum*
Mature seeds rugose or pitted ; leaves never parted or divided :
Seeds up to 3 mm. long and 2 mm. broad ; leaves petiolate (petiole 0·2–2·5 cm. long),
varying from obovate-elliptic (the lower) to narrowly oblong (the upper), 3·5–
11 cm. long and 1·2–4 cm. broad ; capsule up to 3 cm. long and 1 cm. broad
3. *radiatum*

Seeds up to 1·5 mm. long and 1 mm. broad ; leaves sessile or subsessile (very rarely
the lower ones shortly petiolate ; the petiole then up to 5 mm. long), linear to
linear-lanceolate, 2–12 cm. long and 0·1–1 cm. (lower ones sometimes up to 4 cm.)
broad ; capsule up to 2·4 cm. long and 4 mm. broad ..   ..   4. *angustifolium*

1. S. alatum *Thonning* in Beskr. Guin. Pl. 284 (1827) ; F.T.A. 4, 2 : 559 ; Chev. Bot. 489 ; Berhaut, Fl. Sén.
11, 23, 264 ; F.T.E.A., Pedaliac., 17, fig. 7, 1–3. *S. sabulosum* A. Chev. Bot. 489 (1920), name only. Erect
herb 2–3 ft. high with sulcate 4-angled stems finely glandular when young ; flowers pink or bright red with
spotted throat ; in dry savanna.
Sen.: *Roger* ! *Berhaut* 757. **Mali:** Timbuktu (fl. & fr. June, July) *Chev.* 1293 ! *Hagerup* 117 ! Diendéla
(Mar.) *Chev.* 624 ! Fenadjie (fl. & fr. June) *Davey* 124 ! **Ghana:** *Thonning* (photo) ! Burufo Plain (fl. &
fr. Dec.) *Adams* 4442 ! Bongo to Nangodi (fr. Nov.) *Morton* A3805 ! **N.Nig.:** Borgu *Barter* 119 ! Kontagora (fr. Dec.) *Dalz.* 151 ! Sokoto (Nov.) *Moiser* 12 ! Alkalire *Lely* 645 ! Chad *Gaillard* ! Daura (fl. & fr.
Nov.) *Sampson* 11 ! [**Br.**]**Cam.:** Bama, Dikwa Dist. (fl. & fr. Nov.) *McClintock* 31 ! Also in Sudan, Ethiopa,
E. Africa to Mozambique, S. Rhodesia, Bechuanaland and S.W. Africa. (See Appendix, p. 447.)

---

[1] This is apparently a separate publication, which was subsequently published in Caillaud's " Voyage à
Méroé", vol. 4 (1827), as an appendix. The plate no. in the whole work is No. LXII and the reference to the genus
is accordingly 4 : 368 (1827).
[2] This does not constitute a valid description, as no proper generic description was given at the time of
publication.

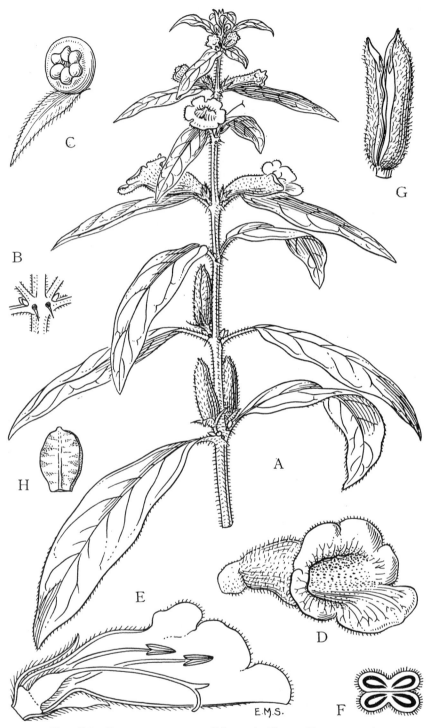

Fig. 297.—Sesamum radiatum *Schum. & Thonn.* (Pedaliaceae).

A, habit, × ⅔. B, internode showing glands, × 2. C, gland, × 20. D, corolla, × 2. E,
longitudinal section of flower, × 2. F, transverse section of ovary (diagrammatic). G, fruit,
× 1¼. H, seed, × 6. A–F drawn from *Meikle* 361. G–H from *Heudelot* 547.

2. **S. indicum** *Linn.* Sp. Pl. 634 (1753) ; F.T.A. 4, 2 : 558 ; Chev. Bot. 488, 489 (incl. vars.) ; F.T.E.A., Pedaliac. 17, fig. 7, 4–8 ; Fl. Males., ser. 1, 4 : 217 ; Berhaut Fl. Sén. 67. Erect, simple or branched, herb, 1–6 ft. high ; flowers whitish or pink and purple-tinged and spotted ; cultivated. Beniseed.
Sen.: *Heudelot* ! Galam (fl. & fr. Dec.) *Roger* 140 ! *Berhaut* 1813. **Mali:** Binkola (fr. Aug.) *Thomas* 1776 ! Tassakant (fl. & fr. Aug.) *Chev.* 3322 ! **S.L.:** Momaba (fr. Nov.) *Thomas* 4596 ! Bumban (Sept.) *Thomas* 1996 ! Mano *Thomas* 1002 ! Njala (fl. & fr. Aug.) *Deighton* 1972 ! Pehala (Oct.) *Fischer* 34 ! **Lib.:** Monrovia (Nov.) *Cooper* 133 ! Kakatown *Whyte* ! Firestone Plantation No. 3 (fl. & fr. July) *Linder* 122 ! Firestone Plantation, Dukwai R., *Lely* 622b ! 671 ! **N.Nig.:** Nupe *Barter* 1260 ! Kafanchan (fl. & fr. Dec.) *Sampson* 38 ! 40 ! Kontagora (fl. & fr. Dec.) *Dalz.* 150 ! Biu, Bornu Prov. (fl. & fr. Sept.) *Noble* 21 ! **S.Nig.:** Ibadan (fl. & fr. Jan.) *Smith* 1 ! Cultivated and naturalized in most tropical and subtropical countries. (See Appendix, p. 447.)

3. **S. radiatum** *Schum. & Thonn.* Beskr. Guin. Pl. 282 (1827) ; F.T.A. 4, 2 : 557 ; Chev. Bot. 489 ; Backer in Fl. Males., ser. 1, 4 : 218 ; Berhaut Fl. Sén. 67. *S. talbotii* Wernham Cat. Talb. 73 (1913). *S. caillei* A. Chev. Bot. 488 (1920), name only. Erect herb about 2 ft. high, glandular hairs, with an unpleasant smell ; flowers purplish-pink ; cultivated for its seed.
Sen.: *Heudelot* 547 ! *Berhaut* 1453. **Gam.:** Toniataba (fl. & fr. Oct.) *Agric. Dept.* ! **Guin.** Kouria to Longdery *Caille* in Hb. *Chev.* 14623. Kouroussa (June) *Pobéguin* ! **S.L.:** *Afzelius* ! *Don* ! *Smeathman* ! Sherbro Isl. (fr. Nov.) *Hunter* 21 ! Mahela (fr. Dec.) *Sc. Elliot* 4136 ! Njala (fr. Dec.) *Sc. Elliot* 1794 ! **Lib.:** *Whyte* ! Ganta (May) *Harley* ! Monroviatown (fl. & fr. Aug.) *Baldwin* 9003 ! Brewersville (fl. & fr. Sept.) *Barker* 1429 ! **Iv.C.:** Dabou (fr. July) *de Wilde* 108 ! Sassandra (fr. Mar.) *Leeuwenberg* 3097 ! Port Bouet (fl. & fr. Feb.) *Leeuwenberg* 2713 ! Assakra (fr. Aug.) *de Wit* 7007 ! **Ghana:** Dawa (May) *Howes* 912 ! Legon Hill (fl. & fr. Apr.) *Adams* 4242 ! **Dah.:** *Burton* ! **N.Nig.:** Nupe *Barter* 1202 ! Jebba (fl. & fr. Mar.) *Meikle* 1287 ! Vom *Dent Young* 190a ! **S.Nig.:** Lagos (fl. & fr. June) *Dalz.* 1244 ! 1080 ! Abeokuta *Migeod* ! *Irving* ! Ijebu (fl. & fr. July) *Dalz.* 149 ! Ibadan (fl. & fr. Oct.) *Okafor* FHI 35878 ! Ibadan (fl. & fr. Jan.) *Smith* 3 ! **[Br.]Cam.:** Victoria (Apr.) *Ejiofor* FHI 29334 ! Kumba to Victoria (fl. & fr. Apr.) *Olorunfemi* FHI 30534 ! Bama, Dikwa Dist. (fr. Jan.) *McClintock* 159 ! **F.Po:** *T. Vogel* 23. Throughout tropical Africa. Pantropical weed, undoubtedly of tropical African origin, sometimes cultivated. (See Appendix, p. 449.)

4. **S. angustifolium** (*Oliv.*) *Engl.* Pflanzenw. Ost-Afr. C : 365 (1895) ; F.T.A. 4, 2 : 554 ; F.T.E.A., Pedaliac. 19, fig. 7, 9–16. *S. indicum* var. ? *angustifolium* Oliv. in Trans. Linn. Soc. 29 : 131 (1875). A herb up to 6 ft. high, with pink or mauve flowers about 1 inch long ; sometimes cultivated.
N.Nig.: *Yates* 18a ! 19a ! Zungeru (fl. & fr. Apr.) *Dalz.* 149 ! **S.Nig.:** Lagos *Phillips* 48 ! Onitsha (fl. & fr. Mar.) *Killick* 99 ! Shaki, Oyo Prov. (Apr.) *Sofoluwe* FHI 38160 ! From Nigeria eastwards throughout tropical Africa, in cultivated and waste areas.

5. CERATOTHECA Endl. in Linnæa 7 : 5 (1832) ; F.T.A. 4, 2 : 563.

Stems pubescent ; leaves ovate-deltoid to ovate or linear-oblong, base more or less truncate or hastate, dentate, 3–6 cm. long ; petiole up to 2·5 cm. long ; corolla 1·5–4 cm. long (in one specimen described by Berhaut as var. *grandiflora* 5 cm. long) ; capsule 1·2–2·3 cm. long, pubescent, horns up to 3·5 mm. long .. .. *sesamoides*

C. sesamoides *Endl.* in Linnaea 7 : 5, figs. 1, 2 (1832) ; F.T.A. 4, 2 : 563 ; Chev. Bot. 489–490, incl. var. *baoulensis* A. Chev., name only ; Berhaut Fl. Sén. 67, incl. var. (see note below) ; F.T.E.A. Pedaliac. 14, fig. 6. *Sesamum heudelotii* Stapf in F.T.A. 4, 2 : 552 (1906) ; F.W.T.A., ed. 1, 2 : 244. Erect or suberect herb, branched, ½–3 ft. high, with foxglove-like flowers ¾–1 in. long, solitary in the axils, pink or light red-purple outside, paler with purple streaks inside, sometimes white.
Sen.: *Kohaut* ! *Heudelot* 147 ! *Perrottet* 552 ! *Roger* 141 ! Kouroussa (fl. & fr. Dec.) *Pobéguin* 616 ! Tambacounda *Berhaut* 1730 (the type of var. *grandiflora* Berh.) **Mali:** Sindou *Chev.* 868 ! Tiédiana *Chev.* 1001; San *Chev.* 1024; 3324. Ségou *Chev.* 3324. **Guin.:** Conakry *Chev.* 12154. Kankan *Brossart* in Hb. *Chev.* 15678. **S.L.:** Port Loko (fr. Dec.) *Thomas* 5819 ! Bumbunah (fl. & fr. Oct.) *Thomas* 3281 ! 3402 ! Ronietta (Nov.) *Thomas* 5293 ! Njala (fl. & fr. Dec.) *Deighton* 1783 ! Yetaya (fl. & fr. Sept.) *Thomas* 2389 ! Rokupr (Jan.) *Deighton* 4039 ! **Lib.:** Sanokwele (fr. Sept.) *Baldwin* 9527 ! Abuji (fl. & fr. Mar.) *Patterson* ! Ashanti *Kitson* ! Tamale (fl. & fr. June) *Saunders* 9 ! (Nov.) *Sampson* 2 ! *Lloyd Williams* 531 ! Gambaga Cliff (June) *Harris* ! **Togo Rep.:** Lorro, Missodu *Kersting* A374 ! **N.Nig.:** *Yates* 39 ! Ilorin (Mar.) *Lamb* 40 ! Bida Town (fr. July) *Onochie* FHI 35369 ! Maifoni *Parsons* ! Lokoja & Kontagora (fl. & fr. May) *Dalz.* 148 ! Sokoto (Sept.) *Moiser* 105 ! Fodama (fl. & fr. Jan.) *Moiser* 180 ! 212 ! **S.Nig.:** Ibadan (fl. & fr. Jan.) *Smith* 2 ! *Meikle* 998 ! Olokemeji (May) *Foster* 304 ! Throughout tropical Africa. (See Appendix, p. 446.)
[Berhaut's var. *grandiflora* (l.c., 1954), French descr. only, has "corollas up to 5 cm. long and a flat U-shaped capsule with two truncate horns". Not seen, but almost certainly only an occasional form.—H. H.]

159. ACANTHACEAE

By H. Heine

Herbaceous or climbing, rarely somewhat shrubby. Leaves opposite, often with distinct cystoliths ; stipules absent. Flowers hermaphrodite, zygomorphic, often with conspicuous bracts. Calyx-segments or lobes 4 or 5, imbricate or valvate; rarely the calyx reduced to a ring. Corolla gamopetalous, 2-lipped or sometimes 1-lipped, lobes imbricate or contorted. Stamens 4, didynamous or 2, inserted on the corolla-tube and alternate with its lobes ; filaments free amongst themselves or partially connate in pairs ; anthers 2-celled or 1-celled by reduction, cells confluent or separated, sometimes one much smaller than the other, opening lengthwise. Disk present. Ovary superior, sessile on the disk, 2-celled. Capsule mostly elastically dehiscent from the apex downwards, the valves recurved and leaving the central axis. Seeds mostly with indurated funicle ; endosperm rarely present ; embryo large.

Recognised by the usually herbaceous habit, opposite leaves and the peculiar capsule.
Corolla-lobes contorted in bud :
 Stamens 4 :
  Anthers 2-celled :
   Anther-cells not tailed at base :
    Ovules superimposed in each cell or rarely 1 :
     Ovules 3 or more in each cell :
      Corolla 2-lipped ; retinacula curved or papilliform ; seeds usually with hygro-
       scopic hairs ; herbs of various habit, sometimes heterophyllous, nearly always
       growing in wet habitats   ..   .. 1. **Hygrophila**
      Corolla not 2-lipped :
       Leaves in each pair equal- or subequal-sized :
        Flowers solitary or in small cymes, axillary :
         Small marsh-herb ; leaves up to 2 cm. long and 12 mm. broad; corolla 9–
          13 mm. long (*H. abyssinica*)..   ..   .. 1. **Hygrophila**
         Tall herbs ; leaves up to 6 cm. long and 3 cm. broad; corolla up to 4 cm. long
                                                                      2. **Ruellia**
        Flowers three to very many in a terminal spike or head, or inflorescence
         elongated and loose :
         Calyx divided into subequal segments, the lower part sometimes forming a
          little tube :
          Corolla-tube linear-cylindric almost to the small mouth, corolla-lobes 5,
           subreflexed, more or less at one side   ..   .. 3. **Eremomastax**
          Corolla-tube infundibuliform, gradually widened upwards into an oblique
           mouth, corolla-lobes more or less regularly arranged   4. **Heteradelphia**
         Calyx with two segments free to the base and three united at least one third
          of their length, usually more than three quarters ..   5. **Dischistocalyx**
        Leaves in each pair unequal-sized   ..   .. 6. **Endosiphon**
     Ovules 1–2 in each cell :
      Valves of the placenta not rising elastically from the base of the capsule :
       Stigma with one linear arm and one minute or obsolete ; seeds with many
        hygroscopic hairs   ..   .. 7. **Acanthopale**
       Stigma capitate, obscurely 2-lobed ; seeds glabrous   ..   .. 8. **Whitfieldia**
      Valves of the placenta rising elastically from the base of the ripe capsule and
       throwing out the seeds   ..   .. 9. **Phaulopsis**
     Ovules 2, and collateral in each cell :
      Fruit dry, beaked, 2-celled, dehiscent   ..   .. 10. **Thunbergia**
      Fruit a drupe, i.e. fleshy, 1-celled, indehiscent   ..   ..   .. 11. **Mendoncia**
   Anther-cells tailed at base :
    Ovules 3 or more in each cell ; anther-cells 6   ..   ..   .. 12. **Mimulopsis**
    Ovules 2 in each cell ; anthers 2 :
     Flowers solitary, opposite in the leaf-axils ; corolla about 5 cm. long ; anther-
      cells without spurs ; calyx tubular, up to 3·5 cm. long   13. **Satanocrater**
     Flowers mostly axillary, solitary or clustered in small condensed, rarely lax,
      cymes ; corolla up to 3 cm. long ; anthers pointed or spurred at base ; calyx
      never longer than 15 mm., not tubular   ..   .. 14. **Dyschoriste**
  Anthers 1-celled   ..   .. 15. **Physacanthus**
 Stamens 2 :
  Ovules 3 or more in each cell ; corolla 2-lipped :
   Flowers paniculate ; seeds with large curved retinacula   .. 16. **Brillantaisia**
   Flowers verticillate ; seeds with minute conical retinacula   .. 1. **Hygrophila**
  Ovules 1–2 in each cell ; corolla not 2-lipped   ..   .. 17. **Lankesteria**
Corolla-lobes imbricate in bud :
 *Stamens 4 :
  Anthers 1-celled ; corolla 1-lipped :
   Calyx-segments 5 :
    Calyx-segments all similar, the adaxia 1-nerved   ..   .. 18. **Sclerochiton**
    Calyx-segments not all similar, the adaxial larger, 2-nerved and often 2-toothed at
     the top..   ..   .. 19. **Crossandra**
   Calyx-segments 4 :
    Bracts or leaves spinous :
     Seeds covered with rope-like hair bundles, which unroll when moistened
                                                                      20. **Blepharis**
     Seeds glabrous   ..   .. 21. **Acanthus**
    Bracts and leaves not spinous   ..   .. 22. **Crossandrella**
  Anthers (at least of one pair of stamens) 2-celled ; corolla 2-lipped :
   Anther-cells tailed   ..   .. 23. **Asystasia**

Anther-cells not tailed :
Shrubs or undershrubs ; capsules always with retinacula ; bracteoles absent :
Flowers in terminal, elongated many-flowered racemes ; capsule compressed
globose-ellipsoid .. .. 24. **Thomandersia**
Flowers in short, bracteate, sometimes unilateral spikes, not always terminal, but
also axillary, basal etc. ; capsules elongate (not known in No. 26) :
Leaves triplinerve, acicular or ericaceous .. .. 25. **Lepidagathis**
Leaves with 5–10 pairs of lateral nerves, never triplinerve, acicular or ericaceous :
Corolla inflated upwards, mouth oblique, abaxial stamens with 2-celled anthers,
one anther-cell below the other, smaller ; infloresence terminal, loose-
racemose ; minutely hairy ; leaves green on both sides .. 26. **Filetia**
Corolla-mouth funnel-shaped, hardly oblique, abaxial stamens with 1-celled
anthers ; inflorescence of dense axillary heads with long white hairs ; leaves
with thick white tomentum beneath .. .. .. 27. **Neuracanthus**
Herbs or creepers, often prostrate and rooting at the nodes; bracteoles present,
longer than the calyx :
Leaves with cystoliths ; capsules with retinacula and 2 ovules in each cell
28. **Afrofittonia**
Leaves without cystoliths ; capsules with papilliform, not curved retinacula,
many-seeded.. .. .. 29. **Staurogyne**
*Stamens 2 :
Anthers 2-celled :
Anther-cells at equal or nearly equal levels, the cells not or (in No. 34) slightly tailed :
Ovules numerous in each ovary-cell ; calyx-segments 4 :
Leaves alternate, subradical, crenate (in some forms lyrate) to subentire
30. **Elytraria**
Leaves opposite, not subradical, entire .. .. 31. **Nelsonia**
Ovules 1–2 in each cell :
Bracteoles not present :
Calyx-segments 4 .. .. 32. **Barleria**
Calyx-segments 5 .. .. 33. **Pseuderanthemum**
Bracteoles present :
Stigma shortly bifid :
Flowers spicate, spikes thyrsoid or solitary .. .. .. 34. **Adhatoda**
Flowers paniculate or subracemose :
Staminodes (2) present .. .. 35. **Graptophyllum**
Staminodes absent .. .. 36. **Schaueria**
Stigma undivided :
Flowers in a terminal spike up to 15 cm. long, corolla up to 12 mm. long, whitish
37. **Chlamydocardia**
Flowers solitary, axillary, corolla up to 4 cm. long .. .. 38. **Anisotes**
Anther-cells at unequal levels :
Anther-cells not tailed :
Placentas not rising elastically in fruit and not throwing out the seeds :
Seeds smooth .. .. 39. **Peristrophe**
Seeds tubercular-scabrous :
Corolla-tube thick-cylindric or inflated in the upper half (not linear)
40. **Isoglossa**
Corolla-tube linear-cylindric, regular (not inflated) .. 41. **Rhinacanthus**
Placentas rising elastically in fruit and throwing out the seeds .. 42. **Dicliptera**
Anther-cells tailed at the base :
Placentas not rising elastically in fruit and not throwing out the seeds :
Capsule 4-seeded ; seeds rough or tubercled .. .. .. 43. **Justicia**
Capsule 2-seeded ; seeds smooth .. .. 44. **Monechma**
Placentas rising elastically in fruit and throwing out the seeds .. 45. **Rungia**
Anthers 1-celled :
Flowers appearing as though enclosed between two opposite bracts which are often
connate into a tube .. .. 46. **Hypoëstes**
Flowers not appearing as though enclosed between two opposite bracts :
Herb up to 1 m. high ; flowers purple to pale-pink, in rather dense, spike-like
terminal inflorescences ; corolla-tube 2–2·5 cm. long ; capsule about 1 cm. long
47. **Brachystephanus**
Shrubby plants with woody branches, up to 4 m. high :
Shrub up to 1·5 m. high ; flowers bright red, in a terminal spike-like inflorescence,
composed of shortly stalked cymules ; corolla-tube 2·5 cm. long ; capsule up
to 4 cm. long .. .. 48. **Ruspolia**
Shrub up to 4 m. high ; flowers mauve to whitish, in a very loose panicle with
100–200 flowers, corolla-tube very short, 2–3 mm. long ; capsule about 1·5 cm.
long .. .. 49. **Oreacanthus**

1. HYGROPHILA R. Br.—F.T.A. 5 : 30 ; Benoist in Bull. Soc. Bot. Fr. 60 : 330 (1913).

Synnema Benth. (Afr. spp.)—F.T.A. 5 : 29 ; Benoist l.c. (1913) ; F.W.T.A., ed. 1, 2 : 248. *Kita* A. Chev. in Rev. Bot. Appliq. 30 : 266 (1950).

Corolla not 2-lipped, up to 1 cm. long, tube nearly linear-cylindric ; segments about
  2 mm. long ; stamens nearly equal ; capsules about 6 mm. long, with about 8 seeds ;
  leaves usually about 15 mm. long and 8 mm. broad, sometimes the lower ones much
  longer (African " *Hemigraphis* " species)   ..   ..   **1.** *abyssinica*
Corolla markedly 2-lipped :
  Inflorescence laxly cymose, axillary, its branches few-flowered ; leaves sessile, usually
    linear to narrowly subacute, sometimes (in No. 4) with large obtuse teeth along the
    margins (Sect. *Nomaphila* ; *Kita* A. Chev.):
      Corolla 2 cm. long ; leaves lanceolate-elliptic, acuminate, up to 10 cm. long and
        1·5 cm. broad, minutely and closely puberulous beneath ; calyx-segments linear,
        unequal, up to 1 cm. long   ..   ..   **2.** *laevis*
      Corolla 4–11 mm. long ; leaves up to 9 cm. long and 1 cm. broad, but usually much
        smaller :
        Leaves broadly lanceolate, up to 2 cm. long, obtuse, sometimes the lower ones
          obscurely toothed ; calyx segments linear-lanceolate, at flowering time about
          3 mm. long, slightly accrescent in fruit ; corolla up to 4·5 mm. long ; capsule
          5 mm. long, with short, but well distinguishable retinaculas, containing 10–14
          small seeds (about 0·5 mm. long)   ..   ..   **3.** *stagnalis*
        Leaves linear to linear-lanceolate, up to 8 cm. long, acute, usually entire, but some-
          times strongly toothed with up to 6 teeth on each side up to 7 mm. long ; calyx-
          segments linear-filiform, the posticous 8–9(–10) mm., the 4 others 5–7 mm. long ;
          corolla up to 1 mm. long ; capsule 4–5 mm. long with papilliform retinacula,
          containing 40–50 very small sand-grain like seeds (about 0·3 mm. long)   **4.** *borellii*
  Inflorescences subspicate or in axillary clusters :
    Capsules with 4–8 seeds; leaves elongate-oblanceolate to linear-oblanceolate, up to
      25 cm. long and 3·5 cm. broad, more or less setose-pilose on the upper surface, with
      inconspicuous rod-like cystoliths beneath ; bracts and calyx-segments setose with
      white hairs (Sect. *Asteracantha*) :
      Flower-whorls with stiff spines up to 4 cm. long, bracts lanceolate ; fruits with up
        to 8 seeds   ..   ..   **5.** *auriculata*
      Flower-whorls without spines, bracts much resembling the ordinary leaves but
        smaller, with a large dilated base which sheathes the flowers or fruits ; fruits with
        4 seeds ..   ..   **6.** *niokoloensis*
    Capsules with 8 to numerous seeds ; leaves never longer than 12 cm. :
      Capsules with very short papilliform (not curved) retinacula (Afr. "*Synnema*" spp.):
        Stamens 2 ; stems glandular-puberulous ; leaves of two forms, the lower, if sub-
          merged, sometimes pinnatifid, the upper sessile, linear-oblanceolate, 2–5 cm. long,
          about 6 mm. broad, puberulous ; flowers at the nodes ; bracts foliaceous, up
          to 1 cm. long, pilose ; corolla-tube 5–6 mm. long ; style pubescent   **7.** *africana*
        Stamens 4 ; corolla with an extremely short or wanting upper-lip, flowers in axillary
          clusters ; leaves sessile, narrowly lanceolate, about 4–5 cm. long and 1 cm. broad,
          very variable from entire to remotely toothed or very deeply serrate-pinnatifid ;
          corolla-tube 3 mm. long ; style pubescent   ..   ..   ..   **8.** *brevituba*
      Capsules with markedly curved retinacula (Sect. *Physichilus, Polyechma* and
        Hygrophila) :
        Inflorescence continuous, spike-like ; bracts and calyx-segments markedly pectinate-
          ciliate, linear ; corolla about 1·5 cm. long ; leaves linear, sometimes slightly
          dentate or even pinnatifid, up to 5 cm. long, with rod-like cystoliths on the upper
          surface   ..   ..   **9.** *senegalensis*
        Inflorescence not spike-like, the flowers in interrupted axillary clusters :
          Plants entirely glabrous except for a few occasional caducous bristles around the
            nodes and some cilia on the calyx-lobes ; leaves linear-lanceolate, sessile, up to
            6 cm. long and 8 mm. broad   ..   ..   **10.** *pobeguinii*
          Plants pubescent to setose-pilose ; leaves lanceolate or oblanceolate, petiolate
            up to 12 cm. long and 3 cm. broad :
            Calyx and bracts densely setose-ciliate with stiff white hairs :
              Flowers large, about 1·5 cm. long, few in each cluster ..   ..   **9.** *senegalensis*
              Flowers smaller, numerous in each cluster ; flower-clusters 1·5 cm. diam. ;
                calyx-segments 5–6 mm. long ; leaves linear, up to 5 cm. long, with rod-like
                cystoliths on the upper surface ; fruit 8 mm. long ..   ..   **11.** *barbata*
            Calyx and bracts softly pilose (not markedly setose) ; flowers very few or several
              at each node :
              Small spreading plants, much-branched ; leaves up to 2·5 cm. long and 5 mm.
                broad, often remotely and obtusely toothed (not absolutely entire, as stated
                in F.T.A. 5 : 35) ; corolla about 10 mm. long ; calyx divided to a little below
                the middle, up to 9 mm. long ..   ..   **12.** *micrantha*

Stout erect plants, unbranched or little branched ; leaves up to 7 cm. long and
2·5 cm. broad, always entire ; corolla 1·2–2 cm. long ; calyx divided nearly
to base :
  Leaves and inflorescences softly and shortly pubescent :
    Pubescence velvety, intermixed with longer hairs ; leaves up to 12 cm. long
    and 2·5 cm. broad, rarely heterophyllous, i.e. submerged leaves
    pinnatifid ; calyx-segments obtuse ; corolla about 2 cm. long   13. odora
    Pubescence not velvety, hairs more or less equal, small and rather stiff-
    bristly ; leaves up to 4·5 cm. long and 8 mm. broad ; calyx-segments
    acute ; corolla 12–13 mm. long .. ..   14. chevalieri
  Leaves shortly setose-pilose only on the nerves beneath, oblanceolate, nervose,
  4–8 cm. long, up to 3 cm. broad ; bracts much shorter than the calyx-
  segments, the latter linear-filiform ; corolla 2 cm. long ..   15. uliginosa

1. **H. abyssinica** (*Hochst. ex Nees*) *T. Anders.* in J. Linn. Soc. 7: 22 (1863); Heine in Kew Bull. 16: 174 (1962).
Hemigraphis abyssinica (Hochst. ex Nees) C. B. Cl. in F.T.A. 5 : 58 (1899) ; Morton in J. W. Afr. Sci.
Assoc. 3 : 175 (1957), partly (*Morton* 8830). *Polyechma abyssinicum* Hochst. ex Nees in DC. Prod. 11 :
83 (1847). (?) *Synnema abyssinicum* (Hochst. ex Nees) Bremek. in Verh. Ned. Akad. Wetensch., Afd.
Natuurk., ser. 2, 41, 1 : 136 (1944). *Hemigraphis schweinfurthii* (S. Moore) C. B. Cl. in F.T.A. 5 : 58,
partly (*Schweinfurth* 2708 ; 2799) ; Schnell in Mém. I.F.A.N. 18 : 13 (1952). *Cardanthera africana*
var. *schweinfurthii* S. Moore in J. Bot. 18 : 7 (1880), partly (*Schweinfurth* 2708 ; 2799). (?) *Synnema
schweinfurthii* (S. Moore) Bremek. l.c. 141 (1944), partly (*Schweinfurth* 2708 ; 2799). A very variable
herb in large specimens up to 1 ft. tall, but mostly much smaller, glabrescent or hairy, with dense
axillary clusters and small purplish-blue flowers ; in wet or muddy places.
  Ghana: Tumu (fl. & fr. Mar.) Jos waterworks *Schnell* 9622. Near Toto, Jos to
Bauchi (fr. Apr.) *Morton* K402 ! [**Br.**]**Cam.:** Dikwa (fl. & fr. Mar.) *McClintock* 14 ! Throughout tropical
Africa.
  [An extremely variable marsh-plant, described and treated under numerous names in the genera
Cardanthera, Hemigraphis, Hygrophila, Polyechma and *Synnema*. Its systematic position is not yet
satisfactorily elucidated, but it is certainly a member of Hygrophilineae, and not Strobilanthineae ; see
Bremekamp in Acta Bot. Neerland. 4 : 647 (1955). The synonymy as given above contains only the names
(and corresponding synonyms) which have been used by previous authors for plants from the region of the
F.W.T.A. ; it may be completed by further studies on the African " *Hemigraphis* "-species.—H.H.]

2. **H. laevis** (*Nees*) *Lindau* in E. & P. Pflanzenfam. 4, 3B : 297 (1895) ; F.T.A. 5 : 36 ; Benoist l.c. 332.
Nomaphila laevis Nees in DC. Prod. 11 : 85 (1847) ; T. Anders. in J. Linn. Soc. 7 : 21 (1863) ; Berhaut
Fl. Sén. 94. *Kita laevis* (Nees) A. Chev. in Rev. Bot. Appliq. 30 : 266 (1950). Erect herb about 1 ft. high,
becoming woody at the base, rooting at the lower nodes, with pale mauve flowers ; beside streams.
  Sen.: *Heudelot* 197 ! *Berhaut* 1468. **Mali:** Kéniéba (fl. & fr. Mar.) *Roberty* 17048 ! **Ghana:** base of Gambaga
Scarp (Dec.) *Morton* A1377 !

3. **H. stagnalis** *Benoist* in Mém. Soc. Bot. Fr. 2, 8 : 277 (1917). (?) *Kita gracilis* A. Chev. l.c. 267 (1950). Low
herb with long creeping stems rooting at the nodes ; flowers pale mauve ; in temporarily dry river beds
and marshy places.
  Guin.: Kollangui *Chev.* 12229 ! [(?) **Mali:** " in R. Bakoy, 30 km. W. of Kita " *Chev.* s.n. (1950)].
  [*Kita gracilis* is " probably only an ecological form " of *H. laevis*, according to Chevalier *l.c.*]

4. **H. borellii** (*Lindau*) *Heine* in Kew Bull. 16: 176 (1962). *Brillantaisia borellii* Lindau in Engl. Bot. Jahrb. 33:
186 (1902). *Synnema borellii* (Lindau) Benoist in Bull. Soc. Bot. Fr. 60 : 330 (1913). *S. diffusum*
J. K. Morton in J. W. Afr. Sci. Assoc. 2 : 67 (1956), as *diffusa*. A prostrate, sometimes creeping plant,
very bushy, rooting at the nodes and with ascending inflorescences ; flowers violet with pale brownish
tube ; in marshy grassland.
  Ghana: Nassia Swamp (fl. & fr. Dec.) *Vigne* FH 4672 ! Pong Tamale (Nov.) *Akpabla* 430 ! Bungari
(fl. & fr. Dec.) *Morton* GC 25078 ! Nyankpala to Tamale (fl. & fr. Dec.) *Morton* GC 9936 ! Tamale (fl. &
fr. Dec.) *Morton* GC 9846 ! Bimbila (fl. & fr. Dec.) *Morton* A1427 ! **Dah.:** Paonignan (fl. & fr. Nov.) *Poisson*
34 !

5. **H. auriculata** (*Schumach.*) *Heine* l.c. 172 (1962), for full synonymy. *Barleria auriculata* Schumach.
Beskr. Guin. Pl. 285 (1827). *H. longifolia* (Linn.) Kurz (1870)—Berhaut Fl. Sén. 75, not of Nees (1847).
Asteracantha longifolia (Linn.) Nees (1832)—Robyns Fl. Sperm. Parc Nat. Alb. 2 : 274 (1947). *Barleria
longifolia* Linn. (1759). *Hygrophila spinosa* T. Anders. (1860)—F.T.A. 5 : 31 ; Benoist in Bull. Soc.
Bot. Fr. 60 : 330 ; F.W.T.A., ed. 1, 2 : 247 ; Chev. Bot. 492. A stout erect herb of marshy places, 2–3 ft.
or up to 6 ft. or more, with 4-angled usually bristly stem, and bluish or purple flowers 1–1¼ in. long in
dense whorls with about 6 strong spines.
  Sen.: *Heudelot* 180 ! *Berhaut* 491. M'bidgem *Chev.* 2800. Matam *Chev.* 2807. **Gam.:** Boteler ! *Brown-
Lester* 15 ! *Brooks* 27 ! Genieri (fl. & fr. Feb.) *Fox* 62 ! Jarra East (fl. & fr. Nov.) *Frith* 262 ! **Mali:**
Goumal (fl. & fr. Oct.) *Davey* 059 ! **Port.G.:** Buruntuma *Esp. Santo* 374. Sucujaque (Nov.) *Esp. Santo*
3145 ! Prabis, Bissau (fr. Feb.) *Esp. Santo* 1806 ! **Ghana:** Cape Coast Castle *T. Vogel* ! Accra *Irvine* 682 !
Achimota F.R. (Apr.) *Morton* GC 6166 ! Amisano (Oct.) *Andoh* 5080 ! Babianiba (fl. & fr. Oct.) *Howes*
991 ! **Togo Rep.:** Lomé *Warnecke* 227 ! **Dah.:** Zagnanado, Abbo *Chev.* 22962. Savé *Chev.* 23582. **N.Nig.:**
Nupe *Barter* 773 ! Katagum *Dalz.* 102 ! Kabba *Parsons* ! Naraguta (May) *Lely* 30 ! Gajibo, Bornu
(Nov.) *Johnston* N60 ! **S.Nig.:** Lagos *Barton* 64 ! Ikom *Rosevear* ! [**Br.**]**Cam.:** Gurum, Vogel Peak area
(fl. & fr. Nov.) *Hepper* 1253 ! Dikwa (Nov.) *McClintock* 15 ! Widespread in the Old World tropics. (See
Appendix, p. 451.)

6. **H. niokoloensis** *Berhaut* in Bull. Soc. Bot. Fr. 101 : 374 (1955) ; Fl. Sén. 92 (1954), French descr. only.
Erect or ascending herb about 1 ft. high, with fleshy stem only very slightly tinged purple ; flowers pale
blue-mauve with fine mauve striations and on the lower lip, and 2 white processes in throat, upper lip
pale blue-mauve without noticeable striations, anthers beneath upper lip and same blue-mauve colour.
  Sen.: Niokolo-Koba (fr. Jan.-Apr.) *Berhaut* 869 ! 4630 ! **N.Nig.:** Kagara, Minna Div. (Oct.) *Hepper*
981 ! Dogon Kurmi, Jemaa Dist. (fr. Nov.) *Keay* FHI 37232 ! Anara F.R., Zaria (fr. Nov.) *Keay* FHI
5472 ! 28122 ! Jos, 4,000 ft. (Oct.) *Hepper* 1014 !

7. **H. africana** (*T. Anders.*) *Heine* l.c. 176 (1962). *Adenosma africana* T. Anders. in J. Linn. Soc. 7 : 21
(1863). *Cardanthera africana* (T. Anders.) Benth. (1875). *C. africana* var. *schweinfurthii* S. Moore in J.
Bot. 18 : 7 (1880), partly (*Schweinfurth* 2764). *Synnema africanum* (T. Anders.) O. Ktze. Rev. Gen.
Pl. 2 : 500 (1891) ; Benoist l.c. ; F.T.A. 5 : 29 ; F.W.T.A., ed. 1, 2 : 248. *Hemigraphis abyssinica* of
J. K. Morton in J. W. Afr. Sci. Assoc. 3 : 177, partly (*Morton* A1297), not of (Hochst. ex Nees) T. Anders.
A branched herb of swamps with white 2-lipped flowers.
  Mali: Kara (Dec.) *Davey* 191 ! Dogo (Apr.) *Davey* 564 ! Saredina (May) *Davey* 135 (heterophyllous) !
Ghana: N. of Pusiga, Bawku (fr. Dec.) *Morton* A1297 ! **N.Nig.:** Jebba *Barter* 751 ! Sokoto (Nov.) *Lely*
117 ! [**Br.**] **Cam.:** Bama, Dikwa Div. (fl. & fr. Dec.) *McClintock* 35 ! Also in Cameroun.

8. **H. brevituba** (*Burkill*) *Heine* l.c. 176 (1962). *Synnema brevitubum* Burkill in F.T.A. 5 : 30 (1899) ; Benoist
l.c. ; F.W.T.A., ed. 1, 2 : 248. *Cardanthera parviflora* Turrill in Kew Bull. 1914 : 82. *C. breviflora*

(Burkill) Turrill l.c. (1914). A branched hairy herb attaining 2–3 ft. in height; flowers white or yellow and heavily brown-veined ; amongst grass in swamps.
Iv.C.: Mankona to Marahouē (Jan.) *de Wit* 7967 ! Mankona (fl. & fr. Nov.) *de Wilde* 965 ! **Ghana:** Afram Plains (Dec.) *Irvine* 886 ! Atabubu (Dec.) *Darko* 368 ! Ejura (Dec.) *Morton* GC 9493 ! Salaga *Krause* ! Krakye, Volta R. *Krause* ! **N.Nig.:** Abinsi *Dalz.* 720 ! [**Br.]Cam.:** Daksami, Vogel Peak area (fl. & fr. Dec.) *Hepper* 1557 !

9. **H. senegalensis** (*Nees*) *T. Anders.* in J. Linn. Soc. 7 : 22 (1863) ; F.T.A. 5 : 34 ; Benoist l.c. 332 ; Chev. Bot. 492 ; F.W.T.A., ed. 1, 2 : 247, excl. syn. ; Berhaut Fl. Sén. 91. *Physichilus senegalensis* Nees in DC. Prod. 11 : 81 (1847), incl. forma β *macer* Nees l.c. 82 (1847). Herb, erect, branched, with 4-angled glabrous stems, spongy at the base when in damp ground, and the lower leaves become pinnatifid, like those of *Myriophyllum*, if submerged ; flowers ¾ in. long, violet-purple.
Sen.: *Heudelot* 139 ! *Roger* 32 ! *Leprieur. Berhaut* 601. Matam *Chev.* 2808. **Gam.:** *Brown-Lester* 212 ! *Duke* 15 ! **Mali:** *Scaëtta* 3275 ! Koulikoro *Chev.* 2790 ; 2791. Dogo (fl. & fr. Apr.) *Davey* 563 ! **Port.G.:** Bissoram (Nov.) *Esp. Santo* 2408 ! Peluba, Bissau (Dec.) *Esp. Santo* 7611 ! **Guin.:** Kindia (fl. & fr. May) *Jac.-Fél.* 1640 ! **S.L.:** Mamaka (Feb.) *Glanville* 165 ! Yomboro, Konta (fl. & fr. Apr.) *Hepper* 2653 ! **Iv.C.:** Séguéla to Mankono (Nov.) *de Wilde* 929 ! Séguéla (Nov.) *de Wilde* 907 ! **Ghana:** Afram Plains (Dec.) *Irvine* 619 ! Sampa, 1,000 ft. (Dec.) *Vigne* FH 3489 ! Dormaa-Ahenkro (Dec.) *Adams* 3014 ! Nangodi (Dec.) *Adams & Akpabla* 4285 ! **N.Nig.:** Katagum Dist. *Dalz.* 103 ! Dabeti to Damaturu (fl. & fr. Nov.) *Noble* A18 !
[Forms with pinnatifid leaves have been known since the original description of Nees (1847) (e.g. *Leprieur* 31), but there is no mention of those in F.T.A., F.W.T.A. etc.]

10. **H. pobeguinii** *Benoist* in Notulae Syst. 2 : 339 (1913), and Bull. Soc. Bot. Fr. 60 : 331 ; Chev. in Rev. Bot. Appliq. 30 : 266 (1950). *H. vanderystii* S. Moore in J. Bot. 58 : 46 (1920). Erect herb up to 1 ft. high, with crimson stems ; flowers pale blue.
Mali: Bamako to Koulouba (fl. & fr. Dec.) *Monod* ! Sotuba, Bamako (fl. & fr. Jan.) *Raynal* 5404 ! **Guin.:** Kouroussa (Nov.) *Pobéguin* 1108 ! **Port.G.:** Bissabanca, Bissau (fr. Jan.) *Esp. Santo* 1602 ! **N.Nig.:** Bukuru to Barakin Ladi (Nov.) *Keay* FHI 21442 ! Also in Ubangi-Shari, Congo and in E. Africa.

11. **H. barbata** (*Nees*) *T. Anders.* in J. Linn. Soc. 7 : 22 (1863) ; F.T.A. 5 : 32 ; Benoist in Bull. Soc. Bot. Fr. 60 : 332. *Physichilus barbatus* Nees in DC. Prod. 11 : 82 (1847). Erect much-branched annual herb, with 4-angled, glabrous stems (sometimes slightly hairy on the nodes) and pink, occasionally white or magenta flowers ; in marshy places and rice fields.
Sen.: *Heudelot* 573 ! **Mali:** Kita (Oct.) *Jaeger* 3339 ! **Port.G.:** Antula, Bissau (Nov.) *Esp. Santo* 1381 ! **Guin.:** Iles Tristao (fl. & fr. Nov.) *Jac.-Fél.* 7341 ! Friguiagbé *Chillou* 791 ! 1018 ! **S.L.:** *Garrett* ! Talia (fr. Feb.) *Thomas* 8865 ! Bonti, near Samaia (fl. & fr. Dec.) *Deighton* 5963 ! Ruka (Dec.) *Sc. Elliot* 4329 ! Rowankiliboli (Nov.) *Jordan* 828 ! **Lib.:** Monrovia (Dec.) *Baldwin* 10493 !

12. **H. micrantha** (*Nees*) *T. Anders.* l.c. 247 : 22 (1863) ; F.T.A. 5 : 35 ; Berhaut Fl. Sén. 91 ; Benoist l.c. 331. *Polyechma micranthum* Nees in DC. Prod. 11 : 83 (1847). *Hygrophila senegalensis* of Hutch. & Dalz. F.W.T.A., ed. 1, 2 : 247, partly, not of (Nees) T. Anders. *H. tumbuctuensis* A. Chev. in Bull. Mus. Hist. Nat., sér. 2, 4 : 586 (1932). *H. chevalieri* " Benoist MS ex A. Chev." in Rev. Bot. Appliq. 30 : 265 (1950), partly (*de Wailly* 5356), not of Benoist (1917). A small, much-branched spreading plant, up to 6–8 in. high, stems hispid ; flowers pale bluish, 1–6 in false whorls in the leaf-axils.
Sen.: *Perrottet* 32 ; 477. Dagama *Leprieur.* Niokolo-Koba *Berhaut* 539. Saouor *Leprieur* 5 ! **Mali:** Gao (fl. & fr. Mar.-Apr.) *de Wailly* 4676 ! 4995 ! 5356 ! Bourem to Bamba, Timbuktu *Chev.* ! Diafarabé (Jan.) *Davey* 4101 ! Sotuba, Bamako (fl. & fr. Feb.) *Raynal* 5476 !

13. **H. odora** (*Nees*) *T. Anders.* l.c. 247 : 22 (1863) ; F.T.A. 5 : 33 ; Benoist l.c. 332 ; F.W.T.A., ed. 1, 2 : 247, excl. syn. *H. acutisepala* of Chev. Bot. 492, not of Burkill (as to *Chev.* 2793) ; Berhaut Fl. Sén. 265. *Polyechma odorum* Nees in DC. Prod. 11 : 83 (1847). An aromatic viscid herb, up to 2–3 ft. high, woody at the base, with pink, purple-lilac (or sometimes white) flowers with a musk-like scent.
Sen.: Kountaour *Berhaut* 1338. **Port.G.:** Chitole, Saltimbo (Feb.) *Esp. Santo* 2668 ! Corubal, Cossita (May) *Esp. Santo* 1239 ! **Guin.:** *Heudelot* 807 ! *Paroisse* 5 ! *Farmar* 301 ! Chute du Tinkisso *Pobéguin* 985 ! Kolenté R. *Chillou* 1043 ! **S.L.:** Freetown *Burbridge* 505 ! Komoh country (fl. & fr. Mar.) *Burbridge* 504 ! Mofari Jau *Sc. Elliot* 4723 ! *Deighton* 4006 ! Njala (Jan.) *Deighton* 2960 ! **Lib.:** Kailahun (fl. & fr. Feb.) *Bequaert* 58 ! Baila (fr. May) *Harley* 1422 !

14. **H. chevalieri** *Benoist* in Mém. Soc. Bot. Fr. 2, 8 : 278 (1917) ; Chev. in Rev. Bot. Appliq. 30 : 265 (1950), partly (*Chev.* 637). *Dyschoriste perrottetii* O. Ktze. " var." of Chev. Bot. 494. *H. acutisepala* of Chev. Bot. 492 (*Chev.* 2793), not of Burkill (1899). *H. odora* of F.W.T.A., ed. 1, 2 : 247, partly, (*Chev.* 2793), not of T. Anders. Very similar to the last, but in all its parts smaller and more slender, indument less dense, composed of subequal hairs, and flowers about half the size.
Sen.: Sédhiou to Tambanaba (Feb.) *Chev.* 2793 ! **Mali:** Niagalétoumanina *Chev.* 637 !
[Note: Chevalier's description (1950) is based partly on characters of *H. micrantha* (Nees) T. Anders. (see under No. 10).]

15. **H. uliginosa** *S. Moore* in J. Bot. 18 : 197 (1880) ; F.T.A. 5 : 32; Chev. Bot. 493 ; F.W.T.A., ed. 1, 2 : 608. *H. teuczii* Lindau (1894) ; F.T.A. 5 : 33 ; F.W.T.A., ed. 1, 2 : 247. *H. sereti* ; De Wild. (1910). An erect herb up to 2 ft. high, with angled, hispid stems, becoming glabrescent, with mauve flowers about ¾ in. long.
Togo Rep.: *Büttner* 329 ! **N.Nig.:** Panyam (fl. & fr. Sept.) *Lely* 726 ! Vom, 3,000–4,500 ft. (Apr.) *Morton* K339 ! *Dent Young* 196 ! 197 ! Also in Ubangi-Shari, Chad, Cameroun, Congo and Angola.

2. RUELLIA Linn.—F.T.A. 5 : 44. *Stenochista* Bremek. in Engl. Bot. Jahrb. 73 : 147 (1943).

Calyx-lobes linear-subulate, about 5 mm. long ; corolla 2·5 cm. long, the linear-cylindric portion up to 1 cm. long ; leaves ovate-elliptic to lanceolate, sub-acute, 3–4 cm. long, 1–2 cm. broad, setulose-pubescent ; stamens 4 ; anther-cells not spurred at the base ; style pubescent, one style-arm suppressed ; fruit 1·5 cm. long, club-shaped, glabrous . . . . 1. *praetermissa*

Calyx-lobes filiform 6–7 mm. long ; corolla about 5 cm. long, inflated in the upper third, the linear-cylindric portion 3 cm. long ; leaves broadly ovate-lanceolate to elliptic, acutely acuminate, 5–10 cm. long, 2–5 cm. broad ; stamens, style and fruit as above
2. *togoensis*

1. **R. praetermissa** *Schweinf. ex Lindau* in Engl. Bot. Jahrb. 20 : 15 (1894) ; F.T.A. 5 : 45 ; Chev. Etud Fl. Afr. Centr. 1 : 232 (1913) ; F. W. Andr. Fl. Sudan 3 : 187 (1956). *R. patula* of F.T.A. 5 : 45, partly (W. African specimens) of F.W.T.A., ed. 1, 2 : 246, not of Jacq. *R. patula* Jacq. " forma " of Chev. Bot. 494 ; Berhaut Fl. Sén. 91. A decumbent undershrub with white or blue-purple flowers, the corolla pubescent outside, 1 in. or more long.
Sen.: *Richard* ! *Berhaut* 1336. **Mali:** Gao (fl. & fr. Sept.) *de Wailly* 4766 ! **S.L.:** *Brass* ! **Iv.C.:** Guidéko to Zozro (Oct.) *Chev.* 19072 ! Gaoloubré to Soubré *Chev.* 17788 ! **Ghana:** Accra (Nov.) *Brown* ! Achimota (fl. Sept., fr. July, Sept.) *Akpabla* 578 ! *Morton* A2 ! Accra Plains (July) *Deighton* 3386 Nungua (June) *Rose Innes* GC 30073 ! **Dah.:** Ouémé (Feb.) *Gironcourt* 239 !

[Note : As the taxonomy of the genus *Ruellia* Linn. in Africa is still uncertain and much in need of a revision, this name for the plants concerned may be altered in future. The application of Jacquin's *R. patula* to the W. African specimens does not seem to be justified, and these specimens, named since C. B. Clarke's treatment of Acanthaceae for F.T.A. as " *Ruellia patula* Jacq.", are practically conspecific with *R. praetermissa* Lindau, a Sudan species, which has also been recorded from Middle Shari—H.H.]

2. **R. togoensis** (*Lindau*) *Heine* in Kew Bull. 16 181 (1962). *Dischistocalyx togoensis* Lindau in Engl. Bot. Jahrb. 33 : 188 (1902) ; F.W.T.A , ed. 1, 2 : 246. *Stenochista togoensis* (Lindau) Bremek. in Engl. Bot. Jahrb. 73 : 148 (1943). *R. ardeicollis* Benoist in Mém. Soc. Bot. Fr. 2, 8 : 278 (1917) ; Chev. Bot. 494 ; F.W.T.A., ed. 1, 2 : 246. A slightly woody herb 1 ft. or more high, with white flowers about 2 in. long more or less like the last but with a longer tube.
> **Iv.C.:** Orodougou (June) *Chev.* 21816 ! Morénou (Nov.) *Chev.* 22436 ! **Ghana:** Agogo (Apr.) *Adams* 2635 ! Sunyani to Berekum (Dec.) *Adams* 2776 ! Koforidua, Begoro (Nov.) *Morton* A33 ! Ejura Scarp GC 9798 ! **Togo Rep.:** Lomé *Warnecke* 264 !

3. **EREMOMASTAX** Lindau in Engl. Bot. Jahrb. 20 : 8 (1894). *Paulowilhelmia* Hochst. in Flora 27, 1, Beil. 4 (after June 1844) ; F.T.A. 5 : 52 ; F.W.T.A., ed. 1, 2 : 246, not *Paulowilhelmia* Hochst. in Flora 27 : 17 (Jan. 1844).

Coarse herb up to 2 m. high ; leaves irregularly toothed or lobed, ovate, truncate or subcordate at the base, 5–13 cm. long, sparsely pubescent or nearly glabrous ; petiole 2·5–6·5 cm. long ; flowers blue, in lax divided subscorpioid cymes 2–7 cm. long ; corolla-tube 2·5–3 cm. long, linear, lobes elliptic, 12 mm. long ; capsule about 15 mm. long, usually 4–8-seeded .. .. *polysperma*

E. polysperma (*Benth.*) *Dandy* in F. W. Andr. Fl. Sudan 3 : 174 (1956). *Paulowilhelmia polysperma* Benth. in Fl. Nigrit. 479 (1849) ; F.T.A. 5 : 52 ; F.W.T.A., ed. 1, 2 : 246 ; Chev. Bot. 494. *P. sclerochiton* (S. Moore) Lindau (1893)—F.T.A. 5 : 53 ; F.W.T.A., ed. 1, 2 : 246. *P. pubescens* Lindau in Mém. Soc. Bot. Fr. 2, 8 : 49 (1907) ; Chev. Bot. 494. *Ruellia sclerochiton* S. Moore (1880). *Paulowilhelmia togoensis* Lindau (1893)—F.T.A. 5 : 53. A stout erect branched herb with long-petioled broad leaves and lilac or bluish-purple flowers 1½–2 in. long with darker spots and streaks on the corolla-lobes.
> **Guin.:** *Scaëtta* 3369 ! Timbo *Caille* in *Hb. Chev.* 14671 ! Diaguissa, 4,000 ft. (fl. & fr. Nov.) *Jac.-Fél.* 2078 ! *Chev.* 12920 ; 13468. Kindia *Chev.* 13120bis ; 13224bis. Ditinn to Diaguissa *Chev.* 13843bis. **S.L.:** Winwood *Reade* ! York Pass (fr. Dec.) *Deighton* 3334 ! Sugar Loaf Mt. (fr. Mar.) *Dalz.* 932 ! *Sc. Elliot* 3990b ! 4024 ! Boge (Dec.) *Marmo* 121 ! **Lib.:** Kakatown *Whyte* ! Totokwelli (Oct.) *Linder* 1296 ! Zuie (Nov.) *Baldwin* 10246a ! **Ghana:** Aburi Hills (Nov.) *Johnson* 237 ! 446 ! Amedzofe (Mar.) *Morton* 6496 ! *de Wit* 614 ! Asuboi (fl. & fr. Nov.) *Morton* 8002 ! Kumasi (fl. & fr. Nov.) *Vigne* FH 4082 ! **Togo Rep.:** Sokodé-Basari *Kersting* 1721 ! Misahöhe *Mildbr.* 7234 ! Seggebach *Büttner* 342 ! **N.Nig.:** Jos *Batten-Poole* 142 ! Kogin Giri, Jos (Oct.) *Wimbush* 18 ! Vom, 4,000 ft. (fl. & fr. Oct.) *Hepper* 1152 ! **S.Nig.:** Ibadan (fr. Jan.) *Meikle* 976 ! Aboabam (Dec.) *Keay* FHI 28226 ! Cross R. (fl. & fr. Jan.) *Johnston* ! Oban *Talbot* 10 ! [**Br.]Cam.:** Cam. Mt. *Mann* 1259 ! Buea, 3,000 ft. (Dec.) *Maitland* 136 ! Near Vogel Peak, Adamawa (Dec.) *Hepper* 1522 ! Widespread in tropical Africa. (See Appendix, p. 452.)

4. **HETERADELPHIA** Lindau in Engl. Bot. Jahrb. 17 : 108 (1893) ; Exell Cat. S. Tomé 261 (1944).

Subshrub or coarse herb up to 1 m. (?) high ; leaves broadly obovate-acuminate, 10–13 cm. long and 4–5·5 cm. broad, cordate at the base, crenate, sparsely pubescent or nearly glabrous ; petiole 3–4 cm. long ; flowers in a dense compound inflorescence, composed of very many axillary 5–8-flowered cymes up to 4 cm. long ; corolla-tube funnel-shaped, up to 3·5 cm. long, lobes more or less equal, 12 mm. long ; capsule about 15 mm. long, 8–12-seeded .. .. *paulojaegeria*

H. paulojaegeria *Heine* in Bull. I.F.A.N. 23 : 1028 (1961). An undershrub or coarse herb up to about 3 ft. high ; flowers purplish-red.
> **Guin.:** Forékaria, Faranah (fr. Jan.) *Chev.* 20521 ! Beyla (Oct.) *Jac.-Fél.* 2199 ! Banian to Bambaya (fl. & fr. Oct.) *Jaeger* 2103 !

5. **DISCHISTOCALYX** T. Anders. ex Benth.—F.T.A. 5 : 60, as *Distichocalyx* ; Bremekamp in Engl. Bot. Jahrb. 73 : 134 (1943).

Leaves with 6–7 pairs of lateral nerves, narrowly obovate, acutely cuneate at base, acuminate, 10–15 cm. long, 3–7 cm. broad, with numerous short rod-like cystoliths on both surfaces ; spikes few-flowered ; bracts early deciduous ; calyx-segments linear-oblanceolate, acute, 2 cm. long ; corolla 6 cm. long ; fruit 2 cm. long, narrow, glabrous .. .. 1. *thunbergiiflorus*
Leaves with 10 pairs of lateral nerves, otherwise as above ; bracts persistent, 2·5 cm. long ; calyx-segments scarcely 1·5 cm. long .. .. .. 2. *obanensis*

1. **D. thunbergiiflorus** (*T. Anders.*) *Benth. ex C.B.Cl.* in F.T.A. 5 : 62 (1899) ; Bremekamp l.c. 135. *Ruellia thunbergiaeflora* T. Anders. in J. Linn. Soc. 7 : 24 (1863). *Dischistocalyx ruellioides* S. Moore in Cat. Talb. b. 76 (1913). A herbaceous undershrub several feet high, woody below, glabrous except on the young parts, with spikes of violet-purple flowers 1½–2 in. long.
> **S.Nig.:** Oban *Talbot* 1527 ! (Mar.) *Richards* 5142 ! Okarara, Calabar (fl. & fr. May) *Ujor* FHI 30824 ! Boshi-Okwangwo F.R., Ogoja (fl. & fr. May) *Latilo* FHI 30941 ! [**Br.]Cam.:** Mamfe to Bamenda (fl. & fr. Mar.) *Richards* 5234 ! **F.Po:** 1,300–2,000 ft. *Mann* 316 ! 1446 !

2. **D. obanensis** *S. Moore* in Cat. Talb. 77 (1913). An undershrub like the last with pale violet flowers.
> **S. Nig.:** Oban *Talbot* 73 ! 1485 !
> [Note : *D. staudtii* Bremek. l.c. 141 (1943) (*Staudt* 875, Johann-Albrechtshöhe) may be conspecific with this species, but the taxonomic position could not be decided without seeing type material—H.H.]

6. **ENDOSIPHON** T. Anders. ex Benth.—F.T.A. 5 : 49.

Stems long-pilose ; leaves anisophyllous, one sometimes very small, ovate to oblong-elliptic, acuminate, rounded and very unequal-sided at the base, 6–10 cm. long, 2·5–5 cm. broad, marked with linear cystoliths on both surfaces, thinly pilose on the nerves beneath ; flowers solitary ; pedicels up to 4 cm. long, pilose or pubescent ; calyx-lobes slightly unequal, linear, 2·5–3 cm. long, 1-nerved, pilose ; corolla-tube up to 5 cm. long, narrow, limb about 4 cm. diam., with rounded lobes ; anthers 4, deep in the corolla-tube, linear ; fruit linear, 2 cm. long, valves grooved in the middle

primuloides

E. primuloides *T. Anders. ex Benth.* in Benth. & Hook. f. Gen. Pl. 2 : 1086 (1876) ; F.T.A. 5 : 49 ; Chev. Bot. 496, incl. var. *hirsuta* Benoist ex A. Chev. (1920), name only. *E. obliquus* C. B. Cl. in F.T.A. 5 : 50 (1899). Herb or small half-woody undershrub in forest, 1–3 ft. high, clothed with multicellular hairs, with pale blue or lilac slender-tubular flowers with spreading limb.
S.L.: Damawulo, Tunkia (fr. Dec.) *Deighton* 3825 ! Gola Forest (Nov.) *Deighton* 370 ! **Lib.:** Monrovia *Whyte* ! Gbanga *Linder* 388 ! Mt. Bili (fl. & fr. Dec.) *Barker* 1169 ! Bangee, Boporo Dist. (Nov.) *Baldwin* 10354 ! Vahon, Kolahun Dist. (Nov.) *Baldwin* 10209 ! **Iv.C.:** Accrédiou *Chev.* 17069 ; 17117. Fort Binger *Chev.* 19522. Nékaougnié to Grabo *Chev.* 19594. Taté *Chev.* 19799. Yapo (Mar.) *Roberty* 15476 ! Abidjan (fl. & fr. Jan.) *Leeuwenberg* 2334 ! **Ghana:** Kibbi Hills (fr. Dec.) *Johnson* 262 ! Bompata (Dec.) *Dalz.* 151 ! Aniwase (fr. Oct.) *Fishlock* 59 ! Amentia *Irvine* 546 ! **S.Nig.:** Oban *Talbot* 979 ! Budeng to Ndeokoro (Nov.) *Rosevear* C27 ! Afi River F.R., Ogoja (Dec.) *Keay* FHI 28190 ! 28237 ! Calabar (fr. Feb.) *Onochie* FHI 36324x ! **F.Po:** (Nov., Dec.) *Mann* 571 ! *Guinea* 739 ! Also in Rio Muni and Gabon.
[Very hairy forms occur in Ivory Coast (= var. *hirsuta* Benoist ex A. Chev.), but it is impossible to regard them as of any taxonomic value—H.H.]

7. ACANTHOPALE C. B. Cl. in F.T.A. 5 : 62 (1899) ; Bremekamp in Engl. Bot. Jahrb. 73 : 142 (1943).

Plants up to 3 m. high, nearly glabrous ; leaves very unequal in each pair, the larger obovate, long-acuminate, attenuated at base into the winged petiole, up to 20 cm. long and 8 cm. broad ; racemes axillary, up to 5 cm. long, with about 4 flowers ; corolla widely funnel-shaped, about 3 cm. long ; calyx-lobes about 1 cm. long ; filaments glabrescent[1]   ..   ..   *decempedalis*

A. decempedalis *C. B. Cl.* in F.T.A. 5 : 63 (1899) ; Bremekamp l.c. 143, 148. *A. laxiflora* (Lindau) C. B. Cl. in F.T.A. 5 : 36 (1899), partly (*Preuss* 947) ; F.W.T.A., ed. 1, 2 : 248, partly (*Preuss* 947). *Dischistocalyx laxiflorus* Lindau in Engl. Bot. Jahrb. 20 : 13 (1894), partly (*Preuss* 947). *Acanthopale cameronensis* Bremek. l.c. 145 (1943). A soft-wooded shrub with funnel-shaped white flowers.
S.Nig.: Obudu Plateau, 5,500 ft. (Aug.) *Stone* 65 ! 75 ! [**Br.**]**Cam.:** Buea *Preuss* 947 ! Mimbia (Mar.) *Brenan* 9365 ! Cam. Mt., 3 mls. W. of Mann's Spring (Dec.) *Morton* K859 ! Bamenda *Ledermann* 5845. Konga Mts. *Ledermann* 6029. Lakom, 6,000 ft. (fl. & fr. May) *Maitland* 1535 ![2] **F.Po:** 7,000 ft. *Mann* 2347 ! 3747 ! El Pico, 7,000 ft. (fr. Dec.) *Boughey* 117 ! Biao, 6,500 ft. (Sept.) *Melville* 464 ! Peak of St. Isabel, above Basilé, 3,600–4,400 ft. *Mildbr.* 6301.

8. WHITFIELDIA Hook.—F.T.A. 5 : 65.

Corolla-tube very slender, about 5 cm. long, elongated, abruptly expanded into the limb, indumentum variable, glabrescent or more or less pubescent ; lobes oblong-elliptic, 2 cm. long ; calyx-segments broadly linear, 3 cm. long ; glandular-pilose ; leaves more or less elliptic, long-acuminate, acutely cuneate at base, 6–18 cm. long, 3–6 cm. broad, minutely pustulate on the upper surface ; bracts ovate, about 1·5 cm. long   ..   ..   1. *elongata*
Corolla-tube gradually expanded from the base, about 2·4 cm. long :
  Bracteoles broadly and obliquely obovate, about 0·8–1 cm. long ; flowers rather few :
    Corolla 3–3·5 cm. long ; leaves oblong-lanceolate to ovate-elliptic, acutely acuminate, acute at base, 10–18 cm. long, 3–7 cm. broad, pubescent on the midrib beneath ; fruit 2 cm. long   ..   ..   2. *lateritia*
    Corolla 2 cm. long, very broad ; inflorescence rather elongated ; leaves as above, but midrib glabrous   ..   ..   3. *preussii*
  Bracteoles lanceolate to narrowly elliptic, 1–1·5 cm. long ; flowers rather numerous ; corolla 3 cm. long ; leaves oblong-lanceolate to ovate-elliptic, acutely acuminate, acute at base, 10–18 cm. long, 3–7 cm. broad, glabrous ; fruit 2 cm. long
4. colorata

1. **W. elongata** (*P. Beauv.*) *De Wild. & Th. Dur.* in Bull. Soc. Bot. Belg. 38 : 110 (1899) ; F.T.A. 5 : 66. *Ruellia elongata* P. Beauv. Fl. Oware 1 : 45, t. 26 (1806). *Whitfieldia longifolia* T. Anders. (1863)—F.T.A. 5 : 66 ; F.W.T.A., ed. 1, 2 : 248, incl. var. *perglabra* (C. B. Cl.) Hutch. & Dalz. (1931). *W. perglabra* C. B. Cl. in F.T.A. 5 : 66 (1899). *W. subviridis* C. B. Cl. in F.T.A. 5 : 66 (1899). Nearly glabrous undershrub, erect, several feet high, or straggling and half-scandent, with pith in the stems, constricted nodes, and conspicuous white flowers 2–3 in. long, with petaloid calyx, in terminal panicles.
S.Nig.: Ikorodu, Lagos (Dec.) *Onochie* FHI 26694 ! Ilaro F.R., Abeokuta (Dec.) *Onochie* FHI 31888 ! Igbessa *Millen* 141 ! Ibazo (Nov.) *Thomas* 2003 ! Oban *Talbot* 992 ! 2009 ! [**Br.**]**Cam.:** Kumba (Dec.) *Binuyo & Daramola* FHI 35074 ! **F.Po:** (Jan.) *Barter* 198 ! 2069 ! *T. Vogel* 187 ! 242 ! *Mann* 198 ! Extends to Angola, Sudan and Tanganyika. (See Appendix, p. 453.)
2. **W. lateritia** *Hook.* in Bot. Mag. 71 : t. 4155 (1845) ; F.T.A. 5 : 67. A forest undershrub about 10 ft. high, with red or salmon flowers 1–1½ in. long in terminal pubescent racemes.
Guin.: Sérédou *Bouquet* ! **S.L.:** *Whitfield* ! *Sc. Elliot* 3976 ! Jagba Hill (Sept.) *Melville & Hooker* 631 ! Lumbaraya (Feb.) *Sc. Elliot* 4933 ! Kambui Hills South F.R. (Dec.) *Small* 860 ! Batkanu (Jan.) *Deighton*

---

[1]The key character " filaments glabrous " used by C. B. Clarke in F.T.A. for *A. decempedalis* is erroneous ; *Mann* 3747 has densely hirsute filaments, and *Mann* 2347 glabrescent ones (" glabrous except for the few hairs near the tips " in C. B. Clarke's original MS. !), and therefore *A. cameronensis* Bremek. is clearly conspecific with *A. decempedalis* C. B. Cl.
[2] *Maitland* 1535 is a specimen covered with a rather dense indumentum. Similar plants from the Parc National Albert (Congo) have been named " *A. pubescens* (Lindau) C. B. Cl. " (Robyns Fl. Parc Nat. Alb. 2 : 285), but the original description of *Dischistocalyx pubescens* Lindau ex Engl. (1895), the basionym of this species, differs very much from them.

2861! **Lib.**: Kakatown *Whyte*! Gbanga (Sept.) *Linder* 440! Gletown (July) *Baldwin* 6776! Tappita (Aug.) *Baldwin* 9031! Karamadhun (Nov.) *Baldwin* 10195! **Iv.C.**: Taï (Mar.) *Leeuwenberg* 3030! Man (Nov.) *de Wilde* 847! Taï to Tabou (Aug.) *Boughey* GC 14916!
3. **W. preussii** (*Lindau*) *C. B. Cl.* in F.T.A. 5 : 67 (1899). *Stylarthropus preussii* Lindau (1894)—E. & P. Pflanzenfam. 4, 3 : 306, fig. 123, A, B, C (1895). A nearly glabrous undershrub with yellow flowers scarcely 1 in. long in a terminal raceme.
   [**Br.**]**Cam.**: Barombi (Sept.) *Preuss* 516! Also in Cameroun.
4. **W. colorata** *C. B. Cl. ex Stapf* in Johnston Lib. 2 : 640 (1906). A glabrous undershrub about 10 ft. high, with short dense panicles of bright red or reddish-purple flowers 1–1¾ in. long, the bracts turning brick-red.
   S.L.: Pujehun (Feb.) *Thomas* 8451! 8663! *Deighton* 363! Gola Forest (Mar.) *Small* 548! **Lib.**: Monrovia *Whyte*! Kakatown *Whyte*! *Cooper* 34! Farmington R. (Dec.) *Bequaert* 10! Dukwia R. (fr. Feb.) *Cooper* 152! Sinoe Basin *Whyte*! **Iv.C.**: Béyo (Dec.) *Leeuwenberg* 2223! Niapidou (fl. & fr. Feb.) *Leeuwenberg* 2475! 2757!

9. PHAULOPSIS Willd. Sp. Pl. 3 : 342 (1800). *Nom. cons.*, '*Phaylopsis*' ; corr. Spreng. Anleit. ed. 2, 1 : 422 (1817) ; F.T.A. 5 : 82.

Abaxial calyx-lobes broadly spathulate-oblanceolate, adaxial ones asymmetric, oblong-falcate, black towards the base, their adjacent margins straight, very close together, both hispid-ciliate ; inflorescence strobilate, short, sessile or subsessile, bracts ovate-orbicular, thin and reticulate, ciliate ; leaves long-petiolate, ovate to ovate-lanceolate, up to 12 cm. long and 5 cm. broad, crenate except towards the base   1. *falcisepala*
Abaxial calyx-lobes linear or subulate-lanceolate, adaxial ones symmetric :
Leaves small, at most about 5 cm. long and 2·5 cm. broad, rounded at apex ; flower-clusters ellipsoid, not strobiliform ; bracts ciliate with long rather stiff hairs
                                                                      2. *silvestris*
Leaves much larger, acuminate, up to 10 cm. long and 4 cm. broad, shortly pubescent ; flower-clusters strobiliform :
   Bracts ovate-orbicular, about 1 cm. long, always entire, long-ciliate, nervose ; leaves up to 8 cm. long and 4 cm. broad ; adaxial calyx-lobes narrowly linear-ligulate, about 5 mm. long, not dilated at the acute tip ; corolla 7–10 mm. long
                                                                      3. *imbricata*
   Bracts suborbicular to ovate-lanceolate, 1·5–2 cm. long ; leaves up to 10 cm. long and 3·5 cm. broad, in large specimens sometimes crenulate ; adaxial calyx-lobes spathulate, up to 8 mm. long, dilated at the rounded tip ; corolla up to 16 mm. long   ..   ..   4. *barteri*

1. **P. falcisepala** *C. B. Cl.* in F.T.A. 5 : 84 (1899), as *Phaylopsis* ; Berhaut Fl. Sén. 93 ; F. W. Andr. Fl. Sudan 3 : 185 ; Aké Assi, Contrib. 1: 200. Herb or weak undershrub up to 4 ft. high, erect or decumbent, with dense ovoid or short-cylindric spikes of small flowers red or white with purple markings.
   Sen.: *Berhaut* 720. **Guin.**: *Farmar* 311! **S.L.**: Kumrabai (fl. & fr. Dec.-Jan.) *Thomas* 6792! 6798! 6833! Kenema (fl. & fr. Jan.) *Thomas* 7661! 7739! 7912! Senehun, Kamgai (Nov.) *Cole* 55! Gegbwema (Nov.) *Deighton* 400! **Lib.**: Boporo (Nov.) *Baldwin* 10381! Monrovia (Nov.) *Barker* 1458! *Linder* 1414! Kakatown (Sept.) *Whyte*! Peahtah (fr. Oct.) *Linder* 1103! **Iv.C.**: Mankona (Nov.) *de Wilde* 967! Beoumi (Dec.) *Lowe*! **Ghana**: Amoma (fr. Jan.) *Vigne* FH 3550! Agogo (fr. Dec.) *Irvine* 581! Abetifi, Kwahu (fr. Dec.) *Irvine* 1656! Kibi (fl. & fr. Dec.) *Morton* GC 8122! Bamboi to Wenchi (Dec.) *Adams & Akpabla* GC 4505! **Togo Rep.**: Misahöhe (Nov.) *Mildbr.* 7226! **N.Nig.**: Ilorin *Clarke* 2! Nupe *Barter* 913 (partly)! Kontagora (fl. & fr. Nov.) *Dalz.* 186! Damisa, Zaria Prov. (Nov.) *Keay & Onochie* FHI 21658! Anara F.R., Zaria Prov. (Oct.) *Keay* FHI 20114! **S.Nig.**: Lagos *Millson* 127! Ilaro F.R. (Dec.) *Onochie* FHI 31897! Idanre (Jan.) *Brenan & Jones* 8702! 8725! Onitsha Olona (Oct.) *Maitland* 1842! **F.Po**: Bahia de Venus (Dec.) *Guinea* 294! (See Appendix, p. 452.)
2. **P. silvestris** (*Lindau*) Lindau in E. & P. Pflanzenfam. Nachtr. 2–4 : 305 (1897). *Micranthus silvestris* Lindau in Engl. Bot. Jahrb. 17 : 107 (1893). *Micranthus microphyllus* O. Ktze. Rev. Gen. Pl. 2 : 493 (1891), name only. *Phaylopsis microphylla* C. B. Cl. in F.T.A. 5 : 85 (1899) ; F.W.T.A., ed. 1, 2 : 249. Habit of the last, with smaller leaves and looser flower-clusters.
   Ghana: Assin-Yan-Kumasi *Cummins* 71! near Kumasi (Dec.) *Morton* A123! Kibi (Dec.) *Morton* GC 8121! Bosumkese F.R. (Dec.) *Adams* 2910! 5256! **S.Nig.**: Oban *Talbot* 994! Usonigbe F.R. (Nov.) *Keay & Onochie* FHI 21502! [**Br.**]**Cam.**: Buea, 3,000–4,000 ft. (Dec.) *Maitland* 121! *Migeod* 145! Mimba to Buea (Dec.) *Akpabla* GC 10523! Also in Cameroun and Gabon.
   [" *Phaylopsis microphylla* T. Anders." is a MS name, quoted in synonymy by O. Kuntze (1891), and accepted for this species by C. B. Clarke in F.T.A. l.c. (1899).]
3. **P. imbricata** (*Forsk.*) *Sweet* Hort. Brit., ed. 1, 327 (1827) ; Alston in Trimen, Handb. Fl. Ceylon 6, Suppl. : 225 ; Milne-Redhead in Mem. N.Y. Bot. Gard. 9 : 22 (1954). *Ruellia imbricata* Forsk. Fl. Aeg.-Arab. Descr. Pl. 113 (1775). *Aetheilema imbricata* (Forsk.) R. Br. (1810). *Micranthus oppositifolius* Wendl. (1798). *Phaulopsis oppositifolia* (Wendl.) Lindau (1897)—Chev. Bot. 495. *Phaylopsis parviflora* Willd. Sp. Pl. 3 : 342 (1801). Bot. Mag. 60 : t. 2433 ; F.T.A. 5 : 83 ; Chev. Bot. 495 ; F.W.T.A., ed. 1, 2 : 249 ; Berhaut Fl. Sén. 93 ; Schnell in Mém. I.F.A.N. 22 : 517 (1952) ; Aké Assi, Contrib. 1: 201. *P. talbotii* S. Moore in Cat. Talb. 78 (1913). A more or less hairy and slightly viscid herb or small undershrub up to 2 ft. high, with small white or purplish flowers in ovoid or short-cylindric spikes.
   Sen.: *Heudelot* 7! 8! 585! *Leprieur*! *Berhaut* 1201. Niayes *Chev.* 2802. **Gam.**: *Brown-Lester* 11! Genieri (fr. Feb.) *Fox* 70! **Mali**: Tabacco *Chev.* 140. Kita (Nov.) *Jaeger* 3613! **Port.G.**: Nhambanhana *Esp. Santo* 310. Pessubé (Dec.) *Esp. Santo* 1095! **S.L.**: Freetown (Dec.) *Jaeger* 594! **Iv.C.**: Abengourou *Aké Assi* IA 2666. Taï *Miège & Aké Assi* IA 2716. **S.Nig.**: Oban *Talbot* 977! Also in Ethiopia, E. Africa and Nyasaland. (See Appendix, p. 452.)
4. **P. barteri** (*T. Anders.*) *Lindau* in E. & P. Pflanzenfam. Nachtr. 2–4 : 305 (1897). *Phaylopsis barteri* T. Anders. in J. Linn. Soc. 7 : 27 (1862) ; F.T.A. 5 : 86 ; F.W.T.A., ed. 1, 2: 249; Aké Assi, Contrib. 1 : 200. A weak undershrub 2–4 ft. high in woodland undergrowth etc., with white flowers ⅔ in. long, well exserted from the bracts, in dense strobilate spikes 1–3 in. long and 1 in. broad.
   Mali: Kita Massif (Sept.-Oct.) *Jaeger* 62! 2466! **Port.G.**: Boma *Esp. Santo* 322. **S.L.**: *Afzelius. Brown.* Musaia (fl. & fr. Dec.) *Deighton* 4416! 4497! Bintumane Peak 3,200–5,400 ft. (fl. & fr. Jan.) *T. S. Jones* 126! 157! **Iv.C.**: Ourossantiakara (Nov.) *Leeuwenberg* 2001! Kouroumkourounga Forest, Séguela to Mankono (Nov.) *de Wilde* 953! *Aké Assi* 4599*bis*. **Ghana**: Techiman (Dec.) *Adams & Akpabla* 4496! Banda Ravine, Wenchi (Dec.) *Morton* GC 25123! 25164! Gambaga Scarp (Dec.) *Adams* 2235! Dutukpene (Dec.) *Adams* GC 4539! Kintampo (Dec.) *Morton* A1171! Ejura Scarp (Jan.) *Morton* A1553! **Togo Rep.**: *Baumann* 364. Atakpame (Nov.) *Mildbr.* 7461! Bismarckburg (Nov.) *Büttner* 336! *Kling* 193! **N.Nig.**:

Nupe *Barter* 913! Lokoja (Oct.) *Parsons*! Toro (Oct.) *Lely* 683! Damisa, Zaria Prov. (Jan.) *Keay & Onochie* FHI 21542! Jos (Oct.) *Hepper* 1097! 1180! **S.Nig.**: Onitsha *Barter* 845! Ikwette Plateau, Obudu Div. (Dec.) *Savory & Keay* FHI 25155! [**Br.**]**Cam.**: Gangumi, Gashaka Dist. (fl. & fr. Dec.) *Latilo & Daramola* FHI 28829! Bellel and Maisamari, 4,600 ft., Mambila Plateau (Jan.) *Hepper* 1673! 1692! Vogel Peak, 4,000 ft., Adamawa (Nov.) *Hepper* 1343! A tropical African highland species ; also in Cameroun, Gabon, Congo, and tropical E. Africa.

[An intermediate between this and *P. falcisepala* C. B. Cl. is *Adams & Akpabla* 4496! (Ghana : Techiman) ; Jaeger's specimens from Kita Mts. are too young for certainty and may be intermediates too. Generally, the distinction between *P. falcisepala* and *P. barteri* is not always easy and *P. falcisepala* may be an infraspecific taxon of the latter.]

10. THUNBERGIA Retz. Phys. Saellsk. Handl. 1 : 163 (1776), *nom. cons.* ; F.T.A. 5 : 8 ; Benoist in Notulae Syst. 2 : 287 (1912–13), and 11 : 144 (1944).

Petioles winged, 1–2 cm. long ; leaves ovate-triangular, sagittate or hastate at base, 3·5–7 cm. long, 2·5–4·5 cm. broad, pubescent, entire or undulate-dentate ; flowers axillary, solitary ; pedicels up to 6 cm. long ; bracteoles ovate, subcordate, about 2 cm. long ; corolla-tube 2 cm. long, limb oblique, 1·5 cm. long ; fruit 2·5 cm. long

1. *alata*

Petioles not winged :
Climbing herbs :
Flowers 2–3 together between two sessile leaves on a peduncle-like branchlet ; leaves very broadly ovate, cordate, acutely acuminate, 7–9 cm. long, up to 6 cm. broad, pubescent on the nerves beneath, repand-dentate ; petiole up to 6 cm. long ; bracteoles ovate-elliptic, mucronate, 2 cm. long ; corolla 6 cm. long   2. *fasciculata*
Flowers solitary or sometimes (in No. 5) in a few-flowered axillary inflorescence up to 15 cm. long :
Leaves entire, or at most undulate :
Leaves lanceolate, sometimes with a tooth on each side at the base, rounded at the base, 4–10 cm. long, acute, minutely setulose ; bracteoles lanceolate, 1·5 cm. long ; corolla 2·5 cm. long ; fruit 1·5 cm. long, pubescent   ..   3. *laevis*
Leaves ovate or triangular :
Leaves acutely mucronate, cordate at base, up to 6 cm. long and 4 cm. broad, glabrous ; petiole pubescent ; flowers solitary ; bracteoles ovate-lanceolate, 1 cm. long ; corolla 2 cm. long ; fruit 1·5 cm. long, softly puberulous

4. *cynanchifolia*

Leaves gradually acuminate, rounded-subcordate at base, about 17 cm. long and 8 cm. broad, prominently trinerved at base, glabrous ; petiole glabrous ; bracteoles narrowly ovate, 1·5 cm. long, venose ; calyx lobed ; corolla-tube 4 cm. long   ..   ..   5. *togoensis*
Leaves lobulate or lobed, widely cordate at base, lobes triangular, setose-pilose above; petiole slender ; flowers solitary ; pedicel up to 5 cm. long ; bracteoles ovate, 2·5 cm. long, slightly setulose outside ; calyx obsolete, truncate ; corolla 8–10 cm. long, broadly tubular ..   ..   6. *chrysops*
Shrubs or woody climbers :
Bracteoles densely covered outside with very small stellate hairs, ovate, rounded at apex, 3 cm. long and 2·5 cm. broad ; corolla hispid outside near base, about 7 cm. long ; leaves broadly elliptic, acuminate, shortly cuneate at base, about 12 cm. long and 7 cm. broad, nearly glabrous   ..   ..   7. *rufescens*
Bracteoles glabrous :
Calyx about 1 cm. long ; leaves oblong to elliptic, acuminate, subacute at base, 10–15 cm. long, up to 8 cm. broad, glabrous, entire ; bracteoles 2 cm. long ; corolla 6–7 cm. long ..   ..   8. *vogeliana*
Calyx at most 3 mm. long ; leaves small, ovate to broadly lanceolate, acute, 3–4·5 cm. long, 1·5–2 cm. broad, subentire ; bracteoles soon deciduous ; corolla 6 cm. long

9. *erecta*

1. **T. alata** *Boj. ex Sims* in Bot. Mag. 52 : t. 2591 (1825) ; F.T.A. 5 : 16 ; Milne-Redhead in Mem. N.Y. Bot. Gard. 9, 1 : 19 (1954) ; Bremekamp in Verh. K. Nederl. Akad. Wetensch., Afd. Natuurk. sect. 2, 50 : 41 (1955) ; Aké Assi, Contrib. 1: 202. A softly pubescent herbaceous twiner, with pale to dark yellow flowers (sometimes white) usually red or dark purple in the throat.
S.L.: *Reade*! Freetown (Feb., Aug.) *Johnston* 59! *Melville & Hooker* 349! Regent (Dec.) *Sc. Elliot* 4141a! Njala (July) *Deighton* 455! **Lib.**: Sinoe Basin (fr. Jan.) *Johnston*! Monrovia (Nov.) *Linder* 1565! **Iv.C.**: Abidjan *Aké Assi* IA 925! **N.Nig.**: (fr. Sept.) *Lely* P711! This is the oldest name available within this aggregate species, which is a native of E. and S. Africa ; frequently cultivated and often naturalized in all the warmer parts of the world.
2. **T. fasciculata** *Lindau* in Engl. Bot. Jahrb. 17 : 97 (1893) ; F.T.A. 5 : 15. Climber with sparingly pilose stems ; flowers completely hidden by two leaf-like sessile bracts until full-out, corolla yellow or purplish with yellowish tube.
S.Nig.: Aponmu, Ondo (Oct.) *Keay* FHI 16248! Awba Hills F.R. (Oct.) *Jones* FHI 7081! Etomi to British Obokum, Ogoja Prov. (Dec.) *Keay* FHI 28276! [**Br.**]**Cam.**: Bambili, Bamenda (Aug.) *Ujor* FHI 29983! Also in Congo, Uganda and Sudan.
3. **T. laevis** *Nees* in Wallich Pl. As. Rar. 3 : 77 (1832) ; Bremekamp l.c. 51. *T. fragrans* var. *laevis* (Nees) C. B. Cl. in Hook. f., Fl. Brit. Ind. 4 : 391 (1884). *T. fragrans* of F.W.T.A., ed. 1, 2 : 250 (1931), not strictly of Roxb. (1795). A slender twiner, sparingly hairy with whitish flowers 1½ in. long ; cultivated and more or less naturalized.
S.L.: *Lane-Poole* 278! Freetown (May) *Deighton* 2689! Regent (Oct.) *Jordan* 948! *Sc. Elliot* 4141b! A native of tropical Asia, now widespread in the tropics.

Fig. 298.—Thunbergia erecta (*Benth.*) *T. Anders.* (Acanthaceae).
A, pistil. B, anther. C, ovary.

[The specimens cited above have pubescent ovaries, but one gathering from Ghana (Achimota *Morton* GC 25399) has glabrous ovaries and therefore appears to be the true *T. fragrans* Roxb., also a tropical Asian plant.]

4. **T. cynanchifolia** Benth. in Fl. Nigrit. 475 (1849) ; F.T.A. 5 : 20 ; Chev. Bot. 490. A slender climbing plant, on trees, etc., with pure white flowers ¾–1 in. long, the stems either glabrous or spreading-pilose.
Guin.: Mt. Ntogon, 3,100 ft., *Chev.* 20927. Oua, upper Cavally *Chev.* 21312 ! S.L.: Njala (Aug.) *Deighton* 5579 ! Musaia (Aug.)*Jordan* 493 ! (fr. Mar.) *Deighton* 5392 ! Segbwema (Dec.) *Deighton* 3464 ! Warantamba (Oct.) *Small* 377 ! Lib.: Firestone Plantations, Dukwai R., Monrovia (Oct.-Nov.) *Cooper* 10 ! Gbanga (Sept.) *Linder* 481 ! Firestone Cavalla R. Plantation, Maryland Co. (June) *Baldwin* 6021 ! Iv.C.: Taï *Leeuwenberg* 3033 ! Sago (fl. & fr. Mar.) *Leeuwenberg* 3096 ! Ghana: Aburi (Oct.) *Brown* 346 ! 375 ! Kwadaso (Dec.) *Darko* 646 ! New Tafo (Sept.) *Lovi* 3916 ! Amuni *Vigne* FH 1271 (partly) ! Amedzofe (fl. & fr. June) *Morton* A3423 ! S.Nig.: *Thomas* 1146 ! Niger R. (= Quorra) *T. Vogel* 7 ! Eket *Talbot* 3016 ! Oban *Talbot* 988 ! [Br.]Cam.: Mamfe (Mar.) *Morton* GC 7141 ! Also in Cameroun.

5. **T. togoensis** Lindau in Engl. Bot. Jahrb. 22: 112 (1895); F.T.A. 5: 22 ; Aké Assi, Contrib. 1: 205. A glabrous climber with blue flowers 2 in. long.
Iv.C.: Adiopodoumé (cult., fr. Aug.) *Aké Assi* 271 ! 5571 ! Ghana: Amedzofe, 2,000 ft. (Jan., Mar., Apr.) *Irvine* 3361 ! *Hall* 1754 ! *Morton* GC 6503 ! A2204 ! Togo Rep.: Misahöhe, near Aundjowe-Avatime, 2,000 ft. (July) *Baumann* 74 ! Lib.: Olokemeji F.R. (fl. & fr. Aug., Sept.) *Onochie* FHI 40265 ! *Chizea & Odukwe* FHI 33912 ! [Br.]Cam.: Aba-ajia, Wum Dist. (July) *Uior* FHI 30452 ! Also in Cameroun.

6. **T. chrysops** Hook. Bot. Mag. 70 : t. 4119 (1844) ; F.T.A. 5 : 21 ; Chev. Bot. 490 ; F.W.T.A., ed. 1, 2 : 250 and Aké Assi, Contrib. 1: 203, excl. syn. *T. talbotiae* S. Moore and *Talbot* 3394. *T. geraniifolia* Benth. (1849)—F.T.A. 5 : 21 ; Chev. Bot. 491. *T. subnymphaeifolia* Lindau in Engl. Bot. Jahrb. 38 : 67 (1905). A slender climber, pilose on the young stems, glabrescent, with beautiful violet-purple flowers 3–4 in. long, with yellow throat. Often cultivated as an ornamental.
Guin.: Kouria *Chev.* 14972 ! Fouta Djalon *Chev.* 18247. S.L.: *Don* ! *Burbridge* ! Hill Station, Freetown (fl. & fr. Oct.) *Deighton* 2169 ! Makump (Sept.) *Deighton* 1288 ! Balima (Sept.) *Deighton* 3030 ! Rowalla (July) *Thomas* 1190 ! Bo *Gardner* 38 ! Lib.: (Nov.) *Bunting* 101 ! Kassa Ta (Sept.) *Linder* 823 ! Iv.C.: Oroumba Boca (Aug.) *de Wit* 7646 ! Rocher d'Issia (fl. & fr. Aug.) *de Wilde* 430 ! Ghana: *Vigne* FH 3121 ! Bou *Vigne* FH 1583 ! Kibbi *Gould* ! Maliato to Tonogo *Morton* A3475 ! Mpraeso Scarp (fr. Nov.) *Morton* A2757 ! Togo Rep.: *Kersting* 585. [Br.]Cam.: Ikom to Mamfe (Dec.) *Morton* K674 ! (See Appendix, p. 453.)

7. **T. rufescens** Lindau in Engl. Bot. Jahrb. 17 : 96 (1893) ; F.T.A. 5 : 10 ; F.W.T.A., ed. 1, 2 : 250. A climbing shrub about 20 ft. high, rusty-stellate-tomentose on the young parts with white flowers 2½–3 in. long ; fruit unknown ; in forest.
S.Nig.: Akure F.R., Ondo (Oct., Nov.) *Latilo & Edwin* FHI 15282 ! *Keay* FHI 25477 ! Aiyetoro Owenna, Ondo (Nov.) *Onochie* FHI 34246 ! [Br.]Cam.: Barombi-ba-Mbu to Kake (Sept.) *Preuss* 432 !
[This species has a quite abnormal position within the genus *Thunbergia* : it shows some striking similarities with the genus *Pseudocalyx* (entire leaves ; bracteoles which are covered with rusty-brown stellate hairs outside ; stellate hairs outside at the base of the corolla-tube). The floral characters, however, do not permit its inclusion in that genus ; fruits are required.—H.H.]

8. **T. vogeliana** Benth. in Fl. Nigrit. 476 (1849) ; H. Ross in Neubert's Garten-Mag. 51 : 57, t. 2 (1898) ; F.T.A 5 : 10 ; Bremekamp l.c. 36 (1955). *T. kamerunensis* Lindau in Engl. Bot. Jahrb. 17 : 97 (1893). A branching shrub, up to 15 ft. or more high, with flowers 2–3 in. long, pale or dark violet-purple with white tube. Sometimes cultivated as an ornamental.
Ghana: Anyaboui to Jaketi (cult. in Achimota) (Oct.) *Morton* A35 ! S.Nig.: Ohe, Benin (Dec.) *Keay* FHI 28303 ! Benin F.R. (Nov.) *Obomanu* FHI 26930 ! Iseneki, Benin (Apr.) *Farquhar* 199 ! Oban *Talbot* 387 ! [Br.]Cam.: Barombi (Aug., Sept.) *Preuss* 387 ! *Njia* FHI 8416 ! Bova, Kumba *Olorunfemi* FHI 30776 ! F.Po: (Nov.) *T. Vogel* 147 ! *Mann* 557 ! 558 ! Conception (Sept.) *Melville* 446 ! (See Appendix, p. 453.)
[The specimens from Ghana and Benin Prov. have toothed leaves and pointed bracts.]

9. **T. erecta** (Benth.) T. Anders. in J. Linn. Soc. 7 : 18 (1863) ; F.T.A. 5 : 12 ; Chev. Bot. 491 ; Bremekamp l.c. 37 (1955) ; Mangenot in Ic. Pl. Afr. I.F.A.N. 4 : 96 (1957). *Meyenia erecta* Benth. in Fl. Nigrit. 476 (1849) ; Bot. Mag. 83 : t. 5013 (1857). A branching shrub, up to 15 ft. or more high with flowers 2–3 in. long, pale or dark violet-purple with white tube ; often cultivated as an ornamental.
Port.G.: Pobresa (fr. Feb.) *Esp. Santo* 2377 ! Guin.: Conakry *Chev.* 25685. S.L.: Sherboro Isl. *Sc. Elliot* 5856 ! Njala (fr. Nov.) *Deighton* 4924 ! Lib.: Ganta (Sept.) *Baldwin* 9304 ! Iv.C.: Adiopodoumé (cult.) *Aké Assi.* Oellé (Sept.) *Aké Assi.* Ghana: Aburi *Chev.* 13830. N.Nig.: Sokoto (Nov.) *Moiser* 71 ! Mada Hills *Hepburn* ! Vom (Oct.) *Dent Young* 203a ! *Hepper* 1159a ! S.Nig.: Lagos (Feb.) *Millen* 190 ! Okomu F.R., Benin (Sept., Dec.) *Brenan* 8511 ! *Iriah* FHI 23089 ! Akilla F.R., Ijebu (Oct.) *Enwioglon* FHI 37951 ! Owo *Burton* 2 ! Calabar *Williams* 54 ! [Br.]Cam.: Bamenda, 4,700 ft. (Feb.) *Migeod* 499 ! Gangumi, Adamawa (Dec.) *Latilo & Daramola* FHI 28912 !

T. grandiflora (Roxb. ex Rottl.) Roxb. and *T. laurifolia* Lindl. are introduced and cultivated as ornamentals (for the taxonomy of these two Asiatic species see Bremekamp l.c., pp. 45–49, 1955). *T. grandiflora* was found " in thick bush near Awa " S. Nigeria (*Talbot* 3394) and described as *T. talbotiae* by S. Moore in J. Bot. 52 : 31 (1914). Talbot's plant was unfortunately considered by Hutch. & Dalz. (F.W.T.A., ed. 1, 2 : 250) as conspecific with the indigenous *T. chrysops* Hook. and the name *T. talbotiae* S. Moore therefore wrongly added to the synonymy of *T. chrysops* Hook. (see above). *T. grandiflora* (Roxb. ex Rottl.) Roxb. is apparently naturalized in Calabar : " creeping over steep slopes of bluffs ; white flower," 8 Feb. 1958, *Stubbings* 104 ! It seems probable that this gathering has a connection with Talbot's, which would mean that the species has been established in Calabar Prov. for about 50 years.

11. MENDONCIA Vell. ex Vand. Fl. Lusit. 43, t. 3, f. 22 (1788) ; Benoist in Bull. Mus. Hist. Nat. 31 : 386 (1925), Notulae Syst. 11 : 139 (1944). *Monachochlamys* Bak. (1883). *Afromendoncia* Gilg ex Lindau (1893)—F.T.A. 5 : 6 ; Turrill in Kew Bull. 1919 : 407 ; F.W.T.A., ed. 1, 2 : 250.

Plants entirely glabrous ; leaves oblanceolate, acuminate, slightly cordate at base, up to 9 cm. long and 5·5 cm. broad, with about 3–4 pairs of lateral nerves ; petioles up to 3 cm. long ; flowers unknown ; bracts purplish, broadly ovate cordate, up to 1·5 cm. long and broad, acuminate ; fruit a pointed drupe about 8 mm. long (when dry), somewhat angled   ..   ..   1b. *gilgiana* var. *tisserantii*
Plants hairy ; leaves becoming glabrescent :
  Stems 4-angled, villous when young ; leaves ovate-oblong, shortly acuminate, 5–8 cm. long, up to 4 cm. broad, pilose beneath, with 5–6 pairs of lateral nerves ; petiole up to 3 cm. long ; flowers axillary, fasciculate ; pedicels about 2 cm. long ; calyx short ; corolla-tube 2 cm. long ; anthers hairy at the base
                                                                1a. *gilgiana* var. *gilgiana*

Stems terete, thinly pubescent at the bracts ; leaves broadly elliptic, slightly pointed, rounded at base, 10–14 cm. long, 5–8 cm. broad, pilose on the 2–3 pairs of lateral nerves beneath ; petiole 2 cm. long ; pedicels 2 cm. long, thinly setose ; bracteoles broadly ovate-elliptic, 3-nerved from the base, setose-pilose ; calyx an undulate rim ; corolla-tube 1·3 cm. long :
Indumentum composed of ordinary hairs up to 3 mm. long and some small branched (but not stellate) hairs   ..   ..   2. *iodioides*
Indumentum composed exclusively of small stellate hairs ; corolla white
3. *combretoides*

1. **M. gilgiana** (*Lindau*) *Benoist* in Bull. Soc. Bot. Fr. 85 : 679 (1939) ; Notulae Syst. 11 : 143 (1943). *Afromendoncia gilgiana* Lindau (1894)—F.T.A. 5 : 7 ; F.W.T.A., ed. 1, 2 : 250. *Monachochlamys gilgiana* (Lindau) S. Moore in J. Bot. 67 : 227 (1929).
1a. **M. gilgiana** (*Lindau*) *Benoist* var. **gilgiana**. A slender twiner with white flowers ¾ in. long enclosed in broad bracteoles in axillary clusters of 2–4.
   [**Br.**]**Cam.**: Barombi *Preuss* 481 ! Also in Cameroun.
1b. **M. gilgiana** (*Lindau*) *Benoist* var. **tisserantii** *Benoist* Bull. Soc. Bot. Fr. 85 : 679 (1939) ; Aké Assi, Contrib. 1: 205. Climber with purplish bracts, vegetative parts similar to *Calycobolus heudelotii* (Convolvulaceae) but with opposite leaves which are slightly cordate at the base.
   Lib.: Nekabozu (fr. Oct.) *Baldwin* 9971 ! **Iv.C.**: Danané to N'Zo (fr. Aug.) *Aké Assi* 3271 ; 5424 ! Also in Ubangi-Shari and Congo.
2. **M. iodioides** (*S. Moore*) *Heine* in Kew Bull. 16: 180 (1962). *Afromendoncia iodioides* S. Moore in Cat. Talb. 74, 139 (1913) ; F.W.T.A., ed. 1, 2 : 250. *Monachochlamys iodioides* (S. Moore) S. Moore in J. Bot. 67 : 227 (1929). A climbing undershrub, sparsely rusty-hairy, with broad leaves and whitish flowers nearly 1 in. long few in axillary clusters.
   S.Nig.: Oban *Talbot* 388 !
3. **M. combretoides** (*A. Chev.*) *Benoist* in Notulae Syst. 11 : 143 (1943). *Thunbergia combretoides* A. Chev. Bot. 490 (1920) ; F.W.T.A., ed. 1, 2 : 250. A lofty woody climber with arm-thick stem having adventitious roots at the base, and white flowers in fascicles on the old wood.
   Lib.: Gletown, Webo Dist. (July) *Baldwin* 6764 ! **Iv.C.**: Grabo (Aug.) *Chev.* 19745 ! 60 km. N.of Sassandra (Feb.) *Leeuwenberg* 2820 ! Nékaougnié to Grabo (Aug.) *Chev.* 19600 ! Banco *Hallé* 796 !
   [Note : *Thunbergia combretoides* A. Chev. (1920) is not only a name, as stated by Index Kewensis, by Hutchinson & Dalziel, and Benoist ; it is accompanied by a short descriptive note which indicates very clearly the diagnostic characters.]

12. MIMULOPSIS Schweinf.—F.T.A. 5 : 54.

Stems glabrous ; leaves very long-petiolate, ovate-triangular, cordate at base, 8–14 cm. long, 5·5–8·5 cm. broad, doubly and coarsely dentate, lateral nerves 6–7 pairs, sparingly pubescent, with very minute rod-like cystoliths on the upper surface ; panicle lax, large, leafy, pubescent ; bracteoles linear ; calyx densely viscid-pilose ; lobes linear ; corolla 2–2·5 cm. long, subregular, glabrous, widely funnel-shaped ; stamens 4 ; fruit beaked, pubescent   ..   ..   *solmsii*

M. solmsii *Schweinf.* in Verh. Zool.-Bot. Ges. Wien 18 : 677 (1868) ; F.T.A. 5 : 55 ; Milne-Redhead in Mem. N.Y. Bot. Gard. 9 : 21 (1954). *M. violacea* Lindau (1893)—F.T.A. 5 : 55 ; F.W.T.A., ed. 1, 2 : 252. A coarse herb or undershrub up to 12 ft. high, with 4-angled pithy stems and viscid-hairy inflorescence of lilac or pale violet flowers 1 in. long.
   [**Br.**]**Cam.**: Buea, 3,000–7,800 ft. (fl. & fr. May) *Preuss* 814 ! 1056 ! *Maitland* 1314 ! *Mann* 1260 ! Mba Kokeka Mt., Bamenda (Jan.) *Keay* FHI 28403 ! Bambili, Bemenda (Aug.) *Uior* FHI 29971 ! **F.Po**: 7,000 ft. (fl. & fr. Mar.) *Mann* 2348 ! Moka 4,000–7,000 ft. (fl.Sept., fr. Dec.) *Boughey* 87 ! 158 ! 172 ! 194 ! *Wrigley* 565 ! (See Appendix, p. 452.)
   [A poor specimen from Sierra Leone (Bintumane Peak, 5,800 ft. (Jan.) *T. S. Jones* 105 !) is doubtfully this species.]

13. SATANOCRATER Schweinf.—F.T.A. 5 : 68.

Young branches densely light-brownish pubescent ; mature leaves subsessile, ovate-lanceolate, up to 10 cm. long and 4·5 cm. broad, glabrous, but with small sunken yellowish glands on both sides ; flowers solitary, opposite in the leaf-axils ; pedicels 5–8 mm. long, with two linear bracteoles up to 12 mm. long ; calyx tubular, up to 3·5 cm. long ; corolla large, tube about 4·5 cm. long, cylindric at the base for a length of about 2 cm., 4 mm. diam. ; throat 10–12 cm. diam., with 5 rounded, subequal lobes ; stamens 4, included, filaments sparsely pilose, anther-cells 2, muticous ; ovary with 2 ovules in each cell ; fruit oblong, up to 22 mm. long, glabrous, but covered with small yellowish scale-like glands ; seeds discoid   ..   ..   ..   *berhautii*

S. berhautii *Benoist* in Bull. Soc. Bot. Fr. 99 : 325 (1953) ; Berhaut Fl. Sén. 89. Herb with axillary mauve foxglove like flowers up to 2½ in. long.
   Guin.: Koundara (fl. & fr. Jan.-Dec.) *Berhaut* 1863 ; 4157 ; 4158 !
   [Very close to *S. fellatensis* Schweinf. (1868) from Galabat and Somaliland, only differing in the much larger bracteoles, calyx-tubes and capsules—H.H.]

14. DYSCHORISTE Nees—F.T.A. 5 : 71.

Flowers (at least the lower ones) in stalked cymes ; calyx glandular-pubescent, 1·5 cm. long, segments united to the middle, subulate-lanceolate ; corolla 2·5 cm. long, softly pubescent ; leaves oblanceolate, 3–4 cm. long, shortly pubescent on the nerves beneath   ..   ..   1. *pedicellata*
Flowers in axillary clusters or subsolitary :
Flowers 1–2 in each leaf-axil, shortly pedicellate ; calyx densely glandular-pubescent, 2 cm. long, segments united to the middle, subulate ; corolla 3 cm. long ; leaves oblanceolate, subacute, 4–7 cm. long, up to 2 cm. broad, minutely pubescent
2. *heudelotiana*

Flowers 3 or more in each leaf-axil, sessile or subsessile ; calyx 6–8 mm. long, covered with cystoliths but not pubescent ; lobes filiform-subulate ; corolla about 8 mm. long ; leaves oblanceolate, up to 10 cm. long and 1·5 cm. broad, glabrous ; fruit 1 cm. long   ..   ..   **3. *perrottetii***

1. **D. pedicellata** *C. B. Cl.* in F.T.A. 5 : 75 (1899) ; Monod in Mém. I.F.A.N. 18 : 13 (1952). *Hygrophila chariensis* Lindau in Mém. Soc. Bot. Fr. 2, 8 : 48 (1908). *Dyschoriste radicans* of F.W.T.A., ed. 1, 2 : 252, partly (*Lely* 483, 548), not of Nees (1847). A low shrublet, shortly pubescent, with pinkish or light purplish flowers 1 in. long.
   Gam.: (July) *Ozanne* 8 ! **Mali:** Koumana (fl. & fr. Feb.) *Chev.* 367 ! **N.Nig.:** Naraguta (fl. & fr. Aug.) *Lely* 483 ! 548 ! Kogigiri, 4,100 ft., Jos (fl. & fr. Oct.) *Hepper* 1058 ! Zaria (fl. & fr. Oct.) *Keay* FHI 21419 ! Wusasa, Zaria (Aug.) *Keay* FHI 28029 ! Faskari, Katsina Prov. (fl. & fr. Nov.) *Daggash* FHI 35029 ! Also in Ubangi-Shari.

2. **D. heudelotiana** (*Nees*) *O. Ktze.* Rev. Gen. Pl. 2 : 486 (1891) ; F.T.A. 5 : 87 ; Berhaut Fl. Sén. 91, 95. *Calophanes heudelotianus* Nees in DC. Prod. 11 : 112 (1847). Half-woody herb or undershrub, shortly pubescent, with purple flowers 1 in. long.
   Sen.: Bondou *Heudelot* 18 ; 144 ! 190 ! *Berhaut* 807. Vodouli (Jan.) *Roberty* 16645 ! **Mali:** Kita (Dec.) *Jaeger* 3750 !
   [*D. heudelotiana* and *D. pedicellata* C. B. Cl. are doubtfully distinct and perhaps conspecific.—H.H.]

3. **D. perrottetii** (*Nees*) *O. Ktze.* l.c. 486 (1891) ; F.T.A. 5 : 72 ; Chev. Bot. 492 ; Berhaut Fl. Sén. 93 ; Aké Assi, Contrib. **1**: 205. *Calophanes perrottetii* Nees in DC. Prod. 11: 111 (1847). Herb or undershrub, branched, 1–3 ft. high, woody below, sometimes decumbent and rooting at the lower nodes, with rather small flowers ½ in. long, pink or purplish or white with purple markings.
   Sen.: *Perrottet* 28 ; 515. *Leprieur. Roger* 136 ! *Berhaut* 1543. **Port.G.:** Canjambarim (fl. & fr. Mar.) *Esp. Santo* 2439 ! **S.L.:** Musaia (Dec.) *Deighton* 4561 ! Sasseni (Jan.) *Sc. Elliot* 4534 ! Wallia (Jan.) *Sc. Elliot* 4634 ! Mano (Mar.) *Thomas* 10398 ! Njala (fr. Feb.) *Deighton* 1073 ! **Lib.:** Baila *Harley* 1421 ! **Iv.C.:** Tiassale (fl. & fr. Dec.) *Leeuwenberg* 2139 ! Adiopodoumé *Aké Assi* IA 5503 ! **Ghana:** Paliba (fr. Dec.) *Adams & Akpabla* 4112 ! Port Tamale (fr. Mar.) *Morton* GC 8768 ! Damongo (fr. Aug.) *Harris* ! Akuse (fr. Apr.) *Morton* A3654 ! Kete Krachi (fr. Dec.) *Adams* 4591 ! **Togo Rep.:** *Baumann* 410 ! **N.Nig.:** Kwaki (fl. & fr. Nov.) *Onochie & Keay* FHI 21737 ! Borgu *Barter* 820 ! (fr. Feb.) *Onyeagocha* FHI 7631 ! Kontagora (fr. Dec.) *Dalz.* 210 ! Fodama (fl. & fr. Jan.) *Moiser* 182 ! **S.Nig.:** Lagos (Mar.) *Millen* 99 ! Ibadan to Abeokuta (fr. Jan.) *Schlechter* 13035 ! Ibadan (fl. & fr. Jan., Dec.) *Umana* FHI 26758 ! Oyo (fl. & fr. Feb.) *Brenan* 8962 ! [**Br.]Cam.:** Belo, 5,000 ft., Bamenda *Newberry & Etim* FHI 184 ! Gangumi to Garbabi, Adamawa (fl. & fr. Nov.) *Latilo & Daramola* FHI 28781 ! Extends to E. Sudan and Ethiopia. (See Appendix, p. 450.)

Excluded species.

D. radicans *Nees* in DC. Prod. 11 : 106 (1847) ; F.T.A. 5 : 73. Not recorded from our area, but perhaps overlooked ; the citations in F.W.T.A., ed. 1, 2 : 252 belong to *D. pedicellata* (*Lely* 483, 548). *Chev.* 367, cited also under *D. radicans*, in F.W.T.A., ed. 1, 2 : 252, is a syntype of *Hygrophila chariensis* Lindau (1908).

15. PHYSACANTHUS Benth.—F.T.A. 5 : 57 ; Rendle & Britten in J. Bot. 47 : 377 (1909) ; N. Hallé in Adansonia, n. sér. 1 : 343 (1962, dated 1961).

Corolla-lobes widely bilobed, about 7 mm. long ; tube 3·5 cm. long, very slender, bent and pubescent towards apex ; leaves elliptic, rounded at both ends, 3–5 cm. long, 2–2·5 cm. broad, thinly pilose above, pilose beneath mainly on the nerves ; calyx 1·5 cm. long, straight-tubular, never inflated, striate-pilose with rather long hyaline hairs   ..   ..   **1. *nematosiphon***

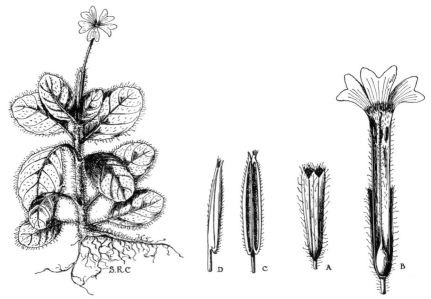

Fig. 299.—Physacanthus nematosiphon (*Lindau*) Rendle & Britten (Acanthaceae).
A, calyx.   B, flower in longitudinal section.   C and D, stamens.

Corolla-lobes entire or nearly so :
Leaves broadly elliptic, rounded at both ends, sometimes remotely and obtusely toothed, 3–5 cm. long, 2·3–5 cm. broad, otherwise as above ; calyx up to 3·5 cm. long, slightly inflated, very shortly pubescent ; corolla-tube up to 5 cm. long, villous at the mouth ; lobes broadly obovate, 1·3 cm. long, 1 cm. broad .. 2. *talbotii*
Leaves broadly oblanceolate, rounded at apex, narrowed at base, sometimes obtusely crenate, 8–14 cm. long, 3–5 cm. broad, thinly pilose on both surfaces ; petiole 1 cm. long, densely pilose ; flowers up to 4 in each inflorescence ; calyx 2 cm. long, conspicuously inflated, pubescent with up to 2 mm. long hyaline hairs ; corolla-tube 4 cm. long, bent and pubescent towards apex ; lobes obovate, 1 cm. long
3. *batanganus*

1. **P. nematosiphon** (*Lindau*) *Rendle & Britten* l.c. 378 (1909) ; Hallé l.c. 343. *Haselhoffia nematosiphon* Lindau in Engl. Bot. Jahrb. 43 : 351 (1909). *P. cylindricus* C. B. Cl. in F.T.A. 5 : 58 (1899), Hallé l.c. 345, and *Haselhoffia cylindrica* (C. B. Cl.) Lindau l.c. 352 (1909), partly (4 specimens of *Mann* 1669, Gabon). Creeping herb rooting at the nodes, densely hairy, the leaves with few nerves prominent beneath, and usually solitary narrow-tubular white flowers 1½ in. long.
Lib.: Gola Forest (June) *Bunting*! Firestone Plantations, Dukwai R. (Oct.-Nov.) *Cooper* 17! Duo, Sinoe Co. (Mar.) *Baldwin* 11353! Jabroke, Webo Dist. (July) *Baldwin* 6482! Zuie, Boporo Dist. (Nov.) *Baldwin* 10241a! **Iv.C.:** Mt. Copé, Cavally Basin (July) *Chev.* 19670! Tiapleu Forest (Nov.) *de Wilde* 904! Also in Gabon.
2. **P. talbotii** *S. Moore* in Cat. Talb. 75 (1913) ; Hallé l.c. 345. A pubescent creeping herb like the last, with slender-tubed solitary flowers 2 in. long, blue-violet with darker throat.
S.Nig.: Oban *Talbot* 972! Stubbs Creek F.R., Eket (Jan.-May) *Keay* FHI 37718a! Onochie FHI 33170!
3. **P. batanganus** (*G. Braun & K. Schum.*) *Lindau* in Schlechter Westafr. Kautschuk-Exped. 315 (1900) ; Rendle & Britten in J. Bot. 47: 378 (1909), for full synonymy ; Hallé l.c., t. 1, 6–8. *Ruellia batangana* G. Braun & K. Schum. Mitt. Deutsch. Schutzgeb. 2 : 173 (1889). *Physacanthus inflatus* C. B. Cl. in F.T.A. 5 : 57 (1899) ; F.W.T.A., ed. 1, 2 : 253 ; Chev. Bot. 495. *P. cylindricus* C. B. Cl. and *Haselhoffia cylindrica* (C. B. Cl.) Lindau in Engl. Bot. Jahrb. 43 : 352 (1909), Hallé l.c. 345, partly (*Mann* 1670 ; 3 specimens of 1669, Gabon). Stems half-woody, erect, 1–2 ft. high, the leaves sometimes marbled with white or purple-tinge, with white or pale blue-violet flowers.
Lib.: Bumbumi-Moala (Nov.) *Linder* 1340! Webo (June) *Baldwin* 6088! Diebla, Webo Dist. (July) *Baldwin* 6297! Duo, Sinoe (Mar.) *Baldwin* 11353a! Nana Kru *Baldwin* 11590! **Iv.C.:** Mid. Sassandra to Mid. Cavally *Chev.* 19255bis. Kéeta *Chev.* 19373. Fort Binger to Toula *Chev.* 19540. Toula to Nékaougnié *Chev.* 19570. Mt. Copé *Chev.* 19691 ; 19705. **S.Nig.:** Eket *Talbot* 3301! Stubbs Creek F.R., Eket (Jan.) *Keay* FHI 37718b! Also in Cameroun and Gabon.

16. BRILLANTAISIA P. Beauv.—F.T.A. 5 : 37 ; Benoist in Bull. Soc. Bot. Fr. 60 : 334 (1913).

Lamina lanceolate, with ascending lateral nerves, gradually acuminate, acute at base, 7–10 cm. long, 1–2·5 cm. broad, with very numerous rod-like cystoliths beneath, serrulate ; petiole about 1 cm. long ; panicle small, few-flowered, about 10 cm. long ; calyx-segments unequal, the longest 6 mm. long, glandular ; corolla 2 cm. long
1. *lancifolia*
Lamina more or less broadly ovate :
Lamina entire or very obscurely crenulate :
Lamina rounded at base, variable in size, broadly ovate, acuminate, pilose on both surfaces, with conspicuous rod-like cystoliths on the upper surface ; petiole not or only slightly winged, pilose ; panicle lax ; calyx-segments linear, 8–10 mm. long, with a few stipitate glands ; corolla about 3 cm. long, the tube about ½ as long as the lips ; fruit glabrous, linear, 3–3·5 cm. long, many-seeded .. 2. *lamium*
Lamina broadly rounded-cuneate into the winged petiole, about 16 cm. long, 7–8 cm. broad, otherwise as above ; inflorescence spike-like ; calyx-segments filiform, plumose .. .. 3. *madagascariensis*
Lamina toothed, mostly cuneate at base into the winged petiole :
Petiole very shortly or scarcely winged towards the apex, up to 10 cm. long ; leaves broadly ovate, rounded to truncate at base, shortly acuminate, 5–15 cm. long, up to 10 cm. broad, very thin, with numerous rod-like cystoliths prominent on the upper surface, obtusely repand-dentate ; panicle many-flowered ; calyx-segments 1 cm. long, glandular-pubescent ; corolla 2·5 cm. long ; fruit 2 cm. long, very shortly pubescent .. .. 4. *vogeliana*
Petiole more or less winged in the upper half, the lamina cuneate at base :
Lamina gradually narrowed into the petiole :
Branches of inflorescence markedly zigzag ; petiole winged to the base ; calyx-segments linear-spathulate .. .. 5. *patula*
Branches of inflorescence dichotomous ; petiole not winged to the base ; calyx-segments subfiliform .. .. 6. *owariensis*
Lamina abruptly narrowed into the winged petiole, rounded or truncate at the base:
Petiole broadly winged, long-pilose to nearly glabrous, wings dentate ; leaves very broadly ovate, abruptly acuminate, truncate or cordate at base, about 15–18 cm. long and 12 cm. broad, coarsely serrate, pilose on both surfaces ; calyx-segments about 1 cm. long ; corolla 3–4 cm. long .. .. .. 7. *nitens*
Petiole not winged, tomentose, about 2·5 cm. long ; leaves broadly ovate, broadly acuminate, rounded at base, obtusely dentate, 8–15 cm. long, 5–10 cm. broad, shortly setulose, especially on the nerves beneath ; calyx-segments linear, densely glandular-pilose, up to 1·5 cm. long ; corolla 4 cm. long .. .. 8. *bauchiensis*

1. **B. lancifolia** *Lindau* in Engl. Bot. Jahrb. 17 : 98 (1893) ; F.T.A. 5 : 40 ; Benoist l.c. 335. *B. talbotii* S. Moore in Cat. Talb. 75 (1913) ; F.W.T.A., ed. 1, 2 : 254. A rather scrambling or suberect herb, the stems soon glabrous, the inflorescence glandular-pubescent, with blue flowers ¾ in. or more long.
 S.Nig.: Oban *Talbot* 1270 ! 2000 ! Also in Gabon.

2. **B. lamium** (*Nees*) *Benth.* in Fl. Nigrit. 477 (1849) ; F.T.A. 5 : 38 ; Benoist l.c. 334 ; Chev. Bot. 493 ; Exell Cat. S. Tomé 260 ; Monod in Mém. I.F.A.N. 18 : 13 (1952). *Leucorhaphis lamium* Nees in Prod. 11 : 97 (1847). A coarse herb, erect branched 2–4 ft. high, more or less pilose at first, with blue- or violet-purple flowers 1–1¼ in. long in a glandular-pubescent panicle ; in damp shady places.

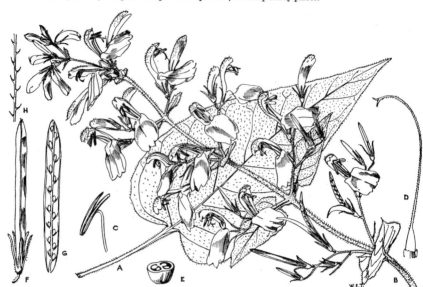

Fig. 300.—Brillantaisia lamium (*Nees*) *Benth.* (Acanthaceae).

A, leaf. B, inflorescence. C, anther. D, pistil. E, cross-section of ovary. F, fruit. G, one valve of fruit. H, axis of fruit.

 Guin.: Kouria *Caille* in *Hb. Chev.* 14713. Fouta Djalon *Chev.* 18365 ; 18397 ; 18697 ; 18848. **S.L.:** Mt. Aureole, Freetown (Mar., Aug.) *Dalz.* 994 ! *Lane-Poole* 392 ! Fiaama, Nomo (Oct.) *Deighton* 81 ! Njala (Oct.) *Deighton* 1419 ! Madina (Sept.) *Adames* 73 ! **Lib.:** Cape Palmas (fl. & fr. July) *Ansell* ! Bakratown (Oct.) *Linder* 849 ! White Plains (fl. & fr. Sept.) *Barker* 1418 ! Harper (fl. & fr. June) *Baldwin* 5993 ! Browntown (Aug.) *Baldwin* 7070 ! **Iv.C.:** Bingerville *Chev.* 16056. Potou to Alépé *Chev.* 17379 ; 17403. Nékaougnié to Grabo *Chev.* 19592. Yapo (Oct.) *Leeuwenberg* 1808 ! Boka (fl. & fr. Aug.) *Schnell* 6513 ! **Ghana:** *Burton & Cameron* ! Axim (Sept.) *Morton* ! Assuantsi road (fl. & fr. Sept.) *Dalz.* 149 ! **Togo Rep.:** Misahöhe (fl. & fr. July) *Baumann* 67 ! **N.Nig.:** Naraguta (fl. & fr. July) *Lely* 469 ! Vom *Dent Young* 194 ! (fl. & fr. Oct.) *Hepper* 1151 ! Jos *Monod* 9619. **S.Nig.:** Okomu F.R., Benin (fl. & fr. Sept.) *Iriah* FHI 23095 ! Ibadan (Jan.) *Meikle* 953 ! Ilaro F.R. (fl. & fr. Nov.) *W. MacGregor* 333 ! Sapoba (fl. & fr. Nov.) *Meikle* 590 ! Obudu Plateau (Aug.) *Stone* 63 ! 86 ! Oban *Talbot* 981 ! **[Br.]Cam.:** Mungo Ndaw, Kumba Div. (fl. & fr. Oct.) *Dundas* FHI 15323 ! Bamenda (June) *Maitland* 1619 ! Also in Cameroun, Congo and Uganda. (See Appendix, p. 450.)

3. **B. madagascariensis** *T. Anders. ex Lindau* in Engl. Bot. Jahrb. 17 : 103 (1893) ; F.T.A. 5 : 43 ; Benoist l.c. 334 ; Chev. Bot. 493. *B. verruculosa* Lindau l.c. 22 : 113 (1895) ; F.T.A. 5 : 43 ; Benoist l.c. 334. Like the last but in W. Africa usually with whitish flowers 1–1½ in. long in a spike with glandular hairs.
 Guin.: Timbikounda (fl. & fr. Jan.) *Chev.* 20607 ! **F.Po:** El Pico, 4,300 ft. (Dec.) *Boughey* 170 ! Moka, 4,600 ft. (Sept.) *Wrigley* 643 ! Also in E. Africa and Madagascar.

4. **B. vogeliana** (*Nees*) *Benth.* in Fl. Nigrit. 477 (1849) ; F.T.A. 5 : 40 ; Benoist l.c. 335 ; Exell Cat. S. Tomé 260. *Leucorhaphis vogeliana* Nees in DC. Prod. 11 : 97 (1847). *Brillantaisia preussii* Lindau in Engl. Bot. Jahrb. 17 : 100 (1893). Habit of the last and with similar violet-purple flowers about 1 in. long.
 Ghana: Begoro (Nov.) *Morton* A36 ! Kibi (fl. & fr. Dec.) *Morton* GC 8123 ! **S.Nig.:** Oban *Talbot* ! **[Br.] Cam.:** Victoria (fl. & fr. Oct.) *Maitland* 757 ! Buea, 3,000 ft. (fl. & fr. Sept., fr. Oct.) *Preuss* 998 ! *Maitland* 110 ! Barombi (fl. & fr. June) *Preuss* 320 ! **F.Po:** (fl. & fr. Nov.) *T. Vogel* 179 ! *Mann* 36 ! Monte Balea (fl. & fr. Dec.) *Guinea* 482 ! Moka, 4,000 ft. (fl. & fr. Dec.) *Boughey* 7 ! Also in Cameroun and S. Tomé.

5. **B. patula** *T. Anders.* in J. Linn. Soc. 7 : 21 (1863) ; F.T.A. 5 : 41 ; Benoist l.c. 335 ; Exell Cat. S. Tomé 260 ; Monod l.c. (1952). A stout plant up to 6–10 ft. high, sometimes several inches in circumference at base of stem, with comparatively large violet-purple flowers, the upper lip yellowish and purple-spotted.
 Togo Rep.: *Kersting* A108 ! **N.Nig.:** *Yates* 35 ! Lokoja (Mar.) *Shaw* 26 ! Jos, 3,600–4,000 ft. (Oct., Dec.) *Lamb* 66 ! *Hepper* 1094 ! Pedong *Monod* 9620. **S.Nig.:** Lagos (cult.) *Dalz.* 1136 ! **[Br.]Cam.:** Jamtari, Gashaka Dist. (fl. & fr. Dec.) *Latilo & Daramola* FHI 28926 ! Also in Cameroun ; extending to Angola and Uganda. (See Appendix, p. 450.)

6. **B. owariensis** *P. Beauv.* Fl. Oware. 2 : 68, t. 100 (1818) ; F.T.A. 5 : 40 ; Benoist l.c. 335. Erect up to 10–12 ft. high, stems glandular-pubescent at first, later glabrous, with purple flowers 1–1½ in. long.
 S.Nig.: Awka *Thomas* 102 ! 114 ! Umuahia (Dec.) *Carpenter* 312 ! Ibadan to Lagos (fl. & fr. Jan.) *Hambler* 106 ! below Koloishe Mt., Ogoja Prov. (Dec.) *Savory & Keay* FHI 25056 ! **[Br.]Cam.:** Cam. Mt. (Dec., Jan.) *Mann* 1959 ! *Dunlap* 25 ! Buea, 3,000–3,800 ft. (Feb.) *Dalz.* 8180 ! (fl. & fr. Jan.) *Maitland* 199 ! 207 ! *Preuss* 600 ! *Hutch. & Metcalfe* 122 ! Maisamari, 4,600 ft., Mambila Plateau (fl. & fr. Jan.) *Hepper* 1651 ! Vogel Peak, 3,850 ft., Adamawa (fl. & fr. Nov.) *Hepper* 1332 !

7. **B. nitens** *Lindau* in Engl. Bot. Jahrb. 17 : 102 (1893) ; F.T.A. 5 : 41 ; Benoist l.c. 336. *B. salviiflora* Lindau l.c. 101 (1893) ; F.T.A. 5 : 41. *B. leonensis* Burkill in F.T.A. 5 : 41 (1899) ; Benoist l.c. 335 ; Chev. Bot. 493 ; F.W.T.A., ed. 1, 2: 254. Aké Assi, Contrib. 1: 206. Stems up to 9 ft. high, glandular-pubescent becoming glabrous, with blue-purple flowers 1½–2 in. long.

Guin.: Kindia *Chev.* 13362. Ditinn *Chev.* 13557. Forekaria *Chev.* 20526. **S.L.:** Regent to Sugar Loaf Mt. (Dec.) *Sc. Elliot* 3990*a*! *Winwood Reade*! Bandajuma *Burbridge* 473! Gegbwema (Nov.) *Deighton* 356! Nyago (Jan.) *Smythe* 240! **Lib.:** Karmadhun (fl. & fr. Nov.) *Baldwin* 10182! Genna Tanyehun (Dec.) *Baldwin* 10778! Bilipia (Dec.) *Harley* 1631! **Iv.C.:** Morenou *Chev.* B22421. Man to Danane (Nov.) *de Wilde* 850! 859! Mt. Tonkui, 1,500 ft., Danane (fl. Nov., fr. Mar.) *Aké Assi* 5502! *Leeuwenberg* 2948! **Ghana:** Oda (Oct.) *Fishlock* 55! Koforidua (Nov.) *Dalz.* 148! Adankrono (Aug.) *Chipp* 560! Assin-Yan-Kumasi (fl. & fr. Jan.) *Cummins* 245! Manso (fl. & fr. Aug.) *Howes* 957! **Togo Rep.:** Misahöhe (Nov.) *Mildbr.* 7353! **S.Nig.:** Ogwashi (Nov.) *Thomas* 2034! Benin *Dennett* 34! Mamu F.R., Onitsha (Jan.) *Burtt* B16! Iso Bendiga to Bendiga Afi, Ogoja (fl. & fr. Dec.) *Keay* FHI 28264! Oban *Talbot* 993! **[Br.] Cam.:** Buea, 6,000 ft. (Dec., Feb.) *Preuss* 847! *Deistel* 106! Mayuko (Feb.) *Dalz.* 8233! Mbalanga F.R. (Jan.) *Binuyo & Daramola* FHI 35484! Bamenda, 6,700 ft. (fl. & fr. Jan.) *Daramola* FHI 40574! Also in Cameroun. (See Appendix, p. 450.)

8. **B. bauchiensis** *Hutch. & Dalz.* F.W.T.A., ed. 1, 2 : 254 (1931) ; Monod in Mém. I.F.A.N. 18 : 13 (1952). (?) *B. fulva* Lindau in Engl. Bot. Jahrb. 24 : 313 (1897). Herbaceous perennial up to 6–8 ft. high, densely tomentose and glandular-pubescent, with small hair-like cystoliths on the upper surface 2 in. long.
N.Nig.: Vom (fl. & fr. Feb.) *Dent Young* 195! *McClintock* 222! Jos Plateau (fl. & fr. Jan.) *Lely* P129! *Batten-Poole* 209! Ropp (Dec.) *MacGregor* 402! **[Br.]Cam.:** Bamenda, 4,000–6,000 ft. (fl. & fr. Jan.) *Migeod* 359! *Keay* FHI 28347! Kumbo (Feb.) *Hepper* 1972! Mandanga, Mambila Plateau (fr. Jan.) *Latilo & Daramola* FHI 28992!
[A species of doubtful position, but at the present not identifiable with another described taxon. Most likely only a highland form of *B. owariensis*. If it should prove to be specifically distinct and found to be conspecific with *B. fulva* (1897) from Cameroun the latter name would have priority over *B. bauchiensis* (1931). *B. fulva*, an inadequately described species based on fragmentary material, was not accounted for in F.T.A. (1899) and the type was not available for this revision.—H.H.]

17. LANKESTERIA Lindl.—F.T.A. 5 : 69.

Bracts ovate :
　Calyx 5 mm. long, not visible in the spike ; bracts nervose, 2–3 cm. long ; leaves obovate-oblanceolate, acuminate, 10–20 cm. long, 4–7 cm. broad, with small hair-like cystoliths conspicuous on the upper surface :
　　Corolla-tube about 3·5–5 cm. long, slender, shortly pubescent ; bracts glabrous except the margin　..　..　..　..　..　..　　1. *elegans*
　　Corolla-tube 1·5–3 cm. long ; bracts pubescent　..　..　　2. *brevior*
　Calyx over 1 cm. long, conspicuous amongst the bracts ; bracts as above, but more obovate ; leaves more or less as above ; corolla-tube about 2·5 cm. long, shortly pilose　..　..　..　..　..　..　..　..　　3. *barteri*
Bracts linear, densely ciliate and glandular :
　Inflorescence spicate, longer than broad ; branchlets densely pilose ; leaves obovate-oblanceolate, 7–12 cm. long, 3–5 cm. broad, hardly acuminate, narrowed from well above middle to the base, with small hair-like cystoliths on the upper surface ; calyx-segments 1·5 cm. long ; corolla a little longer than the calyx　4. *thyrsoidea*
　Inflorescence depressed-globose ; branchlets, leaves, etc., as above, but leaves acuminate; calyx-segments 2 cm. long; corolla considerably longer than the calyx; fruit 1 cm. long, glabrous　..　..　..　..　..　..　　5. *hispida*

1. **L. elegans** (*P. Beauv.*) *T. Anders.* in J. Linn. Soc. 7 : 33 (1863), partly (excl. *Barter* 3381) ; F.T.A. 5 : 70; F.W.T.A., ed. 1, 2: 255, partly (excl. Chev. Bot. 495 and Iv. C. specimens) ; Mangenot in Ic. Pl. Afr. I.F.A.N. 4 : 87 (1957). *Justicia elegans* P. Beauv. Fl. Oware. 1 : 84, t. 50 (1806). Undershrub up to 4 ft. high, in forest undergrowth etc., with spikes 2–7 in. long of orange- or pinkish-tinged slender tubular flowers 1½ in. long enclosed in broad smooth but ciliate bracts.
S.L.: Kambui F.R. (fl. & fr. Feb.) *Lane-Poole* 339! Kingohun (Jan.) *Deighton* 3866! **Lib.:** Ganta (Feb.) *Harley* 877! **Iv.C.:** Sassandra (Jan.) *Leeuwenberg* 2867! Biguale to Gouané (May) *Chev.* 21507! Tingouéla to Assikasso *Chev.* 22573! **Ghana:** Wiawso (Dec.) *Andoh* 5606! Bechema (Jan.) *Foggie* FH 4443! Bosum-kese F.R. (Dec.) *Adams* 2884! Mampong Scarp (Jan.) *Morton* A1538! **Dah.:** Sakété (Jan.) *Chev.* 22810! **S.Nig.:** Lagos (Jan.-Apr.) *Millen* 103! 120! *Moloney* 6! *Rowland*! Okomu F.R., Benin (Dec.) *Brenan* 8515! Amahor F.R. (fl. & fr. Feb.) *Ongunseitan* 41704! Sapoba (Jan.-Mar.) *Kennedy* 542! Shasha F.R., Ife (Dec.) *Omiyale* FHI 44061! **[Br.]Cam.:** R. Kindong, S. Bakundu (Jan.) *Binuyo & Daramola* FHI 35061! Mamfe (Dec.-Feb.) *Keay* FHI 28319! *Maitland* 1155! *Hepper* 2210! Extends to Congo, S.E. Sudan and Uganda. (See Appendix, p. 451.)

2. **L. brevior** *C. B. Cl.* in J. Linn. Soc. 37 : 110 (1905) ; Aké Assi, Contrib. 1: 207. *L. elegans* of Chev. Bot. 495 (*Chev.* 16567*bis*, 16646, 19964), not of (P. Beauv.) T. Anders. Habit of the last but flowers pure white but yellow throat, corolla-tube usually considerably shorter, and bracts smaller, pubescent and often turning bronze-tinged when old.
S.L.: Blama (Dec.) *Deighton* 312! Kpaku (Apr.) *Deighton* 1609! Fairo (Mar.) *Deighton* 4095! Gardohun *Deighton* 1596! Gorahun (Feb.-Mar.) *Maitland* 467! 473! Gola Forest (Dec.) *King* 83*b*! **Lib.:** Monrovia *Dinklage* 2917! 2955! *Whyte*! Firestone Plantation, Du R. (Dec.) *Bequaert* 3! Dukwia R. (fl. & fr. Feb.) *Cooper* 160! Kakatown *Whyte*! Robertsport *Baldwin* 10926! **Iv.C.:** Bouroukrou (Dec.) *Chev.* 16567*bis*! 16646! Tabou to Bèriby *Chev.* 19964! Banco (July) *de Wit* 7096! 7761! Téke (Aug.) *de Wit* 7097! Mange-not & Aké Assi IA 108! Sassandra (Jan., Aug.) *Leeuwenberg* 2521! *de Wilde* 346! Angédédou, Abidjan (Nov.) *Leeuwenberg* 1861! **Ghana:** Wiawso (Dec.) *Andoh* FH 5603! Akim (Dec.) *Johnson* 261! *Irvine* 1816! Bompata (Dec.) *Dalz.* 150! Kumasi (Feb.) *Vigne* FH 1833! Akota (Dec.) *Darko* 1064! **[Br.]Cam.:** Tiko (Jan.) *Dunlap* 173! Johann-Albrechtshöhe (=Kumba) *Staudt* 538! S. Bakundu F.R., Kumba (Jan.) *Keay* FHI 28561! (See Appendix, p. 451.)

3. **L. barteri** *Hook. f.* Bot. Mag. 91 : t. 5533 (1865) ; F.T.A. 5 : 70. Undershrub 2–4 ft. high, pubescent, with slender tubular flowers 1¼ in. long, colour deep orange or reddish, in dense spike-like panicles 2–4 in. long.
S.Nig.: Abeokuta *Barter* 3381! W. of Okomu F.R. (Feb.) *Lomax* FHI 9055! Benin (Feb.) *Jones* FHI 9197! Benin to Agbor (Jan.) *Keay* FHI 37355! Calabar (Feb., Mar.) *Mann* 2313! *Latilo* FHI 41324! **[Br.]Cam.:** Bafia, Victoria (Feb.) *Keay* FHI 37519! Also in Cameroon.

4. **L. thyrsoidea** *S. Moore* in Cat. Talb. 77 (1913). A hairy undershrub, 1 ft. or more high, with slender flowers 1–1¼ in. long, white, lilac or pinkish with yellow throat in a dense glandular-hairy narrow panicle ; in forest.
S.Nig.: Ibadan South F.R. (Jan.) *Keay* FHI 19805! Mamu F.R. (Jan.) *Burtt* 13! Orem, Calabar (Jan.) *Onochie & Okafor* FHI 36024! Oban *Talbot* 1646! *Richards* 5192! Cross R. (Dec.) *Holland* 229! **[Br.]Cam.:** Mamfe (Jan.) *Maitland* 1151!

5. L. hispida (*Willd.*) *T. Anders.* in J. Linn. Soc. 7 : 32 (1863) ; Schnell in Mém. I.F.A.N. 22 : 517 ; Aké Assi, Contrib. 1 : 207. *Justicia hispida* Willd. Sp. Pl., ed. 2, 1 : 84 (1797). *Lankesteria parviflora* Lindl. in Bot. Reg. 31, Misc. 86 (1845) and 32 : t. 12 (1846) ; F.T.A. 5 : 70 ; Chev. Bot. 495. Pubescent undershrub 1–2 ft. high (sometimes decumbent) in forest undergrowth, with slender tubular flowers 1¼ in. long, lemon-yellow or pure white, but turning yellow, in dense heads.
Guin.: Conakry *Maclaud* ! Kotouma *Paroisse* 68 ! Yogoronia *Pobéguin* 922 ! Nzérékoré (fl. & fr. Jan.) *Schnell* 4213 ! Kindia *Chev.* 13093. **S.L.:** Leicester (May) *Barter* ! York Pass *Deighton* 3487 ! Kessewe (Feb.) *MacDonald* 2 ! Free-town (Mar.) *Dalz.* 972 ! Wallia, Benna (Jan.) *Sc. Elliot* 4452 ! **Lib.:** (Jan.) *Dinklage* 3340 ! Belefanai (Dec.) *Baldwin* 10553 ! Betandu (Mar.) *Bequaert* 149 ! Ganta (Feb.-Apr.) *Harley* 528 ! 868 ! **Iv.C.:** Danané to Tulépleu *Aké Assi* IA 4729. Also in Cameroun.

18. SCLEROCHITON Harvey—F.T.A. 5 : 109.

Calyx-segments subequal, one much broader than the other, broadly lanceolate, about 4 cm. long and 1·3 cm. broad, many-nerved, obtuse, the others very acutely pointed ; fruit 2·5 cm. long ; leaves oblong-elliptic, broadly acuminate, 15–20 cm. long, 5–8 cm. broad, very thin, glabrous .. .. 1. *preussii*
Calyx-segments subequal in length, one somewhat broader, narrowly subulate-lanceolate, 2·5 cm. long ; fruit 1·5 cm. long ; leaves ovate to lanceolate, broadly acuminate, 4–12 cm. long, up to 5 cm. broad, subcoriaceous, pubescent on the midrib beneath ; corolla 3 cm. long, limb of 5 united segments in one plane, lobes about 5 mm. long ; stamens 4 ; anthers 1-celled, pubescent .. .. 2. *vogelii*

1. S. preussii (*Lindau*) *C. B. Cl.* in F.T.A. 5 : 110 (1899). *Pseudoblepharis preussii* Lindau in Engl. Bot. Jahrb. 20 : 34 (1894). A nearly glabrous shrub with fairly large leaves and short flower-spikes with shining striate papery bracts and calyx.
[Br.]Cam.: Buea (fr. Oct.) *Preuss* 1073 !
2. S. vogelii (*Nees*) *T. Anders.* in J. Linn. Soc. 7: 37 (1863); F.T.A. 5: 111; Aké Assi, Contrib. 1: 209. *Isacanthus vogelii* Nees in DC. Prod. 11 : 279 (1847). A nearly glabrous shrub with dark smooth foliage and white flowers (turning red-brown on drying) about 1 in. long in abbreviated spikes with striate dry calyx and bracts.
Guin.: Macenta (fl. & fr. Sept.) *Jac.-Fél.* .1203! **S.L.:** Makump (fl. & fr. July) *Thomas* 930 ! Bumbuna (Aug.) *Deighton* 1202 ! Matotoka (July) *Thomas* 1278 ! **Lib.:** Cape Palmas *Ansell* ! *T. Vogel* 54 ! Monrovia (July-Aug.) *Baldwin* 13007 ! *Linder* 2 ! *Dinklage* 2823 ! 2833 ! Baila (Oct.) *Harley* 1367 ! **Iv.C.:** N. Donei to Oroumba-Boca (fl. Oct.) *de Wilde* 674 ! Kassa *Mangenot & Aké Assi* IA 2695. Canal d'Assinié *Mangenot, Miège & Aké Assi* IA 3148. **Ghana:** Asientien (July) *Chipp* 287 !

19. CROSSANDRA Salisb.—F.T.A. 5 : 112. *Stenandriopsis* S. Moore in J. Bot. 44 : 153 (1906) ; Benoist in Bull. Mus. Hist. Nat., sér. 2, 15 : 231 (1943), W. Afr. spp.

Short herbs ; inflorescences terminal :
Leaves gradually attenuated into the short petiole, oblanceolate, rounded-triangular at apex, up to 25 cm. long and 6 cm. broad, puberulous beneath ; spikes with a hairy peduncle, broadly oblong-cylindric, with densely overlapping very broadly ovate spinose-dentate pubescent bracts ; corolla-tube slender, about 3·5 cm. long, softly pubescent .. .. 1. *flava*
Leaves rounded to subcordate at base :
Leaves oblong, rounded at base, 3–6 cm. long, 1·5–2·5 cm. broad, setulose above, pubescent on the nerves beneath ; petiole 8 mm. long, pubescent ; spikes small and 2–3-flowered .. .. 2. *buntingii*
Leaves rounded or subcordate at base, obovate-elliptic, rounded at apex, 6–12 cm. long, 3–6 cm. broad, pubescent on the nerves beneath ; spikes sessile, very narrowly cylindrical, up to 12 cm. long, with ovate-lanceolate striate shortly toothed bracts :
Bracts up to 6 mm. long and 2·5 mm. broad, irregularly toothed ; calyx-segments 13–14 mm. long and 1·5 mm. broad .. .. 3. *thomensis*
Bracts up to 25 mm. long (usually not exceeding 18 mm.) ; calyx-segments 6–10 mm. long and 1·5–2 mm. broad :
Bracts acuminate, sharply pointed, up to 5 mm. broad, teeth almost only in the upper third and rather regular, small .. .. 4. *guineensis*
Bracts truncate, without a point, up to 7 mm. broad, teeth along the upper two thirds, very irregular and rather large .. .. 5. *talbotii*
Shrubs up to 1 m. high ; inflorescences axillary :
Leaves pseudo-verticillate, usually 4 per node; petioles about 2 cm. long, lamina entire, ovate, acute, up to 7 cm. long and 4 cm. broad, with 6 pairs of lateral nerves ; inflorescences on long spreading, shortly pubescent peduncles, flowering portion 5 cm. long ; bracts ovate-elliptic, 12 mm. long and 7 mm. broad, puberulous with a thinly ciliate margin, bracteoles linear, 10 mm. long and 1 mm. broad ; calyx up to 8 mm. long ; corolla about 2 cm. long and 2·4 cm. diam. ; capsule oblong-ellipsoid, about 1 cm. long, shortly apiculate, flattened, with four flattened seeds .. 6. *massaica*
Leaves opposite, petioles up to 2 cm. long, lamina lanceolate-elliptic, 15·5–17·5 cm. long and 4·5–5 cm. broad, lateral nerves 11–12 pairs ; inflorescences on peduncles up to 4·5 cm. long, inflorescence up to 5·5 cm. long, few-flowered ; bracts lanceolate, 15 mm. long and 5·5 mm. broad, puberulous ; bracteoles 14 mm. long and 1·5–2 mm. broad ; calyx up to 11 mm. long ; corolla-tube up to 3·5 cm. long 7. *obanensis*

1. C. flava Hook. Bot. Mag. 79 : t. 4710 (1853) ; F.T.A. 5 : 113 ; Aké Assi, Contrib. 1: 210 (1961). A short-stemmed herb woody at the root, with rather long leaves and slender-tubular yellow flowers 1½ in. long in a very dense bracteate almost spiny spike 1–2 in. long ; wet places in forest.

FIG. 301.—CROSSANDRA FLAVA *Hook.* (ACANTHACEAE).

A, calyx and style.   B, ovary.   C, longitudinal section of ovary.   D, seed.   E, cross-section of fruit.   F, fruit.

S.L.: *Unwin & Smythe* 24! York Pass (Dec.) *Lane-Poole* 426! *Deighton* 3344! Colonial F.R. (fl. & fr. Jan.) *King-Church* 11! **Iv.C.:** Kassa *Mangenot, Miège & Aké Assi* IA 904. Mudjika *Mangenot, Miège & Aké Assi* IA 1195. **Ghana:** Ebene, Kwahu (Jan.) *Chipp* 635! Agogo (Dec.) *Chipp* 611! Kunkerentumi (Dec.) *Johnson* 271! Oboum (Dec.) *Vigne* FH 1485! **Togo Rep.:** Misahöhe (fl. & fr. Nov.) *Mildbr.* 7347! **S.Nig.:** *Foster* 183! Lagos (Mar.) *Millen* 122! Akure F.R. (fl. & fr. Dec.) *Okereke* FHI 26882! Okelife, Ondo (Nov.) *Onochie* FHI 34349! Omo F.R. (Dec.) *Tamajong* FHI 20274! Idanre F.R. (Dec.) *Keay* FHI 37270!

2. **C. buntingii** *S. Moore* in J. Bot. 49: 321 (1911); Aké Assi, Contrib. 1: 209. A low herb with pale blue or white flowers with two purple dots, about 2 in. long ; moist places in forest.
  S.L.: *Glanville* 392! Dama Chiefdom (Nov.) *Fisher & Deighton* 78! Bobu (Dec.) *Deighton* 3810! Masandrov (June) *Marmo* 163! Kenema (fl. & fr. Sept.) *Deighton* 4030! Pendembu (July-Aug.) *Dawe* 530! **Lib.:** Gola Forest *Bunting*! *Linder*! Karmadhun (Nov.) *Baldwin* 10205! Bilimu (May, Sept.) *Harley* 1354! 1542a! **Iv.C.:** Yakassé to Adzopé (Dec.) *Chev.* 22666! Mt. Tonkui, 3,000–5,000 ft. (Nov., Apr., Aug.) *de Wilde* 917! *Aké Assi* IA 4870! *Mangenot & Miège* IA 5414!

3. **C. thomensis** *Milne-Redhead* in Kew Bull. 1935 : 280 ; Exell Cat. S. Tomé 262. Low herb with purple flower ; brown stem ; leaves white beneath.
  [Br.]**Cam.:** bank of R. Metschum, Wum, Bamenda Prov. (June) *Ujor* FHI 29253! The only mainland record of this species described from S. Tomé.

4. **C. guineensis** *Nees* in DC. Prod. 11 : 281 (1847) ; Bot. Mag. 104 : t. 6346 (1878) ; F.T.A. 5 : 117 ; F.W.T.A., ed. 1, 2 : 260, partly (excl. syn. *C. talbotii* S. Moore, and *Talbot* 80) ; Milne-Redhead in Kew Bull. 1935 : 281. *Stenandriopsis guineensis* (Nees) Benoist in Bull. Mus. Hist. Nat., sér. 2, 15 : 235 (1943) ; Aké Assi, Contrib. 1: 210. *C. elatior* S. Moore Cat. Talb. 79 (1913). Stem densely short-hairy about 6 in. high, with leaves prominently nerved beneath and often vinous-tinged, and narrow spike 2–6 in. long of pale purplish or white flowers 1 in. long ; ground herb on forest floor.
  S.L.: Njagbema (Sept.) *Deighton* 2232! Seli R. (Oct.) *Thomas* 3103! **Lib.:** Yratoke (fl. & fr. July) *Baldwin* 6251! **Iv.C.:** Sassandra *Aké Assi* IA 3179. **Ghana:** Bomfa *Cansdale* 4427! Kibbi Hills (Dec.) *Johnson* 488! Ekosu (July) *Darko* 915! Joacri Tafo (July) *Darko* 906! Nyiraheri Hills (fl. & fr. Nov.) *Morton* A2809! **S.Nig.:** Akure F.R. (fl. & fr. Oct.) *Keay* FHI 21578! Ubiaja F.R. (Aug.) *Onochie* FHI 33258! Owena R. (Aug.) *Jones* FHI 19528! Ekeji-Ipetu F.R. (fr. Dec.) *Onochie* FHI 5240! Oban *Talbot* 101! 134! 383! [Br.]**Cam.:** Northern Korup F.R. (July) *Olorunfemi* FHI 30661! S. Bakundu F.R., Kumba (fr. Nov.) *Ejiofor* FHI 8461! Nkam-Wum F.R., 2,000 ft., Bamenda (Aug.) *Savory* UCI 321! Wum (Aug.) *Savory* UCI 316! **F.Po:** 2,000 ft. *Mann* 50! Also in Cameroun, Gabon, the Congos and Uganda. (See Appendix, p. 450.)

5. **C. talbotii** *S. Moore* in Cat. Talb. 78 (1913) ; Milne-Redhead l.c. 281. *C. guineensis* of F.W.T.A., ed. 1, 2 : 260, partly (*Talbot* 1026) not of Nees. Herb with stout purplish ascending stems ; leaves thickly papery, deep green and slightly glossy above, very pale grey-green with prominent venation beneath ; bracts pale green, corolla white. Very similar to No. 4, but easily distinguished by the key characters.
  S.Nig.: Okomu F.R., Benin (fr. Dec.) *Brenan* FHI 8463! Afi River F.R., Ogoja (Dec.) *Keay* FHI 28245! Oban *Talbot* 1026!

6. **C. massaica** *Mildbr.* in J. Arn. Arb. 11 : 54 (1930) ; J. K. Morton in J. W. Afr. Sci. Assoc. 2 : 145 (1956). A straggling shrub with bright salmon-pink flowers ; undershrub on rocky hills (Morton).
  Ghana: Shai Hills (Dec., Feb.) *Morton* GC 8088! A3862! Also in Congo, Kenya and Tanganyika.

7. **C. obanensis** *Heine* in Kew Bull. 16 : 170 (1962). A semi-shrubby plant up to 18 in. high with a few-flowered axillary inflorescence up to 2½ in. long, corolla primrose yellow outside, deep yellow inside ; on wet rocks in moist forest.
  S.Nig.: Oban, near Oban Rock (Mar.) *Richards* 5190! *Coombe* 187!

20. BLEPHARIS Juss.—F.T.A. 5 : 94.

Leaves obovate or broadly oblanceolate, rounded to acutely acuminate, about 4 in a whorl, up to 6 cm. long and 2 cm. broad, scabrid above, sparingly setulose beneath and on the margin ; inflorescence 1–1·5 cm. long, often in pairs or clusters ; bracts

broadly obovate, with long stiff bristles on the margins, bristles with short reflexed
hairs ; corolla 1-lipped, 1·3 cm. long   ..   ..   1. *maderaspatensis*
Leaves linear, up to 12 cm. long and 2 cm. broad ; inflorescences solitary, strobilate,
often in the forks of branches, 1·5–8 cm. long :
Inflorescences very reduced (one-flowered branches), up to 2·5 cm. long and 8 mm.
broad, in appearance much like certain grass-spikelets; both leaves and bracts entire,
sharply pointed but never with teeth or spines along the margins, with no bracteoles;
corolla very shortly 3-lipped, 1·7 cm. long   ..   ..   2. *glumacea*
Inflorescences few to many-flowered, 3–8 cm. long ; bracts broadly ovate, spinous-
toothed, ribbed ; corolla 1-lipped, about 2 cm. long, striate   **3. *linariifolia***

1. **B. maderaspatensis** (*Linn.*) *Heyne ex Roth* Nov. Pl. Sp. Ind. Or. 320 (1821); Berhaut Fl. Sén. 62. *Acanthus maderaspatensis* Linn. (1763). *Blepharis boerhaviaefolia* Pers. (1807)—F.T.A. 5 : 96, as *B. boerhaaviaefolia* ; Chev. Bot. 496. *B. molluginifolia* of Chev. Bot. 496, not of Pers. (1807). Suffrutescent herb, usually procumbent, coarsely pubescent on the young parts, with whorled leaves and white flowers ½ in. long in short finely-spiny spikes ; a weed of roadsides, waste places and forest margins in savanna.
   Sen.: *Heudelot* 509 ! *Berhaut* 528. **Mali:** Koulikoro *Chev.* 2783. San *Chev.* 2784 ! Fakoroba (Sept.) *Davey* 038 ! **Port.G.:** Pussubé *Esp. Santo* 1094. **Guin.:** Kita (Sept.) *Jaeger* 2644 ! **S.L.:** Hill Station, Freetown (Dec.) *Deighton* 472 ! Sasseni (fl. & fr. Jan.) *Sc. Elliot* 4525 ! Roruks (fl. & fr. Nov.) *Thomas* 5719 ! Kasawe F.R. (fl. & fr. Dec.) *King* 37b ! 144b ! Kumrabai (fr. Dec.-Jan.) *Thomas* 6900 ! **Ghana:** Burufo (fr. Dec.) *Adams* 4410 ! Achimota F.R. (fr. Aug.) *Akpabla* 853 ! Legon to Aburi (fr. May) *Boughey* GC 10392 ! **N.Nig.:** Sokoto (fr. Sept.) *Dalz.* 362 ! Kano (fr. Nov.) *Hagerup* 619 ! **S.Nig.:** Lagos (fl. & fr. Mar.) *Millen* 50 ! Ibadan (fl. & fr. Nov.) *Meikle* 664 ! Abeokuta (fr. Nov.) *Ejiofor & Olorunfemi* FHI 30783 ! In tropical Africa generally, extending to India.

2. **B. glumacea** *S. Moore* in J. Bot. 18 : 232 (1880) ; F.T.A. 5 : 97. A small herb about 6 in. high with slender stems ; in savanna.
   N.Nig.: Anara F.R., Zaria (Sept.) *Olorunfemi* FHI 24398 ! Jos *Batten-Poole* 120 ! Also in Angola, Congo and tropical E. Africa.

3. **B. linariifolia** *Pers.* Syn. 2 : 180 (1806) ; F.T.A. 5 : 100 ; Chev. Bot. 496 ; F.W.T.A., ed. 1, 2 : 260 ; Monod. Contr. Etud. Sahara Occ. 2 : 113 (1939) ; Berhaut Fl. Sén. 2: 71; F.W. Andr. Fl. Pl. Sudan 3: 171, fig. 47. *B. passargei* Lindau Engl. Bot. Jahrb. 22 : 117 (1895). A herb with wiry branches often close to the ground, and prickly bracts ; flowers blue with white throat.
   Maur.: Pharaon-Adrar *Charles* in Hb. Chev. 25451 ! Edderoum *Charles* in Hb. Chev. 28806. **Sen.:** *Berhaut* 552. *Heudelot* 179. *Lelièvre.* **Mali:** Goundam *Chev.* 2789. Labezenga (fr. Sept.) *Hagerup* 440 ! Dioura (Aug.-Nov.) *Davey* 048 ! Kourbita (fr. Jan.) *Davey* 247 ! **Iv.C.:** *Poisson* ! **N.Nig.:** Bimasa, Sokota (fr. Jan.) *Bornu & Jibirinjia* FHI 29051 ! Nupe *Barter* 716 ! Kontagora (fl. & fr. Nov.) *Dalz.* 183 ! Jos Plateau (fr. Nov.) *Lely* 636 ! Yola (fl. & fr. Aug.) *Dalz.* 142 ! **[Br.]Cam.:** Gurum, Adamawa (fl. & fr. Nov.) *Hepper* 1273 ! Extends to Sudan ; apparently also in S. Africa and S.W. Asia. (See Appendix, p. 450.)
   [The closely related *B. persica* (Burm. f.) O. Ktze. has a similar distribution but is apparently not represented in its strict sense in our area. The genus is much in need of critical revision and monographic treatment.]

21. ACANTHUS Linn.—F.T.A. 5 : 105.

Stout erect plants, glabrescent or hirsute in the upper parts ; leaves shortly petiolate,
obovate to oblanceolate, often deeply pinnately lobed, rarely only dentate, spinous-
toothed, up to 20 cm. long and 10 cm. broad, scabrid or thinly pilose ; spikes up to
20 cm. long.

Bracts broadly elliptic-ovate, 2–2·5 cm. long, mostly glabrescent or very sparsely
puberulous, spinously toothed with about 10–12 teeth, the largest always in the
middle, bracteoles with up to 5 teeth ; upper calyx-segment 2·5–3 cm. long, the
lowest one nearly equal ; corolla rose or pink (sometimes pinkish-white) ; style-
base hairy   ..   ..   1. *montanus*
Bracts broadly ovate, 2–2·5 cm. long, mostly conspicuously puberulous or hairy, spinously
toothed, with about 6–8 teeth, the largest pair always just beyond the terminal
tooth ; bracteoles entire, very rarely with 1(–2) small teeth ; upper calyx-segment
2·5–3 cm. long, the lowest one always only about half that size ; corolla white or pale
yellow with greenish yellow marked veins (very rarely light pinkish) ; style entirely
glabrous   ..   ..   2. *guineensis*

1. **A. montanus** (*Nees*) *T. Anders.* in J. Linn. Soc. 7 : 37 (1863) ; Bot. Mag. 91 : t. 5516 (1865) ; F.T.A. 5 : 107 ; F.W.T.A., ed. 1, 2 : 260, partly (specimens from N.Nig.-F.Po, fig. 268). *Cheilopsis montana* Nees in DC. Prod. 11 : 272 (1847). *A. barteri* T. Anders. l.c. (1863) ; F.T.A. 5 : 108. Plant up to 6 ft. high, stems woody, sparsely branched ; leaves dark glossy green above, paler beneath, papery, scabrid bracts and calyx-lobes whitish with green veins ; in high forest.
   Dah.: Sakété (Jan.) *Chev.* 22834 ! **N.Nig.:** Patti Lokoja (Nov.) *Dalz.* 138 ! Vom, 3,000–4,500 ft., Jos Plateau *Dent Young* 191 ! *McClintock* 226 ! **S.Nig.:** Okomu F.R., Benin *Brenan* 5863 ! Ibadan F.R. (fl. & fr. Nov.) *Chizea* FHI 23965 ! Onitsha *Barter* 1301 ! *Thomas* 1665 ! Oban *Talbot* 1027 ! 2006 ! **[Br.]Cam.:** Victoria (fl. & fr. Nov.) *Maitland* 774 ! Buea, 3,000 ft. (fl. Dec., Jan., fr. Mar.) *Maitland* 123 ! 460 ! *Migeod* 249 ! *Dunlap* 124 ! Vogel Peak, 2,000 ft., Adamawa (Nov.) *Hepper* 1467 ! **F.Po:** (Nov., Dec.) *T. Vogel* 239 ! *Mann* 639 ! Also in Cameroun, Ubangi-Shari, Gabon, Principe, Congo and Angola ; plants known as *A. montanus* to the east of these countries may belong to different taxa. (See Appendix, p. 449.)

2. **A. guineensis** *Heine & P. Taylor* in Kew Bull. 16: 161, fig. 1, 1-3 (1962). *A. montanus* of F.W.T.A., ed. 1, 2: 206 , partly ; Chev. Bot. 496 ; Aké Assi, Contrib. 1: 211, partly (specimens W. of Togo), not of T. Anders. An erect almost unbranched plant with fleshy stem up to 6 ft. high; in wet forests. Very similar to the preceding species, and only distinguishable by the key characters.
   Guin.: Timbo *Chev.* 12427 ! Longuery *Caille* in Hb. Chev. 14782 ! Bouloukountou *Caille* in Hb. Chev. 14863 ! Bambaya *Pobéguin* 50 ! Dalaba (Nov.) *Jac.-Fél.* 568 ! *Schnell* 2223 ! **S.L.:** Bumbuna (Oct.) *Thomas* 3464 ! Gorahun (Nov.) *Deighton* 381 ! Kabala (Dec.) *Deighton* 2937 ! Koflu Mt. (fl. & fr. Jan.) *Sc. Elliot* 4619 ! Mopalma (fl. & fr. Jan.) *Adames* 21 ! **Lib.:** Kakatown (Sept.) *Whyte* ! Karmadhun (Nov.) *Baldwin* 10189 ! Ganta (Nov., Dec.) *Barker* 1133 ! *Harley* 775 ! **Ghana:** Johnson 275 ! Abetifi (Jan.) *Irvine* 1680 ! Begoro, 1,500 ft. (Jan.) *Plumpton* 90 ! Kibi (Nov.) *Morton* GC 8015 ! Pusa Pusa Ravine, Kibi Hills (Jan.) *Morton* GC 6382 ! Amedzofe (fl. & fr. Feb.) *Irvine* 162 ! Mpraeso Scarp (fl. & fr. Dec.) *Morton* A2421 !

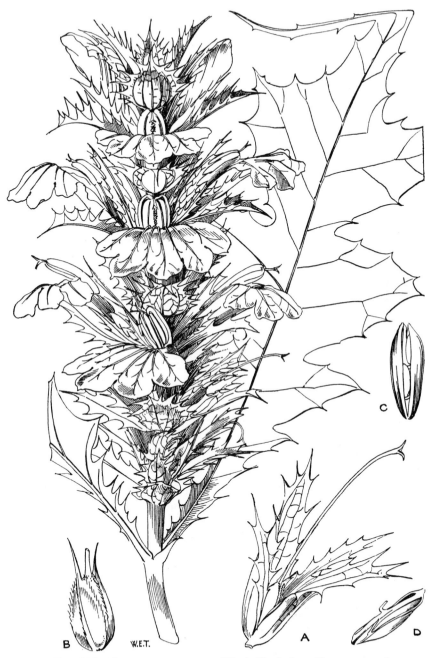

W.E.T.

Fig. 302.—Acanthus montanus (*Nees*) *T. Anders.* (Acanthaceae).
A, bracts and pistil. B, calyx and ovary. C, fruit. D, one valve of fruit.

22. CROSSANDRELLA C. B. Cl. in Kew Bull. 1906 : 251.

Stems puberulous ; leaves oblong-lanceolate, gradually acuminate, acutely cuneate at base, 10–18 cm. long, 4–6 cm. broad, with 12–15 pairs of spreading arcuate nerves, minutely punctate beneath ; petiole 2–3 cm. long ; spike about 10 cm. long ; bracts ovate, acutely acuminate, 8 mm. long, strongly nerved, glabrous ; bracteoles obovate-elliptic, 1 cm. long, 5-nerved ; calyx deeply 4-lobed ; corolla nearly 1·5 cm. long

dusenii

C. dusenii (*Lindau*) *S. Moore* in Cat. Talb. 74 (1913). *Pseudoblepharis dusenii* Lindau in Engl. Bot. Jahrb. 20 : 34 (1894). *Acanthus dusenii* (Lindau) C. B. Cl. in F.T.A. 5 : 108 (1899). *Crossandrella laxispica* C. B. Cl. in Kew Bull. 1906 : 251. An erect undershrub with rather long thin pale green leaves and bracteate flowers nearly ½ in. long in a terminal spike.
S. Nig. : Oban *Talbot* 1269 ! [Br.]**Cam.** : *Dusen* 348a. Korup F.R., Kumba (fl. & fr. June) *Olorunfemi* FHI 30643 ! Bamenda (fl. & fr. July) *Ujor* FHI 29276 ! Extends to Rio Muni and Uganda.

23. ASYSTASIA Blume—F.T.A. 5 : 130.

Leaves very large, 20–30 cm. long, 5–12 cm. broad, elongate-obovate, caudate-acuminate, acute at base, with 7–9 pairs of prominent nerves, glabrous ; inflorescence spike-like, with rather distant clusters of flowers, up to 30 cm. long ; calyx-segments subulate, about 4 mm. long ; corolla 3–5 cm. long, blackish when dry ; style glabrous, thread-like and persistent .. .. 1. *macrophylla*
Leaves much smaller than above, up to 15 cm. long and 6 cm. broad :
Corolla about 5 cm. long :
Corolla narrowly cylindric in the lower third, rather suddenly and narrowly bell-shaped in the upper two-thirds ; calyx-segments about 1 cm. long ; leaves oblong to obovate-elliptic, acuminate, acute at base, 8–16 cm. long, 3·5–5·5 cm. broad, glabrous .. .. 2. *scandens*
Corolla gradually widened to near the top ; calyx-segments linear, 7 mm. long, glandular-puberulous ; leaves narrowly elliptic, gradually acuminate, acute at base, 8–15 cm. long, 3·5–5·5 cm. broad, glabrous .. .. .. 3. *vogeliana*
Corolla up to 3 cm. long :
Calyx-segments about 1 cm. long ; corolla-tube 15–18 mm. long, glabrous outside :
Calyx-segments broadly linear (2 mm. broad), 1 cm. long, hairy to glabrescent, but not glandular ; inflorescences elongate, about 5 cm. long ; petiole 3–10 cm. long ; leaves ovate to oblong-elliptic, broadly acuminate, rounded to cuneate at base, 5–7 cm. long, 2·5–4 cm. broad, glabrous or nearly so .. .. 4. *calycina*
Calyx-segments narrowly linear (1–1·5 mm. broad), 9 mm. long, glandular-pilose ; inflorescences short, about 2 cm. long, bracts 5 mm. long, 1 mm. broad ; petiole 5–30 mm. long ; leaves elliptic-ovate, cuspidate, rounded to cuneate at base, 7–12 cm. long, 15–55 mm. broad, slightly hirsute on the veins on both sides

5. *glandulifera*

Calyx-segments 5 mm. long or less, subulate :
Calyx-segments 5 mm. long ; leaves ovate or ovate-triangular, broadly pointed, rounded or truncate at base, 4–7 cm. long, 2·5–3·5 cm. broad, pubescent on the nerves ; petiole pubescent ; corolla-tube 1·5 cm. long or more, shortly pubescent with descending hairs ; fruit 2 cm. long, softly puberulous .. 6. *gangetica*
Calyx-segments 1–2 mm. long ; leaves oblong- to obovate-elliptic, obtusely acuminate, bluntly cuneate at base, 6–15 cm. long, 2–6 cm. broad, subglabrous on the nerves beneath ; corolla-tube 6–8 mm. long, minutely puberulous ; fruit 2 cm. long, minutely puberulous .. .. 7. *decipiens*

1. **A. macrophylla** (*T. Anders.*) *Lindau* in E. & P. Pflanzenfam. 4, 3B : 326 (1895) ; F.T.A. 5 : 134. *Dicentranthera macrophylla* T. Anders. in J. Linn. Soc. 7 : 52 (1863). A soft-wooded shrub up to 10 ft. high, with strong spikes of white or purple-tinged flowers 1½ in. long.
   S.Nig.: Oban *Talbot* 2010 ! Akor to Orem, Oban F.R. (Jan.) *Onochie* FHI 36079x ! Akorkwa *Thompson* 14 ! [Br.]**Cam.**: S. Bakunda F.R., Kumba (Mar., Apr.) *Olorunfemi* FHI 30545 ! *Brenan* 9293 ! 9293a ! Ikom to Mamfe, between Cross R. and Mun-Aya R. (Dec.) *Keay* FHI 28320 ! Johann-Albrechtshöhe (=Kumba) *Staudt* 470 ! Kembong F.R. (Mar.) *Onochie* FHI 32054 ! **F.Po**: (Dec.) *Mann* 13 ! Also in Cameroun and Gabon.
2. **A. scandens** (*Lindley*) *Hook.* Bot. Mag. 75 : t. 4449 (1849) ; F.T.A. 5 : 133. *Henfreya scandens* Lindley Bot. Reg. 33 : t. 31 (1847). A straggling under-shrub, sometimes low and subherbaceous, with pale purple or white and purple flowers 2 in. long in a terminal somewhat viscid inflorescence about 3 in. long ; sometimes cultivated as an ornamental.
   Guin.: Kindia (Jan.) *Pobéguin* 1377 ! Fouta Djalon (Mar.) *Chev.* 12587 ! **S.L.**: *Don* ! *Afzelius* ! *Smeathmann* ! (May) *Barter* ! *Thomas* 9011 ! 9382 ! 9418 ! 9537 ! Kangahun (fl. & fr. Feb.) *Deighton* 6036 ! York Pass (Feb.) *Deighton* 3486 ! Guma (fl. & fr. Mar.) *Hepper & Pyne* 2509 ! **Lib.**: (Feb.) *Harley* 861a ! Kakatown (fl. & fr. Sept.) *Whyte* 56 ! Also in Congo (Beni) ; *fide* Mildbr. in Wiss. Erg. 1907–08 : 300 (1914).
   [Note : The Smeathmann specimen in the British Museum bears the synonym note " *Ruellia acuminata* Afzelius ", and the same name occurs in Afzelius' own handwriting on Afzelius' specimen.]
3. **A. vogeliana** *Benth.* in Fl. Nigrit. 479 (1849) ; F.T.A. 5 : 133 ; Chev. Bot. 499. *A. longituba* Lindau in Engl. Bot. Jahrb. 22 : 118 (1895) ; F.T.A. 5: 133. A weak more or less straggling undershrub, with tubular pale lilac or bluish flowers 2–2½ in. long in a terminal compound lax inflorescence.
   S.L.: Vaama, Koya (Dec.) *Deighton* 3837 ! Gola Forest (Jan.) *King* 12 ! Malema (Nov.) *Deighton* 454 ! **Lib.**: Kakatown (Sept.) *Whyte* ! Bili Mt. (Jan.) *Harley* 1322 ! Genna Tanyehun (Dec.) *Baldwin* 10774 ! **Iv.C.**: Makougnié *Chev.* 16956 ! **Iv. C.**: Kassigué (Jan.) *Chev.* 16992 *ter* ! 17048 ! **Dah.**: Pedjile to Pobé (Feb.) *Chev.* 22922 ! Guébo *Chev.* 17211 ! Potou to Alépé *Chev.* 17392 ! Andédédou Forest, near Abidjan (Feb.) *Leeuwenberg* 2683 ! **Ghana**: Tafo (Nov.) *Pyne* 184 ! Begoro (Nov.) *Morton* A32 ! Goaso (Feb.) *Andoh* 5109 ! Akim Hills (Dec.) *Johnson* 265 ! Bibiami to Munasi (Dec.) *Adams* 1937 ! Pramkese (Jan.) *Morton*

A 2965 ! Assin-Yan-Kumasi *Cummins* 79 ! **S.Nig.**: Aponmu, Ondo (Nov., Dec.) *Trochain & Keay* FHI 25607 ! *Keay* FHI 25531 ! Angiama *Barter* 2096 ! Okomu F.R., Benin (fl. & fr. Dec., Jan.) *Brenan* 8576 ! 8763 ! Degema *Talbot* 3653 ! Eket *Talbot* 3008 ! Aboabam, Ogoja (Dec.) *Keay* FHI 28224 ! [**Br.**]**Cam.**: Victoria (Jan.) *Maitland* 314 ! Cam. Mt., 3,000 ft. (Dec.) *Mann* 1955 ! *Preuss* 1363 ! Lisongo, 4,000 ft. *Maitland* ! Bokosso (fl. & fr. Jan.) *Keay* FHI 37363 ! **F.Po**: (Dec.) *T. Vogel* 221 ! *Mann* 47 ! Also in Cameroun, Ubangi-Shari, Gabon and Congo.

4. **A. calycina** *Benth.* in Fl. Nigrit. 478 (1849). *A. coromandeliana* of F.T.A. 5 : 131, partly (syn.) *A. buettneri* Lindau (1894)—F.T.A. 5 : 132 ; *Chev.* Bot. 499 ; F.W.T.A., ed. 1, 2 : 257 ; Aké Assi, Contrib. 1: 208. *A. dryadum* S. Moore in Cat. Talb. 87 (1913). Herb, erect or half-straggling, 1–2 ft. or more high, slightly pubescent, with tubular flowers inflated above, ¾–1 in. or more long, usually white with red-purple markings, in one-sided racemes.

Guin.: Kouria (fl. & fr. Nov.) *Caille* in *Hb. Chev.* 14872 ! Nzo (fr. Oct.) *Schnell* 3858 ! **S.L.**: Leicester (Dec.) *Sc. Elliot* 3895 (partly, mixed with No. 6) ! Makeni (fl. & fr. Nov.) *Tindall* 55a ! Mattru to Gbangbama (fl. & fr. Nov.) *Deighton* 2342 ! Kaballa (fl. & fr. Sept.) *Thomas* 2226 ! Gberia Fotombu (Oct.) *Small* 414 ! **Lib.**: Monrovia (fl. Nov., fl. & fr. Apr.) *Barker* 1263 ! *Linder* 1531 ! *Whyte* ! Grand Bassa *T. Vogel* 87 ! Gbanga (Jan.) *Daniel* 100 ! **Iv. C.**: Makougnie (Jan.) *Chev.* 16962 ! Grabo (fl. & fr. Aug.) *Chev.* 19747 ! Lamé (fl. & fr. Oct.) *Leeuwenberg* 1737 ! **Ghana**: Accra (fl. & fr. Jan.) *Dalz.* 152 ! Asafo road junction (fr. Oct.) *Morton* GC 7836 ! Kumasi (fl. Sept., Nov., fr. Nov.) *Darko* 339 ! 636 ! Top of Bame Pass (Oct.) *Morton* GC 9301 ! **Togo Rep.**: Misahöhe (fl. & fr. Nov.) *Mildbr.* 7321 ! Depongo *Büttner* 260 ! **S.Nig.**: Aguku Dist. *Thomas* 738 ! 1689 ! Ibadan (Feb.) *Meikle* 1134 ! Mamu River F.R., Onitsha *E. W. Jones* 2 ! Eket *Talbot* 3114 ! Oban *Talbot* 991 ! 991a ! Also in Cameroun. (See Appendix, p. 449.)

5. **A. glandulifera** *Lindau* in Engl. Bot. Jahrb. 57 : 23 (1920). Ascending herb 1–2 m. high, with terminal inflorescences of white flowers with mauve mottling on lower lip.

[**Br.**]**Cam.**: Mayo Ini Plateau, 4,800 ft., Vogel Peak, Adamawa (Dec.) *Hepper* 1585 ! Also in Cameroun.

6. **A. gangetica** (*Linn.*) *T. Anders.* in Thwaites Enum. Pl. Zeyl. 235 (1860) ; J. Linn. Soc. 7: 52 (1863) ; Chev. Bot. 499 ; F.W.T.A., ed. 1, 2 : 257 ; Backer Onkruidfl. Javan. Suikerrietgr. 668 (1931) ; Robyns Fl. Sperm. Parc. Nat. Alb. 2 : 294 ; Benoist Bol. Soc. Brot. sér. 2, 24: 39 (1950) ; E. C. Leonard in Contrib. U.S. Nat. Herb. 31 : 288 (1953) ; Berhaut Fl. Sén. 87 ; F. W. Andr. Fl. Pl. Sudan 3 : 166. *Justicia gangetica* Linn. (1759). *A. coromandeliana* Nees (1832)—F.T.A. 5 : 131, partly (excl. syn. *A. calycina* Benth. (1849)) ; Chev. Bot. 499 ; C. B. Cl. in Fl. Brit. Ind. 4 : 493, and in Journ. As. Soc. Beng. 74 : 667 (1908), incl. syn. *A. parvula* of Aké Assi, Contrib. 1: 209, t. 16 (1961), not of C. B. Cl. (1900). Herb, 1–4 ft. high, very similar to *A. calycina* Benth. but easily distinguishable from this species by the much smaller calyx.

Sen.: *Farmar* 75 ! *Berhaut* 731 ! Niayes *Chev.* 2801. Dakar *Chev.* 2804. **Gam.**: Genieri (July) *Fox* 135 ! *Ingram* ! *Skues* ! **Mali**: Kita (Aug.) *Jaeger* 22a ! **Port.G.**: Boi, Bissau *Baptista* 30. Pussubé, Bissau *Esp. Santo* 1006 ! **Guin.**: Macenta (Oct.) *Baldwin* 9809 ! **S.L.**: Leicester (fl. & fr. Dec.) *Sc. Elliot* 3895 (partly, mixed with No. 4) ! Freetown (fl. & fr. Dec.) *Deighton* 1537 ! Mahera (fl. & fr. May) *Deighton* 5675 ! Kasewe F.R. (Dec.) *King* 11 ! Rokupr (fl. & fr. Nov.) *Jordan* 698 ! **Lib.**: Cape Palmas (July) *T. Vogel* 56 ! Kakatown (Sept.) *Whyte* ! Suen (Nov.) *Linder* 1391 ! Suah Koko (Sept.) *Linder* 373 ! Gbau (Sept.) *Baldwin* 9386 ! **Iv.C.**: Bingerville *Chev.* 16008. Adiopodoumé (fl. & fr. Oct.) *Roberty* 15324 ! **Ghana**: Cape Coast *T. Vogel*!9 ! Accra *Irvine* 673 ! Kumasi (fl. & fr. Aug.) *Darko* 697 ! Atwabo *Fishlock* 42/1931 ! Yendi (fl. & fr. Apr.) *Williams* 168 ! **Togo Rep.**: Lomé *Warnecke* 272 ! **N.Nig.**: Jebba *Barter* ! Mongu (July) *Lely* 402 ! Lokoja (fl. & fr. Sept., Oct.) *Dalz.* 135 ! *Parsons* 30 ! Vom *Dent Young* 192 ! Jos (fl. & fr. June) *Lely* 843 ! P171 ! **S.Nig.**: Lagos *Hagerup* 843 ! Benin (fl. & fr. May) *Umana* FHI 29108 ! Ibadan (Feb.) *Meikle* 1133 ! Odo-Ona Bridge, Ibadan (fl. & fr. Sept.) *Oladoyinbo* FHI 24264 ! Ikom *Rosevear* 20/29 ! [**Br.**]**Cam.**: *Preuss* 1115 ! Mamfe (fl. & fr. Dec.) *Migeod* 264 ! Sawyer's Camp, Kumba Dist. (fl. & fr. Apr.) *Olorunfemi* FHI 30552 ! Gashaka, Adamawa (fl. & fr. Dec.) *Latilo & Daramola* FHI 28963 ! **F.Po**: *Mann* 265 ! A very complex aggregate species. Widespread throughout the Old World Tropics, introduced into tropical America. (See Appendix, p. 449.)

7. **A. decipiens** *Heine* in Kew Bull. 16 : 169 (1962). *A. lindauiana* Hutch. & Dalz. F.W.T.A., ed. 1, 2 : 257 (1931), cited specimens only (*Talbot* 380, 985), otherwise = *Filetia africana* Lindau, see under this species (p. 417). *Filetia africana* of S. Moore in Cat. Talb. 140 (1913), not of Lindau (1894). Erect herb, about 80 cm. high ; stems dark green ; leaves dark above, paler beneath with dull crimson midrib and petiole ; calyx green ; corolla pale greenish-cream with faint dull crimson markings on the throat ; in swampy areas of rain forest.

S.L.: Njala (July) *Deighton* 764 ! Kasawe F.R. (fr. Jan.) *King* 205b ! Lowoma (fl. & fr. Oct.) *Deighton* 56 ! Gola Forest (Dec.) *King* 104b ! **Lib.**: Medina, Bumbuma (Oct.) *Linder* 1309 ! Du R. (Aug.) *Linder* 213 ! 12 miles up R. Cess (fr. Mar.) *Baldwin* 11246 ! Yila, St. John R. (Aug.) *Baldwin* 9127 ! Peahtah (Oct.) *Bequaert* in *Hb. Linder* 1047 ! **Ghana**: Mankrong R., Afram (fl. & fr. Apr.) *Morton* A563 ! **S.Nig.**: Oban *Talbot* 380 ! 985 ! Oban F.R. (Jan.) *Keay* FHI 37732 !

24. THOMANDERSIA Baill. (1891)—F.T.A. 5 : 119 ; Bremekamp in Réc. trav. bot. néerl. 39 : 166 (1942). *Scytanthus* T. Anders. ex Benth. & Hook. f. (1876), not *Skytanthus* Meyen (1834).

A shrub ; branches terete ; glabrous ; leaves markedly anisophyllous, the larger oblong-elliptic, subcaudate-acuminate, unequal-sided at base, 8–15 cm. long, 3–6·5 cm. broad, with 3–5 pairs of lateral nerves, pubescent in their axils ; petiole up to 4 cm. long ; smaller leaves ovate ; racemes axillary, very slender, up to 30 cm. long ; calyx-leathery, 2·5 mm. long ; corolla 1·5 cm. long ; stamens 4 ; fruit ovate-orbicular, 1·5 cm. long   ..   ..   *laurifolia*

T. laurifolia (*T. Anders. ex Benth.*) *Baill.* Hist. Pl. 10 : 456 (1891) ; F.T.A. 5 : 120 ; Aké Assi, Contrib. 1 : 208. *Scytanthus laurifolius* T. Anders. ex Benth. in Benth. & Hook. f. Gen. Pl. 2 : 1093 (1876), and in Hook. Ic. Pl. 13 : t. 1209 (1877). A nearly glabrous shrub up to 15 ft. high with shining foliage, and long narrow racemes 6–12 in. long of flowers ¼ in. long, white or red and with red-purple markings. **Lib.**: Nyaake, Webo Dist. (June) *Baldwin* 6151 ! 11264 ! **Iv.C.**: Sassandra (Aug.) *de Wilde* 323 ! **S.Nig.**: Calabar (Feb., May) *Mann* 2321 ! *Kalbreyer* 204 ! *Holland* 87 ! Orukin to Unyene, Calabar (May) *Onochie* FHI 33179 ! Umon-Ndealichi F.R., Calabar (May) *Onochie* FHI 33225 ! Oban (fr. May) *Talbot* 381 ! Ujor FHI 31785 ! [**Br.**]**Cam.**: Barombi *Preuss* 339 ! Victoria to Kumba (Apr.) *Hutch. & Metcalfe* 145 ! S. Bakundu F.R., Kumba Dist. (Mar.) *Brenan* 9314 ! (fl. Jan., Apr., fr. Aug.) *Binuyo & Daramola* FHI 35167 ! FHI 35168 ! *Ejiofor* FHI 29345 ! *Olorunfemi* FHI 30702 ! *Onochie* FHI 30857 ! This species is in need of further taxonomic study and after such work other taxa may be recognized. In the circumscription of the original authors, *T. laurifolia* also occurs in Cameroun, Gabon and Congo. The plants from Liberia differ slightly in indumentum and density of the inflorescence.

25. LEPIDAGATHIS Willd.—F.T.A. 5 : 120.

Inflorescences terminal, up to 5 cm. long ; bracts broadly lanceolate, long-bristle-pointed, strongly nerved, about 7 mm. long ; leaves lanceolate to narrowly oblanceo-

late, slightly pubescent on the nerves beneath ; corolla nearly 1 cm. long ; adaxial
    stamens 2- or 1-celled ; fruit shorter than the bracts .. ..   1. *alopecuroides*
Inflorescences axillary or at the base of the plant near the ground, sometimes both
    types of inflorescence on the same plant :
Inflorescence elongate, oblong-cylindric to finger-like, longer than broad :
    Inflorescences (in the specimens known at present) always near the ground, finger-
        like to elongate-strobilate, up to 5 cm. long and (at the top) 1·5 cm. diam., ciliate
        on the margins but otherwise entirely glabrous, upper bracts sharply pointed,
        points erect ; leaves acicular, up to 1·5 cm. long .. .. ..   2. *felicis*
Inflorescences oblong-cylindric, nearly always in the upper parts of the plant, 1–5 cm.
    long and 1·5–2·5 cm. broad, formed by silvery-silky imbricate bracts with reflexed
    or curved points :
    Leaves glabrous, linear, 4–5 cm. long, 1-nerved ; inflorescence 3 cm. long, 1·5 cm.
        diam. ; bracts ovate, long-subulate-acuminate, about 1 cm. long, villous
                                                                            3. *sericea*
    Leaves pubescent :
    Stems softly villous ; leaves linear, prominently 3-nerved, appressed-villous-pilose,
        6–7 cm. long ; inflorescence sessile, 3 cm. long ; bracts lanceolate, with very
        long bristly tips ; juvenile forms .. ..   4. *heudelotiana*
    Stems thinly pubescent, otherwise plants as above ; leaves linear, 3-nerved,
        appressed-pubescent ; inflorescences about 4 cm. long ; bracts long-pilose, with
        curved slender tips ; mature and fruiting forms .. .. 4. *heudelotiana*
Inflorescence globose or subglobose, about as broad as long, sometimes (in Nos. 10
    and 11) ellipsoid :
    Leaves long-acicular to filiform, up to 5 cm. long and 2 mm. broad :
    Inflorescences strobilate to capituliform, 4–4·5 cm. diam. at the top of creeping
        stems up to 5 cm. long .. ..   5. *capituliformis*
    Inflorescences sessile, never strobilate, occurring at the base of the plants as well as
        on the stems ; leaves glabrous :
    Inner bracts with a terete subulate acumen, outer ovate-lanceolate, thin, pubescent ;
        inflorescences densely crowded .. ..   6. *fimbriata*
    Inner bracts with a flat acumen, outer ovate-lanceolate, glabrous outside, up to
        1·3 cm. long ; inflorescences in a sessile cluster .. ..   7. *chariensis*
    Leaves not acicular or filiform ; bracts acute or with spine-like tips :
    Leaves with lateral nerves or 3-nerved, very variable, up to 10 cm. long and 1·5 cm.
        broad ; plants never ericaceous in habit :
    Bracts villous-plumose, with subulate points ; leaves linear to elliptic-lanceolate,
        with lateral nerves, sometimes 3-nerved, with anastomosing lateral nerves ;
        extremely variable plants ; inflorescences often confined to the base of the plant,
        those sometimes occurring in the upper parts with longer and deflexed bracts,
        indumentum often yellowish .. ..   8. *collina*
    Bracts glabrous, usually ciliate on the margins (never villous-plumose), with long
        subulate points, up to 1·5 cm. long ; leaves linear, very acute, up to 17 cm. long,
        strongly 3-nerved, with prominent linear cystoliths spreading from the midrib
        at right angles .. ..   9. *anobrya*
    Leaves small, subulate, with no (or very inconspicuous parallel) lateral nerves, up
        to 2 cm. long and 1–2 mm. broad (in the vegetative parts), plants therefore often
        ericaceous in habit :
    Inflorescences 1–3 together, 1·5–3 cm. long, about 2 cm. diam., truncate at the apex ;
        the outer bracts ½–1 cm. broad, truncate with a rigid point ; bracteoles and
        sepals oblong, truncate and tomentose at the tip ; leaves of the erect stems
        ericoid, 4–10 mm. long, glabrous .. .. ..   10. *pobeguinii*
    Inflorescences often proliferous, forming a globose cluster ; bracteoles and sepals
        with a long linear ciliate tip ; leaves of the erect stems 1 cm. long 11. *chevalieri*

1. **L. alopecuroides** (*Vahl*) *R. Br. ex Griseb.* Fl. Brit. W. Ind. 453 (1861), excl. syn. *Adenosma chenopodifolia*
(Poir.) Sprengel (1825) ; Benoist in Notulae Syst. 2 : 18 (1911) ; Chev. Bot. 497. *Ruellia alopecuroidea*
Vahl Eclog. Am. 2 : 49 (1796). *Teliostachya alopecuroidea* (Vahl) Nees in DC. Prod. 11 : 262 (1847).
L. laguroidea (Nees) T. Anders. (1863)—F.T.A. 5 : 128 ; Chev. Bot. 498 ; F.W.T.A., ed. 1, 2 : 256 ;
Berhaut Fl. Sén. 88. *Teliostachya laguroidea* Nees (1847). *L. hyssopifolia* (Benth.) T. Anders. (1863)—
F.T.A. 5 : 128. *Teliostachya hyssopifolia* Benth. (1849). Decumbent slender branching herb, woody
below, rooting at the lower nodes, with small pink or purplish flowers ⅓ in. long in rather dense almost
bristly pseudo-spikes ; inflorescences very variable in length and density.
    Sen.: Niokolo-Koba *Berhaut* 1541. **Mali:** Kita *Chev.* 99. **Guin.:** Faranna *Chev.* 20443. Forékaria *Chev.*
20528. Macenta, 2,000–2,500 ft. (fr. Oct.) *Baldwin* 9756 ! **S.L.:** *Afzelius* ! *Don* ! *Smeathmann* ! *Sc. Elliot*
3902 ! Kambia, Magbema (fl. & fr. Dec.-Jan.) *Jordan* 761 ! *Sc. Elliot* 4376 ! Wallia (fl. & fr. Jan.) *Sc.
Elliot* 4744 ! Makump (fl. & fr. July) *Thomas* 936 ! Njala (fr. Jan.) *Dalz.* 8077 ! Kennema *Deighton* 427 !
Lib.: Tawata, Boporo Dist. (Nov.) *Baldwin* 10311 ! Baila (fl. & fr. Feb.) *Harley* 1470a ! Accra *T. Vogel* !
Iv.C.: Tiassalé (fl. & fr. Dec.) *Leeuwenberg* 2140 ! Bingerville *Chev.* 16079. Makougnié *Chev.* 16978*bis*.
Tébo *Chev.* 19393. Grabo *Chev.* 19748. Taï (fl. & fr. Aug.) *Boughey* GC 14942 ! **Ghana:** Apaaso, Volta
gap (fr. Feb.) *Morton* A119 ! Afram Mankrong F.R. (Jan.) *Adams* 5020 ! Mankrong, Kwahu (Jan.) *Adams*
5059 ! Pramkese (fl. & fr. Jan.) *de Wit & Morton* A2954 ! Nsawkaw (fl. & fr. Dec.) *Adams* 3157 ! **N.Nig.:**
Oly R. *Barter* 746 ! Naraguta (fl. & fr. May) *Lely* 36 ! Kwakuti, Niger Prov. (fl. & fr. Dec.) *Keay* FHI
37307 ! **S.Nig.:** Olokemeji (Dec.) *Meikle* 921 ! Iseyin (fl. & fr. Feb.) *Brenan & Keay* 8963 ! Oban *Talbot*
984 ! Kwa Falls (Mar.-Dec.) *Brenan* 9239 ! Bendiga Ayuk, Ogoja (fl. & fr. Dec.) *Keay* FHI 28167 ! *Coombe*
164 ! **[Br.]Cam.:** Bamenda, 3,500 ft. (fl. & fr. May) *Maitland* 1689 ! Widespread in the tropics.

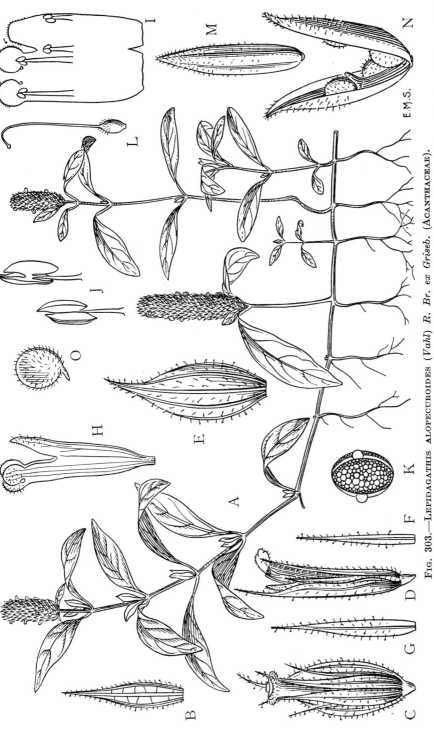

Fig. 303.—LEPIDAGATHIS ALOPECUROIDES (*Vahl*) *R. Br. ex Griseb.* (ACANTHACEAE).

A, habit, × ⅔. Drawn from *Brenan & Keay* 8963. B, floral bract, × 6. C, flower (frontal view), × 6. D, flower (lateral view), × 6. E, posterior sepal, × 6. F, lateral sepal, × 6. G, anterior sepal, × 6. H, corolla, × 6. I, corolla opened out, × 6. J, stamens in two aspects, × 40. K, pollen grain, × 600. L, pistil, × 6. M, fruit, × 12. N, opened fruit, × 12. O, seed, × 18. Drawn from *Thomas* 6780.

415

2. **L. felicis** *Benoist* in Bull. Soc. Bot. Fr. 85 : 682 (1939), as *felixii*. " Close to *L. pobeguini* Hua " (Benoist) and probably an occasional (juvenile ?) form of that species.
Guin.: near Mamou (Feb.) *Jac.-Fél.* 774 ! Only known from the type collection.

3. **L. sericea** *Benoist* Notulae Syst. 2 : 154 (1911) ; Chev. Bot. 499, excl. var. *hirta* Benoist ; Berhaut Fl. Sén. 81. Erect woody undershrub with narrow leaves and pink-purple flowers in somewhat 4-sided spikes ¾–1¼ in. long.
Sen.: Galam *Leprieur* ! Bakel to Fangalla *Carrey* 33 ! Bakel *Collin* 77. Niokolo-Koba *Berhaut* 1227 ; 4508. Bandia *Trochain* 1226 ! **Mali:** Balani *Chev.* 144. **Guin.:** Col de Sokotoro (fr. Dec.) *Pobéguin* 1458 ! Pied col Sita (Oct.) *Adam* 12687 ! Dinguiraye *Jac.-Fél.* 1483 ! Timbo *Maclaud* 21. Kadé *Pobéguin*.

4. **L. heudelotiana** *Nees* in DC. Prod. 11 : 254 (1847) ; F.T.A. 5 : 126; Benoist l.c. 21 : 155 ; Chev. Bot. 498 ; Berhaut Fl. Sén. 80. *L. sericea* var. *hirta* Benoist l.c. 155 (1911) ; Chev. Bot. 499. *L. medusae* of F.W.T.A., ed. 1, 2 : 256, not of S. Moore. Erect woody undershrub with narrow velvety hairy leaves and whitish-hairy flower-spikes 1–1½ in. long, often 2 together, with tail-tips of the bracts becoming deflexed.
Sen.: Rio Nunez *Heudelot* 666 ! Tambacounda *Berhaut* 1288. **Gam.:** *Hayes* 574 ! **Mali:** Kita (fl. & fr. Oct.) *Jaeger* 2912 ! Mossi *Chev.* 24630 ! Koulikoro (Oct.) *Chev.* 2773 ! **Port.G.:** Madina de Boé (fr. Dec.) *Esp. Santo* 2356 ! 2852 ! Gabu, Pirada (Sept.) *Esp. Santo* 2786 ! **Guin.:** Kouroussa *Pobéguin* 912 ! Mamou to Dalaba (fr. Sept.) *Chev.* 34703 ! Boumalol, Madina Tossekré (Oct.) *Adam* 12562 ! Diaguissa *Chev.* 18786 ! Kindia *Jac.-Fél.* 1842 ! **Ghana:** Kayoro (Oct.) *Vigne* FH 4615 ! (See Appendix, p. 452.)
[Juvenile forms of this species (Chev. 2773, 2774), originally correctly identified by Lindau as *L. heudelotiana* Nees, have been described as *L. sericea* var. *hirta* Benoist (1911).]

5. **L. capituliformis** *Benoist* l.c. 19 (1911) ; Berhaut Fl. Sén. 81 : 91. A herb with mauve flowers ; in appearance very similar to species Nos. 2, 10 and 11.
Sen.: Niokolo-Koba *Berhaut* 1494 ; 4455. *Raynal* 6777 ! Tantaba (May) *Trochain* 3490 ! **Guin.:** Coniaguin (Feb.) *Pobéguin* 2144 !

6. **L. fimbriata** *C. B. Cl.* in F.T.A. 5 : 125 (1899) ; Benoist l.c. 21 ; Chev. Bot. 498. *L. anobrya* Nees var. *angustissima* Nees in DC. Prod. 11 : 255 (1847). A slender half-woody herb, nearly glabrous, with very narrow leaves and short dense inflorescences 1–1½ in. across.
Sen.: Fouta Djalon *Heudelot* 679 ! **Port.G.:** Boé, Bilonco (fl. & fr. Dec.-Jan.) *Esp. Santo* 2868 ! **Guin.:** Kolen *Maclaud* 149. Bambaya *Pobéguin* 761 ! Timbo to Conakry *Pobéguin* 767. Kindia (fl. & fr. Aug.) *Pobéguin* 1356 ! *Jac.-Fél.* 364 ! Mamou (fr. Oct.) *Adam* 12643 !

7. **L. chariensis** *Benoist* in Mém. Soc. Bot. Fr. 2, 8 : 189 (before Oct. 1912). *L. acicularis* Turrill in Kew Bull. 1912 : 361 (25 Oct. 1912). Wiry, terete, glabrous stems, with very narrow leaves 1–2 in. long and yellow flowers ½–⅝ in. long in short congested inflorescences either basal or axillary.
N.Nig.: Patti Lokoja (Oct.) *Dalz.* 139 ! Also in Ubangi-Shari.
[Most likely only an extreme form of the following.]

8. **L. collina** (*Endl.*) *Milne-Redhead* in Kew Bull. 8 : 119 (1953). *Russegera collina* Endl. Iconogr. Gen. Pl., t. 94 (1838) ; Nov. Stirp. 38 (1839) ; F. W. Andr. Fl. Pl. Sudan 3 : 182. *L. radicalis* Hochst. ex Nees (1847)—F.T.A. 5 : 123 ; Berhaut Fl. Sén. 93. *L. radicalis* var. *acrantha* Benoist ex Tisserant Mém. Inst. Etud. Centrafric. 2 : 14 (1950), French descr. ; Monod in Mém. I.F.A.N. 18 : 14 (1952). *L. radicalis* var. *caulispica* Benoist, var. *elata* Benoist, var. *polyneura* Benoist ex Tisserant l.c. (1950). *L. mollis* T. Anders. (1863)—F.T.A. 5 : 126 (1899) ; Benoist in Notulae Syst. 2 : 22 ; F.W.T.A., ed. 1, 2 : 256. *L. diversa* C. B. Cl. in F.T.A. 5 : 126 (1899). *L. schweinfurthii* Lindau (1894)—F.T.A. 5 : 123. *L. reticulata* Benoist in Notulae Syst. 2 : 22 (1911). (?) *L. garuensis* Lindau in Engl. Bot. Jahrb. 49 : 402 (1913). (?) *L. petrophila* Lindau l.c. 403 (1913). An extremely polymorphic coarse woody herb, 1–2 ft. high, with ribbed pubescent stems, and yellow purple-spotted flowers ⅝ in. long in very dense heads 1–2 in. across either axillary or basal ; growth forms much affected by grassland-burning and other ecological factors.
Sen.: *Berhaut* 803. **Mali:** Kita (Oct.) *Jaeger* 3037 ! **Port.G.:** Bafata, Uhambauha (Nov.) *Esp. Santo* 2817 ! **Guin.:** Kouroussa *Pobéguin* 916. Diendou *Pobéguin* 1515 ! Siguiri (Nov.) *Jac.-Fél.* 514 ! Timbo *Maclaud* 263 ! Dalaba Plateau *Chev.* 20276 ! **Ghana:** Paliba (Dec.) *Adams & Akpabla* 4106 ! Prang (Sept.) *Darko* 991 ! Damongo Scarp *Morton* GC 25001 ! Kete Krachi to Atebubu *Adams* 4614 ! Tongo *Adams & Akpabla* 4315 ! Banda Ravine (Dec.) *Morton* GC 25191 ! **N.Nig.:** Nupe *Barter* 955 ! Niger *Baikie* ! Jos, 3,900 ft. (Oct.) *Hepper* 1025 ! *Batten-Poole* 341 ! Lapai to Paiko, Niger Prov. (Aug.) *Onochie* FHI 35434 ! Yola *Dalz.* 303 ! **[Br.]Cam.:** Vogel Peak, 4,000 ft., Adamawa (Nov., Dec.) *Hepper* 1361! 2712 ! Also in Cameroun, Ubangi-Shari, Sudan, Ethiopia and Uganda.

9. **L. anobrya** *Nees* in DC. Prod. 11 : 255 (1847), excl. var. *angustissima* Nees l.c. (1847) ; F.T.A. 5 : 124 ; Chev. Bot. 498 ; Benoist in Notulae Syst. 2 : 19 ; Berhaut Fl. Sén. 81 : 93. *L. fimbriata* of Chev. Bot. 498, partly (*Chev.* 24539), not of C.B. Cl. Stems slender, glabrous, angled, with long narrow leaves and dense flower-heads 1–1½ in. diam., mostly basal or with one at an upper node.
Sen.: *Heudelot* 204 ! Koulikoro *Chev.* 3326 ! *Berhaut* 1281 ! 4482. **Mali:** Fodébouyou, Kita (Aug.) *Jaeger* 2314 ! Kita (fl. & fr. Nov.) *Jaeger* 3622 ! Ravin Gaïbala, Kita (Sept.) *Jaeger* 2515 ! Sanga, near Bandigara *Ganay* ! **Guin.:** Dalaba to Diaguissa (Sept.) *Chev.* 18578. **Iv.C.:** Sific *Hallé* 128 ! **Ghana:** Berekum to Sampa *Adams* 5304 ! Sawla junction on Tamale road (Sept.) *Rose Innes* GC 30185 ! Bagabaga, Tamale (Nov.) *Darko* 435 ! Kete Krachi to Dutukpene *Adams* 4628 ! Kete Krachi *Morton* A1489 ! **Niger:** *Marchal* 80 ! Fada to Koupéla, Gourma *Chev.* 24539 !

10. **L. pobeguinii** *Hua* in Bull. Soc. Bot. Fr. 50 : 578, t. 18 (1904), and in Pobéguin Ess. Fl. Guin. Fr. 125, t. 27*bis* (1906) ; Chev. Bot. 499. *Un-named plant* of A. L. Smith in J. Linn. Soc. 30 : 155–157, t. 8 (1894). A very variable, heath-like plant, 12–18 in. high with a woody rootstock ; flowers yellowish.
Mali: *Scaëtta* 3172 ! Dougoura *Chev.* 335 ! 356. **Guin.:** Fouta Djalon, Kouroussa *Pobéguin* 209 ! *Sc. Elliot* 5386 ! Beyla *Chev.* 20879 !

11. **L. chevalieri** *Benoist* l.c. 20 (1911) ; Chev. Bot. 498. Habit of the last, with inflorescence-heads more confluent and often in dense clusters.
Mali: Diaragouéla *Chev.* 481 ! **Guin.:** Coniaguin to Boussoura *Pobéguin* 2145 !

Note : The species Nos. 2, 5, 10, 11 are very closely allied and probably (ecological ?) forms of one polymorphic aggregate species. The accepted distinctions are artificial and based upon herbarium material only.

Doubtful species.

L. dahomeyensis *Benoist* in Mém. Soc. Bot. Fr. 2, 8 : 190 (1910) ; Chev. Bot. 498, as *dahomensis* ; F.W.T.A., ed. 1, 2 : 256.
Dah.: Farfa to Toukountouna, Mts. Atacora, 1,500 ft. *Chev.* 24072 !
The species is only known from the type-collection and the holotype consists of a poor specimen with partly burned inflorescences, which may well represent an occasional form of *L. collina* occurring after burning. See also note under No. 7.

Additional species.

L. scabra *C. B. Cl.* in F.T.A. 5 : 129 (1899), an Angolan and E. African species, was found by Dr. J. K. Morton in Ghana on the top of Damongo Scarp, in savanna woodland (fl. & fr. Dec.) *Morton* A120 !

26. FILETIA Miq.—F.T.A. 5 : 136.

Shrub, about 1 m. high, glabrous, with simple round stem ; leaves ovate, up to 18 cm. long and 7 cm. broad, with truncate or often slightly cordate base ; petioles 0·5–2 cm.

long ; inflorescences terminal, racemose, up to 8 cm. long ; bracts filiform, 2–4 mm. long, bracteoles similar, up to 2 mm. long ; flowers in 1–3-flowered dichasial partial inflorescences, peduncles and pedicels short, up to 4 mm. long ; calyx about 4 mm. long, glandular, with 5 equal, filiform lobes ; corolla about 9 mm. long, 1·5 mm. diam., with minute glandular hairs outside ; upper lip 5 mm. long, 2 mm. broad, with two lobes, lower lip 4 mm. diam. ; stamens 4, of different length, the outer ones 3 mm., the inner ones 2 mm. long, filaments of the two different stamens close together and apparently united, anther-cells muticous ; ovules 2 in each cell ; fruit unknown

africana

F. africana *Lindau* in Engl. Bot. Jahrb. 20 : 41 (1894) ; F.T.A. 5 : 136 (not of S. Moore in Cat. Talb. 140, nor syn. *F. africana* of F.W.T.A., ed. 1, 2 : 257, excl. distrib.). *Asystasia lindauiana* Hutch. & Dalz. F.W.T.A., ed. 1, 2 : 257 (1931), excl. cited specimens (*Talbot* 380, 985). A shrub about 3 ft. high with a simple slender green stem ; flowers pale purplish pink ; in deep shade in forest.
[Br.]**Cam.**: S. Bakundu F.R., Kumba (Mar.) *Brenan* 9412 ! Also in Cameroun and Gabon.

[A very rare plant, confused in Cat. Talb. and F.W.T.A., ed. 1, with an at that time undescribed *Asystasia* sp. (*A. decipiens*, see p. 413 above).]

27. NEURACANTHUS Nees—F.T.A. 5 : 137.

Plants about 2 ft. high, from a short woody rootstock, stems and leaves beneath white stellate-tomentose ; leaves broadly elliptic, acute, up to 10 cm. long and 6·5 cm. broad, with about 10 pairs of lateral nerves, entire, subsessile or with a petiole up to 5 mm. long ; flowers in dense axillary heads up to 5 cm. in diam., made up of few-flowered condensed unilateral spikes with 3 mm. long hyaline hairs ; bracts 9–12 mm. long, ovate, acute, soft, mucronate with almost deflexed tips, dark brown, parallel-nerved ; calyx 9 mm. long, with long-acuminate lobes ; corolla about 1 cm. long, pilose ; capsule glabrous, flattened, spindle-like, quadrangular and with sharp tips on both ends, about 14 mm. long, 4 mm. broad and 2 mm. diam., with 2–4 lentil-like seeds about 3 mm. diam. with fine white-silky hygroscopic hairs ..   ..   *niveus*

N. niveus *S. Moore* in J. Bot. 18 : 37 (1880) ; F.T.A. 5 : 139 ; Benoist in Notulae Syst. 2 : 145 (1911) ; A. Chev. in Etud. Fl. Afr. Centr. Fr. 1 : 235 (1913) ; F. W. Andr. Fl. Pl. Sud. 3 : 184. General appearance like *Lepidagathis heudelotiana*, but very different from all W. African *Lepidagathis* by its broad leaves with pinnate venation of many lateral nerves and white tomentum beneath ; flowers violet.
Guin.: Dinguiraye, Bocaria (fr. Dec.) *Jac.-Fél.* 1447 ! Also in Cameroun, Ubangi-Shari, Sudan and E. Africa.

28. AFROFITTONIA Lindau in Engl. Bot. Jahrb. 49 : 406 (Mar. 1913) ; Rendle in Cat. Talb. IX (Apr. 1913). *Talbotia* S. Moore in Cat. Talb. 80, t. 11, fig. 1–6 (Apr. 1913).

Herb with procumbent stems rooting at the nodes ; branches pubescent ; leaves ovate-elliptic, rounded at both ends, 3–5 cm. long, 2·5–3·5 cm. broad, densely pubescent beneath ; petioles pubescent ; spikes axillary, pedunculate, 2–4 cm. long ; bracts obovate, 1 cm. long ; bracteoles spathulate-oblanceolate, 8–10 mm. long ; corolla-tube 1 cm. long, 2-lipped, the posterior lip bidentate, the anterior lip 3-lobed ; stamens 4, the anthers of the 2 shorter ones 1-celled ..   ..   ..   *silvestris*

A. silvestris *Lindau* l.c. 407, fig. 2 (1913) ; F.W.T.A., ed. 1, 2 : 608. *Talbotia radicans* S. Moore l.c. (1913) ; F.W.T.A., ed. 1, 2 : 257. A procumbent herb trailing on moist ground on the forest floor ; flowers whitish with mauve markings.
S.Nig.: Oban *Talbot* 971 ! Calabar to Mamfe (Jan., Nov.) *Onochie* FHI 36025 ! *Baldwin* 13761 ! Cross R. North F.R. (Dec.) *Keay* FHI 28149 ! **F.Po**: Monte Balea (Dec.) *Guinea* 362 ! Bokoko (Oct.) *Mildbr.* 6830. Also in Cameroun (Ndoungué *Ledermann* 6369.)

The related *Fittonia argyroneura* Coem., a native of tropical America, is cultivated as an ornamental, e.g. in Ghana (Legon (Nov.) *Morton* A2819 !) and in S. Nigeria (Ibadan (Oct.) *Keay* FHI 37677 !)

29. STAUROGYNE Wall. Pl. As. Rar. 2 : 80, t. 186 (1831) ; Bremekamp in Reinwardtia 3 : 163 (1955) ; Heine in Kew Bull. 16 : 181 (1962). *Ebermaiera* Nees (1832). *Zenkerina* Engler (1897)—F.T.A. 4, 2 : 262. (?)*Neozenkerina* Mildbr. (1921). *Staurogynopsis* Mangenot & Aké Assi in Bull. Jard. Bot. Brux. 29 : 27 (1959).

Erect plants, little-branched ; inflorescence spicate, many-flowered, up to 10 cm. long ; leaves elliptic, up to 15 cm. long and 6 cm. broad, cuneate at base, acute at apex

1. *kamerunensis*

Small creeping herbs, much-branched ; inflorescence capitate, few-flowered, up to 1·5 cm. diam. ; leaves ovate to ovate-lanceolate, up to 6 cm. long and 3·5 cm. broad, truncate at base, very obtuse at apex :
  Leaves up to 6 cm. long and 3·5 cm. broad, hairs distributed over the whole surface as well as on the nerves ; calyx-lobes free from near the base   ..   2. *paludosa*
  Leaves up to 2 cm. long and 1·5 cm. broad, hairs on the surface nearly confined to the nerves, with very few also in the lower margins ; calyx-lobes united in the lower half   ..   ..   ..   3. *capitata*

1. **S. kamerunensis** (*Engl.*) *Benoist* in Notulae Syst. 2 : 290 (1913) ; Mangenot & Aké Assi l.c. 29 : 36 (1959). *Zenkerina kamerunensis* Engl. Bot. Jahrb. 23 : 498, t. 10, fig. A–F (1897) ; F.T.A. 4, 2 : 262 (in note) ; Cat. Talb. 139. (?) *Neozenkerina bicolor* Mildbr. (1921). Erect herb up to 18 in. high ; flowers creamy-white, up to ½ in. long in rain forest.

S.Nig.: Oban *Talbot* 1369! 2004! Oban F.R. (Jan.) *Onochie* FHI 36145*x*! Orem, Calabar (Jan.) *Onochie* FHI 36063*x*! Aboabam, Ikom (Dec.) *Keay* FHI 28210! Also in Cameroun.

2. **S. paludosa** (*Mangenot & Aké Assi*) *Heine* in Kew Bull. 16 : 183 (1962). *S. capitata* E. A. Bruce in Kew Bull. 1935 : 284, partly (*Chipp* 154). *Staurogynopsis paludosa* Mangenot & Aké Assi l.c. 29, t. 2 (caption reversed with t. 3)(1959); Aké Assi, Contrib. 1 : 213. *Geophila hirsuta* of Chev. Bot. 345, partly (*Chev.* 15381; 19554), not of Benth. Creeping herb; leaves pale green above, very pale beneath; calyx pale green, corolla and stamens white; in rain forest.

Lib.: Nyaake, Webo Dist. (June) *Baldwin* 6100! Diebla, Webo Dist. (July) *Baldwin* 6291! Duo, Sinoe (Mar.) *Baldwin* 11353*b*! **Iv.C.:** La Mamba *Mangenot & Aké Assi* 4597. Yapo (Sept.) *Mangenot & Aké Assi* 4595! Fort-Binger to Toula *Chev.* 19554! Abidjan, Dabou *Chev.* 15381! Guéde (Dec.) *Leeuwenberg* 2447! **Ghana:** Asamang, Axim Dist. (Apr.) *Chipp* 154! Also in Cameroun (*Letouzey* 3642).

3. **S. capitata** *E. A. Bruce* in Kew Bull. 1935 : 284, mainly (*Bequaert* 1141; *Cook* 139); Heine l.c. 183. *Staurogynopsis maiana* Mangenot & Aké Assi l.c. 33, t. 3 (caption reversed with t. 2). *Staurogynopsis capitata* (E. A. Bruce) Mangenot & Aké Assi, Contrib. 1 : 212(1961). *Geophila hirsuta* of Chev. Bot. 345, partly (*Chev.* 19267; 19500; 19555) not of Benth. Very similar to the last, with which it often grows, but much more slender and delicate in all its parts.

S.L.: Masakio (June) *Marmo* 46! **Lib.:** (June) *Cook* 139! Fayapulu (Oct.) *Bequaert* 1141! **Iv.C.:** Mid. Sassandra to Mid. Cavally (July) *Chev.* 19267! Fort-Binger to Mt. Niénokué (July) *Chev.* 19500! Fort-Binger to Toula (July) *Chev.* 19555! Tiapleu Forest (May) *Mangenot & Aké Assi* 4596! Béyo (Jan.) *Leeuwenberg* 2452!

---

30. **ELYTRARIA** L. C. Rich. in Michx. Fl. Bor. Am. 1 : 8 (1803), *nom. cons.*; F.T.A. 5 : 27; Bremekamp in Reinwardtia 3 : 249 (1955); J. K. Morton in Rev. Biologia 1 : 49 (1956); Johri & Singh in Bot. Notiser 112 : 227 (1959).

Plants simple, erect, with a solitary, usually stalked rosette; leaves broadly oblanceolate to obovate, obtusely triangular-pointed, attenuated to base, sometimes lyrate or obtusely and remotely crenate, very shortly petiolate, up to 15 cm. long and 5 cm. broad, very thinly pilose on the nerves; inflorescence a slender pedunculate spike up to 15 cm. long; bracts broadly ovate, bristly-acuminate, about 5 mm. long, green with paler shortly ciliate margins; fruit 4–5 mm. long, glabrous   ..   1. *marginata*

Plants diffusely branched, prostrate, with many small rosettes of entire leaves rarely exceeding 7 cm.; all parts smaller than the preceding, but otherwise similar
         2. *maritima*

1. **E. marginata** *Vahl* Enum. Pl. 1 : 108 (1804); Milne-Redhead in Exell Suppl. Cat. S. Tomé 37 (1956); Morton l.c. 51. *E. crenata* of F.T.A. 5 : 27 (1899), excl. syn. F.T.A. 5 : 509 (1900), Cat. Talb. 139, not of Vahl (1805). *E. acaulis* of F.W.T.A., ed. 1, 2 : 261 (1931), not of (Linn. f.) Lindau (1897). *E. squamosa* Lindau in Anal. fis.-geogr. Costa Rica 8 : 299 (1895), partly (Afr. specimens), and in Schlechter Westafr. Kautschuk-Exp. 314; Wiss. Ergebn. 1907–08 : 291; Chev. Bot. 491, partly (excl. *Chev.* 16074), not *Verbena squamosa* Jacq. (1797). *E. lyrata* Vahl l.c. 106 (1804); Morton l.c. 54, fig. 3. *E. acaulis* var. *lyrata* (Vahl) Bremek. l.c. 251 (1955). Stem up to 9 in. high, with small white or bluish flowers ¼–⅓ in. long in narrow stiff spikes with sharp-pointed bracts.

Guin.: Badabou valley (fr. Oct.) *Schnell* 7600! **S.L.:** Newton (fl. & fr. Nov.) *Deighton* 1485! York Pass, 1,300 ft. (fr. Nov.) *T. S. Jones* 183! Kamalu (fr. May) *Thomas* 394! Gegbwema (fr. Nov.) *Deighton* 402! Lowoma, Nomo (fl. & fr. Oct.) *Fisher* 47! **Lib.:** Grand Bassa *T. Vogel* 93! Sinkor (fl. & fr. July) *Barker* 1384! Soplima (Nov.) *Baldwin* 10023! 10025! Bobei Mt. (Sept.) *Baldwin* 12503! Ganta (June) *Baldwin* 6102! **Iv.C.:** Divo (fr. May) *Aké Assi* 4363! Tiapleu (fr. Sept.) *Aké Assi* 3276! Amitioro (fr. Dec.) *Aké Assi* 4486! Guédeyo, Soubré (fl. & fr. Dec.) *Leeuwenberg* 2163! **Ghana:** Cape Coast *Brass*! *Farmar* 519*a*! Obuasi (fl. & fr. June) *Andoh* 4216! Amentia *Irvine* 444! Volta R., near Jaketi (fl. & fr. Jan.) *Morton* GC 8021! Afram Mankrong F.R. (fl. & fr. Dec.) *Adams* 4936! **N.Nig.:** Lokoja (fl. & fr. Oct.) *Dalz.* 198! Sanga River F.R. (fr. Nov.) *Keay* FHI 22265! **S.Nig.:** Lagos *Millen* 24! *Dalz.* 1140! Aguku Dist. *Thomas* 723! Onitsha *Barter* 1302! Degema (fr. Sept.) *Holland* 138! Oban *Talbot* 386! [**Br.]Cam.:** Man o'war Bay *Schlechter* 12386! Johann-Albrechtshöhe *Staudt* 453! Bakosi, Kumba (fl. & fr. May) *Olorunfemi* FHI 30572! Aba-ajia, Bamenda (fr. July) *Ujor* FHI 30457! **F.Po:** *T. Vogel* 75! (fr. Dec.) *Mann* 26! *Guinea* 677! Also in Cameroun, Gabon, Congo and (partly) S. E. Africa. (See Appendix, p. 450.)
[A polymorphic species which is accepted here in a broad sense. The reviser feels quite unable to distinguish specimens with lyrate leaves in specific rank.—H.H.]

2. **E. maritima** *J. K. Morton* l.c. 54, fig. 1 (1956). *E. squamosa* of Chev. Bot. 491, partly (*Chev.* 16074), not of (Jacq.) Lindau. A prostrate perennial herb; flowers pale mauve; "only known from under coconut palms near the coast when this plant forms large patches in the almost bare sand " (Morton l.c. 57). **Iv.C.:** Abidjan *Boughey* GC 14539! *Leeuwenberg* 1789! Anguédédou (fl. & fr. Oct.) *Mangenot & Aké Assi* 3792! Bingerville (fr. Dec.) *Chev.* 16074! **Ghana:** Princesstown *Morton* A326! GC 8451. Ancobra R. *Cudjoe* 123. Axim (fr. Feb.) *Morton* A1631!
[A local race of *E. marginata*, described in specific rank.—H.H.]

---

31. **NELSONIA** R. Br.—F.T.A. 5 : 28; Bremekamp in Reinwardtia 3 : 247 (1955).

A small herb with decumbent villous branches; lower leaves long-petiolate, upper becoming subsessile, elliptic to ovate, entire, up to 4 cm. long, pubescent or often woolly; flowers in cylindric villous spikes up to 10 cm. long; bracts imbricate, broadly ovate, about 6 mm. long; calyx-lobes 4, lanceolate, unequal, the largest often shortly bilobed; fruits a little longer than the bracts   ..   *canescens*

N. canescens (*Lam.*) *Spreng.* Syst. Veg. 1 : 42 (1825); F.W.T.A., ed. 1, 2 : 608; Milne-Redhead in Mem. N.Y. Bot. Gard. 9, 1 : 19(1954); Bremekamp l.c. 248. *Justicia canescens* Lam. Tab. Encycl. Méth. Bot. 1 : 41 (1791). *Nelsonia campestris* R. Br. (1810)—F.T.A. 5 : 28; Chev. Bot. 491; F.W.T.A., ed. 1, 2 : 261. Softly pubescent prostrate branching herb with erect spikes of small pink or purple flowers.

Sen.: *Heudelot* 48! *Boivin*! Gorée *Ealmy*! *Farmar* 60! **Gam.:** *Brown-Lester*! *Saunders* 8! **Mali:** Fetodie (fr. Apr.) *Davey* 565! Tilembeya (Mar.) *Davey* 97! **Guin.** (?): Kano *Pobéguin* 627! **S.L.:** Kamalu (fr. May) *Thomas* 397! Freetown *Sc. Elliot* 5905! *Deighton* 239! Magbile (Dec.) *Thomas* 6037! Daru (fr. Mar.) *Deighton* 4100! **Lib.:** *Dinklage* 3356! Beiden *Baldwin* 10269! Monrovia (fl. & fr. Sept.) *Wrigley* 714! *Iv.C.:* Nambonkaha (fr. Nov.) *Leeuwenberg* 2059! Davo (Aug.) *de Wilde* 374! Sassandra (Jan.) *Leeuwenberg* 2554! **Ghana:** R. Kakum (fr. Mar.) *Irvine* 1594! Krobo (fl. & fr. May) *Morton*! Dawa (fl. & fr. Dec.) *Thomas* D32! Assin-Yan-Kumasi *Cummins* 191! Atwabo *Fishlock* 50/1931! Navrongo (fr. Feb.) *Vigne* FH 4479! **N.Nig.:** Nupe *Barter* 1159! Kontagora *Dalz.* 155! Sokoto (fr. Nov.) *Lely* 141! Vom *Dent Young* 202! Naraguta (Nov.) *Lely* 707! **S.Nig.:** Ibadan (fr. Jan.) *Meikle* 956! Gambari F.R. (fl. & fr. Mar.) *Hepper* 2283! Utanga, Obudu Div. (fr. Dec.) *Savory & Keay* FHI 25164! Oban *Talbot*!

Calabar (fl. & fr. Feb.) *Onochie* FHI 25149! 36341! [**Br.**]**Cam.**: *Preuss*! Buea, 3,000 ft. (fr. Dec.) *Maitland* 119! Mbalange, Kumba (Jan.) *Binuyo & Daramola* FHI 35495! Bamenda, 4,000 ft. (fr. June) *Maitland* 1699! Near Vogel Peak, Adamawa (Nov.) *Hepper* 1433! **F.Po**: *Barter* 2046! *Mann* 253! 1869! Widespread in the tropics. (See Appendix, p. 452.)

32. **BARLERIA** Linn.—F.T.A. 5 : 140.

Plants, especially young leaves and inflorescences, covered with both simple and stellate hairs ; stellate hairs mostly composed of one very long hair with a dense tuft of much smaller hairs at its base :

Flowers in a terminal spike-like inflorescence up to 15 cm. long ; leaves linear-lanceolate, acute, up to 17 cm. long and 23 mm. broad, with about 5 pairs of lateral nerves which are, like the midrib, covered with stellate hairs beneath ; outer sepals lanceolate, up to 15 mm. long and 4·5 mm. broad, adaxial ones usually slightly emarginate, like all other parts of the inflorescence hairy-glandulose, of the same texture as the leaves (i.e. not becoming dry and papery in flowering and fruiting state)   ..   ..   1. *maclaudii*

Flowers axillary and terminal ; leaves elliptic-lanceolate, acuminate, cuneate at base, up to 10 cm. long and 3·5 cm. broad (but usually only about 5 cm. long and 2 cm. broad) ; stellate hairs particularly conspicuous on the younger parts of the inflorescences ; outer calyx-segments subequal, broadly ovate, up to 1·5 cm. long and 1 cm. broad, sharply acuminate, becoming dry and papery in flowering and fruiting state, finely and irregularly spinulose-dentate, covered along the margins and nerves on the outer surface with stellate hairs but glabrous along the nerves within, and on the inner surface only covered with sparsely distributed minute simple hairs between the veins ..   ..   2. *asterotricha*

Plants only covered with simple hairs :

Branches without interpetiolar spines :

Bracts with spiny teeth on the margin, strongly venose, broadly ovate, about 2·5 cm. long and 2 cm. broad ; corolla 5 cm. long ; leaves narrowly obovate, acutely acuminate, about 15 cm. long and 5 cm. broad, with rod-like cystoliths on the upper surface, appressed-pilose on the nerves beneath ; corolla yellow

       3. *oenotheroides*

Bracts entire ; corolla white, blue, purplish, or (particularly on drying) greyish-violet :

Leaves glabrous beneath, ovate-elliptic, shortly and acutely acuminate, rounded or subcordate at base, 6–9 cm. long, 3–4·5 cm. broad ; petioles (of the lower leaves) 1·5–2·5 cm. long, upper leaves subsessile ; inflorescence substrobilate ; adaxial calyx-segment broadly obovate-orbicular, shortly ciliate, up to 2 cm. long and 15 mm. broad ; corolla 5 cm. long, pubescent ; stamens long-exserted

       4. *brownii*

Leaves pilose, pubescent or villose beneath ; leaf-base always cuneate :

Flowers very few not forming a strobilus ; corolla 3–4 cm. long :

Stems covered with minute grey appressed indumentum ; leaves narrow-lanceolate, 1–3·5 cm. long, 0·5–1 cm. broad, finely grey-pubescent in young leaves (older leaves may be glabrescent and larger) ; the 2 outer sepals ovate about 1·5 cm. long and 7 mm. broad ; fruits 1-seeded, beaked   5. *bonifacei*

Stems more or less setose with brown hairs ; leaves elliptic, 3·5–15 cm. long, 2–5 cm. broad, setose mainly on the nerves beneath ; the 2 outer sepals broadly ovate up to 1·8 cm. long and 1·3 cm. broad, setose, becoming slightly transparent or papery and minutely and sparsely dentate in fruit, often purplish-tinged ; fruits 2-seeded, not beaked   ..   ..   6. *ruellioides*

Flowers in substrobilate, inflorescence-like clusters ; the two outer sepals lanceolate to ovate-lanceolate, 13–17 mm. long, not becoming transparent nor papery nor denticulate nor purplish-tinged in fruiting specimens ; corolla 4 cm. long :

Indumentum pilose, not densely villose, hairs spreading ; adaxial sepal narrowly lanceolate, up to 15 mm. long and 4 mm. broad, abaxial sepal of about the same size, at its apex narrowly 2-cleft for about $\frac{1}{4}$ (or sometimes more) of its length ; corolla white, sometimes with a few purple markings ..   ..   7. *opaca*

Indumentum densely villose, sometimes slightly silky-shiny, hairs straight ; adaxial sepal broadly ovate-lanceolate, up to 17 mm. long and 8 mm. broad, abaxial sepal of about the same size (sometimes even larger), at its apex broadly 2-cleft for about $\frac{1}{3}$ of its length ; corolla blue-purple ..   ..   8. *villosa*

Branches with interpetiolar spines :

Spines spinulose-dentate, slightly winged (but often nearly simple and with no, or very small, teeth) ; plants softly pubescent ; leaves oblanceolate, up to 4 cm. long and 1·5 cm. broad, loosely pilose, drying blackish, with a sharp tip ; flowers solitary in the upper axils, blue ; corolla 2·5 cm. long, glabrous outside ; fruit oblanceolate, 2 cm. long   ..   ..   9. *bornuensis*

Spines rigid, acicular, simple ; plants nearly glabrous ; leaves more or less elliptic, narrowed at both ends, usually slightly larger than in No. 8, glabrous, drying light greenish or greenish-grey, with a sharp tip ; flowers solitary and axillary, yellow-orange, more or less crowded at the tips of branches and forming an inflorescence, the upper leaves and bracts often becoming spine-like ; corolla up to 3 cm. long, glabrous or with very few hairs outside ; fruit oblanceolate, 1·5 cm. long

10. *eranthemoides*

1. **B. maclaudii** *Benoist* in Notulae Syst. 2 : 156 (1911). A stout erect annual (?) herb, unbranched, up to 5 ft. high with very long willow-like leaves ; flowers pale mauve, about 1 in. long, funnel-shaped, about 1½ in. across.
  Guin.: Saman valley (Oct.) *Maclaud* 209 ! Fouta Djalon *Unknown collector* (*Hb.* Paris) ! Dinguiraye (fl. & fr. Dec.) *Jac.-Fél.* 1430 !

2. **B. asterotricha** *Benoist* l.c. 155 (1911). Subshrub, up to 3 ft. high, very similar to *B. ruellioides* (No. 6), but easily distinguishable by its indumentum ; flowers handsome violet with a narrow corolla-tube up to ¾ in. long, corolla if fully opened, saucer-shaped, ¾ in. across ; preferably in shady places (Jac.-Fél.).
  Guin.: Dinguiraye (fl. & fr. Dec.) *Maclaud* 24 ! *Jac.-Fél.* 1484 !

FIG. 304.—BARLERIA OENOTHEROIDES *Dum. Cours.* (ACANTHACEAE).

A, calyx and style.   B, corolla laid open.   C, pistil.   C', longitudinal section of ovary.   D, cross-section of fruit.   E, seed.

3. **B. oenotheroides** *Dum. Cours.* Botan. Cultiv., ed. 2, 2 : 561 (1811). *B. flava* Jacq. f. Eclog. Pl. rar. 1 : 67, t. 46 (1813), excl. syn. ; l.c. 153 ; Bot. Mag. 70 : t. 4113 (1844), excl. syn. *B. mitis* Ker ; F.T.A. 5 : 155 ; Chev. Bot. 497 ; F.W.T.A., ed. 1, 2 : 262 ; Benoist in Bol. Soc. Brot., sér. 2, 24 : 39. *Eranthemum flavum* Willd. (1813), name only. *Barleria senegalensis* Nees (1847). A woody undershrub erect at first then often straggling, the stems covered with stiff yellowish hairs ; the flowers nearly 2 in. long, yellow turning mauve in drying, in dense bracteate spikes ; sometimes cultivated as an ornamental.
  Mali: Moussaia *Chev.* 397. Guienguénia *Chev.* 437. **Port.G. :** Tébé *Esp. Santo* 566. Bissalanca, Bissau (Jan.) *Esp. Santo* 2240 ! **Guin. :** *Heudelot* 644 ! *Farmar* 260 ! Kindia *Chev.* 13011. Soya to Koulounadala *Chev.* 20378. Santa to Timbo *Chev.* 12612 ; 12812*bis*. **S.L.:** Sugar Loaf Mt., 2,000 ft. (Feb.) *Tindall* 35 ! Mofadu (Jan.) *Sc. Elliot* 4423 ! Loma Mts. (Jan.) *Jaeger* 3914 ! *Glanville* 391 ! *Thomas* 8254 ! Kabala (Dec.) *Deighton* 2936 ! **Iv.C.:** Bouroukrou *Chev.* 16779 ; 16859. Sassandra (Jan.) *Leeuwenberg* 2534 ! 2609 ! **Ghana:** Banka *Irvine* 474 ! Ofin Headwaters F.R. (Jan.) *Andoh* 4288 ! Abranja-Oda (Jan.) *Darko* 598 ! Goaso (Feb.) *Andoh* 5108 ! Kibbi Hills (Jan–Feb.) *Morton* GC 8394 ! 25329 ! **S.Nig.:** Omo (formerly part Shasha) F.R. (Feb.) *Jones & Onochie* FHI 14726 ! 16741 !

4. **B. brownii** *S. Moore* in J. Bot. 46 : 73 (1908). *B. talbotii* S. Moore in Cat. Talb. 86 (1913). A straggling shrub with smooth branches and blue flowers about 2 in. long.
  Ghana: Sunyani to Chira, N. Ashanti (Dec.) *Morton* A2826 ! Dormaa to Berekum (Dec.) *Morton* A2647 ! **S.Nig.:** Owo (Jan.) *Burton* FHI 30798 ! Akure F.R., Ondo (Dec.) *Keay & Trochain* FHI 25603 ! Onipanu, Ondo (Dec.) *Onochie* FHI 5235 ! Oban *Talbot* 1396 ! Also in Ubangi-Shari, Gabon, Congo, Uganda and Angola.

5. **B. bonifacei** *Benoist* in Bull. Soc. Bot. Fr. 99 : 325 (1953). *B. hereroensis* Engl. var. *charlesii* Benoist in Mem. Soc. Bot. Fr. 8 : 279 (1917) ; Chev. Bot. 497. Herb 1–2 ft. high ; flowers bluish or mauve ; in dry savanna.
  Maur.: Oumoulsouitigat (Oct.) *Charles* in *Hb. Chev.* 28813 ! Assaba (fl. & fr. Jan.) *Boniface* 107 !
  [This is very near the aggregate species *B. lancifolia* T. Anders. (in J. Linn. Soc. 7 : 28 (1863) ; Obermeijer in Ann. Transvaal Mus. 16 : 147 (1933), for synonymy and taxonomic discussion. *B. alata* S. Moore (1880)—F.T.A. 5 : 158), which is widespread in S. Africa and Angola ; but the incomplete knowledge of the taxonomy of the representatives of this group outside the area covered by Miss Obermeijer's revision does not permit the inclusion of *B. bonifacei* in its synonymy—H.H.]

6. **B. ruellioides** *T. Anders.* in J. Linn. Soc. 7 : 30 (1863) ; F.T.A. 5 : 165. Woody herbaceous straggling plant with pale blue sub-campanulate flowers 1–1½ in. long ; in riverine forest in savanna country.
  Mali: Bamako (Sept.) *Jaeger* 3/1943 ! Siriguibougou (Nov.) *Raynal* 5193 ! Sikasso (fl. & fr. Dec., Apr.) *Vuillet* 504 ! *Chev.* 761 ! **Guin.:** Kouroussa (Dec.) *Pobéguin* 606 ! Siguiri (Jan.) *Jac.-Fél.* 1524 ! **Iv.C.:** Bobo (Dec.) *Aubrév.* 2225 ! **Ghana:** Banda Ravine (fr. Dec.) *Morton* GC 25157 ! Dutukpene (Dec.) *Adams* 4530 ! **N.Nig.:** Nupe *Barter* 932 ! Mokwa *Meikle* 1028 ! Patti Lokoja (Nov.) *Dalz.* 141 ! Bebi, by R. Ata (fr. Dec.) *Savory & Keay* FHI 25009 ! Anara F.R., Zaria (Nov.) *Keay* FHI 21691 ! [**Br.**] **Cam.:** Labati Serti, Adamawa (Dec.) *Latilo & Daramola* FHI 28949 ! Vogel Peak, 3,850 ft., Adamawa (Nov.) *Hepper* 1331 ! Also in Ubangi-Shari.

7. **B. opaca** (*Vahl*) *Nees* in DC. Prod. 11 : 230 (1847) ; F.T.A. 5 : 163 ; Chev. Bot. 497. *Justicia opaca* Vahl Enum. 1 : 133 (1805). A straggling undershrub ; flowers white, 1½ in. long.
Iv.C.: Bouroukrou *Chev.* 16779 ; 16859. Anoumaba (Nov.) *Chev.* 22401 ! **Ghana:** Accra *Don* ! Cape Coast *Brass* ! Akroful *Cummins* 30 ! Alafango *Krause* ! Krobo (Nov.) *Irvine* 1701 ! Aboma F.R. (Nov.) *Vigne* FH 3446 ! Aburi Scarp (Nov.) *Morton* GC 6133 ! **Togo Rep.:** Lomé *Warnecke* 20 ! **Dah.:** Sakété (Jan.) *Chev.* 22835. **S.Nig.:** Lagos (Feb., Mar.) *Millen* 46 ! 179 ! *Rowland* ! Gambari F.R. (Nov., Dec.) *Onochie* FHI 34948 ! 35346 ! Ibadan (Jan.) *Newberry & Etim* 181 ! *Meikle* 874 ! Olokemeji to Alabata (fl. & fr. Nov.) *Onochie* FHI 8135 ! Utanga, Obudu Div. (Dec.) *Savory & Keay* FHI 25146 ! **[Br.]Cam.:** Buea, 3,000 ft. (Dec., Jan.) *Maitland* 139 ! 142 ! 272 ! *Dunlap* 161 ! *Preuss* 714. Maisamari, 4,600 ft., Mambila Plateau (Jan.) *Hepper* 1679 ! Also in Cameroun. (See Appendix, p. 450.)

8. **B. villosa** *S. Moore* in J. Bot. 18 : 267 (1880) ; F.T.A. 5 : 164. A woody climbing shrub with villous stems and foliage, and blue-purple flowers 1½ in. long.
N.Nig.: Jos Plateau (fl. & fr. Jan.) *Lely* P114 ! *Batten-Poole* 206 ! Vom (Jan.) *Dent Young* 193 ! Ropp (Dec.) *W. D. MacGregor* 406 ! Pankshin, 4,000 ft. (Dec.) *Coombe* 61 ! **[Br.]Cam.:** Vogel Peak, 4,900 ft., Adamawa (Dec.) *Hepper* 1578 ! Also in Cameroun and Angola.

[*B. opaca* and *B. villosa* are very similar and, particularly in herbarium specimens, sometimes difficult to distinguish (see note in F.T.A. l.c.). The differentiation based on the size and shape of the adaxial calyx-segments as indicated by Clarke in F.T.A. 5 : 144, key ; 163, 164, is very unreliable because these organs are rapidly accrescent after the corollas have fully opened ; the measurements—taken from a full series of dissections of herbarium material of our region—are therefore very different from those given in F.T.A. The isotype-specimens of *B. villosa* (*Welw.* 5070, 5071) which were apparently used by Clarke for his description of *B. villosa* in F.T.A. are very young and have therefore much smaller calyx-segments than usual ; in fact, *B. villosa* has, in all fully flowering specimens seen from W. Africa for this revision, larger outer calyx-segments than *B. opaca*—H.H.]

9. **B. bornuensis** *S. Moore* in Macleod, Chiefs & Cities Centr. Afr. 304 (1912) ; Cat. Talb. 85. Low, branched, woody undershrub of savanna, with spines ½–¾ in. long, and blue flowers 1 in. long.
N.Nig.: Kontagora & Sokoto Prov. (fl. & fr. Dec.) *Dalz.* 359 ! Keana (Aug.) *Hepburn* 53 ! Dutsin Makurdi (fr. June) *Elderry* FHI 16352 ! 16372 ! L. Chad Dist. *Talbot* 1009 ! Also in Cameroun.

10. **B. eranthemoides** *R. Br. ex C. B. Cl.* in F.T.A. 5 : 147 (1899). *B. eranthemoides* R. Br. in Salt, Voy. Abyss. Append. IV : LXIV (1813), name only. Low prickly semi-shrub up to 75 cm. high with spines up to 1 in. long, leaves bright green, smooth ; flowers pale orange, corolla-tube about 1 in. long.
N.Nig.: Wusasa, Zaria (Aug.) *Keay* FHI 28030 ! **[Br.]Cam.:** Gwoza, 3,500 ft., Dikwa Div. (Jan.) *McClintock* 149 ! Also in Sudan, Ethiopia, Somaliland, E. Africa and Mozambique.

B. cristata Linn. and *B. prionitis* Linn., both natives of tropical Asia, are cultivated as ornamentals.

Additional species.

B. schmittii *Benoist* in Notulae Syst. 3 : 218 (1916) ; Chev. Bot. 497. A prickly undershrub up to 1 ft. high ; flowers white.
Maur.: Bir Moghreïn *Schmitt* in *Hb. Chev.* 28571 ! Chareb El Ba'ïr (fl. & fr. Feb.) *Chudeau* ! Recorded apparently just outside our area and included here to draw attention to the fact that it may occur within it. The two gatherings cited (the only specimens hitherto known) belong to a complex group which is mainly represented in S. Africa and for which the following name seems today the most reliable : *B. rigida* Nees in DC. Prod. 11 : 242 (1847) ; P. G. Meyer in Mitt. Bot. Staatssammlg. München 2 : 384 (1957), for synonymy and taxonomic discussion. *B. irritans* var. *ilicina* T. Anders. in J. Linn. Soc. 7 : 28 (1863). The taxonomy of these plants is still in need of further study.

33. PSEUDERANTHEMUM Radlk. in Sitzb. math. phys. Cl. k. bayer. Akad. Wiss. 13 : 282 (1883) ; Milne-Redhead in Kew Bull. 1936 : 255. *Eranthemum* of F.T.A. 5 : 169, partly, and of F.W.T.A., ed. 1, 2 : 262, partly, not of Linn.

Capsule normally 2-seeded, slightly pubescent outside ; flowers brick to red-scarlet, with a darker centre ; inflorescence finely glandular ; seeds finely muricate, especially on the inner face ..   .. 1. *dispermum*
Capsule normally 4-seeded :
Corolla distinctly bilabiate, stamens exserted ; corolla-tube scarcely inflated above ; capsule often more than 3 cm. long ; nodes of the inflorescence many-flowered ; shrubs up to 4 m. high ..   .. 2. *ludovicianum*
Corolla not distinctly bilabiate, stamens included ; inflorescence rather slender with few flowers at each node ; corolla somewhat inflated above ; capsule less than 3 cm. long ; herbs, sometimes woody at the base, up to 20 cm. high .. 3. *tunicatum*

1. **P. dispermum** *Milne-Redhead* in Kew Bull. 1936 : 267, fig. 6, 4. *Eranthemum hypocrateriforme* of Cat. Talb. 140 (1913) ; F.W.T.A., ed. 1, 2 : 262 ; not of Roem. & Schult. Erect perennial herb or shrubby, up to 2½ ft. high ; corolla brick-red outside, bright coral-red inside, about 1½ in. long.
S.Nig.: Oban *Talbot* 1552 ! Okuni, Cross R. (Jan.) *Holland* 172 ! Agoi Efut-Ukwop Eyere *Rosevear* 30/31 ! **[Br.]Cam.:** Meandja *Schlechter* 12681 ! Babenga to Owe, Victoria (fl. & fr. Nov.) *Keay* FHI 37544 ! Also in Cameroun. (See Appendix, p. 453.)

2. **P. ludovicianum** (*Büttner*) *Lindau* in E. & P. Pflanzenfam. 4, 3b : 310 (1895) and in Schlechter, Westafr. Kautschuk.-Exped. 316 ; F.W.T.A., ed. 1, 2 : 608 ; Milne-Redhead l.c. 263, fig. 6, 1. *Eranthemum ludovicianum* Büttner (1890)—F.T.A. 5 : 172 ; F.W.T.A., ed. 1, 2 : 262. A forest undershrub up to 12 ft. high, with narrow tubular flowers white or purplish-tinged, 1½ in. or more long forming a spike-like panicle 4–10 in. long.
Lib.: (fl. & fr. Jan.) *Harley* 1063 ! **Ghana:** Pusa Pusa Ravine, Kibbi Hills (Jan.) *Morton* GC 6383 ! Kibbi, 2,370 ft. (Dec.) *Morton* GC 8129 ! Mampong Scarp (fl. & fr. Jan.) *Morton* A1537 ! Kwahu Tafo to Mankrong (fr. Apr.) *Morton* A673 ! **S.Nig.:** Oban *Talbot* 379 ! 1390 ! 1398 ! 1423 ! 1437 ! Ikwette to Balegete, Ogoja (Dec.) *Savory & Keay* FHI 25199 ! R. Ata, below Koloishe, 4,000 ft. (Dec.) *Savory & Keay* FHI 25046 ! Boje to Afi River F.R. (Dec.) *Keay* FHI 28244 ! **[Br.]Cam.:** Victoria *Preuss* 598 ! Buea 2,600–3,500 ft. (Jan., Feb.) *Schlechter* 12853 ! *Dalz.* 8210 ! *Maitland* 196 ! Ngusi to Mafura (Jan.) *Schlechter* 12905 ! Johann-Albrechtshöhe (= Kumba) *Staudt* 537 ! Barombi (Feb.) *Binuyo & Daramola* FHI 35551 ! **F.Po:** *Mann* 1437 ! Also in Cameroun.

3. **P. tunicatum** (*Afzel.*) *Milne-Redhead* l.c. 264, fig. 6, 2 (1936). *Justicia tunicata* Afzel. (1814). *Eranthemum nigritianum* T. Anders. (1863)—F.T.A. 5 : 171 ; F.W.T.A., ed. 1, 2 : 262. *Pseuderanthemum nigritanum* (T. Anders.) Radlk. (1883—Chev. Bot. 500 ; F.W.T.A., ed. 1, 2 : 608. *Eranthemum lindaui* C. B. Cl. in F.T.A. 5 : 173 (1899). A woody herb or sometimes of the forest, 1–2 ft. high with pale bluish or lilac or nearly white flowers, spotted on the lip, 1–1½ in. long including the corolla-lobes and slender tube.
S.L.: *Afzelius* ! Freetown (Jan.) *Dalz.* 948 ! Kukuna (Jan.) *Sc. Elliot* 4692 ! Njala (Feb.) *Deighton* 1079 ! 1756 ! Port Loko *Glanville* 260 ! Loma Mts., 2,600–3,000 ft. (Jan.) *Jaeger* 3916 ! 4085 ! **Lib.:** Nana Kru (Mar.) *Baldwin* 11588 ! **Iv.C.:** Makougnié *Chev.* 16992 ! 17188 ! Kassigué *Chev.* 17048 ! Zaranou *Chev.*

17632! Nékaougnié to Grabo *Chev.* 19603! Mt. Copé *Chev.* 19698! Sassandra (Jan.) *Leeuwenberg* 2608! **Ghana:** Banka *Irvine* 492! Assin Yan to Kumasi *Cummins* 237! Abetifi, Kwahu (Jan.) *Irvine* 1665! Sunyani (Jan.) *Chipp* 74! N'Kwanta (Jan.) *Chipp* 64! Koforidua Hills (fl. & fr. Dec.) *Johnston* 440! **Togo Rep.:** *Baumann* 375! **Dah.:** near Sakété (Jan.) *Chev.* 22864! 22899! **S.Nig.:** Kradu Lagoon, Lagos *Barter* 3300! Okomu F.R., Benin (fl. & fr. Dec.-Feb.) *Brenan* 8618! 8833!a! 9003a1b! Usonigbe F.R. (fl. & fr. Feb.) *Onochie* FHI 27699! Omo F.R. (fr. Apr.) *Onochie* FHI 15516! Eket *Talbot*! Oban *Talbot* 1356; 1392! [**Br.**]**Cam.:** Victoria *Preuss* 1108! 1366! Johann-Albrechtshöhe (= Kumba) *Staudt* 561! Mombo, S. Bakosi *Olorunfemi* FHI 30574! Lus, Nkambe Div. (fl. & fr. Feb.) *Hepper* 1880! **F.Po:** *Mann* 156! Extends to Angola. (See Appendix, p. 453.)

34. ADHATODA Medik.—F.T.A. 5 : 221.[1]

Plants persistently brownish-tomentose and hairy, particularly stem, petioles, inflorescences and leaves beneath ; leaves broadly elliptic-lanceolate, up to 35 cm. long and 11 cm. broad, with up to 17 pairs of lateral nerves ; inflorescences axillary and subterminal, very short (up to 3·5 cm. long) and compact, strobiliform in appearance, densely hairy ; corolla about 2 cm. long   ..   .. 1. *camerunensis*
Plants not persistently hairy or tomentose (sometimes the younger parts pubescent or even brownish-tomentose, but always becoming more or less glabrous) ; inflorescences terminal (but sometimes with some additional axillary branches in the upper leaf-axils), elongate, never strobiliform in appearance, paniculate or spiciform ; always (in normal specimens) much longer :
    Lateral nerves about 15 pairs, puberulous beneath ; leaves broadly elliptic, acuminate, shortly decurrent on the petiole, 17–25 cm. long, 8–10 cm. broad ; petiole up to 4 cm. long ; . panicle large, with stiff ascending branches ; bracts leafy ; bracteoles obovate-oblanceolate, pubescent ; calyx-segments oblong-lanceolate, 1 cm. long, shortly pubescent ; ovary pubescent ; corolla 2 cm. long ..   .. 2. *robusta*
    Lateral nerves up to 10 pairs :
        Inflorescences lax, almost branched, paniculate or compound-racemose :
        Bracts narrowly lanceolate, about 5 mm. long ; leaves broadly elliptic or obovate-elliptic, acuminate, up to 20 cm. long and 10 cm. broad, glabrous, lateral nerves in 6–8 pairs ; calyx 7 mm. long, lobes about 5 mm. long and 1·5 mm. broad at base ..   .. 4. *maculata*
        Bracts ovate to orbicular, sometimes leafy :
        Partial inflorescence elongate, sometimes with zigzag branches, flowers in rather remote clusters with large orbicular leafy bracts of 15 cm. diam. ; leaves elliptic, rather abruptly acuminate, about 15 cm. long and 8 cm. broad, glabrous ; lateral nerves about 6 pairs ; calyx-segments lanceolate, about 8 mm. long
                                                           3. *buchholzii*
        Partial inflorescences straight, often branched, flowers more or less racemosely arranged at the terminal branchlets in the axils of ovate, shortly apiculate bracts 9–14 mm. long and 6–8 mm. broad ; leaves narrowly obovate-lanceolate, up to 30 cm. long and 8 cm. broad, glabrous, lateral nerves about 8 pairs ; calyx-segments lanceolate, about 7 mm. long and 3 mm. broad at base   5. *guineensis*
    Inflorescence a dense spike :
        Bracts linear-lanceolate ; spikes slender, about 15 cm. long ; leaves obovate, auriculate at the broadly cuneate base, 14–18 cm. long, 5–8 cm. broad, minutely pubescent beneath, with 9–10 pairs of lateral nerves ; corolla 1 cm. long 6. *tristis*
        Bracts suborbicular, overlapping, strongly nervose, about 1·3 cm. long, softly puberulous ; spikes strobiliform, elongated, up to 25 cm. long, pedunculate ; leaves elliptic-obovate, shortly acuminate, broadly cuneate at base, up to 25 cm. long and 12 cm. broad, shortly pubescent on nerves and veins beneath
                                                          7. *orbicularis*

1. **A. camerunensis** *Heine* in Kew Bull. 16 : 165, fig. 2 (1962). *Duvernoia robusta* Lindau in Schlechter, Westafr-Kautschuk-Exp. 317 (1900), as to *Schlechter* 12928, not *Adhatoda robusta* C. B. Cl. Erect, little-branched shrub up to 6 ft. high with brown-pilose stem and petioles ; leaves glabrescent above, softly hairy beneath ; corolla pure white or white with crimson markings ; in lowland rain forest.
    [**Br.**]**Cam.:** Bova to Kuke Bova, Kumba (fl. & fr. Jan.) *Keay* FHI 37391! Also in Cameroun.
2. **A. robusta** *C. B. Cl.* in F.T.A. 5 : 223 (June 1900) ; Aké Assi Contrib. 1 : 214. *Duvernoia robusta* (C. B. Cl.) Lindau in Schlechter, Westafr. Kautschuk-Exp. 317 (not before Dec. 1900) excl. *Schlechter* 12928. A shrub or small tree up to 15 ft. high, brownish-velvety tomentose on the young parts, with large shining leaves and white flowers ¾ in. long in a large erect panicle ; in rain forest.
    Iv.C.: 30 km. E. of Gagnoa (fr. Dec.) *Leeuwenberg* 2154! Oumé (Nov.) *Aké Assi* 5505! **Ghana:** Begoro, Akim (Jan.) *Irvine* 1158! *Darko* 956! Akwapim Hills (Oct.) *Johnson* 793! Akparafe (Nov.) *Morton* A4041! **S.Nig.:** Inguobazuwa, Benin (Nov.) *Onochie, Latilo & Binuyo* FHI 40300! Degema *Talbot* 3646! [**Br.**]**Cam.:** Buea, 3,500 ft. (Dec.) *Migeod* 248! N. slopes of Cam. Mt. 4,700 ft. (Feb.) *Keay* FHI 37498! Bonakanda, 2,200 ft. (fr. Dec.) *Maitland* 1211! **F.Po:** 3,000 ft. (Dec.) *Mann* 634! Also in Cameroun.
3. **A. buchholzii** (*Lindau*) *S. Moore* in Cat. Talb. 74, 140 (1913). *Duvernoia buchholzii* Lindau in Engl. Bot. Jahrb. 20 : 43 (1894). *A. maculata* of F.T.A. 5 : 223, partly (syn. *D. buchholzii*), not of (T. Anders.) C. B. Cl. A climbing shrub finely rusty-pubescent on the young parts, becoming glabrous, with fairly large leaves and white flowers about ¾ in. long in interrupted spikes forming a panicle ; fruits 1 in. long.
    S.Nig.: Oban *Talbot* 372! 1025! 2076! Uwet-Odot F.R., Calabar (Jan.) *Jones* FHI 6859! Also in Cameroun.
    (See Appendix, p. 449.)
4. **A. maculata** *C. B. Cl.* in F.T.A. 5 : 223 (1900), partly ; Aké Assi, Contrib. 1: 213, excl. syn. *Duvernoia buchholzii* Lindau. *Justicia maculata* T. Anders. in J. Linn. Soc. 7 : 38 (1863), not of Lodd. (1822), nor of Perrottet (1825). *Asystasia trichotogyne* Lindau in Engl. Bot. Jahrb. 33 : 190 (1902). A climbing shrub like the last, with interrupted simple spikes of flowers 1 in. long, white with purple spots.

[1] See footnote under *Justicia* on p. 426.

Iv.C.: Sakiré (Jan.) *Mangenot & Aké Assi* 2790 ! **Ghana:** Busua Bay *Morton* A1805 ! **S.Nig.:** Oban *Talbot* 1405 ! **[Br.]Cam.:** S. Bakundu F.R., Kumba (Jan.) *Binuyo & Daramola* FHI 35498 ! **F.Po:** (Jan.) *Mann* 202 ! Also in Cameroun.

5. **A. guineensis** *Heine* in Kew Bull. 16 : 167, fig. 3 (1962). Tall herb or undershrub 3–4 ft. high, with yellow-green flowers with dull purple stripes on the upper lip and purple markings on the lower lip ; in undergrowth of forest.
   Guin.: Nzérékoré (Sept.) *Jac.-Fél.* 1098 ! Nimba Mts. (Oct.) *Schnell* 3821 ! **S.L.:** Joru (fr. Jan.) *Deighton* 3857 ! Giewahun (Dec.) *Deighton* 3833 ! Lalahun, Gaura (Dec.) *King* 84*b* ! Gola Forest (Jan.) *King* 4 ! **Lib.:** Harbel (Dec.) *Bequaert* 2 ! Gbanga (Sept.) *Linder* 441 ! Soplima (Nov.) *Baldwin* 10018*a* ! *Dinklage* 3360 ! **Iv.C.:** Tiapleu (Oct.) *Aké Assi* 5670 !

6. **A. tristis** *Nees* in DC. Prod. 11 : 404 (1847) ; F.T.A. 5 : 223 ; Cat. Talb. 140. *A. auriculata* S. Moore in Cat. Talb. 87, 140 (1913), not of Nees (1847). Herbaceous undershrub 2–4 ft. high, with 4-angled branches brownish-pubescent when young, with purple flowers ¾ in. long in spikes interrupted at the base.
   S.Nig.: Oban *Talbot* 50 ! 996 ! 2011 ! Calabar to Mamfe (Jan.) *Onochie & Okafor* FHI 36052*x* ! **F.Po:** (Jan.) *Mann* 165 ! (Nov.) *T. Vogel* 161 ! S. Isabel to S. Carlos (fr. Dec.) *Guinea* 766 ! Also in Cameroun.

7. **A. orbicularis** *(Lindau)* C. B. Cl. in F.T.A. 5 : 222 (1900) ; Cat. Talb. 140. *Duvernoia orbicularis* Lindau in Engl. Bot. Jahrb. 22 : 123 (1895). A shrub about 12 ft. high, tawny-pubescent on the young parts, with rather large leaves and flowers ½ in. long in long close spikes with strongly-nerved bracts.
   S.Nig.: Oban *Talbot* 975 ! **F.Po:** *Mann* 1435 ! Also in Cameroun.

35. GRAPTOPHYLLUM Nees—F.T.A. 5 : 241.

Leaves oblong-elliptic, usually variegated, sometimes copper-coloured, very shortly and broadly acuminate, acute at base, 9–15 cm. long, 5–7 cm. broad, with 7–8 pairs of lateral nerves, glabrous ; petiole about 5 mm. long ; racemes several-flowered ; pedicels with 3–4 small bracts at the base, about 5 mm. long ; calyx-lobes lanceolate, 3–5 mm. long ; corolla glabrous, 3·5 cm. long, limb very oblique ; stamens 2, exserted, anther-cells equal .. .. 1. *pictum*
Leaves more or less as above but normally green with longer acumen and long petioles ; calyx and corolla minutely glandular .. .. 2. *glandulosum*

1. **G. pictum** *(Linn.)* *Griffith* Notulae 4 : 139 (1854). *Justicia picta* Linn. Sp. Pl. ed. 2, 1 : 21 (1763) ; Bot. Reg. 15 : t. 1227 (1829). *G. hortense* Nees (1832)—C. B. Cl. in Fl. Brit. Ind. 4 : 545 ; F.T.A. 5 : 241, partly (excl. *Staudt* " 4551 "). A glabrous shrub with variegated leaves, often copper-coloured (see note below) and red-purple tubular flowers 1–1½ in. long in narrow panicles.
   Guin.: Boa Entrada (cult., Aug.) *Chev.* 13492 ! **S.L.:** (cult., May) *Barter* ! Kitchom (Dec.) *Sc. Elliot* 4326 ! Freetown (cult., Feb.) *Dalz.* 1015 ! Magbile (Dec.) *Thomas* 6379 ! **Ghana:** Kokorantumi (cult., June) *Deighton* 3391 ! Planted in gardens as an ornamental ; native country uncertain, but certainly Indo-Pacific, sometimes known as " Joseph's Coat " or " Caricature Plant ". Plants with normal green leaves have been described from Java under the name *G. hortense* var. *viride* Hasskarl (in Cat. Hort. Bogor. alt. 150, 1844), but they are apparently not known from Africa ; they may be very near the next species. Plants with copper-coloured leaves are known as **G. pictum** var. **lurido-sanguineum** *(Sims)* Bremek. & Backer in Bekn. Fl. Java 9B : 71 (1949) (syn. *Justicia picta* β *lurido-sanguinea* Sims Bot. Mag. 44 : t. 1870 (1816[7?]) ; Blume Bijdr. 14 : 784 (1826), as var. ; *Graptophyllum hortense* var. *rubrum* Hassk. l.c. (1844). *G. hortense* var. *lurido-sanguineum* Chittenden, R.H.S. Dict. Gard. 2 : 921 (1951) ; and in Bot. Mag. Index 114 (1956)).

2. **G. glandulosum** *Turrill* in Kew Bull. 1912 : 331 ; Cat. Talb. 140. *G. pictum* Lindau (sic.) in Schlechter Westafr. Kautschuk-Exp. 316 (1900), not of (Linn.) Griffith. *G. hortense* of F.T.A. 5 : 241, partly (*Staudt* " 4551 "), not of Nees. A shrub with 4-angled, nearly glabrous branches, normal green leaves, and red-purple flowers 1 in. or more long.
   S.Nig.: Oban *Talbot* 377 ! Oban F.R. (Jan.) *Onochie* FHI 36142 ! Ifunkpa-Atakom, Calabar (Dec.) *Holland* 216 ! Obutong, Calabar (Mar.) *Onochie* FHI 7708 ! **[Br.]Cam.:** Bangwe *Conrau* 52 ! Johann-Albrechtshöhe (=Kumba) (June) *Staudt* 455 ! Nyasoso, 2,600 ft. *Schlechter* 12894 ! Also in Cameroun.
   A relative of these plants, *Odontonema cuspidatum* (Nees) O. Ktze. (*Thyrsacanthus cuspidatus* Nees) from Mexico, is cultivated as an ornamental. It is sometimes wrongly named "*Jacobinia coccinea*". This species is regarded by some botanists as conspecific with the C. American *O. strictum* (Nees) O. Ktze. Specimen seen from S. Nigeria, Ibadan (cult., Jan.) *Keay* FHI 37330 !

36. SCHAUERIA Nees Ind. Sem. Hort. Ratisb. 1838, *nom. cons.* ; F.T.A. 5 : 242.

Branchlets glabrous ; leaves ovate, rather abruptly acuminate, cordate at base, 8–10 cm. long, 3·5–6·5 cm. broad, covered with numerous minute rod-like cystoliths ; petiole up to 7 cm. long, pubescent ; inflorescence spike-like, cylindric, up to 9 cm. long ; bracts filiform, 1–1·5 cm. long, with stipitate glands towards the apex ; calyx-lobes subulate, 6 mm. long ; corolla 1·5 cm. long ; stamens 2 ; anther-cells 2, rounded at the base, pubescent at the tip .. .. *populifolia*

S. populifolia C. B. Cl. in F.T.A. 5 : 242 (1900) ; Cat. Talb. 140. Herbaceous undershrub 4–5 ft. high, with red-purple flowers ½–¾ in. long in a dense cylindrical panicle.
   S.Nig.: Cross River North F.R. (Dec.) *Keay* FHI 28157 ! Afi River F.R. (Dec.) *Keay* FHI 28238 ! Oban *Talbot* 993 ! **F.Po:** *Mann* 1426 ! Clarence Peak (Nov.) *T. Vogel* ! Also in Cameroun.

37. CHLAMYDOCARDIA Lindau—F.T.A. 5 : 234.

Branches with a line of hairs up the side ; leaves narrowly ovate-elliptic, gradually acuminate, cuneate at base, 6–8 cm. long, 2·5–3·5 cm. broad, with numerous rod-like cystoliths more conspicuous beneath ; spikes terminal, up to 15 cm. long ; bracts very variable, from long-apiculate subulate and emarginate-cordate at the apex to broadly spathulate to subrhomboid, leafy, up to 1 cm. long and 5 mm. broad ; calyx about 5 mm. long ; corolla up to 12 mm. long ; stamens 2, anther-cells muticous ; fruit 1 cm. long ; seeds tuberculate .. .. *buettneri*

C. buettneri *Lindau* in Engl. Bot. Jahrb. 20 : 139 (1894) ; F.T.A. 5 : 234 ; Aké Assi, Contrib. 1 : 214. *C. subrhomboidea* Lindau l.c. 22 : 119 (1895) ; F.T.A. 5 : 235. Slender, branched, woody herb, 12–18 in. high, creeping at the base, with narrow-cylindric 2-lipped whitish flowers ⅓ in. long in narrow spike-like panicles ; wet places in high forest.

Iv.C.: Baba to Jumbley, San Pédro (Feb.) *Aké Assi* 2858 ! **S.Nig.:** Okomu F.R., Benin (fl. & fr. Dec.) *Brenan* 8455 ! Shasha F.R., Ijebu Prov. (Mar.) *Richards* 3224 ! Oban *Talbot* 1374 ! Oban F.R. (fr. Feb.) *Onochie* FHI 36297 ! British Obokum, Ikom (fl. & fr. Dec.) *Keay* FHI 28278 ! **[Br.]Cam.:** Victoria *Preuss* 1383. Victoria to Bimbia *Preuss* 1309. S. Bakundu F.R., Kumba (Mar.) *Brenan* 9317 ! Extends to Gabon.

38. ANISOTES Nees in DC. Prod. 11 : 424 (1847), *nom. cons.*; F.T.A. 5 : 226.

Stems puberulous ; leaves oblong-oblanceolate, obtuse at apex, 5–8 cm. long, 2–4 cm. broad, entire, shortly pubescent on the midrib ; flowers solitary, axillary ; bracteoles 2 cm. long, oblong ; corolla 4 cm. long, 2-lipped nearly to the base, tube 1 cm. long, glabrous   ..   ..   *guineensis*

A. guineensis *Lindau* in Mém. Soc. Bot. Fr. 2, 8 : 53 (1908) ; Chev. Bot. 501. Erect shrub or small tree up to 13 ft. high, with laurel-like leaves dark green and glossy above ; flowers dark red ; in wet ground by streams.
Guin.: Diaguissa to Timbo, 3,300–4,000 ft. (Apr.) *Chev.* 13444 ! Télimélé (fr. Jan.) *Roberty* 16454 ! Kindia to Forécariah, Benna Massif (June) *Jac.-Fél.* 1758 ! Pita (July) *Pobéguin* 2136 ! Kinkou Falls, Pita (fl. & fr. Oct.) *Jac.-Fél.* 1962 !

39. PERISTROPHE Nees—F.T.A. 5 : 242.

Stems sparingly setose-pilose ; leaves subequal ovate-lanceolate, acute, abruptly narrowed into the winged petiole, up to 10 cm. long and 4 cm. broad, with conspicuous rod-like cystoliths on both sides ; flowers very numerous in a lax leafy panicle ; pedicels angular and scabrid ; calyx-segments linear-lanceolate, very acute, 8 mm. long ; corolla 1·5 cm. long ; stamens 2, anthers 2-celled ; fruit 1 cm. long, pubescent *bicalyculata*

P. bicalyculata (*Retz.*) *Nees* in Wall. Pl. As. Rar. 3 : 113 (1832) ; F.T.A. 5 : 242, 514 ; Chev. Bot. 501 ; Schnell in Icones Pl. Afr. (I.F.A.N.) 1 : t. 20 (1953), for full syn. ; Berhaut Fl. Sén. 83, 85 ; Milne-Redhead in Mem. N.Y. Bot. Gard. 9 : 28 (1954). *Dianthera bicalyculata* Retz. in Sv. Vetensk. Handl. 36 : 297, t. 9 (1775). *Justicia bicalyculata* (Retz.) Vahl (1791). *Peristrophe pilosa* Turrill in Kew Bull. 1921 : 394. Erect branched herb, 2–4 ft. high, with stems several-angled, and narrow pink or pale purple flowers ½–¾ in. long.
Sen.: *Perrottet* 633 ! *Roger* ! *Farmar* 18 ! Thiès (fl. & fr. Dec.) *Chev.* 2785 ! 2786. Dakar (fr. Feb.) *Caille* ! Richard Tol (fl. & fr. Feb.) *Roberty* 16855 ! Kaolak (Nov.) *Berhaut* 415 ! **Gam.:** *Boteler* ! **Mali:** Gao (fl. & fr. Sept.) *Hagerup* 363 ! Tenadjie (Nov.) *Davey* 152 ! **Niger:** Kolo to Marona (fr. Oct.) *Vaillant* 571 ! **N.Nig.:** Panyam, 4,500 ft. *Lely* 415 ! Borgu *Barter* 767 ! Sokoto *Lely* 159 ! *Moiser* 19 ! Kuka, Bornu (fl. & fr. Feb.) *E. Vogel* 34 ! In Cape Verde Islands, throughout tropical Africa from Mauritania to S. Africa, and Arabia. (See Appendix, p. 452.)

40. ISOGLOSSA Oerst.—F.T.A. 5 : 227.

Leaves rounded at base, subequal in each pair, ovate, acuminate, up to 7 cm. long and 3·5 cm. broad, with small rod-like cystoliths very conspicuous on the nerves ; petiole slender, up to 3 cm. long, the upper leaves becoming sessile ; panicle slender, few-flowered ; calyx-lobes 2·5 mm. long, fringed with gland-tipped hairs ; corolla 2 cm. long ; stamens 2 ; anther-cells widely separated ; fruit 1·5 cm. long   1. *glandulifera*
Leaves cuneate at base, subequal, ovate, acuminate, up to 8 cm. long and 4·5 cm. broad, with very short rod-like cystoliths on the upper surface, pubescent on the nerves beneath ; petiole pilose with multicellular hairs ; panicle as above   2. *nervosa*

1. **I. glandulifera** *Lindau* in Engl. Bot. Jahrb. 20 : 54 (1894) ; F.T.A. 5 : 229. Erect, slender-branched herb 4–10 ft. high, with very thin leaves and light red flowers ¾ in. long in a lax slightly glandular panicle.
S.Nig.: Obudu Plateau, 5,500 ft., Ogoja Prov. (Aug.) *Stone* 78 ! **[Br.]Cam.:** Victoria (Mar.) *Brenan & E. W. Jones* 9355 ! 9555 ! Mimbia, 3,900 ft. *Preuss* 1062 ! Cam. Mt., 7,000 ft. (Mar.) *Kalbreyer* 132 ! (fl. & fr. Nov.) *Mann* 1972 ! *Maitland* 1022 ! Bafut-Ngemba F.R., 7,000 ft. (fl. & fr. Feb.) *Hepper* 2172 ! **F.Po:** 4,000 ft. (Nov.) *Mann* 579 ! El Pico, 5,000 ft. (Dec.) *Boughey* 180 ! S.W. of Lago Bianco, 2,000 ft. (Sept.) *Melville* 659 !

2. **I. nervosa** *C. B. Cl.* in F.T.A. 5 : 229 (1900). Stems fairly stout, ribbed, rather densely glandular-hairy on the young parts, 5 ft. high, with coarser strongly-nerved leaves and white flowers, turning reddish.
[Br.]Cam.: Cam. Mt., 7,000–8,000 ft. (Dec.) *Mann* 2009 !

Imperfectly known species.

I. sp. A.: [Br.]Cam.: Nkom-Wum F.R. (fl. & fr. Jan.) *Lightbody* FHI 26290 !
I. sp. B.: F.Po: Picode S. Isabel (fr. Mar.) *Guinea* 2953 !
I. sp. C.: S.L.: Bitumane Peak (fl. & fr. Jan.) *T. S. Jones* 117 !
Single gatherings of doubtful position.

41. RHINACANTHUS Nees—F.T.A. 5 : 224.

Branches pubescent, leaves equal to very unequal-sized in each pair, ovate-lanceolate to lanceolate, with numerous small root-like cystoliths on both surfaces, panicles axillary and terminal ; pedicels and linear calyx-segments glandular-puberulous ; corolla-tube slender, tubular, 1·5 cm. long ; stamens 2 ; anthers shortly exserted ; fruit 1·5 cm. long, softly pubescent ; seeds finely warted :
Leaves elliptic-lanceolate to ovate, gradually acuminate-cuspidate, cuneate at the base, up to 16 cm. long and 5·5 cm. broad ; petiole up to 1 cm. long (but upper leaves sometimes subsessile) ; corolla-tube 1·5–2·5 cm. long, white to whitish-yellow, sometimes pale mauve   ..   ..   1a. *virens* var. *virens*
Leaves elliptic, strongly obtuse or rounded at apex, rounded to subcordate at base, up to 10 cm. long and 4·5 cm. broad, subsessile (petiole never exceeding 3 mm.) ; corolla-tube 10–15 mm. long, white, lower lip with purple dots
       1b. *virens* var. *obtusifolius*

1. **R. virens** (*Nees*) *Milne-Redhead* in Exell Suppl. Cat. S. Tomé 37 (1956), (q.v. for synonymy etc.). *Lepto-stachya virens* Nees in DC. Prod. 9 : 378 (1847). *Rhinacanthus subcaudatus* C. B. Cl. in F.T.A. 5 : 225 (1900). *R. communis* of F.T.A. 5 : 224, 514, partly (W. Afr. specimens), of F.W.T.A., ed. 1, 2 : 266, of Aké Assi, Contrib. 1 : 220, not of Nees (1832).

1a. **R. virens** (*Nees*) *Milne-Redhead* var. **virens**—Heine in Kew Bull. 16 : 180. Herb 1–3 ft. high or weak undershrub woody below, with slender white or purplish flowers about 1 in. long ; variable in habit.
Guin.: Paroisse 1 ! Bafing *Adam* 12653 ! **S.L.:** Makump (Jan.) *Glanville* 138 ! Kasawe F.R. (Dec.) King 40b ! Njala (Mar.) *Deighton* 4715 ! Bafi R. (Feb.) *Macdonald* 2 ! Kangahun, Gandima *Deighton* 6024 ! Wallia (Feb.) *Sc. Elliot* 4276 ! **Lib.:** Kakatown (Sept.) *Whyte* 99 ! Genna Tanyehun & Mecca (Dec.) *Baldwin* 10739 ! 10800 ! Ba, Mano R. *Baldwin* 10725 ! **Iv.C.:** Anguédédou *Miège & Aké Assi* IA 1242. Zaranou (Mar.) *Chev.* 17622 ! *Leeuwenberg* 3046 ! **Ghana:** Agogo (fl. & fr. Dec.) *Irvine* 574 ! Begoro, Akim, 1,500 ft. (Jan.) *Irvine* 1171 ! Mt. Ejuanema (Dec.) *Adams* 5140 ! Afram Mankrong F.R. (fl. & fr. Dec.) *Adams* 4937 ! **S.Nig.:** Idanre (Jan.) *Brenan & Keay* 8695 ! Benin (fl. & fr. Feb., Mar.) Onochie FHI 19749 ! *Morton* K453 ! Eket *Talbot* 3131 ! Oban *Talbot* 1365 ! **[Br.]Cam.:** Victoria (Jan., Feb.) *Keay* FHI 28673 ! *Dalz.* 8200 ! Banga, Kumba (fl. & fr. Jan., Mar.) *Binuyo & Daramola* FHI 35166 ! *Brenan* 9406 ! *Olorunfemi* FHI 30511 ! Also in Cameroun, Gabon, Congo, S. Tomé and Principe. (See Appendix, p. 453.)
   [There is a very hairy-pilose form, occurring in Ghana (*Irvine* 1171, *Adams* 5140), but it seems impossible to distinguish such aberrant specimens in infraspecific rank.]

1b. **R. virens** var. **obtusifolius** *Heine* in Kew Bull. 16 : 180 (1962). *R. tenuipes* (S. Moore) Aké Assi, Contrib. 1 : 221, t. 17 (1961), partly (specimens cited from Iv. C.), not *Justicia tenuipes* S. Moore (1913). Small herb ; leaves dark green above, paler beneath ; inflorescences up to 2 in. long ; on forest floor.
Lib.: Ganta (fl. & fr. Jan.) *Harley* 817 ! **Iv.C.:** Niapidou (Jan.) *Leeuwenberg* 2417 ! Abengourou (Jan.-Feb.) *Hallé* 4867 ! *Aké Assi* 5515 ! Koleahinou (Jan.) *Aké Assi* 4866 ! Sangouine *Aké Assi* 5517 ! **Ghana:** Begoro (Jan.) *Morton* A2672 ! Doubtfully recorded also from Guinée (Schnell in Mém. I.F.A.N. 22 : 518 (1952)) ; *fide* Aké Assi *l.c.*

42. **DICLIPTERA** Juss. in Ann. Mus. Hist. Nat. Paris 9 : 267 (1807), *nom. cons.* ; F.T.A. 5 : 256.

Bracteoles with sharp spiny tips, acuminate, narrowly oblanceolate to elliptic, villous-ciliate, sometimes glabrescent, up to 1 cm. long ; leaves long-petiolate, ovate, broadly acuminate, up to 6 cm. long and 3 cm. broad, with prominent rod-like cystoliths on the lower surface ..   ..   **1. *verticillata***
Bracteoles not spiny at the tips (but often sharply acuminate) :
  The bracteoles suborbicular to broadly obovate, sometimes nearly as broad as long :
    Bracteoles densely fringed with long soft white hairs, suborbicular, 6–8 mm. long ; leaves narrowly obovate, broadly acuminate, cuneate at base, 3–8 cm. long and 1·5–3 cm. broad, with rod-like cystoliths ; corolla up to 2·5 cm. long   **2. *elliotii***
    Bracteoles fringed with hairs, broadly obovate, 1–2 cm. long, strongly nerved ; leaves on long slender petioles, narrowly obovate, broadly acuminate, 8–12 cm. long, 2–4 cm. broad, with about 6 pairs of lateral nerves ; corolla not exceeding 1·5 cm. ..   ..   **3. *obanensis***
  The bracteoles oblong-elliptic to oblanceolate :
    Bracteoles oblong-elliptic, not narrowed to base, 5–6 mm. long, with prominent green veins ; spikes lax ; leaves ovate-elliptic, abruptly acuminate, 4–6 cm. long, 2–3·5 cm. broad, with minute hair-like cystoliths ; petiole about 2 cm. long   **4. *alternans***

    Bracteoles oblanceolate, narrowed to base :
    Bracteoles villous ; leaves lanceolate, 5–6 cm. long, 1–2 cm. broad ; flowers crowded   **5. *adusta***

    Bracteoles glabrescent :
    Small decumbent herb ; leaves oblanceolate, up to 2 cm. long and 8 mm. broad ; flowers in terminal, spike-like inflorescences ; bracteoles lanceolate, up to 5 mm. long and 1 mm. broad   ..   ..   **6. *laxispica***
    Stout upright herb ; leaves elliptic, up to 8 cm. long and 3·5 cm. broad ; cymes axillary, spikelets nearly all pedicelled ; bracts up to 15 mm. long and 5 mm. broad ..   ..   **7. *laxata***

1. **D. verticillata** (*Forsk.*) *C. Christens.* in Dansk Bot. Ark. 4, 3 : 11 (1922) ; Aké Assi, Contrib. 1 : 215. *Dianthera verticillata* Forsk. Fl. Aeg. Arab. Descr. pl. 9 (1775). *Justicia cuspidata* Vahl (1791). *J. umbellata* Vahl (1804). *Dicliptera umbellata* (Vahl) Juss. (1807)—F.T.A. 5 : 259 ; F.W.T.A., ed. 1, 2 : 264. *D. micrantha* Nees (1832)—F.T.A. 5 : 258 ; F.W.T.A., ed. 1, 2 : 264. *D. ocimoides* (Lam.) Juss. l.c. (1807) ; Berhaut Fl. Sén. 92. *Justicia ocymoides* Lam. (1784). *D. maculata* var. *senegambica* Nees in DC. Prod. 11 : 485 (1847). *D. senegambica* (Nees) Benoist ex Tisserant in Mém. Inst. Fr. Centrafric. 2 : 15 (1950). Stems several-angled almost woody below, decumbent, 1–2 ft. long, with red flowers in whorls running into terminal spikes.
Sen.: *Roger* 42 ! *Brunner* ! *Berhaut* 697. **Gam.:** Genieri (Feb.) *Fox* 69 ! **S.L.:** Wellington (fl. & fr. Feb.) *Deighton* 4994 ! Gbanbama (Mar.) *Thomas* 8924 ! 9432 ! **Iv.C.:** Bouaké (Jan.) *Aké Assi* 4551 ! **Ghana:** Kumasi (Jan.) *Darko* 370 ! Dormaa-Ahenkro (Dec.) *Adams* 3024 ! Zuarungu (Nov.) *Akpabla* 403 ! Nangodi (Dec.) *Morton* A1253 ! Gambaga (fr. Dec.) *Morton* A1368 ! **Niger:** Marona to Zinder (Dec.) *Vaillant* 621/2 ! **N.Nig.:** Kano (fl. & fr. Dec.) *Hagerup* 663 ! **S.Nig.:** Lagos *Barter* 2176 ! Ibadan (Jan.) *de Wit* 7785 ! Aguku Dist. *Thomas* 1394 ! Eket *Talbot* 3202 ! **[Br.]Cam.:** Victoria (fl. & fr. Feb.) *Maitland* 978 ! Also in Cape Verde Islands, Congo, Angola, E. Africa and Arabia.

2. **D. elliotii** *C. B. Cl.* in F.T.A. 5 : 258 (1900) ; Aké Assi l.c. 215. *D. talbotii* S. Moore (1914). *D. silvicola* Lindau (1911). *D. heterostegia* of Chev. Bot. 501, not (?) of Presl ex Nees (see note below). *D. insignis* Mildbr. (1935). Decumbent, 1 ft. or more long, slightly angled stems, with small white flowers ½ in. long in axillary and terminal whorls of broad-bracteate spikelets.
Mali: Bamacora (fl. & fr. Feb.) *Chev.* 502 ! **S.L.:** *Thomas* 8793 ! Wallia (fl. & fr. Jan.) *Sc. Elliot* 4632 ! Njala (fl. & fr. Jan.-Mar.) *Deighton* 1101 ! 1806 ! Moselelo (fr. Nov.) *Deighton* 2373 ! Baoma (fr. Apr.) *Deighton* 3169 ! **Lib.:** Mao R., Nyandamolahum (fr. Feb.) *Bequaert* 75. **Iv.C.:** Dakpadoo (fl. & fr. Feb.) *Leeuwenberg* 2863 ! Taï (fl. & fr. Mar.) *Leeuwenberg* 3035 ! Aké Assi IA 3094. **Ghana:** Cape Coast (Dec.) *Hall* 1676 ! Axim (fr. Feb.) *Irvine* 2370 ! Akuse (Oct.) *Morton* P1752 ! Dormaa

(fr. Dec.) *Adams* 2991! Tain II F.R. (fr. Dec.) *Morton* A1029! **N.Nig.:** *Thornewill* 26! **S.Nig.:** Eket *Talbot* 3217! Balinge, Obudu Div. (fl. & fr. Dec.) *Savory & Keay* FHI 25165! A very variable and widespread tropical African weed described under many names ; owing to the imperfect taxonomic knowledge of the genus its true distribution is not properly known.

[Plants here called *D. elliotii* C. B. Cl. are only doubtfully distinct (mainly by the usually smaller flowers) from those which have been described as *D. heterostegia* Presl ex Nees in DC. Prod. 11 : 478 (1847) ; Fl. Cap. 5 : 90 ; Chev. Bot. 501. (*D. heterostegia* (E. Mey.) Presl in Abh. k. böhm. Ges. Wiss. V. Folge, 3 : 525 (1845), name only. *Justicia heterostegia* E. Mey. in Drège, Zwei Pflanzengeogr. Docum. 152, 159, 195 (1843), name only.)—Described from S. Africa. *Chev.* 502 was named by Lindau as " *D. heterostegia* Nees "].

3. **D. obanensis** *S. Moore* in Cat. Talb. 90 (1913). A slender herb 1–2 ft. high, rooting at the lower nodes, with green foliaceous bracts and small white flowers.
   S.L.: Lalakum to Gbwema, Gaura (fr. Dec.) *King* 103b! **Ghana:** Accra (fl. & fr. Jan.) *Morton* GC 25346! Kibi (fr. Dec.) *Morton* GC 8213! E. Akim (fr. Mar.) *Johnson* 717! Obomeng, Kwahu (Dec.) *Adams* 5170! **S.Nig.:** Oban (fr. Feb.) *Talbot* 1363! Onochie FHI 36314!
   [This seems to be doubtfully distinct from *D. chinensis* (Linn.) Juss. (syn. *Justicia sexangularis* Forsk. (1775), not of Linn. (1753) ; C. Christens. in Dansk Bot. Ark. 4, 3 : 10 (1912) ; *J. chinensis* Linn. (1753)).]

4. **D. alternans** *Lindau* in Engl. Bot. Jahrb. 20 : 47 (1894) ; F.T.A. 5 : 258. A slender herb up to 1 ft. high, slightly pubescent, with white flowers and green-veined pointed bracts.
   [Br.]Cam.: Buea (fr. Jan.) *Preuss* 604! Also in Cameroun.

5. **D. adusta** *Lindau* in Mém. Soc. Bot. Fr. 2, 8 : 51 (1908). *D. leyi* Hutch. & Dalz. F.W.T.A., ed. 1, 2 : 265 (1931). Erect, branched herb, woody below, with several-angled stems nearly glabrous, and many flowers ½–¾ in. long, pinkish purple with spots and darker lower lip ; in savanna, flowering after fires.
   N.Nig.: Naraguta (Nov.) *Lely* 712! Birnin Gwari (fl. & fr. Mar.) *Meikle* 1349! Utachu to Madengyen, Niger Prov. (Jan.) *Meikle* 1071! Jos Plateau (Jan.) *Lely* P91! Sanga River F.R., Jemaa (Apr.) *Keay & Jones* FHI 37623! **S.Nig.:** Oyo F.R. (fl. & fr. Feb.) *Keay* FHI 22513! Also in Ubangi-Shari.
   [This is only slightly different from *D. pumila* (Lindau) Dandy in Mem. N.Y. Bot. Gard. 9, 1 : 27 (1954), see discussion there.]

6. **D. laxispica** *Lindau* in Engl. Bot. Jahrb. 30 : 113 (1901), and in Schlechter Westafr. Kautschuk-Exp. 316 (1900), name only. A decumbent herb rooting at the lower nodes, with small leaves and small white flowers ¼ in. long.
   Iv.C.: Sassandra *Aké Assi* 4869! Taï (fr. Jan.) *Mangenot & Aké Assi* 3090! **Ghana:** Mpameso F.R., W. Ashanti (Dec.) *Adams* 2970! **S.Nig.:** Oban *Talbot*! Also in Cameroun. [This may be only a form of No. 4, but with the material at present available it is inadvisable to add it to the synonymy of that species.]

7. **D. laxata** *C. B. Cl.* in F.T.A. 5 : 258 (1900) ; Mildbraed in Notizbl. Bot. Gart. Berl. 9 : 507 (1926). Scrambling shrub with dark green stems and pale green leaf undersurfaces ; bracts usually dark tinged ; flowers white ; a typical highland species.
   [Br.]Cam.: Buea (Jan.) *Deistel* 489. Bamenda, 5,000 ft. (fl. Jan., fr. Apr.) *Maitland*! *Keay & Lightbody* FHI 28476! Bafut-Ngemba F.R. (fr. Feb.) *Hepper* 2193! *Tiku* FHI 22163! Bali (fr. Feb.) *Conrau* 18. *Rudin* in Hb. *Hepper* 2864! **F.Po:** Balachá to Ureka (fr. Feb.) *Guinea* 2298! Also in Congo and E. Africa.

Doubtful species.

D. hyalina of Berhaut Fl. Sén. 92, not of Nees (1847). " *Sp.*" of Chev. Sudania 1 : 44. Villous-hairy plants resembling vegetatively *D. verticillata*, but they do not seem to be a form of it and they are certainly not conspecific with *D. hyalina* Nees (in DC. Prod. 11 : 484 (1847)) described from Comoro Islands. Further collections and monographic study are necessary to elucidate their taxonomic position.
Sen.: Tanma *Berhaut* 1304. N'diaye bop *Chev.* 2809! Ntiaye (fl. & fr. May) *Raynal* 7053!

43. JUSTICIA Linn.[1]—F.T.A. 5 : 179.

Bracts white, transparent, reticulate with greenish margins and veins, ovate-acute, up to 12 mm. long and 7 mm. broad ; inflorescences terminal, strobilate ; leaves ovate, up to 7 cm. long and 4 cm. broad   ..   ..   1. *betonica*
Bracts never white and reticulate with greenish margins and veins :
  Flowers axillary ; stems pubescent to glabrous ; leaves very variable, from linear to ovate, subsessile to long-petiolate, acute or acuminate, up to 12 cm. long and 4 cm. broad, setulose to nearly glabrous ; calyx-lobes lanceolate, acuminate, more or less ciliate ; corolla up to 2·5 cm. long ; pubescent outside ; fruit 1·5 cm. long, slightly pubescent or glabrous   ..   ..   2. *insularis*
  Flowers spicate racemose or paniculate :
    Flowers all to one side of the short slender spikes ; leaves linear to lanceolate, acute, 3·5–6 cm. long, up to 1 cm. broad, glabrous or nearly so, thin ; calyx-segments subulate-lanceolate, 3·5 mm. long, with narrow hyaline margins ; fruit 1 cm. long, glabrous   ..   ..   3. *anselliana*
    Flowers not all to one side of the inflorescence :
      Inflorescence spicate :
        Spikes interrupted, the lower internodes 1 cm. long ; flowers in small dense cymules ; leaves broadly obovate, acuminate, cuneate at base, 10–18 cm. long, 5–8 cm. broad, glabrous, with about 5 pairs of lateral nerves ; petiole 2–5 cm. long ; corolla 1 cm. long ; stamens exserted   ..   ..   4. *nigerica*
        Spikes more or less continuous :
          Spikes terminal, with lanceolate, linear to spathulate bracts, stout undershrubs (No. 7 sometimes herbaceous) :
            Flowers 2·5 cm. long, very deeply 2-lipped, opening about 5 cm. wide, blue, inflorescence sometimes bearing small lateral cymes ; up to 3 m. high, with sparsely pubescent ovate leaves up to 12 cm. long and 6 cm. broad 5. *preussii*
            Flowers much smaller, 9–22 mm. long, not deeply lipped, opening about 1 cm. wide, inflorescence never with lateral cymes :

[1] The type species of *Justicia* Linn. (1753) is, according to A. S. Hitchcock and M. L. Green in Intern. Bot. Congr. Cambridge 1930, Propos. by Brit. Botanists, p. 116, No. 26 (1929), *J. hyssopifolia* Linn. (1753), not *J. adhatoda* Linn. (1753), the type species of *Adhatoda* Medik. (1790).

Bracts spathulate, about 1 cm. long, at the top about 2 mm. broad and in the upper half densely covered with glandular hairs 2 mm. long ; corolla about 2·2 cm. long, bluish ; petioles up to 2·5 cm. long ; leaves elliptic-oblong, 5–15 cm. long and 2·5–7·5 cm. broad, velvety, with about 7–9 pairs of lateral nerves .. .. 6. *hepperi*

Bracts oblanceolate, narrowed to the base, about 1 cm. long, ciliate with eglandular hairs ; corolla about 2 cm. long, yellow ; petioles up to 1·7 cm. long; leaves like the last, hairy (not velvety), up to 12 cm. long and 5 cm. broad 7. *flava*

Spikes lateral, 1–2 cm. long, on slender peduncles, with broadly obovate nervose bracts about 5 mm. long ; leaves ovate, long-petiolate, rounded at apex, up to 5 cm. long and 3 cm. broad, very thin and glabrescent ; fruits 3 mm. long ; a creeping slender herb .. .. 8. *tenella*

Inflorescence paniculate :

Panicles axillary, few-flowered, terminal and very loose, compound with capillary branches :

Plants pubescent or nearly glabrous ; sepals up to 4 mm. long ; corolla up to 12 mm. long ; seeds covered with long scabrous tubercles .. 9. *glabra*

Plants pubescent ; sepals about 8 mm. long ; corolla up to 10 mm. long ; seeds minutely warted .. .. 10. *tenuipes*

Panicles terminal, usually 15–25 cm. long, subpyramidal (in some forms small and compact), rich-flowered ; herbs about 1·5 m. high ; seeds with blunt tubercles ; bracts linear, subulate or sometimes leafy ; leaves long-petiolate :

Ovary and capsule glabrous .. .. 11. *laxa*

Ovary and capsule hairy .. .. 12. *extensa*

1. **J. betonica** *Linn.* Sp. Pl. 15 (1753) ; Sp. Pl. ed. 2, 1 : 21 (1762)[1]; F.T.A. 5 : 184. A rather variable herb, sometimes becoming shrubby, with white flowers, the tube slightly violet in the throat.
Mali: Kita (Dec.) *Jaeger* 3788 ! Throughout the Old World tropics.
[This species is cited by C. B. Clarke in F.T.A. from Sierra Leone, but there is no reference in the corresponding literature nor a specimen at Kew. Clarke attributes his indication to Lindau ; I was unable to find out where Lindau published the note to which Clarke referred.—H.H.]

2. **J. insularis** *T. Anders.* in J. Linn. Soc. 7 : 40 (1863) ; F.T.A. 5 : 195 ; Mildbraed in Notizbl. Bot. Gart. Berl. 9: 501 (1926); Aké Assi, Contrib 1: 217. *Adhatoda diffusa* Benth. in Fl. Nigrit. 483 (1849), not *Justicia diffusa* Willd. (1797). *J. galeopsis* T. Anders. ex C. B. Cl. in F.T.A. 5 : 196 (1900) ; Lindau in Mildbr. Wiss. Egebn. 1907–08 : 309 (1914) ; Chev. Bot. 500 ; Benoist in Bol. Soc. Brot., sér. 2a, 24 : 39. *Siphonoglossa macleodiae* S. Moore in Macleod, Chiefs & Cities Centr. Afr. 304 (1912) ; Cat. Talb. 84. A straggling extremely variable and widespread herb 1–3 ft. high, rooting below, with bright pink, red-purple or sometimes yellowish flowers with purple markings ¾ in. long, 1–3 in the axils.
Sen.: *Berhaut* 446. *Farmar* 45a ! **Gam.:** *Skues* ! Kuntaur *Ruxton* 157 ! **Guin.:** Fouta Djalon (Jan.) *Maclaud* 312 ! Mamou (Sept.) *Dalz.* 8390 ! *Schnell* 6792 ! **Port.G.:** *Esp. Santo* 820. Calicunda to Paunca (fl. & fr. Sept.) *Esp. Santo* 2788 ! **S.L.:** Regent (fl. & fr. Dec.) *Sc. Elliot* 4149 ! Sugar Loaf Mt. (fl. & fr. Dec.) *Sc. Elliot* 3974 ! Ronietta (fl. & fr. Nov.) *Thomas* 5589 ! Masactaba (fl. & fr. Oct.) *Glanville* 33 ! Musaia (fl. & fr. Dec.) *Deighton* 4529 ! **Lib.:** Kakatown & Monrovia *Whyte* ! **Ghana:** Bukrum (Feb.) *A. S. Thomas* D122 ! Tano-Offin F.R. (fl. & fr. Jan.) *Vigne* FH 1012 ! Kpedsu (Jan.) *Howes* 1079 ! Kwahu (Nov.) *Irvine* 1712 ! Dzolo Kpuita (fr. Nov.) *Morton* GC 9350 ! Kumasi (fl. & fr. Dec.) *Morton* GC 9683 ! **Dah.:** *Burton* ! **N.Nig.:** *Yates* 26 ! *Millson* 40 ! Nupe *Barter* 1038 ! Aguji, Ilorin *Thornton* ! **S.Nig.:** Lagos *Rowland* ! *Millen* 169 ! Abeokuta *Irving* 65 ! Eket *Talbot* ! Oban *Talbot* 983 ! [Br.]**Cam.:** S. Bakundu F.R., Kumba (fl. & fr. Jan.) *Binuyo & Daramola* FHI 35196 ! **F.Po:** (fl. & fr. Nov., Dec.) *T. Vogel* ! *Guinea* 756 ! Mamfe (fl. & fr. Dec.) *Morton* K675 ! Throughout tropical Africa ; a very complex species and, like the whole genus, much in need of a monographic revision. (See Appendix, p. 451.)
[This is, as accepted here, a complex group of plants which often appear heterogenous and rather different in habit, morphological characters and other features. The oldest name for it seems to be *J. lithospermifolia* Jacq. (1797), but without monographic revision of the whole group it is undesirable to change in this Revised Edition the traditional taxonomic concept for only a small part of the area inhabited by this very widespread Old World species—H.H.]

3. **J. anselliana** (*Nees*) *T. Anders.* in J. Linn. Soc. 7 : 44 (1883) ; F.T.A. 5 : 208. *Adhatoda anselliana* Nees in DC. Prod. 11 : 403 (1847). A nearly glabrous herb, erect or decumbent, 1–2 ft., with small white and purple-streaked flowers ⅓ in. long, few in short axillary one-sided inflorescences.
Mali: " Fleuve blanc " *Peney* ! **Guin.:** Agoué *Ménager* ! **Lib.:** Cape Palmas *Ansell* ! **Ghana:** Kpong-Akuse (fl. & fr. June) *Irvine* 1772 ! Accra *Unknown Coll.* (Hb. Paris) ! Shai Plains (May) *Irvine* 2674 ! Senya Beraku (fr. May) *Morton* GC 9214 ! Togo Gap (Sept.) *Mahoux* 408 ! **Togo Rep.:** Toblekoré (fl. & fr. Aug., Sept.) *Morton* K324 ! Naraguta & 423 ! **Dah.:** Zagnanado, L. Azri (Feb.) *Chev.* 23036 ! **N.Nig.:** Vom (fr. Apr.) *Morton* K324 ! Naraguta & Jos (fl. & fr. Apr.) *Lely* 571 ! P232 ! **S.Nig.:** Nkisi, Onitsha (fl. & fr. May) *Killick* 118 ! Abo (fl. & fr. Aug.) *T. Vogel* 14 ! Otumoye L., Onitsha (Mar.) *Onochie* FHI 35753 ! Throughout tropical Africa.

4. **J. nigerica** *S. Moore* Cat. Talb. 81 (1913). A woody undershrub with fairly stout puberulous branches flowers ⅓ in. long in spikes up to 9 in. or more long, simple or slightly compound below.
S.Nig.: Oban *Talbot* 995 ! 1308 ! 2008 ! " Central Prov." *Rosevear* C19 ! (See Appendix, p. 451.)

5. **J. preussii** (*Lindau*) *C. B. Cl.* in F.T.A. 5 : 204 (1900). *Salviacanthus preusii* Lindau in Engl. Bot. Jahrb. 20 : 75 (1894). *Brillantaisia majestica* Wernham in J. Bot. 54 : 229 (1916). A puberulous shrub up to 10 ft. high ; the large clear blue flowers in shape much resemble those of *Brillantaisia* or Mexican Salvias, capsule up to ¾ in. long.
[Br.]**Cam.:** Cam. Mt., 2,000 ft. *Mann* 1298 ! Buea, 3,000–5,500 ft. (fl. & fr. Jan., Dec.) *Preuss* 675 ! *Bates* 817 ! *Migeod* 124 ! Mimbia, 4,000 ft. (Mar.) *Brenan & E. W. Jones* 9336 ! Also in Cameroon.

6. **J. hepperi** *Heine* in Kew Bull. 16 : 178, fig. 4 (1962). An erect perennial herb about 5 ft. high in large clumps ; flowers pale blue with darker blue veins ; in hilly savanna.
[Br.]**Cam.:** Dakemi, Vogel Peak, 3,200 ft., Adamawa (fl. & fr. Nov.) *Hepper* 1393 !

7. **J. flava** (*Forsk.*) *Vahl* Symb. Bot. 2 : 15 (1791) ; F.T.A. 5 : 190 ; Chev. Bot. 500. *Dianthera americana* α *flava* Forsk. Fl. Aeg.-Arab. Desc. Pl. 9 (1775). *D. flava* (Forsk.) Vahl (1790). *Justicia palustris* of Oliv. in Trans. Linn. Soc., ser. 2, 2 : 345 ; Lindau in E. & P. Pflanzenf. 4, 3B : 349 (*Baumann* 164), not of (Hochst.) T. Anders. Herb, erect or straggling 1–4 ft., with pubescent sulcate stems and yellow dark-streaked flowers ⅓–½ in. long in pubescent spikes 2–6 in. or more long.
Sen.: Niayes *Chev.* 2799. **Guin.:** Iles de Los (fr. Nov.) *Schnell* 2073 ! Kouria *Caille* in *Hb. Chev.* 14720 !

---

[1] This is the citation given by C. B. Clarke in F.T.A. l.c.

14908 ; 14910*bis*. Nzo (fr. Oct.) *Schnell* 3859 ! Nzérékoré (fl. & fr. Oct.) *Baldwin* 9737 ! **S.L.**: Binkolo (Aug.) *Deighton* 1278 ! Gbap (fl. & fr. Oct.) *Jordan* 627 ! Rokupr (fl. & fr. Dec.) *Jordan* 747 ! Kasawe F.R. (fr. Dec.) *King* 57*b* ! Musaia (Dec.) *Deighton* 4450 ! **Lib.**: Gbanga (fl. & fr. Oct.) *Linder* 1264 ! Yratoke (July) *Baldwin* 6244 ! **Iv.C.**: Bingerville (fr. Dec.) *Chev.* 15568 ! 16039 ! *de Wilde* 500 ! Nekaougnié to Grabo *Chev.* 19591. Adiopodoumé (fl. & fr. Oct.) *Leeuwenberg* 1722 ! Grand Bassam (fl. & fr. Sept.) *de Wit* 7110 ! Cape Coast (July) *T. Vogel* ! **Ghana**: Swedru (fl. Feb.) *Dalz.* 8278 ! Dunkwa (fl. & fr. Aug.) *Sampeney* 711 ! Apampam (fr. Jan.) *Morton* GC 6403 ! Red Volta R. (fr. Dec.) *Adams & Akpabla* 4292 ! Tamale (fr. Dec.) *Morton* GC 6187 ! **Togo Rep.**: *Baumann* 164. *Büttner* 258 ! Lomé *Warnecke* 443 ! Talessou (fr. Oct.) *Gironcourt* 258 ! **Dah.**: Niaouli to Allada (fr. Mar.) *Chev.* 23141 ! **N.Nig.**: Quorra (= Niger) (Sept.) *T. Vogel* 167 ! Benue R. *T. Vogel* ! Patti Lokoja (fr. Oct.) *Dalz.* 136 ! Alkaline (fl. & fr. Nov.) *Lely* 646 ! Damisa, Zaria Prov. (Nov.) *Keay & Onochie* FHI 21720 ! **S.Nig.**: Lagos (fr. Mar.) *Millen* 37 ! Orosun, Idanre Hills (fr. Oct.) *Keay* FHI 25509 ! Aguku Dist. *Thomas* 769 ! Onitsha *Barter* 619 ! 1779 ! Degema *Talbot* 3659 ! [**Br.**]**Cam.**: Vogel Peak, Adamawa (fl. & fr. Nov.) *Hepper* 1367 ! Throughout tropical Africa. (See Appendix, p. 451.)

8. **J. tenella** (*Nees*) *T. Anders.* in J. Linn. Soc. 7 : 4 (1863) ; F.T.A. 5 : 187 (for full synonymy) ; *Chev. Bot.* 501 ; Berhaut Fl. Sén. 95. *Rostellaria tenella* Nees in DC. Prod. 11 : 369 (1847). *Rungia baumannii* Lindau in Engl. Bot. Jahrb. 22 : 120 (1895). A slender nearly glabrous herb, erect 6–12 in. high, creeping and rooting at the lower nodes, with small white and lilac or pink-tinged flowers in short 4-sided spikes with pale green broad bracts.

Sen.: *Heudelot* 327 ! Tambacounda *Berhaut* 1287 ! Badi (fr. Nov.) *Berhaut* 1930 ! **Mali**: Kita (fr. Oct.) *Jaeger* 3445 ! **Guin.**: Kouria (fr. Dec.) *Caille* in *Hb. Chev.* 14640 ! Bouria (fl. & fr. Nov.) *Caille* in *Hb. Chev.* 14956 ! *Farmar* 190 ! **Port.G.**: Brene, Bissau (fr. Feb.) *Esp. Santo* 1731 ! **S.L.**: Njala (fl. & fr. Sept.) *Deighton* 2081 ! Makump (fr. Aug.) *Deighton* 1375 ! Kambui Hills (fl. & fr. Dec.) *Small* 847 ! Rokupr (fl. & fr. Oct.) *Jordan* 653 ! Kabala (fl. & fr. Oct.) *Deighton* 4194 ! **Lib.**: Gbanga (fl. & fr. Sept.) *Linder* 570 ! Gbau (fl. & fr. Sept.) *Baldwin* 9417 ! Tawata (fr. Nov.) *Baldwin* 10320 ! 10321 ! Yratoke (fl. & fr. July) *Baldwin* 6263 ! **Iv.C.**: Bingerville *Chev.* 15587 ! Alépé *Chev.* 17474 ! Grabo *Chev.* 19748 ! Banco (fl. & fr. July) *de Wit* 7109 ! Tonkoui (fr. Aug.) *Boughey* 18332 ! Taï to Tabou (fl. & fr. Aug.) *Boughey* GC 14948 ! **Ghana**: Aburi *Brown* 384 ! Apampam (fl. & fr. Jan.) *Morton* GC 6401 ! **N.Nig.**: Kontagora (fr. Dec.) *Dalz.* 185 ! **S.Nig.**: Sapoba *Kennedy* 1767 ! 2699 ! Usonigbe F.R., Benin (fr. Oct.) *Keay* FHI 21580 ! Owan, Benin (fr. Nov.) *Latilo* FHI 35357 ! Idanre (fr. Jan.) *Brenan* 8700 ! Oban *Talbot* 378 ! Afi River F.R., Ikom (Dec.) *Keay* FHI 28246 ! [**Br.**]**Cam.**: Victoria (fr. Nov.) *Maitland* 776 ! Lus, Nkambe Div. (fl. & fr. Feb.) *Hepper* 1868 ! Mayo Selbe, 1,600 ft., Gashaka Dist. (fl. & fr. Jan.) *Hepper* 1636 ! Vogel Peak, 2,500 ft., Adamawa (fl. & fr. Nov.) *Hepper* 1469 ! **F.Po**: El Pico (fl. & fr. Dec.) *Boughey* 181 ! Throughout tropical Africa.

9. **J. glabra** *Koenig ex Roxb.* Fl. Ind., ed. Carey & Wall., 1 : 132 (1820) ; *Hort. Beng.* 4 (1814), name only ; F.T.A. 5 : 208 ; Milne-Redhead in Mem. N.Y. Bot. Gard. 9 : 24 (1954). Erect or straggling herb about 6 ft. high ; flowers greenish-yellow.

Iv.C.: Béoumi to Toumodi-Sakasso *Aké Assi* IA 3450. **N.Nig.**: Kwakuti, Niger Prov. (fl. & fr. Dec.) *Keay* FHI 37308 ! Lokoja (Oct.) *Parsons* L87 ! [**Br.**]**Cam.**: Vogel Peak, 4,800 ft., Adamawa (fl. & fr. Dec.) *Hepper* 1584 ! Throughout tropical Africa, and India.

10. **J. tenuipes** *S. Moore* in Cat. Talb. 82 (1913). *Rhinacanthus tenuipes* (S. Moore) *Aké Assi*, Contrib. 1 : 221 (1961), excl. t. 17 and Iv. C. specimens. A short-stemmed herb, creeping and rooting below, densely pubescent, with wrinkled leaves and narrow flowers $\frac{1}{2}$–$\frac{2}{3}$ in. long in a glandular-pubescent loose panicle, sometimes reduced to 1 or 2 flowers.

S.Nig.: Oban *Talbot* 1327 ! 1483 ! Only known from the type collections.

11. **J. laxa** *T. Anders.* in J. Linn. Soc. 7 : 43 (1863) ; F.T.A. 5 : 205 ; F.W.T.A., ed. 1, 2 : 266, partly (*T. Vogel* 144, *Preuss* 1350). *Adhatoda paniculata* Benth. (1849). *Duvernoya paniculata* (Benth.) Lindau (1894), not *Justicia paniculata* Burm. f. (1768), nor Forsk. (1775). Herbaceous undershrub 4–5 ft. or more high or more or less straggling, with fairly large leaves and more or less dense terminal panicles or spike-like inflorescences of greenish-white flowers $\frac{1}{2}$–$\frac{2}{3}$ in. long.

S.Nig.: Oban *Talbot* 70 ! 102 ! 978 ! [**Br.**]**Cam.**: *Preuss* 1350 ! Bafia, Victoria (Feb.) *Keay* FHI 37534 ! Kumba, crater lake (fr. Dec.) *Morton* K818 ! **F.Po**: (Nov.–Dec.) *T. Vogel* 144 ! *Mann* 63 ! Also in Cameroun and Gabon. (See Appendix, p. 451.)

12. **J. extensa** *T. Anders.* in J. Linn. Soc. 7 : 44 (1863) ; F.T.A. 5 : 206. *J. talbotii* S. Moore in Cat. Talb. 83 (1913). *J. thyrsiflora* S. Moore l.c. 84 (1913) ; F.W.T.A., ed. 1, 2 : 266, not of Roxb. (1820). *J. laxa* of F.W.T.A., ed. 1, 2 : 266, partly (excl. *T. Vogel* 144, *Preuss* 1350), not of T. Anders. Herbaceous undershrub up to 10 ft. high, sometimes climbing, densely pubescent on the young stems, with small pale green pink-purple-spotted flowers $\frac{1}{4}$ in. long in short-branched tawny-pubescent terminal inflorescences up to 1 ft. long.

Guin.: Konkouré Mt. (Nov.) *Pobéguin* 1479 ! Kouria (Nov.) *Caille* in *Hb. Chev.* 14709 ! **S.L.**: Kwaoma (fr. Jan.) *Deighton* 3883 ! Male R. ferry (Dec.) *Deighton* 3567 ! Kenema (fl. & fr. Jan., Mar.) *Thomas* 7483 ! *Lane-Poole* 444 ! Bumbuna (Oct.) *Thomas* 3933 ! Giewahun (Nov.) *Deighton* 358 ! **Lib.**: Monrovia (Oct.) *Dinklage* 2916 ! *Whyte* ! Kakatown *Whyte* ! Moala (Nov.) *Linder* 1376 ! Peahtah (Dec.) *Baldwin* 10613 ! Pronyon *Bunting* 64 ! **Iv.C.**: Morénou (Nov.) *Chev.* B22425 ! Singrobo (Nov.) *Aké Assi* 4459 ! Kassa (Oct.) *Aké Assi* 5504 ! **Ghana**: Shai Plains (Jan.) *Johnson* 582 ! Agogo (fl. & fr. Dec.) *Chipp* 610 ! Banka S. (Sept.) *Vigne* FH 1373 ! Apampam (fl. & fr. Nov.) *Vigne* FH 4274 ! Ankasa *Taylor* FH 5446 ! Wudidi, Togo Plateau (fl. & fr. Nov.) *Morton* A3445 ! **Togo Rep.**: Misahöhe (Nov.) *Mildbr.* 7274 ! **Dah.**: Sakété (Jan.) *Chev.* 22812 ! Sakété to Pedjilé (fr. Feb.) *Chev.* 22893 ! **S.Nig.**: Eppah, Lagos *Barter* 3301 ! 3396 ! Etemi to Etemi Odo, Ijebu (Oct.) *Tamajong* FHI 20991 ! Degema *Talbot* 3654 ! Eket *Talbot* 1425 ! Oban *Talbot* 976 ! 1371 ! 1408 ! Aboabam, Ikom (Dec.) *Keay* FHI 28211 ! Iso Bendiga to Afi River F.R. (Dec.) *Keay* FHI 20991 ! [**Br.**]**Cam.**: Mamfe (fl. & fr. Dec.) *Morton* K681 ! Also in Cameroun, Gabon, Ubangi-Shari and Congo.

44. MONECHMA Hochst.—F.T.A. 5 : 212.

Bracts with long white bristles, lanceolate, leafy, about 1·5 cm. long ; leaves linear or narrowly linear-lanceolate, up to 12 cm. long and 1·5 cm. broad, more or less pilose ; spikes short ; calyx-segments sub-equal, linear, hispid ; corolla 1·5 cm. long ; fruit 1 cm. long   ..   ..   1. *ciliatum*
Bracts without stiff white bristles, elliptic-oblong to lanceolate, leaves elliptic to oblong-lanceolate :
Leaves lanceolate, usually long-cuspidate, acuminate, up to 10 cm. long and 3 cm. broad ; pilose on the nerves beneath ; spikes up to 6 cm. long ; bracts acute, sometimes nearly glabrous, with pale margins, up to 12 mm. long and 4 mm. broad   2. *depauperatum*
Leaves oblong-lanceolate, obtusely acuminate or rounded at apex, 3–3·5 cm. long and 0·8–1 cm. broad, puberulous, spikes up to 10 cm. long ; bracts puberulous, mostly obtuse, 7–10 mm. long and 2–3 mm. broad (but in forms sometimes much larger), with no pale margins   ..   ..   3. *ndellense*

1. **M. ciliatum** (*Jacq.*) *Milne-Redhead* in Kew Bull. 1934 : 304 ; l.c. 5 : 381 (1951) (both for full synonymy). *Justicia ciliata* Jacq. Hort. Vind. 2, 47, t. 104 (1772). *Monechma hispidum* Hochst. (1841)—F.T.A. 5 : 213 ; F.W.T.A., ed. 1, 2 : 266 ; Benoist in Bol. Soc. Brot., sér. 2, 24 : 39 ; Aké Assi, Contrib. 1 : 219 ; Berhaut Fl. Sén. 93. *Justicia togoensis* Lindau (1894). *J. buettneri* Lindau (1894). *Hygrophila lutea* T. Anders. (1863)— F.T.A. 5 : 32 ; F.W.T.A., ed. 1, 2 : 247. Herb 1–2 ft. high, coarsely whitish-hairy, woody below, with 2-lipped cream-white flowers purple- and orange-streaked ; in savanna and fields.
Sen.: *Perrottet* 639 ! *Heudelot* 167 ! 505 ! *Berhaut* 551. Thiès to Soussoune (Sept.) *Trochain* 1580 ! **Gam.:** *Brown-Lester* 251 ! Uyambai forest, Kombo Central *Frith* 238 ! **Mali:** San (Sept.) *Chev.* 2792 ! Dioura *Davey* 025 ! **Port.G.:** *Esp. Santo* 834. **Guin.:** Koumi *Pobéguin* 4193 ! Mamou *Jac.-Fél.* 2198 ! **S.L.:** Boyaboya (fr. Feb.) *Sc. Elliot* 4588 ! **Iv.C.:** Morenou (fr. Dec.) *Chev.* 22498 ! Ferkéssédougou *Aubrév.* 2557 ! Mankono, near R. Manakone (fr. Nov.) *de Wilde* 964 ! Namboukara to Ouronautiakara (fr. Nov.) *Leeuwenberg* 1981 ! Bouaké (fl. & fr. Nov.) *Leeuwenberg* 2068 ! **Ghana:** Banda to Luenji (fr. Dec.) *Morton* GC 25119 ! GC 25253 ! Paliba (fr. Dec.) *Adams & Akpabla* 4117 ! Tate to Daboya (fr. Dec.) *Morton* GC 6207 ! Damongo (fl. & fr. Nov.) *Rose Innes* GC 30189 ! **Togo Rep.:** Kitschenke (Aug.) *Büttner* 148 ! Misahöhe (fl. & fr. Oct.-Nov.) *Baumann* 326 ! *Mildbr.* 7236 ! *Schroeder* 99 ! 104 (partly, mixed with *Hypoestes strobilifera* var. *tisserantii*) ! **Dah.:** Agouagon to Savé (fr. May) *Chev.* 23556 ! **Niger:** Zinder (fr. Dec.) *Vaillant* 899 ! Kolo (fr. Jan.) *Vaillant* 668 ! Niamey (fr. Oct.) *Hagerup* 468 ! **N.Nig.:** Zaria (Aug., Sept.) *Keay* FHI 28016 ! Olorunfemi FHI 24357 ! Dogon Kurmi, Jemaa Div. (Nov.) *Keay* FHI 21038 ! Katagum *Dalz.* 101 ! Jos Plateau (July) *Lely* P342 ! Katsina (fr. & galls Dec.-Jan.) *Kennedy* 3019 ! **S.Nig.:** Lagos (fr. Sept.) *Dodd* 426 ! Onitsha *Barter* 1163 ! (galled) [**Br.**]**Cam.:** Dikwa (fr. Dec.) *McClintock* 52 ! Throughout tropical Africa. (See Appendix, p. 452.)

2. **M. depauperatum** (*T. Anders.*) *C. B. Cl.* in F.T.A. 5 : 217 (1900) ; Aké Assi, Contrib. 1 : 218. *Justicia depauperata* T. Anders. in J. Linn. Soc. 7 : 40 (1863). *J. barteri* T. Anders. l.c. 39 (1863). *Monechma scabridum* (S. Moore) C. B. Cl. in F.T.A. 5 : 217 (1900); Berhaut Fl. Sén. 88. *Justicia scabrida* S. Moore (1880). *Monechma marginatum* (Lindau) C. B. Cl. in F.T.A. 5 : 217 (1900). *Nicoteba marginata* Lindau (1895). An erect, very variable, branched undershrub several ft. high (in grass) with pubescent angled stems, and white purple-streaked flowers ½ in. long ; in wet grassland.
Sen.: *Berhaut* 1679. Niokolo-Koba (fr. Jan.) *Berhaut* 4594 ! **Mali:** Bamako (Sept.) *Waterlot* 1312 ! 1369 ! **Guin.:** Kouroussa (fl. & fr. Sept., Oct.) *Pobéguin* 448 ! 570 ! 1834*bis* ! Mamou to Timbo (fl. & fr. Oct.) *Jac.-Fél.* 1897 ! Loura Mt., Mali (fr. Oct.) *Jac.-Fél.* 1977 ! Diaguissa to Boulivel (Sept.) *Chev.* 18601 ! Kouria to Trébéleya (fr. Sept.) *Chev.* 18233 ! **S.L.:** Musaia (fl. Dec., Feb., fr. Mar.) *Deighton* 4197 ! 4449 ! 5434 ! Lomaburu (fr. Feb.) *Sc. Elliot* 5035 ! **Iv.C.:** Séguéla (fl. & fr. Aug.) *Boughey* GC 18483 ! Ouronautiakara (Nov.) *Leeuwenberg* 1998 ! **Ghana:** Accra (fr. Dec.) *Morton* A2838 ! Afram Plains (fr. Jan.) *Irvine* 621 ! Ejura (fr. Dec.) *Morton* GC 9478 ! Mampong (fl. & fr. Nov.) *Vigne* FH 3428 ! **Togo Rep.:** Schroeder 99 ! *Büttner* 202. **Dah.:** Atacora Mts. *Chev.* 24092 ! **N.Nig.:** Naraguta (fl. & fr. Oct.-Dec.) *Coombe* 39 ! *Hepper* 1166 ! Jos Plateau *Batten-Poole* 174 ! Kurmin Damisa, Jemaa (fl. & fr. Nov.) *Keay & Onochie* FHI 21543 ! **S.Nig.:** Lagos *Dalz.* 1243 ! Sapoba *Ejiofor* FHI 24617 ! Benin F.R. (Nov.) *Obomanu & Ajayi* FHI 26929 ! Olokemeji (Nov.) *Onochie* FHI 8122 ! *Obaseki* FHI 23839 ! Onitsha *Barter* 592 ! 1300 ! Also in Cameroun and in Angola.

3. **M. ndellense** (*Lindau*) *Miège & Heine* in Ann. Fac. Sci. Univ. Dakar 6 : 121, fig. 1 (1961). *Justicia ndelleniss* Lindau in Mém. Soc. Bot. Fr. 2, 8 : 52 (1907) ; Chev. in Etud. Fl. Afr. Centr. fr. 1 : 238 (1913). A small shrub to 2 ft. high, often with a thick woody stem, greyish pubescent stems and leaves ; in fissures of sandstone rocks of high plateaux, etc.
Sen.: Banharé, Niokolo-Koba (fr. Dec.) *J. & A. Raynal* 6951 ! **Mali:** *Scaëtta* 3335 ! Bandiagara, Sanga (fr. Jan.) *Ganay* 123 ! Kita *Jaeger* 1 ! Sikasso to Bodo *Maire* ! **Guin.:** Dinguiraye, Mafadié to Tamba (fr. Dec.) *Jac.-Fél.* 1463 ! **Ghana:** Aboma F.R., Ashanti (fl. & fr. Nov.) *Vigne* FH 3443 ! Also in Ubangi-Shari and Sudan.

45. RUNGIA Nees—F.T.A. 5 : 252.

Bracts dimorphic (sterile bracts elliptic-lanceolate, fertile bracts cordate, with emarginate apex and a sharp acumen in the sinus), but dimorphism sometimes not very conspicuous :
Bracts small, up to 9 mm. long and in the broadest part 8 mm. broad (at the base 2 mm. broad), with a well-marked hyaline, sometimes pinkish margin up to 1·5 mm. broad, ciliate with multicellular hairs .. .. .. 1. *paucinervia*
Bracts 10–15 mm. long and 6–8 mm. broad, with obscure, or sometimes lacking, hyaline margins, ciliate with few-cellular hairs .. .. .. 2. *dimorpha*
Bracts uniform, with a green middle portion and broad pale hyaline margins :
Bracts rounded or emarginate at apex :
Bracts woolly-tomentose outside, about 8 mm. long ; inflorescence strobilate, 5–6 cm. long, about 1·5 cm. diam. ; leaves lanceolate, 2 cm. long, shortly pubescent
3. *eriostachya*
Bracts thinly setose outside and long-ciliate, 8–10 mm. long ; inflorescence up to 4 cm. long ; leaves broadly oblanceolate, acuminate, acute at base, 8–14 cm. long, thinly pubescent .. .. 4. *paxiana*
Bracts more or less pointed :
Bracts about 5 mm. long, minutely setulose outside ; inflorescence 2–3 cm. long ; leaves narrowly obovate, obtusely pointed, 3–6 cm. long, 1–2·5 cm. broad, glabrous, with 3–4 pairs of lateral nerves .. .. .. 5. *congoensis*
Bracts 1·5 cm. long, ovate or ovate-lanceolate, slightly pubescent outside ; inflorescence up to 12 cm. long ; leaves obovate or elliptic and pointed at both ends, up to 18 cm. long and 7 cm. broad, with 8–10 pairs of lateral nerves ; petiole long
6. *grandis*

1. **R. paucinervia** (*T. Anders. ex C. B. Cl.*) *Heine* in Kew Bull. 16 : 181 (1962). *Justicia paucinervia* T. Anders. ex C. B. Cl. in F.T.A. 5 : 186 (1899). *Rungia obcordata* Lindau in Engl. Bot. Jahrb. 38 : 71 (1905). Slender herb up to 12 in. high, rooting at lower nodes ; stems dark red ; bracts pale green with purple edges.
S.Nig.: Oban F.R. (Jan.) *Onochie & Olorunfemi* FHI 36124*x* ! Also in Cameroun and Gabon.

2. **R. dimorpha** *S. Moore* in Cat. Talb. 89 (1913). Erect branched herb or undershrub, stems pubescent at first, soon glabrous, drying blackish, with greenish flowers about ½ in. long in spikes with half-imbricate bracts.
S.Nig.: Oban *Talbot* 1528 ! [Only known from the type collection and perhaps conspecific with the preceding species—H.H.]

3. **R. eriostachya** *Hua* in Bull. Mus. Hist. Nat. 11 : 62 (1905). *R. pobeguinii* Hutch. & Dalz. F.W.T.A., ed. 1, 2 : 267 (1931). Undershrub with pubescent woody stems and leaves, and woolly spikes of small flowers hidden in closely imbricate membranous-edged bracts.

Port.G.: Boé, Madina (Jan.) *Esp. Santo* 2366! **Guin.:** Kouroussa (Sept.) *Pobéguin* 447! Friguiagbé to Bambaya (fl. & fr. Dec.) *Pobéguin* 36! Fouta Djalon (Sept.) *Chev.* 18443! *Schnell* 7021!

4. **R. paxiana** (*Lindau*) *C. B. Cl.* in F.T.A. 5 : 253 (1900). *Justicia paxiana* Lindau in Engl. Bot. Jahrb. 20 : 63 (1894). Herbaceous, branched, 4–6 ft. high in forest undergrowth ; more or less pilose, woody below, with light-purple flowers less than ½ in. long in spikes with thin papery bracts.
Guin.: Macenta, 2,000–2,500 ft. (Sept., Oct.) *Baldwin* 9783! *Jac.-Fél.* 1178! Nimba Mts. (Sept.) *Schnell* 3566! **Iv.C.:** Port Bouet to Grand Bassam (fr. Dec.) *de Wilde* 990! Mt. Tonkui (Sept., Oct.) *Schnell* 1711! *Hallé* 131! **S.Nig.:** Sonkwala area, Ogoja Prov. *Savory & Keay* FHI 25041! **[Br.]Cam.:** Cam. Mt., 3,000–4,000 ft. (Nov.) *Mann* 1968! *Dunlap* 10! Buea, 4,000 ft. (Nov., Jan.) *Migeod* 18! *Maitland* 258! below Hut I (fr. Dec.) *Morton* K701! **F.Po:** 3,000–4,000 ft. (Dec.) *Mann* 588! *Boughey* 71! Also in Cameroun.

5. **R. congoensis** *C. B. Cl.* in F.T.A. 5 : 254 (1900). A half-woody herb about 1 ft. high, with shortly pubescent branches, and very small flowers in short spikes with whitish-edged bracts.
S.Nig.: Olokemeji F.R. *Meikle* 920! Oban *Talbot* 1391! **[Br.]Cam.:** Victoria Bot. Gard. (Jan., Feb.) *Keay* FHI 28660! *Morton* K901! Extends to Congo.

6. **R. grandis** *T. Anders.* in J. Linn. Soc. 7 : 46 (1863) ; F.T.A. 5 : 252 ; Benoist in Bol. Soc. Brot., sér. 2a, 24 : 39. *Justicia grandis* (T. Anders.) Lindau in Schlechter Westafr. Kautschuk-Exp. 317 (1900) ; Chev. Bot. 500. Herbaceous undershrub up to 10 ft. high, woody below, with white, purple-veined flowers ¾ in. long in conspicuous spikes with broad green bracts with whitish papery edges.
Mali: Moussaia (fr. Feb.) *Chev.* 459! **Port.G.:** Prábis, Bissau (Feb.) *Esp. Santo* 2242! Mansaba *Esp. Santo* 762. **Guin.:** Farmar 208! Kindia *Chev.* 13145! *Jac.-Fél.* 2168! Mamou (Jan.) *Chev.* 20376! **S.L.:** Mt. Horton (fr. Jan.) *T. S. Jones* 301! Sugar Loaf Mt. (fl. & fr. Dec.) *Sc. Elliot* 3964! Kasawe F.R. (Dec.) *King* 56b! Mayogbo (fl. & fr. Feb.) *Marmo* 180! Kofiu Mt. (Jan.) *Sc. Elliot* 4616! **Lib.:** Sanokwele (May) *Harley* 1432! **Ghana:** Krachi Dist. (Nov.) *Morton* A4074! **Dah:** Pobé to Adjaouéré, Zagnanado (Feb.) *Chev.* 22934! **S.Nig.:** Ibadan (fl. & fr. Jan.) *Meikle* 931! Benin (fr. Nov.) *Keay* FHI 25589! Okomu F.R., Benin (fl. & fr. Feb.) *Richards* 5069! Idanre (fl. & fr. Jan.) *Brenan* 8674! Omo F. R., Ijebu (Feb.) *Jones & Onochie* FHI 16717! Oban *Talbot* 1394! Extends to Congo and Uganda. (See Appendix, p. 453.)

46. HYPOËSTES Soland. ex R. Br.—F.T.A. 5 : 244 ; Benoist in Notulae Syst. 10 : 241 (1942).

Flowers in slender more or less unilateral spikes or substrobiliform :
　Bracts densely long yellow setose-ciliate ; inflorescence strobiliform, terminal :
　　Bracts linear-lanceolate with a linear tail 2–5 mm. long ; inflorescences up to 4 cm. long and 2 cm. broad ; leaves linear-lanceolate, acute, 3–6 cm. long, 0·5–1 cm. broad ; corolla about 1·3 cm. long ; fruit 6 mm. long, glabrous .. .. 1. *cancellata*
　　Bracts broadly ovate, acuminate, but without tail, closely imbricate ; inflorescences up to 3 cm. long and 1·5 cm. broad ; leaves linear-oblong, base often rounded, up to 14 cm. long and 1·5 cm. broad ; corolla about 1·7 mm. long ; fruit up to 7 mm. long .. .. 2. *strobilifera* var. *tisserantii*
　Bracts not yellow setose-ciliate ; inflorescence more or less spicate :
　　Upper lip of the corolla rhomboid, sometimes marked with purple, corolla white, often cream-coloured ; external bracts of the involucre united up to the middle, sometimes marked with purple ; an extremely variable and polymorphic plant
　　　　　　　　　　　　　　　　　　　　　　　　　　　　　　　3. *verticillaris*
　　Upper lip of the corolla regularly ovate, oblong to linear, corolla almost rose or purplish ; external bracts of the involucre only united at the base　　4. *rosea*
Flowers in whorls or clusters :
　Bracts with long slender tails at the apex, slightly pubescent, up to 1·5 cm. long ; corolla-tube 1·5 cm. long ; stamens long-exserted ; leaves ovate-elliptic, gradually acuminate, broadly cuneate at base, up to 15 cm. long and 5 cm. broad, thinly pilose on the midrib beneath .. .. 5. *aristata*
　Bracts without tails at the apex:
　　Inflorescences more or less sessile ; floral leaves small, lanceolate, acuminate, up to 8 mm. long, often with long slender hairs in the upper part　3. *verticillaris*
　　Inflorescences often apparently terminal on lateral branches, 1·2–5 cm. long ; floral leaves obovate to oblanceolate, up to 1 cm. long and 4 mm. broad, shortly setose-pilose ; leaves broadly ovate, acuminate, 3–5 cm. long and 1·5–3 cm. broad, setose-pilose on both surfaces ; petiole slender, up to 3 cm. long :
　　　Floral leaves rounded at apex, about 10 mm. long and 4 mm. broad ; flowers almost whitish, upper lip sometimes spotted with purple .. .. 6. *triflora*
　　　Floral leaves markedly acuminate, usually 12 mm. long and 3 mm. broad ; flowers pinkish, magenta or mauve .. .. 7. *consanguinea*

1. **H. cancellata** *Nees* in DC. Prod. 11 : 505 (1847) ; F.T.A. 5 : 246 ; Chev. Bot. 501 ; Benoist l.c. 242, Bol. Soc. Brot., sér. 2, 24 : 39 ; Berhaut Fl. Sén. 85. Herb 1–3 ft. high, with slightly pubescent sulcate stems and flowers ⅝ in. long, bright red, red- or blue-purple.
Sen.: Niokolo-Koba *Berhaut* 1441. **Port.G.:** Cacine *Esp. Santo* 722. **Guin.:** Boulivel to Dalaba *Chev.* 18673bis. Timbo (fr. Nov.) *Maclaud* 26! Kouroussa *Pobéguin* 1837! **S.L.:** Leicester Peak (fl. & fr. Dec.) *Sc. Elliot* 3843! Kasawe F.R. (Dec.) *King* 20b! Musaia (Dec.) *Deighton* 4448! Wallia (fl. & fr. Feb.) *Sc. Elliot* 4265! Kumrabai (Dec.) *Thomas* 6747! 6832! 7093! Yalamba (fr. Jan.) *Jaeger* 3948! **Iv.C.:** Touba (Nov.) *Aké Assi* 2814! **[Br.]Cam.:** Kwagiri to Gangumi, Gashaka Dist. (fr. Dec.) *Latilo & Daramola* FHI 28787! Vogel Peak, 3,850 ft., Adamawa (Nov.) *Hepper* 1336! Also in Congo.

2. **H. strobilifera** *S. Moore* in J. Bot. 18 : 40 (1880) ; F.T.A. 5 : 248 ; Benoist l.c. 242 (1942). Var. *strobilifera* occurs in Congo, Sudan, and tropical E. Africa.
H. strobilifera var. **tisserantii** *Benoist* l.c. (1942) ; Monod in Mém. I.F.A.N. 18 : 13 (1952). An erect herb up to 3 ft. high ; flowers bright mauve, in strobiliform inflorescences ; in open savanna.
Togo Rep.: Sokodé-Basari *Schroeder* 104 (partly, mixed with *Monechma hispidum*)! **N.Nig.:** Dogon Kurmi, Jemaa Div. (fl. & fr. Nov.) *Keay* FHI 37222! Naraguta & Jos (Sept.) *Lely* 569! Kwa

Plantations, Plateau Prov. (Nov.) *Keay* FHI 21029 ! Damisa, Jemaa (fl. & fr. Nov.) *Keay & Onochie* FHI 21653 ! **[Br.]Cam.:** Ndop to Kumbo road (fr. Dec.) *Boughey* GC 11179 ! Vogel Peak, Adamawa (fl. & fr. Nov.) *Hepper* 1411 ! Also in Ubangi-Shari and Congo.

3. **H. verticillaris** (*Linn. f.*) *Soland. ex Roem. & Schult.* Syst. Veg. 1 : 140 (1817) ; F.T.A. 5 : 250 ; Chev. Bot. 502 ; Benoist l.c. 247 ; Milne-Redhead in Mem. N.Y. Bot. Gard. 9, 1 : 26 (1954) ; Berhaut Fl. Sén. 22. *Justicia verticillaris* Linn. f. (1781). *Hypoestes preussii* Lindau (1894)—F.T.A. 5 : 251. *H. violaceo-tincta* Lindau (1897)—F.T.A. 5 : 251 ; F.W.T.A., ed. 1, 2 : 268. *H. forskalei* (Vahl) Soland. ex Roem. & Schult. (1817). *Justicia forskalei* Vahl (1790). *Hypoestes paniculata* (Forsk.) Schweinf. (1912)—Christensen in Dansk Bot. Arkiv 4 : 10 (1922) ; Robyns in Fl. Parc. Nat. Alb. 2 : 298. *Justicia paniculata* Forsk. (1775), not of Burm. f. (1768). *Hypoestes verticillaris* R. Br. var. *forskalei* (R. Br.) Benoist, var. *hildebrandtii* (Lindau) Benoist, var. *mollis* (T. Anders.) Benoist, var. *violaceo-tincta* (Lindau) Benoist, l.c. 246–247 (1942).[1] Erect tufted herb 2–3 ft. high ; in open places in various habitats.
Sen.: *Heudelot* 552 ! " Ravin des Voleurs " *Berhaut* 1587. **Mali:** Badinko *Chev.* 114. Bâlani *Chev.* 147. Djenne *Chev.* 2788. Kara (Dec.) *Davey* 194 ! **Port.G.:** Suzana, Praira Varela (fr. Mar.) *Esp. Santo* 2247 ! **Guin.:** Mali (Jan.) *Roberty* 16573 ! Santa to Timbo *Chev.* 12591. **S.L.:** Yiraia (Jan.) *Glanville* 482 ! Gbwema (Dec.) *King* 102b ! Bumbuna (fr. Jan.) *Deighton* 3599 ! Senge, Tunkia (Jan.) *Deighton* 3864 ! Mamansu (Feb.) *Deighton* 4063 ! Bafodeya (Apr.) *Sc. Elliot* ! **Iv.C.:** Oroumba-Boca (Aug.) *de Wit* 5692 ! Bemouni (Jan.) *de Wit* 7974 ! **Ghana:** Kpeve (Nov.) *Adams & Akpabla* 4003 ! Bamboi to Wenchi (Dec.) *Adams & Akpabla* 4508 ! Krepi Plains (Jan.) *Johnson* 543 ! Agogo (Dec.) *Irvine* 575 ! Kaasi, Kumasi (Feb.) *Darko* 514 ! **Togo Rep.:** Misahöhe (Apr., Nov.) *Baumann* 476 ! Mildbr. 7315 ! Alafayo *Krause* ! **N.Nig.:** Kontagora (Oct.) *Dalz.* 187 ! Naraguta (Nov.) *Lely* 702 ! Vom, 3,000–4,500 ft. *Dent Young* 198 ! Jos (Nov.) *Keay* FHI 21745 ! **S.Nig.:** Lagos (Mar.) *Millen* 54 ! Awka Dist. *Thomas* 28 ! Owo *Burton* 6 ! Eket *Talbot* ! Oban *Talbot* 1373 ! Cross R. *Johnston* ! **[Br.]Cam.:** Cam. Mt., Buea *Preuss* 755 ! Bafut-Ngemba F.R. (Jan.-Feb.) *Tiku* FHI 22161 ! *Keay & Lightbody* FHI 28371 ! Bamenda, 6,500–8,000 ft. (Jan.) *Migeod* 339 ! 414 ! Kumbo (Feb.) *Hepper* 1978 ! Vogel Peak, Adamawa (Nov.) *Hepper* 1459 ! Widespread in Africa, and in Arabia. (See Appendix, p. 451.)

4. **H. rosea** *P. Beauv.* Fl. Oware 2 : 66, t. 100 (1818) ; F.T.A. 5 : 248 (1900), partly (excl. syn. *H. consanguinea* Lindau, *Preuss* 599) ; F.W.T.A., ed. 1, 2 : 268, partly (excl. *Preuss* 599) ; Milne-Redhead in Kew Bull. 1940 : 64. *H. barteri* T. Anders. (1863)—F.T.A. 5 : 248. F.W.T.A., ed. 1, 2 : 268. *H. aristata* var. *barteri* (T. Anders.) Benoist l.c. 244 (1942). *H. triticea* Lindau (1894). *H. talbotii* S. Moore in Cat. Talb. 88 (1913). Erect herb 1½–3 ft. high ; gregarious in forest.
S.Nig.: Lagos (fl. & fr. Mar.) *Millen* 35 ! Barter 3285 ! Benin to Sapele (Jan.) *Latilo* FHI 27334 ! Ojogba to Uku, Benin (Jan.) *Okon* FHI 22751 ! Ubulubu (Dec.) *Thomas* 2093 ! Idumaye (Dec.) *Thomas* 2117 ! Oban *Talbot* 2005 ! 2020 ! Calabar (Feb.) *Onochie* FHI 36338 ! **[Br.]Cam.:** Victoria (Feb.) *Dalz.* 8220 ! Buea, 3,000 ft. (Jan.) *Maitland* 235 ! *Dunlap* 130 ! Crater L., Kumba *Morton* K819 ! Oku, 7,000 ft. (Feb.) *Hepper* 2718 !

5. **H. aristata** (*Vahl*) *Soland. ex Roem. & Schult.* Syst. Veg. 1 : 140 (1817) ; F.T.A. 5 : 245 ; Benoist l.c. 243 ; Milne-Redhead in Mem. N.Y. Bot. Gard. 9 : 26 (1954). *Justicia aristata* Vahl Symb. bot. 2 : 2 (1791). *Hypoestes insularis* T. Anders. (1863)—F.T.A. 5 : 246. *H. aristata* var. *insularis* (T. Anders.) Benoist l.c. 244 (1942). *H. staudtii* Lindau (1895)—F.T.A. 5 : 246. *H. aristata* var. *staudtii* (Lindau) Benoist l.c. 244 (1942). Erect herb about 3 ft. high ; flowers pale mauve with darker markings ; in forest.
N.Nig.: Dogon Kurmi, Jemaa Div. (Nov.) *Keay & Onochie* FHI 21536 ! Ropp (Dec.) *W. D. MacGregor* FHI 405 ! **S.Nig.:** Oban *Talbot* 1409 ! 2019 ! Old Ndebiji, Calabar (fr. Feb.) *Aninze* FHI 15414 ! Unya, Sonkwala, Obudu Div. (Dec.) *Savory & Keay* FHI 25018 ! Cross R. (Dec.) *Holland* 182 ! **[Br.]Cam.:** Cam. Mt., 2,500 ft. (Dec.) *Mann* 1951 ! Buea, 3,000 ft. (Dec.) *Maitland* 143 ! Bakundu F.R., Kumba (Jan.) *Binuyo & Daramola* FHI 35069 ! Bamenda *Morton* K162 ! **F.Po:** (Jan.) *Mann* 179 ! Moka, 4,000–5,000 ft. (Jan.) *Exell* 779 !

6. **H. triflora** (*Forsk.*) *Roem. & Schult.* Syst. Veg. 1 : 141 (1817) ; F.T.A. 5 : 247 ; Chev. Bot. 502 ; Benoist l.c. 243 ; Milne-Redhead l.c. 27. *Justicia triflora* Forsk. Fl. Aegypt.-Arab. Descr. Pl. 4 (1775). Straggling herb 2–3 ft. high, laxly branched, sometimes with viscid hairs ; in montane forest.
Guin.: Diaguissa Plateau *Chev.* 12680. **Ghana:** Shiare (fl. & fr. Apr.) *Hall* 1433 ! **[Br.]Cam.:** Cam. Mt., 3,000–7,000 ft. (fl. & fr. Feb.-Mar.) *Mann* 1979 ! *Preuss* 745 ! *Maitland* 257 ! 826 ! *Brenan & Onochie* FHI 9397 ! Gembu, 4,500 ft., Mambila Plateau (Jan.) *Hepper* 1834 ! **F.Po:** Clarence Peak, 5,000 ft. (Dec.) *Mann* 580 ! Moka, 4,000–4,500 ft. (Feb.) *Exell* 867 ! near Pico Biao (Sept.) *Wrigley* 591 ! Extends to Ethiopia, E. Africa and Angola.

7. **H. consanguinea** Lindau in Engl. Bot. Jahrb. 20 : 50 (1894). *H. rosea* of F.T.A. 5 : 248, partly (excl. *Beauvois, Don*), not of P. Beauv. ; Milne-Redhead in Kew Bull. 1940 : 64. Erect, laxly branched herb, 2–3 ft. high ; flowers pale pink or whitish with mauve markings on lip ; in montane forest.
Togo Rep.: Bismarckburg (Oct., Nov.) *Büttner* 315 ! *Kling* 1889 ! **S.Nig.:** Sonkwala area, 3,800–4,000 ft., Obudu Div. (Dec.) *Savory & Keay* FHI 25039 ! 25195 ! **[Br.]Cam.:** Buea (Jan.) *Preuss* 599 ! Lakom, 6,000 ft., Bamenda (fl. & fr. Apr.) *Maitland* 1687 ! Mbakokeka Mt., 7,000 ft. (Jan.) *Keay* FHI 28393 ! Kumbo (fl. & fr. Jan., Feb.) *Keay* FHI 28452 ! *Hepper* 2841 ! Oku, 6,000 ft. (Feb.) *Hepper* 2032 !

47. BRACHYSTEPHANUS Nees—F.T.A. 5 : 177.

Branches glabrescent ; leaves subequal, ovate, broadly acuminate, very shortly cuneate at base, about 10 cm. long and 6 cm. broad, ciliate, with numerous very minute short cystoliths beneath ; spikes sessile, dense, up to 7 cm. long ; bracts with very long ciliate tails ; calyx-segments about 1·5 cm. long, linear-subulate ; corolla 3 cm. long ; stamens 2, anthers 1-celled ; fruit 1 cm. long .. .. *longiflorus*

B. **longiflorus** *Lindau* in Engl. Bot. Jahrb. 20 : 53 (1894) ; F.T.A. 5 : 178. Herbaceous undershrub 3 ft. high, with slender purple flowers nearly 1 in. long in dense spike-like panicles.
S.Nig.: Ikwette Plateau, 5,500 ft., Obudu Div. (Aug.) *Stone* 77 ! **[Br.]Cam.:** Buea *Preuss* 890 ! Cam. Mt., grassland 1,000 ft., Hut II (fr. Dec.) *Gregory* 229 ! **F.Po:** 3,000 ft. (fl. & fr. Dec.) *Mann* 589 ! Lago Biao, 4,000 ft. (Sept.) *Wrigley* 671 ! Monte Balea (fr. Dec.) *Guinea* 446 !

48. RUSPOLIA Lindau in E. & P. Pflanzenfam. 4, 3B : 354 (1895) ; Milne-Redhead in Kew Bull. 1936 : 268. *Eranthemum* of F.T.A. 5 : 169, partly ; and of F.W.T.A., ed. 1, 2 : 262, partly, not of Linn.

Flowers in shortly stalked cymules ; calyx-lobes broadly subulate-lanceolate ; corolla-tube 2·5 cm. long ; anthers exserted, 1-celled ; leaves ovate or elliptic, broadly acuminate, 5–12 cm. long, 3–5 cm. broad, nearly glabrous ; fruit 4 cm. long, stoutly club-shaped .. .. *hypocrateriformis*

R. **hypocrateriformis** (*Vahl*) *Milne-Redhead* in Kew Bull. 1936 : 270 ; Benoist in Bol. Soc. Brot., sér. 2, 24 : 39. *Justicia hypocrateriformis* Vahl Enum. 1 : 165 (1805). *Eranthemum hypocrateriforme* (Vahl) Roem. & Schult. (1817)—F.T.A. 5 : 171 ; F.W.T.A., ed. 1, 2 : 262. *Pseuderanthemum hypocrateriforme* (Vahl)

[1] am quite unable to distinguish properly these taxa accepted by Benoist in varietal rank.—H.H.

Radlk. (1883)—F.W.T.A., ed. 1, 2 : 608 ; Berhaut Fl. Sén. 116. *P. decurrens* of Chev. Bot. 500, not of Radlk. A shrubby plant 3–4 ft. high, with stiff pithy angled branches, and slender tubular flowers 1½ in. long coral-red with a darker centre forming a rather dense linear panicle 2–7 in. long.
Sen.: Dakar (Dec.) *Vigne* FH 1773 ! *Berhaut* 1151. Casamance, Carabane (Jan.) *Chev.* 2795 ! **Mali**: Kita (fl. & fr. Oct.) *Jaeger* 2964 ! 3353 ! **Port.G.**: Bissau *Esp. Santo* 1224. Cusselinta (fr. Dec.) *Esp. Santo* 3174 ! **Guin.**: Kindia to Forécariah (fl. & fr. Oct.) *Jac.-Fél.* 2113 ! **S.L.**: *T. Vogel* 159 ! *Smeathmann* ! Kent (Nov.) *Tindall* 11 ! *Deighton* 5536 ! Picket Hill, 2,300 ft. (Nov.) *Jones* 207 ! Sugar Loaf Mt., 1,000 ft. (Aug.) *Melville & Hooker* 360 ! Wallia (Jan.) *Sc. Elliot* 4451 ! **Ghana**: *Farmar* 469 ! Accra (Jan.-Mar.) *Deighton* 617 ! *Dalz.* 72 ! Achimota (Apr., Dec.) *Bally* 12 ! *Andoh* 4471 ! Aburi (Nov.) *Howes* 1015 ! *Williams* 288 ! (fl. & fr. Jan.) *de Wit & Morton* A2857 ! **Togo Rep.**: Lomé *Warnecke* 386 ! 387 ! 388 ! **N.Nig.**: Wana, Plateau Prov. (fl. & fr. Dec.) *Hepburn* 150 ! **S.Nig.**: Eruwa rocks, Oyo *Rowland* ! *Hoskyns-Abrahall* FHI 25657 ! **[Br.]Cam.**: Gashaka Dist. (Nov.) *Latilo & Daramola* FHI 28729 ! Vogel Peak, 4,800 ft., Adamawa (Dec.) *Hepper* 1588 ! Also in Congo and Angola. Only the type variety occurs in our area ; var. *australis* Milne-Redhead occurs in Transvaal.

49. OREACANTHUS Benth.—F.T.A. 5 : 176.

Branches shortly pubescent, with long internodes ; leaves subequal in each pair, elliptic, narrowed at both ends, long-acuminate, 15–20 cm. long, 4–7 cm. broad, thin, pubescent on the numerous lateral nerves beneath ; petiole pubescent ; panicle large and pyramidal ; calyx-lobes linear, 4 mm. long, viscid ; corolla-2-lipped, about 1·3 cm. long, the posticous subentire, the anticous 3-lobed ; stamens 2, long-exserted ; anthers 1-celled ; fruit 1·5 cm. long   ..   ..   *mannii*

O. mannii *Benth.* l.c. (1876) ; F.T.A. 5 : 177. Shrub 8–12 ft. or more high, with stems constricted and brittle above the nodes, and numerous white or light purple flowers ⅓ in. long.
[Br.]Cam.: Cam. Mt., 7,000 ft. (fl. & fr. Dec.) *Mann* 1259 ! 8,000 ft. (fl. Dec., fl. & fr. Feb.) *Johnston* 64 ! 65 ! *Kalbreyer* 132 ! *Preuss* 1078 ! *Dunlap* 50 ! *Keay* FHI 3703 ! Mimbia, 5,000 ft. (fr. Mar.) *Brenan* 9339 ! **F.Po**: El Pico, 7,000 ft. (Dec.) *Boughey* 118 ! 195 ! Also in Cameroun.

160. VERBENACEAE

By H. Huber[1], F. N. Hepper[2] and R. D. Meikle[3]

Herbaceous or woody, often with quadrangular branchlets. Leaves usually opposite or whorled, simple or compound ; stipules absent. Flowers hermaphrodite, zygomorphic. Calyx 4–5-lobed or toothed, persistent. Corolla gamopetalous, tubular, 4–5-lobed, lobes imbricate. Stamens on the corolla, 4 or rarely 2 or 5 ; anthers 2-celled, cells often divergent, opening lengthwise. Ovary superior, 2–8-celled, often 4-celled. Style terminal, simple. Fruit a drupe or berry. Seeds with straight embryo and scanty or no endosperm.

Perfect stamens 2, staminodes 2 ; flowers in slender spikes, more or less immersed in the soft swollen rhachis ; herbs or undershrubs   ..   ..   1. **Stachytarpheta**
Perfect stamens 4, often didynamous :
  Flowers spicate or racemose ; leaves always simple ; herbs or undershrubs :
    Ovary 4-celled :
      Calyx in fruit open, not inflated ; pyrenes 4   ..   ..   2. **Verbena**
      Calyx closed over the fruit as a small bladder ; pyrenes 2   ..   ..   3. **Priva**
    Ovary 2-celled ; fruit of 1 or 2, 1-seeded pyrenes :
      Epicarp fleshy ; corolla-tube usually much exceeding calyx   ..   4. **Lantana**
      Epicarp dry :
        Trailing herbs rooting at the nodes ; corolla-tube only a little longer than the calyx
                                                               5. **Phyla**
        Erect herbs or undershrubs:
          Lower stems not corky ; corolla-tube only a little longer than the calyx ; savanna herbs or subshrubs   ..   ..   6. **Lippia**
          Lower stems with soft corky bark ; corolla-tube twice as long as calyx ; undershrub in semi-desert   ..   ..   7. **Chascanum**
  Flowers cymose ; shrubs or trees :
    Fruit dry ; leaves always simple :
      Pyrenes 4, 1-celled ; shrubs   ..   ..   8. **Premna**
      Pyrene 1, 4-celled ; mostly climbing shrubs   ..   ..   9. **Clerodendrum**
    Fruit a fleshy drupe ; leaves usually digitately compound ; mostly trees   10. **Vitex**

The following genera have been introduced into our area : *Gmelina arborea* Roxb. and Teak, *Tectona grandis* Linn. f. (both from tropical Asia), as plantation trees ; *Duranta repens* Linn. (tropical America), *Holmskioldia sanguinea* Retz. (India) and *Petrea volubilis* Linn. (tropical America), as ornamental shrubs.

1. STACHYTARPHETA Vahl—F.T.A. 5 : 283 ; Brenan in Kew Bull. 5 : 223 (1950).[4]

Calyx, when seen from the anticous [out-]side, showing 4 prominent teeth at apex, the two central ones equalling in length or rather shorter than the laterals ; leaves broad,

[1] Genera 9, 10.
[2] Genera 1–3, 7, 8.
[3] Genera 4–6.
[4] The taxonomy and nomenclature of this account are adapted directly from the excellent revision of the W. African species by J. P. M. Brenan, to whom acknowledgement is made.

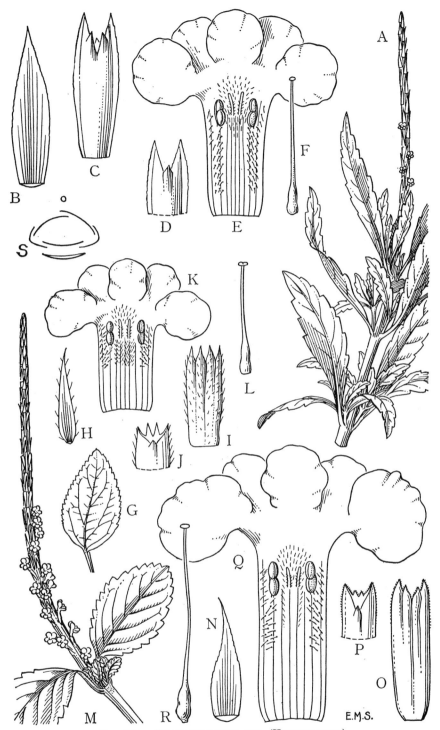

FIG. 305.—STACHYTARPHETA SPP. (VERBENACEAE).

S. *angustifolia* (Mill.) Vahl—A, habit, × ⅔. B, floral bract. C, calyx, anticous side. D, apex
of calyx, posticous side. E, corolla opened out. F, pistil. All × 6. Drawn from
Deighton 4172.
S. *cayennensis* (L. C. Rich.) Schau.—G, leaf, × ⅔. H, floral bract. I, calyx, anticous side.
J, apex of calyx, posticous side. K, corolla opened out. L, pistil. All × 6. Drawn from
Deighton 4291.
S. *indica* (Linn.) Vahl—M, inflorescence, × ⅔. N, floral bract. O, calyx, anticous side. P,
apex of calyx, posticous side. Q, corolla opened out. R, pistil. All × 6. Drawn from
Deighton 4173.
S, floral diagram suitable for each species.

ovate to elliptic or oblong-elliptic, teeth on the well-developed stem leaves rather close
and numerous (10–)11–18 on each margin :

Plant shrubby, with woody stems at least below ; inflorescences up to 20 cm. long in
fruit, almost always sparsely but distinctly pubescent ; calyx 5 mm. long; corolla-tube
4–5 mm. long, lobes 1·5 mm. long ; leaves 2–8 cm. long, 1–3·5 cm. broad, marginal
teeth rather small   ..   ..   1. *cayennensis*

Plant herbaceous ; inflorescences up to 40 cm. long in fruit, glabrous ; calyx 6·5 cm.
long ; corolla-tube about 10 mm. long, lobes about 3 mm. long ; leaves 4–11 cm.
long, 2–4·5 cm. broad, marginal teeth rather large   ..   ..   ..   2. *indica*

Calyx, when seen from the anticous [out-]side, appearing bifid at apex, the 2 central
teeth reduced to small inconspicuous denticles on the side of much larger lateral
teeth ; inflorescences up to 25 cm. long in fruit ; leaves usually narrow, oblanceolate
to linear-oblanceolate, 3–8(–13) cm. long, 0·5–2(–6) cm. broad, with 4–9(–10) large
distinct teeth on each margin ; plant herbaceous, rather fleshy   3. *angustifolia*

1. **S. cayennensis** (*L. C. Rich.*) *Schau.* in DC. Prod. 11 : 562 (1847) (as *Stachytarpha*) ; Danser in Ann. Jard.
Bot. Buitenz. 40 : 2 (1929) ; Brenan in Kew Bull. 5 : 223 (for other refs. and synonymy). *Verbena
cayennensis* L. C. Rich. (1792). *Stachytarpheta indica* of F.T.A. 5 : 284, partly (*Sc. Elliot* 4162), not of
(Linn.). Vahl. *S. jamaicensis* of F.W.T.A., ed. 1, 2 : 277, partly (*Sc. Elliot* 4162, *Dawodu* 364, *Dennett* 502),
not of (Linn.) Vahl. A slender-branched undershrub 3–6 ft. high, with narrow spreading spikes ; flowers
white or pale blue with white centre ; roadsides and waste places.
S.L.: *Sc. Elliot* 4162! Freetown (June) *Deighton* 2716! Leicester (Mar.) *Hepper* 2484! Regent (Oct.)
Jordan 947! Kenema (Apr.) *Deighton* 4291! Dambye, Nongowa (Feb.) *Bakshi* 21! **Lib.:** Monrovia (June)
Barker 1341! **Ghana:** Benso, Tarkwa Dist. (June) *Andoh* FHI 5526! **S.Nig.:** Lagos (June) *Dennett*
502! *Dawodu* 364! Idanre, Ondo Prov. (Jan.) *Brenan* 8709! Ibadan *Meikle* 516! Olokemeji F.R. (Aug.)
Keay FHI 37152! Eket Dist. *Talbot*! A native of tropical America now widely introduced into other
parts of the tropics.

2. **S. indica** (*Linn.*) *Vahl* Enum. Pl. 1 : 206 (1805) ; F.T.A. 5 : 284, mostly ; Danser l.c. 5 ; Chev. Bot. 503 ;
Brenan l.c. 225. *Verbena indica* Linn. (1762). *Stachytarpheta jamaicensis* of F.W.T.A., ed. 1, 2 : 277, partly
(*T. Vogel* 30, *Mann* 89, *Sc. Elliot* 3833, *Johnston*). A well-branched herb 2–3 ft. high with very long narrow
spikes ; flowers deep blue with white centre ; a weed.
S.L.: *Sc. Elliot* 3833! No. 2 River F.R. (Apr.) *Hepper* 2525! Makumri (June) *Thomas* 532! Njala (Oct.)
Deighton 2545! Rokupr (Feb.) *Deighton* 4173! **Lib.:** *Johnston*! Monrovia (May, Nov.) *Barker* 1315!
Linder 1567! **Iv.C.:** *fide* Chev. l.c. **Ghana:** *Chipp* 48! Ankobra Junction (Feb.) *Kitson* 1028! Cape Coast
T. Vogel 30! **S.Nig.:** Ilesha to Akure (Oct.) *Keay* FHI 25455! Usonigbe F.R., Sapoba (Nov.) *Meikle* 612!
Oban *Talbot* 307! Orem, Calabar (Jan.) *Onochie & Daramola* FHI 36180z! Old Calabar (Jan.) *Holland*
15! **F.Po:** (Dec.) *Mann* 89! Also in E. Africa and tropical Asia. (See Appendix, p. 456.)

3. **S. angustifolia** (*Mill.*) *Vahl* Enum. Pl. 1 : 205 (1805) ; F.T.A. 5 : 284 (1900) ; Chev. Bot. 503 ; F.W.T.A.,
ed. 1, 2 : 277 ; Brenan l.c. 226 ; Berhaut Fl. Sén. 72. *Verbena angustifolia* Mill. (1768). A herb about 1 ft.
(sometimes up to 4 ft.) high, simple or slightly branched, often rather succulent ; flowers pale blue with or
without a white centre ; in moist places.
Sen.: *Perrottet* 650! *Berhaut* 789. **Gam.:** *Hayes* 549! **Mali:** Kabarah (Aug.) *Chev.* 1322! Timbuktu
(Aug.) *Hagerup* 257! Bamako (Apr.) *Hagerup* 32! Niger R., E. of Macina *Lean* 17! Dogo (fl. & fr. Apr.)
Davey 567! **Port.G.:** Povoacão, Buba *Esp. Santo* 1241! Buba (June) *Esp. Santo* 2489! Madina, Boé
(June) *Esp. Santo* 2934! **Guin.:** Kouroussa (June) *Pobéguin* 298! **S.L.:** Bulom (Feb.) *T. S. Jones* 400!
Rokupr (Feb.) *Deighton* 4172! Kabala (fl. & fr. Sept.) *Thomas* 2181! **Iv.C.:** Alépé to Malamalasso *Chev.*
17508. Daloa (Aug.) *Schnell* 5953! **Ghana:** Achimota (Apr.) *Akpabla* 18! Prampram *Robertson* 38!
Nungua, Accra Plains (May, June) *Ankrah* GC 20146! 20186! Otisu ferry (Apr.) *Goodall* GC 16628! Salaga
Krause! **Togo Rep.:** Lomé *Warnecke* 259! Sokodé *Schroeder* 121! **Dah.:** *Burton*! **Niger:** Gao (Aug.)
Popov 99! **N.Nig.:** Aguji, Ilorin *Thornton*! Jebba *Barter*! Nupe *Barter* 1200! Gwari Hill near Minna
(Dec.) *Meikle* 702! Lemme (May) *Lely* 142! Gindiri, Jos Plateau (Oct.) *Hepper* 1149! **S.Nig.:** Lagos
Dawodu 224! Ogun R., Olokemeji (Mar.) *Meikle* 1245! Nkisi Bridge, Onitsha (May) *Killick* 117! Otu
Enenia, Onitsha (June) *Onochie* FHI 35858! [Br.]**Cam.:** Rio del Rey *Johnston*! Gwoza, Dikwa Div. (Nov.)
McClintock 62! Scattered in other parts of tropical Africa ; also in tropical America.

2. VERBENA Linn.—F.T.A. 5 : 286.

A densely pubescent annual with several stems arising from near the base and branched
above ; leaves deltoid in outline, bipinnatifid, 1–4 cm. long, 5–15 mm. broad ;
inflorescences terminal and axillary, dense at first, elongating considerably in fruit ;
calyx 2 mm. long, pilose ; corolla-tube 3–4 mm. long, lobes 1 mm. long ; fruits 4-
celled, oblong, 2 mm. long   ..   ..   *supina*

V. supina *Linn.* Sp. Pl. 21 (1753) ; F.T.A. 5 : 286 ; Berhaut Fl. Sén. 65. A small annual herb 2–12 ins. high,
appearing grey-green ; flowers very small, lilac ; in seasonally flooded ground.
Maur.: Aftout es Sahéli Region (fl. & fr. Apr.) *Popov* 15! **Sen.:** Richard Toll *Berhaut* 554. **Mali:** Koubita
(fl. & fr. Feb.) *Davey* 440! Also in Sudan, Ethiopia, and around the Mediterranean.

V. tenera Spreng., a native of S. America, has been introduced into Nigeria (Obubra Dist., FHI 43964) ;
a creeping herb with very dissected leaves and white or violet flowers.
V. officinalis Linn. is reported by De Miré & Gillet (in J. Agric. Trop. 3 : 721 (1956)) from Aïr, Niger
Republic, on the northern limit of our area.

3. PRIVA Adans.—F.T.A. 5 : 285 ; Moldenke in Fedde Rep. 41 : 1–76 (1936).

A perennial herb 15 (–100) cm. high ; stems acutely 4-angled, slightly pubescent
especially at the nodes ; leaves ovate, acute or acuminate at apex, truncate at base,
1·4 (–14·5) cm. long, 1–2 (–8·5) cm. broad, margins coarsely serrate, with scattered
white bulbous-based hairs above, more or less pubescent beneath ; petioles 0·5–3 cm.
long ; inflorescences terminating the stems and branchlets, racemose with rather
widely spaced flowers ; calyx tubular-campanulate, about 2 mm. long, inflating in
fruit and about 6 mm. long, densely pubescent ; corolla 2-lipped, 5-lobed, 2–3 mm.
diam., fruits with 2 pyrenes, each with 2 seeds   ..   ..   ..   *lappulacea*

P. lappulacea (*Linn.*) *Pers.* Syn. Pl. 2 : 139 (1806) ; Moldenke l.c. 24. A perennial weed with spreading and decumbent or erect stems, markedly 4-angled ; flowers small, the corolla violet with paler stripes.
Ghana: Asebu, near Cape Coast (fl. & fr. Apr.) *Hall* 1891 ! A native of the warmer parts of America and the West Indies ; introduced into Java and possibly elsewhere, but not previously recorded from Africa.

4. LANTANA Linn.—F.T.A. 5 : 275.

Bracts narrow, linear ; flowers showy, white, yellow, red, orange or pink ; corolla much longer than subtending bract ; foliage turning blackish on drying :
  Leaves, petioles and peduncles pilose or strigose, not or very sparsely glandular ; stems usually armed with short recurved prickles .. .. ..   1. *camara*
  Leaves, petioles and peduncles densely glandular ; stems unarmed
                2. *glandulosissima*
Bracts broader, ovate or ovate-lanceolate ; flowers not showy ; foliage not turning black on drying :
  Corolla distinctly longer than the subtending bract, 3–5 mm. wide at apex ; peduncle usually long and slender, longer than the fruiting spike :
    Leaves (except on the weakest growths) in threes ; lower floral bracts ovate-lanceolate, gradually narrowing to a slender acumen ; flowers pink or mauve   3. *trifolia*
    Leaves opposite ; lower floral bracts broadly ovate with a short cuspidate apex ; flowers white .. ..   4. *viburnoides*
  Corolla scarcely longer than subtending bract, 1–2 mm. wide at apex ; peduncle usually shorter than fruiting spike ; bracts of fruiting spike broadly ovate ; flowers usually magenta or purple .. ..   5. *rhodesiensis*

1. **L. camara** *Linn.* Sp. Pl. 2 : 627 (1753) ; F.T.A. 5 : 275 ; Chev. Bot. 502. *L. antidotalis* Thonning (1827)—F.T.A. 5 : 276. Erect or spreading much-branched, square-stemmed shrub, usually with short prickles (var. *aculeata* (Linn.) Moldenke) and showy flowers in convex heads 1–2 in. across.
Sen.: Lelièvre. **Port.G.:** Balama (fl. & fr. Apr.) *Esp. Santo* 1917 ! **S.L.:** *Barter* ! *Kirk* 44 ! Freetown (Jan.) *Johnston* 1 ! Magbile (Dec.) *Thomas* 6420 ! Nyanyahun (Apr.) *Deighton* 1613 ! Sumbuya (Apr.) *Deighton* 1693 ! Makump (May) *Deighton* 1704 ! **Lib.:** Monrovia *Johnston* ! **Iv.C.:** Adiopodoumé (Oct.) *Leeuwenberg* 1723 ! 2122 ! **Ghana:** *Thonning.* Cape Coast (July) *T. Vogel* 69 ! Accra (Mar., Oct.) *Deighton* 599 ! *Morton* GC 6005 ! Dawa (May) *Howes* 906 ! Kumasi (Aug.) *Darko* 698 ! Asabotchway (Nov.) *Rose Innes* GC 30092 ! **Togo Rep.:** Lomé *Warnecke* 23 ! *Mildbr.* 7474 ! **S.Nig.:** Lagos *Barter* 3318 ! *Moloney* ! Ogwashi (fl. Nov.) *Thomas* 2033 ! Ezi (Feb.) *Thomas* 2334 ! Ibadan (fl. & fr. Jan.) *Meikle* 1125 ! Port Harcourt (Oct.) *Stubbings* 62 ! Calabar (Jan.) *Holland* 18 ! [**Br.**]**Cam.:** Gwoza, Dikwa Div. (Jan.) *McClintock* 152 ! Cultivated throughout the tropics and commonly found as an escape, or a relic of cultivation. (See Appendix, p. 455.)

2. **L. glandulosissima** *Hayek* in Fedde Rep. 2 : 161 (1906) ; Moldenke, Resumé of Verbenac. 138 (1958). Very similar to the last but with densely glandular stems, leaves etc. and without prickles ; flowers yellow or orange.
Mali: near Djimal, Douentza (fr. Oct.) *Jaeger* 5522 ! **N.Nig.:** Naraguta, Jos Plateau (fl. & fr. July) *Lely* 540 ! 547 ! A native of Mexico.

3. **L. trifolia** *Linn.* Sp. Pl. 2 : 626 (1753) ; F.T.A. 5 : 277. *L. mearnsii* Moldenke in Phytologia 1 : 421 (1940). A woody herb or subshrub, to about 6 ft. high, sparsely branched, with hairy stems and rugose scabrid leaves in threes ; flowers pink or mauve ; fruits purple.
S. Nig.: Lagos *Rowland* ! A native of tropical America now widely naturalized in the Old World and a pest in some areas, but apparently rare in our area.

4. **L. viburnoides** (*Forsk.*) *Vahl* Symb. Bot. 1 : 45 (1790) ; F.T.A. 5 : 276, partly. *L. salvifolia* of F.T.A. 5 : 276, partly ; F.W.T.A., ed. 1, 2 : 269, partly, not of Jacq. Aromatic straggling subshrub or woody herb to about 3 ft. high ; peduncles long, usually exceeding the leaves ; flowers relatively large, white.
N.Nig.: Yola (fl. & fr. July) *Dalz.* 114 ! Widely distributed in the drier areas of tropical and subtropical Africa, also in Arabia and eastwards to India.

5. **L. rhodesiensis** *Moldenke* in Phytologia 3 : 269 (1950). *L. viburnoides* of F.T.A. 5 : 276, partly, not of (Forsk.) Vahl. *L. salvifolia* of F.T.A. 5 : 276, partly ; F.W.T.A., ed. 1, 2 : 269, partly, not of Jacq. *L. mearnsii* Moldenke var. *latibracteolata* Moldenke in Phytologia 2 : 313 (1947). A woody herb of open grassland, 1–6 ft. high with leaves in threes or occasionally opposite ; peduncles short ; flowers minute usually magenta or purple ; fruits crimson or purple.
Port.G.: Pirada (June) *Esp. Santo* 3023 ! Piche to Paiama (fl. & fr. Oct.) *Esp. Santo* 3472 ! **Guin.:** Timbo (fl. & fr. Oct.) *Jac.-Fél.* 1876 ! **S.L.:** Yetaya (Sept.) *Thomas* 2438 ! Kanya (fl. Oct.) *Thomas* 303 ! Bintumane (fl. Jan.) *Glanville* 461 ! Musaia (fl. Apr.) *Deighton* 4589 ! **Ghana:** Accra Plains (fl. & fr. Mar.) *Howes* 1132 ! Nchenenchene to Kwahu Tafo (fl. & fr. May) *Kitson* 1151 ! Damongo (fl. & fr. July) *Andoh* 5205 ! Prang (fl. & fr. Sept.) *Darko* WACRI 989 ! Salaga *Krause* ! *Dalz.* 56 ! **N.Nig.:** Nupe *Barter* 1286 ! Lokoja (Oct.) *Parsons* 11 ! 122 ! Jos Plateau *Batten-Poole* 177 ! 318 ! Anara F.R., Zaria (May) *Keay* FHI 22938 ! Abinsi (Mar.) *Dalz.* 695 ! Biu, Bornu (fl. & fr. Aug.) *Noble* 14 ! **S.Nig.:** Lagos *Dawodu* in Hb. MacGregor 182 ! Olokemeji (Apr.) *Savory* UCI 87 ! Widely distributed in the savanna regions of tropical Africa. (See Appendix, p. 455.)
  [The following specimens should probably be assigned to this species though differing somewhat in their rather larger, paler (pink or mauve) flowers and in their smaller floral bracts: **Ghana:** Assuantsi Road (fl. & fr. Jan.) *Fishlock* 27 ! Adaiso (fl. & fr. Nov.) *Morton* 6147 ! Amedzofe (fl. Nov.) *Morton* GC 9383 ! **N.Nig.:** Jos Plateau (July) *Lely* P358 ! Pankshin (fl. & fr. July) *Lely* 431 ! **S.Nig.:** Shaki, Oyo (fl. & fr. Nov.) *Olorunfemi* FHI 40306 ! [**Br.**]**Cam.:** Vogel Peak, Adamawa (fl. & fr. Dec.) *Hepper* 1536 ! Hybrids occur in the genus, and it is just possible that some of these aberrant specimens may be of hybrid origin, a possibility which should be borne in mind by collectors—R.D.M.]

5. PHYLA Lour. Fl. Cochinch. 66 (1790) ; Greene in Pittonia 4 : 48 (1899).

Rootstock woody ; stems slender herbaceous, rooting at the nodes ; leaves opposite, oblanceolate or obovate, 1·5–4 cm. long, 0·5–3 cm. broad, with an entire cuneate base and an acute serrate apex, glabrous or thinly clothed with appressed medifixed hairs ; inflorescence a long-stalked, dense, bracteate cylindrical spike ; calyx dorsiventrally flattened, cleft almost to base abaxially and to about the middle adaxially ; corolla small, unequally 4-lobed ; stamens 4, didynamous ; stigma obliquely capitate ; pyrenes 2 .. ..   *nodiflora*

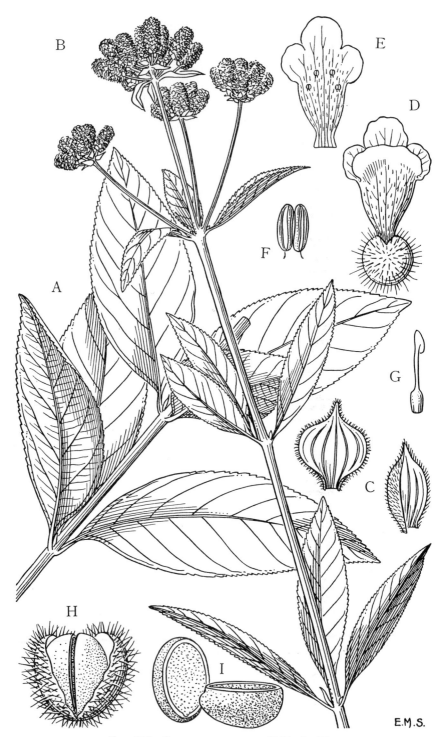

B

E

D

A

F

G

C

H

I

E.M.S.

Fig. 306.—Lippia multiflora *Moldenke* (Verbenaceae).

A, leaves, × ½. B, inflorescence, × ⅔. C, floral bracts, × 10. D, flower, × 12. E, part of corolla showing insertion of stamens, × 12. F, stamen, × 100. G, style, × 12. H, fruit, × 20. I, seeds, × 20. A, H, I are drawn from *Meikle* 722. B–G from *Meikle* 1231.

P. nodiflora (*Linn.*) *Greene* in Pittonia 4 : 46 (1899). *Verbena nodiflora* Linn. (1753). *Lippia nodiflora* (Linn.) Michx. (1803)—F.T.A. 5 : 279 ; F.W.T.A., ed. 1, 2 : 270. A prostrate creeping herb of moist places ; leaves and bracts often tinged purple ; flowers mauve or pink.
Sen.: *Heudelot*! Dakar (May) *Baldwin* 5738! L. Retbo, near Cape Verde (Apr.) *Giovanetti*! **Mali:** Grand Daouna (fl. Aug.) *Chev.* 2752! Modou to Berivem (Sept.) *Chev.* 10155! Timbuktu (Aug.) *Hagerup* 254! side of L. Faguibini (Oct.) *Monod* 4047! **Port.G.:** Bananto to Conjabarim (June) *Esp. Santo* 2396! **Lib.:** Grand Cess (Mar.) *Baldwin* 11621! **Ghana:** Busua Bay (Mar.) *Morton* A305! Pui, Keta Dist. (July) *Akpabla* 40! Keta (Mar.) *Morton* GC 6533! **N.Nig.:** Sokoto (July) *Dalz.* 366! Katsina to Daura (Apr.) *Meikle* 1361! Kalkala (fl. Mar.) *Gwynn* 101! S. of Yo, Bornu (fl. & fr. Dec.) *Elliott* 147! Baga, Seyoram (Oct.) *Golding* 24! Bindawa (Aug.) *Grove* 8! **S.Nig.:** Lagos (fl. & fr. Sept.) *Stubbings* 8! Widespread in the tropics and subtropics. (See Appendix, p. 456.)

6. LIPPIA Linn.—F.T.A. 5 : 278, partly.

Flower-heads in spreading, much-branched corymbose cymes :
  Upper surface of leaves smooth to the touch, venation obscure ; stems subglabrous or sparsely appressed-pubescent ; apex of floral bracts very obtuse, shortly mucronate

                                                   1. *multiflora*
  Upper surface of leaves rough to the touch, venation prominently rugose-reticulate ; stems usually distinctly appressed-pubescent ; apex of floral bracts acute or cuspidate .. .. 2. *rugosa*
Flower-heads in narrow, elongate terminal and axillary verticels or spikes :
  Floral bracts obtuse or shortly cuspidate ; leaves smooth or slightly scabridulous ; stems subglabrous or sparsely appressed-pubescent .. .. .. 3. *chevalieri*
  Floral bracts acute ; leaves strongly scabrid ; stems densely pubescent or pilose

                                                   4. *savoryi*

1. **L. multiflora** *Moldenke* in Phytologia 3 : 168 (1949). *L. adoensis* of F.T.A. 5 : 280, partly ; F.W.T.A., ed. 1, 2, partly, not of Hochst. Robust woody perennial up to 12 ft. high with large oblong-lanceolate bluish-green leaves ; pleasantly aromatic flowers ; small, whitish in branched inflorescences ; in savanna.
Guin.: Boulivel, Mamou *Mission Pharm. A.O.F.* 2! Kindougou *Martine* 188! **S.L.:** R. Scarcies, Wallia *Sc. Elliot* 4262! Duunia, Talla Hills (Feb.) *Sc. Elliot* 4878! Madina (fl. & fr. Aug.) *Smythe* 121! Kamaramka (Feb.) *Deighton* 4089! Kabala (fl. & fr. Jan.) *King* 190b! Falamba (fl. & fr. Jan.) *Jaeger* 3944! **Iv.C.:** 30 km. N. of Bouaké (Nov.) *Leeuwenberg* 2066! **Ghana:** Krepi Plains (Jan.) *Johnson* 542! Esuosa (Feb.) *Vigne* FH 1813! Kpere (Jan.) *Irvine* 1743! Salaga (fl. & fr. June) *Vigne* FH 3881! Ejura (Dec.) *Morton* GC 9551! Tamale (Dec.) *Morton* GC 9911! Apepesu F.R., Hohoe (Nov.) *Enti* FH 6848! **Togo Rep.:** Misahöhe (Nov.) *Mildbr.* 7238! **Dah.:** *Poisson*! **N.Nig.:** Borgu *Barter* 768! Minna (Dec.) *Meikle* 722! Bosso, Minna (Feb.) *Meikle* 1231! N. of Bukuru (Nov.) *Keay* FHI 21040! Biu, Bornu (Nov.) *Noble* 40! **S.Nig.:** Lagos *Millen* 143! Olokemeji *Foster* 133! Awba Hills F.R. (Dec.) *Keay & Jones* FHI 14810! Igboho, Oyo (Feb.) *Keay* FHI 23433! Akure (Oct.) *Latilo* FHI 27331! Also in Rio Muni, Gabon, Congo, Sudan and Ubangi-Shari. (See Appendix, p. 455.)
2. **L. rugosa** *A. Chev.* in Mém. Soc. Bot. Fr. 2, 8 : 191 (1912) ; Chev. Bot. 503. *L. adoensis* of F.W.T.A., ed. 1, 2 : 270, partly, not of Hochst. *L. nigeriensis* Moldenke in Phytologia 3 : 272 (1950). Similar to the last in general habit, but with scabrid leaves and prominent leaf-venation.
Guin.: Tagania, Faranah (fl. & fr. Jan.) *Chev.* 20410! **N.Nig.:** Funtua (fl. & fr. Dec.) *Meikle* 824! Anara F.R., Zaria (Nov.) *Latilo* FHI 37974! Naraguta & Jos (fl. Sept.–Dec.) *Lely* 582! *Monod* 9361! 9872! *Coombe* 9! Pankshin (Dec.) *MacGregor* 429! Gindiri (Oct.) *Hepper* 1092! Sokoto *Lely* 115! **S.Nig.:** Aku Rock (Nov.) *Hambler* 585! [**Br.**]**Cam.:** Bamenda (May) *Maitland* 1697! 10 km. N. of Bafut (fl. & fr. Jan.) *Keay* FHI 28522! Gangumi, Adamawa (Dec.) *Latilo & Daramola* FHI 28795! Gurum, Vogel Peak area (Nov.) *Hepper* 1269!
3. **L. chevalieri** *Moldenke* in Phytologia 2 : 313 (1947). *L. adoensis* of F.T.A. 5 : 280, partly ; F.W.T.A., ed. 1, 2 : 270, partly, not of Hochst. An erect aromatic subshrub to about 3 ft. high ; floral bracts covered with whitish or silvery-grey tomentum often becoming bright yellow on drying ; flowers minute, whitish with a yellow throat.
Sen.: *Heudelot* 103! M'Bidjem *Thierry* 227! Sanyako, Toubacouta (Oct.) *Monod* 8681! **Gam.:** *Ingram*! *Brown-Lester* 4! 28! *Dawe* 30! *Saunders* 108! Genieri (Feb.) *Fox* 48! **Mali:** Toukota (Dec.) *Chev.* 67! **Port.G.:** Sonaco to Gabu (Dec.) *Esp. Santo* 2341! Nova Lamego to Canjufa (Oct.) *Esp. Santo* 3552! **Guin.:** Timbo, Kouria etc. *Caille* in *Hb. Chev.* 14772! Boulivel, Mamou (Nov.) *Mission Pharm. A.O.F.* 3/1934! Ditinn (Oct.) *Schnell* 7411! **S.L.:** Wallia, Scarcies (Feb.) *Sc. Elliot* 4285! **N.Nig.:** Zaria (fl. & fr. Dec.) *Meikle* 289! Samaru (Aug.) *Keay* FHI 28012! Gwaram, Farin Dutse F.R. (Nov.) *Onwudinjoh* FHI 24037!
4. **L. savoryi** *Meikle* in Kew Bull. 17 : 173 (1963). *L. multiflora* Moldenke var. *pubescens* Moldenke in Phytologia 3 : 271 (1950). *L. africana* Moldenke var. *sessilis* l.c. 271 (1950). *L. nigeriensis* Moldenke var. *brevipedunculata* Moldenke l.c. 272 (1950). *L. adoensis* of F.W.T.A., ed. 1, 2 : 270, partly, not of Hochst. Erect woody perennial or subshrub to about 5 ft. high ; leaves greyish-green, very scabrid ; floral bracts pale green, densely white villose ; flowers white with a yellow throat.
Ghana: Bole (fl. Nov.) *Harris*! **N.Nig.:** Jos and Naraguta (May–Jan.) *Lely* 69! 241! 320! P40! Shere Hills (Dec.) *Coombe* 112! Kafenkey to Jos (Apr.) *Gregory* 281! Kontagora (Oct.) *Dalz.* 177! Ukata to Kontagora (fl. & fr. Jan.) *Meikle* 1059! Birnin Gwari (secondary growth, fl. Mar.) *Meikle* 1344! Kurra Falls *Savory* UCI 108! Assob (fr. Feb.) *McClintock* 244!

7. CHASCANUM E. Mey. Comm. Pl. Afr. Austr. 275 (1837) ; Moldenke in Fedde Rep. 45 : 114 (1938). *Bouchea* of F.T.A. 5 : 281, not of Cham.

Small shrub or perennial herb ; lower stems with soft corky bark, branchlets densely puberulous, hairs about 1 mm. long ; leaves ovate, broadly elliptic or suborbicular, obtuse at apex, truncate or broadly cuneate at base, 0·5–4·5 cm. long, 0·5–3 cm. broad, coarsely dentate, puberulous on both sides, venation prominent beneath ; petioles usually about 1 cm. long ; inflorescences terminal-spikes 3–35 cm. long ; calyx about 6 mm. long, less than twice as long as bract ; corolla-tube about 12 mm. long, lobes 5, subequal, 2 mm. diam. ; fruits dry, included in calyx-tube, linear, dividing into two pyrenes .. .. *marrubiifolium*

C. marrubiifolium *Fenzl ex Walp.* Repert. 4 : 38 (1845) ; Moldenke l.c. 135 ; Gillett in Kew Bull. 10 : 133 (1955). *Bouchea marrubiifolia* (Fenzl ex Walp.) Schau. in DC. Prod. 11 : 558 (1847) ; F.T.A. 5 : 282 ; De Miré & Gillet in J. Agric. Trop. 3 : 720. A branched herb or small shrub up to 3 ft. high, with white-pubescent foliage ; flowers white in narrow spikes ; in semi-desert.

Maur.: Oualata to Néma *Jumelle*. **Guin.** (?): Upper Senegal & Niger *Buchanan*. **Mali**: Gao (fl. & fr. Sept.) *Hagerup* 345 ! **Niger**: Agadez, Aïr Mts. *Petit-Lagrange* 60 ; 66. Taraouadji massif, Aïr *Gillet* 1057 ; 1066. Elmiki, Aïr *De Mire* N2–96. Distributed across N. tropical Africa to Arabia and Pakistan.

8. PREMNA Linn.—F.T.A. 5 : 287.

Leaves narrowed to the base :
 Lamina pilose-pubescent beneath :
  Leaves broadly obovate, acutely acuminate, up to 20 cm. long and 10 cm. broad, with 8–10 pairs of lateral nerves ; corymbs terminal, many-flowered ; calyx pubescent, enlarging and nervose in fruit ; fruit obovoid, 1 cm. long.. ..   1. *hispida*
  Leaves elongate-obovate-oblanceolate, long and acutely acuminate, 25–30 cm. long, 6–9 cm. broad, thin, denticulate in the upper half, pilose especially on the midrib beneath, with 10–14 pairs of lateral nerves ; petiole 1 cm. long, hirsute ; cyme small and subsessile, rusty-pilose except the pedicels ; calyx very small, glabrous   2. *grandifolia*
 Lamina glabrous beneath, oblong, acutely acuminate, 7–8 cm. long, 2·5–3 cm. broad, with about 4 pairs of lateral nerves ; petiole 1·5–2 cm. long ; corymbs small, few-flowered .. .. 3. *milnei*
Leaves rounded to cordate at base:
 Lamina ovate, usually in whorls of 4 ; calyx truncate or slightly bilobed :
  Leaves entire, 9–14 cm. long, 6–12 cm. broad, glabrous or finely pubescent mainly on the nerves beneath ; petioles 3–6 cm. long, glabrous or pubescent ; inflorescence glabrous or finely and densely pubescent, terminal, becoming lax and expansive, about 20 cm. diam., with numerous small flowers .. .. 4. *angolensis*
  Leaves coarsely crenate, 7–11 cm. long, 4–7·5 cm. broad, softly pubescent beneath ; petioles 2–3 cm. long, pubescent ; inflorescence finely and densely pubescent, terminal, pyramidal, about 12 cm. long and 10 cm. across the base, with numerous small flowers .. .. 5. *quadrifolia*
 Lamina broadly oblong to oblong-elliptic, rounded (or sometimes slightly cordate) at base, 5–9 cm. long, 2–6 cm. broad, glabrous to softly pubescent beneath ; frequently with reflexed arrested axillary branches terminating in a pair of leaves and a small inflorescence about 1 cm. diam. ; principal inflorescence terminal, a rather dense corymb about 5 cm. diam. ; calyx with small teeth .. .. 6. *lucens*

1. **P. hispida** *Benth.* in Fl. Nigrit. 485 (1849) ; F.T.A. 5 : 290 ; Chev. Bot. 504 ; Berhaut Fl. Sén. 109, 119 ; Aubrév. Fl. For. C. Iv., ed. 2, 3 : 236, t. 339. A straggling shrub or small tree up to 20 ft. high ; flowers greenish-white, fruits orange ; often in secondary forest.
 Sen.: *Leprieur* ! *Heudelot* 61 ! Koulaye, Casamance *Chev.* 2770. **Port.G.:** Bissau (fl. & fr. Mar.) *Esp. Santo* 1157 ! 1843 ! 1870 ! **Guin.:** *Heudelot* 761 ! Conakry *Chev.* 15408. Kindia *Chev.* 12791 ; 13383. Mamou (Mar.) *Dalz.* 8401 ! *Farmar* 302 ! **S.L.:** Sugar Loaf Mt. (fr. Mar., May) *Barter* ! *Hepper* 2490 ! Njala (Sept.) *Deighton* 2113 ! Mabonto (Feb.) *King-Church* 17 ! Bonganema (fr. May) *Deighton* 5948 ! Sasseni, Scarcies (Jan.) *Sc. Elliot* 4521 ! **Lib.:** Sodu (Jan.) *Bequaert* 49 ! Sinoe Basin *Whyte* ! Kakatown *Whyte* ! Gbanga (fr. Sept.) *Linder* 489 ! Buchanan (Mar.) *Baldwin* 11171 ! Suacoco (fr. Feb.) *Konneh* 126 ! **Iv.C.:** Bingerville *Chev.* 15408 ; 15560. Bingerville to Potou lagoon *Chev.* 20086. Diamancrou, Nzi *Croux* in *Hb. Chev.* 20129. Mt. Tonkoui, Man (fr. Mar.) *Leeuwenberg* 2935 ! **Ghana:** Nsemre F.R., near Wenchi (Dec.) *Adams* 5369 ! Twindorase, Kwahu Plateau (May) *Kitson* 1192 ! Aboma F.R. (Jan.) *Andoh* FH 5452 ! Amedzofe (fr. Feb.) *Irvine* 171 ! (See Appendix, p. 456.)
2. **P. grandifolia** *A. Meeuse* in Blumea 5 : 72 (1942). *P. macrophylla* A. Chev. ex Hutch. & Dalz. F.W.T.A., ed. 1, 2 : 272 (1931) ; Chev. Bot. 505, name only ; not of Wall. ex Schau. (1847), nor of H. J. Lam (1919). A shrub 2–3 ft. high, densely rusty-pilose on the branchlets and midribs, with long leaves and small white flowers.
 Iv.C.: Kéeta, Mid. Cavally (July) *Chev.* 19366. Grabo, Cavally Basin *Chev.* 19654.
3. **P. milnei** *Bak.* in F.T.A. 5 : 291 (1900). A shrub with glabrous branchlets and small white flowers in a pubescent slender-branched corymbose panicle.
 S.Nig.: Calabar *Robb* ! **F.Po:** *Milne* !
4. **P. angolensis** *Gürke* in Engl. Bot. Jahrb. 18 : 165 (1893). *P. zenkeri* Gürke l.c. 33 : 292 (1903) ; F.W.T.A., ed. 1, 2 : 272 ; Berhaut Fl. Sén. 100, 119. A small tree up to 40 ft. high or (?) climbing shrub, flowers white, very small in copious panicles.
 Sen.: " gallery forest " *Berhaut* 596. **Guin.:** Timbo *Pobéguin* 741 ! **S.L.:** Bumbuna (fr. Oct.) *Thomas* 3171 ! Pujehun (July) *Deighton* 5802 ! Sumbuya, near Mattru (June) *Deighton* 4774 ! **Ghana:** Komenda, Cape Coast (July) *Andoh* 5737 ! **Togo Rep.:** Basari (July) *Kersting* I1121 ! **N.Nig.:** summit of Gwalor Hill, Niger Prov. (June) *Onochie, Latilo & Okeke* FHI 40176 ! **S.Nig.:** Ibadan South F.R. (Apr.) *Keay* FHI 22818 ! Iyamoyong, Obubra Dist. (fr. Aug.) *Adebusuyi* FHI 43976 ! [**Br.]Cam.:** Buea, 3,200 ft. (fr. May) *Maitland* 661 ! *Ainsley* 283 ! Su, 3,000–4,000 ft., Bamenda (May) *Maitland* ! Widespread in tropical Africa.
 [The Sierra Leone specimens and *Keay* FHI 22818 are cited with caution since they are described as climbing shrubs, and their inflorescences are usually smaller than typical *P. angolensis*. The diagnostic character of bilobed calyx attributed to *P. zenkeri* is only visible in young buds and it cannot be maintained as a distinct species ; amounts of indumentum are very variable—F.N.H.]
5. **P. quadrifolia** *Schum. & Thonn.* Beskr. Guin. Pl. 275 (1827) ; F.T.A. 5 : 289 ; Chev. Bot. 505. *Gaertnera ferruginea* A. Chev. Bot. 444, name only. A shrub up to 10 ft. high, aromatic ; flowers white ; in thickets.
 Port.G.: Catio, N'emberem to Quéfu (June) *Esp. Santo* 2078 ! **Iv.C.:** Mt. Niénokoué, Mid. Cavally (July) *Chev.* 19457 ! Orodougou, Siñé to Séguéla *Chev.* 21819. **Ghana:** Accra (June) *Dalz.* 144 ! (Mar.) *Deighton* 594 ! Nungua, Accra Plains (May) *Rose Innes* GC 30063 ! Legon Hill (Dec.) *Adams* 3597 ! Achimota (Feb.) *Dalz.* 8263 ! Aburi Hill (June) *Williams* 296 ! **S.Nig.:** *T. Vogel* ! Okpauani *A. E. Kitson* ! Abeokuta *Barter* 3358 ! (See Appendix, p. 456.)
6. **P. lucens** *A. Chev.* in Mém. Soc. Bot. Fr. 2, 8 : 192 (1912) ; Chev. Bot. 505. *P. quadrifolia* of F.W.T.A., ed. 1, 2 : 272, partly (syn. and Chev. 22392). *P. gracilis* A. Chev. l.c. 191 (1912) ; Chev. Bot. 504. A climber with clusters of white flowers and red fruits.
 Guin.: Comoe waterfalls, near Banfora (Sept.) *Jaeger* 5258 ! **S.L.:** Nganyama (fr. Nov.) *Deighton* 5264 ! **Lib.:** Mecca, Boporo Dist. (Nov.) *Baldwin* 10415 ! Peahtah (Oct.) *Linder* 916 ! Gbau, Sanokwele Dist. *Baldwin* 9411 ! Totokwelli (fr. Oct.) *Linder* 1284 ! Tappita *Baldwin* 9098 ! **Iv.C.:** Diahbo, Baoulé-Nord (July) *Chev.* 22071 ! Bouaké to Mt. Lémélébou *Fleury* in *Hb. Chev.* 22101 ! Touminiaré to Angouakoukro (Aug.) *Chev.* 22392 ! **Ghana:** Odomi River F.R. (Oct.) *St. C.-Thompson* FH 3637 !

9. CLERODENDRUM Linn.—F.T.A. 5 : 292 ; B. Thomas in Engl. Bot. Jahrb. 68 : 1 (1936).

Calyx-lobes obtuse and semi-orbiculate to semi-elliptic ; corolla-tube zygomorphic, lobes longer than the tube ; fruit 4-lobed, not breaking into pieces at maturity ; inflorescences always terminal :
Leaves sessile, entire or serrate, mostly verticillate, obovate-elliptic to obovate-lanceolate, 7–14 cm. long, 2·5–6 cm. broad, glabrescent or shortly puberulous ; stems erect, subherbaceous .. .. 1. *alatum*
Leaves distinctly petiolate, mostly entire :
Leaves quite glabrous on both sides except for a minute pubescence on the nerves beneath ; leaves thinly membranaceous, ovate, elliptic or oblong, 5–12 cm. long, 3–10 cm. broad ; straggling or climbing shrub .. .. .. 2. *violaceum*
Leaves pubescent or tomentose, at least all over the undersurface ; leaves almost coriaceous and very fragile when dry (probably slightly fleshy when alive) :
Leaves sparsely puberulous and when dry blackish-green on their upper side, puberulous but not tomentose beneath ; leaves opposite, broadly ovate to ovate-lanceolate, 3–12 cm. long, 1–10 cm. broad ; calyx lobed to about ½ or more of its length .. .. 3. *carnosulum*
Leaves thinly tomentellous and when dry to dull brown above, densely tomentellous beneath ; opposite or in whorls of 3, broadly ovate to elliptic, 3–6 cm. long and 2–4 cm. broad ; calyx lobed to about ⅓ of its length .. 4. *tomentellum*
Calyx-lobes acute or acuminate ; corolla-tube actinomorphic or slightly zygomorphic, the lobes shorter than the tube ; fruit breaking into 2 or 4 pieces at maturity :
Fruit breaking into 2 halves at maturity ; inflorescences axillary, pedunculate, few-flowered, becoming spinescent ; corolla-tube about 1·5 mm. long ; introduced erect spiny shrub with rather small ovate, elliptic or oblong leaves, cuneate at base, entire, 1·5–5 cm. long, 5–20 mm. broad, glabrous or almost so ; young branches and peduncles puberulous ; branches solid .. .. .. 5. *aculeatum*
Fruit breaking into 4 quarters at maturity ; inflorescences axillary or terminal, never spinescent :
*Corolla-tube 1·4–12 cm. long ; branches hollow or solid (to p. 440) :
†Calyx-teeth 5–30 mm. long :
Branches deeply sulcate, hollow in the internodes ; calyx 1–1·5 cm. long, lobed almost to the base, segments linear, less than 2 mm. broad ; corolla-tube about 3 cm. long ; leaves deltoid-ovate, acuminate, 6–9 cm. long, 4–6 cm. broad, crenate, long-petiolate, puberulous beneath .. .. 6. *eupatorioides*
Branches not sulcate :
Flowers in branched, ample, corymbose, terminal inflorescences, never clustered ; corolla-tube 1·4–5 cm. long (to p. 440):
Leaf-margin entire, leaves ovate or elliptic ; African species :
Calyx-teeth 5 mm. long or longer, deltoid ; leaves herbaceous, ovate, 6–20 cm. long, 3·5–11 cm. broad, rounded or very often slightly cordate at base, glabrous or frequently minutely puberulous ; lower leaves long-petiolate, petiole 2–8 cm. long ; corolla-tube puberulous .. .. 7. *umbellatum*
Calyx 12 mm. or more long, lobed to ⅔ or more of its length ; corolla-tube up to 1·5 times longer than calyx ; calyx glabrous or hairy :
Leaves, branches and calyx glabrous ; leaves thinly membranaceous, ovate, 6–18 cm. long, 3·5–8 cm. broad ; calyx white, 14–30 mm. long ; corolla crimson .. .. 8. *thomsonae*
Leaves, branches and calyx long-hirsute (hairs 2–5 mm. long) ; leaves ovate, 6–10 cm. long and 2·5–6 cm. broad ; calyx 12–15 mm. long ; white or yellowish with a red centre .. .. 9. *buettneri*
Leaf-margin crenate, dentate or undulate-sinuate, leaves broadly ovate or deltoid, branches solid ; introduced species :
Inflorescence a divaricate panicle ; calyx (3–)6 mm. long ; corolla deep red, its tube 5–7 times longer than calyx ; leaves broadly ovate-cordate, 5–25 cm. long, almost as broad, tomentose beneath ; climbing shrub, 10. *speciosissimum*
Inflorescence a dense head ; calyx 1–1·5 cm. long (including teeth) ; corolla-tube a little longer than the calyx ; leaves deltoid, 7–20 cm. long and almost as broad, shortly pilose, especially beneath ; erect undershrub ; flowers usually double .. .. 11. *japonicum*
Flowers in head-like clusters, peduncled or sessile, terminal or axillary :
Branches hollow in the internodes ; calyx lobed to ¾ of its length or deeper, 12–20 mm. long, the segments lanceolate, 4–8 mm. broad ; corolla-tube 4–12 cm. long :
Bracts linear-filiform ; inflorescences only lateral on old leafless branches ; flowering branches mostly about 1 cm. diam. ; leaves glabrous on both sides, large, elliptic, 15–25 cm. long, 6–12 cm. broad, with 5–9 pairs of lateral nerves, entire or indistinctly and obtusely dentate ; inflorescences more or less sessile, many-flowered .. .. 12. *globuliflorum*

Bracts ovate-acuminate or ovate-lanceolate ; inflorescences terminal or terminal and lateral[1] on young branches ; not cauliflorous :
Leaves with 3–6 pairs of lateral nerves ; corolla-tube 4–8 cm. long ; leaves ovate or elliptic, 7–20 cm. long, 4–10 cm. broad, short- to long-petiolate :
Calyx-segments ciliate or fimbriate on the margins ; young branches patent-pilose or hispid ; leaves scabrid or glabrescent above, pilose beneath especially on the midrib and lateral nerves, rarely on the midrib only ; leaves entire or coarsely sinuate-dentate, shortly cuneate to slightly cordate at base .. .. 13. *capitatum* var. *capitatum*
Calyx-segments quite glabrous, as well as the young branches and leaves ; leaves always entire, shortly cuneate or rounded at base
13a. *capitatum* var. *cephalanthum*
Leaves with 7–10 pairs of lateral nerves ; corolla-tube 8–10 cm. long ; leaves elliptic-oblong, 15–17 cm. long, 5–6 cm. broad, margin entire, glabrous except the sparsely pubescent midrib beneath ; petioles up to 1 cm. long ; young branches densely and shortly pubescent ; calyx segments ciliate
14. *whitfieldii*
Branches solid ; calyx lobed to about ½ its length, 6–14 mm. long ; corolla-tube 1·5–5 cm. long :
Branches, leaves and inflorescences whitish-tomentose when young ; leaves rather small 1·5–7 cm. long, 0·7–4 cm. broad, broadly ovate to ovate-lanceolate, petiolate ; flowers in lateral and terminal, peduncled, head-like clusters ; calyx-lobes lanceolate ; corolla-tube 1·5–2·5 cm. long 15. *acerbianum*
Branches yellowish-brown pubescent ; leaves rather large, ovate, cordate or truncate at base, 10–18 cm. long and 7–12 cm. broad, short- or long-petiolate ; flowers in solitary terminal subcapitate inflorescences ; calyx-lobes deltoid ; corolla-tube 3–5 cm. long .. .. 16. *welwitschii*
†Calyx-teeth up to 4 mm. long :
Corolla-tube 6–12 cm. long ; leaves opposite, broadly lanceolate, acuminate, gradually attenuate to base and scarcely petiolate, 6–14 cm. long, 1·5–5 cm. broad, coarsely dentate ; stem erect, herbaceous .. .. 17. *incisum*
Corolla-tube up to 4 cm. long ; leaves distinctly petiolate :
Calyx-tube 3–5 times longer than the teeth ; teeth usually glabrous ; frequently cauliflorous :
Inflorescence an elongate, leafless panicle, frequently cauliflorous ; rhachis 5–30 cm. long ; corolla-tube up to 2·5 cm. long, (1·2–)2–3 times longer than the calyx ; leaves elliptic or ovate, entire, 8–20 cm. long, 3–10 cm. broad, glabrous
18. *buchholzii*
Inflorescence capitate, with the rhachis abbreviated, 0·5–1(–2) cm. long ; corolla-tube 3–4 cm. long, 5–8 times longer than the calyx ; leaves elliptic, 9–20 cm. long, 4–9 cm. broad, glabrous or minutely puberulous beneath, entire or frequently dentate .. .. 19. *schweinfurthii*
Calyx-tube shorter to about as long as the teeth, if longer than the teeth then distinctly pilose :
Inflorescence a solitary, terminal, densely head-like cluster ; corolla-tube at most 1·5 cm. long :
Corolla-tube 3–4 times longer than the calyx ; calyx-teeth deltoid ; (other characters as above) .. .. 16. *welwitschii*
Corolla-tube 1–1·5 times longer than the calyx ; calyx-teeth lanceolate to subulate :
Calyx 10–15 mm. long, with the teeth ciliate ; leaves deltoid ; flowers usually double ; introduced erect undershrub ; (other characters as below)
11. *japonicum*
Calyx 6–9 mm. long, pilose all over ; leaves broadly ovate or elliptic, all of margin serrate or obscurely dentate, scabrous-pubescent above, yellow-tomentose to greyish-pubescent beneath ; corolla-tube 10–14 mm. long ; native climbing shrub .. .. 20. *sinuatum*
Inflorescence laxly branched or if clustered not contracted into a single head :
Inflorescence on young, leafy branches, terminal or axillary ; calyx-teeth 2–4 mm. long ; leaves elliptic or ovate, rounded, subcordate or subcuneate at base, 6–20 cm. long, 4–10 cm. broad, glabrous on both sides ; petiole 3–20 mm. long .. .. 21. *splendens*
Cauliflorous ; inflorescence leafless, much-branched ; calyx-teeth about 1 mm. long ; corolla-tube about 1·5 cm. long or perhaps longer, puberulous, about 1 mm. broad when pressed ; leaves elliptic, 10–20 cm. long, 4–8 cm. broad, glabrous .. .. 22. *bipindense*
*Corolla-tube up to 1 cm. long ; calyx-teeth not more than 3 mm. long ; branches solid (except No. 28 with hollow internodes) :

---

[1] Lateral inflorescences do occur ,but rarely, and it seems only when the terminal bud has been damaged.

Leaves opposite or subopposite :
  Calyx 5 mm. long or longer, the tube 3–5 times longer than the teeth ; frequently cauliflorous ; leaves glabrous; (see also above)   ..   ..   ..   18. *buchholzii*
  Calyx less than 5 mm. long or if longer then leaves markedly pilose :
    Calyx-lobes as broad as or broader than long, usually spreading ; pedicels slender, mostly more than 5 mm. long ; leaves glabrous :
      Whole plant turning black when dry ; petiole fairly long, more than 3 cm. long, lamina ovate, cordate at base, 6–10 cm. long and almost as broad ; corolla-tube glabrous or nearly so, less than 1 cm. long, usually less than 1 mm. broad when pressed ; corolla-lobes 2–3 mm. long ; inflorescence terminal
                                          23. *melanocrater*
      Plant not turning black when dry ; petioles short, up to about 2 cm. long ; corolla-lobes longer ; corolla-tube 6–10 mm. long ; inflorescence terminal :
        Branches puberulous ; inflorescence elongate, the internodes of the rhachis usually 2 cm. or more long ; corolla-tube 2 mm. or more broad when pressed, densely pilose ; leaves ovate or elliptic, acuminate, 6–14 cm. long, 3–7 cm. broad   ..   ..   24. *thyrsoideum*
        Branches glabrous ; inflorescence short, usually subcorymbose, the internodes of the axis less than 1 cm. long ; corolla-tube less than 1 mm. diam. when pressed, almost glabrous ; leaves ovate or elliptic, acuminate, 5–15 cm. long, 2–7 cm. broad   ..   ..   25. *volubile*
    Calyx-lobes longer than broad, mostly erect ; pedicels short, usually less than 5 mm. long ; inflorescence frequently clustered ; leaves usually pilose, especially on the nerves beneath :
      Calyx and the whole inflorescence hairy all over, 5–7 mm. long ; leaves opposite, 12–18 cm. long, 7–12 cm. broad, broadly ovate, pubescent on both surfaces ; inflorescences composed of few rather large head-like clusters
                                          26. *inaequipetiolatum*
      Calyx minutely puberulous to almost glabrous, 2–4 mm. long ; leaves mostly opposite, rarely in whorls of 3, rounded or sometimes cuneate at base, elliptic to ovate, glabrous above or slightly pubescent on the midrib only, pubescent on the midrib and nerves beneath ; corolla-tube about 5 mm. long
                                          27. *dusenii*
Leaves in whorls of mostly 3 ; inflorescence always terminal :
  Branches hollow in the internodes ; leaves acuminate at apex, usually short-cuneate at base, elliptic, 4–10 cm. long, 2–5 cm. broad, quite glabrous (except sometimes midrib beneath) ; calyx 2–3 mm. long ; corolla-tube very short, 4–5 mm. long ; axis of inflorescence short, branches long and almost horizontally spreading   ..   ..   28. *formicarum*
  Branches solid ; leaves glabrous, puberulous or pilose :
    Flowers in head-like clusters or on almost horizontally spreading branches ; leaves ovate, truncate or mostly subcordate at base, 7–15 cm. long, 4–10 cm. broad, entire, pubescent above ; branches with yellow indumentum
                                          29. *polycephalum*
    Flowers in terminal, subcorymbose, thyrsoid inflorescences ; calyx 2–3 mm. long, teeth very acute, minutely puberulous outside ; corolla-tube 5–8 mm. long ; leaves ovate, acuminate at apex, short-cuneate at base, smooth, 3–14 cm. long, 2–6 cm. broad, glabrous or finely whitish-tomentellous..   ..   30. *glabrum*

1. **C. alatum** *Gürke* in Engl.Bot. Jahrb. 18 : 182 (1893) ; F.T.A. 5 : 311 ; Thomas l.c. 83. *C. fleuryi* A. Chev. in Mém. Soc. Bot. Fr. 2, 8 : 191 (1911) ; Chev. Bot. 508 ; F.W.T.A., ed. 1, 2 : 273 ; Thomas l.c. 83. *C. lelyi* Hutch. & Dalz. in Kew Bull. 1921 : 395. An erect subherbaceous plant, 2–4 ft. high ; flowers blue-purple with the upper corolla-lobes white ; inflorescence a thyrsoid panicle.
   Iv.C.: Langouassou to Fétékro, Boualé-Nord (July) *Chev.* 22160 ! **N.Nig.**: Ropp (July) *Keay* FHI 37099 ! Vom, Jos Plateau *Dent Young* 205 ! Mongu, 4,300 ft. (fr. July) *Lely* 384 ! Kilba country (July) *Dalz.* 119 ! Biu, Bornu Prov. (fl. & fr. Sept.) *Noble* 18 ! Also Cameroun, Congo, Egypt, Sudan and Ethiopia.
2. **C. violaceum** *Gürke* l.c. 28 : 303 (1900) ; F.T.A. 5 : 520 ; Thomas l.c. 82. *C. kalbreyeri* Bak. in F.T.A. 5 : 311 (1900). *C. noiroti* A. Chev. Bot. 508, name only. A straggling or climbing shrub with almost glabrous leaves and conspicuous violet, violet and white or greenish flowers up to 1 in. across ; inflorescence a terminal panicle.
   Guin.: Labé *Chev.* 12393 ! Mamou to Labé (Mar.) *Jac.-Fél.* 796 ! Dalaba (Mar.) *Dalz.* 8403 ! Macenta (Oct.) *Baldwin* 9770 ! **S.L.**: Hill Station *Smythe* 105 ! Koinadugu (fr. Apr.) *Glanville* 198 ! Kondembaia to Dalakuru (Apr.) *Deighton* 5071 ! N. Kabala (Apr.) *Roberty* 17254 ! Dodo (Apr.) *Deighton* 3905 ! **Iv.C.**: Lower Comoé (Mar.) *Chev.* 17566 ! **Ghana**: Manso (fl. & fr. Apr.) *Scholes* 269 ! Bobiri F.R. (fr. Mar.) *Andoh* 4954 ! Begoro, Akim (fl. & fr. Apr.) *Irvine* 1353 ! Mampong Scarp (Mar.) *Vigne* FH 1066 ! Asamankese (fl. & fr. Mar.) *Vigne* FH 4375 ! **S.Nig.**: Lagos *Rowland* ! Gambari *W. D. MacGregor* 589 ! Owo F.R., Ijebu (May) *Tamajong* FHI 23260 ! Ibadan South F.R. (Apr.) *Keay* FHI 22541 ! Iyamoyong F.R., Obubra Dist. (Apr.) *Binuyo* FHI 41222 ! **[Br.]Cam.**: Bolifambo (fl. & fr. Mar.) *Maitland* 554 ! Kumba (Apr.) *Hutch. & Metcalfe* 159 ! Cam. Mt. (July) *Kalbreyer* 94 ! Also in Cameroun, Congo and N. Rhodesia. (See Appendix, p. 455.)
3. **C. carnosulum** *Bak.* in F.T.A. 5 : 311, 520 (1900) ; Thomas l.c. 81. Erect or climbing shrub with drooping branches ; calyx crimson, corolla light blue.
   [Br.]Cam.: Johann-Albrechtshöhe (= Kumba) *Staudt* 872. Bamenda (fl. Mar., Apr., fr. Apr.) *Onochie* FHI 34885 ! *Ujor* FHI 30023 ! Bamenda Nkwe (Mar.) *Daramola* FHI 40539 ! Bambui (Apr.) *Maitland* 1648 ! Also in Cameroun, Congo and Angola.
4. **C. tomentellum** *Hutch. & Dalz.* F.W.T.A., ed. 1, 2 : 272, 273 (1931) ; Thomas l.c. 83. Erect shrub or small,

tortuous tree with the young parts densely tomentose ; flowers large, up to 1 in. across, with 4 white and 1 mauve corolla-lobes ; inflorescence a terminal panicle.

N.Nig.: Vom, Jos Plateau *Dent Young* 205a ! Jos Plateau (May) *Lely* P264 ! Naraguta F.R. (June) *Kennedy* 7255 ! Heipang Tabo, 4,300 ft. (fl. & fr. Jan.) *Wimbush* FHI 41824 !

5. **C. aculeatum** (*Linn.*) *Griseb.* Fl. Brit. W. Ind. 500 (1864) ; Thomas l.c. 90 ; Berhaut Fl. Sén. 97, 98, 109. *Volkameria aculeata* Linn. (1753). Erect shrub ; branches with small opposite conical thorns ; flowers white, in few-flowered inflorescences from the upper nodes.

Sen.: *Farmar* ! near Hann beach, Dakar (May) *Baldwin* 5743 ! St. Louis (cult.) *Brunner* 36 ! Manulea Rufisque *Chev.* ! **Port.G.:** Cacheu (cult., Oct.) *Esp. Santo* 2276 ! A native of the W. Indies ; introduced into Africa.

6. **C. eupatorioides** *Bak.* in F.T.A. 5 : 295 (1900) ; Thomas l.c. 55. Stem erect, herbaceous, 4–6 ft. high, glandulose-pilose ; flowers white, in dense terminal cymes.

[Br.]Cam.: Cam. Mt., 2,000 ft. (Dec.) *Mann* 1295 !

[This plant, the fruits of which are unknown, must be regarded as doubtfully belonging to *Clerodendrum.*]

7. **C. umbellatum** *Poir.* Encycl. Méth. Bot. 5 : 166 (Jan. 1804). *C. scandens* P. Beauv. Fl. Oware 1 : 52, t. 62 (Sept.-Oct. 1804) ; F.T.A. 5 : 304 ; F.W.T.A., ed. 1, 2 : 274 ; Thomas l.c. 58 (incl. vars. *asperi-folium* B. Thomas l.c. 59 (1936), and *speciosum* (Teysm. & Binn.) B. Thomas l.c. 58 (1936)) ; Chev. Bot. 508. A climbing shrub up to 20 ft. high, or sometimes suberect ; flowers white, sometimes with a pink throat, calyx green below and white above.

Sen.: *Heudelot* 519 ! **Guin.:** *Farmar* 177 ! Santa to Timbo *Chev.* 12812 *ter* ! Bowali *Caille* in *Hb. Chev.* 14705. **S.L.:** Freetown (fl. & fr. Feb.) *Dalz.* 978 ! New England (fl. Dec.) *Small* 810 ! Kambia (Dec.) *Deighton* 894 ! Roruks (fr. Nov.) *Thomas* 5687 ! Musaia (Nov.) *Deighton* 5962 ! **Lib.:** Monrovia (Nov.) *Okeke* 62 ! Cape Palmas (July) *T. Vogel* 53 ! Moala-Suen (Nov.) *Linder* 1382 ! Zigida, Vonjama Dist. (Oct.) *Baldwin* 10020a ! Ganta (May) *Harley* ! **Iv.C.:** Bingerville *Chev.* 16069. Makougnié *Chev.* 16992 *bis.* Bouroukrou *Chev.* 16748 ; 16798. Adiopodoumé (Oct.) *Leeuwenberg* 1774 ! Niapidou (Jan.) *Leeuwenberg* 2476 ! **Ghana:** Dixcove (fr. Jan.) *Deakin* 67 ! Aburi Scarp (fl. & fr. Jan.) *Morton* GC 6343 ! Kumasi (Jan.) *Vigne* FH 1798 ! Akroso (Aug.) *Howes* 952 ! Takasi to Jaware (Dec.) *Adams* 2160 ! **Togo Rep.:** *Baumann* 373. Misahöhe *Busse* 3414. Agome Mts. *Busse* 3436. Atakpame *Doering* 226. **S.Nig.:** Lagos (fl. & fr. July) *Dalz.* 1403 ! Sapoba *Thompson* 3 ! Mt. Koloishe, 4,800 ft. (Dec.) *Savory & Keay* FHI 25126 ! Oban *Talbot* ! Opoba (Sept.) *Holland* 143 ! **[Br.]Cam.:** Victoria (fl. & fr. Apr.) *Dundas* FHI 20501 ! 20503 ! Eyang, Mamfe Dist. (Dec.) *Tamajong* FHI 22135 ! Bali to Bamoa, Bamenda (fr. Feb.) *Migeod* 503 ! 505 ! *Maitland* ! Kakara, 4,900 ft., Mambila Plateau (Jan.) *Hepper* 1799 ! Gangumi, Gashaka Dist. (Dec.) *Latilo & Daramola* FHI 28796 ! **F.Po:** *Ansell* ! (Dec.) *Mann* 83 ! near Clarence (Nov.) *Buchholz* ! *Barter* 2058 ! *T. Vogel* 98 ! Musola (Jan.) *Guinea* 975 ! Also in Cameroun, Ubangi-Shari, Uganda, Tanganyika and Congo. (See Appendix, p. 454.)

[Note : B. Thomas (l.c.) indicates *C. congensis* Engl. (Bot. Jahrb. 8 : 65 (1887)) for Sierra Leone (*Afzelius* s.n., *Bastian* s.n.). *C. congensis* is said to differ from *C. umbellatum* : 1) in having the glandular spots of the undersurface of the leaves shining golden, a feature which, however, may also be observed exceptionally in *C. umbellatum* : 2) in having the branches grey-brown, and 3) in the almost glabrous corolla tube. I have not seen any West African material with these characters—H. Huber.]

FIG. 307.—CLERODENDRUM THOMSONAE *Balf. f.* (VERBENACEAE).

A, anther. B, pistil. C, cross-section of ovary.

8. **C. thomsonae** *Balf.f.* in Trans. Bot. Soc. Edinb. 7 : 265, t. 7 and 580, t. 16 (1863) ; F.T.A. 5 : 303 ; Thomas l.c. 60. An almost glabrous climbing shrub, with a large white calyx and bright crimson flowers.

Sen.: Dakar *Chev.* 3437 ! **S.L.:** Njala (cult. Dec.) *Deighton* 5282 ! **Ghana:** Achimota (cult., Dec.) *Irvine* 1651 ! **N.Nig.:** Koton Karifa (fr. Dec.) *Elliott* 219 ! Lokoja (July) *Parsons* L159 ! Oduape, Kabba (fr. Oct.) *Parsons* L111 ! **S.Nig.:** Ifon, Ondo Prov. (June) *Foster* 200 ! Sapoba *Kennedy* 426 ! Angiama *Barter* 2093 ! Eket *Talbot* 3037 ! Calabar (Feb.) *Mann* 2327 ! (cult.) *Thomson* ! Umon-Ndealichi, Itu Dist. (June) *Ujor* FHI 27995 ! **[Br.]Cam.:** *Buchholz. Lips.* Buea *Deistel* 547. Also in Cameroun. (See Appendix, p. 455.)

9. **C. buettneri** *Gürke* in Engl. Bot. Jahrb. 18 : 174 (1893) ; F.T.A. 5 : 302 ; Thomas l.c. 61. A shrub, probably climbing, with leaves and branches brown-hispid ; flowers white or yellowish with a red centre.

S.Nig.: Afi River F.R., Ikom Dist. (May) *Latilo* FHI 31813 ! Also in Cameroun, Rio Muni and Gabon.

10. **C. speciosissimum** *Paxt.* Mag. Bot. 3 : 217 (1837). *C. fallax* Lindl. Bot. Reg. 1844 : sub. t. 19 ; Thomas l.c. 59. *C. greyi* Bak. in F.T.A. 5 : 308 (1900) ; F.W.T.A., ed. 1, 2 : 274. An introduced decorative climber with blood-red flowers in ample panicles.
    W. Africa ; without precise locality, 1860, *Grey* ! **S.L.**: Njala (cult., Apr.) *Deighton* 2494 ! **Ghana:** Achimota (cult. fl. & fr. Mar.) *Asamany* 27 ! **S.Nig.**: Lagos (cult., Mar.) *Lugard* ! A native of Java.

11. **C. japonicum** (*Thunb.*) *Sweet* Hort. Brit., ed. 1, 322 (1827). *Volkameria japonica* Thunb. Fl. Jap. 255 (1784). *Volkmannia japonica* Jacq. (1797–1801). *Clerodendrum fragrans* Vent. Jard. Malm. 2, t. 70 (1804), (in syn. under *Volkameria fragrans* Vent.). A small erect subherbaceous shrub with white or pink flowers in dense terminal heads ; introduced and occasionally naturalised.
    S.L.: Bathurst, Freetown (cult., June) *Deighton* 2710 ! Bahama, Bo Dist. (Feb.) *Marmo* 184 ! Momaia (cult., Aug.) *Adames* 65 ! Baoma (cult., Apr.) *Small* 904 ! **Ghana:** Aburi (cult., June) *Irvine* 34 ! **S.Nig.**: Oban *Talbot* ! [Br.]**Cam.**: Victoria (Jan.) *Ndi* FHI 50351 ! A native of China and Japan ; the double-flowered form is extensively grown in warmer countries.

12. **C. globuliflorum** *B. Thomas* l.c. 63, 99 (1936). A shrub about 6 ft. high, glabrous on branches and leaves, with thick globose inflorescences on the older wood ; flowers white ; calyx violet to brownish-purple.
    S.Nig.: Sapoba *Mutch* FHI 21853 ! Okomu F.R., Benin (Jan.) *Brenan* 8829 ! Onochie FHI 40414 ! Oban *Talbot* ! Obutong, Oban F.R. (Mar.) *Onyeagocha* FHI 7707 ! [Br.]**Cam.**: Rio del Rey *Johnston* ! Cam. Mt. *Weberbauer* 58. Bangwe *Conrau* 249. Bafia (Feb.) *Keay* FHI 37536 ! Bamenda (Jan.) *Daramola* FHI 40612 ! **F.Po:** St. Isabel Peak *Mildbr.* 6345. Also in Cameroun.

13. **C. capitatum** (*Willd.*) *Schum. & Thonn.* var. **capitatum**—Beskr. Guin. Pl. 61 (1827) ; F.T.A. 5 : 305 ; Chev. Bot. 507 ; Thomas l.c. 64 ; Berhaut Fl. Sén. 107. *Volkameria capitata* Willd. Sp. Pl. 3 : 384 (1800). *C. obanense* Wernham in Cat. Talb. 91 (1913). *C. talbotii* Wernham l.c. (1913). *C. capitatum* var. *talbotii* (Wernham) B. Thomas l.c. 63, 65 (1936). *C. capitatum* var. *conglobatum* B. Thomas l.c. (1936). *C. frutectorum* S. Moore in J. Bot. 57 : 249 (1919), partly. Erect or scrambling shrub with long petiolar thorns and white flowers in globose terminal heads.
    Sen.: *Berhaut* 423 ; 3157. M'Bidjem *Thierry* 4 ! **Gam.**: *Brooks* 8 ! *Ingram* ! *Hayes* 502 ! Nyumbai Forest *Frith* 125 ! Grand Daouna *Chev.* 2751. Fodibougou, Kita Massif (Oct.) *Jaeger* 11 ! **Mali:** Kangala *Chev.* 832bis. Dio *Chev.* 2573. **Port.G.**: Suzana to Casselol, S. Domingos (Aug.) *Esp. Santo* 3088 ! **Guin.:** Timbo *Jac.-Fél.* 1880 ! Douné *Caille* in *Hb. Chev.* 14678 ! **S.L.**: Dansogoria (Jan.) *King* 156b Yakala (Sept.) *Thomas* 2379 ! Kabala *Thomas* 2233 ! Dogoloia, near Musaia (Sept.) *Miszewski* 47 ! R. Maleki (Nov.) *Marmo* 289 ! **Lib.:** Busi (Dec.) *Harley* 792 ! **Iv.C.:** Mankono *Chev.* 21938. **Ghana:** Accra (July) *Dalz.* 143 ! Achimota (fl. & fr. July) *Irvine* 748 ! Aquapim *T. Vogel* ! Dodowa (Feb.) *Irvine* 1973 ! Tafo (July) *Darko* WACRI 890 ! **Togo Rep.**: Sokode *Kersting* A65. Difale *Kersting* A191 ! **N.Nig.:** Lokoja *Barter* ! Naraguta (Aug.) *Lely* 497 ! Pankshin, 5,100 ft. (July) *Lely* 435 ! Katagum (Nov.) *Dalz.* 107 ! Biu (Aug.) *Noble* 6 ! **S.Nig.**: Lagos (Oct.) *Punch* 42 ! Ibadan (Aug.) *Newberry* 60 ! Owo F.R. (July) *Onochie* FHI 33354 ! Onitsha *Barter* 342 ! Degema *Talbot* ! Oban *Talbot* 341 ! [Br.]**Cam.**: Cam. Mt., 3,000 ft. (Dec.) *Mann* 1975 ! Also in Cameroun, Congo, Angola, Egypt and E. African from Sudan to N. Rhodesia and Nyasaland. (See Appendix, p. 454.)

13a. **C. capitatum** var. **cephalanthum** (*Oliv.*) *H. Huber* in Kew Bull. 16 : 174 (1963). *C. cephalanthum* Oliv. in Hook. Ic. Pl. 16, t. 1550 (1887) ; Thomas l.c. 66.
    S.L.: Mano (Sept.) *Deighton* 3790 ! Taiama (Aug.) *Deighton* 3523 ! Gbonge (Sept.) *Deighton* 6113 ! Yetaya (Sept.) *Thomas* 2296 ! **Lib.:** Ganta (Sept.) *Baldwin* 9303 ! Bilimu (Aug.) *Harley* 1453 ! **Iv.C.:** Baléko (Sept.) *Aké Assi* 3196 ! **Ghana:** Suhum (Sept.) *Deighton* 3417 ! Ampunyase (Aug.) *Andoh* 4237 ! Also in Kenya and Tanganyika.

14. **C. whitfieldii** *Seem.* in Bonplandia 10 : 250 (1862) ; Thomas l.c. 64.
    S.L.: *Whitfield* ! This species is known only from a single specimen.

15. **C. acerbianum** (*Vis.*) *Benth. & Hook.* Gen. Pl. 2 : 1156 (1876) ; F.T.A. 5 : 295 ; Thomas l.c. 89 ; Berhaut Fl. Sén. 98, 100, 119. *Volkameria acerbiana* Vis. Ic. Pl. Aegypt. Nub. 23, t. 4 (1836). Erect or scrambling shrub, densely pubescent to tomentose ; flowers white in lateral and terminal, long-peduncled clusters.
    Gam.: *Pirie* 46 ! Ouassadou *Berhaut* 794. **Port.G.**: Piche to Buruntuma (July) *Esp. Santo* 2707 ! Widespread through Egypt, Sudan, and the drier parts of E. Africa from Somaliland to Tanganyika.

16. **C. welwitschii** *Gürke* in Engl. Bot. Jahrb. 18 : 174 (1893) ; F.W.T.A., ed. 1, 2 : 273. A climbing shrub with large leaves and brown-pubescent branches ; flowers white in dense terminal inflorescences.
    S.Nig.: Eket *Talbot* 3033 ! Ikwette Plateau, 5,200 ft., Obudu Div. (Dec.) *Savory & Keay* FHI 25189 ! Also in Cameroun, Congo and Angola.

17. **C. incisum** *Klotzsch* in Peters Reise Mossamb. Bot. 257 (1862) ; F.T.A. 5 : 307 ; Thomas l.c. 78. An erect, subherbaceous plant with very long-tubular flowers in rather few-flowered, dense terminal inflorescences.
    S.Nig.: Benin *Kennedy* 1420 ! Sapoba *Kennedy* 2640 ! Distributed from Somalia to Mozambique. probably only introduced into our area.

18. **C. buchholzii** *Gürke* in Engl. Bot. Jahrb. 18 : 175 (1893) ; F.T.A. 5 : 301 ; Thomas l.c. 69. *C. preussii* Gürke l.c. 175 (1893) ; F.T.A. 5 : 302 ; Thomas l.c. 69. *C. schifferi* A. Chev. Bot. 509, name only. A glabrous woody climber up to 30 ft. high with petiolar thorns ; frequently cauliflorous ; flowers white, fragrant ; fruits red, with the calyx enlarged, white.
    Guin.: Dalaba (Sept.) *Schnell* 6871 ! Kouria (Sept., Oct.) *Caille* in *Hb. Chev.* 14899 ; 15049 ! **S.L.**: Bumba (Sept.) *Thomas* 1921 ! Gbangbama (fr. Nov.) *Deighton* 2358 ! Rokupr (Sept.) *Jordan* 124 ! Njala (fr. Oct.) *Deighton* 2530 ! Loma Mts. (Sept.) *Jaeger* 1634 ! **Lib.:** Du R. (July, Aug.) *Linder* 184 ! 252 ! Gbanga (Sept.) *Linder* 730 ! Browntown, Tchien Dist. (Aug.) *Baldwin* 7082 ! **Iv.C.:** Kéeta (July) *Chev.* 19320. Bingerville *Chev.* 16009 ; 16010 ; 20150. **Ghana:** Aburi (July–Sept.) *Johnson* 169 ! *Gould* ! Kibi (Sept.) *Vigne* FH 4256 ! Akwapim (Aug.) *Johnson* 771 ! Tafo (Aug.) *Darko* WACRI 946 ! **S.Nig.**: Lagos *Punch* ! Olode Oshonde, Ibadan (Oct.) *Moses & Jonathan* FHI 19181 ! Ama Ekpelima (Sept.) *Jones* FHI 6721 ! Ukpon F.R., Obubra Dist. (July) *Latilo* FHI 31876 ! Oban *Talbot* ! [Br.]**Cam.**: Cam. Mt., 3,000–6,000 ft. (fl. & fr. Jan., fr. Apr.) *Maitland* 311 ! Hutch. & Metcalfe 5 ! *Keay* FHI 28580 ! S. Bakundu F.R., Kumba (Apr.) *Olorunfemi* FHI 30520 ! Likomba (Oct.) *Mildbr.* 10507 ! Mamfe (Mar.) *Onochie* FHI 34838 ! Bamenda to Bafut (Aug.) *Savory* UCI 322 ! **F.Po:** (fr. Nov.) *T. Vogel* 157 ! Moka, 4,000 ft., Plateau area (Dec.) *Boughey* 62 ! Mt. Balea (Dec.) *Guinea* 424 ! Musola (Sept.) *Wrigley* 684 ! Also in Cameroun, Gabon, E. Africa, N. Rhodesia and Angola. (See Appendix, p. 454.)
    [Note : The corolla is frequently transformed by insect activity into a gall ; thus the corolla, normally with a very narrow tube, becomes enlarged, the tube broadly cylindric-campanulate, about 2 cm. long and 1·5 cm. in diameter, the lobes semi-elliptic, 5–7 mm. long, the corolla resembling a flower of *Digitalis purpurea*.]

19. **C. schweinfurthii** *Gürke* in Engl. Bot. Jahrb. 18 : 177 (1893) ; F.T.A. 5 : 296 ; Thomas l.c. 70. *C. bakeri* Gürke l.c. 175 (1893) ; F.T.A. 5 : 296 ; Chev. Bot. 507. *C. schweinfurthii* var. *bakeri* (Gürke) B. Thomas l.c. 71 (1936). Small shrub, 2–7 ft. high with a creeping rhizome ; inflorescences capitate, mostly produced from the rhizome ; flowers white, fragrant.
    S.L.: Freetown (cult., May) *Deighton* 2900 ! Lane-Poole 68 ! Musaia and Falaba (Mar.) *Sc. Elliot* 5201 ! 5309 ! Gaahun (Apr.) *Deighton* 3933 ! **Iv.C.:** Mid. Sassandra to Mid. Cavally (July) *Chev.* 19237. [Br.]**Cam.**: Kimbi Bridge, Wum Div. (Feb.) *Hepper* 1924 ! Also in Cameroun, Ubangi-Shari, Uganda, Congo, Egypt, Sudan, Tanganyika and Angola.

20. **C. sinuatum** *Hook.* in Bot. Mag. 72 : t. 4255 (1846) ; F.T.A. 5 : 295 ; Chev. Bot. 509 ; Thomas l.c. 74 ; Berhaut Fl. Sén. 108 (1954), incl. var. *aureum* Berhaut, French descr. only. A climbing shrub, often with dense yellowish-brown indumentum ; flowers white, in dense, terminal heads.
    Sen.: Ouassadou *Berhaut* 799 ; 3037 ; 3088 ; 3089. **Gam.**: Genieri (July) *Fox* 175 ! Kuntaur *Ruxton*

139! **Guin.:** Timbo *Chev.* 12533. Timbo to Faranna *Chev.* 13328. **S.L.:** *Whitfield!* foot of Da Bulen, Loma Mts. (fr. Aug.) *Jaeger* 1317! Yetaya (Sept.) *Thomas* 2452! **Ghana:** Mampong Scarp (June) *Morton!* *Vigne* FH 1133! Akrodum (July) *Vigne* FH 4415! Dawa (May) *Howes* 908! Kotanfo (Apr.) *Andoh* FH 5483! Koforidua to Aburi (Apr.) *Morton* A504! **S.Nig.:** Lagos *Foster* 102! *Millen* 151! Abeokuta *Harrison* 2!

21. **C. splendens** *G. Don* in Edinb. Phil. Journ. 11 : 349 (1824) ; F.T.A. 5 : 300 ; Chev. Bot. 509 ; Thomas l.c. 91 ; Berhaut Fl. Sén. 97. A climbing, glabrous shrub, mostly fairly low, with deep red flowers in corymbose inflorescences ; fruits black and shining ; in forest.
Sen.: *Farmar* 149! Mampalogou, Casamance (Feb.) *Chev.* 2763! **Port.G.:** Pussubé, Bissau (Apr.) *Esp. Santo* 1169! Brene, Bissau (Jan.) *Esp. Santo* 1715! **Guin.:** Conakry (Feb.) *Chev.* 12157! Kindia *Chev.* 12792. **S.L.:** *Don!* Freetown (Jan.) *Dalz.* 935! Taiama (Mar.) *Deighton* 3371! Kambui Hills (fl. & fr. Mar.) *Small* 520! Kenema (fr. Apr.) *Jordan* 2052! **Lib.:** Monrovia (Nov.) *Linder* 1411! Kailahun (Feb.) *Bequaert* 62! Brewersville (June) *Barker* 1203! 1332! Jaurazon, Sinoe (fr. Mar.) *Baldwin* 11444! Barclayville (Mar.) *Baldwin* 11120! **Iv.C.:** Bouroukrou *Chev.* 16548 ; 16570. Accrédou *Chev.* 17064. Yaou, Sanvi *Chev.* 17781. Abouabou (Jan.) *Leeuwenberg* 2396! **Ghana:** W. Afao Hills F.R., Awaso (Feb.) *Darko* 758! **N.Nig.:** Ankpa Bassa (Feb.) *Lamb* 24! **S.Nig.:** Lagos (Dec.) *Hagerup* 756! Ibadan (Feb.) *Meikle* 1135! Okomu F.R., Benin (Jan.) *Brenan* 8758! Asaba (Jan.) *Chizea* FHI 8252! Onitsha *Barter!* Calabar (Apr.) *Espley* 6! **[Br.]Cam.:** Bali to Bamoa, Bamenda Prov. (Feb.) *Migeod* 504! Victoria to Kumba (fl. & fr. Apr.) *Hutch. & Metcalfe* 143! Buea to Kumba (Feb.) *Dundas* FHI 13986! S. Bakundu F.R., Kumba (Jan.) *Binuyo & Daramola* FHI 35159! Also in Cameroun, Rio Muni, Ubangi-Shari, Gabon, Angola and Congo. (See Appendix, p. 454.)

22. **C. bipindense** *Gürke* in Engl. Bot. Jahrb. 28 : 296 (1900) ; F.T.A. 5 : 516 ; Thomas l.c. 72. A slender liane to 20 ft. high with very ample, much-branched inflorescences from the leafless part of the stem near the ground ; flowers yellow or cream.
S.Nig.: Afi River F.R., Ikom Dist. (Dec.) *Keay* FHI 28182! Oban *Talbot* 1530 ; 2058! **[Br.]Cam.:** Victoria (Nov.) *Maitland* 783! *Preuss* 1358! Likomba (Nov.) *Mildbr.* 10706! S. Bakundu F.R., Kumba (fr. Jan.) *Binuyo & Daramola* FHI 35082! **F.Po:** (Dec.) *Mann* 71! Also in Cameroun.

23. **C. melanocrater** *Gürke* l.c. 18 : 180 (1893) ; F.T.A. 5 : 299 ; Thomas l.c. 72. A small climbing shrub ; calyx white and flowers brownish-yellow, fragrant, uniformly black, when dry.
[Br.]Cam.: Likomba (Oct.) *Mildbr.* 10510! **F.Po:** Conception (Sept.) *Melville* 445! Also in Cameroun, Congo, Uganda and Tanganyika.

24. **C. thyrsoideum** *Gürke* l.c. 28 : 293 (May 1900) ; F.T.A. 5 : 516 ; Thomas l.c. 73, not of Bak. (June 1900). *C. streptocaulon* Hutch. & Dalz. F.W.T.A., ed. 1, 2 : 273, 274 (1931). Climbing shrub 3–60 ft. high with puberulous branches and thyrsoid inflorescences ; flowers greenish, yellowish or white.
Gam.: Genieri (fr. Feb.) *Fox* 71! **Port.G.:** *Esp. Santo* 275! Nova Lamego to Canjufa (fl. & fr. Oct.) *Esp. Santo* 3548! Bangacia, Bafata (fr. Jan.) *Esp. Santo* 2872! **Guin.:** Pita (Jan.) *Jac.-Fél.* 741! **S.L.:** Wellington (fr. Feb.) *Deighton* 5007! Njala (Dec.) *Deighton* 2576! Funkadeh (fl. & fr. Jan.) *Jordan* 186! Rokupr (fr. Jan.) *Deighton* 2954! Kitchom (Jan.) *Deighton* 962! **Lib.:** Monrovia (fl. & fr. Dec.) *Okeke* 76! *Dinklage* 2981. **S.Nig.:** Apapa (Nov.) *Keay* FHI 22452! Ikoyi Plains, Lagos (Dec.) *Dalz.* 1249! Igboedun, Ogun R., Ikeja Dist. (Dec.) *Onochie* FHI 26690! Also in Cameroun, Ubangi-Shari and the Congo.

25. **C. volubile** *P. Beauv.* Fl. Oware 1 : 52, t. 32 (1804) ; F.T.A. 5 : 297 ; Chev. Bot. 509 ; Thomas l.c. 72; Berhaut Fl. Sén. 56. Climbing shrub up to 10 ft. high, with petiolar-thorns ; glabrous except the inflorescences ; flowers in dense, subcorymbose, mostly terminal inflorescences.
Sen.: Badi *Berhaut* 1611. **Mali:** Bafaga (Apr.) *Chev.* 709! **Guin.:** Kindia *Chev.* 12794 ; 13370. Bilima *Caille* in Hb. *Chev.* 14739. Konkouré *Caille* in Hb. *Chev.* 14857. Nzérékoré (Sept.) *Jac.-Fél.* 1156! **S.L.:** *Don!* Waterloo to York Pass (Jan.) *Tindall* 27! Port Loko (Dec.) *Thomas* 6574! Njala (Feb.) *Dalz.* 8109! Gbinti (Jan.) *Deighton* 5901! Kambia (Jan.) *Sc. Elliot* 4424! **Lib.:** Kle, Boporo Dist. (Dec.) *Baldwin* 10631! Peahtah, Salala Dist. (Dec.) *Baldwin* 10580! Robertsport (Dec.) *Baldwin* 10959! **Iv.C.:** Bouroukrou *Chev.* 16893. Makougnié *Chev.* 16980. Bliéron *Chev.* 19901. Tabou to Bériby *Chev.* 19976. Abouabou (Jan.) *Leeuwenberg* 2345! *Irvine* (Feb.) *Irvine* 2293! Kwahu Prasu (Feb.) *Vigne* FH 1627! *Irvine* 1683! Tafo (Nov.) *Robertson* K8! Ho *Schröder* 13. Amedzofe (fr. Feb.) *Irvine* 161! **Togo Rep.:** *Schröder* 189. Nedéme *Busse* 3448. Lavanyo *Baumann* 370. **N.Nig.:** Gudi, S. Plateau Div. *Thornewill* 60! **S.Nig.:** Lagos, Eppah *Barter* 3275! Gambari F.R. (Mar.) *Hepper* 2288! Ibadan (Dec.) *Meikle* 928! *Onochie* FHI 34138! Okomu F.R., Benin (Dec.) *Brenan* 8510! Port Harcourt (Nov.) *Stubbings* 187! Eket *Talbot!* **F.Po:** *Mann* 71. Also in Cameroun, Rio Muni, Gabon, Congo and Angola. (See Appendix, p. 455.)

26. **C. inaequipetiolatum** *Good* in J. Bot. 68, Suppl. 2 : 141 (1930) ; Thomas l.c. 75. Scrambling shrub up to about 10 ft. high, with yellowish-brown indumentum and white flowers ; similar and closely akin to *C. sinuatum.*
[Br.]Cam.: Metchen R., Bamenda (Aug.) *Savory* UCI 305! Wum to Bamenda (July) *Ujor* FHI 30489! Also in Cameroun and Angola.

27. **C. duseni** *Gürke* in Engl. Bot. Jahrb. 28 : 293 (1900) ; F.T.A. 5 : 518 ; Thomas l.c. 76. *C. barteri* Bak. in F.T.A. 5 : 298 (1900). *Premna macrosiphon* Bak. in F.T.A. 5 : 290 (1900). *Clerodendrum macrosiphon* (Bak.) Pieper in Engl. Bot. Jahrb. 62, Beibl. 141 (as 142) : 80 (1928), not of Hook. *C. intermedium* B. Thomas l.c. 76 (1936). *C. premnoides* Meeuse in Blumea 5 : 74 (1942). A scrambling shrub with the young branches and the petioles brownish-puberulous ; flowers small, white, in branches, mostly sub-corymbose thyrsi.
S.Nig.: Nun R. (Sept.) *Mann* 486! Brass *Barter* 51! Okomu F.R., Benin (Feb.) *Brenan* 9016! Oban *Talbot* 1491! Eket *Talbot* 3212! Calabar *Thomson* 28! **[Br.]Cam.:** Tiko (Feb.) *Maitland* 980! Bolifambo (Mar.) *Maitland* 531! Also in S. Tomé, Cameroun and Angola.

28. **C. formicarum** *Gürke* in Engl. Bot. Jahrb. 18 : 179 (1893) ; F.T.A. 5 : 297 ; Thomas l.c. 74 (incl. var. *sulcatum* B. Thomas l.c. 74 (1936)). A climbing shrub with minute flowers and horizontally branched inflorescences.
Guin.: near Macenta (Sept.) *Jac.-Fél.* 1175! **Lib.:** *Harley* 1237! **Iv.C.:** Mt. Tonkoui (Aug.) *Aké Assi* 5423! **S.Nig.:** Sapoba *Kennedy* 1408! Ugo, Benin (Aug.) *Okeke & Emwiogbon* FHI 30142! Eket *Talbot* 3077! Calabar (July) *Holland* 53! **[Br.]Cam.:** Ngel Nyaki, 5,400 ft., Mambila Plateau (Jan.) *Hepper* 1726! Also in Cameroun, Rio Muni, Gabon, Congo, Angola, Egypt, Sudan, Uganda, Tanganyika and N. Rhodesia. (See Appendix, p. 454.)

29. **C. polycephalum** *Bak.* in Kew Bull. 1895 : 116 ; F.T.A. 5 : 300 ; Thomas l.c. 75 (as *C. polycephalatum*); Chev. Bot. 508. Climbing or erect shrub up to 15 ft. high ; leaves and branches yellow- or brown-pilose; flowers small, white, arranged in headlike clusters.
Guin.: Nzérékoré (May) *Roberty* 18024! Nzérékoré to Macenta (Oct.) *Baldwin* 9741! Macenta (fr. Oct.) *Baldwin* 9774! **S.L.:** Falaba to Musaia (Sept.) *Small* 329! Gberia Fotombu (Sept.) *Small* 295! Kayima *Dawe* 525! **Iv.C.:** Dialakoro to Kénégoué *Chev.* 21976. Bouaké *Chev.* 22141. Man (Aug.) *Boughey* GC 18401! **Ghana:** Mampong, Ashanti (Apr., July) *Darko* 688! *Vigne* FH 1921! Techeri, Ashanti (June) *Chipp* 475! Nsuta (May) *Vigne* FH 1730! Bame Pass (Dec.) *Adams* 4487! **Togo Rep.:** Misahöhe *Mildbr.* 7325. Basari *Kersting* 135. Losso *Kersting* 463. Paratau *Büttner* 606. **N.Nig.:** (May to Oct.) *Thornewill* 77! Bonu F.R., Gwari Dist. (June) *Onochie* FHI 40213! **S.Nig.:** Lagos *Millen* 12! 153! 157! *Phillips* 39! Ogbomosho *Rowland!* *Keay* FHI 19190! Old Oyo F.R. (Sept.) *Latilo* FHI 23549! Also in Cameroun. (See Appendix, p. 454.)

30. **C. glabrum** *E. Mey.* Comm. Pl. Afr. Austr. 273 (1837) ; F.T.A. 5 : 297 ; Thomas l.c. 76. A glabrous to

minutely tomentellous shrub with small, white flowers in dense, subcorymbose inflorescences ; introduced into our area as a hedge plant.
S.L.: Freetown (cult., June) *Deighton* 4623 ! Newton (cult., Nov.) *Deighton* 3349 ! From Mombasa southwards to the eastern Cape Province ; also in Angola.

10. VITEX Linn.—F.T.A. 5 : 315 ; Pieper in Engl. Bot. Jahrb. 62, Beibl. 141 (as " 142 ") :
1–91 (1928) ; Moldenke in Phytologia 5 & 6 (1955–58).

Cymes arranged in terminal or terminal and axillary panicles ; fruit globose, up to about
5 mm. diam. ; leaves 5-foliolate, leaflets broadly obovate to elliptic, acuminate,
puberulous beneath on the nerves, otherwise glabrous and densely dotted with yellow
glands ; central leaflets with petiolule 5–18 cm. long, 3·5–9 cm. broad ; calyx truncate,
with obsolete teeth ; corolla with a right-angle bend, about twice as long as the calyx
1. *thyrsiflora*

Cymes axillary or axillary and terminal, arranged in compound dichasia ; fruits obovoid
to sub-globose, 1 cm. or more long :
Calyx and pedicels minutely puberulous (hairs shorter than ⅓ of the diam. of the pedicel);
leaves glabrous beneath or pubescent only on the nerves, 5–7-foliolate :
Leaflets crenate or dentate, at least in the apical ⅓, acuminate ; central leaflet with
9–10 pairs of lateral nerves ; ovary glabrous :
Calyx and pedicels with short, appressed indumentum ; flowers about 6 mm. long ;
leaflets obovate-cuneate to oblanceolate, shortly acuminate, the middle one
2·5–10 cm. long, 1–3 cm. broad .. .. 2. *micrantha*
Calyx and pedicels with short, patent indumentum ; flowers about 7 mm. long ;
leaflets obovate to obovate-oblong, mostly long-acuminate, the middle ones
8–25 cm. long, 2·5–8 cm. broad .. .. 3. *oxycuspis*
Leaflets entire :
Peduncle 10–20 cm. long ; ovary glabrous, with sessile glands ; corolla less than
1 cm. long :
Middle leaflets with 4–8 pairs of lateral nerves, leaflets obovate-cuneate, abruptly
acuminate at apex, narrowly cuneate at base, central ones 12–25 cm. long,
4·5–10 cm. broad ; calyx 3–5 mm. long .. .. 4. *lehmbachii*
Middle leaflets with 12–20 pairs of lateral nerves, leaflets elliptic, elliptic-oblong or
broadly lanceolate, acuminate at apex, shortly cuneate at base, central ones
8–18 cm. long, 3–6 cm. broad ; calyx about 1–1·5 mm. long .. 5. *rivularis*
Peduncle 0·5–8 cm. long ; ovary densely pilose at apex with erect hairs ; corolla
mostly 1–2 cm. long :
Leaflets obtuse or broadly deltoid or rarely shortly apiculate at apex, obovate to
elliptic, the middle ones 5–18 (rarely 20) cm. long including the petiolule, 4–10 cm.
broad, with the petiolule 1–2·5 cm. long ; peduncle 2–8 cm. long ; calyx 3–4 mm.
long ; inflorescences congested .. .. 6. *doniana*
Leaflets abruptly short- to long-acuminate at apex, obovate-cuneate, the middle
ones usually 13–40 cm. long, 6–20 cm. broad, very gradually and narrowly
attenuated into a usually very short petiolule, rarely exceeding 1 cm. in length ;
peduncle 0·5–5 cm. long .. .. 7. *grandifolia*
Calyx and pedicels distinctly pilose to tomentose (hairs at least half as long as, mostly
longer than, the diameter of the pedicel) :
Ovary glabrous or with very few solitary hairs on the top, with sessile glands ; calyx-
teeth 0·4–1·5 mm. long ; peduncles 4–16 cm. long :
Leaves 5(–7) foliolate, middle leaflet mostly with 10–12 pairs of lateral nerves ;
calyx 1–2 mm. long, actinomorphic :
Inflorescence and petioles glabrous or glabrescent ; leaflets glabrous or glabrescent
beneath except on the nerves ; indumentum (distinct only in very young parts)
greyish to yellowish-white (other characters as above) .. .. 5. *rivularis*
Inflorescence, petioles, undersurface of the leaves and young branches densely
villous with long rust-red to orange-brown hairs ; peduncles 4–16 cm. long ;
leaflets obovate, the middle one 9–16 cm. long, 4–7·5 cm. broad, usually with
10–16 pairs of lateral nerves .. .. 8. *phaeotricha*
Leaves 1–3(–5)-foliolate ; calyx 2·5 mm. long, bilabiate ; indumentum greyish-
white to pale yellow ; lateral nerves (when leaves compound of the middle leaflet)
5–12(–14) pairs :
Leaves all 3(–5) foliolate, obovate, obovate-oblong to lanceolate, entire or crenate
on the margin, at first tomentose beneath, later frequently glabrescent except
on the nerves ; middle leaflet 1·5–4 times as long as broad, 5–15 cm. long and 2·5–
9 cm. broad, with 8–12(–14) pairs of lateral nerves ; calyx often irregularly
pilose .. .. 9. *madiensis*
Leaves in the floral region undivided or with a few ternate ones among them ;
simple leaves broadly ovate to ovate-acuminate, persistently tomentose beneath,
margins entire, in compound leaves the middle leaflet obovate ; leaf 1–2 times as
long as broad, 5–14 cm. long and 3·5–10 cm. broad, with 5–10 pairs of lateral
nerves ; calyx densely tomentose with flexuous hairs .. .. 10. *simplicifolia*
Ovary densely pilose with erect hairs ; calyx 4–5 mm. long, actinomorphic, with the

teeth mostly 2–3 mm. long ; peduncles wanting or up to 6 cm. long ; middle leaflet
with 8–20 pairs of lateral nerves :
Peduncles 0–2 cm. long ; fruit glabrous when mature except at the apex ; inflorescence and young leaves frequently with a dark greyish-yellow pubescence ; leaves
5–7 foliolate, leaflets long-acuminate, obovate to elliptic, entire or often crenate,
pubescent beneath, the central one 5–14 cm. long and 2–5 cm. broad, with 10–12
pairs of lateral nerves   ..   ..   11. *ferruginea*
Peduncles 2–6 cm. long ; fruit yellowish-puberulous all over even when ripe ;
inflorescence and undersurface of leaves persistently tomentose with greyish-white
to pale-yellow indumentum ; leaves 3– 5-foliolate, leaflets obtuse or triangular at
apex, obovate or elliptic, entire or frequently crenate, central ones 5–12 cm. long
and 2·5–5 cm. broad, with 8–14 pairs of lateral nerves   ..   .. 12. *chrysocarpa*

1. **V. thyrsiflora** *Bak.* in Kew Bull. 1895 : 152 ; F.T.A. 5 : 319 ; Pieper l.c. 54 ; Aubrév. Fl. For. C. Iv., ed.
2, 3 : 233 ; Moldenke in Phytologia 6 : 152. *V. staudtii* Gürke in Engl. Bot. Jahrb. 33 : 299 (1904).
V. obanensis Wernham in Cat. Talb. 92 (1913). *V. agraria* A. Chev. Bot. 505, name only. An undershrub or small tree with glabrous branches, 5-foliolate leaves and small white flowers in terminal panicles ;
in forest.
Guin.: Faranna *Chev.* 13199. Kaba to Upper Mamou *Chev.* 13267. Macenta (Apr.) *Jac.-Fél.* 852 ! Dyeke
(fr. Oct.) *Baldwin* 9669 ! **S.L.:** Binkolo (fr. Aug.) *Thomas* 1692 ! Bumba (Sept.) *Thomas* 1953 ! Segbwema
(July) *Deighton* 3747 ! **Lib.:** Vonjama (fr. Oct.) *Baldwin* 9945 ! Ganta (fr. Oct.) *Harley* ! Sanokwele
(fr. Sept.) *Baldwin* 9510 ! Suacoco, Gbanga (May) *Konneh* 175 ! Nyaake, Webo (June) *Baldwin* 6172 !
Iv.C.: Atteou *Chev.* 17055 ; 17140 ! Kéeta *Chev.* 19340. Taté *Chev.* 19805. **Togo Rep.:** *Baumann* 564.
N.Nig.: Dogon Kurmi, Jemaa Div. *Killick* 67 ! **S.Nig.:** W. Lagos *Rowland* ! Abeokuta *Harrison* 5 !
Aponmu, Akure Dist. (fr. July) *Symington* FHI 5052 ! Ojogba-Ugun F.R., Ishan Dist. (June) *Olorunfemi*
FHI 38057 ! Oban *Talbot* ! **[Br.]Cam.:** Bulejambo, Buea (Mar.) *Maitland* 565 ! Talanganji, Kumba
Dist. (June) *Olorunfemi* FHI 30608 ! Bamenda to Kentu (June) *Maitland* 1577 ! Tinaberi, Wum Dist.
(fr. July) *Ujor* FHI 29288 ! Extends to Congo.
2. **V. micrantha** *Gürke* in Engl. Bot. Jahrb. 18 : 170 (1894) ; F.T.A. 5 : 324 ; Pieper l.c. 56 ; Aubrév. Fl.
For. C. Iv., ed. 2, 3 : 232, t. 336, 1–3 ; Chev. Bot. 507 ; Moldenke in Phytologia 5 : 473. *V. longe-
acuminata* A. Chev. Bot. 506, name only. Forest tree, 15–80 ft. high, with fairly small, mostly 5-foliolate
leaves and small white flowers in poor, long-pedunculate cymes.
Guin.: Macenta (fl. & fr. Apr.) *Jac.-Fél.* 845 ! **S.L.:** Bagroo R. (Apr.) *Mann* 860*bis* ! Newton (fr. June)
Deighton 3010 ! Njala (May) *Deighton* 658 ! Falaba (Apr.) *Aylmer* 46 ! **Lib.:** Dukwia R. (fr. Apr.) *Cooper*
70 ! Du R. (fr. Aug.) *Linder* 279 ! Tubman Bridge (Mar.) *Barker* 1230 ! Gbanga (fr. Dec.) *Baldwin*
10531 ! Nyaake, Webo (fr. June) *Baldwin* 6114 ! **Iv.C.:** Abidjan (fr. July) *Aubrév.* 54 ! *Chev.* 15409 !
Alépé *Chev.* 16229. Ahiamé *Chev.* 17808. Mt. Momi, Dans Massif *Aubrév.* 907 ! 1171 ! Taï (fr. Aug.)
Schnell 6082 ! **Ghana:** Ankobra Junction *Kitson* 1001 ! Kumasi *Cummins* 74 ! Axim (fl. & fr. Feb.)
Irvine 2128 ! Simpa, near Tarkwa (Feb.) *Vigne* FH 2798 ! *Kinloch* FH 3234 ! According to Moldenke
(l.c. 474) also in Cameroun. (See Appendix, p. 458.)
3. **V. oxycuspis** *Bak.* in F.T.A. 5 : 326 (1900) ; Pieper l.c. 56 ; Moldenke l.c. 6 : 31 ; Aubrév. Fl. For. C. Iv.,
ed. 2, 3 : 232, t. 336, 4. Forest tree 15–35 ft. high, similar to the above.
S.L.: *Unwin & Smythe* 37 ! Gola Forest *Small* 613 ! **Lib.:** Dukwia R. *Cooper* 321 ! **Iv.C.:** Taï *Boughey*
GC 14858 ! **S.Nig.:** Sapoba *A. F. Ross* 234 ! Abeokuta *Jones & Keay* FHI 14236 ! Calabar R. (Feb.)
Mann 2243 ! Oban F.R. (Jan.) *Onochie* FHI 36163 ! Oban *Talbot* 2061*bis* ! Also in Angola. (See Appendix,
p. 458.)
4. **V. lehmbachii** *Gürke* in Engl. Bot. Jahrb. 33 : 297 (1904) ; Pieper l.c. 56 ; Moldenke l.c. 5 : 435. Shrub
or small tree up to 30 ft. high with glabrous branches and leaves, and long-peduncled cymes of pale pink or
white flowers.
[Br.]Cam.: Barike to Manya, Kumba (June) *Olorunfemi* FHI 30629 ! Buea *Lehmbach* 11. *Reder* 687 ;
1087 ; 1384. Bum, Bamenda (June) *Maitland* 1725 ! Also in Cameroun.
5. **V. rivularis** *Gürke* in Engl. Bot. Jahrb. 33 : 297 (1904) ; Pieper l.c. 56 ; Moldenke l.c. 6 : 114 ; Aubrév.
l.c. 233, t. 336, 5–7. *V. cilio-foliolata* A. Chev. Bot. 506, name only. Forest tree 30–40 ft. high with long-
petioled, 5–7 foliolate leaves and very small white flowers tinged with purple in rich, long-stalked cymes.
Lib.: Diebla, Webo Dist. (fr. July) *Baldwin* 6285 ! Jabroke, Webo Dist. (fr. July) *Baldwin* 6491 ! **Iv.C.:**
Guidéko *Chev.* 19097 ! **Ghana:** Dunkwa *Vigne* FH 895 ! Agogo (Apr.) *Vigne* FH 1094 ! *Irvine* 951 !
Adiembra (May) *Vigne* FH 865 ! **S.Nig.:** *Kennedy* 910 ! Afi River F.R., Ogoja Prov. (May) *Jones &
Onochie* FHI 18760 ! **[Br.]Cam.:** Likomba (fr. Oct.) *Mildbr.* 10535 ! Also in Cameroun and Congo. (See
Appendix, p. 458.)
6. **V. doniana** *Sweet* Hort. Brit., ed. 1 : 323 (1827) ; F.T.A. 5 : 323 ; Pieper l.c. 64 ; Moldenke l.c. 5 : 322,
not of Hiern (1900). *V. umbrosa* G. Don ex Sabine (1824) not of Swartz (1788). *V. cuneata* Schum. &
Thonn. Beskr. Guin. Pl. 289 (1827) ; F.T.A. 5 : 328 ; Chev. Bot. 506 ; Aubrév. Fl. For. Soud.-Guin. 504,
t. 113, 1–2 ; Fl. For. C. Iv., ed. 2, 3 : 230, t. 335, fig. 1–2 ; Berhaut Fl. Sén. 21. *V. cienkowskii* Kotschy
& Peyr. (1867)—F.T.A. 5 : 328. Tree 30–60 ft. high, with glabrous branches and 5-foliolate, coriaceous
leaves ; savanna woodland and open country.
Sen.: Casamance *Heudelot* 379 ! Floup-Fedyan *Chev.* 2757. Tambanaba *Chev.* 2758. **Mali:** San (Mar.)
Davey 500/102 ! Sékoba, Kita Massif (fr. Nov.) *Jaeger* 17 ! Mopti *Chev.* 2769 ! Diafarabé *Chev.* 2766.
Négala *Chev.* 164*bis*. **Port.G.** Brene, Bissau (Feb.) *Esp. Santo* 1735 ! **Guin.:** Kouroussa *Pobéguin* 682 !
Bobengali, Kouria (Dec.) *Caille* in *Hb. Chev.* 14799 ! Faranna (fl. & fr. Mar.) *Sc. Elliot* 5211 ! Kollangui
Chev. 12866. **S.L.:** Jigaya *Thomas* 2824 ! Yisai (Apr.) *Lane-Poole* 1 ! Batkanu *Deighton* 3841 ! Musaia
(Mar.) *Miszewski* 33 ! Tabe (fr. Sept.) *Deighton* 3052 ! **Iv.C.:** Dimbokro *Aubrév.* 428*bis*. Baoulé *Aubrév.*
806. Touba *Aubrév.* 1244. **Ghana:** Achimota *Irvine* 194 ! Ho to Adidome *Ankrah* GC 20347 ! Ejura (fr.
Aug.) *Chipp* 727 ! 744 ! Kumawo, Ashanti (fr. June) *Chipp* 463 ! Yeji (fl. & fr. Apr.) *Saunders* 3 ! **Togo
Rep.:** Lomé *Warnecke* 156 ! **Dah.:** *Poisson.* **N.Nig.:** Nupe *Barter* 1108 ! Agaie (Apr.) *Yates* 59 ! Lafia,
Benue Prov. (Feb.) *Jones* FHI 675 ! Kargi, Zaria Prov. (Mar.) *Lafia* FHI 7758 ! Jos Plateau (Jan.) *Lely*
P134 ! **S.Nig.:** Lagos *Millen* 118 ! Ogun R., Oyo to Iseyin (Feb.) *Brenan* 8956 ! Ori (Feb.) *A. F. Ross*
R71 ! Onitsha *D. F. Chesters* 190 ! Millikin Hill, Enugu (Mar.) *Hepper* 2225 ! **[Br.]Cam.:** Bamenda
(Feb.) *Daramola* FHI 40485 ! Bum, Wum Div. (Feb.) *Hepper* 1921 ! Mai Idoanu, Gashaki Dist. (Feb.)
Latilo & Daramola FHI 34485 ! Jamtari, Gashaka Dist. (fr. Dec.) *Latilo & Daramola* FHI 28927 !
Bama, Dikwa Div. (Jan.) *McClintock* 175 ! Widespread in tropical Africa extending to the Comoro
Islands (See Appendix, p. 458.).
7. **V. grandifolia** *Gürke* in Engl. Bot. Jahrb. 18 : 169 (1894) ; F.T.A. 5 : 324 ; Moldenke l.c. 5 : 385, 404 ;
Aubrév. Fl. For. C. Iv., ed. 2, 3 : 234, t. 337B ; and var. *bipindensis* (Gürke) Pieper l.c. 73 (1928). *V.
bipindensis* Gürke in Engl. Bot. Jahrb. 33 : 295 (1904). *V. lutea* A. Chev. Bot. 506 (1920), name only.
A small tree 20–30 ft. high with large 5-foliolate leaves and pale-yellowish to brownish-yellow flowers in
short-peduncled cymes at the base of the leaves ; in high or secondary forest.
S.L.: Bagroo R. (Apr.) *Mann* 880 ! Luseniya, Samu *Sc. Elliot* 4327 ! Gbangbama (fr. Nov.) *Deighton*
2386 ! Baiima (fr. Sept.) *Deighton* 3078 ! Gola Forest (fr. May) *Small* 696 ! **Lib.:** Monrovia (fr. June)
Baldwin 5908 ! Taninewa (Mar.) *Bequaert* 152 ! Sinoe Basin *Whyte* ! Peahtah (fr. Oct.) *Linder* 966 !

Fig. 308.—Vitex doniana *Sweet* (Verbenaceae).

A, flower.   B, longitudinal section of flower.   C, same of ovary.   D, fruit.   E, cross-section of ovary.

Nyaake, Webo (June) *Baldwin* 6171 ! **Iv.C.:** Dabou (fr. Apr.) *Roberty* 15559 ! *Chev.* 15470. Adiopodoumé (fr. Nov.) *Leeuwenberg* 1922 ! Bingerville *Chev.* 17275. Soubré *Chev.* 19091. **Ghana:** Ankobra R., Axim (Mar.) *Morton* A379 ! Akwapim (Mar.) *Murphy* 676 ! Benso *Andoh* FH 5488 ! Kumasi (Feb.) *Vigne* FH 1041 ! Nkwatia to Mpraeso, Kwahu Plateau (fr. May) *Kitson* 1182 ! **S.Nig.:** Lagos *Barter* 2180 ! Nun R. *Barter* 2098 ! Sapoba (Feb.) *Onochie* FHI 27686 ! Ibadan (fl. & fr. Aug.) *Keay* FHI 25362 ! Oban *Talbot* 2057 ! [Br.]**Cam.:** Victoria (Feb.) *Maitland* 361 ! Also in Cameroun, Rio Muni and Gabon. (See Appendix, p. 457.)

8. **V. phaeotricha** *Mildbr. ex Pieper* in Fedde Rep. 26 : 163 (1929) ; Moldenke l.c. 6 : 57. *V. monroviana* Pieper l.c. (1929) ; Moldenke l.c. 118. *V. rufa* A. Chev. ex Hutch. & Dalz. F.W.T.A., ed. 1, 2 : 275 (1931) ; Chev. Bot. 501, name only ; Cavaco in Bull. Mus. Hist. Nat. sér. 2, 27 : 91 (1955) ; Aubrév. Fl. For. C. Iv., ed. 2, 3 : 232, t. 335, 4–6. A forest tree 40–45 ft. high with long abundant, rust-coloured indumentum ; flowers small, white with purple.
S.L.: *Edwardson* 221 ! Kambui *Wallace* SLFD 74 ! **Lib.:** Harbel (Apr.) *Harley* 1480 ! Monrovia *Dinklage* 2194. White Plains (June) *Dinklage* 3058 ! Dukwai R. (fr. Mar.) *Cooper* 67 ! Palilah, Gbanga Dist. (fr. Aug.) *Baldwin* 9155 ! **Iv.C.:** Grabo, Cavally *Chev.* 19712. *Aubrév.* 4076. Dakpadou F.R. *Aubrév.* 95. Also in Cameroun. (See Appendix, p. 458.)

9. **V. madiensis** *Oliv.* in Trans. Linn. Soc. 29 : 35, t. 131 (1890) ; F.T.A. 5 : 322 ; Pieper in Engl. Bot. Jahrb. 62, Beibl. 141 (as 142) : 61 ; Moldenke l.c. 5 : 451 ; Aubrév. Fl. For. Soud.-Guin. 504, 507, t. 115, 1–2. *V. camporum* Büttn. (1890)–F.T.A. 5 ; 323 ; F.W.T.A., ed. 1, 2 : 276, partly, and Chev. Bot. 505, partly (excl. *Chev.* 12467, which is *Harrisonia abyssinica* Oliv.). *V. barbata* Planch. ex Bak. in F.T.A. 5 : 323 (1900) ; Pieper l.c. 61 ; F.W.T.A. ed. 1, 2 : 276 ; Berhaut Fl. Sén 21 ; Aubrév. l.c. t. 115, 4. *V. pobeguini* Aubrév. l.c. 506, 507, t. 115, 3 (1950), French. descr. only. Shrub or small tree with the young parts densely pubescent ; flowers pilose, yellowish and blue-purple in long-peduncled cymes ; in savanna.
Sen.: *Perrottet* 658 ! *Heudelot* 30 ! *Berhaut* 106. **Gam.:** *Ingram* ! Marremba, Central Div. (fr. July) *Frith* 77 ! Sukuko *Dalz.* 8061 ! **Mali:** Koulikoro *Vuillet* 618. Medinani *Chev.* 510*bis* ; 511. **Port.G.:** Gabu, Pirada (fr. June) *Esp. Santo* 3030 ! Bissau (fr. Feb.) *Esp. Santo* 1142 ! Buba (fr. May) *Esp. Santo* 2484 ! **Guin.:** Banancoro *Chev.* 496. Kadé *Pobéguin* 2007. Dinguiraye *Maclaud* 64. **S.L.:** Mt. Gonkwi, Talla Hills (Feb.) *Sc. Elliot* 4881 ! Laminaia (fr. Apr.) *Thomas* 152 ! 160 ! Falaba (Mar., Apr.) *Sc. Elliot* 5189 ! *Deighton* 5430 ! **Iv.C.:** Dimbokro (July) *Serv. For.* 428*bis* ! **Ghana:** Bujan to Sekaii *Kitson* 689 ! Gurum-bele to Bantala (Apr.) *Kitson* 835 ! Tumu (Apr.) *Vigne* FH 3786 ! **N.Nig.:** Jos Plateau *Lely* 14 ! In savanna areas of Cameroun, Gabon, Ubangi-Shari, Sudan, Uganda, Congo, N. Rhodesia, Mozambique and Angola. (See Appendix, p. 456.)

10. **V. simplicifolia** *Oliv.* in Trans. Linn. Soc. 29 : 113 (1875) ; F.T.A. 5 : 320 ; Pieper l.c. 64, incl. var. *vogelii* (Bak.) Pieper (1928) ; Moldenke l.c. 6 : 133. *V. diversifolia* Bak. in F.T.A. 5 : 323 (1900) ; Chev. Bot. 506 ; F.W.T.A., ed. 1, 2 : 276 ; Aubrév. Fl. For. Soud.-Guin. 506, 507, t. 114, 1–2, not of Kurz (1870). *V. schweinfurthii* Bak. l.c. 322 (1900). *V. vogelii* Bak. l.c. 319 (1900) ; Aubrév. l.c. t. 114, 5. (?) *V. cordata* Aubrév. l.c. 506 (1950), French descr. A small tree or shrub with dense, pale indumentum and mauve flowers ; in savanna.
Mali: Sansanding *Chev.* 2767. Bandiagara *de Ganay* 22. Bobo-Dioulasso *Aubrév.* 1868. **Iv.C.:** *Aubrév.* 1967–1969. Ouangolo *Aubrév.* 1394, Ferkéssédougou *Aubrév.* 1540. Dimbroko *Aubrév.* 428. **Dah.:** Savalou *Aubrév.* 46*d* ; 57*d*. **N.Nig.:** Sokoto (fl. & fr. Mar.) *Lely* 849 ! Kontagora (Jan.) *Dalz.* 176 ! *Meikle* 1070 ! Jos Plateau (fr. Mar.) *Lely* P197 ! Vom *Dent Young* 206 ! Nupe *Barter* 1644 ! Makurdi South (Feb.) *Trueblood* FHI 4319 ! **S.Nig.:** Yoruba country *Barter* 1096 ! [Br.]**Cam.:** Serti to Goje, Gashaka Dist. (Feb.) *Latilo & Daramola* FHI 34490 ! Also in Cameroun, Uganda and extending to Egypt and Sudan. (See Appendix, p. 457.)

11. **V. ferruginea** *Schum. & Thonn.* Beskr. Guin. Pl. 62 (1827) ; F.T.A. 5 : 324 ; Pieper l.c. 70. *V. fosteri* Wright in Kew Bull. 1908 : 437 ; Pieper l.c. 68 ; F.W.T.A., ed. 1, 2 : 276 ; Aubrév. Fl. For. C. Iv., ed. 2, 3 : 233, t. 337 A ; Moldenke l.c. 5 : 334. Tree 15–50 ft. high with ochre indumentum on inflore-scences and young parts ; flowers pink to purple, sometimes almost cauliflorous ; in closed forest.
Port.G.: Catio, Guebil (May) *Esp. Santo* 2059 ! Canchungo, Empacaca *Esp. Santo* 1948 ! **Guin.:** Dyeke (fr. Oct.) *Baldwin* 9678 ! **S.L.:** Mano (Apr.) *Deighton* 3128 ! Njala (fr. Apr., July) *Deighton* 3520 ! 3534 ! Baiima (fr. Sept.) *Deighton* 3077 ! Tabe (fl. Aug.) *Deighton* 6085 ! **Lib.:** Gbau, Sanokwele Dist. (fr. Sept.) *Baldwin* 9430 ! **Iv.C.:** Rasso *Aubrév.* 154 ! Agboville *Aubrév.* 421 ! Man *Aubrév.* 938 ! **Ghana:** Cape Coast

(fr. July) *Hall* 1509 ! Kumasi (Apr.) *Vigne* FH 1893 ! Kwahu Praso (fr. June) *Vigne* FH 1761 ! Banka (Mar.) *Vigne* FH 1876 ! Ashanti (Mar.) *Thompson* 37 ! **S.Nig.:** Lagos *Foster* 34 ! Ikoyi Plains (Apr.) *Dalz.* 1246 ! *Wright* R135 ! " E. Prov." (Mar.) *Unwin* 9 ! Itu, Cross R. (Apr.) *Ainslie* 129 ! Aro Chuku, Calabar Prov. (Apr.) *Ainslie* 139 ! Also in Congo. (See Appendix, p. 457.)

12. **V. chrysocarpa** *Planch. ex Benth.* in Fl. Nigrit. 486 (1849) ; F.T.A. 5 : 325 ; Chev. Bot. 506 ; Pieper l.c. 68 ; Moldenke l.c. 5 : 269. *V. pseudochrysocarpa* Pieper in Fedde Rep. 26 : 164 (1929). Small shrub or spreading tree, especially on banks of rivers ; indumentum mostly pale yellow, flowers violet, in peduncled cymes.

Mali: Nyamina to Koulikoro *Chev.* 2756. Koulikoro *Vuillet* 611. Bamako *Waterlot* 1116. *Jaeger* 6 ! Mopti to Djenné *Chev.* 2765. **Guin.:** Kouroussa *Pobéguin* 232. Siguiri *Chev.* 2768. **Iv.C.:** Black Volta *Serv. For.* 2408. Diébougou *Serv. For.* 2736. **Ghana:** Yeji (Apr.) *Vigne* FH 1688! Bugiyenga to Loggoda (May) *Kitson* 688 ! Kete Krachi (Dec.) *Adams* 4567 ! Tamale (Apr.) *Williams* 150 ! Yendi (Apr.) *Dalz.* 2 ! **Togo Rep.:** Sansanné Mango *Aubrév.* 78d. **Niger:** Gourma *Chev.* 24394. **N.Nig.:** Nupe *Barter* 1214 ! 1651 ! Jebba (fr. June) *Onochie* FHI 18669 ! Lokoja *Barter* 388 ! Badeggi-Lapai, Lapai Dist. (fr. June) *Onochie* FHI 40236 ! Yola *Dalz.* 115 ! **S.Nig.:** " Quorra " (= R. Niger) *T. Vogel* 142.! Olokemeji (Jan.) *Kennedy* 58 ! *Obaseki* FHI 23825 ! (See Appendix, p. 456.)

An ornamental shrub from the Mediterranean region, *V. agnus-castus* L., is occasionally cultivated in our area.

Imperfectly known species.

V. milnei *Pieper* in Engl. Bot. Jahrb. 62, Beibl. 141 (as 142) : 71 (1928) ; Moldenke in Phytologia 5 : 475 (1957). *V. divaricata* Bak. in F.T.A. 5 : 327 (1900), not of Sweet (1788). *V. cienkowskii* of F.W.T.A., ed. 1, 2 : 276, partly (syn. & *Milne*).

F.Po: *Milne* ! Further material is required in order to decide the status of this plant. It differs from *V. doniana* in having a smaller calyx (2 mm. long) and a laxer inflorescence, and when better known it may prove to be recognisable as a distinct infraspecific taxon of that species. Pieper also cites *Isert*, and *Warnecke* 156a from Togo.

V. sp. A. Trees with large digitate leaves and rather membranaceous leaflets.

S.Nig.: Okomu F.R., Benin *Brenan* 9188 ! [**Br.**]**Cam.:** Bambuko F.R., Kumbo *Olorunfemi* FHI 30751 ! Flowers are required to decide if these are distinct from *V. grandifolia* or whether they are referrable to the Cameroun species *V. zenkeri* Gürke (in Engl. Bot. Jahrb. 33 : 293 (1903)).

V. sp. B. A tree close to *V. rivularis* from which it differs by having the leaves more scabridulous on both sides, obtuse or shortly deltoid at the apex, and the central leaflet with 9–10 pairs of lateral nerves.

[**Br.**]**Cam.:** S. Bakundu F.R., Kumba (fr. Aug.) *Olorunfemi* FHI 30712 !

161. AVICENNIACEAE

By F. N. Hepper

Shrubs or trees, branches terete. Leaves opposite and decussate, simple, entire ; stipules absent. Inflorescences axillary or terminal ; flowers sessile, hermaphrodite, actinomorphic. Calyx very deeply 5-lobed, lobes imbricate, subtended by a pseudo-involucre composed of a scale-like bractlet and 2 alternate scale-like prophylla. Corolla gamopetalous, campanulate-rotate, actinomorphic, deeply 4-lobed. Stamens 4, inserted in throat of the corolla-tube, equal or subdidynamous. Ovary superior, with 2 united carpels, placenta free, central. Style shortly divided. Ovules 4, pendulous. Fruit dehiscent by 2 valves, usually only 1 seed develops. Seed with 2 cotyledons, embryo viviparous.

A monogeneric family, formerly in *Verbenaceae*, inhabiting the maritime regions of the tropics and subtropics around the world and forming characteristic mangrove thickets.

AVICENNIA Linn.—F.T.A. 5 : 287 ; Moldenke in Phytologia 7 : 123, 179, 259 (1960).

Branchlets puberulous ; leaves lanceolate to oblong-lanceolate, or rarely narrowly elliptic, subacute, acute at the base, up to 18 cm. long and 8 cm. broad, minutely and closely pitted above, with a close grey scurf beneath ; lateral nerves 10–12 pairs, looped and forming an intra-marginal nerve, raised on both surfaces ; flowers crowded at the end of short peduncles, these often forming a small panicle ; bracts imbricate, ovate, shortly pointed, about 4 mm. long, tomentellous ; fruit obliquely ellipsoid, 2·5 cm. long, beaked, tomentellous ..   ..   *africana*

A. africana *P. Beauv.* Fl. Oware 1 : 80, t. 47 (1809) ; F.T.A. 5 : 331 ; Chev. Bot. 510 ; Exell Cat. S. Tomé 265 ; De Sousa in Anais Jun. Inv. Col. 3 : 52 (1949) ; Moldenke in Phytologia 7 : 146 (1960). *A. nitida* of F.W.T.A., ed. 1, 2 : 270 ; Berhaut Fl. Sén. 113, 115 ; Aubrév. Fl. For. C. Iv. 3 : 234, t. 338, not of Jacq. A shrub or tree up to 50 ft. high with pneumatophores covering the ground round about but stilt roots absent ; flowers white with yellow centre ; in mangrove thickets. The black mangrove.

Maur.: St. Louis (Feb.) *Döllinger* 73, partly ! **Sen.:** Joual (fr. Jan.) *Döllinger* 73, partly ! Hann beach (May) *Baldwin* 5753 ! Ruflsque *Chev.* 2759. Ile de Jafal *Leprieur.* Carabane *Chev.* 2760 ! **Gam.:** Brumen ferry (Feb.) *Frith* 34 ! Kuntaur *Ruxton* ! Cape St. Mary (Mar.) *Pitt* 693 ! Koto stream area (May) *Fox* 106 ! **Port.G.:** Bissau *Esp. Santo* 1219 ; 1796 ! **Guin.:** *Debeaux.* **S.L.:** *Don.* 168 ! Yele, Turtle Is. *Deighton* 2362 ! opposite York Isl., Bonthe Dist. (Jan.) *T. S. Jones* 410 ! Waterloo (Jan.) *Lane-Poole* 320 ! Mambolo (Apr.) *Glanville* 246 ! Kumrabai (Dec.) *Thomas* 7070 ! **Lib.:** *Dinklage* 1910. Grand Bassa (fl. & fr. July) *T. Vogel* 101 ! Bushrod Isl. (fl. & fr. Aug.) *Baldwin* 13050 ! Sinoe Basin *Whyte* ! **Iv.C.:** Bliéron (May) 19908. **Ghana:** Elmina (fl. & fr. Apr.) *Morton* A494 ! Accra (Oct.) *Johnson* 984 ! *Irvine* 754 ! Ada (Jan.) de Wit & Morton A2971 ! Dwira Akyinim (fl. & fr. Jan.) *Andoh* 5604 ! Bushua (fl. & fr. Apr.) *Chipp* 175 ! **Togo Rep.:** Lomé *Warnecke* 63. **S.Nig.:** Lagos (fl. & fr. Feb.) *MacGregor* 341 ! Forcados (fl. & fr. July) *Unwin* 56 ! Niger Delta *Barter* 44 ! Bamuso (fr. Oct.) *Rosevear* 16 ! Eket Dist. *Talbot* ! [**Br.**]**Cam.:** Victoria (fr. July) *Maitland* 30 ! **F.Po:** *Mann* 231 ! *Milne* ! Extends along the coast of western Africa to Cabinda and Congo ; also in S. Tomé. If this plant is regarded as conspecific with similar ones in America, as does Compère in Taxon 12 : 150–152 (1963), the correct name appears to be *A. germinans* (Linn.) Linn.

Fig. 309.—Avicennia africana *P. Beauv.* (Avicenniaceae).

A, flower. B, corolla laid open. C, stamens. D, pistil. E, fruit. F, fruit in longitudinal
section. G, radicle.

162. LABIATAE

By J. K. Morton

Herbaceous or rarely woody, often odoriferous. Stems usually quadrangular. Leaves opposite or whorled, simple ; stipules absent. Flowers hermaphrodite, zygomorphic, rarely almost actinomorphic, axillary, whorled, racemose or paniculate. Calyx persistent, of 5 variously united sepals, often 2-lipped. Corolla gamopetalous, hypogynous, tubular ; lobes 4–5, imbricate, often forming 2 lips or rarely 1 lip. Stamens inserted in the corolla-tube, 4 or 2 ; anthers 2-celled, cells often divergent, opening lengthwise. Ovary superior, of 2 deeply lobed carpels, the style (gynobasic) arising from the inner base of the lobes ; stigmas mostly bifid. Ovules 4 in each ovary, erect. Fruit of 4 achene-like nutlets, free or cohering in pairs. Seeds with usually straight embryo without endosperm, or the latter very scanty.

A large cosmopolitan family of herbs or undershrubs, recognised by the usually 4-angled stems, opposite (very rarely alternate) leaves, commonly glandular and odoriferous, deeply lobed ovary with gynobasic style, and by the fruit of 4 nutlets.

Calyx with all the teeth equal, or nearly equal, in size and shape without a distinct lip :
  Lower lobe of the corolla considerably larger than the upper :
    Lower lip of corolla flat or very nearly so :
      Lower lip of corolla 3-lobed, with central lobe larger and emarginate ; calyx campanulate ; flowers whorled ..   ..   **26. Stachys**
      Lower lip of corolla entire ; mature calyx ventricose, tube curved ; flowers in axillary panicles of dichotomous cymes ..   ..   ..   **15. Homalocheilos**
    Lower lip of corolla keeled :
      Calyx tube bent near the top   ..   ..   **16. Leocus**
      Calyx tube straight :
        Teeth of calyx very long and spiny   ..   ..   ..   **13. Pycnostachys**
        Teeth of calyx lanceolate, acuminate or triangular, not spiny :
          Mature calyx up to 4 mm. long ; slender delicate plants   **18. Englerastrum**
          Mature calyx 8–10 mm. long ; large, stout erect plant   **19. Isodictyophorus**
  Lower lobe of the corolla not much larger than the upper :
    Mature calyx dry ; stamens 4 :
      Calyx teeth 4   ..   ..   **5. Geniosporum**
      Calyx teeth 5 :
        Calyx teeth subulate   ..   ..   **21. Hyptis**
        Calyx teeth not subulate, either triangular, lanceolate, or lanceolate-acuminate :
          Calyx with teeth as long as, or longer than, the tube   ..   ..   **20. Neohyptis**
          Calyx with teeth much shorter than the tube :
            Calyx teeth narrow, lanceolate, long-acuminate   ..   ..   **22. Satureja**
            Calyx teeth broad, triangular, not acuminate   ..   **25. Achyrospermum**
    Mature calyx forming an orange berry about 8 mm. diam. ; stamens 2 10. **Hoslundia**
Calyx with teeth of different size and shape, or bilabiate, or truncate or (in No. 11) nearly so :
  Calyx circumscissile in fruit :
    Leaves opposite ; rootstock annual and fibrous or perennial and creeping
                                                             11. Aeolanthus
    Leaves alternate ; rootstock perennial, tuberous, not creeping   ..   **12. Icomum**
  Calyx not circumscissile :
    Calyx with 8–10 teeth :
      Upper and lower lips of corolla about equal in length ; middle lobe of lower lip larger than the lateral ones   ..   ..   **27. Leucas**
      Upper lip of corolla much longer than the lower ; middle lobe of lower lip scarcely longer than the lateral ones   ..   ..   **28. Leonotis**
    Calyx with up to 5 teeth :
      Calyx bladder-like in fruit ..   ..   **29. Tinnea**
      Calyx not bladder-like in fruit :
        *Calyx distinctly 2-lipped :
          Inflorescences of dense, globose, terminal heads, frequently with large coloured bracts   ..   ..   **9. Haumaniastrum**
          Inflorescences not as above :
            Corolla markedly zygomorphic, with lower lip deeply concave :
              Inflorescence with 2 flowers at each node   ..   ..   ..   **24. Scutellaria**
              Inflorescence with many flowers at each node   ..   ..   **17. Solenostemon**

Corolla not markedly zygomorphic ; lower lip not deeply concave :
 Calyx-tube much longer than broad ; flowers dark pink **3. Endostemon**
 Calyx-tube about as long as broad ; flowers more or less white **4. Platostoma**
*Calyx not 2-lipped, though frequently with one lobe larger than the others and
 forming a single lip :
 Calyx with 3 broadly triangular or ovate lobes :
  Inflorescences in rather lax racemes ; stamens 2 .. .. .. **23. Salvia**
  Inflorescences dense, conical or oblong ; stamens 4 .. **25. Achyrospermum**
 Calyx not as above, at least some of the teeth lanceolate-acuminate or subulate :
  Upper calyx-lobe oblong or lanceolate :
   Upper calyx-lobe longer than the others .. .. .. **16. Leocus**
   Upper calyx-lobe not longer than the others .. .. **22. Satureja**
  Upper calyx-lobe broadly ovate or triangular :
   Calyx with dense fringe of short white cilia between the teeth .. **2. Becium**
   Calyx not as above:
    Inflorescence with large, pink, ovate or rotund bracts enveloping the flowers
                     8. Hemizygia
    Inflorescence not as above :
     Lower lip of corolla much enlarged and keeled :
      Calyx with median teeth deltoid or obsolete .. **17. Solenostemon**
      Calyx with median teeth well developed and acuminate **14. Plectranthus**
     Lower lip of corolla not enlarged or keeled :
      Mature calyx up to 2 mm. long, upper lip not or slightly decurrent on the
       tube .. .. **6. Basilicum**
      Mature calyx 5 mm. or more long, upper lip decurrent on the tube :
       Stigma capitate .. .. **7. Orthosiphon**
       Stigma deeply bifid .. .. **1. Ocimum**

Besides the above, one or more "species" of *Mentha* (? *M.* × *gentilis* Linn.) are occasionally cultivated as herbs but apparently do not flower in these latitudes.

1. OCIMUM Linn.—F.T.A. 5 : 334.

Calyx twice as long as broad when viewed from above ; perennials :
 Calyx dull, densely woolly ; spikes dense, whorls continuous except at base ; calyx
  7 mm. long in fruit, upper lip ovate, apiculate, lower teeth small, deltoid ; leaves
  pubescent, ovate to ovate-lanceolate, acutely acuminate, 7–10 cm. long, 2–4 cm.
  broad ; petiole 1–3 cm. long .. .. 1. *suave*
 Calyx glossy, glabrous or nearly so ; spikes rather lax :
  Leaves linear, about 3 cm. long and 3 mm. broad, or less, glabrous with a slightly wavy
   margin, gland-dotted ; stems stiff, erect, glabrous, leafy ; inflorescence a lax
   terminal raceme, bracts minute ; pedicels about half as long as the fruiting calyx,
   arcuate, puberulous ; fruiting calyx 8 mm. long, tube twice as long as the teeth,
   thinly gland-dotted ; upper lip recurved, deltoid ; lower teeth ovate, subulate
                     2. *irvinei*

  Leaves ovate :
   Calyx about 6 mm. long in fruit ; leaves 6–12 cm. long and 3 cm. broad, ovate to
    obovate, cuneate at base, acutely acuminate, subglabrous beneath but gland-
    pitted and slightly pubescent on the nerves ; whorls only continuous in flower,
    interrupted in fruit ; pedicels puberulous ; calyx 2-lipped, upper lip broadly ovate
    and mucronate, lower lip oblong with 2 teeth .. .. .. 3. *gratissimum*
   Calyx about 10 mm. long in fruit ; leaves 2–5 cm. long ovate, acute, petiolate,
    pubescent beneath ; spikes long, lax, often branched at base ; bracts bright-
    coloured ; calyx with upper lip ovate, obtuse, lower lip with 2 large teeth ; stamens
    long-exserted .. .. 4. *lamiifolium*
Calyx orbicular when viewed from above ; leaves at most shallowly serrate ; annuals ;
 stems square ; leaves distinctly petiolate, thin, ovate to obovate, cuneate at base,
 acuminate ; inflorescence lax, elongate ; calyx with upper lobe orbicular and much
 longer than the very short tube, lower lobes deltoid-mucronate, slightly exceeding the
 upper ; stamens slightly exserted :
  Mature calyx about 6 mm. diam. ; corolla large, twice as long as the calyx or more ;
   plant bushy, 50 cm. or more in height, usually subglabrous .. 5. *basilicum*
  Mature calyx about 4 mm. diam. ; corolla small, hardly exceeding the calyx; plant
   smaller, 30–45 cm. high, pubescent .. .. 6. *canum*

1. **O. suave** *Willd.* Enum. Pl. Hort. Berol. 629 (1809) ; F.T.A. 5 : 338. *O. trichodon* Bak. ex Gürke (1895)—
F.T.A. 5 : 338 ; F.W.T.A., ed. 1, 2 : 286. *O. dalabaense* A. Chev. (1909)—Chev. Bot. 511. *O. caillei* A.
Chev. Bot. 511, name only. A branched, erect, pubescent, aromatic shrub reaching several feet in height
with dense spikes of small greenish-white flowers.
 Guin.: Fouta Djalon 3,000–3,500 ft. (Sept.) *Chev.* 18382! 18637! Labé (Mar.) *Chev.* 12382! Mali to
Kédougou (Apr.) *Pitot* T51–1180! Dalaba (Oct.) *Adam* 12583! Niguelandé (Mar.) *Adam* 11632! **N.Nig.:**
Jos Plateau (Oct.) *Lely* P798! Ropp *Morton* A2043! [Br.]**Cam.:** Mambila Plateau (Jan.) *Hepper* 1689!
Also in Cameroun, Congo and E. Africa, and widespread in tropical Asia.
2. **O. irvinei** *J. K. Morton* in J. Linn. Soc. 58 : 266, t. 1 (1962). An erect herb about 18 in. high ; stems

several from a perennial woody rootstock with thick roots ; flowers pink rarely white ; in seasonally marshy savanna.

Ghana: N.W. Ashanti : near Nwereme (May) *Irvine* 2498! New Drobo (Nov.-Apr.) *Harris*! *Morton* A3876! A3877! Dwinim (fr. Dec.) *Adams* 5294! Adamsu (Apr.) *Morton* A3238! W. of Bimbila, Dagomba Dist. (Jan., Mar.) *Hepper & Morton* A3096! Akpabla 1853!

3. **O. gratissimum** *Linn.* Sp. Pl. 2: 1197 (1753) ; Chev. Bot. 511 (as var. *macrophyllum* Briq.) ; Morton l.c. 232 (1962). *O. viride* Willd. (1809)—F.T.A. 5 : 337 ; Chev. Bot. 512 ; F.W.T.A., ed. 1, 2 : 285. *O. guineense* Schum. & Thonn. (1827). Similar to *O. suave* but subglabrous ; usually occurring around villages. The Tea Bush.

Gam.: Albredaar (Nov.) *Gardiner* 37! **Port.G.:** Gidade, Bissau (Oct.) *Esp. Santo* 945! **Guin.:** Korse (Mar.) *Adam* 11706! Bcyla (Mar.) *Chev.* 20901! **S.L.:** Kuntara (June) *Thomas* 417! **Lib.:** Ganta *Harley*! **Iv.C.:** Abidjan *Chev.* 15386 ; 15410. Tiebissou to Languira (July) *Chev.* 22199. **Ghana:** Suhum (Dec.) *Morton* GC 6335! Akuse (Oct.) *Morton* A12! **Dah.:** Dassa-Zoume (May) *Chev.* 23627. **N.Nig.:** Abinsi (Mar.) *Dalz.* 773! **S.Nig.:** Sapoba *Kennedy* 1771! [**Br.**]**Cam.:** Victoria (Apr.) *Schlechter* 12570! Buea (Dec., Apr.) *Morton* K646! K836! **F.Po:** (Oct.) *T. Vogel* 4! Probably originally introduced from Asia, now widespread in tropical Africa. (See Appendix, p. 462.)

4. **O. lamiifolium** *Hochst. ex Benth.* in DC. Prod. 12 : 37 (1848) ; F.T.A. 5 : 346. An erect, hairy perennial several feet in height ; flowers white, in long lax racemes.

[**Br.**]**Cam.:** Bamenda, 5,000 ft. (May, July) *Maitland* 1438! *Ujor* FHI 30460! Extends to Ethiopia and Tanganyika.

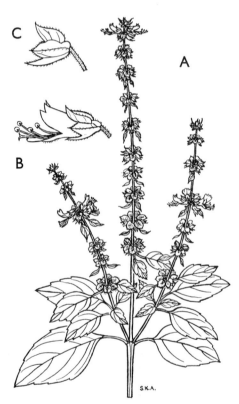

Fig. 310.—Ocimum basilicum *Linn.* (Labiatae).

A, leafy shoot with inflorescence.
B, flower.   C, accrescent calyx.

5. **O. basilicum** *Linn.* Sp. Pl. 2 : 597 (1753) ; F.T.A. 5 : 336. *O. americanum* Linn. (1763)–F.W.T.A., ed. 1, 2 : 285, partly. *O. lanceolatum* Schum. & Thonn. (1827). *O. dichotomum* Hochst. ex Benth. (1848)—F.T.A. 5 : 346. *O. menthaefolium* of Chev. Bot. 512, not of Hochst. ex Benth. A stout, bushy, aromatic herb, with white flowers in loose racemes ; commonly cultivated for culinary purposes ; tetraploid.

Sen.: Tamboukane (Dec.) *Chev.* 2780! **Mali:** Dogo (Apr.) *Davey* 609! **Port.G.:** Bafata (Aug.) *Esp. Santo* 3291! **S.L.:** (Oct.) *Thomas* 2652! 9071! Mt. Aureol, Freetown (Aug.) *Melville* 97! **Lib.:** Peahtah (Oct.) *Linder* 1057! **Ghana:** Legon (Nov.) *Morton* A981! Denu (May) *Morton* A2201! **N.Nig.:** Zungeru (June) *Dalz.* 137! **S.Nig.:** Lagos (May) *Denton* 58! **F.Po:** (Dec.) *Guinea* 563! Probably introduced from Asia, now widespread in the tropics. (See Appendix, p. 462.)

[This species is not always readily distinguishable from the next.]

6. **O. canum** *Sims* Bot. Mag. 51 : t. 2452 (1823) ; F.T.A. 5 : 337 ; Chev. Bot. 511. *O. americanum* of F.W.T.A. ed. 1, 2 : 285, partly. *O. hispidulum* Schum. & Thonn. (1827). *O. thymoides* Bak. in F.T.A. 5 : 344 (1900). Similar to *O. basilicum* but smaller in all its parts and having a more pungent odour ; commonly cultivated for medicinal purposes ; perhaps not native ; diploid.

Maur.: Tichoten (July) *Charles* in *Hb. Chev.* 28696. Niemelane (July) *Charles* in *Hb. Chev.* 28713. **S.L.:** Kanno (July-Aug.) *Dawe* 523! **Lib.:** Gletown (July) *Baldwin* 6794! **Iv.C.:** *fide* Chev. *l.c.* **Ghana:** Achimota (Feb.) *Morton* GC 25350! Tumu (Mar.) GC 8884! **Togo Rep.:** Lomé *Warnecke* 269! **Dah.:** Dassa-Zoume (May) *Chev.* 23622. Atacora Mts. (June) *Chev.* 24124. **N.Nig.:** Jos Plateau (July) *Lely* P525! Maiduguri (Oct.) *Noble* A13! **S.Nig.:** Oban (Feb.) *Talbot* 228! [**Br.**]**Cam.:** Gurum, Adamawa (Nov.) *Hepper* 1293! **F.Po:** (Oct.) *T. Vogel* 32! Widespread in the tropics. (See Appendix, p. 462.)

2. BECIUM Lindl. Bot. Reg. 28 : Misc. 42 (1842).

Whole plant glabrous, thinly puberulous to rather densely puberulous ; leaves linear to narrowly elliptic or obovate-lanceolate, subsessile or shortly petiolate, thick and prominently veined, acute to obtuse, tapered to base, 3–8 cm. long, 5–20 mm. broad, more or less crenate towards apex, inflorescence terminal, dense when young, elongating and interrupted in flower and fruit ; pedicels very short, stout, puberulous ; calyx accrescent, whitish-pubescent when young, mature calyx subglabrous to pubescent 8–10 mm. long, upper lip ovate, shortly apiculate, lower teeth up to 2 mm. long, subulate, upcurved ; corolla 7–14 mm. long, distinctly 2-lipped, puberulous ; stamens long-exserted .. .. *obovatum*

B. obovatum (*E. Mey.*) *N. E. Br.* in Fl. Cap. 5, 1 : 230 (1910). *Ocimum obovatum* E. Mey. (1837). *Ocimum knyanum* of F.T.A. 5 : 346 ; of F.W.T.A., ed. 1, 2 : 286, not of Vatke. *O. knyanum* var. *astephanum* Bak. in F.T.A. 5 : 346 (1900). *O. odontopetalum* C. H. Wright in Kew Bull. 1907 : 54. *O. affine* of Chev. Bot. 510, not of Hochst. *Becium knyanum* (Vatke) G. Tayl. in J. Bot. 69, Suppl. 2, 146 (1931), partly. Rootstock swollen, woody, perennial, producing erect, branched or unbranched flowering shoots 6 in.–2 ft. high, bearing compact or elongate inflorescences of attractive pink (rarely white) flowers with very long stamens. Very variable in habit, leaf shape, pubescence and inflorescence but intermediates appear to connect all the W. African forms.
   Guin.: *Farmar* 289 ! Kéméné (Mar.) *Chev.* 12935. Timbo to Ditinn (Sept.) *Chev.* 18400. Labé (Apr.) *Adam* 11816 ! Fouta Djalon (fr. June) *Adam* 14496 ! 14666 ! **Ghana:** Bimbila (Jan.-Mar.) *Hepper & Morton* A3069 ! *Morton* A3740 ! Akpabla 1857 ! Shiare to Chilinga, Transvolta (fr. May) *Morton* A3985 ! **N.Nig.:** *Kennedy* 2908 ! Vom (Apr.) *Morton* K347 ! K392 ! K398 ! Iyatawa, Zaria (July) *Keay* FHI 25945 ! Ropp, 4,600 ft., Jos Plateau (Apr., July) *Lely* 444 ! *Morton* K372 ! Nupe *Barter* 1283 ! Sokoto (Nov.) *Lely* 113 ! **[Br.]Cam.:** Binka, Nkambe Div. (Feb.) *Hepper* 1886 ! Widespread in tropical Africa.

3. ENDOSTEMON N. E. Br. in Fl. Cap. 5, 1 : 295 (1910) ; Ashby in J. Bot. 74 : 121 (1936).

Leaves linear-lanceolate, distantly serrate, 3–6 cm. long, 0·5–1 cm. broad, shortly pubescent ; racemes composed of distinct whorls with leafy bracts ; axis, pedicels and calyx setulose-pubescent with white hairs ; calyx narrowly tubular, the upper lip suborbicular with recurved margin .. .. *tereticaulis*

E. tereticaulis (*Poir.*) *M. Ashby* in J. Bot. 74 : 129 (1936). *Ocimum tereticaule* Poir. (1810)—F.T.A. 5 : 347 ; F.W.T.A., ed. 1, 2 : 285 ; Chev. Bot. 512. *O. thonningii* Schumach. (1827). A low bushy herb up to about 1 ft. high, woody below, strongly aromatic, with magenta flowers.
   Sen.: *Heudelot* 168 ! *Perrottet* 670. **Mali:** Dioura (Aug.) *Davey* 83 ! **Iv.C.:** Ouahigouya (Aug.) *Chev.* 24774 ; 24780. **Ghana:** Legon (Nov.) *Morton* GC 6004 ! Nungua, Accra (Oct.) *Morton* GC 7887 ! Accra (May) *Dalz.* 26 ! *Irvine* 658 ! Extends to Ethiopia, Somaliland, E. Africa, S. Rhodesia, Bechuanaland and Transvaal.

4. PLATOSTOMA P. Beauv. Fl. Oware 2 : 61 (1818) ; F.T.A. 5 : 349, as *Platystoma*.

A slender erect annual ; stems slightly pubescent between the rounded angles ; leaves long-petiolate, ovate, about 5 cm. long and 2·5 cm. broad, serrate except for the broad cuneate base, glabrous or nearly so ; racemes numerous, axillary, rather loose, many-flowered ; bracts persistent, broadly ovate to obovate, foliaceous, nearly as long as the whorl, often reflexed and coloured ; pedicels short, puberulous   *africanum*

P. africanum *P. Beauv.* l.c. t. 95, fig. 2 (1818) ; F.T.A. 5 : 349 ; Chev. Bot. 513. *Ocimum sylvaticum* Thonning (1827). *Platostoma djalonense* A. Chev. Bot. 513, name only. A branched slender herb up to about 2 ft. high, with slender racemes of very small flowers ; corolla white, spotted with mauve ; in damp places.
   Sen.: Niokolo-Koba (Oct.) *Adam* 15576 ! **Mali:** Koulikoro (Oct.) *Chev.* **Port.G.:** Fulacunda (Sept.) *Esp. Santo* 2185 ! Bafata (Sept.) *Esp. Santo* 2766 ! **Guin.:** Gangan, Kindia (Apr.) *Adam* 11938 ! Tossékré, Fouta Djalon (Oct.) *Adam* 12753 ! **S.L.:** Sefadu (Nov.) *Deighton* 4664 ! Freetown (Sept.) *Hepper* 914 ! **Lib.:** Cape Palmas (Mar.) *Hale* 20 ! Gbanga (Aug.) *Freeman* 3 ! **Iv.C.:** Nieki (Oct.) *Roberty* 12416 ! Adiopodoumé (July) *Boughey* GC 14512 ! **Ghana:** Nwereme *Morton* ! Accra to Nsawam (Sept.) *Morton* A10 ! Kpandu (Apr.) *Asamang* 166 ! Axim (Mar.) *Morton* GC 8443 ! Togo Rep.: Lomé *Warnecke* 440 ! **N.Nig.:** Abinsi (Aug.) *Dalz.* 795 ! **S.Nig.:** Onitsha (Sept.) *Onochie* FHI 33447 ! **[Br.]Cam.:** above Buea (Mar.) *Morton* GC 6742 ! K454 ! Gurum, Adamawa (Nov.) *Hepper* 1381 ! **F.Po:** (Jan.) *Guinea* 1084 ! Estrada (Jan.) *Akpabla* GC 11486 ! Widespread in tropical Africa and in India. (See Appendix, p. 463.)

5. GENIOSPORUM Wall. ex Benth.—F.T.A. 5 : 350.

Stems stout, woody, grooved, densely ferrugineous-pubescent ; leaves subrotund to broadly lanceolate, slightly acuminate, usually acute, rounded at base, shortly petiolate, crenulate, 2–5 cm. long, 1–3 cm. broad, pubescent above, densely ferrugineous-pubescent beneath, coriaceous ; inflorescences several, dense, cylindrical, 2·5–10 cm. long, 1–2 cm. broad ; bracts broadly ovate, exceeding calyx, lower ones larger, white ; corolla twice as long as calyx, lobes subequal, not concave, slightly bilabiate ; anthers and stigma exserted ; calyx about 4 mm. long when mature, tubular with 4 teeth, the lower one much broader, obtuse or emarginate and incurved or reflexed .. .. *rotundifolium*

G. rotundifolium *Briq.* in Engl. Bot. Jahrb. 19 : 163 (1894) ; F.T.A. 5 : 351. *G. paludosum* Bak. in F.T.A. 5 : 352 (1900) ; F.W.T.A., ed. 1, 2 : 286 : Robyns Fl. Sperm. Parc Nat. Alb. 2 : 194, t. 18. *Ocimum konianense* A. Chev. in Mém. Soc. Bot. Fr. 2, 8 : 198 (1912) ; Chev. Bot. 511. *O. paludosum* (Bak.) Roberty in Bull. I.F.A.N. 16, 1 : 329 (1954). A stout, erect, perennial, several feet in height, with a densely pubescent stem and conspicuous white or mauve-tinged bracts ; in damp grassland at high altitudes.
   S.L.: Bintumane, 4,500 ft. (Jan.) *Glanville* 468 ! **Iv.C.:** Beyla (Mar.) *Chev.* 20869 ! **N.Nig.:** Bukuru to Hepham, 4,300 ft. (Aug.) *Lely* 345 ! Vom, 3,000–4,500 ft. *Dent Young* 210 ! **S.Nig.:** Ikwette Plateau, 4,000–5,600 ft., Obudu (Aug.-Dec.) *Keay & Savory* FHI 25035 ! *Stone* 52 ! **[Br.]Cam.:** Bamenda, 5,800–6,000 ft. (Apr.) *Maitland* 1668 ! *Morton* K283 ! Mambila Plateau (Jan.) *Hepper* 1665 ! Vogel Peak, Adamawa (Nov.) *Hepper* 1366 ! Also in Congo, Angola and E. Africa.

6. BASILICUM Moench Suppl. Meth. Pl. 143 (1802).

Stems quadrangular, glabrous ; leaves long-petiolate, broadly ovate, cuneate and entire
at base, long- and gradually-acuminate, up to 6 cm. long and 3 cm. broad, margin
crenate-serrate, closely and minutely gland-dotted ; racemes slender, axillary,
whorls about 6-flowered, rather lax ; pedicels short, puberulous, articulating at the
top ; bracts minute ; calyx gland-dotted .. .. 1. *polystachyon*

B. **polystachyon** (*Linn.*) *Moench* l.c. (1802) ; F.W.T.A., ed. 1, 2 : 608. *Ocimum polystachyon* Linn. (1771).
Moschosma polystachyon (Linn.) Benth. (1831)—F.T.A. 5 : 353 ; Chev. Bot. 515 ; F.W.T.A., ed. 1, 2 :
287. *Ocimum dimidiatum* Schum. & Thonn. (1827). A branched, erect, subglabrous herb, 1–3 ft. high,
with very small, pale mauve, almost white flowers ; in damp places.
Iv.C.: Assinie (Apr.) *Chev.* 17871 ; 17874 ; 17882. Gottoro to Diahbo, Baoule-Nord (July) *Chev.* 22035.
Ghana: Accra to Nsawam (July) *Morton* GC 9324 ! Tumu, N.R. (May) *Morton* GC 7571 ! Cape Coast
(July) *T. Vogel* 44 ! **Dah.:** Abbo to Massé (Feb.) *Chev.* 22988. Abomey (Feb.) *Chev.* 23189. Aguégué
(Mar.) *Chev.* 23320. **N.Nig.:** Nupe *Barter* 1157 ! Jebba (Dec.) *Hagerup* 678 ! Katagum (Nov.) *Dalz.* 111 !
S.Nig.: Lagos *Rowland* ! *Dalz.* 1402 ! Ogun River F.R., Lagos (Mar.) *Hepper* 2261 ! Idanre, Ondo Prov.
(Jan.) *Brenan & Jones* 8735 ! Widespread in the Old World tropics. (See Appendix, p. 462.)

7. ORTHOSIPHON Benth.—F.T.A. 5 : 365.

Root tuberous, woody, perennial ; stems stiff, erect ; calyx 9–10 mm. long when mature,
usually bright pink, setose, upper lip reflexed, deltoid and slightly mucronate, lower
teeth ovate with a long acumen, tube twice as long as the teeth ; corolla 1 cm. long
or more, conspicuous ; leaves 5–8 cm. long, including the petiole, 1–3 cm. broad,
cuneate at base, rounded-obtuse at apex, lanceolate, unevenly crenate-serrate on
margins, thick, subglabrous to pubescent, glaucous and distinctly veined beneath ;
inflorescence elongate, many-flowered with 4–6 flowers per whorl   1. *rubicundus*
Root fibrous ; calyx 5–7 mm. long, either greenish or dark purple ; leaves small, less
than 5 cm. long including the petiole :
Calyx 5 mm. long, pale green ; upper lip ovate, obtuse, lower teeth ovate, subulate ;
leaves up to 3·5 cm. long including the long petiole, and 2 cm. broad, but usually
much smaller, lanceolate, subacute to obtuse, crenate-serrate ; stems square,
recurved-pubescent ; corolla small, pale greenish .. .. 2. *incisus*
Calyx 7 mm. long, usually tinged with dark purple ; upper lip deltoid, apiculate,
teeth ovate, subulate ; leaves up to 5 cm. long and 4 cm. broad, broadly ovate-
triangular, long-petiolate, coarsely serrate, glabrous except for a few appressed hairs
on the veins, minutely glandular-punctate beneath ; stems square, subglabrous,
slender ; inflorescence elongate, lax, thinly pubescent, whorls about 6-flowered ;
corolla small, pale violet .. .. 3. *suffrutescens*

1. **O. rubicundus** (*D. Don*) *Benth.* in Wall. Pl. As. Rar. 2 : 14 (1831). *Plectranthus rubicundus* D. Don (1825).
Orthosiphon salagensis Bak. in F.T.A. 5 : 368 (1900) ; F.W.T.A., ed. 1, 2 : 286, partly. *O. coloratus* Vatke
(1881)—F.T.A. 5 : 372. *O. xylorrhizus* Briq. (1917)—Chev. Bot. 515. *Ocimum affine* var. *bafingensis*
A. Chev. Bot. 510, name only. An attractive erect aromatic perennial herb about 1½–2 ft. high ; flowers
white or pink ; inflorescence usually pink ; calyx accrescent ; common in savanna woodland.
   Mali: Bama to Samandini (June) *Chev.* 944. Iv.C.: junction of R. Sassandra and R. Bafing (May) *Chev.*
21776 *bis* ! **Ghana:** Salaga (Aug.) *Morton* GC 7191 ! *Krause* ! Banguon, Lawra (May) *Vigne* FH 3816 !
Yendi (May) *Akpabla* 516 ! **Dah.:** Pobégou to Birni, Atacora Mts. (June) *Chev.* 23944 ! **N.Nig.:** S.W. of
Wamba (Apr.) *Morton* K419 ! Naraguta (May) *Hepper* 678 ! Zaria (May) *Keay* FHI 25796 ! Dutsin
Makurdi (June) *McElderry* FHI 16377 ! Also in Congo, E. and S. tropical Africa and in India.
2. **O. incisus** *A. Chev.* in Mém. Soc. Bot. Fr. 2, 8 : 199 (1912) ; Chev. Bot. 515. *O. salagensis* of Chev. Bot.
515 ; of F.W.T.A., ed. 1, 2 : 286, partly ; not of Bak. A small, annual, pubescent, bushy, aromatic herb,
under 12 in. high.
   Mali: Ouagadougou to Ouahigouya (Aug.) *Chev.* 24698. **Ghana :** Accra (Dec.) *Morton* GC 6135 ! Prampram (Oct.) *Morton* GC 6083 ! Accra to Ada (June) *Morton* GC 9237 ! N. of Nangodi (Dec.) *Morton* A1261 !
Dah.: Somba, Atacora Mts. (June) *Chev.* 24120 !
3. **O. suffrutescens** (*Thonning*) *J. K. Morton* in J. Linn. Soc. 58 : 266 (1962). *Ocimum suffrutescens* Thonning
in Kongel. Dansk. Vid. Selsk. 4 : 330 & Rettelser (1829). *O. thoningii* Thonning Beskr. Guin. Pl. 269
(1827), not *O. thonningii* Schum. & Thonn. l.c. 265 (1927). *Orthosiphon australis* Vatke (1876)—F.T.A. 5 :
373. *O. rabaiensis* var. *parvifolia* S. Moore in J. Bot. 44 : 26 (1906). A slender straggling subglabrous
perennial herb of deciduous and dry forest and thicket, frequently purple-tinged on the stem, leaves and
inflorescence.
   Ghana: Mampong Scarp, Akwapim (June) *Morton* GC 9325 ! Afram Mankrong F.R. (Dec.) *Adams* 4935 !
Swedru (Feb.) *Dalz.* 8279 ! Asafo road junction, Accra to Cape Coast (Mar.) *Morton* A251 ! Fume, Trans-
volta (Nov.) *Morton* A959 ! **S.Nig.:** (Oct.) *Foster* 348 ! Also in Congo and widespread in E. Africa.
   [Very closely allied to, and probably only racially distinct from *O. glabratus* Benth. in India.]

8. HEMIZYGIA (Benth.) Briq. in E. & P. Pflanzenfam. 4, 3A : 368 (1897) ; Ashby in J. Bot. 73 : 312 (1935). *Bouetia* A. Chev. (1912).

Leaves linear-lanceolate, 5–15 cm. long, 1–2·5 cm. broad, narrowly acute at apex,
gradually cuneate at base, sessile or shortly petiolate, hairy on both surfaces ; racemes
long, lax, hairy ; flowers 4–6 in a whorl ; pedicels short ; upper bracts large, narrowly
ovate-lanceolate, reddish, membranous ; calyx 7–8 mm. long in fruit, upper lip
orbicular, lower teeth long and subulate ; corolla twice as long as the calyx, 2-lipped ;
stamens long-exserted ; whole plant covered densely with long, straight hairs
                                                                    1. *bracteosa*
Leaves ovate to ovate-lanceolate, 2·5–5 cm. long, 1–2·5 cm. broad, broadly acute at
apex, rounded to cuneate at base, sessile or nearly so, pubescent ; racemes shorter
and denser, hairy ; flowers 6 in a whorl ; pedicels short, stout, pubescent ; upper
bracts large ovate-orbicular, reddish, membranous ; calyx about 10 mm. long in

fruit, upper lip broadly ovate, lower teeth long and subulate ; corolla and stamens as above ; whole plant covered fairly densely with crispate hairs . . 2. *welwitschii*

1. **H. bracteosa** (*Benth.*) *Briq.* in Ann. Jard. Gen. 2 : 248 (1898) ; Ashby in J. Bot. 73 : 318, 350 (q.v. for syn.). *Ocimum bracteosum* Benth. (1832). *Orthosiphon bracteosus* (Benth.) Bak. in F.T.A. 5 : 375 (1900) ; F.W.T.A. ed. 1, 2 : 286. *Hemizygia nigritiana* S. Moore (1909). *Bouetia ocymoides* A. Chev. (1912)—Chev. Bot. 513. An erect greyish pubescent annual up to 3 ft. high, with pale pink flowers and clusters of pink bracts at the top of the inflorescence ; in marshy grassland.
    Sen.: Heudelot 396 ! **Dah.:** Agouagon (May) *Chev.* 23546 ! Pobégou to Birni, Atacora Mts. (June) *Chev.* 23945 ! **N.Nig.:** Nupe *Barter* 946 ! Alagbede, Ilorin (Feb.) *Ejiofor* FHI 19807a ! Yola (May) *Dalz.* 108 ! **S.Nig.:** "W. Prov." *Kitson* ! [Br.]**Cam.:** Alantika Mts., Adamawa (Dec.) *Hepper* 1606 ! Widespread in tropical and S. Africa. (See Appendix, p. 463.)

2. **H. welwitschii** (*Rolfe*) *M. Ashby* in J. Bot. 73 : 318, 350 (1935). *Orthosiphon welwitschii* Rolfe (1893)— F.T.A. 5 : 376. *O. bracteosus* of F.W.T.A., ed. 1, 2 : 286, partly. A bushy, aromatic perennial, up to 2 ft. high, with white flowers and terminal clusters of pink bracts ; growing in clumps in dry stony grassland.
    N.Nig.: Kafanchan (May) *Jones* FHI 14106 ! Zaria (Apr.) *Dalz.* 285 ! Jos (Jan., Mar.) *Hill* 19 ! *Lely* P112 ! *Batten-Poole* 379 ! Vom (Feb.) *McClintock* 198 ! Gindiri, Pankshin Div. (Oct.) *Hepper* 1130 ! [Br.]**Cam.:** Bamenda (Apr.-Mar.) *Maitland* 1599 ! *Morton* K190 ! Ndop to Bamenda (Dec.) *Adams* GC 11252 ! Ndop (Dec.) *Boughey* GC 11141 ! Also in Ubangi-Shari, Congo, Angola and S. Rhodesia.

9. HAUMANIASTRUM Duvign. & Plancke in Biol. Jaarb. 27 : 222 (1959) ; Morton in J. Linn. Soc. 58 : 239 (1962). *Acrocephalus* Benth. (partly)—F.T.A. 5 : 354 ; F.W.T.A., ed. 1, 2 : 288.

Leaves broad, ovate to lanceolate or broadly oblanceolate :
   Leaves sessile, ovate, rounded at base, acutely acuminate, sharply serrate, 2–4 cm. long, 1–2 cm. broad, thick ; flower-heads numerous, corymbose ; bracts ovate-reniform, broadly apiculate, densely tomentose outside with glabrescent margins
1. *alboviride*
Leaves distinctly petiolate, ovate-lanceolate to oblanceolate, cuneate at base, subacute, serrulate, 3–6 cm. long, 1–2 cm. broad, thinly pilose on both surfaces ; flower-heads few or subsolitary, small, globose, subtended by normal, non-coloured leaves
2. *galeopsifolium*
Leaves narrow, linear to linear-lanceolate or narrowly oblanceolate :
   Stem and leaves glabrous or nearly so ; leaves erecto-patent, narrow, sessile, 3·5–5 cm. long, 3–5 mm. broad, obscurely dentate, densely gland-dotted on both surfaces, with a few hairs on the veins beneath ; stem wiry, square, purple ; flower-heads few, rather large ; outer bracts coloured, with a long cusp ; inner bracts broadly ovate, lanate . . . . 3. *quarrei*
   Stem and leaves densely pubescent ; leaves spreading ; flower-heads numerous :
      Stem with long spreading pubescence ; leaves linear to linear-lanceolate, acute, up to 10 cm. long and 1 cm. broad, usually less, obscurely dentate, thinly to densely long-pubescent and gland-dotted on both surfaces ; outer bracts frequently with a white or coloured crenate margin and a short to medium cusp ; inner bracts broadly ovate, lanate throughout . . . . 4. *lilacinum*
      Stem with closely appressed short pubescence :
         Flower-heads small ; inner floral bracts with subglabrous brown margin, lanate below ; outer bracts with narrow base, without coloured crenate margin ; leaves oblanceolate, acute, subentire, 7 cm. long and 1 cm. broad or less, densely silvery appressed-pubescent especially on the veins beneath . . . . 5. *buettneri*
         Flower-heads medium to large ; inner floral bracts lanate throughout ; outer bracts usually with a broad crenate white or coloured base and a medium to short cusp ; leaves linear to linear-lanceolate, acute, more or less dentate, up to 12 cm. long and 2 cm. broad, but usually much less, shortly appressed-pubescent at least on the veins and gland-dotted on both surfaces. . . . . . 6. *caeruleum*

1. **H. alboviride** (*Hutch.*) *Duvign. & Plancke* in Biol. Jaarb. 27 : 224 (1959). *Acrocephalus alboviridis* Hutch. in Kew Bull. 1921 : 396 ; F.W.T.A. ed. 1, 2 : 287. An erect woody plant, 3–4 ft. high, with dark purple stems and pale violet flowers.
    Ghana: Shiare to Chilinga, 2,000 ft., Transvolta (fl. Nov., fr. Apr.) *Hall* 1441 ! *Morton* A3959 ! A3960 ! A4101 ! **N.Nig.:** Jos Plateau (Oct.) *Lely* P847 ! Vom, 3,000–4,500 ft. *Dent Young* 207 ! Ropp (Aug.) *Lely* 456 ! [Br.]**Cam.:** Ndop, 5,000 ft., Bamenda (Dec.–Apr.) *Adams* 1513 ! *Maitland* 1522 ! Vogel Peak (Nov.) *Hepper* 1421 !

2. **H. galeopsifolium** (*Bak.*) *Duvign. & Plancke* l.c. 225 (1959). *Acrocephalus galeopsifolius* Bak. in F.T.A. 5 : 356 (1900). *A. ramosissimus* A. Chev. in Journ. de Bot., sér. 2, 2 : 120 (1909) ; Chev. Bot. 514 ; F.W.T.A., ed. 1, 2 : 288. A slender erect, branched, pilose herb, up to about 2 ft. high, with pale mauve flowers in small heads about ⅓ in. diam.
    Guin.: Dalaba-Diaguissa Plateau, 3,000–4,000 ft. (Oct.) *Chev.* 18698 ! 18898 ! 18899 ! Niguelandé (Mar.) *Adam* 11652 ! **N.Nig.:** Jos Plateau (Sept.-Nov.) *Lely* 577 ! P746 ! P809 ! Vom *Dent Young* 208 ! Anara F.R., Zaria (Oct.) *Keay* FHI 5471 ! Widespread in tropical Africa.

3. **H. quarrei** (*Robyns & Lebrun*) *J. K. Morton* in J. Linn. Soc. 58 : 266 (1962). *Acrocephalus quarrei* Robyns & Lebrun in Ann. Soc. Sci. Brux. sér. B, 48 : Mém. 196 (1928). A stiffly erect, little-branched, subglabrous herb, up to 4 ft. high, with pale violet flowers and conspicuous floral bracts ; in marshy savanna.
    Ghana: W. of Bimbila (Jan., Dec.) *Morton* GC 6273 ! A1416 ! Also in Congo.

4. **H. lilacinum** (*Oliv.*) *J. K. Morton* l.c. 266 (1962). *Acrocephalus lilacinus* Oliv. in Trans. Linn. Soc. 29 : 135, t. 134 (1875) ; F.T.A. 5 : 359 ; F.W.T.A., ed. 1, 2 : 288. *A. polytrichus* Bak. in F.T.A. 5 : 358 (1900). *A. centratheroides* Bak. in F.T.A. 5 : 356 (1900). *A. crinitus* Briq. in Mém. Soc. Bot. Fr. 2, 8 : 195 (1912) ; Chev. Bot. 514. *A. sordidus* Briq. l.c. (1912) ; Chev. Bot. 514. An erect, branched often bushy, pilose herb, 1–4 ft. high, with violet flowers in pubescent heads about ⅓ in. diam. ; in damp savanna areas.
    Sen.: Niokolo-Koba (Oct.) *Adam* 15882 ! Tambanaba, Casamance (Feb.) *Chev.* **Mali:** Koulikoro *Chev.* 2781 ! **Port.G.:** Gabu (Oct.) *Esp. Santo* 3511 ! 3536 ! **S.L.:** *Glanville* 378 ! **Iv.C.:** Bouaké *Leeuwenberg* 2071 ! **Ghana:** Tali, Tamale (Dec.) *Morton* GC 9929 ! Tumu, N.R. (May) *Morton* GC 7524 ! Pong Tamale

FIG. 311.—HAUMANIASTRUM ALBOVIRIDE (*Hutch.*) *Duvign. & Plancke* (LABIATAE).
A, flower.   B, stamen.   C, pistil.

(Nov.) *Akpabla* 376! Kpeve (Jan.) *Irvine* 1745! Bimbila (Dec.) *Morton* A1418! **N.Nig.**: Bukuru to Barakin Ladi, Plateau Prov. (Nov.) *Keay* FHI 21444! Jos (Apr.) *Morton* K352! Nupe *Barter*! **S.Nig.**: Emiworo, Ilorin (Jan.) *Ujor* FHI 31605! Also in E. Africa. (See Appendix, p. 458.)

5. **H. buettneri** (*Gürke*) *J. K. Morton* l.c. 267 (1962). *Acrocephalus buettneri* Gürke in Engl. Bot. Jahrb. 19 : 198 (1894); F.T.A. 5 : 360; Chev. Bot. 514; F.W.T.A., ed. 1, 2 : 288. A stout, erect, woody, silvery-pubescent plant, growing in clumps, up to 4 ft. high; flowers pale violet; in upland savanna.
Sen.: Niokolo-Koba (Oct.) *Adam* 15957! **Port.G.**: Bafata (Nov.) *Esp. Santo* 2853! Gabu (Oct.-Dec.) *Esp. Santo* 3512! 3767! **Guin.**: Kissidougou (Dec.) *Martine* 185! Madina Tossékéré *Adam* 12526! **S.L.**: Musaia (Dec.) *Deighton* 4521! Bambaye to Keressadji (Jan.) *Jaeger* 3884! **Ghana**: Amedzofe (Nov.) *Morton* GC 9404! *De Wit & Morton* A2878! **Togo Rep.**: Bismarckburg *Büttner* 304! **Dah.**: Kétou, Zagnanado (Feb.) *Chev.* 23001; 23043. [**Br.]Cam.**: Ndop, 5,400 ft., Bamenda (Dec.) *Boughey* GC 10477!
   [Plants from the W. of our area have narrower leaves with a thinner pubescence and smaller flower-heads.]

6. **H. caeruleum** (*Oliv.*) *J. K. Morton* l.c. 267 (1962). *Acrocephalus caeruleus* Oliv. in Trans. Linn. Soc. 19 : 135, t. 133 (1875); F.T.A. 5 : 359. *A. heudelotii* Briq. (1894)—F.T.A. 5 : 361; Chev. Bot. 514; F.W.T.A., ed. 1, 2 : 288. *A. lagoensis* Bak. (1895)—F.T.A. 5 : 360; Chev. Bot. 514; F.W.T.A., ed. 1, 2 : 288. A stout, erect, branched, often bushy, shortly pubescent herb, with pale violet flowers in villous heads and usually with conspicuous whitish bracts; in damp savanna.
Sen.: Tambanaba, Casamance (Feb.) *Chev.* 2775. Niokolo-Koba (Oct.) *Adam* 15912! **Mali**: Koundian (Feb.) *Chev.* 419! Sotuba to Bamako (Dec.) *Adam* 11273! **Guin.**: Rio Nunez *Heudelot*! Madina Tossékéré (Mar.) *Adam* 11620! Erimakuna (Mar.) *Sc. Elliot* 5262! **Iv.C.**: Aboumaba (Nov.) *Chev.* 22390. **Ghana**: Attabubu (Dec.) *Morton* GC 6185! Gonja (Dec.) *Morton* GC 25043! Yendi to Zabzugu (Dec.) *Morton* A1470! Dzolo Kpuita (Jan.) *De Wit & Morton* A2911! **S.Nig.**: Western Lagos *Rowland*! Olokemeji to Eruwa (Nov.) *Latilo* FHI 26747! [**Br.]Cam.**: Bamenda (Jan.) *Migeod* 306! 423! Ndop, 5,400 ft. (Dec.) *Boughey* GC 10477! Wum (Dec.) *Boughey* s.n.! GC 10957! Vogel Peak (Nov.) *Hepper* 1440! Also in Congo and E. Africa.
   [Plants intermediate between two or more of the above species often occur, probably as the result of hybridity—J.K.M.]

10. HOSLUNDIA Vahl—F.T.A. 5 : 377.

Young stems puberulous; leaves paired or ternate narrowly ovate-lanceolate, gradually acuminate, cuneate at base, about 10 cm. long and 4 cm. broad, crenulate-serrulate, puberulous on the nerves above and sometimes sub-scabrid, minutely glandular beneath; panicle terminal, many-flowered, puberulous; pedicels slender, articulate at the top; calyx very small in flower, campanulate, regular, with 5 small teeth, puberulous, enlarging in fruit, fleshy and berry-like. .   . .   .   *opposita*

H. opposita *Vahl* Enum. Pl. 1 : 212 (1805); F.T.A. 5 : 377, incl. var. *verticillata* (Vahl) Bak.; Chev. Bot. 515. *H. oppositifolia* P. Beauv. Fl. Oware 1 : 53, t. 33 (1805); Chev. Bot. 516. *H. verticillata* Vahl (1805)—Chev. Bot. 516. An erect or half-climbing, odorous shrub, up to 10 or 15 ft. high, with small greenish-cream flowers in copious panicles and orange-yellow succulent fruits ¼ in. across.
Sen.: *Heudelot* 41! **Gam.**: *Skues*! Genieri (July) *Fox* 168! **Mali**: Guiri (May) *Chev.* 884. **Port.G.**: Bissau (Mar.) *Esp. Santo* 1869! **Guin.**: Kaba (May) *Chev.* 13169. Kouria to Longuery (Aug.) *Caille* in *Hb. Chev.* 14624. **S.L.**: Koinadugu (Aug.) *Dawe* 500! **Lib.**: Jabroke (July) *Baldwin* 6470! **Iv.C.**: Bouroukrou (Dec.) *Chev.* 16601. Makougnie (Jan.) *Chev.* 16995. Zaranou (Mar.) *Chev.* 17633. **Ghana**: Burufu, Lawra (May) *Morton* GC 7639! Bomase *A. S. Thomas* D195! Dixcove (Mar.) *Morton* A287! Djodje (July) *Akpabla* 47! **Togo Rep.**: Lomé *Warnecke* 388! **Niger**: Fada to Koupela (July) *Chev.* 24527. **N.Nig.**: Jos Plateau (May) *Lely* P276! Cece F.R., Bida Div. (June) *E. W. Jones* 260! **S.Nig.**: Ibadan (Aug.) *Newberry* 56! [**Br.]Cam.**: below Buea (Jan.) *Morton* K919! **F.Po**: *Mann* 264! Widespread in tropical Africa, S. Africa and Madagascar. (See Appendix, p. 460.)

11. AEOLANTHUS Mart.—F.T.A. 5 : 388.

Leaves long-petiolate, spathulate-oblong or spathulate-lanceolate, broadly cuneate at base, rounded at apex, up to 6 cm. long and 2·5 cm. broad, including the rather long petiole, subentire, minutely glandular-punctate ; cymules forming a lax leafy panicle, the lateral branches trifurcate, branches slender, spicate up to 6 cm. long ; calyx hardened, cupular   ..   ..   *1. pubescens*
Leaves sessile, narrower than above :
   A rather small herb with a prostrate woody perennial stock ; leaves about 2 cm. long and 7 mm. broad, acute, entire, thick, villose ; panicle 5–15 cm. long, little-branched, spikes 1–2·5 cm. long, rhachis and small lanceolate bracts pubescent ; calyx pubescent, truncate   ..   ..   *2. repens*
   A robust erect much-branched annual ; leaves about 5 cm. long and 1·5 cm. broad, obtuse, entire or obscurely crenate, thin, subglabrous ; panicle large, much-branched, spikes 2–5 cm. long, rhachis and very small narrowly-oblong bracts densely villose ; calyx hardened, disk-like   ..   ..   *3. heliotropioides*

1. **A. pubescens** *Benth.* in DC. Prod. 12 : 80 (1848) ; F.T.A. 5 : 394 ; Chev. Bot. 521. *A. buettneri* Gürke (1894)—F.T.A. 5 : 396. *A. chevalieri* Briq. in Mém. Soc. Bot. Fr. 2, 8 : 197 (1912) ; Chev. Bot. 521. *A. pubescens* var. *nuda* A. Chev. l.c. 201 (1912). *Solenostemon pubescens* A. Chev. Bot. 518, name only. An erect, branched, strongly aromatic, slightly fleshy annual, 1–3 ft. high, with small violet sessile flowers ; sometimes cultivated.
Mali: Koulikoro (Oct.) *Chev.* 2778! **Guin.:** Kouroussa *Pobéguin*! Fouta Djalon (Sept.) *Chev.* 18408! Madina Tossékéré, 3,000 ft. (Oct.) *Adam* 12557! **S.L.:** Kanya (Oct.) *Thomas* 3044! **Lib.:** Gbau (Sept.) *Baldwin* 9444! **Ghana:** Jaketi to Ajena (Oct.) *Morton* A13! Kintampo (Dec.) *Morton* A1146! Togo Plateau F.R. (Oct.) *St. C.-Thompson* FH 1515! **Togo Rep.:** Bismarckburg (fl. & fr. Sept.) *Büttner* 250! **N.Nig.:** Patti Lokoja *T. Vogel* 174! Jebba *Barter*! Fuka, Minna (Oct.) *Hepper* 962! Yola *Dalz.* 302! **S.Nig.:** Akure F.R. (Nov.) *Ajayi* FHI 26920! Abeokuta *Irving* 112! (See Appendix, p. 458.)

2. **A. repens** *Oliv.* in Trans. Linn. Soc. 29 : 137, t. 136, fig. A (1875) ; F.T.A. 5 : 395. A small creeping herb, with erect flowering stems bearing a few small bluish flowers about ¼ in. long ; in moist rocky situations and exposed rock-crevices at high elevations.
[Br.]Cam.: Bamenda, 4,000–7,000 ft. (Aug.-Feb.) *Migeod* 495! *Savory* UCI 377! *Ujor* FHI 29962! 30222! *Maitland* 1642! Vogel Peak, Adamawa (Nov.) *Hepper* 1356! Also in Sudan and E. Africa.

3. **A. heliotropioides** *Oliv.* l.c. 137, t. 82 (1875) ; F.T.A. 5 : 393. An erect much-branched, robust aromatic annual, about 2 ft. high, with dense panicles of small violet flowers.
N.Nig.: cult. in Munchi farms, R. Benue *Sampson* 35! **[Br.]Cam.:** Vogel Peak, Adamawa (Nov.) *Hepper* 1420! Also in Sudan and E. Africa. (See Appendix, p. 458.)

12. ICOMUM Hua—F.T.A. 5 : 387.

Stems short, slender, simple or sparingly branched towards the top, subterete and softly pubescent when young ; leaves alternate, linear, entire, up to 3·5 cm. long, sessile, ascending, glandular-punctate ; spikes short, dense, forming a small terminal panicle ; bracts broadly linear ; calyx campanulate, pubescent, teeth minute ; corolla-tube cylindrical, pubescent, much exceeding the calyx, lobes small, the lower shallowly saccate   ..   ..   *1. paradoxum*
Stems very short, softly tomentose when young ; young leaves oblong, with small glandular teeth, softly tomentellous ; bracts subulate-linear   ..   *2. gambicola*

1. **I. paradoxum** *Hua* in Bull. Mus. Hist. Nat. 3 : 329 (1897) ; F.T.A. 5 : 387 ; Chev. Bot. 513. *I. foutadjalonensis* De Wild. (1928). *Aeolanthus foutadjalonense* (De Wild.) De Wild. (1928). Stems ¼ in. long ; corolla white with rose-spotted lip.
Guin.: Moussaia (June) *Pobéguin* 705! Fouta Djalon (May) *Chev.* 12750! Timbo (May, July) *Miquel* 43! *Pobéguin* 738! Sineya and Toumanea (June) *Pobéguin* 709!

2. **I. gambicola** *A. Chev. ex Hutch. & Dalz.* F.W.T.A., ed. 1, 2 : 282 (1931) ; Chev. Bot. 513 (1920), name only. Plant only of a few inches high from a perennial base, with ovoid, fragrant tuber 2–2½ in. long ; corolla white.
Guin.: at sources of the Gambia R., 3,300 ft., Fouta Djalon (Apr.) *Chev.* 13573!
[Perhaps not distinct from *I. paradoxum* Hua—J.K.M.]

13. PYCNOSTACHYS Hook.—F.T.A. 5 : 379 ; Perkins in Notizbl. Bot. Gart. Berl. 8 : 63 (1921) ; E. A. Bruce in Kew Bull. 1939 : 563.

Leaves ovate-lanceolate to lanceolate, distinctly crenate-serrate ; shrubs ; spines on calyx sharp to the touch :
   Mature calyx densely tomentose, teeth slender, short, about 2·5 mm. long, pale brown ; spikes narrow, about 1·5 cm. broad and 2·5–5 cm. long ; leaves about 10 cm. long and 4 cm. broad, cuneate at base into a long petiole about ⅓ length of the lamina, thin, shortly pubescent on both surfaces, glandular-punctate beneath ..   *1. eminii*
   Mature calyx shortly and rather thinly pubescent, teeth stout, about 5 mm. long, dark reddish-brown ; spikes stout, 2·5 cm. or more broad, 3–5 cm. long ; leaves similar to last species but thicker, densely pubescent and with a shorter petiole   *2. meyeri*
Leaves linear to linear-lanceolate, serrulate ; herbs, sometimes woody ; spines on calyx not very sharp to the touch :
   Leaves small, up to 3 cm. long and 2 mm. broad, pseudo-whorled and numerous ; stems strongly sulcate   ..   ..   *3. pallide-caerulea*
   Leaves large, up to 10 cm. long and 15 mm. broad, sessile, opposite or in whorls of 3, glandular punctate beneath :
     Leaves thin, flaccid, opposite with smaller leafy axillary shoots, thinly scabrid up to 10 cm. long and 6 mm. broad ; spikes on long slender peduncles, ovoid, 2 cm. diam.,

with small reflexed bracts at base ; stems slender with many spreading branches ;
corolla about 14 mm. long .. .. 4. *schweinfurthii*
Leaves thick, ascending, 3 in each whorl, 4–10 cm. long, 4–15 mm. broad, appressed-
puberulous on veins beneath, otherwise glabrous ; spikes solitary, terminal on short
stout peduncle, conical, about 3 cm. diam. with spreading leafy bracts at base ;
stems stout, erect, branches few, ascending ; corolla about 13 mm. long
5. *reticulata*

1. **P. eminii** *Gürke* in Engl. Bot. Jahrb. 22 : 145 (1895) ; F.T.A. 5 : 385 ; E. A. Bruce in Kew Bull. 1939 :
589. A pubescent aromatic shrub several feet high with cylindrical spikes of pale blue flowers.
[Br.]**Cam.**: Bambui, 6,300 ft., Bamenda (Dec.) *Adams* 11304 ! Bamenda, 4,300–5,000 ft. (fl. Feb.,
May, fr. Mar.) *Migeod* 475 ! *Morton* K105 ! K137 ! Vogel Peak, Adamawa (Nov.) *Hepper* 1489 ! Also in
Congo and E. Africa.

2. **P. meyeri** *Gürke* in Engl. Hochgebirgsfl. Afr. 362 (1892) ; F.T.A. 5 : 384 ; E. A. Bruce l.c. 591. *P. volkensii*
Gürke (1895)—F.T.A. 5 : 384 ; F.W.T.A., ed. 1, 2 : 288. *P. bowalensis* A. Chev. in Journ. de Bot., sér. 2,
2 : 126 (1909) ; Chev. Bot. 521. *P.* aff. *bowalensis* A. Chev. Bot. 521. *Coleus urticaefolius* (Hook. f.)
Roberty in Bull. I.F.A.N. 16, 1 : 330 (1954), partly. Similar to the last species but with stouter spikes
of bluish-purple flowers.
Guin.: Fouta Djalon, 3,500 ft. (fl. Sept.-Oct., fr. Jan.) *Chev.* 18544 ! 18591 ! 18707 ! 18853 ! 20345 !
Schnell 7081 ! Labé (Jan.) *Roberty* 16402 ! **S.Nig.**: Koloishe Mt., Obudu Div. (Dec.) *Keay & Savory*
FHI 25075 ! [Br.]**Cam.**: Cam. Mt., 8,100 ft. (Jan.) *Keay* FHI 28598 ! *Boughey* GC 12685 ! Bamenda,
7,500 ft. (Jan.) *Keay* FHI 28405 ! Nchan, Bamenda (fl. & fr. June) *Maitland* 1563 ! **F.Po:** Clarence Peak,
7,000 ft. (Dec.) *Boughey* GC 12685 ! Also in Congo and E. Africa.

3. **P. pallide-caerulea** *Perkins* in Notizbl. Bot. Gart. Berl. 8 : 67 (1921) ; E. A. Bruce l.c. 581.
[Br.]**Cam.**: Kumbo, Banso Mts., 6,000 ft. (Oct.) *Ledermann* 5738 !

4. **P. schweinfurthii** *Briq.* in Engl. Bot. Jahrb. 19 : 191 (1894) ; F.T.A. 5 : 380 ; E. A. Bruce l.c. 586. *P.
togoensis* Perkins (1921)—F.W.T.A., ed. 1, 2 : 288. *Coleus urticaefolius* (Hook. f.) Roberty in Bull. I.F.A.N.
16, 1 : 330 (1954), partly. A slender fast-growing often bushy annual herb up to 4 ft. high ; in marshy
savanna.
Ghana: Dubwesein *Chipp* 621 ! Afram Plains *Irvine* 890 ! Yeji (Nov.) *Morton* A2704 ! Ejura (Dec.)
Morton GC 9490 ! 9825 ! Yendi to Zabzugu (Dec.) *Morton* A1456 ! Kete Krachi to R. Volta (Jan.) *Baumann*
404 ! Krakye *Krause* ! Also in Sudan and Tanganyika.

5. **P. reticulata** (*E. Mey.*) *Benth.* in DC. Prod. 12 : 83 (1848) ; F.T.A. 5 : 382 ; E. A. Bruce l.c. 584. *Echino-
stachys reticulata* E. Mey. (1837). An erect half-woody, perennial herb with short spikes of blue flowers ;
in marshy grassland.
[Br.]**Cam.**: Gurum, Vogel Peak area, Adamawa (Nov.) *Hepper* 1600 ! Also in East and Central Africa.

14. PLECTRANTHUS L'Hérit. Stirp Nov. 1 : 84 (1784) ; Morton in J. Linn. Soc. 58 : 242 (1962). *Nom. cons.*

*Mature calyx small, 2–5 mm. long :
Flowers in sessile fascicles :
Inflorescence a dense broad panicle with lateral branches about 15 cm. long and 1 cm.
broad ; mature calyx 3 mm. long, upper lip triangular, slightly broader and shorter
than the other lanceolate teeth ; corolla 6 mm. long ; leaves broadly ovate, about
6 cm. long and 5·5 cm. broad, truncate, obtuse, deeply and coarsely crenate with a
dense soft pubescence on both surfaces .. .. 1. *cyaneus*
Inflorescence racemose :
Mature calyx 3–4 mm. long, upper lip ovate, acute, median and lower teeth slightly
longer, narrowly lanceolate-acuminate ; corolla up to 5 mm. long ; inflorescence
branched below, up to 15 cm. long, 1 cm. broad ; pedicels 4–5 mm. long, unbranched ; leaves broadly ovate-triangular, 4 cm. long and broad, truncate to
slightly cordate, long-petiolate, coarsely crenate, pubescent on both surfaces ;
stems with a long soft pubescence .. .. 2. *hallii*
Mature calyx 5–6 mm. long ; corolla up to 1 cm. long ; inflorescence up to 35 cm.
long and 2–2·5 cm. broad ; pedicels 10 mm. long ; leaves 2·5–5 cm. long and
broad, truncate to abruptly cuneate, subglabrous to pubescent ; stems glabrous
to shortly pubescent .. .. 3. *assurgens*
Flowers not fascicled ; inflorescence a panicle with the lateral branches dichotomous
or racemose :
Lateral branches of inflorescence with first node dichotomous, racemose above ;
inflorescence lax, up to 7 cm. broad ; mature calyx 5 mm. long, upper lip ovate,
teeth slightly exceeding it ; corolla about 1 cm. long ; leaves ovate, twice as long
as broad, 3·5–7 cm. long, finely crenate, rounded cuneate at base into a short petiole
up to ½ as long as the lamina, subglabrous .. .. .. 4. *peulhorum*
Lateral branches of inflorescence racemose :
Stems pubescent, slender ; leaves lanceolate, acute at apex, gradually cuneate at
base, sessile or shortly petiolate, 0·5–6 cm. long, a third as broad, crenate, thinly
pubescent ; mature calyx 3–4 mm. long, upper lip triangular-acute, teeth subulate,
subequal ; corolla 8 mm. long .. .. 5. *tenuicaulis*
Stems covered with narrowly lanceolate brown scales 4–5 mm. long, stout ; leaves
ovate to ovate-lanceolate, acute, more or less acuminate, decurrent into a long
petiole, up to 16 cm. long and 8 cm. broad, usually less, coarsely crenate, subglabrous
above, appressed-puberulous on veins beneath ; inflorescence a fairly dense terminal
panicle about 20 cm. long and 6 cm. broad ; pedicels very slender, about 3 mm.
long, each with a minute bract ; mature calyx similar to No. 5 ; corolla about
1 cm. long .. .. 6. *harrisii*

*Mature calyx large, 5–20 mm. long :
  Leafless at time of flowering :
    Large woody undershrub ; infloresence a very lax ample racemose panicle with ultimate
      branches 15–30 cm. long or smaller and on the older wood ; mature calyx 2 cm. long,
      teeth very unequal, lower long and subulate ; corolla 2 cm. long ; leaves narrowly
      obovate-elliptic, acuminate, gradually cuneate into a short petiole, 15–25 cm. long
      and 6–10 cm. broad, pubescent on the veins, finely serrate   ..   7. *insignis*
    Perennial herb with edible tuberous roots ; inflorescence a much-branched panicle ;
      mature calyx 8 mm. long ; corolla 1·5 cm. long ; leaves oblong-lanceolate, sessile,
      rounded at apex, entire to somewhat crenate, setulose on the veins beneath
                                                                                8. *esculentus*

Flowering stems leafy :
  Leaves cordate, not decurrent on the long petiole :
    Flowers fascicled on the main axis or on a very short common peduncle ; inflores-
      cence little-branched ; leaves ovate, acutely acuminate, up to 11 cm. long and
      9 cm. broad, evenly and coarsely crenate, crenations broader than deep, glandular-
      punctate and with long white articulate hairs on the veins of both surfaces ;
      mature calyx 8 mm. long, with a long white pubescence ; upper lip ovate, acute,
      lower teeth lanceolate acuminate, exceeding upper lip, middle teeth similar but a
      little shorter ; whole plant densely pubescent with long white jointed more or less
      glandular hairs   ..   ..   9. *kamerunensis*
    Flowers in pedunculate cymes on the main axis ; inflorescence a much-branched
      lax pyramidal panicle ; branches of cymes slender, elongate, pubescent, straight ;
      leaves ovate, acutely acuminate, up to 15 cm. long and 13 cm. broad, with very
      uneven double or treble rather small crenations, glandular-punctate, subglabrous
      or pubescent on the veins ; mature calyx 9 mm. long, upper lip ovate acute, lower
      teeth lanceolate, long-acuminate, much exceeding upper lip, middle teeth similar
      but shorter, glabrous or minutely puberulous on the veins, gland-dotted ; whole
      plant densely pubescent with short glandular hairs   ..   ..   10. *glandulosus*
  Leaves cuneate to truncate, decurrent on the petiole :
    Plants producing fusiform, densely brown-villose bulbils in the axils of the inflores-
      cence and branches ; leaves shortly petiolate, elliptic-lanceolate, cuneate at base,
      long-acuminate, about 7 cm. long and 3 cm. broad, coarsely crenate, rusty-
      pubescent on veins beneath, subglabrous above ; inflorescences axillary and
      terminal, flowers in sessile fascicles ; pedicels about 7 mm. long, puberulous ;
      mature calyx glandular, about 7 mm. long, upper lip ovate, acute, teeth narrowly
      lanceolate, subequal, ¼ to ⅓ of the length of the tube ; corolla 1 cm. long, pubescent
                                                                                11. *luteus*

Plant without bulbils in axils of branches :
  Leaves sessile or nearly so, ovate to ovate-lanceolate, obtuse to subacute, serrate-
    crenate ; stems angular ; inflorescence of several whorls, dense in flower,
    elongate in fruit ; calyx with ovate upper lip and narrowly lanceolate upcurved
    teeth, pubescent and gland-dotted, exceeding the pedicels :
      Corolla about 7 mm. long ; mature calyx 7–8 mm. long ; leaves broadly ovate,
        2–4 cm. long, 1·5–3 cm. broad, shortly petiolate ; plant subglabrous to pubescent
        with short woolly hairs   ..   ..   ..   12. *punctatus* subsp. *punctatus*
      Corolla about 1·5 cm. long ; mature calyx 8–10 mm. long ; leaves ovate-lanceolate,
        up to 7·5 cm. long and 4 cm. broad, sessile ; whole plant with a dense long
        woolly indumentum   ..   ..   ..   12a. *punctatus* subsp. *lanatus*
  Leaves distinctly petiolate :
    Mature calyx 5–6 mm. long (for other characters see above)   3. *assurgens*
    Mature calyx 8–10 mm. long :
      Leaves nearly as long as broad, ovate-rotund, abruptly cuneate, coarsely
        crenate, thin, 4–6 cm. long, glabrous above, thinly pubescent on the veins
        beneath ; petiole about as long as lamina ; inflorescence racemose, lax,
        simple ; mature calyx about 8 mm. long, lower teeth long-acuminate, exceed-
        ing the others ; corolla 1–1·5 cm. long, blue   ..   13. *dissitiflorus*
      Leaves at least twice as long as broad, ovate-lanceolate, up to 16 cm. long and
        7 cm. broad, gradually cuneate and decurrent on a long petiole, crenate-
        serrate, thin, subglabrous except for a short. brown pubescence on the veins ;
        inflorescence elongate, racemose, lax ; pedicels shorter than the mature calyx ;
        mature calyx 9–10 mm. long, upper lip orbicular, teeth long-acuminate ;
        corolla 10–15 mm. long, yellow   ..   ..   14. *decurrens*

1. **P. cyaneus** *Gürke* in Engl. Bot. Jahrb. 19 : 208 (1894) ; F.T.A. 5 : 409. A densely pubescent bushy herb
   with woody stems and copious panicles of small violet flowers.
   [Br.]Cam.: Jada, Adamawa (Nov.) *Hepper* 1362 ! Also in Uganda, Kenya and Tanganyika.
2. **P. hallii** *J. K. Morton* in J. Linn. Soc. 58 : 267 (1962). An erect annual branched delicate herb up to
   about 1 ft. high with loose spikes of very pale blue flowers ; under trees in rocky savanna.
   Ghana: Gambaga Scarp (fl. Sept., fr. Nov.) *Hall* 790 ! *Morton* A2749 ! A3791 !
3. **P. assurgens** (*Bak.*) *J. K. Morton* l.c. 267 (1962). *Coleus assurgens* Bak. in F.T.A. 5 : 428 (1900). A much-
   branched, rather slender herb with long straggling stems thin subglabrous leaves and many-flowered
   whorls of bluish-purple flowers.

S.Nig.: Mt. Koloishe, 5,000 ft., Obudu Div. (Dec.) *Keay & Savory* FHI 25067! **[Br.]Cam.:** path to Mann's Spring, 3,800 ft., Cam. Mt. (Dec.) *Adams* 1671! *Morton* K835! L. Oku, Bamenda (Jan.) *Keay & Lightbody* FHI 28508! **F.Po:** (Dec.) *Monod* 10427! Also in E. Africa.

4. **P. peulhorum** (*A. Chev.*) *J. K. Morton* l.c. 268 (1962). *Coleus peulhorum* A. Chev. in Journ. de Bot. 12 : 123 (1909) ; Bot. 519. An erect annual herb 1–1½ ft. high, with a rather slender, branched, thinly puberulous stem and pale blue flowers ⅓ in. long.
    Guin.: Fouta Djalon (Oct.) *Chev.* 18424! 18566! 18895! **N.Nig.:** Jos Plateau (Sept.) *Lely* P741! *Batten-Poole* 167! **[Br.]Cam.:** Bafut-Ngemba F.R., 5,000–6,500 ft., Bamenda (Sept.-Nov.) *Lightbody* FHI 26263! *Savory* UCI 379!

5. **P. tenuicaulis** (*Hook. f.*) *J. K. Morton* l.c. 268 (1962). *Coleus tenuicaulis* Hook. f. in J. Linn. Soc. 7 : 211 (1864) ; F.T.A. 5 : 443. *Plectranthus minimus* Gürke (1894)—F.T.A. 5 : 403 ; F.W.T.A., ed. 1, 2 : 289. A usually small erect, branched, slender pubescent herb 2–36 in. high with lax panicles of small blue flowers.
    [Br.]Cam.: Cam. Mt., 6,000–7,500 ft. (fl. Sept.-Dec., fr. Nov.-Jan.) *Mann* 1939! *Preuss* 1019! *Morton* K529! K716! K741! K846! Bamenda Dist. (fr. Dec.) *Boughey* GC 10585!

6   **P. harrisii** *J. K. Morton* l.c. 267, t. 5 (1962). A stout, erect unbranched herb to 8 ft. high with perennial root-stock and narrow panicles of yellow flowers heavily suffused with purple ; in hill savanna and woodland at the forest edge.
    Ghana: Togo Plateau, 2,000 ft. (fl. Dec.-Jan.) *Morton* A3528! A3830! *Adams* 1842!

7. **P. insignis** *Hook. f.* in J. Linn. Soc. 7 : 210 (1864). F.T.A. 5 : 404. A large shrub 10–15 ft. high, with soft wood and inflorescences in a very loose panicle or raceme ; flowers yellow suffused with purple ; usually leafless at time of flowering ; in montane forest.
    [Br.]Cam.: Cam. Mt., 6,000–7,000 ft. (Jan.-Dec.) *Preuss* 991! *Dunlap* 4! 51! *Mann* 1257! *Morton* K497! K697! Bafut-Ngemba F.R., Bamenda (Feb.) *Hepper* 2162!

8. **P. esculentus** *N. E. Br.* in Kew Bull. 1894 : 12. *Coleus esculentus* (N. E. Br.) G. Tayl. (1931). *C. dazo* A. Chev. (1904)—Chev. & Perrot. in Vég. Util. Afr. Trop. Fr. 1 : 126 (1905) ; F.W.T.A., ed. 1, 2 : 292. *C. langouassiensis* A. Chev. (1905)—Chev. Bot. 519. Perennial with herbaceous erect stems, rather coarsely pilose with whitish hairs ; flowers yellow ½–⅔ in. long ; root elongated-tuberous ; often cultivated.
    Mali: Tangaye to Lozitenga, Mossi (July) *Chev.* 24579. **N.Nig.:** Bukuru, Jos (Jan.) *Keay & Russell* FHI 28448! **[Br.]Cam.:** Kumbo (Feb.) *Hepper* 1975! Also in Angola, Ubangi-Shari, Congo and extending to Angola and Natal. (See Appendix, p. 459.)

9. **P. kamerunensis** *Gürke* in Engl. Bot. Jahrb. 19 : 202 (1894) ; F.T.A. 5 : 408 ; F.W.T.A., ed. 1, 2 : 289. *P. mannii* Bak. in F.T.A. 5 : 408 (1900) ; F.W.T.A., ed. 1, 2 : 289. A straggling woolly herb with erect flowering branches some 3 ft. in height ; flowers violet.
    S.Nig.: Ikwette Plateau, 5,200 ft., Obudu Div. (Dec.) *Keay & Savory* FHI 25253! **[Br.]Cam.:** Cam. Mt., 3,000–4,000 ft. (Nov., Dec.) *Mann* 1947! *Migeod* 65! *Morton* K500! K705! path to Mann's Spring, 5,600 ft. (Dec.) *Adams* 1665! *Morton* K642! Bamenda (Jan., Mar.) *Migeod* 415! *Boughey* GC 10533! GC 10575! *Morton* K43! Bafut-Ngemba F.R. (Feb.) *Hepper* 2145!

10. **P. glandulosus** *Hook. f.* in J. Linn. Soc. 6 : 17 (1861) ; F.T.A. 5 : 411. *P. hylophilus* Gürke (1894)—F.T.A. 5 : 413. *P. urticoides* Bak. in F.T.A. 5 : 412 (1900) ; F.W.T.A., ed. 1, 2 : 289. *P. almamii* A. Chev. (1909)—Chev. Bot. 516. *C. laxiflorus* (Benth.) Roberty in Bull. I.F.A.N. 16, 1 : 331 (1954), partly. A coarse scrambling to erect glandular strongly aromatic herb up to 10 ft. high with copious loose panicles of violet flowers ; in openings in the montane forest and amongst scrub.
    Mali: Nianguere (Mar.) *Adam* 11606! Madina Tossékéré (Mar.) *Adam* 11722! **Guin.:** Fouta Djalon (Sept.) *Chev.* 18609! **N.Nig.:** Jos Plateau (Sept.) *Lely* P722! **S.Nig.:** Mt. Koloishe, 4,800 ft. Obudu Div. (Dec.) *Keay & Savory* FHI 25122! **[Br.]Cam.:** Cam. Mt., 5,000 ft. *Morton* GC 6908! K946! K719! K893! Mann's Spring, 7,200 ft. (Dec.-Mar.) *Preuss* 815! *Mann* 2029! Bamenda (Jan., Mar.) *Morton* K185! *Keay & Lightbody* FHI 28379! Maisamari, 4,600 ft., Mambila Plateau (Jan.) *Hepper* 1680! Vogel Peak, 4,900 ft. (Dec.) *Hepper* 1561! **F.Po:** (Dec., Mar.) *Guinea* 2966! *Mann* 318! *Monod* 10429! *Boughey* 11! 178! (See Appendix, p. 463.)

11. **P. luteus** *Gürke* in Engl. Bot. Jahrb. 28 : 468 (1900) ; F.T.A. 5 : 524. *Coleus luteus* (Gürke) Staner (1934) *Coleus entebbensis* S. Moore (1906). A branched woody herb, about 3 ft. high, with orange-yellow flowers 1 cm. long.
    Lib.: Mt. Wologwisi, 4,500 ft., Pandonai (Dec.) *Bequaert* 105! Also in Congo and E. Africa.

12. **P. punctatus** *L'Hérit.* subsp. **punctatus**—Stirp. Nov. 87, t. 42 (1784). *Coleus glandulosus* Hook. f. in J. Linn. Soc. 7 : 211 (1864) ; F.T.A. 5 : 425 ; F.W.T.A., ed. 1, 2 : 292. An erect, glandular-pubescent herb about 1 ft. high ; flowers bluish-purple, small ; in montane grassland.
    [Br.]Cam.: Cam. Mt., 2,700–7,000 ft. (Nov., Dec.) *Mann* 1301! 1988! Mann's Spring, 7,500 ft. (Apr., Dec.) *Boughey* GC 10691! *Morton* GC 6796! K594! **F.Po:** summit grassland (July) *Boughey* GC 10968!

12a. **P. punctatus** subsp. **lanatus** *J. K. Morton* in J. Linn. Soc. 58 : 268, t. 14 (1962). Larger than the type subspecies, with large showy, royal-blue flowers.
    [Br.]Cam.: Bamenda Prov.: Bambui, 6,300 ft. (Dec.) *Adams* GC 11322! Bamenda, 6,000 ft. (June) *Maitland* 1724! Ndu, Banso (Sept.) *Savory* UCI 386! ridge N. of L. Oku (Jan.) *Keay & Lightbody* FHI 28464!
    [Similar to the E. African *Coleus edulis* Vatke which has tuberous roots and is subglabrous.]

13. **P. dissitiflorus** (*Gürke*) *J. K. Morton* l.c. 267 (1962). *Coleus dissitiflorus* Gürke in Engl. Bot. Jahrb. 19 : 217 (1894) ; F.T.A. 5 : 444 ; F.W.T.A., ed. 1, 2 : 292. A perennial herb with glandular-pubescent stems up to 3 or 4 ft. long and blue-purple flowers ¾ in. long.
    [Br.]Cam.: Buea to Bimbia (fl. & fr. Oct.) *Preuss* 1055!

14. **P. decurrens** (*Gürke*) *J. K. Morton* l.c. 267 (1962). *Coleus decurrens* Gürke l.c. 215 (1894) ; F.T.A. 5 : 427 ; F.W.T.A., ed. 1, 2 : 292. *C. elatus* Bak. in F.T.A. 5 : 427 (1900). A herb up to 6 ft. high, with pubescent stems and large leaves : flowers either red or yellowish, about ½ in. long.
    S.Nig.: Oban *Talbot* 369! **[Br.]Cam.:** near Buea, 3,000 ft. (fl. & fr. Sept.) *Preuss* 948! Northern Korup F.R., Kumba (July) *Olorunfemi* FHI 30688! **F.Po:** (Dec.) *Mann* 584! Extends to Gabon.

15. HOMALOCHEILOS J. K. Morton in J. Linn. Soc. 58 : 268 (1962).

Inflorescence a panicle of many-flowered dichotomous cymes borne in the axils of leafy bracts ; stems straggling, square, pubescent ; leaves ovate, rounded to cuneate at base, acute at apex, up to 7 cm. long and 4 cm. broad, crenate, thinly pubescent and dark-green above, densely pubescent, gland-dotted and pale-green beneath, distinctly petiolate ; calyx densely pubescent, 4 mm. long, tube declinate, ventricose, teeth subequal, short, triangular ; corolla 5 mm. long, upper lip very small, recurved, lower lip small, almost flat, deflexed in flower   ..   ..   ..   *ramosissimus*

H. ramosissimus (*Hook. f.*) *J. K. Morton* l.c. 268, t. 6 (1962). *Plectranthus ramosissimus* Hook. f. in J. Linn. Soc. 6 : 17 (1862) ; F.T.A. 5 : 418 : F.W.T.A., ed. 1, 2 : 289. *Coleus ramosissimus* (*Hook. f.*) Roberty in Bull. I.F.A.N. 16, 1 : 330 (1954), partly. A straggling to erect herb, with sharply angled pilose stems up to 12 ft. long and copious glandular-pubescent inflorescences of small white flowers with mauve marks in the throat of the corolla.
S.L.: Loma Mt. *Jaeger* 1628! **S.Nig.:** Koloishe Mt., 5,000 ft., Obudu Div. (Dec.) *Keay & Savory* FHI

25074! [Br.]Cam.: Cam. Mt., 7,000–7,500 ft. (Jan., Apr.) *Dunlap* 213! *Mann* 1320! *Morton* GC 7099! Bamenda, 7,000 ft. (Jan., Mar.) *Morton* K138! *Keay & Lightbody* FHI 28367! Mambila Plateau (Jan.) *Hepper* 1675! Vogel Peak, Adamawa (Nov.) *Hepper* 1551! **F.Po:** (Mar., Dec.) *Monod* 10430! *Guinea* 1266! 1979! Also on the Sudan-Uganda border and in S. Rhodesia.

16. LEOCUS A. Chev. in Journ. de Bot., sér. 2, 2 : 125 (1909), incl. *Briquetastrum* Robyns & Lebrun in Ann. Soc. Sci. Brux., sér. B, 49, Mém. 102 (1929).

Erect more or less unbranched herb with quadrangular pubescent stems bearing a single terminal inflorescence, leafless above the middle ; leaves lyrate-oblanceolate, crenate above and entire below the middle, acute, more or less auricled on the stem, about 10 cm. long, 3–4 cm. broad, densely crispate-pubescent especially on the veins, setose above ; inflorescence dense, cylindrical about 6 cm. long and 1·7 cm. broad, bracts small, inconspicuous, lanceolate, caducous ; calyx purple, pubescent, prominently veined, 6 mm. long in fruit, tube bulbous, rotund, equalling the 4 lanceolate shorter teeth, upper tooth larger, narrowly lanceolate, acuminate ; corolla pubescent, 8 mm. long   ..   ..   *1. lyratus*

Shrubby plants with woody stems ; leaves lanceolate to ovate, distinctly petiolate ; inflorescences several to numerous, terminal on the branches, cylindrical with conspicuous ovate to rotund apiculate or acuminate bracts at least when in bud, bracts often caducous in flower :

Leaves oblong-lanceolate slightly cuneate at base, about 15 cm. long and 4 cm. broad, undulate-crenate, puberulous above, softly tomentellous and closely reticulate beneath ; petiole 1–2 cm. long ; stems grooved, softly tomentellous ; inflorescence of 3 long-pedunculate cylindrical spikes 6–10 cm. long, 1·5–2·5 cm. broad, bracteate in bud ; calyx campanulate, with the adaxial lip large, oblong-elliptic, forming a hood over the other 4 small ovate teeth, nervose ; corolla 6 mm. long, tomentellous   ..   ..   *2. africanus*

Leaves ovate :

Bracts persistent, membranous, 1–2 cm. diam., rotund, apiculate ; calyx 6 mm. long, teeth narrowly lanceolate, acuminate, shortly scabrid-pubescent, stems stout woody, terete, tomentellous when young ; leaves ovate, cuneate and petiolate at base, obtuse at apex, about 8 cm. long and 4·5 cm. broad, thinly pubescent above, tomentellous and gland-dotted beneath ; plant leafless when in flower ..   ..   *3. membranaceus*

Bracts caducous at time of flowering, not membranous, ovate-lanceolate, acuminate ; calyx prominently veined, about 4·5 mm. long in fruit, tube longer than broad, exceeding the long-lanceolate upper tooth, other teeth short and triangular ; corolla pubescent ; leaves ovate to ovate-triangular acutely acuminate or obtuse, rounded at base, distinctly petiolate, about 8 cm. long and 5 cm. broad, coarsely crenate, finely tomentellous on both surfaces, whitish beneath :

Corolla 7–10 mm. long ; calyx rather thinly and shortly pubescent   ..   *4. caillei*
Corolla about 15 mm. long ; calyx villose-tomentose   ..   ..   *5. pobeguinii*

1. L. lyratus *A. Chev.* in Journ. de Bot., sér. 2, 2 : 126 (1909) ; Chev. Bot. 520. *Coleus lyratus* (A. Chev.) Roberty in Bull. I.F.A.N. 16, 1 : 330 (1954). An erect perennial herb, about 3 ft. high, with dense terminal spike-like inflorescences of bluish-purple flowers.
   Guin.: Dalaba-Diaguissa Plateau, 3,000–4,000 ft. (Sept.-Oct.) *Chev.* 18824! Tinka Mt., 4,000 ft., Dalaba (Sept.) *Chev.* **S.L.:** Loma Mt. (Sept.) *Jaeger* 1879! Warantamoa (Oct.) *Small* 348! Musaia (Sept.) *Small* 245! Falaba (Sept.) *Deighton* 5168! Also in N. & S. Rhodesia.

2. L. africanus *(Bak. ex Sc. Elliot) J. K. Morton* in J. Linn. Soc. 58 : 270 (1962). *Anisochilus africanus* Bak. ex Sc. Elliot in J. Linn. Soc. 30 : 94 (1 Feb. 1894) ; F.T.A. 5 : 446 ; Chev. Bot. 521 ; F.W.T.A., ed. 1, 2 : 292. *A. engleri* Briq. in Engl. Bot. Jahrb. 19 : 190 (13 Apr. 1894). *Geniosporum congoense* Gürke (1900), name only. *Briquetastrum africanum* (Bak. ex Sc. Elliot) Robyns and Lebrun (1929). *Coleus africanus* (Bak. ex Sc. Elliot) Roberty l.c. (1954). A woody perennial herb of savanna swamps, 4–5 ft. or more in height, with pale ochrey-pubescent, with pale blue flowers arranged in dense cylindrical spikes.
   Guin: Télimélé (Apr.) *Pitot* T51–873! Dalaba *fide* Roberty. Faranna (Jan.) *Chev.* 20411! 20442! **S.L.:** Lomaburn (Feb.) *Sc. Elliot* 5033! near Koinadugu (Dec.) *Glanville* 457! Kiridoy to Kamaro (Jan.) *Jaeger* 3897! **N.Nig.:** Kontagora (Nov.) *Dalz.* 133! *Meikle* 1087! Vom *Dent Young* 213! Rafin Bauna North F.R., Jos (Oct.) *Hepper* 1187! **S.Nig.:** Koloishe Mt., 5,000 ft., Ogoja Prov. (Dec.) *Keay & Savory* FHI 25134! [Br.]Cam.: Buca (May) *Mildbr.* 9483! Mai-Idoanu, Adamawa (Jan.) *Latilo & Daramola* FHI 28984! Mambila Plateau (Jan.) *Hepper* 1649! Also in Cameroun, Ubangi-Shari, Chad, Congo and Uganda.

3. L. membranaceus *J. K. Morton* l.c. 270, t. 7 (1962). An aromatic bush several feet high, with thick, rather succulent stems, bearing many terminal spike-like inflorescences of pale blue flowers with large whitish or purple bracts ; in rock crevices in savanna country.
   Guin.: Mali (fl. Jan., fr. Apr.) *Adam* 12174! *Roberty* 16533! Mt. Loura, 4,700 ft., Fouta Djalon (fr. June) *Adam* 14526! **Iv.C.:** Pouake, Mt. Lemelebou *Chev.* 22103! **Ghana:** Kintampo (fl. Dec.-Jan., leaf Apr.-June) *Morton* A2585! *Akpabla* 1942! **N.Nig.:** near Zaria *Keay* FHI 28025! *Meikle* 1141! Anara F.R., Zaria (May) *Keay* FHI 22922! Naraguta Hills (Dec.) *Coombe* 8! Ropp (leaf May-July) *Keay & King* FHI 37103! **S.Nig.:** Carter's Peak, Idanre (fl. Jan., leaf Oct.) *Brenan* 8639! *Keay* FHI 22651!

4. L. caillei *(A. Chev. ex Hutch. & Dalz.) J. K. Morton* l.c. 270 (1962). *Coleus caillei* A. Chev. ex Hutch. & Dalz. F.W.T.A., ed. 1, 2 : 292 (1931) ; Chev. Bot. 518, name only. A stout erect undershrub, 2–3 ft. high with dense oblong spikes of bright blue flowers.
   Guin.: Longuery (Dec.) *Caille* in Hb. *Chev.* 14864! **S.L.:** Bintumane, 5,000 ft. (Jan.) *Glanville* 477! **N.Nig.:** Naraguta (Oct.) *Hepper* 1170! Sara Hills, Bauchi Div. (Oct.) *Hepper* 1128! Kadun Peak, Pankshin Div. (Oct.) *Wimbush* 17! [Br.]Cam.: Vogel Peak, Adamawa (Nov.) *Hepper* 1553!

5. L. pobeguinii *(Hutch. & Dalz.) J. K. Morton* l.c. 270 (1962). *Coleus pobeguinii* Hutch. & Dalz. F.W.T.A., ed. 1, 2 : 292 (1931). A stout undershrub with woody stems ¼ in. thick and 3–6 ft. high ; flowers rich bluish-purple in dense oblong heads about 1 in. long, forming a stiff, branched panicle.
   Guin.: Kouroussa (Oct.) *Pobéguin* 556! 996! Mali to Kédougou, Fouta Djalon (fr. June) *Adam* 14443! **S.L.:** *Glanville* 385!

17. SOLENOSTEMON Thonning—F.T.A. 5 : 420 ; Morton in J. Linn. Soc. 58 : 251 (1962).

Median teeth of calyx deltoid, not acuminate :

Inflorescence a terminal umbellate cyme of about 4–6 short racemes 3–4 cm. long ; calyx 6 mm. long, upper lip ovate-rotund, decurrent, lower lip exceeding upper with 2 very long mucronate curved teeth, median lobes lanceolate blunt short ; stems creeping, pubescent ; leaves long-petiolate, broadly ovate, 2–3 cm. long and broad, obtuse, cuneate, coarsely crenate, thin, setose-pubescent   ..   ..   1. *repens*

Inflorescence pseudo-racemose, paniculate or spicate, not umbellate :

Plant perennial producing numerous underground edible tubers 1–6 cm. long ; inflorescence narrow, about 1 cm. broad, elongate and lax ; mature calyx 2–3 mm. long, upper lip ovate, lower lip oblong with 2 small teeth, median teeth very small and triangular ; corolla 4–8 mm. long ; leaves thick, half-succulent, ovate, up to 6 cm. long and 4 cm. broad, shallowly crenate, cuneate into a petiole, shortly pubescent, gland-dotted beneath ..   ..   2. *rotundifolius*

Plant annual not producing underground tubers, roots fibrous ; erect or creeping herb, subglabrous to long-pubescent ; inflorescence lax, interrupted, elongate, up to 30 cm. long in fruit and 1–2·5 cm. broad ; corolla 7–15 mm. long ; mature calyx 3–6 mm. long, upper lip broadly ovate, lower lip strap-shaped with 2 sharp teeth, exceeding upper, median lobes deltoid-rotund ; leaves thin to somewhat fleshy, linear-lanceolate to ovate, truncate to cuneate, 1·5–8 cm. long and 1–5 cm. broad, crenate, long-petiolate   ..   ..   3. *latifolius*

Median teeth of calyx lanceolate, acute or acuminate :

Pedicels much exceeding mature calyx, often branched :

Lower surface of leaves obscured by a dense even felt of hairs, upper surface with dense tufts of hairs obscuring all but the veins ; leaves broadly triangular-ovate, 3–4 cm. long, cuneate at base ; inflorescence elongate, lax, simple, about 30 cm. long, few-flowered, cymes stalked, distant ; mature calyx 4 mm. long, lower lip ovate, emarginate ; corolla 10–14 mm. long   ..   ..   ..   4. *koualensis*

Lower surface of leaves pubescent to glabrous, not obscured by hairs :

Leaves lanceolate, somewhat fleshy, 2–8 cm. long, ½–3 cm. broad, acute, gradually cuneate at base, almost sessile, dentate except towards base, thinly puberulous above and on the veins beneath ; annual erect bushy herb ; inflorescence lax, 1–15 cm. long, 3–5 cm. broad ; mature calyx globose, about 5 mm. diam., tube very short, upper lip very broadly ovate, lower emarginate ; corolla about 7 mm. long

5. *chevalieri*

Leaves ovate ; perennial herbs :

Small herb 10–30 cm. high with decumbent stem bases and thick fleshy leaves ; leaves 1–2·5 cm. long, 7–15 mm. broad, triangular, bluntly acute, coarsely crenate, with a broad cuneate base and a short petiole, gland-dotted and pubescent on the veins beneath, thinly pubescent above ; inflorescence 5–15 cm. long, rather dense, fascicles of flowers sessile, pedicels simple ; mature calyx 3–5 mm. long, pubescent, both lips rotund, upper shortly apiculate ; corolla 1–1·5 cm. long   ..   ..   6. *decumbens*

Large climbing or bushy herbs ; leaves thin to fleshy, completely subglabrous to densely pubescent on the veins beneath, ovate, 4–15 cm. long, crenate, acutely acuminate, long-petiolate, cuneate at base ; inflorescence rather dense, copiously flowered, 3–4 cm. broad and up to 25 cm. or more long in fruit ; calyx 4–5 mm. long, puberulous, upper lip ovate, lower slightly emarginate ; corolla 1·5 cm. long

7. *mannii*

Pedicels about equalling the mature calyx :

Leaves linear, about 8 cm. long and 5 mm. broad, sessile, dentate, scabrid above, appressed-pubescent on the veins beneath and gland-dotted ; stems wiry, 4-angled, thinly appressed-pubescent ; inflorescence simple, elongate, lax ; bracts ovate, acuminate, about 3 mm. long, caducous ; pedicels puberulous, shorter than the mature calyx ; mature calyx inflated, thin and papery, about 5 mm. long and 3·7 mm. broad, thinly puberulous and gland-dotted ; corolla 6–8 mm. long

8. *linearifolius*

Leaves lanceolate, ovate, or rotund :

Leaves rugose, densely pubescent, especially on the veins beneath, thick ; perennials ; stems rounded quadrangular, tomentose :

Erect bushy woody herb 1 m. or more high ; leaves ovate, acute, subsessile ; inflorescence dense, long and spicate, subcontiguous even in fruit, 15–40 cm. long, 1·5 cm. broad, bracts ovate-acuminate about 5 mm. long, reflexed in flower, persistent until fruit is mature ; pedicels often branched at base ; mature calyx 4–4·5 mm. long, pubescent and gland-dotted, lips divergent, upper ovate slightly apiculate, lower oblong emarginate ; corolla 10–12 mm. long, lower lobe 5 mm. broad   ..   ..   9. *graniticola*

Stems decumbent forming a mat over rocks ; flowering shoots erect 30–40 cm. high ; leaves ovate to ovate-rotund, obtuse, about 3 cm. long and 2 cm. broad, petiolate ;

inflorescence whitish-pubescent when young, dense in flower but elongate and interrupted in fruit, 10–30 cm. long, 1 cm. broad, bracts slightly broader than above and caducous as the flower opens ; pedicels simple ; mature calyx as above but 3–3·5 mm. long ; corolla about 7 mm. long, lower lobe 3 mm. broad
11c. *monostachyus* subsp. *marrubiifolius*

Leaves not rugose :
Small creeping herb with decumbent branched stems pubescent on the angles ; leaves ovate-triangular, acutely acuminate, 1·5–3 cm. long, 1–2·5 cm. broad, coarsely crenate, with purple blotch in centre, pubescent on both surfaces, cuneate into a petiole almost equal to the lamina ; inflorescence fairly dense, elongate, 10–25 cm. long, up to 2 cm. broad ; mature calyx 3 mm. long, reticulate, lower lip oblong-rotund, median teeth small and acute, upper lip exceeding lower, ovate ; corolla about 7 mm. long. .. .. **10.** *minor*

More or less erect, sometimes bushy herbs, not creeping :
Inflorescence dense in flower, subcontiguous in fruit, with large persistent broadly ovate-acuminate bracts which are reflexed in flower, about 1 cm. long and broad or larger ; corolla 10–15 mm. long, lower lip about 9 mm. long and 3 mm. broad ; mature calyx 6 mm. long ; stems stout woody, erect, shortly crispate pubescent ; leaves coriaceous, ovate 8–12 cm. long, 4·5–7 cm. broad, cuneate into a short petiole 1–2 cm. long, finely and thinly appressed-puberulous above, densely and shortly crispate-pubescent on the veins beneath
11b. *monostachyus* subsp. *lateriticola*

Inflorescence rather dense in flower but elongate and interrupted in fruit, bracts about 5 mm. long, lanceolate, usually caducous in flower ; corolla 5–10 mm. long, lower lip up to 6 mm. long and 2 mm. broad ; mature calyx 4–5 mm. long ; stems slender or stout and fleshy, erect and bushy or decumbent at base ; leaves thin to somewhat succulent, petiole exceeding half length of lamina :
Annual herb with rather fleshy acutely angled stems ; leaves broadly ovate, usually obtuse, sharply cuneate and long-petiolate at base, up to 10 cm. long and 8·5 cm. broad, thin, coarsely crenate, shortly pubescent above and on the veins beneath ; inflorescence 5–40 cm. long, 1·5–1·7 cm. broad in fruit ; mature calyx 4–5 mm. long, pubescent ; corolla about 5 mm. long
11. *monostachyus* subsp. *monostachyus*

Perennial herb with small tuberous stem-base ; leaves lanceolate to ovate-lanceolate up to 9 cm. long and 5 cm. broad, bluntly acute, abruptly cuneate into petiole as long as the lamina, somewhat fleshy, subglabrous apart from a minute pubescence on the veins ; inflorescence very elongate and lax, 20–50 cm. long, 1–2 cm. broad, whorls distant ; corolla about 9 mm. long
11a. *monostachyus* subsp. *perennis*

1. **S. repens** (*Gürke*) *J. K. Morton* in J. Linn. Soc. 58 : 272 (1962). *Coleus repens* in Engl. Bot. Jahrb. 19 : 213 (1894) ; F.T.A. 5 : 439 ; F.W.T.A., ed. 1, 2 : 292. *C. carnosus* A. Chev. in Journ. de Bot., sér. 2, 2 : 125 (1909), incl. var. *lamiifolius* A. Chev. Bot. 518. A small creeping herb a few inches high, with lax cymes of greenish-white flowers ; in dense wet forest particularly on rocks and decaying tree-trunks. **Guin.:** Fassakoidou (Feb.) *Chev.* 20806 ! Nzo Mts. (Mar.) *Chev.* 21077 ! **S.L.:** Baoma (Dec.) *Deighton* 3806 ! **Lib.:** Ganta to Sanokwele (Sept.) *Baldwin* 13262 ! **Iv.C.:** Davo to Gaouloubré (May) *Chev.* 17985 *bis* ! Moyen-Cavally (June–July) *Chev.* 19221 ! 19254 ! 19403 ! **Ghana:** Begoro waterfall (Nov.) *Morton* A43 ! Akim (Dec.) *Johnson* 787 ! Puso Puso Ravine, Kibbi (Dec.) *Morton* GC 8216 ! **[Br.]Cam.:** near Buea, 4,000 ft. *Preuss* 949 ! **F.Po:** Moka (Dec.) *Boughey* 75 !
2. **S. rotundifolius** (*Poir.*) *J. K. Morton* l.c. 272 (1962). *Germanea rotundifolia* Poir. (1812). *Coleus rotundi-folius* (Poir.) A. Chev. & E. Perrot. in Vég. Util. Afr. Trop. Fr. 1 : 101, 119 (1905) ; Chev. Bot. 520, incl. var. *nigra* A. Chev. *C. dysentericus* Bak. in Kew Bull. 1894 : 10 ; F.T.A. 5 : 437 ; F.W.T.A., ed. 1, 2 : 292. *C. salagensis* Gürke (1894). *Plectranthus coppinii* Cornu (1901). *Coleus pallidiflorus* A. Chev. (1909)—Chev. Bot. 519. An erect nearly glabrous half-succulent herb with square stems, branched at base, tuberous roots and tiny blue flowers 4–5 mm. long ; cultivated in the drier savanna areas. **Sen.:** Casamance (Feb.) *Chev.* 2813 ! **Guin.:** Kouria (Nov.) *Caille* in Hb. *Chev.* 14878 ! Camayenne (June) *Chev.* 4418 ! Kouria (Nov.) *Chev.* 14838 ; 14858. Conakry (Oct.) *Chev.* 18908 ! **Iv.C.:** Upper Sassandra (May) *Chev.* 21532 ! **Ghana:** *Johnson* 794 ! *Director of Agric.* 1912 ! Salaga, cult. (Oct.) *Morton* ! **N.Nig.:** (Dec.) *Lamb* 67 ! Lokoja (Nov.) *Dalz.* 103 ! **S.Nig.:** Yoruba country *Barter* 846 ! (See Appendix, p. 459.)
3. **S. latifolius** (*Hochst. ex Benth.*) *J. K. Morton* l.c. 271 (1962). *Coleus latifolius* Hochst. ex Benth. in DC. Prod. 12 ; 74 (1848) ; F.T.A. 5 : 437. *C. copiosiflorus* Briq. (1917)—Chev. Bot. 519. *C. phymatodes* Briq. (1917)—Chev. Bot. 520. *C. splendidus* A. Chev. in Journ. de Bot., sér. 2, 2 : 122 (1909). *C. djalo-nensis* A. Chev. l.c. 123 (1909). *C. peulhorum* var. *violacea* A. Chev. l.c. 124 (1909) ; Chev. Bot. 519, 520. *C. laxiflorus* (Benth.) Roberty in Bull. I.F.A.N. 16, 1 : 331 (1954), partly. *Plectranthus bongensis* Baker in F.T.A. 5 : 410 (1900). An erect, or rarely creeping, branched annual herb up to 3 ft. high with cylindrical inflorescences of violet flowers ; amongst rocks. Very variable and probably containing several taxa. **Mali:** Koulikoro (Oct.) *Chev.* ! **Guin.:** Bilima (Sept.) *Caille* in Hb. *Chev.* 15042 ! Fouta Djalon (Sept.-Oct.) *Chev.* 18274 ! 18297 ! 18327 ! 18895 ! *Schnell* 7019 ! Kita massif (Oct.) *Jaeger* 32 ! **S.L.:** Kabala, 2,000 ft. (Apr.) *Glanville* 254 ! 359 ! Falaba (Sept.) *Deighton* 5167 ! Warantamba Fotombu (Oct.) *Small* 369 ! 381 ! Musaia (Aug.) *Deighton* 5130 ! **Iv.C.:** Nimba Mts., 2,000 ft. (Aug.) *Boughey* GC 18145 ! *Roberty* 3108 ! *Schnell* 6238 ! **Ghana:** Banda, N.W. Ashanti (Nov.) *Harris* ! Gambaga Scarp (fl. Sept., fr. Nov.) *Hall* 789 ! *Morton* ! Bongo Hills, Bolgatanga (fr. Nov.) *Morton* ! **N.Nig.:** Naraguta to Jos (fl. Sept., fr. Oct.) *Lely* 575 ! *Hepper* 1076 ! **[Br.]Cam.:** Bamenda *Ujor* FHI 29985 ! Widespread in tropical Africa.
4. **S. koualensis** (*A. Chev. ex Hutch. & Dalz.*) *J. K. Morton* l.c. 270 (1962). *Coleus koualensis* A. Chev. Bot. ex Hutch. & Dalz. F.W.T.A., ed. 1, 2 : 291 (1931) ; Chev. Bot. 519, name only. A slender erect plant with woody stems 2–3 ft. high and blue flowers. **Iv.C.:** Mt. Dourou, summit 3,400 ft., Upper Sassandra (May) *Fleury* in Hb. *Chev.* 21715 !
5. **S. chevalieri** *Briq.* in Mém. Soc. Bot. Fr. 2, 8 : 286 (1917) ; Chev. Bot. 517. *Coleus nigericus* A. Chev. Bot.

519, name only. An erect, branched annual herb about 1 ft. high with broad inflorescences of small pale violet flowers.

Mali: Niokolo-Koba (fr. Oct.) *Adam* 15608 ! Koulikoro (Oct.) *Chev.* 2779 !

6. **S. decumbens** (*Hook. f.*) *Bak.* in F.T.A. 5 : 421 (1900). *Plectranthus decumbens* Hook. f. (1864). A small aromatic herb with ascending flowering shoots and decumbent stems ; flowers pale blue.
   [**Br.**]**Cam.:** Cam. Mt., 7,000–8,000 ft. (Nov.-Jan.) *Johnston* 79 ! *Maitland* 923 ! *Mann* 2002 ! *Boughey* GC 10624 ! *Morton* K740 ! K583 ! Bamenda, in bamboo thicket, 7,500 ft. (fr. Apr.) *Morton* K218 !

7. **S. mannii** (*Hook. f.*) *Bak.* in F.T.A. 5 : 422 (1900). *Coleus mannii* Hook. f. (1864). *Solenostemon cymosus* Bak. in F.T.A. 5 : 422 (1900) ; F.W.T.A., ed. 1, 2 : 290. A herbaceous or somewhat woody perennial growing in wet forest or amongst scrub at higher altitudes, with climbing or erect and bushy square stems and long, rather dense racemes of rich bluish-purple flowers.
   S.L.: Linbaia, near Falaba (Oct.) *Small* 429 ! **Lib.:** Sanokwele (Sept.) *Baldwin* 9517 ! **Ghana:** Begoro (Nov.) *Morton* A45 ! Kibbi (Dec.) *Morton* GC 8217 ! **S.Nig.:** Sonkwala, 1,000 ft., Obudu (Dec.) *Carpenter* 713 ! Ikom (Dec.) *Keay* FHI 28175 ! [**Br.**]**Cam.:** Cam. Mt., 1,000–4,500 ft. (Nov.-Dec.) *Mann* 1251 ! 1967 ! *Morton* K891 ! *Maitland* 137 ! Bamenda, 3,600–6,500 ft. (fl. Dec., fr. Apr.) *Boughey* GC 1126 ! *Morton* K240 ! *Keay & Lightbody* FHI 28364 ! **F.Po:** *Boughey* !
   [The erect bushy forms represented by *Morton* K240 and *Keay & Lightbody* FHI 28364 may be distinct.]

8. **S. linearifolius** *J. K. Morton* in J. Linn. Soc. 58 : 271, photo. 2 (1962). An erect annual herb about 1 ft. high with whorls of violet flowers in lax erect racemes.
   Guin.: Diendiou (Mar.) *Pobéguin* 1920 !

9. **S. graniticola** *A. Chev.* in Journ. de Bot., sér. 2, 2 : 121 (1909) ; Chev. Bot. 517. *S. monostachyus* var. *graniticola* (A. Chev.) Brenan in Kew Bull. 5 : 231 (1950). A densely whitish pubescent rather fleshy shrub about 3 ft. high with showy dense inflorescences of large purple flowers ; on granite rocks.
   Lib.: Zwedru (Aug.) *Baldwin* 7068 ! **Iv.C.:** Mt. Nienokue (July) *Chev.* 19469 ! Issa, Daloa to Abidjan (Dec.) *Boughey* GC 13598 ! GC 14708 !

10. **S. minor** *J. K. Morton* l.c. 271 (1962). A creeping herb, rooting at the nodes and tuberous at the base ; flowering stems 6–18 in. high ; flowers purple ; on wet granite rocks.
    N.Nig.: W. Kagoro Hills, 3,700 ft., Jos Plateau (Aug.) *Killick* 15 ! **S.Nig.:** Mt. Koloishe, 5,500 ft., Obudu Div. (Dec.) *Keay & Savory* FHI 25086 !

11. **S. monostachyus** (*P. Beauv.*) *Briq.* subsp. **monostachyus**—in E. & P. Pflanzenfam. 5, 3A : 359 (1897) ; Chev. Bot. 517 ; Morton in Bull. Soc. Bot. Fr. 104 : 164 (1957) (fig. citations reversed). *Ocimum* ? *monostachyum* P. Beauv. Fl. Oware 2 : 60, t. 95, fig. 1 (1818). *Solenostemon ocymoides* Schum. & Thonn. (1827)—F.T.A. 5 : 420 ; Chev. Bot. 518 ; F.W.T.A., ed. 1, 2 : 290. *S. ocymoides* var. *monostachyus* (P. Beauv.) Bak. in F.T.A. 5 : 421 (1900). An erect, branched, pubescent, annual weedy herb with long rather dense inflorescences of small, pale violet flowers ; frequently with a purple blotch in the centre of the leaves. Diploid.

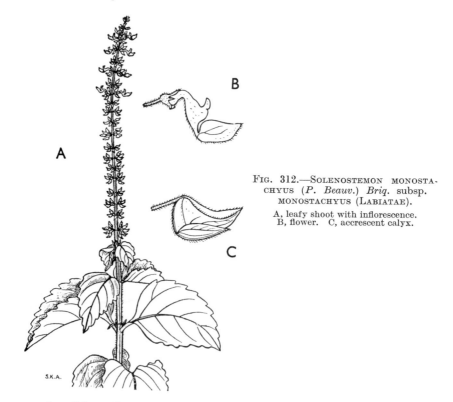

FIG. 312.—SOLENOSTEMON MONOSTA-
CHYUS (*P. Beauv.*) *Briq.* subsp.
MONOSTACHYUS (LABIATAE).

A, leafy shoot with inflorescence.
B, flower. C, accrescent calyx.

Sen.: *Unknown collector* ! **Mali:** Penia (May) *Chev.* 825. Mossi (Aug.) *Chev.* 24621. **Port.G.:** Gabu (Oct.) *Esp. Santo* 3403 ! **Guin.:** Conakry (Oct.) *Adam* 12579 ! Kaba (May) *Chev.* 13169. Longuery *Caille* in *Hb. Chev.* 14624. **S.L.:** Heddles Farm, Freetown *Sc. Elliot* 3908 ! **Lib.:** Monrovia *Barker* 1459 ! Cape Palmas (Aug.) *Hale* 122 ! Buchanan (Nov.) *Adam* 16038 ! **Iv.C.:** Abidjan *Chev.* 15359 ; 15484 ; 15567 ; 15598. **Ghana:** Aburi (Sept.) *Morton* GC 9274 ! Amedzofe (Oct.) *Morton* A1996 ! Achimota (Mar.) *Morton* A1944 ! **Dah.:** Cotonou (July) *Chev.* 4493. **N.Nig.:** Lokoja *Parsons* 29 ! **S.Nig.:** Lagos

(Sept.) *Dawodu* 27! Ibadan *Newberry* 64! [**Br.**]**Cam.**: Victoria (Dec.) *Morton* K690! Kumba *Hutch.* & *Metcalfe* 156! **F.Po**: *Mann* 404! Widespread in tropical Africa. (See Appendix, p. 463.)

11a. **S. monostachyus** subsp. **perennis** *J. K. Morton* l.c. 272 (1962). An erect, sometimes branched, slender, thinly pubescent herb with somewhat fleshy leaves and very long lax inflorescences ; flowers rich purple ; whole plant usually purplish, perennating by means of the fleshy swollen stem base ; in rocky savanna. Tetraploid.

    Iv.C.: Sindou *Adam* 18173! **Ghana**: Damongo Scarp (fl. Sept., fr. Nov.) *Rose Innes* 366! *Harris*! *Morton* A1242! **N.Nig.**: Naraguta *Lely* 540! Jos Plateau *Summerhayes* 11! Kilba country, Yola *Dalz*. 101! Abinsi *Dalz.* 799! **S.Nig.**: Gawu Hills, Niger Prov. *Onochie* FHI 35722!
    [Specimens apparently intermediate with the type subspecies and subsp. *perennis* occur.]

11b. **S. monostachyus** subsp. **lateriticola** (*A. Chev.*) *J. K. Morton* l.c. 271 (1962). *S. lateriticola* A. Chev. in Journ. de Bot. 1909: 121; *Chev.* Bot. 517. *Coleus lateriticola* A. Chev. Bot. 519, name only. A stout erect herb some 3 ft. high with dense spikes of purple flowers and purplish stems and leaves.
    Guin.: Faranna *Chev.* 13410. Ditinn 2,600 ft. (fr. Apr.) *Chev.* 13553! ut Foa Djalon (fl. Sept.-Oct.) *Chev.* 18231! 18254! 18300! 18381! 18884! 18893! **S.L.**: Mt. Loma (bud July) *Jaeger*!

11c. **S. monostachyus** subsp. **marrubiifolius** (*Brenan*) *J. K. Morton* l.c. 272, photo. 1 (1962). *S. monostachyus* var. *marrubiifolius* Brenan in Kew Bull. 5 : 231 (1950). *S. graniticola* J. K. Morton in Bull. Soc. Bot. Fr. 104: 164 (1957), partly, with fig. (fig. citations reversed). *S. gouanensis* A. Chev. Bot. 517, name only. A densely hairy, fleshy herb with perennial creeping branched woody stems and erect loose inflorescences of medium-sized rich purple flowers ; on flat rocks. Diploid.
    Iv.C.: Cavally (Apr.) *Fleury* in *Hb. Chev.* 21324! **Ghana**: Kwahu-Tafo (Apr.) *Morton* A689! A1693! *Irvine* 1720! Kumasi (Dec.) *Morton* GC 9688! Trumeso, Chira to Wenchi (Dec.) *Adams* GC 5110! **S.Nig.**: Carter Peak, Idanre Hills (Oct.) *Keay* FHI 22585! Ado Rock (Oct.-Nov.) *Hambler* 510! *Keay* FHI 37181!
    [Plants intermediate with the type subspecies occasionally occur in rocky places—e.g. *Morton* A3526, from Togo Plateau, Ghana, and *Keay* FHI 22561, from rock-outcrop near Ondo, S. Nigeria.]

18. ENGLERASTRUM Briq.—F.T.A. 5 : 445 ; Hutch. & Dandy in Kew Bull. 1926 : 479.

Plant creeping, perennial ; leaves persistent, ovate, cuneate at base, obtuse at apex, about 2·5 cm. long and 1·5 cm. broad, obscurely crenate, very thin, thinly pubescent on the veins, petiolate ; stems slender, pilose ; inflorescences axillary on long slender peduncles, racemose ; calyx pubescent, broadly campanulate, 3 mm. long, teeth short and triangular, equal ; pedicels short ..   ..   ..   1. *schweinfurthii*
Plant erect, annual ; leaves falling before time of flowering, lanceolate, cuneate at base, dentate, sessile, up to 10 cm. long and 3 cm. broad, thinly pilose chiefly on the veins :
Pedicels very long and slender, several times as long as the calyx ; inflorescence lax
                                                   2. *gracillimum*
Pedicels about equal to the calyx ; inflorescence dense   ..   ..   3. *nigericum*

1. **E. schweinfurthii** *Briq.* in Engl. Bot. Jahrb. 19 : 178, t. 3, fig. A (1894) ; F.T.A. 5 : 445. *E. djalonense* A. Chev. (1909)—F.W.T.A., ed. 1, 2 : 290. *E. hutchinsonianum* Alston (1926)—F.W.T.A., ed. 1, 2 : 290. *Coleus englerastrum* Roberty in Bull I.F.A.N. 16, 1 : 330 (1954), partly. A slender creeping herb with small, pale blue flowers ; usually in marshy savanna.
    Sen.: Niokola-Koba (Jan.) *Berhaut* 4620! **Mali**: Bamako, R. Niger (Dec.) *Adam* 11188! Diaka (Nov.) *Monod* ! **Guin.**: Fouta Djalon (Sept.) *Chev.* 18495! **Ghana**: foot of Ejura Scarp (Dec.-Jan.) *Morton* A1554! GC 9598! Wenchi to Sampa (Dec.) *Morton* A2603! Yendi to Zabzugu (Dec.) *Morton* A1462! **Togo Rep.**: Sokode *Schröder* 66. **N.Nig.**: Patti Lokoja (Nov.) *Dalz.* 105! Also in Sudan and Congo.

2. **E. gracillimum** *Th. C. E. Fries* in Notizbl. Bot. Gart. Berl. 9 : 69, t. 1b, 2a (1924) ; Milne-Redhead in Kew Bull. 5 : 380 (q.v. for syn.). *Plectranthus tenuis* Hutch. & Dandy (1926)— F.W.T.A., ed. 1, 2 : 289. *P. ramosissimus* of Chev. Bot. 516, not of Hook. f. *Coleus ramosissimus* (Hook. f.) Roberty l.c. (1954), partly. An erect, slender, branched, annual herb with racemes or panicles of small blue flowers borne on long slender pedicels, leafless when in fruit.
    Mali: Po, Upper Volta (Nov.) *Darko* 458! **S.L.** Musaia (Dec.) *Deighton* 4411! 4473! Soremodou (Jan.) *Chev.* 20567! **Ghana**: Damongo Scarp (Dec.) *Morton* GC 25068! Gambaga (Dec.) *Morton* A1391! Dutukpene (Dec.) *Adams* 4544! **N.Nig.**: Nupe *Barter*! Mokwa, Niger Prov. (Nov.) *Mutch* FHI 27965! Wana, Mada Hills (May) *Hepburn* 92! Vom (Nov.) *Morton*! [**Br.**]**Cam.**: Gangumi, Adamawa (Dec.) *Latilo & Daramola* FHI 28855! Jada (Nov.) *Hepper* 1302!

3. **E. nigericum** *Alston* in Kew Bull. 1926 : 298 ; F.W.T.A., ed. 1, 2 : 290. *Plectranthus denudatus* A. Chev. ex Hutch. & Dalz. F.W.T.A., ed. 1, 2 : 289 (1931) ; Chev. Bot. 516 (1920), name only. *Coleus englerastrum* Roberty in Bull. I.F.A.N. 16, 1 : 330 (1954), partly. An erect, slender, branched herb, about 1 ft. or more high, with small bluish flowers in more or less compressed one-sided racemes.
    Sen.: Floup-Fedyan, Casamance (Jan.) *Chev.* 2815! **Guin.**: Kondika (Jan.) *Roberty* 16736! **Port.G.**: Pussube (Dec.) *Esp. Santo* 1087! Bafata (Sept.) *Esp. Santo* 3388! **S.L.**: Njala (Jan.) *Deighton* 1569! **N.Nig.**: Katagum (Oct.) *Dalz.* 109! Naraguta (Oct.) *Hepper* 1005! Vom (Oct.) *Hepper* 1156!

19. ISODICTYOPHORUS Briq. in Mém. Soc. Bot. Fr. 2, 8 : 285 (1917).

Leaves lanceolate, up to 15 cm. or more long, petiolate, crenulate, glabrous above, pubescent and glandular-punctate beneath ; inflorescence lax, spreading ; pedicels slender, puberulous, crowded at the nodes, often branched, 2–3 cm. long in fruit ; mature calyx broadly campanulate, 8–10 mm. long, actinomorphic, teeth triangular, about equalling the tube ; corolla 17 mm. long ..   ..   ..   *reticulatus*

I. **reticulatus** (*A. Chev.*) *J. K. Morton* in J. Linn. Soc. 58 : 272 (1962). *Coleus reticulatus* A. Chev. in Mém. Soc. Bot. Fr. 2, 8 : 200 (1912) ; Chev. Bot. 520. *Isodictyophorus chevalieri* Briq. l.c. (1917) ; Chev. Bot. 524. *Coleus casamanicus* A. Chev. ex Hutch. & Dalz. F.W.T.A., ed. 1, 2 : 291 (1931) ; Chev. Bot. 518, name only. An erect undershrub with quadrangular stems, whitish-woolly when young, and large bright-blue flowers in racemes up to 1 ft. long.
    Sen.: Tambanaba (Feb.) *Chev.* 2777 ; s.n.! **Port.G.**: Mancrossé, Bafata (Jan.) *Esp. Santo* 433! Bafata (Nov.-Dec.) *Esp. Santo* 2343! 2822! **Guin.**: Farana, between Kaba and Tonfili (Jan.) *Chev.* 20393! **N.Nig.**: Jos Plateau (Sept.) *Lely* P718! **S.Nig.**: Balinge to Ikwette, Obudu Div. (Dec.) *Keay & Savory* FHI 25154!

20. NEOHYPTIS J. K. Morton in J. Linn. Soc. 58 : 272 (1962).

Stem slender erect, little branched, sharply quadrangular, grooved, shortly pubescent ; leaves oblong to ovate-lanceolate, subsessile, subacute, obscurely crenate, up to 3 cm.

long and 1 cm. broad, glabrous, minutely appressed-puberulous on veins beneath, veins arising from base, curved, prominent ; inflorescences terminal and lateral, sessile or shortly stalked, oblong, dense, spike-like racemes 1–3 cm. long, about 8 mm. broad ; bracts small ; mature calyx 2·5–3 mm. long with 5 subequal lanceolate teeth, finely pubescent ; corolla twice as long as calyx　　..　　　..　　　　*paniculata*

N. paniculata (*Bak.*) *J. K. Morton* l.c. 273, t. 10 (1962). *Geniosporum paniculatum* Bak. in F.T.A. 5 : 351 (1900) ; F.W.T.A., ed. 1, 2 : 286. *Plectranthus guerkei* Briq. (1904). *Hyptis quadrialata* A. Chev. Bot. 522, name only. A slender, erect perennial herb, 1–3 ft. high ; corolla small, white with purple dots ; growing in savanna swamps.
Port.G.: Bafata (Dec.) *Esp. Santo* 3802 ! Farim (Nov.) *Esp. Santo* 3639 ! **Guin.:** Mansonia, Kissi (Feb.) *Chev.* 20684 ! **S.L.:** Falaba (Mar.) *Sc. Elliot* 5100 ! **Ghana:** Sampa, N. W. Ashanti (Apr.) *Morton* A2619 ! **N.Nig.:** Kontagora (Nov.) *Dalz.* 134 ! Vom *Dent Young* 211 ! Rafin Bauna F.R., Jos (Oct.) *Hepper* 1183 ! **S.Nig.:** Ikwette Plateau, 5,000 ft., Obudu Div. (Dec.) *Keay & Savory* FHI 25175 ! **[Br.]Cam.:** Bamenda (June) *Maitland* 1382 ! Mambila Plateau (Jan.) *Hepper* 1668 ! Also in Angola and E. Africa.

21. HYPTIS Jacq.—F.T.A. 5 : 447 ; Epling in Kew Bull. 1936 : 278. *Nom. cons.*

Inflorescence terminal :

Inflorescence a dense cylindrical or ovoid-cylindrical spike up to 9 cm. long ; bracts linear-filiform, ciliate ; mature calyx 5 mm. long, strongly 10-ribbed, pubescent, but lacking conspicuous tufts of white hairs between the subulate teeth ; leaves lanceolate, acute, up to 8 cm. long and 3 cm. broad, closely gland-dotted beneath
　　　　　　　　　　　　　　　　　　　　　　　　　　　　　　1. *spicigera*
Inflorescence much-branched, not leafy, flowers secund, numerous ; mature calyx 3 mm. long, 10-ribbed, pubescent, with conspicuous tufts of white hairs between the subulate teeth ; leaves ovate-elliptic, up to 8 cm. long and 4 cm. broad, toothed, softly tomentose, long-petiolate　　..　　　..　　　..　　　..　　　..　　2. *pectinata*
Inflorescence of axillary heads of few-flowered cymes :
Inflorescence lax, consisting of few-flowered axillary cymes ; calyx broadly campanu-late, strongly 10-ribbed, accrescent in fruit, about 11 mm. long when mature, teeth subequal, subulate ; leaves broadly ovate, rounded at base, 4–5 cm. long, up to 4 cm. broad, pubescent above, white-tomentose especially on the veins beneath, very long-petiolate　　..　　　..　　　..　　　..　　　..　　　..　　3. *suaveolens*
Inflorescence dense, globose, many-flowered ; calyx narrow and tubular :
Mature calyx 7 mm. long, tube truncate at apex, teeth linear-filiform ; leaves ovate-lanceolate to ovate, cuneate at base, up to 5 cm. long and 2 cm. broad, crenate, densely gland-dotted beneath and setose on the nerves, petiolate　　4. *atrorubens*
Mature calyx 4–5 mm. long, tube not truncate, the subulate-triangular teeth with a V-shaped sinus between ; leaves lanceolate, cuneate at base, up to 11 cm. long and 3·5 cm. broad, but usually smaller, serrate, subglabrous, gland-dotted beneath, shortly petiolate　　..　　　..　　　..　　　..　　　..　　　..　　5. *lanceolata*

1. **H. spicigera** *Lam.* Encycl. Méth. Bot. 3 : 185 (1789) ; F.T.A. 5 : 448 ; Chev. Bot. 523 ; Epling in Kew Bull. 1936 : 278. A tall, erect, aromatic, rather scabrid herb ; flowers very small, corolla white with mauve marks on the lip ; a weed of roadsides and cultivated land, often occurring in damp places.
Sen.: *Perrottet* 676 ! **Gam.:** Basse and Kudang (Mar.) *Todd* 15 ! Genieri (Feb.) *Fox* 47 ! **Mali:** Garango, Upper Volta (Oct.) *Grost* ! Koulikoro (Oct.) *Chev.* 2782 ! Kara (Oct.) *Davey* 188 ! **Port.G.:** Bafata (Nov.) *Esp. Santo* 2967 ! **Guin.:** Bordo, Kankan (Apr.) *Adam* 12132 ! **S.L.:** Musaia (Dec.) *Deighton* 4451 ! **Ghana:** Gonja (Dec.) *Morton* GC 25029 ! Yendi (Dec.) *Adams & Akpabla* 4079 ! **Niger:** Niamey (Oct.) *Hagerup* 548 ! **N.Nig.:** Kontagora (Nov.) *Dalz.* 136 ! **[Br.]Cam.:** Jada, Adamawa (Nov.) *Hepper* 1223 ! Perhaps a native of Brazil, but now widespread in Africa and Asia. (See Appendix, p. 460.)
2. **H. pectinata** (*Linn.*) *Poit.* in Ann. Mus. Paris 7 : 474, t. 30 (1806) ; F.T.A. 5 : 448 ; Chev. Bot. 522 ; Epling l.c. *Nepeta pectinata* Linn. (1759). *Bystropogon coarctatus* Schum. & Thonn. (1827). A tall, erect, aromatic, woody herb with a dense, much-branched inflorescence of very small flowers ; corolla pale yellowish-green with mauve marks on the lip.
Guin.: Niguelandé (Mar.) *Adam* 11631 ! **S.L.:** Kukuna (Jan.) *Sc. Elliot* 4667 ! **Iv.C.:** Toumodi (Dec.) *Boughey* GC 14423 ! **Ghana:** Gonja, N.T. (Dec.) *Morton* GC 9943 ! Weija, Accra (Nov.) *Morton* A996 ! Butre, Western Region (Mar.) *Morton* A435 ! Kete Krachi (Apr.) *Morton* GC 9127 ! **Togo Rep.:** Lomé *Warnecke* 16 ! **Dah.:** Zagnanado (Feb.) *Chev.* 22983. Lahama (Feb.) *Chev.* 23257. **N.Nig.:** Naraguta (Aug.) *Lely* 551 ! **S.Nig.:** Lagos *Dawodu* 207 ! **[Br.]Cam.:** *Preuss* 1100 ! A weed native in tropical America, now in tropical Africa, Madagascar and Asia. (See Appendix, p. 460.)
3. **H. suaveolens** *Poit.* l.c. 472, t. 29 (1806) ; F.T.A. 5 : 449 ; Chev. Bot. 523 ; Epling l.c. A stout bushy, strongly aromatic, pubescent herb with large ¼ in. blue flowers ; a weed of roadside and cultivated land.
Sen.: *Farmar* 156 ! **Gam.:** (July) *Brooks* 22 ! **Port.G.:** Bissau, Pelum (Jan.) *Esp. Santo* 1614 ! **Guin.:** Conakry *Chev.* 12064 ! *Debeaux* 142 ! **S.L.:** Mafwe (Apr.) *Deighton* 1691 ! Freetown (Sept.) *Hepper* 916 ! **Lib.:** Monrovia (Nov.) *Linder* 1572 ! Cape Palms (Apr.) *Todd* 33 ! **Ghana:** Achimota (Feb., Oct.) *Morton* GC 25383 ! A14 ! Winneba (May) *Adams* 311 ! **N.Nig.:** Bichikki (July) *Lely* 183 ! **S.Nig.:** Lagos (July) *Dalz.* 1366 ! Native in tropical America, now widespread in tropical Africa and Asia and in Queensland. (See Appendix, p. 461.)
4. **H. atrorubens** *Poit.* l.c. 466, t. 27 f. 3 (1806) ; F.T.A. 5 : 447 ; Chev. Bot. 522 ; *H. atrorubensa* var. *african* Epling l.c. 279 (1936). A straggling, aromatic herb with globose, terminal and axillary heads ¾ in. diam. of small flowers ; corolla white with mauve marks on the lip.
Guin.: *Heudelot* 922 ! Dalaba-Diaguissa Plateau (Oct.) *Chev.* 18891 ; 20214. Macenta (Oct.) *Baldwin* 9808 ! **S.L.:** Kitchom (Dec.) *Sc. Elliot* 4343 ! Musaia (Dec.) *Deighton* 4447 ! Yetaya (Sept.) *Thomas* 2329 ! **Lib.:** Cape Palmas (Dec.) *Hale* 70a ! Gbau (Sept.) *Baldwin* 9448 ! **Iv.C.:** Grabo, Cavally basin (July) *Chev.* 19619. Mt. Oroumba Boka, Toumodi (Dec.) *Boughey* 14418 ! **Ghana:** Essiama (May, Dec.) *Morton* A2118 ! A2526 ! Also in tropical America.
[Epling named our West African plant var. *africana* because of its slightly larger calyx, but it differs little from the American plant.]
5. **H. lanceolata** *Poir.* in Lam. Encycl. Méth. Bot., Suppl. 3 : 114 (1813) ; Epling l.c. 279. *H. brevipes* of F.T.A. 5 : 447 ; Chev. Bot. 522 ; F.W.T.A., ed. 1, 2 : 284, not of Poit. *H. lanceifolia* Thonning (1827)—Epling l.c. 280. An erect, branched, aromatic herb of damp places, with axillary globose inflorescences ¼ in. across of very small flowers ; corolla white with pale mauve marks on the lip.

Sen.: *Perrottet* 673! **Gam.**: Kuntaur *Ruxton* 128! **Mali**: Dendela (Mar.) *Chev.* 632. **Guin.**: Kindia (Mar.) *Chev.* 13074. Fouta Djalon (Sept.) *Chev.* 18251 ; 18587. **S.L.**: Gbundapi (Mar.) *MacDonald* 5! Rokupr (Apr.) *Hepper* 2666! **Lib.**: Suacoco (Jan.) *Daniel* 87! Cape Palmas (Dec.) *Hale* 70! **Iv.C.**: Grand Bassam (Aug.) *Jaeger* 4850! Guitry (Dec.) *Boughey* GC 13532! **Ghana**: Kpong to Akuse (Oct.) *Morton* A11! Essiama (Dec.) *Morton* A2465! Yendi (Apr.) *Morton* GC 9075! **Dah.**: Porto-Novo (Mar.) *Chev.* 23307. Atacora Mts. (June) *Chev.* 24121. **N.Nig.**: Abinsi (Apr.) *Dalz.* 697! **S.Nig.**: Lagos (May) *Dalz.* 1245! *Dawodu* 316! **[Br.]Cam.**: Victoria (Sept.) *Boughey* X1175! Kumba to Mamfe (Jan.) *Morton*! Mamfe (Jan.) *Keay* FHI 28547! **F.Po**: (June) *Barter*! (Dec.) *Monod* 10489! Apparently native in our area ; widespread in tropical Africa and America.

22. SATUREJA Linn. Sp. Pl. 2 : 567 (1753) ; Brenan in Mem. N.Y. Bot. Gard. 9 : 45 (1954).

Older leaves 4–10 mm. long, ovate, rounded at base with several ascending nerves, scabrid above, dotted with large glands beneath, subsessile, entire ; flowers few together, in axillary whorls ; pedicels very short ; calyx tubular, 4 mm. long, strongly 12–15-ribbed, minutely setulose, teeth subequal, subulate, straight ; corolla twice as long as calyx ; stamens 4, anthers 2-celled ; a small woody herb with perennial base ; stems pubescent with long and short recurved hairs   ..   ..   1. *punctata*
Older leaves 15–35 mm. long, crenate :
  Inflorescence of distant axillary whorls ; a small slender, straggling herb branched from the base ; stems softly pubescent and pilose ; leaves subsessile, broadly ovate, rounded at base, up to 2 cm. long and nearly as broad, setose on the nerves beneath, crenulate ; flowers about 6 in a whorl ; pedicels up to 5 mm. long, pilose ; calyx broadly tubular, about 12-nerved, 6–8 mm. long in fruit, teeth lanceolate, acuminate, subequal, pilose, curved upwards ; corolla nearly twice as long as the calyx
                                                                   2. *pseudosimensis*
  Inflorescence of many, dense, terminal, broad, sessile spikes, up to about 5 cm. long, rarely more ; a stout, erect, branched, woody herb ; branchlets very leafy, bristly-pubescent ; leaves ovate-rotund, on the main shoots about 3·5 cm. long and on the terminal shoots about 1·5 cm. long, rounded or slightly cordate at base, sessile, crenate, strongly nerved beneath ; mature calyx 4–5 mm. long, 13–15-ribbed, teeth subulate, short, equal, straight, shortly setose-pubescent ; corolla shortly exserted from calyx   ..   ..   3. *robusta*

1. **S. punctata** (*Benth.*) *Briq.* in E. & P. Pflanzenfam. 4, 3A : 299 (1896) ; Brenan in Mem. N.Y. Bot. Gard. 9 : 45. *Micromeria punctata* Benth. (1834). *M. biflora* of F.T.A. 5 : 452 ; of F.W.T.A., ed. 1, 2 : 278, not of (Buch.-Ham. ex D. Don) Benth. *M. purtschelleri* Gürke (1891)—E. & K. Walther in Mitt. Thur. Bot. Ges. 1, 4 : 10 (1957). An erect or spreading heath-like plant with wiry stems up to 1 ft. or more high ; flowers small, corolla rich pink ; growing in montane grassland.
  [Br.]Cam.: Cam. Mt., 6,000–13,000 ft. (Dec.-Apr.) *Dundas* FHI 20358! *Morton* GC 6866! K746! *Mann* 1292! 1981! Bamenda, above 5,000 ft. (Dec.-Apr.) *Richards* 5296! *Morton* K18! 253! Jakiri (Feb.) *Hepper* 2071! Mambila Plateau (Jan.) *Hepper* 1662! Extends to Sudan, Congo, E. and S. Africa.
2. **S. pseudosimensis** Brenan *l.c.* 50 (1954). *Calamintha simensis* of F.T.A. 5 : 455 ; of F.W.T.A., ed. 1, 2 : 282 ; of Robyns Fl. Sperm. Parc Nat. Alb. 2 : 163 ; not of Benth. A much-branched, perennial, weak straggling herb, 6 in. to 2 ft. high, with mauve flowers about ¼ in. long ; the calyx frequently purple tinged ; in montane grassland and thicket.
  [Br.]Cam.: Cam. Mt., 6,000–13,000 ft. (Dec.-Apr.) *Dunlap* 241! *Dalz.* 8309! *Mann* 1292! 1981! *Morton* GC 6884! 7002! K588! *Plumptre* 230! Bamenda, above 6,000 ft. (Dec.-Apr.) *Adams* GC 11307! *Morton* K226! K256! *Maitland* 1392! 1487! **F.Po**: 8,500 ft. (Mar., Dec.) *Mann* 617! *Guinea*! Also in the Congo, Sudan, E. Africa and Nyasaland.
3. **S. robusta** (*Hook. f.*) Brenan *l.c.* 48 (1954). *Nepeta robusta* Hook. f. (1864)—F.T.A. 5 : 460 ; F.W.T.A., ed. 1, 2 : 280. An erect, branched, robust, aromatic perennial, 3–4 ft. or more high, with small flowers about ¼ in. long ; corolla white with mauve marks on the lip ; in montane grassland and scrub.
  [Br.]Cam.: Cam. Mt., above 6,000 ft. (Dec.-Apr.) *Dundas* FHI 20357! *Morton* GC 6937! 6998! *Plumptre* 222! Bamenda, above 6,000 ft. (Dec.-April.) *Ujor* FHI 30021! *Keay & Lightbody* FHI 28354! *Morton* K156! Ndu, Nkambe Div. (Feb.) *Hepper* 1941! 1942! (See Appendix p. 462.)
  [*Thomson* s.n. at Kew labelled " Old Calabar " presumably came from the slopes of Cam. Mt. on the Calabar side.]

23. SALVIA Linn.—F.T.A. 5 : 456.

Calyx truncate, not prominently lobed, densely white tomentellous, about 6 mm. long in fruit ; corolla blue, about 15 mm. long, tube 7 mm. long ; stamens not exserted ; leaves lanceolate, about 5 cm. long and 2 cm. broad, acute, cuneate and long-petiolate at base, serrate, puberulous on the veins, gland-dotted beneath ; stems sulcate puberulous, woody at base   ..   ..   1. *farinacea*
Calyx distinctly 3-lobed, shortly pubescent ; corolla usually red, sometimes pink, yellow or white :
  Calyx scarlet, inflated, large, 15 mm. long in flower, lobes ovate, mucronate ; leaves ovate, acuminate, cuneate and long-petiolate at base, serrate, about 7 cm. long and 4 cm. broad, glabrous ; stems sulcate, glabrous, woody below ; corolla about 25 mm. long ; stamens exserted   ..   ..   2. *splendens*
  Calyx green, not inflated, about 8 mm. long in flower, lobes ovate, acute not mucronate ; leaves ovate, obtuse or acute, truncate at base, long-petiolate, 3–6 cm. long and 2–4 cm. broad, puberulous above, pubescent on the nerves beneath, serrate-crenate ; stems sulcate, pubescent with long and short hairs ; corolla about 23 mm. long ; stamens long-exserted   ..   ..   3. *coccinea*

1. **S. farinacea** Benth. Lab. Gen. & Sp. 274 (1833). A perennial, bushy herb, several feet high, with blue flowers; commonly cultivated and occasionally occurring as a garden throwout, etc.
  S.L.: Njala (Oct.)*Deighton* 2423! **Ghana**: Legon (Dec.) *Morton* A2645! Native of tropical America.

2. **S. splendens** *Ker-Gawl.* in Bot. Reg. 8 : t. 687 (1823). An erect, perennial, bushy herb several feet high, with scarlet flowers and calyx.

　　Ghana: Achimota *Irvine* 454 ! (Jan.) *Akpabla* 1907 ! Amedzofe *fide* Morton. Aburi (July) *Sampram* 121 ! **S.Nig.**: Ibadan *fide* Morton. A native of Brazil ; commonly cultivated and occasionally escaping in our area.

3. **S. coccinea** *Buc'hoz ex Ettl.* Salvia 23 (1777) ; F.T.A. 5 : 459 ; G. Taylor in Exell Cat. S. Tomé 267. An erect bushy herb, up to 2 ft. high, with a green calyx and scarlet or more rarely yellow, pink or white flowers ; commonly cultivated and frequently occurring as a weed on waste ground etc.

　　Mali: Ségou *fide* Roberty. **S.L.**: Freetown (Oct.) *Deighton* 1977 ! **Ghana**: Achimota *Anoff* ! *Akwei* 2 ! *Attoh* 43 ! *Morton* A2446 ! Kumasi *Kyei* ! Tamale *fide* Roberty. A native of tropical America.

24. SCUTELLARIA Linn.—F.T.A. 5 : 461.

Inflorescence subglabrous to shortly pubescent, racemose, flowers paired at the nodes, with small leafy bracts ; stems shortly pubescent ; leaves ovate to lanceolate, shortly petiolate, small at time of flowering but up to 7 cm. long when mature, rounded-cordate at base, crenulate, puberulous on the nerves and slightly gland-dotted ; pedicels short and pubescent ; mature calyx shortly campanulate, about 1 cm. long, upper and lower lips broadly deltoid ; corolla about 1·7 cm. long, pubescent, not glandular. .　　..　　..　　..　　..　　..　　..　　..　　1. *paucifolia*

Inflorescence densely tomentose and glandular-pubescent, racemose ; stems woody, pubescent ; leaves full-grown at time of flowering, ovate-lanceolate, rounded or sub-cordate at base, crenate, about 5 cm. long and 2·5 cm. broad, petiolate, glandular-punctate, beneath ; corolla about 1·3 cm. long, pubescent　　..　　2. *violascens*

1. **S. paucifolia** *Bak.* in Kew Bull. 1895 : 292 ; F.T.A. 5 : 462. *S. briquetii* A. Chev. in Mém. Soc. Bot. Fr. 2, 8 : 198 (1912) ; Chev. Bot. 523. Rootstock of fusiform tubers producing erect, 4-angled stems up to 1 ft. high, with dark purple flowers appearing after grass-burning ; the main leafy shoots develop later in the season and are straggling.

　　Guin.: Kindia (Aug.) *Pobéguin* 1520 ! Dabola (Mar.) *Pobéguin* 986 ! Fouta Djalon (Apr.) *Chev.* 12648 ! **S.L.**: Bintumane (Jan., Feb.) *Glanville* 476 ! *Jaeger* 4285 ! **Ghana**: Banda, Wenchi Dist. (Mar.) *Morton* GC 8612 ! A3273 ! A1094 ! *Asare* ! **N.Nig.**: Jos Plateau (Apr.) *Lely* P245 ! S.W. of Jos (Dec.) *Monod* 9762 ! Widely distributed in E. Africa.

2. **S. violascens** *Gürke* in Engl. Bot. Jahrb. 30 : 392 (1901). A shrub about 3 ft. high with stiff racemes of purple flowers.

　　[Br.]Cam.: Bamenda, 5,500 ft. (Jan., Aug.) *Migeod* 445 ! *Ujor* FHI 29991 ! Also in Cameroun and E. Africa.

25. ACHYROSPERMUM Blume—F.T.A. 5 : 463 ; Perkins in Notizbl. Bot. Gart. Berl. 8 : 78 (1921) : E. A. Bruce in Kew Bull. 1936 : 47.

Calyx irregular with 5 unequal teeth, 3 of which are partially united to form a broad lobe :

Mature calyx 4–5 mm. long, ciliate, with 2 subacute teeth and a broad lobe with 3 very short obtuse triangular teeth, divided to near middle on one side ; corolla 3–4 mm. long ; inflorescences terminal and axillary 5 cm. long and 1 cm. broad or less ; leaves ovate, shortly acuminate at apex, cuneate at base, up to 8 cm. long and 5 cm. broad, crenate ; petiole about 1·5 cm. long; whole plant covered with a long soft pubescence　　1. *dasytrichum*

Mature calyx about 7 mm. long, minutely puberulous, margin thinly ciliate, with 2 narrow triangular lobes and one broad lobe having 3 acute teeth ; inflorescences lateral, about 2 cm. long ; stem tomentose ; leaves narrowly obovate-elliptic, acuminate, gradually narrowed to base, up to 15 cm. long and 6 cm. broad, crenate, subglabrous except for fine appressed pubescence on the veins beneath ; petiole up to 1·5 cm. long　　..　　..　　..　　..　　..　　..　　2. *schlechteri*

Calyx regular with 5 equal teeth, 8–10 mm. long when mature :

Calyx teeth as broad as long, ciliate around margin, green ; inflorescence terminal 3–5 (–10) cm. long, 2 cm. broad ; bracts broadly ovate, ciliate ; leaves narrowly obovate-elliptic, acuminate at apex, cuneate and entire at base, 10–18 cm. long and up to 8 cm. broad, crenate, appressed-pubescent on nerves beneath ; petiole 1–3 cm. long　　3. *oblongifolium*

Calyx teeth twice as long as broad ; leaves ovate, sharply acuminate, rounded at base, cuneate, about 15 cm. long and 8 cm. broad, crenate-serrate ; petiole up to 5 cm. long :

Stem and leaves with long spreading pubescence ; inflorescences terminal and lateral, elongate, conical, terminal inflorescences 8–12 cm. long and 1·5 cm. broad, lateral smaller ; bracts broadly ovate, about 8 mm. long, green, pubescent with densely ciliate margin, deep pink ; corolla exceeding calyx, pubescent, magenta with white throat ; stamens exserted　　..　　..　　..　　..　　..　　4. *africanum*

Stem and leaves thinly appressed-puberulous ; inflorescences shorter and broader, 3–6 cm. long and 1–2 cm. broad (without the corolla) ; bracts almost orbicular, abruptly apiculate, membranous ; bracts, calyx and corolla dark red

　　　　　　　　　　　　　　　　　　　　　　　　　　　　　　5. *erythrobotrys*

1. **A. dasytrichum** *Perkins* in Notizbl. Bot. Gart. Berl. 8 : 81 (1921). A small woody perennial with a creeping rhizome and erect aerial shoots ; corolla cream ; occurring in forest farms.

　　Ghana: Kade Agric. Sta. (Dec.) *Morton* A3639 ! Also in Cameroun.

2. **A. schlechteri** *Gürke* in Engl. Bot. Jahrb. 36 : 127 (1905) ; E. A. Bruce in Kew Bull. 1936 : 53. A woody perennial, with lateral inflorescences usually on the one year old shoots.

　　S.Nig.: Oban *Talbot* ! **[Br.]Cam.**: Buea, 2,500 ft. *Schlechter* 12850 !

3. **A. oblongifolium** *Bak.* in F.T.A. 5 : 464 (1900) ; Bruce l.c. 55 ; G. Taylor in Excell Cat. S. Tomé 266. "*Ocimum sassandrae* A. Chev. Bot. 512, name only. An erect, little-branched undershrub, 1–2 ft. high, with tomentose stems and dense terminal spikes of greenish-white flowers ; in forest and bamboo thicket.
Guin.: Fassakoidou to Késséridou (Feb.) *Chev.* 20814 ! Ntongon Mt., 3,150 ft. (Mar.) *Chev.* 20926 ! **S.L.:** Kenema (Jan.) *Thomas* 7572 ! York Pass (Dec.) *Deighton* 3310 ! **Lib.:** Vahum, W. Prov. (Nov.) *Baldwin* 10212 ! **Iv.C.:** Mid. Sassandra & Mid. Cavally (June-July) *Chev.* 19244 ! Oroumba Boka (Dec.) *Schnell* ! *Boughey* GC 14438 ! **Ghana:** Apapam (Dec.) *Morton* GC 8161 ! Mpraeso (Apr.) *Morton* A690 ! Bosumkese (Dec.) *Adams* GC 5147 ! **S.Nig.:** Aboabam, Ikom Div. (Dec.) *Keay* FHI 28195 ! [Br.]**Cam.:** Bamenda, 6,000–7,000 ft. (Apr.) *Morton* K211 ! *Adams* GC 11222 ! **F.Po:** (June, Dec.) *Barter* 1697 ! *Boughey* 164 !
Also in S. Tomé, Cameroun and Cabinda.

4. **A. africanum** *Hook. f. ex Bak.* in F.T.A. 5 : 465 (1900). *A. schimperi* Perkins l.c. 79 (1921), partly. *Ocimum peulhorum* A. Chev. Bot. 512, name only. *Achyrospermum peulhorum* E. A. Bruce l.c. 59 (1936). An erect, herbaceous undershrub up to about 8 ft. high ; with dense spikes of purplish or purple-marked flowers.
Guin.: Dalaba to Mamou (Jan.) *Chev.* 20347 ! Dalaba (Jan.) *Roberty* 16478 ! **N.Nig.:** Sanga River F.R., Zaria Prov. (Nov.) *Keay* FHI 37240 ! [Br.]**Cam.:** Cam. Mt., 2,000–4,500 ft. (Dec.-Jan.) *Morton* K700 ! *Adams* GC 11294 ! *Keay* FHI 28650 ! *Mann* 1297 ! *Mann* 1949 ! Buea (Jan.,Mar.) *Maitland* 267 ! 546 ! Mbakokeke Mt., Bamenda (Mar.) *Richards* 5326 ! Abonando *Rudatis* !

5. **A. erythrobotrys** Perkins in Notizbl. Bot. Gart. Berl. 8 : 80 (1921) ; Bruce l.c. 56. A much-branched shrub, up to about 6 ft. high, with dark red flowers and inflorescences.
S.Nig.: Ikwette Plateau, 5,200 ft., Obudu Div. (Dec.) *Keay & Savory* FHI 25211 ! [Br.]**Cam.:** near L. Bambulue, 6,200 ft., Bamenda (Jan.) *Keay & Lightbody* FHI 28363 ! Also in Cameroun.
[Note : *A. erythrobotrys* may only be a form of *A. africanum. Keay & Savory* FHI 25024 ! from Ijua, Ogoja Prov., S. Nigeria, is intermediate—J.K.M.]

26. STACHYS Linn.—F.T.A. 5 : 465.

Stout erect or scrambling herbs ; calyx 7–10 mm. long ; corolla about 10 mm. long :
  Leaves abruptly cuneate at base into a petiole 1 cm. or more long, ovate, acute at apex, about 5·5 cm. long and 4·5 cm. broad, unevenly and rather coarsely crenate-serrate, softly pubescent above, densely ferrugineous-hairy on veins beneath ; stems softly and densely pubescent, stout, square ; inflorescences terminal, contiguous, with numerous, crowded many-flowered whorls about 3·5 cm. diam. in fruit forming a dense pyramidal spike, with small leafy bracts which are densely pubescent with long eglandular and short glandular hairs ; bracteoles small, subulate, pubescent ; mature calyx 1 cm. long, campanulate, teeth lanceolate, ½ as long as the tube, densely pubescent ; corolla about 1 cm. long   ..   ..   1. *pyramidalis*
  Leaves cordate at base, ovate-triangular, acute at apex, up to 5·5 cm. long and 4·5 cm. broad but usually less, evenly crenate, pilose above, whitish-pubescent especially on the veins beneath ; petiole as long as the lamina, pilose ; flowers in distant several-flowered whorls up to 2 cm. diam. in fruit, with leafy bracts ; stems with stiff prickly bristles and short glandular hairs ; mature calyx about 7 mm. long, campanulate, teeth lanceolate, ⅓ to ½ as long as the tube, densely pubescent ; corolla about 1 cm. long   ..   ..   2. *aculeolata*
Small annual, erect, branched herb 10–25 cm. high ; stem hirsute ; leaves 1·5–3 cm. long, ovate, truncate or cordate, crenate-serrate, petiolate ; inflorescence a very lax spike, much interrupted below ; whorls 2–6-flowered with foliose bracts which are smaller and subsessile above ; calyx 4–6 mm. long, tubular-campanulate, hirsute with triangular mucronate teeth ; corolla 6–7 mm. long   ..   ..   3. *arvensis*

1. **S. pyramidalis** *J. K. Morton* in J. Linn. Soc. 58 : 273 (1962). A stout, erect, branched herb with square grooved, densely hairy stems, and dense many-flowered, terminal, inflorescences.
N.Nig.: Lokoja (Oct.) *Dalz.* 104 !
2. **S. aculeolata** *Hook. f.* in J. Linn. Soc. 6 : 18 (1861) ; F.T.A. 5 : 466 ; Th. C. E. Fries in Notizbl. Bot. Gart. Berl. 8 : 629, fig. 1 & 2, incl. var. *camerunensis* Th. C. E. Fries (1923). A perennial prickly herb with straggling, scrambling, rather slender stems and axillary whorls of flowers in the upper parts ; corolla white or pale pink with dark pink markings ; in openings in montane forest and in gullys in the grassland.
[Br.]**Cam.:** Cam. Mt. (Jan.-Dec.) *Maitland* 926 ! *Johnston* 71 ! *Keay* FHI 28600 ! *Morton* GC 6707 ! 7003 ! *Adams* GC 11779 ! Bamenda (Dec.-Apr.) *Maitland* 1654 ! *Richards* 5320 ! 5323 ! *Morton* K102 ! K259 ! K306 ! *Mann* 319 ! *Monod* 68 ! Also in the mountainous parts of Cameroun, Congo, Ethiopia, Uganda and Kenya.
3. **S. arvensis** (*Linn.*) *Linn.* Sp. Pl., ed. 2, 2 : 814 (1763) ; Chev. Bot. 523. *Glecoma arvensis* Linn. Sp. Pl. 578 (1753). A weed of cultivated land ; flowers pale purple.
Mali: Goundam, Timbuktu *Chev.* 2776 ! Also occurring in N. Africa and widespread in Europe.
[A specimen in the British Museum labelled " Senegal 1848 " probably belongs to another species but is too poor for identification.]

27. LEUCAS R. Br.—F.T.A. 5 : 472.

Longest tooth of the calyx on the adaxial (upper) side ; calyx pubescent ; leaves petiolate, ovate-lanceolate, with a broad, blunt acumen, up to 9 cm. long and 4 cm. broad, coarsely dentate, pubescent on both surfaces, glandular-punctate beneath ; flowers in dense axillary whorls ; stems pubescent   ..   ..   1. *martinicensis*
Longest tooth of the calyx on the abaxial (lower) side :
  Calyx strongly deflexed towards the tip, teeth short triangular mucronate, shortly pubescent ; flowers in dense axillary whorls with numerous linear-subulate bracteoles about 8 mm. long ; leaves lanceolate, cuneate at base, serrate, with an entire, acute tip, petiolate, pubescent on both surfaces ..   ..   2. *deflexa*
  Calyx straight or slightly curved, densely ciliate, teeth narrowly lanceolate, subulate :
    Leaves elliptic to elliptic-lanceolate, up to 2·3 cm. long and 1 cm. broad, crenate, more or less obtuse, tapered to base, thick, with dense stiff appressed hairs above and on the veins beneath, gland-dotted beneath ; stems pilose, more or less round ; flowers in one, rarely 2, dense, globose whorls, terminal on each branch ; mature

calyx 8–10 mm. long ; bracteoles about 7 mm. long, linear, slightly spathulate,
ciliate ..   ..   **3.** *oligocephala* subsp. *oligocephala*
Leaves linear-lanceolate to lanceolate, usually exceeding 2·5 cm. in length ; inflorescence with several distinct whorls :
Stout branched erect pubescent to woolly herb ; leaves up to 8 cm. long and 2 cm.
broad, coriaceous, prominently veined beneath ; calyx about 12 mm. long
                                            3a. *oligocephala* subsp. *bowalensis*
Slender low much-branched thinly pubescent herb ; leaves up to 5 cm. long and
6 mm. broad, neither coriaceous nor prominently veined ; calyx 8–10 mm. long
                                            3b. *oligocephala* subsp. *tenuifolia*

1. **L. martinicensis** (*Jacq.*) *Ait. f.* in Ait. Hort. Kew., ed. 2, 3 : 409 (1811) ; F.T.A. 5 : 479 ; Chev. Bot. 524·
   Clinopodium martinicense Jacq. (1760)—Select. Stirp. Amer. Hist. 173, t. 177, fig. 75 (1763). *Phlomis
   mollis* Schum. & Thonn. (1827). An erect, branched, annual aromatic herb usually about 1½–2 ft. high,
   with very small white flowers in dense axillary whorls.
   Sen.: Tambacounda (Oct.) *Roberty* 10122 ! **Gam.:** (Jan.) *Ingram* ! **Mali:** Dioura (Oct.) *Davey* 114 !
   Guin.: Soumbalako to Boulivel (Sept.) *Chev.* 18625. **S.L.:** Musaia (Dec.) *Deighton* 4469 ! **Iv.C.:** Nimba
   (Apr.) *Schnell* 5312 ! **Ghana:** Akra on coast (Dec.) *Morton* A2541 ! Lumbunga, Tamale (Dec.) *Morton*
   GC 6253 ! GC 6234 ! Yendi (Dec.) *Irvine* 2710 ! **Niger :** Niamey *Hagerup* 477 ! **N.Nig.:** Kontagora (Nov.)
   Dalz. 135 ! **S.Nig.:** Oke Ado, Oyo Prov. (Nov.) *Onochie* FHI 14506 ! **F.Po:** Moka (Dec.) *Boughey* 10 ! A
   weed of the tropics. (See Appendix, p. 461.)
2. **L. deflexa** *Hook f.* in J. Linn. Soc. 7 : 213 (1864) ; F.T.A. 5 : 487. *L. decurvata* Bak. ex Hiern (1900)—
   F.T.A. 5 : 481. A long straggling, aromatic herb with globose, axillary whorls of small white flowers.
   Ghana: Aduamoa, 1,500 ft., Kwahu (Jan.) *Irvine* 1728 ! Begoro (Jan.) *Morton* A2683 ! Vane (July)
   Harris ! **[Br.]Cam.:** Cam. Mt., 4,000–7,000 ft. (Mar., Nov.) *Mann* 1976 ! *Morton* GC 6694 ! Bamenda
   (Jan.-May) *Migeod* 399 ! *Morton* K42 ! K51 ! Nkambe (Feb.) *Hepper* 1893 ! **F.Po:**
   4,000–5,000 ft. (Feb.) *Exell* 857 ! *Boughey* GC 10803 ! Also in Angola. (See Appendix, p. 461.)
3. **L. oligocephala** *Hook. f.* subsp. **oligocephala**—in J. Linn. Soc. 7 : 213 (1864) ; F.T.A. 5 : 486. A perennial
   aromatic herb with slender, branched stems 1–4 ft. high, and with dense, terminal, or subterminal, whorls
   of white flowers ; buds densely hairy and purple-tinged.
   [Br.]Cam.: Cam. Mt., above 7,000 ft. (Nov.-Apr.) *Keay* FHI 28583 ! *Mann* 1220 ! 1987 ! *Morton* 7076 !
   7085 ! *Maitland* 446 ! Bamenda, 6,000 ft. (Apr.-May) *Maitland* 1050 *bis* ! 1452 ! Also in E. Africa and
   Nyasaland.
3a. **L. oligocephala** subsp. **bowalensis** (*A. Chev.*) *J. K. Morton* in J. Linn. Soc. 58 : 273 (1962). *L. bowalensis* A. Chev.
   in Journ. de Bot., sér. 2, 2 : 127 (1909) ; Chev. Bot. 524. A stout erect branched woody herb up to 3 ft.
   high, with several dense whorls of white flowers in each branch ; buds densely hairy and purple-tinged.
   Guin.: Dalaba, 3,000–4,000 ft. (Oct.) *Chev.* 18750 ! 20175 ! **N.Nig.:** Jos Plateau (Sept.) *Lely* P715 !
   [Br.]Cam.: Bamenda, 5,000–7,000 ft. (Sept.-Dec.) *Savory* UCI 396 ! *Adams* GC 11088 ! 11266 !
3b. **L. oligocephala** subsp. **tenuifolia** *J. K. Morton* l.c. (1962). A slender, branched, low, woody herb about
   9–12 ins. high with dense whorls of white flowers ; buds densely hairy and purple-tinged.
   N.Nig.: Jos Plateau (May, Oct.-Dec.) *Lely* 62 ! P263 ! *MacGregor* 428 ! *Kennedy* FHI 7271 ! *Hepper*
   1018 !
   [Note: intermediate plants occur where the subspecies grow together—J.K.M.]

28. LEONOTIS (Pers.) R. Br.—F.T.A. 5 : 490.

Corolla orange-brown, villous ; stems tomentellous ; leaves long-petiolate, ovate-
triangular, broadly cuneate at base, the largest about 12 cm. long and 6 cm. broad,
softly puberulous on both surfaces, coarsely serrate ; flowers in globose whorls at the
upper nodes ; bracts crowded at base, linear, subulate, stiff, about 1 cm. long ; pedicels
very short ; calyx 1·5 cm. long, shortly pubescent, teeth subulate, stiff, equal
                                            1. *nepetifolia* var. *nepetifolia*
Corolla cream, otherwise as above ..   ..   ..   **1a.** *nepetifolia* var. *africana*

1. **L. nepetifolia** (*Linn.*) *Ait. f.* var. **nepetifolia**—in Ait. Hort. Kew., ed. 2 : 409 (1811) ; Chev. Bot. 524.
   Phlomis nepetaefolia Linn. (1753). A robust herb 4–8 ft. high, mainly found around dwellings and some-
   times cultivated, but not common.
   Guin.: Moussaia (Nov.) *Schnell* 2078 ! **S.L.:** (June) *T. Vogel* 94 ! Sherbro *Sc. Elliot* 5765 ! Makene (Apr.)
   Deighton 5500 ! Mano (Oct.) *Deighton* 2542 ! **Lib.:** Kakatown *Whyte* ! **Iv.C.:** Tabou to Bériby (Aug.)
   Chev. 199974. **N.Nig.:** Dakida, Bornu Prov. (Oct.) *Daggash* FHI 24869 ! **[Br.]Cam.:** Tiko (Feb.) *Maitland* !
   Buea (Mar.) *Morton* K661 ! Kumba (May, June) *Olorunfemi* FHI 30592 ! *Rosevear* Cam. 87/37 ! Wide-
   spread in the tropics. (See Appendix, p. 461.)
1a. **L. nepetifolia** var. **africana** (*P. Beauv.*) *J. K. Morton* in J. Linn. Soc. 58 : 275 (1962). *Phlomis africana*
   P. Beauv. Fl. Oware 2 : 82, t. 111 (1819). *P. pallida* Schum. & Thonn. (1827). *Leonotis pallida* (Schum.
   & Thonn.) Benth. (1834)—F.T.A. 5 : 491. *L. africana* (P. Beauv.) Briq. in E. & P. Pflanzenfam. 4, 3A :
   246 (1896) ; Chev. Bot. 524. A robust herb, 2–5 ft. high, with 4-angled, pale, softly pubescent stems and
   cream, tomentose flowers 1 in. long in dense whorls, with spine-tipped calyx teeth and bracts ; common
   near dwellings and sometimes cultivated.
   Sen.: (Mar.) *Perrottet* 678 ! (Dec.) *Monod* ! **Mali:** Kobale (Feb.) *Chev.* 350. Macina *Roberty* 801 ! Legan
   (Oct.) *Roberty* 3042 ! **Lib.:** Ganta (May) *Harley* ! Cape Palmas (July) *T. Vogel* 36 ! **Ghana:** Axim road
   (Mar.) *Morton* GC 8500 ! Banda, N.W. Ashanti (Nov.) *Harris* ! Krobo (Mar.) *Gould* ! **Togo Rep.:** Misahöhe
   Baumann 47 ! Lomé *Warnecke* 288 ! **Niger :** Niamey (Oct.) *Hagerup* 541 ! **N.Nig.:** Borgu *Barter* 1053 !
   Katagum *Dalz.* 110 ! Kogigiri, Jos (Oct.) *Hepper* 1069 ! **S.Nig.:** Lagos *Rowland* ! Oban *Talbot* 2064 !
   [Br.]Cam.: Kumba *Boughey* ! Extends to the Congo, E. Africa and Nyasaland. (See Appendix, p. 461.)

29. TINNEA Kotschy & Peyr.—F.T.A. 5 : 496 ; Robyns in
Bull. Jard. Bot. Brux. 8 : 161 (1930).

Leaves sessile or subsessile, ovate-orbicular to broadly ovate, mucronate, entire, up to
5 cm. long, tomentellous or pubescent beneath, densely glandular on both surfaces ;
stems tomentellous ; upper bracteate leaves markedly mucronate, small, about 1 cm.
long ; calyx with 2 entire orbicular lips about 7 mm. long, at length splitting to the
base, tomentellous outside ; corolla-tube pubescent, longer than the calyx ; calyx
much enlarged and bladder like in fruit, thick and brittle, about 11 mm. diam.
                                                                            1. *barteri*

Fig. 313.—Leonotis nepetifolia (*Linn.*) *Ait.* f. var. africana (*P. Beauv.*) *J. K. Morton* (Labiatae).

A, flower.   B, stamen.   C, ovary.

471

S.R.C

Fig. 314.—Tinnea barteri *Gürke* (Labiatae).

A, flower. B, corolla laid open. C, pistil. D, ovary. E, stamen. F, calyx in fruit. G, same in section showing fruit. H, nutlet with part of hairs removed. I, hair from nutlet.

472

Leaves petiolate, elliptic-lanceolate, apiculate, entire, about 3 cm. long and 1·5 cm. broad, glabrous except on the midrib, densely glandular-punctate ; stems tomentellous ; calyx as above but thinly pubescent, larger and membranous in fruit, about 2 cm. diam. .. ..   2. *aethiopica*

1. **T. barteri** *Gürke* in Engl. Bot. Jahrb. 28 : 314 (1900) ; F.T.A. 5 : 499 ; Chev. Bot. 524 ; Robyns l.c. 192. An erect, little-branched half-woody herb with round stems up to 5 ft. high, and lax terminal spikes of purple flowers ½ in. long.
Mali: Bougouni *Roberty* 3225 ! Bamako to Zougouni (Sept.) *Jaeger* 5130 ! **Guin.:** Kouroussa (Aug.) *Pobéguin* 331 ! **Ghana:** Prang (June) *Vigne* FH 3910 ! Gonja (fr. Dec.) *Morton* GC 25006 ! Wa (June) *Adams* 860 ! Walewale to Gambaga (Nov.) *Morton* A2716 ! **N.Nig.:** Nupe *Barter* 971 ! 1261 ! Jebba to Bida (Aug.) *Onochie* FHI 35101 ! banks of Gurara R. *Elliott* 207 ! Zungeru (Aug.) *Dalz.* 180 ! Jos Plateau (Sept.) *Lely* 638 ! Also in Congo. (See Appendix, p. 464.)

2. **T. aethiopica** *Kotschy & Peyr.* Plant. Tinn. 25, t. 11 (1867) ; F.T.A. 5 : 497 ; Bot. Mag. tt. 5637 & 6744 ; Robyns l.c. 182. A shrub, up to 8–10 ft. high, with dusky purple flowers ½ in. long.
Mali: Bamako (Jan.) *Roberty* 10382 ! **N.Nig.:** Katagum *Dalz.* 108 ! Kachia to Kaduna (Nov.) *Meikle* 1208 ! Jos Plateau (fl. May, fr. Dec.) *Monod* 9761 ! *Lely* P256 ! *Batten-Poole* 199 ! Zelau, 3,200 ft. *Lely* 112 ! Sara Hills, Bauchi Div. (Oct.) *Hepper* 1118 ! [Br.]**Cam.:** Vogel Peak (Nov.) *Hepper* 1371 ! Extends to the Congo, Sudan, E. Africa and Nyasaland. (See Appendix, p. 464.)

KEY TO THE FAMILIES OF DICOTYLEDONS

By J. E. Dandy

(This key is specially designed to cover plants represented in this Flora. The basic arrangement is that of Hutchinson, *The Families of Flowering Plants*, ed. 2, vol. I, *Dicotyledons*, Oxford, 1959.)

Gynoecium composed of 2 or more free or almost free carpels with separate styles and stigmas :
Gynoecium composed of 1 carpel or of 2 or more united carpels with free or united styles and stigmas, or if carpels free below then the styles or stigmas united into 1 :
  Ovules 2 or more in the gynoecium, attached to the wall or walls of the ovary (placentation parietal) :
    Ovary superior :
    Ovary inferior or semi-inferior :
  Ovules 1 or more in the gynoecium, if more than 1 then attached to the central axis or the base or apex of the ovary :
    Ovary superior :
    Ovary inferior or semi-inferior :

Group 1

Leaves opposite, simple ; stamens and carpels as many as the petals ; herbs, often succulent .. .. **24. Crassulaceae** (1 : 114)
Leaves alternate:
  Sepals 6–18 in 2 or more series, free or the inner ones united ; flowers dioecious, small ; leaves simple or sometimes 3-foliolate ; woody climbers or sometimes erect shrubs or small trees .. .. **12. Menispermaceae** (1 : 66)
  Sepals 5 or fewer in 1 series, free or united, sometimes forming a spathaceous calyx :
    Leaves all or mostly compound with 3 or more leaflets :
      Stamens numerous, more than twice as many as the petals ; carpels indefinite in number, often numerous ; leaves with stipules or sheathing at the base ; leaflets toothed or lobed :
        Shrubs, erect or scrambling, with prickles ; sepals united below ; petals and stamens perigynous, inserted at the mouth of the calyx-tube ; carpels with 2 ovules, drupaceous in fruit .. .. **87. Rosaceae** (1 : 423)
        Herbs, without prickles ; sepals free ; petals and stamens hypogynous, not inserted on the calyx ; carpels with 1 ovule, achenial in fruit **9. Ranunculaceae** (1 : 62)
      Stamens as many or twice as many as the petals ; carpels 2–5 ; leaves without stipules, not sheathing at the base ; leaflets entire or crenate ; trees or shrubs, sometimes climbing :
        Flowers bisexual ; stamens twice as many as the petals ; leaves pinnate or 1–3-foliolate .. .. **122. Connaraceae** (1 : 739)
        Flowers unisexual or polygamous ; stamens as many as the petals ; leaves pinnate :
          Leaflets dotted with pellucid glands ; carpels with 2 ovules ; branches armed with prickles .. .. **114. Rutaceae** (1 : 683)
          Leaflets without pellucid glands ; carpels with 1 ovule ; branches unarmed **115. Simaroubaceae** (1 : 689)

Leaves simple or 1-foliolate, sometimes lobed or deeply divided :
  Sepals valvate or forming a spathaceous calyx ; trees or shrubs, sometimes climbing :
    Leaves without stipules ; petals 6 in 2 series, or 6 or 4 or 3 in 1 series ; anthers
    commonly with a prolongation of the connective above the thecae
                                  4. Annonaceae (1 : 34)
    Leaves with stipules ; petals 5 in 1 series ; anthers without a prolongation of the
    connective above the thecae :
      Flowers bisexual, with an androgynophore bearing the stamens and carpels at its
      apex ; stamens accompanied on the inside by 5 petaloid staminodes
                                  75. Sterculiaceae (1 : 310)
      Flowers unisexual, without an androgynophore ; stamens not accompanied by
      petaloid staminodes .. .. **74. Tiliaceae** (1 : 300)
  Sepals imbricate, not forming a spathaceous calyx :
    Herbs ; leaves palmately lobed or deeply divided ; flowers zygomorphic ; sepals
    petaloid, the median (adaxial) one spurred at the base ; upper (adaxial) pair of
    petals spurred at the base, the spurs fitting into the calyx-spur
                                9. Ranunculaceae (1 : 62)
    Trees or shrubs, sometimes climbing ; leaves entire or toothed ; flowers actino-
    morphic ; sepals and petals not spurred :
      Stamens twice as many as the petals ; carpels with 2 ovules
                                122. Connaraceae (1 : 739)
      Stamens numerous, more than twice as many as the petals ; carpels with several
      (more than 2) ovules .. .. **49. Dilleniaceae** (1 : 179)

Group 2

Leaves opposite or the lower ones peltate ; corolla-lobes 4–5, as many as the sepals or
calyx-lobes ; corolla-tube often longer than the lobes ; herbs or shrubs, often
succulent .. .. **24. Crassulaceae** (1 : 114)
Leaves alternate, not peltate ; corolla-lobes 6, twice as many as the sepals ; corolla-
tube much shorter than the lobes ; trees or shrubs, sometimes climbing
                                4. Annonaceae (1 : 34)

Group 3

Flowers zygomorphic, unisexual, the perianth produced unilaterally into an elongated
lobe ; anthers opening by upcurving valves ; carpels sunk in the receptacle ; trees or
shrubs .. .. **5. Monimiaceae** (1 : 54)
Flowers actinomorphic ; anthers opening by slits ; carpels not sunk in the receptacle :
  Stamens united into a column or synandrium ; flowers unisexual :
    Leaves with stipules ; sepals or calyx-lobes 3–8 in 1 series ; trees or shrubs
                              75. Sterculiaceae (1 : 310)
    Leaves without stipules ; sepals 6–12 in 2 or more series ; woody climbers
                              12. Menispermaceae (1 : 66)

  Stamens free :
    Sepals united below ; stamens perigynous, inserted at the mouth of the calyx-tube ;
    leaves all or mostly compound or palmately lobed, with stipules ; shrubs or herbs,
    sometimes scrambling with prickles .. .. .. **87. Rosaceae** (1 : 423)
    Sepals free ; stamens hypogynous, not inserted on the calyx :
      Leaves all or mostly compound ; sepals more or less petaloid ; herbs or woody
      climbers or trailers .. .. **9. Ranunculaceae** (1 : 62)
      Leaves simple ; sepals not petaloid :
        Herbs with opposite or subopposite leaves ; flowers bisexual or unisexual
                              30. Molluginaceae (1 : 133)
        Woody climbers with alternate leaves ; flowers dioecious :
          Sepals 6–18 in 2 or more series ; stamens 3–6 ; flowers in clustered cymes
                              12. Menispermaceae (1 : 66)
         Sepals 5 in 1 series ; stamens 10–15 ; flowers in elongated racemes
                              35. Phytolaccaceae (1 : 142)

Group 4

Gynoecium composed of 1 carpel, thus with only 1 placenta in the ovary :
  Leaves 2-pinnate, with stipules or stipular spines :
    Petals valvate, sometimes coherent above the base ; flowers in heads or spikes or
    spike-like racemes ; trees or shrubs or herbs, sometimes climbing or aquatic
                                90. Mimosaceae (1 : 484)
    Petals imbricate ; flowers in lax or dense racemes or spikes, sometimes paniculate ;
    trees or shrubs, sometimes climbing .. .. **89. Caesalpiniaceae** (1 : 439)
  Leaves simple or 1-pinnate or 1–3-foliolate :
    Flowers zygomorphic ; leaves simple or compound ; trees or shrubs or herbs, some-
    times climbing:

Corolla papilionaceous; petals 5, the uppermost (adaxial) one outside the others in bud, the lower (abaxial) pair forming a carina enclosing the stamens and ovary
91. **Papilionaceae** (1 : 505)
Corolla not papilionaceous ; petals 1–6, the uppermost (adaxial) one often inside the others in bud, the lower ones not forming a carina 89. **Caesalpiniaceae** (1 : 439)
Flowers actinomorphic :
Leaves with stipules :
Stamens as many as and opposite to the petals, the filaments united below into a tube ; ovary with 2 ovules ; leaves simple, toothed ; herbs or shrublets
75. **Sterculiaceae** (1 : 310)
Stamens more numerous than the petals, or as many as the petals and alternating with them, free or the filaments united towards the base ; ovary with 2 or more ovules ; leaves simple or pinnate or 2-foliolate ; trees or shrubs
89. **Caesalpiniaceae** (1 : 439)
Leaves without stipules ; trees or shrubs, sometimes climbing :
Flowers bisexual; stamens as many as the petals, not accompanied by staminodes ; leaves pinnate or 3-foliolate, the leaflets dotted with pellucid glands ; branches often armed with prickles .. .. 114. **Rutaceae** (1 : 683)
Flowers unisexual ; stamens twice as many as the petals, or as many as the petals and alternating with staminodes ; leaves pinnate or 1–3-foliolate, without pellucid glands ; branches unarmed .. .. 122. **Connaraceae** (1 : 739)
Gynoecium composed of 2 or more united carpels, thus with 2 or more placentas in the ovary :
Stamens 6, tetradynamous (the outer 2 shorter than the inner 4) ; sepals 4 ; petals 4 or fewer ; ovary divided by a false septum connecting the 2 placentas ; herbs, sometimes aquatic .. .. 19. **Cruciferae** (1 : 96)
Stamens 2 or more, if 6 then not tetradynamous :
Leaves all or mostly compound :
Leaves 1–3-pinnate ; trees or shrubs :
Leaves 2–3-pinnate ; flowers zygomorphic ; stamens as many as the petals, free, alternating with staminodes .. .. .. 18. **Moringaceae** (1 : 95)
Leaves 1-pinnate ; flowers actinomorphic ; stamens twice as many as the petals, the filaments united into a tube, not accompanied by staminodes
118. **Meliaceae** (1 : 697)
Leaves digitate, or mostly digitate and some 1-foliolate :
Styles 3, free or united at the base ; ovary subsessile ; stamens as many as the petals, surrounded by a fimbriate corona ; flowers actinomorphic ; climbing shrubs with tendrils .. .. 58. **Passifloraceae** (1 : 199)
Style 1 or stigma sessile ; ovary often borne on an elongated gynophore ; stamens numerous or few, sometimes accompanied by staminodes but not surrounded by a fimbriate corona ; flowers actinomorphic or zygomorphic ; trees or shrubs or herbs, sometimes climbing, without tendrils .. 17. **Capparidaceae** (1 : 86)
Leaves simple (sometimes lobed) or absent :
Flowers unisexual :
Leaves opposite ; stamens numerous, united into 5 spathulate bundles opposite to the petals ; trees .. .. 72. **Guttiferae** (1 : 290)
Leaves alternate :
Petals in the male flowers united into an elongated tube, free in the female flowers ; stamens twice as many as the corolla-lobes, inserted at the mouth of the corolla-tube ; trees with palmately lobed leaves .. .. 61. **Caricaeae** (1 : 220)
Petals free in both the male and female flowers :
Stamens more numerous than the petals ; trees or shrubs, sometimes climbing :
Ovary borne on an elongated gynophore ; sepals 4 ; branches armed with stipular spines .. .. 17. **Capparidaceae** (1 : 86)
Ovary sessile or shortly stipitate ; sepals 2–5 ; branches unarmed
53. **Flacourtiaceae** (1 : 185)
Stamens as many as the petals :
Petals smaller than the sepals ; leaves with 1–2 glands at the top of the petiole, the lamina sometimes palmately lobed ; herbaceous or woody climbers with tendrils .. .. 58. **Passifloraceae** (1 : 199)
Petals larger than the sepals ; leaves without glands at the top of the petiole, the lamina unlobed ; trees or shrubs, without tendrils
50. **Pittosporaceae** (1 : 181)
Flowers bisexual :
Stamens as many as the petals :
Herbaceous or woody climbers with tendrils ; flowers with a fimbriate corona outside the stamens ; leaves entire or 3-lobed .. 58. **Passifloraceae** (1 : 199)
Plants without tendrils ; flowers with or without a corona :
Style 1, elongated and unbranched :
Stamens alternating with petaloid staminodes, surrounded by an outer series of

filiform staminodes ; leaves with pectinate stipules ; herbs
                63. Ochnaceae (1 : 221)

Stamens not accompanied by staminodes :
  Connective of the anthers prolonged above the thecae into an appendage ; leaves with stipules ; flowers actinomorphic or zygomorphic ; trees or shrubs or herbs   ..   ..   **20. Violaceae** (1 : 98)
  Connective of the anthers not prolonged above the thecae ; leaves without stipules ; flowers actinomorphic ; trees or shrubs **50. Pittosporaceae** (1 : 181)

Styles 3–5, free or united below, sometimes branched :
  Leaves bearing numerous sticky stipitate glands ; styles 3, deeply 2-branched ; insectivorous herbs   ..   ..   **26. Droseraceae** (1 : 120)
  Leaves without sticky stipitate glands ; styles unbranched ; plants not insectivorous :
    Flowers with a hairy fimbriate corona outside the stamens ; stigmas capitate ; trees or shrubs   ..   ..   **58. Passifloraceae** (1 : 199)
    Flowers without a corona but sometimes with a hairy scale at the base of each petal ; stigmas not capitate :
      Herbs ; sepals united below into a tube ; seeds arillate at the base ; leaves often pinnately lobed   ..   ..   **16. Turneraceae** (1 : 85)
      Trees ; sepals free ; seeds not arillate ; leaves unlobed
                53. Flacourtiaceae (1 : 185)

Stamens more numerous than the petals :
  Flowers zygomorphic :
    Sepals 5–6 ; petals at least the upper (adaxial) ones laciniate ; fruit a capsule gaping at the apex ; leaves entire or 3-lobed ; herbs with flowers in dense terminal racemes   ..   ..   ..   **21. Resedaceae** (1 : 107)
    Sepals 4 ; petals not laciniate ; fruit indehiscent or dehiscent throughout its length by 2 valves ; leaves entire or sometimes absent ; herbs or shrubs
                17. Capparidaceae (1 : 86)

  Flowers actinomorphic :
    Leaves opposite ; herbs :
      Sepals free ; stamens numerous ; styles 2–4, free ; leaves dotted with minute pellucid glands   ..   ..   **71. Hypericaceae** (1 : 286)
      Sepals united below into a tube ; stamens 6 ; style 1, 2–4-branched at the apex ; leaves without pellucid glands   ..   **56. Frankeniaceae** (1 : 198)
    Leaves alternate :
      Anthers opening at the top by 1–2 pores or short pore-like slits :
        Leaves palmately lobed ; seeds covered with long cottony hairs ; shrubs or shrublets   ..   ..   **52. Cochlospermaceae** (1 : 183)
        Leaves not palmately lobed ; seeds without long hairs ; trees or shrubs :
          Anthers horseshoe-shaped, opening at the top of the bend ; sepals 5, caducous, leaving large persistent glands at the base of the flower ; fruit a 2-valved bristly capsule   ..   ..   **51. Bixaceae** (1 : 183)
          Anthers straight, opening at the apex ; sepals 3 ; fruit a 5–8-valved smooth or prickly capsule ..   ..   **53. Flacourtiaceae** (1 : 185)
      Anthers opening by longitudinal slits :
        Leaves pinnately lobed ; sepals 2–3, horned at the apex, caducous ; fruit a capsule dehiscent by short valves at the top ; prickly herbs
                15. Papaveraceae (1 : 84)

        Leaves not pinnately lobed :
          Ovary borne on an elongated gynophore ; trees or shrubs, often climbing or scrambling with stipular spines ..   ..   **17. Capparidaceae** (1 : 86)
          Ovary sessile or subsessile :
            Herbs ; stamens 6 ; flowers in racemes ..   **77. Capparidaceae** (1 : 86)
            Trees or shrubs, sometimes climbing ; stamens 10 or more :
              Stamens 10, twice as many as the petals ; seeds few, large, disciform, with an annular wing, developing in the open capsule ; leaves often 2-hooked at the apex, or sometimes reduced to the midrib bearing stipitate glands ; climbing shrubs ..   ..   **54. Dioncophyllaceae** (1 : 191)
              Stamens numerous, more than twice as many as the petals ; seeds unwinged, developing within the closed capsular or indehiscent fruit ; leaves without apical hooks, not reduced to the midrib ; erect shrubs or trees :
                Stamens arranged in bundles opposite to the petals ; petals twice as many as the sepals ; styles 3–4 ..   **55. Samydaceae** (1 : 194)
                Stamens not arranged in bundles ; petals as many as the sepals or more numerous ; styles 1–5 :
                  Flowers with a fimbriate or hairy corona outside the stamens
                       58. Passifloraceae (1 : 199)
                  Flowers without a corona but sometimes with a hairy scale at the base of each petal   ..   ..   ..   **53. Flacourtiaceae** (1 : 185)

Group 5

Gynoecium composed of 1 carpel, thus with only 1 placenta in the ovary ; leaves compound, with stipules or stipular spines ; flowers in heads or spikes or spike-like racemes :

Leaves 3-foliolate ; corolla zygomorphic, papilionaceous, with imbricate lobes, the uppermost (adaxial) lobe outside the others in bud, the lowest (abaxial) pair of lobes forming a carina enclosing the stamens and ovary ; herbs

91. Papilionaceae (1 : 505)

Leaves 2-pinnate ; corolla actinomorphic, with valvate lobes ; trees or shrubs or herbs, sometimes climbing .. .. 90. Mimosaceae (1 : 484)

Gynoecium composed of 2 or more united carpels, thus with 2 or more placentas in the ovary ; leaves simple or 1-pinnate, without stipules, sometimes reduced to scales :

Flowers zygomorphic ; stamens fewer than the corolla-lobes, inserted on the corolla-tube :

Leaves reduced to scales ; fleshy herbs parasitic on roots ; stamens 4

154. Orobanchaceae (2 : 374)

Leaves (or solitary leaf) well developed ; plants not parasitic :

Trees ; leaves pinnate ; stamens 4 ; fruit large, subcylindric, indehiscent

157. Bignoniaceae (2 : 383)

Herbs, sometimes epiphytic ; leaves (or solitary leaf) simple ; stamens 2 or 4 ; fruit a capsule .. .. 156. Gesneriaceae (2 : 381)

Flowers actinomorphic ; stamens as many as the corolla-lobes or more numerous :

Flowers unisexual, the petals united into a tube in the male flowers but often free in the female flowers ; leaves palmately lobed ; trees .. .. 61. Caricaceae (1 : 220)

Flowers bisexual ; leaves not palmately lobed :

Stamens more numerous than the corolla-lobes, not inserted on the corolla; corolla-lobes twice as many as the sepals or calyx-lobes or more numerous ; trees or shrubs with alternate leaves :

Corolla-lobes 6 in 1 series or in 2 unequal series ; anthers subsessile, with a broad prolongation of the connective above the thecae ; ovary with numerous ovules and sessile radiating stigmas .. .. 4. Annonaceae (1 : 34)

Corolla-lobes more than 10 in 3–4 irregular series ; anthers borne on elongated filaments, without a prolongation of the connective above the thecae ; ovary with 4 ovules and a 2-branched style .. 128. Hoplestigmataceae (2 : 15)

Stamens as many as the corolla-lobes, inserted on the corolla-tube ; corolla-lobes as many as the sepals or calyx-lobes :

Leaves alternate, the lamina cordate-orbicular ; corolla-lobes induplicate-valvate ; aquatic herbs .. .. 141. Menyanthaceae (2 : 302)

Leaves opposite, sometimes reduced to scales ; corolla-lobes contorted ; plants terrestrial :

Trees or shrubs, often climbing, sometimes with tendrils ; fruit a large berry

134. Apocynaceae (2 : 51)

Herbs, sometimes saprophytic with the leaves reduced to scales ; fruit a septicidal capsule .. .. 140. Gentianaceae (2 : 297)

Group 6

Leaves opposite ; fruit a circumscissile capsule ; herbs   31. Ficoidaceae (1 : 135)

Leaves alternate ; fruit not circumscissile :

Gynoecium composed of 1 carpel, thus with only 1 placenta in the ovary ; leaves pinnate or 1–3-foliolate, with stipules ; trees or shrubs   89. Caesalpiniaceae (1 : 439)

Gynoecium composed of 2 or more united carpels, thus with 2 or more placentas in the ovary :

Herbs ; leaves pinnate ; fruit a 2-valved silique .. .. 19. Cruciferae (1 : 96)

Trees or shrubs ; leaves simple :

Perianth absent ; flowers in catkins, unisexual ; seeds with a basal tuft of long fine hairs .. .. 93. Salicaceae (1 : 588)

Perianth present ; flowers not in catkins ; seeds without a tuft of long fine hairs :

Ovary borne on an elongated gynophore ; stigma sessile or terminating a short style .. .. 17. Capparidaceae (1 : 86)

Ovary sessile :

Style 1 ; sepals more or less united below ; stamens perigynous, inserted at the mouth of the calyx-tube ; flowers bisexual ; branches unarmed

55. Samydaceae (1 : 194)

Styles 2–6, free or united at the base ; sepals free ; stamens hypogynous, not inserted on the calyx ; flowers bisexual or unisexual ; branches often armed with spines .. .. 53. Flacourtiaceae (1 : 185)

Group 7

Flowers unisexual ; stamens 3–5, the anthers straight or flexuous or folded ; climbing or trailing herbs or shrubs with tendrils ; leaves often palmately lobed or divided
59. **Cucurbitaceae** (1 : 204)

Flowers bisexual ; stamens with straight anthers ; plants without tendrils :
Leaves absent or reduced to scales ; succulent shrubs, sometimes epiphytic
62. **Cactaceae** (1 : 221)

Leaves present, well developed, not reduced to scales :
Aquatic herbs ; leaves peltate, cordate, long-petiolate ; flowers solitary, long-pedunc10·late, large and conspicuous ; petals and stamens numerous ; ovary multilocular
11. **Nymphaeaceae** (1 : 65)

Trees or shrubs ; leaves not peltate ; flowers grouped in inflorescences ; petals 5–8, persistent and often accrescent in fruit ; stamens arranged singly or in bundles opposite to the petals ; ovary 1-locular ..   ..   55. **Samydaceae** (1 : 194)

Group 8

Flowers unisexual ; leaves alternate, often lobed or divided or sometimes digitate or pedate ; stamens 3–5, the anthers straight or curved or flexuous or folded ; climbing or trailing herbs or shrubs with tendrils ..   ..   59. **Cucurbitaceae** (1 : 204)
Flowers bisexual ; leaves opposite or verticillate, simple, entire, with interpetiolar stipules ; stamens with straight anthers ; trees or shrubs, sometimes climbing but without tendrils   ..   ..   137. **Rubiaceae** (2 : 104)

Group 9

Herbs or shrubs, often climbing, with alternate simple leaves ; perianth with an elongated tube inflated below, the mouth 2–6-lobed or produced unilaterally into a single lobe ; stamens 6–24, the anthers adnate to the style   ..   13. **Aristolochiaceae** (1 : 77)

Group 10

Ovary 1-locular, sometimes septate towards the base :
Leaves compound :
Flowers bisexual ; leaves 1–2-pinnate, with stipules :
Herbs ; corolla papilionaceous, the uppermost (adaxial) petal outside the others in bud, the lowest pair forming a carina enclosing the stamens and ovary ; stamens diadelphous, the uppermost (adaxial) one free, the filaments of the others united into a sheath ; leaves 1-pinnate ..   ..   91. **Papilionaceae** (1 : 505)
Trees ; corolla not papilionaceous, the uppermost (adaxial) petal inside the others in bud, the lower ones not forming a carina ; stamens free or the filaments shortly united at the base ; leaves 1–2-pinnate ..   89. **Caesalpiniaceae** (1 : 439)
Flowers unisexual ; leaves 1-pinnate or 3-foliolate, without stipules ; trees or shrubs, sometimes climbing :
Ovary with 2 ovules ; leaflets dotted with pellucid glands 114. **Rutaceae** (1 : 683)
Ovary with 1 ovule ; leaflets without pellucid glands 121. **Anacardiaceae** (1 : 726)
Leaves simple :
Leaves opposite or verticillate :
Herbs ; ovary with 3 or more ovules on a free-basal or free-central placenta :
Leaves with stipules   ..   ..   29. **Caryophyllaceae** (1 : 129)
Leaves without stipules :
Petals and stamens perigynous, inserted on the calyx-tube ; style 1
47. **Lythraceae** (1 : 163)
Petals and stamens hypogynous or subhypogynous, not inserted on the calyx ; styles 2–6   ..   ..   29. **Caryophyllaceae** (1 : 129)
Trees or shrubs ; ovary with 1–2 ovules :
Flowers unisexual or polygamous ; styles 3, free or united at the base ; stamens as many as and alternating with the petals or twice as many, free
121. **Anacardiaceae** (1 : 726)

Flowers bisexual ; style 1 :
Leaves with stipules ; stamens as many as and opposite to the petals, free ; ovules 1–2, erect from the base of the ovary ..   112. **Rhamnaceae** (1 : 667)
Leaves without stipules ; stamens numerous, united into 5 bundles fused to form a 5-lobed tube surrounding the ovary ; ovule 1, pendulous from the apex of the ovary ..   ..   71. **Hypericaceae** (1 : 286)
Leaves alternate :
Leaves peltate or subpeltate ; woody climbers with small dioecious flowers
12. **Menispermaceae** (1 : 66)

Leaves not peltate :
Stamens as many as and opposite to the petals ; trees or shrubs, sometimes climbing :

Stamens alternating with as many or fewer staminodes :
　Petals imbricate ; staminodes much longer than the petals ; ovary with 6–8
　　ovules pendulous from the apex of a free-central axis ; styles 3–4, short
　　　　　　　　　　　　　　　　　　　　107. **Medusandraceae** (1 : 652)
　Petals valvate ; staminodes shorter than the petals ; ovary with 3 ovules
　　pendulous from the apex of a free-basal placenta ; style 1
　　　　　　　　　　　　　　　　　　　　　　104. **Olacaceae** (1 : 644)
Stamens not accompanied by staminodes :
　Leaves with stipules ; ovary with 1–2 basal ovules　112. **Rhamnaceae** (1 : 667)
　Leaves without stipules :
　　Petals imbricate ; flowers unisexual or polygamous　130. **Myrsinaceae** (2 : 30)
　　Petals valvate ; flowers bisexual :
　　　Stamens united into a tube surrounding the ovary ; anthers opening by down-
　　　　curving valves ; ovary with 2–3 ovules pendulous from the apex of a free-
　　　　basal placenta ; fruit surrounded by the persistent accrescent calyx
　　　　　　　　　　　　　　　　　　　　　　104. **Olacaceae** (1 : 644)
　　　Stamens free ; anthers opening by longitudinal slits ; ovary with 1 ovule
　　　　pendulous from the apex of a free-basal placenta ; fruit not surrounded by a
　　　　persistent accrescent calyx　..　..　..　..　106. **Opiliaceae** (1 : 651)
Stamens as many as and alternating with the petals, or more numerous or fewer :
　Ovary with 1 ovule :
　　Perianth zygomorphic, the lowest (abaxial) petal or pair of petals forming a
　　　carina enclosing the stamens and ovary :
　　　Herbs ; leaves with stipules ; carina composed of 2 coherent petals ; fruit not
　　　　winged ..　..　..　..　..　..　91. **Papilionaceae** (1 : 505)
　　　Trees or shrubs, sometimes climbing ; leaves without stipules ; carina composed
　　　　of 1 petal ; fruit with an elongated subterminal wing
　　　　　　　　　　　　　　　　　　　　　22. **Polygalaceae** (1 : 108)
　　Perianth actinomorphic, the petals not forming a carina :
　　　Leaves with stipules ; ovary with 3 subsessile stigmas ; herbs
　　　　　　　　　　　　　　　　　　　　　34. **Illecebraceae** (1 : 142)
　　　Leaves without stipules ; ovary with 1–3 styles :
　　　　Flowers bisexual, in involucrate heads or solitary or few together in the axils
　　　　　of the leaves ; petals and stamens perigynous, inserted high up on the
　　　　　elongated calyx-tube ; stamens twice as many as the petals, not accom-
　　　　　panied by staminodes ; herbs or shrubs or trees
　　　　　　　　　　　　　　　　　　　　46. **Thymelaeaceae** (1 : 171)
　　　　Flowers unisexual or polygamous, in lax or dense panicles ; petals and stamens
　　　　　hypogynous, not inserted on the calyx ; stamens as many or twice as many
　　　　　as the petals, or reduced to 1–2 and accompanied by staminodes ; trees or
　　　　　shrubs　..　..　..　..　..　..　121. **Anacardiaceae** (1 : 726)
　Ovary with 2 or more ovules :
　　Leaves minute, scale-like ; ovary with 3 short styles and numerous ovules on
　　　basal placentas ; trees or shrubs　..　..　57. **Tamaricaceae** (1 : 198)
　　Leaves well developed, not scale-like ; ovary with 1–3 styles :
　　　Petals valvate ; trees or shrubs, sometimes climbing :
　　　　Stamens as many as the petals, not accompanied by staminodes ; ovules 2,
　　　　　pendulous from the apex of the ovary ; flowers bisexual or unisexual
　　　　　　　　　　　　　　　　　　　　102. **Icacinaceae** (1 : 636)
　　　　Stamens more numerous than the petals, or fewer than the petals and accom-
　　　　　panied by staminodes ; ovules 2 or more on a basal or free-basal placenta ;
　　　　　flowers bisexual :
　　　　　Ovules 6 on a basal placenta ; stamens twice as many as the petals ; flowers
　　　　　　solitary or paired in the axils of the leaves　131. **Styracaceae** (2 : 33)
　　　　　Ovules 2–4, pendulous from the apex of a free-basal placenta ; stamens twice
　　　　　　as many as the petals, or fewer and accompanied by staminodes ; flowers
　　　　　　in axillary racemes　..　..　..　..　104. **Olacaceae** (1 : 644)
　　　Petals imbricate or contorted :
　　　　Herbs with fleshy leaves ; sepals 2, free ; flowers in lax terminal racemes or
　　　　　panicles　..　..　..　..　..　32. **Portulacaceae** (1 : 136)
　　　　Trees or shrubs, sometimes climbing ; sepals 5 or more, free or united below :
　　　　　Styles 3 ; ovules 6, pendulous from the apex of a free-central axis ; flowers in
　　　　　　axillary spikes　..　..　..　..　107. **Medusandraceae** (1 : 652)
　　　　　Style 1 ; ovules 2 or more, basal :
　　　　　　Sepals free, persistent and accrescent in fruit, 2 becoming wing-like and one
　　　　　　　of these much larger than the other ; petals and stamens hypogynous, not
　　　　　　　inserted on the calyx ; ovary central, with a terminal style and numerous
　　　　　　　ovules　..　..　..　..　..　..　63. **Ochnaceae** (1 : 221)
　　　　　　Sepals united below, not becoming wing-like in fruit ; petals and stamens
　　　　　　　perigynous, inserted at the mouth of the calyx-tube ; ovary central or

inserted laterally on the calyx-tube, with a terminal or basal style and 2 ovules .. ..   **87. Rosaceae** (1 : 423)

Ovary 2- or more-locular, completely or almost completely septate :

Stamens as many as and opposite to the petals :

Filaments united at least below into a tube or cup and sometimes alternating with staminodes :

Leaves with stipules ; anthers opening by longitudinal slits ; fruit a capsule, not surrounded by a persistent calyx ; shrubs or herbs, sometimes climbing
75. Sterculiaceae (1 : 310)

Leaves without stipules ; anthers opening by downcurving valves ; fruit a drupe surrounded by the persistent accrescent calyx ; trees or shrubs
104. Olacaceae (1 : 644)

Filaments free, sometimes very short or absent, not alternating with staminodes :

Ovary with 4–5 ovules in each loculus ; fruit a loculicidal capsule ; herbs
75. Sterculiaceae (1 : 310)

Ovary with 1–2 ovules in each loculus ; fruit a berry or drupe :

Ovules 2 in each loculus ; inflorescences leaf-opposed, sometimes pseudo-axillary ; leaves simple (sometimes palmately lobed) or digitate or pedate, sometimes caducous ; herbs or shrubs, often climbing or trailing with tendrils
113. Ampelidaceae (1 : 672)

Ovule 1 in each loculus ; inflorescences axillary ; leaves simple, undivided ; trees or shrubs without tendrils :

Leaves with stipules or stipular spines ; ovules erect from the base of the loculi ; fruit not enclosed by an accrescent receptacle   **112. Rhamnaceae** (1 : 667)

Leaves without stipules or stipular spines ; ovules pendulous from the apex of the loculi or central axis ; fruit wholly or partly enclosed by the accrescent receptacle .. .. **104. Olacaceae** (1 : 644)

Stamens as many as and alternating with the petals, or more numerous or fewer :

Leaves compound with 2 or more leaflets :

Inflorescences bearing tendrils ; leaves pinnate or biternate ; climbing shrubs or herbs .. .. **119. Sapindaceae** (1 : 709)

Inflorescences without tendrils :

Anthers 1-thecous ; leaves digitate ; trees with large flowers
76. Bombacaceae (1 : 332)

Anthers 2-thecous :

Styles 3–5 :

Herbs ; flowers bisexual ; petals contorted ; ovary with 1 or more ovules in each loculus ; fruit capsular ; leaves pinnate or 3-foliolate
40. Oxalidaceae (1 : 157)

Trees or shrubs ; flowers unisexual or polygamous ; petals imbricate or valvate ; ovary with 1 ovule in each loculus ; fruit drupaceous ; leaves pinnate
121. Anacardiaceae (1 : 726)

Style 1 or stigma sessile :

Leaves opposite, pinnate or 2–3-foliolate, with stipules or stipular spines ; herbs or shrublets ; stamens twice as many as the petals
85. Zygophyllaceae (1 : 361)

Leaves alternate ; trees or shrubs, sometimes climbing :

Leaves with intrapetiolar stipules, pinnate ; flowers zygomorphic, in racemes ; stamens 4–6 .. .. **120. Melianthaceae** (1 : 725)

Leaves without stipules ; flowers actinomorphic or zygomorphic :

Filaments united into an entire or lobed tube, the anthers included within the tube or exserted from it ; leaves 1–3-pinnate   **118. Meliaceae** (1 : 697)

Filaments free :

Leaflets dotted with pellucid glands ; leaves pinnate or digitately 3–5-foliolate .. .. **114. Rutaceae** (1 : 683)

Leaflets without pellucid glands :

Ovary with 2 ovules in each loculus :

Leaves paripinnate ; flowers unisexual ..   **119. Sapindaceae** (1 : 709)

Leaves imparipinnate or 3-foliolate ; flowers bisexual or unisexual
117. Burseraceae (1 : 694)

Ovary with 1 ovule in each loculus :

Ovules erect or ascending from the middle or below the middle of the loculi ; leaves paripinnate or 2–3-foliolate ; flowers unisexual
119. Sapindaceae (1 : 709)

Ovules pendulous from the apex or upper part of the loculi :

Leaves 2-foliolate ; branches often armed with straight spines ; flowers bisexual .. .. **85. Zygophyllaceae** (1 : 361)

Leaves 3- or more-foliolate ; flowers bisexual or unisexual :

Filaments each with a hairy scale-like appendage at the base ; branches unarmed or armed with hooked prickles **115. Simaroubaceae** (1 : 689)

Filaments without a scale-like appendage ; branches unarmed :
Petals 5, hairy ; stamens 10, with hairy filaments
            115. **Simaroubaceae** (1 : 689)
Petals 4, glabrous ; stamens 8, with glabrous filaments
            121. **Anacardiaceae** (1 : 726)
Leaves simple or 1-foliolate, sometimes lobed or deeply divided :
Flowers strongly zygomorphic, the median sepal petaloid and produced dorsally into a spur ; median petal free ; lateral petals united into 2 pairs, the pairs free from each other or sometimes coherent ; stamens 5, the anthers united round the ovary ; fruit an explosively dehiscent capsule ; herbs ..   41. **Balsaminaceae** (1 : 159)
Flowers actinomorphic or zygomorphic, the sepals not spurred ; petals all free :
Leaves opposite or verticillate, not fasciculate :
Stamens more than twice as many as the petals :
Leaves with free or interpetiolar stipules :
Trees or shrubs ; petals laciniate at the apex ; stamens 15–45, the filaments not united into bundles   ..   ..   ..   70. **Rhizophoraceae** (1 : 281)
Herbs ; petals not laciniate ; stamens 15, the filaments united below into 5 bundles of 3   ..   ..   39. **Geraniaceae** (1 : 156)
Leaves without stipules :
Petals and stamens perigynous, inserted on the calyx-tube ; stamens free ; leaves without glands ; herbs or shrublets   .. 42. **Lythraceae** (1 : 163)
Petals and stamens hypogynous, not inserted on the calyx ; stamens free or united into 2–5 bundles ; leaves often dotted or streaked with pellucid or opaque glands :
Styles 4–5, free or united in the lower part ; trees or shrubs or herbs ; flowers bisexual   ..   ..   71. **Hypericaceae** (1 : 286)
Style 1 or stigma sessile ; trees or shrubs :
Fruit a capsule ; style elongated, the stigma small and subentire ; flowers bisexual ; leaves sessile   ..   ..   71. **Hypericaceae** (1 : 286)
Fruit a berry ; style elongated or short or absent, the stigma disciform or 2–5-lobed ; flowers bisexual or unisexual ; leaves petiolate or sometimes subsessile   ..   ..   72. **Guttiferae** (1 : 290)
Stamens twice as many as the petals or fewer :
Sepals united below into a tube :
Petals and stamens inserted on the calyx-tube ; ovary with numerous ovules in each loculus ; herbs or shrubs or trees ; leaves without stipules :
Anthers opening by an apical pore ; connective often appendaged below the anther ; leaves with 3 or more parallel longitudinal nerves
            68. **Melastomataceae** (1 : 245)
Anthers opening by longitudinal slits ; connective unappendaged ; leaves without parallel longitudinal nerves   ..   .. 42. **Lythraceae** (1 : 163)
Petals and stamens not inserted on the calyx ; ovary with 2–4 ovules in each loculus ; trees or shrubs, sometimes climbing :
Stamens 10, twice as many as the petals, the filaments free or united into a tube ; leaves with interpetiolar stipules ..   70. **Rhizophoraceae** (1 : 281)
Stamens 3, fewer than the petals, the filaments free ; leaves with or without small free stipules   ..   ..   100. **Celastraceae** (1 : 623)
Sepals free or almost free :
Ovary with 1 ovule in each fertile loculus (sometimes 1 or 2 of the loculi empty) :
Shrubs, often climbing ; ovary 3-locular, sometimes with 1 or 2 empty loculi ; styles 3, elongated ; fruit winged, indehiscent or separating into 2–3 cocci
            78. **Malpighiaceae** (1 : 350)
Herbs ; ovary 2-locular, without empty loculi ; styles 2, short ; fruit not winged, separating into 2 cocci   ..   30. **Molluginaceae** (1 : 133)
Ovary with 2 or more ovules in each loculus :
Leaves palmately lobed or deeply divided, with stipules ; ovary beaked, the beak surmounted by a 5-branched style ; herbs   39. **Geraniaceae** (1 : 156)
Leaves not lobed or divided ; ovary unbeaked :
Styles 3–5, free :
Leaves with stipules ; sepals entire ; ovules numerous in each loculus ; herbs or shrublets, sometimes aquatic ..   28. **Elatinaceae** (1 : 127)
Leaves without stipules ; sepals toothed or lobed at the apex ; ovules 2 in each loculus ; annual herbs   ..   ..   84. **Linaceae** (1 : 358)
Style 1 or stigma sessile :
Stamens as many as or fewer than the petals ; trees or shrubs, often climbing   ..   ..   100. **Celastraceae** (1 : 623)
Stamens twice as many as the petals :
Trees with flowers in short terminal panicles ; leaves with interpetiolar stipules ; ovary 2-locular with a 2-branched style ; seeds with a fibrous pectinate aril   ..   ..   ..   83. **Ctenolophonaceae** (1 : 357)

Herbs or small shrublets with solitary axillary flowers ; leaves with free
stipules or stipular spines ; ovary 4–5-locular with an unbranched style ;
seeds without an aril   ..   ..   ..   85. **Zygophyllaceae** (1 : 361)

Leaves alternate, sometimes fasciculate :
Ovary with 1 ovule in each fertile loculus (sometimes 1 or 2 of the loculi empty) :
Flowers unisexual or polygamous :
Leaves without stipules ; petals 4 ; shrubs with small flowers in narrow spike-
like or raceme-like panicles ..   ..   ..   119. **Sapindaceae** (1 : 709)
Leaves with stipules ; petals 5–6 or sometimes reduced to 3 :
Calyx cupular, open in bud ; style-branches 3–4, entire ; fruit a drupe ;
stamens twice as many as the petals ; trees   101. **Pandaceae** (1 : 634)
Calyx composed of 5 sepals or splitting into 2–5 valvate lobes ; styles or style-
branches 2 or more, lobed or divided ; fruit a capsule ; stamens as many as
the petals or more numerous ; trees or shrubs or herbs, sometimes climbing
86. **Euphorbiaceae** (1 : 364)

Flowers bisexual :
Anthers 1-thecous ; stamens numerous, the filaments more or less united into a
tube ; sepals valvate, with or without an epicalyx of bracteoles ; herbs or
shrublets, often with stellate hairs ..   ..   ..   77. **Malvaceae** (1 : 335)
Anthers 2-thecous :
Leaves with stipules ; trees or shrubs :
Petals and stamens perigynous, inserted at the mouth of the calyx-tube ;
ovary lateral, inserted on one side of the calyx-tube ; style arising laterally
from the base of the ovary   ..   ..   87. **Rosaceae** (1 : 423)
Petals and stamens hypogynous, not inserted on the calyx ; ovary central ;
style or styles central or terminal on the ovary :
Ovary deeply lobed, with 1 style, the carpels separating in fruit ; stamens
free, twice as many as the petals or more numerous
63. **Ochnaceae** (1 : 221)
Ovary entire, with 1 or 3 styles, the carpels not separating in fruit ; stamens
free or the filaments united below :
Filaments united below into a tube or cup, twice as many as the petals ;
styles or style-branches 3 ; ovary with 1 fertile loculus and 2 empty
loculi   ..   ..   81. **Erythroxylaceae** (1 : 355)
Filaments free ; style 1, unbranched ; ovary with 2–6 fertile loculi :
Shrubs with spiny branches ; leaves mostly fasciculate ; stipules not
leaving annular scars round the nodes ; stamens twice as many as the
petals or more numerous   ..   ..   85. **Zygophyllaceae** (1 : 361)
Trees without spines ; leaves not fasciculate ; stipules leaving annular
scars round the nodes ; stamens twice as many as the petals
116. **Irvingiaceae** (1 : 692)
Leaves without stipules :
Herbs ; sepals 4 ; petals 4 ; stamens 6 ; leaves (at least the basal ones)
more or less pinnately divided   ..   ..   19. **Cruciferae** (1 : 96)
Trees or shrubs, sometimes climbing ; sepals 4–6 ; petals 4–10 ; stamens 8
or more ; leaves entire or crenulate :
Styles 2–5 :
Stamens numerous, united into 5 bundles opposite to the petals ; styles 5 ;
fruit a berry, not winged   ..   ..   71. **Hypericaceae** (1 : 286)
Stamens twice as many as the petals, free ; styles 2–3 ; fruit winged
78. **Malpighiaceae** (1 : 350)
Style 1 :
Petals 8–10, twice as many as the sepals ; stamens as many as and alter-
nating with the petals ..   ..   ..   46. **Thymelaeaceae** (1 : 171)
Petals 4–6, as many as the sepals :
Petals valvate ; stamens twice as many as the petals or more numerous ;
ovary 3–4-locular, sometimes incompletely septate at the top
104. **Olacaceae** (1 : 644)
Petals imbricate ; stamens twice as many as the petals ; ovary 4–5-
locular, completely septate :
Leaves dotted with pellucid glands ; petals glabrous, gland-dotted ;
filaments equal   ..   ..   ..   114. **Rutaceae** (1 : 683)
Leaves without pellucid glands ; petals hairy outside, not gland-dotted ;
filaments unequal   ..   ..   ..   79. **Humiriaceae** (1 : 354)
Ovary with 2 or more ovules in each loculus :
Stamens as many as or fewer than the petals :
Ovary 5-locular :
Sepals valvate, surrounded by an epicalyx of 3 bracteoles ; stamens alter-
nating with liguliform staminodes ; herbs or shrublets with stellate hairs
75. **Sterculiaceae** (1 : 310)

Sepals imbricate, without an epicalyx ; stamens not accompanied by staminodes ; trees or shrubs :
Anthers opening by apical pores ; ovules numerous in each loculus ; seeds
winged, without an aril   ..   ..   ..   **63. Ochnaceae** (1 : 221)
Anthers opening by longitudinal slits ; ovules 2 in each loculus ; seeds
with an aril, not winged ..   ..   ..   **80. Ixonanthaceae** (1 : 355)
Ovary 2–4-locular :
Styles 2–3, free or united at the base, often lobed or branched ; flowers uni-
sexual ; trees or shrubs   ..   ..   ..   **86. Euphorbiaceae** (1 : 364)
Style 1 or stigmas sessile ; flowers bisexual or unisexual :
Herbs ; sepals valvate ; fruit covered with hooked prickles
                                          74. Tiliaceae (1 : 300)
Trees or shrubs, sometimes climbing ; sepals imbricate ; fruit without hooked
prickles :
Stamens 3, fewer than the petals ; ovules 2 or more in each loculus
                                        100. Celastraceae (1 : 623)
Stamens 4–5, as many as the petals ; ovules 2 in each loculus :
Ovules erect from the base of the loculi ..   **100. Celastraceae** (1 : 623)
Ovules pendulous from the apex or near the apex of the loculi :
Style absent, the stigmas sessile ; flowers unisexual ; petals short and
scale-like, entire ; fruit a capsule   ..   **86. Euphorbiaceae** (1 : 364)
Style present, elongated ; flowers bisexual or unisexual ; petals 2-lobed
or sometimes entire ; fruit a drupe ..   **88. Chailletiaceae** (1 : 433)
Stamens more numerous than the petals :
Leaves without stipules ; trees or shrubs, sometimes climbing :
Ovules 2 in each loculus :
Filaments united into a tube ; petals imbricate ; fruit a capsule
                                        118. Meliaceae (1 : 697)
Filaments free or shortly united at the base ; petals valvate :
Stamens twice as many as the petals ; calyx 4-lobed ; fruit drupaceous ;
flowers bisexual or unisexual ..   ..   **117. Burseraceae** (1 : 694)
Stamens numerous, more than twice as many as the petals ; calyx cupular ;
fruit a capsule or drupe ; flowers bisexual   **73. Scytopetalaceae** (1 : 299)
Ovules numerous in each loculus :
Leaves dotted with pellucid glands ; branches often armed with axillary
spines   ..   ..   **114. Rutaceae** (1 : 683)
Leaves without pellucid glands ; branches unarmed :
Petals valvate ; calyx cupular ; stamens numerous ; anthers opening by
apical pore-like slits ; flowers bisexual ..   **73. Scytopetalaceae** (1 : 299)
Petals imbricate, with thick scale-like bases ; calyx composed of 5 free
sepals ; stamens 9–13 ; anthers opening by longitudinal slits ; flowers
polygamous   ..   ..   ..   **105. Pentadiplandraceae** (1 : 649)
Leaves with stipules :
Sepals imbricate ; ovules 2 in each loculus ; trees or shrubs, sometimes
climbing :
Flowers unisexual, in short sessile catkin-like racemes or fascicles ; petals
imbricate ; petioles jointed at the top ..   **82. Lepidobotryaceae** (1 : 356)
Flowers bisexual ; petals contorted ; petioles not jointed :
Stamens numerous, more than twice as many as the petals ; fruit indehiscent,
surrounded by the 5 persistent accrescent wing-like sepals
                                        65. Dipterocarpaceae (1 : 234)
Stamens twice as many as the petals ; fruit not surrounded by persistent
wing-like sepals :
Styles 3–5 ; petals fugacious ; fruit a drupe ; branches often armed with
hooked spines   ..   ..   **84. Linaceae** (1 : 358)
Style 1 ; petals persistent ; fruit a septicidal capsule ; branches unarmed
                                        80. Ixonanthaceae (1 : 355)
Sepals or calyx-lobes valvate, sometimes with an epicalyx of bracteoles ;
ovules 2 or more in each loculus :
Anthers 1-thecous ; filaments united into a tube surrounding the ovary ;
calyx with or without an epicalyx ; trees or shrubs or herbs
                                        77. Malvaceae (1 : 335)
Anthers 2-thecous :
Stamens all fertile, not accompanied by staminodes, free or the filaments
united below into a short tube or into bundles opposite to the petals ;
trees or shrubs or herbs ..   ..   ..   **74. Tiliaceae** (1 : 300)
Stamens accompanied by staminodes :
Staminodes petaloid or liguliform or filiform, inserted inside or in the same
whorl as the stamens and often united with them to form a tube or cup ;

stamens often grouped into bundles of 2–4 alternating with the staminodes ; trees or shrubs ..   ..   ..   **75. Sterculiaceae** (1 : 310)
Staminodes filiform, inserted outside the stamens and free from them ;
stamens not grouped into bundles ; herbs or shrubs
74. Tiliaceae (1 : 300)

Group 11

Ovary 1-locular, sometimes septate towards the base :
Leaves peltate or subpeltate ; flowers dioecious, the males with united petals, the
females with 1–3 free petals ; stamens united into a synandrium ; woody climbers
12. Menispermaceae (1 : 66)
Leaves not peltate ; flowers bisexual or sometimes dioecious, all with united petals ;
stamens free or with united filaments :
Ovary with 1 ovule :
Stamens twice as many as the calyx-lobes, inserted on the elongated calyx-tube ;
corolla ring-like, inserted at the mouth of the calyx-tube ; ovule pendulous from
the apex of the ovary ; shrubs ..   ..   ..   **46. Thymelaeaceae** (1 : 171)
Stamens as many as the calyx-lobes, not inserted on the calyx ; corolla 4–5-lobed ;
ovule arising from the base of the ovary :
Leaves alternate ; stamens opposite to the corolla-lobes ; calyx covered outside
with stalked glands ; climbing shrubs ..   ..   **143. Plumbaginaceae** (2 : 306)
Leaves opposite ; stamens alternating with the corolla-lobes ; calyx without
stalked glands ; trees or shrubs   ..   ..   **103. Salvadoraceae** (1 : 644)
Ovary with 2 or more ovules :
Stamens fewer than the corolla-lobes ; flowers more or less zygomorphic :
Ovules 2 ; calyx enclosed by 2 opposite ovate bracteoles ; climbing plants with
opposite leaves ; stamens 4   ..   ..   ..   **159. Acanthaceae** (2 : 391)
Ovules numerous ; calyx not enclosed by bracteoles ; herbs, often aquatic or sub-
aquatic or sometimes epiphytic :
Stamens 2 ; corolla-tube spurred ; leaves undivided or much divided and often
bearing insectivorous bladders   ..   ..   **155. Lentibulariaceae** (2 : 375)
Stamens 4 ; corolla-tube not spurred ; leaves undivided, without bladders
153. Scrophulariaceae (2 : 352)
Stamens as many as the corolla-lobes ; flowers actinomorphic or slightly zygo-
morphic :
Stamens opposite to the corolla-lobes :
Trees or shrublets ; fruit a berry ; flowers in fascicles of 3 or more
130. Myrsinaceae (2 : 30)
Herbs ; fruit a capsule ; flowers solitary or in fascicles of 2–3
142. Primulaceae (2 : 303)
Stamens alternating with the corolla-lobes :
Ovules 4, pendulous from the apex of a 4-winged free-basal placenta ; fruit a
compressed 2-valved capsule ; flowers bisexual ; corolla-lobes and stamens 4 ;
leaves opposite ; littoral trees or shrubs   ..   **161. Avicenniaceae** (2 : 448)
Ovules 2, pendulous from the apex of the ovary ; fruit a drupe ; flowers bisexual
or unisexual ; corolla-lobes and stamens 3–5 ; leaves alternate or opposite ;
trees or shrubs, sometimes climbing ..   ..   **102. Icacinaceae** (1 : 636)
Ovary 2- or more-locular, completely or almost completely septate or the loculi free :
Corolla-lobes numerous (12 or more) ; trees or shrubs, sometimes climbing or epiphytic :
Leaves opposite ; corolla-lobes contorted ; stamens as many as the corolla-lobes, the
filaments united below into a tube or cup ; ovary with numerous ovules in each
loculus   ..   ..   **132. Loganiaceae** (2 : 34)
Leaves alternate ; corolla-lobes imbricate or valvate :
Ovary with numerous ovules in each loculus ; corolla-lobes in 1 series, much shorter
than the broad-campanulate many-ribbed tube ; stamens numerous, not accom-
panied by staminodes ; branches with foliaceous wings   **67. Lecythidaceae** (1 : 241)
Ovary with 1 ovule in each loculus ; corolla-lobes imbricate in 2 or more series, as
long as or longer than the tube ; stamens as many as and opposite to the inner
corolla-lobes and alternating with staminodes ; branches not winged
129. Sapotaceae (2 : 16)
Corolla-lobes fewer than 12 :
Stamens more numerous than the corolla-lobes :
Leaves with stipules ; trees or shrubs or herbs, sometimes climbing :
Flowers bisexual ; anthers 1-thecous ; filaments united into a tube surrounding
the ovary ; leaves simple, sometimes palmately lobed or divided
77. Malvaceae (1 : 335)
Flowers unisexual ; anthers 2-thecous ; filaments free or some or all united ;
leaves simple (often palmately lobed) or digitate   **86. Euphorbiaceae** (1 : 364)
Leaves without stipules :
Flowers zygomorphic, the lowest (abaxial) corolla-lobe forming a carina bearing

32

the stamens and enclosing the ovary ; herbs or shrublets ; ovary 2-locular with
1 ovule in each loculus   ..   ..   **22. Polygalaceae** (1 : 108)
Flowers actinomorphic, the corolla not forming a carina ; trees or shrubs :
  Anthers opening by apical pores or pore-like slits ; ovary with more than 2
    ovules in each loculus ; leaves alternate or opposite or verticillate, sometimes
    small and heath-like   ..   ..   **126. Ericaceae** (2 : 1)
  Anthers opening by longitudinal slits ; ovary with 1–2 ovules in each loculus ;
    leaves alternate :
    Leaves pinnate or 1–3-foliolate ; flowers bisexual ; corolla-lobes valvate ;
      stamens twice as many as the corolla-lobes, united into a tube
          118. Meliaceae (1 : 697)
    Leaves simple ; flowers unisexual or sometimes bisexual ; corolla-lobes imbricate
      or contorted ; stamens free or the filaments united in pairs or at the base :
      Corolla-lobes imbricate ; stamens arranged in groups opposite to the corolla-
        lobes and alternating with staminodes ; style 1, unbranched
          129. Sapotaceae (2 : 16)
      Corolla-lobes contorted ; stamens sometimes paired but not arranged in groups
        alternating with staminodes ; styles (or style-branches) 2–5
          127. Ebenaceae (2 : 2)
Stamens as many as or fewer than the corolla-lobes :
  Stamens fewer than the corolla-lobes :
    Ovary with numerous (more than 4) ovules in each loculus :
      Leaves pinnate ; stamens 4 ; fruit a loculicidal capsule with winged seeds ; trees
      or shrubs ..   ..   **157. Bignoniaceae** (2 : 383)
      Leaves simple or digitate, sometimes deeply divided or reduced to scales :
        Stamen 1 ; climbing shrubs with opposite entire leaves ; fruit a septicidal
        capsule with winged seeds   ..   ..   ..   **132. Loganiaceae** (2 : 34)
        Stamens 2–4 :
        Ovary completely or incompletely 4-chambered, each of the 2 original loculi
          becoming divided into 2 by a false septum ; flowers solitary or cymose in the
          axils of the leaves ; stamens 4 ; herbs ..   **158. Pedaliaceae** (2 : 388)
        Ovary 2-locular, the loculi not divided by false septa :
          Ovules arranged in 1–2 series on each placenta ; fruit a loculicidal capsule, the
          seeds often borne on hardened hook-like funicles (retinacula) ; leaves oppo-
          site or sometimes alternate ; herbs or shrublets **159. Acanthaceae** (2 : 391)
          Ovules arranged in more than 2 series on each placenta ; fruit a septicidal or
          loculicidal capsule, the seeds not borne on hardened hook-like funicles :
          Corolla-lobes very short, alternating with claviform processes ; stamens 2 ;
            leaves alternate ; herbs ..   ..   **151. Solanaceae** (2 : 325)
          Corolla-lobes not alternating with claviform processes ; stamens 2–4 ; leaves
            alternate or opposite or verticillate, sometimes reduced to scales ; herbs or
            shrublets, sometimes aquatic or parasitic on roots
            153. Scrophulariaceae (2 : 352)
  Ovary with 1–4 ovules in each loculus :
    Peduncles adnate to the petioles of the subtending leaves ; leaves alternate,
      simple, with stipules ; stamens 2–3 ; trees .. **88. Chailletiaceae** (1 : 433)
    Peduncles free from the leaves ; leaves without stipules :
      Perianth actinomorphic ; stamens 2 ; leaves simple or 3-foliolate, opposite or
      sometimes ternate ; trees or shrubs, sometimes climbing   **133. Oleaceae** (2 : 47)
      Perianth zygomorphic :
      Ovary more or less deeply 4-lobed, the style gynobasic ; fruit separating into
        4 nutlets (or fewer by abortion) ; leaves simple, opposite or verticillate or
        sometimes alternate ; herbs or shrubs, often aromatic   **162. Labiatae** (2 : 450)
      Ovary not deeply 4-lobed, the style not gynobasic :
        Fruit a loculicidal capsule, the seeds often borne on hardened hook-like
        funicles (retinacula) ; leaves simple, opposite ; herbs or shrubs, sometimes
        climbing   ..   ..   **159. Acanthaceae** (2 : 391)
        Fruit indehiscent or separating into 2–4 pyrenes or cocci, the seeds not borne on
        hardened hook-like funicles :
        Flowers solitary, axillary ; fruit indehiscent, armed with 4 lateral spines ;
          stamens 4 ; leaves simple, opposite ; herbs   **158. Pedaliaceae** (2 : 388)
        Flowers grouped in inflorescences ; fruit drupaceous or separating into 2–4
          pyrenes or cocci, without spines ; stamens 4 or 2 ; leaves simple or
          digitate, opposite or verticillate ; herbs or shrubs or trees, sometimes
          climbing ..   ..   **160. Verbenaceae** (2 : 432)
Stamens as many as the corolla-lobes :
  Leaves absent or reduced to scales :
    Slender twining parasitic herbs ; corolla-tube with a series of fimbriate scales
    below and opposite to the stamens ; ovary with 2 ovules in each loculus ; fruit
    a capsule ..   ..   **152. Convolvulaceae** (2 : 335)

Shrubs (sometimes climbing) or succulent plants, not parasitic ; corolla-tube with a 1–2-seriate corona above the stamens or adnate to the outside of the united filaments ; ovary with numerous ovules in each loculus ; fruit formed of 2 separated follicular carpels (or 1 by abortion)   136. **Asclepiadaceae** (2 : 85)
Leaves present, well developed, sometimes appearing after the flowers :
Leaves opposite or verticillate, not all radical :
Stamens hypogynous, not inserted on the corolla, the anthers opening by apical pore-like slits ; heath-like shrubs with small verticillate leaves
                                                  126. **Ericaceae** (2 : 1)
Stamens inserted on the corolla-tube ; plants not heath-like :
Ovary with 1–2 ovules in each loculus :
Style 2-branched, the branches again 2-branched ; trees or shrubs :
Calyx-lobes longer than the tube ; ovary 2-locular with 2 ovules in each loculus ; fruit a compressed 2-lobed loculicidal capsule
                                                  132. **Loganiaceae** (2 : 34)
Calyx-lobes shorter than the tube ; ovary 4-locular with 1 ovule in each loculus ; fruit a drupe   ..   ..   ..   150. **Boraginaceae** (2 : 317)
Style unbranched or with 2 simple branches :
Corolla-lobes valvate ; leaves with interpetiolar stipules ; trees or shrubs
                                                  137. **Rubiaceae** (2 : 104)
Corolla-lobes imbricate or contorted ; leaves without stipules :
Corolla-lobes 4, imbricate ; herbs or shrubs or trees, sometimes climbing
                                                  160. **Verbenaceae** (2 : 432)
Corolla-lobes 5, contorted :
Herbs, often hispid with bulbous-based hairs ; anthers hairy, connivent into a twisted cone protruding from the corolla-tube ; ovary 4-locular
                                                  150. **Boraginaceae** (2 : 317)
Trees or shrubs ; anthers glabrous, included in the corolla-tube, not connivent into a twisted cone   ..   134. **Apocynaceae** (2 : 51)
Ovary with 3 or more ovules in each loculus :
Ovary-loculi free or almost free, connected above by the style ; fruit formed of 2 or more separated carpels (or 1 by abortion) :
Pollen granular, liberated freely, without pollen-carriers ; corolla-tube without a corona or sometimes with scales or appendages at the mouth or above the stamens ; fruiting carpels follicular or baccate ; seeds with or without hairs, sometimes arillate ; trees or shrubs, sometimes climbing
                                                  134. **Apocynaceae** (2 : 51)
Pollen granular or waxy, transported by 5 pollen carriers at first resting on the style ; corolla-tube with a 1–2-seriate corona above the stamens or adnate to the outside of the filaments ; fruiting carpels follicular ; seeds with an apical tuft of silky hairs ; herbs or shrubs, often climbing :
Pollen agglutinated into waxy masses (pollinia) attached in pairs or fours to the pollen-carriers ; filaments united into a tube or ring
                                                  136. **Asclepiadaceae** (2 : 85)
Pollen granular, in tetrads, transported on glandular spathulate carriers ; filaments free   ..   ..   ..   135. **Periplocaceae** (2 : 80)
Ovary-loculi completely united ; fruit a capsule or berry :
Corolla-lobes valvate ; trees or shrubs or herbs, sometimes climbing, the branches often bearing spines or hooked tendrils
                                                  132. **Loganiaceae** (2 : 34)
Corolla-lobes imbricate or contorted :
Corolla-lobes imbricate :
Herbs ; flowers solitary or paired in the axils of the leaves
                                                  153. **Scrophulariaceae** (2 : 352)
Trees or shrubs ; flowers in cymes or cymose panicles
                                                  132. **Loganiaceae** (2 : 34)
Corolla-lobes contorted :
Herbs ; fruit a septicidal capsule   ..   140. **Gentianaceae** (2 : 297)
Trees or shrubs, sometimes climbing or epiphytic ; fruit a berry :
Corolla-lobes and stamens 6–11; filaments united below into a tube or cup   ..   ..   132. **Loganiaceae** (2 : 34)
Corolla-lobes and stamens 5 ; filaments free   134. **Apocynaceae** (2 : 51)
Leaves alternate, sometimes all radical :
Leaves all radical ; corolla-lobes and stamens 4 ; fruit a circumscissile capsule ; scapigerous herbs with small flowers in spikes   144. **Plantaginaceae** (2 : 306)
Leaves not all radical :
Ovary with numerous ovules in each loculus :
Corolla-lobes contorted ; fruit formed of 2 separated follicular carpels (or 1 by abortion) ; shrubs   ..   ..   134. **Apocynaceae** (2 : 51)
Corolla-lobes imbricate or valvate or plicate ; fruit a capsule or berry :

                                                  32—2

Styles 2 ; corolla-lobes imbricate ; fruit a capsule ; herbs
                                    149. Hydrophyllaceae (2 : 316)
Style 1 :
   Corolla-lobes plicate or valvate ; herbs or shrubs or trees, sometimes
     climbing   ..   ..   **151. Solanaceae** (2 : 325)
   Corolla-lobes imbricate :
     Flowers subsessile in long scorpioid cymes ; fruit a circumscissile capsule
       enclosed in the persistent accrescent calyx ; herbs
                                 151. Solanaceae (2 : 325)
     Flowers solitary or paired in the axils of the leaves, pedicellate ; fruit not
       circumscissile :
       Spiny shrubs ; fruit a berry   ..   ..   **151. Solanaceae** (2 : 325)
       Herbs or shrublets, not spiny ; fruit a capsule
                             153. Scrophulariaceae (2 : 352)
Ovary with 1–2 ovules in each loculus :
   Petals united below into a sheath open on the upper (adaxial) side and bearing
     the stamens, the filaments also united below into a sheath ; flowers zygo-
     morphic ; trees or shrubs, sometimes climbing   **22. Polygalaceae** (1 : 108)
   Petals united at least at the base into a tube ; flowers actinomorphic or slightly
     zygomorphic :
     Stamens opposite to the corolla-lobes ; fruit a berry ; trees or shrubs :
       Leaves 2–3-pinnate ; corolla-lobes valvate ; filaments united into a lobed
         tube   ..   ..   **113. Ampelidaceae** (1 : 672)
       Leaves simple ; corolla-lobes imbricate ; filaments free, sometimes alter-
         nating with staminodes   ..   ..   ..   **129. Sapotaceae** (2 : 16)
     Stamens alternating with the corolla-lobes :
     Flowers unisexual ; style absent, the stigma sessile ; trees
                             99. Aquifoliaceae (1 : 623)
     Flowers bisexual ; styles 1–2, sometimes branched :
       Ovules pendulous from the apex of the loculi ; leaves with stipules ; trees
         or shrubs, sometimes climbing   ..   **88. Chailletiaceae** (1 : 433)
       Ovules erect from the base of the loculi ; leaves without stipules :
       Styles 2, free, sometimes 2-branched :
         Corolla-lobes and stamens 4 ; leaves toothed or pinnately lobed ;
           annual herbs   ..   ..   ..   **150. Boraginaceae** (2 : 317)
         Corolla-lobes and stamens 5 ; leaves entire ; herbs or shrubs, sometimes
           climbing   ..   ..   ..   **152. Convolvulaceae** (2 : 335)
       Style 1, sometimes with 2 or 4 branches :
       Fruit separating into 2–4 nutlets ; style terminal or gynobasic ; herbs or
         shrublets   ..   ..   ..   **150. Boraginaceae** (2 : 317)
       Fruit not separating into nutlets ; style terminal :
         Corolla-lobes imbricate ; sepals united at least at the base ; fruit
          drupaceous ; trees or shrubs   ..   **150. Boraginaceae** (2 : 317)
         Corolla-lobes plicate or valvate ; sepals free ; fruit a capsule or some-
          times indehiscent ; herbs or shrubs, often trailing or climbing
                             152. Convolvulaceae (2 : 335)

Group 12

Ovary 1- or more-locular with 2 or more ovules in each loculus :
   Plants aquatic, moss-like or liverwort-like in habit, growing on rocks ; flowers small
     and inconspicuous, with 1–2 stamens ; ovary with numerous ovules ; fruit a capsule
                             27. Podostemaceae (1 : 122)
   Plants terrestrial or sometimes aquatic but not moss-like or liverwort-like in habit :
     Leaves compound ; trees or shrubs with unisexual or polygamous flowers ; ovary
       2- or more-locular :
       Leaves pinnate ; ovary with 2 ovules in each loculus   **119. Sapindaceae** (1 : 709)
       Leaves digitate :
       Leaves alternate ; stamens united into a column ; ovules numerous (more than 2)
         in each loculus ; fruit formed of 3 or more separated follicular carpels (or fewer by
         abortion)   ..   ..   **75. Sterculiaceae** (1 : 310)
       Leaves opposite ; stamens free ; ovules 2 in each loculus ; fruit a woody loculicidal
         capsule   ..   ..   **86. Euphorbiaceae** (1 : 364)
     Leaves simple, sometimes lobed :
     Leaves all or some opposite or verticillate, sometimes all radical :
       Flowers unisexual ; woody plants :
       Climbing shrubs with tendrils ; perianth-segments valvate ; stamens as many as
         and alternating with the perianth-segments ; ovary 1-locular with 2 pendulous
         apical ovules   ..   ..   **102. Icacinaceae** (1 : 636)
       Shrubs or trees without tendrils ; perianth-segments imbricate ; stamens 6, 2

opposite to the outer perianth-segments, 4 in pairs opposite to the inner perianth-segments ; ovary 3-locular with 2 ovules in each loculus 92. **Buxaceae** (1 : 587)

Flowers bisexual ; herbs, sometimes aquatic :
  Style 1 :
    Leaves with stipules ; sepals free ; stamens hypogynous ; style with 3 stigmatic
    branches   ..   .. 29. **Caryophyllaceae** (1 : 129)
    Leaves without stipules ; sepals united into a tube ; stamens or single stamen
    perigynous, inserted on the calyx-tube ; style unbranched
                .42. **Lythraceae** (1 : 163)

  Styles 2–5 :
    Ovary 1-locular with a free-basal or free-central placenta ; leaves without
    stipules .. .. 29. **Caryophyllaceae** (1 : 129)
    Ovary 2-5-locular with axile placentas ; leaves with or without stipules :
      Sepals united below ; stamens perigynous, inserted on the calyx-tube ; fruit a
      circumscissile capsule .. .. .. 31. **Ficoidaceae** (1 : 135)
      Sepals free ; stamens hypogynous or subhypogynous ; fruit a loculicidal
      capsule .. .. 30. **Molluginaceae** (1 : 133)

Leaves alternate, not all radical :
  Ovary 1-locular :
    Stamens numerous (more than 10) ; style arising from the base of the ovary ;
    flowers polygamous, in panicles ; trees .. .. 87. **Rosaceae** (1 : 423)
    Stamens 3–8 ; style or sessile stigma terminal or lateral, not arising from the base
    of the ovary ; flowers in racemes or spikes or fascicles :
      Herbs, sometimes climbing ; flowers bisexual ; fruit a circumscissile capsule
              37. **Amaranthaceae** (1 : 145)
      Trees or shrubs, sometimes climbing ; flowers unisexual ; fruit indehiscent, often
      drupaceous :
        Leaves with stipules ; perianth-segments imbricate ; stamens as many as and
        opposite to the perianth-segments or more numerous
              86. **Euphorbiaceae** (1 : 364)
        Leaves without stipules ; perianth-segments valvate ; stamens as many as and
        alternating with the perianth-segments .. 102. **Icacinaceae** (1 : 636)
  Ovary 2- or more-locular :
    Flowers bisexual :
      Ovary borne on an elongated gynophore ; stamens numerous (more than 10) ;
      fruit indehiscent, unwinged ; flowers solitary, axillary, pedunculate ; shrubs
      or herbs .. .. 17. **Capparidaceae** (1 : 86)
      Ovary sessile or subsessile ; stamens 5–8 ; fruit a septicidal capsule with 2 or
      more longitudinal membranous wings ; flowers in panicles or racemes ; shrubs
      or trees .. .. 119. **Sapindaceae** (1 : 709)
    Flowers unisexual or polygamous :
      Filaments completely united into a column :
        Calyx-segments valvate ; anthers 5 or more ; fruit formed of 3 or more separated
        follicular or indehiscent carpels (or fewer by abortion) ; trees or shrubs
              75. **Sterculiaceae** (1 : 310)
        Calyx-segments imbricate ; anthers 2–3 ; fruit a capsule ; herbs or shrubs
              86. **Euphorbiaceae** (1 : 364)
      Filaments all or some free at least above :
        Stamens 10 or more ; trees or shrubs :
          Stigmas terminating 2 or more styles ; branches often armed with axillary
          spines .. .. 53. **Flacourtiaceae** (1 : 185)
          Stigmas sessile or subsessile, 2 or more, entire or 2-lobed ; branches without
          spines .. .. 86. **Euphorbiaceae** (1 : 364)
        Stamens fewer than 10 :
          Leaves with stipules or stipular spines ; styles 2–4, free or united below, or
          stigmas sessile ; fruit an unwinged capsule or indehiscent or separating into
          2 winged cocci ; trees or shrubs or herbs, sometimes climbing
              86. **Euphorbiaceae** (1 : 364)
          Leaves without stipules ; style 1 ; fruit a septicidal capsule with 2 or more
          longitudinal membranous wings ; shrubs or trees 119. **Sapindaceae** (1 : 709)

Ovary 1- or more-locular with 1 ovule in each loculus :
  Ovary 2- or more-locular :
    Flowers bisexual :
      Trees or shrubs ; sepals united into a tube ; stamens twice as many as the calyx-lobes, perigynous, inserted on the calyx-tube ; fruit a drupe
              46. **Thymelaeaceae** (1 : 171)
      Herbs ; sepals free ; stamens hypogynous or subhypogynous ; fruit separating into
      2 cocci sometimes winged along the back .. 30. **Molluginaceae** (1 : 133)
    Flowers unisexual or polygamous, sometimes grouped in involucrate inflorescences
    simulating bisexual or male flowers :

Male flowers reduced to a single stamen with or without a minute vestigial calyx :
    Flowers solitary or paired (1 male and 1 female) in the axils of the leaves ; ovary
       4-locular ; leaves opposite ; aquatic or subaquatic herbs
                                      45. **Haloragaceae** (1 : 171)
    Flowers grouped in involucrate inflorescences, the involucres either cupular or
       composed of free bracts and containing a number of male flowers with or without
       a female flower ; ovary 2–4-locular ; leaves alternate or opposite, sometimes
       reduced to scales or stipular spines ; terrestrial herbs or shrubs or trees, sometimes
       succulent   ..   ..   86. **Euphorbiaceae** (1 : 364)
Male flowers with 2 or more stamens and with a calyx of 2 or more free or united
    sepals :
    Leaves pinnate ; trees or shrubs   ..   ..   ..   119. **Sapindaceae** (1 : 709)
    Leaves simple or digitately 3-foliolate :
    Filaments completely united into a column bearing 8–20 anthers ; trees with
       simple or 3-foliolate leaves   ..   ..   ..   86. **Euphorbiaceae** (1 : 364)
    Filaments all or some free at least above ; leaves simple, sometimes lobed or deeply
       divided :
       Stamens 2–5 ; trees or shrubs or herbs, sometimes climbing
                               86. **Euphorbiaceae** (1 : 364)
       Stamens 6 or more :
       Branches armed with axillary spines ; fruit a berry ; shrubs or trees
                           53. **Flacourtiaceae** (1 : 185)
       Branches without spines :
         Leaves with stipules ; ovary 2–4-locular ; styles or stigmas 1–4, often branched
         or lobed ; fruit a capsule, sometimes tardily dehiscent or indehiscent ;
         trees or shrubs or herbs, sometimes climbing   86. **Euphorbiaceae** (1 : 364)
        Leaves without stipules ; ovary 5–7-locular ; styles 5–7, not branched or
         lobed ; fruit a berry ; climbing shrubs with flowers in elongated racemes
                            35. **Phytolaccaceae** (1 : 142)
Ovary 1-locular :
  Leaves absent or reduced to scales ; flowers spicate :
  Slender twining parasitic plants ; flowers not immersed in the rhachis of the spike ;
    stamens 6–9, accompanied by staminodes ; anthers opening by upcurving valves
                               6. **Lauraceae** (1 : 56)
  Succulent salt-marsh herbs with articulated branches, not parasitic ; flowers
    immersed in the rhachis of the spike ; stamens 1–2 ; anthers opening by longi-
    tudinal slits ..   ..   36. **Chenopodiaceae** (1 : 143)
  Leaves present, not reduced to scales :
  Leaves with stipules or with a sheathing or annular base surrounding the stem :
    Leaves 1–4-pinnate or 2-foliolate :
    Herbs ; leaves 2–4-pinnate ; fruit a small stipitate achene, not winged
                             9. **Ranunculaceae** (1 : 62)
    Trees ; leaves 1-pinnate or 2-foliolate ; fruit large, sessile or stipitate, woody or
    winged   ..   ..   89. **Caesalpiniaceae** (1 : 439)
    Leaves simple or digitate with 3 or more leaflets :
    Ovule pendulous from the apex or near the apex of the ovary ; flowers unisexual
       or polygamous :
       Leaves all or mostly digitate ; male flowers in elongated panicles, female flowers
         spicate and enclosed by bracts ; annual herbs   98. **Cannabinaceae** (1 : 623)
       Leaves simple :
       Flowers densely spicate or capitate, or arranged on or in a flat or cupular recep-
         tacle or inside a hollow almost closed receptacle (fig), the females sometimes
         solitary on a bracteate receptacle or sometimes immersed in the tissue of the
         receptacle ; perianth present or absent ; trees or shrubs or herbs, sometimes
         epiphytic   ..   ..   96. **Moraceae** (1 : 593)
       Flowers solitary or fasciculate or in cymes or racemes or panicles ; perianth
         present ; trees or shrubs :
         Style 2-branched with simple or divided stigmas ; fruit a drupe or a
          membranous-winged samara   ..   ..   ..   95. **Ulmaceae** (1 : 591)
         Style short, unbranched ; fruit a capsule or indehiscent, not winged
                            86. **Euphorbiaceae** (1 : 364)
  Ovule arising from the base or near the base of the ovary :
    Leaves opposite :
    Flowers unisexual ; stamens as many as and opposite to the calyx-segments or
    reduced to 1 ; leaves marked with cystoliths ; herbs or shrubs
                            97. **Urticaceae** (1 : 616)
    Flowers bisexual ; stamens as many as and alternating with the calyx-segments ;
    leaves small, fleshy, not marked with cystoliths ; herbs
                            31. **Ficoidaceae** (1 : 135)

Leaves alternate :
  Perianth absent ; flowers minute, in dense spikes ; shrubs, sometimes climbing
                            14. **Piperaceae** (1 : 81)
  Perianth present, sometimes vestigial in female flowers :
    Styles or style-branches 2-3 ; stamens 5 or more ; leaves often with a basal
      sheath (ocrea) or annulus surrounding the stem ; herbs or shrubs or trees,
      sometimes climbing or aquatic .. .. 33. **Polygonaceae** (1 : 137)
    Style 1, unbranched, or stigma sessile ; stamens 1-8 :
      Flowers bisexual, in spike-like racemes ; stamens as many as and alternating
        with the calyx-segments or more numerous ; herbs or shrublets
                            35. **Phytolaccaceae** (1 : 142)
      Flowers unisexual ; stamens as many as and opposite to the calyx-segments or
        reduced to 1 :
        Stamens or single stamen erect in bud ; flowers dioecious, the males densely
          crowded in paniculate spikes or heads, the females in pedunculate heads ;
          leaves simple or digitately compound ; trees or shrubs
                            96. **Moraceae** (1 : 593)
        Stamens inflexed in bud ; flowers monoecious or dioecious, cymose, the
          cymes often dense or head-like and sometimes enclosed in an involucre ;
          leaves simple, sometimes pinnately lobed or palmately 3-lobed, often
          marked with cystoliths ; herbs or shrubs, sometimes climbing or epiphytic
                            97. **Urticaceae** (1 : 616)
Leaves without stipules, the base not sheathing or surrounding the stem :
  Submerged aquatic herbs ; leaves verticillate, deeply dichotomously divided into
    linear or filiform segments ; flowers unisexual, solitary and sessile in the axils
    of the leaves .. .. 10. **Ceratophyllaceae** (1 : 65)
  Plants terrestrial ; leaves alternate or opposite or verticillate, sometimes pinnately
    lobed but not dichotomously divided :
    Perianth absent ; flowers in spikes :
      Trees ; flowers unisexual ; stamens 3-12 ; styles 2, free or shortly united below ;
        leaves alternate, mostly toothed .. .. .. 94. **Myricaceae** (1 : 589)
      Herbs, sometimes climbing or epiphytic ; flowers bisexual ; stamens 2 ; style
        absent, the stigma sessile ; leaves alternate or opposite or verticillate, entire
                            14. **Piperaceae** (1 : 81)

    Perianth present :
      Stamens united into a column ; seed arillate ; flowers dioecious, in small solitary
        or clustered (sometimes paniculate) umbels or heads ; trees or climbing shrubs
                            8. **Myristicaceae** (1 : 60)
      Stamens free or the filaments united at the base into a cup ; seed not arillate :
        Anthers opening by upcurving valves ; calyx with 6-8 lobes in 2 series ; stamens
          6-12 in 2-3 series, accompanied on the inside by staminodes ; trees or shrubs
                            6. **Lauraceae** (1 : 56)
        Anthers opening by longitudinal slits ; calyx with 3-6 segments or lobes in
          1 series :
          Stamens more numerous than the calyx-segments or -lobes :
            Ovule arising from the base of the ovary ; stamens fewer than twice as many
              as the calyx-lobes ; calyx tubular, petaloid, the lower portion persistent
              round the fruit and often bearing glands on the outside ; herbs or shrubs,
              sometimes climbing .. .. .. 47. **Nyctaginaceae** (1 : 176)
            Ovule pendulous from the apex of the ovary ; stamens twice as many as the
              calyx-segments or -lobes or more numerous :
              Flowers bisexual ; sepals united into an elongated tube ; stamens inserted
                on the calyx-tube ; leaves alternate or opposite ; shrubs or herbs, some-
                times climbing .. .. 46. **Thymelaeaceae** (1 : 171)
              Flowers unisexual ; sepals shortly united at the base ; stamens not inserted
                on the calyx ; leaves opposite or subopposite ; shrubs or trees
                            5. **Monimiaceae** (1 : 54)
          Stamens as many as or fewer than the calyx-segments or -lobes, sometimes
            alternating with staminodes :
            Leaves opposite :
              Calyx petaloid, the sepals united into a tube constricted above the ovary,
                the lower portion persistent round the fruit and often bearing glands on
                the outside ; stamens not accompanied by staminodes ; herbs, sometimes
                climbing .. .. 47. **Nyctaginaceae** (1 : 176)
              Calyx dry and scarious, the sepals free or shortly united at the base, not
                forming a tube constricted above the ovary ; stamens often alternating
                with staminodes ; herbs or shrubs, sometimes climbing
                            37. **Amaranthaceae** (1 : 145)

            Leaves alternate :
              Calyx with 3-4 valvate segments or lobes :

Trees or shrubs ; flowers bisexual, in large bracteate heads or in elongated
  spikes ; stamens inserted on the calyx-segments   48. **Proteaceae** (1 : 178)
Herbs, sometimes climbing ; flowers unisexual or polygamous, in small
  dense or lax cymes in the axils of the leaves ; stamens not inserted on
  the calyx ..   ..   97. **Urticaceae** (1 : 616)
Calyx with 3–5 imbricate segments or lobes :
  Stamens 4, as many as and alternating with the calyx-segments ; calyx
    zygomorphic ; herbs with flowers in racemes   35. **Phytolaccaceae** (1 : 142)
  Stamens 1–5, as many as and opposite to the calyx-segments or -lobes or
    fewer ; calyx actinomorphic :
    Sepals united below into a tube with 2 adnate bracteoles outside ; twining
      herbs with flowers in axillary pedunculate spikes
                                          38. **Basellaceae** (1 : 155)
    Sepals free or shortly united below, without adnate bracteoles ; herbs or
      shrubs, not twining :
      Calyx dry and more or less scarious ; stamens sometimes alternating with
        staminodes ; leaves entire   ..   ..   37. **Amaranthaceae** (1 : 145)
      Calyx herbaceous ; stamens not accompanied by staminodes ; leaves
        entire or toothed or pinnately lobed   36. **Chenopodiaceae** (1 : 143)

Group 13

Parasitic shrubs growing on branches of other shrubs or trees ; ovule scarcely distinguishable from the surrounding tissue of the ovary ; calyx truncate or shortly lobed ;
  stamens as many as and inserted on the petals ; leaves simple, entire
                                          109. **Loranthaceae** (1 : 658)
Plants not parasitic ; ovule or ovules clearly distinguishable within the ovary :
  Ovary 1-locular :
    Ovule 1 :
      Leaves digitate with 3–5 leaflets ; anthers opening by upcurving valves ; stamens
        as many as the petals ; fruit with 2–4 broad lateral wings ; climbing shrubs
                                          7. **Hernandiaceae** (1 : 58)
      Leaves simple, sometimes palmately lobed ; anthers or single anther opening by
        longitudinal slits :
        Leaves opposite ; flowers zygomorphic, the median (adaxial) sepal with 2 super-
          posed spurs ; stamen 1 ; fruit indehiscent, crowned by the persistent accrescent
          sepals, 2 much larger than the others and wing-like ; trees
                                          23. **Vochysiaceae** (1 : 114)
        Leaves alternate ; flowers actinomorphic ; stamens as many as or more numerous
          than the petals :
          Petals contorted ; fruit indehiscent, crowned by the persistent accrescent wing-
            like sepals ; leaves equal-sided ; climbing shrubs with hooked spines
                                          64. **Ancistrocladaceae** (1 : 233)
          Petals valvate ; fruit a drupe, not crowned by accrescent wing-like sepals ; leaves
            often unequal-sided ; trees without spines   ..   123. **Alangiaceae** (1 : 749)
    Ovules 2 or more :
      Style 1, unbranched ; trees or shrubs, sometimes climbing :
        Ovules pendulous from the apex or near the apex of the ovary ; fruit indehiscent,
          often with 2–6 longitudinal wings or angles ; flowers in racemes or spikes or heads ;
          leaves opposite or verticillate or alternate   ..   69. **Combretaceae** (1 : 264)
        Ovules borne on a free-basal or free-central placenta ; fruit a berry, not winged or
          angled ; flowers in lax or dense cymes or fascicles or umbels, sometimes pani-
          culate ; leaves opposite ..   ..   ..   68. **Melastomataceae** (1 : 245)
      Styles or style-branches 2–6 ; herbs or shrublets :
        Ovules numerous on 1 or more free-basal placentas ; sepals 2, often deciduous ;
          fruit a circumscissile capsule   ..   ..   ..   32. **Portulacaceae** (1 : 136)
        Ovules pendulous from the apex of the ovary or on pendulous apical placentas ;
          sepals or calyx-lobes 4–5 ; fruit not circumscissile :
          Ovules numerous on 2–3 pendulous placentas ; flowers bisexual, in axillary pairs ;
            petals and stamens 5 ; leaves opposite   ..   25. **Saxifragaceae** (1 : 119)
          Ovules 4, pendulous from the apex of the ovary ; flowers bisexual or unisexual, in
            axillary clusters ; petals and stamens 4 ; leaves opposite or alternate
                                          45. **Haloragaceae** (1 : 171)
  Ovary 2- or more-locular, septate up to or almost up to the apex :
    Ovary with 2 or more ovules in each loculus :
      Leaves with free or interpetiolar stipules :
        Leaves alternate, sometimes all radical, often unequal-sided ; flowers unisexual ;
          stamens numerous, more than twice as many as the petals ; ovules numerous in
          each loculus ; herbs or shrubs, sometimes climbing or epiphytic, not viviparous
                                          60. **Begoniaceae** (1 : 216)

Leaves opposite ; flowers bisexual ; stamens twice as many as the petals ; ovules 2 in each loculus ; viviparous littoral trees or shrubs   70. **Rhizophoraceae** (1 : 281)

Leaves without stipules :

Anthers opening by a terminal pore ; connective often appendaged below the anther ; leaves with 3 or more longitudinal parallel nerves, opposite, sometimes all radical ; herbs or shrubs, sometimes climbing or epiphytic
68. **Melastomataceae** (1 : 245)

Anthers opening by longitudinal slits ; connective unappendaged ; leaves without parallel longitudinal nerves :

Stamens as many or twice as many as the petals ; herbs or shrubs, sometimes aquatic   ..   ..   43. **Onagraceae** (1 : 166)

Stamens numerous, more than twice as many as the petals :

Leaves alternate ; fruit with 4 longitudinal membranous wings ; trees
67. **Lecythidaceae** (1 : 241)

Leaves opposite ; fruit a berry, unwinged ; trees or shrubs or sometimes herbs
66. **Myrtaceae** (1 : 235)

Ovary with 1 ovule in each loculus :

Stamens twice as many as the petals :

Annual herbs ; leaves toothed or pinnately lobed ; flowers solitary, axillary ; fruit a prickly capsule ..   ..   87. **Rosaceae** (1 : 423)

Trees or shrubs ; leaves entire ; flowers in solitary or aggregated spikes ; fruit a drupe, not prickly   ..   ..   70. **Rhizophoraceae** (1 : 281)

Stamens as many as the petals :

Stamens opposite to the petals :

Ovules erect from the base of the loculi ; leaves with stipules ; fruit unwinged or with 1–3 wings, sometimes separating into 3 cocci ; trees or shrubs, sometimes climbing with or without tendrils   ..   ..   112. **Rhamnaceae** (1 : 667)

Ovules pendulous from the apex or near the apex of the loculi ; leaves without stipules ; fruit drupaceous, not winged ; trees   104. **Olacaceae** (1 : 644)

Stamens alternating with the petals :

Flowers solitary in the axils of the leaves ; floating aquatic herbs ; leaves alternate, rosulate, with a more or less inflated petiole ; fruit large, indehiscent, with hard endocarp and armed with 2 horns   ..   ..   44. **Trapaceae** (1 : 170)

Flowers grouped in inflorescences ; plants not floating on water :

Fruit separating into 2 cocci ; herbs or sometimes shrubs or trees ; flowers in simple or compound umbels, sometimes capitate or spicate ; leaves often much divided or compound ..   ..   125. **Umbelliferae** (1 : 751)

Fruit drupaceous, not separating into cocci ; trees or shrubs, sometimes climbing :

Leaves pinnate or digitate ; flowers in umbels or racemes or spikes or heads, sometimes paniculate   ..   ..   ..   124. **Araliaceae** (1 : 750)

Leaves simple, often unequal-sided ; flowers in axillary cymes
123. **Alangiaceae** (1 : 749)

Group 14

Parasitic shrubs growing on branches of other shrubs or trees ; ovule scarcely distinguishable from the surrounding tissue of the ovary ; calyx truncate or lobed ; stamens as many as and inserted on the corolla-lobes ; leaves simple, entire
109. **Loranthaceae** (1 : 658)

Plants not parasitic ; ovule or ovules clearly distinguishable within the ovary :

Ovary 1-locular :

Ovule 1 ; flowers actinomorphic or zygomorphic, in involucrate heads ; fruit indehiscent, often crowned by the persistent calyx forming a pappus of bristles or scales :

Anthers united into a tube surrounding the style ; ovule erect from the base of the ovary ; herbs or shrubs or trees, sometimes climbing ; corolla of the outer (ray) flowers often differing from that of the inner (disk) flowers 139. **Compositae** (2 : 225)

Anthers free ; ovule pendulous from the apex of the ovary ; herbs with opposite leaves   ..   ..   138. **Dipsacaceae** (2 : 223)

Ovules numerous or several on 1 or more free-basal placentas ; flowers actinomorphic, not in involucrate heads :

Calyx composed of 2 often deciduous sepals ; stamens more numerous than the corolla-lobes ; fruit a circumscissile capsule ; herbs   32. **Portulacaceae** (1 : 136)

Calyx 4–5-lobed ; stamens as many as and opposite to the corolla-lobes ; fruit not circumscissile :

Herbs ; flowers bisexual, in terminal racemes ; fruit a 5-valved capsule
142. **Primulaceae** (2 : 303)

Trees or shrubs, sometimes climbing ; flowers unisexual or polygamous, in axillary racemes or panicles ; fruit a berry   ..   ..   130. **Myrsinaceae** (2 : 30)

Ovary 2- or more-locular :

Climbing herbs with tendrils ; flowers unisexual ; stamens 3, with flexuous anthers
59. **Cucurbitaceae** (1 : 204)

Plants without tendrils ; flowers bisexual or unisexual ; anthers not flexuous :
Corolla forming a deciduous cap (calyptra) ; stamens very numerous ; trees or shrubs with gland-dotted leaves .. .. 66. **Myrtaceae** (1 : 235)
Corolla not forming a deciduous cap :
Leaves opposite or verticillate, with interpetiolar or intrapetiolar stipules, the stipules sometimes leaf-like ; trees or shrubs or herbs, sometimes climbing
137. **Rubiaceae** (2 : 104)
Leaves alternate or sometimes opposite, without stipules :
Flowers with a 2-seriate laciniate corona outside the stamens ; corolla many-ribbed, the ribs ending in short teeth ; trees or shrubs, sometimes climbing
67. **Lecythidaceae** (1 : 241)
Flowers without a corona ; corolla 5-lobed :
Corolla zygomorphic, split down the upper (adaxial) side :
Anthers united into a tube surrounding the style ; ovules numerous in each loculus ; fruit a capsule ; herbs .. .. 147. **Lobeliaceae** (2 : 311)
Anthers free ; ovule 1 in each loculus ; fruit a drupe ; littoral shrubs or herbs
148. **Goodeniaceae** (2 : 315)
Corolla actinomorphic, not split down one side ; herbs :
Corolla-lobes imbricate ; fruit a circumscissile capsule ; flowers in terminal spikes .. .. 145. **Sphenocleaceae** (2 : 307)
Corolla-lobes valvate ; fruit a capsule dehiscent by apical valves ; flowers in cymes or panicles .. .. 146. **Campanulaceae** (2 : 309)

Group 15

Ovary 2- or more-locular with numerous clearly distinguishable ovules in each loculus ; plants with well-developed leaves :
Flowers unisexual ; perianth composed of 2 or more free petaloid segments ; stamens numerous, the anthers free ; leaves with stipules, often unequal-sided ; herbs or shrubs, sometimes climbing or epiphytic .. .. 60. **Begoniaceae** (1 : 216)
Flowers bisexual ; perianth with an elongated tube inflated below, the mouth 2–6-lobed or produced unilaterally into a single lobe; stamens 6–24, the anthers adnate to the style ; leaves without stipules ; herbs or shrubs, often climbing
13. **Aristolochiaceae** (1 : 77)
Ovary 1-locular, the ovule or ovules sometimes scarcely distinguishable from the surrounding tissue ; plants with or without well-developed leaves :
Leaves reduced to scales ; plants parasitic on roots :
Flowers unisexual, in involucrate heads ; ovule 1, scarcely distinguishable from the surrounding tissue of the ovary ; fleshy herbs with unbranched flowering stems densely covered with scale-leaves .. .. 111. **Balanophoraceae** (1 : 666)
Flowers bisexual, solitary or clustered or in short spikes ; ovules 2–4, pendulous from the apex of a slender free-basal placenta ; herbs or shrublets with slender branches
110. **Santalaceae** (1 : 665)
Leaves well developed, not scale-like :
Parasitic shrubs growing on branches of other shrubs or trees ; ovule or ovules scarcely distinguishable from the surrounding tissue of the ovary ; flowers unisexual ; anthers as many as and sessile on the perianth-segments ; leaves opposite
109. **Loranthaceae** (1 : 658)
Plants terrestrial, not parasitic ; ovule or ovules clearly distinguishable within the ovary :
Ovule 1 :
Flowers unisexual, densely capitate or arranged on or in a flat or cupular receptacle, the females sometimes solitary in the receptacle ; anthers opening by longitudinal slits ; leaves with stipules ; trees or shrubs or herbs 96. **Moraceae** (1 : 593)
Flowers bisexual or unisexual, in cymes or panicles ; anthers opening by up-curving valves ; leaves without stipules :
Stamens 9 ; anthers 4-valved ; perianth-segments 6 ; trees
6. **Lauraceae** (1 : 56)
Stamens 3–7 ; anthers 2-valved ; perianth-segments 4–10 ; trees or climbing shrubs .. .. 7. **Hernandiaceae** (1 : 58)
Ovules 2 or more :
Stamens 6–10, more numerous than the perianth-segments or -lobes ; flowers in racemes or spikes or heads ; fruit indehiscent, often with 2–5 longitudinal wings ; trees or shrubs .. .. 69. **Combretaceae** (1 : 264)
Stamens 4–5, as many as the perianth-segments or -lobes :
Styles 4 ; stamens 4 ; ovules 4, pendulous from the apex of the ovary ; herbs or shrublets with flowers in axillary clusters .. 45. **Haloragaceae** (1 : 171)
Style 1 with an entire or lobed stigma ; stamens 5 :
Stigma large, with radiating lobes ; stamens opposite to the perianth-segments ; ovules 3–4, pendulous from the apex of the ovary at the top of a slender free-

central axis ; flowers with stellate hairs on the outside of the perianth and ovary, in racemes or spikes or panicles ; trees   108. **Octoknemataceae** (1 : 656)

Stigma small, entire ; stamens alternating with the perianth-lobes ; ovules 2, pendulous from the apex of the ovary ; flowers without stellate hairs, in dense heads; maritime shrubs   ..   ..   ..   69. **Combretaceae** (1 : 264)

ADDITIONS AND CORRECTIONS

Since the preparation of Volume II the following additions and corrections have been noted. Unless otherwise stated the information originates from the reviser of the family concerned.

22.   **SYNSEPALUM**
Add under *Imperfectly known species:*

S. aubrevillei (*Pellegr.*) *Aubrév. & Pellegr.* in Notulae Syst. 16 : 263 (1961). *Sideroxylon aubrevillei* Pellegr. in Bull. Soc. Bot. Fr. 78 : 682 (1932) ; Aubrév. Fl. For. C. Iv. ed. 1, 3 : 106, t. 283, 4–6 ; ed. 2, 3 : 128, t. 293, 6–8. **Iv. C.**: Massa Mé *Aubrév.* 133 ! Alépé (Mar.) *Aké Assi* 4166 ! Toumanguié *Aké Assi* 4857 ! This species was inadequately known and imperfectly described at the time of the revision. Its generic position was doubtful and it was therefore not possible to include it in one of the accepted genera.

37.   **STRYCHNOS**
6. **S. innocua**—The two varieties **innocua** and **pubescens** cannot now be maintained as distinct. The reference and citations under the two varieties on p. 41 should be combined and the supposed distinctions in the key on p. 38 deleted. The species should be called simply *S. innocua.*

Three species from Liberia and Ivory Coast are to be described (Acta Bot. Neerl. ined. (1964)). One is related to *S. camptoneura* Gilg (No. 27), but it has leaves dull beneath ; one to *S. floribunda* Gilg (No. 25), but tendrils are paired, fruit globose and branches are not lenticillate ; and one to *S. longicaudata* Gilg (No. 24), but it differs in its indumentum and sepals.

49.   **OLEA**
O. hochstetteri—**N. Nig.**: Panshanu, Bauchi Dist. (July) *A. F. A. Lamb* FHI 3165 ! It has been recorded by Wimbush and White from Jos Plateau.—Ed.

116.   **SCHUMANNIOPHYTON**
2. **S. problematicum**—add **Ghana** : Neung F.R. (fr. Oct.) **Enti** FH 7535 !

169.   **SABICEA**
N. Hallé in Adansonia 3 : 168–177 (1963), as well as maintaining *Stipularia* P. Beauv. distinct from *Sabicea*, places several species of *Sabicea* with 2(–3) locular ovaries in the genera *Pseudosabicea* N. Hallé and *Ecpoma* K. Schum. The following species occurring in our area are affected : **S. medusula** K. Schum. ex Wernham = *P. medusula* (K. Schum. ex Wernham) N. Hallé ; **S. pedicellata** Wernham = *P. pedicellata* (Wernham) N. Hallé ; **S. floribunda** K. Schum. = *P. floribunda* (K. Schum.) N. Hallé ; **S. gigantistipula** K. Schum. = *E. gigantistipula* (K. Schum.) N. Hallé. Detailed re-appraisal of these proposals has not been possible before going to press.

271.   **AEDESIA**
Add: **A. spectabilis** *Mattf.* in Engl. Bot. Jahrb. 59, Beibl. 133 : 11 (1924). A stout erect herb 1–2ft. high, easily distinguished from the other species by the long white hairs. [Br.] **Cam.**: Balibansin, Ndop Plain, 3,800ft. (May) *Brunt* 447 ! Mbiami to Mbaw Plain road (June) *Brunt* 739 ! Also in Cameroun.—Ed.

293.   **LACTUCA**
5. **L. tuberosa** *A. Chev.*—substitute **L. praevia** C. D. Adams **nom. nov.** *L. tuberosa* A. Chev. in Mém. Soc. Bot. Fr. 2, 8 : 261 (1917), not of Jacq. (1772).

327.   **NICOTIANA**
N. rustica—add **Iv. C.**: Bouaflé (Sept.) *Aké Assi* 6639 !

336.   **CONVOLVULACEAE**
Add : **Argyreia nervosa** (Burm. f.) Bojer (syn. *A. speciosa* (Linn. f.) Sweet) a native of India, is cultivated at Ouagadougou, Upper Volta.

342.   **MERREMIA**
8. **M. pterygocaulos**—add *Ipomoea hydrosmifolia* A. Chev. Bot. (1920), name only. **Guin.**: Faranna (Jan.) *Chev.* 20397.

348–350. **IPOMOEA**
14. **Ipomoea verbascoidea**—add **Guin.**: Timbo to Ditinn (fr. Sept.) *Chev.* 18456 ! 18508 !
17. **I. stenobasis**—add **Iv. C.**: Séguéla (fl. & fr. Oct.) *Aké Assi* 6662 !
24. **I. triloba**—add syn. *I. webbii* Coutinho (1914)—A. Chev. in Rev. Bot. Appliq. 15 : 918 (1935).

385.   **BIGNONIACEAE**
Add to the list of ornamentals : **Cybistax donnell-smithii** (*Rose*) *Seibert* (C. America).

397.   **EREMOMASTAX**
E. polysperma—add **Iv. C.**: Fourounou to Attienkro (fr. Jan.) *Aké Assi* 4549

416. **LEPIDAGATHIS**
9. **L. anobrya**—add **Iv. C.:** Bowal near Séguéla (Sept.) *Aké Assi* 3351 !

431. **BRACHYSTEPHANUS**
Add as *Imperfectly known species:*

1. **B.** sp. cf. **nemoralis** *S. Moore* in **J**. Bot. 68, Suppl. 2 : 136 (1930). **Iv. C.:** Mt. Tonkoui (Nov.) *Aké Assi* 5501 ! 7031 !
2. **B.** sp. Single gathering of doubtful position. **Ghana :** Bosumkese F. R., Ashanti (fl. & fr. Dec.) *Adams* 5259 !

INDEX TO VOLUME II

(Synonyms and misapplied names are shown in italics ; they may be located by means of the number of the genus and species shown in brackets.)

34

Saba.
senegalensis (*A. DC.*) *Pichon* 61.
senegalensis var. glabriflora (*Hua*) *Pichon* 61.
thompsonii (*A. Chev.*) *Pichon* 61.
Sabicea *Aubl.* 169.
adamsii Hepper (41 : 9) 160.
africana (*P. Beauv.*) *Hepper* 173, fig. 236.
arachnoidea *Hutch. & Dalz.* 174.
barteri Wernham (49 : 14) 172.
bracteolata *Wernham* 172.
brevipes *Wernham* 173.
brunnea Wernham (49 : 31) 174.
calycina *Benth.* 172, fig. 235.
capitellata *Benth.* 174.
cordata *Hutch. & Dalz.* 172.
cordata *Hutch. & Dalz.* (sp. nr.) 172.
discolor *Stapf* 173.
discolor var. β *laxothyrsa* Wernham (49 : 26) 173.
discolor of Chev. (49 : 12) 172.
efulenensis (*Hutch.*) *Hepper* 173.
elliptica (*Schweinf. ex Hiern*) *Hepper* 174.
ferruginea (*G. Don*) *Benth.* 174.
ferruginea var. *lasiocalyx* (Stapf) Wernham (49 : 33) 174.
floribunda *K. Schum.* 172.
gabonica (*Hiern*) *Hepper* 172.
geophiloides *Wernham* 173.
gigantistipula *K. Schum.* 173.
globifera Hutch. & Dalz. (49 : 20) 173.
harleyae *Hepper* 172.
johnstonii *K. Schum. ex Wernham* 174.
lanuginosa *Wernham* 174.
lasiocalyx Stapf (49 : 33) 174.
liberica *Hepper* 172.
linderi (Hutch. & Dalz.) Bremek. (62 :) 190.
loxothyrsus K. Schum. & Dinkl. ex Stapf (49 : 26) 173.
medusula *K. Schum. ex Wernham* 174.
neglecta *Hepper* 172.
pedicellata *Wernham* 174.
pilosa *Hiern* 172.
robbii *Wernham* 173.
rosea *A. C. Hoyle* 172.
salmonea of Chev. (49 : 33) 174.
schaeferi *Wernham* 172.
schumanniana of Wernham, partly (49 : 12) 172.
segregata of Chev., partly (49 : 5) 172.
speciosa *K. Schum.* 172.
talbotii *Wernham* 173.
tchapensis K. Krause (49 : 18) 173.
urbaniana *Wernham* 172.
urceolata *Hepper* 172.
venosa *Benth.* 172.
vogelii *Benth.* 173.
xanthotricha *Wernham* 172.
Sacosperma *G. Tayl.* 213.
paniculatum (*Benth.*) *G. Tayl.* 213.
parviflorum (*Benth.*) *G. Tayl.* 213.
Saintpaulia ionantha *Wendl.* 382.
Salvia *Linn.* 467.
coccinea *Buc'hoz ex Ettl.* 468.
farinacea *Benth.* 467.
splendens *Ker-Gawl.* 468.
Salviacanthus preussii Lindau (43 : 5) 427.
Samolus *Linn.* 304.
valerandi *Linn.* 304.

Sapota cerasifera Welw. (23 : 2) 30.
mammosa Gaertn. (21 : 6) 27.
Sapotaceae 16.
Sarcocephalus Afzel. ex Sabine (46 :) 163.
badi Aubrév. (46 : 3) 164.
diderrichii De Wild. & Th. Dur. (46 : 3) 164.
diderrichii of ed. 1, partly (46 : 2) 163.
esculentus Afzel. ex Sabine (46 : 1) 163.
nervosus Hutch. & Dalz. (46 : 4) 164.
pobeguini Pobéguin ex Pellegr. (46 : 2) 163.
russeggeri Kotschy ex Schweinf. (46 : 1) 163.
sambucinus K. Schum. (46 : 1) 163.
sassandrae A. Chev. 163.
trillesii Pierre (46 : 3) 164.
vanderguchti De Wild. (46 : 4) 164.
Sarcostemma *R. Br.* 93.
viminale (*Linn.*) R. Br. 93.
Saritaea magnifica (*Sprague ex van Steenis*) *Dugand* 385.
Satanocrater *Schweinf.* 403.
berhautii *Benoist* 403.
Satureja *Linn.* 467.
pseudosimensis *Brenan* 467.
punctata (*Benth.*) *Briq.* 467.
robusta (*Hook. f.*) *Brenan* 467.
Scabiosa Linn., partly (2 :) 223.
succisa Linn. (2 :) 223.
Scaevola *Linn.* 315.
lobelia Murr. 315.
plumieri (*Linn.*) *Vahl* 315.
sericea Vahl 315.
Schaueria *Nees* 423.
populifolia *C. B. Cl.* 423.
Schinziella *Gilg* 300.
tetragona (*Schinz*) *Gilg* 300.
Schizocolea *Bremek.* 190.
linderi (*Hutch. & Dalz.*) *Bremek.* 190.
Schizoglossum angustissimum K. Schum. (13 : 1) 92.
angustissimum of ed. 1, partly (13 : 2) 92.
glanvillei Hutch. & Dalz. (23 : 9) 96.
interruptum (E. Mey.) Schltr. (13 : 2) 92.
Schrebera *Roxb.* 49.
arborea *A. Chev.* 49.
chevalieri Hutch. & Dalz. (3 :) 49.
golungensis of Thompson (3 :) 49.
Schultesia *Mart.* 299.
stenophylla *Mart.* 299.
stenophylla var. latifolia *Mart. ex Prog.* 300.
Schumanniophyton *Harms* 116.
klaineanum Pierre ex A. Chev. (10 : 1) 116.
magnificum (*K. Schum.*) *Harms* 116.
problematicum (*A. Chev.*) *Aubrév.* 116.
Schwenckia *D. van Royen ex Linn.* 327.
americana *Linn.* 327, fig. 279.
Schwenkia Linn. (3 :) 327.
guineensis Schum. & Thonn. (3 :) 327.
hirta of C. H. Wright (3 :) 327.
Sclerocarpus *Jacq. f.* 235.
africanus *Jacq. ex Murr.* 235.
Sclerochiton *Harvey* 408.
preussii (*Lindau*) *C. B. Cl.* 408.
vogelii (*Nees*) *T. Anders.* 408.